NOBLE
GASES

					2	4.0026
					He	
					Helium	
					0	
					$1s^2$	

	3A	4A	5A	6A	7A	
	5 10.811	6 12.0112	7 14.0067	8 15.9994	9 18.9984	10 20.183
	B	C	N	O	F	Ne
	Boron	Carbon	Nitrogen	Oxygen	Fluorine	Neon
	1,2	±4,2	−3,2,3,4,5	−2	−1	0
	$1s^22s^22p^1$	$1s^22s^22p^2$	$1s^22s^22p^3$	$1s^22s^22p^4$	$1s^22s^22p^5$	$1s^22s^22p^6$
	13 26.9815	14 28.086	15 30.9738	16 32.064	17 35.453	18 39.948
	Al	Si	P	S	Cl	Ar
	Aluminum	Silicon	Phosphorus	Sulfur	Chlorine	Argon
	3	4	±3,4,5	−2,4,6	±1,3,5,7	0
	$(Ne)3s^23p^1$	$(Ne)3s^23p^2$	$(Ne)3s^23p^3$	$(Ne)3s^23p^4$	$(Ne)3s^23p^5$	$(Ne)3s^23p^6$

	1B	2B						
71	29 63.54	30 65.37	31 69.72	32 72.59	33 74.922	34 78.96	35 79.909	36 83.80
	Cu	Zn	Ga	Ge	As	Se	Bromine	Kr
	Copper	Zinc	Gallium	Germanium	Arsenic	Selenium	Br	Krypton
	1,2	2	3	2,4	±3,5	−2,4,6	±1,5	0
	$(Ar)3d^{10}4s^1$	$(Ar)3d^{10}4s^2$	$(Ar)3d^{10}4s^24p^1$	$(Ar)3d^{10}4s^24p^2$	$(Ar)3d^{10}4s^24p^3$	$(Ar)3d^{10}4s^24p^4$	$(Ar)3d^{10}4s^24p^5$	$(Ar)3d^{10}4s^24p^6$
5.4	47 107.870	48 112.40	49 114.82	50 118.69	51 121.75	52 127.60	53 126.904	54 131.30
	Ag	Cd	In	Sn	Sb	Te	I	Xe
	Silver	Cadmium	Indium	Tin	Antimony	Tellurium	Iodine	Xenon
	1	2	3	2,4	±3,5	−2,4,6	±1,5,7	0
	$(Kr)4d^{10}5s^1$	$(Kr)4d^{10}5s^2$	$(Kr)4d^{10}5s^25p^1$	$(Kr)4d^{10}5s^25p^2$	$(Kr)4d^{10}5s^25p^3$	$(Kr)4d^{10}5s^25p^4$	$(Kr)4d^{10}5s^25p^5$	$(Kr)4d^{10}5s^25p^6$
09	79 196.967	80 200.59	81 204.37	82 207.19	83 208.980	84 (210)	85 (210)	86 (222)
	Au	Hg	Tl	Pb	Bi	Po	At	Rn
	Gold	Mercury	Thallium	Lead	Bismuth	Polonium	Astatine	Radon
	1,3	2	1,3	2,4	±3,5	−2,4,6	±1,5,7	0(+4 +6)
	$(Xe)4f^{14}5d^{10}6s^1$	$(Xe)4f^{14}5d^{10}6s^2$	$(Xe)4f^{14}5d^{10}6s^26p^1$	$(Xe)4f^{14}5d^{10}6s^26p^2$	$(Xe)4f^{14}5d^{10}6s^26p^3$	$(Xe)4f^{14}5d^{10}6s^26p^4$	$(Xe)4f^{14}5d^{10}6s^26p^5$	$(Xe)4f^{14}5d^{10}6s^26p^6$

LANTHANIDES

157.25	65 158.924	66 162.50	67 164.930	68 167.26	69 168.934	70 173.04	71 174.97
Gd	Tb	Dy	Ho	Er	Tm	Yb	Lu
Gadolinium	Terbium	Dysprosium	Holmium	Erbium	Thulium	Ytterbium	Lutetium
3	3,4	3	3	3	2,3	2,3	3
$(Xe)4f^75d^16s^2$	$(Xe)4f^95d^06s^2$	$(Xe)4f^{10}5d^06s^2$	$(Xe)4f^{11}5d^06s^2$	$(Xe)4f^{12}5d^06s^2$	$(Xe)4f^{13}5d^06s^2$	$(Xe)4f^{14}5d^06s^2$	$(Xe)4f^{14}5d^16s^2$
(245)	97 247	98 (251)	99 (252)	100 (253)	101 (256)	102 (253)	103 (257)
Cm	Bk	Cf	Es	Fm	Md	No	Lw
Curium	Berkelium	Californium	Einsteinium	Fermium	Mendelevium	Nobelium	Lawrencium
3	3,4	3	(3)	(3)	(3)	(3)	(3)
$(Rn)5f^76d^17s^2$	$(Rn)5f^86d^17s^2$	$(Rn)5f^{10}6d^07s^2$	$(Rn)5f^{10}6d^17s^2$	$(Rn)5f^{11}6d^17s^2$	$(Rn)5f^{13}6d^17s^2$	$(Rn)5f^{14}6d^07s^2$	$(Rn)5f^{14}6d^17s^2$

ACTINIDES

$^{12}C = 12.0000$.

HANDBOOK

OF

ENGINEERING FUNDAMENTALS

Prepared by a Staff of Specialists
Under the Editorship of

THE LATE OVID W. ESHBACH
and MOTT SOUDERS

THIRD EDITION

WILEY ENGINEERING
HANDBOOK SERIES

JOHN WILEY & SONS
New York · London · Sydney · Toronto

WILEY ENGINEERING HANDBOOK SERIES

**HANDBOOK OF ENGINEERING FUNDA-
MENTALS.** Third Edition. Edited by
MOTT SOUDERS and OVID W. ESHBACH.

**KENT'S MECHANICAL ENGINEERS' HAND-
BOOK.** Twelfth Edition.
DESIGN AND PRODUCTION. Edited by
COLIN. CARMICHAEL.
POWER. Edited by J. K. SALISBURY.

ELECTRICAL ENGINEERS' HANDBOOK.
Fourth Edition.
ELECTRIC POWER. Edited by HAROLD
PENDER and WILLIAM A. DEL MAR.
ELECTRIC COMMUNICATION AND
ELECTRONICS. Edited by HAROLD PEN-
DER and KNOX MCILWAIN.

MINING ENGINEERS' HANDBOOK. Third
Edition. Edited by the late ROBERT PEELE.

HANDBOOK OF MINERAL DRESSING.
ORES AND INDUSTRIAL MINERALS.
Edited by ARTHUR F. TAGGART.

Library of Congress Cataloging in Publication Data

Souders, Mott, 1904-1974
  Handbook of engineering fundamentals.

  (Wiley engineering handbook series)
  First-2d editions edited by O. W. Eshbach.
1. Engineering—Handbooks, manuals, etc. I. Eshbach,
Ovid Wallace, 1893–1958, ed. Handbook of engineering
fundamentals. II. Title.

TA151.S58 1974   620'.002'02   74-7467
ISBN 0-471-24553-4

Printed in the United States of America

10 9 8 7 6 5 4 3 2 1

CONTRIBUTORS

Chandler H. Barkelew, Ph.D., Senior Research Engineer, Shell Development Company, Houston, Texas.—*Automatic Control.*

John L. Barnes, S.M., A.M., Ph.D., Professor of Engineering, University of California, Los Angeles.—*Mathematical and Physical Tables* and *Mathematics.*

R. R. Bate, Ph.D., Vice Dean USAF Academy, Colorado.—*Foundation of Orbital Mechanics.*

W. C. Bauman, M.M.E., Associate Professor of Aeronautics, USAF Academy, Colorado.—*Atmospheric Structure.*

John Beek, Ph.D., Physical Chemist, Oakland, California.—*The Elements of Fortran.*

J. J. Beoddy, M.S., Instructor of Aeronautics, USAF Academy, Colorado.—*Viscous Flows.*

J. G. Brainerd, Sc.D., Professor, Moore School of Electrical Engineering, University of Pennsylvania, Philadelphia.—*Tables of Conversion Factors—Units of Weights and Measures, Gages.*

George S. Cherniak, Anderson-Nichols & Company, Boston, Massachusetts.—*Dynamic Stresses, Working Stresses,* and *Testing of Materials.*

J. T. Clay, M.A., Assistant Professor of Aeronautics, USAF Academy, Colorado—*Hypersonic Flow.*

D. H. Daley, S.M., Professor and Head, Dept. of Aeronautics, USAF Academy, Colorado.—*Steady One-Dimensional Gas Dynamics.*

Merle E. Dowd, B.S. in M.E., Assistant Professor of Industrial Relations, Northwestern Technological Institute, Evanston, Illinois.—*Symbols and Abbreviations.*

Ovid W. Eshbach, E.E., D.Sc., D.Engr., Late Dean, Northwestern Technological Institute, Evanston, Illinois.—*Mathematical and Physical Tables* and *Mechanics of Deformable Bodies.*

D. Finkleman, Ph.D., Assistant Professor of Aeronautics, USAF Academy, Colorado.—*Aerodynamics of Bodies, Real Gases.*

C. A. Forbrich, Jr., Ph.D., Associate Professor of Aeronautics, USAF Academy, Colorado.—*Kinetic Theory of Matter, Radiation Gas Dynamics, Magnetogasdynamics.*

T. J. Forster, Ph.D., Assistant Professor of Aeronautics, USAF Academy, Colorado.—*Flight Dynamics.*

J. D. Hines, M.S., Instructor of Aeronautics, USAF Academy, Colorado.—*Foundations of Orbital Mechanics.*

Paul J. Kiefer, A.B., M.E., Professor of Mechanical Engineering, U. S. Naval Postgraduate School, Annapolis, Maryland.—*Engineering Thermodynamics.*

John M. Lessells, B.Sc., President, Lessells and Associates, Inc., Consulting Engineers, Boston, Massachusetts, and Associate Professor of Mechanical Engineering, Massachusetts Institute of Technology, Cambridge.—*Dynamic Stresses, Working Stresses* and *Testing of Materials.*

R. B. Lindsay, Ph.D., Professor of Theoretical Physics and Chairman of the Department of Physics, Brown University, Providence, Rhode Island.—*Acoustics.*

Joseph E. Love, Jr., Ph.D., Structural Engineer, General Electric Company, Richland, Washington.—*Reinforced Concrete.*

John A. M. Lyon, Ph.D., Professor of Electrical Engineering, University of Michigan, Ann Arbor, Michigan.—*Electromagnetics and Circuits.*

R. D. Marker, M.S., Chief, Aeronautics Laboratory, USAF Academy, Colorado.—*Experimental Methods.*

R. W. Milling, M.S., Associate Professor of Aeronautics, USAF Academy, Colorado.—*Experimental Methods.*

C. Dean Newman, Ph.D., Associate Professor of Engineering, San Jose State University, San Jose, California.—*Chemistry.*

L. M. Nicolai, Ph.D., Associate Professor of Aeronautics, USAF Academy, Colorado.—*Kinetic Theory of Matter, Conservation Equations of Continuum Flow, Aerodynamics of Wings, Viscous Flows, Aerodynamics of Wing-Body Combinations.*

iii

R. D. Oliver, M.S., Supervisor, External Aerodynamics, Aerophysics Laboratory, North American Aviation, Inc., Downey, California.—*Propulsion.*

B. W. Parkinson, Ph.D., Associate Professor and Deputy Head, Department of Astronautics and Computer Science, USAF Academy, Colorado.—*Inertial Navigation.*

A. E. Preyss, Sc.D., Associate Professor of Astronautics, USAF Academy, Colorado.— *Foundation of Orbital Mechanics.*

Janvier M. Rice, E.E., Supervisor of Engineering Planning, Fairchild Engine Division, Fairchild Engine and Airplane Corporation, Farmingdale, New York.—*Mechanics of Rigid Bodies.*

Warren M. Rohsenow, Ph.D., Professor of Mechanical Engineering, Massachusetts Institute of Technology, Cambridge.—*Heat Transfer.*

J. C. Ruth, Sc.D., Associate Professor of Mathematics, USAF Academy, Colorado.— *Inertial Navigation.*

Sidney C. Singer, Jr., M.S. Consulting Engineer, Piedmont, California.—*Properties of Materials.*

Mott Souders, Ph.D., D.Sc., Chemical Engineer, Piedmont, California.—*Nuclear Reactions* and *Engineering Economy.*

Victor L. Streeter, Sc.D., Research Professor of Mechanics, Illinois Institute of Technology, Chicago.—*Mechanics of Incompressible Fluids.*

Milton C. Stuart, M.E., Professor of Mechanical Engineering, Lehigh University, Bethlehem, Pennsylvania.—*Engineering Thermodynamics.*

G. E. Thompson, M.S., Assistant Professor of Aeronautics, USAF Academy, Colorado.— *Flight Dynamics.*

Joseph Tusinski, Ph.D., Director of Research, Coburn Optical Industries, Inc., Colonial Heights, Virginia, formerly of Old Dominion College, Norfolk, Virginia.—*Electronics.*

Ernst Weber, Dr. Phil., Dr. Techn., Ing., Chairman Division of Engineering, National Research Council, Washington, D. C.—*Physical Units and Standards* and *Radiation and Light.*

R. E. Willes, Ph.D., Tenure Associate Professor of Aeronautics, USAF Academy, Colorado.—*Flight Dynamics, Modern Control Theory.*

J. P. Wittry, M.S., Associate Professor of Astronautics, Department of Astronautics and Computer Science, USAF Academy, Colorado.—*Inertial Navigation.*

PREFACE TO THE THIRD EDITION

The Third Edition of the *Handbook of Engineering Fundamentals* has been prepared to incorporate new knowledge, and to correct errors and ambiguities in the previous editions. Every section has been carefully revised with these goals in mind. New material has been added to Mathematical and Physical Tables, to Mathematics (including an article on Elements of Fortran), to Physical Units and Standards, and to Radiation, Light, and Acoustics.

Two decades have passed since the Second Edition was published. Changes and advances during this period have required complete rewriting of the sections on Aeronautics and Chemistry. Expanding horizons have shown the need for new sections on Astronautics, Heat Transfer, Electronics, Automatic Control, and Engineering Economy.

The former two sections on Properties of Materials have been combined into a single section. In order to provide space for more charts and tables, all textual material has been omitted from Properties of Materials. The section on Engineering Law has been omitted from the Third Edition.

MOTT SOUDERS

February 1974

CANONS OF ETHICS OF ENGINEERS

Fundamental Principles of Professional Engineering Ethics

The Engineer, to uphold and advance the honor and dignity of the engineering profession and in keeping with high standards of ethical conduct:

I. Will be honest and impartial, and will serve with devotion his employer, his clients, and the public;

II. Will strive to increase the competence and prestige of the engineering profession;

III. Will use his knowledge and skill for the advancement of human welfare.

Relations With the Public

1.1 The Engineer will have proper regard for the safety, health and welfare of the public in the performance of his professional duties.

1.2 He will endeavor to extend public knowledge and appreciation of engineering and its achievements, and will oppose any untrue, unsupported, or exaggerated statements regarding engineering.

1.3 He will be dignified and modest in explaining his work and merit, will ever uphold the honor and dignity of his profession, and will refrain from self-laudatory advertising.

1.4 He will express an opinion on an engineering subject only when it is founded on adequate knowledre and honest conviction.

1.5 He will preface any ex parte statements, criticisms, or arguments that he may issue by clearly indicating on whose behalf they are made.

Relations With Employers and Clients

2.1 The Engineer will act in professional matters as a faithful agent or trustee for each employer or client.

2.2 He will act fairly and justly toward vendors and contractors, and will not accept from vendors or contractors, any commissions or allowances, directly or indirectly.

2.3 He will inform his employer or client if he is financially interested in any vendor or contractor, or in any invention, machine, or apparatus, which is involved in a project or work of his employer or client. He will not allow such interest to affect his decisions regarding engineering services which he may be called upon to perform.

2.4 He will indicate to his employer or client the adverse consequences to be expected if his engineering judgment is over-ruled.

2.5 He will undertake only those engineering assignments for which he is qualified. He will engage or advise his employer or client to engage specialists and will cooperate with them whenever his employer's or client's interests are served best by such an arrangement.

2.6 He will not disclose information concerning the business affairs or technical processes of any present or former employer or client without his consent.

2.7 He will not accept compensation—financial or otherwise—from more than one party for the same service, or for other services pertaining to the same work, without the consent of all interested parties.

2.8 The employed engineer will engage in supplementary employment or consulting practice only with the consent of his employer.

Relations With Engineers

3.1 The Engineer will take care that credit for engineering work is given to those to whom credit is properly due.

3.2 He will provide a prospective engineering employee with complete information on working conditions and his proposed status of employment, and after employment will keep him informed of any changes in them.

3.3 He will uphold the principle of appropriate and adequate compensation for those engaged in engineering work, including those in subordinate capacities.

3.4 He will endeavor to provide opportunity for the professional development and advancement of engineers in his employ or under his supervision.

3.5 He will not injure maliciously the professional reputation, prospects, or practice of another engineer. However, if he has proof that another engineer has been unethical, illegal, or unfair in his practice, he should so advise the proper authority.

3.6 He will not compete unfairly with another engineer.

3.7 He will not invite or submit price proposals for professional services, which require creative intellectual effort, on a basis that constitutes competition on price alone. Due regard should be given to all professional aspects of the engagement.

3.8 He will cooperate in advancing the engineering profession by interchanging information and experience with other engineers and students, and by contributing to public communication media, to the efforts of engineering and scientific societies and schools.

Approved by Engineer's Council for Professional Development, September 30, 1963.

GENERAL TABLE OF CONTENTS

Detailed tables of contents are given at the beginnning of each section. An alphabetical index appears following Section 16.

SECTION 1

MATHEMATICAL AND PHYSICAL TABLES

REVISED AND EXTENDED BY
MOTT SOUDERS

FROM THE SECOND EDITION BY
O. W. ESHBACH, J. L. BARNES, J. G. BRAINERD, AND M. E. DOWD

SYMBOLS AND ABBREVIATIONS

Table 1. Greek Alphabet

A	α	Alpha	H	η	Eta	N	ν	Nu	T	τ	Tau
B	β	Beta	Θ	ϑ θ	Theta	Ξ	ξ	Xi	Υ	υ	Upsilon
Γ	γ	Gamma	I	ι	Iota	O	o	Omicron	Φ	ϕ	Phi
Δ	δ	Delta	K	κ	Kappa	Π	π	Pi	X	χ	Chi
E	ϵ	Epsilon	Λ	λ	Lambda	P	ρ	Rho	Ψ	ψ	Psi
Z	ζ	Zeta	M	μ	Mu	Σ	σ s	Sigma	Ω	ω	Omega

Table 2. Symbols for Mathematical Operations

REFERENCES: American Standard for Mathematical Symbols (AESC Report Z10f). Mathematical Association of America (National Committee Report, 1923, Chapter 8). American Standard Mathematical Symbols (ASA Z10f—1928), American Standards Association, New York.

Addition and Subtraction

$a + b$, a plus b
$a - b$, a minus b
$a \pm b$, a plus or minus b
$a \mp b$, a minus or plus b

Multiplication and Division

$a \times b$, or $a \cdot b$, or ab, a times b
$a \div b$, or $\frac{a}{b}$, or a/b, a divided by b

Symbols of Aggregation

() parentheses
[] brackets
{} braces
— vinculum

Equalities and Inequalities

$a = b$, a equals b
$a \approx b$. a approximately equals b
$a \neq b$, a is not equal to b
$a > b$, a is greater than b
$a < b$, a is less than b
$a \geqq b$, a equals or is greater than b
$a \leqq b$, a is less than or equals b
$a \equiv b$, a is identical to b
$a \rightarrow b$, or $a \doteq b$, a approaches b as a limit

Proportion

$a/b = c/d$, or $a : b :: c : d$, a is to b as c is to d
$a \propto b$, a varies directly as b
%, per cent

Powers and Roots

a^2, a squared
a^n, a raised to the nth power
$\sqrt[3]{a}$, cube root of a
$\sqrt[n]{a}$, or $a^{1/n}$, nth root of a
a^{-n}, $1/a^n$
$3.14 \times 10^4 = 31,400$
$3.14 \times 10^{-4} = 0.000314$

Miscellaneous

$a!$, $= 1 \cdot 2 \cdot 3 \ldots a$, factorial a
$P(n, r) = n(n - 1)(n - 2) \ldots (n - r + 1)$
$C(n, r) = \frac{P(n, r)}{r!} = \binom{n}{r} =$ binomial coefficients
$|a| =$ absolute value of a
i (or j) $= \sqrt{-1}$, imaginary unit
$\pi = 3.1416$, ratio of the circumference to the diameter of a circle
∞, infinity

Plane Geometry

$\angle$, angle
$\triangle$, triangle
$\parallel$, parallel
$\perp$, perpendicular
$\odot$, circle
$\square$, parallelogram
$\therefore$, therefore
$\circ$ $'$ $''$, degree, minute, second
$'$ $''$, feet, inches

Logarithms

$\log a = \log_{10} a$, common logarithm of a or log of a to the base 10
$\ln a = \log_e a$, natural logarithm of a or log of a to the base e ($e = 2.718$)
$\log^{-1} a$, number whose log is a

Trigonometry

sin, cos, tan
cosec or csc, sec, cot or ctn } trigonometric functions
vers, covers
$\sin^{-1}$, $\cos^{-1}$, etc., inverse of the functions

Analytic Geometry

x, y, z; ξ, η, ζ, rectangular coordinates
ρ, s, intrinsic coordinates
ρ, radius of curvature
s, length of arc
r, θ, polar coordinates
ψ, angle from radius vector to tangent
r, θ, ϕ, spherical coordinates
θ, co-latitude
ϕ, longitude
r, θ, z, cylindrical coordinates
e, eccentricity in conics
p, semi latus rectum in conics
$l = \cos \alpha$, $m = \cos \beta$, $n = \cos \gamma$, direction cosines

Calculus

$y = f(x)$, y is a function of x
$y' = f'(x) = \frac{dy}{dx} = D_x y$, derivative of $y = f(x)$ with respect to x
$y'' = f''(x) = \frac{d(y')}{dx} = D^2_x y = \frac{d^2 y}{dx^2}$, second derivative of $y = f(x)$ with respect to x
$u = f(x, y)$, u is a function of x and y
$u_x = f_x(x, y) = D_x(u) = \frac{\partial u}{\partial x}$, partial derivative of $u = f(x, y)$ with respect to x
$u_{xy} = f_{xy}(x, y) = D_y(D_x u) = \frac{\partial^2 u}{\partial y \partial x}$, second partial derivative of $u = f(x, y)$ with respect to x and y
Δy, increment of y
dy, differential of y
δy, variation of y
$\sum_{i=a}^{b}$, summation over i from a to b
$\lim_{x \to a} (y) = b$, $y \to b$ as $x \to a$
$\int$, integral of
$\int_a^b$, definite integral of

Vector Analysis

i, j, k, unit vectors along the axes (right-hand system)
$a \cdot b = (ab) = Sab$, scalar product of a and
$a \times b = [ab] = Vab$, vector product of a b
Vectors are indicated in print by bold type.

Table 3. Abbreviations * for Scientific and Engineering Terms

NOTE: This list of abbreviations is adapted from the recommendations of the American Standards Association. (See ASA Z10.1—1941.)

absolute	abs
acre	spell out
acre-foot	acre-ft
air horsepower	air hp
alternating-current (as adjective)	a-c
ampere	amp
ampere-hour	amp-hr
amplitude, an elliptic function	am.
Angstrom unit	A
antilogarithm	antilog
atmosphere	atm
atomic weight	at. wt
average	avg
avoirdupois	avdp
azimuth	az or α
barometer	bar.
barrel	bbl
Baumé	Bé
board feet (feet board measure)	fbm
boiler pressure	spell out
boiling point	bp
brake horsepower	bhp
brake horsepower-hour	bhp-hr
Brinell hardness number	Bhn
British thermal unit †	Btu or B
bushel	bu
calorie	cal
candle	c
candle-hour	c-hr
candlepower	cp
cent	c or ¢
center to center	c to c
centigram	cg
centiliter	cl
centimeter	cm
centimeter-gram-second (system)	cgs
chemical	chem
chemically pure	cp
circular	cir
circular mils	cir mils
coefficient	coef
cologarithm	colog
concentrate	conc
conductivity	cond
constant	const
continental horsepower	cont hp
cord	cd
cosecant	csc
cosine	cos
cosine of the amplitude, an elliptic function	cn
cost, insurance, and freight	cif
cotangent	cot
coulomb	spell out
counter electromotive force	cemf

cubic	cu
cubic centimeter	
cu cm, cm³ (liquid, meaning milliliter, ml)	
cubic foot	cu ft
cubic feet per minute	cfm
cubic feet per second	cfs
cubic inch	cu in.
cubic meter	cu m or m³
cubic micron	cu μ or cu mu or μ^3
cubic millimeter	cu mm or mm³
cubic yard	cu yd
current density	spell out
cycles per second	spell out or c
cylinder	cyl
day	spell out
decibel	db
degree ‡	deg or °
degree centigrade	C
degree Fahrenheit	F
degree Kelvin	K
degree Rankine	R
delta amplitude, an elliptic function	dn
diameter	diam
direct-current (as adjective)	d-c
dollar	$
dozen	doz
dram	dr
efficiency	eff
electric	elec
electromotive force	emf
elevation	el
equation	eq
external	ext
farad	spell out or f
feet board measure (board feet)	fbm
feet per minute	fpm
feet per second	fps
fluid	fl
foot	ft
foot-candle	ft-c
foot-Lambert	ft-L
foot-pound	ft-lb
foot-pound-second (system)	fps
foot-second (see cubic feet per second)	
franc	fr
free aboard ship	spell out
free alongside ship	spell out
free on board	fob
freezing point	fp
frequency	spell out
fusion point	fnp
gallon	gal

* These forms are recommended for readers whose familiarity with the terms used makes possible a maximum of abbreviations. For other classes of readers editors may wish to use less contracted combinations made up from this list. For example, the list gives the abbreviation of the term "feet per second" as "fps." To some readers ft per sec will be more easily understood.

† Abbreviation recommended by the A.S.M.E. Power Test Codes Committee. B = 1 Btu, kB = 1000 Btu, mB = 1,000,000 Btu. The A.S.H.&V.E. recommends the use of Mb = 1000 Btu and Mbh = 1000 Btu per hr.

‡ There are circumstances under which one or the other of these forms is preferred. In general the sign ° is used where space conditions make it necessary, as in tabular matter, and when abbreviations are cumbersome, as in some angular measurements, i.e., 59° 23′ 42″. In the interest of simplicity and clarity the Committee has recommended that the abbreviation for the temperature scale, F, C, K, etc., always be included in expressions for numerical temperatures, but, wherever feasible, the abbreviation or "degree" be omitted; as 69 F.

Table 3. Abbreviations for Scientific and Engineering Terms—*Continued*

gallons per minute	gpm
gallons per second	gps
grain	spell out
gram	g
gram-calorie	g-cal
greatest common divisor	gcd
haversine	hav
hectare	ha
henry	h
high-pressure (adjective)	h-p
hogshead	hhd
horsepower	hp
horsepower-hour	hp-hr
hour	hr
hour (in astronomical tables)	h
hundred	C
hundredweight (112 lb)	cwt
hyperbolic cosine	cosh
hyperbolic sine	sinh
hyperbolic tangent	tanh
inch	in.
inch-pound	in-lb
inches per second	ips
indicated horsepower	ihp
indicated horsepower-hour	ihp-hr
inside diameter	ID
intermediate-pressure (adjective)	i-p
internal	int
joule	j
kilocalorie	kcal
kilocycles per second	kc
kilogram	kg
kilogram-calorie	kg-cal
kilogram-meter	kg-m
kilograms per cubic meter	kg per cu m or kg/m^3
kilograms per second	kgps
kiloliter	kl
kilometer	km
kilometers per second	kmps
kilovolt	kv
kilovolt-ampere	kva
kilowatt	kw
kilowatthour	kwhr
lambert	L
latitude	lat or ϕ
least common multiple	lcm
linear foot	lin ft
liquid	liq
lira	spell out
liter	l
logarithm (common)	log
logarithm (natural)	log$_e$ or ln
longitude	long. or λ
low-pressure (as adjective)	l-p
lumen	lm
lumen-hour	lm-hr
lumens per watt	lpw
mass	spell out
mathematics (ical)	math
maximum	max
mean effective pressure	mep

mean horizontal candlepower	mhcp
megacycle	spell out
megohm	spell out
melting point	mp
meter	m
meter-kilogram	m-kg
mho	spell out
microampere	μa or mu a
microfarad	μf
microinch	μin.
microfarad	μμf
micromicron	μμ or mu mu
micron	μ or mu
microvolt	μv
microwatt	μw or mu w
mile	spell out
miles per hour	mph
miles per hour per second	mphps
milliampere	ma
milligram	mg
millihenry	mh
millilambert	mL
milliliter	ml
millimeter	mm
millimicron	mμ or m mu
million	spell out
million gallons per day	mgd
millivolt	mv
minimum	min
minute	min
minute (angular measure)	
minute (time)(in astronomical tables)	m
mole	spell out
molecular weight	mol. wt
month	spell out
National Electrical Code	NEC
ohm	spell out or Ω
ohm-centimeter	ohm-cm
ounce	oz
ounce-foot	oz-ft
ounce-inch	oz-in.
outside diameter	OD
parts per million	ppm
peck	pk
penny (pence)	d
pennyweight	dwt
per	(See Fundamental Rules)
peso	spell out
pint	pt
potential	spell out
potential difference	spell out
pound	lb
pound-foot	lb-ft
pound-inch	lb-in.
pound sterling	£
pounds per brake horsepower-hour	lb per bhp-hr
pounds per cubic foot	lb per cu ft
pounds per square foot	psf
pounds per square inch	psi
pounds per square inch absolute	psia
power factor	spell out or pf
quart	q

Table 3.   Abbreviations for Scientific and Engineering Terms—*Continued*

radian	spell out
reactive kilovolt-ampere	kvar
reactive volt-ampere	var
revolutions per minute	rpm
revolutions per second	rps
rod	spell out
root mean square	rms
secant	sec
second	sec
second (angular measure)	"
second-foot (see cubic feet per second)	
second (time)(in astronomical tables)	s
shaft horsepower	shp
shilling	s
sine	sin
sine of the amplitude, an elliptic function	sn
specific gravity	sp gr
specific heat	sp ht
spherical candle power	scp
square	sq
square centimeter	sq cm or cm^2
square foot	sq ft
square inch	sq in.
square kilometer	sq km or km^2
square meter	sq m or m^2
square micron	sq μ or sq mu or μ^2
square millimeter	sq mm or mm^2
square root of mean square	rms
standard	std
stere	s
tangent	tan
temperature	temp
tensile strength	ts
thousand	M
thousand foot-pounds	kip-ft
thousand pound	kip
ton	spell out
ton-mile	spell out
versed sine	vers
volt	v
volt-ampere	va
volt-coulomb	spell out
watt	w
watthour	whr
watts per candle	wpc
week	spell out
weight	wt
yard	yd
year	yr

Table 4.   Symbols for Physical Quantities

NOTE: The most frequently used American Standard and Tentative Standard Symbols are included in this table. References to sources, which are publications of the American Standards Association and its successors, American National Standards Institute, are indicated by the numbers in parentheses following the names of the quantities. The numbers correspond to the following:

(1) American Standard Letter Symbols for Aeronautical Sciences, ASA, Y10.7—1954.
(2) American Institute of Electrical Engineers Standards, ASA, Z10.5—1949.
(3) American Standard Symbols for Heat and Thermodynamics, ASA, Z10.4—1943.
(4) American Standard Symbols for Hydraulics, ASA Y10.2—1958.
(5) American Standard Letter Symbols for Mechanics of Solid Bodies, ASA, Y10.8—1962.
(6) American Standard Letter Symbols for Chemical Engineering, ASA, Y10.12—1953.

Where possible, capital letters denote total quantities and small letters denote specific quantities, or quantities per unit.

NAME OF QUANTITY	SYMBOL
Absorption factor (3)	α
Acceleration	
angular (1) (4) (5)	α
linear, general (3) (4) (5) (6)	a
Acceleration due to gravity	
general (1) (3) (4) (5) (6)	g
International Adopted Standard (3) (6)	g_0
local (3)	g_L
gravitational conversion factor (3) (6)	g_c
Activity (6)	a
Activity coefficient, molal basis (6)	γ
Adiabatic factor (3)	X
Admittance (2)	Y
Advanced ratio of propeller (1)	J
Altitude (1)	h, z
Amplitude (5)	A
Angle (6)	α
Angle (1)	$\beta\phi$
blade angle (1)	β
effective helix (1)	ψ
dihedral (1)	Γ
helical angle of advance (1)	ϕ
of attack (1)	α
of downwash (1)	ϵ
of radiation	θ
of sideslip (1)	β
of sidewash (1)	σ
solid (3) (6)	ω

Table 4. Symbols for Physical Quantities—*Continued*

Name of Quantity	Symbol
Angular	
acceleration (1) (5)	α
displacements (1)	δ
frequency (2)	ω
momentum (1)	H
velocity (1) (2) (3) (4) (5)	ω
Area (1) (3) (4) (5) (6)	A
Area * (1) (3) (6)	S
Aspect ratio (1)	A, $\mathcal{R}$
Atomic weight (3)	A
Attack, angle of (1)	α
Attenuation (2)	a
Axes	
of aircraft (left-handed)	
earthbound coordinate system (1)	x, y, z
lateral (1)	Y
longitudinal (1)	X
normal (1)	Z
Bazin's coefficient of roughness (4)	m
Blade width (propellers) (1)	b
Boundary layer thickness (1)	δ
Breadth (3) (5) (6)	b
Capacitance, capacity (2)	C
Capacitivity (2)	ϵ
of evacuated space (2)	ϵ_v
relative (2)	ϵ_r
Charge, electric or quantity of electricity (2)	Q
Charge density	
line density of charge (2)	λ
surface density of charge (2)	σ
volume density of charge (2)	ρ
Chézy's coefficient (4)	C
Chord length (1)	c
Circular frequency $(2\pi f)$ (5)	ω
Circulation, strength of single vortex (1)	Γ
Coefficient	
absolute (1)	C
general (4)	C
of contraction (4)	C_c
of discharge (6)	C
of discharge (4)	C_Q
of energy per unit weight in $C_e \dfrac{V^2}{29}$ (4)	C_e
of flow (Chézy) (4)	C
of friction (Weisbach-Darcy) (4)	f
of friction (5)	μ, f
of momentum per unit weight in $C_m \dfrac{V}{g}$ (4)	C_m
of roughness (Bazin) (4)	m
of roughness (Kutter and Manning) (4)	n
of heat transfer overall	V
of velocity (4)	C_v
Compressibility factor (6)	z
Concentrated load (5)	F, P, Q
Concentration (1) (3)	C, c
Concentration, volumetric (6)	c
Concentration factor, stress (5)	K
Conductance	
electrical (2)	G
thermal (3) (6)	$\dfrac{1}{R}$
per unit area (3) (6)	$\dfrac{1}{RA}$
Conductivity	
electrical (2)	γ, σ
equivalent (2)	Λ
thermal (1) (3)	k

* Use with appropriate subscript.

Table 4. Symbols for Physical Quantities—*Continued*

Name of Quantity	Symbol
Contraction, coefficient of (4)	C_c
Correlation coefficient (1)	R
Coupling coefficient (2)	k
Critical state or indicating critical value (subscript) (3)	c
Current † (2)	I

Damping	
coefficient (5)	c
constant or coefficient (2)	δ
factor (1)	λ
Deflection (5)	δ
of beam, maximum (5)	δ
Density (1) (3) (4) (5) (6)	ρ
relative to standard air density (1)	σ
Depth (5)	h
Depth (6)	y
of flow, channels (4)	y
Diameter (1) (3) (4) (5) (6)	D
Dielectric constant (2)	ϵ
Difference between values (3) (6)	Δ
Difference of potential † ‡ (2)	E, e
Diffusion coefficient	
diffusivity (3)	D_v
diffusivity (1)	α
Diffusivity, thermal (6)	α
of vapor (6)	D_v
Discharge	
coefficient of (4)	C_q
coefficient of (6)	C
rate of; or flow (4)	Q
per unit width (4)	q
Displacement, electric (2)	D
Distance	
from center of gravity to center of pressure of horizontal tail surface (1)	f
linear (3) (5)	s
Drag, absolute coefficient of (1)	D
Dynamic (or impact) pressure (1)	q
Eccentricity of application of load (5)	e
Efficiency (1) (2) (3) (4) (5) (6)	η
Elastance (2)	S
mutual (2)	S_m, S_{rc}
self (2)	S, S_{cc}
Elasticity	
bulk modulus, of liquids (4)	K
kinematic $\dfrac{K}{\rho}$ (4)	e
modulus of (5)	E
Elastivity (2)	σ
Electric potential † ‡ (2)	E, e
Electricity, quantity of (2)	Q, q
Electromotive force † (2)	E, e
Electronic charge, absolute value (2)	e
Electrostatic flux (2)	ψ
Elevation	
above datum (4)	Z
above stream bed (4)	Z_0
Elongation, total (5)	δ
Emissivity, total (1) (3) (6)	ϵ
Energy (2)	W
work total (3) (4) (6)	E
Energy (1)	E
internal; intrinsic (1) § (3) (6)	U, u
kinetic (5)	E_k, T
per unit time (power) (4)	P
potential (5)	E_p, V

† Where distinctions between maximum, instantaneous, effective (root-mean-square), and average values are necessary, E_m, I_m, P_m are recommended for maximum values; e, i, p for instantaneous values; E, I for effective (rms) values; and P for average value.

‡ Where a distinction between electromotive force and difference of electric potential is desirable, the symbols E, e, and V, v, respectively, may be used.

Table 4. Symbols for Physical Quantities—*Continued*

Name of Quantity	Symbol
Enthalpy (1) § (3)	H or h
Enthalpy (6)	H
of dry saturated vapor (3)	h_g
of saturated liquid (3)	h_f
per unit weight (1) (6)	h
Entropy (1) (6)	S, s
Error signal (1)	ϵ
Expansion, exponent of polytropic (1) (3)	n
cubical, thermal coefficient (3) (6)	β
linear, thermal coefficient (3) (5) (6)	α
Factor of safety (5)	N
Film thickness, effective (6)	B
Flow rate (3) (6)	w
in pounds per unit of time (3)	w
volumetric (6)	q
Fluidity (6)	$\dfrac{1}{\mu}$
Flux	
density	
magnetic (2)	B
displacement (2)	ψ
magnetic (2)	Φ
Force (2)	F
Force (1) (4)	F
electromotive † (2)	E, e
magnetomotive (2)	$M, \mathfrak{F}$
moment of (5)	M
normal (1)	N
shearing force in beam section (5)	V
total load (3) (6)	F
Forces or loads, concentrated	P, Q, F
Fraction	
by volume (6)	x_v
by weight (6)	x_w
Free energy	
Gibbs (6)	G
Helmholtz (6)	A
Frequency (1) (2) (3) (5)	f
circular ($2\pi f$) (5)	ω
of radiant energy	ν
reduced (flutter) (1)	k
rotational (2)	n
Frequency, angular (2)	ω
Friction	
coefficient of sliding (5)	f, μ
factor used in expressing pipe-loss (4) (6)	f
in energy balance (6)	F
Fugacity (3) (6)	f
Gas constant (1) (3) (6)	R
Gibbs' function, total potential function § (3) (6)	G, g
Gyration, radius of (5)	k
Head	
atmospheric (4)	h_a
lost * (4)	h
potential (4)	h_{pz}
pressure (4)	h_p
velocity (4)	h_v
Heat	
content; enthalpy § (3)	H, h
content of dry saturated vapor; enthalpy of dry saturated vapor (3)	h_g
content of saturated liquid; enthalpy of saturated liquid (3)	h_f

* Use with appropriate subscript.

† Where distinctions between maximum, instantaneous, effective (root-mean-square), and average values are necessary, E_m, I_m, P_m are recommended for maximum values; e, i, p for instantaneous values; E, I for effective (rms) values; and P for average value.

§ In each instance upper-case italics may be used optionally for values in general, or per mole. Molal values may have subscript M. Lower-case italics are to be used for specific values (per pound, gram, liter, etc.). Molecular values may be represented by lower-case italics or by lower-case italics with subscript m.

Table 4. Symbols for Physical Quantities—*Continued*

NAME OF QUANTITY	SYMBOL
equivalent of work (3)	$\frac{1}{J}$
flow rate (3)	q
across a boundary surface (1)	h
latent, of evaporation (6)	λ, h_{fg}
mechanical equivalent of (3)	J
of vaporization at constant pressure § (3)	$H_{fg}, \lambda,$ or h_{fg}
specific, at constant pressure (1) (3)	c_p
specific, at constant volume (1) (3)	c_v
ratio of specific heats (1)	γ
ratio of specific (3)	$\gamma, \kappa,$ or k
transfer, overall coefficient of (1) (3) (6)	U
transfer, surface coefficient of (3) (6)	h
Height (3) (5)	h
crest, weirs (4)	z
Helix, effective angle (1)	ϕ
Helmholtz' free energy; internal potential function § (3) (6)	A, a
Humidity (6)	H
density of water vapor; weight of water vapor per unit of volume of space (3)	ρ_H
density of water vapor at saturation (3)	ρ_s
enthalpy of the mixture minus the enthalpy of the liquid at the temperature of adiabatic saturation; Carrier sigma function (3)	h_Σ
humid volume, volume of mixture per unit of weight of dry air (3)	v_H
partial pressure of water vapor (3)	p_H
percentage humidity by weight (3)	w_H/w_s
relative humidity; ratio of an actual partial pressure of water vapor in air to the saturation partial pressure (3) (6)	H_R
saturation pressure of water vapor (3)	p_s
saturation weight of water vapor per unit of weight of dry air (3)	H_s, w_s
weight of water vapor per unit of weight of dry air (3)	H, w_H
Hydraulic radius (3) (6)	R_H
mean in a reach (4)	R_m
of cross-sectional area (4)	R
Hydraulic slope (4)	S_w
Impedance (2)	Z
Impulse (1)	I
Inductance (2)	L
magnetic (2)	B
mutual (2)	L_m
self (2)	L, L_{cc}
Inertia, moment of (5)	I
Inertia, moment of	
polar (5)	J
rectangular (1) (5)	I
product moment of (5)	I_{xy}
Intensity	
electric (2)	E, K
magnetic (2)	H
Isentropic factor (1)	X
Joule-Thomson's coefficient (3)	μ
Kutter's coefficient of roughness (4)	n
Length (3) (4) (5) (6)	L
Length (1) (5)	l
Lift (1)	L
Linear expansion, coefficient (5)	α
Linear velocity (5)	v
Load	
concentrated (5)	F, P, Q
eccentricity of application of (5)	e
factor (1)	n
per unit distance (5)	w, q
total (5)	W, P
Mach	
angle (1)	μ
number (1)	M

§ In each instance upper-case italics may be used optionally for values in general, or per mole. Molal values may have subscript M. Lower-case italics are to be used for specific values (per pound, gram, iter, etc.). Molecular values may be represented by lower-case italics or by lower-case italics with ubscript m.

Table 4. Symbols for Physical Quantities—*Continued*

NAME OF QUANTITY	SYMBOL
Magnetic	
flux (2)	Φ
intensity (2)	H
Magnetomotive force (2)	$M, \mathfrak{F}$
Mass (1) (3) (4) (5) (6)	m
flow rate (1) (6)	w
velocity (1) (6)	G
Mean free path (1)	λ
Mechanical equivalent of heat (1) (3) (6)	J
Microscale (turbulence) (1)	λ
Modulus	
bulk, of elasticity of liquids	K
of elasticity (1) (5)	E
of elasticity in shear (5)	G
section (5)	Z
shear (1)	G
Molecular weight (3) (6)	M
Moment	
electric (2)	p
magnetic (2)	m
of any area about a given axis, statical (5)	Q
of force, including bending moment (1) (5)	M
of inertia, polar (1) (5)	J
rectangular (1) (5) (6)	I
Mutual inductance (2)	L_m
Neutral axis, distance to extreme fiber (5)	c
Nozzle divergence factor (1)	λ
Number in general (1) (3) (4) (5) (6)	N
of conductors or turns (2)	N
of moles, pound-moles, kilogram-moles, etc. (3)	n
of phases (2)	m
of poles (2)	p
of revolutions per unit of time (1) (3) (5)	n
Perimeter, wetted, of a sectional area (4)	P
Period (2) (5)	T
Permeability	
magnetic (2)	μ
of evacuated space (2)	μ_c
relative (2)	μ_r
Permeance (2)	$\mathcal{P}, \Lambda$
Permittivity (2)	ϵ
Phase	
angle (2)	ϕ
constant (2)	β
displacement (5)	ϕ
Pitch, geometric (1)	p
Planck constant (3)	h
Poisson ratio (5)	μ, ν
Poisson ratio (1)	μ
Polarization, magnetic (2)	B_i
Pole strength (2)	m
Potential	
electric †‡ (2)	V
function (1)	ϕ
function, internal; Helmholtz' free energy (3)	A, a
function, total; Gibbs' function (3)	G, g
magnetic (2)	$M, \mathfrak{F}$
magnetic vector (2)	A
retarded vector (2)	A_r
Power (4) (5) (6)	P
Power	
active (2)	P
apparent (2)	S
factor (2)	F_p
reactive (2)	Q

† Where distinctions between maximum, instantaneous, effective (root-mean-square), and average values are necessary, E_m, I_m, P_m are recommended for maximum values; e, i, p for instantaneous values; E, I for effective (rms) values; and P for average value.

‡ Where a distinction between electromotive force and difference of electric potential is desirable, the symbols E, e, and V, v, respectively, may be used.

Table 4. Symbols for Physical Quantities—*Continued*

NAME OF QUANTITY	SYMBOL
Power (1) (3) (4) (6)	P
Pressure	
dynamic (1)	q
intensity; force per unit area (1) (3) (4) (5) (6)	p
relative (1)	δ
saturation of water vapor (3)	p_s
Propagation constant (2)	γ
Poynting vector (2)	Π
Q factor of a reactor (2)	Q
Quality of vapor (3)	x
Quantity	
of electricity (2)	Q
of heat per unit mass or unit weight (1)	q
of heat per unit time (1) (3) (6)	q
of matter (6)	W
total, of a fluid, water, gas, heat (by volume) (1) (3) (6)	Q
Radiant density (3)	u
Radiant energy (3)	U
Radiant flux (3)	Φ
density (3)	W
Radiant intensity (3)	J
Radiation, intensity of (6)	N
Radii (5)	r, R
Radius (1) (3) (4) (6)	r
of gyration (1) (5)	k
Range (1)	R
Reactance (2)	X
capacitive (2)	X_c
inductive (2)	X_L
mutual (2)	X_m, X_{rc}
self (2)	X, X_{cc}
Reactive factor (2)	F_q
Recovery factor (1)	η_r
Reduced frequency (flutter) (1)	k
Reflection factor (3)	ρ
Reluctance (2)	$\mathcal{R}$
Reluctivity (2)	ν
Resistance	
electrical (2)	R
temperature coefficient (2)	α
thermal (3) (6)	R
per unit area (3)	RA
Resistivity	
electrical (2)	ρ
thermal (3)	$\dfrac{1}{k}$
Revolutions per unit time (1) (2) (3) (5)	n
Reynolds' number (1) (4)	R
Richness; equivalence ratio (combustion) (1)	R
Rotation	
rate of (6)	n
speed of (2)	n
Safety factor (5)	N
saturation pressure of water vapor (3)	p_s
Section modulus (5)	Z
Self-inductance (2)	L, L_{cc}
Set of control surfaces, angle of * (1)	δ
Shape factor (3)	S
Shearing force in beam section (5)	V
Slip (2)	s
Slope	
of channel bed (4)	S_o
of cuts and embankments (4)	s
of energy grade line (4)	S
of hydraulic grade line (4)	S_w
of lift curve (1)	a
Solidity, propellers (1)	σ
Span (1)	b
effectiveness (1)	e

* Use with appropriate subscript.

Table 4.   Symbols for Physical Quantities—*Continued*

NAME OF QUANTITY	SYMBOL
Specific	
gravity (3)	G
heat (3) (6)	c
heat at constant pressure (1) (3) (6)	c_p
molar (3)	C_p
heat at constant volume (1) (3) (6)	c_v
molar (3)	C_v
heats, ratio (1) (3)	γ
heats, ratio (3) (6)	κ or k
volume (3) (6)	v
weight (3) (4)	γ
Speed	
linear (3)	V, v, u
of rotation (2)	n
Spring constant (5)	k
Stefan-Boltzmann constant (3) (6)	σ
Strain	
normal (1) (5)	ϵ
shear (1) (5)	γ
Stream function (1)	ψ
Stress	
concentration factor (5)	K
normal (5)	σ, s
normal (1)	σ
shear (1)	τ
shear (5)	τ, s_s
Supercompressibility factor (3)	z
Surface coefficient of heat transfer	h
Surface per unit volume (6)	a
Surface tension (3) (4) (6)	σ
kinematic $\dfrac{\sigma}{\rho}$ (4)	ω
Susceptance (2)	B
Susceptibility	
dielectric (2)	η
magnetic (2)	κ
Sweepback angle (1)	Λ
Taper ratio (1)	λ
Temperature	
absolute ∥ (F abs or K) (1) (2) (3) (4) (5) (6)	T or Θ
ordinary ∥ (F or C) (1) (2) (3) (4) (5) (6)	t or θ
ratio (1)	θ
Thermal	
conductance (3)	$\dfrac{1}{R}$
per unit area; "unit conductance" (3) (6)	$\dfrac{1}{RA}$
conductivity (1) (3)	k
diffusivity (3)	α
resistance (3) (6)	R
resistance of unit area (3)	RA
resistivity (3)	$\dfrac{1}{k}$
transfer factor (6)	j
transmission (3)	q
Thickness (1) (3)	d or t
Thickness (4)	t
Thickness (5)	h
Thrust	
stream (1)	F
propeller (1)	T
Time (1) (2) (4) (5) (6)	t
Time ¶ (1) (3) (6)	t or τ
Time constant (2)	τ
Torque (1)	Q
Torque (5)	T or M

∥ θ and Θ in sources (2) and (3) only. θ is preferable only when t is used for time in the same discussion. Θ is preferable only when θ is used for ordinary temperature.

¶ τ should be used only when t is used for ordinary temperature in the same discussion.

Table 4. Symbols for Physical Quantities—*Continued*

NAME OF QUANTITY	SYMBOL
Transmission	
factor (3)	τ
thermal (3)	q
Turbulence exchange, coefficient (1)	ϵ
Turbulence scale (1)	L
Vaporization, heat of, at constant pressure (3)	H_{fg}, h_{fg}, λ
Velocity (1) (3) (4)	V
Velocity (3)	V or v
acoustic (6)	V_a
angular (1) (2) (3) (4) (5) (6)	ω
average (6)	V
Belanger critical (4)	V_c
components in x, y, z directions, respectively (1) (4)	u, v, w
linear (5)	v
local (6)	u
mass, mass-flow, per unit cross-sectional area, per unit time (1) (3)	G
mean (Q/A) (4)	V
of light (2) (3)	c
of sound (1)	a, c
of uniform flow (4)	V_o
of wave celerity (4)	c
relative (3)	v
temporal means of components (4)	$\bar{u}, \bar{v}, \bar{w}$
Vibration constant (2)	p
Viscosity	
absolute; coefficient of (1) (3) (4) (6)	μ
kinematic (1) (3) (4) (6)	ν
relative (to absolute viscosity of water) (3). (6)	$\dfrac{\mu}{\mu_w}$
relative kinematic (3)	use $\dfrac{\nu}{\nu_w}$
Voltage † (2)	E, e
Volume (1) (4) (5) (6)	V
molar (3)	V, V_m
specific (1) (3)	v
total (3)	V, V_L
Volume rate; discharge by volume, fluid rate of flow by volume (3)	q, Q
Wavelength (1) (2) (3) (5)	λ
constant (2)	β
Weight	
molecular (3)	M
per unit time per unit area of cross section; "mass velocity" (6)	G
per unit volume (4) (5)	γ
rate; per unit of power; for unit of time (3)	w
rate of flow per unit of breadth (6)	Γ
specific, with g_c (3)	γ
total (1) (3) (4) (5) (6)	W
Weirs	
crest height (4)	z
crest length (4)	b
degree of submergence (4)	N
Wetted perimeter (3) (4)	L_p
Width (same as breadth) (5) (6)	b
of stream bed (4)	b
Width, channel surface (4)	b_w
Wing setting angle of (angle between the wing chord and the thrust line) (1)	i_w
Work (2) (4) (5) (6)	W
external (6)	W_e
heat equivalent of (3)	$\dfrac{1}{J}$ or A
per unit weight (3)	w, w_k

† Where distinctions between maximum, instantaneous, effective (root-mean-square), and average values are necessary, E_m, I_m, P_m are recommended for maximum values; e, i, p for instantaneous values; E, I for effective (rms) values; and P for average value.

MATHEMATICAL TABLES

Table 5. Certain Constants Containing e and π

$e = 2.7182818285$ $M = \log_{10} e = 0.4342944819$
$\pi = 3.1415926536$ $M^{-1} = \log_e 10 = 2.3025850930$

e^n	Value	Logarithm	$n\pi$	Value	Logarithm	π/n	Value	Logarithm
e	2.718282	0.43294	π	3.141593	0.497150	$\pi/2$	1.570780	0.196120
e^{-1}	0.367879	$\overline{1}$.565706	2π	6.283185	0.798180	$\pi/3$	1.047198	0.020029
e^2	7.389057	0.868589	3π	9.424778	0.974271	$\pi/4$	0.785398	$\overline{1}$.895090
e^{-2}	0.135335	$\overline{1}$.131411	4π	12.566371	1.099210	$\pi/180$	0.017453*	$\overline{2}$.241877
$e^{\frac{1}{2}}$	1.648721	0.217147	5π	15.707963	1.196120			

	Reciprocals of π			Powers of π			Roots of π	
n/π	Value	Logarithm	$\pi^{\pm n}$	Value	Logarithm	$\pi^{\pm 1/n}$	Value	Logarithm
$1/\pi$	0.318310	$\overline{1}$.502850	π^2	9.869604	0.994300	$\sqrt{\pi}$	1.772454	0.248575
$2/\pi$	0.636620	$\overline{1}$.803880	$1/\pi^2$	0.101321	$\overline{1}$.005700	$1/\sqrt{\pi}$	0.564190	$\overline{1}$.751425
$3/\pi$	0.954930	$\overline{1}$.979971	π^3	31.006277	1.491450	$\sqrt[3]{\pi}$	1.464592	0.165717
$180/\pi$	57.295780†	1.758123	$1/\pi^3$	0.032252	$\overline{2}$.508550	$1/\sqrt[3]{\pi}$	0.682784	$\overline{1}$.834283

* Number of radians per degree.   † Number of degrees per radian.

Table 6. Factorials

n	$n! = 1 \cdot 2 \cdot 3 \ldots n$	$1/n!$	n	$n! = 1 \cdot 2 \cdot 3 \ldots n$	$1/n!$
1	1	1.	11	$399{,}168 \times 10^2$	0.250521×10^{-7}
2	2	0.5	12	$479{,}002 \times 10^3$	$.208768 \times 10^{-8}$
3	6	.166667	13	$622{,}702 \times 10^4$	$.160590 \times 10^{-9}$
4	24	$.416667 \times 10^{-1}$	14	$871{,}783 \times 10^5$	$.114707 \times 10^{-10}$
5	120	$.833333 \times 10^{-2}$	15	$130{,}767 \times 10^7$	$.764716 \times 10^{-12}$
6	720	$.138889 \times 10^{-2}$	16	$209{,}228 \times 10^8$	$.477948 \times 10^{-13}$
7	5,040	$.198413 \times 10^{-3}$	17	$355{,}687 \times 10^9$	$.281146 \times 10^{-14}$
8	40,320	$.248016 \times 10^{-4}$	18	$640{,}237 \times 10^{10}$	$.156192 \times 10^{-15}$
9	362,880	$.275573 \times 10^{-5}$	19	$121{,}645 \times 10^{12}$	$.822064 \times 10^{-17}$
10	3,628,800	$.275573 \times 10^{-6}$	20	$243{,}290 \times 10^{12}$	$.411032 \times 10^{-19}$

Table 7

Decimal Equivalents, Squares, Cubes, Square Roots, Cube Roots, Three-halves Powers, Fifth Roots, Reciprocals, Circumference and Area of Circles

Number, N		N^2	N^3	$\sqrt{N}$	$\sqrt[3]{N}$	$N^{3/2}$	$\sqrt[5]{N}$	$\frac{1}{N}$	Circle ($N = D$)	
Fraction	Decimal								Circum.	Area
1/64	.015625	0.000244	.381×10⁻⁵	.1250	.2500	.00195	.4353	64.0	.04909	.00019
1/32	.03125	.000977	.305×10⁻⁴	.1768	.3150	.00552	.5000	32.0	.09818	.00077
3/64	.046875	.002197	.103×10⁻³	.2165	.3606	.01015	.5422	18.8235	.14726	.00173
1/16	.0625	.003906	.244×10⁻³	.2500	.3969	.01563	.5744	16.0	.19635	.00307
5/64	.078125	.006104	.477×10⁻³	.2795	.4275	.02184	.6006	12.80	.24544	.00479
3/32	.09375	.008789	.824×10⁻³	.3062	.4543	.02871	.6229	10.6667	.29452	.00690
	.10	.010	.00100	.3162	.4642	.03162	.6310	10.0	.31416	.00785
7/64	.109375	.01196	.01308	.3307	.4782	.03617	.6424	9.1429	.34361	.00939
1/8	.125	.01563	.001953	.3536	.5000	.04419	.6598	8.0	.39270	.01227
9/64	.140625	.01978	.002782	.3750	.5200	.05273	.6755	7.1111	.44179	.01554
5/32	.15625	.02441	.003814	.3953	.5386	.06176	.6899	6.40	.49087	.01917
11/64	.171875	.02954	.005077	.4146	.5560	.07126	.7031	5.8182	.53996	.02320
3/16	.1875	.03516	.006592	.4330	.5724	.08119	.7155	5.3333	.58905	.02761
	.20	.040	.0080	.4472	.5848	.08944	.7248	5.0	.62832	.03142
13/64	.203125	.04126	.008381	.4507	.5878	.09155	.7270	4.9231	.63814	.03241
7/32	.21875	.04785	.01047	.4677	.6025	.10231	.7379	4.5714	.68722	.03758
15/64	.234375	.05493	.01287	.4841	.6166	.11347	.7481	4.2667	.73631	.04314
1/4	.250	.0625	.01563	.5000	.6300	.12500	.7579	4.0	.78540	.04909
17/64	.265625	.07056	.01874	.5154	.6428	.13690	.7671	3.7647	.83448	.05542
9/32	.28125	.07910	.02225	.5303	.6552	.14916	.7759	3.5556	.88357	.06213
19/64	.296875	.08813	.02616	.5449	.6671	.16176	.7844	3.3684	.93266	.06922
	.30	.090	.0270	.5477	.6694	.16432	.7860	3.3333	.94248	.07069
5/16	.3125	.09766	.03052	.5590	.6786	.17469	.7925	3.2000	.98175	.07670
21/64	.328125	.10767	.03533	.5728	.6897	.18796	.8002	3.0476	1.0308	.08456
11/32	.34375	.11816	.04062	.5863	.7005	.20154	.8077	2.9091	1.0799	.09281
23/64	.359375	.12915	.04641	.5995	.7110	.21544	.8149	2.7826	1.1290	.10143
3/8	.375	.14063	.05273	.6124	.7211	.22964	.8219	2.6667	1.1781	.11045
25/64	.390625	.15259	.05961	.6250	.7310	.24414	.8286	2.5600	1.2272	.11984
	.40	.16	.0640	.6325	.7368	.25298	.8326	2.50	1.2566	.12566
13/32	.40625	.16504	.06705	.6374	.7406	.25894	.8351	2.4615	1.2763	.12962
27/64	.421875	.17798	.07508	.6495	.7500	.27402	.8415	2.3704	1.3254	.13979
7/16	.4375	.19141	.08374	.6614	.7592	.28938	.8476	2.2857	1.3744	.15033
29/64	.453125	.20532	.09304	.6732	.7681	.30502	.8536	2.2069	1.4235	.16126
15/32	.46875	.21973	.10300	.6847	.7768	.32093	.8594	2.1333	1.4726	.17257
31/64	.484375	.23462	.11364	.6960	.7854	.33711	.8650	2.0645	1.5217	.18427
1/2	.50	.2500	.12500	.7071	.7937	.35355	.8706	2.0	1.5708	.19635
33/64	.515625	.26587	.13709	.7181	.8019	.37025	.8759	1.9394	1.6199	.20881
17/32	.53125	.28223	.14993	.7289	.8099	.38721	.8812	1.8824	1.6690	.22166
35/64	.546875	.29907	.16355	.7395	.8178	.40442	.8863	1.8286	1.7181	.23489
9/16	.5625	.31641	.17798	.7500	.8255	.42188	.8913	1.7778	1.7671	.24850
37/64	.578125	.33423	.19323	.7604	.8331	.43957	.8962	1.7297	1.8162	.26250
19/32	.59375	.35254	.20932	.7706	.8405	.45751	.9010	1.6842	1.8653	.27688
	.60	.3600	.21600	.7746	.8434	.46476	.9029	1.6667	1.8850	.28274
39/64	.609375	.37134	.22628	.7806	.8478	.47569	.9057	1.6410	1.9144	.29165
5/8	.625	.39063	.24414	.7906	.8550	.49410	.9103	1.6000	1.9635	.30680
41/64	.640625	.41040	.26291	.8004	.8621	.51275	.9148	1.5610	2.0126	.32233
21/32	.65625	.43066	.28262	.8101	.8690	.53162	.9192	1.5238	2.0617	.33824
43/64	.671875	.45142	.30330	.8197	.8759	.55072	.9235	1.4884	2.1108	.35454
11/16	.6875	.47266	.32495	.8297	.8826	.57005	.9278	1.4545	2.1598	.37122
	.70	.4900	.34300	.8367	.8879	.58566	.9312	1.4286	2.1991	.38485
45/64	.703125	.49438	.34761	.8385	.8892	.58959	.9320	1.4222	2.2089	.38829
23/32	.71875	.51660	.37131	.8478	.8958	.60935	.9361	1.3913	2.2580	.40574
47/64	.734375	.53931	.39605	.8570	.9022	.62933	.9401	1.3617	2.3071	.42357
3/4	.750	.56250	.42188	.8660	.9086	.64952	.9441	1.3333	2.3562	.44179
49/64	.765625	.58618	.44879	.8750	.9148	.66992	.9480	1.3061	2.4053	.46038
25/32	.78125	.61035	.47684	.8839	.9210	.69053	.9518	1.2800	2.4544	.47937
51/64	.796875	.63501	.50602	.8927	.9271	.71135	.9556	1.2549	2.5035	.49874
	.80	.6400	.51200	.8944	.9283	.71554	.9564	1.2500	2.5133	.50265
13/16	.8125	.66016	.53638	.9014	.9331	.73238	.9593	1.2308	2.5525	.51849
53/64	.828125	.68579	.56792	.9100	.9391	.75361	.9630	1.2075	2.6016	.53862
27/32	.84375	.71191	.60067	.9186	.9449	.77503	.9666	1.1852	2.6507	.55914
55/64	.859375	.73853	.63467	.9270	.9507	.79666	.9702	1.1636	2.6998	.58004
7/8	.875	.76563	.66992	.9354	.9565	.81849	.9737	1.1429	2.7489	.60132
57/64	.890625	.79321	.70645	.9437	.9621	.84051	.9771	1.1228	2.7980	.62299
	.90	.81000	.72900	.9487	.9655	.85435	.9792	1.1111	2.8274	.63617
29/32	.90625	.82129	.74429	.9520	.9677	.86272	.9805	1.1034	2.8471	.64504
59/64	.921875	.84985	.78346	.9601	.9733	.88513	.9839	1.0847	2.8962	.66747
15/16	.9375	.87891	.82398	.9683	.9787	.90773	.9872	1.0667	2.9452	.69029
61/64	.953125	.90845	.86587	.9763	.9841	.93053	.9905	1.0492	2.9943	.71349
31/32	.96875	.93848	.90915	.9843	.9895	.95349	.9937	1.0323	3.0434	.73708
63/64	.984375	.96899	.95385	.9922	.9948	.97666	.9969	1.0159	3.0925	.76104

Table 7 —*Continued*

N	N^2	N^3	$\sqrt{N}$	$\sqrt[3]{N}$	$N^{3/2}$	$\sqrt[5]{N}$	$\frac{1}{N}$	Circle ($N = D$) Circum.	Circle ($N = D$) Area
1.	1.0000	1.0000	1.0000	1.0000	1.0000	1.0000	1.0000000	3.1416	.7854
1.125	1.2656	1.4238	1.0606	1.0400	1.1932	1.0238	.8888888	3.5343	.9940
1.25	1.5625	1.9531	1.1180	1.0772	1.3975	1.0456	.80000000	3.9270	1.2272
1.375	1.8906	2.5996	1.1726	1.1120	1.6123	1.0658	.72727272	4.3197	1.4849
1.5	2.25	3.3750	1.2247	1.1447	1.8371	1.0845	.66666666	4.7124	1.7671
1.625	2.6406	4.2910	1.2748	1.1757	2.0715	1.1020	.61538462	5.1051	2.0739
1.75	3.0625	5.3594	1.3229	1.2051	2.3150	1.1186	.57142857	5.4978	2.4053
1.875	3.5156	6.5918	1.3693	1.2331	2.5675	1.1340	.53333333	5.8905	2.7612
2.	4.0000	8.0000	1.4142	1.2599	2.8284	1.1487	.50000000	6.2832	3.1416
2.125	4.5156	9.5957	1.4577	1.2856	3.0977	1.1627	.47058823	6.6759	3.5466
2.25	5.0625	11.3906	1.5000	1.3104	3.3750	1.1761	.44444444	7.0686	3.9761
2.375	5.6406	13.3965	1.5411	1.3342	3.6601	1.1889	.42105263	7.4613	4.4301
2.5	6.2500	15.6250	1.5811	1.3572	3.9529	1.2011	.40000000	7.8540	4.9087
2.625	6.8906	18.0879	1.6202	1.3795	4.2530	1.2129	.38095231	8.2467	5.4119
2.75	7.5625	20.7969	1.6583	1.4011	4.5604	1.2242	.36363636	8.6394	5.9396
2.875	8.2656	23.7637	1.6956	1.4219	4.8748	1.2352	.34782609	9.0321	6.4918
3.	9.000	27.0000	1.7321	1.4422	5.1962	1.2457	.33333333	9.4248	7.0686
3.125	9.7656	30.5176	1.7678	1.4620	5.5243	1.2559	.32000000	9.8175	7.6699
3.25	10.5625	34.3281	1.8028	1.4813	5.8590	1.2658	.30769231	10.2102	8.2958
3.375	11.3906	38.4434	1.8371	1.5000	6.2003	1.2754	.29629629	10.6029	8.9462
3.5	12.2500	42.8750	1.8708	1.5183	6.5479	1.2847	.28571429	10.9956	9.6211
3.625	13.1406	47.6348	1.9039	1.5362	6.9018	1.2938	.27586207	11.3883	10.3206
3.75	14.0625	52.7344	1.9365	1.5536	7.2619	1.3026	.26666666	11.7810	11.0447
3.875	15.0156	58.1856	1.9685	1.5707	7.6279	1.3112	.25806452	12.1737	11.7932
4.	16.0000	64.0000	2.0000	1.5874	8.0000	1.3195	.25000000	12.5664	12.5664
4.125	17.0156	70.1895	2.0310	1.6038	8.3779	1.3277	.24242424	12.9591	13.3640
4.25	18.0625	76.7656	2.0616	1.6198	8.7616	1.3356	.23529412	13.3518	14.1863
4.375	19.1406	83.7402	2.0916	1.6355	9.1510	1.3434	.22857143	13.7445	15.0330
4.5	20.2500	91.1250	2.1213	1.6510	9.5460	1.3510	.22222222	14.1372	15.9043
4.625	21.3906	98.9317	2.1506	1.6661	9.9465	1.3584	.21621622	14.5299	16.8001
4.75	22.5625	107.1719	2.1795	1.6810	10.3524	1.3656	.21052632	14.9226	17.7205
4.875	23.7656	115.8574	2.2079	1.6956	10.7637	1.3728	.20512821	15.3153	18.6655
5.	25.0000	125.0000	2.2361	1.7100	11.1803	1.3799	.20000000	15.7080	19.6350
5.125	26.2656	134.6113	2.2638	1.7241	11.6022	1.3866	.19512195	16.1006	20.6289
5.25	27.5625	144.7031	2.2913	1.7380	12.0293	1.3933	.19047619	16.4933	21.6475
5.375	28.8906	155.2871	2.3184	1.7517	12.4614	1.3998	.18604651	16.8860	22.6906
5.5	30.2500	166.3750	2.3452	1.7652	12.8987	1.4063	.18181818	17.2787	23.7583
5.625	31.6406	177.9785	2.3727	1.7784	13.3409	1.4126	.17777777	17.6714	24.8505
5.75	33.0625	190.1094	2.3979	1.7915	13.7880	1.4188	.17391304	18.0641	25.9672
5.875	34.5156	202.7793	2.4238	1.8044	14.2400	1.4250	.17021277	18.4568	27.1085
6.	36.0000	216.0000	2.4495	1.8171	14.6969	1.4310	.16666666	18.8495	28.2743
6.125	37.5156	229.7832	2.4749	1.8297	15.1586	1.4369	.16326531	19.2422	29.4647
6.25	39.0625	244.1406	2.5000	1.8420	15.6250	1.4427	.16000000	19.6349	30.6796
6.375	40.6406	259.0840	2.5249	1.8542	16.0961	1.4484	.15686275	20.0276	31.9190
6.5	42.2500	274.6250	2.5495	1.8663	16.5718	1.4542	.15384615	20.4203	33.1831
6.625	43.8906	290.7754	2.5739	1.8781	17.0522	1.4596	.15094339	20.8130	34.4716
6.75	45.5625	307.5469	2.5981	1.8899	17.5370	1.4651	.14814815	21.2057	35.7847
6.875	47.2656	324.9512	2.6220	1.9015	18.0264	1.4705	.14545454	21.5984	37.1223
7.	49.0000	343.0000	2.6458	1.9129	18.5203	1.4758	.14285714	21.9911	38.4845
7.125	50.7656	361.7051	2.6693	1.9243	19.0186	1.4810	.14035088	22.3838	39.8712
7.25	52.5625	381.0781	2.6926	1.9354	19.5212	1.4862	.13793103	22.7765	41.2825
7.375	54.3906	401.1309	2.7157	1.9465	20.0283	1.4913	.13559322	23.1692	42.7183
7.5	56.2500	421.8750	2.7386	1.9574	20.5396	1.4963	.13333333	23.5619	44.1786
7.625	58.1406	443.3223	2.7613	1.9683	21.0552	1.5012	.13114764	23.9546	45.6633
7.75	60.0625	465.4844	2.7839	1.9789	21.5751	1.5061	.12903226	24.3473	47.1730
7.875	62.0156	488.3731	2.8063	1.9895	22.0992	1.5110	.12698413	24.7400	48.7069
8.	64.0000	512.0000	2.8284	2.0000	22.6274	1.5157	.12500000	25.1327	50.2655
8.125	66.0156	536.3770	2.8504	2.0104	23.1598	1.5204	.12307692	25.5254	51.8485
8.25	68.0625	561.5156	2.8723	2.0206	23.6963	1.5251	.12121212	25.9181	53.4562
8.375	70.1406	587.4278	2.8940	2.0308	24.2369	1.5297	.11940298	26.3108	55.0883
8.5	72.2500	614.1250	2.9155	2.0408	24.7816	1.5342	.11764706	26.7035	56.7450
8.625	74.3906	641.6192	2.9368	2.0508	25.3301	1.5387	.11594203	27.0962	58.4262
8.75	76.5625	669.9219	2.9580	2.0606	25.8828	1.5431	.11428571	27.4889	60.1320
8.875	78.7656	699.0450	2.9791	2.0704	26.4394	1.5475	.11267605	27.8816	61.8623
9.	81.0000	729.0000	3.0000	2.0801	27.0000	1.5518	.11111111	28.2743	63.6172
9.125	83.2656	759.7989	3.0207	2.0897	27.5645	1.5561	.10958904	28.6670	65.3966
9.25	85.5625	791.4531	3.0414	2.0992	28.1328	1.5604	.10810811	29.0597	67.2006
9.375	87.8906	823.9746	3.0619	2.1086	28.7050	1.5646	.10666666	29.4524	69.0291
9.5	90.2500	857.3750	3.0822	2.1179	29.2810	1.5687	.10526316	29.8451	70.8822
9.625	92.6406	891.6660	3.1024	2.1272	29.8608	1.5728	.10389610	30.2378	72.7597
9.75	95.0625	926.8594	3.1225	2.1363	30.4444	1.5769	.10256410	30.6305	74.6619
9.875	97.5156	962.9668	3.1425	2.1454	31.0317	1.5809	.10126582	31.0232	76.5886

Table 7 —*Continued*

N	N^2	N^3	$\sqrt{N}$	$\sqrt[3]{N}$	$N^{3/2}$	$\sqrt[5]{N}$	$\dfrac{1}{N}$	Circle ($N = D$) Circum.	Area
10	100	1000	3.1623	2.1544	31.623	1.5849	.10000000	31.4159	78.5398
11	121	1331	3.3166	2.2240	36.483	1.6154	.09090909	34.5575	95.0332
12	144	1728	3.4641	2.2894	41.569	1.6438	.08333333	37.6991	113.0973
13	169	2197	3.6056	2.3513	46.873	1.6703	.07692308	40.8407	132.7323
14	196	2744	3.7417	2.4101	52.384	1.6953	.07142857	43.9823	153.9380
15	225	3375	3.8730	2.4662	58.095	1.7188	.06666667	47.1239	176.7146
16	256	4096	4.0000	2.5198	64.000	1.7411	.06250000	50.2654	201.0619
17	289	4913	4.1231	2.5713	70.093	1.7623	.05882353	53.4070	226.9801
18	324	5832	4.2426	2.6207	76.367	1.7826	.05555556	56.5486	254.4690
19	361	6859	4.3589	2.6684	82.819	1.8020	.05263158	59.6902	283.5287
20	400	8000	4.4721	2.7144	89.442	1.8206	.05000000	62.8318	314.1593
21	441	9261	4.5826	2.7589	96.235	1.8384	.04761905	65.9734	346.3606
22	484	10648	4.6904	2.8020	103.19	1.8556	.04545455	69.1150	380.1327
23	529	12167	4.7958	2.8439	110.30	1.8722	.04347826	72.2566	415.4756
24	576	13824	4.8990	2.8845	117.58	1.8882	.04166667	75.3982	452.3893
25	625	15625	5.0000	2.9240	125.00	1.9037	.04000000	78.5398	490.8739
26	676	17576	5.0990	2.9625	132.57	1.9186	.03846154	81.6813	530.9292
27	729	19683	5.1962	3.0000	140.30	1.9332	.03703704	84.8229	572.5553
28	784	21952	5.2915	3.0366	148.16	1.9473	.03571429	87.9645	615.7522
29	841	24389	5.3852	3.0723	156.17	1.9610	.03448276	91.1061	660.5198
30	900	27000	5.4772	3.1072	164.32	1.9744	.03333333	94.2477	706.8583
31	961	29791	5.5678	3.1414	172.60	1.9873	.03225806	97.3893	754.7676
32	1024	32768	5.6569	3.1748	181.02	2.0000	.03125000	100.5309	804.2477
33	1089	35937	5.7446	3.2075	189.57	2.0123	.03030303	103.6725	855.2986
34	1156	39304	5.8310	3.2396	198.25	2.0244	.02941176	106.8141	907.9203
35	1225	42875	5.9161	3.2711	207.06	2.0362	.02857143	109.9557	962.1127
36	1296	46656	6.0000	3.3019	216.00	2.0477	.02777778	113.0972	1017.8760
37	1369	50653	6.0828	3.3322	225.06	2.0589	.02702703	116.2388	1075.2101
38	1444	54872	6.1644	3.3620	234.25	2.0699	.02631579	119.3804	1134.1149
39	1521	59319	6.2450	3.3912	243.56	2.0807	.02564103	122.5220	1194.5906
40	1600	64000	6.3246	3.4200	252.98	2.0913	.02500000	125.6636	1256.6371
41	1681	68921	6.4031	3.4482	262.53	2.1016	.02439024	128.8052	1320.2543
42	1764	74088	6.4807	3.4760	272.19	2.1118	.02380952	131.9468	1385.4424
43	1849	79507	6.5574	3.5034	281.97	2.1218	.02325581	135.0884	1452.2012
44	1936	85184	6.6332	3.5303	291.86	2.1315	.02272727	138.2300	1520.5308
45	2025	91125	6.7082	3.5569	301.87	2.1411	.02222222	141.3716	1590.4313
46	2116	97336	6.7823	3.5830	311.99	2.1506	.02173913	144.5131	1661.9025
47	2209	103823	6.8557	3.6088	322.22	2.1598	.02127660	147.6547	1734.9445
48	2304	110592	6.9282	3.6342	332.55	2.1689	.02083333	150.7963	1809.5574
49	2401	117649	7.0000	3.6593	343.00	2.1779	.02040816	153.9379	1885.7410
50	2500	125000	7.0711	3.6840	353.55	2.1867	.02000000	157.0795	1963.500
51	2601	132651	7.1414	3.7084	364.21	2.1954	.01960784	160.2211	2042.820
52	2704	140608	7.2111	3.7325	374.98	2.2039	.01923077	163.3627	2123.716
53	2809	148877	7.2801	3.7563	385.85	2.2124	.01886792	166.5043	2206.183
54	2916	157464	7.3485	3.7798	396.82	2.2206	.01851852	169.6459	2290.221
55	3025	166375	7.4162	3.8030	407.89	2.2288	.01818182	172.7875	2375.829
56	3136	175616	7.4833	3.8259	419.07	2.2369	.01785714	175.9290	2463.008
57	3249	185193	7.5498	3.8485	430.35	2.2448	.01754386	179.0706	2551.758
58	3364	195112	7.6158	3.8709	441.72	2.2526	.01724138	182.2122	2642.079
59	3481	205379	7.6811	3.8930	453.19	2.2603	.01694915	185.3538	2733.970
60	3600	216000	7.7460	3.9149	464.76	2.2679	.01666667	188.4954	2827.433
61	3721	226981	7.8102	3.9365	476.43	2.2755	.01639344	191.6370	2922.466
62	3844	238328	7.8740	3.9579	488.19	2.2829	.01612903	194.7786	3019.070
63	3969	250047	7.9373	3.9791	500.05	2.2902	.01587302	197.9202	3117.245
64	4096	262144	8.0000	4.0000	512.00	2.2974	.01562500	201.0618	3216.990
65	4225	274625	8.0623	4.0207	524.05	2.3045	.01538462	204.2034	3318.307
66	4356	287496	8.1240	4.0412	536.19	2.3116	.01515152	207.3449	3421.194
67	4489	300763	8.1854	4.0615	548.42	2.3186	.01492537	210.4865	3525.652
68	4624	314432	8.2462	4.0817	560.74	2.3254	.01470588	213.6281	3631.680
69	4761	328509	8.3066	4.1016	573.16	2.3322	.01449275	216.7697	3739.280
70	4900	343000	8.3666	4.1213	585.66	2.3389	.01428571	219.9113	3848.450
71	5041	357911	8.4261	4.1408	598.26	2.3456	.01408451	223.0529	3959.191
72	5184	373248	8.4853	4.1602	610.94	2.3522	.01388889	226.1945	4071.503
73	5329	389017	8.5440	4.1793	623.71	2.3587	.01369863	229.3361	4185.386
74	5476	405224	8.6023	4.1983	636.57	2.3651	.01351351	232.4777	4300.839
75	5625	421875	8.6603	4.2172	649.52	2.3714	.01333333	235.6193	4417.864
76	5776	438976	8.7178	4.2358	662.55	2.3777	.01315789	238.7608	4536.459
77	5929	456533	8.7750	4.2543	675.68	2.3840	.01298701	241.9024	4656.625
78	6084	474552	8.8318	4.2727	688.88	2.3901	.01282051	245.0440	4778.361
79	6241	493039	8.8882	4.2908	702.17	2.3962	.01265823	248.1856	4901.669

Table 7—*Continued*

N	N^2	N^3	$\sqrt{N}$	$\sqrt[3]{N}$	$N^{3/2}$	$\sqrt[5]{N}$	$\dfrac{1}{N}$	Circle ($N = D$) Circum.	Area
80	6400	512000	8.9443	4.3089	715.54	2.4022	.01250000	251.327	5026.547
81	6561	531441	9.0000	4.3267	729.00	2.4082	.01234568	254.469	5152.998
82	6724	551368	9.0554	4.3445	742.54	2.4141	.01219512	257.610	5281.016
83	6889	571787	9.1104	4.3621	756.17	2.4200	.01204819	260.752	5410.607
84	7056	592704	9.1652	4.3795	769.88	2.4258	.01190476	263.894	5541.770
85	7225	614125	9.2195	4.3968	783.66	2.4315	.01176471	267.035	5674.501
86	7396	636056	9.2736	4.4140	797.53	2.4372	.01162791	270.177	5808.805
87	7569	658503	9.3274	4.4310	811.49	2.4429	.01149425	273.318	5944.679
88	7744	681472	9.3808	4.4480	825.52	2.4485	.01136364	276.460	6082.124
89	7921	704969	9.4340	4.4647	839.63	2.4540	.01123596	279.602	6221.138
90	8100	729000	9.4868	4.4814	853.82	2.4595	.01111111	282.743	6361.725
91	8281	753571	9.5394	4.4979	868.09	2.4650	.01098901	285.885	6503.882
92	8464	778688	9.5917	4.5144	882.44	2.4705	.01086957	289.026	6647.610
93	8649	804357	9.6437	4.5307	896.86	2.4758	.01075269	292.168	6792.909
94	8836	830584	9.6954	4.5468	911.36	2.4810	.01063830	295.309	6939.778
95	9025	857375	9.7468	4.5629	925.95	2.4863	.01052632	298.451	7088.219
96	9216	884736	9.7980	4.5789	940.61	2.4915	.01041667	301.593	7238.230
97	9409	912673	9.8489	4.5947	955.34	2.4966	.01030928	304.734	7389.812
98	9604	941192	9.8995	4.6104	970.15	2.5018	.01020408	307.876	7542.962
99	9801	970299	9.9499	4.6261	985.04	2.5069	.01010101	311.017	7697.688
100	10000	1000000	10.0000	4.6416	1000.0	2.5119	.01000000	314.159	7853.982
101	10201	1030301	10.0499	4.6570	1015.0	2.5169	.00990099	317.301	8011.85
102	10404	1061208	10.0995	4.6723	1030.1	2.5219	.00980392	320.442	8171.28
103	10609	1092727	10.1489	4.6875	1045.3	2.5268	.00970874	323.584	8332.29
104	10816	1124864	10.1980	4.7027	1060.6	2.5317	.00961538	326.725	8494.87
105	11025	1157625	10.2470	4.7177	1075.9	2.5365	.00952381	329.867	8659.01
106	11236	1191016	10.2956	4.7326	1091.3	2.5413	.00943396	333.009	8824.73
107	11449	1225043	10.3441	4.7475	1106.8	2.5461	.00934579	336.150	8992.02
108	11664	1259712	10.3923	4.7622	1122.4	2.5509	.00925926	339.292	9160.88
109	11881	1295029	10.4403	4.7769	1138.0	2.5556	.00917431	342.433	9331.32
110	12100	1331000	10.4881	4.7914	1153.7	2.5602	.00909091	345.575	9503.32
111	12321	1367631	10.5357	4.8059	1169.5	2.5649	.00900901	348.716	9676.89
112	12544	1404928	10.5830	4.8203	1185.3	2.5695	.00892857	351.858	9852.03
113	12769	1442897	10.6301	4.8346	1201.2	2.5740	.00884956	355.000	10028.75
114	12996	1481544	10.6771	4.8488	1217.2	2.5786	.00877193	358.141	10207.03
115	13225	1520875	10.7238	4.8629	1233.2	2.5831	.03869565	361.283	10386.89
116	13456	1560896	10.7703	4.8770	1249.4	2.5876	.00862069	364.424	10568.32
117	13689	1601613	10.8167	4.8910	1265.5	2.5920	.00854701	367.566	10751.31
118	13924	1643032	10.8628	4.9049	1281.8	2.5964	.00847458	370.708	10935.88
119	14161	1685159	10.9087	4.9187	1298.1	2.6008	.00840336	373.849	11122.02
120	14400	1728000	10.9545	4.9324	1314.5	2.6052	.00833333	376.991	11309.73
121	14641	1771561	11.0000	4.9461	1331.0	2.6095	.00826446	380.132	11499.01
122	14884	1815848	11.0454	4.9597	1347.5	2.6138	.00819672	383.274	11689.86
123	15129	1860867	11.0905	4.9732	1364.1	2.6181	.00813008	386.416	11882.29
124	15376	1906624	11.1355	4.9866	1380.8	2.6223	.00806452	389.557	12076.28
125	15625	1953125	11.1803	5.0000	1397.5	2.6265	.00800000	392.699	12271.84
126	15876	2000376	11.2250	5.0133	1414.4	2.6307	.00793651	395.840	12468.98
127	16129	2048383	11.2694	5.0265	1431.2	2.6349	.00787402	398.982	12667.68
128	16384	2097152	11.3137	5.0397	1448.2	2.6390	.00781250	402.124	12867.96
129	16641	2146689	11.3578	5.0528	1465.2	2.6431	.00775194	405.265	13069.81
130	16900	2197000	11.4018	5.0658	1482.2	2.6472	.00769231	408.407	13273.23
131	17161	2248091	11.4455	5.0788	1499.4	2.6513	.00763359	411.548	13478.22
132	17424	2299968	11.4891	5.0916	1516.6	2.6553	.00757576	414.690	13684.77
133	17689	2352637	11.5326	5.1045	1533.8	2.6593	.00751880	417.831	13892.91
134	17956	2406104	11.5758	5.1172	1551.2	2.6633	.00746269	420.973	14102.61
135	18225	2460375	11.6190	5.1299	1568.6	2.6673	.00740741	424.115	14313.88
136	18496	2515456	11.6619	5.1426	1586.0	2.6712	.00735294	427.256	14526.72
137	18769	2571353	11.7047	5.1551	1603.6	2.6751	.00729927	430.398	14741.14
138	19044	2628072	11.7473	5.1676	1621.1	2.6790	.00724638	433.539	14957.12
139	19321	2685619	11.7898	5.1801	1638.8	2.6829	.00719424	436.681	15174.67
140	19600	2744000	11.8322	5.1925	1656.5	2.6867	.00714286	439.823	15393.80
141	19881	2803221	11.8743	5.2048	1674.3	2.6906	.00709220	442.964	15614.50
142	20164	2863288	11.9164	5.2171	1692.1	2.6944	.00704225	446.106	15836.77
143	20449	2924207	11.9583	5.2293	1710.0	2.6981	.00699301	449.247	16060.60
144	20736	2985984	12.0000	5.2415	1728.0	2.7019	.00694444	452.389	16286.01
145	21025	3048625	12.0416	5.2536	1746.0	2.7057	.00689655	455.531	16512.99
146	21316	3112136	12.0830	5.2656	1764.1	2.7094	.00684932	458.672	16741.54
147	21609	3176523	12.1244	5.2776	1782.2	2.7131	.00680272	461.814	16971.67
148	21904	3241792	12.1655	5.2896	1800.5	2.7168	.00675676	464.955	17203.36
149	22201	3307949	12.2066	5.3015	1818.8	2.7204	.00671141	468.097	17436.62

Table 7 —*Continued*

N	N^2	N^3	$\sqrt{N}$	$\sqrt[3]{N}$	$N^{3/2}$	$\sqrt[5]{N}$	$\dfrac{1}{N}$	Circle ($N = D$) Circum.	Circle ($N = D$) Area
150	22500	3375000	12.2474	5.3133	1837.1	2.7241	.00666667	471.239	17671.46
151	22801	3442951	12.2882	5.3251	1855.5	2.7277	.00662252	474.380	17907.86
152	23104	3511808	12.3288	5.3368	1874.0	2.7314	.00657895	477.522	18145.84
153	23409	3581577	12.3693	5.3485	1892.5	2.7349	.00653595	480.663	18385.38
154	23716	3652264	12.4097	5.3601	1911.1	2.7385	.00649351	483.805	18626.50
155	24025	3723875	12.4499	5.3717	1929.7	2.7420	.00645161	486.946	18869.19
156	24336	3796416	12.4900	5.3832	1948.4	2.7455	.00641026	490.088	19113.45
157	24649	3869893	12.5300	5.3947	1967.2	2.7490	.00636943	493.230	19359.28
158	24964	3944312	12.5698	5.4061	1986.0	2.7525	.00632911	496.371	19606.68
159	25281	4019679	12.6095	5.4175	2004.9	2.7560	.00628931	499.513	19855.65
160	25600	4096000	12.6491	5.4288	2023.9	2.7595	.00625000	502.654	20106.19
161	25921	4173281	12.6886	5.4401	2042.9	2.7629	.00621118	505.796	20358.30
162	26244	4251528	12.7279	5.4514	2061.9	2.7663	.00617284	508.938	20611.99
163	26569	4330747	12.7671	5.4626	2081.0	2.7697	.00613497	512.079	20867.24
164	26896	4410944	12.8062	5.4737	2100.2	2.7731	.00609756	515.221	21124.06
165	27225	4492125	12.8452	5.4848	2119.5	2.7765	.00606061	518.362	21382.46
166	27556	4574296	12.8841	5.4959	2138.8	2.7799	.00602410	521.504	21642.43
167	27889	4657463	12.9228	5.5069	2158.1	2.7832	.00598802	524.646	21903.96
168	28224	4741632	12.9615	5.5178	2177.5	2.7865	.00595238	527.787	22167.07
169	28561	4826809	13.0000	5.5288	2197.0	2.7898	.00591716	530.929	22431.75
170	28900	4913000	13.0384	5.5397	2216.5	2.7931	.00588235	534.070	22698.00
171	29241	5000211	13.0767	5.5505	2236.1	2.7964	.00584795	537.212	22965.82
172	29584	5088448	13.1149	5.5613	2255.8	2.7997	.00581395	540.353	23235.21
173	29929	5177717	13.1529	5.5721	2275.5	2.8029	.00578035	543.495	23506.18
174	30276	5268024	13.1909	5.5828	2295.2	2.8061	.00574713	546.637	23778.71
175	30625	5359375	13.2288	5.5934	2315.0	2.8094	.00571429	549.778	24052.81
176	30976	5451776	13.2665	5.6041	2334.9	2.8126	.00568182	552.920	24328.49
177	31329	5545233	13.3041	5.6147	2354.8	2.8158	.00564972	556.061	24605.73
178	31684	5639752	13.3417	5.6252	2374.8	2.8189	.00561798	559.203	24884.55
179	32041	5735339	13.3791	5.6357	2394.9	2.8221	.00558659	562.345	25164.94
180	32400	5832000	13.4164	5.6462	2415.0	2.8252	.00555556	565.486	25446.90
181	32761	5929741	13.4536	5.6567	2435.1	2.8284	.00552486	568.628	25730.42
182	33124	6028568	13.4907	5.6671	2455.3	2.8315	.00549451	571.769	26015.52
183	33489	6128487	13.5277	5.6774	2475.6	2.8346	.00546448	574.911	26302.19
184	33856	6229504	13.5647	5.6877	2495.9	2.8377	.00543478	578.053	26590.43
185	34225	6331625	13.6015	5.6980	2516.3	2.8408	.00540541	581.194	26880.25
186	34596	6434856	13.6382	5.7083	2536.7	2.8438	.00537634	584.336	27171.63
187	34969	6539203	13.6748	5.7185	2557.2	2.8469	.00534759	587.477	27464.58
188	35344	6644672	13.7113	5.7287	2577.7	2.8499	.00531915	590.619	27759.11
189	35721	6751269	13.7477	5.7388	2598.3	2.8529	.00529101	593.761	28055.20
190	36100	6859000	13.7840	5.7489	2619.0	2.8560	.00526316	596.902	28352.87
191	36481	6967871	13.8203	5.7590	2639.7	2.8590	.00523560	600.044	28652.10
192	36864	7077888	13.8564	5.7690	2660.4	2.8619	.00520833	603.185	28952.91
193	37249	7189057	13.8924	5.7790	2681.2	2.8649	.00518135	606.327	29255.29
194	37636	7301384	13.9284	5.7890	2702.1	2.8679	.00515464	609.468	29559.24
195	38025	7414875	13.9642	5.7989	2723.0	2.8708	.00512821	612.610	29864.76
196	38416	7529536	14.0000	5.8088	2744.0	2.8738	.00510204	615.752	30171.85
197	38809	7645373	14.0357	5.8186	2765.0	2.8767	.00507614	618.893	30480.51
198	39204	7762392	14.0712	5.8285	2786.1	2.8796	.00505051	622.035	30790.74
199	39601	7880599	14.1067	5.8383	2807.2	2.8825	.00502513	625.176	31102.55
200	40000	8000000	14.1421	5.8480	2828.4	2.8854	.00500000	628.318	31415.93
201	40401	8120601	14.1774	5.8578	2849.7	2.8883	.00497512	631.460	31730.87
202	40804	8242408	14.2127	5.8675	2871.0	2.8911	.00495050	634.601	32047.39
203	41209	8365427	14.2478	5.8771	2892.3	2.8940	.00492611	637.743	32365.47
204	41616	8489664	14.2829	5.8868	2913.7	2.8968	.00490196	640.884	32685.13
205	42025	8615125	14.3178	5.8964	2935.2	2.8997	.00487805	644.026	33006.36
206	42436	8741816	14.3527	5.9059	2956.7	2.9025	.00485437	647.168	33329.16
207	42849	8869743	14.3875	5.9155	2978.2	2.9053	.00483092	650.309	33653.53
208	43264	8998912	14.4222	5.9250	2999.8	2.9081	.00480769	653.451	33979.47
209	43681	9129329	14.4568	5.9345	3021.5	2.9109	.00478469	656.592	34306.98
210	44100	9261000	14.4914	5.9439	3043.2	2.9137	.00476190	659.734	34636.06
211	44521	9393931	14.5258	5.9533	3065.0	2.9165	.00473934	662.875	34966.71
212	44944	9528128	14.5602	5.9627	3086.8	2.9192	.00471698	666.017	35298.94
213	45369	9663597	14.5945	5.9721	3108.7	2.9220	.00469484	669.159	35632.73
214	45796	9800344	14.6287	5.9814	3130.6	2.9247	.00467290	672.300	35968.09
215	46225	9938375	14.6629	5.9907	3152.5	2.9274	.00465116	675.442	36305.03
216	46656	10077696	14.6969	6.0000	3174.5	2.9302	.00462963	678.583	36643.54
217	47089	10218313	14.7309	6.0092	3196.6	2.9329	.00460829	681.725	36983.61
218	47524	10360232	14.7648	6.0185	3218.7	2.9356	.00458716	684.867	37325.26
219	47961	10503459	14.7986	6.0277	3240.9	2.9383	.00456621	688.008	37668.48

Table 7 —*Continued*

N	N^2	N^3	$\sqrt{N}$	$\sqrt[3]{N}$	$N^{3/2}$	$\sqrt[5]{N}$	$\dfrac{1}{N}$	Circle ($N = D$) Circum.	Circle ($N = D$) Area
220	48400	10648000	14.8324	6.0368	3263.1	2.9409	.00454545	691.150	38013.27
221	48841	10793861	14.8661	6.0459	3285.4	2.9436	.00452489	694.291	38359.63
222	49284	10941048	14.8997	6.0550	3307.7	2.9463	.00450450	697.433	38707.56
223	49729	11089567	14.9332	6.0641	3330.1	2.9489	.00448430	700.575	39057.07
224	50176	11239424	14.9666	6.0732	3352.5	2.9516	.00446429	703.716	39408.14
225	50625	11390625	15.0000	6.0822	3375.0	2.9542	.00444444	706.858	39760.78
226	51076	11543176	15.0333	6.0912	3397.5	2.9568	.00442478	709.999	40115.00
227	51529	11697083	15.0665	6.1002	3420.1	2.9594	.00440529	713.141	40470.78
228	51984	11852352	15.0997	6.1091	3442.7	2.9620	.00438596	716.283	40828.14
229	52441	12008989	15.1327	6.1180	3465.4	2.9646	.00436681	719.424	41187.07
230	52900	12167000	15.1658	6.1269	3488.1	2.9672	.00434783	722.566	41547.56
231	53361	12326391	15.1987	6.1358	3510.9	2.9698	.00432900	725.707	41909.63
232	53824	12487168	15.2315	6.1446	3533.7	2.9723	.00431034	728.849	42273.27
233	54289	12649337	15.2643	6.1534	3556.6	2.9749	.00429185	731.990	42638.48
234	54756	12812904	15.2971	6.1622	3579.5	2.9774	.00427350	735.132	43005.26
235	55225	12977875	15.3297	6.1710	3602.5	2.9800	.00425532	738.274	43373.61
236	55696	13144256	15.3623	6.1797	3625.5	2.9825	.00423729	741.415	43743.54
237	56169	13312053	15.3948	6.1885	3648.6	2.9850	.00421941	744.557	44115.03
238	56644	13481272	15.4272	6.1972	3671.7	2.9875	.00420168	747.698	44488.09
239	57121	13651919	15.4596	6.2058	3694.8	2.9900	.00418410	750.840	44862.73
240	57600	13824000	15.4919	6.2145	3718.0	2.9925	.00416667	753.982	45238.93
241	58081	13997521	15.5242	6.2231	3741.3	2.9950	.00414938	757.123	45616.71
242	58564	14172488	15.5563	6.2317	3764.6	2.9975	.00413223	760.265	45996.06
243	59049	14348907	15.5885	6.2403	3788.0	3.0000	.00411523	763.406	46376.98
244	59536	14526784	15.6205	6.2488	3811.4	3.0025	.00409836	766.548	46759.47
245	60025	14706125	15.6525	6.2573	3834.9	3.0049	.00408163	769.690	47143.52
246	60516	14886936	15.6844	6.2658	3858.4	3.0074	.00406504	772.831	47529.16
247	61009	15069223	15.7162	6.2743	3881.9	3.0098	.00404858	775.973	47916.36
248	61504	15252992	15.7480	6.2828	3905.5	3.0122	.00403226	779.114	48305.13
249	62001	15438249	15.7797	6.2912	3929.2	3.0147	.00401606	782.256	48695.47
250	62500	15625000	15.8114	6.2996	3952.9	3.0171	.00400000	785.398	49087.39
251	63001	15813251	15.8430	6.3080	3976.6	3.0195	.00398406	788.539	49480.87
252	63504	16003008	15.8745	6.3164	4000.4	3.0219	.00396825	791.681	49875.92
253	64009	16194277	15.9060	6.3247	4024.2	3.0243	.00395257	794.822	50272.55
254	64516	16387064	15.9374	6.3330	4048.1	3.0267	.00393701	797.964	50670.75
255	65025	16581375	15.9687	6.3413	4072.0	3.0291	.00392157	801.105	51070.52
256	65536	16777216	16.0000	6.3496	4096.0	3.0314	.00390625	804.247	51471.85
257	66049	16974593	16.0312	6.3579	4120.0	3.0338	.00389105	807.389	51874.76
258	66564	17173512	16.0624	6.3661	4144.1	3.0362	.00387597	810.530	52279.24
259	67081	17373979	16.0935	6.3743	4168.2	3.0385	.00386100	813.672	52685.29
260	67600	17576000	16.1245	6.3825	4192.4	3.0418	.00384615	816.813	53092.92
261	68121	17779581	16.1555	6.3907	4216.6	3.0432	.00383142	819.955	53502.11
262	68644	17984728	16.1864	6.3988	4240.8	3.0455	.00381679	823.097	53912.87
263	69169	18191447	16.2173	6.4070	4265.1	3.0478	.00380228	826.238	54325.21
264	69696	18399744	16.2481	6.4151	4289.5	3.0501	.00378788	829.380	54739.11
265	70225	18609625	16.2788	6.4232	4313.9	3.0524	.00377358	832.521	55154.59
266	70756	18821096	16.3095	6.4312	4338.3	3.0547	.00375940	835.663	55571.63
267	71289	19034163	16.3401	6.4393	4362.8	3.0570	.00374532	838.805	55990.25
268	71824	19248832	16.3707	6.4473	4387.3	3.0593	.00373134	841.946	56410.44
269	72361	19465109	16.4012	6.4553	4411.9	3.0616	.00371747	845.088	56832.20
270	72900	19683000	16.4317	6.4633	4436.5	3.0639	.00370370	848.229	57255.53
271	73441	19902511	16.4621	6.4713	4461.2	3.0662	.00369004	851.371	57680.43
272	73984	20123648	16.4924	6.4792	4485.9	3.0684	.00367647	854.512	58106.90
273	74529	20346417	16.5227	6.4872	4510.7	3.0707	.00366300	857.654	58534.94
274	75076	20570824	16.5529	6.4951	4535.5	3.0729	.00364964	860.796	58964.55
275	75625	20796875	16.5831	6.5030	4560.4	3.0752	.00363636	863.937	59395.74
276	76176	21024576	16.6132	6.5108	4585.3	3.0774	.00362319	867.079	59828.49
277	76729	21253933	16.6433	6.5187	4610.2	3.0796	.00361011	870.220	60262.82
278	77284	21484952	16.6733	6.5265	4635.2	3.0818	.00359712	873.362	60698.71
279	77841	21717639	16.7033	6.5343	4660.2	3.0840	.00358423	876.504	61136.18
280	78400	21952000	16.7332	6.5421	4685.3	3.0863	.00357143	879.645	61575.22
281	78961	22188041	16.7631	6.5499	4710.4	3.0885	.00355872	882.787	62015.82
282	79524	22425768	16.7929	6.5577	4735.6	3.0907	.00354610	885.928	62458.00
283	80089	22665187	16.8226	6.5654	4760.8	3.0928	.00353357	889.070	62901.75
284	80656	22906304	16.8523	6.5731	4786.0	3.0950	.00352113	892.212	63347.07
285	81225	23149125	16.8819	6.5808	4811.3	3.0972	.00350877	895.353	63793.97
286	81796	23393656	16.9115	6.5885	4836.7	3.0994	.00349650	898.495	64242.43
287	82369	23639903	16.9411	6.5962	4862.1	3.1015	.00348432	901.636	64692.46
288	82944	23887872	16.9706	6.6039	4887.5	3.1037	.00347222	904.778	65144.07
289	83521	24137569	17.0000	6.6115	4913.0	3.1058	.00346021	907.920	65597.24

Table 7 —*Continued*

N	N^2	N^3	$\sqrt{N}$	$\sqrt[3]{N}$	$N^{3/2}$	$\sqrt[5]{N}$	$\frac{1}{N}$	Circle ($N = D$) Circum.	Area
290	**84100**	**24389000**	**17.0294**	**6.6191**	**4938.5**	**3.1080**	**.00344828**	**911.061**	**66051.99**
291	84681	24642171	17.0587	6.6267	4964.1	3.1101	.00343643	914.203	66508.30
292	85264	24897088	17.0880	6.6343	4989.7	3.1123	.00342466	917.344	66966.19
293	85849	25153757	17.1172	6.6419	5015.4	3.1144	.00341297	920.486	67425.65
294	86436	25412184	17.1464	6.6494	5041.1	3.1165	.00340136	923.627	67886.68
295	87025	25672375	17.1756	6.6569	5066.8	3.1186	.00338983	926.769	68349.28
296	87616	25934336	17.2047	6.6644	5092.6	3.1207	.00337838	929.911	68813.45
297	88209	26198073	17.2337	6.6719	5118.4	3.1228	.00336700	933.052	69279.19
298	88804	26463592	17.2627	6.6794	5144.3	3.1249	.00335570	936.194	69746.50
299	89401	26730899	17.2916	6.6869	5170.2	3.1270	.00334448	939.335	70215.38
300	**90000**	**27000000**	**17.3205**	**6.6943**	**5196.2**	**3.1291**	**.00333333**	**942.477**	**70685.83**
301	90601	27270901	17.3494	6.7018	5222.2	3.1312	.00332226	945.619	71157.86
302	91204	27543608	17.3781	6.7092	5248.2	3.1333	.00331126	948.760	71631.45
303	91809	27818127	17.4069	6.7166	5274.3	3.1354	.00330033	951.902	72106.62
304	92416	28094464	17.4356	6.7240	5300.4	3.1374	.00328947	955.043	72583.36
305	93025	28372625	17.4642	6.7313	5326.6	3.1395	.00327869	958.185	73061.66
306	93636	28652616	17.4929	6.7387	5352.8	3.1416	.00326797	961.327	73541.54
307	94249	28934443	17.5214	6.7460	5379.1	3.1436	.00325733	964.468	74022.99
308	94864	29218112	17.5499	6.7533	5405.4	3.1456	.00324675	967.610	74506.01
309	95481	29503629	17.5784	6.7606	5431.7	3.1477	.00323625	970.751	74990.60
310	**96100**	**29791000**	**17.6068**	**6.7679**	**5458.1**	**3.1497**	**.00322581**	**973.893**	**75476.76**
311	96721	30080231	17.6352	6.7752	5484.5	3.1518	.00321543	977.034	75964.50
312	97344	30371328	17.6635	6.7824	5511.0	3.1538	.00320513	980.176	76453.80
313	97969	30664297	17.6918	6.7897	5537.5	3.1558	.00319489	983.318	76944.67
314	98596	30959144	17.7200	6.7969	5564.1	3.1578	.00318471	986.459	77437.12
315	99225	31255875	17.7482	6.8041	5590.7	3.1598	.00317460	989.601	77931.13
316	99856	31554496	17.7764	6.8113	5617.3	3.1618	.00316456	992.742	78426.72
317	100489	31855013	17.8045	6.8185	5644.0	3.1638	.00315457	995.884	78923.88
318	101124	32157432	17.8326	6.8256	5670.7	3.1658	.00314465	999.026	79422.60
319	101761	32461759	17.8606	6.8328	5697.5	3.1678	.00313480	1002.167	79922.90
320	**102400**	**32768000**	**17.8885**	**6.8399**	**5724.3**	**3.1698**	**.00312500**	**1005.309**	**80424.77**
321	103041	33076161	17.9165	6.8470	5751.2	3.1718	.00311526	1008.450	80928.21
322	103684	33386248	17.9444	6.8541	5778.1	3.1737	.00310559	1011.592	81433.22
323	104329	33698267	17.9722	6.8612	5805.0	3.1757	.00309598	1014.734	81939.80
324	104976	34012224	18.0000	6.8683	5832.0	3.1777	.00308642	1017.875	82447.96
325	105625	34328125	18.0278	6.8753	5859.0	3.1796	.00307692	1021.017	82957.68
326	106276	34645976	18.0555	6.8824	5886.1	3.1816	.00306748	1024.158	83468.97
327	106929	34965783	18.0831	6.8894	5913.2	3.1835	.00305810	1027.300	83981.84
328	107584	35287552	18.1108	6.8964	5940.3	3.1855	.00304878	1030.442	84496.28
329	108241	35611289	18.1384	6.9034	5967.5	3.1874	.00303951	1033.583	85012.28
330	**108900**	**35937000**	**18.1659**	**6.9104**	**5994.7**	**3.1894**	**.00303030**	**1036.725**	**85529.86**
331	109561	36264691	18.1934	6.9174	6022.0	3.1913	.00302115	1039.866	86049.01
332	110224	36594368	18.2209	6.9244	6049.3	3.1932	.00301205	1043.008	86569.73
333	110889	36926037	18.2483	6.9313	6076.7	3.1951	.00300300	1046.149	87092.02
334	111556	37259704	18.2757	6.9382	6104.1	3.1970	.00299401	1049.291	87615.88
335	112225	37595375	18.3030	6.9451	6131.5	3.1989	.00298507	1052.433	88141.31
336	112896	37933056	18.3303	6.9521	6159.0	3.2009	.00297619	1055.574	88668.31
337	113569	38272753	18.3576	6.9589	6186.5	3.2028	.00296736	1058.716	89196.88
338	114244	38614472	18.3848	6.9658	6214.1	3.2047	.00295858	1061.857	89727.03
339	114921	38958219	18.4120	6.9727	6241.7	3.2066	.00294985	1064.999	90258.74
340	**115600**	**39304000**	**18.4391**	**6.9795**	**6269.3**	**3.2085**	**.00294118**	**1068.141**	**90792.03**
341	116281	39651821	18.4662	6.9864	6297.0	3.2103	.00293255	1071.282	91326.88
342	116964	40001688	18.4932	6.9932	6324.7	3.2122	.00292398	1074.424	91863.31
343	117649	40353607	18.5203	7.0000	6352.4	3.2141	.00291545	1077.565	92401.31
344	118336	40707584	18.5472	7.0068	6380.2	3.2160	.00290698	1080.707	92940.88
345	119025	41063625	18.5742	7.0136	6408.1	3.2178	.00289855	1083.849	93482.02
346	119716	41421736	18.6011	7.0203	6436.0	3.2197	.00289017	1086.990	94024.73
347	120409	41781923	18.6279	7.0271	6463.9	3.2216	.00288184	1090.132	94569.01
348	121104	42144192	18.6548	7.0338	6491.9	3.2234	.00287356	1093.273	95114.86
349	121801	42508549	18.6815	7.0406	6519.9	3.2253	.00286533	1096.415	95662.28
350	**122500**	**42875000**	**18.7083**	**7.0473**	**6547.9**	**3.2271**	**.00285714**	**1099.557**	**96211.28**
351	123201	43243551	18.7350	7.0540	6576.0	3.2289	.00284900	1102.698	96761.84
352	123904	43614208	18.7617	7.0607	6604.1	3.2308	.00284091	1105.840	97313.97
353	124609	43986977	18.7883	7.0674	6632.3	3.2326	.00283286	1108.981	97867.68
354	125316	44361864	18.8149	7.0740	6660.5	3.2345	.00282486	1112.123	98422.96
355	126025	44738875	18.8414	7.0807	6688.7	3.2363	.00281690	1115.264	98979.80
356	126736	45118016	18.8680	7.0873	6717.0	3.2381	.00280899	1118.406	99538.22
357	127449	45499293	18.8944	7.0940	6745.3	3.2399	.00280112	1121.548	100098.21
358	128164	45882712	18.9209	7.1006	6773.7	3.2417	.00279330	1124.689	100659.77
359	128881	46268279	18.9473	7.1072	6802.1	3.2435	.00278552	1127.831	101222.90

Table 7 —*Continued*

N	N^2	N^3	$\sqrt{N}$	$\sqrt[3]{N}$	$N^{3/2}$	$\sqrt[5]{N}$	$\dfrac{1}{N}$	Circle ($N = D$) Circum.	Area
360	129600	46656000	18.9737	7.1138	6830.5	3.2453	.00277778	1130.972	101787.60
361	130321	47045881	19.0000	7.1204	6859.0	3.2471	.00277008	1134.114	102353.87
362	131044	47437928	19.0263	7.1269	6887.5	3.2489	.00276243	1137.256	102921.72
363	131769	47832147	19.0526	7.1335	6916.1	3.2507	.00275482	1140.397	103491.13
364	132496	48228544	19.0788	7.1400	6944.7	3.2525	.00274725	1143.539	104062.12
365	133225	48627125	19.1050	7.1466	6973.3	3.2543	.00273973	1146.680	104634.67
366	133956	49027896	19.1311	7.1531	7002.0	3.2561	.00273224	1149.822	105208.80
367	134689	49430863	19.1572	7.1596	7030.7	3.2579	.00272480	1152.964	105784.49
368	135424	49836032	19.1833	7.1661	7059.5	3.2597	.00271739	1156.105	106361.76
369	136161	50243409	19.2094	7.1726	7088.3	3.2614	.00271003	1159.247	106940.60
370	136900	50653000	19.2354	7.1791	7117.1	3.2632	.00270270	1162.388	107521.01
371	137641	51064811	19.2614	7.1855	7146.0	3.2650	.00269542	1165.530	108102.99
372	138384	51478848	19.2873	7.1920	7174.9	3.2668	.00268817	1168.671	108686.54
373	139129	51895117	19.3132	7.1984	7203.9	3.2685	.00268097	1171.813	109271.66
374	139876	52313624	19.3391	7.2048	7232.8	3.2702	.00267380	1174.955	109858.35
375	140625	52734375	19.3649	7.2112	7261.8	3.2719	.00266667	1178.096	110446.62
376	141376	53157376	19.3907	7.2177	7290.9	3.2737	.00265957	1181.238	111036.45
377	142129	53582633	19.4165	7.2240	7320.0	3.2754	.00265252	1184.379	111627.86
378	142884	54010152	19.4422	7.2304	7349.2	3.2772	.00264550	1187.521	112220.83
379	143641	54439939	19.4679	7.2368	7378.4	3.2789	.00263852	1190.663	112815.38
380	144400	54872000	19.4936	7.2432	7407.6	3.2807	.00263158	1193.804	113411.49
381	145161	55306341	19.5192	7.2495	7436.8	3.2824	.00262467	1196.946	114009.18
382	145924	55742968	19.5448	7.2558	7466.1	3.2841	.00261780	1200.087	114608.44
383	146689	56181887	19.5704	7.2622	7495.4	3.2858	.00261097	1203.229	115209.27
384	147456	56623104	19.5959	7.2685	7524.8	3.2875	.00260417	1206.371	115811.67
385	148225	57066625	19.6214	7.2748	7554.2	3.2892	.00259740	1209.512	116415.64
386	148996	57512456	19.6469	7.2811	7583.7	3.2909	.00259067	1212.654	117021.18
387	149769	57960603	19.6723	7.2874	7613.2	3.2926	.00258398	1215.795	117628.30
388	150544	58411072	19.6977	7.2936	7642.7	3.2943	.00257732	1218.937	118236.98
389	151321	58863869	19.7231	7.2999	7672.3	3.2960	.00257069	1222.079	118847.24
390	152100	59319000	19.7484	7.3061	7701.9	3.2977	.00256410	1225.220	119459.06
391	152881	59776471	19.7737	7.3124	7731.5	3.2994	.00255754	1228.362	120072.46
392	153664	60236288	19.7990	7.3186	7761.2	3.3011	.00255102	1231.503	120687.42
393	154449	60698457	19.8242	7.3248	7790.9	3.3028	.00254453	1234.645	121303.96
394	155236	61162984	19.8494	7.3310	7820.7	3.3045	.00253807	1237.786	121922.07
395	156025	61629875	19.8746	7.3372	7850.5	3.3061	.00253165	1240.928	122541.75
396	156816	62099136	19.8997	7.3434	7880.3	3.3078	.00252525	1244.070	123163.00
397	157609	62570773	19.9249	7.3496	7910.2	3.3095	.00251889	1247.211	123785.82
398	158404	63044792	19.9499	7.3558	7940.1	3.3111	.00251256	1250.353	124410.21
399	159201	63521199	19.9750	7.3619	7970.0	3.3128	.00250627	1253.494	125036.17
400	160000	64000000	20.0000	7.3681	8000.0	3.3145	.00250000	1256.636	125663.71
401	160801	64481201	20.0250	7.3742	8030.0	3.3161	.00249377	1259.778	126292.81
402	161604	64964808	20.0499	7.3803	8061.1	3.3178	.00248756	1262.919	126923.48
403	162409	65450827	20.0749	7.3864	8090.2	3.3194	.00248139	1266.061	127555.73
404	163216	65939264	20.0998	7.3925	8120.3	3.3211	.00247525	1269.202	128189.55
405	164025	66430125	20.1246	7.3986	8150.5	3.3227	.00246914	1272.344	128824.93
406	164836	66923416	20.1494	7.4047	8180.7	3.3243	.00246305	1275.486	129461.89
407	165649	67419143	20.1742	7.4108	8210.9	3.3260	.00245700	1278.627	130100.42
408	166464	67917312	20.1990	7.4169	8241.2	3.3276	.00245098	1281.769	130740.52
409	167281	68417929	20.2237	7.4229	8271.5	3.3292	.00244499	1284.910	131382.19
410	168100	68921000	20.2485	7.4290	8301.9	3.3308	.00243902	1288.052	132025.43
411	168921	69426531	20.2731	7.4350	8332.3	3.3325	.00243309	1291.193	132670.24
412	169744	69934528	20.2978	7.4410	8362.7	3.3341	.00242718	1294.335	133316.63
413	170569	70444997	20.3224	7.4470	8393.2	3.3357	.00242131	1297.477	133964.58
414	171396	70957944	20.3470	7.4530	8423.7	3.3373	.00241546	1300.618	134614.10
415	172225	71473375	20.3715	7.4590	8454.2	3.3390	.00240964	1303.760	135265.20
416	173056	71991296	20.3961	7.4650	8484.8	3.3406	.00240385	1306.901	135917.86
417	173889	72511713	20.4206	7.4710	8515.4	3.3422	.00239808	1310.043	136572.10
418	174724	73034632	20.4450	7.4770	8546.0	3.3438	.00239234	1313.185	137227.91
419	175561	73560059	20.4695	7.4829	8576.7	3.3454	.00238663	1316.326	137885.29
420	176400	74088000	20.4939	7.4889	8607.4	3.3470	.00238095	1319.468	138544.24
421	177241	74618461	20.5183	7.4948	8638.2	3.3485	.00237530	1322.609	139204.76
422	178084	75151448	20.5426	7.5007	8669.0	3.3501	.00236967	1325.751	139866.85
423	178929	75686967	20.5670	7.5067	8699.8	3.3517	.00236407	1328.893	140530.51
424	179776	76225024	20.5913	7.5126	8730.7	3.3533	.00235849	1332.034	141195.74
425	180625	76765625	20.6155	7.5185	8761.6	3.3559	.00235294	1335.176	141862.54
426	181476	77308776	20.6398	7.5244	8792.5	3.3564	.00234742	1338.317	142530.92
427	182329	77854483	20.6640	7.5302	8823.5	3.3580	.00234192	1341.459	143200.86
428	183184	78402752	20.6882	7.5361	8854.6	3.3596	.00233645	1344.601	143872.38
429	184041	78953589	20.7123	7.5420	8885.6	3.3612	.00233100	1347.742	144545.46

Table 7 —Continued

N	N^2	N^3	$\sqrt{N}$	$\sqrt[3]{N}$	$N^{3/2}$	$\sqrt[5]{N}$	$\dfrac{1}{N}$	Circle ($N = D$) Circum.	Area
430	184900	79507000	20.7364	7.5478	8916.7	3.3627	.00232558	1350.884	145220.12
431	185761	80062991	20.7605	7.5537	8947.8	3.3643	.00232019	1354.025	145896.35
432	186624	80621568	20.7846	7.5595	8979.0	3.3659	.00231481	1357.167	146574.15
433	187489	81182737	20.8087	7.5654	9010.1	3.3674	.00230947	1360.308	147253.52
434	188356	81746504	20.8327	7.5712	9041.4	3.3690	.00230415	1363.450	147934.46
435	189225	82312875	20.8567	7.5770	9072.7	3.3705	.00229885	1366.592	148616.97
436	190096	82881856	20.8806	7.5828	9104.0	3.3720	.00229358	1369.733	149301.05
437	190969	83453453	20.9045	7.5886	9135.3	3.3736	.00228833	1372.875	149986.70
438	191844	84027672	20.9284	7.5944	9166.7	3.3752	.00228311	1376.016	150673.92
439	192721	84604519	20.9523	7.6001	9198.1	3.3767	.00227790	1379.158	151362.72
440	193600	85184000	20.9762	7.6059	9229.5	3.3783	.00227273	1382.300	152053.08
441	194481	85766121	21.0000	7.6117	9261.0	3.3798	.00226757	1385.441	152745.02
442	195364	86350888	21.0238	7.6174	9292.5	3.3813	.00226244	1388.583	153438.53
443	196249	86938307	21.0476	7.6232	9324.1	3.3828	.00225734	1391.724	154133.60
444	197136	87528384	21.0713	7.6289	9355.7	3.3844	.00225225	1394.866	154830.25
445	198025	88121125	21.0950	7.6346	9387.3	3.3859	.00224719	1398.008	155528.47
446	198916	88716536	21.1187	7.6403	9419.0	3.3874	.00224215	1401.149	156228.26
447	199809	89314623	21.1424	7.6460	9450.7	3.3889	.00223714	1404.291	156929.62
448	200704	89915392	21.1660	7.6517	9482.4	3.3904	.00223214	1407.432	157632.55
449	201601	90518849	21.1896	7.6574	9514.2	3.3919	.00222717	1410.574	158337.05
450	202500	91125000	21.2132	7.6631	9546.0	3.3935	.00222222	1413.716	159043.13
451	203401	91733851	21.2368	7.6688	9577.8	3.3950	.00221729	1416.857	159750.77
452	204304	92345408	21.2603	7.6744	9609.6	3.3965	.00221239	1419.999	160459.99
453	205209	92959677	21.2838	7.6801	9641.5	3.3980	.00220751	1423.140	161170.77
454	206116	93576664	21.3073	7.6857	9673.5	3.3995	.00220264	1426.282	161883.13
455	207025	94196375	21.3307	7.6914	9705.5	3.4010	.00219780	1429.423	162597.05
456	207936	94818816	21.3542	7.6970	9737.5	3.4025	.00219298	1432.565	163312.55
457	208849	95443993	21.3776	7.7026	9769.5	3.4039	.00218818	1435.707	164029.62
458	209764	96071912	21.4009	7.7082	9801.6	3.4054	.00218341	1438.848	164748.26
459	210681	96702579	21.4243	7.7138	9833.8	3.4069	.00217865	1441.990	165468.47
460	211600	97336000	21.4476	7.7194	9865.9	3.4084	.00217391	1445.131	166190.25
461	212521	97972181	21.4709	7.7250	9898.1	3.4199	.00216920	1448.273	166913.60
462	213444	98611128	21.4942	7.7306	9930.3	3.4113	.00216450	1451.415	167638.52
463	214369	99252847	21.5174	7.7362	9962.6	3.4128	.00215983	1454.556	168365.02
464	215296	99897344	21.5407	7.7418	9994.8	3.4143	.00215517	1457.698	169093.08
465	216225	100544625	21.5639	7.7473	10027.	3.4158	.00215054	1460.839	169822.72
466	217156	101194696	21.5870	7.7529	10060.	3.4173	.00214592	1463.981	170553.92
467	218089	101847563	21.6102	7.7584	10092.	3.4187	.00214133	1467.123	171286.70
468	219024	102503232	21.6333	7.7639	10124.	3.4202	.00213675	1470.264	172021.05
469	219961	103161709	21.6564	7.7695	10157.	3.4217	.00213220	1473.406	172756.96
470	220900	103823000	21.6795	7.7750	10189.	3.4231	.00212766	1476.547	173494.45
471	221841	104487111	21.7025	7.7805	10222.	3.4246	.00212314	1479.689	174233.51
472	222784	105154048	21.7256	7.7860	10255.	3.4260	.00211864	1482.830	174974.14
473	223729	105823817	21.7486	7.7915	10287.	3.4275	.00211416	1485.972	175716.34
474	224676	106496424	21.7715	7.7970	10320.	3.4289	.00210970	1489.114	176460.12
475	225625	107171875	21.7945	7.8025	10352.	3.4304	.00210526	1492.255	177205.46
476	226576	107850176	21.8174	7.8079	10385.	3.4318	.00210084	1495.397	177952.37
477	227529	108531333	21.8403	7.8134	10418.	3.4332	.00209644	1498.538	178700.86
478	228484	109215352	21.8632	7.8188	10450.	3.4347	.00209205	1501.680	179450.91
479	229441	109902239	21.8861	7.8243	10483.	3.4361	.00208768	1504.822	180202.54
480	230400	110592000	21.9089	7.8297	10516.	3.4375	.00208333	1507.963	180955.74
481	231361	111284641	21.9317	7.8352	10549.	3.4390	.00207900	1511.105	181710.50
482	232324	111980168	21.9545	7.8406	10582.	3.4404	.00207469	1514.246	182466.84
483	233289	112678587	21.9773	7.8460	10615.	3.4418	.00207039	1517.388	183224.75
484	234256	113379904	22.0000	7.8514	10648.	3.4433	.00206612	1520.530	183984.23
485	235225	114084125	22.0227	7.8568	10681.	3.4447	.00206186	1523.671	184745.28
486	236196	114791256	22.0454	7.8622	10714.	3.4461	.00205761	1526.813	185507.90
487	237169	115501303	22.0681	7.8676	10747.	3.4475	.00205339	1529.954	186272.10
488	238144	116214272	22.0907	7.8730	10780.	3.4489	.00204918	1533.096	187037.86
489	239121	116930169	22.1133	7.8784	10813.	3.4504	.00204499	1536.238	187805.19
490	240100	117649000	22.1359	7.8837	10847.	3.4518	.00204082	1539.379	188574.10
491	241081	118370771	22.1585	7.8891	10880.	3.4532	.00203666	1542.521	189344.57
492	242064	119095488	22.1811	7.8944	10913.	3.4546	.00203252	1545.662	190116.62
493	243049	119823157	22.2036	7.8998	10946.	3.4560	.00202840	1548.804	190890.24
494	244036	120553784	22.2261	7.9051	10980.	3.4574	.00202429	1551.945	191665.43
495	245025	121287375	22.2486	7.9105	11013.	3.4588	.00202020	1555.087	192442.18
496	246016	122023936	22.2711	7.9158	11046.	3.4602	.00201613	1558.229	193220.51
497	247009	122763473	22.2935	7.9211	11080.	3.4616	.00201207	1561.370	194000.41
498	248004	123505992	22.3159	7.9264	11113.	3.4630	.00200803	1564.512	194781.89
499	249001	124251499	22.3383	7.9317	11147.	3.4643	.00200401	1567.653	195564.93

Table 7 —Continued

N	N^2	N^3	$\sqrt{N}$	$\sqrt[3]{N}$	$N^{3/2}$	$\sqrt[5]{N}$	$\frac{1}{N}$	Circum.	Area
500	250000	125000000	22.3607	7.9370	11180	3.4657	.00200000	1570.795	196349.54
501	251001	125751501	22.3830	7.9423	11214	3.4671	.00199601	1573.937	197135.72
502	252004	126506008	22.4054	7.9476	11247	3.4685	.00199203	1577.078	197923.48
503	253009	127263527	22.4277	7.9528	11281	3.4699	.00198807	1580.220	198712.80
504	254016	128024064	22.4499	7.9581	11315	3.4713	.00198413	1583.361	199503.70
505	255025	128787625	22.4722	7.9634	11348	3.4726	.00198020	1586.503	200296.17
506	256036	129554216	22.4944	7.9686	11382	3.4740	.00197628	1589.645	201090.20
507	257049	130323843	22.5167	7.9739	11416	3.4754	.00197239	1592.786	201885.81
508	258064	131096512	22.5389	7.9791	11450	3.4768	.00196850	1595.928	202682.99
509	259081	131872229	22.5610	7.9843	11484	3.4781	.00196464	1599.069	203481.74
510	260100	132651000	22.5832	7.9896	11517	3.4795	.00196078	1602.211	204282.06
511	261121	133432831	22.6053	7.9948	11551	3.4808	.00195695	1605.352	205083.95
512	262144	134217728	22.6274	8.0000	11585	3.4822	.00195313	1608.494	205887.42
513	263169	135005697	22.6495	8.0052	11619	3.4836	.00194932	1611.636	206692.45
514	264196	135796744	22.6716	8.0104	11653	3.4849	.00194553	1614.777	207499.05
515	265225	136590875	22.6936	8.0156	11687	3.4863	.00194175	1617.919	208307.23
516	266256	137388096	22.7156	8.0208	11721	3.4876	.00193798	1621.060	209116.97
517	267289	138188413	22.7376	8.0260	11755	3.4890	.00193424	1624.202	209928.29
518	268324	138991832	22.7596	8.0311	11789	3.4904	.00193050	1627.344	210741.18
519	269361	139798359	22.7816	8.0363	11824	3.4917	.00192678	1630.485	211555.63
520	270400	140608000	22.8035	8.0415	11858	3.4930	.00192308	1633.627	212371.66
521	271441	141420761	22.8254	8.0466	11892	3.4944	.00191939	1636.768	213189.26
522	272484	142236648	22.8473	8.0517	11926	3.4957	.00191571	1639.910	214008.43
523	273529	143055667	22.8692	8.0569	11960	3.4970	.00191205	1643.052	214829.17
524	274576	143877824	22.8910	8.0620	11995	3.4984	.00190840	1646.193	215651.49
525	275625	144703125	22.9129	8.0671	12029	3.4997	.00190476	1649.335	216475.37
526	276676	145531576	22.9347	8.0723	12064	3.5010	.00190114	1652.476	217300.82
527	277729	146363183	22.9565	8.0774	12098	3.5024	.00189753	1655.618	218127.85
528	278784	147197952	22.9783	8.0825	12133	3.5037	.00189394	1658.760	218956.44
529	279841	148035889	23.0000	8.0876	12167	3.5050	.00189036	1661.901	219786.61
530	280900	148877000	23.0217	8.0927	12202	3.5064	.00188679	1665.043	220618.34
531	281961	149721291	23.0434	8.0978	12236	3.5077	.00188324	1668.184	221451.65
532	283024	150568768	23.0651	8.1028	12271	3.5090	.00187970	1671.326	222286.53
533	284089	151419437	23.0868	8.1079	12305	3.5103	.00187617	1674.467	223122.98
534	285156	152273304	23.1084	8.1130	12340	3.5116	.00187266	1677.609	223961.00
535	286225	153130375	23.1301	8.1180	12375	3.5130	.00186916	1680.751	224800.59
536	287296	153990656	23.1517	8.1231	12410	3.5143	.00186567	1683.892	225641.75
537	288369	154854153	23.1733	8.1281	12444	3.5156	.00186220	1687.034	226484.48
538	289444	155720872	23.1948	8.1332	12479	3.5169	.00185874	1690.175	227328.79
539	290521	156590819	23.2164	8.1382	12514	3.5182	.00185529	1693.317	228174.66
540	291600	157464000	23.2379	8.1433	12549	3.5195	.00185185	1696.459	229022.10
541	292681	158340421	23.2594	8.1483	12583	3.5208	.00184843	1699.600	229871.12
542	293764	159220088	23.2809	8.1533	12618	3.5221	.00184502	1702.742	230721.71
543	294849	160103007	23.3024	8.1583	12653	3.5234	.00184162	1705.883	231573.86
544	295936	160989184	23.3238	8.1633	12688	3.5247	.00183824	1709.025	232427.59
545	297025	161878625	23.3452	8.1683	12723	3.5260	.00183486	1712.167	233282.89
546	298116	162771336	23.3666	8.1733	12758	3.5273	.00183150	1715.308	234139.76
547	299209	163667323	23.3880	8.1783	12793	3.5286	.00182815	1718.450	234998.20
548	300304	164566592	23.4094	8.1833	12828	3.5299	.00182482	1721.591	235858.21
549	301401	165469149	23.4307	8.1882	12863	3.5311	.00182149	1724.733	236719.79
550	302500	166375000	23.4521	8.1932	12899	3.5324	.00181818	1727.875	237582.94
551	303601	167284151	23.4734	8.1982	12934	3.5337	.00181488	1731.016	238447.67
552	304704	168196608	23.4947	8.2031	12969	3.5350	.00181159	1734.158	239313.96
553	305809	169112377	23.5160	8.2081	13004	3.5363	.00180832	1737.299	240181.83
554	306916	170031464	23.5372	8.2130	13040	3.5376	.00180505	1740.441	241051.26
555	308025	170953875	23.5584	8.2180	13075	3.5388	.00180180	1743.582	241922.27
556	309136	171879616	23.5797	8.2229	13110	3.5401	.00179856	1746.724	242794.85
557	310249	172808693	23.6008	8.2278	13146	3.5414	.00179533	1749.866	243668.99
558	311364	173741112	23.6220	8.2327	13181	3.5426	.00179211	1753.007	244544.71
559	312481	174676879	23.6432	8.2377	13217	3.5439	.00178891	1756.149	245422.00
560	313600	175616000	23.6643	8.2426	13252	3.5451	.00178571	1759.290	246300.86
561	314721	176558481	23.6854	8.2475	13288	3.5464	.00178253	1762.432	247181.30
562	315844	177504328	23.7065	8.2524	13323	3.5477	.00177936	1765.574	248063.30
563	316969	178453547	23.7276	8.2573	13359	3.5490	.00177620	1768.715	248946.87
564	318096	179406144	23.7487	8.2621	13394	3.5502	.00177305	1771.857	249832.01
565	319225	180362125	23.7697	8.2670	13430	3.5515	.00176991	1774.998	250718.73
566	320356	181321496	23.7908	8.2719	13466	3.5527	.00176678	1778.140	251607.01
567	321489	182284263	23.8118	8.2768	13501	3.5540	.00176367	1781.282	252496.87
568	322624	183250432	23.8328	8.2816	13537	3.5553	.00176056	1784.423	253388.30
569	323761	184220009	23.8537	8.2865	13573	3.5565	.00175747	1787.565	254281.29

Table 7 —*Continued*

N	N^2	N^3	$\sqrt{N}$	$\sqrt[3]{N}$	$N^{3/2}$	$\sqrt[5]{N}$	$\dfrac{1}{N}$	Circle ($N = D$) Circum.	Area
570	324900	185193000	23.8747	8.2913	13609	3.5577	.00175439	1790.706	255175.86
571	326041	186169411	23.8956	8.2962	13644	3.5590	.00175131	1793.848	256072.00
572	327184	187149248	23.9165	8.3010	13680	3.5602	.00174825	1796.989	256969.71
573	328329	188132517	23.9374	8.3059	13716	3.5615	.00174520	1800.131	257868.99
574	329476	189119224	23.9583	8.3107	13752	3.5627	.00174216	1803.273	258769.85
575	330625	190109375	23.9792	8.3155	13788	3.5640	.00173913	1806.414	259672.27
576	331776	191102976	24.0000	8.3203	13824	3.5652	.00173611	1809.556	260576.26
577	332929	192100033	24.0208	8.3251	13860	3.5664	.00173310	1812.697	261481.83
578	334084	193100552	24.0416	8.3300	13896	3.5677	.00173010	1815.839	262388.96
579	335241	194104539	24.0624	8.3348	13932	3.5689	.00172712	1818.981	263297.67
580	336400	195112000	24.0832	8.3396	13968	3.5702	.00172414	1822.122	264207.94
581	337561	196122941	24.1039	8.3443	14004	3.5714	.00172117	1825.264	265119.79
582	338724	197137368	24.1247	8.3491	14040	3.5726	.00171821	1828.405	266033.21
583	339889	198155287	24.1454	8.3539	14077	3.5738	.00171527	1831.547	266948.20
584	341056	199176704	24.1661	8.3587	14113	3.5751	.00171233	1834.689	267864.76
585	342225	200201625	24.1868	8.3634	14149	3.5763	.00170940	1837.830	268782.89
586	343396	201230056	24.2074	8.3682	14186	3.5775	.00170648	1840.972	269702.59
587	344569	202262003	24.2281	8.3730	14222	3.5787	.00170358	1844.113	270623.86
588	345744	203297472	24.2487	8.3777	14258	3.5799	.00170068	1847.255	271546.70
589	346921	204336469	24.2693	8.3825	14295	3.5812	.00169779	1850.397	272471.12
590	348100	205379000	24.2899	8.3872	14331	3.5824	.00169492	1853.538	273397.10
591	349281	206425071	24.3105	8.3919	14368	3.5836	.00169205	1856.680	274324.66
592	350464	207474688	24.3311	8.3967	14404	3.5848	.00168919	1859.821	275253.78
593	351649	208527857	24.3516	8.4014	14440	3.5860	.00168634	1862.963	276184.48
594	352836	209584584	24.3721	8.4061	14477	3.5872	.00168350	1866.104	277116.75
595	354025	210644875	24.3926	8.4108	14514	3.5884	.00168067	1869.246	278050.58
596	355216	211708736	24.4131	8.4155	14550	3.5896	.00167785	1872.388	278985.99
597	356409	212776173	24.4336	8.4202	14587	3.5908	.00167504	1875.529	279922.97
598	357604	213847192	24.4540	8.4249	14624	3.5920	.00167224	1878.671	280861.52
599	358801	214921799	24.4745	8.4296	14660	3.5932	.00166945	1881.812	281801.65
600	360000	216000000	24.4949	8.4343	14697	3.5944	.00166667	1884.954	282743.34
601	361201	217081801	24.5153	8.4390	14734	3.5956	.00166389	1888.096	283686.60
602	362404	218167208	24.5357	8.4437	14770	3.5968	.00166113	1891.237	284631.44
603	363609	219256227	24.5561	8.4484	14807	3.5980	.00165837	1894.379	285577.84
604	364816	220348864	24.5764	8.4530	14844	3.5992	.00165563	1897.520	286525.82
605	366025	221445125	24.5967	8.4577	14881	3.6004	.00165289	1900.662	287475.36
606	367236	222545016	24.6171	8.4623	14918	3.6016	.00165017	1903.804	288426.48
607	368449	223648543	24.6374	8.4670	14955	3.6028	.00164745	1906.945	289379.17
608	369664	224755712	24.6577	8.4716	14992	3.6040	.00164474	1910.087	290333.43
609	370881	225866529	24.6779	8.4763	15029	3.6052	.00164204	1913.228	291289.26
610	372100	226981000	24.6982	8.4809	15066	3.6063	.00163934	1916.370	292246.66
611	373321	228099131	24.7184	8.4856	15103	3.6075	.00163666	1919.511	293205.63
612	374544	229220928	24.7386	8.4902	15140	3.6087	.00163399	1922.653	294166.17
613	375769	230346397	24.7588	8.4948	15177	3.6099	.00163132	1925.795	295128.28
614	376996	231475544	24.7790	8.4994	15214	3.6111	.00162866	1928.936	296091.97
615	378225	232608375	24.7992	8.5040	15252	3.6122	.00162602	1932.078	297057.22
616	379456	233744896	24.8193	8.5086	15289	3.6134	.00162338	1935.219	298024.05
617	380689	234885113	24.8395	8.5132	15326	3.6146	.00162075	1938.361	298992.44
618	381924	236029032	24.8596	8.5178	15363	3.6158	.00161812	1941.503	299962.41
619	383161	237176659	24.8797	8.5224	15400	3.6169	.00161551	1944.644	300933.95
620	384400	238328000	24.8998	8.5270	15437	3.6181	.00161290	1947.786	301907.05
621	385641	239483061	24.9199	8.5316	15475	3.6192	.00161031	1950.927	302881.73
622	386884	240641848	24.9399	8.5362	15513	3.6204	.00160772	1954.069	303857.98
623	388129	241804367	24.9600	8.5408	15550	3.6216	.00160514	1957.211	304835.80
624	389376	242970624	24.9800	8.5453	15588	3.6227	.00160256	1960.352	305815.20
625	390625	244140625	25.0000	8.5499	15625	3.6239	.00160000	1963.494	306796.16
626	391876	245314376	25.0200	8.5544	15663	3.6250	.00159744	1966.635	307778.69
627	393129	246491883	25.0400	8.5590	15700	3.6262	.00159490	1969.777	308762.79
628	394384	247673152	25.0599	8.5635	15738	3.6274	.00159236	1972.919	309748.47
629	395641	248858189	25.0799	8.5681	15775	3.6285	.00158983	1976.060	310735.71
630	396900	250047000	25.0998	8.5726	15813	3.6297	.00158730	1979.202	311724.53
631	398161	251239591	25.1197	8.5772	15850	3.6309	.00158479	1982.343	312714.92
632	399424	252435968	25.1396	8.5817	15888	3.6320	.00158228	1985.485	313706.88
633	400689	253636137	25.1595	8.5862	15926	3.6331	.00157978	1988.627	314700.40
634	401956	254840104	25.1794	8.5907	15964	3.6343	.00157729	1991.768	315695.50
635	403225	256047875	25.1992	8.5952	16002	3.6354	.00157480	1994.910	316692.17
636	404496	257259456	25.2190	8.5997	16040	3.6366	.00157233	1998.051	317690.42
637	405769	258474853	25.2389	8.6043	16077	3.6377	.00156986	2001.193	318690.23
638	407044	259694072	25.2587	8.6088	16115	3.6389	.00156740	2004.334	319691.61
639	408321	260917119	25.2784	8.6132	16153	3.6400	.00156495	2007.476	320694.56

Table 7 —Continued

N	N^2	N^3	$\sqrt{N}$	$\sqrt[3]{N}$	$N^{3/2}$	$\sqrt[5]{N}$	$\frac{1}{N}$	Circle ($N = D$) Circum.	Area
640	409600	262144000	25.2982	8.6177	16191	3.6411	.00156250	2010.618	321699.09
641	410881	263374721	25.3180	8.6222	16229	3.6423	.00156006	2013.759	322705.18
642	412164	264609288	25.3377	8.6267	16267	3.6435	.00155763	2016.901	323712.85
643	413449	265847707	25.3574	8.6312	16305	3.6446	.00155521	2020.042	324722.09
644	414736	267089984	25.3772	8.6357	16343	3.6457	.00155280	2023.184	325732.89
645	416025	268336125	25.3969	8.6401	16381	3.6468	.00155039	2026.326	326745.27
646	417316	269586136	25.4165	8.6446	16419	3.6479	.00154799	2029.467	327759.22
647	418609	270840023	25.4362	8.6490	16457	3.6499	.00154560	2032.609	328774.74
648	419904	272097792	25.4558	8.6535	16495	3.6502	.00154321	2035.750	329791.83
649	421201	273359449	25.4755	8.6579	16534	3.6513	.00154083	2038.892	330810.49
650	422500	274625000	25.4951	8.6624	16572	3.6524	.00153846	2042.034	331830.72
651	423801	275894451	25.5147	8.6668	16610	3.6536	.00153610	2045.175	332852.53
652	425104	277167808	25.5343	8.6713	16648	3.6547	.00153374	2048.317	333875.90
653	426409	278445077	25.5539	8.6757	16687	3.6558	.00153139	2051.458	334900.85
654	427716	279726264	25.5734	8.6801	16725	3.6569	.00152905	2054.600	335927.36
655	429025	281011375	25.5930	8.6845	16764	3.6580	.00152672	2057.741	336955.45
656	430336	282300416	25.6125	8.6890	16802	3.6592	.00152439	2060.883	337985.10
657	431649	283593393	25.6320	8.6934	16840	3.6603	.00152207	2064.025	339016.33
658	432964	284890312	25.6515	8.6978	16879	3.6614	.00151976	2067.166	340049.13
659	434281	286191179	25.6710	8.7022	16917	3.6625	.00151745	2070.308	341083.50
660	435600	287496000	25.6905	8.7066	16956	3.6636	.00151515	2073.449	342119.44
661	436921	288804781	25.7099	8.7110	16994	3.6647	.00151286	2076.591	343156.95
662	438244	290117528	25.7294	8.7154	17033	3.6658	.00151057	2079.733	344196.03
663	439569	291434247	25.7488	8.7198	17071	3.6669	.00150830	2082.874	345236.69
664	440896	292754944	25.7682	8.7241	17110	3.6680	.00150602	2086.016	346278.91
665	442225	294079625	25.7876	8.7285	17149	3.6691	.00150376	2089.157	347322.70
666	443556	295408296	25.8070	8.7329	17187	3.6702	.00150150	2092.299	348368.07
667	444889	296740963	25.8263	8.7373	17226	3.6713	.00149925	2095.441	349415.00
668	446224	298077632	25.8457	8.7416	17265	3.6724	.00149701	2098.582	350463.51
669	447561	299418309	25.8650	8.7460	17304	3.6735	.00149477	2101.724	351513.59
670	448900	300763000	25.8844	8.7503	17343	3.6746	.00149254	2104.865	352565.24
671	450241	302111711	25.9037	8.7547	17381	3.6757	.00149031	2108.007	353618.45
672	451584	303464448	25.9230	8.7590	17420	3.6768	.00148810	2111.148	354673.24
673	452929	304821217	25.9422	8.7634	17459	3.6779	.00148588	2114.290	355729.60
674	454276	306182024	25.9615	8.7677	17498	3.6790	.00148368	2117.432	356787.54
675	455625	307546875	25.9808	8.7721	17537	3.6801	.00148148	2120.573	357847.04
676	456976	308915776	26.0000	8.7764	17576	3.6812	.00147929	2123.715	358908.11
677	458329	310288733	26.0192	8.7807	17615	3.6823	.00147710	2126.856	359970.75
678	459684	311665752	26.0384	8.7850	17654	3.6834	.00147493	2129.998	361034.97
679	461041	313046839	26.0576	8.7893	17693	3.6845	.00147275	2133.140	362100.75
680	462400	314432000	26.0768	8.7937	17732	3.6856	.00147059	2136.281	363168.11
681	463761	315821241	26.0960	8.7980	17771	3.6866	.00146843	2139.423	364237.04
682	465124	317214568	26.1151	8.8023	17810	3.6877	.00146628	2142.564	365307.54
683	466489	318611987	26.1343	8.8066	17850	3.6888	.00146413	2145.706	366379.60
684	467856	320013504	26.1534	8.8109	17889	3.6899	.00146199	2148.848	367453.24
685	469225	321419125	26.1725	8.8152	17928	3.6909	.00145985	2151.989	368528.45
686	470596	322828856	26.1916	8.8194	17967	3.6920	.00145773	2155.131	369605.23
687	471969	324242703	26.2107	8.8237	18007	3.6931	.00145560	2158.272	370683.59
688	473344	325660672	26.2298	8.8280	18046	3.6942	.00145349	2161.414	371763.51
689	474721	327082769	26.2488	8.8323	18085	3.6953	.00145138	2164.556	372845.00
690	476100	328509000	26.2679	8.8366	18125	3.6963	.00144928	2167.697	373928.07
691	477481	329933371	26.2869	8.8408	18164	3.6974	.00144718	2170.839	375012.70
692	478864	331373888	26.3059	8.8451	18204	3.6985	.00144509	2173.980	376098.91
693	480249	332812557	26.3249	8.8493	18243	3.6995	.00144300	2177.122	377186.68
694	481636	334255384	26.3439	8.8536	18283	3.7006	.00144092	2180.263	378276.03
695	483025	335702375	26.3629	8.8578	18322	3.7016	.00143885	2183.405	379366.95
696	484416	337153536	26.3818	8.8621	18362	3.7027	.00143678	2186.547	380459.44
697	485809	338608873	26.4008	8.8663	18401	3.7038	.00143472	2189.688	381553.50
698	487204	340068392	26.4197	8.8706	18441	3.7049	.00143266	2192.830	382649.13
699	488601	341532099	26.4386	8.8748	18480	3.7059	.00143062	2195.971	383746.33
700	490000	343000000	26.4575	8.8790	18520	3.7070	.00142857	2199.113	384845.10
701	491401	344472101	26.4764	8.8833	18560	3.7080	.00142653	2202.255	385945.44
702	492804	345948408	26.4953	8.8875	18600	3.7091	.00142450	2205.396	387047.36
703	494209	347428927	26.5141	8.8917	18640	3.7101	.00142248	2208.538	388150.84
704	495616	348913664	26.5330	8.8959	18679	3.7112	.00142045	2211.679	389255.90
705	497025	350402625	26.5518	8.9001	18719	3.7123	.00141844	2214.821	390362.52
706	498436	351895816	26.5707	8.9043	18759	3.7133	.00141643	2217.963	391470.72
707	499849	353393243	26.5895	8.9085	18799	3.7144	.00141443	2221.104	392580.49
708	501264	354894912	26.6083	8.9127	18839	3.7154	.00141243	2224.246	393691.82
709	502681	356400829	26.6271	8.9169	18879	3.7165	.00141044	2227.387	394804.73

Table 7 —Continued

N	N^2	N^3	$\sqrt{N}$	$\sqrt[3]{N}$	$N^{3/2}$	$\sqrt[5]{N}$	$\dfrac{1}{N}$	Circle ($N = D$) Circum.	Area
710	504100	357911000	26.6458	8.9211	18919	3.7175	.00140845	2230.529	395919.21
711	505521	359425431	26.6646	8.9253	18959	3.7185	.00140647	2233.670	397035.26
712	506944	360944128	26.6833	8.9295	18999	3.7196	.00140449	2236.812	398152.89
713	508369	362467097	26.7021	8.9337	19039	3.7206	.00140252	2239.954	399272.08
714	509796	363994344	26.7208	8.9378	19079	3.7217	.00140056	2243.095	400392.84
715	511225	365525875	26.7395	8.9420	19119	3.7227	.00139860	2246.237	401515.18
716	512656	367061696	26.7582	8.9462	19159	3.7238	.00139665	2249.378	402639.08
717	514089	368601813	26.7769	8.9503	19199	3.7248	.00139470	2252.520	403764.56
718	515524	370146232	26.7955	8.9545	19239	3.7258	.00139276	2255.662	404891.60
719	516961	371694959	26.8142	8.9587	19280	3.7269	.00139082	2258.803	406020.22
720	518400	373248000	26.8328	8.9628	19320	3.7279	.00138889	2261.945	407150.41
721	519841	374805361	26.8514	8.9670	19360	3.7290	.00138696	2265.086	408282.17
722	521284	376367048	26.8701	8.9711	19400	3.7300	.00138504	2268.228	409415.50
723	522729	377933067	26.8887	8.9752	19440	3.7310	.00138313	2271.370	410550.40
724	524176	379503424	26.9072	8.9794	19481	3.7321	.00138122	2274.511	411686.87
725	525625	381078125	26.9258	8.9835	19521	3.7331	.00137931	2277.653	412824.91
726	527076	382657176	26.9444	8.9876	19562	3.7341	.00137741	2280.794	413964.52
727	528529	384240583	26.9629	8.9918	19602	3.7351	.00137552	2283.936	415105.71
728	529984	385828352	26.9815	8.9959	19643	3.7362	.00137363	2287.078	416248.46
729	531441	387420489	27.0000	9.0000	19683	3.7372	.00137174	2290.219	417392.79
730	532900	389017000	27.0185	9.0041	19724	3.7382	.00136986	2293.361	418538.68
731	534361	390617891	27.0370	9.0082	19764	3.7392	.00136799	2296.502	419686.15
732	535824	392223168	27.0555	9.0123	19805	3.7403	.00136612	2299.644	420835.19
733	537289	393832837	27.0740	9.0164	19845	3.7413	.00136426	2302.785	421985.79
734	538756	395446904	27.0924	9.0205	19886	3.7423	.00136240	2305.927	423137.97
735	540225	397065375	27.1109	9.0246	19927	3.7433	.00136054	2309.069	424291.72
736	541696	398688256	27.1293	9.0287	19967	3.7443	.00135870	2312.210	425447.04
737	543169	400315553	27.1477	9.0328	20008	3.7454	.00135685	2315.352	426603.94
738	544644	401947272	27.1662	9.0369	20049	3.7464	.00135501	2318.493	427762.40
739	546121	403583419	27.1846	9.0410	20090	3.7474	.00135318	2321.635	428922.43
740	547600	405224000	27.2029	9.0450	20130	3.7484	.00135135	2324.777	430084.03
741	549081	406869021	27.2213	9.0491	20171	3.7494	.00134953	2327.918	431247.21
742	550564	408518488	27.2397	9.0532	20212	3.7504	.00134771	2331.060	432411.95
743	552049	410172407	27.2580	9.0572	20253	3.7514	.00134590	2334.201	433578.27
744	553536	411830784	27.2764	9.0613	20294	3.7524	.00134409	2337.343	434746.16
745	555025	413493625	27.2947	9.0654	20335	3.7534	.00134228	2340.485	435915.62
746	556516	415160936	27.3130	9.0694	20376	3.7545	.00134048	2343.626	437086.64
747	558009	416832723	27.3313	9.0735	20417	3.7555	.00133869	2346.768	438259.24
748	559504	418508992	27.3496	9.0775	20458	3.7565	.00133690	2349.909	439433.41
749	561001	420189749	27.3679	9.0816	20499	3.7575	.00133511	2353.051	440609.16
750	562500	421875000	27.3861	9.0856	20540	3.7585	.00133333	2356.193	441786.47
751	564001	423564751	27.4044	9.0896	20581	3.7595	.00133156	2359.334	442965.35
752	565504	425259008	27.4226	9.0937	20622	3.7605	.00132979	2362.476	444145.80
753	567009	426957777	27.4408	9.0977	20663	3.7615	.00132802	2365.617	445327.83
754	568516	428661064	27.4591	9.1017	20704	3.7625	.00132626	2368.759	446511.42
755	570025	430368875	27.4773	9.1057	20745	3.7635	.00132450	2371.900	447696.59
756	571536	432081216	27.4955	9.1098	20787	3.7645	.00132275	2375.042	448883.32
757	573049	433798093	27.5136	9.1138	20828	3.7655	.00132100	2378.184	450071.63
758	574564	435519512	27.5318	9.1178	20869	3.7665	.00131926	2381.325	451261.51
759	576081	437245479	27.5500	9.1218	20910	3.7675	.00131752	2384.467	452452.96
760	577600	438976000	27.5681	9.1258	20952	3.7685	.00131579	2387.608	453645.98
761	579121	440711081	27.5862	9.1298	20993	3.7694	.00131406	2390.750	454840.57
762	580644	442450728	27.6043	9.1338	21035	3.7704	.00131234	2393.892	456036.73
763	582169	444194947	27.6225	9.1378	21076	3.7714	.00131062	2397.033	457234.46
764	583696	445943744	27.6405	9.1418	21117	3.7724	.00130890	2400.175	458433.77
765	585225	447697125	27.6586	9.1458	21159	3.7734	.00130719	2403.316	459634.64
766	586756	449455096	27.6767	9.1498	21200	3.7744	.00130548	2406.458	460837.08
767	588289	451217663	27.6948	9.1537	21242	3.7754	.00130378	2409.600	462041.10
768	589824	452984832	27.7128	9.1577	21283	3.7764	.00130208	2412.741	463246.69
769	591361	454756609	27.7308	9.1617	21325	3.7774	.00130039	2415.883	464453.84
770	592900	456533000	27.7489	9.1657	21367	3.7784	.00129870	2419.024	465662.57
771	594441	458314011	27.7669	9.1696	21408	3.7793	.00129702	2422.166	466872.87
772	595984	460099648	27.7849	9.1736	21450	3.7803	.00129534	2425.307	468084.74
773	597529	461889917	27.8029	9.1775	21492	3.7813	.00129366	2428.449	469298.18
774	599076	463684824	27.8209	9.1815	21533	3.7822	.00129199	2431.591	470513.19
775	600625	465484375	27.8388	9.1855	21575	3.7832	.00129032	2434.732	471729.77
776	602176	467288576	27.8568	9.1894	21617	3.7842	.00128866	2437.874	472947.92
777	603729	469097433	27.8747	9.1933	21658	3.7852	.00128700	2441.015	474167.65
778	605284	470910952	27.8927	9.1973	21700	3.7861	.00128535	2444.157	475388.94
779	606841	472729139	27.9106	9.2012	21742	3.7871	.00128370	2447.299	476611.81

Table 7 —*Continued*

N	N^2	N^3	$\sqrt{N}$	$\sqrt[3]{N}$	$N^{3/2}$	$\sqrt[5]{N}$	$\dfrac{1}{N}$	Circle ($N = D$) Circum.	Area
780	608400	474552000	27.9285	9.2052	21784	3.7881	.00128205	2450.440	477836.24
781	609961	476379541	27.9464	9.2091	21826	3.7890	.00128041	2453.582	479062.25
782	611524	478211768	27.9643	9.2130	21868	3.7900	.00127877	2456.723	480289.83
783	613089	480048687	27.9821	9.2170	21910	3.7910	.00127714	2459.865	481518.97
784	614656	481890304	28.0000	9.2209	21952	3.7920	.00127551	2463.007	482749.69
785	616225	483736625	28.0179	9.2248	21994	3.7929	.00127389	2466.148	483981.98
786	617796	485587656	28.0357	9.2287	22036	3.7939	.00127226	2469.290	485215.84
787	619369	487443403	28.0535	9.2326	22078	3.7949	.00127065	2472.431	486451.28
788	620944	489303872	28.0713	9.2365	22120	3.7959	.00126904	2475.573	487688.28
789	622521	491169069	28.0891	9.2404	22162	3.7969	.00126743	2478.715	488926.85
790	624100	493039000	28.1069	9.2443	22205	3.7978	.00126582	2481.856	490166.99
791	625681	494913671	28.1247	9.2482	22247	3.7987	.00126422	2484.998	491408.71
792	627264	496793088	28.1425	9.2521	22289	3.7997	.00126263	2488.139	492651.99
793	628849	498677257	28.1603	9.2560	22331	3.8006	.00126103	2491.281	493896.85
794	630436	500566184	28.1780	9.2599	22373	3.8016	.00125945	2494.422	495143.28
795	632025	502459875	28.1957	9.2638	22416	3.8025	.00125786	2497.564	496391.27
796	633616	504358336	28.2135	9.2677	22458	3.8035	.00125628	2500.706	497640.84
797	635209	506261573	28.2312	9.2716	22500	3.8044	.00125471	2503.847	498891.98
798	636804	508169592	28.2489	9.2754	22543	3.8054	.00125313	2506.989	500144.69
799	638401	510082399	28.2666	9.2793	22585	3.8064	.00125156	2510.130	501398.97
800	640000	512000000	28.2843	9.2832	22627	3.8073	.00125000	2513.272	502654.82
801	641601	513922401	28.3019	9.2870	22670	3.8083	.00124844	2516.414	503912.25
802	643204	515849608	28.3196	9.2909	22712	3.8092	.00124688	2519.555	505171.24
803	644809	517781627	28.3373	9.2948	22755	3.8102	.00124533	2522.697	506431.80
804	646416	519718464	28.3549	9.2986	22797	3.8111	.00124378	2525.838	507693.94
805	648025	521660125	28.3725	9.3025	22840	3.8121	.00124224	2528.980	508957.64
806	649636	523606616	28.3901	9.3063	22883	3.8130	.00124069	2532.122	510222.92
807	651249	525557943	28.4077	9.3102	22925	3.8139	.00123916	2535.263	511489.77
808	652864	527514112	28.4253	9.3140	22968	3.8149	.00123762	2538.405	512758.19
809	654481	529475129	28.4429	9.3179	23010	3.8158	.00123609	2541.546	514028.18
810	656100	531441000	28.4605	9.3217	23053	3.8168	.00123457	2544.688	515299.74
811	657721	533411731	28.4781	9.3255	23096	3.8177	.00123305	2547.829	516572.87
812	659344	535387328	28.4956	9.3294	23138	3.8186	.00123153	2550.971	517847.57
813	660969	537367797	28.5132	9.3332	23181	3.8196	.00123001	2554.113	519123.84
814	662596	539353144	28.5307	9.3370	23224	3.8205	.00122850	2557.254	520401.68
815	664225	541343375	28.5482	9.3408	23267	3.8215	.00122699	2560.396	521681.10
816	665856	543338496	28.5657	9.3447	23310	3.8224	.00122549	2563.537	522962.08
817	667489	545338513	28.5832	9.3485	23352	3.8234	.00122399	2566.679	524244.63
818	669124	547343432	28.6007	9.3523	23395	3.8243	.00122249	2569.821	525528.76
819	670761	549353259	28.6182	9.3561	23438	3.8252	.00122100	2572.962	526814.46
820	672400	551368000	28.6356	9.3599	23481	3.8262	.00121951	2576.104	528101.73
821	674041	553387661	28.6531	9.3637	23524	3.8271	.00121803	2579.245	529390.56
822	675684	555412248	28.6705	9.3675	23567	3.8280	.00121655	2582.387	530680.97
823	677329	557441767	28.6880	9.3713	23610	3.8290	.00121507	2585.529	531972.95
824	678976	559476224	28.7054	9.3751	23653	3.8299	.00121359	2588.670	533266.50
825	680625	561515625	28.7228	9.3789	23696	3.8308	.00121212	2591.812	534561.62
826	682276	563559976	28.7402	9.3827	23740	3.8317	.00121065	2594.953	535858.32
827	683929	565609283	28.7576	9.3865	23783	3.8327	.00120919	2598.095	537156.58
828	685584	567663552	28.7750	9.3902	23826	3.8336	.00120773	2601.237	538456.41
829	687241	569722789	28.7924	9.3940	23869	3.8345	.00120627	2604.378	539757.82
830	688900	571787000	28.8097	9.3978	23912	3.8355	.00120482	2607.520	541060.79
831	690561	573856191	28.8271	9.4016	23955	3.8364	.00120337	2610.661	542365.34
832	692224	575930368	28.8444	9.4053	23999	3.8373	.00120192	2613.803	543671.46
833	693889	578009537	28.8617	9.4091	24042	3.8382	.00120048	2616.944	544979.15
834	695556	580093704	28.8791	9.4129	24085	3.8391	.00119904	2620.086	546288.40
835	697225	582182875	28.8964	9.4166	24128	3.8401	.00119760	2623.228	547599.23
836	698896	584277056	28.9137	9.4204	24172	3.8410	.00119617	2626.369	548911.63
837	700569	586376253	28.9310	9.4241	24215	3.8419	.00119474	2629.511	550225.61
838	702244	588480472	28.9482	9.4279	24259	3.8428	.00119332	2632.652	551541.15
839	703921	590589719	28.9655	9.4316	24302	3.8437	.00119190	2635.794	552858.26
840	705600	592704000	28.9828	9.4354	24346	3.8446	.00119048	2638.936	554176.94
841	707281	594823321	29.0000	9.4391	24389	3.8456	.00118906	2642.077	555497.20
842	708964	596947688	29.0172	9.4429	24432	3.8465	.00118765	2645.219	556819.02
843	710649	599077107	29.0345	9.4466	24476	3.8474	.00118624	2648.360	558142.42
844	712336	601211584	29.0517	9.4503	24520	3.8483	.00118483	2651.502	559467.39
845	714025	603351125	29.0689	9.4541	24563	3.8492	.00118343	2654.644	560793.92
846	715716	605495736	29.0861	9.4578	24607	3.8501	.00118203	2657.785	562122.03
847	717409	607645423	29.1033	9.4615	24650	3.8510	.00118064	2660.927	563451.71
848	719104	609800192	29.1204	9.4652	24694	3.8519	.00117925	2664.068	564782.96
849	720801	611960049	29.1376	9.4690	24738	3.8528	.00117786	2667.210	566115.78

Table 7 —*Continued*

N	N^2	N^3	$\sqrt{N}$	$\sqrt[3]{N}$	$N^{3/2}$	$\sqrt[5]{N}$	$\dfrac{1}{N}$	Circle ($N=D$) Circum.	Area
850	722500	614125000	29.1548	9.4727	24782	3.8558	.00117647	2670.352	567450.17
851	724201	616295051	29.1719	9.4764	24825	3.8547	.00117509	2673.493	568786.14
852	725904	618470208	29.1890	9.4801	24869	3.8556	.00117371	2676.635	570123.67
853	727609	620650477	29.2062	9.4838	24913	3.8565	.00117233	2679.776	571462.77
854	729316	622835864	29.2233	9.4875	24957	3.8574	.00117096	2682.918	572803.45
855	731025	625026375	29.2404	9.4912	25000	3.8582	.00116959	2686.059	574145.69
856	732736	627222016	29.2575	9.4949	25044	3.8592	.00116822	2689.201	575489.51
857	734449	629422793	29.2746	9.4986	25088	3.8601	.00116686	2692.343	576834.90
858	736164	631628712	29.2916	9.5023	25132	3.8610	.00116550	2695.484	578181.85
859	737881	633839779	29.3087	9.5060	25176	3.8619	.00116414	2698.626	579530.38
860	739600	636056000	29.3258	9.5097	25220	3.8628	.00116279	2701.767	580880.48
861	741321	638277381	29.3428	9.5134	25264	3.8637	.00116144	2704.909	582232.15
862	743044	640503928	29.3598	9.5171	25308	3.8646	.00116009	2708.051	583585.39
863	744769	642735647	29.3769	9.5207	25352	3.8655	.00115875	2711.192	584940.20
864	746496	644972544	29.3939	9.5244	25396	3.8664	.00115741	2714.334	586296.59
865	748225	647214625	29.4109	9.5281	25440	3.8673	.00115607	2717.475	587654.54
866	749956	649461896	29.4279	9.5317	25485	3.8682	.00115473	2720.617	589014.07
867	751689	651714363	29.4449	9.5354	25529	3.8691	.00115340	2723.759	590375.16
868	753424	653972032	29.4618	9.5391	25573	3.8700	.00115207	2726.900	591737.83
869	755161	656234909	29.4788	9.5427	25617	3.8708	.00115075	2730.042	593102.06
870	756900	658503000	29.4958	9.5464	25661	3.8717	.00114943	2733.183	594467.87
871	758641	660776311	29.5127	9.5501	25706	3.8726	.00114811	2736.325	595835.25
872	760384	663054848	29.5296	9.5537	25750	3.8735	.00114679	2739.466	597204.20
873	762129	665338617	29.5466	9.5574	25794	3.8744	.00114548	2742.608	598574.72
874	763876	667627624	29.5635	9.5610	25839	3.8753	.00114416	2745.750	599946.81
875	765625	669921875	29.5804	9.5647	25883	3.8762	.00114286	2748.891	601320.47
876	767376	672221376	29.5973	9.5683	25927	3.8771	.00114155	2752.033	602695.70
877	769129	674526133	29.6142	9.5719	25972	3.8780	.00114025	2755.174	604072.50
878	770884	676836152	29.6311	9.5756	26016	3.8789	.00113895	2758.316	605450.88
879	772641	679151439	29.6479	9.5792	26061	3.8797	.00113766	2761.458	606830.82
880	774400	681472000	29.6648	9.5828	26105	3.8806	.00113636	2764.599	608212.34
881	776161	683797841	29.6816	9.5865	26150	3.8815	.00113507	2767.741	609595.42
882	777924	686128968	29.6985	9.5901	26194	3.8823	.00113379	2770.882	610980.08
883	779689	688465387	29.7153	9.5937	26239	3.8832	.00113250	2774.024	612366.31
884	781456	690807104	29.7321	9.5973	26283	3.8841	.00113122	2777.166	613754.11
885	783225	693154125	29.7489	9.6010	26328	3.8850	.00112994	2780.307	615143.48
886	784996	695506456	29.7658	9.6046	26373	3.8859	.00112867	2783.449	616534.42
887	786769	697864103	29.7825	9.6082	26417	3.8868	.00112740	2786.590	617926.93
888	788544	700227072	29.7993	9.6118	26462	3.8877	.00112613	2789.732	619321.01
889	790321	702595369	29.8161	9.6154	26507	3.8885	.00112486	2792.874	620716.66
890	792100	704969000	29.8329	9.6190	26551	3.8894	.00112360	2796.015	622113.89
891	793881	707347971	29.8496	9.6226	26596	3.8902	.00112233	2799.157	623512.68
892	795664	709732288	29.8664	9.6262	26641	3.8911	.00112108	2802.298	624913.04
893	797449	712121957	29.8831	9.6298	26686	3.8920	.00111982	2805.440	626314.98
894	799236	714516984	29.8998	9.6334	26730	3.8929	.00111857	2808.581	627718.49
895	801025	716917375	29.9166	9.6370	26775	3.8937	.00111732	2811.723	629123.56
896	802816	719325136	29.9333	9.6406	26820	3.8946	.00111607	2814.865	630530.21
897	804609	721734273	29.9500	9.6442	26865	3.8955	.00111483	2818.006	631938.43
898	806404	724150792	29.9666	9.6477	26910	3.8963	.00111359	2821.148	633348.22
899	808201	726572699	29.9833	9.6513	26955	3.8972	.00111235	2824.289	634759.58
900	810000	729000000	30.0000	9.6549	27000	3.8981	.00111111	2827.431	636172.51
901	811801	731432701	30.0167	9.6585	27045	3.8989	.00110988	2830.573	637587.01
902	813604	733870808	30.0333	9.6620	27090	3.8998	.00110865	2833.714	639003.09
903	815409	736314327	30.0500	9.6656	27135	3.9007	.00110742	2836.856	640420.73
904	817216	738763264	30.0666	9.6692	27180	3.9015	.00110619	2839.997	641839.95
905	819025	741217625	30.0832	9.6727	27225	3.9024	.00110497	2843.139	643260.73
906	820836	743677416	30.0998	9.6763	27270	3.9032	.00110375	2846.281	644683.09
907	822649	746142643	30.1164	9.6799	27316	3.9041	.00110254	2849.422	646107.01
908	824464	748613312	30.1330	9.6834	27361	3.9050	.00110132	2852.564	647532.51
909	826281	751089429	30.1496	9.6870	27406	3.9059	.00110011	2855.705	648959.58
910	828100	753571000	30.1662	9.6905	27451	3.9067	.00109890	2858.847	650388.22
911	829921	756058031	30.1828	9.6941	27497	3.9076	.00109769	2861.988	651818.43
912	831744	758550528	30.1993	9.6976	27542	3.9084	.00109649	2865.130	653250.21
913	833569	761048497	30.2159	9.7012	27587	3.9093	.00109529	2868.272	654683.56
914	835396	763551944	30.2324	9.7047	27632	3.9101	.00109409	2871.413	656118.48
915	837225	766060875	30.2490	9.7082	27678	3.9110	.00109290	2874.555	657554.98
916	839056	768575296	30.2655	9.7118	27723	3.9118	.00109170	2877.696	658993.04
917	840889	771095213	30.2820	9.7153	27769	3.9127	.00109051	2880.838	660432.68
918	842724	773620632	30.2985	9.7188	27814	3.9135	.00108932	2883.980	661873.88
919	844561	776151559	30.3150	9.7224	27859	3.9144	.00108814	2887.121	663316.66

Table 7 —Continued

N	N^2	N^3	$\sqrt{N}$	$\sqrt[3]{N}$	$N^{3/2}$	$\sqrt[5]{N}$	$\frac{1}{N}$	Circle ($N = D$)	
								Circum.	Area
920	**846400**	**778688000**	30.3315	9.7259	27905	3.9153	.00108696	2890.263	664761.01
921	848241	781229961	30.3480	9.7294	27950	3.9161	.00108578	2893.404	666206.92
922	850084	783777448	30.3645	9.7329	27996	3.9169	.00108460	2896.546	667654.41
923	851929	786330467	30.3809	9.7364	28042	3.9178	.00108342	2899.688	669103.47
924	853776	788889024	30.3974	9.7400	28087	3.9186	.00108225	2902.829	670554.10
925	855625	791453125	30.4138	9.7435	28133	3.9194	.00108108	2905.971	672006.30
926	857476	794022776	30.4302	9.7470	28179	3.9203	.00107991	2909.112	673460.08
927	859329	796597983	30.4467	9.7505	28224	3.9212	.00107875	2912.254	674915.42
928	861184	799178752	30.4631	9.7540	28270	3.9220	.00107759	2915.396	676372.33
929	863041	801765089	30.4795	9.7575	28315	3.9229	.00107643	2918.537	677830.82
930	**864900**	**804357000**	**30.4959**	**9.7610**	**28361**	**3.9237**	**.00107527**	**2921.679**	**679290.87**
931	866761	806954491	30.5123	9.7645	28407	3.9246	.00107411	2924.820	680752.50
932	868624	809557568	30.5287	9.7680	28453	3.9254	.00107296	2927.962	682215.69
933	870489	812166237	30.5450	9.7715	28499	3.9262	.00107181	2931.103	683680.46
934	872356	814780504	30.5614	9.7750	28544	3.9271	.00107066	2934.245	685146.80
935	874225	817400375	30.5778	9.7785	28590	3.9279	.00106952	2937.387	686614.71
936	876096	820025856	30.5941	9.7819	28636	3.9288	.00106838	2940.528	688084.19
937	877969	822656953	30.6105	9.7854	28682	3.9296	.00106724	2943.670	689555.24
938	879844	825293672	30.6268	9.7889	28728	3.9304	.00106610	2946.811	691027.86
939	881721	827936019	30.6431	9.7924	28774	3.9313	.00106496	2949.953	692502.05
940	**883600**	**830584000**	**30.6594**	**9.7959**	**28820**	**3.9521**	**.00106383**	**2953.095**	**693977.82**
941	885481	833237621	30.6757	9.7993	28866	3.9329	.00106270	2956.236	695455.15
942	887364	835896888	30.6920	9.8028	28912	3.9338	.00106157	2959.378	696934.06
943	889249	838561807	30.7083	9.8063	28958	3.9346	.00106045	2962.519	698414.53
944	891136	841232384	30.7246	9.8097	29004	3.9354	.00105932	2965.661	699896.58
945	893025	843908625	30.7409	9.8132	29050	3.9363	.00105820	2968.803	701380.19
946	894916	846590536	30.7571	9.8167	29096	3.9371	.00105708	2971.944	702865.38
947	896809	849278123	30.7734	9.8201	29142	3.9379	.00105597	2975.086	704352.14
948	898704	851971392	30.7896	9.8236	29189	3.9388	.00105485	2978.227	705840.47
949	900601	854670349	30.8058	9.8270	29235	3.9396	.00105374	2981.369	707330.37
950	**902500**	**857375000**	**30.8221**	**9.8305**	**29281**	**3.9404**	**.00105263**	**2984.511**	**708821.84**
951	904401	860085351	30.8383	9.8339	29327	3.9413	.00105152	2987.652	710314.88
952	906304	862801408	30.8545	9.8374	29374	3.9421	.00105042	2990.794	711809.50
953	908209	865523177	30.8707	9.8408	29420	3.9429	.00104932	2993.935	713305.68
954	910116	868250664	30.8869	9.8443	29466	3.9438	.00104822	2997.077	714803.43
955	912025	870983875	30.9031	9.8477	29513	3.9446	.00104712	3000.218	716302.76
956	913936	873722816	30.9192	9.8511	29559	3.9454	.00104603	3003.360	717803.66
957	915849	876467493	30.9354	9.8546	29605	3.9462	.00104493	3006.502	719306.12
958	917764	879217912	30.9516	9.8580	29652	3.9471	.00104384	3009.643	720810.16
959	919681	881974079	30.9677	9.8614	29698	3.9479	.00104275	3012.785	722315.77
960	**921600**	**884736000**	**30.9839**	**9.8648**	**29745**	**3.9487**	**.00104167**	**3015.926**	**723822.95**
961	923521	887503681	31.0000	9.8683	29791	3.9495	.00104058	3019.068	725331.70
962	925444	890277128	31.0161	9.8717	29838	3.9503	.00103950	3022.210	726842.02
963	927369	893056347	31.0322	9.8751	29884	3.9512	.00103842	3025.351	728353.91
964	929296	895841344	31.0483	9.8785	29931	3.9520	.00103734	3028.493	729867.37
965	931225	898632125	31.0644	9.8819	29977	3.9528	.00103627	3031.634	731382.40
966	933156	901428696	31.0805	9.8854	30024	3.9536	.00103520	3034.776	732899.01
967	935089	904231063	31.0966	9.8888	30070	3.9544	.00103413	3037.918	734417.18
968	937024	907039232	31.1127	9.8922	30117	3.9553	.00103306	3041.059	735936.93
969	938961	909853209	31.1288	9.8956	30164	3.9561	.00103199	3044.201	737458.24
970	**940900**	**912673000**	**31.1448**	**9.8990**	**30210**	**3.9569**	**.00103093**	**3047.342**	**738981.13**
971	942841	915498611	31.1609	9.9024	30257	3.9577	.00102987	3050.484	740505.59
972	944784	918330048	31.1769	9.9058	30304	3.9585	.00102881	3053.625	742031.62
973	946729	921167317	31.1929	9.9092	30351	3.9593	.00102775	3056.767	743559.22
974	948676	924010424	31.2090	9.9126	30398	3.9602	.00102669	3059.909	745088.39
975	950625	926859375	31.2250	9.9160	30444	3.9610	.00102564	3063.050	746619.13
976	952576	929714176	31.2410	9.9194	30491	3.9618	.00102459	3066.192	748151.44
977	954529	932574833	31.2570	9.9227	30538	3.9626	.00102354	3069.333	749685.32
978	956484	935441352	31.2730	9.9261	30585	3.9634	.00102249	3072.475	751220.78
979	958441	938313739	31.2890	9.9295	30632	3.9642	.00102145	3075.617	752757.80
980	**960400**	**941192000**	**31.3050**	**9.9329**	**30679**	**3.9650**	**.00102041**	**3078.758**	**754296.40**
981	962361	944076141	31.3209	9.9363	30726	3.9658	.00101937	3081.900	755836.59
982	964324	946966168	31.3369	9.9396	30773	3.9666	.00101833	3085.041	757378.30
983	966289	949862087	31.3528	9.9430	30820	3.9674	.00101729	3088.183	758921.61
984	968256	952763904	31.3688	9.9464	30867	3.9682	.00101626	3091.325	760466.48
985	970225	955671625	31.3847	9.9497	30914	3.9691	.00101523	3094.466	762012.93
986	972196	958585256	31.4006	9.9531	30961	3.9699	.00101420	3097.608	763560.95
987	974169	961504803	31.4166	9.9565	31008	3.9707	.00101317	3100.749	765110.54
988	976144	964430272	31.4325	9.9598	31055	3.9715	.00101215	3103.891	766661.70
989	978121	967361669	31.4484	9.9632	31102	3.9723	.00101112	3107.033	768214.44

Table 7 —Continued

N	N^2	N^3	$\sqrt{N}$	$\sqrt[3]{N}$	$N^{3/2}$	$\sqrt[5]{N}$	$\frac{1}{N}$	Circle ($N = D$) Circum.	Area
990	980100	970299000	31.4643	9.9666	31150	3.9731	.00101010	3110.174	769768.74
991	982081	973242271	31.4802	9.9699	31197	3.9739	.00100908	3113.316	771324.61
992	984064	976191488	31.4960	9.9733	31244	3.9747	.00100806	3116.457	772882.06
993	986049	979146657	31.5119	9.9766	31291	3.9755	.00100705	3119.599	774441.07
994	988036	982107784	31.5278	9.9800	31339	3.9763	.00100604	3122.740	776001.66
995	990025	985074875	31.5436	9.9833	31386	3.9771	.00100503	3125.882	777563.82
996	992016	988047936	31.5595	9.9866	31433	3.9779	.00100402	3129.024	779127.54
997	994009	991026973	31.5753	9.9900	31480	3.9787	.00100301	3132.165	780692.84
998	996004	994011992	31.5911	9.9933	31528	3.9795	.00100200	3135.307	782259.71
999	998001	997002999	31.6070	9.9967	31575	3.9803	.00100100	3138.448	783828.15
1000	1000000	1000000000	31.6228	10.0000	31623	3.9811	.00100000	3141.593	785398.16

Table 8.   Inches to Decimals of a Foot

In.	Ft.	In.	Ft.	In.	Ft.	In.	Ft.	In.	Ft.	In.	Ft.	In.	Ft.
1/16	.0052	5/16	.0260	9/16	.0469	13/16	.0677	1	.0833	5	.4167	9	.7500
1/8	.0104	3/8	.0313	5/8	.0521	7/8	.0729	2	.1667	6	.5000	10	.8333
3/16	.0156	7/16	.0364	11/16	.0573	15/16	.0781	3	.2500	7	.5833	11	.9167
1/4	.0208	1/2	.0417	3/4	.0625	1	.0833	4	.3333	8	.6667	12	1.0000

Table 9.   Circular Arcs, Chords, and Segments

Central Angle in Degrees	Arc $\frac{}{R}$	Height $\frac{}{R}$	Chord $\frac{}{R}$	Height $\frac{}{Chord}$	Area $\frac{}{R^2}$	Central Angle in Degrees	Arc $\frac{}{R}$	Height $\frac{}{R}$	Chord $\frac{}{R}$	Height $\frac{}{Chord}$	Area $\frac{}{R^2}$
1	0.0175	0.0000	0.0175	0.0022	0.00000	31	0.5411	0.0364	0.5345	0.0680	0.01301
2	.0349	.0002	.0349	.0044	.00000	32	.5585	.0387	.5513	.0703	.01429
3	.0524	.0003	.0524	.0066	.00001	33	.5760	.0412	.5680	.0725	.01566
4	.0698	.0006	.0698	.0087	.00003	34	.5934	.0437	.5847	.0747	.01711
5	.0873	.0010	.0872	.0109	.00006	35	.6109	.0463	.6014	.0770	.01864
6	.1047	.0014	.1047	.0131	.00010	36	.6283	.0489	.6180	.0792	.02027
7	.1222	.0019	.1221	.0153	.00015	37	.6458	.0517	.6346	.0814	.02198
8	.1396	.0024	.1395	.0175	.00023	38	.6632	.0545	.6511	.0837	.02378
9	.1571	.0031	.1569	.0196	.00032	39	.6807	.0574	.6676	.0859	.02568
10	.1745	.0038	.1743	.0218	.00044	40	.6981	.0603	.6840	.0882	.02767
11	.1920	.0046	.1917	.0240	.00059	41	.7156	.0633	.7004	.0904	.02976
12	.2094	.0055	.2091	.0262	.00076	42	.7330	.0664	.7167	.0927	.03195
13	.2269	.0064	.2264	.0284	.00097	43	.7505	.0696	.7330	.0949	.03425
14	.2443	.0075	.2437	.0306	.00121	44	.7679	.0728	.7492	.0972	.03664
15	.2618	.0086	.2611	.0328	.00149	45	.7854	.0761	.7654	.0995	.03915
16	.2793	.0097	.2783	.0350	.00181	46	.8029	.0795	.7815	.1017	.04176
17	.2967	.0110	.2956	.0372	.00217	47	.8203	.0829	.7975	.1040	.04448
18	.3142	.0123	.3129	.0394	.00257	48	.8378	.0865	.8135	.1063	.04731
19	.3316	.0137	.3301	.0415	.00302	49	.8552	.0900	.8294	.1086	.05025
20	.3491	.0152	.3473	.0437	.00352	50	.8727	.0937	.8452	.1108	.05331
21	.3665	.0167	.3645	.0459	.00408	51	.8901	.0974	.8610	.1131	.05649
22	.3840	.0184	.3816	.0481	.00468	52	.9076	.1012	.8767	.1154	.05978
23	.4014	.0201	.3987	.0503	.00535	53	.9250	.1051	.8924	.1177	.06319
24	.4189	.0219	.4158	.0526	.00607	54	.9425	.1090	.9080	.1200	.06673
25	.4363	.0237	.4329	.0548	.00686	55	.9599	.1130	.9235	.1223	.07039
26	.4538	.0256	.4499	.0570	.00771	56	.9774	.1171	.9389	.1247	.07417
27	.4712	.0276	.4669	.0592	.00862	57	.9948	.1212	.9543	.1270	.07808
28	.4887	.0297	.4838	.0614	.00961	58	1.0123	.1254	.9696	.1293	.08212
29	.5061	.0319	.5008	.0636	.01067	59	1.0297	.1296	.9848	.1316	.08629
30	.5236	.0341	.5176	.0658	.01180	60	1.0472	.1340	1.0000	.1340	.09059

Table 9 —Continued

Central Angle in Degrees	Arc R	Height R	Chord R	Height Chord	Area R²	Central Angle in Degrees	Arc R	Height R	Chord R	Height Chord	Area R²
61	1.0647	.1384	1.015	.1363	.09502	121	2.1118	.5076	1.741	.2916	.62734
62	1.0821	.1428	1.030	.1387	.09958	122	2.1293	.5152	1.749	.2945	.64063
63	1.0996	.1474	1.045	.1410	.10428	123	2.1468	.5228	1.758	.2975	.65404
64	1.1170	.1520	1.060	.1434	.10911	124	2.1642	.5305	1.766	.3004	.66759
65	1.1345	.1566	1.075	.1457	.11408	125	2.1817	.5383	1.774	.3034	.68125
66	1.1519	.1613	1.089	.1481	.11919	126	2.1991	.5460	1.782	.3064	.69505
67	1.1694	.1661	1.104	.1505	.12443	127	2.2166	.5538	1.790	.3094	.70897
68	1.1868	.1710	1.118	.1529	.12982	128	2.2340	.5616	1.798	.3124	.72301
69	1.2043	.1759	1.133	.1553	.13535	129	2.2515	.5695	1.805	.3155	.73716
70	1.2217	.1808	1.147	.1576	.14102	130	2.2689	.5774	1.813	.3185	.75143
71	1.2392	.1859	1.161	.1601	.14683	131	2.2864	.5853	1.820	.3216	.76584
72	1.2566	.1910	1.176	.1625	.15279	132	2.3038	.5933	1.827	.3247	.78034
73	1.2741	.1961	1.190	.1649	.15889	133	2.3213	.6013	1.834	.3278	.79497
74	1.2915	.2014	1.204	.1673	.16514	134	2.3387	.6093	1.841	.3309	.80970
75	1.3090	.2066	1.218	.1697	.17154	135	2.3562	.6173	1.848	.3341	.82454
76	1.3265	.2120	1.231	.1722	.17808	136	2.3736	.6254	1.854	.3373	.83949
77	1.3439	.2174	1.245	.1746	.18477	137	2.3911	.6335	1.861	.3404	.85455
78	1.3614	.2229	1.259	.1771	.19160	138	2.4086	.6416	1.867	.3436	.86971
79	1.3788	.2284	1.272	.1795	.19859	139	2.4260	.6498	1.873	.3469	.88497
80	1.3963	.2340	1.286	.1820	.20573	140	2.4435	.6580	1.879	.3501	.90034
81	1.4137	.2396	1.299	.1845	.21301	141	2.4609	.6662	1.885	.3534	.91580
82	1.4312	.2453	1.312	.1869	.22045	142	2.4784	.6744	1.891	.3566	.93135
83	1.4486	.2510	1.325	.1894	.22804	143	2.4958	.6827	1.897	.3599	.94700
84	1.4661	.2569	1.338	.1919	.23578	144	2.5133	.6910	1.902	.3633	.96274
85	1.4835	.2627	1.351	.1944	.24367	145	2.5307	.6993	1.907	.3666	.97858
86	1.5010	.2686	1.364	.1970	.25171	146	2.5482	.7076	1.913	.3700	.99449
87	1.5184	.2746	1.377	.1995	.25990	147	2.5656	.7160	1.918	.3734	1.0105
88	1.5359	.2807	1.389	.2020	.26825	148	2.5831	.7244	1.923	.3768	1.0266
89	1.5533	.2867	1.402	.2046	.27675	149	2.6005	.7328	1.927	.3802	1.0428
90	1.5708	.2929	1.414	.2071	.28540	150	2.6180	.7412	1.932	.3837	1.0590
91	1.5882	.2991	1.427	.2097	.29420	151	2.6354	.7496	1.936	.3871	1.0753
92	1.6057	.3053	1.439	.2122	.30316	152	2.6529	.7581	1.941	.3906	1.0917
93	1.6232	.3116	1.451	.2148	.31226	153	2.6704	.7666	1.945	.3942	1.1082
94	1.6406	.3180	1.463	.2174	.32152	154	2.6878	.7750	1.949	.3977	1.1247
95	1.6581	.3244	1.475	.2200	.33093	155	2.7053	.7836	1.953	.4013	1.1413
96	1.6755	.3309	1.486	.2226	.34050	156	2.7227	.7921	1.956	.4049	1.1580
97	1.6930	.3374	1.498	.2252	.35021	157	2.7402	.8006	1.960	.4085	1.1747
98	1.7104	.3439	1.509	.2279	.36008	158	2.7576	.8092	1.963	.4122	1.1915
99	1.7279	.3506	1.521	.2305	.37009	159	2.7751	.8178	1.967	.4158	1.2084
100	1.7453	.3572	1.532	.2332	.38026	160	2.7925	.8264	1.970	.4195	1.2253
101	1.7628	.3639	1.543	.2358	.39058	161	2.8100	.8350	1.973	.4233	1.2422
102	1.7802	.3707	1.554	.2385	.40104	162	2.8274	.8436	1.975	.4270	1.2592
103	1.7977	.3775	1.565	.2412	.41166	163	2.8449	.8522	1.978	.4308	1.2763
104	1.8151	.3843	1.576	.2439	.42242	164	2.8623	.8608	1.981	.4346	1.2934
105	1.8326	.3912	1.587	.2466	.43333	165	2.8798	.8695	1.983	.4385	1.3105
106	1.8500	.3982	1.597	.2493	.44439	166	2.8972	.8781	1.985	.4424	1.3277
107	1.8675	.4052	1.608	.2520	.45560	167	2.9147	.8868	1.987	.4463	1.3449
108	1.8850	.4122	1.618	.2548	.46695	168	2.9322	.8955	1.989	.4502	1.3621
109	1.9024	.4193	1.628	.2575	.47844	169	2.9496	.9042	1.991	.4542	1.3794
110	1.9199	.4264	1.638	.2603	.49008	170	2.9671	.9128	1.992	.4582	1.3967
111	1.9373	.4336	1.648	.2631	.50187	171	2.9845	.9215	1.994	.4622	1.4140
112	1.9548	.4408	1.658	.2659	.51379	172	3.0020	.9302	1.995	.4663	1.4314
113	1.9722	.4481	1.668	.2687	.52586	173	3.0194	.9390	1.996	.4704	1.4488
114	1.9897	.4554	1.677	.2715	.53807	174	3.0369	.9477	1.997	.4745	1.4662
115	2.0071	.4627	1.687	.2743	.55041	175	3.0543	.9564	1.998	.4786	1.4836
116	2.0246	.4701	1.696	.2772	.56289	176	3.0718	.9651	1.999	.4828	1.5010
117	2.0420	.4775	1.705	.2800	.57551	177	3.0892	.9738	1.999	.4871	1.5184
118	2.0595	.4850	1.714	.2829	.58827	178	3.1067	.9825	2.000	.4914	1.5359
119	2.0769	.4925	1.723	.2858	.60116	179	3.1241	.9913	2.000	.4957	1.5533
120	2.0944	.5000	1.732	.2887	.61418	180	3.1416	1.0000	2.000	.5000	1.5708

Table 10.   Common Logarithms of Numbers

Mantissas in Six Decimal Places

The common logarithm of a number is the index of the power to which the base 10 must be raised in order to equal the number.

The common logarithm of every positive number not an integral power of 10 consists of an *integral* and a *decimal part*. The integral part or whole number is called the *characteristic* and may be either *positive or negative*. The decimal or fractional part is a *positive* number called the *mantissa* and is the same for all numbers which have the same sequential digits.

The characteristic of the logarithm of any positive number greater than one is positive and is one less than the number of digits before the decimal point.

The characteristic of the logarithm of any positive number less than one is negative and is one more than the number of ciphers immediately after the decimal point.

A negative number or number less than zero has no real logarithm.

Examples. $\mathrm{Log}_{10}\ 25400. = 4.404834$ $\mathrm{Log}_{10}\ 0.0254 = \bar{2}.404834$ or $8.404834 - 10$

The two systems of logarithms in general use are the common or Briggsian logarithms, introduced in 1615 by Henry Briggs, a contemporary of John Napier, the inventor of logarithms, and the natural or less appropriately termed Napierian or hyperbolic logarithms, which developed somewhat accidentally from Napier's original work. The latter have a base denoted by e, an irrational number, which is:

$$\mathrm{Lim}_{u=\infty}\left(1 + \frac{1}{u}\right)^{u} = 1 + 1 + \frac{1}{2!} + \frac{1}{3!} + \frac{1}{4!} + \ldots = 2.7182818$$

To obtain the natural logarithm, the common logarithm is multiplied by $\log_e 10$ which is 2.302585, or $\log_e N = 2.302585 \log_{10} N$.

N	0	1	2	3	4	5	6	7	8	9
0		000000	301030	477121	602060	698970	778151	845098	903090	954243
1	000000	041393	079181	113943	146128	176091	204120	230449	255273	278754
2	301030	322219	342423	361728	380211	397940	414973	431364	447158	462398
3	477121	491362	505150	518514	531479	544068	556303	568202	579784	591065
4	602060	612784	623249	633468	643453	653213	662758	672098	681241	690196
5	698970	707570	716003	724276	732394	740363	748188	755875	763428	770852
6	778151	785330	792392	799341	806180	812913	819544	826075	832509	838849
7	845098	851258	857332	863323	869232	875061	880814	886491	892095	897627
8	903090	908485	913814	919078	924279	929419	934498	939519	944483	949390
9	954243	959041	963788	968483	973128	977724	982271	986772	991226	995635
10	000000	004321	008600	012837	017033	021189	025306	029384	033424	037426
1	041393	045323	049218	053078	056905	060698	064458	068186	071882	075547
2	079181	082785	086360	089905	093422	096910	100371	103804	107210	110590
3	113943	117271	120574	123852	127105	130334	133539	136721	139879	143015
4	146128	149219	152288	155336	158362	161368	164353	167317	170262	173186
5	176091	178977	181844	184691	187521	190332	193125	195900	198657	201397
6	204120	206826	209515	212188	214844	217484	220108	222716	225309	227887
7	230449	232996	235528	238046	240549	243038	245513	247973	250420	252853
8	255273	257679	260071	262451	264818	267172	269513	271842	274158	276462
9	278754	281033	283301	285557	287802	290035	292256	294466	296665	298853
20	301030	303196	305351	307496	309630	311754	313867	315970	318063	320146
1	322219	324282	326336	328380	330414	332438	334454	336460	338456	340444
2	342423	344392	346353	348305	350248	352183	354108	356026	357935	359835
3	361728	363612	365488	367356	369216	371068	372912	374748	376577	378398
4	380211	382017	383815	385606	387390	389166	390935	392697	394452	396199
5	397940	399674	401401	403121	404834	406540	408240	409933	411620	413300
6	414973	416641	418301	419956	421604	423246	424882	426511	428135	429752
7	431364	432969	434569	436163	437751	439333	440909	442480	444045	445604
8	447158	448706	450249	451786	453318	454845	456366	457882	459392	460898
9	462398	863893	465383	466868	468347	469822	471292	472756	474216	475671
30	477121	478566	480007	481443	482874	484300	485721	487138	488551	489958
1	491362	492760	494155	495544	496930	498311	499687	501059	502427	503791
2	505150	506505	507856	509203	510545	511883	513218	514548	515874	517196
3	518514	519828	521138	522444	523746	525045	526339	527630	528917	530200
4	531479	532754	534026	535294	536558	537819	539076	540329	541579	542825
5	544068	545307	546543	547775	549003	550228	551450	552668	553883	555094

N	0	1	2	3	4	5	6	7	8	9
5	544068	545307	546543	547775	549003	550228	551450	552668	553883	555094
6	556303	557507	558709	559907	561101	562293	563481	564666	565848	567026
7	568202	569374	570543	571709	572872	574031	575188	576341	577492	578639
8	579784	580925	582063	583199	584331	585461	586587	587711	588832	589950
9	591065	592177	593286	594393	595496	596597	597695	598791	599883	600973
40	602060	603144	604226	605305	606381	607455	608526	609594	610660	611723
1	612784	613842	614897	615950	617000	618048	619093	620136	621176	622214
2	623249	624232	625312	626340	627366	628389	629410	630428	631444	632457
3	633468	634477	635484	636488	637490	638489	639486	640481	641474	642465
4	643453	644439	645422	646404	647383	648360	649335	650308	651278	652246
5	653213	654177	655138	656098	657056	658011	658965	659916	660865	661713
6	662758	663701	664642	665581	666518	667453	668386	669317	670246	671173
7	672098	673021	673942	674861	675778	676694	677607	678518	679428	680336
8	681241	682145	683047	683947	684845	685742	686636	687529	688420	689309
9	690196	691081	691965	692847	693727	694605	695482	696356	697229	698100
50	698970	699838	700704	701568	702431	703291	704151	705008	705864	706718
1	707570	708421	709270	710117	710963	711807	712650	713491	714330	715167
2	716003	716838	717671	718502	719331	720159	720986	721811	722634	723456
3	724276	725095	725912	726727	727541	728354	729165	729974	730782	731589
4	732394	733197	733999	734800	735599	736397	737193	737987	738781	739572
5	740363	741152	741939	742725	743510	744293	745075	745855	746634	747412
6	748188	748963	749736	750508	751279	752048	752816	753583	754348	755112
7	755875	756636	757396	758155	758912	759668	760422	761176	761928	762679
8	763428	764176	764923	765669	766413	767156	767898	768638	769377	770115
9	770852	771587	772322	773055	773786	774517	775246	775974	776701	777427
60	778151	778874	779596	780317	781037	781755	782473	783189	783904	784617
1	785330	786041	786751	787460	788168	788875	789581	790285	790988	791691
2	792392	793092	793790	794488	795185	795880	796574	797268	797960	798651
3	799341	800029	800717	801404	802089	802774	803457	804139	804821	805501
4	806180	806858	807535	808211	808886	809560	810233	810904	811575	812245
5	812913	813581	814248	814913	815578	816241	816904	817565	818226	818885
6	819544	820201	820858	821514	822168	822822	823474	824126	824776	825426
7	826075	826723	827369	828015	828660	829304	829947	830589	831230	831870
8	832509	833147	833784	834421	835056	835691	836324	836957	837588	838219
9	838849	839478	840106	840733	841359	841985	842609	843233	843856	844477
70	845098	845718	846337	846955	847573	848189	848805	849419	850033	850646
1	851258	851870	852480	853090	853698	854306	854913	855519	856124	856729
2	857332	857935	858537	859138	859739	860338	860937	861534	862131	862728
3	863323	863917	864511	865104	865696	866287	866878	867467	868056	868644
4	869232	869818	870404	870989	871573	872156	872739	873321	873902	874482
5	875061	875640	876218	876795	877371	877947	878522	879096	879669	880242
6	880814	881385	881955	882525	883093	883661	884229	884795	885361	885926
7	886491	887054	887617	888179	888741	889302	889862	890421	890980	891537
8	892095	892651	893207	893762	894316	894870	895423	895975	896526	897077
9	897627	898176	898725	899273	899821	900367	900913	901458	902003	902547
80	903090	903633	904174	904716	905256	905796	906335	906874	907411	907949
1	908485	909021	909556	910091	910624	911158	911690	912222	912753	913284
2	913814	914343	914872	915400	915927	916454	916980	917506	918030	918555
3	919078	919601	920123	920645	921166	921686	922206	922725	923244	923762
4	924279	924796	925312	925828	926342	926857	927370	927883	928396	928908
5	929419	929930	930440	930949	931458	931966	932474	932981	933487	933993
6	934498	935003	935507	936011	936514	937016	937518	938019	938520	939020
7	939519	940018	940516	941014	941511	942008	942504	943000	943495	943989
8	944483	944976	945469	945961	946452	946943	947434	947924	948413	948902
9	949390	949878	950365	950851	951338	951823	952308	952792	953276	953760
90	954243	954725	955207	955688	956168	956649	957128	957607	958086	958564
1	959041	959518	959995	960471	960946	961421	961895	962369	962843	963316
2	963788	964260	964731	965202	965672	966142	966611	967080	967548	968016
3	968483	968950	969416	969882	970347	970812	971276	971740	972203	972666
4	973128	973590	974051	974512	974972	975432	975891	976350	976808	977266
5	977724	978181	978637	979093	979548	980003	980458	980912	981366	981819
6	982271	982723	983175	983626	984077	984527	984977	985426	985875	986324
7	986772	987219	987666	988113	988559	989005	989450	989895	990339	990783
8	991226	991669	992111	992554	992995	993436	993877	994317	994757	995196
9	995635	996074	996512	996949	997386	997823	998259	998695	999131	999565
100	000000	000434	000868	001301	001734	002166	002598	003029	003461	003891

N	0	1	2	3	4	5	6	7	8	9	Diff.
100	000000	000434	000868	001301	001734	002166	002598	003029	003461	003891	432
1	004321	004751	005181	005609	006038	006466	006894	007321	007748	008174	428
2	008600	009026	009451	009876	010300	010724	011147	011570	011993	012415	424
3	012837	013259	013680	014100	014521	014940	015360	015779	016197	016616	420
4	017033	017451	017868	018284	018700	019116	019532	019947	020361	020775	416
5	021189	021603	022016	022428	022841	023252	023664	024075	024486	024896	412
6	025306	025715	026125	026533	026942	027350	027757	028164	028571	028978	408
7	029384	029789	030195	030600	031004	031408	031812	032216	032619	033021	404
8	033424	033826	034227	034628	035029	035430	035830	036230	036629	037028	400
9	037426	037825	038223	038620	039017	039414	039811	040207	040602	040998	397
110	041393	041787	042182	042576	042969	043362	043755	044148	044540	044932	393
1	045323	045714	046105	046495	046885	047275	047664	048053	048442	048830	390
2	049218	049606	049993	050380	050766	051153	051538	051924	052309	052694	386
3	053078	053463	053846	054230	054613	054996	055378	055760	056142	056524	383
4	056905	057286	057666	058046	058426	058805	059185	059563	059942	060320	379
5	060698	061075	061452	061829	062206	062582	062958	063333	063709	064083	376
6	064458	064832	065206	065580	065953	066326	066699	067071	067443	067815	373
7	068186	068557	068928	069298	069668	070038	070407	070776	071145	071514	370
8	071882	072250	072617	072985	073352	073718	074085	074451	074816	075182	366
9	075547	075912	076276	076640	077004	077368	077731	078094	078457	078819	363
120	079181	079543	079904	080266	080626	080987	081347	081707	082067	082426	360

PROPORTIONAL PARTS

Diff.	1	2	3	4	5	6	7	8	9
434	43.4	86.8	130.2	173.6	217.0	260.4	303.8	347.2	390.6
432	43.2	86.4	129.6	172.8	216.0	259.2	302.4	345.6	388.8
430	43.0	86.0	129.0	172.0	215.0	258.0	301.0	344.0	387.0
428	42.8	85.6	128.4	171.2	214.0	256.8	299.6	342.4	385.2
426	42.6	85.2	127.8	170.4	213.0	255.6	298.2	340.8	383.4
424	42.4	84.8	127.2	169.6	212.0	254.4	296.8	339.2	381.6
422	42.2	84.4	126.6	168.8	211.0	253.2	295.4	337.6	379.8
420	42.0	84.0	126.0	168.0	210.0	252.0	294.0	336.0	378.0
418	41.8	83.6	125.4	167.2	209.0	250.8	292.6	334.4	376.2
416	41.6	83.2	124.8	166.4	208.0	249.6	291.2	332.8	374.4
414	41.4	82.8	124.2	165.6	207.0	248.4	289.8	331.2	372.6
412	41.2	82.4	123.6	164.8	206.0	247.2	288.4	329.6	370.8
410	41.0	82.0	123.0	164.0	205.0	246.0	287.0	328.0	369.0
408	40.8	81.6	122.4	163.2	204.0	244.8	285.6	326.4	367.2
406	40.6	81.2	121.8	162.4	203.0	243.6	284.2	324.8	365.4
404	40.4	80.8	121.2	161.6	202.0	242.4	282.8	323.2	363.6
402	40.2	80.4	120.6	160.8	201.0	241.2	281.4	321.6	361.8
400	40.0	80.0	120.0	160.0	200.0	240.0	280.0	320.0	360.0
398	39.8	79.6	119.4	159.2	199.0	238.8	278.6	318.4	358.2
396	39.6	79.2	118.8	158.4	198.0	237.6	277.2	316.8	356.4
394	39.4	78.8	118.2	157.6	197.0	236.4	275.8	315.2	354.6
392	39.2	78.4	117.6	156.8	196.0	235.2	274.4	313.6	352.8
390	39.0	78.0	117.0	156.0	195.0	234.0	273.0	312.0	351.0
388	38.8	77.6	116.4	155.2	194.0	232.8	271.6	310.4	349.2
386	38.6	77.2	115.8	154.4	193.0	231.6	270.2	308.8	347.4
384	38.4	76.8	115.2	153.6	192.0	230.4	268.8	307.2	345.6
382	38.2	76.4	114.6	152.8	191.0	229.2	267.4	305.6	343.8
380	38.0	76.0	114.0	152.0	190.0	228.0	266.0	304.0	342.0
378	37.8	75.6	113.4	151.2	189.0	226.8	264.6	302.4	340.2
376	37.6	75.2	112.8	150.4	188.0	225.6	263.2	300.8	338.4
374	37.4	74.8	112.2	149.6	187.0	224.4	261.8	299.2	336.6
372	37.2	74.4	111.6	148.8	186.0	223.2	260.4	297.6	334.8
370	37.0	74.0	111.0	148.0	185.0	222.0	259.0	296.0	333.0

N	0	1	2	3	4	5	6	7	8	9	Diff.
120	079181	079543	079904	080266	080626	080987	081347	081707	082067	082426	360
1	082785	083144	083503	083861	084219	084576	084934	085291	085647	086004	357
2	086360	086716	087071	087426	087781	088136	088490	088845	089198	089552	355
3	089905	090258	090611	090963	091315	091667	092018	092370	092721	093071	352
4	093422	093772	094122	094471	094820	095169	095518	095866	096215	096562	349
5	096910	097257	097604	097951	098298	098644	098990	099335	099681	100026	346
6	100371	100715	101059	101403	101747	102091	102434	102777	103119	103462	343
7	103804	104146	104487	104828	105169	105510	105851	106191	106531	106871	341
8	107210	107549	107888	108227	108565	108903	109241	109579	109916	110253	338
9	110590	110926	111263	111599	111934	112270	112605	112940	113275	113609	335
130	113943	114277	114611	114944	115278	115611	115943	116276	116608	116940	333
1	117271	117603	117937	118265	118595	118926	119256	119586	119915	120245	330
2	120574	120903	121231	121560	121888	122216	122544	122871	123198	123525	328
3	123852	124178	124504	124830	125156	125481	125806	126131	126456	126781	325
4	127105	127429	127753	128076	128399	128722	129045	129368	129690	130012	323
5	130334	130655	130977	131298	131619	131939	132260	132580	132900	133219	321
6	133539	133858	134177	134496	134814	135133	135451	135769	136086	136403	318
7	136721	137037	137354	137671	137987	138303	138618	138934	139249	139564	316
8	139879	140194	140508	140822	141136	141450	141763	142076	142389	142702	314
9	143015	143327	143639	143951	144263	144574	144885	145196	145507	145818	311
140	146128	146438	146748	147058	147367	147676	147985	148294	148603	148911	309

PROPORTIONAL PARTS

Diff.	1	2	3	4	5	6	7	8	9
370	37.0	74.0	111.0	148.0	185.0	222.0	259.0	296.0	333.0
368	36.8	73.6	110.4	147.2	184.0	220.8	257.6	294.4	331.2
366	36.6	73.2	109.8	146.4	183.0	219.6	256.2	292.8	329.4
36	36.4	72.8	109.2	145.6	182.0	218.4	254.8	291.2	327.6
362	36.2	72.4	108.6	144.8	181.0	217.2	253.4	289.6	325.8
360	36.0	72.0	108.0	144.0	180.0	216.0	252.0	288.0	324.0
358	35.8	71.6	107.4	143.2	179.0	214.8	250.6	286.4	322.2
356	35.6	71.2	106.8	142.4	178.0	213.6	249.2	284.8	320.4
354	35.4	70.8	106.2	141.6	177.0	212.4	247.8	283.2	318.6
352	35.2	70.4	105.6	140.8	176.0	211.2	246.4	281.6	316.8
350	35.0	70.0	105.0	140.0	175.0	210.0	245.0	280.0	315.0
348	34.8	69.6	104.4	139.2	174.0	208.8	243.6	278.4	313.2
346	34.6	69.2	103.8	138.4	173.0	207.6	242.2	276.8	311.4
344	34.4	68.8	103.2	137.6	172.0	206.4	240.8	275.2	309.6
342	34.2	68.4	102.6	136.8	171.0	205.2	239.4	273.6	307.8
340	34.0	68.0	102.0	136.0	170.0	204.0	238.0	272.0	306.0
338	33.8	67.6	101.4	135.2	169.0	202.8	236.6	270.4	304.2
336	33.6	67.2	100.8	134.4	168.0	201.6	235.2	268.8	302.4
334	33.4	66.8	100.2	133.6	167.0	200.4	233.8	267.2	300.6
332	33.2	66.4	99.6	132.8	166.0	199.2	232.4	265.6	298.8
330	33.0	66.0	99.0	132.0	165.0	198.0	231.0	264.0	297.0
328	32.8	65.6	98.4	131.2	164.0	196.8	229.6	262.4	295.2
326	32.6	65.2	97.8	130.4	163.0	195.6	228.2	260.8	293.4
324	32.4	64.8	97.2	129.6	162.0	194.4	226.8	259.2	291.6
322	32.2	64.4	96.6	128.8	161.0	193.2	225.4	257.6	289.8
320	32.0	64.0	96.0	128.0	160.0	192.0	224.0	256.0	288.0
318	31.8	63.6	95.4	127.2	159.0	190.8	222.6	254.4	286.2
316	31.6	63.2	94.8	126.4	158.0	189.6	221.2	252.8	284.4
314	31.4	62.8	94.2	125.6	157.0	188.4	219.8	251.2	282.6
312	31.2	62.4	93.6	124.8	156.0	187.2	218.4	249.6	280.8
310	31.0	62.0	93.0	124.0	155.0	186.0	217.0	248.0	279.0
308	30.8	61.6	92.4	123.2	154.0	184.8	215.6	246.4	277.2

N	0	1	2	3	4	5	6	7	8	9	Diff.
140	146128	146438	146748	147058	147367	147676	147985	148294	148603	148911	309
1	149219	149527	149835	150142	150449	150756	151063	151370	151676	151982	307
2	152288	152594	152900	153205	153510	153815	154120	154424	154728	155032	305
3	155336	155640	155943	156246	156549	156852	157154	157457	157759	158061	303
4	158362	158664	158965	159266	159567	159868	160168	160469	160769	161068	301
5	161368	161667	161967	162266	162564	162863	163161	163460	163758	164055	299
6	164353	164650	164947	165244	165541	165838	166134	166430	166726	167022	297
7	167317	167613	167908	168203	168497	168792	169086	169380	169674	169968	295
8	170262	170555	170848	171141	171434	171726	172019	172311	172603	172895	293
9	173186	173478	173769	174060	174351	174641	174932	175222	175512	175802	291
150	176091	176381	176670	176959	177248	177536	177825	178113	178401	178689	289
1	178977	179264	179552	179839	180126	180413	180699	180986	181272	181558	287
2	181844	182129	182415	182700	182985	183270	183555	183839	184123	184407	285
3	184691	184975	185259	185542	185825	186108	186391	186674	186956	187239	283
4	187521	187803	188084	188366	188647	188928	189209	189490	189771	190051	281
5	190332	190612	190892	191171	191451	191730	192010	192289	192567	192846	279
6	193125	193403	193681	193959	194237	194514	194792	195069	195346	195623	278
7	195900	196176	196453	196729	197005	197281	197556	197832	198107	198382	276
8	198657	198932	199206	199481	199755	200029	200303	200577	200850	201124	274
9	201397	201670	201943	202216	202488	202761	203033	203305	203577	203848	272
160	204120	204391	204663	204934	205204	205475	205746	206016	206286	206556	271
1	206826	207096	207365	207634	207904	208173	208441	208710	208979	209247	269
2	209515	209783	210051	210319	210586	210853	211121	211388	211654	211921	267
3	212188	212454	212720	212986	213252	213518	213783	214049	214314	214579	266
4	214844	215109	215373	215638	215902	216166	216430	216694	216957	217221	264
5	217484	217747	218010	218273	218536	218798	219060	219323	219585	219846	262
6	220108	220370	220631	220892	221153	221414	221675	221936	222196	222456	261
7	222716	222976	223236	223496	223755	224015	224274	224533	224792	225051	259
8	225309	225568	225826	226084	226342	226600	226858	227115	227372	227630	258
9	227887	228144	228400	228657	228913	229170	229426	229682	229938	230193	256
170	230449	230704	230960	231215	231470	231724	231979	232234	232488	232742	255

PROPORTIONAL PARTS

Diff.	1	2	3	4	5	6	7	8	9
310	31.0	62.0	93.0	124.0	155.0	186.0	217.0	248.0	279.0
308	30.8	61.6	92.4	123.2	154.0	184.8	215.6	246.4	277.2
306	30.6	61.2	91.8	122.4	153.0	183.6	214.2	244.8	275.4
304	30.4	60.8	91.2	121.6	152.0	182.4	212.8	243.2	273.6
302	30.2	60.4	90.6	120.8	151.0	181.2	211.4	241.6	271.8
300	30.0	60.0	90.0	120.0	150.0	180.0	210.0	240.0	270.0
298	29.8	59.6	89.4	119.2	149.0	178.8	208.6	238.4	268.2
296	29.6	59.2	88.8	118.4	148.0	177.6	207.2	236.8	266.4
294	29.4	58.8	88.2	117.6	147.0	176.4	205.8	235.2	264.6
292	29.2	58.4	87.6	116.8	146.0	175.2	204.4	233.6	262.8
290	29.0	58.0	87.0	116.0	145.0	174.0	203.0	232.0	261.0
288	28.8	57.6	86.4	115.2	144.0	172.8	201.6	230.4	259.2
286	28.6	57.2	85.8	114.4	143.0	171.6	200.2	228.8	257.4
284	28.4	56.8	85.2	113.6	142.0	170.4	198.8	227.2	255.6
282	28.2	56.4	84.6	112.8	141.0	169.2	197.4	225.6	253.8
280	28.0	56.0	84.0	112.0	140.0	168.0	196.0	224.0	252.0
278	27.8	55.6	83.4	111.2	139.0	166.8	194.6	222.4	250.2
276	27.6	55.2	82.8	110.4	138.0	165.6	193.2	220.8	248.4
274	27.4	54.8	82.2	109.6	137.0	164.4	191.8	219.2	246.6
272	27.2	54.4	81.6	108.8	136.0	163.2	190.4	217.6	244.8
270	27.0	54.0	81.0	108.0	135.0	162.0	189.0	216.0	243.0
268	26.8	53.6	80.4	107.2	134.0	160.8	187.6	214.4	241.2
266	26.6	53.2	79.8	106.4	133.0	159.6	186.2	212.8	239.4
264	26.4	52.8	79.2	105.6	132.0	158.4	184.8	211.2	237.6

N	0	1	2	3	4	5	6	7	8	9	Diff.
170	230449	230704	230960	231215	231470	231724	231979	232234	232488	232742	255
1	232996	233250	233504	233757	234011	234264	234517	234770	235023	235276	253
2	235528	235781	236033	236285	236537	236789	237041	237292	237544	237795	252
3	238046	238297	238548	238799	239049	239299	239550	239800	240050	240300	250
4	240549	240799	241048	241297	241546	241795	242044	242293	242541	242790	249
5	243038	243286	243534	243782	244030	244277	244525	244772	245019	245266	248
6	245513	245759	246006	246252	246499	246745	246991	247237	247482	247728	246
7	247973	248219	248464	248709	248954	249198	249443	249687	249932	250176	245
8	250420	250664	250908	251151	251395	251638	251881	252125	252368	252610	243
9	252853	253096	253338	253580	253822	254064	254306	254548	254790	255031	242
180	255273	255514	255755	255996	256237	256477	256718	256958	257198	257439	241
1	257679	257918	258158	258398	258637	258877	259116	259355	259594	259833	239
2	260071	260310	260548	260787	261025	261263	261501	261739	261976	262214	238
3	262451	262688	262925	263162	263399	263636	263873	264109	264346	264582	237
4	264818	265054	265290	265525	265761	265996	266232	266467	266702	266937	235
5	267172	267406	267641	267875	268110	268344	268578	268812	269046	269279	234
6	269513	269746	269980	270213	270446	270679	270912	271144	271377	271609	233
7	271842	272074	272306	272538	272770	273001	273233	273464	273696	273927	232
8	274158	274389	274620	274850	275081	275311	275542	275772	276002	276232	230
9	276462	276692	276921	277151	277380	277609	277838	278067	278296	278525	229
190	278754	278982	279211	279439	279667	279895	280123	280351	280578	280806	228
1	281033	281261	281488	281715	281942	282169	282396	282622	282849	283075	227
2	283301	283527	283753	283979	284205	284431	284656	284882	285107	285332	226
3	285557	285782	286007	286232	286456	286681	286905	287130	287354	287578	225
4	287802	288026	288249	288473	288696	288920	289143	289366	289589	289812	223
5	290035	290257	290480	290702	290925	291147	291369	291591	291813	292034	222
6	292256	292478	292699	292920	293141	293363	293584	293804	294025	294246	221
7	294466	294687	294907	295127	295347	295567	295787	296007	296226	296446	220
8	296665	296884	297104	297323	297542	297761	297979	298198	298416	298635	219
9	298853	299071	299289	299507	299725	299943	300161	300378	300595	300813	218
200	301030	301247	301464	301681	301898	302114	302331	302547	302764	302980	217

PROPORTIONAL PARTS

Diff.	1	2	3	4	5	6	7	8	9
262	26.2	52.4	78.6	104.8	131.0	157.2	183.4	209.6	235.8
260	26.0	52.0	78.0	104.0	130.0	156.0	182.0	208.0	234.0
258	25.8	51.6	77.4	103.2	129.0	154.8	180.6	206.4	232.2
256	25.6	51.2	76.8	102.4	128.0	153.6	179.2	204.8	230.4
254	25.4	50.8	76.2	101.6	127.0	152.4	177.8	203.2	228.6
252	25.2	50.4	75.6	100.8	126.0	151.2	176.4	201.6	226.8
250	25.0	50.0	75.0	100.0	125.0	150.0	175.0	200.0	225.0
248	24.8	49.6	74.4	99.2	124.0	148.8	173.6	198.4	223.2
246	24.6	49.2	73.8	98.4	123.0	147.6	172.2	196.8	221.4
244	24.4	48.8	73.2	97.6	122.0	146.4	170.8	195.2	219.6
242	24.2	48.4	72.6	96.8	121.0	145.2	169.4	193.6	217.8
240	24.0	48.0	72.0	96.0	120.0	144.0	168.0	192.0	216.0
238	23.8	47.6	71.4	95.2	119.0	142.8	166.6	190.4	214.2
236	23.6	47.2	70.8	94.4	118.0	141.6	165.2	188.8	212.4
234	23.4	46.8	70.2	93.6	117.0	140.4	163.8	187.2	210.6
232	23.2	46.4	69.6	92.8	116.0	139.2	162.4	185.6	208.8
230	23.0	46.0	69.0	92.0	115.0	138.0	161.0	184.0	207.0
228	22.8	45.6	68.4	91.2	114.0	136.8	159.6	182.4	205.2
226	22.6	45.2	67.8	90.4	113.0	135.6	158.2	180.8	203.4
224	22.4	44.8	67.2	89.6	112.0	134.4	156.8	179.2	201.6
222	22.2	44.4	66.6	88.8	111.0	133.2	155 4	177.6	199.8
220	22.0	44.0	66.0	88.0	110.0	132.0	154.0	176.0	198.0
218	21.8	43.6	65.4	87.2	109.0	130.8	152.6	174.4	196.2
216	21.6	43.2	64.8	86.4	108.0	129.6	151.2	172.8	194.4

N	0	1	2	3	4	5	6	7	8	9	Diff.
200	301030	301247	301464	301681	301898	302114	302331	302547	302764	302980	217
1	303196	303412	303628	303844	304059	·304275	304491	304706	304921	305136	216
2	305351	305566	305781	305996	306211	306425	306639	306854	307068	307282	215
3	307496	307710	307924	308137	308351	308564	308778	308991	309204	309417	213
4	309630	309843	310056	310268	310481	310693	310906	311118	311330	311542	212
5	311754	311966	312177	312389	312600	312812	313023	313234	313445	313656	211
6	313867	314078	314289	314499	314710	314920	315130	315340	315551	315760	210
7	315970	316180	316390	316599	316809	317018	317227	317436	317646	317854	209
8	318063	318272	318481	318689	318898	319106	319314	319522	319730	319938	208
9	320146	320354	320562	320769	320977	321184	321391	321598	321805	322012	207
210	322219	322426	322633	322839	323046	323252	323458	323665	323871	324077	206
1	324282	324488	324694	324899	325105	325310	325516	325721	325926	326131	205
2	326336	326541	326745	326950	327155	327359	327563	327767	327972	328176	204
3	328380	328583	328787	328991	329194	329398	329601	329805	330008	330211	203
4	330414	330617	330819	331022	331225	331427	331630	331832	332034	332236	202
5	332438	332640	332842	333044	333246	333447	333649	333850	334051	334253	
6	334454	334655	334856	335057	335257	335458	335658	335859	336059	336260	201
7	336460	336660	336860	337060	337260	337459	337659	337858	338058	338257	200
8	338456	338656	338855	339054	339253	339451	339650	339849	340047	340246	199
9	340444	340642	340841	341039	341237	341435	341632	341830	342028	342225	198
220	342423	342620	342817	343014	343212	343409	343606	343802	343999	344196	197
1	344392	344589	344785	344981	345178	345374	345570	345766	345962	346157	196
2	346353	346549	346744	346939	347135	347330	347525	347720	347915	348110	195
3	348305	348500	348694	348889	349083	349278	349472	349666	349860	350054	194
4	350248	350442	350636	350829	351023	351216	351410	351603	351796	351989	193
5	352183	352375	352568	352761	352954	353147	353339	353532	353724	353916	
6	354108	354301	354493	354685	354876	355068	355260	355452	355643	355834	192
7	356026	356217	356408	356599	356790	356981	357172	357363	357554	357744	191
8	357935	358125	358316	358506	358696	358886	359076	359266	359456	359646	190
9	359835	360025	360215	360404	360593	360783	360972	361161	361350	361539	189
230	361728	361917	362105	362294	362482	362671	362859	363048	363236	363424	188
1	363612	363800	363988	364176	364363	364551	364739	364926	365113	365301	
2	365488	365675	365862	366049	366236	366423	366610	366796	366983	367169	187
3	367356	367542	367729	367915	368101	368287	368473	368659	368845	369030	186
4	369216	369401	369587	369772	369958	370143	370328	370513	370698	370883	185
5	371068	371253	371437	371622	371806	371991	372175	372360	372544	372728	184
6	372912	373096	373280	373464	373647	373831	374015	374198	374382	374565	
7	374748	374932	375115	375298	375481	375664	375846	376029	376212	376394	183
8	376577	376759	▸376942	377124	377306	377488	377670	377852	378034	378216	182
9	378398	378580	378761	378943	379124	379306	379487	379668	379849	380030	181
240	380211	380392	380573	380754	380934	381115	381296	381476	381656	381837	

PROPORTIONAL PARTS

Diff.	1	2	3	4	5	6	7	8	9
216	21.6	43.2	64.8	86.4	108.0	129.6	151.2	172.8	194.4
214	21.4	42.8	64.2	85.6	107.0	128.4	149.8	171.2	192.6
212	21.2	42.4	63 6	84.8	106.0	127.2	148.4	169.6	190.8
210	21.0	42.0	63.0	84.0	105.0	126.0	147.0	168.0	189.0
208	20.8	41.6	62.4	83.2	104.0	124.8	145.6	166.4	187.2
206	20.6	41.2	61.8	82.4	103.0	123.6	144.2	164.8	185.4
204	20.4	40.8	61.2	81.6	102.0	122.4	142.8	163.2	183.6
202	20.2	40.4	60.6	80.8	101.0	121.2	141.4	161.6	181.8
200	20.0	40.0	60.0	80.0	100.0	120.0	140.0	160.0	180.0
198	19.8	39.6	59.4	79.2	99.0	118.8	138.6	158.4	178.2
196	19.6	39.2	58.8	78.4	98.0	117.6	137.2	156.8	176.4
194	19.4	38.8	58.2	77.6	97.0	116.4	135.8	155.2	174.6
192	19.2	38.4	57.6	76.8	96.0	115.2	134.4	153.6	172.8
190	19.0	38.0	57.0	76.0	95.0	114.0	133.0	152.0	171.0
188	18.8	37.6	56.4	75.2	94.0	112.8	131.6	150.4	169.2
186	18.6	37.2	55.8	74.4	93.0	111.6	130.2	148.8	167.4

N	0	1	2	3	4	5	6	7	8	9	Diff.
240	**380211**	**380392**	**380573**	**380754**	**380934**	**381115**	**381296**	**381476**	**381656**	**381837**	181
1	382017	382197	382377	382557	382737	382917	383097	383277	383456	383636	180
2	383815	383995	384174	384353	384533	384712	384891	385070	385249	385428	179
3	385606	385785	385964	386142	386321	386499	386678	386856	387034	387212	178
4	387390	387568	387746	387924	388101	388279	388456	388634	388811	388989	
5	389166	389343	389520	389698	389875	390051	390228	390405	390582	390759	177
6	390935	391112	391288	391464	391641	391817	391993	392169	392345	392521	176
7	392697	392873	393048	393224	393400	393575	393751	393926	394101	394277	
8	394452	394627	394802	394977	395152	395326	395501	395676	395850	396025	175
9	396199	396374	396548	396722	396896	397071	397245	397419	397592	397766	174
250	**397940**	**398114**	**398287**	**398461**	**398634**	**398808**	**398981**	**399154**	**399328**	**399501**	173
1	399674	399847	400020	400192	400365	400538	400711	400883	401056	401228	
2	401401	401573	401745	401917	402089	402261	402433	402605	402777	402949	172
3	403121	403292	403464	403635	403807	403978	404149	404320	404492	404663	171
4	404834	405005	405176	405346	405517	405688	405858	406029	406199	406370	
5	406540	406710	406881	407051	407221	407391	407561	407731	407901	408070	170
6	408240	408410	408579	408749	408918	409087	409257	409426	409595	409764	169
7	409933	410102	410271	410440	410609	410777	410946	411114	411283	411451	
8	411620	411788	411956	412124	412293	412461	412629	412796	412964	413132	168
9	413300	413467	413635	413803	413970	414137	414305	414472	414639	414806	167
260	**414973**	**415140**	**415307**	**415474**	**415641**	**415808**	**415974**	**416141**	**416308**	**416474**	
1	416641	416807	416973	417139	417306	417472	417638	417804	417970	418135	166
2	418301	418467	418633	418798	418964	419129	419295	419460	419625	419791	165
3	419956	420121	420286	420451	420616	420781	420945	421110	421275	421439	
4	421604	421768	421933	422097	422261	422426	422590	422754	422918	423082	164
5	423246	423410	423574	423737	423901	424065	424228	424392	424555	424718	
6	424882	425045	425208	425371	425534	425697	425860	426023	426186	426349	163
7	426511	426674	426836	426999	427161	427324	427486	427648	427811	427973	162
8	428135	428297	428459	428621	428783	428944	429106	429268	429429	429591	
9	429752	429914	430075	430236	430398	430559	430720	430881	431042	431203	161
270	**431364**	**431525**	**431685**	**431846**	**432007**	**432167**	**432328**	**432488**	**432649**	**432809**	
1	432969	433130	433290	433450	433610	433770	433930	434090	434249	434409	160
2	434569	434729	434888	435048	435207	435367	435526	435685	435844	436004	159
3	436163	436322	436481	436640	436799	436957	437116	437275	437433	437592	
4	437751	437909	438067	438226	438384	438542	438701	438859	439017	439175	158
5	439333	439491	439648	439806	439964	440122	440279	440437	440594	440752	
6	440909	441066	441224	441381	441538	441695	441852	442009	442166	442323	157
7	442480	442637	442793	442950	443106	443263	443419	443576	443732	443889	
8	444045	444201	444357	444513	444669	444825	444981	445137	445293	445449	156
9	445604	445760	445915	446071	446226	446382	446537	446692	446848	447003	155
280	**447158**	**447313**	**447468**	**447623**	**447778**	**447933**	**448088**	**448242**	**448397**	**448552**	

PROPORTIONAL PARTS

Diff.	1	2	3	4	5	6	7	8	9
184	18.4	36.8	55.2	73.6	92.0	110.4	128.8	147.2	165.6
182	18.2	36.4	54.6	72.8	91.0	109.2	127.4	145.6	163.8
180	18.0	36.0	54.0	72.0	90.0	108.0	126.0	144.0	162.0
178	17.8	35.6	53.4	71.2	89.0	106.8	124.6	142.4	160.2
176	17.6	35.2	52.8	70.4	88.0	105.6	123.2	140.8	158.4
174	17.4	34.8	52.2	69.6	87.0	104.4	121.8	139.2	156.6
172	17.2	34.4	51.6	68.8	86.0	103.2	120.4	137.6	154.8
170	17.0	34.0	51.0	68.0	85.0	102.0	119.0	136.0	153.0
168	16.8	33.6	50.4	67.2	84.0	100.8	117.6	134.4	151.2
166	16.6	33.2	49.8	66.4	83.0	99.6	116.2	132.8	149.4
164	16.4	32.8	49.2	65.6	82.0	98.4	114.8	131.2	147.6
162	16.2	32.4	48.6	64.8	81.0	97.2	113.4	129.6	145.8
160	16.0	32.0	48.0	64.0	80.0	96.0	112.0	128.0	144.0
158	15.8	31.6	47.4	63.2	79.0	94.8	110.6	126.4	142.2
156	15.6	31.2	46.8	62.4	78.0	93.6	109.2	124.8	140.4
154	15.4	30.8	46.2	61.6	77.0	92.4	107.8	123.2	138.6

N	0	1	2	3	4	5	6	7	8	9	Diff.
280	**447158**	**447313**	**447468**	**447623**	**447778**	**447933**	**448088**	**448242**	**448397**	**448552**	
1	448706	448861	449015	449170	449324	449478	449633	449787	449941	450095	154
2	450249	450403	450557	450711	450865	451018	451172	451326	451479	451633	
3	451786	451940	452093	452247	452400	452553	452706	452859	453012	453165	153
4	453318	453471	453624	453777	453930	454082	454235	454387	454540	454692	
5	454845	454997	455150	455302	455454	455606	455758	455910	456062	456214	152
6	456366	456518	456670	456821	456973	457125	457276	457428	457579	457731	
7	457882	458033	458184	458336	458487	458638	458789	458940	459091	459242	151
8	459392	459543	459694	459845	459995	460146	460296	460447	460597	460748	
9	460898	461048	461198	461348	461499	461649	461799	461948	462098	462248	150
290	**462398**	**462548**	**462697**	**462847**	**462997**	**463146**	**463296**	**463445**	**463594**	**463744**	
1	463893	464042	464191	464340	464490	464639	464788	464936	465085	465234	149
2	465383	465532	465680	465829	465977	466126	466274	466423	466571	466719	
3	466868	467016	467164	467312	467460	467608	467756	467904	468052	468200	148
4	468347	468495	468643	468790	468938	469085	469233	469380	469527	469675	
5	469822	469969	470116	470263	470410	470557	470704	470851	470998	471145	147
6	471292	471438	471585	471732	471878	472025	472171	472318	472464	472610	146
7	472756	472903	473049	473195	473341	473487	473633	473779	473925	474071	
8	474216	474362	474508	474653	474799	474944	475090	475235	475381	475526	
9	475671	475816	475962	476107	476252	476397	476542	476687	476832	476976	145
300	**477121**	**477266**	**477411**	**477555**	**477700**	**477844**	**477989**	**478133**	**478278**	**478422**	
1	478566	478711	478855	478999	479143	479287	479431	479575	479719	479863	144
2	480007	480151	480294	480438	480582	480725	480869	481012	481156	481299	
3	481443	481586	481729	481872	482016	482159	482302	482445	482588	482731	143
4	482874	483016	483159	483302	483445	483587	483730	483872	484015	484157	
5	484300	484442	484585	484727	484869	485011	485153	485295	485437	485579	142
6	485721	485863	486005	486147	486289	486430	486572	486714	486855	486997	
7	487138	487280	487421	487563	487704	487845	487986	488127	488269	488410	141
8	488551	488692	488833	488974	489114	489255	489396	489537	489677	489818	
9	489958	490099	490239	490380	490520	490661	490801	490941	491081	491222	140
310	**491362**	**491502**	**491642**	**491782**	**491922**	**492062**	**492201**	**492341**	**492481**	**492621**	
1	492760	492900	493040	493179	493319	493458	493597	493737	493876	494015	139
2	494155	494294	494433	494572	494711	494850	494989	495128	495267	495406	
3	495544	495683	495822	495960	496099	496238	496376	496515	496653	496791	
4	496930	497068	497206	497344	497483	497621	497759	497897	498035	498173	138
5	498311	498448	498586	498724	498862	498999	499137	499275	499412	499550	
6	499687	499824	499962	500099	500236	500374	500511	500648	500785	500922	137
7	501059	501196	501333	501470	501607	501744	501880	502017	502154	502291	
8	502427	502564	502700	502837	502973	503109	503246	503382	503518	503655	136
9	503791	503927	504063	504199	504335	504471	504607	504743	504878	505014	
320	**505150**	**505286**	**505421**	**505557**	**505693**	**505828**	**505964**	**506099**	**506234**	**506370**	
1	506505	506640	506776	506911	507046	507181	507316	507451	507586	507721	135
2	507856	507991	508126	508260	508395	508530	508664	508799	508934	509068	
3	509203	509337	509471	509606	509740	509874	510009	510143	510277	510411	134
4	510545	510679	510813	510947	511081	511215	511349	511482	511616	511750	
5	511883	512017	512151	512284	512418	512551	512684	512818	512951	513084	133
6	513218	513351	513484	513617	513750	513883	514016	514149	514282	514415	
7	514548	514681	514813	514946	515079	515211	515344	515476	515609	515741	
8	515874	516006	516139	516271	516403	516535	516668	516800	516932	517064	132
9	517196	517328	517460	517592	517724	517855	517987	518119	518251	518382	
330	**518514**	**518646**	**518777**	**518909**	**519040**	**519171**	**519303**	**519434**	**519566**	**519697**	131

PROPORTIONAL PARTS

Diff.	1	2	3	4	5	6	7	8	9
154	15.4	30.8	46.2	61.6	77.0	92.4	107.8	123.2	138.6
152	15.2	30.4	45.6	60.8	76.0	91.2	106.4	121.6	136.8
150	15.0	30.0	45.0	60.0	75.0	90.0	105.0	120.0	135.0
148	14.8	29.6	44.4	59.2	74.0	88.8	103.6	118.4	133.2
146	14.6	29.2	43.8	58.4	73.0	87.6	102.2	116.8	131.4
144	14.4	28.8	43.2	57.6	72.0	86.4	100.8	115.2	129.6
142	14.2	28.4	42.6	56.8	71.0	85.2	99.4	113.6	127.8
140	14.0	28.0	42.0	56.0	70.0	84.0	98.0	112.0	126.0
138	13.8	27.6	41.4	55.2	69.0	82.8	96.6	110.4	124.2
136	13.6	27.2	40.8	54.4	68.0	81.6	95.2	108.8	122.4

N	0	1	2	3	4	5	6	7	8	9	Diff.
330	518514	518646	518777	518909	519040	519171	519303	519434	519566	519697	
1	519828	519959	520090	520221	520353	520484	520615	520745	520876	521007	
2	521138	521269	521400	521530	521661	521792	521922	522053	522183	522314	
3	522444	522575	522705	522835	522966	523096	523226	523356	523486	523616	130
4	523746	523876	524006	524136	524266	524396	524526	524656	524785	524915	
5	525045	525174	525304	525434	525563	525693	525822	525951	526081	526210	129
6	526339	526469	526598	526727	526856	526985	527114	527243	527372	527501	
7	527630	527759	527888	528016	528145	528274	528402	528531	528660	528788	
8	528917	529045	529174	529302	529430	529559	529687	529815	529943	530072	128
9	530200	530328	530456	530584	530712	530840	530968	531096	531223	531351	
340	531479	531607	531734	531862	531990	532117	532245	532372	532500	532627	
1	532754	532882	533009	533136	533264	533391	533518	533645	533772	533899	127
2	534026	534153	534280	534407	534534	534661	534787	534914	535041	535167	
3	535294	535421	535547	535674	535800	535927	536053	536180	536306	536432	126
4	536558	536685	536811	536937	537063	537189	537315	537441	537567	537693	
5	537819	537945	538071	538197	538322	538448	538574	538699	538825	538951	
6	539076	539202	539327	539452	539578	539703	539829	539954	540079	540204	125
7	540329	540455	540580	540705	540830	540955	541080	541205	541330	541454	
8	541579	541704	541829	541953	542078	542203	542327	542452	542576	542701	
9	542825	542950	543074	543199	543323	543447	543571	543696	543820	543944	124
350	544068	544192	544316	544440	544564	544688	544812	544936	545060	545183	
1	545307	545431	545555	545678	545802	545925	546049	546172	546296	546419	
2	546543	546666	546789	546913	547036	547159	547282	547405	547529	547652	
3	547775	547898	548021	548144	548267	548389	548512	548635	548758	548881	123
4	549003	549126	549249	549371	549494	549616	549739	549861	549984	550106	
5	550228	550351	550473	550595	550717	550840	550962	551084	551206	551328	122
6	551450	551572	551694	551816	551938	552060	552181	552303	552425	552547	
7	552668	552790	552911	553033	553155	553276	553398	553519	553640	553762	121
8	553883	554004	554126	554247	554368	554489	554610	554731	554852	554973	
9	555094	555215	555336	555457	555578	555699	555820	555940	556061	556182	
360	556303	556423	556544	556664	556785	556905	557026	557146	557267	557387	120
1	557507	557627	557748	557868	557988	558108	558228	558349	558469	558589	
2	558709	558829	558948	559068	559188	559308	559428	559548	559667	559787	
3	559907	560026	560146	560265	560385	560504	560624	560743	560863	560982	119
4	561101	561221	561340	561459	561578	561698	561817	561936	562055	562174	
5	562293	562412	562531	562650	562769	562887	563006	563125	563244	563362	
6	563481	563600	563718	563837	563955	564074	564192	564311	564429	564548	
7	564666	564784	564903	565021	565139	565257	565376	565494	565612	565730	118
8	565848	565966	566084	566202	566320	566437	566555	566673	566791	566909	
9	567026	567144	567262	567379	567497	567614	567732	567849	567967	568084	
370	568202	568319	568436	568554	568671	568788	568905	569023	569140	569257	117
1	569374	569491	569608	569725	569842	569959	570076	570193	570309	570426	
2	570543	570660	570776	570893	571010	571126	571243	571359	571476	571592	
3	571709	571825	571942	572058	572174	572291	572407	572523	572639	572755	116
4	572872	572988	573104	573220	573336	573452	573568	573684	573800	573915	
5	574031	574147	574263	574379	574494	574610	574726	574841	574957	575072	
6	575188	575303	575419	575534	575650	575765	575880	575996	576111	576226	115
7	576341	576457	576572	576687	576802	576917	577032	577147	577262	577377	
8	577492	577607	577722	577836	577951	578066	578181	578295	578410	578525	
9	578639	578754	578868	578983	579097	579212	579326	579441	579555	579669	114
380	579784	579898	580012	580126	580241	580355	580469	580583	580697	580811	

PROPORTIONAL PARTS

Diff.	1	2	3	4	5	6	7	8	9
134	13.4	26.8	40.2	53.6	67.0	80.4	93.8	107.2	120.6
132	13.2	26.4	39.6	52.8	66.0	79.2	92.4	105.6	118.8
130	13.0	26.0	39.0	52.0	65.0	78.0	91.0	104.0	117.0
128	12.8	25.6	38.4	51.2	64.0	76.8	89.6	102.4	115.2
126	12.6	25.2	37.8	50.4	63.0	75.6	88.2	100.8	113.4
124	12.4	24.8	37.2	49.6	62.0	74.4	86.8	99.2	111.6
122	12.2	24.4	36.6	48.8	61.0	73.2	85.4	97.6	109.8
120	12.0	24.0	36.0	48.0	60.0	72.0	84.0	96.0	108.0
118	11.8	23.6	35.4	47.2	59.0	70.8	82.6	94.4	106.2
116	11.6	23.2	34.8	46.4	58.0	69.6	81.2	92.8	104.4
114	11.4	22.8	34.2	45.6	57.0	68.4	79.8	91.2	102.6

N	0	1	2	3	4	5	6	7	8	9	Diff.
380	579784	579898	580012	580126	580241	580355	580469	580583	580697	580811	114
1	580925	581039	581153	581267	581381	581495	581608	581722	581836	581950	
2	582063	582177	582291	582404	582518	582631	582745	582858	582972	583085	
3	583199	583312	583426	583539	583652	583765	583879	583992	584105	584218	
4	584331	584444	584557	584670	584783	584896	585009	585122	585235	585348	113
5	585461	585574	585686	585799	585912	586024	586137	586250	586362	586475	
6	586587	586700	586812	586925	587037	587149	587262	587374	587486	587599	
7	587711	587823	587935	588047	588160	588272	588384	588496	588608	588720	112
8	588832	588944	589056	589167	589279	589391	589503	589615	589726	589838	
9	589950	590061	590173	590284	590396	590507	590619	590730	590842	590953	
390	591065	591176	591287	591399	591510	591621	591732	591843	591955	592066	
1	592177	592288	592399	592510	592621	592732	592843	592954	593064	593175	111
2	593286	593397	593508	593618	593729	593840	593950	594061	594171	594282	
3	594393	594503	594614	594724	594834	594945	595055	595165	595276	595386	
4	595496	595606	595717	595827	595937	596047	596157	596267	596377	596487	
5	596597	596707	596817	596927	597037	597146	597256	597366	597476	597586	110
6	597695	597805	597914	598024	598134	598243	598353	598462	598572	598681	
7	598791	598900	599009	599119	599228	599337	599446	599556	599665	599774	
8	599883	599992	600101	600210	600319	600428	600537	600646	600755	600864	109
9	600973	601082	601191	601299	601408	601517	601625	601734	601843	601951	
400	602060	602169	602277	602386	602494	602603	602711	602819	602928	603036	
1	603144	603253	603361	603469	603577	603686	603794	603902	604010	604118	108
2	604226	604334	604442	604550	604658	604766	604874	604982	605089	605197	
3	605305	605413	605521	605628	605736	605844	605951	606059	606166	606274	
4	606381	606489	606596	606704	606811	606919	607026	607133	607241	607348	
5	607455	607562	607669	607777	607884	607991	608098	608205	608312	608419	107
6	608526	608633	608740	608847	608954	609061	609167	609274	609381	609488	
7	609594	609701	609808	609914	610021	610128	610234	610341	610447	610554	
8	610660	610767	610873	610979	611086	611192	611298	611405	611511	611617	106
9	611723	611829	611936	612042	612148	612254	612360	612466	612572	612678	
410	612784	612890	612996	613102	613207	613313	613419	613525	613630	613736	
1	613842	613947	614053	614159	614264	614370	614475	614581	614686	614792	105
2	614897	615003	615108	615213	615319	615424	615529	615634	615740	615845	
3	615950	616055	616160	616265	616370	616476	616581	616686	616790	616895	
4	617000	617105	617210	617315	617420	617525	617629	617734	617839	617943	
5	618048	618153	618257	618362	618466	618571	618676	618780	618884	618989	104
6	619093	619198	619302	619406	619511	619615	619719	619824	619928	620032	
7	620136	620240	620344	620448	620552	620656	620760	620864	620968	621072	104
8	621176	621280	621384	621488	621592	621695	621799	621903	622007	622110	
9	622214	622318	622421	622525	622628	622732	622835	622939	623042	623146	
420	623249	623353	623456	623559	623663	623766	623869	623973	624076	624179	
1	624282	624385	624488	624591	624695	624798	624901	625004	625107	625210	103
2	625312	625415	625518	625621	625724	625827	625929	626032	626135	626238	
3	626340	626443	626546	626648	626751	626853	626956	627058	627161	627263	
4	627366	627468	627571	627673	627775	627878	627980	628082	628185	628287	
5	628389	628491	628593	628695	628797	628900	629002	629104	629206	629308	102
6	629410	629512	629613	629715	629817	629919	630021	630123	630224	630326	
7	630428	630530	630631	630733	630835	630936	631038	631139	631241	631342	
8	631444	631545	631647	631748	631849	631951	632052	632153	632255	632356	
9	632457	632559	632660	632761	632862	632963	633064	633165	633266	633367	
430	633468	633569	633670	633771	633872	633973	634074	634175	634276	634376	101

PROPORTIONAL PARTS

Diff.	1	2	3	4	5	6	7	8	9
114	11.4	22.8	34.2	45.6	57.0	68.4	79.8	91.2	102.6
112	11.2	22.4	33.6	44.8	56.0	67.2	78.4	89.6	100.8
110	11.0	22.0	33.0	44.0	55.0	66.0	77.0	88.0	99.0
108	10.8	21.6	32.4	43.2	54.0	64.8	75.6	86.4	97.2
106	10.6	21.2	31.8	42.4	53.0	63.6	74.2	84.8	95.4
104	10.4	20.8	31.2	41.6	52.0	62.4	72.8	83.2	93.6
102	10.2	20.4	30.6	40.8	51.0	61.2	71.4	81.6	91.8

N	0	1	2	3	4	5	6	7	8	9	Diff.
430	633468	633569	633670	633771	633872	633973	634074	634175	634276	634376	
1	634477	634578	634679	634779	634880	634981	635081	635182	635283	635383	
2	635484	635584	635685	635785	635886	635986	636087	636187	636287	636388	
3	636488	636588	636688	636789	636889	636989	637089	637189	637290	637390	
4	637490	637590	637690	637790	637890	637990	638090	638190	638290	638389	100
5	638489	638589	638689	638789	638888	638988	639088	639188	639287	639387	
6	639486	639586	639686	639785	639885	639984	640084	640183	640283	640382	
7	640481	640581	640680	640779	640879	640978	641077	641177	641276	641375	
8	641474	641573	641672	641771	641871	641970	642069	642168	642267	642366	
9	642465	642563	642662	642761	642860	642959	643058	643156	643255	643354	99
440	643453	643551	643650	643749	643847	643946	644044	644143	644242	644340	
1	644439	644537	644636	644734	644832	644931	645029	645127	645226	645324	
2	645422	645521	645619	645717	645815	645913	646011	646110	646208	646306	
3	646404	646502	646600	646698	646796	646894	646992	647089	647187	647285	98
4	647383	647481	647579	647676	647774	647872	647969	648067	648165	648262	
5	648360	648458	648555	648653	648750	648848	648945	649043	649140	649237	
6	649335	649432	649530	649627	649724	649821	649919	650016	650113	650210	
7	650308	650405	650502	650599	650696	650793	650890	650987	651084	651181	
8	651278	651375	651472	651569	651666	651762	651859	651956	652053	652150	97
9	652246	652343	652440	652536	652633	652730	652826	652923	653019	653116	
450	653213	653309	653405	653502	653598	653695	653791	653888	653984	654080	
1	654177	654273	654369	654465	654562	654658	654754	654850	654946	655042	
2	655138	655235	655331	655427	655523	655619	655715	655810	655906	656002	96
3	656098	656194	656290	656386	656482	656577	656673	656769	656864	656960	
4	657056	657152	657247	657343	657438	657534	657629	657725	657820	657916	
5	658011	658107	658202	658298	658393	658488	658584	658679	658774	658870	
6	658965	659060	659155	659250	659346	659441	659536	659631	659726	659821	
7	659916	660011	660106	660201	660296	660391	660486	660581	660676	660771	95
8	660865	660960	661055	661150	661245	661339	661434	661529	661623	661718	
9	661813	661907	662002	662096	662191	662286	662380	662475	662569	662663	
460	662758	662852	662947	663041	663135	663230	663324	663418	663512	663607	
1	663701	663795	663889	663983	664078	664172	664266	664360	664454	664548	
2	664642	664736	664830	664924	665018	665112	665206	665299	665393	665487	94
3	665581	665675	665769	665862	665956	666050	666143	666237	666331	666424	
4	666518	666612	666705	666799	666892	666986	667079	667173	667266	667360	
5	667453	667546	667640	667733	667826	667920	668013	668106	668199	668293	
6	668386	668479	668572	668665	668759	668852	668945	669038	669131	669224	
7	669317	669410	669503	669596	669689	669782	669875	669967	670060	670153	93
8	670246	670339	670431	670524	670617	670710	670802	670895	670988	671080	
9	671173	671265	671358	671451	671543	671636	671728	671821	671913	672005	
470	672098	672190	672283	672375	672467	672560	672652	672744	672836	672929	
1	673021	673113	673205	673297	673390	673482	673574	673666	673758	673850	
2	673942	674034	674126	674218	674310	674402	674494	674586	674677	674769	92
3	674861	674953	675045	675137	675228	675320	675412	675503	675595	675687	
4	675778	675870	675962	676053	676145	676236	676328	676419	676511	676602	
5	676694	676785	676876	676968	677059	677151	677242	677333	677424	677516	
6	677607	677698	677789	677881	677972	678063	678154	678245	678336	678427	
7	678518	678609	678700	678791	678882	678973	679064	679155	679246	679337	91
8	679428	679519	679610	679700	679791	679882	679973	680063	680154	680245	
9	680336	680426	680517	680607	680698	680789	680879	680970	681060	681151	
480	681241	681332	681422	681513	681603	681693	681784	681874	681964	682055	

PROPORTIONAL PARTS

Diff.	1	2	3	4	5	6	7	8	9
102	10.2	20.4	30.6	40.8	51.0	61.2	71.4	81.6	91.8
100	10.0	20.0	30.0	40.0	50.0	60.0	70.0	80.0	90.0
98	9.8	19.6	29.4	39.2	49.0	58.8	68.6	78.4	88.2
96	9.6	19.2	28.8	38.4	48.0	57.6	67.2	76.8	86.4
94	9.4	18.8	28.2	37.6	47.0	56.4	65.8	75.2	84.6
92	9.2	18.4	27.6	36.8	46.0	55.2	64.4	73.6	82.8
90	9.0	18.0	27.0	36.0	45.0	54.0	63.0	72.0	81.0

N	0	1	2	3	4	5	6	7	8	9	Diff.
480	**681241**	**681332**	**681422**	**681513**	**681603**	**681693**	**681784**	**681874**	**681964**	**682055**	
1	682145	682235	682326	682416	682506	682596	682686	682777	682867	682957	
2	683047	683137	683227	683317	683407	683497	683587	683677	683767	683857	90
3	683947	684037	684127	684217	684307	684396	684486	684576	684666	684756	
4	684845	684935	685025	685114	685204	685294	685383	685473	685563	685652	
5	685742	685831	685921	686010	686100	686189	686279	686368	686458	686547	
6	686636	686726	686815	686904	686994	687083	687172	687261	687351	687440	
7	687529	687618	687707	687796	687886	687975	688064	688153	688242	688331	
8	688420	688509	688598	688687	688776	688865	688953	689042	689131	689220	89
9	689309	689398	689486	689575	689664	689753	689841	689930	690019	690107	
490	**690196**	**690285**	**690373**	**690462**	**690550**	**690639**	**690728**	**690816**	**690905**	**690993**	
1	691081	691170	691258	691347	691435	691524	691612	691700	691789	691877	
2	691965	692053	692142	692230	692318	692406	692494	692583	692671	692759	
3	692847	692935	693023	693111	693199	693287	693375	693463	693551	693639	88
4	693727	693815	693903	693991	694078	694166	694254	694342	694430	694517	
5	694605	694693	694781	694868	694956	695044	695131	695219	695307	695394	
6	695482	695569	695657	695744	695832	695919	696007	696094	696182	696269	
7	696356	696444	696531	696618	696706	696793	696880	696968	697055	697142	
8	697229	697317	697404	697491	697578	697665	697752	697839	697926	698014	87
9	698100	698188	698275	698362	698449	698535	698622	698709	698796	698883	
500	**698970**	**699057**	**699144**	**699231**	**699317**	**699404**	**699491**	**699578**	**699664**	**699751**	
1	699838	699924	700011	700098	700184	700271	700358	700444	700531	700617	
2	700704	700790	700877	700963	701050	701136	701222	701309	701395	701482	
3	701568	701654	701741	701827	701913	701999	702086	702172	702258	702344	
4	702431	702517	702603	702689	702775	702861	702947	703033	703119	703205	
5	703291	703377	703463	703549	703635	703721	703807	703893	703979	704065	86
6	704151	704236	704322	704408	704494	704579	704665	704751	704837	704922	
7	705008	705094	705179	705265	705350	705436	705522	705607	705693	705778	
8	705864	705949	706035	706120	706206	706291	706376	706462	706547	706632	
9	706718	706803	706888	706974	707059	707144	707229	707315	707400	707485	
510	**707570**	**707655**	**707740**	**707826**	**707911**	**707996**	**708081**	**708166**	**708251**	**708336**	
1	708421	708506	708591	708676	708761	708846	708931	709015	709100	709185	85
2	709270	709355	709440	709524	709609	709694	709779	709863	709948	710033	
3	710117	710202	710287	710371	710456	710540	710625	710710	710794	710879	
4	710963	711048	711132	711217	711301	711385	711470	711554	711639	711723	
5	711807	711892	711976	712060	712144	712229	712313	712397	712481	712566	
6	712650	712734	712818	712902	712986	713070	713154	713238	713323	713407	
7	713491	713575	713659	713742	713826	713910	713994	714078	714162	714246	
8	714330	714414	714497	714581	714665	714749	714833	714916	715000	715084	84
9	715167	715251	715335	715418	715502	715586	715669	715753	715836	715920	
520	**716003**	**716087**	**716170**	**716254**	**716337**	**716421**	**716504**	**716588**	**716671**	**716754**	
1	716838	716921	717004	717088	717171	717254	717338	717421	717504	717587	
2	717671	717754	717837	717920	718003	718086	718169	718253	718336	718419	83
3	718502	718585	718668	718751	718834	718917	719000	719083	719165	719248	
4	719331	719414	719497	719580	719663	719745	719828	719911	719994	720077	
5	720159	720242	720325	720407	720490	720573	720655	720738	720821	720903	
6	720986	721068	721151	721233	721316	721398	721481	721563	721646	721728	
7	721811	721893	721975	722058	722140	722222	722305	722387	722469	722552	
8	722634	722716	722798	722881	722963	723045	723127	723209	723291	723374	
9	723456	723538	723620	723702	723784	723866	723948	724030	724112	724194	82
530	**724276**	**724358**	**724440**	**724522**	**724604**	**724685**	**724767**	**724849**	**724931**	**725013**	

PROPORTIONAL PARTS

Diff.	1	2	3	4	5	6	7	8	9
90	9.0	18.0	27.0	36.0	45.0	54.0	63.0	72.0	81.0
88	8.8	17.6	26.4	35.2	44.0	52.8	61.6	70.4	79.2
86	8.6	17.2	25.8	34.4	43.0	51.6	60.2	68.8	77.4
84	8.4	16.8	25.2	33.6	42.0	50.4	58.8	67.2	75.6
82	8.2	16.4	24.6	32.8	41.0	49.2	57.4	65.6	73.8

N	0	1	2	3	4	5	6	7	8	9	Diff.
530	724276	724358	724440	724522	724604	724685	724767	724849	724931	725013	
1	725095	725176	725258	725340	725422	725503	725585	725667	725748	725830	
2	725912	725993	726075	726156	726238	726320	726401	726483	726564	726646	
3	726727	726809	726890	726972	727053	727134	727216	727297	727379	727460	
4	727541	727623	727704	727785	727866	727948	728029	728110	728191	728273	
5	728354	728435	728516	728597	728678	728759	728841	728922	729003	729084	
6	729165	729246	729327	729408	729489	729570	729651	729732	729813	729893	81
7	729974	730055	730136	730217	730298	730378	730459	730540	730621	730702	
8	730782	730863	730944	731024	731105	731186	731266	731347	731428	731508	
9	731589	731669	731750	731830	731911	731991	732072	732152	732233	732313	
540	732394	732474	732555	732635	732715	732796	732876	732956	733037	733117	
1	733197	733278	733358	733438	733518	733598	733679	733759	733839	733919	
2	733999	734079	734160	734240	734320	734400	734480	734560	734640	734720	80
3	734800	734880	734960	735040	735120	735200	735279	735359	735439	735519	
4	735599	735679	735759	735838	735918	735998	736078	736157	736237	736317	
5	736397	736476	736556	736635	736715	736795	736874	736954	737034	737113	
6	737193	737272	737352	737431	737511	737590	737670	737749	737829	737908	
7	737987	738067	738146	738225	738305	738384	738463	738543	738622	738701	
8	738781	738860	738939	739018	739097	739177	739236	739335	739414	739493	
9	739572	739651	739731	739810	739889	739968	740047	740126	740205	740284	79
550	740363	740442	740521	740600	740678	740757	740836	740915	740994	741073	
1	741152	741230	741309	741388	741467	741546	741624	741703	741782	741860	
2	741939	742018	742096	742175	742254	742332	742411	742489	742568	742647	
3	742725	742804	742882	742961	743039	743118	743196	743275	743353	743431	
4	743510	743588	743667	743745	743823	743902	743980	744058	744136	744215	
5	744293	744371	744449	744528	744606	744684	744762	744840	744919	744997	
6	745075	745153	745231	745309	745387	745465	745543	745621	745699	745777	78
7	745855	745933	746011	746089	746167	746245	746323	746401	746479	746556	
8	746634	746712	746790	746868	746945	747023	747101	747179	747256	747334	
9	747412	747489	747567	747645	747722	747800	747878	747955	748033	748110	
560	748188	748266	748343	748421	748498	748576	748653	748731	748808	748885	
1	748963	749040	749118	749195	749272	749350	749427	749504	749582	749659	
2	749736	749814	749891	749968	750045	750123	750200	750277	750354	750431	
3	750508	750586	750663	750740	750817	750894	750971	751048	751125	751202	
4	751279	751356	751433	751510	751587	751664	751741	751818	751895	751972	77
5	752048	752125	752202	752279	752356	752433	752509	752586	752663	752740	
6	752816	752893	752970	753047	753123	753200	753277	753353	753430	753506	
7	753583	753660	753736	753813	753889	753966	754042	754119	754195	754272	
8	754348	754425	754501	754578	754654	754730	754807	754883	754960	755036	
9	755112	755189	755265	755341	755417	755494	755570	755646	755722	755799	
570	755875	755951	756027	756103	756180	756256	756332	756408	756484	756560	
1	756636	756712	756788	756864	756940	757016	757092	757168	757244	757320	76
2	757396	757472	757548	757624	757700	757775	757851	757927	758003	758079	
3	758155	758230	758306	758382	758458	758533	758609	758685	758761	758836	
4	758912	758988	759063	759139	759214	759290	759366	759441	759517	759592	
5	759668	759743	759819	759894	759970	760045	760121	760196	760272	760347	
6	760422	760498	760573	760649	760724	760799	760875	760950	761025	761101	
7	761176	761251	761326	761402	761477	761552	761627	761702	761778	761853	
8	761928	762003	762078	762153	762228	762303	762378	762453	762529	762604	75
9	762679	762754	762829	762904	762978	763053	763128	763203	763278	763353	
580	763428	763503	763578	763653	763727	763802	763877	763952	764027	764101	

PROPORTIONAL PARTS

Diff.	1	2	3	4	5	6	7	8	9
82	8.2	16.4	24.6	32.8	41.0	49.2	57.4	65.6	73.8
80	8.0	16.0	24.0	32.0	40.0	48.0	56.0	64.0	72.0
78	7.8	15.6	23.4	31.2	39.0	46.8	54.6	62.4	70.2
76	7.6	15.2	22.8	30.4	38.0	45.6	53.2	60.8	68.4
74	7.4	14.8	22.2	29.6	37.0	44.4	51.8	59.2	66.6

N	0	1	2	3	4	5	6	7	8	9	Diff.
580	763428	763503	763578	763653	763727	763802	763877	763952	764027	764101	
1	764176	764251	764326	764400	764475	764550	764624	764699	764774	764848	
2	764923	764998	765072	765147	765221	765296	765370	765445	765520	765594	
3	765669	765743	765818	765892	765966	766041	766115	766190	766264	766338	
4	766413	766487	766562	766636	766710	766785	766859	766933	767007	767082	
5	767156	767230	767304	767379	767453	767527	767601	767675	767749	767823	
6	767898	767972	768046	768120	768194	768268	768342	768416	768490	768564	74
7	768638	768712	768786	768860	768934	769008	769082	769156	769230	769303	
8	769377	769451	769525	769599	769673	769746	769820	769894	769968	770042	
9	770115	770189	770263	770336	770410	770484	770557	770631	770705	770778	
590	770852	770926	770999	771073	771146	771220	771293	771367	771440	771514	
1	771587	771661	771734	771808	771881	771955	772028	772102	772175	772248	
2	772322	772395	772468	772542	772615	772688	772762	772835	772908	772981	
3	773055	773128	773201	773274	773348	773421	773494	773567	773640	773713	
4	773786	773860	773933	774006	774079	774152	774225	774298	774371	774444	73
5	774517	774590	774663	774736	774809	774882	774955	775028	775100	775173	
6	775246	775319	775392	775465	775538	775610	775683	775756	775829	775902	
7	775974	776047	776120	776193	776265	776338	776411	776483	776556	776629	
8	776701	776774	776846	776919	776992	777064	777137	777209	777282	777354	
9	777427	777499	777572	777644	777717	777789	777862	777934	778006	778079	
600	778151	778224	778296	778368	778441	778513	778585	778658	778730	778802	
1	778874	778947	779019	779091	779163	779236	779308	779380	779452	779524	
2	779596	779669	779741	779813	779885	779957	780029	780101	780173	780245	
3	780317	780389	780461	780533	780605	780677	780749	780821	780893	780965	72
4	781037	781109	781181	781253	781324	781396	781468	781540	781612	781684	
5	781755	781827	781899	781971	782042	782114	782186	782258	782329	782401	
6	782473	782544	782616	782688	782759	782831	782902	782974	783046	783117	
7	783189	783260	783332	783403	783475	783546	783618	783689	783761	783832	
8	783904	783975	784046	784118	784189	784261	784332	784403	784475	784546	
9	784617	784689	784760	784831	784902	784974	785045	785116	785187	785259	
610	785330	785401	785472	785543	785615	785686	785757	785828	785899	785970	
1	786041	786112	786183	786254	786325	786396	786467	786538	786609	786680	71
2	786751	786822	786893	786964	787035	787106	787177	787248	787319	787390	
3	787460	787531	787602	787672	787744	787815	787885	787956	788027	788098	
4	788168	788239	788310	788381	788451	788522	788593	788663	788734	788804	
5	788875	788946	789016	789087	789157	789228	789299	789369	789440	789510	
6	789581	789651	789722	789792	789863	789933	790004	790074	790144	790215	
7	790285	790356	790426	790496	790567	790637	790707	790778	790848	790918	
8	790988	791059	791129	791199	791269	791340	791410	791480	791550	791620	
9	791691	791761	791831	791901	791971	792041	792111	792181	792252	792322	
620	792392	792462	792532	792602	792672	792742	792812	792882	792952	793022	70
1	793092	793162	793231	793301	793371	793441	793511	793581	793651	793721	
2	793790	793860	793930	794000	794070	794139	794209	794279	794349	794418	
3	794488	794558	794627	794697	794767	794836	794906	794976	795045	795115	
4	795185	795254	795324	795393	795463	795532	795602	795672	795741	795811	
5	795880	795949	796019	796088	796158	796227	796297	796366	796436	796505	
6	796574	796644	796713	796782	796852	796921	796990	797060	797129	797198	
7	797268	797337	797406	797475	797545	797614	797683	797752	797821	797890	
8	797960	798029	798098	798167	798236	798305	798374	798443	798513	798582	
9	798651	798720	798789	798858	798927	798996	799065	799134	799203	799272	69
630	799341	799409	799478	799547	799616	799685	799754	799823	799892	799961	

PROPORTIONAL PARTS

Diff.	1	2	3	4	5	6	7	8	9
76	7.6	15.2	22.8	30.4	38.0	45.6	53.2	60.8	68.4
74	7.4	14.8	22.2	29.6	37.0	44.4	51.8	59.2	66.6
72	7.2	14.4	21.6	28.8	36.0	43.2	50.4	57.6	64.8
70	7.0	14.0	21.0	28.0	35.0	42.0	49.0	56.0	63.0
68	6.8	13.6	20.4	27.2	34.0	40.8	47.6	54.4	61.2

N	0	1	2	3	4	5	6	7	8	9	Diff.
630	799341	799409	799478	799547	799616	799685	799754	799823	799892	799961	
1	800029	800098	800167	800236	800305	800373	800442	800511	800580	800648	
2	800717	800786	800854	800923	800992	801061	801129	801198	801266	801335	
3	801404	801472	801541	801609	801678	801747	801815	801884	801952	802021	
4	802089	802158	802226	802295	802363	802432	802500	802568	802637	802705	
5	802774	802842	802910	802979	803047	803116	803184	803252	803321	803389	
6	803457	803525	803594	803662	803730	803798	803867	803935	804003	804071	
7	804139	804208	804276	804344	804412	804480	804548	804616	804685	804753	
8	804821	804889	804957	805025	805093	805161	805229	805297	805365	805433	68
9	805501	805569	805637	805705	805773	805841	805908	805976	806044	806112	
640	806180	806248	806316	806384	806451	806519	806587	806655	806723	806790	
1	806858	806926	806994	807061	807129	807197	807264	807332	807400	807467	
2	807535	807603	807670	807738	807806	807873	807941	808008	808076	808143	
3	808211	808279	808346	808414	808481	808549	808616	808684	808751	808818	
4	808886	808953	809021	809088	809156	809223	809290	809358	809425	809492	
5	809560	809627	809694	809762	809829	809896	809964	810031	810098	810165	
6	810233	810300	810367	810434	810501	810569	810636	810703	810770	810837	
7	810904	810971	811039	811106	811173	811240	811307	811374	811441	811508	67
8	811575	811642	811709	811776	811843	811910	811977	812044	812111	812178	
9	812245	812312	812379	812445	812512	812579	812646	812713	812780	812847	
650	812913	812980	813047	813114	813181	813247	813314	813381	813448	813514	
1	813581	813648	813714	813781	813848	813914	813981	814048	814114	814181	
2	814248	814314	814381	814447	814514	814581	814647	814714	814780	814847	
3	814913	814980	815046	815113	815179	815246	815312	815378	815445	815511	
4	815578	815644	815711	815777	815843	815910	815976	816042	816109	816175	
5	816241	816308	816374	816440	816506	816573	816639	816705	816771	816838	
6	816904	816970	817036	817102	817169	817235	817301	817367	817433	817499	
7	817565	817631	817698	817764	817830	817896	817962	818028	818094	818160	
8	818226	818292	818358	818424	818490	818556	818622	818688	818754	818820	66
9	818885	818951	819017	819083	819149	819215	819281	819346	819412	819478	
660	819544	819610	819676	819741	819807	819873	819939	820004	820070	820136	
1	820201	820267	820333	820399	820464	820530	820595	820661	820727	820792	
2	820858	820924	820989	821055	821120	821186	821251	821317	821382	821448	
3	821514	821579	821645	821710	821775	821841	821906	821972	822037	822103	
4	822168	822233	822299	822364	822430	822495	822560	822626	822691	822756	
5	822822	822887	822952	823018	823083	823148	823213	823279	823344	823409	
6	823474	823539	823605	823670	823735	823800	823865	823930	823996	824061	
7	824126	824191	824256	824321	824386	824451	824516	824581	824646	824711	
8	824776	824841	824906	824971	825036	825101	825166	825231	825296	825361	65
9	825426	825491	825556	825621	825686	825751	825815	825880	825945	826010	
670	826075	826140	826204	826269	826334	826399	826464	826528	826593	826658	
1	826723	826787	826852	826917	826981	827046	827111	827175	827240	827305	
2	827369	827434	827499	827563	827628	827692	827757	827821	827886	827951	
3	828015	828080	828144	828209	828273	828338	828402	828467	828531	828595	
4	828660	828724	828789	828853	828918	828982	829046	829111	829175	829239	
5	829304	829368	829432	829497	829561	829625	829690	829754	829818	829882	
6	829947	830011	830075	830139	830204	830268	830332	830396	830460	830525	
7	830589	830653	830717	830781	830845	830909	830973	831037	831102	831166	
8	831231	831294	831358	831422	831486	831550	831614	831678	831742	831806	64
9	831870	831934	831998	832062	832126	832189	832253	832317	832381	832445	
680	832509	832573	832637	832700	832764	832828	832892	832956	833020	833083	

PROPORTIONAL PARTS

Diff.	1	2	3	4	5	6	7	8	9
70	7.0	14.0	21.0	28.0	35.0	42.0	49.0	56.0	63.0
68	6.8	13.6	20.4	27.2	34.0	40.8	47.6	54.4	61.2
66	6.6	13.2	19.8	26.4	33.0	39.6	46.2	52.8	59.4
64	6.4	12.8	19.2	25.6	32.0	38.4	44.8	51.2	57.6
62	6.2	12.4	18.6	24.8	31.0	37.2	43.4	49.6	55.8

N	0	1	2	3	4	5	6	7	8	9	Diff.
680	832509	832573	832637	832700	832764	832828	832892	832956	833020	833083	
1	833147	833211	833275	833338	833402	833466	833530	833593	833657	833721	
2	833784	833848	833912	833975	834039	834103	834166	834230	834294	834357	
3	834421	834484	834548	834611	834675	834739	834802	834866	834929	834993	
4	835056	835120	835183	835247	835310	835373	835437	835500	835564	835627	
5	835691	835754	835817	835881	835944	836007	836071	836134	836197	836261	
6	836324	836387	836451	836514	836577	836641	836704	836767	836830	836894	
7	836957	837020	837083	837146	837210	837273	837336	837399	837462	837525	
8	837588	837652	837715	837778	837841	837904	837967	838030	838093	838156	63
9	838219	838282	838345	838408	838471	838534	838597	838660	838723	838786	
690	838849	838912	838975	839038	839101	839164	839227	839289	839352	839415	
1	839478	839541	839604	839667	839729	839792	839855	839918	839981	840043	
2	840106	840169	840232	840294	840357	840420	840482	840545	840608	840671	
3	840733	840796	840859	840921	840984	841046	841109	841172	841234	841297	
4	841359	841422	841485	841547	841610	841672	841735	841797	841860	841922	
5	841985	842047	842110	842172	842235	842297	842360	842422	842484	842547	
6	842609	842672	842734	842796	842859	842921	842983	843046	843108	843170	
7	843233	843295	843357	843420	843482	843544	843606	843669	843731	843793	
8	843855	843918	843980	844042	844104	844166	844229	844291	844353	844415	
9	844477	844539	844601	844664	844726	844788	844850	844912	844974	845036	
700	845098	845160	845222	845284	845346	845408	845470	845532	845594	845656	62
1	845718	845780	845842	845904	845966	846028	846090	846151	846213	846275	
2	846337	846399	846461	846523	846585	846646	846708	846770	846832	846894	
3	846955	847017	847079	847141	847202	847264	847326	847388	847449	847511	
4	847573	847634	847696	847758	847819	847881	847943	848004	848066	848128	
5	848189	848251	848312	848374	848435	848497	848559	848620	848682	848743	
6	848805	848866	848928	848989	849051	849112	849174	849235	849297	849358	
7	849419	849481	849542	849604	849665	849726	849788	849849	849911	849972	
8	850033	850095	850156	850217	850279	850340	850401	850462	850524	850585	
9	850646	850707	850769	850830	850891	850952	851014	851075	851136	851197	
710	851258	851320	851381	851442	851503	851564	851625	851686	851747	851809	61
1	851870	851931	851992	852053	852114	852175	852236	852297	852358	852419	
2	852480	852541	852602	852663	852724	852785	852846	852907	852968	853029	
3	853090	853150	853211	853272	853333	853394	853455	853516	853577	853637	
4	853698	853759	853820	853881	853941	854002	854063	854124	854185	854245	
5	854306	854367	854428	854488	854549	854610	854670	854731	854792	854852	
6	854913	854974	855034	855095	855156	855216	855277	855337	855398	855459	
7	855519	855580	855640	855701	855761	855822	855882	855943	856003	856064	
8	856124	856185	856245	856306	856366	856427	856487	856548	856608	856668	
9	856729	856789	856850	856910	856970	857031	857091	857152	857212	857272	
720	857332	857393	857453	857513	857574	857634	857694	857755	857815	857875	
1	857935	857995	858056	858116	858176	858236	858297	858357	858417	858477	
2	858537	858597	858657	858718	858778	858838	858898	858958	859018	859078	
3	859138	859198	859258	859318	859379	859439	859499	859559	859619	859679	60
4	859739	859799	859859	859919	859978	860038	860098	860158	860218	860278	
5	860338	860398	860458	860518	860578	860637	860697	860757	860817	860877	
6	860937	860996	861056	861116	861176	861236	861295	861355	861415	861475	
7	861534	861594	861654	861714	861773	861833	861893	861952	862012	862072	
8	862131	862191	862251	862310	862370	862430	862489	862549	862608	862668	
9	862728	862787	862847	862906	862966	863025	863085	863144	863204	863263	
730	863323	863382	863442	863501	863561	863620	863680	863739	863799	863858	

PROPORTIONAL PARTS

Diff.	1	2	3	4	5	6	7	8	9
64	6.4	12.8	19.2	25.6	32.0	38.4	44.8	51.2	57.6
62	6.2	12.4	18.6	24.8	31.0	37.2	43.4	49.6	55.8
60	6.0	12.0	18.0	24.0	30.0	36.0	42.0	48.0	54.0
58	5.8	11.6	17.4	23.2	29.0	34.8	40.6	46.4	52.2

N	0	1	2	3	4	5	6	7	8	9	Diff.
730	863323	863382	863442	863501	863561	863620	863680	863739	863799	863858	
1	863917	863977	864036	864096	864155	864214	864274	864333	864392	864452	
2	864511	864570	864630	864689	864748	864808	864867	864926	864985	865045	
3	865104	865163	865222	865282	865341	865400	865459	865519	865578	865637	
4	865696	865755	865814	865874	865933	865992	866051	866110	866169	866228	
5	866287	866346	866405	866465	866524	866583	866642	866701	866760	866819	
6	866878	866937	866996	867055	867114	867173	867232	867291	867350	867409	59
7	867467	867526	867585	867644	867703	867762	867821	867880	867939	867998	
8	868056	868115	868174	868233	868292	868350	868409	868468	868527	868586	
9	868644	868703	868762	868821	868879	868938	868997	869056	869114	869173	
740	869232	869290	869349	869408	869466	869525	869584	869642	869701	869760	
1	869818	869877	869935	869994	870053	870111	870170	870228	870287	870345	
2	870404	870462	870521	870579	870638	870696	870755	870813	870872	870930	
3	870989	871047	871106	871164	871223	871281	871339	871398	871456	871515	
4	871573	871631	871690	871748	871806	871865	871923	871981	872040	872098	
5	872156	872215	872273	872331	872389	872448	872506	872564	872622	872681	
6	872739	872797	872855	872913	872972	873030	873088	873146	873204	873262	
7	873321	873379	873437	873495	873553	873611	873669	873727	873785	873844	
8	873902	873960	874018	874076	874134	874192	874250	874308	874366	874424	58
9	874482	874540	874598	874656	874714	874772	874830	874888	874945	875003	
750	875061	875119	875177	875235	875293	875351	875409	875466	875524	875582	
1	875640	875698	875756	875813	875871	875929	875987	876045	876102	876160	
2	876218	876276	876333	876391	876449	876507	876564	876622	876680	876737	
3	876795	876853	876910	876968	877026	877083	877141	877199	877256	877314	
4	877371	877429	877487	877544	877602	877659	877717	877774	877832	877889	
5	877947	878004	878062	878119	878177	878234	878292	878349	878407	878464	
6	878522	878579	878637	878694	878752	878809	878866	878924	878981	879039	
7	879096	879153	879211	879268	879325	879383	879440	879497	879555	879612	
8	879669	879726	879784	879841	879898	879956	880013	880070	880127	880185	
9	880242	880299	880356	880413	880471	880528	880585	880642	880699	880756	
760	880814	880871	880928	880985	881042	881099	881156	881213	881271	881328	
1	881385	881442	881499	881556	881613	881670	881727	881784	881841	881898	
2	881955	882012	882069	882126	882183	882240	882297	882354	882411	882468	57
3	882525	882581	882638	882695	882752	882809	882866	882923	882980	883037	
4	883093	883150	883207	883264	883321	883377	883434	883491	883548	883605	
5	883661	883718	883775	883832	883888	883945	884002	884059	884115	884172	
6	884229	884285	884342	884399	884455	884512	884569	884625	884682	884739	
7	884795	884852	884909	884965	885022	885078	885135	885192	885248	885305	
8	885361	885418	885474	885531	885587	885644	885700	885757	885813	885870	
9	885926	885983	886039	886096	886152	886209	886265	886321	886378	886434	
770	886491	886547	886604	886660	886716	886773	886829	886885	886942	886998	
1	887054	887111	887167	887223	887280	887336	887392	887449	887505	887561	
2	887617	887674	887730	887786	887842	887898	887955	888011	888067	888123	
3	888179	888236	888292	888348	888404	888460	888516	888573	888629	888685	
4	888741	888797	888853	888909	888965	889021	889077	889134	889190	889246	
5	889302	889358	889414	889470	889526	889582	889638	889694	889750	889806	56
6	889862	889918	889974	890030	890086	890141	890197	890253	890309	890365	
7	890421	890477	890533	890589	890645	890700	890756	890812	890868	890924	
8	890980	891035	891091	891147	891203	891259	891314	891370	891426	891482	
9	891537	891593	891649	891705	891760	891816	891872	891928	891983	892039	
780	892095	892150	892206	892262	892317	892373	892429	892484	892540	892595	

PROPORTIONAL PARTS

Diff.	1	2	3	4	5	6	7	8	9
60	6.0	12.0	18.0	24.0	30.0	36.0	42.0	48.0	54.0
58	5.8	11.6	17.4	23.2	29.0	34.8	40.6	46.4	52.2
56	5.6	11.2	16.8	22.4	28.0	33.6	39.2	44.8	50.4
54	5.4	10.8	16.2	21.6	27.0	32.4	37.8	43.2	48.6

N	0	1	2	3	4	5	6	7	8	9	Diff.
780	892095	892150	892206	892262	892317	892373	892429	892484	892540	892595	
1	892651	892707	892762	892818	892873	892929	892985	893040	893096	893151	
2	893207	893262	893318	893373	893429	893484	893540	893595	893651	893706	
3	893762	893817	893873	893928	893984	894039	894094	894150	894205	894261	
4	894316	894371	894427	894482	894538	894593	894648	894704	894759	894814	
5	894870	894925	894980	895036	895091	895146	895201	895257	895312	895367	
6	895423	895478	895533	895588	895644	895699	895754	895809	895864	895920	
7	895975	896030	896085	896140	896195	896251	896306	896361	896416	896471	
8	896526	896581	896636	896692	896747	896802	896857	896912	896967	897022	
9	897077	897132	897187	897242	897297	897352	897407	897462	897517	897572	
790	897627	897682	897737	897792	897847	897902	897957	898012	898067	898122	55
1	898176	898231	898286	898341	898396	898451	898506	898561	898615	898670	
2	898725	898780	898835	898890	898944	898999	899054	899109	899164	899218	
3	899273	899328	899383	899437	899492	899547	899602	899656	899711	899766	
4	899821	899875	899930	899985	900039	900094	900149	900203	900258	900312	
5	900367	900422	900476	900531	900586	900640	900695	900749	900804	900859	
6	900913	900968	901022	901077	901131	901186	901240	901295	901349	901404	
7	901458	901513	901567	901622	901676	901731	901785	901840	901894	901948	
8	902003	902057	902112	902166	902221	902275	902329	902384	902438	902492	
9	902547	902601	902655	902710	902764	902818	902873	902927	902981	903036	
800	903090	903144	903199	903253	903307	903361	903416	903470	903524	903578	
1	903633	903687	903741	903795	903849	903904	903958	904012	904066	904120	
2	904174	904229	904283	904337	904391	904445	904499	904553	904607	904661	
3	904716	904770	904824	904878	904932	904986	905040	905094	905148	905202	
4	905256	905310	905364	905418	905472	905526	905580	905634	905688	905742	54
5	905796	905850	905904	905958	906012	906066	906119	906173	906227	906281	
6	906335	906389	906443	906497	906551	906604	906658	906712	906766	906820	
7	906874	906927	906981	907035	907089	907143	907196	907250	907304	907358	
8	907411	907465	907519	907573	907626	907680	907734	907787	907841	907895	
9	907949	908002	908056	908110	908163	908217	908270	908324	908378	908431	
810	908485	908539	908592	908646	908699	908753	908807	908860	908914	908967	
1	909021	909074	909128	909181	909235	909289	909342	909396	909449	909503	
2	909556	909610	909663	909716	909770	909823	909877	909930	909984	910037	
3	910091	910144	910197	910251	910304	910358	910411	910464	910518	910571	
4	910624	910678	910731	910784	910838	910891	910944	910998	911051	911104	
5	911158	911211	911264	911317	911371	911424	911477	911530	911584	911637	
6	911690	911743	911797	911850	911903	911956	912009	912063	912116	912169	
7	912222	912275	912328	912381	912435	912488	912541	912594	912647	912700	
8	912753	912806	912859	912913	912966	913019	913072	913125	913178	913231	
9	913284	913337	913390	913443	913496	913549	913602	913655	913708	913761	53
820	913814	913867	913920	913973	914026	914079	914132	914184	914237	914290	
1	914343	914396	914449	914502	914555	914608	914660	914713	914766	914819	
2	914872	914925	914977	915030	915083	915136	915189	915241	915294	915347	
3	915400	915453	915505	915558	915611	915664	915716	915769	915822	915875	
4	915927	915980	916033	916085	916138	916191	916243	916296	916349	916401	
5	916454	916507	916559	916612	916664	916717	916770	916822	916875	916927	
6	916980	917033	917085	917138	917190	917243	917295	917348	917400	917453	
7	917506	917558	917611	917663	917716	917768	917820	917873	917925	917978	
8	918030	918083	918135	918188	918240	918293	918345	918397	918450	918502	
9	918555	918607	918659	918712	918764	918816	918869	918921	918973	919026	
830	919078	919130	919183	919235	919287	919340	919392	919444	919496	919549	
1	919601	919653	919706	919758	919810	919862	919914	919967	920019	920071	
2	920123	920176	920228	920280	920332	920384	920436	920489	920541	920593	
3	920645	920697	920749	920801	920853	920906	920958	921010	921062	921114	52
4	921166	921218	921270	921322	921374	921426	921478	921530	921582	921634	
5	921686	921738	921790	921842	921894	921946	921998	922050	922102	922154	

PROPORTIONAL PARTS

Diff.	1	2	3	4	5	6	7	8	9
56	5.6	11.2	16.8	22.4	28.0	33.6	39.2	44.8	50.4
54	5.4	10.8	16.2	21.6	27.0	32.4	37.8	43.2	48.6
52	5.2	10.4	15.6	20.8	26.0	31.2	36.4	41.6	46.8

N	0	1	2	3	4	5	6	7	8	9	Diff.
835	921686	921738	921790	921842	921894	921946	921998	922050	922102	922154	
6	922206	922258	922310	922362	922414	922466	922518	922570	922622	922674	
7	922725	922777	922829	922881	922933	922985	923037	923089	923140	923192	
8	923244	923296	923348	923399	923451	923503	923555	923607	923658	923710	
9	923762	923814	923865	923917	923969	924021	924072	924124	924176	924228	
840	924279	924331	924383	924434	924486	924538	924589	924641	924693	924744	
1	924796	924848	924899	924951	925003	925054	925106	925157	925209	925261	
2	925312	925364	925415	925467	925518	925570	925621	925673	925725	925776	
3	925828	925879	925931	925982	926034	926085	926137	926188	926240	926291	
4	926342	926394	926445	926497	926548	926600	926651	926702	926754	926805	
5	926857	926908	926959	927011	927062	927114	927165	927216	927268	927319	
6	927370	927422	927473	927524	927576	927627	927678	927730	927781	927832	
7	927883	927935	927986	928037	928088	928140	928191	928242	928293	928345	
8	928396	928447	928498	928549	928601	928652	928703	928754	928805	928857	
9	928908	928959	929010	929061	929112	929163	929215	929266	929317	929368	
850	929419	929470	929521	929572	929623	929674	929725	929776	929827	929879	
1	929930	929981	930032	930083	930134	930185	930236	930287	930338	930389	51
2	930440	930491	930542	930592	930643	930694	930745	930796	930847	930898	
3	930949	931000	931051	931102	931153	931204	931254	931305	931356	931407	
4	931458	931509	931560	931610	931661	931712	931763	931814	931865	931915	
5	931966	932017	932068	932118	932169	932220	932271	932322	932372	932423	
6	932474	932524	932575	932626	932677	932727	932778	932829	932879	932930	
7	932981	933031	933082	933133	933183	933234	933285	933335	933386	933437	
8	933487	933538	933589	933639	933690	933740	933791	933841	933892	933943	
9	933993	934044	934094	934145	934195	934246	934296	934347	934397	934448	
860	934498	934549	934599	934650	934700	934751	934801	934852	934902	934953	
1	935003	935054	935104	935154	935205	935255	935306	935356	935406	935457	
2	935507	935558	935608	935658	935709	935759	935809	935860	935910	935960	
3	936011	936061	936111	936162	936212	936262	936313	936363	936413	936463	
4	936514	936564	936614	936665	936715	936765	936815	936865	936916	936966	
5	937016	937066	937116	937167	937217	937267	937317	937367	937418	937468	
6	937518	937568	937618	937668	937718	937769	937819	937869	937919	937969	
7	938019	938069	938119	938169	938219	938269	938320	938370	938420	938470	50
8	938520	938570	938620	938670	938720	938770	938820	938870	938920	938970	
9	939020	939070	939120	939170	939220	939270	939320	939369	939419	939469	
870	939519	939569	939619	939669	939719	939769	939819	939869	939918	939968	
1	940018	940068	940118	940168	940218	940267	940317	940367	940417	940467	
2	940516	940566	940616	940666	940716	940765	940815	940865	940915	940964	
3	941014	941064	941114	941163	941213	941263	941313	941362	941412	941462	
4	941511	941561	941611	941660	941710	941760	941809	941859	941909	941958	
5	942008	942058	942107	942157	942207	942256	942306	942355	942405	942455	
6	942504	942554	942603	942653	942702	942752	942801	942851	942901	942950	
7	943000	943049	943099	943148	943198	943247	943297	943346	943396	943445	
8	943495	943544	943593	943643	943692	943742	943791	943841	943890	943939	
9	943989	944038	944088	944137	944186	944236	944285	944335	944384	944433	
880	944483	944532	944581	944631	944680	944729	944779	944828	944877	944927	
1	944976	945025	945074	945124	945173	945222	945272	945321	945370	945419	
2	945469	945518	945567	945616	945665	945715	945764	945813	945862	945912	
3	945961	946010	946059	946108	946157	946207	946256	946305	946354	946403	
4	946452	946501	946551	946600	946649	946698	946747	946796	946845	946894	
5	946943	946992	947041	947090	947140	947189	947238	947287	947336	947385	
6	947434	947483	947532	947581	947630	947679	947728	947777	947826	947875	49
7	947924	947973	948022	948070	948119	948168	948217	948266	948315	948364	
8	948413	948462	948511	948560	948608	948657	948706	948755	948804	948853	
9	948902	948951	948999	949048	949097	949146	949195	949244	949292	949341	
890	949390	949439	949488	949536	949585	949634	949683	949731	949780	949829	

PROPORTIONAL PARTS

Diff.	1	2	3	4	5	6	7	8	9
52	5.2	10.4	15.6	20.8	26.0	31.2	36.4	41.6	46.8
50	5.0	10.0	15.0	20.0	25.0	30.0	35.0	40.0	45.0
48	4.8	9.6	14.4	19.2	24.0	28.8	33.6	38.4	43.2

N	0	1	2	3	4	5	6	7	8	9	Diff.
890	**949390**	**949439**	**949488**	**949536**	**949585**	**949634**	**949683**	**949731**	**949780**	**949829**	
1	949878	949926	949975	950024	950073	950121	950170	950219	950267	950316	
2	950365	950414	950462	950511	950560	950608	950657	950706	950754	950803	
3	950851	950900	950949	950997	951046	951095	951143	951192	951240	951289	
4	951338	951386	951435	951483	951532	951580	951629	951677	951726	95177	
5	951823	951872	951920	951969	952017	952066	952114	952163	952211	9522(	
6	952308	952356	952405	952453	952502	952550	952599	952647	952696	95274-	
7	952792	952841	952889	952938	952986	953034	953083	953131	953180	953228	
8	953276	953325	953373	953421	953470	953518	953566	953615	953663	953711	
9	953760	953808	953856	953905	953953	954001	954049	954098	954146	954194	
900	**954243**	**954291**	**954339**	**954387**	**954435**	**954484**	**954532**	**954580**	**954628**	**954677**	
1	954725	954773	954821	954869	954918	954966	955014	955062	955110	955158	
2	955207	955255	955303	955351	955399	955447	955495	955543	955592	955640	
3	955688	955736	955784	955832	955880	955928	955976	956024	956072	956120	
4	956168	956216	956265	956313	956361	956409	956457	956505	956553	956601	
5	956649	956697	956745	956793	956840	956888	956936	956984	957032	957080	48
6	957128	957176	957224	957272	957320	957368	957416	957464	957512	957559	
7	957607	957655	957703	957751	957799	957847	957894	957942	957990	958038	
8	958086	958134	958181	958229	958277	958325	958373	958421	958468	958516	
9	958564	958612	958659	958707	958755	958803	958850	958898	958946	958994	
910	**959041**	**959089**	**959137**	**959185**	**959232**	**959280**	**959328**	**959375**	**959423**	**959471**	
1	959518	959566	959614	959661	959709	959757	959804	959852	959900	959947	
2	959995	960042	960090	960138	960185	960233	960280	960328	960376	960423	
3	960471	960518	960566	960613	960661	960709	960756	960804	960851	960899	
4	960946	960994	961041	961089	961136	961184	961231	961279	961326	961374	
5	961421	961469	961516	961563	961611	961658	961706	961753	961801	961848	
6	961895	961943	961990	962038	962085	962132	962180	962227	962275	962322	
7	962369	962417	962464	962511	962559	962606	962653	962701	962748	962795	
8	962843	962890	962937	962985	963032	963079	963126	963174	963221	963268	
9	963316	963363	963410	963457	963504	963552	963599	963646	963693	963741	
920	**963788**	**963835**	**963882**	**963929**	**963977**	**964024**	**964071**	**964118**	**964165**	**964212**	
1	964260	964307	964354	964401	964448	964495	964542	964590	964637	964684	
2	964731	964778	964825	964872	964919	964966	965013	965061	965108	965155	
3	965202	965249	965296	965343	965390	965437	965484	965531	965578	965625	
4	965672	965719	965766	965813	965860	965907	965954	966001	966048	966095	47
5	966142	966189	966236	966283	966329	966376	966423	966470	966517	966564	
6	966611	966658	966705	966752	966799	966845	966892	966939	966986	967033	
7	967080	967127	967173	967220	967267	967314	967361	967408	967454	967501	
8	967548	967595	967642	967688	967735	967782	967829	967875	967922	967969	
9	968016	968062	968109	968156	968203	968249	968296	968343	968390	968436	
930	**968483**	**968530**	**968576**	**968623**	**968670**	**968716**	**968763**	**968810**	**968856**	**968903**	
1	968950	968996	969043	969090	969136	969183	969229	969276	969323	969369	
2	969416	969463	969509	969556	969602	969649	969695	969742	969789	969835	
3	969882	969928	969975	970021	970068	970114	970161	970207	970254	970300	
4	970347	970393	970440	970486	970533	970579	970626	970672	970719	970765	
5	970812	970858	970904	970951	970997	971044	971090	971137	971183	971229	
6	971276	971322	971369	971415	971461	971508	971554	971601	971647	971693	
7	971740	971786	971832	971879	971925	971971	972018	972064	972110	972157	
8	972203	972249	972295	972342	972388	972434	972481	972527	972573	972619	
9	972666	972712	972758	972804	972851	972897	972943	972989	973035	973082	
940	**973128**	**973174**	**973220**	**973266**	**973313**	**973359**	**973405**	**973451**	**973497**	**973543**	
1	973590	973636	973682	973728	973774	973820	973866	973913	973959	974005	
2	974051	974097	974143	974189	974235	974281	974327	974374	974420	974466	
3	974512	974558	974604	974650	974696	974742	974788	974834	974880	974926	
4	974972	975018	975064	975110	975156	975202	975248	975294	975340	975386	46
5	975432	975478	975524	975570	975616	975662	975707	975753	975799	975845	

PROPORTIONAL PARTS

Diff.	1	2	3	4	5	6	7	8	9
50	5.0	10.0	15.0	20.0	25.0	30.0	35.0	40.0	45.0
48	4.8	9.6	14.4	19.2	24.0	28.8	33.6	38.4	43.2
46	4.6	9.2	13.8	18.4	23.0	27.6	32.2	36.8	41.4

N	0	1	2	3	4	5	6	7	8	9	Diff.
945	975432	975478	975524	975570	975616	975662	975707	975753	975799	975845	
6	975891	975937	975983	976029	976075	976121	976167	976212	976258	976304	
7	976350	976396	976442	976488	976533	976579	976625	976671	976717	976763	
8	976808	976854	976900	976946	976992	977037	977083	977129	977175	977220	
9	977266	977312	977358	977403	977449	977495	977541	977586	977632	977678	
950	977724	977769	977815	977861	977906	977952	977998	978043	978089	978135	
1	978181	978226	978272	978317	978363	978409	978454	978500	978546	978591	
2	978637	978683	978728	978774	978819	978865	978911	978956	979002	979047	
3	979093	979138	979184	979230	979275	979321	979366	979412	979457	979503	
4	979548	979594	979639	979685	979730	979776	979821	979867	979912	979958	
5	980003	980049	980094	980140	980185	980231	980276	980322	980367	980412	
6	980458	980503	980549	980594	980640	980685	980730	980776	980821	980867	
7	980912	980957	981003	981048	981093	981139	981184	981229	981275	981320	
8	981366	981411	981456	981501	981547	981592	981637	981683	981728	981773	
9	981819	981864	981909	981954	982000	982045	982090	982135	982181	982226	
960	982271	982316	982362	982407	982452	982497	982543	982588	982633	982678	
1	982723	982769	982814	982859	982904	982949	982994	983040	983085	983130	
2	983175	983220	983265	983310	983356	983401	983446	983491	983536	983581	
3	983626	983671	983716	983762	983807	983852	983897	983942	983987	984032	
4	984077	984122	984167	984212	984257	984302	984347	984392	984437	984482	45
5	984527	984572	984617	984662	984707	984752	984797	984842	984887	984932	
6	984977	985022	985067	985112	985157	985202	985247	985292	985337	985382	
7	985426	985471	985516	985561	985606	985651	985696	985741	985786	985830	
8	985875	985920	985965	986010	986055	986100	986144	986189	986234	986279	
9	986324	986369	986413	986458	986503	986548	986593	986637	986682	986727	
970	986772	986817	986861	986906	986951	986996	987040	987085	987130	987175	
1	987219	987264	987309	987353	987398	987443	987488	987532	987577	987622	
2	987666	987711	987756	987800	987845	987890	987934	987979	988024	988068	
3	988113	988157	988202	988247	988291	988336	988381	988425	988470	988514	
4	988559	988604	988648	988693	988737	988782	988826	988871	988916	988960	
5	989005	989049	989094	989138	989183	989227	989272	989316	989361	989405	
6	989450	989494	989539	989583	989628	989672	989717	989761	989806	989850	
7	989895	989939	989983	990028	990072	990117	990161	990206	990250	990294	
8	990339	990383	990428	990472	990516	990561	990605	990650	990694	990738	
9	990783	990827	990871	990916	990960	991004	991049	991093	991137	991182	
980	991226	991270	991315	991359	991403	991448	991492	991536	991580	991625	
1	991669	991713	991758	991802	991846	991890	991935	991979	992023	992067	
2	992111	992156	992200	992244	992288	992333	992377	992421	992465	992509	
3	992554	992598	992642	992686	992730	992774	992819	992863	992907	992951	
4	992995	993039	993083	993127	993172	993216	993260	993304	993348	993392	
5	993436	993480	993524	993568	993613	993657	993701	993745	993789	993833	
6	993877	993921	993965	994009	994053	994097	994141	994185	994229	994273	
7	994317	994361	994405	994449	994493	994537	994581	994625	994669	994713	44
8	994757	994801	994845	994889	994933	994977	995021	995065	995108	995152	
9	995196	995240	995284	995328	995372	995416	995460	995504	995547	995591	
990	995635	995679	995723	995767	995811	995854	995898	995942	995986	996030	
1	996074	996117	996161	996205	996249	996293	996337	996380	996424	996468	
2	996512	996555	996599	996643	996687	996731	996774	996818	996862	996906	
3	996949	996993	997037	997080	997124	997168	997212	997255	997299	997343	
4	997386	997430	997474	997517	997561	997605	997648	997692	997736	997779	
5	997823	997867	997910	997954	997998	998041	998085	998129	998172	998216	
6	998259	998303	998347	998390	998434	998477	998521	998564	998608	998652	
7	998695	998739	998782	998826	998869	998913	998956	999000	999043	999087	
8	999131	999174	999218	999261	999305	999348	999392	999435	999479	999522	
9	999565	999609	999652	999696	999739	999783	999826	999870	999913	999957	
1000	000000	000043	000087	000130	000174	000217	000260	000304	000347	000391	43

PROPORTIONAL PARTS

Diff.	1	2	3	4	5	6	7	8	9
46	4.6	9.2	13.8	18.4	23.0	27.6	32.2	36.8	41.4
44	4.4	8.8	13.2	17.6	22.0	26.4	30.8	35.2	39.6
42	4.2	8.4	12.6	16.8	21.0	25.2	29.4	33.6	37.8

Table 11.　Natural (Napierian) Logarithms of Numbers

The natural logarithm of a number is the index of the power to which the base e ($= 2.7182818$) must be raised in order to equal the number.

EXAMPLE: $\log_e 4.12 = \ln 4.12 = 1.4159$.

The table gives the natural logarithms of numbers from 1.00 to 9.99 directly, and permits the finding of the logarithms of numbers outside of that range by the addition or subtraction of the natural logarithms of powers of 10.

EXAMPLES: $\log_e 679. = \log_e 6.79 + \log_e 10^2 = 1.9155 + 4.6052 = 6.5207.$

$\log_e .0679 = \log_e 6.79 - \log_e 10^2 = 1.9155 - 4.6052 = -2.6897.$

Natural Logarithms of Powers of 10

$\log_e 10 = 2.302\ 585$　　$\log_e 10^4 = 9.210\ 340$　　$\log_e 10^7 = 16.118\ 096$
$\log_e 10^2 = 4.605\ 170$　　$\log_e 10^5 = 11.512\ 925$　　$\log_e 10^8 = 18.420\ 681$
$\log_e 10^3 = 6.907\ 755$　　$\log_e 10^6 = 13.815\ 511$　　$\log_e 10^9 = 20.723\ 266$

To obtain the common logarithm, the natural logarithm is multiplied by $\log_{10} e$, which is $0.434\ 294$, or $\log_{10} N = 0.434\ 294\ \log_e N$.

A negative number or number less than zero has no real logarithm.

N	0	1	2	3	4	5	6	7	8	9
1.0	0.0000	0.0100	0.0198	0.0296	0.0392	0.0488	0.0583	0.0677	0.0770	0.0862
1.1	0.0953	0.1044	0.1133	0.1222	0.1310	0.1398	0.1484	0.1570	0.1655	0.1740
1.2	0.1823	0.1906	0.1989	0.2070	0.2151	0.2231	0.2311	0.2390	0.2469	0.2546
1.3	0.2624	0.2700	0.2776	0.2852	0.2927	0.3001	0.3075	0.3148	0.3221	0.3293
1.4	0.3365	0.3436	0.3507	0.3577	0.3646	0.3716	0.3784	0.3853	0.3920	0.3988
1.5	0.4055	0.4121	0.4187	0.4253	0.4318	0.4383	0.4447	0.4511	0.4574	0.4637
1.6	0.4700	0.4762	0.4824	0.4886	0.4947	0.5008	0.5068	0.5128	0.5188	0.5247
1.7	0.5306	0.5365	0.5423	0.5481	0.5539	0.5596	0.5653	0.5710	0.5766	0.5822
1.8	0.5878	0.5933	0.5988	0.6043	0.6098	0.6152	0.6206	0.6259	0.6313	0.6366
1.9	0.6419	0.6471	0.6523	0.6575	0.6627	0.6678	0.6729	0.6780	0.6831	0.6881
2.0	0.6931	0.6981	0.7031	0.7080	0.7129	0.7178	0.7227	0.7275	0.7324	0.7372
2.1	0.7419	0.7467	0.7514	0.7561	0.7608	0.7655	0.7701	0.7747	0.7793	0.7839
2.2	0.7885	0.7930	0.7975	0.8020	0.8065	0.8109	0.8154	0.8198	0.8242	0.8286
2.3	0.8329	0.8372	0.8416	0.8459	0.8502	0.8544	0.8587	0.8629	0.8671	0.8713
2.4	0.8755	0.8796	0.8838	0.8879	0.8920	0.8961	0.9002	0.9042	0.9083	0.9123
2.5	0.9163	0.9203	0.9243	0.9282	0.9322	0.9361	0.9400	0.9439	0.9478	0.9517
2.6	0.9555	0.9594	0.9632	0.9670	0.9708	0.9746	0.9783	0.9821	0.9858	0.9895
2.7	0.9933	0.9969	1.0006	1.0043	1.0080	1.0116	1.0152	1.0188	1.0225	1.0260
2.8	1.0296	1.0332	1.0367	1.0403	1.0438	1.0473	1.0508	1.0543	1.0578	1.0613
2.9	1.0647	1.0682	1.0716	1.0750	1.0784	1.0818	1.0852	1.0886	1.0919	1.0953
3.0	1.0986	1.1019	1.1053	1.1086	1.1119	1.1151	1.1184	1.1217	1.1249	1.1282
3.1	1.1314	1.1346	1.1378	1.1410	1.1442	1.1474	1.1506	1.1537	1.1569	1.1600
3.2	1.1632	1.1663	1.1694	1.1725	1.1756	1.1787	1.1817	1.1848	1.1878	1.1909
3.3	1.1939	1.1969	1.2000	1.2030	1.2060	1.2090	1.2119	1.2149	1.2179	1.2208
3.4	1.2238	1.2267	1.2296	1.2326	1.2355	1.2384	1.2413	1.2442	1.2470	1.2499
3.5	1.2528	1.2556	1.2585	1.2613	1.2641	1.2669	1.2698	1.2726	1.2754	1.2782
3.6	1.2809	1.2837	1.2865	1.2892	1.2920	1.2947	1.2975	1.3002	1.3029	1.3056
3.7	1.3083	1.3110	1.3137	1.3164	1.3191	1.3218	1.3244	1.3271	1.3297	1.3324
3.8	1.3350	1.3376	1.3403	1.3429	1.3455	1.3481	1.3507	1.3533	1.3558	1.3584
3.9	1.3610	1.3635	1.3661	1.3686	1.3712	1.3737	1.3762	1.3788	1.3813	1.3838
4.0	1.3863	1.3888	1.3913	1.3938	1.3962	1.3987	1.4012	1.4036	1.4061	1.4085
4.1	1.4110	1.4134	1.4159	1.4183	1.4207	1.4231	1.4255	1.4279	1.4303	1.4327
4.2	1.4351	1.4375	1.4398	1.4422	1.4446	1.4469	1.4493	1.4516	1.4540	1.4563
4.3	1.4586	1.4609	1.4633	1.4656	1.4679	1.4702	1.4725	1.4748	1.4770	1.4793
4.4	1.4816	1.4839	1.4861	1.4884	1.4907	1.4929	1.4951	1.4974	1.4996	1.5019
4.5	1.5041	1.5063	1.5085	1.5107	1.5129	1.5151	1.5173	1.5195	1.5217	1.5239
4.6	1.5261	1.5282	1.5304	1.5326	1.5347	1.5369	1.5390	1.5412	1.5433	1.5454
4.7	1.5476	1.5497	1.5518	1.5539	1.5560	1.5581	1.5602	1.5623	1.5644	1.5665
4.8	1.5686	1.5707	1.5728	1.5748	1.5769	1.5790	1.5810	1.5831	1.5851	1.5872
4.9	1.5892	1.5913	1.5933	1.5953	1.5974	1.5994	1.6014	1.6034	1.6054	1.6074

Table 11—*Continued*

N	0	1	2	3	4	5	6	7	8	9
5.0	**1.6094**	**1.6114**	**1.6134**	**1.6154**	**1.6174**	**1.6194**	**1.6214**	**1.6233**	**1.6253**	**1.6273**
5.1	1.6292	1.6312	1.6332	1.6351	1.6371	1.6390	1.6409	1.6429	1.6448	1.6467
5.2	1.6487	1.6506	1.6525	1.6544	1.6563	1.6582	1.6601	1.6620	1.6639	1.6658
5.3	1.6677	1.6696	1.6715	1.6734	1.6752	1.6771	1.6790	1.6808	1.6827	1.6845
5.4	1.6864	1.6882	1.6901	1.6919	1.6938	1.6956	1.6974	1.6993	1.7011	1.7029
5.5	1.7047	1.7066	1.7084	1.7102	1.7120	1.7138	1.7156	1.7174	1.7192	1.7210
5.6	1.7228	1.7246	1.7263	1.7281	1.7299	1.7317	1.7334	1.7352	1.7370	1.7387
5.7	1.7405	1.7422	1.7440	1.7457	1.7475	1.7492	1.7509	1.7527	1.7544	1.7561
5.8	1.7579	1.7596	1.7613	1.7630	1.7647	1.7664	1.7681	1.7699	1.7716	1.7733
5.9	1.7750	1.7766	1.7783	1.7800	1.7817	1.7834	1.7851	1.7867	1.7884	1.7901
6.0	**1.7918**	**1.7934**	**1.7951**	**1.7967**	**1.7984**	**1.8001**	**1.8017**	**1.8034**	**1.8050**	**1.8066**
6.1	1.8083	1.8099	1.8116	1.8132	1.8148	1.8165	1.8181	1.8197	1.8213	1.8229
6.2	1.8245	1.8262	1.8278	1.8294	1.8310	1.8326	1.8342	1.8358	1.8374	1.8390
6.3	1.8405	1.8421	1.8437	1.8453	1.8469	1.8485	1.8500	1.8516	1.8532	1.8547
6.4	1.8563	1.8579	1.8594	1.8610	1.8625	1.8641	1.8656	1.8672	1.8687	1.8703
6.5	1.8718	1.8733	1.8749	1.8764	1.8779	1.8795	1.8810	1.8825	1.8840	1.8856
6.6	1.8871	1.8886	1.8901	1.8916	1.8931	1.8946	1.8961	1.8976	1.8991	1.9006
6.7	1.9021	1.9036	1.9051	1.9066	1.9081	1.9095	1.9110	1.9125	1.9140	1.9155
6.8	1.9169	1.9184	1.9199	1.9213	1.9228	1.9242	1.9257	1.9272	1.9286	1.9301
6.9	1.9315	1.9330	1.9344	1.9359	1.9373	1.9387	1.9402	1.9416	1.9430	1.9445
7.0	**1.9459**	**1.9473**	**1.9488**	**1.9502**	**1.9516**	**1.9530**	**1.9544**	**1.9559**	**1.9573**	**1.9587**
7.1	1.9601	1.9615	1.9629	1.9643	1.9657	1.9671	1.9685	1.9699	1.9713	1.9727
7.2	1.9741	1.9755	1.9769	1.9782	1.9796	1.9810	1.9824	1.9838	1.9851	1.9865
7.3	1.9879	1.9892	1.9906	1.9920	1.9933	1.9947	1.9961	1.9974	1.9988	2.0001
7.4	2.0015	2.0028	2.0042	2.0055	2.0069	2.0082	2.0096	2.0109	2.0122	2.0136
7.5	2.0149	2.0162	2.0176	2.0189	2.0202	2.0215	2.0229	2.0242	2.0255	2.0268
7.6	2.0281	2.0295	2.0308	2.0321	2.0334	2.0347	2.0360	2.0373	2.0386	2.0399
7.7	2.0412	2.0425	2.0438	2.0451	2.0464	2.0477	2.0490	2.0503	2.0516	2.0528
7.8	2.0541	2.0554	2.0567	2.0580	2.0592	2.0605	2.0618	2.0631	2.0643	2.0656
7.9	2.0669	2.0681	2.0694	2.0707	2.0719	2.0732	2.0744	2.0757	2.0769	2.0782
8.0	**2.0794**	**2.0807**	**2.0819**	**2.0832**	**2.0844**	**2.0857**	**2.0869**	**2.0882**	**2.0894**	**2.0906**
8.1	2.0919	2.0931	2.0943	2.0956	2.0968	2.0980	2.0992	2.1005	2.1017	2.1029
8.2	2.1041	2.1054	2.1066	2.1078	2.1090	2.1102	2.1114	2.1126	2.1138	2.1150
8.3	2.1163	2.1175	2.1187	2.1199	2.1211	2.1223	2.1235	2.1247	2.1258	2.1270
8.4	2.1282	2.1294	2.1306	2.1318	2.1330	2.1342	2.1353	2.1365	2.1377	2.1389
8.5	2.1401	2.1412	2.1424	2.1436	2.1448	2.1459	2.1471	2.1483	2.1494	2.1506
8.6	2.1518	2.1529	2.1541	2.1552	2.1564	2.1576	2.1587	2.1599	2.1610	2.1622
8.7	2.1633	2.1645	2.1656	2.1668	2.1679	2.1691	2.1702	2.1713	2.1725	2.1736
8.8	2.1748	2.1759	2.1770	2.1782	2.1793	2.1804	2.1815	2.1827	2.1838	2.1849
8.9	2.1861	2.1872	2.1883	2.1894	2.1905	2.1917	2.1928	2.1939	2.1950	2.1961
9.0	**2.1972**	**2.1983**	**2.1994**	**2.2006**	**2.2017**	**2.2028**	**2.2039**	**2.2050**	**2.2061**	**2.2072**
9.1	2.2083	2.2094	2.2105	2.2116	2.2127	2.2138	2.2148	2.2159	2.2170	2.2181
9.2	2.2192	2.2203	2.2214	2.2225	2.2235	2.2246	2.2257	2.2268	2.2279	2.2289
9.3	2.2300	2.2311	2.2322	2.2332	2.2343	2.2354	2.2364	2.2375	2.2386	2.2396
9.4	2.2407	2.2418	2.2428	2.2439	2.2450	2.2460	2.2471	2.2481	2.2492	2.2502
9.5	2.2513	2.2523	2.2534	2.2544	2.2555	2.2565	2.2576	2.2586	2.2597	2.2607
9.6	2.2618	2.2628	2.2638	2.2649	2.2659	2.2670	2.2680	2.2690	2.2701	2.2711
9.7	2.2721	2.2732	2.2742	2.2752	2.2762	2.2773	2.2783	2.2793	2.2803	2.2814
9.8	2.2824	2.2834	2.2844	2.2854	2.2865	2.2875	2.2885	2.2895	2.2905	2.2915
9.9	2.2925	2.2935	2.2946	2.2956	2.2966	2.2976	2.2986	2.2996	2.3006	2.3016

Table 12.   Values of Degrees, Minutes, and Seconds in Radians

Lengths of Circular Arcs, Radius Unity

Example.   Θ = 30° 20′ 10″
30° = 0.52359878
20′ = 0.00581776
10″ = 0.00004848

Arc length = 0.52946502

Degrees	Radians Arc Length R = 1	Degrees	Radians Arc Length R = 1	Degrees	Radians Arc Length R = 1		Radians Arc Length R = 1	
							Minutes	Seconds
0		60	1.04719755	120	2.09439510	0		
1	0.01745329	61	1.06465084	121	2.11184840	1	0.00029089	0.00000485
2	0.03490659	62	1.08210414	122	2.12930169	2	.00058178	.00000970
3	0.05235988	63	1.09955743	123	2.14675498	3	.00087266	.00001454
4	0.06981317	64	1.11701072	124	2.16420828	4	.00116355	.00001939
5	0.08726646	65	1.13446401	125	2.18166157	5	.00145444	.00002424
6	0.10471976	66	1.15191731	126	2.19911486	6	.00174533	.00002909
7	0.12217305	67	1.16937060	127	2.21656815	7	.00203622	.00003394
8	0.13962634	68	1.18682389	128	2.23402145	8	.00232711	.00003879
9	0.15707963	69	1.20427718	129	2.25147474	9	.00261799	.00004363
10	0.17453293	70	1.22173048	130	2.26892803	10	.00290888	.00004848
11	0.19198622	71	1.23918377	131	2.28638133	11	.00319977	.00005333
12	0.20943951	72	1.25663706	132	2.30383462	12	.00349066	.00005818
13	0.22689280	73	1.27409035	133	2.32128791	13	.00378155	.00006303
14	0.24434610	74	1.29154365	134	2.33874121	14	.00407243	.00006787
15	0.26179939	75	1.30899694	135	2.35619450	15	.00436332	.00007272
16	0.27925268	76	1.32645023	136	2.37364780	16	.00465421	.00007757
17	0.29670597	77	1.34390352	137	2.39110107	17	.00494510	.00008242
18	0.31415927	78	1.36135682	138	2.40855436	18	.00523599	.00008727
19	0.33161256	79	1.37881011	139	2.42600766	19	.00552688	.00009211
20	0.34906585	80	1.39626340	140	2.44346095	20	.00581776	.00009696
21	0.36651914	81	1.41371669	141	2.46091424	21	.00610865	.00010181
22	0.38397244	82	1.43116999	142	2.47836754	22	.00639954	.00010666
23	0.40142573	83	1.44862328	143	2.49582083	23	.00669043	.00011151
24	0.41887902	84	1.46607657	144	2.51327413	24	.00698132	.00011636
25	0.43633231	85	1.48352986	145	2.53072742	25	.00727221	.00012120
26	0.45378561	86	1.50098316	146	2.54818071	26	.00756309	.00012605
27	0.47123890	87	1.51843645	147	2.56563401	27	.00785398	.00013090
28	0.48869219	88	1.53588974	148	2.58308729	28	.00814487	.00013575
29	0.50614548	89	1.55334303	149	2.60054058	29	.00843576	.00014060
30	0.52359878	90	1.57079633	150	2.61799388	30	.00872665	.00014544
31	0.54105207	91	1.58824962	151	2.63544717	31	.00901753	.00015029
32	0.55850536	92	1.60570291	152	2.65290046	32	.00930842	.00015514
33	0.57595865	93	1.62315620	153	2.67035375	33	.00959931	.00015999
34	0.59341195	94	1.64060950	154	2.68780705	34	.00989020	.00016484
35	0.61086524	95	1.65806279	155	2.70526034	35	.01018109	.00016968
36	0.62831853	96	1.67551608	156	2.72271363	36	.01047198	.00017453
37	0.64577182	97	1.69296937	157	2.74016693	37	.01076286	.00017938
38	0.66322512	98	1.71042267	158	2.75762022	38	.01105375	.00018423
39	0.68067841	99	1.72787596	159	2.77507351	39	.01134464	.00018908
40	0.69813170	100	1.74532925	160	2.79252680	40	.01163553	.00019393
41	0.71558499	101	1.76278254	161	2.80998009	41	.01192642	.00019877
42	0.73303829	102	1.78023584	162	2.82743338	42	.01221730	.00020362
43	0.75049158	103	1.79768913	163	2.84488668	43	.01250819	.00020847
44	0.76794487	104	1.81514242	164	2.86233997	44	.01279908	.00021332
45	0.78539816	105	1.83259571	165	2.87979327	45	.01308997	.00021817
46	0.80285146	106	1.85004901	166	2.89724655	46	.01338086	.00022301
47	0.82030475	107	1.86750230	167	2.91469985	47	.01367175	.00022786
48	0.83775804	108	1.88495559	168	2.93215314	48	.01396263	.00023271
49	0.85521133	109	1.90240888	169	2.94960643	49	.01425352	.00023756
50	0.87266463	110	1.91986218	170	2.96705972	50	.01454441	.00024241
51	0.89011792	111	1.93731547	171	2.98451302	51	.01483530	.00024725
52	0.90757121	112	1.95476876	172	3.00196631	52	.01512619	.00025210
53	0.92502450	113	1.97222205	173	3.01941961	53	.01541707	.00025695
54	0.94247780	114	1.98967535	174	3.03687289	54	.01570796	.00026180
55	0.95993109	115	2.00712864	175	3.05432619	55	.01599885	.00026665
56	0.97738438	116	2.02458193	176	3.07177948	56	.01628974	.00027150
57	0.99483767	117	2.04203522	177	3.08923277	57	.01658063	.00027634
58	1.01229097	118	2.05948852	178	3.10668607	58	.01687152	.00028119
59	1.02974426	119	2.07694181	179	3.12413962	59	.01716240	.00028604
				180	3.14159265			

Table 13. Values of Radians in Degrees

Rad.	.00	.01	.02	.03	.04	.05	.06	.07	.08	.09
	Deg	Deg	Deg	Deg	Deg	Deg	Deg	Deg	Deg	Deg
0.0	0.0000	0.5730	1.1459	1.7189	2.2918	2.8648	3.4377	4.0107	4.5837	5.1566
.1	5.7296	6.3025	6.8755	7.4485	8.0214	8.5944	9.1673	9.7403	10.3132	10.8862
.2	11.4591	12.0321	12.6051	13.1780	13.7510	14.3239	14.8969	15.4699	16.0428	16.6158
.3	17.1887	17.7617	18.3346	18.9076	19.4806	20.0535	20.6265	21.1994	21.7724	22.3454
.4	22.9183	23.4913	24.0642	24.6372	25.2101	25.7831	26.3561	26.9290	27.5020	28.0749
.5	28.6479	29.2208	29.7938	30.3668	30.9397	31.1527	32.0856	32.6586	33.2316	33.8045
.6	34.3775	34.9504	35.5234	36.0963	36.6693	37.2423	37.8152	38.3882	38.9611	39.5341
.7	40.1070	40.6800	41.2530	41.8259	42.3989	42.9718	43.5448	44.1178	44.6907	45.2637
.8	45.8366	46.4096	46.9825	47.5555	48.1285	48.7014	49.2744	49.8473	50.4203	50.9932
.9	51.5662	52.1392	52.7121	53.2851	53.8580	54.4310	55.0039	55.5769	56.1499	56.7228

1 Radian = 57.29578 deg | 2 Radians = 114.59156 deg | 3 Radians = 171.88734 deg

Table 14. Decimals of a Degree in Minutes and Seconds

Decimal	.00		.01		.02		.03		.04		.05		.06		.07		.08		.09	
	Min	Sec	Min	Sec	Min	Sec	Min	Sec	Min	Sec	Min	Sec	Min	Sec	Min	Sec	Min	Sec	Min	Sec
0.0	0	0	0	36	1	12	1	48	2	24	3	0	3	36	4	12	4	48	5	24
.1	6	0	6	36	7	12	7	48	8	24	9	0	9	36	10	12	10	48	11	24
.2	12	0	12	36	13	12	13	48	14	24	15	0	15	36	16	12	16	48	17	24
.3	18	0	18	36	19	12	19	48	20	24	21	0	21	36	22	12	22	48	23	24
.4	24	0	24	36	25	12	25	48	26	24	27	0	27	36	28	12	28	48	29	24
.5	30	0	30	36	31	12	31	48	32	24	33	0	33	36	34	12	24	48	35	24
.6	36	0	36	36	37	12	37	48	38	24	39	0	39	36	40	12	40	48	41	24
.7	42	0	42	36	43	12	43	48	44	24	45	0	45	36	46	12	46	48	47	24
.8	48	0	48	36	49	12	49	48	50	24	51	0	51	36	52	12	52	48	53	24
.9	54	0	54	36	55	12	55	48	56	24	57	0	57	36	58	12	58	48	59	24

Table 15. Minutes in Decimals of a Degree

Minutes	0	1	2	3	4	5	6	7	8	9
	Degrees	Degrees	Degrees	Degrees	Degrees	Degrees	Degrees	Degrees	Degrees	Degrees
0	0.00000	0.01667	0.03333	0.05000	0.06667	0.08333	0.10000	0.11667	0.13333	0.15000
10	.16667	.18333	.20000	.21667	.23333	.25000	.26667	.28333	.30000	.31667
20	.33333	.35000	.36667	.38333	.40000	.41667	.43333	.45000	.46667	.48333
30	.50000	.51667	.53333	.55000	.56667	.58333	.60000	.61667	.63333	.65000
40	.66667	.68333	.70000	.71667	.73333	.75000	.76667	.78333	.80000	.81667
50	.83333	.85000	.86667	.88333	.90000	.91667	.93333	.95000	.96667	.98333

Table 16. Seconds in Decimals of a Degree

Seconds	0	1	2	3	4
	Degrees	Degrees	Degrees	Degrees	Degrees
0	0	0.0002778	0.0005555	0.0008333	0.0011111
10	0.0027778	.0030555	.0033333	.0036111	.0038888
20	.0055555	.0058333	.0061111	.0063888	.0066667
30	.0083333	.0086111	.0088888	.0091667	.0094444
40	.0111111	.0113888	.0116667	.0119444	.0122222
50	.0138888	.0141667	.0144444	.0147222	.0150000

Seconds	5	6	7	8	9
	Degrees	Degrees	Degrees	Degrees	Degrees
0	0.0013888	0.0016667	0.0019444	0.0022222	0.0024999
10	.0041667	.0044444	.0047222	.0050000	.0052778
20	.0069444	.0072222	.0075000	.0077778	.0080555
30	.0097222	.0100000	.0102778	.0105555	.0108333
40	.0125000	.0127778	.0130555	.0133333	.0136111
50	.0152778	.0155555	.0158333	.0161111	.0163888

Table 17.   Radii, Logarithms, Offsets for Degrees of Curvature, 0°–50°

Deg. D	Radius R	Logarithm log R	Tan Off. t	Mid. Ord. m	Deg. D	Radius R	Logarithm log R	Tan Off. t	Mid. Ord. m
0° 0′	Infinite	Infinite	0.000	0.000	1° 0′	5729.65	3.758128	0.873	0.218
1	343775.	5.536274	.015	.004	1	5635.72	3.750950	0.887	.222
2	171887.	5.235244	.029	.007	2	5544.83	3.743888	0.902	.225
3	114592.	5.059153	.044	.011	3	5456.82	3.736939	0.916	.229
4	85943.7	4.934214	.058	.015	4	5371.56	3.730100	0.931	.233
5	68754.9	4.837304	.073	.018	5	5288.92	3.723367	0.945	.236
6	57295.8	4.758123	.087	.022	6	5208.79	3.716737	0.960	.240
7	49110.7	4.691176	.102	.025	7	5131.05	3.710206	0.974	.244
8	42971.8	4.633184	.116	.029	8	5055.59	3.703772	0.989	.247
9	38197.2	4.582031	.131	.033	9	4982.33	3.697432	1.004	.251
10	34377.5	4.536274	.145	.036	10	4911.15	3.691183	1.018	.255
11	31252.3	4.494881	.160	.040	11	4841.98	3.685023	1.033	.258
12	28647.8	4.457093	.175	.044	12	4774.74	3.678949	1.047	.262
13	26444.2	4.422331	.189	.047	13	4709.33	3.672959	1.062	.265
14	24555.4	4.390146	.204	.051	14	4645.69	3.667051	1.076	.269
15	22918.3	4.360183	.218	.055	15	4583.75	3.661221	1.091	.273
16	21485.9	4.332154	.233	.058	16	4523.44	3.655469	1.105	.276
17	20222.1	4.305825	.247	.062	17	4464.70	3.649792	1.120	.280
18	19098.6	4.281002	.262	.065	18	4407.46	3.644189	1.134	.284
19	18093.4	4.257521	.276	.069	19	4351.67	3.638656	1.149	.287
20	17188.8	4.235244	.291	.073	20	4297.28	3.633194	1.164	.291
21	16370.2	4.214055	.305	.076	21	4244.23	3.627799	1.178	.295
22	15626.1	4.193852	.320	.080	22	4192.47	3.622470	1.193	.298
23	14946.7	4.174547	.335	.084	23	4141.96	3.617206	1.207	.302
24	14324.0	4.156064	.349	.087	24	4092.66	3.612005	1.222	.305
25	13751.0	4.138335	.364	.091	25	4044.51	3.606866	1.236	.309
26	13222.1	4.121302	.378	.095	26	3997.49	3.601787	1.251	.313
27	12732.4	4.104911	.393	.098	27	3951.54	3.596766	1.265	.316
28	12277.7	4.089117	.407	.102	28	3906.64	3.591803	1.280	.320
29	11854.3	4.073877	.422	.105	29	3862.74	3.586896	1.294	.324
30	11459.2	4.059154	.436	.109	30	3819.83	3.582044	1.309	.327
31	11089.6	4.044914	.451	.113	31	3777.85	3.577245	1.324	.331
32	10743.0	4.031125	.465	.116	32	3736.79	3.572499	1.338	.335
33	10417.5	4.017762	.480	.120	33	3696.61	3.567804	1.353	.338
34	10111.1	4.004797	.495	.124	34	3657.29	3.563160	1.367	.342
35	9822.18	3.992208	.509	.127	35	3618.80	3.558564	1.382	.345
36	9549.34	3.979973	.524	.131	36	3581.10	3.554017	1.396	.349
37	9291.25	3.968074	.538	.135	37	3544.19	3.549517	1.411	.353
38	9046.75	3.956493	.553	.138	38	3508.02	3.545063	1.425	.356
39	8814.78	3.945212	.567	.142	39	3472.59	3.540654	1.440	.360
40	8594.42	3.934216	.582	.145	40	3437.87	3.536289	1.454	.364
41	8384.80	3.923493	.596	.149	41	3403.83	3.531968	1.469	.367
42	8185.16	3.913027	.611	.153	42	3370.46	3.527690	1.483	.371
43	7994.81	3.902808	.625	.156	43	3337.74	3.523453	1.498	.375
44	7813.11	3.892824	.640	.160	44	3305.65	3.519257	1.513	.378
45	7639.49	3.883065	.654	.164	45	3274.17	3.515101	1.527	.382
46	7473.42	3.873519	.669	.167	46	3243.29	3.510985	1.542	.385
47	7314.41	3.864179	.684	.171	47	3212.98	3.506908	1.556	.389
48	7162.03	3.855036	.698	.174	48	3183.23	3.502868	1.571	.393
49	7015.87	3.846082	.713	.178	49	3154.03	3.498866	1.585	.396
50	6875.55	3.837308	.727	.182	50	3125.36	3.494900	1.600	.400
51	6740.74	3.828708	.742	.185	51	3097.20	3.490970	1.614	.404
52	6611.12	3.820275	.756	.189	52	3069.55	3.487075	1.629	.407
53	6486.38	3.812002	.771	.193	53	3042.39	3.483215	1.643	.411
54	6366.26	3.803885	.785	.196	54	3015.71	3.479389	1.658	.414
55	6250.51	3.795916	.800	.200	55	2989.48	3.475596	1.673	.418
56	6138.90	3.788091	.814	.204	56	2963.72	3.471836	1.687	.422
57	6031.20	3.780404	.829	.207	57	2938.39	3.468109	1.702	.425
58	5927.22	3.772851	.844	.211	58	2913.49	3.464413	1.716	.429
59	5826.76	3.765427	.858	.215	59	2889.01	3.460749	1.731	.433
60	5729.65	3.758128	.873	.218	60	2864.93	3.457115	1.745	.436

Table 17. Radii, Logarithms, Offsets for Degrees of Curvature, 0°–50°—*Continued*

Deg. D	Radius R	Logarithm log R	Tan Off. t	Mid. Ord. m	Deg. D	Radius R	Logarithm log R	Tan Off. t	Mid. Ord. m
2° 0′	2864.93	3.457115	1.745	0.436	3° 0′	1910.08	3.281051	2.618	0.654
1	2841.26	3.453511	1.760	.440	1	1899.53	3.278646	2.632	.658
2	2817.97	3.449937	1.774	.444	2	1889.09	3.276253	2.647	.662
3	2795.06	3.446392	1.789	.447	3	1878.77	3.273874	2.661	.665
4	2772.53	3.442876	1.803	.451	4	1868.56	3.271508	2.676	.669
5	2750.35	3.439388	1.818	.454	5	1858.47	3.269155	2.690	.673
6	2728.52	3.435928	1.832	.458	6	1848.48	3.266814	2.705	.676
7	2707.04	3.432495	1.847	.462	7	1838.59	3.264486	2.719	.680
8	2685.89	3.429089	1.862	.465	8	1828.82	3.262170	2.734	.684
9	2665.08	3.425710	1.876	.469	9	1819.14	3.259867	2.749	.687
10	2644.58	3.422356	1.891	.473	10	1809.57	3.257576	2.763	.691
11	2624.39	3.419029	1.905	.476	11	1800.10	3.255296	2.778	.694
12	2604.51	3.415727	1.920	.480	12	1790.73	3.253029	2.792	.698
13	2584.93	3.412449	1.934	.484	13	1781.45	3.250774	2.807	.702
14	2565.65	3.409197	1.949	.487	14	1772.27	3.248530	2.821	.705
15	2546.64	3.405968	1.963	.491	15	1763.18	3.246297	2.836	.709
16	2527.92	3.402763	1.978	.494	16	1754.19	3.244077	2.850	.713
17	2509.47	3.399582	1.992	.498	17	1745.29	3.241867	2.865	.716
18	2491.29	3.396424	2.007	.502	18	1736.48	3.239669	2.879	.720
19	2473.37	3.393289	2.022	.505	19	1727.75	3.237481	2.894	.723
20	2455.70	3.390176	2.036	.509	20	1719.12	3.235305	2.908	.727
21	2438.29	3.387085	2.051	.513	21	1710.57	3.233140	2.923	.731
22	2421.12	3.384016	2.065	.516	22	1702.10	3.230985	2.938	.734
23	2404.19	3.380969	2.080	.520	23	1693.72	3.228841	2.952	.738
24	2387.50	3.377943	2.094	.524	24	1685.42	3.226707	2.967	.742
25	2371.04	3.374938	2.109	.527	25	1677.20	3.224584	2.981	.745
26	2354.80	3.371954	2.123	.531	26	1669.06	3.222472	2.996	.749
27	2338.78	3.368990	2.138	.534	27	1661.00	3.220369	3.010	.753
28	2322.98	3.366046	2.152	.530	28	1653.01	3.218277	3.025	.756
29	2307.39	3.363122	2.167	.542	29	1645.11	3.216195	3.039	.760
30	2292.01	3.360217	2.181	.545	30	1637.28	3.214122	3.054	763
31	2276.84	3.357332	2.196	.549	31	1629.52	3.212060	3.068	.767
32	2261.86	3.354466	2.211	.553	32	1621.84	3.210007	3.083	.771
33	2247.08	3.351618	2.225	.556	33	1614.22	3.207964	3.097	.774
34	2232.49	3.348789	2.240	.560	34	1606.68	3.205930	3.112	.778
35	2218.09	3.345979	2.254	.564	35	1599.21	3.203906	3.127	.782
36	2203.87	3.343187	2.269	.567	36	1591.81	3.201892	3.141	.785
37	2189.84	3.340412	2.283	.571	37	1584.48	3.199886	3.156	.789
38	2175.98	3.337655	2.298	.574	38	1577.21	3.197890	3.170	.793
39	2162.30	3.334916	2.312	.578	39	1570.01	3.195903	3.185	.796
40	2148.79	3.332193	2.327	.582	40	1562.88	3.193925	3.199	.800
41	2135.44	3.329488	2.341	.585	41	1555.81	3.191956	3.214	.803
42	2122.26	3.326799	2.356	.589	42	1548.80	3.189996	3.228	.807
43	2109.24	3.324127	2.371	.593	43	1541.86	3.188045	3.243	.811
44	2096.39	3.321471	2.385	.596	44	1534.98	3.186103	3.257	.814
45	2083.68	3.318832	2.400	.600	45	1528.16	3.184169	3.272	.818
46	2071.13	3.316208	2.414	.604	46	1521.40	3.182244	3.286	.822
47	2058.73	3.313600	2.429	.607	47	1514.70	3.180327	3.301	.825
48	2046.48	3.311008	2.443	.611	48	1508.06	3.178419	3.316	.829
49	2034.37	3.308431	2.458	.614	49	1501.48	3.176519	3.330	.832
50	2022.41	3.305869	2.472	.618	50	1494.95	3.174627	3.345	.836
51	2010.59	3.303323	2.487	.622	51	1488.48	3.172744	3.359	.840
52	1998.90	3.300791	2.501	.625	52	1482.07	3.170868	3.374	.843
53	1987.35	3.298274	2.516	.629	53	1475.71	3.169001	3.388	.847
54	1975.93	3.295771	2.530	.633	54	1469.41	3.167142	3.403	.851
55	1964.64	3.293283	2.545	.636	55	1463.16	3.165291	3.417	.854
56	1953.48	3.290809	2.560	.640	56	1456.96	3.163447	3.432	.858
57	1942.44	3.288349	2.574	.644	57	1450.81	3.161612	3.446	.862
58	1931.53	3.285902	2.589	.647	58	1444.72	3.159784	3.461	.865
59	1920.75	3.283470	2.603	.651	59	1438.68	3.157963	3.475	.869
60	1910.08	3.281051	2.618	.654	60	1432.69	3.156151	3.490	.872

Table 17.   Radii, Logarithms, Offsets for Degrees of Curvature, 0°–50°—*Continued*

Deg. D	Radius R	Logarithm log R	Tan Off. t	Mid. Ord. m	Deg. D	Radius R	Logarithm log R	Tan Off. t	Mid. Ord. m
4° 0′	1432.69	3.156151	3.490	0.872	5° 0′	1146.28	3.059290	4.362	1.091
1	1426.74	3.154346	3.505	0.876	1	1142.47	3.057846	4.376	1.094
2	1420.85	3.152548	3.519	0.880	2	1138.69	3.056407	4.391	1.098
3	1415.01	3.150758	3.534	0.883	3	1134.94	3.054972	4.405	1.102
4	1409.21	3.148975	3.548	0.887	4	1131.21	3.053542	4.420	1.105
5	1403.46	3.147200	3.563	0.891	5	1127.50	3.052116	4.435	1.109
6	1397.76	3.145431	3.577	0.894	6	1123.82	3.050696	4.449	1.112
7	1392.10	3.143670	3.592	0.898	7	1120.16	3.049280	4.464	1.116
8	1386.49	3.141916	3.606	0.902	8	1116.52	3.047868	4.478	1.120
9	1380.92	3.140170	3.621	0.905	9	1112.91	3.046462	4.493	1.123
10	1375.40	3.138430	3.635	0.909	10	1109.33	3.045059	4.507	1.127
11	1369.92	3.136697	3.650	0.912	11	1105.76	3.043662	4.522	1.131
12	1364.49	3.134971	3.664	0.916	12	1102.22	3.042268	4.536	1.134
13	1359.10	3.133251	3.679	0.920	13	1098.70	3.040880	4.551	1.138
14	1353.75	3.131539	3.693	0.923	14	1095.20	3.039495	4.565	1.142
15	1348.45	3.129833	3.708	0.927	15	1091.73	3.038115	4.580	1.146
16	1343.18	3.128134	3.723	0.931	16	1088.28	3.036740	4.594	1.149
17	1337.96	3.126442	3.737	0.934	17	1084.85	3.035368	4.609	1.153
18	1332.77	3.124756	3.752	0.938	18	1081.44	3.034002	4.623	1.157
19	1327.63	3.123077	3.766	0.942	19	1078.05	3.032639	4.638	1.160
20	1322.53	3.121404	3.781	0.945	20	1074.68	3.031281	4.653	1.164
21	1317.46	3.119738	3.795	0.949	21	1071.34	3.029927	4.667	1.168
22	1312.43	3.118078	3.810	0.952	22	1068.01	3.028577	4.682	1.171
23	1307.45	3.116424	3.824	0.956	23	1064.71	3.027231	4.696	1.175
24	1302.50	3.114777	3.839	0.960	24	1061.43	3.025890	4.711	1.179
25	1297.58	3.113136	3.853	0.963	25	1058.16	3.024552	4.725	1.182
26	1292.71	3.111501	3.868	0.967	26	1054.92	3.023219	4.740	1.186
27	1287.87	3.109872	3.882	0.971	27	1051.70	3.021890	4.754	1.190
28	1283.07	3.108249	3.897	0.974	28	1048.49	3.020565	4.769	1.193
29	1278.30	3.106632	3.911	0.978	29	1045.31	3.019244	4.783	1.197
30	1273.57	3.105022	3.926	0.982	30	1042.14	3.017927	4.798	1.200
31	1268.87	3.103417	3.941	0.985	31	1039.00	3.016614	4.812	1.204
32	1264.21	3.101818	3.955	0.989	32	1035.87	3.015305	4.827	1.208
33	1259.58	3.100225	3.970	0.993	33	1032.76	3.013999	4.841	1.211
34	1254.98	3.098638	3.984	0.996	34	1029.67	3.012698	4.856	1.215
35	1250.42	3.097057	3.999	1.000	35	1026.60	3.011401	4.870	1.218
36	1245.89	3.095481	4.013	1.003	36	1023.55	3.010107	4.885	1.222
37	1241.40	3.093912	4.028	1.007	37	1020.51	3.008818	4.900	1.226
38	1236.94	3.092347	4.042	1.011	38	1017.49	3.007532	4.914	1.229
39	1232.51	3.090789	4.057	1.014	39	1014.50	3.006250	4.929	1.233
40	1228.11	3.089236	4.071	1.018	40	1011.51	3.004972	4.943	1.237
41	1223.74	3.087689	4.086	1.022	41	1008.55	3.003698	4.958	1.240
42	1219.40	3.086147	4.100	1.025	42	1005.60	3.002427	4.972	1.244
43	1215.09	3.084610	4.115	1.029	43	1002.67	3.001160	4.987	1.247
44	1210.82	3.083079	4.129	1.032	44	999.762	2.999897	5.001	1.251
45	1206.57	3.081553	4.144	1.036	45	996.867	2.998637	5.016	1.255
46	1202.36	3.080033	4.159	1.040	46	993.988	2.997381	5.030	1.258
47	1198.17	3.078518	4.173	1.043	47	991.126	2.996129	5.045	1.262
48	1194.01	3.077008	4.188	1.047	48	988.280	2.994880	5.059	1.266
49	1189.88	3.075504	4.202	1.051	49	985.451	2.993635	5.074	1.269
50	1185.78	3.074005	4.217	1.054	50	982.638	2.992393	5.088	1.273
51	1181.71	3.072511	4.231	1.058	51	979.840	2.991155	5.103	1.277
52	1177.66	3.071022	4.246	1.062	52	977.060	2.989921	5.117	1.280
53	1173.65	3.069538	4.260	1.065	53	974.294	2.988690	5.132	1.284
54	1169.66	3.068059	4.275	1.069	54	971.544	2.987463	5.146	1.288
55	1165.70	3.066585	4.289	1.073	55	968.810	2.986238	5.161	1.291
56	1161.76	3.065116	4.304	1.076	56	966.091	2.985018	5.175	1.295
57	1157.85	3.063653	4.318	1.080	57	963.387	2.983801	5.190	1.298
58	1153.97	3.062194	4.333	1.083	58	960.698	2.982587	5.205	1.302
59	1150.11	3.060740	4.347	1.088	59	958.025	2.981377	5.219	1.306
60	1146.28	3.059290	4.362	1.091	60	955.366	2.980170	5.234	1.309

Table 17. Radii, Logarithms, Offsets for Degrees of Curvature, 0°–50°—*Continued*

Deg. D	Radius R	Logarithm log R	Tan Off. t	Mid. Ord. m	Deg. D	Radius R	Logarithm log R	Tan Off. t	Mid. Ord. m
6° 0′	955.366	2.980170	5.234	1.309	7° 0′	819.020	2.913295	6.105	1.528
1	952.722	2.978966	5.248	1.313	1	817.077	2.912263	6.119	1.531
2	950.093	2.977766	5.263	1.317	2	815.144	2.911234	6.134	1.535
3	947.478	2.976569	5.277	1.320	3	813.219	2.910208	6.148	1.539
4	944.877	2.975375	5.292	1.324	4	811.303	2.909183	6.163	1.543
5	942.291	2.974185	5.306	1.327	5	809.397	2.908162	6.177	1.546
6	939.719	2.972998	5.321	1.331	6	807.499	2.907142	6.192	1.550
7	937.161	2.971814	5.335	1.335	7	805.611	2.906125	6.206	1.553
8	934.616	2.970633	5.350	1.338	8	803.731	2.905111	6.221	1.557
9	932.086	2.969456	5.364	1.342	9	801.860	2.904098	6.236	1.561
10	929.569	2.968282	5.379	1.346	10	799.997	2.903089	6.250	1.564
11	927.066	2.967111	5.393	1.349	11	798.144	2.902081	6.265	1.568
12	924.576	2.965943	5.408	1.353	12	796.299	2.901076	6.279	1.572
13	922.100	2.964778	5.422	1.356	13	794.462	2.900073	6.294	1.575
14	919.637	2.963616	5.437	1.360	14	792.634	2.899073	6.308	1.579
15	917.187	2.962458	5.451	1.364	15	790.814	2.898074	6.323	1.582
16	914.750	2.961303	5.466	1.368	16	789.003	2.897078	6.337	1.586
17	912.326	2.960150	5.480	1.371	17	787.200	2.896085	6.352	1.590
18	909.915	2.959001	5.495	1.375	18	785.405	2.895094	6.366	1.593
19	907.517	2.957855	5.510	1.378	19	783.618	2.894105	6.381	1.597
20	905.131	2.956711	5.524	1.382	20	781.840	2.893118	6.395	1.600
21	902.758	2.955571	5.539	1.386	21	780.069	2.892133	6.410	1.604
22	900.397	2.954434	5.553	1.389	22	778.307	2.891151	6.424	1.608
23	898.048	2.953300	5.568	1.393	23	776.552	2.890171	6.439	1.611
24	895.712	2.952168	5.582	1.397	24	774.806	2.889193	6.453	1.615
25	893.388	2.951040	5.597	1.400	25	773.067	2.888217	6.468	1.619
26	891.076	2.949915	5.611	1.404	26	771.336	2.887244	6.482	1.623
27	888.776	2.948792	5.626	1.407	27	769.613	2.886272	6.497	1.626
28	886.488	2.947673	5.640	1.411	28	767.897	2.885303	6.511	1.630
29	884.211	2.946556	5.655	1.415	29	766.190	2.884336	6.526	1.633
30	881.946	2.945442	5.669	1.418	30	764.489	2.883371	6.540	1.637
31	879.693	2.944331	5.684	1.422	31	762.797	2.882409	6.555	1.641
32	877.451	2.943223	5.698	1.426	32	761.112	2.881448	6.569	1.644
33	875.221	2.942118	5.713	1.429	33	759.434	2.880490	6.584	1.648
34	873.002	2.941015	5.727	1.433	34	757.764	2.879534	6.598	1.651
35	870.795	2.939916	5.742	1.437	35	756.101	2.878580	6.613	1.655
36	868.598	2.938819	5.756	1.440	36	754.445	2.877627	6.627	1.659
37	866.412	2.937725	5.771	1.444	37	752.796	2.876678	6.642	1.662
38	864.238	2.936633	5.785	1.447	38	751.155	2.875730	6.656	1.666
39	862.075	2.935545	5.800	1.451	39	749.521	2.874784	6.671	1.670
40	859.922	2.934459	5.814	1.455	40	747.894	2.873840	6.685	1.673
41	857.780	2.933376	5.829	1.458	41	746.274	2.872898	6.700	1.677
42	855.648	2.932295	5.844	1.462	42	744.661	2.871959	6.714	1.680
43	853.527	2.931218	5.858	1.466	43	743.055	2.871021	6.729	1.684
44	851.417	2.930142	5.873	1.469	44	741.456	2.870086	6.743	1.688
45	849.317	2.929070	5.887	1.473	45	739.864	2.869152	6.758	1.691
46	847.228	2.928000	5.902	1.476	46	738.279	2.868221	6.773	1.695
47	845.148	2.926933	5.916	1.480	47	736.701	2.867291	6.787	1.699
48	843.080	2.925869	5.931	1.484	48	735.129	2.866363	6.802	1.702
49	841.021	2.924807	5.945	1.487	49	733.564	2.865438	6.816	1.706
50	838.972	2.923747	5.960	1.491	50	732.005	2.864514	6.831	1.710
51	836.933	2.922691	5.974	1.495	51	730.454	2.863593	6.845	1.713
52	834.904	2.921637	5.989	1.498	52	728.909	2.862673	6.860	1.717
53	832.885	2.920585	6.003	1.502	53	727.370	2.861755	6.874	1.720
54	830.876	2.919536	6.018	1.505	54	725.838	2.860840	6.889	1.724
55	828.876	2.918489	6.032	1.510	55	724.312	2.859926	6.903	1.728
56	826.886	2.917446	6.047	1.513	56	722.793	2.859014	6.918	1.731
57	824.905	2.916404	6.061	1.517	57	721.280	2.858104	6.932	1.735
58	822.934	2.915365	6.076	1.520	58	719.774	2.857196	6.947	1.739
59	820.973	2.914329	6.090	1.524	59	718.273	2.856290	6.961	1.742
60	819.020	2.913295	6.105	1.528	60	716.779	2.855385	6.976	1.746

Table 17.   Radii, Logarithms, Offsets for Degrees of Curvature, 0°–50°—*Continued*

Deg. D	Radius R	Logarithm log R	Tan Off. t	Mid. Ord. m	Deg. D	Radius R	Logarithm log R	Tan Off. t	Mid. Ord. m
8° 0′	716.779	2.855385	6.976	1.746	9° 0′	637.275	2.804327	7.846	1.965
1	715.291	2.854483	6.990	1.749	1	636.099	2.803525	7.860	1.968
2	713.810	2.853583	7.005	1.753	2	634.928	2.802724	7.875	1.972
3	712.335	2.852684	7.019	1.756	3	633.761	2.801926	7.889	1.975
4	710.865	2.851787	7.034	1.760	4	632.599	2.801128	7.904	1.979
5	709.402	2.850892	7.048	1.764	5	631.440	2.800332	7.918	1.983
6	707.945	2.849999	7.063	1.768	6	630.286	2.799538	7.933	1.987
7	706.493	2.849108	7.077	1.771	7	629.136	2.798745	7.947	1.990
8	705.048	2.848219	7.092	1.775	8	627.991	2.797953	7.962	1.994
9	703.609	2.847331	7.106	1.778	9	626.849	2.797163	7.976	1.998
10	702.175	2.846445	7.121	1.782	10	625.712	2.796374	7.991	2.001
11	700.748	2.845562	7.135	1.786	11	624.579	2.795587	8.005	2.005
12	699.326	2.844679	7.150	1.790	12	623.450	2.794801	8.020	2.008
13	697.910	2.843799	7.164	1.793	13	622.325	2.794017	8.034	2.012
14	696.499	2.842921	7.179	1.797	14	621.203	2.793234	8.049	2.016
15	695.095	2.842044	7.193	1.801	15	620.087	2.792453	8.063	2.019
16	693.696	2.841169	7.208	1.804	16	618.974	2.791673	8.078	2.023
17	692.302	2.840296	7.222	1.807	17	617.865	2.790894	8.092	2.026
18	690.914	2.839424	7.237	1.811	18	616.760	2.790117	8.107	2.030
19	689.532	2.838555	7.251	1.815	19	615.660	2.789341	8.121	2.034
20	688.156	2.837687	7.266	1.819	20	614.563	2.788566	8.136	2.037
21	686.785	2.836821	7.280	1.822	21	613.470	2.787793	8.150	2.041
22	685.419	2.835956	7.295	1.826	22	612.380	2.787021	8.165	2.045
23	684.059	2.835093	7.309	1.829	23	611.295	2.786251	8.179	2.048
24	682.704	2.834232	7.324	1.833	24	610.214	2.785482	8.194	2.052
25	681.354	2.833373	7.338	1.837	25	609.136	2.784714	8.208	2.056
26	680.010	2.832515	7.353	1.840	26	608.062	2.783948	8.223	2.060
27	678.671	2.831660	7.367	1.844	27	606.992	2.783183	8.237	2.063
28	677.338	2.830805	7.382	1.848	28	605.926	2.782420	8.252	2.066
29	676.008	2.829953	7.396	1.851	29	604.864	2.781657	8.266	2.070
30	674.686	2.829102	7.411	1.855	30	603.805	2.780897	8.281	2.074
31	673.369	2.828253	7.425	1.858	31	602.750	2.780137	8.295	2.077
32	672.056	2.827405	7.440	1.862	32	601.698	2.779379	8.310	2.081
33	670.748	2.826560	7.454	1.866	33	600.651	2.778622	8.324	2.084
34	669.446	2.825715	7.469	1.869	34	599.607	2.777867	8.339	2.088
35	668.148	2.824873	7.483	1.873	35	598.567	2.777112	8.353	2.092
36	666.856	2.824032	7.498	1.877	36	597.530	2.776360	8.368	2.096
37	665.568	2.823193	7.512	1.880	37	596.497	2.775608	8.382	2.099
38	664.286	2.822355	7.527	1.884	38	595.467	2.774858	8.397	2.103
39	663.008	2.821519	7.541	1.888	39	594.441	2.774109	8.411	2.106
40	661.736	2.820685	7.556	1.892	40	593.419	2.773361	8.426	2.110
41	660.468	2.819852	7.570	1.895	41	592.400	2.772615	8.440	2.113
42	659.205	2.819021	7.585	1.899	42	591.384	2.771870	8.455	2.117
43	657.947	2.818191	7.599	1.903	43	590.372	2.771126	8.469	2.121
44	656.694	2.817363	7.614	1.906	44	589.364	2.770383	8.484	2.125
45	655.446	2.816537	7.628	1.910	45	588.359	2.769642	8.498	2.128
46	654.202	2.815712	7.643	1.914	46	587.357	2.768902	8.513	2.132
47	652.963	2.814889	7.657	1.918	47	586.359	2.768164	8.527	2.135
48	651.729	2.814067	7.672	1.921	48	585.364	2.767426	8.542	2.139
49	650.499	2.813247	7.686	1.924	49	584.373	2.766690	8.556	2.142
50	649.274	2.812428	7.701	1.928	50	583.385	2.765955	8.571	2.146
51	648.054	2.811611	7.715	1.932	51	582.400	2.765221	8.585	2.150
52	646.838	2.810796	7.730	1.935	52	581.419	2.764489	8.600	2.154
53	645.627	2.809982	7.744	1.939	53	580.441	2.763758	8.614	2.158
54	644.420	2.809169	7.759	1.943	54	579.466	2.763028	8.629	2.161
55	643.218	2.808358	7.773	1.946	55	578.494	2.762299	8.643	2.165
56	642.021	2.807549	7.788	1.950	56	577.526	2.761572	8.658	2.168
57	640.828	2.806741	7.802	1.953	57	576.561	2.760845	8.672	2.172
58	639.639	2.805935	7.817	1.957	58	575.599	2.760120	8.687	2.175
59	638.455	2.805130	7.831	1.961	59	574.641	2.759397	8.701	2.179
60	637.275	2.804327	7.846	1.965	60	573.686	2.758674	8.716	2.183

Table 17. Radii, Logarithms, Offsets for Degrees of Curvature, 0°–50°—*Continued*

Deg. D	Radius R	Logarithm log R	Tan Off. t	Mid. Ord. m	Deg. D	Radius R	Logarithm log R	Tan Off. t	Mid. Ord. m
10° 0′	573.686	2.758674	8.716	2.183	12° 0′	478.339	2.679735	10.453	2.620
2	571.784	2.757232	8.745	2.190	2	477.018	2.678535	10.482	2.628
4	569.896	2.755796	8.774	2.198	4	475.705	2.677338	10.511	2.635
6	568.020	2.754364	8.803	2.205	6	474.400	2.676145	10.540	2.642
8	566.156	2.752937	8.831	2.212	8	473.102	2.674954	10.569	2.650
10	564.305	2.751514	8.860	2.219	10	471.810	2.673767	10.597	2.657
12	562.466	2.750096	8.889	2.227	12	470.526	2.672584	10.626	2.664
14	560.638	2.748683	8.918	2.234	14	469.249	2.671403	10.655	2.671
16	558.823	2.747274	8.947	2.241	16	467.978	2.670226	10.684	2.679
18	557.019	2.745870	8.976	2.249	18	466.715	2.669052	10.713	2.686
20	555.227	2.744471	9.005	2.256	20	465.459	2.667881	10.742	2.693
22	553.447	2.743076	9.034	2.263	22	464.209	2.666713	10.771	2.701
24	551.678	2.741686	9.063	2.270	24	462.966	2.665549	10.800	2.708
26	549.920	2.740300	9.092	2.278	26	461.729	2.664387	10.829	2.715
28	548.174	2.738918	9.121	2.285	28	460.500	2.663229	10.858	2.722
30	546.438	2.737541	9.150	2.293	30	459.276	2.662074	10.887	2.730
32	544.714	2.736169	9.179	2.300	32	458.060	2.660922	10.916	2.737
34	543.001	2.734800	9.208	2.307	34	456.850	2.659773	10.945	2.744
36	541.298	2.733436	9.237	2.314	36	455.646	2.658628	10.973	2.752
38	539.606	2.732077	9.266	2.321	38	454.449	2.657485	11.002	2.759
40	537.924	2.730721	9.295	2.329	40	453.259	2.656345	11.031	2.766
42	536.253	2.729370	9.324	2.336	42	452.073	2.655208	11.060	2.774
44	534.593	2.728023	9.353	2.343	44	450.894	2.654075	11.089	2.781
46	532.943	2.726681	9.382	2.351	46	449.722	2.652944	11.118	2.788
48	531.303	2.725342	9.411	2.358	48	448.556	2.651816	11.147	2.795
50	529.673	2.724008	9.440	2.365	50	447.395	2.650691	11.176	2.803
52	528.053	2.722677	9.469	2.372	52	446.241	2.649570	11.205	2.810
54	526.443	2.721351	9.498	2.380	54	445.093	2.648451	11.234	2.817
56	524.843	2.720029	9.527	2.387	56	443.951	2.647335	11.263	2.825
58	523.252	2.718711	9.556	2.394	58	442.814	2.646221	11.291	2.832
11° 0′	521.671	2.717397	9.585	2.402	13° 0′	441.684	2.645111	11.320	2.839
2	520.100	2.716087	9.614	2.409	2	440.559	2.644004	11.349	2.846
4	518.539	2.714781	9.642	2.416	4	439.440	2.642899	11.378	2.854
6	516.986	2.713479	9.671	2.423	6	438.326	2.641798	11.407	2.861
8	515.443	2.712181	9.700	2.431	8	437.219	2.640699	11.436	2.868
10	513.909	2.710887	9.729	2.438	10	436.117	2.639603	11.465	2.876
12	512.385	2.709596	9.758	2.445	12	435.020	2.638510	11.494	2.883
14	510.869	2.708310	9.787	2.453	14	433.929	2.637419	11.523	2.890
16	509.363	2.707027	9.816	2.460	16	432.844	2.636331	11.552	2.898
18	507.865	2.705748	9.845	2.467	18	431.764	2.635246	11.580	2.905
20	506.376	2.704473	9.874	2.475	20	430.690	2.634164	11.609	2.912
22	504.896	2.703202	9.903	2.482	22	429.620	2.633085	11.638	2.919
24	503.425	2.701934	9.932	2.489	24	428.557	2.632008	11.667	2.927
26	501.962	2.700671	9.961	2.496	26	427.498	2.630934	11.696	2.934
28	500.507	2.699410	9.990	2.504	28	426.445	2.629863	11.725	2.941
30	499.061	2.698154	10.019	2.511	30	425.396	2.628794	11.754	2.949
32	497.624	2.696901	10.048	2.518	32	424.354	2.627728	11.783	2.956
34	496.195	2.695652	10.077	2.526	34	423.316	2.626665	11.812	2.963
36	494.774	2.694407	10.106	2.533	36	422.283	2.625604	11.840	2.971
38	493.361	2.693165	10.135	2.540	38	421.256	2.624546	11.869	2.978
40	491.956	2.691926	10.164	2.547	40	420.233	2.623490	11.898	2.985
42	490.559	2.690692	10.192	2.555	42	419.215	2.622437	11.927	2.992
44	489.171	2.689460	10.221	2.562	44	418.203	2.621387	11.956	3.000
46	487.790	2.688233	10.250	2.569	46	417.195	2.620339	11.985	3.007
48	486.417	2.687008	10.279	2.577	48	416.192	2.619294	12.014	3.014
50	485.051	2.685788	10.308	2.584	50	415.194	2.618251	12.043	3.022
52	483.694	2.684570	10.337	2.591	52	414.201	2.617211	12.071	3.029
54	482.344	2.683357	10.366	2.598	54	413.212	2.616173	12.100	3.036
56	481.001	2.682146	10.395	2.606	56	412.229	2.615138	12.129	3.044
58	479.666	2.680939	10.424	2.613	58	411.250	2.614106	12.158	3.051
60	478.339	2.679735	10.453	2.620	60	410.275	2.613075	12.187	3.058

Table 17. Radii, Logarithms, Offsets for Degrees of Curvature, 0°–50°—*Continued*

Deg. D	Radius R	Logarithm log R	Tan Off. t	Mid. Ord. m	Deg. D	Radius R	Logarithm log R	Tan Off. t	Mid. Ord. m
14° 0′	410.275	2.613075	12.187	3.058	16° 0′	359.265	2.555415	13.917	3.496
2	409.306	2.612048	12.216	3.065	2	358.523	2.554517	13.946	3.504
4	408.341	2.611023	12.245	3.073	4	357.784	2.553621	13.975	3.511
6	407.380	2.610000	12.274	3.080	6	357.048	2.552727	14.004	3.518
8	406.424	2.608980	12.302	3.087	8	356.315	2.551834	14.033	3.526
10	405.473	2.607962	12.331	3.095	10	355.585	2.550944	14.061	3.533
12	404.526	2.606946	12.360	3.102	12	354.859	2.550055	14.090	3.540
14	403.583	2.605933	12.389	3.109	14	354.135	2.549169	14.119	3.547
16	402.645	2.604923	12.418	3.117	16	353.414	2.548284	14.148	3.555
18	401.712	2.603914	12.447	3.124	18	352.696	2.547401	14.177	3.562
20	400.782	2.602908	12.476	3.131	20	351.981	2.546519	14.205	3.569
22	399.857	2.601905	12.504	3.138	22	351.269	2.545640	14.234	3.577
24	398.937	2.600904	12.533	3.146	24	350.560	2.544762	14.263	3.584
26	398.020	2.599905	12.562	3.153	26	349.854	2.543887	14.292	3.591
28	397.108	2.598908	12.591	3.160	28	349.150	2.543013	14.320	3.599
30	396.200	2.597914	12.620	3.168	30	348.450	2.542140	14.349	3.606
32	395.296	2.596922	12.649	3.175	32	347.752	2.541270	14.378	3.613
34	394.396	2.595933	12.678	3.182	34	347.057	2.540401	14.407	3.621
36	393.501	2.594945	12.706	3.190	36	346.365	2.539535	14.436	3.628
38	392.609	2.593960	12.735	3.197	38	345.676	2.538670	14.464	3.635
40	391.722	2.592978	12.764	3.204	40	344.990	2.537806	14.493	3.643
42	390.838	2.591997	12.793	3.211	42	344.306	2.536945	14.522	3.650
44	389.959	2.591019	12.822	3.219	44	343.625	2.536085	14.551	3.657
46	389.084	2.590043	12.851	3.226	46	342.947	2.535227	14.580	3.664
48	388.212	2.589069	12.880	3.233	48	342.271	2.534370	14.608	3.672
50	387.345	2.588097	12.908	3.241	50	341.598	2.533516	14.637	3.679
52	386.481	2.587128	12.937	3.248	52	340.928	2.532663	14.666	3.686
54	385.621	2.586161	12.966	3.255	54	340.260	2.531811	14.695	3.694
56	384.765	2.585196	12.995	3.263	56	339.595	2.530962	14.723	3.701
58	383.913	2.584233	13.024	3.270	58	338.933	2.530114	14.752	3.708
15° 0′	383.065	2.583272	13.053	3.277	17° 0′	338.273	2.529268	14.781	3.716
2	382.220	2.582314	13.081	3.284	2	337.616	2.528424	14.810	3.723
4	381.380	2.581358	13.110	3.292	4	336.962	2.527581	14.838	3.730
6	380.543	2.580403	13.139	3.299	6	336.310	2.526740	14.867	3.738
8	379.709	2.579451	13.168	3.306	8	335.660	2.525900	14.896	3.745
10	378.880	2.578501	13.197	3.314	10	335.013	2.525062	14.925	3.752
12	378.054	2.577553	13.226	3.321	12	334.369	2.524226	14.954	3.760
14	377.231	2.576608	13.254	3.328	14	333.727	2.523392	14.982	3.767
16	376.412	2.575664	13.283	3.336	16	333.088	2.522559	15.011	3.774
18	375.597	2.574722	13.312	3.343	18	332.451	2.521728	15.040	3.781
20	374.785	2.573783	13.341	3.350	20	331.816	2.520898	15.069	3.789
22	373.977	2.572845	13.370	3.358	22	331.184	2.520070	15.097	3.796
24	373.173	2.571910	13.399	3.365	24	330.555	2.519244	15.126	3.803
26	372.372	2.570977	13.427	3.372	26	329.928	2.518419	15.155	3.811
28	371.574	2.570045	13.456	3.379	28	329.303	2.517596	15.184	3.818
30	370.780	2.569116	13.485	3.387	30	328.689	2.516774	15.212	3.825
32	369.989	2.568189	13.514	3.394	32	328.061	2.515954	15.241	3.833
34	369.202	2.567264	13.543	3.401	34	327.443	2.515136	15.270	3.840
36	368.418	2.566340	13.572	3.409	36	326.828	2.514319	15.299	3.847
38	367.637	2.565419	13.600	3.416	38	326.215	2.513504	15.327	3.855
40	366.859	2.564500	13.629	3.423	40	325.604	2.512690	15.356	3.862
42	366.085	2.563582	13.658	3.431	42	324.996	2.511878	15.385	3.869
44	365.315	2.562667	13.687	3.438	44	324.390	2.511067	15.414	3.877
46	364.547	2.561754	13.716	3.445	46	323.786	2.510258	15.442	3.884
48	363.783	2.560843	13.744	3.452	48	323.184	2.509451	15.471	3.891
50	363.022	2.559933	13.773	3.460	50	322.585	2.508645	15.500	3.899
52	362.264	2.559026	13.802	3.467	52	321.989	2.507840	15.529	3.906
54	361.510	2.558120	13.831	3.474	54	321.394	2.507037	15.557	3.913
56	360.758	2.557216	13.860	3.482	56	320.801	2.506236	15.586	3.920
58	360.010	2.556315	13.889	3.489	58	320.211	2.505436	15.615	3.928
60	359.265	2.555415	13.917	3.496	60	319.623	2.504638	15.643	3.935

Table 17.　Radii, Logarithms, Offsets for Degrees of Curvature, 0°–50°—*Continued*

Deg. D	Radius R	Logarithm log R	Tan Off. t	Mid. Ord. m	Deg. D	Radius R	Logarithm log R	Tan Off. t	Mid. Ord. m
18° 0′	319.623	2.504638	15.643	3.935	20° 0′	287.939	2.459300	17.365	4.374
2	319.037	2.503841	15.672	3.942	10	285.583	2.455733	17.508	4.411
4	318.453	2.503045	15.701	3.950	20	283.267	2.452195	17.651	4.448
6	317.871	2.502251	15.730	3.957	30	280.988	2.448688	17.794	4.484
8	317.292	2.501459	15.758	3.964	40	278.746	2.445209	17.937	4.521
10	316.715	2.500668	15.787	3.972	50	276.541	2.441759	18.081	4.558
12	316.139	2.499879	15.816	3.979	21° 0′	274.370	2.438337	18.224	4.594
14	315.566	2.499091	15.845	3.986	10	272.234	2.434943	18.367	4.631
16	314.993	2.498304	15.873	3.994	20	270.132	2.431576	18.509	4.668
18	314.426	2.497519	15.902	4.001	30	268.062	2.428235	18.652	4.704
					40	266.024	2.424921	18.795	4.741
20	313.860	2.496736	15.931	4.008	50	264.018	2.421633	18.938	4.778
22	313.295	2.495953	15.959	4.016					
24	312.732	2.495173	15.988	4.023	22° 0′	262.042	2.418371	19.081	4.814
26	312.172	2.494393	16.017	4.030	10	260.098	2.415134	19.224	4.851
28	311.613	2.493616	16.046	4.038	20	258.180	2.411922	19.366	4.888
30	311.056	2.492839	16.074	4.045	30	256.292	2.408734	19.509	4.925
32	310.502	2.492064	16.103	4.052	40	254.431	2.405571	19.652	4.961
34	309.949	2.491291	16.132	4.060	50	252.599	2.402431	19.794	4.998
36	309.399	2.490518	16.160	4.067	23° 0′	250.793	2.399315	19.937	5.035
38	308.850	2.489748	16.189	4.074	10	249.013	2.396222	20.079	5.071
					20	247.258	2.393151	20.222	5.108
40	308.303	2.488978	16.218	3.081	30	245.529	2.390103	20.364	5.145
42	307.759	2.488210	16.246	4.089	40	243.825	2.387077	20.507	5.182
44	307.216	2.487444	16.275	4.096	50	242.144	2.384074	20.649	5.218
46	306.675	2.486679	16.304	4.103					
48	306.136	2.485915	16.333	4.111	24° 0′	240.487	2.381091	20.791	5.255
50	305.599	2.485152	16.361	4.118	10	238.853	2.378130	20.933	5.292
52	305.064	2.484391	16.390	4.125	20	237.241	2.375190	21.076	5.329
54	304.531	2.483632	16.419	4.133	30	235.652	2.372270	21.218	5.366
56	304.000	2.482873	16.447	4.140	40	234.084	2.369371	21.360	5.402
58	303.470	2.482116	16.476	4.147	50	232.537	2.366492	21.502	5.439
					25° 0′	231.011	2.363633	21.644	5.476
19° 0′	302.943	2.481361	16.505	4.155	10	229.506	2.360794	21.786	5.513
2	302.417	2.480607	16.533	4.162	20	228.020	2.357974	21.928	5.549
4	301.893	2.479854	16.562	4.169	30	226.555	2.355173	22.070	5.586
6	301.371	2.479102	16.591	4.177	40	225.108	2.352391	22.212	5.623
8	300.851	2.478352	16.620	4.184	50	223.680	2.349627	22.353	5.660
10	300.333	2.477603	16.648	4.191					
12	299.816	2.476855	16.677	4.199	26° 0′	222.271	2.346882	22.495	5.697
14	299.302	2.476109	16.706	4.206	10	220.879	2.344155	22.637	5.734
16	298.789	2.475364	16.734	4.213	20	219.506	2.341446	22.778	5.770
18	298.278	2.474621	16.763	4.221	30	218.150	2.338755	22.920	5.807
					40	216.811	2.336081	23.062	5.844
20	297.768	2.473878	16.792	4.228	50	215.489	2.333424	23.203	5.881
22	297.260	2.473137	16.820	4.235	27° 0′	214.183	2.330785	23.345	5.918
24	296.755	2.472398	16.849	4.243	10	212.893	2.328162	23.486	5.955
26	296.250	2.471659	16.878	4.250	20	211.620	2.325556	23.627	5.992
28	295.748	2.470922	16.906	4.257	30	210.362	2.322967	23.769	6.029
30	295.247	2.470186	16.935	4.265	40	209.119	2.320393	23.910	6.065
32	294.748	2.469452	16.964	4.272	50	207.891	2.317836	24.051	6.102
34	294.251	2.468718	16.992	4.279	28° 0′	206.678	2.315295	24.192	6.139
36	293.756	2.467986	17.021	4.287	10	205.480	2.312769	24.333	6.176
38	293.262	2.467256	17.050	4.294	20	204.296	2.310259	24.474	6.213
40	292.770	2.466526	17.078	4.301	30	203.125	2.307764	24.615	6.250
42	292.279	2.465798	17.107	4.308	40	201.969	2.305285	24.756	6.287
44	291.790	2.465071	17.136	4.316	50	200.826	2.302820	24.897	6.324
46	291.303	2.464345	17.164	4.323	29° 0′	199.696	2.300370	25.038	6.360
48	290.818	2.463621	17.193	4.330	10	198.580	2.297935	25.179	6.398
50	290.334	2.462897	17.222	4.338	20	197.476	2.295515	25.320	6.435
52	289.851	2.462175	17.250	4.345	30	196.385	2.293108	25.460	6.472
54	289.371	2.461455	17.279	4.352	40	195.306	2.290716	25.601	6.509
56	288.892	2.460735	17.308	4.360	50	194.240	2.288338	25.741	6.545
58	288.414	2.460017	17.336	4.367					
60	287.939	2.459300	17.365	4.374	30° 0′	193.185	2.285974	25.882	6.583

Table 17.   Radii, Logarithms, Offsets for Degrees of Curvature, 0°–50°—*Continued*

Deg. D	Radius R	Logarithm log R	Tan Off. t	Mid. Ord. m	Deg. D	Radius R	Logarithm log R	Tan Off. t	Mid. Ord. m
30° 20'	191.111	2.281286	26.163	6.657	38° 30'	151.657	2.180863	32.969	8.479
40	189.083	2.276652	26.443	6.731	39° 0'	149.787	2.175475	33.381	8.592
31° 0'	187.099	2.272071	26.724	6.805	30	147.965	2.170160	33.792	8.704
20	185.158	2.267541	27.004	6.879	40° 0'	146.190	2.164918	34.202	8.816
40	183.258	2.263062	27.284	6.953	30	144.460	2.159747	34.612	8.929
32° 0'	181.398	2.258632	27.564	7.027	41° 0'	142.773	2.154645	35.021	9.041
20	179.577	2.254250	27.843	7.101	30	141.127	2.149610	35.429	9.154
40	177.794	2.249916	28.123	7.175	42° 0'	139.521	2.144641	35.837	9.267
33° 0'	176.047	2.245628	28.402	7.250	30	137.955	2.139736	36.244	9.380
20	174.336	2.241386	28.680	7.324	43° 0'	136.425	2.134895	36.650	9.493
40	172.659	2.237188	28.959	7.398	30	134.932	2.130114	37.056	9.606
34° 0'	171.015	2.233035	29.237	7.473	44° 0'	133.473	2.125395	37.461	9.719
20	169.404	2.228924	29.515	7.547	30	132.049	2.120734	37.865	9.832
40	167.825	2.224855	29.793	7.621	45° 0'	130.656	2.116130	38.268	9.946
35° 0'	166.275	2.220828	30.071	7.696	30	129.296	2.111584	38.671	10.059
20	164.756	2.216842	30.348	7.770	46° 0'	127.965	2.107092	39.073	10.173
40	163.266	2.212895	30.625	7.845	30	126.664	2.102655	39.474	10.286
36° 0'	161.803	2.208988	30.902	7.919	47° 0'	125.392	2.098270	39.875	10.400
20	160.368	2.205119	31.178	7.994	30	124.148	2.093938	40.275	10.514
40	158.960	2.201288	31.454	8.068	48° 0'	122.930	2.089657	40.674	10.628
37° 0'	157.577	2.197494	31.730	8.143	30	121.738	2.085425	41.072	10.742
20	156.220	2.193736	32.006	8.218	49° 0'	120.571	2.081243	41.469	10.856
40	154.887	2.190014	32.282	8.292	30	119.429	2.077109	41.866	10.970
38° 0'	153.578	2.186328	32.557	8.367	50° 0'	118.310	2.073022	42.262	11.085

Table 18.   Tangents and Externals to a 1° Curve

Angle Δ	Tan T	External E	Angle Δ	Tan T	External E	Angle Δ	Tan T	External E
1°	50.00	0.218	5°	250.16	5.459	9°	450.93	17.717
10'	58.34	0.297	10'	258.51	5.829	10'	459.32	18.381
20	66.67	0.388	20	266.86	6.211	20	467.71	19.058
30	75.01	0.491	30	275.21	6.606	30	476.10	19.746
40	83.34	0.606	40	283.57	7.013	40	484.49	20.447
50	91.68	0.733	50	291.92	7.432	50	492.88	21.161
2	100.01	0.873	6	300.28	7.863	10	501.28	21.887
10	108.35	1.024	10	308.64	8.307	10	509.68	22.624
20	116.68	1.188	20	316.99	8.762	20	518.08	23.375
30	125.02	1.364	30	325.35	9.230	30	526.48	24.138
40	133.36	1.552	40	333.71	9.710	40	534.89	24.913
50	141.70	1.752	50	342.08	10.202	50	543.29	25.700
3	150.04	1.964	7	350.44	10.707	11	551.70	26.500
10	158.38	2.188	10	358.81	11.224	10	560.11	27.313
20	166.72	2.425	20	367.17	11.753	20	568.53	28.137
30	175.06	2.674	30	375.54	12.294	30	576.95	28.974
40	183.40	2.934	40	383.91	12.847	40	585.36	29.824
50	191.74	3.207	50	392.28	13.413	50	593.79	30.686
4	200.08	3.492	8	400.66	13.991	12	602.21	31.561
10	208.43	3.790	10	409.03	14.582	10	610.64	32.447
20	216.77	4.099	20	417.41	15.184	20	619.07	33.347
30	225.12	4.421	30	425.79	15.799	30	627.50	34.259
40	233.47	4.755	40	434.17	16.426	40	635.93	35.183
50	241.81	5.100	50	442.55	17.065	50	644.37	36.120

Table 18. Tangents and Externals to a 1° Curve—*Continued*

Angle Δ	Tan T	External E	Angle Δ	Tan T	External E	Angle Δ	Tan T	External E
13°	652.81	37.070	23°	1165.7	117.38	33°	1697.2	246.08
10'	661.25	38.031	10'	1174.4	119.12	10'	1706.3	248.66
20	669.70	39.006	20	1183.1	120.87	20	1715.3	251.26
30	678.15	39.993	30	1191.8	122.63	30	1724.4	253.87
40	686.60	40.992	40	1200.5	124.41	40	1733.5	256.50
50	695.06	42.004	50	1209.2	126.20	50	1742.6	259.14
14	703.51	43.029	24	1217.9	128.00	34	1751.7	261.80
10	711.97	44.066	10	1226.6	129.82	10	1760.8	264.47
20	720.44	45.116	20	1235.3	131.65	20	1770.0	267.16
30	728.90	46.178	30	1244.0	133.50	30	1779.1	269.86
40	737.37	47.253	40	1252.8	135.35	40	1788.2	272.58
50	745.85	48.341	50	1261.5	137.23	50	1797.4	275.31
15	754.32	49.441	25	1270.2	139.11	35	1806.6	278.05
10	762.80	50.554	10	1279.0	141.01	10	1815.7	280.82
20	771.29	51.679	20	1287.7	142.93	20	1824.9	283.60
30	779.77	52.818	30	1296.5	144.85	30	1834.1	286.39
40	788.26	53.969	40	1305.3	146.79	40	1843.3	289.20
50	796.75	55.132	50	1314.0	148.75	50	1852.5	292.02
16	805.25	56.309	26	1322.8	150.71	36	1861.7	294.86
10	813.75	57.498	10	1331.6	152.69	10	1870.9	297.72
20	822.25	58.699	20	1340.4	154.69	20	1880.1	300.59
30	830.76	59.914	30	1349.2	156.70	30	1889.4	303.47
40	839.27	61.141	40	1358.0	158.72	40	1898.6	306.37
50	847.78	62.381	50	1366.8	160.76	50	1907.9	309.29
17	856.30	63.634	27	1375.6	162.81	37	1917.1	312.22
10	864.82	64.900	10	1384.4	164.86	10	1926.4	315.17
20	873.35	66.178	20	1393.2	166.95	20	1935.7	318.13
30	881.88	67.470	30	1402.0	169.04	30	1945.0	321.11
40	890.41	68.774	40	1410.9	171.15	40	1954.3	324.11
50	898.95	70.091	50	1419.7	173.27	50	1963.6	327.12
18	907.49	71.421	28	1428.6	175.41	38	1972.9	330.15
10	916.03	72.764	10	1437.4	177.55	10	1982.2	333.19
20	924.58	74.119	20	1446.3	179.72	20	1991.5	336.25
30	933.13	75.488	30	1455.1	181.89	30	2000.9	339.32
40	941.69	76.869	40	1464.0	184.08	40	2010.2	342.41
50	950.25	78.264	50	1472.9	186.29	50	2019.6	345.52
19	958.81	79.671	29	1481.8	188.51	39	2029.0	348.64
10	967.38	81.092	10	1490.7	190.74	10	2038.4	351.78
20	975.96	82.525	20	1499.6	192.99	20	2047.8	354.94
30	984.53	83.972	30	1508.5	195.25	30	2057.2	358.11
40	993.12	85.431	40	1517.4	197.53	40	2066.6	361.29
50	1001.7	86.904	50	1526.3	199.82	50	2076.0	364.50
20	1010.3	88.389	30	1535.3	202.12	40	2085.4	367.72
10	1018.9	89.888	10	1544.2	204.44	10	2094.9	370.95
20	1027.5	91.399	20	1553.1	206.77	20	2104.3	374.20
30	1036.1	92.924	30	1562.1	209.12	30	2113.8	377.47
40	1044.7	94.462	40	1571.0	211.48	40	2123.3	380.76
50	1053.3	96.013	50	1580.0	213.86	50	2132.7	384.06
21	1061.9	97.577	31	1589.0	216.25	41	2142.2	387.38
10	1070.6	99.155	10	1598.0	218.66	10	2151.7	390.71
20	1079.2	100.75	20	1606.9	221.08	20	2161.2	394.06
30	1087.8	102.35	30	1615.9	223.51	30	2170.8	397.43
40	1096.4	103.97	40	1624.9	225.96	40	2180.3	400.82
50	1105.1	105.60	50	1633.9	228.42	50	2189.9	404.22
22	1113.7	107.24	32	1643.0	230.90	42	2199.4	407.64
10	1122.4	108.90	10	1652.0	233.39	10	2209.0	411.07
20	1131.0	110.57	20	1661.0	235.90	20	2218.6	414.52
30	1139.7	112.25	30	1670.0	238.43	30	2228.1	417.99
40	1148.4	113.95	40	1679.1	240.96	40	2237.7	421.48
50	1157.0	115.66	50	1688.1	243.52	50	2247.3	424.98

Table 18. Tangents and Externals to a 1° Curve—*Continued*

Angle Δ	Tan T	External E	Angle Δ	Tan T	External E	Angle Δ	Tan T	External E
43°	2257.0	428.50	53°	2856.7	672.66	63°	3511.1	990.24
10′	2266.6	432.04	10′	2867.1	677.32	10′	3522.6	996.24
20	2276.2	435.59	20	2877.5	681.99	20	3534.1	1002.3
30	2285.9	439.16	30	2888.0	686.68	30	3545.6	1008.3
40	2295.6	442.75	40	2898.4	691.40	40	3557.2	1014.4
50	2305.2	446.35	50	2908.9	696.13	50	3568.7	1020.5
44	2314.19	449.98	54	2919.4	700.89	64	3580.3	1026.6
10	2324.6	453.62	10	2929.9	705.66	10	3591.9	1032.8
20	2334.3	457.27	20	2940.4	710.46	20	3603.5	1039.0
30	2344.1	460.95	30	2951.0	715.28	30	3615.1	1045.2
40	2353.8	464.64	40	2961.5	720.11	40	3626.8	1051.4
50	2363.5	468.35	50	2972.1	724.97	50	3638.5	1057.7
45	2373.3	472.08	55	2982.7	729.85	65	3650.2	1063.9
10	2383.1	475.82	10	2993.3	734.76	10	3661.9	1070.2
20	2392.8	479.59	20	3003.9	739.68	20	3673.7	1076.6
30	2402.6	483.37	30	3014.5	744.62	30	3685.4	1082.9
40	2412.4	487.17	40	3025.2	749.59	40	3697.2	1089.3
50	2422.3	490.98	50	3035.8	754.57	50	3709.0	1095.7
46	2432.1	494.82	56	3046.5	759.58	66	3720.9	1102.2
10	2441.9	498.67	10	3057.2	764.61	10	3732.7	1108.6
20	2451.8	502.54	20	3067.9	769.66	20	3744.6	1115.1
30	2461.7	506.42	30	3078.7	774.73	30	3756.5	1121.7
40	2471.5	510.33	40	3089.4	779.83	40	3768.5	1128.2
50	2481.4	514.25	50	3100.2	784.94	50	3780.4	1134.8
47	2491.3	518.20	57	3110.9	790.08	67	3792.4	1141.4
10	2501.2	522.16	10	3121.7	795.24	10	3804.4	1148.0
20	2511.2	526.13	20	3132.6	800.42	20	3816.4	1154.7
30	2521.1	530.13	30	3143.4	805.62	30	3828.4	1161.3
40	2531.1	534.15	40	3154.2	810.85	40	3840.5	1168.1
50	2541.0	538.18	50	3165.1	816.10	50	3852.6	1174.8
48	2551.0	542.23	58	3176.0	821.37	68	3864.7	1181.6
10	2561.0	546.30	10	3186.9	826.66	10	3876.8	1188.4
20	2571.0	550.39	20	3197.8	831.98	20	3889.0	1195.2
30	2581.0	554.50	30	3208.8	837.31	30	3901.2	1202.0
40	2591.1	558.63	40	3219.7	842.67	40	3913.4	1208.9
50	2601.1	562.77	50	3230.7	848.06	50	3925.6	1215.8
49	2611.2	566.94	59	3241.7	853.46	69	3937.9	1222.7
10	2621.2	571.12	10	3252.7	858.89	10	3950.2	1229.7
20	2631.3	575.32	20	3263.7	864.34	20	3962.5	1236.7
30	2641.4	579.54	30	3274.8	869.82	30	3974.8	1243.7
40	2651.5	583.78	40	3285.8	875.32	40	3987.2	1250.8
50	2661.6	588.04	50	3296.9	880.84	50	3999.5	1257.9
50	2671.8	592.32	60	3308.0	886.38	70	4011.9	1265.0
10	2681.9	596.62	10	3319.1	891.95	10	4024.4	1272.1
20	2692.1	600.93	20	3330.3	897.54	20	4036.8	1279.3
30	2702.3	605.27	30	3341.4	903.15	30	4049.3	1286.5
40	2712.5	609.62	40	3352.6	908.79	40	4061.8	1293.6
50	2722.7	614.00	50	3363.8	914.45	50	4074.4	1300.9
51	2732.9	618.39	61	3375.0	920.14	71	4086.9	1308.2
10	2743.1	622.81	10	3386.3	925.85	10	4099.5	1315.6
20	2753.4	627.24	20	3397.5	931.58	20	4112.1	1322.9
30	2763.7	631.69	30	3408.8	937.34	30	4124.8	1330.3
40	2773.9	636.17	40	3420.1	943.12	40	4137.4	1337.7
50	2784.2	640.66	50	3431.4	948.92	50	4150.1	1345.1
52	2794.5	645.17	62	3442.7	954.75	72	4162.8	1352.6
10	2804.9	649.70	10	3454.1	960.60	10	4175.6	1360.1
20	2815.2	654.25	20	3465.4	966.48	20	4188.4	1367.6
30	2825.6	658.83	30	3476.8	972.38	30	4201.2	1375.2
40	2835.9	663.42	40	3488.3	978.31	40	4214.0	1382.8
50	2846.3	668.03	50	3499.7	984.27	50	4226.8	1390.4

Table 18. Tangents and Externals to a 1° Curve—*Continued*

Angle Δ	Tan T	External E	Angle Δ	Tan T	External E	Angle Δ	Tan T	External E
73°	4239.7	1398.0	83°	5069.2	1920.5	93°	6037.8	2594.0
10′	4252.6	1405.7	10′	5084.0	1930.4	10′	6055.4	2606.8
20	4265.6	1413.5	20	5099.0	1940.3	20	6073.1	2619.7
30	4278.5	1421.2	30	5113.9	1950.3	30	6090.8	2632.6
40	4291.5	1429.0	40	5128.9	1960.2	40	6108.6	2645.5
50	4304.6	1436.8	50	5143.9	1970.3	50	6126.4	2658.5
74	4317.6	1444.6	84	5159.0	1980.4	94	6144.3	2671.6
10	4330.7	1452.5	10	5174.1	1990.5	10	6162.2	2684.7
20	4343.8	1460.4	20	5189.3	2000.6	20	6180.2	2697.9
30	4356.9	1468.4	30	5204.4	2010.8	30	6198.3	2711.2
40	4370.1	1476.4	40	5219.7	2021.1	40	6216.4	2724.5
50	4383.3	1484.4	50	5234.9	2031.4	50	6234.6	2737.9
75	4396.5	1492.4	85	5250.3	2041.7	95	6252.8	2751.3
10	4409.8	1500.5	10	5265.6	2052.1	10	6271.1	2764.8
20	4423.1	1508.6	20	5281.0	2062.5	20	6289.4	2778.3
30	4436.4	1516.7	30	5296.4	2073.0	30	6307.9	2792.0
40	4449.7	1524.9	40	5311.9	2083.5	40	6326.3	2805.6
50	4463.1	1533.1	50	5327.4	2094.1	50	6344.8	2819.4
76	4476.5	1541.4	86	5343.0	2104.7	96	6363.4	2833.2
10	4489.9	1549.7	10	5358.6	2115.3	10	6382.1	2847.0
20	4503.4	1558.0	20	5374.2	2126.0	20	6400.8	2861.0
30	4516.9	1566.3	30	5389.9	2136.7	30	6419.5	2875.0
40	4530.4	1574.7	40	5405.6	2147.5	40	6438.4	2889.0
50	4544.0	1583.1	50	5421.4	2158.4	50	6457.3	2903.1
77	4557.6	1591.6	87	5437.2	2169.2	97	6476.2	2917.3
10	4571.2	1600.1	10	5453.1	2180.2	10	6495.2	2931.6
20	4584.8	1608.6	20	5469.0	2191.1	20	6514.3	2945.9
30	4598.5	1617.1	30	5484.9	2202.2	30	6533.4	2960.3
40	4612.2	1625.7	40	5500.9	2213.2	40	6552.6	2974.7
50	4626.0	1634.4	50	5517.0	2224.3	50	6571.0	2989.2
78	4639.8	1643.0	88	5533.1	2235.5	98	6591.2	3003.8
10	4653.6	1651.7	10	5549.2	2246.7	10	6610.6	3018.4
20	4667.4	1660.5	20	5565.4	2258.0	20	6630.1	3033.1
30	4681.3	1669.2	30	5581.6	2269.3	30	6649.6	3047.9
40	4695.2	1678.1	40	5597.8	2280.6	40	6669.2	3062.8
50	4709.2	1686.9	50	5614.2	2292.0	50	6688.8	3077.7
79	4723.2	1695.8	89	5630.5	2303.5	99	6708.6	3092.7
10	4737.2	1704.7	10	5646.9	2315.0	10	6728.4	3107.7
20	4751.2	1713.7	20	5663.4	2326.6	20	6748.2	3122.9
30	4765.3	1722.7	30	5679.9	2338.2	30	6768.1	3138.1
40	4779.4	1731.7	40	5696.4	2349.8	40	6788.1	3153.3
50	4793.6	1740.8	50	5713.0	2361.5	50	6808.2	3168.7
80	4807.7	1749.9	90	5729.7	2373.3	100	6828.3	3184.1
10	4822.0	1759.0	10	5746.3	2385.1	10	6848.5	3199.6
20	4836.2	1768.2	20	5763.1	2397.0	20	6868.8	3215.1
30	4850.5	1777.4	30	5779.9	2408.9	30	6889.2	3230.8
40	4864.8	1786.7	40	5796.7	2420.9	40	6909.6	3246.5
50	4879.2	1796.0	50	5813.6	2432.9	50	6930.1	3262.3
81	4893.6	1805.3	91	5830.5	2444.9	101	6950.6	3278.1
10	4908.0	1814.7	10	5847.5	2457.1	10	6971.3	3294.1
20	4922.5	1824.1	20	5864.6	2469.3	20	6992.0	3310.1
30	4937.0	1833.6	30	5881.7	2481.5	30	7012.7	3326.1
40	4951.5	1843.1	40	5898.8	2493.8	40	7033.6	3342.3
50	4966.1	1852.6	50	5916.0	2506.1	50	7054.5	3358.5
82	4980.7	1862.2	92	5933.2	2518.5	102	7075.5	3374.9
10	4995.4	1871.8	10	5950.5	2531.0	10	7096.6	3391.2
20	5010.0	1881.5	20	5967.9	2543.5	20	7117.8	3407.7
30	5024.8	1891.2	30	5985.3	2556.0	30	7139.0	3424.3
40	5039.5	1900.9	40	6002.7	2568.6	40	7160.3	3440.9
50	5054.3	1910.7	50	6020.2	2581.3	50	7181.7	3457.6

Table 18. Tangents and Externals to a 1° Curve—*Continued*

Angle Δ	Tan T	External E	Angle Δ	Tan T	External E	Angle Δ	Tan T	External E
103°	7203.2	3474.4	109°	8032.7	4137.1	115°	8993.8	4934.1
10'	7224.7	3491.3	10'	8057.4	4157.3	10'	9022.7	4958.6
20	7246.3	3508.2	20	8082.3	4177.5	20	9051.7	4983.1
30	7268.0	3525.2	30	8107.3	4197.9	30	9080.9	5007.8
40	7289.8	3542.4	40	8132.3	4218.4	40	9110.3	5032.6
50	7311.7	3559.6	50	8157.5	4239.0	50	9139.8	5057.6
104	7333.6	3576.8	110	8182.8	4259.7	116	9169.4	5082.7
10	7355.6	3594.2	10	8208.2	4280.5	10	9199.1	5107.9
20	7377.8	3611.7	20	8233.7	4301.4	20	9229.0	5133.3
30	7399.9	3629.2	30	8259.3	4322.4	30	9259.0	5158.8
40	7422.2	3646.8	40	8285.0	4343.6	40	9289.2	5184.5
50	7444.5	3664.5	50	8310.8	4364.8	50	9319.5	5210.3
105	7467.0	3682.3	111	8336.7	4386.1	117	9349.9	5236.2
10	7489.6	3700.2	10	8362.7	4407.6	10	9380.5	5262.3
20	7512.2	3718.2	20	8388.9	4429.2	20	9411.3	5288.6
30	7534.9	3736.2	30	8415.1	4450.9	30	9442.2	5315.0
40	7557.7	3754.4	40	8441.5	4472.7	40	9473.2	5341.5
50	7580.5	3772.6	50	8468.0	4494.6	50	9504.4	5368.2
106	7603.5	3791.0	112	8494.6	4516.6	118	9535.7	5395.1
10	7626.6	3809.4	10	8521.3	4538.8	10	9567.2	5422.1
20	7649.7	3827.9	20	8548.1	4561.1	20	9598.9	5449.2
30	7672.9	3846.5	30	8575.0	4583.4	30	9630.7	5476.5
40	7696.3	3865.2	40	8602.1	4606.0	40	9662.6	5504.0
50	7719.7	3884.0	50	8629.3	4628.6	50	9694.7	5531.7
107	7743.2	3902.9	113	8656.6	4651.3	119	9727.0	5559.4
10	7766.8	3921.9	10	8684.0	4674.2	10	9759.4	5587.4
20	7790.5	3940.9	20	8711.5	4697.2	20	9792.0	5615.5
30	7814.3	3960.1	30	8739.2	4720.3	30	9824.8	5643.8
40	7838.1	3979.4	40	8767.0	4743.6	40	9857.7	5672.3
50	7862.1	3998.7	50	8794.9	4766.9	50	9890.8	5700.9
108	7886.2	4018.2	114	8822.9	4790.4	120	9924.0	5729.7
10	7910.4	4037.8	10	8851.0	4814.1	10	9957.5	5758.6
20	7934.6	4057.4	20	8879.3	4837.8	20	9991.0	5787.7
30	7959.0	4077.2	30	8907.7	4861.7	30	10025.0	5817.0
40	7983.5	4097.1	40	8936.3	4885.7	40	10059.0	5846.5
50	8008.0	4117.0	50	8965.0	4909.9	50	10093.0	5876.1

Table 19. Corrections for Tangents and Externals

Angle Δ	For Tangents, add						Angle Δ	For Externals, add					
	5° Curve	10° Curve	15° Curve	20° Curve	25° Curve	30° Curve		5° Curve	10° Curve	15° Curve	20° Curve	25° Curve	30° Curve
10°	0.03	0.06	0.09	0.13	0.16	0.19	10°	0.001	0.003	0.004	0.006	0.007	0.008
20	.06	0.13	0.19	0.26	0.32	0.39	20	.006	.011	0.017	0.022	0.028	0.034
30	.10	0.19	0.29	0.39	0.49	0.59	30	.013	.025	0.038	0.051	0.065	0.078
40	.13	0.26	0.40	0.53	0.67	0.80	40	.023	.046	0.070	0.093	0.117	0.141
50	.17	0.34	0.51	0.68	0.85	1.02	50	.037	.075	0.116	0.151	0.189	0.227
60	.21	0.42	0.63	0.84	1.05	1.27	60	.056	.112	0.168	0.225	0.283	0.340
70	.25	0.51	0.76	1.02	1.28	1.54	70	.080	.159	0.240	0.321	0.403	0.485
80	.30	0.61	0.91	1.22	1.53	1.84	80	.110	.220	0.332	0.445	0.558	0.671
90	.36	0.72	1.09	1.45	1.83	2.20	90	.149	.299	0.450	0.603	0.756	0.910
100	.43	0.86	1.30	1.74	2.18	2.62	100	.200	.401	0.604	0.809	1.015	1.221
110	.51	1.03	1.56	2.08	2.61	3.14	110	.268	.536	0.806	1.082	1.355	1.633
120	.62	1.25	1.93	2.52	3.16	3.81	120	.360	.721	1.086	1.456	1.825	2.197

Table 20. Stadia Reductions
Differences in Elevation for 100 ft Inclined Distance

Min- utes	0°	1°	2°	3°	4°	5°	6°	7°	8°	9°	10°	11°	12°
0	0.00	1.74	3.49	5.23	6.96	8.68	10.40	12.10	13.78	15.45	17.10	18.73	20.34
2	0.06	1.80	3.55	5.28	7.02	8.74	10.45	12.15	13.84	15.51	17.16	18.78	20.39
4	0.12	1.86	3.60	5.34	7.07	8.80	10.51	12.21	13.89	15.56	17.21	18.84	20.44
6	0.17	1.92	3.66	5.40	7.13	8.85	10.57	12.26	13.95	15.62	17.26	18.89	20.50
8	0.23	1.98	3.72	5.46	7.19	8.91	10.62	12.32	14.01	15.67	17.32	18.95	20.55
10	0.29	2.04	3.78	5.52	7.25	8.97	10.68	12.38	14.06	15.73	17.37	19.00	20.60
12	0.35	2.09	3.84	5.57	7.30	9.03	10.74	12.43	14.12	15.78	17.43	19.05	20.66
14	0.41	2.15	3.90	5.63	7.36	9.08	10.79	12.49	14.17	15.84	17.48	19.11	20.71
16	0.47	2.21	3.95	5.69	7.42	9.14	10.85	12.55	14.23	15.89	17.54	19.16	20.76
18	0.52	2.27	4.01	5.75	7.48	9.20	10.91	12.60	14.28	15.95	17.59	19.21	20.81
20	0.58	2.33	4.07	5.80	7.53	9.25	10.96	12.66	14.34	16.00	17.65	19.27	20.87
22	0.64	2.38	4.13	5.86	7.59	9.31	11.02	12.72	14.40	16.06	17.70	19.32	20.92
24	0.70	2.44	4.18	5.92	7.65	9.37	11.08	12.77	14.45	16.11	17.76	19.38	20.97
26	0.76	2.50	4.24	5.98	7.71	9.43	11.13	12.83	14.51	16.17	17.81	19.43	21.03
28	0.81	2.56	4.30	6.04	7.76	9.48	11.19	12.88	14.56	16.22	17.86	19.48	21.08
30	0.87	2.62	4.36	6.09	7.82	9.54	11.25	12.94	14.62	16.28	17.92	19.54	21.13
32	0.93	2.67	4.42	6.15	7.88	9.60	11.30	13.00	14.67	16.33	17.97	19.59	21.18
34	0.99	2.73	4.48	6.21	7.94	9.65	11.36	13.05	14.73	16.39	18.03	19.64	21.24
36	1.05	2.79	4.53	6.27	7.99	9.71	11.42	13.11	14.79	16.44	18.08	19.70	21.29
38	1.11	2.85	4.59	6.33	8.05	9.77	11.47	13.17	14.84	16.50	18.14	19.75	21.34
40	1.16	2.91	4.65	6.38	8.11	9.83	11.53	13.22	14.90	16.55	18.19	19.80	21.39
42	1.22	2.97	4.71	6.44	8.17	9.88	11.59	13.28	14.95	16.61	18.24	19.86	21.45
44	1.28	3.02	4.76	6.50	8.22	9.94	11.64	13.33	15.01	16.66	18.30	19.91	21.50
46	1.34	3.08	4.82	6.56	8.28	10.00	11.70	13.39	15.06	16.72	18.35	19.96	21.55
48	1.40	3.14	4.88	6.61	8.34	10.05	11.76	13.45	15.12	16.77	18.41	20.02	21.60
50	1.45	3.20	4.94	6.67	8.40	10.11	11.81	13.50	15.17	16.83	18.46	20.07	21.66
52	1.51	3.26	4.99	6.73	8.45	10.17	11.87	13.56	15.23	16.88	18.51	20.12	21.71
54	1.57	3.31	5.05	6.79	8.51	10.22	11.93	13.61	15.28	16.94	18.57	20.18	21.76
56	1.63	3.37	5.11	6.84	8.57	10.28	11.98	13.67	15.34	16.99	18.62	20.23	21.81
58	1.69	3.43	5.17	6.90	8.63	10.34	12.04	13.73	15.40	17.05	18.68	20.28	21.87
60	1.74	3.49	5.23	6.96	8.68	10.40	12.10	13.78	15.45	17.10	18.73	20.34	21.92
$f+c$													
.75	0.01	0.02	0.03	0.05	0.06	0.07	0.08	0.10	0.11	0.12	0.14	0.15	0.16
1.00	0.01	0.03	0.04	0.06	0.08	0.09	0.11	0.13	0.15	0.16	0.18	0.20	0.22
1.25	0.02	0.03	0.05	0.08	0.10	0.11	0.14	0.16	0.18	0.21	0.23	0.25	0.27

Corrections to Horizontal Distances

Dis- tance	0°	1°	2°	3°	4°	5°	6°	7°	8°	9°	10°	11°	12°
100	0.0	0.0	0.1	0.3	0.5	0.8	1.1	1.5	1.9	2.5	3.0	3.6	4.3
200	.0	.1	0.2	0.5	1.0	1.5	2.2	3.0	3.9	4.9	6.0	7.3	8.6
300	.0	.1	0.4	0.8	1.5	2.3	3.3	4.5	5.8	7.4	9.1	10.9	13.0
400	.0	.1	0.5	1.1	2.0	3.0	4.4	6.0	7.8	9.8	12.1	14.6	17.3
500	.0	.2	0.6	1.4	2.5	3.8	5.5	7.5	9.7	12.3	15.1	18.2	21.6
600	.0	.2	0.7	1.6	2.9	4.6	6.5	8.9	11.6	14.7	18.1	21.8	25.9
700	.0	.2	0.8	1.9	3.4	5.3	7.6	10.4	13.6	17.2	21.1	25.5	30.2
800	.0	.2	1.0	2.2	3.9	6.1	8.7	11.9	15.5	19.6	24.2	29.1	34.6
900	.0	.3	1.1	2.4	4.4	6.8	9.8	13.4	17.5	22.1	27.2	32.8	38.9
1000	.0	.3	1.2	2.7	4.9	7.6	10.9	14.9	19.4	24.5	30.2	36.4	43.2

Table 20. Stadia Reductions—*Continued*

Differences in Elevation for 100 ft Inclined Distance

Min-utes	13°	14°	15°	16°	17°	18°	19°	20°	21°	22°	23°	24°	25°
0	21.92	23.47	25.00	26.50	27.96	29.39	30.78	32.14	33.46	34.73	35.97	37.16	38.30
2	21.97	23.52	25.05	26.55	28.01	29.44	30.83	32.18	33.50	34.77	36.01	37.20	38.34
4	22.02	23.58	25.10	26.59	28.06	29.48	30.87	32.23	33.54	34.82	36.05	37.23	38.38
6	22.08	23.63	25.15	26.64	28.10	29.53	30.92	32.27	33.59	34.86	36.09	37.37	38.41
8	22.13	23.68	25.20	26.69	28.15	29.58	30.97	32.32	33.63	34.90	36.13	37.31	38.45
10	22.18	23.73	25.25	26.74	28.20	29.62	31.01	32.36	33.67	34.94	36.17	37.35	38.49
12	22.23	23.78	25.30	26.79	28.25	29.67	31.06	32.41	33.72	34.98	36.21	37.39	38.53
14	22.28	23.83	25.35	26.84	28.30	29.72	31.10	32.45	33.76	35.02	36.25	37.43	38.56
16	22.34	23.88	25.40	26.89	28.34	29.76	31.15	32.49	33.80	35.07	36.29	37.47	38.60
18	22.39	23.93	25.45	26.94	28.39	29.81	31.19	32.54	33.84	35.11	36.33	37.51	38.64
20	22.44	23.99	25.50	26.99	28.44	29.86	31.24	32.58	33.89	35.15	36.37	37.54	38.67
22	22.49	24.04	25.55	27.04	28.49	29.90	31.28	32.63	33.93	35.19	36.41	37.58	38.71
24	22.54	24.09	25.60	27.09	28.54	29.95	31.33	32.67	33.97	35.23	36.45	37.62	38.75
26	22.60	24.14	25.65	27.13	28.58	30.00	31.38	32.72	34.01	35.27	36.49	37.66	38.78
28	22.65	24.19	25.70	27.18	28.63	30.04	31.42	32.76	34.06	35.31	36.53	37.70	38.82
30	22.70	24.24	25.75	27.23	28.68	30.09	31.47	32.80	34.10	35.36	36.57	37.74	38.86
32	22.75	24.29	25.80	27.28	28.73	30.14	31.51	32.85	34.14	35.40	36.61	37.77	38.89
34	22.80	24.34	25.85	27.33	28.77	30.19	31.56	32.89	34.18	35.44	36.65	37.81	38.93
36	22.85	24.39	25.90	27.38	28.82	30.23	31.60	32.93	34.23	35.48	36.69	37.85	38.97
38	22.91	24.44	25.95	27.43	28.87	30.28	31.65	32.98	34.27	35.52	36.73	37.89	39.00
40	22.96	24.49	26.00	27.48	28.92	30.32	31.69	33.02	34.31	35.56	36.77	37.93	39.04
42	23.01	24.55	26.05	27.52	28.96	30.37	31.74	33.07	34.35	35.60	36.80	37.96	39.08
44	23.06	24.60	26.10	27.57	29.01	30.41	31.78	33.11	34.40	35.64	36.84	38.00	39.11
46	23.11	24.65	26.15	27.62	29.06	30.46	31.83	33.15	34.44	35.68	36.88	38.04	39.15
48	23.16	24.70	26.20	27.67	29.11	30.51	31.87	33.20	34.48	35.72	36.92	38.08	39.18
50	23.22	24.75	26.25	27.72	29.15	30.55	31.92	33.24	34.52	35.76	36.96	38.11	39.22
52	23.27	24.80	26.30	27.77	29.20	30.60	31.96	33.28	34.57	35.80	37.00	38.15	39.26
54	23.32	24.85	26.35	27.81	29.25	30.65	32.01	33.33	34.61	35.85	37.04	38.19	39.29
56	23.37	24.90	26.40	27.86	29.30	30.69	32.05	33.37	34.65	35.89	37.08	38.23	39.33
58	23.42	24.95	26.45	27.91	29.34	30.74	32.09	33.41	34.69	35.93	37.12	38.26	39.36
60	23.47	25.00	26.50	27.96	29.39	30.78	32.14	33.46	34.73	35.97	37.16	38.30	39.40

f + c													
.75	0.17	0.19	0.20	0.21	0.23	0.24	0.25	0.26	0.27	0.29	0.30	0.31	0.32
1.00	0.23	0.25	0.27	0.28	0.30	0.32	0.33	0.35	0.37	0.38	0.40	0.41	0.43
1.25	0.29	0.31	0.34	0.36	0.38	0.40	0.42	0.44	0.46	0.48	0.50	0.52	0.54

Corrections to Horizontal Distances

Dis-tance	13°	14°	15°	16°	17°	18°	19°	20°	21°	22°	23°	24°	25°
100	5.1	5.9	6.7	7.6	8.5	9.5	10.6	11.7	12.8	14.0	15.3	16.5	17.9
200	10.1	11.7	13.4	15.2	17.1	19.1	21.2	23.4	25.7	28.1	30.5	33.1	35.7
300	15.2	17.6	20.1	22.8	25.6	28.6	31.8	35.1	38.5	42.1	45.8	49.6	53.6
400	20.2	23.4	26.8	30.4	34.2	38.2	42.4	46.8	51.4	56.1	61.1	66.2	71.4
500	25.3	29.3	33.5	38.0	42.7	47.7	53.0	58.5	64.2	70.2	76.4	82.7	89.3
600	30.4	35.1	40.2	45.6	51.3	57.3	63.6	70.2	77.0	84.2	91.6	99.2	107.2
700	35.4	41.0	46.9	53.2	59.8	66.8	74.2	81.9	89.9	98.2	106.9	115.8	125.0
800	40.5	46.8	53.6	60.8	68.4	76.4	84.8	93.6	102.7	112.4	122.2	132.3	142.9
900	45.5	52.7	60.3	68.4	76.9	85.9	95.4	105.3	115.6	126.3	137.4	148.9	160.7
1000	50.6	58.5	67.0	76.0	85.5	95.5	106.0	117.0	128.4	140.3	152.7	165.4	178.6

Table 21. Values and Logarithms of Trigonometric Functions

Decimals	Minutes	Natural Values				Common Logarithms				Minutes	Decimals
		Sin	Cos	Tan	Cot	Sin	Cos	Tan	Cot		
.00	0	.00000	1.00000	.00000	$+\infty$	$-\infty$	10.000000	$-\infty$	$+\infty$	60	1.00
	1	.00029	1.00000	.00029	3437.75	6.463726	.000000	6.463726	13.536274	59	
	2	.00058	1.00000	.00058	1718.87	.764756	.000000	.764756	.235244	58	
.05	3	.00087	1.00000	.00087	1145.92	.940847	.000000	.940847	.059153	57	.95
	4	.00116	1.00000	.00116	859.436	7.065786	.000000	7.065786	12.934214	56	
	5	.00145	1.00000	.00145	687.549	.162696	.000000	.162696	.837304	55	
.10	6	.00175	1.00000	.00175	572.957	.241877	9.999999	.241878	.758122	54	.90
	7	.00204	1.00000	.00204	491.106	.308824	.999999	.308825	.691175	53	
	8	.00233	1.00000	.00233	429.718	.366816	.999999	.366817	.633183	52	
.15	9	.00262	1.00000	.00262	381.971	.417968	.999999	.417970	.582030	51	.85
	10	.00291	1.00000	.00291	343.774	.463726	.999998	.463727	.536273	50	
	11	.00320	.99999	.00320	312.521	.505118	.999998	.505120	.494880	49	
.20	12	.00349	.99999	.00349	286.478	.542906	.999997	.542909	.457091	48	.80
	13	.00378	.99999	.00378	264.441	.577668	.999997	.577672	.422328	47	
	14	.00407	.99999	.00407	245.552	.609853	.999996	.609857	.390143	46	
.25	15	.00436	.99999	.00436	229.182	.639816	.999996	.639820	.360180	45	.75
	16	.00465	.99999	.00465	214.858	.667845	.999995	.667849	.332151	44	
	17	.00495	.99999	.00495	202.219	.694173	.999995	.694179	.305821	43	
.30	18	.00524	.99999	.00524	190.984	.718997	.999994	.719003	.280997	42	.70
	19	.00553	.99998	.00553	180.932	.742478	.999993	.742484	.257516	41	
	20	.00582	.99998	.00582	171.885	.764754	.999993	.764761	.235239	40	
.35	21	.00611	.99998	.00611	163.700	.785943	.999992	.785951	.214049	39	.65
	22	.00640	.99998	.00640	156.259	.806146	.999991	.806155	.193845	38	
	23	.00669	.99998	.00669	149.465	.825451	.999990	.825460	.174540	37	
.40	24	.00698	.99998	.00698	143.237	.843934	.999989	.843944	.156056	36	.60
	25	.00727	.99997	.00727	137.507	.861662	.999989	.861674	.138326	35	
	26	.00756	.99997	.00756	132.219	.878695	.999988	.878708	.121292	34	
.45	27	.00785	.99997	.00785	127.321	.895085	.999987	.895099	.104901	33	.55
	28	.00814	.99997	.00815	122.774	.910879	.999986	.910894	.089106	32	
	29	.00844	.99996	.00844	118.540	.926119	.999985	.926134	.073866	31	
.50	30	.00873	.99996	.00873	114.589	.940842	.999983	.940858	.059142	30	.50
	31	.00902	.99996	.00902	110.892	.955082	.999982	.955100	.044900	29	
	32	.00931	.99996	.00931	107.426	.968870	.999981	.968889	.031111	28	
.55	33	.00960	.99995	.00960	104.171	.982233	.999980	.982253	.017747	27	.45
	34	.00989	.99995	.00989	101.107	.995198	.999979	.995219	.004781	26	
	35	.01018	.99995	.01018	98.2179	8.007787	.999977	8.007809	11.992191	25	
.60	36	.01047	.99995	.01047	95.4895	.020021	.999976	.020044	.979956	24	.40
	37	.01076	.99994	.01076	92.9085	.031919	.999975	.031945	.968055	23	
	38	.01105	.99994	.01105	90.4633	.043501	.999973	.043527	.956473	22	
.65	39	.01134	.99994	.01135	88.1436	.054781	.999972	.054809	.945191	21	.35
	40	.01164	.99993	.01164	85.9398	.065776	.999971	.065806	.934194	20	
	41	.01193	.99993	.01193	83.8435	.076500	.999969	.076531	.923469	19	
.70	42	.01222	.99993	.01222	81.8470	.086965	.999968	.086997	.913003	18	.30
	43	.01251	.99992	.01251	79.9434	.097183	.999966	.097217	.902783	17	
	44	.01280	.99992	.01280	78.1263	.107167	.999964	.107203	.892797	16	
.75	45	.01309	.99991	.01309	76.3900	.116926	.999963	.116963	.883037	15	.25
	46	.01338	.99991	.01338	74.7292	.126471	.999961	.126510	.873490	14	
	47	.01367	.99991	.01367	73.1390	.135810	.999959	.135851	.864149	13	
.80	48	.01396	.99990	.01396	71.6151	.144953	.999958	.144996	.855004	12	.20
	49	.01425	.99990	.01425	70.1533	.153907	.999956	.153952	.846048	11	
	50	.01454	.99989	.01455	68.7501	.162681	.999954	.162727	.837273	10	
.85	51	.01483	.99989	.01484	67.4019	.171280	.999952	.171328	.828672	9	.15
	52	.01513	.99989	.01513	66.1055	.179713	.999950	.179763	.820237	8	
	53	.01542	.99988	.01542	64.8580	.187985	.999948	.188036	.811964	7	
.90	54	.01571	.99988	.01571	63.6567	.196102	.999946	.196156	.803844	6	.10
	55	.01600	.99987	.01600	62.4992	.204070	.999944	.204126	.795874	5	
	56	.01629	.99987	.01629	61.3829	.211895	.999942	.211953	.788047	4	
.95	57	.01658	.99986	.01658	60.3058	.219581	.999940	.219641	.780359	3	.05
	58	.01687	.99986	.01687	59.2659	.227134	.999938	.227195	.772805	2	
	59	.01716	.99985	.01716	58.2612	.234557	.999936	.234621	.765379	1	
1.00	60	.01745	.99985	.01746	57.2900	8.241855	9.999934	8.241921	11.758079	0	.00
Decimals	Minutes	Cos	Sin	Cot	Tan	Cos	Sin	Cot	Tan	Minutes	Decimals
		Natural Values				Common Logarithms					

Decimals	Minutes	Natural Values — Sin	Cos	Tan	Cot	Common Logarithms — Sin	Cos	Tan	Cot	Minutes	Decimals
.00	0	.01745	.99985	.01746	57.2900	8.241855	9.999934	8.241921	11.758079	60	1.00
	1	.01774	.99984	.01775	56.3506	.249033	.999932	.249102	.750898	59	
	2	.01803	.99984	.01804	55.4415	.256094	.999929	.256165	.743835	58	
.05	3	.01832	.99983	.01833	54.5613	.263042	.999927	.263115	.736885	57	.95
	4	.01862	.99983	.01862	53.7086	.269881	.999925	.269956	.730044	56	
	5	.01891	.99982	.01891	52.8821	.276614	.999922	.276691	.723309	55	
.10	6	.01920	.99982	.01920	52.0807	.283243	.999920	.283323	.716677	54	.90
	7	.01949	.99981	.01949	51.3032	.289773	.999918	.289856	.710144	53	
	8	.01978	.99980	.01978	50.5485	.296207	.999915	.296292	.703708	52	
.15	9	.02007	.99980	.02007	49.8157	.302546	.999913	.302634	.697366	51	.85
	10	.02036	.99979	.02036	49.1039	.308794	.999910	.308884	.691116	50	
	11	.02065	.99979	.02066	48.4121	.314954	.999907	.315046	.684954	49	
.20	12	.02094	.99978	.02095	47.7395	.321027	.999905	.321122	.678878	48	.80
	13	.02123	.99977	.02124	47.0853	.327016	.999902	.327114	.672886	47	
	14	.02152	.99977	.02153	46.4489	.332924	.999899	.333025	.666975	46	
.25	15	.02181	.99976	.02182	45.8294	.338753	.999897	.338856	.661144	45	.75
	16	.02211	.99976	.02211	45.2261	.344504	.999894	.344610	.655390	44	
	17	.02240	.99975	.02240	44.6386	.350181	.999891	.350289	.649711	43	
.30	18	.02269	.99974	.02269	44.0661	.355783	.999888	.355895	.644105	42	.70
	19	.02298	.99974	.02298	43.5081	.361315	.999885	.361430	.638570	41	
	20	.02327	.99973	.02328	42.9641	.366777	.999882	.366895	.633105	40	
.35	21	.02356	.99972	.02357	42.4335	.372171	.999879	.372292	.627708	39	.65
	22	.02385	.99972	.02386	41.9158	.377499	.999876	.377622	.622378	38	
	23	.02414	.99971	.02415	41.4106	.382762	.999873	.382889	.617111	37	
.40	24	.02443	.99970	.02444	40.9174	.387962	.999870	.388092	.611908	36	.60
	25	.02472	.99969	.02473	40.4358	.393101	.999867	.393234	.606766	35	
	26	.02501	.99969	.02502	39.9655	.398179	.999864	.398315	.601685	34	
.45	27	.02530	.99968	.02531	39.5059	.403199	.999861	.403338	.596662	33	.55
	28	.02560	.99967	.02560	39.0568	.408161	.999858	.408304	.591696	32	
	29	.02589	.99966	.02589	38.6177	.413068	.999854	.413213	.586787	31	
.50	30	.02618	.99966	.02619	38.1885	.417919	.999851	.418068	.581932	30	.50
	31	.02647	.99965	.02648	37.7686	.422717	.999848	.422869	.577131	29	
	32	.02676	.99964	.02677	37.3579	.427462	.999844	.427618	.572382	28	
.55	33	.02705	.99963	.02706	36.9560	.432156	.999841	.432315	.567685	27	.45
	34	.02734	.99963	.02735	36.5627	.436800	.999838	.436962	.563038	26	
	35	.02763	.99962	.02764	36.1776	.441394	.999834	.441560	.558440	25	
.60	36	.02792	.99961	.02793	35.8006	.445941	.999831	.446110	.553890	24	.40
	37	.02821	.99960	.02822	35.4313	.450440	.999827	.450613	.549387	23	
	38	.02850	.99959	.02851	35.0695	.454893	.999824	.455070	.544930	22	
.65	39	.02879	.99959	.02881	34.7151	.459301	.999820	.459481	.540519	21	.35
	40	.02908	.99958	.02910	34.3678	.463665	.999816	.463849	.536151	20	
	41	.02938	.99957	.02939	34.0273	.467985	.999813	.468172	.531828	19	
.70	42	.02967	.99956	.02968	33.6935	.472253	.999809	.472454	.527546	18	.30
	43	.02996	.99955	.02997	33.3662	.476498	.999805	.476693	.523307	17	
	44	.03025	.99954	.03026	33.0452	.480693	.999801	.480892	.519108	16	
.75	45	.03054	.99953	.03055	32.7303	.484848	.999797	.485050	.514950	15	.25
	46	.03083	.99952	.03084	32.4213	.488963	.999794	.489170	.510830	14	
	47	.03112	.99952	.03114	32.1181	.493040	.999790	.493250	.506750	13	
.80	48	.03141	.99951	.03143	31.8205	.497078	.999786	.497293	.502707	12	.20
	46	.03170	.99950	.03172	31.5284	.501080	.999782	.501298	.498702	11	
	50	.03199	.99949	.03201	31.2416	.505045	.999778	.505267	.494733	10	
.85	51	.03228	.99948	.03230	30.9599	.508974	.999774	.509200	.490800	9	.15
	52	.03257	.99947	.03259	30.6833	.512867	.999769	.513098	.486902	8	
	53	.03286	.99946	.03288	30.4116	.516726	.999765	.516961	.483039	7	
.90	54	.03316	.99945	.03317	30.1446	.520551	.999761	.520790	.479210	6	.10
	55	.03345	.99944	.03346	29.8823	.524343	.999757	.524586	.475414	5	
	56	.03374	.99943	.03376	29.6245	.528102	.999753	.528349	.471651	4	
.95	57	.03403	.99942	.03405	29.3711	.531828	.999748	.532080	.467920	3	.05
	58	.03432	.99941	.03434	29.1220	.535523	.999744	.535779	.464221	2	
	59	.03461	.99940	.03463	28.8771	.539186	.999740	.539447	.460553	1	
1.00	60	.03490	.99939	.03492	28.6363	8.542819	9.999735	8.543084	11.456916	0	.00
		Cos	Sin	Cot	Tan	Cos	Sin	Cot	Tan		
		Natural Values				Common Logarithms				Minutes	Decimals

Decimals	Minutes	Natural Values Sin	Cos	Tan	Cot	Common Logarithms Sin	Cos	Tan	Cot	Minutes	Decimals
.00	0	.03490	.99939	.03492	28.6363	8.542819	9.999735	8.543084	11.456916	60	1.00
	1	.03519	.99938	.03521	28.3994	.546422	.999731	.546691	.453309	59	
	2	.03548	.99937	.03550	28.1664	.549995	.999726	.550268	.449732	58	
.05	3	.03577	.99936	.03579	27.9372	.553539	.999722	.553817	.446183	57	.95
	4	.03606	.99935	.03609	27.7117	.557054	.999717	.557336	.442664	56	
	5	.03635	.99934	.03638	27.4899	.560540	.999713	.560828	.439172	55	
.10	6	.03664	.99933	.03667	27.2715	.563999	.999708	.564291	.435709	54	.90
	7	.03693	.99932	.03696	27.0566	.567431	.999704	.567727	.432273	53	
	8	.03723	.99931	.03725	26.8450	.570836	.999699	.571137	.428863	52	
.15	9	.03752	.99930	.03754	26.6367	.574214	.999694	.574520	.425480	51	.85
	10	.03781	.99929	.03783	26.4316	.577566	.999689	.577877	.422123	50	
	11	.03810	.99927	.03812	26.2296	.580892	.999685	.581208	.418792	49	
.20	12	.03839	.99926	.03842	26.0307	.584193	.999680	.584514	.415486	48	.80
	13	.03868	.99925	.03871	25.8348	.587469	.999675	.587795	.412205	47	
	14	.03897	.99924	.03900	25.6418	.590721	.999670	.591051	.408949	46	
.25	15	.03926	.99923	.03929	25.4517	.593948	.999665	.594260	.405717	45	.75
	16	.03955	.99922	.03958	25.2644	.597152	.999660	.597492	.402508	44	
	17	.03984	.99921	.03987	25.0798	.600332	.999655	.600677	.399323	43	
.30	18	.04013	.99919	.04016	24.8978	.603489	.999650	.603839	.396161	42	.70
	19	.04042	.99918	.04046	24.7185	.606623	.999645	.606978	.393022	41	
	20	.04071	.99917	.04075	24.5418	.609734	.999640	.610094	.389906	40	
.35	21	.04100	.99916	.04104	24.3675	.612823	.999635	.613189	.386811	39	.65
	22	.04129	.99915	.04133	24.1957	.615891	.999629	.616262	.383738	38	
	23	.04159	.99913	.04162	24.0263	.618937	.999624	.619313	.380687	37	
.40	24	.04188	.99912	.04191	23.8593	.621962	.999619	.622343	.377657	36	.60
	25	.04217	.99911	.04220	23.6945	.624965	.999614	.625352	.374648	35	
	26	.04246	.99910	.04250	23.5321	.627942	.999608	.628340	.371660	34	
.45	27	.04275	.99909	.04279	23.3718	.630911	.999603	.631308	.368692	33	.55
	28	.04304	.99907	.04308	23.2137	.633854	.999597	.634256	.365744	32	
	29	.04333	.99906	.04337	23.0577	.636776	.999592	.637184	.362816	31	
.50	30	.04362	.99905	.04366	22.9038	.639680	.999586	.640093	.359907	30	.50
	31	.04391	.99904	.04395	22.7519	.642563	.999581	.642982	.357018	29	
	32	.04420	.99902	.04424	22.6020	.645428	.999575	.645853	.354147	28	
.55	33	.04449	.99901	.04454	22.4541	.648274	.999570	.648704	.351296	27	.45
	34	.04478	.99900	.04483	22.3081	.651102	.999564	.651537	.348463	26	
	35	.04507	.99898	.04512	22.1640	.653911	.999558	.654352	.345648	25	
.60	36	.04536	.99897	.04541	22.0217	.656702	.999553	.657149	.342851	24	.40
	37	.04565	.99896	.04570	21.8813	.659475	.999547	.659928	.340072	23	
	38	.04594	.99894	.04599	21.7426	.662230	.999541	.662689	.337311	22	
.65	39	.04623	.99893	.04628	21.6056	.664968	.999535	.665433	.334567	21	.35
	40	.04653	.99892	.04658	21.4704	.667689	.999529	.668160	.331840	20	
	41	.04682	.99890	.04687	21.3369	.670393	.999524	.670870	.329130	19	
.70	42	.04711	.99889	.04716	21.2049	.673080	.999518	.673563	.326437	18	.30
	43	.04740	.99888	.04745	21.0747	.675751	.999512	.676239	.323761	17	
	44	.04769	.99886	.04774	20.9460	.678405	.999506	.678900	.321100	16	
.75	45	.04798	.99885	.04803	20.8188	.681043	.999500	.681544	.318456	15	.25
	46	.04827	.99883	.04833	20.6932	.683665	.999493	.684172	.315828	14	
	47	.04856	.99882	.04862	20.5691	.686272	.999487	.686784	.313216	13	
.80	48	.04885	.99881	.04891	20.4465	.688863	.999481	.689381	.310619	12	.20
	49	.04914	.99879	.04920	20.3253	.691438	.999475	.691963	.308037	11	
	50	.04943	.99878	.04949	20.2056	.693998	.999469	.694529	.305471	10	
.85	51	.04972	.99876	.04978	20.0872	.696543	.999463	.697081	.302919	9	.15
	52	.05001	.99875	.05007	19.9702	.699073	.999456	.699617	.300383	8	
	53	.05030	.99873	.05037	19.8546	.701589	.999450	.702139	.297861	7	
.90	54	.05059	.99872	.05066	19.7403	.704090	.999443	.704646	.295354	6	.10
	55	.05088	.99870	.05095	19.6273	.706577	.999437	.707140	.292860	5	
	56	.05117	.99869	.05124	19.5156	.709049	.999431	.709618	.290382	4	
.95	57	.05146	.99867	.05153	19.4051	.711507	.999424	.712083	.287917	3	.05
	58	.05175	.99866	.05182	19.2959	.713952	.999418	.714534	.285466	2	
	59	.05205	.99864	.05212	19.1879	.716383	.999411	.716972	.283028	1	
1.00	60	.05234	.99863	.05241	19.0811	8.718800	9.999404	8.719396	11.280604	0	.00

Decimals	Minutes	Cos	Sin	Cot	Tan	Cos	Sin	Cot	Tan	Minutes	Decimals
		Natural Values				Common Logarithms					

87

Decimals	Minutes	Natural Values Sin	Cos	Tan	Cot	Common Logarithms Sin	Cos	Tan	Cot	Minutes	Decimals
.00	0	.05234	.99863	.05241	19.0811	8.718800	9.999404	8.719396	11.280604	60	1.00
	1	.05263	.99861	.05270	18.9755	.721204	.999398	.721806	.278194	59	
	2	.05292	.99860	.05299	18.8711	.723595	.999391	.724204	.275796	58	
.05	3	.05321	.99858	.05328	18.7678	.725972	.999384	.726588	.273412	57	.95
	4	.05350	.99857	.05357	18.6656	.728337	.999378	.728959	.271041	56	
	5	.05379	.99855	.05387	18.5645	.730688	.999371	.731317	.268683	55	
.10	6	.05408	.99854	.05416	18.4645	.733027	.999364	.733663	.266337	54	.90
	7	.05437	.99852	.05445	18.3655	.735354	.999357	.735996	.264004	53	
	8	.05466	.99851	.05474	18.2677	.737667	.999350	.738317	.261683	52	
.15	9	.05495	.99849	.05503	18.1708	.739969	.999343	.740626	.259374	51	.85
	10	.05524	.99847	.05533	18.0750	.742259	.999336	.742922	.257078	50	
	11	.05553	.99846	.05562	17.9802	.744536	.999329	.745207	.254793	49	
.20	12	.05582	.99844	.05591	17.8863	.746802	.999322	.747479	.252521	48	.80
	13	.05611	.99842	.05620	17.7934	.749055	.999315	.749740	.250260	47	
	14	.05640	.99841	.05649	17.7015	.751297	.999308	.751989	.248011	46	
.25	15	.05669	.99839	.05678	17.6106	.753528	.999301	.754227	.245773	45	.75
	16	.05698	.99838	.05708	17.5205	.755747	.999294	.756453	.243547	44	
	17	.05727	.99836	.05737	17.4314	.757955	.999287	.758668	.241332	43	
.30	18	.05756	.99834	.05766	17.3432	.760151	.999279	.760872	.239128	42	.70
	19	.05785	.99833	.05795	17.2558	.762337	.999272	.763065	.236935	41	
	20	.05814	.99831	.05824	17.1693	.764511	.999265	.765246	.234754	40	
.35	21	.05844	.99829	.05854	17.0837	.766675	.999257	.767417	.232583	39	.65
	22	.05873	.99827	.05883	16.9990	.768828	.999250	.769578	.230422	38	
	23	.05902	.99826	.05912	16.9150	.770970	.999242	.771727	.228273	37	
.40	24	.05931	.99824	.05941	16.8319	.773101	.999235	.773866	.226134	36	.60
	25	.05960	.99822	.05970	16.7496	.775223	.999227	.775995	.224005	35	
	26	.05989	.99821	.05999	16.6681	.777333	.999220	.778114	.221886	34	
.45	27	.06018	.99819	.06029	16.5874	.779434	.999212	.780222	.219778	33	.55
	28	.06047	.99817	.06058	16.5075	.781524	.999205	.782320	.217680	32	
	29	.06076	.99815	.06087	16.4283	.783605	.999197	.784408	.215592	31	
.50	30	.06105	.99813	.06116	16.3499	.785675	.999189	.786486	.213514	30	.50
	31	.06134	.99812	.06145	16.2722	.787736	.999181	.788554	.211446	29	
	32	.06163	.99810	.06175	16.1952	.789787	.999174	.790613	.209387	28	
.55	33	.06192	.99808	.06204	16.1190	.791828	.999166	.792662	.207338	27	.45
	34	.06221	.99806	.06233	16.0435	.793859	.999158	.794701	.205299	26	
	35	.06250	.99804	.06262	15.9687	.795881	.999150	.796731	.203269	25	
.60	36	.06279	.99803	.06291	15.8945	.797894	.999142	.798752	.201248	24	.40
	37	.06308	.99801	.06321	15.8211	.799897	.999134	.800763	.199237	23	
	38	.06337	.99799	.06350	15.7483	.801892	.999126	.802765	.197235	22	
.65	39	.06366	.99797	.06379	15.6762	.803876	.999118	.804758	.195242	21	.35
	40	.06395	.99795	.06408	15.6048	.805852	.999110	.806742	.193258	20	
	41	.06424	.99793	.06437	15.5340	.807819	.999102	.808717	.191283	19	
.70	42	.06453	.99792	.06467	15.4638	.809777	.999094	.810683	.189317	18	.30
	43	.06482	.99790	.06496	15.3943	.811726	.999086	.812641	.187359	17	
	44	.06511	.99788	.06525	15.3254	.813667	.999077	.814589	.185411	16	
.75	45	.06540	.99786	.06554	15.2571	.815599	.999069	.816529	.183471	15	.25
	46	.06569	.99784	.06584	15.1893	.817522	.999061	.818461	.181539	14	
	47	.06598	.99782	.06613	15.1222	.819436	.999053	.820384	.179616	13	
.80	48	.06627	.99780	.06642	15.0557	.821343	.999044	.822298	.177702	12	.20
	49	.06656	.99778	.06671	14.9898	.823240	.999036	.824205	.175795	11	
	50	.06685	.99776	.06700	14.9244	.825130	.999027	.826103	.173897	10	
.85	51	.06714	.99774	.06730	14.8596	.827011	.999019	.827992	.172008	9	.15
	52	.06743	.99772	.06759	14.7954	.828884	.999010	.829874	.170126	8	
	53	.06773	.99770	.06788	14.7317	.830749	.999002	.831748	.168252	7	
.90	54	.06802	.99768	.06817	14.6685	.832607	.998993	.833613	.166387	6	.10
	55	.06831	.99766	.06847	14.6059	.834456	.998984	.835471	.164529	5	
	56	.06860	.99764	.06876	14.5438	.836297	.998976	.837321	.162679	4	
.95	57	.06889	.99762	.06905	14.4823	.838130	.998967	.839163	.160837	3	.05
	58	.06918	.99760	.06934	14.4212	.839956	.998958	.840998	.159002	2	
	59	.06947	.99758	.06963	14.3607	.841774	.998950	.842825	.157175	1	
1.00	60	.06976	.99756	.06993	14.3007	8.843585	9.998941	8.844644	11.155356	0	.00

Decimals	Minutes	Cos	Sin	Cot	Tan	Cos	Sin	Cot	Tan	Minutes	Decimals
		Natural Values				Common Logarithms					

86°

Decimals	Minutes	Natural Values				Common Logarithms				Minutes	Decimals
		Sin	Cos	Tan	Cot	Sin	Cos	Tan	Cot		
.00	0	.06976	.99756	.06993	14.3007	8.843585	9.998941	8.844644	11.155356	60	1.00
	1	.07005	.99754	.07022	14.2411	.845387	.998932	.846455	.153545	59	
	2	.07034	.99752	.07051	14.1821	.847183	.998923	.848260	.151740	58	
.05	3	.07063	.99750	.07080	14.1235	.848971	.998914	.850057	.149943	57	.95
	4	.07092	.99748	.07110	14.0655	.850751	.998905	.851846	.148154	56	
	5	.07121	.99746	.07139	14.0079	.852525	.998896	.853628	.146372	55	
.10	6	.07150	.99744	.07168	13.9507	.854291	.998887	.855403	.144597	54	.90
	7	.07179	.99742	.07197	13.8940	.856049	.998878	.857171	.142829	53	
	8	.07208	.99740	.07227	13.8378	.857801	.998869	.858932	.141068	52	
.15	9	.07237	.99738	.07256	13.7821	.859546	.998860	.860686	.139314	51	.85
	10	.07266	.99736	.07285	13.7267	.861283	.998851	.862433	.137567	50	
	11	.07295	.99734	.07314	13.6719	.863014	.998841	.864173	.135827	49	
.20	12	.07324	.99731	.07344	13.6174	.864738	.998832	.865906	.134094	48	.80
	13	.07353	.99729	.07373	13.5634	.866455	.998823	.867632	.132368	47	
	14	.07382	.99727	.07402	13.5098	.868165	.998813	.869351	.130649	46	
.25	15	.07411	.99725	.07431	13.4566	.869868	.998804	.871064	.128936	45	.75
	16	.07440	.99723	.07461	13.4039	.871565	.998795	.872770	.127230	44	
	17	.07469	.99721	.07490	13.3515	.873255	.998785	.874469	.125531	43	
.30	18	.07498	.99719	.07519	13.2996	.874938	.998776	.876162	.123838	42	.70
	19	.07527	.99716	.07548	13.2480	.876615	.998766	.877849	.122151	41	
	20	.07556	.99714	.07578	13.1969	.878285	.998757	.879529	.120471	40	
.35	21	.07585	.99712	.07607	13.1461	.879949	.998747	.881202	.118798	39	.65
	22	.07614	.99710	.07636	13.0958	.881607	.998738	.882869	.117131	38	
	23	.07643	.99708	.07665	13.0458	.883258	.998728	.884530	.115470	37	
.40	24	.07672	.99705	.07695	12.9962	.884903	.998718	.886185	.113815	36	.60
	25	.07701	.99703	.07724	12.9469	.886542	.998708	.887833	.112167	35	
	26	.07730	.99701	.07753	12.8981	.888174	.998699	.889476	.110524	34	
.45	27	.07759	.99699	.07782	12.8496	.889801	.998689	.891112	.108888	33	.55
	28	.07788	.99696	.07812	12.8014	.891421	.998679	.892742	.107258	32.	
	29	.07817	.99694	.07841	12.7536	.893035	.998669	.894366	.105634	31	
.50	30	.07846	.99692	.07870	12.7062	.894643	.998659	.895984	.104016	30	.50
	31	.07875	.99689	.07899	12.6591	.896246	.998649	.897596	.102404	29	
	32	.07904	.99687	.07929	12.6124	.897842	.998639	.899203	.100797	28	
.55	33	.07933	.99685	.07958	12.5660	.899432	.998629	.900803	.099197	27	.45
	34	.07962	.99683	.07987	12.5199	.901017	.998619	.902398	.097602	26	
	35	.07991	.99680	.08017	12.4742	.902596	.998609	.903987	.096013	25	
.60	36	.08020	.99678	.08046	12.4288	.904169	.998599	.905570	.094430	24	.40
	37	.08049	.99676	.08075	12.3838	.905736	.998589	.907147	.092853	23	
	38	.08078	.99673	.08104	12.3390	.907297	.998578	.908719	.091281	22	
.65	39	.08107	.99671	.08134	12.2946	.908853	.998568	.910285	.089715	21	.35
	40	.08136	.99668	.08163	12.2505	.910404	.998558	.911846	.088154	20	
	41	.08165	.99666	.08192	12.2067	.911949	.998548	.913401	.086599	19	
.70	42	.08194	.99664	.08221	12.1632	.913488	.998537	.914951	.085049	18	.30
	43	.08223	.99661	.08251	12.1201	.915022	.998527	.916495	.083505	17	
	44	.08252	.99659	.08280	12.0772	.916550	.998516	.918034	.081966	16	
.75	45	.08281	.99657	.08309	12.0346	.918073	.998506	.919568	.080432	15	.25
	46	.08310	.99654	.08339	11.9923	.919591	.998495	.921096	.078904	14	
	47	.08339	.99652	.08368	11.9504	.921103	.998485	.922619	.077381	13	
.80	48	.08368	.99649	.08397	11.9087	.922610	.998474	.924136	.075864	12	.20
	49	.08397	.99647	.08427	11.8673	.924112	.998464	.925649	.074351	11	
	50	.08426	.99644	.08456	11.8262	.925609	.998453	.927156	.072844	10	
.85	51	.08455	.99642	.08485	11.7853	.927100	.998442	.928658	.071342	9	.15
	52	.08484	.99639	.08514	11.7448	.928587	.998431	.930155	.069845	8	
	53	.08513	.99637	.08544	11.7045	.930068	.998421	.931647	.068353	7	
.90	54	.08542	.99635	.08573	11.6645	.931544	.998410	.933134	.066866	6	.10
	55	.08571	.99632	.08602	11.6248	.933015	.998399	.934616	.065384	5	
	56	.08600	.99630	.08632	11.5853	.934481	.998388	.936093	.063907	4	
.95	57	.08629	.99627	.08661	11.5461	.935942	.998377	.937565	.062435	3	.05
	58	.08658	.99625	.08690	11.5072	.937398	.998366	.939032	.060968	2	
	59	.08687	.99622	.08720	11.4685	.938850	.998355	.940494	.059506	1	
1.00	60	.08716	.99619	.08749	11.4301	8.940296	9.998344	8.941952	11.058048	0	.00
Decimals	Minutes	Cos	Sin	Cot	Tan	Cos	Sin	Cot	Tan	Minutes	Decimals
		Natural Values				Common Logarithms					

85°

Decimals	Minutes	Natural Values Sin	Cos	Tan	Cot	Common Logarithms Sin	Cos	Tan	Cot	Minutes	Decimals
.00	0	.08716	.99619	.08749	11.4301	8.940296	9.998344	8.941952	11.058048	60	1.00
	1	.08745	.99617	.08778	11.3919	.941738	.998333	.943404	.056596	59	
	2	.08774	.99614	.08807	11.3540	.943174	.998322	.944852	.055148	58	
.05	3	.08803	.99612	.08837	11.3163	.944606	.998311	.946295	.053705	57	.95
	4	.08831	.99609	.08866	11.2789	.946034	.998300	.947734	.052266	56	
	5	.08860	.99607	.08895	11.2417	.947456	.998289	.949168	.050832	55	
.10	6	.08889	.99604	.08925	11.2048	.948874	.998277	.950597	.049403	54	.90
	7	.08918	.99602	.08954	11.1681	.950287	.998266	.952021	.047979	53	
	8	.08947	.99599	.08983	11.1316	.951696	.998255	.953441	.046559	52	
.15	9	.08976	.99596	.09013	11.0954	.953100	.998243	.954856	.045144	51	.85
	10	.09005	.99594	.09042	11.0594	.954499	.998232	.956267	.043733	50	
	11	.09034	.99591	.09071	11.0237	.955894	.998220	.957674	.042326	49	
.20	12	.09063	.99588	.09101	10.9882	.957284	.998209	.959075	.040925	48	.80
	13	.09092	.99586	.09130	10.9529	.958670	.998197	.960473	.039527	47	
	14	.09121	.99583	.09159	10.9178	.960052	.998186	.961866	.038134	46	
.25	15	.09150	.99580	.09189	10.8829	.961429	.998174	.963255	.036745	45	.75
	16	.09179	.99578	.09218	10.8483	.962801	.998163	.964639	.035361	44	
	17	.09208	.99575	.09247	10.8139	.964170	.998151	.966019	.033981	43	
.30	18	.09237	.99572	.09277	10.7797	.965534	.998139	.967394	.032606	42	.70
	19	.09266	.99570	.09306	10.7457	.966893	.998128	.968766	.031234	41	
	20	.09295	.99567	.09335	10.7119	.968249	.998116	.970133	.029867	40	
.35	21	.09324	.99564	.09365	10.6783	.969600	.998104	.971496	.028504	39	.65
	22	.09353	.99562	.09394	10.6450	.970947	.998092	.972855	.027145	38	
	23	.09382	.99559	.09423	10.6118	.972289	.998080	.974209	.025791	37	
.40	24	.09411	.99556	.09453	10.5789	.973628	.998068	.975560	.024440	36	.60
	25	.09440	.99553	.09482	10.5462	.974962	.998056	.976906	.023094	35	
	26	.09469	.99551	.09511	10.5136	.976293	.998044	.978248	.021752	34	
.45	27	.09498	.99548	.09541	10.4813	.977619	.998032	.979586	.020414	33	.55
	28	.09527	.99545	.09570	10.4491	.978941	.998020	.980921	.019079	32	
	29	.09556	.99542	.09600	10.4172	.980259	.998008	.982251	.017749	31	
.50	30	.09585	.99540	.09629	10.3854	.981573	.997996	.983577	.016423	30	.50
	31	.09614	.99537	.09658	10.3538	.982883	.997984	.984899	.015101	29	
	32	.09642	.99534	.09688	10.3224	.984189	.997972	.986217	.013783	28	
.55	33	.09671	.99531	.09717	10.2913	.985491	.997959	.987532	.012468	27	.45
	34	.09700	.99528	.09746	10.2602	.986789	.997947	.988842	.011158	26	
	35	.09729	.99526	.09776	10.2294	.988083	.997935	.990149	.009851	25	
.60	36	.09758	.99523	.09805	10.1988	.989374	.997922	.991451	.008549	24	.40
	37	.09787	.99520	.09834	10.1683	.990660	.997910	.992750	.007250	23	
	38	.09816	.99517	.09864	10.1381	.991943	.997897	.994045	.005955	22	
.65	39	.09845	.99514	.09893	10.1080	.993222	.997885	.995337	.004663	21	.35
	40	.09874	.99511	.09923	10.0780	.994497	.997872	.996624	.003376	20	
	41	.09903	.99508	.09952	10.0483	.995768	.997860	.997908	.002092	19	
.70	42	.09932	.99506	.09981	10.0187	.997036	.997847	.999188	.000812	18	.30
	43	.09961	.99503	.10011	9.98931	.998299	.997835	9.000465	10.999535	17	
	44	.09990	.99500	.10040	9.96007	.999560	.997822	.001738	.998262	16	
.75	45	.10019	.99497	.10069	9.93101	9.000816	.997809	.003007	.996993	15	.25
	46	.10048	.99494	.10099	9.90211	.002069	.997797	.004272	.995728	14	
	47	.10077	.99491	.10128	9.87338	.003318	.997784	.005534	.994466	13	
.80	48	.10106	.99488	.10158	9.84482	.004563	.997771	.006792	.993208	12	.20
	49	.10135	.99485	.10187	9.81641	.005805	.997758	.008047	.991953	11	
	50	.10164	.99482	.10216	9.78817	.007044	.997745	.009298	.990702	10	
.85	51	.10192	.99479	.10246	9.76009	.008273	.997732	.010546	.989454	9	.15
	52	.10221	.99476	.10275	9.73217	.009510	.997719	.011790	.988210	8	
	53	.10250	.99473	.10305	9.70441	.010737	.997706	.013031	.986969	7	
.90	54	.10279	.99470	.10334	9.67680	.011962	.997693	.014268	.985732	6	.10
	55	.10308	.99467	.10363	9.64935	.013182	.997680	.015502	.984498	5	
	56	.10337	.99464	.10393	9.62205	.014400	.997667	.016732	.983268	4	
.95	57	.10366	.99461	.10422	9.59490	.015613	.997654	.017959	.982041	3	.05
	58	.10395	.99458	.10452	9.56791	.016824	.997641	.019183	.980817	2	
	59	.10424	.99455	.10481	9.54106	.018031	.997628	.020403	.979597	1	
1.00	60	.10453	.99452	.10510	9.51436	9.019235	9.997614	9.021620	10.978380	0	.00
Decimals	Minutes	Cos	Sin	Cot	Tan	Cos	Sin	Cot	Tan	Minutes	Decimals
		Natural Values				Common Logarithms					

84°

Decimals	Minutes	Natural Values				Common Logarithms				Minutes	Decimals
		Sin	Cos	Tan	Cot	Sin	Cos	Tan	Cot		
.00	0	.10453	.99452	.10510	9.51436	9.019235	9.997614	9.021620	10.978380	60	1.00
	1	.10482	.99449	.10540	9.48781	.020435	.997601	.022834	.977166	59	
	2	.10511	.99446	.10569	9.46141	.021632	.997588	.024044	.975956	58	
.05	3	.10540	.99443	.10599	9.43515	.022825	.997574	.025251	.974749	57	.95
	4	.10569	.99440	.10628	9.40904	.024016	.997561	.026455	.973545	56	
	5	.10597	.99437	.10657	9.38307	.025203	.997547	.027655	.972345	55	
.10	6	.10626	.99434	.10687	9.35724	.026386	.997534	.028852	.971148	54	.90
	7	.10655	.99431	.10716	9.33155	.027567	.997520	.030046	.969954	53	
	8	.10684	.99428	.10746	9.30599	.028744	.997507	.031237	.968763	52	
.15	9	.10713	.99424	.10775	9.28058	.029918	.997493	.032425	.967575	51	.85
	10	.10742	.99421	.10805	9.25530	.031089	.997480	.033609	.966391	50	
	11	.10771	.99418	.10834	9.23016	.032257	.997466	.034791	.965209	49	
.20	12	.10800	.99415	.10863	9.20516	.033421	.997452	.035969	.964031	48	.80
	13	.10829	.99412	.10893	9.18028	.034582	.997439	.037144	.962856	47	
	14	.10858	.99409	.10922	9.15554	.035741	.997425	.038316	.961684	46	
.25	15	.10887	.99406	.10952	9.13093	.036896	.997411	.039485	.960515	45	.75
	16	.10916	.99402	.10981	9.10646	.038048	.997397	.040651	.959349	44	
	17	.10945	.99399	.11011	9.08211	.039197	.997383	.041813	.958187	43	
.30	18	.10973	.99396	.11040	9.05789	.040342	.997369	.042973	.957027	42	.70
	19	.11002	.99393	.11070	9.03379	.041485	.997355	.044130	.955870	41	
	20	.11031	.99390	.11099	9.00983	.042625	.997341	.045284	.954716	40	
.35	21	.11060	.99386	.11128	8.98598	.043762	.997327	.046434	.953566	39	.65
	22	.11089	.99383	.11158	8.96227	.044895	.997313	.047582	.952418	38	
	23	.11118	.99380	.11187	8.93867	.046026	.997299	.048727	.951273	37	
.40	24	.11147	.99377	.11217	8.91520	.047154	.997285	.049869	.950131	36	.60
	25	.11176	.99374	.11246	8.89185	.048279	.997271	.051008	.948992	35	
	26	.11205	.99370	.11276	8.86862	.049400	.997257	.052144	.947856	34	
.45	27	.11234	.99367	.11305	8.84551	.050519	.997242	.053277	.946723	33	.55
	28	.11263	.99364	.11335	8.82252	.051635	.997228	.054407	.945593	32	
	29	.11291	.99360	.11364	8.79964	.052749	.997214	.055535	.944465	31	
.50	30	.11320	.99357	.11394	8.77689	.053859	.997199	.056659	.943341	30	.50
	31	.11349	.99354	.11423	8.75425	.054966	.997185	.057781	.942219	29	
	32	.11378	.99351	.11452	8.73172	.056071	.997170	.058900	.941100	28	
.55	33	.11407	.99347	.11482	8.70931	.057172	.997156	.060016	.939984	27	.45
	34	.11436	.99344	.11511	8.68701	.058271	.997141	.061130	.938870	26	
	35	.11465	.99341	.11541	8.66482	.059367	.997127	.062240	.937760	25	
.60	36	.11494	.99337	.11570	8.64275	.060460	.997112	.063348	.936652	24	.40
	37	.11523	.99334	.11600	8.62078	.061551	.997098	.064453	.935547	23	
	38	.11552	.99331	.11629	8.59893	.062639	.997083	.065556	.934444	22	
.65	39	.11580	.99327	.11659	8.57718	.063724	.997068	.066655	.933345	21	.35
	40	.11609	.99324	.11688	8.55555	.064806	.997053	.067752	.932248	20	
	41	.11638	.99320	.11718	8.53402	.065885	.997039	.068846	.931154	19	
.70	42	.11667	.99317	.11747	8.51259	.066962	.997024	.069938	.930062	18	.30
	43	.11696	.99314	.11777	8.49128	.068036	.997009	.071027	.928973	17	
	44	.11725	.99310	.11806	8.47007	.069107	.996994	.072113	.927887	16	
.75	45	.11754	.99307	.11836	8.44896	.070176	.996979	.073197	.926893	15	.25
	46	.11783	.99303	.11865	8.42795	.071242	.996964	.074278	.925722	14	
	47	.11812	.99300	.11895	8.40705	.072306	.996949	.075356	.924644	13	
.80	48	.11840	.99297	.11924	8.38625	.073366	.996934	.076432	.923568	12	.20
	49	.11869	.99293	.11954	8.36555	.074424	.996919	.077505	.922495	11	
	50	.11898	.99290	.11983	8.34496	.075480	.996904	.078576	.921424	10	
.85	51	.11927	.99286	.12013	8.32446	.076533	.996889	.079644	.920356	9	.15
	52	.11956	.99283	.12042	8.30406	.077583	.996874	.080710	.919290	8	
	53	.11985	.99279	.12072	8.28376	.078631	.996858	.081773	.918227	7	
.90	54	.12014	.99276	.12101	8.26355	.079676	.996843	.082833	.917167	6	.10
	55	.12043	.99272	.12131	8.24345	.080719	.996828	.083891	.916109	5	
	56	.12071	.99269	.12160	8.22344	.081759	.996812	.084947	.915053	4	
.95	57	.12100	.99265	.12190	8.20352	.082797	.996797	.086000	.914000	3	.05
	58	.12129	.99262	.12219	8.18370	.083832	.996782	.087050	.912950	2	
	59	.12158	.99258	.12249	8.16398	.084864	.996766	.088098	.911902	1	
1.00	60	.12187	.99255	.12278	8.14435	9.085894	9.996751	9.089144	10.910856	0	.00
Decimals	Minutes	Cos	Sin	Cot	Tan	Cos	Sin	Cot	Tan	Minutes	Decimals
		Natural Values				Common Logarithms					

Decimals	Minutes	Natural Values				Common Logarithms				Minutes	Decimals
		Sin	Cos	Tan	Cot	Sin	Cos	Tan	Cot		
.00	0	.12187	.99255	.12278	8.14435	9.085894	9.996751	9.089144	10.910856	60	1.00
	1	.12216	.99251	.12308	8.12481	.086922	.996735	.090187	.909813	59	
	2	.12245	.99248	.12338	8.10536	.087947	.996720	.091228	.908772	58	
.05	3	.12274	.99244	.12367	8.08600	.088970	.996704	.092266	.907734	57	.95
	4	.12302	.99240	.12397	8.06674	.089990	.996688	.093302	.906698	56	
	5	.12331	.99237	.12426	8.04756	.091008	.996673	.094336	.905664	55	
.10	6	.12360	.99233	.12456	8.02848	.092024	.996657	.095367	.904633	54	.90
	7	.12389	.99230	.12485	8.00948	.093037	.996641	.096395	.903605	53	
	8	.12418	.99226	.12515	7.99058	.094047	.996625	.097422	.902578	52	
.15	9	.12447	.99222	.12544	7.97176	.095056	.996610	.098446	.901554	51	.85
	10	.12476	.99219	.12574	7.95302	.096062	.996594	.099468	.900532	50	
	11	.12504	.99215	.12603	7.93438	.097065	.996578	.100487	.899513	49	
.20	12	.12533	.99211	.12633	7.91582	.098066	.996562	.101504	.898496	48	.80
	13	.12562	.99208	.12662	7.89734	.099065	.996546	.102519	.897481	47	
	14	.12591	.99204	.12692	7.87895	.100062	.996530	.103532	.896468	46	
.25	15	.12620	.99200	.12722	7.86064	.101056	.996514	.104542	.895458	45	.75
	16	.12649	.99197	.12751	7.84242	.102048	.996498	.105550	.894450	44	
	17	.12678	.99193	.12781	7.82428	.103037	.996482	.106556	.893444	43	
.30	18	.12706	.99189	.12810	7.80622	.104025	.996465	.107559	.892441	42	.70
	19	.12735	.99186	.12840	7.78825	.105010	.996449	.108560	.891440	41	
	20	.12764	.99182	.12869	7.77035	.105992	.996433	.109559	.890441	40	
.35	21	.12793	.99178	.12899	7.75254	.106973	.996417	.110556	.889444	39	.65
	22	.12822	.99175	.12929	7.73480	.107951	.996400	.111551	.888449	38	
	23	.12851	.99171	.12958	7.71715	.108927	.996384	.112543	.887457	37	
.40	24	.12880	.99167	.12988	7.69957	.109901	.996368	.113533	.886467	36	.60
	25	.12908	.99163	.13017	7.68208	.110873	.996351	.114521	.885479	35	
	26	.12937	.99160	.13047	7.66466	.111842	.996335	.115507	.884493	34	
.45	27	.12966	.99156	.13076	7.64732	.112809	.996318	.116491	.883509	33	.55
	28	.12995	.99152	.13106	7.63005	.113774	.996302	.117472	.882528	32	
	29	.13024	.99148	.13136	7.61287	.114737	.996285	.118452	.881548	31	
.50	30	.13053	.99144	.13165	7.59575	.115698	.996269	.119429	.880571	30	.50
	31	.13081	.99141	.13195	7.57872	.116656	.996252	.120404	.879596	29	
	32	.13110	.99137	.13224	7.56176	.117613	.996235	.121377	.878623	28	
.55	33	.13139	.99133	.13254	7.54487	.118567	.996219	.122348	.877652	27	.45
	34	.13168	.99129	.13284	7.52806	.119519	.996202	.123317	.876683	26	
	35	.13197	.99125	.13313	7.51132	.120469	.996185	.124284	.875716	25	
.60	36	.13226	.99122	.13343	7.49465	.121417	.996168	.125249	.874751	24	.40
	37	.13254	.99118	.13372	7.47806	.122362	.996151	.126211	.873789	23	
	38	.13283	.99114	.13402	7.46154	.123306	.996134	.127172	.872828	22	
.65	39	.13312	.99110	.13432	7.44509	.124248	.996117	.128130	.871870	21	.35
	40	.13341	.99106	.13461	7.42871	.125187	.996100	.129087	.870913	20	
	41	.13370	.99102	.13491	7.41240	.126125	.996083	.130041	.869959	19	
.70	42	.13399	.99098	.13521	7.39616	.127060	.996066	.130994	.869006	18	.30
	43	.13427	.99094	.13550	7.37999	.127993	.996049	.131944	.868056	17	
	44	.13456	.99091	.13580	7.36389	.128925	.996032	.132893	.867107	16	
.75	45	.13485	.99087	.13609	7.34786	.129854	.996015	.133839	.866161	15	.25
	46	.13514	.99083	.13639	7.33190	.130781	.995998	.134784	.865216	14	
	47	.13543	.99079	.13669	7.31600	.131706	.995980	.135726	.864274	13	
.80	48	.13572	.99075	.13698	7.30018	.132630	.995963	.136667	.863333	12	.20
	49	.13600	.99071	.13728	7.28442	.133551	.995946	.137605	.862395	11	
	50	.13629	.99067	.13758	7.26873	.134470	.995928	.138542	.861458	10	
.85	51	.13658	.99063	.13787	7.25310	.135387	.995911	.139476	.860524	9	.15
	52	.13687	.99059	.13817	7.23754	.136303	.995894	.140409	.859591	8	
	53	.13716	.99055	.13846	7.22204	.137216	.995876	.141340	.858660	7	
.90	54	.13744	.99051	.13876	7.20661	.138128	.995859	.142269	.857731	6	.10
	55	.13773	.99047	.13906	7.19125	.139037	.995841	.143196	.856804	5	
	56	.13802	.99043	.13935	7.17594	.139944	.995823	.144121	.855879	4	
.95	57	.13831	.99039	.13965	7.16071	.140850	.995806	.145044	.854956	3	.05
	58	.13860	.99035	.13995	7.14553	.141754	.995788	.145966	.854034	2	
	59	.13889	.99031	.14024	7.13042	.142655	.995771	.146885	.853115	1	
1.00	60	.13917	.99027	.14054	7.11537	9.143555	9.995753	9.147803	10.852197	0	.00
Decimals	Minutes	Cos	Sin	Cot	Tan	Cos	Sin	Cot	Tan	Minutes	Decimals
		Natural Values				Common Logarithms					

Decimals	Minutes	Natural Values				Common Logarithms				Minutes	Decimals
		Sin	Cos	Tan	Cot	Sin	Cos	Tan	Cot		
.00	0	.13917	.99027	.14054	7.11537	9.143555	9.995753	9.147803	10.852197	60	1.00
	1	.13946	.99023	.14084	7.10038	.144453	.995735	.148718	.851282	59	
	2	.13975	.99019	.14113	7.08546	.145349	.995717	.149632	.850368	58	
.05	3	.14004	.99015	.14143	7.07059	.146243	.995699	.150544	.849456	57	.95
	4	.14033	.99011	.14173	7.05579	.147136	.995681	.151454	.848546	56	
	5	.14061	.99006	.14202	7.04105	.148026	.995664	.152363	.847637	55	
.10	6	.14090	.99002	.14232	7.02637	.148915	.995646	.153269	.846731	54	.90
	7	.14119	.98998	.14262	7.01174	.149802	.995628	.154174	.845826	53	
	8	.14148	.98994	.14291	6.99718	.150686	.995610	.155077	.844923	52	
.15	9	.14177	.98990	.14321	6.98268	.151569	.995591	.155978	.844022	51	.85
	10	.14205	.98986	.14351	6.96823	.152451	.995573	.156877	.843123	50	
	11	.14234	.98982	.14381	6.95385	.153330	.995555	.157775	.842225	49	
.20	12	.14263	.98978	.14410	6.93952	.154208	.995537	.158671	.841329	48	.80
	13	.14292	.98973	.14440	6.92525	.155083	.995519	.159565	.840435	47	
	14	.14320	.98969	.14470	6.91104	.155957	.995501	.160457	.839543	46	
.25	15	.14349	.98965	.14499	6.89388	.156830	.995482	.161347	.838653	45	.75
	16	.14378	.98961	.14529	6.88278	.157700	.995464	.162236	.837764	44	
	17	.14407	.98957	.14559	6.86874	.158569	.995446	.163123	.836877	43	
.30	18	.14436	.98953	.14588	6.85475	.159435	.995427	.164008	.835992	42	.70
	19	.14464	.98948	.14618	6.84082	.160301	.995409	.164892	.835108	41	
	20	.14493	.98944	.14648	6.82694	.161164	.995390	.165774	.834226	40	
.35	21	.14522	.98940	.14678	6.81312	.162025	.995372	.166654	.833346	39	.65
	22	.14551	.98936	.14707	6.79936	.162885	.995353	.167532	.832468	38	
	23	.14580	.98931	.14737	6.78564	.163743	.995334	.168409	.831591	37	
.40	24	.14608	.98927	.14767	6.77199	.164600	.995316	.169284	.830716	36	.60
	25	.14637	.98923	.14796	6.75838	.165454	.995297	.170157	.829843	35	
	26	.14666	.98919	.14826	6.74483	.166307	.995278	.171029	.828971	34	
.45	27	.14695	.98914	.14856	6.73133	.167159	.995260	.171899	.828101	33	.55
	28	.14723	.98910	.14886	6.71789	.168008	.995241	.172767	.827233	32	
	29	.14752	.98906	.14915	6.70450	.168856	.995222	.173634	.826366	31	
.50	30	.14781	.98902	.14945	6.69116	.169702	.995203	.174499	.825501	30	.50
	31	.14810	.98897	.14975	6.67787	.170547	.995184	.175362	.824638	29	
	32	.14838	.98893	.15005	6.66462	.171389	.995165	.176224	.823776	28	
.55	33	.14867	.98889	.15034	6.65144	.172230	.995146	.177084	.822916	27	.45
	34	.14896	.98884	.15064	6.63831	.173070	.995127	.177942	.822058	26	
	35	.14925	.98880	.15094	6.62523	.173908	.995108	.178799	.821201	25	
.60	36	.14954	.98876	.15124	6.61219	.174744	.995089	.179655	.820345	24	.40
	37	.14982	.98871	.15153	6.59921	.175578	.995070	.180508	.819492	23	
	38	.15011	.98867	.15183	6.58627	.176411	.995051	.181360	.818640	22	
.65	39	.15040	.98863	.15213	6.57339	.177242	.995032	.182211	.817789	21	.35
	40	.15069	.98858	.15243	6.56055	.178072	.995013	.183059	.816941	20	
	41	.15097	.98854	.15272	6.54777	.178900	.994993	.183907	.816093	19	
.70	42	.15126	.98849	.15302	6.53503	.179726	.994974	.184752	.815248	18	.30
	43	.15155	.98845	.15332	6.52234	.180551	.994955	.185597	.814403	17	
	44	.15184	.98841	.15362	6.50970	.181374	.994935	.186439	.813561	16	
.75	45	.15212	.98836	.15391	6.49710	.182196	.994916	.187280	.812720	15	.25
	46	.15241	.98832	.15421	6.48456	.183016	.994896	.188120	.811880	14	
	47	.15270	.98827	.15451	6.47206	.183834	.994877	.188958	.811042	13	
.80	48	.15299	.98823	.15481	6.45961	.184651	.994857	.189794	.810206	12	.20
	49	.15327	.98818	.15511	6.44720	.185466	.994838	.190629	.809371	11	
	50	.15356	.98814	.15540	6.43484	.186280	.994818	.191462	.808538	10	
.85	51	.15385	.98809	.15570	6.42253	.187092	.994798	.192294	.807706	9	.15
	52	.15414	.98805	.15600	6.41026	.187903	.994779	.193124	.806876	8	
	53	.15442	.98800	.15630	6.39804	.188712	.994759	.193953	.806047	7	
.90	54	.15471	.98796	.15660	6.38587	.189519	.994739	.194780	.805220	6	.10
	55	.15500	.98791	.15689	6.37374	.190325	.994720	.195606	.804394	5	
	56	.15529	.98787	.15719	6.36165	.191130	.994700	.196430	.803570	4	
.95	57	.15557	.98782	.15749	6.34961	.191933	.994680	.197253	.802747	3	.05
	58	.15586	.98778	.15779	6.33761	.192734	.994660	.198074	.801926	2	
	59	.15615	.98773	.15809	6.32566	.193534	.994640	.198894	.801106	1	
1.00	60	.15643	.98769	.15838	6.31375	9.194332	9.994620	9.199713	10.800287	0	.00
Decimals	Minutes	Cos	Sin	Cot	Tan	Cos	Sin	Cot	Tan	Minutes	Decimals
		Natural Values				Common Logarithms					

Decimals	Minutes	Natural Values				Common Logarithms				Minutes	Decimals
		Sin	Cos	Tan	Cot	Sin	Cos	Tan	Cot		
.00	0	.15643	.98769	.15838	6.31375	9.194332	9.994620	9.199713	10.800287	60	1.00
	1	.15672	.98764	.15868	6.30189	.195129	.994600	.200529	.799471	59	
	2	.15701	.98760	.15898	6.29007	.195925	.994580	.201345	.798655	58	
.05	3	.15730	.98755	.15928	6.27829	.196719	.994560	.202159	.797841	57	.95
	4	.15758	.98751	.15958	6.26655	.197511	.994540	.202971	.797029	56	
	5	.15787	.98746	.15988	6.25486	.198302	.994519	.203782	.796218	55	
.10	6	.15816	.98741	.16017	6.24321	.199091	.994499	.204592	.795408	54	.90
	7	.15845	.98737	.16047	6.23160	.199879	.994479	.205400	.794600	53	
	8	.15873	.98732	.16077	6.22003	.200666	.994459	.206207	.793793	52	
.15	9	.15902	.98728	.16107	6.20851	.201451	.994438	.207013	.792987	51	.85
	10	.15931	.98723	.16137	6.19703	.202234	.994418	.207817	.792183	50	
	11	.15959	.98718	.16167	6.18559	.203017	.994398	.208619	.791381	49	
.20	12	.15988	.98714	.16196	6.17419	.203797	.994377	.209420	.790580	48	.80
	13	.16017	.98709	.16226	6.16283	.204577	.994357	.210220	.789780	47	
	14	.16046	.98704	.16256	6.15151	.205354	.994336	.211018	.788982	46	
.25	15	.16074	.98700	.16286	6.14023	.206131	.994316	.211815	.788185	45	.75
	16	.16103	.98695	.16316	6.12899	.206906	.994295	.212611	.787389	44	
	17	.16132	.98690	.16346	6.11779	.207679	.994274	.213405	.786595	43	
.30	18	.16160	.98686	.16376	6.10664	.208452	.994254	.214198	.785802	42	.70
	19	.16189	.98681	.16405	6.09552	.209222	.994233	.214989	.785011	41	
	20	.16218	.98676	.16435	6.08444	.209992	.994212	.215780	.784220	40	
.35	21	.16246	.98671	.16465	6.07340	.210760	.994191	.216568	.783432	39	.65
	22	.16275	.98667	.16495	6.06240	.211526	.994171	.217356	.782644	38	
	23	.16304	.98662	.16525	6.05143	.212291	.994150	.218142	.781858	37	
.40	24	.16333	.98657	.16555	6.04051	.213055	.994129	.218926	.781074	36	.60
	25	.16361	.98652	.16585	6.02962	.213818	.994108	.219710	.780290	35	
	26	.16390	.98648	.16615	6.01878	.214579	.994087	.220492	.779508	34	
.45	27	.16419	.98643	.16645	6.00797	.215338	.994066	.221272	.778728	33	.55
	28	.16447	.98638	.16674	5.99720	.216097	.994045	.222052	.777948	32	
	29	.16476	.98633	.16704	5.98646	.216854	.994024	.222830	.777170	31	
.50	30	.16505	.98629	.16734	5.97576	.217609	.994003	.223607	.776393	30	.50
	31	.16533	.98624	.16764	5.96510	.218363	.993982	.224382	.775618	29	
	32	.16562	.98619	.16794	5.95448	.219116	.993960	.225156	.774844	28	
.55	33	.16591	.98614	.16824	5.94390	.219868	.993939	.225929	.774071	27	.45
	34	.16620	.98609	.16854	5.93335	.220618	.993918	.226700	.773300	26	
	35	.16648	.98604	.16884	5.92283	.221367	.993897	.227471	.772529	25	
.60	36	.16677	.98600	.16914	5.91236	.222115	.993875	.228239	.771761	24	.40
	37	.16706	.98595	.16944	5.90191	.222861	.993854	.229007	.770993	23	
	38	.16734	.98590	.16974	5.89151	.223606	.993832	.229773	.770227	22	
.65	39	.16763	.98585	.17004	5.88114	.224349	.993811	.230539	.769461	21	.35
	40	.16792	.98580	.17033	5.87080	.225092	.993789	.231302	.768698	20	
	41	.16820	.98575	.17063	5.86051	.225833	.993768	.232065	.767935	19	
.70	42	.16849	.98570	.17093	5.85024	.226573	.993746	.232826	.767174	18	.30
	43	.16878	.98565	.17123	5.84001	.227311	.993725	.233586	.766414	17	
	44	.16906	.98561	.17153	5.82982	.228048	.993703	.234345	.765655	16	
.75	45	.16935	.98556	.17183	5.81966	.228784	.993681	.235103	.764897	15	.25
	46	.16964	.98551	.17213	5.80953	.229518	.993660	.235859	.764141	14	
	47	.16992	.98546	.17243	5.79944	.230252	.993638	.236614	.763386	13	
.80	48	.17021	.98541	.17273	5.78938	.230984	.993616	.237368	.762632	12	.20
	49	.17050	.98536	.17303	5.77936	.231715	.993594	.238120	.761880	11	
	50	.17078	.98531	.17333	5.76937	.232444	.993572	.238872	.761128	10	
.85	51	.17107	.98526	.17363	5.75941	.233172	.993550	.239622	.760378	9	.15
	52	.17136	.98521	.17393	5.74949	.233899	.993528	.240371	.759629	8	
	53	.17164	.98516	.17423	5.73960	.234625	.993506	.241118	.758882	7	
.90	54	.17193	.98511	.17453	5.72974	.235349	.993484	.241865	.758135	6	.10
	55	.17222	.98506	.17483	5.71992	.236073	.993462	.242610	.757390	5	
	56	.17250	.98501	.17513	5.71013	.236795	.993440	.243354	.756646	4	
.95	57	.17279	.98496	.17543	5.70037	.237515	.993418	.244097	.755903	3	.05
	58	.17308	.98491	.17553	5.69064	.238235	.993396	.244839	.755161	2	
	59	.17336	.98486	.17603	5.68094	.238953	.993374	.245579	.754421	1	
1.00	60	.17365	.98481	.17633	5.67128	9.239670	9.993351	9.246319	10.753681	0	.00

Decimals	Minutes	Cos	Sin	Cot	Tan	Cos	Sin	Cot	Tan	Minutes	Decimals
		Natural Values				Common Logarithms					

84

80°

Decimals	Minutes	Natural Values				Common Logarithms				Minutes	Decimals
		Sin	Cos	Tan	Cot	Sin	Cos	Tan	Cot		
.00	0	.17365	.98481	.17633	5.67128	9.239670	9.993351	9.246319	10.753681	60	1.00
	1	.17393	.98476	.17663	5.66165	.240386	.993329	.247057	.752943	59	
	2	.17422	.98471	.17693	5.65205	.241101	.993307	.247794	.752206	58	
.05	3	.17451	.98466	.17723	5.64248	.241814	.993284	.248530	.751470	57	.95
	4	.17479	.98461	.17753	5.63295	.242526	.993262	.249264	.750736	56	
	5	.17508	.98455	.17783	5.62344	.243237	.993240	.249998	.750002	55	
.10	6	.17537	.98450	.17813	5.61397	.243947	.993217	.250730	.749270	54	.90
	7	.17565	.98445	.17843	5.60452	.244656	.993195	.251461	.748539	53	
	8	.17594	.98440	.17873	5.59511	.245363	.993172	.252191	.747809	52	
.15	9	.17623	.98435	.17903	5.58573	.246069	.993149	.252920	.747080	51	.85
	10	.17651	.98430	.17933	5.57638	.246775	.993127	.253648	.746352	50	
	11	.17680	.98425	.17963	5.56706	.247478	.993104	.254374	.745626	49	
.20	12	.17708	.98420	.17993	5.55777	.248181	.993081	.255100	.744900	48	.80
	13	.17737	.98414	.18023	5.54851	.248883	.993059	.255824	.744176	47	
	14	.17766	.98409	.18053	5.53927	.249583	.993036	.256547	.743453	46	
.25	15	.17794	.98404	.18083	5.53007	.250282	.993013	.257269	.742731	45	.75
	16	.17823	.98399	.18113	5.52090	.250980	.992990	.257990	.742010	44	
	17	.17852	.98394	.18143	5.51176	.251677	.992967	.258710	.741290	43	
.30	18	.17880	.98389	.18173	5.50264	.252373	.992944	.259429	.740571	42	.70
	19	.17909	.98383	.18203	5.49356	.253067	.992921	.260146	.739854	41	
	20	.17937	.98378	.18233	5.48451	.253761	.992898	.260863	.739137	40	
.35	21	.17966	.98373	.18263	5.47548	.254453	.992875	.261578	.738422	39	.65
	22	.17995	.98368	.18293	5.46648	.255144	.992852	.262292	.737708	38	
	23	.18023	.98362	.18323	5.45751	.255834	.992829	.263005	.736995	37	
.40	24	.18052	.98357	.18353	5.44857	.256523	.992806	.263717	.736283	36	.60
	25	.18081	.98352	.18384	5.43966	.257211	.992783	.264428	.735572	35	
	26	.18109	.98347	.18414	5.43077	.257898	.992759	.265138	.734862	34	
.45	27	.18138	.98341	.18444	5.42192	.258583	.992736	.265847	.734153	33	.55
	28	.18166	.98336	.18474	5.41309	.259268	.992713	.266555	.733445	32	
	29	.18195	.98331	.18504	5.40429	.259951	.992690	.267261	.732739	31	
.50	30	.18224	.98325	.18534	5.39552	.260633	.992666	.267967	.732033	30	.50
	31	.18252	.98320	.18564	5.38677	.261314	.992643	.268671	.731329	29	
	32	.18281	.98315	.18594	5.37805	.261994	.992619	.269375	.730625	28	
.55	33	.18309	.98310	.18624	5.36936	.262673	.992596	.270077	.729923	27	.45
	34	.18338	.98304	.18654	5.36070	.263351	.992572	.270779	.729221	26	
	35	.18367	.98299	.18684	5.35206	.264027	.992549	.271479	.728521	25	
.60	36	.18395	.98294	.18714	5.34345	.264703	.992525	.272178	.727822	24	.40
	37	.18424	.98288	.18745	5.33487	.265377	.992501	.272876	.727124	23	
	38	.18452	.98283	.18775	5.32631	.266051	.992478	.273573	.726427	22	
.65	39	.18481	.98277	.18805	5.31778	.266723	.992454	.274269	.725731	21	.35
	40	.18509	.98272	.18835	5.30928	.267395	.992430	.274964	.725036	20	
	41	.18538	.98267	.18865	5.30080	.268065	.992406	.275658	.724342	19	
.70	42	.18567	.98261	.18895	5.29235	.268734	.992382	.276351	.723649	18	.30
	43	.18595	.98256	.18925	5.28393	.269402	.992359	.277043	.722957	17	
	44	.18624	.98250	.18955	5.27553	.270069	.992335	.277734	.722266	16	
.75	45	.18652	.98245	.18986	5.26715	.270735	.992311	.278424	.721576	15	.25
	46	.18681	.98240	.19016	5.25880	.271400	.992287	.279113	.720887	14	
	47	.18710	.98234	.19046	5.25048	.272064	.992263	.279801	.720199	13	
.80	48	.18738	.98229	.19076	5.24218	.272726	.992239	.280488	.719512	12	.20
	49	.18767	.98223	.19106	5.23391	.273388	.992214	.281174	.718826	11	
	50	.18795	.98218	.19136	5.22566	.274049	.992190	.281858	.718142	10	
.85	51	.18824	.98212	.19166	5.21744	.274708	.992166	.282542	.717458	9	.15
	52	.18852	.98207	.19197	5.20925	.275367	.992142	.283225	.716775	8	
	53	.18881	.98201	.19227	5.20107	.276025	.992118	.283907	.716093	7	
.90	54	.18910	.98196	.19257	5.19293	.276681	.992093	.284588	.715412	6	.10
	55	.18938	.98190	.19287	5.18480	.277337	.992069	.285268	.714732	5	
	56	.18967	.98185	.19317	5.17671	.277991	.992044	.285947	.714053	4	
.95	57	.18995	.98179	.19347	5.16863	.278645	.992020	.286624	.713376	3	.05
	58	.19024	.98174	.19378	5.16058	.279297	.991996	.287301	.712699	2	
	59	.19052	.98168	.19408	5.15256	.279948	.991971	.287977	.712023	1	
1.00	60	.19081	.98163	.19438	5.14455	9.280599	9.991947	9.288652	10.711348	0	.00
Decimals	Minutes	Cos	Sin	Cot	Tan	Cos	Sin	Cot	Tan	Minutes	Decimals
		Natural Values				Common Logarithms					

Decimals	Minutes	Natural Values				Common Logarithms				Minutes	Decimals
		Sin	Cos	Tan	Cot	Sin	Cos	Tan	Cot		
.00	0	.19081	.98163	.19438	5.14455	9.280599	9.991947	9.288652	10.711348	60	1.00
	1	.19109	.98157	.19468	5.13658	.281248	.991922	.289326	.710674	59	
	2	.19138	.98152	.19498	5.12862	.281897	.991897	.289999	.710001	58	
.05	3	.19167	.98146	.19529	5.12069	.282544	.991873	.290671	.709329	57	.95
	4	.19195	.98140	.19559	5.11279	.283190	.991848	.291342	.708658	56	
	5	.19224	.98135	.19589	5.10490	.283836	.991823	.292013	.707987	55	
.10	6	.19252	.98129	.19619	5.09704	.284480	.991799	.292682	.707318	54	.90
	7	.19281	.98124	.19649	5.08921	.285124	.991774	.293350	.706650	53	
	8	.19309	.98118	.19680	5.08139	.285766	.991749	.294017	.705983	52	
.15	9	.19338	.98112	.19710	5.07360	.286408	.991724	£94684	.705316	51	.85
	10	.19366	.98107	.19740	5.06584	.287048	.991699	.295349	.704651	50	
	11	.19395	.98101	.19770	5.05809	.287688	.991674	.296013	.703987	49	
.20	12	.19423	.98096	.19801	5.05037	.288326	.991649	.296677	.703323	48	.80
	13	.19452	.98090	.19831	5.04267	.288964	.991624	.297339	.702661	47	
	14	.19481	.98084	.19861	5.03499	.289600	.991599	.298001	.701999	46	
.25	15	.19509	.98079	.19891	5.02734	.290236	.991574	.298662	.701338	45	.75
	16	.19538	.98073	.19921	5.01971	.290870	.991549	.299322	.700678	44	
	17	.19566	.98067	.19952	5.01210	.291504	.991524	.299980	.700020	43	
.30	18	.19595	.98061	.19982	5.00451	.292137	.991498	.300638	.699362	42	.70
	19	.19623	.98056	.20012	4.99695	.292768	.991473	.301295	.698705	41	
	20	.19652	.98050	.20042	4.98940	.293399	.991448	.301951	.698049	40	
.35	21	.19680	.98044	.20073	4.98188	.294029	.991422	.302607	.697393	39	.65
	22	.19709	.98039	.20103	4.97438	.294658	.991397	.303261	.696739	38	
	23	.19737	.98033	.20133	4.96690	.295286	.991372	.303914	.696086	37	
.40	24	.19766	.98027	.20164	4.95945	.295913	.991346	.304567	.695433	36	.60
	25	.19794	.98021	.20194	4.95201	.296539	.991321	.305218	.694782	35	
	26	.19823	.98016	.20224	4.94460	.297164	.991295	.305869	.694131	34	
.45	27	.19851	.98010	.20254	4.93721	.297788	.991270	.306519	.693481	33	.55
	28	.19880	.98004	.20285	4.92984	.298412	.991244	.307168	.692832	32	
	29	.19908	.97998	.20315	4.92249	.299034	.991218	.307816	.692184	31	
.50	30	.19937	.97992	.20345	4.91516	.299655	.991193	.308463	.691537	30	.50
	31	.19965	.97987	.20376	4.90785	.300276	.991167	.309109	.690891	29	
	32	.19994	.97981	.20406	4.90056	.300895	.991141	.309754	.690246	28	
.55	33	.20022	.97975	.20436	4.89330	.301514	.991115	.310399	.689601	27	.45
	34	.20051	.97969	.20466	4.88605	.302132	.991090	.311042	.688958	26	
	35	.20079	.97963	.20497	4.87882	.302748	.991064	.311685	.688315	25	
.60	36	.20108	.97958	.20527	4.87162	.303364	.991038	.312327	.687673	24	.40
	37	.20136	.97952	.20557	4.86444	.303979	.991012	.312968	.687032	23	
	38	.20165	.97946	.20588	4.85727	.304593	.990986	.313608	.686392	22	
.65	39	.20193	.97940	.20618	4.85013	.305207	.990960	.314247	.685753	21	.35
	40	.20222	.97934	.20648	4.84300	.305819	.990934	.314885	.685115	20	
	41	.20250	.97928	.20679	4.83590	.306430	.990908	.315523	.684477	19	
.70	42	.20279	.97922	.20709	4.82882	.307041	.990882	.316159	.683841	18	.30
	43	.20307	.97916	.20739	4.82175	.307650	.990855	.316795	.683205	17	
	44	.20336	.97910	.20770	4.81471	.308259	.990829	.317430	.682570	16	
.75	45	.20364	.97905	.20800	4.80769	.308867	.990803	.318064	.681936	15	.25
	46	.20393	.97899	.20830	4.80068	.309474	.990777	.318697	.681303	14	
	47	.20421	.97893	.20861	4.79370	.310080	.990750	.319330	.680670	13	
.80	48	.20450	.97887	.20891	4.78673	.310685	.990724	.319961	.680039	12	.20
	49	.20478	.97881	.20921	4.77978	.311289	.990697	.320592	.679408	11	
	50	.20507	.97875	.20952	4.77286	.311893	.990671	.321222	.678778	10	
.85	51	.20535	.97869	.20982	4.76595	.312495	.990645	.321851	.678149	9	.15
	52	.20563	.97863	.21013	4.75906	.313097	.990618	.322479	.677521	8	
	53	.20592	.97857	.21043	4.75219	.313698	.990591	.323106	.676894	7	
.90	54	.20620	.97851	.21073	4.74534	.314297	.990565	.323733	.676267	6	.10
	55	.20649	.97845	.21104	4.73851	.314897	.990538	.324358	.675642	5	
	56	.20677	.97839	.21134	4.73170	.315495	.990511	.324983	.675017	4	
.95	57	.20706	.97833	.21164	4.72490	.316092	.990485	.325607	.674393	3	.05
	58	.20734	.97827	.21195	4.71813	.316689	.990458	.326231	.673769	2	
	59	.20763	.97821	.21225	4.71137	.317284	.990431	.326853	.673147	1	
1.00	60	.20791	.97815	.21256	4.70463	9.317879	9.990404	9.327475	10.672525	0	.00

Decimals	Minutes	Cos	Sin	Cot	Tan	Cos	Sin	Cot	Tan	Minutes	Decimals
		Natural Values				Common Logarithms					

78°

Decimals	Minutes	Natural Values				Common Logarithms				Minutes	Decimals
		Sin	Cos	Tan	Cot	Sin	Cos	Tan	Cot		
.00	0	.20791	.97815	.21256	4.70463	9.317879	9.990404	9.327475	10.672525	60	1.00
	1	.20820	.97809	.21286	4.69791	.318473	.990378	.328095	.671905	59	
	2	.20848	.97803	.21316	4.69121	.319066	.990351	.328715	.671285	58	
.05	3	.20877	.97797	.21347	4.68452	.319658	.990324	.329334	.670666	57	.95
	4	.20905	.97791	.21377	4.67786	.320249	.990297	.329953	.670047	56	
	5	.20933	.97784	.21408	4.67121	.320840	.990270	.330570	.669430	55	
.10	6	.20962	.97778	.21438	4.66458	.321430	.990243	.331187	.668813	54	.90
	7	.20990	.97772	.21469	4.65797	.322019	.990215	.331803	.668197	53	
	8	.21019	.97766	.21499	4.65138	.322607	.990188	.332418	.667582	52	
.15	9	.21047	.97760	.21529	4.64480	.323194	.990161	.333033	.666967	51	.85
	10	.21076	.97754	.21560	4.63825	.323780	.990134	.333646	.666354	50	
	11	.21104	.97748	.21590	4.63171	.324366	.990107	.334259	.665741	49	
.20	12	.21132	.97742	.21621	4.62518	.324950	.990079	.334871	.665129	48	.80
	13	.21161	.97735	.21651	4.61868	.325534	.990052	.335482	.664518	47	
	14	.21189	.97729	.21682	4.61219	.326117	.990025	.336093	.663907	46	
.25	15	.21218	.97723	.21712	4.60572	.326700	.989997	.336702	.663298	45	.75
	16	.21246	.97717	.21743	4.59927	.327281	.989970	.337311	.662689	44	
	17	.21275	.97711	.21773	4.59283	.327862	.989942	.337919	.662081	43	
.30	18	.21303	.97705	.21804	4.58641	.328442	.989915	.338527	.661473	42	.70
	19	.21331	.97698	.21834	4.58001	.329021	.989887	.339133	.660867	41	
	20	.21360	.97692	.21864	4.57363	.329599	.989860	.339739	.660261	40	
.35	21	.21388	.97686	.21895	4.56726	.330176	.989832	.340344	.659656	39	.65
	22	.21417	.97680	.21925	4.56091	.330753	.989804	.340948	.659052	38	
	23	.21445	.97673	.21956	4.55458	.331329	.989777	.341552	.658448	37	
.40	24	.21474	.97667	.21986	4.54826	.331903	.989749	.342155	.657845	36	.60
	25	.21502	.97661	.22017	4.54196	.332478	.989721	.342757	.657243	35	
	26	.21530	.97655	.22047	4.53568	.333051	.989693	.343358	.656642	34	
.45	27	.21559	.97648	.22078	4.52941	.333624	.989665	.343958	.656042	33	.55
	28	.21587	.97642	.22108	4.52316	.334195	.989637	.344558	.655442	32	
	29	.21616	.97636	.22139	4.51693	.334767	.989610	.345157	.654843	31	
.50	30	.21644	.97630	.22169	4.51071	.335337	.989582	.345755	.654245	30	.50
	31	.21672	.97623	.22200	4.50451	.335906	.989553	.346353	.653647	29	
	32	.21701	.97617	.22231	4.49832	.336475	.989525	.346949	.653051	28	
.55	33	.21729	.97611	.22261	4.49215	.337043	.989497	.347545	.652455	27	.45
	34	.21758	.97604	.22292	4.48600	.337610	.989469	.348141	.651859	26	
	35	.21786	.97598	.22322	4.47986	.338176	.989441	.348735	.651265	25	
.60	36	.21814	.97592	.22353	4.47374	.338742	.989413	.349329	.650671	24	.40
	37	.21843	.97585	.22383	4.46764	.339307	.989385	.349922	.650078	23	
	38	.21871	.97579	.22414	4.46155	.339871	.989356	.350514	.649486	22	
.65	39	.21899	.97573	.22444	4.45548	.340434	.989328	.351106	.648894	21	.35
	40	.21928	.97566	.22475	4.44942	.340996	.989300	.351697	.648303	20	
	41	.21956	.97560	.22505	4.44338	.341558	.989271	.352287	.647713	19	
.70	42	.21985	.97553	.22536	4.43735	.342119	.989243	.352876	.647124	18	.30
	43	.22013	.97547	.22567	4.43134	.342679	.989214	.353465	.646535	17	
	44	.22041	.97541	.22597	4.42534	.343239	.989186	.354053	.645947	16	
.75	45	.22070	.97534	.22628	4.41936	.343797	.989157	.354640	.645360	15	.25
	46	.22098	.97528	.22658	4.41340	.344355	.989128	.355227	.644773	14	
	47	.22126	.97521	.22689	4.40745	.344912	.989100	.355813	.644187	13	
.80	48	.22155	.97515	.22719	4.40152	.345469	.989071	.355398	.643602	12	.20
	49	.22183	.97508	.22750	4.39560	.346024	.989042	.356982	.643018	11	
	50	.22212	.97502	.22781	4.38969	.346579	.989014	.357566	.642434	10	
.85	51	.22240	.97496	.22811	4.38381	.347134	.988985	.358149	.641851	9	.15
	52	.22268	.97489	.22842	4.37793	.347687	.988956	.358731	.641269	8	
	53	.22297	.97483	.22872	4.37207	.348240	.988927	.359313	.640687	7	
.90	54	.22325	.97476	.22903	4.36623	.348792	.988898	.359893	.640107	6	.10
	55	.22353	.97470	.22934	4.36040	.349343	.988869	.360474	.639526	5	
	56	.22382	.97463	.22964	4.35459	.349893	.988840	.361053	.638947	4	
.95	57	.22410	.97457	.22995	4.34879	.350443	.988811	.361632	.638368	3	.05
	58	.22438	.97450	.23026	4.34300	.350992	.988782	.362210	.637790	2	
	59	.22467	.97444	.23056	4.33723	.351540	.988753	.362787	.637213	1	
1.00	60	.22495	.97437	.23087	4.33148	9.352088	9.988724	9.363364	10.636636	0	.00
Decimals	Minutes	Cos	Sin	Cot	Tan	Cos	Sin	Cot	Tan	Minutes	Decimals
		Natural Values				Common Logarithms					

87

77°

Decimals	Minutes	Natural Values Sin	Cos	Tan	Cot	Common Logarithms Sin	Cos	Tan	Cot	Minutes	Decimals
.00	0	.22495	.97437	.23087	4.33148	9.352088	9.988724	9.363364	10.636636	60	1.00
	1	.22523	.97430	.23117	4.32573	.352635	.988695	.363940	.636060	59	
	2	.22552	.97424	.23148	4.32001	.353181	.988666	.364515	.635485	58	
.05	3	.22580	.97417	.23179	4.31430	.353726	.988636	.365090	.634910	57	.95
	4	.22608	.97411	.23209	4.30860	.354271	.988607	.365664	.634336	56	
	5	.22637	.97404	.23240	4.30291	.354815	.988578	.366237	.633763	55	
.10	6	.22665	.97398	.23271	4.29724	.355358	.988548	.366810	.633190	54	.90
	7	.22693	.97391	.23301	4.29159	.355901	.988519	.367382	.632618	53	
	8	.22722	.97384	.23332	4.28595	.356443	.988489	.367953	.632047	52	
.15	9	.22750	.97378	.23363	4.28032	.356984	.988460	.368524	.631476	51	.85
	10	.22778	.97371	.23393	4.27471	.357524	.988430	.369094	.630906	50	
	11	.22807	.97365	.23424	4.26911	.358064	.988401	.369663	.630337	49	
.20	12	.22835	.97358	.23455	4.26352	.358603	.988371	.370232	.629768	48	.80
	13	.22863	.97351	.23485	4.25795	.359141	.988342	.370799	.629201	47	
	14	.22892	.97345	.23516	4.25239	.359678	.988312	.371367	.628633	46	
.25	15	.22920	.97338	.23547	4.24685	.360215	.988282	.371933	.628067	45	.75
	16	.22948	.97331	.23578	4.24132	.360752	.988252	.372499	.627501	44	
	17	.22977	.97325	.23608	4.23580	.361287	.988223	.373064	.626936	43	
.30	18	.23005	.97318	.23639	4.23030	.361822	.988193	.373629	.626371	42	.70
	19	.23033	.97311	.23670	4.22481	.362356	.988163	.374193	.625807	41	
	20	.23062	.97304	.23700	4.21933	.362889	.988133	.374756	.625244	40	
.35	21	.23090	.97298	.23731	4.21387	.363422	.988103	.375319	.624681	39	.65
	22	.23118	.97291	.23762	4.20842	.363954	.988073	.375881	.624119	38	
	23	.23146	.97284	.23793	4.20298	.364485	.988043	.376442	.623558	37	
.40	24	.23175	.97278	.23823	4.19756	.365016	.988013	.377003	.622997	36	.60
	25	.23203	.97271	.23854	4.19215	.365546	.987983	.377563	.622437	35	
	26	.23231	.97264	.23885	4.18675	.366075	.987953	.378122	.621878	34	
.45	27	.23260	.97257	.23916	4.18137	.366604	.987922	.378681	.621319	33	.55
	28	.23288	.97251	.23946	4.17600	.367131	.987892	.379239	.620761	32	
	29	.23316	.97244	.23977	4.17064	.367659	.987862	.379797	.620203	31	
.50	30	.23345	.97237	.24008	4.16530	.368185	.987832	.380354	.619646	30	.50
	31	.23373	.97230	.24039	4.15997	.368711	.987801	.380910	.619090	29	
	32	.23401	.97223	.24069	4.15465	.369236	.987771	.381466	.618534	28	
.55	33	.23429	.97217	.24100	4.14934	.369761	.987740	.382020	.617980	27	.45
	34	.23458	.97210	.24131	4.14405	.370285	.987710	.382575	.617425	26	
	35	.23486	.97203	.24162	4.13877	.370808	.987679	.383129	.616871	25	
.60	36	.23514	.97196	.24193	4.13350	.371330	.987649	.383682	.616318	24	.40
	37	.23542	.97189	.24223	4.12825	.371852	.987618	.384234	.615766	23	
	38	.23571	.97182	.24254	4.12301	.372373	.987588	.384786	.615214	22	
.65	39	.23599	.97176	.24285	4.11778	.372894	.987557	.385337	.614663	21	.35
	40	.23627	.97169	.24316	4.11256	.373414	.987526	.385888	.614112	20	
	41	.23656	.97162	.24347	4.10736	.373933	.987496	.386438	.613562	19	
.70	42	.23684	.97155	.24377	4.10216	.374452	.987465	.386987	.613013	18	.30
	43	.23712	.97148	.24408	4.09699	.374970	.987434	.387536	.612464	17	
	44	.23740	.97141	.24439	4.09182	.375487	.987403	.388084	.611916	16	
.75	45	.23769	.97134	.24470	4.08666	.376003	.987372	.388631	.611369	15	.25
	46	.23797	.97127	.24501	4.08152	.376519	.987341	.389178	.610822	14	
	47	.23825	.97120	.24532	4.07639	.377035	.987310	.389724	.610276	13	
.80	48	.23853	.97113	.24562	4.07127	.377549	.987279	.390270	.609730	12	.20
	49	.23882	.97105	.24593	4.06616	.378063	.987248	.390815	.609185	11	
	50	.23910	.97100	.24624	4.06107	.378575	.987217	.391360	.608640	10	
.85	51	.23938	.97093	.24655	4.05599	.379089	.987186	.391903	.608097	9	.15
	52	.23966	.97086	.24686	4.05092	.379601	.987155	.392447	.607553	8	
	53	.23995	.97079	.24717	4.04586	.380113	.987124	.392989	.607011	7	
.90	54	.24023	.97072	.24747	4.04081	.380624	.987092	.393531	.606469	6	.10
	55	.24051	.97065	.24778	4.03578	.381134	.987061	.394073	.605927	5	
	56	.24079	.97058	.24809	4.03076	.381643	.987030	.394614	.605386	4	
.95	57	.24108	.97051	.24840	4.02574	.382152	.986998	.395154	.604846	3	.05
	58	.24136	.97044	.24871	4.02074	.382661	.986967	.395694	.604306	2	
	59	.24164	.97037	.24902	4.01576	.383168	.986936	.396233	.603767	1	
1.00	60	.24192	.97030	.24933	4.01078	9.383675	9.986904	9.396771	10.603229	0	.00
Decimals	Minutes	Cos	Sin	Cot	Tan	Cos	Sin	Cot	Tan	Minutes	Decimals
		Natural Values				Common Logarithms					

76°

Decimals	Minutes	Natural Values				Common Logarithms				Minutes	Decimals
		Sin	Cos	Tan	Cot	Sin	Cos	Tan	Cot		
.00	0	.24192	.97030	.24933	4.01078	9.383675	9.986904	9.396771	10.603229	60	1.00
	1	.24220	.97023	.24964	4.00582	.384182	.986873	.397309	.602691	59	
	2	.24249	.97015	.24995	4.00086	.384687	.986841	.397846	.602154	58	
.05	3	.24277	.97008	.25026	3.99592	.385192	.986809	.398383	.601617	57	.95
	4	.24305	.97001	.25056	3.99099	.385697	.986778	.398919	.601081	56	
	5	.24333	.96994	.25087	3.98607	.386201	.986746	.399455	.600545	55	
.10	6	.24362	.96987	.25118	3.98117	.386704	.986714	.399990	.600010	54	.90
	7	.24390	.96980	.25149	3.97627	.387207	.986683	.400524	.599476	53	
	8	.24418	.96973	.25180	3.97139	.387709	.986651	.401058	.598942	52	
.15	9	.24446	.96966	.25211	3.96651	.388210	.986619	.401591	.598409	51	.85
	10	.24474	.96959	.25242	3.96165	.388711	.986587	.402124	.597876	50	
	11	.24503	.96952	.25273	3.95680	.389211	.986555	.402656	.597344	49	
.20	12	.24531	.96945	.25304	3.95196	.389711	.986523	.403187	.596813	48	.80
	13	.24559	.96937	.25335	3.94713	.390210	.986491	.403718	.596282	47	
	14	.24587	.96930	.25366	3.94232	.390708	.986459	.404249	.595751	46	
.25	15	.24615	.96923	.25397	3.93751	.391206	.986427	.404778	.595222	45	.75
	16	.24644	.96916	.25428	3.93271	.391703	.986395	.405308	.594692	44	
	17	.24672	.96909	.25459	3.92793	.392199	.986363	.405836	.594164	43	
.30	18	.24700	.96902	.25490	3.92316	.392695	.986331	.406364	.593636	42	.70
	19	.24728	.96894	.25521	3.91839	.393191	.986299	.406892	.593108	41	
	20	.24756	.96887	.25552	3.91364	.393685	.986266	.407419	.592581	40	
.35	21	.24784	.96880	.25583	3.90890	.394179	.986234	.407945	.592055	39	.65
	22	.24813	.96873	.25614	3.90417	.394673	.986202	.408471	.591529	38	
	23	.24841	.96866	.25645	3.89945	.395166	.986169	.408996	.591004	37	
.40	24	.24869	.96858	.25676	3.89474	.395658	.986137	.409521	.590479	36	.60
	25	.24897	.96851	.25707	3.89004	.396150	.986104	.410045	.589955	35	
	26	.24925	.96844	.25738	3.88536	.396641	.986072	.410569	.589431	34	
.45	27	.24954	.96837	.25769	3.88068	.397132	.986039	.411092	.588908	33	.55
	28	.24982	.96829	.25800	3.87601	.397621	.986007	.411615	.588385	32	
	29	.25010	.96822	.25831	3.87136	.398111	.985974	.412137	.587863	31	
.50	30	.25038	.96815	.25862	3.86671	.398600	.985942	.412658	.587342	30	.50
	31	.25066	.96807	.25893	3.86208	.399088	.985909	.413179	.586821	29	
	32	.25094	.96800	.25924	3.85745	.399575	.985876	.413699	.586301	28	
.55	33	.25122	.96793	.25955	3.85284	.400062	.985843	.414219	.585781	27	.45
	34	.25151	.96786	.25986	3.84824	.400549	.985811	.414738	.585262	26	
	35	.25179	.96778	.26017	3.84364	.401035	.985778	.415257	.584743	25	
.60	36	.25207	.96771	.26048	3.83906	.401520	.985745	.415775	.584225	24	.40
	37	.25235	.96764	.26079	3.83449	.402005	.985712	.416293	.583707	23	
	38	.25263	.96756	.26110	3.82992	.402489	.985679	.416810	.583190	22	
.65	39	.25291	.96749	.26141	3.82537	.402972	.985646	.417326	.582674	21	.35
	40	.25320	.96742	.26172	3.82083	.403455	.985613	.417842	.582158	20	
	41	.25348	.96734	.26203	3.81630	.403938	.985580	.418358	.581642	19	
.70	42	.25376	.96727	.26235	3.81177	.404420	.985547	.418873	.581127	18	.30
	43	.25404	.96719	.26266	3.80726	.404901	.985514	.419387	.580613	17	
	44	.25432	.96712	.26297	3.80276	.405382	.985480	.419901	.580099	16	
.75	45	.25460	.96705	.26328	3.79827	.405862	.985447	.420415	.579585	15	.25
	46	.25488	.96697	.26359	3.79378	.406341	.985414	.420927	.579073	14	
	47	.25516	.96690	.26390	3.78931	.406820	.985381	.421440	.578560	13	
.80	48	.25545	.96682	.26421	3.78485	.407299	.985347	.421952	.578048	12	.20
	49	.25573	.96675	.26452	3.78040	.407777	.985314	.422463	.577537	11	
	50	.25601	.96667	.26483	3.77595	.408254	.985280	.422974	.577026	10	
.85	51	.25629	.96660	.26515	3.77152	.408731	.985247	.423484	.576516	9	.15
	52	.25657	.96653	.26546	3.76709	.409207	.985213	.423993	.576007	8	
	53	.25685	.96645	.26577	3.76268	.409682	.985180	.424503	.575497	7	
.90	54	.25713	.96638	.26608	3.75828	.410157	.985146	.425011	.574989	6	.10
	55	.25741	.96630	.26639	3.75388	.410632	.985113	.425519	.574481	5	
	56	.25769	.96623	.26670	3.74950	.411106	.985079	.426027	.573973	4	
.95	57	.25798	.96615	.26701	3.74512	.411579	.985045	.426534	.573466	3	.05
	58	.25826	.96608	.26733	3.74075	.412052	.985011	.427041	.572959	2	
	59	.25854	.96600	.26764	3.73640	.412524	.984978	.427547	.572453	1	
1.00	60	.25882	.96593	.26795	3.73205	9.412996	9.984944	9.428052	10.571948	0	.00
Decimals	Minutes	Cos	Sin	Cot	Tan	Cos	Sin	Cot	Tan	Minutes	Decimals
		Natural Values				Common Logarithms					

75°

Decimals	Minutes	Natural Values				Common Logarithms				Minutes	Decimals
		Sin	Cos	Tan	Cot	Sin	Cos	Tan	Cot		
.00	0	.25882	.96593	.26795	3.73205	9.412996	9.984944	9.428052	10.571948	60	1.00
	1	.25910	.96585	.26826	3.72771	.413467	.984910	.428558	.571442	59	
	2	.25938	.96578	.26857	3.72338	.413938	.984876	.429062	.570938	58	
.05	3	.25966	.96570	.26888	3.71907	.414408	.984842	.429566	.570434	57	.95
	4	.25994	.96562	.26920	3.71476	.414878	.984808	.430070	.569930	56	
	5	.26022	.96555	.26951	3.71046	.415347	.984774	.430573	.569427	55	
.10	6	.26050	.96547	.26982	3.70616	.415815	.984740	.431075	.568925	54	.90
	7	.26079	.96540	.27013	3.70188	.416283	.984706	.431577	.568423	53	
	8	.26107	.96532	.27044	3.69761	.416751	.984672	.432079	.567921	52	
.15	9	.26135	.96524	.27076	3.69335	.417217	.984638	.432580	.567420	51	.85
	10	.26163	.96517	.27107	3.68909	.417684	.984603	.433080	.566920	50	
	11	.26191	.96509	.27138	3.68485	.418150	.984569	.433580	.566420	49	
.20	12	.26219	.96502	.27169	3.68061	.418615	.984535	.434080	.565920	48	.80
	13	.26247	.96494	.27201	3.67638	.419079	.984500	.434579	.565421	47	
	14	.26275	.96486	.27232	3.67217	.419544	.984466	.435078	.564922	46	
.25	15	.26303	.96479	.27263	3.66796	.420007	.984432	.435576	.564424	45	.75
	16	.26331	.96471	.27294	3.66376	.420470	.984397	.436073	.563927	44	
	17	.26359	.96463	.27326	3.65957	.420933	.984363	.436570	.563430	43	
.30	18	.26387	.96456	.27357	3.65538	.421395	.984328	.437067	.562933	42	.70
	19	.26415	.96448	.27388	3.65121	.421857	.984294	.437563	.562437	41	
	20	.26443	.96440	.27419	3.64705	.422318	.984259	.438059	.561941	40	
.35	21	.26471	.96433	.27451	3.64289	.422778	.984224	.438554	.561446	39	.65
	22	.26500	.96425	.27482	3.63874	.423238	.984190	.439048	.560952	38	
	23	.26528	.96417	.27513	3.63461	.423697	.984155	.439543	.560457	37	
.40	24	.26556	.96410	.27545	3.63048	.424156	.984120	.440036	.559964	36	.60
	25	.26584	.96402	.27576	3.62636	.424615	.984085	.440529	.559471	35	
	26	.26612	.96394	.27607	3.62224	.425073	.984050	.441022	.558978	34	
.45	27	.26640	.96386	.27638	3.61814	.425530	.984015	.441514	.558486	33	.55
	28	.26668	.96379	.27670	3.61405	.425987	.983981	.442006	.557994	32	
	29	.26696	.96371	.27701	3.60996	.426443	.983946	.442497	.557503	31	
.50	30	.26724	.96363	.27732	3.60588	.426899	.983911	.442988	.557012	30	.50
	31	.26752	.96355	.27764	3.60181	.427354	.983875	.443479	.556521	29	
	32	.26780	.96347	.27795	3.59775	.427809	.983840	.443968	.556032	28	
.55	33	.26808	.96340	.27826	3.59370	.428263	.983805	.444458	.555542	27	.45
	34	.26836	.96332	.27858	3.58966	.428717	.983770	.444947	.555053	26	
	35	.26864	.96324	.27889	3.58562	.429170	.983735	.445435	.554565	25	
.60	36	.26892	.96316	.27921	3.58160	.429623	.983700	.445923	.554077	24	.40
	37	.26920	.96308	.27952	3.57758	.430075	.983664	.446411	.553589	23	
	38	.26948	.96301	.27983	3.57357	.430527	.983629	.446898	.553102	22	
.65	39	.26976	.96293	.28015	3.56957	.430978	.983594	.447384	.552616	21	.35
	40	.27004	.96285	.28046	3.56557	.431429	.983558	.447870	.552130	20	
	41	.27032	.96277	.28077	3.56159	.431879	.983523	.448356	.551644	19	
.70	42	.27060	.96269	.28109	3.55761	.432329	.983487	.448841	.551159	18	.30
	43	.27088	.96261	.28140	3.55364	.432778	.983452	.449326	.550674	17	
	44	.27116	.96253	.28172	3.54968	.433226	.983416	.449810	.550190	16	
.75	45	.27144	.96246	.28203	3.54573	.433675	.983381	.450294	.549706	15	.25
	46	.27172	.96238	.28234	3.54179	.434122	.983345	.450777	.549223	14	
	47	.27200	.96230	.28266	3.53785	.434569	.983309	.451260	.548740	13	
.80	48	.27228	.96222	.28297	3.53393	.435016	.983273	.451743	.548257	12	.20
	49	.27256	.96214	.28329	3.53001	.435462	.983238	.452225	.547775	11	
	50	.27284	.96206	.28360	3.52609	.435908	.983202	.452706	.547294	10	
.85	51	.27312	.96198	.28391	3.52219	.436353	.983166	.453187	.546813	9	.15
	52	.27340	.96190	.28423	3.51829	.436798	.983130	.453668	.546332	8	
	53	.27368	.96182	.28454	3.51441	.437242	.983094	.454148	.545852	7	
.90	54	.27396	.96174	.28486	3.51053	.437686	.983058	.454628	.545372	6	.10
	55	.27424	.96166	.28517	3.50666	.438129	.983022	.455107	.544893	5	
	56	.27452	.96158	.28549	3.50279	.438572	.982986	.455586	.544414	4	
.95	57	.27480	.96150	.28580	3.49894	.439014	.982950	.456064	.543936	3	.05
	58	.27508	.96142	.28612	3.49509	.439456	.982914	.456542	.543458	2	
	59	.27536	.96134	.28643	3.49125	.439897	.982878	.457019	.542981	1	
1.00	60	.27564	.96126	.28675	3.48741	9.440338	9.982842	9.457496	10.542504	0	.00

Decimals	Minutes	Cos	Sin	Cot	Tan	Cos	Sin	Cot	Tan	Minutes	Decimals
		Natural Values				Common Logarithms					

74°

Decimals	Minutes	Natural Values				Common Logarithms				Minutes	Decimals
		Sin	Cos	Tan	Cot	Sin	Cos	Tan	Cot		
.00	0	.27564	.96126	.28675	3.48741	9.440338	9.982842	9.457496	10.542504	60	1.00
	1	.27592	.96118	.28706	3.48359	.440778	.982805	.457973	.542027	59	
	2	.27620	.96110	.28738	3.47977	.441218	.982769	.458449	.541551	58	
.05	3	.27648	.96102	.28769	3.47596	.441658	.982733	.458925	.541075	57	.95
	4	.27676	.96094	.28800	3.47216	.442096	.982696	.459400	.540600	56	
	5	.27704	.96086	.28832	3.46837	.442535	.982660	.459875	.540125	55	
.10	6	.27731	.96078	.28864	3.46458	.442973	.982624	.460349	.539651	54	.90
	7	.27759	.96070	.28895	3.46080	.443410	.982587	.460823	.539177	53	
	8	.27787	.96062	.28927	3.45703	.443847	.982551	.461297	.538703	52	
.15	9	.27815	.96054	.28958	3.45327	.444284	.982514	.461770	.538230	51	.85
	10	.27843	.96046	.28990	3.44951	.444720	.982477	.462242	.537758	50	
	11	.27871	.96037	.29021	3.44576	.445155	.982441	.462715	.537285	49	
.20	12	.27899	.96029	.29053	3.44202	.445590	.982404	.463186	.536814	48	.80
	13	.27927	.96021	.29084	3.43829	.446025	.982367	.463658	.536342	47	
	14	.27955	.96013	.29116	3.43456	.446459	.982331	.464128	.535872	46	
.25	15	.27983	.96005	.29147	3.43084	.446893	.982294	.464599	.535401	45	.75
	16	.28011	.95997	.29179	3.42713	.447326	.982257	.465069	.534931	44	
	17	.28039	.95989	.29210	3.42343	.447759	.982220	.465539	.534461	43	
.30	18	.28067	.95981	.29242	3.41973	.448191	.982183	.466008	.533992	42	.70
	19	.28095	.95972	.29274	3.41604	.448623	.982146	.466477	.533523	41	
	20	.28123	.95964	.29305	3.41236	.449054	.982109	.466945	.533055	40	
.35	21	.28150	.95956	.29337	3.40869	.449485	.982072	.467413	.532587	39	.65
	22	.28178	.95948	.29368	3.40502	.449915	.982035	.467880	.532120	38	
	23	.28206	.95940	.29400	3.40136	.450345	.981998	.468347	.531653	37	
.40	24	.28234	.95931	.29432	3.39771	.450775	.981961	.468814	.531186	36	.60
	25	.28262	.95923	.29463	3.39406	.451204	.981924	.469280	.530720	35	
	26	.28290	.95915	.29495	3.39042	.451632	.981886	.469746	.530254	34	
.45	27	.28318	.95907	.29526	3.38679	.452060	.981849	.470211	.529789	33	.55
	28	.28346	.95898	.29558	3.38317	.452488	.981812	.470676	.529324	32	
	29	.28374	.95890	.29590	3.37955	.452915	.981774	.471141	.528859	31	
.50	30	.28402	.95882	.29621	3.37594	.453342	.981737	.471605	.528395	30	.50
	31	.28429	.95874	.29653	3.37234	.453768	.981700	.472069	.527931	29	
	32	.28457	.95865	.29685	3.36875	.454194	.981662	.472532	.527468	28	
.55	33	.28485	.95857	.29716	3.36516	.454619	.981625	.472995	.527005	27	.45
	34	.28513	.95849	.29748	3.36158	.455044	.981587	.473457	.526543	26	
	35	.28541	.95841	.29780	3.35800	.455469	.981549	.473919	.526081	25	
.60	36	.28569	.95832	.29811	3.35443	.455893	.981512	.474381	.525619	24	.40
	37	.28597	.95824	.29843	3.35087	.456316	.981474	.474842	.525158	23	
	38	.28625	.95816	.29875	3.34732	.456739	.981436	.475303	.524697	22	
.65	39	.28652	.95807	.29906	3.34377	.457162	.981399	.475763	.524237	21	.35
	40	.28680	.95799	.29938	3.34023	.457584	.981361	.476223	.523777	20	
	41	.28708	.95791	.29970	3.33670	.458006	.981323	.476683	.523317	19	
.70	42	.28736	.95782	.30001	3.33317	.458427	.981285	.477142	.522858	18	.30
	43	.28764	.95774	.30033	3.32965	.458848	.981247	.477601	.522399	17	
	44	.28792	.95766	.30065	3.32614	.459268	.981209	.478059	.521941	16	
.75	45	.28820	.95757	.30097	3.32264	.459688	.981171	.478517	.521483	15	.25
	46	.28847	.95749	.30128	3.31914	.460108	.981133	.478975	.521025	14	
	47	.28875	.95740	.30160	3.31565	.460527	.981095	.479432	.520568	13	
.80	48	.28903	.95732	.30192	3.31216	.460946	.981057	.479889	.520111	12	.20
	49	.28931	.95724	.30224	3.30868	.461364	.981019	.480345	.519655	11	
	50	.28959	.95715	.30255	3.30521	.461782	.980981	.480801	.519199	10	
.85	51	.28987	.95707	.30287	3.30174	.462199	.980942	.481257	.518743	9	.15
	52	.29015	.95698	.30319	3.29829	.462616	.980904	.481712	.518288	8	
	53	.29042	.95690	.30351	3.29483	.463032	.980866	.482167	.517833	7	
.90	54	.29070	.95681	.30382	3.29139	.463448	.980827	.482621	.517379	6	.10
	55	.29098	.95673	.30414	3.28795	.463864	.980789	.483075	.516925	5	
	56	.29126	.95664	.30446	3.28452	.464279	.980750	.483529	.516471	4	
.95	57	.29154	.95656	.30478	3.28109	.464694	.980712	.483982	.516018	3	.05
	58	.29182	.95647	.30509	3.27767	.465108	.980673	.484435	.515565	2	
	59	.29209	.95639	.30541	3.27426	.465522	.980635	.484887	.515113	1	
1.00	60	.29237	.95630	.30573	3.27085	9.465935	9.980596	9.485339	10.514661	0	.00
Decimals	Minutes	Cos	Sin	Cot	Tan	Cos	Sin	Cot	Tan	Minutes	Decimals
		Natural Values				Common Logarithms					

91

Decimals	Minutes	Natural Values				Common Logarithms				Minutes	Decimals
		Sin	Cos	Tan	Cot	Sin	Cos	Tan	Cot		
.00	0	.29237	.95630	.30573	3.27085	9.465935	9.980596	9.485339	10.514661	60	1.00
	1	.29265	.95622	.30605	3.26745	.466348	.980558	.485791	.514209	59	
	2	.29293	.95613	.30637	3.26406	.466761	.980519	.486242	.513758	58	
.05	3	.29321	.95605	.30669	3.26067	.467173	.980480	.486693	.513307	57	.95
	4	.29348	.95596	.30700	3.25729	.467585	.980442	.487143	.512857	56	
	5	.29376	.95588	.30732	3.25392	.467996	.980403	.487593	.512407	55	
.10	6	.29404	.95579	.30764	3.25055	.468407	.980364	.488043	.511957	54	.90
	7	.29432	.95571	.30796	3.24719	.468817	.980325	.488492	.511508	53	
	8	.29460	.95562	.30828	3.24383	.469227	.980286	.488941	.511059	52	
.15	9	.29487	.95554	.30860	3.24049	.469637	.980247	.489390	.510610	51	.85
	10	.29515	.95545	.30891	3.23714	.470046	.980208	.489838	.510162	50	
	11	.29543	.95536	.30923	3.23381	.470455	.980169	.490286	.509714	49	
.20	12	.29571	.95528	.30955	3.23048	.470863	.980130	.490733	.509267	48	.80
	13	.29599	.95519	.30987	3.22715	.471271	.980091	.491180	.508820	47	
	14	.29626	.95511	.31019	3.22384	.471679	.980052	.491627	.508373	46	
.25	15	.29654	.95502	.31051	3.22053	.472086	.980012	.492073	.507927	45	.75
	16	.29682	.95493	.31083	3.21722	.472492	.979973	.492519	.507481	44	
	17	.29710	.95485	.31115	3.21392	.472898	.979934	.492965	.507035	43	
.30	18	.29737	.95476	.31147	3.21063	.473304	.979895	.493410	.506590	42	.70
	19	.29765	.95467	.31178	3.20734	.473710	.979855	.493854	.506146	41	
	20	.29793	.95459	.31210	3.20406	.474115	.979816	.494299	.505701	40	
.35	21	.29821	.95450	.31242	3.20079	.474519	.979776	.494743	.505257	39	.65
	22	.29849	.95441	.31274	3.19752	.474923	.979737	.495186	.504814	38	
	23	.29876	.95433	.31306	3.19426	.475327	.979697	.495630	.504370	37	
.40	24	.29904	.95424	.31338	3.19100	.475730	.979658	.496073	.503927	36	.60
	25	.29932	.95415	.31370	3.18775	.476133	.979618	.496515	.503485	35	
	26	.29960	.95407	.31402	3.18451	.476536	.979579	.496957	.503043	34	
.45	27	.29987	.95398	.31434	3.18127	.476938	.979539	.497399	.502601	33	.55
	28	.30015	.95389	.31466	3.17804	.477340	.979499	.497841	.502159	32	
	29	.30043	.95380	.31498	3.17481	.477741	.979459	.498282	.501718	31	
.50	30	.30071	.95372	.31530	3.17159	.478142	.979420	.498722	.501278	30	.50
	31	.30098	.95363	.31562	3.16838	.478542	.979380	.499163	.500837	29	
	32	.30126	.95354	.31594	3.16517	.478942	.979340	.499603	.500397	28	
.55	33	.30154	.95345	.31626	3.16197	.479342	.979300	.500042	.499958	27	.45
	34	.30182	.95337	.31658	3.15877	.479741	.979260	.500481	.499519	26	
	35	.30209	.95328	.31690	3.15558	.480140	.979220	.500920	.499080	25	
.60	36	.30237	.95319	.31722	3.15240	.430539	.979180	.501359	.498641	24	.40
	37	.30265	.95310	.31754	3.14922	.480937	.979140	.501797	.498203	23	
	38	.30292	.95301	.31786	3.14605	.481334	.979100	.502235	.497765	22	
.65	39	.30320	.95293	.31818	3.14288	.481731	.979059	.502672	.497328	21	.35
	40	.30348	.95284	.31850	3.13972	.482128	.979019	.503109	.496891	20	
	41	.30376	.95275	.31882	3.13656	.482525	.978979	.503546	.496454	19	
.70	42	.30403	.95266	.31914	3.13341	.482921	.978939	.503982	.496018	18	.30
	43	.30431	.95257	.31946	3.13027	.483316	.978898	.504418	.495582	17	
	44	.30459	.95248	.31978	3.12713	.483712	.978858	.504854	.495146	16	
.75	45	.30486	.95240	.32010	3.12400	.484107	.978817	.505289	.494711	15	.25
	46	.30514	.95231	.32042	3.12087	.484501	.978777	.505724	.494276	14	
	47	.30542	.95222	.32074	3.11775	.484895	.978737	.506159	.493841	13	
.80	48	.30570	.95213	.32106	3.11464	.485289	.978696	.506593	.493407	12	.20
	49	.30597	.95204	.32139	3.11153	.485682	.978655	.507027	.492973	11	
	50	.30625	.95195	.32171	3.10842	.486075	.978615	.507460	.492540	10	
.85	51	.30653	.95186	.32203	3.10532	.486467	.978574	.507893	.492107	9	.15
	52	.30680	.95177	.32235	3.10223	.486860	.978533	.508326	.491674	8	
	53	.30708	.95168	.32267	3.09914	.487251	.978493	.508759	.491241	7	
.90	54	.30736	.95159	.32299	3.09606	.487643	.978452	.509191	.490809	6	.10
	55	.30763	.95150	.32331	3.09298	.488034	.978411	.509622	.490378	5	
	56	.30791	.95142	.32363	3.08991	.488424	.978370	.510054	.489946	4	
.95	57	.30819	.95133	.32396	3.08685	.488814	.978329	.510485	.489515	3	.05
	58	.30846	.95124	.32428	3.08379	.489204	.978288	.510916	.489084	2	
	59	.30874	.95115	.32460	3.08073	.489593	.978247	.511346	.488654	1	
1.00	60	.30902	.95106	.32492	3.07768	9.489982	9.978206	9.511776	10.488224	0	.00

Decimals	Minutes	Cos	Sin	Cot	Tan	Cos	Sin	Cot	Tan	Minutes	Decimals
			Natural Values				Common Logarithms				

72°

Decimals	Minutes	Natural Values				Common Logarithms				Minutes	Decimals
		Sin	Cos	Tan	Cot	Sin	Cos	Tan	Cot		
.00	0	.30902	.95106	.32492	3.07768	9.489982	9.978206	9.511776	10.488224	60	1.00
	1	.30929	.95097	.32524	3.07464	.490371	.978165	.512206	.487794	59	
	2	.30957	.95088	.32556	3.07160	.490759	.978124	.512635	.487365	58	
.05	3	.30985	.95079	.32588	3.06857	.491147	.978083	.513064	.486936	57	.95
	4	.31012	.95070	.32621	3.06554	.491535	.978042	.513493	.486507	56	
	5	.31040	.95061	.32653	3.06252	.491922	.978001	.513921	.486079	55	
.10	6	.31068	.95052	.32685	3.05950	.492308	.977959	.514349	.485651	54	.90
	7	.31095	.95043	.32717	3.05649	.492695	.977918	.514777	.485223	53	
	8	.31123	.95033	.32749	3.05349	.493081	.977877	.515204	.484796	52	
.15	9	.31151	.95024	.32782	3.05049	.493466	.977835	.515631	.484369	51	.85
	10	.31178	.95015	.32814	3.04749	.493851	.977794	.516057	.483943	50	
	11	.31206	.95006	.32846	3.04450	.494236	.977752	.516484	.483516	49	
.20	12	.31233	.94997	.32878	3.04152	.494621	.977711	.516910	.483090	48	.80
	13	.31261	.94988	.32911	3.03854	.495005	.977669	.517335	.482665	47	
	14	.31289	.94979	.32943	3.03556	.495388	.977628	.517761	.482239	46	
.25	15	.31316	.94970	.32975	3.03260	.495772	.977586	.518186	.481814	45	.75
	16	.31344	.94961	.33007	3.02963	.496154	.977544	.518610	.481390	44	
	17	.31372	.94952	.33040	3.02667	.496537	.977503	.519034	.480966	43	
.30	18	.31399	.94943	.33072	3.02372	.496919	.977461	.519458	.480542	42	.70
	19	.31427	.94933	.33104	3.02077	.497301	.977419	.519882	.480118	41	
	20	.31454	.94924	.33136	3.01783	.497682	.977377	.520305	.479695	40	
.35	21	.31482	.94915	.33169	3.01489	.498064	.977335	.520728	.479272	39	.65
	22	.31510	.94906	.33201	3.01196	.498444	.977293	.521151	.478849	38	
	23	.31537	.94897	.33233	3.00903	.498825	.977251	.521573	.478427	37	
.40	24	.31565	.94888	.33266	3.00611	.499204	.977209	.521995	.478005	36	.60
	25	.31593	.94878	.33298	3.00319	.499584	.977167	.522417	.477583	35	
	26	.31620	.94869	.33330	3.00028	.499963	.977125	.522838	.477162	34	
.45	27	.31648	.94860	.33363	2.99738	.500342	.977083	.523259	.476741	33	.55
	28	.31675	.94851	.33395	2.99447	.500721	.977041	.523680	.476320	32	
	29	.31703	.94842	.33427	2.99158	.501099	.976999	.524100	.475900	31	
.50	30	.31730	.94832	.33460	2.98868	.501476	.976957	.524520	.475480	30	.50
	31	.31758	.94823	.33492	2.98580	.501854	.976914	.524940	.475060	29	
	32	.31786	.94814	.33524	2.98292	.502231	.976872	.525359	.474641	28	
.55	33	.31813	.94805	.33557	2.98004	.502607	.976830	.525778	.474222	27	.45
	34	.31841	.94795	.33589	2.97717	.502984	.976787	.526197	.473803	26	
	35	.31868	.94786	.33621	2.97430	.503360	.976745	.526615	.473385	25	
.60	36	.31896	.94777	.33654	2.97144	.503735	.976702	.527033	.472967	24	.40
	37	.31923	.94768	.33686	2.96858	.504110	.976660	.527451	.472549	23	
	38	.31951	.94758	.33718	2.96573	.504485	.976617	.527868	.472132	22	
.65	39	.31979	.94749	.33751	2.96288	.504860	.976574	.528285	.471715	21	.35
	40	.32006	.94740	.33783	2.96004	.505234	.976532	.528702	.471298	20	
	41	.32034	.94730	.33816	2.95721	.505608	.976489	.529119	.470881	19	
.70	42	.32061	.94721	.33848	2.95437	.505981	.976446	.529535	.470465	18	.30
	43	.32089	.94712	.33881	2.95155	.506354	.976404	.529951	.470049	17	
	44	.32116	.94702	.33913	2.94872	.506727	.976361	.530366	.469634	16	
.75	45	.32144	.94693	.33945	2.94591	.507099	.976318	.530781	.469219	15	.25
	46	.32171	.94684	.33978	2.94309	.507471	.976275	.531196	.468804	14	
	47	.32199	.94674	.34010	2.94028	.507843	.976232	.531611	.468389	13	
.80	48	.32227	.94665	.34043	2.93748	.508214	.976189	.532025	.467975	12	.20
	49	.32254	.94656	.34075	2.93468	.508585	.976146	.532439	.467561	11	
	50	.32282	.94646	.34108	2.93189	.508956	.976103	.532853	.467147	10	
.85	51	.32309	.94637	.34140	2.92910	.509326	.976060	.533266	.466734	9	.15
	52	.32337	.94627	.34173	2.92632	.509696	.976017	.533679	.466321	8	
	53	.32364	.94618	.34205	2.92354	.510065	.975974	.534092	.465908	7	
.90	54	.32392	.94609	.34238	2.92076	.510434	.975930	.534504	.465496	6	.10
	55	.32419	.94599	.34270	2.91799	.510803	.975887	.534916	.465084	5	
	56	.32447	.94590	.34303	2.91523	.511172	.975844	.535328	.464672	4	
.95	57	.32474	.94580	.34335	2.91246	.511540	.975800	.535739	.464261	3	.05
	58	.32502	.94571	.34368	2.90971	.511907	.975757	.536150	.463850	2	
	59	.32529	.94561	.34400	2.90696	.512275	.975714	.536561	.463439	1	
1.00	60	.32557	.94552	.34433	2.90421	9.512642	9.975670	9.536972	10.463028	0	.00
Decimals	Minutes	Cos	Sin	Cot	Tan	Cos	Sin	Cot	Tan	Minutes	Decimals
		Natural Values				Common Logarithms					

71°

Decimals	Minutes	Natural Values				Common Logarithms				Minutes	Decimals
		Sin	Cos	Tan	Cot	Sin	Cos	Tan	Cot		
.00	0	.32557	.94552	.34433	2.90421	9.512642	9.975670	9.536972	10.463028	60	1.00
	1	.32584	.94542	.34465	2.90147	.513009	.975627	.537382	.462618	59	
	2	.32612	.94533	.34498	2.89873	.513375	.975583	.537792	.462208	58	
.05	3	.32639	.94523	.34530	2.89600	.513741	.975539	.538202	.461798	57	.95
	4	.32667	.94514	.34563	2.89327	.514107	.975496	.538611	.461389	56	
	5	.32694	.94504	.34596	2.89055	.514472	.975452	.539020	.460980	55	
.10	6	.32722	.94495	.34628	2.88783	.514837	.975408	.539429	.460571	54	.90
	7	.32749	.94485	.34661	2.88511	.515202	.975365	.539837	.460163	53	
	8	.32777	.94476	.34693	2.88240	.515566	.975321	.540245	.459755	52	
.15	9	.32804	.94466	.34726	2.87970	.515930	.975277	.540653	.459347	51	.85
	10	.32832	.94457	.34758	2.87700	.516294	.975233	.541061	.458939	50	
	11	.32859	.94447	.34791	2.87430	.516657	.975189	.541468	.458532	49	
.20	12	.32887	.94438	.34824	2.87161	.517020	.975145	.541875	.458125	48	.80
	13	.32914	.94428	.34856	2.86892	.517382	.975101	.542281	.457719	47	
	14	.32942	.94418	.34889	2.86624	.517745	.975057	.542688	.457312	46	
.25	15	.32969	.94409	.34922	2.86356	.518107	.975013	.543094	.456906	45	.75
	16	.32997	.94399	.34954	2.86089	.518468	.974969	.543499	.456501	44	
	17	.33024	.94390	.34987	2.85822	.518829	.974925	.543905	.456095	43	
.30	18	.33051	.94380	.35020	2.85555	.519190	.974880	.544310	.455690	42	.70
	19	.33079	.94370	.35052	2.85289	.519551	.974836	.544715	.455285	41	
	20	.33106	.94361	.35085	2.85023	.519911	.974792	.545119	.454881	40	
.35	21	.33134	.94351	.35118	2.84758	.520271	.974748	.545524	.454476	39	.65
	22	.33161	.94342	.35150	2.84494	.520631	.974703	.545928	.454072	38	
	23	.33189	.94332	.35183	2.84229	.520990	.974659	.546331	.453669	37	
.40	24	.33216	.94322	.35216	2.83965	.521349	.974614	.546735	.453265	36	.60
	25	.33244	.94313	.35248	2.83702	.521707	.974570	.547138	.452862	35	
	26	.33271	.94303	.35281	2.83439	.522066	.974525	.547540	.452460	34	
.45	27	.33298	.94293	.35314	2.83176	.522424	.974481	.547943	.452057	33	.55
	28	.33326	.94284	.35346	2.82914	.522781	.974436	.548345	.451655	32	
	29	.33353	.94274	.35379	2.82653	.523138	.974391	.548747	.451253	31	
.50	30	.33381	.94264	.35412	2.82391	.523495	.974347	.549149	.450851	30	.50
	31	.33408	.94254	.35445	2.82130	.523852	.974302	.549550	.450450	29	
	32	.33436	.94245	.35477	2.81870	.524208	.974257	.549951	.450049	28	
.55	33	.33463	.94235	.35510	2.81610	.524564	.974212	.550352	.449648	27	.45
	34	.33490	.94225	.35543	2.81350	.524920	.974167	.550752	.449248	26	
	35	.33518	.94215	.35576	2.81091	.525275	.974122	.551153	.448847	25	
.60	36	.33545	.94206	.35608	2.80833	.525630	.974077	.551552	.448448	24	.40
	37	.33573	.94196	.35641	2.80574	.525984	.974032	.551952	.448048	23	
	38	.33600	.94186	.35674	2.80316	.526339	.973987	.552351	.447649	22	
.65	39	.33627	.94176	.35707	2.80059	.526693	.973942	.552750	.447250	21	.35
	40	.33655	.94167	.35740	2.79802	.527046	.973897	.553149	.446851	20	
	41	.33682	.94157	.35772	2.79545	.527400	.973852	.553548	.446452	19	
.70	42	.33710	.94147	.35805	2.79289	.527753	.973807	.553946	.446054	18	.30
	43	.33737	.94137	.35838	2.79033	.528105	.973761	.554344	.445656	17	
	44	.33764	.94127	.35871	2.78778	.528458	.973716	.554741	.445259	16	
.75	45	.33792	.94118	.35904	2.78523	.528810	.973671	.555139	.444861	15	.25
	46	.33819	.94108	.35937	2.78269	.529161	.973625	.555536	.444464	14	
	47	.33846	.94098	.35969	2.78014	.529513	.973580	.555933	.444067	13	
.80	48	.33874	.94088	.36002	2.77761	.529864	.973535	.556329	.443671	12	.20
	49	.33901	.94078	.36035	2.77507	.530215	.973489	.556725	.443275	11	
	50	.33929	.94068	.36068	2.77254	.530565	.973444	.557121	.442879	10	
.85	51	.33956	.94058	.36101	2.77002	.530915	.973398	.557517	.442483	9	.15
	52	.33983	.94049	.36134	2.76750	.531265	.973352	.557913	.442087	8	
	53	.34011	.94039	.36167	2.76498	.531614	.973307	.558308	.441692	7	
.90	54	.34038	.94029	.36199	2.76247	.531963	.973261	.558703	.441297	6	.10
	55	.34065	.94019	.36232	2.75996	.532312	.973215	.559097	.440903	5	
	56	.34093	.94009	.36265	2.75746	.532661	.973169	.559491	.440509	4	
.95	57	.34120	.93999	.36298	2.75496	.533009	.973124	.559885	.440115	3	.05
	58	.34147	.93989	.36331	2.75246	.533357	.973078	.560279	.439721	2	
	59	.34175	.93979	.36364	2.74997	.533704	.973032	.560673	.439327	1	
1.00	60	.34202	.93969	.36397	2.74748	9.534052	9.972986	9.561066	10.438934	0	.00

Decimals	Minutes	Cos	Sin	Cot	Tan	Cos	Sin	Cot	Tan	Minutes	Decimals
		Natural Values				Common Logarithms					

Decimals	Minutes	Natural Values				Common Logarithms				Minutes	Decimals
		Sin	Cos	Tan	Cot	Sin	Cos	Tan	Cot		
.00	0	.34202	.93969	.36397	2.74748	9.534052	9.972986	9.561066	10.438934	60	1.00
	1	.34229	.93959	.36430	2.74499	.534399	.972940	.561459	.438541	59	
	2	.34257	.93949	.36463	2.74251	.534745	.972894	.561851	.438149	58	
.05	3	.34284	.93939	.36496	2.74004	.535092	.972848	.562244	.437756	57	.95
	4	.34311	.93929	.36529	2.73756	.535438	.972802	.562636	.437364	56	
	5	.34339	.93919	.36562	2.73509	.535783	.972755	.563028	.436972	55	
.10	6	.34366	.93909	.36595	2.73263	.536129	.972709	.563419	.436581	54	.90
	7	.34393	.93899	.36628	2.73017	.536474	.972663	.563811	.436189	53	
	8	.34421	.93889	.36661	2.72771	.536818	.972617	.564202	.435798	52	
.15	9	.34448	.93879	.36694	2.72526	.537163	.972570	.564593	.435407	51	.85
	10	.34475	.93869	.36727	2.72281	.537507	.972524	.564983	.435017	50	
	11	.34503	.93859	.36760	2.72036	.537851	.972478	.565373	.434627	49	
.20	12	.34530	.93849	.36793	2.71792	.538194	.972431	.565763	.434237	48	.80
	13	.34557	.93839	.36826	2.71548	.538538	.972385	.566153	.433847	47	
	14	.34584	.93829	.36859	2.71305	.538880	.972338	.566542	.433458	46	
.25	15	.34612	.93819	.36892	2.71062	.539223	.972291	.566932	.433068	45	.75
	16	.34639	.93809	.36925	2.70819	.539565	.972245	.567320	.432680	44	
	17	.34666	.93799	.36958	2.70577	.539907	.972198	.567709	.432291	43	
.30	18	.34694	.93789	.36991	2.70335	.540249	.972151	.568098	.431902	42	.70
	19	.34721	.93779	.37024	2.70094	.540590	.972105	.568486	.431514	41	
	20	.34748	.93769	.37057	2.69853	.540931	.972058	.568873	.431127	40	
.35	21	.34775	.93759	.37090	2.69612	.541272	.972011	.569261	.430739	39	.65
	22	.34803	.93748	.37123	2.69371	.541613	.971964	.569648	.430352	38	
	23	.34830	.93738	.37157	2.69131	.541953	.971917	.570035	.429965	37	
.40	24	.34857	.93728	.37190	2.68892	.542293	.971870	.570422	.429578	36	.60
	25	.34884	.93718	.37223	2.68653	.542632	.971823	.570809	.429191	35	
	26	.34912	.93708	.37256	2.68414	.542971	.971776	.571195	.428805	34	
.45	27	.34939	.93698	.37289	2.68175	.543310	.971729	.571581	.428419	33	.55
	28	.34966	.93688	.37322	2.67937	.543649	.971682	.571967	.428033	32	
	29	.34993	.93677	.37355	2.67700	.543987	.971635	.572352	.427648	31	
.50	30	.35021	.93667	.37388	2.67462	.544325	.971588	.572738	.427262	30	.50
	31	.35048	.93657	.37422	2.67225	.544663	.971540	.573123	.426877	29	
	32	.35075	.93647	.37455	2.66989	.545000	.971493	.573507	.426493	28	
.55	33	.35102	.93637	.37488	2.66752	.545338	.971446	.573892	.426108	27	.45
	34	.35130	.93626	.37521	2.66516	.545674	.971398	.574276	.425724	26	
	35	.35157	.93616	.37554	2.66281	.546011	.971351	.574660	.425340	25	
.60	36	.35184	.93606	.37588	2.66046	.546347	.971303	.575044	.424956	24	.40
	37	.35211	.93596	.37621	2.65811	.546683	.971256	.575427	.424573	23	
	38	.35239	.93585	.37654	2.65576	.547019	.971208	.575810	.424190	22	
.65	39	.35266	.93575	.37687	2.65342	.547354	.971161	.576193	.423807	21	.35
	40	.35293	.93565	.37720	2.65109	.547689	.971113	.576576	.423424	20	
	41	.35320	.93555	.37754	2.64875	.548024	.971066	.576959	.423041	19	
.70	42	.35347	.93544	.37787	2.64642	.548359	.971018	.577341	.422659	18	.30
	43	.35375	.93534	.37820	2.64410	.548693	.970970	.577723	.422277	17	
	44	.35402	.93524	.37853	2.64177	.549027	.970922	.578104	.421896	16	
.75	45	.35429	.93514	.37887	2.63945	.549360	.970874	.578486	.421514	15	.25
	46	.35456	.93503	.37920	2.63714	.549693	.970827	.578867	.421133	14	
	47	.35484	.93493	.37953	2.63483	.550026	.970779	.579248	.420752	13	
.80	48	.35511	.93483	.37986	2.63252	.550359	.970731	.579629	.420371	12	.20
	49	.35538	.93472	.38020	2.63021	.550692	.970683	.580009	.419991	11	
	50	.35565	.93462	.38053	2.62791	.551024	.970635	.580389	.419611	10	
.85	51	.35592	.93452	.38086	2.62561	.551356	.970586	.580769	.419231	9	.15
	52	.35619	.93441	.38120	2.62332	.551687	.970538	.581149	.418851	8	
	53	.35647	.93431	.38153	2.62103	.552018	.970490	.581528	.418472	7	
.90	54	.35674	.93420	.38186	2.61874	.552349	.970442	.581907	.418093	6	.10
	55	.35701	.93410	.38220	2.61646	.552680	.970394	.582286	.417714	5	
	56	.35728	.93400	.38253	2.61418	.553010	.970345	.582665	.417335	4	
.95	57	.35755	.93389	.38286	2.61190	.553341	.970297	.583044	.416956	3	.05
	58	.35782	.93379	.38320	2.60963	.553670	.970249	.583422	.416578	2	
	59	.35810	.93368	.38353	2.60736	.554000	.970200	.583800	.416200	1	
1.00	60	.35837	.93358	.38386	2.60509	9.554329	9.970152	9.584177	10.415823	0	.00
Decimals	Minutes	Cos	Sin	Cot	Tan	Cos	Sin	Cot	Tan	Minutes	Decimals
		Natural Values				Common Logarithms					

69°

Decimals	Minutes	Natural Values				Common Logarithms				Minutes	Decimals
		Sin	Cos	Tan	Cot	Sin	Cos	Tan	Cot		
.00	0	.35837	.93358	.38386	2.60509	9.554329	9.970152	9.584177	10.415823	60	1.00
	1	.35864	.93348	.38420	2.60283	.554658	.970103	.584555	.415445	59	
	2	.35891	.93337	.38453	2.60057	.554987	.970055	.584932	.415068	58	
.05	3	.35918	.93327	.38487	2.59831	.555315	.970006	.585309	.414691	57	.95
	4	.35945	.93316	.38520	2.59606	.555643	.969957	.585686	.414314	56	
	5	.35973	.93306	.38553	2.59381	.555971	.969909	.586062	.413938	55	
.10	6	.36000	.93295	.38587	2.59156	.556299	.969860	.586439	.413561	54	.90
	7	.36027	.93285	.38620	2.58932	.556626	.969811	.586815	.413185	53	
	8	.36054	.93274	.38654	2.58708	.556953	.969762	.587190	.412810	52	
.15	9	.36081	.93264	.38687	2.58484	.557280	.969714	.587566	.412434	51	.85
	10	.36108	.93253	.38721	2.58261	.557606	.969665	.587941	.412059	50	
	11	.36135	.93243	.38754	2.58038	.557932	.969616	.588316	.411684	49	
.20	12	.36162	.93232	.38787	2.57815	.558258	.969567	.588691	.411309	48	.80
	13	.36190	.93222	.38821	2.57593	.558583	.969518	.589066	.410934	47	
	14	.36217	.93211	.38854	2.57371	.558909	.969469	.589440	.410560	46	
.25	15	.36244	.93201	.38888	2.57150	.559234	.969420	.589814	.410186	45	.75
	16	.36271	.93190	.38921	2.56928	.559558	.969370	.590188	.409812	44	
	17	.36298	.93180	.38955	2.56707	.559883	.969321	.590562	.409438	43	
.30	18	.36325	.93169	.38988	2.56487	.560207	.969272	.590935	.409065	42	.70
	19	.36352	.93159	.39022	2.56266	.560531	.969223	.591308	.408692	41	
	20	.36379	.93148	.39055	2.56046	.560855	.969173	.591681	.408319	40	
.35	21	.36406	.93137	.39089	2.55827	.561178	.969124	.592054	.407946	39	.65
	22	.36434	.93127	.39122	2.55608	.561501	.969075	.592426	.407574	38	
	23	.36461	.93116	.39156	2.55389	.561824	.969025	.592799	.407201	37	
.40	24	.36488	.93106	.39190	2.55170	.562146	.968976	.593171	.406829	36	.60
	25	.36515	.93095	.39223	2.54952	.562468	.968926	.593542	.406458	35	
	26	.36542	.93084	.39257	2.54734	.562790	.968877	.593914	.406086	34	
.45	27	.36569	.93074	.39290	2.54516	.563112	.968827	.594285	.405715	33	.55
	28	.36596	.93063	.39324	2.54299	.563433	.968777	.594656	.405344	32	
	29	.36623	.93052	.39357	2.54082	.563755	.968728	.595027	.404973	31	
.50	30	.36650	.93042	.39391	2.53865	.564075	.968678	.595398	.404602	30	.50
	31	.36677	.93031	.39425	2.53648	.564396	.968628	.595768	.404232	29	
	32	.36704	.93020	.39458	2.53432	.564716	.968578	.596138	.403862	28	
.55	33	.36731	.93010	.39492	2.53217	.565036	.968528	.596508	.403492	27	.45
	34	.36758	.92999	.39526	2.53001	.565356	.968479	.596878	.403122	26	
	35	.36785	.92988	.39559	2.52786	.565676	.968429	.597247	.402753	25	
.60	36	.36812	.92978	.39593	2.52571	.565995	.968379	.597616	.402384	24	.40
	37	.36839	.92967	.39626	2.52357	.566314	.968329	.597985	.402015	23	
	38	.36867	.92956	.39660	2.52142	.566632	.968278	.598354	.401646	22	
.65	39	.36894	.92945	.39694	2.51929	.566951	.968228	.598722	.401278	21	.35
	40	.36921	.92935	.39727	2.51715	.567269	.968178	.599091	.400909	20	
	41	.36948	.92924	.39761	2.51502	.567587	.968128	.599459	.400541	19	
.70	42	.36975	.92913	.39795	2.51289	.567904	.968078	.599827	.400173	18	.30
	43	.37002	.92902	.39829	2.51076	.568222	.968027	.600194	.399806	17	
	44	.37029	.92892	.39862	2.50864	.568539	.967977	.600562	.399438	16	
.75	45	.37056	.92881	.39896	2.50652	.568856	.967927	.600929	.399071	15	.25
	46	.37083	.92870	.39930	2.50440	.569172	.967876	.601296	.398704	14	
	47	.37110	.92859	.39963	2.50229	.569488	.967826	.601663	.398337	13	
.80	48	.37137	.92849	.39997	2.50018	.569804	.967775	.602029	.397971	12	.20
	49	.37164	.92838	.40031	2.49807	.570120	.967725	.602395	.397605	11	
	50	.37191	.92827	.40065	2.49597	.570435	.967674	.602761	.397239	10	
.85	51	.37218	.92816	.40098	2.49386	.570751	.967624	.603127	.396873	9	.15
	52	.37245	.92805	.40132	2.49177	.571066	.967573	.603493	.396507	8	
	53	.37272	.92794	.40166	2.48967	.571380	.967522	.603858	.396142	7	
.90	54	.37299	.92784	.40200	2.48758	.571695	.967471	.604223	.395777	6	.10
	55	.37326	.92773	.40234	2.48549	.572009	.967421	.604588	.395412	5	
	56	.37353	.92762	.40267	2.48340	.572323	.967370	.604953	.395047	4	
.95	57	.37380	.92751	.40301	2.48132	.572636	.967319	.605317	.394683	3	.05
	58	.37407	.92740	.40335	2.47924	.572950	.967268	.605682	.394318	2	
	59	.37434	.92729	.40369	2.47716	.573263	.967217	.606046	.393954	1	
1.00	60	.37461	.92718	.40403	2.47509	9.573575	9.967166	9.606410	10.393590	0	.00

Decimals	Minutes	Cos	Sin	Cot	Tan	Cos	Sin	Cot	Tan	Minutes	Decimals
		Natural Values				Common Logarithms					

68°

Decimals	Minutes	Natural Values Sin	Cos	Tan	Cot	Common Logarithms Sin	Cos	Tan	Cot	Minutes	Decimals
.00	0	.37461	.92718	.40403	2.47509	9.573575	9.967166	9.606410	10.393590	60	1.00
	1	.37488	.92707	.40436	2.47302	.573888	.967115	.606773	.393227	59	
	2	.37515	.92697	.40470	2.47095	.574200	.967064	.607137	.392863	58	
.05	3	.37542	.92686	.40504	2.46888	.574512	.967013	.607500	.392500	57	.95
	4	.37569	.92675	.40538	2.46682	.574824	.966961	.607863	.392137	56	
	5	.37595	.92664	.40572	2.46476	.575136	.966910	.608225	.391775	55	
.10	6	.37622	.92653	.40606	2.46270	.575447	.966859	.608588	.391412	54	.90
	7	.37649	.92642	.40640	2.46065	.575758	.966808	.608950	.391050	53	
	8	.37676	.92631	.40674	2.45860	.576069	.966756	.609312	.390688	52	
.15	9	.37703	.92620	.40707	2.45655	.576379	.966705	.609674	.390326	51	.85
	10	.37730	.92609	.40741	2.45451	.576689	.966653	.610036	.389964	50	
	11	.37757	.92598	.40775	2.45246	.576999	.966602	.610397	.389603	49	
.20	12	.37784	.92587	.40809	2.45043	.577309	.966550	.610759	.389241	48	.80
	13	.37811	.92576	.40843	2.44839	.577618	.966499	.611120	.388880	47	
	14	.37838	.92565	.40877	2.44636	.577927	.966447	.611480	.388520	46	
.25	15	.37865	.92554	.40911	2.44433	.578236	.966395	.611841	.388159	45	.75
	16	.37892	.92543	.40945	2.44230	.578545	.966344	.612201	.387799	44	
	17	.37919	.92532	.40979	2.44027	.578853	.966292	.612561	.387439	43	
.30	18	.37946	.92521	.41013	2.43825	.579162	.966240	.612921	.387079	42	.70
	19	.37973	.92510	.41047	2.43623	.579470	.966188	.613281	.386719	41	
	20	.37999	.92499	.41081	2.43422	.579777	.966136	.613641	.386359	40	
.35	21	.38026	.92488	.41115	2.43220	.580085	.966085	.614000	.386000	39	.65
	22	.38053	.92477	.41149	2.43019	.580392	.966033	.614359	.385641	38	
	23	.38080	.92466	.41183	2.42819	.580699	.965981	.614718	.385282	37	
.40	24	.38107	.92455	.41217	2.42618	.581005	.965929	.615077	.384923	36	.60
	25	.38134	.92444	.41251	2.42418	.581312	.965876	.615435	.384565	35	
	26	.38161	.92432	.41285	2.42218	.581618	.965824	.615793	.384207	34	
.45	27	.38188	.92421	.41319	2.42019	.581924	.965772	.616151	.383849	33	.55
	28	.38215	.92410	.41353	2.41819	.582229	.965720	.616509	.383491	32	
	29	.38241	.92399	.41387	2.41620	.582535	.965668	.616867	.383133	31	
.50	30	.38268	.92388	.41421	2.41421	.582840	.965615	.617224	.382776	30	.50
	31	.38295	.92377	.41455	2.41223	.583145	.965563	.617582	.382418	29	
	32	.38322	.92366	.41490	2.41025	.583449	.965511	.617939	.382061	28	
.55	33	.38349	.92355	.41524	2.40827	.583754	.965458	.618295	.381705	27	.45
	34	.38376	.92343	.41558	2.40629	.584058	.965406	.618652	.381348	26	
	35	.38403	.92332	.41592	2.40432	.584361	.965353	.619008	.380992	25	
.60	36	.38430	.92321	.41626	2.40235	.584665	.965301	.619364	.380636	24	.40
	37	.38456	.92310	.41660	2.40038	.584968	.965248	.619720	.380280	23	
	38	.38483	.92299	.41694	2.39841	.585272	.965195	.620076	.379924	22	
.65	39	.38510	.92287	.41728	2.39645	.585574	.965143	.620432	.379568	21	.35
	40	.38537	.92276	.41763	2.39449	.585877	.965090	.620787	.379213	20	
	41	.38564	.92265	.41797	2.39253	.586179	.965037	.621142	.378858	19	
.70	42	.38591	.92254	.41831	2.39058	.586482	.964984	.621497	.378503	18	.30
	43	.38617	.92243	.41865	2.38863	.586783	.964931	.621852	.378148	17	
	44	.38644	.92231	.41899	2.38668	.587085	.964879	.622207	.377793	16	
.75	45	.38671	.92220	.41933	2.38473	.587386	.964826	.622561	.377439	15	.25
	46	.38698	.92209	.41968	2.38279	.587688	.964773	.622915	.377085	14	
	47	.38725	.92198	.42002	2.38084	.587989	.964720	.623269	.376731	13	
.80	48	.38752	.92186	.42036	2.37891	.588289	.964666	.623623	.376377	12	.20
	49	.38778	.92175	.42070	2.37697	.588590	.964613	.623976	.376024	11	
	50	.38805	.92164	.42105	2.37504	.588890	.964560	.624330	.375670	10	
.85	51	.38832	.92152	.42139	2.37311	.589190	.964507	.624683	.375317	9	.15
	52	.38859	.92141	.42173	2.37118	.589489	.964454	.625036	.374964	8	
	53	.38886	.92130	.42207	2.36925	.589789	.964400	.625388	.374612	7	
.90	54	.38912	.92119	.42242	2.36733	.590088	.964347	.625741	.374259	6	.10
	55	.38939	.92107	.42276	2.36541	.590387	.964294	.626093	.373907	5	
	56	.38966	.92096	.42310	2.36349	.590686	.964240	.626445	.373555	4	
.95	57	.38993	.92085	.42345	2.36158	.590984	.964187	.626797	.373203	3	.05
	58	.39020	.92073	.42379	2.35967	.591282	.964133	.627149	.372851	2	
	59	.39046	.92062	.42413	2.35776	.591580	.964080	.627501	.372499	1	
1.00	60	.39073	.92050	.42447	2.35585	9.591878	9.964026	9.627852	10.372148	0	.00

Decimals	Minutes	Cos	Sin	Cot	Tan	Cos	Sin	Cot	Tan	Minutes	Decimals
		Natural Values				Common Logarithms					

67°

Decimals	Minutes	Natural Values				Common Logarithms				Minutes	Decimals
		Sin	Cos	Tan	Cot	Sin	Cos	Tan	Cot		
.00	0	.39073	.92050	.42447	2.35585	9.591878	9.964026	9.627852	10.372148	60	1.00
	1	.39100	.92039	.42482	2.35395	.592176	.963972	.628203	.371797	59	
	2	.39127	.92028	.42516	2.35205	.592473	.963919	.628554	.371446	58	
.05	3	.39153	.92016	.42551	2.35015	.592770	.963865	.628905	.371095	57	.95
	4	.39180	.92005	.42585	2.34825	.593067	.963811	.629255	.370745	56	
	5	.39207	.91994	.42619	2.34636	.593363	.963757	.629606	.370394	55	
.10	6	.39234	.91982	.42654	2.34447	.593659	.963704	.629956	.370044	54	.90
	7	.39260	.91971	.42688	2.34258	.593955	.963650	.630306	.369694	53	
	8	.39287	.91959	.42722	2.34069	.594251	.963596	.630656	.369344	52	
.15	9	.39314	.91948	.42757	2.33881	.594547	.963542	.631005	.368995	51	.85
	10	.39341	.91936	.42791	2.33693	.594842	.963488	.631355	.368645	50	
	11	.39367	.91925	.42826	2.33505	.595137	.963434	.631704	.368296	49	
.20	12	.39394	.91914	.42360	2.33317	.595432	.963379	.632053	.367947	48	.80
	13	.39421	.91902	.42894	2.33130	.595727	.963325	.632402	.367598	47	
	14	.39448	.91891	.42929	2.32943	.596021	.963271	.632750	.367250	46	
.25	15	.39474	.91879	.42963	2.32756	.596315	.963217	.633099	.366901	45	.75
	16	.39501	.91868	.42998	2.32570	.596609	.963163	.633447	.366553	44	
	17	.39528	.91856	.43032	2.32383	.596903	.963108	.633795	.366205	43	
.30	18	.39555	.91845	.43067	2.32197	.597196	.963054	.634143	.365857	42	.70
	19	.39581	.91833	.43101	2.32012	.597490	.962999	.634490	.365510	41	
	20	.39608	.91822	.43136	2.31826	.597783	.962945	.634838	.365162	40	
.35	21	.39635	.91810	.43170	2.31641	.598075	.962890	.635185	.364815	39	.65
	22	.39661	.91799	.43205	2.31456	.598368	.962836	.635532	.364468	38	
	23	.39688	.91787	.43239	2.31271	.598660	.962781	.635879	.364121	37	
.40	24	.39715	.91775	.43274	2.31086	.598952	.962727	.636226	.363774	36	.60
	25	.39741	.91764	.43308	2.30902	.599244	.962672	.636572	.363428	35	
	26	.39768	.91752	.43343	2.30718	.599536	.962617	.636919	.363081	34	
.45	27	.39795	.91741	.43378	2.30534	.599827	.962562	.637265	.362735	33	.55
	28	.39822	.91729	.43412	2.30351	.600118	.962508	.637611	.362389	32	
	29	.39848	.91718	.43447	2.30167	.600409	.962453	.637956	.362044	31	
.50	30	.39875	.91706	.43481	2.29984	.600700	.962398	.638302	.361698	30	.50
	31	.39902	.91694	.43516	2.29801	.600990	.962343	.638647	.361353	29	
	32	.39928	.91683	.43550	2.29619	.601280	.962288	.638992	.361008	28	
.55	33	.39955	.91671	.43585	2.29437	.601570	.962233	.639337	.360663	27	.45
	34	.39982	.91660	.43620	2.29254	.601860	.962178	.639682	.360318	26	
	35	.40008	.91648	.43654	2.29073	.602150	.962123	.640027	.359973	25	
.60	36	.40035	.91636	.43689	2.28891	.602439	.962067	.640371	.359629	24	.40
	37	.40062	.91625	.43724	2.28710	.602728	.962012	.640716	.359284	23	
	38	.40088	.91613	.43758	2.28528	.603017	.961957	.641060	.358940	22	
.65	39	.40115	.91601	.43793	2.28348	.603305	.961902	.641404	.358596	21	.35
	40	.40141	.91590	.43828	2.28167	.603594	.961846	.641747	.358253	20	
	41	.40168	.91578	.43862	2.27987	.603882	.961791	.642091	.357909	19	
.70	42	.40195	.91566	.43897	2.27806	.604170	.961735	.642434	.357566	18	.30
	43	.40221	.91555˙	.43932	2.27626	.604457	.961680	.642777	.357223	17	
	44	.40248	.91543	.43966	2.27447	.604745	.961624	.643120	.356880	16	
.75	45	.40275	.91531	.44001	2.27267	.605032	.961569	.643463	.356537	15	.25
	46	.40301	.91519	.44036	2.27088	.605319	.961513	.643806	.356194	14	
	47	.40328	.91508	.44071	2.26909	.605606	.961458	.644148	.355852	13	
.80	48	.40355	.91496	.44105	2.26730	.605892	.961402	.644490	.355510	12	.20
	49	.40381	.91484	.44140	2.26552	.606179	.961346	.644832	.355168	11	
	50	.40408	.91472	.44175	2.26374	.606465	.961290	.645174	.354826	10	
.85	51	.40434	.91461	.44210	2.26196	.606751	.961235	.645516	.354484	9	.15
	52	.40461	.91449	.44244	2.26018	.607036	.961179	.645857	.354143	8	
	53	.40488	.91437	.44279	2.25840	.607322	.961123	.646199	.353801	7	
.90	54	.40514	.91425	.44314	2.25663	.607607	.961067	.646540	.353460	6	.10
	55	.40541	.91414	.44349	2.25486	.607892	.961011	.646881	.353119	5	
	56	.40567	.91402	.44384	2.25309	.608177	.960955	.647222	.352778	4	
.95	57	.40594	.91390	.44413	2.25132	.608461	.960899	.647562	.352438	3	.05
	58	.40621	.91378	.44453	2.24956	.608745	.960843	.647903	.352097	2	
	59	.40647	.91366	.44488	2.24780	.609029	.960786	.648243	.351757	1	
1.00	60	.40674	.91355	.44523	2.24604	9.609313	9.960730	9.648583	10.351417	0	.00
Decimals	Minutes	Cos	Sin	Cot	Tan	Cos	Sin	Cot	Tan	Minutes	Decimals
		Natural Values				Common Logarithms					

Decimals	Minutes	Natural Values				Common Logarithms				Minutes	Decimals
		Sin	Cos	Tan	Cot	Sin	Cos	Tan	Cot		
.00	0	.40674	.91355	.44523	2.24604	9.609313	9.960730	9.648583	10.351417	60	1.00
	1	.40700	.91343	.44558	2.24428	.609597	.960674	.648923	.351077	59	
	2	.40727	.91331	.44593	2.24252	.609880	.960618	.649263	.350737	58	
.05	3	.40753	.91319	.44627	2.24077	.610164	.960561	.649602	.350398	57	.95
	4	.40780	.91307	.44662	2.23902	.610447	.960505	.649942	.350058	56	
	5	.40806	.91295	.44697	2.23727	.610729	.960448	.650281	.349719	55	
.10	6	.40833	.91283	.44732	2.23553	.611012	.960392	.650620	.349380	54	.90
	7	.40860	.91272	.44767	2.23378	.611294	.960335	.650959	.349041	53	
	8	.40886	.91260	.44802	2.23204	.611576	.960279	.651297	.348703	52	
.15	9	.40913	.91248	.44837	2.23030	.611858	.960222	.651636	.348364	51	.85
	10	.40939	.91236	.44872	2.22857	.612140	.960165	.651974	.348026	50	
	11	.40966	.91224	.44907	2.22683	.612421	.960109	.652312	.347688	49	
.20	12	.40992	.91212	.44942	2.22510	.612702	.960052	.652650	.347350	48	.80
	13	.41019	.91200	.44977	2.22337	.612983	.959995	.652988	.347012	47	
	14	.41045	.91188	.45012	2.22164	.613264	.959938	.653326	.346674	46	
.25	15	.41072	.91176	.45047	2.21992	.613545	.959832	.653663	.346337	45	.75
	16	.41098	.91164	.45082	2.21819	.613825	.959825	.654000	.346000	44	
	17	.41125	.91152	.45117	2.21647	.614105	.959768	.654337	.345663	43	
.30	18	.41151	.91140	.45152	2.21475	.614385	.959711	.654674	.345326	42	.70
	19	.41178	.91128	.45187	2.21304	.614665	.959654	.655011	.344989	41	
	20	.41204	.91116	.45222	2.21132	.614944	.959596	.655348	.344652	40	
.35	21	.41231	.91104	.45257	2.20961	.615223	.959539	.655684	.344316	39	.65
	22	.41257	.91092	.45292	2.20790	.615502	.959482	.656020	.343980	38	
	23	.41284	.91080	.45327	2.20619	.615781	.959425	.656356	.343644	37	
.40	24	.41310	.91068	.45362	2.20449	.616060	.959368	.656692	.343308	36	.60
	25	.41337	.91056	.45397	2.20278	.616338	.959310	.657028	.342972	35	
	26	.41363	.91044	.45432	2.20108	.616616	.959253	.657364	.342636	34	
.45	27	.41390	.91032	.45467	2.19938	.616894	.959195	.657699	.342301	33	.55
	28	.41416	.91020	.45502	2.19769	.617172	.959138	.658034	.341966	32	
	29	.41443	.91008	.45538	2.19599	.617450	.959080	.658369	.341631	31	
.50	30	.41469	.90996	.45573	2.19430	.617727	.959023	.658704	.341296	30	.50
	31	.41496	.90984	.45608	2.19261	.618004	.958965	.659039	.340961	29	
	32	.41522	.90972	.45643	2.19092	.618281	.958908	.659373	.340627	28	
.55	33	.41549	.90960	.45678	2.18923	.618558	.958850	.659708	.340292	27	.45
	34	.41575	.90948	.45713	2.18755	.618834	.958792	.660042	.359958	26	
	35	.41602	.90936	.45748	2.18587	.619110	.958734	.660376	.339624	25	
.60	36	.41628	.90924	.45784	2.18419	.619386	.958677	.660710	.339290	24	.40
	37	.41655	.90911	.45819	2.18251	.619662	.958619	.661043	.338957	23	
	38	.41681	.90899	.45854	2.18084	.619938	.958561	.661377	.338623	22	
.65	39	.41707	.90887	.45889	2.17916	.620213	.958503	.661710	.338290	21	.35
	40	.41734	.90875	.45924	2.17749	.620488	.958445	.662043	.337957	20	
	41	.41760	.90863	.45960	2.17582	.620763	.958387	.662376	.337624	19	
.70	42	.41787	.90851	.45995	2.17416	.621038	.958329	.662709	.337291	18	.30
	43	.41813	.90839	.46030	2.17249	.621313	.958271	.663042	.336958	17	
	44	.41840	.90826	.46065	2.17083	.621587	.958213	.663375	.336625	16	
.75	45	.41866	.90814	.46101	2.16917	.621861	.958154	.663707	.336293	15	.25
	46	.41892	.90802	.46136	2.16751	.622135	.958096	.664039	.335961	14	
	47	.41919	.90790	.46171	2.16585	.622409	.958038	.664371	.335629	13	
.80	48	.41945	.90778	.46206	2.16420	.622682	.957979	.664703	.335297	12	.20
	49	.41972	.90766	.46242	2.16255	.622956	.957921	.665035	.334965	11	
	50	.41998	.90753	.46277	2.16090	.623229	.957863	.665366	.334634	10	
.85	51	.42024	.90741	.46312	2.15925	.623502	.957804	.665698	.334302	9	.15
	52	.42051	.90729	.46348	2.15760	.623774	.957746	.666029	.333971	8	
	53	.42077	.90717	.46383	2.15596	.624047	.957687	.666360	.333640	7	
.90	54	.42104	.90704	.46418	2.15432	.624319	.957628	.666691	.333309	6	.10
	55	.42130	.90692	.46454	2.15268	.624591	.957570	.667021	.332979	5	
	56	.42156	.90680	.46489	2.15104	.624863	.957511	.667352	.332648	4	
.95	57	.42183	.90668	.46525	2.14940	.625135	.957452	.667682	.332318	3	.05
	58	.42209	.90655	.46560	2.14777	.625406	.957393	.668013	.331987	2	
	59	.42235	.90643	.46595	2.14614	.625677	.957335	.668343	.331657	1	
1.00	60	.42262	.90631	.46631	2.14451	9.625948	9.957276	9.668673	10.331327	0	.00
Decimals	Minutes	Cos	Sin	Cot	Tan	Cos	Sin	Cot	Tan	Minutes	Decimals

Natural Values   Common Logarithms

65°

Decimals	Minutes	Natural Values Sin	Cos	Tan	Cot	Common Logarithms Sin	Cos	Tan	Cot	Minutes	Decimals
.00	0	.42262	.90631	.46631	2.14451	9.625948	9.957276	9.668673	10.331327	60	1.00
	1	.42288	.90618	.46666	2.14288	.626219	.957217	.669002	.330998	59	
	2	.42315	.90606	.46702	2.14125	.626490	.957158	.669332	.330668	58	
.05	3	.42341	.90594	.46737	2.13963	.626760	.957099	.669661	.330339	57	.95
	4	.42367	.90582	.46772	2.13801	.627030	.957040	.669991	.330009	56	
	5	.42394	.90569	.46808	2.13639	.627300	.956981	.670320	.329680	55	
.10	6	.42420	.90557	.46843	2.13477	.627570	.956921	.670649	.329351	54	.90
	7	.42446	.90545	.46879	2.13316	.627840	.956862	.670977	.329023	53	
	8	.42473	.90532	.46914	2.13154	.628109	.956803	.671306	.328694	52	
.15	9	.42499	.90520	.46950	2.12993	.628378	.956744	.671635	.328365	51	.85
	10	.42525	.90507	.46985	2.12832	.628647	.956684	.671963	.328037	50	
	11	.42552	.90495	.47021	2.12671	.628916	.956625	.672291	.327709	49	
.20	12	.42578	.90483	.47056	2.12511	.629185	.956566	.672619	.327381	48	.80
	13	.42604	.90470	.47092	2.12350	.629453	.956506	.672947	.327053	47	
	14	.42631	.90458	.47128	2.12190	.629721	.956447	.673274	.326726	46	
.25	15	.42657	.90446	.47163	2.12030	.629989	.956387	.673602	.326398	45	.75
	16	.42683	.90433	.47199	2.11871	.630257	.956327	.673929	.326071	44	
	17	.42709	.90421	.47234	2.11711	.630524	.956268	.674257	.325743	43	
.30	18	.42736	.90408	.47270	2.11552	.630792	.956208	.674584	.325416	42	.70
	19	.42762	.90396	.47305	2.11392	.631059	.956148	.674911	.325089	41	
	20	.42788	.90383	.47341	2.11233	.631326	.956089	.675237	.324763	40	
.35	21	.42815	.90371	.47377	2.11075	.631593	.956029	.675564	.324436	39	.65
	22	.42841	.90358	.47412	2.10916	.631859	.955969	.675890	.324110	38	
	23	.42867	.90346	.47448	2.10758	.632125	.955909	.676217	.323783	37	
.40	24	.42894	.90334	.47483	2.10600	.632392	.955849	.676543	.323457	36	.60
	25	.42920	.90321	.47519	2.10442	.632658	.955789	.676869	.323131	35	
	26	.42946	.90309	.47555	2.10284	.632923	.955729	.677194	.322806	34	
.45	27	.42972	.90296	.47590	2.10126	.633189	.955669	.677520	.322480	33	.55
	28	.42999	.90284	.47626	2.09969	.633454	.955609	.677846	.322154	32	
	29	.43025	.90271	.47662	2.09811	.633719	.955548	.678171	.321829	31	
.50	30	.43051	.90259	.47698	2.09654	.633984	.955488	.678496	.321504	30	.50
	31	.43077	.90246	.47733	2.09498	.634249	.955428	.678821	.321179	29	
	32	.43104	.90233	.47769	2.09341	.634514	.955368	.679146	.320854	28	
.55	33	.43130	.90221	.47805	2.09184	.634778	.955307	.679471	.320529	27	.45
	34	.43156	.90208	.47840	2.09028	.635042	.955247	.679795	.320205	26	
	35	.43182	.90196	.47876	2.08872	.635306	.955186	.680120	.319880	25	
.60	36	.43209	.90183	.47912	2.08716	.635570	.955126	.680444	.319556	24	.40
	37	.43235	.90171	.47948	2.08560	.635834	.955065	.680768	.319232	23	
	38	.43261	.90158	.47984	2.08405	.636097	.955005	.681092	.318908	22	
.65	39	.43287	.90146	.48019	2.08250	.636360	.954944	.681416	.318584	21	.35
	40	.43313	.90133	.48055	2.08094	.636623	.954883	.681740	.318260	20	
	41	.43340	.90120	.48091	2.07939	.636886	.954823	.682063	.317937	19	
.70	42	.43366	.90108	.48127	2.07785	.637148	.954762	.682387	.317613	18	.30
	43	.43392	.90095	.48163	2.07630	.637411	.954701	.682710	.317290	17	
	44	.43418	.90082	.48198	2.07476	.637673	.954640	.683033	.316967	16	
.75	45	.43445	.90070	.48234	2.07321	.637935	.954579	.683356	.316644	15	.25
	46	.43471	.90057	.48270	2.07167	.638197	.954518	.683679	.316321	14	
	47	.43497	.90045	.48306	2.07014	.638458	.954457	.684001	.315999	13	
.80	48	.43523	.90032	.48342	2.06860	.638720	.954396	.684324	.315676	12	.20
	49	.43549	.90019	.48378	2.06706	.638981	.954335	.684646	.315354	11	
	50	.43575	.90007	.48414	2.06553	.639242	.954274	.684968	.315032	10	
.85	51	.43602	.89994	.48450	2.06400	.639503	.954213	.685290	.314710	9	.15
	52	.43628	.89981	.48486	2.06247	.639764	.954152	.685612	.314388	8	
	53	.43654	.89968	.48521	2.06094	.640024	.954090	.685934	.314066	7	
.90	54	.43680	.89956	.48557	2.05942	.640284	.954029	.686255	.313745	6	.10
	55	.43706	.89943	.48593	2.05790	.640544	.953968	.686577	.313423	5	
	56	.43733	.89930	.48629	2.05637	.640804	.953906	.686898	.313102	4	
.95	57	.43759	.89918	.48665	2.05485	.641064	.953845	.687219	.312781	3	.05
	58	.43785	.89905	.48701	2.05333	.641324	.953783	.687540	.312460	2	
	59	.43811	.89892	.48737	2.05182	.641583	.953722	.687861	.312139	1	
1.00	60	.43837	.89879	.48773	2.05030	9.641842	9.953660	9.688182	10.311818	0	.00

Decimals	Minutes	Cos	Sin	Cot	Tan	Cos	Sin	Cot	Tan	Minutes	Decimals
		Natural Values				Common Logarithms					

64°

Decimals	Minutes	Natural Values				Common Logarithms				Minutes	Decimals
		Sin	Cos	Tan	Cot	Sin	Cos	Tan	Cot		
.00	0	.43837	.89879	.48773	2.05030	9.641842	9.953660	9.688182	10.311818	60	1.00
	1	.43863	.89867	.48809	2.04879	.642101	.953599	.688502	.311498	59	
	2	.43889	.89854	.48845	2.04728	.642360	.953537	.688823	.311177	58	
.05	3	.43916	.89841	.48881	2.04577	.642618	.953475	.689143	.310857	57	.95
	4	.43942	.89828	.48917	2.04426	.642877	.953413	.689463	.310537	56	
	5	.43968	.89816	.48953	2.04276	.643135	.953352	.689783	.310217	55	
.10	6	.43994	.89803	.48989	2.04125	.643393	.953290	.690103	.309897	54	.90
	7	.44020	.89790	.49026	2.03975	.643650	.953228	.690423	.309577	53	
	8	.44046	.89777	.49062	2.03825	.643908	.953166	.690742	.309258	52	
.15	9	.44072	.89764	.49098	2.03675	.644165	.953104	.691062	.308938	51	.85
	10	.44098	.89752	.49134	2.03526	.644423	.953042	.691381	.308619	50	
	11	.44124	.89739	.49170	2.03376	.644680	.952980	.691700	.308300	49	
.20	12	.44151	.89726	.49206	2.03227	.644936	.952918	.692019	.307981	48	.80
	13	.44177	.89713	.49242	2.03078	.645193	.952855	.692338	.307662	47	
	14	.44203	.89700	.49278	2.02929	.645450	.952793	.692656	.307344	46	
.25	15	.44229	.89687	.49315	2.02780	.645706	.952731	.692975	.307025	45	.75
	16	.44255	.89674	.49351	2.02631	.645962	.952669	.693293	.306707	44	
	17	.44281	.89662	.49387	2.02483	.646218	.952606	.693612	.306388	43	
.30	18	.44307	.89649	.49423	2.02335	.646474	.952544	.693930	.306070	42	.70
	19	.44333	.89636	.49459	2.02187	.646729	.952481	.694248	.305752	41	
	20	.44359	.89623	.49495	2.02039	.646984	.952419	.694566	.305434	40	
.35	21	.44385	.89610	.49532	2.01891	.647240	.952356	.694883	.305117	39	.65
	22	.44411	.89597	.49568	2.01743	.647494	.952294	.695201	.304799	38	
	23	.44437	.89584	.49604	2.01596	.647749	.952231	.695518	.304482	37	
.40	24	.44464	.89571	.49640	2.01449	.648004	.952168	.695836	.304164	36	.60
	25	.44490	.89558	.49677	2.01302	.648258	.952106	.696153	.303847	35	
	26	.44516	.89545	.49713	2.01155	.648512	.952043	.696470	.303530	34	
.45	27	.44542	.89532	.49749	2.01008	.648766	.951980	.696787	.303213	33	.55
	28	.44568	.89519	.49786	2.00862	.649020	.951917	.697103	.302897	32	
	29	.44594	.89506	.49822	2.00715	.649274	.951854	.697420	.302580	31	
.50	30	.44620	.89493	.49858	2.00569	.649527	.951791	.697736	.302264	30	.50
	31	.44646	.89480	.49894	2.00423	.649781	.951728	.698053	.301947	29	
	32	.44672	.89467	.49931	2.00277	.650034	.951665	.698369	.301631	28	
.55	33	.44698	.89454	.49967	2.00131	.650287	.951602	.698685	.301315	27	.45
	34	.44724	.89441	.50004	1.99986	.650539	.951539	.699001	.300999	26	
	35	.44750	.89428	.50040	1.99841	.650792	.951476	.699316	.300684	25	
.60	36	.44776	.89415	.50076	1.99695	.651044	.951412	.699632	.300368	24	.40
	37	.44802	.89402	.50113	1.99550	.651297	.951349	.699947	.300053	23	
	38	.44828	.89389	.50149	1.99406	.651549	.951286	.700263	.299737	22	
.65	39	.44854	.89376	.50185	1.99261	.651800	.951222	.700578	.299422	21	.35
	40	.44880	.89363	.50222	1.99116	.652052	.951159	.700893	.299107	20	
	41	.44906	.89350	.50258	1.98972	.652304	.951096	.701208	.298792	19	
.70	42	.44932	.89337	.50295	1.98828	.652555	.951032	.701523	.298477	18	.30
	43	.44958	.89324	.50331	1.98684	.652806	.950968	.701837	.298163	17	
	44	.44984	.89311	.50368	1.98540	.653057	.950905	.702152	.297848	16	
.75	45	.45010	.89298	.50404	1.98396	.653308	.950841	.702466	.297534	15	.25
	46	.45036	.89285	.50441	1.98253	.653558	.950778	.702781	.297219	14	
	47	.45062	.89272	.50477	1.98110	.653808	.950714	.703095	.296905	13	
.80	48	.45088	.89259	.50514	1.97966	.654059	.950650	.703409	.296591	12	.20
	49	.45114	.89245	.50550	1.97823	.654309	.950586	.703722	.296278	11	
	50	.45140	.89232	.50587	1.97681	.654558	.950522	.704036	.295964	10	
.85	51	.45166	.89219	.50623	1.97538	.654808	.950458	.704350	.295650	9	.15
	52	.45192	.89206	.50660	1.97395	.655058	.950394	.704663	.295337	8	
	53	.45218	.89193	.50696	1.97253	.655307	.950330	.704976	.295024	7	
.90	54	.45243	.89180	.50733	1.97111	.655556	.950266	.705290	.294710	6	.10
	55	.45269	.89167	.50769	1.96969	.655805	.950202	.705603	.294397	5	
	56	.45295	.89153	.50806	1.96827	.656054	.950138	.705916	.294084	4	
.95	57	.45321	.89140	.50843	1.96685	.656302	.950074	.706228	.293772	3	.05
	57	.45347	.89127	.50879	1.96544	.656551	.950010	.706541	.293459	2	
	59	.45373	.89114	.50916	1.96402	.656799	.949945	.706854	.293146	1	
1.00	60	.45399	.89101	.50953	1.96261	9.657047	9.949881	9.707166	10.292834	0	.00

Decimals	Minutes	Cos	Sin	Cot	Tan	Cos	Sin	Cot	Tan	Minutes	Decimals
		Natural Values				Common Logarithms					

63°

Decimals	Minutes	Natural Values Sin	Cos	Tan	Cot	Common Logarithms Sin	Cos	Tan	Cot	Minutes	Decimals
.00	0	.45399	.89101	.50953	1.96261	9.657047	9.949881	9.707166	10.292834	60	1.00
	1	.45425	.89087	.50989	1.96120	.657295	.949816	.707478	.292522	59	
	2	.45451	.89074	.51026	1.95979	.657542	.949752	.707790	.292210	58	
.05	3	.45477	.89061	.51063	1.95838	.657790	.949688	.708102	.291898	57	.95
	4	.45503	.89048	.51099	1.95698	.658037	.949623	.708414	.291586	56	
	5	.45529	.89035	.51136	1.95557	.658284	.949558	.708726	.291274	55	
.10	6	.45554	.89021	.51173	1.95417	.658531	.949494	.709037	.290963	54	.90
	7	.45580	.89008	.51209	1.95277	.658778	.949429	.709349	.290651	53	
	8	.45606	.88995	.51246	1.95137	.659025	.949364	.709660	.290340	52	
.15	9	.45632	.88981	.51283	1.94997	.659271	.949300	.709971	.290029	51	.85
	10	.45658	.88968	.51319	1.94858	.659517	.949235	.710282	.289718	50	
	11	.45684	.88955	.51356	1.94718	.659763	.949170	.710593	.289407	49	
.20	12	.45710	.88942	.51393	1.94579	.660009	.949105	.710904	.289096	48	.80
	13	.45736	.88928	.51430	1.94440	.660255	.949040	.711215	.288785	47	
	14	.45762	.88915	.51467	1.94301	.660501	.948975	.711525	.288475	46	
.25	15	.45787	.88902	.51503	1.94162	.660746	.948910	.711836	.288164	45	.75
	16	.45813	.88888	.51540	1.94023	.660991	.948845	.712146	.287854	44	
	17	.45839	.88875	.51577	1.93885	.661236	.948780	.712456	.287544	43	
.30	18	.45865	.88862	.51614	1.93746	.661481	.948715	.712766	.287234	42	.70
	19	.45891	.88848	.51651	1.93608	.661726	.948650	.713076	.286924	41	
	20	.45917	.88835	.51688	1.93470	.661970	.948584	.713386	.286614	40	
.35	21	.45942	.88822	.51724	1.93332	.662214	.948519	.713696	.286304	39	.65
	22	.45968	.88808	.51761	1.93195	.662459	.948454	.714005	.285995	38	
	23	.45994	.88795	.51798	1.93057	.662703	.948388	.714314	.285686	37	
.40	24	.46020	.88782	.51835	1.92920	.662946	.948323	.714624	.285376	36	.60
	25	.46046	.88768	.51872	1.92782	.663190	.948257	.714933	.285067	35	
	26	.46072	.88755	.51909	1.92645	.663433	.948192	.715242	.284758	34	
.45	27	.46097	.88741	.51946	1.92508	.663677	.948126	.715551	.284449	33	.55
	28	.46123	.88728	.51983	1.92371	.663920	.948060	.715860	.284140	32	
	29	.46149	.88715	.52020	1.92235	.664163	.947995	.716168	.283832	31	
.50	30	.46175	.88701	.52057	1.92098	.664406	.947929	.716477	.283523	30	.50
	31	.46201	.88688	.52094	1.91962	.664648	.947863	.716785	.283215	29	
	32	.46226	.88674	.52131	1.91826	.664891	.947797	.717093	.282907	28	
.55	33	.46252	.88661	.52168	1.91690	.665133	.947731	.717401	.282599	27	.45
	34	.46278	.88647	.52205	1.91554	.665375	.947665	.717709	.282291	26	
	35	.46304	.88634	.52242	1.91418	.665617	.947600	.718017	.281983	25	
.60	36	.46330	.88620	.52279	1.91282	.665859	.947533	.718325	.281675	24	.40
	37	.46355	.88607	.52316	1.91147	.666100	.947467	.718633	.281367	23	
	38	.46381	.88593	.52353	1.91012	.666342	.947401	.718940	.281060	22	
.65	39	.46407	.88580	.52390	1.90876	.666583	.947335	.719248	.280752	21	.35
	40	.46433	.88566	.52427	1.90741	.666824	.947269	.719555	.280445	20	
	41	.46458	.88553	.52464	1.90607	.667065	.947203	.719862	.280138	19	
.70	42	.46484	.88539	.52501	1.90472	.667305	.947136	.720169	.279831	18	.30
	43	.46510	.88526	.52538	1.90337	.667546	.947070	.720476	.279524	17	
	44	.46536	.88512	.52575	1.90203	.667786	.947004	.720783	.279217	16	
.75	45	.46561	.88499	.52613	1.90068	.668027	.946937	.721089	.278911	15	.25
	46	.46587	.88485	.52650	1.89935	.668267	.946871	.721396	.278604	14	
	47	.46613	.88472	.52687	1.89801	.668506	.946804	.721702	.278298	13	
.80	48	.46639	.88458	.52724	1.89667	.668746	.946738	.722009	.277991	12	.20
	49	.46664	.88445	.52761	1.89533	.668986	.946671	.722315	.277685	11	
	50	.46690	.88431	.52798	1.89400	.669225	.946604	.722621	.277379	10	
.85	51	.46716	.88417	.52836	1.89266	.669464	.946538	.722927	.277073	9	.15
	52	.46742	.88404	.52873	1.89133	.669703	.946471	.723232	.276768	8	
	53	.46767	.88390	.52910	1.89000	.669942	.946404	.723538	.276462	7	
.90	54	.46793	.88377	.52947	1.88867	.670181	.946337	.723844	.276156	6	.10
	55	.46819	.88363	.52985	1.88734	.670419	.946270	.724149	.275851	5	
	56	.46844	.88349	.53022	1.88602	.670658	.946203	.724454	.275546	4	
.95	57	.46870	.88336	.53059	1.88469	.670896	.946136	.724760	.275240	3	.05
	58	.46896	.88322	.53096	1.88337	.671134	.946069	.725065	.274935	2	
	59	.46921	.88308	.53134	1.88205	.671372	.946002	.725370	.274630	1	
1.00	60	.46947	.88295	.53171	1.88073	9.671609	9.945935	9.725674	10.274326	0	.00

Decimals	Minutes	Cos	Sin	Cot	Tan	Cos	Sin	Cot	Tan	Minutes	Decimals
		Natural Values				Common Logarithms					

Decimals	Minutes	Natural Values Sin	Cos	Tan	Cot	Common Logarithms Sin	Cos	Tan	Cot	Minutes	Decimals
.00	0	.46947	.88295	.53171	1.88073	9.671609	9.945935	9.725674	10.274326	60	1.00
	1	.46973	.88281	.53208	1.87941	.671847	.945868	.725979	.274021	59	
	2	.46999	.88267	.53246	1.87809	.672084	.945800	.726284	.273716	58	
.05	3	.47024	.88254	.53283	1.87677	.672321	.945733	.726588	.273412	57	.95
	4	.47050	.88240	.53320	1.87546	.672558	.945666	.726892	.273108	56	
	5	.47076	.88226	.53358	1.87415	.672795	.945598	.727197	.272803	55	
.10	6	.47101	.88213	.53395	1.87283	.673032	.945531	.727501	.272499	54	.90
	7	.47127	.88199	.53432	1.87152	.673268	.945464	.727805	.272195	53	
	8	.47153	.88185	.53470	1.87021	.673505	.945396	.728109	.271891	52	
.15	9	.47178	.88172	.53507	1.86891	.673741	.945328	.728412	.271588	51	.85
	10	.47204	.88158	.53545	1.86760	.673977	.945261	.728716	.271284	50	
	11	.47229	.88144	.53582	1.86630	.674213	.945193	.729020	.270980	49	
.20	12	.47255	.88130	.53620	1.86499	.674448	.945125	.729323	.270677	48	.80
	13	.47281	.88117	.53657	1.86369	.674684	.945058	.729626	.270374	47	
	14	.47306	.88103	.53694	1.86239	.674919	.944990	.729929	.270071	46	
.25	15	.47332	.88089	.53732	1.86109	.675155	.944922	.730233	.269767	45	.75
	16	.47358	.88075	.53769	1.85979	.675390	.944854	.730535	.269465	44	
	17	.47383	.88062	.53807	1.85850	.675624	.944786	.730838	.269162	43	
.30	18	.47409	.88048	.53844	1.85720	.675859	.944718	.731141	.268859	42	.70
	19	.47434	.88034	.53882	1.85591	.676094	.944650	.731444	.268556	41	
	20	.47460	.88020	.53920	1.85462	.676328	.944582	.731746	.268254	40	
.35	21	.47486	.88006	.53957	1.85333	.676562	.944514	.732048	.267952	39	.65
	22	.47511	.87993	.53995	1.85204	.676796	.944446	.732351	.267649	38	
	23	.47537	.87979	.54032	1.85075	.677030	.944377	.732653	.267347	37	
.40	24	.47562	.87965	.54070	1.84946	.677264	.944309	.732955	.267045	36	.60
	25	.47588	.87951	.54107	1.84818	.677498	.944241	.733257	.266743	35	
	26	.47614	.87937	.54145	1.84689	.677731	.944172	.733558	.266442	34	
.45	27	.47639	.87923	.54183	1.84561	.677964	.944104	.733860	.266140	33	.55
	28	.47665	.87909	.54220	1.84433	.678197	.944036	.734162	.265838	32	
	29	.47690	.87896	.54258	1.84305	.678430	.943967	.734463	.265537	31	
.50	30	.47716	.87882	.54296	1.84177	.678663	.943899	.734764	.265236	30	.50
	31	.47741	.87868	.54333	1.84049	.678895	.943830	.735066	.264934	29	
	32	.47767	.87854	.54371	1.83922	.679128	.943761	.735367	.264633	28	
.55	33	.47793	.87840	.54409	1.83794	.679360	.943693	.735668	.264332	27	.45
	34	.47818	.87826	.54446	1.83667	.679592	.943624	.735969	.264031	26	
	35	.47844	.87812	.54484	1.83540	.679824	.943555	.736269	.263731	25	
.60	36	.47869	.87798	.54522	1.83413	.680056	.943486	.736570	.263430	24	.40
	37	.47895	.87784	.54560	1.83286	.680288	.943417	.736870	.263130	23	
	38	.47920	.87770	.54597	1.83159	.680519	.943348	.737171	.262829	22	
.65	39	.47946	.87756	.54635	1.83033	.680750	.943279	.737471	.262529	21	.35
	40	.47971	.87743	.54673	1.82906	.680982	.943210	.737771	.262229	20	
	41	.47997	.87729	.54711	1.82780	.681213	.943141	.738071	.261929	19	
.70	42	.48022	.87715	.54748	1.82654	.681443	.943072	.738371	.261629	18	.30
	43	.48048	.87701	.54786	1.82528	.681674	.943003	.738671	.261329	17	
	44	.48073	.87687	.54824	1.82402	.681905	.942934	.738971	.261029	16	
.75	45	.48099	.87673	.54862	1.82276	.682135	.942864	.739271	.260729	15	.25
	46	.48124	.87659	.54900	1.82150	.682365	.942795	.739570	.260430	14	
	47	.48150	.87645	.54938	1.82025	.682595	.942726	.739870	.260130	13	
.80	48	.48175	.87631	.54975	1.81899	.682825	.942656	.740169	.259831	12	.20
	49	.48201	.87617	.55013	1.81774	.683055	.942587	.740468	.259532	11	
	50	.48226	.87603	.55051	1.81649	.683284	.942517	.740767	.259233	10	
.85	51	.48252	.87589	.55089	1.81524	.683514	.942448	.741066	.258934	9	.15
	52	.48277	.87575	.55127	1.81399	.683743	.942378	.741365	.258635	8	
	53	.48303	.87561	.55165	1.81274	.683972	.942308	.741664	.258336	7	
.90	54	.48328	.87546	.55203	1.81150	.684201	.942239	.741962	.258038	6	.10
	55	.48354	.87532	.55241	1.81025	.684430	.942169	.742261	.257739	5	
	56	.48379	.87518	.55279	1.80901	.684658	.942099	.742559	.257441	4	
.95	57	.48405	.87504	.55317	1.80777	.684887	.942029	.742858	.257142	3	.05
	58	.48430	.87490	.55355	1.80653	.685115	.941959	.743156	.256844	2	
	59	.48456	.87476	.55393	1.80529	.685343	.941889	.743454	.256546	1	
1.00	60	.48481	.87462	.55431	1.80405	9.685571	9.941819	9.743752	10.256248	0	.00

Decimals	Minutes	Cos	Sin	Cot	Tan	Cos	Sin	Cot	Tan	Minutes	Decimals
			Natural Values				Common Logarithms				

Decimals	Minutes	Natural Values Sin	Cos	Tan	Cot	Common Logarithms Sin	Cos	Tan	Cot	Minutes	Decimals
.00	0	.48481	.87462	.55431	1.80405	9.685571	9.941819	9.743752	10.256248	60	1.00
	1	.48506	.87448	.55469	1.80281	.685799	.941749	.744050	.255950	59	
	2	.48532	.87434	.55507	1.80158	.686027	.941679	.744348	.255652	58	
.05	3	.48557	.87420	.55545	1.80034	.686254	.941609	.744645	.255355	57	.95
	4	.48583	.87406	.55583	1.79911	.686482	.941539	.744943	.255057	56	
	5	.48608	.87391	.55621	1.79788	.686709	.941469	.745240	.254760	55	
.10	6	.48634	.87377	.55659	1.79665	.686936	.941398	.745538	.254462	54	.90
	7	.48659	.87363	.55697	1.79542	.687163	.941328	.745835	.254165	53	
	8	.48684	.87349	.55736	1.79419	.687389	.941258	.746132	.253868	52	
.15	9	.48710	.87335	.55774	1.79296	.687616	.941187	.746429	.253571	51	.85
	10	.48735	.87321	.55812	1.79174	.687843	.941117	.746726	.253274	50	
	11	.48761	.87306	.55850	1.79051	.688069	.941046	.747023	.252977	49	
.20	12	.48786	.87292	.55888	1.78929	.688295	.940975	.747319	.252681	48	.80
	13	.48811	.87278	.55926	1.78807	.688521	.940905	.747616	.252384	47	
	14	.48837	.87264	.55964	1.78685	.688747	.940834	.747913	.252087	46	
.25	15	.48862	.87250	.56003	1.78563	.688972	.940763	.748209	.251791	45	.75
	16	.48888	.87235	.56041	1.78441	.689198	.940693	.748505	.251495	44	
	17	.48913	.87221	.56079	1.78319	.689423	.940622	.748801	.251199	43	
.30	18	.48938	.87207	.56117	1.78198	.689648	.940551	.749097	.250903	42	.70
	19	.48964	.87193	.56156	1.78077	.689873	.940480	.749393	.250607	41	
	20	.48989	.87178	.56194	1.77955	.690098	.940409	.749689	.250311	40	
.35	21	.49014	.87164	.56232	1.77834	.690323	.940338	.749985	.250015	39	.65
	22	.49040	.87150	.56270	1.77713	.690548	.940267	.750281	.249719	38	
	23	.49065	.87136	.56309	1.77592	.690772	.940196	.750576	.249424	37	
.40	24	.49090	.87121	.56347	1.77471	.690996	.940125	.750872	.249128	36	.60
	25	.49116	.87107	.56385	1.77351	.691220	.940054	.751167	.248833	35	
	26	.49141	.87093	.56424	1.77230	.691444	.939982	.751462	.248538	34	
.45	27	.49166	.87079	.56462	1.77110	.691668	.939911	.751757	.248243	33	.55
	28	.49192	.87064	.56501	1.76990	.691892	.939840	.752052	.247948	32	
	29	.49217	.87050	.56539	1.76869	.692115	.939768	.752347	.247653	31	
.50	30	.49242	.87036	.56577	1.76749	.692339	.939697	.752642	.247358	30	.50
	31	.49268	.87021	.56616	1.76629	.692562	.939625	.752937	.247063	29	
	32	.49293	.87007	.56654	1.76510	.692785	.939554	.753231	.246769	28	
.55	33	.49318	.86993	.56693	1.76390	.693008	.939482	.753526	.246474	27	.45
	34	.49344	.86978	.56731	1.76271	.693231	.939410	.753820	.246180	26	
	35	.49369	.86964	.56769	1.76151	.693453	.939339	.754115	.245885	25	
.60	36	.49394	.86949	.56808	1.76032	.693676	.939267	.754409	.245591	24	.40
	37	.49419	.86935	.56846	1.75913	.693898	.939195	.754703	.245297	23	
	38	.49445	.86921	.56885	1.75794	.694120	.939123	.754997	.245003	22	
.65	39	.49470	.86906	.56923	1.75675	.694342	.939052	.755291	.244709	21	.35
	40	.49495	.86892	.56962	1.75556	.694564	.938980	.755585	.244415	20	
	41	.49521	.86878	.57000	1.75437	.694786	.938908	.755878	.244122	19	
.70	42	.49546	.86863	.57039	1.75319	.695007	.938836	.756172	.243828	18	.30
	43	.49571	.86849	.57078	1.75200	.695229	.938763	.756465	.243535	17	
	44	.49596	.86834	.57116	1.75082	.695450	.938691	.756759	.243241	16	
.75	45	.49622	.86820	.57155	1.74964	.695671	.938619	.757052	.242948	15	.25
	46	.49647	.86805	.57193	1.74846	.695892	.938547	.757345	.242655	14	
	47	.49672	.86791	.57232	1.74728	.696113	.938475	.757638	.242362	13	
.80	48	.49697	.86777	.57271	1.74610	.696334	.938402	.757931	.242069	12	.20
	49	.49723	.86762	.57309	1.74492	.696554	.938330	.758224	.241776	11	
	50	.49748	.86748	.57348	1.74375	.696775	.938258	.758517	.241483	10	
.85	51	.49773	.86733	.57386	1.74257	.696995	.938185	.758810	.241190	9	.15
	52	.49798	.86719	.57425	1.74140	.697215	.938113	.759102	.240898	8	
	53	.49824	.86704	.57464	1.74022	.697435	.938040	.759395	.240605	7	
.90	54	.49849	.86690	.57503	1.73905	.697654	.937967	.759687	.240313	6	.10
	55	.49874	.86675	.57541	1.73788	.697874	.937895	.759979	.240021	5	
	56	.49899	.86661	.57580	1.73671	.698094	.937822	.760272	.239728	4	
.95	57	.49924	.86646	.57619	1.73555	.698313	.937749	.760564	.239436	3	.05
	58	.49950	.86632	.57657	1.73438	.698532	.937676	.760856	.239144	2	
	59	.49975	.86617	.57696	1.73321	.698751	.937604	.761148	.238852	1	
1.00	60	.50000	.86603	.57735	1.73205	9.698970	9.937531	9.761439	10.238561	0	.00
Decimals	Minutes	Cos	Sin	Cot	Tan	Cos	Sin	Cot	Tan	Minutes	Decimals
		Natural Values				Common Logarithms					

Decimals	Minutes	Natural Values				Common Logarithms				Minutes	Decimals
		Sin	Cos	Tan	Cot	Sin	Cos	Tan	Cot		
.00	0	.50000	.86603	.57735	1.73205	9.698970	9.937531	9.761439	10.238561	60	1.00
	1	.50025	.86588	.57774	1.73089	.699189	.937458	.761731	.238269	59	
	2	.50050	.86573	.57813	1.72973	.699407	.937385	.762023	.237977	58	
.05	3	.50076	.86559	.57851	1.72857	.699626	.937312	.762314	.237686	57	.95
	4	.50101	.86544	.57890	1.72741	.699844	.937238	.762606	.237394	56	
	5	.50126	.86530	.57929	1.72625	.700062	.937165	.762897	.237103	55	
.10	6	.50151	.86515	.57968	1.72509	.700280	.937092	.763188	.236812	54	.90
	7	.50176	.86501	.58007	1.72393	.700498	.937019	.763479	.236521	53	
	8	.50201	.86486	.58046	1.72278	.700716	.936946	.763770	.236230	52	
.15	9	.50227	.86471	.58085	1.72163	.700933	.936872	.764061	.235939	51	.85
	10	.50252	.86457	.58124	1.72047	.701151	.936799	.764352	.235648	50	
	11	.50277	.86442	.58162	1.71932	.701368	.936725	.764643	.235357	49	
.20	12	.50302	.86427	.58201	1.71817	.701585	.936652	.764933	.235067	48	.80
	13	.50327	.86413	.58240	1.71702	.701802	.936578	.765224	.234776	47	
	14	.50352	.86398	.58279	1.71588	.702019	.936505	.765514	.234486	46	
.25	15	.50377	.86384	.58318	1.71473	.702236	.936431	.765805	.234195	45	.75
	16	.50403	.86369	.58357	1.71358	.702452	.936357	.766095	.233905	44	
	17	.50428	.86354	.58396	1.71244	.702669	.936284	.766385	.233615	43	
.30	18	.50453	.86340	.58435	1.71129	.702885	.936210	.766675	.233325	42	.70
	19	.50478	.86325	.58474	1.71015	.703101	.936136	.766965	.233035	41	
	20	.50503	.86310	.58513	1.70901	.703317	.936062	.767255	.232745	40	
.35	21	.50528	.86295	.58552	1.70787	.703533	.935988	.767545	.232455	39	.65
	22	.50553	.86281	.58591	1.70673	.703749	.935914	.767834	.232166	38	
	23	.50578	.86266	.58631	1.70560	.703964	.935840	.768124	.231876	37	
.40	24	.50603	.86251	.58670	1.70446	.704179	.935766	.768414	.231586	36	.60
	25	.50628	.86237	.58709	1.70332	.704395	.935692	.768703	.231297	35	
	26	.50654	.86222	.58748	1.70219	.704610	.935618	.768992	.231008	34	
.45	27	.50679	.86207	.58787	1.70106	.704825	.935543	.769281	.230719	33	.55
	28	.50704	.86192	.58826	1.69992	.705040	.935469	.769571	.230429	32	
	29	.50729	.86178	.58865	1.69879	.705254	.935395	.769860	.230140	31	
.50	30	.50754	.86163	.58905	1.69766	.705469	.935320	.770148	.229852	30	.50
	31	.50779	.86148	.58944	1.69653	.705683	.935246	.770437	.229563	29	
	32	.50804	.86133	.58983	1.69541	.705898	.935171	.770726	.229274	28	
.55	33	.50829	.86119	.59022	1.69428	.706112	.935097	.771015	.228985	27	.45
	34	.50854	.86104	.59061	1.69316	.706326	.935022	.771303	.228697	26	
	35	.50879	.86089	.59101	1.69203	.706539	.934948	.771592	.228408	25	
.60	36	.50904	.86074	.59140	1.69091	.706753	.934873	.771880	.228120	24	.40
	37	.50929	.86059	.59179	1.68979	.706967	.934798	.772168	.227832	23	
	38	.50954	.86045	.59218	1.68866	.707180	.934723	.772457	.227543	22	
.65	39	.50979	.86030	.59258	1.68754	.707393	.934649	.772745	.227255	21	.35
	40	.51004	.86015	.59297	1.68643	.707606	.934574	.773033	.226967	20	
	41	.51029	.86000	.59336	1.68531	.707819	.934499	.773321	.226679	19	
.70	42	.51054	.85985	.59376	1.68419	.708032	.934424	.773608	.226392	18	.30
	43	.51079	.85970	.59415	1.68308	.708245	.934349	.773896	.226104	17	
	44	.51104	.85956	.59454	1.68196	.708458	.934274	.774184	.225816	16	
.75	45	.51129	.85941	.59494	1.68085	.708670	.934199	.774471	.225529	15	.25
	46	.51154	.85926	.59533	1.67974	.708882	.934123	.774759	.225241	14	
	47	.51179	.85911	.59573	1.67863	.709094	.934048	.775046	.224954	13	
.80	48	.51204	.85896	.59612	1.67752	.709306	.933973	.775333	.224667	12	.20
	49	.51229	.85881	.59651	1.67641	.709518	.933898	.775621	.224379	11	
	50	.51254	.85866	.59691	1.67530	.709730	.933822	.775908	.224092	10	
.85	51	.51279	.85851	.59730	1.67419	.709941	.933747	.776195	.223805	9	.15
	52	.51304	.85836	.59770	1.67309	.710153	.933671	.776482	.223518	8	
	53	.51329	.85821	.59809	1.67198	.710364	.933596	.776768	.223232	7	
.90	54	.51354	.85806	.59849	1.67088	.710575	.933520	.777055	.222945	6	.10
	55	.51379	.85792	.59888	1.66978	.710786	.933445	.777342	.222658	5	
	56	.51404	.85777	.59928	1.66867	.710997	.933369	.777628	.222372	4	
.95	57	.51429	.85762	.59967	1.66757	.711208	.933293	.777915	.222085	3	.05
	58	.51454	.85747	.60007	1.66647	.711419	.933217	.778201	.221799	2	
	59	.51479	.85732	.60046	1.66538	.711629	.933141	.778488	.221512	1	
1.00	60	.51504	.85717	.60086	1.66428	9.711839	9.933066	9.778774	10.221226	0	.00
Decimals	Minutes	Cos	Sin	Cot	Tan	Cos	Sin	Cot	Tan	Minutes	Decimals
		Natural Values				Common Logarithms					

105

Decimals	Minutes	Natural Values				Common Logarithms				Minutes	Decimals
		Sin	Cos	Tan	Cot	Sin	Cos	Tan	Cot		
.00	0	.51504	.85717	.60086	1.66428	9.711839	9.933066	9.778774	10.221226	60	1.00
	1	.51529	.85702	.60126	1.66318	.712050	.932990	.779060	.220940	59	
	2	.51554	.85687	.60165	1.66209	.712260	.932914	.779346	.220654	58	
.05	3	.51579	.85672	.60205	1.66099	.712469	.932838	.779632	.220368	57	.95
	4	.51604	.85657	.60245	1.65990	.712679	.932762	.779918	.220082	56	
	5	.51628	.85642	.60284	1.65881	.712889	.932685	.780203	.219797	55	
.10	6	.51653	.85627	.60324	1.65772	.713098	.932609	.780489	.219511	54	.90
	7	.51678	.85612	.60364	1.65663	.713308	.932533	.780775	.219225	53	
	8	.51703	.85597	.60403	1.65554	.713517	.932457	.781060	.218940	52	
.15	9	.51728	.85582	.60443	1.65445	.713726	.932380	.731346	.218654	51	.85
	10	.51753	.85567	.60483	1.65337	.713935	.932304	.781631	.218369	50	
	11	.51778	.85551	.60522	1.65228	.714144	.932228	.781916	.218084	49	
.20	12	.51803	.85536	.60562	1.65120	.714352	.932151	.782201	.217799	48	.80
	13	.51828	.85521	.60602	1.65011	.714561	.932075	.782486	.217514	47	
	14	.51852	.85506	.60642	1.64903	.714769	.931998	.782771	.217229	46	
.25	15	.51877	.85491	.60681	1.64795	.714978	.931921	.783056	.216944	45	.75
	16	.51902	.85476	.60721	1.64687	.715186	.931845	.783341	.216659	44	
	17	.51927	.85461	.60761	1.64579	.715394	.931768	.783626	.216374	43	
.30	18	.51952	.85446	.60801	1.64471	.715602	.931691	.783910	.216090	42	.70
	19	.51977	.85431	.60841	1.64363	.715809	.931614	.784195	.215805	41	
	20	.52002	.85416	.60881	1.64256	.716017	.931537	.784479	.215521	40	
.35	21	.52026	.85401	.60921	1.64148	.716224	.931460	.784764	.215236	39	.65
	22	.52051	.85385	.60960	1.64041	.716432	.931383	.785048	.214952	38	
	23	.52076	.85370	.61000	1.63934	.716639	.931306	.785332	.214668	37	
.40	24	.52101	.85355	.61040	1.63826	.716846	.931229	.785616	.214384	36	.60
	25	.52126	.85340	.61080	1.63719	.717053	.931152	.785900	.214100	35	
	26	.52151	.85325	.61120	1.63612	.717259	.931075	.786184	.213816	34	
.45	27	.52175	.85310	.61160	1.63505	.717466	.930998	.786468	.213532	33	.55
	28	.52200	.85294	.61200	1.63398	.717673	.930921	.786752	.213248	32	
	29	.52225	.85279	.61240	1.63292	.717879	.930843	.787036	.212964	31	
.50	30	.52250	.85264	.61280	1.63185	.718085	.930766	.737319	.212681	30	.50
	31	.52275	.85249	.61320	1.63079	.718291	.930688	.787603	.212397	29	
	32	.52299	.85234	.61360	1.62972	.718497	.930611	.787886	.212114	28	
.55	33	.52324	.85218	.61400	1.62866	.718703	.930533	.788170	.211830	27	.45
	34	.52349	.85203	.61440	1.62760	.718909	.930456	.788453	.211547	26	
	35	.52374	.85188	.61480	1.62654	.719114	.930378	.788736	.211264	25	
.60	36	.52399	.85173	.61520	1.62548	.719320	.930300	.789019	.210981	24	.40
	37	.52423	.85157	.61561	1.62442	.719525	.930223	.789302	.210698	23	
	38	.52448	.85142	.61601	1.62336	.719730	.930145	.789585	.210415	22	
.65	39	.52473	.85127	.61641	1.62230	.719935	.930067	.789868	.210132	21	.35
	40	.52498	.85112	.61681	1.62125	.720140	.929989	.790151	.209849	20	
	41	.52522	.85096	.61721	1.62019	.720345	.929911	.790434	.209566	19	
.70	42	.52547	.85081	.61761	1.61914	.720549	.929833	.790716	.209284	18	.30
	43	.52572	.85066	.61801	1.61808	.720754	.929755	.790999	.209001	17	
	44	.52597	.85051	.61842	1.61703	.720958	.929677	.791281	.208719	16	
.75	45	.52621	.85035	.61882	1.61598	.721162	.929599	.791563	.208437	15	.25
	46	.52646	.85020	.61922	1.61493	.721366	.929521	.791846	.208154	14	
	47	.52671	.85005	.61962	1.61388	.721570	.929442	.792128	.207872	13	
.80	48	.52696	.84989	.62003	1.61283	.721774	.929364	.792410	.207590	12	.20
	49	.52720	.84974	.62043	1.61179	.721978	.929286	.792692	.207308	11	
	50	.52745	.84959	.62083	1.61074	.722181	.929207	.792974	.207026	10	
.85	51	.52770	.84943	.62124	1.60970	.722385	.929129	.793256	.206744	9	.15
	52	.52794	.84928	.62164	1.60865	.722588	.929050	.793538	.206462	8	
	53	.52819	.84913	.62204	1.60761	.722791	.928972	.793819	.206181	7	
.90	54	.52844	.84897	.62245	1.60657	.722994	.928893	.794101	.205899	6	.10
	55	.52869	.84882	.62285	1.60553	.723197	.928815	.794383	.205617	5	
	56	.52893	.84866	.62325	1.60449	.723400	.928736	.794664	.205336	4	
.95	57	.52918	.84851	.62366	1.60345	.723603	.928657	.794946	.205054	3	.05
	58	.52943	.84836	.62406	1.60241	.723805	.928578	.795227	.204773	2	
	59	.52967	.84820	.62446	1.60137	.724007	.928499	.795508	.204492	1	
1.00	60	.52992	.84805	.62487	1.60033	9.724210	9.928420	9.795789	10.204211	0	.00

Decimals	Minutes	Cos	Sin	Cot	Tan	Cos	Sin	Cot	Tan	Minutes	Decimals
		Natural Values				Common Logarithms					

Decimals	Minutes	Sin	Cos	Tan	Cot	Sin	Cos	Tan	Cot	Minutes	Decimals
		Natural Values				**Common Logarithms**					
.00	0	.52992	.84805	.62487	1.60033	9.724210	9.928420	9.795789	10.204211	60	1.00
	1	.53017	.84789	.62527	1.59930	.724412	.928342	.796070	.203930	59	
	2	.53041	.84774	.62568	1.59826	.724614	.928263	.796351	.203649	58	
.05	3	.53066	.84759	.62608	1.59723	.724816	.928183	.796632	.203368	57	.95
	4	.53091	.84743	.62649	1.59620	.725017	.928104	.796913	.203087	56	
	5	.53115	.84728	.62689	1.59517	.725219	.928025	.797194	.202806	55	
.10	6	.53140	.84712	.62730	1.59414	.725420	.927946	.797474	.202526	54	.90
	7	.53164	.84697	.62770	1.59311	.725622	.927867	.797755	.202245	53	
	8	.53189	.84681	.62811	1.59208	.725823	.927787	.798036	.201964	52	
.15	9	.53214	.84666	.62852	1.59105	.726024	.927708	.798316	.201684	51	.85
	10	.53238	.84650	.62892	1.59002	.726225	.927629	.798596	.201404	50	
	11	.53263	.84635	.62933	1.58900	.726426	.927549	.798877	.201123	49	
.20	12	.53288	.84619	.62973	1.58797	.726626	.927470	.799157	.200843	48	.80
	13	.53312	.84604	.63014	1.58695	.726827	.927390	.799437	.200563	47	
	14	.53337	.84588	.63055	1.58593	.727027	.927310	.799717	.200283	46	
.25	15	.53361	.84573	.63095	1.58490	.727228	.927231	.799997	.200000	45	.75
	16	.53386	.84557	.63136	1.58388	.727428	.927151	.800277	.199723	44	
	17	.53411	.84542	.63177	1.58286	.727628	.927071	.800557	.199443	43	
.30	18	.53435	.84526	.63217	1.58184	.727828	.926991	.800836	.199164	42	.70
	19	.53460	.84511	.63258	1.58083	.728027	.926911	.801116	.198884	41	
	20	.53484	.84495	.63299	1.57981	.728227	.926831	.801396	.198604	40	
.35	21	.53509	.84480	.63340	1.57879	.728427	.926751	.801675	.198325	39	.65
	22	.53534	.84464	.63380	1.57778	.728626	.926671	.801955	.198045	38	
	23	.53558	.84448	.63421	1.57676	.728825	.926591	.802234	.197766	37	
.40	24	.53583	.84433	.63462	1.57575	.729024	.926511	.802513	.197487	36	.60
	25	.53607	.84417	.63503	1.57474	.729223	.926431	.802792	.197208	35	
	26	.53632	.84402	.63544	1.57372	.729422	.926351	.803072	.196928	34	
.45	27	.53656	.84386	.63584	1.57271	.729621	.926270	.803351	.196649	33	.55
	28	.53681	.84370	.63625	1.57170	.729820	.926190	.803630	.196370	32	
	29	.53705	.84355	.63666	1.57069	.730018	.926110	.803909	.196091	31	
.50	30	.53730	.84339	.63707	1.56969	.730217	.926029	.804187	.195813	30	.50
	31	.53754	.84324	.63748	1.56868	.730415	.925949	.804466	.195534	29	
	32	.53779	.84308	.63789	1.56767	.730613	.925868	.804745	.195255	28	
.55	33	.53804	.84292	.63830	1.56667	.730811	.925788	.805023	.194977	27	.45
	34	.53828	.84277	.63871	1.56566	.731009	.925707	.805302	.194698	26	
	35	.53853	.84261	.63912	1.56466	.731206	.925626	.805580	.194420	25	
.60	36	.53877	.84245	.63953	1.56366	.731404	.925545	.805859	.194141	24	.40
	37	.53902	.84230	.63994	1.56265	.731602	.925465	.806137	.193863	23	
	38	.53926	.84214	.64035	1.56165	.731799	.925384	.806415	.193585	22	
.65	39	.53951	.84198	.64076	1.56065	.731996	.925303	.806693	.193307	21	.35
	40	.53975	.84182	.64117	1.55966	.732193	.925222	.806971	.193029	20	
	41	.54000	.84167	.64158	1.55866	.732390	.925141	.807249	.192751	19	
.70	42	.54024	.84151	.64199	1.55766	.732587	.925060	.807527	.192473	18	.30
	43	.54049	.84135	.64240	1.55666	.732784	.924979	.807805	.192195	17	
	44	.54073	.84120	.64281	1.55567	.732980	.924897	.808083	.191917	16	
.75	45	.54097	.84104	.64322	1.55467	.733177	.924816	.808361	.191639	15	.25
	46	.54122	.84088	.64363	1.55368	.733373	.924735	.808638	.191362	14	
	47	.54146	.84072	.64404	1.55269	.733569	.924654	.808916	.191084	13	
.80	48	.54171	.84057	.64446	1.55170	.733765	.924572	.809193	.190807	12	.20
	49	.54195	.84041	.64487	1.55071	.733961	.924491	.809471	.190529	11	
	50	.54220	.84025	.64528	1.54972	.734157	.924409	.809748	.190252	10	
.85	51	.54244	.84009	.64569	1.54873	.734353	.924328	.810025	.189975	9	.15
	52	.54269	.83994	.64610	1.54774	.734549	.924246	.810302	.189698	8	
	53	.54293	.83978	.64652	1.54675	.734744	.924164	.810580	.189420	7	
.90	54	.54317	.83962	.64693	1.54576	.734939	.924083	.810857	.189143	6	.10
	55	.54342	.83946	.64734	1.54478	.735135	.924001	.811134	.188866	5	
	56	.54366	.83930	.64775	1.54379	.735330	.923919	.811410	.188590	4	
.95	57	.54391	.83915	.64817	1.54281	.735525	.923837	.811687	.188313	3	.05
	58	.54415	.83899	.64858	1.54183	.735719	.923755	.811964	.188036	2	
	59	.54440	.83883	.64899	1.54085	.735914	.923673	.812241	.187759	1	
1.00	60	.54464	.83867	.64941	1.53986	9.736109	9.923591	9.812517	10.187483	0	.00
Decimals	Minutes	Cos	Sin	Cot	Tan	Cos	Sin	Cot	Tan	Minutes	Decimals
		Natural Values				**Common Logarithms**					

57°

Decimals	Minutes	Natural Values Sin	Cos	Tan	Cot	Common Logarithms Sin	Cos	Tan	Cot	Minutes	Decimals
.00	0	.54464	.83867	.64941	1.53986	9.736109	9.923591	9.812517	10.187483	60	1.00
	1	.54488	.83851	.64982	1.53888	.736303	.923509	.812794	.187206	59	
	2	.54513	.83835	.65024	1.53791	.736498	.923427	.813070	.186930	58	
.05	3	.54537	.83819	.65065	1.53693	.736692	.923345	.813347	.186653	57	.95
	4	.54561	.83804	.65106	1.53595	.736886	.923263	.813623	.186377	56	
	5	.54586	.83788	.65148	1.53497	.737080	.923181	.813899	.186101	55	
.10	6	.54610	.83772	.65189	1.53400	.737274	.923098	.814176	.185824	54	.90
	7	.54635	.83756	.65231	1.53302	.737467	.923016	.814452	.185548	53	
	8	.54659	.83740	.65272	1.53205	.737661	.922933	.814728	.185272	52	
.15	9	.54683	.83724	.65314	1.53107	.737855	.922851	.815004	.184996	51	.85
	10	.54708	.83708	.65355	1.53010	.738048	.922768	.815280	.184720	50	
	11	.54732	.83692	.65397	1.52913	.738241	.922686	.815555	.184445	49	
.20	12	.54756	.83676	.65438	1.52816	.738434	.922603	.815831	.184169	48	.80
	13	.54781	.83660	.65480	1.52719	.738627	.922520	.816107	.183893	47	
	14	.54805	.83645	.65521	1.52622	.738820	.922438	.816382	.183618	46	
.25	15	.54829	.83629	.65563	1.52525	.739013	.922355	.816658	.183342	45	.75
	16	.54854	.83613	.65604	1.52429	.739206	.922272	.816933	.183067	44	
	17	.54878	.83597	.65646	1.52332	.739398	.922189	.817209	.182791	43	
.30	18	.54902	.83581	.65688	1.52235	.739590	.922106	.817484	.182516	42	.70
	19	.54927	.83565	.65729	1.52139	.739783	.922023	.817759	.182241	41	
	20	.54951	.83549	.65771	1.52043	.739975	.921940	.818035	.181965	40	
.35	21	.54975	.83533	.65813	1.51946	.740167	.921857	.818310	.181690	39	.65
	22	.54999	.83517	.65854	1.51850	.740359	.921774	.818585	.181415	38	
	23	.55024	.83501	.65896	1.51754	.740550	.921691	.818860	.181140	37	
.40	24	.55048	.83485	.65938	1.51658	.740742	.921607	.819135	.180865	36	.60
	25	.55072	.83469	.65980	1.51562	.740934	.921524	.819410	.180590	35	
	26	.55097	.83453	.66021	1.51466	.741125	.921441	.819684	.180316	34	
.45	27	.55121	.83437	.66063	1.51370	.741316	.921357	.819959	.180041	33	.55
	28	.55145	.83421	.66105	1.51275	.741508	.921274	.820234	.179766	32	
	29	.55169	.83405	.66147	1.51179	.741699	.921190	.820508	.179492	31	
.50	30	.55194	.83389	.66189	1.51084	.741889	.921107	.820783	.179217	30	.50
	31	.55218	.83373	.66230	1.50988	.742080	.921023	.821057	.178943	29	
	32	.55242	.83356	.66272	1.50893	.742271	.920939	.821332	.178668	28	
.55	33	.55266	.83340	.66314	1.50797	.742462	.920856	.821606	.178394	27	.45
	34	.55291	.83324	.66356	1.50702	.742652	.920772	.821880	.178120	26	
	35	.55315	.83308	.66398	1.50607	.742842	.920688	.822154	.177846	25	
.60	36	.55339	.83292	.66440	1.50512	.743033	.920604	.822429	.177571	24	.40
	37	.55363	.83276	.66482	1.50417	.743223	.920520	.822703	.177297	23	
	38	.55388	.83260	.66524	1.50322	.743413	.920436	.822977	.177023	22	
.65	39	.55412	.83244	.66566	1.50228	.743602	.920352	.823251	.176749	21	.35
	40	.55436	.83228	.66608	1.50133	.743792	.920268	.823524	.176476	20	
	41	.55460	.83212	.66650	1.50038	.743982	.920184	.823798	.176202	19	
.70	42	.55484	.83195	.66692	1.49944	.744171	.920099	.824072	.175928	18	.30
	43	.55509	.83179	.66734	1.49849	.744361	.920015	.824345	.175655	17	
	44	.55533	.83163	.66776	1.49755	.744550	.919931	.824619	.175381	16	
.75	45	.55557	.83147	.66818	1.49661	.744739	.919846	.824893	.175107	15	.25
	46	.55581	.83131	.66860	1.49566	.744928	.919762	.825166	.174834	14	
	47	.55605	.83115	.66902	1.49472	.745117	.919677	.825439	.174561	13	
.80	48	.55630	.83098	.66944	1.49378	.745306	.919593	.825713	.174287	12	.20
	49	.55654	.83082	.66986	1.49284	.745494	.919508	.825986	.174014	11	
	50	.55678	.83066	.67028	1.49190	.745683	.919424	.826259	.173741	10	
.85	51	.55702	.83050	.67071	1.49097	.745871	.919339	.826532	.173468	9	.15
	52	.55726	.83034	.67113	1.49003	.746060	.919254	.826805	.173195	8	
	53	.55750	.83017	.67155	1.48909	.746248	.919169	.827078	.172922	7	
.90	54	.55775	.83001	.67197	1.48816	.746436	.919085	.827351	.172649	6	.10
	55	.55799	.82985	.67239	1.48722	.746624	.919000	.827624	.172376	5	
	56	.55823	.82969	.67282	1.48629	.746812	.918915	.827897	.172103	4	
.95	57	.55847	.82953	.67324	1.48536	.746999	.918830	.828170	.171830	3	.05
	58	.55871	.82936	.67366	1.48442	.747187	.918745	.828442	.171558	2	
	59	.55895	.82920	.67409	1.48349	.747373	.918659	.828715	.171285	1	
1.00	60	.55919	.82904	.67451	1.48256	9.747562	9.918574	9.828987	10.171013	0	.00
Decimals	Minutes	Cos	Sin	Cot	Tan	Cos	Sin	Cot	Tan	Minutes	Decimals
		Natural Values				Common Logarithms					

56°

Decimals	Minutes	Natural Values Sin	Cos	Tan	Cot	Common Logarithms Sin	Cos	Tan	Cot	Minutes	Decimals
.00	0	.55919	.82904	.67451	1.48256	9.747562	9.918574	9.828987	10.171013	60	1.00
	1	.55943	.82887	.67493	1.48163	.747749	.918489	.829260	.170740	59	
	2	.55968	.82871	.67536	1.48070	.747936	.918404	.829532	.170468	58	
.05	3	.55992	.82855	.67578	1.47977	.748123	.918318	.829805	.170195	57	.95
	4	.56016	.82839	.67620	1.47885	.748310	.918233	.830077	.169923	56	
	5	.56040	.82822	.67663	1.47792	.748497	.918147	.830349	.169651	55	
.10	6	.56064	.82806	.67705	1.47699	.748683	.918062	.830621	.169379	54	.90
	7	.56088	.82790	.67748	1.47607	.748870	.917976	.830893	.169107	53	
	8	.56112	.82773	.67790	1.47514	.749056	.917891	.831165	.168835	52	
.15	9	.56136	.82757	.67832	1.47422	.749243	.917805	.831437	.168563	51	.85
	10	.56160	.82741	.67875	1.47330	.749429	.917719	.831709	.168291	50	
	11	.56184	.82724	.67917	1.47238	.749615	.917634	.831981	.168019	49	
.20	12	.56208	.82708	.67960	1.47146	.749801	.917548	.832253	.167747	48	.80
	13	.56232	.82692	.68002	1.47053	.749987	.917462	.832525	.167475	47	
	14	.56256	.82675	.68045	1.46962	.750172	.917376	.832796	.167204	46	
.25	15	.56280	.82659	.68088	1.46870	.750358	.917290	.833068	.166932	45	.75
	16	.56305	.82643	.68130	1.46778	.750543	.917204	.833339	.166661	44	
	17	.56329	.82626	.68173	1.46686	.750729	.917118	.833611	.166389	43	
.30	18	.56353	.82610	.68215	1.46595	.750914	.917032	.833882	.166118	42	.70
	19	.56377	.82593	.68258	1.46503	.751099	.916946	.834154	.165846	41	
	20	.56401	.82577	.68301	1.46411	.751284	.916859	.834425	.165575	40	
.35	21	.56425	.82561	.68343	1.46320	.751469	.916773	.834696	.165304	39	.65
	22	.56449	.82544	.68386	1.46229	.751654	.916687	.834967	.165033	38	
	23	.56473	.82528	.68429	1.46137	.751839	.916600	.835238	.164762	37	
.40	24	.56497	.82511	.68471	1.46046	.752023	.916514	.835509	.164491	36	.60
	25	.56521	.82495	.68514	1.45955	.752208	.916427	.835780	.164220	35	
	26	.56545	.82478	.68557	1.45864	.752392	.916341	.836051	.163949	34	
.45	27	.56569	.82462	.68600	1.45773	.752576	.916254	.836322	.163678	33	.55
	28	.56593	.82446	.68642	1.45682	.752760	.916167	.836593	.163407	32	
	29	.56617	.82429	.68685	1.45592	.752944	.916081	.836864	.163136	31	
.50	30	.56641	.82413	.68728	1.45501	.753128	.915994	.837134	.162866	30	.50
	31	.56665	.82396	.68771	1.45410	.753312	.915907	.837405	.162595	29	
	32	.56689	.82380	.68814	1.45320	.753495	.915820	.837675	.162325	28	
.55	33	.56713	.82363	.68857	1.45229	.753679	.915733	.837946	.162054	27	.45
	34	.56736	.82347	.68900	1.45139	.753862	.915646	.838216	.161784	26	
	35	.56760	.82330	.68942	1.45049	.754046	.915559	.838487	.161513	25	
.60	36	.56784	.82314	.68985	1.44958	.754229	.915472	.838757	.161243	24	.40
	37	.56808	.82297	.69028	1.44868	.754412	.915385	.839027	.160973	23	
	38	.56832	.82281	.69071	1.44778	.754595	.915297	.839297	.160703	22	
.65	39	.56856	.82264	.69114	1.44688	.754778	.915210	.839566	.160432	21	.35
	40	.56880	.82248	.69157	1.44598	.754960	.915123	.839838	.160162	20	
	41	.56904	.82231	.69200	1.44508	.755143	.915035	.840108	.159892	19	
.70	42	.56928	.82214	.69243	1.44418	.755326	.914948	.840378	.159622	18	.30
	43	.56952	.82198	.69286	1.44329	.755508	.914860	.840648	.159352	17	
	44	.56976	.82181	.69329	1.44239	.755690	.914773	.840917	.159083	16	
.75	45	.57000	.82165	.69372	1.44149	.755872	.914685	.841187	.158813	15	.25
	46	.57024	.82148	.69416	1.44060	.756054	.914598	.841457	.158543	14	
	47	.57047	.82132	.69459	1.43970	.756236	.914510	.841727	.158273	13	
.80	48	.57071	.82115	.69502	1.43881	.756418	.914422	.841996	.158004	12	.20
	49	.57095	.82098	.69545	1.43792	.756600	.914334	.842266	.157734	11	
	50	.57119	.82082	.69588	1.43703	.756782	.914246	.842535	.157465	10	
.85	51	.57143	.82065	.69631	1.43614	.756963	.914158	.842805	.157195	9	.15
	52	.57167	.82048	.69675	1.43525	.757144	.914070	.843074	.156926	8	
	53	.57191	.82032	.69718	1.43436	.757326	.913982	.843343	.156657	7	
.90	54	.57215	.82015	.69761	1.43347	.757507	.913894	.843612	.156388	6	.10
	55	.57238	.81999	.69804	1.43258	.757688	.913806	.843882	.156118	5	
	56	.57262	.81982	.69847	1.43169	.757869	.913718	.844151	.155849	4	
.95	57	.57286	.81965	.69891	1.43080	.758050	.913630	.844420	.155580	3	.05
	58	.57310	.81949	.69934	1.42992	.758230	.913541	.844689	.155311	2	
	59	.57334	.81932	.69977	1.42903	.758411	.913453	.844958	.155042	1	
1.00	60	.57358	.81915	.70021	1.42815	9.758591	9.913365	9.845227	10.154773	0	.00

Decimals	Minutes	Cos	Sin	Cot	Tan	Cos	Sin	Cot	Tan	Minutes	Decimals
		Natural Values				Common Logarithms					

55°

Decimals	Minutes	Natural Values				Common Logarithms				Minutes	Decimals
		Sin	Cos	Tan	Cot	Sin	Cos	Tan	Cot		
.00	0	.57358	.81915	.70021	1.42815	9.758591	9.913365	9.845227	10.154773	60	1.00
	1	.57381	.81899	.70064	1.42726	.758772	.913276	.845496	.154504	59	
	2	.57405	.81882	.70170	1.42638	.758952	.913187	.845764	.154236	58	
.05	3	.57429	.81865	.70151	1.42550	.759132	.913099	.846033	.153967	57	.95
	4	.57453	.81848	.70194	1.42462	.759312	.913010	.846302	.153698	56	
	5	.57477	.81832	.70238	1.42374	.759492	.912922	.846570	.153430	55	
.10	6	.57501	.81815	.70281	1.42286	.759672	.912833	.846839	.153161	54	.90
	7	.57524	.81798	.70325	1.42198	.759852	.912744	.847108	.152892	53	
	8	.57548	.81782	.70368	1.42110	.760031	.912655	.847376	.152624	52	
.15	9	.57572	.81765	.70412	1.42022	.760211	.912566	.847644	.152356	51	.85
	10	.57596	.81748	.70455	1.41934	.760390	.912477	.847913	.152087	50	
	11	.57619	.81731	.70499	1.41847	.760569	.912388	.848181	.151819	49	
.20	12	.57643	.81714	.70542	1.41759	.760748	.912299	.848449	.151551	48	.80
	13	.57667	.81698	.70586	1.41672	.760927	.912210	.848717	.151283	47	
	14	.57691	.81681	.70629	1.41584	.761106	.912121	.848986	.151014	46	
.25	15	.57715	.81664	.70673	1.41497	.761285	.912031	.849254	.150746	45	.75
	16	.57738	.81647	.70717	1.41409	.761464	.911942	.849522	.150478	44	
	17	.57762	.81631	.70760	1.41322	.761642	.911853	.849790	.150210	43	
.30	18	.57786	.81614	.70804	1.41235	.761821	.911763	.850057	.149943	42	.70
	19	.57810	.81597	.70848	1.41148	.761999	.911674	.850325	.149675	41	
	20	.57833	.81580	.70891	1.41061	.762177	.911584	.850593	.149407	40	
.35	21	.57857	.81563	.70935	1.40974	.762356	.911495	.850861	.149139	39	.65
	22	.57881	.81546	.70979	1.40887	.762534	.911405	.851129	.148871	38	
	23	.57904	.81530	.71023	1.40800	.762712	.911315	.851396	.148604	37	
.40	24	.57928	.81513	.71066	1.40714	.762889	.911226	.851664	.148336	36	.60
	25	.57952	.81496	.71110	1.40627	.763067	.911136	.851931	.148069	35	
	26	.57976	.81479	.71154	1.40540	.763245	.911046	.852199	.147801	34	
.45	27	.57999	.81462	.71198	1.40454	.763422	.910956	.852466	.147534	33	.55
	28	.58023	.81445	.71242	1.40367	.763600	.910866	.852733	.147267	32	
	29	.58047	.81428	.71285	1.40281	.763777	.910776	.853001	.146999	31	
.50	30	.58070	.81412	.71329	1.40195	.763954	.910686	.853268	.146732	30	.50
	31	.58094	.81395	.71373	1.40109	.764131	.910596	.853535	.146465	29	
	32	.58118	.81378	.71417	1.40022	.764308	.910506	.853802	.146198	28	
.55	33	.58141	.81361	.71461	1.39936	.764485	.910415	.854069	.145931	27	.45
	34	.58165	.81344	.71505	1.39850	.764662	.910325	.854336	.145664	26	
	35	.58189	.81327	.71549	1.39764	.764838	.910235	.854603	.145397	25	
.60	36	.58212	.81310	.71593	1.39679	.765015	.910144	.854870	.145130	24	.40
	37	.58236	.81293	.71637	1.39593	.765191	.910054	.855137	.144863	23	
	38	.58260	.81276	.71681	1.39507	.765367	.909963	.855404	.144596	22	
.65	39	.58283	.81259	.71725	1.39421	.765544	.909873	.855671	.144329	21	.35
	40	.58307	.81242	.71769	1.39336	.765720	.909782	.855938	.144062	20	
	41	.58330	.81225	.71813	1.39250	.765896	.909691	.856204	.143796	19	
.70	42	.58354	.81208	.71857	1.39165	.766072	.909601	.856471	.143529	18	.30
	43	.58378	.81191	.71901	1.39079	.766247	.909510	.856737	.143263	17	
	44	.58401	.81174	.71946	1.38994	.766423	.909419	.857004	.142996	16	
.75	45	.58425	.81157	.71990	1.38909	.766598	.909328	.857270	.142730	15	.25
	46	.58449	.81140	.72034	1.38824	.766774	.909237	.857537	.142463	14	
	47	.58472	.81123	.72078	1.38738	.766949	.909146	.857803	.142197	13	
.80	48	.58496	.81106	.72122	1.38653	.767124	.909055	.858069	.141931	12	.20
	49	.58519	.81089	.72167	1.38568	.767300	.908964	.858336	.141664	11	
	50	.58543	.81072	.72211	1.38484	.767475	.908873	.858602	.141398	10	
.85	51	.58567	.81055	.72255	1.38399	.767649	.908781	.858868	.141132	9	.15
	52	.58590	.81038	.72299	1.38314	.767824	.908690	.859134	.140866	8	
	53	.58614	.81021	.72344	1.38229	.767999	.908599	.859400	.140600	7	
.90	54	.58637	.81004	.72388	1.38145	.768173	.908507	.859666	.140334	6	.10
	55	.58661	.80987	.72432	1.38060	.768348	.908416	.859932	.140068	5	
	56	.58684	.80970	.72477	1.37976	.768522	.908324	.860198	.139802	4	
.95	57	.58708	.80953	.72521	1.37891	.768697	.908233	.860464	.139536	3	.05
	58	.58731	.80936	.72565	1.37807	.768871	.908141	.860730	.139270	2	
	59	.58755	.80919	.72610	1.37722	.769045	.908049	.860995	.139005	1	
1.00	60	.58779	.80902	.72654	1.37638	9.769219	9.907958	9.861261	10.138739	0	.00

Decimals	Minutes	Cos	Sin	Cot	Tan	Cos	Sin	Cot	Tan	Minutes	Decimals
		Natural Values				Common Logarithms					

110

54°

Decimals	Minutes	Natural Values				Common Logarithms				Minutes	Decimals
		Sin	Cos	Tan	Cot	Sin	Cos	Tan	Cot		
.00	0	.58779	.80902	.72654	1.37638	9.769219	9.907958	9.861261	10.138739	60	1.00
	1	.58802	.80885	.72699	1.37554	.769393	.907866	.861527	.138473	59	
	2	.58826	.80867	.72743	1.37470	.769566	.907774	.861792	.138208	58	
.05	3	.58849	.80850	.72788	1.37386	.769740	.907682	.862058	.137942	57	.95
	4	.58873	.80833	.72832	1.37302	.769913	.907590	.862323	.137677	56	
	5	.58896	.80816	.72877	1.37218	.770087	.907498	.862589	.137411	55	
.10	6	.58920	.80799	.72921	1.37134	.770260	.907406	.862854	.137146	54	.90
	7	.58943	.80782	.72966	1.37050	.770433	.907314	.863119	.136881	53	
	8	.58967	.80765	.73010	1.36967	.770606	.907222	.863385	.136615	52	
.15	9	.58990	.80748	.73055	1.36883	.770779	.907129	.863650	.136350	51	.85
	10	.59014	.80730	.73100	1.36800	.770952	.907037	.863915	.136085	50	
	11	.59037	.80713	.73144	1.36716	.771125	.906945	.864180	.135820	49	
.20	12	.59061	.80696	.73189	1.36633	.771298	.906852	.864445	.135555	48	.80
	13	.59084	.80679	.73234	1.36549	.771470	.906760	.864710	.135290	47	
	14	.59108	.80662	.73278	1.36466	.771643	.906667	.864975	.135025	46	
.25	15	.59131	.80644	.73323	1.36383	.771815	.906575	.865240	.134760	45	.75
	16	.59154	.80627	.73368	1.36300	.771987	.906482	.865505	.134495	44	
	17	.59178	.80610	.73413	1.36217	.772159	.906389	.865770	.134230	43	
.30	18	.59201	.80593	.73457	1.36134	.772331	.906296	.866035	.133965	42	.70
	19	.59225	.80576	.73502	1.36051	.772503	.906204	.866300	.133700	41	
	20	.59248	.80558	.73547	1.35968	.772675	.906111	.866564	.133436	40	
.35	21	.59272	.80541	.73592	1.35885	.772847	.906018	.866829	.133171	39	.65
	22	.59295	.80524	.73637	1.35802	.773018	.905925	.867094	.132906	38	
	23	.59318	.80507	.73681	1.35719	.773190	.905832	.867358	.132642	37	
.40	24	.59342	.80489	.73726	1.35637	.773361	.905739	.867623	.132377	36	.60
	25	.59365	.80472	.73771	1.35554	.773533	.905645	.867887	.132113	35	
	26	.59389	.80455	.73816	1.35472	.773704	.905552	.868152	.131848	34	
45	27	.59412	.80438	.73861	1.35389	.773875	.905459	.868416	.131584	33	.55
	28	.59436	.80420	.73906	1.35307	.774046	.905366	.868680	.131320	32	
	29	.59459	.80403	.73951	1.35224	.774217	.905272	.868945	.131055	31	
.50	30	.59482	.80386	.73996	1.35142	.774388	.905179	.869209	.130791	30	.50
	31	.59506	.80368	.74041	1.35060	.774558	.905085	.869473	.130527	29	
	32	.59529	.80351	.74086	1.34978	.774729	.904992	.869737	.130263	28	
.55	33	.59552	.80334	.74131	1.34896	.774899	.904898	.870001	.129999	27	.45
	34	.59576	.80316	.74176	1.34814	.775070	.904804	.870265	.129735	26	
	35	.59599	.80299	.74221	1.34732	.775240	.904711	.870529	.129471	25	
.60	36	.59622	.80282	.74267	1.34650	.775410	.904617	.870793	.129207	24	.40
	37	.59646	.80264	.74312	1.34568	.775580	.904523	.871057	.128943	23	
	38	.59669	.80247	.74357	1.34487	.775750	.904429	.871321	.128679	22	
.65	39	.59693	.80230	.74402	1.34405	.775920	.904335	.871585	.128415	21	.35
	40	.59716	.80212	.74447	1.34323	.776090	.904241	.871849	.128151	20	
	41	.59739	.80195	.74492	1.34242	.776259	.904147	.872112	.127888	19	
.70	42	.59763	.80178	.74538	1.34160	.776429	.904053	.872376	.127624	18	.30
	43	.59786	.80160	.74583	1.34079	.776598	.903959	.872640	.127360	17	
	44	.59809	.80143	.74628	1.33998	.776768	.903864	.872903	.127097	16	
.75	45	.59832	.80125	.74674	1.33916	.776937	.903770	.873167	.126833	15	.25
	46	.59856	.80108	.74719	1.33835	.777106	.903676	.873430	.126570	14	
	47	.59879	.80091	.74764	1.33754	.777275	.903581	.873694	.126306	13	
.80	48	.59902	.80073	.74810	1.33673	.777444	.903487	.873957	.126043	12	.20
	49	.59926	.80056	.74855	1.33592	.777613	.903392	.874220	.125780	11	
	50	.59949	.80038	.74900	1.33511	.777781	.903298	.874484	.125516	10	
.85	51	.59972	.80021	.74946	1.33430	.777950	.903203	.874747	.125253	9	.15
	52	.59995	.80003	.74991	1.33349	.778119	.903108	.875010	.124990	8	
	53	.60019	.79986	.75037	1.33268	.778287	.903014	.875273	.124727	7	
.90	54	.60042	.79968	.75082	1.33187	.778455	.902919	.875537	.124463	6	.10
	55	.60065	.79951	.75128	1.33107	.778624	.902824	.875800	.124200	5	
	56	.60089	.79934	.75173	1.33026	.778792	.902729	.876063	.123937	4	
.95	57	.60112	.79916	.75219	1.32946	.778960	.902634	.876326	.123674	3	.05
	58	.60135	.79899	.75264	1.32865	.779128	.902539	.876589	.123411	2	
	59	.60158	.79881	.75310	1.32785	.779295	.902444	.876852	.123148	1	
1.00	60	.60182	.79864	.75355	1.32704	9.779463	9.902349	9.877114	10.122886	0	.00
Decimals	Minutes	Cos	Sin	Cot	Tan	Cos	Sin	Cot	Tan	Minutes	Decimals
		Natural Values				Common Logarithms					

111

Decimals	Minutes	Natural Values				Common Logarithms				Minutes	Decimals
		Sin	Cos	Tan	Cot	Sin	Cos	Tan	Cot		
.00	0	.60182	.79864	.75355	1.32704	9.779463	9.902349	9.877114	10.122886	60	1.00
	1	.60205	.79846	.75401	1.32624	.779631	.902253	.877377	.122623	59	
	2	.60228	.79829	.75447	1.32544	.779798	.902158	.877640	.122360	58	
.05	3	.60251	.79811	.75492	1.32464	.779966	.902063	.877903	.122097	57	.95
	4	.60274	.79793	.75538	1.32384	.780133	.901967	.878165	.121835	56	
	5	.60298	.79776	.75584	1.32304	.780300	.901872	.878428	.121572	55	
.10	6	.60321	.79758	.75629	1.32224	.780467	.901776	.878691	.121309	54	.90
	7	.60344	.79741	.75675	1.32144	.780634	.901681	.878953	.121047	53	
	8	.60367	.79723	.75721	1.32064	.780801	.901585	.879216	.120784	52	
.15	9	.60390	.79706	.75767	1.31984	.780968	.901490	.879478	.120522	51	.85
	10	.60414	.79688	.75812	1.31904	.781134	.901394	.879741	.120259	50	
	11	.60437	.79671	.75858	1.31825	.781301	.901298	.880003	.119997	49	
.20	12	.60460	.79653	.75904	1.31745	.781468	.901202	.880265	.119735	48	.80
	13	.60483	.79635	.75950	1.31666	.781634	.901106	.880528	.119472	47	
	14	.60506	.79618	.75996	1.31586	.781800	.901010	.880790	.119210	46	
.25	15	.60529	.79600	.76042	1.31507	.781966	.900914	.881052	.118948	45	.75
	16	.60553	.79583	.76088	1.31427	.782132	.900818	.881314	.118686	44	
	17	.60576	.79565	.76134	1.31348	.782298	.900722	.881577	.118423	43	
.30	18	.60599	.79547	.76180	1.31269	.782464	.900626	.881839	.118161	42	.70
	19	.60622	.79530	.76226	1.31190	.782630	.900529	.882101	.117899	41	
	20	.60645	.79512	.76272	1.31110	.782796	.900433	.882363	.117637	40	
.35	21	.60668	.79494	.76318	1.31031	.782961	.900337	.882625	.117375	39	.65
	22	.60691	.79477	.76364	1.30952	.783127	.900240	.882887	.117113	38	
	23	.60714	.79459	.76410	1.30873	.783292	.900144	.883148	.116852	37	
.40	24	.60738	.79441	.76456	1.30795	.783458	.900047	.883410	.116590	36	.60
	25	.60761	.79424	.76502	1.30716	.783623	.899951	.883672	.116328	35	
	26	.60784	.79406	.76548	1.30637	.783788	.899854	.883934	.116066	34	
.45	27	.60807	.79388	.76594	1.30558	.783953	.899757	.884196	.115804	33	.55
	28	.60830	.79371	.76640	1.30480	.784118	.899660	.884457	.115543	32	
	29	.60853	.79353	.76686	1.30401	.784282	.899564	.884719	.115281	31	
.50	30	.60876	.79335	.76733	1.30323	.784447	.899467	.884980	.115020	30	.50
	31	.60899	.79318	.76779	1.30244	.784612	.899370	.885242	.114758	29	
	32	.60922	.79300	.76825	1.30166	.784776	.899273	.885504	.114496	28	
.55	33	.60945	.79282	.76871	1.30087	.784941	.899176	.885765	.114235	27	.45
	34	.60968	.79264	.76918	1.30009	.785105	.899078	.886026	.113974	26	
	35	.60991	.79247	.76964	1.29931	.785269	.898981	.886288	.113712	25	
.60	36	.61015	.79229	.77010	1.29853	.785433	.898884	.886549	.113451	24	.40
	37	.61038	.79211	.77057	1.29775	.785597	.898787	.886811	.113189	23	
	38	.61061	.79193	.77103	1.29696	.785761	.898689	.887072	.112928	22	
.65	39	.61084	.79176	.77149	1.29618	.785925	.898592	.887333	.112667	21	.35
	40	.61107	.79158	.77196	1.29541	.786089	.898494	.887594	.112406	20	
	41	.61130	.79140	.77242	1.29463	.786252	.898397	.887855	.112145	19	
.70	42	.61153	.79122	.77289	1.29385	.786416	.898299	.888116	.111884	18	.30
	43	.61176	.79105	.77335	1.29307	.786579	.898202	.888378	.111622	17	
	44	.61199	.79087	.77382	1.29229	.786742	.898104	.888639	.111361	16	
.75	45	.61222	.79069	.77428	1.29152	.786906	.898006	.888900	.111100	15	.25
	46	.61245	.79051	.77475	1.29074	.787069	.897908	.889161	.110839	14	
	47	.61268	.79033	.77521	1.28997	.787232	.897810	.889421	.110579	13	
.80	48	.61291	.79016	.77568	1.28919	.787395	.897712	.889682	.110318	12	.20
	49	.61314	.78998	.77615	1.28842	.787557	.897614	.889943	.110057	11	
	50	.61337	.78980	.77661	1.28764	.787720	.897516	.890204	.109796	10	
.85	51	.61360	.78962	.77708	1.28687	.787883	.897418	.890465	.109535	9	.15
	52	.61383	.78944	.77754	1.28610	.788045	.897320	.890725	.109275	8	
	53	.61406	.78926	.77801	1.28533	.788208	.897222	.890986	.109014	7	
.90	54	.61429	.78908	.77848	1.28456	.788370	.897123	.891247	.108753	6	.10
	55	.61451	.78891	.77895	1.28379	.788532	.897025	.891507	.108493	5	
	56	.61474	.78873	.77941	1.28302	.788694	.896926	.891768	.108232	4	
.95	57	.61497	.78855	.77988	1.28225	.788856	.896828	.892028	.107972	3	.05
	58	.61520	.78837	.78035	1.28148	.789018	.896729	.892289	.107711	2	
	59	.61543	.78819	.78082	1.28071	.789180	.896631	.892549	.107451	1	
1.00	60	.61566	.78801	.78129	1.27994	9.789342	9.896532	9.892810	10.107190	0	.00
Decimals	Minutes	Cos	Sin	Cot	Tan	Cos	Sin	Cot	Tan	Minutes	Decimals
		Natural Values				Common Logarithms					

52°

Decimals	Minutes	Natural Values				Common Logarithms				Minutes	Decimals
		Sin	Cos	Tan	Cot	Sin	Cos	Tan	Cot		
.00	0	.61566	.78801	.78129	1.27994	9.789342	9.896532	9.892810	10.107190	60	1.00
	1	.61589	.78783	.78175	1.27917	.789504	.896433	.893070	.106930	59	
	2	.61612	.78765	.78222	1.27841	.789665	.896335	.893331	.106669	58	
.05	3	.61635	.78747	.78269	1.27764	.789827	.896236	.893591	.106409	57	.95
	4	.61658	.78729	.78316	1.27688	.789988	.896137	.893851	.106149	56	
	5	.61681	.78711	.78363	1.27611	.790149	.896038	.894111	.105889	55	
.10	6	.61704	.78694	.78410	1.27535	.790310	.895939	.894372	.105628	54	.90
	7	.61726	.78676	.78457	1.27458	.790471	.895840	.894632	.105368	53	
	8	.61749	.78658	.78504	1.27382	.790632	.895741	.894892	.105108	52	
.15	9	.61772	.78640	.78551	1.27306	.790793	.895641	.895152	.104848	51	.85
	10	.61795	.78622	.78598	1.27230	.790954	.895542	.895412	.104588	50	
	11	.61818	.78604	.78645	1.27153	.791115	.895443	.895672	.104328	49	
.20	12	.61841	.78586	.78692	1.27077	.791275	.895343	.895932	.104068	48	.80
	13	.61864	.78568	.78739	1.27001	.791436	.895244	.896192	.103808	47	
	14	.61887	.78550	.78786	1.26925	.791596	.895145	.896452	.103548	46	
.25	15	.61909	.78532	.78834	1.26849	.791757	.895045	.896712	.103288	45	.75
	16	.61932	.78514	.78881	1.26774	.791917	.894945	.896971	.103029	44	
	17	.61955	.78496	.78928	1.26698	.792077	.894846	.897231	.102769	43	
.30	18	.61978	.78478	.78975	1.26622	.792237	.894746	.897491	.102509	42	.70
	19	.62001	.78460	.79022	1.26546	.792397	.894646	.897751	.102249	41	
	20	.62024	.78442	.79070	1.26471	.792557	.894546	.898010	.101990	40	
.35	21	.62046	.78424	.79117	1.26395	.792716	.894446	.898270	.101730	39	.65
	22	.62069	.78405	.79164	1.26319	.792876	.894346	.898530	.101470	38	
	23	.62092	.78387	.79212	1.26244	.793035	.894246	.898789	.101211	37	
.40	24	.62115	.78369	.79259	1.26169	.793195	.894146	.899049	.100951	36	.60
	25	.62138	.78351	.79306	1.26093	.793354	.894046	.899308	.100692	35	
	26	.62160	.78333	.79354	1.26018	.793514	.893946	.899568	.100432	34	
.45	27	.62183	.78315	.79401	1.25943	.793673	.893846	.899827	.100173	33	.55
	28	.62206	.78297	.79449	1.25867	.793832	.893745	.900087	.099913	32	
	29	.62229	.78279	.79496	1.25792	.793991	.893645	.900346	.099654	31	
.50	30	.62251	.78261	.79544	1.25717	.794150	.893544	.900605	.099395	30	.50
	31	.62274	.78243	.79591	1.25642	.794308	.893444	.900864	.099136	29	
	32	.62297	.78225	.79639	1.25567	.794467	.893343	.901124	.098876	28	
.55	33	.62320	.78206	.79686	1.25492	.794626	.893243	.901383	.098617	27	.45
	34	.62342	.78188	.79734	1.25417	.794784	.893142	.901642	.098358	26	
	35	.62365	.78170	.79781	1.25343	.794942	.893041	.901901	.098099	25	
.60	36	.62388	.78152	.79829	1.25268	.795101	.892940	.902160	.097840	24	.40
	37	.62411	.78134	.79877	1.25193	.795259	.892839	.902420	.097580	23	
	38	.62433	.78116	.79924	1.25118	.795417	.892739	.902679	.097321	22	
.65	39	.62456	.78098	.79972	1.25044	.795575	.892638	.902938	.097062	21	.35
	40	.62479	.78079	.80020	1.24969	.795733	.892536	.903197	.096803	20	
	41	.62502	.78061	.80067	1.24895	.795891	.892435	.903456	.096544	19	
.70	42	.62524	.78043	.80115	1.24820	.796049	.892334	.903714	.096286	18	.30
	43	.62547	.78025	.80163	1.24746	.796206	.892233	.903973	.096027	17	
	44	.62570	.78007	.80211	1.24672	.796364	.892132	.904232	.095768	16	
.75	45	.62592	.77988	.80258	1.24597	.796521	.892030	.904491	.095509	15	.25
	46	.62615	.77970	.80306	1.24523	.796679	.891929	.904750	.095250	14	
	47	.62638	.77952	.80354	1.24449	.796836	.891827	.905008	.094992	13	
.80	48	.62660	.77934	.80402	1.24375	.796993	.891726	.905267	.094733	12	.20
	49	.62683	.77916	.80450	1.24301	.797150	.891624	.905526	.094474	11	
	50	.62706	.77897	.80498	1.24227	.797307	.891523	.905785	.094215	10	
.85	51	.62728	.77879	.80546	1.24153	.797464	.891421	.906043	.093957	9	.15
	52	.62751	.77861	.80594	1.24079	.797621	.891319	.906302	.093698	8	
	53	.62774	.77843	.80642	1.24005	.797777	.891217	.906560	.093440	7	
.90	54	.62796	.77824	.80690	1.23931	.797934	.891115	.906819	.093181	6	.10
	55	.62819	.77806	.80738	1.23858	.798091	.891013	.907077	.092923	5	
	56	.62842	.77788	.80786	1.23784	.798247	.890911	.907336	.092664	4	
.95	57	.62864	.77769	.80834	1.23710	.798403	.890809	.907594	.092406	3	.05
	58	.62887	.77751	.80882	1.23637	.798560	.890707	.907853	.092147	2	
	59	.62909	.77733	.80930	1.23563	.798716	.890605	.908111	.091889	1	
1.00	60	.62932	.77715	.80978	1.23490	9.798872	9.890503	9.908369	10.091631	0	.00
Decimals	Minutes	Cos	Sin	Cot	Tan	Cos	Sin	Cot	Tan	Minutes	Decimals
		Natural Values				Common Logarithms					

Decimals	Minutes	Natural Values				Common Logarithms				Minutes	Decimals
		Sin	Cos	Tan	Cot	Sin	Cos	Tan	Cot		
.00	0	.62932	.77715	.80978	1.23490	9.798872	9.890503	9.908369	10.091631	60	1.00
	1	.62955	.77696	.81027	1.23416	.799028	.890400	.908628	.091372	59	
	2	.62977	.77678	.81075	1.23343	.799184	.890298	.908886	.091114	58	
.05	3	.63000	.77660	.81123	1.23270	.799399	.890195	.909144	.090856	57	.95
	4	.63022	.77641	.81171	1.23196	.799495	.890093	.909402	.090598	56	
	5	.63045	.77623	.81220	1.23123	.799651	.889990	.909660	.090340	55	
.10	6	.63068	.77605	.81268	1.23050	.799806	.889888	.909918	.090082	54	.90
	7	.63090	.77586	.81316	1.22977	.799962	.889785	.910177	.089823	53	
	8	.63113	.77568	.81364	1.22904	.800117	.889682	.910435	.089565	52	
.15	9	.63135	.77550	.81413	1.22831	.800272	.889579	.910693	.089307	51	.85
	10	.63158	.77531	.81461	1.22758	.800427	.839477	.910951	.089049	50	
	11	.63180	.77513	.81510	1.22685	.800582	.889374	.911209	.088791	49	
.20	12	.63203	.77494	.81553	1.22612	.800737	.839271	.911467	.088533	48	.80
	13	.63225	.77476	.81606	1.22539	.800892	.889168	.911725	.088275	47	
	14	.63248	.77458	.81655	1.22467	.801047	.889064	.911982	.088018	46	
.25	15	.63271	.77439	.81703	1.22394	.801201	.888961	.912240	.087760	45	.75
	16	.63293	.77421	.81752	1.22321	.801356	.888858	.912498	.087502	44	
	17	.63316	.77402	.81800	1.22249	.801511	.888755	.912756	.087244	43	
.30	18	.63338	.77384	.81849	1.22176	.801665	.888651	.913014	.086986	42	.70
	19	.63361	.77366	.81898	1.22104	.801819	.888548	.913271	.086729	41	
	20	.63383	.77347	.81946	1.22031	.801973	.888444	.913529	.086471	40	
.35	21	.63406	.77329	.81995	1.21959	.802128	.888341	.913787	.086213	39	.65
	22	.63428	.77310	.82044	1.21886	.802282	.888237	.914044	.085956	38	
	23	.63451	.77292	.82092	1.21814	.802436	.888134	.914302	.085698	37	
.40	24	.63473	.77273	.82141	1.21742	.802589	.888030	.914560	.085440	36	.60
	25	.63496	.77255	.82190	1.21670	.802743	.887926	.914817	.085183	35	
	26	.63518	.77236	.82238	1.21598	.802897	.887822	.915075	.084925	34	
.45	27	.63540	.77218	.82287	1.21526	.803050	.887718	.915332	.084668	33	.55
	28	.63563	.77199	.82336	1.21454	.803204	.887614	.915590	.084410	32	
	29	.63585	.77181	.82385	1.21382	.803357	.887510	.915847	.084153	31	
.50	30	.63608	.77162	.82434	1.21310	.803511	.887406	.916104	.083896	30	.50
	31	.63630	.77144	.82483	1.21238	.803664	.887302	.916362	.083638	29	
	32	.63653	.77125	.82531	1.21166	.803817	.887198	.916619	.083381	28	
.55	33	.63675	.77107	.82580	1.21094	.803970	.887093	.916877	.083123	27	.45
	34	.63698	.77088	.82629	1.21023	.804123	.886989	.917134	.082866	26	
	35	.63720	.77070	.82678	1.20951	.804276	.886885	.917391	.082609	25	
.60	36	.63742	.77051	.82727	1.20879	.804428	.886780	.917648	.082352	24	.40
	37	.63765	.77033	.82776	1.20808	.804581	.886676	.917906	.032094	23	
	38	.63787	.77014	.82825	1.20736	.804734	.886571	.918163	.081837	22	
.65	39	.63810	.76996	.82874	1.20665	.804886	.886466	.918420	.081580	21	.35
	40	.63832	.76977	.82923	1.20593	.805039	.886362	.918677	.081323	20	
	41	.63854	.76959	.82972	1.20522	.805191	.886257	.918934	.081066	19	
.70	42	.63877	.76940	.83022	1.20451	.805343	.886152	.919191	.080809	18	.30
	43	.63899	.76921	.83071	1.20379	.805495	.886047	.919448	.080552	17	
	44	.63922	.76903	.83120	1.20308	.805647	.885942	.919705	.080295	16	
.75	45	.63944	.76884	.83169	1.20237	.805799	.885837	.919962	.080038	15	.25
	46	.63966	.76866	.83218	1.20166	.805951	.885732	.920219	.079781	14	
	47	.63989	.76847	.83268	1.20095	.806103	.885627	.920476	.079524	13	
.80	48	.64011	.76828	.83317	1.20024	.806254	.885522	.920733	.079267	12	.20
	49	.64033	.76810	.83366	1.19953	.806406	.885416	.920990	.079010	11	
	50	.64056	.76791	.83415	1.19882	.806557	.885311	.921247	.078753	10	
.85	51	.64078	.76772	.83465	1.19811	.806709	.885205	.921503	.078497	9	.15
	52	.64100	.76754	.83514	1.19740	.806860	.885100	.921760	.078240	8	
	53	.64123	.76735	.83564	1.19669	.807011	.884994	.922017	.077983	7	
.90	54	.64145	.76717	.83613	1.19599	.807163	.884889	.922274	.077726	6	.10
	55	.64167	.76698	.83662	1.19528	.807314	.884783	.922530	.077470	5	
	56	.64190	.76679	.83712	1.19457	.807465	.884677	.922787	.077213	4	
.95	57	.64212	.76661	.83761	1.19387	.807615	.884572	.923044	.076956	3	.05
	58	.64234	.76642	.83811	1.19316	.807766	.884466	.923300	.076700	2	
	59	.64256	.76623	.83860	1.19246	.807917	.884360	.923557	.076443	1	
1.00	60	.64279	.76604	.83910	1.19175	9.808067	9.884254	9.923814	10.076186	0	.00
Decimals	Minutes	Cos	Sin	Cot	Tan	Cos	[Sin	Cot	Tan	Minutes	Decimals
		Natural Values				Common Logarithms					

Decimals	Minutes	Natural Values				Common Logarithms				Minutes	Decimals
		Sin	Cos	Tan	Cot	Sin	Cos	Tan	Cot		
.00	0	.64279	.76604	.83910	1.19175	9.808067	9.884254	9.923814	10.076186	60	1.00
	1	.64301	.76586	.83960	1.19105	.808218	.884148	.924070	.075930	59	
	2	.64323	.76567	.84009	1.19035	.808368	.884042	.924327	.075673	58	
.05	3	.64346	.76548	.84059	1.18964	.808519	.883936	.924583	.075417	57	.95
	4	.64368	.76530	.84108	1.18894	.808669	.883829	.924840	.075160	56	
	5	.64390	.76511	.84158	1.18824	.808819	.883723	.925096	.074904	55	
.10	6	.64412	.76492	.84208	1.18754	.808969	.883617	.925352	.074648	54	.90
	7	.64435	.76473	.84258	1.18684	.809119	.883510	.925609	.074391	53	
	8	.64457	.76455	.84307	1.18614	.809269	.883404	.925865	.074135	52	
.15	9	.64479	.76436	.84357	1.18544	.809419	.883297	.926122	.073878	51	.85
	10	.64501	.76417	.84407	1.18474	.809569	.883191	.926378	.073622	50	
	11	.64524	.76398	.84457	1.18404	.809718	.883084	.926634	.073366	49	
.20	12	.64546	.76380	.84507	1.18334	.809868	.882977	.926890	.073110	48	.80
	13	.64568	.76361	.84556	1.18264	.810017	.882871	.927147	.072853	47	
	14	.64590	.76342	.84606	1.18194	.810167	.882764	.927403	.072597	46	
.25	15	.64612	.76323	.84656	1.18125	.810316	.882657	.927659	.072341	45	.75
	16	.64635	.76304	.84706	1.18055	.810465	.882550	.927915	.072085	44	
	17	.64657	.76286	.84756	1.17986	.810614	.882443	.928171	.071829	43	
.30	18	.64679	.76267	.84806	1.17916	.810763	.882336	.928427	.071573	42	.70
	19	.64701	.76248	.84856	1.17846	.810912	.882229	.928684	.071316	41	
	20	.64723	.76229	.84906	1.17777	.811061	.882121	.928940	.071060	40	
.35	21	.64746	.76210	.84956	1.17708	.811210	.882014	.929196	.070804	39	.65
	22	.64768	.76192	.85006	1.17638	.811358	.881907	.929452	.070548	38	
	23	.64790	.76173	.85057	1.17569	.811507	.881799	.929708	.070292	37	
.40	24	.64812	.76154	.85107	1.17500	.811655	.881692	.929964	.070036	36	.60
	25	.64834	.76135	.85157	1.17430	.811804	.881584	.930220	.069780	35	
	26	.64856	.76116	.85207	1.17361	.811952	.881477	.930475	.069525	34	
.45	27	.64879	.76097	.85257	1.17292	.812100	.881369	.930731	.069269	33	.55
	28	.64901	.76078	.85308	1.17223	.812248	.881261	.930987	.069013	32	
	29	.64923	.76059	.85358	1.17154	.812396	.881153	.931243	.068757	31	
.50	30	.64945	.76041	.85408	1.17085	.812544	.881046	.931499	.068501	30	.50
	31	.64967	.76022	.85458	1.17016	.812692	.880938	.931755	.068245	29	
	32	.64989	.76003	.85509	1.16947	.812840	.880830	.932010	.067990	28	
.55	33	.65011	.75984	.85559	1.16878	.812988	.880722	.932266	.067734	27	.45
	34	.65033	.75965	.85609	1.16809	.813135	.880613	.932522	.067478	26	
	35	.65055	.75946	.85660	1.16741	.813283	.880505	.932778	.067222	25	
.60	36	.65077	.75927	.85710	1.16672	.813430	.880397	.933033	.066967	24	.40
	37	.65100	.75908	.85761	1.16603	.813578	.880289	.933289	.066711	23	
	38	.65122	.75889	.85811	1.16535	.813725	.880180	.933545	.066455	22	
.65	39	.65144	.75870	.85862	1.16466	.813872	.880072	.933800	.066200	21	.35
	40	.65166	.75851	.85912	1.16398	.814019	.879963	.934056	.065944	20	
	41	.65188	.75832	.85963	1.16329	.814166	.879855	.934311	.065689	19	
.70	42	.65210	.75813	.86014	1.16261	.814313	.879746	.934567	.065433	18	.30
	43	.65232	.75794	.86064	1.16192	.814460	.879637	.934822	.065178	17	
	44	.65254	.75775	.86115	1.16124	.814607	.879529	.935078	.064922	16	
.75	45	.65276	.75756	.86166	1.16056	.814753	.879420	.935333	.064667	15	.25
	46	.65298	.75738	.86216	1.15987	.814900	.879311	.935589	.064411	14	
	47	.65320	.75719	.86267	1.15919	.815046	.879202	.935844	.064156	13	
.80	48	.65342	.75700	.86318	1.15851	.815193	.879093	.936100	.063900	12	.20
	49	.65364	.75680	.86368	1.15783	.815339	.878984	.936355	.063645	11	
	50	.65386	.75661	.86419	1.15715	.815485	.878875	.936611	.063389	10	
.85	51	.65408	.75642	.86470	1.15647	.815632	.878766	.936866	.063134	9	.15
	52	.65430	.75623	.86521	1.15579	.815778	.878656	.937121	.062879	8	
	53	.65452	.75604	.86572	1.15511	.815924	.878547	.937377	.062623	7	
.90	54	.65474	.75585	.86623	1.15443	.816069	.878438	.937632	.062368	6	.10
	55	.65496	.75566	.86674	1.15375	.816215	.878328	.937887	.062113	5	
	56	.65518	.75547	.86725	1.15308	.816361	.878219	.938142	.061858	4	
.95	57	.65540	.75528	.86776	1.15240	.816507	.878109	.938398	.061602	3	.05
	58	.65562	.75509	.86827	1.15172	.816652	.877999	.938653	.061347	2	
	59	.65584	.75490	.86878	1.15104	.816798	.877890	.938908	.061092	1	
1.00	60	.65606	.75471	.86929	1.15037	9.816943	9.877780	9.939163	10.060837	0	.00
Decimals	Minutes	Cos	Sin	Cot	Tan	Cos	Sin	Cot	Tan	Minutes	Decimals
		Natural Values				Common Logarithms					

Decimals	Minutes	Natural Values				Common Logarithms				Minutes	Decimals
		Sin	Cos	Tan	Cot	Sin	Cos	Tan	Cot		
.00	0	.65606	.75471	.86929	1.15037	9.816943	9.877780	9.939163	10.060837	60	1.00
	1	.65628	.75452	.86980	1.14969	.817088	.877670	.939418	.060582	59	
	2	.65650	.75433	.87031	1.14902	.817233	.877560	.939673	.060327	58	
.05	3	.65672	.75414	.87082	1.14834	.817379	.877450	.939928	.060072	57	.95
	4	.65694	.75395	.87133	1.14767	.817524	.877340	.940183	.059817	56	
	5	.65716	.75375	.87184	1.14699	.817668	.877230	.940439	.059561	55	
.10	6	.65738	.75356	.87236	1.14632	.817813	.877120	.940694	.059306	54	.90
	7	.65759	.75337	.87287	1.14565	.817958	.877010	.940949	.059051	53	
	8	.65781	.75318	.87338	1.14498	.818103	.876899	.941204	.058796	52	
.15	9	.65803	.75299	.87389	1.14430	.818247	.876789	.941459	.058541	51	.85
	10	.65825	.75280	.87441	1.14363	.818392	.876678	.941713	.058287	50	
	11	.65847	.75261	.87492	1.14296	.818536	.876568	.941968	.058032	49	
.20	12	.65869	.75241	.87543	1.14229	.818681	.876457	.942223	.057777	48	.80
	13	.65891	.75222	.87595	1.14162	.818825	.876347	.942478	.057522	47	
	14	.65913	.75203	.87646	1.14095	.818969	.876236	.942733	.057267	46	
.25	15	.65935	.75184	.87698	1.14028	.819113	.876125	.942988	.057012	45	.75
	16	.65956	.75165	.87749	1.13961	.819257	.876014	.943243	.056757	44	
	17	.65978	.75146	.87801	1.13894	.819401	.875904	.943498	.056502	43	
.30	18	.66000	.75126	.87852	1.13828	.819545	.875793	.943752	.056248	42	.70
	19	.66022	.75107	.87904	1.13761	.819689	.875682	.944007	.055993	41	
	20	.66044	.75088	.87955	1.13694	.819832	.875571	.944262	.055738	40	
.35	21	.66066	.75069	.88007	1.13627	.819976	.875459	.944517	.055483	39	.65
	22	.66088	.75050	.88059	1.13561	.820120	.875348	.944771	.055229	38	
	23	.66109	.75030	.88110	1.13494	.820263	.875237	.945026	.054974	37	
.40	24	.66131	.75011	.88162	1.13428	.820406	.875126	.945281	.054719	36	.60
	25	.66153	.74992	.88214	1.13361	.820550	.875014	.945535	.054465	35	
	26	.66175	.74973	.88265	1.13295	.820693	.874903	.945790	.054210	34	
.45	27	.66197	.74953	.88317	1.13228	.820836	.874791	.946045	.053955	33	.55
	28	.66218	.74934	.88369	1.13162	.820979	.874680	.946299	.053701	32	
	29	.66240	.74915	.88421	1.13096	.821122	.874568	.946554	.053446	31	
.50	30	.66262	.74896	.88473	1.13029	.821265	.874456	.946808	.053192	30	.50
	31	.66284	.74876	.88524	1.12963	.821407	.874344	.947063	.052937	29	
	32	.66306	.74857	.88576	1.12897	.821550	.874232	.947318	.052682	28	
.55	33	.66327	.74838	.88628	1.12831	.821693	.874121	.947572	.052428	27	.45
	34	.66349	.74818	.88680	1.12765	.821835	.874009	.947827	.052173	26	
	35	.66371	.74799	.88732	1.12699	.821977	.873896	.948081	.051919	25	
.60	36	.66393	.74780	.88784	1.12633	.822120	.873784	.948335	.051665	24	.40
	37	.66414	.74760	.88836	1.12567	.822262	.873672	.948590	.051410	23	
	38	.66436	.74741	.88888	1.12501	.822404	.873560	.948844	.051156	22	
.65	39	.66458	.74722	.88940	1.12435	.822546	.873448	.949099	.050901	21	.35
	40	.66480	.74703	.88992	1.12369	.822688	.873335	.949353	.050647	20	
	41	.66501	.74683	.89045	1.12303	.822830	.873223	.949608	.050392	19	
.70	42	.66523	.74664	.89097	1.12238	.822972	.873110	.949862	.050138	18	.30
	43	.66545	.74644	.89149	1.12172	.823114	.872998	.950116	.049884	17	
	44	.66566	.74625	.89201	1.12106	.823255	.872885	.950371	.049629	16	
.75	45	.66588	.74606	.89253	1.12041	.823397	.872772	.950625	.049375	15	.25
	46	.66610	.74586	.89306	1.11975	.823539	.872659	.950879	.049121	14	
	47	.66632	.74567	.89358	1.11909	.823680	.872547	.951133	.048867	13	
.80	48	.66653	.74548	.89410	1.11844	.823821	.872434	.951388	.048612	12	.20
	49	.66675	.74528	.89463	1.11778	.823963	.872321	.951642	.048358	11	
	50	.66697	.74509	.89515	1.11713	.824104	.872208	.951896	.048104	10	
.85	51	.66718	.74489	.89567	1.11648	.824245	.872095	.952150	.047850	9	.15
	52	.66740	.74470	.89620	1.11582	.824386	.871981	.952405	.047595	8	
	53	.66762	.74451	.89672	1.11517	.824527	.871868	.952659	.047341	7	
.90	54	.66783	.74431	.89725	1.11452	.824668	.871755	.952913	.047087	6	.10
	55	.66805	.74412	.89777	1.11387	.824808	.871641	.953167	.046833	5	
	56	.66827	.74392	.89830	1.11321	.824949	.871528	.953421	.046579	4	
.95	57	.66848	.74373	.89883	1.11256	.825090	.871414	.953675	.046325	3	.05
	58	.66870	.74353	.89935	1.11191	.825230	.871301	.953929	.046071	2	
	59	.66891	.74334	.89988	1.11126	.825371	.871187	.954183	.045817	1	
1.00	60	.66913	.74314	.90040	1.11061	9.825511	9.871073	9.954437	10.045563	0	.00
		Cos	Sin	Cot	Tan	Cos	Sin	Cot	Tan	Minutes	Decimals
Decimals	Minutes	Natural Values				Common Logarithms					

48°

Decimals	Minutes	Sin	Cos	Tan	Cot	Sin	Cos	Tan	Cot	Minutes	Decimals
		Natural Values				**Common Logarithms**					
00	0	.66913	.74314	.90040	1.11061	9.825511	9.871073	9.954437	10.045563	60	1.00
	1	.66935	.74295	.90093	1.10996	.825651	.870960	.954691	.045309	59	
	2	.66956	.74276	.90146	1.10931	.825791	.870846	.954946	.045054	58	
.05	3	.66978	.74256	.90199	1.10867	.825931	.870732	.955200	.044800	57	.95
	4	.66999	.74237	.90251	1.10802	.826071	.870618	.955454	.044546	56	
	5	.67021	.74217	.90304	1.10737	.826211	.870504	.955708	.044292	55	
.10	6	.67043	.74198	.90357	1.10672	.826351	.870390	.955961	.044039	54	.90
	7	.67064	.74178	.90410	1.10607	.826491	.870276	.956215	.043785	53	
	8	.67086	.74159	.90463	1.10543	.826631	.870161	.956469	.043531	52	
.15	9	.67107	.74139	.90516	1.10478	.826770	.870047	.956723	.043277	51	.85
	10	.67129	.74120	.90569	1.10414	.826910	.869933	.956977	.043023	50	
	11	.67151	.74100	.90621	1.10349	.827049	.869818	.957231	.042769	49	
.20	12	.67172	.74080	.90674	1.10285	.827189	.869704	.957485	.042515	48	.80
	13	.67194	.74061	.90727	1.10220	.827328	.869589	.957739	.042261	47	
	14	.67215	.74041	.90781	1.10156	.827467	.869474	.957993	.042007	46	
.25	15	.67237	.74022	.90834	1.10091	.827606	.869360	.958247	.041753	45	.75
	16	.67258	.74002	.90887	1.10027	.827745	.869245	.958500	.041500	44	
	17	.67280	.73983	.90940	1.09963	.827884	.869130	.958754	.041246	43	
.30	18	.67301	.73963	.90993	1.09899	.828023	.869015	.959008	.040992	42	.70
	19	.67323	.73944	.91046	1.09834	.828162	.868900	.959262	.040738	41	
	20	.67344	.73924	.91099	1.09770	.828301	.868785	.959516	.040484	40	
.35	21	.67366	.73904	.91153	1.09706	.828439	.868670	.959769	.040231	39	.65
	22	.67387	.73885	.91206	1.09642	.828578	.868555	.960023	.039977	38	
	23	.67409	.73865	.91259	1.09578	.828716	.868440	.960277	.039723	37	
.40	24	.67430	.73846	.91313	1.09514	.828855	.868324	.960530	.039470	36	.60
	25	.67452	.73826	.91366	1.09450	.828993	.868209	.960784	.039216	35	
	26	.67473	.73806	.91419	1.09386	.829131	.868093	.961038	.038962	34	
.45	27	.67495	.73787	.91473	1.09322	.829269	.867978	.961292	.038708	33	.55
	28	.67516	.73767	.91526	1.09258	.829407	.867862	.961545	.038455	32	
	29	.67538	.73747	.91580	1.09195	.829545	.867747	.961799	.038201	31	
.50	30	.67559	.73728	.91633	1.09131	.829683	.867631	.962052	.037948	30	.50
	31	.67580	.73708	.91687	1.09067	.829821	.867515	.962306	.037694	29	
	32	.67602	.73688	.91740	1.09003	.829959	.867399	.962560	.037440	28	
.55	33	.67623	.73669	.91794	1.08940	.830097	.867283	.962813	.037187	27	.45
	34	.67645	.73649	.91847	1.08876	.830234	.867167	.963067	.036933	26	
	35	.67666	.73629	.91901	1.08813	.830372	.867051	.963320	.036680	25	
.60	36	.67688	.73610	.91955	1.08749	.830509	.866935	.963574	.036426	24	.40
	37	.67709	.73590	.92008	1.08686	.830646	.866819	.963828	.036172	23	
	38	.67730	.73570	.92062	1.08622	.830784	.866703	.964081	.035919	22	
.65	39	.67752	.73551	.92116	1.08559	.830921	.866586	.964335	.035665	21	.35
	40	.67773	.73531	.92170	1.08496	.831058	.866470	.964588	.035412	20	
	41	.67795	.73511	.92224	1.08432	.831195	.866353	.964842	.035158	19	
.70	42	.67816	.73491	.92277	1.08369	.831332	.866237	.965095	.034905	18	.30
	43	.67837	.73472	.92331	1.08306	.831469	.866120	.965349	.034651	17	
	44	.67859	.73452	.92385	1.08243	.831606	.866004	.965602	.034398	16	
.75	45	.67880	.73432	.92439	1.08179	.831742	.865887	.965855	.034145	15	.25
	46	.67901	.73413	.92493	1.08116	.831879	.865770	.966109	.033891	14	
	47	.67923	.73393	.92547	1.08053	.832015	.865653	.966362	.033638	13	
.80	48	.67944	.73373	.92601	1.07990	.832152	.865536	.966616	.033384	12	.20
	49	.67965	.73353	.92655	1.07927	.832288	.865419	.966869	.033131	11	
	50	.67987	.73333	.92709	1.07864	.832425	.865302	.967123	.032877	10	
.85	51	.68008	.73314	.92763	1.07801	.832561	.865185	.967376	.032624	9	.15
	52	.68029	.73294	.92817	1.07738	.832697	.865068	.967629	.032371	8	
	53	.68051	.73274	.92872	1.07676	.832833	.864950	.967883	.032117	7	
.90	54	.68072	.73254	.92926	1.07613	.832969	.864833	.968136	.031864	6	.10
	55	.68093	.73234	.92980	1.07550	.833105	.864716	.968389	.031611	5	
	56	.68115	.73215	.93034	1.07487	.833241	.864598	.968643	.031357	4	
.95	57	.68136	.73195	.93088	1.07425	.833377	.864481	.968896	.031104	3	.05
	58	.68157	.73175	.93143	1.07362	.833512	.864363	.969149	.030851	2	
	59	.68179	.73155	.93197	1.07299	.833648	.864245	.969403	.030597	1	
1.00	60	.68200	.73135	.93252	1.07237	9.833783	9.864127	9.969656	10.030344	0	.00
Decimals	Minutes	Cos	Sin	Cot	Tan	Cos	Sin	Cot	Tan	Minutes	Decimals
		Natural Values				**Common Logarithms**					

47°

Decimals	Minutes	Natural Values				Common Logarithms				Minutes	Decimals
		Sin	Cos	Tan	Cot	Sin	Cos	Tan	Cot		
.00	0	.68200	.73135	.93252	1.07237	9.833783	9.864127	9.969656	10.030344	60	1.00
	1	.68221	.73116	.93306	1.07174	.833919	.864010	.969909	.030091	59	
	2	.68242	.73096	.93360	1.07112	.834054	.863892	.970162	.029838	58	
.05	3	.68264	.73076	.93415	1.07049	.834189	.863774	.970416	.029584	57	.95
	4	.68285	.73056	.93469	1.06987	.834325	.863656	.970669	.029331	56	
	5	.68306	.73036	.93524	1.06925	.834460	.863538	.970922	.029078	55	
.10	6	.68327	.73016	.93578	1.06862	.834595	.863419	.971175	.028825	54	.90
	7	.68349	.72996	.93633	1.06800	.834730	.863301	.971429	.028571	53	
	8	.68370	.72976	.93688	1.06738	.834865	.863183	.971682	.028318	52	
.15	9	.68391	.72957	.93742	1.06676	.834999	.863064	.971935	.028065	51	.85
	10	.68412	.72937	.93797	1.06613	.835134	.862946	.972188	.027812	50	
	11	.68434	.72917	.93852	1.06551	.835269	.862827	.972441	.027559	49	
.20	12	.68455	.72897	.93906	1.06489	.835403	.862709	.972695	.027305	48	.80
	13	.68476	.72877	.93961	1.06427	.835538	.862590	.972948	.027052	47	
	14	.68497	.72857	.94016	1.06365	.835672	.862471	.973201	.026799	46	
.25	15	.68518	.72837	.94071	1.06303	.835807	.862353	.973454	.026546	45	.75
	16	.68539	.72817	.94125	1.06241	.835941	.862234	.973707	.026293	44	
	17	.68561	.72797	.94180	1.06179	.836075	.862115	.973960	.026040	43	
.30	18	.68582	.72777	.94235	1.06117	.836209	.861996	.974213	.025787	42	.70
	19	.68603	.72757	.94290	1.06056	.836343	.861877	.974466	.025534	41	
	20	.68624	.72737	.94345	1.05994	.836477	.861757	.974720	.025280	40	
.35	21	.68645	.72717	.94400	1.05932	.836611	.861638	.974973	.025027	39	.65
	22	.68666	.72697	.94455	1.05870	.836745	.861519	.975226	.024774	38	
	23	.68688	.72677	.94510	1.05809	.836878	.861400	.975479	.024521	37	
.40	24	.68709	.72657	.94565	1.05747	.837012	.861280	.975732	.024268	36	.60
	25	.68730	.72637	.94620	1.05685	.837146	.861161	.975985	.024015	35	
	26	.68751	.72617	.94676	1.05624	.837279	.861041	.976238	.023762	34	
.45	27	.68772	.72597	.94731	1.05562	.837412	.860922	.976491	.023509	33	.55
	28	.68793	.72577	.94786	1.05501	.837546	.860802	.976744	.023256	32	
	29	.68814	.72557	.94841	1.05439	.837679	.860682	.976997	.023003	31	
.50	30	.68835	.72537	.94896	1.05378	.837812	.860562	.977250	.022750	30	.50
	31	.68857	.72517	.94952	1.05317	.837945	.860442	.977503	.022497	29	
	32	.68878	.72497	.95007	1.05255	.838078	.860322	.977756	.022244	28	
.55	33	.68899	.72477	.95062	1.05194	.838211	.860202	.978009	.021991	27	.45
	34	.68920	.72457	.95118	1.05133	.838344	.860082	.978262	.021738	26	
	35	.68941	.72437	.95173	1.05072	.838477	.859962	.978515	.021485	25	
.60	36	.68962	.72417	.95229	1.05010	.838610	.859842	.978768	.021232	24	.40
	37	.68983	.72397	.95284	1.04949	.838742	.859721	.979021	.020979	23	
	38	.69004	.72377	.95340	1.04888	.838875	.859601	.979274	.020726	22	
.65	39	.69025	.72357	.95395	1.04827	.839007	.859480	.979527	.020473	21	.35
	40	.69046	.72337	.95451	1.04766	.839140	.859360	.979780	.020220	20	
	41	.69067	.72317	.95506	1.04705	.839272	.859239	.980033	.019967	19	
.70	42	.69088	.72297	.95562	1.04644	.839404	.859119	.980286	.019714	18	.30
	43	.69109	.72277	.95618	1.04583	.839536	.858998	.980538	.019462	17	
	44	.69130	.72257	.95673	1.04522	.839668	.858877	.980791	.019209	16	
.75	45	.69151	.72236	.95729	1.04461	.839800	.858756	.981044	.018956	15	.25
	46	.69172	.72216	.95785	1.04401	.839932	.858635	.981297	.018703	14	
	47	.69193	.72196	.95841	1.04340	.840064	.858514	.981550	.018450	13	
.80	48	.69214	.72176	.95897	1.04279	.840196	.858393	.981803	.018197	12	.20
	49	.69235	.72156	.95952	1.04218	.840328	.858272	.982056	.017944	11	
	50	.69256	.72136	.96008	1.04158	.840459	.858151	.982309	.017691	10	
.85	51	.69277	.72116	.96064	1.04097	.840591	.858029	.982562	.017438	9	.15
	52	.69298	.72095	.96120	1.04036	.840722	.857908	.982814	.017186	8	
	53	.69319	.72075	.96176	1.03976	.840854	.857786	.983067	.016933	7	
.90	54	.69340	.72055	.96232	1.03915	.840985	.857665	.983320	.016680	6	.10
	55	.69361	.72035	.96288	1.03855	.841116	.857543	.983573	.016427	5	
	56	.69382	.72015	.96344	1.03794	.841247	.857422	.983826	.016174	4	
.95	57	.69403	.71995	.96400	1.03734	.841378	.857300	.984079	.015921	3	.05
	58	.69424	.71974	.96457	1.03674	.841509	.857178	.984332	.015668	2	
	59	.69445	.71954	.96513	1.03613	.841640	.857056	.984584	.015416	1	
1.00	60	.69466	.71934	.96569	1.03553	9.841771	9.856934	9.984837	10.015163	0	.00
Decimals	Minutes	Cos	Sin	Cot	Tan	Cos	Sin	Cot	Tan	Minutes	Decimals
			Natural Values				Common Logarithms				

46°

Decimals	Minutes	Natural Values				Common Logarithms				Minutes	Decimals
		Sin	Cos	Tan	Cot	Sin	Cos	Tan	Cot		
.00	0	.69466	.71934	.96569	1.03553	9.841771	9.856934	9.984837	10.015163	60	1.00
	1	.69487	.71914	.96625	1.03493	.841902	.856812	.985090	.014910	59	
	2	.69508	.71894	.96681	1.03433	.842033	.856690	.985343	.014657	58	
.05	3	.69529	.71873	.96738	1.03372	.842163	.856568	.985596	.014404	57	.95
	4	.69549	.71853	.96794	1.03312	.842294	.856446	.985848	.014152	56	
	5	.69570	.71833	.96850	1.03252	.842424	.856323	.986101	.013899	55	
.10	6	.69591	.71813	.96907	1.03192	.842555	.856201	.986354	.013646	54	.90
	7	.69612	.71792	.96963	1.03132	.842685	.856078	.986607	.013393	53	
	8	.69633	.71772	.97020	1.03072	.842815	.855956	.986860	.013140	52	
.15	9	.69654	.71752	.97076	1.03012	.842946	.855833	.987112	.012888	51	.85
	10	.69675	.71732	.97133	1.02952	.843076	.855711	.987365	.012635	50	
	11	.69696	.71711	.97189	1.02892	.843206	.855588	.987618	.012382	49	
.20	12	.69717	.71691	.97246	1.02832	.843336	.855465	.987871	.012129	48	.80
	13	.69737	.71671	.97302	1.02772	.843466	.855342	.988123	.011877	47	
	14	.69758	.71650	.97359	1.02713	.843595	.855219	.988376	.011624	46	
.25	15	.69779	.71630	.97416	1.02653	.843725	.855096	.988629	.011371	45	.75
	16	.69800	.71610	.97472	1.02593	.843855	.854973	.988882	.011118	44	
	17	.69821	.71590	.97529	1.02533	.843984	.854850	.989134	.010866	43	
.30	18	.69842	.71569	.97586	1.02474	.844114	.854727	.989387	.010613	42	.70
	19	.69862	.71549	.97643	1.02414	.844243	.854603	.989640	.010360	41	
	20	.69883	.71529	.97700	1.02355	.844372	.854480	.989893	.010107	40	
.35	21	.69904	.71508	.97756	1.02295	.844502	.854356	.990145	.009855	39	.65
	22	.69925	.71488	.97813	1.02236	.844631	.854233	.990398	.009602	38	
	23	.69946	.71468	.97870	1.02176	.844760	.854109	.990651	.009349	37	
.40	24	.69966	.71447	.97927	1.02117	.844889	.853986	.990903	.009097	36	.60
	25	.69987	.71427	.97984	1.02057	.845018	.853862	.991156	.008844	35	
	26	.70008	.71407	.98041	1.01998	.845147	.853738	.991409	.008591	34	
.45	27	.70029	.71386	.98098	1.01939	.845276	.853614	.991662	.008338	33	.55
	28	.70049	.71366	.98155	1.01879	.845405	.853490	.991914	.008086	32	
	29	.70070	.71345	.98213	1.01820	.845533	.853366	.992167	.007833	31	
.50	30	.70091	.71325	.98270	1.01761	.845662	.853242	.992420	.007580	30	.50
	31	.70112	.71305	.98327	1.01702	.845790	.853118	.992672	.007328	29	
	32	.70132	.71284	.98384	1.01642	.845919	.852994	.992925	.007075	28	
.55	33	.70153	.71264	.98441	1.01583	.846047	.852869	.993178	.006822	27	.45
	34	.70174	.71243	.98499	1.01524	.846175	.852745	.993431	.006569	26	
	35	.70195	.71223	.98556	1.01465	.846304	.852620	.993683	.006317	25	
.60	36	.70215	.71203	.98613	1.01406	.846432	.852496	.993936	.006064	24	.40
	37	.70236	.71182	.98671	1.01347	.846560	.852371	.994189	.005811	23	
	38	.70257	.71162	.98728	1.01288	.846688	.852247	.994441	.005559	22	
.65	39	.70277	.71141	.98786	1.01229	.846816	.852122	.994694	.005306	21	.35
	40	.70298	.71121	.98843	1.01170	.846944	.851997	.994947	.005053	20	
	41	.70319	.71100	.98901	1.01112	.847071	.851872	.995199	.004801	19	
.70	42	.70339	.71080	.98958	1.01053	.847199	.851747	.995452	.004548	18	.30
	43	.70360	.71059	.99016	1.00994	.847327	.851622	.995705	.004295	17	
	44	.70381	.71039	.99073	1.00935	.847454	.851497	.995957	.004043	16	
.75	45	.70401	.71019	.99131	1.00876	.847582	.851372	.996210	.003790	15	.25
	46	.70422	.70998	.99189	1.00818	.847709	.851246	.996463	.003537	14	
	47	.70443	.70978	.99247	1.00759	.847836	.851121	.996715	.003285	13	
.80	48	.70463	.70957	.99304	1.00701	.847964	.850996	.996968	.003032	12	.20
	49	.70484	.70937	.99362	1.00642	.848091	.850870	.997221	.002779	11	
	50	.70505	.70916	.99420	1.00583	.848218	.850745	.997473	.002527	10	
.85	51	.70525	.70896	.99478	1.00525	.848345	.850619	.997726	.002274	9	.15
	52	.70546	.70875	.99536	1.00467	.848472	.850493	.997979	.002021	8	
	53	.70567	.70855	.99594	1.00408	.848599	.850368	.998231	.001769	7	
.90	54	.70587	.70834	.99652	1.00350	.848726	.850242	.998484	.001516	6	.10
	55	.70608	.70813	.99710	1.00291	.848852	.850116	.998737	.001263	5	
	56	.70628	.70793	.99768	1.00233	.848979	.849990	.998989	.001011	4	
.95	57	.70649	.70772	.99826	1.00175	.849106	.849864	.999242	.000758	3	.05
	58	.70670	.70752	.99884	1.00116	.849232	.849738	.999495	.000505	2	
	59	.70690	.70731	.99942	1.00058	.849359	.849611	.999747	.000253	1	
1.00	60	.70711	.70711	1.00000	1.00000	9.849485	9.849485	10.000000	10.000000	0	.00
Decimals	Minutes	Cos	Sin	Cot	Tan	Cos	Sin	Cot	Tan	Minutes	Decimals
		Natural Values				Common Logarithms					

119

45°

Table 22. Values and Logarithms of Exponentials and Hyperbolic Functions

The following tables give values of e^x, e^{-x}, sinh x, cosh x, and tanh x for values of x from 0.00 to 6.00 in intervals of 0.01.

To facilitate computations involving multiplication, the common logarithms of e^x, sinh x, cosh x, and tanh x are also given.

For values of x greater than 6 : e^x may be computed from the relationship $e^x = \log^{-1}$ $(x \log_{10} e) = \log^{-1} 0.43429x$; e^{-x} approaches zero; sinh x and cosh x are approximately equal and become $0.5\ e^x$; and tanh x and coth x have values approximately equal to unity.

Where more accurate values of the exponentials and functions are required they may be computed from the following relationships.

$$e = 2.71828\ 18285 \qquad\qquad \frac{1}{e} = 0.36787\ 94412$$

$$M = \log_{10} e = 0.43429\ 44819 \qquad\qquad \frac{1}{M} = \log_e 10 = 2.30258\ 50930$$

$$e^x = \log^{-1} Mx \qquad\qquad e^{-x} = \log^{-1} - Mx$$

$$\text{Sinh } x = \frac{e^x - e^{-x}}{2} \qquad \text{cosh } x = \frac{e^x + e^{-x}}{2} \qquad \text{tanh } x = \frac{e^x - e^{-x}}{e^x + e^{-x}}$$

$$\text{csch } x = \frac{1}{\text{sinh } x} \qquad \text{sech } x = \frac{1}{\text{cosh } x} \qquad \text{coth } x = \frac{1}{\text{tanh } x}$$

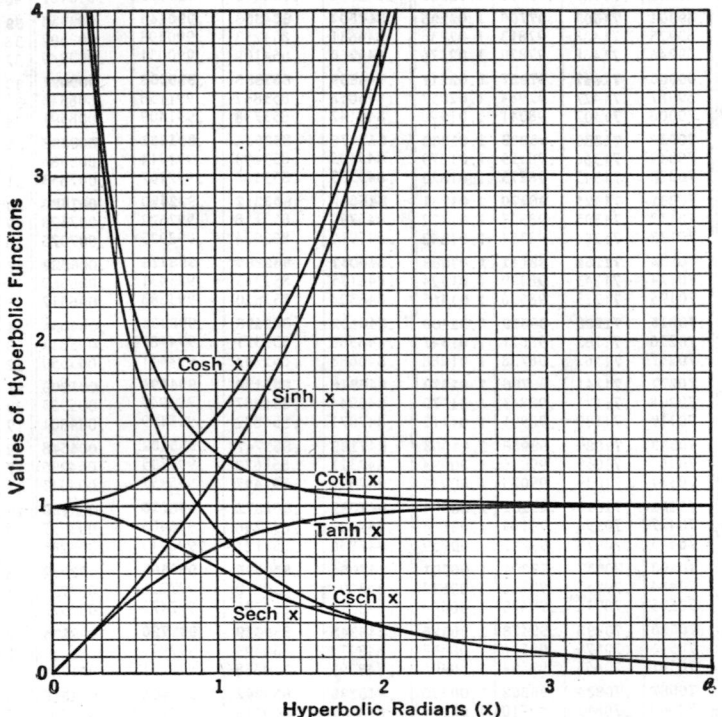

Chart of the Hyperbolic Functions.

x	Natural Values					Common Logarithms			
	e^x	e^{-x}	Sinh x	Cosh x	Tanh x	e^x	Sinh x	Cosh x	Tanh x
0.00	1.0000	1.0000	0.0000	1.0000	.00000	0.00000	− ∞	0.00000	− ∞
0.01	1.0101	.99005	0.0100	1.0001	.01000	.00434	2̄.00001	.00002	3̄.99999
0.02	1.0202	.98020	0.0200	1.0002	.02000	.00869	.30106	.00009	2̄.30097
0.03	1.0305	.97045	0.0300	1.0005	.02999	.01303	.47719	.00020	.47699
0.04	1.0408	.96079	0.0400	1.0008	.03998	.01737	.60218	.00035	.60183
0.05	1.0513	.95123	0.0500	1.0013	.04996	.02171	.69915	.00054	.69861
0.06	1.0618	.94176	0.0600	1.0018	.05993	.02606	.77841	.00078	.77763
0.07	1.0725	.93239	0.0701	1.0025	.06989	.03040	.84545	.00106	.84439
0.08	1.0833	.92312	0.0801	1.0032	.07983	.03474	.90355	.00139	.90216
0.09	1.0942	.91393	0.0901	1.0041	.08976	.03909	.95483	.00176	.95307
0.10	1.1052	.90484	0.1002	1.0050	.09967	0.04343	1̄.00072	0.00217	2̄.99856
0.11	1.1163	.89583	0.1102	1.0061	.10956	.04777	.04227	.00262	1̄.03965
0.12	1.1275	.88692	0.1203	1.0072	.11943	.05212	.08022	.00312	.07710
0.13	1.1388	.87810	0.1304	1.0085	.12927	.05646	.11517	.00366	.11151
0.14	1.1503	.86936	0.1405	1.0098	.13909	.06080	.14755	.00424	.14330
0.15	1.1618	.86071	0.1506	1.0113	.14889	.06514	.17772	.00487	.17285
0.16	1.1735	.85214	0.1607	1.0128	.15865	.06949	.20597	.00554	.20044
0.17	1.1853	.84366	0.1708	1.0145	.16838	.07383	.23254	.00625	.22629
0.18	1.1972	.83527	0.1810	1.0162	.17808	.07817	.25762	.00700	.25062
0.19	1.2092	.82696	0.1911	1.0181	.18775	.08252	.28136	.00779	.27357
0.20	1.2214	.81873	0.2013	1.0201	.19738	0.08686	1̄.30392	0.00863	1̄.29529
0.21	1.2337	.81058	0.2115	1.0221	.20697	.09120	.32541	.00951	.31590
0.22	1.2461	.80252	0.2218	1.0243	.21652	.09554	.34592	.01043	.33549
0.23	1.2586	.79453	0.2320	1.0266	.22603	.09989	.36555	.01139	.35416
0.24	1.2712	.78663	0.2423	1.0289	.23550	.10423	.38437	.01239	.37198
0.25	1.2840	.77880	0.2526	1.0314	.24492	.10857	.40245	.01343	.38902
0.26	1.2969	.77105	0.2629	1.0340	.25430	.11292	.41986	.01452	.40534
0.27	1.3100	.76338	0.2733	1.0367	.26362	.11726	.43663	.01564	.42099
0.28	1.3231	.75578	0.2837	1.0395	.27291	.12160	.45282	.01681	.43601
0.29	1.3364	.74826	0.2941	1.0423	.28213	.12595	.46847	.01801	.45046
0.30	1.3499	.74082	0.3045	1.0453	.29131	0.13029	1̄.48362	0.01926	1̄.46436
0.31	1.3634	.73345	0.3150	1.0484	.30044	.13463	.49830	.02054	.47775
0.32	1.3771	.72615	0.3255	1.0516	.30951	.13897	.51254	.02107	.49067
0.33	1.3910	.71892	0.3360	1.0549	.31852	.14332	.52637	.02323	.50314
0.34	1.4049	.71177	0.3466	1.0584	.32748	.14766	.53981	.02463	.51518
0.35	1.4191	.70469	0.3572	1.0619	.33638	.15200	.55290	.02607	.52682
0.36	1.4333	.69768	0.3678	1.0655	.34521	.15635	.56564	.02755	.53809
0.37	1.4477	.69073	0.3785	1.0692	.35399	.16069	.57807	.02907	.54899
0.38	1.4623	.68386	0.3892	1.0731	.36271	.16503	.59019	.03063	.55956
0.39	1.4770	.67706	0.4000	1.0770	.37136	.16937	.60202	.03222	.56980
0.40	1.4918	.67032	0.4108	1.0811	.37995	0.17372	1̄.61358	0.03385	1̄.57973
0.41	1.5068	.66365	0.4216	1.0852	.38847	.17806	.62488	.03552	.58936
0.42	1.5220	.65705	0.4325	1.0895	.39693	.18240	.63594	.03723	.59871
0.43	1.5373	.65051	0.4434	1.0939	.40532	.18675	.64677	.03897	.60780
0.44	1.5527	.64404	0.4543	1.0984	.41364	.19109	.65738	.04075	.61663
0.45	1.5683	.63763	0.4653	1.1030	.42190	.19543	.66777	.04256	.62521
0.46	1.5841	.63128	0.4764	1.1077	.43008	.19978	.67797	.04441	.63355
0.47	1.6000	.62500	0.4875	1.1125	.43820	.20412	.68797	.04630	.64167
0.48	1.6161	.61878	0.4986	1.1174	.44624	.20846	.69779	.04822	.64957
0.49	1.6323	.61263	0.5098	1.1225	.45422	.21280	.70744	.05018	.65726
0.50	1.6487	.60653	0.5211	1.1276	.46212	0.21715	1̄.71692	0.05217	1̄.66475
0.51	1.6653	.60050	0.5324	1.1329	.46995	.22149	.72624	.05419	.67205
0.52	1.6820	.59452	0.5438	1.1383	.47770	.22583	.73540	.05625	.67916
0.53	1.6989	.58860	0.5552	1.1438	.48538	.23018	.74442	.05834	.68608
0.54	1.7160	.58275	0.5666	1.1494	.49299	.23452	.75330	.06046	.69284
0.55	1.7333	.57695	0.5782	1.1551	.50052	.23886	.76204	.06262	.69942
0.56	1.7507	.57121	0.5897	1.1609	.50798	.24320	.77065	.06481	.70584
0.57	1.7683	.56553	0.6014	1.1669	.51536	.24755	.77914	.06703	.71211
0.58	1.7860	.55990	0.6131	1.1730	.52267	.25189	.78751	.06929	.71822
0.59	1.8040	.55433	0.6248	1.1792	.52990	.25623	.79576	.07157	.72419
0.60	1.8221	.54881	0.6367	1.1855	.53705	0.26058	1̄.80390	0.07389	1̄.73001

x	Natural Values					Common Logarithms			
	e^x	e^{-x}	Sinh x	Cosh x	Tanh x	e^x	Sinh x	Cosh x	Tanh x
0.60	**1.8221**	**.54881**	**0.6367**	**1.1855**	**.53705**	**0.26058**	**$\bar{1}$.80390**	**0.07389**	**$\bar{1}$.73001**
0.61	1.8404	.54335	0.6485	1.1919	.54413	.26492	.81194	.07624	.73570
0.62	1.8589	.53794	0.6605	1.1984	.55113	.26926	.81987	.07861	.74125
0.63	1.8776	.53259	0.6725	1.2051	.55805	.27361	.82770	.08102	.74667
0.64	1.8965	.52729	0.6846	1.2119	.56490	.27795	.83543	.08346	.75197
0.65	1.9155	.52205	0.6967	1.2188	.57167	.28229	.84308	.08593	.75715
0.66	1.9348	.51685	0.7090	1.2258	.57836	.28663	.85063	.08843	.76220
0.67	1.9542	.51171	0.7213	1.2330	.58498	.29098	.85809	.09095	.76714
0.68	1.9739	.50662	0.7336	1.2402	.59152	.29532	.86548	.09351	.77197
0.69	1.9937	.50158	0.7461	1.2476	.59798	.29966	.87278	.09609	.77669
0.70	**2.0138**	**.49659**	**0.7586**	**1.2552**	**.60437**	**0.30401**	**$\bar{1}$.88000**	**0.09870**	**$\bar{1}$.78130**
0.71	2.0340	.49164	0.7712	1.2628	.61068	.30835	.88715	.10134	.78581
0.72	2.0544	.48675	0.7838	1.2706	.61691	.31269	.89423	.10401	.79022
0.73	2.0751	.48191	0.7966	1.2785	.62307	.31704	.90123	.10670	.79453
0.74	2.0959	.47711	0.8094	1.2865	.62915	.32138	.90817	.10942	.79875
0.75	2.1170	.47237	0.8223	1.2947	.63515	.32572	.91504	.11216	.80288
0.76	2.1383	.46767	0.8353	1.3030	.64108	.33006	.92185	.11493	.80691
0.77	2.1598	.46301	0.8484	1.3114	.64693	.33441	.92859	.11773	.81086
0.78	2.1815	.45841	0.8615	1.3199	.65271	.33875	.93527	.12055	.81472
0.79	2.2034	.45384	0.8748	1.3286	.65841	.34309	.94190	.12340	.81850
0.80	**2.2255**	**.44933**	**0.8881**	**1.3374**	**.66404**	**0.34744**	**$\bar{1}$.94846**	**0.12627**	**1.82219**
0.81	2.2479	.44486	0.9015	1.3464	.66959	.35178	.95498	.12917	.82581
0.82	2.2705	.44043	0.9150	1.3555	.67507	.35612	.96144	.13209	.82935
0.83	2.2933	.43605	0.9286	1.3647	.68048	.36046	.96784	.13503	.83281
0.84	2.3164	.43171	0.9423	1.3740	.68581	.36481	.97420	.13800	.83620
0.85	2.3396	.42741	0.9561	1.3835	.69107	.36915	.98051	.14099	.83952
0.86	2.3632	.42316	0.9700	1.3932	.69626	.37349	.98677	.14400	.84277
0.87	2.3869	.41895	0.9840	1.4029	.70137	.37784	.99299	.14704	.84595
0.88	2.4109	.41478	0.9981	1.4128	.70642	.38218	.99916	.15009	.84906
0.89	2.4351	.41066	1.0122	1.4229	.71139	.38652	0.00528	.15317	.85211
0.90	**2.4596**	**.40657**	**1.0265**	**1.4331**	**.71630**	**0.39087**	**0.01137**	**0.15627**	**$\bar{1}$.85509**
0.91	2.4843	.40252	1.0409	1.4434	.72113	.39521	.01741	.15939	.85801
0.92	2.5093	.39852	1.0554	1.4539	.72590	.39955	.02341	.16254	.86088
0.93	2.5345	.39455	1.0700	1.4645	.73059	.40389	.02937	.16570	.86368
0.94	2.5600	.39063	1.0847	1.4753	.73522	.40824	.03530	.16888	.86642
0.95	2.5857	.38674	1.0995	1.4862	.73978	.41258	.04119	.17208	.86910
0.96	2.6117	.38289	1.1144	1.4973	.74428	.41692	.04704	.17531	.87173
0.97	2.6379	.37908	1.1294	1.5085	.74870	.42127	.05286	.17855	.87431
0.98	2.6645	.37531	1.1446	1.5199	.75307	.42561	.05864	.18181	.87683
0.99	2.6912	.37158	1.1598	1.5314	.75736	.42995	.06439	.18509	.87930
1.00	**2.7183**	**.36788**	**1.1752**	**1.5431**	**.76159**	**0.43429**	**0.07011**	**0.18839**	**$\bar{1}$.88172**
1.01	2.7456	.36422	1.1907	1.5549	.76576	.43864	.07580	.19171	.88409
1.02	2.7732	.36059	1.2063	1.5669	.76987	.44298	.08146	.19504	.88642
1.03	2.8011	.35701	1.2220	1.5790	.77391	.44732	.08708	.19839	.88869
1.04	2.8292	.35345	1.2379	1.5913	.77789	.45167	.09268	.20176	.89092
1.05	2.8577	.34994	1.2539	1.6038	.78181	.45601	.09825	.20515	.89310
1.06	2.8864	.34646	1.2700	1.6164	.78566	.46035	.10379	.20855	.89524
1.07	2.9154	.34301	1.2862	1.6292	.78946	.46470	.10930	.21197	.89733
1.08	2.9447	.33960	1.3025	1.6421	.79320	.46904	.11479	.21541	.89938
1.09	2.9743	.33622	1.3190	1.6552	.79688	.47338	.12025	.21886	.90139
1.10	**3.0042**	**.33287**	**1.3356**	**1.6685**	**.80050**	**0.47772**	**0.12569**	**0.22233**	**$\bar{1}$.90336**
1.11	3.0344	.32956	1.3524	1.6820	.80406	.48207	.13111	.22582	.90529
1.12	3.0649	.32628	1.3693	1.6956	.80757	.48641	.13649	.22931	.90718
1.13	3.0957	.32303	1.3863	1.7093	.81102	.49075	.14186	.23283	.90903
1.14	3.1268	.31982	1.4035	1.7233	.81441	.49510	.14720	.23636	.91085
1.15	3.1582	.31664	1.4208	1.7374	.81775	.49944	.15253	.23990	.91262
1.16	3.1899	.31349	1.4382	1.7517	.82104	.50378	.15783	.24346	.91436
1.17	3.2220	.31037	1.4558	1.7662	.82427	.50812	.16311	.24703	.91607
1.18	3.2544	.30728	1.4735	1.7808	.82745	.51247	.16836	.25062	.91774
1.19	3.2871	.30422	1.4914	1.7957	.83058	.51681	.17360	.25422	.91938
1.20	**3.3201**	**.30119**	**1.5095**	**1.8107**	**.83365**	**0.52115**	**0.11882**	**0.25784**	**$\bar{1}$.92099**

	Natural Values					Common Logarithms			
x	e^x	e^{-x}	Sinh x	Cosh x	Tanh x	e^x	Sinh x	Cosh x	Tanh x
1.20	**3.3201**	**.30119**	**1.5095**	**1.8107**	**.83365**	**0.52115**	**0.17882**	**0.25784**	**$\bar{1}$.92099**
1.21	3.3535	.29820	1.5276	1.8258	.83668	.52550	.18402	.26146	.92256
1.22	3.3872	.29523	1.5460	1.8412	.83965	.52984	.18920	.26510	.92410
1.23	3.4212	.29229	1.5645	1.8568	.84258	.53418	.19437	.26876	.92561
1.24	3.4556	.28938	1.5831	1.8725	.84546	.53853	.19951	.27242	.92709
1.25	3.4903	.28650	1.6019	1.8884	.84828	.54287	.20464	.27610	.92854
1.26	3.5254	.28365	1.6209	1.9045	.85106	.54721	.20975	.27979	.92996
1.27	3.5609	.28083	1.6400	1.9208	.85380	.55155	.21485	.28349	.93135
1.28	3.5966	.27804	1.6593	1.9373	.85648	.55590	.21993	.28721	.93272
1.29	3.6328	.27527	1.6788	1.9540	.85913	.56024	.22499	.29093	.93406
1.30	**3.6693**	**.27253**	**1.6984**	**1.9709**	**.86172**	**0.56458**	**0.23004**	**0.29467**	**$\bar{1}$.93537**
1.31	3.7062	.26982	1.7182	1.9880	.86428	.56893	.23507	.29842	.93665
1.32	3.7434	.26714	1.7381	2.0053	.86678	.57327	.24009	.30217	.93791
1.33	3.7810	.26448	1.7583	2.0228	.86925	.57761	.24509	.30594	.93914
1.34	3.8190	.26185	1.7786	2.0404	.87167	.58195	.25008	.30972	.94035
1.35	3.8574	.25924	1.7991	2.0583	.87405	.58630	.25505	.31352	.94154
1.36	3.8962	.25666	1.8198	2.0764	.87639	.59064	.26002	.31732	.94270
1.37	3.9354	.25411	1.8406	2.0947	.87869	.59498	.26496	.32113	.94384
1.38	3.9749	.25158	1.8617	2.1132	.88095	.59933	.26990	.32495	.94495
1.39	4.0149	.24908	1.8829	2.1320	.88317	.60367	.27482	.32878	.94604
1.40	**4.0552**	**.24660**	**1.9043**	**2.1509**	**.88535**	**0.60801**	**0.27974**	**0.33262**	**$\bar{1}$.94712**
1.41	4.0960	.24414	1.9259	2.1700	.88749	.61236	.28464	.33647	.94817
1.42	4.1371	.24171	1.9477	2.1894	.88960	.61670	.28952	.34033	.94919
1.43	4.1787	.23931	1.9697	2.2090	.89167	.62104	.29440	.34420	.95020
1.44	4.2207	.23693	1.9919	2.2288	.89370	.62538	.29926	.34807	.95119
1.45	4.2631	.23457	2.0143	2.2488	.89569	.62973	.30412	.35196	.95216
1.46	4.3060	.23224	2.0369	2.2691	.89765	.63407	.30896	.35585	.95311
1.47	4.3492	.22993	2.0597	2.2896	.89958	.63841	.31379	.35976	.95404
1.48	4.3929	.22764	2.0827	2.3103	.90147	.64276	.31862	.36367	.95495
1.49	4.4371	.22537	2.1059	2.3312	.90332	.64710	.32343	.36759	.95584
1.50	**4.4817**	**.22313**	**2.1293**	**2.3524**	**.90515**	**0.65144**	**0.32823**	**0.37151**	**$\bar{1}$.95672**
1.51	4.5267	.22091	2.1529	2.3738	.90694	.65578	.33303	.37545	.95758
1.52	4.5722	.21871	2.1768	2.3955	.90870	.66013	.33781	.37939	.95842
1.53	4.6182	.21654	2.2008	2.4174	.91042	.66447	.34258	.38334	.95924
1.54	4.6646	.21438	2.2251	2.4395	.91212	.66881	.34735	.38730	.96005
1.55	4.7115	.21225	2.2496	2.4619	.91379	.67316	.35211	.39126	.96084
1.56	4.7588	.21014	2.2743	2.4845	.91542	.67750	.35686	.39524	.96162
1.57	4.8066	.20805	2.2993	2.5073	.91703	.68184	.36160	.39921	.96238
1.58	4.8550	.20598	2.3245	2.5305	.91860	.68619	.36633	.40320	.96313
1.59	4.9037	.20393	2.3499	2.5538	.92015	.69053	.37105	.40719	.96386
1.60	**4.9530**	**.20190**	**2.3756**	**2.5775**	**.92167**	**0.69487**	**0.37577**	**0.41119**	**$\bar{1}$.96457**
1.61	5.0028	.19989	2.4015	2.6013	.92316	.69921	.38048	.41520	.96528
1.62	5.0531	.19790	2.4276	2.6255	.92462	.70356	.38518	.41921	.96597
1.63	5.1039	.19593	2.4540	2.6499	.92606	.70790	.38987	.42323	.96664
1.64	5.1552	.19398	2.4806	2.6746	.92747	.71224	.39456	.42725	.96730
1.65	5.2070	.19205	2.5075	2.6995	.92886	.71659	.39923	.43129	.96795
1.66	5.2593	.19014	2.5346	2.7247	.93022	.72093	.40391	.43532	.96858
1.67	5.3122	.18825	2.5620	2.7502	.93155	.72527	.40857	.43937	.96921
1.68	5.3656	.18637	2.5896	2.7760	.93286	.72961	.41323	.44341	.96982
1.69	5.4195	.18452	2.6175	2.8020	.93415	.73396	.41788	.44747	.97042
1.70	**5.4739**	**.18268**	**2.6456**	**2.8283**	**.93541**	**0.73830**	**0.42253**	**0.45153**	**$\bar{1}$.97100**
1.71	5.5290	.18087	2.6740	2.8549	.93665	.74264	.42717	.45559	.97158
1.72	5.5845	.17907	2.7027	2.8818	.93786	.74699	.43180	.45966	.97214
1.73	5.6407	.17728	2.7317	2.9090	.93906	.75133	.43643	.46374	.97269
1.74	5.6973	.17552	2.7609	2.9364	.94023	.75567	.44105	.46782	.97323
1.75	5.7546	.17377	2.7904	2.9642	.94138	.76002	.44567	.47191	.97376
1.76	5.8124	.17204	2.8202	2.9922	.94250	.76436	.45028	.47600	.97428
1.77	5.8709	.17033	2.8503	3.0206	.94361	.76870	.45488	.48009	.97479
1.78	5.9299	.16864	2.8806	3.0492	.94470	.77304	.45948	.48419	.97529
1.79	5.9895	.16696	2.9112	3.0782	.94576	.77739	.46408	.48830	.97578
1.80	**6.0496**	**.16530**	**2.9422**	**3.1075**	**.94681**	**0.78173**	**0.46867**	**0.49241**	**$\bar{1}$.97626**

x	Natural Values					Common Logarithms			
	e^x	e^{-x}	Sinh x	Cosh x	Tanh x	e^x	Sinh x	Cosh x	Tanh x
1.80	**6.0496**	**.16530**	**2.9422**	**3.1075**	**.94681**	**0.78173**	**0.46867**	**0.49241**	**1̄.97626**
1.81	6.1104	.16365	2.9734	3.1371	.94783	.78607	.47325	.49652	.97673
1.82	6.1719	.16203	3.0049	3.1669	.94884	.79042	.47783	.50064	.97719
1.83	6.2339	.16041	3.0367	3.1972	.94983	.79476	.48241	.50476	.97764
1.84	6.2965	.15882	3.0689	3.2277	.95080	.79910	.48698	.50889	.97809
1.85	6.3598	.15724	3.1013	3.2585	.95175	.80344	.49154	.51302	.97852
1.86	6.4237	.15567	3.1340	3.2897	.95268	.80779	.49610	.51716	.97895
1.87	6.4883	.15412	3.1671	3.3212	.95359	.81213	.50066	.52130	.97936
1.88	6.5535	.15259	3.2005	3.3530	.95449	.81647	.50521	.52544	.97977
1.89	6.6194	.15107	3.2341	3.3852	.95537	.82082	.50976	.52959	.98017
1.90	**6.6859**	**.14957**	**3.2682**	**3.4177**	**.95624**	**0.82516**	**0.51430**	**0.53374**	**1̄.98057**
1.91	6.7531	.14808	3.3025	3.4506	.95709	.82950	.51884	.53789	.98095
1.92	6.8210	.14661	3.3372	3.4838	.95792	.83385	.52338	.54205	.98133
1.93	6.8895	.14515	3.3722	3.5173	.95873	.83819	.52791	.54621	.98170
1.94	6.9588	.14370	3.4075	3.5512	.95953	.84253	.53244	.55038	.98206
1.95	7.0287	.14227	3.4432	3.5855	.96032	.84687	.53696	.55455	.98242
1.96	7.0993	.14086	3.4792	3.6201	.96109	.85122	.54148	.55872	.98272
1.97	7.1707	.13946	3.5156	3.6551	.96185	.85556	.54600	.56290	.98311
1.98	7.2427	.13807	3.5523	3.6904	.96259	.85990	.55051	.56707	.98344
1.99	7.3155	.13670	3.5894	3.7261	.96331	.86425	.55502	.57126	.98377
2.00	**7.3891**	**.13534**	**3.6269**	**3.7622**	**.96403**	**0.86859**	**0.55953**	**0.57544**	**1̄.98409**
2.01	7.4633	.13399	3.6647	3.7987	.96473	.87293	.56403	.57963	.98440
2.02	7.5383	.13266	3.7028	3.8355	.96541	.87727	.56853	.58382	.98471
2.03	7.6141	.13134	3.7414	3.8727	.96609	.88162	.57303	.58802	.98502
2.04	7.6906	.13003	3.7803	3.9103	.96675	.88596	.57753	.59221	.98531
2.05	7.7679	.12873	3.8196	3.9483	.96740	.89030	.58202	.59641	.98560
2.06	7.8460	.12745	3.8593	3.9867	.96803	.89465	.58650	.60061	.98589
2.07	7.9248	.12619	3.8993	4.0255	.96865	.89899	.59099	.60482	.98617
2.08	8.0045	.12493	3.9398	4.0647	.96926	.90333	.59547	.60903	.98644
2.09	8.0849	.12369	3.9806	4.1043	.96986	.90768	.59995	.61324	.98671
2.10	**8.1662**	**.12246**	**4.0219**	**4.1443**	**.97045**	**0.91202**	**0.60443**	**0.61745**	**1̄.98697**
2.11	8.2482	.12124	4.0635	4.1847	.97103	.91636	.60890	.62167	.98723
2.12	8.3311	.12003	4.1056	4.2256	.97159	.92070	.61337	.62589	.98748
2.13	8.4149	.11884	4.1480	4.2669	.97215	.92505	.61784	.63011	.98773
2.14	8.4994	.11765	4.1909	4.3085	.97269	.92939	.62231	.63433	.98798
2.15	8.5849	.11648	4.2342	4.3507	.97323	.93373	.62677	.63856	.98821
2.16	8.6711	.11533	4.2779	4.3932	.97375	.93808	.63123	.64278	.98845
2.17	8.7583	.11418	4.3221	4.4362	.97426	.94242	.63569	.64701	.98868
2.18	8.8463	.11304	4.3666	4.4797	.97477	.94676	.64015	.65125	.98890
2.19	8.9352	.11192	4.4116	4.5236	.97526	.95110	.64460	.65548	.98912
2.20	**9.0250**	**.11080**	**4.4571**	**4.5679**	**.97574**	**0.95545**	**0.64905**	**0.65972**	**1̄.98934**
2.21	9.1157	.10970	4.5030	4.6127	.97622	.95979	.65350	.66396	.98955
2.22	9.2073	.10861	4.5494	4.6580	.97668	.96413	.65795	.66820	.98975
2.23	9.2999	.10753	4.5962	4.7037	.97714	.96848	.66240	.67244	.98996
2.24	9.3933	.10646	4.6434	4.7499	.97759	.97282	.66684	.67668	.99016
2.25	9.4877	.10540	4.6912	4.7966	.97803	.97716	.67128	.68093	.99035
2.26	9.5831	.10435	4.7394	4.8437	.97846	.98151	.67572	.68518	.99054
2.27	9.6794	.10331	4.7880	4.8914	.97888	.98585	.68016	.68943	.99073
2.28	9.7767	.10228	4.8372	4.9395	.97929	.99019	.68459	.69368	.99091
2.29	9.8749	.10127	4.8868	4.9881	.97970	.99453	.68903	.69794	.99109
2.30	**9.9742**	**.10026**	**4.9370**	**5.0372**	**.98010**	**0.99888**	**0.69346**	**0.70219**	**1̄.99127**
2.31	10.074	.09926	4.9876	5.0868	.98049	1.00322	.69789	.70645	.99144
2.32	10.176	.09827	5.0387	5.1370	.98087	.00756	.70232	.71071	.99161
2.33	10.278	.09730	5.0903	5.1876	.98124	.01191	.70675	.71497	.99178
2.34	10.381	.09633	5.1425	5.2388	.98161	.01625	.71117	.71923	.99194
2.35	10.486	.09537	5.1951	5.2905	.98197	.02059	.71559	.72349	.99210
2.36	10.591	.09442	5.2483	5.3427	.98233	.02493	.72002	.72776	.99226
2.37	10.697	.09348	5.3020	5.3954	.98267	.02928	.72444	.73203	.99241
2.38	10.805	.09255	5.3562	5.4487	.98301	.03362	.72885	.73630	.99256
2.39	10.913	.09163	5.4109	5.5026	.98335	.03796	.73327	.74056	.99271
2.40	**11.023**	**.09072**	**5.4662**	**5.5569**	**.98367**	**1.04231**	**0.73769**	**0.74484**	**1̄.99285**

x	Natural Values					Common Logarithms			
	e^x	e^{-x}	Sinh x	Cosh x	Tanh x	e^x	Sinh x	Cosh x	Tanh x
2.40	**11.023**	**.09072**	**5.4662**	**5.5569**	**.98367**	**1.04231**	**0.73769**	**0.74484**	**$\overline{1}$.99285**
2.41	11.134	.08982	5.5221	5.6119	.98400	.04665	.74210	.74911	.99299
2.42	11.246	.08892	5.5785	5.6674	.98431	.05099	.74652	.75338	.99313
2.43	11.359	.08804	5.6354	5.7235	.98462	.05534	.75093	.75766	.99327
2.44	11.473	.08716	5.6929	5.7801	.98492	.05968	.75534	.76194	.99340
2.45	11.588	.08629	5.7510	5.8373	.98522	.06402	.75975	.76621	.99353
2.46	11.705	.08543	5.8097	5.8951	.98551	.06836	.76415	.77049	.99366
2.47	11.822	.08458	5.8689	5.9535	.98579	.07271	.76856	.77477	.99379
2.48	11.941	.08374	5.9288	6.0125	.98607	.07705	.77296	.77906	.99391
2.49	12.061	.08291	5.9892	6.0721	.98635	.08139	.77737	.78334	.99403
2.50	**12.182**	**.08208**	**6.0502**	**6.1323**	**.98661**	**1.08574**	**0.78177**	**0.78762**	**$\overline{1}$.99415**
2.51	12.305	.08127	6.1118	6.1931	.98688	.09008	.78617	.79191	.99426
2.52	12.429	.08046	6.1741	6.2545	.98714	.09442	.79057	.79619	.99438
2.53	12.554	.07966	6.2369	6.3166	.98739	.09877	.79497	.80048	.99449
2.54	12.680	.07887	6.3004	6.3793	.98764	.10311	.79937	.80477	.99460
2.55	12.807	.07808	6.3645	6.4426	.98788	.10745	.80377	.80906	.99470
2.56	12.936	.07730	6.4293	6.5066	.98812	.11179	.80816	.81335	.99481
2.57	13.066	.07654	6.4946	6.5712	.98835	.11614	.81256	.81764	.99491
2.58	13.197	.07577	6.5607	6.6365	.98858	.12048	.81695	.82194	.99501
2.59	13.330	.07502	6.6274	6.7024	.98881	.12482	.82134	.82623	.99511
2.60	**13.464**	**.07427**	**6.6947**	**6.7690**	**.98903**	**1.12917**	**0.82573**	**0.83052**	**$\overline{1}$.99521**
2.61	13.599	.07353	6.7628	6.8363	.98924	.13351	.83012	.83482	.99530
2.62	13.736	.07280	6.8315	6.9043	.98946	.13785	.83451	.83912	.99540
2.63	13.874	.07208	6.9008	6.9729	.98966	.14219	.83890	.84341	.99549
2.64	14.013	.07136	6.9709	7.0423	.98987	.14654	.84329	.84771	.99558
2.65	14.154	.07065	7.0417	7.1123	.99007	.15088	.84768	.85201	.99566
2.66	14.296	.06995	7.1132	7.1831	.99026	.15522	.85206	.85631	.99575
2.67	14.440	.06925	7.1854	7.2546	.99045	.15957	.85645	.86061	.99583
2.68	14.585	.06856	7.2583	7.3268	.99064	.16391	.86083	.86492	.99592
2.69	14.732	.06788	7.3319	7.3998	.99083	.16825	.86522	.86922	.99600
2.70	**14.880**	**.06721**	**7.4063**	**7.4735**	**.99101**	**1.17260**	**0.86960**	**0.87352**	**$\overline{1}$.99608**
2.71	15.029	.06654	7.4814	7.5479	.99118	.17694	.87398	.87783	.99615
2.72	15.180	.06587	7.5572	7.6231	.99136	.18128	.87836	.88213	.99623
2.73	15.333	.06522	7.6338	7.6991	.99153	.18562	.88274	.88644	.99631
2.74	15.487	.06457	7.7112	7.7758	.99170	.18997	.88712	.89074	.99638
2.75	15.643	.06393	7.7894	7.8533	.99186	.19431	.89150	.89505	.99645
2.76	15.800	.06329	7.8683	7.9316	.99202	.19865	.89588	.89936	.99652
2.77	15.959	.06266	7.9480	8.0106	.99218	.20300	.90026	.90367	.99659
2.78	16.119	.06204	8.0285	8.0905	.99233	.20734	.90463	.90798	.99666
2.79	16.281	.06142	8.1098	8.1712	.99248	.21168	.90901	.91229	.99672
2.80	**16.445**	**.06081**	**8.1919**	**8.2527**	**.99263**	**1.21602**	**0.91339**	**0.91660**	**$\overline{1}$.99679**
2.81	16.610	.06020	8.2749	8.3351	.99278	.22037	.91776	.92091	.99685
2.82	16.777	.05961	8.3586	8.4182	.99292	.22471	.92213	.92522	.99691
2.83	16.945	.05901	8.4432	8.5022	.99306	.22905	.92651	.92953	.99698
2.84	17.116	.05843	8.5287	8.5871	.99320	.23340	.93088	.93385	.99704
2.85	17.288	.05784	8.6150	8.6728	.99333	.23774	.93525	.93816	.99709
2.86	17.462	.05727	8.7021	8.7594	.99346	.24208	.93963	.94247	.99715
2.87	17.637	.05670	8.7902	8.8469	.99359	.24643	.94400	.94679	.99721
2.88	17.814	.05613	8.8791	8.9352	.99372	.25077	.94837	.95110	.99726
2.89	17.993	.05558	8.9689	9.0244	.99384	.25511	.95274	.95542	.99732
2.90	**18.174**	**.05502**	**9.0596**	**9.1146**	**.99396**	**1.25945**	**0.95711**	**0.95974**	**$\overline{1}$.99737**
2.91	18.357	.05448	9.1512	9.2056	.99408	.26380	.96148	.96405	.99742
2.92	18.541	.05393	9.2437	9.2976	.99420	.26814	.96584	.96837	.99747
2.93	18.728	.05340	9.3371	9.3905	.99441	.27248	.97021	.97269	.99752
2.94	18.916	.05287	9.4315	9.4844	.99443	.27683	.97458	.97701	.99757
2.95	19.106	.05234	9.5268	9.5791	.99454	.28117	.97895	.98133	.99762
2.96	19.298	.05182	9.6231	9.6749	.99464	.28551	.98331	.98565	.99767
2.97	19.492	.05130	9.7203	9.7716	.99475	.28985	.98768	.98997	.99771
2.98	19.688	.05079	9.8185	9.8693	.99485	.29420	.99205	.99429	.99776
2.99	19.886	.05029	9.9177	9.9680	.99496	.29854	.99641	.99861	.99780
3.00	**20.086**	**.04979**	**10.018**	**10.068**	**.99505**	**1.30288**	**1.00078**	**1.00293**	**$\overline{1}$.99785**

x	Natural Values					Common Logarithms			
	e^x	e^{-x}	Sinh x	Cosh x	Tanh x	e^x	Sinh x	Cosh x	Tanh x
3.00	20.086	.04979	10.018	10.068	.99505	1.30288	1.00078	1.00293	$\bar{1}$.99785
3.01	20.287	.04929	10.119	10.168	.99515	.30723	.00514	.00725	.99789
3.02	20.491	.04880	10.221	10.270	.99525	.31157	.00950	.01157	.99793
3.03	20.697	.04832	10.325	10.373	.99534	.31591	.01387	.01589	.99797
3.04	20.905	.04783	10.429	10.477	.99543	.32026	.01823	.02022	.99801
3.05	21.115	.04736	10.534	10.581	.99552	.32460	.02259	.02454	.99805
3.06	21.328	.04689	10.640	10.687	.99561	.32894	.02696	.02886	.99809
3.07	21.542	.04642	10.748	10.794	.99570	.33328	.03132	.03319	.99813
3.08	21.758	.04596	10.856	10.902	.99578	.33763	.03568	.03751	.99817
3.09	21.977	.04550	10.966	11.011	.99587	.34197	.04004	.04184	.99820
3.10	22.198	.04505	11.077	11.122	.99595	1.34631	1.04440	1.04616	$\bar{1}$.99824
3.11	22.421	.04460	11.188	11.233	.99603	.35066	.04876	.05049	.99827
3.12	22.646	.04416	11.301	11.345	.99611	.35500	.05312	.05481	.99831
3.13	22.874	.04372	11.415	11.459	.99618	.35934	.05748	.05914	.99834
3.14	23.104	.04328	11.530	11.574	.99626	.36368	.06184	.06347	.99837
3.15	23.336	.04285	11.647	11.689	.99633	.36803	.06620	.06779	.99841
3.16	23.571	.04243	11.764	11.807	.99641	.37237	.07056	.07212	.99844
3.17	23.807	.04200	11.883	11.925	.99648	.37671	.07492	.07645	.99847
3.18	24.047	.04159	12.003	12.044	.99655	.38106	.07927	.08078	.99850
3.19	24.288	.04117	12.124	12.165	.99662	.38540	.08363	.08510	.99853
3.20	24.533	.04076	12.246	12.287	.99668	1.38974	1.08799	1.08943	$\bar{1}$.99856
3.21	24.779	.04036	12.369	12.410	.99675	.39409	.09235	.09376	.99859
3.22	25.028	.03996	12.494	12.534	.99681	.39843	.09670	.09809	.99861
3.23	25.280	.03956	12.620	12.660	.99688	.40277	.10106	.10242	.99864
3.24	25.534	.03916	12.747	12.786	.99694	.40711	.10542	.10675	.99867
3.25	25.790	.03877	12.876	12.915	.99700	.41146	.10977	.11108	.99869
3.26	26.050	.03839	13.006	13.044	.99706	.41580	.11413	.11541	.99872
3.27	26.311	.03801	13.137	13.175	.99712	.42014	.11849	.11974	.99875
3.28	26.576	.03763	13.269	13.307	.99717	.42449	.12284	.12407	.99877
3.29	26.843	.03725	13.403	13.440	.99723	.42883	.12720	.12840	.99879
3.30	27.113	.03688	13.538	13.575	.99728	1.43317	1.13155	1.13273	$\bar{1}$.99882
3.31	27.385	.03652	13.674	13.711	.99734	43751	.13591	.13706	.99884
3.32	27.660	.03615	13.812	13.848	.99739	.44186	.14026	.14139	.99886
3.33	27.938	.03579	13.951	13.987	.99744	.44620	.14461	.14573	.99889
3.34	28.219	.03544	14.092	14.127	.99749	.45054	.14897	.15006	.99891
3.35	28.503	.03508	14.234	14.269	.99754	.45489	.15332	.15439	.99893
3.36	28.789	.03474	14.377	14.412	.99759	.45923	.15768	.15872	.99895
3.37	29.079	.03439	14.522	14.556	.99764	.46357	.16203	.16306	.99897
3.38	29.371	.03405	14.668	14.702	.99768	.46792	.16638	.16739	.99899
3.39	29.666	.03371	14.816	14.850	.99773	.47226	.17073	.17172	.99901
3.40	29.964	.03337	14.965	14.999	.99777	1.47660	1.17509	1.17605	$\bar{1}$.99903
3.41	30.265	.03304-	15.116	15.149	.99782	.48094	.17944	.18039	.99905
3.42	30.569	.03271	15.268	15.301	.99786	.48529	.18379	.18472	.99907
3.43	30.877	.03239	15.422	15.455	.99790	.48963	.18814	.18906	.99909
3.44	31.187	.03206	15.577	15.610	.99795	.49397	.19250	.19339	.99911
3.45	31.500	.03175	15.734	15.766	.99799	.49832	.19685	.19772	.99912
3.46	31.817	.03143	15.893	15.924	.99803	.50266	.20120	.20206	.99914
3.47	32.137	.03112	16.053	16.084	.99807	.50700	.20555	.20639	.99916
3.48	32.460	.03081	16.215	16.245	.99810	.51134	.20990	.21073	.99918
3.49	32.786	.03050	16.378	16.408	.99814	.51569	.21425	.21506	.99919
3.50	33.115	.03020	16.543	16.573	.99818	1.52003	1.21860	1.21940	$\bar{1}$.99921
3.51	33.448	.02990	16.709	16.739	.99821	.52437	.22296	.22373	.99922
3.52	33.784	.02960	16.877	16.907	.99825	.52872	.22731	.22807	.99924
3.53	34.124	.02930	17.047	17.077	.99828	.53306	.23166	.23240	.99925
3.54	34.467	.02901	17.219	17.248	.99832	.53740	.23601	.23674	.99927
3.55	34.813	.02872	17.392	17.421	.99835	.54175	.24036	.24107	.99928
3.56	35.163	.02844	17.567	17.596	.99838	.54609	.24471	.24541	.99930
3.57	35.517	.02816	17.744	17.772	.99842	.55043	.24906	.24975	.99931
3.58	35.874	.02788	17.923	17.951	.99845	.55477	.25341	.25408	.99933
3.59	36.234	.02760	18.103	18.131	.99848	.55912	.25776	.25842	.99934
3.60	36.598	.02732	18.285	18.313	.99851	1.56346	1.26211	1.26275	$\bar{1}$.99935

x	Natural Values					Common Logarithms			
	e^x	e^{-x}	Sinh x	Cosh x	Tanh x	e^x	Sinh x	Cosh x	Tanh x
3.60	36.598	.02732	18.285	18.313	.99851	1.56346	1.26211	1.26275	1̄.99935
3.61	36.966	.02705	18.470	18.497	.99854	.56780	.26646	.26709	.99936
3.62	37.338	.02678	18.655	18.682	.99857	.57215	.27080	.27143	.99938
3.63	37.713	.02652	18.843	18.870	.99859	.57649	.27515	.27576	.99939
3.64	38.092	.02625	19.033	19.059	.99862	.58083	.27950	.28010	.99940
3.65	38.475	.02599	19.224	19.250	.99865	.58517	.28385	.28444	.99941
3.66	38.861	.02573	19.418	19.444	.99868	.58952	.28820	.28878	.99942
3.67	39.252	.02548	19.613	19.639	.99870	.59386	.29255	.29311	.99944
3.68	39.646	.02522	19.811	19.836	.99873	.59820	.29690	.29745	.99945
3.69	40.045	.02497	20.010	20.035	.99875	.60255	.30125	.30179	.99946
3.70	40.447	.02472	20.211	20.236	.99878	1.60689	1.30559	1.30612	1̄.99947
3.71	40.854	.02448	20.415	20.439	.99880	.61123	.30994	.31046	.99948
3.72	41.264	.02423	20.620	20.644	.99883	.61558	.31429	.31480	.99949
3.73	41.679	.02399	20.828	20.852	.99885	.61992	.31864	.31914	.99950
3.74	42.098	.02375	21.037	21.061	.99887	.62426	.32299	.32348	.99951
3.75	42.521	.02352	21.249	21.272	.99889	.62860	.32733	.32781	.99952
3.76	42.948	.02328	21.463	21.486	.99892	.63295	.33168	.33215	.99953
3.77	43.380	.02305	21.679	21.702	.99894	.63729	.33603	.33649	.99954
3.78	43.816	.02282	21.897	21.919	.99896	.64163	.34038	.34083	.99955
3.79	44.256	.02260	22.117	22.140	.99898	.64598	.34472	.34517	.99956
3.80	44.701	.02237	22.339	22.362	.99900	1.65032	1.34907	1.34951	1̄.99957
3.81	45.150	.02215	22.564	22.586	.99902	.65466	.35342	.35384	.99957
3.82	45.604	.02193	22.791	22.813	.99904	.65900	.35777	.35818	.99958
3.83	46.063	.02171	23.020	23.042	.99906	.66335	.36211	.36252	.99959
3.84	46.525	.02149	23.252	23.274	.99908	.66769	.36646	.36686	.99960
3.85	46.993	.02128	23.486	23.507	.99909	.67203	.37081	.37120	.99961
3.86	47.465	.02107	23.722	23.743	.99911	.67638	.37515	.37554	.99961
3.87	47.942	.02086	23.961	23.982	.99913	.68072	.37950	.37988	.99962
3.88	48.424	.02065	24.202	24.222	.99915	.68506	.38385	.38422	.99963
3.89	48.911	.02045	24.445	24.466	.99916	.68941	.38819	.38856	.99964
3.90	49.402	.02024	24.691	24.711	.99918	1.69375	1.39254	1.39290	1̄.99964
3.91	49.899	.02004	24.939	24.960	.99920	.69809	.39689	.39724	.99965
3.92	50.400	.01984	25.190	25.210	.99921	.70243	.40123	.40158	.99966
3.93	50.907	.01964	25.444	25.463	.99923	.70678	.40558	.40591	.99966
3.94	51.419	.01945	25.700	25.719	.99924	.71112	.40993	.41025	.99967
3.95	51.935	.01925	25.958	25.977	.99926	.71546	.41427	.41459	.99968
3.96	52.457	.01906	26.219	26.238	.99927	.71981	.41862	.41893	.99968
3.97	52.985	.01887	26.483	26.502	.99929	.72415	.42296	.42327	.99969
3.98	53.517	.01869	26.749	26.768	.99930	.72849	.42731	.42761	.99970
3.99	54.055	.01850	27.018	27.037	.99932	.73284	.43166	.43195	.99970
4.00	54.598	.01832	27.290	27.308	.99933	1.73718	1.43600	1.43629	1̄.99971
4.01	55.147	.01813	27.564	27.583	.99934	.74152	.44035	.44063	.99971
4.02	55.701	.01795	27.842	27.860	.99936	.74586	.44469	.44497	.99972
4.03	56.261	.01777	28.122	28.139	.99937	.75021	.44904	.44931	.99973
4.04	56.826	.01760	28.404	28.422	.99938	.75455	.45339	.45365	.99973
4.05	57.397	.01742	28.690	28.707	.99939	.75889	.45773	.45799	.99974
4.06	57.974	.01725	28.979	28.996	.99941	.76324	.46208	.46233	.99974
4.07	58.557	.01708	29.270	29.287	.99942	.76758	.46642	.46668	.99975
4.08	59.145	.01691	29.564	29.581	.99943	.77192	.47077	.47102	.99975
4.09	59.740	.01674	29.862	29.878	.99944	.77626	.47511	.47536	.99976
4.10	60.340	.01657	30.162	30.178	.99945	1.78061	1.47946	1.47970	1̄.99976
4.11	60.947	.01641	30.465	30.482	.99946	.78495	.48380	.48404	.99977
4.12	61.559	.01624	30.772	30.788	.99947	.78929	.48815	.48838	.99977
4.13	62.178	.01608	31.081	31.097	.99948	.79364	.49249	.49272	.99978
4.14	62.803	.01592	31.393	31.409	.99949	.79798	.49684	.49706	.99978
4.15	63.434	.01576	31.709	31.725	.99950	.80232	.50118	.50140	.99978
4.16	64.072	.01561	32.028	32.044	.99951	.80667	.50553	.50574	.99979
4.17	64.715	.01545	32.350	32.365	.99952	.81101	.50987	.51008	.99979
4.18	65.366	.01530	32.675	32.691	.99953	.81535	.51422	.51442	.99980
4.19	66.023	.01515	33.004	33.019	.99954	.81969	.51856	.51876	.99980
4.20	66.686	.01500	33.336	33.351	.99955	1.82404	1.52291	1.52310	1̄.99980

x	Natural Values					Common Logarithms			
	e^x	e^{-x}	Sinh x	Cosh x	Tanh x	e^x	Sinh x	Cosh x	Tanh x
4.20	**66.686**	**.01500**	**33.336**	**33.351**	**.99955**	**1.82404**	**1.52291**	**1.52310**	**$\bar{1}$.99980**
4.21	67.357	.01485	33.671	33.686	.99956	.82838	.52725	.52745	.99981
4.22	68.033	.01470	34.009	34.024	.99957	.83272	.53160	.53179	.99981
4.23	68.717	.01455	34.351	34.366	.99958	.83707	.53594	.53613	.99982
4.24	69.408	.01441	34.697	34.711	.99958	.84141	.54029	.54047	.99982
4.25	70.105	.01426	35.046	35.060	.99959	.84575	.54463	.54481	.99982
4.26	70.810	.01412	35.398	35.412	.99960	.85009	.54898	.54915	.99983
4.27	71.522	.01398	35.754	35.768	.99961	.85444	.55332	.55349	.99983
4.28	72.240	.01384	36.113	36.127	.99962	.85878	.55767	.55783	.99983
4.29	72.966	.01370	36.476	36.490	.99962	.86312	.56201	.56217	.99984
4.30	**73.700**	**.01357**	**36.843**	**36.857**	**.99963**	**1.86747**	**1.56636**	**1.56652**	**$\bar{1}$.99984**
4.31	74.440	.01343	37.214	37.227	.99964	.87181	.57070	.57086	.99984
4.32	75.189	.01330	37.588	37.601	.99965	.87615	.57505	.57520	.99985
4.33	75.944	.01317	37.966	37.979	.99965	.88050	.57939	.57954	.99985
4.34	76.708	.01304	38.347	38.360	.99966	.88484	.58373	.58388	.99985
4.35	77.478	.01291	38.733	38.746	.99967	.88918	.58808	.58822	.99986
4.36	78.257	.01278	39.122	39.135	.99967	.89352	.59242	.59256	.99986
4.37	79.044	.01265	39.515	39.528	.99968	.89787	.59677	.59691	.99986
4.38	79.838	.01253	39.913	39.925	.99969	.90221	.60111	.60125	.99986
4.39	80.640	.01240	40.314	40.326	.99969	.90655	.60546	.60559	.99987
4.40	**81.451**	**.01228**	**40.719**	**40.732**	**.99970**	**1.91090**	**1.60980**	**1.60993**	**$\bar{1}$.99987**
4.41	82.269	.01216	41.129	41.141	.99970	.91524	.61414	.61427	.99987
4.42	83.096	.01203	41.542	41.554	.99971	.91958	.61849	.61861	.99987
4.43	83.931	.01191	41.960	41.972	.99972	.92392	.62283	.62296	.99988
4.44	84.775	.01180	42.382	42.393	.99972	.92827	.62718	.62730	.99988
4.45	85.627	.01168	42.808	42.819	.99973	.93261	.63152	.63164	.99988
4.46	86.488	.01156	43.238	43.250	.99973	.93695	.63587	.63598	.99988
4.47	87.357	.01145	43.673	43.684	.99974	.94130	.64021	.64032	.99989
4.48	88.235	.01133	44.112	44.123	.99974	.94564	.64455	.64467	.99989
4.49	89.121	.01122	44.555	44.566	.99975	.94998	.64890	.64901	.99989
4.50	**90.017**	**.01111**	**45.003**	**45.014**	**.99975**	**1.95433**	**1.65324**	**1.65335**	**$\bar{1}$.99989**
4.51	90.922	.01100	45.455	45.466	.99976	.95867	.65759	.65769	.99989
4.52	91.836	.01089	45.912	45.923	.99976	.96301	.66193	.66203	.99990
4.53	92.759	.01078	46.374	46.385	.99977	.96735	.66627	.66637	.99990
4.54	93.691	.01067	46.840	46.851	.99977	.97170	.67062	.67072	.99990
4.55	94.632	.01057	47.311	47.321	.99978	.97604	.67496	.67506	.99990
4.56	95.583	.01046	47.787	47.797	.99978	.98038	.67931	.67940	.99990
4.57	96.544	.01036	48.267	48.277	.99979	.98473	.68365	.68374	.99991
4.58	97.514	.01025	48.752	48.762	.99979	.98907	.68799	.68808	.99991
4.59	98.494	.01015	49.242	49.252	.99979	.99341	.69234	.69243	.99991
4.60	**99.484**	**.01005**	**49.737**	**49.747**	**.99980**	**1.99775**	**1.69668**	**1.69677**	**$\bar{1}$.99991**
4.61	100.48	.00995	50.237	50.247	.99980	2.00210	.70102	.70111	.99991
4.62	101.49	.00985	50.742	50.752	.99981	.00644	.70537	.70545	.99992
4.63	102.51	.00975	51.252	51.262	.99981	.01078	.70971	.70979	.99992
4.64	103.54	.00966	51.767	51.777	.99981	.01513	.71406	.71414	.99992
4.65	104.58	.00956	52.288	52.297	.99982	.01947	.71840	.71848	.99992
4.66	105.64	.00947	52.813	52.823	.99982	.02381	.72274	.72282	.99992
4.67	106.70	.00937	53.344	53.354	.99982	.02816	.72709	.72716	.99992
4.68	107.77	.00928	53.880	53.890	.99983	.03250	.73143	.73151	.99993
4.69	108.85	.00919	54.422	54.431	.99983	.03684	.73577	.73585	.99993
4.70	**109.95**	**.00910**	**54.969**	**54.978**	**.99983**	**2.04118**	**1.74012**	**1.74019**	**$\bar{1}$.99993**
4.71	111.05	.00900	55.522	55.531	.99984	.04553	.74446	.74453	.99993
4.72	112.17	.00892	56.080	56.089	.99984	.04987	.74881	.74887	.99993
4.73	113.30	.00883	56.643	56.652	.99984	.05421	.75315	.75322	.99993
4.74	114.43	.00874	57.213	57.222	.99985	.05856	.75749	.75756	.99993
4.75	115.58	.00865	57.788	57.796	.99985	.06290	.76184	.76190	.99993
4.76	116.75	.00857	58.369	58.377	.99985	.06724	.76618	.76624	.99994
4.77	117.92	.00848	58.955	58.964	.99986	.07158	.77052	.77059	.99994
4.78	119.10	.00840	59.548	59.556	.99986	.07593	.77487	.77493	.99994
4.79	120.30	.00831	60.147	60.155	.99986	.08027	.77921	.77927	.99994
4.80	**121.51**	**.00823**	**60.751**	**60.759**	**.99985**	**2.08461**	**1.78355**	**1.78361**	**$\bar{1}$.99994**

x	Natural Values					Common Logarithms			
	e^x	e^{-x}	Sinh x	Cosh x	Tanh x	e^x	Sinh x	Cosh x	Tanh x
4.80	121.51	.00823	60.751	60.760	.99986	2.08461	1.78355	1.78361	$\bar{1}$.99994
4.81	122.73	.00815	61.362	61.370	.99987	.08896	.78790	.78796	.99994
4.82	123.97	.00807	61.979	61.987	.99987	.09330	.79224	.79230	.99994
4.83	125.21	.00799	62.601	62.609	.99987	.09764	.79658	.79664	.99994
4.84	126.47	.00791	63.231	63.239	.99987	.10199	.80093	.80098	.99995
4.85	127.74	.00783	63.866	63.874	.99988	.10633	.80527	.80532	.99995
4.86	129.02	.00775	64.508	64.516	.99988	.11067	.80962	.80967	.99995
4.87	130.32	.00767	65.157	65.164	.99988	.11501	.81396	.81401	.99995
4.88	131.63	.00760	65.812	65.819	.99988	.11936	.81830	.81835	.99995
4.89	132.95	.00752	66.473	66.481	.99989	.12370	.82265	.82269	.99995
4.90	134.29	.00745	67.141	67.149	.99989	2.12804	1.82699	1.82704	$\bar{1}$.99995
4.91	135.64	.00737	67.816	67.823	.99989	.13239	.83133	.83138	.99995
4.92	137.00	.00730	68.498	68.505	.99989	.13673	.83568	.83572	.99995
4.93	138.38	.00723	69.186	69.193	.99990	.14107	.84002	.84006	.99995
4.94	139.77	.00715	69.882	69.889	.99990	.14541	.84436	.84441	.99996
4.95	141.17	.00708	70.584	70.591	.99990	.14976	.84871	.84875	.99996
4.96	142.59	.00701	71.293	71.300	.99990	.15410	.85305	.85309	.99996
4.97	144.03	.00694	72.010	72.017	.99990	.15844	.85739	.85743	.99996
4.98	145.47	.00687	72.734	72.741	.99991	.16279	.86174	.86178	.99996
4.99	146.94	.00681	73.465	73.472	.99991	.16713	.86608	.86612	.99996
5.00	148.41	.00674	74.203	74.210	.99991	2.17147	1.87042	1.87046	$\bar{1}$.99996
5.01	149.90	.00667	74.949	74.956	.99991	.17582	.87477	.87480	.99996
5.02	151.41	.00660	75.702	75.710	.99991	.18016	.87911	.87915	.99996
5.03	152.93	.00654	76.463	76.470	.99991	.18450	.88345	.88349	.99996
5.04	154.47	.00647	77.232	77.238	.99992	.18884	.88780	.88783	.99996
5.05	156.02	.00641	78.008	78.014	.99992	.19319	.89214	.89217	.99996
5.06	157.59	.00635	78.792	78.798	.99992	.19753	.89648	.89652	.99997
5.07	159.17	.00628	79.584	79.590	.99992	.20187	.90083	.90086	.99997
5.08	160.77	.00622	80.384	80.390	.99992	.20622	.90517	.90520	.99997
5.09	162.39	.00616	81.192	81.198	.99992	.21056	.90951	.90955	.99997
5.10	164.02	.00610	82.008	82.014	.99993	2.21490	1.91386	1.91389	$\bar{1}$.99997
5.11	165.67	.00604	82.832	82.838	.99993	.21924	.91820	.91823	.99997
5.12	167.34	.00598	83.665	83.671	.99993	.22359	.92254	.92257	.99997
5.13	169.02	.00592	84.506	84.512	.99993	.22793	.92689	.92692	.99997
5.14	170.72	.00586	85.355	85.361	.99993	.23227	.93123	.93126	.99997
5.15	172.43	.00580	86.213	86.219	.99993	.23662	.93557	.93560	.99997
5.16	174.16	.00574	87.079	87.085	.99993	.24096	.93992	.93994	.99997
5.17	175.91	.00568	87.955	87.960	.99994	.24530	.94426	.94429	.99997
5.18	177.68	.00563	88.839	88.844	.99994	.24965	.94860	.94863	.99997
5.19	179.47	.00557	89.732	89.737	.99994	.25399	.95294	.95297	.99997
5.20	181.27	.00552	90.633	90.639	.99994	2.25833	1.95729	1.95731	$\bar{1}$.99997
5.21	183.09	.00546	91.544	91.550	.99994	.26267	.96163	.96166	.99997
5.22	184.93	.00541	92.464	92.470	.99994	.26702	.96597	.96600	.99997
5.23	186.79	.00535	93.394	93.399	.99994	.27136	.97032	.97034	.99998
5.24	188.67	.00530	94.332	94.338	.99994	.27570	.97466	.97469	.99998
5.25	190.57	.00525	95.281	95.286	.99994	.28005	.97900	.97903	.99998
5.26	192.48	.00520	96.238	96.243	.99995	.28439	.98335	.98337	.99998
5.27	194.42	.00514	97.205	97.211	.99995	.28873	.98769	.98771	.99998
5.28	196.37	.00509	98.182	98.188	.99995	.29307	.99203	.99206	.99998
5.29	198.34	.00504	99.169	99.174	.99995	.29742	.99638	.99640	.99998
5.30	200.34	.00499	100.17	100.17	.99995	2.30176	2.00072	2.00074	$\bar{1}$.99998
5.31	202.35	.00494	101.17	101.18	.99995	.30610	.00506	.00508	.99998
5.32	204.38	.00489	102.19	102.19	.99995	.31045	.00941	.00943	.99998
5.33	206.44	.00484	103.22	103.22	.99995	.31479	.01375	.01377	.99998
5.34	208.51	.00480	104.25	104.26	.99995	.31913	.01809	.01811	.99998
5.35	210.61	.00475	105.30	105.31	.99995	.32348	.02244	.02246	.99998
5.36	212.72	.00470	106.36	106.36	.99996	.32782	.02678	.02680	.99998
5.37	214.86	.00465	107.43	107.43	.99996	.33216	.03112	.03114	.99998
5.38	217.02	.00461	108.51	108.51	.99996	.33650	.03547	.03548	.99998
5.39	219.20	.00456	109.60	109.60	.99996	.34085	.03981	.03983	.99998
5.40	221.41	.00452	110.70	110.71	.99996	2.34519	2.04415	2.04417	$\bar{1}$.99998

x	Natural Values					Common Logarithms			
	e^x	e^{-x}	Sinh x	Cosh x	Tanh x	e^x	Sinh x	Cosh x	Tanh x
5.40	221.41	.00452	110.70	110.71	.99996	2.34519	2.04415	2.04417	1.99998
5.41	223.63	.00447	111.81	111.82	.99996	.34953	.04849	.04851	.99998
5.42	225.88	.00443	112.94	112.94	.99996	.35388	.05284	.05285	.99998
5.43	228.15	.00438	114.07	114.08	.99996	.35822	.05718	.05720	.99998
5.44	230.44	.00434	115.22	115.22	.99996	.36256	.06152	.06154	.99998
5.45	232.76	.00430	116.38	116.38	.99996	.36690	.06587	.06588	.99998
5.46	235.10	.00425	117.55	117.55	.99996	.37125	.07021	.07023	.99998
5.47	237.46	.00421	118.73	118.73	.99996	.37559	.07455	.07457	.99998
5.48	239.85	.00417	119.92	119.93	.99997	.37993	.07890	.07891	.99998
5.49	242.26	.00413	121.13	121.13	.99997	.38428	.08324	.08325	.99999
5.50	244.69	.00409	122.34	122.35	.99997	2.38862	2.08758	2.08760	1.99999
5.51	247.15	.00405	123.57	123.58	.99997	.39296	.09193	.09194	.99999
5.52	249.64	.00401	124.82	124.82	.99997	.39731	.09627	.09628	.99999
5.53	252.14	.00397	126.07	126.07	.99997	.40165	.10061	.10063	.99999
5.54	254.68	.00393	127.34	127.34	.99997	.40599	.10495	.10497	.99999
5.55	257.24	.00389	128.62	128.62	.99997	.41033	.10930	.10931	.99999
5.56	259.82	.00385	129.91	129.91	.99997	.41468	.11364	.11365	.99999
5.57	262.43	.00381	131.22	131.22	.99997	.41902	.11798	.11800	.99999
5.58	265.07	.00377	132.53	132.54	.99997	.42336	.12233	.12234	.99999
5.59	267.74	.00374	133.87	133.87	.99997	.42771	.12667	.12668	.99999
5.60	270.43	.00370	135.21	135.22	.99997	2.43205	2.13101	2.13103	1.99999
5.61	273.14	.00366	136.57	136.57	.99997	.43639	.13536	.13537	.99999
5.62	275.89	.00362	137.94	137.95	.99997	.44074	.13970	.13971	.99999
5.63	278.66	.00359	139.33	139.33	.99997	.44508	.14404	.14405	.99999
5.64	281.46	.00355	140.73	140.73	.99997	.44942	.14839	.14840	.99999
5.65	284.29	.00352	142.14	142.15	.99998	.45376	.15273	.15274	.99999
5.66	287.15	.00348	143.57	143.58	.99998	.45811	.15707	.15708	.99999
5.67	290.03	.00345	145.02	145.02	.99998	.46245	.16141	.16142	.99999
5.68	292.95	.00341	146.47	146.48	.99998	.46679	.16576	.16577	.99999
5.69	295.89	.00338	147.95	147.95	.99998	.47114	.17010	.17011	.99999
5.70	298.87	.00335	149.43	149.44	.99998	2.47548	2.17444	2.17445	1.99999
5.71	301.87	.00331	150.93	150.94	.99998	.47982	.17879	.17880	.99999
5.72	304.90	.00328	152.45	152.45	.99998	.48416	.18313	.18314	.99999
5.73	307.97	.00325	153.98	153.99	.99998	.48851	.18747	.18748	.99999
5.74	311.06	.00321	155.53	155.53	.99998	.49285	.19182	.19182	.99999
5.75	314.19	.00318	157.09	157.10	.99998	.49719	.19616	.19617	.99999
5.76	317.35	.00315	158.67	158.68	.99998	.50154	.20050	.20051	.99999
5.77	320.54	.00312	160.27	160.27	.99998	.50588	.20484	.20485	.99999
5.78	323.76	.00309	161.88	161.88	.99998	.51022	.20919	.20920	.99999
5.79	327.01	.00306	163.51	163.51	.99998	.51457	.21353	.21354	.99999
5.80	330.30	.00303	165.15	165.15	.99998	2.51891	2.21787	2.21788	1.99999
5.81	333.62	.00300	166.81	166.81	.99998	.52325	.22222	.22222	.99999
5.82	336.97	.00297	168.48	168.49	.99998	.52759	.22656	.22657	.99999
5.83	340.36	.00294	170.18	170.18	.99998	.53194	.23090	.23091	.99999
5.84	343.78	.00291	171.89	171.89	.99998	.53628	.23525	.23525	.99999
5.85	347.23	.00288	173.62	173.62	.99998	.54062	.23959	.23960	.99999
5.86	350.72	.00285	175.36	175.36	.99998	.54497	.24393	.24394	.99999
5.87	354.25	.00282	177.12	177.13	.99998	.54931	.24828	.24828	.99999
5.88	357.81	.00279	178.90	178.91	.99998	.55365	.25262	.25262	.99999
5.89	361.41	.00277	180.70	180.70	.99998	.55799	.25696	.25697	.99999
5.90	365.04	.00274	182.52	182.52	.99998	2.56234	2.26130	2.26131	1.99999
5.91	368.71	.00271	184.35	184.35	.99999	.56668	.26565	.26565	.99999
5.92	372.41	.00269	186.20	186.21	.99999	.57102	.26999	.27000	.99999
5.93	376.15	.00266	188.08	188.08	.99999	.57537	.27433	.27434	.99999
5.94	379.93	.00263	189.97	189.97	.99999	.57971	.27868	.27868	.99999
5.95	383.75	.00261	191.88	191.88	.99999	.58405	.28302	.28303	.99999
5.96	387.61	.00258	193.80	193.81	.99999	.58840	.28736	.28737	.99999
5.97	391.51	.00255	195.75	195.75	.99999	.59274	.29171	.29171	.99999
5.98	395.44	.00253	197.72	197.72	.99999	.59708	.29605	.29605	.99999
5.99	399.41	.00250	199.71	199.71	.99999	.60142	.30039	.30040	.99999
6.00	403.43	.00248	201.71	201.72	.99999	2.60577	2.30473	2.30474	1.99999

Table 23. Table of Integrals

Elementary Indefinite Integrals

1. $\int a\,dx = ax$

2. $\int (u + v + w + \ldots)\,dx = \int u\,dx + \int v\,dx + \int w\,dx + \ldots$

3. $\int u\,dv = uv - \int v\,du$, integration by parts

4. $\int f(x)\,dx = \int f[\phi(y)]\phi'(y)\,dy$, $x = \phi(y)$, change of variable

5. $\int x^n\,dx = \dfrac{x^{n+1}}{n+1}$, $(n \neq -1)$

6. $\int \dfrac{dx}{x} = \log_e x + c = \log_e c_1 x$, $[\log_e x = \log_e (-x) + (2k+1)\pi i]$

7. $\int e^{ax}\,dx = \dfrac{1}{a}e^{ax}$

8. $\int a^x\,dx = \dfrac{a^x}{\log_e a}$

9. $\int a^x \log_e a\,dx = a^x$

10. $\int \sin ax\,dx = -\dfrac{1}{a}\cos ax$

11. $\int \cos ax\,dx = \dfrac{1}{a}\sin ax$

12. $\int \tan ax\,dx = -\dfrac{1}{a}\log_e \cos ax = \dfrac{1}{a}\log_e \sec ax$

13. $\int \cot ax\,dx = \dfrac{1}{a}\log_e \sin ax = -\dfrac{1}{a}\log_e \csc ax$

14. $\int \sec ax\,dx = \dfrac{1}{a}\log_e (\sec ax + \tan ax) = \dfrac{1}{a}\log_e \tan \left(\dfrac{ax}{2} + \dfrac{\pi}{4}\right)$

15. $\int \csc ax\,dx = \dfrac{1}{a}\log_e (\csc ax - \cot ax) = \dfrac{1}{a}\log_e \tan \dfrac{ax}{2}$

16. $\int \dfrac{dx}{\sqrt{a^2 - x^2}} = \sin^{-1}\dfrac{x}{a} = -\cos^{-1}\dfrac{x}{a}$ $(x^2 < a^2)$

17. $\int \dfrac{dx}{a^2 + x^2} = \dfrac{1}{a}\tan^{-1}\dfrac{x}{a} = -\dfrac{1}{a}\cot^{-1}\dfrac{x}{a}$

18. $\int \sinh ax\,dx = \dfrac{1}{a}\cosh ax$

19. $\int \cosh ax\,dx = \dfrac{1}{a}\sinh ax$

20. $\int \tanh ax\,dx = \dfrac{1}{a}\log_e (\cosh ax)$

21. $\int \coth ax\,dx = \dfrac{1}{a}\log_e (\sinh ax)$

22. $\int \operatorname{sech} ax\,dx = \dfrac{1}{a}\sin^{-1}(\tanh ax) = \dfrac{1}{a}\tan^{-1}(\sinh ax)$

23. $\int \operatorname{csch} ax\,dx = \dfrac{1}{a}\log_e \left(\tanh \dfrac{ax}{2}\right)$

24. $\int \sin^2 ax\,dx = \dfrac{1}{2}x - \dfrac{1}{2a}\sin ax \cos ax = \dfrac{1}{2}x - \dfrac{1}{4a}\sin 2ax$

25. $\int \cos^2 ax\,dx = \dfrac{1}{2}x + \dfrac{1}{2a}\sin ax \cos ax = \dfrac{1}{2}x + \dfrac{1}{4a}\sin 2ax$

26. $\int \tan^2 ax\,dx = \dfrac{1}{a}\tan ax - x$

27. $\int \cot^2 ax\,dx = -\dfrac{1}{a}\cot ax - x$

Elementary Indefinite Integrals—*Continued*

28. $\int \sec^2 ax \, dx = \dfrac{1}{a} \tan ax$

29. $\int \csc^2 ax \, dx = -\dfrac{1}{a} \cot ax$

30. $\int \sin^{-1} ax \, dx = x \sin^{-1} ax + \dfrac{1}{a} \sqrt{1 - a^2 x^2}$

31. $\int \cos^{-1} ax \, dx = x \cos^{-1} ax - \dfrac{1}{a} \sqrt{1 - a^2 x^2}$

32. $\int \tan^{-1} ax \, dx = x \tan^{-1} ax - \dfrac{1}{2a} \log_e (1 + a^2 x^2)$

33. $\int \cot^{-1} ax \, dx = x \cot^{-1} ax + \dfrac{1}{2a} \log_e (1 + a^2 x^2)$

34. $\int \sec^{-1} ax \, dx = x \sec^{-1} ax - \dfrac{1}{a} \log_e (ax + \sqrt{a^2 x^2 - 1})$

35. $\int \csc^{-1} ax \, dx = x \csc^{-1} ax + \dfrac{1}{a} \log_e (ax + \sqrt{a^2 x^2 - 1})$

Integrals Involving $(ax + b)$

36. $\int (ax + b)^n \, dx = \dfrac{1}{a(n + 1)} (ax + b)^{n+1} \quad (n \neq -1)$

37. $\int \dfrac{dx}{ax + b} = \dfrac{1}{a} \log_e (ax + b)$

38. $\int x(ax + b)^n \, dx = \dfrac{1}{a^2(n + 2)} (ax + b)^{n+2} - \dfrac{b}{a^2(n + 1)} (ax + b)^{n+1} \quad (n \neq -1, -2)$

39. $\int \dfrac{x \, dx}{ax + b} = \dfrac{x}{a} - \dfrac{b}{a^2} \log_e (ax + b)$

40. $\int \dfrac{x \, dx}{(ax + b)^2} = \dfrac{b}{a^2(ax + b)} + \dfrac{1}{a^2} \log_e (ax + b)$

41. $\int \dfrac{x^2 \, dx}{ax + b} = \dfrac{1}{a^3} \left[\dfrac{1}{2}(ax + b)^2 - 2b(ax + b) + b^2 \log_e (ax + b) \right]$

42. $\int \dfrac{x^2 \, dx}{(ax + b)^2} = \dfrac{1}{a^3} \left[(ax + b) - 2b \log_e (ax + b) - \dfrac{b^2}{ax + b} \right]$

43. $\int \dfrac{x^2 \, dx}{(ax + b)^3} = \dfrac{1}{a^3} \left[\log_e (ax + b) + \dfrac{2b}{ax + b} - \dfrac{b^2}{2(ax + b)^2} \right]$

44. $\int \dfrac{dx}{x(ax + b)} = \dfrac{1}{b} \log_e \dfrac{x}{ax + b}$

45. $\int \dfrac{dx}{x^2 (ax + b)} = -\dfrac{1}{bx} + \dfrac{a}{b^2} \log_e \dfrac{ax + b}{x}$

46. $\int \dfrac{dx}{x(ax + b)^2} = \dfrac{1}{b(ax + b)} - \dfrac{1}{b^2} \log_e \dfrac{ax + b}{x}$

47. $\int \dfrac{dx}{x^2 (ax + b)^2} = -\dfrac{b + 2 ax}{b^2 x(ax + b)} + \dfrac{2a}{b^3} \log_e \dfrac{ax + b}{x}$

48. $\int \dfrac{dx}{x \sqrt{ax + b}} = \dfrac{1}{\sqrt{b}} \log_e \dfrac{\sqrt{ax + b} - \sqrt{b}}{\sqrt{ax + b} + \sqrt{b}} \quad$ (*b* positive)

49. $\int \dfrac{dx}{x \sqrt{ax + b}} = \dfrac{2}{\sqrt{-b}} \tan^{-1} \sqrt{\dfrac{ax + b}{-b}} \quad$ (*b* negative)

50. $\int \dfrac{\sqrt{ax + b}}{x} \, dx = 2 \sqrt{ax + b} + \sqrt{b} \log_e \dfrac{\sqrt{ax + b} - \sqrt{b}}{\sqrt{ax + b} + \sqrt{b}} \quad$ (*b* positive)

51. $\int \dfrac{\sqrt{ax + b}}{x} \, dx = 2 \sqrt{ax + b} - 2 \sqrt{-b} \tan^{-1} \sqrt{\dfrac{ax + b}{-b}} \quad$ (*b* negative)

Integrals Involving $(ax + b)$—Continued

52. $\displaystyle\int \frac{dx}{x^2 \sqrt{ax + b}} = -\frac{\sqrt{ax + b}}{bx} - \frac{a}{2b\sqrt{b}} \log_e \frac{\sqrt{ax + b} - \sqrt{b}}{\sqrt{ax + b} + \sqrt{b}}$ (b positive)

53. $\displaystyle\int \frac{dx}{x^2 \sqrt{ax + b}} = -\frac{\sqrt{ax + b}}{bx} - \frac{a}{b\sqrt{-b}} \tan^{-1}\sqrt{\frac{ax + b}{-b}}$ (b negative)

54. $\displaystyle\int \frac{ax + b}{fx + g} \, dx = \frac{ax}{f} + \frac{bf - cg}{f^2} \log_e (fx + g)$

55. $\displaystyle\int \frac{dx}{(ax + b)(fx + g)} = \frac{1}{bf - ag} \log_e \left(\frac{fx + g}{ax + b}\right)$ ($ag \neq bf$)

56. $\displaystyle\int \frac{x \, dx}{(ax + b)(fx + g)} = \frac{1}{bf - ag}\left[\frac{b}{a} \log_e (ax + b) - \frac{g}{f} \log_e (fx + g)\right]$ ($ag \neq bf$)

57. $\displaystyle\int \frac{dx}{(ax + b)^2(fx + g)} = \frac{1}{bf - ag}\left(\frac{1}{ax + b} + \frac{f}{bf - ag} \log_e \frac{fx + g}{ax + b}\right)$ ($ag \neq bf$)

Integrals Involving $(ax^n + b)$

58. $\displaystyle\int (ax^2 + b)^n x \, dx = \frac{1}{2a} \frac{(ax^2 + b)^{n+1}}{n + 1}$ ($n \neq -1$)

59. $\displaystyle\int \frac{dx}{ax^2 + b} = \frac{1}{\sqrt{ab}} \tan^{-1}\left(x\sqrt{\frac{a}{b}}\right)$ (a and b positive)

60. $\displaystyle\int \frac{dx}{ax^2 + b} = \frac{1}{2\sqrt{-ab}} \log_e \frac{x\sqrt{a} - \sqrt{-b}}{x\sqrt{a} + \sqrt{-b}}$ (a positive, b negative)

$\displaystyle\qquad\qquad = \frac{1}{2\sqrt{-ab}} \log_e \frac{\sqrt{b} + x\sqrt{-a}}{\sqrt{b} - x\sqrt{-a}}$ (a negative, b positive)

61. $\displaystyle\int \frac{dx}{x(ax^2 + b)} = \frac{1}{2b} \log_e \frac{x^2}{ax^2 + b}$

62. $\displaystyle\int \frac{dx}{(ax^2 + b)^n} = \frac{1}{2(n - 1)b} \frac{x}{(ax^2 + b)^{n-1}} + \frac{2n - 3}{2(n - 1)b}\int \frac{dx}{(ax^2 + b)^{n-1}}$ (n integer > 1)

63. $\displaystyle\int \frac{x^2 \, dx}{ax^2 + b} = \frac{x}{a} - \frac{b}{a}\int \frac{dx}{ax^2 + b}$

64. $\displaystyle\int \frac{x^2 \, dx}{(ax^2 + b)^n} = -\frac{1}{2(n - 1)a} \frac{x}{(ax^2 + b)^{n-1}} + \frac{1}{2(n - 1)a}\int \frac{dx}{(ax^2 + b)^{n-1}}$ (n integer > 1)

65. $\displaystyle\int \frac{dx}{x^2 (ax^2 + b)^n} = \frac{1}{b}\int \frac{dx}{x^2 (ax^2 + b)^{n-1}} - \frac{a}{b}\int \frac{dx}{(ax^2 + b)^n}$ (n = positive integer)

66. $\displaystyle\int \sqrt{ax^2 + b} \, dx = \frac{x}{2} \sqrt{ax^2 + b} + \frac{b}{2\sqrt{a}} \log_e \frac{x\sqrt{a} + \sqrt{ax^2 + b}}{\sqrt{b}}$ (a positive)

67. $\displaystyle\int \sqrt{ax^2 + b} \, dx = \frac{x}{2} \sqrt{ax^2 + b} + \frac{b}{2\sqrt{-a}} \sin^{-1}\left(x\sqrt{-\frac{a}{b}}\right)$ (a negative)

68. $\displaystyle\int \frac{dx}{\sqrt{ax^2 + b}} = \frac{1}{\sqrt{a}} \log_e (x\sqrt{a} + \sqrt{ax^2 + b})$ (a positive)

69. $\displaystyle\int \frac{dx}{\sqrt{ax^2 + b}} = \frac{1}{\sqrt{-a}} \sin^{-1}\left(x\sqrt{-\frac{a}{b}}\right)$ (a negative)

70. $\displaystyle\int \frac{x \, dx}{\sqrt{ax^2 + b}} = \frac{1}{a} \sqrt{ax^2 + b}$

71. $\displaystyle\int \frac{\sqrt{ax^2 + b}}{x} \, dx = \sqrt{ax^2 + b} + \sqrt{b} \log_e \frac{\sqrt{ax^2 + b} - \sqrt{b}}{x}$ (b positive)

72. $\displaystyle\int \frac{\sqrt{ax^2 + b}}{x} \, dx = \sqrt{ax^2 + b} - \sqrt{-b} \tan^{-1} \frac{\sqrt{ax^2 + b}}{\sqrt{-b}}$ (b negative)

73. $\displaystyle\int x \sqrt{ax^2 + b} \, dx = \frac{1}{3a} (ax^2 + b)^{3/2}$

Integrals Involving $(ax^n + b)$—Continued

74. $\int x^2 \sqrt{ax^2+b}\, dx = \dfrac{x}{4a}(ax^2+b)^{3/2} - \dfrac{bx}{8a}\sqrt{ax^2+b} - \dfrac{b^2}{8a\sqrt{a}}\log_e (x\sqrt{a}+\sqrt{ax^2+b})$
(a positive)

75. $\int x^2 \sqrt{ax^2+b}\, dx = \dfrac{x}{4a}(ax^2+b)^{3/2} - \dfrac{bx}{8a}\sqrt{ax^2+b} - \dfrac{b^2}{8a\sqrt{-a}}\sin^{-1}\left(x\sqrt{\dfrac{-a}{b}}\right)$
(a negative)

76. $\int \dfrac{dx}{x\sqrt{ax^2+b}} = \dfrac{1}{\sqrt{b}}\log_e \dfrac{\sqrt{ax^2+b}-\sqrt{b}}{x}$ (b positive)

77. $\int \dfrac{dx}{x\sqrt{ax^2+b}} = \dfrac{1}{\sqrt{-b}}\sec^{-1}\left(x\sqrt{-\dfrac{a}{b}}\right)$ (b negative)

78. $\int \dfrac{x^2\, dx}{\sqrt{ax^2+b}} = \dfrac{x}{2a}\sqrt{ax^2+b} - \dfrac{b}{2a\sqrt{a}}\log_e (x\sqrt{a}+\sqrt{ax^2+b})$ (a positive)

79. $\int \dfrac{x^2\, dx}{\sqrt{ax^2+b}} = \dfrac{x}{2a}\sqrt{ax^2+b} - \dfrac{b}{2a\sqrt{-a}}\sin^{-1}\left(x\sqrt{-\dfrac{a}{b}}\right)$ (a negative)

80. $\int \dfrac{\sqrt{ax^2+b}}{x^2}\, dx = -\dfrac{\sqrt{ax^2+b}}{x} + \sqrt{a}\log_e (x\sqrt{a}+\sqrt{ax^2+b})$ (a positive)

81. $\int \dfrac{\sqrt{ax^2+b}}{x^2}\, dx = -\dfrac{\sqrt{ax^2+b}}{x} - \sqrt{-a}\sin^{-1}\left(x\sqrt{-\dfrac{a}{b}}\right)$ (a negative)

82. $\int \dfrac{dx}{x(ax^n+b)} = \dfrac{1}{bn}\log_e \dfrac{x^n}{ax^n+b}$

83. $\int \dfrac{dx}{x\sqrt{ax^n+b}} = \dfrac{1}{n\sqrt{b}}\log_e \dfrac{\sqrt{ax^n+b}-\sqrt{b}}{\sqrt{ax^n+b}+\sqrt{b}}$ (b positive)

84. $\int \dfrac{dx}{x\sqrt{ax^n+b}} = \dfrac{2}{n\sqrt{-b}}\sec^{-1}\sqrt{-\dfrac{ax^n}{b}}$ (b negative)

Integrals Involving $(ax^2 + bx + d)$

85. $\int \dfrac{dx}{ax^2+bx+d} = \dfrac{1}{\sqrt{b^2-4ad}}\log_e \dfrac{2ax+b-\sqrt{b^2-4ad}}{2ax+b+\sqrt{b^2-4ad}}$ $(b^2 > 4ad)$

86. $\int \dfrac{dx}{ax^2+bx+d} = \dfrac{2}{\sqrt{4ad-b^2}}\tan^{-1}\dfrac{2ax+b}{\sqrt{4ad-b^2}}$ $(b^2 < 4ad)$

87. $\int \dfrac{dx}{ax^2+bx+d} = -\dfrac{2}{2ax+b}$ $(b^2 = 4ad)$

88. $\int \dfrac{dx}{\sqrt{ax^2+bx+d}} = \dfrac{1}{\sqrt{a}}\log_e \left(2ax+b+2\sqrt{a(ax^2+bx+d)}\right)$ (a positive)

89. $\int \dfrac{dx}{\sqrt{ax^2+bx+d}} = \dfrac{1}{\sqrt{-a}}\sin^{-1}\dfrac{-2ax-b}{\sqrt{b^2-4ad}}$ (a negative)

90. $\int \dfrac{x\, dx}{ax^2+bx+d} = \dfrac{1}{2a}\log_e (ax^2+bx+d) - \dfrac{b}{2a}\int \dfrac{dx}{ax^2+bx+d}$

91. $\int \dfrac{x\, dx}{\sqrt{ax^2+bx+d}} = \dfrac{\sqrt{ax^2+bx+d}}{a} - \dfrac{b}{2a}\int \dfrac{dx}{\sqrt{ax^2+bx+d}}$

92. $\int \dfrac{dx}{x\sqrt{ax^2+bx+d}} = -\dfrac{1}{\sqrt{d}}\log_e \left(\dfrac{\sqrt{ax^2+bx+d}+\sqrt{d}}{x} + \dfrac{b}{2\sqrt{d}}\right)$ (d positive)

93. $\int \dfrac{dx}{x\sqrt{ax^2+bx+d}} = \dfrac{1}{\sqrt{-d}}\sin^{-1}\dfrac{bx+2d}{x\sqrt{b^2-4ad}}$ (d negative)

94. $\int \dfrac{dx}{x\sqrt{ax^2+bx}} = -\dfrac{2}{bx}\sqrt{ax^2+bx}$

Integrals Involving $(ax^2 + bx + d)$—*Continued*

95. $\displaystyle\int \sqrt{ax^2 + bx + d}\; dx = \frac{2\,ax + b}{4a}\sqrt{ax^2 + bx + d} + \frac{4\,ad - b^2}{8a}\int \frac{dx}{\sqrt{ax^2 + bx + d}}$

96. $\displaystyle\int x\sqrt{ax^2 + bx + d}\; dx = \frac{(ax^2 + bx + d)^{3/2}}{3a} - \frac{b}{2a}\int \sqrt{ax^2 + bx + d}\; dx$

Integrals Involving $\sin^n ax$

97. $\displaystyle\int \sin^3 ax\; dx = -\frac{1}{a}\cos ax + \frac{1}{3a}\cos^3 ax$

98. $\displaystyle\int \sin^4 ax\; dx = \frac{3}{8}x - \frac{1}{4a}\sin 2\,ax + \frac{1}{32\,a}\sin 4\,ax$

99. $\displaystyle\int \sin^n ax\; dx = -\frac{\sin^{n-1} ax\cos ax}{na} + \frac{n-1}{n}\int \sin^{n-2} ax\; dx$ $(n = \text{positive integer})$

100. $\displaystyle\int x\sin ax\; dx = \frac{\sin ax}{a^2} - \frac{x\cos ax}{a}$

101. $\displaystyle\int x^2 \sin ax\; dx = \frac{2x}{a^2}\sin ax - \left(\frac{x^2}{a} - \frac{2}{a^3}\right)\cos ax$

102. $\displaystyle\int x^3 \sin ax\; dx = \left(\frac{3x^2}{a^2} - \frac{6}{a^4}\right)\sin ax - \left(\frac{x^3}{a} - \frac{6x}{a^3}\right)\cos ax$

103. $\displaystyle\int x^n \sin ax\; dx = -\frac{x^n}{a}\cos ax + \frac{n}{a}\int x^{n-1}\cos ax\; dx$ $(n > 0)$

104. $\displaystyle\int \frac{\sin ax}{x^n}\; dx = -\frac{1}{n-1}\frac{\sin ax}{x^{n-1}} + \frac{a}{n-1}\int \frac{\cos ax}{x^{n-1}}\; dx$

105. $\displaystyle\int \frac{dx}{\sin^n ax} = -\frac{1}{a(n-1)}\frac{\cos ax}{\sin^{n-1} ax} + \frac{n-2}{n-1}\int \frac{dx}{\sin^{n-2} ax}$ $(n\ \text{integer} > 1)$

106. $\displaystyle\int \frac{x\,dx}{\sin^2 ax} = -\frac{x}{a}\cot ax + \frac{1}{a^2}\log_e \sin ax$

107. $\displaystyle\int \frac{dx}{1 + \sin ax} = -\frac{1}{a}\tan\left(\frac{\pi}{4} - \frac{ax}{2}\right)$

108. $\displaystyle\int \frac{dx}{1 - \sin ax} = \frac{1}{a}\cot\left(\frac{\pi}{4} - \frac{ax}{2}\right)$

109. $\displaystyle\int \frac{x\,dx}{1 + \sin ax} = -\frac{x}{a}\tan\left(\frac{\pi}{4} - \frac{ax}{2}\right) + \frac{2}{a^2}\log_e \cos\left(\frac{\pi}{4} - \frac{ax}{2}\right)$

110. $\displaystyle\int \frac{x\,dx}{1 - \sin ax} = \frac{x}{a}\cot\left(\frac{\pi}{4} - \frac{ax}{2}\right) + \frac{2}{a^2}\log_e \sin\left(\frac{\pi}{4} - \frac{ax}{2}\right)$

111. $\displaystyle\int \frac{dx}{b + d\sin ax} = \frac{-2}{a\sqrt{b^2 - d^2}}\tan^{-1}\left[\sqrt{\frac{b - d}{b + d}}\tan\left(\frac{\pi}{4} - \frac{ax}{2}\right)\right]$ $(b^2 > d^2)$

112. $\displaystyle\int \frac{dx}{b + d\sin ax} = \frac{-1}{a\sqrt{d^2 - b^2}}\log_e \frac{d + b\sin ax + \sqrt{d^2 - b^2}\cos ax}{b + d\sin ax}$ $(d^2 > b^2)$

113. $\displaystyle\int \sin ax \sin bx\; dx = \frac{\sin(a - b)x}{2(a - b)} - \frac{\sin(a + b)x}{2(a + b)}$ $(a^2 \neq b^2)$

Integrals Involving $\cos^n ax$

114. $\displaystyle\int \cos^3 ax\; dx = \frac{1}{a}\sin ax - \frac{1}{3a}\sin^3 ax$

115. $\displaystyle\int \cos^4 ax\; dx = \frac{3}{8}x + \frac{1}{4a}\sin 2\,ax + \frac{1}{32a}\sin 4\,ax$

116. $\displaystyle\int \cos^n ax\; dx = \frac{\cos^{n-1} ax\sin ax}{na} + \frac{n-1}{n}\int \cos^{n-2} ax\; dx$ $(n = \text{positive integer})$

117. $\displaystyle\int x\cos ax\; dx = \frac{\cos ax}{a^2} + \frac{x\sin ax}{a}$

Integrals Involving $\cos^n ax$—*Continued*

118. $\int x^2 \cos ax \, dx = \dfrac{2x}{a^2} \cos ax + \left(\dfrac{x^2}{a} - \dfrac{2}{a^3} \right) \sin ax$

119. $\int x^3 \cos ax \, dx = \left(\dfrac{3x^2}{a^2} - \dfrac{6}{a^4} \right) \cos ax + \left(\dfrac{x^3}{a} - \dfrac{6x}{a^3} \right) \sin ax$

120. $\int x^n \cos ax \, dx = \dfrac{x^n \sin ax}{a} - \dfrac{n}{a} \int x^{n-1} \sin ax \, dx \quad (n > 0)$

121. $\int \dfrac{\cos ax}{x^n} \, dx = - \dfrac{1}{n-1} \dfrac{\cos ax}{x^{n-1}} - \dfrac{a}{n-1} \int \dfrac{\sin ax}{x^{n-1}} \, dx$

122. $\int \dfrac{dx}{\cos^n ax} = \dfrac{1}{a(n-1)} \dfrac{\sin ax}{\cos^{n-1} ax} + \dfrac{n-2}{n-1} \int \dfrac{dx}{\cos^{n-2} ax} \quad (n \text{ integer} > 1)$

123. $\int \dfrac{x \, dx}{\cos^2 ax} = \dfrac{x}{a} \tan ax + \dfrac{1}{a^2} \log_e \cos ax$

124. $\int \dfrac{dx}{1 + \cos ax} = \dfrac{1}{a} \tan \dfrac{ax}{2}$

125. $\int \dfrac{dx}{1 - \cos ax} = - \dfrac{1}{a} \cot \dfrac{ax}{2}$

126. $\int \dfrac{x \, dx}{1 + \cos ax} = \dfrac{x}{a} \tan \dfrac{ax}{2} + \dfrac{2}{a^2} \log_e \cos \dfrac{ax}{2}$

127. $\int \dfrac{x \, dx}{1 - \cos ax} = - \dfrac{x}{a} \cot \dfrac{ax}{2} + \dfrac{2}{a^2} \log_e \sin \dfrac{ax}{2}$

128. $\int \dfrac{dx}{b + d \cos ax} = \dfrac{2}{a \sqrt{b^2 - d^2}} \tan^{-1} \left(\sqrt{\dfrac{b-d}{b+d}} \tan \dfrac{ax}{2} \right) \quad (b^2 > d^2)$

129. $\int \dfrac{dx}{b + d \cos ax} = \dfrac{1}{a \sqrt{d^2 - b^2}} \log_e \dfrac{d + b \cos ax + \sqrt{d^2 - b^2} \sin ax}{b + d \cos ax} \quad (d^2 > b^2)$

130. $\int \cos ax \cos bx \, dx = \dfrac{\sin (a-b)x}{2(a-b)} + \dfrac{\sin (a+b)x}{2(a+b)} \quad (a^2 \neq b^2)$

Integrals Involving $\sin^n ax, \cos^n ax$

131. $\int \sin ax \cos bx \, dx = - \dfrac{1}{2} \left[\dfrac{\cos (a-b)x}{a-b} + \dfrac{\cos (a+b)x}{a+b} \right] \quad (a^2 \neq b^2)$

132. $\int \sin^2 ax \cos^2 ax \, dx = \dfrac{x}{8} - \dfrac{\sin 4ax}{32a}$

133. $\int \sin^n ax \cos ax \, dx = \dfrac{1}{a(n+1)} \sin^{n+1} ax \quad (n \neq -1)$

134. $\int \sin ax \cos^n ax \, dx = - \dfrac{1}{a(n+1)} \cos^{n+1} ax \quad (n \neq -1)$

135. $\int \sin^n ax \cos^m ax \, dx = - \dfrac{\sin^{n-1} ax \cos^{m+1} ax}{a(n+m)} + \dfrac{n-1}{n+m} \int \sin^{n-2} ax \cos^m ax \, dx \quad (m, n \text{ pos})$

136. $\int \dfrac{\sin^n ax}{\cos^m ax} \, dx = \dfrac{\sin^{n+1} ax}{a(m-1) \cos^{m-1} ax} - \dfrac{n-m+2}{m-1} \int \dfrac{\sin^n ax}{\cos^{m-2} ax} \, dx \quad (m, n \text{ pos}, m \neq 1)$

137. $\int \dfrac{\cos^m ax}{\sin^n ax} \, dx = \dfrac{-\cos^{m+1} ax}{a(n-1) \sin^{n-1} ax} + \dfrac{n-m-2}{(n-1)} \int \dfrac{\cos^m ax}{\sin^{n-2} ax} \, dx \quad (m, n \text{ pos}, n \neq 1)$

138. $\int \dfrac{dx}{\sin ax \cos ax} = \dfrac{1}{a} \log_e \tan ax$

139. $\int \dfrac{dx}{b \sin ax + d \cos ax} = \dfrac{1}{a \sqrt{b^2 + d^2}} \log_e \tan \tfrac{1}{2} \left(ax + \tan^{-1} \dfrac{d}{b} \right)$

140. $\int \dfrac{\sin ax}{b + d \cos ax} \, dx = - \dfrac{1}{ad} \log_e (b + d \cos ax)$

141. $\int \dfrac{\cos ax}{b + d \sin ax} \, dx = \dfrac{1}{ad} \log_e (b + d \sin ax)$

Integrals Involving $\tan^n ax$, $\cot^n ax$, $\sec^n ax$, $\csc^n ax$

142. $\displaystyle\int \tan^n ax\, dx = \frac{1}{a(n-1)} \tan^{n-1} ax - \int \tan^{n-2} ax\, dx$ (n integer > 1)

143. $\displaystyle\int \cot^n ax\, dx = -\frac{1}{a(n-1)} \cot^{n-1} ax - \int \cot^{n-2} ax\, dx$ (n integer > 1)

144. $\displaystyle\int \sec^n ax\, dx = \frac{1}{a(n-1)} \frac{\sin ax}{\cos^{n-1} ax} + \frac{n-2}{n-1} \int \sec^{n-2} ax\, dx$ (n integer > 1)

145. $\displaystyle\int \csc^n ax\, dx = -\frac{1}{a(n-1)} \frac{\cos ax}{\sin^{n-1} ax} + \frac{n-2}{n-1} \int \csc^{n-2} ax\, dx$ (n integer > 1)

146. $\displaystyle\int \frac{dx}{b + d \tan ax} = \frac{1}{b^2 + d^2}\left[bx + \frac{d}{a} \log_e (b \cos ax + d \sin ax) \right]$

147. $\displaystyle\int \frac{dx}{\sqrt{b + d \tan^2 ax}} = \frac{1}{a\sqrt{b-d}} \sin^{-1}\left[\sqrt{\frac{b-d}{b}} \sin ax \right]$ (b pos, $b^2 > d^2$)

148. $\displaystyle\int \tan ax \sec ax\, dx = \frac{1}{a} \sec ax$

149. $\displaystyle\int \tan^n ax \sec^2 ax\, dx = \frac{1}{a(n+1)} \tan^{n+1} ax$ ($n \ne -1$)

150. $\displaystyle\int \frac{\sec^2 ax\, dx}{\tan ax} = \frac{1}{a} \log_e \tan ax$

151. $\displaystyle\int \cot ax \csc ax\, dx = -\frac{1}{a} \csc ax$

152. $\displaystyle\int \cot^n ax \csc^2 ax\, dx = -\frac{1}{a(n+1)} \cot^{n+1} ax$ ($n \ne -1$)

153. $\displaystyle\int \frac{\csc^2 ax}{\cot ax}\, dx = -\frac{1}{a} \log_e \cot ax$

Integrals Involving b^{ax}, e^{ax}, $\sin bx$, $\cos bx$

154. $\displaystyle\int x b^{ax}\, dx = \frac{x b^{ax}}{a \log_e b} - \frac{b^{ax}}{a^2(\log_e b)^2}$

155. $\displaystyle\int x e^{ax}\, dx = \frac{e^{ax}}{a^2} (ax - 1)$

156. $\displaystyle\int x^n b^{ax}\, dx = \frac{x^n b^{ax}}{a \log_e b} - \frac{n}{a \log_e b} \int x^{n-1} b^{ax}\, dx$ (n positive)

157. $\displaystyle\int x^n e^{ax}\, dx = \frac{1}{a} x^n e^{ax} - \frac{n}{a} \int x^{n-1} e^{ax}\, dx$ (n positive)

158. $\displaystyle\int \frac{dx}{b + de^{ax}} = \frac{1}{ab}\left[ax - \log_e (b + de^{ax}) \right]$

159. $\displaystyle\int \frac{e^{ax}\, dx}{b + de^{ax}} = \frac{1}{ad} \log_e (b + de^{ax})$

160. $\displaystyle\int \frac{dx}{be^{ax} + de^{-ax}} = \frac{1}{a\sqrt{bd}} \tan^{-1}\left(e^{ax} \sqrt{\frac{b}{d}} \right)$ (b and d positive)

161. $\displaystyle\int \frac{e^{ax}}{x}\, dx = \log_e x + ax + \frac{(ax)^2}{2 \cdot 2!} + \frac{(ax)^3}{3 \cdot 3!} + \dots$

162. $\displaystyle\int \frac{e^{ax}}{x^n}\, dx = \frac{1}{n-1}\left(-\frac{e^{ax}}{x^{n-1}} + a \int \frac{e^{ax}}{x^{n-1}}\, dx \right)$ (n integer > 1)

163. $\displaystyle\int e^{ax} \sin bx\, dx = \frac{e^{ax}}{a^2 + b^2} (a \sin bx - b \cos bx)$

164. $\displaystyle\int e^{ax} \cos bx\, dx = \frac{e^{ax}}{a^2 + b^2} (a \cos bx + b \sin bx)$

165. $\displaystyle\int x e^{ax} \sin bx\, dx = \frac{x e^{ax}}{a^2 + b^2} (a \sin bx - b \cos bx)$
$$- \frac{e^{ax}}{(a^2 + b^2)^2} [(a^2 - b^2) \sin bx - 2ab \cos bx]$$

166. $\displaystyle\int x e^{ax} \cos bx\, dx = \frac{x e^{ax}}{a^2 + b^2} (a \cos bx + b \sin bx)$
$$- \frac{e^{ax}}{(a^2 + b^2)^2}\left[(a^2 - b^2) \cos bx + 2ab \sin bx \right]$$

Integrals Involving log$_e$ ax

167. $\int \log_e ax \, dx = x \log_e ax - x$

168. $\int (\log_e ax)^n \, dx = x(\log_e ax)^n - n(\log_e ax)^{n-1} \, dx$ (n positive)

169. $\int x^n \log_e ax \, dx = x^{n+1} \left(\dfrac{\log_e ax}{n+1} \right) - \dfrac{1}{(n+1)^2}$, $n \neq -1$

170. $\int \dfrac{(\log_e ax)^n}{x} \, dx = \dfrac{(\log_e ax)^{n+1}}{n+1}$, $n \neq -1$

171. $\int \dfrac{dx}{x \log_e ax} = \log_e(\log_e x)$

172. $\int \dfrac{dx}{\log_e ax} = \dfrac{1}{a} \left[\log_e (\log_e ax) + \log_e ax + \dfrac{(\log_e ax)^2}{2 \cdot 2!} + \dots \right]$

173. $\int x^m (\log_e ax)^n \, dx = \dfrac{x^{m+1}(\log_e ax)^n}{m+1} - \dfrac{n}{m+1} \int x^m (\log_e ax)^{n-1} \, dx$, $m, n \neq 1$

174. $\int \dfrac{x^m \, dx}{(\log_e ax)^n} = - \dfrac{x^{m+1}}{(n-1)(\log_e ax)^{n-1}} + \dfrac{m+1}{n-1} \int \dfrac{x^m \, dx}{(\log_e ax)^{n-1}}$

Some Definite Integrals

1. $\int_0^a \sqrt{a^2 - x^2} \, dx = \dfrac{\pi a^2}{4}$

2. $\int_0^a \sqrt{2ax - x^2} \, dx = \dfrac{\pi a^2}{4}$

3. $\int_0^\infty \dfrac{dx}{a + bx^2} = \dfrac{\pi}{2\sqrt{ab}}$ (a and b positive)

4. $\int_0^{\sqrt{a/b}} \dfrac{dx}{a + bx^2} \, dx = \int_{\sqrt{a/b}}^\infty \dfrac{dx}{a + bx^2} = \dfrac{\pi}{4\sqrt{ab}}$ (a and b positive)

5. $\int_0^{\sqrt{a/b}} \dfrac{dx}{\sqrt{a - bx^2}} = \dfrac{\pi}{2\sqrt{b}}$ (a and b positive)

6. $\int_0^\infty \dfrac{\sin bx}{x} \, dx = \dfrac{\pi}{2}$ $(b > 0)$ $= 0$ $(b = 0)$ $= -\dfrac{\pi}{2}$ $(b < 0)$

7. $\int_0^\infty \dfrac{\tan x}{x} \, dx = \dfrac{\pi}{2}$

8. $\int_0^{\pi/2} \sin^{2n+1} x \, dx = \int_0^{\pi/2} \cos^{2n+1} x \, dx = \dfrac{2 \cdot 4 \cdot 6 \cdot \ldots \cdot 2n}{3 \cdot 5 \cdot 7 \cdot \ldots \cdot (2n+1)}$. $(n > 0)$

9. $\int_0^{\pi/2} \sin^{2n} x \, dx = \int_0^{\pi/2} \cos^{2n} x \, dx = \dfrac{1 \cdot 3 \cdot 5 \cdot \ldots \cdot (2n-1)}{2 \cdot 4 \cdot 6 \cdot \ldots \cdot 2n} \cdot \dfrac{\pi}{2}$ $(n > 0)$

10. $\int_0^\pi \sin ax \sin bx \, dx = \int_0^\pi \cos ax \cos bx \, dx = 0$ $(a \neq b)$

11. $\int_0^\pi \sin^2 ax \, dx = \int_0^\pi \cos^2 ax \, dx = \dfrac{\pi}{2}$

12. $\int_0^{\pi/2} \log_e \cos x \, dx = \int_0^{\pi/2} \log_e \sin x \, dx = -\dfrac{\pi}{2} \log_e 2$

13. $\int_0^\infty e^{-ax^2} \, dx = \dfrac{1}{2} \sqrt{\dfrac{\pi}{a}}$

14. $\int_0^\infty x^n e^{-ax} \, dx = \dfrac{n!}{a^{n+1}}$ $(a > 0, \; n = 1, 2, 3, \dots)$

15. $\int_0^1 \dfrac{\log_e x}{1-x} \, dx = -\dfrac{\pi^2}{6}$

16. $\int_0^1 \dfrac{\log_e x}{1+x} \, dx = -\dfrac{\pi^2}{12}$

17. $\int_0^1 \dfrac{\log_e x}{1-x^2} \, dx = \dfrac{\pi^2}{8}$

Table 24. Haversines

$$\text{hav } \theta = \tfrac{1}{2}\,\text{vers } \theta = \tfrac{1}{2}\,(1 - \cos \theta) = \sin^2 \tfrac{1}{2}\theta$$
$$\text{hav } (-\theta) = \text{have } \theta$$
$$\text{hav } (180° - \theta) = \text{hav } (180° + \theta) = 1 - \text{hav } \theta$$

Characteristics of the logarithms are omitted.

θ°	Value	Log	θ°	Value	Log	θ°	Value	Log	θ°	Value	Log
0	.00000	—	50	.17861	.25190	100	.58682	.76851	150	.93301	.96989
1	.00008	.88168	51	.18534	.26797	101	.59540	.77481	151	.93731	.97188
2	.00030	.48371	52	.19217	.28368	102	.60396	.78101	152	.94147	.97381
3	.00069	.83584	53	.19909	.29905	103	.61248	.78709	153	.94550	.97566
4	.00122	.08564	54	.20611	.31409	104	.62096	.79306	154	.94940	.97745
5	.00190	.27936	55	.21321	.32281	105	.62941	.79893	155	.95315	.97916
6	.00274	.43760	56	.22040	.34322	106	.63782	.80470	156	.95677	.98081
7	.00373	.57135	57	.22768	.35733	107	.64619	.81036	157	.96025	.98239
8	.00487	.68717	58	.23504	.37114	108	.65451	.81592	158	.96359	.98389
9	.00616	.78929	59	.24248	.38468	109	.66278	.82137	159	.96679	.98533
10	.00760	.88059	60	.25000	.39794	110	.67101	.82673	160	.96985	.98670
11	.00919	.96315	61	.25760	.41094	111	.67918	.83199	161	.97276	.98801
12	.01093	.03847	62	.26526	.42368	112	.68730	.83715	162	.97553	.98924
13	.01281	.10772	63	.27300	.43617	113	.69537	.84221	163	.97815	.99041
14	.01485	.17179	64	.28081	.44842	114	.70337	.84718	164	.98063	.99151
15	.01704	.23140	65	.28869	.46043	115	.71131	.85206	165	.98296	.99254
16	.01937	.28711	66	.29663	.47222	116	.71919	.85684	166	.98515	.99350
17	.02185	.33940	67	.30463	.48378	117	.72700	.86153	167	.98719	.99440
18	.02447	.38867	68	.31270	.49512	118	.73474	.86613	168	.98907	.99523
19	.02724	.43522	69	.32082	.50625	119	.74240	.87064	169	.99081	.99599
20	.03015	.47934	70	.32899	.51718	120	.75000	.87506	170	.99240	.99669
21	.03321	.52127	71	.33722	.52791	121	.75752	.87939	171	.99384	.99732
22	.03641	.56120	72	.34549	.53844	122	.76496	.88364	172	.99513	.99788
23	.03975	.59931	73	.35381	.54878	123	.77232	.88780	173	.99627	.99838
24	.04323	.63576	74	.36218	.55893	124	.77960	.89187	174	.99726	.99881
25	.04685	.67067	75	.37060	.56889	125	.78679	.89586	175	.99810	.99917
26	.05060	.70418	76	.37904	.57868	126	.79389	.89976	176	.99878	.99947
27	.05450	.73637	77	.38752	.58830	127	.80091	.90358	177	.99931	.99970
28	.05853	.76735	78	.39604	.59774	128	.80783	.90732	178	.99970	.99987
29	.06269	.79720	79	.40460	.60702	129	.81466	.91098	179	.99992	.99997
30	.06699	.82599	80	.41318	.61613	130	.82139	.91455	180	1.00000	.00000
31	.07142	.85380	81	.42178	.62509	131	.82803	.91805			
32	.07598	.88068	82	.43041	.63389	132	.83457	.92146			
33	.08066	.90668	83	.43907	.64253	133	.84100	.92480			
34	.08548	.93187	84	.44774	.65102	134	.84733	.92805			
35	.09042	.95628	85	.45642	.65937	135	.85355	.93123			
36	.09549	.97996	86	.46512	.66757	136	.85967	.93433			
37	.10068	.00295	87	.47383	.67562	137	.86568	.93736			
38	.10599	.02528	88	.48255	.68354	138	.87157	.94030			
39	.11143	.04699	89	.49127	.69132	139	.87735	.94318			
40	.11698	.06810	90	.50000	.69897	140	.88302	.94597			
41	.12265	.08865	91	.50873	.70648	141	.88857	.94869			
42	.12843	.10866	92	.51745	.71387	142	.89401	.95134			
43	.13432	.12815	93	.52617	.72112	143	.89932	.95391			
44	.14033	.14715	94	.53488	.72825	144	.90451	.95641			
45	.14645	.16568	95	.54358	.73526	145	.90958	.95884			
46	.15267	.18376	96	.55226	.74215	146	.91452	.96119			
47	.15900	.20140	97	.56093	.74891	147	.91934	.96347			
48	.16543	.21863	98	.56959	.75556	148	.92402	.96568			
49	.17197	.23545	99	.57822	.76209	149	.92858	.96782			

Table 25. Complete Elliptic Integrals

$$K = \int_0^{\pi/2} \frac{d\Phi}{\sqrt{1-k^2 \sin^2 \Phi}} = F\left(k, \frac{\pi}{2}\right) \qquad E = \int_0^{\pi/2} \sqrt{1-k^2 \sin^2 \Phi} \cdot d\Phi = E\left(k, \frac{\pi}{2}\right)$$

$\sin^{-1} k$	K	$\log K$	E	$\log E$	$\sin^{-1} k$	K	$\log K$	E	$\log E$
0°	1.5708	0.196120	1.5708	0.196120	45°	1.8541	0.268127	1.3506	0.130541
1	1.5709	0.196153	1.5707	0.196087	46	1.8691	0.271644	1.3418	0.127690
2	1.5713	0.196252	1.5703	0.195988	47	1.8848	0.275267	1.3329	0.124788
3	1.5719	0.196418	1.5697	0.195822	48	1.9011	0.279001	1.3238	0.121836
4	1.5727	0.196649	1.5689	0.195591	49	1.9180	0.282848	1.3147	0.118836
5	1.5738	0.196947	1.5678	0.195293	50	1.9356	0.286811	1.3055	0.115790
6	1.5751	0.197312	1.5665	0.194930	51	1.9539	0.290895	1.2963	0.112698
7	1.5767	0.197743	1.5649	0.194500	52	1.9729	0.295101	1.2870	0.109563
8	1.5785	0.198241	1.5632	0.194004	53	1.9927	0.299435	1.2776	0.106386
9	1.5805	0.198806	1.5611	0.193442	54	2.0133	0.303901	1.2681	0.103169
10	1.5828	0.199438	1.5589	0.192815	55	2.0347	0.308504	1.2587	0.099915
11	1.5854	0.200137	1.5564	0.192121	56	2.0571	0.313247	1.2492	0.096626
12	1.5882	0.200904	1.5537	0.191362	57	2.0804	0.318138	1.2397	0.093303
13	1.5913	0.201740	1.5507	0.190537	58	2.1047	0.323182	1.2301	0.089950
14	1.5946	0.202643	1.5476	0.189646	59	2.1300	0.328384	1.2206	0.086569
15	1.5981	0.203615	1.5442	0.188690	60	2.1565	0.333753	1.2111	0.083164
16	1.6020	0.204657	1.5405	0.187668	61	2.1842	0.339295	1.2015	0.079738
17	1.6061	0.205768	1.5367	0.186581	62	2.2132	0.345020	1.1920	0.076293
18	1.6105	0.206948	1.5326	0.185428	63	2.2435	0.350936	1.1826	0.072834
19	1.6151	0.208200	1.5283	0.184210	64	2.2754	0.357053	1.1732	0.069364
20	1.6200	0.209522	1.5238	0.182928	65	2.3088	0.363384	1.1638	0.065889
21	1.6252	0.210916	1.5191	0.181580	66	2.3439	0.369940	1.1545	0.062412
22	1.6307	0.212382	1.5141	0.180168	67	2.3809	0.376736	1.1453	0.058937
23	1.6365	0.213921	1.5090	0.178691	68	2.4198	0.383787	1.1362	0.055472
24	1.6426	0.215533	1.5037	0.177150	69	2.4610	0.391112	1.1272	0.052020
25	1.6490	0.217219	1.4981	0.175545	70	2.5046	0.398730	1.1184	0.048589
26	1.6557	0.218981	1.4924	0.173876	71	2.5507	0.406665	1.1096	0.045183
27	1.6627	0.220818	1.4864	0.172144	72	2.5998	0.414943	1.1011	0.041812
28	1.6701	0.222732	1.4803	0.170348	73	2.6521	0.423596	1.0927	0.038481
29	1.6777	0.224723	1.4740	0.168489	74	2.7081	0.432660	1.0844	0.035200
30	1.6858	0.226793	1.4675	0.166567	75	2.7681	0.442176	1.0764	0.031976
31	1.6941	0.228943	1.4608	0.164583	76	2.8327	0.452196	1.0686	0.028819
32	1.7028	0.231173	1.4539	0.162537	77	2.9026	0.462782	1.0611	0.025740
33	1.7119	0.233485	1.4469	0.160429	78	2.9786	0.474008	1.0538	0.022749
34	1.7214	0.235880	1.4397	0.158261	79	3.0617	0.485967	1.0468	0.019858
35	1.7312	0.238359	1.4323	0.156031	80	3.1534	0.498777	1.0401	0.017081
36	1.7415	0.240923	1.4248	0.153742	81	3.2553	0.512591	1.0338	0.014432
37	1.7522	0.243575	1.4171	0.151393	82	3.3699	0.527613	1.0278	0.011927
38	1.7633	0.246315	1.4092	0.148985	83	3.5004	0.544120	1.0223	0.009584
39	1.7748	0.249146	1.4013	0.146519	84	3.6519	0.562514	1.0172	0.007422
40	1.7868	0.252068	1.3931	0.143995	85	3.8317	0.583396	1.0127	0.005465
41	1.7992	0.255085	1.3849	0.141414	86	4.0528	0.607751	1.0086	0.003740
42	1.8122	0.258197	1.3765	0.138778	87	4.3387	0.637355	1.0053	0.002278
43	1.8256	0.261406	1.3680	0.136086	88	4.7427	0.676027	1.0026	0.001121
44	1.8396	0.264716	1.3594	0.133340	89	5.4349	0.735192	1.0008	0.000326
					90	∞	∞	1.0000	0.000000

Table 25. Complete Elliptic Integrals—Continued

$\sin^{-1} k$	K	$\log K$	$\sin^{-1} k$	K	$\log K$	$\sin^{-1} k$	K	$\log K$
89 20	5.840	0.76641	89 40	6.533	0.81511	89 50	7.226	0.85890
89 22	5.891	0.77019	89 41	6.584	0.81849	89 51	7.332	0.86522
89 24	5.946	0.77422	89 42	6.639	0.82210	89 52	7.449	0.87210
89 26	6.003	0.77837	89 43	6.696	0.82582	89 53	7.583	0.87984
89 28	6.063	0.78269	89 44	6.756	0.82969	89 54	7.737	0.88857
89 30	6.128	0.78732	89 45	6.821	0.83385	89 55	7.919	0.89867
89 32	6.197	0.79218	89 46	6.890	0.83822	89 56	8.143	0.91078
89 34	6.271	0.79734	89 47	6.964	0.84286	89 57	8.430	0.92583
89 36	6.351	0.80284	89 48	7.044	0.84782	89 58	8.836	0.94626
89 38	6.438	0.80875	89 49	7.131	0.85315	89 59	9.529	0.97905
						90 0	∞	∞

Table 26. Gamma Function†

Values of $\Gamma(n) = \int_0^\infty e^{-x} x^{n-1}\, dx$; $\Gamma(n+1) = n\Gamma(n)$

n	$\Gamma(n)$	n	$\Gamma(n)$	n	$\Gamma(n)$	n	$\Gamma(n)$
1.00	1.00000	1.25	.90640	1.50	.88623	1.75	.91906
1.01	.99433	1.26	.90440	1.51	.88659	1.76	.92137
1.02	.98884	1.27	.90250	1.52	.88704	1.77	.92376
1.03	.98355	1.28	.90072	1.53	.88757	1.78	.92623
1.04	.97844	1.29	.89904	1.54	.88818	1.79	.92877
1.05	.97350	1.30	.89747	1.55	.88887	1.80	.93138
1.06	.96874	1.31	.89600	1.56	.88964	1.81	.93408
1.07	.96415	1.32	.89464	1.57	.89049	1.82	.93685
1.08	.95973	1.33	.89338	1.58	.89142	1.83	.93969
1.09	.95546	1.34	.89222	1.59	.89243	1.84	.94261
1.10	.95135	1.35	.89115	1.60	.89352	1.85	.94561
1.11	.94739	1.36	.89018	1.61	.89468	1.86	.94869
1.12	.94359	1.37	.88931	1.62	89592	1.87	.95184
1.13	.93993	1.38	.88854	1.63	.89724	1.88	.95507
1.14	.93642	1.39	.88785	1.64	.89864	1.89	.95838
1.15	.93304	1.40	.88726	1.65	.90012	1.90	.96177
1.16	.92980	1.41	.88676	1.66	.90167	1.91	.96523
1.17	.92670	1.42	.88636	1.67	.90330	1.92	.96878
1.18	.92373	1.43	.88604	1.68	.90500	1.93	.97240
1.19	.92088	1.44	.88580	1.69	.90678	1.94	.97610
1.20	.91817	1.45	.88565	1.70	.90864	1.95	.97988
1.21	.91558	1.46	.88560	1.71	.91057	1.96	.98374
1.22	.91311	1.47	.88563	1.72	.91258	1.97	.98768
1.23	.91075	1.48	.88575	1.73	.91466	1.98	.99171
1.24	.90852	1.49	.88595	1.74	.91683	1.99	.99581
						2.00	1.00000

For large positive integers, Stirling's formula gives an approximation in which the relative error decreases as n increases:

$$\Gamma(n+1) = (2\pi n)^{1/2} \left(\frac{n}{e}\right)^n$$

† From *CRC Standard Mathematical Tables*, Chemical Rubber Publishing Co., 12th ed., 1959. Used by permission.

Table 27. Fractional Powers of Numbers

N	Power of N								
	0.1	0.2	0.3	0.4	0.5	0.6	0.7	0.8	0.9
0.01	0.6310	.3981	0.2512	0.1585	0.1000	0.0631	0.0398	0.0251	0.0159
.02	.6762	.4573	.3093	.2091	.1414	.0956	.0647	.0437	.0296
.03	.7042	.4959	.3492	.2460	.1732	.1220	.0859	.0605	.0426
.04	.7248	.5253	.3807	.2760	.2000	.1450	.1051	.0762	.0552
.05	.7411	.5493	.4071	.3017	.2236	.1657	.1228	.0910	.0675
.06	.7548	.5697	.4300	.3245	.2449	.1849	.1395	.1053	.0795
.07	.7665	.5875	.4503	.3452	.2646	.2028	.1554	.1192	.0913
.08	.7768	.6034	.4687	.3641	.2828	.2197	.1707	.1326	.1030
.09	.7860	.6178	.4856	.3817	.3000	.2358	.1853	.1457	.1145
.10	.7943	.6310	.5012	.3981	.3162	.2512	.1995	.1585	.1259
.11	.8019	.6431	.5157	.4136	.3317	.2660	.2133	.1711	.1372
.12	.8089	.6544	.5294	.4282	.3464	.2802	.2267	.1834	.1483
.13	.8154	.6650	.5422	.4422	.3606	.2940	.2398	.1955	.1594
.14	.8215	.6749	.5544	.4555	.3742	.3074	.2525	.2074	.1704
.15	.8272	.6843	.5660	.4682	.3873	.3204	.2650	.2192	.1813
.16	.8326	.6931	.5771	.4805	.4000	.3330	.2773	.2308	.1922
.17	.8376	.7016	.5877	.4922	.4123	.3454	.2893	.2423	.2030
.18	.8424	.7097	.5978	.5036	.4243	.3574	.3011	.2536	.2137
.19	.8470	.7174	.6076	.5146	.4359	.3692	.3127	.2649	.2243
.20	.8513	.7248	.6170	.5253	.4472	.3807	.3241	.2760	.2349
.21	.8555	.7319	.6261	.5357	.4583	.3920	.3354	.2869	.2455
.22	.8595	.7387	.6349	.5457	.4690	.4031	.3465	.2978	.2560
.23	.8633	.7453	.6435	.5555	.4796	.4140	.3575	.3086	.2664
.24	.8670	.7517	.6517	.5650	.4899	.4248	.3683	.3193	.2768
.25	.8706	.7579	.6598	.5744	.5000	.4353	.3789	.3299	.2872
.26	.8740	.7638	.6676	.5834	.5099	.4456	.3895	.3404	.2975
.27	.8773	.7696	.6752	.5923	.5196	.4559	.3999	.3508	.3078
.28	.8805	.7752	.6826	.6010	.5292	.4659	.4102	.3612	.3180
.29	.8836	.7807	.6898	.6095	.5385	.4758	.4204	.3715	.3282
.30	.8866	.7860	.6969	.6178	.5477	.4856	.4305	.3817	.3384
.31	.8895	.7912	.7037	.6260	.5568	.4952	.4405	.3918	.3485
.32	.8923	.7962	.7105	.6340	.5757	.5048	.4504	.4019	.3586
.33	.8951	.8011	.7171	.6418	.5745	.5142	.4602	.4119	.3687
.34	.8977	.8059	.7235	.6495	.5831	.5235	.4699	.4219	.3787
.35	.9003	.8106	.7298	.6571	.5916	.5327	.4796	.4318	.3887
.36	.9029	.8152	.7360	.6645	.6000	.5417	.4891	.4416	.3987
.37	.9054	.8197	.7421	.6719	.6083	.5507	.4986	.4514	.4087
.38	.9078	.8241	.7481	.6791	.6164	.5596	.5080	.4611	.4186
.39	.9101	.8284	.7539	.6862	.6245	.5684	.5173	.4708	.4285
.40	.9124	.8326	.7597	.6932	.6325	.5771	.5266	.4805	.4384
.41	.9147	.8367	.7653	.7000	.6403	.5857	.5357	.4900	.4482
.42	.9169	.8407	.7709	.7068	.6481	.5942	.5449	.4996	.4581
.43	.9191	.8447	.7763	.7135	.6558	.6027	.5539	.5091	.4679
.44	.9212	.8486	.7817	.7201	.6633	.6110	.5629	.5185	.4777
.45	.9233	.8524	.7870	.7266	.6708	.6193	.5718	.5279	.4874
.46	.9253	.8562	.7922	.7330	.6782	.6276	.5807	.5373	.4971
.47	.9273	.8598	.7973	.7393	.6856	.6357	.5895	.5466	.5069
.48	.9292	.8635	.8023	.7456	.6928	.6438	.5982	.5559	.5166
.49	.9312	.8670	.8073	.7518	.7000	.6518	.6069	.5651	.5262

Table 27. Fractional Powers of Numbers—*Continued*

N	\multicolumn{9}{Power of N}								
	0.1	0.2	0.3	0.4	0.5	0.6	0.7	0.8	0.9
0.50	.9330	.8706	.8123	.7579	.7071	.6598	.6156	.5743	.5359
.51	.9349	.8740	.8171	.7639	.7141	.6676	.6242	.5835	.5455
.52	.9367	.8774	.8219	.7698	.7211	.6755	.6327	.5927	.5551
.53	.9385	.8808	.8266	.7757	.7280	.6832	.6412	.6018	.5647
.54	.9402	.8841	.8312	.7816	.7349	.6909	.6497	.6108	.5743
.55	.9420	.8873	.8358	.7873	.7416	.6986	.6580	.6199	.5839
.56	.9437	.8905	.8403	.7930	.7843	.7062	.6664	.6289	.5934
.57	.9453	.8937	.8448	.7986	.7550	.7137	.6747	.6378	.6030
.58	.9470	.8968	.8492	.8042	.7616	.7212	.6830	.6467	.6125
.59	.9486	.8999	.8536	.8098	.7681	.7286	.6912	.6557	.6220
.60	.9502	.9029	.8579	.8152	.7746	.7360	.6994	.6645	.6315
.61	.9518	.9059	.8622	.8206	.7810	.7434	.7075	.6734	.6409
.62	.9533	.9088	.8664	.8260	.7864	.7506	.7156	.6822	.6504
.63	.9549	.9117	.8706	.8313	.7937	.7579	.7237	.6910	.6598
.64	.9564	.9146	.8747	.8365	.8000	.7651	.7317	.6998	.6692
.65	.9578	.9175	.8788	.8417	.8062	.7722	.7397	.7085	.6786
.66	.9593	.9203	.8828	.8469	.8124	.7793	.7476	.7172	.6880
.67	.9608	.9230	.8868	.8520	.8185	.7864	.7555	.7259	.6974
.68	.9622	.9258	.8907	.8571	.8246	.7934	.7634	.7345	.7067
.69	.9636	.9285	.8947	.8621	.8307	.8004	.7713	.7342	.7161
.70	.9650	.9312	.8985	.8670	.8367	.8074	.7791	.7518	.7254
.71	.9663	.9338	.9024	.8720	.8426	.8143	.7868	.7603	.7347
.72	.9677	.9364	.9062	.8769	.8485	.8211	.7946	.7689	.7441
.73	.9690	.9390	.9099	.8817	.8544	.8279	.8023	.7774	.7533
.74	.9703	.9416	.9136	.8865	.8602	.8347	.8100	.7859	.7626
.75	.9716	.9441	.9173	.8913	.8660	.8415	.8176	.7944	.7719
.76	.9729	.9466	.9210	.8960	.8718	.8482	.8252	.8029	.7811
.77	.9742	.9491	.9246	.9007	.8775	.8549	.8328	.8113	.7904
.78	.9755	.9515	.9282	.9054	.8832	.8615	.8404	.8197	.7996
.79	.9767	.9540	.9317	.9100	.8888	.8681	.8479	.8281	.8088
.80	.9779	.9564	.9353	.9146	.8944	.8747	.8554	.8365	.8181
.81	.9792	.9587	.9388	.9192	.9000	.8812	.8629	.8449	.8273
.82	.9804	.9611	.9422	.9237	.9055	.8877	.8703	.8532	.8364
.83	.9816	.9634	.9456	.9282	.9110	.8942	.8777	.8615	.8456
.84	.9827	.9657	.9490	.9326	.9165	.9007	.8851	.8698	.8548
.85	.9839	.9680	.9524	.9371	.9220	.9071	.8925	.8781	.8639
.86	.9850	.9703	.9558	.9415	.9274	.9135	.8998	.8863	.8731
.87	.9862	.9725	.9591	.9458	.9327	.9198	.9071	.8946	.8822
.88	.9873	.9748	.9624	.9502	.9381	.9262	.9144	.9028	.8913
.89	.9884	.9770	.9656	.9545	.9434	.9325	.9217	.9110	.9004
.90	.9895	.9792	.9689	.9587	.9487	.9387	.9289	.9192	.9095
.91	.9906	.9813	.9721	.9630	.9539	.9450	.9361	.9273	.9186
.92	.9917	.9835	.9753	.9672	.9592	.9512	.9433	.9355	.9277
.93	.9928	.9856	.9785	.9714	.9644	.9574	.9505	.9436	.9368
.94	.9938	.9877	.9816	.9756	.9695	.9636	.9576	.9517	.9458
.95	.9949	.9898	.9847	.9797	.9747	.9697	.9647	.9598	.9549
.96	.9959	.9919	.9878	.9838	.9798	.9758	.9718	.9679	.9639
.97	.9970	.9939	.9909	.9879	.9849	.9819	.9789	.9759	.9730
.98	.9980	.9960	.9940	.9920	.9900	.9880	.9860	.9840	.9820
.99	.9990	.9980	.9970	.9960	.9950	.9940	.9930	.9920	.9910

Table 28. Higher Powers of Numbers

N	N^4	N^5	N^6	N^7	N^8	N^9
1	1	1	1	1	1	1
2	16	32	64	128	256	512
3	81	243	729	2187	6561	19683
4	256	1024	4096	16384	65536	262144
5	625	3125	15625	78125	390625	1953125
6	1296	7776	46656	279936	1679616	10077696
7	2401	16807	117649	823543	5764801	40353607
8	4096	32768	262144	2097152	16777216	134217728
9	6561	59049	531441	4782969	43046721	387420489
					$\times 10^8$	$\times 10^9$
10	10000	100000	1000000	10000000	1.000000	1.000000
11	14641	161051	1771561	19487171	2.143589	2.357948
12	20736	248832	2985984	35831808	4.299817	5.159780
13	28561	371293	4826809	62748517	8.157307	10.604499
14	38416	537824	7529536	105413504	14.757891	20.661047
15	50625	759375	11390625	170859375	25.628906	38.443359
16	65536	1048576	16777216	268435456	42.949673	68.719477
17	83521	1419857	24137569	410338673	69.757574	118.587876
18	104976	1889568	34012224	612220032	110.199606	198.359291
19	130321	2476099	47045881	893871739	169.835630	322.687697
				$\times 10^9$	$\times 10^{10}$	$\times 10^{11}$
20	160000	3200000	64000000	1.280000	2.560000	5.120000
21	194481	4084101	85766121	1.801089	3.782286	7.942800
22	234256	5153632	113379904	2.494358	5.487587	12.072692
23	279841	6436343	148035889	3.404825	7.831099	18.011527
24	331776	7962624	191102976	4.586471	11.007531	26.418075
25	390625	9765625	244140625	6.103516	15.258789	38.146973
26	456976	11881376	308915776	8.031810	20.882706	54.295037
27	531441	14348907	387420489	10.460353	28.242954	76.255975
28	614656	17210368	481890304	13.492929	37.780200	105.784559
29	707281	20511149	594823321	17.249876	50.024641	145.071460
			$\times 10^8$	$\times 10^{10}$	$\times 10^{11}$	$\times 10^{13}$
30	810000	24300000	7.290000	2.187000	6.561000	1.968300
31	923521	28629151	8.875037	2.751261	8.528910	2.643962
32	1048576	33554432	10.737418	3.435974	10.995116	3.518437
33	1185921	39135393	12.914680	4.261844	14.064086	4.641148
34	1336336	45435424	15.448044	5.252335	17.857939	6.071699
35	1500625	52521875	18.382656	6.433930	22.518754	7.881564
36	1679616	60466176	21.767823	7.836416	28.211099	10.155996
37	1874161	69343957	25.657264	9.493188	35.124795	12.996174
38	2085136	79235168	30.109364	11.441558	43.477921	16.521610
39	2313441	90224199	35.187438	13.723101	53.520093	20.872836
			$\times 10^9$	$\times 10^{10}$	$\times 10^{12}$	$\times 10^{14}$
40	2560000	102400000	4.096000	16.384000	6.553600	2.621440
41	2825761	115856201	4.750104	19.475427	7.984925	3.273819
42	3111696	130691232	5.489032	23.053933	9.682652	4.066714
43	3418801	147008443	6.321363	27.181861	11.688200	5.025926
44	3748096	164916224	7.256314	31.927781	14.048224	6.181218
45	4100625	184528125	8.303766	37.366945	16.815125	7.566806
46	4477456	205962976	9.474297	43.581766	20.047612	9.221902
47	4879681	229345007	10.779215	50.662312	23.811287	11.191305
48	5308416	254803968	12.230590	58.706834	28.179280	13.526055
49	5764801	282475249	13.841287	67.822307	33.232931	16.284136
50	6250000	312500000	15.625000	78.125000	39.062500	19.531250

Reproduced by permission from *CRC Standard Mathematical Tables*, Cleveland Ohio, Chemical Rubber Publishing Co., 12th ed., 1959.

Table 29. Bessel Functions

$J_0(x)$ and $J_1(x)$

x	$J_0(x)$	$J_1(x)$	x	$J_0(x)$	$J_1(x)$	x	$J_0(x)$	$J_1(x)$
0.0	1.0000	0.0000	3.0	−0.2601	0.3391	6.0	0.1506	−0.2767
0.1	0.9975	0.0499	3.1	−0.2921	0.3009	6.1	0.1773	−0.2559
0.2	0.9900	0.0995	3.2	−0.3202	0.2613	6.2	0.2017	−0.2329
0.3	0.9776	0.1483	3.3	−0.3443	0.2207	6.3	0.2238	−0.2081
0.4	0.9604	0.1960	3.4	−0.3643	0.1792	6.4	0.2433	−0.1816
0.5	0.9385	0.2423	3.5	−0.3801	0.1374	6.5	0.2601	−0.1538
0.6	0.9120	0.2867	3.6	−0.3918	0.0955	6.6	0.2740	−0.1250
0.7	0.8812	0.3290	3.7	−0.3992	0.0538	6.7	0.2851	−0.0953
0.8	0.8463	0.3668	3.8	−0.4026	0.0128	6.8	0.2931	−0.0652
0.9	0.8075	0.4059	3.9	−0.4018	−0.0272	6.9	0.2981	−0.0349
1.0	0.7652	0.4401	4.0	−0.3971	−0.0660	7.0	0.3001	−0.0047
1.1	0.7196	0.4709	4.1	−0.3887	−0.1033	7.1	0.2991	0.0252
1.2	0.6711	0.4983	4.2	−0.3766	−0.1386	7.2	0.2951	0.0543
1.3	0.6201	0.5220	4.3	−0.3610	−0.1719	7.3	0.2882	0.0826
1.4	0.5669	0.5419	4.4	−0.3423	−0.2028	7.4	0.2786	0.1096
1.5	0.5118	0.5579	4.5	−0.3205	−0.2311	7.5	0.2663	0.1352
1.6	0.4554	0.5699	4.6	−0.2961	−0.2566	7.6	0.2516	0.1592
1.7	0.3980	0.5778	4.7	−0.2693	−0.2791	7.7	0.2346	0.1813
1.8	0.3400	0.5815	4.8	−0.2404	−0.2985	7.8	0.2154	0.2014
1.9	0.2818	0.5812	4.9	−0.2097	−0.3147	7.9	0.1944	0.2192
2.0	0.2239	0.5767	5.0	−0.1776	−0.3276	8.0	0.1717	0.2346
2.1	0.1666	0.5683	5.1	−0.1443	−0.3371	8.1	0.1475	0.2476
2.2	0.1104	0.5560	5.2	−0.1103	−0.3432	8.2	0.1222	0.2580
2.3	0.0555	0.5399	5.3	−0.0758	−0.3460	8.3	0.0960	0.2657
2.4	0.0025	0.5202	5.4	−0.0412	−0.3453	8.4	0.0692	0.2708
2.5	−0.0484	0.4971	5.5	−0.0068	−0.3414	8.5	0.0419	0.2731
2.6	−0.0968	0.4708	5.6	0.0270	−0.3343	8.6	0.0146	0.2728
2.7	−0.1424	0.4416	5.7	0.0599	−0.3241	8.7	−0.0125	0.2697
2.8	−0.1850	0.4097	5.8	0.0917	−0.3110	8.8	−0.0392	0.2641
2.9	−0.2243	0.3754	5.9	0.1220	−0.2951	8.9	−0.0653	0.2559

$J_1(x) = 0$ for $x = 0, 3.832, 7.016, 10.173, 13.324, \cdots$
$J_0(x) = 0$ for $x = 2.405, 5.520, 8.654, 11.792, \cdots$

$Y_0(x)$ and $Y_1(x)$

x	$Y_0(x)$	$Y_1(x)$	x	$Y_0(x)$	$Y_1(x)$	x	$Y_0(x)$	$Y_1(x)$
0.0	$(-\infty)$	$(-\infty)$	2.5	0.498	0.146	5.0	−0.309	0.148
0.5	−0.445	−1.471	3.0	0.377	0.325	5.5	−0.340	−0.024
1.0	0.088	−0.781	3.5	0.189	0.410	6.0	−0.288	−0.175
1.5	0.382	−0.412	4.0	−0.017	0.398	6.5	−0.173	−0.274
2.0	0.510	−0.107	4.5	−0.195	0.301	7.0	−0.026	−0.303

TABLES OF CONVERSION FACTORS, UNITS OF WEIGHTS AND MEASURES

By J. G. Brainerd

Table 30. Temperature Conversion

$$^\circ F = (^\circ C \times 9/5) + 32 = (^\circ C + 40) \times 9/5 - 40$$
$$^\circ C = (F - 32) \times 5/9 = (^\circ F + 40) \times 5/9 - 40$$
$$^\circ R = {}^\circ F + 459.69$$
$$^\circ K = {}^\circ C + 273.16$$

Interpolation differences

°C	Temp	°F	°C	Temp	°F
0.5556	1	1.8	3.3334	6	10.8
1.1111	2	3.6	3.8889	7	12.6
1.6667	3	5.4	4.4445	8	14.4
2.2222	4	7.2	5.0000	9	16.2
2.7778	5	9.0	5.5556	10	18.0

°C	Temp	°F	°C	Temp	°F	°C	Temp	°F
−206.67	−340		−123.33	−190	−310	−40.00	−40	−40
−201.11	−330		−117.78	−180	−292	−34.44	−30	−22
−195.56	−320		−112.22	−170	−274	−28.89	−20	−4
−190.00	−310		−106.67	−160	−256	−23.35	−10	14
−184.44	−300		−101.11	−150	−238	−17.78	0	32
−178.89	−290		−95.56	−140	−220	−17.22	1	33.8
−173.33	−280		−90.0	−130	−202	−16.67	2	35.6
−167.78	−270	−454	−84.44	−120	−184	−16.11	3	37.4
−162.22	−260	−436	−78.89	−110	−166	−15.56	4	39.2
−156.67	−250	−418	−73.33	−100	−148	−15.00	5	41.0
−151.11	−240	−400	−67.78	−90	−130	−14.44	6	42.8
−145.56	−230	−382	−62.22	−80	−121	−13.89	7	44.6
−140.00	−220	−364	−56.67	−70	−94	−13.33	8	46.4
−134.44	−210	−346	−51.11	−60	−76	−12.78	9	48.2
−128.89	−200	−328	−45.56	−50	−58	−12.22	10	50.0

Table 30. Temperature Conversion—*Continued*

°C	Temp	°F	°C	Temp	°F	°C	Temp	°F
−11.67	11	51.8	18.89	66	150.8	154.44	310	590
−11.11	12	53.6	19.44	67	152.6	160.00	320	608
−10.56	13	55.4	20.00	68	154.4	165.56	330	626
−10.00	14	57.2	20.56	69	156.2	171.11	340	644
−9.44	15	59.0	21.11	70	158.0	176.67	350	662
−8.89	16	60.8	21.67	71	159.8	182.22	360	680
−8.33	17	62.6	22.22	72	161.6	187.78	370	698
−7.78	18	64.4	22.78	73	163.4	193.33	380	716
−7.22	19	66.2	23.33	74	165.2	198.89	390	734
−6.67	20	68.0	23.89	75	167.0	204.44	400	752
−6.11	21	69.8	24.44	76	168.8	210.00	410	770
−5.56	22	71.6	25.00	77	170.6	215.55	420	788
−5.00	23	73.4	25.56	78	172.4	221.11	430	806
−4.44	24	75.2	26.11	79	174.2	226.66	440	824
−3.89	25	77.0	26.67	80	176.0	232.22	450	842
−3.33	26	78.8	27.22	81	177.8	237.77	460	860
−2.78	27	80.6	27.78	82	179.6	243.33	470	878
−2.22	28	82.4	28.33	83	181.4	248.88	480	896
−1.67	29	84.2	28.89	84	183.2	254.44	490	914
−1.11	30	86.0	29.44	85	185.0	260.00	500	932
−0.56	31	87.8	30.00	86	186.8	315.6	600	1112
0	32	89.6	30.56	87	188.6	371.1	700	1292
0.56	33	91.4	31.11	88	190.4	426.7	800	1472
1.11	34	93.2	31.67	89	192.2	482.2	900	1652
1.67	35	95.0	32.22	90	194.0	537.8	1000	1832
2.22	36	96.8	32.78	91	195.8	593.3	1100	2012
2.78	37	98.6	33.33	92	197.6	648.9	1200	2192
3.33	38	100.4	33.89	93	199.4	704.4	1300	2372
3.89	39	102.2	34.44	94	201.2	760.0	1400	2552
4.44	40	104.0	35.00	95	203.0	815.6	1500	2732
5.00	41	105.8	35.56	96	204.8	871.1	1600	2912
5.56	42	107.6	36.11	97	206.6	926.7	1700	3092
6.11	43	109.4	36.67	98	208.4	982.2	1800	3272
6.67	44	111.2	37.22	99	210.2	1038	1900	3452
7.22	45	113.0	37.78	100	212.0	1093	2000	3632
7.78	46	114.8	43.33	110	230	1149	2100	3812
8.33	47	116.6	48.89	120	248	1204	2200	3992
8.89	48	118.4	54.44	130	266	1260	2300	4172
9.44	49	120.2	60.00	140	284	1316	2400	4352
10.00	50	122.0	65.56	150	302	1371	2500	4532
10.56	51	123.8	71.11	160	320	1427	2600	4712
11.11	52	125.6	76.67	170	338	1482	2700	4892
11.67	53	127.4	82.22	180	356	1538	2800	5072
12.22	54	129.2	87.78	190	374	1593	2900	5252
12.78	55	131.0	93.33	200	392	1649	3000	5432
13.33	56	132.8	100.00	212	413.6	1704	3100	5612
13.89	57	134.6	104.44	220	428	1760	3200	5792
14.44	58	136.4	110.00	230	446	1816	3300	5972
15.00	59	138.2	115.56	240	464	1871	3400	6152
15.56	60	140.0	121.11	250	482	1927	3500	6332
16.11	61	141.8	126.67	260	500	1982	3600	6512
16.67	62	143.6	132.22	270	518	2038	3700	6692
17.22	63	145.4	137.78	280	536	2093	3800	6872
17.78	64	147.2	143.33	290	554	2149	3900	7052
18.33	65	149.0	148.89	300	572	2205	4000	7232

Table 31. Length [L]

to Obtain \ Multiply Number of → by	Centimeters	Feet	Inches	Kilometers	Nautical miles	Meters	Mils	Miles	Millimeters	Yards
Centimeters	1	30.48	2.540	10^5	1.853×10^5	100	2.540×10^{-3}	1.609×10^5	0.1	91.44
Feet	3.281×10^{-2}	1	8.333×10^{-2}	3281	6080.27	3.281	8.333×10^{-5}	5280	3.281×10^{-3}	3
Inches	0.3937	12	1	3.937×10^4	7.296×10^4	39.37	0.001	6.336×10^4	3.937×10^{-2}	36
Kilometers	10^{-5}	3.048×10^{-4}	2.540×10^{-5}	1	1.853	0.001	2.540×10^{-8}	1.609	10^{-6}	9.144×10^{-4}
Nautical miles		1.645×10^{-4}		0.5396	1	5.396×10^{-4}		0.8684		4.934×10^{-4}
Meters	0.01	0.3048	2.540×10^{-2}	1000	1853	1		1609	0.001	0.9144
Mils	393.7	1.2×10^4	1000	3.937×10^7		3.937×10^4	1		39.37	3.6×10^4
Miles	6.214×10^{-6}	1.894×10^{-4}	1.578×10^{-5}	0.6214	1.1516	6.214×10^{-4}		1	6.214×10^{-7}	5.682×10^{-4}
Millimeters	10	304.8	25.40	10^6		1000	2.540×10^{-2}		1	914.4
Yards	1.094×10^{-2}	0.3333	2.778×10^{-2}	1094	2027	1.094	2.778×10^{-5}	1760	1.094×10^{-3}	1

Metric Multiples

10^6 microns $= 10^3$ millimeters $= 10^2$ centimeters $= 10$ decimeters $= 1$ meter
$= 10^{-1}$ dekameter $= 10^{-2}$ hectometer $= 10^{-3}$ kilometer $= 10^{-4}$ myriameter
$= 10^{-6}$ megameter $= 10^{10}$ Angstrom Units.

Land Measure

7.92 inches = 1 link
25 links   = 1 rod = 16.5 feet = 5.5 yards (1 rod = 1 pole = 1 perch)
4 rods   = 1 chain (Gunther's) = 66 feet = 22 yards = 100 links
10 chains = 1 furlong = 660 feet = 220 yards = 1000 links = 40 rods
8 furlongs = 1 mile = 5280 feet = 1760 yards = 8000 links = 320 rods = 80 chains

Ropes and Cables

2 yards = 1 fathom   120 fathoms = 1 cable's length

Nautical Measure

6080.27 feet = 1 nautical mile = 1.15156 statute miles
3 nautical miles = 1 league (U. S.)   3 statute miles = 1 league (Gr. Britain)

(NOTE. A nautical mile is the length of a minute of longitude of the earth at the equator at sea level. The British Admiralty uses the round figure of 6080 feet. The word " knot " is used to denote " nautical miles per hour.")

Miscellaneous

3 inches = 1 palm   9 inches = 1 span
4 inches = 1 hand   2 1/2 feet = 1 military pace

Table 32. Area [L^2]

to Obtain ↓ \ Multiply Number of →	Acres	Circular mils	Square centimeters	Square feet	Square inches	Square kilometers	Square meters	Square miles	Square millimeters	Square yards
Acres	1			2.296×10^{-5}		247.1	2.471×10^{-4}	640		2.066×10^{-4}
Circular mils		1	1.973×10^{5}	1.833×10^{8}	1.273×10^{6}		1.973×10^{9}		1973	
Square centimeters	5.067×10^{-6}	1		929.0	6.452	10^{10}	10^{4}	2.590×10^{10}	0.01	8361
Square feet	4.356×10^{4}		1.076×10^{-3}	1	6.944×10^{-3}	1.076×10^{7}	10.76	2.788×10^{7}	1.076×10^{-5}	9
Square inches	6,272,640	7.854×10^{-7}	0.1550	144	1	1.550×10^{9}	1550	4.015×10^{9}	1.550×10^{-3}	1296
Square kilometers	4.047×10^{-3}		10^{-10}	9.290×10^{-8}	6.452×10^{-10}	1	10^{-6}	2.590	10^{-12}	8.361×10^{-7}
Square meters	4047		0.0001	9.290×10^{-2}	6.452×10^{-4}	10^{6}	1	2.590×10^{6}	10^{-6}	0.8361
Square miles	1.562×10^{-3}		3.861×10^{-11}	3.587×10^{-8}		0.3861	3.861×10^{-7}	1	3.861×10^{-12}	3.228×10^{-7}
Square millimeters	5.067×10^{-4}		100	9.290×10^{4}	645.2	10^{12}	10^{6}		1	8.361×10^{5}
Square yards	4840		1.196×10^{-4}	0.1111	7.716×10^{-4}	1.196×10^{6}	1.196	3.098×10^{6}	1.196×10^{-6}	1

Land Measure

$30 \tfrac{1}{4}$ square yards = 1 square rod = $272 \tfrac{1}{4}$ square feet
16 square rods = 1 square chain = 484 square yards = 4356 square feet
$2 \tfrac{1}{2}$ square chains = 1 rood = 40 square rods = 1210 square yards
4 roods = 1 acre = 10 square chains = 160 square rods
640 acres = 1 square mile = 2560 roods = 102,400 square rods
1 section of land = 1 square mile; 1 quarter section = 160 acres

Architect's Measure

100 square feet = 1 square

Circular Inch and Circular Mil

A circular inch is the area of a circle 1 inch in diameter = 0.7854 square inch
1 square inch = 1.2732 circular inches
A circular mil is the area of a circle 1 mil (or 0.001 inch) in diameter = 0.7854 square mil
1 square mil = 1.2732 circular mils
1 circular inch = 10^{6} circular mils = 0.7854×10^{6} square mils
1 square inch = 1.2732×10^{6} circular mils = 10^{6} square mils

Metric Multiples

1 square meter = 1 centiare = 10^{-2} are = 10^{-4} hectare
= 10^{-6} square kilometer = 10^{-8} square myriameter

Table 33. Volume [L^3]

to Obtain ↓ \ Multiply Number of → by	Bushels (dry)	Cubic centimeters	Cubic feet	Cubic inches	Cubic meters	Cubic yards	Gallons (liquid)	Liters	Pints (liquid)	Quarts (liquid)
Bushels (dry)	1		0.8036	4.651×10^{-4}	28.38			2.838×10^{-2}		
Cubic centimeters	3.524×10^4	1	2.832×10^4	16.39	10^6	7.646×10^5	3785	1000	473.2	946.4
Cubic feet	1.2445	3.531×10^{-5}	1	5.787×10^{-4}	35.31	27	0.1337	3.531×10^{-2}	1.671×10^{-2}	3.342×10^{-2}
Cubic inches	2150.4	6.102×10^{-2}	1728	1	6.102×10^4	46,656	231	61.02	28.87	57.75
Cubic meters	3.524×10^{-2}	10^{-6}	2.832×10^{-2}	1.639×10^{-5}	1	0.7646	3.785×10^{-3}	0.001	4.732×10^{-4}	9.464×10^{-4}
Cubic yards		1.308×10^{-6}	3.704×10^{-2}	2.143×10^{-5}	1.308	1	4.951×10^{-3}	1.308×10^{-3}	6.189×10^{-4}	1.238×10^{-3}
Gallons (liquid)		2.642×10^{-4}	7.481	4.329×10^{-3}	264.2	202.0	1	0.2642	0.125	0.25
Liters	35.24	0.001	28.32	1.639×10^{-2}	1000	764.6	3.785	1	0.4732	0.9464
Pints (liquid)		2.113×10^{-3}	59.84	3.463×10^{-2}	2113	1616	8	2.113	1	2
Quarts (liquid)......		1.057×10^{-3}	29.92	1.732×10^{-2}	1057	807.9	4	1.057	0.5	1

Metric Multiples

10 milliliters	= 1 centiliter	= 0.338 fluid ounce
10 centiliters	= 1 deciliter	= 0.845 liquid gill
10 deciliters	= 1 liter	= 1.0567 liquid quarts
10 liters	= 1 dekaliter	= 2.6417 liquid gallons
10 dekaliters	= 1 hectoliter	= 2.8375 U. S. bushels
10 hectoliters	= 1 kiloliter (or stere)	= 28.375 U. S. bushels

Cubic Measure

1 cord of wood = a pile cut 4 feet long, piled 4 feet high and 8 feet on the ground = 128 cubic feet

1 perch of stone = a quantity 1 $1/2$ feet thick, 1 foot high and 16 $1/2$ feet long = 24 $3/4$ cubic feet

(NOTE.—A perch of stone is, however, often computed differently in different localities; thus, in most if not all of the States and Territories west of the Mississippi, stonemasons figure rubble by the perch of 16 $1/2$ cubic feet. In Philadelphia, 22 cubic feet are called a perch. In Chicago, stone is measured by the cord of 100 cubic feet. Check should be made against local practice.)

Board Measure

In board measure, boards are assumed to be one inch in thickness. Therefore, feet board measure of a stick of square timber = length in feet × breadth in feet × thickness in inches.

Shipping Measure

For register tonnage or measurement of the entire internal capacity of a vessel, it is arbitrarily assumed, to facilitate computation, that:

100 cubic feet = 1 register ton

For the measurement of cargo:

40 cubic feet = 1 U. S. shipping ton = 32.143 U. S. bushels

42 cubic feet = 1 British shipping ton = 32.703 Imperial bushels

Dry Measure

One U. S. Winchester bushel contains 1.2445 cubic feet or 2150.42 cubic inches. It holds 77.601 pounds distilled water at 62° F.

(NOTE.—The above is a *struck* bushel. A *heaped* bushel in general equals 1 1/4 struck bushels, although for apples and pears it contains 1.2731 struck bushels = 2737.72 cubic inches.)

One U. S. gallon (dry measure) = 1/8 bushel and contains 268.8 cubic inches.

(NOTE.—This is not a legal U. S. *dry measure* and therefore is given for comparison only.)

One British Imperial bushel contains 1.2843 cubic feet or 2219.36 cubic inches. It holds 80 pounds distilled water at 62° F.

One British Imperial gallon = 1/8 Imperial bushel and contains 277.42 cubic inches.

1 Winchester bushel = 0.9694 Imperial bushel

1 Imperial bushel = 1.032 Winchester bushels

Same relations as above maintain for gallons (dry measure)

(NOTE.—1 U. S. gallon (dry) = 1.164 U. S. gallons (liquid)).

U. S. Units

2 pints	= 1 quart	=	67.2 cubic inches
4 quarts	= 1 gallon * = 8 pints	=	268.8 cubic inches
2 gallons*	= 1 peck = 16 pints = 8 quarts	=	537.6 cubic inches
4 pecks	= 1 bushel = 64 pints = 32 quarts = 8 gallons*	=	2150.42 cubic inches

1 cubic foot contains 6.428 gallons (dry measure)*

Liquid Measure

One U. S. gallon (liquid measure) contains 231 cubic inches. It holds 8.336 pounds distilled water at 62° F.

One British Imperial gallon contains 277.42 cubic inches. It holds 10 pounds distilled water at 62° F.

1 U. S. gallon (liquid) = 0.8327 Imperial gallon

1 Imperial gallon   = 1.201 U. S. gallons (liquid)

(NOTE.—1 U. S. gallon (liquid) = 0.8594 U. S. gallon (dry)).

U. S. Units

4 gills	= 1 pint	= 16 fluid ounces
2 pints	= 1 quart = 8 gills	= 32 fluid ounces
4 quarts	= 1 gallon = 32 gills = 8 pints	= 128 fluid ounces

1 cubic foot contains 7.4805 gallons (liquid measure)

Apothecaries' Fluid Measure

60 minims = 1 fluid drachm.   8 drachms = 1 fluid ounce

In the U. S. a fluid ounce is the 128th part of a U. S. gallon, or 1.805 cu in. or 29.58 cu cm. It contains 455.8 grains of water at 62° F. In Great Britain the fluid ounce is 1.732 cu in. and contains 1 ounce avoirdupois (or 437.5 grains) of water at 62° F.

---

* The gallon is not a U. S. legal dry measure.

Table 34.   Plane Angle [*No Dimensions*]

to Obtain ↓ / Multiply Number of → by	Degrees	Minutes	Quadrants	Radians *	Revolutions * (Circumferences)	Seconds
Degrees	1	1.667×10^{-2}	90	57.30	360	2.778×10^{-4}
Minutes	60	1	5400	3438	2.16×10^{4}	1.667×10^{-2}
Quadrants	1.111×10^{-2}	1.852×10^{-4}	1	0.6366	4	3.087×10^{-6}
Radians *	1.745×10^{-2}	2.909×10^{-4}	1.571	1	6.283	4.848×10^{-6}
Revolutions * (Circumferences)	2.778×10^{-3}	4.630×10^{-5}	0.25	0.1591	1	7.716×10^{-7}
Seconds	3600	60	3.24×10^{5}	2.063×10^{5}	1.296×10^{6}	1

* 2π radians = 1 circumference = 360 degrees by definition.

Table 35.   Solid Angle [*No Dimensions*]

to Obtain ↓ / Multiply Number of → by	Hemispheres	Spheres *	Spherical right angles	Steradians †
Hemispheres	1	2	0.25	0.1592
Spheres *	0.5	1	0.125	7.958×10^{-2}
Spherical right angles	4	8	1	0.6366
Steradians †	6.283	12.57	1.571	1

* A sphere is the total solid angle about a point.   † 4π steradians = 1 sphere by definition.

Table 36.   Time [*T*]

to Obtain ↓ / Multiply Number of → by	Days	Hours	Minutes	Months (average)*	Seconds	Weeks
Days	1	4.167×10^{-2}	6.944×10^{-4}	30.42	1.157×10^{-5}	7
Hours	24	1	1.667×10^{-2}	730.0	2.778×10^{-4}	168
Minutes	1440	60	1	4.380×10^{4}	1.667×10^{-2}	1.008×10^{4}
Months (average) *	3.288×10^{-2}	1.370×10^{-3}	2.283×10^{-5}	1	3.806×10^{-7}	0.2302
Seconds	8.64×10^{4}	3600	60	2.628×10^{6}	1	6.048×10^{5}
Weeks	0.1429	5.952×10^{-3}	9.921×10^{-5}	4.344	1.654×10^{-6}	1

* One common year = 365 days; one leap year = 366 days; one average month = $\frac{1}{12}$ of a common year.

Table 37. Linear Velocity $[LT^{-1}]$

Multiply Number of → / to Obtain ↓	Centimeters per second	Feet per minute	Feet per second	Kilometers per hour	Kilometers per minute	Knots *	Meters per minute	Meters per second	Miles per hour	Miles per minute
Centimeters per second	1	0.5080	30.48	27.78	1667	51 48	1.667	100	44.70	2682
Feet per minute	1.969	1	60	54.68	3281	101.3	3.281	196.8	88	5280
Feet per second	3.281×10^{-2}	1.667×10^{-2}	1	0.9113	54.68	1.689	5.468×10^{-2}	3.281	1.467	88
Kilometers per hour	0.036	1.829×10^{-2}	1.097	1	60	1.853	0.06	3.6	1.609	96.54
Kilometers per minute	0.0006	3.048×10^{-4}	1.829×10^{-2}	1.667×10^{-2}	1	3.088×10^{-2}	0.001	0.06	2.682×10^{-2}	1.609
Knots *	1.943×10^{-2}	9.868×10^{-3}	0.5921	0.5396	32.38	1	3.238×10^{-2}	1.943	0.8684	52.10
Meters per minute	0.6	0.3048	18.29	16.67	1000	30.88	1	60	26.82	1609
Meters per second	0.01	5.080×10^{-3}	0.3048	0.2778	16.67	0.5148	1.667×10^{-2}	1	0.4470	26.82
Miles per hour	2.237×10^{-2}	1.136×10^{-2}	0.6818	0.6214	37.28	1.152	3.728×10^{-2}	2.237	1	60
Miles per minute	3.728×10^{-4}	1.892×10^{-4}	1.136×10^{-2}	1.036×10^{-2}	0.6214	1.919×10^{-2}	6.214×10^{-4}	3.728×10^{-2}	1.667×10^{-2}	1

* Nautical miles per hour.

The Miner's Inch
(Used in Measuring Flow of Water)

An Act of the California legislature, May 23, 1901, makes the standard miner's inch 1.5 cu ft per minute, measured through any aperture or orifice.

The term Miner's Inch is more or less indefinite, for the reason that California water companies do not all use the same head above the center of the aperture, and the inch varies from 1.36 to 1.73 cu ft per minute, but the most common measurement is through an aperture 2 in. high and whatever length is required, and through a plank 1 1/4 in. thick. The lower edge of the aperture should be 2 in. above the bottom of the measuring-box, and the plank 5 in. high above the aperture, thus making a 6-in. head above the center of the stream. Each square inch of this opening represents a miner's inch, which is equal to a flow of 1.5 cu ft per minute.

Table 38.   Angular Velocity $[T^{-1}]$

Multiply Number of → / to Obtain by	Degrees per second	Radians per second	Revolutions per minute	Revolutions per second
Degrees per second	1	57.30	6	360
Radians per second	1.745×10^{-2}	1	0.1047	6.283
Revolutions per minute	0.1667	9.549	1	60
Revolutions per second	2.778×10^{-3}	0.1592	1.667×10^{-2}	1

Table 39.   Linear Acceleration * $[LT^{-2}]$

Multiply Number of → / to Obtain by	Centimeters per second per second	Feet per second per second	Kilometers per hour per second	Meters per second per second	Miles per hour per second
Centimeters per second per second	1	30.48	27.78	100	44.70
Feet per second per second	3.281×10^{-2}	1	0.9113	3.281	1.467
Kilometers per hour per second	0.036	1.097	1	3.6	1.609
Meters per second per second	0.01	0.3048	0.2778	1	0.4470
Miles per hour per second	2.237×10^{-2}	0.6818	0.6214	2.237	1

* The (standard) acceleration due to gravity $(g_0) = 980.7$ cm per sec per sec, $= 32.17$ feet per sec per sec $= 35.30$ km per hour per sec $= 9.807$ meters per sec per sec $= 21.94$ miles per hour per sec.

Table 40.   Angular Acceleration $[T^{-2}]$

Multiply Number of → / to Obtain by	Radians per second per second	Revolutions per minute per minute	Revolutions per minute per second	Revolutions per second per second
Radians per second per second	1	1.745×10^{-3}	0.1047	6.283
Revolutions per minute per minute	573.0	1	60	3600
Revolutions per minute per second	9.549	1.667×10^{-2}	1	60
Revolutions per second per second	0.1592	2.778×10^{-4}	1.667×10^{-2}	1

Table 41. Mass [M] and Weight *

to Obtain ↓ / Multiply Number of → by	Grains	Grams	Kilograms	Milligrams	Ounces †	Pounds †	Tons (long)	Tons (metric)	Tons (short)
Grains	1	15.43	1.543×10^4	1.543×10^{-2}	437.5	7000			
Grams	6.481×10^{-2}	1	1000	0.001	28.35	453.6	1.016×10^6	10^6	9.072×10^5
Kilograms	6.481×10^{-5}	0.001	1	10^{-6}	2.835×10^{-2}	0.4536	1016	1000	907.2
Milligrams	64.81	1000	10^6	1	2.835×10^4	4.536×10^5	1.016×10^9	10^9	9.072×10^8
Ounces †	2.286×10^{-3}	3.527×10^{-2}	35.27	3.527×10^{-5}	1	16	3.584×10^4	3.527×10^4	3.2×10^4
Pounds †	1.429×10^{-4}	2.205×10^{-3}	2.205	2.205×10^{-6}	6.250×10^{-2}	1	2240	2205	2000
Tons (long)		9.842×10^{-7}	9.842×10^{-4}	9.842×10^{-10}	2.790×10^{-5}	4.464×10^{-4}	1	0.9842	0.8929
Tons (metric)		10^{-6}	0.001	10^{-9}	2.835×10^{-5}	4.536×10^{-4}	1.016	1	0.9072
Tons (short)		1.102×10^{-6}	1.102×10^{-3}	1.102×10^{-9}	3.125×10^{-5}	0.0005	1.120	1.102	1

* These same conversion factors apply to the *gravitational* units of force having the corresponding names. The dimensions of these units when used as gravitational units of force are MLT^{-2}; see table for *Force*.
† Avoirdupois pounds and ounces.

Metric Multiples

10^6 micrograms $= 10^3$ milligrams $= 10^2$ centigrams $= 10$ decigrams $= 1$ gram $\simeq 10^{-1}$ dekagram $= 10^{-2}$ hectogram $= 10^{-3}$ kilogram $= 10^{-4}$ myriagram $= 10^{-6}$ megagram

Avoirdupois Weight
(Used Commercially)

27.343 grains	= 1 drachm
16 drachms	= 1 ounce (oz) = 437.5 grains
16 ounces	= 1 pound (lb) = 7000 grains
28 pounds	= 1 quarter (qr)
4 quarters	= 1 hundredweight (cwt) = 112 pounds
20 hundredweight	= 1 gross or long ton *
2000 pounds	= 1 net or short ton

(* Note.—The long ton is used by the U. S. custom-houses in collecting duties upon foreign goods. It is also used in freighting coal and selling it wholesale.)

14 pounds = 1 stone; 100 pounds = 1 quintal

Troy Weight
(Used in weighing gold or silver)

24 grains	= 1 pennyweight (dwt)
20 pennyweights	= 1 ounce (oz) = 480 grains
12 ounces	= 1 pound (lb) = 5760 grains

The grain is the same in Avoirdupois, Troy and Apothecaries' weights. A carat, for weighing diamonds = 3.086 grains = 0.200 gram. (International Standard, 1913.)

1 pound troy = .8229 pound avoirdupois
1 pound avoirdupois = 1.2153 pounds troy

Apothecaries' Weight

(Used in compounding medicines)

20 grains   = 1 scruple (Θ)
3 scruples = 1 drachm ($\mathfrak{Z}$) = 60 grains
8 drachms = 1 ounce ($\mathfrak{Z}$) = 480 grains
12 ounces = 1 pound (lb) = 5760 grains

The grain is the same in Avoirdupois, Troy and Apothecaries' weights.

1 pound apothecaries = 0.82286 pound avoirdupois
1 pound avoirdupois = 1.2153 pounds apothecaries

Table 42.   Density or Mass per Unit Volume $[ML^{-3}]$

to Obtain ↓ / Multiply Number of → by	Grams per cubic centimeter	Kilograms per cubic meter	Pounds per cubic foot	Pounds per cubic inch
Grams per cubic centimeter	1	0.001	1.602×10^{-2}	27.68
Kilograms per cubic meter	1000	1	16.02	2.768×10^4
Pounds per cubic foot	62.43	6.243×10^{-2}	1	1728
Pounds per cubic inch	3.613×10^{-2}	3.613×10^{-5}	5.787×10^{-4}	1
Pounds per mil foot *	3.405×10^{-7}	3.405×10^{-10}	5.456×10^{-9}	9.425×10^{-6}

* Unit of volume is a volume one foot long and one circular mil in cross-section area.

Table 43.   Force * $[MLT^{-2}]$ or $[F]$

to Obtain ↓ / Multiply Number of → by	Dynes	Grams	Joules per cm	Joules per meter	Kilograms	Pounds	Poundals
Dynes	1	980.7	10^7	10^5	9.807×10^5	4.448×10^5	1.383×10^4
Grams	1.020×10^{-3}	1	1.020×10^4	102.0	1000	453.6	14.10
Joules per cm	10^{-7}	9.807×10^{-5}	1	.01	9.807×10^{-2}	4.448×10^{-2}	1.383×10^{-3}
Newtons, or joules per meter	10^{-5}	9.807×10^{-3}	100	1	9.807	4.448	0.1383
Kilograms	1.020×10^{-6}	0.001	10.20	0.1020	1	0.4536	1.410×10^{-2}
Pounds	2.248×10^{-6}	2.205×10^{-3}	22.48	0.2248	2.205	1	3.108×10^{-2}
Poundals	7.233×10^{-5}	7.093×10^{-2}	723.3	7.233	70.93	32.17	1

* Conversion factors between absolute and gravitational units apply only under standard acceleration due to gravity conditions.   (See Sec. 3.)

Table 44. Pressure or Force per Unit Area $[ML^{-1}T^{-2}]$ or $[FL^{-2}]$

Multiply Number of → / to Obtain ↓	Atmospheres *	Baryes or dynes per square centimeter †	Centimeters of mercury at 0°C ‡	Inches of mercury at 0°C ‡	Inches of water at 4°C	Kilograms per square meter §	Pounds per square foot	Pounds per square inch	Tons (short) per square foot	Newtons per square meter
Atmospheres *	1	9.869×10^{-7}	1.316×10^{-2}	3.342×10^{-2}	2.458×10^{-3}	9.678×10^{-5}	4.725×10^{-4}	6.804×10^{-2}	0.9450	9.869×10^{-6}
Baryes or dynes per square centimeter	1.013×10^{6}	1	1.333×10^{4}	3.386×10^{4}	2.491×10^{-3}	98.07	478.8	6.895×10^{4}	9.576×10^{5}	10
Centimeters of mercury at 0°C ‡	76.00	7.501×10^{-5}	1	2.540	0.1868	7.356×10^{-3}	3.591×10^{-2}	5.171	71.83	7.501×10^{-4}
Inches of mercury at 0°C ‡	29.92	2.953×10^{-5}	0.3937	1	7.355×10^{-2}	2.896×10^{-3}	1.414×10^{-2}	2.036	28.28	2.953×10^{-4}
Inches of water at 4°C	406.8	4.015×10^{-4}	5.354	13.60	1	3.937×10^{-2}	0.1922	27.68	384.5	4.015×10^{-8}
Kilograms per square meter §	1.033×10^{4}	1.020×10^{-2}	136.0	345.3	25.40	1	4.882	703.1	9765	0.1020
Pounds per square foot	2117	2.089×10^{-3}	27.85	70.73	5.204	0.2048	1	144	2000	2.089×10^{-2}
Pounds per square inch	14.70	1.450×10^{-5}	0.1934	0.4912	3.613×10^{-2}	1.422×10^{-3}	6.944×10^{-3}	1	13.89	1.450×10^{-4}
Tons (short) per square foot	1.058	1.044×10^{-6}	1.392×10^{-2}	3.536×10^{-2}	2.601×10^{-3}	1.024×10^{-4}	0.0005	0.072	1	1.044×10^{-5}
Newtons per square meter	1.013×10^{5}	10^{-1}	1.333×10^{3}	3.386×10^{3}	2.491×10^{-4}	9.807	47.88	6.895×10^{3}	9.576×10^{4}	1

* Definition: One atmosphere (standard) = 76 cm of mercury at 0 C.

‡ To convert height h of a column of mercury at t degrees Centigrade to the equivalent height h_0 at 0 C use $h_0 = h \left\{ 1 - \dfrac{(m - l)t}{1 + mt} \right\}$ where $m = 0.0001818$ and $l = 18.4 \times 10^{-6}$ if the scale is engraved on brass; $l = 8.5 \times 10^{-6}$ if on glass. This assumes the scale is correct at 0 C; for other cases (any liquid) see *International Critical Tables*, Vol 1, 68.

§ 1 gram per sq cm = 10 kilograms per sq m.

Table 45. Torque or Moment of Force $[ML^2T^{-2}]$ or $[FL]$ *

Multiply Number of → / to Obtain ↓	Dyne-centimeters	Gram-centimeters	Kilogram-meters	Pound-feet	Newton-meter
Dyne-centimeters	1	980.7	9.807×10^{7}	1.356×10^{7}	10^{7}
Gram-centimeters	1.020×10^{-3}	1	10^{5}	1.383×10^{4}	1.020×10^{4}
Kilogram-meters	1.020×10^{-8}	10^{-5}	1	0.1383	0.1020
Pound-feet	7.376×10^{-8}	7.233×10^{-5}	7.233	1	0.7376
Newton-meter	10^{-7}	9.807×10^{-4}	9.807	1.356	1

* Same dimensions as energy.

Table 46. Moment of Inertia $[ML^2]$

to Obtain ↓ / Multiply Number of → by	Gram-centimeters squared	Kilogram-meters squared	Pound-inches squared	Pound-feet squared	Slug-feet squared
Gram-centimeters squared	1	10^7	2.9266×10^3	4.21434×10^5	1.3559×10^7
Kilogram-meters squared	10^{-7}	1	2.9266×10^{-4}	4.21434×10^{-2}	1.3559
Pound-inches squared	3.4169×10^{-4}	3.4169×10^3	1	144	4.63304×10^3
Pound-feet squared	2.37285×10^{-6}	23.7285	6.944×10^{-3}	1	32.1739
Slug-feet squared	7.37507×10^{-8}	0.737507	2.15841×10^{-4}	3.10811×10^{-2}	1

Table 47. Energy, Work and Heat * $[ML^2T^{-2}]$ or $[FL]$

to Obtain ↓ / Multiply Number of → by	British thermal units †	Centimeter-grams	Ergs or centimeter-dynes	Foot-pounds	Horsepower-hours	Joules ‡ or watt-seconds	Kilogram-calories †	Kilowatt-hours	Meter-kilograms	Watt-hours
British thermal units †	1	9.297×10^{-8}	9.480×10^{-11}	1.285×10^{-3}	2545	9.480×10^{-4}	3.969	3413	9.297×10^{-3}	3.413
Centimeter-grams	1.076×10^7	1	1.020×10^{-3}	1.383×10^4	2.737×10^{10}	1.020×10^4	4.269×10^7	3.671×10^{10}	10^5	3.671×10^7
Ergs or centimeter-dynes	1.055×10^{10}	980.7	1	1.356×10^7	2.684×10^{13}	10^7	4.186×10^{10}	3.6×10^{13}	9.807×10^7	3.6×10^{10}
Foot-pounds	778.0	7.233×10^{-5}	7.367×10^{-8}	1	1.98×10^6	0.7376	3087	2.655×10^6	7.233	2655
Horsepower-hours	3.929×10^{-4}	3.654×10^{-11}	3.722×10^{-14}	5.050×10^{-7}	1	3.722×10^{-7}	1.559×10^{-3}	1.341	3.653×10^{-6}	1.341×10^{-3}
Joules ‡ or watt-seconds	1054.8	9.807×10^{-5}	10^{-7}	1.356	2.684×10^6	1	4186	3.6×10^6	9.807	3600
Kilogram-calories †	0.2520	2.343×10^{-8}	2.389×10^{-11}	3.239×10^{-4}	641.3	2.389×10^{-4}	1	860.0	2.343×10^{-3}	0.8600
Kilowatt-hours	2.930×10^{-4}	2.724×10^{-11}	2.778×10^{-14}	3.766×10^{-7}	0.7457	2.778×10^{-7}	1.163×10^{-3}	1	2.724×10^{-6}	0.001
Meter-kilograms	107.6	10^{-5}	1.020×10^{-8}	0.1383	2.737×10^5	0.1020	426.9	3.671×10^5	1	367.1
Watt-hours	0.2930	2.724×10^{-8}	2.778×10^{-11}	3.766×10^{-4}	745.7	2.778×10^{-4}	1.163	1000	2.724×10^{-3}	1

* See note at the bottom of Table 48.

† Mean calorie and Btu used throughout. One gram-calorie = 0.001 kilogram-calorie; one Ostwald calorie = 0.1 kilogram-calorie.

The IT cal, 1000 international steam-table calories, has been defined as the 1/860th part of the international kilowatthour (see *Mechanical Engineering*, Nov., 1935, p. 710). Its value is very nearly equal to the mean kilogram-calorie, 1 IT cal = 1.00037 kilogram-calories (mean). 1 Btu = 251.996 IT cal.

‡ Absolute joule, defined as 10^7 ergs. The international joule, based on the international ohm and ampere, equals 1.0003 absolute joules.

Table 48. Power or Rate of Doing Work * $[ML^2T^{-3}]$ or $[FLT^{-1}]$

to Obtain ↓ Multiply Number of →	British thermal units per minute	Ergs per second	Foot-pounds per minute	Foot-pounds per second	Horsepower *	Kilogram-calories per minute	Kilowatts	Metric horsepower	Watts
British thermal units per minute	1	5.689×10^{-9}	1.285×10^{-3}	7.712×10^{-2}	42.41	3.969	56.89	41.83	5.689×10^{-2}
Ergs per second	1.758×10^{8}	1	2.259×10^{5}	1.356×10^{7}	7.457×10^{9}	6.977×10^{8}	10^{10}	7.355×10^{9}	10^{7}
Foot-pounds per minute	778.0	4.426×10^{-6}	1	60	3.3×10^{4}	3087	4.426×10^{4}	3.255×10^{4}	44.26
Foot-pounds per second	12.97	7.376×10^{-8}	1.667×10^{-2}	1	550	51.44	737.6	542.5	0.7376
Horsepower *	2.357×10^{-2}	1.341×10^{-10}	3.030×10^{-5}	1.818×10^{-3}	1	9.355×10^{-2}	1.341	0.9863	1.341×10^{-3}
Kilogram-calories per minute	0.2520	1.433×10^{-9}	3.239×10^{-4}	1.943×10^{-2}	10.69	1	14.33	10.54	1.433×10^{-2}
Kilowatts	1.758×10^{-2}	10^{-10}	2.260×10^{-5}	1.356×10^{-3}	0.7457	6.977×10^{-2}	1	0.7355	10^{-3}
Metric horsepower	2.390×10^{-2}	1.360×10^{-10}	3.072×10^{-5}	1.843×10^{-3}	1.014	9.485×10^{-2}	1.360	1	1.360×10^{-3}
Watts	17.58	10^{-7}	2.260×10^{-2}	1.356	745.7	69.77	1000	735.5	1

1 Cheval-vapeur = 75 kilogram-meters per second
1 Poncelet = 100 kilogram-meters per second

* The "horsepower" used in these tables is equal to 550 foot-pounds per second by definition. Other definitions are one horsepower equals 746 watts (U. S. and Great Britain) and one horsepower equals 736 watts (continental Europe). Neither of these latter definitions is equivalent to the first; the "horsepowers" defined in these latter definitions are widely used in the rating of electrical machinery.

Table 49. Quantity of Electricity and Dielectric Flux $[Q]$

to Obtain ↓ Multiply Number of →	Abcoulombs	Ampere-hours	Coulombs	Faradays	Stat-coulombs
Abcoulombs	1	360	0.1	9649	3.335×10^{-11}
Ampere-hours	2.778×10^{-3}	1	2.778×10^{-4}	26.80	9.259×10^{-14}
Coulombs	10	3600	1	9.649×10^{4}	3.335×10^{-10}
Faradays	1.036×10^{-4}	3.731×10^{-2}	1.036×10^{-5}	1	3.457×10^{-15}
Statcoulombs	2.998×10^{10}	1.080×10^{13}	2.998×10^{9}	2.893×10^{14}	1

Table 50.   Charge per Unit Area and Electric Flux Density $[QL^{-2}]$

Multiply Number of → to Obtain ↓	Abcoulombs per square centimeter *	Coulombs per square centimeter	Coulombs per square inch	Statcoulombs per square centimeter	Coulombs per square meter
Abcoulombs per square centi-centimeter	1	0.1	1.550×10^{-2}	3.335×10^{-11}	10^{-5}
Coulombs per square centimeter	10	1	0.1550	3.335×10^{-10}	10^{-4}
Coulombs per square inch	64.52	6.452	1	2.151×10^{-9}	6.452×10^{-4}
Statcoulombs per square centimeter	2.998×10^{10}	2.998×10^{9}	4.647×10^{8}	1	2.998×10^{5}
Coulombs per square meter	10^{5}	10^{4}	1550	3.335×10^{-6}	1

Table 51.   Electric Current $[QT^{-1}]$

Multiply Number of → to Obtain ↓	Abamperes	Amperes	Statamperes
Abamperes	1	0.1	3.335×10^{-11}
Amperes	10	1	3.335×10^{-10}
Statamperes	2.998×10^{10}	2.998×10^{9}	1

Table 52.   Current Density $[QT^{-1}L^{-2}]$

Multiply Number of → to Obtain ↓	Abamperes per square centimeter	Amperes per square centimeter	Amperes per square inch	Statamperes per square centimeter	Amperes per square meter
Abamperes per square centimeter	1	0.1	1.550×10^{-2}	3.335×10^{-11}	10^{-5}
Amperes per square centimeter	10	1	0.1550	3.335×10^{-10}	10^{-4}
Amperes per square inch	64.52	6.452	1	2.151×10^{-9}	6.452×10^{-4}
Statamperes per square centimeter	2.998×10^{10}	2.998×10^{9}	4.647×10^{8}	1	2.998×10^{5}
Amperes per square meter	10^{5}	10^{4}	1550	3.335×10^{-6}	1

Table 53. Electric Potential and Electromotive Force $[MQ^{-1}L^2T^{-2}]$ or $[FQ^{-1}L]$

Multiply Number of → / to Obtain ↓ by	Abvolts	Microvolts	Millivolts	Statvolts	Volts
Abvolts	1	100	10^5	2.998×10^{10}	10^8
Microvolts	0.01	1	1000	2.998×10^8	10^6
Millivolts	10^{-5}	0.001	1	2.998×10^5	1000
Statvolts	3.335×10^{-11}	3.335×10^{-9}	3.335×10^{-6}	1	3.335×10^{-3}
Volts	10^{-8}	10^{-6}	0.001	299.8	1

Table 54. Electric Field Intensity and Potential Gradient $[MQ^{-1}LT^{-2}]$ or $[FQ^{-1}]$

Multiply Number of → / to Obtain ↓ by	Abvolts per centimeter	Microvolts per meter	Millivolts per meter	Statvolts per centimeter	Volts per centimeter	Kilovolts per centimeter	Volts per inch	Volts per mil	Volts per meter
Abvolts per centimeter	1	1	1000	2.998×10^{10}	10^8	10^{11}	3.937×10^7	3.937×10^{10}	10^9
Microvolts per meter	1	1	1000	2.998×10^{10}	10^8	10^{11}	3.937×10^7	3.937×10^{10}	10^6
Millivolts per meter	0.001	0.001	1	2.998×10^7	10^5	10^8	3.937×10^4	3.937×10^7	1000
Statvolts per centimeter	3.335×10^{-11}	3.335×10^{-11}	3.335×10^{-8}	1	3.335×10^{-3}	3.335	1.313×10^{-3}	1.313	3.335×10^{-5}
Volts per centimeter	10^{-8}	10^{-8}	10^{-5}	299.8	1	1000	0.3937	393.7	10^{-2}
Kilovolts per centimeter	10^{-11}	10^{-11}	10^{-8}	0.2998	0.001	1	3.937×10^{-4}	0.3937	10^{-5}
Volts per inch	2.540×10^{-8}	2.540×10^{-8}	2.540×10^{-5}	761.6	2.540	2540	1	1000	2.540×10^{-2}
Volts per mil	2.540×10^{-11}	2.540×10^{-11}	2.540×10^{-8}	0.7616	2.540×10^{-3}	2.540	0.001	1	2.540×10^{-5}
Volts per meter	10^{-6}	10^{-6}	10^{-3}	2.998×10^4	100	10^5	39.37	3.937×10^4	1

Table 55.   Electric Resistance $[MQ^{-2}L^2T^{-1}]$ or $[FQ^{-2}LT]$

Multiply Number of → / to Obtain ↓ by	Abohms	Megohms	Microhms	Ohms	Statohms
Abohms	1	10^{15}	1000	10^9	8.988×10^{20}
Megohms	10^{-15}	1	10^{-12}	10^{-6}	8.988×10^5
Microhms	0.001	10^{12}	1	10^6	8.988×10^{17}
Ohms	10^{-9}	10^6	10^{-6}	1	8.988×10^{11}
Statohms	1.112×10^{-21}	1.112×10^{-6}	1.112×10^{-18}	1.112×10^{-12}	1

Electrical Conductance $[F^{-1}QL^{-1}T^{-1}]$

1 mho = 1 ohm^{-1} = 10^{-6} megmho = 10^6 micromho

Table 56.   Electric Resistivity * $[MQ^{-2}L^3T^{-1}]$ or $[FQ^{-2}L^2T]$

Multiply Number of → / to Obtain ↓ by	Abohm-centimeters	Microhm-centimeters	Microhm-inches	Ohms (mil, foot)	Ohms (meter, gram) †	Ohm-meters
Abohm-centimeters	1	1000	2540	166.2	$\dfrac{10^5}{\delta}$	10^{11}
Microhm-centimeters	0.001	1	2.540	0.1662	$\dfrac{100}{\delta}$	10^8
Microhm-inches	3.937×10^{-4}	0.3937	1	6.545×10^{-2}	$\dfrac{39.37}{\delta}$	3.937×10^7
Ohms (mil, foot)	6.015×10^{-3}	6.015	15.28	1	$\dfrac{601.5}{\delta}$	6.015×10^8
Ohms (meter, gram) †	$10^{-5}\delta$	0.01δ	$2.540 \times 10^{-2}\delta$	$1.662 \times 10^{-3}\delta$	1	$10^{-6}\delta$
Ohm-meters	10^{-11}	10^{-8}	2.540×10^{-8}	1.662×10^{-9}	$\dfrac{10^{-6}}{\delta}$	1

* In this table δ is density in grams per cm.^{3} The following names, corresponding respectively to those at the tops of columns, are sometimes used: abohms per cm cube; microhms per cm cube; microhms per inch cube; ohms per mil-foot; ohms per meter-gram.   The first four columns are headed by units of *volume* resistivity, the last by a unit of *mass* resistivity.   The dimensions of the latter are $Q^{-2}L^6T^{-1}$; not these given in the heading of the table.

† One ohm (meter, gram) = 5710 ohms (mile, pound).

Table 57. Electric Conductivity * $[M^{-1}Q^2L^{-3}T]$ or $[F^{-1}Q^2L^{-2}T^{-1}]$

Multiply Number of → / to Obtain ↓	Abmhos per cm	Mhos (mil, foot)	Mhos (meter, gram)	Micromhos per cm	Micromhos per inch	Mhos per meter
Abmhos per cm	1	6.015×10^{-3}	$10^{-5}\delta$	0.001	3.937×10^{-4}	10^{-11}
Mhos (mil, foot)	166.2	1	$1.662 \times 10^{-3}\delta$	0.1662	6.524×10^{-2}	1.662×10^{-9}
Mhos (meter, gram)	$10^5/\delta$	$601.5/\delta$	1	$100/\delta$	$39.37/\delta$	$10^{-6}/\delta$
Micromhos per cm	1000	6.015	0.01δ	1	0.3937	10^{-8}
Micromhos per inch	2540	15.28	$2.540 \times 10^{-2}\delta$	2.540	1	2.54×10^{-8}
Mhos per meter	10^{11}	6.015×10^8	$10^6\delta$	10^8	3.937×10^7	1

* See footnote of Table 56, Electric Resistivity. Names sometimes used are abmho per cm cube, mho per mil-foot, etc. Dimensions of mass conductivity are $Q^2L^{-6}T$.

Table 58. Capacitance $[M^{-1}Q^2L^{-2}T^2]$ or $[F^{-1}Q^2L^{-1}]$

Multiply Number of → / to Obtain ↓	Abfarads	Farads	Microfarads	Statfarads
Abfarads	1	10^{-9}	10^{-15}	1.112×10^{-21}
Farads	10^9	1	10^{-6}	1.112×10^{-12}
Microfarads	10^{15}	10^6	1	1.112×10^{-6}
Statfarads	8.988×10^{20}	8.988×10^{11}	8.988×10^5	1

Table 59. Inductance $[MQ^{-2}L^2]$ or $[FQ^{-2}LT^2]$

to Obtain ↓ \ Multiply Number of →	Abhenries *	Henries	Microhenries	Millihenries	Stathenries
Abhenries *	1	10^9	1000	10^6	8.988×10^{20}
Henries	10^{-9}	1	10^{-6}	0.001	8.988×10^{11}
Microhenries	0.001	10^6	1	1000	8.988×10^{17}
Millihenries	10^{-6}	1000	0.001	1	8.988×10^{14}
Stathenries	1.112×10^{-21}	1.112×10^{-12}	1.112×10^{-18}	1.112×10^{-15}	1

* An abhenry is sometimes called a "centimeter."

Table 60. Magnetic Flux $[MQ^{-1}L^2T^{-1}]$ or $[FQ^{-1}LT]$

to Obtain ↓ \ Multiply Number of →	Kilolines	Maxwells (or lines)	Webers
Kilolines	1	0.001	10^5
Maxwells (or lines)	1000	1	10^8
Webers	10^{-5}	10^{-8}	1

Table 61. Magnetic Flux Density $[MQ^{-1}T^{-1}]$ or $[FQ^{-1}L^{-1}T]$

to Obtain ↓ \ Multiply Number of →	Gausses (or lines per square centimeter)	Lines per square inch	Webers per square centimeter	Webers per square inch	Webers per square meter
Gausses (or lines per square centimeter)	1	0.1550	10^8	1.550×10^7	10^4
Lines per square inch	6.452	1	6.452×10^8	10^8	6.452×10^4
Webers per square centimeter	10^{-8}	1.550×10^{-9}	1	0.1550	10^{-4}
Webers per square inch	6.452×10^{-8}	10^{-8}	6.452	1	6.452×10^{-4}
Webers per square meter	10^{-4}	1.550×10^{-5}	10^4	1550	1

Table 62. Magnetic Potential and Magnetomotive Force $[QT^{-1}]$

to Obtain ↓ \\ Multiply Number of → by	Abampere-turns	Ampere-turns	Gilberts
Abampere-turns	1	0.1	7.958×10^{-2}
Ampere-turns	10	1	0.7958
Gilberts	12.57	1.257	1

Table 63. Magnetic Field Intensity, Potential Gradient, and Magnetizing Force $[QL^{-1}T^{-1}]$

to Obtain ↓ \\ Multiply Number of → by	Abampere-turns per centimeter	Ampere-turns per centimeter	Ampere-turns per inch	Oersteds (gilberts per centimeter)	Ampere-turns per meter
Abampere-turns per centimeter	1	0.1	3.937×10^{-2}	7.958×10^{-2}	10^{-3}
Ampere-turns per centimeter	10	1	0.3937	0.7958	10^{-2}
Ampere-turns per inch	25.40	2.540	1	2.021	2.54×10^{-2}
Oersteds (gilberts per centimeter)	12.57	1.257	0.4950	1	1.257×10^{-2}
Ampere-turns per meter	10^3	10^2	39.37	79.58	1

Table 64. Specific Heat $[L^2T^{-2}t^{-1}]$

(t = temperature)

To change specific heat in gram-calories per gram per degree Centigrade to the units given in any line of the following table, multiply by the factor in the last column.

Unit of Heat or Energy	Unit of Mass	Temperature Scale *	Factor
Gram-calories	Gram	Centigrade	1
Kilogram-calories	Kilogram	Centigrade	1
British thermal units	Pound	Centigrade	1.800
British thermal units	Pound	Fahrenheit	1.000
Joules	Gram	Centigrade	4.186
Joules	Pound	Fahrenheit	1055
Kilowatt-hours	Kilogram	Centigrade	1.163×10^{-3}
Kilowatt-hours	Pound	Fahrenheit	2.930×10^{-4}

* Temperature conversion formulas:

$$t_c = \text{temperature in Centigrade degrees}$$
$$t_f = \text{temperature in Fahrenheit degrees}$$
$$1\,F = 5/9\ C$$
$$t_c = 5/9\ (t_f - 32)$$
$$t_f = 9/5\ t_c + 32$$

Table 65. Thermal Conductivity $[MLT^{-3}t^{-1}]$ and Thermal Resistivity $[M^{-1}L^{-1}T^3t]$

(t = temperature)

To convert thermal conductivity, in gram-calories transmitted per second from one face of a cube 1 cm on edge to the opposite face per degree centigrade temperature difference between these faces, to the units given in any line of the following table, multiply by the factor in the last column.

To convert thermal conductivity in any unit given to any other unit multiply the number of original units by a factor obtained by dividing the factor in the last column for the final unit by the factor for the original unit.

To convert thermal resistivity, in degrees centigrade between one face of a cube 1 cm on edge and the opposite face per gram-calories transmitted per second between these faces, to the units given in any line of the following table, divide by the factor in the last column.

To convert thermal resistivity in any given unit to any other unit multiply the number of the original units by a factor obtained by dividing the factor in the last column for the original unit by the factor for the final unit.

Surface emission resistance in thermal ohms per square centimeter is derived from degrees Fahrenheit per Btu per hour per square foot by multiplying the number of the latter units by 1761.

Heat	Units of Area	Thickness	Time	Temperature Scale	Factor
Gram-calories................	cm^2	cm	second	Centigrade	1.
Kilogram-calories............	m^2	cm	hour	Centigrade	3.6×10^4
British thermal units.........	ft^2	inch	hour	Fahrenheit	2903
Joules *.....................	cm^2	cm	second	Centigrade	4.186
Joules.......................	ft^2	inch	second	Fahrenheit	850.6
Kilowatt-hours..............	m^2	cm	hour	Centigrade	41.86
Kilowatt-hours...............	ft^2	inch	hour	Fahrenheit	0.8506

* Thermal resistances in these units are known as *thermal ohms*.

Table 66. Light

Multiply Number of → by / to Obtain ↓	Inter-national candles	Hefners	10-cp pentanes	Carcels	Bougie deci-males	English candles	German candles
International candles	1.00	0.90	10.0	9.61	1.00	1.04	1.055
Hefners	1.11	1.00	11.1	10.66	1.11	1.154	1.17
10-cp pentanes	0.10	0.09	1.00	0.96	0.10	0.104	0.105
Carcels	0.104	0.094	1.04	1.00	0.104	0.1	0.109
Bougie decimales	1.00	0.90	10.0	9.61	1.00	1.04	1.055
English candles	0.96	0.864	9.6	9.24	0.96	1.00	1.02
German candles	0.95	0.855	9.5	9.19	0.95	0.98	1.00

International System of Units

Candela, the unit of luminous intensity, is defined such that the luminance of a blackbody radiator at the freezing temperature of platinum is 60 candelas per square centimeter. The former candle is abandoned.   Same as National Bureau of Standards candle.

Lumen is the unit of luminous flux.

Lumen per unit area is the unit of luminous flux density.   The former footcandle is abandoned.

Lux is the unit of luminous flux density = 1 lumen per square meter

1 footcandle = 10.764 lux or lumens per square meter
1 footlambert = 3.4263 candelas per square meter

Table 67. Specific Gravity Conversions

$$°Be = 145 - \frac{145}{sp\ gr} \text{ (heavier than } H_2O\text{)} \qquad °Be = \frac{140}{sp\ gr} - 130 \text{ (lighter than } H_2O\text{)}$$

$$°Tw = \frac{sp\ gr\ 60°/60°F - 1}{0.005} \qquad °API = \frac{141.5}{sp\ gr} - 131.5$$

Sp gr 60°/60°	°Bé.	°A.P.I.	Lb/gal at 60°F, wt in air	Lb/ft³ at 60°F, wt in air	Sp gr 60°/60°	°Bé.	°A.P.I.	Lb/gal at 60°F, wt in air	Lb/ft³ at 60°F wt in air
0.600	103.33	104.33	4.9929	37.350	0.900	25.76	25.72	7.4944	56.062
.605	101.40	102.38	5.0346	37.662	.905	24.70	24.85	7.5361	56.374
.610	99.51	100.47	5.0763	37.973	.910	23.85	23.99	7.5777	56.685
.615	97.64	98.58	5.1180	38.285	.915	23.01	23.14	7.6194	56.997
.620	95.81	96.73	5.1597	38.597	.920	22.17	22.30	7.6612	57.410
.625	94.00	94.90	5.2014	39.910	.925	21.35	21.47	7.7029	57.622
.630	92.22	93.10	5.2431	39.222	.930	20.54	20.65	7.7446	57.934
.635	90.47	91.33	5.2848	39.534	.935	19.73	19.84	8.7863	58.246
.640	88.75	89.59	5.3265	39.845	.940	18.94	19.03	7.8280	58.557
.645	87.05	87.88	5.3682	40.157	.945	18.15	18.24	7.8697	58.869
.650	85.38	86.19	5.4098	40.468	.950	17.37	17.45	7.9114	59.181
.655	83.74	84.53	5.4515	40.780	.955	16.60	16.67	7.9531	59.493
.660	82.12	82.89	5.4932	41.092	.960	15.83	15.90	7.9947	59.805
.665	80.53	81.28	5.5349	41.404	.965	15.08	15.13	8.0364	60.117
.670	78.96	79.69	5.5766	41.716	.970	14.33	14.38	8.0780	60.428
.675	77.41	78.13	5.6183	42.028	.975	13.59	13.63	8.1197	60.740
.680	75.88	76.59	5.6600	42.340	.980	12.86	12.89	8.1615	61.052
.685	74.38	75.07	5.7017	42.652	.985	12.13	12.15	8.2032	61.364
.690	72.90	73.57	5.7434	42.963	.990	11.41	11.43	8.2449	61.676
.695	71.44	72.10	5.7851	43.275	.995	10.70	10.71	8.2866	61.988
					1.000	10.00	10.00	8.3283	62.300
.700	70.00	70.64	5.8268	43.587					
.705	68.50	69.21	5.8685	43.899				Lb/gal at 60°F, wt in air	Lb/ft³ at 60°F, wt in air
.710	67.18	67.80	5.9101	44.211					
.715	65.80	66.40	5.9518	44.523	Sp gr 60°/60°	°Bé.	°Tw,		
.720	64.44	65.03	5.9935	44.834					
.725	63.10	63.67	6.0352	45.146	1.005	0.72	1	8.3700	62.612
.730	61.78	62.34	6.0769	45.458	1.010	1.44	2	8.4117	62.924
.735	60.48	61.02	6.1186	45.770	1.015	2.14	3	8.4534	63.236
.740	59.19	59.72	6.1603	46.082	1.020	2.84	4	8.4950	63.547
.745	57.92	58.43	6.2020	46.394	1.025	3.54	5	8.5367	63.859
.750	56.67	57.17	6.2437	46.706	1.030	4.22	6	8.5784	64.171
.755	55.43	55.92	6.2854	47.018	1.035	4.90	7	8.6201	64.483
.760	54.21	54.68	6.3271	47.330	1.040	5.58	8	8.6618	64.795
.765	53.01	53.47	6.3688	47.642	1.045	6.24	9	8.7035	65.107
.770	51.82	52.27	6.4104	47.953	1.050	6.91	10	8.7452	65.419
.775	50.65	51.08	6.4521	47.265	1.055	7.56	11	8.7869	65.731
.780	49.49	49.91	6.4938	48.577	1.060	8.21	12	8.8286	66.042
.785	48.34	48.75	6.5355	48.889	1.065	8.85	13	8.8703	66.354
.790	47.22	47.61	6.5772	49.201	1.070	9.49	14	8.9120	66.666
.795	46.10	46.49	6.6189	49.513	1.075	10.12	15	8.9537	66.978
.800	45.00	45.38	6.6606	49.825	1.080	10.74	16	8.9954	67.290
.805	43.91	44.28	6.7023	50.137	1.085	11.36	17	9.0371	67.602
.810	42.84	43.19	6.7440	50.448	1.090	11.97	18	9.0787	67.914
.815	41.78	42.12	6.7857	50.760	1.095	12.58	19	9.1204	68.226
.820	40.73	41.06	6.8274	51.072	1.100	13.18	20	9.1621	58.537
.825	39.70	40.02	6.8691	51.384	1.105	13.78	21	9.2038	68.849
.830	38.67	38.98	6.9108	51.696	1.110	14.37	22	9.2455	69.161
.835	37.66	37.96	6.9525	52.008	1.115	14.96	23	9.2872	69.473
.840	36.67	36.95	6.9941	52.320	1.120	15.54	24	9.3289	69.785
.845	35.68	35.96	7.0358	52.632	1.125	16.11	25	9.3706	70.097
.850	34.71	34.97	7.0775	52.943	1.130	16.68	26	9.4123	70.409
.855	33.74	34.00	7.1192	53.225	1.135	17.25	27	9.4540	70.721
.860	32.79	33.03	7.1609	53.567	1.140	17.81	28	9.4957	71.032
.865	31.85	32.08	7.2026	53.879	1.145	18.36	29	9.5374	71.344
.870	30.92	31.14	7.2443	54.191	1.150	18.91	30	9.5790	71.656
.875	30.00	30.21	7.2860	54.503	1.155	19.46	31	9.6207	71.968
.880	29.09	29.30	7.3277	54.815	1.160	20.00	32	9.6624	72.280
.885	28.19	28.38	7.3694	55.127	1.165	20.54	33	9.7041	72.592
.890	27.30	27.49	7.4111	55.438	1.170	21.07	34	9.7458	72.904
.895	26.42	26.60	8.4528	55.750	1.175	21.60	35	9.7875	73.216

Table 67. Specific Gravity Conversions—*Continued*

Sp gr 60°/60°	°Bé.	°Tw.	Lb/gal at 60°F, wt in air	Lb/ft³ at 60°F, wt in air	Sp gr 60°/60°	°Bé.	°Tw.	Lb/gal at 60°F, wt in air	Lb/ft³ at 60°F, wt in air
1.180	22.12	36	9.8292	73.528	1.51	48.97	102	12.581	94.11
1.185	22.64	37	9.8709	73.840	1.52	49.61	104	12.644	94.79
1.190	23.15	38	9.9126	74.151	1.53	50.23	106	12.748	95.36
1.195	23.66	39	9.9543	74.463	1.54	50.84	108	12.831	95.98
1.200	24.17	40	9.9960	74.775	1.55	51.45	110	12.914	96.61
1.205	24.67	41	10.0377	75.087	1.56	52.05	112	12.998	97.23
1.210	25.17	42	10.0793	75.399	1.57	52.64	114	13.081	97.85
1.215	25.66	43	10.1210	75.711	1.58	53.23	116	13.165	98.48
1.220	26.15	44	10.1627	76.022	1.59	53.81	118	13.248	99.10
1.225	26.63	45	10.2044	76.334	1.60	54.38	120	13.331	99.73
1.230	27.11	46	10.2461	76.646	1.61	54.94	122	13.415	100.35
1.235	27.59	47	10.2878	76.958	1.62	55.49	124	13.498	100.97
1.240	28.06	48	10.3295	77.270	1.63	56.04	126	13.582	101.60
1.245	28.53	49	10.3712	77.582	1.64	56.59	128	13.665	102.22
1.250	29.00	50	10.4129	77.894	1.65	57.12	130	13.748	102.84
1.255	29.46	51	10.4546	78.206	1.66	57.65	132	13.832	103.47
1.260	29.92	52	10.4963	78.518	1.67	58.17	134	13.915	104.09
1.265	30.38	53	10.5380	78.830	1.68	58.69	136	13.998	104.72
1.270	30.83	54	10.5797	79.141	1.69	59.20	138	14.082	105.34
1.275	31.27	55	10.6214	79.453	1.70	59.71	140	14.165	105.96
1.280	31.72	56	10.6630	79.765	1.71	60.20	142	14.249	106.59
1.285	32.16	57	10.7047	80.077	1.72	60.70	144	14.332	107.21
1.290	32.60	58	10.7464	80.389	1.73	61.18	146	14.415	107.83
1.295	33.03	59	10.7881	80.701	1.74	61.67	148	14.499	108.46
1.300	33.46	60	10.8298	81.013	1.75	62.14	150	14.582	109.08
1.305	33.89	61	10.8715	81.325	1.76	62.61	152	14.665	109.71
1.310	34.31	62	10.9132	81.636	1.77	63.08	154	14.749	110.32
1.315	34.73	63	10.9549	81.948	1.78	63.54	156	14.832	110.95
1.320	35.15	64	10.9966	82.260	1.79	63.99	158	14.916	111.58
1.325	35.57	65	11.0383	82.572	1.80	64.44	160	14.999	112.20
1.330	35.98	66	11.0800	82.884	1.81	64.89	162	15.082	112.82
1.335	36.39	67	11.1217	83.196	0.82	65.33	164	15.166	113.45
1.340	36.79	68	11.1634	83.508	1.83	65.77	166	15.249	114.07
1.345	37.19	69	11.2051	83.820	1.84	66.20	168	15.333	114.70
1.530	37.59	70	11.2467	84.131	1.85	66.62	070	15.416	115.31
1.355	37.99	71	11.2884	84.443	1.86	67.04	172	15.499	115.94
1.360	38.38	72	11.3301	84.755	1.87	67.46	174	15.583	116.56
1.365	38.77	73	11.3718	95.067	1.88	67.87	176	15.666	117.19
1.370	39.16	74	11.4135	85.379	1.89	68.28	178	15.750	117.81
1.375	39.55	75	11.4552	85.691	1.90	68.68	180	15.832	118.43
1.380	39.93	76	11.4969	86.003	1.91	69.08	182	15.916	119.06
1.385	40.31	77	11.5386	86.315	1.92	69.48	184	16.000	119.68
1.390	40.68	78	11.5803	86.626	1.93	69.87	186	16.083	120.31
1.395	41.06	79	11.6220	86.938	1.94	70.26	188	16.166	120.93
1.400	41.43	80	11.6637	87.250	1.95	70.64	190	16.250	121.56
1.405	41.80	81	11.7054	87.562	1.96	71.02	192	16.333	122.18
1.410	42.16	82	11.7471	87.874	1.97	71.40	194	16.417	122.80
1.415	42.53	83	11.7888	88.186	1.98	71.77	196	16.500	123.43
1.420	42.89	84	11.8304	88.498	1.99	72.14	198	16.583	124.05
1.425	43.25	85	11.8721	88.810	2.00	72.50	200	16.667	124.68
1.430	43.60	86	11.9138	89.121					
1.435	43.95	87	11.9555	89.433					
1.440	44.31	88	11.9972	89.745					
1.445	44.65	89	12.0389	90.057					
1.450	45.00	90	12.0806	90.369					
1.455	45.34	91	12.1223	90.681					
1.460	45.68	92	12.1640	90.993					
1.465	46.02	93	12.2057	91.305					
1.470	46.36	94	12.2473	91.616					
1.475	46.69	95	12.2890	91.928					
1.480	47.03	96	12.3307	92.240					
1.485	47.36	97	12.3724	92.552					
1.490	47.68	98	12.4141	92.864					
1.495	48.01	99	12.4558	93.176					
1.500	48.33	100	12.4975	93.488					

GAGES

Table 68a. United States Standard Gage * for Sheet and Plate Iron and Steel, and Its Extension †

Gage No.	Weight per square foot		Weight per square meter	Approximate thickness			
				Wrought iron 480 lb/ft³		Steel and open-hearth iron 489.6 lb/ft³	
	Ounces	Pounds	kg	Inch	mm	Inch	mm
0000000	320	20.00	97.65	0.500	12.70	0.490	12.45
000000	300	18.75	91.55	.469	11.91	.460	11.67
00000	280	17.50	85.44	.438	11.11	.429	10.90
0000	260	16.25	79.34	.406	10.32	.398	10.12
000	240	15.00	73.24	.375	9.52	.368	9.34
00	220	13.75	67.13	.344	8.73	.337	8.56
0	200	12.50	61.03	.312	7.94	.306	7.78
1	180	11.25	54.93	.2812	7.14	.2757	7.00
2	170	10.62	51.88	.2656	6.75	.2604	6.62
3	160	10.00	48.82	.2500	6.35	.2451	6.23
4	150	9.375	45.77	.2344	5.95	.2298	5.84
5	140	8.750	42.72	.2188	5.56	.2145	5.45
6	130	8.125	39.67	.2031	5.16	.1991	5.06
7	120	7.500	36.62	.1875	4.76	.1838	4.67
8	110	6.875	33.57	.1719	4.37	.1685	4.28
9	100	6.250	30.52	.1562	3.97	.1532	3.89
10	90	5.625	27.46	.1406	3.57	.1379	3.50
11	80	5.000	24.41	.1250	3.18	.1225	3.11
12	70	4.375	21.36	.1094	2.778	.1072	2.724
13	60	3.750	18.31	.0938	2.381	.0919	2.335
14	50	3.125	15.26	.0781	1.984	.0766	1.946
15	45	2.812	13.73	.0703	1.786	.0689	1.751
16	40	2.500	12.21	.0625	1.588	.0613	1.557
17	36	2.250	10.99	.0562	1.429	.0551	1.400
18	32	2.000	9.765	.0500	1.270	.0490	1.245
19	28	1.750	8.544	.0438	1.111	.0429	1.090
20	24	1.500	7.324	.0375	.952	.0368	.934
21	22	1.375	6.713	.0344	.873	.0337	.856
22	20	1.250	6.103	.0312	.794	.0306	.778
23	18	1.125	5.493	.0281	.714	.0276	.700
24	16	1.000	4.882	.0250	.635	.0245	.623
25	14	.8750	4.272	.0219	.556	.0214	.545
26	12	.7500	3.662	.0188	.476	.0184	.467
27	11	.6875	3.357	.0172	.437	.0169	.428
28	10	.6250	3.052	.0156	.397	.0153	.389
29	9	.5625	2.746	.0141	.357	.0138	.350
30	8	.5000	2.441	.0125	.318	.0123	.311
31	7	.4375	2.136	.0109	.278	.0107	.272
32	6 1/2	.4062	1.983	.0102	.258	.0100	.253
33	6	.3750	1.831	.0094	.238	.0092	.233
34	5 1/2	.3438	1.678	.0086	.218	.0084	.214
35	5	.3125	1.526	.0078	.198	.0077	.195
36	4 1/2	.2812	1.373	.0070	.179	.0069	.175
37	4 1/4	.2656	1.297	.0066	.169	.0065	.165
38	4	.2500	1.221	.0062	.159	.0061	.156
39	3 3/4	.2344	1.144	.0059	.149	.0057	.146
40	3 1/2	.2188	1.068	.0055	.139	.0054	.136
41	3 3/8	.2109	1.030	.0053	.134	.0052	.131
42	3 1/4	.2031	.9917	.0051	.129	.0050	.126
43	3 1/8	.1953	.9536	.0049	.124	.0048	.122
44	3	.1875	.9155	.0047	.119	.0046	.117

* For the Galvanized Sheet Gage, add 2.5 ounces to the weight per square foot as given in the table. Gage numbers below 8 and above 34 are not used in the Galvanized Sheet Gage.

† Gage numbers greater than 38 were not in the standard as set up by law, but are in general use.

Table 68b. American Wire Gage—Weights of Copper, Aluminum and Brass Sheets and Plates

Gage No.	Thickness		Approximate weight * per sq ft in lb		
	Inch	mm	Copper	Aluminum	Commercial (high) brass
0000...............	0.4600	11.68	21.27	6.49	20.27
000................	.4096	10.40	18.94	5.78	18.05
00.................	.3648	9.266	16.87	5.14	16.07
0..................	.3249	8.252	15.03	4.58	14.32
1..................	.2893	7.348	13.38	4.08	12.75
2..................	.2576	6.544	11.91	3.632	11.35
3..................	.2294	5.827	10.61	3.234	10.11
4..................	.2043	5.189	9.45	2.880	9.00
5..................	.1819	4.621	8.41	2.565	8.01
6..................	.1620	4.115	7.49	2.284	7.14
7..................	.1443	3.665	6.67	2.034	6.36
8..................	.1285	3.264	5.94	1.812	5.66
9..................	.1144	2.906	5.29	1.613	5.04
10.................	.1019	2.588	4.713	1.437	4.490
11.................	.0907	2.305	4.195	1.279	3.996
12.................	.0808	2.053	3.737	1.139	3.560
13.................	.0720	1.828	3.330	1.015	3.172
14.................	.0641	1.628	2.965	0.904	2.824
15.................	.0571	1.450	2.641	.805	2.516
16.................	.0508	1.291	2.349	.716	2.238
17.................	.0453	1.150	2.095	.639	1.996
18.................	.0403	1.024	1.864	.568	1.776
19.................	0.0359	0.9116	1.660	.506	1.582
20.................	.0320	.8118	1.480	.451	1.410
21.................	.0285	.7230	1.318	.402	1.256
22.................	.0253	.6438	1.170	.3567	1.115
23.................	.0226	.5733	1.045	.3186	0.996
24.................	.0201	.5106	0.930	.2834	.886
25.................	.0179	.4547	.828	.2524	.789
26.................	.0159	.4049	.735	.2242	.701
27.................	.0142	.3606	.657	.2002	.626
28.................	.0126	.3211	.583	.1776	.555
29.................	.0113	.2859	.523	.1593	.498
30.................	.0100	.2546	.4625	.1410	.4406
31.................	.00893	.2268	.4130	.1259	.3935
32.................	.00795	.2019	.3677	.1121	.3503
33.................	.00708	.1798	.3274	.0998	.3119
34.................	.00630	.1601	.2914	.0888	.2776
35.................	.00561	.1426	.2595	.0791	.2472
36.................	.00500	.1270	.2312	.0705	.2203
37.................	.00445	.1131	.2058	.0627	.1961
38.................	.00397	.1007	.1836	.0560	.1749
39.................	.00353	.0897	.1633	.0498	.1555
40.................	.00314	.0799	.1452	.0443	.1383

* Assumed specific gravities or densities in grams per cubic centimeter; Copper, 8.89; Aluminum, 2.71; brass, 8.47.

Wire Gages

The sizes of wires having a diameter less than $1/2$ inch are usually stated in terms of certain arbitrary scales called "gages." The size or gage number of a solid wire refers to the cross section of the wire perpendicular to its length; the size or gage number of a stranded wire refers to the total cross section of the constituent wires, irrespective of the pitch of the spiraling. Larger wires are usually described in terms of their area expressed in circular mils. A circular mil is the area of a circle 1 mil in diameter, and the area of any circle in circular mils is equal to the square of its diameter in mils.

Table 69. Comparison of Wire Gage Diameters in Mils
(Bureau of Standards, Circulars No. 31 and No. 67)

Gage No.	American wire gage (B. & S.)	Steel wire gage	Birmingham wire gage (Stubs')	Old English wire gage (London)	Stubs' steel wire gage	(British) Standard wire gage	Metric gage *	Gage No.
7–0	....	490.0	....	....	....	500	....	7–0
6–0	....	461.5	....	....	....	464	....	6–0
5–0	....	430.5	....	....	....	432	....	5–0
4–0	460	393.8	454	454	....	400	....	4–0
3–0	410	362.5	425	425	....	372	....	3–0
2–0	365	331.0	380	380	....	348	....	2–0
0	325	306.5	340	340	....	324	....	0
1	289	283.0	300	300	227	300	3.94	1
2	258	262.5	284	284	219	276	7.87	2
3	229	243.7	259	259	212	252	11.8	3
4	204	225.3	238	238	207	232	15.7	4
5	182	207.0	220	220	204	212	19.7	5
6	162	192.0	203	203	201	192	23.6	6
7	144	177.0	180	180	199	176	27.6	7
8	128	162.0	165	165	197	160	31.5	8
9	114	148.3	148	148	194	144	35.4	9
10	102	135.0	134	134	191	128	39.4	10
11	91	120.5	120	120	188	116	....	11
12	81	105.5	109	109	185	104	47.2	12
13	72	91.5	95	95	182	92	....	13
14	64	80.0	83	83	180	80	55.1	14
15	57	72.0	72	72	178	72	....	15
16	51	62.5	65	65	175	64	63.0	16
17	45	54.0	58	58	172	56	....	17
18	40	47.5	49	49	168	48	70.9	18
19	36	41.0	42	42	164	40	....	19
20	32	34.8	35	35	161	36	78.7	20
21	28.5	31.7	32	31.5	157	32	....	21
22	25.3	28.6	28	29.5	155	28	....	22
23	22.6	25.8	25	27.0	153	24	....	23
24	20.1	23.0	22	25.0	151	22	....	24
25	17.9	20.4	20	23.0	148	20	98.4	25
26	15.9	18.1	18	20.5	146	18	....	26
27	14.2	17.3	16	18.75	143	16.4	....	27
28	12.6	16.2	14	16.50	139	14.8	....	28
29	11.3	15.0	13	15.50	134	13.6	....	29
30	10.0	14.0	12	13.75	127	12.4	118	30
31	8.9	13.2	10	12.25	120	11.6	....	31
32	8.0	12.8	9	11.25	115	10.8	....	32
33	7.1	11.8	8	10.25	112	10.0	....	33
34	6.3	10.4	7	9.50	110	9.2	....	34
35	5.6	9.5	5	9.00	108	8.4	138	35
36	5.0	9.0	4	7.50	106	7.6	....	36
37	4.5	8.5	....	6.50	103	6.8	....	37
38	4.0	8.0	....	5.75	101	6.0	....	38
39	3.5	7.5	....	5.00	99	5.2	....	39
40	3.1	7.0	....	4.50	97	4.8	157	40
41	....	6.6	....	....	95	4.4	....	41
42	....	6.2	....	....	92	4.0	....	42
43	....	6.0	....	....	88	3.6	....	43
44	....	5.8	....	....	85	3.2	....	44
45	....	5.5	....	....	81	2.8	177	45
46	....	5.2	....	....	79	2.4	....	46
47	....	5.0	....	....	77	2.0	....	47
48	....	4.8	....	....	75	1.6	....	48
49	....	4.6	....	....	72	1.2	....	49
50	....	4.4	....	....	69	1.0	197	50

* For diameters corresponding to metric gage numbers, 1.2, 1.4, 1.6, 1.8, 2.5, 3.5, and 4.5, divide those of 12, 14, etc., by ten.

STANDARD STRUCTURAL SIZES—STEEL

Steel Sections. Tables 70 to 77 give the dimensions, weights, and properties of *rolled steel* structural sections, including wide flange sections, American standard beams, channels, angles, tees, and zees. The values for the various structural forms, taken from the fifth edition, 1949, of *Steel Construction*, by the kind permission of the publisher, the American Institute of-Steel Construction, give the section specifications required in designing steel structures. The theory of design is covered in Section 5.

Most of the sections can be supplied promptly by the Bethlehem, Carnegie, and Illinois Steel Company mills. Owing to variations in the rolling practice of the different mills, their products are not identical, although their divergence from the values given in the tables is practically negligible. For standardization, only the lesser values are given, and therefore they are on the side of safety.

Further information on sections listed in the tables, together with information on other products and on the requirements for placing orders, may be gathered from mill catalogs.

Table 70. Properties of Wide-Flange Sections

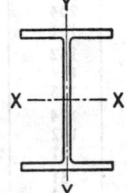

Nominal Size	Weight per Foot	Area	Depth	Flange Width	Flange Thickness	Web Thickness	Axis X–X I	Axis X–X S	Axis X–X r	Axis Y–Y I	Axis Y–Y S	Axis Y–Y r
in.	lb	in.^{2}	in.	in.	in.	in.	in.^{4}	in.^{3}	in.	in.^{4}	in.^{3}	in.
36×16 1/2	300	88.17	36.72	16.655	1.680	0.945	20290.2	1105.1	15.17	1225.2	147.1	3.73
	280	82.32	36.50	16.595	1.570	.885	18819.3	1031.2	15.12	1127.5	135.9	3.70
	260	76.56	36.24	16.555	1.440	.845	17233.8	951.1	15.00	1020.6	123.3	3.65
	245	72.03	36.06	16.512	1.350	.802	16092.2	892.5	14.95	944.7	114.4	3.62
	230	67.73	35.88	16.475	1.260	.765	14988.4	835.5	14.88	870.9	105.7	3.59
36×12	194	57.11	36.48	12.117	1.260	.770	12103.4	663.6	14.56	355.4	58.7	2.49
	182	53.54	36.32	12.072	1.180	.725	11281.5	621.2	14.52	327.7	54.3	2.47
	170	49.98	36.16	12.027	1.100	.680	10470.0	579.1	14.47	300.6	50.0	2.45
	160	47.09	36.00	12.000	1.020	.653	9738.8	541.0	14.38	275.4	45.9	2.42
	150	44.16	35.84	11.972	0.940	.625	9012.1	502.9	14.29	250.4	41.8	2.38
33×15 3/4	240	70.52	33.50	15.865	1.400	.830	13585.1	811.1	13.88	874.3	110.2	3.52
	220	64.73	33.25	15.810	1.275	.775	12312.1	740.6	13.79	782.4	99.0	3.48
	200	58.79	33.00	15.750	1.150	.715	11048.2	669.6	13.71	691.7	87.8	3.43
33×11 1/2	152	44.71	33.50	11.565	1.055	.635	8147.6	486.4	13.50	256.1	44.3	2.39
	141	41.51	33.31	11.535	0.960	.605	7442.2	446.8	13.39	229.7	39.8	2.35
	130	38.26	33.10	11.510	0.855	.580	6699.0	404.8	13.23	201.4	35.0	2.29
30×15	210	61.78	30.38	15.105	1.315	.775	9872.4	649.9	12.64	707.9	93.7	3.38
	190	55.90	30.12	15.040	1.185	.710	8825.9	586.1	12.57	624.6	83.1	3.34
	172	50.65	29.88	14.985	1.065	.655	7891.5	528.2	12.48	550.1	73.4	3.30
30×10 1/2	132	38.83	30.30	10.551	1.000	.615	5753.1	379.7	12.17	185.0	35.1	2.18
	124	36.45	30.16	10.521	0.930	.585	5347.1	354.6	12.11	169.7	32.3	2.16
	116	34.13	30.00	10.500	0.850	.564	4919.1	327.9	12.00	153.2	29.2	2.12
	108	31.77	29.82	10.484	0.760	.548	4461.0	299.2	11.85	135.1	25.8	2.06
27×14	177	52.10	27.31	14.090	1.190	.725	6728.6	492.8	11.36	518.9	73.7	3.15
	160	47.04	27.08	14.023	1.075	.658	6018.6	444.5	11.31	458.0	65.3	3.12
	145	42.68	26.88	13.965	0.975	.600	5414.3	402.9	11.26	406.9	58.3	3.09
27×10	114	33.53	27.28	10.070	0.932	.570	4080.5	299.2	11.03	149.6	29.7	2.11
	102	30.01	27.07	10.018	0.827	.518	3604.1	266.3	10.96	129.5	25.9	2.08
	94	27.65	26.91	9.990	0.747	.490	3266.7	242.8	10.87	115.1	23.0	2.04
24×14	160	47.04	24.72	14.091	1.135	.656	5110.3	413.5	10.42	492.6	69.9	3.23
	145	42.62	24.49	14.043	1.020	.608	4561.0	372.5	10.34	434.3	61.8	3.19
	130	38.21	24.25	14.000	0.900	.565	4009.5	330.7	10.24	375.2	53.6	3.13

Table 70. Properties of Wide-Flange Sections—*Continued*

Nominal Size	Weight per Foot	Area	Depth	Flange Width	Flange Thickness	Web Thickness	Axis X-X I	Axis X-X S	Axis X-X r	Axis Y-Y I	Axis Y-Y S	Axis Y-Y r
in.	lb	in.²	in.	in.	in.	in.	in.⁴	in.³	in.	in.⁴	in.³	in.
24×12	120	35.29	24.31	12.088	0.930	0.556	3635.3	299.1	10.15	254.0	42.0	2.68
	110	32.36	24.16	12.042	0.855	0.510	3315.0	274.4	10.12	229.1	38.0	2.66
	100	29.43	24.00	12.000	0.775	0.468	2987.3	248.9	10.08	203.5	33.9	2.63
24×9	94	27.63	24.29	9.061	0.872	0.516	2683.0	220.9	9.85	102.2	22.6	1.92
	84	24.71	24.09	9.015	0.772	0.470	2364.3	196.3	9.78	88.3	19.6	1.89
	76	22.37	23.91	8.985	0.682	0.440	2096.4	175.4	9.68	76.5	17.0	1.85
21×13	142	41.76	21.46	13.132	1.095	0.659	3403.1	317.2	9.03	385.9	58.8	3.04
	127	37.34	21.24	13.061	0.985	0.588	3017.2	284.1	8.99	338.6	51.8	3.01
	112	32.93	21.00	13.000	0.865	0.527	2620.6	249.6	8.92	289.7	44.6	2.96
21×9	96	28.21	21.14	9.038	0.935	0.575	2088.9	197.6	8.60	109.3	24.2	1.97
	82	24.10	20.86	8.962	0.795	0.499	1752.4	168.0	8.53	89.6	20.0	1.93
21×8 1/4	73	21.46	21.24	8.295	0.740	0.455	1 00.3	150.7	8.64	66.2	16.0	1.76
	68	20.02	21.13	8.270	0.685	0.430	1478.3	139.9	8.59	60.4	14.6	1.74
	62	18.23	20.99	8.240	0.615	0.400	1326.8	126.4	8.53	53.1	12.9	1.71
18×11 3/4	114	33.51	18.48	11.833	0.991	0.595	2033.8	220.1	7.79	255.6	43.2	2.76
	105	30.86	18.32	11.792	0.911	0.554	1852.5	202.2	7.75	231.0	39.2	2.73
	96	28.22	18.16	11.750	0.831	0.512	1674.7	184.4	7.70	206.8	35.2	2.71
18×8 3/4	85	24.97	18.32	8.838	0.911	0.526	1429.9	156.1	7.57	99.4	22.5	2.00
	77	22.63	18.16	8.787	0.831	0.475	1286.8	141.7	7.54	88.6	20.2	1.98
	70	20.56	18.00	8.750	0.751	0.438	1153.9	128.2	7.49	78.5	17.9	1.95
	64	18.80	17.87	8.715	0.686	0.403	1045.8	117.0	7.46	70.3	16.1	1.93
18×7 1/2	60	17.64	18.25	7.558	0.695	0.416	984.0	107.8	7.47	47.1	12.5	1.63
	55	16.19	18.12	7.532	0.630	0.390	889.9	98.2	7.41	42.0	11.1	1.61
	50	14.71	18.00	7.500	0.570	0.358	800.6	89.0	7.38	37.2	9.9	1.59
16×11 1/2	96	28.22	16.32	11.533	0.875	0.535	1355.1	166.1	6.93	207.2	35.9	2.71
	88	25.87	16.16	11.502	0.795	0.504	1222.6	151.3	6.87	185.2	32.2	2.67
16×8 1/2	78	22.92	16.32	8.586	0.875	0.529	1042.6	127.8	6.74	87.5	20.4	1.95
	71	20.86	16.16	8.543	0.795	0.486	936.9	115.9	6.70	77.9	18.2	1.93
	64	18.80	16.00	8.500	0.715	0.443	833.8	104.2	6.66	68.4	16.1	1.91
	58	17.04	15.86	8.464	0.645	0.407	746.4	94.1	6.62	60.5	14.3	1.88
16×7	50	14.70	16.25	7.073	0.628	0.380	655.4	80.7	6.68	34.8	9.8	1.54
	45	13.24	16.12	7.039	0.563	0.346	583.3	72.4	6.64	30.5	8.7	1.52
	40	11.77	16.00	7.000	0.503	0.307	515.5	64.4	6.62	26.5	7.6	1.50
	36	10.59	15.85	6.992	0.428	0.299	446.3	56.3	6.49	22.1	6.3	1.45
14×16	426	125.25	18.69	16.695	3.033	1.875	6610.3	707.4	7.26	2359.5	282.7	4.34
	398	116.98	18.31	16.590	2.843	1.770	6013.7	656.9	7.17	2169.7	261.6	4.31
	370	108.78	17.94	16.475	2.658	1.655	5454.2	608.1	7.08	1986.0	241.1	4.27
	342	100.59	17.56	16.365	2.468	1.545	4911.5	559.4	6.99	1806.9	220.8	4.24
	314	92.30	17.19	16.235	2.283	1.415	4399.4	511.9	6.90	1631.4	201.0	4.20
	287	84.37	16.81	16.130	2.093	1.310	3912.1	465.5	6.81	1466.5	181.8	4.17
	264	77.63	16.50	16.025	1.938	1.205	3526.0	427.4	6.74	1331.2	166.1	4.14
	246	72.33	16.25	15.945	1.813	1.125	3228.9	397.4	6.68	1226.6	153.9	4.12
	237	69.69	16.12	15.910	1.748	1.090	3080.9	382.2	6.65	1174.8	147.7	4.11
	228	67.06	16.00	15.865	1.688	1.045	2942.4	367.8	6.62	1124.8	141.8	4.10
	219	64.36	15.87	15.825	1.623	1.005	2798.2	352.6	6.59	1073.2	135.6	4.08
	211	62.07	15.75	15.800	1.563	0.980	2671.4	339.2	6.56	1028.6	130.2	4.07
	202	59.39	15.63	15.750	1.503	0.930	2538.8	324.9	6.54	979.7	124.4	4.06
	193	56.73	15.50	15.710	1.438	0.890	2402.4	310.0	6.51	930.1	118.4	4.05
	184	54.07	15.38	15.660	1.378	0.840	2274.8	295.8	6.49	882.7	112.7	4.04
	176	51.73	15.25	15.640	1.313	0.820	2149.6	281.9	6.45	837.9	107.1	4.02
	167	49.09	15.12	15.600	1.248	0.780	2020.8	267.3	6.42	790.2	101.3	4.01
	158	46.47	15.00	15.550	1.188	0.730	1900.6	253.4	6.40	745.0	95.8	4.00
	150	44.08	14.88	15.515	1.128	0.695	1786.9	240.2	6.37	702.5	90.6	3.99
	142	41.85	14.75	15.500	1.063	0.680	1672.2	226.7	6.32	660.1	85.2	3.97
	320 *	94.12	16.81	16.710	2.093	1.890	4141.7	492.8	6.63	1635.1	195.7	4.17
14×14 1/2	136	39.98	14.75	14.740	1.063	0.660	1593.0	216.0	6.31	567.7	77.0	3.77
	127	37.33	14.62	14.690	0.998	0.610	1476.7	202.0	6.29	527.6	71.8	3.76
	119	34.99	14.50	14.650	0.938	0.570	1373.1	189.4	6.26	491.8	67.1	3.75
	111	32.65	14.37	14.620	0.873	0.540	1266.5	176.3	6.23	454.9	62.2	3.73
	103	30.26	14.25	14.575	0.813	0.495	1165.8	163.6	6.21	419.7	57.6	3.72
	95	27.94	14.12	14.545	0.748	0.465	1063.5	150.6	6.17	383.7	52.8	3.71
	87	25.56	14.00	14.500	0.688	0.420	966.9	138.1	6.15	349.7	48.2	3.70

* Column core section.

Table 70. Properties of Wide Flange Sections—*Continued*

Nominal Size	Weight per Foot	Area	Depth	Flange Width	Flange Thickness	Web Thickness	Axis X-X I	Axis X-X S	Axis X-X r	Axis Y-Y I	Axis Y-Y S	Axis Y-Y r
in.	lb	in.^{2}	in.	in.	in.	in.	in.^{4}	in.^{3}	in.	in.^{4}	in.^{3}	in.
14×12	84	24.71	14.18	12.023	0.778	0.451	928.4	130.9	6.13	225.5	37.5	3.02
	78	22.94	14.06	12.000	0.718	0.428	851.2	121.1	6.09	206.9	34.5	3.00
14×10	74	21.76	14.19	10.072	0.783	0.450	796.8	112.3	6.05	133.5	26.5	2.48
	68	20.00	14.06	10.040	0.718	0.418	724.1	103.0	6.02	121.2	24.1	2.46
	61	17.94	13.91	10.000	0.643	0.378	641.5	92.2	5.98	107.3	21.5	2.45
14×8	53	15.59	13.94	8.062	0.658	0.370	542.1	77.8	5.90	57.5	14.3	1.92
	48	14.11	13.81	8.031	0.593	0.339	484.9	70.2	5.86	51.3	12.8	1.91
	43	12.65	13.68	8.000	0.528	0.308	429.0	62.7	5.82	45.1	11.3	1.89
14×6 3/4	38	11.17	14.12	6.776	0.513	0.313	385.3	54.6	5.87	24.6	7.3	1.49
	34	10.00	14.00	6.750	0.453	0.287	339.2	48.5	5.83	21.3	6.3	1.46
	30	8.81	13.86	6.733	0.383	0.270	289.6	41.8	5.73	17.5	5.2	1.41
12×12	190	55.86	14.38	12.670	1.736	1.060	1892.5	263.2	5.82	589.7	93.1	3.25
	161	47.38	13.88	12.515	1.486	0.905	1541.8	222.2	5.70	486.2	77.7	3.20
	133	39.11	13.38	12.365	1.236	0.755	1221.2	182.5	5.59	389.9	63.1	3.16
	120	35.31	13.12	12.320	1.106	0.710	1071.7	163.4	5.51	345.1	56.0	3.13
	106	31.19	12.88	12.230	0.986	0.620	930.7	144.5	5.46	300.9	49.2	3.11
	99	29.09	12.75	12.190	0.921	0.580	858.5	134.7	5.43	278.2	45.7	3.09
	92	27.06	12.62	12.155	0.856	0.545	788.9	125.0	5.40	256.4	42.2	3.08
	85	24.98	12.50	12.105	0.796	0.495	723.3	115.7	5.38	235.5	38.9	3.07
	79	23.22	12.38	12.080	0.736	0.470	663.0	107.1	5.34	216.4	35.8	3.05
	72	21.16	12.25	12.040	0.671	0.430	597.4	97.5	5.31	195.3	32.4	3.04
	65	19.11	12.12	12.000	0.606	0.390	533.4	88.0	5.28	174.6	29.1	3.02
12×10	58	17.06	12.19	10.014	0.641	0.359	476.1	78.1	5.28	107.4	21.4	2.51
	53	15.59	12.06	10.000	0.576	0.345	426.2	70.7	5.23	96.1	19.2	2.48
12×8	50	14.71	12.19	8.077	0.641	0.371	394.5	64.7	5.18	56.4	14.0	1.96
	45	13.24	12.06	8.042	0.576	0.336	350.8	58.2	5.15	50.0	12.4	1.94
	40	11.77	11.94	8.000	0.516	0.294	310.1	51.9	5.13	44.1	11.0	1.94
12×6 1/2	36	10.59	12.24	6.565	0.540	0.305	280.8	45.9	5.15	23.7	7.2	1.50
	31	9.12	12.09	6.525	0.465	0.265	238.4	39.4	5.11	19.8	6.1	1.47
	27	7.97	11.95	6.500	0.400	0.240	204.1	34.1	5.06	16.6	5.1	1.44
10×10	112	32.92	11.38	10.415	1.248	0.755	718.7	126.3	4.67	235.4	45.2	2.67
	100	29.43	11.12	10.345	1.118	0.685	625.0	112.4	4.61	206.6	39.9	2.65
	89	26.19	10.88	10.275	0.998	0.615	542.4	99.7	4.55	180.6	35.2	2.63
	77	22.67	10.62	10.195	0.868	0.535	457.2	86.1	4.49	153.4	30.1	2.60
	72	21.18	10.50	10.170	0.808	0.510	420.7	80.1	4.46	141.8	27.9	2.59
	66	19.41	10.38	10.117	0.748	0.457	382.5	73.7	4.44	129.2	25.5	2.58
	60	17.66	10.25	10.075	0.683	0.415	343.7	67.1	4.41	116.5	23.1	2.57
	54	15.88	10.12	10.028	0.618	0.368	305.7	60.4	4.39	103.9	20.7	2.56
	49	14.40	10.00	10.000	0.558	0.340	272.9	54.6	4.35	93.0	18.6	2.54
10×8	45	13.24	10.12	8.022	0.618	0.350	248.6	49.1	4.33	53.2	13.3	2.00
	39	11.48	9.94	7.990	0.528	0.318	209.7	42.2	4.27	44.9	11.2	1.98
	33	9.71	9.75	7.964	0.433	0.292	170.9	35.0	4.20	36.5	9.2	1.94
10×5 3/4	29	8.53	10.22	5.799	0.500	0.289	157.3	30.8	4.29	15.2	5.2	1.34
	25	7.35	10.08	5.762	0.430	0.252	133.2	26.4	4.26	12.7	4.4	1.31
	21	6.19	9.90	5.750	0.340	0.240	106.3	21.5	4.14	9.7	3.4	1.25
8×8	67	19.70	9.00	8.287	0.933	0.575	271.8	60.4	3.71	88.6	21.4	2.12
	58	17.06	8.75	8.222	0.808	0.510	227.3	52.0	3.65	74.9	18.2	2.10
	48	14.11	8.50	8.117	0.683	0.405	183.7	43.2	3.61	60.9	15.0	2.08
	40	11.76	8.25	8.077	0.558	0.365	146.3	35.5	3.53	49.0	12.1	2.04
	35	10.30	8.12	8.027	0.493	0.315	126.5	31.1	3.50	42.5	10.6	2.03
	31	9.12	8.00	8.000	0.433	0.288	109.7	27.4	3.47	37.0	9.2	2.01
8×6 1/2	28	8.23	8.06	6.540	0.463	0.285	97.8	24.3	3.45	21.6	6.6	1.62
	24	7.06	7.93	6.500	0.398	0.245	82.5	20.8	3.42	18.2	5.6	1.61
8×5 1/4	20	5.88	8.14	5.268	0.378	0.248	69.2	17.0	3.43	8.5	3.2	1.20
	17	5.00	8.00	5.250	0.308	0.230	56.4	14.1	3.36	6.7	2.6	1.16

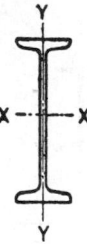

Table 71. Properties of American Standard Beams

Nominal Size	Weight per Foot	Area	Depth	Flange Width	Flange Thickness	Web Thickness	Axis X-X I	Axis X-X S	Axis X-X r	Axis Y-Y I	Axis Y-Y S	Axis Y-Y r
in.	lb	in.^{2}	in.	in.	in.	in.	in.^{4}	in.^{3}	in.	in.^{4}	in.^{3}	in.
24 × 7 7/8	120.0	35.13	24.00	8.048	1.102	0.798	3010.8	250.9	9.26	84.9	21.1	1.56
	105.9	30.98	24.00	7.875	1.102	.625	2811.5	234.3	9.53	78.9	20.0	1.60
24 × 7	100.0	29.25	24.00	7.247	0.871	.747	2371.8	197.6	9.05	48.4	13.4	1.29
	90.0	26.30	24.00	7.124	0.871	.624	2230.1	185.8	9.21	45.5	12.8	1.32
	79.9	23.33	24.00	7.000	0.871	.500	2087.2	173.9	9.46	42.9	12.2	1.36
20 × 7	95.0	27.74	20.00	7.200	0.916	.800	1599.7	160.0	7.59	50.5	14.0	1.35
	85.0	24.80	20.00	7.053	0.916	.653	1501.7	150.2	7.78	47.0	13.3	1.38
20 × 6 1/4	75.0	21.90	20.00	6.391	0.789	.641	1263.5	126.3	7.60	30.1	9.4	1.17
	65.4	19.08	20.00	6.250	0.789	.500	1169.5	116.9	7.83	27.9	8.9	1.21
18 × 6	70.0	20.46	18.00	6.251	0.691	.711	917.5	101.9	6.70	24.5	7.8	1.09
	54.7	15.94	18.00	6.000	0.691	.460	795.5	88.4	7.07	21.2	7.1	1.15
15 × 5 1/2	50.0	14.59	15.00	5.640	0.622	.550	481.1	64.2	5.74	16.0	5.7	1.05
	42.9	12.49	15.00	5.500	0.622	.410	441.8	58.9	5.95	14.6	5.3	1.08
12 × 5 1/4	50.0	14.57	12.00	5.477	0.659	.687	301.6	50.3	4.55	16.0	5.8	1.05
	40.8	11.84	12.00	5.250	0.659	.460	268.9	44.8	4.77	13.8	5.3	1.08
12 × 5	35.0	10.20	12.00	5.078	0.544	.428	227.0	37.8	4.72	10.0	3.9	0.99
	31.8	9.26	12.00	5.000	0.544	.350	215.8	36.0	4.83	9.5	3.8	1.01
10 × 4 5/8	35.0	10.22	10.00	4.944	0.491	.594	145.8	29.2	3.78	8.5	3.4	0.91
	25.4	7.38	10.00	4.660	0.491	.310	122.1	24.4	4.07	6.9	3.0	0.97
8 × 4	23.0	6.71	8.00	4.171	0.425	.441	64.2	16.0	3.09	4.4	2.1	0.81
	18.4	5.34	8.00	4.000	0.425	.270	56.9	14.2	3.26	3.8	1.9	0.84
7 × 3 5/8	20.0	5.83	7.00	3.860	0.392	.450	41.9	12.0	2.68	3.1	1.6	0.74
	15.3	4.43	7.00	3.660	0.392	.250	36.2	10.4	2.86	2.7	1.5	0.78
6 × 3 3/8	17.25	5.02	6.00	3.565	0.359	.465	26.0	8.7	2.28	2.3	1.3	0.68
	12.5	3.61	6.00	3.330	0.359	.230	21.8	7.3	2.46	1.8	1.1	0.72
5 × 3	14.75	4.29	5.00	3.284	0.326	.494	15.0	6.0	1.87	1.7	1.0	0.63
	10.0	2.87	5.00	3.000	0.326	.210	12.1	4.8	2.05	1.2	0.82	0.65
4 × 2 5/8	9.5	2.76	4.00	2.796	0.293	.326	6.7	3.3	1.56	0.91	0.65	0.58
	7.7	2.21	4.00	2.660	0.293	.190	6.0	3.0	1.64	0.77	0.58	0.59
3 × 2 3/8	7.5	2.17	3.00	2.509	0.260	.349	2.9	1.9	1.15	0.59	0.47	0.52
	5.7	1.64	3.00	2.330	0.260	.170	2.5	1.7	1.23	0.46	0.40	0.53

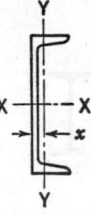

Table 72. Properties of American Standard Channels

Nominal Size	Weight per Foot	Area	Depth	Flange		Web Thickness	Axis X-X			Axis Y-Y			
				Width	Average Thickness		I	S	r	I	S	r	x
in.	lb	in.^{2}	in.	in.	in.	in.	in.^{4}	in.^{3}	in.	in.^{4}	in.^{3}	in.	in.
18×4 *	58.0	16.98	18.00	4.200	0.625	0.700	670.7	74.5	6.29	18.5	5.6	1.04	0.88
	51.9	15.18	18.00	4.100	.625	.600	622.1	69.1	6.40	17.1	5.3	1.06	.87
	45.8	13.38	18.00	4.000	.625	.500	573.5	63.7	6.55	15.8	5.1	1.09	.89
	42.7	12.48	18.00	3.950	.625	.450	549.2	61.0	6.64	15.0	4.9	1.10	.90
$15 \times 3\,3/8$	50.0	14.64	15.00	3.716	.650	.716	401.4	53.6	5.24	11.2	3.8	0.87	.80
	40.0	11.70	15.00	3.520	.650	.520	346.3	46.2	5.44	9.3	3.4	0.89	.78
	33.9	9.90	15.00	3.400	.650	.400	312.6	41.7	5.62	8.2	3.2	0.91	.79
12×3	30.0	8.79	12.00	3.170	.501	.510	161.2	26.9	4.28	5.2	2.1	0.77	.68
	25.0	7.32	12.00	3.047	.501	.387	143.5	23.9	4.43	4.5	1.9	0.79	.68
	20.7	6.03	12.00	2.940	.501	.280	128.1	21.4	4.61	3.9	1.7	0.81	.70
$10 \times 2\,5/8$	30.0	8.80	10.00	3.033	.436	.673	103.0	20.6	3.42	4.0	1.7	0.67	.65
	25.0	7.33	10.00	2.886	.436	.526	90.7	18.1	3.52	3.4	1.5	0.68	.62
	20.0	5.86	10.00	2.739	.436	.379	78.5	15.7	3.66	2.8	1.3	0.70	.61
	15.3	4.47	10.00	2.600	.436	.240	66.9	13.4	3.87	2.3	1.2	0.72	.64
$9 \times 2\,1/2$	20.0	5.86	9.00	2.648	.413	.448	60.6	13.5	3.22	2.4	1.2	0.65	.59
	15.0	4.39	9.00	2.485	.413	.285	50.7	11.3	3.40	1.9	1.0	0.67	.59
	13.4	3.89	9.00	2.430	.413	.230	47.3	10.5	3.49	1.8	0.97	0.67	.61
$8 \times 2\,1/4$	18.75	5.49	8.00	2.527	.390	.487	43.7	10.9	2.82	2.0	1.0	0.60	.57
	13.75	4.02	8.00	2.343	.390	.303	35.8	9.0	2.99	1.5	0.86	0.62	.56
	11.5	3.36	8.00	2.260	.390	.220	32.3	8.1	3.10	1.3	0.79	0.63	.58
$7 \times 2\,1/8$	14.75	4.32	7.00	2.299	.366	.419	27.1	7.7	2.51	1.4	0.79	0.57	.53
	12.25	3.58	7.00	2.194	.366	.314	24.1	6.9	2.59	1.2	0.71	0.58	.53
	9.8	2.85	7.00	2.090	.366	.210	21.1	6.0	2.72	0.98	0.63	0.59	.55
6×2	13.0	3.81	6.00	2.157	.343	.437	17.3	5.8	2.13	1.1	0.65	0.53	.52
	10.5	3.07	6.00	2.034	.343	.314	15.1	5.0	2.22	0.87	0.57	0.53	.50
	8.2	2.39	6.00	1.920	.343	.200	13.0	4.3	2.34	0.70	0.50	0.54	.52
$5 \times 1\,3/4$	9.0	2.63	5.00	1.885	.320	.325	8.8	3.5	1.83	0.64	0.45	0.49	.48
	6.7	1.95	5.00	1.750	.320	.190	7.4	3.0	1.95	0.48	0.38	0.50	.49
$4 \times 1\,5/8$	7.25	2.12	4.00	1.720	.296	.320	4.5	2.3	1.47	0.44	0.35	0.46	.46
	5.4	1.56	4.00	1.580	.296	.180	3.8	1.9	1.56	0.32	0.29	0.45	.46
$3 \times 1\,1/2$	6.0	1.75	3.00	1.596	.273	.356	2.1	1.4	1.08	0.31	0.27	0.42	.46
	5.0	1.46	3.00	1.498	.273	.258	1.8	1.2	1.12	0.25	0.24	0.41	.44
	4.1	1.19	3.00	1.410	.273	.170	1.6	1.1	1.17	0.20	0.21	0.41	.44

* Car and Shipbuilding Channel; not an American Standard.

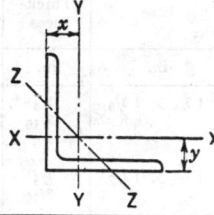

Table 73. Properties of Angles with Equal Legs

Size	Thickness	Weight per Foot	Area	Axis X–X and Axis Y–Y				Axis Z–Z
				I	S	r	x or y	r
in.	in.	lb	in.^{2}	in.^{4}	in.^{3}	in.	in.	in.
8 × 8	1 1/8	56.9	16.73	98.0	17.5	2.42	2.41	1.56
	1	51.0	15.00	89.0	15.8	2.44	2.37	1.56
	7/8	45.0	13.23	79.6	14.0	2.45	2.32	1.57
	3/4	38.9	11.44	69.7	12.2	2.47	2.28	1.57
	5/8	32.7	9.61	59.4	10.3	2.49	2.23	1.58
	9/16	29.6	8.68	54.1	9.3	2.50	2.21	1.58
	1/2	26.4	7.75	48.6	8.4	2.50	2.19	1.59
6 × 6	1	37.4	11.00	35.5	8.6	1.80	1.86	1.17
	7/8	33.1	9.73	31.9	7.6	1.81	1.82	1.17
	3/4	28.7	8.44	28.2	6.7	1.83	1.78	1.17
	5/8	24.2	7.11	24.2	5.7	1.84	1.73	1.18
	9/16	21.9	6.43	22.1	5.1	1.85	1.71	1.18
	1/2	19.6	5.75	19.9	4.6	1.86	1.68	1.18
	7/16	17.2	5.06	17.7	4.1	1.87	1.66	1.19
	3/8	14.9	4.36	15.4	3.5	1.88	1.64	1.19
	5/16	12.5	3.66	13.0	3.0	1.89	1.61	1.19
5 × 5	7/8	27.2	7.98	17.8	5.2	1.49	1.57	0.97
	3/4	23.6	6.94	15.7	4.5	1.51	1.52	0.97
	5/8	20.0	5.86	13.6	3.9	1.52	1.48	0.98
	1/2	16.2	4.75	11.3	3.2	1.54	1.43	0.98
	7/16	14.3	4.18	10.0	2.8	1.55	1.41	0.98
	3/8	12.3	3.61	8.7	2.4	1.56	1.39	0.99
	5/16	10.3	3.03	7.4	2.0	1.57	1.37	0.99
4 × 4	.3/4	18.5	5.44	7.7	2.8	1.19	1.27	0.78
	5/8	15.7	4.61	6.7	2.4	1.20	1.23	0.78
	1/2	12.8	3.75	5.6	2.0	1.22	1.18	0.78
	7/16	11.3	3.31	5.0	1.8	1.23	1.16	0.78
	3/8	9.8	2.86	4.4	1.5	1.23	1.14	0.79
	5/16	8.2	2.40	3.7	1.3	1.24	1.12	0.79
	1/4	6.6	1.94	3.0	1.1	1.25	1.09	0.80
3 1/2 × 3 1/2	1/2	11.1	3.25	3.6	1.5	1.06	1.06	0.68
	7/16	9.8	2.87	3.3	1.3	1.07	1.04	0.68
	3/8	8.5	2.48	2.9	1.2	1.07	1.01	0.69
	5/16	7.2	2.09	2.5	0.98	1.08	0.99	0.69
	1/4	5.8	1.69	2.0	0.79	1.09	0.97	0.69
3 × 3	1/2	9.4	2.75	2.2	1.1	0.90	0.93	0.58
	7/16	8.3	2.43	2.0	0.95	0.91	0.91	0.58
	3/8	7.2	2.11	1.8	0.83	0.91	0.89	0.58
	5/16	6.1	1.78	1.5	0.71	0.92	0.87	0.59
	1/4	4.9	1.44	1.2	0.58	0.93	0.84	0.59
	3/16	3.71	1.09	0.96	0.44	0.94	0.82	0.59
2 1/2 × 2 1/2	1/2	7.7	2.25	1.2	0.72	0.74	0.81	0.49
	3/8	5.9	1.73	0.98	0.57	0.75	0.76	0.49
	5/16	5.0	1.47	0.85	0.48	0.76	0.74	0.49
	1/4	4.1	1.19	0.70	0.39	0.77	0.72	0.49
	3/16	3.07	0.90	0.55	0.30	0.78	0.69	0.49
2 × 2	3/8	4.7	1.36	0.48	0.35	0.59	0.64	0.39
	5/16	3.92	1.15	0.42	0.30	0.60	0.61	0.39
	1/4	3.19	0.94	0.35	0.25	0.61	0.59	0.39
	3/16	2.44	0.71	0.27	0.19	0.62	0.57	0.39
	1/8	1.65	0.48	0.19	0.13	0.63	0.55	0.40

Table 73. Properties of Angles with Equal Legs—*Continued*

Size	Thickness	Weight per Foot	Area	Axis X–X and Axis Y–Y				Axis Z–Z
				I	S	r	x or y	r
in.	in.	lb	in.^{2}	in.^{4}	in.^{3}	in.	in.	in.
$1\,3/4 \times 1\,3/4$	1/4	2.77	0.81	0.23	0.19	0.53	0.53	0.34
	3/16	2.12	.62	.18	.14	.54	.51	.34
	1/8	1.44	.42	.13	.10	.55	.48	.35
$1\,1/2 \times 1\,1/2$	1/4	2.34	.69	.14	.13	.45	.47	.29
	3/16	1.80	.53	.11	.10	.46	.44	.29
	1/8	1.23	.36	.08	.07	.47	.42	.30
$1\,1/4 \times 1\,1/4$	1/4	1.92	.56	.08	.09	.37	.40	.24
	3/16	1.48	.43	.06	.07	.38	.38	.24
	1/8	1.01	.30	.04	.05	.38	.36	.25
1×1	1/4	1.49	.44	.04	.06	.29	.34	.20
	3/16	1.16	.34	.03	.04	.30	.32	.19
	1/8	0.80	.23	.02	.03	.30	.30	.20

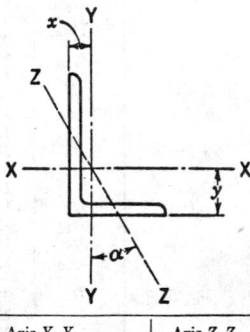

Table 74. Properties of Angles with Unequal Legs

Size	Thickness	Weight per Foot	Area	Axis X–X				Axis Y–Y				Axis Z–Z	
				I	S	r	y	I	S	r	x	r	Tan α
in.	in.	lb	in.^{2}	in.^{4}	in.^{3}	in.	in.	in.^{4}	in.^{3}	in.	in.	in.	
9×4	1	40.8	12.00	97.0	17.6	2.84	3.50	12.0	4.0	1.00	1.00	0.83	0.203
	7/8	36.1	10.61	86.8	15.7	2.86	3.45	10.8	3.6	1.01	0.95	0.84	.208
	3/4	31.3	9.19	76.1	13.6	2.88	3.41	9.6	3.1	1.02	0.91	0.84	.212
	5/8	26.3	7.73	64.9	11.5	2.90	3.36	8.3	2.6	1.04	0.86	0.85	.216
	9/16	23.8	7.00	59.1	10.4	2.91	3.33	7.6	2.4	1.04	0.83	0.85	.218
	1/2	21.3	6.25	53.2	9.3	2.92	3.31	6.9	2.2	1.05	0.81	0.85	.220
8×6	1	44.2	13.00	80.8	15.1	2.49	2.65	38.8	8.9	1.73	1.65	1.28	.543
	7/8	39.1	11.48	72.3	13.4	2.51	2.61	34.9	7.9	1.74	1.61	1.28	.547
	3/4	33.8	9.94	63.4	11.7	2.53	2.56	30.7	6.9	1.76	1.56	1.29	.551
	5/8	28.5	8.36	54.1	9.9	2.54	2.52	26.3	5.9	1.77	1.52	1.29	.554
	9/16	25.7	7.56	49.3	9.0	2.55	2.50	24.0	5.3	1.78	1.50	1.30	.556
	1/2	23.0	6.75	44.3	8.0	2.56	2.47	21.7	4.8	1.79	1.47	1.30	.558
	7/16	20.2	5.93	39.2	7.1	2.57	2.45	19.3	4.2	1.80	1.45	1.31	.560
8×4	1	37.4	11.00	69.6	14.1	2.52	3.05	11.6	3.9	1.03	1.05	0.85	.247
	7/8	33.1	9.73	62.5	12.5	2.53	3.00	10.5	3.5	1.04	1.00	0.85	.253
	3/4	28.7	8.44	54.9	10.9	2.55	2.95	9.4	3.1	1.05	0.95	0.85	.258
	5/8	24.2	7.11	46.9	9.2	2.57	2.91	8.1	2.6	1.07	0.91	0.86	.262
	9/16	21.9	6.43	42.8	8.4	2.58	2.88	7.4	2.4	1.07	0.88	0.86	.265
	1/2	19.6	5.75	38.5	7.5	2.59	2.86	6.7	2.2	1.08	0.86	0.86	.267
	7/16	17.2	5.06	34.1	6.6	2.60	2.83	6.0	1.9	1.09	0.83	0.87	.269
7×4	7/8	30.2	8.86	42.9	9.7	2.20	2.55	10.2	3.5	1.07	1.05	0.86	.318
	3/4	26.2	7.69	37.8	8.4	2.22	2.51	9.1	3.0	1.09	1.01	0.86	.324
	5/8	22.1	6.48	32.4	7.1	2.24	2.46	7.8	2.6	1.10	0.96	0.86	.329
	9/16	20.0	5.87	29.6	6.5	2.24	2.44	7.2	2.4	1.11	0.94	0.87	.332
	1/2	17.9	5.25	26.7	5.8	2.25	2.42	6.5	2.1	1.11	0.92	0.87	.335
	7/16	15.8	4.62	23.7	5.1	2.26	2.39	5.8	1.9	1.12	0.89	0.88	.337
	3/8	13.6	3.98	20.6	4.4	2.27	2.37	5.1	1.6	1.13	0.87	0.88	.339
6×4	7/8	27.2	7.98	27.7	7.2	1.86	2.12	9.8	3.4	1.11	1.12	0.86	.421
	3/4	23.6	6.94	24.5	6.3	1.88	2.08	8.7	3.0	1.12	1.08	0.86	.428
	5/8	20.0	5.86	21.1	5.3	1.90	2.03	7.5	2.5	1.13	1.03	0.86	.435
	9/16	18.1	5.31	19.3	4.8	1.90	2.01	6.9	2.3	1.14	1.01	0.87	.438
	1/2	16.2	4.75	17.4	4.3	1.91	1.99	6.3	2.1	1.15	0.99	0.87	.440
	7/16	14.3	4.18	15.5	3.8	1.92	1.96	5.6	1.9	1.16	0.96	0.87	.443
	3/8	12.3	3.61	13.5	3.3	1.93	1.94	4.9	1.6	1.17	0.94	0.88	.446
	5/16	10.3	3.03	11.4	2.8	1.94	1.92	4.2	1.4	1.17	0.92	0.88	.449

Table 74. Properties of Angles with Unequal Legs—*Continued*

Size	Thickness	Weight per Foot	Area	Axis X–X				Axis Y–Y				Axis Z–Z	
				I	S	r	y	I	S	r	x	r	Tan α
in.	in.	lb	in.^{2}	in.^{4}	in.^{3}	in.	in.	in.^{4}	in.^{3}	in.	in.	in.	
6 × 3½	1/2	15.3	4.50	16.6	4.2	1.92	2.08	4.3	1.6	0.97	0.83	0.76	0.344
	3/8	11.7	3.42	12.9	3.2	1.94	2.04	3.3	1.2	0.99	0.79	.77	.350
	5/16	9.8	2.87	10.9	2.7	1.95	2.01	2.9	1.0	1.00	0.76	.77	.352
	1/4	7.9	2.31	8.9	2.2	1.96	1.99	2.3	0.85	1.01	0.74	.78	.355
5 × 3½	3/4	19.8	5.81	13.9	4.3	1.55	1.75	5.6	2.2	0.98	1.00	.75	.464
	5/8	16.8	4.92	12.0	3.7	1.56	1.70	4.8	1.9	0.99	0.95	.75	.472
	1/2	13.6	4.00	10.0	3.0	1.58	1.66	4.1	1.6	1.01	0.91	.75	.479
	7/16	12.0	3.53	8.9	2.6	1.59	1.63	3.6	1.4	1.01	0.88	.76	.482
	3/8	10.4	3.05	7.8	2.3	1.60	1.61	3.2	1.2	1.02	0.86	.76	.486
	5/16	8.7	2.56	6.6	1.9	1.61	1.59	2.7	1.0	1.03	0.84	.76	.489
	1/4	7.0	2.06	5.4	1.6	1.61	1.56	2.2	0.83	1.04	0.81	.76	.492
5 × 3	1/2	12.8	3.75	9.5	2.9	1.59	1.75	2.6	1.1	0.83	0.75	.65	.357
	7/16	11.3	3.31	8.4	2.6	1.60	1.73	2.3	1.0	0.84	0.73	.65	.361
	3/8	9.8	2.86	7.4	2.2	1.61	1.70	2.0	0.89	0.84	0.70	.65	.364
	5/16	8.2	2.40	6.3	1.9	1.61	1.68	1.8	0.75	0.85	0.68	.66	.368
	1/4	6.6	1.94	5.1	1.5	1.62	1.66	1.4	0.61	0.86	0.66	.66	.371
4 × 3½	5/8	14.7	4.30	6.4	2.4	1.22	1.29	4.5	1.8	1.03	1.04	.72	.745
	1/2	11.9	3.50	5.3	1.9	1.23	1.25	3.8	1.5	1.04	1.00	.72	.750
	7/16	10.6	3.09	4.8	1.7	1.24	1.23	3.4	1.4	1.05	0.98	.72	.753
	3/8	9.1	2.67	4.2	1.5	1.25	1.21	3.0	1.2	1.06	0.96	.73	.755
	5/16	7.7	2.25	3.6	1.3	1.26	1.18	2.6	1.0	1.07	0.93	.73	.757
	1/4	6.2	1.81	2.9	1.0	1.27	1.16	2.1	0.81	1.07	0.91	.73	.759
4 × 3	5/8	13.6	3.98	6.0	2.3	1.23	1.37	2.9	1.4	0.85	0.87	.64	.534
	1/2	11.1	3.25	5.1	1.9	1.25	1.33	2.4	1.1	0.86	0.83	.64	.543
	7/16	9.8	2.87	4.5	1.7	1.25	1.30	2.2	1.0	0.87	0.80	.64	.547
	3/8	8.5	2.48	4.0	1.5	1.26	1.28	1.9	0.87	0.88	0.78	.64	.551
	5/16	7.2	2.09	3.4	1.2	1.27	1.26	1.7	0.73	0.89	0.76	.65	.554
	1/4	5.8	1.69	2.8	1.0	1.28	1.24	1.4	0.60	0.90	0.74	.65	.558
3½ × 3	1/2	10.2	3.00	3.5	1.5	1.07	1.13	2.3	1.1	0.88	0.88	.62	.714
	7/16	9.1	2.65	3.1	1.3	1.08	1.10	2.1	0.98	0.89	0.85	.62	.718
	3/8	7.9	2.30	2.7	1.1	1.09	1.08	1.9	0.85	0.90	0.83	.62	.721
	5/16	6.6	1.93	2.3	0.95	1.10	1.06	1.6	0.72	0.90	0.81	.63	.724
	1/4	5.4	1.56	1.9	0.78	1.11	1.04	1.3	0.59	0.91	0.79	.63	.727
3½ × 2½	1/2	9.4	2.75	3.2	1.4	1.09	1.20	1.4	0.76	0.70	0.70	.53	.486
	7/16	8.3	2.43	2.9	1.3	1.09	1.18	1.2	0.68	0.71	0.68	.54	.491
	3/8	7.2	2.11	2.6	1.1	1.10	1.16	1.1	0.59	0.72	0.66	.54	.496
	5/16	6.1	1.78	2.2	0.93	1.11	1.14	0.94	0.50	0.73	0.64	.54	.501
	1/4	4.9	1.44	1.8	0.75	1.12	1.11	0.78	0.41	0.74	0.61	.54	.506
3 × 2½	1/2	8.5	2.50	2.1	1.0	0.91	1.00	1.3	0.74	0.72	0.75	.52	.667
	7/16	7.6	2.21	1.9	0.93	0.92	0.98	1.2	0.66	0.73	0.73	.52	.672
	3/8	6.6	1.92	1.7	0.81	0.93	0.96	1.0	0.58	0.74	0.71	.52	.676
	5/16	5.6	1.62	1.4	0.69	0.94	0.93	0.90	0.49	0.74	0.68	.53	.680
	1/4	4.5	1.31	1.2	0.56	0.95	0.91	0.74	0.40	0.75	0.66	.53	.684
3 × 2	1/2	7.7	2.25	1.9	1.0	0.92	1.08	0.67	0.47	0.55	0.58	.43	.414
	7/16	6.8	2.00	1.7	0.89	0.93	1.06	0.61	0.42	0.55	0.56	.43	.421
	3/8	5.9	1.73	1.5	0.78	0.94	1.04	0.54	0.37	0.56	0.54	.43	.428
	5/16	5.0	1.47	1.3	0.66	0.95	1.02	0.47	0.32	0.57	0.52	.43	.435
	1/4	4.1	1.19	1.1	0.54	0.95	0.99	0.39	0.26	0.57	0.49	.43	.440
	3/16	3.07	0.90	0.84	0.41	0.97	0.97	0.31	0.20	0.58	0.47	.44	.446
2½ × 2	3/8	5.3	1.55	0.91	0.55	0.77	0.83	0.51	0.36	0.58	0.58	.42	.614
	5/16	4.5	1.31	0.79	0.47	0.78	0.81	0.45	0.31	0.58	0.56	.42	.620
	1/4	3.62	1.06	0.65	0.38	0.78	0.79	0.37	0.25	0.59	0.54	.42	.626
	3/16	2.75	0.81	0.51	0.29	0.79	0.76	0.29	0.20	0.60	0.51	.43	.631
2½ × 1½	3/8	4.7	1.36	0.82	0.52	0.78	0.92	0.22	0.20	0.40	0.42	.32	.340
	5/16	3.92	1.15	0.71	0.44	0.79	0.90	0.19	0.17	0.41	0.40	.32	.349
	1/4	3.19	0.94	0.59	0.36	0.79	0.88	0.16	0.14	0.41	0.38	.32	.357
	3/16	2.44	0.72	0.46	0.28	0.80	0.85	0.13	0.11	0.42	0.35	.33	.364
2 × 1½	1/4	2.77	0.81	0.32	0.24	0.62	0.66	0.15	0.14	0.43	0.41	.32	.543
	3/16	2.12	0.62	0.25	0.18	0.63	0.64	0.12	0.11	0.44	0.39	.32	.551
	1/8	1.44	0.42	0.17	0.13	0.64	0.62	0.09	0.08	0.45	0.37	.33	.558
1¾ × 1¼	1/4	2.34	0.69	0.20	0.18	0.54	0.60	0.09	0.10	0.35	0.35	.27	.486
	3/16	1.80	0.53	0.16	0.14	0.55	0.58	0.07	0.08	0.36	0.33	.27	.496
	1/8	1.23	0.36	0.11	0.09	0.56	0.56	0.05	0.05	0.37	0.31	.27	.506

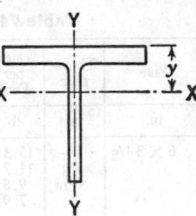

Table 75. Properties and Dimensions of Tees

Section Number	Weight per Foot	Area	Depth of Tee	Flange Width	Flange Average Thickness	Stem Thickness	Axis X–X I	S	r	y	Axis Y–Y I	S	r
	lb	in.²	in.	in.	in.	in.	in.⁴	in.³	in.	in.	in.⁴	in.³	in.
ST 18 Wᶠ *	150	44.09	18.36	16.655	1.680	0.945	1222.7	85.9	5.27	4.13	612.6	73.6	3.73
	140	41.16	18.25	16.595	1.570	.885	1133.3	79.9	5.25	4.07	563.7	67.9	3.70
	130	38.28	18.12	16.555	1.440	.845	1059.2	75.4	5.26	4.07	510.3	61.6	3.65
	122.5	36.01	18.03	16.512	1.350	.802	994.3	71.1	5.25	4.04	472.3	57.2	3.62
	115	33.86	17.94	16.475	1.260	.765	935.8	67.2	5.26	4.02	435.5	52.9	3.59
ST 18 Wᶠ	97	28.56	18.24	12.117	1.260	.770	904.0	67.3	5.63	4.81	177.7	29.3	2.49
	91	26.77	18.16	12.072	1.180	.725	844.0	63.0	5.61	4.77	163.9	27.1	2.47
	85	24.99	18.08	12.027	1.100	.680	784.7	58.8	5.60	4.74	150.3	25.0	2.45
	80	23.54	18.00	12.000	1.020	.653	741.0	56.0	5.61	4.76	137.7	22.9	2.42
	75	22.08	17.92	11.972	0.940	.625	696.7	53.0	5.62	4.79	125.2	20.9	2.38
ST 16 Wᶠ	120	35.26	16.75	15.865	1.400	.830	822.5	63.2	4.83	3.73	437.2	55.1	3.52
	110	32.36	16.63	15.810	1.275	.775	754.1	58.4	4.83	3.71	391.2	49.5	3.48
	100	29.40	16.50	15.750	1.150	.715	683.6	53.3	4.82	3.67	345.8	43.9	3.43
ST 16 Wᶠ	76	22.35	16.75	11.565	1.055	.635	591.9	47.4	5.15	4.26	128.1	22.1	2.39
	70.5	20.76	16.66	11.535	0.960	.603	551.8	44.7	5.16	4.30	114.9	19.9	2.35
	65	19.13	16.55	11.510	0.855	.580	513.0	42.1	5.18	4.37	100.7	17.5	2.29
ST 15 Wᶠ	105	30.89	15.19	15.105	1.315	.775	578.0	48.7	4.33	3.31	354.0	46.9	3.38
	95	27.95	15.06	15.040	1.185	.710	520.4	44.1	4.31	3.26	312.3	41.5	3.34
	86	25.32	14.94	14.985	1.065	.655	471.0	40.2	4.31	3.23	275.1	36.7	3.30
ST 15 Wᶠ	66	19.41	15.15	10.551	1.000	.615	420.7	37.4	4.66	3.90	92.5	17.5	2.18
	62	18.22	15.08	10.521	0.930	.585	394.8	35.3	4.65	3.90	84.8	16.1	2.16
	58.0	17.07	15.00	10.500	0.850	.564	371.8	33.6	4.67	3.94	76.6	14.6	2.12
	54.0	15.88	14.91	10.484	0.760	.548	349.5	32.1	4.69	4.03	67.6	12.9	2.06
ST 13 Wᶠ	88.5	26.05	13.66	14.090	1.190	.725	391.8	36.7	3.88	2.97	259.4	36.8	3.16
	80	23.72	13.54	14.023	1.075	.658	351.4	33.1	3.87	2.91	229.0	32.7	3.12
	72.5	21.34	13.44	13.965	0.975	.600	316.3	29.9	3.85	2.85	203.5	29.1	3.09
ST 13 Wᶠ	57	16.77	13.64	10.070	0.932	.570	288.9	28.3	4.15	3.42	74.8	14.9	2.11
	51	15.01	13.53	10.018	0.827	.518	257.7	25.4	4.14	3.39	64.8	12.9	2.08
	47	13.83	13.45	9.990	0.747	.490	238.5	23.7	4.15	3.41	57.5	11.5	2.04
ST 12 Wᶠ	80	23.54	12.36	14.091	1.135	.656	271.6	27.6	3.40	2.51	246.3	35.0	3.23
	72.5	21.31	12.24	14.043	1.020	.608	246.2	25.2	3.40	2.48	217.1	30.9	3.19
	65	19.11	12.13	14.000	0.900	.565	222.6	23.1	3.41	2.47	187.6	26.8	3.13
ST 12 Wᶠ	60	17.64	12.16	12.088	0.930	.556	213.6	22.4	3.48	2.62	127.0	21.0	2.68
	55	16.18	12.08	12.042	0.855	.510	195.2	20.5	3.47	2.57	114.5	19.0	2.66
	50	14.71	12.00	12.000	0.775	.468	176.7	18.7	3.46	2.54	101.8	17.0	2.63
ST 12 Wᶠ	47	13.81	12.15	9.061	0.872	.516	185.9	20.3	3.67	2.99	51.1	11.3	1.92
	42	12.35	12.04	9.015	0.772	.470	165.9	18.3	3.66	2.97	44.2	9.8	1.89
	38	11.18	11.95	8.985	0.682	.440	151.1	16.9	3.68	3.00	38.3	8.5	1.85
ST 10 Wᶠ	71	20.88	10.73	13.132	1.095	.659	177.3	20.8	2.91	2.18	193.0	29.4	3.04
	63.5	18.67	10.62	13.061	0.985	.588	155.8	18.3	2.89	2.11	169.3	25.9	3.01
	56	16.47	10.50	13.000	0.865	.527	136.4	16.2	2.88	2.06	144.8	22.3	2.96

* Wᶠ indicates structural tee cut from wide-flange section.

Table 75. Properties and Dimensions of Tees—*Continued*

Section Number	Weight per Foot	Area	Depth of Tee	Flange Width	Flange Average Thickness	Stem Thickness	Axis X-X I	S	r	y	Axis Y-Y I	S	r
	lb	in.^{2}	in.	in.	in.	in.	in.^{4}	in.^{3}	in.	in.	in.^{4}	in.^{3}	in.
ST 10 WF *	48	14.11	10.57	9.038	0.935	0.575	137.1	17.1	3.11	2.55	54.7	12.1	1.97
	41	12.05	10.43	8.962	0.795	.499	115.4	14.5	3.09	2.48	44.8	10.0	1.93
ST 10 WF	36.5	10.73	10.62	8.295	0.740	.455	110.2	13.7	3.21	2.60	33.1	7.98	1.76
	34	10.01	10.57	8.270	0.685	.430	102.8	12.9	3.20	2.59	30.2	7.30	1.74
	31	9.12	10.49	8.240	0.615	.400	93.7	11.9	3.21	2.59	26.6	6.45	1.71
ST 9 WF	57	16.77	9.24	11.833	0.991	.595	102.6	13.9	2.47	1.85	127.8	21.6	2.76
	52.5	15.43	9.16	11.792	0.911	.554	93.9	12.8	2.47	1.82	115.5	19.6	2.73
	48	14.11	9.08	11.750	0.831	.512	85.3	11.7	2.46	1.78	103.4	17.6	2.71
ST 9 WF	42.5	12.49	9.16	8.838	0.911	.526	84.4	11.9	2.60	2.05	49.7	11.3	2.00
	38.5	11.32	9.08	8.787	0.831	.475	75.3	10.6	2.58	1.99	44.3	10.1	1.98
	35	10.28	9.00	8.750	0.751	.438	68.1	9.67	2.57	1.96	39.2	8.97	1.95
	32	9.40	8.94	8.715	0.686	.403	61.8	8.82	2.56	1.93	35.2	8.07	1.93
ST 9 WF	30	8.82	9.12	7.558	0.695	.416	64.8	9.32	2.71	2.17	23.5	6.23	1.63
	27.5	8.09	9.06	7.532	0.630	.390	59.6	8.63	2.71	2.16	21.0	5.57	1.61
	25	7.35	9.00	7.500	0.570	.358	53.9	7.85	2.71	2.14	18.6	4.96	1.59
ST 8 WF	48	14.11	8.16	11.533	0.875	.535	64.7	9.82	2.14	1.57	103.6	18.0	2.71
	44	12.94	8.08	11.502	0.795	.504	59.5	9.11	2.14	1.55	92.6	16.1	2.67
ST 8 WF	39	11.46	8.16	8.586	0.875	.529	60.0	9.45	2.28	1.81	43.8	10.2	1.95
	35.5	10.43	8.08	8.543	0.795	.486	54.0	8.57	2.28	1.77	38.9	9.11	1.93
	32	9.40	8.00	8.500	0.715	.443	48.3	7.71	2.27	1.73	34.2	8.05	1.91
	29	8.52	7.93	8.464	0.645	.407	43.6	7.00	2.26	1.70	30.2	7.14	1.88
ST 8 WF	25	7.35	8.13	7.073	0.628	.380	42.2	6.77	2.40	1.89	17.4	4.92	1.54
	22.5	6.62	8.06	7.039	0.563	.346	37.8	6.10	2.39	1.87	15.2	4.33	1.52
	20	5.88	8.00	7.000	0.503	.307	33.2	5.37	2.37	1.82	13.3	3.79	1.50
	18	5.30	7.93	6.992	0.428	.299	30.7	5.10	2.41	1.90	11.1	3.17	1.45
ST 7 WF	105.5	31.04	7.88	15.800	1.563	.980	102.2	16.2	1.81	1.57	514.3	65.1	4.07
	101	29.70	7.82	15.750	1.503	.930	95.7	15.2	1.80	1.53	489.8	62.2	4.06
	96.5	28.36	7.75	15.710	1.438	.890	90.1	14.4	1.78	1.49	465.1	59.2	4.05
	92	27.04	7.69	15.660	1.378	.840	83.9	13.4	1.76	1.45	441.4	56.4	4.04
	88	25.87	7.63	15.640	1.313	.820	80.2	12.9	1.76	1.42	418.9	53.6	4.02
	83.5	24.55	7.56	15.600	1.248	.780	75.0	12.1	1.75	1.39	395.1	50.7	4.01
	79	23.24	7.50	15.550	1.188	.730	69.3	11.3	1.73	1.34	372.5	47.9	4.00
	75	22.04	7.44	15.515	1.128	.695	64.9	10.6	1.72	1.31	351.3	45.3	3.99
	71	20.92	7.38	15.500	1.063	.680	62.1	10.2	1.72	1.29	330.1	42.6	3.97
ST 7 WF	68	19.99	7.38	14.740	1.063	.660	60.0	9.89	1.73	1.31	283.9	38.5	3.77
	63.5	18.67	7.31	14.690	0.998	.610	54.7	9.04	1.71	1.26	263.8	35.9	3.76
	59.5	17.49	7.25	14.650	0.938	.570	50.4	8.36	1.70	1.22	245.9	33.6	3.75
	55.5	16.33	7.19	14.620	0.873	.540	46.7	7.80	1.69	1.19	227.4	31.1	3.73
	51.5	15.13	7.13	14.575	0.813	.495	42.4	7.10	1.67	1.15	209.9	28.8	3.72
	47.5	13.97	7.06	14.545	0.748	.465	39.1	6.58	1.67	1.12	191.9	26.4	3.71
	43.5	12.78	7.00	14.5	0.688	.420	34.9	5.88	1.65	1.08	174.8	24.1	3.70
ST 7 WF	42	12.36	7.09	12.023	0.778	.451	37.4	6.36	1.74	1.21	112.7	18.8	3.02
	39	11.47	7.03	12.000	0.718	.428	34.8	5.96	1.74	1.19	103.5	17.2	3.00
ST 7 WF	37	10.88	7.10	10.072	0.783	.450	36.1	6.26	1.82	1.32	66.7	13.3	2.48
	34	10.00	7.03	10.040	0.718	.418	33.0	5.74	1.81	1.29	60.6	12.1	2.46
	30.5	8.97	6.96	10.000	0.643	.378	29.2	5.13	1.80	1.25	53.6	10.7	2.45
ST 7 WF	26.5	7.79	6.97	8.062	0.658	.370	27.7	4.95	1.88	1.38	28.8	7.14	1.92
	24	7.06	6.91	8.031	0.593	.339	24.9	4.49	1.88	1.35	25.6	6.38	1.91
	21.5	6.32	6.84	8.000	0.528	.308	22.2	4.02	1.87	1.33	22.6	5.64	1.89

* WF indicates structural tee cut from wide-flange section.

Table 75. Properties and Dimensions of Tees—*Continued*

Section Number	Weight per Foot	Area	Depth of Tee	Flange		Stem Thickness	Axis X–X				Axis Y–Y		
				Width	Average Thickness		I	S	r	y	I	S	r
	lb	in.²	in.	in.	in.	in.	in.⁴	in.³	in.	in.	in.⁴	in.³	in.
ST 7 WF *	19	5.59	7.06	6.776	0.513	0.313	23.5	4.27	2.05	1.56	12.3	3.64	1.49
	17	5.06	7.00	6.750	0.453	.287	21.1	3.86	2.05	1.55	10.6	3.15	1.46
	15	4.41	6.93	6.733	0.383	.270	19.0	3.55	2.08	1.59	8.77	2.61	1.41
ST 6 WF	80.5	23.69	6.94	12.515	1.486	.905	62.6	11.5	1.63	1.47	243.1	38.9	3.20
	66.5	19.56	6.69	12.365	1.236	.755	48.4	9.03	1.57	1.33	195.0	31.5	3.16
	60	17.65	6.56	12.320	1.106	.710	43.4	8.22	1.57	1.28	172.5	28.0	3.13
	53	15.59	6.44	12.230	0.986	.620	36.7	7.01	1.53	1.20	150.4	24.6	3.11
	49.5	14.54	6.38	12.190	0.921	.580	33.7	6.46	1.52	1.16	139.1	22.8	3.09
	46	13.53	6.31	12.155	0.856	.545	31.0	5.98	1.51	1.13	128.2	21.1	3.08
	42.5	12.49	6.25	12.105	0.796	.495	27.8	5.38	1.49	1.08	117.7	19.5	3.07
	39.5	11.61	6.19	12.080	0.736	.470	25.8	5.02	1.48	1.06	108.2	17.9	3.05
	36	10.58	6.13	12.040	0.671	.430	23.1	4.53	1.48	1.02	97.6	16.2	3.04
	32.5	9.55	6.06	12.000	0.606	.390	20.6	4.06	1.47	0.98	87.3	14.6	3.02
ST 6 WF	29	8.53	6.10	10.014	0.641	.359	19.0	3.75	1.49	1.03	53.7	10.7	2.51
	26.5	7.80	6.03	10.000	0.576	.345	17.7	3.54	1.51	1.02	48.0	9.60	2.48
ST 6 WF	25	7.36	6.10	8.077	0.641	.371	18.7	3.80	1.60	1.17	28.2	6.98	1.96
	22.5	6.62	6.03	8.042	0.576	.336	16.6	3.40	1.59	1.13	25.0	6.20	1.94
	20	5.89	5.97	8.000	0.516	.294	14.4	2.94	1.56	1.08	22.0	5.50	1.94
ST 6 WF	18	5.29	6.12	6.565	0.540	.305	15.3	3.14	1.70	J.26	11.9	3.62	1.50
	15.5	4.56	6.04	6.525	0.465	.265	13.0	2.69	1.69	1.22	9.9	3.04	1.47
	13.5	3.98	5.98	6.500	0.400	.240	11.4	2.39	1.69	1.21	8.3	2.55	1.44
ST 6 WF	7	2.07	5.96	3.970	0.224	.200	7.66	1.83	1.92	1.76	1.13	0.57	0.74
ST 6 I †	25	7.29	6.00	5.477	0.660	.687	25.2	6.05	1.85	1.84	7.85	2.87	1.03
	20.4	5.92	6.00	5.250	0.660	.460	18.8	4.26	1.77	1.57	6.77	2.58	1.06
ST 6 I	17.5	5.10	6.00	5.078	0.544	.428	17.2	3.95	1.83	1.65	4.93	1.94	0.98
	15.9	4.63	6.00	5.000	0.544	.350	14.9	3.31	1.78	1.51	4.68	1.87	1.00
ST 5 I	17.5	5.11	5.00	4.944	0.491	.594	12.5	3.63	1.56	1.56	4.18	1.69	0.90
	12.7	3.69	5.00	4.660	0.491	.310	7.81	2.05	1.45	1.20	3.39	1.46	0.95
ST 4 I	11.5	3.36	4.00	4.171	0.425	.441	5.03	1.77	1.22	1.15	2.15	1.03	0.80
	9.2	2.67	4.00	4.000	0.425	.270	3.50	1.14	1.14	0.94	1.86	0.93	0.83
ST 3.5 I	10	2.92	3.50	3.860	0.392	.450	3.36	1.36	1.07	1.04	1.58	0.82	0.73
	7.65	2.22	3.50	3.660	0.392	.250	2.18	0.81	0.99	0.81	1.32	0.72	0.77
ST 3 I	8.625	2.51	3.00	3.565	0.359	.465	2.13	1.02	0.92	0.91	1.15	0.65	0.67
	6.25	1.81	3.00	3.330	0.359	.230	1.27	0.55	0.83	0.69	0.93	0.56	0.71
ST 5 WF	56	16.46	5.69	10.415	1.248	.755	28.8	6.42	1.32	1.21	117.7	22.6	2.67
	50	14.72	5.56	10.345	1.118	.685	24.8	5.62	1.30	1.14	103.3	20.0	2.65
	44.5	13.09	5.44	10.275	0.998	.615	21.3	4.88	1.28	1.07	90.3	17.6	2.63
	38.5	11.33	5.31	10.195	0.868	.535	17.7	4.10	1.25	1.00	76.7	15.1	2.60
	36	10.59	5.25	10.170	0.808	.510	16.4	3.83	1.24	0.97	70.9	13.9	2.59
	33	9.70	5.19	10.117	0.748	.457	14.5	3.39	1.22	0.92	64.6	12.8	2.58
	30	8.83	5.13	10.075	0.683	.415	12.8	3.02	1.21	0.88	58.2	11.6	2.57
	27	7.94	5.06	10.028	0.618	.368	11.2	2.64	1.18	0.84	51.95	10.4	2.56
	24.5	7.20	5.00	10.000	0.558	.340	10.1	2.40	1.18	0.81	46.5	9.30	2.54
ST 5 WF	22.5	6.62	5.06	8.022	0.618	.350	10.3	2.48	1.25	0.91	26.6	6.63	2.00
	19.5	5.74	4.97	7.990	0.528	.318	8.96	2.19	1.25	0.88	22.5	5.62	1.98
	16.5	4.85	4.88	7.964	0.433	.292	7.80	1.95	1.27	0.88	18.2	4.58	1.94

* WF indicates structural tee cut from wide-flange section.
† I indicates structural tee cut from standard beam section.

Table 75. Properties and Dimensions of Tees—*Continued*

Section Number	Weight per Foot	Area	Depth of Tee	Flange Width	Flange Average Thickness	Stem Thickness	Axis X-X I	S	r	y	Axis Y-Y I	S	r
	lb	in.²	in.	in.	in.	in.	in.⁴	in.³	in.	in.	in.⁴	in.³	in.
ST 5 WF *	14.5	4.27	5.11	5.799	0.500	0.289	8.38	2.07	1.40	1.05	7.61	2.62	1.34
	12.5	3.67	5.04	5.762	.430	.252	7.12	1.77	1.39	1.02	6.34	2.20	1.31
	10.5	3.10	4.95	5.750	.340	.240	6.3	1.62	1.43	1.06	4.87	1.69	1.25
ST 4 WF	33.5	9.85	4.50	8.287	.933	.575	10.94	3.07	1.05	0.94	44.3	10.7	2.12
	29	8.53	4.38	8.222	.808	.510	9.11	2.60	1.03	0.87	37.5	9.10	2.10
	24	7.06	4.25	8.117	.683	.405	6.92	2.00	0.99	0.78	30.45	7.50	2.08
	20	5.88	4.13	8.077	.558	.365	5.80	1.71	0.99	0.74	24.5	6.05	2.04
	17.5	5.15	4.06	8.027	.493	.315	4.88	1.45	0.97	0.69	21.25	5.30	2.03
	15.5	4.56	4.00	8.000	.433	.288	4.31	1.30	0.97	0.67	18.5	4.60	2.01
ST 4 WF	14	4.11	4.03	6.540	.463	.285	4.22	1.28	1.01	0.73	10.8	3.30	1.62
	12	3.53	3.97	6.500	.398	.245	3.53	1.08	1.00	0.70	9.10	2.80	1.61
ST 4 WF	10	2.94	4.07	5.268	.378	.248	3.66	1.13	1.12	0.83	4.25	1.61	1.20
	8.5	2.50	4.00	5.250	.308	.230	3.21	1.01	1.13	0.84	3.36	1.28	1.16

Nominal Size	Weight per Foot	Area	Dimensions Depth	Width of flange	Minimum thickness Flange	Stem	Axis X-X I	S	r	y	Axis Y-Y I	S	r
in.	lb	in.²	in.	in.	in.	in.	in.⁴	in.³	in	in.	in.⁴	in.³	in.
5 × 3 1/8	13.6	4.00	3 1/8	5	1/2	13/32	2.7	1.1	0.82	0.76	5.2	2.1	1.14
5 × 3	11.5	3.37	3	5	3/8	13/32	2.4	1.1	0.84	0.76	3.9	1.6	1.10
4 × 4 1/2	11.2	3.29	4 1/2	4	3/8	3/8	6.3	2.0	1.39	1.31	2.1	1.1	0.80
4 × 4	13.5	3.97	4	4	1/2	1/2	5.7	2.0	1.20	1.18	2.8	1.4	0.84
4 × 3	9.2	2.68	3	4	3/8	3/8	2.0	0.90	0.86	0.78	2.1	1.1	0.89
4 × 2 1/2	8.5	2.48	2 1/2	4	3/8	3/8	1.2	0.62	0.69	0.62	2.1	1.0	0.92
3 × 3	7.8	2.29	3	3	3/8	3/8	1.84	0.86	0.89	0.88	0.89	0.60	0.63
3 × 3	6.7	1.97	3	3	5/16	5/16	1.61	0.74	0.90	0.85	0.75	0.50	0.62
3 × 2 1/2	6.1	1.77	2 1/2	3	5/16	5/16	0.94	0.51	0.73	0.68	0.75	0.50	0.65
2 1/2 × 2 1/2	6.4	1.87	2 1/2	2 1/2	3/8	3/8	1.0	0.59	0.74	0.76	0.52	0.42	0.53
2 1/2 × 2 1/2	4.6	1.33	2 1/2	2 1/2	1/4	1/4	0.74	0.42	0.75	0.71	0.34	0.27	0.51
2 1/4 × 2 1/4	4.1	1.19	2 1/4	2 1/4	1/4	1/4	0.52	0.32	0.66	0.65	0.25	0.22	0.46
2 × 2	4.3	1.26	2	2	5/16	5/16	0.44	0.31	0.59	0.61	0.23	0.23	0.43
2 × 2	3.56	1.05	2	2	1/4	1/4	0.37	0.26	0.59	0.59	0.18	0.18	0.42

* WF indicates structural tee cut from wide-flange section.

Table 76. Properties and Dimensions of Zees

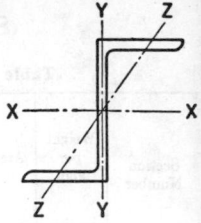

Nominal Size	Weight per Foot	Area	Dimensions			Axis X–X			Axis Y–Y			Axis Z–Z
			Depth	Width of Flange	Thickness	I	S	r	I	S	r	r
in.	lb	in.^{2}	in.	in.	in.	in.^{4}	in.^{3}	in.	in.^{4}	in.^{3}	in.	in.
6 × 3 1/2	21.1	6.19	6 1/8	3 5/8	1/2	34.4	11.2	2.36	12.9	3.8	1.44	0.84
	15.7	4.59	6	3 1/2	3/8	25.3	8.4	2.35	9.1	2.8	1.41	.83
5 × 3 1/4	17.9	5.25	5	3 1/4	1/2	19.2	7.7	1.91	9.1	3.0	1.31	.74
	16.4	4.81	5 1/8	3 3/8	7/16	19.1	7.4	1.99	9.2	2.9	1.38	.77
	14.0	4.10	5 1/16	3 5/16	3/8	16.2	6.4	1.99	7.7	2.5	1.37	.76
	11.6	3.40	5	3 1/4	5/16	13.4	5.3	1.98	6.2	2.0	1.35	.75
4 × 3 1/16	15.9	4.66	4 1/16	3 1/8	1/2	11.2	5.5	1.55	8.0	2.8	1.31	.67
	12.5	3.66	4 1/8	3 3/16	3/8	9.6	4.7	1.62	6.8	2.3	1.36	.69
	10.3	3.03	4 1/16	3 1/8	5/16	7.9	3.9	1.62	5.5	1.8	1.34	.68
	8.2	2.41	4	3 1/16	1/4	6.3	3.1	1.62	4.2	1.4	1.33	.67
3 × 2 11/16	12.6	3.69	3	2 11/16	1/2	4.6	3.1	1.12	4.9	2.0	1.15	.53
	9.8	2.86	3	2 11/16	3/8	3.9	2.6	1.16	3.9	1.6	1.17	.54
	6.7	1.97	3	2 11/16	1/4	2.9	1.9	1.21	2.8	1.1	1.19	.55

Tees and zees are seldom used as structural framing members. When so used they are generally employed on short spans in flexure. These tables list a few selected sizes, the range of whose section moduli will cover all ordinary conditions. For sizes not listed, the catalogs of the respective rolling mills should be consulted.

Table 77. Properties and Dimensions of H Bearing Piles

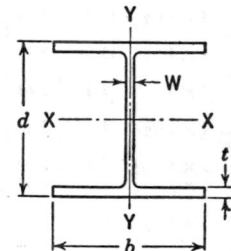

Section Number and Nominal Size	Weight per Foot	Area A	Depth d	Flange		Web Thickness W	Axis X–X			Axis Y–Y		
				Width b	Thickness t		I	S	r	I'	S'	r'
	lb	in.^{2}	in.	in.	in.	in.	in.^{4}	in.^{3}	in.	in.^{4}	in.^{3}	in.
BP 14	117	34.44	14.234	14.885	0.805	0.805	1228.5	172.6	5.97	443.1	59.5	3.59
14 × 14 1/2	102	30.01	14.032	14.784	.704	.704	1055.1	150.4	5.93	379.6	51.3	3.56
	89	26.19	13.856	14.696	.616	.616	909.1	131.2	5.89	326.2	44.4	3.53
	73	21.46	13.636	14.586	.506	.506	733.1	107.5	5.85	261.9	35.9	3.49
BP 12	74	21.76	12.122	12.217	.607	.607	566.5	93.5	5.10	184.7	30.2	2.91
12 × 12	53	15.58	11.780	12.046	.436	.436	394.8	67.0	5.03	127.3	21.2	2.86
BP 10	57	16.76	10.012	10.224	.564	.564	294.7	58.9	4.19	100.6	19.7	2.45
10 × 10	42	12.35	9.720	10.078	.418	.418	210.8	43.4	4.13	71.4	14.2	2.40
BP 8 } 8 × 8 }	36	10.60	8.026	8.158	.446	.446	119.8	29.9	3.36	40.4	9.9	1.95

Table 78. Square and Round Bars *

Size, in.	Square Weight, lb	Square Area, sq in.	Round Weight, lb	Round Area, sq in.	Size, in.	Square Weight, lb	Square Area, sq in.	Round Weight, lb	Round Area, sq in.
0					4	54.40	16.000	42.73	12.566
1/16	0.013	0.0039	0.010	0.0031	1/16	56.11	16.504	44.07	12.962
1/8	.053	.0156	.042	.0123	1/8	57.85	17.016	45.44	13.364
3/16	.120	.0352	.094	.0276	3/16	59.62	17.535	46.83	13.772
1/4	.213	.0625	.167	.0491	1/4	61.41	18.063	48.23	14.186
5/16	.332	.0977	.261	.0767	5/16	63.23	18.598	49.66	14.607
3/8	.478	.1406	.376	.1105	3/8	65.08	19.141	51.11	15.033
7/16	.651	.1914	.511	.1503	7/16	66.95	19.691	52.58	15.466
1/2	.850	.2500	.668	.1963	1/2	68.85	20.250	54.07	15.904
9/16	1.076	.3164	.845	.2485	9/16	70.78	20.816	55.59	16.349
5/8	1.328	.3906	1.043	.3068	5/8	72.73	21.391	57.12	16.800
11/16	1.607	.4727	1.262	.3712	11/16	74.71	21.973	58.67	17.257
3/4	1.913	.5625	1.502	.4418	3/4	76.71	22.563	60.25	17.721
13/16	2.245	.6602	1.763	.5185	13/16	78.74	23.160	61.85	18.190
7/8	2.603	.7656	2.044	.6013	7/8	80.80	23.766	63.46	18.665
15/16	2.988	.8789	2.347	.6903	15/16	82.89	24.379	65.10	19.147
1	3.400	1.0000	2.670	.7854	5	85.00	25.000	66.76	19.635
1/16	3.838	1.1289	3.015	.8866	1/16	87.14	25.629	68.44	20.129
1/8	4.303	1.2656	3.380	.9940	1/8	89.30	26.266	70.14	20.629
3/16	4.795	1.4102	3.766	1.1075	3/16	91.49	26.910	71.86	21.135
1/4	5.313	1.5625	4.172	1.2272	1/4	93.71	27.563	73.60	21.648
5/16	5.857	1.7227	4.600	1.3530	5/16	95.96	28.223	75.36	22.166
3/8	6.428	1.8906	5.049	1.4849	3/8	98.23	28.891	77.15	22.691
7/16	7.026	2.0664	5.518	1.6230	7/16	100.53	29.566	78.95	23.221
1/2	7.650	2.2500	6.008	1.7671	1/2	102.85	30.250	80.78	23.758
9/16	8.301	2.4414	6.519	1.9175	9/16	105.20	30.941	82.62	24.301
5/8	8.978	2.6406	7.051	2.0739	5/8	107.58	31.641	84.49	24.850
11/16	9.682	2.8477	7.604	2.2365	11/16	109.98	32.348	86.38	25.406
3/4	10.413	3.0625	8.178	2.4053	3/4	112.41	33.063	88.29	25.967
13/16	11.170	3.2852	8.773	2.5802	13/16	114.87	33.785	90.22	26.535
7/8	11.953	3.5156	9.388	2.7612	7/8	117.35	34.516	92.17	27.109
15/16	12.763	3.7539	10.024	2.9483	15/16	119.86	35.254	94.14	27.688
2	13.600	4.0000	10.681	3.1416	6	122.40	36.000	96.13	28.274
1/16	14.463	4.2539	11.359	3.3410	1/16	124.96	36.754	98.15	28.866
1/8	15.353	4.5156	12.058	3.5466	1/8	127.55	37.516	100.18	29.465
3/16	16.270	4.7852	12.778	3.7583	3/16	130.17	38.285	102.23	30.069
1/4	17.213	5.0625	13.519	3.9761	1/4	132.81	39.063	104.31	30.680
5/16	18.182	5.3477	14.280	4.2000	5/16	135.48	39.848	106.41	31.296
3/8	19.178	5.6406	15.062	4.4301	3/8	138.18	40.641	108.53	31.919
7/16	20.201	5.9414	15.866	4.6664	7/16	140.90	41.441	110.66	32.548
1/2	21.250	6.2500	16.690	4.9087	1/2	143.65	42.250	112.82	33.183
9/16	22.326	6.5664	17.534	5.1572	9/16	146.43	43.066	115.00	33.824
5/8	23.428	6.8906	18.400	5.4119	5/8	149.23	43.891	117.20	34.472
11/16	24.557	7.2227	19.287	5.6727	11/16	152.06	44.723	119.43	35.125
3/4	25.713	7.5625	20.195	5.9396	3/4	154.91	45.563	121.67	35.785
13/16	26.895	7.9102	21.123	6.2126	13/16	157.79	46.410	123.93	36.450
7/8	28.103	8.2656	22.072	6.4918	7/8	160.70	47.266	126.22	37.122
15/16	29.338	8.6289	23.042	6.7771	15/16	163.64	48.129	128.52	37.800
3	30.60	9.000	24.03	7.069	7	166.60	49.000	130.85	38.485
1/16	31.89	9.379	25.05	7.366	1/16	169.59	49.879	133.19	39.175
1/8	33.20	9.766	26.08	7.670	1/8	172.60	50.766	135.56	39.871
3/16	34.54	10.160	27.13	7.980	3/16	175.64	51.660	137.95	40.574
1/4	35.91	10.563	28.21	8.296	1/4	178.71	52.563	140.36	41.282
5/16	37.31	10.973	29.30	8.618	5/16	181.81	53.473	142.79	41.997
3/8	38.73	11.391	30.42	8.946	3/8	184.93	54.391	145.24	42.718
7/16	40.18	11.816	31.55	9.281	7/16	188.07	55.316	147.71	43.445
1/2	41.65	12.250	32.71	9.621	1/2	191.25	56.250	150.21	44.179
9/16	43.15	12.691	33.89	9.968	9/16	194.45	57.191	152.72	44.918
5/8	44.68	13.141	35.09	10.321	5/8	197.68	58.141	155.26	45.664
11/16	46.23	13.598	36.31	10.680	11/16	200.93	59.098	157.81	46.415
3/4	47.81	14.063	37.55	11.045	3/4	204.21	60.063	160.39	47.173
13/16	49.42	14.535	38.81	11.416	13/16	207.52	61.035	162.99	47.937
7/8	51.05	15.016	40.10	11.793	7/8	210.85	62.016	165.60	48.707
15/16	52.71	15.504	41.40	12.177	15/16	214.21	63.004	168.24	49.483
4	54.40	16.000	42.73	12.566	8	217.60	64.000	170.90	50.265

* One cubic inch of rolled steel is assumed to weigh 0.2833 pound.

Table 79. Pipe

(Steel Construction, 1949, A.I.S.C.)

Nom. Diam., in.	Outside Diam., in.	Inside Diam., in.	Thickness, in.	Weight per Foot, lb — Plain Ends	Weight per Foot, lb — Thread and Coupling	Threads per inch	Couplings Outside Diam., in.	Length, in.	Weight, lb	I, in.⁴	A, in.²	k, in.
						Schedule 40ST						
1/8	0.405	0.269	0.068	0.24	0.25	27	0.562	7/8	0.03	0.001	0.072	0.12
1/4	0.540	0.364	0.088	0.42	0.43	18	0.685	1	0.04	0.003	0.125	0.16
3/8	0.675	0.493	0.091	0.57	0.57	18	0.848	1 1/8	0.07	0.007	0.167	0.21
1/2	0.840	0.622	0.109	0.85	0.85	14	1.024	1 3/8	0.12	0.017	0.250	0.26
3/4	1.050	0.824	0.113	1.13	1.13	14	1.281	1 5/8	0.21	0.037	0.333	0.33
1	1.315	1.049	0.133	1.68	1.68	11 1/2	1.576	1 7/8	0.35	0.087	0.494	0.42
1 1/4	1.660	1.380	0.140	2.27	2.28	11 1/2	1.950	2 1/8	0.55	0.195	0.669	0.54
1 1/2	1.900	1.610	0.145	2.72	2.73	11 1/2	2.218	2 3/8	0.76	0.310	0.799	0.62
2	2.375	2.067	0.154	3.65	3.68	11 1/2	2.760	2 5/8	1.23	0.666	1.075	0.79
2 1/2	2.875	2.469	0.203	5.79	5.82	8	3.276	2 7/8	1.76	1.530	1.704	0.95
3	3.500	3.068	0.216	7.58	7.62	8	3.948	3 1/8	2.55	3.017	2.228	1.16
3 1/2	4.000	3.548	0.226	9.11	9.20	8	4.591	3 5/8	4.33	4.788	2.680	1.34
4	4.500	4.026	0.237	10.79	10.89	8	5.091	3 5/8	5.41	7.233	3.174	1.51
5	5.563	5.047	0.258	14.62	14.81	8	6.296	4 1/8	9.16	15.16	4.300	1.88
6	6.625	6.065	0.280	18.97	19.19	8	7.358	4 1/8	10.82	28.14	5.581	2.25
8	8.625	8.071	0.277	24.70	25.00	8	9.420	4 5/8	15.84	63.35	7.265	2.95
8	8.625	7.981	0.322	28.55	28.81	8	9.420	4 5/8	15.84	72.49	8.399	2.94
10	10.750	10.192	0.279	31.20	32.00	8	11.721	6 1/8	33.92	125.4	9.178	3.70
10	10.750	10.136	0.307	34.24	35.00	8	11.721	6 1/8	33.92	137.4	10.07	3.69
10	10.750	10.020	0.365	40.48	41.13	8	11.721	6 1/8	33.92	160.7	11.91	3.67
12	12.750	12.090	0.330	43.77	45.00	8	13.958	6 1/8	48.27	248.5	12.88	4.39
12	12.750	12.000	0.375	49.56	50.71	8	13.958	6 1/8	48.27	279.3	14.38	4.38
						Schedule 80XS						
1/8	0.405	0.215	0.095	0.31	0.32	27	0.582	1 1/8	0.05	0.001	0.093	0.12
1/4	0.540	0.302	0.119	0.54	0.54	18	0.724	1 3/8	0.07	0.004	0.157	0.16
3/8	0.675	0.423	0.126	0.74	0.75	18	0.898	1 5/8	0.13	0.009	0.217	0.20
1/2	0.840	0.546	0.147	1.09	1.10	14	1.085	1 7/8	0.22	0.020	0.320	0.25
3/4	1.050	0.742	0.154	1.47	1.49	14	1.316	2 1/8	0.33	0.045	0.433	0.32
1	1.315	0.957	0.179	2.17	2.20	11 1/2	1.575	2 3/8	0.47	0.106	0.639	0.41
1 1/4	1.660	1.278	0.191	3.00	3.05	11 1/2	2.054	2 7/8	1.04	0.242	0.881	0.52
1 1/2	1.900	1.500	0.200	3.63	3.69	11 1/2	2.294	2 7/8	1.17	0.391	1.068	0.61
2	2.375	1.939	0.218	5.02	5.13	11 1/2	2.870	3 5/8	2.17	0.868	1.477	0.77
2 1/2	2.875	2.323	0.276	7.66	7.83	8	3.389	4 1/8	3.43	1.924	2.254	0.92
3	3.500	2.900	0.300	10.25	10.46	8	4.014	4 1/8	4.13	3.894	3.016	1.14
3 1/2	4.000	3.364	0.318	12.51	12.82	8	4.628	4 5/8	6.29	6.280	3.678	1.31
4	4.500	3.826	0.337	14.98	15.39	8	5.233	4 5/8	8.16	9.610	4.407	1.48
5	5.563	4.813	0.375	20.78	21.42	8	6.420	5 1/8	12.87	20.67	6.112	1.84
6	6.625	5.761	0.432	28.57	29.33	8	7.482	5 1/8	15.18	40.49	8.405	2.20
·8	8.625	7.625	0.500	43.39	44.72	8	9.596	6 1/8	26.63	105.7	12.76	2.88
10	10.750	9.750	0.500	54.74	56.94	8	11.958	6 5/8	44.16	211.9	16.10	3.63
12	12.750	11.750	0.500	65.42	68.02	8	13.958	6 5/8	51.99	361.5	19.24	4.34
						Schedule XX						
1/2	0.840	0.252	0.294	1.71	1.73	14	1.085	1 7/8	0.22	0.024	0.504	0.22
3/4	1.050	0.434	0.308	2.44	2.46	14	1.316	2 1/8	0.33	0.058	0.718	0.28
1	1.315	0.599	0.358	3.66	3.68	11 1/2	1.575	2 3/8	0.47	0.140	1.076	0.36
1 1/4	1.660	0.896	0.382	5.21	5.27	11 1/2	2.054	2 7/8	1.04	0.341	1.534	0.47
1 1/2	1.900	1.100	0.400	6.41	6.47	11 1/2	2.294	2 7/8	1.17	0.568	1.885	0.55
2	2.375	1.503	0.436	9.03	9.14	11 1/2	2.870	3 5/8	2.17	1.311	2.656	0.70
2 1/2	2.875	1.771	0.552	13.70	13.87	8	3.389	4 1/8	3.43	2.871	4.028	0.84
3	3.500	2.300	0.600	18.58	18.79	8	4.014	4 1/8	4.13	5.992	5.466	1.05
3 1/2	4.000	2.728	0.636	22.85	23.16	8	4.628	4 5/8	6.29	9.848	6.721	1.21
4	4.500	3.152	0.674	27.54	27.95	8	5.233	4 5/8	8.16	15.28	8.101	1.37
5	5.563	4.063	0.750	38.55	39.20	8	6.420	5 1/8	12.87	33.64	11.34	1.72
6	6.625	4.897	0.864	53.16	53.92	8	7.482	5 1/8	15.18	66.33	15.64	2.06
8	8.625	6.875	0.875	72.42	73.76	8	9.596	6 1/8	26.63	162.0	21.30	2.76

Large O. D. Pipe

Pipe 14″ and larger is sold by actual O. S. diameter and thickness.
Sizes 14″, 15″, and 16″ are available regularly in thicknesses varying by 1/16″ from 1/4″ to 1″, inclusive.
All pipe is furnished random length unless otherwise ordered, viz: 12 to 22 ft with privilege of furnishing 5 per cent in 6 to 12 ft lengths. Pipe railing is most economically detailed with slip joints and random lengths between couplings.

STANDARD STRUCTURAL SIZES—ALUMINUM

Aluminum Sections. Tables 80 to 86 give the dimensions, weights, and properties of extruded or rolled aluminum alloy structural sections. The values for the various structural forms were taken from *Alcoa Structural Handbook*, 1948, sixth edition, by the kind permission of the Aluminum Company of America. The tables include section specifications necessary to design structures. The theory of design is covered in Section 5.

All values indicated in the tables were computed on the basis of the nominal dimensions shown. Fillets and roundings have been included throughout all calculations except those for the torsion factor, J. On the profiles shown, the X–X axis represents the axis of maximum moment of inertia; the Y–Y axis represents the axis of minimum moment of inertia. The maximum X–X axis and the minimum Y–Y axis are indicated for sections having an axis of symmetry. Axis Z–Z is the axis of least moment of inertia for unsymmetrical sections. Each of the above axes is the neutral axis for flexure in a plane at right angles to the axes.

The weights given are for 14S * alloy. The weights based on other alloys may be found as follows:

For 3S,* multiply by 0.980   For 24S,* multiply by 0.990
For 4S,* and 61S multiply by 0.970   For 52S,* multiply by 0.960

Only *standard* shapes most generally used are included in the following tables. Further information on sections listed in the tables, together with information on sections not listed, may be obtained from catalogs of the various mills producing the sections. ALCOA aluminum standard structural shapes are indicated by an asterisk (*) at the head of the column. These standard shapes may ordinarily be obtained with a minimum of delay and usually at lower prices than non-standard shapes.

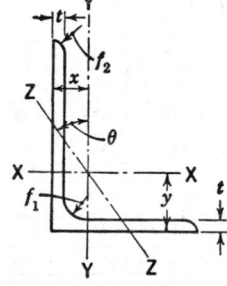

Table 80.   Properties of Angles

Elements of Sections

All dimensions in inches. S = section modulus in in.^{3}
Weight in pounds per foot. r = radius of gyration in inches.
Area in square inches. J = torsion factor in in.^{4}
I = moment of inertia in in.^{4}

Size	Legs	$1/2 \times 1/2$	$5/8 \times 5/8$		$3/4 \times 3/8$	$3/4 \times 3/4$			
	t	$1/16$	$3/32$	$1/8$	$3/32$	$1/16$	$3/32$	$1/8$ *	$3/16$ *
Weight		0.071	0.134	0.169	0.120	0.108	0.159	0.207	0.297
Area		0.059	0.111	0.140	0.099	0.089	0.132	0.171	0.246
	f_1	$1/16$	$1/8$	$1/16$	$1/8$	$1/8$	$1/8$	$1/8$	$1/8$
	f_2	$1/32$	$3/64$	$1/16$	$3/64$	$3/32$	$3/32$	$3/32$	$3/32$
Axis	I	0.0013	0.0037	0.0046	0.0054	0.0043	0.0063	0.0082	0.0112
X–X	S	0.0038	0.0084	0.0109	0.0114	0.0079	0.0118	0.0157	0.0224
	r	0.150	0.183	0.182	0.232	0.220	0.219	0.219	0.214
	y	0.146	0.187	0.199	0.279	0.199	0.214	0.227	0.251
Axis	I	0.0013	0.0037	0.0046	0.0009	0.0043	0.0063	0.0082	0.0112
Y–Y	S	0.0038	0.0084	0.0109	0.0031	0.0079	0.0118	0.0157	0.0224
	r	0.150	0.183	0.182	0.094	0.220	0.219	0.219	0.214
	x	0.146	0.187	0.199	0.098	0.199	0.214	0.227	0.251
Axis	θ	45° 0'	45° 0'	45° 0'	13° 44'	45° 0'	45° 0'	45° 0'	45° 0'
Z–Z	I	0.0006	0.0015	0.0020	0.0006	0.0018	0.0026	0.0034	0.0049
	r	0.097	0.117	0.120	0.077	0.142	0.141	0.141	0.141
	J	0.00008	0.00034	0.00081	0.00031	0.00012	0.00041	0.00098	0.0033

* Alloy designation of the Aluminum Company of America.

Table 80. Properties of Angles—*Continued*

Size	Legs	1 × 5/8		1 × 3/4		1 × 1			
	t	1/8	1/4	1/8	1/16	3/32 *	1/8 *	3/16 *	1/4 *
Weight		0.226	0.418	0.245	0.147	0.216	0.283	0.411	0.529
Area		0.187	0.345	0.202	0.122	0.178	0.234	0.339	0.437
f_1		1/16	1/8	1/16	1/16	1/8	1/8	1/8	1/8
f_2		1/16	1/8	1/16	1/32	3/32	3/32	3/32	3/32
Axis X–X	I	0.0181	0.0306	0.0194	0.0118	0.0161	0.0208	0.0291	0.0361
	S	0.0279	0.0507	0.0288	0.0162	0.0223	0.0293	0.0424	0.0544
	r	0.312	0.298	0.309	0.311	0.301	0.298	0.293	0.287
	y	0.351	0.396	0.329	0.271	0.276	0.290	0.314	0.336
Axis Y–Y	I	0.0054	0.0089	0.0092	0.0118	0.0161	0.0208	0.0291	0.0361
	S	0.0117	0.0215	0.0169	0.0162	0.0223	0.0293	0.0424	0.0544
	r	0.170	0.161	0.214	0.311	0.301	0.298	0.293	0.287
	x	0.165	0.210	0.205	0.271	0.276	0.290	0.314	0.336
Axis Z–Z	θ	20° 34'	9° 08'	28° 23'	45° 0'	45° 0'	45° 0'	45° 0'	45° 0'
	I	0.0033	0.0084	0.0051	0.0048	0.0066	0.0085	0.0124	0.0162
	r	0.132	0.156	0.158	0.199	0.193	0.191	0.191	0.193
J		0.00106	0.00846	0.00114	0.00016	0.00055	0.00130	0.00439	0.01042

Size	Legs	1 1/8 × 1 1/8	1 1/4 × 3/4	1 1/4 × 1		1 1/4 × 1 1/4			
	t	1/8	3/32 *	1/8 *	3/32	1/8 *	3/16 *	1/4 *	5/16
Weight		0.32	0.22	0.32	0.28	0.36	0.53	0.68	0.83
Area		0.27	0.18	0.27	0.23	0.30	0.43	0.56	0.68
f_1		3/16	3/32	1/8	3/32	3/32	3/32	3/32	3/32
f_2		1/8	3/64	1/16	3/64	1/8	1/8	1/8	1/8
Axis X–X	I	0.030	0.029	0.040	0.033	0.042	0.059	0.074	0.088
	S	0.037	0.035	0.047	0.036	0.046	0.068	0.087	0.106
	r	0.33	0.40	0.39	0.38	0.37	0.37	0.36	0.36
	y	0.32	0.42	0.39	0.34	0.35	0.37	0.40	0.42
Axis Y–Y	I	0.030	0.008	0.023	0.033	0.042	0.059	0.074	0.088
	S	0.037	0.014	0.031	0.036	0.046	0.068	0.087	0.106
	r	0.33	0.21	0.29	0.38	0.37	0.37	0.36	0.36
	x	0.32	0.17	0.27	0.34	0.35	0.37	0.40	0.42
Axis Z–Z	θ	45° 0'	19° 47'	31° 51'	45° 0'	45° 0'	45° 0'	45° 0'	45° 0'
	I	0.012	0.005	0.012	0.014	0.017	0.025	0.032	0.040
	r	0.21	0.16	0.21	0.24	0.24	0.24	0.24	0.24
J		0.0015	0.0005	0.0015	0.0007	0.0016	0.0055	0.013	0.025

Size	Legs	1 1/2 × 3/4		1 1/2 × 7/8	1 1/2 × 1		1 1/2 × 1 1/4		
	t	1/8 *	3/16 *	3/16	5/32 *	1/4 *	•1/8 *	3/16 *	1/4 *
Weight		0.32	0.47	0.50	0.45	0.68	0.40	0.58	0.76
Area		0.27	0.39	0.41	0.37	0.56	0.33	0.48	0.63
f_1		1/8	1/8	1/8	5/32	3/16	3/16	3/16	3/16
f_2		1/16	3/32	3/32	5/64	1/8	1/8	1/8	1/8
Axis X–X	I	0.061	0.085	0.090	0.080	0.117	0.070	0.100	0.127
	S	0.064	0.091	0.093	0.081	0.122	0.066	0.097	0.126
	r	0.48	0.47	0.47	0.47	0.46	0.46	0.46	0.45
	y	0.54	0.57	0.54	0.50	0.53	0.44	0.47	0.49
Axis Y–Y	I	0.010	0.014	0.022	0.027	0.040	0.044	0.063	0.079
	S	0.018	0.025	0.034	0.036	0.057	0.047	0.069	0.090
	r	0.20	0.19	0.23	0.27	0.27	0.37	0.36	0.36
	x	0.17	0.19	0.23	0.26	0.29	0.32	0.35	0.37
Axis Z–Z	θ	14° 25'	13° 45'	18° 08'	23° 02'	22° 23'	33° 59'	33° 53'	33° 36'
	I	0.007	0.009	0.014	0.015	0.025	0.022	0.032	0.041
	r	0.16	0.16	0.18	0.20	0.21	0.26	0.26	0.26
J		0.0015	0.0049	0.0052	0.0032	0.0130	0.0018	0.0060	0.0143

Table 80. Properties of Angles—*Continued*

Size / Legs	1 1/2 × 1 1/2						1 5/8 × 1 1/4	1 3/4 × 1 1/8
t	3/32	1/8 *	3/16 *	1/4 *	5/16	3/8	1/8	3/16
Weight	0.33	0.44	0.64	0.83	1.02	1.19	0.41	0.61
Area	0.28	0.36	0.53	0.69	0.84	0.99	0.34	0.51
f_1	3/16	3/16	3/16	3/16	3/16	3/16	1/8	3/16
f_2	1/8	1/8	1/8	1/8	1/8	1/8	1/8	3/32
Axis X-X I	0.058	0.074	0.107	0.135	0.161	0.184	0.087	0.152
S	0.053	0.068	0.100	0.130	0.158	0.185	0.077	0.132
r	0.46	0.45	0.45	0.44	0.44	0.43	0.51	0.55
y	0.40	0.41	0.44	0.46	0.48	0.51	0.50	0.59
Axis Y-Y I	0.058	0.074	0.107	0.135	0.161	0.184	0.045	0.049
S	0.053	0.068	0.100	0.130	0.158	0.185	0.048	0.058
r	0.46	0.45	0.45	0.44	0.44	0.43	0.36	0.31
x	0.40	0.41	0.44	0.46	0.48	0.51	0.31	0.29
Axis Z-Z θ	45° 0'	45° 0'	45° 0'	45° 0'	45° 0'	45° 0'	29° 55'	21° 47'
I	0.024	0.031	0.044	0.057	0.070	0.083	0.024	0.029
r	0.30	0.29	0.29	0.29	0.29	0.29	0.26	0.24
J	0.0008	0.0020	0.0066	0.016	0.031	0.053	0.0019	0.0063

Size / Legs	1 3/4 × 1 1/4			1 3/4 × 1 3/4				
t	1/8 *	3/16 *	1/4 *	3/32	1/8 *	3/16 *	1/4 *	5/16 *
Weight	0.44	0.64	0.83	0.39	0.51	0.75	0.98	1.21
Area	0.36	0.53	0.69	0.32	0.42	0.62	0.81	1.00
f_1	3/16	3/16	3/16	3/32	3/16	3/16	3/16	3/16
f_2	1/8	1/8	1/8	3/64	1/8	1/8	1/8	1/8
Axis X-X I	0.108	0.156	0.199	0.096	0.121	0.174	0.223	0.266
S	0.090	0.132	0.172	0.075	0.094	0.139	0.181	0.221
r	0.55	0.54	0.54	0.55	0.53	0.53	0.52	0.52
y	0.54	0.57	0.60	0.47	0.47	0.50	0.52	0.55
Axis Y-Y I	0.046	0.066	0.083	0.096	0.121	0.174	0.223	0.266
S	0.048	0.071	0.092	0.075	0.094	0.139	0.181	0.221
r	0.36	0.35	0.35	0.55	0.53	0.53	0.52	0.52
x	0.30	0.32	0.35	0.47	0.47	0.50	0.52	0.55
Axis Z-Z θ	26° 22'	26° 08'	25° 47'	45° 0'	45° 0'	45° 0'	45° 0'	45° 0'
I	0.026	0.037	0.048	0.039	0.050	0.072	0.093	0.113
r	0.27	0.26	0.26	0.35	0.34	0.34	0.34	0.34
J	0.0020	0.0066	0.016	0.0010	0.0023	0.0077	0.018	0.036

Size / Legs	2 × 1 1/4			2 × 1 3/8	2 × 1 1/2				
t	1/8	3/16	1/4	1/4	1/8 *	3/16 *	1/4 *	5/16	3/8 *
Weight	0.47	0.70	0.91	0.95	0.51	0.75	0.98	1.21	1.42
Area	0.39	0.58	0.75	0.79	0.42	0.62	0.81	1.00	1.17
f_1	3/16	3/16	3/16	1/4	3/16	3/16	3/16	3/16	3/16
f_2	1/8	1/8	1/8	1/8	1/8	1/8	1/8	1/8	1/8
Axis X-X I	0.158	0.228	0.291	0.302	0.17	0.24	0.31	0.37	0.43
S	0.117	0.172	0.224	0.228	0.12	0.18	0.23	0.28	0.33
r	0.63	0.63	0.62	0.62	0.63	0.62	0.62	0.61	0.60
y	0.65	0.68	0.70	0.68	0.60	0.63	0.66	0.68	0.70
Axis Y-Y I	0.047	0.068	0.086	0.114	0.08	0.12	0.15	0.18	0.20
S	0.049	0.072	0.093	0.113	0.07	0.10	0.14	0.16	0.19
r	0.35	0.34	0.34	0.38	0.44	0.43	0.43	0.42	0.41
x	0.28	0.31	0.33	0.37	0.36	0.38	0.41	0.43	0.45
Axis Z-Z θ	21° 05'	20° 52'	20° 31'	24° 14'	28° 44'	28° 36'	28° 20'	28° 0'	27° 37'
I	0.028	0.041	0.053	0.067	0.04	0.06	0.08	0.10	0.12
r	0.27	0.27	0.26	0.29	0.32	0.32	0.32	0.32	0.32
J	0.0021	0.0071	0.017	0.018	0.0023	0.0077	0.018	0.036	0.062

Table 80. Properties of Angles—*Continued*

Size	Legs	2 × 1 3/4	2 × 2						2 1/4 × 1 1/2
	t	1/4	1/8 *	3/16 *	1/4 *	5/16 *	3/8 *	1/4
Weight		1.07	0.59	0.87	1.14	1.40	1.65	1.07
Area		0.88	0.49	0.72	0.94	1.16	1.37	0.88
f_1		1/4	1/4	1/4	1/4	1/4	1/4	1/4
f_2		1/8	1/8	1/8	1/8	1/8	1/8	1/8
Axis X–X	I	0.33	0.18	0.27	0.34	0.41	0.47	0.43
	S	0.24	0.13	0.19	0.24	0.30	0.35	0.29
	r	0.61	0.61	0.61	0.60	0.60	0.59	0.70
	y	0.62	0.53	0.56	0.58	0.61	0.63	0.76
Axis Y–Y	I	0.23	0.18	0.27	0.34	0.41	0.47	0.15
	S	0.18	0.13	0.19	0.24	0.30	0.35	0.14
	r	0.51	0.61	0.61	0.60	0.60	0.59	0.42
	x	0.49	0.53	0.56	0.58	0.61	0.63	0.39
Axis Z–Z	θ	34° 44′	45° 0′	45° 0′	45° 0′	45° 0′	45° 0′	23° 09′
	I	0.11	0.08	0.11	0.14	0.17	0.20	0.09
	r	0.36	0.40	0.39	0.39	0.39	0.39	0.32
	J	0.020	0.0026	0.0088	0.021	0.041	0.070	0.020

Size	Legs	2 1/2 × 1 1/4		2 1/2 × 1 1/2				2 1/2 × 2	
	t	1/8	3/16	1/8	3/16 *	1/4 *	5/16 *	1/8 *	3/16 *
Weight		0.55	0.82	0.59	0.87	1.14	1.40	0.67	0.99
Area		0.46	0.68	0.49	0.72	0.94	1.16	0.55	0.82
f_1		3/16	1/4	1/4	1/4	1/4	3/16	1/4	1/4
f_2		3/32	3/32	1/8	1/8	1/8	1/8	1/8	1/8
Axis X–X	I	0.30	0.43	0.31	0.46	0.59	0.71	0.34	0.50
	S	0.18	0.27	0.19	0.27	0.36	0.44	0.19	0.29
	r	0.81	0.80	0.80	0.79	0.79	0.78	0.79	0.78
	y	0.87	0.89	0.80	0.84	0.86	0.89	0.72	0.75
Axis Y–Y	I	0.05	0.07	0.09	0.12	0.16	0.19	0.20	0.29
	S	0.05	0.08	0.07	0.11	0.14	0.17	0.13	0.19
	r	0.34	0.33	0.42	0.41	0.41	0.40	0.60	0.59
	x	0.25	0.28	0.32	0.35	0.37	0.39	0.48	0.51
Axis Z–Z	θ	14° 45′	14° 29′	19° 40′	19° 38′	19° 25′	19° 06′	31° 57′	31° 58′
	I	0.03	0.05	0.05	0.08	0.10	0.12	0.10	0.15
	r	0.27	0.27	0.33	0.32	0.32	0.32	0.43	0.42
	J	0.0024	0.0082	0.0026	0.0088	0.021	0.041	0.0029	0.010

Size	Legs	2 1/2 × 2 (Cont.)			2 1/2 × 2 1/2					
	t	1/4 *	5/16 *	3/8 *	1/8 *	3/16 *	1/4 *	5/16 *	3/8 *	7/16
Weight		1.29	1.59	1.88	0.75	1.10	1.45	1.78	2.11	2.42
Area		1.07	1.32	1.55	0.62	0.91	1.19	1.47	1.74	2.00
f_1		1/4	1/4	1/4	1/4	1/4	1/4	1/4	1/4	1/4
f_2		1/8	1/8	1/8	1/8	1/8	1/8	1/8	1/8	1/8
Axis X–X	I	0.65	0.78	0.91	0.37	0.54	0.69	0.84	0.98	1.10
	S	0.38	0.46	0.54	0.20	0.30	0.39	0.48	0.56	0.64
	r	0.78	0.77	0.76	0.77	0.77	0.77	0.76	0.75	0.74
	y	0.78	0.80	0.83	0.65	0.68	0.71	0.73	0.76	0.78
Axis Y–Y	I	0.37	0.44	0.51	0.37	0.54	0.69	0.84	0.98	1.10
	S	0.25	0.30	0.36	0.20	0.30	0.39	0.48	0.56	0.64
	r	0.58	0.58	0.57	0.77	0.77	0.77	0.76	0.75	0.74
	x	0.53	0.55	0.58	0.65	0.68	0.71	0.73	0.76	0.78
Axis Z–Z	θ	31° 51′	31° 41′	31° 28′	45° 0′	45° 0′	45° 0′	45° 0′	45° 0′	45° 0′
	I	0.19	0.23	0.27	0.15	0.22	0.29	0.35	0.41	0.47
	r	0.42	0.42	0.42	0.50	0.49	0.49	0.49	0.48	0.48
	J	0.023	0.046	0.079	0.0033	0.011	0.026	0.051	0.088	0.140

Table 80. Properties of Angles—*Continued*

Size	Legs	3 × 1 1/2	3 × 2					3 × 2 1/2		
	t	1/4	3/16 *	1/4 *	5/16 *	3/8 *	7/16 *	1/4 *	5/16 *	3/8 *
Weight		1.30	1.10	1.44	1.78	2.11	2.42	1.58	1.95	2.32
Area		1.08	0.91	1.19	1.47	1.74	2.00	1.31	1.62	1.92
f_1		5/16	5/16	5/16	5/16	5/16	5/16	5/16	5/16	5/16
f_2		1/8	3/16	3/16	3/16	3/16	3/16	1/4	1/4	1/4
Axis X-X	I	0.98	0.82	1.06	1.29	1.51	1.71	1.12	1.37	1.60
	S	0.51	0.40	0.52	0.65	0.76	0.88	0.53	0.66	0.78
	r	0.95	0.95	0.94	0.94	0.93	0.92	0.92	0.92	0.91
	y	1.08	0.94	0.97	1.00	1.03	1.05	0.89	0.92	0.94
Axis Y-Y	I	0.16	0.29	0.38	0.45	0.53	0.59	0.70	0.86	1.00
	S	0.14	0.19	0.25	0.30	0.36	0.41	0.38	0.47	0.55
	r	0.39	0.56	0.56	0.56	0.55	0.55	0.73	0.73	0.72
	x	0.34	0.46	0.48	0.51	0.53	0.56	0.64	0.67	0.69
Axis Z-Z	θ	14° 22'	23° 25'	23° 22'	23° 13'	23° 0'	22° 44'	34° 03'	34° 0'	33° 54'
	I	0.11	0.17	0.22	0.27	0.31	0.36	0.35	0.43	0.51
	r	0.32	0.43	0.43	0.43	0.42	0.42	0.52	0.51	0.51
J		0.023	0.011	0.026	0.051	0.088	0.140	0.029	0.056	0.097

Size	Legs	3 × 3						3 1/2 × 2 1/2		
	t	3/16 *	1/4 *	5/16 *	3/8 *	7/16 *	1/2 *	1/4 *	5/16 *	3/8 *
Weight		1.33	1.73	2.14	2.55	2.94	3.32	1.73	2.14	2.55
Area		1.10	1.43	1.77	2.10	2.43	2.74	1.43	1.77	2.10
f_1		5/16	5/16	5/16	5/16	5/16	5/16	5/16	5/16	5/16
f_2		1/4	1/4	1/4	1/4	1/4	1/4	1/4	1/4	1/4
Axis X-X	I	0.93	1.18	1.45	1.70	1.94	2.16	1.73	2.12	2.49
	S	0.42	0.54	0.67	0.80	0.92	1.04	0.72	0.89	1.06
	r	0.92	0.91	0.91	0.90	0.89	0.89	1.10	1.09	1.09
	y	0.80	0.82	0.85	0.87	0.90	0.92	1.09	1.12	1.14
Axis Y-Y	I	0.93	1.18	1.45	1.70	1.94	2.16	0.73	0.89	1.05
	S	0.42	0.54	0.67	0.80	0.92	1.04	0.38	0.48	0.57
	r	0.92	0.91	0.91	0.90	0.89	0.89	0.71	0.71	0.71
	x	0.80	0.82	0.85	0.87	0.90	0.92	0.60	0.62	0.65
Axis Z-Z	θ	45° 0'	45° 0'	45° 0'	45° 0'	45° 0'	45° 0'	26° 23'	26° 18'	26° 10'
	I	0.38	0.49	0.60	0.70	0.81	0.91	0.41	0.50	0.59
	r	0.59	0.58	0.58	0.58	0.58	0.58	0.53	0.53	0.53
J		0.013	0.031	0.061	0.105	0.167	0.250	0.031	0.061	0.105

Size	Legs	3 1/2 × 2 1/2 (Cont.)	3 1/2 × 3				3 1/2 × 3 1/2			
	t	1/2 *	1/4 *	5/16 *	3/8 *	1/2 *	1/4 *	5/16 *	3/8 *	1/2 *
Weight		3.32	1.89	2.34	2.78	3.63	2.05	2.53	3.01	3.94
Area		2.74	1.57	1.94	2.30	3.00	1.69	2.09	2.49	3.25
f_1		5/16	3/8	3/8	3/8	3/8	3/8	3/8	3/8	3/8
f_2		1/4	1/4	1/4	1/4	1/4	1/4	1/4	1/4	1/4
Axis X-X	I	3.17	1.84	2.26	2.65	3.38	1.93	2.37	2.79	3.56
	S	1.37	0.74	0.92	1.09	1.42	0.76	0.94	1.11	1.45
	r	1.08	1.08	1.08	1.07	1.06	1.07	1.06	1.06	1.05
	y	1.19	1.01	1.04	1.06	1.11	0.94	0.97	1.00	1.05
Axis Y-Y	I	1.32	1.28	1.52	1.79	2.27	1.93	2.37	2.79	3.56
	S	0.73	0.57	0.69	0.82	1.06	0.76	0.94	1.11	1.45
	r	0.69	0.90	0.89	0.88	0.87	1.07	1.06	1.06	1.05
	x	0.70	0.76	0.79	0.82	0.86	0.94	0.97	1.00	1.05
Axis Z-Z	θ	25° 48'	36° 13'	35° 40'	35° 37'	35° 26'	45° 0'	45° 0'	45° 0'	45° 0'
	I	0.76	0.63	0.74	0.87	1.13	0.80	0.98	1.15	1.49
	r	0.53	0.63	0.62	0.62	0.61	0.69	0.68	0.68	0.68
J		0.250	0.034	0.066	0.114	0.271	0.036	0.071	0.123	0.292

Table 80. Properties of Angles—*Continued*

Size	Legs	4 × 3							4 × 3 1/2	
	t	1/4 *	5/16 *	3/8 *	7/16 *	1/2 *	9/16	5/8 *	3/8 *	1/2 *
Weight		2.05	2.53	3.01	3.48	3.94	4.39	4.83	3.22	4.22
Area		1.69	2.09	2.49	2.87	3.25	3.62	3.99	2.66	3.49
f_1		3/8	3/8	3/8	3/8	3/8	3/8	3/8	3/8	3/8
f_2		1/4	1/4	1/4	1/4	1/4	1/4	1/4	5/16	5/16
Axis	I	2.68	3.29	3.88	4.43	4.96	5.47	5.95	4.02	5.17
X–X	S	0.96	1.19	1.42	1.63	1.85	2.05	2.25	1.43	1.87
	r	1.26	1.25	1.25	1.24	1.24	1.23	1.22	1.23	1.22
	y	1.21	1.24	1.26	1.29	1.31	1.34	1.36	1.18	1.23
Axis	I	1.29	1.58	1.86	2.12	2.36	2.60	2.82	2.85	3.67
Y–Y	S	0.56	0.70	0.83	0.96	1.08	1.20	1.32	1.11	1.46
	r	0.87	0.87	0.86	0.86	0.85	0.85	0.84	1.04	1.03
	x	0.72	0.74	0.77	0.79	0.82	0.84	0.86	0.94	0.99
Axis	Θ	28° 42'	28° 40'	28° 35'	28° 28'	28° 20'	28° 11'	28° 0'	36° 53'	36° 46'
Z–Z	I	0.70	0.85	1.01	1.15	1.30	1.44	1.58	1.36	1.77
	r	0.64	0.64	0.64	0.63	0.63	0.63	0.63	0.72	0.71
J		0.036	0.071	0.123	0.195	0.292	0.415	0.570	0.132	0.313

Size	Legs	4 × 4								
	t	1/4 *	5/16 *	3/8 *	7/16 *	1/2 *	9/16 *	5/8 *	11/16 *	3/4 *
Weight		2.35	2.91	3.46	4.01	4.54	5.07	5.58	6.09	6.58
Area		1.94	2.41	2.86	3.31	3.75	4.19	4.61	5.03	5.44
f_1		3/8	3/8	3/8	3/8	3/8	3/8	3/8	3/8	3/8
f_2		1/4	1/4	1/4	1/4	1/4	1/4	1/4	1/4	1/4
Axis	I	2.94	3.61	4.26	4.87	5.46	6.02	6.56	7.08	7.57
X–X	S	1.00	1.24	1.48	1.71	1.93	2.15	2.36	2.57	2.77
	r	1.23	1.23	1.22	1.21	1.21	1.20	1.19	1.19	1.18
	y	1.07	1.10	1.12	1.15	1.17	1.20	1.22	1.24	1.26
Axis	I	2.94	3.61	4.26	4.87	5.46	6.02	6.56	7.08	7.57
Y–Y	S	1.00	1.24	1.48	1.71	1.93	2.15	2.36	2.57	2.77
	r	1.23	1.23	1.22	1.21	1.21	1.20	1.19	1.19	1.18
	x	1.07	1.10	1.12	1.15	1.17	1.20	1.22	1.24	1.26
Axis	Θ	45° 0'	45° 0'	45° 0'	45° 0'	45° 0'	45° 0'	45° 0'	45° 0'	45° 0'
Z–Z	I	1.21	1.48	1.75	2.01	2.26	2.51	2.76	3.00	3.25
	r	0.79	0.78	0.78	0.78	0.78	0.77	0.77	0.77	0.77
J		0.042	0.081	0.141	0.223	0.333	0.475	0.651	0.867	1.125

Size	Legs	5 × 2 1/2	5 × 3		5 × 3 1/2					
	t	1/2	3/8 *	1/2 *	5/16 *	3/8 *	7/16 *	1/2 *	9/16	5/8 *
Weight		4.24	3.45	4.52	3.09	3.69	4.27	4.84	5.40	5.95
Area		3.50	2.85	3.73	2.56	3.05	3.53	4.00	4.46	4.92
f_1		3/8	3/8	3/8	7/16	7/16	7/16	7/16	7/16	7/16
f_2		1/4	5/16	5/16	5/16	5/16	5/16	5/16	5/16	5/16
Axis	I	8.74	7.15	9.24	6.39	7.56	8.69	9.77	10.82	11.82
X–X	S	2.77	2.15	2.83	1.85	2.21	2.56	2.90	3.24	3.56
	r	1.58	1.59	1.57	1.58	1.58	1.57	1.56	1.56	1.55
	y	1.84	1.68	1.73	1.55	1.58	1.61	1.63	1.66	1.68
Axis	I	1.46	1.93	2.48	2.58	3.04	3.49	3.91	4.32	4.70
Y–Y	S	0.77	0.84	1.10	0.96	1.15	1.33	1.50	1.67	1.84
	r	0.64	0.82	0.81	1.00	1.00	0.99	0.99	0.98	0.98
	x	0.60	0.69	0.74	0.81	0.84	0.87	0.89	0.92	0.94
Axis	Θ	14° 12'	19° 40'	19° 26'	25° 33'	25° 32'	25° 27'	25° 22'	25° 15'	25° 07'
Z–Z	I	0.96	1.17	1.52	1.45	1.71	1.97	2.22	2.46	2.70
	r	0.52	0.64	0.64	0.75	0.75	0.75	0.74	0.74	0.74
J		0.313	0.141	0.333	0.086	0.149	0.237	0.354	0.504	0.692

Table 80. Properties of Angles—*Continued*

Size	Legs	5 × 3½ (Cont.)	5 × 5				6 × 3½			
	t	3/4	3/8 *	7/16 *	1/2 *	5/8 *	5/16 *	3/8 *	1/2 *	5/8
Weight		7.025	4.36	5.05	5.74	7.08	3.49	4.15	5.46	6.73
Area		5.806	3.60	4.18	4.74	5.85	2.88	3.43	4.51	5.56
f_1		1/2	1/2	1/2	1/2	1/2	1/2	1/2	1/2	1/2
f_2		3/8	3/8	3/8	3/8	3/8	5/16	5/16	5/16	5/16
Axis X–X	I	13.619	8.37	9.65	10.89	13.22	10.64	12.60	16.34	19.83
	S	4.161	2.30	2.67	3.03	3.73	2.64	3.15	4.14	5.09
	r	1.532	1.52	1.52	1.52	1.50	1.92	1.92	1.90	1.89
	y	1.727	1.36	1.38	1.41	1.46	1.97	2.00	2.06	2.11
Axis Y–Y	I	5.371	8.37	9.65	10.89	13.22	2.70	3.19	4.11	4.94
	S	2.135	2.30	2.67	3.03	3.73	0.98	1.17	1.53	1.88
	r	0.962	1.52	1.52	1.52	1.50	0.97	0.96	0.95	0.94
	x	0.984	1.36	1.38	1.41	1.46	0.74	0.77	0.82	0.87
Axis Z–Z	θ	24° 45′	45° 0′	45° 0′	45° 0′	45° 0′	18° 52′	18° 51′	18° 42′	18° 28′
	I	3.142	3.44	3.96	4.47	5.47	1.65	1.95	2.52	3.07
	r	0.736	0.98	0.97	0.97	0.97	0.76	0.75	0.75	0.74
J		1.195	0.176	0.279	0.417	0.814	0.097	0.167	0.396	0.773

Size	Legs	6 × 4						
	t	3/8 *	7/16 *	1/2 *	9/16 *	5/8 *	11/16	3/4 *
Weight		4.36	5.05	5.74	6.42	7.08	7.74	8.39
Area		3.60	4.18	4.74	5.30	5.85	6.40	6.93
f_1		1/2	1/2	1/2	1/2	1/2	1/2	1/2
f_2		3/8	3/8	3/8	3/8	3/8	3/8	3/8
Axis X–X	I	13.02	15.02	16.95	18.82	20.63	22.39	24.08
	S	3.17	3.69	4.19	4.69	5.17	5.64	6.11
	r	1.90	1.90	1.89	1.88	1.88	1.87	1.86
	y	1.90	1.93	1.96	1.98	2.01	2.03	2.06
Axis Y–Y	I	4.63	5.34	6.01	6.65	7.27	7.86	8.43
	S	1.50	1.74	1.98	2.21	2.44	2.66	2.87
	r	1.13	1.13	1.13	1.12	1.11	1.11	1.10
	x	0.91	0.94	0.97	0.99	1.02	1.04	1.07
Axis Z–Z	θ	23° 33′	23° 31′	23° 27′	23° 22′	23° 16′	23° 10′	23° 02′
	I	2.67	3.07	3.47	3.86	4.24	4.61	4.98
	r	0.86	0.86	0.86	0.85	0.85	0.85	0.85
J		0.176	0.279	0.417	0.593	0.814	1.08	1.41

Size	Legs	6 × 6						
	t	3/8 *	7/16 *	1/2 *	9/16	5/8 *	11/16	3/4
Weight		5.27	6.11	6.95	7.78	8.59	9.40	10.20
Area		4.35	5.05	5.74	6.43	7.10	7.77	8.43
f_1		1/2	1/2	1/2	1/2	1/2	1/2	1/2
f_2		3/8	3/8	3/8	3/8	3/8	3/8	3/8
Axis X–X	I	14.85	17.15	19.38	21.54	23.64	25.67	27.64
	S	3.38	3.93	4.46	4.99	5.51	6.02	6.52
	r	1.85	1.84	1.84	1.83	1.82	1.82	1.81
	y	1.60	1.63	1.66	1.68	1.71	1.73	1.76
Axis Y–Y	I	14.85	17.15	19.38	21.54	23.64	25.67	27.64
	S	3.38	3.93	4.46	4.99	5.51	6.02	6.52
	r	1.85	1.84	1.84	1.83	1.82	1.82	1.81
	x	1.60	1.63	1.66	1.68	1.71	1.73	1.76
Axis Z–Z	θ	45° 0′	45° 0′	45° 0′	45° 0′	45° 0′	45° 0′	45° 0′
	I	6.07	7.01	7.92	8.82	9.70	10.57	11.43
	r	1.18	1.18	1.17	1.17	1.17	1.17	1.16
J		0.211	0.335	0.500	0.712	0.977	1.30	1.69

Table 80.   Properties of Angles—*Continued*

Size	Legs	8 × 6			8 × 8		
	t	5/8 *	11/16 *	3/4 *	1/2 *	3/4 *	1 *
Weight		10.129	11.07	12.016	9.41	13.87	18.18
Area		8.371	9.15	9.931	7.77	11.46	15.02
f_1		1/2	1/2	1/2	5/8	5/8	5/8
f_2		5/16	3/8	3/8	3/8	3/8	3/8
Axis	I	53.571	57.99	62.603	47.74	68.86	88.11
X–X	S	9.737	10.58	11.474	8.16	11.99	15.60
	r	2.530	2.52	2.511	2.48	2.45	2.42
	y	2.498	2.52	2.544	2.15	2.26	2.35
Axis	I	25.939	27.98	30.150	47.74	68.86	88.11
Y–Y	S	5.770	6.25	6.774	8.16	11.99	15.60
	r	1.760	1.75	1.742	2.48	2.45	2.42
	x	1.504	1.53	1.549	2.15	2.26	2.35
Axis	Θ	28° 52′	28° 47′	28° 43′	45° 0′	45° 0′	45° 0′
Z–Z	I	13.891	15.02	16.236	19.51	28.20	36.46
	r	1.288	1.28	1.279	1.58	1.57	1.56
J		1.139	1.516	1.969	0.667	2.25	5.33

Table 81. Properties of Standard I-Beams

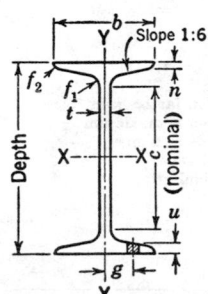

Elements of Sections

All dimensions in inches.
Weight in pounds per foot.
Area in square inches.
I = moment of inertia in in.[4]
S = section modulus in in.[3]
r = radius of gyration in inches.

J = torsion factor in in.[4]
Rivet given is maximum allowable in flange.
g = usual gage.
u = nominal grip.

Size	Depth	3			4		5			6	
	t	0.170*	0.251	0.349*	0.190*	0.326*	0.210*	0.347	0.494*	0.230*	0.343*
Weight		2.02	2.31	2.67	2.72	3.38	3.53	4.36	5.25	4.43	5.25
Area		1.67	1.91	2.21	2.25	2.79	2.92	3.60	4.34	3.66	4.34
b		2.330	2.411	2.509	2.660	2.796	3.000	3.137	3.284	3.330	3.443
n		0.170	0.170	0.170	0.190	0.190	0.210	0.210	0.210	0.230	0.230
f_1		0.27	0.27	0.27	0.29	0.29	0.31	0.31	0.31	0.33	0.33
f_2		0.10	0.10	0.10	0.11	0.11	0.13	0.13	0.13	0.14	0.14
c		1 3/4	1 3/4	1 3/4	2 3/4	2 3/4	3 1/2	3 1/2	3 1/2	4 1/2	4 1/2
Axis X-X	I	2.52	2.71	2.93	6.06	6.79	12.26	13.69	15.22	22.08	24.11
	S	1.68	1.80	1.95	3.03	3.39	4.90	5.48	6.09	7.36	8.04
	r	1.23	1.19	1.15	1.64	1.56	2.05	1.95	1.87	2.46	2.36
Axis Y-Y	I	0.46	0.51	0.59	0.76	0.90	1.21	1.41	1.66	1.82	2.04
	S	0.39	0.42	0.47	0.57	0.65	0.81	0.90	1.01	1.09	1.19
	r	0.52	0.52	0.52	0.58	0.57	0.64	0.63	0.62	0.71	0.69
Rivet Data	Diam.	3/8	3/8	3/8	1/2	1/2	1/2	1/2	1/2	5/8	5/8
	g	3/4	3/4	3/4	3/4	3/4	7/8	7/8	7/8	1	1
	u	5/16	5/16	5/16	5/16	5/16	3/8	3/8	3/8	3/8	3/8
J		0.045	0.061	0.093	0.074	0.12	0.12	0.19	0.33	0.17	0.24

Size	Depth	6 (Cont.)	7	8		9	10		12	
	t	0.465	0.345*	0.270*	0.532*	0.290	0.310*	0.594	0.350*	0.565
Weight		6.13	6.23	6.53	9.07	7.72	9.01	12.45	11.31	16.01
Area		5.07	5.15	5.40	7.49	6.38	7.45	10.29	9.35	13.23
b		3.565	3.755	4.000	4.262	4.330	4.660	4.944	5.000	5.355
n		0.230	0.250	0.270	0.270	0.290	0.310	0.310	0.350	0.460
f_1		0.33	0.35	0.37	0.37	0.39	0.41	0.41	0.45	0.56
f_2		0.14	0.15	0.16	0.16	0.17	0.19	0.19	0.21	0.28
c		4 1/2	5 1/4	6 1/4	6 1/4	7	8	8	9 3/4	9 1/4
Axis X-X	I	26.31	39.40	57.55	68.73	85.90	123.39	147.06	218.13	287.27
	S	8.77	11.26	14.39	17.18	19.09	24.68	29.41	36.35	47.88
	r	2.28	2.77	3.27	3.03	3.67	4.07	3.78	4.83	4.66
Axis Y-Y	I	2.31	2.88	3.73	4.66	5.09	6.78	8.36	9.35	14.50
	S	1.30	1.53	1.86	2.19	2.35	2.91	3.38	3.74	5.42
	r	0.68	0.75	0.83	0.79	0.89	0.95	0.90	1.00	1.05
Rivet Data	Diam.	5/8	5/8	3/4	3/4	3/4	3/4	3/4	3/4	3/4
	g	1	1 1/8	1 1/8	1 1/8	1 1/4	1 3/8	1 3/8	1 1/2	1 1/2
	u	3/8	3/8	7/16	1/2	1/2	1/2	1/2	9/16	3/4
J		0.38	0.32	0.34	0.75	0.46	0.62	1.31	0.92	2.19

Table 82.   Properties of Standard Channels

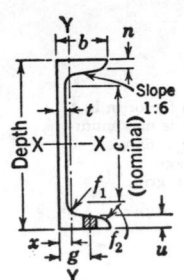

Elements of Sections

All dimensions in inches.
Weight in pounds per foot.
Area in square inches.
I = moment of inertia in in.[4]
S = section modulus in in.[3]
r = radius of gyration in inches.

J = torsion factor in in.[4]
Rivet given is maximum
allowable in flange.
g = usual gage.
u = nominal grip.

Size		3					4			5
Depth										
t		0.170*	0.187	0.258*	0.320	0.356*	0.180*	0.247*	0.320*	0.190*
Weight		1.46	1.52	1.78	2.00	2.13	1.90	2.22	2.58	2.38
Area		1.21	1.26	1.47	1.66	1.76	1.57	1.84	2.13	1.97
b		1.410	1.427	1.498	1.560	1.596	1.580	1.647	1.720	1.750
n		0.170	0.170	0.170	0.170	0.170	0.180	0.180	0.180	0.190
f_1		0.27	0.27	0.27	0.27	0.27	0.28	0.28	0.28	0.29
f_2		0.10	0.10	0.10	0.10	0.10	0.11	0.11	0.11	0.11
c		1 3/4	1 3/4	1 3/4	1 3/4	1 3/4	2 3/4	2 3/4	2 3/4	3 3/4
Axis X–X	I	1.66	1.69	1.85	1.99	2.07	3.83	4.19	4.58	7.49
	S	1.10	1.13	1.24	1.33	1.38	1.92	2.10	2.29	3.00
	r	1.17	1.16	1.12	1.10	1.08	1.56	1.51	1.47	1.95
Axis Y–Y	I	0.20	0.21	0.25	0.28	0.31	0.32	0.37	0.43	0.48
	S	0.20	0.21	0.23	0.25	0.27	0.28	0.31	0.34	0.38
	r	0.40	0.41	0.41	0.41	0.42	0.45	0.45	0.45	0.49
	x	0.44	0.44	0.44	0.45	0.46	0.46	0.45	0.46	0.48
Rivet Data	Diam.	1/2	1/2	1/2	1/2	1/2	1/2	1/2	1/2	1/2
	g	7/8	7/8	7/8	7/8	7/8	1	1	1	1 1/8
	u	1/4	1/4	1/4	1/4	1/4	5/16	5/16	5/16	5/16
J		0.031	0.033	0.047	0.066	0.080	0.045	0.062	0.090	0.064

Size		5 (Cont.)			6				7	
Depth										
t		0.225	0.325*	0.472*	0.200*	0.225*	0.314*	0.437*	0.230*	0.314*
Weight		2.59	3.20	4.09	2.91	3.09	3.73	4.63	3.64	4.36
Area		2.14	2.64	3.38	2.40	2.55	3.09	3.82	3.01	3.60
b		1.785	1.885	2.032	1.920	1.945	2.034	2.157	2.110	2.194
n		0.190	0.190	0.190	0.200	0.200	0.200	0.200	0.210	0.210
f_1		0.29	0.29	0.29	0.30	0.30	0.30	0.30	0.31	0.31
f_2		0.11	0.11	0.11	0.12	0.12	0.12	0.12	0.13	0.13
c		3 3/4	3 3/4	3 3/4	4 1/2	4 1/2	4 1/2	4 1/2	5 1/2	5 1/2
Axis X–X	I	7.86	8.90	10.43	13.12	13.57	15.18	17.39	21.84	24.24
	S	3.14	3.56	4.17	4.37	4.52	5.06	5.80	6.24	6.93
	r	1.91	1.83	1.76	2.34	2.31	2.22	2.13	2.69	2.60
Axis Y–Y	I	0.52	0.63	0.81	0.69	0.73	0.87	1.05	1.01	1.17
	S	0.40	0.45	0.53	0.49	0.51	0.56	0.64	0.64	0.70
	r	0.49	0.49	0.49	0.54	0.54	0.53	0.52	0.58	0.57
	x	0.48	0.48	0.51	0.51	0.51	0.50	0.51	0.54	0.52
Rivet Data	Diam.	1/2	1/2	1/2	5/8	5/8	5/8	5/8	5/8	5/8
	g	1 1/8	1 1/8	1 1/8	1 1/8	1 1/8	1 1/8	1 3/8	1 1/4	1 1/4
	u	5/16	5/16	5/16	5/16	5/16	3/8	3/8	7/16	3/8
J		0.074	0.12	0.25	0.088	0.097	0.14	0.26	0.13	0.18

Table 82. Properties of Standard Channels—*Continued*

Size		7 (Cont.)		8					9	
	Depth t	0.419*	0.524	0.250*	0.303*	0.395*	0.487*	0.520	0.230*	0.448*
Weight		5.24	6.13	4.38	4.89	5.78	6.67	6.99	4.74	7.11
Area		4.33	5.07	3.62	4.04	4.78	5.51	5.78	3.91	5.88
b		2.299	2.404	2.290	2.343	2.435	2.527	2.560	2.430	2.648
n		0.210	0.210	0.220	0.220	0.220	0.220	0.220	0.230	0.230
f_1		0.31	0.31	0.32	0.32	0.32	0.32	0.32	0.33	0.33
f_2		0.13	0.13	0.13	0.13	0.13	0.13	0.13	0.14	0.14
c		5 1/2	5 1/2	6 1/2	6 1/4	6 1/4	6 1/4	6 1/4	7 1/4	7 1/4
Axis X–X	I	27.24	30.25	33.85	36.11	40.04	43.96	45.37	47.68	60.92
	S	7.78	8.64	8.46	9.03	10.01	10.99	11.34	10.60	13.54
	r	2.51	2.44	3.06	2.99	2.90	2.82	2.80	3.49	3.22
Axis Y–Y	I	1.38	1.59	1.40	1.53	1.75	1.98	2.07	1.75	2.42
	S	0.78	0.86	0.81	0.85	0.93	1.01	1.04	0.96	1.17
	r	0.56	0.56	0.62	0.61	0.61	0.60	0.60	0.67	0.64
	x	0.53	0.55	0.56	0.55	0.55	0.57	0.57	0.60	0.58
Rivet Data	Diam.	5/8	5/8	3/4	3/4	3/4	3/4	3/4	3/4	3/4
	g	1 1/4	1 1/2	1 3/8	1 3/8	1 1/2	1 1/2	1 1/2	1 3/8	1 1/2
	u	7/16	7/16	3/8	3/8	7/16	7/16	7/16	7/16	1/2
J		0.29	0.47	0.17	0.21	0.32	0.47	0.55	0.20	0.47

Size		9 (Cont.)	10		12				15	
	Depth t	0.612	0.240*	0.526*	0.300*	0.387*	0.510*	0.632	0.400*	0.716*
Weight		8.90	5.43	8.89	7.63	8.89	10.67	12.45	12.05	17.78
Area		7.35	4.49	7.35	6.30	7.35	8.82	10.29	9.96	14.70
b		2.812	2.600	2.886	2.960	3.047	3.170	3.292	3.400	3.716
n		0.230	0.240	0.240	0.280	0.280	0.280	0.280	0.400	0.400
f_1		0.33	0.34	0.34	0.38	0.38	0.38	0.38	0.500	0.500
f_2		0.14	0.14	0.14	0.17	0.17	0.17	0.17	0.240	0.240
c		7 1/4	8 1/4	8 1/4	10	10	10	10	12 3/8	12 3/8
Axis X–X	I	70.89	67.37	91.20	131.84	144.37	162.08	179.65	314.76	403.64
	S	15.75	13.47	18.24	21.97	24.06	27.01	29.94	41.97	53.82
	r	3.11	3.87	3.52	4.57	4.43	4.29	4.18	5.62	5.24
Axis Y–Y	I	2.94	2.28	3.36	3.99	4.47	5.14	5.82	9.63	12.53
	S	1.34	1.16	1.48	1.76	1.89	2.06	2.24	3.11	4.30
	r	0.63	0.71	0.68	0.80	0.78	0.76	0.75	0.90	0.92
	x	0.61	0.63	0.62	0.69	0.67	0.67	0.69	0.79	0.80
Rivet Data	Diam.	3/4	3/4	3/4	7/8	7/8	7/8	7/8	1	1
	g	1 1/2	1 1/2	1 3/4	1 3/4	1 3/4	1 3/4	2	2	2
	u	1/2	7/16	1/2	1/2	1/2	1/2	5/8	5/8	5/8
J		0.92	0.25	0.75	0.46	0.61	0.95	1.48	1.17	2.89

Table 83. Properties of Wide-Flange Beams

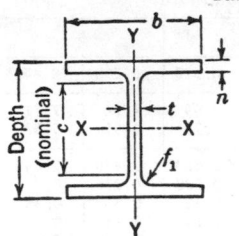

Elements of Sections

All dimensions in inches.
Weight in pounds per foot.
Area in square inches.
I = moment of inertia in in.[4]

S = section modulus in in.[3]
r = radius of gyration in inches.
J = torsion factor in in.[4]

Nominal Size		6×4	6×6	8×5	8×7	8×8	$10 \times 5\,3/4$
Actual Depth		6.00 *	6.00 *	8.00 *	8.00 *	8.00 *	9.90 *
t		0.230	0.240	0.230	0.245	0.288	0.240
Weight Area		4.28 3.54	5.56 4.59	6.07 5.02	8.56 7.08	11.04 9.12	7.51 6.21
b n f_1 c		4.00 0.279 0.250 4 7/8	6.00 0.269 0.250 4 7/8	5.25 0.308 0.320 6 3/4	6.50 0.398 0.400 6 3/8	8.00 0.433 0.400 6 3/8	5.75 0.340 0.312 8 1/2
Axis X–X	I S r	21.75 7.25 2.48	30.17 10.06 2.56	56.73 14.18 3.36	84.15 21.04 3.45	109.66 27.41 3.47	106.74 21.56 4.15
Axis Y–Y	I S r	2.98 1.49 0.92	9.69 3.23 1.45	7.44 2.83 1.22	18.23 5.61 1.61	36.97 9.24 2.01	10.77 3.75 1.32
J		0.082	0.106	0.135	0.312	0.497	0.196

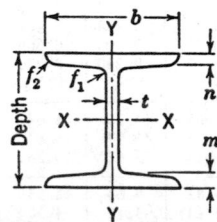

Table 84. Properties of H-Beams

Elements of Sections

All dimensions in inches.
Weight in pounds per foot.
Area in square inches.
I = moment of inertia in in.[4]

S = section modulus in in.[3]
r = radius of gyration in inches.
J = torsion factor in in.[4]

Size	Depth	4	5	6			8		
	t	0.313*	0.313*	0.250*	0.313	0.438	0.313*	0.375	0.500*
Weight Area		4.85 4.00	6.63 5.48	8.04 6.64	8.49 7.02	9.40 7.77	11.51 9.52	12.11 10.01	13.32 11.01
b m n f_1 f_2		4.000 0.453 0.290 0.313 0.145	5.000 0.503 0.330 0.313 0.165	5.938 0.542 0.360 0.313 0.180	6.000 0.542 0.360 0.313 0.180	6.125 0.542 0.360 0.313 0.180	7.938 0.560 0.358 0.313 0.179	8.000 0.560 0.358 0.313 0.179	8.125 0.560 0.358 0.313 0.179
Axis X–X	I S r	10.72 5.36 1.64	23.82 9.53 2.08	44.06 14.69 2.58	45.19 15.06 2.54	47.44 15.81 2.47	112.94 28.23 3.45	115.58 28.90 3.40	120.92 30.23 3.31
Axis Y–Y	I S r	3.56 1.78 0.94	7.82 3.13 1.19	14.18 4.77 1.46	14.65 4.88 1.44	15.65 5.11 1.42	34.15 8.60 1.89	35.01 8.75 1.87	36.79 9.06 1.83
J		0.22	0.34	0.45	0.50	0.62	0.68	0.75	0.96

Table 85. Properties of Tees

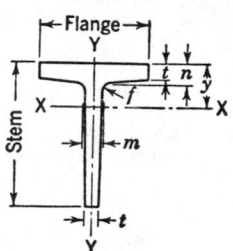

Elements of Sections

All dimensions in inches. I = moment of inertia in in.[4]
Weight in pounds per foot. S = section modulus in in.[3]
Area in square inches. r = radius of gyration in inches.

Size	Flange	1	1½					2		2¼
	Stem	1	1¼	1¼	1½	1½	2	2*	2	2¼*
	t	1/8	1/8	3/16	3/16	1/4	3/16	1/4	5/16	1/4
	Weight	0.323	0.451	0.633	0.704	0.895	0.884	1.29	1.55	1.47
	Area	0.267	0.373	0.523	0.581	0.740	0.730	1.07	1.28	1.21
	m	5/32	5/32	7/32	7/32	9/32	1/4	5/16	3/8	5/16
	n	5/32	5/32	7/32	7/32	9/32	1/4	5/16	3/8	5/16
	f	1/8	1/8	1/8	3/16	3/16	3/16	1/4	1/4	1/4
Axis X-X	I	0.023	0.049	0.067	0.114	0.142	0.269	0.37	0.43	0.53
	S	0.032	0.053	0.075	0.108	0.137	0.195	0.26	0.31	0.33
	r	0.293	0.363	0.359	0.443	0.438	0.606	0.59	0.58	0.66
	y	0.292	0.326	0.352	0.437	0.464	0.624	0.58	0.61	0.64
Axis Y-Y	I	0.011	0.038	0.056	0.056	0.075	0.060	0.18	0.23	0.26
	S	0.023	0.051	0.075	0.075	0.100	0.080	0.18	0.23	0.23
	r	0.206	0.319	0.328	0.312	0.319	0.286	0.41	0.42	0.46

Size	Flange	2½			3		4				4½	5
	Stem	1¼	2½*	3	2½	3*	2	3	4*	5	5	3	3
	t	3/16	5/16	5/16	5/16	3/8	3/8	5/16	3/8	3/8	1/2	5/16	3/8
	Weight	1.03	1.97	2.17	2.19	2.79	2.78	2.84	3.85	4.34	5.56	3.04	4.14
	Area	0.85	1.62	1.80	1.81	2.31	2.30	2.34	3.18	3.59	4.60	2.52	3.42
	m	5/16	3/8	3/8	3/8	7/16	7/16	3/8	7/16	7/16	9/16	3/8	5/8
	n	9/32	3/8	3/8	3/8	7/16	7/16	3/8	7/16	7/16	9/16	3/8	7/16
	f	3/16	1/4	1/4	5/16	5/16	1/4	3/8	1/2	1/2	1/2	3/8	3/8
Axis X-X	I	0.08	0.89	1.49	0.94	1.83	0.60	1.72	4.56	8.56	10.84	1.78	2.37
	S	0.09	0.50	0.72	0.51	0.86	0.40	0.77	1.58	2.43	3.14	0.78	1.06
	r	0.31	0.74	0.91	0.72	0.89	0.51	0.86	1.20	1.54	1.54	0.84	0.83
	y	0.30	0.73	0.92	0.68	0.88	0.48	0.75	1.11	1.48	1.54	0.71	0.76
Axis Y-Y	I	0.28	0.44	0.44	0.75	0.90	2.10	1.77	2.12	2.13	2.83	2.52	4.13
	S	0.22	0.35	0.35	0.50	0.60	1.05	0.89	1.06	1.06	1.42	1.12	1.65
	r	0.57	0.52	0.50	0.65	0.63	0.96	0.87	0.82	0.77	0.79	1.00	1.10

Table 86.　Properties of Zees

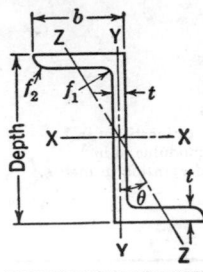

Elements of Sections

All dimensions in inches.
Weight in pounds per foot.
Area in square inches.
I = moment of inertia in in.[4]

S = section modulus in in.[3]
r = radius of gyration in inches.
J = torsion factor in in.[4]

Size	Nominal Depth	1 3/4	2	2 3/8	3		4	
	t	3/16	3/16	3/16	1/4 *	3/8 *	1/4 *	5/16 *
Weight		1.116	0.946	1.031	2.40	3.48	2.93	3.68
Area		0.922	0.782	0.852	1.98	2.87	2.42	3.04
Actual Depth		1 3/4	2	2 3/8	3	3	4	4 1/16
b		1 3/4	1 1/4	1 1/4	2 11/16	2 11/16	3 1/16	3 1/8
f_1		3/16	3/16	3/16	5/16	5/16	5/16	5/16
f_2		1/8	1/8	1/8	1/4	1/4	1/4	1/4
Axis X–X	I	0.446	0.458	0.694	2.89	3.86	6.32	7.97
	S	0.510	0.458	0.584	1.92	2.57	3.16	3.92
	r	0.695	0.765	0.902	1.21	1.16	1.62	1.62
Axis Y–Y	I	0.551	0.186	0.186	2.64	3.76	4.01	5.24
	S	0.333	0.161	0.161	1.03	1.50	1.36	1.76
	r	0.773	0.488	0.467	1.15	1.14	1.29	1.31
Axis Z–Z	θ	48° 49'	29° 12'	23° 12'	43° 24'	44° 31'	36° 47'	37° 24'
	I	0.101	0.063	0.082	0.59	0.82	1.08	1.39
	r	0.331	0.283	0.310	0.54	0.53	0.67	0.68
J		0.012	0.010	0.011	0.044	0.15	0.053	0.10

Size	Nominal Depth	4 (Cont.)			5			6
	t	3/8 *	7/16	9/16	5/16	3/8 *	1/2 *	11/16
Weight		4.44	4.92	6.40	4.13	4.98	6.37	10.00
Area		3.67	4.06	5.29	3.41	4.12	5.27	8.27
Actual Depth		4 1/8	4	4 1/8	5	5 1/16	5	6 1/8
b		3 3/16	3 1/16	3 3/16	3 1/4	.3 5/16	3 1/4	3 5/8
f_1		5/16	5/16	5/16	5/16	5/16	5/16	5/16
f_2		1/4	1/4	1/4	1/4	1/4	1/4	1/4
Axis X–X	I	9.66	9.68	12.76	13.41	16.23	19.23	43.24
	S	4.68	4.84	6.19	5.36	6.41	7.69	14.12
	r	1.62	1.54	1.55	1.98	1.99	1.91	2.29
Axis Y–Y	I	6.54	6.53	9.05	5.94	7.40	8.82	16.07
	S	2.18	2.30	3.11	1.92	2.37	2.94	4.90
	r	1.33	1.27	1.31	1.32	1.34	1.29	1.39
Axis Z–Z	θ	37° 55'	37° 50'	38° 41'	30° 40'	31° 08'	31° 09'	27° 44'
	I	1.72	1.74	2.41	1.89	2.33	2.82	5.70
	r	0.68	0.66	0.68	0.74	0.75	0.73	0.83
J		0.18	0.28	0.62	0.12	0.21	0.48	1.45

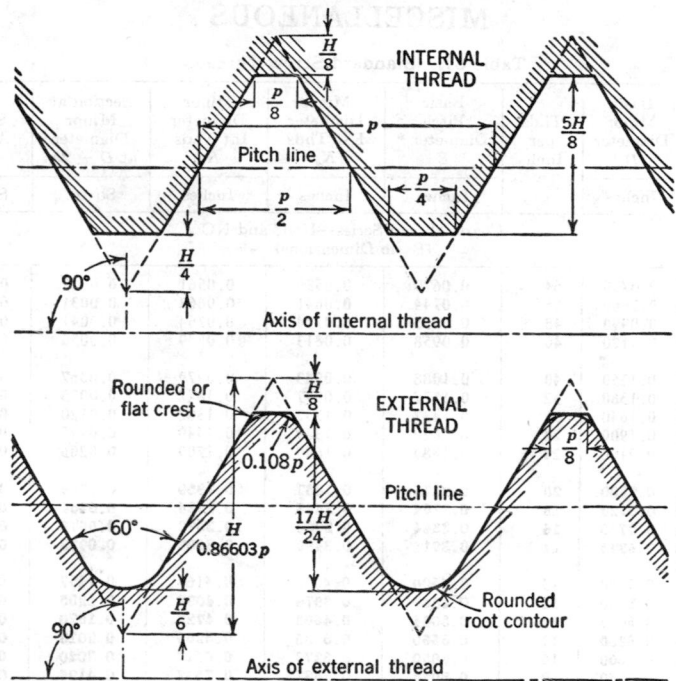

Standard Screw Threads

The Unified and American Screw Threads included in Table 87 are taken from the publication of the American Standards Association, ASA B1.1—1949. The *coarse-thread series* is the former United States Standard Series. It is recommended for general use in engineering work where conditions do not require the use of a fine thread. The *fine-thread series* is the former "Regular Screw Thread Series" established by the Society of Automotive Engineers. The *fine-thread series* is recommended for general use in automotive and aircraft work, and where special conditions require a fine thread. The *extra-fine-thread series* is the same as the former SAE fine series and the present SAE extra-fine series. It is used particularly in aircraft and aeronautical equipment where (1) thin-walled material is to be threaded; (2) thread depth of nuts clearing ferrules, coupling flanges, etc., must be held to a minimum; and (3) a maximum practicable number of threads is required within a given thread length.

The method of designating a screw thread is by the use of the initial letters of the thread series, preceded by the nominal size (diameter in inches or the screw number) and number of threads per inch, all in Arabic numerals, and followed by the classification designation, with or without the pitch diameter tolerances or limits of size. An example of an external thread designation and its meaning is given below:

Example. 1/4″—20 UNC—2A

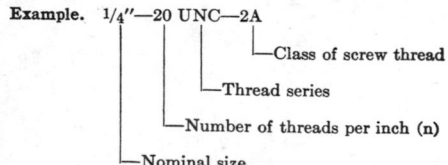

        └─Class of screw thread

      └─Thread series

    └─Number of threads per inch (n)

  └─Nominal size

A left-hand thread must be identified by the letters "LH" following the class designation. If no such designation is used, the thread is assumed to be right hand.

Classes of thread are distinguished from each other by the amounts of tolerance and allowance specified in ASA B1.1—1949.

MISCELLANEOUS

Table 87.   Standard Screw Threads

Sizes	Basic Major Diameter D	Thds per Inch n	Basic Pitch Diameter * E	Minor Diameter Ext Thds K_s	Minor Diameter Int Thds K_n	Section at Minor Diameter at $D - 2h_b$	Stress Area †
	Inches		Inches	Inches	Inches	Sq In.	Sq In.

Coarse-thread Series—UNC and NC
(Basic Dimensions)

Sizes	D	n	E	K_s	K_n	Section	Stress Area
1 (.073)	0.0730	64	0.0629	0.0538	0.0561	0.0022	0.0026
2 (.086)	0.0860	56	0.0744	0.0641	0.0667	0.0031	0.0036
3 (.099)	0.0990	48	0.0855	0.0734	0.0764	0.0041	0.0048
4 (.112)	0.1120	40	0.0958	0.0813	0.0849	0.0050	0.0060
5 (.125)	0.1250	40	0.1088	0.0943	0.0979	0.0067	0.0079
6 (.138)	0.1380	32	0.1177	0.0997	0.1042	0.0075	0.0090
8 (.164)	0.1640	32	0.1437	0.1257	0.1302	0.0120	0.0139
10 (.190)	0.1900	24	0.1629	0.1389	0.1449	0.0145	0.0174
12 (.216)	0.2160	24	0.1889	0.1649	0.1709	0.0206	0.0240
1/4	**0.2500**	**20**	**0.2175**	**0.1887**	**0.1959**	**0.0269**	**0.0317**
5/16	**0.3125**	**18**	**0.2764**	**0.2443**	**0.2524**	**0.0454**	**0.0522**
3/8	**0.3750**	**16**	**0.3344**	**0.2983**	**0.3073**	**0.0678**	**0.0773**
7/16	**0.4375**	**14**	**0.3911**	**0.3499**	**0.3602**	**0.0933**	**0.1060**
1/2	0.5000	13	0.4500	0.4056	0.4167	0.1257	0.1416
1/2	**0.5000**	**12**	**0.4459**	**0.3978**	**0.4098**	**0.1205**	**0.1374**
9/16	**0.5625**	**12**	**0.5084**	**0.4603**	**0.4723**	**0.1620**	**0.1816**
5/8	**0.6250**	**11**	**0.5660**	**0.5135**	**0.5266**	**0.2018**	**0.2256**
3/4	**0.7500**	**10**	**0.6850**	**0.6273**	**0.6417**	**0.3020**	**0.3340**
7/8	**0.8750**	**9**	**0.8028**	**0.7387**	**0.7547**	**0.4193**	**0.4612**
1	1.0000	8	0.9188	0.8466	0.8647	0.5510	0.6051
1 1/8	**1.1250**	**7**	**1.0322**	**0.9497**	**0.9704**	**0.6931**	**0.7627**
1 1/4	**1.2500**	**7**	**1.1572**	**1.0747**	**1.0954**	**0.8898**	**0.9684**
1 3/8	**1.3750**	**6**	**1.2667**	**1.1705**	**1.1946**	**1.0541**	**1.1538**
1 1/2	**1.5000**	**6**	**1.3917**	**1.2955**	**1.3196**	**1.2938**	**1.4041**
1 3/4	**1.7500**	**5**	**1.6201**	**1.5046**	**1.5335**	**1.7441**	**1.8983**
2	**2.0000**	**4 1/2**	**1.8557**	**1.7274**	**1.7594**	**2.3001**	**2.4971**
2 1/4	**2.2500**	**4 1/2**	**2.1057**	**1.9774**	**2.0094**	**3.0212**	**3.2464**
2 1/2	**2.5000**	**4**	**2.3376**	**2.1933**	**2.2294**	**3.7161**	**3.9976**
2 3/4	**2.7500**	**4**	**2.5876**	**2.4433**	**2.4794**	**4.6194**	**4.9326**
3	**3.0000**	**4**	**2.8376**	**2.6933**	**2.7294**	**5.6209**	**5.9659**
3 1/4	**3.2500**	**4**	**3.0876**	**2.9433**	**2.9794**	**6.7205**	**7.0992**
3 1/2	**3.5000**	**4**	**3.3376**	**3.1933**	**3.2294**	**7.9183**	**8.3268**
3 3/4	**3.7500**	**4**	**3.5876**	**3.4433**	**3.4794**	**9.2143**	**9.6546**
4	**4.0000**	**4**	**3.8376**	**3.6933**	**3.7294**	**10.6084**	**11.0805**

* British: Effective Diameter.
† The stress area is the assumed area of an externally threaded part which is used for the purpose of computing the tensile strength.
Bold type indicates Unified threads—UNF.

Table 87. Standard Screw Threads—*Continued*

Sizes	Basic Major Diameter D Inches	Thds per Inch n	Basic Pitch Diameter * E Inches	Minor Diameter Ext Thds K_s Inches	Minor Diameter Int Thds K_n Inches	Section at Minor Diameter at $D-2h_b$ Sq In.	Stress Area † Sq In.
			Fine-Thread Series—UNF and NF (Basic Dimensions)				
0 (.060)	0.0600	80	0.0519	0.0447	0.0465	0.0015	0.0018
1 (.073)	0.0730	72	0.0640	0.0560	0.0580	0.0024	0.0027
2 (.086)	0.0860	64	0.0759	0.0668	0.0691	0.0034	0.0039
3 (.099)	0.0990	56	0.0874	0.0771	0.0797	0.0045	0.0052
4 (.112)	0.1120	48	0.0985	0.0864	0.0894	0.0057	0.0065
5 (.125)	0.1250	44	0.1102	0.0971	0.1004	0.0072	0.0082
6 (.138)	0.1380	40	0.1218	0.1073	0.1109	0.0087	0.0101
8 (.164)	0.1640	36	0.1460	0.1299	0.1339	0.0128	0.0146
10 (.190)	0.1900	32	0.1697	0.1517	0.1562	0.0175	0.0199
12 (.216)	0.2160	28	0.1928	0.1722	0.1773	0.0226	0.0257
1/4	**0.2500**	**28**	**0.2268**	**0.2062**	**0.2113**	**0.0326**	**0.0362**
5/16	**0.3125**	**24**	**0.2854**	**0.2614**	**0.2674**	**0.0524**	**0.0579**
3/8	**0.3750**	**24**	**0.3479**	**0.3239**	**0.3299**	**0.0809**	**0.0876**
7/16	**0.4375**	**20**	**0.4050**	**0.3762**	**0.3834**	**0.1090**	**0.1185**
1/2	**0.5000**	**20**	**0.4675**	**0.4387**	**0.4459**	**0.1486**	**0.1597**
9/16	**0.5625**	**18**	**0.5264**	**0.4943**	**0.5024**	**0.1888**	**0.2026**
5/8	**0.6250**	**18**	**0.5889**	**0.5568**	**0.5649**	**0.2400**	**0.2555**
3/4	**0.7500**	**16**	**0.7094**	**0.6733**	**0.6823**	**0.3513**	**0.3724**
7/8	**0.8750**	**14**	**0.8286**	**0.7874**	**0.7977**	**0.4805**	**0.5088**
1	**1.0000**	**12**	**0.9459**	**0.8978**	**0.9098**	**0.6245**	**0.6624**
1 1/8	**1.1250**	**12**	**1.0709**	**1.0228**	**1.0348**	**0.8118**	**0.8549**
1 1/4	**1.2500**	**12**	**1.1959**	**1.1478**	**1.1598**	**1.0237**	**1.0721**
1 3/8	**1.3750**	**12**	**1.3209**	**1.2728**	**1.2848**	**1.2602**	**1.3137**
1 1/2	**1.5000**	**12**	**1.4459**	**1.3978**	**1.4098**	**1.5212**	**1.5799**
			Extra-Fine-Thread Series—NEF (Basic Dimensions)				
12 (.216)	0.2160	32	0.1957	0.1777	0.1822	0.0242	0.0269
1/4	0.2500	32	0.2297	0.2117	0.2162	0.0344	0.0377
5/16	0.3125	32	0.2922	0.2742	0.2787	0.0581	0.0622
3/8	0.3750	32	0.3547	0.3367	0.3412	0.0878	0.0929
7/16	0.4375	28	0.4143	0.3937	0.3988	0.1201	0.1270
1/2	0.5000	28	0.4768	0.4562	0.4613	0.1616	0.1695
9/16	0.5625	24	0.5354	0.5114	0.5174	0.2030	0.2134
5/8	0.6250	24	0.5979	0.5739	0.5799	0.2560	0.2676
11/16	0.6875	24	0.6604	0.6364	0.6424	0.3151	0.3280
3/4	0.7500	20	0.7175	0.6887	0.6959	0.3685	0.3855
13/16	0.8125	20	0.7800	0.7512	0.7584	0.4388	0.4573
7/8	0.8750	20	0.8425	0.8137	0.8209	0.5153	0.5352
15/16	0.9375	20	0.9050	0.8762	0.8834	0.5979	0.6194
1	1.0000	20	0.9675	0.9387	0.9459	0.6866	0.7095
1 1/16	1.0625	18	1.0264	0.9943	1.0024	0.7702	0.7973
1 1/8	1.1250	18	1.0889	1.0568	1.0649	0.8705	0.8993
1 3/16	1.1875	18	1.1514	1.1193	1.1274	0.9770	1.0074
1 1/4	1.2500	18	1.2139	1.1818	1.1899	1.0895	1.1216
1 5/16	1.3125	18	1.2764	1.2443	1.2524	1.2082	1.2420
1 3/8	1.3750	18	1.3389	1.3068	1.3149	1.3330	1.3684
1 7/16	1.4375	18	1.4014	1.3693	1.3774	1.4640	1.5010
1 1/2	1.5000	18	1.4639	1.4318	1.4399	1.6011	1.6397
1 9/16	1.5625	18	1.5264	1.4943	1.5024	1.7444	1.7846
1 5/8	1.6250	18	1.5889	1.5568	1.5649	1.8937	1.9357
1 11/16	1.6875	18	1.6514	1.6193	1.6274	2.0493	2.0929
1 3/4	1.7500	16	1.7094	1.6733	1.6823	2.1873	2.2382
2	2.0000	16	1.9594	1.9233	1.9323	2.8917	2.9501

* British: Effective Diameter.

† The stress area is the assumed area of an externally threaded part which is used for the purpose of computing the tensile strength.

Bold type indicates Unified threads—UNC.

Table 88. ASA * Standard Bolts and Nuts

Nominal Size	Across Flats	Across Square Corners	Across Hex Corners	Thickness Unfinished	Thickness Semi-finished
		Regular Bolt Heads			
1/4	3/8	0.498	0.413	11/64	5/32
5/16	1/2	0.665	0.552	13/64	3/16
3/8	9/16	0.747	0.620	1/4	15/64
7/16	5/8	0.828	0.687	19/64	9/32
1/2	3/4	0.995	0.826	21/64	19/64
9/16	7/8	1.163	0.966	3/8	11/32
5/8	15/16	1.244	1.033	27/64	25/64
3/4	1 1/8	1.494	1.240	1/2	15/32
7/8	1 5/16	1.742	1.447	19/32	9/16
1	1 1/2	1.991	1.653	21/32	19/32
1 1/8	1 11/16	2.239	1.859	3/4	11/16
1 1/4	1 7/8	2.489	2.066	27/32	25/32
1 3/8	2 1/16	2.738	2.273	29/32	27/32
1 1/2	2 1/4	2.986	2.480	1	15/16
1 5/8	2 7/16	3.235	2.686	1 3/32	1 1/32
1 3/4	2 5/8	3.485	2.893	1 5/32	1 3/32
1 7/8	2 13/16	3.733	3.100	1 1/4	1 3/16
2	3	3.982	3.306	1 11/32	1 7/32
2 1/4	3 3/8	4.479	3.719	1 1/2	1 3/8
2 1/2	3 3/4	4.977	4.133	1 21/32	1 17/32
2 3/4	4 1/8	5.476	4.546	1 53/64	1 11/16
3	4 1/2	5.973	4.959	2	1 7/8
		Heavy Bolt Heads			
1/2	7/8	1.167	0.969	7/16	13/32
9/16	15/16	1.249	1.037	15/32	7/16
5/8	1 1/16	1.416	1.175	17/32	1/2
3/4	1 1/4	1.665	1.383	5/8	19/32
7/8	1 7/16	1.914	1.589	23/32	11/16
1	1 5/8	2.162	1.796	13/16	3/4
1 1/8	1 13/16	2.411	2.002	29/32	27/32
1 1/4	2	2.661	2.209	1	15/16
1 3/8	2 3/16	2.909	2.416	1 3/32	1 1/32
1 1/2	2 3/8	3.158	2.622	1 3/16	1 1/8
1 5/8	2 9/16	3.406	2.828	1 9/32	1 7/32
1 3/4	2 3/4	3.655	3.036	1 3/8	1 5/16
1 7/8	2 15/16	3.905	3.242	1 15/32	1 13/32
2	3 1/8	4.153	3.449	1 9/16	1 7/16
2 1/4	3 1/2	4.652	3.862	1 3/4	1 5/8
2 1/2	3 7/8	5.149	4.275	1 15/16	1 13/16
2 3/4	4 1/4	5.646	4.688	2 1/8	2
3	4 5/8	6.144	5.102	2 5/16	2 3/16

Nominal Size	Width Across Flats	Width Across Corners		Thickness Unfinished		Thickness Semi-finished	
		Square	Hex	Regular nuts	Regular jam nuts	Regular nuts	Regular jam nuts
			Regular Nuts and Regular Jam Nuts				
1/4	7/16	0.584	0.484	7/32	5/32	13/64	9/64
5/16	9/16	0.751	0.624	17/64	3/16	1/4	11/64
3/8	5/8	0.832	0.691	21/64	7/32	5/16	13/64
7/16	3/4	1.000	0.830	3/8	1/4	23/64	15/64
1/2	13/16	1.082	0.898	7/16	5/16	27/64	19/64
9/16	7/8	1.163	0.966	1/2	11/32	31/64	21/64
5/8	1	1.330	1.104	35/64	3/8	17/32	23/64
3/4	1 1/8	1.494	1.240	21/32	7/16	41/64	27/64
7/8	1 5/16	1.742	1.447	49/64	1/2	3/4	31/64
1	1 1/2	1.991	1.653	7/8	9/16	55/64	35/64
1 1/8	1 11/16	2.239	1.859	1	5/8	31/32	39/64
1 1/4	1 7/8	2.489	2.066	1 3/32	3/4	1 1/16	23/32
1 3/8	2 1/16	2.738	2.273	1 13/64	13/16	1 11/64	25/32
1 1/2	2 1/4	2.986	2.480	1 5/16	7/8	1 9/32	27/32
1 5/8	2 7/16	3.235	2.686	1 27/64	15/16	1 25/64	29/32
1 3/4	2 5/8	3.485	2.893	1 17/32	1	1 1/2	31/32
1 7/8	2 13/16	3.733	3.100	1 41/64	1 1/16	1 39/64	1 1/32
2	3	3.982	3.306	1 3/4	1 1/8	1 23/32	1 3/32
2 1/4	3 3/8	4.479	3.719	1 31/32	1 1/4	1 59/64	1 13/64
2 1/2	3 3/4	4.977	4.133	2 3/16	1 1/2	2 9/64	1 29/64
2 3/4	4 1/8	5.476	4.546	2 13/32	1 5/8	2 23/64	1 37/64
3	4 1/2	5.973	4.959	2 5/8	1 3/4	2 37/64	1 45/64

* American Standards Association, Bulletin ASA B18.2—1941.

Table 88. ASA Standard Bolts and Nuts—Continued

Nominal Size	Width Across Flats	Width Across Corners		Thickness Unfinished		Thickness Semi-finished	
		Square	Hex	Heavy nuts	Heavy jam nuts	Heavy nuts	Heavy jam nuts
			Heavy Nuts and Heavy Jam Nuts				
1/4	1/2	0.670	0.556	1/4	3/16	15/64	11/64
5/16	19/32	0.794	0.659	5/16	7/32	19/64	13/64
3/8	11/16	0.919	0.763	3/8	1/4	23/64	15/64
7/16	25/32	1.042	0.865	7/16	9/32	27/64	17/64
1/2	7/8	1.167	0.969	1/2	5/16	31/64	19/64
9/16	15/16	1.249	1.037	9/16	11/32	35/64	21/64
5/8	1 1/16	1.416	1.175	5/8	3/8	39/64	23/64
3/4	1 1/4	1.665	1.382	3/4	7/16	47/64	27/64
7/8	1 7/16	1.914	1.589	7/8	1/2	55/64	31/64
1	1 5/8	2.162	1.796	1	9/16	63/64	35/64
1 1/8	1 13/16	2.411	2.002	1 1/8	5/8	1 7/64	39/64
1 1/4	2	2.661	2.209	1 1/4	3/4	1 7/32	23/32
1 3/8	2 3/16	2.909	2.416	1 3/8	13/16	1 11/32	25/32
1 1/2	2 3/8	3.158	2.622	1 1/2	7/8	1 15/32	27/32
1 5/8	2 9/16	3.406	2.828	1 5/8	15/16	1 19/32	29/32
1 3/4	2 3/4	3.656	3.035	1 3/4	1	1 23/32	31/32
1 7/8	2 15/16	3.905	3.242	1 7/8	1 1/16	1 27/32	1 1/32
2	3 1/8	4.153	3.449	2	1 1/8	1 31/32	1 3/32
2 1/4	3 1/2	4.652	3.862	2 1/4	1 1/4	2 13/64	1 13/64
2 1/2	3 7/8	5.149	4.275	2 1/2	1 1/2	2 29/64	1 29/64
2 3/4	4 1/4	5.646	4.688	2 3/4	1 5/8	2 45/64	1 37/64
3	4 5/8	6.144	5.102	3	1 3/4	2 61/64	1 45/64
3 1/4	5	6.643	5.515	3 1/4	1 7/8	3 3/16	1 13/16
3 1/2	5 3/8	7.140	5.928	3 1/2	2	3 7/16	1 15/16
3 3/4	5 3/4	7.637	6.341	3 3/4	2 1/8	3 11/16	2 1/16
4	6 1/8	8.135	6.755	4	2 1/4	3 15/16	2 3/16

Nominal Size	Regular Slotted Nuts Semi-finished			Heavy Slotted Nuts Semi-finished			Slot	
	Width		Thickness	Width		Thickness	Width	Depth
	Across flats	Across corners		Across flats	Across corners			
1/4	7/16	0.485	13/64	1/2	0.556	15/64	5/64	3/32
5/16	9/16	0.624	1/4	19/32	0.659	19/64	3/32	3/32
3/8	5/8	0.691	5/16	11/16	0.763	23/64	1/8	1/8
7/16	3/4	0.830	23/64	25/32	0.865	27/64	1/8	5/32
1/2	13/16	0.898	27/64	7/8	0.969	31/64	5/32	5/32
9/16	7/8	0.966	31/64	15/16	1.037	35/64	5/32	3/16
5/8	1	1.104	17/32	1 1/16	1.175	39/64	3/16	7/32
3/4	1 1/8	1.240	41/64	1 1/4	1.382	47/64	3/16	1/4
7/8	1 5/16	1.447	3/4	1 7/16	1.589	55/64	3/16	1/4
1	1 1/2	1.653	55/64	1 5/8	1.796	63/64	1/4	9/32
1 1/8	1 11/16	1.859	31/32	1 13/16	2.002	1 7/64	1/4	11/32
1 1/4	1 7/8	2.066	1 1/16	2	2.209	1 7/32	5/16	3/8
1 3/8	2 1/16	2.273	1 11/64	2 3/16	2.416	1 11/32	5/16	3/8
1 1/2	2 1/4	2.480	1 9/32	2 3/8	2.622	1 15/32	3/8	7/16
1 5/8	2 7/16	2.686	1 25/64	2 9/16	2.828	1 19/32	3/8	7/16
1 3/4	2 5/8	2.893	1 1/2	2 3/4	3.035	1 23/32	7/16	1/2
1 7/8	2 13/16	3.100	1 39/64	2 15/16	3.242	1 27/32	7/16	9/16
2	3	3.306	1 23/32	3 1/8	3.449	1 31/32	7/16	9/16
2 1/4	3 3/8	3.719	1 59/64	3 1/2	3.862	2 13/64	7/16	9/16
2 1/2	3 3/4	4.133	2 9/64	3 7/8	4.275	2 29/64	9/16	11/16
2 3/4	4 1/8	4.546	2 23/64	4 1/4	4.688	2 45/64	9/16	11/16
3	4 1/2	4.959	2 37/64	4 5/8	5.102	2 61/64	5/8	3/4

Table 89. Properties of American Standard Yard Lumber and Timber Sizes

Nominal Size in inches	American Standard Dressed Size, inches	Area of Section, A, sq in.	Weight per lin ft,* W, lb	Moment of Inertia, I, in.^{4}	Section-Modulus, S, in.^{3}
2 × 4	1 5/8 × 3 5/8	5.89	1.6	6.45	3.56
2 × 6	1 5/8 × 5 5/8	9.14	2.5	24.10	8.57
2 × 8	1 5/8 × 7 1/2	12.19	3.4	57.13	15.32
2 × 10	1 5/8 × 9 1/2	15.44	4.3	116.09	24.44
2 × 12	1 5/8 × 11 1/2	18.69	5.2	205.94	35.82
2 × 14	1 5/8 × 13 1/2	23.62	6.5	333.15	49.36
2 × 16	1 5/8 × 15 1/2	25.18	7.0	504.24	65.07
2 × 18	1 5/8 × 17 1/2	28.43	7.9	725.71	82.94
2 × 20	1 5/8 × 19 1/2	31.69	8.8	1,004.05	102.98
3 × 4	2 5/8 × 3 5/8	9.51	2.6	10.42	5.75
3 × 6	2 5/8 × 5 5/8	14.76	4.2	38.93	13.84
3 × 8	2 5/8 × 7 1/2	19.68	5.7	92.28	24.60
3 × 10	2 5/8 × 9 1/2	24.93	7.2	187.55	39.48
3 × 12	2 5/8 × 11 1/2	30.18	8.8	332.69	57.86
3 × 14	2 5/8 × 13 1/2	35.43	10.3	538.21	79.73
3 × 16	2 5/8 × 15 1/2	40.68	11.3	814.60	105.11
3 × 18	2 5/8 × 17 1/2	45.94	12.8	1,172.36	133.98
3 × 20	2 5/8 × 19 1/2	51.19	14.21	1,622.00	166.36
4 × 4	3 5/8 × 3 5/8	13.14	3.6	14.38	7.94
4 × 6	3 5/8 × 5 5/8	20.39	5.7	53.76	19.11
4 × 8	3 5/8 × 7 1/2	27.18	7.5	127.44	33.98
4 × 10	3 5/8 × 9 1/2	34.43	9.6	258.99	54.52
4 × 12	3 5/8 × 11 1/2	41.68	11.6	459.42	79.90
4 × 14	3 5/8 × 13 1/2	48.93	13.6	743.23	110.11
4 × 16	3 5/8 × 15 1/2	56.18	15.6	1,124.90	145.15
4 × 18	3 5/8 × 17 1/2	63.43	17.6	1,618.96	185.02
4 × 20	3 5/8 × 19 1/2	70.69	19.6	2,239.88	229.73
6 × 6	5 1/2 × 5 1/2	30.25	8.4	76.25	27.73
6 × 8	5 1/2 × 7 1/2	41.25	11.4	193.35	51.56
6 × 10	5 1/2 × 9 1/2	52.25	14.5	329.96	82.73
6 × 12	5 1/2 × 11 1/2	63.25	17.5	697.06	121.23
6 × 14	5 1/2 × 13 1/2	74.25	20.6	1,127.66	167.06
6 × 16	5 1/2 × 15 1/2	85.25	23.6	1,706.76	220.22
6 × 18	5 1/2 × 17 1/2	96.25	26.7	2,456.36	280.73
6 × 20	5 1/2 × 19 1/2	107.25	29.8	3,398.46	348.53
6 × 22	5 1/2 × 21 1/2	118.25	32.8	4,555.05	423.76
8 × 8	7 1/2 × 7 1/2	56.25	15.6	263.67	70.31
8 × 10	7 1/2 × 9 1/2	71.25	19.8	535.85	112.81
8 × 12	7 1/2 × 11 1/2	86.25	23.9	950.55	165.31
8 × 14	7 1/2 × 13 1/2	101.25	28.0	1,537.73	227.81
8 × 16	7 1/2 × 15 1/2	116.25	32.0	2,327.42	300.31
8 × 18	7 1/2 × 17 1/2	131.25	36.4	3,349.60	382.81
8 × 20	7 1/2 × 19 1/2	146.25	40.6	4,634.30	475.31
8 × 22	7 1/2 × 21 1/2	161.25	44.8	6,211.48	577.81
8 × 24	7 1/2 × 23 1/2	176.25	48.9	8,111.17	690.31
10 × 10	9 1/2 × 9 1/2	90.25	25.0	678.75	142.89
10 × 12	9 1/2 × 11 1/2	109.25	30.3	1,204.01	209.39
10 × 14	9 1/2 × 13 1/2	128.25	35.6	1,947.78	288.56
10 × 16	9 1/2 × 15 1/2	147.25	40.9	2,948.04	380.39
10 × 18	9 1/2 × 17 1/2	166.25	46.1	4,242.80	484.89

* Based on assumed average weight of 40 lb per cu ft

Table 89. Properties of American Standard Yard Lumber and Timber Sizes—*Continued*

Nominal Size in inches	American Standard Dressed Size, inches	Area of Section, A, sq in.	Weight per lin ft,* W, lb	Moment of Inertia, I, in.^{4}	Section-Modulus, S, in.^{3}
10 × 20	9 1/2 × 19 1/2	185.25	51.4	5,870.05	602.06
10 × 22	9 1/2 × 21 1/2	204.25	56.7	7,867.81	731.89
10 × 24	9 1/2 × 23 1/2	223.25	62.0	10,274.06	874.39
10 × 26	9 1/2 × 25 1/2	242.25	67.3	13,126.81	1029.56
10 × 28	9 1/2 × 27 1/2	261.25	72.5	16,465.24	1197.39
10 × 30	9 1/2 × 29 1/2	280.25	77.8	20,323.79	1377.89
12 × 12	11 1/2 × 11 1/2	132.25	36.7	1,457.50	253.47
12 × 14	11 1/2 × 13 1/2	155.25	43.1	2,357.85	349.31
12 × 16	11 1/2 × 15 1/2	178.25	49.5	3,568.70	460.48
12 × 18	11 1/2 × 17 1/2	201.25	55.9	5,136.49	586.98
12 × 20	11 1/2 × 19 1/2	224.25	62.3	7,105.90	728.81
12 × 22	11 1/2 × 21 1/2	247.25	68.7	9,524.24	885.98
12 × 24	11 1/2 × 23 1/2	270.25	75.0	12,437.08	1058.47
12 × 26	11 1/2 × 25 1/2	293.25	81.4	15,890.42	1246.31
12 × 28	11 1/2 × 27 1/2	316.25	87.8	19,932.58	1449.47
12 × 30	11 1/2 × 29 1/2	339.25	94.2	24,602.61	1667.97
14 × 14	13 1/2 × 13 1/2	182.25	50.6	2,767.92	410.06
14 × 16	13 1/2 × 15 1/2	209.25	58.1	4,189.36	540.56
14 × 18	13 1/2 × 17 1/2	236.25	65.6	6,029.29	689.06
14 × 20	13 1/2 × 19 1/2	263.25	73.1	8,341.73	855.56
14 × 22	13 1/2 × 21 1/2	290.25	80.6	11,180.67	1040.06
14 × 24	13 1/2 × 23 1/2	317.25	88.1	14,600.10	1242.56
14 × 26	13 1/2 × 25 1/2	344.25	95.6	18,654.04	1463.06
14 × 28	13 1/2 × 27 1/2	371.25	103.1	23,398.73	1701.56
14 × 30	13 1/2 × 29 1/2	398.25	110.6	28,881.42	1958.06
16 × 16	15 1/2 × 15 1/2	240.25	66.7	4,809.98	620.64
16 × 18	15 1/2 × 17 1/2	271.25	75.3	6,922.49	791.14
16 × 20	15 1/2 × 19 1/2	302.25	83.9	9,577.50	982.31
16 × 22	15 1.2 × 21 1/2	333.25	92.5	12,837.00	1194.14
16 × 24	15 1/2 × 23 1/2	3ۅ4.25	101.2	16,763.00	1426.64
16 × 26	15 1/2 × 25 1/2	395.25	109.8	21,417.50	1679.81
16 × 28	15 1/2 × 27 1/2	426.25	118.4	26,863.78	1953.64
16 × 30	15 1/2 × 29 1/2	457.25	127.0	33,159.98	2248.14
18 × 18	17 1/2 × 17 1/2	3.6.25	85.0	7,815.73	893.23
18 × 20	17 1/2 × 19 1/2	341.25	94.8	0,813.33	1109.06
18 × 22	17 1/2 × 21 1/2	376.25	104.5	14,493.43	1348.23
18 × 24	17 1/2 × 23 1/2	411.25	114.2	18,926.02	1610.72
18 × 26	17 1/2 × 25 1/2	446.25	123.9	24,181.11	1896.56
18 × 28	17 1/2 × 27 1/2	481.25	133.7	30,331.62	2205.72
18 × 30	17 1/2 × 29 1/2	516.25	143.4	37,438.79	2538.22
20 × 20	19 1/2 × 19 1/2	380.25	105.6	12,049.49	1235.81
20 × 22	19 1/2 × 21 1/2	419.25	116.4	16,149.86	1502.31
20 × 24	19 1/2 × 23 1/2	458.25	127.3	21,089.04	1794.81
20 × 26	19 1/2 × 25 1/2	497.25	138.1	26,944.73	2113.31
20 × 28	19 /2 × 27 1/2	536.25	148.9	33,798.17	2457.81
20 × 30	19 1/2 × 29 1/2	575.25	159.8	41,717.61	2828.31
24 × 24	23 1/2 × 23 1/2	552.25	153.4	25,414.96	2162.97
24 × 26	23 1/2 × 25 1/2	599.25	166.4	32,471.80	2546.81
24 × 28	23 1/2 × 27 1/2	646.25	179.5	40,731.06	2916.97
24 × 30	23 1/2 × 29 1/2	693.25	192.5	50,274.98	3408.47

* Based on assumed average weight of 40 lb per cu ft

Table 90. Dimensions of Ferrous Pipe

Nominal pipe size, in.	Outside diam., in.	Schedule No.	Wall thickness, in.	Inside diam., in.	Cross-sectional area		Circumference, ft., or surface, sq. ft./ft. of length		Capacity at 1 ft./sec. velocity		Weight of plain-end pipe, lb./ft.
					Metal, sq. in.	Flow, sq. ft.	Outside	Inside	U.S. gal./min.	Lb./hr. water	
⅛	0.405	10S	.049	.307	.055	.00051	.106	.0804	0.231	115.5	0.19
		40ST, 40S	.068	.269	.072	.00040	.106	.0705	.179	89.5	.24
		80XS, 80S	.095	.215	.093	.00025	.106	.0563	.113	56.5	.31
¼	0.540	10S	.065	.410	.097	.00092	.141	.107	.412	206.5	.33
		40ST, 40S	.088	.364	.125	.00072	.141	.095	.323	161.5	.42
		80XS, 80S	.119	.302	.157	.00050	.141	.079	.224	112.0	.54
⅜	0.675	10S	.065	.545	.125	.00162	.177	.143	.727	363.5	.42
		40ST, 40S	.091	.493	.167	.00133	.177	.129	.596	298.0	.57
		80XS, 80S	.126	.423	.217	.00098	.177	.111	.440	220.0	.74
½	0.840	5S	.065	.710	.158	.00275	.220	.186	1.234	617.0	.54
		10S	.083	.674	.197	.00248	.220	.176	1.112	556.0	.67
		40ST, 40S	.109	.622	.250	.00211	.220	.163	0.945	472.0	.85
		80XS, 80S	.147	.546	.320	.00163	.220	.143	0.730	365.0	1.09
		160	.188	.464	.385	.00117	.220	.122	0.527	263.5	1.31
		XX	.294	.252	.504	.00035	.220	.066	0.155	77.5	1.71
¾	1.050	5S	.065	.920	.201	.00461	.275	.241	2.072	1036.0	0.69
		10S	.083	.884	.252	.00426	.275	.231	1.903	951.5	0.86
		40ST, 40S	.113	.824	.333	.00371	.275	.216	1.665	832.5	1.13
		80XS, 80S	.154	.742	.433	.00300	.275	.194	1.345	672.5	1.47
		160	.219	.612	.572	.00204	.275	.160	0.917	458.5	1.94
		XX	.308	.434	.718	.00103	.275	.114	0.461	230.5	2.44
1	1.315	5S	.065	1.185	.255	.00768	.344	.310	3.449	1725	0.87
		10S	.109	1.097	.413	.00656	.344	.287	2.946	1473	1.40
		40ST, 40S	.133	1.049	.494	.00600	.344	.275	2.690	1345	1.68
		80XS, 80S	.179	0.957	.639	.00499	.344	.250	2.240	1120	2.17
		160	.250	0.815	.836	.00362	.344	.213	1.625	812.5	2.84
		XX	.358	0.599	1.076	.00196	.344	.157	0.878	439.0	3.66
1¼	1.660	5S	.065	1.530	0.326	.01277	.435	.401	5.73	2865	1.11
		10S	.109	1.442	0.531	.01134	.435	.378	5.09	2545	1.81
		40ST, 40S	.140	1.380	0.668	.01040	.435	.361	4.57	2285	2.27
		80XS, 80S	.191	1.278	0.881	.00891	.435	.335	3.99	1995	3.00
		160	.250	1.160	1.107	.00734	.435	.304	3.29	1645	3.76
		XX	.382	0.896	1.534	.00438	.435	.235	1.97	985	5.21
1½	1.900	5S	.065	1.770	0.375	.01709	.497	.463	7.67	3835	1.28
		10S	.109	1.682	0.614	.01543	.497	.440	6.94	3465	2.09
		40ST, 40S	.145	1.610	0.800	.01414	.497	.421	6.34	3170	2.72
		80XS, 80S	.200	1.500	1.069	.01225	.497	.393	5.49	2745	3.63
		160	.281	1.338	1.429	.00976	.497	.350	4.38	2190	4.86
		XX	.400	1.100	1.885	.00660	.497	.288	2.96	1480	6.41
2	2.375	5S	.065	2.245	0.472	.02749	.622	.588	12.34	6170	1.61
		10S	.109	2.157	0.776	.02538	.622	.565	11.39	5695	2.64
		40ST, 40S	.154	2.067	1.075	.02330	.622	.541	10.45	5225	3.65
		80ST, 80S	.218	1.939	1.477	.02050	.622	.508	9.20	4600	5.02
		160	.344	1.687	2.195	.01552	.622	.436	6.97	3485	7.46
		XX	.436	1.503	2.656	.01232	.622	.393	5.53	2765	9.03
2½	2.875	5S	.083	2.709	0.728	.04003	.753	.709	17.97	8985	2.48
		10S	.120	2.635	1.039	.03787	.753	.690	17.00	8500	3.53
		40ST, 40S	.203	2.469	1.704	.03322	.753	.647	14.92	7460	5.79
		80XS, 80S	.276	2.323	2.254	.02942	.753	.608	13.20	6600	7.66
		160	.375	2.125	2.945	.02463	.753	.556	11.07	5535	10.01
		XX	.552	1.771	4.028	.01711	.753	.464	7.68	3840	13.70
3	3.500	5S	.083	3.334	0.891	.06063	.916	.873	27.21	13,605	3.03
		10S	.120	3.260	1.274	.05796	.916	.853	26.02	13,010	4.33
		40ST, 40S	.216	3.068	2.228	.05130	.916	.803	23.00	11,500	7.58
		80XS, 80S	.300	2.900	3.016	.04587	.916	.759	20.55	10,275	10.25
		160	.438	2.624	4.213	.03755	.916	.687	16.86	8430	14.31
		XX	.600	2.300	5.466	.02885	.916	.602	12.95	6475	18.58
3½	4.0	5S	.083	3.834	1.021	.08017	1.047	1.004	35.98	17,990	3.48
		10S	.120	3.760	1.463	.07711	1.047	.984	34.61	17,305	4.97
		40ST, 40S	.226	3.548	2.680	.06870	1.047	.929	30.80	15,400	9.11
		80XS, 80S	.318	3.364	3.678	.06170	1.047	.881	27.70	13,850	12.51
4	4.5	5S	.083	4.334	1.152	.10245	1.178	1.135	46.0	23,000	3.92
		10S	.120	4.260	1.651	.09898	1.178	1.115	44.4	22,200	5.61
		40ST, 40S	.237	4.026	3.17	.08840	1.178	1.054	39.6	19,800	10.79
		80XS, 80S	.337	3.826	4.41	.07986	1.178	1.002	35.8	17,900	14.98
		120	.438	3.624	5.58	.07170	1.178	0.949	32.2	16,100	19.01
		160	.531	3.438	6.62	.06647	1.178	0.900	28.9	14,450	22.52
		XX	.674	3.152	8.10	.05419	1.178	0.825	24.3	12,150	27.54
5	5.563	5S	.109	5.345	1.87	.1558	1.456	1.399	69.9	34,950	6.36
		10S	.134	5.295	2.29	.1529	1.456	1.386	68.6	34,300	7.77
		40ST, 40S	.258	5.047	4.30	.1390	1.456	1.321	62.3	31,150	14.62
		80XS, 80S	.375	4.813	6.11	.1263	1.456	1.260	57.7	28,850	20.78
		120	.500	4.563	7.95	.1136	1.456	1.195	51.0	25,500	27.04
		160	.625	4.313	9.70	.1015	1.456	1.129	45.5	22,750	32.96
		XX	.750	4.063	11.34	.0900	1.456	1.064	40.4	20,200	38.55

Table 90. Dimensions of Ferrous Pipe—*Continued*

Nominal pipe size, in.	Outside diam., in.	Schedule No.	Wall thickness, in.	Inside diam., in.	Cross-sectional area		Circumference, ft., or surface, sq. ft./ft. of length		Capacity at 1 ft./sec. velocity		Weight of plain-end pipe, lb./ft.
					Metal, sq. in.	Flow, sq. ft.	Outside	Inside	U.S. gal./min.	Lb./hr. water	
6	6.625	5S	0.109	6.407	2.23	0.2239	1.734	1.677	100.5	50,250	7.60
		10S	.134	6.357	2.73	.2204	1.734	1.664	98.9	49,450	9.29
		40ST, 40S	.280	6.065	5.58	.2006	1.734	1.588	90.0	45,000	18.97
		80XS, 80S	.432	5.761	8.40	.1810	1.734	1.508	81.1	40,550	28.57
		120	.562	5.501	10.70	.1650	1.734	1.440	73.9	36,950	36.42
		160	.719	5.187	13.34	.1467	1.734	1.358	65.9	32,950	45.34
		XX	.864	4.897	15.64	.1308	1.734	1.282	58.7	29,350	53.16
8	8.625	5S	.109	8.407	2.915	.3855	2.258	2.201	173.0	86,500	9.93
		10S	148	8.329	3.941	.3784	2.258	2.180	169.8	84,900	13.40
		20	.250	8.125	6.578	.3601	2.258	2.127	161.5	80,750	22.36
		30	.277	8.071	7.260	.3553	2.258	2.113	159.4	79,700	24.70
		40ST, 40S	.322	7.981	8.396	.3474	2.258	2.089	155.7	77,850	28.55
		60	.406	7.813	10.48	.3329	2.258	2.045	149.4	74,700	35.66
		80XS, 80S	.500	7.625	12.76	.3171	2.258	1.996	142.3	71,150	43.39
		100	.594	7.437	14.99	.3017	2.258	1.947	135.4	67,700	50.93
		120	.719	7.187	17.86	.2817	2.258	1.882	126.4	63,200	60.69
		140	.812	7.001	19.93	.2673	2.258	1.833	120.0	60,000	67.79
		XX	.875	6.875	21.30	.2578	2.258	1.800	115.7	57,850	72.42
		160	.906	6.813	21.97	.2532	2.258	1.784	113.5	56,750	74.71
10	10.75	5S	.134	10.842	4.47	.5993	2.814	2.744	269.0	134,500	15.23
		10S	.165	10.420	5.49	.5922	2.814	2.728	265.8	132,900	18.70
		20	.250	10.250	8.25	.5731	2.814	2.685	257.0	128,500	28.04
		30	.307	10.136	10.07	.5603	2.814	2.655	252.0	126,000	34.24
		40ST, 40S	.365	10.020	11.91	.5475	2.814	2.620	246.0	123,000	40.48
		80S, 60XS	.500	9.750	16.10	.5185	2.814	2.550	233.0	116,500	54.74
		80	.594	9.562	18.95	.4987	2.814	2.503	223.4	111,700	64.40
		100	.719	9.312	22.66	.4729	2.814	2.438	212.3	106,150	77.00
		120	.844	9.062	26.27	.4479	2.814	2.372	201.0	100,500	89.27
		140, XX	1.000	8.750	30.63	.4176	2.814	2.291	188.0	94,000	104.13
		160	1.125	8.500	34.02	.3941	2.814	2.225	177.0	88,500	115.65
12	12.75	5S	0.156	12.438	6.17	.8438	3.338	3.26	378.7	189,350	22.22
		10S	0.180	12.390	7.11	.8373	3.338	3.24	375.8	187,900	24.20
		20	0.250	12.250	9.82	.8185	3.338	3.21	367.0	183,500	33.38
		30	0.330	12.090	12.88	.7972	3.338	3.17	358.0	179,000	43.77
		ST, 40S	0.375	12.000	14.58	.7854	3.338	3.14	352.5	176,250	49.56
		40	0.406	11.938	15.74	.7773	3.338	3.13	349.0	174,500	53.56
		XS, 80S	0.500	11.750	19.24	.7530	3.338	3.08	338.0	169,000	65.42
		60	0.562	11.626	21.52	.7372	3.338	3.04	331.0	165,500	73.22
		80	0.688	11.374	26.07	.7056	3.338	2.98	316.7	158,350	88.57
		100	0.844	11.062	31.57	.6674	3.338	2.90	299.6	149,800	107.29
		120, XX	1.000	10.750	36.91	.6303	3.338	2.81	283.0	141,500	125.49
		140	1.125	10.500	41.09	.6013	3.338	2.75	270.0	135,000	139.68
		160	1.312	10.126	47.14	.5592	3.338	2.65	251.0	125,500	160.33
14	14	5S	0.156	13.688	6.78	1.0219	3.665	3.58	459	229,500	22.76
		10S	0.188	13.624	8.16	1.0125	3.665	3.57	454	227,000	27.70
		10	0.250	13.500	10.80	0.9940	3.665	3.53	446	223,000	36.71
		20	0.312	13.376	13.42	0.9750	3.665	3.50	438	219,000	45.68
		30, ST	0.375	13.250	16.05	0.9575	3.665	3.47	430	215,000	54.57
		40	0.438	13.124	18.66	0.9397	3.665	3.44	422	211,000	63.37
		XS	0.500	13.000	21.21	0.9218	3.665	3.40	414	207,000	72.09
		60	0.594	12.812	25.02	0.8957	3.665	3.35	402	201,000	85.01
		80	0.750	12.500	31.22	0.8522	3.665	3.27	382	191,000	106.13
		100	0.938	12.124	38.49	0.8017	3.665	3.17	360	180,000	130.79
		120	1.094	11.812	44.36	0.7610	3.665	3.09	342	171,000	150.76
		140	1.250	11.500	50.07	0.7213	3.665	3.01	324	162,000	170.22
		160	1.406	11.188	55.63	0.6827	3.665	2.93	306	153,000	189.12
16	16	5S	0.165	15.670	8.18	1.3393	4.189	4.10	601	300,500	27.87
		10S	0.188	15.624	9.34	1.3314	4.189	4.09	598	299,000	31.62
		10	0.250	15.500	12.37	1.3104	4.189	4.06	587	293,500	42.05
		20	0.312	15.376	15.38	1.2895	4.189	4.03	578	289,000	52.36
		30, ST	0.375	15.250	18.41	1.2680	4.189	3.99	568	284,000	62.58
		40, XS	0.500	15.000	24.35	1.2272	4.189	3.93	550	275,000	82.77
		60	0.656	14.688	31.62	1.1766	4.189	3.85	528	264,000	107.54
		80	0.844	14.312	40.19	1.1171	4.189	3.75	501	250,500	136.58
		100	1.031	13.938	48.48	1.0596	4.189	3.65	474	237,000	164.86
		120	1.219	13.562	56.61	1.0032	4.189	3.55	450	225,000	192.40
		140	1.438	13.124	65.79	0.9394	4.189	3.44	422	211,000	223.57
		160	1.594	12.812	72.14	0.8953	4.189	3.35	402	201,000	245.72
18	18	5S	0.165	17.670	9.25	1.7029	4.712	4.63	764	382,000	31.32
		10S	0.188	17.624	10.52	1.6941	4.712	4.61	760	379,400	35.48
		10	0.250	17.500	13.94	1.6703	4.712	4.58	750	375,000	47.39
		20	0.312	17.376	17.34	1.6468	4.712	4.55	739	369,500	59.03
		ST	0.375	17.250	20.76	1.6230	4.712	4.52	728	364,000	70.59
		30	0.438	17.124	24.16	1.5993	4.712	4.48	718	359,000	82.06
		XS	0.500	17.000	27.49	1.5763	4.712	4.45	707	353,500	93.45
		40	0.562	16.876	30.79	1.5533	4.712	4.42	697	348,500	104.76
		60	0.750	16.500	40.64	1.4849	4.712	4.32	666	333,000	138.17
		80	0.938	16.124	50.28	1.4180	4.712	4.22	636	318,000	170.75
		100	1.156	15.688	61.17	1.3423	4.712	4.11	602	301,000	208.00
		120	1.375	15.250	71.82	1.2684	4.712	3.99	569	284,500	244.14
		140	1.562	14.876	80.66	1.2070	4.712	3.89	540	270,000	274.30
		160	1.781	14.438	90.75	1.1370	4.712	3.78	510	255,000	308.55

Table 90.   Dimensions of Ferrous Pipe—*Continued*

Nominal pipe size, in.	Outside diam., in.	Schedule No.	Wall thickness, in.	Inside diam., in.	Cross-sectional area		Circumference, ft., or surface, sq. ft./ft. of length		Capacity at 1 ft./sec. velocity		Weight of plain-end pipe, lb./ft.
					Metal, sq. in.	Flow, sq. ft.	Outside	Inside	U.S. gal./min.	Lb./hr. water	
20	20	5S	0.188	19.624	11.70	2.1004	5.236	5.14	943	471,500	39.76
		10S	.218	19.564	13.55	2.0878	5.236	5.12	937	467,500	45.98
		10	.250	19.500	15.51	2.0740	5.236	5.11	930	465,000	52.73
		20, ST	.375	19.250	23.12	2.0211	5.236	5.04	902	451,000	78.60
		30, XS	.500	19.000	30.63	1.9689	5.236	4.97	883	441,500	104.13
		40	.594	18.812	36.21	1.9302	5.236	4.92	866	433,000	123.06
		60	.812	18.376	48.95	1.8417	5.236	4.81	826	413,000	166.50
		80	1.031	17.938	61.44	1.7550	5.236	4.70	787	393,500	208.92
		100	1.281	17.438	75.33	1.6585	5.236	4.57	744	372,000	256.15
		120	1.500	17.000	87.18	1.5763	5.236	4.45	707	353,500	296.37
		140	1.750	16.500	100.3	1.4849	5.236	4.32	665	332,500	341.10
		160	1.969	16.062	111.5	1.4071	5.236	4.21	632	316,000	379.14
24	24	5S	0.218	23.564	16.29	3.0285	6.283	6.17	1359	679,500	55.08
		10, 10S	0.250	23.500	18.65	3.012	6.283	6.15	1350	675,000	63.41
		20, ST	0.375	23.250	27.83	2.948	6.283	6.09	1325	662,500	94.62
		XS	0.500	23.000	36.90	2.885	6.283	6.02	1295	642,500	125.49
		30	0.562	22.876	41.39	2.854	6.283	5.99	1281	640,500	140.80
		40	0.688	22.624	50.39	2.792	6.283	5.92	1253	626,500	171.17
		60	0.969	22.062	70.11	2.655	6.283	5.78	1192	596,000	238.29
		80	1.219	21.562	87.24	2.536	6.283	5.64	1138	569,000	296.53
		100	1.531	20.938	108.1	2.391	6.283	5.48	1073	536,500	367.45
		120	1.812	20.376	126.3	2.264	6.283	5.33	1016	508,000	429.50
		140	2.062	19.876	142.1	2.155	6.283	5.20	965	482,500	483.24
		160	2.344	19.312	159.5	2.034	6.283	5.06	913	456,500	542.09
30	30	5S	0.250	29.500	23.37	4.746	7.854	7.72	2130	1,065,000	79.43
		10, 10S	0.312	29.376	29.10	4.707	7.854	7.69	2110	1,055,000	99.08
		ST	0.375	29.250	34.90	4.666	7.854	7.66	2094	1,048,000	118.65
		20, XS	0.500	29.000	46.34	4.587	7.854	7.59	2055	1,027,500	157.53
		30	0.625	28.750	57.68	4.508	7.854	7.53	2020	1,010,000	196.08

5S, 10S, and 40S are taken from A.S.A. B16.19, "Stainless Steel Pipe."   ST = standard wall, XS = extra strong wall, XX = double extra strong wall are all taken from Table 4 of A. S.A. B16.10, "Wrought-steel and Wrought Iron Pipe."   Wrought-iron pipe has slightly thicker walls, approximately 3 per cent, but the same weight per foot, because of lower density.   10, 20, 30, 40, 60, 80, 100, 120, 140, and 160 are taken from Table 2 of A.S.A. B16.10, "Wrought Steel and Wrought Iron Pipe," and apply to steel pipe only.   Decimal thicknesses for respective pipe sizes represent their nominal or average wall dimensions.   Mill tolerances as high as 12½ per cent are permitted.

Plain-end pipe is produced by a square cut.   Pipe is also shipped from the mills threaded, with a threaded coupling on one end, or with the ends beveled for welding, or grooved or sized for patented couplings.   Weights per foot for threaded and coupled pipe are slightly greater because of the weight of the coupling, but it is not available larger than 12 in., or lighter than Schedule 30 sizes 8 through 12 in., or Schedule 40 6 in. and smaller.

From *Chemical Engineer's Handbook*, New York, McGraw-Hill, 4th ed., 1963.   Used by permission.

SECTION 2

MATHEMATICS

BY

JOHN L. BARNES*

* This section is a revision of the second edition, which was written by J. L. Barnes and M. S. Barnes.

MATHEMATICS

The names of Greek letters are found in Table 1, Section 1; standard mathematical symbols in Table 2, Section 1; and abbreviations for engineering terms in Table 3, Section 1.

ARITHMETIC

1. ROMAN NUMERALS

Roman Notation uses seven letters and a bar; a letter with a bar placed over it represents a thousand times as much as it does without the bar. The letters and rules for combining them to represent numbers are as follows:

I.	V	X	L	C	D	M	$\overline{\text{L}}$
1	5	10	50	100	500	1000	50,000

Rule 1. If no letter precedes a letter of greater value, add the numbers represented by the letters.

Example. XXX represents 30; VI represents 6.

Rule 2. If a letter precedes a letter of greater value, subtract the smaller from the greater; add the remainder or remainders thus obtained to the numbers represented by the other letters.

Example. IV represents 4; XL represents 40; CXLV represents 145.

Other Illustrations:

IX	XIII	XIV	LV	XLII	XCVI	MDCI	$\overline{\text{IV}}$CCXL
9	13	14	55	42	96	1601	4240

2. ROOTS OF NUMBERS

Roots can be found by use of tables (p. 16),   logarithms (p. 220), or a slide rule (p. 221).

To find an nth root by arithmetic, use a method indicated by the binomial theorem expansion of $(a + b)^n$.

$$(a + b)^n = a^n + na^{n-1}b + \frac{n(n-1)}{2} a^{n-2}b^2 + \frac{n(n-1)(n-2)}{3 \cdot 2} a^{n-3}b^3 + \cdots + b^n$$
$$= a^n + bD$$

in which $D = na^{n-1} + \frac{n(n-1)}{2} a^{n-2}b + \cdots + b^{n-1}$.

(1) Point off the given number into periods of n figures each, starting at the decimal point and going both ways.

(2) Find the largest nth power in the left-hand period and use its root as the first digit of the result. Subtract this nth power from the left-hand period and bring down the next period.

(3) Use the quantity D, in which a is 10 times the first digit since the first digit occupies a higher place than the second, as the divisor to obtain the second digit b. As a trial divisor to estimate b, use the first term in D, since it is the largest. Multiply D by b, subtract, and bring down the next period.

(4) To get the next digit use 10 times the first two digits as a and proceed as before.

Examples.

1. Square root.

$$3'02'.98'06'52' \mid 17.406+$$

	1
$D = 2a + b = 27$	202
	189
344	1398
	1376
34806	220652
	208836

2. Cube root.

$$158'252'.632'929 \mid 54.09$$

$5^3 =$		125
Trial divisor $= 3a^2 = 3 \times 50^2 =$	7500	33252
$3ab = 3 \times 50 \times 4 =$	600	
$b^2 = 4^2 =$	16	
$D = 3a^2 + 3ab + b^2 =$	8116	32464
$3 \times 5400^2 = $	87480000	788632929
$3 \times 5400 \times 9 =$	145800	
$9^2 =$	81	
	87625881	788632929

3. APPROXIMATE COMPUTATION

Standard Notation. $N = a \cdot 10^b$, N is a given number; $1 \leqq a < 10$, the figures in a being the *significant figures* in N; b is an integer, positive or negative or zero.

Example. If $N = 2,953,000$, in which the first five figures are significant, then $N = 2.9530 \times 10^6$.

A number is *rounded* to contain fewer significant figures by dropping figures from the right-hand side. If the figures dropped amount to more than $1/2$ in the last figure kept, this last figure is increased by 1. If the figures dropped amount to $1/2$, the last figure may or may not be increased.

Since the last significant figure used in making a measurement, an estimate, etc., is not exact, but is usually the nearer of two consecutive figures, an approximate number may represent any value in a range from $1/2$ less in its last significant figure to $1/2$ more. The *absolute error* in an approximate number may be as much as $1/2$ in the last significant figure.

Example. If $N = 2.9530 \times 10^6$ is an approximate number, then $2.95295 \times 10^6 \leqq N \leqq 2.95305 \times 10^6$. The absolute error is between -0.00005×10^6 and 0.00005×10^6.

The size of the absolute error depends on the location of the decimal point.

The *relative error* is the ratio of the absolute error to the number. Its size depends on the number of significant figures.

Example. The relative error in the preceding example is at most $\dfrac{0.00005 \times 10^6}{2.9530 \times 10^6}$, or about 1 in 60,000; the percentage error is at most $100 \times \dfrac{0.00005}{2.9530}$, or less than 0.002%.

In the result of a computation with approximate numbers some figures on the right are doubtful and should be rounded off. In slide-rule computations and to some extent in computations done with tables, rounding is done automatically. It is always possible, by using the bounds of the ranges which approximate numbers represent, to compute exactly the bounds of the range in which the result lies, and then round off the uncertain figures.

Example. Divide the approximate number 536 by the approximate number 217.4.

	At least	*At most*
$536/217.4 = 2.47-$	$535.5/217.45 = 2.46+$	$536.5/217.35 = 2.47-$

In the quotient the third figure may be in error. It is useless to carry the division further.

The following rules usually give the largest number of significant figures that it is reasonable to keep.

Addition and Subtraction. Keep as the last significant figure in the result the figure in the last full column. The absolute accuracy of the result is determined by the least absolutely accurate number.

Example.

$$\begin{array}{r} 2.953\text{xx} \\ 0.8942\text{x} \\ 0.06483 \\ \hline 3.912\text{xx} \end{array}$$

Multiplication, Division, Powers, and Roots. Keep no more significant figures in the result than the fewest in any number involved. The relative accuracy of the result is determined by that of the least relatively accurate number. Shortcuts as shown in the examples may be used.

Examples.

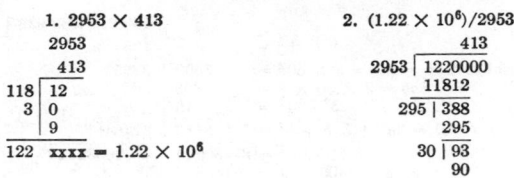

In intermediate results keep one additional figure.

If there is much difference in the relative accuracy, that is, the number of significant figures, of the numbers involved in a computation, round all of them to one more significant figure than the least accurate number has. This procedure may introduce a small error in the last figure kept in the result. A three-digit number beginning with 8 or 9 has about the same relative accuracy as a four-digit number beginning with 1.

Use of Tables. In using a table to find the value of a function corresponding to an approximate value of an argument, it is usually advisable to retain no more significant figures in the function than there are in the argument, although the accuracy of the function varies considerably, depending inversely on the slope of the curve representing the function. However, there is no need for many-place tables if the values of the argument are known only to a few significant figures.

Examples. $1/_{52} = 0.019$; $\cos 61.3° = 0.877$; $\log 3.74 = 0.573$.

To investigate the behavior of the error for any given function, the differential approximation is useful. If $y = f(x)$, then $dy = f'(x)\,dx$ approximates the absolute error, and $dy/y = f'(x)\,dx/f(x)$ the relative error.

For particular approximate values of the arguments, the bounds of the ranges of the functions can be found directly from a table with arguments given to one additional place.

4. INTERPOLATION

Gregory-Newton Interpolation Formula. Let $f(x)$ be a tabulated function of the argument x, Δx the constant difference between values of x for which the function is tabulated, and p a proper fraction. To find $f(x + p\Delta x)$, use the formula

$$f(x + p\Delta x) = f(x) + p\cdot\Delta f + {_pC_2}\cdot\Delta_2 f + {_pC_3}\cdot\Delta_3 f + \cdots$$

in which ${_pC_r} = \dfrac{p(p-1)\cdots(p-r+1)}{r!}$ and $\Delta_r f = r$th functional difference.

Binomial coefficients for interpolation:

p	$_pC_2$	$_pC_3$	$_pC_4$	$_pC_5$	p	$_pC_2$	$_pC_3$	$_pC_4$	$_pC_5$
0.1	−0.0450	0.0285	−0.0207	0.0161	0.6	−0.1200	0.0560	−0.0336	0.0228
0.2	−0.0800	0.0480	−0.0336	0.0255	0.7	−0.1050	0.0455	−0.0262	0.0173
0.3	−0.1050	0.0595	−0.0402	0.0297	0.8	−0.0800	0.0320	−0.0176	0.0113
0.4	−0.1200	0.0640	−0.0416	0.0300	0.9	−0.0450	0.0165	−0.0087	0.0054
0.5	−0.1250	0.0625	−0.0391	0.0273					

In ordinary linear interpolation the first two terms of the formula are used.

Example. Find $\sqrt{15.4}$.

x	$f(x) = \sqrt{x}$	Δf	$\Delta_2 f$	$\Delta_3 f$
15	3.8730			
		0.1270		
16	4.0000		−0.0039	
		0.1231		0.0003
17	4.1231		−0.0036	
		0.1195		
18	4.2426			

$$\Delta x = 1, \quad p = 0.4$$

$$f(15 + 0.4 \times 1) = 3.8730 + 0.4 \times 0.1270 + 0.1200 \times 0.0039 + 0.0640 \times 0.0003$$

$$= 3.9243$$

ALGEBRA

5. NUMBERS

Classification.

(a) *Real* (*positive* and *negative*).
  (1) *Rational*, expressible as the quotient of two integers.
    (i) *Integers*, as $-1, 2, 53$.
    (ii) *Fractions*, as $3/4, -5/2$.
  (2) *Irrational*, not expressible as the quotient of two integers, as $\sqrt{2}$, π.

(b) *Imaginary*, a product of a real number and the *imaginary unit* $i (= \sqrt{-1})$. Electrical engineers use j to avoid confusion with i for current. Example: $\sqrt{-2} = \sqrt{2}i$.

(c) *Complex*, a sum of a real number and an imaginary number, as $a + bi$ (a and b real), $-3 + 0.5i$. A real number may be regarded as a complex number in which $b = 0$, and an imaginary number as one in which $a = 0$.

The Absolute Value of

(a) a *real number* is the number itself if the number is positive, and the number with its sign changed if it is negative, as, for example, $|3| = |-3| = 3$;

(b) a *complex number* $a + bi$ is $\sqrt{a^2 + b^2}$, as, for example, $|-3 + 0.5i| = \sqrt{9 - 1/4} = 3.04+$.

6. IDENTITIES

Powers

1. $(-a)^n = a^n$, if n is even.
2. $(-a)^n = -a^n$, if n is odd.
3. $a^m \cdot a^n = a^{m+n}$.

4. $\dfrac{a^m}{a^n} = a^{m-n}$.

5. $(ab)^n = a^n b^n$.

6. $\left(\dfrac{a}{b}\right)^n = \dfrac{a^n}{b^n} = \left(\dfrac{b}{a}\right)^{-n} = \dfrac{b^{-n}}{a^{-n}}$.

7. $a^{-n} = \left(\dfrac{1}{a}\right)^n = \dfrac{1}{a^n}$.

8. $(a^m)^n = a^{mn}$.
9. $a^0 = 1$; $0^n = 0$; 0^0 is meaningless.

Roots

1. $\sqrt[n]{a} = a^{1/n}$.
2. $(\sqrt[n]{a})^n = \sqrt[n]{a^n} = a$.
3. $\sqrt[n]{ab} = \sqrt[n]{a}\,\sqrt[n]{b}$.
4. $\sqrt[n]{\dfrac{a}{b}} = \dfrac{\sqrt[n]{a}}{\sqrt[n]{b}}$.
5. $\sqrt[m]{a}\,\sqrt[n]{a} = a^{(1/m)+(1/n)} = \sqrt[mn]{a^{m+n}}$.
6. $\sqrt[m]{a^n} = (\sqrt[m]{a})^n = a^{n/m}$.
7. $\sqrt[m]{\sqrt[n]{a}} = \sqrt[mn]{a} = \sqrt[n]{\sqrt[m]{a}} = (a^{1/m})^{1/n} = a^{1/mn}$.
8. $\sqrt{a} + \sqrt{b} = \sqrt{a + b + 2\sqrt{ab}}$.

Products

1. $(a \pm b)^2 = a^2 \pm 2ab + b^2$.
2. $(a + b)(a - b) = a^2 - b^2$.
3. $(a + b + c)^2 = a^2 + b^2 + c^2 + 2ab + 2ac + 2bc$.
4. $(a \pm b)^3 = a^3 \pm 3a^2b + 3ab^2 \pm b^3$.
5. $a^3 \pm b^3 = (a \pm b)(a^2 \mp ab + b^2)$.

Quotients

1. $(a^n - b^n)/(a - b) = a^{n-1} + a^{n-2}b + a^{n-3}b^2 + \cdots + ab^{n-2} + b^{n-1}$, if $a \neq b$.
2. $(a^n + b^n)/(a + b) = a^{n-1} - a^{n-2}b + a^{n-3}b^2 - \cdots - ab^{n-2} + b^{n-1}$, if n is odd.
3. $(a^n - b^n)/(a + b) = a^{n-1} - a^{n-2}b + a^{n-3}b^2 - \cdots + ab^{n-2} - b^{n-1}$, if n is even.

Fractions

Signs. $\dfrac{a}{b} = \dfrac{-a}{-b} = -\dfrac{-a}{b} = -\dfrac{a}{-b}$.

Addition and Subtraction. $\dfrac{a}{c} \pm \dfrac{b}{d} = \dfrac{ad \pm bc}{cd}, \dfrac{a}{c} \pm \dfrac{b}{c} = \dfrac{a \pm b}{c}, \dfrac{a}{c} \pm \dfrac{a}{d} = \dfrac{a(d \pm c)}{cd}$,

$\dfrac{a}{def} + \dfrac{b}{e^3g} - \dfrac{c}{df^2} = \dfrac{ae^2fg + bdf^2 - ce^3g}{de^3f^2g}$.

Multiplication. $\dfrac{a}{b} \times \dfrac{c}{d} = \dfrac{ac}{bd}, \dfrac{a}{b} = \dfrac{ac}{bc}$.

Division. $\dfrac{a}{b} \Big/ \dfrac{c}{d} = \dfrac{a}{b} \times \dfrac{d}{c} = \dfrac{ad}{bc}, \dfrac{a}{b} = \dfrac{a}{c} \Big/ \dfrac{b}{c}$.

Series

1. $1 + 2 + 3 + 4 + \cdots + (n - 1) + n = \dfrac{n(n + 1)}{2}$.

2. $p + (p + 1) + (p + 2) + \cdots + (q - 1) + q = \dfrac{(q + p)(q - p + 1)}{2}$.

3. $2 + 4 + 6 + 8 + \cdots + (2n - 2) + 2n = n(n + 1)$.
4. $1 + 3 + 5 + 7 + \cdots + (2n - 3) + (2n - 1) = n^2$.

5. $1^2 + 2^2 + 3^2 + 4^2 + \cdots + (n - 1)^2 + n^2 = \dfrac{n(n + 1)(2n + 1)}{6}$.

6. $1^3 + 2^3 + 3^3 + 4^3 + \cdots + (n - 1)^3 + n^3 = \dfrac{n^2(n + 1)^2}{4}$.

7. $1^4 + 2^4 + 3^4 + 4^4 + \cdots + (n - 1)^4 + n^4 = \dfrac{n}{30}(n + 1)(2n + 1)(3n^2 + 3n - 1)$.

7. BINOMIAL THEOREM

$$(a \pm b)^n = a^n \pm na^{n-1}b + \frac{n(n - 1)}{1 \cdot 2} a^{n-2}b^2 \pm \frac{n(n - 1)(n - 2)}{1 \cdot 2 \cdot 3} a^{n-3}b^3 + \cdots$$

$$+ (\pm 1)^r \frac{n(n - 1) \cdots (n - r + 1)}{r!} a^{n-r}b^r + \cdots,$$

in which the last term shown is the $(r + 1)$th; $r!$, called r *factorial*, equals $1 \cdot 2 \cdot 3 \cdots (r - 1) \cdot r$; and $0! = 1$.

If n is a positive integer, the series is finite; it has $(n + 1)$ terms, the last being b^n; and it holds for all values of a and b. If n is fractional or negative, the series is infinite; it converges only for $|b| < |a|$ (see p. 306).

The coefficients n, $\dfrac{n(n - 1)}{2!}$, $\dfrac{n(n - 1)(n - 2)}{3!}$, $\cdots$ are called *binomial coefficients*.

For brevity the coefficient $\dfrac{n(n - 1) \cdots (n - r + 1)}{r!}$ of the $(r + 1)$th term is written $\dbinom{n}{r}$ or $_nC_r$. If n is a positive integer, the coefficients of the rth term from the beginning and the rth from the end are equal.

For any value of n, and $-1 < x < 1$:

$$(1 \pm x)^n = 1 \pm nx + \frac{n(n-1)}{1 \cdot 2} x^2 \pm \frac{n(n-1)(n-2)}{1 \cdot 2 \cdot 3} x^3 + \frac{n(n-1)(n-2)(n-3)}{1 \cdot 2 \cdot 3 \cdot 4} x^4 \pm \cdots$$

$$\frac{1}{1 \pm x} = (1 \pm x)^{-1} = 1 \mp x + x^2 \mp x^3 + x^4 \mp x^5 + \cdots$$

$$\sqrt{1 \pm x} = (1 \pm x)^{\frac{1}{2}} = 1 \pm \frac{1}{2} x - \frac{1}{2 \cdot 4} x^2 \pm \frac{1 \cdot 3}{2 \cdot 4 \cdot 6} x^3 - \frac{1 \cdot 3 \cdot 5}{2 \cdot 4 \cdot 6 \cdot 8} x^4 \pm \frac{1 \cdot 3 \cdot 5 \cdot 7}{2 \cdot 4 \cdot 6 \cdot 8 \cdot 10} x^5 -.$$

$$\frac{1}{\sqrt{1 \pm x}} = (1 \pm x)^{-\frac{1}{2}} = 1 \mp \frac{1}{2} x + \frac{1 \cdot 3}{2 \cdot 4} x^2 \mp \frac{1 \cdot 3 \cdot 5}{2 \cdot 4 \cdot 6} x^3 + \cdots$$

8. APPROXIMATE FORMULAS

(a) If $|x|$ and $|y|$ are small compared with 1:

   1. $(1 \pm x)^2 = 1 \pm 2x$.

   2. $(1 \pm x)^{\frac{1}{2}} = 1 \pm \dfrac{x}{2}$.

   3. $\dfrac{1}{1 \pm x} = 1 \mp x$.

   4. $(1 + x)(1 + y) = 1 + x + y$.

   5. $(1 + x)(1 - y) = 1 + x - y$.

   6. $e^x = 1 + x + \dfrac{x^2}{2}$ (where $e = 2.71828$).

   7. $\log_e (1 \pm x) = \pm x - \dfrac{x^2}{2} \pm \dfrac{x^3}{3}$.   (Last term often may be omitted.)

   8. $\log_e \left(\dfrac{1 + x}{1 - x} \right) = 2 \left(x + \dfrac{x^3}{3} + \dfrac{x^5}{5} \right)$.

(b) If $|x|$ is small compared with a and $a > 0$:

   9. $a^x = 1 + x \log_e a + \dfrac{x^2}{2} (\log_e a)^2$.   (Last term often may be omitted.)

(c) If a and b are nearly equal and both > 0:

   10. $\sqrt{ab} = \dfrac{a + b}{2}$.

(d) If b is small compared with a and both > 0:

   11. $\sqrt{a^2 \pm b} = a \pm \dfrac{b}{2a}$.

   12. $\sqrt{a^3 \pm b} = a \pm \dfrac{b}{3a^2}$.

   13. $\sqrt{a^2 + b^2} = 0.960a + 0.398b$. This is within 4 per cent of the true value if $a > b$. A closer approximation is $\sqrt{a^2 + b^2} = 0.9938a + 0.0703b + 0.3567 \dfrac{b^2}{a}$.

   14. $\sqrt{a^2 + b^2 + c^2} = 0.939a + 0.389b + 0.297c$. This is within 6 per cent of the true value if $a > b > c$. For instance, for the numbers 43, 42, and 41, the error < 5.2 per cent.

(e) If $|x|$ is less than $\dfrac{\pi}{18}$:

   15. $\sin x = x - \dfrac{x^3}{6}$.

   16. $\cos x = 1 - \dfrac{x^2}{2}$.   (Last term often may be omitted.)

   17. $\tan x = x + \dfrac{x^3}{3}$.

(Note: If $x = 8° = \dfrac{8\pi}{180} = 0.13963$, $\sin x = x - \frac{1}{6} x^3 = 0.13918$, which is one unit in error in the fifth decimal place. If the absolute value of the angle is less than $5°$, the values of x and $\sin x$ do not differ more than one unit in the fourth decimal place.)

(*f*) If $|y|$ is less than $\dfrac{\pi}{36}$ and small compared with $|x|$:

    18. $\sin(x \pm y) = \sin x \pm y \cos x$.
    19. $\cos(x \pm y) = \cos x \mp y \sin x$.

    20. $\tan(x \pm y) = \tan x \pm \dfrac{y}{\cos^2 x}$.

(*g*) If $|n| > 1$:

    21. $e^{1/n} = 1 + \dfrac{1}{n - 0.5}$.

    22. $e^{-1/n} = 1 - \dfrac{1}{n + 0.5}$.

(*h*) As $n \to \infty$:

    23. $\dfrac{1 + 2 + 3 + 4 + 5 \cdots + n}{n^2} \to \dfrac{1}{2}$.

    24. $\dfrac{1 + 2^2 + 3^2 + 4^2 + \cdots + n^2}{n^3} \to \dfrac{1}{3}$.

    25. $\dfrac{1 + 2^3 + 3^3 + 4^3 + \cdots + n^3}{n^4} \to \dfrac{1}{4}$.

9. INEQUALITIES

Laws of Inequalities for positive quantities:

(*a*) If $a > b$, then: $a + c > b + c$ $b < a$

                   $a - c > b - c$ $c - a < c - b$

                   $ac > bc$ $-ca < -cb$

                   $\dfrac{a}{c} > \dfrac{b}{c}$ $\dfrac{c}{a} < \dfrac{c}{b}$

Corollary: If $a - c > b$, then $a > b + c$.

(*b*) If $a > b$ and $c > d$, then: $a + c > b + d$; $ac > bd$; but $a - c$ may be $>$ or $=$ or $< b - d$; a/c may be $>$ or $=$ or $< b/d$.

10. RATIO AND PROPORTION

Laws of Ratio and Proportion

(*a*) If $\dfrac{a}{b} = \dfrac{c}{d}$ then: $\dfrac{a}{c} = \dfrac{b}{d}$; $ad = bc$; $\dfrac{ma + nb}{pa + qb} = \dfrac{mc + nd}{pc + qd}$; $\left(\dfrac{a}{b}\right)^n = \left(\dfrac{c}{d}\right)^n$. If also

$\dfrac{e}{f} = \dfrac{g}{h}$, then: $\dfrac{ae}{bf} = \dfrac{cg}{dh}$.

(*b*) If $\dfrac{a}{b} = \dfrac{c}{d} = \dfrac{e}{f} = \cdots$, then: $\dfrac{a}{b} = \dfrac{c}{d} = \dfrac{e}{f} = \cdots = \dfrac{pa + qc + re + \cdots}{pb + qd + rf + \cdots}$.

Variation.

If $y = kx$, y varies directly as x; i.e., y is directly proportional to x.

If $y = \dfrac{k}{x}$, y varies inversely as x; i.e., y is inversely proportional to x.

If $y = kxz$, y varies jointly as x and z.

If $y = k\dfrac{x}{z}$, y varies directly as x and inversely as z.

The constant k is called the proportionality factor.

ALBEGRA 219

11. PROGRESSIONS

An Arithmetic Progression is a sequence in which the *difference d* of any two consecutive terms is a constant. If n = number of terms, a = first term, l = last term, s = sum of n terms, then $l = a + (n - 1)d$, and $s = \dfrac{n}{2}(a + l)$. The *arithmetic mean A* of two quantities m, n is the quantity which placed between them makes with them an arithmetic progression; $A = (m + n)/2$.

Example. Given the series $3 + 5 + 7 + \cdots$ to 10 terms. Here $n = 10$, $a = 3$, $d = 2$; hence $l = 3 + (10 - 1) \times 2 = 21$, and $s = (10/2)(3 + 21) = 120$.

A Geometric Progression is a sequence in which the *ratio r* of any two consecutive terms is a constant. If n = number of terms, a = first term, l = last term, s = sum of n terms, then $l = ar^{n-1}$, $s = (rl - a)/(r - 1) = a(1 - r^n)/(1 - r)$. The *geometric mean G* of two quantities m, n is the quantity which placed between them makes with them a geometric progression; $G = \sqrt{mn}$.

Example. Given the series $3 + 6 + 12 + \cdots$ to 6 terms. Here $n = 6$, $a = 3$, $r = 2$; hence $l = 3 \times 2^{6-1} = 96$, and $s = (2 \times 96 - 3)/(2 - 1)$, or $= 3(1 - 2^6)/(1 - 2) = 189$.

If $|r| < 1$, then, as $n \to \infty$, $s \to a/(1 - r)$.

Example. Given the infinite series $1/2 + 1/4 + 1/8 + \cdots$. Here $a = 1/2$ and $r = 1/2$; hence $s \to (1/2)/(1 - 1/2) = 1$ as $n \to \infty$.

A Harmonic Progression is a sequence in which the reciprocals of the terms form an arithmetic progression. The *harmonic mean H* of two quantities m, n is the quantity which placed between them makes with them a harmonic progression; $H = 2mn/(m + n)$.

Relation among the Arithmetic, Geometric, and Harmonic Means of two quantities; $G^2 = AH$.

12. PARTIAL FRACTIONS

A *proper* algebraic fraction is one in which the numerator is of lower degree than the denominator. An improper fraction can be changed to the sum of a polynomial and a proper fraction by dividing the numerator by the denominator.

A proper fraction can be resolved into *partial fractions*, the denominators of which are factors, prime to each other, of the denominator of the given fraction.

Case 1. The denominator can be factored into real linear factors, P, Q, R, $\cdots$, all different. Let

$$\frac{\text{Num}}{PQR \cdots} = \frac{A}{P} + \frac{B}{Q} + \frac{C}{R} + \cdots$$

Example.

$$\frac{6x^2 - x + 1}{x^3 - x} = \frac{A}{x} + \frac{B}{x - 1} + \frac{C}{x + 1}$$

Clearing of fractions,

$$6x^2 - x + 1 = A(x - 1)(x + 1) + Bx(x + 1) + Cx(x - 1) \tag{1}$$

(a) Substitution method. Letting $x = 0$, $A = -1$; $x = 1$, $B = 3$; $x = -1$, $C = 4$. Then

$$\frac{6x^2 - x + 1}{x^3 - x} = -\frac{1}{x} + \frac{3}{x - 1} + \frac{4}{x + 1}$$

(b) Method of undetermined coefficients. Rewriting (1),

$$6x^2 - x + 1 = (A + B + C)x^2 + (B - C)x - A$$

Equating coefficients of like powers of x, $A + B + C = 6$, $B - C = -1$, $-A = 1$. Solving this system of equations, $A = -1$, $B = 3$, $C = 4$.

Case 2. The denominator can be factored into real linear factors, P, Q, $\cdots$, one or more repeated. Let

$$\frac{\text{Num}}{P^2Q^3} = \frac{A}{P} + \frac{B}{P^2} + \frac{C}{Q} + \frac{D}{Q^2} + \frac{E}{Q^3} + \cdots$$

Example.

$$\frac{x + 1}{x(x - 1)^3} = \frac{A}{x} + \frac{B}{x - 1} + \frac{C}{(x - 1)^2} + \frac{D}{(x - 1)^3}$$

Clearing of fractions,

$$x + 1 = A(x - 1)^3 + Bx(x - 1)^2 + Cx(x - 1) + Dx$$

A and D can be found by substituting $x = 0$ and $x = 1$. After inserting these numerical values for A and D, B and C can be found by the method of undetermined coefficients.

Case 3. The denominator can be factored into quadratic factors, $P, Q, \cdots$, all different, which cannot be factored into real linear factors. Let

$$\frac{\text{Num}}{PQ \cdots} = \frac{Ax + B}{P} + \frac{Cx + D}{Q} + \cdots$$

Example.

$$\frac{3x^2 - 2}{(x^2 + x + 1)(x + 1)} = \frac{Ax + B}{x^2 + x + 1} + \frac{C}{x + 1}$$

Clearing of fractions,

$$3x^2 - 2 = (Ax + B)(x + 1) + C(x^2 + x + 1)$$
$$= (A + C)x^2 + (A + B + C)x + (B + C)$$

Use the method of undetermined coefficients to find A, B, C.

Case 4. The denominator can be factored into quadratic factors, $P, Q, \cdots$, one or more repeated, which cannot be factored into real linear factors. Let

$$\frac{\text{Num}}{P^2 Q^3 \cdots} = \frac{Ax + B}{P} + \frac{Cx + D}{P^2} + \frac{Ex + F}{Q} + \frac{Gx + H}{Q^2} + \frac{Ix + J}{Q^3} + \cdots$$

Example.

$$\frac{5x^2 - 4x + 16}{(x - 3)(x^2 - x + 1)^2} = \frac{A}{x - 3} + \frac{Bx + C}{x^2 - x + 1} + \frac{Dx + E}{(x^2 - x + 1)^2}$$

Clearing of fractions,

$$5x^2 - 4x + 16 = A(x^2 - x + 1)^2 + (Bx + C)(x - 3)(x^2 - x + 1) + (Dx + E)(x - 3)$$

Find A by substituting $x = 3$. Then use the method of undetermined coefficients to find B, C, D, E.

13. LOGARITHMS

If $N = b^x$, then x is the *logarithm* of the number N to the *base b*. For computation, *common*, or *Briggs*, logarithms to the base 10 (abbreviated $\log_{10}$ or log) are used. For theoretical work involving calculus, *natural*, or *Naperian*, logarithms to the irrational base $e = 2.71828 \cdots$ (abbreviated ln, $\log_e$, or log) are used. The relation between logarithms of the two systems is:

$$\log_e n = \log_{10} n / \log_{10} e = \log_{10} n / 0.4343 = 2.303 \log_{10} n$$

The integral part of a common logarithm, called the *characteristic*, may be positive, negative, or zero. The decimal part, called the *mantissa* and given in tables, is always positive.

To find the common logarithm of a number, first find the mantissa from Table 10, Section 1, disregarding the decimal point of the number. Then from the location of the decimal point find the characteristic as follows. If the number is greater than 1, the characteristic is positive or zero. It is one less than the number of figures preceding the decimal point. For a number expressed in standard notation the characteristic is the exponent of 10.

Examples. $\log 6.54 = 0.8156$, $\log 6540 = \log (6.54 \times 10^3) = 3.8156$.

If the number is less than 1, the characteristic is negative and is numerically one greater than the number of zeros immediately following the decimal point. To avoid having a negative integral part and a positive decimal part, the characteristic is written as a difference.

Examples. $\log\ 0.654 = \log\ (6.54 \times 10^{-1}) = \bar{1}.8156 = 9.8156 - 10$, $\log\ 0.000654 = \log (6.54 \times 10^{-4}) = \bar{4}.8156 = 6.8156 - 10$.

To find a number whose logarithm is given, each of the above steps is reversed.

The cologarithm of a number is the logarithm of its reciprocal. Hence, colog $N = \log 1/N = \log 1 - \log N = -\log N$.

Use of Logarithms in Computation.

To multiply a and b	$\log ab = \log a + \log b$
To divide a by b	$\log a/b = \log a - \log b$
To raise a to the nth power	$\log a^n = n \log a$
To find the nth root of a	$\log a^{1/n} = (1/n) \log a$

Examples. (1) $68.31 \times 0.2754 = 18.81$.

$$\log 68.31 = 1.8345$$
$$\log 0.2754 = 9.4400 - 10$$
$$\overline{11.2745 - 10} = 1.2745 = \log 18.81$$

(2) $0.6831^{1.53} = 0.5582$.

$$\log 0.6831 = 9.8345 - 10$$
$$1.53 \times (9.8345 - 10) = 15.0468 - 15.3$$

To subtract 15.3 from 15.0468, add 10 to 15.0468 and subtract 10 from it.

$$25.0468 - 10$$
$$15.3$$
$$\overline{9.7468 - 10} = \log 0.5582$$

(3) $\sqrt[5]{0.6831} = 0.9266$.

$$\log 0.6831 = 9.8345 - 10$$

$$\frac{49.8345 - 50}{5} = 9.9669 - 10 = \log 0.9266$$

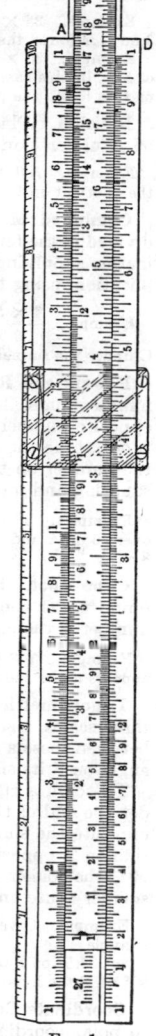

To solve a simple exponential equation of the form $a^x = b$, equate the logarithms of the two sides of the equation:

$x \log a = \log b$, from which $x = \dfrac{\log b}{\log a}$, and $\log x = \log (\log b) - \log (\log a)$

Example. $0.6831^x = 27.54$.

$$x = \frac{\log 27.54}{\log 0.6831} = \frac{1.4400}{9.8345 - 10} = \frac{1.4400}{-0.1655} = -8.701$$

Theory of the Slide Rule

A Slide Rule is an instrument for mechanical computation based on the use of logarithms. With it the operations of multiplication, division, and the finding of powers and roots can be performed rapidly and with an accuracy sufficient for most engineering work. With a good 10-inch Mannheim rule the results obtained are accurate to approximately $1/10$ of 1 per cent.

The Simple Mannheim Rule consists of a fixed and a sliding part both of which are ruled with logarithmic scales, i.e., with divisions at distances equal to the logarithms of the numbers marking the division points. Since the logarithm of the product of two numbers is the sum of the logarithms of the numbers and the logarithm of the quotient of two numbers is the difference of their logarithms, multiplication or division of the numbers can be effected by moving the slide to the right or left to add or subtract the logarithms. The scales on the fixed part of the rule are called the A and D scales, and those on the slide the B and C scales. The A and B scales are each divided into two parts, each part being a half-size reproduction of the C and D scale. A "runner," which consists of a glass plate with a fine vertical line on it, is used to facilitate some of the operations. Since 0 is the logarithm of 1 the numbering on each scale begins with the Fig. 1. In using the scale, the position of the decimal point is disregarded.

Fig. 1

Use of the Simple Mannheim Slide Rule

The following examples illustrate the use of the slide rule.

Proportion. Set the first term of a proportion on the C scale opposite the second term on the D scale; then opposite the third term on the C scale read the fourth term on the D scale.

Example. Find the fourth term in the proportion $12 : 21 :: 30 : x$. Move the slide to the right until 12 on C is opposite 21 on D, then opposite 30 on C read x on D, $= 52.5$.

Multiplication. Set 1 on the C scale opposite one of the factors on D. Under the other factor on the C scale read the product on the D scale.

Example. 25 × 3. Move the slide to the right until the left 1 on C is opposite 25 on D. Under 3 on C is found the product on D, = 75.

Example. 25 × 5. If the slide is moved to the right as above, the 5 on C will be beyond the end of D. In this case move the slide to the left until the right 1 on C is opposite 25 on D. Under 5 on C is found the product on D, = 125.

Division. Place the divisor on C opposite the dividend on D, and the quotient is found on D under 1 on C.

Example. 750 ÷ 25. Move the slide until 25 on C is opposite 750 on D. Under 1 on C is found the quotient on D, = 30.

Combined Multiplication and Division. Arrange the numbers to be multiplied and divided in the form of a fraction with one more factor in the numerator than in the denominator, supplying the factor 1 if necessary. Then perform alternate division and multiplication, using the runner to indicate partial results.

Example. $\dfrac{4 \times 5 \times 8}{3 \times 6} = 8.9$ nearly. Set 3 on C over 4 on D, set runner to 5 on C, then set 6 on C under the runner, and read under 8 on C the result 8.9— on D.

Powers and Roots. The numbers on scales A and B are the squares of the opposite numbers on scales C and D, and the numbers on scales C and D the square roots of the opposite numbers on scales A and B. In extracting square roots, if the number of digits is odd, take the number on the left-hand scale of A; if the number of digits is even, take the number on the right-hand scale of A. To cube a number, perform the operations of squaring and multiplication.

Examples. $4^2 = 16$. Set the runner over 4 on D and read 16 on A. $\sqrt{16} = 4$. Set the runner over 16 on A and read 4 on D. $2^3 = 8$. Set 1 on C over 2 on D, and above 2 on B read the result 8 on A.

Cube Root. Set the runner over the number on A; then move the slide until there is found under the runner on B the same number which is found under 1 on C on D; this number is the cube root desired.

Example. $\sqrt[3]{8} = 2$. Set the runner over 8 on A, move the slide along until the same number appears under the runner on B and under 1 on C on D; this is the number 2.

Trigonometric Computations. On the under side of the slide (which is reversible) are three scales, a scale of natural sines marked S, a scale of natural tangents marked T, and between these a scale of equal parts. To use these scales, turn the slide over. Opposite an angle on S its sine is found on A, and opposite an angle on T is found its tangent on D. To solve a right triangle in which the two arms are given, use the T and D scales. Set the end of the T scale opposite the longer arm on D and find the smaller angle on T opposite the shorter arm on D. In all other cases use the A and S scales. Set a given side on A opposite its opposite angle on S. With this setting the hypotenuse is opposite 90°, the longer arm opposite the larger acute angle, and the shorter arm opposite the smaller acute angle.

Example. Given, shorter arm = 24, opposite angle = 35.5°. Find the remaining parts of the triangle.
Set 35.5° on S opposite 24 on A. Opposite 54.5° on S read the longer arm 33.6 on A and opposite 90° on S the hypotenuse 41.3 on A.

Coordinate Conversion; Log Log Scales. Scales for easy conversion from rectangular to polar coordinates, and also scales for solution of problems involving logarithmic and exponential equations are available on certain slide rules. For detailed instructions in their use consult the *Manual of The Log Log Vector Slide Rule*, published by Keuffel and Esser Co.

14. EQUATIONS

The equation $f(x) = a_0 x^n + a_1 x^{n-1} + a_2 x^{n-2} + \cdots + a_n = 0$, a_i real, is a *polynomial equation* of *degree n* in one variable.
For $n = 1$, the equation $f(x) = ax + b = 0$ is *linear*. It has one root $x_1 = -b/a$.

Quadratic Equation

For $n = 2$, the equation $f(x) = ax^2 + bx + c = 0$ is *quadratic*. It has two roots, both real or both complex, given by the formulas

$$x_1, x_2 = \frac{-b \pm \sqrt{b^2 - 4ac}}{2a} = \frac{2c}{-b \mp \sqrt{b^2 - 4ac}}$$

To avoid loss of precision if $\sqrt{b^2 - 4ac}$ and $|b|$ are nearly equal, use the form that does not involve the difference.

If the quantity $b^2 - 4ac$, called the *discriminant*, is > 0, the roots are real and unequal; if it $= 0$, the roots are real and equal; if it is < 0, the roots are complex.

Cubic Equation

For $n = 3$, the equation $f(x) = a_0x^3 + a_1x^2 + a_2x + a_3 = 0$ is *cubic*. It has three roots, all real or one real and two complex.

Algebraic Solution. Write the equation in the form $ax^3 + 3bx^2 + 3cx + d = 0$. Let

$$q = ac - b^2 \quad \text{and} \quad r = \tfrac{1}{2}(3abc - a^2d) - b^3$$

Also let $\quad s_1 = (r + \sqrt{q^3 + r^2})^{\frac{1}{3}} \quad \text{and} \quad s_2 = (r - \sqrt{q^3 + r^2})^{\frac{1}{3}}$

Then the roots are:

$$x_1 = [(s_1 + s_2) - b] \div a$$

$$x_2 = \left[-\frac{1}{2}(s_1 + s_2) + \frac{\sqrt{-3}}{2}(s_1 - s_2) - b \right] \div a$$

$$x_3 = \left[-\frac{1}{2}(s_1 + s_2) - \frac{\sqrt{-3}}{2}(s_1 - s_2) - b \right] \div a$$

If $q^3 + r^2 > 0$, there are one real and two complex roots. If $q^3 + r^2 = 0$, there are three real roots of which at least two are equal. If $q^3 + r^2 < 0$, there are three real roots but the numerical solution leads to finding the cube roots of complex quantities. In such a case the trigonometric solution is employed.

Example. Given the equation $x^3 + 12x^2 + 45x + 54 = 0$.
Here $a = 1$, $b = 4$, $c = 15$, $d = 54$. $q = 15 - 16 = -1$; $r = \tfrac{1}{2}(180 - 54) - 64 = -1$; $q^3 + r^2 = -1 + 1 = 0$, $s_1 = s_2 = (-1)^{\frac{1}{3}} = -1$. $s_1 + s_2 = -2$; $s_1 - s_2 = 0$.
Hence the roots are $x_1 = (-2 - 4) = -6$; $x_2 = x_3 = [-\tfrac{1}{2}(-2) - 4] = -3$.

Trigonometric Solution. Write the equation in the form $ax^3 + 3bx^2 + 3cx + d = 0$. Let $q = ac - b^2$ and $r = \tfrac{1}{2}(3abc - a^2d) - b^3$ (as in algebraic solution). Then the roots are:

$$x_1 = (y_1 - b) \div a$$

$$x_2 = (y_2 - b) \div a$$

$$x_3 = (y_3 - b) \div a$$

where y_1, y_2, and y_3 have the following values (upper of alternative signs being used when r is $+$ and the lower when r is $-$):

Case 1. If q is $-$ and $q^3 + r^2 \leqq 0$:

$$y_1 = \pm 2\sqrt{-q} \cos\left[\frac{1}{3}\cos^{-1}\frac{\pm r}{\sqrt{-q^3}} \right]$$

$$y_2 = \pm 2\sqrt{-q} \cos\left[\frac{1}{3}\cos^{-1}\frac{\pm r}{\sqrt{-q^3}} + \frac{2\pi}{3} \right]$$

$$y_3 = \pm 2\sqrt{-q} \cos\left[\frac{1}{3}\cos^{-1}\frac{\pm r}{\sqrt{-q^3}} + \frac{4\pi}{3} \right]$$

Case 2. If q is $-$ and $q^3 + r^2 \geqq 0$:

$$y_1 = \pm 2\sqrt{-q} \cosh\left[\frac{1}{3}\cosh^{-1}\frac{\pm r}{\sqrt{-q^3}} \right]$$

$$y_2 = \mp \sqrt{-q} \cosh\left[\frac{1}{3}\cosh^{-1}\frac{\pm r}{\sqrt{-q^3}} \right] + i\sqrt{-3q}\sinh\left[\frac{1}{3}\cosh^{-1}\frac{\pm r}{\sqrt{-q^3}} \right]$$

$$y_3 = \mp \sqrt{-q} \cosh\left[\frac{1}{3}\cosh^{-1}\frac{\pm r}{\sqrt{-q^3}} \right] - i\sqrt{-3q}\sinh\left[\frac{1}{3}\cosh^{-1}\frac{\pm r}{\sqrt{-q^3}} \right]$$

Case 3. If q is $+$:

$$y_1 = \pm 2\sqrt{q} \sinh\left[\frac{1}{3}\sinh^{-1}\frac{\pm r}{\sqrt{q^3}}\right]$$

$$y_2 = \mp\sqrt{q} \sinh\left[\frac{1}{3}\sinh^{-1}\frac{\pm r}{\sqrt{q^3}}\right] + i\sqrt{3q} \cosh\left[\frac{1}{3}\sinh^{-1}\frac{\pm r}{\sqrt{q^3}}\right]$$

$$y_3 = \mp\sqrt{q} \sinh\left[\frac{1}{3}\sinh^{-1}\frac{\pm r}{\sqrt{q^3}}\right] - i\sqrt{3q} \cosh\left[\frac{1}{3}\sinh^{-1}\frac{\pm r}{\sqrt{q^3}}\right]$$

Example. Given the equation $x^3 + 6x^2 - 9x - 54 = 0$.
Here $a = 1$, $b = 2$, $c = -3$, $d = -54$; $q = -3 - 4 = -7$; $r = 1/2\,(-18 + 54) - 8 = 10$; $q^3 + r^2 = -343 + 100 = -243$. Note that q is $-$; $q^3 + r^2 < 0$; r is $+$. Therefore use *Case 1* with upper signs.

$$y_1 = 2\sqrt{7}\cos\left[\frac{1}{3}\cos^{-1}\frac{10}{\sqrt{343}}\right] = 2\sqrt{7}\cos 19.1° = 5$$

Hence, one root is $x_1 = 5 - 2 = 3$. The other roots can be similarly determined.

Quartic Equation

For $n = 4$, the equation $f(x) = a_0x^4 + a_1x^3 + a_2x^2 + a_3x + a_4 = 0$ is *quartic*. It has four roots, all real, all complex, or two real and two complex.
To solve, first divide the equation by a_0 to put it in the form $x^4 + ax^3 + bx^2 + cx + d = 0$. Find any real root y_1 of the cubic equation:

$$8y^3 - 4by^2 + 2(ac - 4d)y - [c^2 + d(a^2 - 4b)] = 0$$

Then the four roots of the quartic equation are given by the roots of the two quadratic equations:

$$x^2 + \left[\frac{a}{2} + \sqrt{\frac{a^2}{4} + 2y_1 - b}\right]x + (y_1 + \sqrt{y_1{}^2 - d}) = 0$$

$$x^2 + \left[\frac{a}{2} - \sqrt{\frac{a^2}{4} + 2y_1 - b}\right]x + (y_1 - \sqrt{y_1{}^2 - d}) = 0$$

Nth Degree Equation

Properties of $f(x) = a_0x^n + a_1x^{n-1} + \cdots + a_n = 0$. a_n's are real.
1. *Remainder Theorem.* If $f(x)$ is divided by $(x - r)$ until a remainder independent of x is obtained, this remainder is equal to $f(r)$, the value of $f(x)$ for $x = r$.
2. *Factor Theorem.* If, and only if, $(x - r)$ is a factor of $f(x)$, then $f(r) = 0$.
3. The equation $f(x) = 0$ has n roots, not necessarily distinct. Complex roots occur in conjugate pairs, $a + bi$ and $a - bi$. If n is odd there is at least one real root.
4. The sum of the roots is $-a_1/a_0$, the sum of the products of the roots taken two at a time is a_2/a_0, the sum of the products of the roots taken three at a time is $-a_3/a_0$, etc. The product of all the roots is $(-1)^n a_n/a_0$.
5. If the a_i are integers and p/q is a rational root of $f(x) = 0$, reduced to its lowest terms, then p is a divisor of a_n and q of a_0. If a_0 is 1, the rational roots are integers.
6. If x is replaced by (a) y/m, (b) $-y$, (c) $y + h$, the roots of the resulting equation $\phi(y) = 0$ are (a) m times, (b) the negatives of, (c) less by h than the corresponding roots of $f(x) = 0$.
7. *Descartes' Rule of Signs.* A *variation* of sign occurs in $f(x) = 0$ if two consecutive terms have unlike signs. The number of positive roots is either equal to the number of variations of sign or is less by a positive even integer. For negative roots apply the rule to $f(-x) = 0$.
8. If, for two real numbers a and b, $f(a)$ and $f(b)$ have opposite signs, there is an odd number of roots between a and b.
9. If k is the exponent of the first term with a negative coefficient and G the greatest of the absolute values of the negative coefficients, then an upper bound of the real roots is $1 + \sqrt[n-k]{G/a_0}$.
10. *Sturm's Theorem.* Let the equation $f(x) = 0$ have no multiple roots. With $f_0 = f(x)$ and $f_1 = f'(x)$, form the sequence $f_0, f_1, f_2, \cdots, f_n$ as follows:

$$f_0 = q_1f_1 - f_2, \quad f_1 = q_2f_2 - f_3, \quad f_2 = q_3f_3 - f_4, \quad \cdots, \quad f_{n-2} = q_{n-1}f_{n-1} - f_n$$

At any step, a function f_i may be multiplied by a positive number to avoid fractions. Let a and b be real numbers, $a < b$, such that $f(a) \neq 0$, $f(b) \neq 0$, and let $V(a)$ be the number of variations of sign in the non-zero members of the sequence $f_0(a)$, $f_i(a)$, $\cdots$, f_n. Then the number of real roots between a and b is $V(a) - V(b)$.

If $f(x) = 0$ has multiple roots, the sequence terminates with the function f_m, $m < n$, when $f_{m-1} = q_m f_m$. For this sequence, $V(a) - V(b)$ is the number of distinct real roots between a and b.

Examples. (1) Locate the real roots of $x^3 - 7x - 7 = 0$.

$x =$	-2	-1	0	1	2	3	4
$f_0 = x^3 - 7x - 7$	$-$	$-$	$-$	$-$	$-$	$-$	$+$
$f_1 = 3x^2 - 7$	$+$	$-$	$-$	$-$	$+$	$+$	$+$
$f_2 = 2x + 3$	$-$	$+$	$+$	$+$	$+$	$+$	$+$
$f_3 = 1$	$+$	$+$	$+$	$+$	$+$	$+$	$+$
$V(x) =$	3	1	1	1	1	1	0

$3x - 9/2 \qquad x$

$$2x + 3 \overline{\smash{\big)}\ 3x^2 \qquad\quad - 7} \qquad x^3 - 7x - 7$$
$$\underline{6x^2 \qquad - 14} \qquad \overline{\smash{\big)}\ 3x^3 - 21x - 21}$$
$$6x^2 + 9x \qquad\qquad 3x^3 - 7x$$
$$\underline{-9x - 14} \qquad\quad -14x - 21$$
$$-9x - 27/2 \qquad\qquad 2x + 3 = f^2$$
$$\underline{-1/2}$$
$$1 = f_3$$

$V(-2) - V(-1) = 2 \qquad V(3) - V(4) = 1$
$-2 < r_1, r_2 < -1 \qquad\quad 3 < r_3 < 4$

(2) Locate the real roots of $4x^3 - 3x - 1 = 0$.

$x =$	-1	0	1	2
$f_0 = 4x^3 - 3x - 1$	$-$	$-$	0	$+$
$f_1 = 3(4x^2 - 1)$	$+$	$-$		$+$
$f_2 = 2x + 1$	$-$	$+$		$+$
$V(x) =$	2	1		0

$2x - 1 \qquad x$

$$2x + 1 \overline{\smash{\big)}\ 4x^2 - 1} \qquad 4x^3 - 3x - 1$$
$$\underline{4x^2 + 2x} \qquad\quad 4x^3 - x$$
$$-2x - 1 \qquad -2x - 1$$
$$\underline{-2x - 1} \qquad\quad 2x + 1 = f_2$$

$V(-1) - V(0) = 1 \quad V(0) - V(2) = 1$
$-1 < r_1 < 0 \qquad\qquad 0 < r_2 < 2$
r_1 can be found to be a double root.

Synthetic Division. To divide a polynomial $f(x)$ by $(x - a)$, proceed as in the example. Divide $f(x) = 4x^3 - 7x + 1$ by $x + 2$. Arrange the coefficients in order of descending powers of x, supplying zeros for missing powers. Place a ($= -2$) to the left. Bring down the first coefficient, multiply it by a, and add the product to the next coefficient. Multiply the sum by a, add the product to the next coefficient, and continue thus.

$$
\begin{array}{r|rrrrr}
-2 & 4 & + \;0 & - \;7 & + \;1 \\
   &   & - \;8 & + 16 & - 18 \\
\hline
   & 4 & - \;8 & + \;9 & \big| - 17
\end{array}
$$

The last number is the remainder. It is the value of the polynomial $f(x) = 4x^3 - 7x + 1$ for $x = -2$, or $f(-2) = -17$. The other numbers in the last line are the coefficients of the quotient $4x^2 - 8x + 9$, a polynomial of one degree less than the dividend.

Rational Roots. Possible integral and fractional roots can be found by Property 5, and tested by synthetic division. If a rational root r is found, then the remaining roots are roots of $q(x) \equiv f(x)/(x - r) = 0$.

Irrational Roots. (1) Horner's Method consists of diminishing a root repeatedly toward zero and adding together the amounts by which it is diminished. This sum approximates the original root. The method is explained by an example.

A root of $x^3 + 4x - 7 = 0$ is located between the successive integers 1 and 2, graphically or by synthetic division, using Property 8. First, the roots are diminished by 1 (Property 6-c) to give an equation $f(y + 1) \equiv \phi(y) = 0$, which has a root between 0 and 1. The method of obtaining the coefficients of $\phi(y)$ by use of successive synthetic divisions is illustrated. The remainders are the required coefficients. The root between 0 and 1 of $\phi(y) = 0$ is then located between successive tenths. Since its value is small, the last two terms set equal to 0 suffice to estimate that it is between 0.2 and 0.3. Next, diminish the roots by 0.2 to obtain an equation with a root between 0 and 0.1. To check that the root was between 0.2 and 0.3, note that the first remainder, which is the value of $\phi(0.2)$, remains negative when $\phi(y)$ is divided by $y - 0.2$, and that the remainder would be found to be positive if $\phi(y)$ were divided by $y - 0.3$. Repeat the process, using the last two terms to estimate that the root of the new equation is between 0.05 and 0.06, and then diminish by 0.05. At the next stage it is frequently possible to estimate two more figures by using the last two terms.

$$
\begin{array}{rrrrr}
1 & + \;0 & + \;4 & - \;7 & | \,1 \\
  & + \;1 & + \;1 & + \;5 & \underline{} \\
\hline
1 & + \;1 & + \;5 & - \;2 & \\
  & + \;1 & + \;2 & & \\
\hline
1 & + \;2 & + \;7 & & \\
  & + \;1 & & & \\
\hline
1 & + \;3 & + \;7 & - \;2 & | \,0.2 \\
  & + \;0.2 & + \;0.64 & + \;1.528 & \underline{} \\
\hline
1 & + \;3.2 & + \;7.64 & - \;0.472 & \\
  & + \;0.2 & + \;0.68 & & \\
\hline
1 & + \;3.4 & + \;8.32 & & \\
  & + \;0.2 & & & \\
\hline
1 & + \;3.6 & + \;8.32 & - \;0.472 & | \,0.05 \\
  & + \;0.05 & + \;0.1825 & + \;0.425125 & \\
\hline
1 & + \;3.65 & + \;8.5025 & - \;0.046875 & \\
  & + \;0.05 & + \;0.185 & & \\
\hline
1 & + \;3.70 & + \;8.6875 & & \\
  & + \;0.05 & & & \\
\hline
1 & + \;3.75 & & & \\
\end{array}
$$

$$8.6875x - 0.046875 = 0$$

$$x = 0.0054-$$

The root is 1.2554−

To find a *negative* irrational root $-r$ by Horner's method, replace x in $f(x) = 0$ by $-y$, find the positive root r of $\phi(y) = f(-y) = 0$, and change its sign.

(2) Newton's Method can be used to find a root of either an algebraic or a transcendental equation. The root is first located graphically between α and β, $f(\alpha)$ and $f(\beta)$ having un-

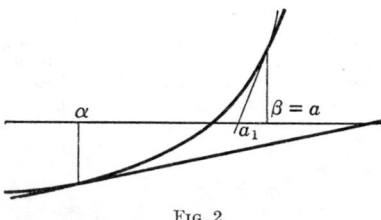

FIG. 2

like signs (Fig. 2). Assume that there is no maximum, minimum, or inflection point in the interval (α, β), that is, that neither $f'(x)$ nor $f''(x)$ equals zero for any point in (α, β). Take as a first approximation a the end point α or β for which $f(x)$ and $f''(x)$ have the same sign, that is, if the curve is concave up, take the end point at which $f(x)$ is positive, and, if concave down, the end point at which $f(x)$ is negative. The point $a_1 = a - f(a)/f'(a)$, at which the tangent to the curve at $[a, f(a)]$ intersects the x axis, is between a and the root. Then,

by using a_1 instead of a, a still better approximation a_2 is obtained, etc If the end point for which $f(x)$ and $f''(x)$ have opposite signs were used, it could happen that the approximation obtained would be better than a_1, but it might be much worse since the tangent would not cross the x axis between the end point used and the root (Fig. 2).

Example. Find the real root of $x^3 + 4x - 7 = 0$.

$$f(x) = x^3 + 4x - 7$$

$$f'(x) = 3x^2 + 4$$

$$f''(x) = 6x$$

Graphically (Fig. 3), $\alpha = 1.2$, $\beta = 1.3$. Since $f(1.2) = -0.472$ and $f(1.3) = 0.397$, and $f''(x)$ is positive in the interval, then $a = 1.3$.

$$a_1 = a - \frac{f(a)}{f'(a)} = 1.3 - \frac{0.397}{9.07} = 1.3 - 0.044 = 1.256$$

$$a_2 = 1.256 - \frac{0.005385}{8.7326} = 1.256 - 0.00062 = 1.25538$$

If Newton's method of using the tangent is not applicable, either because of the presence of a maximum, minimum, or inflection point, or because of difficulty in finding $f'(x)$, the interpolation method using the chord joining $[\alpha, f(\alpha)]$ and $[\beta, f(\beta)]$ can be used. The chord crosses the x axis at $a = \alpha - f(\alpha)(\beta - \alpha)/[f(\beta) - f(\alpha)]$, a better approximation than either α or β. Note that this formula differs from Newton's only in having the difference quotient, which is the slope of the chord, in place of the derivative, which is the slope of the tangent. To get a still better approximation, repeat the procedure, using as one end point a and as the other either α or β, chosen so that $f(x)$ has opposite signs at the end points of the new interval.

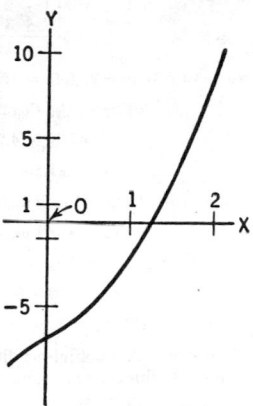

Fig. 3

(3) **The Graphical Method of Solution** can be used to solve any kind of equation if it gives sufficient accuracy. To solve the equation $f(x) = 0$, graph the function $y = f(x)$. The x coordinates of the points at which the graph intersects the x axis are roots of $f(x) = 0$. Another method is to set $f(x)$ equal to any convenient difference $f_1(x) - f_2(x)$, and graph the functions $y = f_1(x)$ and $y = f_2(x)$ on the same axes. The x coordinates of the points of intersection of the two graphs are real roots of $f(x) = 0$.

(4) **Graeffe's Method for Real and Complex Roots.** Let $x_1, x_2, \cdots, x_n$ be the roots of the equation $a_0 x^n + a_1 x^{n-1} + \cdots + a_n = 0$, arranged in descending order of absolute values. Form a sequence of equations such that the roots of each are the negatives of the squares of the roots of the preceding equation. Using the negatives of the squares gives more uniform formulas.

Let A_i be a coefficient of the equation being formed, and a_i a coefficient of the preceding equation.

$$A_0 = a_0 = 1$$
$$A_1 = a_1^2 - 2a_0 a_2 = a_1^2 - 2a_2$$
$$A_2 = a_2^2 - 2a_1 a_3 + 2a_4$$
$$A_3 = a_3^2 - 2a_2 a_4 + 2a_1 a_5 - 2a_6$$
$$\cdots \cdots \cdots \cdots \cdots \cdots \cdots$$
$$A_{n-1} = a_{n-1}^2 - 2a_{n-2} a_n$$
$$A_n = a_n^2$$

Each coefficient is the sum of the square of the preceding and twice the product of all pairs of equidistant coefficients in the preceding equation, taken with alternately minus and plus signs. Missing coefficients are zero. The process is ended when further steps do not affect the non-fluctuating coefficients to the accuracy desired in the roots.

As the successive equations are formed, various cases arise depending on the behavior of the coefficients. Among them are:

Case 1. Each coefficient approaches the square of the preceding. The roots are real and unequal in absolute value. Let A_i be a coefficient of the equation whose roots are $-x_1^p, -x_2^p, \cdots, -x_n^p$. Then, approximately, $x_1 = \pm \sqrt[p]{A_1}$, $x_2 = \pm \sqrt[p]{A_2/A_1}$, $\cdots$, $x_n = \pm \sqrt[p]{A_n/A_{n-1}}$. The signs of the roots are determined by substitution in the original equation. It is usually sufficient to find successive integers between which a root is located.

Example. $f(x) = x^3 - 2x^2 - 5x + 4 = 0$.

	x^3	x^2	x	x^0
1st	1	-2	-5	4
	1	4	25	16
		10	16	
2nd	1	14	41	16
	1	196	1.681×10^3	256
		-82	-0.448×10^3	
4th	1	1.14×10^2	1.233×10^3	256
	1	1.300×10^4	1.520×10^6	6.554×10^4
		-0.247×10^4	-0.058×10^6	
8th	1	1.053×10^4	1.462×10^6	6.554×10^4
	1	1.109×10^8	2.137×10^{12}	4.295×10^9
		-0.029×10^8	-0.001×10^{12}	
16th	1	1.080×10^8	2.136×10^{12}	4.295×10^9

$$\log x_1 = 1/16 \log 1.080 \times 10^8 = 1/16 \times 8.0334 = 0.5021$$
$$x_1 = \pm 3.177$$

Using synthetic substitution,

$$3 \,|\, \begin{array}{r} 1 - 2 - 5 + 4 \\ + 3 + 3 - 6 \\ \hline 1 + 1 - 2 - 2 \end{array} \qquad\qquad 4 \,|\, \begin{array}{r} 1 - 2 - 5 + 4 \\ + 4 + 8 + 12 \\ \hline 1 + 2 + 3 + 16 \end{array}$$

we have $f(3) = -2$, $f(4) = 16$. Therefore there is a root between 3 and 4, and $x_1 = 3.177$.

$$\log x_2 = {}^1\!/_{16}\,(\log 2.136 \times 10^{12} - \log 1.080 \times 10^8) = {}^1\!/_{16}\,(12.3296 - 8.0334)$$

$$= {}^1\!/_{16} \times 4.2962 = 0.2685$$

$$x_2 = \pm 1.856$$

Using synthetic substitution, $f(-2) = -2$, $f(-1) = 6$. Therefore $x_2 = -1.856$.

$$\log x_3 = {}^1\!/_{16}\,(\log 4.295 \times 10^9 - \log 2.136 \times 10^{12}) = {}^1\!/_{16}\,(9.6330 - 12.3296)$$

$$= {}^1\!/_{16}\,(157.3034 - 160) = 9.8315 - 10$$

$$x_3 = \pm 0.678$$

Since $x_1 + x_2 + x_3 = 2$, $x_3 = 0.678$.

Case 2. A coefficient fluctuates in sign. There is a pair of complex roots. If the sign of A_i fluctuates, then $x_i = u + iv$ and $x_{i+1} = u - iv$ are complex. Let $r^2 = u^2 + v^2$. Then $r^2 = \sqrt[p]{A_{i+1}/A_{i-1}}$, $2u = -a_1 - $ (sum of real roots), $v = \sqrt{r^2 - u^2}$.

Example. $f(x) = x^4 - 2x^3 - 4x^2 + 5x - 7 = 0$.

	x^4	x^3	x^2	x	x^0
1st	1	-2	-4	5	-7
	1	4	16	25	49
		8	20	-56	
			-14		
2nd	1	12	22	-31	49
	1	144	484	961	2401
		-44	744	-2156	
			98		
4th	1	100	1326	-1195	2401
	1	1.0000×10^4	1.758×10^6	1.428×10^6	5.765×10^6
		-0.2652×10^4	0.239×10^6	-6.367×10^6	
			0.005×10^6		
8th	1	7.348×10^3	2.002×10^6	-4.939×10^6	5.765×10^6
	1	5.399×10^7	4.008×10^{12}	2.439×10^{13}	3.324×10^{13}
		-0.400×10^7	0.073×10^{12}	-2.308×10^{13}	
			*		
16th	1	4.999×10^7	4.081×10^{12}	1.31×10^{12}	3.324×10^{13}
	1	2.499×10^{15}	1.665×10^{25}	0.017×10^{26}	1.105×10^{27}
		-0.001×10^{15}	*	-2.713×10^{26}	
			*		
32nd	1	2.498×10^{15}	1.665×10^{25}	-2.696×10^{26}	1.105×10^{27}

Since the sign of A_3 fluctuates, x_3 and x_4 are complex.

$$x_1 = \pm \sqrt[32]{2.498 \times 10^{15}} = \pm 3.028 \qquad\qquad f(3) = -,\, f(4) = +$$

$$\frac{\log(2.498 \times 10^{15})}{32} = \frac{15.3976}{32} = 0.4812 \qquad\qquad x_1 = 3.028$$

$$x_2 = \pm \sqrt[32]{\frac{1.665 \times 10^{25}}{2.498 \times 10^{15}}} = \pm 2.028 \qquad\qquad f(-3) = +,\, f(-2) = -$$

$$\begin{array}{r} 25.2214 \\ 15.3976 \\ \hline 9.8238 \end{array} \Big/ 32 = 0.3070 \qquad\qquad x_2 = -2.028$$

$$r^2 = \sqrt[32]{\frac{1.105 \times 10^{27}}{1.665 \times 10^{25}}} = 1.140 \qquad\qquad u = \frac{2 - (3.028 - 2.028)}{2} = 0.500$$

$$\begin{array}{r} 27.0434 \\ 25.2214 \\ \hline 1.8220 \end{array} \Big/ 32 = 0.05694 \qquad\qquad v = \sqrt{1.140 - 0.250} = \sqrt{0.890} = 0.943$$

$$x_3,\, x_4 = 0.5 \pm 0.943i$$

If, for a fourth-degree equation, alternate coefficients, that is, the second and fourth, fluctuate in sign, all four roots are complex. Let the roots be $u_1 \pm iv_1$, $u_2 \pm iv_2$. Then $r_1^2 = \sqrt[p]{A_2}$, $r_2^2 = \sqrt[p]{A_4/A_2}$, $2(u_1 + u_2) = -a_1$, $2(r_2^2u_1 + r_1^2u_2) = -a_3$.

Example. $f(x) = x^4 - 3x^3 - x^2 + 4x + 14 = 0$.

	x^4	x^3	x^2	x	x^0
1st	1	-3	-1	4	14
	1	9	1	16	196
		2	24	28	
			28		
2nd	1	11	53	44	196
	1	121	2809	1936	38416
		-106	-968	-20776	
			392		
4th	1	15	2.233×10^3	-1.884×10^4	3.842×10^4
	1	0.225×10^3	4.986×10^6	3.549×10^8	1.476×10^9
		-4.466×10^3	0.565×10^6	-1.716×10^8	
			0.077×10^6		
8th	1	-4.241×10^3	5.628×10^6	1.833×10^8	1.476×10^9
	1	1.799×10^7	3.1674×10^{13}	3.360×10^{16}	2.178×10^{18}
		-1.126×10^7	0.1555×10^{13}	-1.661×10^{16}	
			0.0003×10^{13}		
16th	1	0.675×10^7	3.323×10^{13}	1.699×10^{16}	2.178×10^{18}

Since A_1 and A_3 fluctuate in sign, there are four complex roots.

$$r_1^2 = \sqrt[16]{3.323 \times 10^{13}} = 7.000$$

$$\frac{\log (3.323 \times 10^{13})}{16} = \frac{13.5215}{16} = 0.8451$$

$$r_2^2 = \sqrt[16]{\frac{2.178 \times 10^{18}}{3.323 \times 10^{13}}} = 2.000$$

$$\begin{array}{r} 18.3380 \\ 13.5215 \\ \hline 4.8165 \end{array}$$

$$\frac{4.8165}{16} = 0.3010$$

$2(u_1 + u_2) = 3$

$2(2u_1 + 7u_2) = -4$

$u_2 = -1$

$u_1 = 2.5$

$v_2 = \sqrt{r_2^2 - u_2^2} = \sqrt{2 - 1} = 1$

$v_1 = \sqrt{r_1^2 - u_1^2} = \sqrt{7 - 6.25}$

$\qquad\qquad = \sqrt{0.75} = 0.866$

$x_1, x_2 = 2.5 \pm 0.866i$

$x_3, x_4 = -1 \pm i$

Case 3. A coefficient approaches one-half the square of the preceding. There is a double real root, or there are two real roots of equal absolute value. If A_i approaches one-half the square of the preceding coefficient, then $|x_i| = |x_{i+1}| = \sqrt[2p]{A_{i+1}/A_{i-1}}$.

Example. $f(x) = x^3 + 2.20x^2 - 2.95x + 0.80 = 0$.

	x^3	x^2	x	x^0
1st	1	2.20	-2.95	0.80
	1	4.84	8.703	0.64
		5.90	-3.52	
2nd	1	10.74	5.183	0.64
	1	1.1535×10^2	2.686×10	0.4096
		-0.1037×10^2	-1.375×10	
4th	1	1.050×10^2	1.311×10	0.4096
	1	1.1025×10^4	1.719×10^2	0.1678
		-0.0026×10^4	-0.860×10^2	
8th	1	1.100×10^4	0.859×10^2	0.1678

Since A_2 approaches one-half the square of the preceding coefficient, $|x_2| = |x_3|$.

$$x_1 = \pm \sqrt[8]{1.100 \times 10^4} = \pm 3.20$$

$$\frac{\log (1.100 \times 10^4)}{8} = \frac{4.0414}{8} = 0.5052$$

$$|x_2| = |x_3| = \sqrt[16]{\frac{0.1678}{1.100 \times 10^4}} = 0.50$$

$$\begin{array}{r} 9.2248 - 10 \\ 4.0414 \\ \hline 155.1834 - 160 \end{array}$$

$$\frac{155.1834 - 160}{16} = 9.6990 - 10$$

$f(-4) = -, f(-3) = +$

$x_1 = -3.20$

$f(0.5) = 0, f(-0.5) \ne 0$

$x_2 = x_3 = 0.50$

For a more complete treatment of Graeffe's method, including other cases, see Doherty and Keller, *Mathematics of Modern Engineering*, John Wiley and Sons, 1936; or Whittaker and Robinson, *The Calculus of Observations*, Blackie and Son, Ltd., 1942.

15. MATRICES AND DETERMINANTS

Definitions

1. A *matrix* is a system of mn quantities, called *elements*, arranged in a rectangular array of m rows and n columns.

$$A = \begin{pmatrix} a_{11} & a_{12} & \cdots & a_{1n} \\ a_{21} & a_{22} & \cdots & a_{2n} \\ \cdots & \cdots & \cdots & \cdots \\ a_{m1} & a_{m2} & \cdots & a_{mn} \end{pmatrix} = \begin{Vmatrix} a_{11} & a_{12} & \cdots & a_{1n} \\ a_{21} & a_{22} & \cdots & a_{2n} \\ \cdots & \cdots & \cdots & \cdots \\ a_{m1} & a_{m2} & \cdots & a_{mn} \end{Vmatrix} = (a_{ij}) = \| a_{ij} \|, \quad \begin{array}{l} i = 1, \cdots, m \\ j = 1, \cdots, n \end{array}$$

2. If $m = n$, then A is a *square* matrix of *order* n.

3. Two matrices are *equal* if and *only if* they have the same number of rows and of columns, and corresponding elements are equal.

4. Two matrices are *transposes* (sometimes called *conjugates*) of each other, if either is obtained from the other by interchanging rows and columns.

5. The *complex conjugate* of a matrix (a_{ij}) with complex elements is the matrix $(\bar{a}_{ij})$. See p. 347.

6. A matrix is *symmetric* if it is equal to its transpose, that is, if $a_{ij} = a_{ji}$; $i, j = 1, \cdots, n$.

7. A matrix is *skew-symmetric*, or *anti-symmetric*, if $a_{ij} = -a_{ji}$; $i, j = 1, \cdots, n$. The diagonal elements $a_{ii} = 0$.

8. A matrix all of whose elements are zero is a *zero matrix*.

9. If the non-diagonal elements a_{ij}, $i \neq j$, of a square matrix A are all zero, then A is a *diagonal matrix*. If, furthermore, the diagonal elements are all equal, the matrix is a *scalar matrix;* if they are all 1, it is an *identity* or *unit matrix*, denoted by I.

10. The *determinant* $| A |$ of a square matrix (a_{ij}); $i, j = 1, \cdots, n$, is the sum of the $n!$ products $a_{1r_1} a_{2r_2} \cdots a_{nr_n}$, in which $r_1, r_2, \cdots, r_n$ is a permutation of $1, 2, \cdots, n$, and the sign of each product is $+$ or $-$ according as the permutation is obtained from $1, 2, \cdots, n$ by an even or an odd number of interchanges of two numbers.

Symbols used are

$$| A | = \begin{vmatrix} a_{11} & a_{12} & \cdots & a_{1n} \\ a_{21} & a_{22} & \cdots & a_{2n} \\ \cdots & \cdots & \cdots & \cdots \\ a_{n1} & a_{n2} & \cdots & a_{nn} \end{vmatrix} = | a_{ij} |, \quad i, j = 1, \cdots, n$$

11. A square matrix (a_{ij}) is *singular* if its determinant $| a_{ij} |$ is zero.

12. The determinants of the square submatrices of any matrix A, obtained by striking out certain rows or columns, or both, are called the *determinants* or *minors* of A. A matrix is of *rank r* if it has at least one r-rowed determinant which is not zero, while all its determinants of order higher than r are zero. The *nullity d* of a square matrix of order n is $d = n - r$. The zero matrix is of rank 0.

13. The *minor* D_{ij} of the element a_{ij} of a square matrix is the determinant of the submatrix obtained by striking out the row and column in which a_{ij} lies. The *cofactor* A_{ij} of the element a_{ij} is $(-1)^{i+j} D_{ij}$. A *principal minor* is the minor obtained by striking out the same rows as columns.

14. The *inverse* of the square matrix A is

$$A^{-1} = \begin{pmatrix} \dfrac{A_{11}}{| A |} & \cdots & \dfrac{A_{n1}}{| A |} \\ \cdots & \cdots & \cdots \\ \dfrac{A_{1n}}{| A |} & \cdots & \dfrac{A_{nn}}{| A |} \end{pmatrix}$$

$$A A^{-1} = A^{-1} A = I$$

15. The *adjoint* of A is

$$\text{Adj } A = \begin{pmatrix} A_{11} & \cdots & A_{n1} \\ \cdots & \cdots & \cdots \\ A_{1n} & \cdots & A_{nn} \end{pmatrix}$$

16. *Elementary transformations* of a matrix are

(1) The interchange of two rows or of two columns.

(2) The addition to the elements of a row (or column) of any constant multiple of the corresponding elements of another row (or column).

(3) The multiplication of each element of a row (or column) by any non-zero constant.

17. Two $m \times n$ matrices A and B are *equivalent* if it is possible to pass from one to the other by a finite number of elementary transformations.

(1) The matrices A and B are equivalent if and only if there exist two non-singular square matrices E and F, having m and n rows respectively, such that $EAF = B$.

(2) The matrices A and B are equivalent if and only if they have the same rank.

Matrix Operations

1. Addition and Subtraction. The sum or difference of two matrices (a_{ij}) and (b_{ij}) is the matrix $(a_{ij} \pm b_{ij})$, $i = 1, \cdots, m; j = 1, \cdots, n$.

2. Scalar Multiplication. The product of the scalar k and the matrix (a_{ij}) is the matrix (ka_{ij}).

3. Matrix Multiplication. The product (p_{ik}), $i = 1, \cdots, m; k = 1, \cdots, q$, of two matrices (a_{ij}), $i = 1, \cdots, m; j = 1, \cdots, n$, and (b_{jk}), $j = 1, \cdots, n; k = 1, \cdots, q$, is the matrix whose elements are

$$p_{ik} = \sum_{j=1}^{n} a_{ij}b_{jk} = a_{i1}b_{1k} + a_{i2}b_{2k} + \cdots + a_{in}b_{nk}$$

The element in the ith row and kth column of the product is the sum of the n products of the n elements of the ith row of (a_{ij}) by the corresponding n elements of the kth column of (b_{jk}).

Example.

$$\begin{pmatrix} a_{11} & a_{12} \\ a_{21} & a_{22} \end{pmatrix}\begin{pmatrix} b_{11} & b_{12} & b_{13} \\ b_{21} & b_{22} & b_{23} \end{pmatrix} = \begin{pmatrix} a_{11}b_{11} + a_{12}b_{21} & a_{11}b_{12} + a_{12}b_{22} & a_{11}b_{13} + a_{12}b_{23} \\ a_{21}b_{11} + a_{22}b_{21} & a_{21}b_{12} + a_{22}b_{22} & a_{21}b_{13} + a_{22}b_{23} \end{pmatrix}$$

All the laws of ordinary algebra hold for the addition and subtraction of matrices and for scalar multiplication.

Multiplication of matrices is not in general commutative, but it is associative and distributive.

If the product of two or more matrices is zero, it does not follow that one of the factors is zero. The factors are *divisors of zero*.

Example.

$$\begin{pmatrix} a & 0 \\ b & 0 \end{pmatrix}\begin{pmatrix} 0 & 0 \\ c & d \end{pmatrix} = \begin{pmatrix} 0 & 0 \\ 0 & 0 \end{pmatrix}$$

Linear Dependence

1. The quantities $l_1, l_2, \cdots, l_n$ are *linearly dependent* if there exist constants $c_1, c_2, \cdots c_n$, not all zero, such that

$$c_1 l_1 + c_2 l_2 + \cdots + c_n l_n = 0$$

If no such constants exist, the quantities are *linearly independent*.

2. The linear functions

$$l_i = a_{i1}x_1 + a_{i2}x_2 + \cdots + a_{in}x_n, \quad i = 1, 2, \cdots, m$$

are *linearly dependent* if and only if the matrix of the coefficients is of rank $r < m$. Exactly r of the l_i form a linearly independent set.

3. For $m > n$, any set of m linear functions are linearly dependent.

Consistency of Equations

1. The system of homogeneous linear equations

$$a_{i1}x_1 + a_{i2}x_2 + \cdots + a_{in}x_n = 0, \quad i = 1, 2, \cdots, m$$

has solutions not all zero if the rank r of the matrix (a_{ij}) is less than n.

If $m < n$, there always exist solutions not all zero. If $m = n$, there exist solutions not all zero if $|a_{ij}| = 0$.

If r of the equations are so selected that their matrix is of rank r, they determine uniquely r of the variables as homogeneous linear functions of the remaining $n - r$ variables. A solution of the system is obtained by assigning arbitrary values to the $n - r$ variables and finding the corresponding values of the r variables.

2. The system of linear equations

$$a_{i1}x_1 + a_{i2}x_2 + \cdots + a_{in}x_n = k_i, \quad i = 1, 2, \cdots, m$$

is consistent if and only if the *augmented* matrix derived from (a_{ij}) by annexing the column $k_1, \cdots, k_m$ has the same rank r as (a_{ij}).

As in the case of a system of homogeneous linear equations, r of the variables can be expressed in terms of the remaining $n - r$ variables.

Linear Transformations

1. If a linear transformation

$$x'_i = a_{i1}x_1 + a_{i2}x_2 + \cdots + a_{in}x_n, \quad i = 1, 2, \cdots, n$$

with matrix (a_{ij}) transforms the variables x_i into the variables x'_i, and a linear transformation

$$x''_i = b_{i1}x'_1 + b_{i2}x'_2 + \cdots + b_{in}x'_n, \quad i = 1, 2, \cdots, n$$

with matrix (b_{ij}) transforms the variables x'_i into the variables x''_i, then the linear transformation with matrix $(b_{ij})(a_{ij})$ transforms the variables x_i into the variables x''_i directly.

2. A real *orthogonal* transformation is a linear transformation of the variables x_i into the variables x'_i such that

$$\sum_{i=1}^{n} x_i^2 = \sum_{i=1}^{n} x'_i{}^2$$

A transformation is orthogonal if and only if the transpose of its matrix is the inverse of its matrix.

3. A *unitary* transformation is a linear transformation of the variables x_i into the variables x'_i such that

$$\sum_{i=1}^{n} x_i \bar{x}_i = \sum_{i=1}^{n} x'_i \bar{x}'_i$$

A transformation is unitary if and only if the transpose of the conjugate of its matrix is the inverse of its matrix.

Quadratic Forms

A *quadratic form* in n variables is

$$\begin{aligned}
\sum_{i,j=1}^{n} a_{ij}x_i x_j = {} & a_{11}x_1^2 + a_{12}x_1 x_2 + \cdots + a_{1n}x_1 x_n \\
& + a_{21}x_2 x_1 + a_{22}x_2^2 + \cdots + a_{2n}x_2 x_n \\
& \cdots \cdots \cdots \cdots \cdots \cdots \\
& + a_{n1}x_n x_1 + a_{n2}x_n x_2 + \cdots + a_{nn}x_n^2
\end{aligned}$$

in which $a_{ji} = a_{ij}$. The symmetric matrix (a_{ij}) of the coefficients is the *matrix* of the quadratic form and the rank of (a_{ij}) is the *rank* of the quadratic form.

A real quadratic form of rank r can be reduced by a real non-singular linear transformation to the *normal form*

$$x_1^2 + \cdots + x_p^2 - x_{p+1}^2 - \cdots - x_r^2$$

in which the *index* p is uniquely determined.

If $p = r$, a quadratic form is *positive*, and, if $p = 0$, it is *negative*. If, furthermore, $r = n$, both are *definite*. A quadratic form is positive definite if and only if the determinant and all the principal minors of its matrix are positive.

A method of reducing a quadratic form to its normal form is illustrated.

Example.

$$q = 3x^2 - 4y^2 - z^2 + 4xy - 2xz + 4yz$$

$$q = \left\{\begin{array}{l} 3x^2 + 2xy - xz \\ +2xy - 4y^2 + 2yz \\ -xz + 2yz - z^2 \end{array}\right\} = \frac{1}{3}(3x + 2y - z)^2 + q_1, \text{ in which the quantity in parentheses is obtained by factoring } x \text{ out of the first row}$$

$$= \frac{1}{3}(9x^2 + 4y^2 + z^2 + 12xy - 6xz - 4yz) + q_1$$

$$q_1 = -\frac{4}{3}y^2 - \frac{1}{3}z^2 + \frac{4}{3}yz - 4y^2 + 4yz - z^2$$

$$= \left\{\begin{array}{l} -\frac{16}{3}y^2 + \frac{8}{3}yz \\ +\frac{8}{3}yz - \frac{4}{3}z^2 \end{array}\right\} = -\frac{3}{16}\left(-\frac{16}{3}y + \frac{8}{3}z\right)^2 + q_2$$

$$q_2 = 0$$

The transformation

$$x' = 3x + 2y - z$$
$$y' = -\frac{16}{3}y + \frac{8}{3}z$$
$$z' = z$$

reduces q to $\frac{1}{3}x'^2 - \frac{3}{16}y'^2$.

The transformation

$$x'' = \sqrt{3}\,x'$$

$$y'' = \frac{4}{\sqrt{3}}y'$$

$$z'' = z'$$

further reduces q to the normal form $x''^2 - y''^2$ of rank 2 and index 1.

Expressing x, y, z in terms of x'', y'', z''. the real non-singular linear transformation which reduces q to the normal form is

$$x = \frac{\sqrt{3}}{3}x'' + \frac{1}{2\sqrt{3}}y''$$

$$y = \qquad -\frac{\sqrt{3}}{4}y'' + \frac{1}{2}z''$$

$$z = \qquad\qquad z''$$

Hermitian Forms

A *Hermitian form* in n variables is

$$\sum_{i,j=1}^{n} a_{ij}x_i\bar{x}_j, \quad a_{ji} = \bar{a}_{ij}$$

The matrix (a_{ij}) is a *Hermitian matrix*. Its transpose is equal to its conjugate. The rank of (a_{ij}) is the *rank* of the Hermitian form.

A Hermitian form of rank r can be reduced by a non-singular linear transformation to the *normal form*

$$x_1\bar{x}_1 + \cdots + x_p\bar{x}_p - x_{p+1}\bar{x}_{p+1} - \cdots - x_r\bar{x}_r$$

in which the *index* p is uniquely determined.

If $p = r$, the Hermitian form is *positive*, and, if $p = 0$, it is *negative*. If, furthermore, $r = n$, both are *definite*.

Determinants

Second- and third-order determinants are formed from their square symbols by taking diagonal products, down from left to right being positive and up negative.

$$\begin{vmatrix} a_{11} & a_{12} \\ a_{21} & a_{22} \end{vmatrix} = a_{11}a_{22} - a_{21}a_{12}$$

$$\begin{vmatrix} a_{11} & a_{12} & a_{13} \\ a_{21} & a_{22} & a_{23} \\ a_{31} & a_{32} & a_{33} \end{vmatrix} = a_{11}a_{22}a_{33} + a_{12}a_{23}a_{31} + a_{13}a_{32}a_{21} \\ - a_{31}a_{22}a_{13} - a_{32}a_{23}a_{11} - a_{33}a_{12}a_{21}$$

Third and higher order determinants are formed by selecting any row or column and taking the sum of the products of each element and its cofactor. This process is continued until second- or third-order cofactors are reached.

$$\begin{vmatrix} a_{11} & a_{12} & a_{13} \\ a_{21} & a_{22} & a_{23} \\ a_{31} & a_{32} & a_{33} \end{vmatrix} = a_{11}\begin{vmatrix} a_{22} & a_{23} \\ a_{32} & a_{33} \end{vmatrix} - a_{21}\begin{vmatrix} a_{12} & a_{13} \\ a_{32} & a_{33} \end{vmatrix} + a_{31}\begin{vmatrix} a_{12} & a_{13} \\ a_{22} & a_{23} \end{vmatrix}$$

The determinant of a matrix A is
(1) Zero, if two rows or two columns of A have proportional elements.
(2) Unchanged, if
    (a) The rows and columns of A are interchanged.
    (b) To each element of a row or column of A is added a constant multiple of the corresponding element of another row or column.
(3) Changed in sign, if two rows or two columns of A are interchanged.
(4) Multiplied by c, if each element of any row or column of A is multiplied by c.
(5) The sum of the determinants of two matrices B and C, if A, B, and C have all the same elements, except that in one row or column each element of A is the sum of the corresponding elements of B and C.

Example.

$$\begin{vmatrix} 2 & 9 & 9 & 4 \\ 2 & -3 & 12 & 8 \\ 4 & 8 & 3 & -5 \\ 1 & 2 & 6 & 4 \end{vmatrix} = \begin{vmatrix} 2 & 5 & 9 & 4 \\ 2 & -7 & 12 & 8 \\ 4 & 0 & 3 & -5 \\ 1 & 0 & 6 & 4 \end{vmatrix} = 3\begin{vmatrix} 2 & 5 & 3 & 4 \\ 2 & -7 & 4 & 8 \\ 4 & 0 & 1 & -5 \\ 1 & 0 & 2 & 4 \end{vmatrix}$$

Multiply 1st column   Factor 3 out of
by -2 and add to 2nd.   the 3rd column.

$$= 3\times(-5)\begin{vmatrix} 2 & 4 & 8 \\ 4 & 1 & -5 \\ 1 & 2 & 4 \end{vmatrix} + 3\times(-7)\begin{vmatrix} 2 & 3 & 4 \\ 4 & 1 & -5 \\ 1 & 2 & 4 \end{vmatrix} = \qquad 0 \qquad -21\begin{vmatrix} 1 & 1 & 0 \\ 4 & 1 & -5 \\ 1 & 2 & 4 \end{vmatrix}$$

Expand according to 2nd column.   1st and 3rd   Subtract 3rd
    rows proportional.   row from 1st.

$$= -21\begin{vmatrix} 1 & -5 \\ 2 & 4 \end{vmatrix} - (-21)\begin{vmatrix} 4 & -5 \\ 1 & 4 \end{vmatrix} = -21[(4+10)-(16+5)] = +147$$

Expand according to 1st row.

16. SYSTEMS OF EQUATIONS

Linear Systems

Homogeneous. $a_{i1}x_1 + \cdots + a_{in}x_n = 0$, $i = 1, \cdots, m$. Let r = rank of (a_{ij}).

	For $m = n$,	
$r = n$	$\mid a_{ij} \mid \ne 0$	One solution, $x_1 = \cdots = x_n = 0$.
$r < n$	$\mid a_{ij} \mid = 0$	Infinite number of solutions.

Non-homogeneous. $a_{i1}x_1 + \cdots + a_{in}x_n = k_i$, $i = 1, \cdots, m$.

Let $a = (a_{ij})$, an $m \times n$ matrix,

$$k = \text{augmented matrix} = \begin{pmatrix} a_{11} & \cdots & a_{1n}k_1 \\ \cdot & \cdots & \cdot \\ a_{m1} & \cdots & a_{mn}k_m \end{pmatrix}, \text{ an } m \times (n+1) \text{ matrix,}$$

r_a = rank of a,

r_k = rank of k.

	For $m = n$,	
$r_a = r_k$		Consistent.
(a) $r_a = r_k = n$	$\mid a_{ij} \mid \ne 0$	Independent. One solution.
(b) $r_a = r_k < n$	$\mid a_{ij} \mid = 0$	Dependent. Infinite number of solutions.
$r_a < r_k$		Inconsistent. No solution.

Methods of Solution

Elimination is a practical method of solution for a system of two or three linear equations in as many variables.

Examples. 1. By addition and subtraction.

Solve:
$$\begin{cases} 2x + y + 3z = 9 & (1) \\ x - 2y + z = -2 & (2) \\ 3x + 2y + 2z = 7 & (3) \end{cases}$$

$(2) + (3)$ gives: $\quad 4x + 3z = 5 \quad$ (4)

$2 \times (1) + (2)$ gives: $\quad 5x + 7z = 16 \quad$ (5)

$5 \times (4) - 4 \times (5)$ gives: $\quad -13z = -39 \quad$ or $\quad z = 3$

Putting $z = 3$ in (4) or (5): $x = -1$

Then from (1), (2), or (3): $y = 2$

2. By substitution.

Solve:
$$\begin{cases} x + 2y - z = 5 & (1) \\ x - y = 2 & (2) \\ 2x + z = 1 & (3) \end{cases}$$

From (2), $y = x - 2$, and from (3), $z = -2x + 1$. Substituting for y and z in (1), $x - 2x - 4 + 2x - 1 = 5$, from which $x = 2$. Then $y = 2 - 2 = 0$, $z = -4 + 1 = -3$.

Determinants can be used to solve a system of n non-homogeneous linear equations in n variables for which $|a_{ij}| \neq 0$. To solve for x_j, form a fraction, the denominator of which is the determinant $|a_{ij}|$ and the numerator the determinant obtained from $|a_{ij}|$ by replacing its jth column by the constants k_i.

Example.

Solve:
$$\begin{cases} 2x + y + 3z = 9 \\ x - 2y + z = -2 \\ 3x + 2y + 2z = 7 \end{cases}$$

$$x = \frac{\begin{vmatrix} 9 & 1 & 3 \\ -2 & -2 & 1 \\ 7 & 2 & 2 \end{vmatrix}}{\begin{vmatrix} 2 & 1 & 3 \\ 1 & -2 & 1 \\ 3 & 2 & 2 \end{vmatrix}} = \frac{\begin{vmatrix} 9 & 1 & 3 \\ -2 & -2 & 1 \\ 5 & 0 & 3 \end{vmatrix}}{\begin{vmatrix} 2 & 1 & 3 \\ 1 & -2 & 1 \\ 4 & 0 & 3 \end{vmatrix}} = \frac{\begin{vmatrix} 9 & 1 & 3 \\ 16 & 0 & 7 \\ 5 & 0 & 3 \end{vmatrix}}{\begin{vmatrix} 2 & 1 & 3 \\ 5 & 0 & 7 \\ 4 & 0 & 3 \end{vmatrix}} = \frac{-(48 - 35)}{-(15 - 28)} = -1$$

Miscellaneous Systems

To be solvable a system of equations must have as many independent equations as variables.

A system of two polynomial equations of degrees m and n has mn solutions, real or complex. For systems in general no statement can be made regarding the number of solutions.

The Graphical Method of Solution is a general one for systems of two equations in two variables. It consists of graphing both equations on the same axes and reading the pairs of coordinates of the points of intersection of the graphs as solutions of the system. This method gives real solutions only.

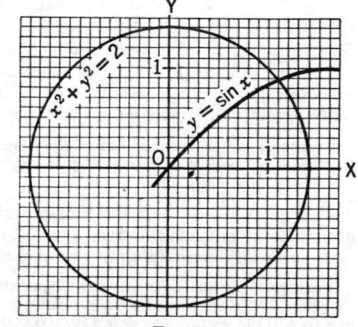

Example.

Solve:
$$\begin{cases} y = \sin x \\ x^2 + y^2 = 2 \end{cases}$$

Solution from the graph, $x = 1.1$, $y = 0.9$.

From symmetry, $x = -1.1$, $y = -0.9$, is also a solution.

Fig. 4

The Method of Elimination of Variables is a general one which can be applied to systems composed of any kinds of equations, algebraic or transcendental. However, except in fairly simple cases, practical difficulties are frequently encountered.

Example.

Solve:
$$\begin{cases} y = \sin x \\ x^2 + y^2 = 2 \end{cases}$$

Squaring both sides of the first equation and subtracting it from the second to eliminate y, $x^2 = 2 - \sin^2 x$. This equation can be solved by Newton's method. Extraneous solutions introduced by squaring can be eliminated by reference to the graph.

There are numerous devices for eliminating variables in special systems. For example, to solve the system of *two general quadratics*

$$a_1x^2 + b_1xy + c_1y^2 + d_1x + e_1y + f_1 = 0 \qquad (1)$$

$$a_2x^2 + b_2xy + c_2y^2 + d_2x + e_2y + f_2 = 0 \qquad (2)$$

eliminate x^2 by multiplying (1) by a_2 and (2) by a_1 and subtracting, solve the resulting equation for x, substitute this expression in either of the given equations, and clear of fractions. The resulting fourth-degree equation in y can be solved by Horner's method. In a similar manner y could have been eliminated instead of x.

17. PERMUTATIONS AND COMBINATIONS

Fundamental Principle. If in a sequence of s events the first event can occur in n_1 ways, the second in n_2, $\cdots$, the sth in n_s, then the number of different ways in which the sequence can occur is $n_1n_2 \cdots n_s$.

A Permutation of n objects taken r at a time is an arrangement of any r objects selected from the n objects. The number of permutations of n objects taken r at a time is

$$_nP_r = n(n-1)(n-2) \cdots (n-r+1) = \frac{n!}{(n-r)!}$$

In particular, $_nP_1 = n$, $_nP_n = n!$.

Cyclic permutations, $_nP_r^c = \frac{n!}{r(n-r)!}$, $_nP_n^c = (n-1)!$.

Distinguishable permutations, if the n objects are divided into s sets each containing n_i objects which are alike, $n = n_1 + n_2 + \cdots + n_s$, $_nP_n = \frac{n!}{n_1!n_2! \cdots n_s!}$.

A Combination of n objects taken r at a time is an unarranged selection of any r of the n objects. The number of combinations of n objects taken r at a time is

$$_nC_r = \frac{_nP_r}{r!} = \frac{n!}{r!(n-r)!} = {_nC_{n-r}}$$

In particular, $_nC_1 = n$, $_nC_n = 1$.
Combinations taken any number at a time, $_nC_1 + {_nC_2} + \cdots + {_nC_n} = 2^n - 1$.

18. PROBABILITY

If, in a set M of m events which are mutually exclusive and equally likely, one event will occur, and if in the set M there is a subset N of n events ($n \leqq m$), then the *a priori probability* p that the event which will occur is one of the subset N is n/m. The probability q that the event which will occur does not belong to N is $1 - n/m$.

Example. If the probability of drawing one of the 4 aces from a deck of 52 cards is to be found, then $m = 52$, $n = 4$, and $p = 4/52 = 1/13$. The probability of drawing a card that is not an ace is $q = 1 - 1/13 = 12/13$.

If, out of a large number r of observations in which a given event might or might not occur, the event has occurred s times, then a useful approximate value of the *experimental*, or *a posteriori, probability* of the occurrence of the event under the same conditions is s/r.

Example. From the American Experience Mortality Table, out of 100,000 persons living at age 10 years 749 died within a year. Here $r = 100,000$, $s = 749$, and the probability that a person of age 10 will die within a year is $749/100,000$.

If p is the probability of receiving an amount A, then the *expectation* is pA.
Addition Rule (either or). The probability that any one of several mutually exclusive events will occur is the sum of their separate probabilities.

Example. The probability of drawing an ace from a deck of cards is $1/13$, and the probability of drawing a king is the same. Then the probability of drawing either an ace or a king is $1/13 + 1/13 = 2/13$.

Multiplication Rule (both and). (*a*) The probability that two (or more) independent events will both (or all) occur is the product of their separate probabilities.

(*b*) If p_1 is the probability that an event will occur, and if, after it has occurred, p_2 is the probability that another event will occur, then the probability that both will occur in the given order is p_1p_2. This rule can be extended to more than 2 events.

Example. (*a*) The probability of drawing an ace from a deck of cards is $1/13$, and the probability of drawing a king from another deck is $1/13$. Then the probability that an ace will be drawn from the first deck and a king from the second is $1/13 \cdot 1/13 = 1/169$. (*b*) After an ace has been drawn from a deck of cards, the probability of drawing a king is $4/51$. If 2 cards are drawn in succession without the first being replaced, the probability that the first is an ace and the second a king is $1/13 \cdot 4/51 = 4/663$.

Repeated Trials. If p is the probability that an event will occur in a single trial, then the probability that it will occur exactly s times in r trials is the *binomial* or *Bernoulli* distribution function

$$_rC_sp^s(1 - p)^{r-s}$$

The probability that it will occur at least s times is

$$p^r + {}_rC_{r-1}p^{r-1}(1 - p) + {}_rC_{r-2}p^{r-2}(1 - p)^2 + \cdots + {}_rC_sp^s(1 - p)^{r-s}$$

Example. If 5 cards are drawn, one from each of 5 decks, the probability that exactly 3 will be aces is $_5C_3(1/13)^3(12/13)^2$. The probability that at least 3 will be aces is $(1/13)^5 + {}_5C_4(1/13)^4(12/13) + {}_5C_3(1/13)^3(12/13)^2$.

SET ALGEBRA

19. SETS

A **set** is a collection of objects called **elements** which are distinguished by a particular characteristic. Examples are: a set of engineers, a set of integers, a set of points. Element e belongs to set S is written $e \epsilon S$. If not, $e \slashed{\epsilon} S$. A set can be denoted by including the listed elements, or merely by a typical element, in curly brackets: $\{2, 4, 6\}$; $\{e_1, e_2\}$, $\{e\}$. A set with no elements is called the **null set**, and is denoted by $\varnothing$. A set with one element e_1 is denoted by $\{e_1\}$; and to avoid a paradox in logic, these two ideas must be kept distinct.

Two sets S_1, S_2 may be compared as follows. If every element of set S_1 is also an element of S_2, then S_1 is contained in S_2. This is written $S_1 \subset S_2$, and is read "S_1 is **contained in** S_2," or "S_1 is a **subset** of S_2." If, in addition, $S_2 \subset S_1$, then their relation is written $S_1 = S_2$. On the other hand, if S_4 has at least one element not contained in S_3, but $S_3 \subset S_4$, S_3 is a **proper subset** of S_4. If S_5 can contain all the elements of S_6, this can be stressed by writing $S_5 \subseteq S_6$. Evidently $\varnothing \subset S$, for every set S.

If S, called the **space**, is the largest set concerned in a particular discussion, all the other sets are subsets of S. Thus set $\alpha \subset S$. The **complement** of α, α^c, **with respect to space S** is the set of elements in S which are not elements of α.

Binary Operations for sets. The **union**, $S_a \cup S_b$, of sets S_a and S_b is the set of elements in S_a or S_b or in both. Note that union differs from the idea of sum since in the union the common elements are counted only once. The **intersection**, $S_a \cap S_b$, of sets S_a and S_b is the set of elements in both S_1 and S_2.

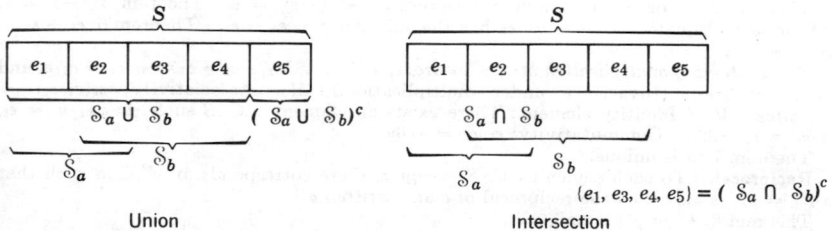

Union   Intersection

Boolean Algebra has as one representation the following:

Set Algebra. Let S_a, S_b, S_c have their elements in space S.

Unicity. Unique union $S_a \cup S_b \subset S$. Unique intersection $S_a \cap S_b \subset S$.

Commutativity. $S_a \cup S_b = S_b \cup S_a$, $S_a \cap S_b = S_b \cap S_a$.

Associativity. $S_a \cup (S_b \cup S_c) = (S_a \cup S_b) \cup S_c$, $S_a \cap (S_b \cap S_c) = (S_a \cap S_b) \cap S_c$.

Distributivity. $S_a \cup (S_b \cap S_c) = (S_a \cup S_b) \cap (S_a \cup S_c)$, $S_a \cap (S_b \cup S_c)$
$= (S_a \cap S_b) \cup (S_a \cap S_c)$.

Idempotency. $S_a \cup S_a = S_a$, $S_a \cap S_a = S_a$.

Space. $S_a \cup S = S$, $S_a \cap S = S_a$; **Null Set** $S_a \cup \varnothing = S_a$, $S_a \cap \varnothing = \varnothing$.

Subset. $\varnothing \subset S_a \subset S$, $S_a \subset (S_a \cup S_b)$, $(S_a \cap S_b) \subset S_a$, $S_a \subset S_b \Rightarrow S_a \cup S_b = S_b$, and $S_a \cap S_b = S_a$.

Complement. To $S_a \subset S$ there corresponds unique $S_a{}^c \subset S$. $S_a \cup S_a{}^c = S$, $S_a \cap S_a{}^c = \varnothing$.

de Morgan's Relations. $(S_a \cup S_b)^c = S_a{}^c \cap S_b{}^c$, $(S_a \cap S_b)^c = S_a{}^c \cup S_b{}^c$.

Invariant under the **duality transformation,** $\cup \leftrightarrow \cap$, $\subset \leftrightarrow \supset$, $S \leftrightarrow \varnothing$, are all the preceding relations.

20. GROUPS

A *group* is a system composed of a *set of elements* $\{a\}$ and a *rule of combination* of any two of them to form a product, such that

1. The *product* of any ordered pair of elements and the square of each element are elements of the set.

2. The *associative* law holds.

3. The set contains an *identity* element I such that $Ia = aI = a$ for any element a of the set.

4. For any element a of the set there is in the set an *inverse* a^{-1} such that $aa^{-1} = a^{-1}a = I$.

If, in addition,

5. The *commutative* law holds, the group is *commutative* or *Abelian*.

The *order* of a group is the number n of elements in the group.

21. RINGS, INTEGRAL DOMAINS, FIELDS

Rings. Space S consists of a set of elements e_1, e_2, e_3, $\cdots$. These elements are compared for **equality** and **order,** and combined by the operations of **addition and multiplication.** These terms in **bold face** are partially defined by the following sets of assumptions.

Equality is a term from logic, and means that if two expressions have this relation, then one may be substituted for the other.

Assumptions of equality. E_1. Unicity: either $e_1 = e_2$ or $e_1 \neq e_2$. E_2. Reflexivity: $e_1 = e_1$. E_3. Symmetry: $e_1 = e_2 \Rightarrow e_2 = e_1$. E_4. Transitivity: $e_1 = e_2$, $e_2 = e_3 \Rightarrow e_1 = e_3$.

Assumptions of addition A_1. Closure: $e_1 + e_2 \subset S$. A_2. $e_1 = e_2 \Rightarrow e_1 + e_3 = e_2 + e_3$ and $e_3 + e_1 = e_3 + e_2$. (Invariance under addition.) A_3. Associativity: $e_1 + (e_2 + e_3) = (e_1 + e_2) + e_3$. A_4. Identity Element: There exists an element $z \subset S$, such that $e_1 + z = e_1$, $z + e_1 = e_1$. A_5. Commutativity: $e_1 + e_2 = e_2 + e_1$.

Theorem 1: z is unique.

Negative. To each $e \subset S$, there corresponds an $e' \subset S$ such that $e + e' = z$. e' is called the negative of e and written $-e$.

Theorem 2: e' or $-e$ is unique. Theorem 3: $-(-e) = e$. Theorem 4: $-z = z$. Theorem 5: Equation $x + e_1 = e_2$ has the solution $x = e_2 - e_1$. Theorem 6: $e_1 + e_3 = e_1 \Rightarrow e_3 = z$.

Assumptions of multiplication M_1. Closure: $e_1 \cdot e_2 \subset S$. M_2. $e_1 = e_2 \Rightarrow e_1 \cdot e_3 = e_2 \cdot e_3$ and $e_3 \cdot e_1 = e_3 \cdot e_2$. (Invariance under multiplication.) M_3. Associativity: $e_1(e_2 \cdot e_3) = (e_1 \cdot e_2)e_3$. M_4. Identity element: There exists an element $u \subset S$ such that $e_1 \cdot u = e_1$, $u \cdot e_1 = e_1$. M_5. Commutativity: $e_1 \cdot e_2 = e_2 \cdot e_1$.

Theorem 7: u is unique.

Reciprocal. To each element $e \subset S$ except z, there corresponds an $e'' \subset S$ such that $e \cdot e'' = u$. e'' is called the reciprocal of e and written e^{-1}.

Theorem 8: e'' or e^{-1} is unique.

M_7. Distributivity: $e_1(e_2 + e_3) = e_1 \cdot e_2 + e_1 \cdot e_3$. Theorem 9: $e \cdot z = z$. Theorem 10: $e_1(-e_2) = -(e_1 \cdot e_2) = (-e_1)e_2$. Theorem 11: $(-e_1)(-e_2) = e_1 \cdot e_2$. Theorem 12: If S contains an element besides z then it is $u \neq z$. Theorem 13: $e_1 \cdot e_2 = z \Rightarrow$ either $e_1 = z$ or $e_2 = z$.

A **ring** is a space S having at least two elements for which assumptions E_1 ro E_4, A_1 to A_6, M_1 to M_5 and M_7 hold. An example is a residue system modulo 4.

An **integral domain** is a ring for which, as an assumption, Theorem 13 holds. An example is the set of all integers.

A **field** is an integral domain for which M_6 holds. An example of a field is the set of algebraic numbers.

Assumptions of (linear) order. O_1. (Contains E_1.) If e_1, $e_2 \subset S$, then either $e_1 < e_2$, $e_1 = e_2$, or $e_2 < e_1$. O_2. $e_1 < e_2 \Rightarrow e_1 + e_3 < e_2 + e_3$. (Invariance under addition.) O_3. Transitivity: $e_1 < e_2$, $e_2 < e_3 \Rightarrow e_1 < e_3$.

Negative. If $e_1 < z$, then e_1 is called **negative.**

Positive. If $z < e_2$, then e_2 is called **positive.**

O_4. $z < e_2$ $z < e_3 \Rightarrow z < e_2 \cdot e_3$.

An **ordered integral domain** is an integral domain for which O_1 to O_4 hold. An example is the set of all integers.

An **ordered field** is an ordered integral domain for which M_6 holds. An example is the set of all rational numbers.

If an additional order assumption, O_5, known as the Dedekind assumption—see a book on Real Analysis—be included, then the space S for which assumptions E_1 to E_4, A_1 to A_6, M_1 to M_7, and O_1 to O_5 hold is called the real number space. An example is the set of real numbers. Here z is denoted 0, and u is denoted 1. Another example is the set of points on the real line.

STATISTICS AND PROBABILITY

22. FREQUENCY DISTRIBUTIONS OF ONE VARIABLE

Definitions

A *frequency distribution* of statistical data consisting of N values of a variable x is a tabulation by intervals, called *classes*, showing the number f_i, called the *frequency* or *weight*, in each class; $N = \Sigma f_i$.

The mid value x_i of a class is the *class mark*.

For equal classes, the *class interval* is $c = x_{i+1} - x_i$.

The *cumulative frequency*, cum f, at any class is the sum of the frequencies of all classes up to and including the given class.

Graphs

Frequency Polygon. Plot the points (x_i, f_i) and draw a broken line through them.

Histogram. Draw a set of rectangles, using as bases intervals representing the classes marked off on a straight line, and using altitudes proportional to the frequencies.

Frequency Curve. Draw a continuous curve approximating a frequency polygon, or such that the region under the curve approximates a histogram. As the class interval c is taken smaller and the total frequency N larger, the approximation becomes better.

An Ogive is a graph of cumulative frequencies.

2. MATHEMATICS

Averages

Arithmetic Mean: A.M. $= \bar{x} = \dfrac{1}{N} \displaystyle\sum_{i=1}^{k} f_i x_i$, in which $N = \displaystyle\sum_{i=1}^{k} f_i$.

Geometric Mean: G.M. $= (x_1^{f_1} \cdot x_2^{f_2} \cdots x_k^{f_k})^{\frac{1}{N}}$; log G.M. $= \dfrac{1}{N} \displaystyle\sum_{i=1}^{k} f_i \log x_i$.

Harmonic Mean: H.M. $= \dfrac{N}{\displaystyle\sum_{i=1}^{k} \dfrac{f_i}{x_i}}$.

Root Mean Square: R.M.S. $= \sqrt{\dfrac{\displaystyle\sum_{i=1}^{k} f_i x_i^2}{N}}$.

Median: (a) for continuously varying data, the value of x for which cum $f = N/2$;
(b) for discrete data, the value of x such that there is an equal number of values larger and smaller; for N odd, $N = 2k - 1$, the median is x_k; for N even, $N = 2k$, the median may be taken as $1/2\,(x_k + x_{k+1})$.
Mode: the value of x which occurs most frequently.

Moments

1. About x_0.

In x units: $\nu_r = \dfrac{1}{N} \displaystyle\sum_{i=1}^{k} f_i (x_i - x_0)^r$, $r = 0, 1, \cdots$.

If $x_0 = 0$, $\nu_1 = \bar{x}$ which is the *arithmetic mean*.

In u units: $\nu_r = \dfrac{1}{N} \displaystyle\sum_{i=1}^{k} f_i u_i^r$, $r = 0, 1, \cdots$, $u = \dfrac{x - x_0}{c}$, $c =$ class interval.

2. About the *mean*.

In x units: $\mu_r = \dfrac{1}{N} \displaystyle\sum_{i=1}^{k} f_i (x_i - \bar{x})^r$, $r = 0, 1, \cdots$, $\bar{x} = \nu_1$ in x units.

In u units: $\mu_r = \dfrac{1}{N} \displaystyle\sum_{i=1}^{k} f_i (u_i - \bar{u})^r$, $r = 0, 1, \cdots$, $\bar{u} = \nu_1$ in u units.

In either x or u units, the μ's as functions of the ν's:

$$\mu_0 = 1$$
$$\mu_1 = 0$$
$$\mu_2 = \nu_2 - \nu_1^2$$
$$\mu_3 = \nu_3 - 3\nu_1\nu_2 + 2\nu_1^3$$
$$\mu_4 = \nu_4 - 4\nu_1\nu_3 + 6\nu_1^2\nu_2 - 3\nu_1^4$$
$$\mu_r \text{ (in } x \text{ units)} = c^r \mu_r \text{ (in } u \text{ units)}, \ r = 0, 1, \cdots.$$

In x units, μ_2 is the *variance*; $\sqrt{\mu_2}$ is the *standard deviation* σ.
Both are used as measures of dispersion. To compute σ,

$$\sigma = c \sqrt{\dfrac{\displaystyle\sum_{i=1}^{k} f_i u_i^2}{N} - \bar{u}^2}$$

Probable error = 0.6745σ.

3. In *standard* (deviation) *units:*

$$\alpha_1 = 0$$

$$\alpha_2 = 1$$

$$\alpha_3 = \frac{\mu_3}{\sigma^3}, \text{ a measure of } skewness$$

$$\alpha_4 = \frac{\mu_4}{\sigma^4}, \text{ a measure of } kurtosis$$

The Moment Generating Function, or arbitrary-range inverse real Laplace transform, is

$$M(\theta) = \int_a^b e^{\theta x} f(x) \, dx$$

The rth moment is

$$\mu_r = \frac{d^r M}{d\theta^r}\bigg|_{\theta=0}, \quad r = 0, 1, 2, \cdots$$

M-Tiles

The rth *quartile* Q_r is the value of x for which $\dfrac{\text{cum } f}{N} = \dfrac{r}{4}$.

The rth *percentile* P_r is the value of x for which $\dfrac{\text{cum } f}{N} = \dfrac{r}{100}$.

For $r = 10s$, $P_r = D_s$, the sth *decile*.

Other Measures of Shape

Dispersion. 1. *Range* of x, the difference between the largest and the smallest values of x.

2. *Mean deviation* $= \dfrac{1}{N} \sum_{i=1}^{k} f_i |x_i - \bar{x}|$.

3. *Semi-interquartile range*, or *quartile deviation*, $Q = \dfrac{|Q_3 - Q_1|}{2}$.

Skewness. 1. *Quartile coefficient of skewness* $= \dfrac{Q_3 - 2Q_2 + Q_1}{Q}$.

Statistical Hypotheses

A hypothesis concerning one or more statistical distribution parameters is a *statistical hypothesis*. A *test* of such a hypothesis is a procedure leading to a decision to accept or reject the hypothesis. The *significance level* is the probability value below which a hypothesis is rejected.

A *type 1 error* is made if the hypothesis is correct but the test rejects the hypothesis A *type 2 error* is made if the hypothesis is false but the test accepts the hypothesis.

If the variable x has a distribution function $f(x; \theta)$, with parameter θ, then the *likelihood* function, that is, distribution function of a random sample of size n, is $P(\theta) = f(x_1; \theta)f(x_2; \theta) \cdots f(x_n; \theta)$. The use of $P_{\max}(\theta)$ in the estimation of population parameters is the *method of maximum likelihood*. It often consists of solving $\dfrac{dP}{d\theta} = 0$ for θ.

Random Sampling

A set $x_1, x_2, \cdots, x_n$ of values of x with distribution function $f(x)$ is a *sample of size* n drawn from the population described by $f(x)$. If repeated samples of size n drawn from the population have the x_r's independently distributed in the probability sense and each x_r has the same distribution as the population, then the sampling is *random*.

Normal and Non-normal Distributions

The normal distribution function in analytic and tabular form is found on pages 245 to 247.

A linear combination of independent normal variables is normally distributed.

The *Poisson distribution*, $P(x) = \dfrac{e^{-m}m^x}{x!}$, is the limit approached by the binomial distribution (p. 2-32), if the probability p that an event will occur in a single trial approaches zero and the number of trials r becomes infinite in such a way that $rp = m$ remains constant.

If m is the mean of a non-normal distribution of x, σ the standard deviation, and if the moment generating function exists, then the variable $(\bar{x} - m)n^{1/2}/\sigma$, in which $\bar{x}$ is the mean of a sample of size n, has a distribution which approaches the normal distribution as $n \to \infty$.

Non-parametric methods are those which do not involve the estimation of parameters of a distribution function. Tchebycheff's inequality (p. 2-38) provides non-parametric tests for the validity of hypotheses. It leads to the *law of large numbers*. Let p be the probability of an event occurring in one trial, and p_n the ratio of the number of occurrences in n trials to the number n. The probability that $|p_n - p| > \epsilon$ is $\leq \dfrac{pq}{n\epsilon}$; this can be made arbitrarily small, however small ϵ is, by taking n large enough. The ratio p_n converges *stochastically* to the probability p.

Two numbers L_1, L_2 between which a large fraction of a population is expected to lie are *tolerance limits*. If z is the fraction of the population of a variable with a continuous distribution that lies between the extreme values of a random sample of size n from this population, then the distribution of z is $f(z) = n(n-1)z^{n-2}(1-z)$.

Statistical Control of Production Processes

A chart on which percentage defective in a sample is graphed as a function of output time can be used for control of an industrial process. Horizontal lines are drawn through the mean m and the controls $m \pm 3\dfrac{\sigma}{n^{1/2}}$. The behavior of the graph with respect to these control lines is used as an error signal in a feedback system which controls the process. If the graph goes out of the band bounded by the control lines, the process is stopped until the trouble is located and removed.

23. CORRELATION

To discover whether there is a simple relation between two variables, corresponding pairs of values are used as coordinates to plot the points of a *scatter diagram*. The simplest relation exists if the scatter diagram can be approximated more or less closely by a straight line.

The Least Square Straight Line, that is, the line which minimizes the sum of the squares of the y deviations of the points, is

$$\hat{y} - \bar{y} = M(x - \bar{x})$$

in which

$$M = \frac{\Sigma(x - \bar{x})y}{\Sigma(x - \bar{x})^2}$$

(x, y) is a plotted point, and $(x, \hat{y})$ is a point on this *line of regression* of y on x. The *correlation coefficient*

$$r = \pm\left[1 - \frac{\Sigma(y - \hat{y})^2}{\Sigma(y - \bar{y})^2}\right]^{1/2}$$

is a measure of the usefulness of the regression line. If $r = 0$, the line is useless; if $r = \pm 1$, the line gives a perfect estimate. The percentage of the variance of y that has been accounted for by y's relation to x is equal to r^2.

A Polynomial of Degree $n - 1$ can be passed through n points (x_i, y_i). The method of doing this by *divided differences* is as follows:

Example. Find the polynomial through (1, 5), (3, 11), (4, 31), (6, 3).

Using the first three values of x, assume the polynomial to be of the form $y = a_1 + a_2(x - 1) + a_3(x - 1)(x - 3) + a_4(x - 1)(x - 3)(x - 4)$. The a_i are the last four numbers in the top diagonal of the following table.

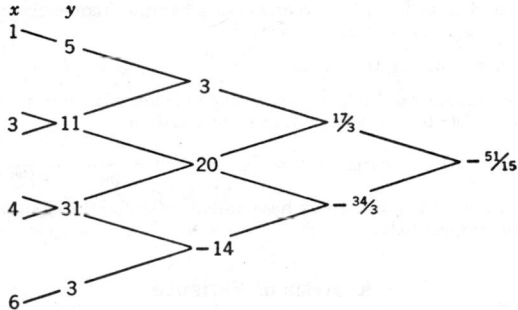

To form the table, put the given (x_i, y_i) in the first two columns. To find a number in any other column, divide the difference of the two numbers just above and below it immediately to the left by the difference of the x's in the two diagonals through it. The polynomial is $y = 5 + 3(x - 1) + 17/3\,(x - 1)(x - 3) - 51/15\,(x - 1)(x - 3)(x - 4)$.

A Power Formula $y = ax^n$ fits well if the points (x_i, y_i) lie approximately on a straight line when plotted on logarithmic (log scales on both horizontal and vertical axes) graph paper. To find a and n use two of the points (x_1, y_1) and (x_2, y_2), preferably far apart.

$$n = \frac{\log y_2 - \log y_1}{\log x_2 \quad \log x_1}$$

$$\log a = \log y_1 - n \log x_1$$

An Exponential Formula $y = ae^{nx}$ fits well if the points (x_i, y_i) lie approximately on a straight line when plotted on semilogarithmic (log scale on vertical axis) graph paper, To find a and n use two of the points (x_1, y_1) and (x_2, y_2), preferably far apart.

$$n = \frac{\ln y_1 - \ln y_2}{x_1 - x_2} = 2.3026 \frac{\log y_1 - \log y_2}{x_1 - x_2}$$

$$\ln a = \ln y_1 - nx_1 \quad \text{or} \quad \log a = \log y_1 - 0.4343nx_1$$

24. STATISTICAL ESTIMATION BY SMALL SAMPLES

A statistic is an *unbiased estimate* of a population parameter if its expected value is equal to the population parameter.

In the problem of estimating a population parameter, such as the mean or variance, the interval within which $c\%$ of the sample parameter values lies is the $c\%$ *confidence interval* for the parameter.

The χ^2 Distribution Function for ν *degrees of freedom* is

$$f(\chi^2) = \frac{1}{2^{\nu/2}\Gamma\left(\dfrac{\nu}{2}\right)} (\chi^2)^{(\nu-2)/2} e^{-\chi^2/2}$$

and its moment generating function is

$$M(\theta) = (1 - 2\theta)^{-\nu/2}$$

The sum of the squares of n random sample values of x has a χ^2 distribution with n degrees of freedom if x has a normal distribution with zero mean and unit variance.

Binomial Index of Dispersion is

$$\chi^2 = \sum_{r=1}^{k} \frac{(x_r - \bar{x})^2}{\bar{x}\left(1 - \dfrac{\bar{x}}{n}\right)}$$

For p small and n large, this reduces to the *Poisson index of dispersion*

$$\sum_{r=1}^{k} \frac{(x_r - \bar{x})^2}{\bar{x}}$$

These indices are used to test the hypothesis that k sample frequencies x_r came from the same binomial or Poisson population, respectively.

Student's t Distribution for the variable $t = \dfrac{u\nu^{\frac{1}{2}}}{v}$ is $f(t) = c\left(1 + \dfrac{t^2}{\nu}\right)^{-(\nu+1)/2}$, ν degrees of freedom, c constant; if u has a normal distribution with zero mean and unit variance and v^2 has a χ^2 distribution with ν degrees of freedom.

The F Distribution for the variable $F = \dfrac{u/\nu_1}{v/\nu_2}$ is $f(F) = \dfrac{cF^{(\nu_1-2)/2}}{(\nu_2 + \nu_1 F)^{(\nu_1+\nu_2)/2}}$, ν_1 and ν_2 degrees of freedom, c constant, if u and v have independent χ^2 distributions with ν_1 and ν_2 degrees of freedom, respectively.

Analysis of Variance

Experimental error is the variation in the basic variable remaining after the effects of controlled variables have been removed (p. 245). The *analysis of variance* means the resolution of the basic sum of squares into the component which measures the part of the variation being tested and the component which measures the experimental error.

25. STATISTICAL DESIGN OF EXPERIMENTS

To get *valid* conclusions from an experiment, there is need for proper control of the other variables besides those being investigated, and also for sufficiently large and random samples.

Sampling Inspection. To make an inspection *efficient*, the cost and usually the amount of sampling should be minimized.

It is a common practice in industry for a consumer to accept or reject a lot on the basis of a sample drawn from the lot. There is a maximum fraction of defectives that the consumer will tolerate. This is the *lot tolerance fraction defective* p_t. A random sample of n pieces is selected from a lot of N pieces. The maximum allowable number of defective pieces in an acceptable sample is c. *Single sampling* means: (1) Inspect a sample of n pieces. (2) Accept the lot if the number of defective pieces is c or less; otherwise inspect the remainder of the lot. (3) Replace all defective pieces found by non-defective.

The *consumer's risk*, that is, the probability that a consumer will accept a lot of quality lower than p_t is

$$P_c = \sum_{x=0}^{c} \frac{\dbinom{Np_t}{x}\dbinom{N - Np_t}{n - x}}{\dbinom{N}{n}} \tag{1}$$

If a producer has standardized his quality at a fractional value $\bar{p}$, the *process average fraction defective*, then the *producer's risk*, that is, the probability that a lot of his will be erroneously rejected, is

$$P_p = 1 - \sum_{x=0}^{c} \frac{\dbinom{N\bar{p}}{x}\dbinom{N - N\bar{p}}{n - x}}{\dbinom{N}{n}} \tag{2}$$

These two risks correspond to errors of type 2 and type 1, respectively.

The average number of pieces inspected per lot for single sampling is $I = n + (N - n)P_p$. The amount of inspection and ordinarily the cost is minimized by finding the pair of values of n and c which satisfy (1) for an assigned value of P_c and minimize I.

Sequential Analysis. An improvement on the fixed-size sampling methods described above results in greater efficiency if the inspection can be conducted on an accumulation-of-information basis. Such sequential methods operate on successive terms of a sequence of observations as they are received. They involve two steps: 1. to accept or reject the hypothesis under test, 2. to continue taking additional observations if the hypothesis is rejected.

For a more extensive treatment of the elementary theory of statistics see *Introduction to Mathematical Statistics* by P. G. Hoel, John Wiley & Sons (1947).

26. PRECISION OF MEASUREMENTS

Observations and Errors

The Error of an Observation is $e_i = m_i - m$, $i = 1, 2, \cdots, n$, where the m_i are the observed values, the e_i the errors, and m the mean value, that is, the arithmetic mean of a very large number (theoretically infinite) of observations.

In a large number of measurements *random errors* are as often negative as positive and have little effect on the arithmetic mean. All other errors are classed as *systematic*. If due to the same cause, they affect the mean in the same sense and give it a definite bias.

Best Estimate and Measured Value. If all systematic errors have been eliminated, it is possible to consider the sample of individual repeated measurements of a quantity with a view to securing the "best" estimate of the mean value m and assessing the degree of reproducibility which has been obtained. The final result will then be expressed in the form $E \pm L$, where E is the best estimate of m, and L the characteristic limit of variation associated with a certain risk. Not merely E, but the entire result $E \pm L$ is the value measured.

The Arithmetic Mean. If a large number of measurements have been made to determine directly the mean m of a certain quantity, all measurements having been made with equal skill and care, the best estimate of m from a sample of n is the arithmetic mean $\overline{m}$ of the measurements in the sample,

$$\overline{m} = \frac{1}{n} \sum_{i=1}^{n} m_i$$

Standard Deviation is the *root-mean-square* of the deviations e_i of a set of observations from the mean,

$$\sigma = \left(\frac{1}{n} \sum_{i=1}^{n} e_i^2 \right)^{\frac{1}{2}}$$

Since neither the mean m nor the errors of observation e_i are ordinarily known, the deviations from the arithmetic mean, or the *residuals*, $x_i = m_i - \overline{m}$, $i = 1, 2, \cdots, n$, will be referred to as errors. Likewise, for σ the unbiased value

$$\sigma = (n - 1)^{-\frac{1}{2}} \left[\sum_{i=1}^{n} (m_i - \overline{m})^2 \right]^{\frac{1}{2}} = (n - 1)^{-\frac{1}{2}} \left(\sum_{i=1}^{n} e_i^2 \right)^{\frac{1}{2}}$$

will be used, in which n is replaced by $n - 1$ since one degree of freedom is lost by using $\overline{m}$ instead of m, $\overline{m}$ being related to the m_i.

The Normal Distribution

Relative Frequency of Errors. The *Gauss-Laplace*, or *normal*, distribution of frequency of errors is (Fig. 1)

$$y = \frac{1}{\sigma \sqrt{2\pi}} e^{-x^2/2\sigma^2}$$

or

$$y = \frac{1}{\sqrt{\pi}} h e^{-h^2 x^2}$$

where $2h^2\sigma^2 = 1$, or $h = 1/(\sqrt{2}\sigma)$, and y represents the proportionate number of errors of value x. The area under the curve is unity. The dotted curve is also an error distribution curve with a greater value

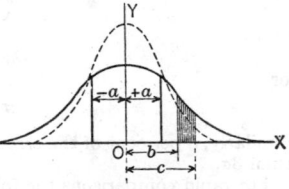

Fig. 1

of the precision index h which measures the concentration of observations about their mean.

Probability. The fraction of the total number of errors whose values lie between $x = -a$ and $x = a$ is

$$P = \frac{h}{\sqrt{\pi}} \int_{-a}^{+a} e^{-h^2 x^2}\, dx = \frac{2}{\sqrt{\pi}} \int_0^{ha} e^{-h^2 x^2}\, d(hx) \tag{3}$$

that is, P is the probability of an error x having a value between $-a$ and a. Similarly, the shaded area represents the probability of errors between b and c.

Table 1. Values of $P = \dfrac{2}{\sqrt{\pi}} \displaystyle\int_0^{ha} e^{-h^2 x^2}\, d(hx)$

ha*	0	1	2	3	4	5	6	7	8	9
0.0		.01128	.02256	.03384	.04511	.05637	.06762	.07886	.09008	.10128
.1	.11246	.12362	.13476	.14587	.15695	.16800	.17901	.18999	.20094	.21184
.2	.22270	.23352	.24430	.25502	.26570	.27633	.28690	.29742	.30788	.31828
.3	.32863	.33891	.34913	.35928	.36936	.37938	.38933	.39921	.40901	.41874
.4	.42839	.43797	.44747	.45689	.46623	.47548	.48466	.49375	.50275	.51167
.5	.52050	.52924	.53790	.54646	.55494	.56332	.57162	.57982	.58792	.59594
.6	.60386	.61168	.61941	.62705	.63459	.64203	.64938	.65663	.66378	.67084
.7	.67780	.68467	.69143	.69810	.70468	.71116	.71754	.72382	.73001	.73610
.8	.74210	.74800	.75381	.75952	.76514	.77067	.77610	.78144	.78669	.79184
.9	.79691	.80188	.80677	.81156	.81627	.82089	.82542	.82987	.83423	.83851
1.0	.84270	.84681	.85084	.85478	.85865	.86244	.86614	.86977	.87333	.87680
1.1	.88021	.88353	.88679	.88997	.89308	.89612	.89910	.90200	.90484	.90761
1.2	.91031	.91296	.91553	.91805	.92051	.92290	.92524	.92751	.92973	.93190
1.3	.93401	.93606	.93807	.94002	.94191	.94376	.94556	.94731	.94902	.95067
1.4	.95229	.95385	.95538	.95686	.95830	.95970	.96105	.96237	.96365	.96490
1.5	.96611	.96728	.96841	.96952	.97059	.97162	.97263	.97360	.97455	.97546
1.6	.97635	.97721	.97804	.97884	.97962	.98038	.98110	.98181	.98249	.98315
1.7	.98379	.98441	.98500	.98558	.98613	.98667	.98719	.98769	.98817	.98864
1.8	.98909	.98952	.98994	.99035	.99074	.99111	.99147	.99182	.99216	.99248
1.9	.99279	.99309	.99338	.99366	.99392	.99418	.99443	.99466	.99489	.99511
2.0	.99532	.99552	.99572	.99591	.99609	.99626	.99642	.99658	.99673	.99688
2.1	.99702	.99715	.99728	.99741	.99753	.99764	.99775	.99785	.99795	.99805
2.2	.99814	.99822	.99831	.99839	.99846	.99854	.99861	.99867	.99874	.99880
2.3	.99886	.99891	.99897	.99902	.99906	.99911	.99915	.99920	.99924	.99928
2.4	.99931	.99935	.99938	.99941	.99944	.99947	.99950	.99952	.99955	.99957
2.5	.99959	.99961	.99963	.99965	.99967	.99969	.99971	.99972	.99974	.99975
2.6	.99976	.99978	.99979	.99980	.99981	.99982	.99983	.99984	.99985	.99986
2.7	.99987	.99987	.99988	.99989	.99989	.99990	.99991	.99991	.99992	.99992
2.8	.99992	.99993	.99993	.99994	.99994	.99994	.99995	.99995	.99995	.99996
2.9	.99996	.99996	.99996	.99997	.99997	.99997	.99997	.99997	.99997	.99998
3.0	.99998	1.0000	1.0000	1.0000						

$$* \; ha = 0.47694\,\frac{a}{r} = \frac{1}{\sqrt{2}}\,\frac{a}{\sigma}$$

Probable Error. Results of measurements are sometimes expressed in the form $E \pm r$, where r is the *probable error* of a single observation and is defined as the number which the actual error may with equal probability be greater or less than. From (3)

$$\frac{2}{\sqrt{\pi}} \int_0^{hr} e^{-h^2 x^2}\, d(hx) = 0.50$$

and

$$hr = 0.47694$$

or

$$r = 0.4769 \times \sqrt{2}\sigma = 0.6745\sigma$$

Similarly, 5 per cent of the errors x are greater than 2σ, and less than 1 per cent are greater than 3σ.

For rapid comparisons the following approximate formula due to Peters is useful:

$$r \approx 0.8453\,[n(n-1)]^{-\frac{1}{2}} \sum_{i=1}^{n} |x_i|$$

The Standard Deviation of the Arithmetic Mean, $\sigma_{\overline{m}}$, as calculated from data, is related to the standard deviation, σ, by the formula

$$\sigma_{\overline{m}} = n^{-\frac{1}{2}}\sigma = [n(n-1)]^{-\frac{1}{2}}\left(\sum_{i=1}^{n} x_i^2\right)^{\frac{1}{2}}$$

From this formula and Tables 1 and 2 the limits corresponding to given risks can be determined as indicated above. It is evident that the stability of the mean increases with n, i.e., the effect of the erratic behavior of single cases decreases with increase of n.

Table 2. Values of Functions of n and $(n-1)$

Factors for Computing Actual and Approximate Values of r and $r_{\overline{m}}$

n	$\dfrac{0.6745}{\sqrt{n-1}}$	$\dfrac{0.6745}{\sqrt{n(n-1)}}$	$\dfrac{0.8453}{\sqrt{n(n-1)}}$	$\dfrac{0.8453}{n\sqrt{n-1}}$	n	$\dfrac{0.6745}{\sqrt{n-1}}$	$\dfrac{0.6745}{\sqrt{n(n-1)}}$	$\dfrac{0.8453}{\sqrt{n(n-1)}}$	$\dfrac{0.8453}{n\sqrt{n-1}}$
1					51	0.0954	0.0134	0.0167	0.0023
2	0.6745	0.4769	0.5978	0.4227	52	.0944	.0131	.0164	.0023
3	.4769	.2754	.3451	.1993	53	.0935	.0128	.0161	.0022
4	.3894	.1947	.2440	.1220	54	.0926	.0126	.0158	.0022
5	.3372	.1508	.1890	.0845	55	.0918	.0124	.0155	.0021
6	.3016	.1231	.1543	.0630	56	.0909	.0122	.0152	.0020
7	.2754	.1041	.1304	.0493	57	.0901	.0119	.0150	.0020
8	.2549	.0901	.1130	.0399	58	.0893	.0117	.0147	.0019
9	.2385	.0795	.0996	.0332	59	.0886	.0115	.0145	.0019
10	0.2248	0.0711	0.0891	0.0282	60	0.0878	0.0113	0.0142	0.0018
11	.2133	.0643	.0806	.0243	61	.0871	.0111	.0140	.0018
12	.2034	.0587	.0736	.0212	62	.0864	.0110	.0137	.0017
13	.1947	.0540	.0677	.0188	63	.0857	.0108	.0135	.0017
14	.1871	.0500	.0627	.0167	64	.0850	.0106	.0133	.0017
15	.1803	.0465	.0583	.0151	65	.0843	.0105	.0131	.0016
16	.1742	.0435	.0546	.0136	66	.0837	.0103	.0129	.0016
17	.1686	.0409	.0513	.0124	67	.0830	.0101	.0127	.0016
18	.1636	.0386	.0483	.0114	68	.0824	.0100	.0125	.0015
19	.1590	.0365	.0457	.0105	69	.0818	.0098	.0123	.0015
20	0.1547	0.0346	0.0434	0.0097	70	0.0812	0.0097	0.0122	0.0015
21	.1508	.0329	.0412	.0090	71	.0806	.0096	.0120	.0014
22	.1472	.0314	.0393	.0084	72	.0800	.0094	.0118	.0014
23	.1438	.0300	.0376	.0078	73	.0795	.0093	.0117	.0014
24	.1406	.0287	.0360	.0073	74	.0789	.0092	.0115	.0013
25	.1377	.0275	.0345	.0069	75	.0784	.0091	.0113	.0013
26	.1349	.0265	.0332	.0065	76	.0779	.0089	.0112	.0013
27	.1323	.0255	.0319	.0061	77	.0774	.0088	.0111	.0013
28	.1298	.0245	.0307	.0058	78	.0769	.0087	.0109	.0012
29	.1275	.0237	.0297	.0055	79	.0764	.0086	.0108	.0012
30	0.1252	0.0229	0.0287	0.0052	80	0.0759	0.0085	0.0106	0.0012
31	.1231	.0221	.0277	.0050	81	.0754	.0084	.0105	.0012
32	.1211	.0214	.0268	.0047	82	.0749	.0083	.0104	.0011
33	.1192	.0208	.0260	.0045	83	.0745	.0082	.0102	.0011
34	.1174	.0201	.0252	.0043	84	.0740	.0081	.0101	.0011
35	.1157	.0196	.0245	.0041	85	.0736	.0080	.0100	.0011
36	.1140	.0190	.0238	.0040	86	.0732	.0079	.0099	.0011
37	.1124	.0185	.0232	.0038	87	.0727	.0078	.0098	.0010
38	.1109	.0180	.0225	.0037	88	.0723	.0077	.0097	.0010
39	.1094	.0175	.0220	.0035	89	.0719	.0076	.0096	.0010
40	0.1080	0.0171	0.0214	0.0034	90	0.0715	0.0075	0.0094	0.0010
41	.1066	.0167	.0209	.0033	91	.0711	.0075	.0093	.0010
42	.1053	.0163	.0204	.0031	92	.0707	.0074	.0092	.0010
43	.1041	.0159	.0199	.0030	93	.0703	.0073	.0091	.0009
44	.1029	.0155	.0194	.0029	94	.0699	.0072	.0090	.0009
45	.1017	.0152	.0190	.0028	95	.0696	.0071	.0089	.0009
46	.1005	.0148	.0186	.0027	96	.0692	.0071	.0089	.0009
47	.0994	.0145	.0182	.0027	97	.0688	.0070	.0088	.0009
48	.0984	.0142	.0178	.0026	98	.0685	.0069	.0087	.0009
49	.0974	.0139	.0174	.0025	99	.0681	.0068	.0086	.0009
50	0.0964	0.0136	0.0171	0.0024	100	0.0678	0.0068	0.0085	0.0009

The **Probable Error of the Arithmetic Mean** as calculated from data, $r_{\overline{m}}$, is then given by

$$r_{\overline{m}} = 0.6745[n(n-1)]^{-\frac{1}{2}}\left(\sum_{i=1}^{n} x_i^2\right)^{\frac{1}{2}}$$

and Peters' formula for the approximate value is:

$$r_{\overline{m}} \approx 0.8453[n^2(n-1)]^{-\frac{1}{2}}\sum_{i=1}^{n} |x_i|$$

Example. The following are ten measurements, m_i, of the length of a base line. Below are given the values of the residuals, x_i, and their squares. m_i: 455.35, 455.35, 455.20, 455.05, 455.75, 455.40, 455.10, 455.30, 455.50, 455.30.
Arithmetic mean, $\overline{m}$ = 455.330.
x_i: 0.02, 0.02, −0.13, −0.28, 0.42, 0.07, −0.23, −0.03, −0.17, −0.03.
x_i^2: 0.0004, 0.0004, 0.0169, 0.0784, 0.1764, 0.0049, 0.0529, 0.0009, 0.0289, 0.0009.

Hence: $\sum_{i=1}^{10} x_i^2 = 0.3610$, and $\sum_{i=1}^{10} |x_i| = 1.40$.

So by the standard formulas, $r = 0.6745(9)^{-\frac{1}{2}}(0.3610)^{\frac{1}{2}} = 0.13$, $r_{\overline{m}} = (10)^{-\frac{1}{2}}r = 0.042$. By the approximate formulas, $r \approx 0.8453(90)^{-\frac{1}{2}}(1.40) = 0.12$, $r_{\overline{m}} \approx 0.039$.
For the best estimate of the base line, the result is 455.330 with probable error ± 0.042 (using result given by standard formula), usually written 455.330 ± 0.042. In any considerable number of observations it should be the case, as it is here, that half of the residuals are less than the probable error.

Rounded Numbers. It can be shown that the standard deviation, σ, of a *rounded number* (p. 213) due to rounding is $\sigma = 0.2887w$, where w is a unit in the last place retained. Consequently, the probable error of a rounded number due to rounding is:

$$r = 0.6745 \times 0.2887w = 0.1947w$$

Weighted Observations. Sometimes, notwithstanding the care with which observations are taken, there are reasons for believing that certain observations are better than others. In such cases the observations are given different weights, that is, are counted different numbers of times, the weights or numbers expressing their relative practical worth. If there are n weighted observations, m_i, with weights, p_i, these being made directly on the same quantity, then the best estimate of the mean value m of the quantity is the *weighted arithmetic mean* $\overline{m}$ of the sample,

$$\overline{m} \equiv \sum_{i=1}^{n} p_i m_i \Big/ \sum_{i=1}^{n} p_i$$

For the set of weighted observations we have

$$r = 0.6745(n-1)^{-\frac{1}{2}}\left(\sum_{i=1}^{n} p_i x_i^2\right)^{\frac{1}{2}}$$

as the probable error of an observation of unit weight, and

$$r_{\overline{m}} = 0.6745\left[(n-1)\sum_{i=1}^{n} p_i\right]^{-\frac{1}{2}}\left(\sum_{i=1}^{n} p_i x_i^2\right)^{\frac{1}{2}}$$

as the probable error of the arithmetic mean of weighted items, in which

$$x_i \equiv m_i - \sum_{i=1}^{n} p_i m_i \Big/ \sum_{i=1}^{n} p_i$$

Example. Let six observations on the same quantity be made, with weights, p_i, the sum of these weights being 21 (see tabulation below). The sum of the weighted observations, $\sum_{i=1}^{6} p_i m_i$, is 3741.36.

The best estimate of the value of m for the observed quantity is $\overline{m}$ = 3741.36/21 = 178.16. Subtracting this from each m_i gives the residuals x_i. The sum of the weighted squares of the residuals.

$\sum_{i=1}^{6} p_i x_i^2$, is 62.95. Then the preceding formulas give the probable error of an observation of weight unity as $r = 2.39$ and the probable error of the weighted mean as $r_{\overline{m}} = 0.52$. The final result then is 178.16 ± 0.52.

p_i:	5	4	1	4	3	4
m_i:	178.26	176.30	181.06	177.95	176.20	180.85
$p_i m_i$:	891.30	705.20	181.06	711.80	528.60	723.40
x_i:	.10	1.86	2.90	.21	1.96	2.69
$x_i{}^2$:	.010	3.460	8.410	.441	3.842	7.230
$p_i x_i{}^2$:	.05	13.84	8.41	.18	11.53	28.94

Probable Error in a Result Calculated from the Means of Several Observed Quantities. Let Z be a sum of n means of observed independent quantities, each taken with a plus or a minus sign. Then, if r_j, $j = 1, 2, \cdots, n$, are the probable errors in these means,

the probable error in Z is $\left(\sum_{j=1}^{n} r_j{}^2 \right)^{1/2}$.

Let $Z = Az$, where z is the mean of an observed quantity with probable error r, and A an exact number. Then the probable error in Z is Ar.

Let Z be any differentiable function of the means of independently observed quantities z_j with probable errors r_j. Then the probable error in Z is

$$\left[\sum_{j=1}^{m} \left(\frac{\partial Z}{\partial z_j} \right)^2 r_j{}^2 \right]^{1/2}$$

For example, if $Z = z_1 z_2$, the probable error in Z is

$$(z_1{}^2 r_2{}^2 + z_2{}^2 r_1{}^2)^{1/2}$$

Conditions of Applicability. The theory underlying the foregoing development depends upon the following assumptions: (1) The sample consists of a large number of observations. (2) The observations have been made with equal care and skill so that: (2.1) there are approximately an equal number of readings above and below the mean (except in the case of weighted items), and (2.2) the individual deviations from the mean are small in most cases, and (2.3) the number of deviations diminishes rapidly as their size increases.

The extent to which the observed data satisfy the above assumptions is a measure of the extent to which we are justified in using the Gauss error distribution curve, which is consistent with the statement that $\overline{m}$ is the best estimate of the mean value m, and which leads to the factor 0.6745 used in computing probable error. Even if we were not justified in assuming the Gaussian distribution of errors, the arithmetic mean still remains the best estimate we have for m. Therefore, there is little difficulty in this regard, especially since "errors" appear to follow the Gaussian distribution as closely as any other we know. Our difficu'ties enter in connection with the factor 0.6745 and the accuracy of the σ, as estimated from the data.

If the number of observations n in a sample is small the estimate of the standard deviation of the possible infinity of observations with mean m is itself subject to considerable error. For example, for $n = 3$ the standard error of the standard dev ation is as large as the standard deviation itself, and hence the probable error calculated from $r = 0.6745\sigma$ would not be very reliable. Table 3 * will illustrate this. The second and third columns give the probability that the probable error of a single observation should be out 20 and 50 per cent, respectively.

Table 3. Combination of Observations

n	20 per cent	50 per cent	n	20 per cent	50 per cent
5	0.64	0.24	30	0.12	0.00014
10	.40	.034	40	.076	8×10^{-6}
15	.29	.008	50	.047	6×10^{-7}
20	.21	.0002	100	.0050	

From this table it is clear that with 10 observations the odds are only 3 to 2 that the calculated probable error is within 20 per cent of the correct value, and about 30 to 1 that it is within 50 per cent of the correct value. Of course, the probable error of the mean will be correspondingly out.

The use of Table 3 is quite legitimate for $100 < n$, and for $30 < n < 100$ the tables may be used provided σ is multiplied by $(n - 3)^{-1/2}$. For $n < 30$, a rough estimate can

* David Brunt, *The Combination of Observations*, Cambridge University Press (1917).

be obtained from the fact that the percentage of cases lying outside the range, $m \pm k\sigma$, is $< 100k^{-2}$ for $1 < k$. A striking property of this inequality due to Tchebycheff is that it is *non-parametric*, which means independent of the nature of the distribution assumed.

GEOMETRY

27. GEOMETRIC CONCEPTS

Plane Angles

A Degree (°) is $1/360$ of a revolution (or *perigon*) and is divided into 60 units called *minutes* (′) which in turn are divided into 60 units called *seconds* (″).

A Radian is a *central angle* which intercepts a *circular arc* equal to its *radius*. One radian, therefore, equals $360/2\pi$ degrees or $57.295779513°$, and $1° = 0.017453293$ radian.

An Angle of 90° is a *right angle*, and the lines that form it are *perpendicular*. An angle less than a right angle is *acute*. An angle greater than a right angle but less than 180° is *obtuse*. If the sum of two angles equals 90°, they are *complementary* to each other, and if their sum is 180°, *supplementary* to each other.

Polygons

A Polygon, or *plane rectilinear figure*, is a closed broken line.

A Triangle is a polygon of three sides. It is *isosceles* if two sides (and their opposite angles) are equal; it is *equilateral* if all three sides (and all three angles) are equal.

A Quadrilateral is a polygon of four sides. This classification includes the *trapezium* having no two sides parallel; the *trapezoid* having two opposite sides parallel (*isosceles trapezoid* if the non-parallel sides are equal); and the *parallelogram* having both pairs of opposite sides parallel and equal. The parallelogram includes the *rhomboid* having no right angles and, in general, adjacent sides not equal; the *rhombus* having no right angles but all sides equal; the *rectangle* having only right angles and, in general, adjacent sides not equal; and the *square* having only right angles, and all sides equal.

Similar Polygons have their respective angles equal and their corresponding sides proportional.

A Regular Polygon has all sides equal and all angles equal. An *equilateral triangle* and a *square* are regular polygons.

Other Polygons classified according to number of sides are: (5) *pentagon;* (6) *hexagon;* (7) *heptagon;* (8) *octagon;* (9) *enneagon* or *nonagon;* (10) *decagon;* (12) *dodecagon.* Two regular polygons of the same number of sides are *similar*.

Properties of Triangles

General Triangle. The sum of the angles equals $180°$. $\angle XAB$ (Fig. 1) is an *exterior angle* of $\triangle ABC$ and equals the sum of the opposite *interior angles* (i.e., $\angle XAB = \angle B + \angle C$). A *median* of a triangle is a line joining a vertex to the mid-point of the opposite side. The three medians meet at the *center of gravity*, G, and G trisects each median (i.e., $AG = \frac{2}{3} AD$, etc.). *Bisectors of angles* of a triangle (Fig. 2) meet in a point M equidistant from all sides. M is the center of the *inscribed circle* (tangent to all sides),

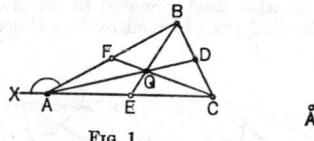
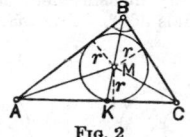
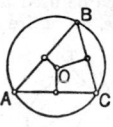

FIG. 1   FIG. 2   FIG. 3

or the *incenter* of the triangle. An angle bisector divides the opposite side into segments proportional to the adjacent sides of the angle (i.e., $AK/KC = AB/BC$, etc.). An *altitude* of a triangle is a perpendicular from a vertex to the opposite side. The three altitudes meet in a point called the *orthocenter*. The *perpendicular bisectors* of the sides of a triangle (Fig. 3) meet in a point O equidistant from all vertices. O is the center of the *circumscribed circle* (passing through all vertices), or the *circumcenter* of the triangle. The longest side of a triangle is opposite the largest angle, and vice versa. The line joining the mid-points of two sides of a triangle is parallel to the third side and half its length. If two triangles are mutually equiangular, they are *similar*, and their corresponding sides are proportional.

Orthogonal Projection. In Figs. 4 and 5, AE is the orthogonal projection of AB on AC, BE being perpendicular to AC. The square of the side opposite an acute angle equals the sum of the squares of the other two sides diminished by twice the product of one of those sides by the orthogonal projection of the other side upon it. In Fig. 4,

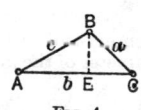
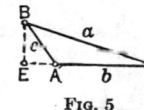
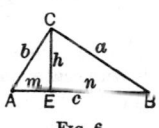

FIG. 4   FIG. 5   FIG. 6

$a^2 = b^2 + c^2 - 2b \cdot AE$. The square of the side opposite an obtuse angle equals the sum of the squares of the other two sides increased by twice the product of one of those sides by the orthogonal projection of the other side upon it. In Fig. 5, $a^2 = b^2 + c^2 + 2b \cdot AE$.

Right Triangle. In Fig. 6, let h be the *altitude* drawn from the vertex of right angle C to the *hypotenuse* c. Then $\angle A + \angle B = 90°$; $c^2 = a^2 + b^2$; $h^2 = mn$; $b^2 = cm$; $a^2 = cn$; (median from C) $= c/2$.

Isosceles Triangle. Two sides are equal and their opposite angles are equal. If a straight line from the vertex at which the equal sides meet bisects the base, it also bisects the angle at the vertex and is perpendicular to the base.

Circles

A **Circle** is a closed plane curve, all the points of which are equidistant from a *center* point. A *chord* is a straight line joining two points on a curve, that is, joining the extremities of an *arc*. A *segment* of a curve is the part of its plane included between a concave arc and its chord. An angle *intercepts* an arc cut off by its sides; the arc *subtends* the angle. A *central angle* of a circle is one whose vertex is at the center and whose sides are two radii. A *sector* of a circle is the part of its plane which is included between an arc and two radii drawn to its extremities. A *secant* of a circle is a straight line intersecting it in two points. Parallel secants (or tangents) intercept equal arcs. A tangent line meets a circle in only one point and is perpendicular to the radius to that point. If a radius is perpendicular to a chord, it bisects both the chord and the arc intercepted by the chord. If two circles are tangent to each other, the line of centers passes through the point of contact; if the circles intersect, the line of centers bisects the common chord at right angles. In Fig. 8, the product of linear segments AC and AE equals the product of linear segments AB and AF. In Fig. 9, the product of the whole secant AB and its external segment AE equals the product of the whole secant AC and its external segment AF. In Fig. 10, the product of the whole secant AD and its external segment AC equals the square of tangent AB (or AE). Also $\angle ABE = \angle AEB$.

Angle Measurement. Considering the arc of a circle to be expressed in terms of the central angle which it subtends, the arc may be said to contain a certain number of degrees and hence be used to express the measurement of other angles related to the circle. On this basis, an entire circle equals 360°. The *inscribed angle* formed by two chords inter-

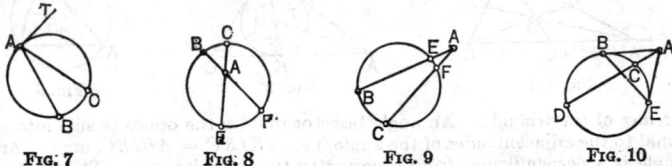

FIG. 7   FIG. 8   FIG. 9   FIG. 10

secting on a circle equals half the arc intercepted by it. Thus, in Fig. 7, $\angle BAC = \frac{1}{2}$ arc BC. An angle inscribed in a semicircle is a right angle. The angle formed by a tangent to a circle and a chord having one extremity at the point of contact equals half the arc intercepted by the chord. In Fig. 7, $\angle BAT = \frac{1}{2}$ arc BCA. The angle formed by two chords intersecting within a circle equals half the sum of the intercepted arcs. In Fig. 8, $\angle BAC$ (or $\angle EAF$) $= \frac{1}{2}$ (arc BC + arc EF). The angle formed by two secants, or two tangents, or a secant and a tangent, intersecting outside a circle, equals half the difference of the intercepted arcs. In Fig. 9, $\angle BAC = \frac{1}{2}$ (arc BC − arc EF). In Fig. 10, $\angle BAE = \frac{1}{2}$ (arc BDE − arc BCE), and $\angle BAD = \frac{1}{2}$ (arc BD − arc BC).

Coaxal Systems

Types. 1. A set of non-intersecting circles having collinear centers and orthogonal to a given circle with center also collinear. The end points of the diameter of the given circle

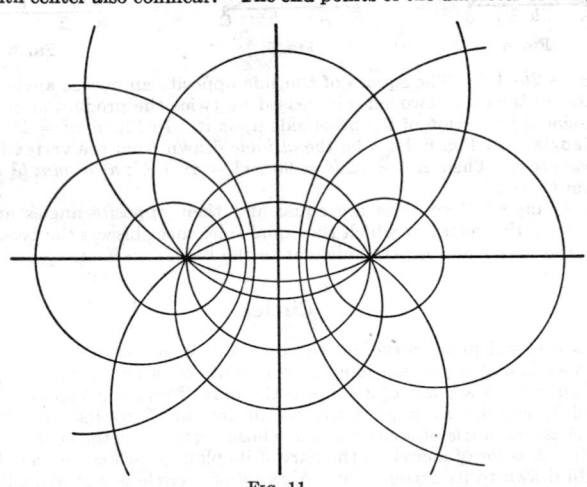

FIG. 11

on the line of centers are the limiting points of the system (Fig. 11, centers on horizontal line).

2. A set of circles through two given points (Fig. 11, centers on vertical line).
3. A set of circles with a common point of tangency.
4. A set of concentric circles.
5. A set of concurrent lines.
6. A set of parallel lines.

Conjugate Systems. Two coaxal systems whose members are mutually orthogonal are *conjugate*. A conjugate pair may consist of (1) a system of Type 1 and one of Type 2, with the limiting points of one the common points of the other (Fig. 11); (2) two systems of Type 3; (3) a system of Type 4 and one of Type 5; (4) two systems of Type 6.

Inversion

If the point O is the center of a circle c of radius r, if P and P' are collinear with O, and if $OP \cdot OP' = r^2$, then P and P' are *inverse* to each other with respect to the circle c (Fig. 12). The point O is the *center of inversion*.

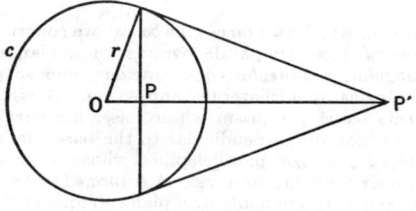

FIG. 12

The inverse of a circle not passing through the center of inversion is a circle, the inverse of a circle through the center is a straight line not through the center, and the inverse of a straight line through the center is itself.

Two intersecting curves invert into curves intersecting at the same angle.

Non-planar Angles

A Dihedral Angle is the opening between two intersecting planes. In Fig. 13, $P-BD-Q$ is a dihedral angle of which the two planes are the *faces* and their line of intersection BD is the *edge*. A *plane angle* which measures a dihedral angle is an angle formed by two lines, one in each face, drawn perpendicular to the edge at the same point (as $\angle ABC$). A *right dihedral angle* is one whose plane angle is a right angle. Through a given line oblique or parallel to a given plane, one and only one plane can be passed perpendicular

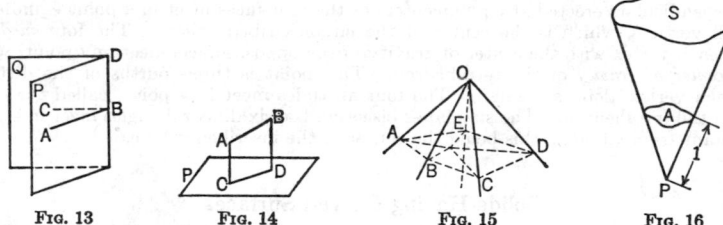

FIG. 13   FIG. 14   FIG. 15   FIG. 16

to the given plane. The line of intersection CD (Fig. 14) is the *orthogonal projection* of line AB upon plane P. The *angle between a line and a plane* is the angle that the line (produced if necessary) makes with its orthogonal projection on the plane. This angle is the least angle which the line makes with any line in the plane.

A Polyhedral Angle is the opening of three or more planes which meet in a common point. In Fig. 15, $O-ABCDE$ is a polyhedral angle of which the intersections of the planes, as OA, OB, etc., are the *edges;* the portions of the planes lying between the edges are the *faces;* and the common point O is the *vertex.* Angles formed by adjacent edges, as angles AOB, BOC, etc., are *face angles.* A polyhedral angle is called a *trihedral angle* if it has three faces; a *tetrahedral angle* if it has four faces; and so on.

A Solid Angle measures the opening between surfaces, either planar or non-planar, which meet in a common point. The polyhedral angle is a special case. In Fig. 16 the solid angle at any point P, subtended by any surface S, is equal numerically to the portion A of the surface of a sphere of unit radius which is cut out by a conical surface with vertex at P and having the boundary of S for base. The *unit solid angle* is the *steradian* and equals the central solid angle which intercepts a spherical area (of any shape) equal to the (radius)^2. The total solid angle about a point equals 4π steradians.

A **Spherical Angle** is the opening between two arcs of great circles drawn on a sphere from the same point (*vertex*), and is measured by the plane angle formed by tangents to its sides at its vertex. If the planes of the great circles are perpendicular, the angle is a *right spherical angle.*

Polyhedrons

A **Polyhedron** is a convex closed surface consisting of parts of four or more planes, called its *faces;* its faces intersect in straight lines, called its *edges;* its edges at points, called its *vertices.*

A **Prism** is a polyhedron of which two faces (the *bases*) are congruent polygons in parallel planes and the other (*lateral*) faces are parallelograms whose planes intersect in the *lateral edges.* Prisms are *triangular, rectangular, quadrangular,* and so on, according as their bases are triangles, rectangles, quadrilaterals, and so on. A *right prism* has its lateral edges perpendicular to its bases. A prism whose bases are parallelograms is a *parallepiped;* if in addition the edges are perpendicular to the bases, it is a *right parallelepiped.* A *rectangular parallelepiped* is a *right* parallelepiped whose bases are rectangles. A *cube* is a parallelepiped whose six faces are squares. A *truncated prism* is that part of a prism included between a base and a section made by a plane oblique to the base. A *right section* of a prism is a section made by a plane which cuts all the lateral edges perpendicularly.

A **Prismatoid** is a polyhedron of which two faces (the *bases*) are polygons in parallel planes and the other (*lateral*) faces are triangles or trapezoids with one side common with one base and the opposite vertex or side common with the other base.

A **Pyramid** is a polyhedron of which one face (the *base*) is a polygon and the other (*lateral*) faces are triangles meeting in a common point called the *vertex* of the pyramid and intersecting one another in its *lateral edges.* Pyramids are *triangular, quadrangular,* and so on, according as their bases are triangles, quadrilaterals, and so on. A *regular pyramid* (or *right pyramid*) has for its base a regular polygon whose center coincides with the foot of the perpendicular dropped from the vertex to the base. A *frustum of a pyramid* is the portion of a pyramid included between its base and a section parallel to the base. If the section is not parallel to the base, a *truncated pyramid* results.

A **Regular Polyhedron** has all faces formed of congruent regular polygons and all polyhedral angles equal. The only regular polyhedrons possible are the five types discussed in the mensuration table, p. 2-58.

A **Tetrahedron** is a polyhedron of four faces. It may be described also as a triangular pyramid, and any one of its four triangular faces may be considered as the base. The four perpendiculars erected at circumcenters of the four faces meet in a point equidistant from all vertices, which is the center of the circumscribed sphere. The four *medians,* joining each vertex with the center of gravity of the opposite face, meet in a point, which is the *center of gravity* of the tetrahedron. This point is three-fourths of the distance from each vertex along a median. The four altitudes meet in a point, called the *orthocenter* of the tetrahedron. The six planes bisecting the six dihedral angles meet in a point equidistant from all faces, this being the center of the inscribed sphere.

Solids Having Curved Surfaces

A **Cylinder** is a solid bounded by two parallel plane surfaces (the *bases*) and a cylindrical *lateral* surface. A *cylindrical surface* is a surface generated by the movement of a straight line (the *generatrix*) which constantly is parallel to a fixed straight line and touches a fixed curve (the *directrix*) not in the plane of the fixed straight line. The generatrix in any position is an *element* of the cylindrical surface. A *circular cylinder* is one having circular bases. A *right cylinder* is one whose elements are perpendicular to its bases. A *truncated cylinder* is the part of a cylinder included between a base and a section made by a plane oblique to the base. A *right section* of a cylinder is a section made by a plane which cuts all the elements perpendicularly.

A **Cone** is a solid bounded by a conic *lateral* surface and a plane (the *base*) which cuts all the elements of the conic surface. A *conic surface* is a surface generated by the movement of a straight line (the *generatrix*) which constantly touches a fixed plane curve (the *directrix*) and passes through a fixed point (the *vertex*) not in the plane of the fixed curve. The generatrix in any position is an *element* of the conic surface. A *circular cone* is one having a circular base. A *right cone* is a circular cone whose center of the base coincides with the foot of the perpendicular dropped from the vertex to the base. A *frustum of a cone* is the portion of a cone included between its base and a section parallel to the base.

A **Sphere** is a solid bounded by a surface all points of which are equidistant from a point within called the *center*. Every plane section of a sphere is a circle. This circle is a *great circle* if its plane passes through the center of the sphere; otherwise, it is a *small circle*. *Poles* of such a circle are the extremities of the diameter of the sphere which is perpendicular to the plane of the circle. Through two points on a spherical surface, not extremities of a diameter, one great circle can be passed. The shortest line that can be drawn on the surface of a sphere between two such points is an arc of a great circle less than a semicircumference joining those points. If two spherical surfaces intersect, their line of intersection is a circle whose plane is perpendicular to the line of centers, and whose center lies on this line.

A **Spherical Sector** is the portion of a sphere generated by the revolution of a circular sector about a diameter of the circle of which the sector is a part. A *hemisphere* is half of a sphere.

A **Spherical Segment** is the portion of a sphere contained between two parallel plane sections (the *bases*), one of which may be tangent to the sphere (in which case there is only one base). The term "segment" also is applied in an analogous manner to various solids of revolution, the planes in such cases being perpendicular to an axis. A *zone* is the portion of a spherical surface included between two parallel planes.

A **Spherical Polygon** is a figure on a spherical surface bounded by three or more arcs of great circles. The sum of the angles of a spherical triangle (polygon of three sides) is greater than two right angles and less than six right angles.

Other Solids appearing in the mensuration table on pp. 2-58 to 2-61, if not sufficiently defined by their figures, may be found discussed in the section on analytic geometry.

28. MENSURATION

Perimeters of similar figures are proportional to their respective linear dimensions, areas to the squares of their linear dimensions, and volumes of similar solids to the cubes of their linear dimensions.

Table 1. Mensuration Formulas

Approximate Decimal Equivalents (for reference):

$\pi = 3.1416$	$\dfrac{1}{\pi} = 0.318$	$\sqrt{2} = 1.414$
$\dfrac{\pi}{2} = 1.5708$	$\dfrac{1}{2\pi} = 0.159$	$\sqrt{3} = 1.732$
$\dfrac{\pi}{4} = 0.7854$	$\dfrac{1}{4\pi} = 0.080$	$\dfrac{1}{\sqrt{2}} = 0.707$
$\dfrac{\pi}{180} = 0.01745$	$\dfrac{180}{\pi} = 57.296$	$\dfrac{1}{\sqrt{3}} = 0.577$
$\dfrac{\pi}{360} = 0.00873$	$\dfrac{360}{\pi} = 114.592$	

2. MATHEMATICS

1a. Plane Rectilinear Figures

Notation. Lines, a, b, c, $\cdots$; angles, α, β, γ, $\cdots$; altitude (perpendicular height), h; side, l; diagonals, d, d_1, $\cdots$; perimeter, p; radius of inscribed circle, r; radius of circumscribed circle, R; area, A.

1. Right Triangle

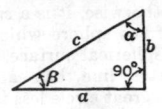

(One angle 90°)

$$p = a + b + c;\ c^2 = a^2 + b^2;$$

$$A = \frac{ab}{2} = \frac{a^2}{2}\tan\beta = \frac{c^2}{4}\sin 2\beta = \frac{c^2}{4}\sin 2\alpha.$$

For additional formulas, see *General Triangle* below, and also trigonometry.

2. General Triangle (and Equilateral Triangle)

For General Triangle:

$$p = a + b + c. \quad \text{Let } s = \tfrac{1}{2}(a + b + c).$$

$$r = \frac{\sqrt{s(s-a)(s-b)(s-c)}}{s}\ ; \quad R = \frac{a}{2\sin\alpha} = \frac{abc}{4rs}\ ;$$

$$A = \frac{ah}{2} = \frac{ab}{2}\sin\gamma = \frac{b^2 \sin\gamma\sin\alpha}{2\sin\beta} = rs = \frac{abc}{4R}.$$

Length of median to side $c = \tfrac{1}{2}\sqrt{2(a^2 + b^2) - c^2}$.

Length of bisector of angle $\gamma = \dfrac{\sqrt{ab[(a+b)^2 - c^2]}}{a+b}$.

For Equilateral Triangle ($a = b = c = l$ and $\alpha = \beta = \gamma = 60°$):

(Equal sides and equal angles)

$$p = 3l,\ r = \frac{l}{2\sqrt{3}}\ ;\quad R = \frac{l}{\sqrt{3}} = 2r;$$

$$h = \frac{l\sqrt{3}}{2}\ ;\ l = \frac{2h}{\sqrt{3}}\ ;\ A = \frac{l^2\sqrt{3}}{4}.$$

For additional formulas, see trigonometry.

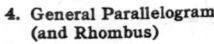

3. Rectangle (and Square)

For Rectangle:

$$p = 2(a + b);\ d = \sqrt{a^2 + b^2};\ A = ab.$$

For Square ($a = b = l$):

$$p = 4l;\ d = l\sqrt{2};\ l = \frac{d}{\sqrt{2}}\ ;\ A = l^2 = \frac{d^2}{2}.$$

4. General Parallelogram (and Rhombus)

For General Parallelogram (Rhomboid):

(Opposite sides parallel)

$$p = 2(a + b);\ d_1 = \sqrt{a^2 + b^2 - 2ab\cos\gamma};$$

$$d_2 = \sqrt{a^2 + b^2 + 2ab\cos\gamma};\ d_1{}^2 + d_2{}^2 = 2(a^2 + b^2);$$

$$A = ah = ab\sin\gamma.$$

For Rhombus ($a = b = l$):

(Opposite sides parallel and all sides equal)

$$p = 4l;\ d_1 = 2l\sin\frac{\gamma}{2}\ ;\ d_2 = 2l\cos\frac{\gamma}{2}\ ;\ d_1{}^2 + d_2{}^2 = 4l^2;$$

$$d_1 d_2 = 2l^2\sin\gamma;\ A = lh = l^2\sin\gamma = \frac{d_1 d_2}{2}.$$

5. General Trapezoid (and Isosceles Trapezoid)

Let mid-line bisecting non-parallel sides $= m$. Then $m = \dfrac{a+b}{2}$.

For General Trapezoid:

(Only one pair of opposite sides parallel)

$$p = a + b + c + d;\ A = \frac{(a+b)h}{2} = mh.$$

For Isosceles Trapezoid ($d = c$):

(Non-parallel sides equal)

$$A = \frac{(a+b)h}{2} = mh = \frac{(a+b)c\sin\gamma}{2}$$

$$= (a - c\cos\gamma)c\sin\gamma = (b + c\cos\gamma)c\sin\gamma.$$

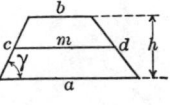

1a. Plane Rectilinear Figures—*Continued*

6. General Quadrilateral (Trapezium)

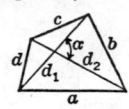

(No sides parallel)

$p = a + b + c + d$

$A = 1/2\, d_1 d_2 \sin \alpha$ = sum of areas of the two triangles formed by either diagonal and the four sides.

7. Quadrilateral Inscribed in Circle

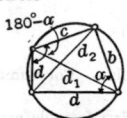

(Sum of opposite angles = 180°)

$ac + bd = d_1 d_2$.

Let $s = 1/2\,(a + b + c + d) = \dfrac{p}{2}$ and α = angle between sides a and b.

$A = \sqrt{(s - a)(s - b)(s - c)(s - d)} = 1/2\,(ab + cd) \sin \alpha$.

8. Regular Polygon (and General Polygon)

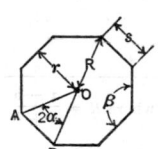

For Regular Polygon:

(Equal sides and equal angles)

Let n = number of sides.

Central angle $= 2\alpha = \dfrac{2\pi}{n}$ radians;

Vertex angle $= \beta = \dfrac{(n - 2)}{n}\,\pi$ radians.

$p = ns$; $s = 2r \tan \alpha = 2R \sin \alpha$;

$r = \dfrac{s}{2} \cot \alpha$; $R = \dfrac{s}{2} \csc \alpha$;

$A = \dfrac{nsr}{2} = nr^2 \tan \alpha = \dfrac{nR^2}{2} \sin 2\alpha = \dfrac{ns^2}{4} \cot \alpha$ = sum of areas of the n equal triangles such as OAB.

For General Polygon:

A = sum of areas of constituent triangles into which it can be divided.

1b. Plane Curvilinear Figures

Notation. Lines, a, b, $\cdots$; radius, r; diameter, d; perimeter, p; circumference, c; central angle n radians, θ; arc, s; chord of arc s, l; chord of half arc $s/2$, l'; rise, h; area, A.

9. Circle (and Circular Arc)

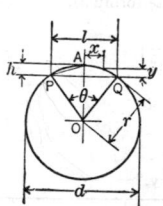

For Circle:

$d = 2r$; $c = 2\pi r = \pi d$; $A = \pi r^2 = \dfrac{\pi d^2}{4} = \dfrac{c^2}{4\pi}$.

For Circular Arc:

Let arc $PAQ = s$; and chord $PA = l'$. Then, $s = r\theta = \dfrac{d\theta}{2}$; $s = \dfrac{8l' - l}{3}$. (The latter equation is Huygens' approximate formula. For θ small; error is very small; for $\theta = 120°$. error is about 0.25%; for $\theta = 180°$, error is less than 1.25%.)

$l = 2r \sin \dfrac{\theta}{2}$; $l = 2\sqrt{2hr - h^2}$ (approximate formula)

$r = \dfrac{s}{\theta} = \dfrac{l}{2 \sin \dfrac{\theta}{2}}$; $r = \dfrac{4h^2 + l^2}{8h}$ (approximate formula)

$h = r \mp \sqrt{r^2 - \dfrac{l^2}{4}}$ ($-$ if $\theta \leq 180°$; $+$ if $\theta \geq 180°$) $= r\left(1 - \cos \dfrac{\theta}{2}\right)$

$= r \,\text{versin}\, \dfrac{\theta}{2} = 2r \sin^2 \dfrac{\theta}{4} = \dfrac{l}{2} \tan \dfrac{\theta}{4} = r + y - \sqrt{r^2 - x^2}$.

Side ordinate $y = h - r + \sqrt{r^2 - x^2}$.

1b.   Plane Curvilinear Figures—*Continued*

10. Circular Sector (and Semicircle)

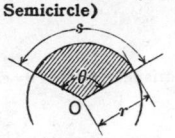

For Circular Sector:

$$A = \frac{\theta r^2}{2} = \frac{sr}{2}.$$

For Semicircle:

$$A = \frac{\pi r^2}{2}.$$

11. Circular Segment

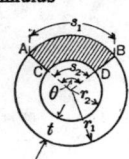

$$A = \frac{r^2}{2}(\theta - \sin \theta)$$

$$= 1/2\,[sr \mp l(r - h)]\,(- \text{ if } h \leqq r;\ + \text{ if } h \geqq r).$$

$$A = \frac{2lh}{3} \text{ or } \frac{h}{15}(8l' + 6l). \quad \text{(Approximate formulas. For } h \text{ small com-}$$

pared with r, error is very small; for $h = \dfrac{r}{4}$, first formula errs about 3.5% and second less than 1.0%.)

12. Annulus

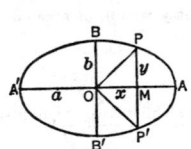

(Region between two concentric circles)

$$A = \pi(r_1{}^2 - r_2{}^2) = \pi(r_1 + r_2)(r_1 - r_2);$$

$$A \text{ of sector } ABCD = \frac{\theta}{2}(r_1{}^2 - r_2{}^2) = \frac{\theta}{2}(r_1 + r_2)(r_1 - r_2)$$

$$= \frac{t}{2}(s_1 + s_2).$$

13. Ellipse

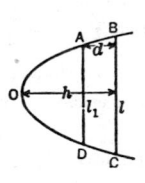

$$p = \pi(a + b)\left(1 + \frac{R^2}{4} + \frac{R^4}{64} + \frac{R^4}{256} + \cdots\right) \text{ where } R = \frac{a - b}{a + b}.$$

$$p = \pi(a + b)\frac{64 - 3R^4}{64 - 16R^2} \text{ (approximate formula).}$$

$$A = \pi ab; \ A \text{ of quadrant } AOB = \frac{\pi ab}{4};$$

$$A \text{ of sector } AOP = \frac{ab}{2}\cos^{-1}\frac{x}{a}; \ A \text{ of sector } POB = \frac{ab}{2}\sin^{-1}\frac{x}{a};$$

$$A \text{ of section } BPP'B' = xy + ab\sin^{-1}\frac{x}{a};$$

$$A \text{ of segment } PAP'P = -xy + ab\cos^{-1}\frac{x}{a}.$$

For additional formulas, see analytic geometry.

14. Parabola

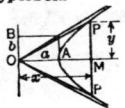

$$\text{Arc } BOC = s = 1/2\sqrt{l^2 + 16h^2} + \frac{l^2}{8h}\log_e\frac{4h + \sqrt{l^2 + 16h^2}}{l}.$$

Let $R = \dfrac{h}{l}.$ Then,

$$s = l\left(1 + \frac{8R^2}{3} - \frac{32R^4}{5} + \cdots\right) \text{ (approximate formula).}$$

$$d = \frac{h}{l^2}(l^2 - l_1{}^2); \ l_1 = l\sqrt{\frac{h - d}{h}}; \ h = \frac{dl^2}{l^2 - l_1{}^2};$$

$$A \text{ of segment } BOC = \frac{2hl}{3};$$

$$A \text{ of section } ABCD = \frac{2}{3}d\left(\frac{l^3 - l_1{}^3}{l^2 - l_1{}^2}\right).$$

For additional formulas, see analytic geometry.

15. Hyperbola

$$A \text{ of figure } OPAP'O = ab\log_e\left(\frac{x}{a} + \frac{y}{b}\right) = ab\cosh^{-1}\frac{x}{a};$$

$$A \text{ of segment } PAP' = xy - ab\log_e\left(\frac{x}{a} + \frac{y}{b}\right) = xy - ab\cosh^{-1}\frac{x}{a}.$$

For additional formulas, see analytic geometry.

1b. Plane Curvilinear Figures—*Continued*

16. Cycloid

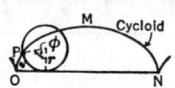

Arc $OP = s = 4r \left(1 - \cos \dfrac{\phi}{2} \right)$; arc $OMN = 8r$;

A under curve $OMN = 3\pi r^2$.

For additional formulas, see analytic geometry.

17. Epicycloid

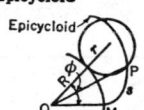

Arc $MP = s = \dfrac{4r}{R} (R + r) \left(1 - \cos \dfrac{R\phi}{2r} \right)$;

Area $MOP = A = \dfrac{r}{2R} (R + r)(R + 2r) \left(\dfrac{R\phi}{r} - \sin \dfrac{R\phi}{r} \right).$

For additional formulas, see analytic geometry.

18. Hypocycloid

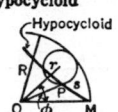

Arc $MP = s = \dfrac{4r}{R} (R - r) \left(1 - \cos \dfrac{R\phi}{2r} \right)$;

Area $MOP = A = \dfrac{r}{2R} (R - r)(R - 2r) \left(\dfrac{R\phi}{r} - \sin \dfrac{R\phi}{r} \right).$

For additional formulas, see analytic geometry.

19. Catenary

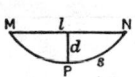

If d is small compared with l:

Arc $MPN = s = l \left[1 + \dfrac{2}{3} \left(\dfrac{2d}{l} \right)^2 \right]$ (approximately).

For additional formulas, see analytic geometry.

20. Helix (a skew curve)

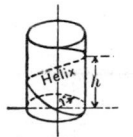

Let length of helix $= s$; radius of coil ($=$ radius of cylinder in figure) $= r$; distance advanced in one revolution $=$ pitch $= h$; and number of revolutions $= n$. Then,

$$s = n\sqrt{(2\pi r)^2 + h^2}.$$

21. Spiral of Archimedes

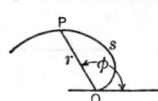

Let $a = \dfrac{r}{\phi}$. Then,

Arc $OP = s = \dfrac{a}{2} [\phi\sqrt{1 + \phi^2} + \log_e (\phi + \sqrt{1 + \phi^2})].$

For additional formulas, see analytic geometry.

22. Irregular Figure

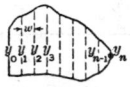

Divide the figure into an *even* number, n, of strips by means of $(n + 1)$ ordinates, y_i, spaced equal distances, w. The area can then be determined approximately by any of the following formulas, which are presented in the order of usual increasing approach to accuracy. In any of the first three cases, the greater the number of strips used, the more nearly accurate will be the result.

(Approximate Formulas)

Trapezoidal Rule......... $A = w \left[\dfrac{y_0 + y_n}{2} + y_1 + y_2 + \cdots + y_{n-1} \right]$;

Durand's Rule............ $A = w[0.4(y_0 + y_n) + 1.1(y_1 + y_{n-1}) + y_2 + y_3 + \cdots + y_{n-2}];$

Simpson's Rule........... $A = \dfrac{w}{3} [(y_0 + y_n) + 4(y_1 + y_3 + \cdots + y_{n-1}) +$
(n *must* be even)

$2(y_2 + y_4 + \cdots + y_{n-2})];$

Weddle's Rule............ $A = \dfrac{3w}{10} [5(y_1 + y_5) + 6y_3 + y_0 + y_2 + y_4 + y_6].$
(for 6 strips only)

Areas of irregular regions can often be determined more quickly by such methods as plotting on squared paper and counting the squares; graphical coordinate representation (see analytic geometry); or use of a planimeter.

1c. Solids Having Plane Surfaces

Notation. Lines, a, b, c, $\cdots$; altitude (perpendicular height), h; slant height, s; perimeter of base, p_b or p_B; perimeter of a right section, p_r; area of base, A_b or A_B; area of a right section, A_r; total area of lateral surfaces, A_l; total area of all surfaces, A_t; volume, V.

---

23. Wedge (and Right Triangular Prism)

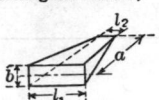

For Wedge:

(Narrow-side rectangular); $V = \dfrac{ab}{6}(2l_1 + l_2)$.

For Right Triangular Prism (or wedge having parallel triangular bases perpendicular to sides) $l_2 = l_1 = l$:

$$V = \frac{abl}{2}.$$

---

24. Rectangular Prism (or Rectangular Parallelepiped) (and Cube)

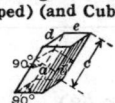

For Rectangular Prism or Rectangular Parallelepiped:

$A_l = 2c(a + b)$; $A_t = 2(de + ac + bc)$;

$V = A_r c = abc$.

For Cube (letting $b = c = a$):

$A_t = 6a^2$; $V = a^3$; Diagonal $= a\sqrt{3}$.

---

25. General Prism

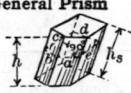

$A_l = hp_b = sp_r = s(a + b + \cdots + n)$;

$V = hA_b = sA_r$.

---

26. General Truncated Prism (and Truncated Triangular Prism)

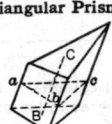

For General Truncated Prism:

$V = A_r \cdot$ (length of line BC joining centers of gravity of bases).

For Truncated Triangular Prism:

$$V = \frac{A_r}{3}(a + b + c).$$

---

27. Prismatoid

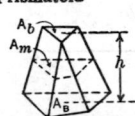

Let area of mid-section $= A_m$.

$$V = \frac{h}{6}(A_B + A_b + 4A_m).$$

---

28. Right Regular Pyramid (and Frustum of Right Regular Pyramid)

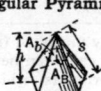

For Right Regular Pyramid:

$$A_l = \frac{spB}{2}; \quad V = \frac{hA_B}{3}.$$

For Frustum of Right Regular Pyramid:

$$A_l = \frac{s}{2}(p_B + p_b); \quad V = \frac{h}{3}(A_B + A_b + \sqrt{A_B A_b}).$$

---

29. General Pyramid (and Frustum of Pyramid)

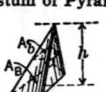

For General Pyramid:

$$V = \frac{hA_B}{3}.$$

For Frustum of General Pyramid:

$$V = \frac{h}{3}(A_B + A_b + \sqrt{A_B A_b}).$$

---

30. Regular Polyhedrons

Tetrahedron   Cube   Octahedron

Dodecahedron   Icosahedron

Let edge $= a$, and radius of inscribed sphere $= r$. Then,

$r = \dfrac{3V}{A_t}$, and:

Number of Faces	Form of Faces	Total Area A_t	Volume V
4	Equilateral triangle	$1.7321a^2$	$0.1179a^3$
6	Square	$6.0000a^2$	$1.0000a^3$
8	Equilateral triangle	$3.4641a^2$	$0.4714a^3$
12	Regular pentagon	$20.6457a^2$	$7.6631a^3$
20	Equilateral triangle	$8.6603a^2$	$2.1817a^3$

(Factors shown only to four decimal places.)

1d. Solids Having Curved Surfaces

Notation. Lines, a, b, c, $\cdots$; altitude (perpendicular height), h, h_1, $\cdots$; slant height, s; radius, r; perimeter of base, p_b; perimeter of a right section, p_r; angle in radians, ϕ; arc, s; chord of segment, l; rise, h; area of base, A_b or A_B; area of a right section, A_r; total area of convex surface, A_l; total area of all surfaces, A_t; volume, V.

---

31. Right Circular Cylinder (and Truncated Right Circular Cylinder)

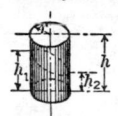

For Right Circular Cylinder:

$A_l = 2\pi rh$; $A_t = 2\pi r(r + h)$;

$V = \pi r^2 h$.

For Truncated Right Circular Cylinder:

$$A_l = \pi r(h_1 + h_2); \ A_t = \pi r\left[h_1 + h_2 + r + \sqrt{r^2 + \left(\frac{h_1 - h_2}{2}\right)^2}\right];$$

$$V = \frac{\pi r^2}{2}(h_1 + h_2).$$

---

32. Ungula (Wedge) of Right Circular Cylinder

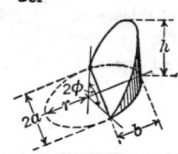

$$A_l = \frac{2rh}{b}[a + (b - r)\phi];$$

$$V = \frac{h}{3b}[a(3r^2 - a^2) + 3r^2(b - r)\phi]$$

$$= \frac{hr^3}{b}\left[\sin\phi - \frac{\sin^3\phi}{3} - \phi\cos\phi\right].$$

For Semicircular Base (letting $a = b = r$):

$$A_l = 2rh; \ V = \frac{2r^2h}{3}.$$

---

33. General Cylinder

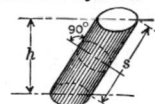

$A_l = p_b h = p_r s$;

$V = A_b h = A_r s$.

---

34. Right Circular Cone (and Frustum of Right Circular Cone)

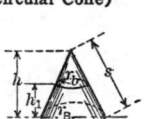

For Right Circular Cone:

$A_l = \pi r_B s = \pi r_B\sqrt{r_B^2 + h^2}$; $A_t = \pi r_B(r_B + s)$;

$V = \frac{\pi r_B^2 h}{3}$.

For Frustum of Right Circular Cone:

$s = \sqrt{h_1^2 + (r_B - r_b)^2}$; $A_l = \pi s(r_B + r_b)$;

$V = \frac{\pi h_1}{3}(r_B^2 + r_b^2 + r_B r_b)$.

---

35. General Cone (and Frustum of General Cone)

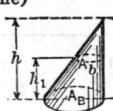

For General Cone:

$$V = \frac{A_B h}{3}.$$

For Frustum of General Cone:

$$V = \frac{h_1}{3}(A_B + A_b + \sqrt{A_B A_b}).$$

---

36. Sphere

Let diameter $= d$.

$A_t = 4\pi r^2 = \pi d^2$;

$V = \frac{4\pi r^3}{3} = \frac{\pi d^3}{6}$.

---

37. Spherical Sector (and Hemisphere)

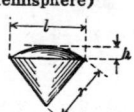

For Spherical Sector:

$$A_t = \frac{\pi r}{2}(4h + l); \ V = \frac{2\pi r^2 h}{3}.$$

For Hemisphere $\left(\text{letting } h = \frac{l}{2} = r\right)$:

$$A_t = 3\pi r^2; \ V = \frac{2\pi r^3}{3}.$$

1d. Solids Having Curved Surfaces—*Continued*

38. Spherical Zone (and Spherical Segment)

For Spherical Zone Bounded by Two Planes:

$$A_l = 2\pi rh; \quad A_t = \frac{\pi}{4}(8rh + a^2 + b^2).$$

For Spherical Zone Bounded by One Plane ($b = 0$):

$$A_l = 2\pi rh = \frac{\pi}{4}(4h^2 + a^2);$$

$$A_t = \frac{\pi}{4}(8rh + a^2) = \frac{\pi}{2}(2h^2 + a^2).$$

For Spherical Segment with Two Bases:

$$V = \frac{\pi h}{24}(3a^2 + 3b^2 + 4h^2).$$

For Spherical Segment with One Base ($b = 0$):

$$V = \frac{\pi h}{24}(3a^2 + 4h^2) = \pi h^2\left(r - \frac{h}{3}\right).$$

---

39. Spherical Polygon (and Spherical Triangle)

For Spherical Polygon:

Let sum of angles in radians $= \theta$ and number of sides $= n$.

$$A = [\theta - (n - 2)\pi]r^2$$

(The quantity $[\theta - (n - 2)\pi]$ is called "spherical excess.")

For Spherical Triangle ($n = 3$):

$$A = (\theta - \pi)r^2$$

For additional formulas, see trigonometry.

---

40. Torus

$$A_l = 4\pi^2 Rr;$$
$$V = 2\pi^2 Rr^2.$$

---

41. Ellipsoid (and Spheroids)

For Ellipsoid:

$$V = \frac{4}{3}\pi abc.$$

For Prolate Spheroid:

Let $c = b$ and $\dfrac{\sqrt{a^2 - b^2}}{a} = e$.

$$A_t = 2\pi b^2 + 2\pi ab\frac{\sin^{-1}e}{e}; \quad V = \frac{4}{3}\pi ab^2.$$

For Oblate Spheroid:

Let $c \doteq a$ and $\dfrac{\sqrt{a^2 - b^2}}{a} = e$.

$$A_t = 2\pi a^2 + \frac{\pi b^2}{e}\ln\left(\frac{1 + e}{1 - e}\right); \quad V = \frac{4}{3}\pi a^2 b.$$

---

42. Paraboloid of Revolution

$$A_l \text{ of segment } DOC = \frac{2\pi l}{3h^2}\left[\left(\frac{l^2}{16} + h^2\right)^{3/2} - \left(\frac{l}{4}\right)^3\right].$$

For Paraboloidal Segment with Two Bases:

$$V \text{ of } ABCD = \frac{\pi d}{8}(l^2 + l_1^2).$$

For Paraboloidal Segment with One Base ($l_1 = 0$ and $d = h$):

$$V \text{ of } DOC = \frac{\pi h l^2}{8}.$$

---

43. Hyperboloid of Revolution

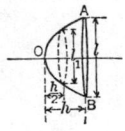

$$V \text{ of segment } AOB = \frac{\pi h}{24}(l^2 + 4l_1^2).$$

1d. Solids Having Curved Surfaces—*Continued*

44. Surface and Solid of Revolution 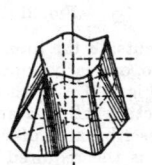 Area = A	Let perpendicular distance from axis to center of gravity (*G*) of curve (or surface) = *r*. Curve (or surface) must not cross axis. Then, *Area of Surface* generated by curve revolving about axis: $A_l = 2\pi rs.$ *Volume of Solid* generated by surface revolving about axis: $V = 2\pi rA.$
45. Irregular Solid	One of the following methods can often be employed to determine the volume of an irregular solid with a reasonable approach to accuracy: (*a*) Divide the solid into prisms, cylinders, etc., and sum their individual volumes. (*b*) Divide one surface into triangles, after replacing curved lines by straight ones and curved surfaces by plane ones. Then multiply the area of each triangle by the mean depth of the section beneath it (which generally approximates the average of the depths at its corners). Sum the volumes thus obtained. (*c*) If two surfaces are parallel, replace any curved lateral surfaces by plane surfaces best suited to the contour and then employ the prismatoidal formula.

29. CONSTRUCTIONS

Lines

1. To draw a line parallel to a given line.
Case 1. At a given distance from the given line (Fig. 17*a*).
With the given distance as radius and with any centers *m* and *n* on the given line *AB*, describe arcs *xy* and *zw*, respectively. Draw *CD* touching these arcs. *CD* is the required parallel line.
Case 2. Through a given point (Fig. 17*b*). Let *C* be the given point and *D* be any point on the given line *AB*. Draw *CD*. With equal radii draw arcs *bf* and *ce* with *D* and *C*, respectively, as centers. With radius equal to chord *bf* and with *c* as center draw an arc cutting arc *ce* at *E*. *CE* is the required parallel line.
2. To bisect a given line (Fig. 18). Let *AB* be the given line. With any radius greater than 0.5 *AB* describe two arcs with *A* and *B* as centers. The line *CD*, through points of intersection of the arcs, is the perpendicular bisector of the given line.

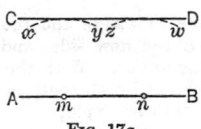

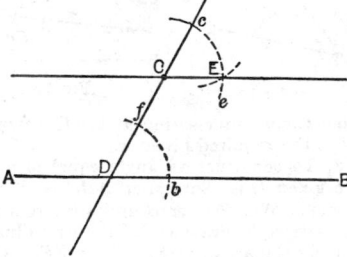

Fig. 17*a*

Fig. 17*b*

3. To divide a given line into a given number of equal parts (Fig. 19). Let *AB* be the given line and let the number of equal parts be five. Draw line *AC* at any convenient angle with *AB*, and step off with dividers five equal lengths from *A* to *b*. Connect *b* with *B*, and draw parallels to *Bb* through the other points in *AC*. The intersections of these parallels with *AB* determine the required equal parts on the given line.
4. To divide a given line into segments proportional to a number of given unequal parts. Follow the same procedure as under 3 above except make the lengths on *AC* equal to (or proportional to) the lengths of the given unequal parts.
5. To erect a perpendicular to a given line at a given point in the line.
Case 1. Point *C* is at or near the middle of the line *AB* (Fig. 20). With *C* as center, describe arcs of equal radii intersecting *AB* at *a* and *b*. With *a* and *b* as centers, and any radius greater than *Ca*, describe arcs intersecting at *D*. *CD* is the required perpendicular.

Case 2. Point C is at or near the extremity of the line AB (Fig. 21). With any point O, as center, and radius OC, describe an arc intersecting AB at a. Extend aO to intersect the arc at D. CD is the required perpendicular.

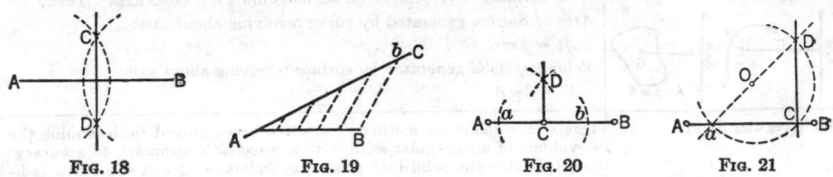

FIG. 18   FIG. 19   FIG. 20   FIG. 21

6. To erect a perpendicular to a given line through a given point outside the line.

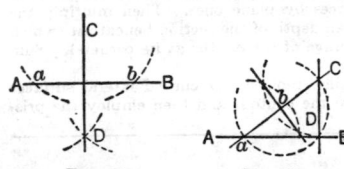

FIG. 22   FIG. 23

Case 1. Point C is opposite, or nearly opposite, the middle of the line AB (Fig. 22). With C as center, describe an arc intersecting AB at a and b. With a and b as centers, describe arcs of equal radii intersecting at D. CD is the required perpendicular.

Case 2. Point C is opposite, or nearly opposite, the extremity of the line AB (Fig. 23). Through C, draw any line intersecting AB at a. Divide line Ca into two equal parts, ab and bC (method given above). With b as center, and radius bC, describe an arc intersecting AB at D. CD is the required perpendicular.

Angles

7. To bisect a given angle.
Case 1. Vertex B is accessible (Fig. 24). Let ABC be the given angle. With B as center, and a large radius, describe an arc intersecting AB and BC at a and c respectively. With a and c as centers, describe arcs of equal radii intersecting at D. DB is the required bisector.
Case 2. The vertex is inaccessible (Fig. 25). Let the given angle be that between lines AB and BC. Draw lines ab and bc parallel to the given lines, and at equal distances

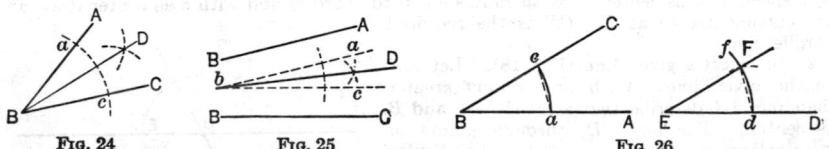

FIG. 24   FIG. 25   FIG. 26

from them, intersecting at b. Construct Db bisecting angle abc (method given above). Db is the required bisector.

8. To construct an angle equal to a given angle if one new side and the new vertex are given (Fig. 26). Let ABC be the given angle; DE the new side; and E the new vertex. With center B and a convenient radius, describe arc ac. With the same radius and center E, draw arc df. With radius equal to chord ac and with center d draw an arc cutting the arc df at F. Draw EF. Then DEF is the required angle.

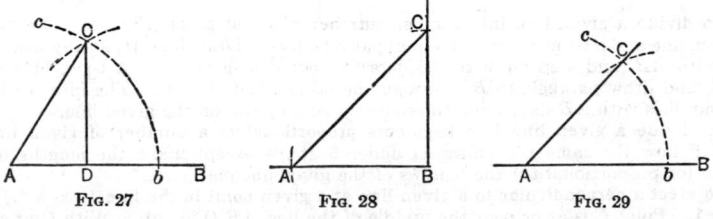

FIG. 27   FIG. 28   FIG. 29

9. To construct angles of 60° and 30° (Fig. 27). About any point A on a line AB, describe with a convenient radius the arc bc. From b, using an equal radius, describe an arc cutting the former one at C. Draw AC, and drop a perpendicular CD from C to line AB. Then CAD is a 60° angle and ACD is a 30° angle.

10. To construct an angle of 45° (Fig. 28). Set off any distance AB; draw BC perpendicular and equal to AB; and join CA. Angles CAB and ACB are each 45°.

11. To draw a line making a given angle with a given line (Fig. 29). Let AB be the given line. With A as the center and with as large a radius as convenient, describe arc bc. Determine from Table 9, Section 1, the length of chord to radius one, corresponding to the given angle. Multiply this chord by the length of Ab, and with the product as a new radius and b as a center, describe an arc cutting bc at C. Draw AC. This line makes the required angle with AB.

Circles

12. To describe through two given points an arc of a circle having a given radius (Fig. 30). Let A and B be the given points. With the given radius, and these points as centers, describe arcs cutting each other at C. From C, with the same radius, describe arc AB, which is the required arc.

13. To bisect a given arc of a circle. Draw the perpendicular bisector of the chord of the arc. The point in which this bisector meets the arc is the required mid-point.

14. To locate the center of a given circle or circular arc (Fig. 31). Select three points, A, B, C, on the circle (or arc), located well apart. Draw chords AB and BC and erect their perpendicular bisectors. The point O, where the bisectors intersect, is the required center.

15. To draw a circle through three given points not in the same straight line.

Case 1. Radius small and center accessible (Fig. 31). Let A, B, C, be the given points. Draw lines AB and BC and erect their perpendicular bisectors. From point O,

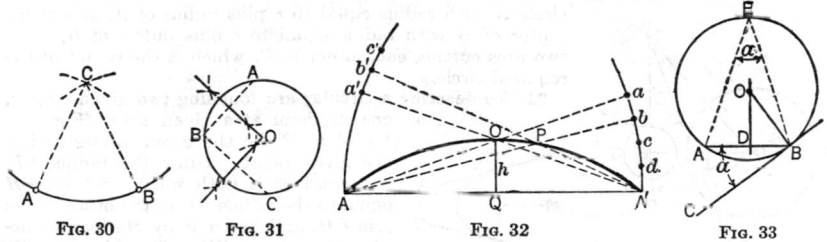

| FIG. 30 | FIG. 31 | FIG. 32 | FIG. 33 |

where the bisectors intersect, describe a circle of radius OA which is the required circle.

Case 2. Radius very long or center inaccessible (Fig. 32). Let A, O, A', be the given points (O not necessarily mid-point of AOA'). Draw arcs Aa' and $A'a$ with centers at A' and A respectively; extend AO to determine a and $A'O$ to determine a'; point off from a on aA' equal parts ab, bc, etc.; lay off $a'b'$, $b'c'$, etc., equal to ab; join A with any point as b and A' with the corresponding point b'; the intersection P of these joining lines is a point on the required circle.

16. To lay out a circular arc without locating the center of the circle, having given the chord and the rise (Fig. 32). Let AA' be the chord and QO the rise. (In this case, O is mid-point of AOA'.) The arc can be constructed through the points A, O, A', as under 15, Case 2, above.

17. To construct, upon a given chord, a circle in which a given angle can be inscribed (Fig. 33). Let AB be the given chord, and α the given angle. Construct angle $A'BC$ equal to angle α. Bisect line AB by the perpendicular at D. Draw a perpendicular to BC from point B. With O, the point of intersection of the perpendiculars, as center, and OB as radius, describe a circle. The angle AEB, with vertex E located anywhere on the arc AEB, equals α, and therefore the circle just drawn is the one required.

18. To draw a tangent to a given circle through a given point.

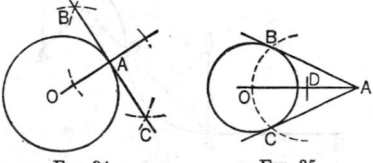

| FIG. 34 | FIG. 35 |

Case 1. Point A is on the circle (Fig. 34). Draw radius OA. Through A, perpendicular to OA, draw BAC, the required tangent.

Case 2. Point A is outside the circle (Fig. 35). Two tangents can be drawn. Join O and A. Bisect OA at D, and with D as center and DO as radius, describe an arc intersecting the given circle at B and C. BA and CA are the required tangents.

19. To draw a common tangent to two given circles. Let the circles have centers O and O' and corresponding radii r and r' $(r > r')$.

Case 1. Common internal tangents (when circles do not intersect) (Fig. 36). Construct a circle having the same center O as the larger circle and a radius equal to the sum of the radii of the given circles $(r + r')$. Construct a tangent $O'P$ from center O' of the smaller circle to this circle. Construct $O'N$ perpendicular to this tangent. Draw OP.

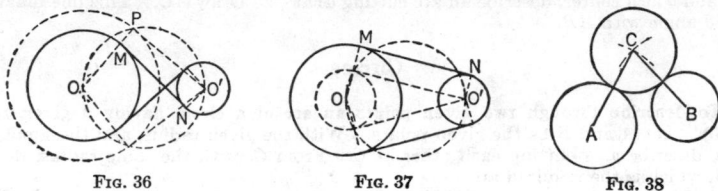

FIG. 36　　　　　　　FIG. 37　　　　　　　FIG. 38

The line MN joining the extremities of the radii OM and $O'N$ is a common tangent. The figure shows two such common internal tangents.

Case 2. Common external tangents (Fig. 37). Construct a circle having the same center O as the larger circle and radius equal to the difference of the radii $(r - r')$. Construct a tangent to this circle from the center of the smaller circle. The line joining the extremities M, N, of the radii of the given circles perpendicular to this tangent is a required common tangent. There are two such tangents.

20. To draw a circle with a given radius that will be tangent to two given circles (Fig. 38). Let r be the given radius and A and B the given circles. About center of circle A with radius equal to r plus radius of A, and about center of B with radius equal to r plus radius of B, draw two arcs cutting each other in C, which is the center of the required circle.

21. To describe a circular arc touching two given circles, one of them at a given point (Fig. 39). Let AB, FG be the given circles and F the given point. Draw the radius EF, and produce it both ways. Set off FH equal to the radius AC of the other circle; join CH, and bisect it by the perpendicular LT, cutting EF at T. About center T, with radius TF, describe arc FA as required.

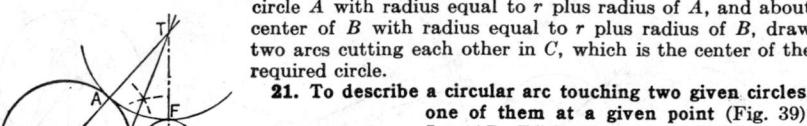

FIG. 39　　　　　　　FIG. 40

22. To draw a circular arc that will be tangent to two given lines inclined to one another, one tangential point being given (Fig. 40). Let AB and CD be the given lines and E the given point. Draw the line GH, bisecting the angle formed by AB and CD. From E draw EF at right angles to AB; then F, its intersection with GH, is the center of the required circular arc.

23. To connect two given parallel lines by a reversed curve composed of two circular arcs of equal radius, the curve being tangent to the lines at given points (Fig. 41). Let AD and BE be the given lines and A and B the given points. Join A and B, and bisect the connecting line at C. Bisect CA and CB by perpendiculars. At A and B erect

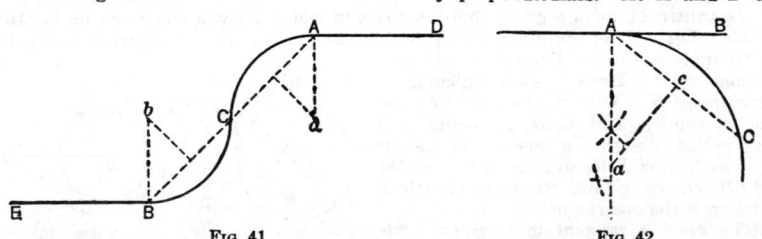

FIG. 41　　　　　　　FIG. 42

perpendiculars to the given lines, and the intersections a and b are the centers of the arcs composing the required curve.

24. To describe a circular arc which will be tangent to a given line at a given point, and pass through another given point outside the line (Fig. 42). Let AB be the given line, A the given point on the line, and C the given point outside it. Draw from A a line perpendicular to the given line. Connect A and C by a straight line, and bisect this line by the perpendicular ca. The point a where these two perpendiculars intersect is the center of the required circular arc.

25. To draw a circular arc joining two given relatively inclined lines, tangent to the lines, and passing through a given point on the line bisecting their included angle (Fig. 43). Let AB and DE be the given lines and F the given point on the line FC which bisects their included angle. Through F draw DA at right angles to FC; bisect the angles A and D by lines intersecting at C, and about C as a center, with radius CF, draw the arc HFG required.

26. To draw a series of circles between two given relatively inclined lines, touching the lines, and touching each other (Fig. 44). Let AB and CD be the given lines. Bisect their included angle by the line NO. From a point P in this line draw the perpendicular PB to the line AB, and on P describe the circle BD, touching the given lines and cutting

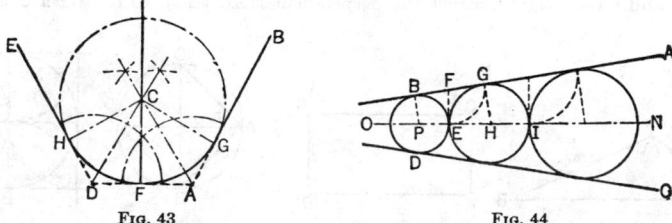

FIG. 43   FIG. 44

the center line at E. From E draw EF perpendicular to the center line, cutting AB at F; and about F as a center describe an arc EG, cutting AB at G. Draw GH parallel to BP, giving H, the center of the next circle, to be described with the radius HE; and so on for the next circle IN.

27. To circumscribe a circle about a given triangle (Fig. 45). Construct perpendicular bisectors of two sides. Their point of intersection O is the center (*circumcenter*) of the required circle.

28. To inscribe a circle in a given triangle (Fig. 46). Draw bisectors of two angles,

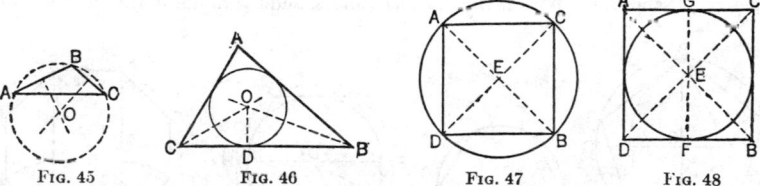

FIG. 45   FIG. 46   FIG. 47   FIG. 48

intersecting in O (*incenter*). From O draw OD perpendicular to BC. Then the circle with center O and radius OD is the required circle.

29. To circumscribe a circle about a given square (Fig. 47). Let $ACBD$ be the given square. Draw diagonals AB and CD of the square, intersecting at E. On center E, with radius AE, describe the required circle. The same procedure can be used for circumscribing a circle about a given rectangle.

30. To inscribe a circle in a given square (Fig. 48). Let $ACBD$ be the given square. Draw diagonals AB and CD of the square, intersecting at E. Drop a perpendicular EF from E to one side. On center E, with radius EF, describe the required circle.

31. To circumscribe a circle about a given regular polygon.

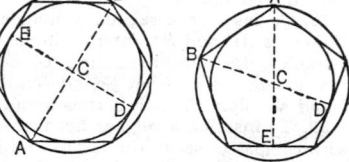

FIG. 49   FIG. 50

Case 1. The polygon has an even number of sides (Fig. 49). Draw a diagonal AB joining two opposite vertices. Bisect the diagonal by a perpendicular line DE, which is another diagonal or a line bisecting two opposite sides, depending upon whether the number of sides is, or is not, divisible by 4. With the midpoint C as the center, and radius CA, describe the required circle.

Case 2. The polygon has an odd number of sides (Fig. 50). Bisect two of the sides at D and E by the perpendicular lines DB and EA which pass through the respective opposite vertices and intersect at a point C. With C as the center, and radius CA, describe the required circle.

32. To inscribe a circle in a given regular polygon (Figs. 49, 50). Locate the center, C, as in 31 above. With C as center, and radius CD, describe the required circle.

Polygons

33. To construct a triangle on a given base, the lengths of the sides being given (Fig. 51). Let AB be the given base and a, b, the given lengths of sides. With A and B as centers, and b and a as respective radii, describe arcs intersecting at C. Draw AC and BC to complete the required triangle.

34. To construct a rectangle of given base and given height (Fig. 52). Let AB be the base and c the height. Erect the perpendicular AC equal to c. With C and B as

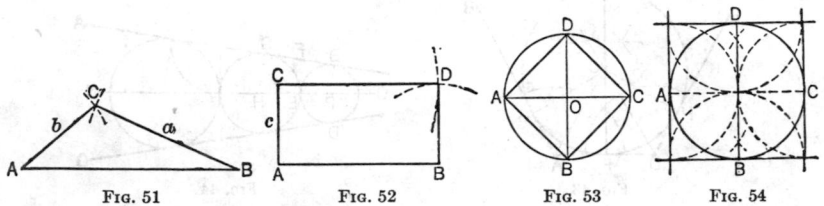

FIG. 51   FIG. 52   FIG. 53   FIG. 54

centers, and AB and c as respective radii, describe arcs intersecting at D. Draw BD and CD to complete the required rectangle.

35. To construct a square with a given diagonal (Fig. 53). Let AC be the given diagonal. Draw a circle on AC as diameter and erect the diameter BD perpendicular to AC. Then $ABCD$ is the required square.

36. To inscribe a square in a given circle (Fig. 53). Draw perpendicular diameters AC and BD. Their extremities are the vertices of an inscribed square.

37. To circumscribe a square about a given circle (Fig. 54). Draw perpendicular diameters AC and BD. With A, B, C, D, as centers, and the radius of the circle as radius,

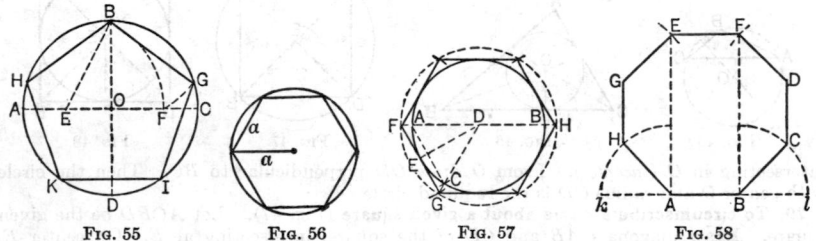

FIG. 55   FIG. 56   FIG. 57   FIG. 58

describe the four semicircular arcs shown. Their outer intersections are the vertices of the required square.

38. To inscribe a regular pentagon in a given circle (Fig. 55). Draw perpendicular diameters AC and BD intersecting at O. Bisect AO at E and with E as center, and EB as radius, draw an arc cutting AC at F. With B as center and BF as radius, draw an arc cutting the circle at G and H; also with the same radius, step around the circle to I and K. Join the points thus found to form the pentagon.

39. To inscribe a regular hexagon in a given circle (Fig. 56). Step around the circle with compasses set to the radius and join consecutive divisions thus marked off.

40. To circumscribe a regular hexagon about a given circle (Fig. 57). Draw a diameter ADB and with center A and radius AD, describe an arc cutting the circle at C. Draw AC and bisect it with the radius DE. Through E, draw FG parallel to AC, cutting diameter AB extended at F. With center D and radius DF, describe the circumscribing circle FH; and within this circle inscribe a regular hexagon as under 39 above. This hexagon circumscribes the given circle, as required.

41. To construct a regular hexagon having a side of given length (Fig. 56). Draw a circle with radius equal to the given length of side and inscribe a regular hexagon (see 39 above).

42. To construct a regular octagon having a side of given length (Fig. 58). Let AB be the given side. Produce AB in both directions, and draw perpendiculars AE and BF. Bisect the external angles at A and B by the lines AH and BC, making them equal to AB. Draw CD and HG parallel to AE, and equal to AB; from the centers G, D, with the radius AB, draw arcs cutting the perpendiculars at E, F, and draw EF to complete the octagon.

43. To inscribe a regular octagon in a given circle (Fig. 59). Draw perpendicular diameters AC and BD. Bisect arcs AB, BC, etc., and join Ae, eB, etc., to form the octagon.

44. To inscribe a regular octagon in a given square (Fig. 60). Draw diagonals of the given square, intersecting at O. With A, B, C, D, as centers, and AO as radius, describe

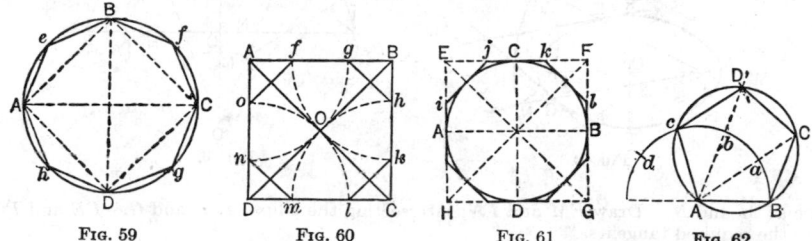

| FIG. 59 | FIG. 60 | FIG. 61 | FIG. 62 |

arcs cutting the sides of the square at gn, fk, hm, and ol. Join the points thus found to form the octagon.

45. To circumscribe a regular octagon about a given circle (Fig. 61). Describe a square about the given circle. Draw perpendiculars ij, kl, etc., to the diagonals of the squares, touching the circle. Then ij, jk, kl, etc., form the octagon.

46. To describe a regular polygon of any given number of sides when one side is given (Fig. 62). Let AB be the given side and let the number of sides be five. Produce the line AB, and with A as center and AB as radius, describe a semicircle. Divide this into as many equal parts as there are to be sides of the polygon—in this case, five. Draw lines from A through the division points a, b, and c (omitting the last). With B and c as centers, and AB as radius, cut Aa at C and Ab at D. Draw cD, DC, and CB, to complete the polygon.

47. To inscribe a regular polygon of a given number of sides in a given circle. Determine the central angle subtended by any side by dividing 360° by the number of sides. Lay off this angle successively round the center of the circle by means of a protractor. The radii thus drawn intersect the circle at vertices of the required polygon.

Ellipse

An **Ellipse** is a curve for which the sum of the distances of any point on it from two fixed points (the *foci*) is constant.

48. To describe an ellipse for which the axes are given (Fig. 63). Let AB be the *major* and RS the *minor* axis $(AB > RS)$. With O as center, and OB and OR as radii, describe circles. From O draw any radial line intersecting the circles at M and N. Through M draw a line parallel to OR, and through N a line parallel to OB. These lines intersect at H, a point on the ellipse. Repeat the construction to obtain other points.

49. To locate the foci of an ellipse, having given the axes (Fig. 63). With R as center, and radius equal to AO, describe arcs intersecting AB at F and F', the required foci.

50. To describe an ellipse mechanically, having given an axis and the foci (Fig. 63). A cord of length equal to the major axis is pinned or fixed at its ends to the foci F and F'. With a pencil inside the loop,

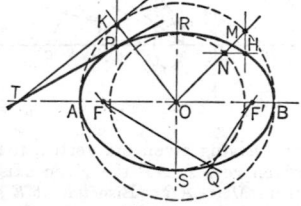

FIG. 63

keeping the cord taut so as to guide the pencil point, trace the outline of the ellipse (Q represents the pencil point and length FQF' the cord). If the minor axis RS is given rather than the major axis AB, the length AB (for the cord) is readily determined as $FR + RF'$.

51. To draw a tangent to a given ellipse through a given point.

Case 1. Point P is on the curve (Fig. 63). With O as center, and OB as radius, describe a circle. Through P draw a line parallel to OR, intersecting the circle at K.

Through K draw a tangent to the circle, intersecting the major axis at T. PT is the required tangent.

Case 2. Point P is not on the curve (Fig. 64). With P as center, and radius PF', describe an arc. With F as center, and radius AB, describe an arc intersecting the first

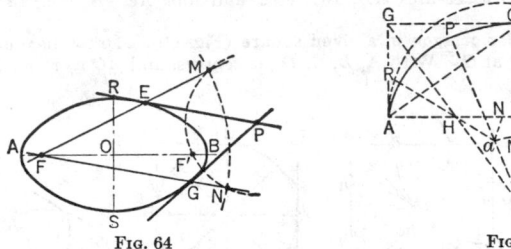

FIG. 64   FIG. 65

arc at M and N. Draw FM and FN, intersecting the ellipse at E and G. PE and PG are the required tangents.

52. To describe an ellipse approximately by means of circular arcs of three radii (Fig. 65). On the major axis AB draw the rectangle BG of altitude equal to half the minor axis, OC; to the diagonal AC draw the perpendicular GHD; set off OK equal to OC, and describe a semicircle on AK; produce OC to L; set off OM equal to CL, and from D describe an arc with radius DM; from A, with radius OL, draw an arc cutting AB at N; from H, with radius HN, draw an arc cutting arc ab at a. Thus the five centers H, a, D, b, H', are found, from which the arcs AR, RP, PQ, QS, SB, are described. The part of the ellipse below axis AB can be constructed in like manner.

Parabola

A **Parabola** is a curve for which the distance of any point on it from a fixed line (the *directrix*) is equal to its distance from a fixed point (the *focus*). For a general discussion of its properties, see the section on analytic geometry.

53. To describe a parabola for which the vertex, the axis, and a point of the curve are given (Fig. 66). Let A be the given vertex, AB the given axis, and M the given point. Construct the rectangle $ABMC$. Divide MC and CA into the same number of equal parts

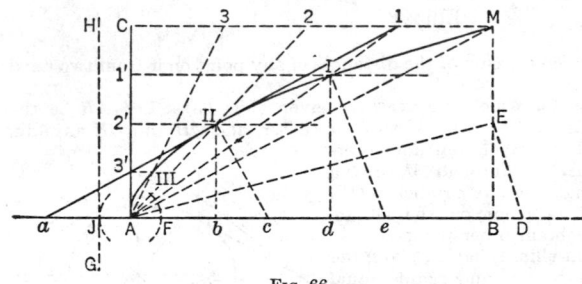

FIG. 66

(say four), numbering the divisions consecutively in the manner shown. Connect $A1$, $A2$, and $A3$. Through $1'$, $2'$, $3'$, draw parallels to the axis AB. The intersections I, II, and III, of these lines are points on the required curve. A similar construction below the axis will give the other symmetric branch of the curve.

54. To locate the focus and directrix of a parabola, having given the vertex, the axis, and a point of the curve (Fig. 66). Let A be the given vertex, AB the given axis, and M the given point. Drop the perpendicular MB from M to AB. Bisect it at E and draw AE. Draw ED perpendicular to AE at E and intersecting the axis at D. With A as center and BD as radius, describe arcs cutting the axis at F and J. Then F is the focus, and the line GH, perpendicular to the axis through J, is the directrix.

55. To describe a parabola mechanically, having given the focus and directrix (Fig. 67). Let F be the given focus and EN the given directrix. Place a straight-edge on the directrix EN, and apply to it a square, LEG. Fasten to the end G one end of a cord equal in length to the edge EG, and attach the other end to the focus F; slide the square along the straight-edge, holding the cord taut against the edge of the square by a pencil D, by which the parabolic curve is described.

56. To draw a tangent to a given parabola through a given point.

Case 1. The point is on the curve (Fig. 66). Let II be the given point. Drop a perpendicular from II to the axis, cutting it at b. Make Aa equal to Ab. Then a line through a and II is the required tangent. The line IIc perpendicular to the tangent at II is the *normal* at that point; bc is the *subnormal*. All subnormals of a given parabola are equal to the distance from the directrix to the focus and hence equal to each other. Thus the subnormal at I is de equal to bc, where d is the foot of the perpendicular dropped from I. The tangent at I can be drawn as a perpendicular to Ie through I.

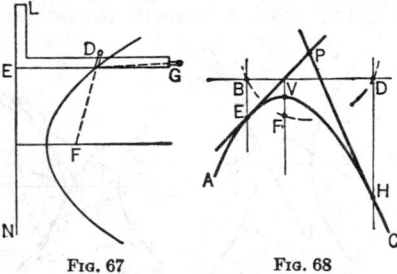

FIG. 67   FIG. 68

Case 2. The point is off the curve (on the convex side) (Fig. 68). Let P be the given point and F the focus of the parabola. With P as center, and PF as radius, draw arcs intersecting the directrix at B and D. Through B and D draw lines parallel to the axis, intersecting the parabola at E and H. PE and PH are the required tangents.

Hyperbola

A **Hyperbola** is a curve for which the difference of the distances of any point on it from two fixed points (the *foci*) is constant. It has two distinct branches.

57. To describe a hyperbola for which the foci and the difference of the focal radii are given (Fig. 69). Let F and F' be the given foci and AOB the given difference of the

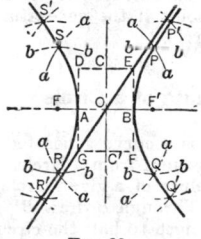

FIG. 69

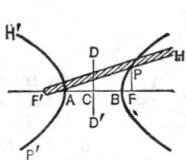

FIG. 70

focal radii. Lay out AOB (the transverse axis) so that $AF = F'B$ and $AO = OB$. A and B are points on the required curve. With centers F and F', and any radius greater than FB or $F'A$, describe arcs aa. With the same centers, and radius equal to the difference between the first radius and the transverse axis AOB, describe arcs bb, intersecting arcs aa at P, Q, R, and S, points on the required curve. Repeat the construction for additional points.

Make $BC = BC' = OF = OF'$, and construct the rectangle $DEFG$; CC' is the conjugate axis. The diagonals DF and EG, produced, are called *asymptotes*. The hyperbola is tangent to its asymptotes at infinity.

58. To locate the foci of a hyperbola, having given the axes (Fig. 69). With O as center and radius equal to BC, describe arcs intersecting AB extended, at F and F', the required foci.

59. To describe a hyperbola mechanically, having given the foci and the difference of the focal radii (Fig. 70). Let F and F' be the given foci and AB the given distance of focal radii. Using a ruler longer than the distance $F'F$, fasten one of its extremities at the focus F'. At the other extremity H attach a cord of such a length that the length of the ruler shall exceed the length of the cord by the given distance AB. Attach the other extremity of the cord at the focus F. Press a pencil P against the ruler, and keep the cord constantly taut while the ruler is turned around F' as a center. The point of the pencil will describe one branch of the curve, and the other can be obtained in like manner.

60. To draw a tangent to a given hyperbola through a given point.

Case 1. Point P is on the curve (Fig. 71). Draw lines connecting P with the foci. Bisect the angle $F'PF$. The bisecting line TP is the required tangent.

Case 2. Point P is off the curve on the convex side (Fig. 72). With P as center and radius PF', describe an arc. With F as center, and radius AB, describe an arc intersecting the first arc at M and N. Produce lines FM and FN to intersect the curve at E and G. PE and PG are the required tangents.

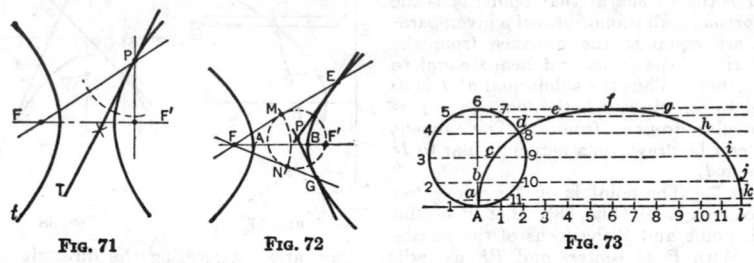

FIG. 71   FIG. 72   FIG. 73

Cycloid

A **Cycloid** is a curve generated by a point on a circle rolling on a straight line.

61. To describe a cycloid for which the generating circle is given (Fig. 73). Let A be the generating point. Divide the circumference of the generating circle into an even number of equal arcs, as $A1$, 1–2, etc., and set off the rectified arcs on the base. Through the points 1, 2, 3, etc., on the circle, draw horizontal lines, and on them set off distances $1a = A1$, $2b = A2$, $3c = A3$, etc. The points A, a, b, c, etc., are points of the cycloid.

An **Epicycloid** is a curve generated by a point on one circle rolling on the *outside* of another circle. A **Hypocycloid** is a curve generated by the point if the generating circle rolls on the *inside* of the second circle.

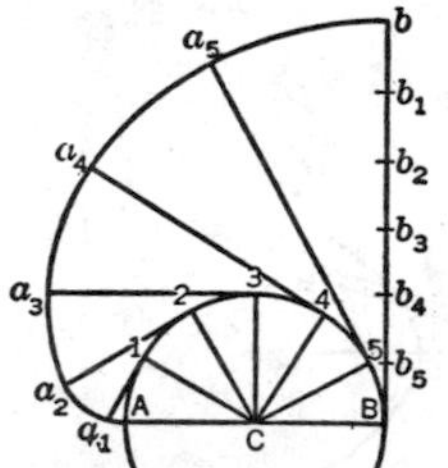

Involute of a Circle

An **Involute of a Circle** is a curve generated by the free end of a taut string as it is unwound from a circle.

62. To describe an involute of a given circle (Fig. 74). Let AB be the given circle. Through B draw Bb perpendicular to AB. Make Bb equal in length to half the circumference of the circle. Divide Bb and the semi-circumference into the same number of equal parts, say six. From each point of division 1, 2, 3, etc., of the circumference, draw lines to the center C of

FIG. 74

the circle. Then draw $1a_1$ perpendicular to $C1$; $2a_2$ perpendicular to $C2$; and so on. Make $1a_1$ equal to bb_1; $2a_2$ equal to bb_2; $3a_3$ equal to bb_3; and so on. Join the points A, a_1, a_2, a_3, etc., by a curve; this curve is the required involute.

TRIGONOMETRY

30. CIRCULAR FUNCTIONS OF PLANE ANGLES

Definitions and Values

Trigonometric Functions. The angle α in Fig. 1 is measured in degrees or radians, as defined in Art. 27. The ratio of any two of the quantities x, y, or r determines the extent of the opening between the lines OP and OX. Since these ratios are functions of the angle they may be used to measure or construct it. The definitions and terms used to designate the functions are as follows:

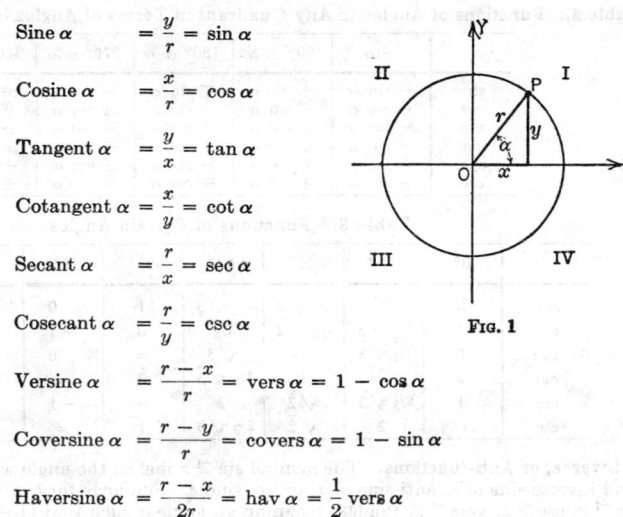

$$\text{Sine } \alpha \quad = \frac{y}{r} = \sin \alpha$$

$$\text{Cosine } \alpha \quad = \frac{x}{r} = \cos \alpha$$

$$\text{Tangent } \alpha \quad = \frac{y}{x} = \tan \alpha$$

$$\text{Cotangent } \alpha = \frac{x}{y} = \cot \alpha$$

$$\text{Secant } \alpha \quad = \frac{r}{x} = \sec \alpha$$

$$\text{Cosecant } \alpha \quad = \frac{r}{y} = \csc \alpha$$

$$\text{Versine } \alpha \quad = \frac{r - x}{r} = \text{vers } \alpha = 1 - \cos \alpha$$

$$\text{Coversine } \alpha = \frac{r - y}{r} = \text{covers } \alpha = 1 - \sin \alpha$$

$$\text{Haversine } \alpha = \frac{r - x}{2r} = \text{hav } \alpha = \frac{1}{2} \text{vers } \alpha$$

FIG. 1

Positive and Negative Values. An angle α (Fig. 1), if measured in a *counterclockwise* direction, is said to be *positive;* if measured *clockwise, negative.* Following the convention that x is positive if measured along OX to the right of the OY axis and negative if measured to the left, and similarly, y is positive if measured along OY above the OX axis and negative if measured below, the signs of the trigonometric functions are different for angles in the quadrants I, II, III, and IV.

Table 1.   Signs of Trigonometric Functions

Quadrant	sin	cos	tan	cot	sec	csc
I	+	+	+	+	+	+
II	+	−	−	−	−	+
III	−	−	+	+	−	−
IV	−	+	−	−	+	−

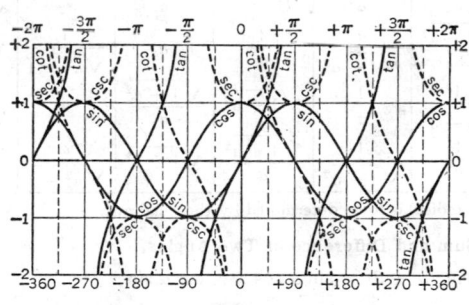

FIG. 2

Values of Trigonometric Functions are periodic, the period of the sin, cos, sec, csc being 2π radians, and that of the tan and cot, π radians. For example, in Fig. 2 (n an integer)

$$\sin (\alpha + 2\pi n) = \sin \alpha$$

$$\tan (\alpha + \pi n) = \tan \alpha$$

Table 2. Functions of Angles in Any Quadrant in Terms of Angles in the First Quadrant

	$-\alpha$	$90° \pm \alpha$	$180° \pm \alpha$	$270° \pm \alpha$	$360° \pm \alpha$
sin	$-\sin \alpha$	$+\cos \alpha$	$\mp \sin \alpha$	$-\cos \alpha$	$\pm \sin \alpha$
cos	$+\cos \alpha$	$\mp \sin \alpha$	$-\cos \alpha$	$\pm \sin \alpha$	$+\cos \alpha$
tan	$-\tan \alpha$	$\mp \cot \alpha$	$\pm \tan \alpha$	$\mp \cot \alpha$	$\pm \tan \alpha$
cot	$-\cot \alpha$	$\mp \tan \alpha$	$\pm \cot \alpha$	$\mp \tan \alpha$	$\pm \cot \alpha$
sec	$+\sec \alpha$	$\mp \csc \alpha$	$-\sec \alpha$	$\pm \csc \alpha$	$+\sec \alpha$
csc	$-\csc \alpha$	$+\sec \alpha$	$\mp \csc \alpha$	$-\sec \alpha$	$\pm \csc \alpha$

Table 3. Functions of Certain Angles

	0°	30°	45°	60°	90°	180°	270°	360°
sin	0	$1/2$	$1/2\sqrt{2}$	$1/2\sqrt{3}$	1	0	-1	0
cos	1	$1/2\sqrt{3}$	$1/2\sqrt{2}$	$1/2$	0	-1	0	1
tan	0	$1/3\sqrt{3}$	1	$\sqrt{3}$	∞	0	∞	0
cot	∞	$\sqrt{3}$	1	$1/3\sqrt{3}$	0	∞	0	∞
sec	1	$2/3\sqrt{3}$	$\sqrt{2}$	2	∞	-1	∞	1
csc	∞	2	$\sqrt{2}$	$2/3\sqrt{3}$	1	∞	-1	∞

Inverse, or Anti-functions. The symbol $\sin^{-1} x$ means the angle whose sine is x, and is read inverse sine of x, anti-sine of x, or arc sine x. Similarly for $\cos^{-1} x$, $\tan^{-1} x$, $\cot^{-1} x$, $\sec^{-1} x$, $\csc^{-1} x$, $\mathrm{vers}^{-1} x$, the last meaning an angle α such that $(1 - \cos \alpha) = x$. While the direct functions (sine, etc.) are single valued, the indirect are many valued; thus $\sin 30° = 0.5$, but $\sin^{-1} 0.5 = 30°,\ 150°,\ \cdots$.

Functional Relations Identities

Table 4. Functions of an Angle in Terms of Each of the Others

	$\sin \alpha = a$	$\cos \alpha = a$	$\tan \alpha = a$	$\cot \alpha = a$	$\sec \alpha = a$	$\csc \alpha = a$
sin	a	$\sqrt{1-a^2}$	$\dfrac{a}{\sqrt{1+a^2}}$	$\dfrac{1}{\sqrt{1+a^2}}$	$\dfrac{\sqrt{a^2-1}}{a}$	$\dfrac{1}{a}$
cos	$\sqrt{1-a^2}$	a	$\dfrac{1}{\sqrt{1+a^2}}$	$\dfrac{a}{\sqrt{1+a^2}}$	$\dfrac{1}{a}$	$\dfrac{\sqrt{a^2-1}}{a}$
tan	$\dfrac{a}{\sqrt{1-a^2}}$	$\dfrac{\sqrt{1-a^2}}{a}$	a	$\dfrac{1}{a}$	$\sqrt{a^2-1}$	$\dfrac{1}{\sqrt{a^2-1}}$
cot	$\dfrac{\sqrt{1-a^2}}{a}$	$\dfrac{a}{\sqrt{1-a^2}}$	$\dfrac{1}{a}$	a	$\dfrac{1}{\sqrt{a^2-1}}$	$\sqrt{a^2-1}$
sec	$\dfrac{1}{\sqrt{1-a^2}}$	$\dfrac{1}{a}$	$\sqrt{1+a^2}$	$\dfrac{\sqrt{1+a^2}}{a}$	a	$\dfrac{a}{\sqrt{a^2-1}}$
csc	$\dfrac{1}{a}$	$\dfrac{1}{\sqrt{1-a^2}}$	$\dfrac{\sqrt{1+a^2}}{a}$	$\sqrt{1+a^2}$	$\dfrac{a}{\sqrt{a^2-1}}$	a

Note: The sign of the radical is to be determined by the quadrant.

Functions of the Sum and Difference of Two Angles.

$$\sin (\alpha \pm \beta) = \sin \alpha \cos \beta \pm \cos \alpha \sin \beta$$
$$\cos (\alpha \pm \beta) = \cos \alpha \cos \beta \mp \sin \alpha \sin \beta$$
$$\tan (\alpha \pm \beta) = (\tan \alpha \pm \tan \beta)/(1 \mp \tan \alpha \tan \beta)$$
$$\cot (\alpha \pm \beta) = (\cot \beta \cot \alpha \mp 1)/(\cot \beta \pm \cot \alpha)$$

If x is small, say 3° or 4°, then the following are close approximations, in which the quantity x is to be expressed in radians (1° = 0.01745 radian).

$$\sin \alpha \approx \alpha, \quad \cos \alpha \approx 1, \quad \tan \alpha \approx \alpha$$
$$\sin (\alpha \pm x) \approx \sin \alpha \pm x \cos \alpha, \quad \cos (\alpha \pm x) \approx \cos \alpha \mp x \sin \alpha$$

Functions of Half Angles.

$$\sin \tfrac{1}{2}\alpha = \sqrt{\tfrac{1}{2}(1 - \cos\alpha)} = \tfrac{1}{2}\sqrt{1 + \sin\alpha} - \tfrac{1}{2}\sqrt{1 - \sin\alpha}$$

$$\cos \tfrac{1}{2}\alpha = \sqrt{\tfrac{1}{2}(1 + \cos\alpha)} = \tfrac{1}{2}\sqrt{1 + \sin\alpha} + \tfrac{1}{2}\sqrt{1 - \sin\alpha}$$

$$\tan \tfrac{1}{2}\alpha = \sqrt{(1 - \cos\alpha)/(1 + \cos\alpha)} = (1 - \cos\alpha)/\sin\alpha = \sin\alpha/(1 + \cos\alpha)$$

$$\cot \tfrac{1}{2}\alpha = \sqrt{(1 + \cos\alpha)/(1 - \cos\alpha)} = (1 + \cos\alpha)/\sin\alpha = \sin\alpha/(1 - \cos\alpha)$$

Functions of Multiples of Angles.

$$\sin 2\alpha = 2 \sin\alpha \cos\alpha$$

$$\tan 2\alpha = 2 \tan\alpha/(1 - \tan^2\alpha)$$

$$\cos 2\alpha = \cos^2\alpha - \sin^2\alpha = 2\cos^2\alpha - 1 = 1 - 2\sin^2\alpha$$

$$\cot 2\alpha = (\cot^2\alpha - 1)/2 \cot\alpha$$

$$\sin 3\alpha = 3 \sin\alpha - 4\sin^3\alpha$$

$$\cos 3\alpha = 4\cos^3\alpha - 3\cos\alpha$$

$$\sin 4\alpha = 8\cos^3\alpha \sin\alpha - 4\cos\alpha \sin\alpha$$

$$\cos 4\alpha = 8\cos^4\alpha - 8\cos^2\alpha + 1$$

$$\sin n\alpha = 2 \sin (n - 1)\alpha \cos\alpha - \sin (n - 2)\alpha$$

$$= n \sin\alpha \cos^{n-1}\alpha - {}_nC_3 \sin^3\alpha \cos^{n-3}\alpha + {}_nC_5 \sin^5\alpha \cos^{n-5}\alpha - \cdots$$

$$\cos n\alpha = 2 \cos (n - 1)\alpha \cos\alpha - \cos (n - 2)\alpha$$

$$= \cos^n\alpha - {}_nC_2 \sin^2\alpha \cos^{n-2}\alpha + {}_nC_4 \sin^4\alpha \cos^{n-4}\alpha - \cdots$$

(For ${}_nC_r$, see p. 2–26.)

Products and Powers of Functions.

$$\sin\alpha \sin\beta = \tfrac{1}{2}\cos (\alpha - \beta) - \tfrac{1}{2}\cos (\alpha + \beta)$$

$$\cos\alpha \cos\beta = \tfrac{1}{2}\cos (\alpha - \beta) + \tfrac{1}{2}\cos (\alpha + \beta)$$

$$\sin\alpha \cos\beta = \tfrac{1}{2}\sin (\alpha - \beta) + \tfrac{1}{2}\sin (\alpha + \beta)$$

$$\tan\alpha \cot\alpha = \sin\alpha \csc\alpha = \cos\alpha \sec\alpha = 1$$

$$\sin^2\alpha = \tfrac{1}{2}(1 - \cos 2\alpha); \quad \cos^2\alpha = \tfrac{1}{2}(1 + \cos 2\alpha)$$

$$\sin^3\alpha = \tfrac{1}{4}(3 \sin\alpha - \sin 3\alpha); \quad \cos^3\alpha = \tfrac{1}{4}(3 \cos\alpha + \cos 3\alpha)$$

$$\sin^4\alpha = \tfrac{1}{8}(3 - 4 \cos 2\alpha + \cos 4\alpha); \quad \cos^4\alpha = \tfrac{1}{8}(3 + 4 \cos 2\alpha + \cos 4\alpha)$$

$$\sin^5\alpha = \tfrac{1}{16}(10 \sin\alpha - 5 \sin 3\alpha + \sin 5\alpha)$$

$$\sin^6\alpha = \tfrac{1}{32}(10 - 15 \cos 2\alpha + 6 \cos 4\alpha - \cos 6\alpha)$$

$$\cos^5\alpha = \tfrac{1}{16}(10 \cos\alpha + 5 \cos 3\alpha + \cos 5\alpha)$$

$$\cos^6\alpha = \tfrac{1}{32}(10 + 15 \cos 2\alpha + 6 \cos 4\alpha + \cos 6\alpha)$$

Sums and Differences of Functions.

$$\sin\alpha + \sin\beta = 2 \sin \tfrac{1}{2}(\alpha + \beta) \cos \tfrac{1}{2}(\alpha - \beta)$$

$$\sin\alpha - \sin\beta = 2 \cos \tfrac{1}{2}(\alpha + \beta) \sin \tfrac{1}{2}(\alpha - \beta)$$

$$\cos\alpha + \cos\beta = 2 \cos \tfrac{1}{2}(\alpha + \beta) \cos \tfrac{1}{2}(\alpha - \beta)$$

$$\cos\alpha - \cos\beta = -2 \sin \tfrac{1}{2}(\alpha + \beta) \sin \tfrac{1}{2}(\alpha - \beta)$$

$$\tan\alpha + \tan\beta = \frac{\sin (\alpha + \beta)}{\cos\alpha \cos\beta}; \quad \cot\alpha + \cot\beta = \frac{\sin (\alpha + \beta)}{\sin\alpha \sin\beta}$$

$$\tan\alpha - \tan\beta = \frac{\sin (\alpha - \beta)}{\cos\alpha \cos\beta}; \quad \cot\alpha - \cot\beta = -\frac{\sin (\alpha - \beta)}{\sin\alpha \sin\beta}$$

$$\sin^2\alpha - \sin^2\beta = \sin (\alpha + \beta) \sin (\alpha - \beta)$$

$$\cos^2\alpha - \cos^2\beta = -\sin (\alpha + \beta) \sin (\alpha - \beta)$$

$$\cos^2\alpha - \sin^2\beta = \cos (\alpha + \beta) \cos (\alpha - \beta)$$

Anti-Trigonometric or Inverse Functional Relations. In the following formulas the periodic constant is omitted.

$$\sin^{-1} x = -\sin^{-1}(-x) = \frac{\pi}{2} - \cos^{-1} x = \cos^{-1}\sqrt{1-x^2} = \tan^{-1}\frac{x}{\sqrt{1-x^2}}$$

$$= \cot^{-1}\frac{\sqrt{1-x^2}}{x} = \csc^{-1}\frac{1}{x} = \sec^{-1}\frac{1}{\sqrt{1-x^2}}$$

$$\cos^{-1} x = \pi - \cos^{-1}(-x) = \frac{\pi}{2} - \sin^{-1} x = \tfrac{1}{2}\cos^{-1}(2x^2-1) = \sin^{-1}\sqrt{1-x^2}$$

$$= \tan^{-1}\frac{\sqrt{1-x^2}}{x} = \cot^{-1}\frac{x}{\sqrt{1-x^2}} = \sec^{-1}\frac{1}{x} = \csc^{-1}\frac{1}{\sqrt{1-x^2}}$$

$$\tan^{-1} x = -\tan^{-1}(-x) = \frac{\pi}{2} - \cot^{-1} x = \sin^{-1}\frac{x}{\sqrt{1+x^2}} = \cos^{-1}\frac{1}{\sqrt{1+x^2}} = \cot^{-1}\frac{1}{x}$$

$$= \sec^{-1}\sqrt{1+x^2} = \csc^{-1}\frac{\sqrt{1+x^2}}{x}$$

$$\cot^{-1} x = \tan^{-1}\frac{1}{x}; \quad \sec^{-1} x = \cos^{-1}\frac{1}{x}; \quad \csc^{-1} x = \sin^{-1}\frac{1}{x}$$

$$\sin^{-1} x \pm \sin^{-1} y = \sin^{-1}\{x\sqrt{1-y^2} \pm y\sqrt{1-x^2}\}$$

$$\cos^{-1} x \pm \cos^{-1} y = \cos^{-1}\{xy \mp \sqrt{(1-x^2)(1-y^2)}\}$$

$$\sin^{-1} x \pm \cos^{-1} y = \sin^{-1}\{xy \pm \sqrt{(1-x^2)(1-y^2)}\} = \cos^{-1}\{y\sqrt{1-x^2} \mp x\sqrt{1-y^2}\}$$

$$\tan^{-1} x \pm \tan^{-1} y = \tan^{-1}\frac{x \pm y}{1 \mp xy}$$

$$\tan^{-1} x \pm \cot^{-1} y = \tan^{-1}\frac{xy \pm 1}{y \mp x} = \cot^{-1}\frac{y \mp x}{xy \pm 1}$$

31. SOLUTION OF TRIANGLES

Relations between Angles and Sides of Plane Triangles. Let a, b, c = sides of triangle; α, β, γ = angles opposite, a, b, c, respectively; A = area of triangle; $s = \tfrac{1}{2}(a+b+c)$; r = radius of inscribed circle (Fig. 3).

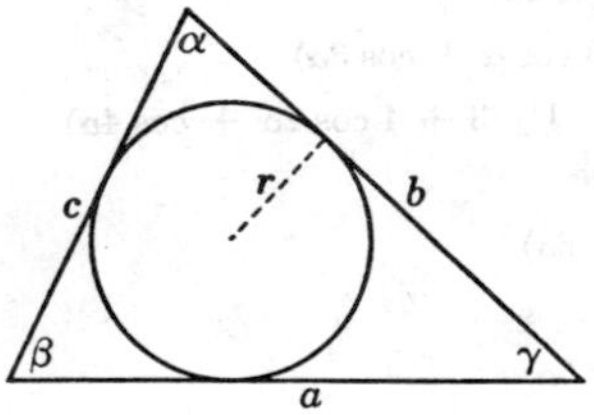

Fig. 3

$$\frac{a}{\sin\alpha} = \frac{b}{\sin\beta} = \frac{c}{\sin\gamma} \quad \text{(Law of Sines)}$$

$$a^2 = b^2 + c^2 - 2bc\cos\alpha \quad \text{(Law of Cosines)}$$

$$\frac{a-b}{a+b} = \frac{\tan\tfrac{1}{2}(\alpha-\beta)}{\tan\tfrac{1}{2}(\alpha+\beta)} \quad \text{(Law of Tangents)}$$

$$\alpha + \beta + \gamma = 180°$$

$$a = b\cos\gamma + c\cos\beta; \quad b = c\cos\alpha + a\cos\gamma; \quad c = a\cos\beta + b\cos\alpha$$

$$A = \sqrt{s(s-a)(s-b)(s-c)}$$

$$\sin\alpha = \frac{2}{bc}A; \quad \sin\beta = \frac{2}{ca}A; \quad \sin\gamma = \frac{2}{ab}A$$

$$\sin\frac{\alpha}{2} = \sqrt{\frac{(s-b)(s-c)}{bc}}; \quad \sin\frac{\beta}{2} = \sqrt{\frac{(s-c)(s-a)}{ca}}; \quad \sin\frac{\gamma}{2} = \sqrt{\frac{(s-a)(s-b)}{ab}}$$

$$\cos\frac{\alpha}{2} = \sqrt{\frac{s(s-a)}{bc}}; \quad \cos\frac{\beta}{2} = \sqrt{\frac{s(s-b)}{ca}}; \quad \cos\frac{\gamma}{2} = \sqrt{\frac{s(s-c)}{ab}}$$

$$\tan\frac{\alpha}{2} = \sqrt{\frac{(s-b)(s-c)}{s(s-a)}}; \quad \tan\frac{\beta}{2} = \sqrt{\frac{(s-c)(s-a)}{s(s-b)}}; \quad \tan\frac{\gamma}{2} = \sqrt{\frac{(s-a)(s-b)}{s(s-c)}}$$

Solution of Plane Oblique Triangles.
Given a, b, c. (If logarithms are to be used, use 1.)

1. $r = \sqrt{\dfrac{(s-a)(s-b)(s-c)}{s}}$; $A = \sqrt{s(s-a)(s-b)(s-c)} = rs$;

$\tan\dfrac{\alpha}{2} = \dfrac{r}{s-a}$; $\tan\dfrac{\beta}{2} = \dfrac{r}{s-b}$; $\tan\dfrac{\gamma}{2} = \dfrac{r}{s-c}$.

2. $\cos\alpha = \dfrac{b^2+c^2-a^2}{2bc}$; $\cos\beta = \dfrac{a^2+c^2-b^2}{2ac}$;

$\cos\gamma = \dfrac{a^2+b^2-c^2}{2ab}$, or $\gamma = 180° - (\alpha + \beta)$.

Given a, b, α.

$\sin\beta = \dfrac{b\sin\alpha}{a}$ (if $a > b$, $\beta < \dfrac{\pi}{2}$ and has only one value; if $b > a$, β has two values,

β_1 and $\beta_2 = 180° - \beta_1$); $\gamma = 180° - (\alpha + \beta)$; $c = \dfrac{a\sin\gamma}{\sin\alpha}$; $A = 1/2\,ab\sin\gamma$.

Given a, α, β.

$b = \dfrac{a\sin\beta}{\sin\alpha}$; $\gamma = 180° - (\alpha + \beta)$; $c = \dfrac{a\sin\gamma}{\sin\alpha}$; $A = 1/2\,ab\sin\gamma$.

Given a, b, γ. (If logarithms are to be used, use 1.)

1. $\tan 1/2\,(\alpha - \beta) = \dfrac{a-b}{a+b}\cot 1/2\,\gamma$; $1/2\,(\alpha + \beta) = 90° - 1/2\,\gamma$; $c = \dfrac{a\sin\gamma}{\sin\alpha}$;

$A = 1/2\,ab\sin\gamma$.

2. $c = \sqrt{a^2 + b^2 - 2ab\cos\gamma}$; $\sin\alpha = \dfrac{a\sin\gamma}{c}$; $\beta = 180° - (\alpha + \gamma)$.

3. $\tan\alpha = \dfrac{a\sin\gamma}{b - a\cos\gamma}$; $\beta = 180° - (\alpha + \gamma)$; $c = \dfrac{a\sin\gamma}{\sin\alpha}$.

Mollweide's Check Formulas.

1. $\dfrac{a-b}{c} = \dfrac{\sin 1/2\,(\alpha - \beta)}{\cos 1/2\,\gamma}$.

2. $\dfrac{a+b}{c} = \dfrac{\cos 1/2\,(\alpha - \beta)}{\sin 1/2\,\gamma}$.

Solution of Plane Right Triangles. Let $\gamma = 90°$ and c be the hypotenuse. Given any two sides or one side and an acute angle α.

$$a = \sqrt{c^2 - b^2} = \sqrt{(c+b)(c-b)} = b\tan\alpha = c\sin\alpha$$

$$b = \sqrt{c^2 - a^2} = \sqrt{(c+a)(c-a)} = \dfrac{a}{\tan\alpha} = c\cos\alpha$$

$$c = \sqrt{a^2 + b^2} = \dfrac{a}{\sin\alpha} = \dfrac{b}{\cos\alpha}$$

$$\alpha = \sin^{-1}\dfrac{a}{c} = \cos^{-1}\dfrac{b}{c} = \tan^{-1}\dfrac{a}{b}\,; \beta = 90° - \alpha$$

$$A = \dfrac{ab}{2} = \dfrac{a^2}{2\tan\alpha} = \dfrac{b^2\tan\alpha}{2} = \dfrac{c^2\sin 2\alpha}{4}$$

32. SPHERICAL TRIGONOMETRY

Spherical Trigonometry. Let O be the center of the sphere and a, b, c the sides of a triangle on the surface with opposite angles α, β, γ, respectively, the sides being measured by the angle subtended at the center of the sphere. Let $s = 1/2\,(a + b + c)$,

$\sigma = \frac{1}{2}(\alpha + \beta + \gamma)$, $E = \alpha + \beta + \gamma - 180°$, the spherical excess. The following formulas are valid usually only for triangles of which the sides and angles are all between $0°$ and $180°$. To each such triangle there is a polar triangle, whose sides are $180° - \alpha$, $180° - \beta$, $180° - \gamma$, and whose angles are $180° - a$, $180° - b$, $180° - c$.

General Formulas.

$$\frac{\sin a}{\sin \alpha} = \frac{\sin b}{\sin \beta} = \frac{\sin c}{\sin \gamma} \quad \text{(Law of Sines)}$$

$$\cos a = \cos b \cos c + \sin b \sin c \cos \alpha \quad \text{(Law of Cosines)}$$

$$\cos \alpha = - \cos \beta \cos \gamma + \sin \beta \sin \gamma \cos a \quad \text{(Law of Cosines)}$$

$$\cos a \sin b = \sin a \cos b \cos \gamma + \sin c \cos \alpha$$

$$\cot a \sin b = \sin \gamma \cot \alpha + \cos \gamma \cos b$$

$$\cos \alpha \sin \beta = \sin \gamma \cos a - \sin \alpha \cos \beta \cos c$$

$$\cot \alpha \sin \beta = \sin c \cot a - \cos c \cos \beta$$

$$\sin \frac{a}{2} = \sqrt{\frac{- \cos \sigma \cos (\sigma - \alpha)}{\sin \beta \sin \gamma}} \; ; \quad \sin \frac{\alpha}{2} = \sqrt{\frac{\sin (s - b) \sin (s - c)}{\sin b \sin c}}$$

$$\cos \frac{a}{2} = \sqrt{\frac{\cos (\sigma - \beta) \cos (\sigma - \gamma)}{\sin \beta \sin \gamma}} \; ; \quad \cos \frac{\alpha}{2} = \sqrt{\frac{\sin s \sin (s - a)}{\sin b \sin c}}$$

$$\tan \frac{a}{2} = \sqrt{\frac{- \cos \sigma \cos (\sigma - \alpha)}{\cos (\sigma - \beta) \cos (\sigma - \gamma)}} \; ; \quad \tan \frac{\alpha}{2} = \sqrt{\frac{\sin (s - b) \sin (s - c)}{\sin s \sin (s - a)}}$$

$$\tan \frac{E}{4} = \sqrt{\tan \frac{s}{2} \tan \frac{(s - a)}{2} \tan \frac{(s - b)}{2} \tan \frac{(s - c)}{2}} \; ; \quad \cot \frac{E}{2} = \frac{\cot \frac{a}{2} \cot \frac{b}{2} + \cos \gamma}{\sin \gamma}$$

$$\tan \left(\frac{a + b}{2}\right) = \frac{\cos \left(\frac{\alpha - \beta}{2}\right)}{\cos \left(\frac{\alpha + \beta}{2}\right)} \tan \frac{c}{2} \; ; \quad \tan \left(\frac{a - b}{2}\right) = \frac{\sin \left(\frac{\alpha - \beta}{2}\right)}{\sin \left(\frac{\alpha + \beta}{2}\right)} \tan \frac{c}{2}$$

$$\tan \left(\frac{\alpha + \beta}{2}\right) = \frac{\cos \left(\frac{a - b}{2}\right)}{\cos \left(\frac{a + b}{2}\right)} \cot \frac{\gamma}{2} \; ; \quad \tan \left(\frac{\alpha - \beta}{2}\right) = \frac{\sin \left(\frac{a - b}{2}\right)}{\sin \left(\frac{a + b}{2}\right)} \cot \frac{\gamma}{2}$$

$$\cos \left(\frac{\alpha + \beta}{2}\right) \cos \frac{c}{2} = \cos \left(\frac{a + b}{2}\right) \sin \frac{\gamma}{2} \; ; \quad \sin \left(\frac{\alpha + \beta}{2}\right) \cos \frac{c}{2} = \cos \left(\frac{a - b}{2}\right) \cos \frac{\gamma}{2}$$

$$\cos \left(\frac{\alpha - \beta}{2}\right) \sin \frac{c}{2} = \sin \left(\frac{a + b}{2}\right) \sin \frac{\gamma}{2} \; ; \quad \sin \left(\frac{\alpha - \beta}{2}\right) \sin \frac{c}{2} = \sin \left(\frac{a - b}{2}\right) \cos \frac{\gamma}{2}$$

The Right Spherical Triangle.

Let $\gamma = 90°$ and c be the hypotenuse.

$$\cos c = \cos a \cos b = \cot \alpha \cot \beta; \quad \cos a = \frac{\cos \alpha}{\sin \beta} \; ; \quad \cos b = \frac{\cos \beta}{\sin \alpha} \; ; \quad \sin \alpha = \frac{\sin a}{\sin c}$$

$$\cos \alpha = \frac{\tan b}{\tan c} \; ; \quad \tan \alpha = \frac{\tan a}{\sin b}$$

33. HYPERBOLIC TRIGONOMETRY

Hyperbolic Angles are defined in a manner similar to circular angles but with reference to an *equilateral hyperbola*. The comparative relations are shown in Figs. 4 and 5. A *circular angle* is a central angle measured in radians by the ratio s/r or the ratio $2A/r^2$, where A is the area of the sector included by the angle α and the arc s (Fig. 4). For the *hyperbola* the radius ρ is not constant and only the value of the *differential hyperbolic angle* $d\theta$ is defined by the ratio ds/ρ. Thus, $\theta = \int ds/\rho = 2A/a^2$, where A represents the shaded area in Fig. 5. If both s and ρ are measured in the same units the angle is expressed in *hyperbolic radians*.

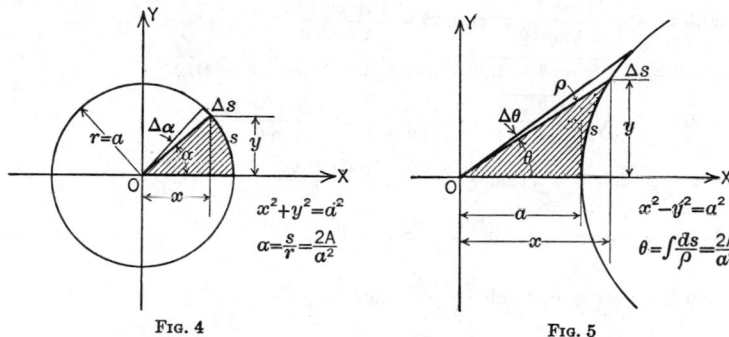

$$x^2 + y^2 = a^2$$
$$\alpha = \frac{s}{r} = \frac{2A}{a^2}$$

$$x^2 - y^2 = a^2$$
$$\theta = \int \frac{ds}{\rho} = \frac{2A}{a^2}$$

FIG. 4   FIG. 5

Hyperbolic Functions are defined by ratios similar to those defining functions of circular angles and also named similarly. Their names and abbreviations are:

$$\text{Hyperbolic sine } \theta \quad = \frac{y}{a} = \sinh \theta$$

$$\text{Hyperbolic cosine } \theta \quad = \frac{x}{a} = \cosh \theta$$

$$\text{Hyperbolic tangent } \theta \quad = \frac{y}{x} = \tanh \theta$$

$$\text{Hyperbolic cotangent } \theta = \frac{x}{y} = \coth \theta$$

$$\text{Hyperbolic secant } \theta \quad = \frac{a}{x} = \operatorname{sech} \theta$$

$$\text{Hyperbolic cosecant } \theta \quad = \frac{a}{y} = \operatorname{csch} \theta$$

Values and Exponential Equivalents. The values of hyperbolic functions may be computed from their exponential equivalents. The graphs are shown in Fig. 6. Values for increments of 0.01 radian are given in Section 1, Table 18.

$$\sinh \theta = \frac{e^\theta - e^{-\theta}}{2} ; \quad \cosh \theta = \frac{e^\theta + e^{-\theta}}{2} ; \quad \tanh \theta = \frac{e^\theta - e^{-\theta}}{e^\theta + e^{-\theta}}$$

If θ is extremely small, $\sinh \theta \approx \theta$, $\cosh \theta \approx 1$, and $\tanh \theta \approx \theta$. For large values of θ, $\sinh \theta \approx \cosh \theta$, and $\tanh \theta \approx \coth \theta \approx 1$.

Fundamental Identities.

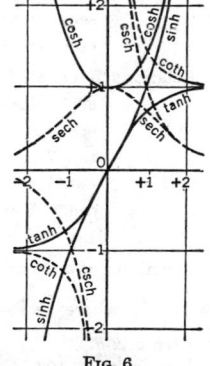

$$\operatorname{csch} \theta = \frac{1}{\sinh \theta} ; \quad \operatorname{sech} \theta = \frac{1}{\cosh \theta} ; \quad \coth \theta = \frac{1}{\tanh \theta}$$

$$\cosh^2 \theta - \sinh^2 \theta = 1; \quad \operatorname{sech}^2 \theta = 1 - \tanh^2 \theta;$$
$$\operatorname{csch}^2 \theta = \coth^2 \theta - 1$$

$$\cosh \theta + \sinh \theta = e^\theta; \quad \cosh \theta - \sinh \theta = e^{-\theta}$$

$$\sinh (-\theta) = -\sinh \theta; \quad \cosh (-\theta) = \cosh \theta$$

$$\tanh (-\theta) = -\tanh \theta; \quad \coth (-\theta) = -\coth \theta$$

$$\sinh (\theta_1 \pm \theta_2) = \sinh \theta_1 \cosh \theta_2 \pm \cosh \theta_1 \sinh \theta_2$$

$$\cosh (\theta_1 \pm \theta_2) = \cosh \theta_1 \cosh \theta_2 \pm \sinh \theta_1 \sinh \theta_2$$

$$\tanh (\theta_1 \pm \theta_2) = \frac{\tanh \theta_1 \pm \tanh \theta_2}{1 \pm \tanh \theta_1 \tanh \theta_2} ; \quad \coth (\theta_1 \pm \theta_2) = \frac{1 \pm \coth \theta_1 \coth \theta_2}{\coth \theta_1 \pm \coth \theta_2}$$

$$\sinh 2\theta = 2 \sinh \theta \cosh \theta = \frac{2 \tanh \theta}{1 - \tanh^2 \theta}$$

FIG. 6

$$\cosh 2\theta = \sinh^2 \theta + \cosh^2 \theta = 1 + 2\sinh^2 \theta = 2\cosh^2 \theta - 1 = \frac{1 + \tanh^2 \theta}{1 - \tanh^2 \theta}$$

$$\tanh 2\theta = \frac{2\tanh \theta}{1 + \tanh^2 \theta}; \quad \coth 2\theta = \frac{1 + \coth^2 \theta}{2\coth \theta}$$

$$\sinh \theta/2 = \sqrt{(\cosh \theta - 1)/2}; \quad \cosh \theta/2 = \sqrt{(\cosh \theta + 1)/2}$$

$$\tanh \theta/2 = \sqrt{\frac{\cosh \theta - 1}{\cosh \theta + 1}} = \frac{\sinh \theta}{\cosh \theta + 1} = \frac{\cosh \theta - 1}{\sinh \theta}$$

$$\sinh \theta_1 \pm \sinh \theta_2 = 2\sinh \frac{(\theta_1 \pm \theta_2)}{2}\cosh \frac{(\theta_1 \mp \theta_2)}{2}$$

$$\cosh \theta_1 + \cosh \theta_2 = 2\cosh \frac{(\theta_1 + \theta_2)}{2}\cosh \frac{(\theta_1 - \theta_2)}{2}$$

$$\cosh \theta_1 - \cosh \theta_2 = 2\sinh \frac{(\theta_1 + \theta_2)}{2}\sinh \frac{(\theta_1 - \theta_2)}{2}$$

$$\tanh \theta_1 \pm \tanh \theta_2 = \frac{\sinh (\theta_1 \pm \theta_2)}{\cosh \theta_1 \cosh \theta_2}$$

$$(\cosh \theta \pm \sinh \theta)^n = \cosh n\theta \pm \sinh n\theta$$

Anti-hyperbolic or Inverse Functions. The inverse hyperbolic sine of u is written: $\sinh^{-1} u$. Values of the inverse functions may be computed from their logarithmic equivalents.

$$\sinh^{-1} u = \log_e (u + \sqrt{u^2 + 1}); \quad \cosh^{-1} u = \log_e (u + \sqrt{u^2 - 1})$$

$$\tanh^{-1} u = \tfrac{1}{2}\log_e \frac{1 + u}{1 - u}; \quad \coth^{-1} u = \tfrac{1}{2}\log_e \frac{u + 1}{u - 1}$$

34. FUNCTIONS OF IMAGINARY AND COMPLEX ANGLES

Relations of Hyperbolic to Circular Functions. By comparison of the exponential equivalents of hyperbolic and circular functions the following identities are established $(i = \sqrt{-1})$:

$\sin \alpha = -i\sinh i\alpha$	$\sinh \beta = -i\sin i\beta$
$\cos \alpha = \cosh i\alpha$	$\cosh \beta = \cos i\beta$
$\tan \alpha = -i\tanh i\alpha$	$\tanh \beta = -i\tan i\beta$
$\cot \alpha = i\coth i\alpha$	$\coth \beta = i\cot i\beta$
$\sec \alpha = \text{sech}\, i\alpha$	$\text{sech}\, \beta = \sec i\beta$
$\csc \alpha = i\,\text{csch}\, i\alpha$	$\text{csch}\, \beta = i\csc i\beta$

Relations between Inverse Functions.

$\sin^{-1} A = -i\sinh^{-1} iA$	$\sinh^{-1} B = -i\sin^{-1} iB$
$\cos^{-1} A = -i\cosh^{-1} A$	$\cosh^{-1} B = i\cos^{-1} B$
$\tan^{-1} A = -i\tanh^{-1} iA$	$\tanh^{-1} B = -i\tan^{-1} iB$
$\cot^{-1} A = i\coth^{-1} iA$	$\coth^{-1} B = i\cot^{-1} iB$
$\sec^{-1} A = -i\,\text{sech}^{-1} A$	$\text{sech}^{-1} B = i\sec^{-1} B$
$\csc^{-1} A = i\,\text{csch}^{-1} iA$	$\text{csch}^{-1} B = i\csc^{-1} iB$

Functions of a Complex Angle. In complex notation $c = a + ib = |c|(\cos \theta + i\sin \theta)$ $= |c|e^{i\theta}$, where $|c| = \sqrt{a^2 + b^2}$, $i = \sqrt{-1}$, and $\theta = \tan^{-1}\frac{b}{a}$. $|c|e^{i\theta}$ is frequently written $c \angle \theta$.

$\text{Log}_e |c|e^{i\theta} = \log|c| + i(\theta + 2k\pi)$ and is infinitely many valued. By its principal part will be understood $\log_e |c| + i\theta$. Some convenient identities are:

$$\log_e 1 = 0; \quad \log_e (-1) = i\pi; \quad \log_e i = i\frac{\pi}{2}; \quad \log_e (-i) = i\frac{3\pi}{2}$$

$$(\cos \theta \pm i\sin \theta)^n = \cos n\theta \pm i\sin n\theta; \quad \sqrt[n]{\cos \theta \pm i\sin \theta} = \cos \frac{\theta + 2\pi k}{n} \pm i\sin \frac{\theta + 2\pi k}{n}$$

The use of complex angles occurs frequently in electric circuit problems where it is often necessary to express the functions of them as a complex number.

$$\sin (\alpha \pm i\beta) = \sin \alpha \cosh \beta \pm i \cos \alpha \sinh \beta = \sqrt{\cosh^2 \beta - \cos^2 \alpha}\, e^{\pm i\theta}$$

where $\theta = \tan^{-1} \cot \alpha \tanh \beta$.

$$\cos (\alpha \pm i\beta) = \cos \alpha \cosh \beta \mp i \sin \alpha \sinh \beta = \sqrt{\cosh^2 \beta - \sin^2 \alpha}\, e^{\pm i\nu}$$

where $\theta = \tan^{-1} \tan \alpha \tanh \beta$.

$$\sinh (\alpha \pm i\beta) = \sinh \alpha \cos \beta \pm i \cosh \alpha \sin \beta$$
$$= \sqrt{\sinh^2 \alpha + \sin^2 \beta}\, e^{\pm i\theta} = \sqrt{\cosh^2 \alpha - \cos^2 \beta}\, e^{\pm i\theta}$$

where $\theta = \tan^{-1} \coth \alpha \tan \beta$.

$$\cosh (\alpha \pm i\beta) = \cosh \alpha \cos \beta \pm i \sinh \alpha \sin \beta$$
$$= \sqrt{\sinh^2 \alpha + \cos^2 \beta}\, e^{\pm i\theta} = \sqrt{\cosh^2 \alpha - \sin^2 \beta}\, e^{\pm i\theta}$$

where $\theta = \tan^{-1} \tanh \alpha \tan \beta$.

$$\tan (\alpha \pm i\beta) = \frac{\sin 2\alpha \pm i \sinh 2\beta}{\cos 2\alpha + \cosh 2\beta}\; ; \quad \tanh (\alpha \pm i\beta) = \frac{\sinh 2\alpha \pm i \sin 2\beta}{\cosh 2\alpha + \cos 2\beta}$$

The hyperbolic sine and cosine have the period $2\pi i$; the hyperbolic tangent has the period πi.

$$\sinh (\alpha + 2k\pi i) = \sinh \alpha; \quad \cosh (\alpha + 2k\pi i) = \cosh \alpha$$
$$\tanh (\alpha + k\pi i) = \tanh \alpha; \quad \coth (\alpha + k\pi i) = \coth \alpha$$

Inverse Functions of Complex Numbers.

$$\sin^{-1} (A \pm iB) = \sin^{-1} \left[\frac{\sqrt{B^2 + (1 + A)^2} - \sqrt{B^2 + (1 - A)^2}}{2} \right]$$

$$\pm i \cosh^{-1} \left[\frac{\sqrt{B^2 + (1 + A)^2} + \sqrt{B^2 + (1 - A)^2}}{2} \right]$$

$$\cos^{-1} (A \pm iB) = \cos^{-1} \left[\frac{\sqrt{B^2 + (1 + A)^2} - \sqrt{B^2 + (1 - A)^2}}{2} \right]$$

$$\mp i \cosh^{-1} \left[\frac{\sqrt{B^2 + (1 + A)^2} + \sqrt{B^2 + (1 - A)^2}}{2} \right]$$

$$\tan^{-1} (A \pm iB) = \left[\frac{\pi - \tan^{-1}\dfrac{A}{\pm B - 1} + \tan^{-1}\dfrac{A}{\pm B + 1}}{2} \right]$$

$$\pm i\, {}^1\!/_4 \log_e \frac{A^2 + (1 \pm B)^2}{A^2 + (1 \mp B)^2}$$

$$\sinh^{-1} (A \pm iB) = \cosh^{-1} \left[\frac{\sqrt{A^2 + (1 + B)^2} + \sqrt{A^2 + (1 - B)^2}}{2} \right]$$

$$\pm i \sin^{-1} \left[\frac{\sqrt{A^2 + (1 + B)^2} - \sqrt{A^2 + (1 - B)^2}}{2} \right]$$

$$\cosh^{-1} (A \pm iB) = \cosh^{-1} \left[\frac{\sqrt{B^2 + (1 + A)^2} + \sqrt{B^2 + (1 - A)^2}}{2} \right]$$

$$\pm i \cos^{-1} \left[\frac{\sqrt{B^2 + (1 + A)^2} - \sqrt{B^2 + (1 - A)^2}}{2} \right]$$

$$\tanh^{-1} (A \pm iB) = {}^1\!/_2 \tanh^{-1} \frac{2A}{1 + A^2 + B^2} + i\, {}^1\!/_2 \tan^{-1} \frac{\pm 2B}{1 - A^2 - B^2}$$

PLANE ANALYTIC GEOMETRY

35. POINT AND LINE

Coordinates. The position of a point P_1 in a plane is determined if its distance and direction from each of two lines or axes OX and OY, which are perpendicular to each other, are known. The distances x and y (Fig. 1) perpendicular to the axes are called the *cartesian* or *rectangular coordinates* of the point. The directions to the right of OY and

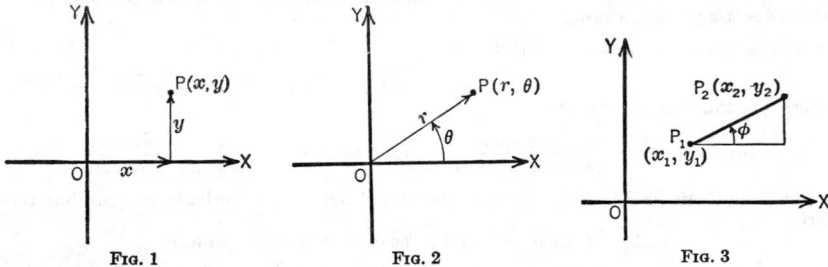

| FIG. 1 | FIG. 2 | FIG. 3 |

above OX are called *positive*, and opposite directions *negative*. The point O of intersection of OY and OX is called the *origin*.

The position of a point P is also given by its radial distance r from the origin and the angle θ between the radius r and the horizontal axis OX (Fig. 2). These coordinates r, θ, are called *polar coordinates*.

The distance s between two points $P_1(x_1, y_1)$ and $P_2(x_2, y_2)$ on a straight line is

$$s = \sqrt{(x_2 - x_1)^2 + (y_2 - y_1)^2} \tag{1}$$

In polar coordinates the distance s between $P_1(r_1, \theta_1)$ and $P_2(r_2, \theta_2)$ is

$$s = \sqrt{r_1{}^2 + r_2{}^2 - 2r_1r_2 \cos(\theta_2 - \theta_1)} \tag{2}$$

The slope m of the line P_1P_2 is defined as the tangent of the angle ϕ, which the line makes with OX.

$$m = \tan \phi = \frac{y_2 - y_1}{x_2 - x_1} \tag{3}$$

To divide the segment P_1P_2 in the ratio c_1/c_2, internally or externally,

$$x = \frac{c_2 x_1 \pm c_1 x_2}{c_2 \pm c_1}, \quad y = \frac{c_2 y_1 \pm c_1 y_2}{c_2 \pm c_1}$$

The midpoint of P_1P_2 is

$$x = 1/2\,(x_1 + x_2), \quad y = 1/2\,(y_1 + y_2)$$

The Equation of a Straight Line in cartesian coordinates is of the first degree and is expressed as follows:

$$Ax + By + C = 0 \tag{4}$$

where A, B, and C are constants.

Other forms of the equation are:

$$y = mx + b \tag{5}$$

where m is the slope and b is the y intercept;

$$y - y_1 = m(x - x_1) \tag{6}$$

where m is the slope and (x_1, y_1) is a point on the line;

$$\frac{x - x_1}{y - y_1} = \frac{x_1 - x_2}{y_1 - y_2} \tag{7}$$

where (x_1, y_1) and (x_2, y_2) are two points on the line;

$$\frac{x}{a} + \frac{y}{b} = 1 \tag{8}$$

where a and b are the x and y intercepts, respectively;

$$x \cos \alpha + y \sin \alpha - p = 0 \qquad (9)$$

where α is the angle between OX and the perpendicular from the origin to the line and p is the length of the perpendicular (Fig. 4). This is called the *perpendicular* form and is obtained by dividing the general form $Ax + By + C = 0$ by $\pm\sqrt{A^2 + B^2}$. The sign before the radical is taken opposite to that of C if $C \neq 0$ and the same as that of B if $C = 0$.

Equations of lines parallel to the x and y axes, respectively, are

$$y = k, \quad x = k \qquad (10)$$

The Perpendicular Distance of a Point $P_1(x_1, y_1)$ (Fig. 4) from the line $Ax + By + C = 0$ is

$$p_1 = \frac{Ax_1 + By_1 + C}{\pm\sqrt{A^2 + B^2}} \qquad (11)$$

FIG. 4

where the sign before the radical is opposite to that of C if $C \neq 0$, and the same as B if $C = 0$.

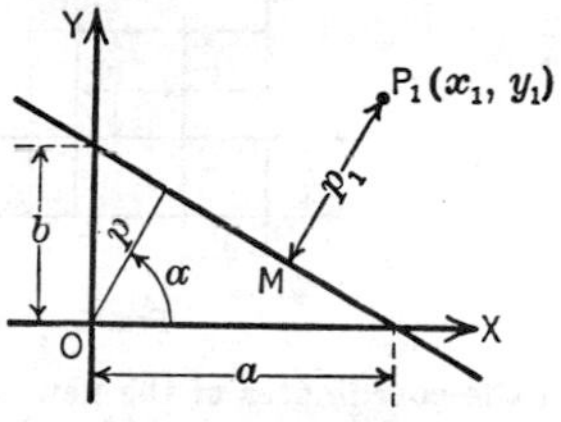

Parallel Lines. The two lines $y = m_1x + b_1$, $y = m_2x + b_2$ are parallel if $m_1 = m_2$. For the form $A_1x + B_1y + C_1 = 0$, $A_2x + B_2y + C_2 = 0$, the lines are parallel if

$$\frac{A_1}{A_2} = \frac{B_1}{B_2} \qquad (12)$$

The equation of a line through the point (x_1, y_1) and parallel to the line $Ax + By + C = 0$ is

$$A(x - x_1) + B(y - y_1) = 0 \qquad (13)$$

Perpendicular Lines. The two lines $y = m_1x + b_1$ and $y = m_2x + b_2$ are perpendicular if

$$m_1 = -\frac{1}{m_2} \qquad (14)$$

For the form $A_1x + B_1y + C_1 = 0$, $A_2x + B_2y + C_2 = 0$, the lines are perpendicular if

$$A_1A_2 + B_1B_2 = 0 \qquad (15)$$

The equation of a line through the point (x_1, y_1) perpendicular to the line $Ax + By + C = 0$ is

$$B(x - x_1) - A(y - y_1) = 0 \qquad (16)$$

Intersecting Lines. Let $A_1x + B_1y + C_1 = 0$ and $A_2x + B_2y + C_2 = 0$ be the equations of two intersecting lines and λ an arbitrary real number. Then

$$(A_1x + B_1y + C_1) + \lambda(A_2x + B_2y + C_2) = 0 \qquad (17)$$

represents the system of lines through the point of intersection.

The three lines $A_1x + B_1y + C_1 = 0$, $A_2x + B_2y + C_2 = 0$, $A_3x + B_3y + C_3 = 0$ meet in a point if

$$\begin{vmatrix} A_1 & B_1 & C_1 \\ A_2 & B_2 & C_2 \\ A_3 & B_3 & C_3 \end{vmatrix} = 0 \qquad (18)$$

The Angle θ between Two Lines with equations $A_1x + B_1y + C_1 = 0$ and $A_2x + B_2y + C_2 = 0$ can be found from

$$\sin\theta = \frac{A_1B_2 - A_2B_1}{\sqrt{(A_1^2 + B_1^2)(A_2^2 + B_2^2)}}, \quad \cos\theta = \frac{A_1A_2 + B_1B_2}{\sqrt{(A_1^2 + B_1^2)(A_2^2 + B_2^2)}}$$

$$\tan\theta = \frac{A_1B_2 - A_2B_1}{A_1A_2 + B_1B_2} \qquad (19)$$

The signs of $\tan\theta$ and $\cos\theta$ determine whether the acute or obtuse angle is meant. If the equations are in the form $y = m_1x + b_1$, $y = m_2x + b_2$, then

$$\sin\theta = \frac{m_2 - m_1}{\sqrt{(1 + m_1^2)(1 + m_2^2)}}, \quad \cos\theta = \frac{1 + m_1m_2}{\sqrt{(1 + m_1^2)(1 + m_2^2)}}, \quad \tan\theta = \frac{m_2 - m_1}{1 + m_1m_2} \qquad (20)$$

36. TRANSFORMATION OF COORDINATES

Change of Origin O to O'. Let (x, y) denote the coordinates of a point P with respect to the old axes, and (x', y') the coordinates with respect to the new axes (Fig. 5). Then,

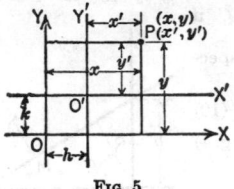 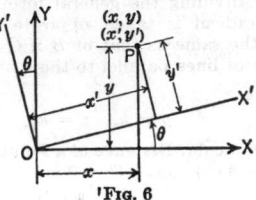

FIG. 5   FIG. 6

if the coordinates of the new origin O' with respect to the old axes are $x = h$, $y = k$, the relations between the old and the new coordinates under transformation are

$$\left. \begin{array}{l} x = x' + h \\ y = y' + k \end{array} \right\} \tag{21}$$

Rotation of Axes about the Origin. Let θ (Fig. 6) be the angle through which the axes are rotated. Then

$$\left. \begin{array}{l} x = x' \cos\theta - y' \sin\theta \\ y = x' \sin\theta + y' \cos\theta \end{array} \right\} \tag{22}$$

If the axes are both translated and rotated,

$$\left. \begin{array}{l} x = x' \cos\theta - y' \sin\theta + h \\ y = x' \sin\theta + y' \cos\theta + k \end{array} \right\} \tag{23}$$

Coordinate Transformation. The relations between the rectangular coordinates x, y and the polar coordinates r, θ are

$$x = r \cos\theta, \quad y = r \sin\theta, \quad r = \sqrt{x^2 + y^2}, \quad \theta = \tan^{-1}\frac{y}{x} \tag{24}$$

37. CONIC SECTIONS

A **Conic Section** is a curve traced by a point P moving in a plane so that the distance PF of the point from a fixed point (*focus*) is in constant ratio to the distance PM of the point from a fixed line (*directrix*) in the plane of the curve. The ratio, $e = \dfrac{PF}{PM}$, is called the *eccentricity*. If $e < 1$, the curve is an *ellipse; $e = 1$, a *parabola*; $e > 1$, a *hyperbola*; and $e = 0$, a *circle*, which is a special case of an ellipse.

The Circle. The equation is

$$(x - x_0)^2 + (y - y_0)^2 = r^2 \tag{25}$$

where (x_0, y_0) is the center and r the radius. If the center is at the origin,

$$x^2 + y^2 = r^2 \tag{26}$$

Another form is

$$x^2 + y^2 + 2gx + 2fy + c = 0 \tag{27}$$

FIG. 7

with center $(-g, -f)$ and radius $\sqrt{g^2 + f^2 - c}$.

The equation of the tangent to (27) at a point $P_1(x_1, y_1)$ is

$$xx_1 + yy_1 + g(x + x_1) + f(y + y_1) + c = 0 \tag{28}$$

The Ellipse (Fig. 7). The equation is

$$\frac{(x - x_0)^2}{a^2} + \frac{(y - y_0)^2}{b^2} = 1 \tag{29}$$

where (x_0, y_0) is the center, a = semi-major axis, b = semi-minor axis. In Fig. 7, (x_0, y_0) = $(0, 0)$.

Coordinates of *foci* are $F_1 = (-ae, 0)$, $F_2 = (ae, 0)$; $e^2 = \dfrac{(F_1P)^2}{(MP)^2} = 1 - \dfrac{b^2}{a^2} < 1$; and the *directrices* are the lines $x = -\dfrac{a}{e}$, $x = \dfrac{a}{e}$.

The chord LL' through F is called the *latus rectum* and has the length $\dfrac{2b^2}{a} = 2a(1 - e^2)$.
If P_1 is any point on the ellipse, $F_1P_1 = a - ex_1$, $F_2P_1 = a + ex_1$, and $F_1P_1 + F_2P_1 = 2a$ (a constant).

The *area* of the ellipse with semi-axes a and b is

$$A = \pi ab \tag{30}$$

The equation of the *tangent* to the ellipse (Fig. 7) at the point (x_1, y_1) is

$$\frac{xx_1}{a^2} + \frac{yy_1}{b^2} = 1 \tag{31}$$

the equation of the tangent with slope m is

$$y = mx \pm \sqrt{a^2m^2 + b^2} \tag{32}$$

The equation of the *normal* to the ellipse at the point (x_1, y_1) is

$$a^2y_1(x - x_1) - b^2x_1(y - y_1) = 0 \tag{33}$$

Conjugate Diameters. A line through the center of an ellipse is a *diameter;* if the slopes m and m' of the two diameters $y = mx$ and $y = m'x$ are such that $mm' = -\dfrac{b^2}{a^2}$ each diameter bisects all chords parallel to the other and the diameters are called *conjugate*.

Other Forms of the Equation of the Ellipse.

$$\frac{x^2}{a^2} + \frac{y^2}{a^2(1 - e^2)} = 1 \tag{34}$$

$$ax^2 + by^2 + 2gx + 2fy + c = 0 \tag{35}$$

If a, b, and $\left(\dfrac{g^2}{a} + \dfrac{f^2}{b}\right) - c$ have the same sign, (35) is an ellipse whose axes are parallel to the coordinate axes.

The parametric form is

$$x = a \cos\phi, \quad y = b \sin\phi \tag{36}$$

The Hyperbola (Fig. 8). The equation is

$$\frac{(x - x_0)^2}{a^2} - \frac{(y - y_0)^2}{b^2} = 1 \tag{37}$$

where (x_0, y_0) is the center, $AA' = 2a$ is the transverse axis, $BB' = 2b$ is the conjugate axis. In Fig. 8, $(x_0, y_0) = (0, 0)$.

$e^2 = \dfrac{(F_1P)^2}{(PM)^2} = 1 + \dfrac{b^2}{a^2} > 1$; the coordinates of the *foci*, $F_1 = (-ae, 0)$, $F_2 = (ae, 0)$; and the *directrices* are the lines $x = -\dfrac{a}{e}$, $x = \dfrac{a}{e}$.

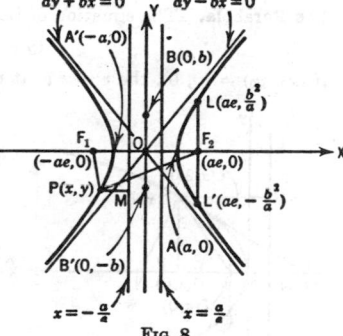

Fig. 8

The chord LL' through F is called the *latus rectum* and has the length $\dfrac{2b^2}{a} = 2a(e^2 - 1)$.
If P_1 is any point on the curve, $F_1P_1 = ex_1 - a$, $F_2P_1 = ex_1 + a$, and $|F_2P_1 - F_1P_1| = 2a$ (a constant).

The equation of the *tangent* to the hyperbola (Fig. 8) at the point (x_1, y_1) is

$$\frac{xx_1}{a^2} - \frac{yy_1}{b^2} = 1 \tag{38}$$

The equation of the tangent whose slope is m is

$$y = mx \pm \sqrt{a^2m^2 - b^2} \tag{39}$$

The equation of the *normal* to the hyperbola at the point (x_1, y_1) is

$$a^2y_1(x - x_1) + b^2x_1(y - y_1) = 0 \tag{40}$$

Conjugate Hyperbolas and Diameters. The two hyperbolas $\dfrac{x^2}{a^2} - \dfrac{y^2}{b^2} = 1$ and $\dfrac{y^2}{b^2} - \dfrac{x^2}{a^2} = 1$ are conjugate. The transverse axis of each is the conjugate axis of the other.

If the slopes of the two lines $y = mx$ and $y = m_1x$ through the center O are connected by the relation $mm_1 = \dfrac{b^2}{a^2}$, each of these lines bisects all chords of the hyperbola which are parallel to the other line. Two such lines are called *conjugate diameters*. The equation of the hyperbola referred to its conjugate diameters as oblique axes is

$$\frac{x'^2}{a_1^2} - \frac{y'^2}{b_1^2} = 1 \qquad (41)$$

where $2a_1$ and $2b_1$ are the conjugate axes.

The Asymptotes. The lines $y = \dfrac{b}{a}x$ and $y = -\dfrac{b}{a}x$ are the *asymptotes* of the hyperbola $\dfrac{x^2}{a^2} - \dfrac{y^2}{b^2} = 1$. The asymptotes are two tangents whose points of contact with the curve are at an infinite distance from the center. The equation of the hyperbola when referred to its asymptotes as oblique axes is

$$4x'y' = a^2 + b^2 \qquad (42)$$

If $a = b$, the asymptotes are the perpendicular lines $y = x$, $y = -x$; the corresponding hyperbola

$$x^2 - y^2 = a^2 \qquad (43)$$

is called the *rectangular* or *equilateral hyperbola*.

Other Forms of the Equation of the Hyperbola.

$$\frac{x^2}{a^2} - \frac{y^2}{a^2(e^2 - 1)} = 1 \qquad (44)$$

$$ax^2 + by^2 + 2gx + 2fy + c = 0 \qquad (45)$$

If a and b have unlike signs, (45) is a hyperbola with axes parallel to the coordinate axes. The parametric form is

$$x = a \sec \phi, \quad y = a \tan \phi \qquad (46)$$

The Parabola. The equation of the parabola is

$$(y - y_0)^2 = 4a(x - x_0) \qquad (47)$$

If $(x_0, y_0) = (0, 0)$, the *vertex* is at the origin (Fig. 9); the *focus F* is on OX, called the axis of the parabola, and. has the coordinates $(a, 0)$; and the *directrix* is $x = -a$. The chord LL' through F is the *latus rectum* and has the length $4a$. The *eccentricity* $e = \dfrac{FP}{PM} = 1$.

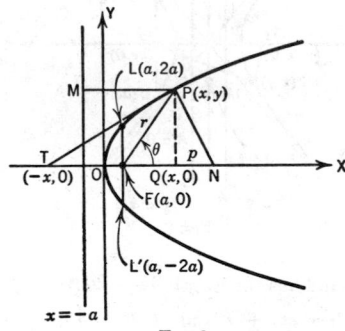

The *tangent* to the parabola $y^2 = 4ax$ at the point (x_1, y_1) is

$$yy_1 = 2a(x + x_1) \qquad (48)$$

The equation of the tangent whose slope is m is

$$y = mx + \frac{a}{m} \qquad (49)$$

The *normal* to the parabola at the point (x_1, y_1) is

$$2a(y - y_1) + y_1(x - x_1) = 0 \qquad (50)$$

FIG. 9

A *diameter* of the curve is a straight line parallel to the axis. It bisects all chords parallel to the tangent at the point where the diameter meets the parabola.

If P_1T is tangent to the curve at (x_1, y_1), then $TQ = 2x_1$ is the *subtangent*, and $QN = 2a$ (a constant) is the *subnormal*, where P_1N is perpendicular to P_1T.

The equation of the form $y^2 + 2gx + 2fy + c = 0$, where $g \neq 0$, is a parabola whose axis is parallel to OX; and the equation $x^2 + 2gx + 2fy + c = 0$, where $f \neq 0$, is a parabola whose axis is parallel to OY.

The parabola referred to the tangents at the extremities of its latus rectum as axes of coordinates is

$$x^{1/2} \pm y^{1/2} = b^{1/2} \qquad (51)$$

where b is the distance from the origin to each point of tangency.

Polar Equations of the Conics. If e is the eccentricity, if the directrix is vertical, if the focus is at a distance p to the right or left of it, respectively, and if the polar origin is taken at the focus, the polar equation is

$$r = \frac{ep}{1 \mp e \cos \theta}, \quad \text{for ellipse, hyperbola, or parabola} \qquad (52)$$

$$r = \frac{a(1 - e^2)}{1 \mp e \cos \theta}, \quad \text{for ellipse or circle} \qquad (53)$$

$$r = \frac{a(e^2 - 1)}{1 \mp e \cos \theta}, \quad \text{for hyperbola} \qquad (54)$$

If the directrix is horizontal and the focus is at a distance p above or below it, respectively, the polar equation is

$$r = \frac{ep}{1 \mp e \sin \theta}, \quad \text{for ellipse, hyperbola, or parabola} \qquad (55)$$

$$r = \frac{a(1 - e^2)}{1 \mp e \sin \theta}, \quad \text{for ellipse or circle} \qquad (56)$$

$$r = \frac{a(e^2 - 1)}{1 \mp e \sin \theta}, \quad \text{for hyperbola} \qquad (57)$$

The General Equation of a Conic Section has the form

$$ax^2 + 2hxy + by^2 + 2gx + 2fy + c = 0 \qquad (58)$$

Let

$$D = \begin{vmatrix} a & h & g \\ h & b & f \\ g & f & c \end{vmatrix}; \quad d = \begin{vmatrix} a & h \\ h & b \end{vmatrix}; \quad \delta = a + b \qquad (59)$$

Then the following is a classification of conic sections.

1. A parabola for $d = 0$, $D \neq 0$.
2. Two parallel lines (possibly coincident or imaginary) for $d = 0$, $D = 0$.
3. An ellipse for $d > 0$, $\delta D < 0$.
4. No locus (imaginary ellipse) for $d > 0$, $\delta D > 0$.
5. Point ellipse for $d > 0$, $D = 0$.
6. A hyperbola for $d < 0$, $D \neq 0$.
7. Two intersecting lines for $d < 0$, $D = 0$.

Let $A + B = a + b$, $AB = ab - h^2 = d$, and $A - B$ have the same sign as h. Let $c' = D/d$; then the equation of the conic referred to its axes is

$$\frac{x^2}{-\dfrac{c'}{A}} + \frac{y^2}{-\dfrac{c'}{B}} = 1 \qquad (60)$$

To find the center (x_0, y_0) of the conic solve the equations

$$\left. \begin{array}{l} ax_0 + hy_0 + g = 0 \\ hx_0 + by_0 + f = 0 \end{array} \right\} \qquad (61)$$

To remove the term in xy from (50), rotate the axes about the origin through an angle θ such that $\tan 2\theta = -\dfrac{2h}{a - b}$.

38. HIGHER PLANE CURVES

Plane Curves. The point (x, y) describes a plane curve if x and y are continuous functions of a variable t (parameter), as $x = x(t)$, $y = y(t)$. The elimination of t from the two equations gives $F(x, y) = 0$ or in explicit form $y = f(x)$. The angle τ which a tangent to the curve makes with OX can be found from

$$\sin \tau = \frac{dy}{ds}, \quad \cos \tau = \frac{dx}{ds}, \quad \tan \tau = \frac{dy}{dx} = y' \tag{62}$$

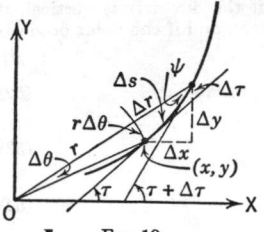

Fig. 10

where ds is the element of arc length:

$$ds = \sqrt{dx^2 + dy^2} = \sqrt{1 + y'^2}\, dx \tag{63}$$

In polar coordinates,

$$ds = \sqrt{dr^2 + r^2\, d\theta^2} = \sqrt{\left(\frac{dr}{d\theta}\right)^2 \theta\rho + r^2} \tag{64}$$

From Fig. 10, it may be seen that

$$\sin \psi = \frac{r\, d\theta}{ds}, \quad \cos \psi = \frac{dr}{ds}, \quad \tan \psi = \frac{r\, d\theta}{dr} \tag{65}$$

The equation of the *tangent* to the curve $F(x, y) = 0$ at the point (x_1, y_1) is

$$\left(\frac{\partial F}{\partial x}\right)_{x=x_1, y=y_1} (x - x_1) + \left(\frac{\partial F}{\partial y}\right)_{x=x_1, y=y_1} (y - y_1) = 0 \tag{66}$$

The equation of the *normal* to the curve $F(x, y) = 0$ at the point (x_1, y_1) is

$$\left(\frac{\partial F}{\partial y}\right)_{x=x_1, y=y_1} (x - x_1) - \left(\frac{\partial F}{\partial x}\right)_{x=x_1, y=y_1} (y - y_1) = 0 \tag{67}$$

The equation of the *tangent* to the curve $y = f(x)$ at the point (x_1, y_1) is

$$y - y_1 = \left(\frac{dy}{dx}\right)_{x=x_1} (x - x_1) \tag{68}$$

The equation of the *normal* to the curve $y = f(x)$ at the point (x_1, y_1) is

$$y - y_1 = -\frac{1}{\left(\dfrac{dy}{dx}\right)_{x=x_1}} (x - x_1) \tag{69}$$

The *radius of curvature* of the curve at the point (x, y) is

$$\rho = \frac{ds}{d\tau} = \frac{\left[1 + \left(\dfrac{dy}{dx}\right)^2\right]^{3/2}}{\dfrac{d^2 y}{dx^2}} = \frac{[1 + y'^2]^{3/2}}{y''} \tag{70}$$

The reciprocal $1/\rho$ is called the *curvature of the curve* at (x, y).

The coordinates (x_0, y_0) of the center of curvature for the point (x, y) on the curve (the center of the circle of curvature tangent to the curve at (x, y) and of radius ρ) are

$$\left. \begin{array}{l} x_0 = x - \rho\,\dfrac{dy}{ds} = x - y'\,\dfrac{[1 + y'^2]}{y''} \\[3mm] y_0 = y + \rho\,\dfrac{dx}{ds} = y + \dfrac{[1 + y'^2]}{y''} \end{array} \right\} \tag{71}$$

A curve has a *singular point* if simultaneously,

$$F(x, y) = 0, \quad \frac{\partial F}{\partial x} = 0, \quad \frac{\partial F}{\partial y} = 0 \tag{72}$$

Let

$$D = \left(\frac{\partial^2 F}{\partial x\, \partial y}\right)^2 - \frac{\partial^2 F}{\partial x^2}\frac{\partial^2 F}{\partial y^2} \tag{73}$$

Then for $D > 0$, the curve has a *double point* with two real different tangents. For $D = 0$, the curve has a *cusp* with two coincident tangents. For $D < 0$, the curve has an *isolated point* with no real tangent.

Semicubic, or Neil's, Parabola

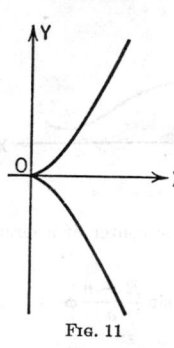

FIG. 11

$$y^2 = ax^3$$

Logarithmic Curve

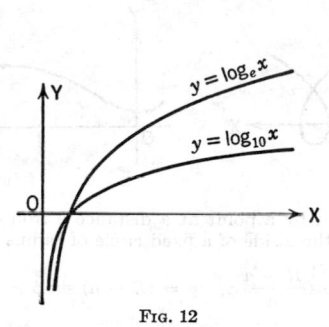

FIG. 12

$$y = \log_b x$$

Exponential Curve

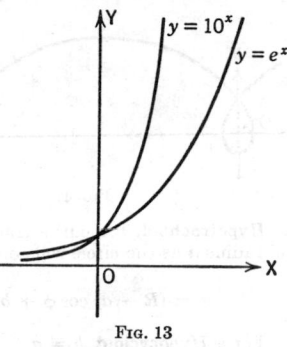

FIG. 13

$$y = b^x$$

Catenary

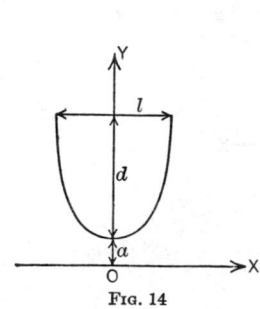

FIG. 14

$$y = \frac{a}{2}(e^{x/a} + e^{-x/a}) = a \cosh \frac{x}{a}$$

For l large compared with d,

$$s \approx l\left[1 + \frac{2}{3}\left(\frac{2d}{l}\right)^2\right]$$

Damped Wave

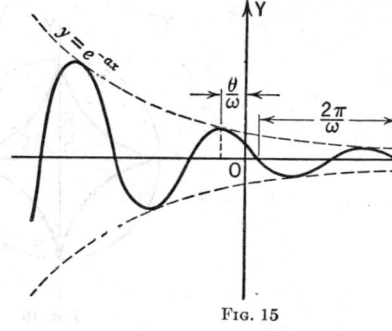

FIG. 15

$$y = e^{-ax} \cos(\omega x + \theta)$$

Trochoid. A curve traced by a point at a distance b from the center of a circle of radius a as the circle rolls on a straight line.

$$x = a\phi - b \sin\phi, \quad y = a - b \cos\phi$$

Cycloid

$$a = b$$

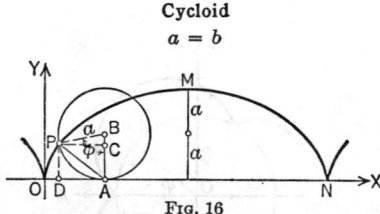

FIG. 16

$$x = a(\phi - \sin\phi), \quad y = a(1 - \cos\phi)$$

$$x = a \cos^{-1}\frac{a - y}{a} \pm \sqrt{(2a - y)y}$$

For one arch, arc length $= 8a$, area $= 3\pi a^2$

Prolate Cycloid
$a < b$

Curtate Cycloid
$a > b$

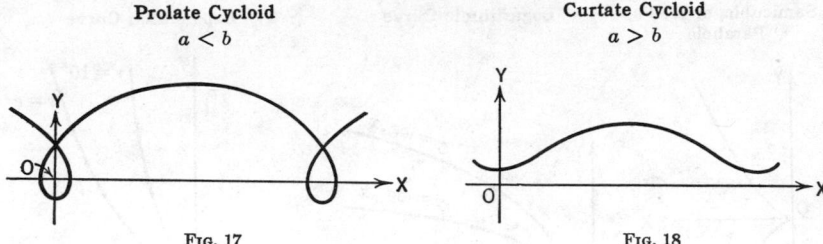

FIG. 17   FIG. 18

Hypotrochoid. A curve traced by a point at a distance b from the center of a circle of radius a as the circle rolls on the inside of a fixed circle of radius R.

$$x = (R - a)\cos\phi + b\cos\frac{R - a}{a}\phi, \quad y = (R - a)\sin\phi - b\sin\frac{R - a}{a}\phi$$

For a Hypocycloid, $b = a$.

Hypocycloid of Four Cusps, or Astroid

$$b = a = \tfrac{1}{4}R$$

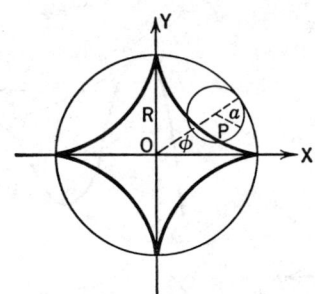

FIG. 19

$$x = R\cos^3\phi, \quad y = R\sin^3\phi$$

$$x^{\frac{2}{3}} + y^{\frac{2}{3}} = R^{\frac{2}{3}}$$

Epitrochoid. A curve traced by a point at a distance b from the center of a circle of radius a as the circle rolls on the outside of a fixed circle of radius R.

$$x = (R + a)\cos\phi - b\cos\frac{R + a}{a}\phi, \quad y = (R + a)\sin\phi - b\sin\frac{R + a}{a}\phi$$

Epicycloid
$$b = a$$

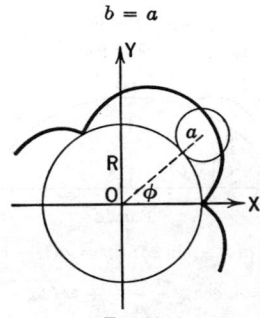

FIG. 20

PLANE ANALYTIC GEOMETRY

Limaçon of Pascal

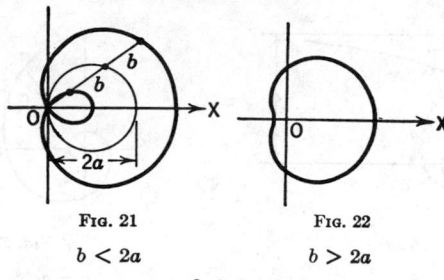

FIG. 21   FIG. 22

$b < 2a$ $b > 2a$

$$r = b + 2a \cos \theta$$

Other forms of the right-hand side of the equation, $b + 2a \sin \theta$, $b - 2a \cos \theta$, $b - 2a \sin \theta$, give curves rotated through 1, 2, 3 right angles, respectively.

Cardioid

Limaçon in which $b = 2a$

Epicycloid in which $R = a$

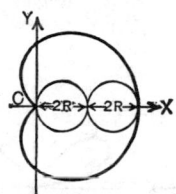

FIG. 23

$$r = 2a(1 + \cos \theta)$$
$$(x^2 + y^2 - 2ax)^2 = 4a^2(x^2 + y^2)$$

Involute of a Circle

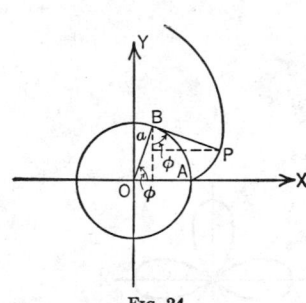

FIG. 24

$$x = a(\cos \phi + \phi \sin \phi)$$
$$y = a(\sin \phi - \phi \cos \phi)$$
$$\theta = \sqrt{r^2/a^2 - 1} - \tan^{-1}\sqrt{r^2/a^2 - 1}$$

Spiral traced by the end of a taut string unwinding from a circle.

Spiral of Archimedes

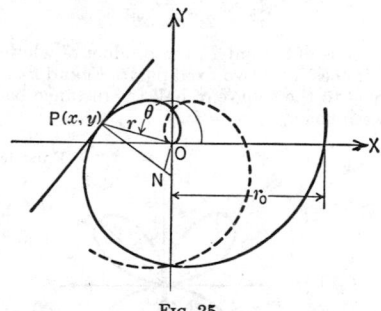

FIG. 25

$$r = a\theta$$

Polar subnormal $ON = a$

Length of arc $OP = s = \frac{1}{2} a(\theta\sqrt{1 + \theta^2} + \sinh^{-1}\theta)$

For many turns, $s \approx \frac{1}{2} a\theta^2$

Hyperbolic, or Reciprocal, Spiral

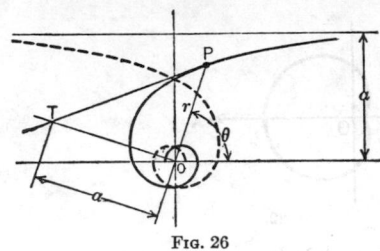

FIG. 26

$$r\theta = a$$

Polar subtangent $OT = -a$

As $\theta \to \infty$, $r \to 0$. The curve winds an indefinite number of times around the origin.

As $\theta \to 0$, $r \to \infty$. The curve has an asymptote parallel to the polar axis at a distance a.

Logarithmic, or Equiangular, Spiral

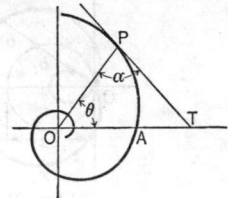

FIG. 27

$$r = ae^{m\theta}, \quad m > 0$$

or

$$\ln \frac{r}{a} = m\theta$$

The tangent to the curve at any point makes a constant angle $\alpha (= \cot^{-1} m)$ with the radius vector.

As $\theta \to -\infty$, $r \to 0$. The curve winds an indefinite number of times around the origin.

Lemniscate of Bernoulli

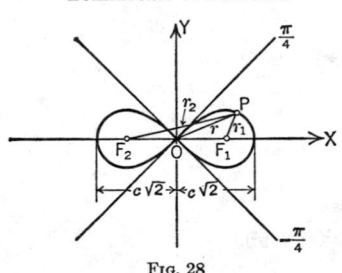

FIG. 28

$$(x^2 + y^2)^2 + 2c^2(y^2 - x^2) = 0$$

$$r^2 = 2c^2 \cos 2\theta$$

Locus of a point P, the product of whose distances from two fixed points F_1 and F_2 is equal to the square of half the distance between them, $r_1 \cdot r_2 = c^2$.

Three-leaved Roses

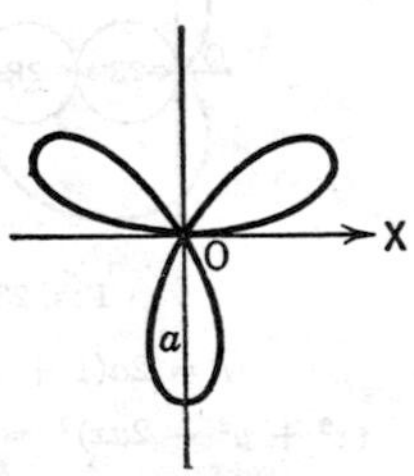

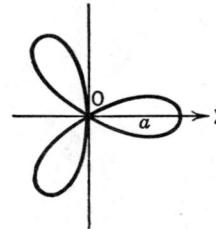

FIG. 29

$$r = a \sin 3\theta$$

FIG. 30

$$r = a \cos 3\theta \text{ }^{.}$$

Four-leaved Roses

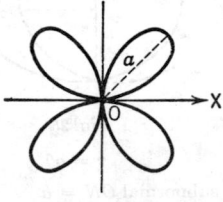

FIG. 31

$$r = a \sin 2\theta$$

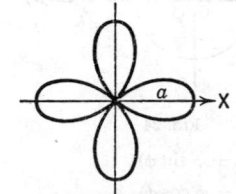

FIG. 32

$$r = a \cos 2\theta$$

The roses, $r = a \sin n\theta$ and $r = a \cos n\theta$, have, for n even, $2n$ leaves; for n odd, n leaves.

Cissoid of Diocles

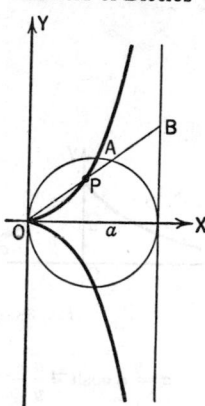

Fig. 33

$$y^2 = \frac{x^3}{a - x}$$

$$r = a(\sec \theta - \cos \theta)$$

Locus of point P such that $OP = AB$.

Strophoid

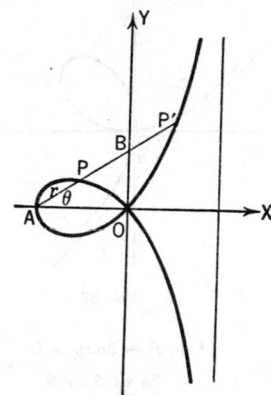

Fig. 34

$$y^2 = \frac{x^2(a + x)}{a - x}$$

$$r = a(\sec \theta - \tan \theta)$$

If the line AB rotates about A, intersecting the y axis at B, and if $PB = BP' = OB$, the locus of P and P' is the strophoid.

Conchoid of Nicomedes

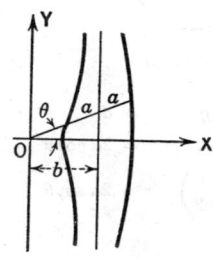

Fig. 35

$$(x^2 + y^2)(x - b)^2 = a^2 x^2$$

$$r = b \sec \theta - a$$

Witch of Agnesi

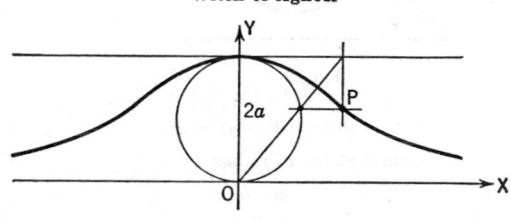

Fig. 36

$$y = \frac{8a^3}{x^2 + 4a^2}$$

$$x = 2a \tan \phi$$

$$y = 2a \cos^2 \phi$$

Folium of Descartes

Tractrix

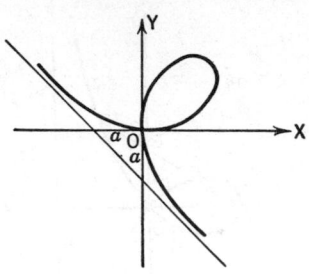

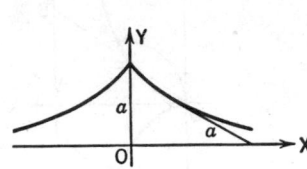

FIG. 37

FIG. 38

$$x^3 + y^3 - 3axy = 0$$

$$r = \frac{3a \sin\theta \cos\theta}{\sin^3\theta + \cos^3\theta}$$

$$x = a \cosh^{-1}\frac{a}{y} - \sqrt{a^2 - y^2}$$

Locus of one end P of tangent line of length a as the other end Q is moved along the x axis.

Circles in Polar Coordinates

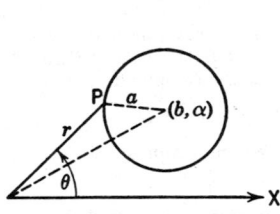

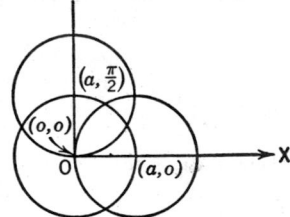

FIG. 39

FIG. 40

$$r^2 + b^2 - 2rb\cos(\theta - \alpha) = a^2$$

Center at (b, α), radius a

Center $(0, 0)$ $r = a$

Center $(a, 0)$ $r = 2a\cos\theta$

Center $\left(a, \frac{\pi}{2}\right)$ $r = 2a\sin\theta$

Frequency-modulated Wave

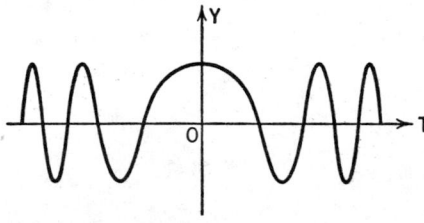

FIG. 41

$$y = k\cos[\phi(t)], \quad \text{instantaneous frequency} = \Omega(t) = \frac{d\phi}{dt}$$

In Fig. 41, $y = \cos\frac{\pi}{2}t^2$, $\Omega(t) = \pi t$.

SOLID ANALYTIC GEOMETRY

39. COORDINATE SYSTEMS

Right-hand Rectangular (Fig. 1). The position of a point $P(x, y, z)$ is fixed by its distances x, y, z from the mutually perpendicular planes yz, xz, and xy, respectively.

Spherical, or Polar (Fig. 2). The position of a point $P(r, \theta, \phi)$ is fixed by its distance from a given point O, the origin, and its direction from O, determined by the angles θ and ϕ.

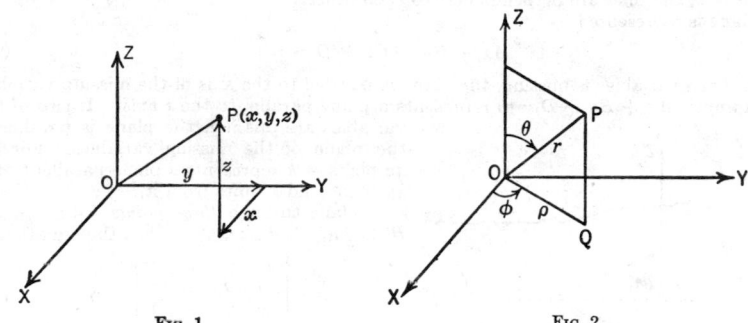

FIG. 1   FIG. 2

Cylindrical (Fig. 2). The position of a point $P(\rho, \phi, z)$ is fixed by its distance z from a given plane and the polar coordinates (ρ, ϕ) of the projection Q of P on the given plane.

Relations among coordinates of the three systems.

$$x = r \sin \theta \cos \phi = \rho \cos \phi \tag{1}$$

$$y = r \sin \theta \sin \phi = \rho \sin \phi \tag{2}$$

$$z = r \cos \theta \tag{3}$$

$$\rho = \sqrt{x^2 + y^2} = r \sin \theta \tag{4}$$

$$\phi = \tan^{-1} \frac{y}{x} \tag{5}$$

$$r = \sqrt{x^2 + y^2 + z^2} = \sqrt{\rho^2 + z^2} \tag{6}$$

$$\theta = \tan^{-1} \frac{\sqrt{x^2 + y^2}}{z} = \tan^{-1} \frac{\rho}{z} \tag{7}$$

40. POINT, LINE, AND PLANE

The Euclidean Distance between Two Points $P_1(x_1, y_1, z_1)$ and $P_2(x_2, y_2, z_2)$ is

$$s = \sqrt{(x_2 - x_1)^2 + (y_2 - y_1)^2 + (z_2 - z_1)^2} \tag{8}$$

To Divide the segment P_1P_2 in the ratio c_1/c_2, internally or externally,

$$x = \frac{c_2 x_1 \pm c_1 x_2}{c_2 \pm c_1}, \quad y = \frac{c_2 y_1 \pm c_1 y_2}{c_2 \pm c_1}, \quad z = \frac{c_2 z_1 \pm c_1 z_2}{c_2 \pm c_1} \tag{9}$$

The *midpoint* of P_1P_2 is

$$x = \frac{x_1 + x_2}{2}, \quad y = \frac{y_1 + y_2}{2}, \quad z = \frac{z_1 + z_2}{2} \tag{10}$$

The Angles α, β, γ which the line P_1P_2 makes with the coordinate directions x, y, z, respectively, are the *direction angles* of P_1P_2. The cosines

$$\cos \alpha = \frac{x_2 - x_1}{s}, \quad \cos \beta = \frac{y_2 - y_1}{s}, \quad \cos \gamma = \frac{z_2 - z_1}{s} \tag{11}$$

are the *direction cosines* of P_1P_2, and

$$\cos^2\alpha + \cos^2\beta + \cos^2\gamma = 1 \tag{12}$$

If $l : m : n = \cos\alpha : \cos\beta : \cos\gamma$, then

$$\cos\alpha = \frac{l}{\sqrt{l^2 + m^2 + n^2}}, \quad \cos\beta = \frac{m}{\sqrt{l^2 + m^2 + n^2}}, \quad \cos\gamma = \frac{n}{\sqrt{l^2 + m^2 + n^2}} \tag{13}$$

The Angle θ between Two Lines in terms of their direction angles α_1, β_1, γ_1, and α_2, β_2, γ_2 is obtained from

$$\cos\theta = \cos\alpha_1 \cos\alpha_2 + \cos\beta_1 \cos\beta_2 + \cos\gamma_1 \cos\gamma_2 \tag{14}$$

If $\cos\theta = 0$, the lines are perpendicular to each other.

A Plane is represented by

$$Ax + By + Cz + D = 0 \tag{15}$$

If one of the variables is missing, the plane is parallel to the axis of the missing variable. For example, $Ax + By + D = 0$ represents a plane parallel to the z axis. If two of the variables are missing, the plane is parallel to the plane of the missing variables. For example, $z = k$ represents a plane parallel to the xy plane and k units from it.

A plane through *three points* $P_1(x_1, y_1, z_1)$, $P_2(x_2, y_2, z_2)$, $P_3(x_3, y_3, z_3)$ has the equation

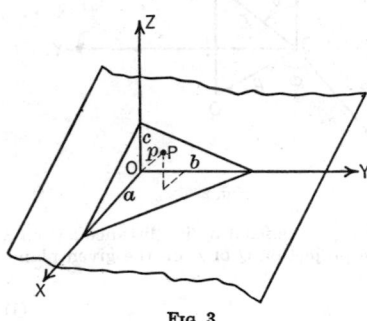

$$\begin{vmatrix} x & y & z & 1 \\ x_1 & y_1 & z_1 & 1 \\ x_2 & y_2 & z_2 & 1 \\ x_3 & y_3 & z_3 & 1 \end{vmatrix} = 0 \tag{16}$$

The equation of a plane whose x, y, z *intercepts* are respectively a, b, c (Fig. 3) is

$$\frac{x}{a} + \frac{y}{b} + \frac{z}{c} = 1 \tag{17}$$

FIG. 3

The *perpendicular* form of the equation of a plane, where $OP = p$ is the perpendicular distance of the plane from the origin O and has the direction angles α, β, γ, is

$$x\cos\alpha + y\cos\beta + z\cos\gamma - p = 0 \tag{18}$$

To bring the general form $Ax + By + Cz + D = 0$ into the perpendicular form, divide it by $\pm\sqrt{A^2 + B^2 + C^2}$, where the sign before the radical is opposite to that of D.

The coefficients A, B, C are proportional to the direction cosines λ, μ, ν of a line perpendicular to the plane. Therefore,

$$A(x - x_1) + B(y - y_1) + C(z - z_1) = 0 \tag{19}$$

is a plane through $P_1(x_1, y_1, z_1)$ and perpendicular to a line with direction cosines λ, μ, ν proportional to A, B, C.

The Perpendicular Distance of a Point P_1 from a Plane $Ax + By + Cz + D = 0$ is given by

$$PP_1 = \frac{Ax_1 + By_1 + Cz_1 + D}{\pm\sqrt{A^2 + B^2 + C^2}} \tag{20}$$

where the sign before the radical is opposite to that of D.

Parallel Planes. Two planes $A_1x + B_1y + C_1z + D_1 = 0$ and $A_2x + B_2y + C_2z + D_2 = 0$ are parallel if $A_1 : B_1 : C_1 = A_2 : B_2 : C_2$.

$$A(x - x_1) + B(y - y_1) + C(z - z_1) = 0 \tag{21}$$

is a plane through the point $P_1(x_1, y_1, z_1)$ and parallel to the plane $Ax + By + Cz + D = 0$

The Angle θ between Two Planes $Ax + By + Cz + D = 0$ and $A_1x + B_1y + C_1z + D_1 = 0$ is the angle between two intersecting lines, each perpendicular to one of the planes:

$$\cos\theta = \frac{AA_1 + BB_1 + CC_1}{\pm\sqrt{(A^2 + B^2 + C^2)(A_1^2 + B_1^2 + C_1^2)}} \tag{22}$$

The two planes are perpendicular if $AA_1 + BB_1 + CC_1 = 0$.

Four Points, $P_k(x_k, y_k, z_k)$ ($k = 1, 2, 3, 4$), lie in the same plane if

$$\begin{vmatrix} 1 & x_1 & y_1 & z_1 \\ 1 & x_2 & y_2 & z_2 \\ 1 & x_3 & y_3 & z_3 \\ 1 & x_4 & y_4 & z_4 \end{vmatrix} = 0 \tag{23}$$

Four Planes, $A_k x + B_k y + C_k z + D_k = 0$ ($k = 1, 2, 3, 4$), pass through the same point if

$$\begin{vmatrix} A_1 & B_1 & C_1 & D_1 \\ A_2 & B_2 & C_2 & D_2 \\ A_3 & B_3 & C_3 & D_3 \\ A_4 & B_4 & C_4 & D_4 \end{vmatrix} = 0 \tag{24}$$

A Straight Line is represented as the intersection of two planes by two first-degree equations

$$\left. \begin{aligned} A_1 x + B_1 y + C_1 z + D_1 = 0 \\ A_2 x + B_2 y + C_2 z + D_2 = 0 \end{aligned} \right\} \tag{25}$$

The three planes through the line perpendicular to the coordinate planes are its *projecting planes*. The equation of the xy projecting plane is found by eliminating z between the two given equations, etc. The line can be represented by any two of its projecting planes, for example,

$$\left. \begin{aligned} y = m_1 x + b_1 \\ z = m_2 x + b_2 \end{aligned} \right\} \tag{26}$$

If the line goes through a *point* $P_1(x_1, y_1, z_1)$ and has the *direction angles* α, β, γ, then

$$\frac{x - x_1}{\cos \alpha} = \frac{y - y_1}{\cos \beta} = \frac{z - z_1}{\cos \gamma} \tag{27}$$

and $m_1 = \dfrac{\cos \beta}{\cos \alpha}$, $m_2 = \dfrac{\cos \gamma}{\cos \alpha}$

The equations of a line through *two points* (x_1, y_1, z_1) and (x_2, y_2, z_2) are

$$\frac{x - x_1}{x_2 - x_1} = \frac{y - y_1}{y_2 - y_1} = \frac{z - z_1}{z_2 - z_1} \tag{28}$$

A line through a *point* P_1 perpendicular to a plane $Ax + By + Cz + D = 0$ has the equations

$$\frac{x - x_1}{A} = \frac{y - y_1}{B} = \frac{z - z_1}{C} \tag{29}$$

Line of Intersection of Two Planes. The direction cosines λ, μ, ν of the line of intersection of two planes $Ax + By + Cz + D = 0$ and $A_1 x + B_1 y + C_1 z + D_1 = 0$ are found from the ratios

$$\lambda : \mu : \nu = \begin{vmatrix} B & C \\ B_1 & C_1 \end{vmatrix} : \begin{vmatrix} C & A \\ C_1 & A_1 \end{vmatrix} : \begin{vmatrix} A & B \\ A_1 & B_1 \end{vmatrix} \tag{30}$$

41. TRANSFORMATION OF COORDINATES

Changing the Origin. Let the coordinates of a point P with respect to the original axes be x, y, z and with respect to the new axes x', y', z'. For a parallel displacement of the axes with x_0, y_0, z_0 the coordinates of the new origin

$$x = x_0 + x', \quad y = y_0 + y', \quad z = z_0 + z' \tag{31}$$

Rotation of the Axes about the Origin. Let the cosines of the angles of the new axes x', y', z', with the x axis be λ_1, μ_1, ν_1, with the y axis be λ_2, μ_2, ν_2, with the z axis be λ_3, μ_3, ν_3. Then

$$\left. \begin{aligned} x = \lambda_1 x' + \mu_1 y' + \nu_1 z' \\ y = \lambda_2 x' + \mu_2 y' + \nu_2 z' \\ z = \lambda_3 x' + \mu_3 y' + \nu_3 z' \end{aligned} \right. \qquad \left. \begin{aligned} x' = \lambda_1 x + \lambda_2 y + \lambda_3 z \\ y' = \mu_1 x + \mu_2 y + \mu_3 z \\ z' = \nu_1 x + \nu_2 y + \nu_3 z \end{aligned} \right\} \tag{32}$$

The following relations exist:

(1) $\lambda_1^2 + \mu_1^2 + \nu_1^2 = 1$
$\lambda_2^2 + \mu_2^2 + \nu_2^2 = 1$
$\lambda_3^2 + \mu_3^2 + \nu_3^2 = 1$

(2) $\lambda_1^2 + \lambda_2^2 + \lambda_3^2 = 1$
$\mu_1^2 + \mu_2^2 + \mu_3^2 = 1$
$\nu_1^2 + \nu_2^2 + \nu_3^2 = 1$

(3) $\lambda_1\lambda_2 + \mu_1\mu_2 + \nu_1\nu_2 = 0$
$\lambda_2\lambda_3 + \mu_2\mu_3 + \nu_2\nu_3 = 0$
$\lambda_3\lambda_1 + \mu_3\mu_1 + \nu_3\nu_1 = 0$

(4) $\lambda_1\mu_1 + \lambda_2\mu_2 + \lambda_3\mu_3 = 0$
$\mu_1\nu_1 + \mu_2\nu_2 + \mu_3\nu_3 = 0$
$\nu_1\lambda_1 + \nu_2\lambda_2 + \nu_3\lambda_3 = 0$

(5) $\lambda_1 = \mu_2\nu_3 - \nu_2\mu_3$
$\mu_1 = \nu_2\lambda_3 - \lambda_2\nu_3$
$\nu_1 = \lambda_2\mu_3 - \mu_2\lambda_3$

(6) $\lambda_2 = \nu_1\mu_3 - \mu_1\nu_3$
$\mu_2 = \lambda_1\nu_3 - \nu_1\lambda_3$
$\nu_2 = \mu_1\lambda_3 - \lambda_1\mu_3$

(7) $\lambda_3 = \mu_1\nu_2 - \nu_1\mu_2$
$\mu_3 = \nu_1\lambda_2 - \lambda_1\nu_2$
$\nu_3 = \lambda_1\mu_2 - \mu_1\lambda_2$

(8) $\begin{vmatrix} \lambda_1 & \mu_1 & \nu_1 \\ \lambda_2 & \mu_2 & \nu_2 \\ \lambda_3 & \mu_3 & \nu_3 \end{vmatrix} = 1$

For a combination of displacement and rotation, apply the corresponding equations simultaneously.

42. QUADRIC SURFACES

The General Form of the Equation of a Surface of the Second Degree is

$$F(x, y, z) \equiv a_{11}x^2 + 2a_{12}xy + 2a_{13}xz + a_{22}y^2 + 2a_{23}yz + a_{33}z^2 + 2a_{14}x + 2a_{24}y$$
$$+ 2a_{34}z + a_{44} = 0 \quad (33)$$

where the a_{ik} are constants, and $a_{ik} = a_{ki}$, i.e., $a_{12} = a_{21}$, etc. Let

$$D = \begin{vmatrix} a_{11} & a_{12} & a_{13} & a_{14} \\ a_{21} & a_{22} & a_{23} & a_{24} \\ a_{31} & a_{32} & a_{33} & a_{34} \\ a_{41} & a_{42} & a_{43} & a_{44} \end{vmatrix}, \quad d = \begin{vmatrix} a_{11} & a_{12} & a_{13} \\ a_{21} & a_{22} & a_{23} \\ a_{31} & a_{32} & a_{33} \end{vmatrix}$$

Let $I \equiv a_{11} + a_{22} + a_{33}$ and $J \equiv a_{22}a_{33} + a_{33}a_{11} + a_{11}a_{22} - a_{23}^2 - a_{13}^2 - a_{12}^2$. D, d, I, and J are invariant under coordinate transformation. The following is a classification of the quadratic surfaces, so far as they are real and do not degenerate into curves in one plane:

Ellipsoid, for $D < 0$, $Id > 0$, $J > 0$.
Hyperboloid of two sheets, for $D < 0$, Id and J not both > 0.
Hyperboloid of one sheet, for $D > 0$, Id and J not both > 0.
Cone, for $D = 0$, $d \neq 0$, Id and J not both > 0.
Elliptic paraboloid, for $D < 0$, $d = 0$, $J > 0$.
Hyperbolic paraboloid, for $D > 0$, $d = 0$, $J < 0$.
Cylinder, for $D = 0$, $d = 0$.

Ellipsoid and Hyperboloids. Consider the center of the quadric as the origin and the principal axes of the quadric as the orthogonal coordinate axes. Then

$$\frac{x^2}{a^2} + \frac{y^2}{b^2} + \frac{z^2}{c^2} = 1 \text{ is an } \textit{ellipsoid} \text{ (Fig. 4)} \quad (34)$$

$$\frac{x^2}{a^2} + \frac{y^2}{b^2} - \frac{z^2}{c^2} = 1 \text{ is a } \textit{hyperboloid of one sheet} \text{ (Fig. 5)} \quad (35)$$

$$\frac{x^2}{a^2} + \frac{y^2}{b^2} - \frac{z^2}{c^2} = -1 \text{ is a } \textit{hyperboloid of two sheets} \text{ (Fig. 6)} \quad (36)$$

where a, b, c are the semi-axes.

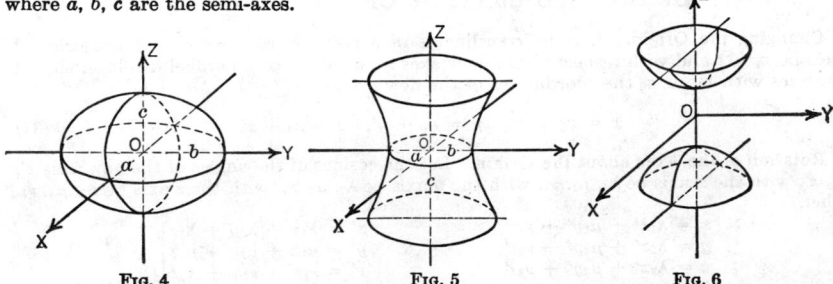

FIG. 4　　　　　　FIG. 5　　　　　　FIG. 6

The length of the semi-axis is found from

$$a^2 = -\frac{D}{\lambda_1 d}, \quad b^2 = -\frac{D}{\lambda_2 d}, \quad c^2 = -\frac{D}{\lambda_3 d} \tag{37}$$

where λ_1, λ_2, λ_3 are the real roots of the following cubic equation:

$$\begin{vmatrix} a_{11} - \lambda & a_{12} & a_{13} \\ a_{12} & a_{22} - \lambda & a_{23} \\ a_{13} & a_{23} & a_{33} - \lambda \end{vmatrix} = 0 \tag{38}$$

Cone. The equation

$$ax^2 + by^2 + cz^2 + 2hxy + 2gxz + 2fyz = 0 \tag{39}$$

represents a cone with vertex at the origin. If the cross-section of the cone is an ellipse with axes $2a$ and $2b$, whose plane is parallel to the xy plane and at a distance c from the origin, then the equation of the cone with vertex at the origin is

$$\frac{x^2}{a^2} + \frac{y^2}{b^2} - \frac{z^2}{c^2} = 0 \tag{40}$$

If $a = b$, the cross-section is circular, and the cone is a cone of revolution.
Sphere. An equation of the form

$$x^2 + y^2 + z^2 + ax + by + cz + d = 0 \tag{41}$$

represents a sphere with radius

$$r = \tfrac{1}{2}\sqrt{a^2 + b^2 + c^2 - 4d} \tag{42}$$

and center

$$x_0 = -\tfrac{1}{2}a, \quad y_0 = -\tfrac{1}{2}b, \quad z_0 = -\tfrac{1}{2}c \tag{43}$$

If (x_0, y_0, z_0) are the coordinates of the center and r is the radius, then the equation of the sphere is

$$(x - x_0)^2 + (y - y_0)^2 + (z - z_0)^2 = r^2 \tag{44}$$

If $x_0 = 0$, $y_0 = 0$, $z_0 = 0$, then the equation is

$$x^2 + y^2 + z^2 = r^2 \tag{45}$$

Paraboloids. The equation

$$\frac{x^2}{a^2} + \frac{y^2}{b^2} = 2cz \tag{46}$$

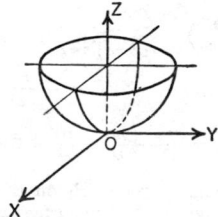

FIG. 7

represents an *elliptic paraboloid* (Fig. 7).
If $a = b$, the equation is of the form

$$x^2 + y^2 = 2cz, \text{ a paraboloid of revolution} \tag{47}$$

The equation

$$\frac{x^2}{a^2} - \frac{y^2}{b^2} = 2cz \text{ represents a } hyperbolic\ paraboloid \text{ (Fig. 8)} \tag{48}$$

Cylinder. The equation of a cylinder perpendicular to the yz, xz, or xy plane is the same as the equation of a section of the cylinder in the corresponding plane. Thus

$$\frac{x^2}{a^2} + \frac{y^2}{b^2} = 1 \tag{49}$$

$$\frac{x^2}{a^2} - \frac{y^2}{b^2} = 1 \tag{50}$$

$$y^2 = 4ax \tag{51}$$

are *elliptic, hyperbolic,* and *parabolic* cylinders respectively with elements or generators parallel to OZ.

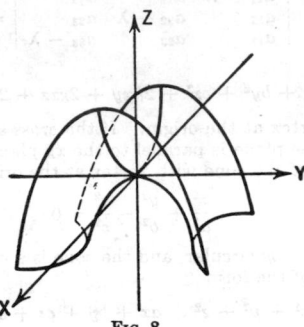

Fig. 8

Tangent Plane. The equation of the tangent plane to any quadric

$$F(x, y, z) \equiv a_{11}x^2 + 2a_{12}xy + 2a_{13}xz + a_{22}y^2 + 2a_{23}yz + a_{33}z^2 + 2a_{14}x$$
$$+ 2a_{24}y + 2a_{34}z + a_{44} = 0 \tag{52}$$

at the point (x_1, y_1, z_1) is

$$\left(\frac{\partial F}{\partial x}\right)_{x=x_1, y=y_1, z=z_1} (x - x_1) + \left(\frac{\partial F}{\partial y}\right)_{x=x_1, y=y_1, z=z_1} (y - y_1)$$
$$+ \left(\frac{\partial F}{\partial z}\right)_{x=x_1, y=y_1, z=z_1} (z - z_1) = 0 \tag{53}$$

Example. Find the tangent plane to the hyperboloid of one sheet at point (x_1, y_1, z_1).

Given $\dfrac{x^2}{a^2} + \dfrac{y^2}{b^2} - \dfrac{z^2}{c^2} = 1$. Then

$$\left(\frac{\partial F}{\partial x}\right)_{x=x_1, y=y_1, z=z_1} (x - x_1) + \left(\frac{\partial F}{\partial y}\right)_{x=x_1, y=y_1, z=z_1} (y - y_1) + \left(\frac{\partial F}{\partial z}\right)_{x=x_1, y=y_1, z=z_1} (z - z_1)$$
$$= \frac{2x_1(x - x_1)}{a^2} + \frac{2y_1(y - y_1)}{b^2} - \frac{2z_1(z - z_1)}{c^2} = 0$$

$\dfrac{xx_1}{a^2} + \dfrac{yy_1}{b^2} - \dfrac{zz_1}{c^2} - \dfrac{x_1{}^2}{a^2} - \dfrac{y_1{}^2}{b^2} + \dfrac{z_1{}^2}{c^2} = \dfrac{xx_1}{a^2} + \dfrac{yy_1}{b^2} - \dfrac{zz_1}{c^2} - 1 = 0$ is the tangent plane.

The Normal. The line through a point P_1 on a surface and perpendicular to the tangent plane at P_1 is called the *normal* to the surface at P_1.

The equations of the normal to the surface $F(x, y, z) = 0$ at the point (x_1, y_1, z_1) are

$$\frac{x - x_1}{\left(\dfrac{\partial F}{\partial x}\right)_{x=x_1, y=y_1, z=z_1}} = \frac{y - y_1}{\left(\dfrac{\partial F}{\partial y}\right)_{x=x_1, y=y_1, z=z_1}} = \frac{z - z_1}{\left(\dfrac{\partial F}{\partial z}\right)_{x=x_1, y=y_1, z=z_1}} \tag{54}$$

DIFFERENTIAL CALCULUS

43. FUNCTIONS AND DERIVATIVES

Function. If two variables x and y are so related that to each value of x in a given domain there corresponds a value of y, then y is a *function* of x in that domain. The variable x is the *independent* variable and y the *dependent* variable. The symbols $F(x)$,

$f(x)$, $\phi(x)$, etc., are used to represent functions of x; the symbol $f(a)$ represents the value of $f(x)$ for $x = a$.

Limit, Derivative, Differential. The function $f(x)$ approaches the limit 1 as x approaches a if the difference $|f(x) - 1|$ can be made arbitrarily small for all values of x except a within a sufficiently small interval with a as midpoint. In symbols, $\lim\limits_{x \to a} f(x) = 1$.

The symbols $\lim\limits_{x \to a} f(x) = \infty$ or $\lim\limits_{x \to a} f(x) = -\infty$ mean that, for all values of x except a within a sufficiently small interval with a as midpoint, the values of $f(x)$ can be made arbitrarily large positively or negatively, respectively.

The symbols $\lim\limits_{x \to \infty} f(x) = 1$ or $\lim\limits_{x \to -\infty} f(x) = 1$ mean that the difference $|f(x) - 1|$ can be made arbitrarily small for all values of x sufficiently large positively or negatively, respectively.

A change in x is called an *increment* of x and is denoted by Δx. The corresponding change in y is denoted by Δy. If

$$\lim_{\Delta x \to 0} \frac{f(x + \Delta x) - f(x)}{\Delta x}$$

exists, it is called the *derivative* of y with respect to x and is denoted by $\frac{dy}{dx}$, $f'(x)$, or $D_x y$.

The geometric interpretation of $f'(x)$ is

$$f'(x) = \frac{dy}{dx} = \tan \theta \tag{1}$$

or $f'(x)$ is equal to the slope of the tangent to the curve $y = f(x)$ at the point $P(x, y)$ (Fig. 1).

$$\frac{RS}{PR} = \lim_{PR \to 0} \frac{RQ}{PR} = \lim_{\Delta x \to 0} \frac{\Delta y}{\Delta x} = \lim_{\Delta x \to 0} \frac{f(x + \Delta x) - f(x)}{\Delta x} = \frac{dy}{dx} = f'(x) = \tan \theta \tag{2}$$

The differentials of x and y, respectively, are

$$dx = \Delta x$$
$$dy = f'(x)\, dx$$

Continuity. A function is *continuous* at $x = b$ if it has a definite value at b and approaches that value as a limit whenever x approaches b as a limit. The notion of continuity at a point suggests that the graph of the function can be drawn without lifting pencil from paper at the point. The analytic conditions that $f(x)$ be continuous at b are that $f(b)$ have a definite value and that for an arbitrarily small positive number ϵ there exist a $\delta(\epsilon)$ such that

$$|f(x) - f(b)| < \epsilon$$

for all values of x for which $|x - b| < \delta(\epsilon)$ (3)

A function which is continuous at each point of an interval is said to be continuous in that interval. An example of a continuous function is $f(x) = x^2$. The function

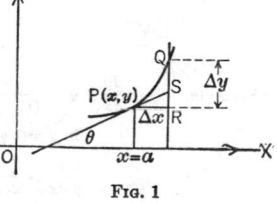

Fɪɢ. 1

$\phi(x) = \dfrac{1}{x - a}$ is continuous for all values of x except $x = a$, at which point it becomes infinite. Every differentiable function is continuous, although the reverse is not always true.

If, in the above definition of continuity, the number δ can be chosen the same for all points in the interval, the function is said to be *uniformly continuous* in that interval.

Derivatives of Higher Order. The *derivative* of the *first derivative* of y with respect to x is called the *second derivative* of y with respect to x and is denoted by

$$\frac{d}{dx}\left(\frac{dy}{dx}\right) = \frac{d^2 y}{dx^2} = f''(x) = D_x^2 y \tag{4}$$

By successive differentiations the nth derivative

$$\frac{d^n y}{dx^n} = f^{(n)}(x) = D_x^n y \tag{5}$$

is obtained.

The nth differential of y is denoted by

$$d^n y = f^{(n)}(x)\, dx^n \tag{6}$$

Parametric Differentiation. To find the derivatives of y with respect to x if $y = y(t)$ and $x = x(t)$:

$$y' = \frac{dy}{dx} = \frac{\dfrac{dy}{dt}}{\dfrac{dx}{dt}} \tag{7}$$

$$y'' = \frac{d^2y}{dx^2} = \frac{\dfrac{dy'}{dt}}{\dfrac{dx}{dt}} \tag{8}$$

$$y^{(n)} = \frac{d^ny}{dx^n} = \frac{\dfrac{dy^{(n-1)}}{dt}}{\dfrac{dx}{dt}} \tag{9}$$

Example. Find the derivatives of y with respect to x for the ellipse $x = a \cos t$, $y = b \sin t$.

$$y' = \frac{dy}{dx} = \frac{b \cos t}{-a \sin t} = -\frac{b}{a} \cot t$$

$$y'' = \frac{dy'}{dx} = \frac{\dfrac{b}{a} \csc^2 t}{-a \sin t} = -\frac{b}{a^2} \csc^3 t$$

$$y''' = \frac{dy''}{dx} = \frac{\dfrac{3b}{a^2} \csc^3 t \cot t}{-a \sin t} = -\frac{3b}{a^3} \csc^4 t \cot t$$

Logarithmic Differentiation for Products and Quotients. If

$$y = \frac{u^l v^m}{w^n} \tag{10}$$

take the logarithms of both sides before differentiating.

$$\ln y = l \ln u + m \ln v - n \ln w \tag{11}$$

$$\frac{1}{y}\frac{dy}{dx} = \frac{l}{u}\frac{du}{dx} + \frac{m}{v}\frac{dv}{dx} - \frac{n}{w}\frac{dw}{dx} \tag{12}$$

$$\frac{dy}{dx} = y\left(\frac{l}{u}\frac{du}{dx} + \frac{m}{v}\frac{dv}{dx} - \frac{n}{w}\frac{dw}{dx}\right) \tag{13}$$

Example. Find $\dfrac{dy}{dx}$ if $y = \dfrac{\sqrt{x^2 - 25}}{(x - 1)^3(x + 5)^2}$

$$\ln y = 1/2 \ln (x^2 - 25) - 3 \ln (x - 1) - 2 \ln (x + 5)$$

$$\frac{1}{y}\frac{dy}{dx} = \frac{2x}{2(x^2 - 25)} - \frac{3}{x - 1} - \frac{2}{x + 5}$$

$$\frac{dy}{dx} = \frac{y(-4x^2 + 11x + 65)}{(x^2 - 25)(x - 1)}$$

Mean Value Theorem. If $f(x)$ is single valued, continuous in the interval $a \leqq x \leqq b$, and has a derivative for all values of x between a and b, then there is a value $x = \xi$, $a < \xi < b$, such that

$$f(b) - f(a) = (b - a)f'(\xi) \tag{14}$$

Another form is

$$f(x + h) = f(x) + hf'(x + \theta h), \quad 0 < \theta < 1 \tag{15}$$

Indeterminate Forms

If a function $f(x)$ for $x = a$ (where a can also be ∞) has no determined value but appears in one of the meaningless forms

$$\frac{0}{0}, \quad \frac{\infty}{\infty}, \quad 0 \cdot \infty, \quad \infty - \infty, \quad 0^0, \quad \infty^0, \quad 0^\infty, \quad 1^\infty$$

DIFFERENTIAL CALCULUS

303

then it may happen that the $\lim f(x)$ has a definite value. For the determination of this limiting value, if it exists, the following rules can be used:

$\dfrac{0}{0}$. If $f(x) = \dfrac{\phi(x)}{\psi(x)}$, $\phi(a) = 0$, and $\psi(a) = 0$, then

$$\lim_{x \to a} f(x) = \lim_{x \to a} \frac{\phi'(x)}{\psi'(x)} \qquad \text{(L'Hospital's rule)} \quad (16)$$

If, however, $\phi'(a) = 0$ and $\psi'(a) = 0$, the rule is applied again, with the result

$$\lim_{x \to a} \frac{\phi(x)}{\psi(x)} = \lim_{\xi \to a} \frac{\phi'(\xi)}{\psi'(\xi)} = \frac{\phi''(a)}{\psi''(a)} \qquad (17)$$

unless $\phi''(a) = 0$ and $\psi''(a) = 0$. In this case, the rule is applied again, etc.

Example. Find the value of $\dfrac{\sin x}{x}$ for $x = 0$.

$$\lim_{x \to 0} \frac{\sin x}{x} = \lim_{x \to 0} \frac{\cos x}{1} = 1$$

$\dfrac{\infty}{\infty}$. If $f(x) = \dfrac{\phi(x)}{\psi(x)}$, $\phi(a) = \infty$, and $\psi(a) = \infty$, then

$$\lim_{x \to a} \frac{\phi(x)}{\psi(x)} = \lim_{x \to a} \frac{\phi'(x)}{\psi'(x)} \qquad (18)$$

as before.

$0 \cdot \infty$. If $f(x) = \phi(x) \cdot \psi(x)$, $\phi(a) = 0$, and $\psi(a) = \infty$, then place $\dfrac{1}{\psi(x)} = \omega(x)$ and obtain the previous case $\dfrac{0}{0}$.

$\infty - \infty$. If $f(x) = \phi(x) - \psi(x)$, $\phi(a) = \infty$, and $\psi(a) = \infty$, then place $\phi(x) = \dfrac{1}{u(x)}$. $\psi(x) = \dfrac{1}{v(x)}$ and obtain

$$f(x) = \frac{v(x) - u(x)}{u(x)v(x)} \qquad (19)$$

which takes the form $\dfrac{0}{0}$.

$0°, \infty°, 0^\infty, 1^\infty$. An expression of the type $[\psi(x)]^{\phi(x)}$ may, for $x = a$, give rise to the forms $0°, \infty°, 0^\infty, 1^\infty$.

Such an expression may be reduced to a type $\dfrac{0}{0}$ or $\dfrac{\infty}{\infty}$ by the use of logarithms. Thus,

$$\left. \begin{array}{l} u = [\psi(x)]^{\phi(x)} \\ \log_e u = \phi(x) \cdot \log_e \psi(x) \end{array} \right\} \qquad (20)$$

If $\lim_{x \to a} \phi(x) \cdot \log_e \psi(x)$ can be found by the previous methods, the limit approached by u can be found.

Example. $u = (1 - x)^{1/x}$ for $x = 0$.

$$\log_e u = \frac{\log_e (1 - x)}{x}$$

$$\lim_{x \to 0} \frac{\log_e (1 - x)}{x} = \lim_{x \to 0} \frac{\dfrac{-1}{1 - x}}{1} = -1$$

Therefore $\lim_{x \to 0} \log_e u = -1$ and $\lim_{x \to 0} u = e^{-1}$.

44. DIFFERENTIATION FORMULAS

Table 1. Differentiation Formulas

Let u, v, w, $\cdots$ be functions of x; a and n be constants; and e be the base of the natural or Napierian logarithms.   Then $e = 2.7183^-$.

$$\frac{d}{dx} a = 0$$

$$\frac{d}{dx}(u + v + w + \cdots) = \frac{du}{dx} + \frac{dv}{dx} + \frac{dw}{dx} + \cdots$$

$$\frac{d}{dx} au = a\frac{du}{dx}$$

$$\frac{d}{dx} uv = u\frac{dv}{dx} + v\frac{du}{dx}$$

$$\frac{d}{dx}(uvw\cdots) = \left(\frac{1}{u}\frac{du}{dx} + \frac{1}{v}\frac{dv}{dx} + \frac{1}{w}\frac{dw}{dx} + \cdots\right)(uvw\cdots)$$

$$\frac{d}{dx}\left(\frac{u}{v}\right) = \frac{v\dfrac{du}{dx} - u\dfrac{dv}{dx}}{v^2}$$

$$\frac{d}{dx} u^n = nu^{n-1}\frac{du}{dx}$$

$$\frac{d}{dx} \log_e u = \frac{1}{u}\frac{du}{dx}$$

$$\frac{d}{dx} \log_{10} u = \frac{1}{u}\frac{du}{dx}\log_{10} e = (0.4343)\frac{1}{u}\frac{du}{dx}$$

$$\frac{d}{dx} e^u = e^u\frac{du}{dx}$$

$$\frac{d}{dx} u^v = vu^{v-1}\frac{du}{dx} + u^v\frac{dv}{dx}\log_e u$$

$$\frac{d}{dx} f(u) = \frac{df(u)}{du}\cdot\frac{du}{dx}$$

$$\frac{d^2 f(u)}{dx^2} = \frac{df(u)}{du}\cdot\frac{d^2 u}{dx^2} + \frac{d^2 f(u)}{du^2}\left(\frac{du}{dx}\right)^2$$

$$\frac{d}{dx} \sin u = \cos u\frac{du}{dx}$$

$$\frac{d}{dx} \cos u = -\sin u\frac{du}{dx}$$

$$\frac{d}{dx} \tan u = \sec^2 u\frac{du}{dx}$$

$$\frac{d}{dx} \cot u = -\csc^2 u\frac{du}{dx}$$

$$\frac{d}{dx} \sec u = \sec u\tan u\frac{du}{dx}$$

$$\frac{d}{dx} \csc u = -\csc u\cot u\frac{du}{dx}$$

$$\frac{d}{dx} \sin^{-1} u = \frac{1}{\sqrt{1 - u^2}}\frac{du}{dx}\left(-\frac{\pi}{2} \leqq \sin^{-1} u \leqq \frac{\pi}{2}\right)$$

$$\frac{d}{dx} \cos^{-1} u = -\frac{1}{\sqrt{1 - u^2}}\frac{du}{dx}\ (0 \leqq \cos^{-1} u \leqq \pi)$$

$$\frac{d}{dx} \tan^{-1} u = \frac{1}{1 + u^2}\frac{du}{dx}$$

$$\frac{d}{dx} \cot^{-1} u = -\frac{1}{1 + u^2}\frac{du}{dx}$$

$$\frac{d}{dx} \sec^{-1} u = \frac{1}{u\sqrt{u^2 - 1}}\frac{du}{dx}\ *$$

$$\frac{d}{dx} \csc^{-1} u = -\frac{1}{u\sqrt{u^2 - 1}}\frac{du}{dx}\ *$$

$$\frac{d}{dx} \sinh u = \cosh u\frac{du}{dx}$$

$$\frac{d}{dx} \cosh u = \sinh u\frac{du}{dx}$$

$$\frac{d}{dx} \tanh u = \operatorname{sech}^2 u\frac{du}{dx}$$

$$\frac{d}{dx} \coth u = -\operatorname{csch}^2 u\frac{du}{dx}$$

$$\frac{d}{dx} \operatorname{sech} u = -\operatorname{sech} u\tanh u\frac{du}{dx}$$

$$\frac{d}{dx} \operatorname{csch} u = -\operatorname{csch} u\coth u\frac{du}{dx}$$

$$\frac{d}{dx} \sinh^{-1} u = \frac{1}{\sqrt{u^2 + 1}}\frac{du}{dx}$$

$$\frac{d}{dx} \cosh^{-1} u = \frac{1}{\sqrt{u^2 - 1}}\frac{du}{dx}$$

$$\frac{d}{dx} \tanh^{-1} u = \frac{1}{1 - u^2}\frac{du}{dx}$$

$$\frac{d}{dx} \coth^{-1} u = \frac{1}{1 - u^2}\frac{du}{dx}$$

$$\frac{d}{dx} \operatorname{sech}^{-1} u = -\frac{1}{u\sqrt{1 - u^2}}\frac{du}{dx}$$

$$\frac{d}{dx} \operatorname{csch}^{-1} u = -\frac{1}{u\sqrt{u^2 + 1}}\frac{du}{dx}$$

* For angles in the first and third quadrants. Use the opposite sign in the second and fourth quadrants.

45. PARTIAL DERIVATIVES

Functions of Two Variables.   If three variables $f(x, y)$, x, y are so related that to each pair of values of x and y in a given domain there corresponds a value of $f(x, y)$, then $f(x, y)$ is a function of x and y in that domain.   If x is considered as the only variable while y is taken as constant, then the derivative of $f(x, y)$ with respect to x is called the *partial derivative* of f with respect to x and is denoted by

$$\frac{\partial f}{\partial x} = f_x = \lim_{\Delta x \to 0}\frac{f(x + \Delta x, y) - f(x, y)}{\Delta x} \qquad (21)$$

Likewise, the partial derivative of f with respect to y is obtained by considering x to be constant while y varies:

$$\frac{\partial f}{\partial y} = f_y = \lim_{\Delta y \to 0} \frac{f(x, y + \Delta y) - f(x, y)}{\Delta y} \tag{22}$$

If $\dfrac{\partial f}{\partial x}$ and $\dfrac{\partial f}{\partial y}$ are again differentiable, the partial derivatives of the second order may be found.

$$\frac{\partial}{\partial x}\left(\frac{\partial f}{\partial x}\right) = \frac{\partial^2 f}{\partial x^2} = f_{xx} \qquad \frac{\partial}{\partial y}\left(\frac{\partial f}{\partial y}\right) = \frac{\partial^2 f}{\partial y^2} = f_{yy}$$

$$\frac{\partial}{\partial x}\left(\frac{\partial f}{\partial y}\right) = \frac{\partial^2 f}{\partial x\, \partial y} = f_{yx} \qquad \frac{\partial}{\partial y}\left(\frac{\partial f}{\partial x}\right) = \frac{\partial^2 f}{\partial y\, \partial x} = f_{xy} \tag{23}$$

If the derivatives in question are continuous, the order of differentiation is immaterial, that is,

$$\frac{\partial^2 f}{\partial y\, \partial x} = \frac{\partial^2 f}{\partial x\, \partial y} \tag{24}$$

Similarly, the third and higher partial derivatives of $f(x, y)$ may be found. The third partial derivatives, if continuous, are four in number:

$$\frac{\partial}{\partial x}\left(\frac{\partial^2 f}{\partial x^2}\right) = \frac{\partial^3 f}{\partial x^3} \qquad \frac{\partial}{\partial x}\left(\frac{\partial^2 f}{\partial y^2}\right) = \frac{\partial}{\partial y}\left(\frac{\partial^2 f}{\partial x\, \partial y}\right) = \frac{\partial^2}{\partial y^2}\left(\frac{\partial f}{\partial x}\right) = \frac{\partial^3 f}{\partial x\, \partial y^2}$$

$$\frac{\partial}{\partial y}\left(\frac{\partial^2 f}{\partial y^2}\right) = \frac{\partial^3 f}{\partial y^3} \qquad \frac{\partial}{\partial y}\left(\frac{\partial^2 f}{\partial x^2}\right) = \frac{\partial}{\partial x}\left(\frac{\partial^2 f}{\partial x\, \partial y}\right) = \frac{\partial^2}{\partial x^2}\left(\frac{\partial f}{\partial y}\right) = \frac{\partial^3 f}{\partial x^2\, \partial y} \tag{25}$$

Functions of N Variables. The formulas above may be generalized to the case where f is a function of more than two variables, that is, there corresponds a value of $f(x, y, z, \cdots)$ to every set of values of x, y, z, $\cdots$.

If the increments Δx, Δy, Δz, $\cdots$ are assigned to x, y, z, $\cdots$ in $f(x, y, z, \cdots)$, the *total increment* of f is

$$\Delta f = f(x + \Delta x, y + \Delta y, z + \Delta z, \cdots) - f(x, y, z, \cdots) \tag{26}$$

The *total differential* of f is

$$df = \frac{\partial f}{\partial x}\, dx + \frac{\partial f}{\partial y}\, dy + \frac{\partial f}{\partial z}\, dz + \cdots \tag{27}$$

The *second total differential* of f is

$$d^2 f = \frac{\partial^2 f}{\partial x^2}(dx)^2 + \frac{\partial^2 f}{\partial y^2}(dy)^2 + \frac{\partial^2 f}{\partial z^2}(dz)^2 + \cdots + 2\frac{\partial^2 f}{\partial x\, \partial y}\, dx\, dy + \cdots \tag{28}$$

In general,

$$d^n f = \left(\frac{\partial}{\partial x}\, dx + \frac{\partial}{\partial y}\, dy + \frac{\partial}{\partial z}\, dz + \cdots\right)^n f(x, y, z, \cdots) \tag{29}$$

Exact Differential. In order for the expression $P(x, y)\, dx + Q(x, y)\, dy$ to be the *exact* or *complete* differential of a function of two variables, it is necessary and sufficient that

$$\frac{\partial Q}{\partial x} = \frac{\partial P}{\partial y} \quad \text{(integrability condition)} \tag{30}$$

For three variables, $P\, dx + Q\, dy + R\, dz$, the corresponding conditions are

$$\frac{\partial Q}{\partial z} = \frac{\partial R}{\partial y}, \quad \frac{\partial R}{\partial x} = \frac{\partial P}{\partial z}, \quad \frac{\partial P}{\partial y} = \frac{\partial Q}{\partial x} \tag{31}$$

Differentiation of Composite Functions. If $u = f(x, y, z, \cdots w)$, and x, y, z, $\cdots w$ are functions of a single variable t, then

$$\frac{du}{dt} = \frac{\partial u}{\partial x}\frac{dx}{dt} + \frac{\partial u}{\partial y}\frac{dy}{dt} + \cdots + \frac{\partial u}{\partial w}\frac{dw}{dt} \tag{32}$$

which is the total derivative of u with respect to t.

Example. Given: $u = x^2 + y^2 + 3xy$, $x = t^2$, $y = \dfrac{1}{t}$.

Then

$$\frac{dx}{dt} = 2t, \quad \frac{dy}{dt} = -\frac{1}{t^2}$$

and

$$\frac{du}{dt} = \left(2t^2 + \frac{3}{t}\right)2t - \left(\frac{2}{t} + 3t^2\right)\frac{1}{t^2}$$

The equation reduces to

$$\frac{du}{dt} = 4t^3 + 3 - \frac{2}{t^3}$$

which expresses the rate of change of u with respect to t as a function of t.

Implicit Functions. The equation $F(x, y) = 0$ defines y as an *implicit* function of x, and x as an implicit function of y. If the equation is solved for y in terms of x, $y = f(x)$, then y is called an *explicit* function of x.

Example. Implicit function: $F(x,y) = x^2 + y^2 - r^2 = 0$. Explicit function: $y = \pm\sqrt{r^2 - x^2}$.

To find $\dfrac{dy}{dx}$, either differentiate $y = f(x)$ or use

$$\frac{dy}{dx} = -\frac{\dfrac{\partial F}{\partial x}}{\dfrac{\partial F}{\partial y}} \qquad \left(\frac{\partial F}{\partial y} \neq 0\right) \qquad (33)$$

$$\frac{d^2y}{dx^2} = -\frac{\dfrac{\partial^2 F}{\partial x^2}\left(\dfrac{\partial F}{\partial y}\right)^2 - 2\dfrac{\partial^2 F}{\partial x \partial y}\dfrac{\partial F}{\partial x}\dfrac{\partial F}{\partial y} + \dfrac{\partial^2 F}{\partial y^2}\left(\dfrac{\partial F}{\partial x}\right)^2}{\left(\dfrac{\partial F}{\partial y}\right)^3} \qquad \left(\frac{\partial F}{\partial y} \neq 0\right) \qquad (34)$$

46. INFINITE SERIES

Let $a_1, a_2, \cdots, a_n, \cdots$ be a sequence of numbers formed according to some rule. The indicated sum

$$\sum_{n=1}^{\infty} a_n = a_1 + a_2 + \cdots + a_n + \cdots \qquad (35)$$

is called an *infinite series*. Let $s_n = a_1 + a_2 + \cdots + a_n$. If the partial sums s_n approach a limit S as $n \to \infty$, then the series is *convergent* and S is the *sum* or *value* of the series. A series which is not convergent is *divergent*.

If the series of absolute values $|a_1| + |a_2| + \cdots + |a_n| + \cdots$ is convergent, then the series (35) is *absolutely convergent*. A series which converges, but not absolutely, is *conditionally convergent*. The sum of an absolutely convergent series is not changed by rearrangement of its terms.

Tests for Convergence

Comparison Test for series of positive terms. If there is a convergent series of positive terms $c_1 + c_2 + \cdots + c_n + \cdots$, such that $a_n \leq c_n$ for every n from some term on, then the series (35) converges. If there is a divergent series of positive terms $d_1 + d_2 + \cdots + d_n + \cdots$, such that $a_n \geq d_n$ for every n from some term on, then the series (35) diverges. Two useful comparison series are

1. The *geometric* series $a + ar + ar^2 + \cdots + ar^{n-1} + \cdots$, which converges for $|r| < 1$ and diverges for $|r| \geq 1$.

2. The p-series $1 + \dfrac{1}{2^p} + \dfrac{1}{3^p} + \cdots + \dfrac{1}{n^p} + \cdots$, which converges for $p > 1$ and diverges for $p \leq 1$.

Ratio Test. Let

$$L = \lim_{n \to \infty} \left|\frac{a_{n+1}}{a_n}\right| \qquad (36)$$

If $L < 1$, the series (35) converges absolutely; if L does not exist or if $L > 1$, the series (35) diverges; if $L = 1$, the test fails.

Examples.

$$(1) \quad 10 + \frac{10^2}{2!} + \frac{10^3}{3!} + \cdots + \frac{10^n}{n!} + \cdots$$

Since $L = \lim_{n \to \infty} \dfrac{\dfrac{10^{n+1}}{(n+1)!}}{\dfrac{10^n}{n!}} = \lim_{n \to \infty} \dfrac{10}{n+1} = 0$, the series converges.

$$(2) \quad \frac{1+1}{1+3} + \frac{(1+1)(2+1)}{(1+3)(2+3)} + \cdots + \frac{(1+1)(2+1)\cdots(n+1)}{(1+3)(2+3)\cdots(n+3)} + \cdots$$

Since $L = \lim_{n \to \infty} \dfrac{(n+1)+1}{(n+1)+3} = 1$, the test fails. Raabe's test can be used. See eq. (38).

Root Test. Let

$$L = \lim_{n \to \infty} |a_n|^{1/n} \tag{37}$$

If $L < 1$, the series (35) converges; if $L > 1$, the series (35) diverges; if $L = 1$, the test fails.

Example.

$$1 + \frac{1}{(\log 2)^2} + \frac{1}{(\log 3)^3} + \cdots + \frac{1}{(\log n)^n} + \cdots$$

Since $L = \lim_{n \to \infty} \dfrac{1}{\log n} = 0$, the series converges.

Integral Test. Let $f(n) = a_n$. If $f(x)$ is a positive non-increasing function of x for $x > k$, then the series (35) converges or diverges with the improper integral $\displaystyle\int_k^\infty f(x)\,dx$.

Example.

$$1 + \frac{1}{2(\log 2)^3} + \frac{1}{3(\log 3)^3} + \cdots + \frac{1}{n(\log n)^3} + \cdots$$

Then

$$f(x) = \frac{1}{x(\log x)^3} \text{ for } x \geq 2$$

and

$$\int_2^\infty \frac{dx}{x(\log x)^3} = \lim_{n \to \infty} \frac{1}{2}\left(\frac{1}{(\log 2)^2} - \frac{1}{(\log n)^2}\right) = \frac{1}{2(\log 2)^2}$$

Since the integral is convergent, the series is also.

Raabe's Test. Let

$$L = \lim_{n \to \infty} n\left(\frac{a_n}{a_{n+1}} - 1\right) \tag{38}$$

If $L > 1$, the series (35) converges; if $L < 1$, the series (35) diverges; if $L = 1$, the test fails.

Example.

$$\frac{1+1}{1+3} + \frac{(1+1)(2+1)}{(1+3)(2+3)} + \cdots + \frac{(1+1)(2+1)\cdots(n+1)}{(1+3)(2+3)\cdots(n+3)} + \cdots$$

Since $L = \lim_{n \to \infty} n\left(\dfrac{(n+1)+3}{(n+1)+1} - 1\right) = \lim_{n \to \infty} \dfrac{2n}{n+2} = 2 > 1$, the series converges.

Convergence of an Alternating Series. A series

$$a_1 - a_2 + a_3 - + \cdots + (-1)^{n+1}a_n + \cdots \tag{39}$$

in which the terms are alternately positive and negative is an *alternating series*. If, from some term on, $|a_{n+1}| \leq |a_n|$ and $a_n \to 0$ as $n \to \infty$, the series converges. The sum of the first n terms differs numerically from the sum of the series by less than $|a_{n+1}|$.

Series of Functions

A *power series* is a series of the form

$$\sum_{n=0}^{\infty} a_n x^n = a_0 + a_1 x + a_2 x^2 + \cdots + a_n x^n + \cdots \tag{40}$$

If $\lim\limits_{n \to \infty} \left| \dfrac{a_{n-1}}{a_n} \right| = r$, the power series converges absolutely for all values of x in the interval $-r < x < r$. For $|x| = r$, it is necessary to use one of the convergence tests for series of numerical terms:

Example.

$$1 - \frac{x}{1 \cdot 2} + \frac{x^2}{2 \cdot 2^2} - \frac{x^3}{3 \cdot 2^3} + \cdots + (-1)^n \frac{x^n}{n \cdot 2^n} + \cdots$$

Since $\lim\limits_{n \to \infty} \dfrac{n \cdot 2^n}{(n-1)2^{n-1}} = 2$, the interval of convergence is $-2 < x < 2$. For $x = 2$, the series is a convergent alternating series. For $x = -2$, it is a divergent p-series.

Taylor's Series. If $f(x)$ has continuous derivatives in the neighborhood of a point $x = a$, then

$$f(x) = f(a) + \frac{f'(a)}{1!}(x-a) + \frac{f''(a)}{2!}(x-a)^2 + \cdots + \frac{f^{(n-1)}(a)}{(n-1)!}(x-a)^{n-1} + \cdots \quad (41)$$

with the **remainder** after n terms

$$R_n = \frac{f^{(n)}(\xi)}{n!}(x-a)^n, \quad \xi = a + \theta(x-a), \quad 0 < \theta < 1 \quad (42)$$

Another form of Taylor's series is

$$f(x+h) = f(x) + \frac{h}{1!}f'(x) + \frac{h^2}{2!}f''(x) + \cdots + \frac{h^{n-1}}{(n-1)!}f^{(n-1)}(x) + \cdots \quad (43)$$

with the remainder after n terms

$$R_n = \frac{h^n}{n!}f^{(n)}(\xi), \quad \xi = x + \theta h, \quad 0 < \theta < 1 \quad (44)$$

Maclaurin's Series. If $a = 0$ in equation (41),

$$f(x) = f(0) + \frac{f'(0)}{1!}x + \frac{f''(0)}{2!}x^2 + \cdots + \frac{f^{(n-1)}(0)}{(n-1)!}x^{n-1} + \cdots \quad (45)$$

with the remainder after n terms

$$R_n = \frac{f^{(n)}(\xi)}{n!}x^n, \quad \xi = \theta x, \quad 0 < \theta < 1 \quad (46)$$

A Taylor's or Maclaurin's series represents a function in an interval if and only if $R_n \to 0$ as $n \to \infty$.

Example. Expand e^{ax} in powers of x.

$$f(x) = e^{ax}, f'(x) = ae^{ax}, f''(x) = a^2e^{ax}, f'''(x) = a^3e^{ax}, \cdots$$

$$f(0) = 1, f'(0) = a, f''(0) = a^2, f'''(0) = a^3, \cdots$$

$$f(x) = e^{ax} = 1 + \frac{a}{1!}x + \frac{a^2}{2!}x^2 + \frac{a^3}{3!}x^3 + \cdots$$

Since $\lim\limits_{n \to \infty} \dfrac{\dfrac{a^{n-1}}{(n-1)!}}{\dfrac{a^n}{n!}} = \lim\limits_{n \to \infty} \dfrac{n}{a} = \infty$, the series converges for all values of x.

Taylor's Series for Two Variables.

$$f(x+h, y+k) = f(x, y) + \frac{1}{1!}\left(h\frac{\partial}{\partial x} + k\frac{\partial}{\partial y}\right)f(x, y)$$

$$+ \frac{1}{2!}\left(h\frac{\partial}{\partial x} + k\frac{\partial}{\partial y}\right)^2 f(x, y) + \cdots + \frac{1}{(n-1)!}\left(h\frac{\partial}{\partial x} + k\frac{\partial}{\partial y}\right)^{n-1}f(x, y) + \cdots \quad (47)$$

with the remainder

$$R_n = \frac{1}{n!}\left(h\frac{\partial}{\partial x} + k\frac{\partial}{\partial y}\right)^n f(x + \theta h, y + \theta k), \quad 0 < \theta < 1 \quad (48)$$

Fourier Series. If $f(x)$ is of bounded variation over an interval of length $2l$, that is, if it can be expressed as the difference of two non-decreasing or non-increasing bounded functions. then

$$f(x) = \frac{a_0}{2} + \sum_{n=1}^{\infty} \left(a_n \cos \frac{n\pi x}{l} + b_n \sin \frac{n\pi x}{l} \right)$$

$$= \frac{a_0}{2} + a_1 \cos \frac{\pi x}{l} + a_2 \cos \frac{2\pi x}{l} + \cdots + b_1 \sin \frac{\pi x}{l} + b_2 \sin \frac{2\pi x}{l} + \cdots \quad (49)$$

in which

$$a_n = \frac{1}{l} \int_k^{k+2l} f(x) \cos \frac{n\pi x}{l} \, dx, \quad b_n = \frac{1}{l} \int_k^{k+2l} f(x) \sin \frac{n\pi x}{l} \, dx, \quad n = 0, 1, 2, \cdots \quad (50)$$

In exponential form

$$f(x) = \sum_{n=-\infty}^{\infty} c_n e^{\frac{in\pi x}{l}}, \quad c_n = \frac{1}{2l} \int_k^{k+2l} f(x) e^{\frac{-in\pi x}{l}} \, dx, \quad n = \cdots, -2, -1, 0, 1, 2, \cdots \quad (51)$$

At a point of discontinuity a Fourier series gives the value at the midpoint of the jump.

Example. Expand e^x in the interval 0 to 2π.

$$a_0 = \frac{1}{\pi} \int_0^{2\pi} e^x \, dx = \frac{1}{\pi} (e^{2\pi} - 1), \quad a_n = \frac{1}{\pi} \int_0^{2\pi} e^x \cos nx \, dx = \frac{e^{2\pi} - 1}{\pi(n^2 + 1)}, \quad b_n = -\frac{n(e^{2\pi} - 1)}{\pi(n^2 + 1)}$$

Hence

$$e^x = \frac{1}{\pi} (e^{2\pi} - 1) \left[\frac{1}{2} + \frac{1}{1^2 + 1} \cos x + \frac{1}{2^2 + 1} \cos 2x + \frac{1}{3^2 + 1} \cos 3x + \cdots \right]$$

$$- \frac{1}{\pi} (e^{2\pi} - 1) \left[\frac{1}{1^2 + 1} \sin x + \frac{2}{2^2 + 1} \sin 2x + \frac{3}{3^2 + 1} \sin 3x + \cdots \right]$$

The expansion is valid only in the interval from 0 to 2π; outside that interval the series repeats itself owing to the periodic property of $\sin nx$ and $\cos nx$.

Fourier Series for Even or Odd Functions.
If $f(-x) = f(x)$, it is an *even* function. Then

$$a_n = \frac{2}{l} \int_0^l f(x) \cos \frac{n\pi x}{l} \, dx, \quad n = 0, 1, 2, \cdots$$

$$b_n = 0 \quad (52)$$

If $f(-x) = -f(x)$, it is an *odd* function. Then

$$a_n = 0$$

$$b_n = \frac{2}{l} \int_0^l f(x) \sin \frac{n\pi x}{l} \, dx, \quad n = 0, 1, 2, \cdots \quad (53)$$

Example. Expand $f(x) = x$ in a cosine series in the interval $(0, \pi)$.
Here

$$\frac{1}{2} a_0 = \frac{1}{\pi} \int_0^{\pi} x \, dx = \frac{\pi}{2}$$

$$a_n = \frac{2}{\pi} \int_0^{\pi} x \cos nx \, dx = \frac{2}{\pi} \left\{ \left[\frac{x \sin nx}{n} \right]_0^{\pi} - \int_0^{\pi} \frac{\sin nx}{n} \, dx \right\}$$

$$= \frac{2}{\pi} \left[\frac{1}{n^2} \cos nx \right]_0^{\pi} = \frac{2}{\pi n^2} (\cos n\pi - 1)$$

Therefore

$$x = \frac{\pi}{2} - \frac{4}{\pi} \left[\cos x + \frac{\cos 3x}{3^2} + \frac{\cos 5x}{5^2} + \cdots \right] \quad (0 < x < \pi)$$

If $x = 0$, the sum of the series is 0; if $x = \pi$, the sum of the series is π.

Uniform Convergence.
Let $R_n(x)$ be the remainder after n terms of the series of functions

$$\sum_{n=1}^{\infty} u_n(x) = u_1(x) + u_2(x) + \cdots + u_n(x) + \cdots \quad (54)$$

The series is uniformly convergent in the interval $a \leq x \leq b$ if, for any $\epsilon > 0$, there exists an N, dependent on ϵ but not on x, such that $|R_n(x)| < \epsilon$ for $n > N$.

If a power series converges in the interval $-r < x < r$, then it converges uniformly in any interval within this interval.

The sum of a uniformly convergent series of continuous functions is also a continuous function.

Weierstrass M-Test. If $\sum_{n=1}^{\infty} u_n(x)$ is a series of functions defined in an interval, if

$\sum_{n=1}^{\infty} M_n$ is a series of positive constants, and if $|\, u_n(x)\,| \leqq M_n$ for all values of x in the

interval, then $\sum_{n=1}^{\infty} u_n(x)$ is absolutely and uniformly convergent in the interval.

Operations. *Term by term differentiation.* If $f(x) = \sum_{n=1}^{\infty} u_n(x)$ is a convergent series

of differentiable functions in an interval, and if $\sum_{n=1}^{\infty} u'_n(x)$ is a series of continuous functions

which converges uniformly in the interval, then $\sum_{n=1}^{\infty} u'_n(x) = f'(x)$.

Term by term integration. If $f(x) = \sum_{n=1}^{\infty} u_n(x)$ converges uniformly in an interval, and

if a and x are any two values in the interval, then $\sum_{n=1}^{\infty} \int_a^x u_n(t)\, dt$ converges to $\int_a^x f(t)\, dt$.
It converges uniformly with respect to x for each fixed value of a.

Two *power series* can be *added, subtracted,* or *multiplied* term by term, and the result is a power series which converges when both of the first do and represents the *sum, difference,* or *product,* respectively, of the two series. The *product* of two power series $\sum_{n=0}^{\infty} a_n x^n$

and $\sum_{n=0}^{\infty} b_n x^n$ is

$$a_0 b_0 + (a_0 b_1 + a_1 b_0)x + (a_0 b_2 + a_1 b_1 + a_2 b_0)x^2 + \cdots$$
$$+ (a_0 b_n + a_1 b_{n-1} + \cdots + a_n b_0)x^n + \cdots \quad (55)$$

The *quotient* of two convergent power series $\sum_{n=0}^{\infty} a_n x^n$ and $\sum_{n=0}^{\infty} b_n x^x$, $b_0 \neq 0$, is

$$\sum_{n=0}^{\infty} q_n x^n = \frac{a_0}{b_0} + \frac{a_1 b_0 - a_0 b_1}{b_0^2} x + \frac{a_2 b_0^2 - a_1 b_0 b_1 + a_0 b_1^2 - a_0 b_0 b_2}{b_0^3} x^2 + \cdots \quad (56)$$

The interval of convergence of the quotient series must be determined. To obtain q_3, q_4, $\cdots$, solve the equations

$$a_0 = q_0 b_0$$
$$a_1 = q_0 b_1 + q_1 b_0$$
$$a_2 = q_0 b_2 + q_1 b_1 + q_2 b_0$$
$$\cdot\ \cdot\ \cdot\ \cdot\ \cdot\ \cdot\ \cdot\ \cdot\ \cdot\ \cdot\ \cdot\ \cdot\ \cdot\ \cdot$$
$$a_n = q_0 b_n + q_1 b_{n-1} + \cdots + q_n b_0 \quad (57)$$

Table 2.   Functions Expanded in Series

$$(\log = \log_e)$$

$$(a + x)^n = a^n + n a^{n-1} x + \frac{n(n-1)}{2!} a^{n-2} x^2 + \frac{n(n-1)(n-2)}{3!} a^{n-3} x^3 + \cdots \qquad (x^2 < a^2)$$

$$e^x = 1 + x + \frac{x^2}{2!} + \frac{x^3}{3!} + \frac{x^4}{4!} + \cdots \qquad (-\infty < x < \infty)$$

$$a^x = 1 + x \log a + \frac{(x \log a)^2}{2!} + \frac{(x \log a)^3}{3!} + \cdots \qquad (-\infty < x < \infty)$$

Table 2. Functions Expanded in Series (*Continued*)

$$e^{-x^2} = 1 - x^2 + \frac{x^4}{2!} - \frac{x^6}{3!} + \frac{x^8}{4!} - \cdots \qquad (-\infty < x < \infty)$$

$$e^{\sin x} = 1 + x + \frac{x^2}{2!} - \frac{3x^4}{4!} - \frac{8x^5}{5!} - \frac{3x^6}{6!} + \frac{56x^7}{7!} + \cdots \qquad (-\infty < x < \infty)$$

$$e^{\cos x} = e\left(1 - \frac{x^2}{2!} + \frac{4x^4}{4!} - \frac{31x^6}{6!} + \cdots\right) \qquad (-\infty < x < \infty)$$

$$e^{\tan x} = 1 + x + \frac{x^2}{2!} + \frac{3x^3}{3!} + \frac{9x^4}{4!} + \frac{37x^5}{5!} + \cdots \qquad \left(-\frac{\pi}{2} < x < \frac{\pi}{2}\right)$$

$$\log x = \frac{x-1}{x} + \frac{1}{2}\left(\frac{x-1}{x}\right)^2 + \frac{1}{3}\left(\frac{x-1}{x}\right)^3 + \cdots \qquad \left(x > \frac{1}{2}\right)$$

$$\log x = 2\left[\frac{x-1}{x+1} + \frac{1}{3}\left(\frac{x-1}{x+1}\right)^3 + \frac{1}{5}\left(\frac{x-1}{x+1}\right)^5 + \cdots\right] \qquad (x > 0)$$

$$\log(1+x) = x - \frac{x^2}{2} + \frac{x^3}{3} - \frac{x^4}{4} + \cdots \qquad (-1 < x < 1)$$

$$\log\left(\frac{1+x}{1-x}\right) = 2\left[x + \frac{x^3}{3} + \frac{x^5}{5} + \frac{x^7}{7} + \cdots\right] \qquad (-1 < x < 1)$$

$$\log\left(\frac{x+1}{x-1}\right) = 2\left[\frac{1}{x} + \frac{1}{3x^3} + \frac{1}{5x^5} + \cdots\right] \qquad (x^2 > 1)$$

$$\log \sin x = \log x - \frac{x^2}{6} - \frac{x^4}{180} - \frac{x^6}{2835} - \cdots \qquad (-\pi < x < \pi)$$

$$\log \cos x = -\frac{x^2}{2} - \frac{x^4}{12} - \frac{x^6}{45} - \frac{17x^8}{2520} - \cdots \qquad \left(-\frac{\pi}{2} < x < \frac{\pi}{2}\right)$$

$$\log \tan x = \log x + \frac{x^2}{3} + \frac{7x^4}{90} + \frac{62x^6}{2835} + \cdots \qquad \left(-\frac{\pi}{2} < x < \frac{\pi}{2}\right)$$

$$\sin x = x - \frac{x^3}{3!} + \frac{x^5}{5!} - \frac{x^7}{7!} + \cdots \qquad (-\infty < x < \infty)$$

$$\cos x = 1 - \frac{x^2}{2!} + \frac{x^4}{4!} - \frac{x^6}{6!} + \cdots \qquad (-\infty < x < \infty)$$

$$\tan x = x + \frac{x^3}{3} + \frac{2x^5}{15} + \frac{17x^7}{315} + \frac{62x^9}{2835} + \cdots \qquad \left(-\frac{\pi}{2} < x < \frac{\pi}{2}\right)$$

$$\cot x = \frac{1}{x} - \frac{x}{3} - \frac{x^3}{45} - \frac{2x^5}{945} - \frac{x^7}{4725} - \cdots \qquad (-\pi < x < \pi)$$

$$\sec x = 1 + \frac{x^2}{2!} + \frac{5x^4}{4!} + \frac{61x^6}{6!} + \cdots \qquad \left(-\frac{\pi}{2} < x < \frac{\pi}{2}\right)$$

$$\csc x = \frac{1}{x} + \frac{x}{3!} + \frac{7x^3}{3\cdot5!} + \frac{31x^5}{3\cdot7!} + \cdots \qquad (-\pi < x < \pi)$$

$$\sin^{-1} x = x + \frac{x^3}{2\cdot3} + \frac{3x^5}{2\cdot4\cdot5} + \frac{3\cdot5x^7}{2\cdot4\cdot6\cdot7} + \cdots \qquad (-1 \leqq x \leqq 1)$$

$$\cos^{-1} x = \frac{\pi}{2} - \sin^{-1} x$$

$$\tan^{-1} x = \frac{\pi}{2} - \frac{1}{x} + \frac{1}{3x^3} - \frac{1}{5x^5} + \cdots \qquad (x^2 \geqq 1)$$

$$= x - \frac{x^3}{3} + \frac{x^5}{5} - \frac{x^7}{7} + \cdots \qquad (-1 \leqq x \leqq 1)$$

$$\cot^{-1} x = \frac{\pi}{2} - \tan^{-1} x$$

$$\sec^{-1} x = \frac{\pi}{2} - \frac{1}{x} - \frac{1}{2\cdot3x^3} - \frac{3}{2\cdot4\cdot5x^5} - \frac{3\cdot5}{2\cdot4\cdot6\cdot7x^7} - \cdots \qquad (x^2 > 1)$$

$$\csc^{-1} x = \frac{\pi}{2} - \sec^{-1} x$$

Table 2. Functions Expanded in Series (*Continued*)

$$\sinh x = x + \frac{x^3}{3!} + \frac{x^5}{5!} + \frac{x^7}{7!} + \cdots \qquad (-\infty < x < \infty)$$

$$\cosh x = 1 + \frac{x^2}{2!} + \frac{x^4}{4!} + \frac{x^6}{6!} + \frac{x^8}{8!} + \cdots \qquad (-\infty < x < \infty)$$

$$\tanh x = x - \frac{x^3}{3} + \frac{2x^5}{15} - \frac{17x^7}{315} + \cdots \qquad \left(-\frac{\pi}{2} < x < \frac{\pi}{2}\right)$$

$$\coth x = \frac{1}{x} + \frac{x}{3} - \frac{x^3}{45} + \frac{2x^5}{945} - \frac{x^7}{4725} + \cdots \qquad (-\pi < x < \pi)$$

$$\operatorname{sech} x = 1 - \frac{x^2}{2!} + \frac{5x^4}{4!} - \frac{61x^6}{6!} + \frac{1385x^8}{8!} - \cdots \qquad \left(-\frac{\pi}{2} < x < \frac{\pi}{2}\right)$$

$$\operatorname{csch} x = \frac{1}{x} - \frac{x}{6} + \frac{7x^3}{360} - \frac{31x^5}{15,120} + \cdots \qquad (-\pi < x < \pi)$$

$$\sinh^{-1} x = x - \frac{x^3}{2 \cdot 3} + \frac{3x^5}{2 \cdot 4 \cdot 5} - \frac{3 \cdot 5 x^7}{2 \cdot 4 \cdot 6 \cdot 7} + \cdots \qquad (-1 < x < 1)$$

$$\sinh^{-1} x = \log 2x + \frac{1}{2 \cdot 2x^2} - \frac{3}{2 \cdot 4 \cdot 4x^4} + \frac{3 \cdot 5}{2 \cdot 4 \cdot 6 \cdot x^6} + \cdots \qquad (x^2 > 1)$$

$$\cosh^{-1} x = \pm \left(\log 2x - \frac{1}{2 \cdot 2x^2} - \frac{1 \cdot 3}{2 \cdot 4 \cdot 4x^4} - \frac{1 \cdot 3 \cdot 5}{2 \cdot 4 \cdot 6x^6} - \cdots \right) \qquad (x > 1)$$

$$\tanh^{-1} x = x + \frac{x^3}{3} + \frac{x^5}{5} + \frac{x^7}{7} + \cdots \qquad (-1 < x < 1)$$

$$\coth^{-1} x = \frac{1}{x} + \frac{1}{3x^3} + \frac{1}{5x^5} + \frac{1}{7x^7} + \cdots \qquad (x^2 > 1)$$

$$\operatorname{sech}^{-1} x = \pm \left(\log \frac{2}{x} - \frac{1}{2 \cdot 2} x^2 - \frac{1 \cdot 3}{2 \cdot 4 \cdot 4} x - \frac{1 \cdot 3 \cdot 5}{2 \cdot 4 \cdot 6 \cdot 6} x^6 - \cdots \right) \qquad (0 < x < 1)$$

$$\operatorname{csch}^{-1} x = \frac{1}{x} - \frac{1}{2 \cdot 3x^3} + \frac{3}{2 \cdot 4 \cdot 5x^5} - \frac{3 \cdot 5}{2 \cdot 4 \cdot 6 \cdot 7x^7} + \cdots \qquad (x^2 > 1)$$

47. MAXIMA AND MINIMA

Function of One Variable. A function $f(x)$ has a relative *maximum* (*minimum*) at a point $x = a$ if at every point in some neighborhood of $x = a$ the values of $f(x)$ are all less (greater) than $f(a)$. Either a maximum or a minimum is an *extreme*. If the derivative exists at a relative extreme, it must be zero, that is, the tangent must be parallel to the

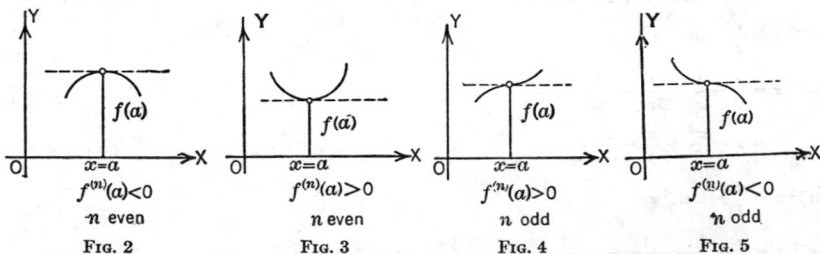

$f^{(n)}(a) < 0$	$f^{(n)}(a) > 0$	$f^{(n)}(a) > 0$	$f^{(n)}(a) < 0$
n even	*n* even	*n* odd	*n* odd
Fig. 2	Fig. 3	Fig. 4	Fig. 5

x axis. To locate possible extreme points solve the equation $f'(x) = 0$. A solution $x = a$ gives a maximum (minimum) value of $f(x)$ if and only if the derivative is positive (negative) for $x < a$ and negative (positive) for $x > a$. If the derivative does not change sign, $x = a$ gives a *point of inflection*.

A solution $x = a$ can be tested also by using the higher derivatives of $f(x)$. Let $f^{(n)}(x)$ be the first derivative which does not equal zero for $x = a$, $f^{(n)}(a) \neq 0$. If n is even, there is a maximum (Fig. 2) if $f^{(n)}(a) < 0$, and a minimum (Fig. 3) if $f^{(n)}(a) > 0$. If n is odd, there is a point of inflection (Fig. 4 and Fig. 5). In many problems physical considerations make testing unnecessary.

Example. A piece of wire of length 30 in. is bent into a rectangle. Find the maximum area. Let $x =$ the base, then $1/2 (30 - 2x) =$ the altitude.
The area, $A = x(15 - x) = 15x - x^2$.
For a maximum or minimum, $\dfrac{dA}{dx} = 15 - 2x = 0$, $x = 7.5$.
Then $A = 7.5(15 - 7.5) = 56.25$ sq in.
To find whether the area is maximum or minimum, $\dfrac{d^2A}{dx^2} = -2$ which is less than 0, and therefore the area 56.25 sq in. is a maximum.

Function of Two or More Variables. A function $f(x, y)$ has a relative *maximum* (*minimum*) at a point $(x, y) = (a, b)$ if at every point in some neighborhood of (a, b) the values of $f(x, y)$ are all less (greater) than $f(a, b)$. If the first partial derivatives exist at a relative extreme, it is necessary that

$$\frac{\partial f}{\partial x} = \frac{\partial f}{\partial y} = 0 \tag{58}$$

If, furthermore, at the point (a, b)

$$\frac{\partial^2 f}{\partial x^2} \frac{\partial^2 f}{\partial y^2} - \left(\frac{\partial^2 f}{\partial x\,\partial y} \right)^2 > 0 \tag{59}$$

then $f(a, b)$ is an extreme value, which is a maximum if

$$\frac{\partial^2 f}{\partial x^2} < 0 \quad \text{(and consequently } \frac{\partial^2 f}{\partial y^2} < 0) \tag{60}$$

and a minimum if

$$\frac{\partial^2 f}{\partial x^2} > 0 \tag{61}$$

If

$$\frac{\partial^2 f}{\partial x^2} \frac{\partial^2 f}{\partial y^2} - \left(\frac{\partial^2 f}{\partial x\,\partial y} \right)^2 < 0 \tag{62}$$

then $f(x, y)$ does not have an extreme value, but has a saddle point. If

$$\frac{\partial^2 f}{\partial x^2} \frac{\partial^2 f}{\partial y^2} - \left(\frac{\partial^2 f}{\partial x\,\partial y} \right)^2 = 0 \tag{63}$$

the test gives no information.

For a function of several variables $f(x, y, z, \cdots)$, necessary conditions for an extreme value are

$$\frac{\partial f}{\partial x} = \frac{\partial f}{\partial y} = \frac{\partial f}{\partial z} = \cdots = 0 \tag{64}$$

INTEGRAL CALCULUS

48. INTEGRATION

The Operation of Integration is the inverse of differentiation. If $F'(x) = f(x)$, then $F(x)$ is an *indefinite integral* of $f(x)$. Any other indefinite integral of $f(x)$ differs from $F(x)$ at most by an additive constant. In symbols, $F(x) = \int f(x)\,dx + c$; or, more precisely, $F(x) = \int_a^x f(\xi)\,d\xi + c$, where a and c are constants (see Definite Integrals below). The function $f(x)$ is the *integrand*.

The integrals of many functions are given in Table 23, p. 1-131. Table 1 is a convenient short table of fundamental integrals. The constant of integration is usually omitted in tables.

2. MATHEMATICS

Table 1. Fundamental Integrals

1. $\int u^n \, du = \dfrac{u^{n+1}}{n+1}.$

2. $\int \dfrac{du}{u} = \log u.$

3. $\int a^u \, du = \dfrac{a^u}{\log a}.$

4. $\int e^u \, du = e^u.$

5. $\int \cos u \, du = \sin u.$

6. $\int \sin u \, du = -\cos u.$

7. $\int \sec^2 u \, du = \tan u.$

8. $\int \csc^2 u \, du = -\cot u.$

9. $\int \sec u \tan u \, du = \sec u.$

10. $\int \csc u \cot u \, du = -\csc u.$

11. $\int \tan u \, du = \log \sec u$

    $= -\log \cos u.$

12. $\int \cot u \, du = \log \sin u$

    $= -\log \csc u.$

13. $\int \sec u \, du = \log (\sec u + \tan u)$

    $= \log \tan \left(\dfrac{\pi}{4} + \dfrac{u}{2} \right).$

14. $\int \csc u \, du = \log(\csc u - \cot u)$

    $= \log \tan \dfrac{u}{2}.$

15. $\int \dfrac{du}{u^2 + a^2} = \dfrac{1}{a} \tan^{-1} \dfrac{u}{a},$

    $= -\dfrac{1}{a} \cot^{-1} \dfrac{u}{a}.$

16. $\int \dfrac{du}{u^2 - a^2} = \dfrac{1}{2a} \log \dfrac{u-a}{u+a},$ $(u^2 > a^2)$

    $= \dfrac{1}{2a} \log \dfrac{a-u}{a+u}$ $(u^2 < a^2)$

    $= -\dfrac{1}{a} \tanh^{-1} \dfrac{u}{a}$ $(u^2 < a^2)$

    $= -\dfrac{1}{a} \coth^{-1} \dfrac{u}{a}$ $(u^2 > a^2).$

17. $\int \dfrac{du}{\sqrt{a^2 - u^2}} = \sin^{-1} \dfrac{u}{a},$

    $= -\cos^{-1} \dfrac{u}{a}.$

18. $\int \dfrac{du}{\sqrt{u^2 \pm a^2}} = \log (u + \sqrt{u^2 \pm a^2})$

    $\int \dfrac{du}{\sqrt{u^2 + a^2}} = \sinh^{-1} \dfrac{u}{a}$

    $\int \dfrac{du}{\sqrt{u^2 - a^2}} = \cosh^{-1} \dfrac{u}{a}.$

19. $\int \dfrac{du}{u\sqrt{u^2 - a^2}} = \dfrac{1}{a} \sec^{-1} \dfrac{u}{a},$

    $= -\dfrac{1}{a} \csc^{-1} \dfrac{u}{a}.$

20. $\int \dfrac{du}{\sqrt{2au - u^2}} = \text{vers}^{-1} \dfrac{u}{a}$

    $= \cos^{-1} \left(1 - \dfrac{u}{a} \right).$

21. $\int \sinh u \, du = \cosh u.$

22. $\int \cosh u \, du = \sinh u.$

23. $\int \tanh u \, du = \log \cosh u.$

24. $\int \coth u \, du = \log \sinh u.$

25. $\int \text{sech } u \, du = 2 \tan^{-1} e^u.$

26. $\int \text{csch } u \, du = \log \tanh \dfrac{u}{2}.$

In general, with the exception of the square root of polynomials of the second degree, integrals containing fractional powers of polynomials above the first degree cannot be integrated in terms of the elementary integral forms.

Elliptic Integrals. An elliptic integral has the form

$$\int R[x, \sqrt{f(x)}\,] \, dx \tag{1}$$

where R represents a rational function and $f(x) = a + bx + cx^2 + dx^3 + ex^4$, an algebraic function of the third or fourth degree.

Elliptic Integral of the First Kind.

$$F(\phi, k) = \int_0^\phi \dfrac{d\theta}{\sqrt{1 - k^2 \sin^2 \theta}} = \int_0^x \dfrac{d\xi}{\sqrt{(1 - \xi^2)(1 - k^2\xi^2)}}, \quad x = \sin \phi, \, k^2 < 1 \tag{2}$$

Elliptic Integral of the Second Kind.

$$E(\phi, k) = \int_0^\phi \sqrt{1 - k^2 \sin^2 \theta} \, d\theta = \int_0^x \dfrac{\sqrt{1 - k^2\xi^2}}{\sqrt{1 - \xi^2}} \, d\xi, \quad x = \sin \phi, \, k^2 < 1 \tag{3}$$

Elliptic Integral of the Third Kind.

$$\Pi(\phi, n, k) = \int_0^\phi \dfrac{d\theta}{(1 + n \sin^2 \theta)\sqrt{1 - k^2 \sin^2 \theta}}$$

$$= \int_0^x \frac{d\xi}{(1 + n\xi^2)\sqrt{(1 - \xi^2)(1 - k^2\xi^2)}}, \quad x = \sin\phi, \ k^2 < 1 \quad (4)$$

The "complete" integrals are

$$K = F\left(\frac{\pi}{2}, k\right) = \frac{\pi}{2}\left[1 + \left(\frac{1}{2}\right)^2 k^2 + \left(\frac{3}{2\cdot4}\right)^2 k^4 + \left(\frac{3\cdot5}{2\cdot4\cdot6}\right)^3 k^6 + \cdots\right] \quad (5)$$

$$E = E\left(\frac{\pi}{2}, k\right) = \frac{\pi}{2}\left[1 - \left(\frac{1}{2^2}\right) k^2 - \left(\frac{3}{2^2\cdot4^2}\right) k^4 - \left(\frac{3^2\cdot5}{2^2\cdot4^2\cdot6^2}\right) k^6 - \cdots\right] \quad (6)$$

$$K' = F\left(\frac{\pi}{2}, \sqrt{1 - k^2}\right), \quad E' = E\left(\frac{\pi}{2}, \sqrt{1 - k^2}\right) \quad (7)$$

They are connected by the relation

$$KE' + EK' - KK' = \frac{\pi}{2} \quad (8)$$

The inverse function of $u = F(\phi, k)$ is $\phi = $ am u, (am = amplitude)

$$x \equiv \sin\phi \equiv \text{sn } u = u - (1 + k^2)\frac{u^3}{3!} + (1 + 14k^2 + k^4)\frac{u^5}{5!} - \cdots \quad (9)$$

$$\cos\phi \equiv \text{cn } u = 1 - \frac{u^2}{2!} + (1 + 4k^2)\frac{u^4}{4!} - (1 + 44k^2 + 16k^4)\frac{u^6}{6!} + \cdots \quad (10)$$

$$\sqrt{1 - k^2x^2} \equiv \Delta\phi \equiv \text{dn } u = 1 - k^2\frac{u^2}{2!} + k^2(4 + k^2)\frac{u^4}{4!} - k^2(16 + 44k^2 + k^4)\frac{u^6}{6!} + \cdots \quad (11)$$

Methods of Integration

Integration by Parts. The formula

$$\int u\, dv = uv - \int v\, du, \ u \text{ and } v \text{ functions of } x. \quad (12)$$

is useful in integrating a product, if factors of the product are a function of x and the derivative of another function of x.

Example. To find $\int x \sin x\, dx$, let $u = x$, $dv = \sin x\, dx$. Then $du = dx$, $v = -\cos x$, and $\int x \sin x\, dx = -x \cos x + \int \cos x\, dx = -x \cos x + \sin x + c$.

Integration of Rational Fractions. If the quotient of two polynomials

$$R(x) = \frac{P_n(x)}{P_d(x)} \quad (13)$$

is not a proper fraction, that is, if the degree of the numerator is not less than that of the denominator, $R(x)$ can be changed, by dividing as indicated, to the sum of a polynomial, which is immediately integrable, and a proper fraction. If the proper fraction cannot be integrated by reference to Table 23, p. 131, use the methods of Art. 12, p. 219, to resolve it, if possible, into partial fractions. These can be integrated from the table.

Irrational Functions can sometimes be put into integrable forms by rationalizing them by a change of variable.

Form	Substitution
$\int[(ax + b)^{p/q}]\, dx$	let $ax + b = y^q$
$\int[(ax + b)^{p/q}(ax + b)^{r/s}]\, dx$	let $ax + b = y^n$, where n is the L.C.M. of q, s
$\int[x, \sqrt{x^3 + ax + b}]\, dx$	let $\sqrt{x^2 + ax + b} = y - x$
$\int[x, \sqrt{-x^2 + ax + b}]\, dx$	let $\sqrt{-x^2 + ax + b} = \sqrt{(\alpha - x)(\beta + x)}$
	$= (\alpha - x)y$ or $= (\beta + x)y$
$\int[\sin x, \cos x]\, dx$	let $\tan \dfrac{x}{2} = y$
$\int[x, \sqrt{a^2 - x^2}]\, dx$	let $x = a \sin y$
$\int[x, \sqrt{x^2 - a^2}]\, dx$	let $x = a \sec y$ or $x = a \cosh y$
$\int[x, \sqrt{x^2 + a^2}]\, dx$	let $x = a \tan y$ or $x = a \sinh y$

49. DEFINITE INTEGRALS

The Definite Integral of $f(x)$ from a to b is

$$\int_a^b f(x)\, dx = \lim_{\substack{n \to \infty \\ \max \Delta x_\nu \to 0}} \sum_{\nu=1}^n f(\xi_\nu)\, \Delta x_\nu \tag{14}$$

in which the interval $a \leqq x \leqq b$ is divided into n arbitrary parts Δx_ν, $\nu = 1, 2, \cdots, n$, and ξ_ν is an arbitrary point in Δx_ν. A sufficient condition that this integral exists is that $f(x)$ be continuous. However, it is necessary and sufficient only that $f(x)$ be bounded and that its points of discontinuity form a set of Lebesgue measure 0. A set of points is of Lebesgue measure 0 if the points can be enclosed in a set of intervals I_ν, $\nu = 1, 2, 3,$

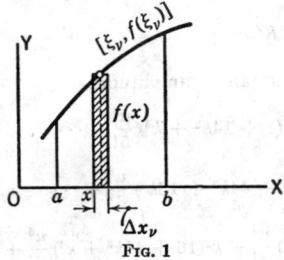

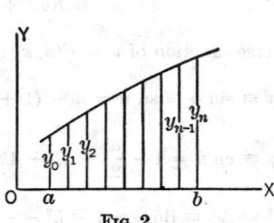

FIG. 1   FIG. 2

$\cdots$, finite or infinite in number, such that, for any $\epsilon > 0$, the sum of the lengths of the I_ν is $< \epsilon$.

A definite integral of a given function over a given interval is a number. The geometric interpretation of the definite integral of $f(x)$, $f(x) \geqq 0$, is the area bounded by the curve $y = f(x)$, the x axis, and the ordinates at $x = a$ and $x = b$.

To evaluate a definite integral, if $F(x) = \int f(x)\, dx$, then $\int_a^b f(x)\, dx = F(b) - F(a)$.

Example.

$$\int_3^5 x^2\, dx = \frac{x^3}{3}\Big|_3^5 = \frac{125}{3} - 9 = \frac{98}{3}$$

Some Fundamental Theorems.

$$\frac{d}{dx}\int_a^x f(\xi)\, d\xi = f(x)$$

$$\int_a^b cf(x)\, dx = c\int_a^b f(x)\, dx$$

$$\int_a^b [f_1(x) + f_2(x) + \cdots + f_n(x)]\, dx = \int_a^b f_1(x)\, dx + \int_a^b f_2(x)\, dx + \cdots + \int_a^b f_n(x)\, dx$$

$$\int_a^b f(x)\, dx = -\int_b^a f(x)\, dx$$

$$\int_a^b f(x)\, dx = \int_a^c f(x)\, dx + \int_c^b f(x)\, dx, \quad a \leqq c \leqq b$$

$$\int_a^b f(x)\, dx = (b - a)f(\xi)$$

for some ξ such that $a \leqq \xi \leqq b$ (Mean value theorem).

Simpson's Rule for Approximate Integration. To evaluate $\int_a^b f(x)\, dx$ approximately, divide the interval from a to b into an even number n of equal parts with end points $x_0 = a$, $x_1, \cdots, x_n = b$ and let $y_i = f(x_i)$ (Fig. 2). Then

$$\int_a^b f(x)\, dx \approx \frac{b-a}{3n}(y_0 + 4y_1 + 2y_2 + 4y_3 + 2y_4 + \cdots + 4y_{n-1} + y_n) \tag{15}$$

Improper Integrals. *If one limit is infinite,*

$$\int_a^\infty f(x)\, dx = \lim_{b \to \infty} \int_a^b f(x)\, dx \qquad (16)$$

The integral exists, or converges, if there is a number $k > 1$ and a number M independent of x such that $x^k | f(x) | < M$ for arbitrarily large values of x. If $x | f(x) | > m$, an arbitrary positive number, for sufficiently large values of x, the interval diverges.

Example. The integral $\int_0^\infty \dfrac{x\, dx}{(x + x^2)^{\frac{3}{2}}}$ exists, since, for $k = 2$ and $M = 1$, $x^2 \left| \dfrac{x}{(x + x^2)^{\frac{3}{2}}} \right| = \left(\dfrac{x^2}{x + x^2} \right)^{\frac{3}{2}} < 1$, no matter how large the value of x.

If the integrand is *infinite* at the upper limit

$$\int_a^b f(x)\, dx = \lim_{\epsilon \to 0} \int_a^{b - \epsilon} f(x)\, dx, \quad 0 < \epsilon < (b - a) \qquad (17)$$

The integral exists if there is a number $k < 1$ and a number M independent of x such that $(b - x)^k | f(x) | < M$ for $a \leqq x < b$. If there is a number $k \geqq 1$ and a number m such that $(b - x)^k | f(x) | > m$ for $a \leqq x < b$, the integral diverges.

Example. The integral $\int_0^1 \dfrac{dx}{1 - x}$ diverges, since, for $k = 1$ and $m = 1/2$, $\dfrac{1 - x}{1 - x} = 1 > 1/2$.

If the integrand is infinite at the lower limit, the tests are analogous. If the integrand is infinite at an intermediate point, use the point to divide the interval into two subintervals and apply the above tests.

Multiple Integrals. Let $f(x, y)$ be defined in the region R of the xy plane. Divide R into subregions $\Delta R_1, \Delta R_2, \cdots, \Delta R_n$ of areas $\Delta A_1, \Delta A_2, \cdots, \Delta A_n$. Let (ξ_i, η_i) be any point in ΔR_i. If the sum $\sum_{i=1}^n f(\xi_i, \eta_i)\, \Delta A_i$ has a limit as $n > \infty$ and the maximum diameter of the subregions ΔR_i approaches 0, then

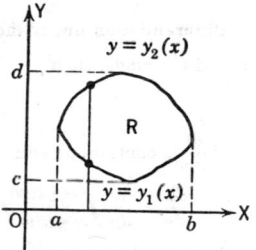

$$\int_R f(x, y)\, dA = \lim_{n \to \infty} \sum_{i=1}^n f(\xi_i, \eta_i)\, \Delta A_i \qquad (18)$$

The double integral is evaluated by two successive single integrations, first with respect to y holding x constant, between variable limits of integration, and then with respect to x between constant limits (Fig. 3). If $f(x, y)$ is continuous, the order of integration can be reversed.

Fig. 3

$$\int_R f(x, y)\, dA = \int_a^b \int_{y_1(x)}^{y_2(x)} f(x, y)\, dy\, dx = \int_c^d \int_{x_1(y)}^{x_2(y)} f(x, y)\, dx\, dy \qquad (19)$$

In polar coordinates,

$$\int_R F(r, \theta)\, dA = \int_\alpha^\beta \int_{r_1(\theta)}^{r_2(\theta)} F(r, \theta) r\, dr\, d\theta = \int_k^l \int_{\theta_1(r)}^{\theta_2(r)} F(r, \theta) r\, d\theta\, dr \qquad (20)$$

The analogous triple integrals are evaluated by three single integrations. In rectangular coordinates,

$$\int_R f(x, y, z)\, dV = \iiint f(x, y, z)\, dx\, dy\, dz \qquad (21)$$

In spherical coordinates,

$$\int_R F(r, \theta, \phi)\, dV = \iiint F(r, \theta, \phi) r^2 \sin \theta\, dr\, d\theta\, d\phi \qquad (22)$$

In cylindrical coordinates,

$$\int_R G(\rho, \phi, z)\, dV = \iiint G(\rho, \phi, z) \rho\, d\rho\, d\phi\, dz \qquad (23)$$

Integrals Containing a Parameter. If $f(x, y)$ is a continuous function of x and y in the closed rectangle $x_0 \leqq x \leqq x_1$, $y_0 \leqq y \leqq y_1$, and if $f(x, y)$ is integrated with respect to x, with y regarded as fixed and called a *parameter*, then

$$\int_{x_0}^{x_1} f(x, y)\, dx = \phi(y) \tag{24}$$

is a continuous function of y. Geometrically, the function $f(x, y)$ may be plotted as a surface $z = f(x, y)$. Then the value of $\phi(y_i)$ is the area of the section under the surface made by the plane $y = y_i$ (Fig. 4). If the limits of integration are continuous functions of y instead of constants, then $\phi(y)$ is continuous.

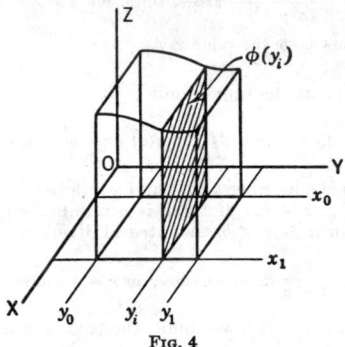

Fig. 4

Differentiation under the Integral Sign. If $\dfrac{\partial f}{\partial y}$ is a continuous function of x and y in a closed rectangle, then

$$\frac{d\phi}{dy} = \int_{x_0}^{x_1} \frac{\partial f(x, y)}{\partial y}\, dx \tag{25}$$

If $\dfrac{\partial f}{\partial y}$ is continuous and the limits of integration are differentiable functions of y, then

$$\frac{d\phi}{dy} = \frac{d}{dy} \int_{x_0 = g_0(y)}^{x_1 = g_1(y)} f(x, y)\, dx = \int_{g_0(y)}^{g_1(y)} \frac{\partial f(x, y)}{\partial y}\, dx - f(g_0, y)\frac{dg_0}{dy} + f(g_1, y)\frac{dg_1}{dy} \tag{26}$$

If $f(x)$ is integrable in the interval $a \leqq x \leqq b$ and continuous at a point within the interval, then at that point the function $F(x) = \int_a^x f(\xi)\, d\xi$ has a derivative $F'(x) = f(x)$.

Uniform Convergence and Change of Order of Integration. The improper integral

$$\phi(y) = \int_{x_0}^{\infty} f(x, y)\, dx \tag{27}$$

converges uniformly in y in the interval $y_0 \leqq y \leqq y_1$, if for any $\epsilon > 0$ there exists an L, dependent on ϵ but not on y, such that

$$\left| \int_{l}^{\infty} f(x_1 y)\, dx \right| < \epsilon \quad \text{for } l \geqq L \tag{28}$$

If $\int_{x_0}^{\infty} f(x, y)\, dx$ is uniformly convergent for $y_0 \leqq y \leqq y_1$, then

$$\int_{y_0}^{y_1} \int_{x_0}^{\infty} f(x, y)\, dx\, dy = \int_{x_0}^{\infty} \int_{y_0}^{y_1} f(x, y)\, dy\, dx \tag{29}$$

Stieltjes Integral. If $f(x)$ and $\phi(x)$ are defined in the interval (a, b), the Stieltjes integral of $f(x)$ with respect to $\phi(x)$ is

$$\int_a^b f(x)\, d\phi(x) = \lim_{\substack{n \to \infty \\ \max \Delta x_v \to 0}} \sum_{v=1}^{n} f(\xi_v)[\phi(x_v) - \phi(x_{v-1})] \tag{30}$$

in which the interval (a, b) is divided into n arbitrary parts $\Delta x_v = x_v - x_{v-1}$ by the points $x_0 = a, x_1, \cdots, x_n = b$, and ξ_v is an arbitrary point in Δx_v. This limit exists if $f(x)$ is continuous and $\phi(x)$ is of bounded variation, that is, can be expressed as the difference of two non-increasing or two non-decreasing bounded functions. However, it is not necessary that $f(x)$ be continuous, but only that the variation of $\phi(x)$ over the set of points of discontinuity of $f(x)$ be zero.

Lebesgue Integral. Let S be a set of points in the interval (a, b), and $C(S)$ the complement of S, that is, the set of all the points of (a, b) that do not belong to S. Enclose the points of S in a set of intervals I_ν, $\nu = 1, 2, 3,$ $\cdots$, finite or infinite in number, and let the sum of the lengths of the I_ν be L. The greatest lower bound of all possible values of L is the exterior measure $\overline{m}(S)$ of S. The interior measure of S is $m(S) = (b - a) - \overline{m}[C(S)]$. If $\overline{m}(S) = m(S)$, the set S is measurable and its *measure* is $m(S) = \overline{m}(S)$.

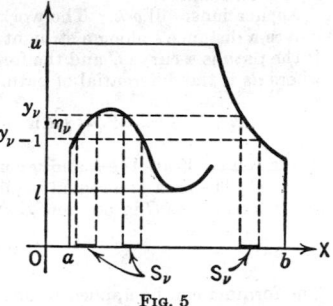

A function $f(x)$ defined in the interval (a, b) is *measurable* if the set of points x for which $y_0 \leqq f(x) < y_1$ is measurable for any values of y_0 and y_1.

Let u and l be the upper and lower bounds of a measurable function $f(x)$ defined in the interval (a, b) (Fig. 5). Divide the interval (u, l) into n arbitrary subintervals Δy_ν by the points $y_0 = 1, y_1,$ $\cdots, y_n = u$. Let S_ν be the set of points for which $y_{\nu-1} \leqq f(x) < y_\nu$, and η_ν any point in the interval Δy_ν. Then the Lebesgue integral of $f(x)$ in the interval (a, b) is

$$\int_a^b f(x)\, dx = \lim_{\substack{n \to \infty \\ \text{max } \Delta y_\nu \to 0}} \sum_{\nu=1}^n \eta_\nu \cdot m(S_\nu) \tag{31}$$

If the Riemann integral in the interval (a, b), defined on p. 316, exists, the Lebesgue integral does also, and the two are equal, but not conversely.

50. LINE, SURFACE, AND VOLUME INTEGRALS

Line Integrals. Let $P(x, y)$ and $Q(x, y)$ be functions continuous at all points of a continuous curve C joining the points A and B in the xy plane. Divide the curve C into n

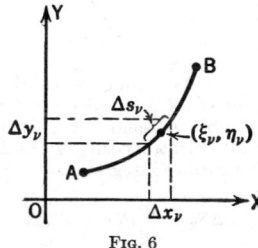

Fig. 6

arbitrary parts Δs_ν by the points (x_ν, y_ν), let (ξ_ν, η_ν) be an arbitrary point on Δs_ν, and let Δx_ν and Δy_ν be the projections of ΔS_ν on the x and the y axes (Fig. 6). The line integral is

$$\int_A^B [P(x, y)\, dx + Q(x, y)\, dy] = \lim_{\substack{n \to \infty \\ \text{max } \Delta x_\nu, \Delta y_\nu \to 0}} \sum_{\nu=1}^n [P(\xi_\nu, \eta_\nu)\, \Delta x_\nu + Q(\xi_\nu, \eta_\nu)\, \Delta y_\nu] \tag{32}$$

If the equation of the curve C is $y = f(x)$, $x = \phi(y)$, or the parametric equations $x = x(t)$, $y = y(t)$, the line integral can be evaluated as a definite integral in the one variable x, y, or t, respectively.

Example. Find the value of $\int_{0,0}^{1,3} [y^2\, dx + (xy - x^2)\, dy]$ along the paths (a) $y = 3x$, (b) $y^2 = 9x$.

(a) Substitute $y = 3x$, $dy = 3\, dx$ and obtain $\int_0^1 [9x^2 + (3x^2 - x^2)3]\, dx = \int_0^1 15x^2\, dx = 5$.

(b) Substitute $y^2 = 9x$, $2y\, dy = 9\, dx$ and obtain $\int_0^3 \left[\frac{2}{9}y^3 + \left(\frac{y^3}{9} - \frac{y^4}{81}\right)\right] dy = \left[\frac{1}{12}y^4 - \frac{y^5}{405}\right]_0^3$

$= 6\,3/20.$

A line integral in the xyz space

$$\int_A^B [P(x, y, z)\, dx + Q(x, y, z)\, dy + R(x, y, z)\, dz] \qquad (33)$$

is defined similarly.

Applications. *Work.* The work done by a constant force F acting on a particle which moves a distance s along a straight line inclined at an angle θ to the force is $W = Fs \cos \theta$. If the path is a curve C and the force variable, the differential of work is $dW = F \cos \theta\, ds$, where ds is the differential of path. Then

$$W = \int dW = \int_C F \cos \theta\, ds = \int_C (X\, dx + Y\, dy) \qquad (34)$$

where X and Y are the x and y components of F (Fig. 7).

Area. The area of a region bounded by a closed curve C such that a line parallel to the x or y axis meets C in no more than two points is

$$A = \tfrac{1}{2} \int_C (x\, dy - y\, dx) \qquad (35)$$

The formula can be applied to any region that can be divided by a finite number of lines into regions satisfying the above condition.

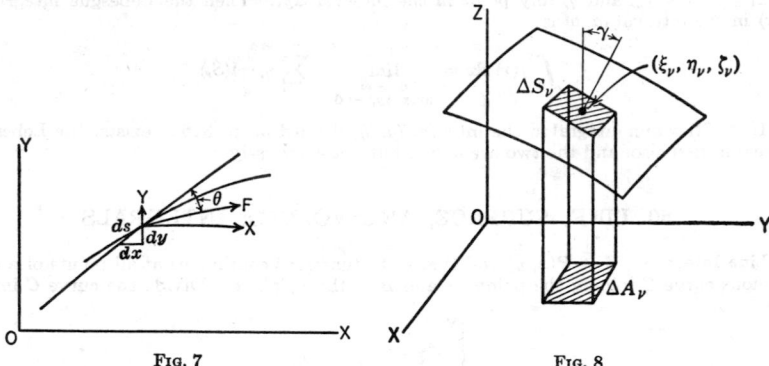

FIG. 7   FIG. 8

Surface Integrals. Let $P(x, y, z)$ be a function continuous at all points of a region S (bounded by a simple closed curve) of a surface $z = f(x, y)$, which has a continuously turning tangent plane except possibly at isolated points or lines. Let A be the projection of S on the xy plane. Divide S into arbitrary subregions ΔS_ν and let $(\xi_\nu, \eta_\nu, \zeta_\nu)$ be an arbitrary point in ΔS_ν (Fig. 8). The surface integral is

$$\lim_{\substack{n \to \infty \\ \max \text{diam } \Delta S_\nu \to 0}} \sum_{\nu=1}^n P(\xi_\nu, \eta_\nu, \zeta_\nu)\, \Delta S_\nu = \int_S P(x, y, z)\, dS$$

$$= \iint_A P(x, y, z) \sqrt{1 + \left(\frac{\partial z}{\partial x}\right)^2 + \left(\frac{\partial z}{\partial y}\right)^2}\, dx\, dy \qquad (36)$$

If α, β, γ are the direction angles of the normal to S, the form of the surface integral analogous to the line integral above is

$$\iint_A (P\, dy\, dz + Q\, dz\, dx + R\, dx\, dy) = \int_S (P \cos \alpha + Q \cos \beta + R \cos \gamma)\, dS \qquad (37)$$

Green's Theorem. Let $P(x, y)$ and $Q(x, y)$ be continuous functions with continuous partial derivatives $\dfrac{\partial P}{\partial y}$ and $\dfrac{\partial Q}{\partial x}$ in a simply connected region R bounded by a simple closed curve C. Then

$$\iint_R \left(\frac{\partial Q}{\partial x} - \frac{\partial P}{\partial y}\right) dx\, dy = \int_C (P\, dx + Q\, dy) \qquad (38)$$

A region is *simply connected* if any closed curve in the region can be shrunk to a point without passing outside the region.

Stokes' Theorem. Let $P(x, y, z)$, $Q(x, y, z)$, $R(x, y, z)$ be continuous functions with continuous first partial derivatives, S a region (bounded by a simple closed curve C) of a surface $z = f(x, y)$, continuous with continuous first partial derivatives. Then

$$\iint_S \left[\left(\frac{\partial R}{\partial y} - \frac{\partial Q}{\partial z} \right) dy\, dz + \left(\frac{\partial P}{\partial z} - \frac{\partial R}{\partial x} \right) dz\, dx + \left(\frac{\partial Q}{\partial x} - \frac{\partial P}{\partial y} \right) dx\, dy \right]$$

$$= \int_C (P\, dx + Q\, dy + R\, dz) \quad (39)$$

The signs are such that an observer standing on the surface with head in the direction of the normal will see the integration around C taken in the positive direction.

Divergence, or Gauss's, Theorem. Let $P(x, y, z)$, $Q(x, y, z)$, $R(x, y, z)$ be continuous functions with continuous first partial derivatives. Let V be a region in the xyz space bounded by a closed surface S with a continuously turning tangent plane except possibly at isolated points or lines. Then

$$\iiint_V \left(\frac{\partial P}{\partial x} + \frac{\partial Q}{\partial y} + \frac{\partial R}{\partial z} \right) dx\, dy\, dz = \iint_S (P\, dy\, dz + Q\, dz\, dx + R\, dx\, dy) \quad (40)$$

Example.

Evaluate $\iint (x\, dy\, dz + y\, dz\, dx + z\, dx\, dy)$ over the cylinder

$$x^2 + y^2 = a^2, \quad z = \pm b$$

Since

$$P = x, \quad Q = y, \quad R = z; \quad \frac{\partial P}{\partial x} = 1, \quad \frac{\partial Q}{\partial y} = 1, \quad \frac{\partial R}{\partial z} = 1$$

and

$$\int_{-a}^{a} \int_{-\sqrt{a^2 - x^2}}^{\sqrt{a^2 - x^2}} \int_{-b}^{+b} 3\, dz\, dy\, dx = 6\pi a^2 b$$

Independence of Path and Exact Differential. Under the conditions of Green's theorem, the following statements are equivalent:

1. $\int_C (P\, dx + Q\, dy) = 0$ for any closed curve C in the region R.

2. The value of $\int_{(a,b)}^{(\xi, \eta)} (P\, dx + Q\, dy)$ is independent of the curve connecting (a, b) and (ξ, η), any points in R.

3. $\frac{\partial P}{\partial y} = \frac{\partial Q}{\partial x}$ at all points of R.

4. There exists a function $F(x, y)$ such that $dF = P\, dx + Q\, dy$.

Under the conditions of Stokes' theorem, the corresponding statements for three dimensions are:

1. $\int_C (P\, dx + Q\, dy + R\, dz) = 0$ for any closed curve C in the region S.

2. The value of $\int_{(a,b,c)}^{(\xi, \eta, \zeta)} (P\, dx + Q\, dy + R\, dz)$ is independent of the curve connecting (a, b, c) and (ξ, η, ζ), any points in S.

3. $\frac{\partial P}{\partial y} = \frac{\partial Q}{\partial x}, \frac{\partial Q}{\partial z} = \frac{\partial R}{\partial y}, \frac{\partial R}{\partial x} = \frac{\partial P}{\partial z}$ at all points of S.

4. There exists a function $F(x, y, z)$ such that $dF = P\, dx + Q\, dy + R\, dz$.

51. APPLICATIONS OF INTEGRATION

Length of Arc of a Curve. The length s of the arc of a plane curve $y = f(x)$ from the point (a, b) to the point (c, d) is

$$s = \int_a^c \sqrt{1 + \left(\frac{dy}{dx} \right)^2}\, dx = \int_b^d \sqrt{1 + \left(\frac{dx}{dy} \right)^2}\, dy \quad (41)$$

If the equation of the curve is in polar coordinates, $r = f(\theta)$, then the length of the arc from the point (r_1, θ_1) to the point (r_2, θ_2) is

$$s = \int_{\theta_1}^{\theta_2} \sqrt{r^2 + \left(\frac{dr}{d\theta} \right)^2}\, d\theta = \int_{r_1}^{r_2} \sqrt{1 + r^2 \left(\frac{d\theta}{dr} \right)^2}\, dr \quad (42)$$

If the curve is in three dimensions, represented by the equations $y = f_1(x)$, $z = f_2(x)$, the length of arc from $x_1 = a$ to $x_2 = b$ is

$$s = \int_a^b \sqrt{1 + \left(\frac{dy}{dx}\right)^2 + \left(\frac{dz}{dx}\right)^2} \, dx \tag{43}$$

Plane Area. The area bounded by the curve $y = f(x)$, the x axis, and the ordinates at $x = a$, $x = b$ is

$$A = \int_a^b f(x) \, dx \tag{44}$$

where y has the same sign for all values of x between a and b.

In polar coordinates, the area bounded by the curve $r = f(\theta)$, and the two radii $\theta = \alpha$, $\theta = \beta$ (Fig. 9) is

$$A = \frac{1}{2} \int_\alpha^\beta r^2 \, d\theta \tag{45}$$

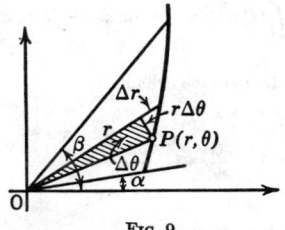

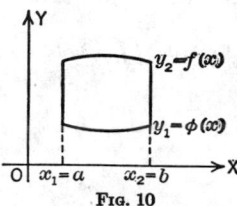

Fig. 9   Fig. 10

In rectangular coordinates, if the area is bounded by the two curves $y_2 = f(x)$, $y_1 = \phi(x)$, and the lines $x_2 = b$, $x_1 = a$ (Fig. 10), then

$$A = \int_a^b dx \int_{\phi(x)}^{f(x)} dy \tag{46}$$

If the area is bounded by the two curves $x_2 = \psi(y)$, $x_1 = \xi(y)$, and the lines $y_2 = d$, $y_1 = c$, then

$$A = \int_c^d dy \int_{\xi(y)}^{\psi(y)} dx \tag{47}$$

If expressed in polar coordinates, the area by double integration is

$$A = \int_{\theta_1}^{\theta_2} d\theta \int_{r_1=f_1(\theta)}^{r_2=f_2(\theta)} r \, dr \quad \text{or} \quad \int_{r_1}^{r_2} r \, dr \int_{\theta_1=\phi_1(r)}^{\theta_2=\phi_2(r)} d\theta \tag{48}$$

Area of a Surface of Revolution. The area of the surface of a solid of revolution generated by revolving the curve $y = f(x)$ between $x = a$ and $x = b$:

about the x axis is

$$2\pi \int_a^b y \sqrt{1 + \left(\frac{dy}{dx}\right)^2} \, dx \tag{49}$$

about the y axis is

$$2\pi \int_c^d x \sqrt{1 + \left(\frac{dx}{dy}\right)^2} \, dy \tag{50}$$

where $c = f(a)$ and $d = f(b)$.

Volume. By triple integration:

Rectangular coordinates $V = \iiint dx \, dy \, dz \tag{51}$

Spherical coordinates $V = \iiint r^2 \sin\theta \, d\theta \, d\phi \, dr \tag{52}$

Cylindrical coordinates $V = \iiint \rho \, d\rho \, d\phi \, dz \tag{53}$

(the limits of integration to be supplied)

Volume of a Solid of Revolution. The volume of a solid of revolution generated by revolving the region bounded by the x axis and the curve $y = f(x)$ between $x = a$ and $x = b$:

about the x axis is $\pi \int_a^b y^2 \, dx \tag{54a}$

about the y axis is $\pi \int_c^d x^2 \, dy \tag{54b}$

where $c = f(a)$ and $d = f(b)$.

Surfaces. If the equation of a surface is written in the parametric form $x = f_1(u, v)$, $y = f_2(u, v)$, $z = f_3(u, v)$, the length of arc of a curve $u = u(t)$, $v = v(t)$ on the surface is

$$s = \int \sqrt{E\left(\frac{du}{dt}\right)^2 + 2F\frac{du}{dt}\frac{dv}{dt} + G\left(\frac{dv}{dt}\right)^2}\, dt \tag{55}$$

The area S of a region on the surface is

$$S = \iint \sqrt{EG - F^2}\, du\, dv \tag{56}$$

where

$$E = \left(\frac{\partial x}{\partial u}\right)^2 + \left(\frac{\partial y}{\partial u}\right)^2 + \left(\frac{\partial z}{\partial u}\right)^2$$

$$F = \frac{\partial x}{\partial u}\frac{\partial x}{\partial v} + \frac{\partial y}{\partial u}\frac{\partial y}{\partial v} + \frac{\partial z}{\partial u}\frac{\partial z}{\partial v}$$

$$G = \left(\frac{\partial x}{\partial v}\right)^2 + \left(\frac{\partial y}{\partial v}\right)^2 + \left(\frac{\partial z}{\partial v}\right)^2$$

If the equation of the surface is written as $x = u$, $y = v$, $z = f(u, v) = f(x, y)$,

the arc length $\quad s = \int \sqrt{(1 + p^2)\left(\frac{dx}{dt}\right)^2 + 2pq\frac{dx}{dt}\frac{dy}{dt} + (1 + q^2)\left(\frac{dy}{dt}\right)^2}\, dt \tag{57}$

the area $\quad S = \iint \sqrt{1 + p^2 + q^2}\, dx\, dy$, where $p = \dfrac{\partial z}{\partial x}$, $q = \dfrac{\partial z}{\partial y}$ $\tag{58}$

(the limits of integration to be supplied)

Moment. The moment of a mass m

about the yz plane, $\qquad M_{yz} = \int x\, dm;$

about the xz plane, $\qquad M_{xz} = \int y\, dm;$ $\qquad\qquad$ (59)

about the xy plane, $\qquad M_{xy} = \int z\, dm$

(the limits of integration to be supplied)

Center of Gravity. The coordinates of the center of gravity of a mass m are

$$x = \frac{\int x\, dm}{\int dm}, \quad y = \frac{\int y\, dm}{\int dm}, \quad z = \frac{\int z\, dm}{\int dm} \tag{60}$$

(the limits of integration to be supplied)

Moment of Inertia. The moments of inertia, I, are

for a plane curve about the x axis, $\qquad I_x = \int y^2\, ds$

for a plane curve about the y axis, $\qquad I_y = \int x^2\, ds$ $\qquad$ (61)

for a plane curve about the origin, $\qquad I_0 = \int (x^2 + y^2)\, ds$

for a plane area about the x axis, $\qquad I_x = \int y^2\, dA$

for a plane area about the y axis, $\qquad I_y = \int x^2\, dA$ $\qquad$ (62)

for a plane area about the origin, $\qquad I_0 = \int (x^2 + y^2)\, dA$

for a solid of mass m about the yz plane, $\quad I_{yz} = \int x^2\, dm$

for a solid of mass m about the xz plane, $\quad I_{xz} = \int y^2\, dm$

for a solid of mass m about the xy plane, $\quad I_{xy} = \int z^2\, dm$ $\qquad$ (63)

for a solid of mass m about the x axis, $\quad I_x = I_{xz} + I_{xy}$, etc.

(the limits of integration to be supplied)

Fluid Pressure. The total force F against a plane surface vertical to the surface of the liquid and between the depths a and b is

$$F = \int_{y=a}^{y=b} \rho y \, dA = \int_a^b \rho y x \, dy \tag{64}$$

where ρ is the weight of the liquid per unit volume and y is the depth beneath the surface of the liquid of a horizontal element of area dA. Usually, $dA = x \, dy$, where x is the width of the vertical surface expressed as a function of y.

Center of Pressure. The depth $\bar{y}$ of the center of pressure against surface vertical to the surface of the liquid and between the depths a and b is

$$\bar{y} = \frac{\int_{y=a}^{y=b} \rho y^2 \, dA}{\int_{y=a}^{y=b} \rho y \, dA} \tag{65}$$

Work. The work W done in moving a particle from $s = a$ to $s = b$ against a force whose component expressed as a function of s in the direction of motion is $F(s)$ is

$$W = \int_{s=a}^{s=b} F(s) \, ds \tag{66}$$

DIFFERENTIAL EQUATIONS

52. DEFINITIONS

A **Differential Equation** is an equation containing an unknown function of a set of variables and its derivatives. If the equation has derivatives with respect to one variable only, it is an *ordinary differential equation*, otherwise it is a *partial differential equation*.

Examples.

$$\frac{d^2y}{dx^2} + k^2 y = 0 \tag{1}$$

$$\frac{d^2y}{dx^2} = \sqrt{1 + y^2 + \frac{dy}{dx}} \tag{2}$$

$$y \frac{\partial^2 z}{\partial x^2} + zx \frac{\partial^2 z}{\partial x \, \partial y} - \frac{\partial z}{\partial y} = xyz \tag{3}$$

$$y - x \frac{dy}{dx} + 3 \frac{dx}{dy} = 0 \tag{4}$$

Equations (1), (2), (4) are ordinary differential equations, and (3) is a partial differential equation.

The Order of a differential equation is the order of the highest derivative involved. Thus in (1), (2), (3), the order is two; in (4), the order is one.

The Degree of a differential equation is the exponent of the highest order appearing in the equation after it is rationalized and cleared of fractions with respect to the derivatives. The degree of (1), (3), (4) is one; that of (2) is two.

A **Solution or Integral** of a differential equation is a relation among the variables which satisfies the equation identically.

A **General Solution** of an ordinary differential equation of the nth order is one that contains n independent constants. Thus, $y = \sin x + c$ is a general solution of the equation $\frac{dy}{dx} = \cos x$.

A **Particular Solution** is one that is derivable from a general solution by assigning fixed values to the arbitrary constants. Thus, $y_1 = \sin x$, $y_2 = \sin x + 4$ are two particular solutions of the above equation.

53. FIRST-ORDER EQUATIONS

Separation of Variables. A differential equation of the first order

$$f\left(x, y, \frac{dy}{dx}\right) = 0 \tag{5}$$

can be brought into the form

$$P(x, y)\, dx + Q(x, y)\, dy = 0 \tag{6}$$

For the special case where P is a function of x only and Q a function of y only,

$$P(x)\, dx + Q(y)\, dy = 0 \tag{7}$$

the variables are separated. The solution is

$$\int P(x)\, dx + \int Q(y)\, dy = c \tag{8}$$

Example. Solve, $\dfrac{dy}{dx} = -\dfrac{x}{y}$.

This can be written as $x\, dx + y\, dy = 0$ and has the solution $\int x\, dx + \int y\, dy = \dfrac{x^2}{2} + \dfrac{y^2}{2} = c$. If $c = \dfrac{r^2}{2}$, then $x^2 + y^2 = r^2$, a set of concentric circles. There are an infinite number of solutions depending on the value of r. Through each point in the plane there passes one circle and only one.

Homogeneous Equations. A function $f(x, y)$ is homogeneous of the nth degree in x and y, if $f(kx, ky) = k^n f(x, y)$. An equation

$$P(x, y)\, dx + Q(x, y)\, dy = 0 \tag{9}$$

is homogeneous if the functions $P(x, y)$ and $Q(x, y)$ are homogeneous in x and y. By substituting $y = vx$, the variables can be separated.

Example. Solve $(x^2 + y^2)\, dx - 2xy\, dy = 0$.
This is of the form $P(x, y)\, dx + Q(x, y)\, dy = 0$, where P and Q are homogeneous functions of the second degree. Making the substitution $y = vx$, the equation becomes $(1 + v^2)\, dx - 2v(x\, dv + v\, dx) = 0$. Separating variables,

$$\frac{dx}{x} - \frac{2v}{1 - v^2}\, dv = 0$$

Integrating, $\log_e x(1 - v^2) = \log_e c$; replacing $v = \dfrac{y}{x}$, $\log\left(1 - \dfrac{y^2}{x^2}\right) x = \log_e c$; and taking exponentials $x^2 - y^2 = cx$.

Linear Differential Equation. The differential equation

$$\frac{dy}{dx} + P(x)y = Q(x) \tag{10}$$

in which y and $\dfrac{dy}{dx}$ appear only in the first degree, and P and Q are functions of x, is a *linear equation of the first order.* This has the general solution

$$y = e^{-\int P(x)\, dx}\left[\int Q(x)e^{\int P(x)\, dx}\, dx + c\right] \tag{11}$$

Example. An equation in the theory of electric networks is $L\dfrac{di}{dt} + Ri = E$, where i is the current, L the inductance (a constant), R the resistance (a constant), and E the electromotive force, a function of time or constant. If $E = E(t)$

$$i = e^{-(R/L)t}\left[\int \frac{E}{L}e^{(R/L)t}\, dt + c\right]$$

If E is constant and if $i = 0$ at $t = 0$, then $i = \dfrac{E}{R}(1 - e^{-(R/L)t})$.

The Bernoulli Equation is

$$\frac{dy}{dx} + P(x)y = Q(x)y^n \tag{12}$$

in which $n \neq 1$. By making the substitution $z = y^{1-n}$, a linear equation is obtained and the general solution is

$$y = e^{-\int P(x)\, dx}\left[(1 - n)\int e^{(1-n)\int P(x)\, dx}\, Q(x)\, dx + c\right]^{\frac{1}{1-n}} \tag{13}$$

Example. Solve the equation $\dfrac{dy}{dx} - xy = xy^2$.

Substitute $z = y^{-1}$ and obtain $\dfrac{dz}{dx} + xz = -x$. The general integral is

$$z = ce^{-x^2/2} - 1 \quad \text{or} \quad y = \frac{1}{ce^{-x^2/2} - 1}$$

Exact Differential Equation. The equation

$$P(x, y)\, dx + Q(x, y)\, dy = 0 \tag{14}$$

is an *exact differential equation* if its left side is an exact differential

$$du = P\, dx + Q\, dy \tag{15}$$

that is, if $\dfrac{\partial P}{\partial y} = \dfrac{\partial Q}{\partial x}$. Then,

$$\int P\, dx + \int \left[Q - \frac{\partial \int P\, dx}{\partial y} \right] dy = c \tag{16}$$

is a solution.

Example. Solve $(x^2 - 4xy - y^2)\, dx + (y^2 - 2xy - 2x^2)\, dy = 0$.

This is an exact equation because $\dfrac{\partial P}{\partial y} = -4x - 2y = \dfrac{\partial Q}{\partial x}$.

$$\int (x^2 - 4xy - y^2)\, dx = \frac{x^3}{3} - 2x^2 y - xy^2$$

$$\int [(y^2 - 2xy - 2x^2) - (-2x^2 - 2xy)]\, dy = \frac{y^3}{3}$$

The general solution is

$$\frac{x^3}{3} - 2x^2 y - xy^2 + \frac{y^3}{3} = c$$

Integrating Factor. If the left member of the differential equation $P(x,\ y)\, dx + Q(x, y)\, dy = 0$ is not an exact differential, look for a factor $v(x, y)$ such that $du = v(P\, dx + Q\, dy)$ is an exact differential. Such an *integrating factor* satisfies the equation

$$Q \frac{\partial v}{\partial x} - P \frac{\partial v}{\partial y} + \left(\frac{\partial Q}{\partial x} - \frac{\partial P}{\partial y} \right) v = 0 \tag{17}$$

Example. The equation $(xy^2 - y^3)\, dx + (1 - xy^2)\, dy = 0$ when multiplied by $v = \dfrac{1}{y^2}$ becomes $(x - y)\, dx + \left(\dfrac{1}{y^2} - x \right) dy = 0$, of which the left side $du = (x - y)\, dx + \left(\dfrac{1}{y^2} - x \right) dy$ is an exact differential since $\dfrac{\partial P}{\partial y} = \dfrac{\partial Q}{\partial x}$. The integration gives $u = \dfrac{x^2}{2} - xy - \dfrac{1}{y}$. The general solution is $u = c$ or $x^2 y - 2xy^2 - 2cy - 2 = 0$.

Riccati's Equation is

$$\frac{dy}{dx} + P(x)y^2 + Q(x)y + R(x) = 0 \tag{18}$$

If a particular integral y_1 is known, place $y = y_1 + \dfrac{1}{z}$ and obtain a linear equation in z.

54. SECOND-ORDER EQUATIONS

The differential equation

$$F\left(x,\ y,\ \frac{dy}{dx},\ \frac{d^2 y}{dx^2} \right) = 0 \tag{19}$$

is of the *second order*. If some of these variables are missing there is a straightforward method of solution.

Case 1. With y and $\dfrac{dy}{dx}$ missing.

$$\frac{d^2 y}{dx^2} = f(x) \tag{20}$$

This has the solution

$$y = \int dx \int f(x)\, dx + cx + c_1 \tag{21}$$

Case 2. With x and $\dfrac{dy}{dx}$ missing.

$$\frac{d^2y}{dx^2} = f(y) \tag{22}$$

Multiply both sides by $2\dfrac{dy}{dx}$ and obtain

$$x = \int \frac{dy}{\sqrt{c + 2\int f(y)\,dy}} + c_1 \tag{23}$$

as a solution.

Case 3. With x and y missing.

$$\frac{d^2y}{dx^2} = f\left(\frac{dy}{dx}\right) \tag{24}$$

Place

$$\frac{dy}{dx} = p, \quad \frac{d^2y}{dx^2} = \frac{dp}{dx} \tag{25}$$

Then

$$x = \int \frac{dp}{f(p)} + c$$

Solve for p, replace p by $\dfrac{dy}{dx}$, and solve the resulting first order equation.

Example. The differential equation of the catenary is $a\dfrac{d^2y}{dx^2} = \sqrt{1 + \left(\dfrac{dy}{dx}\right)^2}$.

Let $p = \dfrac{dy}{dx}$, then $a\dfrac{dp}{dx} = \sqrt{1 + p^2}$. By separating variables, $\dfrac{dp}{\sqrt{1 + p^2}} = \dfrac{dx}{a}$ which has the

solution $\sinh^{-1} p = \dfrac{x + c}{a}$, or $p = \dfrac{dy}{dx} = \sinh \dfrac{x + c}{a}$. Integrating this latter, $y = a \cosh \dfrac{x + c}{a} + c_1$.

Case 4. With y missing.

$$\frac{d^2y}{dx^2} = f\left(\frac{dy}{dx}, x\right) \tag{26}$$

Place $\dfrac{dy}{dx} = p$ and obtain the first order equation $\dfrac{dp}{dx} = f(p, x)$. If this can be solved for p, then

$$y = \int p(x)\,dx + c \tag{27}$$

Case 5. With x missing.

$$\frac{d^2y}{dx^2} = f\left(\frac{dy}{dx}, y\right) \tag{28}$$

Place $\dfrac{dy}{dx} = p$ and obtain the first order equation $p\dfrac{dp}{dy} = f(p, y)$. If this can be solved for p, then

$$x = \int \frac{dy}{p(y)} + c \tag{29}$$

55. BESSEL FUNCTIONS

Wherever the mathematics of problems having circular or cylindrical symmetry appears, it is usually appropriate to consider the solutions of Friedrich Wilhelm Bessel's differential equation (29a). Such applications include radiation from a cylindrical antenna, eddy current losses in a cylindrical wire, and sinusoidal angle modulations including phase and frequency modulation.

$$x^2 \frac{d^2y}{dx^2} + x \frac{dy}{dx} + (x^2 - n^2)y = 0 \tag{29a}$$

where n is real, possibly integral or fractional, or complex, and the solution $y(x)$ is said to be of the **first-kind** and denoted $J_n(x)$ for $0 \leq n$ an integer. Tables of $J_0(x)$ and $J_1(x)$ are available on pp. 145. Graphs of these are shown in Fig. 1.

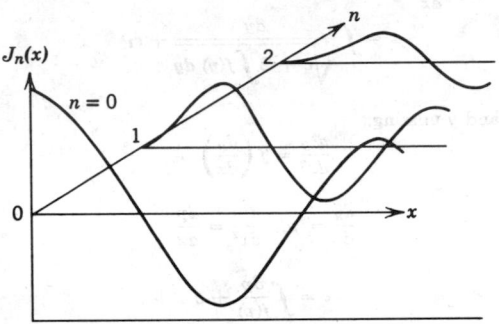

FIG. 1. Bessel functions of first kind.

Bessel functions $J_n(x)$ are **almost periodic functions** which for increasing x have a zero-crossing half "period" approaching π from below. A sequence of these functions can be used to construct an orthogonal series much in the same way that periodic functions, sine and cosine waves make up a Fourier series.

For an extensive set of tables of Bessel functions of many types, see Jahnke, Emde, and Lösch, *Tables of Higher Functions*, Stuttgart, B. G. Teubner Verlag, 1960.

56. LINEAR EQUATIONS

General Theorem. The differential equation

$$\frac{d^n y}{dx^n} + P_1(x)\frac{d^{n-1}y}{dx^{n-1}} + \cdots + P_{n-1}(x)\frac{dy}{dx} + P_n(x)y = F(x) \tag{30}$$

is called the general nth order linear differential equation. If $F(x) = 0$, the equation is *homogeneous*; otherwise it is *non-homogeneous*. If $\phi(x)$ is a solution of the non-homogeneous equation and $y_1, y_2, \cdots, y_n$ are linearly independent solutions of the homogeneous equation, then the *general solution* of (30) is

$$y = c_1 y_1 + c_2 y_2 + \cdots + c_n y_n + \phi(x) \tag{31}$$

The part $\phi(x)$ is called the *particular integral*, and the part $c_1 y_1 + \cdots + c_n y_n$ is the *complementary function*.

Homogeneous Differential Equation with Constant Coefficients.

$$\frac{d^n y}{dx^n} + a_1\frac{d^{n-1}y}{dx^{n-1}} + \cdots + a_{n-1}\frac{dy}{dx} + a_n y = 0 \tag{32}$$

A solution of this equation is

$$y_k = ce^{r_k x} \tag{33}$$

if r_k is a root of the algebraic equation

$$r^n + a_1 r^{n-1} + \cdots + a_{n-1}r + a_n = 0 \tag{34}$$

If all the n roots $r_1, r_2, \cdots, r_n$ of (34) are different, then

$$y = c_1 e^{r_1 x} + c_2 e^{r_2 x} + \cdots + c_n e^{r_n x} \qquad (35)$$

is a general solution of (32). If k of the roots are equal, $r_1 = r_2 = \cdots = r_k$ while r_{k+1}, $\cdots, r_n$ are different, then

$$y = (c_1 + c_2 x + \cdots + c_k x^{k-1}) e^{r_1 x} + c_{k+1} e^{r_{k+1} x} + \cdots + c_n e^{r_n x} \qquad (36)$$

is a general solution. If $r_1 = p + iq$, $r_2 = p - iq$ are conjugate complex roots of (34), then

$$c_1 e^{r_1 x} + c_2 e^{r_2 x} = e^{px}(C_1 \cos qx + C_2 \sin qx) \qquad (37)$$

Example 1. $\dfrac{d^2 y}{dx^2} + 13 \dfrac{dy}{dx} + 40y = 0$ has the solution $y = c_1 e^{-5x} + c_2 e^{-8x}$.

Example 2. $\dfrac{d^2 y}{dx^2} + 6 \dfrac{dy}{dx} + 34y = 0$ has the solution $y = (c_1 \cos 5x + c_2 \sin 5x) e^{-3x}$.

Non-homogeneous Differential Equation with Constant Coefficients.

$$\frac{d^n y}{dx^n} + a_1 \frac{d^{n-1} y}{dx^{n-1}} + \cdots + a_{n-1} \frac{dy}{dx} + a_n y = F(x) \qquad (38)$$

The complementary function is found as above. To find the particular integral, replace

$$\frac{dy}{dx} \text{ by } D, \quad \frac{d^2 y}{dx^2} \text{ by } D^2, \quad \cdots, \quad \frac{d^n y}{dx^n} \text{ by } D^n \qquad (39)$$

$$P(D)y = (D^n + a_1 D^{n-1} + \cdots + a_{n-1} D + a_n) y = F(x) \qquad (40)$$

Particular integrals y_p, in which B_i, A, B are undetermined coefficients, to be determined by substituting y_p in (40) and equating coefficients of like terms, are:

(a) If $F(x) = x^n + b_1 x^{n-1} + \cdots + b_{n-1} x + b_n$, then $y_p = x^n + B_1 x^{n-1} + \cdots + B_{n-1} x + B_n$. If D^m is a factor of $P(D)$, then $y_p = (x^n + B_1 x^{n-1} + \cdots + B_{n-1} x + B_n) x^m$.

(b) If $F(x) = b \sin ax$ or $b \cos ax$, then $y_p = A \sin ax + B \cos ax$. If $(D^2 + a^2)^m$ is a factor of $P(D)$, then $y_p = (A \sin ax + B \cos ax) x^m$.

(c) If $F(x) = c e^{ax}$, then $y_p = A e^{ax}$. If $(D - a)^m$ is a factor of $P(D)$, then $y_p = x^m A e^{ax}$.

(d) If $F(x) = g(x) e^{ax}$, place $y_p = e^{ax} w$ in (38), divide out e^{ax}, and solve the equation for w_p as a function of x.

(e) If $F(x)$ is the sum of a number of the above functions, then y_p is the sum of the particular integrals corresponding to each of the functions.

(f) If $F(x)$ is not of the type (e), try the method of Laplace transformation, p. 2-131.

Example. $\dfrac{d^2 y}{dx^2} + 4y = x^2 + \cos x$ can be written as $(D^2 + 4)y = (D + 2i)(D - 2i) = x^2 + \cos x$.

By (37), the complementary function is $y = c_1 \cos 2x + c_2 \sin 2x$.
For a particular integral take $y_p = ax^2 + bx + c + f \sin x + g \cos x$ [by (a), (b), (e)].

Then $\dfrac{d^2 y_p}{dx^2} = 2a - f \sin x - g \cos x$, and substituting in the original equation

$$\frac{d^2 y_p}{dx^2} + 4y_p = 2a - f \sin x - g \cos x + 4ax^2 + 4bx + 4c + 4f \sin x + 4g \cos x = x^2 + \cos x$$

Equating coefficients, $a = 1/4$, $b = 0$, $c = -1/8$, $f = 0$, $g = 1/3$ and the general solution is $y = c_1 \cos 2x + c_2 \sin 2x + \dfrac{x^2}{4} - 1/8 + 1/3 \cos x$.

Euler's Homogeneous Equation.

$$x^n \frac{d^n y}{dx^n} + a x^{n-1} \frac{d^{n-1} y}{dx^{n-1}} + \cdots + a_{n-1} x \frac{dy}{dx} + a_n y = 0 \qquad (41)$$

Place $x = e^t$, and since

$$x \frac{dy}{dx} = \frac{dy}{dt}, \quad x^2 \frac{d^2 y}{dx^2} = \left[\frac{d}{dt} \left(\frac{d}{dt} - 1 \right) \right] y, \quad x^3 \frac{d^3 y}{dx^3} = \left[\frac{d}{dt} \left(\frac{d}{dt} - 1 \right) \left(\frac{d}{dt} - 2 \right) \right] y, \quad \cdots \qquad (42)$$

(41) is transformed into a linear homogeneous differential equation with constant coefficients.

Depression of Order. If a particular integral of a linear homogeneous differential equation is known, the order of the equation can be lowered. If y_1 is a particular integral of

$$\frac{d^n y}{dx^n} + P_1(x)\frac{d^{n-1}y}{dx^{n-1}} + \cdots + P_{n-1}(x)\frac{dy}{dx} + P_n(x) = 0 \tag{43}$$

substitute $y = y_1 z$. The coefficient of z will be zero, and then by placing $\dfrac{dz}{dx} = u$, the equation is reduced to the $(n-1)$st order.

Example. Given $\dfrac{d^2 y}{dx^2} + p(x)\dfrac{dy}{dx} + q(x)y = 0$ and y_1, a particular integral of this equation.

Let $y = y_1 z$, then $\dfrac{dy}{dx} = y_1\dfrac{dz}{dx} + z\dfrac{dy_1}{dx}$, $\dfrac{d^2 y}{dx^2} = y_1\dfrac{d^2 z}{dx^2} + 2\dfrac{dy_1}{dx}\dfrac{dz}{dx} + z\dfrac{d^2 y_1}{dx^2}$.

Substituting in the original equation

$$y_1\frac{d^2 z}{dx^2} + 2\frac{dy_1}{dx}\frac{dz}{dx} + z\frac{d^2 y_1}{dx^2} + p\left[y_1\frac{dz}{dx} + z\frac{dy_1}{dx}\right] + qy_1 z = 0$$

and since the coefficient of z is zero, this reduces to

$$y_1\frac{d^2 z}{dx^2} + \left(2\frac{dy_1}{dx} + py_1\right)\frac{dz}{dx} = 0. \quad \text{Writing } \frac{dz}{dx} = u, \quad \frac{du}{u} + \left(2\frac{dy_1}{dx} + py_1\right)\frac{dx}{y_1} = 0$$

By integrating,

$$\log_e u + \int p\,dx + \log_e y_1^2 = \log_e c \quad \text{or} \quad u = \frac{c}{y_1^2}e^{-\int p\,dx}$$

Another integration gives z. Then

$$y = y_1\int\frac{c}{y_1^2}e^{-\int p\,dx}\,dx + c_1$$

Systems of Linear Differential Equations with Constant Coefficients. For a system of n linear equations with constant coefficients in n dependent variables and one independent variable t, the symbolic algebraic method of solution may be used. If $n = 2$,

$$\left. \begin{array}{l} (D^n + a_1 D^{n-1} + \cdots + a_n)x + (D^m + b_1 D^{m-1} + \cdots + b_m)y = R(t) \\ (D^p + c_1 D^{p-1} + \cdots + c_p)x + (D^q + d_1 D^{q-1} + \cdots + d_q)y = S(t) \end{array} \right\} \tag{44}$$

where $D = \dfrac{d}{dt}$. The equations may be written as

$$P_1(D)x + Q_1(D)y = R, \quad P_2(D)x + Q_2(D)y = S \tag{45}$$

Treating these as algebraic equations, eliminate either x or y and solve the equation thus obtained.

Example. Solve the system,

$$(1)\ \frac{dx}{dt} + \frac{dy}{dt} + 2x + y = 0, \quad (2)\ \frac{dy}{dt} + 5x + 3y = 0$$

By using the symbol D these equations can be written

$$(D+2)x + (D+1)y = 0, \quad 5x + (D+3)y = 0$$

Eliminating x, $(D^2+1)y = 0$. From (37) (1) this has the solution $y = c_1\cos t + c_2\sin t$. Substituting this in (2), $x = -\dfrac{3c_1 + c_2}{5}\cos t + \dfrac{c_1 - 3c_2}{5}\sin t$.

57. PARTIAL DIFFERENTIAL EQUATIONS

First Order

Definition. If $x_1, x_2, \cdots, x_n$ are n independent variables, $z = z(x_1, x_2, \cdots, x_n)$ the dependent variable, and if

$$\frac{\partial z}{\partial x_1} = p_1, \quad \cdots, \quad \frac{\partial z}{\partial x_n} = p_n \tag{46}$$

then

$$F(x_1, x_2, \cdots, x_n, z, p_1, p_2, \cdots, p_n) = 0 \tag{47}$$

is a partial differential equation of the first order. An equation

$$f(x_1, x_2, \cdots, x_n, z, c_1, \cdots, c_n) = 0 \tag{48}$$

with n independent constants is a *complete integral* of (47) if the elimination of the constants by partial differentiation gives the differential equation (47).

Example.

$$F = z^2 \left[\left(\frac{\partial z}{\partial x} \right)^2 + \left(\frac{\partial z}{\partial y} \right)^2 + 1 \right] - c^2 = 0$$

Then

$$f = (x - h)^2 + (y - k)^2 + z^2 - c^2 = 0$$

is a solution, since by differentiating it with respect to x and y

$$(x - h) + z \frac{\partial z}{\partial x} = 0$$

$$(y - k) + z \frac{\partial z}{\partial y} = 0$$

and substituting the values of $x - h$, $y - k$ from the last two equations in f, expression F is obtained.

If the eliminant obtained by eliminating $c_1, \cdots, c_n$ from the equations $f = 0$, $\dfrac{\partial f}{\partial c_1} = 0$, $\cdots$, $\dfrac{\partial f}{\partial c_n} = 0$ satisfies the differential equation, it is a *singular solution*. This differs from a particular integral in that it is usually not obtainable from the complete integral by giving particular values to the constants.

Suppose that the equation $F\left(x, y, z, \dfrac{\partial z}{\partial x}, \dfrac{\partial z}{\partial y} \right) = 0$ has the complete integral $f(x, y, z, a, b) = 0$. Let one of the constants $b = \phi(a)$; then $f[x, y, z, a, \phi(a)] = 0$. The *general integral* is the set of solutions found by eliminating a between $f[x, y, z, a, \phi(a)] = 0$ and $\dfrac{d\phi}{da} = 0$, for all choices of ϕ.

Linear Differential Equations.

$$P(x, y, z)p + Q(x, y, z)q = R(x, y, z), \quad \text{in which } p = \frac{\partial z}{\partial x}, \quad q = \frac{\partial z}{\partial y} \tag{49}$$

is a linear partial differential equation. From the system of ordinary equations $\dfrac{dx}{P} = \dfrac{dy}{Q} = \dfrac{dz}{R}$, the two independent solutions $u(x, y, z) = c_1$, $v(x, y, z) = c_2$ are obtained. Then $\Phi(u, v) = 0$, where Φ is an arbitrary function, is the *general solution* of $Pp + Qq = R$.

Example 1. Given $xp + yq = z$. The system $\dfrac{dx}{x} = \dfrac{dy}{y} = \dfrac{dz}{z}$ has the solution $u = \dfrac{y}{x} = c_1, v = \dfrac{z}{x} = c_2$. Then the general solution is $\Phi(u, v) = \Phi\left(\dfrac{y}{x}, \dfrac{z}{x} \right) = 0$.

Example 2. Given $(ny - mz)p + (lz - nx)q = mx - ly$.

From $\dfrac{dx}{ny - mz} = \dfrac{dy}{lz - nx} = \dfrac{dz}{mx - ly}$, by using the multipliers l, m, n, and adding the fraction $\dfrac{l\,dx + m\,dy + n\,dz}{0}$ is obtained. Therefore $l\,dx + m\,dy + n\,dz = 0$. This has the solution $lx + my + nz = c_1$. Similarly, $x\,dx + y\,dy + z\,dz = 0$, or $x^2 + y^2 + z^2 = c_2$. Then the general solution is $\Phi(x^2 + y^2 + z^2, lx + my + nz) = 0$.

General Method of Solution. Given $F(x, y, z, p, q) = 0$, the partial differential equation to be solved. Since z is a function of x and y, it follows that $dz = p\,dx + q\,dy$. If another relation can be found among x, y, z, p, q, such as $f(x, y, z, p, q) = 0$; then p and q can be eliminated. The solution of the ordinary differential equation thus formed, involving x, y, z, will satisfy the given equation, $F(x, y, z, p, q) = 0$. The unknown function f must satisfy the following linear partial differential equation:

$$\frac{\partial F}{\partial p} \frac{\partial f}{\partial x} + \frac{\partial F}{\partial q} \frac{df}{\partial y} + \left(p \frac{\partial F}{\partial p} + q \frac{\partial F}{\partial q} \right) \frac{\partial f}{\partial z} - \left(\frac{\partial F}{\partial x} + p \frac{\partial F}{\partial z} \right) \frac{\partial f}{\partial p} - \left(\frac{\partial F}{\partial y} + q \frac{\partial F}{\partial z} \right) \frac{\partial f}{\partial q} = 0 \tag{50}$$

which is satisfied by any of the solutions of the system

$$\frac{\partial x}{\dfrac{\partial F}{\partial p}} = \frac{\partial y}{\dfrac{\partial F}{\partial q}} = \frac{dz}{p\dfrac{\partial F}{\partial p} + q\dfrac{\partial F}{\partial q}} = \frac{-dp}{\dfrac{\partial F}{\partial x} + p\dfrac{\partial F}{\partial z}} = \frac{-dq}{\dfrac{\partial F}{\partial y} + q\dfrac{\partial F}{\partial z}} \tag{51}$$

Example. Solve $p(q^2 + 1) + (b - z)q = 0$. Here the equations above reduce to

$$\frac{dp}{pq} = \frac{dq}{q^2} = \frac{dz}{3pq^2 + p + (b - z)q} = \frac{dx}{q^2 + 1} = \frac{dy}{-z + b + 2pq}$$

The third fraction, by virtue of the given equation, reduces to $\dfrac{dz}{2pq^2}$. From the first two fractions, by integration, $q = cp$. This and the original equation determine the values of p and q, namely, $p = \dfrac{\sqrt{c_1(z - b) - 1}}{c_1}$, $q = \sqrt{c_1(z - b) - 1}$. Substitution of these values in $dz = p\,dx + q\,dy$ gives $dz = \left(\dfrac{dx}{c_1} + dy\right)\sqrt{c_1(z - b) - 1}$. In this equation the variables are separable; this on integration gives the complete integral $2\sqrt{c_1(z - b) + 1} = x + c_1 y + c_2$. There is no singular solution. In the above work, had another pair of ratios been chosen, say $\dfrac{dq}{q^2} = \dfrac{dx}{q^2 + 1}$, another complete integral would have been obtained, namely,

$$(z - b)\left\{\frac{x + k_1}{2} - \sqrt{\left(\frac{x + k_1}{2}\right)^2 + 1}\right\} + y + k_2 = 0$$

Second Order

Definitions. A linear partial differential equation of the second order with two independent variables is of the form

$$L = Ar + 2Bs + Ct + Dp + Eq + Fz = f(x, y) \tag{52}$$

where

$$r = \frac{\partial^2 z}{\partial x^2}, \quad s = \frac{\partial^2 z}{\partial x\,\partial y}, \quad t = \frac{\partial^2 z}{\partial y^2}, \quad p = \frac{\partial z}{\partial x}, \quad q = \frac{\partial z}{\partial y}$$

The coefficients $A, \cdots, F$ are real continuous functions of the real variables x and y. Let $\xi = \xi(x, y)$, $\eta = \eta(x, y)$ be two solutions of the following homogeneous partial differential equation of the first order:

$$Ap^2 + 2Bpq + Cq^2 = 0 \tag{53}$$

If $B^2 - AC = 0$, the homogeneous form of (52), $L = 0$, is called the *parabolic* type, and has the normal form

$$\frac{\partial^2 z}{\partial \xi^2} + a\frac{\partial z}{\partial \xi} + b\frac{\partial z}{\partial \eta} + cz = 0 \tag{54}$$

where a, b, c are functions of ξ and η. An example is the equation of heat flow, $\dfrac{\partial u}{\partial t} = a^2\dfrac{\partial^2 u}{\partial t^2}$ where $u = u(x, t)$ is the temperature, t is the time, a^2 is constant. If $B^2 - AC > 0$ in (53), the homogeneous form of (52) is the *hyperbolic* type which has as its two normal forms

$$\frac{\partial^2 z}{\partial \xi\,\partial \eta} + a\frac{\partial z}{\partial \xi} + b\frac{\partial z}{\partial \eta} + cz = 0 \tag{55}$$

$$\frac{\partial^2 z}{\partial \xi^2} - \frac{\partial^2 z}{\partial \eta^2} + a\frac{\partial z}{\partial \xi} + b\frac{\partial z}{\partial \eta} + cz = 0 \tag{56}$$

An example is the equation of a vibrating string $\dfrac{\partial^2 z}{\partial t^2} = a^2\dfrac{\partial^2 z}{\partial x^2}$, where z is the transverse displacement of a point on the string, with abscissa x at time t and a^2 is constant. If $B^2 - AC < 0$, the equation is of the *elliptic* type which has the normal form

$$\frac{\partial^2 z}{\partial \xi^2} + \frac{\partial^2 z}{\partial z^2} + a\frac{\partial z}{\partial \xi} + b\frac{\partial z}{\partial \eta} + cz = 0 \tag{57}$$

An example is Laplace's equation $\dfrac{\partial^2 z}{\partial \xi^2} + \dfrac{\partial^2 z}{\partial \eta^2} = 0$, usually written $\nabla^2 z = 0$. The two solutions of (53) are real in the hyperbolic case and conjugate complex in the elliptic case.

That is, in the latter case, $\xi = \frac{1}{2}(\alpha + i\beta)$, $\eta = \frac{1}{2}(\alpha - i\beta)$, where α and β are real, and

$$\frac{\partial^2 z}{\partial \xi \, \partial \eta} = \frac{1}{4}\left(\frac{\partial^2 z}{\partial \alpha^2} + \frac{\partial^2 z}{\partial \beta^2}\right).$$

As in ordinary linear equations, the whole solution consists of the complementary function and the particular integral. Also if $z = z_1$, $z = z_2$, $\cdots$, $z = z_n$ are solutions of the homogeneous equation (52), $L = 0$, then $z = c_1 z_1 + c_2 z_2 + \cdots + c_n z_n$ is again a solution.

Equations Linear in the Second Derivatives. The general type of second order equation linear in the second derivatives may be written in the form

$$Ar + Bs + Ct = V \tag{58}$$

where A, B, C, V are functions of x, y, z, p, q. From the equations

$$A \, dy^2 - B \, dx \, dy + C \, dx^2 = 0 \tag{59}$$

$$A \, dp \, dy + C \, dq \, dx - V \, dx \, dy = 0 \tag{60}$$

$$p \, dx + q \, dy = dz \tag{61}$$

it may be possible to derive either one or two relations between x, y, z, p, q, called intermediary integrals, and from these to deduce the solution of (58). To obtain an intermediary integral, resolve (59), supposing the left member not a perfect square, into the two equations $dy - n_1 \, dx = 0$, $dy - n_2 \, dx = 0$. From the first of these and from (60), combined, if necessary, with (61), obtain the two integrals $u_1(x, y, z, p, q) = a$, $v_1(x, y, z, p, q) = b$; then $u_1 = f_1(v_1)$, where f_1 is an arbitrary function, is now an intermediary integral. In the same way, from $dy - n_2 \, dx = 0$, obtain another pair of integrals $u_2 = a_1$, $v_2 = b_1$; then $u_2 = f_2(v_2)$ is an intermediary integral. For the final integral, if $n_1 = n_2$, the intermediary integral may be integrated. If $n_1 \neq n_2$, solve the two intermediary integrals for p and q, substitute in $p \, dx + q \, dy = dz$, and integrate for the solution.

Example. Solve $\qquad\qquad r^2 - a^2 t = 0 \tag{58a}$
the equation for a vibrating string.
The auxiliary equations are

$$dy - a \, dx = 0, \quad dy + a \, dx = 0, \quad \text{and} \quad dp \, dy - a^2 \, dx \, dq = 0 \tag{60a}$$

Hence $y + ax = c_1$, $y - ax = c_2$. Combining $y + ax = c_1$ with (60a), $dp + a \, dq = 0$ is obtained, whereupon $p + aq = c_3 = f_1(y + ax)$. Combining $y - ax = c_1$ with (60a), $dp - a \, dq = 0$ is obtained, whereupon $p - aq = c_4 = f_2(y - ax)$.

Solving for p and q, $p = \frac{1}{2}[f_1(y + ax) + f_2(y - ax)]$, $q = \frac{1}{2a}[f_1(y + ax) - f_2(y - ax)]$. Substituting these in $p \, dx + q \, dy = dz$, $dz = \frac{1}{2a}[f_1(y + a)(dy + a \, dx) - f_2(y - ax)(dy - a \, dx)]$, which is an exact differential. Integration gives $z = \phi(y + ax) + \psi(y - ax)$.

Homogeneous Equation with Constant Coefficients.

$$\frac{\partial^2 z}{\partial x^2} + A_1 \frac{\partial^2 z}{\partial x \, \partial y} + A_2 \frac{\partial^2 z}{\partial y^2} = 0 \tag{62}$$

This equation is equivalent to

$$\left(\frac{\partial}{\partial x} - m_1 \frac{\partial}{\partial y}\right)\left(\frac{\partial}{\partial x} - m_2 \frac{\partial}{\partial y}\right) z = 0 \tag{63}$$

where m_1 and m_2 are roots of the auxiliary equation $X^2 + A_1 X + A_2 = 0$. The general solution of (63) is

$$z = f_1(y + m_1 x) + f_2(y + m_2 x) \tag{64}$$

Example. Solve: $8 \dfrac{\partial^2 z}{\partial x^2} + 2 \dfrac{\partial^2 z}{\partial x \, \partial y} - 15 \dfrac{\partial^2 z}{\partial y^2} = 0$.

The auxiliary equation is $8X^2 + 2X - 15 = (2X + 3)(4X - 5) = 0$. Hence $m_1 = -3/2$, $m_2 = 5/4$. The general solution is $z = f_1(2y - 3x) + f_2(4y + 5x)$.

If the auxiliary equation has multiple factors, the general solution is $z = f_1(y + m_1 x) + x f_2(y + m_1 x)$.

Example. Solve: $\dfrac{\partial^2 z}{\partial x^2} + 6 \dfrac{\partial^2 z}{\partial x \, \partial y} + 9 \dfrac{\partial^2 z}{\partial y^2} = 0$.

The auxiliary equation is $X^2 + 6X + 9 = (X + 3)(X + 3) = 0$. The general solution is $z = f_1(y - 3x) + x f_2(y - 3x)$.

If the coefficients in equation (62) are real, the complex roots of the auxiliary equation occur in conjugate pairs. Then the general solution will have the form

$$z = f(y + \alpha x + i\beta x) + g(y + \alpha x - i\beta x)$$

Example. Solve: $\dfrac{\partial^2 z}{\partial x^2} - 2 \dfrac{\partial^2 z}{\partial x \, \partial y} + 2 \dfrac{\partial^2 z}{\partial y^2} = 0$.

The auxiliary equation is $X^2 - 2X + 2 = 0$ and $m = 1 \pm i$. The general solution is $z = f(y + x + ix) + g(y + x - ix)$, which can be written as $z = f_1(y + x + ix) + f_1(y + x - ix) + i[g_1(y + x + ix) - g_1(y + x - ix)]$, where f_1 and g_1 are any twice differentiable real functions. If, in particular, $f_1 = \cos u$ and $g_1 = e^u$, it can be shown that $z = 2 \cos (x + y) \cosh x - 2e^{x+y} \sin x$.

Method of Separation of Variables. As an example of this method, the solution will be given to Laplace's equation

$$\nabla^2 u = \frac{\partial^2 u}{\partial x^2} + \frac{\partial^2 u}{\partial y^2} = 0 \tag{65}$$

Assume that

$$u = X(x) \cdot Y(y) \tag{66}$$

where X is a function of x only, and Y a function of y only. By substitution and dividing by $X \cdot Y$, (65) becomes

$$\frac{1}{X} \frac{d^2 X}{dx^2} = - \frac{1}{Y} \frac{d^2 Y}{dy^2} \tag{67}$$

Since the left side does not contain y, the right side does not contain x, and the two sides are equal, they must equal a constant, say $-k^2$.

$$\frac{1}{X} \frac{d^2 X}{dx^2} = -k^2, \quad \frac{1}{Y} \frac{d^2 Y}{dy^2} = k^2 \tag{67a}$$

The solutions of these homogeneous linear differential equations with constant coefficients are

$$X = c_1 \cos kx + c_2 \sin kx, \quad Y = c_3 e^{ky} + c_4 e^{-ky} \tag{68}$$

Hence, from (66),

$$\begin{aligned} u &= (c_1 \cos kx + c_2 \sin kx)(c_3 e^{ky} + c_4 e^{-ky}) \\ &= e^{ky}(k_1 \cos kx + k_2 \sin kx) + e^{-ky}(k_3 \cos kx + k_4 \sin kx) \end{aligned} \tag{69}$$

Since (65) is linear, the sum of any number of solutions is again a solution. An infinite number of solutions may be taken provided the series converges and may be differentiated term by term. Then

$$u = \sum_{n=0}^{\infty} [e^{ky}(A_n \cos kx + B_n \sin kx) + e^{-ky}(D_n \cos kx + E_n \sin kx)] \tag{70}$$

is a solution of (65). The coefficients of (70) are determined by using the series as a Fourier series to fit the boundary conditions.

Functions which satisfy Laplace's equation are *harmonic*. In polar coordinates (65) becomes,

$$\nabla^2 u = \frac{\partial^2 u}{\partial r^2} + \frac{1}{r^2} \frac{\partial^2 u}{\partial \theta^2} + \frac{1}{r} \frac{\partial u}{\partial r} = 0 \tag{71}$$

In three dimensions, Laplace's equation in rectangular coordinates is

$$\nabla^2 u = \frac{\partial^2 u}{\partial x^2} + \frac{\partial^2 u}{\partial y^2} + \frac{\partial^2 u}{\partial z^2} = 0 \tag{72}$$

In cylindrical coordinates,

$$\nabla^2 u = \frac{\partial^2 u}{\partial \rho^2} + \frac{1}{\rho} \frac{\partial u}{\partial \rho} + \frac{1}{\rho^2} \frac{\partial^2 u}{\partial \phi^2} + \frac{\partial^2 u}{\partial z^2} = 0 \tag{73}$$

In spherical coordinates,

$$\nabla^2 u = \frac{1}{r^2} \frac{\partial}{\partial r}\left(r^2 \frac{\partial u}{\partial r}\right) + \frac{1}{r^2 \sin^2 \theta} \frac{\partial^2 u}{\partial \phi^2} + \frac{1}{r^2 \sin \theta} \frac{\partial}{\partial \theta}\left(\sin \theta \frac{\partial u}{\partial \theta}\right) \tag{74}$$

LAPLACE TRANSFORMATION
By J. L. Barnes

58. TRANSFORMATION PRINCIPLES

The Laplace and Fourier transformation methods and the Heaviside operational calculus are in essence different aspects of the same method. This method simplifies the solving of linear constant-coefficient integrodifferential equations and convolution type integral equations. For brevity the conditions under which the steps of the method may be

validly applied will be omitted.   Hence the correctness of a final result should be checked
in each case by showing that the formal solution satisfies the given equation and conditions.

1. Direct Laplace Transformation.   Let t be a real variable, s a complex variable (p.
348), $f(t)$ a real function of t which equals zero for $t < 0$, $F(s)$ a function of s, and e
the base of the natural logarithms.   If the Lebesgue integral

$$\int_0^\infty e^{-st} f(t)\, dt = F(s) \tag{1}$$

then $F(s)$ is the *direct Laplace transform* of $f(t)$; in simpler notation

$$\mathcal{L}[f(t)] = F(s) \tag{2}$$

2. Inverse Laplace Transformation.   Under certain conditions the direct transformation can be inverted, giving as one explicit representation

$$\frac{1}{2\pi i} \int_{c-i\infty}^{c+i\infty} e^{ts} F(s)\, ds \ (=) f(t) \tag{3}$$

in which c is a real constant chosen so that the path of integration lies to the right of all
the singularities of $F(s)$, and $(=)$ means equals except possibly for a set of values of t of
measure zero (p. 336).   If this relation holds, then $f(t)$ is the *inverse Laplace transform*
of $F(s)$.   In simpler notation the transformation is written

$$\mathcal{L}^{-1}[F(s)]\, (=) f(t) \tag{4}$$

3. Transformation of nth Derivative.   If $\mathcal{L}[f(t)] = F(s)$, then

$$\mathcal{L}\left[\frac{d^n f(t)}{dt^n}\right] = s^n F(s) - \sum_{k=0}^{n-1} f^{(k)}(0+) \cdot s^{n-1-k} \tag{5}$$

where $f^{(2)}(0+)$ means $\dfrac{d^2 f(t)}{dt^2}$ evaluated for $t \to 0$, and $f^{(0)}(0+)$ means $f(0+)$, and $n = 1, 2,$
$3, \cdots$.

4. Transformation of nth Integral.   If $\mathcal{L}[f(t)] = F(s)$, then

$$\mathcal{L}\left[\overbrace{\int\!\!\int \cdots \int}^{n} f(t)\, dt\right] = s^{-n} F(s) \ | \ \sum_{k=-1}^{-n} f^{(k)}(0+) \cdot s^{-n-1-k} \tag{6}$$

where $n = 1, 2, 3, \cdots$.   For example, $f^{(-2)}(0+)$ means $\int\!\!\int f(t)\, dt\, dt$ evaluated for $t \to 0$.

5. Inverse Transformation of Product.   If

$$\mathcal{L}^{-1}[F_1(s)] = f_1(t), \quad \mathcal{L}^{-1}[F_2(s)] = f_2(t) \tag{7}$$

then

$$\mathcal{L}^{-1}[F_1(s) \cdot F_2(s)] = \int_0^t f_1(t - \lambda) \cdot f_2(\lambda)\, d\lambda \tag{8}$$

6. Linear Transformations $\mathcal{L}$ and $\mathcal{L}^{-1}$.   Let k_1, k_2 be real constants.   Then

$$\mathcal{L}[k_1 f_1(t) + k_2 f_2(t)] = k_1 \mathcal{L}[f_1(t)] + k_2 \mathcal{L}[f_2(t)] \tag{9}$$

and

$$\mathcal{L}^{-1}[k_1 F_1(s) + k_2 F_2(s)] = k_1 \mathcal{L}^{-1}[F_1(s)] + k_2 \mathcal{L}^{-1}[F_2(s)] \tag{10}$$

59. PROCEDURE

To illustrate the application of the rules of procedure the following simple initial-value
problem will be solved.   Given the equation

$$k_1 \frac{dy(t)}{dt} + k_2 y(t) + k_3 \int y(t)\, dt = u(t)$$

and initial values $y(0)$, $y^{(-1)}(0)$ where $u(t) = 0$ for $t < 0$, and 1 for $0 < t$, and k_1, k_2, k_3
are real constants.   Assume that $y(t)$ has a Laplace transform $Y(s)$, that is, $\mathcal{L}[y(t)] = Y(s)$.

Step A.   Find the Laplace transform of the equation to be solved and express it in
terms of the transform of the unknown function.
Thus,

$$\mathcal{L}\left[k_1 \frac{dy(t)}{dt} + k_2 y(t) + k_3 \int y(t)\, dt\right] = \mathcal{L}[u(t)]$$

By **(9)** this becomes

$$k_1 \mathcal{L}\left[\frac{dy(t)}{dt}\right] + k_2 \mathcal{L}[y(t)] + k_3 \mathcal{L}\left[\int y(t)\, dt\right] = \mathcal{L}[u(t)]$$

By **(5)** and **(6)** and the given initial conditions of the problem the equation becomes

$$k_1[sY(s) - y(0)] + k_2 Y(s) + k_3[s^{-1}Y(s) + y^{(-1)}(0)\cdot s^{-1}] = \mathcal{L}[u(t)]$$

Step B. Solve the resulting equation for the transform of the unknown function. Thus,

$$Y(s) = \frac{\mathcal{L}[u(t)] + k_1 y(0) - y^{(-1)}(0)\cdot s^{-1}}{k_1 s + k_2 + k_3 s^{-1}}$$

Step C. Evaluate the direct transform of the given function (right member) in the original equation. Since

$$\mathcal{L}[u(t)] = \frac{1}{s}$$

$$Y(s) = \frac{k_1 y(0)\cdot s - y^{(-1)}(0) + 1}{k_1 s^2 + k_2 s + k_3}$$

Step D. Obtain the solution of the problem by evaluating the inverse Laplace transform of the function obtained by the preceding steps.

One way to carry out Step **D** is to find the inverse transform from the table of Laplace transforms in Art. 60. To use the table, the denominator of the fraction should be factored.

$$y(t) = \mathcal{L}^{-1}[Y(s)] = \mathcal{L}^{-1}\left[\frac{k_1 y(0)\cdot s - y^{(-1)}(0) + 1}{k_1 s^2 + k_2 s + k_3}\right]$$

$$= \mathcal{L}^{-1}\left[\frac{k_1 y(0)\cdot s - y^{(-1)}(0) + 1}{k_1(s + K_1)(s + K_2)}\right]$$

in which

$$K_1 \equiv \frac{k_2}{2k_1} - \frac{1}{2k_1}(k_2{}^2 - 4k_1 k_3)^{\frac{1}{2}}$$

$$K_2 \equiv \frac{k_2}{2k_1} + \frac{1}{2k_1}(k_2{}^2 - 4k_1 k_3)^{\frac{1}{2}}$$

To find the result it is necessary to distinguish between two cases.

Case 1: If $K_1 \neq K_2$,

$$y(t) = \{[k_1 y(0)K_1 + y^{(-1)}(0) - 1]e^{-K_1 t} - [k_1 y(0)K_2 + y^{(-1)}(0) - 1]e^{-K_2 t}\}/[k_1(K_1 - K_2)]$$

for $0 < t$, and $= 0$ for $t < 0$.

Case 2: If $K_1 = K_2 = K$, then $K = \dfrac{k_2}{2k_1}$, and

$$y(t) = \mathcal{L}^{-1}\left[\frac{k_1 y(0)\cdot s - y^{(-1)}(0) + 1}{k_1(s + K)^2}\right]$$

From the table,

$$y(t) = \{k_1 y(0)e^{-Kt} - [y^{(-1)}(0) - 1 + k_1 y(0)K]te^{-Kt}\}/k_1$$

for $0 < t$, and $= 0$ for $t < 0$.

The solutions can be shown to satisfy the original equation and initial conditions.

The use of Step **C** can be avoided by using Steps **E, F,** and **G** in place of Steps **C** and **D** in the following way.

Step E. Factor the transform of the unknown function obtained by Step **B,** and evaluate the inverse Laplace transform of each factor.

Note. The inverse transform of a rational fraction can be found only if it is a proper fraction.

Thus,

$$Y(s) = \frac{k_1 y(0)\cdot s - y^{(-1)}(0)}{k_1 s^2 + k_2 s + k_3} + \frac{s\mathcal{L}[u(t)]}{k_1 s^2 + k_2 s + k_3}$$

Let

$$y_1(t) \equiv \mathcal{L}^{-1}\left[\frac{k_1 y(0) \cdot s - y^{(-1)}(0)}{k_1(s + K_1)(s + K_2)}\right] = \{[k_1(y)(0)K_1 + y^{(-1)}(0)]e^{-K_1 t}$$
$$- [k_1 y(0)K_2 + y^{(-1)}(0)]e^{-K_2 t}\}/[k_1(K_1 - K_2)]$$

for $0 < t$, and $= 0$ for $t < 0$.   Also

$$\mathcal{L}^{-1}\left[\frac{s}{k_1(s + K_1)(s + K_2)}\right] = (K_1 e^{-K_1 t} - K_2 e^{-K_2 t})/[k_1(K_1 - K_2)]$$

for $0 < t$, and $= 0$ for $t < 0$.   Finally, $\mathcal{L}^{-1}\{\mathcal{L}[u(t)]\} = u(t)$.

Step F.   Use **5** to find the inverse transform of the product.
Thus, by **6** and Step **F**,

$$y(t) = y_1(t) + [k_1(K_1 - K_2)]^{-1}\int_0^t [K_1 e^{-K_1(t-\tau)} - K_2 e^{-K_2(t-\tau)}]u(\tau)\,d\tau$$

Step G.   Evaluate the (convolution) integral arising from Step **F**.   Thus,

$$y(t) = y_1(t) + [k_1(K_1 - K_2)]^{-1}(e^{-K_2 t} - e^{-K_1 t})$$

for $0 < t$, and $= 0$ for $t < 0$.

For the particular problem treated above it is much simpler to use Steps **C** and **D** than Steps **E**, **F**, and **G**.   However, for a more complicated right member of the original equation it could happen that Step **G** would be easier to carry out than Step **C**, in which case the second method (**A**, **B**, **E**, **F**, **G**) should be used rather than the first (**A**, **B**, **C**, **D**).

One physical representation of the initial-value problem which we have used for illustration is the problem of finding the current response of a series electric circuit containing constant lumped inductance, resistance, and capacitance to an applied electromotive force $u(t)$, with an initial current in the inductance and an initial charge on the condenser.

The complete method (of which only a part has been given above) is not restricted in its field of application to linear equations with constant coefficients, but the solution of this type of equation is most simplified.

60. TRANSFORM PAIRS

The following tables of Laplace transforms are applicable in the solution of ordinary integrodifferential and difference equations.   These tables are, by permission, from *Transients in Linear Systems*, by M. F. Gardner and J. L. Barnes, John Wiley and Sons, New York, 1942.

Unilateral Laplace Operation-Transform Pairs

Name	$f(t)$ $0 \le t$	$F(s)$
Linearity	$af(t)$ a is a constant or a variable independent of t and s. $f_1(t) \pm f_2(t)$	$aF(s)$ $F_1(s) \pm F_2(s)$
Real differentiation	$\dfrac{df(t)}{dt} \triangleq f'(t)$	$sF(s) - f(0+)$
Multiplication by s	$f'(t)$ if $f(0+) = 0$.	$sF(s)$
Real integration	$\int f(t)dt \triangleq f^{(-1)}(t)$	$\dfrac{F(s)}{s} + \dfrac{f^{(-1)}(0+)}{s}$
Division by s	$\int_0^t f(t)dt \triangleq f^{(-1)}(t) - f^{(-1)}(0+)$	$\dfrac{F(s)}{s}$
Scale change	$f\left(\dfrac{t}{a}\right)$ a is a positive constant or a positive variable independent of t and s.	$aF(as)$
Complex multiplication	$\int_0^t f_1(t - \tau)f_2(\tau)d\tau \triangleq f_1(t) * f_2(t)$	$F_1(s)F_2(s)$
Real translation	$f(t - a)$ if $f(t - a) = 0, 0 < t < a$ $f(t + a)$ if $f(t + a) = 0, -a < t < 0$ a is a non-negative real number.	$e^{-as}F(s)$ $e^{as}F(s)$

Operation-Transform Pairs—Continued

Name	$f(t)$ $0 \leq t$	$F(s)$
Complex translation	$e^{-at}f(t)$ $e^{at}f(t)$ a is a complex number with non-negative real part.	$F(s+a)$ $F(s-a)$
Second independent variable	$\lim\limits_{a \to a_0} f(t,a)$ a is a second variable independent of t and s.	$\lim\limits_{a \to a_0} F(s,a)$
Differentiation with respect to second independent variable	$\dfrac{\partial}{\partial a} f(t,a)$ a is a second variable independent of t and s.	$\dfrac{\partial}{\partial a} F(s,a)$
Final value	$\lim\limits_{t \to \infty} f(t) = \lim\limits_{s \to 0} sF(s)$ if $sF(s)$ is analytic on the axis of imaginaries and in the right half-plane.	
Initial value	$\lim\limits_{t \to 0} f(t) = \lim\limits_{s \to \infty} sF(s)$	
Complex differentiation	$tf(t)$	$-\dfrac{d}{ds} F(s)$
Complex integration	$\dfrac{1}{t} f(t)$	$\displaystyle\int_s^\infty F(s)ds$
Integration with respect to second independent variable	$\displaystyle\int_{a_0}^a f(t,a)da$ a is a second variable independent of t and s.	$\displaystyle\int_{a_0}^a F(s,a)da$
Real multiplication	$f_1(t)f_2(t)$	$\dfrac{1}{2\pi j} \displaystyle\int_{c_1-j\infty}^{c_2+j\infty} F_1(s-w)F_2(w)dw \triangleq F_1(s) \circledast F_2(s),$ $\max(\sigma_{a1}, \sigma_{a2}, \sigma_{a1}+\sigma_{a2}) < \sigma, \sigma_{a2} < c_2 < \sigma - \sigma_{a2}$ $\displaystyle\sum_{k=1}^q \dfrac{A_1(s_k)}{B'_1(s_k)} F_2(s-s_k)$ if $F_1(s) \triangleq \dfrac{A_1(s)}{B_1(s)}$ is rational algebraic fraction having only first-order poles. $\displaystyle\sum_{k=1}^n \sum_{j=1}^{m_k} \dfrac{(-1)^{m_k-j}K_{kj}}{(m_k-j)!} \left[\dfrac{d^{m_k-j}}{ds^{m_k-j}} F_2(s) \right]_{s=s-s_k}$ $K_{kj} \triangleq \dfrac{1}{(j-1)!} \left[\dfrac{d^{j-1}}{ds^{j-1}} (s-s_k)^{m_k}F_1(s) \right]_{s-s_k}$ if $F_1(s)$ is rational algebraic fraction having multiple-order poles.

Unilateral Laplace Function-Transform Pairs*

$F(s)$	$f(t)$	$0 \leq t$
$\dfrac{A(s)}{B(s)}$ Rational proper fraction; first-order poles only.	$\displaystyle\sum_{k=1}^{q} \frac{A(s_k)}{B'(s_k)} e^{s_k t}$	
$\dfrac{A(s)}{B(s)}$ Rational proper fraction; higher-order poles, general case.	$\displaystyle\sum_{k=1}^{n}\sum_{j=1}^{m_k} \frac{K_{kj}}{(m_k - j)!} t^{m_k - j} e^{s_k t}$ $K_{kj} \triangleq \dfrac{1}{(j-1)!}\left[\dfrac{d^{j-1}}{ds^{j-1}}\dfrac{(s-s_k)^{m_k}A(s)}{B(s)}\right]_{s=s_k}$ $B(s) \triangleq (s-s_1)^{m_1}(s-s_2)^{m_2}\cdots(s-s_k)^{m_k}\cdots(s-s_n)^{m_n}$	$m_1 + m_2 + \cdots + m_n = q$
1	$u_1(t) \triangleq \lim_{a\to 0}\dfrac{u(t)-u(t-a)}{a}$, unit impulse at $t=0$	
s	$u_2(t) \triangleq \lim_{a\to 0}\dfrac{u(t)-2u(t-a)+u(t-2a)}{a^2}$, unit doublet impulse at $t=0$	
$\dfrac{1}{s}$	1, or $u(t)$, unit step at $t=0$	
$\dfrac{1}{s+\alpha}$	$e^{-\alpha t}$	
$\dfrac{1}{(s+\alpha)(s+\gamma)}$	$\dfrac{e^{-\alpha t}-e^{-\gamma t}}{\gamma-\alpha}$	
$\dfrac{s+a_0}{(s+\alpha)(s+\gamma)}$	$\dfrac{(a_0-\alpha)e^{-\alpha t}-(a_0-\gamma)e^{-\gamma t}}{\gamma-\alpha}$	

* α, β, γ, δ, and λ are real numbers.

$F(s)$	$f(t)$
$\dfrac{1}{s(s+\alpha)(s+\gamma)}$	$\dfrac{1}{\alpha\gamma}+\dfrac{\gamma e^{-\alpha t}-\alpha e^{-\gamma t}}{\alpha\gamma(\alpha-\gamma)}$
$\dfrac{s+a_0}{s(s+\alpha)(s+\gamma)}$	$\dfrac{a_0}{\alpha\gamma}+\dfrac{a_0-\alpha}{\alpha(\alpha-\gamma)}e^{-\alpha t}+\dfrac{a_0-\gamma}{\gamma(\gamma-\alpha)}e^{-\gamma t}$
$\dfrac{s^2+a_1 s+a_0}{s(s+\alpha)(s+\gamma)}$	$\dfrac{a_0}{\alpha\gamma}+\dfrac{\alpha^2-a_1\alpha+a_0}{\alpha(\alpha-\gamma)}e^{-\alpha t}-\dfrac{\gamma^2-a_1\gamma+a_0}{\gamma(\alpha-\gamma)}e^{-\gamma t}$
$\dfrac{1}{(s+\alpha)(s+\gamma)(s+\delta)}$	$\dfrac{e^{-\alpha t}}{(\gamma-\alpha)(\delta-\alpha)}+\dfrac{e^{-\gamma t}}{(\alpha-\gamma)(\delta-\gamma)}+\dfrac{e^{-\delta t}}{(\alpha-\delta)(\gamma-\delta)}$
$\dfrac{s+a_0}{(s+\alpha)(s+\gamma)(s+\delta)}$	$\dfrac{a_0-\alpha}{(\gamma-\alpha)(\delta-\alpha)}e^{-\alpha t}+\dfrac{a_0-\gamma}{(\alpha-\gamma)(\delta-\gamma)}e^{-\gamma t}+\dfrac{a_0-\delta}{(\alpha-\delta)(\gamma-\delta)}e^{-\delta t}$
$\dfrac{s^2+a_1 s+a_0}{(s+\alpha)(s+\gamma)(s+\delta)}$	$\dfrac{\alpha^2-a_1\alpha+a_0}{(\gamma-\alpha)(\delta-\alpha)}e^{-\alpha t}+\dfrac{\gamma^2-a_1\gamma+a_0}{(\alpha-\gamma)(\delta-\gamma)}e^{-\gamma t}+\dfrac{\delta^2-a_1\delta+a_0}{(\alpha-\delta)(\gamma-\delta)}e^{-\delta t}$
$\dfrac{1}{s^2+\beta^2}$	$\dfrac{1}{\beta}\sin\beta t$
$\dfrac{1}{s^2-\beta^2}$	$\dfrac{1}{\beta}\sinh\beta t$
$\dfrac{s}{s^2+\beta^2}$	$\cos\beta t$
$\dfrac{s}{s^2-\beta^2}$	$\cosh\beta t$

Function-Transform Pairs—Continued

$F(s)$	$f(t)$	$0 \leq t$
$\dfrac{s + a_0}{s^2 + \beta^2}$	$\dfrac{1}{\beta} (a_0 + \beta^2)^{1/2} \sin (\beta t + \psi)$ $\psi \triangleq \tan^{-1} \dfrac{\beta}{a_0}$	
$\dfrac{1}{s(s^2 + \beta^2)}$	$\dfrac{1}{\beta^2} (1 - \cos \beta t)$	
$\dfrac{s + a_0}{s(s^2 + \beta^2)}$	$\dfrac{a_0}{\beta^2} - \dfrac{(a_0{}^2 + \beta^2)^{1/2}}{\beta^2} \cos (\beta t + \psi)$ $\psi \triangleq \tan^{-1} \dfrac{\beta}{a_0}$	
$\dfrac{s^2 + a_1 s + a_0}{s(s^2 + \beta^2)}$	$\dfrac{a_0}{\beta^2} - \dfrac{[(a_0 - \beta^2)^2 + a_1{}^2\beta^2]^{1/2}}{\beta^2} \cos (\beta t + \psi)$ $\psi \triangleq \tan^{-1} \dfrac{a_1\beta}{a_0 - \beta^2}$	
$\dfrac{s + a_0}{(s + \alpha)(s^2 + \beta^2)}$	$\dfrac{a_0 - \alpha}{\alpha^2 + \beta^2} e^{-\alpha t} + \dfrac{1}{\beta} \left[\dfrac{a_0{}^2 + \beta^2}{\alpha^2 + \beta^2} \right]^{1/2} \sin (\beta t + \psi)$ $\psi \triangleq \tan^{-1} \dfrac{\beta}{a_0} - \tan^{-1} \dfrac{\beta}{\alpha}$	
$\dfrac{s^2 + a_1 s + a_0}{(s + \alpha)(s^2 + \beta^2)}$	$\dfrac{\alpha^2 - a_1\alpha + a_0}{\alpha^2 + \beta^2} e^{-\alpha t} + \dfrac{1}{\beta} \left[\dfrac{(a_0 - \beta^2)^2 + a_1{}^2\beta^2}{\alpha^2 + \beta^2} \right]^{1/2} \sin (\beta t + \psi)$ $\psi \triangleq \tan^{-1} \dfrac{a_1\beta}{a_0 - \beta^2} - \tan^{-1} \dfrac{\beta}{\alpha}$	
$\dfrac{s + a_0}{s(s + \alpha)(s^2 + \beta^2)}$	$\dfrac{a_0}{\alpha\beta^2} + \dfrac{\alpha - a_0}{\alpha(\alpha^2 + \beta^2)} e^{-\alpha t} - \dfrac{1}{\beta^2} \left[\dfrac{a_0{}^2 + \beta^2}{\alpha^2 + \beta^2} \right]^{1/2} \cos (\beta t + \psi)$ $\psi \triangleq \tan^{-1} \dfrac{\beta}{a_0} - \tan^{-1} \dfrac{\beta}{\alpha}$	
$\dfrac{s^2 + a_1 s + a_0}{s(s + \alpha)(s^2 + \beta^2)}$	$\dfrac{a_0}{\alpha\beta^2} - \dfrac{\alpha^2 - a_1\alpha + a_0}{\alpha(\alpha^2 + \beta^2)} e^{-\alpha t}$ $\quad - \dfrac{1}{\beta^2} \left[\dfrac{(a_0 - \beta^2)^2 + a_1{}^2\beta^2}{\alpha^2 + \beta^2} \right]^{1/2} \cos (\beta t + \psi)$ $\psi \triangleq \tan^{-1} \dfrac{a_1\beta}{a_0 - \beta^2} - \tan^{-1} \dfrac{\beta}{\alpha}$	
$\dfrac{s^2 + a_1 s + a_0}{(s + \alpha)(s + \gamma)(s^2 + \beta^2)}$	$\dfrac{\alpha^2 - a_1\alpha + a_0}{(\gamma - \alpha)(\alpha^2 + \beta^2)} e^{-\alpha t} + \dfrac{\gamma^2 - a_1\gamma + a_0}{(\alpha - \gamma)(\gamma^2 + \beta^2)} e^{-\gamma t}$ $\quad + \dfrac{1}{\beta} \left[\dfrac{(a_0 - \beta^2)^2 + a_1{}^2\beta^2}{(\alpha^2 + \beta^2)(\gamma^2 + \beta^2)} \right]^{1/2} \sin (\beta t + \psi)$ $\psi \triangleq \tan^{-1} \dfrac{a_1\beta}{a_0 - \beta^2} - \tan^{-1} \dfrac{\beta}{\alpha} - \tan^{-1} \dfrac{\beta}{\gamma}$	
$\dfrac{s^3 + a_2 s^2 + a_1 s + a_0}{(s + \alpha)(s + \gamma)(s^2 + \beta^2)}$	$\dfrac{-\alpha^3 + a_2\alpha^2 - a_1\alpha - a_0}{(\gamma - \alpha)(\alpha^2 + \beta^2)} e^{-\alpha t} + \dfrac{-\gamma^3 + a_2\gamma^2 - a_1\gamma + a_0}{(\alpha - \gamma)(\gamma^2 + \beta^2)} e^{-\gamma t}$ $\quad + \dfrac{1}{\beta} \left[\dfrac{(a_0 - a_2\beta^2)^2 + \beta^2(a_1 - \beta^2)^2}{(\alpha^2 + \beta^2)(\gamma^2 + \beta^2)} \right]^{1/2} \sin (\beta t + \psi)$ $\psi \triangleq \tan^{-1} \dfrac{\beta(a_1 - \beta^2)}{a_0 - a_2\beta^2} - \tan^{-1} \dfrac{\beta}{\alpha} - \tan^{-1} \dfrac{\beta}{\gamma}$	
$\dfrac{s}{(s^2 + \beta^2)(s^2 + \lambda^2)}$	$\dfrac{\cos \beta t - \cos \lambda t}{\lambda^2 - \beta^2}$	
$\dfrac{s}{[s^2 + (\beta + \lambda)^2][s^2 + (\beta - \lambda)^2]}$	$\dfrac{1}{2\lambda\beta} \sin \lambda t \cdot \sin \beta t$	

Function-Transform Pairs—Continued

$F(s)$	$f(t)$	$0 \leq t$
$\dfrac{s^2 + a_1 s + a_0}{(s^2 + \beta^2)(s^2 + \lambda^2)}$	$\dfrac{[(a_0 - \beta^2)^2 + a_1^2\beta^2]^{1/2}}{\beta(\lambda^2 - \beta^2)} \sin(\beta t + \psi_1)$ $+ \dfrac{[(a_0 - \lambda^2)^2 + a_1^2\lambda^2]^{1/2}}{\lambda(\beta^2 - \lambda^2)} \sin(\lambda t + \psi_2)$ $\psi_1 \triangleq \tan^{-1} \dfrac{a_1\beta}{a_0 - \beta^2};\ \psi_2 \triangleq \tan^{-1} \dfrac{a_1\lambda}{a_0 - \lambda^2}$	
$\dfrac{s^3 + a_2 s^2 + a_1 s + a_0}{(s^2 + \beta^2)(s^2 + \lambda^2)}$	$\dfrac{[(a_0 - a_2\beta^2)^2 + \beta^2(a_1 - \beta^2)^2]^{1/2}}{\beta(\lambda^2 - \beta^2)} \sin(\beta t + \psi_1)$ $+ \dfrac{[(a_0 - a_2\lambda^2)^2 + \lambda^2(a_1 - \lambda^2)^2]^{1/2}}{\lambda(\beta^2 - \lambda^2)} \sin \lambda t + \psi_2)$ $\psi_1 \triangleq \tan^{-1} \dfrac{\beta(a_1 - \beta^2)}{a_0 - a_2\beta^2};\ \psi_2 \triangleq \tan^{-1} \dfrac{\lambda(a_1 - \lambda^2)}{a_0 - a_2\lambda^2}$	
$\dfrac{1}{(s + \alpha)^2 + \beta^2}$	$\dfrac{1}{\beta} e^{-\alpha t} \sin \beta t$	
$\dfrac{s + a_0}{(s + \alpha)^2 + \beta^2}$	$\dfrac{1}{\beta}[(a_0 - \alpha)^2 + \beta^2]^{1/2} e^{-\alpha t} \sin(\beta t + \psi)$ $\psi \triangleq \tan^{-1} \dfrac{\beta}{a_0 - \alpha}$	
$\dfrac{s + \alpha}{(s + \alpha)^2 + \beta^2}$	$e^{-\alpha t} \cos \beta t$	
$\dfrac{1}{s[(s + \alpha)^2 + \beta^2]}$	$\dfrac{1}{\beta_0^2} + \dfrac{1}{\beta_0\beta} e^{-\alpha t} \sin(\beta t - \psi)$ $\psi \triangleq \tan^{-1} \dfrac{\beta}{-\alpha}$ $\beta_0^2 \triangleq \alpha^2 + \beta^2$	
$\dfrac{s + a_0}{s[(s + \alpha)^2 + \beta^2]}$	$\dfrac{a_0}{\beta_0^2} + \dfrac{1}{\beta\beta_0}[(a_0 - \alpha)^2 + \beta^2]^{1/2} e^{-\alpha t} \sin(\beta t + \psi)$ $\psi \triangleq \tan^{-1} \dfrac{\beta}{a_0 - \alpha} - \tan^{-1} \dfrac{\beta}{-\alpha}$ $\beta_0^2 \triangleq \alpha^2 + \beta^2$	
$\dfrac{s^2 + a_1 s + a_0}{s[(s + \alpha)^2 + \beta^2]}$	$\dfrac{a_0}{\beta_0^2} + \dfrac{1}{\beta\beta_0}[(\alpha^2 - \beta^2 - a_1\alpha + a_0)^2 + \beta^2(a_1 - 2\alpha)^2]^{1/2} e^{-\alpha t} \sin(\beta t + \psi)$ $\psi \triangleq \tan^{-1} \dfrac{\beta(a_1 - 2\alpha)}{\alpha^2 - \beta^2 - a_1\alpha + a_0} - \tan^{-1} \dfrac{\beta}{-\alpha}$ $\beta_0^2 \triangleq \beta^2 + \alpha^2$	
$\dfrac{1}{(s + \gamma)[(s + \alpha)^2 + \beta^2]}$	$\dfrac{e^{-\gamma t}}{(\gamma - \alpha)^2 + \beta^2} + \dfrac{1}{\beta[(\gamma - \alpha)^2 + \beta^2]^{1/2}} e^{-\alpha t} \sin(\beta t - \psi)$ $\psi \triangleq \tan^{-1} \dfrac{\beta}{\gamma - \alpha}$	
$\dfrac{s + a_0}{(s + \gamma)[(s + \alpha)^2 + \beta^2]}$	$\dfrac{a_0 - \gamma}{(\alpha - \gamma)^2 + \beta^2} e^{-\gamma t} + \dfrac{1}{\beta}\left[\dfrac{(a_0 - \alpha)^2 + \beta^2}{(\gamma - \alpha)^2 + \beta^2}\right]^{1/2} e^{-\alpha t} \sin(\beta t + \psi)$ $\psi \triangleq \tan^{-1} \dfrac{\beta}{a_0 - \alpha} - \tan^{-1} \dfrac{\beta}{\gamma - \alpha}$	
$\dfrac{s^2 + a_1 s + a_0}{(s + \gamma)[(s + \alpha)^2 + \beta^2]}$	$\dfrac{\gamma^2 - a_1\gamma + a_0}{(\alpha - \gamma)^2 + \beta^2} e^{-\gamma t}$ $+ \dfrac{1}{\beta}\left[\dfrac{(\alpha^2 - \beta^2 - a_1\alpha + a_0)^2 + \beta^2(a_1 - 2\alpha)^2}{(\gamma - \alpha)^2 + \beta^2}\right]^{1/2} e^{-\alpha t} \sin(\beta t + \psi)$ $\psi \triangleq \tan^{-1} \dfrac{\beta(a_1 - 2\alpha)}{\alpha^2 - \beta^2 - a_1\alpha + a_0} - \tan^{-1} \dfrac{\beta}{\gamma - \alpha}$	
$\dfrac{1}{s(s + \gamma)[(s + \alpha)^2 + \beta^2]}$	$\dfrac{1}{\gamma\beta_0^2} - \dfrac{1}{\gamma[(\alpha - \gamma)^2 + \beta^2]} e^{-\gamma t}$ $+ \dfrac{1}{\beta\beta_0[(\gamma - \alpha)^2 + \beta^2]^{1/2}} e^{-\alpha t} \sin(\beta t - \psi)$ $\gamma \triangleq \tan^{-1} \dfrac{\beta}{-\alpha} + \tan^{-1} \dfrac{\beta}{\gamma - \alpha}$ $\beta_0^2 \triangleq \alpha^2 + \beta^2$	

Function-Transform Pairs—Continued

$F(s)$	$f(t)$ $0 \leqq t$
$\dfrac{s + a_0}{s(s + \gamma)[(s + \alpha)^2 + \beta^2]}$	$\dfrac{a_0}{\gamma\beta_0{}^2} + \dfrac{\gamma - a_0}{\gamma[(\alpha - \gamma)^2 + \beta^2]} e^{-\gamma t}$ $+ \dfrac{1}{\beta\beta_0} \left[\dfrac{(a_0 - \alpha)^2 + \beta^2}{(\gamma - \alpha)^2 + \beta^2} \right]^{1/2} e^{-\alpha t} \sin(\beta t + \psi)$ $\psi \triangleq \tan^{-1} \dfrac{\beta}{a_0 - \alpha} - \tan^{-1} \dfrac{\beta}{\gamma - \alpha} - \tan^{-1} \dfrac{\beta}{-\alpha}$ $\beta_0{}^2 \triangleq \alpha^2 + \beta^2$
$\dfrac{s^2 + a_1 s + a_0}{(s + \gamma)(s + \delta)[(s + \alpha)^2 + \beta^2]}$	$\dfrac{\gamma^2 - a_1\gamma + a_0}{(\delta - \gamma)[(\alpha - \gamma)^2 + \beta^2]} e^{-\gamma t} + \dfrac{\delta^2 - a_1\delta + a_0}{(\gamma - \delta)[(\alpha - \delta)^2 + \beta^2]} e^{-\delta t}$ $+ \dfrac{1}{\beta} \left\{ \dfrac{(\alpha^2 - \beta^2 - a_1\alpha + a_0)^2 + \beta^2(a_1 - 2\alpha)^2}{[(\delta - \alpha)^2 + \beta^2][(\gamma - \alpha)^2 + \beta^2]} \right\}^{1/2} e^{-\alpha t} \sin(\beta t + \psi)$ $\psi \triangleq \tan^{-1} \dfrac{\beta(a_1 - 2\alpha)}{\alpha^2 - \beta^2 - a_1\alpha + a_0} - \tan^{-1} \dfrac{\beta}{\gamma - \alpha} - \tan^{-1} \dfrac{\beta}{\delta - \alpha}$
$\dfrac{1}{(s^2 + \lambda^2)[(s + \alpha)^2 + \beta^2]}$	$\dfrac{1}{[(\beta_0{}^2 - \lambda^2)^2 + 4\alpha^2\lambda^2]^{1/2}} \left[\dfrac{1}{\lambda} \sin(\lambda t - \psi_1) + \dfrac{1}{\beta} e^{-\alpha t} \sin(\beta t - \psi_2) \right]$ $\psi_1 \triangleq \tan^{-1} \dfrac{2\alpha\lambda}{\beta_0{}^2 - \lambda^2}; \; \psi_2 \triangleq \dfrac{-2\alpha\beta}{\alpha^2 - \beta^2 + \lambda^2}; \; \beta_0{}^2 \triangleq \alpha^2 + \beta^2$
$\dfrac{s + a_0}{(s^2 + \lambda^2)[(s + \alpha)^2 + \beta^2]}$	$\dfrac{1}{\lambda} \left[\dfrac{a_0{}^2 + \lambda^2}{(\beta_0{}^2 - \lambda^2)^2 + 4\alpha^2\lambda^2} \right]^{1/2} \sin(\lambda t + \psi_1)$ $+ \dfrac{1}{\beta} \left[\dfrac{(a_0 - \alpha)^2 + \beta^2}{(\beta_0{}^2 - \lambda^2)^2 + 4\alpha^2\lambda^2} \right]^{1/2} e^{-\alpha t} \sin(\beta t + \psi_2)$ $\psi_1 \triangleq \tan^{-1} \dfrac{\lambda}{a_0} - \tan^{-1} \dfrac{2\alpha\lambda}{\beta_0{}^2 - \lambda^2};$ $\psi_2 \triangleq \tan^{-1} \dfrac{\beta}{a_0 - \alpha} - \tan^{-1} \dfrac{-2\alpha\beta}{\alpha^2 - \beta^2 + \lambda^2}$ $\beta^2 \triangleq \alpha^2 + \beta_0{}^2$
$\dfrac{s^2 + a_1 s + a_0}{(s^2 + \lambda^2)[(s + \alpha)^2 + \beta^2]}$	$\dfrac{1}{\lambda} \left[\dfrac{(a_0 - \lambda^2)^2 + a_1{}^2\lambda^2}{(\beta_0{}^2 - \lambda^2)^2 + 4\alpha^2\lambda^2} \right]^{1/2} \sin(\lambda t + \psi_1)$ $+ \dfrac{1}{\beta} \left[\dfrac{(\alpha^2 - \beta^2 - a_1\alpha + a_0)^2 + \beta^2(a_1 - 2\alpha)^2}{(\beta_0{}^2 - \lambda^2)^2 + 4\alpha^2\lambda^2} \right]^{1/2} e^{-\alpha t} \sin(\beta t + \psi_2)$ $\psi_1 \triangleq \tan^{-1} \dfrac{a_1\lambda}{a_0 - \lambda^2} - \tan^{-1} \dfrac{2\alpha\lambda}{\beta_0{}^2 - \lambda^2}$ $\psi_2 \triangleq \tan^{-1} \dfrac{\beta(a_1 - 2\alpha)}{\alpha^2 - \beta^2 - a_1\alpha + a_0} - \tan^{-1} \dfrac{-2\alpha\beta}{\alpha^2 - \beta^2 + \lambda^2}$ $\beta_0{}^2 \triangleq \alpha^2 + \beta^2$
$\dfrac{s + a_0}{(s + \gamma)(s^2 + \lambda^2)[(s + \alpha)^2 + \beta^2]}$	$\dfrac{a_0 - \gamma}{(\lambda^2 + \gamma^2)[(\alpha - \gamma)^2 + \beta^2]} e^{-\gamma t}$ $+ \dfrac{1}{\lambda} \left\{ \dfrac{a_0{}^2 + \lambda^2}{(\gamma^2 + \lambda^2)[(\beta_0{}^2 - \lambda^2)^2 + 4\alpha^2\lambda^2]} \right\}^{1/2} \sin(\lambda t + \psi_1)$ $+ \dfrac{1}{\beta} \left\{ \dfrac{(a_0 - \alpha)^2 + \beta^2}{[\gamma - \alpha)^2 + \beta^2][(\beta_0{}^2 - \lambda^2)^2 + 4\alpha^2\lambda^2]} \right\}^{1/2} e^{-\alpha t} \sin(\beta t + \psi_2)$ $\psi_1 \triangleq \tan^{-1} \dfrac{\lambda}{a_0} - \tan^{-1} \dfrac{\lambda}{\gamma} - \tan^{-1} \dfrac{2\alpha\lambda}{\beta_0{}^2 - \lambda^2}$ $\gamma_2 \triangleq \tan^{-1} \dfrac{\beta}{a_0 - \alpha} - \tan^{-1} \dfrac{\beta}{\gamma - \alpha} - \tan^{-1} \dfrac{-2\alpha\beta}{\alpha^2 - \beta^2 + \lambda^2}$ $\beta_0{}^2 \triangleq \alpha^2 + \beta^2$
$\dfrac{1}{s^2}$	t
$\dfrac{1}{s^n}$	$\dfrac{1}{(n - 1)!} t^{n-1}$ n is a positive integer.
$\dfrac{1}{(s + \alpha)s^2}$	$\dfrac{e^{-\alpha t} + \alpha t - 1}{\alpha^2}$

Function-Transform Pairs—Continued

$F(s)$	$f(t)$ $\qquad 0 \le t$
$\dfrac{s + a_0}{(s + \alpha)s^2}$	$\dfrac{a_0 - \alpha}{\alpha^2} e^{-\alpha t} + \dfrac{a_0}{\alpha} t + \dfrac{\alpha - a_0}{\alpha^2}$
$\dfrac{s^2 + a_1 s + a_0}{(s + \alpha)s^2}$	$\dfrac{\alpha^2 - a_1 \alpha + a_0}{\alpha^2} e^{-\alpha t} + \dfrac{a_0}{\alpha} t + \dfrac{a_1 \alpha - a_0}{\alpha^2}$
$\dfrac{1}{(s + \alpha)^2}$	$t e^{-\alpha t}$
$\dfrac{s + a_0}{(s + \alpha)^2}$	$[(a_0 - \alpha)t + 1]e^{-\alpha t}$
$\dfrac{1}{(s + \alpha)^n}$	$\dfrac{1}{(n - 1)!} t^{n-1} e^{-\alpha t} \qquad n \text{ is a positive integer.}$
$\dfrac{s^n}{(s + \alpha)^{n+1}}$	$e^{-\alpha t} \displaystyle\sum_{k=0}^{n} \dfrac{n!(-\alpha)^k}{(n - k)!(k!)^2} t^k \qquad n \text{ is a non-negative integer.}$
$\dfrac{1}{s(s + \alpha)^2}$	$\dfrac{1 - (1 + \alpha t)e^{-\alpha t}}{\alpha^2}$
$\dfrac{s + a_0}{s(s + \alpha)^2}$	$\dfrac{a_0}{\alpha^2} + \left(\dfrac{\alpha - a_0}{\alpha} t - \dfrac{a_0}{\alpha^2} \right) e^{-\alpha t}$
$\dfrac{s^2 + a_1 s + a_0}{s(s + \alpha)^2}$	$\dfrac{a_0}{\alpha^2} + \left(\dfrac{a_1 \alpha - a_0 - \alpha^2}{\alpha} t + \dfrac{\alpha^2 - a_0}{\alpha^2} \right) e^{-\alpha t}$
$\dfrac{1}{(s + \gamma)(s + \alpha)^2}$	$\dfrac{1}{(\gamma - \alpha)^2} e^{-\gamma t} + \dfrac{(\gamma - \alpha)t - 1}{(\gamma - \alpha)^2} e^{-\alpha t}$
$\dfrac{s + a_0}{(s + \gamma)(s + \alpha)^2}$	$\dfrac{a_0 - \gamma}{(\alpha - \gamma)^2} e^{-\gamma t} + \left[\dfrac{a_0 - \alpha}{\gamma - \alpha} t + \dfrac{\gamma - a_0}{(\gamma - \alpha)^2} \right] e^{-\alpha t}$
$\dfrac{s^2 + a_1 s + a_0}{(s + \gamma)(s + \alpha)^2}$	$\dfrac{\gamma^2 - a_1 \gamma + a_0}{(\alpha - \gamma)^2} e^{-\gamma t}$ $\qquad + \left[\dfrac{\alpha^2 - a_1 \alpha + a_0}{\gamma - \alpha} t + \dfrac{\alpha^2 - 2\alpha\gamma + a_1\gamma - a_0}{(\gamma - \alpha)^2} \right] e^{-\alpha t}$
$\dfrac{s + a_0}{(s + \gamma)(s + \alpha)^3}$	$\dfrac{a_0 - \gamma}{(\alpha - \gamma)^3} e^{-\gamma t} + \left[\dfrac{a_0 - \alpha}{2(\gamma - \alpha)} t^2 + \dfrac{\gamma - a_0}{(\gamma - \alpha)^2} t + \dfrac{a_0 - \gamma}{(\gamma - \alpha)^3} \right] e^{-\alpha t}$
$\dfrac{s + a_0}{s(s + \gamma)(s + \alpha)^2}$	$\dfrac{a_0}{\gamma\alpha^2} + \dfrac{\gamma - a_0}{\gamma(\alpha - \gamma)^2} e^{-\gamma t} + \left[\dfrac{a_0 - \alpha}{\alpha(\alpha - \gamma)} t + \dfrac{2a_0\alpha - \alpha^2 - a_0\gamma}{\alpha^2(\alpha - \gamma)^2} \right] e^{-\alpha t}$
$\dfrac{s^2 + a_1 s + a_0}{s(s + \gamma)(s + \alpha)^2}$	$\dfrac{a_0}{\gamma\alpha^2} - \dfrac{\gamma^2 - a_1\gamma + a_0}{\gamma(\alpha - \gamma)^2} e^{-\gamma t}$ $\qquad + \left[\dfrac{\alpha^2 - a_1 \alpha + a_0}{\alpha(\alpha - \gamma)} t + \dfrac{(\gamma - a_1)\alpha^2 + (2\alpha - \gamma)a_0}{\alpha^2(\alpha - \gamma)^2} \right] e^{-\alpha t}$
$\dfrac{s + a_0}{(s + \gamma)(s + \delta)(s + \alpha)^2}$	$\dfrac{a_0 - \gamma}{(\delta - \gamma)(\alpha - \gamma)^2} e^{-\gamma t} + \dfrac{a_0 - \delta}{(\gamma - \delta)(\alpha - \delta)^2} e^{-\delta t}$ $\qquad + \left[\dfrac{a_0 - \alpha}{(\gamma - \alpha)(\delta - \alpha)} t + \dfrac{2a_0\alpha - \alpha^2 - a_0(\gamma + \delta) + \gamma\delta}{(\gamma - \alpha)^2(\delta - \alpha)^2} \right] e^{-\alpha t}$
$\dfrac{s + a_0}{(s + \alpha)(s + \gamma)s^2}$	$\dfrac{a_0 - \alpha}{\alpha^2(\gamma - \alpha)} e^{-\alpha t} + \dfrac{a_0 - \gamma}{\gamma^2(\alpha - \gamma)} e^{\gamma t} + \dfrac{a_0}{\alpha\gamma} t + \dfrac{\alpha\gamma - a_0(\alpha + \gamma)}{\alpha^2\gamma^2}$
$\dfrac{s^2 + a_1 s + a_0}{(s + \alpha)(s + \gamma)s^2}$	$\dfrac{\alpha^2 - a_1\alpha + a_0}{\alpha^2(\gamma - \alpha)} e^{-\alpha t} + \dfrac{\gamma^2 - a_1\gamma + a_0}{\gamma^2(\alpha - \gamma)} e^{-\gamma t} + \dfrac{a_0}{\alpha\gamma} t$ $\qquad\qquad + \dfrac{a_1\alpha\gamma - a_0(\alpha + \gamma)}{\alpha^2\gamma^2}$
$\dfrac{s^2 + a_1 s + a_0}{(s + \alpha)^2 s^2}$	$\left[\dfrac{\alpha^2 - a_1\alpha + a_0}{\alpha^2} t + \dfrac{2a_0 - a_1\alpha}{\alpha^3} \right] e^{-\alpha t} + \dfrac{a_0}{\alpha^2} t + \dfrac{a_1\alpha - 2a_0}{\alpha^3}$
$\dfrac{s + a_0}{(s + \alpha)^2(s + \gamma)^2}$	$\left[\dfrac{a_0 - \alpha}{(\gamma - \alpha)^2} t + \dfrac{\alpha + \gamma - 2a_0}{(\gamma - \alpha)^3} \right] e^{-\alpha t}$ $\qquad\qquad + \left[\dfrac{a_0 - \gamma}{(\alpha - \gamma)^2} t + \dfrac{\alpha + \gamma - 2a_0}{(\alpha - \gamma)^3} \right] e^{-\gamma t}$

Function-Transform Pairs—Continued

$F(s)$	$f(t)$ $0 \leq t$
$\dfrac{s^2 + a_1 s + a_0}{(s + \alpha)^2 (s + \gamma)^2}$	$\left[\dfrac{\alpha^2 - a_1\alpha + a_0}{(\gamma - \alpha)^2} t + \dfrac{a_1(\alpha + \gamma) - 2(\alpha\gamma + a_0)}{(\gamma - \alpha)^3} \right] e^{-\alpha t}$ $\quad + \left[\dfrac{\gamma^2 - a_1\gamma + a_0}{(\gamma - \alpha)^2} t - \dfrac{a_1(\alpha + \gamma) - 2(\alpha\gamma + a_0)}{(\gamma - \alpha)^3} \right] e^{-\gamma t}$
$\dfrac{s^2 + a_1 s + a_0}{(s + \alpha)^3 s^2}$	$\left(\dfrac{\alpha^2 - a_1\alpha + a_0}{2\alpha^2} t^2 + \dfrac{-a_1\alpha + 2a_0}{\alpha^3} t + \dfrac{-a_1\alpha + 3a_0}{\alpha^4} \right) e^{-\alpha t}$ $\qquad\qquad\qquad\qquad + \dfrac{a_0}{\alpha^2} t + \dfrac{a_1\alpha - 3a_0}{\alpha^4}$
$\dfrac{1}{(s^2 + \beta^2) s^2}$	$\dfrac{1}{\beta^2} t - \dfrac{1}{\beta^3} \sin \beta t$
$\dfrac{1}{(s^2 - \beta^2) s^2}$	$\dfrac{1}{\beta^3} \sinh \beta t - \dfrac{1}{\beta^2} t$
$\dfrac{s + a_0}{(s^2 + \beta^2) s^2}$	$\dfrac{a_0}{\beta^2} t + \dfrac{1}{\beta^2} - \dfrac{1}{\beta^3} (a_0{}^2 + \beta^2)^{1/2} \sin (\beta t + \psi)$ $\psi \triangleq \tan^{-1} \dfrac{\beta}{a_0}$
$\dfrac{s^2 + a_1 s + a_0}{(s^2 + \beta^2) s^2}$	$\dfrac{a_0}{\beta^2} t + \dfrac{a_1}{\beta^2} - \dfrac{1}{\beta^3} [(a_0 - \beta^2)^2 + a_1{}^2\beta^2]^{1/2} \sin (\beta t + \psi)$ $\psi \triangleq \tan^{-1} \dfrac{a_1\beta}{a_0 - \beta^2}$
$\dfrac{1}{(s^2 + \beta^2) s^3}$	$\dfrac{1}{\beta^4} (\cos \beta t - 1) + \dfrac{1}{2\beta^2} t^2$
$\dfrac{1}{(s^2 - \beta^2) s^3}$	$\dfrac{1}{\beta^4} (\cosh \beta t - 1) - \dfrac{1}{2\beta^2} t^2$
$\dfrac{1}{(s^2 + \beta^2)(s + \alpha)^2}$	$\dfrac{1}{\beta(\alpha^2 + \beta^2)} \sin (\beta t - \psi) + \left[\dfrac{1}{\alpha^2 + \beta^2} t + \dfrac{2\alpha}{(\alpha^2 + \beta^2)^2} \right] e^{-\alpha t}$ $\psi \triangleq 2 \tan^{-1} \dfrac{\beta}{\alpha}$
$\dfrac{s + a_0}{(s^2 + \beta^2)(s + \alpha)^2}$	$\dfrac{(a_0{}^2 + \beta^2)^{1/2}}{\beta(\alpha^2 + \beta^2)} \sin (\beta t + \psi) + \left[\dfrac{a_0 - \alpha}{\alpha^2 + \beta^2} t + \dfrac{2a_0\alpha + \beta^2 - \alpha^2}{(\alpha^2 + \beta^2)^2} \right] e^{-\alpha t}$ $\psi \triangleq \tan^{-1} \dfrac{\beta}{a_0} - 2 \tan^{-1} \dfrac{\beta}{\alpha}$
$\dfrac{s^2 + a_1 s + a_0}{(s^2 + \beta^2)(s + \alpha)^2}$	$\dfrac{[(a_0 - \beta^2)^2 + a_1{}^2\beta^2]^{1/2}}{\beta(\alpha^2 + \beta^2)} \sin (\beta t + \psi)$ $+ \left[\dfrac{\alpha^2 - a_1\alpha + a_0}{\alpha^2 \beta^2} t + \dfrac{a_1(\beta^2 - \alpha^2) + 2\alpha(a_0 - \beta^2)}{(\alpha^2 + \beta^2)^2} \right] e^{-\alpha t}$ $\psi \triangleq \tan^{-1} \dfrac{a_1\beta}{a_0 - \beta^2} - 2 \tan^{-1} \dfrac{\beta}{\alpha}$
$\dfrac{s + a_0}{s(s^2 + \beta^2)(s + \alpha)^2}$	$\dfrac{a_0}{\beta^2\alpha^2} - \dfrac{(a_0{}^2 + \beta^2)^{1/2}}{\beta^2(\alpha^2 + \beta^2)} \cos (\beta t + \psi)$ $+ \left[\dfrac{\alpha - a_0}{\alpha(\alpha^2 + \beta^2)} t + \dfrac{2\alpha^3 - 3a_0\alpha^2 - a_0\beta^2}{\alpha^2(\alpha^2 + \beta^2)^2} \right] e^{-\alpha t}$ $\psi \triangleq \tan^{-1} \dfrac{\beta}{a_0} - 2 \tan^{-1} \dfrac{\beta}{\alpha}$
$\dfrac{s^2 + a_1 s + a_0}{(s + \gamma)(s^2 + \beta^2)(s + \alpha)^2}$	$\dfrac{\gamma^2 - a_1\gamma + a_0}{(\gamma^2 + \beta^2)(\alpha - \gamma)^2} e^{-\gamma t} + \dfrac{[(a_0 - \beta^2)^2 + a^2\beta^2]^{1/2}}{\beta(\gamma^2 + \beta^2)^{1/2}(\alpha^2 + \beta^2)} \sin (\beta t + \psi)$ $+ \dfrac{\alpha^2 - a_1\alpha + a_0}{(\gamma - \alpha)(\alpha^2 + \beta^2)} t e^{-\alpha t}$ $+ \dfrac{(\gamma - \alpha)(\alpha^2 + \beta^2)(a_1 - 2\alpha) - (\alpha^2 - a_1\alpha + a_0)(3\alpha^2 + \beta^2 - 2\alpha\gamma)}{(\gamma - \alpha)^2(\alpha^2 + \beta^2)^2}$ $\times e^{-\alpha t}$ $\psi \triangleq \tan^{-1} \dfrac{a_1\beta}{a_0 - \beta^2} - \tan^{-1} \dfrac{\beta}{\gamma} - 2 \tan^{-1} \dfrac{\beta}{\alpha}$

Function-Transform Pairs—Continued

$F(s)$	$f(t)$ $\qquad$ $0 \leq t$
$\dfrac{1}{(s^2 + \beta^2)^2}$	$\dfrac{1}{2\beta^3}(\sin \beta t - \beta t \cos \beta t)$
$\dfrac{s}{(s^2 + \beta^2)^2}$	$\dfrac{1}{2\beta} t \sin \beta t$
$\dfrac{s^2}{(s^2 + \beta^2)^2}$	$\dfrac{1}{2\beta}(\sin \beta t + \beta t \cos \beta t)$
$\dfrac{s^2 - \beta^2}{(s^2 + \beta^2)^2}$	$t \cos \beta t$
$\dfrac{1}{s(s^2 + \beta^2)^2}$	$\dfrac{1}{\beta^4}(1 - \cos \beta t) - \dfrac{1}{2\beta^3} t \sin \beta t$
$\dfrac{s^2 + a_1 s + a_0}{s(s^2 + \beta^2)^2}$	$\dfrac{a_0}{\beta^4} - \dfrac{[(a_0 - \beta^2)^2 + a_1{}^2\beta^2]^{1/2}}{2\beta^3} t \sin (\beta t + \psi_1)$ $\qquad\qquad - \dfrac{(4a_0{}^2 + a_1{}^2\beta^2)^{1/2}}{2\beta^4} \cos (\beta t + \psi_2)$ $\psi_1 \triangleq \tan^{-1} \dfrac{a_1 \beta}{a_0 - \beta^2}$ $\psi_2 \triangleq \tan^{-1} \dfrac{a_1 \beta}{2 a_0}$
$\dfrac{1}{[(s + \alpha)^2 + \beta^2] s^2}$	$\dfrac{1}{\beta_0{}^2}\left[t - \dfrac{2\alpha}{\beta_0{}^2} + \dfrac{1}{\beta} e^{-\alpha t} \sin (\beta t - \psi) \right]$ $\psi \triangleq 2 \tan^{-1} \dfrac{\beta}{-\alpha}$ $\beta_0{}^2 \triangleq \alpha^2 + \beta^2$
$\dfrac{1}{(s + \gamma)^2 [(s + \alpha)^2 + \beta^2]}$	$\dfrac{1}{(\alpha - \gamma)^2 + \beta^2}\left[te^{-\gamma t} + \dfrac{2(\gamma - \alpha)}{(\alpha - \gamma)^2 + \beta^2} e^{-\gamma t}\right.$ $\qquad\qquad\left. + \dfrac{1}{\beta} e^{-\alpha t} \sin (\beta t - \psi) \right]$ $\psi \triangleq 2 \tan^{-1} \dfrac{\beta}{\gamma - \alpha}$
$\dfrac{s^2 + a_1 s + a_0}{(s + \gamma)^2 [(s + \alpha)^2 + \beta^2]}$	$\dfrac{\gamma^2 - a_1\gamma + a_0}{(\alpha - \gamma)^2 + \beta^2} te^{-\gamma t}$ $+ \dfrac{[(\alpha - \gamma)^2 + \beta^2](a_1 - 2\gamma) - 2(\alpha - \gamma)(\gamma^2 - a_1\gamma + a_0)}{[(\alpha - \gamma)^2 + \beta^2]^2} e^{-\gamma t}$ $+ \dfrac{[(\alpha^2 - \beta^2 - a_1\alpha + a_0)^2 + \beta^2(a_1 - 2\alpha)^2]^{1/2}}{\beta[(\gamma - \alpha)^2 + \beta^2]} e^{-\alpha t} \sin (\beta t + \psi)$ $\psi \triangleq \tan^{-1} \dfrac{\beta(a_1 - 2\alpha)}{\alpha^2 - \beta^2 - a_1\alpha + a_0} - 2 \tan^{-1} \dfrac{\beta}{\gamma - \alpha}$
$\dfrac{1}{[(s + \alpha)^2 + \beta^2]^2}$	$\dfrac{1}{2\beta^3} e^{-\alpha t}(\sin \beta t - \beta t \cos \beta t)$
$\dfrac{s + \alpha}{[(s + \alpha)^2 + \beta^2]^2}$	$\dfrac{1}{2\beta} te^{-\alpha t} \sin \beta t$
$\dfrac{s^2 + a_0}{[(s + \alpha)^2 + \beta^2]^2}$	$\dfrac{\beta_0{}^2 + a_0}{2\beta^3} e^{-\alpha t} \sin \beta t - \dfrac{[(\alpha^2 - \beta^2 + a_0)^2 + 4\alpha^2\beta^2]^{1/2}}{2\beta^2} te^{-\alpha t} \cos (\beta t + \psi)$ $\psi \triangleq \tan^{-1} \dfrac{-2\alpha\beta}{\alpha^2 - \beta^2 + a_0}; \ \beta_0{}^2 \triangleq \alpha^2 + \beta^2$
$\dfrac{(s + \alpha)^2 - \beta^2}{[(s + \alpha)^2 + \beta^2]^2}$	$te^{-\alpha t} \cos \beta t$
$\tan^{-1} \dfrac{\beta}{s}$	$\dfrac{\sin \beta t}{t}$
$\ln \dfrac{s + \beta}{s + \alpha}$	$\dfrac{e^{-\alpha t} - e^{-\beta t}}{t}$

Function-Transform Pairs—Continued

$F(s)$	$f(t)$	$0 \leqq t$
$e^{s^2/4a}\ \mathrm{cerf}\ \dfrac{s}{2\sqrt{a}}$	$2\sqrt{\dfrac{a}{\pi}}\,e^{-at^2}$ $\mathrm{cerf}\ y \triangleq 1 - \mathrm{erf}\ y \triangleq 1 - \dfrac{2}{\sqrt{\pi}}\displaystyle\int_0^y e^{-z^2}dx$	
$\dfrac{1}{\sqrt{s^2+\alpha^2}}$	$J_0(\alpha t)$	
$\dfrac{1}{\sqrt{s^2+\alpha^2}(\sqrt{s^2+\alpha^2}+s)}$	$\dfrac{1}{\alpha}J_1(\alpha t)$	
$\dfrac{1}{\sqrt{s^2+\alpha^2}(\sqrt{s^2+\alpha^2}+s)^n}$	$\dfrac{1}{\alpha^n}J_n(\alpha t)$	n is a non-negative integer.
$\dfrac{1}{s\sqrt{s^2+\alpha^2}(\sqrt{s^2+\alpha^2}+s)^n}$	$\dfrac{1}{\alpha^n}\displaystyle\int_0^t J_n(\alpha t)dt$	n is a non-negative integer.
$\dfrac{1}{\sqrt{s^2+\alpha^2}+s}$	$\dfrac{1}{\alpha}\dfrac{J_1(\alpha t)}{t}$	
$\dfrac{1}{(\sqrt{s^2+\alpha^2}+s)^n}$	$\dfrac{n}{\alpha^n}\dfrac{J_n(\alpha t)}{t}$	n is a positive integer.
$\dfrac{1}{s(\sqrt{s^2+\alpha^2}+s)^n}$	$\dfrac{n}{\alpha^n}\displaystyle\int_0^t \dfrac{J_n(\alpha t)}{t}dt$	n is a positive integer.
$\dfrac{1}{s}e^{-as}$	$u(t-a)$	
$\dfrac{1}{s^2}e^{-as}$	$(t-a)u(t-a)$	
$\left(\dfrac{a}{s}+\dfrac{1}{s^2}\right)e^{-as}$	$tu(t-a)$	
$\left(\dfrac{2}{s^3}+\dfrac{2a}{s^2}+\dfrac{a^2}{s}\right)e^{-as}$	$t^2u(t-a)$	
$\dfrac{1}{s}(e^{-as}-e^{-bs})$ $a<b$	$u(t-a)-u(t-b)$	
$\left(\dfrac{1-e^{-s}}{s}\right)^2$	$\begin{cases} t \\ 2-t \\ 0 \end{cases}$	$\begin{array}{l}0<t<1\\1<t<2\\2<t\end{array}$
$\left(\dfrac{1-e^{-s}}{s}\right)^3$	$\begin{cases} 0.5t^2 \\ 0.75-(t-1.5)^2 \\ 0.5(t-3)^2 \\ 0 \end{cases}$	$\begin{array}{l}0<t<1\\1<t<2\\2<t<3\\3<t\end{array}$
$\dfrac{1}{s^2}(1-e^{-s})$	$\begin{cases} t \\ 1 \end{cases}$	$\begin{array}{l}0<t<1\\1<t\end{array}$
$\dfrac{1}{s^3}(1-e^{-s})^2$	$\begin{cases} 0.5t^2 \\ 1-0.5(t-2)^2 \\ 1 \end{cases}$	$\begin{array}{l}0<t<0\\1<t<2\\2<t\end{array}$
$\dfrac{1}{s(1+e^{-s})}$	$\displaystyle\sum_{k=0}^{\infty}(-1)^k u(t-k)$	

Function-Transform Pairs—Concluded

$F(s)$	$f(t)$	$0 \leqq t$
$\dfrac{1}{s \sinh s}$	$2 \displaystyle\sum_{k=0}^{\infty} u(t - 2k - 1)$	
$\dfrac{1}{s \cosh s}$	$2 \displaystyle\sum_{k=0}^{\infty} (-1)^k u(t - 2k - 1)$	
$\dfrac{1}{s} \tanh s$	$u(t) + 2 \displaystyle\sum_{k=1}^{\infty} (-1)^k u(t - 2k)$ $\text{or} \displaystyle\sum_{k=0}^{\infty} (-1)^k u(t - 2k) u(2k + 2 - t)$	
$\dfrac{e^s - s - 1}{s^2(e^s - 1)}$	$t - \displaystyle\sum_{k=1}^{\infty} u(t - k)$ $\text{or} \displaystyle\sum_{k=0}^{\infty} (t - k) u(t - k) u(k + 1 - t)$	

COMPLEX ANALYSIS

61. COMPLEX NUMBERS

A *complex number* A is a combination of two real numbers a_1, a_2 in the ordered pair $(a_1, a_2) = A = a_1 + ia_2$, where $i = (-1)^{\frac{1}{2}}$. Real and imaginary numbers are special cases of complex numbers obtained by placing $(a_1, 0) = a_1$, $(0, a_2) = ia_2$.

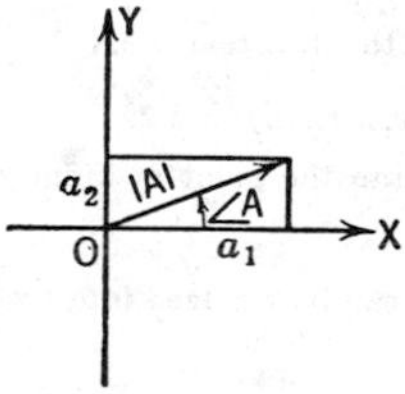

FIG. 1

1. If $a_1 + ia_2 = 0$, then $a_1 = 0$, $a_2 = 0$.
2. If $a_1 + ia_2 = b_1 + ib_2$, then $a_1 = b_1$, $a_2 = b_2$.
3. $a_1 + ia_2$ and $a_1 - ia_2$ are *conjugate* complex numbers. The complex conjugate of A is $\overline{A}$ or $A*$.
4. $A + B = (a_1 + ia_2) + (b_1 + ib_2) = (a_1 + b_1) + i(a_2 + b_2)$.
5. $a_1 + ia_2 = |A|(\cos \angle A + i \sin \angle A) = |A| e^{i \angle A}$.
   $a_1 - ia_2 = |A|(\cos \angle A - i \sin \angle A) = |A| e^{-i \angle A}$.
   where $|A| = \sqrt{a_1{}^2 + a_2{}^2}$, $\sin \angle A = \dfrac{a_2}{|A|}$, $\cos \angle A = \dfrac{a_1}{|A|}$, $|A|$ is the *absolute value (modulus)*, and $\angle A$ is the *angle* of A.
6. $AB = (a_1 + ia_2)(b_1 + ib_2) = (a_1 b_1 - a_2 b_2) + i(a_2 b_1 + a_1 b_2) = |A| |B| e^{i(\angle A + \angle B)}$.
7. $A\overline{A} = (a_1 + ia_2)(a_1 - ia_2) = a_1{}^2 + a_2{}^2 = |A|^2$.
8. $\dfrac{A}{B} = \dfrac{a_1 + ia_2}{b_1 + ib_2} = \dfrac{(a_1 + ia_2)(b_1 - ib_2)}{(b_1 + ib_2)(b_1 - ib_2)} = \dfrac{a_1 b_1 + a_2 b_2}{b_1{}^2 + b_2{}^2} + i \dfrac{a_2 b_1 - a_1 b_2}{b_1{}^2 + b_2{}^2}$
   $= \dfrac{|A|}{|B|} e^{i(\angle A - \angle B)}$.
9. $A^n = (a_1 + ia_2)^n = [|A|(\cos \angle A + i \sin \angle A)]^n = |A|^n e^{in \angle A}$
   $= |A|^n (\cos n \angle A + i \sin n \angle A)$.
   $\overline{A}^n = (a_1 - ia_2)^n = [|A|(\cos \angle A - i \sin \angle A)]^n = |A|^n e^{-in \angle A}$
   $= |A|^n (\cos n \angle A - i \sin n \angle A)$.

10. $\sqrt[n]{A} = \sqrt[n]{a_1 + ia_2} = \sqrt[n]{|A|} \left(\cos \dfrac{\angle A + 2k\pi}{n} + i \sin \dfrac{\angle A + 2k\pi}{n} \right)$

$= \sqrt[n]{|A|} \, e^{i \frac{\angle A + 2k\pi}{n}}$,

where k is an integer. For $k = 0, 1, 2, \cdots, n - 1$, all of the n roots are obtained.

62. COMPLEX VARIABLES

Analytic Functions of a Complex Variable. A function $w = f(z)$, $z = x + iy$, which has a derivative

$$\frac{df}{dz} = f'(z) = \lim_{h \to 0} \frac{f(z + h) - f(z)}{h}$$

at a point z independent of the manner of approach of $z + h$ to z is *analytic* at z and may be expanded in a convergent power series there. A function which is analytic at every point of a region is *analytic in the region*. If $f(z) = w = u(x, y) + iv(x, y)$ and $f(z)$ is analytic at z, then the Cauchy-Riemann differential equations

$$\frac{\partial u}{\partial x} = \frac{\partial v}{\partial y}, \quad \frac{\partial u}{\partial y} = -\frac{\partial v}{\partial x}$$

hold at z. If in a neighborhood of a point z these four partial derivatives exist and are continuous and if the Cauchy-Riemann equations hold, then $f(z)$ is analytic at z. The functions u and v satisfy Laplace's equation

$$\frac{\partial^2 \phi}{\partial x^2} + \frac{\partial^2 \phi}{\partial y^2} = 0$$

Examples of analytic functions are: z, $\dfrac{1}{z}$, e^z, and $\sin z$. An example of a non-analytic function is $w = x - iy$.

Conformal Mapping. The function $w = f(z)$, analytic in a region R_z of the z plane, *conformally* maps each point in R_z on a point of the w plane in the region R_w if $f'(z) \neq 0$ at all points of R_z. This mapping is also *isogonal*, that is, the angle between two curves starting at z_0 is equal to the angle between their mapped curves starting at w_0.

Examples. 1. $w = z + b$, b complex, is a *translation* of magnitude $|b|$ in the direction $\angle b$.

2. $w = az$, a complex, is a *rotation* through $\angle a$ and a *magnification* by $|a|$.

3. $w = az + b$, a, b complex, the *integral linear transformation*, is a combination of 1 and 2.

4. $w = \dfrac{1}{z}$, the *inversion transformation*, carries the origin of the z plane into the point at infinity in the *enlarged* w plane.

5. $w = \dfrac{az + b}{cz + d}$, $ad - bc \neq 0$, the *general linear* or *bilinear transformation*, can be resolved into two linear integral and one inversion transformations.

Integrals of Analytic Functions. If $\dfrac{dF(z)}{dz} = f(z)$ in a simply connected region R_z, then $F(z) = \int f(z) \, dz$ is analytic throughout R_z. If $f_1(z)$, $f_2(z)$ are analytic in R_z and the path of integration is in R_z, then

1. $\displaystyle\int_{z_0}^{z_0} f_1(z) \, dz = 0$

2. $\displaystyle\int_{z_0}^{z_1} [k_1 f_1(z) + k_2 f_2(z)] \, dz = k_1 \int_{z_0}^{z_1} f_1(z) \, dz + k_2 \int_{z_0}^{z_1} f_2(z) \, dz$

3. $\displaystyle\int_{z_1}^{z_0} f_1(z) \, dz = - \int_{z_0}^{z_1} f_1(z) \, dz$

4. $\displaystyle\int_{z_0}^{z_1} f_1(z) \, dz + \int_{z_1}^{z_2} f_1(z) \, dz = \int_{z_0}^{z_2} f_1(z) \, dz$

Cauchy's Integral Theorem. If $f(z)$ is analytic and single-valued on and within a simple closed contour C, then $\displaystyle\int_C f(z) \, dz = 0$.

A *contour* is a continuous curve made up of a finite number of elementary arcs.

If $f(z)$ is continuous on a simple closed contour C and analytic in the region bounded by C, then Cauchy's integral formula

$$f(z) = \frac{1}{2\pi i} \int_C \frac{f(\zeta)}{\zeta - z} d\zeta$$

holds; also

$$f^{(n)}(z) = \frac{n!}{2\pi i} \int_C \frac{f(\zeta)}{(\zeta - z)^{n+1}} d\zeta$$

Laurent Series. A function $f(z)$ has a *zero of order* n at z_1 if it can be put in the form $f(z) = (z - z_1)^n f_1(z)$, n a positive integer, $f_1(z_1) \neq 0$. A function $f(z)$ has a *pole of order* n at z_1 if it can be put in the form $f(z) = \dfrac{f_2(z)}{(z - z_1)^n}$, n a positive integer, $f_2(z_1) \neq 0$.

If $f(z)$ is analytic in a ring between and on two concentric circles C_1 and C_2 with radii R_1 and R_2, $R_1 < R_2$, and center z_1, then the *Laurent series*

$$f(z) = \sum_{n=-\infty}^{\infty} c_n(z - z_1)^n, \quad R_1 < |z - z_1| < R_2$$

is convergent everywhere in the ring, and

$$c_n = \frac{1}{2\pi i} \int_C \frac{f(z)}{(z - z_1)^{n+1}} dz$$

C is circle $|z - z_1| = r$, $R_1 < r < R_2$.

If a single-valued analytic function $f(z)$ is expanded in a Laurent series in the neighborhood of an isolated singularity z_1, then the *residue* of $f(z)$ at z_1 is

$$c_{-1} = \frac{1}{2\pi i} \int_C f(z) \, dz$$

C is any circle with center at z_1 which excludes all other singularities of $f(z)$.

A function is *meromorphic* in a region if it is analytic in the region except for a finite number of poles. If $f(z)$ is analytic on and inside a contour C, except for a finite number of poles, and has no zeros on C, then

$$\frac{1}{2\pi i} \int_C \frac{f'(z)}{f(z)} \, dz = N - P$$

N the total order of the zeros and P the total order of the poles within the contour.

VECTOR ANALYSIS

63. VECTOR ALGEBRA

A *scalar* is a quantity which has magnitude, such as mass, density, temperature. A *vector* is a quantity which has magnitude and direction, such as force, velocity, acceleration. A vector may be represented geometrically by an oriented line segment.

Two vectors A and B are equal if they have the same magnitude and direction. A vector may be displaced parallel to itself provided it retains the same magnitude and direction. A vector having the same magnitude but direction opposite to that of A is the negative of A and is written $-A$. If A is a vector of magnitude, or length, a, then $|A| = a$. A vector parallel to A but with magnitude equal to the reciprocal of the magnitude of A is written $A^{-1} = 1/A$. A unit vector $\dfrac{A}{|A|}$ ($A \neq 0$) has the direction of A and magnitude 1.

The Sum of two vectors A and B is $A + B$ (Fig. 1). Similarly, the sum of three or more vectors can be found by adding them end to end.

The sum of A and $-B$ is $A - B$ (Fig. 1), the *difference* of two vectors.

Let A, B, C be vectors and p, q scalars.

$pA = Ap$, a vector p times as long as A with the same direction as A if p is positive and opposite if p is negative.

$$(p + q)A = pA + qA; \quad p(A + B) = pA + pB$$

$$A + B = B + A$$

$$A + (B + C) = (A + B) + C$$

$|A + B| \leqq |A| + |B|$, where the equality sign holds only for A parallel to B.

Rectangular Coordinates. Figure 2 shows a right-hand coordinate system. Let i, j, k be unit vectors with the directions OX, OY, OZ, respectively. The vector R with initial point O and end point $P(x, y, z)$ can be expressed as the sum of its components

$$R = ix + jy + kz$$

If $A = ia_1 + ja_2 + ka_3$ and $B = ib_1 + jb_2 + kb_3$, then

$$A + B = i(a_1 + b_1) + j(a_2 + b_2) + k(a_3 + b_3)$$

The **Scalar, Inner,** or **Dot Product** of two vectors A and B is $A \cdot B = |A| |B| \cos \theta$ (Fig. 3).

$$A \cdot B = B \cdot A$$

$$A \cdot (B + C) = A \cdot B + A \cdot C$$

$$A \cdot A = A^2 = |A|^2$$

$$i \cdot i = j \cdot j = k \cdot k = 1$$

$$i \cdot j = j \cdot k = k \cdot i = 0$$

If $A \cdot B = 0$, then either $A = 0$, $B = 0$, or A is perpendicular to B.

If $A = ia_1 + ja_2 + ka_3$ and $B = ib_1 + jb_2 + kb_3$, then $A \cdot B = a_1b_1 + a_2b_2 + a_3b_3$.

The **Vector, Restricted Outer,** or **Cross Product** of two vectors A and B is $A \times B = C$, where C is perpendicular to the plane of A and B with the magnitude $|C| = |A| |B| \sin \theta$

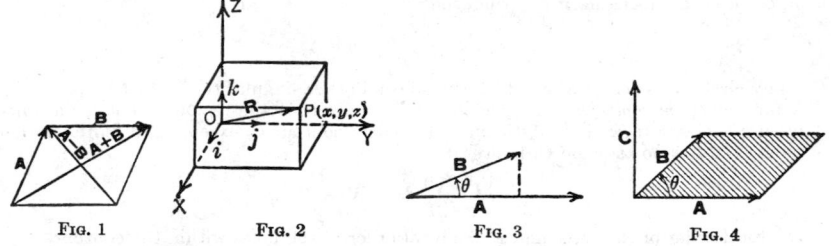

FIG. 1   FIG. 2   FIG. 3   FIG. 4

(the area of the parallelogram made by A and B, Fig. 4) and so directed that a right-hand rotation of less than $180°$ carries A into B.

$$A \times B = -B \times A$$

$$A \times (B + C) = A \times B + A \times C$$

$$(B + C) \times A = B \times A + C \times A$$

$$i \times i = j \times j = k \times k = 0$$

$$i \times j = k = -j \times i$$

$$j \times k = i = -k \times j$$

$$k \times i = j = -i \times k$$

If $A \times B = 0$, then either $A = 0$, $B = 0$, or A is parallel to B.

If $A = ia_1 + ja_2 + ka_3$ and $B = ib_1 + jb_2 + kb_3$, then $A \times B = \begin{vmatrix} i & j & k \\ a_1 & a_2 & a_3 \\ b_1 & b_2 & b_3 \end{vmatrix}$

If $A = ia_1 + ja_2 + ka_3$, $B = ib_1 + jb_2 + kb_3$, $C = ic_1 + jc_2 + kc_3$, then the scalar triple product

$$A \cdot (B \times C) = (A \times B) \cdot C = B \cdot (C \times A) = (A B C) = \begin{vmatrix} a_1 & a_2 & a_3 \\ b_1 & b_2 & b_3 \\ c_1 & c_2 & c_3 \end{vmatrix}$$

and is equal to the volume of a parallelepiped whose three determining edges are A, B, C.

$$(A \times B) \times C = (A \cdot C)B - (B \cdot C)A = -C \times (A \times B)$$

$$(A \times B) \cdot (C \times D) = (A \cdot C)(B \cdot D) - (A \cdot D)(B \cdot C)$$

64. DIFFERENTIATION AND INTEGRATION OF VECTORS

Differentiation. A vector function of one or more scalar variables is called a *variable vector* or *field vector*. The derivative is

$$\frac{dF}{dt} = F'(t) = \lim_{\Delta t \to 0} \frac{F(t + \Delta t) - F(t)}{\Delta t} = \lim_{\Delta t \to 0} \frac{\Delta F}{\Delta t} \quad \text{(Fig. 5)}$$

If the length of F remains unaltered, then $F \cdot dF = 0$. If the direction of F remains unaltered, then $F \times dF = 0$.

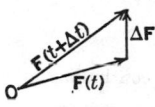

FIG. 5

$$d(A + B) = dA + dB$$
$$d(A \cdot B) = A \cdot dB + B \cdot dA$$
$$d(A \times B) = dA \times B + A \times dB = A \times dB - B \times dA$$
$$d(A \cdot B \cdot C) = A \cdot B \cdot dC + B \cdot C \cdot dA + C \cdot A \cdot dB$$

The Derivative Operators.

$$\nabla = \text{del} = i \frac{\partial}{\partial x} + j \frac{\partial}{\partial y} + k \frac{\partial}{\partial z}$$

$$\nabla^2 = \text{Laplacian} = \frac{\partial^2}{\partial x^2} + \frac{\partial^2}{\partial y^2} + \frac{\partial^2}{\partial z^2}$$

If V is a scalar function, then

$$\nabla V = \text{grad } V = i \frac{\partial V}{\partial x} + j \frac{\partial V}{\partial y} + k \frac{\partial V}{\partial z}$$

If A is a vector function with components A_x, A_y, A_z, then

$$\nabla \cdot A = \text{div } A = \frac{\partial A_x}{\partial x} + \frac{\partial A_y}{\partial y} + \frac{\partial A_z}{\partial z}$$

$$\nabla \times A = \text{curl } A = \text{rot } A = \begin{vmatrix} i & j & k \\ \frac{\partial}{\partial x} & \frac{\partial}{\partial y} & \frac{\partial}{\partial z} \\ A_x & A_y & A_z \end{vmatrix}$$

Formulas for Differentiation. Let U and V be scalar functions and A and B be vector functions of x, y, z. Then:

$$\nabla(U + V) = \nabla U + \nabla V; \quad \nabla \cdot (A + B) = \nabla \cdot A + \nabla \cdot B; \quad \nabla \times (A + B) = \nabla \times A + \nabla \times B$$

$$\nabla(UV) = V\nabla U + U\nabla V; \quad \nabla \cdot (UA) = U\nabla \cdot A + A \cdot \nabla U$$

$$\nabla \times (UA) = \nabla U \times A + U\nabla \times A$$

$$\nabla \cdot (A \times B) = B \cdot \nabla \times A - A \cdot \nabla \times B$$

$$\nabla(A \cdot B) = A \cdot \nabla B + B \cdot \nabla A + A \times (\nabla \times B) + B \times (\nabla \times A)$$

$$\nabla \times (A \times B) = B \cdot \nabla A - A \cdot \nabla B + A(\nabla \cdot B) - B(\nabla \cdot A)$$

$$\nabla \times (\nabla \times A) = \nabla(\nabla \cdot A) - \nabla^2 A$$

$$\nabla \cdot (\nabla \times A) = 0$$

$$\nabla \times (\nabla U) = 0$$

If $R = ix + jy + kz$ (Fig. 2), then

$$\nabla \cdot R = 3; \quad \nabla \times R = 0; \quad A \cdot \nabla |R| = |A|; \quad \nabla \cdot \frac{1}{|R|} = -\frac{R}{|R|^3}; \quad \nabla^2 \frac{1}{|R|} = 0$$

Integration. The line integral of a vector F along a curve AB denotes the integral of the tangential component of the vector along the curve; thus

$$\int_A^B F \cdot dR = \int_A^B |F_c| \, ds \quad \text{(Fig. 6)}$$

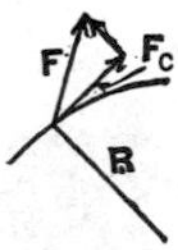

where $dR = i \, dx + j \, dy + k \, dz$.

If $F = \nabla U$ is the gradient of a single-valued continuous function $U(x, y, z)$ the line integral of F depends only on the end points. Conversely, if $F(x, y, z)$ is continuous and $\int_C F \cdot dR = 0$ for any closed path C in a three-dimensional region, there is a function $U(x, y, z)$ such that $F = \nabla U$.

FIG. 6

65. THEOREMS AND FORMULAS

Let n be the vector of unit length perpendicular to a surface at a point P and extending on the positive side (the outward normal); dS, the element of surface, and dv, the element of volume.

The Divergence (Gauss) Theorem. If a field vector F and its first derivatives are continuous at all points in a region of volume v bounded by a closed elementary surface S, then

$$\iint_S F \cdot n \, dS = \iiint_v \nabla \cdot F \, dv$$

Stokes' Theorem. If a field vector F and its first derivatives are continuous at all points in a region of area S bounded by a closed curve C, then

$$\iint_S \nabla \times F \cdot n \, dS = \int_C F \cdot dR$$

Green's Theorem. Under the conditions of the divergence theorem,

$$\iint_S n \cdot U \nabla V \, dS = \iiint_v U \nabla^2 V \, dv + \iiint_v (\nabla U \cdot \nabla V) \, dv$$

$$\iint_S n \cdot (U \nabla V - V \nabla U) \, dS = \iiint_v (U \nabla^2 V - V \nabla^2 U) \, dv$$

Cylindrical Coordinates.

$$x = r \cos \theta, \quad y = r \sin \theta, \quad z = z$$

The element of volume $dv = r \, dr \, d\theta \, dz$. The unit vectors u_r, u_θ, u_z are perpendicular to each other.

$$\text{grad } V = \nabla V = \frac{\partial V}{\partial r} u_r + \frac{1}{r} \frac{\partial V}{\partial \theta} u_\theta + \frac{\partial V}{\partial z} u_z$$

$$\text{div } F = \nabla \cdot F = \frac{1}{r} \frac{\partial}{\partial r} (r F_r) + \frac{1}{r} \frac{\partial}{\partial \theta} (F_\theta) + \frac{\partial}{\partial z} (F_z)$$

$$\text{curl } F = \nabla \times F = \begin{vmatrix} \dfrac{u_r}{r} & u_\theta & \dfrac{u_z}{r} \\ \dfrac{\partial}{\partial r} & \dfrac{\partial}{\partial \theta} & \dfrac{\partial}{\partial z} \\ F_r & r F_\theta & F_z \end{vmatrix}$$

$$\nabla^2 V = \frac{1}{r} \frac{\partial V}{\partial r} + \frac{\partial^2 V}{\partial r^2} + \frac{1}{r^2} \frac{\partial^2 V}{\partial \theta^2} + \frac{\partial^2 V}{\partial z^2}$$

Spherical Coordinates.

$$x = r \cos \phi \sin \theta, \quad y = r \sin \phi \sin \theta, \quad z = r \cos \theta$$

The unit vectors u_r, u_ϕ, u_θ are perpendicular to each other.

$$\text{grad } V = \nabla V = \frac{\partial V}{\partial r} u_r + \frac{1}{r \sin \theta} \frac{\partial V}{\partial \phi} u_\phi + \frac{1}{r} \frac{\partial V}{\partial \theta} u_\theta$$

$$\text{div } F = \nabla \cdot F = \frac{1}{r^2} \frac{\partial}{\partial r} (r^2 F_r) + \frac{1}{r \sin \theta} \frac{\partial F_\phi}{\partial \phi} + \frac{1}{r \sin \theta} \frac{\partial}{\partial \theta} (\sin \theta F_\theta)$$

$$\text{curl } F = \nabla \times F = \begin{vmatrix} \dfrac{u_r}{r^2 \sin \theta} & \dfrac{u_\theta}{r \sin \theta} & \dfrac{u_\phi}{r} \\ \dfrac{\partial}{\partial r} & \dfrac{\partial}{\partial \theta} & \dfrac{\partial}{\partial \phi} \\ F_r & r F_\theta & r \sin \theta F_\phi \end{vmatrix}$$

$$\nabla^2 V = \frac{1}{r^2} \frac{\partial}{\partial r} \left(r^2 \frac{\partial V}{\partial r} \right) + \frac{1}{r^2 \sin^2 \theta} \frac{\partial^2 V}{\partial \phi^2} + \frac{1}{r^2 \sin \theta} \frac{\partial}{\partial \theta} \left(\sin \theta \frac{\partial V}{\partial \theta} \right)$$

Solid Angle. The lines joining the point P to points of a surface S generate a solid angle. If a is the area intercepted by these lines on a sphere of center P with radius r, then

$$\omega = \frac{a}{r^2}$$

is the measure of the solid angle. If S is a surface that does not pass through P, $\cos \theta$ is nowhere zero, and n is everywhere continuous on the surface, then

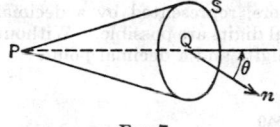

$$\omega = \int_s \frac{R \cdot n \, da}{r^3}$$

FIG. 7

where $R = PQ$. If S forms the complete boundary of a three-dimensional region, the total solid angle subtended by S at P is zero if P lies outside the region, and 4π if P lies inside the region.

THE ELEMENTS OF FORTRAN*

By John Beek

66. FORTRAN LANGUAGE

In an effort to standardize computer languages, an international group of computer users developed Algol algorithmic language, a language for expressive algorithms, and Cobol, a common business-oriented language. Universal acceptance of these languages has been slow.

Algol is patterned after the IBM-developed language called Fortran (formula translator). Due to the wide usage of IBM equipment, Fortran has gained widespread acceptance throughout the industry, Fortran compilers being available for most medium- and large-sized computers. Several Fortran languages have been developed in recent years: Fortran I, Fortran II, Fortran IV, and Fortran V. Although Fortran II is used extensively, the newer Fortran IV will be described here because it is much more powerful.

Fortran Characters

The Fortran language is alphanumeric; that is, variable identifiers (names) may be constructed from combinations of alphabetic and numeric characters. The characters utilized in the Fortran language are found on the standard typewriter keyboard. They include the alphabetic characters A to Z, the numeric characters 0 to 9, and nine special characters used for arithmetic operations—asterisk, comma, dot, equal sign, plus and minus signs, right and left parenthesis, and slash. Computation may be performed with real or complex numbers in either the integer or floating-point mode, but not with mixtures of these in a single expression.

Constants in Fortran Language

Constants as well as variables may appear in any given arithmetic expression. Five kinds of constants are permissible in Fortran IV: (1) fixed-point, (2) real constant, that is, single-precision floating-point, (3) double-precision floating-point, (4) complex, and (5) logical. Fixed-point and real numbers have been defined. A double-precision floating-point number uses the equivalent of two words in memory so that the number of digits carried is doubled.

Fixed-point (integer) constants are whole numbers with no decimal fraction. The terms integer and fixed-point are used synonymously. Integers, in Fortran IV, utilize the entire 36-bit word; from 1 to 11 decimal digits are permitted. This provides a range for integer constants from $-(2^{35} - 1)$ to $+(2^{35} - 1)$.** An exception occurs when the

* Adapted by permission from M. P. Moyle, *Introduction to Computers for Engineers*, John Wiley and Sons, New York, 1967.

** In Fortran II only 17 bits are used so the range is reduced to $-131{,}071$ to $+131{,}071$.

integer is used as a subscript or statement number; then the maximum value is 32,768, the total number of words in memory. Several examples of integer constants are shown in Example 1.

Example 1. Integer Constants.

3	002
18	1
1000571	274938405738517

Real constants are single-precision floating-point numbers, represented by a decimal number with or without exponent; up to 9 significant decimal digits are possible. Without the exponent a real constant is written with one or more digits and a decimal point.

Example 2. Real Constants without Exponents.

3.1	-10038.549
$+16.$	5.38
.002	$-48.$

A real constant with exponent consists of a mantissa, which can have up to 9 significant decimal digits, followed by the letter E and the exponent; the exponent can have up to two digits; at least one digit, which may be zero, must follow the letter E.

Example 3. Real Constants with Exponents.

Constants with Exponents	Numeric Significance
$3.0125E - 2$	3.0125×10^{-2}
$2.0E1$	2.0×10^{1}
$5.1E3$	5.1×10^{3}
$3.27E2$	3.27×10^{2}
$-4.3876E5$	-4.3876×10^{5}

Double-precision constants are double-precision floating-point numbers represented by a decimal number with or without exponent; up to 17 significant decimal digits are permissible. The letter D precedes the exponent. When the constant contains fewer than 10 digits it must be written with the exponent.

Example 4. Double-Precision Constants without Exponents.

$$40296311590.$$
$$10048573190572239.$$
$$8.136024911$$

Example 5. Double-Precision Constants with Exponents.

Constants with Exponents	Numeric Significance
$13.1D0$	13.1
$1.26D5$	1.26×10^{5}
$3.68D - 3$	3.68×10^{-3}
$-8.2D2$	-8.2×10^{2}

The range of double-precision constants is from 10^{-29} to 10^{38}.

Complex constants are numbers that have real and imaginary parts, expressed algebraically,

$$Z = A + Bi$$

They are represented in Fortran by an ordered pair of real numbers separated by a comma and enclosed within parentheses. The first number is the real part and the second is the complex part.

Example 6. Complex Constants.

Complex Constant	Significance
$(1.6, - 2.10)$	$1.6 - 2.10i$
$(8.5, 0.0)$	8.5
$(0.0, 3.25)$	$3.25i$
$(3.28E2, 1.6E5)$	$3.28 \times 10^{2} + 1.6 \times 10^{5}i$
$(-1.4, -1.5)$	$-1.4 - 1.5i$

The parentheses must be used regardless of the way in which the complex number appears in an expression.

Logical constants are used in decision-making expressions. When the decision is based on whether a situation is true or false, it is a logical decision. Fortran IV provides for such decisions with two logical constants:

.TRUE.

.FALSE.

Fortran Variable Identifiers

A variable identifier is a word that refers to a place in the memory where a number represented by that variable is stored. The variables in Fortran may contain up to six alphabetic or numeric characters, subject to the restriction that the first character must be alphabetic. Five types of variable identifiers are permitted; integer, real, double-precision, complex, and logical. The type is specified explicitly by the type statement or implicitly, in the special case of real and integer variables, by the first character of the variable name.

Integer variables are variables that take on fixed-point values with the same restrictions as for fixed-point constants. The general form of the integer type statement is

$$\text{INTEGER } A(\alpha_i),\ B(\alpha_i),\ \ldots$$

where A, B, ... are variable names and the α_i are 1-, 2-, or 3-integer constants and/or integer variables.

Example 7. Integer Variables—Specification by Type Statement.

INTEGER AMY, L, DELTA (3, 1)
INTEGER, ABCDE
INTEGER MIKE

The integer variable may be specified implicitly by using the letters I, J, K, L, M, or N for the first character of the integer variable name.

Example 8. Integer Variable—Implicit Specification.

I12359	NO4
IA1B3	M
JIM	MAYBE
LABCDE	KAT

Real variables are variables that take on floating-point values with the same restrictions as floating-point constants. Real variables are specified by the type statement

$$\text{REAL } A(\alpha_i),\ B(\alpha_i),\ \ldots$$

Example 9. Real Variables—Specification by Type Statement.

REAL ALPHA, BAKER, KAT
REAL IAB

The real variable is specified implicitly by using any letter other than I, J, K, L, M, or N for the first character of the real-variable name.

Example 10. Real Variable—Implicit Specification.

AX	TEMP
A14	PRESS
SAD	BETA
ALPHA	VOLI

Examples 7 and 9 show that the type statement overrides the implicit mode of specification; the variable AMY is a real-variable name but the type statement INTEGER causes AMY to exist as an integer variable in memory.

Double-precision variables are variables that take on double-precision floating-point values subject to the same restrictions as double-precision constants. The general form of the double-precision type statement is

$$\text{DOUBLE PRECISION } A(\alpha_i),\ B(\alpha_i),\ \ldots$$

Example 11. Double-Precision Variable—Specification by Type Statement.

DOUBLE PRECISION GAMA, VOL
DOUBLE PRECISION PRESS
DOUBLE PRECISION DONNER
DOUBLE PRECISION JACK

Complex and logical variables are specified in a similar manner:

$$\text{COMPLEX } A(\alpha_i),\ B(\alpha_i),\ \dots$$
$$\text{LOGICAL } A(\alpha_i),\ B(\alpha_i),\ \dots$$

Example 12. Complex and Logical Variables—Specification by Type Statement.

$$\text{COMPLEX P, Q}$$
$$\text{LOGICAL A, SO, ILL}$$

67. FORTRAN STATEMENTS

A program is carried out by Fortran statements, which are instructions to the computer to carry out an action. A Fortran program consists of a series of such statments, the Fortran language itself providing certain statements necessary for execution of the program. The other statements must be constructed by the programmer.

The statements are entered into the computer via punched cards. A card contains 80 columns; only the first 72 are used in Fortran. A typical card is shown in Fig. 1. Columns 1 through 5 are used for statement numbers, which can be any number less than 32,768: and no two statements may have the same number. It is preferable to number all cards, using columns 73 through 80; the advantage of numbering sequentially is readily apparent if an unnumbered deck of cards is dropped to the floor.

Column 6 is used for continuation of a statement to the next card. The first card in the sequence representing the statement must contain a zero or blank in column 6, the second and subsequent cards have some number different from zero in column 6. Up to nineteen cards may be used for a single statement.

Columns 7 through 72 are used for the Fortran statement. The remaining columns, 73 through 80, are not recognized by the Fortran processor.

COMMENT statement: C. The simplest Fortran statement is the COMMENT statement, as shown in Fig. 2. It may be the title of the program, a discussion of the variables and their units, a code, and so on. Any Fortran character may be used in the comment. In order to identify the statement as a comment, a C is punched in column 1 of the card. No computations are carried out as a result of a COMMENT statement.

DIMENSION statement. The DIMENSION statement,

$$\text{DIMENSION } A(\alpha_i),\ B(\alpha_i),\ \dots$$

is a statement telling the computer to reserve space for the variables A, B, and any others that may be included in the DIMENSION statement. The α_i are the number of locations reserved in one-, two-, or three-dimensional arrays.

The DIMENSION statement provides a means for establishing subscripted variables. By subscripting a variable, it is possible to use that variable identifier to represent many variables simply by changing the index. This can save considerable programming effort.

C	THIS IS A COMMENT CARD. IT CONTAINS A C IN COLUMN ONE. WHATEVER IS
C	PUNCHED IN COLUMNS 7 THROUGH 72 IS PRINTED WHEN THE SOURCE PROGRAM
C	IS COMPILED. TITLES, NOTES, INSTRUCTIONS, ETC., MAY BE PRINTED. FOR
C	EXAMPLE, CONSIDER THE FOLLOWING
C	PROGRAM TO DETERMINE THE OPTIMUM NUMBER IN A MULTIUNIT EVAPORATOR.
C	TEMPERATURES ARE IN DEGREES F. MASS UNITS ARE IN POUNDS PER HOUR.
C	CONCENTRATIONS ARE IN POUNDS PER CUBIC FOOT.

Fig. 2. The COMMENT statement.

C	THE STATEMENT WHICH FOLLOWS IS A DIMENSION STATEMENT
	DIMENSIONTEMP(50)

Fig. 3. Dimension statement for an one-dimensional array.

For example, to sum a set of 100 numbers we could write SUM = SUM + X(1); the initial value being set equal to zero. The first time the statement is executed we will have X_1 in the location identified as SUM. By increasing 1 to 1 + 1 we can proceed to add all 100 numbers with a simple loop.

The set of X_i numbers is called an array; in this case it is a one-dimensional array.

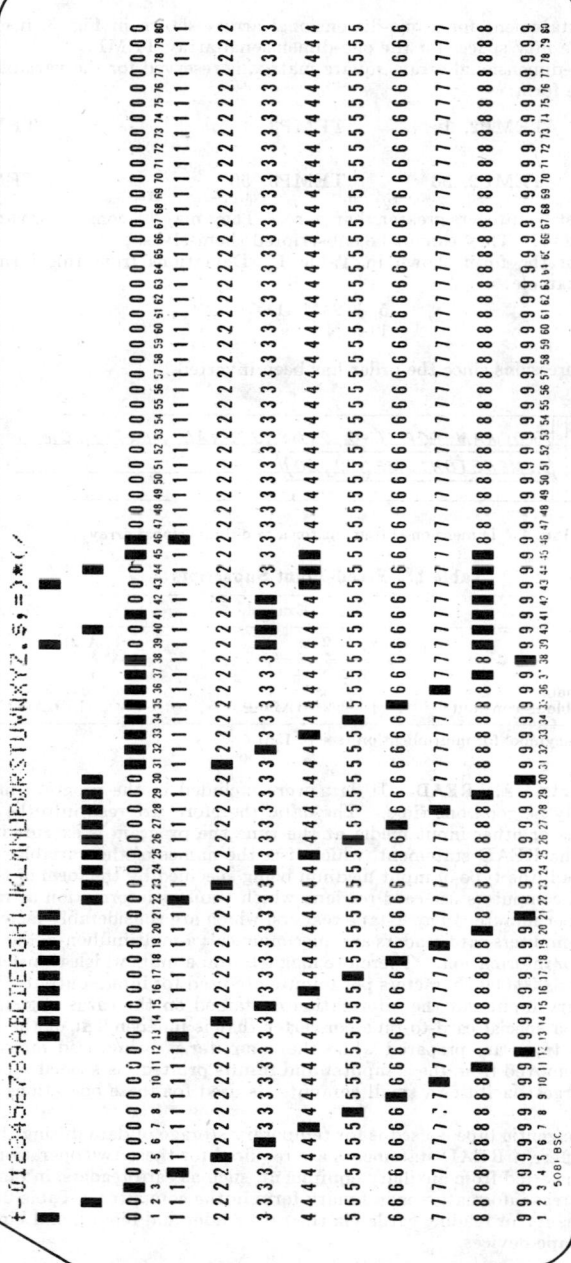

Fig 1.   80-column Fortran card.

The Fortran DIMENSION statement provides for one-, two-, or three-dimensional arrays.

The dimension statement for a one-dimensional array, shown in Fig. 3, instructs the computer to reserve fifty spaces for the one-dimensional array TEMP.

In Fig. 4, a two-dimensional array, square matrix, is reserved for the variable TEMP. The matrix has the form:

TEMP1, 1	TEMP2, 1	TEMP3, 1	⋯	TEMP50, 1
⋮	⋮	⋮	⋮	⋮
TEMP1, 50	TEMP2, 50	TEMP3, 50	⋯	TEMP50, 50

The subscripts must be integers greater than zero. They may be constants, variables, or arithmetic expressions. They cannot be subscripted themselves.

Subscripts are of the form shown in Table 1. Deviations from this form are not permitted. For example:

$$5 + 2*MAD$$
$$1 + N$$

are both illegal expressions since the order has been inverted.

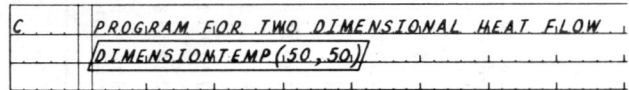

FIG. 4.   Dimension statement for a two-dimensional array.

Table 1.   Fixed-Point Subscripts

Form	Example	Fortran
Constant	2	A(2)
Variable	J	A(J)
Variable ± constant	N + 1	Y(N + 1)
Constant times a variable[a]	3*J	BET(3*J)
Constant times variable ± constant	2*MAD ± 5	ALPHA(2*MAD ± 5)

[a] The asterisk is the symbol for multiplication (see p. 156).

INPUT statements.   A.   READ.   If data were included in the program proper, the program would only be good one time.   The data, therefore, are read into the computer memory from cards or other input media at the time the program is executed.   This is accomplished by the READ statement, which lists the names of the variables for which values are to be read, the type of input medium being specified by the form of the READ statement.   Small computers use card readers, which transfer information at rates up to 80,000 characters per minute.   Paper-tape readers, which are considerably faster, are also used.   On large computers card readers are uneconomical; a multimillion-dollar computer cannot wait idly for information.   Therefore magnetic-tape units which transfer information at rates up to 6,000,000 characters per minute are used for input–output.   Cards are still used as primary input but the information contained on the cards is transferred to magnetic tape via a special card-to-tape converter that is independent of the computer. In this way input tapes are prepared while the computer is being used for computing; output tapes are removed from the computer and results printed via special tape printers. On some of the largest facilities a small computer is used for these operations instead of special equipment.

A second way magentic tape is used is for temporary storage of data during the solution of a problem.   Different READ statements are required for these two operations because information is transferred from auxiliary equipment, such as card readers, in binary-coded decimal form, whereas information is in binary form in the computer.   Table 2 shows the form of the statements for reading cards via the card reader and binary, and binary-coded decimal data via tape devices.

```
C      PROGRAM FOR TWO-DIMENSIONAL HEAT FLOW
C      TEMPERATURES ARE IN DEGREES RANKINE
       DIMENSIONTEMP(50,50)
       READ20,Q
```

Fig. 5. Card READ statement: reading card input.

Table 2. Form of the READ Statements

Type of Input Media	READ Statement
Cards	READ(n) A, B, C, ...
BCD records	READ(i,n) A, B, C, ...
Binary records	READ(i) A, B, C, ...

In Table 2, which will be discussed in Section B, n refers to the Fortran statement containing the format specifications, and i, an unsigned integer constant or integer variable, refers to a particular input device such as a magnetic- or paper-tape reader. In the normal 7090 code, $i = 1, 2, 3$, and 4 are tape units for input or output of binary data; $i = 5$ is for BCD input; $i = 6$ is for BCD output; $i = 7$ is for binary output.

The READ statement illustrated in Fig. 5 calls for the variable Q to be read from a card. The format of Q must be specified in statement 20 of the program.

The READ statement in Fig. 6 calls for the variable Q to be read from an input mag netic tape that had been prepared from cards. The tape is mounted on tape reader number 8. Format of Q is specified by statement 20 of the program.

```
C      PROGRAM FOR TWO-DIMENSIONAL HEAT FLOW
C      TEMPERATURES ARE IN DEGREES RANKINE
       DIMENSIONTEMP(50,50)
       READ(8,20)Q
```

Fig. 6. Binary-coded decimal (BCD) READ statement.

The READ statement in Fig. 7 calls for the variable Q to be read from temporary storage onto magnetic tape. The tape is mounted on tape reader number 1.

B. FORMAT statements. Information used externally to the computer is usually in decimal form if it is a number, or in alphameric form if it is a variable name. Because this information must be stored and handled internally as binary bits, a conversion step is required.

The FORMAT statement specifies how the data is to be converted from the input record to memory or from memory to output record. The input and/or output records contain "fields of information"; a field is a group of information that is a separate entity, such as a row of 5 digits. Input and/or output records may contain many such fields. The information contained in the FORMAT statement specifies the field structure; letters A, D, E, F, H, I, O, L, or X are used to designate the type of information and method of conversion as shown in Table 3. The same letters in Table 3 signify the following:

a is a 1-digit integer that specifies the number of characters in the A-field; $1 \leqslant a \leqslant 6$.

abc is a 1-, 2-, or 3-digit integer that specifies the number of spaces in the field (char-acters + blanks) required; $1 \leqslant abc \leqslant 132$.

dd is a 2-digit integer that specifies the number of digits to the right of the decimal point in a double-precision floating-point number; $00 \leqslant dd \leqslant 17$.

e is a 1-digit integer that specifies the number of digits to the right of the decimal point in real numbers; $0 < e \leqslant 9$.

hh denotes the contents of the abc spaces immediately following H.

n is an unsigned integer that specifies the number of fields to follow.

```
C      PROGRAM FOR TWO-DIMENSIONAL HEAT FLOW
C      TEMPERATURES ARE IN DEGREES RANKINE
       DIMENSIONTEMP(50,50)
       READ(1)Q
```

FIG. 7. Binary READ statement.

Table 3. Field Specification Format Codes

Code	Type of Field	Form	Example
A	Nonnumerical characters	nAa	A4
D	Double-precision with exponent	$nDabc.dd$	3D21.16
E	Real with exponent	$nEabC.e$	E8.1
F	Real without exponent	$nFabc.e$	F10.6
H	Alphanumeric (Hollerith)	$abcHhh...hh$	8HVALUE IS
I	Decimal integer	$nIabc$	312
O	Octal integer	$nOabc$	504
L	Logical	$nLabc$	L6
X	Skip	$abcX$	15X

Some FORMAT statements applications follow.

1. A-field. The A-field statement provides for reading character information in and out of storage. The information may be stored in memory locations whose names are any of the five types of variables. Once the characters have been read into memory they may be moved, provided they are moved to locations of the same mode; that is, an integer variable may be moved to a location containing another integer variable.

We begin with the READ statement, identified as statement 2, which calls for read-in of the name according to FORMAT statement 21. The letter A in the format statement causes the characters IRON to be read into memory and stored as a BCD number. The word, which has only four characters, is shifted left, left-justified, and the remainder of

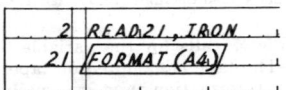

```
   2  READ21,IRON
  21  FORMAT (A4)
```

FIG. 8. Nonnumerical characters; A-conversion format.

```
C      A REGRESSION ANALYSIS
       DIMENSION DOTTER(200),RESULT(200)
       DOUBLE PRECISION DOTTER
       READ2,DOTTER
   2   FORMAT(D26.14)
```

FIG. 9. Double precision: D-conversion format.

the word is filled out with blanks. On output only the leftmost characters of the word are transmitted; that is, only nonblank characters. When the word is too large (greater than six characters), only the six rightmost characters are retained and transmitted.

2. D-field. Double-precision numbers are read in and out of memory by the D-conversion format. When the D-conversion is specified, the variable involved must have a double-precision name.

In Fig. 9 the READ statement calls for a read-in of the double-precision variable DOTTER according to FORMAT statement 2. The format of the number is a 26-column field with 14 places to the right of the decimal. The number might appear on an input card as follows:

$$bbbbb - 0.14571823140000Db'03$$

where b indicates a blank character, and b' is blank or plus if the exponent is positive and minus if negative.

3. E-field. Real numbers with exponent are read in or out by specifying the E-conversion format. Variables must be identified by real names when an E-conversion is specified.

In Fig. 10 the FORMAT statement sets up a 12-column field for a single-precision

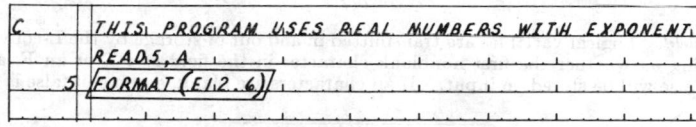

FIG. 10.   Real number with exponent; E-conversion format.

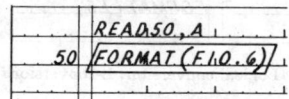

FIG. 11.   Real number without exponent; F-conversion format.

variable with 6 digits to the right of the decimal.   The input data might appear on a card as follows:

0.130625E-02

4. F-field. Real numbers without exponent are read in and out by the F-conversion format.   The information transmitted must have a real-variable name.

In Fig.11 a 10-column field is specified with 6 digits to the right of the decimal.   The input data might appear as follows:

307.254100

5. H-field. For transmitting alphanumeric information such as program titles an H-conversion must be specified.   H-field conversions are not given names in memory so the information cannot be moved or manipulated in storage like the A-conversions.   However, up to 132 characters can be handled in a single field.

The PRINT statement shown below causes THE TRIANGLE IS NOT A RIGHT TRIANGLE to be printed out as specified by FORMAT statement 20.

20 FORMAT (37H THE TRIANGLE IS NOT A RIGHT TRIANGLE)

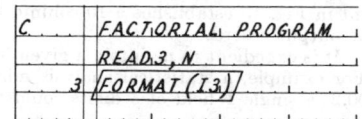

FIG. 12.   Integer numbers: I-conversion format.

6. I-field. Transmission of integer data in and out of storage is accomplished by the I-conversion.   The information transmitted must have an integer-variable name.

The FORMAT statement in Fig. 12 sets up a 3-column field.   The input data might appear as:

*b*12

7. O-field. Octal numbers are transferred by specifying the O-conversion.   The information transmitted may have any variable name.

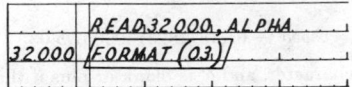

```
       READ,32000,,ALPHA
32000  FORMAT (O3,)
```

FIG. 13. Octal numbers; O-conversion format.

The FORMAT statement in Fig. 13 sets up a 3-column field. The printed output might appear as follows:

$$777$$

8. L-field. Logical variables are transmitted in and out of storage by the L-conversion. Depending on whether the first nonblank character in the field is a T or an F, a value true or false will be stored on input. If all characters are blank, a value of false is stored.

```
       LOGICAL A
   1   READ2,A
   2   FORMAT(L6,)
```

FIG. 14. Logical conversion: L-conversion format.

```
C      PROGRAM FOR N FACTORIAL
       READ2,N
   2   FORMAT(I2)
          .
          .
          .
       PRINT20,,MFAC
  20   FORMAT (15X, 5HMFAC=I10)/
```

FIG. 15. Skipping spaces; X-conversion format.

A value of true or false in storage will cause $abc - 1$ blanks, followed by T or F, respectively, to be written out.

9. X-field. The X-field statement provides for spacing of information on output records and spacing and/or deletion of information on input. The number preceding X specifies the number of blank characters in the input or output record. On input, n spaces of the input record are skipped regardless of whether there is information; n blanks are then inserted in the output record.

The FORMAT statement in Fig. 15 establishes a 10-column field beginning 15 spaces in from the left edge of the output record.

10. Repetition of a field. It is expedient to maintain a given field width when handling large amounts of data. For example, a statistical analysis might involve 100 numbers ranging from 3.18 to 10000.2; a single F-field of 8 digits could then be used to transmit all of the numbers.

The FORMAT statement in Fig. 16 establishes a 10-column field with 6 decimals to

```
C      PROGRAM FOR TWO-DIMENSIONAL HEAT FLOW
C      TEMPERATURES ARE IN DEGREES RANKINE
       DIMENSIONTEMP(50,50)
       READ20,Q,,V,AVGK,TO,,CP
  20   FORMAT (5F10.6)/
```

FIG. 16. Repetition of fields.

```
C      OPTIMUM THICKNESS OF PIPE INSULATION
C      BASIS OF 8000 HOURS OPERATION PER YEAR
       INTEGER HO, HVAP
       READ4, N, ROP, PIPTK, TSTM, HO, HVAP
    4  FORMAT (I2, F10.6, E10.10, F10.6, I2, I4)
```

Fig. 17. Successive-field format.

the right of the decimal and specifies that 5 such fields are to follow. Any number between −99.999999 and 999.999999 can be transmitted.

11. Successive fields. Several different field specifications can be incorporated in a single FORMAT statement. The specifications for successive fields are separated by commas. E, F, and I fields are established by the single FORMAT statement in Fig. 17.

12. Multiple groups of fields. If several sets of calculations are performed, it is usually desirable to identify the set number, then to print out several results.

The READ statement in Fig. 18 calls for the reading of N from the first card; then the values of T[I] and P[I] are read from each successive card. The value of N determines how many cards are called for. The slash appearing in format statement 3 is discussed in the next section.

13. Blank lines on printout. The preparation of output records usually requires spacing title headings, subtitles, and tabular data. This is accomplished by advancing the paper in the line printer without printing, that is, creating blank lines. A slash, /, in a FORMAT statement advances the line printer one line. If *n* slashes appear at the end of a FORMAT statement, *n* records are skipped on input and *n* − 1 blank lines occur on output. If *n* slashes appear anywhere else in the FORMAT statement *n* − 1 records are skipped on

```
C      LEAST SQUARES FIT
       DIMENSION T (50), P (50)
       READ3, N, (T (I), P (I), I = 1, N)
    3  FORMAT (I3 / (2F10.0) )
```

Fig. 18. Multiple-data read-in.

input and *n* − 1 blank lines occur on output.

In Fig. 18 the slash marks the end of the first record which contains only N: the card reader then takes the next data from a following card. In Fig. 19 the double slash yields one blank line between the two records. The first slash marks the end of the first record; note that the FORMAT statement is written on three cards.

```
C      CHANGE MAKING PROBLEM
       DIMENSION DENOM (20), KDENOM (20)
         .
         .
    1  1X, 44HBROKEN DOWN INTO BILLS AND CHANGE AS FOLLOWS //
    2  21X, 6HNUMBER, 10X, 7HDOLLARS
```

Fig. 19

C. OUTPUT statements: PRINT, PUNCH, and WRITE. Output from the computer may be via the PRINT, PUNCH, or WRITE statements. The general form of these statements is shown in Table 4. Notations n and i have the same meanings they have in the READ statements.

Table 4.   Output Statements

Type of Output	Form of Output Statement
Line printer	PRINT n, A, B, C, ...
Cards	PUNCH n,
BCD record	WRITE(i,n)
Binary record	WRITE(i)

1. PRINT. When a printed record is desired, a PRINT statement must be included in the program. The PRINT statement calls for a printout of specified information via the line printer or typewriter, according to the FORMAT statement.

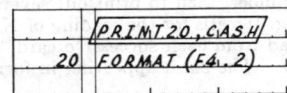

FIG. 20.   PRINT statement format.

The PRINT statement in Fig. 20 causes the printout of the variable CASH as a floating-point number without exponent (signified by the letter F in the format) of up to 3 digits with 2 decimals, 2.00, for example.

2. PUNCH. It may be desirable at times to punch intermediate or final results onto cards, for future read-in or for listing and sorting on auxiliary equipment. The PUNCH statement causes the computer to punch the output information on a card. The first-named variable is punched first and so on; each variable named is punched on a single card. A FORMAT statement controls the format on the card.

As a result of the PUNCH statement in Fig. 21, the variable ALPHA is punched on a card; the format is a floating-point number with two decimals in a 5-column field. The second integer variable JACK is then punched in a 2-column field on the next card.

3. WRITE. To use magnetic tape as output medium it is necessary to use the WRITE instructions. The WRITE statement instructs the computer to write information on magnetic tape. Two forms of the WRITE statement are permitted; WRITE(i,n) and WRITE(i). The WRITE(i,n) statement is used to transfer information from the computer to magnetic tape for printing on a tape printer. Successive records are written on tape according to the format specification in statement number n until all variables named have been written. The WRITE(i) statement is used to write on tape for temporary storage; one logical record containing all words specified is transmitted to tape.

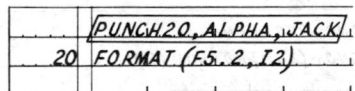

FIG. 21.   Card-punching format.

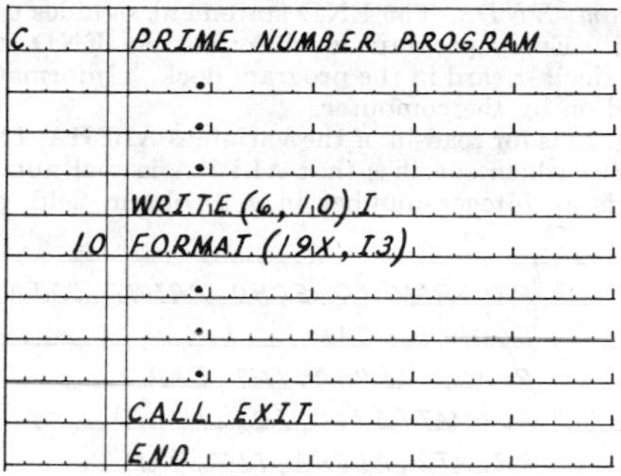

```
C       PRIME  NUMBER  PROGRAM
        •
        •
        •
        WRITE (6, 10) I
10      FORMAT (19X, I3)
        •
        •
        •
        CALL  EXIT
        END
```

FIG. 22. Multiple-record output to magnetic tape;
WRITE(i,n) statement format.

The WRITE statement in Fig. 22 causes the variable I to be written on magnetic tape for ultimate printing on the tape printer. The tape specified is mounted on tape unit 6. Format of the information transmitted must be specified in program statement number 10.

The WRITE statement in Fig. 23 transfers 50 values of the variable M onto one logical tape record. The information will ultimately be read back directly to the computer by a READ(i) statement; the tape is mounted on tape unit number 6. The information cannot be printed as it is not in the form required by the printer.

4. Tape operations. The use of magnetic tape requires some other statements for identification and control. The end of each record must be marked. It is frequently

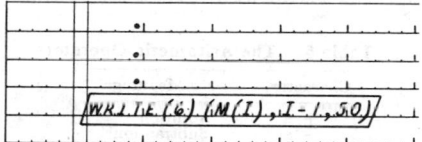

```
        •
        •
        •
        WRITE (6) (M(I), I = 1, 50)
```

FIG. 23. Single logical record output to magnetic
tape; WRITE(i) statement format.

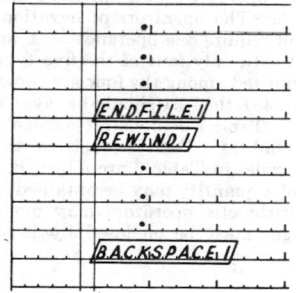

```
        •
        •
        ENDFILE 1
        REWIND 1
        •
        •
        BACKSPACE 1
```

FIG. 24. Tape command format.

necessary to refer back to data or information on a section of tape previously read and tapes must be rewound before a new program is computed. Statements that control these operations are ENDFILE n, REWIND n, and BACKSPACE n.

The ENDFILE statement in Fig. 24 causes an endfile mark to be written on tape 1; the REWIND statement causes tape 1 to be rewound; and the BACKSPACE statement causes tape unit 1 to backspace one record.

5. Ending the program: *E N D.* The END statement signifies to the computer that the program is completed. Every program must contain an END statement, and the card containing it must be the last card in the program deck. Information following the END statement is not acted on by the computer.

The program in Fig. 25 is for read-in of the variables ALPHA, INT, and CAP according to the format statement which specifies that ALPHA ia real number with exponent in a 5-column field, INT is an integer number in a 2-column field, and CAP is a complex

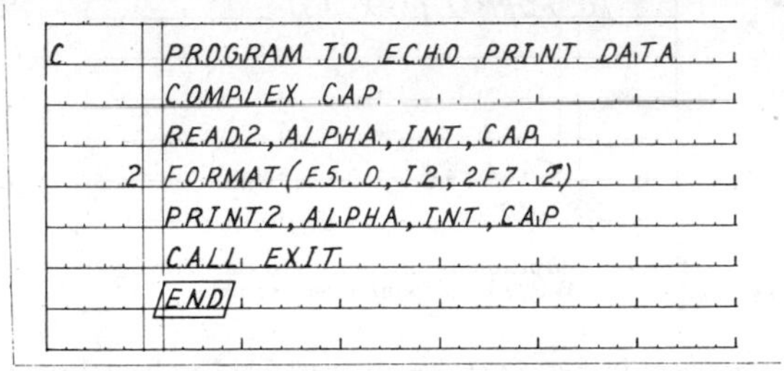

```
C          PROGRAM TO ECHO PRINT DATA
           COMPLEX CAP
           READ2, ALPHA, INT, CAP
        2  FORMAT (E5.0, I2, 2F7.2)
           PRINT2, ALPHA, INT, CAP
           CALL EXIT
           END
```

Fig. 25.   End-of-program statement for temperature rise of a liquid.

number in two 7-column fields. After the variables have been read in they are echo-printed (printed back) and the program ends.

Arithmetic Operations in Fortran

Arithmetic expressions. Numerical computations are specified by arithmetic expressions. An arithmetic expression in Fortran looks very much like the familiar algebraic one, except that special characters are used. These are identified in Table 5.

Table 5.   The Arithmetic Operators

Character	Meaning
+	Addition
−	Subtraction
*	Multiplication
/	Division
**	Raise to power

These operators are binary operators: two operands are required. For example, we must write $X + y$ or $a - b$. The operators of negation (sign reversal) and absolute value are unary operators and require one operand; $-X$ and $|X|$, respectively. Arithmetic operators may be used with only four of the five types of constants and variables; the various combinations permitted among the four are shown in Table 6. In interpreting the exponential operator in Table 6 the base should be taken as the column variable and the exponent as the row variable. Table 7 demonstrates some possible arithmetic operations. The K designates integer numbers. Double-precision and complex numbers follow a similar pattern, provided the rules in Table 6 are followed.

Negation. The negative of a quantity may be obtained by preceding it with the unary minus operator. Because arithmetic operators may not be written side by side, the quantity, with preceding sign, must be enclosed by brackets. The negative of A is written $(-A)$.

Table 6.   Operations Valid in Combination with Various Constants and Variables

	Integer	Real	Double Precision	Complex
Integer	+, −, *, /, **	None	None	None
Real	**	+, −, *, /, **	+, −, *, /, **	+, −, *, /
Double precision	**	+, −, *, /, **	+, −, *, /, **	None
Complex	*	+, −, *, /	None	+, −, *, /

Table 7.   Arithmetic Operations

Arithmetic Operation	Write the Expression	
	Real	Integer
Add B to A	A + B	KA + KB
Subtract B from A	A − B	KA − KB
Multiply A by B	A*B	KA*KB
Divide A by B	A/B	KA/KB
Raise A to power M	A**M	KA**M

Example 13.   Arithmetic Expressions.

(1) $A \times B + C - D/E^{**}X$ means $(A)(B) + C - \dfrac{D}{E^z}$

(2) $A/B^*C/D$ means $\dfrac{AC}{BD}$

(3) R^*T/P means $\dfrac{RT}{P}$

Parentheses for Clarity or Priority

Parentheses are used to clarify the sequence of operations desired and/or to override the rule of precedence.   For example, $(A + B)^2$ would be written $(A+B)^{**}2$.   Multiple use of parentheses is permissible; extra parentheses cause no harm and the beginner would do well to use them freely.   There must be an equal number of left and right parentheses in a Fortran statement.

Example 14.   Use of Parentheses.   The expression for the cubic equation

$$AX^3 + BX^2 + CX + D$$

may be written either

or

(a) $A^*X^{**}3 + B^*X^{**}2 + C^*X + D$

(b) $((((A^*X)^*X) + (B^*X) + C)^*X + D)$

Of the two ways of expressing the cubic equation, the latter is the preferred method when machine time is important because it takes less time to multiply X by X than to raise X to the power 2.   Subroutines are required to carry out the exponential; the time required to call and interpret the subroutine is much greater than the time required for multiplication.

The Substitution Statement

The substitution statement uses the equal sign to assign a value to a given variable identifier.   Whenever the equal sign appears in a program it means that the quantity on the left-hand side of the equal sign in replaced by the quantity on the right-hand side of the equal sign.   The statement

$$Y = X$$

means that the contents of the memory location identified by the variable Y is replaced by the contents of the memory location identified by the variable X, which may be any variable identifier, arithmetic expression, and so forth.

Example 15.   Use of the Substitution Statement.

(a) $CP = ALPHA + BETA^*T + GAMA^*T^*T$

means replace the current contents of the location that represents the variable CP by the sum of the current contents of ALPHA, BETA, and GAMA summed according to the arithmetic expression given.

(b) $P = RT/(V - B) - A/V^{**}2$

means to compute the value of $RT/(V - b) - a/V^2$, then to replace the current value of the location containing the variable P by this result.

(c) $I = I + 1$

means to compute the value of $I + 1$ using the current value of I, and then to place this result in the location called I; in other words, to increase I by 1.

Integer, real, and double-precision variables may be connected via the substitution statement to integer, real, and double-precision expressions in any combination. Complex and logical variables may only be connected to complex and logical expressions respectively.

Example 16.   Write a Program to Calculate the Temperature Rise of a Liquid on Addition of Heat.*
The mass, w, is 100 lb; the heat added, Q, is 1000 Btu; and the specific heat, C_p, is 1.0 Btu/lb°F. The initial temperature is 100°F.
Solution: The equation for the change in temperature of the liquid is:

$$Q = wC_p(T_2 - T_1)$$

Solving for T_2 we obtain

$$T_2 = Q/wC_p + T_1$$

The flow chart and program are given in Fig. 26.
The first statement in the program is the COMMENT statement giving the title of the program; no calculation is performed. It has not been numbered but it could have been if desired. Statement number 1 is the READ statement calling for the variables Q, w, T_1, and C_p to be read into memory from a single card. The data format on the card is 4 numbers with 2 decimals in a field that can have up to 10 columns, as specified in statement number 2. In statement 3 a calculation is called for on the right-hand side of the equal sign (substituion statement) followed by substitution of the result of the calculation for the Variable T_2. Statement 4 calls for a printout of the variable T_2 according to FORMAT statement 5. The program is terminated by statement 6.

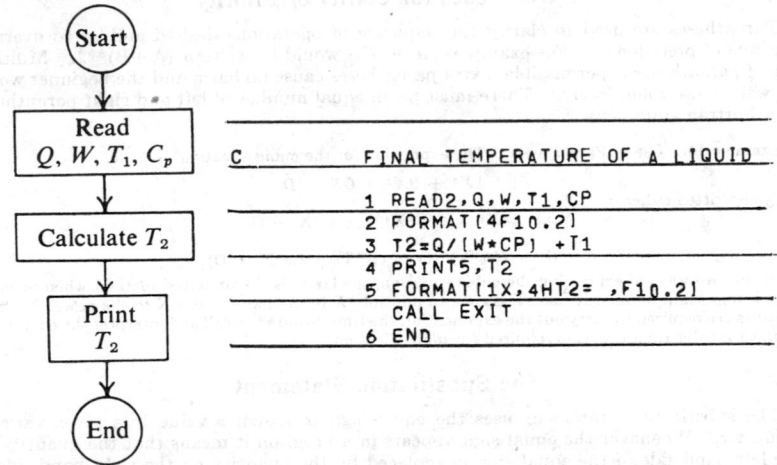

```
C              FINAL TEMPERATURE OF A LIQUID
1  READ2,Q,W,T1,CP
2  FORMAT(4F10.2)
3  T2=Q/(W*CP) +T1
4  PRINT5,T2
5  FORMAT(1X,4HT2= ,F10.2)
   CALL EXIT
6  END
```

Fig. 26.   Flow chart and Fortran program for temperature rise of a liquid.

Output Record:

```
GO GO

T2=      200.00
EXIT CALLED.
```

Example 17.   Write a Program to Calculate the Time to Fill a Tank.
Solution: The time to fill a tank is the volume of the tank divided by the rate of filling.   The volume is $V = \pi R^2 H$, where R is the radius and H is the height of the tank.   The time to fill is:

$$\text{Hours} = \pi R^2 H/Q$$

where Q is the filling rate in ft³/hour.   Let us run the solution for the case where

$$R = 8.00 \text{ ft}$$
$$H = 20.00 \text{ ft}$$
$$Q = 50.00 \text{ gpm}$$

There are 7.48 gallons per cubic foot.   The flow chart and program are given in Fig. 27.

* The compiler records in this and subsequent example problems are taken from the line printers.

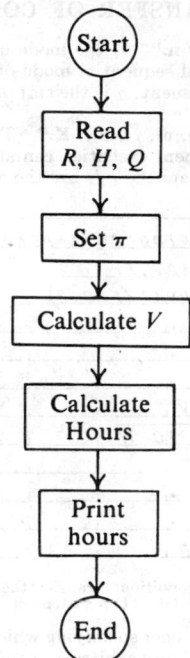

```
C     TIME TO FILL A TANK
      1 READ2,R,H,Q
      2 FORMAT(3F10.2)
      3 PI=3.1416
      4 V=PI*R*R*H
      5 HOURS=V/((Q*60.)/7.48)
      6 PRINT7,HOURS
      7 FORMAT(1X,20HTIME TO FILL TANK IS,F10.2,5HHOURS)
        CALL EXIT
      8 END
```

FIG. 27. Flow chart and program for time required to fill a tank.

The first statement in the program is a COMMENT statement with the title of the program; it is not numbered. Statement number 1 is a READ statement calling for a read-in of the variables, R radius of tank), H (height of tank), and Q (flow rate into tank) according to FORMAT statement numbered 2. The format calls for 3 floating-point numbers with 2 decimals in a field of up to ten columns to be read from a single card. In statement 3, the value of π is set, V is calculated in statement 4 and the time to fill the tank in 5; note the use of brackets for clarity. Statement 6 calls for printout of the variable HOURS according to FORMAT statement 7. The format called for skips one space, prints the heading TIME TO FILL TANK IS, and follows with a floating-point number in a ten-column field; the title HOURS follows the number. The format titles are obtained by using the H specification.

Output Record:

```
   GO GO
─────────────────────────────────────────────────
   TIME TO FILL TANK IS      10.03HOURS
   EXIT CALLED.
─────────────────────────────────────────────────
```

68. TRANSFER OF CONTROL

Unconditional transfer "GO TO n." If the mode of operation must be interrupted, a transfer of control from the normal sequential mode of program statement execution can be achieved via the GO TO n statement; n is the statement number punched in columns 1 through 5, as seen in Fig. 28.

Computed transfer "GO TO (n_1, n_2, ..., n_j), K." Transfer of control from the normal sequential mode of program statement execution can also be accomplished by the Fortran statement GO TO ($n_1 \cdots n_j$), K where the n_j's are the statement numbers punched in col-

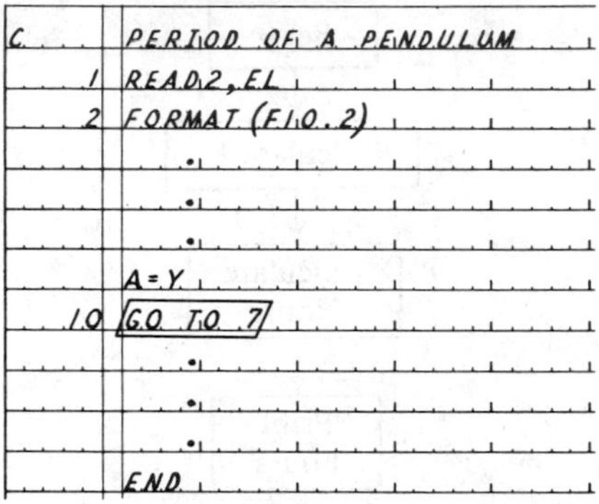

FIG. 28.   Unconditional transfer; the unconditional
GO TO n statement.

umns 1 to 5.   The index K is an integer specifying which of the statement numbers n_j is to be taken currently.   By varying K any arbitrary sequence of statement execution is possible.   In the following example, control is transferred to the statement numbered 15, 45, 60, 200, or 350 according as K is 1, 2, 3, 4, or 5.

$$\text{GO TO } (15,\ 45,\ 60,\ 200,\ 350),\ K$$

Conditional transfer "IF (result)α,β,γ." The conditional transfer statement is one of the more powerful statements available to the programmer; it makes decision-making possible.   In Fortran conditional transfer is accomplished by the IF statement.   Transfer to one of three different program statements is possible depending on the result of the operation within the parentheses:

1. If the result is a negative quantity, transfer to α.
2. If the result is zero, transfer to β.
3. If the result is a positive number, transfer to γ where α, β, and γ are program statement numbers.

In Fig. 29 the quantity $|Y - A|$ is subtracted from 0.005.   If the result is:

1. Negative: the machine goes to statement labeled 9.
2. Zero: the machine goes to statement labeled 11.
3. Positive: the machine goes to statement labeled 11.

The IF statement accomplishes a three-way branch if α, β, and γ are all different values (Fig. 30a): a two-way branch if only two values are different (Figs. 30b and c); and no branch if $\alpha = \beta = \gamma$ (Fig. 30d).   Usually the IF statement is used as a two-way branch to answer a question yes or no.   The possibilities available to the programmer are illustrated schematically in Fig. 30.

```
C       PERIOD OF A PENDULUM
      1 READ2,EL
           .
           .
           .
      7 Y=0.5*(A+X/A)
      8 IF(0.005-ABS(Y-A))9,11,11
      9 A=Y
           .
           .
           .
        END
```

FIG. 29. Conditional transfer; the IF statement.

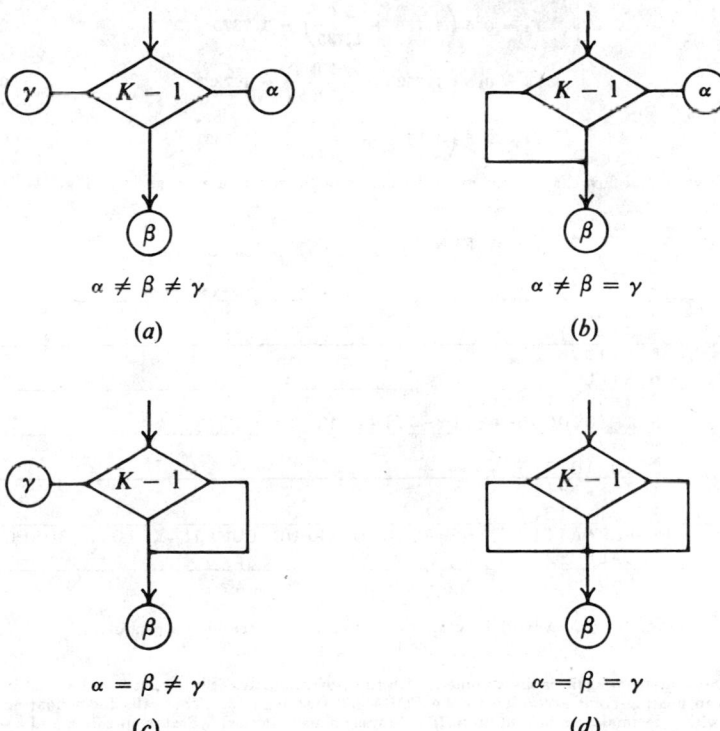

$\alpha \neq \beta \neq \gamma$

(a)

$\alpha \neq \beta = \gamma$

(b)

$\alpha = \beta \neq \gamma$

(c)

$\alpha = \beta = \gamma$

(d)

FIG. 30. Possible branching with the IF statement.

Example 18. Write a Program for the Period of a Pendulum.

Solution: The period of a pendulum is given by the equation

$$T = 2\pi\sqrt{L/g}$$

where T = The period in seconds

L = The length of the pendulum = 3 ft

g = 32.2 ft/sec².

We note that the calculation of the period requires the square root of L/g. The square root of a number X is found by continued iteration of the statement

$$Y = \tfrac{1}{2}(A + X/A)$$

until the difference between successive values of Y is less than or equal to some allowed value δ, called the tolerance.

An arbitary initial value is chosen for A, and Y is calculated; this value of Y is substituted for A and a new value of Y is calculated. When the absolute value of $Y - A$ is less than or equal to δ, the iteration is stopped. If the the difference is greater than δ, the iteration is continued. For example, to calculate $\sqrt{3}$, choose A = 1.8. Then:

$$\hspace{20em} Y - A$$

$$Y_1 = 0.5\left(1.8 + \frac{3.0}{1.8}\right) = 1.735 \hspace{8em} 0.065$$

$$Y_2 = 0.5\left(1.735 + \frac{3.0}{1.735}\right) = 1.7325 \hspace{7em} 0.0025$$

$$Y_3 = 0.5\left(1.7325 + \frac{3.0}{1.7325}\right) = 1.7321 \hspace{6.5em} 0.0004$$

$$Y_4 = 0.5\left(1.7321 + \frac{3.0}{1.7321}\right) = 1.7321 \hspace{6.5em} 0.0000$$

The program and flow chart to calculate the period of the pendulum are given in Fig. 31.

```
       C        PERIOD OF A PENDULUM
 1      1   READ2,EL
 2      2   FORMAT(F10.2)
 3      4   G=32.2
 4      3   PI=3.1416
 5      5   A=10.0
 6      6   X=EL/G
 7      7   Y=0.5*(A+X/A)
10      8   IF(0.0005-ABS(Y-A))9,11,11
11      9   A=Y
12     10   GO TO 7
13     11   T=2.*PI*Y
14     12   PRINT13,T
15     13   FORMAT(1X,21HPERIOD OF PENDULUM IS, F10.2,3HSEC)
16          CALL EXIT
17          END
```

Fig. 31. *(Continued)* Program for calculating the period of a pendulum.

The first statement again is the comment with the program title. Statement 1 is the read-in for L which is in floating-point according to the FORMAT statement 2. This calls for a floating-point number with 2 decimals in a field of up to 10 columns on a single card. Statements 3, 4, and 5 set the values of π, G, and A. Statement 6 is the calculation of L/g; in 7 the trial root is obtained. Statement 8 is the IF statement where a comparison is made to determine whether the error is within the allowed limits. In 9, Y is substituted for A and in 10 the unconditional transfer to 7 is called for; this is the loop for iteration. In 11, T is calculated and in 12 T is printed out according to the format in 13, which calls for the title, floating-point number, and seconds.

Output record.

```
      GO GO

      PERIOD OF PENDULUM IS        1.92SEC
      EXIT CALLED.
```

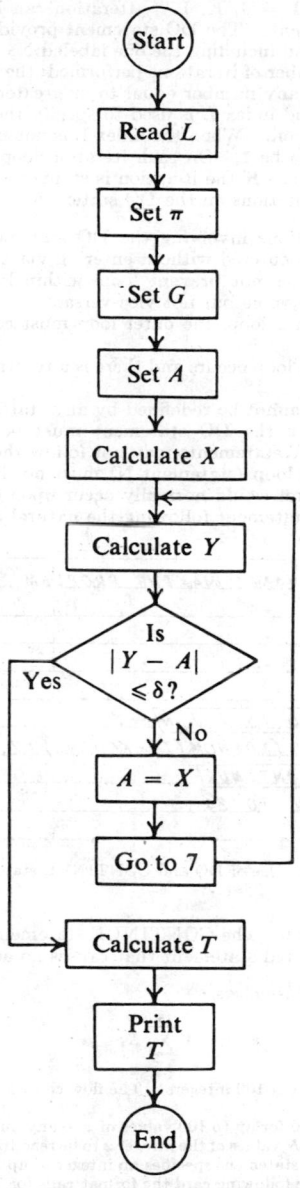

FIG. 31.   Flow chart.

Iteration control: "DO N I = J, K, L." Iteration can be accomplished in another way, by using the DO statement. The DO statement provides for a repetition of all the statements following, up to and including the one labeled N. The indexes I, J, K, and L provide for control of the number of iterations performed: the index J sets the initial value of the index I, which can be any number equal to or greater than 1. Index K sets the total number of iterations; and index L is used to specify the amount that index I is to be incremented on each iteration. When the index L is not specified, the computer automatically assumes its value to be 1. On each iteration loop the value in I is compared with the value in K; if 1 exceeds K the iteration is stopped.

There are a number of restrictions on the DO statement:

1. Iteration or loop operations involving the DO statement must begin from a DO statement; a loop may not be entered without entering via a DO statement unless it is a re-entry. This restriction does not prevent loops within loops; transfer from an inner loop to an outer loop is permissible but not vice-versa.

2. If there is a loop within a loop, the outer loop must contain all statements within th inner loop.

    a. When a transfer out of a loop occurs and there is a return without any index changes, re-entry is permissible.

3. Indices I, J, K, and L cannot be redefined by any statement within the loop.

4. The statement following the DO statement must be an action statement, thus DIMENSION and FORMAT statements may not follow the DO statement directly.

5. The last statement of a loop (statement N) must not involve a transfer of control. Whenever a transfer of control would normally occur upon exit from a loop, the loop is ended with a CONTINUE statement following the natural ending.

F<small>IG</small>. 32. Use of DO and CONTINUE statements.

The CONTINUE statement. The CONTINUE statement is a do-nothing statement. It is used to provide a numbered statement that causes no action.

Example 19. Summation of N Numbers.

$$Y \sum_{i=1}^{N} X_i$$

Let it be required to find the sum of 100 integers. The flow chart and Fortran program are given in Fig. 33.

In Example 19 space is reserved for up to 100 values of x so any number N up to 100 can be added. The READ statement causes the N values of the integer x to be read from cards according to the format in statement 8. The FORMAT statement specifies an integer of up to 3 digits for N on a single card (indicated by the slash). On the following card the format calls for 10 integers of p to 4 digits each. Statement 10, which follows the format declaration, is an initializing statement; it sets the value of the partial sum at zero to start. The DO statement causes a repeated execution of statement 14, N times. Each time the statement is executed, the index I is incremented by 1 (index L is omitted) causing the addition of another value of x to the partial sum. Exit from the iteration loop takes place when the N numbers have been added. A printout of the integer sum ISUM is accomplished according to the statement 18. The 7H activates the circuitry to print out the title "ISUM =" in the first 7 columns of the printout record; the integer sum follows in a 5-digit field.

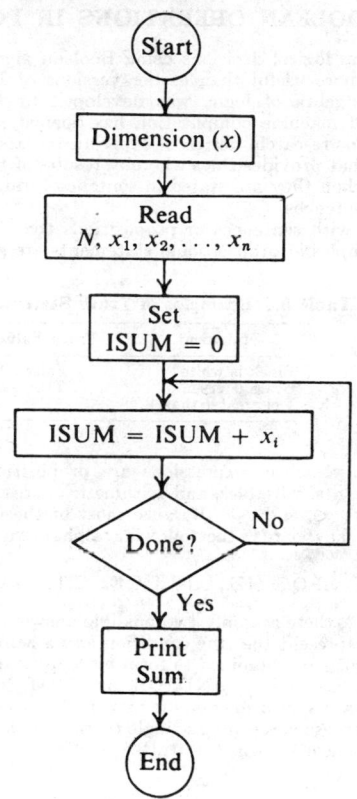

```
C            SUMMATION OF X(I)

    4 DIMENSION IX(100)
    6 READ 8, N, (IX(I), I=1,N)
    8 FORMAT (I3/10I4)
   10 ISUM=0
   12 DO 14 I=1,N
   14 ISUM=ISUM+IX(I)
   18 PRINT 20, ISUM
   20 FORMAT (7H ISUM= I5)
      CALL EXIT
   22 END
```

FIG. 33. Flow chart and program for summation of numbers.

69. BOOLEAN OPERATIONS IN FORTRAN

The ability to perform logical decisions using Boolean algebra makes the Fortran IV compiler considerably more useful than earlier versions of Fortran or other compilers. Boolean algebra, the algebra of logic, was developed by George Boole[1] and Ernest Schroeder.[9] High-speed machine computation has opened the way for ever-increasing applications in operations research, design of experiments, and other areas.

Logic is the science that provides rules whereby results of thinking or reasoning can be tested for correctness when they are stated in sentence form. Logical symbols are used for constructing such sentences.

Symbolic logic deals with sentences or propositions that lead only to a true or false conclusion. Some examples of propositional statements are given in Table 8.

Table 8. Examples of Truth Statements

Statement	Truth Value
Black is white.	False
Grass is green.	True
4 is greater than 8.	False

Boolean expressions. Boolean expressions are propositional statements formed by connecting logical constants, variables, and arithmetic expressions via the basic relational operators $=$, $\neq$, $>$, $\geqslant$, $<$, and $\leqslant$. Because most of these symbols are not found on computer input-output keyboards, the following alphameric representations have been chosen:

.EQ., .NE., .GT., .GE., .LT., .LE.

If we ask, is $X = Y$?, there are only two possible answers: true or false. In the computer we let a one bit represent the true condition and a zero bit represent the false one.

Boolean expressions may be combined to form more involved propositional statements. This requires the Boolean operators AND, OR, and NOT (negation). Operators AND and OR are binary operators used to combine two truth statements, whereas the operator NOT is a unary operator associated with a single truth statement. When the truth values of the basic propositions are known, the truth value of combinations may be defined as follows:

1. .NOT.a is a proposition that has a value of true when the logical expression a is false; it has the value false when a is true.

2. a.AND.b is a combination of two logical expressions a and b that has the value of true if a and b are both true; it has a value of false if either proposition a or b is false.

3. a.OR.b is a combination of the two logical expressions a and b that has the value of true if either a or b is true and has the value of false only if a and b are both false.

The logical operator NOT must be followed by a logical expression; the Boolean operators AND and OR must be preceded and followed by logical expressions.

Examples of the use of the Boolean and relational operators are given in the following expressions:

ALPHA. LE. BETA. OR. GAMA. EQ. DELTA
X. GT. Y + 25.

The first expression is true if ALPHA is less than or equal to BETA, or GAMA is equal to DELTA, or both. The second is true if X is greater than $Y + 25$.

The relational operators cannot be combined with complex or logical variables in any combination. They may be combined with integer, real, and double-precision variables as shown in Table 9.

Table 9. Valid Combinations of Integer, Real, and Double-Precision Variables and Constants with Relational Operators

	Integer	Real	Double-Precision
Integer	Valid	Invalid	Invalid
Real	Invalid	Valid	Valid
Double-Precision	Invalid	Valid	Valid

Rules of Precedence

Rank of operators. The sequence in which the terms in a Fortran expression are evaluated is established by ranking the various operators and by rules of precedence. Each operator must be assigned an order of precedence. Consider the expression:

$$Y = A + B^x C/D + E/F - G$$

Each term must be available before we can proceed with the summation; therefore exponentiation, multiplication, and division must have higher rank than addition and subtraction. It makes no difference in the answer in what order we perform additions and subtractions; these operators therefore must have equal rank. Examination of the expression for Y further reveals that multiplication and division must also have equal rank, because we may multiply B^x by C then divide by D, or vice-versa.

The unary negative sign is the only arithmetic operator left to consider. For example, there is a difference between $(-A)^n$ and $-(A^n)$. Obviously then we must perform the negation before we raise a number to a power.

The Boolean and relational operators must also be ranked. The first question is, which ranks highest: arithmetic, Boolean, or relational operators? To answer this consider the expression

$$X + Y.LE.Z.AND.W.GT.A$$

We observe immediately that it is necessary to evaluate the arithmetic expression $X + Y$ before we can proceed to ask the logical question, is $X + Y$ less than or equal to Z? Therefore the arithmetic operators rank higher than the Boolean or relational operators. Now we must decide which ranks higher, the Boolean or the relational operators. Before deciding whether the AND is true or false, we need the results of the relational expressions on either side. Thus the relational operators must be ranked higher than the Boolean ones. The final ranking of operators and relations is given in Table 10. When two operators have equal rank the leftmost one is done first.

Table 10. Ranking of Operator from High to Low

— (unary)
** function ref.
*, /
+, — (binary)
.EQ., .NE., .GT., .GE., .LT., .LE.
.AND.
.OR.
.NOT.

70. SUBROUTINES

Thousands of people write computer programs every day. Inevitably many of these programs utilize the same basic functions: absolute values, square roots, polynomial fits, and so forth. The subroutine has been developed to reduce this duplication of programming effort.

There are several classes of subroutines in Fortran IV: the arithmetic statement functions, built-in or predefined functions, function subprograms, and subroutine subprograms.

The arithmetic statement function, predefined function, and function subprograms are always single-valued and return a single result. They may be written directly into the program and referenced by an arithmetic expression that contains the function name.

The subroutine subprogram, on the other hand, cannot be written into the program; it must be referenced by a CALL statement. Moreover, it may return multiple results.

Subroutines are named the same way as Fortran variable identifiers; that is, 1 to 6 alphanumeric characters, the first of which must be alphabetic.

The arithmetic function is specified by the name of the function or by a type statement. The type of function subprogram is also determined by the function name if it is an integer or a real function, or by preceding the function statement by a type name. The type of a predefined function is automatically specified in the Fortran processor.

The arithmetic function. The functions, which are defined by a single arithmetic statement written by the programmer, apply only to the source program that defines them. The general form is:

Function Name $(A, B, ...)$ = Arithmetic Expression where $A, B, ...$ are arguments of the function. Neither the function argument nor the arithmetic expression may contain subscripted variables.   In Fig. 34 ALPHA is the function name; pressure and volume are the arguments.

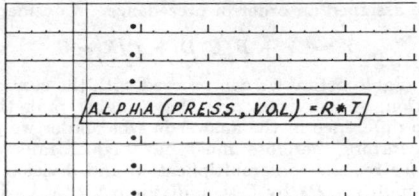

FIG. 34.   Arithmetic function statement.

The predefined function.   Over thirty predefined functions are available in Fortran IV.   These functions may be evaluated at any point in the program simply by calling for them by the appropriate special identifier.   For example, to call for the absolute value of a real variable X, simple include the expression ABS(X) at the appropriate place in the program.   When the computer comes to this instruction, it will exit from the main program, proceed to the subroutine called ABS and evaluate the absolute value of X.   The computer then returns to the main program and continues.   A number of the predefined functions available in Fortran IV are included in Table 11 and illustrated in Fig. 35.

Table 11.   Some Predefined Functions in the Fortran IV Processor

Operation	Number Argu- ments	Type Argument	Type Function	Name
Absolute value	1	Integer	Integer	IABS(X)
		Real	Real	ABS(X)
		Double-precision	Double-precision	DABS(X)
Choose largest value	2	Integer	Integer	MAXO(X)
MAX (ARGI,		Integer	Real	AMAXO(X)
ARG2)		Real	Integer	MAXI(X)
		Real	Real	AMAXI(X)
Choose minimum value:		Replace MAX by MIN in 2		
MIN (ARG 1, ARG 2)				
Change integer variable to real	1	Integer	Real	FLOAT(X)
Change real variable to integer	1	Real	Integer	FIX(X)
Obtain real part of complex argument	1	Complex	Real	REAL(X)
Obtain imaginary part of complex argument	1	Complex	Real	AIMAG(X)
Express two real arguments in complex form	2	Real	Complex	CMPLX(X)
Obtain conjugate of complex argument	1	Complex	Complex	CONJG(X)

```
C       PERIOD OF A PENDULUM
        •
        •
        •
        Y=0.5*(A+X/A)
        IF(0.005-ABS(Y-A) 9,11,11
        A=Y
        •
        •
        •
        END
```

FIG. 35.   Program illustrating the use of predefined functions.

Function subprograms. Various commonly used mathematical subroutines are provided in Fortran IV. These are called the function subprograms. Although they are constructed differently from the predefined functions, they are used in the same way by the programmer; a function subprogram may be called by using its name as the operand in an arithmetic statement.

The function subprogram is a special type of subroutine defined by the statement

$$\text{FUNCTION (A, B, ..., Z)}$$

followed by a set of statements to perform X; the name X must appear at least once on the left-hand side of an arithmetic statement or in an input statement to provide a linkage with the main program.

The termination of a function subprogram is via the RETURN statement, which returns the computation to the next statement in the main program. A subprogram deck follows the main program deck but precedes the END statement.

The function name must be single-valued and unsubscripted; it cannot appear in a DIMENSION statement in the subprogram or in any other program that uses the function. Other function statements and subroutine subprograms may not be included as statements in a function subprogram.

The arguments of a function name may be considered dummy variables to be replaced at the time of execution. The actual arguments, which must have the same number and type of variables as the dummy arguments, may be constants, subscripted or nonsubscripted variables, arithmetic or logical expressions, or the name of a function or subroutine subprogram. There are about thirty-five function subprograms available in Fortran IV. Some of the more useful ones are included in Table 12 and illustrated in Figs. 36 and 37. For evaluation of the trigonometric functions in this table, the angle X must be in radians.

Table 12. Some Function Subprograms Available in Fortran IV

Fortran Identifier

Function	Real Argument and Function	Double-Precision Argument and Function	Complex Argument and Function
Log$_e$ of X	ALOG(X)	DLOG(X)	CLOG(X)
Log$_{10}$ of X	ALOG10(X)	DLOG10(X)	CLOG10(X)
Sine of X	SIN(X)	DSIN(X)	CSIN(X)
Cosine of X	COS(X)	DCOS(X)	CCOS(X)
Arctangent of X	ATAN(X)	DATAN(X)	—
Square root of X	SQRT(X)	DSQRT(X)	CSQRT(X)

```
C       PROGRAM TO CALCULATE THE SQUARE ROOT
        READ 2, X
      2 FORMAT (F10.6)
      3 SQRTX = SQRT(X)
        PRINT 4, SQRTX
      4 FORMAT (2HX=F10.6, 6HSQRTX=F10.6)
```

FIG. 36. Use of the function subprogram for the square root of a number.

Example 20. The Log Mean Temperature Difference for a Double-Pipe Heat Exchanger. The log mean temperature difference (Δt_m) is the driving force for heat transfer between two fluid streams in a countercurrent-flow heat exchanger and is defined by:

$$\Delta t_m = \frac{(T_1 - T_4) - (T_2 - T_3)}{2.303 \log (T_1 - T_4)/(T_2 - T_3)}$$

Write the program and obtain a solution for the case where Stream I is condensing steam at atmospheric pressure and

$$T_1 = 212°F; \quad T_2 = 212°F$$
$$T_3 = 50°F; \quad T_4 = 75°F$$

```
        FUNCTION SIMP(I)
        COMMON N,A,B,C
        DIMENSION Y(20),X(20)
  10    FORMAT(F10.4)
             .
             .
             .
        SIMP=(YZERO + Y(N) + 4.*YODD + 2.*YEVEN)*H/3.
        PRINT 15,A,B,YODD,YEVEN,SIMP
  15    FORMAT(5F10.4)
        RETURN
```

FIG. 37.   Format of a function subprogram.

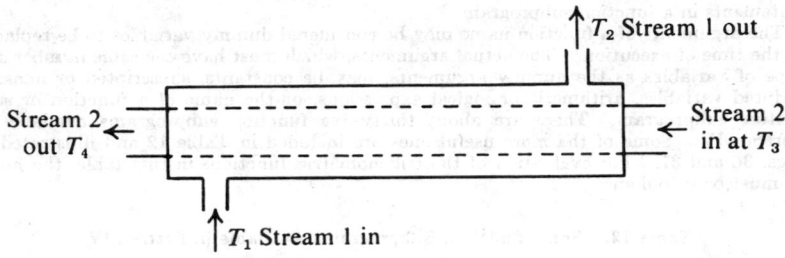

FIG. 38.   Schematic plan of heat exchanger.

Solution: The program is written using the predefined function for the logarithm as shown in Fig. 39.

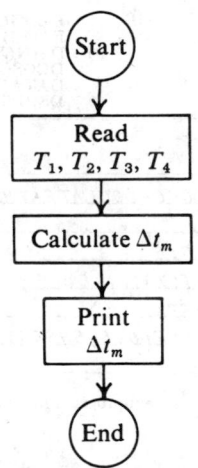

```
  C      LOG MEAN TEMPERATURE DIFFERENCE
  1    1 READ2,T1,T2,T3,T4
  2    2 FORMAT(4F10.2)
  3    3 DELTAM=(((T1-T4)-(T2-T3))/(2.303*ALOG10((T1-T4)/(T2-T3))))
  4    4 PRINT5,DELTAM
  5    5 FORMAT(1X,14HDELTA T MEAN =, F10.2,5HDEG F)
  6      CALL EXIT
  7      END
```

FIG. 39.   Flow chart and Fortran program for log mean temperature differences.

The first statement in the program is the COMMENT statement giving the title of the program. In the next statement, labeled statement 1, the temperatures of the two streams are read in from a single card according to the FORMAT statement, which is labeled 2. Statement 3 is the statement where the calculation is performed using the log function and the substitution statement. A printout of Δt_m is called for in statement 4, with format specified in 5. The format is the title obtained by the 14-column Hollerith field; the floating-point number with 2 decimals in the 10-column field; and the label DEG F obtained with the 5-column Hollerith field. The last statement ends the program.

Output record:

```
OBJECT PROGRAM IS BEING ENTERED INTO STORAGE.
DELTA T MEAN =      53.60DEG F
```

Example 21. Despinning a Satellite. Missiles and projectiles are given an angular spin to stabilize the motion in flight. The Pioneer III lunar space probe was given an initial spin rate of 400 rpm. In order to assure proper operation of the instruments, it was necessary to reduce the spin rate to 5.5 rpm after the probe was in orbit.

A simple technique for reducing the spin rate is based on the exchange of momentum with a small mass m. The mass is attached to a cord wound about the spinning body as shown in Fig. 40. When

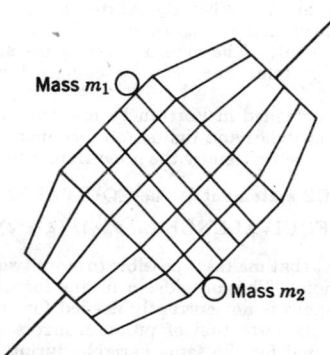

Fig. 40. Arrangement for despinning a satellite.

the mass m is released the cord unwinds and the angular speed of the satellite is reduced. When the cord is completely unwound, it is released and the cord and mass fly off in space.

An analysis of the momentum exchange between the two bodies yields the following expression for the length of cord required to reduce an initial spin rate w_0 to some arbitrary lower value w_f:

$$l_f = R \sqrt{C \frac{w_0 - w_f}{w_0 + w_f}}$$

where R = Diameter of the satellite
  I = Moment of inertia
  C = $I/mR^2 + 1$
  l_f = Length of cord

By using two cords, each with half the mass, the symmetry of the system is maintained.

Data for pioneer space probe:

$$gI = 92 \text{ lb in.}^2$$
$$R = 5 \text{ in.}$$
$$m = 0.2 \text{ oz}$$
$$w_0 = 400 \text{ rpm}$$
$$w_f = 5.5 \text{ rpm}$$

Subroutine subprograms. There are many functions that are of general utility but are not used so frequently as the ones given in Table 12. To provide for these in the basic

```
       C       DESPINNING A SATELLITE
               REAL LF,I


               READ4, R,W,WO,WF,I
       4 FORMAT(5F10.4)


               C=I/((W/16.)*R*R)+1.
               LF=R*SQRT(C*(WO-WF)/(WO+WF))


               PRINT8, WO,WF,LF
       8 FORMAT(1X,4HWO=  ,F10.4/1X,4HWF=  ,F10.4/


          11X,4HLF= F10.4)
               CALL EXIT
               END
```

Fig. 41. Fortran program for despinning a satellite.

Fortran compiler would use too much memory capacity. To accommodate these functions, the library subroutine package where the program is stored on cards or on magnetic tape has been developed. As shown in Fig. 42, the program is called in at the time of execution by the CALL statement. When the subroutine is stored on cards, it is necessary to include the subroutine deck in the program deck; if it is on magnetic tape, all that is required is the CALL statement. The main feature of the subprogram is that it makes available a self-contained program that can be compiled independently of the main program.

Subroutine subprograms provided in Fortran IV are primarily for testing the status of the machine, that is, sense switches and lights, job termination, and core dumps. Other subroutine subprograms may be written by the programmer to accommodate special needs.

EQUIVALENCE statement. The EQUIVALENCE statement

$$\text{EQUIVALENCE } (\alpha, \beta, \gamma), (\delta, \epsilon, \psi)$$

is a nonexecutable statement that makes it possible to store two or more variable identifiers in the same memory location. This provides a means for utilizing memory space by a second variable when the space is not currently needed for the variable it was originally reserved for. It also permits correction of program errors such as when two different variable names have been used for the same variable during the coding step.

```
C    REGRESSION ANALYSIS
     DIMENSIONMATRIX(500),MATROX(500)
        .
        .
        .
     CALLALPHA(MATRIX(5),MATROX(I))
        .
        .
     END
```

Fig. 42. Calling in a subroutine program.

FIG. 43. Use of the EQUIVALENCE statement.

Each pair of parentheses in the statement contains the names of two or more variables that are to be stored in the same location. α, β, γ, . . . must be variable names; they cannot be constants.

In Fig. 43 the variables $T(5)$. ALPHA, ZETA, and DAD are all stored in the same location; variables T(25) and BAKER are stored in a second location.

The EQUIVALENCE statement may be placed anywhere in the main program.

COMMON statement. The COMMON statement

$$\text{COMMON } \alpha, \beta, \gamma$$

is similar to the EQUIVALENCE statement, but it permits data-storage sharing between programs whereas the EQUIVALENCE statement permits sharing only within a program. For example, subroutine variable identifiers are independent of the main program and therefore the same name appearing in the main program and the subroutine is not necessarily taken to be the same; two different locations will be assigned. The order of storage in COMMON, however, depends only on the order in which the identifiers appear in the COMMON statement in the particular subprogram or in the main program. Thus a primary function of the COMMON statement is to communicate values from one program to another, under the same or a different name.

If a substantial amount of programming is to be done, it is desirable to refer to the FORTRAN manual appropriate for the computer installation that is to be used.

BIBLIOGRAPHY

1. BOOLE, G., *An Investigation of the Laws of Thought*, New York, Dover, 1951.
2. GRABBE, E. M. (Ed.), *Automation in Business and Industry*, New York, Wiley, 1957.
3. LEEDS, H. D., and WEINBERG, G. M., *Computer Programming Fundamentals*, New York, McGraw-Hill, 1961.
4. McCORMICK, E. M., *Digital Computer Primer*, New York, McGraw-Hill, 1949.
5. McCRACKEN, D. D., *A Guide to Fortran Programming*, New York, Wiley, 1961.
6. McCRACKEN, D. D., *A Guide to Cobol Programming*, New York, Wiley, 1963.
7. MURPHY, J. S., *Basis of Digital Computers*, Vols. I, II, and III, New York, John F. Rider, 1958.
8. ORGANIK, E. I., *A Fortran Primer*, Reading, Mass., Addison-Wesley, 1963.
9. SCHROEDER, E., *Algebra der Logik*, Vol. 1, Leipzig, Verlagsgesellschaft, 1890.

SECTION 3

PHYSICAL UNITS AND STANDARDS

BY

ERNST WEBER

PHYSICAL UNITS AND STANDARDS

DIMENSION SYSTEMS

1. PHYSICAL QUANTITIES AND THEIR RELATIONS

Mathematics is concerned with relations between numerical quantities, either constant or varying in a specified manner over a specified range of values. The numerical values are unique, absolute, and the same all over the world, being the expression of a fundamental perception of the mind. Any *mathematical equation* defines the values of one numerical quantity, known as the dependent, in terms of constants and one or more other numerical quantities, known as the independent variables, as for example

$$z = y^2 + 3x + 4, \quad y = c \cdot \int_0^x \frac{x^2}{\cos x}\, dx \tag{1}$$

Physics, comprising the knowledge of inanimate nature and her laws, is concerned fundamentally with the measuring of the various quantities founded or created by definition, as for example, *length, mass, electric charge.* In order to specify a *physical quantity* it is not sufficient to state merely a number. The value of a physical quantity can be determined only by comparison of the sample with a known amount of the same quantity, which process is *measuring.* The reference amount is called a unit, and the result of any measurement must be a statement of " how many times the sample was found to contain the reference amount." Thus a physical quantity Q naturally appears to be the product of a numerical value N and a unit U,

$$Q = N \cdot U \tag{2}$$

as for example: The length of a particular rod is 3.5 ft, or the rod is 3 1/2 times the length of 1 ft. Obviously, the reproduction of a unit must be possible at any time in order to facilitate correct measurements. This is being done by means of the " standards," which are simply a set of fundamental unit quantities kept under normalized conditions in order to preserve their values as accurately as facilities permit.

Any physical relation must be the result of a more or less obvious measurement, so that equations in physics are not merely numerical relations, but express dependences between physical quantities. Mathematics does not know "standards"; physics cannot be without "standards." The fact that physics often uses the methods of mathematics must not lead to the identification of the two sciences; it is merely an overlapping in the border regions.

Relations between Units. A unit is a particular amount of the physical quantity to be measured, defined in terms of a standard.* The choice of a unit depends on convenience, facility of reproduction, and easy subdivision so as to obtain smaller units if desired. The value of a physical quantity Q must be independent of the units used, so that for two different units of the same type

$$Q = N_1 \cdot U_1 = N_2 \cdot U_2 \tag{3}$$

The size of the unit and the numerical value of the quantity are inversely related: the larger the unit the smaller the number of units.

A unit relation is an equation between two different units of the same type

$$U_1 = N_{12} \cdot U_2 \tag{4}$$

and serves to convert from one unit U_1 to a different one U_2. The conversion is achieved by replacing U_1, taken as a factor, by its equivalent according to equation 4 so that

$$Q = N_1 \cdot U_1 = N_1 \cdot (N_{12} \cdot U_2) = (N_1 \cdot N_{12}) \cdot U_2 \tag{5}$$

As example, express the length 3.5 ft in centimeters. The unit relation is 1 ft = 30.5 cm,

---

* S. W. Stratton, *Electric Units and Standards,* Circular No. 60 of the Bureau of Standards, 1920.

and therefore $l = 3.5$ ft $= 3.5 \times (30.5$ cm$) = 106.75$ cm. No error is possible if this rule is followed properly.

Physical Equations. Relations between physical quantities are usually given in the form of equations. It is always possible, by the proper use of unit relations (see previous paragraph), to express each side in the same units. Since units are to be considered as factors, they may be canceled and a numerical identity must result. This fact always can be used to check the proper numerical relations and the consistency of the units used.

There are two fundamental types of physical equations:

The Mathematical Definition of a physical quantity determines a new quantity uniquely in terms of known quantities. An example is Newton's definition of mass by $f = m \cdot a$, where f is the force and a the acceleration of a moving body. If f and a are measured, m can be computed as a physical quantity with numerical value $\dfrac{N(f)}{N(a)}$ and unit $\dfrac{U(f)}{U(a)} = U(m)$. A definition should be in agreement with all the other known relations in a particular field of science; it can only be of restricted value if it contradicts other relations (see later the "absolute" electric systems).

The Statement of Proportionality defines one physical quantity as linearly depending on a combination of other, known quantities. It is always the result of an experimental investigation. An example is Newton's law of the gravitational force $f = k \dfrac{m_1 \, m_2}{r^2}$ where m_1 and m_2 are the two masses, r their center distance, and k the proportionality factor. In the case of a proportionality it is permissible to choose arbitrary units for all measurable physical quantities involved and to use the equation as a definition of the proportionality constant which, in general, will be a physical constant with numerical value and unit. In the example the value of k would be

$$\frac{N(f) \cdot N(r^2)}{N(m_1) \cdot N(m_2)} \times \frac{U(f) \cdot U(r^2)}{U(m_1) \cdot U(m_2)} = N(k) \cdot U(k).$$

Most of the fundamental laws of physics are statements of proportionalities, leading to universal physical constants, as for instance the gravitational constant k, the Planck constant h, the gas constant R, the absolute permeability of free space μ_v and the absolute dielectric constant of free space ε_v. It may be observed that each branch of physics is represented by at least one fundamental proportionality constant.*

Derived Physical Quantities are, in general, the result of mathematical definitions. The units of derived quantities are expressed from the combinations of the units used in the definition. All proportionality constants are ordinarily considered as derived physical quantities.

Fundamental Physical Quantities. The physical quantities, arbitrarily chosen to define new quantities or derived quantities, are called fundamental physical quantities. Their number may vary according to needs and convenience. There is no possibility to designate any physical quantity as absolutely fundamental, or a priori fundamental. Quantities which appear to be fundamental in some one special field may be derived quantities in some other field.

2. DIMENSIONS AND DIMENSION SYSTEMS

Definition of Dimension. To choose a unit for a physical quantity one has an infinity of possibilities. The numerous units of length which were in use about 100 years ago present a good practical illustration. Yet all these units have in common the quality of being a distinct length and not, for example, a volume. It is convenient to state this fact by representing with the notation $[L]$ any unit of length whatsoever. The measurement of a physical quantity Q, therefore, leads to the statement

$$Q = N \cdot [Q] \tag{6}$$

where N is a numeric denoting the number of general units $[Q]$ which constitute the total quantity Q. According to Fourier,† who first introduced this concept into the literature, $[Q]$ is called the "dimension" of the quantity Q. Be it clearly understood that dimension is simply the expression of a general unit and therefore a characteristic peculiarity of physical quantities, not occurring in mathematics. Each new physical quantity gives

---

* In the second report of the British Association for the Advancement of Science, Committee on Standards (1873), J. Clerk Maxwell made a statement in connection with the form of Coulomb's law for magnetic poles, p. 63, about the "absurdity" of a proportionality factor, which aroused very pointed remarks by O. Heaviside, in his *Electromagnetic Theory*, Vol. 1, p. 118, 1893.

† J. B. J. Fourier, *Théorie analytique de la chaleur*, Paris, 1822.

rise to a new "dimension" as for instance time $[T]$, force $[F]$, mass $[M]$, and so on.   There are as many dimensions, or general units, as there are kinds of physical quantities.

Derived Dimensions.   Many physical quantities have been introduced by mathematical definition.   Velocity, for example, is defined as $v = \dfrac{ds}{dt}$, where s is the length of the path measured from a definite origin and t is the time.   A possible expression for the dimension of velocity would be $[V]$.   It is customary and convenient, however, to make use of the mathematical definition which is but the rule for the measurement of velocity, and to express the dimension in terms of the more familiar dimensions of length and time as a derived dimension $[V] = [L]/[T] = [L] [T]^{-1}$.   (Read: velocity is of $(+ 1)$ dimension in length and $(- 1)$ dimension in time.)   The use of mathematical definitions, leading to derived dimensions of a composite nature, reduces the number of symbols.   Thus the measurement of volume, if scientifically conducted, gives $[Vol] = [L]^3$, or in words, "volume is of $(+ 3)$ dimensions in length $[L]$."

Proportionality Constants of physics have, in general, *derived dimensions*, as they are defined by the corresponding physical equations.   To assign a special dimension to a proportionality constant would mean to give it preference over a physical quantity and evidently would be illogical.   This seemingly obvious fact has recently been recognized internationally, when the International Electrotechnical Commission (I.E.C.), after decades of discussion, at its meeting in 1930 at Oslo adopted the viewpoint that absolute magnetic permeability is a physical quantity with dimension and not a pure numeric.*   For the absolute dielectric constant this view had been expressed much earlier, and for other proportionality constants this problem never had arisen.

Fundamental Dimensions.   The more familiar dimensions used to express derived dimensions are referred to as fundamental dimensions.   It is advantageous to use as few of these fundamental dimensions as possible, not because the physical relations become simpler or clearer, but merely as a matter of economy in symbols.   In fact, any dimension can be chosen to be a fundamental dimension in a particular field and a derived dimension in some other field of physics.   No fundamental dimension can be made a starting point of natural philosophy.

Dimensional Equations.   Since a physical equation constitutes in fact two equations, one for the units and one for the numerics, one can disregard the numerical factors entirely and write the general units or dimensions only, arriving thus at a dimensional equation.   For instance, the law of gravitation would read $[F] = [k] [M]^2 [L]^{-2}$, using $[F]$, $[k]$, $[M]$, $[L]$ as dimensions for force, gravitation constant, mass, and length, respectively.   From this dimensional equation a derived dimension can be obtained for any quantity involved.   Conversely, dimensional equations are used to check the correctness of physical relations, if all dimensions can be made to cancel.   Finally, the validity of dimensional equations leads to the method of dimensional analysis.   (See Arts. 15 to 18.)

A **Set of Fundamental Dimensions** is any group of fundamental dimensions, convenient and useful to express all the physical quantities of a particular field in terms of derived dimensions.   The number of fundamental dimensions to make a set may vary according to the field of application.   Whether or not a set of fundamental dimensions can be used beyond the field for which it was originally intended will depend upon its suitability as a dimension system.   (See next paragraph.)   In no case should it be used where it can lead to confusion.

A set of fundamental dimensions is *incomplete* when the number of fundamental dimensions composing it is less than the number required for a dimension system.   Incomplete sets of fundamental dimensions should not be used outside the very restricted field for which they are defined; they necessarily would lead to confusing relations.

A **Dimension System** is composed of the smallest number of fundamental dimensions which will form a consistent and complete set for a field of science.   Since each relation between physical quantities can be split up into one relation of numerics and another one of dimensions (as general units), it is possible to combine all known relations of dimensions.   In setting up these relations, all proportionality factors must be taken as physical quantities.   If there are m independent relations known, $(m + p)$ dimensions may be involved, of which m dimensions can be expressed by any p "fundamental" dimensions chosen arbitrarily.

This set of p "fundamental" dimensions is then called a dimension system.   From the theory of numbers, therefore, it is known that one generally has a choice of $\dbinom{m + p}{p}$ possible dimension systems.   Thus, if $p = 3$, $m = 3$, then one has $\dbinom{6}{3} = 20$ different

---

possibilities. A necessary condition, however, is that *each* independent relation involve at least $(p + 1)$ dimensions. If this is not the case, then the number of possible dimension systems is less, so that $\left(\dfrac{m + p}{p}\right)$ indicates the *upper* limit.

Any dimension system chosen in the described manner is consistent, as well as correct, and never leads to ambiguity with respect to the expression of physical quantities. Complete dimension systems in mechanics must have three, in thermodynamics four,* and in electromagnetism four † fundamental dimensions. It seems, according to present knowledge, that five fundamental dimensions suffice for the entire range of physics, namely, the three fundamental dimensions of mechanics, an additional one for thermodynamics, and another additional one for electromagnetism.

All the known dimension systems use length $[L]$ and time $[T]$ as primary fundamental dimensions, adding various fundamental dimensions from the available physical quantities of the fields of physics. The choice of $[L]$ and $[T]$ reduces at once the maximum number of possible dimension systems to $\left(\dfrac{m + p - 2}{p - 2}\right)$.

Why Dimension Systems? Since the proper choice of units is the ultimate goal of any critical analysis of physical quantities the question may be asked: Why is it necessary to discuss dimension systems? The answer is that each physical quantity may be measured by an infinite variety of units but has only one dimension, within a given dimension system. The process of deciding upon the fundamental dimensions before fixing the units within the scope of the fundamental dimensions is, therefore, essentially a matter of economy and logic.

3. THE DIMENSION SYSTEMS OF MECHANICS

Three Fundamental Dimensions are necessary to form a complete mechanical dimension system. With length and time, $[L]$ and $[T]$, as a basis, only a single additional independent dimension is required. One can choose from mass $[M]$, force $[F]$, power $[P]$, energy $[E]$, gravitation constant $[k]$, and others. Only four alternatives have come into use. Table 1 shows the dimensional forms for the more important mechanical quantities in all four dimension systems.

Table 1. Dimensions of Mechanical Quantities

Symbol (according to A.S.A.)	Quantity	Dynamical or Physical System	Gravitational or Technical System	Energetical System	Astrophysical System
l	Length	$[L]$	$[L]$	$[L]$	$[L]$
t	Time	$[T]$	$[T]$	$[T]$	$[T]$
v	Speed, velocity	$[L][T]^{-1}$	$[L][T]^{-1}$	$[L][T]^{-1}$	$[L][T]^{-1}$
m	Mass	$[M]$	$[F][L]^{-1}[T]^2$	$[E][L]^{-2}[T]^2$	$[L]^3[T]^{-2}[k]^{-1}$
f	Force	$[M][L][T]^{-2}$	$[F]$	$[E][L]^{-1}$	$[L]^4[T]^{-4}[k]^{-1}$
p	Pressure	$[M][L]^{-1}[T]^{-2}$	$[F][L]^{-2}$	$[E][L]^{-3}$	$[L]^2[T]^{-4}[k]^{-1}$
M	Momentum	$[M][L][T]^{-1}$	$[F][T]$	$[E][L]^{-1}[T]$	$[L]^4[T]^{-3}[k]^{-1}$
E	Energy	$[M][L]^2[T]^{-2}$	$[F][L]$	$[E]$	$[L]^5[T]^{-4}[k]^{-1}$
P	Power	$[M][L]^2[T]^{-3}$	$[F][L][T]^{-1}$	$[E][T]^{-1}$	$[L]^5[T]^{-5}[k]^{-1}$
T	Torque	$[M][L]^2[T]^{-2}$	$[F][L]$	$[E]$	$[L]^5[T]^{-4}[k]^{-1}$
k	Gravitation constant	$[M]^{-1}[L]^3[T]^{-2}$	$[F]^{-1}[L]^4[T]^{-4}$	$[E]^{-1}[L]^5[T]^{-4}$	$[k]$

The **Dynamical or Physical Dimension System** has as fundamental dimensions length, $[L]$, time $[T]$, and mass $[M]$. It is the most widely used system in physics and is often found even in engineering. The advantage is that any standard of mass can be preserved and copied with relative ease. Comparison of masses at various locations can be made with the ordinary balance; the local constant of gravity has no influence upon the result. This dimension system has often been referred to as the "absolute" system, but on account of inconsistent usage in the literature, it will generally be referred to herein as the physical dimension system.

The **Gravitational or Technical Dimension System** has as fundamental dimensions length $[L]$, time $[T]$, and force $[F]$. It is of wide use in all engineering branches, although

---

* For first reference to this fact, see Fourier's *Théorie analytique de la chaleur*, Paris, 1822.

† For first reference to this fact, see A. W. Rücker, *Proc. Phys. Soc. London*, 1888, Vol. 10, p. 37; *Phil. Mag.*, February, 1889, Vol. 27, p. 104.

often mixed with the physical dimension system. The advantage for the field of engineering is obvious, taking into account the prevalence of force, stress, and pressure computations.* All quantities related to force have particularly simple dimensional forms. The disadvantage is the dependence of the usual force measurements upon the local value of the constant of gravity.† Standards of force are expressed in terms of the weight of a standard mass and, therefore, are indirectly based upon the physical dimension system.

The Energetical Dimension System ‡ has as fundamental dimensions length $[L]$, time $[T]$, and energy $[E]$. The dimensional forms of most of the important quantities are simpler, and this system should appeal to the scientist because of its close relation to the universal quantity energy. The system is not in use in mechanics; it has, however, gained prominence in heat. (See Art. 4.) Its disadvantage is the fact that no substantial standard of energy can be preserved with which results of measurements can readily be compared, as is the case with mass or force.

The Astrophysical Dimension System has as fundamental dimensions length $[L]$, time $[T]$, and the gravitational constant $[k]$. Here is a system which admits a proportionality constant to the role of a fundamental quantity with a fundamental dimension which is not advisable. The temptation exists to assign arbitrarily not a fundamental but a definite dimension and, in fact, astrophysics uses $k = 1$, which means that the numerical value of k is chosen as unity and its dimensions as zero. The dimensions of the other quantities in this special case are obtained from the last column in Table 1 if the factor with $[k]$ is left off on account of $[k] = [L]^\circ[T]^\circ = 1$. This reduces the dimension system to an *incomplete set of dimensions* with very restricted use.

4. THE DIMENSION SYSTEMS OF HEAT

Four Fundamental Dimensions are necessary to form a complete heat dimension system.§ Obviously the simplest extension to four dimensions is to add a thermal fundamental dimension to the three fundamental dimensions of mechanics. As the additional thermal dimension that of temperature $[\theta]$ has been chosen universally. Thus, three heat dimension systems are obtained according to the three first-named mechanical dimension systems from Table 1. No extension of the astrophysical dimension system is known. Table 2 lists the dimensional forms for the more important thermal quantities in all three dimension systems.

Table 2. Dimensions of Thermal Quantities

Symbol (according to ASA)	Quantity	Thermophysical System	Thermotechnical System	Energetical System
	3 fundamental mechanical dimensions	$[L], [T], [M]$	$[L], [T], [F]$	$[L], [T], [E]$
θ	Temperature	$[\theta]$	$[\theta]$	$[\theta]$
H	Quantity of heat	$[M][L]^2[T]^{-2}$	$[F][L]$	$[E]$
c	Thermal capacity	$[L]^2[T]^{-2}[\theta]^{-1}$	$[L]^2[T]^{-2}[\theta]^{-1}$	$[L]^2[T]^{-2}[\theta]^{-1}$
k	Thermal conductivity	$[M][L][T]^{-3}[\theta]^{-1}$	$[F][T]^{-1}[\theta]^{-1}$	$[E][L]^{-1}[T]^{-1}[\theta]^{-1}$
U	Transmittance	$[M][T]^{-3}[\theta]^{-1}$	$[F][L]^{-1}[T]^{-1}[\theta]^{-1}$	$[E][L]^{-2}[T]^{-1}[\theta]^{-1}$
s	Entropy	$[M][L]^2[T]^{-2}[\theta]^{-1}$	$[F][L][\theta]^{-1}$	$[E][\theta]^{-1}$

The Thermophysical Dimension System is the extension of the dynamical or physical mechanical dimension system by adding temperature $[\theta]$ as the fourth fundamental dimension. It is the most widely used system and has all the advantages of the physical dimension system of mechanics. This system is sometimes reduced to an incomplete set of "absolute dimensions" by arbitrarily assuming temperature as a pure numeric. The dimensions in this case can be taken from Table 2 if $[\theta] = 1$ is introduced in the thermophysical system. The objections to such a procedure are obviously the same as in reducing the astrophysical system to less than the required number of fundamental dimensions, and the same argumentation can be used.

The Thermotechnical Dimension System is the extension of the gravitational or technical mechanical dimension system by adding temperature as the fourth fundamental

---

* F. A. Brooks, *J. Eng. Educ.*, Vol. 25, p. 240, 1934.

† Wm. L. DeBaufre, *Proc. Soc. Promotion Eng. Educ.*, Vol. 28, p. 655, 1928.

‡ W. Ostwald, *Berichte der Gesellschaft der Wissenschaften*, Leipzig, 1891, Vol. 43, p. 277.

§ J. B. J. Fourier. *Théorie analytique de la chaleur*, Paris. 1822.

dimension. Its use is rather restricted although it is the natural system to be used in engineering.

The Energetical Dimension System is the extension of the corresponding mechanical dimension system and has been proposed by W. Ostwald. It has not been used to any extent.

5. THE DIMENSION SYSTEMS OF ELECTROMAGNETISM

Four Fundamental Dimensions are necessary to form a complete dimension system for the field of electromagnetism exclusive of thermal phenomena.[*] Obviously the simplest extension to four dimensions is to add an electromagnetic fundamental dimension to the three fundamental dimensions of mechanics. In general, one can choose from among all electromagnetic quantities; in particular, one would probably prefer to choose the dimension of a quantity which comes nearest to the fundamental concept of electricity. Length and time, forming the fundamental background of sensual perception, and mass or force as the fundamental representative of mechanical inertia, demand a companion of equal basic character.

There are three fundamental electromagnetic experiments which involve all the basic quantities of electromagnetism. The two independent experimental laws forming the base of electrostatics on one side and electrodynamics on the other side are Coulomb's law

$$F_e = k_e \frac{Q_1 Q_2}{r^2} \tag{1}$$

and Ampère's law

$$F_m = k_m \frac{I_1 I_2}{r} \cdot l \tag{2}$$

expressing the force actions between two electric charges Q_1 and Q_2 at rest, and between two parallel electric currents I_1 and I_2, respectively. The charges are assumed to reside on two small spheres of comparatively large center distance r; the currents are assumed to flow in thin wires of comparatively large center distance r and of parallel length l.[†] The constants k_e and k_m are the two proportionality factors. As current is merely displacement of charge, the additional relation $I = \dfrac{dQ}{dt}$ must hold, which is the expression of a basic concept.

The third experimental law which connects the purely electric with the magnetic phenomena is Faraday's law of induction

$$V = -k_i \frac{d\phi}{dt} \tag{3}$$

where V is the induced voltage (emf), ϕ the magnetic flux linked with the conductor in which V is produced, and k_i another proportionality factor.

The variety of electromagnetic quantities leads one to expect a corresponding variety of dimension systems. Table 3 shows the dimensions of the more important electromagnetic quantities in eight different dimension systems of varying prominence and varying logical foundation.

The Natural Electrical Dimension Systems

The Electrophysical Dimension System [‡] is the extension of the dynamical or physical mechanical dimension system by adding electric charge [Q] as the fourth fundamental dimension. (See column 1 of Table 3.) This system has been gaining favor among the recent authors on electromagnetic theory and seems destined to be accepted universally.[§]

The Electrotechnical Dimension System [‡] is the extension of the gravitational or technical mechanical dimension system by adding electric charge [Q] as the fourth fundamental dimension. (See column 2 of Table 3.)

[*] A. W. Rücker, *Phil. Mag.*, Vol. 27, p. 104, Feb. 1889.

[†] A. M. Ampère, *Ann. de chimie et de physique*, Vol. 15, pp. 59 and 170, 1820.

[‡] E. Weber, *Trans. A.I.E.E.*, Vol. 51, p. 728, 1932; J. Wallot, *E.T.Z.*, Vol. 43, pp. 1329 and 1372, 1922; E. Brylinski, *Rev. gén. de l'élec.*, Vol. 30, p. 781, 1931.

[§] See particularly J. A. Stratton, *Electromagnetic Theory*, McGraw-Hill, 1941; R. W. P. King, *Electromagnetic Engineering*, McGraw-Hill, 1945; and S. S. Attwood, *Electric and Magnetic Fields*, John Wiley & Sons, 1949 (3rd ed.).

Table 3. Dimensions of Electromagnetic Quantities

Symbol (according to ASA)	Quantity	Electrophysical System ①	Electrotechnical System ②	Definitive System ③
l	Length	$[L]$	$[L]$	$[L]$
t	Time	$[T]$	$[T]$	$[T]$
m	Mass	$[M]$	$[F][L]^{-1}[T]^2$	$[P][L]^{-2}[T]^3$
f	Force	$[M][L][T]^{-2}$	$[F]$	$[P][L]^{-1}[T]$
E	Energy	$[M][L]^2[T]^{-2}$	$[F][L]$	$[P][T]$
P	Power	$[M][L]^2[T]^{-3}$	$[F][L][T]^{-1}$	$[P]$
Q	Electric charge	$[Q]$	$[Q]$	$[Q]$
Ψ	Displacement flux	$[Q]$	$[Q]$	$[Q]$
D	Displacement	$[Q][L]^{-2}$	$[Q][L]^{-2}$	$[Q][L]^{-2}$
E	El. field intensity	$[M][Q]^{-1}[L][T]^{-2}$	$[F][Q]^{-1}$	$[P][Q]^{-1}[L]^{-1}[T]$
C	Capacitance	$[M]^{-1}[Q]^2[L]^{-2}[T]^2$	$[F]^{-1}[Q]^2[L]^{-1}$	$[P]^{-1}[Q]^2[T]^{-1}$
I	Current	$[Q][T]^{-1}$	$[Q][T]^{-1}$	$[Q][T]^{-1}$
V	Voltage	$[M][Q]^{-1}[L]^2[T]^{-2}$	$[F][Q]^{-1}[L]$	$[P][Q]^{-1}[T]$
R	Resistance	$[M][Q]^{-2}[L]^2[T]^{-1}$	$[F][Q]^{-2}[L][T]$	$[P][Q]^{-2}[T]^2$
Φ	Magnetic flux	$[M][Q]^{-1}[L]^2[T]^{-1}$	$[F][Q]^{-1}[L][T]$	$[P][Q]^{-1}[T]^2$
B	Induction	$[M][Q]^{-1}[T]^{-1}$	$[F][Q]^{-1}[L]^{-1}[T]$	$[P][Q]^{-1}[L]^{-2}[T]^2$
H	Magnetic intensity	$[Q][L]^{-1}[T]^{-1}$	$[Q][L]^{-1}[T]^{-1}$	$[Q][L]^{-1}[T]^{-1}$
L	Inductance	$[M][Q]^{-2}[L]^2$	$[F][Q]^{-2}[L][T]^2$	$[P][Q]^{-2}[T]^3$
$\mathcal{F}$	Magnetomotive force	$[Q][T]^{-1}$	$[Q][T]^{-1}$	$[Q][T]^{-1}$
$\mathcal{R}$	Reluctance	$[M]^{-1}[Q]^2[L]^{-2}$	$[F]^{-1}[Q]^2[L]^{-1}[T]^{-2}$	$[P]^{-1}[Q]^2[T]^{-3}$
ε	Absolute diel. const.	$[M]^{-1}[Q]^2[L]^{-3}[T]^2$	$[F]^{-1}[Q]^2[L]^{-2}$	$[P]^{-1}[Q]^2[L]^{-1}[T]^{-1}$
μ	Absolute permeability	$[M][Q]^{-2}[L]$	$[F][Q]^{-2}[T]^2$	$[P][Q]^{-2}[L]^{-1}[T]^3$

The "Natural" Systems

Symbol (according to ASA)	Quantity	"Practical" System ④	Energetical System ⑤	Electrostatic System ⑥
l	Length	$[L]$	$[L]$	$[L]$
t	Time	$[T]$	$[T]$	$[T]$
m	Mass	$[I]^2[R][L]^{-2}[T]^3$	$[E][L]^{-2}[T]^2$	$[M]$
f	Force	$[I]^2[R][L]^{-1}[T]$	$[E][L]^{-1}$	$[M][L][T]^{-2}$
E	Energy	$[I]^2[R][T]$	$[E]$	$[M][L]^2[T]^{-2}$
P	Power	$[I]^2[R]$	$[E][T]^{-1}$	$[M][L]^2[T]^{-3}$
Q	Electric charge	$[I][T]$	$[E][V]^{-1}$	$[M]^{1/2}[L]^{3/2}[T]^{-1}[k_e]^{-1/2}$
Ψ	Displacement flux	$[I][T]$	$[E][V]^{-1}$	$[M]^{1/2}[L]^{3/2}[T]^{-1}[k_e]^{-1/2}$
D	Displacement	$[I][L]^{-2}[T]$	$[E][V]^{-1}[L]^{-2}$	$[M]^{1/2}[L]^{-1/2}[T]^{-1}[k_e]^{-1/2}$
E	El. field intensity	$[I][R][L]^{-1}$	$[V][L]^{-1}$	$[M]^{1/2}[L]^{-1/2}[T]^{-1}[k_e]^{1/2}$
C	Capacitance	$[R]^{-1}[T]$	$[E][V]^{-2}$	$[L][k_e]^{-1}$
I	Current	$[I]$	$[E][V]^{-1}[T]^{-1}$	$[M]^{1/2}[L]^{3/2}[T]^{-2}[k_e]^{-1/2}$
V	Voltage	$[I][R]$	$[V]$	$[M]^{1/2}[L]^{1/2}[T]^{-1}[k_e]^{1/2}$
R	Resistance	$[R]$	$[E]^{-1}[V]^2[T]$	$[L]^{-1}[T][k_e]$
Φ	Magnetic flux	$[I][R][T]$	$[V][T]$	$[M]^{1/2}[L]^{3/2}[T]^{-1}[k_e]^{1/2}$
B	Induction	$[I][R][L]^{-2}[T]$	$[V][L]^{-2}[T]$	$[M]^{1/2}[L]^{-1/2}[T]^{-1}[k_e]^{1/2}$
H	Magnetic intensity	$[I][L]^{-1}$	$[E][V]^{-1}[L]^{-1}[T]^{-1}$	$[M]^{1/2}[L]^{-1/2}[T]^{-2}[k_e]^{-1/2}$
L	Inductance	$[R][T]$	$[E]^{-1}[V]^2[T]^2$	$[L]^{-1}[T]^2[k_e]$
$\mathcal{F}$	Magnetomotive force	$[I]$	$[E][V]^{-1}[T]^{-1}$	$[M]^{1/2}[L]^{3/2}[T]^{-2}[k_e]^{-1/2}$
$\mathcal{R}$	Reluctance	$[R]^{-1}[T]^{-1}$	$[E][V]^{-2}[T]^{-2}$	$[L][T]^{-2}[k_e]^{-1}$
ε	Absolute diel. const.	$[R]^{-1}[L]^{-1}[T]$	$[E][V]^{-2}[L]^{-1}$	$[k_e]^{-1}$
μ	Absolute permeability	$[R][L]^{-1}[T]$	$[E]^{-1}[V]^2[L]^{-1}[T]^2$	$[L]^{-2}[T]^2[k_e]$

The "Practical" Systems   The "Fractional" Systems

Table 3. Dimensions of Electromagnetic Quantities—*Continued*

Symbol (according to ASA)	Quantity	Electromagnetic System ⑦	Gaussian System ⑧
l	Length	$[L]$	$[L]$
t	Time	$[T]$	$[T]$
m	Mass	$[M]$	$[M]$
f	Force	$[M][L][T]^{-2}$	$[M][L][T]^{-2}$
E	Energy	$[M][L]^2[T]^{-2}$	$[M][L]^2[T]^{-2}$
P	Power	$[M][L]^2[T]^{-3}$	$[M][L]^2[T]^{-3}$
Q	Electric charge	$[M]^{\frac12}[L]^{\frac12}[k_m]^{-\frac12}$	$[M]^{\frac12}[L]^{\frac12}[T]^{-1}[k_e]^{-\frac12}$
Ψ	Displacement flux	$[M]^{\frac12}[L]^{\frac12}[k_m]^{-\frac12}$	$[M]^{\frac12}[L]^{\frac12}[T]^{-1}[k_e]^{-\frac12}$
D	Displacement	$[M]^{\frac12}[L]^{-\frac32}[k_m]^{-\frac12}$	$[M]^{\frac12}[L]^{-\frac32}[T]^{-1}[k_e]^{-\frac12}$
E	El. field intensity	$[M]^{\frac12}[L]^{\frac12}[T]^{-2}[k_m]^{\frac12}$	$[M]^{\frac12}[L]^{-\frac12}[T]^{-1}[k_e]^{\frac12}$
C	Capacitance	$[L]^{-1}[T]^2[k_m]^{-1}$	$[L][k_e]^{-1}$
I	Current	$[M]^{\frac12}[L]^{\frac12}[T]^{-1}[k_m]^{-\frac12}$	$[M]^{\frac12}[L]^{\frac12}[T]^{-2}[k_e]^{-\frac12}$
V	Voltage	$[M]^{\frac12}[L]^{\frac12}[T]^{-2}[k_m]^{\frac12}$	$[M]^{\frac12}[L]^{\frac12}[T]^{-1}[k_e]^{\frac12}$
R	Resistance	$[L][T]^{-1}[k_m]$	$[L]^{-1}[T][k_e]$
Φ	Magnetic flux	$[M]^{\frac12}[L]^{\frac12}[T]^{-1}[k_m]^{\frac12}$	$[M]^{\frac12}[L]^{\frac12}[T]^{-1}[k_m]^{\frac12}$
B	Induction	$[M]^{\frac12}[L]^{-\frac12}[T]^{-1}[k_m]^{\frac12}$	$[M]^{\frac12}[L]^{-\frac12}[T]^{-1}[k_m]^{\frac12}$
H	Magnetic intensity	$[M]^{\frac12}[L]^{-\frac12}[T]^{-1}[k_m]^{-\frac12}$	$[M]^{\frac12}[L]^{-\frac12}[T]^{-1}[k_m]^{-\frac12}$
L	Inductance	$[L][k_m]$	$[L][k_m]$
$\mathfrak{F}$	Magnetomotive force	$[M]^{\frac12}[L]^{\frac12}[T]^{-1}[k_m]^{-\frac12}$	$[M]^{\frac12}[L]^{\frac12}[T]^{-1}[k_m]^{-\frac12}$
$\mathfrak{R}$	Reluctance	$[L]^{-1}[k_m]^{-1}$	$[L]^{-1}[k_m]^{-1}$
ε	Absolute diel. const.	$[L]^{-2}[T]^2[k_m]^{-1}$	$[k_e]^{-1}$
μ	Absolute permeability	$[k_m]$	$[k_m]$

The "Fractional" Systems

Both systems are the natural extensions of existing mechanical systems and use the basic concept of quantity of electricity as the additional fundamental dimension. Although the electron is recognized as a truly elemental quantity,* preference was given for a while to resistance as a fundamental dimension since it was thought to be more readily reproduced. Newer investigations show the resistance of a metal to be a very complicated function of the surrounding conditions and a function of the density of free electrons, which points in the direction of the above natural dimension systems as more satisfactory than systems using electrical resistance as a fundamental quantity.

The Definitive Dimension System † is the extension of an energetical mechanical dimension system (using length, time, and power) by adding electric charge $[Q]$ as the fourth fundamental dimension. (See column 3 of Table 3.) As there are no standards available for power, this system is indirectly based upon the physical system. The choice of power as a fundamental quantity was obviously made in view of the extensive use of the watt as a unit in all fields of physics.

The Fractional Electrical Dimension Systems

The Electrostatic Dimension System (ES-System) is the extension of the dynamical or physical mechanical dimension system, by adding the proportionality constant $[k_e]$ as the fourth fundamental dimension. The most extensive use of this dimension system as a preferred one has been made by E. Bennett ‡ (See column 6 of Table 3.)

The Electromagnetic Dimension System (EM-System) is the extension of the dynamical or physical mechanical dimension system, by adding the proportionality constant $[k_m]$ as the fourth fundamental dimension. (See column 7 of Table 3.)

Both systems are complete dimension systems as they are based upon four fundamental dimensions. But the fact that all the electromagnetic quantities appear as derived frac-

---

* Tolman, *Phys. Rev.*, Vol. 9, p. 237, 1917.

† G. A. Campbell, "A definitive system of units," *Bull. National Research Council*, No. 93, p. 48, 1933.

‡ E. Bennett, "A digest of the relations between the electrical units," *Bull. Univ. Wisconsin*, No. 880, 1917.

tional dimensions should indicate that better systems could be found. Both systems introduce a dissymmetry into the dimensional expressions which is undesirable.

The Symmetrical Dimension System. Taking cognizance of the fact that the two proportionality constants k_e and k_m differ in their dimensions by the factor $[L]^2[T]^{-2}$, which happens to be the square of the dimension of a velocity, the relation

$$[k_e] = [k_m][L]^2[T]^{-2}$$

can be used to express either in the electrostatic system all the magnetic quantities in dimensions of $[k_m]$ rather than $[k_e]$, or in the electromagnetic system all the electric quantities in dimensions of $[k_e]$ rather than $[k_m]$. The last column in Table 3 shows the resulting hybrid system which exhibits a remarkable symmetry in the principal electric and magnetic quantities as far as the three mechanical dimensions are concerned; it might be referred to as the *generalized Gaussian dimension system*. Obviously, in all equations combining electric and magnetic quantities, an arbitrary factor must be introduced in order to balance the dimensions on both sides, and this factor must have the characteristics of a velocity or some power of it. In fact, this velocity was found to be of the same value as the velocity of light in free space and this was taken as an indication of a very fundamental principle. In this dimension system, then, the factor c, denoting the velocity of light in free space, appears repeatedly, as for example in the induction law $V = -\dfrac{1}{c}\dfrac{d\phi}{dt}$, where the induced voltage (emf) is measured electrostatically and the magnetic flux is measured electromagnetically. (See column 8 of Table 3.)

The So-called Absolute Electrical Dimension Systems

The original use of the word "absolute" in connection with electrical measurements goes back to the report of the Committee on Electrical Standards, British Association for the Advancement of Science, 1863, which states:

The word "absolute" in the present sense is used as opposed to the word "relative" and by no means implies that the measurement is accurately made or that the unit implied is of perfect construction; in other words, it does not mean that the measurement or units are absolutely correct but only that the measurement instead of being a simple comparison with an arbitrary standard of the same kind as that measured is made by reference to certain fundamental units of another kind treated as postulates.

As the fundamental units, length, mass, and time, were commonly used, any measurement made in terms of these specific quantities is now customarily called an "absolute" measurement. To indicate the injudicious use of the word absolute it is prefixed with "so-called" in this section.

In connection with the use of dimensions it also became customary to call the dynamical or physical dimension system the "absolute" dimension system. Assigning, then, arbitrarily, zero dimension to the proportionality factors occurring in the basic experimental laws of electromagnetism, the fractional electrical dimension systems (see Table 3) reduce to the so-called "absolute" dimension systems. The additional arbitrary disposal of the dimension of the proportionality constant or constants thus leads within electromagnetism to an incomplete set of fundamental dimensions. According to the three fractional dimension systems, three different sets of dimensions for the electromagnetic quantities suggest themselves and have been used.

The So-called Absolute Electrostatic Dimension System is the electrically incomplete set of the three fundamental dimensions of the physical mechanical dimension system derived from the fractional electrostatic dimension system by arbitrarily assuming k_e as purely numeric. The dimensions for this incomplete set are obtained from Table 3, column 6, by introducing $[k_e] = 1$.

The So-called Absolute Electromagnetic Dimension System is the electrically incomplete set of the three fundamental dimensions of the physical mechanical dimension system derived from the fractional electromagnetic dimension system by arbitrarily assuming k_m as purely numeric. The dimensions for this incomplete set are obtained from Table 3, column 7, by introducing $[k_m] = 1$.

The Gaussian or So-called Absolute Symmetrical Dimension System is the electrically incomplete set of the three fundamental dimensions of the physical mechanical dimension system obtained from a peculiar combination of the fractional electrostatic and electromagnetic dimension systems as shown in Table 3, column 8, by specifying $[k_e] = [k_m] = 1$, i.e., defining both proportionality constants as pure numbers. The remarkable result is that all the electric quantities, including current, are in the electrostatic dimension system, and the magnetic quantities are in the electromagnetic system. This Gaussian set of dimensions has been used in theoretical physics * on the basis that the symmetry in the dimensional expressions of electric and magnetic quantities presumably facilitates the teaching of electromagnetism.

On the Use of the So-called Absolute Electrical Dimension Systems. Probably no other single question has raised so many discussions as the question of the "proper" dimensions of the electromagnetic quantities. As has been stated (Art. 2), the chief advantages of using the concept of physical dimensions are:

1. The number of fundamental dimensions is the same as the number of fundamental units.

2. Physical equations can be checked by dimensional homogeneity.

3. Dimensional analysis and model theory can be readily put on a rational basis (see Arts. 17 and 18).

It is obviously desirable to define fundamental and derived dimensions in such a way that the least amount of confusion arises and that easy communication of research is possible. The definition of complete dimension systems in electromagnetism requires from this point of view four fundamental dimensions as has been pointed out repeatedly in the past.† The arbitrary suppression of the dimension of a proportionality constant, as done in the case of the so-called "absolute" systems, leads to confusion in dimensional analysis,‡ and makes difficult the checking of the dimensional homogeneity of equations.

Recent discussions show a definite tendency towards accepting this fact and particularly to choose electric charge as the fourth fundamental dimension, leading to the electrophysical dimension system (see Table 3, column 1), in which all dimensional expressions become rather simple.§

It may be emphasized, that the question of the proper number of fundamental dimensions or units is entirely divorced from the other question of the definition of electromagnetic units. It is well possible to retain the so-called "absolute" electrostatic units as values with dimensions following from any one of the complete dimension systems based upon so-called absolute measurements.

Various "Practical" Dimension Systems

The fractional dimension systems were the first ones used and may duly be called the "classical" dimension systems as they had been proposed by the classical authors in the art. Their value in promoting first quantitative measurements of electromagnetic quantities cannot be doubted. Yet, the inconvenience in using fractional dimensions led to various proposals of so-called practical dimension systems. The aim of these systems is simplicity, stress upon fundamental concepts, and practicability.

The So-called Practical Dimension System has as fundamental dimensions length [L], time [T], current [I], and resistance [R], based upon the fact that convenient standards are available for the two electrical quantities. The system is used rather extensively in the engineering literature,‖ although not always explicitly stated. The fact that two fundamental electrical dimensions are chosen renders it difficult to combine it with the known mechanical dimension systems, and this is an objectionable feature. (See column 4 of Table 3.)

The Energetical Dimension System is the extension of the mechanical dimension system with the same name by adding voltage [V] as a fourth fundamental dimension. It had been proposed by W. Ostwald ¶ but never has been used in the literature. (See column 5 of Table 3.)

The disadvantage of all the "practical" systems is their departure from existing mechanical dimension systems from which, however, they cannot be entirely dissociated, since the fundamental standards of energy or power must be based upon the standard of mass in order to be internationally available.

---

* For example, H. Hertz, *Electric Waves*, London, 1893; A. G. Webster, *Theory of Electricity and Magnetism*, London, 1897; H. A. Lorentz, *The Theory of Electrons*, New York, 1909; J. H. Jeans, *Electricity and Magnetism*, Cambridge, 1911; O. W. Richardson, *The Electron Theory of Matter*, Cambridge, 1914.

† A. W. Rücker, *Phil. Mag.*, February, 1889, Vol. 27, p. 104; S. W. Stratton, *Electric Units and Standards*, Circ. of the Bureau of Standards, No. 60, 1920, p. 10; J. Wallot, *E.T.Z.*, 1922, pp. 1329, 1372; L. Genillon, *Rev. Gén. de l'élec.*, Vol. 13, p. 173, 1923; and others.

‡ See, for example, P. W. Bridgman, *Dimensional Analysis*, Yale Univ. Press, 1922, p. 12.

§ E. Brylinski, *Rev. Gén. de l'élec.*, Vol. 38, p. 589, 1935, Vol. 39, p. 747, 1935, and Vol. 40, p. 99, 1936; A. Sommerfeld, *Zeits. f. techn. Physik*, Vol. 16, p. 420, 1935, and *Phys. Zeits.*, Vol. 36, p. 814, 1935. See particularly footnote † on p. 3-09.

‖ G. Mie, *Lehrbuch der Elektrizität und des Magnetismus*, Stuttgart, 1910; V. Karapetoff, *The Electric Circuit, The Magnetic Circuit*, New York, 1910.

¶ W. Ostwald, see footnote, p. 390

UNIT SYSTEMS

6. UNITS AND UNIT SYSTEMS

Definition of Units. A physical quantity can be measured only by a comparison with a like quantity. By defining a distinct amount of a physical quantity as a *unit*, i.e., as reference value, any physical quantity of the same kind can be compared with it, and its value is then stated in terms of a ratio number and the unit used. Obviously there are infinite possibilities for choosing a unit of a single physical quantity. *All the possible units* of the same physical quantity *must be related by purely numerical factors* which are used as the expressions for direct comparison. (See also Art. 1.)

Units and Dimensions. The general unit of a physical quantity is defined as its *dimension*. (See Art. 2.) There can be only one dimension for each physical quantity if the units are to be related by numerical factors only; but there are as many dimensions as there are physical quantities. Obviously, the concept of dimension systems, as outlined (see Art. 2), facilitates easy orientation and permits the use of a few fundamental dimensions to express all physical quantities in general units. Likewise, the concept of the fundamental unit permits the expression of all other units in terms of the fundamental units. However, whereas dimensional relations are essentially exponential equations, unit relations necessarily include numerical values in their equations.

Unit Systems. On the basis of a proper dimension system a unit system can be developed by choosing, for each fundamental dimension of the system, a specific unit, desirably related to a fundamental standard or standards. These units are called *fundamental units*; the respective physical quantities are called fundamental quantities. Their number must be the same as the number of fundamental dimensions in the respective dimension system in order to constitute a complete unit system. All other physical quantities are then expressible in terms of the fundamental quantities and their units. Obviously there can be an infinity of unit systems for each complete dimension system; but for international understanding it is desirable to limit the usage to a few unit systems.

Systematic Units are all systematically derived units within a unit system, obtained by replacing the general units, as indicated in the derived dimension of the quantity, by the fundamental units of the system. Thus, in a m-kg-sec physical mechanical unit system, the systematic unit of power with the dimension $[M][L]^2[T]^{-3}$ will be 1 kg m^2 sec^{-3}, which is known as 1 watt. The unit relations between systematic units are, therefore, unitary; i.e., they do not involve numerical factors. Unit systems with only systematically derived units are the ultimate goal in any branch of physics; they are difficult to obtain because the units should all be of convenient practical size.

Derived Units. All units which are not fundamental may be called derived units. Systematic units form the most prominent group of derived units, although all mixed units would come into this same category. Derived units can be the same for several unit systems if the defining fundamental units are the same. In many cases, where systematically derived units are inconvenient, mixed units will be used. In the MKS system "charge density" would have the systematic unit "coulomb per square meter," whereas more convenient is the unit "microcoulomb per square centimeter," a multiple of a mixed unit.

Units and Physical Equations. The form of physical equations depends upon the units employed for the various physical quantities. In general, the simplest forms will result when units of the same system are used exclusively. Two different types are, therefore, distinguished.

Systematic Physical Equations use only systematic units for all physical quantities involved. These equations are independent of the specific unit system applied, their form is the simplest obtainable without numerical factors, and their use should be preferred in all general texts. If a definite dimension system is chosen, it will, in conjunction with the proper number of fundamental units, form a sufficient basis for the correct interpretation of the general equations, written in systematic form.

Non-systematic or Hybrid Physical Equations use units from different unit systems or various multiples of units of the same system. These equations involve additional numerical factors (and numerical factors only, if all incomplete dimension sets are excluded). In all non-systematic equations it is imperative to state the units to be used for the various quantities in order to avoid confusion and misunderstandings. Many examples of non-systematic equations can be found in electromagnetism, where hybrid relations are very common. (See Art. 9.)

Comprehensive Unit Systems are composed of five fundamental units to cover the whole field of physics. The difficulties in designing a comprehensive unit system come from the restriction that all the systematic units should be convenient and practical or at least be made so by simple powers of 10. Many attempts have been made to reach international agreement on a single comprehensive system. Though no formal agreement has been reached as yet, the extended MKS system which uses the coulomb as the fourth fundamental unit is finding widespread acceptance among engineers and physicists. The advantage of such a system is obvious, since all physical equations could be used in their systematic form, introducing all the physical quantities in convenient practical values.

So-called Absolute Unit Systems. During the middle of the nineteenth century it became customary to refer to measurements in terms of the centimeter-gram-second mechanical unit system as "absolute" measurements, and the cgs system, as adopted and recommended by the British Association for the Advancement of Science in 1873, was called the "absolute" unit system. This designation is unwarranted, as no system can claim "absoluteness." The reduction of the fractional electrical dimension systems to the three so-called absolute mechanical dimensions is inadvisable and the resulting absolute unit systems will, therefore, not be considered here. The complete unit systems resulting from the fractional dimension systems, which sometimes are called "complete absolute" systems because of their complete set of fundamental dimensions, will be called here *"theoretical" unit systems.* *

International System of Units (SI). This system of units has been chosen by the Conference Générale BIPM (Sèvres, Paris, 1954 and 1960) to be composed of six fundamental units to cover the whole range of physics. The system has been compiled with the aid of organizations such as the International Standards Organization, the International Commission on Illumination, the International Commission, Bureau of Weights & Measures (BIPM) and others. (See Art. 9.) The International System of units is becoming the common basis of the International Law and one in which all international reports are to be expressed.

The Metric Unit Systems are based upon the international metric standards, namely: the meter and the kilogram, or any decimal multiples thereof. Only the metric systems enter in general into the discussion of comprehensive unit systems.

Metric Multiples and Metric Style. Prefixes used to indicate *multiples* and *submultiples* of various metric units are shown in the following table:

10^{12} = tera T	10^2 = hecto h	10^{-3} = milli m	10^{-15} = femto f
10^9 = giga G	10^1 = deca da	10^{-6} = micro μ	10^{-18} = atto a
10^6 = mega M	10^{-1} = deci d	10^{-9} = nano n	
10^3 = kilo k	10^{-2} = centi c	10^{-12} = pico p	

The English Unit Systems are based upon the English standards, namely: the yard and the pound, or any multiples thereof. See Units and Standards, p. 405.

7. THE METRIC UNIT SYSTEMS OF MECHANICS AND HEAT

The Distinctly Mechanical Unit Systems

There are several mechanical unit systems in use, which are shown in Table 1, with their systematic units (the fundamental units are printed bold face) for the most important quantities. It is significant that only one technical and three physical systems have been developed. The variety of the unit systems is not due to variety in dimension systems, but rather to the choice of various multiples of the same unit quantity.

Table 1.   The Metric Unit Systems of Mechanics

Unit System Author Year Basic Dimension System	CGS France 1799 Physical (Dynamical)	Technical Technical (Gravitational)	MKS G. Giorgi 1902 Physical (Dynamical)	MTS France 1913 Physical (Dynamical)
Quantity	①	②	③	④
Length..............	cm	m	m	m
Time................	sec	sec	sec	sec
Mass................	g	9.81 kg mass	kg	ton
Force...............	dyne	kg force	newton = 10^5 dynes	sthène = 10^8 dynes
Power...............	erg/sec	9.81 watt	watt	kilowatt
Energy..............	erg	kg force-m	joule	kilojoule

* G. Mie, "Elektrodynamik," *Handbuch der experimental Physik*, Vol. 11, 1932.

The CGS (Centimeter-gram-second) System of Units is a dynamical system of units and is based upon the centimeter, gram-mass, and second as fundamental units (column 1 of Table 1). It was first proposed in 1795 and was adopted in France by the French statute of Dec. 10, 1799. The standard to which the length unit refers is the prototype meter; the standard to which the gram has reference is the prototype mass of the kilogram (see Standards). This system was endorsed and recommended for use by the British Association for the Advancement of Science in 1863 and since then has been most widely used internationally, in all sciences. Frequent objections to the cgs system are based on the fact that the actual standards are for multiples of these fundamental units and not for the units themselves; that the derived units for force and energy are inconveniently small for practical purposes; and that the system does not fit with the practical electrical units to form a comprehensive unit system.

The MKS (Meter-kilogram-second) System of Units is a dynamical system of units and is based upon the meter, kilogram-mass, and second (column 3 of Table 1). It was proposed by G. Giorgi * (1902) and adopted by the International Electrotechnical Commission at Paris, 1935. Its only disadvantage is the fact that densities become rather large magnitudes; thus, the density of water is 10^3 kg per cu meter, which is inconvenient in practical computations. On the other hand, the advantages of the system are quite numerous. It is based upon the actual standards of the meter and the kilogram as preserved in Paris; the derived force unit *joule per meter*, or *newton*, is of convenient size in engineering applications; and all the practical electrical units fit in as the natural units to form a comprehensive unit system.

The MTS (Meter-ton-second) System of Units is a dynamical system of units and is based upon the meter, metric ton (mass), and second (column 4 of Table 1). It was proposed in 1913 † and legalized in France in 1919. The units are in some respects better adapted to practical use than those in the cgs system; but it is not suitable for extension into a comprehensive system or as a unit system for physics. Its main advantage is the fact that the density of water under normalized conditions becomes 1 ton per cu meter and thus has the same numerical value as in the cgs system (where it was 1 g per cu cm).

The "Technical" System of Units is a gravitational system of units and is based upon the meter, kilogram (force), and second as fundamental units (column 2 of Table 1). The fundamental unit of force was variable until international agreement determined a standard value of the constant of gravity (see Units and Standards) as 9.80665 m/sec². The advantages in practical design work and in numerical computations more than jusitfy the use of this unit system, yielding very simple dimensional formulas and self-explanatory derived units names.

The Thermal Unit Systems

Any mechanical unit system supplemented by temperature as a fourth fundamental unit constitutes a proper thermal unit system. Table 2 gives a summary of the units most frequently used and their coordination in the various thermal unit systems.

Table 2. The Thermal Units Most Commonly Used

Quantity	Caloric Unit System ①	CGS Thermal Unit System ②	MKS Thermal Unit System ③
Fundamental			
Length..........................	cm	cm	m
Time..........................	sec	sec	sec
Mass..........................	(4.184 gram-seven) ‡	g	kg
Temperature...................	°C	°C	°C
Energy.......................	cal	(erg)	(joule)
Derived			
Thermal conductivity...........	cal/sec °C cm	erg/sec °C cm	watt/°C m
Thermal capacity of body........	cal/°C	erg/°C	joule/°C
Thermal capacity of substance....	cal/sec °C g	erg/sec °C g	watt/°C kg
Transmittance.................	cal/sec °C cm²	erg/sec °C cm²	watt/°C m²
Entropy.......................	cal/°C	erg/°C	joule/°C

* G. Giorgi, *Elec. World*, Vol. 40, pp. 355, 368, 1902.
† See A. E. Kennelly, *Proc. Soc. Prom. Eng. Educ.*, Vol. 19, p. 229, 1928.
‡ Gram-seven is the abbreviation for 10^7 gram as proposed by E. Bennett. See footnote, p. 3-09.

The Caloric Unit System is based on the centimeter, second, calorie, and degree Centigrade (column 1 of Table 2). The peculiar definition of the unit of energy produces inconvenient systematic units of mass and force. For this reason a practical modification is widely used, by employing the non-systematic units of gram and dyne in computations. It is with this system that Fourier founded the theory of dimensions.* The most commonly used quantities and their units are shown in Table 2.

The CGS Thermal Unit System is the extension of the cgs mechanical unit system (column 2, Table 2) by the addition of the metric unit of temperature (degree Centigrade). Although it is a dynamical or physical system, it is widely used in heat engineering because the erg is the unit of energy, which provides a close link to the watt and the practical electrical unit systems. The most commonly used quantities and their units are shown in Table 2.

The MKS Thermal Unit System is the extension of the mks mechanical unit system (column 3, Table 2) by the addition of the metric unit of temperature (degree Centigrade). It is widely used in general engineering on account of the convenient magnitudes of its fundamental units. The most commonly used quantities and their units are shown in Table 2.

8. THE METRIC COMPREHENSIVE UNIT SYSTEMS

Prior to the International System of Units (1960) there existed several comprehensive units systems composed of five fundamental units. More than twelve different comprehensive unit systems have been proposed and at least ten have actually been used in publications. Table 3 shows these systems, their authors, the year of proposal or first use, and the dimension system upon which the unit system was based (the fundamental units are bold-face type). Most of the variety can be accounted for by slight changes in the choice of the multiple of the mass unit or by the choice of the supplementary electrical unit. Fortunately, since the international adoption of the MKS system of mechanical units, an increasing number of authors have used the systems in columns 5, 9 and 10 of Table 3.

The "Theoretical" Comprehensive Unit Systems

The "theoretical" comprehensive unit systems are so called because the definition of the supplementary electrical unit is based not upon a standard, but upon a particular choice of the numerical value of either k_e, the constant in Coulomb's electrostatic force law (1), Art. 5, or k_m, the constant in Ampère's electrodynamic force law (2), Art. 5. Table 4 gives the most commonly used electromagnetic quantities and their units in the major theoretical unit systems.

The CGS Electrostatic Unit System is the extension of the cgs mechanical system of units into a comprehensive system by adding the metric unit of temperature, degree Centigrade, and defining the absolute dielectric constant (permittivity) as 1 statfarad per centimeter (see Units and Standards); see column 2, Table 3. It is a dynamical unit system and was introduced by J. C. Maxwell.† It is frequently used in the field of electrostatics, especially in the form of an incomplete set of dimensions as a so-called absolute system which is obtained if the dimension of the dielectric constant is arbitrarily taken as zero, so that capacity will appear with the dimension "centimeter." In all texts employing the so-called absolute units and giving capacity in "centimeters," replace this unit by statfarad. The most commonly used quantities and their units in the complete cgs electrostatic system are listed in Table 4, first column. The unit system has many nonsystematic units, and so the electromagnetic equations contain numerical factors (Art. 9).

The CGS Electromagnetic Unit System is the extension of the cgs mechanical system of units into a comprehensive system by adding the metric unit of temperature (degree Centigrade) and defining the absolute magnetic permeability as 1 abhenry per centimeter (see Units and Standards); see column 3, Table 3. It is a dynamical unit system and was introduced by J. C. Maxwell,† as an alternative to the cgs electrostatic unit system. It is frequently used in the treatment of magnetic fields, especially in the form of an incomplete set of dimensions as a so-called absolute system, which is obtained if the dimension of the permeability is arbitrarily taken as zero, so that inductance will appear with the dimension "centimeter." In all texts employing so-called absolute units and giving inductance in "centimeters," replace this unit by abhenry. The most commonly used

---

* J. B. J. Fourier, *Théorie analytique de la chaleur*, p. 152, Paris, 1822.

† *Reports of the Committee on Standard Resistance of the British Assoc. for the Advancement of Science*, by F. Jenkin, London, 1873, especially appendix C, second report.

Table 3. The Fundamental Units in the Comprehensive Unit Systems

Unit System	CGS Symmetric	CGS Electrostatic	CGS Electromagnetic	"Rationalized" Electromagnetic	MKS
Author Year	C. F. Gauss 1833 W. Weber 1851	J. C. Maxwell 1863	J. C. Maxwell 1863	O. Heaviside 1892	G. Giorgi 1902
Basic Dimension System	Symmetric	Electrostatic	Electromagnetic	Electromagnetic	Practical
Quantity	①	②	③	④	⑤
Length	cm	cm	cm	cm	m
Time	sec	sec	sec	sec	sec
Mass	g	g	g	g	kg
Force	dyne	dyne	dyne	dyne	joule/m
Power	erg/sec	erg/sec	erg/sec	erg/sec	watt
Energy	erg	erg	erg	erg	joule
Temperature	°C	°C	°C	°C	°C
Electric charge	statcoulomb	statcoulomb	abcoulomb	$\frac{1}{\sqrt{4\pi}}$ abcoulomb	coulomb
Current	statampere	statampere	abampere	$\frac{1}{\sqrt{4\pi}}$ abampere	ampere
Resistance	statohm	statohm	abohm	4π abohm	ohm
Dielectric constant	statfarad/cm	statfarad/cm	abfarad/cm	$\frac{1}{4\pi}$ abfarad/cm	farad/m
Permeability	abhenry/cm	stathenry/cm	abhenry/cm	4πabhenry/cm	henry/m

Unit System Author Year Basic Dimension System	"Rationalized" Symmetric H. A. Lorentz 1909 Symmetric	Practical { V. Karapetoff G. Mie, 1910 Practical	CGSS E. Bennett 1917 Electrostatic	Definitive G. A. Campbell 1933 Definitive	MKSC J. A. Stratton 1941 Electrophysical
Quantity	⑥	⑦	⑧	⑩	⑨
Length	cm	cm	cm	m	m
Time	sec	sec	sec	sec	sec
Mass	g	gram-seven = 10^7 g	gram-seven = 10^7 g	kg	kg
Force	dyne	joule/cm	joule/cm	joule/m	joule/m
Power	erg/sec	watt	watt	watt	watt
Energy	erg	joule	joule	joule	joule
Temperature	°C	°C	°C	°C	°C
Electric charge	$\frac{1}{\sqrt{4\pi}}$ statcoulomb	coulomb	coulomb	coulomb	coulomb
Current	$\frac{1}{\sqrt{4\pi}}$ statampere	ampere	ampere	ampere	ampere
Resistance	4π statohm	ohm	ohm	ohm	ohm
Dielectric constant	$\frac{1}{4\pi}$ statfarad/cm	farad/cm	farad/cm	farad/m	farad/m
Permeability	4π abhenry/cm	henry/cm	henry/cm	henry/m	henry/m

quantities and their units in the complete cgs electromagnetic system are listed in Table 4, second column. The system has many non-systematic units, and so the electromagnetic equations contain numerical factors (Art. 9).

The CGS Symmetric Unit System is a combination of the cgs electrostatic and the cgs electromagnetic unit systems in such manner that all the electrostatic and electric current quantities appear in electrostatic units and all the magnetic quantities in electromagnetic units. In order to achieve this result a new factor was introduced into the fundamental electrical equations (Art. 9), so that both the dielectric constant and the permeability could be defined arbitrarily as one statfarad per centimeter and one abhenry per centimeter, respectively. In this way a certain symmetry in the field equations was obtained. The new factor introduced was found to have the same numerical value as the velocity of

Table 4. The Electrical Units Most Commonly Used

Quantity	CGS Electrostatic Unit System	CGS Electromagnetic Unit System	CGS Rational Unit Systems	CGSS | MKSC Unit Systems	Symbol
Electric charge......	statcoul	abcoul	$\frac{1}{\sqrt{4\pi}} \times$ theor.	coul	Q
Surface charge density	statcoul/cm²	abcoul/cm²	$\frac{1}{\sqrt{4\pi}} \times$ "	coul/cm² | coul/m²	η, σ
Displacement flux...	statcoul	abcoul	$\frac{1}{\sqrt{4\pi}} \times$ "	coul	Ψ
Displacement......	statcoul/cm²	abcoul/cm²	$\frac{1}{\sqrt{4\pi}} \times$ "	coul/cm² | coul/m²	D
Electric field strength	statvolt/cm	abvolt/cm	$\sqrt{4\pi} \times$ "	volt/cm | volt/m	E
Capacity...........	statfarad	abfarad	$\frac{1}{4\pi} \times$ "	farad	C
Dielectric constant (permittivity).....	statfarad/cm	abfarad/cm	$1 \times$ "	farad/cm | farad/m	ε
Current............	statamp	abamp	$\frac{1}{\sqrt{4\pi}} \times$ "	amp	I
Voltage............	statvolt	abvolt	$\sqrt{4\pi} \times$ "	volt	V
Potential..........	statvolt	abvolt	$\sqrt{4\pi} \times$ "	volt	V
Resistance.........	statohm	abohm	$4\pi \times$ "	ohm	R
Resistivity.........	statohm cm	abohm cm	$4\pi \times$ "	ohm cm | ohm m	ρ
Conductance........	statmho	abmho	$\frac{1}{4\pi} \times$ "	mho, siemens	A
Conductivity........	statmho/cm	abmho/cm	$\frac{1}{4\pi} \times$ "	mho/cm | mho/m	γ
Magnetic flux.......	statweber	maxwell	$\sqrt{4\pi} \times$ "	weber	Φ
Induction	statweber/cm²	gauss	$\sqrt{4\pi} \times$ "	weber/cm² | weber/m²	B
Magnetic intensity...	....	oersted	$\frac{1}{\sqrt{4\pi}} \times$ "	amp/cm | amp/m	H
Magnetomotive force	....	gilbert	$\frac{1}{\sqrt{4\pi}} \times$ "	amp-turn	$\mathcal{F}$
Reluctance........	....	....	$\frac{1}{4\pi} \times$ "	....	$\mathcal{R}$
Inductance.........	stathenry	abhenry	$4\pi \times$ "	henry	L
Permeability........	stathenry/cm	abhenry/cm	$1 \times$ "	henry/cm | henry/m	μ

NOTES: 1. For relations between the various units see conversion tables.
2. The unit names of the MKSC units are synonomous with those of the SI system. (See p. 404).

light.* This symmetric system is frequently used in European publications, mostly in the form of an incomplete set of fundamental dimensions as the so-called absolute Gaussian system, which is obtained if the dimensions of both the absolute dielectric constant and the absolute permeability are taken as zero. Capacitance as well as inductance will then appear with the dimension "centimeter," which should be replaced properly by statfarad and abhenry, respectively. The "absolute" modification was the one used by C. F. Gauss and W. Weber in the classical investigations on measurements of electromagnetic quantities. The system has many non-systematic units, and so the electromagnetic equations contain numerical factors (Art. 9).

The Rationalized CGS Electromagnetic Unit System was introduced and vigorously defended by O. Heaviside,† whereas *the rationalized cgs symmetric (Gaussian) unit system* was used by H. A. Lorentz.‡ Both systems are identical with the unrationalized systems

---

* W. Weber and R. Kohlrausch, *Pogg. Annalen*, Vol. 99, 1856.
† O. Heaviside, *Electromagnetic Theory*, London, 1893.
‡ H. A. Lorentz, *The Theory of Electrons*, New York, 1909.

of the same name but use a factor 4π in Coulomb's and Ampère's laws instead of in the less logical positions where the classical theory had used them. As a result the electromagnetic field equations take on symmetrical forms (Art. 9). All the units of the rationalized systems bear ratios of $\sqrt{4\pi}$ or multiples thereof to the cgs units of the same systems as shown in Table 4. This fact has prohibited their use in practical computations, although they are often found in modern treatises on electromagnetism, particularly in the English language. Where the rationalized systems have degenerated into so-called absolute systems, the same considerations apply as in the unrationalized cgs systems.

The QES System of Units. J. C. Maxwell [*] has shown that, on the basis of the centimeter, gram, and second, a unit system can be devised in which again the absolute dielectric constant of free space has the value 1 statfarad per centimeter but in which all the other electromagnetic quantities appear in the units of the so-called practical system. Since the cgs units and the practical units are in general related by high powers of 10 (see conversion tables), the fundamental unit of length should be 10^9 cm (or a quadrant), and that of mass should be 10^{-11} gram (eleventh-gram), whereas the second must be preserved as the unit of time. This system never has been used; it is not logical to let a proportionality factor retain unity as its magnitude and accept impractical fundamental units.

The CGSS System of Units. E. Bennett [†] has shown that, on the basis of the mechanical units centimeter, gram-seven (10^7 grams), and second, a unit system can be devised in which the absolute dielectric constant of free space has the value 1 farad per centimeter and in which all the electromagnetic quantities appear in the convenient practical units; see column 8, Table 3. The large unit for the mass has prevented a more general acceptance of the system.

The MKSC System of Units constitutes the *most fortunate compromise* between the classical "theoretical" unit systems and the practical electrical units; see column 9, Table 3. It is the extension of the MKS system of units (Art. 7) into a comprehensive system by adding the metric unit of temperature (degree Centigrade) and the coulomb as fundamental unit of charge, defining, however, the latter in terms of an "absolute" measurement of force action between currents in free space in accordance with (2), article 5 and stipulating for free space (vacuum) $k_{mv} = 2 \times 10^{-7}$ henry/m. The choice of Ampère's law for the definition of the "theoretical" ampere or coulomb is due to the fact that force measurements with the Kelvin balance can be made more accurately than any other basic electrodynamic measurement, permitting a closer approach to the theoretical standard than is possible by any practical standard. The definition of k_{mv} in Ampère's law was accepted internationally at a meeting of the International Electrotechnical Commission [‡] (I.E.C.) in 1938. It still permits two different definitions of the absolute permeability of free space: $\mu_v = 10^{-7}$ henry/m, leading to the *unrationalized* MKSC system, in which the factor 4π appears in the field equations; and $\mu_v = 4\pi \times 10^{-7}$ henry/m, leading to the *rationalized* MKSC system, in which the factor 4π appears only in the force equations and where expected by spherical geometry. (See Art. 9.) The *rationalized MKSC system of units* has been used with increasing frequency in recent publications [§] and seems destined to be accepted universally.

The Various Practical Comprehensive Unit Systems

The "practical" comprehensive unit systems are so called because the definition of the supplementary electrical unit is based upon a standard considered of the same fundamental nature as the primary standards of length and mass. It is now recognized that electrical standards cannot meet the exacting demands of invariance consistent with the accuracy of precision measurements so that these "practical" unit systems have lost their original attractiveness.

The So-called Practical Unit System is based upon the centimeter and second as fundamental mechanical units, adding the metric unit of temperature (degree Centigrade) and two electrical units, the ampere and the ohm (see column 7, Table 3). It was independently introduced into the literature by V. Karapetoff [||] and G. Mie.[¶] The fact that two electrical

[*] J. C. Maxwell, *Electricity and Magnetism*, London, 1881.

[†] E. Bennett, *Bull. Univ. Wisconsin*, No. 880, 1917.

[‡] A. E. Kennelly, *Elec. Eng.*, Vol. 58, p. 78, 1939.

[§] See particularly J. A. Stratton, *Electromagnetic Theory*, McGraw-Hill, 1941; R. W. P. King, *Electromagnetic Engineering*, McGraw-Hill, 1945; and S. S. Attwood, *Electric and Magnetic Fields*, John Wiley & Sons, 1949 (3rd ed.).

[||] V. Karapetoff, *The Electric Circuit*, New York, 1910; *The Magnetic Circuit*, New York, 1911.

[¶] G. Mie, *Elektrizität und Magnetismus*, Stuttgart, 1910.

Table 5. Various Forms of Writing the Electromagnetic Relations

Relation — Name	Relation — Mathematical	Factor (relation)	CGS Electrostatic	CGS Electromagnetic	CGS Symmetric	Rationalized CGS Symmetric	Unrationalized MKSC	Rationalized MKSC
Coulomb's law in free space	$F = k_{ev}\dfrac{Q_1 Q_2}{r^2}$	$k_{ev} = \dfrac{\alpha}{4\pi\varepsilon_v} =$	1 cm/statfarad	V^2 cm/abfarad	1 cm/statfarad	$\dfrac{1}{4\pi}$ cm/statfarad	$\dfrac{c^2}{10^7}$ m/farad	$\dfrac{c^2}{10^7}$ m/farad
Ampère's law in free space	$F = k_{mv}\dfrac{I_1 I_2}{r}\cdot l$	$k_{mv} = \dfrac{\beta}{2\pi}\cdot\mu_v =$	$\dfrac{2}{V^2}$ stathenry/cm	2 abhenry/cm	$\dfrac{2}{V}$ abhenry/cm	$\dfrac{1}{2\pi V}$ abhenry/cm	2×10^{-7} henry/m	2×10^{-7} henry/m
Absolute dielectric constant-permittivity	$D = \varepsilon\cdot E$	$\varepsilon = \varepsilon_v\cdot\varepsilon_r =$	1 statfarad/cm	$\dfrac{1}{V^2}$ abfarad/cm	1 statfarad/cm	1 statfarad/cm	$\dfrac{10^7}{c^2}$ farad/m	$\dfrac{1}{4\pi}\dfrac{10^7}{c^2}$ farad/m
Absolute magnetic permeability	$B = \mu\cdot H$	$\mu = \mu_v\mu_r$, $\mu_v =$	$\dfrac{1}{V^2}$ stathenry/cm	1 abhenry/cm	1 abhenry/cm	1 abhenry/cm	10^{-7} henry/m	$4\pi\times10^{-7}$ henry/m
Electrostatic induction	$\text{div } D = \alpha\cdot\rho$	$\alpha =$	4π	4π	4π	1	4π	1
Magnetic circuit	$\oint H_s\cdot ds = \beta\cdot I$	$\beta =$	4π	4π	$\dfrac{4\pi}{V}$	$\dfrac{1}{V}$	4π	1
Faraday's induction law	curl $E = -\lambda\cdot\dfrac{\partial B}{\partial t}$ $\left(\text{EMF} = -\lambda\cdot\dfrac{\partial\lambda}{\partial t}\right)$	$\lambda =$	1	1	$\dfrac{1}{V}$	$\dfrac{1}{V}$	1	1
Poynting vector	$S = s\cdot(E\times H)$	$s =$	$\dfrac{1}{4\pi}$	$\dfrac{1}{4\pi}$	$\dfrac{V}{4\pi}$	V	$\dfrac{1}{4\pi}$	1

Other relations: Conservation of electricity $\quad I = \dfrac{dQ}{dt}$.

First Maxwell's relation $\quad$ curl $H = \beta\cdot G + \dfrac{\beta}{\alpha}\cdot\dfrac{\partial D}{\partial t}$.

Specific electric field energy $\quad W_e = \dfrac{\beta}{\alpha}\cdot\dfrac{1}{2}ED$.

Specific magnetic field energy $\quad W_m = s\cdot\lambda\cdot 1/2\, HB$.

Velocity of propagation in free space $\quad c^2 = \dfrac{\alpha}{\beta\lambda}\cdot\dfrac{1}{\varepsilon_v\mu_v}$.

NOTES: 1. The units employed for the electromagnetic quantities are those from Table 4 for each system.

2. If units of various systems are used in one equation, the unit relations can be used for transformations, as for example:

$$\oint H\cdot ds = 4\pi\cdot I \text{ if } H \text{ is in oersteds, and } I \text{ in abamperes.}$$
$$= \frac{4\pi}{10} I \text{ if } H \text{ is in oersteds, and } I \text{ in amperes.}$$

3. The factor V (or a multiple thereof) is a pure number, identical with the magnitude of the velocity of light in free space if expressed in centimeters per second.

units are considered fundamental requires primary standards for these and removes the prototype kilogram mass as a primary standard.

The **MKS System of Units** is a dynamical or physical system and is based upon the meter, kilogram (mass), second, degree centigrade, and ohm as fundamental units. (See column 5, Table 3.) This system was accepted internationally in June, 1935, by the I.E.C. without, however, the originally proposed fundamental electrical quantity ohm; it is called the Giorgi-MKS system after its proponent.* The modification of this system leading to the "theoretical" MKSC system (see above) has found wide application.

The **"Definitive" System of Units** is an energetical system based upon the meter, second, watt, degree centigrade, and coulomb as fundamental units (column 9, Table 3). It combines the advantage of the mks system, the convenient relation between mechanical and electrical practical units, with the selection of charge as a fundamental unit. It has the disadvantage of discarding mass as a primary unit and replacing it by the watt for which no convenient standard is known. This system was proposed by G. A. Campbell † and was aimed at a unique, internationally acceptable, practical unit system.

9. THE INTERNATIONAL SYSTEM OF UNITS (SI)

As previously mentioned, this system of units, composed of six fundamental units has been adopted by the Conference Générale (BIPM Sèvres, Paris, 1954 and 1960) to cover the whole range of physics and one in which all international reports are to be expressed.

The six fundamental units

Length	metre	m
Mass	kilogram	kg
Time	second	s
Intensity of electric current	ampere	A
Thermodynamic temperature	degree Kelvin	°K
Luminous intensity	candela	cd

Derived units

Area	square metre	m^2	
Volume	cubic metre	m^3	
Frequency	hertz	Hz	
Density (mass density)	Kilogram per cubic metre	kg/m^3	
Velocity	metre per second	m/s	
Angular velocity	radian per second	rad/s	
Acceleration	metre per square second	m/s^2	
Angular acceleration	radian per square second	rad/s^2	
Force	newton	N	$kg.m/s^2$
Pressure	newton per square metre	N/m^2	
Kinematic viscosity	square metre per second	m^2/s	
Dynamic viscosity	newton-second per square metre	$N.s/m^2$	
Work, energy, heat (quantity of heat)	joule	J	N.m
Power	watt	W	J/s

Supplementary units

Plane angle	radian	rad	
Solid angle	steradian	sr	
Electric charge	coulomb	C	A.s
Electric potential, potential difference, electromotive force	volt	V	W/A
Electric field strength	volt per metre	V/m	
Resistance (to direct current)	ohm	Ω	V/A
Capacitance	farad	F	A.s/V
Magnetic flux	weber	Wb	V.s
Inductance	henry	H	V.s/A
Magnetic flux density (magnetic induction)	tesla	T	Wb/m^2
Magnetic field strength	ampere per metre	A/m	
Magnetomotive force	ampere	A	
Luminous flux	lumen	lm	cd.sr
Luminance	candela per square metre	cd/m^2	
Illumination	lux	lx	lm/m^2

UNITS AND STANDARDS

Dimensions of Units, see Dimension Systems, page 387.
Fundamental and Derived Units, see Unit Systems, page 396.
Conversion of Units, see Tables 30 to 67, Section 1.

---

* G. Giorgi, *Elec. World*, Vol. 40, pp. 355, 368, 1902.
† G. A. Campbell, *Bull. National Research Council*, No. 93, 1933.

10. LENGTH, MASS, AND TIME

The English Units and Standards

Units of Length. The foot (ft) is the *fundamental* unit of length in the foot-pound-second (fps) system. It equals, by definition, one-third of a *yard* (yd), which is the English legalized *standard* unit of length. The *United States yard* was defined by Act of Congress, July 28, 1866, as 3600/3937 the length of the *meter*. (See Metric System for definitions of metric length.)

In Great Britain, the *Imperial yard* is measured by a bronze bar preserved in the Standards Office, Westminster. Its length, in terms of the *international prototype meter*, is 3600/3937.0113 meter. For engineering purposes, the United States and British *yards* may be considered identical.

As **subunits**, the *inch* (in.) is defined as $1/12$ of one standard foot, and the *mil* as the one-thousandth part of one inch. The *nautical mile* (mi) is defined as one minute of arc on the earth's surface at the equator, whereas the United States mile (U. S. mi. statute) is exactly 5280 ft and practically identical with the British mile.

Unit of Capacity (Dry). The bushel (bu) is the *standard* unit of *dry* capacity. The *Winchester bushel* (U. S. standard) has a volume of 2150.42 cu. in.

In Great Britain, the *Imperial bushel* (bu) is defined as the volume of 80 lb of pure water at 62 F, weighed against brass weights in air at the same temperature as the water and with the barometer at 30 in. Its volume is approximately 2219.36 cu in.

Unit of Capacity (Liquid). The *gallon* (gal) is the *standard* unit of *liquid* capacity. The *United States gallon* has a volume of 231 cu in.

In Great Britain, the *Imperial gallon* is defined as the volume of 10 lb of pure water at 62 F, weighed against brass weights in air at the same temperature as the water and with the barometer at 30 in. Its volume is approximately 277.420 cu in. The Imperial gallon (liquid measure) equals exactly one-eighth of the Imperial bushel (dry measure). As subunits, there are used the quart (qt), which is $1/4$ of the standard gallon, and the pint (pt), which is $1/2$ qt.

Units of Mass. The pound (avoirdupois) (lb avdp) is the *fundamental* unit of mass in the fps system.* It is also the English legalized *standard* unit of mass. The *United States pound* (*avoirdupois*) was defined by Act of Congress, 1866, as 1/2.2046 kg, but since 1895 there has been used, for greater accuracy, a value which agrees with that given by law as far as the latter is given; namely, 453.5924277 grams. This value is now used by the Bureau of Standards as an exact definition and is the basis of the customary United States weights (Circular 47, Bureau of Standards).

In Great Britain, the *Imperial pound* (*avoirdupois*) is the mass of a *platinum cylinder* preserved in the Standards Office, Westminster. Its legal equivalent is 453.59243 grams. For engineering purposes, the United States and British pounds (avoirdupois) may be considered as identical.

Subunits of mass are the grain (gr), defined as $1/7000$ of the standard pound (avoirdupois) and the ounce (avoirdupois) (oz-avdp) which is $1/16$ of the standard pound (avoirdupois). The grain was used as fundamental unit in the so-called foot-grain-second (fgs) system of units prior to 1873.

Weight vs. Mass. Unfortunately, the word "weight" is used in two different senses, viz.: (1) by the layman (as well as loosely by the scientist) to designate a given *mass* or quantity of matter; and (2) by the scientist to designate the *pull* in standard gravitational force units which is exerted by the earth upon a piece of matter. The result of the commercial act of "weighing" a specific quantity is independent of the local gravitational pull of the earth, since both spring scales and balances are calibrated locally by comparison with standard masses.

For a method of determining whether or not a balance has equal arms and also for methods of correction of data obtained on an incorrect balance, see discussion on these subjects under the heading Accurate Weighing Procedure in this section.

Auxiliary Fundamental Units and Their Principal Derived Units are defined and discussed under the sections of this handbook pertaining to the topics to which they apply. In general, however, conversion factors are included in the tables of Section 1.

For an interesting and rather complete history see *British Weights and Measures*, London, 1910, by Sir C. M. Watson.

---

* The slug of mass, which is extensively used by engineers and physicists, is (in the English system) the mass to which an acceleration of one foot per second per second would be given by the application of a one-pound force. Under any gravity conditions, one slug of mass = 32.1739 lb. of mass.

The Metric (or French) Units and Standards

Units of Length. The centimeter (cm) is the *fundamental* unit of length in the cgs system. It equals, by definition, $1/100$ of a *meter* (m). The meter has been standardized by international agreement as 1,650,763.73 times the wave length in vacuum of the unperturbed transition $(2p_{10} - 5d_5)$ of Krypton 86. The *basic* meter for international comparisons is the *international prototype meter* which is the distance, at zero degrees Centigrade, between two lines on a platinum-iridium bar located at the International Bureau of Weights and Measures, at Sèvres, France. This meter is the nearest to a duplicate, ever constructed, of the *original* meter which was constructed and deposited in the Archives of the French Republic in 1799. The meter is very nearly equal to one ten-millionth of the distance, measured at sea level, from the equator to either pole.

An interesting history of the development of the *international prototype meter* (as well as the *international prototype kilogram*—see Unit of Mass, below) is given by Wm. Parry, National Bureau of Standards, in Merriman's Civil Engineers' Handbook as follows:

"The use of the meter as the basis of geodetic surveys had become so general throughout Europe that a conference was called in Paris, France, in 1870, for the purpose of establishing a central bureau where the standards of the different countries could be compared. As a result of this conference an International Bureau of Weights and Measures was established near Paris in 1875, by the concurrent action of the principal nations of the world. One of the first tasks undertaken by the Bureau was the construction of exact copies of the meter and kilogram deposited in the Archives. Thirty-one standard meters of iridio-platinum and forty kilograms of the same alloy were constructed and carefully compared with the standards of the Archives and with one another. This great work was completed in 1889, and the meter and kilogram which agreed most nearly with the original standards were called international prototypes, and were deposited at the International Bureau, where they are maintained today subject to the authority of the International Committee on Weights and Measures. The remaining meters and kilograms were distributed by lot to the different nations which contributed to the support of the Bureau. The United States secured two copies of the meter and two copies of the kilogram, which are in the custody of the Bureau of Standards at Washington. One of the meters, known as No. 27, and one kilogram, No. 20, were selected as the United States standards, while the other meter and kilogram are used as secondary standards. It was the declared intention of the International Committee that the various national prototypes should be returned to the International Bureau at regular intervals for the purpose of recomparing them with the international standards and with one another. In this way all measurements based upon metric standards throughout the world are ultimately referred to the international meter and kilogram."

Unit of Capacity. The liter (l) is the *standard* unit of capacity. It is defined as the volume of one kilogram of pure water at the temperature of maximum density (4 C) under a pressure of 76 cm of mercury. For all practical purposes, the liter may be regarded as the equivalent of the cubic decimeter, although the former is actually slightly greater, in the amount of less than three parts in one hundred thousand.

Unit of Mass. The gram (g) is the *fundamental* unit of mass in the cgs system.* It equals, by definition, $1/1000$ of a *kilogram* (kg), which is the *standard* unit of mass. The *basic* kilogram for international comparisons is the *international prototype kilogram* which is a cylinder of platinum-iridium located at the International Bureau of Weights and Measures, at Sèvres, France. This mass is the nearest to a duplicate, ever constructed, of the *original* kilogram which was constructed and deposited in the Archives of the French Republic in 1799. The latter was made as nearly as possible equal to the mass of a cube of pure water at 4 C, the sides of the cube being one-tenth the length of the original meter.

An interesting history of the development of the *international prototype kilogram* is given above under the discussion headed Units of Length.

Weight vs. Mass. See discussion under this same subheading of The English Units and Standards, p. 405.

Auxiliary Fundamental Units and Their Principal Derived Units are defined and discussed under the sections of this handbook pertaining to the topics to which they apply. In general, however, conversion factors are included in tables 31 to 66 of Section 1.

---

* The slug of mass, which is extensively used by engineers and physicists, is (in the metric system) the mass to which an acceleration of one meter per second per second would be given by the application of a one-kilogram force. Under any gravity conditions, one slug of mass = 9.80665 kg of mass.

The Standard of Time

Unit of Time. (1) The second has been standardized by international agreement as 1/31,556,925.9747 of the Tropical Year at 12 hr, Ephemeris Time, Jan 0 for the year 1900.0. (The above definition has been retained for the time being as an Astronomical Time Standard—the following atomic standard of time interval is 100 times more precise.) The second has been standardized by international agreement as the time taken for 9,192,631,770.0 vibrations of the unperturbed hyperfine transition 4,0—3,0 for the $^2S_{1/2}$ fundamental state of the Caesium 133 atom. The Cs^{133} standard has been adopted provisionally (see resolution 5 of the 12th General Conference of Weights and Measures, BIPM, Sèvres, Paris, Oct. 1964). A more accurate hydrogen maser standard may be available in the near future which is 100 times more accurate than the Cs^{133} standard.

Measures of Time. A *solar day* is measured by the rotation of the earth about its axis, with respect to the sun. In *astronomical computations* and in *nautical time* the day commences at noon, and in the former it is counted throughout the 24 hours. In *civil computations* the day commences at midnight, and is divided into two parts of 12 hours each.

A *solar year* is the time in which the earth makes one revolution around the sun. Its average time, called the *mean solar year*, is 365 days, 5 hours, 48 minutes, and 45.9747 seconds, or nearly 365 ¼ days.

Accurate Weighing Procedure

To Determine Whether a Balance Has Equal Arms. After weighing an article and obtaining equilibrium, interchange the article and the weights. If equilibrium still obtains, the balance is true (assuming friction negligible); if equilibrium does not obtain, the pan which descends is suspended from the longer arm.

To Weigh Correctly on an Incorrect Balance. "Weigh" the article first on one pan and then on the other. The correct weight is the square root of the product of the apparent weights thus obtained.

11. FORCE, ENERGY, AND POWER

Dynamical and Gravitational Units. According to the use of two different dimension and unit systems, the dynamical (or physical, or "absolute") system and the gravitational (or technical) system, two different sets of units of force, energy, power, and derived quantities are defined in both the English and the metric systems. *One dynamical unit of force* produces an acceleration of unity on unit standard mass. The *gravitational unit of force* is defined as that force required to give a unit standard mass an acceleration equal to that produced by the gravitational pull of the earth. As the acceleration due to gravity, g, varies with location and altitude,* the gravitational unit of force is not constant, and, therefore, its relation to the dynamical unit of force will vary. By international agreement, the value $g_0 = 980.665$ cm per sec per sec = 32.1739 ft per sec per sec (British) has been chosen as the standard acceleration of gravity to make invariant the gravitational unit of force.

The English Units

Units of Force. The dynamical or physical unit of force is the *poundal*, defined as the force required to give a mass of one pound an acceleration of one foot per second per second.

The *pound-force* (or weight of the pound mass) is the gravitational or technical unit of force. It is, by definition, the force required to give a mass of one pound an acceleration of 32.1739 ft per sec per sec. If a force is measured by "weighing," the result in pounds weight must be multiplied by g/g_0, the ratio of local to standard acceleration of gravity, in order to obtain the absolute value in pound-force units. For engineering purposes this correction can usually be neglected.

The Unit of Pressure is defined as the unit of force acting upon a unit area. The most commonly employed unit is the *pound* (force) *per square inch*.

Pressure is measured also by the height in inches of the column of water at 4 C (39.1 F)

---

* The variation of g with latitude ϕ and altitude H is given approximately by (ϕ in degrees, H in meters): $g = 978.039 \ (1 + 0.005295 \sin^2 \phi) - 0.000307 \ H$. See *International Critical Tables*, Vol. 1, p. 395.

or of the column of mercury at 32 F which it supports. (See Conversion Tables.)

Units of Work or Energy. The foot-poundal is the physical unit of work or energy and is defined as the work done by a force of one poundal in moving a body through the distance of one foot in the direction of the force.

The *foot-pound (force)* is the technical unit of work or energy and is defined as the work required to raise a mass (or weight) of one pound through a vertical distance of one foot at standard acceleration of gravity g_0. If measurements are made in places where the local value of the acceleration of gravity g is different from g_0, a correction factor g/g_0 must be applied, if the exact value of work or energy is desired.

The *British thermal unit* (Btu) is the quantity of heat required to raise the temperature of a one-pound mass of water either at 39 F (at its maximum density) or at 60 F, and standard pressure, through 1 F. The mean British thermal unit is defined as the $1/180$ part of the heat required to raise the temperature of one pound mass of water from 32 to 212 F at standard pressure. It is obvious that the reference temperature must be indicated with the unit used.

Units of Power. Power is the time rate at which work is done. Its physical unit is the *foot-poundal per second*, its technical units are the *foot-pound (force) per second*, or the *British thermal unit per second*. The *horsepower* (hp or Hp) is defined as 33,000 ft-lb (force) per min or 550 ft-lb (force) per sec.

Units of Torque. Torque is the effectiveness of a force to produce rotation. It is defined as the product of the force and the perpendicular distance from its line of action to the instantaneous center of rotation. Its physical unit is the poundal-foot, and its technical unit the pound (force)-foot. (Note the reversal of force and length units in the designation of the units of torque as compared with the units of energy or work.)

The Metric Units

Units of Force. The dynamical, or physical, unit of force is the *dyne*, defined as the force required to give a mass of one gram an acceleration of one centimeter per second per second.

The *newton* is the MKS unit of force. It is the force required to give a mass of one kilogram an acceleration of one meter per sec per sec.

The *kilogram force* (or weight of the kilogram mass) is the gravitational or technical unit of force. It is, by definition, the force required to give a mass of one kilogram an acceleration of 980.665 cm per sec per sec. If a force is measured by "weighing," the result in kilograms weight must be multiplied by g/g_0, the ratio of local to standard acceleration of gravity, in order to obtain the absolute value in kilogram-force units. For engineering purposes this correction can usually be neglected.

In the electrotechnical system of units the systematic unit of force is defined as the *joule per meter*, based upon the fundamental definition of the joule. (See Metric Units of Energy.)

The Unit of Pressure is defined as the unit of force acting upon a unit of area. The physical unit is the *barye* or the pressure of one dyne upon one square centimeter. A larger unit is the *bar* or 10^6 baryes, which has been adopted internationally. (The use of bar for one dyne per square centimeter is discouraged.)

The *newton per square meter* is the MKS unit of pressure.

The *kilogram force per square meter* is the technical unit of pressure. With respect to correction for local gravity, see Force vs. Weight, above.

Pressure is measured also by the height in centimeters of the column of water at 4 C, or of the column of mercury at 0 C, which it supports. (See Conversion Tables.)

The *normal atmosphere* (at), or the standard atmospheric pressure, is defined as the pressure exerted by a column of 76 cm of mercury at sea level and 0 C at standard acceleration of gravity g_0. It is equal to 1.01321 bars or 1.0332 kg force per sq cm and is used extensively in the engineering literature. Some confusion exists since the unit of 1 kg force per sq cm is occasionally called 1 practical atmosphere.

Units of Work or Energy. The *erg* is the physical or so-called absolute unit of work or energy. It is defined as the work done by a force of one dyne acting through the distance of one centimeter. A larger unit is the *theoretical* or "absolute" *joule* defined as 10^7 ergs; it is a systematic unit in the practical electrical unit systems which is based upon the theoretical unit systems. (See Electrical Units.)

The *international joule* is defined as the energy expended during one second by an electric current of one international ampere flowing through a resistance of one international ohm. (See Electrical Units.) The latest value of the international joule is equal to 1.000165 theoretical joules.*

The *kilowatt-hour* is the practical unit of energy in electrical metering. It is defined as a theoretical or an international unit (see definition of joule above) and is equal to 3600 $\times 10^3$ joules.

The *newton meter* is the MKS unit of work. It is the work done by a force of one newton moving its point of application through one meter in the direction of the force.

The *meter-kilogram force* (commonly referred to as the kilogram-meter) is the technical unit of work or energy. It is defined as the work required to raise the mass (or weight) of one kilogram through a vertical distance of one meter at standard acceleration of gravity g_0. If measurements are made in places where the local value of the acceleration of gravity g is different from g_0, a correction factor g/g_0 has to be applied, if the exact value of work or energy is desired. (See Force vs. Weight.)

The *gram-calorie* or small calorie is the physical unit of heat energy. It is defined as the quantity of heat required to raise the temperature of one gram mass of water either from 14.5 to 15.5 C or from 19.5 to 20.5 C, at standard pressure. The two values are designated as 15 C cal and 20 C cal, respectively. The mean gram-calorie is defined as $1/100$ part of the quantity of heat required to raise the temperature of one gram mass of water from 0 to 100 C at standard pressure. The same definitions apply to kilogram-calorie, or large calorie, if the kilogram mass is used as reference standard mass.

The *Ostwald calorie* is the quantity of heat required to raise the temperature of one gram mass from 0 to 100 C. This unit is frequently used by electrochemists and is equal to 100 mean gram-calories.

The *international kilo-calorie* or international steam-table calorie (I T cal) is defined as the 1/860th part of the international kilowatthour. This new unit avoids any reference to the thermal properties of water and was recommended for international adoption at the first International Steam Table Conference (1929).[†] Its value is very nearly equal to the mean kilo-calorie, 1 I T cal = 1.00037 kilogram-calories (mean).

Units of Power. Power is the time rate at which work is done. Its physical unit is the *erg per second*.

A larger unit is the *theoretical* or "absolute" *watt*, defined as 10^7 ergs per second; it is a systematic unit in the practical electrical unit system which is based upon the theoretical unit systems (see electrical units).

The *international watt* is defined as the power expended by an electric current of one international ampere flowing through a resistance of one international ohm. (See Electrical Units.) The latest value of the international watt is equal to 1.000165 theoretical watts.[*]

The *electrical horsepower* is defined as 746 absolute watts and is commonly used in the United States and in England in rating electrical machinery.

The *meter-kilogram force per second* (commonly referred to as the kilogram-meter per second) is the technical unit of power. The *metric horsepower* is defined as 75 kg-meters per sec and is the most common mechanical unit of power.

Units of Torque. Torque is the effectiveness of a force to produce rotation. It is defined as the product of the force and the perpendicular distance from its line of action to the instantaneous center of rotation. Its physical unit is the dyne-centimeter, and its technical unit the kilogram force meter. (Note the reversal of force and length units in the designation of the units of torque as compared with the units of energy and work.)

12. THERMAL UNITS AND STANDARDS

Temperature

Definition of Temperature. The *temperature* of a body may be defined as its thermal state considered from the standpoint of its ability to communicate heat to other bodies. When two bodies are placed in thermal communication, the one which loses heat to the other is said to be at the higher temperature.

Standard Temperatures. Certain thermal states or "temperatures" may be reproduced and recognized by the fact that definite physical phenomena occur at these temperatures. Such thermal states are called "fixed points," and they may, quite apart from any temperature scale, be specified by the physical phenomena characteristic of those temperatures. The two fundamental fixed points are the ice point and the steam point. The *ice point* is defined as the temperature of melting ice, which is realized experimen-

---

* *Announcement of Changes in Electrical and Photometric Units*, Circular of National Bureau of Standards C459, Washington, D. C., 1947.

† *Mechanical Engineering*, Feb., 1930, pp. 122, 139; Nov., 1935, p. 710.

tally as the temperature at which pure finely divided ice is in equilibrium with pure, air-saturated water, under standard atmospheric pressure. The effect of increased pressure is to lower the freezing point to the extent of 0.007 C per atmosphere.

The *steam point* is defined as the temperature of condensing water vapor at standard atmospheric pressure, and it is realized experimentally by the use of a hypsometer so constructed as to avoid superheat of the vapor around the thermometer, or contamination with air or other impurities. If the desired conditions have been attained, the observed temperature should be independent of the rate of heat supply to the boiler, except as this may affect the pressure within the hypsometer, and of the length of time the hypsometer has been in operation.

Definition of Temperature Scale. The purpose of establishing a temperature scale is to assign a number to every thermal state or temperature, and to provide a means for determining the temperature of any particular body.

A *temperature scale* may be defined by (1) selecting definite numbers for certain fixed points, (2) selecting some physical property of a definite substance which varies with temperature, and (3) selecting a mathematical law expressing temperatures on the scale in question in terms of the selected property of the thermometric substance. For example, on the Centigrade mercury-in-glass scale, the ice and steam points are numbered 0 and 100 respectively, the relative or "apparent" expansion of a volume of mercury enclosed in glass of a definite kind is the property used, and the mathematical relation used to express temperature on this scale is that equal increments of apparent volume of the mercury in this glass correspond to equal increments of temperature. If some other substance is substituted for mercury, or if glass of a different kind is used, another scale is obtained which agrees with it at 0 and 100 but not at other temperatures.

Although, in general, a temperature scale depends on the thermometric substance as well as on the expression for the temperature in terms of some property of this substance, Lord Kelvin has shown that, if the property selected is the availability of energy, the scale so defined is wholly independent of the substance and depends only on the mathematical relation chosen. Any scale so defined is known as a thermodynamic scale.

The Kelvin Temperature Scale. The temperature scale finally chosen by Lord Kelvin is the one on which the temperature interval from the ice point to the steam point is 100° and the ratio of the values of any two temperatures is equal to the ratio of the heat taken in to the heat rejected by a reversible thermodynamic engine working with a source and refrigerator at the higher and lower temperatures respectively. On this scale, which is also known as the absolute thermodynamic scale, the lowest attainable temperature is 0 and the ice point is found experimentally to be 273.16°. The steam point therefore is 373.16° or 100° higher.

The *degree Kelvin* (°K) or degree of absolute temperature is the absolute unit of temperature and is, for practical purposes, identical with the degree Centigrade (°C) of the international temperature scale.

The Thermodynamic Centigrade Scale is derived by subtracting from the Kelvin scale a constant number of the proper magnitude to make the ice point 0°. On this scale, therefore, the ice and steam points are 0 and 100°, respectively, and the so-called absolute zero is −273.16°.

The International Centigrade Scale is a practical representation of the thermodynamic Centigrade scale to such a degree of accuracy as is possible with present-day apparatus and methods. It was adopted at the General Conference on Weights and Measures at Sèvres, France, in 1927 and is subject to revision and amendment as improved and more accurate methods of measurement are evolved.

The unit of temperature on the international scale is the *degree Centigrade* (°C, or °C int) and is very nearly equal to $1/100$ the difference between the temperature of melting ice and the temperature of condensing water vapor under standard atmospheric pressure. (See Metric Units for Pressure.)

The standard of the international temperature scale between −190 and +660 C is deduced from the electrical resistance of a standard platinum resistance thermometer by means of a formula connecting the resistance R_t at any temperature t C within the above range with the resistance R_0 at 0 C. The purity of the platinum of which the thermometer is made should be such that the ratio R_t/R_0 for certain fixed temperatures is within specified limits.*

The degree Centigrade is most widely used in scientific publications and increasingly also in the engineering literature. In many countries in Europe it is the common everyday temperature unit. The subdivision into a hundred degrees of the temperature interval between the ice point and the steam point was first used by Celsius, a German, in 1742;

---

* See also U. S. Bureau of Standards, *Journal of Research*, Vol. 1, p. 636, 1928.

therefore, in the European literature "°C" is read "degree Celsius."

The **Fahrenheit Temperature Scale** subdivides the temperature interval between the ice point and the steam point into 180 parts, one part of which is chosen as the unit of temperature and named *degree Fahrenheit* (°F). The ice point is assigned the value 32 F, so that the steam point has a temperature of 212 F.

The Fahrenheit unit of temperature is in common everyday use in the English-speaking countries. It was first introduced in England about 1665 by the physicist Fahrenheit; the choice of 32 F for the ice point has its explanation in the fact that Fahrenheit chose as zero the lowest temperature attainable by means of a salt-ice mixture.

The **Rankine Absolute Temperature Scale** (°R) is the thermodynamic fahrenheit scale where absolute zero is 0 R (−459.69 F). The ice point is assigned the value 491.69 R and the steam point 671.69 R.

Relations between the Temperature Scales. In Table 1 are shown the interrelations between the various temperature scales in form of equations. X indicates the unknown number of chosen temperature units and t the known number of given temperature units.

Table 1. Relations between the Temperature Scales

$x°$ F =	....	$9/5 (t° K − 273.16) + 32$	$9/5 (t° C) + 32$
$x°$ K =	$5/9 (t° F − 32) + 273.16$	....	$(t° C) + 273.16$
$x°$ C =	$5/9 (t° F − 32)$	$(t° K) − 273.16$	....
$x°$ R =	$(t° F) + 459.699$	$9/5 (t° K)$	$9/5 (t° C) + 491.69$

Quantity of Heat and Some Derived Quantities

Units of Quantity of Heat. Quantity of heat is defined as the energy transferred from one body to another by a thermal process, i.e., by radiation or conduction. The units for the quantity of heat are the *British thermal unit* and the *calorie*, as specific thermal units; and the *erg* and *joule*, as general physical units (see units of energy, metric and English system of units).

Thermal Capacity or Specific Heat of a Substance is the quantity of heat required to produce a unit change in temperature in a unit of mass of the substance. The common English unit is the British thermal unit per degree Fahrenheit per pound mass (Btu per °F per lb); the usual metric unit is the gram-calorie per degree Centigrade per gram mass (cal per °C per g); and the general physical unit used in the scientific literature is the erg per degree Centigrade per gram mass (erg per °C per g). In the technical literature thermal capacity of a substance is often expressed in watt-seconds (or joules) per degree Centigrade per kilogram mass (watt-sec per °C per kg) on account of the easy comparison with other technical units.

The Calorimetric or Water Equivalent is the quantity of heat required to produce a unit change in temperature of a body or system. It is numerically equivalent to the mass of water (in units as involved in the definition of the unit of quantity of heat used) which could be raised a unit temperature by the same total quantity of heat. The thermal capacity is expressed in British thermal units per degree Fahrenheit (Btu per °F), calories per degree Centigrade (cal per °C), or watt-seconds per degree Centigrade (watt-sec per °C).

Thermal Conductivity is the time rate of heat transfer through unit area across unit thickness per unit difference in temperature between the end surfaces. It is measured in British thermal units per second per degree Fahrenheit per inch thickness per square inch cross section (Btu per sec per °F per in. per sq in.), in calories per second per degree Centigrade per centimeter thickness per square centimeter cross section (cal per sec per °C per cm per sq cm), or in watts per degree Centigrade per meter thickness per square meter cross section (watts per °C per m per sq m).

Thermal Transmittance * or surface coefficient of transfer is the time rate of heat emitted by unit area for unit difference in temperature between the surface in question and the surroundings. It is measured in British thermal units per second per degree Fahrenheit per square inch (Btu per sec per °F per sq in.), in calories per second per degree Centigrade per square centimeter (cal per sec per °C per sq cm), or in watts per degree Centigrade per square meter (watts per °C per sq m).

The Joule Equivalent

The **Joule Equivalent** is defined as the number of foot-pounds of energy per Btu. In

---

* A. I. Brown and S. M. Morco, *Introduction to Heat Transfer*, McGraw-Hill, New York, 1942.

Table 2 are given the numerical values for the various energy units used in the English and metric systems.*

Table 2. Relations between Energy Units

	Joules "Absolute"	Foot-pounds (force)	Foot-poundals	Meter-kilogram (force)	Kilowatt-hour "International"
1 British thermal unit (Btu) (mean) =	1055.18	778.26	25,040	107.599	2.93019×10^{-4}
1 gram-calorie (cal) (mean)..... =	4.1873	3.0884	99.366	0.42699	1.16279×10^{-6}
1 International kilocalorie (I T-cal) =	4187.3	3088.4	99,366	426.99	1.16279×10^{-3}
1 Ostwald calorie.............. =	418.73	308.84	9936.6	42.699	1.16279×10^{-4}

13. THE THEORETICAL OR "ABSOLUTE" ELECTRICAL UNITS

The Definitions of the Theoretical, or "Absolute," Units are based upon a particular choice of the numerical value of either k_e, the constant in Coulomb's electrostatic force law (1), Art. 5, or k_m, the constant in Ampère's electrodynamic force law (2), Art. 5. The designation "absolute" units is generally used because of historical tradition (see Art. 5); an interesting account of the history can be found in Glazebrook's *Handbook for Applied Physics*, Vol. II, "Electricity," pp. 211 ff., 1922. Because of the theoretical background of the unit definitions, they have also been designated as "theoretical" units which is in good contradistinction to practical units based upon physical standards.

The Theoretical Electrostatic Units

The Theoretical Electrostatic Units are based upon the cgs system of mechanical units and the choice of the numerical value unity for k_{ev} in Coulomb's law (1), Art. 5. They are frequently referred to as the *cgs electrostatic units*, but no specific unit names are available. In order to avoid the cumbersome writing, for example, one "theoretical electrostatic unit of charge," it had been proposed to use the theoretical "practical" unit names and prefix them with either stat † or E.S.,‡ as for example, statcoulomb, or E.S. coulomb. The first alternative will be used here.

The Absolute Dielectric Constant (Permittivity) of free space is the reciprocal of the Coulomb constant k_{ev} and is chosen as the fourth fundamental quantity in the theoretical electrostatic system of units. Its numerical value is defined as unity, and it is identical with one statfarad per centimeter if use is made of prefixing the corresponding unit of the "practical" series.

The Theoretical Electrostatic Unit of Charge or the statcoulomb is defined as the quantity of electricity which, when concentrated at a point and placed at one centimeter distance from an equal quantity of electricity similarly concentrated, will experience a mechanical force of one dyne in free space. An alternative definition, based upon the concept of field lines, gives the theoretical electrostatic unit of charge as a positive charge from which in free space exactly 4π displacement lines emerge.

The Theoretical Electrostatic Unit of Displacement Flux (Dielectric Flux) is the "line of displacement flux" or $1/4\pi$ of the theoretical electrostatic unit of charge. This definition provides the basis for graphical field mapping in so far as it gives a definite rule for the selection of displacement lines to represent the distribution of the field quantitatively.

The Theoretical Electrostatic Unit of Displacement, or Dielectric Flux Density, is chosen as one displacement line per square centimeter area perpendicular to the direction of the displacement lines. It can be given also as $1/4\pi$ statcoulomb per square centimeter (according to Gauss's law). In isotropic media the displacement has the same direction as the potential gradient, and the surfaces perpendicular to the field lines become the equipotential surfaces; the theoretical electrostatic unit of displacement can then be defined as one displacement line per square centimeter of equipotential surface.

The Theoretical Electrostatic Unit of Electrostatic Potential or the Statvolt is defined

---

* For a rather complete historical account see R. Glazebrook, *Dictionary of Applied Physics*, Macmillan & Co., London, Vol. I, p. 477, 1922.

† Proposed by A. E. Kennelly at the International Electrical Congress in St. Louis, 1904, *Trans.*, Vol. 1, p. 180, and used by C. Hering, *Conversion Tables*, J. Wiley & Sons, 1904.

‡ Proposed by A. E. Caswell, *Science*, Vol. 42, p. 695, 1915, and used by E. Bennett. A digest of the relations between the electrical units, *Bull. Univ. Wisconsin*, No. 880, 1917.

as existing at a point in an electrostatic field, if the work done to bring the theoretical electrostatic unit of charge, or the statcoulomb, from infinity to this point equals one erg. This customary definition implies, however, that the potential vanishes at infinite distances and has, therefore, only restricted validity. As it is fundamentally impossible to give absolute values of potential the use of potential difference and its unit (see below) should be preferred.

The Theoretical Electrostatic Unit of Electrical Potential Difference or Voltage, or the Statvolt, is defined as existing between two points in space if the work done to bring the theoretical electrostatic unit of charge, or the statcoulomb, from the one of these points to the other equals one erg. Potential difference is counted positive in the direction in which a negative quantity of electricity would be moved by the electrostatic field.

The Theoretical Electrostatic Unit of Capacitance or the Statfarad is defined as the capacitance which maintains an electrical potential difference of one statvolt between two conductors charged with equal and opposite electrical charges of one statcoulomb. In the older literature, the cgs electrostatic unit of capacitance is identified with the "centimeter"; this should be replaced by statfarad to avoid confusion. (See CGS Electrostatic Unit System, Art. 8.)

The Theoretical Electrostatic Unit of Electric Potential Gradient, or Field Strength (field intensity), is defined to exist at a point in an electric field, if the mechanical force exerted upon the theoretical electrostatic unit of charge concentrated at this point is equal to one dyne. It is expressed as one statvolt per centimeter.

The Theoretical Electrostatic Unit of Current or the Statampere is defined as the time rate of transfer of the theoretical electrostatic unit of charge and is identical with the statcoulomb per second.

The Theoretical Electrostatic Unit of Electrical Resistance or the Statohm is defined as the resistance of a conductor in which a current of one statampere is produced if a potential difference of one statvolt is applied at its ends.

The Theoretical Electrostatic Unit of Electromotive Force (emf) is defined as equivalent to the theoretical electrostatic unit of potential difference if it produces a current of one statampere in a conductor of one statohm resistance. It is identical with the statvolt but, according to its concept, requires an independent definition.

The Theoretical Electrostatic Unit of Magnetic Intensity is defined as the magnetic intensity at the center of a circle of 4π cm diameter in which a current of one statampere is flowing. This unit is equal to 4π statamperes per centimeter but has no name as the factor 4π excludes the possibility of using the prefixed "practical" unit name.

The Theoretical Electrostatic Unit of Magnetic Flux or the Statweber is defined as the magnetic flux whose time rate of change through a linear conductor loop (linear conductor is used to designate a conductor of infinitely small cross-section) produces in this loop an electromotive force (emf) of one statvolt.

The Theoretical Electrostatic Unit of Magnetic Flux Density, or Induction, is defined as the electrostatic unit of magnetic flux per square centimeter area, or the statweber per square centimeter.

The Absolute Magnetic Permeability of Free Space is defined as the ratio of magnetic induction to the magnetic intensity. Its unit is the stathenry per centimeter, as a derived unit; its value is given in Table 5.

The Theoretical Electrostatic Unit of Inductance or the Stathenry is defined as connected with a conductor loop carrying a steady current of one statampere which produces a magnetic flux of one statweber. A more general definition, applicable to varying fields with non-linear relation between magnetic flux and current, gives the stathenry as connected with a conductor loop in which a time rate of change in the current of one statcoulomb produces a time rate of change in the magnetic flux of one statweber per second.

The Theoretical Electromagnetic Units

The Theoretical Electromagnetic Units are based upon the cgs system of mechanical units and Coulomb's law of mechanical force action between two isolated magnetic quantities m_1 and m_2 (approximately true for very long bar magnets) which must be written

$$F_m = \frac{k_m}{2} \frac{m_1 m_2}{r^2} \tag{1}$$

where k_m is the proportionality constant of Ampère's law (2), Art. 5. for force action between parallel currents which is more basic, and amenable to much more accurate measurement, than (1). The factor $1/2$ appears here because of the three-dimensional character of the field distribution around point magnets as compared with the two-dimensional field of two parallel currents. (See also the factor 2π rather than 4π in corresponding relations in Table 5.) Relation (1) also demonstrates that magnetic *force action* is related to magnetic induction **B** and *not to magnetic intensity* **H**; this has been recognized clearly rather recently.*

The theoretical electromagnetic units are obtained by defining the numerical value of $k_{mv}/2$ (for vacuum) as unity; they are frequently referred to as the cgs electromagnetic units. Only a few specific unit names are available. In order to avoid cumbersome writing, for example, one "theoretical electromagnetic unit of charge," it had been proposed to use the theoretical "practical" unit names and prefix them with either ab- † or E.M.,‡ as for example abcoulomb, or E.M. coulomb. The first alternative will be used here.

The **Absolute Magnetic Permeability** of free space is the value $k_{mv}/2$ in (1) and is chosen as the fourth fundamental quantity in the theoretical electromagnetic system of units. Its numerical value is assumed as unity, and it is identical with one abhenry per centimeter if use is made of prefixing the corresponding unit of the "practical" series.

The **Theoretical Electromagnetic Unit of Magnetic Quantity** is defined as the magnetic quantity which, when concentrated at a point and placed at one centimeter distance from an equal magnetic quantity similarly concentrated, will experience a mechanical force of one dyne in free space. An alternative definition, based upon the concept of magnetic intensity lines, gives the theoretical electromagnetic unit of magnetic quantity as a positive magnetic quantity from which, in free space, exactly 4π magnetic intensity lines emerge.

The **Theoretical Electromagnetic Unit of Magnetic Moment** is defined as the magnetic moment possessed by a magnet formed by two theoretical electromagnetic units of magnetic quantity of opposite sign, concentrated at two points one centimeter apart. As a vector, its positive direction is defined from the negative to the positive magnetic quantity along the center line.

The **Theoretical Electromagnetic Unit of Magnetic Induction (Magnetic Flux Density), or the Gauss,** is defined to exist at a point in a magnetic field, if the mechanical torque exerted upon a magnet with theoretical electromagnetic unit of magnetic moment and directed perpendicular to the magnetic field is equal to one dyne-centimeter. The lines to which the vector of magnetic induction is tangent at every point are called induction lines or magnetic flux lines; on the basis of this flux concept, magnetic induction is identical with magnetic flux density.

The **Theoretical Electromagnetic Unit of Magnetic Flux, or the Maxwell,** is the "field line" or line of magnetic induction. In free space, the theoretical electromagnetic unit of magnetic quantity issues 4π induction lines; the unit of magnetic flux, or the maxwell, is then $1/4\pi$ of the theoretical electromagnetic unit of magnetic quantity times the absolute permeability of free space.

The **Theoretical Electromagnetic Unit of Magnetic Intensity (Magnetizing Force), or the Oersted,** is defined to exist at a point in a magnetic field in free space where one measures a magnetic induction of one gauss.

The **Theoretical Electromagnetic Unit of Current, or the Abampere,** is defined as the current which flows in a circle of one centimeter diameter and produces at the center of this circle a magnetic intensity of one oersted.

The **Theoretical Electromagnetic Unit of Inductance, or the Abhenry,** is defined as connected with a conductor loop in which a time rate of change of one maxwell per second in the magnetic flux produces a time rate of change in the current of one abampere per second. In the older literature, the cgs electromagnetic unit of inductance is identified with the "centimeter"; this should be replaced by a henry to avoid confusion. (See CGS Electromagnetic Unit System.)

The **Theoretical Electromagnetic Unit of Magnetomotive Force (mmf)** is defined as the magnetic driving force produced by a conductor loop carrying a steady current of $1/4\pi$ abamperes; it has the name one gilbert. The concept of magnetomotive force as the driving force in a "magnetic circuit" permits an alternative definition of the gilbert as the magnetomotive force which produces a uniform magnetic intensity of one oersted over a length of one centimeter in the magnetic circuit. Obviously, one gilbert equals one oersted-centimeter.

---

* See particularly G. H. Livens, *The Theory of Electricity*, Cambridge Univ. Press, 1926; A. Sommerfeld, *Physik. Z.*, Vol. 36, p. 814, 1935; J. A. Stratton, *Electromagnetic Theory*, p. 241, McGraw-Hill, New York, 1941.

† Proposed by A. E. Kennelly at the International Electrical Congress in St. Louis, 1904, *Trans.*, Vol. 1, p. 180, and used by C. Hering, *Conversion Tables*, John Wiley & Sons, 1904.

‡ Proposed by A. E. Caswell, *Science*, Vol. 42, p. 695, 1915, and used by E. Bennett. A digest of the relations between the electrical units, *Bull. Univ. Wisconsin*, No. 880, 1917.

The Theoretical Electromagnetic Unit of Magnetostatic Potential is defined as the potential existing at a point in a magnetic field, if the work done to bring the theoretical electromagnetic unit of magnetic quantity from infinity to this point equals one erg. This customary definition implies, however, that the potential vanishes at infinite distances, and the definition has, therefore, only restricted validity. The unit, thus defined, is identical with one gilbert. The difference in magnetostatic potential between any two points is usually called magnetomotive force (mmf).

The Theoretical Electromagnetic Unit of Reluctance is defined as the reluctance of a magnetic circuit in which a magnetomotive force of one gilbert produces a magnetic flux of one maxwell.

The Theoretical Electromagnetic Unit of Electric Charge, or the Abcoulomb, is defined as the quantity of electricity which passes through any section of an electric circuit in one second if the current is one abampere.

The Theoretical Electromagnetic Unit of Displacement Flux (Dielectric Flux) is the "line of displacement flux" or $1/4\pi$ of the theoretical electromagnetic unit of electric charge. This definition provides the basis for graphical field mapping in so far as it gives a definite rule for the selection of displacement lines to represent the character of the field.

The Theoretical Electromagnetic Unit of Displacement, or Dielectric Flux Density, is chosen as one displacement line per square centimeter area perpendicular to the direction of the displacement lines. It can also be given as $1/4\pi$ abcoulombs per square centimeter (according to Gauss's law). In isotropic media the theoretical electromagnetic unit of displacement can be defined as one displacement line per square centimeter of equipotential surface. (See Theoretical Electrostatic Unit of Displacement.)

The Theoretical Electromagnetic Unit of Electrical Potential Difference or Voltage, or the Abvolt, is defined as the potential difference existing between two points in space if the work done in bringing the theoretical electromagnetic unit of charge, or the abcoulomb, from one of these points to the other equals one erg. Potential difference is counted positive in the direction in which a negative quantity of electricity would be moved by the electrostatic field.

The Theoretical Electromagnetic Unit of Capacitance, or the Abfarad, is defined as the capacitance which maintains an electrical potential difference of one abvolt between two conductors charged with equal and opposite electrical quantities of one abcoulomb.

The Theoretical Electromagnetic Unit of Potential Gradient, or Field Strength (field intensity), is defined to exist at a point in an electric field if the mechanical force exerted upon the theoretical electromagnetic unit of charge concentrated at this point is equal to one dyne. It is expressed as one abvolt per centimeter.

The Theoretical Electromagnetic Unit of Resistance, or the Abohm, is defined as the resistance of a conductor in which a current of one abampere is produced if a potential difference of one abvolt is applied at its ends.

The Theoretical Electromagnetic Unit of Electromotive Force (emf) is defined as the electromotive force acting in an electric circuit in which a current of one abampere is flowing and electrical energy is converted into other kinds of energy at the rate of one erg per second. This unit is identical with the abvolt.

The Absolute Dielectric Constant of Free Space is defined as the ratio of displacement to the electric field intensity. Its unit is the abfarad per centimeter, a derived unit; its value is given in Table 5.

The Theoretical Electrodynamic Units

The Theoretical Electrodynamic Units are based upon the cgs system of mechanical units and are therefore frequently referred to as *the cgs electrodynamic units*. In contradistinction to the theoretical electromagnetic units, these units are derived from a significant experimental law, Ampère's experiment on the mechanical force between two parallel currents. (See Table 5.) The units as proposed by Ampère and used by W. Weber differ from the electromagnetic units by factors of 2 and multiples thereof. They can be made to coincide with the theoretical electromagnetic units by proper definition of the fundamental unit of current. Some of the important definitions will be given for this latter case only.

For the *absolute magnetic permeability* of free space see Theoretical Electromagnetic Units.

The Theoretical Electrodynamic Unit of Current, or the Abampere, is defined as the current flowing in a circuit consisting of two infinitely long parallel wires one centimeter apart when the electrodynamic force of repulsion between the two wires is *two* dynes per centimeter length in free space. If the more natural choice of *one* dyne per centimeter length is made, the original proposal of Ampère is obtained and the unit of current becomes $1/\sqrt{2}$ abampere.

The **Theoretical Electrodynamic Unit of Magnetic Induction** is defined as the magnetic induction inducing an electromotive force of one abvolt in a conductor of one-centimeter length and moving with a velocity of one centimeter per second, if the conductor, its velocity, and the magnetic induction are mutually perpendicular. The unit thus defined is called one gauss.

The **Theoretical Electrodynamic Unit of Magnetic Flux, or the Maxwell,** is defined as the magnetic flux represented by a uniform magnetic induction of one gauss over an area of one square centimeter perpendicular to the direction of the magnetic induction.

The **Theoretical Electrodynamic Unit of Magnetic Intensity, or the Oersted,** is defined as the magnetic intensity at the center of a circle of 4π-centimeter diameter in which a current of one abampere is flowing.

All the Other Unit Definitions, which do not pertain to magnetic quantities, are identical with the definitions for the theoretical electromagnetic units.

The Rational Units

The factor 4π occurs frequently in the definition of important units. According to Heaviside, this can be avoided by writing the fundamental electromagnetic equations in a form different from and simpler than that used by the classical authors. The form of the equations is given in Table 5, and the unit definitions, which are based upon the definitions of the theoretical electromagnetic units, lead to the "rationalized" units as shown in Table 4. On account of the convenience many authors use the rationalized forms of the equations but very seldom use the rationalized units. Usually conversion into the "practical" units is preferred for any numerical computations.

14. THE INTERNATIONALLY ADOPTED ELECTRICAL UNITS AND STANDARDS

The International Committee on Weights and Measures decided in October, 1946, at Paris to abandon the so-called "international" practical units based on physical standards (see below) and to adopt, effective January 1, 1948, the so-called "absolute" practical units for international use.

The Adopted "Absolute" Practical Units

By a series of international actions, the "absolute" practical electrical units are defined as exact powers of 10 of corresponding theoretical electrodynamic and electromagnetic units because they are based upon the choice of the proportionality constant in Ampère's law for free space as $k_{mv} = 2 \times 10^{-7}$ henry/m. This leads naturally to the MKSC system of units (column 10, Table 3) which can be either rationalized or unrationalized as shown on p. 3–18; the rationalized system has found increasing favor in recent publications.[*]

The **"Absolute" Practical Unit of Current, or the "Absolute" Ampere,** is defined as the current flowing in a circuit consisting of two very long parallel thin wires spaced 1 m apart in free space if the electrodynamic force action between the wires is 2×10^{-7} newton = 0.02 dyne, per meter length. It is 10^{-1} of the theoretical or "absolute" electrodynamic or electromagnetic unit of current and was adopted internationally in 1881.

The **"Absolute" Practical Unit of Electric Charge, or the "Absolute" Coulomb,** is defined as the quantity of electricity which passes through a cross-sectional surface in one second if the current is one "absolute" ampere. It is 10^{-1} of the theoretical or "absolute" electromagnetic unit of electric charge and was adopted internationally in 1881.

The **"Absolute" Practical Unit of Electric Potential Difference, or the "Absolute" Volt,** is defined as the potential difference existing between two points in space if the work done in bringing an electric charge of one "absolute" coulomb from one of these points to another is equal to one "absolute" joule = 10^7 ergs. It is 10^8 of the theoretical or "absolute" electromagnetic unit of potential difference and was adopted internationally in 1881.

The **"Absolute" Practical Unit of Resistance, or the "Absolute" Ohm,** is defined as the resistance of a conductor in which a current of one "absolute" ampere is produced if a

---

[*] See particularly J. A. Stratton, *Electromagnetic Theory*, McGraw-Hill, 1941; R. W. P. King, *Electromagnetic Engineering*, McGraw-Hill, 1945; and S. S. Attwood, *Electric and Magnetic Fields*, John Wiley & Sons, 1949 (3rd ed.).

potential difference of one "absolute" volt is applied at its ends. It is 10^9 of the theoretical or "absolute" electromagnetic unit of resistance and was adopted internationally in 1881.

The "Absolute" Practical Unit of Magnetic Flux, or the "Absolute" Weber, is defined to be linked with a closed loop of thin wire of total resistance one "absolute" ohm if upon removing the wire loop from the magnetic field a total charge of one "absolute" coulomb is passed through any cross section of the wire. It is 10^8 of the theoretical or "absolute" electromagnetic unit of magnetic flux, the maxwell, and was adopted internationally in 1933.

The "Absolute" Practical Unit of Inductance, or the "Absolute" Henry, is defined as connected with a closed loop of thin wire in which a time rate of change of one "absolute" weber per second in the magnetic flux produces a time rate of change in the current of one "absolute" ampere. It is 10^9 of the theoretical or "absolute" electromagnetic unit of inductance and was adopted internationally in 1893.

The "Absolute" Practical Unit of Capacitance, or the "Absolute" Farad, is defined as the capacitance which maintains an electric potential difference of one "absolute" volt between two conductors charged with equal and opposite electrical quantities of one "coulomb." It is 10^{-9} of the theoretical or "absolute" electromagnetic unit of capacitance and was adopted internationally in 1881.

The "Absolute" Practical Derived Units for most of the other electric and magnetic quantities depend upon a decision with respect to rationalization. Table 5 shows the effect of rationalization upon the important relations of electromagnetism.

The Abandoned "International" Practical Units

The International System of electrical and magnetic units is a system for electrical and magnetic quantities which takes as the four fundamental quantities resistance, current, length, and time. The units of resistance and current are defined by physical standards which were originally aimed to be exact replicas of the "absolute" practical units, namely the "absolute" ampere and the "absolute" ohm. On account of long-range variations in the physical standards, it proved impossible to rely upon them for international use and they recently have been replaced by the "absolute" practical units.

The "international" practical standards are defined as follows. [*]

The International Ohm is the resistance at 0 C of a column of mercury of uniform cross section, having a length of 106.300 cm and a mass of 14.4521 grams.

The International Ampere is defined as the current which will deposit silver at the rate of 0.00111800 gram per second.

From these fundamental units, all other electrical and magnetic units can be defined in a manner similar to the "absolute" practical units. Because of the inconvenience of the silver voltameter as a standard, the various national laboratories actually used a volt, defining its value in terms of the other two standards.

At its latest conference in October, 1946, in Paris, the International Committee on Weights and Measures accepted as the best relations between the "international" and the "absolute" practical units the following: [*]

1 mean "international" ohm = 1.00049 "absolute" ohms
1 mean "international" volt = 1.00034 "absolute" volts

These mean values are the averages of values measured in six different national laboratories. On the basis of these mean values, the specific unit relation for converting "international" units appearing on certificates of the National Bureau of Standards, Washington, D. C., into "absolute" practical units, are as follows:

1 international ampere = 0.999835 absolute ampere
1 international coulomb = 0.999835 absolute coulomb
1 international henry   = 1.000495 absolute henries
1 international farad   = 0.999505 absolute farad
1 international watt   = 1.000165 absolute watts
1 international joule   = 1.000165 absolute joules

---

[*] *Announcement of Changes in Electrical and Photometric Units*, Bur. Standards Circular C459, Washington, D. C., 1947.

Table 3. Fundamental Constants

Constant	Symbol	Value	MKSC	Cgs
Speed of light (in vacuo)	c	2.997 925	$\times 10^8$ m/s	$\times 10^{10}$ cm/s
Charge of electron	e	1.602 1	$\times 10^{-19}$ c	$\times 10^{-20}$ emu
		4.802 98		$\times 10^{-10}$ esu
Rest mass of electron	m_e	9.109 1	$\times 10^{-31}$ kg	$\times 10^{-28}$ g
Electron charge/mass ratio	e/m_e	1.758 796	$\times 10^{11}$ c/kg	$\times 10^7$ emu
		5.272 74		$\times 10^{17}$ esu
Planck constant	h	6.625 6	$\times 10^{-34}$ J-s	$\times 10^{-27}$ erg-s
	$\hbar = \dfrac{h}{2\pi}$	1.054 50	$\times 10^{-34}$ J-s	$\times 10^{-27}$ erg-s
Gravitational constant	G	6.670	$\times 10^{-11}$ Jm kg^{-2}	$\times 10^{-8}$ erg cm g^{-2}
Avogadro constant	N_A	6.022 52	$\times 10^{26}$ kmole^{-1}	$\times 10^{23}$ mole^{-1}
Boltzmann constant	K	1.380 54	$\times 10^{-23}$ J deg^{-1} K	$\times 10^{-16}$ erg deg K^{-1}
Rydberg constant	R_∞	1.097 373	$\times 10^7$ m^{-1}	$\times 10^5$ cm^{-1}
Bohr radius	a_0	5.291 167	$\times 10^{-11}$ m	$\times 10^{-9}$ cm
Gas constant	R	8.206	$\times 10$ litre atm deg K^{-1} kmole^{-1}	$\times 10$ cm^3 atm deg K^{-1} mole^{-1}
		8.314 3	$\times 10^3$ J deg K^{-1} kmole^{-1}	$\times 10^7$ erg deg K^{-1} mole^{-1}
Stefan-Boltzmann	σ	5.669 7	$\times 10^{-8}$ W m^{-2} deg K^{-4}	$\times 10^{-5}$ erg cm^{-2} s^{-1} deg K^{-4}
Bohr magneton	μB	9.273 2	10^{-24} J T^{-1}	10^{-21} erg G^{-1}
Faraday constant	F	9.648 70	$\times 10^7$ C kmole^{-1}	$\times 10^3$ emu
		2.892 61		$\times 10^{14}$ esu
1st radiation constant	$c_1 = (2\pi hc^2)$	3.741 5	$\times 10^{-16}$ W m^2	$\times 10^{-5}$ erg cm^{2}s^{-1}
2nd radiation constant	$c_2 = (ch/K)$	1.438 79	$\times 10^{-2}$ m deg K	$\times 1$ cm deg K

Table 4. Conversion Factors from MKSC System

Unit in MKSC	To cgs Electrostatic		To cgs Electromagnetic		To Symmetrical Gaussian	
Meter	10^2	Centimeter	10^2	Centimeter	10^2	Centimeter
Kilogram mass	10^3	Gram	10^3	Gram	10^3	Gram
Joule	10^7	Erg	10^7	Erg	10^7	Erg
Newton	10^5	Dyne	10^5	Dyne	10^5	Dyne
Ampere	3×10^9	Statampere	10^{-1}	Abampere	3×10^9	Statampere
Ampere-turn	$12\pi \times 10^9$	esu of MMF	$4\pi \times 10^{-1}$	Gilbert	$4\pi \times 10^{-1}$	Gilbert
Ampere-turn per meter	$12\pi \times 10^7$	esu of II	$4\pi \times 10^{-3}$	Oersted	$4\pi \times 10^{-3}$	Oersted
Coulomb	3×10^9	Statcoulomb	10^{-1}	Abcoulomb	9×10^{11}	Statcoulomb
Farad	9×10^{11}	Statfarad	10^{-9}	Abfarad	9×10^{11}	Statfarad
Henry	$1/9 \times 10^{-11}$	Stathenry	10^9	Abhenry	10^9	Abhenry
Mho	9×10^{11}	Statmho	10^{-9}	Abmho	9×10^{11}	Statmho
Ohm	$1/9 \times 10^{-11}$	Statohm	10^9	Abohm	$1/9 \times 10^{-11}$	Statohm
Volt	$\frac{1}{3} \times 10^{-2}$	Statvolt	10^8	Abvolt	$\frac{1}{3} \times 10^{-2}$	Statvolt
Weber	$\frac{1}{3} \times 10^{-2}$	Statweber	10^8	Maxwell	10^8	Maxwell
Weber-turn	$\frac{1}{3} \times 10^{-2}$	Statweber-turn	10^8	Maxwell-turn	10^8	Maxwell-turn
Weber per meter^2	$\frac{1}{3} \times 10^{-6}$	Statweber per centimeter^2	10^4	Gauss	10^4	Gauss

DIMENSIONAL ANALYSIS

15. DIMENSIONAL HOMOGENEITY OF PHYSICAL EQUATIONS

A **Physical Quantity** is defined by unit and numerical value. (See Art. 1.) The designation by a mathematical symbol is, therefore, not a unique determination, and proper care has to be taken in all cases where physical quantities are used in mathematical equations. The ratio of any two like physical quantities, however, represents a unique and absolute value, provided both quantities are given in the same unit. This is the basis of physical measurements and also of fixed unit relations, as, for example, 1 meter = 100 cm.

Physical Equations are relations between physical quantities which, in general, state the balance of certain conceptual quantities as forces, energies, voltage drops, currents, momenta, and so on. Obviously all the terms in a physical equation must have the same resultant dimension if the equation is to have sense. This property of physical equations is called *dimensional homogeneity* and can be checked by employing the method of dimensions as used first by Fourier.‡ (See Arts. 1 and 2.) In physical differential equations the differentials are to be treated as having the same dimension as the finite quantities. To use dimensional relations properly, one must adhere to a complete dimension system and must not use incomplete dimension sets; this is of particular importance in dimensional analysis as shown later.§ Proportionality constants, for example, will in general be physical quantities with numerical values and dimensions.

A **Complete Dimension System** suitable for a particular physical equation can be obtained from the equation itself in the following manner. Suppose z quantities are involved in the equation which is composed of q terms of equal resultant dimensions on account of dimensional homogeneity. Between the z dimensions there are, of course, $(q - 1)$ independent relations at the most, so that $(q - 1)$ dimensions can be expressed in general in terms of the remaining $(z - q + 1)$ dimensions. $(z - q + 1)$ constitutes thus the least number of independent or free dimensions in this particular case.

Style. It is of value and utmost convenience to use a definite *style* in writing physical equations. The simplest and most logical expression appears to be the *unitary homogeneous form*, which, of course, involves dimensional homogeneity and is based upon the assumption that one system of units and one only shall be used for all physical quantities involved. Unitary homogeneous equations are independent of the specific unit system employed and can always be reduced to purely numerical relations since all the units cancel on account of the dimensional homogeneity; consequently they avoid the necessity of furnishing a legend with each equation in order to explain the various units to be used. Very illustrative examples of equations not having unitary homogeneity are given by Table 7 in Art. 12.

Functional forms (like trigonometric or exponential functions) can have only numerical arguments; sin E, where E means energy, or voltage, has no sense because it violates dimensional homogeneity. This can be seen by expanding the function into the equivalent power series

$$\sin E = \frac{E}{1!} - \frac{E^3}{3!} + \frac{E^5}{5!} - + \cdots$$

where each term has different dimensions.

Physical Similarity

Physical Similarity can be defined as existing for two systems which are described by exactly the same differential equations with exactly the same numerical constants. This at once discards the meaning of similarity as an absolute property and brings it into the same order of approximation as the differential equation describing the phenomena. In general, difficulties do not arise from the assumption of similarity as an absolute property, but rather from the application of this concept to approximately similar systems.

Practical Similarity, as wanted for extrapolation (or model) experiments, is a vague expression and of doubtful value unless the degree of approximation is stated, or the variables enumerated with respect to which similarity is desired. Practical similarity, therefore, is not a matter of mathematics; it belongs to the physical concept of a phenomenon, and it is the "classifying process" of the experimenter.* If it is found that two systems can be described approximately by the same differential equation, the same variables and constants, then these two systems are called approximately similar, never exactly similar.

It is possible to formulate similarity conditions for two systems suspected to be physically similar. The same information can be gained from a correctly written differential equation as can be gained from proper dimensional analysis. Never, however, can information be gained about physical constants and universal constants without resorting to tricky semi-mathematics. Physical constants are defined by fundamental relations and are not amenable to any analysis.

16. CLASS PARAMETERS AND SIMILARITY CONDITIONS

A differential equation constituting a relation between an unknown physical quantity y, an independent physical variable x, and certain physical parameters Q_i of the symbolic

---

‡ J. B. J. Fourier, *Théorie analytique de la chaleur*, p. 391, Paris, 1822.

§ P. W. Bridgman, *Dimensional Analysis*, Yale Univ. Press, 1922; also H. L. Langhaar, *Dimensional Analysis and Theory of Models*, Wiley, New York, 1951.

* N. R. Campbell, *Phil. Mag.* (6), Vol. 47, p. 482, 1924.

form

$$D\left(Q_i, y, \frac{dy}{dx}, \frac{d^2y}{dx^2}, \cdots\right) = 0 \tag{1}$$

can be treated as a mathematical differential equation. No definite style of writing the differential equation is required for the mathematical solution. If, however, unitary homogeneity of the differential equation is established, a number of deductions are possible which have profound physical significance.

To have concrete expressions, assume the differential equation to be in the form

$$a_n \frac{d^n y}{dx^n} + a_{n-1} \frac{d^{n-1}y}{dx^{n-1}} + \cdots a_1 \frac{dy}{dx} + a_0 y = F \tag{2}$$

where the coefficients a_k may depend on the parameters Q_i, the main variable y or its derivatives, and the independent variable x, in any manner whatsoever, so that non-linear types of equations are equally well included. If the equation is logically written, unitary homogeneity must prevail. Assuming this, the following deductions can be made:

1. Numerical Parameters. The quotients of any two additive terms in the equation are pure and unique numbers, independent of the special units employed. Since the differential quotients $\frac{d^k y}{dx^k}$ all have dimensions y/x^k, we obtain

$$\frac{a_k y/x^k}{a_\nu y/x^\nu} = \text{number}, \quad \frac{a_k y/x^k}{F} = \text{number} \tag{3}$$

The values of all the numbers will, in general, depend upon the specific characteristic physical constants of the substance studied, the variable x, and the applied disturbance F. Some of these numbers will not contain the variable x explicitly; they are fundamental parameters and could be given proper names (like Reynolds' number), if so desired. They can have "critical" values, which in turn will decide the type of solution or the validity of the special assumptions implied in the differential equation.

2. Dimensionless Products. Since the combinations of physical quantities in equation 3 are dimensionless numbers, their mutual products and quotients, and, in fact, any combination of them, will still be dimensionless. They constitute the "numerical" parameters (if they do not contain x explicitly) and the "numerical" variables (if they contain x explicitly) of the problem and must reappear in the mathematical solution. If desired, they might be called "dimensionless products" Π_ν, and the solution written in symbolical form

$$\psi(\Pi_1, \Pi_2, \Pi_3, \Pi_4, \cdots \Pi_N) = 0 \tag{4}$$

The number N of independent products is equal to $z - k$, where z is the total number of physical quantities involved (counting also the dimensional physical constants) and k the necessary number of fundamental units for the special physical field ($k = 3$ in mechanics, $k = 4$ in thermodynamics and in electromagnetism, and $k = 5$ for the general combination of all fields). In order for equation 4 to present a complete solution, the $\Pi_1 \cdots \Pi_N$ must be truly independent; they must not be mutually deducible by multiplication or division. Of course, it is possible to have ratios of like quantities and other numbers involved in the complete solution, which never can be obtained by discussion.

3. The Class Parameters. Considering any differential equation as characteristic for a "type" of problems, the numerical parameters and variables are then identical with the "typical" or class parameters and variables, which should be chosen for graphical repersentations of typical solutions. The N independent class variables or parameters can be selected arbitrarily, according to convenience, from among the much larger number of possible combinations. It seems advantageous, if possible, to let only one dimensionless product, for example Π_1, contain the variable y and to write explicitly

$$\Pi_1 = \phi(\Pi_2, \Pi_3, \Pi_4, \cdots \Pi_N) \tag{4a}$$

where, of course, ϕ is in general given by the mathematical solution of the differential equation. This form is particularly valuable if a direct solution of the differential equation 2 is impossible. Then, of course, ϕ will not be a known function but equation 4a will serve to bring out the class variables for experimental study.

4. The Similarity Conditions. If all the physical parameters and variables be referred to another system and medium according to

$$F = \phi F', \quad Q_i = q_i Q'_i, \quad x = \xi x', \quad y = \eta y' \tag{5}$$

then all the coefficients in equation 2 will become

$$a_\kappa = \alpha_\kappa a'_\kappa \tag{5a}$$

where α_κ is some combination of the q_i, ξ, and η quantities. Introducing these into the differential equation 2,

$$\left(\alpha_n \frac{\eta}{\xi^n}\right) a'_n \frac{d^n y'}{dx'^n} + \left(\alpha_{n-1} \frac{\eta}{\xi^{n-1}}\right) a'_{n-1} \frac{d^{n-1} y'}{dx'^{n-1}} + \cdots \left(\alpha_1 \frac{\eta}{\xi}\right) a'_1 \frac{dy'}{dx'} + (\alpha_0 \eta) a'_0 y' = \phi F' \quad (6)$$

Obviously the same differential equation will hold, or the two systems will react "similarly," if

$$\left(\frac{\alpha_n}{\xi^n} \cdot \frac{\eta}{\phi}\right) = 1, \quad \left(\frac{\alpha_{n-1}}{\xi^{n-1}} \cdot \frac{\eta}{\phi}\right) = 1, \quad \cdots \left(\frac{\alpha_1}{\xi} \cdot \frac{\eta}{\phi}\right) = 1, \quad \left(\alpha_0 \cdot \frac{\eta}{\phi}\right) = 1 \quad (7)$$

which are the *complete similarity conditions* for two systems with respect to the phenomenon described by differential equation 2. They cannot, in any case, be the absolute similarity condition as no absolute similarity is possible.

5. Critical Values of the Class Parameters. The general form (4a) of the solution in terms of class parameters is subject to the assumptions implied in the differential equation. Only experiments can tell whether or not a differential equation is a satisfactory approximation to the proper description of a phenomenon. In general, any solution will be satisfactory for a certain range of the parameters involved. The limiting values of the class parameters, beyond which the specific approximation becomes poor, are often referred to as *critical* values, because they give a criterion for certain phenomena to occur in a certain manner described by the differential equation. It is, of course, necessary to take into account the simultaneous values of all class parameters characteristic for a phenomenon. This leads, in graphical representation, with the class parameters as coordinates, to *regions* of types of solutions according to the number of variables. All the phenomena that come into these ranges of parameters form a "group" which always can be represented by a single type of solution no matter in which field of physics the phenomena might occur. This fact forms the basis of analogies from other fields, as, for example, electromagnetic phenomena are often likened to flow phenomena of incompressible liquids, if certain conditions of stationarity are satisfied.

Example

Stationary Flow of a Viscous Fluid through a Pipe

To illustrate the above point in the analysis of a differential equation, assume as an example the equation for the stationary flow of a viscous fluid through a pipe

$$\frac{\partial v}{\partial t} = G_z - \frac{1}{\rho} \frac{\partial p}{\partial z} + \frac{\mu}{\rho} \left(\frac{\partial^2}{\partial r^2} + \frac{1}{r} \frac{\partial}{\partial r}\right) v \quad (8)$$

(See A. G. Webster, *Differential Equations of Mathematical Physics*, Chap. 1.) To equation 8 the equations of continuity and compressibility have to be added in order to furnish the relations between velocity, pressure, and density of the fluid; since the additional relations add nothing from the point of view of dimensions they will not be considered here. The symbols in (8) denote, respectively, with their dimensions in the technical dimension system:

G_z body force acting per unit mass, or acceleration $[FM^{-1}] = [LT^{-2}]$.
p hydrodynamic pressure $[FL^{-2}]$.
ρ density $[ML^{-3}] = [FL^{-4}T^2]$.
v the axial velocity $[LT^{-1}]$.
μ viscosity coefficient $[FL^{-2}T]$.
r radial distance involved in the differentiations $[L]$.
t time $[T]$.

The differential equation 8 shows dimensional homogeneity as can be checked by the ordinary method of mechanical dimensions and is also unitary homogeneous as it is independent of the units used. It is, therefore, proper to apply the analysis indicated above.

1. The Numerical Parameters. The four terms of equation 8 have the following resultant dimensions which must be identical:

$$\left[\frac{\partial v}{\partial t}\right] = \left[\frac{v}{T}\right], \quad [G] = [G], \quad \left[\frac{1}{\rho} \frac{\partial p}{\partial z}\right] = \left[\frac{p}{\rho L}\right], \quad \left[\frac{\mu}{\rho} \frac{\partial^2 v}{\partial r^2}\right] = \left[\frac{\mu v}{\rho L^2}\right] \quad (9)$$

Forming the mutual quotients by dividing the first term successively by all the following ones, then the second term by the following ones, and so on, leads to the series of distinct dimensionless terms:

$$\frac{v}{TG}, \frac{\rho L v}{pT}, \frac{\rho L^2}{\mu T}, \frac{L\rho G}{p}, \frac{G\rho L^2}{\mu v}, \frac{pL}{\mu v} \quad (10)$$

Here the last three terms appear as numerical parameters not containing time explicitly. They probably could be used as characteristic numbers; this, however, has not become customary.

Since vT/L is also dimensionless, it can be multiplied into the other dimensionless quantities. Multiplication with the first three numbers in (10) results in

$$\frac{vT}{L} \cdot \frac{v}{TG} = \frac{v^2}{LG} \quad \text{which is Froude's number}$$

$$\frac{vT}{L} \cdot \frac{\rho Lv}{pT} = \frac{\rho v^2}{p} \quad \text{which is the reciprocal of Newton's number}$$

$$\frac{vT}{L} \cdot \frac{\rho L^2}{\mu T} = \frac{\rho Lv}{\mu} \quad \text{which is Reynolds' number}$$
(11)

These are the three most widely known numerical parameters of general hydrodynamics and have been named after their first users. In the literature they are usually derived by philosophical arguments or long-winded dimensional deductions; here they follow naturally from the condition of unitary homogeneity of the fundamental differential equation.

2. The Dimensionless Products. Only four independent dimensionless products are possible, since the number of fundamental units $k = 3$ (only mechanical quantities are involved), the number of physical quantities, however, $N = 7$. Choose as these four products the three characteristic numbers and in addition $\dfrac{vT}{L}$; then there results as general type of solution in terms of dimensionless products

$$\psi \left[\frac{vT}{L}, \frac{v^2}{LG}, \frac{p}{v^2\rho}, \frac{v\rho L}{\mu} \right] = 0$$
(12)

That the four combinations are independent is easily seen from the fact that each contains one variable not present in any other product. Other choices could be made with the same degree of generality; however, any such set of dimensionless numbers must be derivable from the one chosen here by multiplication or division with dimensionless numbers only.

3. The Class Parameters. Since the phenomenon is properly described by the maximum number of independent dimensionless products, these can be chosen as class parameters for a whole series of problems represented by the same differential equation. The three numbers from (11) constitute the most widely used class parameters of hydrodynamics, and any solution can be expressed in terms of them. In conducting experiments they should be chosen as variables. If it is desired to express the solution in explicit form of a principal variable, the dimensional products have to be chosen so that only one contains this variable. Suppose it is desired to solve for p; then the third dimensionless product will be selected as principal class variable

$$\frac{p}{v^2\rho} = \phi \left[\frac{vT}{L}, \frac{v^2}{LG}, \frac{v\rho L}{\mu} \right]$$
(12a)

where, of course, ϕ is not known but can be obtained by experiments. It follows

$$p = (v^2\rho) \cdot \phi \left[\frac{vT}{L}, \frac{v^2}{LG}, \frac{v\rho L}{\mu} \right]$$
(13)

4. The Similarity Conditions. If equation 8 is transformed into a system with primed quantities and the proportionality factors are designated by s with a respective subscript

$$G = s_G G' \qquad \rho = s_\rho \cdot \rho' \qquad r = s_r \cdot r'$$
$$p = s_p \cdot p' \qquad v = s_v \cdot v' \qquad t = s_t \cdot t'$$
$$\mu = s_\mu \cdot \mu'$$
(14)

there follows

$$\left(\frac{s_v}{s_t} \right) \cdot \frac{\partial v'}{\partial t'} = (s_G) \cdot G' - \left(\frac{s_p}{s_\rho s_r} \right) \frac{1}{\rho'} \cdot \frac{\partial p'}{\partial z'} + \left(\frac{s_\mu s_v}{s_\rho s_r^2} \right) \cdot \frac{\mu'}{\rho'} \left(\frac{\partial^2}{\partial r'^2} + \frac{1}{r'} \cdot \frac{\partial}{\partial r'} \right) v'$$

and the three similarity conditions are

$$\left(\frac{s_v}{s_t s_G} \right) = 1, \quad \left(\frac{s_p}{s_\rho s_r s_G} \right) = 1, \quad \left(\frac{s_\mu s_v}{s_\rho s_r^2 s_G} \right) = 1$$
(14a)

which must be satisfied to insure similar flow characteristics of two viscous fluids. If these conditions are satisfied, experimental results on one liquid can be utilized for another liquid without repetition of analysis or experiments. See also Section 6, Article 6.

17. DIMENSIONAL ANALYSIS

In the literature on this subject much controversy is found relating to a proper and convincing basis of dimensional analysis. Mathematical proofs have been offered from principles of logic; however, there is no "principle" of similarity, or similitude, which would present a reasonable mathematical basis for the general dimensional analysis. Fundamentally, dimensional analysis is identical with the analysis of physical equations, and, in particular, with the analysis of physical differential equations. The information from dimensional analysis, therefore, is identical with that obtained from differential equations subject to unitary homogeneity. (See Art. 16.) The advantage of dimensional analysis is the more systematic approach in obtaining the class parameters and the similarity conditions, although it seems advisable to take a complete differential equation as the starting point in any case.

The Π Theorem

If z physical quantities Q_k are known to be involved in a certain physical phenomenon, the mutual dependence, in unitary homogeneous form, must be expressible as a power product as has been shown on various occasions.* For example, the quantity Q_1 takes the form

$$Q_1 = Q_2{}^{\alpha_2} \cdot Q_3{}^{\alpha_3} \cdots Q_z{}^{\alpha_z} \qquad (15)$$

where the α_2, $\alpha_3 \cdots \alpha_z$ are definite and unique values which can be determined by dimensional analysis. This fact is based upon the fundamental and reasonable hypothesis of uniqueness of the results of physical measurements, no matter what units are used in measuring. Thus, for example, if $Q_1 = F$ represents force as expressed by mass, length, and time, it would be found $F = MLT^{-2}$ as a unitary homogeneous relation (see Art. 15) independent of the special units employed on the right-hand side. The unit of F, of course, is fixed by the condition of unitary homogeneity; any arbitrary choice of the unit of F would result in an additional numerical conversion factor.

On account of unitary homogeneity, the combination

$$\frac{Q_1}{Q_2{}^{\alpha_2} \cdot Q_3{}^{\alpha_3} \cdots Q_z{}^{\alpha_z}} = \Pi \qquad (15a)$$

is a numerical quantity and can be designated as a "dimensionless product." The number of possible independent dimensionless products is obviously $(z - k)$, if k is the number of independent fundamental units in the particular field. This result is the same as in the analysis of differential equations.

The most general form of a logical mathematical description of the phenomena studied must be some functional relation between the $(z - k)$ independent dimensionless product

$$\Psi(\Pi_1, \Pi_2, \cdots \Pi_{z-k}) = 0 \qquad (16)$$

This, of course, is but the typical solution of a differential equation in terms of class variables and class parameters, but now the functional form Ψ is not known and can be found only by experiment. The forms (15) and (16) are frequently referred to as the "Π theorem," and it is the object of dimensional analysis to find practical ways of obtaining the form (16) with greatest economy and least chance of erroneous reasoning.

The Method of Lord Rayleigh. Usually it is desired to learn the dependence of one physical quantity upon a number of other physical quantities which are supposed to enter into the problem or experiment. If the total number of quantities is z (including the physical constants), and the principal quantity chosen be Q_1, then (15) expresses the general physical relation for Q_1 according to the Π theorem. In general, the dimensions of the z quantities are not independent, but can be represented in terms of k fundamental dimensions, if the respective complete dimension system is composed of k fundamental dimensions. These can be chosen either from among the $Q_1 \cdots Q_z$ themselves, or from any one of the more common dimension systems suitable for the particular field of physics. (See Arts. 3, 4, 5.) It is essential to use a complete dimension system in order to avoid ambiguity with respect to the dimensions of the physical constants. If the dimensional expressions are introduced into (15), and if then the exponents on both sides are equated there result exactly k conditions for the $(z - 1)$ exponents $\alpha_2 \cdots \alpha_z$, permitting the expression of any k exponents in terms of the remaining $(z - 1 - k)$ indeterminate or "free"

---

* J. L. Riabouchinsky, *L'Aerophile*, 1911; E. Buckingham, *Phys. Rev.*, Vol. 4, p. 345, 1914; F. London, *Phys. Zeits.*, Vol. 23, p. 262, 1922.

exponents.* The final expression gives then the desired physical quantity in terms of indeterminate powers of dimensionless products, but in an explicit form, which is the principal advantage of this method.

Example. To illustrate the method of Lord Rayleigh, assume the same example as treated in Art. 16. The differential equation gives all the parameters and variables that enter into the problem so that the list of physical quantities involved is (see the example in Art. 16):

$$v, t, G, \rho, p, \mu, l$$

Suppose it is desired to find the dependence of v on the other quantities; then the form corresponding to (15) is

$$v = t^{\alpha_1} \cdot G^{\alpha_2} \cdot \rho^{\alpha_3} \cdot p^{\alpha_4} \cdot \mu^{\alpha_5} \cdot l^{\alpha_6} \tag{17}$$

In the technical dimension system of mechanics there are three fundamental dimensions L, T, F. (See Art. 3.) It is convenient, in general, to write the dimensions of the various physical quantities in the form of a table.

Physical Quantities		t	G	ρ	p	μ	l	v
L	Exponents of Fundamental Dimensions	0	$+1$	-4	-2	-2	$+1$	$+1$
T		$+1$	-2	$+2$	0	$+1$	0	-1
F		0	0	$+1$	$+1$	$+1$	0	0

Introducing the dimensions from this table into equation 17 and equating the dimensional exponents for L, T, and F gives the three linear simultaneous equations

$$\left. \begin{array}{r} +\alpha_2 - 4\alpha_3 - 2\alpha_4 - 2\alpha_5 + \alpha_6 = +1 \\ +\alpha_1 - 2\alpha_2 + 2\alpha_3 \qquad\quad + \alpha_5 \qquad = -1 \\ +\alpha_3 + \alpha_4 + \alpha_5 \qquad = 0 \end{array} \right\} \tag{18}$$

Since there are six unknowns, three of the exponents will remain indeterminate. Suppose α_1, α_3, and α_6 are chosen for some reason to be left indeterminate; then from (18)

$$\left. \begin{array}{l} \alpha_2 = +1 + 2\alpha_3 - \alpha_6 \\ \alpha_4 = -1 - 3\alpha_3 + 2\alpha_6 + \alpha_1 \\ \alpha_5 = +1 + 2\alpha_3 - 2\alpha_6 - \alpha_1 \end{array} \right\} \tag{18a}$$

so that the velocity can be written, properly collecting terms with the same exponents,

$$v = \frac{\mu G}{p} \cdot \left[\frac{Tp}{\mu} \right]^{\alpha_1} \cdot \left[\frac{G^2 \rho \mu^2}{p^3} \right]^{\alpha_3} \cdot \left[\frac{Lp^2}{G\mu^2} \right]^{\alpha_6} \tag{19}$$

The advantage of having v explicitly given is obvious. Each one of the three independent dimensionless products is a combination of some of the dimensionless products in (10), as can be shown easily. Instead of using the indeterminate exponents $\alpha_1, \alpha_3, \alpha_6$, any arbitrary function of the dimensionless products would be a more general expression.

$$v = \frac{\mu G}{p} \cdot \psi \left[\frac{Tp}{\mu}, \frac{G^2 \rho \mu^2}{p^3}, \frac{Lp^2}{G\mu^2} \right] \tag{20}$$

In the same way the dependence of p on the other quantities involved could be computed, and the result would be similar to (13); the special form would depend upon the particular choice of the $(z - 1 - k)$ free exponents, but could always be transformed into the form (13) by multiplying the dimensionless products by dimensionless numbers from (10).

The Method of E. Buckingham.† Suppose that the total number of physical quantities involved in a problem is z (including the physical constants), and the number of independent fundamental dimensions, necessary for the particular field of physics to which the problem belongs, is k; then k of the z quantities, which are dimensionally independent, are arbitrarily chosen as principal quantities. If Q_1 to Q_k are these principal quantities and Q_{k+1} to Q_z the secondary quantities, $(z - k)$ independent dimensionless products can be formed by multiplying indeterminate powers of the principal quantities successively with one of the secondary quantities and equating the total dimensions to zero, as, for

---

* Lord Rayleigh, *Nature*, Vol. 95, p. 66, 1915.

† E. Buckingham, *Phys. Rev.*, Vol. 4, p. 345, 1914; *Phil. Mag.* (6), Vol. 42, p. 696, 1921.

example,

$$[\Pi_1] = [Q_1^{\alpha_1} \cdot Q_2^{\alpha_2} \cdots Q_k^{\alpha_k} \cdot Q_{k+1}] = 1 \qquad (21)$$

If now the dimensions of Q_{k+1} in terms of the dimensions of the principal quantities are introduced, k linear relations for the exponents $\alpha_1 \cdots \alpha_k$ result so that all the exponents can be determined. In this way $(z - k)$ independent dimensionless products are obtained directly; their special form depends entirely upon the choice of the principal quantities. The advantage of this method lies in the simple form of the equations which determine the exponents of the principal quantities; it is also possible to solve explicitly for any single quantity, if it is not a principal quantity.

Example. To illustrate the method by E. Buckingham, and to be able to compare it with the other methods, assume the same example as before, with the physical quantities

$$v, t, G, \rho, p, \mu, l$$

In choosing the three principal quantities it must be kept in mind that they will appear in all dimensionless products as factors. Suppose l, ρ, and v be chosen here; using the technical mechanical dimension system, the dimensions of all the quantities are again given in the table on page 424. Grouping the dimensional expressions for the dimensionless product $[l^{\alpha_1}\rho^{\alpha_2}v^{\alpha_3} \cdot t] = 1$ in a table

	l^{α_1}	ρ^{α_2}	v^{α_3}	t
$[L]$	$+\alpha_1$	$-4\alpha_2$	$+\alpha_3$	0
$[T]$	0	$+2\alpha_2$	$-\alpha_3$	$+1$
$[F]$	0	$+\alpha_2$	0	0

gives immediately the resulting equations

$$\left.\begin{array}{l} +\alpha_1 - 4\alpha_2 + \alpha_3 + 0 = 0 \\ +2\alpha_2 - \alpha_3 + 1 = 0 \\ +\alpha_2 \qquad + 0 = 0 \end{array}\right\}$$

from which

$$\alpha_2 = 0, \quad \alpha_3 = +1, \quad \alpha_1 = -1$$

so that the product becomes vT/l. Similar tables with t replaced successively by G, p, and μ, give the other three dimensionless products as lG/v^2 (the reciprocal of Froude's number), $p/\rho v^2$ (Newton's number), and $\mu/l\rho v$ (the reciprocal of Reynolds' number). The general solution, therefore, is a functional relation of the four dimensionless products

$$\psi\left[\frac{vT}{l}, \frac{lG}{v^2}, \frac{p}{\rho v^2}, \frac{\mu}{l\rho v}\right] = 0 \qquad (22)$$

which is identical with the solution obtained in (12) from the analysis of the differential equation.

Note on the Application of Dimensional Analysis

It has been customary to surround dimensional analysis with an air of intuitive invention. The comparison of differential equation analysis and dimensional analysis has shown that both are based upon the condition of unitary homogeneity of physical equations. It is obvious, therefore, that dimensional analysis must lead to the same result as the analysis of differential equations, if it is properly used. The chief advantage of dimensional analysis is the fact that it deals with the class variables and parameters and thus reduces the number of variable quantities from z to $(z - k)$, if z is the number of physical quantities and k the number of fundamental dimensions. It can serve as a guide for experiments and gives the conditions of similarity without necessitating a complete solution of the problem. As a basis for model experiments it is extremely valuable. (See Art. 18.)

In applying dimensional analysis, care has to be taken to include all physical quantities which might be involved in the problem. It seems best to start from an approximate differential equation as the expression of "balance" of forces, or energies, or what not, and to include all dimensional physical constants. If any system of $z - k$ independent dimensional products has been found, other systems can be obtained by combination of two or more numerical parameters into new dimensionless products. It is not possible to replace differential equation analysis by dimensional analysis, since the latter never will give the complete form of the solution of a problem.

Dimensional analysis, as well as differential equations, have certain ranges of validity; if experiments do not check the results of an analysis, it might serve as an indication that the problem is incompletely stated and more physical quantities should be included. Neither differential equations nor dimensional analysis can ever lead to new physical constants; these are defined by the fundamental hypotheses.

18. THEORY OF MODELS

The possibility of similarity between various systems of even wholly unrelated fields suggests the exploration of phenomena by means of "models" of either enlarged or reduced scale, or by means of analogies from a different field. Just as similarity cannot be an absolute quality, but is defined as the approximate state of two systems (see Art. 16), so can a model be only an approximate reproduction of the original scale phenomenon; the approximation can, of course, be quite close over certain ranges of the main variables. It is essential, in the use of models, to state definitely the properties to be modeled and to restrict conclusions to the quantities included in the similarity considerations.

The Scale Factors. If the same differential equation is assumed to hold for two different problems of the same field, then similarity exists between the phenomena described by the differential equation. Any one term in the differential equation for *case 1* may be of the form

$$(Q_a{}^{n_1} \cdot Q_b{}^{n_2} \cdot Q_c{}^{n_3} \cdots) \frac{d^\nu y}{dx^\nu}$$

where the factor in parentheses is the coefficient of the νth differential. For *case 2* this term will become

$$(\overline{Q}_a{}^{n_1} \cdot \overline{Q}_b{}^{n_2} \cdot \overline{Q}_c{}^{n_3} \cdots) \frac{d^\nu \bar{y}}{d\bar{x}^\nu}$$

and the new physical quantities (physical constants or variables) may be related to the former ones by the expressions

$$\left. \begin{array}{l} \overline{Q}_a = s_a Q_a, \quad \overline{Q}_b = s_b Q_b, \quad \overline{Q}_c = s_c Q_c \cdots \\ \bar{y} = s_y \cdot y, \quad \bar{x} = s_x \cdot x \end{array} \right\} \tag{23}$$

Suppose now that *case 2* shall be used as a model for *case 1*. Each physical quantity in the model must then take a definite ratio to the same quantity in the original, and the proportionality factors (23) can be considered as *scale factors*, specifying the relations between the model and the original. Obviously, there are as many scale factors as there are physical quantities. It is not possible to choose all the scale factors arbitrarily.

Model Rules. In order to be permitted to use *case 2* as a model of *case 1*, the conditions of similarity must be satisfied. As shown in Art. 16, the similarity conditions require, using (23) above,

$$(s_a{}^{n_1} \cdot s_b{}^{n_2} \cdot s_c{}^{n_3} \cdots) \frac{s_y}{s_x{}^\nu} = k \tag{24}$$

k is the same constant for all the terms, so that the numerical values in the differential equations for both cases are identical.

If there are q terms in the differential equation, there will be q conditions of the type (24) or $(q - 1)$ mutual relations between the scale factors. These are called *the model rules*, as they specify size, shape, and physical properties of the model (as applied forces, densities, viscosities, and so on). The model rules depend entirely on the differential equation chosen to represent the phenomenon.

Fundamental and Dependent Scale Factors. The scale factors which can be chosen arbitrarily are called *fundamental scale factors*. The number of fundamental scale factors is identical with the number of fundamental dimensions necessary for a complete dimension system in the particular field of physics to which the problem belongs. This can be seen from the general differential equation; if z physical quantities are involved, and q terms make up the differential equation, then $(z - q + 1)$ fundamental dimensions are required. (See Art. 15.) But there are also z scale factors, which have to satisfy, according to (24), $(q - 1)$ relations, so that $(z - q + 1)$ fundamental scale factors remain which can be chosen arbitrarily. All the other scale factors follow by means of the model rules, they are dependent, and cannot be chosen freely.

Model Rules from Dimensional Analysis. Dimensional analysis leads to a description of a physical phenomenon in terms of dimensionless products which are identical with, or can be easily transformed into, the class parameters of the analysis of differential equations. (See Art. 17.) Similarity in terms of dimensional analysis exists only if the dimensionless products have the same numerical values for both the model and the original. If any one of the dimensionless products Π_ν for *case 1* is given by

$$\Pi_\nu = (Q_1{}^{\alpha_1} \cdot Q_2{}^{\alpha_2} \cdots Q_k{}^{\alpha_k}) \cdot Q_{k+\nu}$$

and for *case 2* by

$$\overline{\Pi}_\nu = (\overline{Q}_1{}^{\alpha_1} \cdot \overline{Q}_2{}^{\alpha_2} \cdots \overline{Q}_k{}^{\alpha_k}) \cdot \overline{Q}_{k+\nu}$$

and if there are the scale relations

$$\overline{Q}_1 = s_1 Q_1, \quad \overline{Q}_2 = s_2 Q_2 \cdots \quad \overline{Q}_{k+\nu} = s_{k+\nu} Q_{k+\nu}$$

then $\Pi_\nu = \overline{\Pi}_\nu$ only if

$$(s_1{}^{\alpha_1} s_2{}^{\alpha_2} \cdots s_k{}^{\alpha_k}) \cdot s_{k+\nu} = 1 \tag{25}$$

Since there are $(z - k)$ independent dimensionless products, the z scale factors will be related by $z - k$ relations of the type (25), so that the number of free or fundamental scale factors is k, the same as the number of fundamental dimensions. The relations (25) are called the model rules and are identical with the relations (24). They can be *conveniently obtained* by replacing in the dimensionless products all the physical quantities by their respective scale factors and equating to unity the expressions thus obtained.

Example. In order to illustrate the application of the theory of models, it will be chosen to represent the flow of water through a large pipe by the flow of mercury through a model pipe of one-tenth the diameter of the original. Using dimensional analysis, the problem is the same as treated above in Art. 17. As a problem of hydrodynamics, $k = 3$ fundamental dimensions, and, therefore, $k = 3$ fundamental scale factors are necessary. The ratio of the linear dimensions is given by $s_l = 0.1$; the ratio of the densities is $s_\rho = 13.6$, and the ratio of the viscosities can be assumed approximately as $s_\mu = 1.5$. The four dimensionless products are given by equation 22, so that the scale factors have to satisfy the conditions

$$\frac{s_v s_t}{s_l} = 1, \quad \frac{s_l s_G}{s_v{}^2} = 1, \quad \frac{s_p}{s_\rho s_v{}^2} = 1, \quad \frac{s_\mu}{s_l s_\rho s_v} = 1$$

from which the four dependent scale factors follow as

$$s_v = 1.103, \quad s_G = 12.16, \quad s_p = 16.54, \quad s_t = 0.0908$$

which means that the velocity of the mercury flow must be 1.103 of that of water, and the time scale approximately 0.0908 of that of water, and the force, per unit of mass, acting upon the mercury must be made 12.16 times that for water.

SECTION 4

MECHANICS OF RIGID BODIES

BY

JANVIER M. RICE *

* Section 4 was originally a revision of material in previous handbooks published by John Wiley & Sons, most of which was written by Professors C. H. Burnside and E. R. Maurer for Merriman's Civil Engineers' Handbook and Peele's Mining Engineers' Handbook. The present section was revised by the editor with the assistance of M. F. Spotts, who rewrote the articles on Friction, and E. R. Peck, who prepared the articles on Moving Axes, The Gyroscope, and Generalized Coordinates.

MECHANICS OF RIGID BODIES

1. DEFINITIONS

Mechanics is that branch of *science* which treats of forces and motion.

Statics is that branch of *mechanics* which deals with the equilibrium of forces on bodies *at rest* (or moving at a uniform velocity in a straight line).

Kinematics is that branch of *mechanics* which deals with the motion of bodies without consideration of the character of the bodies or of the influence of forces upon their motion. It considers only concepts of *geometry* and *time*.

Kinetics (or Dynamics) is that branch of *mechanics* which deals with the effect of unbalanced external forces in *modifying the motion* of bodies.

Mass and Weight, in the *gravitational system of units* employed by English engineers, are related by the formula $W = Mg$, where W = weight, M = mass, and g = acceleration due to gravity. For a thorough discussion of these terms, see Section 3, Physical Units and Standards.

Force is that which changes or tends to change the state of rest or motion of a body.

Inertia is that property of a body by virtue of which it tends to continue in the state of rest or motion in which it may be placed, until acted on by some force.

Reaction is that *equal and opposite force* exerted by a body in opposing another force acting upon it.

Newton's Laws of Motion. *First Law.* If a body is at rest, it will remain at rest, or if in motion, it will move uniformly in a straight line, until acted on by some force.

Second Law. If a body is acted on by several forces, it will obey each as though the others did not exist, and this whether the body is at rest or in motion. Change of the motion of a body is proportional to the force and to the time during which the force acts, and is in the same direction as the force.

Third Law. If a force acts to change the state of a body with respect to rest or motion, the body will offer a resistance equal and directly opposed to the force. Or, to every action there is opposed an equal and opposite reaction.

Special Terms such as hydrostatics, aerodynamics, etc., are used to denote the theory of statics as applied to *liquid bodies*, the theory of dynamics as applied to *gaseous bodies*, etc. **Mechanics of materials** considers, in addition to *external forces*, the *internal forces* or *stresses* between molecules of a body. Subjects of these types are covered in other sections of this handbook. The present section on *mechanics* is confined, in general, to the discussion of motion of, and *external forces* applied to, *rigid bodies*.

STATICS

2. GRAPHICAL REPRESENTATION AND CLASSIFICATION OF FORCES

Graphical Representation of Force

A Force is completely specified by its *magnitude, direction,* and *point of application.* The word **sense** as applied to a force refers to one of the two directions along the line of action of the force. The effect of any force applied to a rigid body at rest is the same, no matter where in its own line of action the force is applied. This is known as the principle of the **transmissibility of force.** A force may be represented graphically in magnitude and direction by a straight line drawn parallel to its line of action, the length being proportional to the magnitude of the force; its sense is indicated by an arrowhead placed on the line. The English engineers' unit of force is the pound, or the earth's pull on a mass of 1 lb. A drawing which indicates the lines of action of the various forces acting on a machine or structure

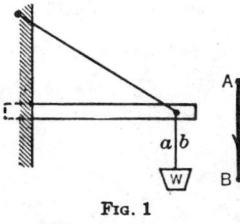

Fig. 1

430

is called a **space diagram**; one in which vectors are drawn to represent the magnitudes and directions of the forces is a **vector diagram**. A force is indicated on a space diagram by two lower-case letters placed on opposite sides of the line of action of the force; the vector, representing its magnitude and direction, by the same capital letters placed at the ends. Thus, in Fig. 1, AB represents the magnitude and direction of the force W, and ab its action line. The vector being read as AB indicates a downward sense; read as BA, an upward sense.

Classification of Systems of Forces

A **System of Forces** consists of any number of forces taken collectively.

Classification of Systems of Forces is made according to the arrangement of their action lines. If the action lines lie in the same plane the system is **coplanar**, otherwise **noncoplanar**. If they pass through the same point the system is **concurrent**, otherwise **nonconcurrent**. If two or more forces have the same action line they are **collinear**. A system of two equal forces, parallel, opposite in sense, and having different action lines is a **couple**. Two or more forces equivalent to a single force are **components** of the single force. **Resolution** is the operation of replacing a single force by a system of components. The single force is the **resultant** of its components. In general, the resultant of a system of forces is the simplest equivalent system. This may be a *single force*, a *single couple*, or a *noncoplanar force and couple* (or *two skewed forces*). When the resultant is a single force the **equilibrant** is a force equal in magnitude, having the same line of action but opposite sense. **Composition** is the operation of replacing a system of forces by its resultant.

3. COMPOSITION AND RESOLUTION OF CONCURRENT FORCES

Composition of Two Concurrent Forces *

Parallelogram Law. If magnitudes, lines of action, and senses of two concurrent forces acting on a rigid body are represented by OA and OB (Fig. 2), the magnitude, line of action, and sense of their resultant are represented by the diagonal OC of the parallelogram $OABC$. The points of application of the forces may be anywhere on the body in the lines OA, OB, and OC, or their extensions. The arrowheads on the lines OA, OB, and OC all point toward or all away from the point of concurrence O.

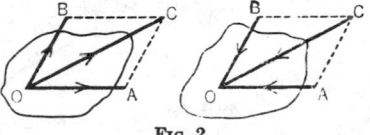

FIG. 2

Triangle Law. This law follows directly from the parallelogram law. If in the triangle ABC (Fig. 3a) AB and BC represent two concurrent forces in magnitude, direction, and sense, AC will represent their resultant in magnitude, direction, and sense; its action line will be ac through the point of concurrency, parallel to AC. It should be noted that the arrowheads on the sides AB and BC are **confluent** (point the same way around) but the arrowhead on AC is not confluent with the others. Also, the point of concurrency need not necessarily be located in or on the body but may be outside it.

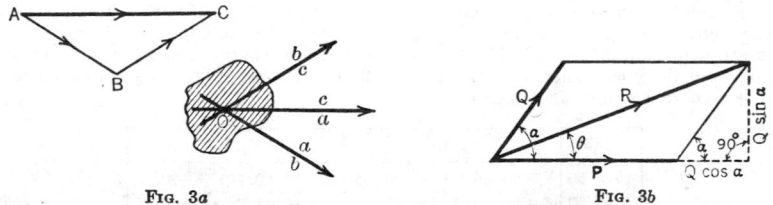

FIG. 3a   FIG. 3b

The resultant may be found **algebraically** thus: In Fig. 3b, let α be the angle between the action lines of forces P and Q, and θ the angle between R and P. Then, $R^2 = P^2 + Q^2 + 2PQ \cos \alpha$, and $\tan \theta = \dfrac{Q \sin \alpha}{P + Q \cos \alpha}$. If $\alpha = 90°$, $R^2 = P^2 + Q^2$. and $\tan \theta = Q/P$.

---

* It is evident that a pair of concurrent forces and their resultant are necessarily *coplanar*.

Resolution into Two Concurrent Forces *

A Force May Be Resolved into an infinite number of pairs of components by constructing different triangles as in Fig. 4a. ´ The action lines of the components must be concurrent at a point on the action line of the force. A common problem is to resolve a force into *rectangular components* (often called resolved parts). In Fig. 4b, AB and BC are a set of rectangular components of the force P. Expressed *algebraically*, $AB = P$ cos α and $BC = P$ sin α. For rectangular components, the resolved part of a force along

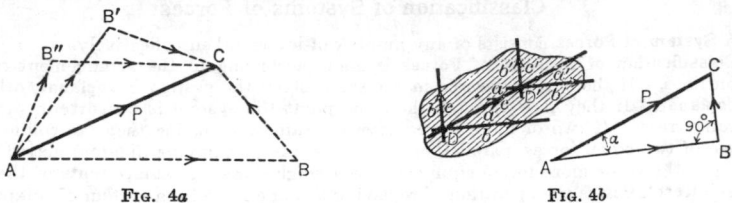

FIG. 4a   FIG. 4b

any line equals the magnitude of the force times the cosine of the angle between the lines of action of the force and its component. Action lines of the components are concurrent on the space diagram at some point on ac, as at D or D'.

Composition of More than Two Coplanar Concurrent Forces †

Graphic Method. In Fig. 5, consider body G acted on by the four forces shown. Construct a **force polygon** as follows: Plot AB parallel to ab, and scale it to represent 60

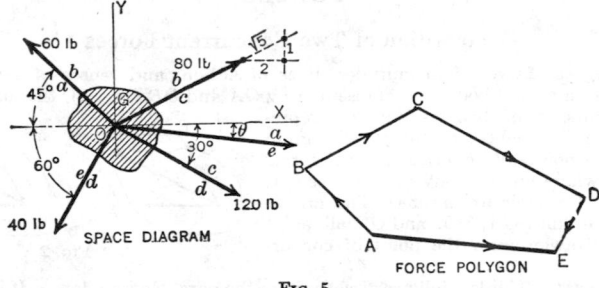

SPACE DIAGRAM   FORCE POLYGON

FIG. 5

lb.; from B plot BC parallel to bc, and scale it to represent 80 lb.; in like manner plot CD and DE, so that the arrows lead *confluently* from A to E. The resultant of the system is AE in magnitude and sense and equals 114 lb. Its action line is ae. The resultant will be the same regardless of the order in which the forces are plotted. Note particularly that the resultant is not confluent with the component forces.

Algebraic Method. Choose rectangular axes OX and OY. Referring to Fig. 5, resolve each force into its x and y components, considering components acting upward or to the right as positive, and those acting downward or to the left as negative. Arrange the results in tabular form, placing the forces in the first column, the x components in the second, and the y components in the third. $\Sigma F_x =$ algebraic sum of x components, and $\Sigma F_y =$ algebraic sum of y components.

F, lb	F_x, lb	F_y, lb
$ab =$ 60	$-60 \times 0.707 = -42.4$	$+60 \times 0.707 = +42.4$
$bc =$ 80	$+80 \times 2/\sqrt{5} = +71.4$	$+80 \times 1/\sqrt{5} = +35.7$
$cd =$ 120	$+120 \times 0.866 = +104$	$-120 \times 0.5 \quad = -60$
$de =$ 40	$-40 \times 0.5 \quad = -20$	$-40 \times 0.866 = -34.6$
	$\Sigma F_x = +113$	$\Sigma F_y = -16.5$

---

* It is evident that a pair of concurrent forces and their resultant are necessarily *coplanar*.
† The composition of two concurrent forces is simply a special case of this more general treatment.

Then $R = \sqrt{\Sigma F_x{}^2 + \Sigma F_y{}^2} = \sqrt{13,041}$
= 114 lb. Sense is downward and to the
right (Fig. 6). Tan $\theta = \dfrac{\Sigma F_y}{\Sigma F_x} = -0.146;$
$\theta = -8°\,20'.$

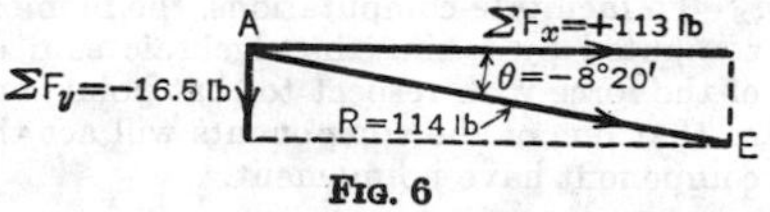

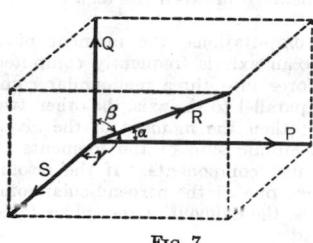

$\Sigma F_x = +113$ lb
$\Sigma F_y = -16.5$ lb
$\theta = -8°20'$
$R = 114$ lb

Fig. 6

Resolution into More than Two Coplanar Concurrent Forces

This type of problem has relatively little practical value. The resolution can be accomplished, however, by employing the force polygon to reverse the process of composition in a manner similar to that employed in the important special case discussed above of resolution into two concurrent forces. An infinite number of force systems is possible for any given number of the component concurrent forces.

Composition of Three Rectangular Noncoplanar Concurrent Forces

Parallelepipedon Law. * Consider the three rectangular forces, P, Q, and S (Fig. 7). On these forces construct to scale a parallelepiped. The resultant of the system is represented in magnitude and direction by the diagonal; its value is $R = \sqrt{P^2 + Q^2 + S^2}$. Its direction cosines with respect to the axes are: $\cos \alpha = P/R$, $\cos \beta = Q/R$, and $\cos \gamma = S/R$.

Fig. 7

Resolution into Three Rectangular Noncoplanar Concurrent Forces

A force F can be resolved into a set of three rectangular noncoplanar concurrent forces by *reversing* the process of composition described in the preceding paragraph. Thus, referring to Fig. 7:

$$P = R \cos \alpha; \quad Q = R \cos \beta; \quad S = R \cos \gamma$$

Composition of Any Number of Concurrent Forces †

This is the most general case of composition of concurrent forces and therefore is applicable also, of course, to the simpler cases previously discussed.

Let the forces be specified with respect to three rectangular axes passing through the point of concurrency: (a) Resolve each force into components along the X, Y, and Z axes; (b) find the algebraic sums of the x, y, and z components, and indicate them by $\Sigma F_x, \Sigma F_y$, and ΣF_z; (c) find the resultant of these three partial resultants by the parallelepipedon law; its value is $R = \sqrt{\Sigma F_x{}^2 + \Sigma F_y{}^2 + \Sigma F_z{}^2}$; its direction angles are $\alpha = \cos^{-1}\dfrac{\Sigma F_x}{R}$; $\beta = \cos^{-1}\dfrac{\Sigma F_y}{R}$; $\gamma = \cos^{-1}\dfrac{\Sigma F_z}{R}$.

Nature of Resultant of Concurrent Forces

The **Resultant** of any system of *concurrent forces* which are not in equilibrium is a *single force*.

4. MOMENTS AND COUPLES

Moment (or Torque) of a Force about a Point

Moment or Torque of a force about a point is the product of the force magnitude and the distance from the point to its action line. This perpendicular distance is called the *arm* of the force, and the point is the *origin or center of moments*. The product is the measure of the rotational tendency of the force. The name of the unit of moment is a combination of the names of force and distance units, as foot-pound, inch-ton, etc. (Some writers use lb-ft as a unit of moment of a force to distinguish from ft-lb as the unit of work or energy, similar distinction being made for the other units.)

---

* The *parallelepipedon law* applies also if the three forces are not rectangular. In such cases, however, it is not practicable to obtain *directly* the value of the resultant either graphically or algebraically. A better method of solution for problems of this nature is given in the second succeeding paragraph.

† As a matter of academic interest, although of little practical value, the resultant of any number of noncoplanar concurrent forces may be represented graphically by extending the principles of the *plane* force polygon to apply to a *space* or *skew* force polygon.

To facilitate computations, the moment of a force with respect to a point is frequently computed by taking the algebraic sum of the moments of two rectangular components of the force with respect to that point; and it will often be convenient to resolve the force so that one of the components will act through the origin of moments, thus making that component have no moment.

Moment (or Torque) of a Force about a Line (or Axis)

At any point on its action line, resolve the force into two rectangular components, one being parallel to the axis. The product of the perpendicular component and the perpen-

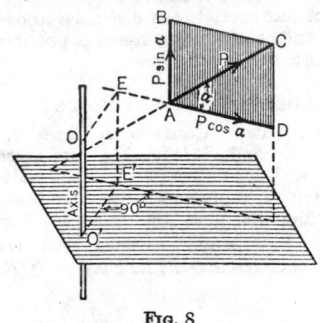

FIG. 8

dicular distance to its line of action from the axis is the *moment* of the given force about the axis. Thus (Fig. 3), P sin α is the component parallel to the axis, and it has no turning effect. All the moment or turning effect is caused by the perpendicular component, and its value is P cos $\alpha \times OE$, OE being the perpendicular distance between the axis and the parallel plane $ABCD$.

To facilitate computations, the moment of a force with respect to an axis is frequently computed by resolving the force into three rectangular components, one being parallel to the axis, the other two perpendicular to it; then the moment of the given force equals the algebraic sum of the moments of the two perpendicular components. If the resolution is made so that one of the perpendicular components cuts the axis, the moment of the given force equals the moment of the other perpendicular component.

Principle of Moments

For a Point. The moment of the resultant of any coplanar forces (not necessarily concurrent) about a point in their plane equals the algebraic sum of the moments of those forces about the point. Thus (Fig. 9), R is the resultant of P and Q, and $R \times r = P \times p - Q \times q$. Moments tending to produce counterclockwise rotation of a body are usually considered positive, and clockwise negative.
Thus, in Fig. 9, the moment of force Q about $O = -Qq$.
It is evident that the moment of a force passing through the origin of moments is zero.

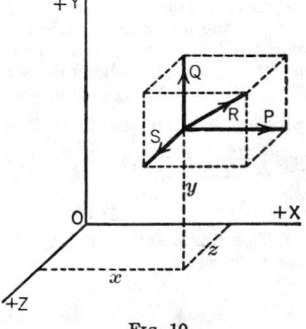

FIG. 9   FIG. 10

For an Axis. The moment of the resultant of any forces (not necessarily either coplanar or concurrent) about a line equals the algebraic sum of the moments of those forces about the line. Thus (Fig. 10), R is the resultant of the three rectangular forces P, Q, and S. Moment of R about axis $X = S \times y - Q \times z$. P contributes nothing to the moment sum, as it is parallel to the axis of moments. Usually in such a case, counterclockwise moment is called positive and clockwise negative, the observer looking toward the origin O, from the positive ends of the axes. Thus, Q has positive moment about axis Z, but negative moment about axis X.
It is evident that the moment of a force parallel to or intersecting the axis of moments is zero.

Couples

Nature of Couples. Two equal and parallel forces of opposite sense are called a *couple*. The tendency of a couple is to produce rotation only. Since a couple has no single resultant, no single force can balance it. To prevent the rotation of a body acted upon by a couple, the application of two other forces is required, forming a second couple.

The *arm* of a couple is the perpendicular distance between the lines of action of the forces. The *moment* of a couple is constant and independent of the origin of moments; it is equal to one of the forces times the arm of the couple. Its sense is positive or negative according as rotational tendency is counterclockwise or clockwise. Couples of equal moments, in the same or parallel planes, are *equivalent* and may be replaced one by the other. Further, the *center of rotation* for a couple may be anywhere in its plane. Hence, a couple may be turned about in its own plane or moved to a parallel plane or replaced by another couple (having an arm of any given length but the same moment) without altering its effect on a rigid body.

Resultant of Couples. The resultant of any number of coplanar couples or of couples in parallel planes is a couple. Its moment and sense equal the algebraic sum of the moments of the component couples.

A couple may be represented by a **vector**. The length of the vector to scale represents the magnitude of the moment; it is drawn perpendicular to the plane of the couple from *any origin*, and an arrow is placed on it to represent the way in which the couple would cause a right-hand screw to advance. The resultant of any number of couples (in oblique or parallel planes) is a couple. The composition is effected by simply adding the vectors representing the couples in a manner analogous to the method of "Composition of Any Number of Concurrent Forces" as previously described in this section. The resultant **vector** defines the resultant couple.

Composition of Single Force and Couple. A *single force and couple in the same plane* (*or parallel planes*) may be composed into *another single force* equal and parallel to the

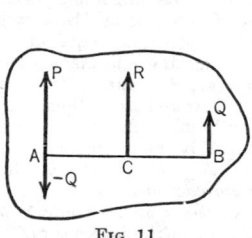

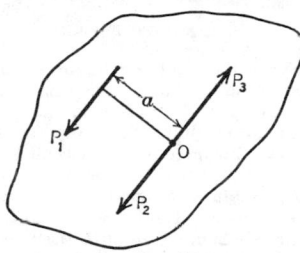

Fig. 11   Fig. 12

original force, at a distance from it equal to the moment of the couple divided by the magnitude of the force and so situated that the moment of the resultant about the point of application of the original force is of the same sign as the moment of the couple. The couple may be brought into the position shown in Fig. 11. The resultant of P, $-Q$, and Q is $R(=P)$ acting in a line through point C so that $(P - Q) \times AC = Q \times BC$. From this it follows that

$$AC = \frac{Q(AC + BC)}{P} = \frac{\text{moment of couple}}{P}$$

Resolution into Single Force through Chosen Point and Couple. A *single force* may be resolved into *another single force acting through a chosen point and a couple* (the new force being equal and parallel to the original force). In Fig. 12, P_1 is the given force and O the chosen point. Through O apply a pair of forces, opposite in sense, equal and parallel to P_1. As P_2 and P_3 balance, no change is produced in the motion of the body due to the addition. P_1 and P_3 constitute a couple of moment $= P \times a$, which is the same as moment of P_1 about O; and P_2 is a force just like P_1, but acting through the chosen point O.

5.   COMPOSITION OF NONCONCURRENT FORCES

Composition of Coplanar Nonparallel Nonconcurrent Forces

Graphic Method (1): String (or Funicular) Polygon. This method involves the resolution of each force into two components in a certain way and the subsequent composition of the components to determine the resultant. Thus, to find the resultant of the three forces ab, bc, and cd acting on the body in Fig. 13, draw a force polygon, as $ABCD$, for the given forces; resolve AB into AO and OB; BC into BO and OC, O having been taken anywhere: all components except the first and last occur in pairs and the forces of each pair are equal and opposite. thus OB and BO, OC and CO, etc.: choose the action lines of these components so that those of any one pair shall be collinear; thus, insert the components of AB at pleasure as at oa and ob, the components of BC so that the component BO shall act in ob, and hence the component OC in oc, etc.; thus the pairs of components consisting of equal, collinear, and opposite forces each balance, leaving only the first and last components AO and OD acting in ao and od; and their resultant (which is also the resultant of the given forces) is AD (magnitude and direction) ad (action line). The point O (Fig. 13) is called the **pole**, lines OA, OB, OC, etc., are **rays**; oa, ob, oc, etc., are **strings**, and all the strings constitute the string (or funicular) **polygon** for the forces. This polygon is also called "link," and "equilibrium polygon," the last being especially appropriate when the given forces are in equilibrium. The object in its construction is to locate one point on the action line of the resultant (or the length of arm to be used with the first and last rays of the force polygon if the resultant is a couple; see example in a later paragraph). The part of the drawing which represents the body and the lines of action of the forces is the *space diagram*; that representing force magnitudes, the *force polygon*. If the force polygon closes, the resultant is in general a couple, the forces of

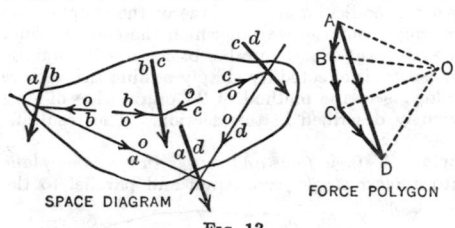

SPACE DIAGRAM

FORCE POLYGON

Fig. 13

the couple acting in the first and last strings of the string polygon, the magnitude of the forces being represented by the corresponding ray. If in addition to the force polygon closing, the first and last strings coincide, the string polygon is closed and the resultant vanishes. If the force polygon does not close, the first and last strings of the string polygon may still happen to have the same line of action but the forces along them will not be equal. The resultant is then the force (in magnitude and direction) required to close the force polygon, and its line of action is that of the coinciding first and last strings.

Graphic Method (2): Triangle Law. If the forces are not parallel and not nearly parallel, find the resultant R_1 of any two forces by the *Triangle Law*, then the resultant R_2 of R_1 and a third force, etc., until all the forces have been compounded. If the forces are parallel or nearly so, the method just explained fails because the lines of action of the several resultants cannot be determined readily on account of inaccessible intersections. In such cases, Graphic Method (1) is the only *graphic* method which can be used.

Algebraic Method. Compute the algebraic sums of the x and y components of the forces (ΣF_x and ΣF_y) and the algebraic sum of the moments of the forces (or of their components if more convenient, the result being the same), with respect to any origin O in their plane (ΣM). Then the resultant $R = \sqrt{(\Sigma F_x)^2 + (\Sigma F_y)^2}$, its angle with the x axis $= \tan^{-1} (\Sigma F_y)/(\Sigma F_x)$, and its arm with respect to O is $a = (\Sigma M)/R$. The general direction of R is apparent from the directions of its components ΣF_x and ΣF_y; a must be measured in such a direction from O that the sign of the moment of R will be the same as that of ΣM. If $\Sigma F_x = \Sigma F_y = 0$, the resultant is in general a couple whose moment $= \Sigma M$. If $\Sigma M = 0$ also, the resultant vanishes.

Composition of Coplanar Parallel Forces *

Graphic Method. If the forces are parallel or nearly so, *Graphic Method* (1) as described under the preceding heading for use with nonparallel forces may be used. *Graphic Method* (2) is not practical.

Using Graphic Method (1), if the force polygon does not close, the resultant is a single force. If it closes, the resultant is in general a couple. As an illustration of the latter, consider the four parallel forces in Fig. 14. The force polygon begins at A and ends at E, the same point; hence the resultant is not a single force (this is called a closed force polygon). Construct the funicular polygon as before. The first and last strings, ao and oe, being parallel, do not intersect.† The forces acting in those lines, AO and OE, being equal and opposite in sense, form a couple. Hence the resultant of the system is a couple whose moment is $AO(=OE) \times MN$. The *sense*, by inspection of the space diagram, is *clockwise*.

---

* Parallel forces are by nature *nonconcurrent*.
† It must be clearly understood that the fact that the forces are parallel and the fact that the first and last strings are parallel in the particular system illustrated in Fig. 14 have no bearing on each other. The first and last strings may or may not be parallel (or coincident) when the forces *either* are or are not parallel depending simply upon whether or not the resultant force vanishes \or happens to coincide in line of action with *both* of the strings).

If in addition to the force polygon closing, the first and last strings coincide, the string polygon is closed and the resultant vanishes.

Algebraic Method. This is a special case of that described under a preceding heading for use with nonparallel forces. The resultant $R = \Sigma F$. Its arm $a = (\Sigma M)/R$ and must be measured in such a direction from the origin of moments that the sign of the moment of R will be the same as that of ΣM.

If $\Sigma F = 0$, the resultant is in general a couple. If $\Sigma M = 0$ also, the resultant vanishes.

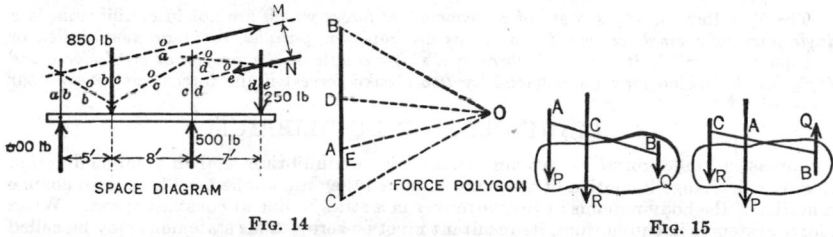

SPACE DIAGRAM   FORCE POLYGON

Fig. 14   Fig. 15

Simple Cases. When only two parallel forces are involved, their resultant (if a force) may be located more quickly by application of the following simple theorem based on Fig. 15. If P and Q act in the same direction, R cuts any line AB internally, and if P and Q are opposite, then externally on the side of the larger force; and in each case the segments of AB are inversely proportional to P and Q; that is, $AC/BC = Q/P$.

Composition of Noncoplanar Parallel Forces *

Graphic Methods are not advantageous in general for use in determining resultants of noncoplanar nonconcurrent force systems. Therefore only the algebraic method will be discussed.

Algebraic Method. Give to the forces F acting in the same direction one sign and to the others the opposite sign; then the resultant $R = \Sigma F$, the sense of R being indicated by the sign of ΣF. Next compute the sums of the moments of the forces with respect to two rectangular axes (x and y, say) perpendicular to the forces; call these sums ΣM_x and ΣM_y, and the arms of R with respect to those axes respectively a_x and a_y; then $a_x = (\Sigma M_x)/R$ and $a_y = (\Sigma M_y)/R$. The signs in these ratios may be disregarded; a_x and a_y have such positions that the moments of R with respect to the x and y axes have the same signs as those of ΣM_x and ΣM_y, respectively. If $\Sigma F = 0$, the resultant in general is a couple which can be determined by finding the resultant of all the forces but one; this resultant and the omitted force constitute the resultant couple. If $\Sigma M_x = \Sigma M_y = 0$ also, the resultant vanishes.

Composition of Noncoplanar Nonparallel Nonconcurrent Forces

Graphic Methods are not advantageous in general for use in determining resultants of noncoplanar nonconcurrent force systems. Therefore only the algebraic method will be discussed.

Algebraic Method. In general, the resultant is not a single force, but in such cases the system can be reduced to a force R acting through any point of the body selected and a couple C, noncoplanar with it; and if desired, R and C can in general be compounded into two noncoplanar forces. To determine R and C: select a set of coordinate axes (x, y, and z) in the body, the origin O being at the selected point referred to; determine the algebraic sums of the x, y, and z components of the given forces (ΣF_x, ΣF_y, and ΣF_z) and the algebraic sums of the moments of the forces with respect to the x, y, and z axes (ΣM_x, ΣM_y, and ΣM_z); ΣF_x, ΣF_y, and ΣF_z are the x, y, and z components of R, and ΣM_x, ΣM_y, and ΣM_z are the moments of the components of C perpendicular to the x, y, and z axes, respectively. They may be represented by vectors parallel to the axes.
$R^2 = (\Sigma F_x)^2 + (\Sigma F_y)^2 + (\Sigma F_z)^2$, and $C^2 = (\Sigma M_x)^2 + (\Sigma M_y)^2 + (\Sigma M_z)^2$; if α_1, α_2, and α_3 denote the angles between R and the x, y, and z axes, and θ_1, θ_2, and θ_3 the angles between the vector representing C and the x, y, and z axes respectively, then

$$\cos \alpha_1 = (\Sigma F_x)/R \qquad \cos \alpha_2 = (\Sigma F_y)/R \qquad \cos \alpha_3 = (\Sigma F_z)/R$$
$$\cos \theta_1 = (\Sigma M_x)/C \qquad \cos \theta_2 = (\Sigma M_y)/C \qquad \cos \theta_3 = (\Sigma M_z)/C$$

The resultant force R and the resultant couple C can be compounded into two forces as follows: take the plane of the couple so that one of the forces of the couple intersects

---

* Parallel forces are by nature *nonconcurrent*.

R; find the resultant of this force and R; this resultant and the other force of C are the two forces sought. In general the final two forces are skewed.

If the plane of C is parallel to R, C and R may be compounded into a single force by the method described in a previous paragraph headed " Composition of Single Force and Couple." If $C = 0$, the resultant is in general a single force; if $R = 0$, the resultant is in general a couple; if $C = R = 0$, the resultant vanishes.

Nature of Resultant of Nonconcurrent Forces

The Resultant of any system of *nonconcurrent forces* which are not in equilibrium is a *single force* or a *single couple* if the forces are *coplanar parallel, coplanar nonparallel,* or *noncoplanar parallel;* it is a *single force* or a *single couple* or a *noncoplanar single force and single couple* (which may be replaced by *two skewed forces*) if the forces are *noncoplanar nonparallel.**

6. PRINCIPLES OF EQUILIBRIUM

Forces in Equilibrium. A system of forces is in equilibrium if their combined action produces no change in motion of the body to which they are applied. There is no change in motion if the body remains at rest or moves in a straight line at constant speed. When a force system is in equilibrium, its resultant must be zero. This statement may be called the *general condition of equilibrium.* It implies both zero force and zero couple.

Table I. Conditions of Equilibrium

Necessary and sufficient independent conditions of equilibrium for the various force systems

System		Algebraical		Graphical	
		No.	Conditions	No.	Conditions
Coplanar	Collinear	1	$\Sigma F = 0$	1	Force polygon closes.
	Concurrent at point O	2	$\Sigma F_x = 0$, $\Sigma F_y = 0$; or, $\Sigma F_x = 0$, $\Sigma M_a = 0$, if x direction is not perpendicular to aO; or, $\Sigma M_a = 0$, $\Sigma M_b = 0$, if aOb is not a straight line.	1	Force polygon closes.
	Parallel	2	$\Sigma F = 0$, $\Sigma M = 0$; or, $\Sigma M_a = 0$, $\Sigma M_b = 0$, if line ab is not parallel to forces.	2	Force and funicular polygons close. (Latter item means that first and last strings coincide.)
	Nonparallel nonconcurrent	3	$\Sigma F_x = 0$, $\Sigma F_y = 0$, $\Sigma M = 0$; or, $\Sigma F_x = 0$, $\Sigma M_a = 0$, $\Sigma M_b = 0$, if x direction is not perpendicular to ab; or, $\Sigma M_a = 0$, $\Sigma M_b = 0$, $\Sigma M_c = 0$, if abc is not a straight line.	2	Force and funicular polygons close. (Latter item means that first and last strings coincide.)
Noncoplanar	Concurrent at point O	3	$\Sigma F_x = 0$, $\Sigma F_y = 0$, $\Sigma F_z = 0$. Combinations of moment and resolution equations can be arranged, but are not common.	2	Force polygon closes. It is warped; hence plan and elevation must show closed. Not commonly used.
	Parallel	3	$\Sigma F_z = 0$, $\Sigma M_x = 0$, $\Sigma M_y = 0$, forces parallel to z axis. Other combinations possible but not common.		Not used.
	Nonparallel nonconcurrent	6	$\Sigma F_x = 0$, $\Sigma F_y = 0$, $\Sigma F_z = 0$, $\Sigma M_x = 0$, $\Sigma M_y = 0$, $\Sigma M_z = 0$. ΣM about every axis $= 0$, and it is often convenient to employ more than three moment equations, instead of using so many resolution equations.		The projection of the system on any plane is in equilibrium, and algebraical or graphical conditions can be used to solve such projected systems.

An oblique system of x, y, z axes may be used for reference provided no two axes are at an angle of 180° with each other. Rectangular axes are preferable generally.

* It is evident that *any force system whatever,* whether *concurrent or nonconcurrent,* even though it includes "sub-systems" coming under all of the various classifications discussed, can be composed into a resultant which is a *single force* or a *single couple* or a *noncoplanar single force and single couple* (or *two skewed forces*).

Body in Equilibrium. A rigid body is in equilibrium if it remains at rest or moves in a straight line at constant speed; i.e., if its state of motion does not change. This condition obtains if all the external forces acting upon it (including those due to pull of gravity, friction, etc.) form a system in equilibrium.

Conditions of Equilibrium

In a problem in *statics* a body is known to be in equilibrium; hence the system composed of all the external forces acting upon it must be in equilibrium. In such a case, tests are not needed to ascertain if equilibrium exists, but they are used to set up relations involving unknown forces, distances, or angles, and the unknown elements are then computed provided their number does not exceed the number of independent equations which may be set up by means of the equilibrium conditions. When the number of unknown elements exceeds the number of independent equations, the problem is said to be statically indeterminate.

Special Conditions. If three forces are in equilibrium they must be coplanar and concurrent or parallel; if concurrent, each force is proportional to the sine of the angle between the other two; if parallel, each force is proportional to the distance between the other two. If a force system is in equilibrium, the resultant of any part must balance the resultant of the other part. It follows that if four coplanar nonconcurrent nonparallel forces are in equilibrium, the resultant of any two is concurrent with the other two.

Stability of Equilibrium. When a body (or collection of bodies) is in equilibrium and the state is such that if when displaced slightly in any way the body returns of itself to its original position, the equilibrium is *stable*; if when displaced slightly the body moves further from its original position, the equilibrium is *unstable*; and if when displaced slightly it remains in that displaced position, the equilibrium is *neutral* or *indifferent*. The body or collection is also said to be stable, unstable, or neutral (or indifferent) under these respective conditions. When the body is stable or unstable, the system of forces is changed by the slight displacement and is no longer in equilibrium; hence the further displacement. Only when the stability is neutral is the equilibrium of the force system undisturbed by a slight displacement of the body.

7. EQUILIBRIUM PROBLEMS

General Principles

(a) It frequently happens that the external force system acting on a body as a whole cannot be solved directly owing to the presence of more unknown elements (forces, distances, and angles) than there are conditions of equilibrium. In such cases, endeavor to separate the original body into simpler parts which will permit solutions. making use of the unknowns thus determined in solving the force systems acting on other sections until the complete solution, if obtainable, has been found.

(b) To facilitate computations, it is desirable if resolution equations are used, resolve perpendicular to one of the unknown forces; if a moment origin is used, to select it on the action line of an unknown force; if moment axes are used, to select them so as to intersect some of the unknown forces.

(c) Assume senses for unknown forces. A plus answer then indicates the sense to have been correctly assumed; a minus answer, incorrect assumption.

(d) When force polygons are used, letter action lines of wholly known forces first and those of the remainder last. Draw the force polygon to the end of the last known vector. Vectors required to close it (remembering that the senses must read confluently from the starting point back to the same point) determine unknown magnitudes and senses and/or lines of action.

Typical Problems

I. **System of Coplanar Concurrent Forces in Equilibrium** *with all forces known except two whose action lines only are known.* The magnitudes and senses of these two forces are to be determined.*

Example. Two smooth cylinders rest upon a 30° plane and against a vertical wall as shown in Fig. 16. Determine all forces acting on each cylinder. (a) The forces involved are 100 lb, 200 lb, P, Q, R, and S, the last four being normal to the surfaces of contact (smooth surfaces). (b) Consider the two cylinders as a single free body. The external force system is 100 lb, 200 lb, P, R, and S (Q_1 and Q_2 are internal). The system is nonconcurrent, so does not come under typical problem I. Consider the large cylinder as a free body. The external force system is 100 lb, Q, R, and S. While

* This is a common problem in the determination of the stresses of a roof or bridge truss.

this system is concurrent, it cannot be solved because there are more than two unknown quantities. Next consider the small cylinder as a free body. The force system is 200 lb, P, and Q, and this is typical problem I.

Algebraic Solution. Choose X and Y directions parallel and perpendicular to the plane.

$\Sigma F_x = 0 = Q_1 \dfrac{\sqrt{60}}{8} - 200 \sin 30°$. Hence $Q_1 = \dfrac{800}{\sqrt{60}} = 103.3$ lb. $\Sigma F_y = 0 = P - 200 \cos 30°$

$- 103.3 \times 2/8$. Hence $P = 199$ lb. Consider the large cylinder as a free body. $Q_1 = Q_2 = 103.3$

lb. Use the same X and Y directions. $\Sigma F_x = 0 = S \cos 30° - 103.3 \times \dfrac{\sqrt{60}}{8} - 100 \sin 30°$.

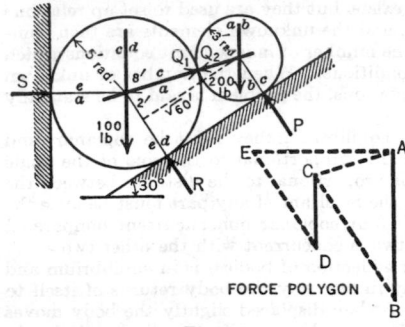

Hence, $S = 173.2$ lb. $\Sigma F_y = 0 = R - 100 \cos 30° - 173.2 \sin 30° + 103.3 \times 2/8$. Hence $R = 147.4$ lb.

Graphic Solution. Discussions (a) and (b) are same as above. The free body is the small cylinder. The force system is ab, bc, and ca, and the polygon is the triangle ABC (Fig. 16). $BC = 199$ lb, $CA = 103.3$ lb. Next, consider the large cylinder as a free body. The force system is ac, cd, de, and ea. Plot the known forces AC and CD. From D draw DE parallel to de, and from A, AE parallel to ae; these lines intersect at E. $DE = 147.4$ lb, and $EA = 173.2$ lb are the magnitudes and senses of the two unknowns.

If the system is concurrent and all the forces are known except one, the two unknown elements will be one angle and one magnitude and sense. Both may be determined by writing

FIG. 16

the equilibrium equations, or by drawing the force polygon.

II. System of Coplanar Parallel Forces in Equilibrium *with all forces known except two whose action lines only are known.* The magnitudes and senses of these two forces are to be determined.

Example. A beam is loaded as shown in Fig. 17 and supported at the points P and Q. Determine the reactions of the supports. Consider the beam as a free body. The external force system consists of the forces 1000 lb, 5000 lb, 2400 lb, P, and Q. This is a coplanar parallel system and is typical problem II.

Algebraic Solution. Assume senses for the reactions.

$$\Sigma M_p = 0 = - 4 \times 1000 - 10 \times 5000 - 14 \times 2400 + 16Q$$

Hence $Q = 5475$ lb; correct sense was assumed.

$$\Sigma M_q = 0 = 2 \times 2400 + 6 \times 5000 + 12 \times 1000 - 16P$$

Hence $P = 2925$ lb; correct sense assumed. As a check, apply a third equilibrium condition, $\Sigma F = 0$.

$$\Sigma F = 0 = - 2925 + 1000 + 5000 + 2400 - 5475$$

Graphic Solution of the same problem involves the construction of a closed force polygon and a closed funicular polygon. On a line of indefinite length parallel to the forces, often called the load line, construct the force poly-
gon by drawing vectors AB, BC, CD to represent 1000 lb, 5000 lb, 2400 lb. Let DE represent Q and EA represent P. The problem is to locate point E. This is done with the aid of the funicular polygon. Draw the rays OA, OB, OC, and OD. Construct the funicular polygon by drawing first the string between the last-lettered unknown force P (i.e, EA) and the first known force AB, and continuing it to the intersection of the last known string with the first-lettered unknown force. The closing line is

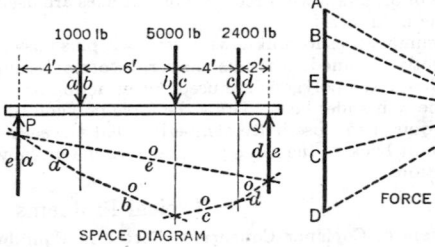

SPACE DIAGRAM

FORCE POLYGON

FIG. 17

oe. Draw the ray OE parallel to oe. The unknown reactions are DE and EA.

For problems of this type, the algebraic solution is preferable to the graphic.

(**NOTE:** This principle cannot be applied to beams having more than two points of support. Such problems require special treatment.)

III. System of Coplanar Nonparallel Nonconcurrent Forces in Equilibrium *with all forces known except two, of which the action line of one and a point in the action line of the* *other are known.* The magnitude and sense of the one and the magnitude, sense, and angular direction of the other are to be determined.*

Example. A roof truss is loaded as in Fig. 18. The left end of the truss rests on a smooth horizontal support. The right end is secured to a wall by means of a pin. Determine the reactions. The external forces acting on the truss are the given loads, the left reaction P (vertical, on account of the smooth support), and the right reaction Q (inclined, through point M). The unknown quantities are the reactions P and Q. This is typical problem III.

Algebraic Solution. Assume P upward.

$\Sigma M_M = 0 = 20,000 \times 18 + 25,000 \times 24 \cos 30° - 36P$; hence, $P = 24,430$ lb; correct sense assumed.

Assume Q upward to the left at angle θ with horizontal.

$\Sigma F_x = 0 = 25,000 \sin 30° - Q \cos \theta$; $\Sigma F_y = 0 = -25,000 \cos 30° - 20,000 + 24,430 + Q \sin \theta$.

Solving simultaneously, $Q = 21,300$ lb, and $\theta = 54°$. Sense and direction were correctly assumed, hence Q acts upward to the left at $54°$ to the horizontal. As a check, apply condition $\Sigma M_P = 0$. $\Sigma M_P = 0 = -25,000 \times 12 \cos 30° - 20,000 \times 18 + 21,300 \times 36 \sin 54°$.

Graphic Solution. ab and bc are the action lines of the given loads, cd of the reaction P and da of the reaction Q. Draw the vectors AB and BC, and a line through C, parallel to cd. Choose a pole and draw the rays. Construct the funicular polygon, drawing oa through M, and draw closing string od from K to M. Draw OD through O parallel to od to intersect CD at D. Draw DA. Vectors CD and DA represent the two unknown forces, $P = 24,430$ lb and $Q = 21,300$ lb. The action line of Q is da, making angle with horizontal $= 54°$.

Special Case. A case coming under the above classification which requires a variation in treatment when employing the graphic method is one in which the action lines of all three forces are known but their magnitudes and senses are unknown. The procedure is in general similar to methods employed before except that, in the graphic solution, two of the unknown forces which are concurrent must be replaced by their unknown resultant acting through their point of concurrency along an unknown action line. After the magnitude and sense of this resultant have been determined (by the method employed in the above example), it is resolved into its two components along the action lines of the two unknown forces which it had replaced. These components represent the magnitudes and senses of this pair of forces.

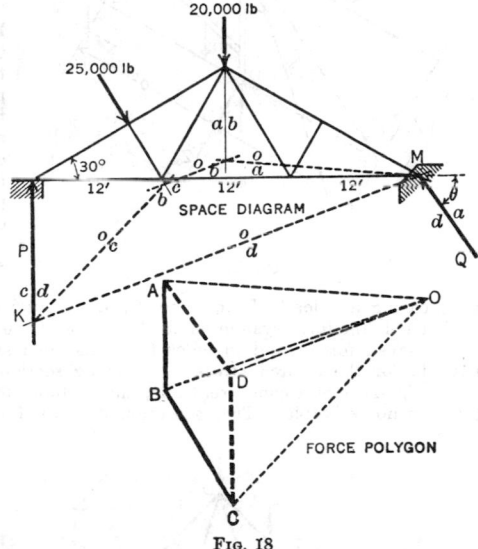

20,000 lb

25,000 lb

SPACE DIAGRAM

FORCE POLYGON

Fig. 18

IV. System of Noncoplanar Nonparallel Nonconcurrent Forces in Equilibrium *with one force completely known and action lines (or a point in the action line) of the others known.* All unknown force magnitudes, action lines and senses are to be determined.

Example. The crane (Fig. 19) is supported by a socket at the foot of the post, at D; is kept from overturning by the backstays AB and AC; and carries a load of 600 lb (E, A, F, G, D, are in the vertical XY plane). Determine the axial components of the reaction on the post at D and the tensions in the backstays. The external forces acting on the post are the load, the reaction at D, and the tensions in the backstays at A. This is typical problem IV. Moment equations are the most convenient to apply for this solution.

$$\Sigma M_{BC} = 0 = 600 \times 40 - 20 D_y; \quad D_y = 1200 \text{ lb}$$

$$\Sigma M_{ZA} = 0 = 600 \times 20 - 16 D_x; \quad D_x = 750 \text{ lb}$$

$$\Sigma M_{XA} = 0 = 600 \times 4 - 16 D_z; \quad D_z = 150 \text{ lb}$$

$$\Sigma M_{XC} = 0 = AB \times \frac{16}{\sqrt{881}} \times 25 - 1200 \times 10 + 600 \times 6; \quad AB = 622 \text{ lb}$$

$$\Sigma M_{XB} = 0 = AC \times \frac{16}{\sqrt{756}} \times 25 + 600 \times 19 - 1200 \times 15; \quad AC = 452 \text{ lb}$$

---

* This is a common problem in the determination of the reactions on a roof truss sustaining wind pressures, the truss being fixed at one end and resting on rollers at the other.

The senses of all forces are as shown in Fig. 19.

The magnitude, action line, and sense of the resultant force on the post at D can be readily determined (if desired) by the Parallelepipedon Law.

Truss Analysis

A **Truss** is a framework * for carrying loads, each *member* of which is subjected only to tension or compression loads. The members are usually pin-jointed with loads applied only at the joints.

The **Stress in a Member** at any section is the force which either of its two parts exerts internally on the other part as a result of the external forces acting on the member. Longitudinal stresses, like external longitudinal forces, may be either tensile or compressive.

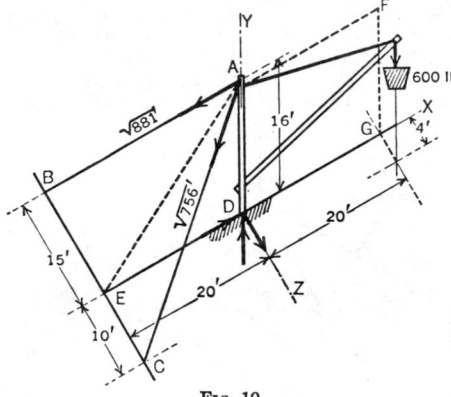

The **Analysis of a Truss** under a given loading condition refers to the determination of the stresses in its members due to the loads.

Analysis by Method of Sections. First, determine the reactions on the truss due to the loads; second, imagine the truss separated into two distinct parts (that is, pass a section through the truss) so that the member under consideration is one of the members cut and so that the system of forces, including stresses, acting on either part of the truss is solvable for the desired stress; third, solve the system.

To Pass the Section, suppose the stress in HI (Fig. 20a) is required, the truss being supported at its ends

FIG. 19

and bearing five loads L and one P, and suppose the reactions determined. Trying section 1–1, the force system on the left part of the truss (Fig. 20b) is a nonconcurrent one of seven forces, and includes four unknown stresses, S_1, S_2, S_3, and S_4; it is not solvable for the desired stress S_1. Trying section 2–2, the force system on the lower part (Fig. 20c) is a concurrent one, and includes four unknown stresses, S_1, S_2, S_5, and S_6; it is not solvable. Trying section 3–3, the force system on the left part (Fig. 20d)

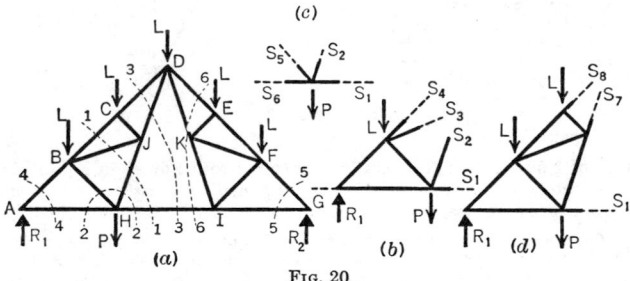

FIG. 20

is nonconcurrent with three unknown stresses, S_1, S_7, and S_8; it is solvable. In some instances different sections may be used, each leading to a solution.

S_1 having been determined, the force system of Fig. 20b becomes solvable, and then, with S_2 also determined, the force system of Fig. 20c may be solved.

Algebraic Solution. Following the general method of procedure outlined above, determine the various stresses by employing algebraic conditions of equilibrium in manners similar to those illustrated heretofore under " Equilibrium Problems."

Graphic Solution. Following the general method of procedure outlined above, determine the various stresses by employing graphic conditions of equilibrium.

In Making the Imaginary Separations of the Truss, care should be taken to cut not more than three members in which the stresses are unknown. It is advantageous to make

---

* *Redundant frames* (i.e., ones having more members than necessary to preserve their shapes under the loading conditions) are not considered in this section, since the stresses in them cannot be determined by elementary static methods.

the separation so that not more than two such members are cut. If this is done, a single force polygon will determine the two unknowns, whereas if three are cut, a force polygon and an equilibrium polygon, or the equivalent, are necessary for determining the three unknowns.

Analysis by Method of Joint Resolution.* Consider the pin at each joint as a body acted upon by forces in equilibrium.

Algebraic Solution. Determine the various stresses by employing algebraic conditions of equilibrium in manners similar to those illustrated heretofore under " Equilibrium Problems."

Graphic Solution. For each joint, draw the force polygon. In doing so, it will be advantageous to represent the forces in the order in which they occur about the joint. A force polygon so drawn will be called a polygon for the joint; and for brevity, if the order taken is clockwise, the polygon will be called a clockwise polygon, and if counterclockwise, it will be called a counterclockwise polygon. If the polygons for all the joints of a truss are drawn separately, the stress in each member will have been represented twice. It is possible to combine the polygons so that it will not be necessary to represent the stress in any member more than once, thus reducing the number of lines to be drawn. Such a combination of force polygons is called a *stress diagram.* Each triangular space in the truss diagram is marked by a small letter; also the space between consecutive action lines of the loads and reactions. Then the two letters on opposite sides of any line serve to designate that line, and the same large letters are used to designate the magnitude of the corresponding force.

To construct a stress diagram for a truss under given loads:

(1) Determine the reactions. (2) Letter the truss diagram as directed. (3) Construct a force polygon for all the external forces applied to the truss (loads and reactions), representing them in the order in which their application points occur about the truss, clockwise or counterclockwise. (4) On the sides of that polygon construct the polygons for all the joints. They must be clockwise or counterclockwise according as the polygon for the loads and reactions was drawn clockwise or counterclockwise. (The first polygon drawn must be for a joint at which only two members are fastened; the joints at the supports are usually such. Next, that joint is considered, and its polygon is drawn, at which not more than two stresses are unknown.)

Example. Fig. 21 represents a roof truss sustaining loads of 600, 1000, 1200, and 1800 lb; the right reaction is 2100 lb, and the left 2500 lb. *ABCDEFA* is a polygon for the loads and reactions, these being represented in the order in which their points of application occur about the truss. The polygon for joint 1 is *FABGF*; the force *BG* acts toward the joint, hence *bg* is under compression, and *GF* acts away from the joint, hence *gf* is in tension. The polygon for joint 2 is *CDEHC*; the force *EH* acts away from the joint, hence *eh* is in tension; and *HC* acts toward the joint, hence *hc* is in compression. The polygon for joint 3 is *HEFGH*; the force *GH* acts away from the joint and hence *gh* is in tension. If the work has been done correctly, *GH* is parallel to *gh*. (In Fig. 21a all the polygons are clockwise, and in Fig. 21b, counterclockwise.)

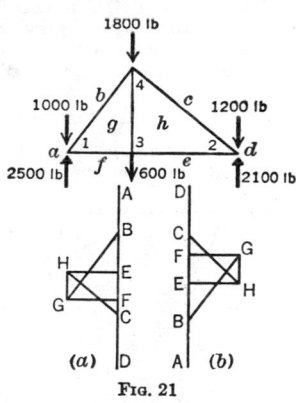

FIG. 21

8. CENTER OF GRAVITY

Definitions

The **Centroid** of a system of parallel forces having fixed application points is the point through which their resultant will always pass regardless of how the forces may be turned, provided they remain parallel.

The **Center of Gravity of a Body** † *or system of bodies* is the *centroid of the forces of gravitation* ‡ acting upon all the particles thereof. Referring the application points of such a force system to a set of coordinate axes, the coordinates of the centroid, or center of gravity (c.g.), are:

$$\bar{x} = (\Sigma F_i \cdot x_i)/\Sigma F_i = \frac{\int x dF}{F}; \quad \bar{y} = (\Sigma F_i \cdot y_i)/\Sigma F_i = \frac{\int y dF}{F}; \quad \bar{z} = (\Sigma F_i \cdot z_i)/\Sigma F_i = \frac{\int z dF}{F}$$

---

* This method is not usually so convenient for determining algebraically the stress in a single specified member.

† Sometimes called *center of mass* or *center of inertia.*

‡ For practical purposes, the forces of gravitation may be considered as parallel.

in which F_i represents the force on (or weight of) one particle and x_i, y_i, z_i are the coordinates of its application point. If a group of bodies is involved, the coordinates of the center of gravity of the group are:

$$\bar{x} = (\Sigma W_i \cdot \bar{x}_i)/\Sigma W_i; \quad \bar{y} = (\Sigma W_i \cdot \bar{y}_i)/\Sigma W_i; \quad \bar{z} = (\Sigma W_i \cdot \bar{z}_i)/\Sigma W_i$$

in which W_i represents the weight of one body and $\bar{x}_i$, $\bar{y}_i$, $\bar{z}_i$ are the coordinates of its center of gravity.* A body (or system of bodies), if supported at its center of gravity, will remain at rest in any position.

The Center of Gravity of Part of a Body may be located by the rule that its moment, with respect to any plane, equals the moment of the whole minus, algebraically, the moment of the remainder.

The Center of Gravity of a Line, Surface, or Volume is that point which would be the center of gravity if the line were replaced by a homogeneous rod of infinitesimal diameter, the surface by a homogeneous plate of infinitesimal thickness, or the volume by a homogeneous body.

Symmetry. Two points are symmetrical with respect to a third point if the line joining the two is bisected by the third. Two points are symmetrical with respect to a line or a plane if the line joining them is perpendicular to the given line or plane and is bisected by it. A body, line, surface, or volume is symmetrical with respect to a point, a line, or a plane if all the points of the body, line, surface, or volume can be paired so that each pair is symmetrical with respect to the point, line, or plane. If a homogeneous body, or a line, surface, or volume is symmetrical with respect to a point, a line, or plane, its center of gravity is at the point, in the line or in the plane.

The Static Moment of a body (having weight), a line (having length), a surface (having area), or a solid † (having volume) with respect to any plane is the product of the weight, length, area, or volume and the distance of the center of gravity of the body, line, surface, or solid from the plane. The static moment of a plane line or plane surface with respect to a straight line in the plane is the product of the length or area and the distance of the center of gravity of the line or surface from the reference line. A static moment is regarded as positive or negative according as the corresponding center of gravity is on the positive or negative side of the reference plane or line.

Determination of Center of Gravity Location

When practicable, determination of center of gravity location by algebraic or integration methods, based on dividing the sum of the moments by the sum of the forces, is

Fig. 22

generally the simplest process. For some bodies of non-homogeneous nature or of very irregular shape, one of the following methods of procedure may be necessary or at least preferable:

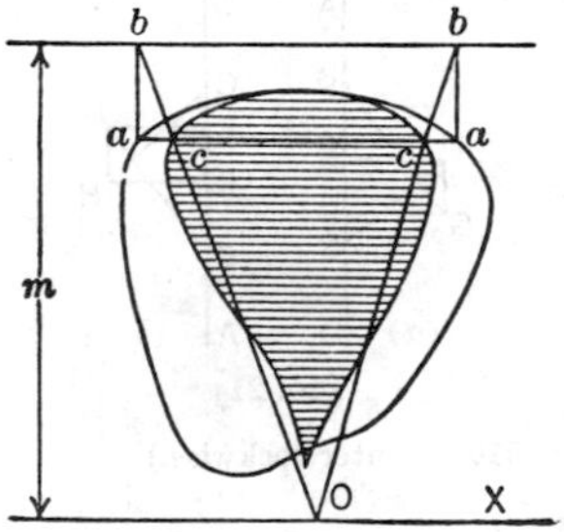

Graphic Method. For application to *plane* figures.‡ Referring to Fig. 22, take a point O and a line bb on opposite sides of the figure at any convenient distance m apart; project any width of the figure parallel to bb as aa on bb, connect the projections bb with O and note the intersections cc; determine other points cc and draw a smooth curve through them as shown; measure the area A' within the curve cc; then $A'm$ is the static moment of the given figure with respect to OX; if A is the area of the given figure and y the distance of its center of gravity from OX, $y = A'm/A$. In a similar way the distance of the center of gravity from a line perpendicular to OX can be determined and its exact position thus definitely located.

Suspension Method. For application to *plane* figures.§ Suspend the body (or a model representing it) from a point near its edge and mark on it the direction of a plumb-line hung from that point. Repeat this operation, using a second suspension point. The center of gravity is at (or behind) the intersection of the two markings.

Weighing Method. Generally applied where location of c.g. in one plane only is required. Determine weight W of the body and then support it on a knife-edge (Fig. 23) and on a point support

---

* As F (in the gravitational system of units) equals W, these symbols may be used interchangeably in the two sets of formulas. Also, any one of the expressions may be read as "the sum of the moments divided by the sum of the forces (or weights)."

† The word "solid" where used herein denotes "that which has volume." Care should be taken to distinguish this from a "body," which has "mass" (as well as "volume"). Some writers use the word "solid" to denote at various times either volume or mass, which is sometimes confusing. In this section, the word "volume" is frequently used even in preference to "solid" to avoid the possibility of confusion with "mass."

‡ Including areas or flat homogeneous bodies of uniform thickness.

§ Including areas or flat bodies of uniform homogeneous thickness.

resting upon a platform scale. Weigh reaction R of the point support and measure horizontal distance a between the point and the knife-edge. Then the horizontal distance from the knife-edge to the center of gravity is $\bar{x} = Ra/W$.

Balancing Method. For general application. Balance the body (or a model representing it) on a straight-edge, marking on the body the vertical plane containing the edge. Repeat for two more balancing positions of the body. The center of gravity is at the point common to the three planes thus determined.

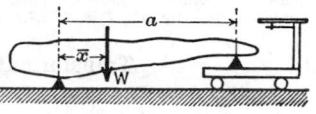

Fɪɢ. 23

9. MOMENT OF INERTIA

Plane Surfaces—Definitions

The **Moment of Inertia of a Plane Surface** (or figure) with respect to (or about) a line (or axis) is the sum of the products obtained by multiplying the area of each element of the surface by the square of its distance from the line.* Letting I_x denote moment of inertia about an X axis:

$$I_x = \int y^2 dA$$

in which A is the total area and y is the perpendicular distance of any element of area dA from the axis. The moment of inertia of a surface is obviously the sum of the moments of inertia of its parts. The moment of inertia of a plane surface is **rectangular** if the axis used is *in* the plane of the area; it is **polar** if the axis is *perpendicular* to the plane of the area.

The **Radius of Gyration of a Plane Surface** with respect to a line is the length whose square multiplied by the area of the surface equals the moment of inertia of the surface with respect to the line. Letting k denote radius of gyration:

$$I = k^2 A \quad \text{or} \quad k = \sqrt{I/A}$$

in which I is the moment of inertia and A the area.

The **Product of Inertia of a Plane Surface** with respect to a pair of coordinate axes in the plane is the algebraic sum of the products obtained by multiplying the area of each element of the surface by its coordinates.* Letting U_{xy} denote product of inertia with respect to X and Y axes:

$$U_{xy} = \int xy dA$$

in which A is the total area and x and y are the coordinates of any element of area dA.

The **Principal Axes of Inertia of a Plane Surface at a Particular Point** in the plane are the two axes about which the moments of inertia are greater and less than for any other axis, through the point in the plane.† The corresponding moments of inertia are called the **principal moments of inertia** of the surface at the point. The principal axes are always at right angles to each other. The product of inertia with respect to them is zero.

The **Customary Engineer's Unit** for both moment and product of inertia of a surface is biquadratic inches (in.⁴).

Determination of Moment of Inertia of Plane Surfaces

When practicable, determination of moment of inertia with respect to an axis by algebraic or integration methods is generally the simplest process. For some surfaces of very irregular shape, the following graphic method of procedure may be necessary or at least preferable:

Graphic Method. Let $aaaa$ (Fig. 24) be the outline and XX' the axis with respect to which the moment of inertia is desired; at any convenient distance m from XX' draw two parallels (but if XX' does not cut the figure, only one parallel, the one on the opposite side of the figure from XX'); draw any line as aa parallel to XX' and project the points aa on the nearer parallel; join the projections bb to any point O in XX', and note the intersections cc on aa; project cc on the same parallel; join the projections dd with O, and note the

Fɪɢ. 24

---

* Moment of inertia is always positive and never zero. Product of inertia may be positive, zero, or negative, depending upon the distribution of the area with respect to the axes. If a surface has an axis of symmetry, its product of inertia with respect to that axis and one perpendicular thereto is zero.

† In certain special cases, as for axes through the point in the center of a circular area, the moment of inertia is the same for any axis and therefore there is no principal axis through that point.

intersections *ee* on *aa*. In a similar manner determine points like *ee* for other widths like *aa*, and connect all points *e* as shown. Then measure the area of the loops *OPO* and *OQO*; denoting this combined area by A'', $I = A'' m^2$. (There will be only one loop if only one parallel *bb* is used.)

Transformation Formulas—Plane Surfaces

Parallel Axes Theorems. Let $I = moment\ of\ inertia$ (either rectangular or polar) of a plane figure with respect to any line or axis, $\bar{I}$ = that with respect to a parallel axis passing through the center of gravity of the figure, d = distance between the axes, k and $\bar{k}$ = the radii of gyration with respect to the same axes respectively, and A = area of the figure; then
$$I = \bar{I} + Ad^2 \quad \text{and} \quad k^2 = \bar{k}^2 + d^2$$

These show that with respect to all parallel axes the moment of inertia and the radius of gyration are least for the one passing through the center of gravity of the figure.

Similarly, let $U = product\ of\ inertia$ of a plane figure with respect to a pair of coordinate axes in the plane, and $\bar{U}$ = that with respect to a parallel pair whose origin is at the center of gravity; $\bar{x}$, $\bar{y}$ the coordinates of the center of gravity referred to the first pair, and A the area of the figure; then $U = \bar{U} + A\bar{x}\bar{y}$.

Relation of Rectangular and Polar Moments of Inertia. Let I_x, I_y, and J_z = the moments of inertia of a plane figure with respect to x, y, and z axes respectively, the axes being *at right angles* to each other and the x and y axes in the plane; and let k_x, k_y, and k_z = the corresponding radii of gyration; then $J_z = I_x + I_y$, $k_z^2 = k_x^2 + k_y^2$.

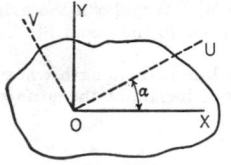

Rotated Axes Theorem. Let XOY and UOV (Fig. 25) be two sets of rectangular coordinate axes with a common origin and in a given plane figure. I_x, I_y, I_u, I_v = moments of inertia of the figure with respect to x, y, u, and v axes respectively; U_{xy} and U_{uv} = its products of inertia with respect to the sets of axes respectively; α = angle through which x axis must be rotated to bring it into u axis, regarded as positive or negative according as the turning is counterclockwise or clockwise. Then $I_u + I_v = I_x + I_y$, and

Fig. 25

$$I_u = I_x \cos^2 \alpha + I_y \sin^2 \alpha - U_{xy} \sin 2\alpha$$
$$U_{uv} = \frac{1}{2}(I_x - I_y) \sin 2\alpha + U_{xy} \cos 2\alpha$$

If OU and OV are *principal axes* (see definition above), $U_{uv} = 0$ and therefore $\tan 2\alpha = 2U_{xy}/(I_y - I_x)$. Hence, the principal axes of a figure at a point can be readily found if the moments of inertia and the product of inertia of the figure with respect to two rectangular axes through the point and in the plane are known.* The *principal moments of inertia* are then I_u from the formula above and I_v from the same formula after replacing α by $\left(\alpha + \dfrac{\pi}{2}\right)$. As a check, $I_u + I_v = I_x + I_y$.

Graphic Transformations—Plane Surfaces

The Inertia Circle is a device for determining *graphically* the moment of inertia of a plane figure with respect to any line of the plane through a given point; and the principal axes and principal moments of inertia for the same point. To construct the circle, it is necessary to know the moments of inertia and the product of inertia with respect to two rectangular axes through the point, in the plane figure. Suppose I_x, I_y, and U_{xy} given for the shaded area in Fig. 26. To convenient scale, plot OX' and OY' to represent I_x and I_y, and $Y'A$ to represent U_{xy} (downward if negative and upward if positive). Center C is mid-

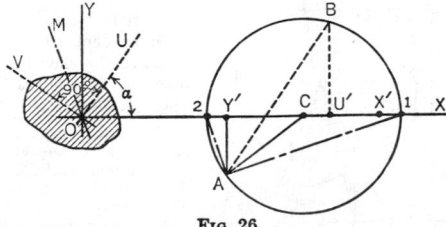

Fig. 26

way between X' and Y'. With CA as radius, describe the inertia circle. To find I_u, draw chord AB parallel to axis OU; draw perpendicular BU'. OU' (to scale) = I_u, and BU' (to scale) = U_{uv}. OM, parallel to $A2$, is axis of least I; and a parallel to $A1$, through O, is axis of greatest I. $O2$ (to scale) is the value of least $I = I_2$; and $O1$, value of greatest $I = I_1$. *Least radius of gyration* for an axis through point $O = \sqrt{I_2/\text{area}}$.

---

* If U_{xy} and $(I_y - I_x)$ are both zero, there is no principal axis through the point.

Bodies—Definitions

The Moment of Inertia of a Body with respect to (or about) a line (or axis) is the sum of the products obtained by multiplying the mass of each elementary part by the square of its distance from the line.† Letting I_x denote moment of inertia about an X axis:

$$I_x = \int y^2 \, dm$$

in which m is the total mass and y is the perpendicular distance of any element of mass dm from the axis.* The moment of inertia of a body is obviously the sum of the moments of inertia of its parts.

The Center of Gyration of a Body with respect to a line is a point at such a distance from the line that, if the entire mass of the body were concentrated there, its moment of inertia would be the same as that of the body.

The Radius of Gyration of a Body with respect to a line is the distance from the center of gyration to the line. Letting k denote radius of gyration:

$$I = k^2 \, m \quad \text{or} \quad k = \sqrt{I/m}$$

in which I is the moment of inertia and m the mass.

The Product of Inertia of a Body with respect to a pair of coordinate planes is the algebraic sum of the products obtained by multiplying the mass of each element of the body by its coordinates with reference to those planes.† Thus with respect to YOZ and ZOX (Fig. 27), ZOX and XOY, and XOY and YOZ planes, the products of inertia are respectively:

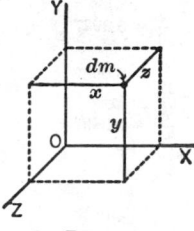

FIG. 27

$$U_{xy} = \int xy\,dm; \quad U_{yz} = \int yz\,dm; \quad U_{zx} = \int zx\,dm$$

The Principal Axes of Inertia of a Body at a Particular Point. The values of moments of inertia of a body for all axes through a given point are in general unequal; for one axis the moment of inertia is greater and for another it is less than for any other axis through the point. These two axes are at right angles, and they together with one at right angles to their plane and passing through the point are **principal axes of inertia of the body at the point**; the corresponding moments of inertia are the **principal moments of inertia of the body at the point**. If the point is the center of gravity of the body, the axes and moments are called *central principal axes* and *central principal moments of inertia*. For a set of principal axes, the three products of inertia, with respect to the **principal planes** determined by them, are zero.

The Customary Engineer's Unit for both moment and product of inertia is slug-ft.[2]

Transformation Formulas—Bodies

Parallel Axes Theorems. Let I = moment of inertia of a body with respect to any line or axis, $\bar{I}$ = that with respect to a parallel axis passing through the center of gravity of the body, d = distance between the axes, k and $\bar{k}$ = the radii of gyration with respect to the same axes respectively, and m = mass of the body; then

$$I = \bar{I} + md^2 \quad \text{and} \quad k^2 = \bar{k}^2 + d^2$$

Rotated Axes Theorems. Let I_x, I_y, and I_z denote the moments of inertia of a body with respect to rectangular axes x, y, and z respectively; U_{xy}, U_{yz}, and U_{zx} its products of inertia with respect to yz and zx planes, and zx and xy planes, and xy and yz planes respectively; I the moment of inertia of the body with respect to a line through the origin of coordinates having direction-angles α, β, and γ; then

$$I = I_x \cos^2\alpha + I_y \cos^2\beta + I_z \cos^2\gamma - 2U_{yz}\cos\beta\cos\gamma - 2U_{zx}\cos\gamma\cos\alpha$$
$$- 2U_{xy}\cos\alpha\cos\beta$$

If $U_{xy} = U_{yz} = 0$, the y axis is a principal axis at the origin.
If $U_{yz} = U_{zx} = 0$, the z axis is a principal axis at the origin.
If $U_{zx} = U_{xy} = 0$, the x axis is a principal axis at the origin.

If a homogeneous body has a plane of symmetry, any perpendicular to the plane is a principal axis of the body at the point where the line pierces the plane. If it has two planes of symmetry at right angles to each other, their intersection is a principal axis at any point of the intersection, the other two being in the planes of symmetry. If it has three planes of symmetry, their lines of intersection are the central principal axes of the body.

---

* Strictly speaking, this is the moment of inertia of the *mass* of a body. If m be replaced by w = weight, the result is the moment of inertia of the *weight* of a body.
† Moment of inertia is always positive and never zero. Product of inertia may be positive, zero, or negative, depending upon the distribution of the mass with respect to the coordinate planes. If a body has a plane of symmetry, its product of inertia with respect to that plane and one perpendicular thereto is zero.

Properties of Various Lines, Surfaces, Volumes, and Bodies

Symbols: I_i = Rectangular Moment of Inertia; k_i = Corresponding Radius of Gyration; J_O = Polar Moment of Inertia about axis through O perpendicular to plane; k_O = Corresponding Radius of Gyration; m = mass = W/g where W = weight and g = acceleration due to gravity. Moments of Inertia of bodies are given in terms of mass. For their values in terms of weight, replace m by W in the formulas.

Decimal Equivalents (*for reference*):

$\pi = 3.1416$	$\dfrac{\pi}{128} = 0.0245$	$\sqrt{10} = 3.162$	$\dfrac{1}{\sqrt{6}} = 0.408$
$\dfrac{\pi}{2} = 1.5708$		$\sqrt{12} = 3.464$	
$\dfrac{\pi}{4} = 0.7854$	$\dfrac{1}{\pi} = 0.318$	$\sqrt{18} = 4.242$	$\dfrac{1}{\sqrt{8}} = 0.354$
	$\sqrt{2} = 1.414$	$\dfrac{1}{\sqrt{2}} = 0.707$	
$\dfrac{\pi}{8} = 0.3927$	$\sqrt{3} = 1.732$		$\dfrac{1}{\sqrt{10}} = 0.316$
$\dfrac{\pi}{32} = 0.0982$	$\sqrt{5} = 2.236$	$\dfrac{1}{\sqrt{3}} = 0.577$	$\dfrac{1}{\sqrt{12}} = 0.289$
	$\sqrt{6} = 2.449$		
$\dfrac{\pi}{64} = 0.0491$	$\sqrt{8} = 2.828$	$\dfrac{1}{\sqrt{5}} = 0.447$	$\dfrac{1}{\sqrt{18}} = 0.238$

Table II.
Lines

Figure	Centroid Location
1. Any Plane Curve	C.G. is at point having coordinates $\bar{x}$, $\bar{y}$, where $\bar{x} = \dfrac{\int x\,ds}{\text{Length}}$ where $ds = \sqrt{1 + \left(\dfrac{dy}{dx}\right)^2}\,dx$ $\bar{y} = \dfrac{\int y\,ds}{\text{Length}}$ where $ds = \sqrt{1 + \left(\dfrac{dx}{dy}\right)^2}\,dy$
2. Circular Arc	C.G. is on axis of symmetry at $\bar{x} = \dfrac{r \sin \alpha}{\alpha} = \dfrac{rc}{s}$. If α is small, distance from C.G. to chord = approx. $\dfrac{2\,h}{3}$. (Error is small even for $\alpha = 45°$) *For semi-circle:* $\bar{x} = \dfrac{2\,r}{\pi}$ *For quadrant:* $\bar{x} = \dfrac{2r\sqrt{2}}{\pi}$, and distance from radius drawn to either end of arc $= \dfrac{2\,r}{\pi}$.

Plane Surfaces

Figure	Centroid Location; Moments of Inertia; Radii of Gyration
3. Any Plane Surface	C.G. is at point having coordinates $\bar{x}$, $\bar{y}$, where $\bar{x} = \dfrac{\iint x\,dxdy}{\text{Area}} = \dfrac{\iint \rho^2 \cos \theta\,d\rho d\theta}{\text{Area}}$; $\bar{y} = \dfrac{\iint y\,dxdy}{\text{Area}} = \dfrac{\iint \rho^2 \sin \theta\,d\rho d\theta}{\text{Area}}$ $I_x = \iint y^2\,dxdy$; $I_y = \iint x^2\,dxdy$; $J_O = \iint \rho^3\,d\rho d\theta = I_x + I_y$ $k_x = \sqrt{\dfrac{I_x}{\text{Area}}}$; $k_y = \sqrt{\dfrac{I_y}{\text{Area}}}$; $k_O = \sqrt{\dfrac{J_O}{\text{Area}}} = \sqrt{\dfrac{I_x + I_y}{\text{Area}}}$

Table II *(Continued)*

Plane Surfaces *(Continued)*

Figure	Centroid Location; Moments of Inertia; Radii of Gyration

4. Triangle

C.G. is at O = intersection of medians.

Perpendicular distance from $a - a = \dfrac{h}{3}$.

$$I_g = \frac{bh^3}{36}; \qquad I_a = \frac{bh^3}{12}; \qquad I_c = \frac{bh^3}{4}$$

$$k_g = \frac{h}{3\sqrt{2}}; \qquad k_a = \frac{h}{\sqrt{6}}; \qquad k_c = \frac{h}{\sqrt{2}}$$

5. Solid Rectangle (or Square)

C.G. is at O = intersection of diagonals.

For Rectangle:

$$I_g = \frac{bh^3}{12}; \qquad I_a = \frac{bh^3}{3}; \qquad I_c = \frac{b^3h^3}{6(b^2 + h^2)}; \qquad J_o = \frac{bh(b^2 + h^2)}{12}$$

$$k_g = \frac{h}{2\sqrt{3}}; \qquad k_a = \frac{h}{\sqrt{3}}; \qquad k_c = \frac{bh}{\sqrt{6(b^2 + h^2)}}; \qquad k_o = \sqrt{\frac{b^2 + h^2}{12}}$$

For Square (letting $b = h = s$):

$$I_g = \frac{s^4}{12}; \qquad I_a = \frac{s^4}{3}; \qquad I_c = \frac{s^4}{12}; \qquad J_o = \frac{s^4}{6}$$

$$k_g = \frac{s}{2\sqrt{3}}; \qquad k_a = \frac{s}{\sqrt{3}}; \qquad k_c = \frac{s}{2\sqrt{3}}; \qquad k_o = \frac{s}{\sqrt{6}}$$

6. Hollow Rectangle (or Square)

C.G. is at O = intersection of diagonals.

For Hollow Rectangle:

$$I_g = \frac{b_1 h_1{}^3 - b_2 h_2{}^3}{12}; \qquad I_a = \frac{b_1 h_1{}^3}{3} - \frac{b_2 h_2 (3h_1{}^2 + h_2{}^2)}{12}$$

$$k_g = \sqrt{\frac{b_1 h_1{}^3 - b_2 h_2{}^3}{12(b_1 h_1 - b_2 h_2)}}; \qquad J_o = \frac{b_1 h_1 (b_1{}^2 + h_1{}^2) - b_2 h_2 (b_2{}^2 + h_2{}^2)}{12}$$

For Hollow Square (letting $b_1 = h_1 = s_1$ and $b_2 = h_2 = s_2$):

$$I_g = \frac{s_1{}^4 - s_2{}^4}{12}; \qquad I_a = \frac{s_1{}^4}{3} - \frac{s_2{}^2(3s_1{}^2 + s_2{}^2)}{12}$$

$$k_g = \sqrt{\frac{s_1{}^2 + s_2{}^2}{12}}; \qquad J_o = \frac{s_1{}^4 - s_2{}^4}{6}$$

(Note: For a diagonal $c - c$, $I_c = I_g$ and $k_c = k_g$)

7. Trapezoid

C.G. is at O, located as shown.

$$I_g = \frac{h^3(B^2 + 4Bb + b^2)}{36(B + b)}; \qquad I_a = \frac{h^3(B + 3b)}{12}$$

$$k_g = \frac{h\sqrt{2(B^2 + 4Bb + b^2)}}{6(B + b)}; \qquad k_a = \frac{h}{\sqrt{6}}\sqrt{\frac{B + 3b}{B + b}}$$

8. Quadrilateral

C.G. is at O, located as follows:

Divide the sides into thirds and construct the parallelogram with sides passing through the third-points as shown. The intersection of the diagonals of this parallelogram is the desired centroid.

9. Regular Polygon

C.G. is at O = geometrical center.

Let $g - g$ be any axis through O and in plane of polygon. Then

$$I_g = \frac{\text{Area} \cdot (6R^2 - a^2)}{24} = \frac{\text{Area} \cdot (12r^2 + a^2)}{48};$$

$$J_o = \frac{\text{Area} \cdot (6R^2 - a^2)}{12} = \frac{\text{Area} \cdot (12r^2 + a^2)}{24}$$

$$k_g = \sqrt{\frac{6R^2 - a^2}{24}} = \sqrt{\frac{12r^2 + a^2}{48}};$$

$$k_o = \sqrt{\frac{6R^2 - a^2}{12}} = \sqrt{\frac{12r^2 + a^2}{24}}$$

Table II *(Continued)*

Plane Surfaces *(Continued)*

Figure	*Centroid Location; Moments of Inertia; Radii of Gyration*
10. Circle	C.G. is at O = geometrical center. $$I_g = \frac{\pi r^4}{4} = \frac{\pi d^4}{64}; \quad J_o = \frac{\pi r^4}{2} = \frac{\pi d^4}{32}$$ $$k_g = \frac{r}{2} = \frac{d}{4}; \quad k_o = \frac{r}{\sqrt{2}} = \frac{d}{\sqrt{8}}$$
11. Circular Sector	C.G. is on axis of symmetry at O. Distance from $a\!-\!a = \dfrac{2r \sin \alpha}{3\alpha}$ $$= \frac{2rc}{3s}$$ A = area = $r^2\alpha$ $$I_g = \frac{Ar^2}{4}\left(1 - \frac{\sin \alpha \cos \alpha}{\alpha}\right); \quad I_a = \frac{Ar^2}{4}\left(1 + \frac{\sin \alpha \cos \alpha}{\alpha}\right)$$ $$k_g = \frac{r}{2}\sqrt{1 - \frac{\sin \alpha \cos \alpha}{\alpha}}; \quad k_a = \frac{r}{2}\sqrt{1 + \frac{\sin \alpha \cos \alpha}{\alpha}}$$
12. Semi-circle	C.G. is on axis of symmetry at O. Distance from $a\!-\!a = \dfrac{4r}{3\pi} = 0.424r$ $$I_g = \frac{d^4(9\pi^2 - 64)}{1152\pi} = \frac{r^4(9\pi^2 - 64)}{72\pi} = 0.1098r^4; \quad I_a = I_b = \frac{\pi d^4}{128} = \frac{\pi r^4}{8};$$ $$J_o = r^4\left(\frac{\pi}{4} - \frac{8}{9\pi}\right) = 0.5025r^4$$ $$k_g = \frac{d\sqrt{9\pi^2 - 64}}{12\pi} = \frac{r\sqrt{9\pi^2 - 64}}{6\pi} = 0.264r; \quad k_a = k_b = \frac{d}{4} = \frac{r}{2};$$ $$k_o = r\sqrt{\frac{1}{2} - \frac{16}{9\pi^2}} = 0.566r$$
13. Circular Segment	C.G. is on axis of symmetry at O. Distance from $a\!-\!a = \dfrac{2r^3 \sin^3 \alpha}{3A}$ $$= \frac{c^3}{12A} \text{ where } A = \text{area} = \frac{r^2(2\alpha - \sin 2\alpha)}{2}.$$ $$I_g = \frac{Ar^2}{4}\left(1 - \frac{2\sin^3 \alpha \cos \alpha}{3(\alpha - \sin \alpha \cos \alpha)}\right); \quad I_a = \frac{Ar^2}{4}\left(1 + \frac{2\sin^3 \alpha \cos \alpha}{(\alpha - \sin \alpha \cos \alpha)}\right)$$ $$k_g = \frac{r}{2}\sqrt{1 - \frac{2\sin^3 \alpha \cos \alpha}{3(\alpha - \sin \alpha \cos \alpha)}}; \quad k_a = \frac{r}{2}\sqrt{1 + \frac{2\sin^3 \alpha \cos \alpha}{(\alpha - \sin \alpha \cos \alpha)}}$$
14. Annulus	C.G. is at O = geometrical center. $$I_g = \frac{\pi(d_1^4 - d_2^4)}{64} = \frac{\pi(r_1^4 - r_2^4)}{4}; \quad J_o = \frac{\pi(d_1^4 - d_2^4)}{32} = \frac{\pi(r_1^4 - r_2^4)}{2}$$ $$k_g = \frac{\sqrt{d_1^2 + d_2^2}}{4} = \frac{\sqrt{r_1^2 + r_2^2}}{2}; \quad k_o = \sqrt{\frac{d_1^2 + d_2^2}{8}} = \sqrt{\frac{r_1^2 + r_2^2}{2}}$$
15. Ellipse	C.G. is at O = geometrical center. *For semi-ellipse ABB′, C.G. is on* OA at distance to right of $c\!-\!c = \dfrac{4a}{3\pi}$. *For quarter-ellipse ABO, C.G. is* at distance to right of $c\!-\!c = \dfrac{4a}{3\pi}$ and at distance above $g\!-\!g = \dfrac{4b}{3\pi}$. A = area = πab $$I_g = \frac{\pi ab^3}{4} = \frac{Ab^2}{4}; \quad I_c = \frac{\pi a^3 b}{4} = \frac{Aa^2}{4}; \quad J_o = \frac{A(a^2 + b^2)}{4}$$ $$k_g = \frac{b}{2}; \qquad k_c = \frac{a}{2}; \qquad k_o = \frac{\sqrt{a^2 + b^2}}{2}$$
16. Parabolic Segment	C.G. is on axis of symmetry at O. Distance from $c\!-\!c = \dfrac{3a}{5}$. $$I_g = \frac{4ab^3}{15}; \qquad I_c = \frac{4a^3 b}{7}$$ $$k_g = \frac{b}{\sqrt{5}} = 0.447b; \quad k_c = a\sqrt{\frac{3}{7}} = 0.654a$$
17. Structural Shapes	See Section 1, Tables 80 to 86.

Table II (*Continued*)

Homogeneous Bodies

(Including Nonplanar Surfaces)

("Body" is to be understood unless "Surface" is indicated.)

Figure	*Centroid Location; Moments of Inertia; Radii of Gyration*

18. Any Surface or Body of Revolution

Let axis of revolution be X axis. Then generating curve is $y = f(x)$.
C.G. is at point having coordinates $\bar{x},\ \bar{y},\ \bar{z}$.

For Surface:

$$\bar{x} = \frac{\displaystyle\int 2\pi xy\, ds}{\displaystyle\int 2\pi y\, ds} = \frac{\displaystyle\int xy \sqrt{1 + \left(\frac{dy}{dx}\right)^2}\, dx}{\displaystyle\int y \sqrt{1 + \left(\frac{dy}{dx}\right)^2}\, dx}; \quad \bar{y} = 0; \quad \bar{z} = 0$$

For Body $\left(\text{letting } \delta = \text{density} = \dfrac{m}{\text{volume}}\right)$:

$$\bar{x} = \frac{\displaystyle\int \pi xy^2\, dx}{\displaystyle\int \pi y^2\, dx}; \quad \bar{y} = 0; \quad \bar{z} = 0.$$

$$I_x = \frac{\pi \delta}{2} \int y^4\, dx; \quad I_y = I_z = \pi \delta \int \left(\frac{y^4}{4} + x^2 y^2\right) dx$$

$$k_x = \sqrt{\frac{I_x}{m}}; \quad k_y = k_z = \sqrt{\frac{I_y}{m}} = \sqrt{\frac{I_z}{m}}$$

For Thin Shell having mass:
C.G. coordinates are same as for surface.

$$I_x = 2\pi \delta \int y^3\, ds = 2\pi \delta \int y^3 \sqrt{1 + \left(\frac{dy}{dx}\right)^2}\, dx$$

$$k_x = \sqrt{\frac{I_x}{m}}$$

19. Thin Straight Rod

C.G. is at O = geometrical center.

$$I_g = \frac{ml^2}{12}; \quad I_b = \frac{ml^2}{3}; \quad I_c = \frac{ml^2 \sin^2 \alpha}{12}; \quad I_d = \frac{ml^2 \sin^2 \alpha}{3}$$

$$k_g = \frac{l}{\sqrt{12}}; \quad k_b = \frac{l}{\sqrt{3}}; \quad k_c = \frac{l \sin \alpha}{\sqrt{12}}; \quad k_d = \frac{l \sin \alpha}{\sqrt{3}}$$

20. Thin Rod Bent into Circular Arc

C.G. is on axis of symmetry at $\bar{x} = \dfrac{r \sin \alpha}{\alpha}$.

$$I_x = \frac{mr^2}{2}\left(1 - \frac{\sin \alpha \cos \alpha}{\alpha}\right); \quad I_y = \frac{mr^2}{2}\left(1 + \frac{\sin \alpha \cos \alpha}{\alpha}\right); \quad I_z = mr^2$$

$$k_x = r\sqrt{\frac{1}{2} - \frac{\sin \alpha \cos \alpha}{2\alpha}}; \quad k_y = r\sqrt{\frac{1}{2} + \frac{\sin \alpha \cos \alpha}{2\alpha}}; \quad k_z = r$$

21. Rectangular Parallelepiped (or Cube)

C.G. is at O = geometrical center.

For Parallelepiped:

$$I_g = \frac{m(b^2 + c^2)}{12}; \quad I_d = \frac{m(a^2 + b^2)}{12}; \quad I_e = \frac{m(4a^2 + b^2)}{12}$$

$$k_g = \sqrt{\frac{b^2 + c^2}{12}}; \quad k_d = \sqrt{\frac{a^2 + b^2}{12}}; \quad k_e = \sqrt{\frac{4a^2 + b^2}{12}}$$

For Cube (letting $a = b = c = s$):

$$I_g = I_d = \frac{ms^2}{6}; \quad I_e = \frac{5ms^2}{12}$$

$$k_g = k_d = \frac{s}{\sqrt{6}}; \quad k_e = s\sqrt{\frac{5}{12}}$$

22. Right Rectangular Pyramid

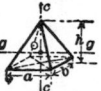

C.G. is on axis of symmetry at O. Distance from base = $\dfrac{h}{4}$.
Drawing g-g axis through O parallel to side a:

$$I_g = \frac{m}{20}\left(b^2 + \frac{3h^2}{4}\right); \quad I_c = \frac{m}{20}(a^2 + b^2)$$

$$k_g = \sqrt{\frac{4b^2 + 3h^2}{80}}; \quad k_c = \sqrt{\frac{a^2 + b^2}{20}}$$

Table II   *(Continued)*
Homogeneous Bodies *(Continued)*

Figure	Centroid Location; Moments of Inertia; Radii of Gyration
23. Pyramid (or Frustum of Pyramid)	*For Surface of Any Pyramid:* C.G. of surface (base excluded) is on line joining apex with centroid of perimeter of base, at a distance two-thirds its length from the apex. *For Body of Any Pyramid:* C.G. of body is on line joining apex with centroid of base, at a distance three-fourths its length from the apex. *For Surface of Frustum of Pyramid having Regular Bases:* Letting R and r be the lengths of sides of the larger and smaller bases respectively, and h the altitude: C.G. of surface (bases excluded) is at distance from larger base $= \dfrac{h(R + 2r)}{3(R + r)}$. *For Body of Frustum of Any Pyramid:* Letting A and a be the areas of the larger and smaller bases, respectively, and h the altitude: C.G. of body is at distance from larger base $= \dfrac{h(A + 2\sqrt{Aa} + 3a)}{4(A + \sqrt{Aa} + a)}$.
24. Right Elliptical Cylinder (or Circular Cylinder)	C.G. is at O = geometrical center. *For Right Elliptical Cylinder:* $$I_g = \frac{m}{12}(3b^2 + h^2); \quad I_c = \frac{m}{4}(a^2 + b^2); \quad I_e = \frac{m}{12}(3r^2 + 4h^2)$$ $$k_g = \sqrt{\frac{3b^2 + h^2}{12}}; \quad k_c = \frac{\sqrt{a^2 + b^2}}{2}; \quad k_e = \sqrt{\frac{3r^2 + 4h^2}{12}}$$ *For Right Circular Cylinder* (letting $a = b = r$): $$I_g = \frac{m}{12}(3r^2 + h^2); \quad I_c = \frac{mr^2}{2}$$ $$k_g = \sqrt{\frac{3r^2 + h^2}{12}}; \quad k_c = \frac{r}{\sqrt{2}}$$
25. Hollow Right Circular Cylinder	C.G. is at O = geometrical center. $$I_g = \frac{m}{12}(3R^2 + 3r^2 + h^2); \quad I_c = \frac{m}{2}(R^2 + r^2); \quad I_e = \frac{m}{12}(3R^2 + 3r^2 + 4h^2)$$ $$k_g = \sqrt{\frac{3R^2 + 3r^2 + h^2}{12}}; \quad k_c = \sqrt{\frac{R^2 + r^2}{2}}; \quad k_e = \sqrt{\frac{3R^2 + 3r^2 + 4h^2}{12}}$$ *For Thin Shell* (radius R): $$I_g = \frac{m}{12}(6R^2 + h^2); \quad I_c = mR^2; \quad I_e = \frac{m}{6}(3R^2 + 2h^2)$$ $$k_g = \sqrt{\frac{6R^2 + h^2}{12}}; \quad k_c = R; \quad k_c = \sqrt{\frac{3R^2 + 2h^2}{6}}$$
26. Right Circular Cone	C.G. is on axis of symmetry at O. Distance from base $= \dfrac{h}{4}$. Drawing g–g axis through O and d–d axis through apex, both parallel to base: $$I_g = \frac{3m}{20}\left(r^2 + \frac{h^2}{4}\right); \quad I_c = \frac{3mr^2}{10}; \quad I_d = \frac{3m}{20}(r^2 + 4h^2)$$ $$k_g = \sqrt{\frac{3}{80}(4r^2 + h^2)}; \quad k_c = \frac{3r}{\sqrt{30}}; \quad k_d = \sqrt{\frac{3}{20}(r^2 + 4h^2)}$$
27. Frustum of Right Circular Cone	C.G. is on axis of symmetry at O. Distance from base $= \dfrac{h(R^2 + 2Rr + 3r^2)}{4(R^2 + Rr + r^2)}$. $$I_c = \frac{3m(R^5 - r^5)}{10(R^3 - r^3)}; \quad k_c = \sqrt{\frac{3(R^5 - r^5)}{10(R^3 - r^3)}}$$

Table II *(Continued)*

Homogeneous Bodies *(Continued)*

Figure	Centroid Location; Moments of Inertia; Radii of Gyration
28. Cone (or Frustum of Cone)	*For Surface of Any Cone:* C.G. of surface (base excluded) is on line joining apex with centroid of perimeter of base, at a distance two-thirds its length from the apex. *For Body of Any Cone:* C.G. of body is on line joining apex with centroid of base, at a distance three-fourths its length from the apex. *For Surface of Frustum of a Circular Cone:* Letting R and r be the radii of the larger and smaller bases, respectively, and h the altitude: C.G. of surface (bases excluded) is at distance from larger base $= \dfrac{h(R+2r)}{3(R+r)}$. *For Body of Frustum of a Circular Cone:* Letting R and r be the radii of the larger and smaller bases, respectively, and h the altitude: C.G. of body is at distance from larger base $= \dfrac{h(R^2+2Rr+3r^2)}{4(R^2+Rr+r^2)}$
29. Thin Circular Lamina	C.G. is at O = geometrical center. $I_g = \dfrac{mr^2}{4}$; $I_c = \dfrac{mr^2}{2}$ (where c–c axis is perpendicular to the plane). $k_g = \dfrac{r}{2}$; $k_o = \dfrac{r}{\sqrt{2}}$
30. Sphere	C.G. is at O = geometrical center. $I_g = \dfrac{2mr^2}{5}$ $k_g = \dfrac{2r}{\sqrt{10}}$
31. Hollow Sphere	C.G. is at O = geometrical center. $I_g = \dfrac{2m}{5}\left(\dfrac{R^5-r^5}{R^3-r^3}\right)$; $k_g = \sqrt{\dfrac{2}{5}\left(\dfrac{R^5-r^5}{R^3-r^3}\right)}$ *For Thin Shell* (radius R): $I_g = \dfrac{2mR^2}{3}$; $k_g = \dfrac{2R}{\sqrt{6}}$
32. Spherical Sector	C.G. is on axis of symmetry at O. Distance from center of sphere $= \dfrac{3(2r-h)}{8}$. $I_g = \dfrac{m}{5}(3rh - h^2)$ $k_g = \sqrt{\dfrac{3rh-h^2}{5}}$
33. Hemisphere	*For Surface:* C.G. is on axis of symmetry at distance from center of sphere $= \dfrac{r}{2}$. *For Body:* C.G. is on axis of symmetry at distance from center of sphere $= \dfrac{3r}{8}$. $I_g = \dfrac{2mr^2}{5}$; $k_g = \dfrac{2r}{\sqrt{10}}$
34. Spherical Segment	C.G. is on axis of symmetry at distance from center of sphere $= \dfrac{3(2r-h)^2}{4(3r-h)}$. $I_g = m\left(r^2 - \dfrac{3rh}{4} + \dfrac{3h^2}{20}\right)\dfrac{2h}{(3r-h)}$ $k_g = \sqrt{\left(r^2 - \dfrac{3rh}{4} + \dfrac{3h^2}{20}\right)\dfrac{2h}{3r-h}}$

Table II *(Continued)*
Homogeneous Bodies *(Continued)*

Figure	Centroid Location; Moments of Inertia; Radii of Gyration
35. Torus	C.G. is at O = geometrical center. $$I_g = \frac{m(4R^2 + 5r^2)}{8}; \quad I_c = \frac{m(4R^2 + 3r^2)}{4}$$ $$k_g = \sqrt{\frac{4R^2 + 5r^2}{8}}; \quad k_c = \frac{\sqrt{4R^2 + 3r^2}}{2}$$
36. Ellipsoid	C.G. is at O = geometrical center. C.G. of *one octant* is at point having coordinates: $$\bar{x} = \frac{3a}{8}; \quad \bar{y} = \frac{3b}{8}; \quad \bar{z} = \frac{3c}{8}$$ *For Complete Ellipsoid:* $$I_x = \frac{m}{5}(b^2 + c^2); \quad I_y = \frac{m}{5}(a^2 + c^2); \quad I_z = \frac{m}{5}(a^2 + b^2)$$ $$k_x = \sqrt{\frac{b^2 + c^2}{5}}; \quad k_y = \sqrt{\frac{a^2 + c^2}{5}}; \quad k_z = \sqrt{\frac{a^2 + b^2}{5}}$$
37. Paraboloid	C.G. is on axis of symmetry at O. Distance from base $= \dfrac{h}{3}$. $$I_g = \frac{mr^2}{3}; \quad I_c = \frac{m}{18}(3r^2 + h^2)$$ $$k_g = \frac{r}{\sqrt{3}}; \quad k_c = \sqrt{\frac{3r^2 + h^2}{18}}$$

KINEMATICS 455

KINEMATICS

10. MOTIONS OF A PARTICLE

Motion of a Particle with respect to other particles or objects is its state of continual changing of position with respect to them.

Rectilinear Motion is motion along a straight path.

Curvilinear Motion is motion along a curved path which may be either planar or skewed.

Displacement of a Particle is its change of position and is a vector quantity. If A is the position of a particle at a time t_1, and B its position at a later time t_2, its displacement in the time interval $t_2 - t_1 = \Delta t$ is the vector AB, no matter whether the path is straight or curved.

Velocity of a Particle is its time rate of displacement (i.e., rate of change of position) and is a vector quantity. **Speed** is the magnitude of velocity without reference to direction or sense.

Acceleration of a Particle is its time rate of change of velocity and is a vector quantity.

11. RECTILINEAR MOTION

Velocity. Let s = distance measured along the path of a particle, s_1 = distance from origin at time t_1, s_2 = distance at a later time t_2, $\Delta s = s_2 - s_1 = displacement$ * in time interval $\Delta t = t_2 - t_1$. Then *average velocity* $= \Delta s/\Delta t$. If the position changes at a uniform rate (which implies no change in sense), actual velocity at any time $= \Delta s/\Delta t$.

For every case, *instantaneous velocity* $= v = \dfrac{ds}{dt} = \displaystyle\lim_{\Delta t \to 0}\left(\dfrac{\Delta s}{\Delta t}\right)$. *Unit of Velocity* is any distance unit divided by any time unit. *Units* commonly used are feet per second and miles per hour.

Acceleration. Let v_1 = velocity of particle at time t_1, v_2 = velocity at a later time t_2, $\Delta v = v_2 - v_1$ = change in velocity in time interval $\Delta t = t_2 - t_1$. Then *average acceleration* $= \Delta v/\Delta t$. If the velocity changes at a uniform rate, the actual acceleration at any time $= \Delta v/\Delta t$. For every case, *instantaneous acceleration* equals

$$a = \frac{dv}{dt} = \frac{d^2s}{dt^2} = \lim_{\Delta t \to 0}\left(\frac{\Delta v}{\Delta t}\right)$$

Unit of acceleration is any velocity unit divided by any time unit. A unit commonly used is feet per second per second (i.e., ft per sec²).

Formulas for Determination of a, v, s, t. If s is given algebraically in terms of t, then v and a may be determined in terms of t by differentiation as indicated above. If a is given algebraically in terms of t, then v and s may be determined in terms of t by integration. Other relations not involving t may be determined by similar methods. The common formulas are:

$$v = \frac{ds}{dt}; \qquad a = \frac{dv}{dt} = \frac{d^2s}{dt^2}; \qquad \frac{a}{v} = \frac{dv}{ds}$$

$$s_2 - s_1 = \int_{t_1}^{t_2} v\, dt; \quad v_2 - v_1 = \int_{t_1}^{t_2} a\, dt; \quad t_2 - t_1 = \int_{s_1}^{s_2} \frac{ds}{v} = \int_{v_1}^{v_2} \frac{dv}{a}; \quad v_2^2 - v_1^2 = 2\int_{s_1}^{s_2} a\, ds$$

For **Uniform Acceleration**, a = constant; $v = at + v_0$; $s = 1/2\,at^2 + v_0 t + s_0$; $v^2 = 2a(s - s_0) + v_0^2$; v_0 being initial velocity and s_0 initial distance.

If algebraic relations between a, v, s, and t are not given but a number of pairs of corresponding values of two of the variables are known, curves may be plotted for the approximate determination of other corresponding pairs of values and of other unknowns, within the range of the data. Such curves are discussed later under " Motion Graphs."

Examples of Rectilinear Motion

Falling Body.† If a body *falls from rest* in a vacuum, $v_0 = 0$, $s_0 = 0$, and $a = g = 32.2$ ft per sec² (approx.). Hence $v = gt = \sqrt{2gs}$; $s = 1/2\,gt^2$. If a body is *projected upward*

---

* The difference in distances along the path equals the displacement *only* when the path is a straight line. (See definition of displacement.)
† Rotation disregarded and body considered as a particle.

at an initial velocity v_0, $a = -g$ and the formulas become $v = -gt + v_0 = \sqrt{-2gs + v_0{}^2}$;
$s = -\frac{1}{2}gt^2 + v_0 t$. Total ascent (to highest position) $= \frac{v_0{}^2}{2g}$, and time required $= \frac{v_0}{g}$.

Crank and Connecting-rod Mechanism. The problem is to find expressions for the velocity and acceleration of any point in the crosshead, as A in Fig. 1. Such a point describes rectilinear motion. Let $c = r/l$, n = revolutions per second (assumed constant), ω = radians of angle described by crank per second, and s = distance of A from its extreme left position, all distances expressed in feet. Then,

$$s = (l + r) - l(1 - c^2 \sin^2 \theta)^{\frac{1}{2}} - r \cos \theta$$

$$v = r\omega\left(\sin \theta + \frac{c \sin 2\theta}{2(1 - c^2 \sin^2 \theta)^{\frac{1}{2}}}\right); \quad a = r\omega^2\left(\cos \theta + \frac{c \cos 2\theta + c^3 \sin^4 \theta}{(1 - c^2 \sin^2 \theta)^{\frac{3}{2}}}\right)$$

The above formulas are exact; close approximations are:

$$s = r(1 - \cos \theta) + \frac{1}{4}cr(1 - \cos 2\theta); \quad v = r\omega(\sin \theta + \frac{1}{2}c \sin 2\theta)$$

$$a = r\omega^2(\cos \theta + c \cos 2\theta)$$

Motion Graphs

Space-time, velocity-time, acceleration-time, velocity-space, acceleration-space curves for a particle are graphs showing the relations between magnitudes of s and t, v and t, a and

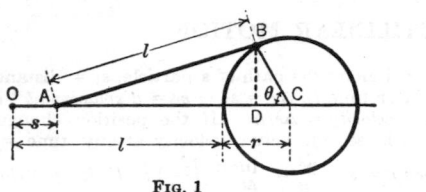

Fig. 1

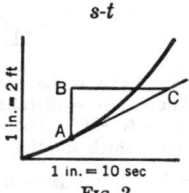

Fig. 2

t, v and s, a and s, respectively. Figs. 2–6 illustrate such graphs but do not correspond to the same motion.

Space-time Diagram. In Fig. 2, the *slope* of the curve at any point represents the magnitude of the velocity. If AB and BC are measured by the s and t scales of the drawing respectively, the slope equals the velocity magnitude; thus if $AB = 0.2$ in. and $BC = 0.4$ in., $v = 0.4/4 = 0.1$ ft per sec.

Velocity-time Diagram. In Fig. 3, the *slope* of the curve at any point represents the magnitude of the acceleration.* If AB and BC are measured by the v and t scales respec-

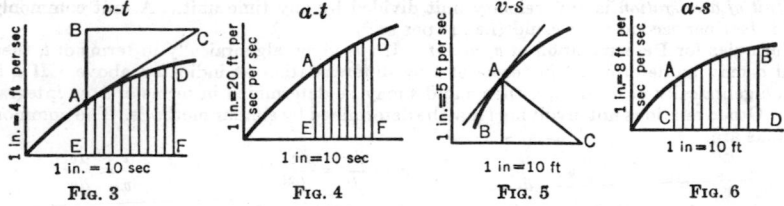

Fig. 3   Fig. 4   Fig. 5   Fig. 6

tively, the slope equals the acceleration magnitude; * thus if $AB = 0.3$ in. and $BC = 0.5$ in., $a = 1.2/5 = 0.24$ ft per sec per sec. The *area* included between any two ordinates (as AE and DF), the curve, and the t axis, represents the displacement † of the moving point in the time EF. If area is below the time axis, it is considered minus. If the area is computed by multiplying its average ordinate measured by the velocity scale (this being the average velocity) by EF measured by the time scale, the product equals the displacement; thus if the average ordinate is 0.35 in., and EF is 0.4 in., the displacement $= 1.4 \times 4 = 5.6$ ft.

Acceleration-time Diagram. In Fig. 4, the *slope* represents the rate at which the acceleration is changing.‡ The *area* (plus above and minus below time axis) included between any two ordinates (as AE and DF), the curve, and the t axis, represents the velocity change in the time EF.‡ Thus if the average ordinate is 0.3 in. and EF is 0.2 in., the velocity change $= 6 \times 2 = 12$ ft per sec.

---

* For curvilinear motion, this is tangential acceleration only.
† For curvilinear motion, this is distance along the path (not displacement).
‡ For rectilinear motion only.

Velocity-space Diagram. In Fig. 5, the subnormal represents the acceleration.* If the length of the subnormal is multiplied by the square of the velocity scale number and the product is divided by the space scale number, the result will equal the acceleration; * thus suppose that the subnormal $BC = 1/3$ in., then $a = (1/3 \times 25)/10 = 0.83$ ft per sec per sec.

Acceleration-space Diagram. In Fig. 6, the area (plus above and minus below space axis) included between two ordinates (as AC and BD), the curve, and the s axis, represents the change in the velocity square. If the area is computed by multiplying the mean ordinate measured by the acceleration scale by CD measured by the space scale, the product times two equals the change in the velocity square; thus if the average ordinate = 0.3 in., and $CD = 0.4$ in., the change = $2.4 \times 4 \times 2 = 19.2$.

Simple Harmonic Motion

Simple Harmonic Motion and Its Motion Graphs have wide application in physics and engineering. If a point P moves in a circular path of radius r at uniform speed, its projection on any diameter has *simple harmonic motion.* The radius r is called the *amplitude.* The *period* is the time required for the projection to go from one end of the diameter to the other and back. The *frequency* is the number of periods per unit time, which makes it the reciprocal of the period. Angle XOP (Fig. 7) (considered as less than 2π radians) is the *phase angle.* The *displacement* at any time is the distance of the point having simple harmonic motion from the center of its path or range.

When $t = 0$, let P be at P_0. ϵ is called the *lead angle* (*lag, if negative*). For simple harmonic motion (*SHM*) of V in the vertical diameter, $y = r \sin(\theta + \epsilon) = r \sin(\omega t + \epsilon)$, in which $\omega = \dfrac{d\theta}{dt} =$ radians per unit time (i.e., 2π times the frequency).

$$v_y = r\omega \cos(\omega t + \epsilon) = \omega x$$
$$a_y = - r\omega^2 \sin(\omega t + \epsilon) = - \omega^2 y$$

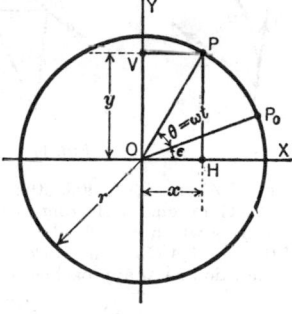

FIG. 7

For *SHM* of H in horizontal diameter:

$$x = r \cos(\theta + \epsilon) = r \cos(\omega t + \epsilon);^3 \; v_x = - r\omega \sin(\omega t + \epsilon) = - \omega y$$
$$a_x = - r\omega^2 \cos(\omega t + \epsilon) = - \omega^2 x$$

If the time is reckoned from the instant when V is in its mid-position, and moving upward, $\epsilon = 0$. The three curves (Fig. 10) OA, $O'B$, and OC are the space-time, velocity-

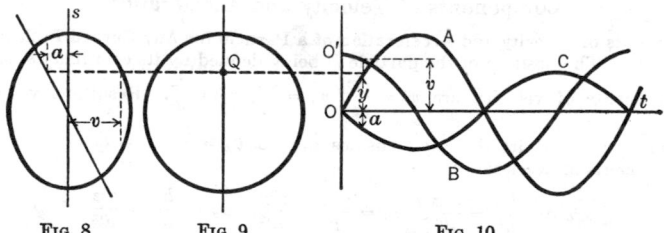

FIG. 8   FIG. 9   FIG. 10

time, and acceleration-time curves, respectively, for one complete period of a simple harmonic motion; $\epsilon = 0$; Ot represents the period; the values of y, v, and a marked are for position Q, shown. In Fig. 8 the curve is the velocity-space curve, and the inclined line the acceleration-space curve. They show how v and a vary with the displacement of the moving point; thus for the position Q, v and a have values as marked.

From the above equations and curves, it will be noted that simple harmonic motion may be defined also as any rectilinear motion in which the acceleration is always directed toward a fixed point in the path and is proportional to the distance between that point and the moving point. †

---

* For curvilinear motion, this is tangential acceleration only.
† A common example of simple harmonic motion is the motion of a weight suspended from an elastic spring.

12. CURVILINEAR MOTION

Velocity. If s is distance measured along the curved path of a particle, then the magnitude of velocity (speed) at any instant $= \dfrac{ds}{dt}$; the linear direction of the velocity is tangent to the path at the instantaneous position of the particle; and the sense of the velocity corresponds to the direction of motion of the particle at the instant.

The velocity vector changes in magnitude and direction. In Fig. 11, let A, B, C represent positions of particle P in its curved path; s, distance along the path; and v_1, v_2,

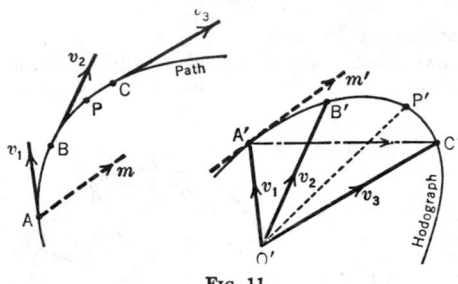

FIG. 11

v_3, velocity vectors at A, B, C. Plot velocity vectors $O'A'$, $O'B'$, $O'C'$, etc., from any origin O' to represent the velocities at A, B, C, etc. The curve $A'B'C'$, drawn through the ends of the vectors, is called a *hodograph* for the motion. For every position of P in its path, there is a corresponding position P' in the hodograph; and P' describes distance s' on the hodograph while P describes distance s on the path. Vector $O'P'$ represents the velocity of P. In time Δt, P moves from A to C, its velocity changes from $O'A'$ to $O'C'$, and the velocity change is $A'C'$.

Acceleration. Referring to the hodograph (Fig. 11), *average acceleration* for interval Δt, during which particle P moves from A to C, is vector $A'C'/\Delta t$, and it has the direction of the chord $A'C'$. The *instantaneous acceleration* of P at $A = a =$ limit of the average acceleration as Δt approaches zero.

$$a = \lim_{\Delta t \to 0} \left(\frac{\text{vector } A'C'}{\Delta t} \right) = \lim_{\Delta t \to 0} = \left(\frac{\text{arc } A'B'C'}{\Delta t} \right) = \frac{ds'}{dt} = \text{speed of } P'$$

on hodograph. The direction of a is along the tangent $A'm'$, and as P' is moving clockwise, the sense is as indicated by arrow at m'. Hence acceleration at A is Am, parallel to $A'M'$ and $= \dfrac{ds'}{dt}$. Its tangential component is $\dfrac{ds}{dt}$, and its normal component is $\dfrac{v^2}{\rho}$, ρ being the radius of curvature at A. *Unit of acceleration* is any velocity unit divided by any time unit.

Components of Velocity and Acceleration

Components of Velocity and Acceleration of a Particle for Any Curved Path (*not necessarily planar*). The position of the particle P being defined by its coordinates, x, y, z, the axial components of velocity are $v_x = \dfrac{dx}{dt}$, $v_y = \dfrac{dy}{dt}$, $v_z = \dfrac{dz}{dt}$. Resultant velocity $v = \sqrt{v_x{}^2 + v_y{}^2 + v_z{}^2}$, and its direction cosines are $\cos \theta_x = \dfrac{v_x}{v}$, $\cos \theta_y = \dfrac{v_y}{v}$, $\cos \theta_z = \dfrac{v_z}{v}$. Axial components of acceleration are:

$$a_x = \frac{dv_x}{dt} = \frac{d^2x}{dt^2}; \quad a_y = \frac{dv_y}{dt} = \frac{d^2y}{dt^2}; \quad a_z = \frac{dv_z}{dt} = \frac{d^2z}{dt^2}$$

Resultant acceleration $a = \sqrt{a_x{}^2 + a_y{}^2 + a_z{}^2}$; and its direction cosines are:

$$\cos \phi_x = \frac{a_x}{a}; \quad \cos \phi_y = \frac{a_y}{a}; \quad \cos \phi_z = \frac{a_z}{a}$$

The tangential and normal components of acceleration are $a_t = \dfrac{dv}{dt} = \dfrac{d^2s}{dt^2}$, and $a_n = \dfrac{v^2}{\rho}$, ρ being the radius of curvature. Resultant acceleration is

$$a = \sqrt{a_t{}^2 + a_n{}^2} = \sqrt{a_x{}^2 + a_y{}^2 + a_z{}^2}$$

If the path is a plane curve, $v_z = 0$ and $a_z = 0$.

The above discussion shows that velocities and accelerations (like forces) may be composed or resolved according to the parallelogram and parallelepipedon laws.

Motion of a Projectile

Projectile * Describing Plane Curvilinear Motion. In the following formulas air resistance is neglected; v_0 = velocity of projection; θ = angle of projection (Fig. 12); x and y = coordinates of the projectile at any time t after projection; v = velocity; v_x and v_y = x and y components respectively of v; r = range on the horizontal plane through O; θ_1 = value of θ for maximum r; h = greatest height attained; and T = time of flight. The path of the projectile, or the trajectory, is a parabola as represented, and a set of parametric equations for it are:

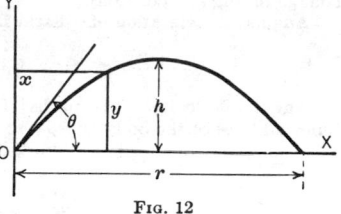

FIG. 12

$$x = v_0 \cos \theta \cdot t, \quad y = v_0 \sin \theta \cdot t - \tfrac{1}{2} g t^2$$

from which
$$y = x \tan \theta - g x^2 / 2 v_0^2 \cos^2 \theta$$

Also:

$$v_x = v_0 \cos \theta; \quad v_y = v_0 \sin \theta - g t; \quad v = \sqrt{v_0^2 - 2gy}; \quad h = \sin^2 \theta \cdot v_0^2 / 2g;$$

$$r = \sin 2\theta \cdot v_0^2 / g; \quad \theta_1 = 45°; \quad T = \frac{2 v_0 \sin \theta}{g}$$

If the direction of projection is horizontal, $\theta = 0$; the equation of the path is $y = -gx^2 / 2v_0^2$; and $x = v_0 t$, and $y = -\tfrac{1}{2} g t^2$.

The fact that the horizontal component of velocity is constant indicates that the hodograph of the motion of a projectile is a straight vertical line.

Motion Graphs

Motion Graphs similar to those previously discussed for rectilinear motion may be constructed for curvilinear motion of a particle. Great care must be exercised, however, in interpreting the significance of slopes, areas, and subnormals when acceleration or distance is involved. In this connection, reference should be made to the footnotes referred to in the previous discussion.

In general, accelerations obtained are tangential components only, while " displacements " must be replaced by " distances along the curve." Thus, in the velocity time graph (Art. 11, Fig. 3), the slope of the curve represents the magnitude of the *tangential component* of the acceleration,[†] while the area under the curve represents the *distance along* the curve.

13. MOTIONS OF A BODY

Translation of a Rigid Body is a motion such that each straight line in it remains fixed in direction. The *paths* of all particles of the body are exactly alike, straight or curved (not necessarily plane curves); the *displacements* of all particles during a given time are the same; the *velocities* of all particles at any instant are the same; and their *accelerations* at any instant are the same. For these reasons, it is customary to use the expressions " velocity of the body " and " acceleration of the body." The motion is described by the same formulas as those previously derived for rectilinear and curvilinear motions of a particle.

Rotation of a Rigid Body is a motion such that one line of the body, or of its extension, remains fixed. The fixed line is the *axis*. The plane through the mass center perpendicular to the axis is the *plane of rotation*.

Plane Motion of a Rigid Body is a motion such that each particle of the body moves in a plane at a constant distance from a fixed plane through the mass center (called the *plane of motion*), while each line of the body parallel to the plane of motion turns through the same angle in the same time interval.

Three-dimensional Motion of a Rigid Body is a term covering all types of motion in three-dimensional space, including pure translation along a skewed curve as a special case. Even in the most general case, any three-dimensional motion of a rigid body may be regarded as consisting of two components: one, a translation equal to that of the mass center, and the other a rotation about some axis through the mass center.

---

* Rotation disregarded and body considered as a particle.
† However, if a velocity-time graph were made for motion along the hodograph of the original motion, the slope of the curve would represent the magnitude of the *total* acceleration of the original particle along the original path.

Angular Displacement of a Rigid Body is the change of angular position of any line in the plane of motion.

Angular Velocity of a Rigid Body is its time rate of angular displacement (i.e., rate of change of angular position).

Angular Acceleration of a Rigid Body is its time rate of change of angular velocity.

14.   ROTATION

Angular Velocity. The paths of all particles are circles with centers on the axis. Since all lines of the body parallel to the plane of rotation sweep out equal angles in equal times, it is customary to describe rotation by the behavior of one radial line. In Fig. 13, let θ be the angle from the x axis to the radial line OP. $\Delta\theta = \theta_2 - \theta_1$ is the *angular displacement* of the body in the time $\Delta t = t_2 - t_1$, and is expressed in any angular unit. *Average angular velocity* =

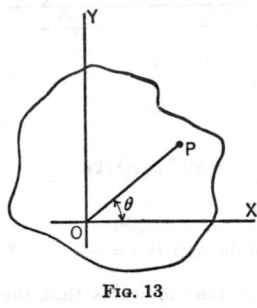

FIG. 13

$\dfrac{\theta_2 - \theta_1}{t_2 - t_1} = \dfrac{\Delta\theta}{\Delta t}$. If the angle changes at a uniform rate, actual velocity at any time = $\Delta\theta/\Delta t$. For every case, the *instantaneous angular velocity* $= \omega = \underset{\Delta t \to 0}{\text{limit}}\left(\dfrac{\Delta\theta}{\Delta t}\right) = \dfrac{d\theta}{dt}$. The sign depends on the numerator of the fraction, or the way in which θ is changing. *The unit of angular velocity* is any angular displacement unit divided by any time unit, such as radians per sec, rev per min, etc.

Angular Acceleration. *Average angular acceleration* $= \dfrac{\omega_2 - \omega_1}{t_2 - t_1} = \dfrac{\Delta\omega}{\Delta t}$. If the angular velocity changes at a uniform rate, actual angular acceleration at any time = $\Delta\omega/\Delta t$. For every case, the *instantaneous angular acceleration* $= \alpha = \underset{\Delta t \to 0}{\text{limit}}\left(\dfrac{\Delta\omega}{\Delta t}\right) = \dfrac{d\omega}{dt} = \dfrac{d^2\theta}{dt^2}$. The sign of α depends on the numerator of the fraction, or on the way in which ω is changing. The *unit of angular acceleration* is any angular velocity unit divided by any time unit, as radians per sec per sec (i.e., radians per sec^2), etc.

Formulas for Determination of a, ω, θ, t. The formulas are exactly analogous to those previously derived for rectilinear motion, a, v, and s being replaced by α, ω and θ, respectively. The formulas are:

$$\omega = \frac{d\theta}{dt}; \qquad\qquad \alpha = \frac{d\omega}{dt} = \frac{d^2\theta}{dt^2}; \qquad\qquad \frac{\alpha}{\omega} = \frac{d\omega}{d\theta}$$

$$\theta_2 - \theta_1 = \int_{t_1}^{t_2}\omega\,dt; \quad \omega_2 - \omega_1 = \int_{t_1}^{t_2}\alpha\,dt; \quad t_2 - t_1 = \int_{\theta_1}^{\theta_2}\frac{d\theta}{w} = \int_{\omega_1}^{\omega_2}\frac{dw}{\alpha}; \quad \omega_2{}^2 - \omega_1{}^2 = 2\int_{\theta_1}^{\theta_2}\alpha\,d\theta$$

Relations between Rectilinear and Angular Velocities and Accelerations. Let ω and α, respectively, be instantaneous angular velocity and acceleration of a rotating body, and v and a the corresponding instantaneous rectilinear velocity and acceleration of a point P of the body located at distance r from the axis of rotation. Then:

$$v = r\omega; \quad a_t = r\alpha; \quad a_n = r\omega^2; \quad a = r\sqrt{\alpha^2 + \omega^4}$$

Sense of v must agree with sense of ω, and sense of a_t with sense of α. Sense of a_n is always toward axis.

Motion Graphs

Motion Graphs analogous to those previously discussed for rectilinear motion may be constructed to show the relations between angular displacement, velocity and acceleration, and time. θ, ω, and α correspond to s, v, and a, respectively.

15.   PLANE MOTION

Any displacement resulting from plane motion may be accomplished by a translation of the body which will bring any one line of it, which is perpendicular to the plane of motion, into final position, followed by a rotation of the body about that line into final position. The necessary amount of translation depends on the line of the body selected as axis of the rotation; the amount of the rotation does not. **The state of motion of a body at any instant may be regarded as consisting of two components, a translational motion and a rotational motion.** Thus a plane motion may be traced by giving the history of the movement of one point of the body (called a base point) in its own curved path, and a descrip tion of the rotation of the body about the selected base point.* The point selected as base should be one for which the motion is readily specified. For a wheel rolling along a straight path, the center would be selected as a base point.

Velocity of Any Point P of the body, at any instant, with respect to a fixed point O, is the vector sum of the velocity of base point A, with respect to O, and of the velocity of P with respect to A due to rotation about A. Thus (Fig. 14) O is the fixed point, A the moving base point, and P any other point of the body at distance r from A; v_1 is velocity of A with respect to O, and $v_2 = r\omega$ is velocity of P with respect to A. Resultant velocity of P with respect to $O = v$; or v_P to $O = v_P$ to $A + v_A$ to O.

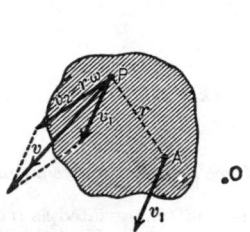

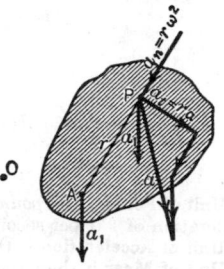

Fig. 14   Fig. 15

Acceleration of Any Point P, with respect to a fixed point O, at any instant, has two components; one is that of the base point A with respect to O, and the other that of P with respect to base A. Acceleration of P with respect to A is rotational, and is conveniently replaced by its tangential and normal components, $a_t = r\alpha$ and $a_n = r\omega^2$. Then resultant acceleration of P, with respect to O, is the vector sum of $r\alpha$, $r\omega^2$, and acceleration of A with respect to O. Thus (Fig. 15) a_1 is acceleration of A to O; and acceleration P to A is resultant of a_t and a_n. Acceleration of P to O is a = vector sum of a_1, a_t, and a_n.

Instantaneous Axis. For a body having plane motion, there is always one point in it (or in its extension), at each instant, for which the velocity with respect to A (Fig. 14) is equal and opposite to velocity of A with respect to O; that is, its velocity is zero at the instant. This point Q is called the *instantaneous* (or *instant*) *center* of rotation, and a line through Q, perpendicular to the plane of motion, is called the *instantaneous axis*. Since Q is at rest for the instant, the resultant velocities of all points at the instant are purely rotational about the instant axis. The instant center is the intersection of two lines drawn from any two points, C and D, in the plane of the motion, perpendicular to their velocities. If the velocity of the point C is known, ω for the body is determined by dividing v_C by the distance of C from Q, or by r_C. The velocity of any other point E is $\omega \times r_E$, perpendicular to the radius r_E.

The position of Q in the body (or in its extension) is continually changing; its locus is a line (usually curved) fixed in the body and moving with it, called the *body centrode*. The locus of the positions of Q in the fixed plane of motion is a line (usually curved) called the *space centrode*. The plane motion may be considered as produced by the rolling, without slipping, of the body centrode upon the space centrode.

Example: Rolling Wheel Describing Plane Motion. A wheel of 6-ft radius rolls along a straight horizontal path, and at a certain instant the point P, 2 ft from center of wheel, is in the position shown in Fig. 16a. At this instant $\omega = 16.75$ radians per sec and $\alpha = 5.6$ radians per sec². Deter-

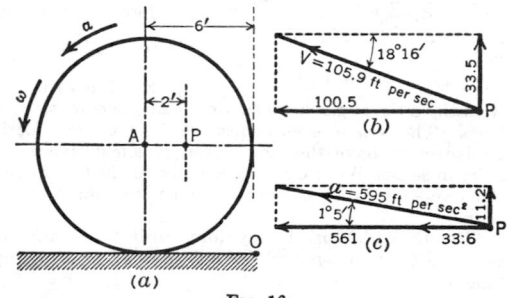

Fig. 16

* To simplify matters, "points" are referred to throughout this and the following discussion but "lines" through the points perpendicular to the plane of motion should be understood. Thus, in Figs. 14–15, the parallel lines through P, A, and O, perpendicular to the plane of motion, move relative to each other.

mine the velocity and acceleration of point P with respect to fixed point O at the specified instant.
Solution: Select center A as base point. From relations between v and a of any point of a rotating
body, and ω and α of the body:

$v_{A \text{ to } O} = r\omega = 6 \times 16.75 = 100.5$ ft per sec, horizontally toward left.
$v_{P \text{ to } A} = r\omega = 2 \times 16.75 = 33.5$ ft per sec, vertically upward.
Therefore, $v_{P \text{ to } O} = 105.9$ ft per sec, upward to left, at $18°$ $16'$ to horizontal (Fig. 16b).
$a_{A \text{ to } O} = r\alpha = 6 \times 5.6 = 33.6$ ft per sec², horizontally toward left.
$a_{tP \text{ to } A} = r\alpha = 2 \times 5.6 = 11.2$ ft per sec², vertically upward.
$a_{nP \text{ to } A} = r w^2 = 2 \times (16.75)^2 = 561$ ft per sec², horizontally toward left.
Therefore, $a_{P \text{ to } O} = 595$ ft per sec², upward to left, at $1°$ $5'$ to horizontal (Fig. 16c).

KINETICS

16. BASIC QUANTITIES

System of Units *

The **Unit of Force** is the *pound force*, defined as the force required to give a slug of mass
an acceleration of 1 ft per second per second.
The **Unit of Acceleration** is the foot per second per second (abbreviated as ft/sec²).
The **Unit of Mass** is the *slug*, defined as 32.1739 lb avoirdupoise. Unlike mass, weight
is not an absolute unit but varies with the local acceleration of gravity. For example,
the weight of a slug of mass on the moon is one-sixth the weight at Seattle, Wash.
The **Relation between Force, Mass, and Acceleration** of a particle, using the above

system of units, is expressed by the formula $F = ma$ $\left(\text{or } F = \dfrac{W}{g} a\right)$, where F = force in

pounds, m = mass in slugs (= weight W in pounds divided by acceleration of gravity in
feet per second per second), and a = acceleration in the direction of the force in feet per
second per second.†

17. DERIVED QUANTITIES AND RELATIONS

Work, Power, and Energy

Work of a Force, if constant, is the product of the force and the effective displacement
of its application point. *Effective displacement* of application point is the component of
the displacement parallel to the force. The body
exerting the force is also said to do work. In
Fig. 1, the work of force F as application point
describes path $AB = F \times AC$. Since $F \cdot (AB \cos \alpha)$
$= (F \cos \alpha) \cdot AB$, the work is also equal to displacement of application point times component of force
parallel to the displacement. *Work of a variable
force* in moving a body through distance $\Delta S =$

FIG. 1

$(s_2 - s_1)$ is $W = \displaystyle\int_{s_1}^{s_2} F \cos \alpha \, ds = \int_{s_1}^{s_2} F_t ds$, in which

F is the variable force, ds is the elementary length
of path, α is angle between force and element ds, and F_t is tangential component of
force. The *sign of work* is positive if force and effective
displacement have the same sense; it is negative if they
differ in sense. Work done by a body against a force is
equal and opposite to work done by the force on the
body.
The **Unit of Work** is any force unit times any distance unit (as foot-pound) or any power unit times any
time unit (as watt-hour).
Work Diagram (Fig. 2). Plot values of F_t as ordinates;
corresponding values of s as abscissas; draw curve AB through
ends of ordinates. Area $ABDC$ times mn equals work, in foot-pounds, done by F_t over distance
$s_2 - s_1$.

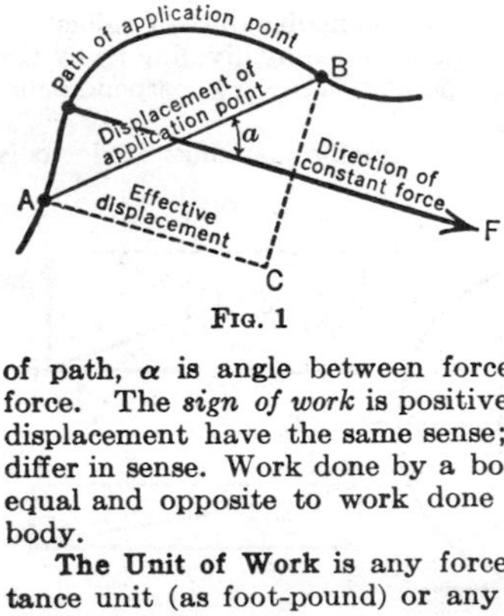

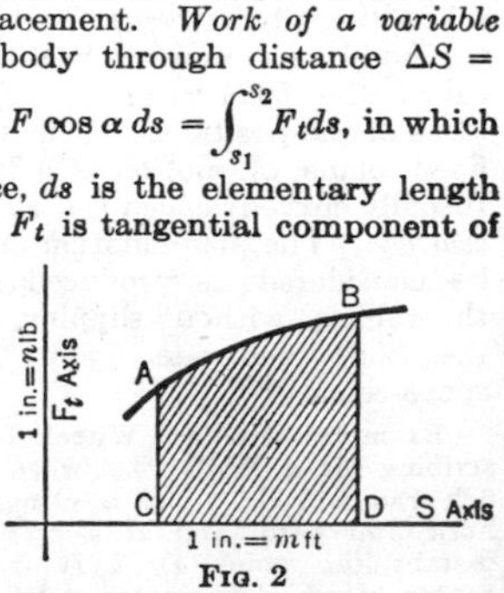

FIG. 2

---

* The units herein defined are those of the English gravitational system. For a complete discussion
of English and metric gravitational and absolute units, see Sec. 3.
† With any system of units, $F = Kma$, where K is a constant. In this system $K = 1$.

Work of Gravity on a body in any motion equals product of weight and change in height of the mass center. *Work of a central force* F (one always directed toward a fixed point), in any displacement of its application point, is $\int_{r_1}^{r_2} F dr$, in which r_2 and r_1 are the distances of the application point from the center at the beginning and end of the displacement.

Work of a Torque T on a rotating body for an angular displacement of $\theta = (\theta_2 - \theta_1)$ radians is $W = \int_{\theta_1}^{\theta_2} T d\theta$. If T is constant, $W = T(\theta_2 - \theta_1)$.

Mechanical Efficiency of a machine is the ratio of useful output to total input of work. Let W_u = useful work performed, W_f = useless work required to overcome friction or air or any other type of resistance, W_a = work applied to the machine. Then $W_a = W_u + W_f$, and Mechanical Efficiency $= \dfrac{W_u}{W_a}$.

Power of a Force is its time rate of doing work. The body exerting the force is also said to have power. Let P = power and W = work. Then instantaneous $P = \dfrac{dW}{dt} = F_t \dfrac{ds}{dt} = F_t v$, where v is instantaneous velocity of application point of force F.

The Unit of Power is any work unit divided by any time unit (as foot-pound per second). One horsepower = 550 ft-lb per sec = 33,000 ft-lb per min = 0.7457 kilowatt.

Power of a Torque at any instant is $P = \dfrac{dW}{dt} = T \dfrac{d\theta}{dt} = T\omega$ where ω is instantaneous angular velocity of the body.

Energy [*] of a *body* (or system of bodies) is the amount of work it can do, by virtue of its motion or position, against forces applied to it, while changing to a standard state.

Potential energy (PE) of a body is that possessed by virtue of its configuration. Thus, a body of weight W, located at a height above the earth's surface such that its mass center can descend h feet, has a potential energy $PE = Wh$.

Kinetic energy (KE) of a body is that possessed by virtue of its velocity, and the standard state is zero velocity. KE of a *body in translation* $= 1/2\, mv^2$. KE of a *rotating body* $= 1/2\, I\omega^2 = 1/2\, mk^2\omega^2$, I, k, and ω being moment of inertia, radius of gyration, and angular velocity, respectively, about axis of rotation. KE of a body having plane motion $= 1/2\, I\omega^2 = 1/2\, mk^2\omega^2 = 1/2\, m\bar{v}^2 + 1/2\, \bar{I}\omega^2$, in which I and k are referred to instantaneous axis, $\bar{v}$ = velocity of mass center, and $\bar{I}$ = moment of inertia about axis through mass center perpendicular to plane of motion. *Unit of energy* is same as unit of work. For KE in foot-pounds, use m in slugs, v in feet per second, ω in radians per second, and k in feet.

Principle of Conservation of Energy. If a body or system of bodies is isolated so that it neither receives nor gives out energy, its total store of energy, all forms included, remains constant; there may be a transfer of energy from one part of the system to another, but the total gain or loss in one part is exactly equivalent to the loss or gain in the remainder. This is the *principle of conservation of energy.*

Principle of Work and Kinetic Energy. Total work of the applied forces acting on any body, or on any system of connected bodies, equals the change in the kinetic energy of the body, or bodies. (This assumes no work converted into non-mechanical types of energy.) Work done $= \Delta KE$. ΔKE in translation $= 1/2\, m(v^2{}_2 - v_1{}^2)$, v_1 and v_2 being initial and final velocities. ΔKE in rotation $= 1/2\, I(\omega^2{}_2 - \omega_1{}^2) = 1/2\, mk^2(\omega_2{}^2 - \omega_1{}^2)$, ω_1 and ω_2 being initial and final angular velocities. In plane motion, change in KE is

$$\Delta KE = 1/2\, I(\omega_2{}^2 - \omega_1{}^2) = 1/2\, mk^2(\omega_2{}^2 - \omega_1{}^2) = 1/2\, m(\bar{v}_2{}^2 - \bar{v}_1{}^2) + 1/2\, \bar{I}(\omega_2{}^2 - \omega_1{}^2)$$

$$= 1/2\, m(\bar{v}_2{}^2 - \bar{v}_1{}^2) + 1/2\, m\bar{k}^2(\omega_2{}^2 - \omega_1{}^2)$$

in which I and k are referred to instantaneous axis, $\bar{I}$ and $\bar{k}$ to a parallel axis through mass center, and $\bar{v}$ is velocity of mass center.

Example. Water falling from a height of 120 ft at the rate of 1000 cu ft per min drives a turbine directly connected to an electric generator at 120 rpm. If the total resisting torque due to friction is 250 lb-ft, and the water leaves the turbine blades with a velocity of 15 ft per sec, find the power developed by the generator.

This is a problem in the conversion of potential energy to work which in turn is converted to useful kinetic energy, wasted kinetic energy, and wasted thermal energy, the total energy of the system of course remaining constant. Assume that 1 cu ft of water weighs 62.5 lb and $g = 32$ ft per sec.[2] In 1 min:

$$\Delta PE = Wh = 1000 \times 62.5 \times 120 = 7,500,000 \text{ ft-lb}$$

$$\text{Wasted } \Delta KE = 1/2\, mv^2 = \frac{1000 \times 62.5 \times 15^2}{2 \times 32} = 219,700 \text{ ft-lb}$$

---

[*] Mechanical energy (which includes potential and kinetic energy) is referred to in this definition. There are other forms of energy such as thermal, chemical, and electrical.

Wasted Friction (Thermal) $\Delta TE = T\theta = 250 \times 2\pi \times 120 = 188{,}500$ ft-lb

Therefore Useful $\Delta KE = \Delta PE -$ Wasted $\Delta KE -$ Wasted ΔTE

$$= 7{,}500{,}000 - 219{,}700 - 188{,}500 = 7{,}091{,}800 \text{ ft-lb}$$

$$P = 7{,}091{,}800/33{,}000 = 215 \text{ hp or } 215 \times 0.7457 = 160 \text{ kw}$$

Impulse, Momentum, and Impact

Linear Impulse of a constant force F for time $t = F \times t$. If the force varies in magnitude and direction, the impulse is computed from axial components of impulse; and the axial components of impulse are found by taking the time integrals of axial components of the force. The three axial component impulses are $\int_{t_1}^{t_2} F_x \, dt,\ \int_{t_1}^{t_2} F_y \, dt,\ \int_{t_1}^{t_2} F_z \, dt$; the resultant of these is the impulse of the force F. Impulse is a *vector quantity*; hence the impulse of the force equals the square root of the sum of the squares of the components. The direction cosines of the resultant vector are determined in the usual manner. *Unit of impulse* is any unit force times any unit time, as pound (force) seconds.

Angular Impulse of a force about a line for a time interval dt is the product of the moment of the force about the line and the time dt. If T represents the moment or torque of the force, the angular impulse for the time interval $(t_2 - t_1)$ is $\int_{t_1}^{t_2} T \, dt$. *Unit of angular impulse* is unit torque times unit time, as pound (force) feet seconds.

Sign of Impulse. Impulse of a force tending to increase velocity of the body to which the force is applied is positive; that which tends to decrease velocity is negative.

Linear Momentum of a *particle* is the product of its mass and velocity. It is a vector quantity and has the *sense* and *direction* of the velocity. *Unit of momentum* is the same as unit of impulse. Linear momentum of a *body* is the resultant, or vector sum, of the momentums of its particles. In any motion the linear momentum of a body is mv, m being mass of the body and $\bar{v}$ the velocity of its mass center.

Angular Momentum of a *particle* about an axis is the moment of its momentum about that axis. In Fig. 3, let $mv =$ momentum of particle P. Resolve the momentum into components parallel and perpendicular to the axis. DE is perpendicular distance from axis to line AP. The angular momentum of $P = mv \cos \alpha \times DE$. The angular momentum of a *body* about an axis is the algebraic sum of the angular momentums of its particles. The angular momentum of a rotating body about the axis of rotation is $I\omega = mk^2\omega$, I and k being moment of inertia and radius of gyration respectively about the axis of rotation, and ω the angular velocity. *Unit of angular momentum* is same as unit of angular impulse.

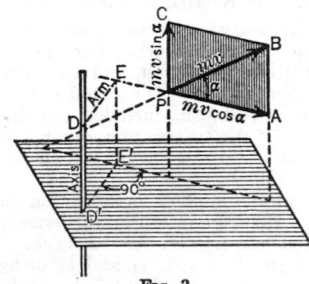

FIG. 3

Principle of Conservation of Linear and Angular Momentum. When no external forces are acting upon a body or system of bodies, the component linear momentum along any line and the angular momentum about any line remain constant; this is the *principle of conservation of linear and angular momentum*.

Principle of Impulse and Momentum. For *linear momentum*, the impulse of the resultant force acting for an infinitesimal time upon a body is equal to the change in linear momentum of its mass center during that time parallel to the direction of the force. Referred to coordinate axes, the change in the component of linear momentum parallel to any axis x for any length of time $t_2 - t_1$ equals the algebraic sum of the components of the impulses of the applied forces parallel to the axis in the same time, or, more briefly,

$\Delta(m\bar{v}_x) = \sum \int_{t_1}^{t_2} F_x dt$.   Similarly, the change in the *angular momentum* about any axis y in the time $t_2 - t_1$ equals the algebraic sum of the angular impulses of the applied forces about the axis in the same time, or, more briefly, $\Delta(I_y\omega) = \sum \int_{t_1}^{t_2} T_y dt$.

Example. A jet of water strikes a concave vessel with a velocity of 80 ft per sec and leaves it with a velocity which has the same magnitude but makes an angle of 120° with the original direction. If the diameter of the jet is 1 in., find the force necessary to hold the vessel in position.

The sustaining force F must bisect the acute angle between the lines representing the original and final velocities. Let the line of action of F (Fig. 4) be taken as the X axis. There is no change in the Y component of momentum. The impulse of the force in the X direction in t seconds = $F \times t$ pound-seconds. The weight of water deflected in t seconds is $W = 80\pi \times 62.5t/576$ pounds. The component of original momentum in the X direction = $-80W$ cos $30°/g$ pound-seconds. The component of final momentum in the X direction = $80W$ cos $30°/g$ pound-seconds. The change in momentum in the X direction = $160W$ cos $30°/g$ = $5W$ cos $30°$ pound-seconds. The fundamental relation gives $F \times t = 5 \times 80\pi \times 62.5 \times$ cos $30° \times t/576$, whence $F = 118$ lb. Observe that the sustaining force F does no mechanical work and that the water suffers no loss of kinetic energy.

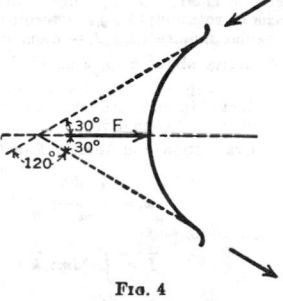

Fio. 4

Impact occurs when two bodies collide. It is *direct* when the motion is perpendicular to the striking surfaces; otherwise it is *oblique*. It is *central* if the forces which the bodies exert on each other are directed along the line joining the mass centers; otherwise it is *eccentric*. In any collision, the forces which the two bodies exert on each other are equal and opposite at each instant; hence the total impulses of these forces during the collision are equal and opposite, and according to the principle of impulse and momentum the changes in the momentums of the bodies produced by the collision must be equal and opposite; or, otherwise stated, the total momentum of the two bodies is unchanged by the collision. Or, for *direct central impact*:

$$m_1 v_1 + m_2 v_2 = m_1 V_1 + m_2 V_2$$

wherein m_1 and m_2 = the masses of the bodies, v_1 and v_2 their velocities before, and V_1 and V_2 their velocities after, the collision; but in numerical substitution, velocities in one direction are given the same sign and those in the other direction the opposite sign.

Experiments on direct central impact of spherical bodies have shown that the relative velocities of spheres after impact are always less than before the impact and that these relative velocities are opposite in direction. The ratio of the relative velocities after impact to that before impact is called *coefficient of restitution*; it seems to depend only on the material of the impinging spheres. For glass the coefficient is $15/16$, for steel and cork $5/9$, ivory $8/9$, wood about $1/2$, clay and putty 0. If e = the coefficient, then

$$(V_1 - V_2) = -e(v_1 - v_2)$$

This equation and the preceding one solved simultaneously show that

$$V_1 = v_1 - \frac{(1+e)m_2}{m_1 + m_2}(v_1 - v_2); \quad V_2 = v_2 - \frac{(1+e)m_1}{m_1 + m_2}(v_2 - v_1)$$

During impact there is, in general, loss of kinetic energy; * the loss is

$$1/2(v_1 - v_2)^2(1 - e^2)m_1 m_2/(m_1 + m_2).$$

Bodies for which $e = 0$ are said to be *inelastic*; and those for which e is nearly 1 are said to be nearly perfectly *elastic*. When a sphere is dropped on a horizontal surface of a large body from a height h, if H = the height of rebound, then $H = e^2 h$. This equation furnishes a means of computing e.

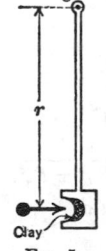

Fig. 5

Example. Ballistic Pendulum. This is a device for determining the velocity of a bullet. The bullet is imbedded in soft material, such as clay, for which $e = 0$. Referring to Fig. 5, let m_1 = mass of bullet, m_2 = mass of pendulum, k = radius of gyration about axis of suspension O, v_1 = velocity of bullet (to be determined), $v_2 = r\omega$ = velocity of bullet after impact, ω = angular velocity of pendulum after impact, and assume pendulum stationary before impact. Then angular momentum of system before impact = $m_1 v_1 r$, and angular momentum of system just after impact = $m_1 r^2 \omega + m_2 k^2 \omega$. Since total momentum of system remains constant, $m_1 v_1 r = m_1 r^2 \omega + m_2 k^2 \omega$. If mass center of pendulum rises to height h, $k^2 \omega^2 = 2gh$. Combining last two equations to eliminate ω and solving for v_1, $v_1 = (m_1 r^2 + m_2 k^2) \sqrt{2gh}/m_1 rk$. The quantities on the right-hand side of this equation are easily determined experimentally.

18. KINEMATIC AND KINETIC FORMULAS

Symbols. s = distance along path of motion; x, y, z = coordinates of any point; $\bar{x}$, $\bar{y}$, $\bar{z}$ = coordinates of mass center; t = time; a = resultant linear acceleration of any point; $\bar{a}$ = resultant

---

* As energy can be neither created nor destroyed, the total energy remains constant. The "lost" kinetic energy is simply converted into other forms, as into work done in distorting the bodies, thermal (heat) energy, etc.

linear acceleration of mass center; $a_{x,y,z}$ = components of resultant acceleration along x, y, z axes; a_t = resultant tangential acceleration of any point; a_n = resultant normal acceleration of any point; v, $\bar{v}$, $v_{x,y,z}$ = linear velocities having corresponding significances; θ = angular displacement; α = angular acceleration; ω = angular velocity; n = revolutions per unit time; r = radius (of curvature); g = acceleration of gravity = 32.2 ft per sec^2 (approx.); m = weight/g = mass; F = resultant force; $F_{x,y,z}$ = components of resultant force along x, y, z axes; F_t = resultant tangential force; F_n = resultant normal force; W = work; Eff = efficiency; P = power; KE = kinetic energy; Imp = linear impulse; Mom = linear momentum; T = resultant torque about axis of rotation; $T_{x,y,z}$ = torques about x, y, z axes; $Ang\ Imp$ = angular impulse; $Ang\ Mom$ = angular momentum; I = moment of inertia (for mass) about axis of rotation; $I_{x,y,z}$ = moments of inertia about x, y, z axes; $\bar{I}$ = moment of inertia about axis through mass center; $k, k_{x,y,z}$, $\bar{k}$ = corresponding radii of gyration; d = distance between axes; U = product of inertia; U_{xy} = product of inertia with respect to YOZ and ZOX planes; (U_{yz}, U_{zx} have corresponding significances); Δ indicates " change in."

Gravitation and Inertia Functions. Mass center has coordinates:

$$\bar{x} = \frac{\int x\,dm}{m}; \qquad \bar{y} = \frac{\int y\,dm}{m}; \qquad \bar{z} = \frac{\int z\,dm}{m}.$$

$$\bar{I} = \int r^2 dm; \quad \bar{k} = \sqrt{\bar{I}/m}; \quad I = \bar{I} + md^2; \quad k = \sqrt{I/m}; \quad k^2 = \bar{k}^2 + d^2;$$

$$U_{xy} = \int xy\,dm; \qquad U_{yz} = \int yz\,dm; \qquad U_{zx} = \int zx\,dm.$$

Translation—(Rectilinear Motion) *

$$v = \frac{ds}{dt}; \qquad a = \frac{dv}{dt} = \frac{d^2s}{dt^2}; \qquad \frac{a}{v} = \frac{dv}{ds};$$

$$\Delta s = \int_{t_1}^{t_2} v\,dt; \qquad \Delta v = \int_{t_1}^{t_2} a\,dt; \qquad \Delta t = \int_{s_1}^{s_2}\frac{ds}{v} = \int_{v_1}^{v_2}\frac{dv}{a}; \qquad \Delta v^2 = 2\int_{s_1}^{s_2} a\,ds;$$

$$F = m\bar{a}; \qquad \Delta W = \int_{s_1}^{s_2} F\,ds; \quad KE = {}^1\!/_2\, m v^2; \qquad \Delta W = \Delta KE;$$

$$P = \frac{dW}{dt} = Fv; \qquad \Delta Imp = \int_{t_1}^{t_2} F\,dt; \quad Mom = m\bar{v}; \qquad \Delta Imp = \Delta Mom.$$

Translation—(Curvilinear Motion) *

$$v = \frac{ds}{dt}; \qquad v_x = \frac{dx}{dt}; \quad v_y = \frac{dy}{dt}; \quad v_z = \frac{dz}{dt}; \qquad v = \sqrt{v_x{}^2 + v_y{}^2 + v_z{}^2};$$

Directional cosines of v are: $\quad \cos\theta_x = \dfrac{v_x}{v}; \quad \cos\theta_y = \dfrac{v_y}{v}; \quad \cos\theta_z = \dfrac{v_z}{v}.$

$$a_x = \frac{dv_x}{dt} = \frac{d^2x}{dt^2}; \qquad a_y = \frac{dv_y}{dt} = \frac{d^2y}{dt^2}; \qquad a_z = \frac{dv_z}{dt} = \frac{d^2z}{dt^2}; \qquad a = \sqrt{a_x{}^2 + a_y{}^2 + a_z{}^2};$$

Directional cosines of a are: $\quad \cos\phi_x = \dfrac{a_x}{a}; \quad \cos\phi_y = \dfrac{a_y}{a}; \quad \cos\phi_z = \dfrac{a_z}{a};$

$$a_t = \frac{dv}{dt} = \frac{d^2s}{dt^2}; \qquad a_n = \frac{v^2}{r}; \qquad a = \sqrt{a_t{}^2 + a_n{}^2}; \qquad \frac{a_t}{v} = \frac{dv}{ds};$$

$$\Delta s = \int_{t_1}^{t_2} v\,dt; \qquad \Delta v = \int_{t_1}^{t_2} a_t\,dt; \quad \Delta t = \int_{s_1}^{s_2}\frac{ds}{v} = \int_{v_1}^{v_2}\frac{dv}{a_t}; \qquad \Delta v^2 = 2\int_{s_1}^{s_2} a_t\,ds;$$

$$F = m\bar{a}; \qquad F_x = m\bar{a}_x; \qquad F_y = m\bar{a}_y; \qquad F_z = m\bar{a}_z; \qquad F = \sqrt{F_x{}^2 + F_y{}^2 + F_z{}^2};$$

$$F_t = m\bar{a}_t; \qquad F_n = m\bar{a}_n; \qquad\qquad\qquad\qquad F = \sqrt{F_t{}^2 + F_n{}^2};$$

$$\Delta W = \int_{s_1}^{s_2} F_t\,ds; \qquad KE = {}^1\!/_2\, m\bar{v}^2; \qquad \Delta W = \Delta KE; \qquad P = \frac{dW}{dt} = F_t v;$$

$$\Delta Imp_x = \int_{t_1}^{t_2} F_x\,dt; \qquad \Delta Imp_y = \int_{t_1}^{t_2} F_y\,dt; \qquad \Delta Imp_z = \int_{t_1}^{t_2} F_z\,dt;$$

$$Imp = \sqrt{Imp_x{}^2 + Imp_y{}^2 + Imp_z{}^2};$$

---

* For a rigid body in translation, accelerations and velocities of all particles are equal. However, $\bar{a}$ and $\bar{v}$ are indicated in certain of the kinetic translation formulas to make them applicable also to non-rigid bodies.

$Mom_x = m\bar{v}_x;$ $Mom_y = m\bar{v}_y;$ $Mom_z = m\bar{v}_z;$

$$Mom = \sqrt{Mom_x{}^2 + Mom_y{}^2 + Mom_z{}^2};$$

$\Delta Imp_x = \Delta Mom_x;$ $\Delta Imp_y = \Delta Mom_y;$ $\Delta Imp_z = \Delta Mom_z;$ $\Delta Imp = \Delta Mom;$

Directional Cosines of $\Delta Imp = \Delta Mom$ are: $\cos \psi_x = \dfrac{\Delta \bar{v}_x}{\Delta \bar{v}};$ $\cos \psi = \dfrac{\Delta \bar{v}_y}{\Delta \bar{v}};$ $\cos \psi_z = \dfrac{\Delta \bar{v}_z}{\Delta \bar{v}}.$

For kinetic formulas applying to a translated *body* for rotation about an axis not fixed in the body or its extension, use formulas applying to " Rotation of a *Particle* " about its axis, considering entire mass of body as concentrated at the mass center.

Rotation †

$$\omega = \frac{d\theta}{dt}; \qquad\qquad \alpha = \frac{d\omega}{dt} = \frac{d^2\theta}{dt^2}; \qquad\qquad \frac{\alpha}{\omega} = \frac{d\omega}{d\theta};$$

$$\Delta\theta = \int_{t_1}^{t_2}\omega\,dt; \qquad \Delta\omega = \int_{t_1}^{t_2}\alpha\,dt; \qquad \Delta t = \int_{\theta_1}^{\theta_2}\frac{d\theta}{\omega} = \int_{\omega_1}^{\omega_2}\frac{d\omega}{\alpha}; \qquad \Delta\omega^2 = 2\int_{\theta_1}^{\theta_2}\alpha\,d\theta.$$

For a "Particle" ($\bar{I}$ *infinitesimal* compared with I) :

$s = r\theta;$ $v = r\omega;$ $a_t = r\alpha;$ $a_n = r\omega^2;$ $a = r\sqrt{\alpha^2 + \omega^4};$

$T = F_t r = mr^2\alpha;$ $F_t = mr\alpha;$ $F_n = mr\omega^2;$ $F = mr\sqrt{\alpha^2 + \omega^4};$

$$\Delta W = \int_{\theta_1}^{\theta_2} T\,d\theta; \qquad KE = {}^1/_2\,mr^2\omega^2; \qquad \Delta W = \Delta KE; \qquad P = \frac{dW}{dt} = T\omega;$$

$\Delta Ang\ Imp = \displaystyle\int_{t_1}^{t_2} T\,dt;$ $Ang\ Mom = mr^2\omega;$ $\Delta Ang\ Imp = \Delta Ang\ Mom.$

For a Body :

$$T = I\alpha = mk^2\alpha; \qquad \Delta W = \int_{\theta_1}^{\theta_2} T\,d\theta; \qquad \Delta W = \Delta KE; \qquad P = \frac{dW}{dt} = T\omega;$$

$$KE = {}^1/_2\,I\omega^2 = {}^1/_2\,mk^2\omega^2;$$

$\Delta Ang\ Imp = \displaystyle\int_{t_1}^{t_2} T\,dt;$ $Ang\ Mom = I\omega = mk^2\omega;$ $\Delta Ang\ Imp = \Delta Ang\ Mom.$

Constrained Rotation *

Plane of rotation fixed above and parallel to horizontal XZ plane; vertical Y axis of rotation not passing through mass center (except as special case).

All previous rotation formulas apply if T is replaced by T_y. Additional formulas are:

For a Particle ($\bar{I}$ *infinitesimal* compared with I) :

$F_x = m\bar{z}\alpha - m\bar{x}\omega^2;$ $F_y = 0;$ $F_z = -m\bar{x}\alpha - m\bar{z}\omega^2;$ $F = \sqrt{F_x^2 + F_y^2 + F_z^2};$

$T_x = F_z\bar{y} = -m\overline{xy}\alpha - m\overline{yz}\omega^2;$ $T_y = F_z\bar{z} = m\bar{z}^2\alpha - m\overline{xz}\omega^2 = mr^2\alpha$

$= -F_z\bar{x} = m\bar{x}^2\alpha + m\overline{xz}\omega^2 = mr^2\alpha;$

$T_z = -F_z\bar{y} = -m\overline{yz}\alpha + m\overline{xy}\omega^2;$ (θ, ω, α positive for counterclockwise rotation facing origin from plus point on axis).

For a Body :

$T_x = -U_{xy}\alpha - U_{yz}\omega^2;$ $T_y = I_y\alpha;$ $T_z = -U_{yz}\alpha + U_{xy}\omega^2$ (Sign convention as above).

Center of Percussion and Center of Oscillation of a pendulum are located at distance from center of suspension $= k^2/\bar{z}$, where $\bar{z}$ is distance from center of suspension to mass center.

Plane and Three-Dimensional Motions

For Translation of Mass Center :

Consider entire mass as concentrated at mass center. Refer motion to set of fixed axes located outside the body. To determine acceleration of mass center, apply formulas for translation.

---

* In obtaining total forces and torques, effect of weight of body (this effect depending on position of plane of rotation) must not be neglected.
† Formulas are for rigid bodies.

For Rotation about Mass Center:

Consider mass center as fixed and resultant of forces as a couple. Refer motion to set of central principal axes. To determine components of angular acceleration about these axes, use formulas:

$$T_x = I_x \alpha_x + (I_z - I_y)\omega_y \omega_z; \quad T_y = I_y \alpha_y + (I_x - I_z)\omega_z \omega_x; \quad T_z = I_z \alpha_z + (I_y - I_x)\omega_x \omega_y$$

For Complete Resultant Motion:

Combine motion of translation of mass center with motion of rotation about mass center.

Work and kinetic energy changes also are equal to the respective sums of the corresponding changes under the above component motions.

19. TRANSLATION

Kinetic Formulas for motion of translation follow directly from the kinematic formulas applying to such motion and the previous discussion on kinetic quantities and their relations. The formulas are summarized in Art. 18 (p. 466) under the heading "Translation." For the solution of a specific problem, careful choice of formulas will often facilitate the computations. As there is no rotation, the resultant force acts through the mass center and there is no couple.

Example. Motion of Parallel Rod of a Locomotive. The problem is to find the forces acting upon the parallel rod when it is in any position with respect to the wheels. The speed of locomotive constant at 60 miles per hour on a level track; driver diameter 5.5 ft; crank length 1 ft; and weight of rod 275 lb. The forces acting on the rod are its weight and the pressures of the crank pins at its ends; the latter are represented (Fig. 6) by their horizontal and vertical components.

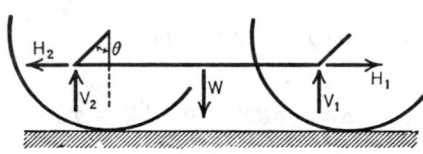

Since the resultant of all these forces acts through the mass center, $V_1 = V_2$; also $2V_1 - 275 = (W/g)a_y = 8.55a_y$ and $H_1 - H_2 = (W/g)a_x = 8.55a_x$. To determine a_x and a_y: The velocity of the center of either crank pin relative to the locomotive is $(88 \times 1)/2.75 = 32$ ft per sec (60 mi per hr = 88 ft per sec), and the relative motion of the pin being circular at constant velocity, the relative acceleration is toward the center of the crank-pin circle at all times and equals $32^2/1 =$

Fig. 6

1024 ft per sec per sec. This is also the absolute acceleration of the crank pin, since the locomotive is assumed to have no acceleration. But the rod has the same acceleration as the crank pin; hence $a_x = 1024 \sin \theta$, and $a_y = 1024 \cos \theta$. Thus $V = 1/2 (8755 \cos \theta + 275)$, and $H_1 - H_2 = 8755 \sin \theta$. In the lowest position of the rod, $\theta = 0$, $a_x = 0$, $a_y = 1024$, $H_1 = H_2$, $V = 1/2 (8755 + 275) = 4515$. In a mid-position when $\theta = 90°$, $a_x = 1024$, $a_y = 0$, $H_1 - H_2 = 8755$, $V = 1/2 (275) = 137.5$. In the highest position, $\theta = 180°$, $a_x = 0$, $a_y = -1024$, $H_1 = H_2$, and $V = 1/2 (275 - 8755) = -4240$, the negative sign meaning that V acts downward on the rod.

20. ROTATION

Kinetic Formulas for motion of rotation follow directly from the kinematic formulas applying to such motion and the previous discussion on kinetic quantities and their relations. The formulas are summarized in Art. 18 (p. 467) under the heading "Rotation." For the solution of a specific problem, careful choice of formulas will often facilitate the computations. As there is no translation, the resultant force is zero but there is a couple.

Example. A Punch is required to exert a force of 100,000 lb through a distance of $1/4$ in., and the work is to be supplied by a flywheel of radius of gyration = 1.5 ft making 120 rpm. Find the weight of the wheel, if the speed is not to be reduced below 100 rpm.

$\omega_1 = 120$ rpm $= 4\pi$ rad per sec, $\omega_2 = 100$ rpm $= (10/3)\pi$ rad per sec. Work done by punch $= 100,000/48$ ft-lb $=$ reduction in KE of flywheel.

Change in $KE = 1/2 \, mk^2 \Delta\omega^2 = W \times 2.25 \, (\omega_1{}^2 - \omega_2{}^2)/64 = W \times 2.25 \, (\omega_1 - \omega_2)(\omega_1 + \omega_2)/64$.

Hence $\dfrac{W \times 2.25}{64} (2\pi/3)(22\pi/3) = \dfrac{100,000}{48}$, whence $W = 1230$ lb = minimum weight of flywheel.

Constrained Rotation

Constrained Rotation refers to rotation of a body about a fixed axis which does not pass through its mass center. Such an axis, since it constrains the motion,* must be held by forces (exerted by bearings) to keep it from shifting position. These bearing reactions depend upon the weight of the body, the manner in which the mass of body is

---

* In certain cases, the "phsyical path" itself constrains the rotation, as the action of the track on a train rounding a curve.

distributed about the axis, the applied forces, the angular velocity ω, and the angular acceleration α. Generally, the resultant of the applied forces for such a body is not a single force, but a single force at a selected origin and a couple. Selecting the origin on the axis of rotation, the axial components of the single force and axial components of the couple are given by the following six equations.*

$$\Sigma F_x = m\bar{z}\alpha - m\bar{x}\omega^2 \qquad \Sigma T_x = -\alpha \int xy\,dm - \omega^2 \int yz\,dm = -U_{xy}\alpha - U_{yz}\omega^2$$

$$\Sigma F_y = 0 \qquad \Sigma T_y = I_y\alpha$$

$$\Sigma F_z = -m\bar{x}\alpha - m\bar{z}\omega^2 \qquad \Sigma T_z = -\alpha \int yz\,dm + \omega^2 \int xy\,dm = -U_{yz}\alpha + U_{xy}\omega^2$$

In these equations, the axis of rotation is fundamentally the y axis; $\bar{x}$, $\bar{y}$, and $\bar{z}$ are the instantaneous coordinates of the mass center; ΣF_x, ΣF_y, ΣF_z are the sums of components of all applied forces in the axial directions; ΣT_x, ΣT_y, ΣT_z are the sums of moments of all applied forces about the axes; and the convention of signs for moments of forces, and senses of θ, ω, and α are that counterclockwise rotation, facing the origin from any plus point on an axis, is positive.

The equations are simultaneous at each instant. They are used more often to determine the forces exerted by the bearings on the axle, than to determine the resultant.

Special Cases (Fig. 7). Choose the x axis through an instantaneous location of the mass center and let XZ be a plane of symmetry of a homogeneous body. The resultant is a single force in the plane of symmetry having the Z component $- m\bar{x}\alpha$ and the X component $- m\bar{x}\omega^2$ acting at point C. $\overline{OC} = k_y^2/\bar{x}$, k_y being radius of gyration about y axis. If $\bar{x} = 0$, the resultant becomes a couple in the XZ plane, of moment $= \Sigma T_y = I_y\alpha$. If $\alpha = 0$ and $\bar{x} \neq 0$, the resultant $= - m\bar{x}\omega^2$, in the sense CO. If $\alpha = 0$ and $\bar{x} = 0$, the resultant vanishes.

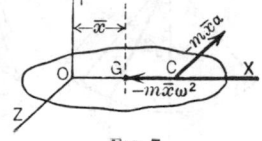

FIG. 7

Centrifugal Force. Let any particle of mass m move in a circular path of radius r about a fixed y axis. The resultant of all forces acting on the particle has a normal component $= mr\omega^2$, and a tangential component $= mr\alpha$. The component $mr\alpha$ increases or decreases the speed of the particle; the component $mr\omega^2$ continually changes the direction of the linear velocity. The resultant of such forces for all the particles of the body is equivalent to the resultant specified by the general equations above. If ω is constant and $\alpha = 0$, the resultant force acting on the particle to make it rotate in its circular path is $mr\omega^2$ toward the axis, and is called centripetal force. Centrifugal force for the particle is equal and opposite to centripetal force, and is exerted by the particle upon its neighboring particles, or upon the axis of rotation. Centrifugal resultant for a body is the resultant of the centrifugal forces of all its particles. Generally, this resultant is not a single force; it may be computed from the general equations by making $\alpha = 0$ and reversing senses of resultant force and couple.

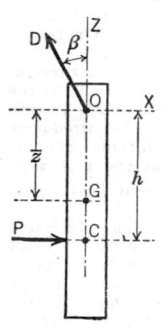

FIG. 8

Center of Percussion. A prismatic bar (Fig. 8) is suspended on a horizontal y axis at O. G is the mass center. If a force P, parallel to x axis, is applied to the body, the axle reaction OD will generally be inclined to the z axis at some angle $\pm\beta$, the angle depending on the distance h of P from the axis of rotation. If $h = k_y^2/\bar{z}$, in which k_y is radius of gyration about y axis, P will cause no x component of axle reaction; that is, β will be zero, and the point C, where the action line of P intersects OG, is the center of percussion. In impact-testing machines, heavy pendulums are used to deliver blows, and proper design requires the striking point to coincide with the center of percussion in order to avoid shock to the axle and detrimental vibration of the pendulum itself.

Examples of Constrained Rotation

A Simple Pendulum consists of a small heavy bob on a light string (Fig. 9).† The forces acting on it are the weight, W, and tension, T. The resultant force along the tangent $= - W \sin \theta$; the resultant force along the normal $= T - W \cos \theta$. The force equations are $Wa_t/g = - W \sin \theta$, $Wa_n/g = T - W \cos \theta$. Since $a_n = l\omega^2$, tension $T = W\left(\cos \theta + \dfrac{l\omega^2}{g}\right)$.

---

* In obtaining total forces and torques, effect of weight of body must not be neglected.
† Radius of gyration of bob about axis through its mass center parallel to axis of rotation is considered negligible compared with radius of its path.

To determine the motion: $a_t/g = l\alpha/g = -\sin\theta$.
The solution of this equation leads to elliptic functions. An approximate solution for small oscillations can be obtained by putting $\sin\theta = \theta$. (Difference between θ and $\sin\theta$ is less than 1 per cent if θ is less than 14°.) The differential equation becomes $\omega \, d\omega/d\theta = - g\theta/l$. If the pendulum is at the end of its swing when $t = 0$, then $\theta = \beta$, $\omega = 0$. Integrating, $\omega^2 = g(\beta^2 - \theta^2)/l$;

$$\omega = \frac{d\theta}{dt} = \pm \sqrt{\frac{g}{l}(\beta^2 - \theta^2)}. \quad \text{Integrating, } \theta = \beta\cos\sqrt{\frac{g}{l}}\,t. \quad \text{Period of oscillation} = 2\pi\sqrt{\frac{l}{g}}.$$

A **Conical Pendulum** * consists of a small heavy bob suspended from a fixed point by a light string so that it can be made to rotate about the vertical axis through the fixed point (Fig. 10). If the bob rotates with constant angular velocity, ω, the quantities ϕ, r, h are constants. Since there is no vertical acceleration, $T\cos\phi = W$. The force acting inward on the bob is $T\sin\phi$. Hence the force equation gives $T\sin\phi = Wa_n/g = Wv^2/gr$, and $\tan\phi = v^2/gr = r\omega^2/g$. Also

$$h = g/\omega^2; \quad T = Wl\omega^2/g; \quad \text{period of one revolution} = \frac{2\pi}{\omega} = 2\pi\sqrt{\frac{h}{g}}.$$

A **Compound (or Physical) Pendulum** is any rigid body suspended from a horizontal axis about which it may rotate under the action of its own weight. The forces acting on the body are its weight, acting downward at G (Fig. 11), and the reaction of the axis at O. Let $\bar{r} = $ distance OG;

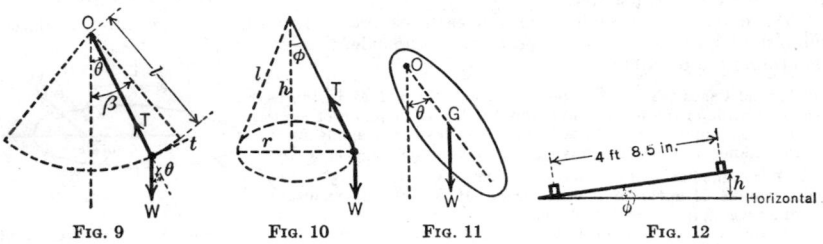

FIG. 9   FIG. 10   FIG. 11   FIG. 12

$k = $ radius of gyration about O. The torque equation gives $Wk^2\alpha/g = - W\bar{r}\sin\theta$, whence $\alpha = - \bar{r}g\sin\theta/k^2$. This is the equation of a simple pendulum (see above) of length $l = k^2/\bar{r}$ called the length of the equivalent simple pendulum. The motion of a compound pendulum is the same as the motion of the equivalent simple pendulum. The point on the compound pendulum located at the distance $k^2/\bar{r}$ from the axis of rotation is called the *center of oscillation*. It coincides with the center of percussion (see above).

Super-Elevation of Outer Rail of a Railroad Track is determined as follows (Fig. 12): Let $r = $ radius of curvature in feet and $v = $ speed in feet per second of car. Then horizontal centrifugal force is Wv^2/gr and vertical force is W, acting through mass center.† For the resultant to be perpendicular to the track and thus impose no side load on the rails, $\tan\phi = v^2/gr$. For small angles, the sine instead of the tangent may be used and, if $h = $ super-elevation of the outer rail in inches, $h = 56.5v^2/gr$.

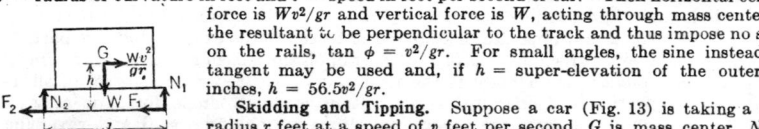

FIG. 13

Skidding and Tipping. Suppose a car (Fig. 13) is taking a curve of radius r feet at a speed of v feet per second, G is mass center, N_1 is the vertical and F_1 the horizontal pressure on the outer wheel, and $f = $ coefficient of friction.† The problem becomes one of statics by introducing $Wv^2/gr = F_1 + F_2 = F = fW$. If $f < v^2/gr$, the car will skid. Suppose $f > v^2/gr$; then $N_1 = W(1/2 + v^2h/dgr)$, $N_2 = W(1/2 - v^2h/dgr)$. The critical speed is $v_1 = \sqrt{dgr/2h}$, when the total weight is borne on the outer wheel. If this critical speed is exceeded, the car will tip over.

Note on Use of Rotation Formulas. In practice, nearly all problems of the nature illustrated by the car problems above are solved by the use of formulas applying to rotation of a *particle* about an axis. It should be realized, however, that the assumption thus made that the mass is concentrated at the mass center is not strictly correct except in the event that the body has a true motion of translation (as exemplified by the motion of the parallel rod of a locomotive). Seldom is this the case in practice as usually every line of the body lying in the plane of motion makes one complete revolution for each revolution of the body about the center of rotation of its path. Therefore the formulas applying to rotation of a *body* are the only ones giving absolutely correct results.

For example, in the problem above on the motion of a simple pendulum, considering the path as that of the mass center, the actual torque $T = Wk^2\alpha/g$, where k is radius of gyration of bob about horizontal axis of rotation through O. Let $\bar{k} = $ radius of gyration of bob about horizontal axis through mass center parallel to axis through O. Then $k^2 = l^2 + \bar{k}^2$, and actual torque $T = W(l^2 + \bar{k}^2)\alpha/g$. But the approximate assumption made in the problem that $F_t = Wl\alpha/g$ gives $T = Wl^2\alpha/g$, which is too small by the amount $W\bar{k}^2\alpha/g$. However, when $\bar{k}$ is very small compared with l, the results obtained are sufficiently near accurate.

---

* The principle of the conical pendulum is employed in the Watt governor for steam engines.
† Radius of gyration of the car about axis through its mass center parallel to axis of rotation is considered negligible compared with radius of its path.

21. PLANE MOTION

Kinetic Formulas for plane motion are a combination of those for motions of translation and rotation. The procedure for solution of a problem is summarized in Art. 18 (p. 467) under the heading "Plane and Three-Dimensional Motions." The general formula given there for determination of angular accelerations reduces to $T_x = I_x\alpha = mk_x{}^2\alpha$, the x axis being perpendicular to the plane of motion and passing through the mass center. The theory forming the basis for the assumptions regarding mass concentration and arrangement of forces is explained below under the general case of "Three-Dimensional Motion."

Example: Wheel on Inclined Plane. In Fig. 14, $\tan \beta = 3/4$, the wheel weighs 100 lb, diameter = 4 ft, radius of gyration = 1.6 ft. (a) Find the acceleration of the center, if the wheel rolls without slipping. (b) Find the least coefficient of friction to prevent slipping. (c) If the coefficient of friction = 0.1, find the acceleration of the center and the number of turns made while the center moves 20 ft.

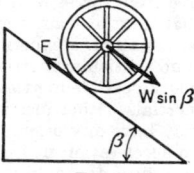

(a) The forces acting to move the wheel are $W \sin \beta = 60$ lb and friction, F. The equation of motion of the center is $100a/32 = 60 - F$. The force acting to turn the wheel is F. The torque equation is $100 \times 1.6 \times 1.6\alpha/32 = 2F$. Since the wheel does not slip, $a = 2\alpha$. Elimination of F and α gives $a = 11.7$ ft per sec².

(b) Friction = min. coeff. of friction × normal pressure, or $F = fW \cos \beta$ = minimum $f \times 80$. From the equation above, $F = 23.4$ lb, whence minimum $f = 0.29$.

FIG. 14

(c) The relation between a and α is not known when the wheel slips. $F = 80 \times 0.1 = 8$ lb. The equation of motion of the center is $100a/32 = 60 - 8 = 52$, whence $a = 16.6$ ft per sec². Distance moved by center, $x = 8.3t^2$. Time to move 20 ft is given by $t^2 = 20/8.3$. Torque equation is $100 \times 1.6 \times 1.6\alpha/32 = 2 \times 8$, whence $\alpha = 2$ rad per sec². The angle turned through, $\theta = t^2 = 20/8.3 = 2.41$ rad = 0.38 revolution.

22. THREE-DIMENSIONAL MOTION

Kinetic Formulas for three-dimensional motion are a combination of those for motions of translation and rotation. The procedure for solution of a problem is summarized in Art. 18 (p. 467) under the heading "Plane and Three-Dimensional Motions."

Any motion of a body may be regarded as consisting of two components: one, a translation equal to that of the mass center, and the other a rotation about some axis through the mass center. These motions may be said to be produced independently by the forces acting on the body; thus (a) the acceleration of the mass center is the same as if the whole mass were concentrated there and acted upon by forces equal in magnitude to, and the same in direction as, the actual external forces; and (b) the angular acceleration is the same as if the mass center were fixed and the actual external forces applied. The reasonableness of this will be seen from the following: imagine each force acting on the body replaced by a force acting at the mass center G and a couple (see p. 435); the resultant of all the forces acting at G is a single force R, and the resultant of all the couples is a single couple C; R cannot turn the body but gives it a motion of translation only, and C cannot move G but merely turns the body about some line through G. In general, C does not cause turning about a line perpendicular to the plane of C, only so if the plane of C is perpendicular to one of the principal central axes of the body. To determine the acceleration of the mass center, take fixed x, y, and z axes outside the body and resolve all external forces F_1, F_2, etc., into x, y, and z components; then

$$\Sigma F_x = m\bar{a}_x \qquad \Sigma F_y = m\bar{a}_y \qquad \Sigma F_z = m\bar{a}_z$$

m denoting the mass of the body. To determine the angular acceleration of the body, take moments of all the forces F_1, F_2, etc., about the three central principal axes; calling the sums of the moments about these axes ΣT_1, ΣT_2, and ΣT_3, the components of the angular acceleration α_1, α_2, and α_3, and the components of the angular velocity ω_1, ω_2, and ω_3, then

$$\Sigma T_1 = I_1\alpha_1 + (I_3 - I_2)\omega_2\omega_3; \quad \Sigma T_2 = I_2\alpha_2 + (I_1 - I_3)\omega_3\omega_1; \quad \Sigma T_3 = I_3\alpha_3 + (I_2 - I_1)\omega_1\omega_2$$

wherein I_1, I_2, and I_3 denote the three central principal moments of inertia of the body. In any motion of a body, the kinetic energy may be computed in two parts: (1) the kinetic energy of the whole body moving with a velocity equal to that of the mass center, and (2) the sum of the kinetic energies of the constituent particles of the body due to their velocities relative to an axis through the mass center.

23. MOVING AXES

A **Frame of Reference** is a set of coordinate axes or coordinate curves with respect to which linear and angular positions, velocities, and accelerations are measured. For brevity, the word *frame* will denote a frame of reference; and a measurement with respect to a frame will be said to be made *in* that frame. A frame may move relative to another frame. In so doing, it has the same freedom of motion as a rigid body.

Forces can be specified and measured in a manner independent of the observer's frame of reference, as for example by spring extensions. It follows that Newton's third law, which is concerned with forces only, holds in all frames: in the mechanical interaction of two bodies, their mutual forces are equal bwt opposite, and in the same line of action.

An Inertial Frame is a frame in which the first law of Newton holds, i.e., that a particle free from forces is unaccelerated. This defines an inertial frame, while experiment shows that inertial frames exist. The second law of Newton, that net force equals mass times acceleration, is experimentally true in all inertial frames and only in inertial frames. There is no particular inertial frame which may be called absolute; but rather there are infinitely many possible inertial frames, and all of them are equivalent. Suppose that a frame R_1 is inertial. Other inertial frames may have any position of their origin in R_1 and their axes may have any orientation in R_1. Another frame R_2 is inertial if and only if its origin has no acceleration in R_1 and in addition its orientation in R_1 is fixed. In other words, R_2 is inertial if and only if its motion in R_1 is at most a pure translation without acceleration. These conditions are necessary and sufficient in order that all accelerations should appear the same in R_2 as in R_1; and they are therefore the conditions that, if R_1 is inertial, R_2 is also. Acceleration in an inertial frame will be called *true acceleration*.

Non-inertial Frames are of practical interest, for all terrestrial experiments are performed in such frames. Any frame fixed in the earth is non-inertial because the earth is both spinning and accelerating relative to an inertial frame. Nevertheless, the observation of phenomena caused by the non-inertial character of an earth-fixed frame is a matter of some delicacy, so that to a certain approximation such a frame may be considered inertial. An earth-fixed frame will be referred to simply as *the earth*.

Effects of the Non-inertial Character of a frame are apparent accelerations or forces not accounted for by actual forces. Let a_x, a_y, and a_z be the components of the apparent acceleration of a point mass m observed in a non-inertial frame S. Let the true acceleration components in the same directions be a'_x, a'_y, and a'_z. Define components of *acceleration difference*, the difference between apparent and true accelerations:

$$g_x \equiv a_x - a'_x; \quad g_y \equiv a_y - a'_y; \quad g_z \equiv a_z - a'_z$$

The acceleration difference involves the motion of S relative to an inertial frame R, but it also involves the position x, y, z and the velocity v_x, v_y, v_z of m in S. Now, if the mass is free from actual forces, it is observed in S to have the acceleration g_x, g_y, g_z; or, if the actual forces on it are specified, it is observed to have this acceleration in addition to that predicted from Newton's second law by the observer in S. If, on the other hand, the observer in S specifies the motion of the particle relative to his frame, he finds it necessary to exert upon m, in addition to the force which he computes from Newton's second law, the force

$$f_x = -mg_x; \quad f_y = -mg_y; \quad f_z = -mg_z$$

The particle m thus exerts a reaction force having no apparent physical cause:

$$r_x = mg_x; \quad r_y = mg_y; \quad r_z = mg_z$$

The following paragraphs classify the possible acceleration differences.

Translational Acceleration of frame S with components a_{0x}, a_{0y}, a_{0z} relative to an inertial frame R produces an acceleration difference

$$g_x = -a_{0x}; \quad g_y = -a_{0y}; \quad g_z = -a_{0z}$$

Thus, in a train which has acceleration a_{0x} along a straight track in the forward direction, a body free to move would have an apparent backward acceleration $g_x = -a_{0x}$. A mass m constrained to remain at rest in the train would exert a backward reaction $-ma_{0x}$.

Centripetal Acceleration causes an acceleration difference when the system S rotates relative to R. If the rotation of S is about its z-axis with an angular speed ω, then a point mass m at x, y, z in S has an acceleration difference $\omega^2 \sqrt{x^2 + y^2}$ directed perpendicularly away from the z-axis. The components of acceleration difference are

$$g_x = \omega^2 x; \quad g_y = \omega^2 y; \quad g_z = 0$$

If m is fixed in S, its reaction force is $r_x = m\omega^2 x$, $r_y = m\omega^2 y$. This is called *centrifugal force*. The rider on a merry-go-round exerts this force on his mount.

Angular Acceleration of S in R produces an acceleration difference. Consider first the case when the axis of rotation remains in a fixed direction, and let this coincide with the z-axis. Let $\alpha = d\omega/dt$ be the rate of change of the magnitude of the angular velocity ω. The resulting acceleration difference for a mass point m at x, y, z is given by $\alpha\sqrt{x^2 + y^2}$ in a direction tangent to the circle of radius $\sqrt{x^2 + y^2}$ about the z-axis. The components of the acceleration difference are

$$g_x = \alpha y; \quad g_y = -\alpha x; \quad g_z = 0$$

If a speck of dust is to cling to a phonograph record as the turntable starts or stops, the frictional force on it must be $-m\alpha y$, $m\alpha x$.

Precessional Acceleration occurs when the frame S is rotating in R in such a way that its angular velocity vector changes direction. This change of direction may be described at a given instant as the rotation of the angular velocity vector about some axis perpendicular to itself with an angular velocity of precession Ω. Let the axes of S and R coincide at a particular instant, let the angular velocity ω of S be directed along positive z, and let the precession be along positive y. The right-handed-screw convention may be used to refer both angular velocities to their axis directions. The components of the acceleration difference for a mass point at x, y, z are then

$$g_x = 0; \quad g_y = \Omega\omega z; \quad g_z = -\Omega\omega y$$

Its magnitude is $\Omega\omega\sqrt{y^2 + z^2}$, and its direction is tangent to a circle about the x-axis. The force required to be exerted on mass m is then

$$f_y = -m\Omega\omega z; \quad f_z = m\Omega\omega y$$

If a collection of particles fixed in S forms a flywheel, there must be exerted on the flywheel a torque about the positive x-axis in order to supply these forces and produce precession. This is the well-known gyroscopic torque.

Coriolis Acceleration is the only difference acceleration which depends on the velocity v_x, v_y, v_z of a point m in a rotating frame S. It occurs when m has a component of velocity in S which is perpendicular to the axis of rotation. Coriolis acceleration is always perpendicular to the velocity of m in S and is independent of the location of m. Let the rotation of S be about the z-axis in the positive sense. The components of the Coriolis acceleration are then

$$g_x = 2\omega v_y; \quad g_y = -2\omega v_x; \quad g_z = 0$$

Coriolis acceleration appears in a body falling freely on the earth, causing it to deviate eastward from the plumbline. If the velocity of fall is v and the latitude of the place is λ, the acceleration difference is $2\omega v \cos \lambda$. As another illustration, let a train run north on a horizontal track with velocity v at the same (north) latitude λ. The train then exerts a Coriolis reaction force eastward, on the track, of magnitude $r = 2\omega v \sin \lambda$.

The General Case may involve all the above types of acceleration simultaneously. These accelerations simply add vectorially.

24. THE GYROSCOPE

A **Gyroscope** is essentially a symmetrical rotor which spins rapidly about its axis. The moment of inertia about the axis is made as large as possible within the limitations of weight and size of the instrument. Gyroscopic phenomena are only those which relate changes of direction of the spin axis to applied torque.

Precession is a term for rotation of the axis direction.

The Angular Momentum of the gyroscope about the spin axis is its basic characteristic quantity. If the moment of inertia about the axis is I_1 and the component of the angular velocity parallel to the axis is ω, the component of angular momentum parallel to the axis is

$$\rho = I_1\omega \tag{1}$$

When precession occurs, the total angular momentum vector is not parallel to the axis; also the value of ω may possibly be affected. In practice, however, the spin is very large compared with precession velocities. It is then a good approximation to consider that the total angular momentum vector is always parallel to the spin axis and has a constant magnitude ρ. This vector will be called $\boldsymbol{\rho}$.

It is necessary to adopt a convention for the sense of $\boldsymbol{\rho}$. The right-handed-screw convention will be adopted for both angular momentum and torque. Thus ρ has the direction in which a right-handed screw spinning with the rotor would advance. Likewise torque is in the direction of advance of a right-handed screw to which it is applied.

The Theory of the gyroscope rests upon the theorem of mechanics that the rate of change of the angular momentum vector equals the applied torque vector. Thus, if L is the vector torque applied to a gyroscope,

$$L = \frac{d\rho}{dt} \tag{2}$$

Since ρ is assumed to have a constant magnitude in a gyroscope problem, $d\rho/dt$ must always be perpendicular to ρ. Thus the gyroscopic torques which can be applied to a

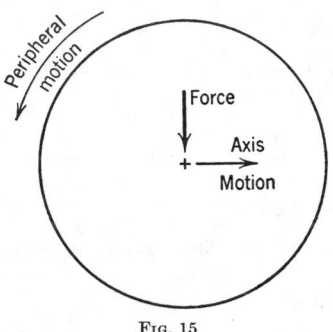

FIG. 15

gyroscope, and hence its reaction torques, are always perpendicular to the axis. If the vector ρ is drawn from a fixed origin, its tip moves always in the direction of the torque vector. Thus, suppose a gyroscope to have its ρ vector pointed towards an observer, and to be mounted so as to be free to precess about its center of mass. See Fig. 15. If the observer exerts a downward force on the end of the axis projecting towards him, the torque direction is horizontally to the right. Hence, this end of the axis precesses horizontally to the right. If the gyroscope is constrained so that horizontal precession is prevented, no downward force (except that required for ordinary angular acceleration) can be exerted on the end of the axis: the axis yields freely. The constraints are then providing torque to produce this downward precession of the end of the axis.

In steady precession about a fixed axis, the tip of the ρ vector drawn from a fixed origin describes a circle about this axis, and the ρ vector sweeps out a half cone about the axis. See Fig. 16. In the simple case where the cone degenerates into a plane, that is, when the ρ vector is perpendicular to the precession axis, the magnitude of the torque is given by

$$L = \rho\Omega = I_1\omega\Omega \tag{3}$$

where Ω is the angular velocity of precession. See Fig. 17. To maintain the precession, the torque axis must rotate with the ρ vector. In the general case of a cone of half angle θ,

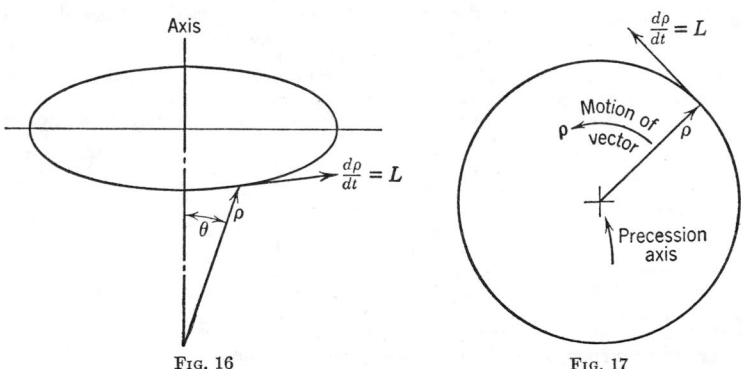

FIG. 16   FIG. 17

the torque is given, to the approximation which has been made, by

$$L = (\rho \sin \theta)\Omega \tag{4}$$

It should be noticed that the axis of precession is always perpendicular to the torque vector.

Example. A gyroscope has a rotor which weighs 8 lb, has a radius of gyration of 3 in., and spins at 3000 rpm. A torque of 2 lb ft is applied, and precession occurs about an axis perpendicular to the spin. Find the velocity of precession.

$$I_1 = mk^2 = {}^{8}\!/_{32} \text{ slugs} \times (1/4 \text{ ft})^2 = {}^{1}\!/_{64} \text{ slug ft}^2$$

$$\omega = 3000 \text{ rpm} = 100\pi \frac{\text{rad}}{\text{sec}}$$

$$\therefore \; \rho = \frac{100\pi}{64} \frac{\text{slug ft}^2}{\text{sec}} = \frac{100\pi}{64} \text{ lb ft sec}$$

$$L = 2 \text{ lb ft}$$

$$\therefore \; \Omega = \frac{2 \text{ lb ft}}{\dfrac{100\pi}{64} \text{ lb ft sec}} = 0.408 \frac{\text{rad}}{\text{sec}}$$

The "Resistance" of a gyroscope to change of its axis direction involves two different types of motion: rotation about the precession axis, and rotation about the torque axis. Suppose a torque L to be applied to a gyroscope which is free to precess. Let L be exerted always about the axis of the angle θ between the spin and precession axes. See Fig. 18. A transient rotation about the torque axis occurs when the torque is first applied. After the transient motion dies out, there is a final steady deflection produced by the torque The gyroscope behaves as if there were a strong spring opposing rotation about the torque axis. The transient motion may be analyzed by means of the general equations for rotation of a body about its mass center.

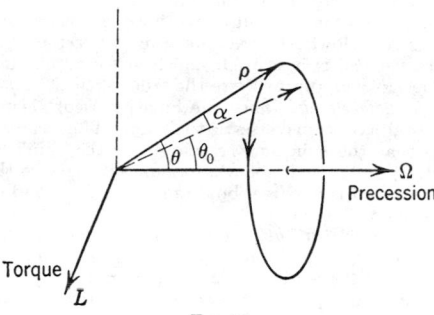

FIG. 18

The result is simple if the increment $\alpha \equiv \theta - \theta_0$ is sufficiently small, θ_0 being the value of θ for no applied torque and no precession. The equation is

$$L = I_2 \frac{d^2\alpha}{dt^2} + \left(\frac{I_1{}^2\omega^2}{I_2} \right) \alpha \tag{5}$$

Here L is torque related by the right-handed-screw rule to positive sense of α, and I_2 is) the moment of inertia of the rotor about a transverse axis. Solutions of (7) lead to simple harmonic motion when L is constant.   When the friction which has been omitted from this equation has damped out the oscillations, the resulting deflection is

$$\alpha_{\text{final}} = \frac{I_2 L}{I_1{}^2\omega^2} \tag{6}$$

Thus the gyroscope acts like a spring whose torsional stiffness is $I_1{}^2\omega^2/I_2 = \rho^2/I_2$.   The stiffness may be very large in practice.

Example. For the gyroscope of the preceding example, find the torsional stiffness, assuming $I_2 = 0.80 I_1$

$$\rho = \frac{100\pi}{64} \text{ lb ft sec} = 4.90 \text{ lb ft sec}$$

$$\rho^2 = 24 \text{ lb}^2 \text{ ft}^2 \text{ sec}^2 \qquad I_2 = \frac{0.80}{64} \text{ slug ft}^2 = \frac{1}{80} \text{ slug ft}^2 = \frac{1}{80} \text{ lb ft sec}^2$$

$$\therefore \; \frac{\rho^2}{I_2} = \frac{24}{\dfrac{1}{80}} \frac{\text{lb ft}}{\text{rad}} = 1920 \frac{\text{lb ft}}{\text{rad}}$$

Here the angle unit of the radian has had to be supplied.

The total deflection of the gyroscope about the precession axis depends on the angular impulse received about the torque axis. If the angle α of Fig. 18 is small, then in the steady state the precession velocity is

$$\Omega = \frac{\rho}{I_2 \sin \theta_0} \alpha \tag{7}$$

If the torque acts for a time Δt, and if $d\alpha/dt$ is the same at the beginning and end of Δt, the resulting deflection about the precession axis is

$$\Delta \psi = \frac{1}{\rho \sin \theta_0} \int_0^{\Delta t} L \, dt$$

$\Delta\psi$ may be made small for a given angular impulse by making ρ large. Thus the gyroscope yields but little in any direction to an applied impulse, provided it is free to precess. In time, however, a steady torque produces an indefinitely large deflection about the precession axis.

Applications of the gyroscope are numerous and important. A few of them are listed below:

1. *Maintenance of a fixed orientation in space for a brief time.* The gyroscope's axis has this property when it is mounted in very friction-free gimbals and when it is balanced to remove gravitational torque. Such a free gyroscope is used to control a torpedo or a rocket.

2. *Indication of vertical in an airplane.* This cannot be done by a free gyroscope for any length of time because of rotation of the earth and motion over the earth's surface. Also the irreducible minimum of friction in the mounting would produce precession after long operation. Small correcting torques are applied to the gyroscope to cause it to precess very slowly towards the *apparent* vertical. In this way the time average of the apparent vertical is indicated; and if accelerations are random and relatively short-lived, this average will approximate the true vertical.

3. *Ship stabilization.* A huge gyroscope is mounted so that it can precess about a horizontal axis transverse to the ship. The precession is limited to a small range of angle so that the spin axis remains near the ship's vertical. The torque resulting from this precession is then about a fore-and-aft axis, so that it can combat rolling of the ship. The precession is not free, but is motor driven, and is controlled by a small gyroscope so as to be most effective.

4. *Bicycle operation* depends on gyroscopic action. Turning is a precession about a vertical axis and requires the torque produced by unbalancing the weight towards the inside of the turn. On the other hand, torque applied to the handlebars about a roughly vertical axis produces precession about a horizontal axis, which is tipping of the bicycle. This assists the rider to control his balance.

5. *Indication of the meridian.* The gyrocompass makes use of the rotation of the earth together with the gravitational field in such a way that the gyroscope axis seeks north.

6. *Indication of rate of turn of an airplane.* The gyroscope is mounted with its axis horizontal and is forced to turn with the airplane. This precession produces a reaction torque which works against a spring and actuates an indicator.

25. GENERALIZED COORDINATES

The Configuration of a mechanical system is that characteristic which is determined by the position and orientation of all its parts. If the system is composed of particles so small relative to the whole that their rotations are irrelevant to the mechanical problem, the configuration of the system may be specified by the cartesian coordinates of all its particles in some inertial frame of reference. In problems of ordinary mechanics, the atoms of matter of which the system is composed may be treated as particles.. The concept of configuration does not involve velocities or accelerations. A mechanical system is solved when its configuration is known as a function of time. From the time-dependence of the configuration, all the kinematic properties, such as velocity and acceleration, may be deduced. Generalized coordinates and Lagrange's equations provide a systematic procedure for solving mechanical systems.

A Displacement of a mechanical system is a change of its configuration.

Constraints are conditions imposed upon a system to limit its possible displacements. For example, a particle may be constrained to remain on a given surface or on a given line. The atoms composing a rigid body may for many purposes be considered to be constrained to remain at fixed distances from one another. A flywheel may be constrained to rotate in fixed bearings. Sometimes constraints change with time, as when a bead is constrained to slide on a moving wire. Such constraints will be called time-dependent. Their motion will be supposed to be given; otherwise, the constraining body must be considered a part of the mechanical system to be solved.

Generalized Coordinates are quantities describing the configuration of a system. They describe the configuration completely if the constraints are not time-dependent; otherwise, the time is needed explicitly, together with the generalized coordinates, to specify the configuration. It is required for what follows that the set of generalized coordinates of a system be so chosen as to contain the least number of quantities necessary to describe the configuration. Thus, for example, the position of three non-collinear points of a rigid body is not a satisfactory set of generalized coordinates for the body: for it requires nine coordinates to locate these points, whereas only six coordinates are required for a rigid body. A true set of generalized coordinates is the location of one point of the body together with a set of three angles, the Euler angles, describing the orientation of the body. A set of generalized coordinates for a mechanical system will be denoted by $q_1, q_2 \cdots, q_j \cdots q_n$.

The **Number of Degrees of Freedom** of a system is equal to the number, n, of generalized coordinates required to specify its configuration, *provided* all n of the increments δq_j are independent. The criterion of independence of the increments δq_j is that any one of them can be specified arbitrarily while all the others are zero. When this is the case, the system is called *holonomic*. This discussion will be limited to holonomic systems. Non-holonomic systems are characterized by the existence of non-integrable relations between infinitesimal increments δq_j so that they are not independent.

Generalized Forces are quantities Q_j such that the work done by the actual forces of the system during an infinitesimal displacement δq_j of only one generalized coordinate is

$$\delta W_j = Q_j \delta q_j \tag{1}$$

The Q_j are ordinarily functions of the generalized coordinates, but they may also involve the time derivatives of the coordinates.

It is important to notice that any actual forces which do no work during possible displacements of the system may be omitted completely from consideration. This kind of force is often present owing to constraints. For example, a weightless, rigid rod connecting two masses may exert forces along its length, but these forces do no work since the length of the rod is constant. Forces which constrain a body to remain in contact with a surface are workless.

A generalized force does not have the dimensions of force unless its corresponding generalized coordinate has the dimensions of length. For example, the generalized force is a torque when its corresponding coordinate is an angle.

The generalized forces of a system can ordinarily be obtained directly by computing the work done by the applied forces for infinitesimal changes of the generalized coordinates. There is a formula for Q_j in terms of the cartesian components of force, X_i, Y_i, and Z_i, acting on the particles of the system whose coordinates are x_i, y_i, and z_i:

$$Q_j = \sum_{i=1}^{N} \left(X_i \frac{\partial x_i}{\partial q_j} + Y_i \frac{\partial y_i}{\partial q_j} + Z_i \frac{\partial z_i}{\partial q_j} \right) \tag{2}$$

This is summed over all N particles of the system.

In the case of moving constraints, the motion of the constraint is not considered in computing the generalized forces. The displacements δq_j of equation 1 are assumed to take place in zero time, i.e., before the constraint has time to change.

The **Kinetic Energy** of a mechanical system may be expressed in terms of the generalized coordinates and their time derivatives. The time appears explicitly only if the constraints are time-dependent. The usual notation for the time derivatives of the generalized coordinates is

$$\dot{q}_j \equiv \frac{dq_j}{dt} \tag{3}$$

The kinetic energy function is then written in general

$$T(q_1, q_2 \cdots q_n; \ \dot{q}_1, \dot{q}_2 \cdots \dot{q}_n; \ t) \tag{4}$$

If the constraints are not time-dependent, this function becomes a homogeneous quadratic form in the $\dot{q}_j$:

$$T = \sum_{j=1}^{n} \sum_{k=1}^{n} A_{jk}(q_1, q_2 \cdots q_n) \dot{q}_j \dot{q}_k \tag{5}$$

Equation 5 is quite easy to prove, starting with the kinetic-energy equation in terms of the ultimate particles of the system

$$T = 1/2 \sum_{l=1}^{N} m_l (\dot{x}_l^2 + \dot{y}_l^2 + \dot{z}_l^2) \tag{6}$$

and using the functional dependence of the cartesian coordinates upon the generalized coordinates in the case of fixed constraints:

$$x_l = x_l(q_1 \cdots q_n) \tag{7}$$

In an actual problem, the kinetic-energy function 4 or 5 is not found by working from equation 6 but rather by inspection of the system and addition of the kinetic energies of its large-scale members.

Lagrange Equations

Lagrange Equations of motion are second-order differential equations in the generalized coordinates, time being the independent variable. They have the same form for any holonomic system, and there are n such equations of this form for a given system:

$$\frac{d}{dt}\frac{\partial T}{\partial \dot{q}_j} - \frac{\partial T}{\partial q_j} = Q_j \qquad j = 1, 2, \cdots n \tag{8}$$

The solution of any holonomic problem is thus systematized. The process involves assigning generalized coordinates; finding the Q_j functions and the T function; use of

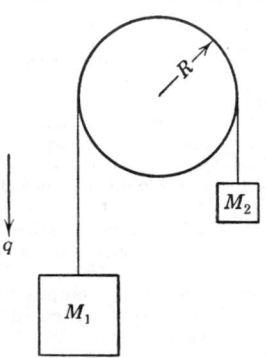

equation 8 to obtain n differential equations; simultaneous solution and integration of the differential equations; and consideration of boundary conditions.

In the use of equation 8, the meaning of the partial derivatives must be clearly understood. In the process of finding $\partial T/\partial q_j$, all other q's besides q_j are considered constants; all the $\dot{q}$'s are considered constant, including $\dot{q}_j$; and, if time appears explicitly, it too is considered constant. Similarly, to find $\partial T/\partial \dot{q}_j$, $\dot{q}_j$ is considered the only variable. It is understood, of course, that T has been expressed in the form of equation 4 to start with.

Two Examples of the use of Lagrange's equations will illustrate most of the ideas involved. Consider first the Atwood's machine shown in Fig. 19. This consists of a frictionless pulley of groove radius R and moment of inertia I, over which is passed a light, inextensible cord tied to two masses m_1 and m_2. The problem is simplest when the masses are constrained by frictionless guides to move only vertically. There is then only one degree of freedom. Let

Fig. 19

the generalized coordinate q be the downward vertical coordinate of m_1. The work done by gravity for a displacement δq is then

$$\delta W = (m_1 g - m_2 g)\,\delta q$$

whence the generalized force is

$$Q = (m_1 - m_2)g$$

The kinetic energy is the sum of the kinetic energies of the pulley and the weights:

$$T = \frac{1}{2} m_1 \dot{q}^2 + \frac{1}{2} m_2 \dot{q}^2 + \frac{1}{2} I \left(\frac{\dot{q}}{R}\right)^2$$

The Lagrange equation is therefore

$$\frac{d}{dt}\left[\left(m_1 + m_2 + \frac{I}{R^2}\right)\dot{q}\right] = (m_1 - m_2)g$$

$$\therefore \frac{d^2 q}{dt^2} = \frac{(m_1 - m_2)g}{m_1 + m_2 + I/R^2}$$

As a second example, consider the problem of a point mass m constrained to move in a plane under a central force $F = kr$ towards the origin, where r is the distance from the origin to m and k is constant. Let us choose polar coordinates r, θ as the generalized coordinates, as in Fig. 20,

$$q_1 = r$$
$$q_2 = \theta$$

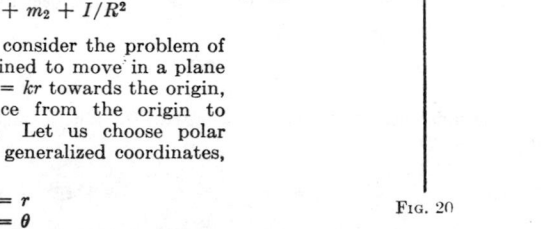

Fig. 20

For displacement δq_1, the force F does work

$$W = -F\,\delta r = -kr\,\delta r = -kq_1\,\delta q_1$$

Thus the generalized force Q_1 is $-kq_1$. No work is done during a rotation $\delta\theta$, so $Q_2 = 0$. The kinetic energy $1/2\ mv^2$ is readily expressed in polar coordinates:

$$T = \frac{1}{2}\ m(\dot{r}^2 + r^2\dot{\theta}^2) = \frac{1}{2}\ m(\dot{q}_1{}^2 + q_1{}^2\dot{q}_2{}^2)$$

The two Lagrange equations of the motion are therefore

$$\frac{d}{dt}(m\dot{q}_1) - mq_1\dot{q}_2{}^2 = -kq_1$$

and

$$\frac{d}{dt}(mq_1{}^2\dot{q}_2) = 0$$

Potential Energy may be used in the case of conservative systems. In such systems, there is a potential energy function V such that, during any displacement of the system, the work done by the force is

$$\delta W = -\delta V \tag{9}$$

Since potential energy is energy of position or configuration, it must be expressible as a function $V(q_1, q_2, \cdots q_n)$ of the generalized coordinates. Then we may write for an infinitesimal displacement

$$\delta V = -\delta W = \left(\frac{\partial V}{\partial q_1}\delta q_1 + \frac{\partial V}{\partial q_2}\delta q_2 + \cdots + \frac{\partial V}{\partial q_n}\delta q_n\right)$$

It follows immediately from definition 1 that the generalized forces are

$$Q_1 = -\frac{\partial V}{\partial q_1} \text{ etc.}$$

or in general

$$Q_j = -\frac{\partial V}{\partial q_j} \tag{10}$$

If the Lagrangian function

$$L(q_1 \cdots q_n; \dot{q}_1 \cdots \dot{q}_n; t) \equiv T(q_1 \cdots q_n; \dot{q}_1 \cdots \dot{q}_n; t) - V(q_1 \cdots q_n) \tag{11}$$

is formed, it may be seen immediately that for a conservative system the Lagrange equation 8 may be written in the compact form

$$\frac{d}{dt}\frac{\partial L}{\partial \dot{q}_j} - \frac{\partial L}{\partial \dot{q}_j} = 0 \qquad j = 1, 2, \cdots n \tag{12}$$

FRICTION
Revised by Merhyle F. Spotts

26. STATIC AND KINETIC FRICTION

A Smooth Surface is one which offers no resistance to the sliding of a body upon it. **A rough surface** does offer resistance to such motion. **The total reaction** (R) (Fig. 1) of the surface of one body upon another body is its resultant force. *Friction* (F) is that component of the total reaction (R) which is tangent to the surface. *Normal reaction* (N) is that component which is normal to the surface.

Static Frictional Force (F) is that friction which opposes motion when there is no slipping. Its value varies as the need for it to prevent motion is developed. *Limiting frictional force* (F') is the value of static frictional force when slipping impends. *Coefficient of static friction* (f) is the ratio F'/N. *Angle of static friction* (ϕ) is defined by $\tan \phi = F'/N = f$. *Angle of repose* is that angle which the surface of one body makes with the horizontal when slipping of another body upon it impends. It applies to the particular rubbing surfaces in contact. It equals the angle of static friction.

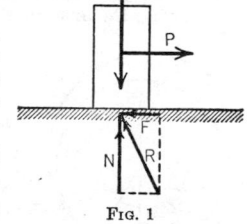

Fig. 1

Kinetic Frictional Force (F_k) opposes motion when one body is slipping on the surface of the other. Its value is usually less than that of the limiting friction. *Coefficient of kinetic friction* (f_k) is the ratio F_k/N. *Angle of kinetic friction* (ϕ_k) is defined by $\tan \phi_k = F_k/N = f_k$.

Laws of Friction for Dry Surfaces:
1. Friction between two given bodies is directly proportional to the pressure; the coefficient of friction is constant for all pressures.

2. The coefficient and amount of friction for given pressures are independent of the area of contact.
3. The coefficient of friction is independent of the relative velocity, although static friction is greater than kinetic friction.

The preceding laws are only approximately true. The coefficient of friction is slightly greater for small pressures upon large areas than for great pressures upon small areas. The coefficient of friction decreases as the speed increases.*

Coefficients of Static and Kinetic Friction are affected by the above laws of friction and also by the characters of the surfaces, the kinds of material, and the nature of any lubricant used. Rough averages for a number of materials and conditions are given in Table 1.†

Table 1. Coefficients of Static and Kinetic Friction

Materials	Con-dition	Sliding Friction ϕ	Sliding Friction f	Static Friction ϕ	Static Friction f
Cast iron on cast iron or bronze......	Wet	$17\,1/4°$	0.31	....	......
Cast iron on cast iron or bronze......	Greaséd	$41/2°- 53/4°$	0.08–0.10	$9°$	0.16
Cast iron on oak (fibers parallel).....	Dry	$163/4°-261/2°$	0.30–0.50	....	......
Cast iron on oak (fibers parallel).....	Wet.	$121/2°$	0.22	$33°$	0.65
Cast iron on oak (fibers parallel).....	Greased	$103/4°$	0.19	....	......
Earth on earth....................		....	......	$14°-45°$	0.25–1.0
Earth on earth (clay)..............	Damp	....	......	$45°$	1.0
Earth on earth (clay)..............	Wet	....	......	$171/4°$	0.31
Hemp-rope on rough wood..........	Dry	$261/2°$	0.50	$261/2°-383/4°$	0.50–0.80
Hemp-rope on polished wood........	Dry	....	......	$181/4°$	0.33
Leather on oak....................	Dry	$163/4°-261/2°$	0.30–0.50	$261/2°-31°$	0.50–0.60
Leather on cast iron...............	Dry	$291/4°$	0.56	$163/4°-261/2°$	0.30–0.50
Oak on oak (fibers parallel)........	Dry	$253/4°$	0.48	$313/4°$	0.62
Oak on oak (fibers crossed).........	Dry	$183/4°$	0.34	$281/4°$	0.54
Oak on oak (fibers crossed).........	Wet	$14°$	0.25	$351/4°$	0.71
Oak on oak (fibers perpendicular)....	Dry	$103/4°$	0.19	$231/4°$	0.43
Steel on ice......................	Dry	....	0.01	$11/2°$	0.027
Steel on steel....................	Dry	Vel. 10 ft per sec	0.09	$81/2°$	0.15
		Vel. 100 ft per sec	0.03		
Stone Masonry on concrete.........	Dry	....	......	$371/4°$	0.76
Stone masonry on undisturbed ground	Dry	....	......	$33°$	0.65
Stone masonry on undisturbed ground	Wet	....	......	$163/4°$	0.30
Wrought iron on wrought iron.......	Dry	$233/4°$	0.44	....	......
Wrought iron on wrought iron.......	Greased	$41/2°-53/4°$	0.08–0.10	$61/2°$	0.11
Wrought iron on cast iron or bronze..	Dry	$101/4°$	0.18	$103/4°$	0.19
Wrought iron on cast iron or bronze..	Greased	....	......	$4°-41/2°$	0.07–0.08

27. AXLE FRICTION AND LUBRICATED SURFACES

Axle Friction, Non-lubricated

Axle friction is the friction that opposes the turning of an axle in its bearing. For a dry bearing, the axle, in Fig. 2, will obviously move to the right. It will climb until a

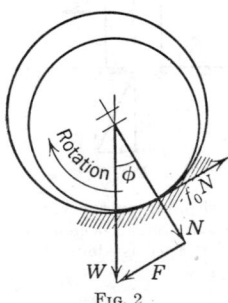

point is reached, at some angle ϕ, where the friction force $f_0 N$ on the shaft or journal is balanced by the tangential component F of the vertical load W, that is $f_0 N = F$. Normal force N is equal to $W \cos \phi$, and f_0 is the coefficient of friction for the surfaces in question. Friction force F is equal to $W \sin \phi$, so that the usual relationship, $f_0 = \tan \phi$, for friction prevails. It is customary, however, to define the coefficient of friction f for a journal bearing as the ratio of load W to the tangential friction force F or $f = F/W = \sin \phi$. The relationship between f_0 and f is then represented by $f = f_0 \cos \phi$.

Friction of Lubricated Surfaces. The friction of lubricated surfaces is characterized by two types of sliding motion:

1. Hydrodynamic or thick-film lubrication.
2. Boundary or thin-film lubrication.

FIG. 2

Hydrodynamic lubrication occurs with a plentiful supply of

---

* Recent experiments have proved also that time of contact affects the coefficient of static friction.
† Hudson's Manual, p. 102.

oil when the thickness of the film is large as compared with the height of the irregularities of the surfaces. Metallic contact or wear does not occur, and the frictional resistance is independent of the kinds of materials composing the bodies.

Although somewhat involved mathematically, the laws for this type of friction are well understood. Figure 3 shows a weightless plate of area A being pushed along at velocity U by force F as it rests on an oil film of thickness h. Newton observed that force F was directly proportional to area A and velocity U and inversely proportional to the film thickness h. This can be written in the form of an equation, $F = \mu U A/h$, where the constant of proportionality μ is called the coefficient of viscosity of the lubricant. It has dimensions of lb sec/in.^{2} or dyne sec/cm^2.

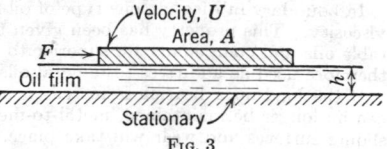

Fig. 3

Example 1. Find the force necessary to maintain a velocity of 20 ft/sec for a 3 in. $\times$ 3 in. plate if the film has a thickness of 0.003 in. A heavy oil of viscosity 0.000005 lb sec/in.^{2} is used.
Solution. Substitution in Newton's equation gives $F = 0.000005 \times 240 \times 9/0.003 = 3.6$ lb.

Journal Friction, Lubricated

Newton's equation can be adapted to the journal bearing of Fig. 4 of radius r and axial length l by substituting $A = 2\pi rl$ and $U = 2\pi rn_s$, where n_s is the revolutions per second. Film thickness h is replaced by radial clearance c. The result,

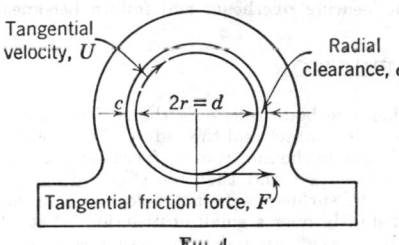

Fig. 4

$$F = 4\pi^2 \mu r^2 h_s \frac{l}{c}$$

is known as *Petroff's equation*. It is valid only for very lightly loaded bearings where the shaft is centrally located in the clearance space. The friction is practically unchanged when the bearing is loaded, and an equation for the coefficient of friction, $f = F/W$, is obtained by dividing both sides of Petroff's equation by the load W.

The result, $f = 2\pi^2 \mu h_s r/pc$, is obtained by the substitution $p = W/2rl$, where p is the load per square inch of projected area of the bearing.

Example 2. Find the coefficient of friction for a 2 in. diameter $\times$ 3 in. long journal bearing carrying a load of 600 lb at a speed of 1200 rpm. Radial clearance is 0.002 in. Viscosity of the lubricant is 0.000003 lb sec/in.^{2}
Solution. Substitution gives

$$f = \frac{2\pi^2\, 0.000003 \times 20 \times 1}{100 \times 0.002} = 0.0059$$

This example shows that the coefficient of friction, under conditions of flooded lubrication, is very much smaller than the value of the coefficient when the surfaces are dry. It should be noted that the coefficient of friction for lubricated surfaces varies directly with the velocity and inversely with the pressure. This is in contrast with the coefficient of friction for dry surfaces, which is independent of both velocity and pressure.

The plate of Fig. 3 could carry a downward load provided that the forward edge would be tipped at a small upward angle. The oil would be drawn into the wedge-shaped opening, and a hydrostatic pressure would be developed in the film sufficient to support the load. When a lubricated journal bearing carries a load, the shaft center will shift sidewise in the direction opposite to that for a dry bearing. The oil film will converge in such a manner as to produce the pressure necessary to carry the load. The mathematical treatment of these phenomena may be found in almost any treatise on lubrication.

Boundary Friction

Boundary friction occurs between surfaces that are slightly oily or greasy but where the oil supply is insufficient to maintain a hydrodynamic film. Although the bodies approach each other closely, the load is carried by the absorbed layers of lubricant attached to the surfaces. The chemical and physical properties of the lubricant and of the materials composing the bodies are now of importance. At the present time, available knowledge for this type of friction is highly specific and cannot be formulated into general laws. It would appear, however, that the friction is relatively independent of the area in contact and the velocity of sliding. The value for the coefficient of friction is usually intermediate between that for dry friction and that for hydrodynamic friction.

In boundary lubrication one type of oil may have less friction than another of the same viscosity. This property has been given the name of *oiliness*. Certain animal and vegetable oils and compounds are superior to mineral oils with respect to oiliness. They are therefore used as additives to mineral oils to decrease the friction.

If the load on a bearing is increased until the absorbed boundary layers of lubricant can no longer be maintained, metal-to-metal contact will occur at the high points of the sliding surfaces and wear will take place. The exact physical and metallurgical process of wear is not clearly understood. The harder metal may gouge off particles of the softer metal. Experimental evidence exists that the surfaces weld together at the points of metallic contact because of the extremely high localized pressures. The welds are broken as soon as made and may account for the major portion of the resistance to the motion. The amount of welding decreases with decrease of solid solubility between the metals of the two bodies. There is thus justification for the long-established rule that the two members of a bearing should be made of dissimilar metals to decrease friction and the possibility of galling and failure. The *running-in* of a bearing is a form of wear process in which an improvement is effected in the surface conditions.

In addition to a scanty oil supply, a bearing operating with boundary friction will be adversely affected by a decrease in the viscosity of the lubricant, a decrease in the speed, or an increase in the pressure. An unfavorable combination of these qualities may cause the coefficient of friction to increase until the bearing overheats and failure becomes imminent.

28. ROLLING FRICTION

Rolling Friction is that friction developed when one body rolls over the surface of another, and depends on the hardness of the surfaces in contact and the radius of the rolling surface. The theory is based on the idea that surfaces are slightly deformed at the place of contact and that the effect of rolling friction is the same as if the surfaces were not deformed and the rolling body passed constantly over a small obstruction. Let P (Fig. 5) be the horizontal force required to overcome the small obstruction B. Then $hP = aW$, and, since h is nearly equal to r, $P = aW/r$ (approximately). *Coefficient of rolling friction* is a, and, as an analogy with definitions of static and kinetic friction coefficients, might be defined as the ratio T_r/W, where $T_r = Pr$ is the torque resisting rolling motion and W is the normal force (in this case, the weight of the body). The coefficient is a linear distance and is usually given in inches. The values of the coefficient of rolling friction given by various investigators are not in close agreement and should be used with caution.

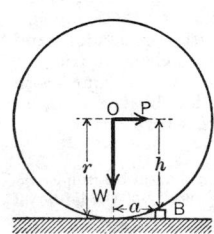

Fig. 5

Coefficients of Rolling Friction. The following are some reported values of coefficients of rolling friction:

Lignum vitae roller on oak track................. 0.019 in.
Elm roller on oak track......................... 0.032 in.
Cast-iron wheel (20-in. diam.) on cast-iron rail..... 0.018–0.019 in.
Railroad wheels (39.4-in. diam.)................. 0.020–0.022 in.
Iron or steel wheels on wood track.............. 0.06–0.10 in.

29. PIVOT FRICTION

Pivot Friction is that friction which opposes the turning of the end of a vertical, or inclined, shaft in its bearing. Some examples of pivot friction, with friction torque and power formulas applying, are shown in the following table.

Table 2. Pivot Friction *

f = coefficient of friction. W = load in pounds.
T = torque of friction about the axis of the shaft.
r = radius in inches. n = revolutions per second.

Type of Pivot	Torque T in Pound-inches	Power P Lost by Friction in Foot-pounds per Second
Shafts and Journals (180° bearing)	$T = fWr$	$P = \dfrac{2\pi n}{12} fWr.$
Flat Pivot	$T = {}^2/_3\, fWr.$	$P = \dfrac{4\pi n}{3 \times 12} fWr$
Collar-bearing	$T = {}^2/_3\, fW\, \dfrac{R^3 - r^3}{R^2 - r^2}$	$P = \dfrac{4\pi n}{3 \times 12} fW\dfrac{R^3 - r^3}{R^2 - r^2}$
Conical Pivot	$T = {}^2/_3\, fW\, \dfrac{r}{\sin \alpha}$	$P = \dfrac{4\pi n\, fWr}{3 \times 12 \sin \alpha}$
Truncated-cone Pivot	$T = {}^2/_3\, fW\, \dfrac{R^3 - r^3}{(R^2 - r^2)\sin \alpha}$	$P = \dfrac{4\pi n\, fW(R^3 - r^3)}{3 \times 12(R^2 - r^2)\sin \alpha}$

* Hudson's Manual, p. 105.

30. BELT FRICTION

Belt or Coil Friction is that friction which opposes the slipping of a belt, rope, brake band, or similar article coiled about a pulley, sheave, post, capstan, or similar device. When power is being transmitted, say by a belt driving a pulley, the tension T_1 on the driving side of the belt is greater than the tension T_2 on the driven side. Neglecting the effect of centrifugal force, which is small at low speeds, the tensions are related by the formula

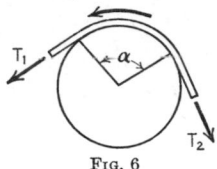

Fig. 6

$T_1/T_2 = e^{f\alpha}$ where $e = 2.718+$ (i.e., base of natural logarithms), f = coefficient of friction between belt and pulley, and α = angle of contact between belt and pulley (Fig. 6). Values of T_1/T_2 for various values of f and α are shown in Table 3. Power transmitted

Table 3.   Maximum Ratio T_1/T_2 (Slipping Impending)

α Radians	Values of f (Coefficient of Friction)								
2π	0.10	0.15	0.20	0.25	0.30	0.35	0.40	0.45	0.50
0.1	1.06	1.1	1.13	1.17	1.21	1.25	1.29	1.33	1.37
0.2	1.13	1.21	1.29	1.37	1.46	1.55	1.65	1.76	1.87
0.3	1.21	1.32	1.45	1.60	1.76	1.93	2.13	2.34	2.57
0.4	1.29	1.46	1.65	1.87	2.12	2.41	2.73	3.10	3.51
0.425	1.31	1.49	1.70	1.95	2.23	2.55	2.91	3.33	3.80
0.45	1.33	1.53	1.76	2.03	2.34	2.69	3.10	3.57	4.11
0.475	1.35	1.56	1.82	2.11	2.45	2.84	3.30	3.83	4.45
0.5	1.37	1.60	1.87	2.19	2.57	3.00	3.51	4.11	4.81
0.525	1.39	1.64	1.93	2.28	2.69	3.17	3.74	4.41	5.20
0.55	1.41	1.68	2.00	2.37	2.82	3.35	3.98	4.74	5.63
0.6	1.46	1.76	2.13	2.57	3.10	3.74	4.52	5.45	6.59
0.7	1.52	1.93	2.41	3.00	3.74	4.66	5.81	7.24	9.02
0.8	1.65	2.13	2.73	3.51	4.52	5.81	7.47	9.60	12.35
0.9	1.76	2.34	3.10	4.11	5.45	7.24	9.60	12.74	16.90
1.0	1.87	2.57	3.51	4.81	6.59	9.02	12.35	16.90	23.14
1.5	2.57	4.11	6.59	10.55	16.90	27.08	43.38	69.49	111.32
2.0	3.51	6.59	12.35	23.14	43.38	81.31	152.40	285.68	535.49
2.5	4.81	10.55	23.14	50.75	111.32	244.15	535.49	1,174.5	2,575.9
3.0	6.59	16.90	43.38	111.32	285.68	733.14	1,881.5	4,828.5	12,391.
3.5	9.02	27.08	81.31	244.15	733.14	2199.9	6,610.7	19,851.	59,608.
4.0	12.35	43.38	152.40	535.49	1881.5	6610.7	23,227.	81,610.	286,744.

is given by $P = (T_1 - T_2)v$, where P is power in foot-pounds per second and v is velocity in feet per second.

Table 4.   Coefficients of Friction for Belts and Pulley Materials *

Belt Material	Pulley Material					
	Iron-Steel	Wood	Paper	Wet Iron	Greasy Iron	Oily Iron
Oak-tanned leather.........	0.25	0.30	0.35	0.20	0.15	0.12
Mineral-tanned leather......	0.40	0.45	0.50	0.35	0.25	0.20
Canvas stitched............	0.20	0.23	0.25	0.15	0.12	0.10
Balata.....................	0.32	0.35	0.40	0.20	....	....
Cotton woven..............	0.22	0.25	0.28	0.15	0.12	0.10
Camel-hair.................	0.35	0.40	0.45	0.25	0.20	0.15
Rubber-friction............	0.30	0.32	0.35	0.18	....	....
Rubber-covered............	0.32	0.35	0.38	0.15	....	....
Rubber on fabric..........	0.35	0.38	0.40	0.20	....	....

* *Machinery,* Vol. 37, 1931, p. 560-A.

BIBLIOGRAPHY

BEER, F. P., and JOHNSON, E. R., *Vector Mechanics for Engineers*, New York, McGraw-Hill, 1964.
DEN HARTOG, J. P., *Mechanics*, New York, Dover, 1961.
FLUGGE, W., *Handbook of Engineering Mechanics*, New York, McGraw-Hill, 1960.
HALLIDAY, D., and RESNICK, R., *Physics*, Part 1, New York, Wiley, 1966.
LEECH, J. W., *Classical Mechanics*, London, Methuen, 1958.
MAXWELL, JAMES CLERK, *Matter and Motion*, New York, Dover, 19—.
RABINOWICZ, E., *Friction and Wear of Materials*, New York, Wiley, 1965.
SEELY, F. B., ENSIGN, N. E., and JONES, P. G., *Analytical Mechanics for Engineers*, New York, Wiley, 5th ed., 1958.
TIMOSHENKO, S. P., and YOUNG, D. H., *Engineering Mechanics*, New York, McGraw-Hill, 4th ed., 1956.

BIBLIOGRAPHY

SECTION 5

MECHANICS OF DEFORMABLE BODIES

REVISED BY

THE EDITOR

IN COLLABORATION WITH JOHN M. LESSELLS, G. S. CHERNIAK,
AND JOSEPH E. LOVE, JR.

FROM THE ORIGINAL SECTION ON MECHANICS OF MATERIALS BY

JASPER O. DRAFFIN AND THEODORE CRANE

SIMPLE STATIC STRESSES

1. TENSION, COMPRESSION, SHEAR

Stress or Total Stress is the resultant internal force that resists change in the size or shape of a body acted on by external forces. A change in size or shape begins when the load is applied and stops when the internal resisting stress holds the external forces in equilibrium. If the external forces acting on the body increase to the extent that the maximum stress that can be developed is unable to balance the external forces, the change in form will increase rapidly and the body will break or rupture. Stresses are measured in the same units as the forces which produce them, i.e., pounds, tons, etc.

Unit Stress or intensity of stress is defined as the stress per unit of area. Its value at any point of a section is the stress on an elementary part of the area, including the point, divided by the elementary area, and generally varies from point to point of the section. When the stress (P) is uniformly distributed over an area (A), the unit stress (σ) at any point of the area is the stress on the area divided by the whole area; i.e., $\sigma = P/A$. Unit stresses are expressed in pounds per square inch, tons per square foot, kilograms per square centimeter, and the like.

Both total stress and unit stress are often referred to simply as "stress." Confusion will be avoided if the units used are stated.

Tensile Stress or Tension is the internal force that resists the action of external forces tending to increase the length of a body. Tension is developed in a bar when the external forces act on it in the directions away from its ends. See Fig. 1. The tendency is to separate the bar into two parts, A and B. To maintain equilibrium each part is acted on at section mn by tensile stresses, σ, whose resultant is equal and opposite in direction to the resultant of the external forces acting at the end of the part considered. If the forces acting on one end of the bar total 1000 lb then the stress on the section mn equals 1000 lb.

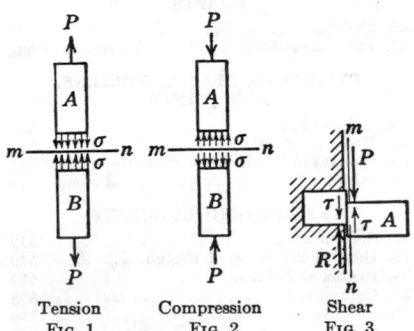

Tension | Compression | Shear
FIG. 1 | FIG. 2 | FIG. 3

Compressive Stress or Compression is the internal force that resists the action of external forces tending to decrease the length of a body. Compression is developed in a bar when external forces act on it in the directions toward its ends. In Fig. 2, the tendency of the external forces is to shorten the bar by pushing any two parts, as A and B, closer together. As long as equilibrium is maintained, the resultant of the compressive stresses acting on either part at section mn is equal and opposite in direction to the resultant of the external forces acting at the end of the part considered.

Shearing Stress or Shear is the internal force acting along a plane between adjacent parts of a body when two equal forces parallel to the plane considered act on each part in opposite directions. The shear resists the tendency of one part to slide over the other part. In Fig. 3, the projecting part of the cantilever beam, A, is acted on by external vertical forces due to the weight of A and any loads that it carries. The resultant of these forces, P, is equal, parallel to, and opposite in general direction to the upward pressure or reaction, R, which acts on the part of the beam embedded in the wall. There exists, therefore, a tendency of the parts to assume the relative positions shown in Fig. 3. An internal force acting along the cross section mn resists the tendency of A to slide vertically downward. This internal force is the shear acting on the cross section. As long as equilibrium is maintained, if P equals 1000 lb, R also equals 1000 lb, and the shear acting on the section equals 1000 lb.

Simple Stress. Tension, compression, and shear are considered singly and in combination in engineering practice. When it is necessary to consider only one of these singly, the case is known as one of simple stress.

488

Normal Stress. A normal stress on a section is one that acts in a direction perpendicular to the section considered. See Fig. 4.

Axial Stress. Axial tension or compression exists in a straight homogeneous bar when the resultant of the applied loads coincides with the axis of the bar. The stress then is distributed uniformly over any section normal to the axis of the bar.

Axial tension is encountered in a vertical bar supported only at its upper end; the tension caused by the weight of the bar and any load carried at its lower end is uniformly distributed over the cross section of the bar. If the resultant of the applied load does not coincide with the axis of the bar, the stress is not uniform over the cross section of the bar. For a bar held in an inclined position, even though subjected to axial force applied at its ends, the weight of the bar itself is an external force acting on it at an angle to its axis. Therefore, this is not simple axial tension, and the stress is not uniform over a cross section of the bar.

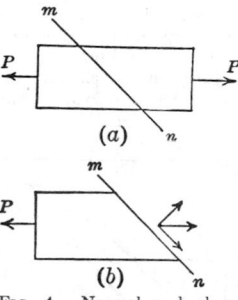

FIG. 4. Normal and shear stress components of resultant stress on section mn

Axial compression exists in a pier or prism of length not exceeding 6 to 8 times the least side or diameter, when the resultant of the load supported acts through the centers of gravity of the end sections. When the length is greater than 6 or 8 times the least side or diameter, the pier is regarded as a column in which some bending has taken place. Then the resultant of the load, though acting through the centers of gravity of the end sections, does not coincide with the axis of the column; the stresses therefore are not uniformly distributed over a cross section and the case is not one of simple axial compression.

2. DEFORMATION UNDER STRESS

Deformation is the amount of the change in the shape of a body caused by the application of external forces. When the external forces cause tension the deformation is the amount that the body is increased in length; when they cause compression, it is the amount that the body is decreased in length; when they cause shear, it is the amount that one part of the body slips over the adjacent part. The deformations accompanying tension, compression, and shear are known, respectively, as elongation, shortening, and detrusion. Deformations are measured by the same units of length used in measuring the linear dimensions of the body.

Unit Deformation, or deformation per unit of length, is determined by dividing the total amount of deformation by the original length of the body before the load causing the deformation was applied. If a steel bar 8 in. long is stretched until it becomes 8.16 in. long the total deformation, in this case elongation, is 0.16 in.; the unit deformation, ϵ, is 0.16 ÷ 8 = 0.02 in. per in. of original length.

The ultimate unit deformation is the unit deformation measured after the body has ruptured; it is usually expressed as a percentage of original length. Thus, if two marks on a bar of hard steel are 2 in. apart before a test and 2.30 in. apart after rupture, the ultimate unit deformation, ϵ, is the difference between these distances divided by the original distance between the two marks: or $\epsilon = 0.30 \div 2.0 = 0.15$ in. per in. $= 15\%$.

Hooke's Law. It has been found by experiment that a body acted on by external forces will deform in proportion to the stress developed as long as the unit stress does not exceed a certain value, which varies for the different materials. This value is the *proportional limit.*

Stress and Strain. The word strain has been frequently used for the internal force in a body, or stress. It is present practice, however, to use the word strain as a synonym for deformation only. Thus the expression "stress and strain" is equivalent to the expression "stress and deformation." Owing to the conflicting meanings of the word strain it has been suggested that it be avoided, the word deformation being used in its place. Physicists usually define strain as the deformation per unit length.

The Proportional Limit is that unit stress at which the unit deformation begins to increase at a faster rate than does the unit stress, or it is the highest unit stress at which the stress is proportional to the deformation. It is determined by noting, on a stress-deformation diagram, the unit stress at which the curve departs from a straight line.

The Elastic Limit is the maximum unit stress to which a material may be subjected and still be able to return to its original form upon the removal of the stress. When stressed beyond the elastic limit a body will return to its original form only partially and thereby acquires a permanent deformation or "set." The determination of the elastic limit logically involves the application and release of a series of increasing loads on the specimen until a set is observed after the release of a load. This procedure is very slow; and, since for many metals experience does not indicate any significant difference between the elastic limit so determined and the proportional limit, the proportional limit is often accepted as equivalent to the elastic limit, and is frequently called the *proportional-elastic limit.* It should be remembered, however, that there is no fundamental relationship between the elastic limit and the proportional limit. In many of the older tests, particularly with timber, the proportional limit was incorrectly reported as the elastic limit.

Johnson's Apparent Elastic Limit. For some materials it is difficult to determine precisely the proportional limit since the point of departure of the stress-deformation curve from a straight line is not well defined. In view of this, J. B. Johnson proposed what

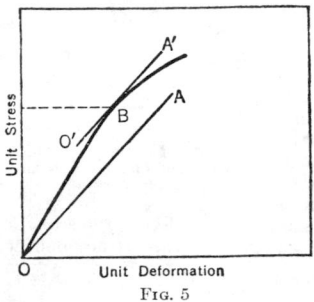

FIG. 5

he called the "apparent elastic limit" as the point on the stress-deformation diagram at which the rate of deformation is 50% greater than at the origin. It is found by drawing a line, OA, Fig. 5, having a slope with respect to the vertical axis 50% greater than the straight-line part of the curve; the unit stress at which this line is tangent to the curve, point B on the line $O'A'$, Fig. 5, is the apparent elastic limit.

The Yield Point of a material is the unit stress at which the deformation first increases markedly without any increase in the applied load. The yield point is always above the proportional limit and true elastic limit. Ductile metals have a well-defined yield point which is practically the same value as the proportional limit. For this reason, and because of the method of commercial testing, the yield point of ductile metals is sometimes reported as the elastic limit, or as the *commercial elastic limit.* The A.S.T.M. Standards, 1949, E6-36, specify that the term yield point shall not be used in connection with material whose stress-deformation curve in the region of yield is a smooth curve of gradual curvature. Only a few materials exhibit a true yield point.

Yield Strength is defined as the unit stress at which a material exhibits a specified limiting permanent set. It is a measure of the useful limit of materials, particularly of those whose stress-deformation curve in the region of yield is a smooth curve of gradual curvature.

The Ultimate Strength, tensile or compressive strength, of a material is the highest unit stress it can sustain before rupturing.

The Point of Rupture, or breaking strength, is the unit stress at which the material tested breaks or ruptures. It is observed in tests on steel to be slightly less than the ultimate strength or maximum stress sustained before rupture because of the large reduction in area before rupture.

Stress-deformation Diagram. The relation between the unit stress and unit deformation for any material tested is shown conveniently by a stress-deformation diagram, in which ordinates to a curve represent unit stresses and abscissas the resulting unit deformations for all values of unit stress to the point of rupture. In a typical tension test on a bar of medium steel, the curve, Fig. 6, is a straight line up to a unit stress of about 35,000 psi, showing a constant ratio of unit stress to unit deformation. The proportional limit, point a, is the unit stress at which the curve deviates from the initial straight line.

At a unit stress slightly higher than the proportional limit, point b, the curve becomes approximately horizontal, showing a rapid increase in deformation without any further increase in the stress; this unit stress is the yield point. For unit stresses beyond the yield point, the curve shows that the rate of deformation increases rapidly. The end of the curve, beyond point c, marks the point of rupture which is shown to be a little lower than the maximum unit stress (ultimate strength) sustained during the test.

The form of the curve obtained from a test will vary according to the material tested, and will be different for compression than for tension. For some materials, like cast iron, concrete, and, frequently, timber, no part of the curve is a straight line.

Modulus of Elasticity in tension or compression is the constant which expresses the ratio of unit stress to unit deformation for all values of unit stress not exceeding the proportional limit of a material. The terms *coefficient of elasticity* and *Young's modulus* are sometimes used to express this ratio.

The deformation caused by any unit stress not exceeding the proportional limit may be computed if the modulus of elasticity is known. Consider a bar of length l and cross-sectional area A acted on by an axial load P which produces a total deformation e. The

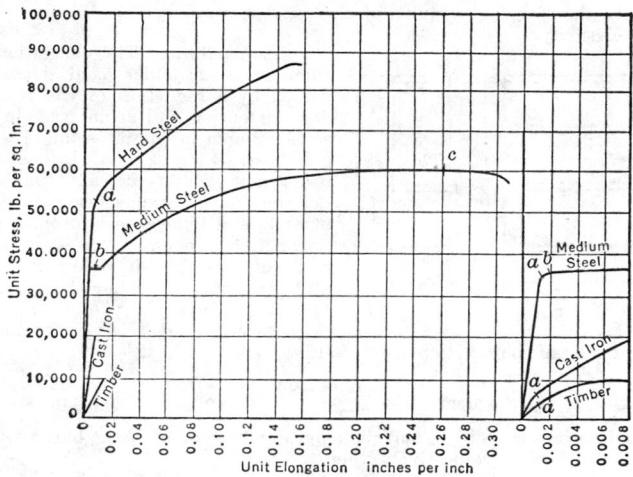

FIG. 6. Stress-deformation diagram

Unit stress is P/A, and the unit deformation is e/l. Let E represent the modulus of elasticity, σ the unit stress, and ϵ the unit deformation; then

$$E = \frac{\text{Unit stress}}{\text{Unit deformation}} = \frac{\sigma}{\epsilon} = \frac{P/A}{e/l} = \frac{Pl}{Ae} \tag{1}$$

or

$$e = \frac{Pl}{AE} = \sigma \frac{l}{E} \tag{2}$$

Since l and e are linear dimensions, ϵ is an abstract number, and E is expressed in the same units as σ, such as pounds per square inch, tons per square foot, or kilograms per square centimeter.

This formula can be used only for unit stresses not greater than the proportional limit because for higher unit stresses the ratio σ/ϵ is not constant. The modulus of elasticity, E, is a measure of the stiffness of a body or its ability to resist deformation within the proportional limit of the material. The greater the modulus of elasticity the less will be the deformation for any unit stress not exceeding the proportional limit.

For any given material, the modulus of elasticity is often the same in tension or compression. Average values for steel and wrought iron are, respectively, 30,000,000 and 25,000,000 psi. Values for other materials are given in Sections 1, 12, and 13. Axial stresses are always implied when the unqualified term "modulus of elasticity" is used.

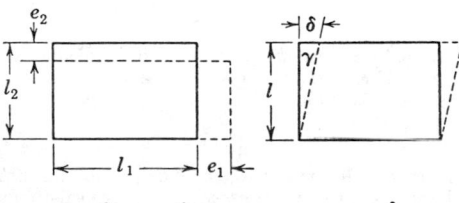

$$\epsilon_1 = \frac{e_1}{l_1}, \epsilon_2 = \frac{e_2}{l_2} \qquad\qquad \gamma = \frac{\delta}{l}$$

(a) Normal strains   (b) Shearing strains

FIG. 7. Strain or unit deformation

Modulus of Elasticity in Shear. The expression for the modulus of elasticity in shear is similar to that for tension or compression; but in shear the deformation is in a direction lying in the plane of shear of the body. The modulus of elasticity in shear is sometimes called the *modulus of rigidity*.

Let G be the shearing modulus of elasticity, τ the unit shearing stress, δ the total lateral deformation or detrusion due to the shear, and γ the unit detrusion; then

$$G = \frac{\text{Unit shearing stress}}{\text{Unit detrusion}} = \frac{\tau}{\gamma} = \frac{P/A}{\delta/l} = \frac{Pl}{\delta A} \tag{3}$$

or

$$\delta = \frac{Pl}{AG} = \frac{\tau l}{G} \tag{4}$$

For metals, the shearing modulus of elasticity is approximately 0.4 times the tensile modulus of elasticity.

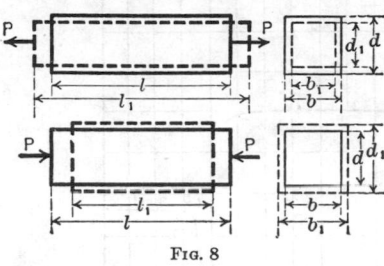

FIG. 8

Poisson's Ratio. The elongation or shortening of a bar under axial stress is accompanied by a reduction of the cross-sectional area in tension and an increase in the cross-sectional area in compression. The two cases are illustrated in Fig. 8. It has been established by experiment that the lateral unit deformation or change per unit diameter or other lateral dimension is proportional to the linear unit elongation or shortening. The ratio of lateral unit deformation to linear unit deformation within the elastic limit is known as *Poisson's ratio* or the "factor of lateral contraction." Average values for the common materials of construction are as follows:

Material	Steel	Wrought Iron	Cast Iron	Brass	Concrete
Poisson's ratio	0.300	0.280	0.270	0.340	0.100

Denoting Poisson's ratio by ν, and the length and diameter of the bar, after stress has been acting, by l_1 and d_1, respectively, then

For tension . . . . $l_1 = (1 + \epsilon)l; \quad d_1 = (1 - \nu\epsilon)d$ (5)

For compression . . . . $l_1 = (1 - \epsilon)l; \quad d_1 = (1 + \nu\epsilon)d$ (6)

in which l and d are the original length and diameter and ϵ the longitudinal unit elongation or shortening as the case may be.

Bulk Modulus. The modulus relating an increase of pressure or unit stress to the corresponding decrease in volume is called the bulk modulus, K.

$$K = - \frac{p}{\Delta V/V} \qquad (7)$$

where p is the pressure and V is the volume.

Relationships. The following relationships exist between the modulus of elasticity, E, the modulus of rigidity G, the bulk modulus of elasticity K, and Poisson's ratio, ν:

$$E = 2G(1 + \nu) \qquad G = \frac{E}{2(1 + \nu)} \qquad \nu = \frac{E - 2G}{2G} \qquad K = \frac{E}{3(1 - 2\nu)} \qquad \nu = \frac{3K - E}{6K}$$

$$(8)$$

True Stress is defined as a ratio of applied axial load to the corresponding cross-sectional area. The units of true stress may be expressed in pounds per square inch, pounds per square foot, etc.

$$\sigma = \frac{P}{A}$$

where σ is the true stress, pounds per square inch; P is the axial load, pounds; and A is the smallest value of cross-sectional area existing under the applied load P, square inches.

True Strain is defined as a function of the original diameter to the instantaneous diameter of the test specimen:

$$q = 2 \log_e \frac{d_0}{d} \text{ in./in.}$$

where q = true strain, inches per inch; d_0 = original diameter of test specimen, inches; and d = instantaneous diameter of test specimen, inches.

True Stress-strain Relationship is obtained when the values of true stress and the corresponding true strain are plotted against each other in the resulting curve (Fig. 9). The slope of the nearly straight line leading from the ultimate load up to fracture is known as the coefficient of strain hardening. It as well as the true tensile strength appear to be related to the other mechanical properties.

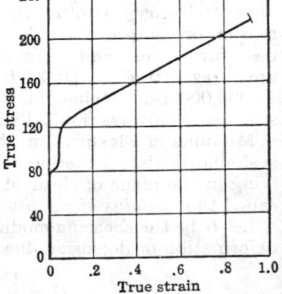

FIG. 9.   True stress-strain relationship

The Allowable Unit Stress or the **allowable working unit stress,** commonly called **allowable stress** or **working stress,** is the maximum unit stress to which it is considered a member may safely be subjected when in service. The term allowable stress is preferable to working stress since the latter is sometimes used to indicate the actual stress in a material when in service. The allowable unit stresses for different materials for various conditions of service are specified by different authorities on the basis of test or experience. In general, for ductile materials, the allowable stress is considerably less than the yield point.

The **Factor of Safety** is the ratio of the ultimate strength of the material to the allowable stress. The term was originated for determining the allowable stress. The ultimate strength of a given material divided by an arbitrary factor of safety, dependent upon material and the use to which it is to be put, gives the allowable stress. In present design practice, it is customary to use allowable stress as specified by recognized authorities or building codes rather than this arbitrary factor of safety. One reason for this decreasing importance of the factor of safety is that it is misleading in that it implies a greater degree of safety than actually exists in a material. For example, a factor of safety of 4 does not mean that a member can carry a load four times as great as that for which it was designed. It should also be clearly understood that, even though each part of a machine is designed with the same factor of safety, the machine as a whole does not have that factor of safety. When one part is stressed beyond the proportional limit, or particularly the yield point, the load or stress distribution may be completely changed throughout the entire machine or structure, and its ability to function may thus be changed though no part has ruptured.

Though no definite rules can be given, if a factor of safety is to be used the following circumstances should be taken into account in its selection:

1. When the ultimate strength of the material is known within narrow limits as for structural steel for which tests of samples have been made, when the load is entirely a steady one of a known amount and there is no reason to fear the deterioration of the metal by corrosion, the lowest factor that should be adopted is 3.

2. When the circumstances of 1 are modified by a portion of the load being variable, as in floors of warehouses, the factor should be not less than 4.

3. When the whole load, or nearly the whole, is apt to be alternately put on and taken off, as in suspension rods of floors of bridges, the factor should be 5 or 6.

4. When the stresses are reversed in direction from tension to compression, as in some bridge diagonals and parts of machines, the factor should be not less than 6.

5. When the piece is subjected to repeated shocks, the factor should be not less than 10.

6. When the piece is subjected to deterioration from corrosion the section should be sufficiently increased to allow for a definite amount of corrosion before the piece will be so far weakened by it as to require removal.

7. When the strength of the material or the amount of the load or both are uncertain, the factor should be increased by an allowance sufficient to cover the amount of the uncertainty.

8. When the strains are of a complex character and of uncertain amount, such as those in the crankshaft of a reversing engine, a very high factor is necessary, possibly even as high as 40.

9. If the property loss caused by failure of the part may be large or if loss of life may result, as in a derrick hoisting materials over a crowded street, the factor should be large.

Table 1. Average Values of Factor of Safety

Material	Steady stress	Variable stress	Shocks	Material	Steady stress	Variable stress	Shocks
Cast iron............	6	10	20	Hard steel.........	5	8	15
Wrought iron.......	4	6	10	Timber............	8	10	15
Structural steel.....	4	6	10	Brick and stone....	15	25	40

3. PHYSICAL PROPERTIES OF MATERIALS OBSERVED IN TESTS

Elasticity. A material that is elastic is able to deform and return to its original shape upon the removal of the load. Regardless of the amount of deformation, the ability to recover its original form is the criterion of elasticity. Steel is elastic within its elastic limit; but for higher unit stresses, causing greater deformations, much of its elasticity is lost, as the material only partially returns to its original form when the stress is removed.

Ductility and Malleability are often used synonymously to indicate the ability of a material to undergo large permanent deformations without rupture. Ductility is commonly thought of as the property which enables a material to be drawn into a wire, whereas malleability is the property which permits it to be beaten or rolled into thin sheets. Ductility is frequently more specifically defined as the ability to undergo large permanent deformations in tension, and malleability, the ability to undergo large permanent deformations in compression.

Plasticity. A material is plastic if the smallest load produces a permanent deformation. A perfectly plastic material is non-elastic and has no ultimate strength in the ordinary meaning of the term. Lead is an example of a plastic material, for a prism tested in compression will deform permanently under a small load and will continue to deform as the load is increased until it flattens out into a thin sheet. Wrought iron and soft steel become plastic when stressed beyond the elastic limit in compression, their behavior resembling that of lead; when stressed beyond the elastic limit in tension they are partly elastic and partly plastic, the degree of plasticity increasing as the ultimate strength is approached.

Brittleness. A material which can be only slightly deformed without rupture is termed brittle. Brittleness is relative, no material being perfectly brittle, that is, capable of no deformation before rupture. Many materials are brittle to a greater or less degree, glass being one of the most brittle of materials. Brittle materials have relatively short stress-deformation curves, since they are capable of only small deformations before rupture. Of the common structural materials, cast iron, brick, and stone are to be considered brittle in comparison with steel.

Brittleness and Plasticity are opposite terms. Materials which have a high degree of plasticity have no brittleness, and they rupture with considerable reduction of area. The reduction of area at rupture may be considered the measure of the plasticity or brittleness of a material, a large reduction of area indicating a high degree of plasticity and little or no reduction of area indicating a high degree of brittleness.

Toughness is the ability to withstand high unit stress together with great unit deformation, without complete fracture. The area under the curve of the stress-deformation diagram, area $OAGH$, or OJK, Fig. 10, is a measure of the toughness of the material. The distinction between ductility and toughness is that ductility deals only with the ability

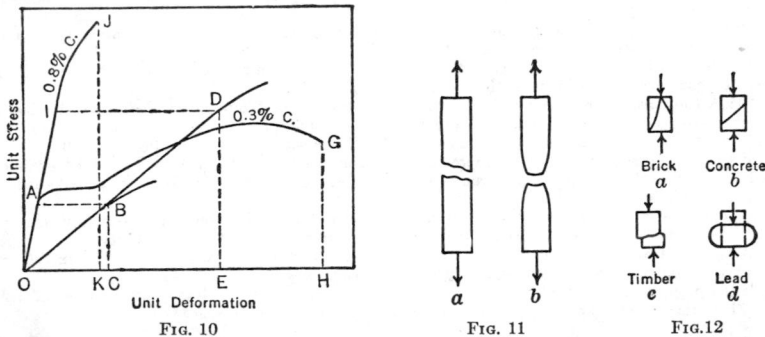

FIG. 10   FIG. 11   FIG. 12

to deform, whereas toughness considers both the ability to deform and the stress developed during deformation.

Stiffness is the ability to resist deformation under stress. The modulus of elasticity is the criterion of the stiffness of a material.

Hardness is the ability to resist very small indentations, abrasion, and plastic deformation. There is no single measure of hardness, as it is not a single property but is a combination of several properties.

Types of Fracture. The materials of engineering have varying degrees of elasticity, ductility, plasticity, and brittleness, which properties are partly indicated by the method in which a test specimen breaks. A bar of brittle material, as cast iron, will rupture in a tension test, as shown in Fig. 11a, in a clean, sharp fracture with very little reduction of cross-sectional area and very little elongation; in a ductile material such as structural steel, the reduction of area and elongation are greater, as shown in Fig. 11b. In compression, a prism of brittle material will break by shearing along oblique planes and the greater the brittleness of the material the more nearly these planes will parallel the direction of the applied force. Figures 12a, b, and c, arranged in order of brittleness, illustrate the type of fracture in prisms of brick, concrete, and timber. Figure 12d represents the deformation of a prism of plastic material, as lead, which flattens out under load without failure.

Creep or Flow of Metals is a phase of plastic or inelastic action. Some solids, as asphalt or paraffin, flow appreciably at room temperatures under extremely small stresses, whereas zinc, lead, and tin show signs of creep at room temperature under moderate stresses. At sufficiently high temperatures, practically all known metals creep under stresses which vary with the temperature; the higher the temperature the lower the stress at which creep takes place. This deformation due to creep continues to increase indefinitely and becomes of extreme importance in members subjected to high temperatures, as parts in turbines, boilers, superheaters, etc. Considerable work has been done on this phase of the strength of materials since about 1919, and many data have been accumulated, but the field has only begun to be explored.

Stress Relaxation. Various types of bolted joints and shrink or press fit assemblies are applications of creep taking place with diminishing stress. This deformation tends to loosen the joint and produce a stress reduction or stress relaxation. The performance of a material to be used under diminishing creep-stress condition is determined by a tensile stress-relaxation test.

Creep Failure. Generally four distinct phases are distinguishable during the course of creep failure. The elapsed time per stage depends on the material, temperature, and stress condition. They are: (a) initial phase, where the total deformation is partially elastic and partially plastic; (b) second phase, where the creep rate decreases with time,

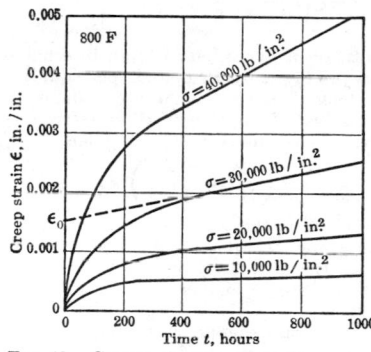

FIG. 13. Curves of creep strain for various stress levels

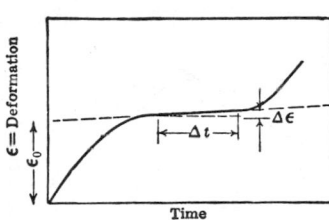

FIG. 14. Method of determining creep rate

indicating the effect of strain hardening; (c) third phase, where the effect of strain hardening is counteracted by the annealing influence of the high temperature which produces a constant or minimum creep rate; (d) final phase, where the creep rate increases until fracture occurs due to the decrease in cross-sectional area of the specimen.

In conducting a conventional creep test, curves of strain as a function of time are obtained for groups of specimens, each specimen in one group being subjected to a different constant stress while all the specimens in the group are tested at one temperature.

In this manner families of curves like those shown in Fig. 13 are obtained. Several methods have been proposed for the interpretation of such data (Ref. 4). Two frequently used expressions of the creep properties of a material can be derived from the data in the following form:

$$C = B\sigma^m$$
$$\epsilon = \epsilon_0 + Ct \tag{9}$$

where C = creep rate; B, m = experimental constants; σ = stress; ϵ = creep strain at any time t; ϵ_0 = zero time strain intercept; and t = time. See Fig. 14.

REFERENCES

1. DEN HARTOG, J. P., *Advanced Strength of Materials*, New York, McGraw-Hill, 1952.
2. HETÉNYI, M., *Handbook of Experimental Stress Analysis*, New York, John Wiley, 1950.
3. MARIN, J., *Strength of Materials*, New York, Macmillan, 1948.
4. MARIN, J., *Mechanical Properties of Materials and Design*, New York, McGraw-Hill, 1942.
5. MAURER, E. R., and WITHEY, M. O., *Strength of Materials*, New York, John Wiley, 1940.
6. SEELY, F. B., *Resistance of Materials*, New York, John Wiley, 4th ed., 1956.
7. SEELY, F. B., and SMITH, J. O., *Advanced Mechanics of Materials*, New York, John Wiley, 2nd ed., 1952.

DYNAMIC STRESSES

By John M. Lessells and G. S. Cherniak

Dynamic Stresses occur where the dimension of time is necessary in defining the loads. They include creep, fatigue, and impact stresses.

Creep Stresses occur when either the load or the deformation progressively varies with time. They are usually associated with non-cyclic phenomena.

Fatigue Stresses occur when the cyclic variation of either load or strain is coincident with respect to time.

Impact Stresses occur from loads which are transient with time. The duration of the load application is of the same order of magnitude as the natural period of vibration of the specimen.

4. WORK AND RESILIENCE

External Work. Let P = axial load, pounds, on a bar, producing an internal stress not exceeding the elastic limit; σ = unit stress produced by P, pounds per square inch; A = cross-sectional area, square inches; l = length of bar, inches; e = deformation, inches; E = modulus of elasticity; W = external work performed on bar, inch-pounds = $1/2\,Pe$. Then

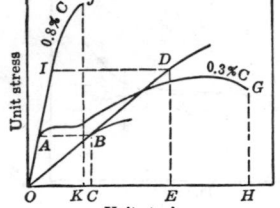

$$W = \frac{1}{2} A\sigma \left(\frac{\sigma l}{E}\right) = \frac{1}{2}\left(\frac{\sigma^2}{E}\right) Al \qquad (1)$$

The factor $(1/2)(\sigma^2/E)$ is the work required per unit volume, the volume being Al. It is represented on the stress-strain diagram by the area ODE or area OBC (Fig. 1), in which DE and BC are ordinates representing the unit stresses considered.

Resilience is the strain energy which may be recovered from a deformed body when the load causing the stress is removed. Within the proportional limit, the resilience is equal to the external work performed in deforming the bar, and may be determined by eq. 1. When σ is equal to the proportional limit, the factor $(1/2)(\sigma^2/E)$ is the *modulus of resilience*, i.e., the measure of capacity of a unit volume of material to store strain energy up to the proportional limit. Average values of the modulus of resilience under tensile stress are given in Table 1.

Fig. 1. Work areas on stress-strain diagram

Table 1. Average Values of Tensile Modulus of Resilience and Toughness

Material	Modulus of Resilience, in.-lb per cu in.	Relative Toughness (Area under Curve of Stress-deformation Diagram)
Gray cast iron	1.2	70
Malleable cast iron	17.4	3,800
Wrought iron	11.6	11,000
Low-carbon steel	15.0	15,700
Medium-carbon steel	34.0	16,300
High-carbon steel	94.0	5,000
Ni-Cr steel, hot-rolled	94.0	44,000
Vanadium steel, 0.98% C, 0.2% V, heat-treated	260.0	22,000
Duralumin, 17 ST	45.0	10,000
Rolled bronze	57.0	15,500
Rolled brass	40.0	10,000
Oak	2.3 *	13 *

* Bending.

The total resilience of a bar is the product of its volume and the modulus of resilience. These formulas for work performed on a bar, and its resilience, do not apply if the unit stress is greater than the proportional limit.

Work Required for Rupture. Since beyond the proportional limit the strains are not proportional to the stresses, $1/2\,P$ does not express the mean value of the force acting. Equation 1, therefore, does not express the work required for strain after the proportional limit of the material has been passed, and cannot express the work required for rupture. The work required per unit volume to produce strains beyond the proportional limit or to cause rupture may be determined from the stress-strain diagram as it is measured by the area included between the axis of abscissas, and the stress-strain curve up to the strain in question, as $OAGH$ or OJK (Fig. 1). This area, however, does not represent the resilience, since part of the work done on the bar is present in the form of hysteresis losses and cannot be recovered.

Damping Capacity (Hysteresis). Observations show that when a tensile load is applied to a bar it does not produce the complete elongation immediately, but there is a definite time lapse which depends on the nature of the material and the magnitude of the stresses involved. In parallel with this it is also noted that, upon unloading, complete recovery of energy does not occur. This phenomenon is variously termed *elastic hysteresis* or, for vibratory stresses, damping. Figure 2 shows a typical hysteresis loop obtained for one cycle of loading. The area of this hysteresis loop, representing the energy dissipated per cycle, is a measure of the damping properties of the material. Although the exact mechanism of damping has not been fully investigated, it has been found that under vibratory conditions the energy dissipated in this manner varies approximately as the cube of the stress.

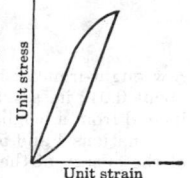

FIG. 2. Hysteresis loop for loading and unloading

5. IMPACT STRESSES IN BARS AND BEAMS

Effect of Sudden Loads. If a sudden load P is applied to a bar, it will cause a deformation el, and the work done by the load will be Pel. Since the external work equals the internal work, $Pel = \sigma^2 Al/2E$, and since $e = \sigma/E$, $P = \sigma A/2$, or $\sigma = 2P/A$. The unit stress and also the unit strain are double those obtained by an equal load applied gradually. However, the bar does not maintain equilibrium at the point of maximum stress and strain. After a series of oscillations in which the surplus energy is dissipated in damping, the bar finally comes to rest with the same strain and stress as that due to the equal static load.

Stress Due to Live Loads. In structural design two loads are considered, the dead load or weight of the structure and the live load or superimposed loads to be carried. The stresses due to the dead load and to the live load are computed separately, each being regarded as a static load. It is obvious that the stress due to the live load may be greatly increased, depending on the suddenness with which the load is applied. It has been shown above that the stress due to a suddenly applied load is double the stress caused by a static load. The term *coefficient of impact* is used extensively in structural engineering to denote the number by which the computed static stress is multiplied to obtain the value of the increased stress assumed to be caused by the suddenness of the application of the live load. If σ = static unit stress computed from the live load, and i = coefficient of impact, then the increase of unit stress due to sudden loading is $i\sigma$, and the total unit stress due to live load is $\sigma + i\sigma$. The value of i has been determined by empirical methods and varies according to different conditions.

In the building codes of most cities, specified floor loadings for buildings include the impact allowance, and no increase is needed for live loads except for special cases of vibration or other unusual conditions. For railroad bridges, the value of i depends upon the proportion of the length of the bridge which is loaded. No increase in the static stress is needed when the mass of the structure, as in monolithic concrete, is great. For machinery and for unusual conditions, such as elevator machinery and its supports, each structure should be considered by itself and the coefficient assumed accordingly. It should be noted that the meaning of the word impact used above differs somewhat from its strict theoretical meaning and as it is used in the next paragraph. The use of the terms impact and coefficient of impact in connection with live load stresses is, however, very general.

Axial Impact on Bars. A load P dropped from a height h onto the end of a vertical bar of cross-sectional area A, rigidly secured at the bottom end, produces in the bar a unit stress which increases from 0 up to σ', with a corresponding total strain increasing from 0 up to e_1. The work done on the bar is $P(h + e_1)$, which, provided no energy is expended in hysteresis losses or in giving velocity to the bar, is equal to the energy $1/2\,\sigma' Ae_1$ stored in the bar; that is,

$$P(h + e_1) = \frac{1}{2}\,\sigma' Ae_1 \tag{2}$$

If e = strain produced by a static load P, within the proportional limit

$$\frac{e}{e_1} = \frac{P/A}{\sigma'} \tag{3}$$

Combining this with eq. 2 gives

$$\sigma' = \sigma + \sigma \sqrt{1 + 2\frac{h}{e}} \qquad (4)$$

$$e_1 = e + e \sqrt{1 + 2\frac{h}{e}} \qquad (5)$$

A wrought-iron bar 1 in. square and 5 ft long under a static load of 5000 lb will be shortened about 0.012 in. assuming no lateral flexure to occur; but, if a weight of 5000 lb drops on its end from a height of 0.048 in., a stress of 20,000 lb will be produced.

Equations 4 and 5 give values of stress and strain that are somewhat high because part of the energy of the applied force is not effective in producing stress but is expended in overcoming the inertia of the bar and in producing local stresses. For light bars they give approximately correct results.

If the bar is horizontal and is struck at one end by a weight P, moving with a velocity V, the strain produced is e_1. Then, as before $\frac{1}{2}\sigma' A e_1 = Ph$. In this case $h = V^2/2g =$ height from which P would have to fall to acquire velocity V ($g =$ acceleration due to gravity $= 32.16$ ft per sec^2). Combining with eq. 3,

$$\sigma' = \sigma \sqrt{2\frac{h}{e}} \qquad (6)$$

$$e_1 = e \sqrt{2\frac{h}{e}} \qquad (7)$$

Impact on Beams. If a weight P falls on a horizontal beam from a height h, producing a maximum deflection y and a maximum unit stress σ' in the extreme fiber, the values of σ' and y are given by

$$\sigma' = \sigma + \sigma \sqrt{1 + 2\frac{h}{y}} \qquad (8)$$

$$y_1 = y + y \sqrt{1 + 2\frac{h}{y}} \qquad (9)$$

where $\sigma =$ extreme fiber unit stress and $y =$ deflection due to P, considered as a static load. The value of σ may be obtained from the flexure formula (p. 513); that of y from the proper formula for deflection under static load.

If a weight P moving horizontally with a velocity V strikes a beam (the ends of which are secured against horizontal movement), the maximum fiber unit stress and the maximum lateral deflection are given by

$$\sigma' = \sigma \sqrt{2\frac{h}{y}} \qquad (10)$$

$$y_1 = y \sqrt{2\frac{h}{y}} \qquad (11)$$

where σ and y are as before and h is height through which P would have to fall to acquire the velocity V. These formulas, like those for axial impact on bars, give results higher than those observed in tests, particularly if the weight of the beam is great.

Rupture from Impact. Rupture may be caused by impact provided the load has the requisite velocity. The above formulas, however, do not apply since they are valid only for stresses within the proportional limit. It has been found that the dynamic properties of a material are dependent on volume, velocity of the applied load, and material condition. If the velocity of the applied load is kept within certain limiting values, the total energy values for static and dynamic conditions are identical. If the velocity is increased, the impact values are considerably reduced. For further information see Ref. 2.

6. STEADY VIBRATORY STRESSES IN BARS AND BEAMS

For steady vibratory stresses the deflection of the bar, or beam, will be increased by the dynamic magnification factor. The relation is given by

$$\delta_{\text{dynamic}} = \delta_{\text{static}} \times \text{Dynamic magnification factor}$$

An example of the calculating procedure for the case of no damping losses is

$$\delta_{\text{dynamic}} = \delta_{\text{static}} \times \frac{1}{1 - (\omega/\omega_n)^2} \qquad (12)$$

where ω is the frequency of oscillation of the load and ω_n the natural frequency of oscillation of the bar. ω_n is determined by

$$\omega_n = \sqrt{\frac{3EIg}{L^3W}} \qquad (13)$$

where E = modulus of elasticity, I = moment of inertia, g = acceleration of gravity, L = length of bar, and W = weight of oscillating load.

7. FATIGUE

Parts Subjected to Repeated Stress. Where parts of a structure or machine are subjected to varying or repeated loads, the ordinary methods of computing stresses and of determining the strength of materials under static load conditions are not satisfactory. In general, more than one-half to ten million repetitions of loading are necessary before the parts come under this classification. Such parts are crankshafts, shafts carrying rotating members (as in motors and generators), turbine blades, valve parts, piston rods, floor beams in elevated railroads, etc.

Nature of Failure. The failure of a ductile member caused by repeated loads is not accompanied by noticeable yielding; the failure is a gradual or progressive fracture. The fracture seems to start at some point in the member at which the stress is much larger than the calculated stress. This high localized stress seems to cause, or at least is accompanied by, a small crack which gradually spreads as the stress is repeated until the whole member ruptures without measurable yielding. The fatigue failure of a ductile material is similar to that of a static failure of a brittle material.

Determination of Endurance Limit. The highest unit stress σ at which a material can be subjected to a very large number of repetitions of load N without evidence of failure is the *endurance limit* of the material. It is usually found for any given material by constructing a σ-N curve for that particular metal from data obtained by tests. A piece of metal is loaded so as to produce a given stress, the load being repeated until failure occurs. Other specimens are then tested at different values of the unit stress until enough points have been obtained to plot a curve. This curve will incline steeply at the higher stresses and gradually flatten out at the lower stresses until, for steel, it approaches asymptotically a limiting stress called the endurance limit. For alloys and materials subjected to corrosive conditions, the σ-N curve continues to slope gradually down; the endurance limit is usually set at that stress which corresponds to approximately 10×10^6 cycles. The most common test is reversed bending by the rotating beam method since this specimen is most easily made and the testing machines are the least costly. Tests are also made in direct tension, direct shear, direct compression, reversed torsion, and in various combinations of them. Typical σ-N curves are shown in Fig. 3.

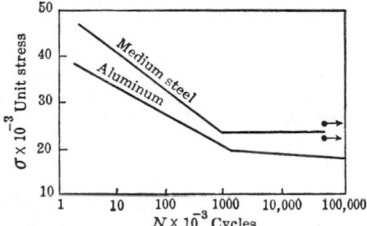

FIG. 3. Typical σ-N curves of fatigue resistance

Effect of Scratches and Defects on Endurance Limit of Rolled or Forged Ferrous Metals. The beginning of rupture under repeated stress usually occurs at a point of discontinuity in the material. The same principle applies to a discontinuity on the outer surface where, in bending or torsion, the stress is greatest. The presence of scratches, sharp

Table 2. Fatigue Endurance Limits

Data from *Bulletin* 124, University of Illinois Engineering Experiment Station.

0.49% C Quenched and Drawn 197 Brinell Steel

Finish	Endurance Limit, psi
High polish, longitudinal	51,000
00 Emery	48,000
Ground	45,000
Smooth turned	43,000
Rough turned	42,000

corners, or grooves reduces the endurance limit an appreciable amount. Moore and Kommers (Ref. 3) obtained the data given in Table 2. The tests were made on bars 0.40 in. in diameter. Horger and Niefert (Ref. 4) have made similar tests which corroborate the earlier work. It has also been found that the effect of scratches, grooves, etc., is not so pronounced in cast iron as in steel.

For keyways on shafts, Thum (Ref. 5) has determined that, for a soft steel of tensile strength 60,000 psi, an end-milled keyway reduced the strength to 20,500 psi, whereas a face-milled keyway reduced the strength to 28,000 psi.

Sharp corners (Ref. 6), oil holes (Ref. 7), press fits and surface rolling (Ref. 8) all produce marked reductions in the strength of the material.

Effect of Range of Stress. The endurance limit as determined by a rotating beam is the endurance limit for a complete reversal of stress from tension to compression, and vice versa. The endurance limit is less for a complete reversal than it is for those stress combinations where the stress goes only from a maximum to zero, or from a maximum to a minimum which is above zero. This is discussed further in article 13 of this section.

Effect of Corrosion. The endurance limit of steels which have been treated (pickling) prior to use in an application of repeated stress with no further corrosion during stressing is affected considerably by roughening of the surface (i.e., production of pits and notches on the surface of the material). It is even more serious to have corrosion occur simultaneously with repeated stress. One explanation advanced is that a pit or notch formed by corrosion is opened up when tension is applied to the piece and it becomes filled with rust or other products of corrosion. When the tension is released and the pit closes down on the corrosion products it contains, these products exert a wedge action to produce cracking. The crack increases the local stress, and matters get progressively worse. Also, repeated stress tends to crack any protective coating of a film of corrosion products and thus allows continued corrosion at the cracks. This phenomenon is clearly outlined in the work of McAdam and Geil (Ref. 9).

Under corrosive conditions, steels have no endurance limit; the σ-N curve never becomes horizontal but continues to drop. Thus no corrodible steel can last indefinitely under corrosive conditions and repeated stress, no matter how small that stress is. (See also Ref. 10.)

The most effective method of protection is to keep the parts subjected to repeated loads from contact with corrosive agents. Where this is not possible, a protective coating of some elastic material such as varnish, cement, or enamel will aid. (See also Section 13.)

Effect of Size. Tests made on specimens 0.3 in. in diameter and larger indicate a reduction of fatigue strength of approximately 10% for an increase in size from 0.3 to 1.0 in. in diameter. In specimens above 1 in. in diameter there was little variation in fatigue strength in geometrically similar specimens without stress concentration. However, with geometrically similar specimens having holes, fillets, and artificial cracks, the smaller specimens gave higher endurance limits than the larger ones (i.e., the strength-reduction factor is lower for the small specimens).

Effect of Speed. There seems to be little change in the fatigue strength for speeds up to 3000 rpm. However, from 3000 to 120,000 rpm, tests have indicated an increase of about 15% in the endurance limit and some increase in life.

Effect of Cold Working. Cold drawing and cold working, as with static strength, increase the fatigue strength. The usefulness of this process is limited because of its tendency to introduce internal strains and minute cracks. Tests have indicated also that by understressing a specimen, that is, subjecting it to a certain safe range of stress before subjecting it to a higher range of stress, a permanent increase of as much as 30% has been obtained in the fatigue strength.

Effect of Temperature. As the material is subjected to temperature above room temperature under repeated stress conditions, the fatigue strength drops. In one set of data reported the endurance limit dropped 10, 37, and 70% for temperature of 300, 400, and 500 F, respectively.

REFERENCES

1. SEELY, F. B., *Resistance of Materials*, New York, John Wiley, 4th ed., 1956.
2. MANN, H. C., articles, *Proc. Am. Soc. Testing Materials*, 1935, 1936, and 1937.
3. MOORE and KOMMERS, *Bull*. 124, Engineering Experiment Station, University of Illinois.
4. HORGER and NIEFERT, *Proc. Am. Soc. Testing Materials*, 1940.
5. THUM, *Trans. North East Coast Inst. Engrs. & Shipbuilders*, 1935.
6. HENDRICKSON, *SAE J.*, 1939.
7. PETERSON and WAHL, *Trans. Am. Soc. Mech. Engrs.*, 1936.
8. BUCKWALTER and HORGER, *Trans. Am. Soc. Metals*, 1937.
9. McADAM, D. J., JR., and GEIL, G. W., *J. Research Natl. Bur. Standards*, 1940.
10. Metallurgical Staff of Battelle Memorial Institute, *Prevention of the Failure of Metals under Repeated Stress*, New York, John Wiley, 1941.

WORKING STRESSES

By John M. Lessells and G. S. Cherniak

8. COMBINED STRESS

Under certain circumstances of loading a body is subjected to a combination of tensile, compressive, and/or shear stresses. For example, a shaft which is simultaneously bent and twisted is subjected to combined stresses, namely, longitudinal tension and compression and torsional shear. For the purposes of analysis it is convenient to reduce such systems of combined stresses to a basic system of stress coordinates known as principal stresses. These stresses act on axes which differ in general from the axes along which the applied stresses are acting and represent the maximum and minimum values of the normal stresses for the particular point considered.

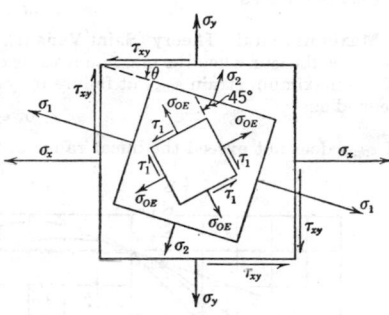

Determination of Principal Stresses. The expressions for the principal stresses in terms of the stresses along the x and y axes are

$$\sigma_1 = \frac{\sigma_x + \sigma_y}{2} + \sqrt{\left(\frac{\sigma_x - \sigma_y}{2}\right)^2 + \tau_{xy}^2} \quad (1)$$

$$\sigma_2 = \frac{\sigma_x + \sigma_y}{2} - \sqrt{\left(\frac{\sigma_x - \sigma_y}{2}\right)^2 + \tau_{xy}^2} \quad (2)$$

$$\tau_1 = \pm \sqrt{\left(\frac{\sigma_x - \sigma_y}{2}\right)^2 + \tau_{xy}^2} \quad (3)$$

FIG. 1. Diagram showing relative orientation of stresses. (Reproduced with modification by permission from *Mechanical Properties of Materials and Design*, by Joseph Marin, McGraw-Hill Book Co.)

where σ_1, σ_2, and τ_1 are the principal stress components and σ_x, σ_y, and τ_{xy} the calculated stress components, all of which are determined at any particular point (Fig. 1).

Graphical Method of Principal Stress Determination—Mohr's Circle. Let the axes x and y be chosen to represent the directions of the applied normal and shearing stresses, respectively (Fig. 2). Lay off to suitable scale distances $OA = \sigma_y$, $OB = \sigma_x$, and $BC = AD = \tau_{xy}$. With point E as a center construct the circle DFC. Then OF and OG are the principal stresses σ_1 and σ_2 respectively, and EC the maximum shear stress τ_1. The inverse also holds; that is, given the principal stresses, σ_x and σ_y can be determined on any plane passing through the point.

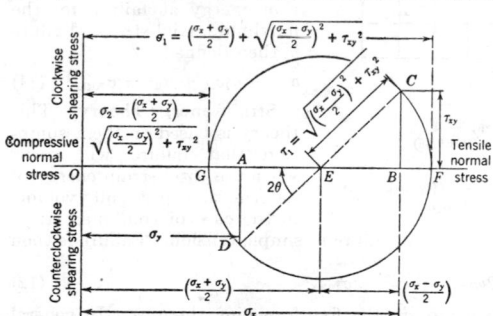

FIG. 2. Mohr's circle used for the determination of the principal stresses. (Reproduced with modification by permission from Joseph Marin, *op. cit.*)

Stress-strain Relations. The linear relation between components of stress and strain is known as *Hooke's law*. This relation for the two-dimensional case can be expressed as

$$\epsilon_x = \frac{1}{E}(\sigma_x - \nu\sigma_y) \quad (4)$$

$$\epsilon_y = \frac{1}{E}(\sigma_y - \nu\sigma_x) \quad (5)$$

$$\gamma_{xy} = \frac{1}{G}\tau_{xy} \quad (6)$$

where σ_x, σ_y, and τ_{xy} are the stress components of a particular point, ν = Poisson's ratio, E = modulus of elasticity, G = modulus of rigidity, and ϵ_x, ϵ_y, and γ_{xy} = strain components.

9. STRENGTH THEORIES

The determination of the magnitudes and directions of the principal stresses and strains and of the maximum shearing stresses is carried out for the purpose of establishing criteria of failure within the material under the anticipated loading conditions. To this end several theories have been advanced to elucidate these criteria. The more noteworthy ones are listed below. The theories are based on the assumption that the principal stresses do not change with time, an assumption that is justified since the applied loads in most cases are synchronous.

Maximum-stress Theory (Rankine's Theory). This theory is based on the assumption that failure will occur when the maximum value of the greatest principal stress reaches the value of the maximum stress σ_{max} at failure in the case of simple axial loading. Failure is then defined as

$$\sigma_1 \text{ or } \sigma_2 = \sigma_{max} \qquad (7)$$

Maximum-strain Theory (Saint Venant). This theory is based on the assumption that failure will occur when the maximum value of the greatest principal strain reaches the value of the maximum strain ϵ_{max} at failure in the case of simple axial loading. Failure is then defined as

$$\epsilon_1 \text{ or } \epsilon_2 = \epsilon_{max} \qquad (8)$$

If ϵ_{max} does not exceed the linear range of the material eq. 8 may be written as

$$\sigma_1 - \nu\sigma_2 = \sigma_{max} \qquad (9)$$

Maximum-shear Theory (Guest). This theory is based on the assumption that failure will occur when the maximum shear stress reaches the value of the maximum shear stress at failure in simple tension. Failure is then defined as

$$\tau_1 = \tau_{max} \qquad (10)$$

Distortion-energy Theory (Hencky-Von Mises) (Shear Energy). This theory is based on the assumption that failure will occur when the distortion energy corresponding to the maximum values of the stress components equals the distortion energy at failure for the maximum axial stress. Failure is then defined as

$$\sigma_1{}^2 - \sigma_1\sigma_2 + \sigma_2{}^2 = \sigma_{max}{}^2 \qquad (11)$$

Strain-energy Theory. This theory is based on the assumption that failure will occur when the total strain energy of deformation per unit volume

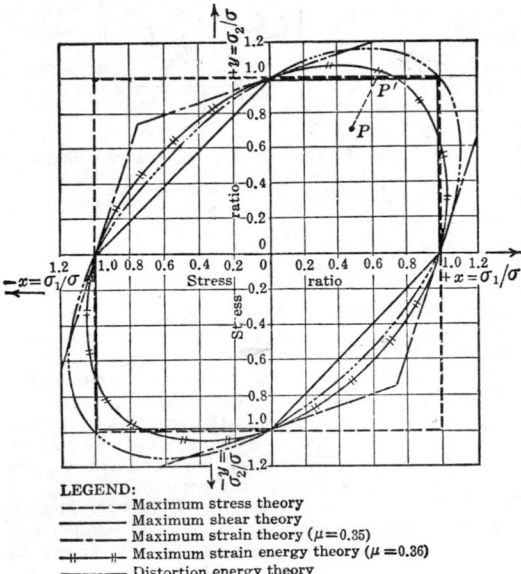

LEGEND:
– – – – – Maximum stress theory
————— Maximum shear theory
– · – · – Maximum strain theory ($\mu = 0.35$)
–‖—‖– Maximum strain energy theory ($\mu = 0.36$)
· · · · · Distortion energy theory

FIG. 3. Comparison of five theories of failure. (Reproduced by permission from Joseph Marin, *op. cit.*)

in the case of combined stress is equal to the strain energy per unit volume at failure in simple tension. Failure is then defined as

$$\sigma_1{}^2 - 2\nu\sigma_1\sigma_2 + \sigma_2{}^2 = \sigma_{max}{}^2 \qquad (12)$$

Comparison of Theories. Figure 3 compares the five foregoing theories. In general the distortion-energy theory is the most satisfactory for ductile materials and the maximum-stress theory for brittle materials. The maximum-shear theory gives conservative results for both ductile and brittle materials. The conditions for yielding, according to the various theories, are given in Table 1, taking $\nu = 0.300$ as for steel.

Table 1. Comparison of Stress Theories

$\tau = \sigma_{yp}$ from the maximum-stress theory
$\tau = 0.77\sigma_{yp}$ from the maximum-strain theory
$\tau = 0.50\sigma_{yp}$ from the maximum-shear theory
$\tau = 0.62\sigma_{yp}$ from the maximum-strain-energy theory

10. STRESS CONCENTRATIONS

The direct design procedure described above assumes no abrupt changes in cross section, discontinuities in the surface or holes, etc., through the member. In most structural parts this is not the case. The stresses produced at these discontinuities are different in magnitude from those calculated by the foregoing methods. The effect of the localized increase in stress, such as that caused by a notch, fillet, hole, or similar *stress raiser*, depends mainly on the type of loading, the geometry of the part, and the material. As a result, it is necessary to consider a stress-concentration factor K_t which is defined by the relationship

$$K_t = \frac{\sigma_{max}}{\sigma_{nominal}} \tag{13}$$

In general σ_{max} will have to be determined by the methods of experimental stress analysis or the theory of elasticity, and $\sigma_{nominal}$ by a simple theory such as $\sigma = P/A$, $\sigma = Mc/I$, $\tau = Tc/J$, without taking into account the variations in stress conditions caused by geometrical discontinuities such as holes, grooves, and fillets. For ductile materials it is not customary to apply stress-concentration factors to members under static loading. For brittle materials, however, stress concentration is serious and should be considered.

Stress-concentration Factors for Fillets, Keyways, Holes, and Shafts. In Table 2 selected stress-concentration factors have been given from a complete table in Ref. 1.

Table 2. Stress-concentration Factors

(Adapted by permission from *Formulas for Stress and Strain* by R. J. Roark, McGraw-Hill Book Co.)

Type	K_t Factors					
Circular hole in plate or rectangular bar	$\frac{h}{a} = 0.67$	0.77	0.91	1.07	1.29	1.56
	$k = 4.37$	3.92	3.61	3.40	3.25	3.16

Square shoulder with fillet for rectangular and circular cross sections in bending	$\frac{r}{d}$ $\frac{h}{r}$	0.05	0.10	0.20	0.27	0.50	1.0
	0.5	1.61	1.49	1.39	1.34	1.22	1.07
	1.0	1.91	1.70	1.48	1.38	1.22	1.08
	1.5	2.00	1.73	1.50	1.39	1.23	1.08
	2.0		1.74	1.52	1.39	1.23	1.09
	3.5		1.76	1.54	1.40	1.23	1.10

11. STATIC WORKING STRESSES

Ductile Materials. For ductile materials the criteria for working stresses are

$$\sigma_w = \frac{\sigma_{yp}}{n} \quad \text{(tension and compression)} \tag{14}$$

$$\tau_w = \frac{1}{2}\frac{\sigma_{yp}}{n} \tag{15}$$

Brittle Materials. For brittle materials the criteria for working stresses are

$$\sigma_w = \frac{\sigma_{ultimate}}{K_t \times n} \quad \text{(tension)} \tag{16}$$

$$\sigma_w = \frac{\sigma_{compressive}}{K_t \times n} \quad \text{(compression)} \tag{17}$$

where K_t is the stress-concentration factor, n is the factor of safety, σ_w and τ_w are working stresses, and σ_{yp} is stress at the yield point.

Working-stress Equations for the Various Theories

Stress Theory

$$\sigma_w = \frac{\sigma_x + \sigma_y}{2} \pm \sqrt{\left(\frac{\sigma_x - \sigma_y}{2}\right)^2 + \tau_{xy}^2} \tag{18}$$

Shear Theory

$$\sigma_w = 2\sqrt{\left(\frac{\sigma_x - \sigma_y}{2}\right)^2 + \tau_{xy}^2} \tag{19}$$

Strain Theory

$$\sigma_w = (1 - \nu)\left(\frac{\sigma_x + \sigma_y}{2}\right) + (1 + \nu)\sqrt{\left(\frac{\sigma_x - \sigma_y}{2}\right)^2 + \tau_{xy}^2} \tag{20}$$

Distortion-energy Theory

$$\sigma_w = \sqrt{\sigma_x^2 - \sigma_x\sigma_y + \sigma_y^2 + 3\tau_{xy}^2} \tag{21}$$

Strain-energy Theory

$$\sigma_w = \sqrt{\sigma_x^2 - 2\nu\sigma_x\sigma_y + \sigma_y^2 + 2(1 + \nu)\tau_{xy}^2} \tag{22}$$

where σ_x, σ_y, τ_{xy} are the stress components of a particular point, ν is Poisson's ratio, and σ_w is working stress.

12. STRENGTH REDUCTION

If the loads acting on a member are cyclic with respect to time, the stress-concentration factor described in article 10 can no longer be applied. It is now necessary to make fatigue tests on specimens to determine the strength-reduction factor K_f:

Definition of K_f:

$$K_f = \frac{\text{Endurance limit without stress concentration}}{\text{Endurance limit with stress concentration}} = \frac{\sigma_e}{\sigma'_e} \tag{23}$$

where σ_e = endurance limit stress. It has been found by test that generally for carbon steel $K_t > K_f$, but for alloy steel $K_t = K_f$. K_f has no meaning unless geometry, size, and material of the specimen are stated.

Notch Sensitivity Index. The notch sensitivity index q is a measure of the degree of agreement between K_f and K_t for a particular specimen, or member, of given size and material containing a stress concentrator of given size and shape:

$$q = \frac{K_f - 1}{K_t - 1} \tag{24}$$

Notch sensitivity varies between zero (where $K_f = 1$) and unity (where $K_f = K_t$). In general q is lower for cast alloys than for high-strength wrought materials.

13. FATIGUE WORKING STRESSES

The application of cyclic stress to a member changes the design calculations owing to a reduction in the maximum allowable stress. (See Article 7.) In this case the strength theories discussed in Article 9 must be modified to take this into account.

Definitions

Stress Cycle. A stress cycle is the smallest section of the stress-time function which is repeated identically and periodically as shown in Fig. 4.

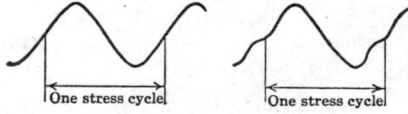

One stress cycle   One stress cycle

FIG. 4. Definition of one stress cycle

Maximum Stress. σ_{max} is the greatest algebraic value of the stress in the stress cycle, being positive for a tensile stress and negative for a compressive stress.

Minimum Stress. σ_{min} is the smallest algebraic value of the stress in the stress cycle, being positive for a tensile stress and negative for a compressive stress.

Range of Stress. σ_r is the algebraic difference between the maximum and minimum stress in one cycle:

$$\sigma_r = \sigma_{max} - \sigma_{min} \tag{25}$$

For most cases of fatigue testing the stress varies about zero stress, but other types of variation may be experienced, as shown in Fig. 5.

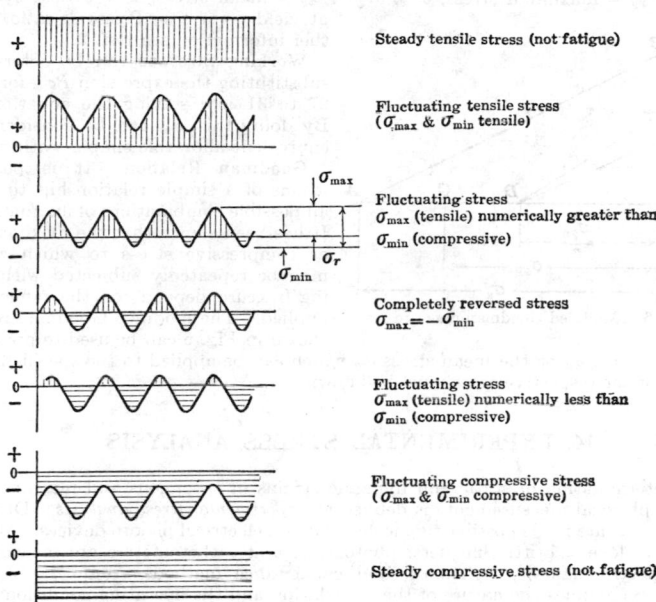

Steady tensile stress (not fatigue)

Fluctuating tensile stress (σ_{max} & σ_{min} tensile)

Fluctuating stress σ_{max} (tensile) numerically greater than σ_{min} (compressive)

Completely reversed stress $\sigma_{max} = -\sigma_{min}$

Fluctuating stress σ_{max} (tensile) numerically less than σ_{min} (compressive)

Fluctuating compressive stress (σ_{max} & σ_{min} compressive)

Steady compressive stress (not fatigue)

FIG. 5. Possible variations of stress with time

Alternating-stress Amplitude (Variable Stress Component). σ_a is half the range of stress $\sigma_a = \sigma_r/2$.

Mean-stress (Steady-stress Component). σ_m is the algebraic mean of the maximum and minimum stress in one cycle:

$$\sigma_m = \frac{\sigma_{max} + \sigma_{min}}{2} \tag{26}$$

Stress Ratio. R is the algebraic ratio of the minimum stress and the maximum stress in one cycle.

Modification of the Strength Theories

Stress Theory

$$\sigma_{max} = (1 - P)\sigma_m \pm \sigma_e \tag{27}$$

where $P = \sigma_e/\sigma_{yp}$. This equation is for normal stresses only where σ_x and σ_y are principal stresses and $\tau_{xy} = 0$.

$$\pm \sigma_e = {}^1\!/_2 \left\{ (\sigma'_x + \sigma'_y) - (1 - P)(\sigma''_x + \sigma''_y) \right.$$
$$\left. \pm \sqrt{[(\sigma'_y - \sigma'_x) - (1 - P)(\sigma''_y - \sigma''_x)]^2 + 4[\tau'_{xy} - (1 - P)\tau''_{xy}]^2} \right\} \tag{28}$$

Shear Theory

$$\pm \sigma_e = \sqrt{[(1 - P)(\sigma''_x - \sigma''_y) - (\sigma'_x - \sigma'_y)]^2 + 4[(1 - P)\tau''_{xy} - \sigma'_{xy}]^2} \tag{29}$$

Distortion-energy Theory

$$\sigma_e = \sqrt{(\sigma'_x)^2 - \sigma'_x\sigma'_y + (\sigma'_y)^2 + 3(\tau'_{xy})^2}$$
$$- (1 - P)\sqrt{(\sigma''_x)^2 - \sigma''_x\sigma''_y + (\sigma''_y)^2 + 3(\tau''_{xy})^2} \tag{30}$$

Strain-energy Theory

$$\sigma_e = \sqrt{(\sigma'_x)^2 - 2\nu\sigma'_x\sigma'_y + (\sigma'_y)^2 + 2(1 + \nu)(\sigma'_{xy})^2}$$

$$- (1 - P)\sqrt{(\sigma''_x)^2 - 2\nu\sigma''_x\sigma''_y + (\sigma''_y)^2 + 2(1 + \nu)(\tau''_{xy})^2} \quad (31)$$

In the foregoing equations σ_e = endurance limit for complete reversal of stress (see Fig. 5), $\sigma'_x, \sigma'_y, \tau'_{xy}$ = maximum stress, $\sigma''_x, \sigma''_y, \tau''_{xy}$ = mean stress, $P = \sigma_e/\sigma_{yp}$, σ_{yp} = stress at yield point, ν = Poisson's ratio. For further information see Ref. 2.

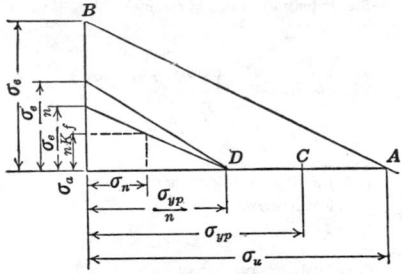

FIG. 6.   Modified Goodman diagram

Working Stresses can be determined by substituting the expression $P\sigma_w$ for σ_e in eqs. 27 to 31 and solving the equation for σ_w. By doing this, σ_w can be considered as an equivalent allowable simple static stress.

Goodman Relation. It is possible by means of a simple relationship to represent all possible combinations of fluctuating stress. It has been shown that the maximum tensile or compressive stress to which a member may be repeatedly subjected without causing fracture depends on the range of stress applied. The modified Goodman relation as shown in Fig. 6 can be used to predict either the cyclic stress σ_a or the mean stress σ_m which can be applied to the specified material. σ_m or σ_a for the respective cases must be known.

14. EXPERIMENTAL STRESS ANALYSIS

The determination of static and dynamic strains in specimens and machine parts by means of physical measurements is defined as *experimental stress analysis*. Direct strain measurements are made possible by mechanical and electrical pickup devices. Mechanical and optical levers, brittle lacquers, photogrids, and cathetometers are mechanical systems. Electromagnetic, resistance, and capacitance methods are forms of electrical systems. Sometimes the nature of the installation and the operating conditions preclude direct strain measurements. Indirect methods such as high-speed photography of body displacements and velocity or acceleration measurements by means of moving coil pickups have proved to be acceptable.

Direct Strain Measurements

Bonded Wire Strain Gage -(Fig. 7) consists of a grid of strain-sensitive wire protected by a paper sheath and cemented to the specimen. A change in the length of the grid due to strains in the specimen changes the resistance of the wire. This change in resistance is readily measured by means of suitable electrical circuits. The gage is suitable for measurement of dynamic as well as static strains.

Mechanical Strain Gage is used to measure static strains in a specimen under load. The instrument employs either the optical or the mechanical lever system to multiply the strain, which then may be read from a suitable scale.

Magnetic Strain Gage is based on the principle of a change in either self or mutual inductance of the gaging element. The simplest type is the variable inductance gage, where the change in an air gap in a magnetic circuit varies the permeance of the circuit, thus giving an indication of the strains produced. See Fig. 8. This gage is used to measure both static and dynamic strains.

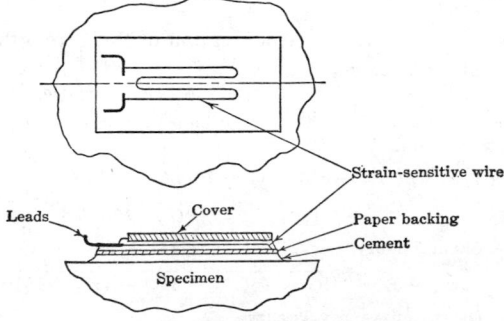

FIG. 7.   Bonded wire strain gage

Capacitance Strain Gage is based on the principle of variation of capacitance by separation of elements, change in area of elements, or change in specific inductive capacity of a dielectric. The first of these methods (Fig. 9) is the most widely used, and the gage is so

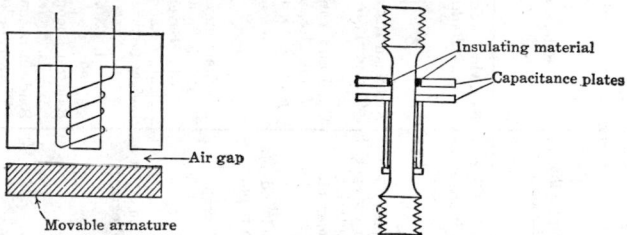

Fɪɢ. 8. Magnetic strain gage   Fɪɢ. 9. Capacitance strain gage

attached to the specimen that the strain produced causes a separation of the two plates. This type of gage is applicable for measuring both static and dynamic strains.

Acoustic Strain Gage employs a taut wire between two points mounted on the specimen, as shown in Fig. 10. The strain produced in the specimen alters the tension in the wire, thus changing its natural frequency. The natural frequency of the wire is determined by a variable frequency oscillator. Resonance is indicated by the output meter of the unit. By suitable methods the strain can be computed. This gage is suitable for measuring static and dynamic strains.

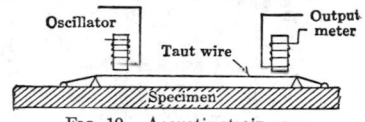

Fɪɢ. 10. Acoustic strain gage

Brittle Lacquer Coatings are very convenient as they graphically present an over-all strain picture, including the principal strain directions. They are revealed by cracks produced in the coating due to the strain in the specimen. The method consists of coating the specimen with lacquer, which becomes brittle on hardening. When the strains in the piece reach a prescribed magnitude, the lacquer cracks. This method, which gives quantitative results accurate to within ±10%, can be used to detect static and dynamic strains in tension or compression.

Photogrid method consists of coating the specimen with a photosensitive film on which a grid system is printed. The dimensional change in the grid system under load is a measure of the strain produced. This method is suitable for measuring permanent static strains only.

Cathetometer method consists of measuring the relative movement of two points on a test specimen under load by means of a microscope with a vernier scale attachment. Both static and dynamic strains can be measured by this method.

Indirect Strain and Displacement Measurements

Displacement Pickups are used to detect the motion or displacement of the specimen. In general they consist of a seismic mass-spring system having suitable dynamic characteristics. The motion of the seismic element is used to produce an electrical signal by any of the electrical methods just described.

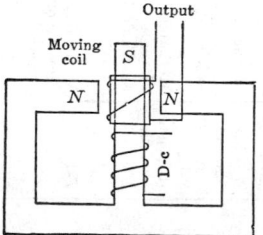

Fɪɢ. 11. Velocity pick-up

Velocity Pickups usually employ the principle of a moving permanent magnet within a coil whose field strength is constant. The voltage induced in the coil is proportional to the relative velocity of the magnet and coil. By mechanical modification, the arrangement can serve as a generator where the voltage output is proportional to the velocity of the moving coil, as shown in Fig. 11. This pickup can be used to measure dynamic strains. By electrically integrating the output, a signal proportional to displacement can be obtained.

Acceleration can be detected by pickups similar to those used for displacement and velocity. The output can be integrated for convenience. In some designs the system consists of a piezoelectric crystal which, by virtue of its high natural frequency, permits the detection of sharp transients and high-frequency accelerations.

Table 3. Characteristics of Strain-measuring Devices

(Reprinted by permission from an article by Karl F. Smith in *Product Engineering*, Jan. 1947)

Name of Device	Type	Principle	Type Strains Best Adapted (in Steels)	Specimen Types	Usual Gage Length, In.	Approx. Smallest Measurable Strain, microinches per inch	Approx. Range of Strains, %	Adaptability to Recording	Reference
"SR-4" Bonded Electric Strain Gage	A-1	Electrical resistance	Elastic	Flats 0.5 × 1 in. or greater. Curved surfaces greater than 1 in. radius on length or 0.5 in. on width	1 (6 to 1/16 in. on other types)	0.5	1 (static loads)	Oscillograph, pen and ink recorders	Baldwin Locomotive Works, *Bulletin* 179
Tuckerman Strain Gage	0.2 in. lozenge	Optical lever	Elastic	Almost any round or flat surface, with suitable attachment	1 or 2	2 on 0.2 in. lozenge	0.25 on 2 in. g. L (without reset)	Not possible, except by special adaptation	American Instrument Company, *Bulletin* 2084
General Motors Gage		Grids and photo-electric cell	Elastic	Any round or flat, with suitable attachment	1/16 to 1/4	2		Oscillograph	Gadd and Van de Grift *Trans. A.S.M.E.*, Vol. 64, p. A-15, 1942.
Interferometer Types		Light wave interference	Elastic	Depending on adaptation		2	About 1	Special adaptation required	
Statham Unbonded Wire Resistance Gages		Electrical resistance	Elastic	Any surface, with suitable attachment	1/2 to 6	4 on 2 in. g. l.	Depending on attachment	Oscillograph	The Statham Laboratories, 9328 Santa Monica Blvd., Beverly Hills, Calif., *Catalog.*
Magnetic Gage	Westinghouse, small size	Magnetic	Elastic	Any round or flat, with suitable attachment	3	5	About 1	Graphic meter or oscillograph	Langer (article), *Experimental Stress Analysis*, Vol. I, No. II, pp. 82–89
Whittemore	10 in. gage length	Mechanical lever	Elastic	Flats of over 10 in. in length	10	10	1	None	Baldwin Locomotive Works, *Bulletin* 167
Huggenberger Tensometer	A	Mechanical lever	Elastic	Any round or flat, with suitable attachment	1	10	1	None	Baldwin Locomotive Works, *Bulletin* 139
Templin	T-1	Mechanical lever (non-averaging)	Elastic	Rounds 0.05–0.505 in. dia., flats 0.01–0.50 in. thick × 0.50 in. wide	2	20 with 1000 magnification	2	Selsyn motors and rotating drum	Baldwin Locomotive Works, *Bulletin* 162

Instrument	Model	Measuring device	Type	Specimen / surface		Sensitivity	Magnification	Special requirements	Reference
Peters	P-1	Mechanical lever (averaging)	Elastic	Rounds 0.505 in. dia.	2	20 with 1000 magnification	2	Selsyn motors and rotating drum	Baldwin Locomotive Works, *Bulletin* 162
Olsen Electronic	YUAAR	Magnetic	Elastic	Rounds or flats, 1/16 to 0.505 in.	2	20 with 1000 magnification	2	Special amplifier and revolving drum	Tinius Olsen Testing Machine Company, *Bulletin* 24
Porter-Lipp		Mechanical lever	Elastic	Any surface, with suitable attachment	1	20	0.5	Special adaptation required	P. L. Porter, 226 Oceano Drive, Los Angeles 24, California
Riehle	Averaging dial type	Dial gage	Elastic and early plastic (averaging)	Rounds or flats, 1/8 to 5/8 in.	2	100	2.5	Special adaptation required	American Machine and Metals, Inc., *Bulletins*
Berry Strain Gage		Mechanical lever	Elastic or plastic	Flats of over 10 in. in length	10	50 on 10 in. g. l.	About 1	None	Davis, Troxell, and Wiskocil
DeForest Scratch Gage		Mechanical trace	Elastic and early plastic	Any flat surface	2	100	2.5	Scratch pattern on special plate	Waugh Laboratories, 420 Lexington Ave., New York, N.Y., *Bulletin on Strain Measurements.*
Amsler		Dial gage	Elastic and early plastic	Rounds 0.02 to 1.5 in. dia. Flats to 1.5 in. wide and to 1.5 in. thick	2 to 10	100 on 2 in. g. l.	8	Special adaptation required	Alfred J. Amsler and Co., Schaffhouse, Switzerland, *Descriptive Bulletin.*
Kenyon-Burns-Young	10 : 1 wedge	Wedge	Plastic	Flats up to 1/8 in. thick and up to 1/2 in. wide	2	500 with 20 magnification	45	Line from wedge around drum	Baldwin Locomotive Works, *Bulletin* 128
Photogrids and Comparator		Microscope comparator	Plastic	Any surface	0.01 to 0.25	Depending on objective and test conditions	About 40	Photography of pattern	O'Haven and Harding (article) *Experimental Stress Analysis*, Vol. II, No. II, pp. 59–70
Micrometer Caliper		Graduated screw	Plastic	Rounds and flats for transverse measurements	0.5	500 on C.505 in. round	No limit	Special adaptation required	....
Stresscoat	ST-103	Brittle coating	Elastic	Any surface	Very small 2	500	0.05 to 0.15	Photography of pattern	Magnaflux Corporation, ST-103 data sheet
Riehle Percentage Gage	2 in. gage length	Vernier	Plastic	Any surface prepared by punch marks		1000	70	Photography of pattern	American Machine and Metals, Inc., *bulletins*
Photoelasticity		Polarized light	Elastic	Transparent plastic model	Very small	Depends on model and testing conditions	Elastic region	Photography of pattern	Frocht, *Photoelasticity*, Vol. I, 1941, Vol. II, 1948, Wiley

Photoelastic Methods involve the construction of a model of the prototype from an isotropic transparent material such as Bakelite, Celluloid, and glass. When these substances are subjected to stress they become doubly refracting, and plane-polarized light on passing through the material is changed into two component mutually perpendicular rays which are parallel and perpendicular to the directions of stress. The emergent rays will be out of phase, depending on the stress and the thickness of the material. A photo-

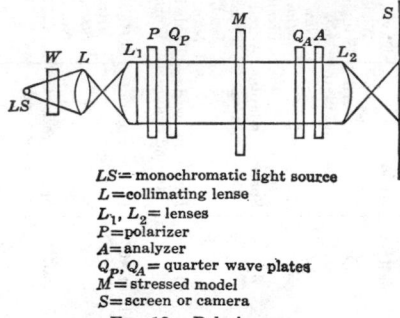

LS = monochromatic light source
L = collimating lense
L_1, L_2 = lenses
P = polarizer
A = analyzer
Q_P, Q_A = quarter wave plates
M = stressed model
S = screen or camera

Fig. 12. Polariscope

graph of the emergent rays gives light and dark fringes indicative of the distribution of the principal stresses. A schematic diagram of one method is shown in Fig. 12. This procedure is used to measure static stress.

Summary of Strain-measuring Devices. Table 3 presents a convenient summary of the more common commercial instruments. Detailed discussion of the various methods may be found in the references cited in the last column of Table 3 and also in Refs. 3 and 4.

REFERENCES

1. Roark, R. J., *Formulas for Stress and Strain*, New York, McGraw-Hill, 4th ed., 1965.
2. Marin, J., *Mechanical Properties of Materials and Design*, New York, McGraw-Hill, 1942.
3. *Proc. Society for Experimental Stress Analysis.*
4. Roberts, H. C., *Mechanical Measurements by Electrical Methods*, Instruments Publishing Co.

BEAMS

15. THEORY OF FLEXURE

Types of Beams. A beam is a bar or structural member subjected to transverse loads that tend to bend it. Usually beams are horizontal bars designed to carry vertical loads, but any structural member acts as a beam if bending is induced by external transverse forces.

A **Simple Beam** (Fig. 1a) is a horizontal member that rests on two supports at the ends of the beam. All parts between the supports have free movement in a vertical plane under the influence of vertical loads.

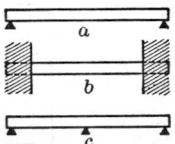

Fig. 1

A **Fixed Beam, Constrained Beam, or Restrained Beam** (Fig. 1b) is one that is rigidly fixed at both ends or rigidly fixed at one end and simply supported at the other.

A **Continuous Beam** (Fig. 1c) is a member resting on more than two supports.

A **Cantilever Beam** (Fig. 1d) is a member with one end projecting beyond the point of support, free to move in a vertical plane under the influence of vertical loads placed between the free end and the support.

Phenomena of Flexure. When a simple beam bends under its own weight, the fibers on the upper or concave side are shortened and the stress acting on them is compression; the fibers on the under or convex side are lengthened and the stress acting on them is tension. In addition to the longitudinal stresses acting on the fibers, shear exists along each cross section, the intensity of the shear being greatest along the sections at the two supports and zero at the middle section.

When a cantilever beam bends under its own weight the fibers on the upper or convex side are lengthened under tensile stresses; the fibers on the under or concave side are shortened under compressive stresses; the shear is greatest along the section at the support and zero at the free end.

The **Neutral Surface** is that horizontal section between the concave and convex surfaces of a loaded beam, where there is no change in the length of the fibers and no tensile or compressive stresses acting upon them.

The **Neutral Axis** is the trace of the neutral surface on any cross section of a beam (Fig. 2).

The **Elastic Curve** of a beam is the curve formed by the intersection of the neutral surface with the side of the beam, it being assumed that the longitudinal stresses on the fibers are within the elastic limit.

Reactions at Supports. The reactions or upward pressures at the points of support are computed by applying the following conditions necessary for equilibrium of a system of vertical forces in the same plane: (1) The algebraic sum of all vertical forces must

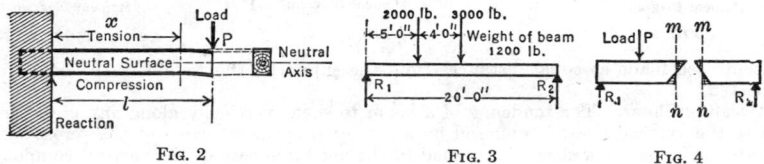

| Fig. 2 | Fig. 3 | Fig. 4 |

equal zero; i.e., the sum of the reactions equals the sum of the downward loads. (2) The algebraic sum of the moments of all the vertical forces must equal zero.

From the first condition it is evident that in a cantilever beam, where there is but one support, the reaction is equal to the sum of all vertical forces acting downward, comprising the weight of the beam and any loads that come upon it. Where there are two supports and the load is uniformly distributed, as is usually true of the weight of the beam itself, or if equal concentrated loads are placed at equal distances from the center of the beam, each support carries one-half of the total load and the reactions are equal. If a beam supported at two points is not uniformly loaded each reaction must be computed separately by applying the second condition for equilibrium. Thus in Fig. 3 let R_1 and R_2 be the two reactions of a simple beam 20 ft long weighing 1200 lb, which carries concentrated loads of 2000 lb and 3000 lb at distances of 5 ft and 9 ft, respectively, from the left support. Taking a center of moments for convenience at the left support, and regarding the weight of the beam as a concentrated load applied at its center of gravity,

$$+2000 \times 5 + 3000 \times 9 + 1200 \times 10 - R_2 \times 20 = 0. \quad R_2 = 2450 \text{ lb}$$

Taking a center of moments at the right support

$$R_1 \times 20 - 2000 \times 15 - 3000 \times 11 - 1200 \times 10 = 0. \quad R_1 = 3750 \text{ lb}$$

$R_1 + R_2 = 6200$ lb, which is equal to the sum of all the loads acting downward.

Conditions of Equilibrium. Imagine a vertical plane cutting the beam, Fig. 4, into two parts at any section mn. The external forces acting on either part, comprising the loads and the reaction, are held in equilibrium by the tensile, compressive, and shearing stresses acting on the fibers of the cross section. From the conditions necessary for static equilibrium of a system of forces in one plane the following fundamental laws are deduced for the stresses at any cross section:

(1) Sum of horizontal tensile stresses = sum of horizontal compressive stresses. (2) Resisting shear = vertical shears. (3) Resisting moment = bending moment.

Vertical Shear. At any cross section of a beam the resultant of the external vertical forces acting on one side of the section is equal and opposite to the resultant of the external vertical forces on the other side of the section. These forces tend to cause the beam to shear vertically along the section. The value of either resultant, which is a measure of the shearing tendency, is known as the vertical shear at the section considered. It is usually computed by finding the algebraic sum of the vertical forces to the left of the section; that is, it is equal to the left reaction minus the sum of the vertical downward forces acting between the left support and the section.

A *Shear Diagram* is a graphic representation of the vertical shear at all cross sections of the beam. Thus in Fig. 5 the ordinates to the line AOB represent to scale the intensity of the vertical shear at the corresponding sections of the simple beam. In this case the vertical shear is greatest at the supports where it is equal to the reactions, and it is zero at the center of the span. In the cantilever beam, Fig. 6, the vertical shear is greatest at the point of support where it is equal to the reaction, and it is zero at the free end. Figure 7 shows graphically the vertical shear on all sections of a simple beam carrying two

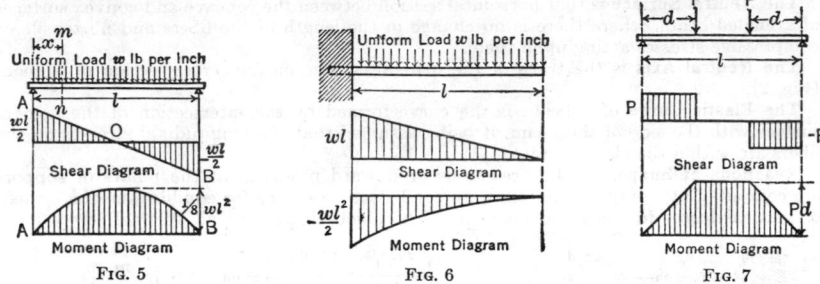

FIG. 5   FIG. 6   FIG. 7

concentrated loads at equal distances from the supports, the weight of the beam being neglected.

Resisting Shear. The tendency of a beam to shear vertically along any cross section due to the vertical shear is opposed by an internal shearing stress at that cross section known as the resisting shear; it is equal to the algebraic sum of the vertical components of all the internal stresses action on the cross section.

If V = vertical shear, pounds; V_r = resisting shear, pounds; τ = average unit shearing stress, pounds per square inch; and A = area of the section, then at any cross section

$$V_r = V = \tau A; \quad \tau = \frac{V}{A} \tag{1}$$

The resisting shear is not uniformly distributed over the cross section, but the intensity varies from zero at the extreme fiber to its maximum value at the neutral axis.

At any point in any cross section the vertical unit shearing stress, pounds per square inch, is

$$\tau = \frac{VA'c'}{It} \tag{2}$$

where V = total vertical shear, pounds, for the section considered; A' = area, square inches, of cross section between a horizontal plane through the point where the shear is being found and the extreme fiber on the same side of the neutral axis; c' = distance, inches, from the neutral axis to the center of gravity of the area A'; I = moment of inertia of the section, inches[4]; t = width of section at plane of shear, inches. For a solid rectangular beam, the maximum value of the unit shearing stress is

$$\tau = \frac{3V}{2A} \tag{3}$$

where A = total area, square inches, of cross section of the beam; for a solid circular beam, the maximum unit shearing stress is

$$\tau = \frac{4V}{3A} \tag{4}$$

Horizontal Shear. If, at any point within a body subject to stress, there exists a unit shearing stress along one plane, there must also be an equal unit shearing stress along a perpendicular plane through that point. For proof of this see Laurson and Cox: *Properties and Mechanics of Materials* (John Wiley, New York), 2nd Ed., 1947. In a beam, at any cross section where there is a vertical shearing force there must be resultant unit shearing stresses acting on the vertical faces of particles which lie at that section. On a horizontal surface of such a particle, there is a unit shearing stress equal to the unit shearing stress on a vertical surface of the particle. Equation 2 therefore also gives the horizontal unit shearing stress at any point on the cross section of a beam.

Bending Moment. The bending moment, or moment, at any cross section of a beam is the algebraic sum of the moments of the external forces acting on either side of the section. It is considered positive when it causes the beam to bend convex downward, hence causing compression in upper fibers and tension in lower fibers of beam. When the bending moment is determined from the forces that lie to the left of the section, it is positive if it acts in a clockwise direction; if determined from forces on the right side, it is positive if it acts in a counterclockwise direction. If the moments of upward forces are given positive signs, and the moments of downward forces negative signs, the bending moment will always have the correct sign, whether determined from the right or left side. The bending moment should be determined for the side for which the calculation will be simplest.

In Fig. 5 let M be the bending moment, pound-inches, at a section mn of a simple beam at a distance x, inches, from the left support, w = weight of the beam per 1 in. of length, and l = length of the beam, inches. Then the reactions are $1/2\, wl$ and $M = 1/2\, wl \cdot x - 1/2\, x \cdot wx$. For the sections at the supports $x = 0$ or l and $M = 0$. For the section at the center of the span $x = 1/2\, l$ and $M = 1/8\, wl^2 = 1/8\, Wl$, where W is the total weight.

A **Moment Diagram** (Figs. 5, 6, and 7) shows the bending moment at all cross sections of a beam. Ordinates to the curve represent to scale the moments at the corresponding cross sections. The curve for a simple beam uniformly loaded is a parabola showing $M = 0$ at the supports and $M = 1/8\, wl^2 = 1/8\, Wl$ at the center, M being in pound-inches.

The **Dangerous Section** is the cross section of a beam where the bending moment is greatest. In a cantilever beam the dangerous section is at the point of support, regardless of the disposition of the loads. In a simple beam it is that section where the vertical shear changes from positive to negative, and may be located graphically by constructing a shear diagram, or numerically by taking the left reaction and subtracting the loads in order from the left until a point is reached where the sum of the loads subtracted equals the reaction. For a simple beam uniformly loaded, the dangerous section is at the center of the span.

Resisting Moment. The tendency to rotate about a point in any cross section of a beam is due to the bending moment at that section; this tendency is resisted by the resisting moment which is the algebraic sum of the moments of all the horizontal stresses with reference to the same point.

Formula for Flexure. If M = bending moment, M_r = resisting moment of the horizontal fiber stresses, σ = unit stress (tensile or compressive) on any fiber, usually that most remote from the neutral surface, and c = distance of that fiber from the neutral surface, then

$$M = M_r = \frac{\sigma I}{c}; \quad \sigma = \frac{Mc}{I} \tag{5}$$

where I is the moment of inertia of the cross section with respect to its neutral axis. If σ is in pounds per square inch, then M must be in pounds-inches, I in inches^4, and c in inches. (For proof of the relation $M = \sigma I/c$, see Seely's *Resistance of Materials*.)

Formula 5 is the basis of the design and investigation of beams. It is true only when the maximum horizontal fiber stress σ does not exceed the proportional limit of the material.

The **Moment of Inertia** is the sum of the products of each elementary area of the cross section multiplied by the square of the distance of that area from the assumed axis of rotation, or

$$I = \Sigma r^2\, \Delta A = \int r^2\, dA \tag{6}$$

in which Σ is the sign of summation, ΔA = an elementary area of the section, and r = distance of ΔA from the axis. It is evident that the moment of inertia is greatest in those sections (such as I-beams) having much of the area concentrated at a distance from the axis. Unless otherwise stated the neutral axis is the axis of rotation considered. I is usually expressed in inches^4.

Modulus of Rupture. To determine the ultimate strength of a material in bending, a beam is loaded to rupture and the maximum load which it carries is noted. The flexure formula, $\sigma = Mc/I$, is true only for stresses within the proportional limit of the material. Therefore, the value, σ, of the rupture strength obtained from the breaking load by this equation is incorrect. However, the equation is used and the nominal value so found is called the *modulus of rupture*, which is a measure of the ultimate load-carrying capacity of the beam. If the strengths in tension and compression are different, the modulus of rupture may be intermediate between the two. It is to be noted that the modulus of rupture does not express the actual stress in the extreme fiber of a beam but is a quantity useful only as a basis of comparison.

The **Section Modulus** is the factor I/c in the flexure formula. It is a measure of the capacity of a section to resist any bending moment to which it may be subjected. The section modulus is expressed in inches^3. For values of I and I/c for simple shapes used as beam sections, see Table 1. Properties of standard structural shapes are given in Section 1, and properties for other geometric sections in Table 2, p. 448.

Table 1. Properties of Sections of Beams *

Section of beam	Moment of inertia, I, inches⁴	Section modulus, I/c, inches³	Radius of gyration, k, inches
	$bd^3/12$	$bd^2/6$	$d/\sqrt{12} = 0.289d$
	$\dfrac{b_1d_1^3 - b_2d_2^3}{12}$	$\dfrac{b_1d_1^3 - b_2d_2^3}{6d_1}$	$\sqrt{\dfrac{b_1d_1^3 - b_2d_2^3}{12(b_1d_1 - b_2d_2)}}$
	$\pi d^4/64$	$\pi d^3/32$	$d/4$
	$\pi(d_1^4 - d_2^4)/64$	$\pi(d_1^4 - d_2^4)/32d_1$	$\sqrt{d_1^2 + d_2^2}/4$
	$bd^3/36$	$bd^2/24$ (min.)	$d/\sqrt{18} = 0.236d$

* See also Table 2, p. 448.

Elastic Deflection of Beams. When a beam bends under load, all points of the elastic curve except those over the supports are deflected from their original positions. The radius of curvature ρ of the elastic curve at any section is expressed as

$$\rho = \frac{EI}{M} \tag{7}$$

where E = the modulus of elasticity of the material, pounds per square inch; I = moment of inertia, inches⁴, of the cross section with reference to its neutral axis; and M = bending moment, pound-inches, at the section considered. Where there is no bending moment ρ is infinity and the curve is a straight line; where M is greatest ρ is smallest and the curvature, therefore, is greatest.

If the elastic curve is referred to a system of coordinate axes in which x represents horizontal distances, y vertical distances, and l distances along the curve, the value of ρ by the aid of the calculus is found to be $d^3l/dx\ d^2y$. Substituting this value in the expression $\rho = EI/M$ and assuming that dx and dl are practically equal there results the following differential equation of the elastic curve which applies to all beams when the elastic limit of the material is not exceeded:

$$EI \frac{d^2y}{dx^2} = M \tag{8}$$

Equation 8 is used to determine the deflection of any point of the elastic curve by regarding the point of support as the origin of the coordinate axis and taking y as the vertical deflection at any point on the curve and x as the horizontal distance from the support to the point considered. The values of E, I, and M are substituted and the expression is integrated twice, giving proper values to the constants of integration, and the deflection y is determined for any point. See Table 3.

Example. The cantilever beam shown in Fig. 2 has a length = l, inches, and carries a load P, pounds, at the free end. It is required to find the deflection of the elastic curve at a point distant x inches, from the support, the weight of the beam being neglected.

The moment $M = -P(l - x)$; substituting in eq. 8, the equation for the elastic curve becomes $EI(d^2y/dx^2) = -Pl + Px$. Integrating and determining the constant of integration by the condition that $dy/dx = 0$ when $x = 0$, there results $EI(dy/dx) = -Plx + \frac{1}{2}Px^2$. Integrating a second time and determining the constant by the condition that $x = 0$ when $y = 0$, there results $EIy = -\frac{1}{2}Plx^2 + \frac{1}{6}Px^3$, which is the equation of the elastic curve. When $x = l$ the value of y, or the deflection in inches at the free end, is found to be $-Pl^3/3EI$.

Deflection Due to Shear. The deflection of a beam as computed by the ordinary formulas is that due to flexural stresses only. In short beams the deflection due to vertical shear is sometimes appreciable and may need to be considered. Because of the non-uniform distribution of the shear over the cross section of the beam, computing the deflection due to shear by exact methods is difficult. It may be approximated by $\delta = M/AG$, where δ = deflection, inches, due to shear; M = bending moment, pound-inches, at the section where the deflection is calculated; G = modulus of elasticity in shear, pounds per square inch; A = area of cross section of beam, square inches (see Seely's *Resistance of Materials*, 3rd ed., p. 164. Swain says (*Strength of Materials*, McGraw-Hill, p. 223) that for a rectangular section the ratio of deflection due to shear, to deflection due to bending, will be less than 5%, so long as the depth of the beam is less than $\frac{1}{8}$ of the length.

16. BEAMS OF UNIFORM CROSS SECTION

Design Procedure. In designing a beam the procedure is: (1) Compute the reactions. (2) Determine the position of the dangerous section and the bending moment at that section. (3) Divide the maximum bending moment (expressed in pound-inches) by the allowable unit stress (expressed in pounds per square inch) to obtain the minimum value of the section modulus. (4) Select a beam section with a section modulus equal to or slightly greater than the section modulus required.

Web Shear. A beam designed in the above manner is safe against rupture of the extreme fibers due to bending in a vertical plane, and usually the cross section will have sufficient area to sustain the shearing stresses with safety. For short beams carrying heavy loads, however, the vertical shear at the supports is large, and it may be necessary to increase the area of the section to keep the unit shearing stress within the limit allowed. For short beams, the average unit shearing stress is computed by $\tau = V/A$, where V = total vertical shear, pounds; and A = area of web, square inches. For allowable average unit shearing stresses in cross section of web of girders and rolled steel beams, as specified by different authorities, see Table 2. For timber beams, see Horizontal Shear in Timber Beams, p. 517.

Miscellaneous Considerations. Other considerations that will influence the choice of section under certain conditions of loading are: (1) The maximum vertical deflection that may be permitted in beams coming in contact with plaster. (2) The danger of failure by sidewise bending in long beams unbraced against lateral deflection. (3) The danger of failure by the buckling of the web of steel beams of short span carrying heavy loads. (4) The danger of failure by horizontal shear, particularly in wooden beams.

Vertical Deflection. When a beam is to be used to support or come in contact with materials like plaster, which may be broken by excessive deflection of the beam, it is usual in practice to select such a beam that the maximum deflection will not be greater than $\frac{1}{360}$ of the span.

It may be shown that for a simple beam, supported at the ends, with a total uniformly distributed load W, pounds, the deflection, inches, is

$$y = 30\frac{\sigma L^2}{Ed} \qquad (9)$$

where σ = allowable fiber unit stress, pounds per square inch; L = span of beam, feet; E = modulus of elasticity, pounds per square inch; d = depth of beam, inches.

If the deflection of a steel beam is to be less than $\frac{1}{360}$ of the span, it may be shown from eq. 9 that, for a maximum allowable fiber stress of 18,000 psi, the limit of span in feet is approximately 1.8 times the depth of the beam in inches.

For the deflection due to the impact of a moving load falling on a beam, see p. 498.

Lateral Deflection. When a beam carries vertical loads, the tensile stresses in the fibers tend to hold the tension flange of the beam in line but the compressive stresses tend to cause lateral bending of the compression flange. To obtain security against failure of long beams by excessive lateral deflection, sidewise bracing of the compression flange should be used or the allowable unit stress for the compression fibers should be reduced. The specifications for steel structures of the American Bridge Company provide that "the lateral unsupported length of beams and girders shall not exceed 40 times the width of

Table 2. Allowable Unit Stresses for Structural Steel as Specified by Different Authorities

All stresses in pounds per square inch. l = length of member; b = width of flange; k = radius of gyration; d = depth of web plate; t = thickness of web plate; all dimensions in inches.

Character of Stress	Report of National Bureau of Standards		A.I.S.C. 1946	A.R.E.A. 1950
	Acceptable Steel [a]	Standard Steel		
Axial tension	16,000	18,000	20,000	18,000
Direct compression	12,500	14,000	20,000 [b]	18,000 [c]
Compression in columns	$16,000 - 60l/k$ Max. 12,500	$18,000 - 70l/k$ Max. 14,000	$17,000 - 0.485\dfrac{l^2}{k^2}$	$17,000 - \dfrac{1}{4}\dfrac{l^2}{k^2}$ [d] $15,000 - \dfrac{1}{3}\dfrac{l^2}{k^2}$ [e]
Fiber stress in flexure in tension or in compression where l/b is 15 or less	16,000	18,000	20,000 [f]	18,000
Compressive fiber stress in flexure for l/b greater than 15 and not exceeding 40	$19,600 - 240l/b$	$22,000 - 270l/b$	$\dfrac{12,000,000}{\dfrac{ld}{bt}}$ [g]	$18,000 - 5l^2/b^2$
Fiber stress on pins	24,000	27,000	30,000	27,000
Bearing on plane-faced or rolled surfaces	24,000	27,000	27,000	27,000
Shear in gross section of web of girders or rolled beams where d/t does not exceed value indicated as Q	$Q = 43$ 10.700	$Q = 43$ 12,000	$Q = 70$ 13,000	$Q = 60$ 11,000
Shear where d/t exceeds the value indicated as Q	$Q = 43$ $13,300 - 62d/t$	$Q = 43$ $15,000 - 70d/t$	$Q = 70$ $\dfrac{64,000,000}{(d/t)^2}$	. . . .
Shear in power-driven rivets or turned bolts	12,000	13,500	15,000	13,500 [h]
Shear in hand-driven rivets or rough bolts	9,000	10,000	10,000	11,000 [i]
Bearing on power-driven rivets, pins, or turned bolts in reamed holes, single shear	24,000	24,000	32,000	27,000 (Rivets) 24,000 (Pins) 20,000 (Bolts)
Bearing on power-driven rivets, pins, or turned bolts in reamed holes, double shear	30,000	30,000	40,000 32,000 40,000	27,000 (Rivets) 24,000 (Pins) 20,000 (Bolts)
Bearing on hand-driven rivets or unfinished bolts, single shear	16,000	16,000	20,000	20,000 [j]
Bearing on hand-driven rivets or unfinished bolts, double shear	20,000	20,000	25,000	20,000 [j]

[a] Steel acceptable to the building official but the origin and physical characteristics of which are not determined.
[b] Plate girder stiffeners, gross section and butt welds, section through throat (crushing).
[c] Plate girder stiffeners, gross section.
[d] Riveted ends.
[e] Pin ends.
[f] In compression ld/bt not in excess of 600.
[g] ld/bt in excess of 600.
[h] Power-driven rivets only.
[i] Hand-driven rivets and turned bolts; rough bolts not acceptable for structural work.
[j] Rivets only.

the compression flange. When the unsupported length l, inches, exceeds 10 times the width b, inches, of the compression flange, the stress per square inch shall not exceed $(19,000 - 300l/b)$." The values specified by other authorities are given in Table 2. See Examples in Beam Design, p. 523.

Buckling of the Web. When a short-span steel beam, carrying heavy concentrated loads, has sufficient web thickness safely to resist shearing stresses, there is still a possibility that the beam is overloaded when the web is considered as a long slender column. The probability of this type of failure is not great if the web thickness is a relatively large percentage of the depth of beam, but it is great if the web is thin and the depth of the beam relatively large. Failure of the web by buckling may be guarded against by increasing its thickness, or by riveting stiffening angles to it. The A.I.S.C. specifications require stiffeners to be placed on the webs of rolled beams and plate girders at the ends, at points of concentrated loads, and at other points where h, the clear distance between flanges in inches, is equal to or greater than $70t$, where t = thickness of web in inches, and v exceeds $64,000,000/(h/t)^2$, where v = the unit shear stress at the section in psi. Where stiffeners are required, the distance between them shall not exceed 84 in. or $11,000t/\sqrt{v}$.

Also, the web over a support may crush, if the end bearing is too short; with a thick web, crushing at the junction of the web and flange may occur, but with a thin web and deep beam the web probably will buckle because of column action before it will crush. Security against this excessive stress in the web may be provided by stiffening angles riveted to the web and bearing against the flanges over the support. Such stiffening angles should be placed either directly over the support or at a distance from the support equal to the depth of the beam. If stiffening angles are not used, the allowable reaction recommended by the A.I.S.C. (A.I.S.C.: *Steel Construction*, 1946) is given by $R = \sigma_b t(a + k)$, where $\sigma_b = 24,000$ psi; R = allowable end reaction, pounds; t = web thickness, inches; a = length of bearing, inches; k = distance from outer face of flange to web toe of fillet, inches.

When the loading reaction exceeds the value of R, the web must be stiffened or additional length of bearing provided. In no case may the loading reaction exceed the maximum web shear V. Lack of proper lateral support for the top flange of the beam at the reaction point so decreases the crippling strength of the web as to render such practice inadmissible. For tests on web buckling, see *Bulls.* 68, 86, and 241, *Eng. Exp. Sta.*, Univ. of Ill.

Horizontal Shear in Timber Beams. (See Horizontal Shear, p. 512). In beams of a homogeneous material which can withstand equally well shearing stresses in any direction, the vertical and horizontal shearing stresses are equally important. In timber, however, the shearing strength along the grain is much less than that perpendicular to the grain, and hence the beams may fail owing to horizontal shear. Short wooden beams should be checked for horizontal shear in order that the allowable unit shearing stress along the grain shall not be exceeded. See Example 4, p. 5-37.

Restrained Beams. A beam is considered to be restrained if one or both ends are not free to rotate. This condition exists if a beam is built into a masonry wall at one or both ends, if it is riveted or otherwise fastened to a column, or if the ends projecting beyond the supports carry loads which tend to prevent the tilting of the ends which would naturally occur as the beam deflects. The shears and moments given in Table 3 for fixed-end conditions are seldom if ever attained, since the restraining elements themselves deform and reduce the magnitude of the restraint. This reduction of restraint decreases the negative moment at the support and increases the positive moment in the central portion of the span. The amount of restraint which exists is a matter which must be judged for each case in the light of the construction used, the rigidity of the connections, and the relative sizes of the connecting members.

Safe Loads on Simple Beams. By substituting in the flexure formula, p. 513, the value of M for a simple beam uniformly loaded as given in Table 3, the expression

$$W = \frac{2}{3}\sigma\frac{S}{L} \tag{10}$$

is obtained, where W = total load, pounds; σ = extreme fiber unit stress, pounds per square inch; S = section modulus, inches^3; L = length of span, feet.

If σ is taken as the maximum allowable fiber unit stress, eq. 10 gives the maximum allowable load on the beam. Most building codes permit a value of $\sigma = 18,000$ psi for quiescent loads. For this value of σ, eq. 10 becomes

$$W = 12,000\frac{S}{L} \tag{11}$$

**Table 3.   Bending Moment, Vertical Shear, and Deflection of Beams of Uniform
Cross Section under Various Conditions of Loading**

P = concentrated loads, lb.
R_1, R_2 = reactions, lb.
w = uniform load per unit of length, lb per in.
W = total uniform load on beam, lb.
l = length of beam, in.
x = distance from support to any section, in.
E = modulus of elasticity, lb per sq in.

I = moment of inertia, in.[4]
V_x = vertical shear at any section, lb.
V = maximum vertical shear, lb.
M_x = bending moment at any section, lb-in.
M = maximum bending moment, lb-in.
y = maximum deflection, in.

SIMPLE BEAM—UNIFORM
LOAD

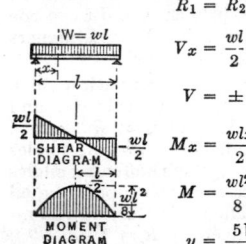

$$R_1 = R_2 = \frac{wl}{2}$$

$$V_x = \frac{wl}{2} - wx$$

$$V = \pm \frac{wl}{2} \left(\text{when} \begin{cases} x = 0 \\ x = l \end{cases} \right)$$

$$M_x = \frac{wlx}{2} - \frac{wx^2}{2}$$

$$M = \frac{wl^2}{8} \left(\text{when } x = \frac{l}{2} \right)$$

$$y = \frac{5Wl^3}{384EI} \quad \begin{array}{c}\text{(at center of}\\ \text{span)}\end{array}$$

SIMPLE BEAM—CONCEN-
TRATED LOAD AT ANY POINT

$R_1 = P(1 - k)$
$R_2 = Pk$
$V_x = R_1$ (when $x < kl$)
$ = R_2$ (when $x > kl$)
$V = P(1 - k)$
$ \text{(when } k < 0.5)$
$ = -Pk$ (when $k > 0.5$)
$M_x = Px(1 - k)$
$ \text{(when } x < kl)$
$ = Pk(l - x)$
$ \text{(when } x > kl)$
$M = Pkl(1 - k)$ (at point of
load)

$$y = \frac{Pl^3}{3EI} (1 - k)$$

$$\times (2/3k - 1/3k^2)^3/2$$

(at $x = l\sqrt{2/_3k - 1/3k^2}$)

SIMPLE BEAM—CONCEN-
TRATED LOAD AT CENTER

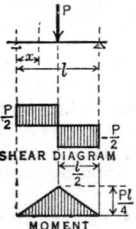

$$R_1 = R_2 = \frac{P}{2}$$

$$V_x = V = \pm \frac{P}{2}$$

$$M_x = \frac{Px}{2}$$

$$M = \frac{Pl}{4} \left(\text{when } x = \frac{l}{2} \right)$$

$$y = \frac{Pl^3}{48EI}$$

(at center of span)

SIMPLE BEAM—TWO EQUAL
CONCENTRATED LOADS AT
EQUAL DISTANCES FROM
SUPPORTS

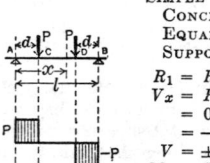

$R_1 = R_2 = P$
$V_x = P$ .... for AC
$ = 0$ .... for CD
$ = -P$ .... for DB
$V = \pm P$
$M_x = Px$ .... for AC
$ = Pd$ .... for CD
$ = P(l - x)$ ...... for DB
$M = Pd$

$$y = \frac{Pd}{24EI} (3l^2 - 4d^2)$$

(at center of span)

SIMPLE BEAM—LOAD IN-
CREASING UNIFORMLY FROM
SUPPORTS TO CENTER OF
SPAN

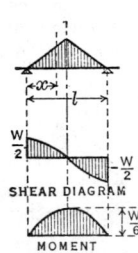

$$R_1 = R_2 = \frac{W}{2}$$

$$V_x = W \left(\frac{1}{2} - \frac{2x^2}{l^2} \right)$$

$$\left(\text{when } x < \frac{l}{2} \right)$$

$$V = \pm \frac{W}{2} \text{ (at supports)}$$

$$M_x = Wx \left(\frac{1}{2} - \frac{2x^2}{3l^2} \right)$$

$$M = \frac{Wl}{6} \text{ (at center of span)}$$

$$y = \frac{Wl^3}{60EI} \quad \begin{array}{c}\text{(at center of}\\ \text{span)}\end{array}$$

CANTILEVER BEAM—LOAD
CONCENTRATED AT FREE END

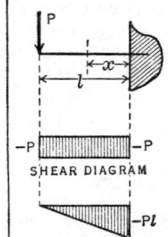

$R = P$

$$V_x = V = -P$$

$$M_x = -P(l - x)$$

$$M = -Pl \text{ (when } x = 0)$$

$$y = \frac{Pl^3}{3EI}$$

Table 3—*Continued*

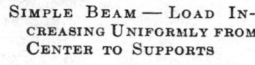

SIMPLE BEAM — LOAD INCREASING UNIFORMLY FROM CENTER TO SUPPORTS	**CANTILEVER BEAM—UNIFORM LOAD**

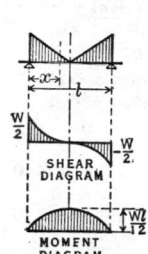

$$R_1 = R_2 = \frac{W}{2}$$

$$V_x = - W\left(\frac{2x}{l} - \frac{2x^2}{l^2} - \frac{1}{2}\right)$$

$$\left(\text{when } x < \frac{l}{2}\right)$$

$$V = \pm \frac{W}{2}$$

$$M_x = Wx\left(\frac{1}{2} - \frac{x}{l} + \frac{2}{3}\frac{x^2}{l^2}\right)$$

$$\left(\text{when } x < \frac{l}{2}\right)$$

$$M = \frac{Wl}{12} \quad \text{(at center of span)}$$

$$y = \frac{3}{320}\frac{Wl^3}{EI} \quad \text{(at center of span)}$$

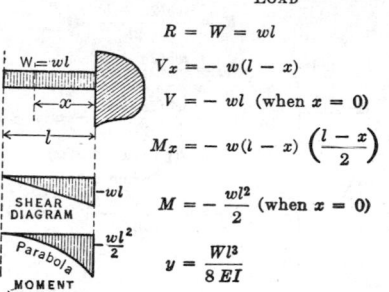

$$R = W = wl$$

$$V_x = - w(l - x)$$

$$V = - wl \quad \text{(when } x = 0)$$

$$M_x = - w(l - x)\left(\frac{l - x}{2}\right)$$

$$M = - \frac{wl^2}{2} \quad \text{(when } x = 0)$$

$$y = \frac{Wl^3}{8\,EI}$$

SIMPLE BEAM — LOAD INCREASING UNIFORMLY FROM ONE SUPPORT TO THE OTHER	**CANTILEVER BEAM—LOAD INCREASING UNIFORMLY FROM FREE END TO SUPPORT**

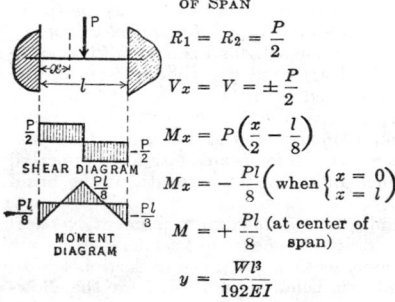

$$R_1 = \frac{W}{3}; \quad R_2 = \frac{2}{3}W$$

$$V_x = W\left(\frac{1}{3} - \frac{x^2}{l^2}\right)$$

$$V = - \frac{2}{3}W \quad \text{(when } x = l)$$

$$M_x = \frac{Wx}{3}\left(1 - \frac{x^2}{l^2}\right)$$

$$M = \frac{2}{9\sqrt{3}}Wl$$

$$\left(\text{when } x = \frac{l}{\sqrt{3}}\right)$$

$$y = \frac{0.01304}{EI}Wl^3$$

$$R = W$$

$$V_x = - W\frac{(l - x)^2}{l^2}$$

$$V = - W \quad \text{(when } x = 0)$$

$$M_x = - \frac{W}{3}\frac{(l - x)^3}{l^2}$$

$$M = - \frac{Wl}{3} \quad \text{(when } x = 0)$$

$$y = \frac{Wl^3}{15EI}$$

FIXED BEAM — CONCENTRATED LOAD AT CENTER OF SPAN	**FIXED BEAM—UNIFORM LOAD**

$$R_1 = R_2 = \frac{P}{2}$$

$$V_x = V = \pm \frac{P}{2}$$

$$M_x = P\left(\frac{x}{2} - \frac{l}{8}\right)$$

$$M_x = - \frac{Pl}{8}\left(\text{when } \begin{cases} x = 0 \\ x = l \end{cases}\right)$$

$$M = + \frac{Pl}{8} \quad \text{(at center of span)}$$

$$y = \frac{Wl^3}{192EI}$$

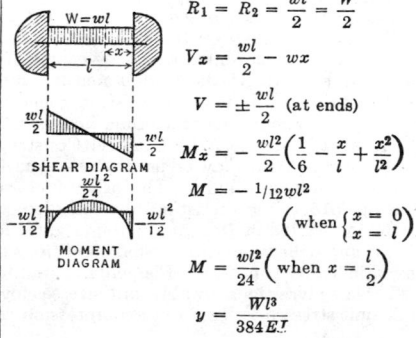

$$R_1 = R_2 = \frac{wl}{2} = \frac{W}{2}$$

$$V_x = \frac{wl}{2} - wx$$

$$V = \pm \frac{wl}{2} \quad \text{(at ends)}$$

$$M_x = - \frac{wl^2}{2}\left(\frac{1}{6} - \frac{x}{l} + \frac{x^2}{l^2}\right)$$

$$M = - 1/12\,wl^2$$

$$\left(\text{when } \begin{cases} x = 0 \\ x = l \end{cases}\right)$$

$$M = \frac{wl^2}{24}\left(\text{when } x = \frac{l}{2}\right)$$

$$y = \frac{Wl^3}{384EI}$$

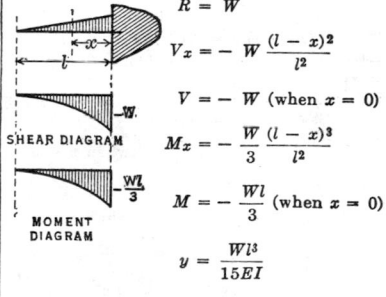

Table 3—*Continued*

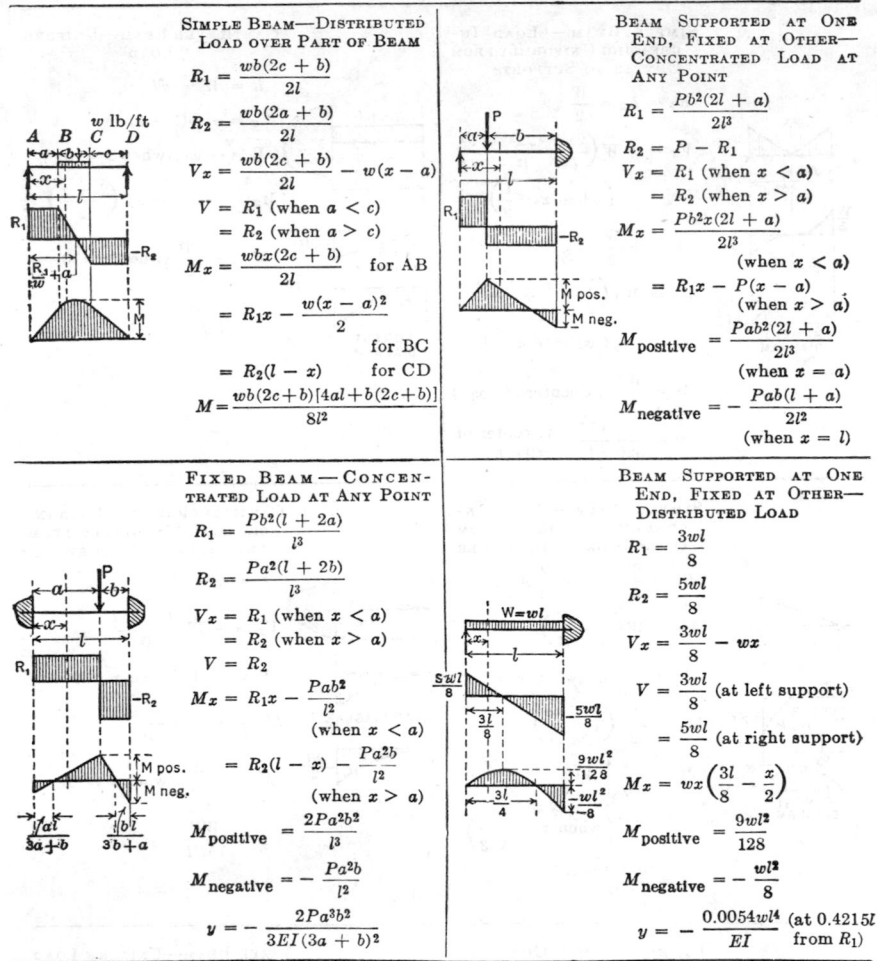

SIMPLE BEAM—DISTRIBUTED LOAD OVER PART OF BEAM

$$R_1 = \frac{wb(2c + b)}{2l}$$

$$R_2 = \frac{wb(2a + b)}{2l}$$

$$V_x = \frac{wb(2c + b)}{2l} - w(x - a)$$

$$V = R_1 \text{ (when } a < c)$$
$$= R_2 \text{ (when } a > c)$$

$$M_x = \frac{wbx(2c + b)}{2l} \quad \text{for } AB$$

$$= R_1 x - \frac{w(x - a)^2}{2}$$
$$\text{for } BC$$

$$= R_2(l - x) \quad \text{for } CD$$

$$M = \frac{wb(2c+b)[4al+b(2c+b)]}{8l^2}$$

BEAM SUPPORTED AT ONE END, FIXED AT OTHER—CONCENTRATED LOAD AT ANY POINT

$$R_1 = \frac{Pb^2(2l + a)}{2l^3}$$

$$R_2 = P - R_1$$

$$V_x = R_1 \text{ (when } x < a)$$
$$= R_2 \text{ (when } x > a)$$

$$M_x = \frac{Pb^2 x(2l + a)}{2l^3}$$
$$\text{(when } x < a)$$
$$= R_1 x - P(x - a)$$
$$\text{(when } x > a)$$

$$M_{\text{positive}} = \frac{Pab^2(2l + a)}{2l^3}$$
$$\text{(when } x = a)$$

$$M_{\text{negative}} = -\frac{Pab(l + a)}{2l^2}$$
$$\text{(when } x = l)$$

FIXED BEAM—CONCENTRATED LOAD AT ANY POINT

$$R_1 = \frac{Pb^2(l + 2a)}{l^3}$$

$$R_2 = \frac{Pa^2(l + 2b)}{l^3}$$

$$V_x = R_1 \text{ (when } x < a)$$
$$= R_2 \text{ (when } x > a)$$
$$V = R_2$$

$$M_x = R_1 x - \frac{Pab^2}{l^2}$$
$$\text{(when } x < a)$$
$$= R_2(l - x) - \frac{Pa^2 b}{l^2}$$
$$\text{(when } x > a)$$

$$M_{\text{positive}} = \frac{2Pa^2 b^2}{l^3}$$

$$M_{\text{negative}} = -\frac{Pa^2 b}{l^2}$$

$$y = -\frac{2Pa^3 b^2}{3EI(3a + b)^2}$$

BEAM SUPPORTED AT ONE END, FIXED AT OTHER—DISTRIBUTED LOAD

$$R_1 = \frac{3wl}{8}$$

$$R_2 = \frac{5wl}{8}$$

$$V_x = \frac{3wl}{8} - wx$$

$$V = \frac{3wl}{8} \text{ (at left support)}$$

$$= \frac{5wl}{8} \text{ (at right support)}$$

$$M_x = wx\left(\frac{3l}{8} - \frac{x}{2}\right)$$

$$M_{\text{positive}} = \frac{9wl^2}{128}$$

$$M_{\text{negative}} = -\frac{wl^2}{8}$$

$$y = -\frac{0.0054wl^4}{EI} \quad \text{(at } 0.4215l \text{ from } R_1)$$

Tables of safe loads, based on eq. 11, for various structural steel beams may be found in the A.I.S.C. Steel Construction, 1946, and in Kent's *Handbook for Mechanical Engineers*.

If the load is concentrated at the center of the span, the safe load is one-half the value given by eq. 11. If the load is neither uniformly distributed nor concentrated at the center of the span, the beam must, in general, be designed as in the problems below. For some special conditions of loading, however, it is possible to find an equivalent uniformly distributed load (see A.I.S.C. Steel Construction, 1946, pp. 366 et seq.), in which case tables of safe loads may be used. The equations above are for beams laterally supported and are for flexure only. The other factors which influence the strength of the beam (see pp. 515 et seq.) must still be considered.

Use of Tables in Design. Table 3 gives formulas for the maximum moment, vertical shear, and deflection under usual conditions of loading. Elements of various sections are given in Table 1, p. 514. Elements of standard steel sections are given in Section 1.

Table 4 gives the allowable unit stresses for timber in bending; in Table 5 are the allowable unit stresses for timber in compression and shear.

Table 4. Allowable Unit Stresses for Timber in Bending *

Recommended by the Forest Products Laboratory, Forest Service, U. S. Dept. of Agriculture.† All values are in pounds per square inch.

Species	Continuously Dry — All Thicknesses		Occasionally Wet but Quickly Dried				More or Less Continuously Damp or Wet			
			4 in. and Thinner		5 in. and Thicker		4 in. and Thinner		5 in. and Thicker	
	Select	Common	Select	Common	Select	Common	Select	Common	Select	Common
Ash, black	1000	800	800	680	900	720	710	600	800	640
" commercial white	1400	1120	1070	910	1200	960	890	760	1000	800
Aspen and large tooth aspen	800	640	580	490	650	520	440	370	500	400
Basswood	800	640	580	490	650	520	440	370	500	400
Beech	1500	1200	1150	980	1300	1040	890	760	1000	800
Birch, paper	900	720	670	570	750	600	530	450	600	480
" yellow and sweet	1500	1200	1150	980	1300	1040	890	760	1000	800
Cedar, Alaska	1100	880	890	760	1000	800	800	680	900	720
" western red	900	720	710	600	800	640	670	570	750	600
" northern and southern white	750	600	580	490	650	520	530	450	600	480
" Port Orford	1100	880	890	760	1000	800	800	680	900	720
Chestnut	950	760	760	650	850	680	620	530	700	560
Cottonwood, eastern and black	800	640	580	490	650	520	530	450	600	480
Cypress, southern	1300	1040	980	830	1100	880	800	680	900	720
Douglas fir (western Wash. and Ore.) ‡	1600	1200	1233	983	1387	1040	948	756	1067	800
Douglas fir (dense) ‡	1750	1400	1349	1147	1517	1213	1037	882	1167	933
" (Rocky Mt.)	1100	880	900	680	900	720	620	530	700	560
Elm, rock	1500	1200	1150	980	1300	1040	890	760	1000	800
" slippery and American	1100	880	800	680	900	720	710	600	800	640
Fir, balsam	900	720	670	570	750	600	530	450	600	480
" commercial white	1100	880	800	680	900	720	710	600	800	640
Gum, red, black, and tupelo	1100	880	800	680	900	720	710	600	800	640
Hemlock, eastern	1100	880	800	680	900	720	710	600	800	640
" western	1300	1040	980	830	1100	880	800	680	900	720
Hickory (true and pecan)	1900	1520	1330	1130	1500	1200	1070	910	1200	960
Larch, western	1200	960	980	830	1100	880	800	680	900	720
Maple, sugar and black	1500	1200	1150	980	1300	1040	890	760	1000	800
" red and silver	1000	800	800	680	900	720	620	530	700	560
Oak, commercial red and white	1400	1120	1070	910	1200	960	890	760	1000	800
Pine, southern yellow ‡		1200		983		1040		756		800
" southern yellow (dense) ‡	1750	1400	1349	1147	1517	1213	1037	882	1167	933
" northern and western white, western yellow, sugar	900	720	710	600	800	640	670	570	750	600
" Norway	1100	880	890	760	1000	800	710	600	800	640
Poplar, yellow	1000	800	800	680	900	720	710	600	800	640
Redwood	1200	960	890	760	1000	800	710	600	800	640
Spruce, red, white and Sitka	1100	880	800	680	900	720	710	600	800	640
" Engelmann	750	600	580	490	650	520	440	370	500	400
Sycamore	1100	880	800	680	900	720	710	600	800	640
Tamarack (eastern)	1200	960	980	830	1100	880	800	680	900	720

* Stress in tension. The working stresses recommended for fiber stress in bending may be safely used for tension parallel to grain.

† American lumber standards. Basic provisions for American lumber standards grades are published by the U. S. Dept. of Commerce, Simplified Practice Recommendation No. 16, Lumber, revised July 1, 1926; specifications for grades conforming to American lumber standards are published in the 1933 Standards A.S.T.M., and in *Amer. Ry. Eng. Assoc. Bull.*, vol. xxx, No. 314, Feb. 1929.

‡ Exact figures given. In order to preserve the exact numerical relations among working stresses for grades involving rate of growth and density requirements, the values for Douglas fir (western Wash. and Ore.) and for southern yellow pine have not been rounded off, as have the values for the other species.

Table 5.   Allowable Unit Stresses for Timber in Compression and Shear

Recommended by the Forest Products Laboratory, U. S. Dept. of Agriculture.*

All values are in pounds per square inch.

Species	Compression ⊥ to Grain, Select and Common Grades			Horizontal Shear ||		Compression || to Grain (Short Columns with Ratio of Length to Least Dimension of 10 or Less)						Average Modulus of Elasticity ‡
	Continuously Dry	Occasionally Wet but Quickly Dried	More or Less Continuously Damp or Wet	Not Varied with Conditions of Exposure		Continuously Dry		Occasionally Wet, but Quickly Dried		More or Less Continuously Damp or Wet		Not Varied with Conditions of Exposure or Grade
				Select	Common	Select	Common	Select	Common	Select	Common	
Ash, black	300	200	150	90	72	650	520	550	440	500	400	1,100,000
" commercial white	500	375	300	125	100	1100	880	1000	800	900	720	1,500,000
Aspen and largetooth aspen	150	125	100	80	64	700	560	550	440	450	360	900,000
Basswood	150	125	100	80	64	700	560	550	440	450	360	900,000
Beech	500	375	300	125	100	1200	960	1100	880	900	720	1,600,000
Birch, paper	200	150	100	80	64	650	520	550	440	450	360	1,000,000
" yellow and sweet	500	375	300	125	100	1200	960	1100	880	900	720	1,600,000
Cedar, Alaska	250	200	150	90	72	800	640	750	600	650	520	1,200,000
" western red	200	150	125	80	64	700	560	700	560	650	520	1,000,000
" northern and southern white	175	140	100	70	56	550	440	500	400	450	360	800,000
Port Orford	250	200	150	90	72	900	720	825	660	750	600	1,200,000
Chestnut	300	200	150	90	72	800	640	700	560	600	480	1,000,000
Cottonwood, eastern and black	150	125	100	80	64	700	560	550	440	450	360	900,000
Cypress, southern	350	250	225	100	80	1100	880	1000	800	800	640	1,200,000
Douglas fir (western Wash. and Ore.) †	§347	§240	§213	90	72	1173	880	1067	800	907	680	1,600,000
Douglas fir (dense) †	379	262	233	105	84	1283	1027	1167	933	992	793	1,600,000
" " (Rocky Mt.)	275	225	200	85	68	800	640	800	640	700	560	1,200,000
Elm, rock	500	375	300	125	100	1200	960	1100	880	900	720	1,300,000
" slippery and Amer.	250	175	125	100	80	800	640	750	600	650	520	1,200,000
Fir, balsam	150	125	100	70	56	700	560	600	480	500	400	1,000,000
" commercial white	300	225	200	70	56	700	560	700	560	600	480	1,100,000
Gum, red, black, and tupelo	300	200	150	100	80	800	640	750	600	650	520	1,200,000
Hemlock, eastern	300	225	200	70	56	700	560	700	560	600	480	1,100,000
" western	300	225	200	75	60	900	720	900	720	800	640	1,400,000
Hickory (true and pecan)	600	400	350	140	112	1500	1200	1200	960	1000	800	1,800,000
Larch, western	325	225	200	100	80	1100	880	1000	800	800	640	1,300,000
Maple, sugar and black	500	375	300	125	100	1200	960	1100	880	900	720	1,600,000
" red and silver	350	250	200	100	80	800	640	700	560	600	480	1,100,000
Oak, commercial red and white	500	375	300	125	100	1000	800	900	720	800	640	1,500,000
Pine, southern yellow †	§	§	§	....	88	.....	880	....	800	....	680	1,600,000
" southern yellow (dense) †	379	262	233	128	103	1283	1027	1167	933	992	793	1,600,000
" northern and western white, western yellow, sugar	250	150	125	85	68	750	600	750	600	650	520	1,000,000
" Norway	300	175	150	85	68	800	640	800	640	700	560	1,200,000
Poplar, yellow	250	150	125	80	64	800	640	700	560	600	480	1,100,000
Redwood	250	150	100	70	56	1000	800	900	720	750	600	1,200,000
Spruce, red, white and Sitka	250	150	125	85	68	800	640	750	600	650	520	1,200,000
" Engelmann	175	140	100	70	56	600	480	550	440	450	360	800,000
Sycamore	300	200	150	80	64	800	640	750	600	650	520	1,200,000
Tamarack (eastern)	300	225	200	95	76	1000	800	900	720	800	640	1,300,000

* See footnote †, Table 4.

† See footnote ‡, Table 4.

‡ The values for modulus of elasticity are average for species, and not safe working stresses. They may be used as given for computing average deflection of beams. To prevent sag in beams, values one-half those given should be used. For safe loads for long columns, values one-third those given should be used.

§ Values are for the Select grade. Working stresses in compression, perpendicular to the grain, for common grades of Douglas fir (western Wash. and Ore.) and southern yellow pine are 325, 225, and 200, respectively, for continuously dry, occasionally wet but quickly dried, and more or less continuously damp or wet conditions.

|| Joint details. The shearing stresses for joint details may be taken for any grades as 50% greater than the horizontal shear values for the Select grade.

Examples in Beam Design: 1. Concentrated Loads. Find the beam section required to carry loads W_1, W_2, W_3, applied at points of a 15-ft span simple beam as follows: W_1 (assumed weight of beam) 40 lb per ft = 600 lb; W_2 = 10,000 lb at 5 ft from left support; W_3 = 8000 lb at 9 ft from left support. Computing the reactions

$$(R_1 \times 15) - (600 \times 7.5) - (10,000 \times 10) - (8000 \times 6) = 0$$
$$R_1 = 10,166 \text{ lb}$$
$$-(R_2 \times 15) + (600 \times 7.5) + (10,000 \times 5) + (8000 \times 9) = 0$$
$$R_2 = 8433 \text{ lb.}$$

The dangerous section is directly under W_2, as at this point the vertical shear passes from positive to negative. The moment at this point is

$$M = (10,166 \times 5) - (5 \times 40 \times 2.5) = 50,330 \text{ lb-ft} = 603,960 \text{ lb-in.}$$

For an allowable unit stress of 20,000 psi, I/c = 603,960/20,000 = 39.20 in.³ = I/c.

In Table 71, Section 1, a section modulus larger than the above is found in a 12-in. 31.8-lb American Standard beam, which is, therefore, selected. If this section is used it may be necessary to support the compression flange by bracing to prevent lateral deflection. If bracing cannot be used a smaller allowable unit stress and, therefore, a larger section may be required. The above section may be checked as follows: The allowable unit stress by the A.I.S.C. specifications (see Table 2) is

$$\sigma = \frac{12,000,000}{ld/bt} = \frac{12,000,000}{180 \times 12/5.00 \times 0.544} = 15,100 \text{ psi}$$

I/c = 603,960/15,100 = 40.0 in.³ Thus the 12-in. 31.8-lb beam which has an I/c of 36.0 in.³ is too small; a 12-in. 40.8 lb, I/c—44.8 in.³, beam will be satisfactory.

2. Uniform Load. Select a beam section required to support a superimposed load of 27,000 lb uniformly distributed over a simple beam with a span of 15 ft, assuming the beam to be braced against lateral deflection.

The beam to be used might be determined directly from a table of safe loads (see Safe Loads, p. 517). Such tables are given in Kent's *Handbook for Mechanical Engineers.* A solution can also be obtained by finding the maximum bending moment as follows:

For a uniformly distributed load, the dangerous section is at the center of the beam. The bending moment at this point (see Table 3) is

$$M = \frac{Wl}{8} = \frac{27,000 \times 15 \times 12}{8} = 607,500 \text{ lb-in.}$$

For an allowable unit stress of 18,000 psi,

$$S = \frac{607,500}{18,000} = 33.75 \text{ in.}^3$$

Referring to Table 71, Section 1, the section modulus of a 12-in. 31.8-lb American Standard beam is 36.0 in.³ The maximum safe load for this beam is

$$W = \frac{2sS}{3L} = \frac{2 \times 18,000 \times 36}{3 \times 15} = 28,800 \text{ lb}$$

The weight of this beam is 31.8 × 15 = 477 lb, making the actual total load 27,477 lb, which is less than the safe load for the beam selected, and hence this beam is satisfactory.

3. Wooden Beams. Design a southern pine girder of common structural grade to carry a load of 9600 lb uniformly distributed over a 16-ft span simple beam in the interior of a building.

$$M = Wl/8 = 9600 \times 16 \times 12/8 = 230,400 \text{ lb-in.}$$

For an allowable unit stress of 1200 psi (see Table 4) I/c = 230,400/1200 = 192 in.³ Referring to Table 1, p. 5-28, the section modulus of a rectangular section is $bd^2/6$. Assume b = 8 in.; then $8d^2/6$ = 192, from which $d = \sqrt{144}$ = 12.0 in. A girder 8 by 12 in. is tentatively selected.

4. Check the horizontal shear in the 8 by 12 wooden girder designed in example 3 above.

Maximum shearing stress (horizontal and vertical) is at the neutral surface over the supports. Using formula 3 for horizontal shear in a solid rectangular beam, p. 512,

$$V = 9600/2 = 4800, \quad A = 8 \times 12$$

$$\tau = \frac{3V}{2A} = \frac{3 \times 4800}{2 \times 96} = 75 \text{ lb per sq in.}$$

According to Table 5, the safe horizontal unit shearing stress for common grade southern yellow pine is 88 lb per sq in. Since the actual horizontal unit shearing stress is less than this, the beam will be satisfactory.

17. BEAMS OF UNIFORM STRENGTH

A **Beam of Uniform Strength** is one in which the dimensions are such that the maximum fiber stress σ is the same throughout the length of the beam. The form of the beam is determined by finding the areas of various cross sections from the flexure formula $M = \sigma I/c$, keeping σ constant and making I/c vary with M. For a rectangular section of width b and depth d, the section modulus I/c = ⅙ bd^2 and, therefore, M = ⅙ σbd^2. By

Table 6.   Rectangular Beams of Uniform Strength *

σ = maximum fiber unit stress, pounds per square inch; E = modulus of elasticity; w = uniform load, pounds per inch; d = depth of beam, inches; b = width of beam, inches; y = maximum deflection, inches. All other dimensions in inches.

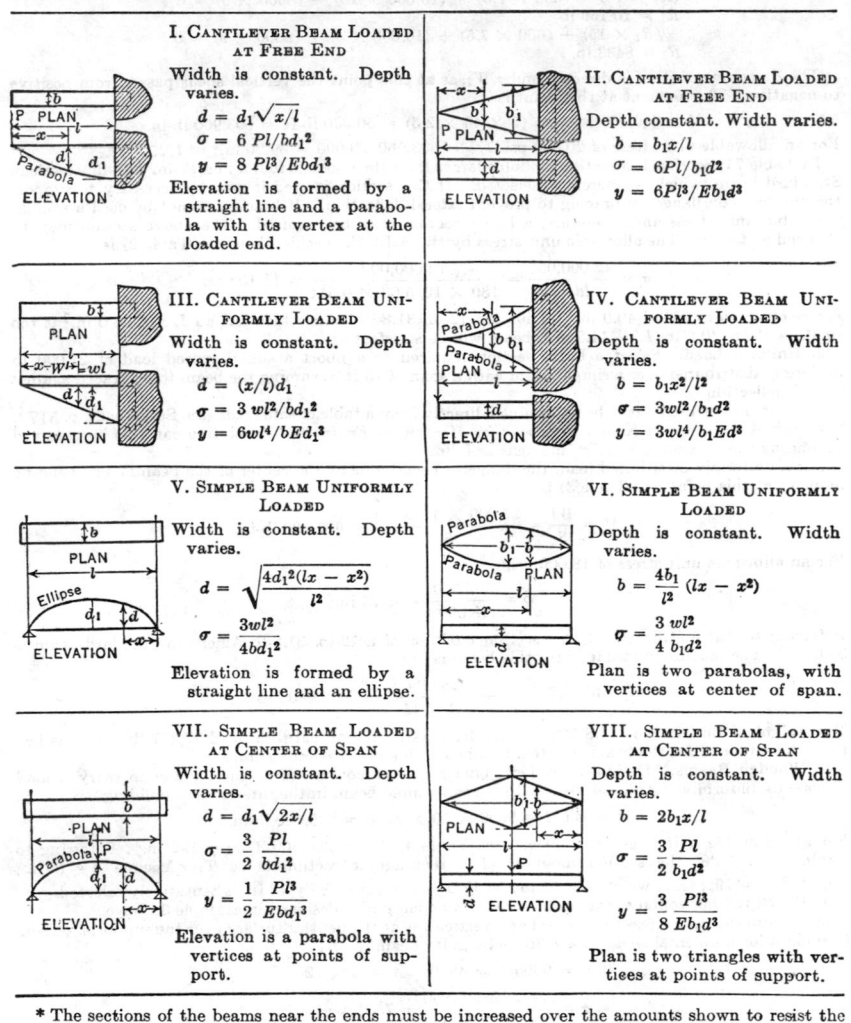

I. Cantilever Beam Loaded at Free End

Width is constant. Depth varies.

$$d = d_1\sqrt{x/l}$$
$$\sigma = 6\,Pl/bd_1{}^2$$
$$y = 8\,Pl^3/Ebd_1{}^3$$

Elevation is formed by a straight line and a parabola with its vertex at the loaded end.

II. Cantilever Beam Loaded at Free End

Depth constant. Width varies.

$$b = b_1x/l$$
$$\sigma = 6Pl/b_1d^2$$
$$y = 6Pl^3/Eb_1d^3$$

III. Cantilever Beam Uniformly Loaded

Width is constant. Depth varies.

$$d = (x/l)d_1$$
$$\sigma = 3\,wl^2/bd_1{}^2$$
$$y = 6wl^4/bEd_1{}^3$$

IV. Cantilever Beam Uniformly Loaded

Depth is constant. Width varies.

$$b = b_1x^2/l^2$$
$$\sigma = 3wl^2/b_1d^2$$
$$y = 3wl^4/b_1Ed^3$$

V. Simple Beam Uniformly Loaded

Width is constant. Depth varies.

$$d = \sqrt{\frac{4d_1{}^2(lx - x^2)}{l^2}}$$
$$\sigma = \frac{3wl^2}{4bd_1{}^2}$$

Elevation is formed by a straight line and an ellipse.

VI. Simple Beam Uniformly Loaded

Depth is constant. Width varies.

$$b = \frac{4b_1}{l^2}(lx - x^2)$$
$$\sigma = \frac{3}{4}\frac{wl^2}{b_1d^2}$$

Plan is two parabolas, with vertices at center of span.

VII. Simple Beam Loaded at Center of Span

Width is constant. Depth varies.

$$d = d_1\sqrt{2x/l}$$
$$\sigma = \frac{3}{2}\frac{Pl}{bd_1{}^2}$$
$$y = \frac{1}{2}\frac{Pl^3}{Ebd_1{}^3}$$

Elevation is a parabola with vertices at points of support.

VIII. Simple Beam Loaded at Center of Span

Depth is constant. Width varies.

$$b = 2b_1x/l$$
$$\sigma = \frac{3}{2}\frac{Pl}{b_1d^2}$$
$$y = \frac{3}{8}\frac{Pl^3}{Eb_1d^3}$$

Plan is two triangles with vertices at points of support.

* The sections of the beams near the ends must be increased over the amounts shown to resist the vertical shear expressed by the formula $\sigma = \frac{3}{2}\,V/A$.

making bd^2 vary with M, the dimensions of the various sections are obtained. Table 6 gives the dimensions b and d, at any section, the maximum fiber unit stress σ, and the maximum deflection y, of some rectangular beams of uniform strength. In this table, the bending moment has been assumed to be the controlling factor. On account of the vertical shear near the ends of the beams, the area of the sections must be increased over that given by an amount necessary to keep the unit shearing stress within the allowable unit shearing stress. The discussion of beams of uniform strength, though of considerable theoretical interest, is of little practical value since the cost of fabrication will offset any economy in the use of the material. A plate girder in a bridge or a building is an approximation in practice to a steel beam of uniform strength.

Table 7. Values of Constant K for Curved Beams

Section	$\frac{R}{c}$	Inside Fiber	Outside Fiber	$\frac{Y_0{}^*}{R}$	Section	$\frac{R}{c}$	Inside Fiber	Outside Fiber	$\frac{Y_0{}^*}{R}$
(two circles)	1.2	3.41	0.54	0.224	(rectangle)	1.2	2.89	0.57	0.305
	1.4	2.40	0.60	0.151		1.4	2.13	0.63	0.204
	1.6	1.96	0.65	0.108		1.6	1.79	0.67	0.149
	1.8	1.75	0.68	0.084		1.8	1.63	0.70	0.112
	2.0	1.62	0.71	0.069		2.0	1.52	0.73	0.090
	3.0	1.33	0.79	0.030		3.0	1.30	0.81	0.041
	4.0	1.23	0.84	0.016		4.0	1.20	0.85	0.021
	6.0	1.14	0.89	0.0070		6.0	1.12	0.90	0.0093
	8.0	1.10	0.91	0.0039		8.0	1.09	0.92	0.0052
	10.0	1.08	0.93	0.0025		10.0	1.07	0.94	0.0033
(trapezoid)	1.2	3.01	0.54	0.336	(trapezoid)	1.2	3.09	0.56	0.336
	1.4	2.18	0.60	0.229		1.4	2.25	0.62	0.229
	1.6	1.87	0.65	0.168		1.6	1.91	0.66	0.168
	1.8	1.69	0.68	0.128		1.8	1.73	0.70	0.128
	2.0	1.58	0.71	0.102		2.0	1.61	0.73	0.102
	3.0	1.33	0.80	0.046		3.0	1.37	0.81	0.046
	4.0	1.23	0.84	0.024		4.0	1.26	0.86	0.024
	6.0	1.13	0.88	0.011		6.0	1.17	0.91	0.011
	8.0	1.10	0.91	0.0060		8.0	1.13	0.94	0.0060
	10.0	1.08	0.93	0.0039		10.0	1.11	0.95	0.0039
(triangle)	1.2	3.14	0.52	0.352	(triangle)	1.2	3.26	0.44	0.361
	1.4	2.29	0.54	0.243		1.4	2.39	0.50	0.251
	1.6	1.93	0.62	0.179		1.6	1.99	0.54	0.186
	1.8	1.74	0.65	0.138		1.8	1.78	0.57	0.144
	2.0	1.61	0.68	0.110		2.0	1.66	0.60	0.116
	3.0	1.34	0.76	0.050		3.0	1.37	0.70	0.052
	4.0	1.24	0.82	0.028		4.0	1.27	0.75	0.020
	6.0	1.15	0.87	0.012		6.0	1.16	0.82	0.013
	8.0	1.12	0.91	0.0060		8.0	1.12	0.86	0.0060
	10.0	1.12	0.93	0.0039		10.0	1.09	0.88	0.0039
(T section)	1.2	3.63	0.58	0.418	(I section)	1.2	3.55	0.67	0.409
	1.4	2.54	0.63	0.299		1.4	2.48	0.72	0.292
	1.6	2.14	0.67	0.229		1.6	2.07	0.76	0.224
	1.8	1.89	0.70	0.183		1.8	1.83	0.78	0.178
	2.0	1.73	0.72	0.149		2.0	1.69	0.80	0.144
	3.0	1.41	0.79	0.069		3.0	1.38	0.86	0.067
	4.0	1.29	0.83	0.040		4.0	1.26	0.89	0.038
	6.0	1.18	0.88	0.018		6.0	1.15	0.92	0.018
	8.0	1.13	0.91	0.010		8.0	1.10	0.94	0.010
	10.0	1.10	0.92	0.0065		10.0	1.08	0.95	0.0065
(I section)	1.2	2.52	0.67	0.408	(H section)	1.2	2.37	0.73	0.453
	1.4	1.90	0.71	0.285		1.4	1.79	0.77	0.319
	1.6	1.63	0.75	0.208		1.6	1.56	0.79	0.236
	1.8	1.50	0.77	0.160		1.8	1.44	0.81	0.183
	2.0	1.41	0.79	0.127		2.0	1.36	0.83	0.147
	3.0	1.23	0.86	0.058		3.0	1.19	0.88	0.067
	4.0	1.16	0.89	0.030		4.0	1.13	0.91	0.036
	6.0	1.10	0.92	0.013		6.0	1.08	0.94	0.016
	8.0	1.07	0.94	0.0076		8.0	1.06	0.95	0.0089
	10.0	1.05	0.95	0.0048		10.0	1.05	0.96	0.0057
(circular tube)	1.2	3.28	0.58	0.269	(box)	1.2	2.63	0.68	0.399
	1.4	2.31	0.64	0.182		1.4	1.97	0.73	0.280
	1.6	1.89	0.68	0.134		1.6	1.66	0.76	0.205
	1.8	1.70	0.71	0.104		1.8	1.51	0.78	0.159
	2.0	1.57	0.73	0.083		2.0	1.43	0.80	0.127
	3.0	1.31	0.81	0.038		3.0	1.23	0.86	0.058
	4.0	1.21	0.85	0.020		4.0	1.15	0.89	0.031
	6.0	1.13	0.90	0.0087		6.0	1.09	0.92	0.014
	8.0	1.10	0.92	0.0049		8.0	1.07	0.94	0.0076
	10.0	1.07	0.93	0.0031		10.0	1.06	0.95	0.0048

* Y_0 is distance from centroidal axis to neutral axis, where beam is subjected to pure bending.

18. CURVED BEAMS

The derivation of the flexure formula, $\sigma = Mc/I$, assumes that the beam is initially straight; therefore, any deviation from this condition introduces an error in the value of the stress. If the curvature is slight the error involved is not large, but in beams with a large amount of curvature, as hooks, chain links, frames of punch presses, etc., the error involved in the use of the ordinary flexure formula is considerable. The effect of the curvature is to increase the stress in the inside and to decrease it on the outside fibers of the beam and to shift the position of the neutral axis from the centroidal axis toward the concave or inner side.

The correct value for the fiber unit stress may be found by introducing a correction factor in the flexure formula, viz., $\sigma = KMc/I$; the factor K depends on the shape of the beam and on the ratio R/c, where R = distance, inches, from the centroidal axis of the section to the center of curvature of the central axis of the unstressed beam; and c = distance, inches, of centroidal axis from the extreme fiber on the inner or concave side. Seely's *Advanced Mechanics of Materials* contains an analysis of curved beams, and Table 7 gives values of K for a number of shapes and ratios of R/c. For slightly different shapes or proportions K may be found by interpolation with a fair degree of approximation.

19. CONTINUOUS BEAMS

As in simple beams, the expressions $M = \sigma I/c$ and $\tau = V/A$ govern the design and investigation of beams resting on more than two supports. For continuous beams, however, the reactions cannot be obtained in the manner described for simple beams. Instead, the bending moments at the various sections must be determined, and from these values the vertical shears at the sections and the reactions at the supports may be derived.

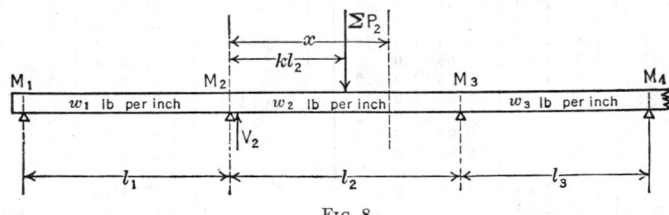

Fig. 8

Consider the second span of length l_2, in., of the continuous beam in Fig. 8. The vertical shear V_x at any section distant x, in., from the left support of the span is equal to the algebraic sum of all the vertical forces on one side of the section. Thus, if V_2 = the vertical shear at a section to the right of, but infinitely close to, the left support, w_2x = the uniform load and ΣP_2 = the sum of the concentrated loads along the distance x, applied at a distance kl_2 from the left support, k being a fraction less than unity, then

$$V_x = V_2 - w_2x - \Sigma P_2 \tag{12}$$

At any section distant x from the left support the bending moment is equal to the algebraic sum of the moments of all forces on one side of the section. If M_2 is the moment, lb-in., at the support to the left,

$$M_x = M_2 + V_2x - (w_2x^2/2) - \Sigma P_2(x - kl_2) \tag{13}$$

Assume that $x = l_2$. Then M_x becomes the moment M_3 at the next support to the right, and the expression may be written

$$V_2l_2 = M_3 - M_2 + (w_2l_2^2/2) + \Sigma P_2(l_2 - kl_2) \tag{14}$$

From the expressions 12, 13, and 14 it is evident that the bending moment M_x and the shear V_x at any section between two consecutive supports may be determined if the bending moments M_2 and M_3 at those supports are known.

To determine the bending moments at the supports an expression known as the *theorem of three moments* is used. This gives the relation between the moments at any three consecutive supports of a beam. For beams with the supports on the same level and uniformly loaded over each span the expression is:

$$M_1l_1 + 2M_2(l_1 + l_2) + M_3l_2 = -\tfrac{1}{4}w_1l_1^3 - \tfrac{1}{4}w_2l_2^3 \tag{15}$$

in which M_1, M_2, and M_3 are the moments at three consecutive supports, l_1 the length between the first and second support, l_2 the length between the second and third support, w_1 the uniform load per lineal unit over the first span, and w_2 the uniform load per lineal unit over the second span. When both spans are of equal length and when the load on each span is the same, $l_1 = l_2$, $w_1 = w_2$, and the above expression reduces to

$$M_1 + 4M_2 + M_3 = -1/2 \, wl^2 \tag{16}$$

which applies to most cases in practice.

Formulas 15 and 16 are used as follows: For any continuous beam of n spans there are $(n + 1)$ supports. Assuming the ends of the beam to be simply supported without any overhang, the moments at the end supports are zero and there are, therefore, $(n - 1)$

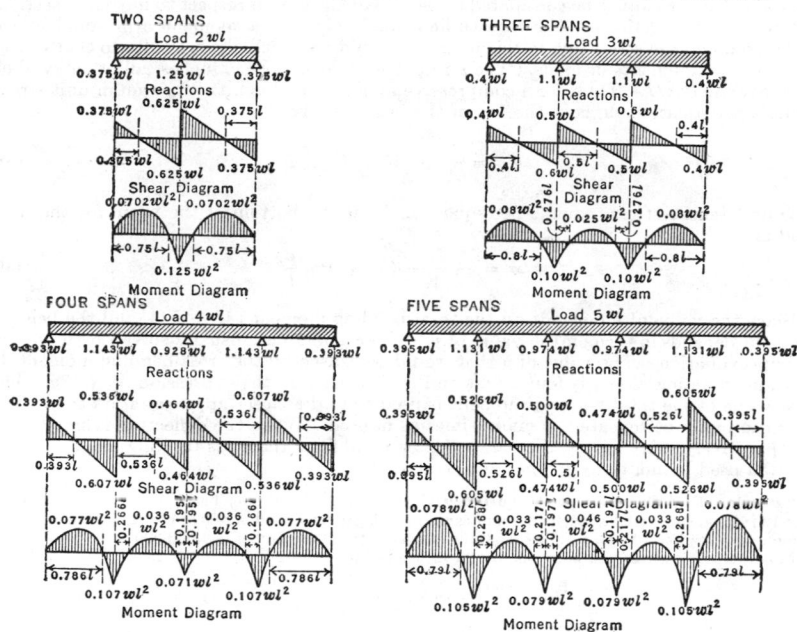

Fig. 9

moments at the other supports to be determined. This may be done by writing $(n - 1)$ equations of the form of 15 or 16 for each support. These equations will contain $(n - 1)$ unknown moments, and their solution will give values of M_1, M_2, M_3, etc., expressed as coefficients of wl^2. The shear V_1 at any support may be determined by substituting values of M_1 and M_2 in formula 14, and the bending moment at any point in any span may be obtained by formula 13. The shear at any point in any span may be determined from formula 12.

Figure 9 gives values and diagrams for the reactions, shears, and moments at all sections of continuous beams uniformly loaded up to five spans. Note that the reaction at any support is equal to the sum of the shears to the right and to the left of that support.

20. AXIALLY END-LOADED BEAMS

Deflection of Beam Negligible. When a beam is subjected to an axial end load as well as to a transverse bending load, the resultant unit stress developed at any point in the beam is the algebraic sum of the unit stresses produced by each of the loads acting independently of each other. If the beam of cross-sectional area A, square inches, is considered so short that its deflection under load may be neglected, the unit stress produced at any cross section of the beam by an axial load P, pounds, is P/A. The maximum unit stress (tensile or compressive), pounds per square inch, due to the transverse or bending load is given by $\sigma = M/(I/c)$ where M, pound-inches, is the maximum bending moment due to this load; I, inches[4], is the moment of inertia of the cross section with respect to the neutral axis; c, inches, is the distance from the neutral axis to the extreme fiber. If the axial load is compressive, the maximum unit stress is compressive, occurring at the top of the beam, and is

$$\sigma = \frac{P}{A} + \frac{Mc}{I} \tag{17}$$

The unit stress on the bottom fiber may be either tensile or compressive according as Mc/I is larger or smaller than P/A. If the longitudinal load P were a tensile load, the maximum unit stress would occur on the bottom fiber and would be a tensile stress.

Deflection of Beam Not Negligible. If the deflection of the beam is not negligible, the load P above cannot be considered to be an axial load with respect to any cross sections except the end sections. The longitudinal load then has a moment arm equal to the deflection, and the stress due to this moment should be added algebraically to that caused by the other moments or loads. The stress due to deflection, y, inches (all other symbols as above), is Pyc/I. Thus, for a compressive longitudinal load, the maximum unit stress, pounds per square inch, is at the top of the beam and is

$$\sigma = \frac{P}{A} + (M + Py)\frac{c}{I} \tag{18}$$

The unit tensile stress, pounds per square inch, at the bottom of the beam for the same load is

$$\sigma = -\frac{P}{A} + (M + Py)\frac{c}{I} \tag{19}$$

Since the value of y depends on the total bending moment $(M + Py)$, and the bending moment depends in turn on the value of y, eqs. 18 and 19 are usually solved by a method of approximation. The value of y that would be caused by the cross-bending moment M, considered acting alone, is found first and is then used in the expression $M + Py$. This new value of the total bending moment is used to find a closer approximation to y. This operation may be repeated as many times as desired. When the deflection is small, as in comparatively short beams, the value of y as found from the cross-bending moment above is often used without any further approximation.

Example. Find the stress at the bottom and at the top of a timber beam 6 in. wide, 8 in. deep, and 12 ft long, supported at each end, and carrying a distributed load of 400 lb per ft, and also an axial compressive load of 10,000 lb. The modulus of elasticity of the timber is 1,600,000. Maximum unit stress will be at the center of the beam.

$$\frac{P}{A} = \frac{10,000}{48} = 208 \text{ psi}$$

$$\frac{Mc}{I} = \frac{400 \times 12 \times 144}{8} \times \frac{6}{6 \times 8^2} = 1350 \text{ psi}$$

See Tables 1 and 3, pp. 5-28, 5-32 to 5-34.

$$y = \frac{5 \times 4800 \times 144^3 \times 12}{384 \times 6 \times 8^3 \times 1,600,000} = 0.456 \text{ in.}$$

$$\frac{Pyc}{I} = \frac{10,000 \times 0.456 \times 6}{6 \times 8^2} = 71 \text{ psi}$$

Total unit compressive stress (top) = 208 + 1350 + 71 = 1629 psi

Total unit tensile stress (bottom) = −208 + 1350 + 71 = 1213 psi.

Here the stress due to deflection is small, as it usually will be in short deep beams.

COLUMNS

21. DEFINITIONS

A **Column** or **Strut** is a bar or structural member under axial compression, which has an unbraced length greater than about 8 or 10 times the least dimension of its cross section. On account of its length, a column cannot be held in a straight line under a load; a slight sidewise bending always occurs, causing flexual stresses in addition to the compressive stresses induced directly by the load. The lateral deflection will be in a direction perpendicular to that axis of the cross section about which the moment of inertia is the least. Thus in Fig. 1A the column will bend in a direction perpendicular to aa; in Fig. 1B it will bend perpendicular to aa or bb; and in Fig. 1C it is apt to bend in any direction.

The **Radius of Gyration** of a section with respect to an axis is equal to the square root of the quotient of the moment of inertia with respect to that axis divided by the area of the section; that is

$$k = \sqrt{\frac{I}{A}}\ ;\ \frac{I}{A} = k^2 \qquad (1)$$

where I is the moment of inertia and A the sectional area. Unless otherwise mentioned, an axis through the center of gravity of the section is the axis considered. As in beams, the moment of inertia is an important factor in the ability of the column to resist bending, but for purposes of computation it is more convenient to use the radius of gyration.

The **Length of a Column** is the distance between points unsupported against lateral deflection.

Slenderness Ratio is the length l divided by the least radius of gyration k, both in inches. For steel, a *short column* is one in which l/k is less than about 20 or 30, and its failure under load is due mainly to direct compression; in a *medium length column*, $l/k = $ about 30 to 175, and it will fail by a combination of direct compression and

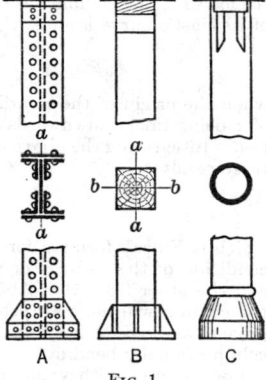

Fig. 1

bending; in a *long column*, l/k is greater than about 175–200, and it will fail mainly by bending. For timber columns these ratios are about 0–30, 30–90, and above 90 respectively. The load which will cause a column to fail decreases as l/k increases. The above ratios apply to round-end columns. If the ends are fixed (see paragraph below), the effective slenderness ratio is one-half that for round-end columns, as the distance between the points of inflection is one-half the total length of the column. For flat ends it is intermediate between the two.

Condition of Ends. As in beams, the conditions of the points of support have an important influence on the ability of the column to resist bending. The various conditions which may exist at the ends of columns are usually divided into four classes. Columns with round ends are such that at the bearing at either end there is perfect freedom of motion, as there would be with a ball-and-socket joint at each end. Columns with hinged ends are such as have perfect freedom of motion at the ends in one plane, as in compression members in bridge trusses where the loads are transmitted through end pins. Columns with flat ends have the bearing surface normal to the axis of the column and of sufficient area to give at least partial fixity to the ends of the columns against lateral deflection. Columns with fixed ends have the ends rigidly secured so that under any load the tangent to the elastic curve at the ends will be parallel to the axis in its original position.

Experiments prove that columns with fixed ends are stronger than columns with either flat, hinged, or round ends, and that columns with round ends are weaker than any of the other types. Columns with hinged ends are equivalent to those with round ends in the plane in which they have free movement; columns with flat ends have a value intermediate between those with fixed ends and those with round ends. It often happens that columns have one end fixed and one end hinged, or various other combinations. Their relative values may be taken as intermediate between those represented by the condition at either end. The extent to which the strength is increased by fixing the ends depends upon the length of the column, fixed ends having a greater effect on long columns than on short ones.

22. COLUMN FORMULAS

There is no exact theoretical formula which gives the strength of a column of any length under an axial load. Formulas involving the use of empirical coefficients have been deduced, however, and they give results which are consistent with the results of tests of full-sized members.

Euler's Formula. In 1757 Euler published his formula, which is based on the assumption that the failure of a column is due solely to the stresses induced by sidewise bending. This assumption is not true for very short columns, which fail mainly by direct compression, nor is it true for columns of medium length such as are usually needed in practice. The failure in such cases is by a combination of direct compression and bending. For columns having a ratio of slenderness greater than 200, Euler's formula is approximately correct and agrees closely with the results of tests.

Let P = axial load, pounds; l = length of column, inches; I = least moment of inertia, inches^4; k = least radius of gyration, inches; E = modulus of elasticity, pounds per square inch; and y = lateral deflection, inches, at any point along the column, that is caused by the load P. If a column has round ends, so that the bending is not restrained, the equation of its elastic curve is

$$EI \frac{dy^2}{dx^2} = -Py \tag{2}$$

when the origin of the coordinate axes is at the top of the column, the positive direction of x being taken downwards and the positive direction of y in the direction of the deflection. Integrating the above expression twice and determining the constants of integration, there results

$$P = \Omega\pi^2 \frac{EI}{l^2} \tag{3}$$

which is Euler's formula for long columns. The factor Ω is a constant depending on the condition of the ends. For round ends $\Omega = 1$; for fixed ends $\Omega = 4$; for one end round and the other fixed $\Omega = 2.05$. P is the load at which, if a slight deflection is produced, the column will not return to its original position; if P is decreased the column will approach its original position, but if P is increased the deflection will increase until the column fails by bending.

For columns with value of l/k less than about 150, Euler's formula gives results distinctly higher than those observed in tests. Euler's formula is now little used except for long members and as a basis for the analysis of the stresses in some types of structural and machine parts. It always gives an *ultimate* and never an allowable load.

Secant Formula. The deflection of the column is used in the Euler formula, but if the load were truly axial it would be impossible to compute the deflection. If the column is assumed to have an initial eccentricity of load of e inches (see Seely, *Resistance of Materials*, 3rd ed., p. 241, for suggested values of e), the equation for the deflection y becomes

$$y_{max} = e\left(\sec \frac{l}{2}\sqrt{\frac{P}{EI}} - 1\right) \tag{4}$$

and the maximum unit compressive stress becomes

$$\sigma = \frac{P}{A}\left(1 + \frac{ec}{k^2}\sec \frac{l}{2}\sqrt{\frac{P}{EI}}\right) \tag{5}$$

where l = length of column, inches; P = total load, pounds; A = area, square inches; I = moment of inertia, inches^4; k = radius of gyration, inches; c = distance from the neutral axis to the most compressed fiber, inches; E = modulus of elasticity, pounds per square inch; I and k are both taken with respect to the axis about which bending takes place. Because the formula contains the secant of the angle ($l/2\sqrt{P/EI}$) it is sometimes called the *secant formula*. It has been suggested by the committee on Steel-column Research (*Trans.*, A.S.C.E., 1933) that the best rational column formula can be constructed on the secant type, though of course it must contain experimental constants.

The secant formula can be used also for columns that are eccentrically loaded if e is taken as the actual eccentricity plus the assumed initial eccentricity.

Parabolic Formula. In 1892, J. B. Johnson suggested a column formula in the form

$$P/A = \sigma - Cl^2/k^2$$

in which σ is the unit stress at the yield point in compression for a ductile material, and C is a constant chosen to make the parabola tangent to Euler's curve.

The values of σ and C can also be adjusted in such a way that the resulting curve closely follows the secant-formula curve over a specified range of values of l/k. The parabolic formula is simpler to apply in design than the secant formula and is more widely used.

Rankine's Formula is a modification of an earlier formula by Gordon and is sometimes called "Gordon's formula." It applies to columns of slenderness ratios of 20 to 200, which are usually required in practice. Rankine's formula is based on the assumption that the maximum unit stress σ, pounds per square inch, occurs in the fiber most remote from the axis on the concave side of the deflected column, and is equal to the sum of the direct unit compressive stress and the unit compressive stress due to bending alone; that is, $\sigma = (P/A) + \sigma_1$ in which P is the load, pounds; A the sectional area, square inches; and σ_1 the unit compressive stress due to bending, pounds per square inch. The value of σ_1 in this equation may be expressed in terms of P by using the flexure formula, since the action is analogous to that of beams. Let l = the length of the column, inches; I = the least moment of inertia, inches[4]; k = corresponding radius of gyration, inches; c = distance from axis to the remotest fiber on the concave side, inches; and y = lateral deflection of column, inches. Substituting in the flexure formula, $\sigma_1 = Mc/I$, and by replacing M by its value Py and I by its value Ak^2, $\sigma_1 = Pyc/(Ak^2)$. Hence the maximum unit compressive stress on the remotest fiber on the concave side is equal to $P/A + Pyc/(Ak^2)$. By analogy with the theory of beams the deflection y varies as l^2/c, so that, by introducing an experimental coefficient ϕ, the expression for maximum unit stress may be written

$$\sigma = \frac{P}{A} + \phi\left(\frac{Pl^2}{Ak^2}\right) = \frac{P}{A}\left[1 + \phi\left(\frac{l}{k}\right)^2\right] \qquad (6)$$

which is Rankine's formula for the investigation of columns. Variations in the value of ϕ have been determined by many experiments on columns of different materials and different conditions of ends. The values in Table 1 are given in Merriman and have been

Table 1. Value of Coefficient ϕ in Rankine's Formula

Material	Both Ends Fixed	Fixed and Round	Both Ends Round
Timber..........	1/3000	1.95/3000	4/3000
Cast iron.........	1/5000	1.95/5000	4/5000
Wrought iron.....	1/36,000	1.95/36,000	4/36,000
Steel...........	1/25,000	1.95/25,000	4/25,000

used extensively in practice in the past. But present practice tends toward the use of the same constant for all end conditions, as tests show that, for columns of the proportions in common use, variations in the material are more important than variations in end conditions. (See Seely, *Resistance of Materials*, 3rd ed., p. 221.) The value to be taken for σ in Rankine's formula is the ultimate compressive stress of the material for rupture, and the allowable compressive unit stress for design. Rankine's formula may be used in the investigation of an existing column by comparing the computed value of σ with the proportional limit and with the ultimate strength of the material, thus determining the factor of safety. Similarly, the safe load P for an existing column may be computed by assuming an allowable unit stress σ.

Ritter's Rational Constant. This is a theoretical value of ϕ such that it connects the curve for the Rankine formula with the curve for the Euler formula. Boyd's *Strength of Materials* gives its value as $\phi = \sigma_u/(\pi^2 E)$, where σ_u = ultimate compressive strength of material, pounds per square inch, and E = modulus of elasticity, pounds per square inch.

Straight-line Formula. The plotted results of actual tests on columns show that the relation between ultimate load and l/k is fairly well represented by a straight line, for columns which fail by flexure of the whole column and not by local collapse. For a value of $l/k = 0$, the average unit stress on the section, P/A, is equal to the ultimate in compression for brittle materials and to the yield point in compression for ductile materials. From the point thus determined the straight line representing the average results of tests is drawn tangent to the curve of Euler's formula for the material. The equation of the straight line is

$$\frac{P}{A} = \sigma - C\frac{l}{k} \qquad (7)$$

in which σ is the unit stress at the ultimate in compression for a brittle material or at the yield point in compression for a ductile material, and C is a constant determined by experiment.

Table 2 gives average values of σ and C for structural steel, cast iron, and wood. The loads given by the straight-line formula using the constants in this table are *ultimate loads*.

For purposes of design the straight-line formula is usually put in a form that gives allowable loads directly. Both the constant σ and the constant C are divided by a factor of safety.

The straight-line formula is not suitable for investigating a column, that is, for determining the values of σ due to given loads, because the term Cl/k is not a function of P/A. It may be used to find the safe load for a given column under a given unit stress, or to design a column for a given load and unit stress. From 1900 to 1930 the straight-line formula probably was used for designing more steel in America than the Rankine formula, but since 1930 the tendency has been towards the use of the Rankine type of formula. See Table 5, p. 535.

Table 2. Ultimate Strength Constants for the Straight-line Column Formula—

$$P/A = \sigma - Cl/k$$

Material	σ psi	C			Limit of l/k		
		Round	Fixed	One End Round One End Fixed	Round	Fixed	One End Round One End Fixed
Structural steel........	35,000	150	75	100	160	320	240
Cast iron.............	34,000 *	175	88	116	90	160	115
Wood...............	5,000 *	40	20	30	75	150	112

* This is less than the ultimate in compression for small specimens of cast iron or wood, but from tests of full-size columns it seems to be the value to be used for full-size castings or timbers which may contain defects.

Eccentric Loads on Short Compression Members. Where a direct push acting on a member does not pass through the centroid but at a distance e, inches, from it, both direct and bending stresses are produced. For short compression members in which column action may be neglected, the direct unit stress is P/A, where P = total load, pounds, and A = area of cross section, square inches. The bending unit stress is Mc/I, where $M = Pe$ is the bending moment, pound-inches; c is the distance, inches, from the centroid to the fiber in which the stress is desired; I = moment of inertia, inches⁴. The total unit stress at any point in the section is $\sigma = P/A + Pec/I$, or $\sigma = P/A(1 + ec/k^2)$, since $I = Ak^2$, where k = radius of gyration, inches.

Eccentric Loads on Columns. Various column formulas must be modified when the loads are not balanced, that is, when the resultant of the loads is not in line with the axis of the column. Let P be the load, pounds, applied at a distance e, inches, from the axis; the bending moment M is Pe. The maximum unit stress σ, pounds per square inch, due to this bending moment alone, is $\sigma = Mc/I = Pec/Ak^2$, where c = distance, inches, from the axis to the most remote fiber on the concave side; A = sectional area, square inches; k = radius of gyration in the direction of the bending inches. This unit stress must be added to the unit stress induced if the resultant load were applied in line along the axis of the column. The modified formulas, expressed in allowable load per unit of area, are:

Modified Rankine's formula for eccentric loads (see p. 531):

$$\frac{P}{A} = \frac{\sigma}{1 + \phi(l/k)^2 + (ec/k^2)} \tag{8}$$

Modified straight-line formula for eccentric loads (see p. 531):

$$\frac{P}{A} = \frac{\sigma - \dfrac{Cl}{k}}{1 + \dfrac{ec}{k^2}} \tag{9}$$

The secant formula, see p. 530, can also be used for columns that are eccentrically loaded if the e in this formula is taken as the actual eccentricity plus the assumed initial eccentricity.

Column Subjected to Transverse or Cross-bending Loads. A compression member that is subjected to cross-bending loads may be considered to be (1) a beam subjected to end thrust as discussed in Article 20 or (2) a column subjected to cross-bending loads, depending on the relative magnitude of the end thrust and cross-bending loads, and on the dimensions of the member. The various column formulas may be modified so as to include the effect of cross-bending loads. In this form they are:

Modified Rankine's formula for transverse loads:

$$\sigma = \frac{P}{A}\left(1 + \phi\frac{l^2}{k^2}\right) + \frac{Pyc}{Ak^2} + \frac{Mc}{Ak^2} \tag{10}$$

Modified straight-line formula for transverse loads:

$$\frac{P}{A} = \left(\sigma - \frac{Cl}{k}\right) - \frac{Pyc}{Ak^2} - \frac{Mc}{Ak^2} \tag{11}$$

Modified secant formula for transverse loads:

$$\sigma = \frac{P}{A}\left(1 + (e + y)\,\frac{c}{k^2}\right)\sec\left(\frac{l}{2k}\right)\sqrt{\frac{P}{AE}} + \frac{Mc}{Ak^2} \qquad (12)$$

In these formulas, σ is the maximum unit stress on the concave side, pounds per square inch; P = the axial end load, pounds; A = cross-sectional area, square inches; M = moment due to cross-bending load, pound-inches; y = deflection due to cross-bending load, inches; k = radius of gyration, inches; l = length of column, inches; e = assumed initial eccentricity, inches; c = distance, inches, from the axis to the most remote fiber on the concave side; C = experimental constant (see Table 2). The modified straight-line formula, like the straight-line formula, cannot be used to solve for σ (see p. 532).

23. WOODEN COLUMNS

Wooden Column Formulas. One of the principal formulas is that formerly used by the A.R.E.A., $P/A = \sigma_1(1 - l/60d)$, where P/A = allowable unit load, pounds per square inch; σ_1 = allowable unit stress in direct compression on short blocks, pounds per square inch; l = length, inches; d = least dimension, inches. This formula is being replaced rapidly by formulas recommended by the A.S.T.M. and A.R.E.A. Committees of these societies, working with the U. S. Forest Products Laboratory, classified timber columns into three groups, as follows (A.S.T.M. Standards, 1937, D245–37):

1. Short Columns. The ratio of unsupported length to least dimension does not exceed 10. For these columns, the allowable unit stress should not be greater than the values given in Table 5, p. 522, under compression parallel to the grain.

2. Intermediate-length Columns. For columns the ratio of whose unsupported length to least dimension is greater than 10, the following formula, of the fourth power parabolic type, shall be used to determine the allowable unit stress until this allowable unit stress is equal to $2/3$ the allowable unit stress for short columns.

$$\frac{P}{A} = \sigma_1\left[1 - \frac{1}{3}\left(\frac{l}{Kd}\right)^4\right] \qquad (13)$$

where P = total load, pounds; A = area, square inches; σ_1 = allowable unit compressive stress parallel to grain, pounds per square inch; see Table 5, p. 522; l = unsupported length, inches; d = least dimension, inches. $K = l/d$ at the point of tangency of the parabolic and Euler curves, at which $P/A = (2/3)s_1$. The value of K for any species and grade is $(\pi/2)\sqrt{E/6\sigma_1}$, where $E =$ modulus of elasticity, pounds per square inch.

3. Long Columns. For columns in which P/A as computed by the formula above is less than $2/3\,\sigma_1$, the following formula of the Euler type, which includes a factor of safety of 3, shall be used:

$$\frac{P}{A} = \frac{\pi^2}{36}\frac{E}{(l/d)^2} \qquad (14)$$

Timber columns should be limited to a ratio of l/d equal to 50. No higher loads are allowed for square-ended columns. The strength of round columns may be considered the same as that of square columns of the same cross-sectional area.

Use of Timber Column Formulas. The values of E (modulus of elasticity) and σ_1 (compression parallel to grain) to be used in the above formulas are given in Table 5, p. 522. Table 3 gives the computed values of K for some common types of timbers.

Table 3. Values of K for Columns of Intermediate Length

A.S.T.M. Standards, 1937, D245–37

Species	Continuously dry		Occasionally wet		Usually wet	
	Select	Common	Select	Common	Select	Common
Cedar, western red.................	24.2	27.1	24.2	27.1	25.1	28.1
" , Port Orford.................	23.4	26.2	24.6	27.4	25.6	28.7
Douglas fir, coast region.............	23.7	27.3	24.9	28.6	27.0	31.1
" " , dense.................	22.6	25.3	23.8	26.5	25.8	28.8
" " , Rocky Mountain region...	24.8	27.8	24.8	27.8	26.5	29.7
Hemlock, west coast.................	25.3	28.3	25.3	28.3	26.8	30.0
Larch, western.....................	22.0	24.6	23.1	25.8	25.8	28.8
Oak, red and white.................	24.8	27.8	26.1	29.3	27.7	31.1
Pine, southern.....................	....	27.3	....	28.6	....	31.1
" , dense.....................	22.6	25.3	23.8	26.5	25.8	28.8
Redwood...........................	22.2	24.8	23.4	26.1	25.6	28.6
Spruce, red, white, Sitka.............	24.8	27.8	25.6	28.7	27.5	30.8

These may be substituted directly in the above formula for intermediate-length columns or may be used in conjunction with Table 4, which gives the strength of columns of intermediate length, expressed as a percentage of strength (σ_1) of short columns. In the tables, the term "continuously dry" refers to interior construction where there is no

Table 4. Strength of Columns of Intermediate Length, Expressed as a Percentage of Strength of Short Columns

A.S.T.M. Standards, 1937, D245–37

Values for the expression $\left[1 - \dfrac{1}{3} \left(\dfrac{l}{Kd} \right)^4 \right]$ in the formula: $\dfrac{P}{A} = \sigma_1 \left[1 - \dfrac{1}{3} \left(\dfrac{l}{Kd} \right)^4 \right]$

K	\multicolumn{20}{c}{Ratio of Length to Least Dimension in Rectangular Timbers, l/d}
	12	13	14	15	16	17	18	19	20	21	22	23	24	25	26	27	28	29	30	31
22	97	96	95	93	91	88	85	81	77	72	67	...	...	...	...	...	...	...	...	...
23	98	97	95	94	92	90	87	84	81	77	72	67	...	...	...	...	...	...	...	...
24	98	97	96	95	93	92	89	87	84	80	76	72	67	...	...	...	...	...	...	...
25	98	98	97	96	94	93	91	89	86	83	80	76	72	67	...	...	...	...	...	...
26	99	98	97	96	95	93	92	91	89	86	83	80	76	72	67	...	...	...	...	...
27	99	98	98	97	96	95	93	92	90	88	85	82	79	74	71	67	...	...	...	...
28	99	98	98	97	96	95	94	93	91	89	87	85	82	79	75	71	67	...	...	...
29	99	99	98	98	97	96	95	94	92	91	89	87	84	82	79	75	71	67	...	...
30	99	99	98	98	97	97	96	95	94	92	90	88	86	84	81	78	75	71	67	...
31	99	99	99	98	98	97	96	95	94	93	92	90	88	86	84	81	78	75	71	67

NOTE: This table can also be used for columns not rectangular, the l/d being equivalent to 0.289l/k, where k is the least radius of gyration of the section.

excessive dampness or humidity; "occasionally wet but quickly dry" refers to bridges, trestles, bleachers and grandstands; "usually wet" refers to timber in contact with the earth or exposed to waves or tide-water.

24. STEEL COLUMNS

Types. Two general types of steel columns are in use: (1) rolled shapes; (2) built-up sections. The rolled shapes have had wide use since the introduction of wide-flanged sections, especially since they may be obtained with large cross-sectional area. They are easily fabricated, accessible for painting, neat in appearance where they are not covered, and convenient in making connections. A disadvantage is the probability that thick sections are of lower-strength material than thin sections owing to the difficulty of adequately rolling the thick material. For the effect of thickness of material on yield point, see *Trans. A.S.C.E.*, vol. xcviii, p. 1377, 1933.

Guiding Principles in Design. The design of steel columns is always a cut-and-try method, as no law governs the relation between area and radius of gyration of the section. A column of given area is selected, and the amount of load that it will carry is computed by the proper formula. If the allowable load so computed is less than that to be carried, a larger column is selected and the load for it computed, the process being repeated until a proper section is found.

A few general principles should guide in proportioning columns. The radius of gyration should be approximately the same in the two directions at right angles to each other; the slenderness ratio of the separate parts of the column should not be greater than that of the column as a whole; the different parts should be adequately connected in order that the column may function as a single unit: the material should be distributed as far as possible from the center line in order to increase the radius of gyration.

Steel Column Formulas. Table 5 gives the more commonly used formulas in past practice. For additional formulas see also Table 2, Article 16. Table 6 gives the maximum values for the slenderness ratio for steel columns as specified by various authorities.

Bureau of Standards Tests of Steel Columns. The collapse of the Quebec Bridge in 1907, due to the failure of a column in the lower chord, brought about a review of existing knowledge of steel columns. In 1913 the A.S.C.E., the A.R.E.A., and the National Bureau of Standards cooperated in a test of columns. These columns were made to represent high-

Table 5. Steel Column Formulas

Organization Recommending or Using Formula	Formula,* Allowable Load, lb per sq in.
New York City Building Code, 1930 Pacific Coast Building Officials Conference Massachusetts Building Officials Conference Philadelphia Building Code, 1929 Wisconsin Building Code American Institute of Steel Construction, 1934	$\dfrac{P}{A} = \dfrac{18,000}{1 + \dfrac{l^2}{18,000k^2}}$ Maximum 15,000 except Mass. B.O.C., which is 13,500
American Bridge Company	$l/k = 0\text{-}60;\ P/A = 13,000$ $l/k = 60\text{-}120;\ P/A = 19,000 - 100l/k$ $l/k = 120\text{-}200;\ P/A = 13,000 - 50l/k$
American Railway Engineering Association	$l/k = 0\text{-}50;\ P/A = 12,500$ $l/k = 50\text{-}150;\ P/A = 15,000 - 50l/k$ $l/k = \text{above } 150;\ P/A = 6.4E/(l/k)^2$
Recommended Building Code for Working Stresses in Building Materials, U. S. Bureau of Standards, 1926	Standard Steel † $P/A = 18,000 - 70l/k$ Maximum 14,000 Acceptable Steel † $P/A = 16,000 - 60l/k$
Chicago Building Code	$P/A = 16,000 - 70l/k$ Maximum 14,000
United States Bureau of Public Roads	Live Load: $\dfrac{P}{A} = \dfrac{16,000}{1 + \dfrac{l^2}{13,500k^2}}$ Dead Load: $\dfrac{P}{A} = \dfrac{24,000}{1 + \dfrac{l^2}{13,500k^2}}$ Maximum is that for $l/k = 40$

* In the formulas given, P = total load on column, pounds; A = cross-sectional area, square inches; l = length of column, inches; k = least radius of gyration, inches; E = modulus of elasticity, pounds per square inch.

† Standard steel is steel which satisfies the A.S.T.M. specifications for Structural Steel for Buildings. Acceptable steel is steel which is acceptable to the building officials but whose origin and properties have not been definitely determined.

Table 6. Maximum Values of Slenderness Ratio (l/k) for Steel Columns

Organization or Authority Recommending or Specifying	Main Members	Secondary Members
American Railway Engineering Association, Railroad bridges...	100	120
New York City Building Code.........................	120	120
United States Bureau of Public Roads.................	120	140
Chicago Building Code...............................	120	150
American Institute of Steel Construction..............		
American Bridge Company...........................		
Philadelphia Building Code...........................	120	200
Wisconsin Building Code.............................		
Pacific Coast Building Officials Conference............		
United States Bureau of Standards....................	160	160
Massachusetts Building Officials Conference...........	160	200

grade practice and covered a wide range of sections, both rolled shapes and built-up sections (*Trans. Am. Soc. C. E.*, vol. lxxxiii, 1919–1920).

The tests showed that, for columns of the proportions commonly used, the effect of the variation in the steel, kinks, initial stresses, and similar defects in the column was more important than the effect of length. They also showed that the thin metal gave definitely higher strength, per unit area, than the thicker metal of the same type of section.

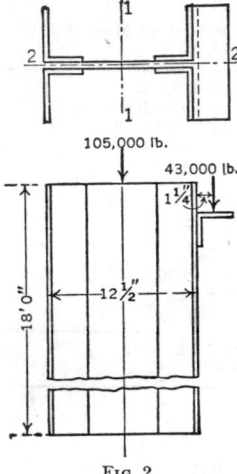

105,000 lb.

43,000 lb.

18' 0"

12½"

1¼"

Fig. 2

Example. Design a steel column 18 ft long to carry an axial load of 105,000 lb and an additional load of 43,000 lb which comes from a crane runway, and which is 1 1/4 in. beyond the outside edge of the column.

The formula used will be the A.I.S.C. formula, 1946, $P/A = 17{,}000 - 0.485l^2/k^2$, and the shape selected will be a plate and four angles, as shown in Fig. 2.

Using 10,000 psi as a preliminary estimate, one plate $12 \times {}^5/_{16}$ in. and four angles $5 \times 3 \times {}^3/_8$ in. will have an area of 15.19 sq in. The safe load for this column will now be found. The moment of inertia I_{1-1} is 406 in.[4], the radius of gyration k_{1-1} is 5.17 in., and the radius of gyration k_{2-2} is 2.13 in.

Considering the safe load with respect to axis 2–2 about which neither load is eccentric,

$$P = 15.19 \left(17{,}000 - 0.485 \frac{(18 \times 12)^2}{2.13^2} \right) = 182{,}400 \text{ lb}$$

Consider now the safe load about axis 1–1, with respect to which the load of 105,000 lb is axial, and the load of 43,000 lb has an eccentricity of 7 1/2 in. The load of 43,000 lb will produce a stress due to its eccentricity, and this stress will reduce by that amount the unit load that the column can carry. The safe load will therefore be

$$P = 15.19 \left(17{,}000 - 0.485 \frac{(18 \times 12)^2}{5.17^2} - \frac{43{,}000 \times 7.5 \times 6.25}{406} \right)$$

$$= 170{,}000 \text{ lb}$$

As the load to be carried is 148,000 lb this column is satisfactory.

25. CAST-IRON COLUMNS

Disadvantages of Cast-iron Columns. Tests made on cast-iron columns show that a large factor of safety should be used when the material is employed in column construction. It is difficult to obtain a practical formula for the strength of a cast-iron column on account of the uncertainty of the quality of the casting, and the danger of hidden defects such as internal stresses due to unequal cooling of the casting, cinder or dirt, blowholes, cold shuts, and cracks on the inner surface, which cannot be discovered by external inspection. Variation in the thickness of the wall due to shifting of the core is another common defect. Brackets which are usually cast on the face of the column to support beams and girders may fail by shearing unless most skillfully designed.

Tests of full-size cast-iron columns made at the Watertown Arsenal and at Phoenixville, Pa., showed that for short columns the strength in compression is much less than the strength of small test specimens of cast iron. For short columns, failure of test columns takes place by shearing on an inclined plane; for long columns, failure takes place by rupture on the convex side. Failure of cast-iron columns is always a sudden, shattering failure.

Advantages of Cast-iron Columns. For buildings of moderate height cast-iron columns are sometimes used instead of wooden columns to save space, and instead of steel columns to save expense. The fact that they are cheap and can be obtained more quickly than steel built-up columns will prevent them from being superseded entirely. Moreover they can be cast in almost any desired shape with lugs and brackets attached for supporting beams or girders.

Design of Cast-iron Columns. Cast-iron columns are usually in the form of a hollow cylinder, as this is the most economical and reliable type. Hollow rectangular sections are permitted by some building codes and are convenient in outer walls where it is necessary to bond into the masonry. They are less reliable than the hollow round type and should not be used if anything else is possible.

The design will usually be governed by building laws, and the following limitations, which are usually specified, should be adhered to: Minimum thickness of wall, 1/2 to 3/4 in.; minimum diameter of column, 6 in.; maximum length, 20 × diameter or 70 × least radius of gyration. When it becomes necessary to use a wall thickness greater than 2 in. or a diameter over 18 in. it will be cheaper and more satisfactory to abandon cast iron and substitute a steel column. Within the above limitations the ratio of length to radius of gyration should be made as small as possible; that is, if a choice may be made of two columns of equal sectional area but of different diameter and, therefore, different wall thicknesses, the one with the larger diameter will be the stronger.

Allowable safe loads on cast-iron columns may be calculated as follows

$$\frac{P}{A} = 9000 - 40\,\frac{l}{k} \tag{15}$$

where P = the safe load, pounds; A = area of section, square inches; l = length, inches; k = least radius of gyration.

SHAFTS

26. DEFINITIONS

Torsional Stress. A bar is under torsional stress when it is held fast at one end and a force acts at the other end to twist the bar. In a round bar, Fig. 1, with a constant force acting, the straight line a–b becomes the helix ad, and a radial line in the cross section, ob, moves to the position od. The angle bad remains constant while the angle bod increases with the length of the bar. Each cross section of the bar tends to shear off the one adjacent to it, and in any cross section the shearing stress at any point is normal to a radial line drawn through the point. Within the shearing proportional limit a radial line of the cross section remains straight after the twisting force has been applied, and the unit shearing stress at any point is proportional to its distance from the axis.

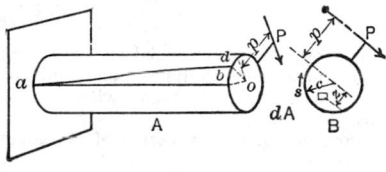

FIG. 1

The **Twisting Moment**, T, is equal to the product of the resultant, P, of the twisting forces, and its distance from the axis, p.

The **Resisting Moment**, T_r, in torsion, is equal to the sum of the moments of the unit shearing stresses acting along a cross section with respect to the axis of the bar. If dA is an elementary area of the section at a distance of z units from the axis of a circular shaft (Fig. 1B), and c is the distance from the axis to the outside of the cross section where the unit shearing stress is τ, then the total shearing force acting on dA is $(\tau z/c)\,dA$, its moment with respect to the axis is $(\tau z^2/c)\,dA$, and the sum of all the moments of the unit shearing stresses on the cross section is $\int (\tau z^2/c)\,dA$. In this expression the factor $\int z^2\,dA$ is the polar moment of inertia of the section with respect to the axis. Denoting this by J, the resisting moment may be written $\tau J/c$.

The **Polar Moment of Inertia** of a surface about an axis through its center of gravity and perpendicular to the surface is the sum of the products obtained by multiplying each elementary area by the square of its distance from the center of gravity of its surface; it is equal to the sum of the moments of inertia taken with respect to two axes in the plane of the surface at right angles to each other passing through the center of gravity. It is represented by J, inches^4. For the cross section of a round shaft,

$$J = (1/32)(\pi d^4) \quad \text{or} \quad 1/2\,\pi r^4 \tag{1}$$

for a hollow shaft

$$J = (1/32)\,[\pi(d^4 - d_1{}^4)] \tag{2}$$

where d is the outside and d_1 the inside diameter, in inches, or

$$J = 1/2\,[\pi(r^4 - r_1{}^4)] \tag{3}$$

where r is the outside and r_1 the inside radius, in inches.

The **Polar Radius of Gyration**, k_p, is also sometimes used in formulas; it is defined as the radius of a circumference along which the entire area of a surface might be concentrated and have the same polar moment of inertia as the distributed area. For a solid circular section,

$$k_p{}^2 = (1/8)d^2 \tag{4}$$

for a hollow circular section,

$$k_p{}^2 = (1/8)(d^2 + d_1{}^2) \tag{5}$$

27. DETERMINATION OF TORSIONAL STRESSES

The Torsion Formula for Round Shafts. The conditions of equilibrium require that the twisting moment, T, be opposed by an equal resisting moment, T_r, so that, for the values of the maximum unit shearing stress, τ, within the proportional limit, the torsion formula for round shafts becomes

$$T_r = T = \frac{\tau J}{c} \tag{6}$$

if τ is in pounds per square inch, then T_r and T must be in pound-inches, J is in inches^4, and c in inches. For solid round shafts having a diameter, d, inches,

$$J = \tfrac{1}{32}\,\pi d^4 \quad \text{and} \quad c = \tfrac{1}{2}\,d \tag{7}$$

and

$$T = \frac{1}{16}\,\pi d^3 \tau \quad \text{or} \quad \tau = \frac{16T}{\pi d^3} \tag{8}$$

For hollow round shafts

$$J = \frac{\pi(d^4 - d_1^4)}{32}, \quad c = \frac{1}{2}\,d \tag{9}$$

and the formula becomes

$$T = \frac{\tau\pi(d^4 - d_1^4)}{16d} \quad \text{or} \quad \tau = \frac{16Td}{\pi(d^4 - d_1^4)} \tag{10}$$

It should be remembered that the torsion formula applies only to solid circular shafts or hollow circular shafts and then only when the load is applied in a plane perpendicular to the axis of the shaft and when the shearing proportional limit of the material is not exceeded.

Shearing Stress in Terms of Horsepower. If the shaft is to be used for the transmission of power, the value of T, pound-inches, in the above formulas becomes $63,030H/n$, where H is the horsepower to be transmitted and $n = $ rpm. The maximum unit shearing stress, pounds per square inch, then, is:

$$\text{For solid round shafts:} \quad \tau = \frac{321,000H}{nd^3} \tag{11}$$

$$\text{For hollow round shafts:} \quad \tau = \frac{321,000Hd}{n(d^4 - d_1^4)} \tag{12}$$

If τ is taken as the allowable unit shearing stress, the diameter, d, inches, necessary to transmit a given horsepower at a given shaft speed can then be determined. It should be remembered, however, that these formulas give the stress due to torsion only, and allowance must be made for any other loads such as the weight of the shaft and pulleys, and tension in belts.

Angle of Twist. When the unit shearing stress τ does not exceed the proportional limit the angle bod (Fig. 1) for a solid round shaft may be computed from the formula

$$\theta = \frac{Tl}{GJ} \tag{13}$$

where $\theta = $ angle, expressed in radians; $l = $ length of shaft, inches; $G = $ shearing modulus of elasticity of the material, pounds per square inch; $T = $ twisting moment, pound-inches. Values of G for different materials are: steel, 12,000,000; wrought iron, 10,000,000; and cast iron, 6,000,000.

When the angle of twist on a section begins to increase in a greater ratio than the twisting moment, it may be assumed that the shearing stress on the outside of the section has reached the proportional limit. The shearing stress at this point may be determined by substituting the twisting moment at this instant in the torsion formula.

Torsion of Non-circular Cross Sections. The analysis of shearing stress distribution along non-circular cross sections of bars under torsion is complex. By drawing two lines at right angles through the center of gravity of a section before twisting, and observing the angular distortion after twisting, it has been found from many experiments that in non-circular sections the shearing unit stresses are not proportional to their distances from the axis. Thus in a rectangular bar there is no shearing stress at the corners of the sections and the stress at the middle of the wide side is greater than at the middle of the narrow

Table 1. Approximate Formulas for Maximum Shearing Stress and Angle of Twist in Members Subjected to Torsion

(From Seely's *Advanced Mechanics of Materials*)

τ = shearing unit stress, pounds per square inch; T = twisting moment, pound-inches; θ = angle of twist, radians, in length, l, inches; J = polar moment of inertia, inches[4]; G = shearing modulus of elasticity, pounds per square inch; A = area of cross section, square inches; y = distance of most remote edge from center of bar, inches. Tests of brittle metal are those made by Bach, *Elastizität u. Festigkeit*, 1924, except for those marked *, which are by Kommers, *Amer. Mach.*, vol. xl, p. 941, 1914.

Shape	Maximum Unit Stress		Angle of Twist
	Mathematical Analysis	Tests of Brittle Material	
	$\tau = \dfrac{16T}{\pi d^3}$		$\theta = \dfrac{Tl}{GJ}$
	$\tau = \dfrac{16Td}{\pi(d^4 - d_1^4)}$		$\theta = \dfrac{32Tl}{\pi(d^4 - d_1^4)G}$
	$\tau = \dfrac{3T}{2\pi r t^2}$		$\theta = \dfrac{3Tl}{2\pi r t^3 G}$
	$\tau = \dfrac{2T}{\pi a b^2}$		$\theta = \dfrac{T(a^2 + b^2)l}{\pi a^3 b^3 G}$
	$\tau = \dfrac{T}{2\pi a_1 b_1^2 k(1 + k)}$ $k = \dfrac{a - a_1}{a_1} = \dfrac{b - b_1}{b_1}$		$\theta = \dfrac{T(a^2 + b^2)l}{4 a_1^3 b_1^3 k G}$
	$\tau = \dfrac{20T}{b^3}$		$\theta = \dfrac{46.2Tl}{b^4 G}$
	$\tau = \dfrac{1.09T}{b^3}$		$\theta = \dfrac{0.967Tl}{b^4 G}$
	$\tau = \dfrac{(3a + 1.8b)T}{a^2 b^2}$	$\tau = \dfrac{9T}{2ab^2}$	$\theta = \dfrac{4\pi^2 TlJ}{A^4 G}$ Error is small where $a > 2b$
	$\tau = \dfrac{T}{2t(a - t)(b - t_1)}$	$\tau = \dfrac{9bT}{2(b^3 a - b^3_1 a_1)}$	$\theta = \dfrac{Tl(at + bt_1 - t^2 - t_1^2)}{2tt_1(a - t)^2(b - t_1)^2 G}$

Table 1—_Continued_

Shape	Maximum Unit Stress		Angle of Twist
	Mathematical Analysis	Tests of Brittle Material	
	$\tau = \dfrac{4.8T}{b^3}$		$\theta = \dfrac{7.11Tl}{b^4G}$
		$\tau = \dfrac{9T}{2t^2(a + 2b_1)}$ $\tau = \dfrac{9T}{2t^2a}$ *	$\theta = \dfrac{36.2TlJ}{A^4G}$ *
		$\tau = \dfrac{9T}{2t^2(a + 2b_1)}$	
		$\tau = \dfrac{9T}{2t^2(a + b - t)}$	
		$\tau = \dfrac{9T}{2t^2(a + b - t)}$	
Any compact section without re-entrant angles.	$\tau = \dfrac{4\pi^2 JTy}{A^4}$		$\theta = \dfrac{kJTl}{A^4G}$ $k = 4\pi^2$ for ellipse. $= 40$ to 42 for rectangles.

side. In an elliptical bar the shearing stress is greater along the flat side than at the round side.

It has been found by tests (Bach, _Elastizität u. Festigkeit;_ and Young, _Bull._ 4, _School of Eng. Research,_ Univ. of Toronto) as well as by mathematical analysis that the torsional resistance of a section made up of a number of rectangular parts is approximately equal to the sum of the resistances of the separate parts. It is on this basis that nearly all the formulas for non-circular sections have been developed. For example, the torsional resistance of an I-beam is approximately equal to the sum of the torsional resistances of the web and the outstanding flanges. In an I-beam in torsion the maximum shearing stress will occur at the middle of the side of the web, except where the flanges are thicker than the web, and then the maximum stress will be at the midpoint of the width of the flange. Re-entrant angles, as those in I-beams and channels, are always a source of weakness in members subjected to torsion. Table 1 gives approximate values of the maximum unit shearing stress τ and the angle of twist θ induced by twisting bars of various cross sections, it being assumed that τ is not greater than the proportional limit and that the modulus of elasticity, G, remains a constant.

Ultimate Strength in Torsion. In a torsion failure, the outer fibers of a section are the first to shear and the rupture extends toward the axis as the twisting is continued. The torsion formula for round shafts has no theoretical basis after the shearing stresses on the outer fibers exceed the proportional limit, as then the stresses along the section are no longer proportional to their distances from the axis. It is convenient, however, to compare the torsional strength of various materials by using the formula to compute values of τ at which rupture takes place. These computed values of the maximum stress sustained before rupture are somewhat higher than the actual maximum stress because of inaccuracy of the simple torsion formula, and the maximum stress computed from the simple formula is known as "modulus of rupture in torsion." Computed values of the modulus of rupture in torsion are found by experiment to be as follows: cast iron, 30,000 psi; wrought iron, 55,000 psi; medium steel, 65,000 psi; timber, 2,000 psi. These computed values of modulus of rupture may be used in the torsion formula to determine the probable twisting moment that will cause rupture of a given round bar or to determine the size of a bar that will be ruptured by a given twisting moment. In design, large factors of safety should be taken, especially when the stress is reversed as in reversing engines and when the torsional stress is combined with other stresses as in shafting.

Effect of Keyway in Shaft. The sharp re-entrant angles in keyways produce high local stresses at these points. On the basis of studies made by hydrodynamical and soap-film methods, the local stress in the corner has been determined as equal to the stress in a similar shaft without a keyway, multiplied by a constant, K, which depends on the radius of the corner. The constants are approximately:

Radius of corner, in.	0.1	0.2	0.3	0.4	0.5	0.6	0.7
K	5.4	3.4	2.7	2.3	2.1	2.0	1.9

Under steady loading these high stresses are probably not very important, as a ductile material allows an adjustment to take place, but if the shaft is subjected to repeated loads a crack may start at the corner.

CYLINDERS, PLATES, ROLLERS, AND JOINTS

28. CYLINDERS

Thin Cylinders under Internal Pressure. A cylinder is regarded as thin when the thickness of the wall is small compared with the diameter. It is assumed that in such cases the tensile stress across a longitudinal section is uniformly distributed over the thickness of the wall. If p = internal pressure, pounds per square inch; l = length of cylinder, inches; t = thickness of wall, inches; d = diameter of cylinder, inches; and σ = tensile stress, pounds per square inch:

$$pdl = 2\sigma tl \quad \text{or} \quad \sigma = \frac{pd}{2t} \qquad (1)$$

For tensile stress across a transverse section

$$\frac{p\pi d^2}{4} = \sigma t\pi d \quad \text{or} \quad \sigma = \frac{pd}{4t} \qquad (2)$$

Formula 2 applies also to the stresses in the walls of a thin hollow sphere, hemisphere, or dome. This analysis does not apply where holes are cut in the cylinder for rivets or for other purposes. Where holes are cut, the tensile stresses must be found by the method used in riveted joints. (See Art. 31.)

Thin Cylinders under External Pressure. It is difficult to derive a rational formula for the stresses in a thin cylinder under external pressure since failure occurs partly by collapse and this is greatly influenced by any slight ellipticity. Tests made by Professor R. T. Stewart (*Trans. A.S.M.E.*, vol. xxvii, p. 730) on lap-welded steel tubes showed that the length was not important provided it was not over 6 times the diameter. For ratios of t/d less than 0.023

$$p = 1000 \left[1 - \sqrt{1 - 1600 \left(\frac{t^2}{d^2} \right)} \right] \qquad (3)$$

where p = collapsing pressure, pounds per square inch; t = thickness of wall, inches; and d = diameter of tube, inches. For ratios of t/d greater than 0.023

$$p = 86,670 \left(\frac{t}{d} \right) - 1386 \qquad (4)$$

Professor A. P. Carmen made tests (*Bull. 5, Eng. Exp. Sta.*, Univ. of Ill., 1906) and found, with ratios of t/d less than 0.025, for cold-drawn seamless steel tubes,

$$p = 50,200,000 \left(\frac{t}{d} \right)^3 \qquad (5)$$

and for seamless brass tubes

$$p = 25,150,000 \left(\frac{t}{d} \right)^3 \qquad (6)$$

For an extended discussion see Strength of Thin Cylindrical Shells under External Pressure by Saunders and Windenburg, and The Collapsing Strength of Steel Tubes by Jasper and Sullivan (*Trans. A.S.M.E.*, Applied Mechanics, September–December, 1931, pp. 207–245).

Thick Cylinders. When the thickness of the shell or wall is relatively large, as in guns, hydraulic machinery piping, and similar installations, the variation in stress from the inner surface to the outer surface is relatively large, and the ordinary formulas for thin wall cylinders are no longer applicable. The maximum stresses can be computed from eqs. 7 and 8

$$\sigma_t = \frac{P_1 r_1^2 - P_2 r_2^2 + (r_1^2 r_2^2 / x^2)(P_1 - P_2)}{r_2^2 - r_1^2} \tag{7a}$$

$$\sigma_r = \frac{P_2 r_2^2 - P_1 r_1^2 + (r_1^2 r_2^2 / x^2)(P_1 - P_2)}{r_2^2 - r_1^2} \tag{7b}$$

With the maximum shear theory the criterion of failure, the maximum shear stress for the case of internal pressure only ($P_2 = 0$, $x = r_1$) is

$$\tau = P_1 \frac{r_2^2}{r_2^2 - r_1^2} \tag{8}$$

where σ_t = tangential stress, σ_r = radial stress, τ = shear stress, P_1 = internal pressure, P_2 = external pressure, r_1 = internal radius, r_2 = external radius, x = radial coordinate of any point in the wall; the minimum and maximum values for x are r_1 and r_2, respectively.

In *shrinkage fits* a hollow cylinder or collar is forced over a cylinder or shaft having an outer diameter slightly greater than the inside diameter of the outer cylinder. This will cause compressive stresses in the inner cylinder and tensile stresses in the outer cylinder. This method is sometimes used in the manufacutre of guns. Timoshenko and Lessells (*Applied Elasticity*) give the equation for the pressure p, pounds per square inch, between the two cylinders as

$$p = \frac{E\delta}{b} \frac{(b^2 - a^2)(c^2 - b^2)}{2b^2(c^2 - a^2)} \tag{9}$$

where E = modulus of elasticity, a = inner radius of inner cylinder, b = outer radius of inner cylinder, c = outer radius of outer cylinder, δ = amount by which the outer radius of the inner cylinder exceeds the inner radius of the outer cylinder. With the pressures known, external pressure on the inner cylinder and internal pressure on the outer cylinder, the stresses can be computed by means of eqs. 7 and 8.

A comparison of the various theories of failure of thick cylinders is given in Seely's *Advanced Mechanics of Materials*.

29. PLATES

Circular Flat Plates. Table 1 gives equations for deflection and stresses in circular plates based mainly on the theoretical analyses developed by Grashof. (See Morley's *Strength of Materials;* Seely's *Advanced Mechanics of Materials;* Timoshenko and Lessels' *Applied Elasticity*.) Experience shows that the strength of flat plates is greater than is indicated by these equations.

Elliptical Flat Plates. An approximate formula for the maximum stress σ in elliptical plates, simply supported at the edge, having a major axis $2a$, a minor axis $2b$, a thickness t, and carrying a distributed load of w pounds per square inch, is

$$\sigma = \frac{(3a - 2b)}{a} \frac{wb^2}{t^2} \tag{10}$$

Square Flat Plates. Where the load is w pounds per square inch and the plate is supported on the four edges, each b inches long, the maximum moment and maximum stress will be along a diagonal. The moment M per unit length of diagonal will be $wb^2/24$, and the unit stress will be

$$\sigma = \frac{wb^2}{4t^2} \tag{11}$$

where t = thickness, inches.

With edges fixed and load distributed the maximum stress at the center of each of the edges is, approximately, $\sigma = 0.20\, wb^2/t^2$. Nichols found in experiments with steel plates

Table 1.   Maximum Stresses and Deflection for Circular Flat Plates

m = Poisson's ratio; P = concentrated load, pounds; w = distributed load, pounds per square inch; r = radius of plate, inches; t = thickness of plate, inches; E = modulus of elasticity.

Type of Load and Support	Maximum Stress at Edge	Maximum Stress at Center	Maximum Deflection at Center, $m = 1/3$
Uniform load, edge simply supported		$\dfrac{3}{8}\dfrac{wr^2}{t^2}(3+m)$	$\dfrac{2}{3}\dfrac{wr^4}{Et^3}$
Uniform load, edge fixed	Radial direction $\dfrac{3}{4}\dfrac{wr^2}{t^2}$ Tangential direction $\dfrac{3}{4}\dfrac{wmr^2}{t^2}$	$\dfrac{3}{8}\dfrac{wr^2}{t^2}(1+m)$	$\dfrac{1}{6}\dfrac{wr^4}{Et^3}$
Concentrated load at center, edge supported		Infinite	$\dfrac{5}{3}\dfrac{Pr^2}{\pi Et^3}$
Concentrated load at center, edge fixed	Radial direction $\dfrac{3P}{2\pi t^2}$ Tangential direction $\dfrac{3}{2}\dfrac{Pm}{\pi t^2}$	Infinite	$\dfrac{2}{3}\dfrac{Pr^2}{\pi Et^3}$
Uniform load, supported on central area of radius r_0		$\dfrac{3}{2t^2}\Big[(1+m)\log_e\dfrac{r}{r_0}$ $+\dfrac{1}{4}(1-m)\Big(1-\dfrac{r_0^2}{r^2}\Big)\Big]$	$\dfrac{3}{16}(1-m)(7+3m)\dfrac{wr^4}{Et^3}$
Load on central area of radius r_0, edge simply supported		$\dfrac{3(1+m)P}{2\pi t^2}\Big[\dfrac{1}{m+1}$ $+\log_e\dfrac{r}{r_0}-\Big(\dfrac{1-m}{1+m}\Big)\dfrac{r_0^2}{4r^2}\Big]$	$\dfrac{5}{3}\dfrac{Pr^2}{\pi Et^3}$
Load on central area of radius r_0, edge fixed		$\dfrac{3(1+m)P}{2\pi t^2}\Big(\log_e\dfrac{r}{r_0}+\dfrac{r_0^2}{4r^2}\Big)$ r must be $> 1.7\,r_0$	$\dfrac{2}{3}\dfrac{!Pr^2}{\pi Et^3}$

a value of

$$\sigma = 0.141\frac{wb^2}{t^2} \tag{12}$$

and Bach found

$$\sigma = 0.19\,\frac{wb^2}{t^2} \tag{13}$$

Rectangular Flat Plates.   For a distributed load w pounds per square inch, with supports along the four sides, and assuming a uniform distribution of moment along a diagonal, Seely gives (*Advanced Mechanics of Materials*) the stress as

$$\sigma = \frac{a^2 b^2 w}{2t^2(a^2 + b^2)} \tag{14}$$

in which a is the length of the long side and b the length of the short side.

For rectangular plates fixed along the four sides, for ductile materials, Bach found experimentally that the coefficient K in the moment equation $M = Kwb^2$ was such that the moment along the fixed edge and along the center was about the same, and that the values of K (see Seely, *Advanced Mechanics of Materials*) for different ratios of the length of the sides a and b were about as follows:

$b/a =$	0	0.2	0.4	0.6	0.8	1.0
$K =$	0.063	0.062	0.056	0.048	0.040	0.032

Timoshenko and Lessells (*Applied Elasticity*) give an analysis which shows that the maximum moment at the center of the plate in strips of unit width through the center of the plate, parallel to the long and short sides respectively, when the sides are simply supported, and the load is w pounds per square inch uniformly distributed over the plate, is given by the equations

$$M_{1\ max} = K_1 w b^2 \tag{15}$$

and

$$M_{2\ max} = K_2 w b^2 \tag{16}$$

the approximate deflection y is given by

$$y_{max} = \frac{D w b^4}{E t^3} \tag{17}$$

where t = thickness, inches; E = modulus of elasticity; K_1, K_2, and D are constants given in Table 2.

Table 2.   Values of Constants D, K_1 and K_2 for Rectangular Plates with Supported Edges and Uniformly Distributed Loads

a = long side, b = short side, Poisson's ratio = 0.3.

a/b	D	K_1	K_2	a/b	D	K_1	K_2
1.0	0.0443	0.0479	0.0479	1.8	0.1017	0.0948	0.0479
1.1	.0530	.0553	.0494	1.9	.1064	.0985	.0471
1.2	.0616	.0626	.0501	2.0	.1106	.1017	.0464
1.3	.0697	.0693	.0504	3.0	.1336	.1189	.0404
1.4	.0770	.0753	.0506	4.0	.1400	.1235	.0384
1.5	.0843	.0812	.0500	5.0	.1416	.1246	.0375
1.6	.0906	.0862	.0493	∞	.1422	.1250	.0375
1.7	.0964	.0908	.0486				

Table 3.   Values of Constants D and K for Rectangular Plates with Fixed Edges and a Uniformly Distributed Load

Poisson's ratio = 0.3

a/b	1.00	1.25	1.50	1.75	2.00	∞
D	0.0138	0.0199	0.0240	0.0264	0.0277	0.0284
K	0.0513	0.0665	0.0757	0.0817	0.0829	0.0833

Table 4.   Formulas for Bending Moments in Rectangular Flat Slabs

Values are per unit width.   Poisson's ratio = 0, a = longer side, b = shorter side, $\alpha = b/a$.

		Moments in Span b		Moments in Span a	
		At center of edge	At center of slab	At center of edge	At center of slab
Rectangular Slabs	Four edges simply supported	0	$\dfrac{1/8\ wb^2}{1 + 2\alpha^3}$	0	$\dfrac{wb^2}{48}(1 + \alpha^2)$
	Span b fixed; span a simply supported	$\dfrac{1/12\ wb^2}{1 + 0.2\ \alpha^4}$	$\dfrac{1/24\ wb^2}{1 + 0.4\ \alpha^4}$	0	$\dfrac{wb^2}{80}(1 + 0.3\ \alpha^2)$
	Span a fixed; span b simply supported	0	$\dfrac{1/8\ wb^2}{1 + 0.8\ \alpha^2 + 6\alpha^4}$	$\dfrac{1/8\ wb^2}{1 + 0.8\ \alpha^4}$	$0.015\ wb^2\left(\dfrac{1 + 3\alpha^2}{1 + \alpha^4}\right)$
	All edges fixed	$\dfrac{1/12\ wb^2}{1 + \alpha^4}$	$\dfrac{1/8\ wb^2}{3 + 4\alpha^4}$	$1/24\ wb^2$	$0.009\ wb^2(1 + 2\alpha^2 - \alpha^4)$
	Elliptical slab with fixed edges; diameters a and b; b/a $= \alpha$	$\dfrac{1/12\ wb^2}{1 + 2/3\ \alpha^2 + \alpha^4}$	$\dfrac{1/24\ wb^2}{1 + 2/3\ \alpha^2 + \alpha^4}$	$\dfrac{1/12\ wb^2\ \alpha^2}{1 + 2/3\ \alpha^2 + \alpha^4}$	$\dfrac{1/24\ wb^2\ \alpha^2}{1 + 2/3\ \alpha^2 + \alpha^4}$

For rectangular plates with the edges fixed, the maximum deflection is given by the same equation as for the plates with supported edges, but with the constant D given in Table 3. The maximum bending moment in a strip of unit width is at the middle of the longer fixed side and is equal to $M_{max} = Kwb^2$; K is given in Table 3, and a and b are the lengths of the long and short sides respectively.

Professor Westergaard in a study of reinforced-concrete slabs supported on columns has computed the moments in medium-thick rectangular and elliptical plates (*Proc. A.C.I*, 1921). These are plates in which there is no appreciable absorption of energy either by the vertical shearing stresses or by the stretching of the middle plane of the plate. By means of a number of simplifying assumptions the values for the moments as given in Table 4 were found.

30. STRESSES IN SOLIDS UNDER PRESSURE

Cylindrical and Spherical Rollers. Approximate equations for areas of contact and pressures have been worked out for a number of cases (Timoshenko and Lessells' *Applied Elasticity*). These are given in Table 5.

Hertz Equations. A general mathematical solution of the problem of the stresses produced by the pressure of one solid upon another, of which cylindrical and spherical rollers are a special case, was made by H. Hertz (English translation in *Miscellaneous Papers* by H. Hertz). The Hertz equations are very complex and deal only with surface stresses. Thomas and Hoersch (*Bull.* 212, *Eng. Exp. Sta.*, University of Illinois, 1930), as a part of an investigation of railroad rails and the development of transverse fissures in them, extended the Hertz equations to a study of the shearing stresses within the solid. They developed methods, too long to be given here, of computing the magnitude and location of the maximum shearing stresses produced in solids under pressure, such as in railroad rails under a wheel load. They have obtained experimental verification of the correctness of their equations by a "strain-etch" method, but point out that the mathematical equations derived do not hold except for thick material and where the ratio of the length of contact to the width of contact is large.

In a cylinder of radius R, inches, under a load P', pounds per inch of length, on a plane, assuming Poisson's ratio as $1/4$ the maximum shearing unit stress $\tau = 677\sqrt{P'/R}$, the width of contact equals $0.000564\sqrt{P'R}$, the depth to the point of maximum shear is $0.000222\sqrt{P'R}$, and the maximum intensity of pressure is $2256\sqrt{P'/R}$. In the case of crossed cylinders, the empirical equation is

$$\tau = \frac{11,750P^{\frac{1}{3}}}{(R_1/R_2)^{0.271}R_2^{\frac{2}{3}}}$$ (18)

where R_1 = radius of larger cylinder; R_2 = radius of smaller cylinder; P = total pressure, pounds, between the two solids; the ratio R_1/R_2 is only true within the limits of 1 and 8. Table 4 gives a number of values computed by the methods of Thomas and Hoersch.

Stribeck made extensive experiments on ball bearings in 1898–1900, and his reports are translated in *Trans. A.S.M.E.*, vol. xxix, p. 421.

Where rollers are moving, as at the end of a truss or girder or with a rolling lift bridge, under relatively high pressures, the plate on which the roller rests is the critical part. The metal of the plate tends to flow longitudinally under the backward and forward motion. Wilson (*Bull.* 191, *Eng. Exp. Sta.*, University of Illinois, 1929) reports on experiments with rollers of the size used in rolling lift bridges, between 116 in. and 476 in. radius, and concludes that the bearing capacity of a plate depends upon its thickness, the tensile strength of the material, and the diameter of the roller. He proposes the formula

$$P = (12,000 + 80D)\left(\frac{p - 13,000}{23,000}\right)$$

where P = safe working load, pounds per inch width of plate; D = diameter of the roller, inches; p = yielding point strength of the material in tension, pound per square inch. The thickness of the plate, inches, should not be less than $(1.0 + 0.004D)$ with D not less than 120 in. Steel bridges usually have one end on rollers to allow for expansion, and the usual allowable load for expansion rollers is $P = 600D$; P = load, pounds per lineal inch of roller, and D = diameter of the roller, inches.

Table 5.   Areas of Contact and Pressures with Two Surfaces in Contact

Poisson's ratio = 0.3; P = load, pounds; P_1 = load per inch of length, pounds; E = modulus of elasticity.

Character of Surfaces	Maximum Pressure, σ, at Center of Contact, lb per sq in.	Radius, r, or Width, b, of Contact Area, in.
Two spheres	$\sigma = 0.616 \sqrt[3]{PE^2 \left(\dfrac{d_1 + d_2}{d_1 d_2}\right)^2}$	$r = 0.881 \sqrt[3]{\dfrac{P}{E}\left(\dfrac{d_1 d_2}{d_1 + d_2}\right)}$
Sphere and plane	$\sigma = 0.616 \sqrt[3]{\dfrac{PE^2}{d^2}}$	$r = 0.881 \sqrt[3]{\dfrac{Pd}{E}}$
Sphere and hollow sphere	$\sigma = 0.616 \sqrt[3]{PE^2 \left(\dfrac{d_2 - d_1}{d_1 d_2}\right)^2}$	$r = 0.881 \sqrt[3]{\dfrac{P}{E}\left(\dfrac{d_1 d_2}{d_2 - d_1}\right)}$
Cylinder and plane	$\sigma = 0.591 \sqrt{\dfrac{P_1 E}{d}}$	$b = 2.15 \sqrt{\dfrac{P_1 d}{E}}$
Two cylinders	$\sigma = 0.591 \sqrt{P_1 E\left(\dfrac{d_1 + d_2}{d_1 d_2}\right)}$	$b = 2.15 \sqrt{\dfrac{P_1}{E}\left(\dfrac{d_1 d_2}{d_1 + d_2}\right)}$

Table 6.   Shearing Stresses in Solids under Pressure

Case	Radius, R_1, in.	Length of Cylinder, in.	Radius, R_2, in.	Load, lb	Area in Contact, sq in.*	Maximum Shearing Stress, lb. per sq in.	Depth below Surface of Point of Maximum Shearing Stress, in.
Crossed cylinders...	16.5	....	14	25,000	0.215	56,500	0.121
Crossed cylinders...	40	....	14	60,000	0.502	59,800	0.179
Cylinder on plane...	16.5	2	Plane	25,000	0.512	18,600	0.100
Cylinder on plane...	40	2	Plane	60,000	1.236	18,600	0.243
Cylinder on plane...	30	....	Plane	39,800 lb per in. length	....	24,700	....
Soleplates Galveston Causeway Bascule Bridge..........	309	....	Plane	112,400 lb per in. length	Width 3.45 in.	12,900	1.355

* Computed by approximate equation.

31. JOINTS

Riveted Joints exist where two or more parts of a structure or machine are connected by rivets. Two types of riveted joints are usually recognized, those in structures or machines, Fig. 1a, and those in internal pressure vessels, Fig. 1b. They are further classified as *butt* joints, Fig. 1c, and *lap* joints, Fig. 1d. The short plates connecting the main plates in Fig. 1c are known as *butt* or *cover* plates and sometimes as *butt straps*.

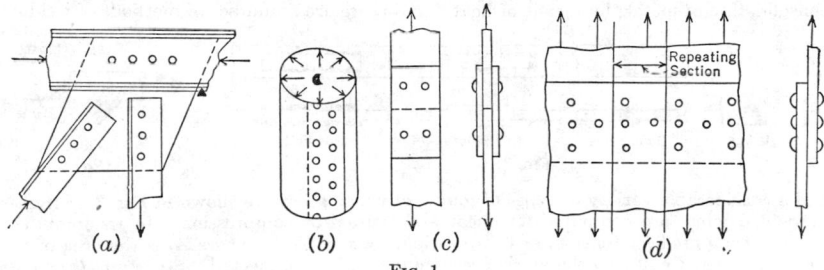

(a)   (b)   (c)   (d)

FIG. 1

Depending upon the number of rows of rivets, joints are known as single, double, and triple riveted. Figure 1d is a triple-riveted joint.

Stresses in Riveted Joints are of three kinds: (1) tensile stresses in the plates or members connected, (2) shearing stresses in the rivets connecting the plates, and (3) bearing stresses in the rivets and in the plates at the surface of contact with the rivets. In computing stresses it is assumed that the rivets are far enough from the end of the plate that the rivets will not tear through the plate and that all stresses are uniformly distributed over the area under consideration. Then the unit stresses, pounds per square inch, will be equal to the load, P, pounds, divided by the area, A, in square inches, or $\sigma = P/A$.

First select a typical portion of the joint and then determine the load on the part selected and the area of each element which is under stress. For structural joints the entire joint is used, while for internal pressure vessels a strip called a *repeating* section is used. Figure 2 shows such a repeating section, enlarged from Fig. 1d. The plate A will tend to tear apart at the section m–n and the plate B at section p–q, since the entire load must be carried across these reduced sections. The area of each plate is the same and the load is the same and therefore the tensile unit stress in each plate is $\sigma_t = P/A_t$, in which $A_t = (b - d)t$, or $\sigma_{t_{m-n}} = P/(b - d)t$. All the rivets being of the same size, each carries one-quarter of the load and therefore the total load carried across the section o–o will be $3P/4$ while the area $A_{t_{o-o}} = (b - 2d)t$, or $\sigma_{t_{o-o}} = 3P/4(b - 2d)t$. The shearing unit stress in the rivets is $\tau = P/A_s$, where A_s is the area of the four rivets or $4\pi d^2/4$, or $\tau = P/\pi d^2$. The bearing unit stress in the plate, or rivet, is $\sigma_b = P/A_b$, where A_b = the bearing area for four rivets, each of which is dt or a total of $4dt$, or $\sigma_b = P/4dt$.

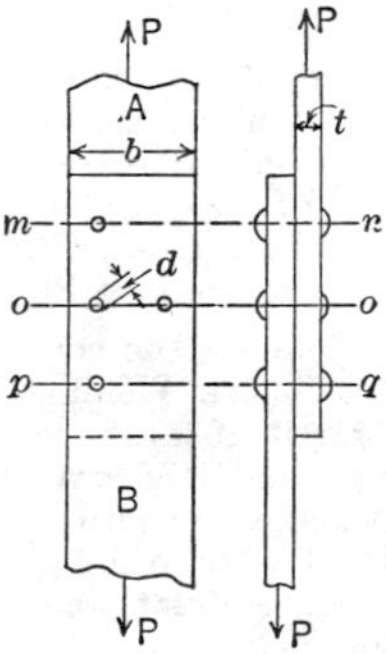

Allowable working stresses vary considerably, those given below representing specifications widely used.

Shearing Strength. ASME boiler code values of τ are 44,000 psi for steel rivets and 38,000 psi for iron rivets. These are ultimate values; with a factor of safety of 5 the working stresses are 8800 psi and 7600 psi, respectively. For structural rivets the working stresses in shear are 10,000 psi for hand-driven rivets and 13,500 psi for power-driven rivets.

For aluminum rivets, ultimate shearing strengths are: 17S-T, 33,000 psi; 24S-T, 42,000 psi (cold-driven); 53S-T, 23,000 psi (cold-driven); 53S-W, 18,000 to 24,000 psi (hot-driven), depending on temperature. Cold-driven rivets are assumed to have had proper quenching treatment immediately before driving.

Bearing or Crushing Strength. ASME boiler code prescribes $\sigma_b = 95,000$ psi for steel plates, corresponding to 19,000 psi working stress with safety factor 5. Allowable crushing stresses in structural joints are 40,000 psi for rivets in double shear, 32,000 psi for rivets in single shear. With aluminum alloys the crushing or bearing strengths are dependent on the plate or shape material and on its heat treatment. Values range from 34,000 psi

FIG. 2

ultimate in soft material to 129,000 psi ultimate in hard material (see Alcoa *Structural Handbook*).

Tearing Strength. A joint may fail by tearing óf the plate between the rivet holes. Resistance to tearing at any section is equal to the net area remaining after the holes have been made times the tensile strength of the plate; ASME code value for σ_t is 55,000 psi ultimate, or 11,000 psi allowable with factor of safety 5. For structural steel the allowable is 20,000 psi.

Welded Joints have come into extensive use in many places in buildings, bridges, tanks, and machinery to replace or supplement riveted joints. Welding is defined as "a localized consolidation of metals by means of heat," and there are a number of methods of welding

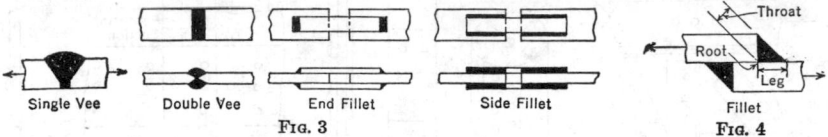

Single Vee   Double Vee   End Fillet   Side Fillet   Fillet

Fig. 3   Fig. 4

and a considerable variety of types of joints, of which a few are shown in Fig. 3. Stresses in welded joints are computed for shear and tension or compression. In *vee* joints the stress is computed for tension, or compression, by $\sigma = P/A_t$, where A_t is the area of the weld, considered equal to the area of the thinnest plate connected. Stress in *fillet* joints is computed for shear, $\tau = P/A_s$, where A_s is equal to the length of all the welds times the thickness at the throat, the throat being measured as indicated in Fig. 4.

The Design Stresses will, of course, vary, depending on the quality of the welds and the conditions of service of the structure. For example, pressure vessels may be built on the basis of factors of safety of 4 or 5, corresponding to an allowable working stress of about 10,000 to 12,000 psi (55,000 psi tensile strength steel), whereas an open storage tank could reasonably employ much higher stresses. The allowable working stresses for welds of various types have been set by engineering codes governing different types of construction.

A list of the commonly used construction codes is given below:

ASME Boiler Construction Code, Section I, Power Boilers; Section III, Locomotive Boilers; Section IV, Heating Boilers; Section V, Miniature Boilers; Section VIII, Unfired Pressure Vessels; Section IX, Qualification of Welding Operators.

API-ASME Code for Unfired Pressure Vessels. Office of Merchant Marine Safety, U. S. Coast Guard. Bureau of Ships, U. S. Navy Department.

ASA Pressure Piping Code. National Board Rules for Repairs by Fusion Welding.

Some of the constructional codes issued by the American Welding Society are: Code for Fusion Welding and Gas Cutting in Building Construction (*J. Am. Welding Soc.*, March, 1935); Specification for the Design, Construction, Alteration, and Repair of Railway and Highway Bridges; Tentative Code for Fusion Welding and Flame Cutting in Machinery Construction; Marine Code for Welding and Gas Cutting, Part D, Rules for the Fusion Welding of Hulls and Hull Parts (*J. Am. Welding Soc.*, February, 1935).

Other Joints and Fasteners. For a more extensive summary of rivets and welded joints and fasteners such as bolts, screws, and nails, see Kent's *Mechanical Engineers' Handbook*, 12th ed.. 1950, Sections 10 and 21.

REINFORCED CONCRETE

By Theodore Crane, Revised by J. E. Love, Jr.

Introduction

Concrete may be reinforced with metal embedded in such a manner that the two materials act together in resisting forces; the principal function of the concrete is to resist compressive stresses, and that of the reinforcement to resist tensile stresses.

The value of combining iron and concrete, to resist respectively tensile and compressive stresses in structural members, was first discovered in France about 1865. Since that time concrete reinforced with steel has become one of the most widely used structural materials for bridges, roads, retaining walls, dams, tanks, and many other types of engineering structures, as well as for buildings. It is almost universally used for multi-story industrial buildings and garages.

The choice between a structural steel or a reinforced concrete skeleton depends on comparative cost, unless the architectural design requires the use of one of these materials. H-section steel cores, extending through four or five lower stories, can be successfully used when column sizes become excessive. For industrial loads, girderless floor systems are often the most satisfactory. They provide better manufacturing facilities and a higher fire-resistance than so-called "mill-construction," or beam-and-slab design. For lighter types of occupancy such as commerical and residential work, ribbed floor systems are more suitable.

The material presented in this section on reinforced concrete is limited to certain common problems encountered in building design. For more detailed information about such features as cost, functional design, construction problems, and other applications of reinforced concrete, the references given at the end of this section should be consulted.

32. SPECIFICATIONS

Concrete. *See* Non-metallic Materials, Section 13.

Reinforcement. Expanded metal, triangular mesh, or wire fabric is often employed in slabs of moderate span, and in walls and partitions. The great bulk of reinforcement used, however, is in the form of bars, either plain or deformed, the deformations aiding the bond between the steel and concrete. Table 1 gives the areas, weights, and perimeters of standard reinforcing bars.

Table 1. Standard A305 Reinforcing Bars

Size	Inches	1/4	3/8	1/2	5/8	3/4	7/8	1			
	Number	2	3	4	5	6	7	8	9	10	11
Diameter, in.		0.250	0.375	0.500	0.625	0.750	0.875	1.000	1.128	1.270	1.410
Area, sq in.		.05	0.11	0.20	0.31	0.44	0.60	0.79	1.00	1.27	1.56
Weight per ft, lb		.167	0.376	0.668	1.043	1.502	2.044	2.670	3.400	4.303	5.313
Perimeter, in.		.786	1.178	1.571	1.963	2.356	2.749	3.142	3.544	3.990	4.430

The bar numbers are based on the number of 1/8 in. increments included in the nominal diameter of the bar.

Bar number 2 is available in plain rounds only. Bars numbered 9, 10 and 11 are round and equivalent in weight and nominal cross-sectional area to the old type 1 in., 1 1/8 in., and 1 1/4 in. square bars.

Deformations on deformed bars shall conform to "Standard Specifications for Minimum Requirements for the Deformations of Deformed Bars for Concrete Reinforcement" (*A.S.T M. Designation:* A305—50T).

The so-called "intermediate grade" of billet steel (see A.S.T.M. Specifications) can generally be used for all purposes where reinforcement is required. Rail steel, rerolled to rods and bars (see A.S.T.M. Specifications for Rail-steel Concrete Reinforcement Bars), is a satisfactory reinforcement in designs that do not require bending of the rods. Wire reinforcement shall conform to the A.S.T.M. Standard Specifications for Cold-drawn Steel Wire for Concrete Reinforcement.

Steel reinforcement should be stored on racks, but need not be protected from the weather unless kept for long periods. A light coating of red rust does not injure the bond between steel and concrete, but heavy scale should be removed by wire brushing.

33. GENERAL PRINCIPLES OF DESIGN

Design. The design of concrete slabs, beams, and girders involves computation of the sectional area of steel required for strength in tension, and of concrete required for strength in compression. Diagonal tension, a function of the shear, is resisted by the concrete alone or with the aid of web reinforcement. Columns or struts supporting concentric loads are designed with a reinforced-concrete section in which each material is assumed to carry a portion of the compression. With comparatively heavy loads, spirals are used to increase the ultimate strength, or structural steel cores may be used to reduce the diameter of the column.

Placing of Reinforcement. All metal reinforcement should be accurately placed, supported at the proper height by concrete or metal chairs, and so secured as to prevent displacement during pouring of concrete. Reinforcement should be cold bent, as indicated on the drawings or steel lists, around a pin whose diameter is four or more times the least dimension of the bar. It should be free from coatings that might destroy or reduce the bond, such as heavy rust, scale, paint, grease, etc. Heating of reinforcement for bending should not be permitted. Sufficient lap should be provided at splices to transfer the stress between bars by bond and shear. Column rods should be lapped at least 20 diameters for deformed steel or 40 diameters for plain steel. Splices should not be made at points of maximum stress.

Insulation of Reinforcement. To protect the steel from corrosion and fire, minimum thickness of concrete should be provided around the reinforcement as follows: floor slabs and walls, $3/4$ in. (some building ordinances require $1 1/2$ in. for walls); beams, girders and columns, $1 1/2$ in. (most building ordinances require 2 in. for girders and columns); footings or other principal structural members where concrete is deposited directly against the ground, 3 in. (most building ordinances require 4 in.); it is also desirable to provide 2-in. protection for all members exposed to the weather, such as spandrel beams and exterior columns.

Contraction Joints. Volumetric changes normally occur when concrete hardens. When concrete is cured in dry air, shrinkage commences immediately after placement, due principally to loss of water by evaporation or absorption. The preliminary shrinkage, occurring while the concrete is plastic, is fairly rapid, particularly in thin members exposed to hot sun and wind. The secondary shrinkage, occurring after the member has assumed definite form, is much slower, and is influenced by atmospheric conditions and the character of mixture. Rich concrete shrinks more than lean; decreasing the water-cement ratio decreases the shrinkage.

Atmospheric temperature changes cause changes in volume during the life of a structure. The coefficient of expansion for concrete through normal temperatures is approximately 0.000,005. Changes in atmospheric moisture cause volumetric changes which may be as great as those due to temperature changes. To meet these conditions, joints must be provided at sufficiently frequent intervals to avoid exceeding the strength of the concrete, or sufficient reinforcement must be used so to distribute deformations caused by initial shrinkage and by temperature and moisture variations that they will not seriously affect the structure.

For buildings not over 300 ft long, the reinforcement required for the principal stresses is usually sufficient to distribute secondary stresses due to shrinkage or expansion. Longer structures should be separated by transverse joints so that free movement of the adjacent parts is possible. In beam-and-slab construction this can be done by erecting two sets of columns supporting parallel floor beams on either side of the joint: a common footing may be used. In girderless construction, contraction joints are placed along the centers of bays, the two sections of which are designed as cantilevers. The break between sections should be complete, and exposed joints should be filled with an elastic joint filler.

Although most damages caused by temperature and moisture changes are traceable to contraction, expansion joints occasionally are necessary where sections of concrete are poured within rigid confines, particularly around the edges of concrete roof fills adjacent to parapet and penthouse walls. Comparatively thin members, particularly in exposed positions, should be lightly reinforced with $3/8$-in. round rods or wire fabric, to distribute deformations due to volumetric change and to prevent cracking. The amount of such reinforcement, of structural grade steel, may be taken as 0.25 per cent of the sectional area of the concrete.

Building Ordinances. Table 2 gives the allowable unit stresses applying to the design of reinforced concrete as now required by the Building Code Requirements for Reinforced Concrete of the American Concrete Institute (ACI 318–51). Unfortunately, there is wide variation in the requirements of different localities, and the engineer should obtain a copy of ordinances controlling his work. Outside of any local jurisdiction, the ACI Building Code is recommended as an excellent standard. Although many municipalities still assume fixed values for concrete mixed in certain specified proportions, it is preferable, when preliminary tests can be made, to compute allowable stresses as percentages of the ultimate compressive strength actually developed at the age of 28 days. The engineering societies have recommended this practice, which is illustrated by data taken from the ACI Building Code.

34. BEAMS AND SLABS

Fundamental Assumptions. A plane cross section of a beam, before bending, remains a plane section after bending; the modulus of elasticity of the concrete in compression remains constant within the assumed working stresses; the distribution of compressive stress in a beam is rectilinear; perfect adhesion exists between concrete and steel; in calculating the moment of resistance of reinforced-concrete beams and slabs, the tensile resistance of the concrete is negligible; the initial stress in the steel caused by contraction

Table 2. Allowable Unit Stresses in Reinforced Concrete Under Static Loads*

Description	For any strength of concrete in accordance with Section 502*	Allowable stresses For strength of concrete shown below			
		$f'_c =$ 2500 psi	$f'_c =$ 3000 psi	$f'_c =$ 4000 psi	$f'_c =$ 5000 psi
Modulus of elasticity..................		29,000,000			
ratio: n		$w^{1.5} 33 \sqrt{f'_c}$			
For concrete weighing 145 lb per cu ft ($=w$) (see Section 1102)*........... n		10	9	8	7
Flexure: f_c					
Extreme fiber stress in compression..... f_c	$0.45f'_c$	1125	1350	1800	2250
Extreme fiber stress in tension in plain concrete footings and walls.......... f_c	$1.6 \sqrt{f'_c}$	80	88	102	113
Shear: v (as a measure of diagonal tension at a distance d from the face of the support)					
Beams with no web reinforcement† ...... v_c	$1.1\sqrt{f'_c}$	55†	60†	70†	78†
Joists with no web reinforcement....... v_c	$1.2\sqrt{f'_c}$	61	66	77	86
Members with vertical or inclined web reinforcement or properly combined bent bars and vertical stirrups....... v	$5\sqrt{f'_c}$	250	274	316	354
Slabs and footings (peripheral shear, Section 1207)†.................... v_c	$2\sqrt{f'_c}$	100†	110†	126†	141†
Bearing: f_c					
On full area.........................	$0.25f'_c$	625	750	1000	1250
On one-third area or less‡............	$0.375f'_c$	938	1125	1500	1875

* From American Concrete Institute Standard 318–63.
† For shear values for lightweight aggregate concrete, see ACI Standard 318–63, Section 1208.
‡ This increase shall be permitted only when the least distance between the edges of the loaded and unloaded areas is a minimum of one-fourth of the parallel side dimension of the loaded area. The allowable bearing stress on a reasonably concentric area greater than one-third, but less than the full areas, shall be interpolated between the values given.

or expansion of the concrete is negligible, except in column design; the ratio of the modulus of elasticity of steel to concrete is expressed by the formula (notation given below):

$$n = \frac{E_s}{1000f'_c} = \frac{30,000}{f'_c} \tag{1}$$

Flexure Formulas for Rectangular Beams and Solid Concrete Slabs. (See Fig. 1.)
The following notation is used: d = effective depth of beam, distance from extreme fibers in compression to center of gravity of tensile reinforcement, inches; b = width of

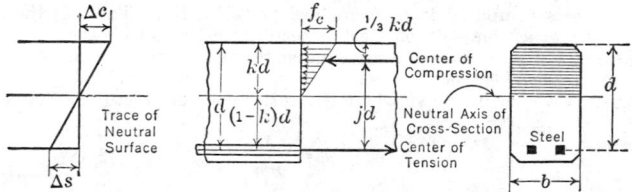

FIG. 1. Diagrams of deformations and stresses

beam or section of slab, inches; k = ratio of distance of neutral axis of cross section from extreme fibers in compression to effective depth of beam; kd = distance of neutral axis from extreme fibers in compression, inches; j = ratio of distance between the center of compression of concrete and center of tension of steel to effective depth of beam, or ratio of arm of resisting couple to d; jd = distance between center of compression in concrete and center of tension in steel, or arm of resisting couple, inches; f_c = compressive unit stress in extreme fibers of concrete, pounds per square inch; f'_c = ultimate crushing unit strength of concrete at age 28 days, pounds per square inch; f_s = tensile unit stress in steel, pounds per square inch; A_s = area of cross section of main tensile reinforcement, square inches; E_s = modulus of elasticity of steel, pounds per square inch; E_c = modulus of elasticity of concrete, pounds per square inch; n = modulus of elasticity of steel divided by modulus of elasticity of concrete, i.e., $n = E_s/E_c$; M = bending moment, inch-pound; p = percentage of reinforcement, or A_s/bd; $K = \frac{1}{2} f_c j k$.

Steel ratio, for balanced reinforcement,

$$p = \frac{1}{2 \dfrac{f_s}{f_c} \left(\dfrac{f_s}{n f_c} + 1 \right)} \tag{2}$$

Ratio of distance of neutral axis of cross section from extreme fibers in compression to effective depth of beam,

$$k = \sqrt{2pn + (pn)^2} - pn \tag{3}$$

Ratio of arm of resisting couple to concrete depth,

$$j = 1 - \frac{k}{3} \tag{4}$$

Concrete depth as controlled by bending moment,

$$d = \sqrt{\frac{M}{\frac{1}{2} f_c j k b}} = \sqrt{\frac{M}{Kb}} \tag{5}$$

Steel area as controlled by bending moment,

$$A_s = \frac{M}{f_s j d} \tag{6a}$$

$$A_s = pbd \tag{6b}$$

Formula 5 determines the depth of a slab or beam to resist a given bending moment without exceeding a specified unit stress in the concrete. If formula 5 is used, $6a$ or $6b$ may be used to determine the steel area. Formula $6a$ is for general use; formula $6b$ determines the sectional area of the reinforcement as a percentage of the concrete section, bd. In Table 3 are given values of k, j, p, and K, as computed by the above formulas.

Table 3.   Formula Factors for Beams and Slabs

f_c, Psi	$f_s = 18,000$ Psi					f_c, Psi	$f_s = 20,000$ Psi				
	n	k	j	p	K		n	k	j	p	K
900	15	0.429	0.857	0.0107	165	900	15	0.403	0.866	0.0091	157
1125	12	.429	.857	.0134	207	1125	12	.403	.866	.0113	196
1350	10	.429	.857	.0161	248	1350	10	.403	.866	.0136	236
1688	8	.429	.857	.0201	310	1688	8	.403	.866	.0170	294

Note: Values of $f_c = 0.45 f'_c$ throughout this table.

If the value of p is computed from formula 2 or taken from Table 3, the result will be to supply the exact amount of steel necessary to "balance" the concrete. This procedure is satisfactory for slabs and rectangular beams, or for other members for which the concrete section has been determined by moment considerations. It is inappropriate for the design of T-beams or others where the entire concrete section is not required by moment.

Fiber stresses,

$$f_s = \frac{M}{A_s j d} = \frac{M}{p j b d^2} \tag{7}$$

$$f_c = \frac{2M}{j k b d^2} = \frac{2 p f_s}{k} \tag{8}$$

Formulas 7 and 8 are for checking fiber stresses of beams already designed.

Flexure Formulas for T-Beams. (See Fig. 2.)   The design of T-beams and T-girders is divided into two classifications: (1) That in which the neutral surface falls either within the flange or at the bottom of it; (2) that in which the neutral surface falls within the web.

In case 1, the design is the same as for rectangular beams, and formulas 2 to 8 apply. The width of flange, however, is used for the value of b, and the steel ratio p is based upon the total area bd.

In case 2, which is more frequent, neglecting the compression resisted by the web, let t = total slab thickness on either side of the web, inches; z = distance from the compression

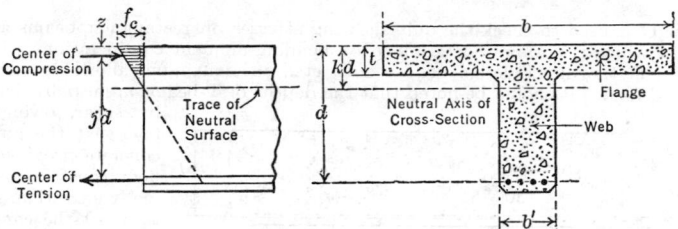

FIG. 2. Stress diagram and cross section of a reinforced-concrete T beam

surface of the beam to the center of compression, inches; b' = width of web, inches; b = allowable width of flange, inches; $p = A_s/b'd$, that is, the steel ratio is based upon the area $b'd$. The remaining notation is the same as for rectangular beams.

Position of neutral axis,

$$kd = \frac{2ndA_s + bt^2}{2nA_s + 2bt} \tag{9}$$

Position of resultant compression,

$$z = \frac{3kd - 2t}{2kd - t} \cdot \frac{t}{3} \tag{10}$$

Arm of resisting couple,

$$jd = d - z \tag{11}$$

Steel area as controlled by bending moment,

$$A_s = \frac{M}{f_s jd} \tag{12}$$

Fiber stresses,

$$f_s = \frac{M}{A_s jd} \tag{13}$$

$$f_c = \frac{Mkd}{btjd(kd - t/2)} = \frac{f_s}{n} \cdot \frac{k}{1-k} \tag{14}$$

Formulas giving approximate results but erring slightly on the safe side are deduced by substituting

$$\left(d - \frac{t}{2}\right) \text{ for } jd \quad \text{and} \quad (1/2) \text{ for } \left(1 - \frac{t}{2kd}\right)$$

This substitution gives:

Extreme fiber stress in concrete,

$$f_c = \frac{2M}{bt\left(d - \dfrac{t}{2}\right)} \tag{15}$$

Steel area,

$$A_s = \frac{M}{f_s\left(d - \dfrac{t}{2}\right)} \tag{16}$$

Shearing Stress. Rectangular beams and slabs should be checked for shear, as a measure of the diagonal tension, and the concrete area increased if the shear is excessive. The depth of T-beams, when economically designed, is generally controlled by shearing stress. Let V = total vertical shear at the section considered, pounds; v = total unit shearing stress at the section, pounds per square inch; $j = 0.875$ (an approximation used for shear computations); b = width of beam for rectangular beams (for T-beams replace b by b' the width of web), inches; d = effective depth of beam, inches. Then,

$$v = \frac{V}{bjd} \tag{17}$$

or

$$d = \frac{V}{bvj} \tag{18}$$

Equation 17 is used to check the unit shearing stresses on rectangular beams and slabs the size of which is normally determined by bending-moment considerations. Equation 18 is used in the direct design of T-beams, etc., the maximum allowable value being substituted for v. It should be noted that the depths of T-beams are usually determined by shear, owing to the fact that the amount of concrete provided by the floor construction is normally more than sufficient to resist the compressive stresses due to moment.

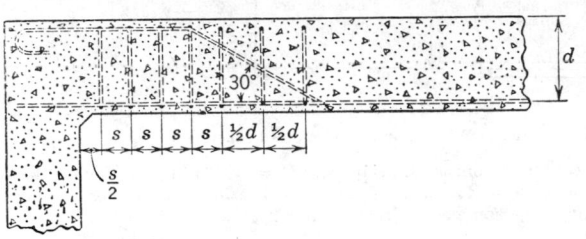

Web Reinforcement. (See Fig. 3.) Stirrups or bent rods are used for web reinforcement in beams and girders of rec-

FIG. 3.   Diagram of a web-reinforcement in a rectangular or T beam   tangular and T-section, as it is more economical to use such reinforcement, and correspondingly higher stresses, than to increase the size of the member.

The design of web reinforcement is as follows: Let v_c = allowable unit shearing stress resisted by the concrete alone, pounds per square inch; A_v = cross-sectional area of stirrups, or of inclined portion of double-bent rods, square inches (in a U-shaped or W-shaped stirrup this is the total cross-sectional area of all legs); f_v = allowable unit tensile stress in stirrups, pounds per square inch; s = spacing of stirrups, inches; remaining notation the same as above.

Then if $v_c bjd$ represents the total shearing stress carried by the concrete,

$$s = \frac{f_v j d A_v}{V - v_c bjd} \tag{19}$$

If it is required that the reinforcement is to be designed to take two-thirds of the total shearing stress, as specified in some codes, the formula becomes

$$s = \frac{3}{2} \frac{f_v j d A_v}{V} \tag{20}$$

These formulas apply to vertical stirrups. Inclined stirrups or inclined bars, forming a part of the main tensile reinforcement, may be designed or checked by the same formulas, by multiplying the value of the shear by the sine of the angle of inclination to the horizontal.

If L = clear span, feet, the distance from the support through which either stirrups or bent bars are required for uniformly loaded beams is given by

$$x = \frac{L}{2} \left(1 - \frac{v_c}{v} \right) \tag{21}$$

Stirrups are usually placed vertically, and are either $3/8$-in. or $1/2$-in. round rods, bent to a U- or W-shape. The maximum spacing for stirrups should be limited to $1/2 \times$ depth of beam; this also is the maximum distance through which the inclined portion of a longitudinal rod, or several rods, bent at the same section, should be considered effective as web reinforcement (see Fig. 3). The diameter of the rod used for stirrups should not exceed $1/50 \times$ depth of beam. Stirrups should encircle the main longitudinal reinforcement, all free ends being anchored by hooks.

Bond. As the tensile stress in the reinforcement is transmitted from the concrete, the rods or bars must have sufficient surface area to avoid exceeding the unit bond stress per square inch. If u = unit bond stress, pounds per square inch, and Σo = sum of perimeters, inches, of all horizontal rods under tensile stress at the section considered,

$$u = \frac{V}{(\Sigma o)jd} \quad \text{or} \quad \Sigma o = \frac{V}{ujd} \tag{22}$$

It is customary to check the bond stress after the design is completed; if the value is excessive, smaller rods should be used and the number increased, since this increases the surface area of the steel. Limiting values for bond stress are usually $0.03f'_c$ for plain rods or bars, and $0.07f'_c$ for deformed types.

Anchorage of Reinforcement. It is necessary to anchor the main longitudinal reinforcement by bond, hooks, or mechanical anchors in or through the supporting member. The ACI Building Code specifies that every reinforcing bar, whether required for positive or negative reinforcement, shall be extended at least 12 diameters beyond the point at which it is no longer needed to resist stress. The maximum tension in any bar must be developed by bond on a sufficiently straight or bent embedment or by other anchorage.

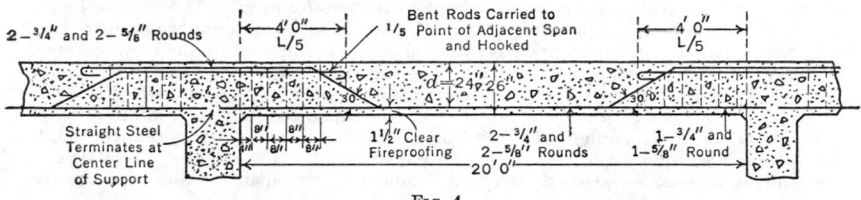

FIG. 4

The anchorage requirements fixed by the ACI Building Code are such that bars which meet the requirements for deformations fixed by A.S.T.M. 305 usually need not be hooked. When hooks are used they may not be assumed to carry a load which would produce a tensile stress in the bar greater than 10,000 psi.

In general, plain bars in tension shall terminate in standard hooks, except that hooks shall not be required on the positive reinforcement at interior supports of continuous members. Figure 4 is an example of a reinforced concrete continuous beam showing a typical arrangement of steel. If the bars meet the requirements of A.S.T.M. 305 they need not be hooked.

Bending Moments and External Shears. Bending moments and vertical shears for continuous and restrained beams as well as for those of single spans simply supported are given on p. 512 and p. 518. These values also apply to reinforced-concrete members subjected to flexure, but for continuous spans they are modified to conform to the conditions of partial restraint existing in concrete frames. Also, to provide for possible variable live loads, the values of the positive bending moments used in concrete design are usually taken larger than the theoretical values.

As the function of the reinforcement is to resist the tensile stresses, the typical arrangement for beams other than cantilever is to place the rods near the bottom face throughout the central portion of the span. For beams simply supported, about two-thirds of the main tensile reinforcement consists of straight rods or bars, the other one-third being bent up at the $1/5$ point of the clear span and anchored into the supports to resist any negative moment resulting from the monolithic construction.

For continuous or restrained beams, it is usual to raise one-half the main tensile reinforcement at the points of inflection, assumed to be the $1/5$ points of the clear span, and, for continuous beams, to carry the reinforcement to the $1/4$ point of the adjoining span. This provides approximately equal areas near the bottom of the beam at mid-span and near the top of the beam over supports, which corresponds to the general design practice of using approximately equal coefficients for both positive and negative moments in the design of continuous members. Raising a portion of the main longitudinal reinforcement also helps to resist diagonal tension and usually permits the omission of one or two stirrups at each end of the beam.

The following equations for determining maximum bending moment are recommended by the ACI Building Code (ACI 318—51) for continuous beams or slabs (the larger of two adjacent spans not exceeding the shorter by more than 20 per cent), under uniform load, where the unit live load does not exceed three times the unit dead load.

Positive moment at center of span:

End spans,

$$M = 1/14 \, wl'^2 \tag{23}$$

Interior spans,

$$M = 1/16 \, wl'^2 \tag{24}$$

Negative moment at exterior face of first interior support:

Two spans,

$$M = \frac{1}{9} wl'^2 \tag{25}$$

More than two spans,

$$M = \frac{1}{10} wl'^2 \tag{26}$$

Negative moment at other faces of interior supports:

$$M = \frac{1}{11} wl'^2 \tag{27}$$

In these equations M = bending moment, foot-pounds; w = uniformly distributed load per foot of beam or per square foot of slab; l' = length of clear span for positive moment and shear, and the average of the two adjacent clear spans for negative moment, in feet.

If the spans are substantially unequal, or subjected to other than uniformly distributed loads, the actual bending moments and shears should be computed by the methods given in the section on Beams, pp. 510 to 528   The computation of moments for continuous girders and other members subjected to concentrated loads may be simplified by determining the moments as if the members were simply supported, and then multiplying by a reducing factor corresponding to the condition of end restraint, $i.e.$, $\frac{4}{5}$ for semi-continuity and $\frac{2}{3}$ for full continuity. The design of continuous members should always provide steel over support at least equal to the amount required in mid-span to resist the negative moment.

The external vertical shear is computed as a measure of the diagonal tension, which, in reinforced-concrete members subject to flexure, is a combination of horizontal and vertical shear. For both simply supported and continuous members, shears are computed as described in the section on Beams, pp.   510 to 528 . For members symmetrically loaded, the maximum shear, occurring at the face of the support, equals $\frac{1}{2}$ load, except as influenced by the conditions of end restraint.

Rectangular Beams. The illustrative problems that follow serve to indicate the application of the theory to the mechanics of solution of individual problems.

Example. Required: to design a rectangular beam, 20-ft clear span, fully continuous at supports, for a uniformly distributed load of 1500 lb per ft, including weight of beam.

Specification data: $f_c = 900$ $(f'_s = 2000)$; $f_s = 18,000$; $n = 15$; $v_c = 60$; maximum vertical shearing stress (v) with web reinforcement $= 120$. *Note:* ACI Building Code (ACI 318—51) allows up to 240 for v but a practical limitation for ordinary design is 120, $(0.06f'_c)$.

By formula 27, $M = wl'^2/11 = (1500 \times 20^2 \times 12)/11 = 655,000$ in.-lb. This is the negative moment at the support.

By formula 5, $d = \sqrt{M/Kb} = \sqrt{655,000/(165 \times 12)} = 18.2$ in., $= 19$ in. (approximately), where breadth b is assumed as 12 in.; K from Table 3.

By formula 6, $A_s = pbd = 0.0107 \times 12 \times 19 = 2.44$ sq in.

Accept a total depth of 21 in. and a width of 12 in., with four number 7 ($\frac{7}{8}$-in. round) bars.

The value of b should be chosen as small as practicable, unless headroom is limited, deep, narrow beams being more economical. Sufficient width should be provided to give a clear space of at least 1 in. between bars and the required fireproofing on the sides. However, the overall depth of a beam or girder should usually be not more than 3 b.

Testing the shear by formula 17, $v = V/bjd = (1500 \times 10)/(12 \times 0.875 \times 19) = 75.2$ psi. As this is more than v_c, the value allowed for plain concrete, and less than the limit for v, 120 psi, the design is acceptable, provided web reinforcement is used. The web reinforcement is computed as follows:

By formula 19, $s = f_v j d A_v/(V - v_c b j d) = 18,000 \times 0.875 \times 19 \times 0.098/(15,000 - 60 \times 12 \times 0.875 \times 19) = 9.7$ in., where A_v = cross-sectional area of the two legs of a U-shaped stirrup, made of a number 2 ($\frac{1}{4}$-in. round) bar.

By formula 21, $x = \frac{L}{2} \left(1 - \frac{v_c}{v}\right) = 20/2(1 - 60/75.2) = 2$ ft approximately.

The first stirrup may be placed 4.5 in. ($\frac{1}{2}$ s) from the face of the support. Three more stirrups, 9 in. apart, provide web reinforcement beyond the section where none is needed.

Testing the bond stress at face of support by formula 22, $u = V/\Sigma o j d = 15,000/4 \times 2.75 \times 0.875 \times 19 = 82$ psi. This value is satisfactory for deformed bars. The arrangement and details of reinforcement shown in Fig. 4 are not the same as the values arrived at in this example. A similar arrangement could be used, however.

Slabs. The design of reinforced concrete one-way slabs is substantially the same as that of rectangular beams. Slabs reinforced principally in one direction only can be used for spans upward to approximately 18 ft, but generally this type of slab construction is reserved for shorter spans.

Large square or rectangular panels with supports at the edges and with no intermediate supports are designed as two-way slabs. In these cases there is principal slab reinforcement running in both span directions. Such slabs should be designed by the methods outlined in the ACI Building Code (ACI 318—51), Art. 709, or by the methods given in any of the standard treatises on reinforced concrete design listed at the end of this section. For a one-way slab the total load on a strip of slab 1 ft wide, spanning between supports, is obtained by multiplying the unit load per square foot by the span length. This strip is then treated as a rectangular beam of width b = 12 in. Web reinforcement is not ordinarily used in slabs.

Example. Required: to design a slab, 7-ft span, fully continuous at supports, to carry a superimposed load of 125 psf.

Specification data: Same as in the example of a rectangular beam. Assume a slab thickness of 4 in., weighing 50 psf; total load per square foot = 125 + 50 = 175 psf.

The ACI Building Code specifies that the negative moment at face of all supports for slabs with spans not exceeding 10 feet shall be $M = 1/12\,wl'^2$ (Art. 701). This value will be used here.

$$M = 1/12\,wl'^2 = 1/12 \times 175 \times 7 \times 7 \times 12 = 8575 \text{ in.-lb}$$

By formula 5, $d = \sqrt{M/Kb} = \sqrt{8575/(165 \times 12)} = 2.08$ in.

Since this value of d is considerably less than the assumed value (d = 3 in.) the concrete is entirely satisfactory in flexure. In this example a total thickness of 4 in. will be taken as the practical minimum. Also, since d = 3 in., the required steel will be computed by formula 6a rather than 6b.

$$A_s = M/(f_s jd) = 8575/(18,000 \times 0.857 \times 3) = 0.19 \text{ sq in.}$$

Satisfactory reinforcement would be number 3 bars (3/8-in. round) 7 in. on centers.

Testing the shear by formula 17, $v = V/(bjd) = 613/(12 \times 0.875 \times 3) = 19.5$ psi. As this is less than the limiting value, v_c, the design is satisfactory.

Testing the bond stress at face of support, by formula 22, $u = V/(\Sigma ojd) = 163/\{(1.71 \times 1.18) \times 0.875 \times 3\} = 116$ psi, where 1.71 = number of bars per foot of width and 1.18 = perimeter of each bar. The value of bond stress is satisfactory for deformed bars.

T-Beams. When a floor system comprising slabs, beams, and girders is designed, the slabs are considered first. These are designed as in the previous example, and the load carried by a T-beam, for which the floor slab acts as a flange, is found by multiplying the unit load per square foot of floor by the area which the beam supports, and adding to the product any other incidental loads and the weight of the stem of the beam.

The loads carried by T-girders, usually a combination of concentrated loads due to beam reactions, and the uniformly distributed weight of the girder itself, are used to

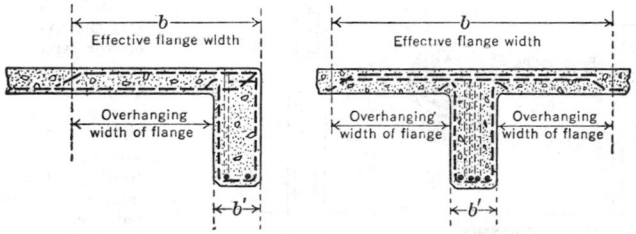

FIG. 5

compute the bending moment, which is then employed in exactly the same way as with uniformly distributed loads.

For purposes of design in computing the sectional area of concrete available to resist compressive stresses, the width of slab which may be considered as acting as the flange of a T-beam should not exceed $1/4$ of the span-length of the beam. See Fig. 5.

The overhanging width on either side of the web should not exceed 8 × slab thickness nor $1/2$ × clear distance to next beam. In angle beams, the overhanging width of flange should be limited to $1/12$ × span or $1/2$ × clear distance to next beam. It is always necessary to pour the floor slab forming the flange of the beam monolithically with the web, and there must be steel in the form of stirrups or bent rods to bond the flange to the web.

The critical section for compression in the concrete of T-beams, continuous over supports, is at the face of the support. As the negative moment acts on the rectangular section of the web, the stress in the concrete at these sections is always higher than through midspan. A choice is open to the designer. The section may be proportioned as a T-beam using the maximum positive moment at the center of the span. At the face of the support and as far out into the span as is necessary the resulting rectangular section must be reinforced with compressive steel as well as with tensile steel for the negative moment. The alternative to this is to consider the rectangular section resisting negative moment and proportion this section as an ordinary rectangular beam. At the section resisting positive moment the stress in the concrete need not be checked, as it must be less than that at the support, as pointed out above. The steel area required may be determined by the approximate formula 16 previously given. The second method is more economical of steel and in an ordinary case is to be favored over the first method. Of course, in either case, the shear and bond requirements must be considered. These are exactly the same as for ordinary rectangular beams.

Reference should be made to one of the standard treatises on reinforced-concrete design for examples of T-beams as used in actual construction.

Ribbed Floor Construction. For floors of apartment houses, hotels, institutional buildings, and others for light occupancy (superimposed loads 30 to 100 psi), the ribbed constructions are particularly suitable. These comprise a series of concrete ribs, 4 or 5 in. wide, separated by terra-cotta blocks, gypsum or concrete units, or metal or wood pans. Rib spacing varies from 16 to 30 in., depending on the width of block or void. A continuous slab, from 2 to 3 in. thick, forms the structural surface of the floor. Plaster is applied directly to the soffits of the blocks used as fillers; under the pans, metal lath is used as a plaster base. All these designs may be used with structural-steel or reinforced-concrete frame. Bridging joists should be used on long or very heavily loaded spans.

As the structural units in such systems are actually small T-beams (the fillers, except in terra-cotta and concrete, acting merely as voids) the design is substantially the same as for T-beams.

35. COLUMNS

The principal types of reinforced-concrete columns are the tied or hooped, spiraled, and cored, the cored being actually a steel column. The main columns of a building should never be less than 12-in. in diameter or in least width; good practice requires a minimum 16-in. diameter for round columns supporting girderless floor construction, and a minimum 14-in. thickness for exterior columns of industrial buildings. The diameter or least width should not be less than $1/15$ the average center-to-center span of the supported bay nor longer than $10 \times$ least lateral dimension of the column, unless the normal load is reduced. Dowels are used between footings and superimposed columns. Splices in vertical reinforcement should be made by lapping the rods a length of at least 20 bar diameters.

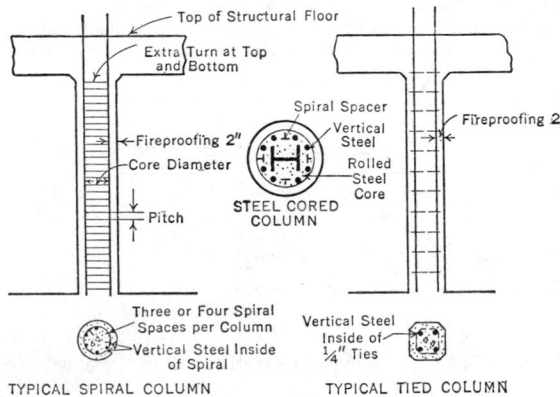

TYPICAL SPIRAL COLUMN   **TYPICAL TIED COLUMN**

Fig. 6

Tied Columns (see Fig. 6) are economical for comparatively light loads, such as those carried in the upper stories of buildings. The ratio P_g of the cross section of area of the vertical reinforcement to effective concrete area, usually taken as the gross sectional area, should not be less than 1 per cent nor should more than 4 per cent be considered in the computations. At least 4 rods, $5/8$-in. diameter, should be placed between 2 and 3 in. from the face of the column, enclosed by $1/4$-in. diameter ties spaced not over 12 in. apart. As both maximum and minimum percentages of steel are controlled by building ordinances, it is usually desirable to increase the column size or to use a richer mixture rather than to add vertical steel to carry load. It is extremely important, however, to have sufficient vertical reinforcement to resist bending, particularly for unsymmetrical loads.

Spiraled Columns (see Fig. 6) are usual for the interiors of buildings designed for heavy loads. Although the minimum amount of vertical reinforcement is required to resist bending stresses, the greatest economy is obtained by using rich mixtures and as large a percentage of spiral as the code or specification will permit. The vertical reinforcement should comprise at least 6 bars of $5/8$-in. minimum diameter, with cross-sectional area not less than 1 per cent nor more than 6 per cent of the core. The spiral is expressed as a percentage of the volume of the enclosed concrete and varies from $1/2$ to 2 per cent (see equation 30 below). The ratio of the percentage of spiral reinforcement to the percentage of vertical reinforcement should not be less than $1/4$. The spiral reinforcement should consist of evenly spaced, continuous spirals held firmly in place by at least three vertical spacer bars. The spacing of the spirals should not be greater than $1/6$ the diameter of the core nor more than 3 in.

Cored Columns (see Fig. 6) are built of steel wide-flange sections enclosed by the minimum of 4 in. of concrete, reinforced with at least 0.5 per cent of vertical reinforcement and 0.5 per cent of spiral. The structural-steel core is then designed on a basis of axial compression, using a stress of 15,000 to 18,000 psi, according to the code or specification. No allowance is made for the concrete casing. The ratio of the unsupported height to the least radius of gyration of the structural-steel section should not exceed 120.

Formulas for Columns. The following formulas apply to the design of tied and spiraled columns. Let A_g = effective cross-sectional area of column, square inches; P_g = ratio of area of vertical steel to sectional area of concrete; P = total safe axial load, pounds, on column whose length $\lessgtr$ 10 × least cross-sectional dimension; n = ratio of modulus of elasticity of steel to modulus of elasticity of concrete; f'_c = ultimate crushing unit strength of concrete at age 28 days, pounds per square inch; f_c = average unit compressive stress, in concrete, pounds per square inch; f = average unit compressive stress, P/A, in the column, pounds per square inch; D = core diameter in spiral columns, inches; a = cross-sectional area of spiral reinforcement steel; s = spacing of spiral reinforcement, i.e., pitch, inches.

For tied columns, the total safe load,

$$P = A_g(0.18f'_c + .80f_sP_g) \tag{28}$$

For spiraled columns, the total safe load,

$$P = A_g(0.225f'_c + f_sP_g) \tag{29}$$

The percentage of spiral reinforcement $= \dfrac{4a}{Ds} \times 100$ (30)

Example of Tied Column Design. Required: to design a square column to carry an axial load of 150,000 lb, including weight of column. For economy, use the minimum amount of vertical reinforcement, $P_g = 1.0$ per cent of the gross cross-sectional area of the column.
Specification data: $f'_c = 2000$ psi; $f_s = 16,000$ psi.
By formula 28, $P = A_g(0.18f'_c + 0.80f_sP_g)$. Substituting the specification data, $P = A_g(0.18 \times 2000 + 0.80 \times 16,000 \times 0.01) = A_g \times 488$. $A_g = P/488 = 150,000/488 = 307$ sq in. This area is given by a 17.5-in. square column, whose area is 306 sq in.
Area of vertical steel $= A_g \times P_g = 307 \times 0.01 = 3.07$ sq in.
Accept eight number 6 bars (3/4 in. round), with 1/4-in. round ties spaced 12 in. on centers. The longitudinal bars are placed 2 in. inside the surface with three on each side of the square.
Note that the amount of steel necessary in the above example could be reduced by using a higher strength concrete.

Example of Spiral Column Design. Required: to design a spiral column to carry an axial load of 425,000 lb including the weight of column. Assume a ratio $P_g = 0.04$ and a spiral ratio $p' = 0.012$.
Specification data: $f'_c = 2500$ psi; $f_s = 16,000$ psi.
By formula 29, $P = A_g(0.225 f'_c + f_sP_g)$. Substituting the specification data, $P = A_g(0.225 \times 2500 + 16,000 \times 0.04) = A_g \times 1202$. $A_g = P/1202 = 425,000/1202 = 353$ sq in. This area is given by a 21.5-in. diameter round column, whose area is 363 sq in.
Area of vertical steel $= A_g \times P_g = 363 \times 0.04 = 13.52$ sq in.
Accept nine number 11 bars. (Eleven number 10 bars would also be satisfactory.) This reinforcement will be placed on a circle whose diameter is 16 in. This will provide a little more than 1 1/2 in. of fire cover outside the spiral steel. Spiral steel, 3/8-in. round, 2-in. pitch. (From equation 30, this corresponds to a spiral reinforcement of 1.20 per cent.) The vertical bars are placed inside the spiral.

36. FOOTINGS

The Function of a Footing is to distribute concentrated loads from walls or columns over a sufficiently large area to bring the pressure within the safe bearing capacity of the soil or rock. Plain concrete footings may be used beneath walls or columns carrying small loads, but reinforced designs are more economical for heavy construction.

Types of Reinforced-concrete Column Footings (see Fig. 7) are: the isolated column footing; the combined or continuous footing, carrying two or more columns and constructed as an inverted beam; the cantilever footing, in which the eccentricity of an exterior footing is resisted by a strap connected with an adjacent interior footing. All types may be used either with or without piles, and the continuous footing may be developed as a mat to cover the entire foundation area. Footings for independent columns are designed in both stepped (Fig. 7a) and pyramidal (Fig. 8c) forms. The use of a pedestal between the footing slab and the base of the superimposed column is of value in reducing the unit compression on the top surface of the footing.

The load used in determining the soil bearing area of a footing is the load used in the design of the column in the story immediately above the footing, plus any live load or dead load in that story, plus the estimated weight of the footing itself. Footing areas should be proportioned for dead load alone or, in industrial buildings, for the dead load plus a fraction of the live load.

The design load, used to determine the thickness and reinforcement of a footing slab, is the load defined above, less the weight of the footing. Individual column footings

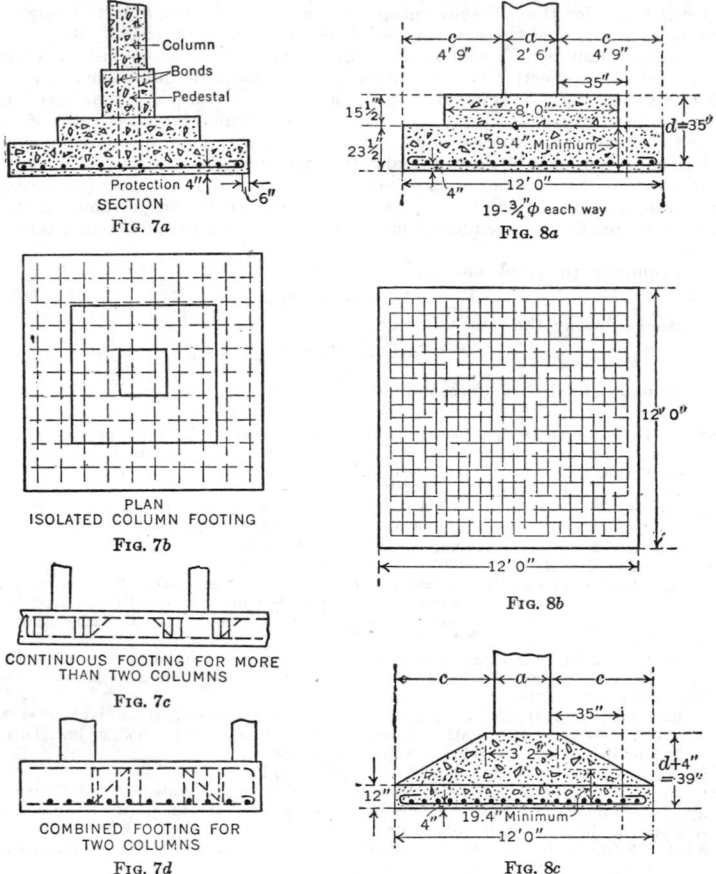

Fig. 7a

PLAN
ISOLATED COLUMN FOOTING
Fig. 7b

CONTINUOUS FOOTING FOR MORE
THAN TWO COLUMNS
Fig. 7c

COMBINED FOOTING FOR
TWO COLUMNS
Fig. 7d

Fig. 8a

Fig. 8b

Fig. 8c

must be centered beneath the column which they support. If more than one column rests on the same footing, the center of gravity of the foundation area must coincide with the center of gravity of the loads.

Reinforced-concrete footings are designed by the same formulas as concrete beams and slabs. The footing must be sufficiently strong to resist the stresses produced by bending moment and diagonal tension, and most building ordinances contain a requirement for "punching shear." A two-way reinforcement, of plain or deformed rods, is used in the slab and dowels to connect the footing slab to the basement column. Pedestals may be used to obtain a uniform height from which to start the first tier of columns.

Isolated footings, preferably square in plan, are generally used unless the proximity of a property line or other obstruction makes a concentric design impossible. Under such conditions, the combined footing, or the continuous wall footing, are both preferable to the cantilever type which is highly indeterminate except when used over piles.

The design of an isolated column footing involves determining the area of the footing slab, the design load, the depth as controlled by punching shear, where this is specified by code, and the steel area as controlled by moment. The diagonal tension, compression in the concrete, and bond stress in the steel are then checked. The requirements of diagonal tension may demand increasing the thickness of the footing slab near its edge, but the concrete stresses are seldom critical if the other considerations are met, and high bond stresses may be avoided by using comparatively small rods and hooking the ends.

Example. Required: to design an isolated footing for a column of 30 in. square, sustaining a load of 525,000 lb, including the weight of the column. Soil pressure, 2 tons per sq ft.

Specification data: $f_c = 650$ psi; $f_s = 16,000$ psi; v limited to 40 psi; punching shear limited to 120 psi; $n = 15$; u limited to 100 psi.

Estimating the weight of the footing at 50,000 lb, area of footing slab = 575,000/4000 = 144 sq ft. Choosing a square footing, 12 × 12 ft, $w = 525,000/144 = 3646$ psf. Depth as controlled by punching shear.

$$d = \frac{(\text{Area of footing} - \text{Area of column}) \times w}{(\text{Perimeter of column}) \times (\text{Allowable punching shear})} = \frac{(144 - 6.25) \times 3646}{4 \times 30 \times 120} = 34.9 \text{ in.}$$

Accept a total depth of 39 in.

If a = width of side of column or pedestal, and c = projection of footing slab, both in feet (see Fig. 8),

$$M = 6w(a + 1.2c)c^2 = 6 \times 3646\{2.5 + (1.20 \times 4.75)\} \times 22.56 = 4,047,000 \text{ lb-in.}$$

By formula 6b, $A_s = M/(f_s jd) = 4,047,000/(16,000 \times 0.875 \times 35) = 8.26$ sq in. Accept nineteen $3/4$-in. rounds.

Thickness (d') of footing slab, as controlled by diagonal tension at distance d from face of column or pedestal (d and d' are both in inches),

$$d' = \frac{(\text{Area outside distance } d)w}{jv(\text{Perimeter at distance } d)} = \frac{[144 - \{2.5 + (35/6)\}^2] \times 3646}{0.875 \times 40 \times \{2.5 + 35/6\} \times 4 \times 12} = 19.4 \text{ in.}$$

By formula 8, minimum width of footing at top, in feet = $b = \dfrac{2M}{jkf_c d^2 \times 12}$

$$M/(1.99 \times f_c \times d^2) = 4,047,000/(1.99 \times 650 \times 1225) = 2.55 \text{ ft.}$$

By formula 22, $u = V/[\Sigma o]jd = \dfrac{(144 - 6.25) \times (3646/4)}{(19 \times 2.35) \times 0.875 \times 35} = 96$ psi, which is acceptable for deformed rods. Fig. 8a shows the footing arranged in a stepped form; Fig. 8c shows the same footing arranged in a pyramidal form.

It should be noted that this example follows the procedure of most building ordinances in considering "punching shear," which usually determines the depth. A somewhat different approach may be had by following the ACI Building Code. The specifications in the above problem are not those of the ACI Building Code.

37. GIRDERLESS OR FLAT SLAB CONSTRUCTION

For building three or more bays wide, where column locations give approximately square bays of equal or nearly equal size, girderless or flat slab construction is the most satisfactory for heavy loads, as are required for industrial occupancy. The advantages

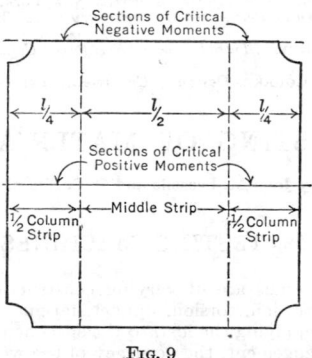

FIG. 9

are a saving in story height, due to the elimination of beams and girders, better lighting facilities, and structural economy. The flared column heads are characteristic of this system, which may be designed either with or without drops over the columns. If depressed panels, usually referred to as drops, are used over the interior columns, half drops are required at the exterior columns. Beams are used to frame around stairwells and elevators, and as lintels connecting exterior columns.

For purposes of establishing the bending moments and the resisting moments of a square panel, the panel is divided into middle and column strips as shown in Fig. 9. In this figure the division is indicated in only one direction, but a similar division is also made at right angles to the strips shown.

The diameter of the column capital is usually taken as 0.225 × average center-to-center span of bay. The minimum width of drop is taken as 0.33 or 0.35 × length of span in that direction; except for heavy loads the shearing stresses around the perimeter of the drop are not excessive, and this requirement normally determines the dimensions of the drop. The offset, or difference in thickness between the slab and drop, varies from 2 to 4 in. It is usually controlled by the sectional area required over the column capital to resist the negative moment on the column strip.

The bending moments used in designing the various sections are derived from empirical data and form a part of the code requirements in all city ordinances. The moments are generally increased by 20 or 25 per cent along the center lines of exterior bays and over the first interior supports. At the wall, or spandrel, the negative moment in the column strip is taken as 80 or 90 per cent and in the middle strip as 50 or 62.5 per cent of the corresponding moments for a normal interior panel. Half bands, containing 50 per cent of the steel required for a column strip, are placed parallel to the outer edge of exterior panels. The spacing of bars is limited to 3 × slab thickness, and the ratio of steel to concrete area in any strip should not be less than 0.0025.

There are two widely used systems of reinforcement, the two-way and the four-way. In the former, bands of steel bars are designed to span from column to column, and a two-way reinforcement is placed in the central rectangular portion of the panel between the bands and running parallel to them. In the four-way system, two diagonal bands supplement the direct bands, and additional reinforcement is added in the form of short bars placed over the lines of support, perpendicular to the steel of the rectangular bands.

Practically every large city has its own individual code governing the design of flat slabs, and when working under any particular jurisdiction, the designer must apply the regulation in the locality. The requirements of the ACI Building Code are typical of these regulations and should be referred to for further information on this type of construction.

BIBLIOGRAPHY ON REINFORCED CONCRETE

AMERICAN CONCRETE INSTITUTE, *Building Code Requirements for Reinforced Concrete* (ACI 318—63), Detroit, 1951.

CAUGHEY, ROBERT, *Reinforced Concrete, Mechanics and Design*, Rev. Ed., New York, D. Van Nostrand, 1946.

DUNHAM, CLARENCE, *The Theory and Practice of Reinforced Concrete*, 3rd Ed., New York, McGraw-Hill, 1953.

HOOL, G. A., and W. S. KINNE, *Reinforced Concrete and Masonry Structures*, 2nd Ed., New York, McGraw-Hill, 1944.

NATIONAL BOARD OF FIRE UNDERWRITERS, *National Building Code*, New York, 1949.

PARKER, HARRY, *Simplified Design of Reinforced Concrete*, New York, John Wiley, 1943.

PEABODY, DEAN, JR., *The Design of Reinforced Concrete Structures*, 3rd Ed., New York, John Wiley, 1968.

SUTHERLAND, HALE, and R. C. REESE, *Introduction to Reinforced Concrete Design*, 2nd Ed., New York, John Wiley, 1943.

URQUHART, L. C., and C. E. O'ROURKE, *Design of Concrete Structures*, 6th Ed., New York, McGraw-Hill, 1958.

TESTING OF MATERIALS

By John M. Lessells and G. S. Sherniak

38. TESTING MACHINES

Machines for testing materials are of varying construction and capacity for testing specimens in tension, compression, torsion, impact, fatigue, and other specific purposes. Usually the mechanism for applying the load to the specimen is independent of the weighing apparatus. By this arrangement, the accuracy of the weighing is not affected by the deformation of the specimen which takes place as the load is applied.

Tensile Testing Machine, Screw Type. This type of machine is in common use in the United States. It may be used for either tension or compression testing. The stress on the test piece is produced through the movement of platforms by screws, which, depending on the direction of turning, will either separate the platforms or draw them toward each other. A compound lever system transmits the platform force to a weighing beam kept in balance by moving the poise. Machines of this type are built by the Tinius Olsen Company in Philadelphia, Riehle Brothers in East Moline, Illinois, and Baldwin Southwark in Philadelphia.

Tensile Testing Machine, Hydraulic Type. A direct-acting hydraulic tensile testing machine has a head moved by hydraulic pressure, intensity of which is read on a pressure gage. The load on the test piece is determined by multiplying the pressure reading by a constant which depends upon the area of the cylinder. These machines are manufactured by Riehle in East Moline, Illinois, and Baldwin Southwark in Philadelphia.

Tension Creep-testing Machines. Since creep testing involves considerable time, most machines are built for multiple testing, and the tests can be carried out at any desired temperature. The test specimens are supported on knife edges at the top and loaded by lever systems. A common way to obtain the creep deformation is by a micrometer microscope.

Torsion-testing Machines. A torsion-testing machine consists essentially of a turning head which applies a torque to a test specimen. This torque is transmitted through a lever system to a hydraulic gage, the reading of which indicates the applied twisting force or torque. The angle of twist is measured directly on the turning head of the machine. The angle through which the specimen twists is recorded for each increment of torque until rupture occurs.

Impact-testing Machines

Drop-weight Machines. The impact is obtained by a hammer of known weight falling from a predetermined height on the specimen. In one type of machine the absorption of kinetic energy by the specimen is determined from the deflection of a spring.

Charpy Pendulum Machine. This machine is used for testing small notched specimens for energy absorption (Fig. 1). The specimen is placed horizontally with the ends sup-

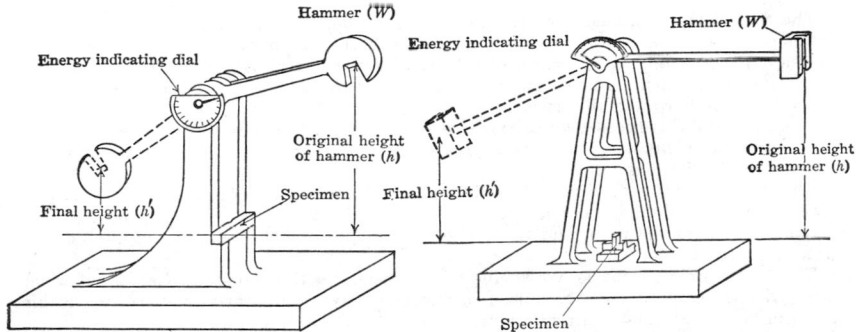

FIG. 1. Charpy impact testing machine   FIG. 2. Izod impact testing machine

ported by an anvil. The hammer strikes the specimen midway between the supports opposite the notch. The energy expended in fracturing the specimen is the product of the weight of the pendulum and the difference between its height of fall and height of rise. This energy is expressed either in foot-pounds or in kilogram-meters.

The Izod Machine is used for testing a notched specimen for energy absorption. The specimen is supported at its lower end in a vertical position as a cantilever (Fig. 2). The pendulum strikes the specimen at a definite distance above the notch on the same side as the notch. The energy used in breaking the specimen is the product of the weight of the pendulum and the difference between its height of fall and height of rise. This energy, expressed in foot-pounds, is known as the Izod value.

The Rotary Hammer Impact Machine is used to determine energy absorption of a test specimen. This is accomplished by means of a flywheel to which is attached a knife that can assume two positions, either flush with the rim of the wheel or projecting from the rim of the wheel. The latter position is required for breaking the test piece. The energy expended by the flywheel in breaking the specimen is measured by a tachometer tube.

The **High-impact Shock-testing Machine** consists of a large hammer hinged to swing in an arc (Fig. 3). It strikes the lower surface of a large anvil table on which equipment to be shock-tested is mounted. This machine is used for determining the vibratory motion and stresses set up in a piece of equipment under impact conditions.

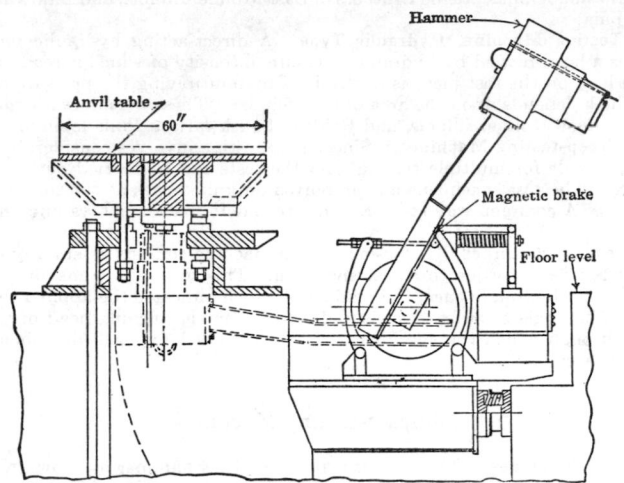

Fɪɢ. 3.   High impact shock-testing machine for medium-weight equipment

Hardness-testing Machines

The **Brinell Hardness-testing Machine** is used to measure the resistance of the material to the penetration of a hardened steel ball subjected to a standard load. The diameter of the indentation produced is measured with the aid of the microscope. A Brinell hardness number can be calculated or read from standard tables, and is used as a comparative measure of the hardness of materials.

The **Rockwell Hardness-testing Machine** is a direct-reading indentation type of machine where the ball or diamond point is impressed into the material. The Rockwell hardness number is read from the dial.

The **Vickers Hardness-testing Machine** follows the indentation principle, where, in this case, a pyramidal diamond indenter and standard load is used. A microscope measures the diagonal of the indentation, and the Vickers hardness number is either computed or read from prepared charts.

The **Shore Scleroscope** is a machine which measures the rebound of a falling weight dropped on the specimen. It is used primarily to give comparative values for checking uniformity of a product in production.

The **Tukon Hardness Machine** is a sensitive indenter type of machine of particular value in measuring hardness on individual grains and of certain small areas of specimens where the hardness of a zone having a certain microstructure is desired.

Fatigue-testing Machines

Fatigue-testing Machines are classified as (a) rotary bending, (b) plane bending, (c) direct stress, and (d) torsional stress.

The **Rotary Bending Machine** is the most widely used and the simplest type of fatigue-testing machine in which the specimen is in the form of a rotating bar. The rotating cantilever or the rotating beam form of loading may be applied. Figure 4 shows one of the modern forms of a rotating cantilever bar. The cantilever bar is end-loaded through a ball-bearing cage. In the rotating beam method the specimen is four-point loaded through ball bearings, as shown in Fig. 5. Machines are built to operate at speeds up to 30,000 rpm.

The Plane Bending Machine subjects the specimen to reversed bending without rotation. The specimen is forced to vibrate by induced magnetism from electromagnets (Fig. 6). Frequencies up to 17,000 cpm have been employed.

The Direct-stress Method subjects the specimen to repeated tensile stress by means of the electromagnetic principle. A typical machine is shown in Fig. 7.

The Torsional-stress Machine applies repeated torsional stress to a specimen by means of inertia or resonant methods. Figure 8 shows a machine of the former type.

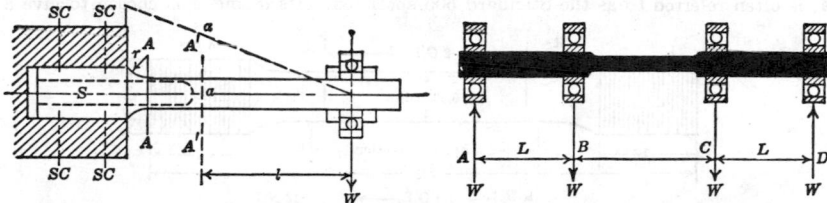

FIG. 4. Rotating cantilever bar fatigue test specimen

FIG. 5. Rotating beam bar fatigue test specimen

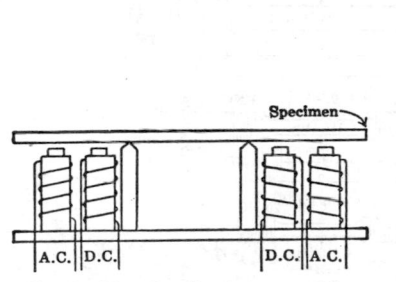

FIG. 6. Plane bending fatigue machine

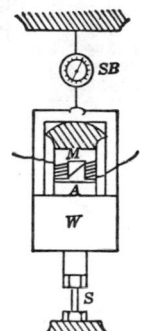

FIG. 7. Direct-stress fatigue machine

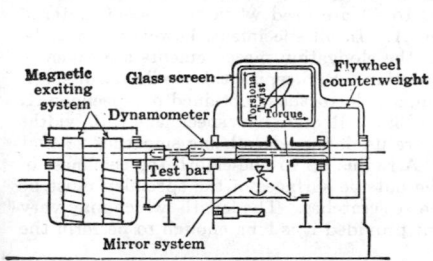

FIG. 8. Inertia torsional fatigue machine

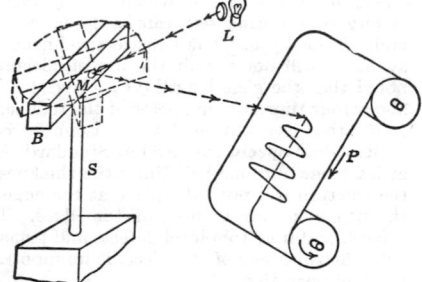

FIG. 9. Damping capacity machine

Damping-testing Machines

One type of machine used to determine the damping capacity of metals (Fig. 9) consists of a cylindrical specimen held rigidly at one end while a torque is applied to the opposite end. When the specimen is released, a record of its vibrations with time is made on sensitized photographic paper.

39. STANDARD TEST PIECES

The fundamental data obtained in a test on material are affected by the method of testing and the size and shape of the specimen. To eliminate variations in results due to these causes, standards have been adopted by the ASTM, the ASME, and various associations of manufacturers. (See ASTM, ASME, etc., Standards for details.)

Specimens for Tension Tests. Figure 10 shows a group of machined test bars in common use for testing steels and some non-ferrous metals. The most common one, specimen A, is often referred to as the Standard 505 specimen. Its diameter is chosen to give a

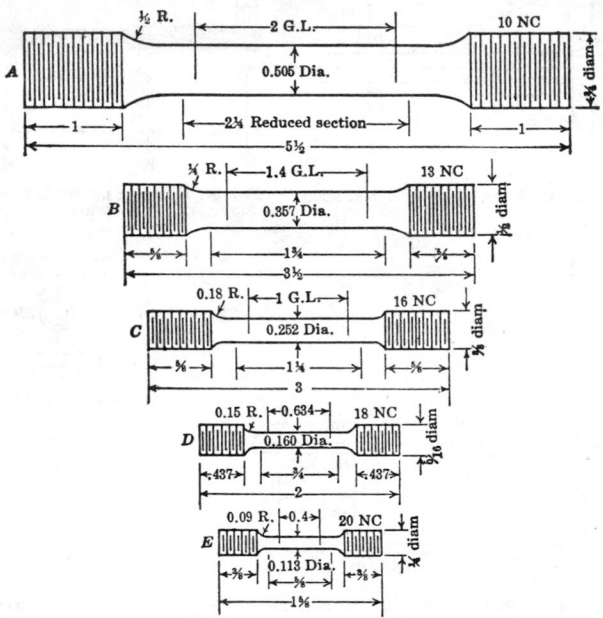

Fig. 10. ASTM Standard Designation E and subsize machined test specimens for cast, rolled, and forged steels, and non-ferrous alloys

0.2-sq-in. area of section which is convenient in that the load in pounds on the specimen at any time during test can easily be converted to stress (pounds per square inch) by multiplying by 5. The subsize specimens (B to E) are used when the size of material available will not permit the use of specimen A. In all specimens, however, it can be noted that the gage length (G.L.) over which the elongation measurements are taken is about four times the diameter of the reduced section. By keeping this similarity of specimens, the data obtained are reasonably comparable to those obtained on specimen A.

Bend-test Specimens (ASTM Standards E16–39). Rectangular specimens with widths at least one and one-half times the thickness are used, the edges being smoothed so that the fracture will not take place at the edge. A radius up to one-eighth the thickness of the specimen may be used on the edges. The outside surfaces of the specimen must be smooth and free from local defects and transverse scratches. The length, which may vary with the thickness of the piece, is unimportant provided it is long enough to perform the bending operation.

Compression-test Specimens (ASTM Tentative Standards E9–46T). It is recommended that standard compression-test specimens be in the form of circular cylinders and that the ends of the specimens shall be plain and normal to the longitudinal axis of the specimen. There are three types of compression-test specimens for metallic materials (short specimens, medium-length specimens, and long specimens). Short specimens are used for compression tests of such metals as bearing metals, which in service are used in the form of a thin plate or shell to carry load perpendicular to the surface. Medium-length specimens are used for determining the general compressive strength properties of metallic materials. Long specimens are best adapted for determining the modulus of elasticity in compression of metallic materials. Suggested dimensions for compression-test specimens for general use are given in Table 1.

Table 1

	Diameter, d, in.	Length, l, in.
Short specimen	$1\,1/8 \pm 0.01$	1 ± 0.05
Medium-length specimens	$\begin{cases} 0.798 \pm 0.01 \\ 1 \pm 0.01 \\ 1\,1/8 \pm 0.01 \end{cases}$	$\begin{array}{l} 2\,3/8 \pm 1/8 \\ 3 \pm 1/8 \\ 3\,3/8 \pm 1/8 \end{array}$
Long specimens	$\begin{cases} 0.798 \pm 0.01 \\ 1\,1/4 \pm 0.01 \end{cases}$	$\begin{array}{l} 6\,3/8 \pm 1/8 \\ 12\,1/2 \text{ min} \end{array}$

Impact-test Specimens (ASTM Tentative Standard E23—41T). The simple-beam (Charpy type) specimen and the cantilever-beam (Izod type) specimen are shown in Figs. 11 and 12, for use in the Charpy type and Izod type impact machines, respectively.

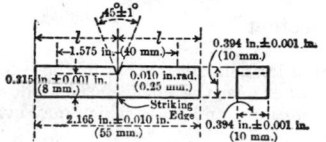

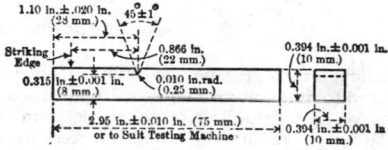

FIG. 11. Simple beam (Charpy) impact-test specimen

FIG. 12. Cantilever beam (Izod) impact-test specimen

Fatigue-test Specimens. No specifications have been drawn up for standard fatigue test bars.

40. NON-DESTRUCTIVE TESTING METHODS

Non-destructive testing is descriptive of test and inspection methods which, when applied to materials for purposes of detecting mechanical defects, variations in metallurgical condition, or composition do not impair the usefulness of the material or part.

Spark Test. The spark test is used to check material for chemical analysis. Since the sparks emitted from material when held to an abrasive wheel have a characteristic color, depending on the elements present in the steel, it is possible to classify unknown material rapidly by recognizing the color pattern produced.

Radiographic Inspection

The x-ray, because of the short wave length, can penetrate opaque materials; the depth of penetration depends upon the power of the x-ray beam. The sensitivity of the process is the ability of the method to reveal defects whose thickness or depth is expressed as a percentage of the metal thickness encountered. In other words, a sensitivity of 2% means that the defect revealed is $0.02t$ in depth, where t = metal thickness.

X-ray Radiography. When the x-ray passes through the material to the opposite side on which a special type sensitive film emulsion has been placed, a shadow picture is obtained wherein changes in density or thickness are revealed by light and dark areas. Voids, cracks, and holes appear as dark areas because they permit more rays to penetrate through them in a given time interval. This method of analysis is useful in checking welds, castings, forgings, plastic parts, and assemblies. Radiography can be used on light alloys, steel, other metals, and plastics. Under careful control a sensitivity of 2% is readily obtained, which meets the requirements of the ASME boiler code. The standard equipment can be used for materials up to one foot thick.

Gamma-ray Radiography. Radium salts emit gamma-rays which are similar to x-rays but have a shorter wavelength, hence a greater penetrating power. Their advantage is that tests can be made without much expense since a capsule of radium salt is all that is required aside from the photographic equipment. The sensitivity is not so high as that of the x-ray for thicknesses less than 2 in. In general the sensitivity is approximately 3%.

Fluoroscopic Inspection. If the x-rays on emerging from the object being tested are allowed to fall on a fluorescent screen, an observer can view the shadow pattern directly. This procedure saves the time and expense of photographing the result. These units are particularly adaptable to conveyor feed systems. A sensitivity range of 5 to 15% is possible.

X-ray Diffraction. The nature and behavior of most substances depend on the arrangement of atoms and molecules in the structure. One of the properties of a material is its diffractive effect produced by x-radiation. Thus, if a beam of x-rays is passed through a crystal the beam will be bent, or redirected, into a series of emergent rays whose separation and intensities are characteristic of the material. With radiographic film a record is obtained of the location and intensity of the diffracted rays. These form a *fingerprint* of the substance, so called, because no two substances have been found to produce identical diffraction patterns. This fingerprint makes possible prediction of the behavior of substances when subjected to various treatments and also identification of what has happened to the substance up to the time of making the diffraction test.

This method is useful in determining the preferred crystal orientation, in deriving specifications for new materials, in discovering whether a steel has been adequately annealed for machining, in determining quality of case hardening, and in giving a quick and exact identification of a material and its prior processing.

Magnetic-particle Inspection

Although radiographic inspection will find deep-seated irregularities, other methods are required to find non-visible, close-lipped discontinuities at or near the surface. The following types provide indications of cracks, semi-cracks, inclusions, and pores which may give rise to fatigue failures by acting as stress raisers.

Magnaflux Inspection. When a magnetic material is magnetized, discontinuities give rise to localized leakage fields which are revealed by the attraction of finely divided magnetic particles sprinkled on the material. There are two types of inspection media. In the wet method, the magnetic particles are prepared in paste form and mixed with oil. This method is more sensitive to minute surface discontinuities. In the dry method, which is more sensitive to subsurface discontinuities, magnetic powder is dusted on the magnetized material. Both direct and alternating currents are used for magnetization. The direct current is useful in finding significant subsurface discontinuities and is used for inspection of welds and castings. Because of *skin effect* the alternating current is limited to surface or very near surface discontinuities and is used for inspection of highly finished machine parts.

Magnaglo Inspection. This method employs fluorescent magnetic particles suspended in oil which give better contrast and extreme sensitivity. It is useful for material of complex contours such as splined shafts and milling cutters.

Zyglo Inspection. This fluorescent penetrant suspended in oil is used to determine discontinuities in non-magnetic materials. Items like ball bearings are more easily inspected by this method. The material is usually dipped into the penetrant, which penetrates the very fine cracks. When viewed under "black" light the discontinuities are revealed.

Supersonic Inspection

Longitudinal sound waves of extremely short wavelength travel through a given material at a velocity corresponding to the density and elastic properties of the substance. Thus a change in velocity or a loss of energy will occur when the sound beam strikes an irregularity such as a crack, separation, inclusion, or blow-hole. By electrical pickup it is possible to detect an irregularity as small as 0.001 in.

Reflectoscope. Internal defects are detected by sending supersonic impulses from a crystal into the material and measuring the time required for these impulses to penetrate the material and be reflected from the opposite side of the defect. The operator views an oscilloscope on which the electrical impulses from the reflected vibrations are impressed. The scanning line indicates the initial impulse, any irregularities, and the back face reflection respectively. The main disadvantage is that the specimens should be finished to about 250 micro-inches rms in order to obtain satisfactory transmission of crystal vibration to the material. It is suitable for metals up to 28 ft thick, but plastics can be tested only to a thickness of 3 in. If plastics have low resin content, the depth is reduced to about $1/2$ in. because the material damps out the sound waves used in the test.

Brush Hypersonic Analyzer. This equipment is best suited to detect defects in plates, strips, and sheets. The ray is not reflected in this type, but pickup equipment is necessary on the opposite face. A pen and ink recorder is used to indicate changes in the sound energy picked up by the receiver. In general this test shows only the existence of the separation but not its nature or depth.

Sperry Supersonic Thruray. This equipment is used to compare production units with a standard unit for soundness of spot welds, the occurrence of laminations in sheet products, the quality of bond in bimetals, and the homogeneity of sintered materials. The principle of operation is similar to that described in the foregoing paragraph except that the units must be placed in a sound-transmitting fluid.

Spectrographic Examination

An accurate analysis of a material can be made rapidly either by spectrography or x-ray spectrometry. It is suitable for checking quality and structure of the material or as an aid in the development of materials and their control in manufacture. The metallic samples can be used as electrodes and an arc, or spark, passed between them. The arc spectrum is then analyzed for the component elements.

Magnetic Analysis Inspection

The magnetic properties of irons and steels vary with their physical, chemical, and metallurgical properties. Therefore with suitable electronic devices it is possible to obtain comparative indications of structure, analysis, internal defects, case depth, and plating thickness. On bars and tubes defects 0.012 in. in depth can be detected.

Cyclograph. A change in structure which results in a significant change in physical properties or any distortion in the crystal lattice will produce some change in the magnetic and electrical properties of the metal. Internal stresses set up during quenching or cold working, residual stresses in shot-peened parts, and near surface variations in structure can be analyzed. The cyclograph does not reveal cracks and surface discontinuities.

For further information see the paper by Rupert Le Grand, Non-destructive Testing Methods, *American Machinist*, May 23, 1946.

SECTION 6

MECHANICS OF INCOMPRESSIBLE FLUIDS

BY

VICTOR L. STREETER

MECHANICS OF INCOMPRESSIBLE FLUIDS

Fluids, consisting of liquids and gases, may be considered incompressible in many situations without appreciable error. Owing to the high resistance which liquids offer to compression, it becomes an important factor only for large or sudden changes in pressure. For small changes in density the flow of gases through closed conduits may be handled as an incompressible fluid of average density. The motion of a body through a gas at low velocity (compared with the velocity of sound) is usually treated as incompressible flow.

1. INTRODUCTION

The subject of incompressible flow is treated from four general viewpoints in this section: as a static fluid; as an ideal (non-viscous) fluid; as a fluid in laminar motion; and as a real fluid in turbulent flow.

Definition of a Fluid

A fluid may be defined as a substance that deforms continuously when subjected to a shear stress, no matter how small that stress may be. Fluids may be defined as *Newtonian* or *non-Newtonian*, depending upon the functional relationship between the applied shear stress and the rate of angular deformation of fluid. The Newtonian fluid has a linear relation, examples of which are the common fluids of low viscosity such as air and water. Most colloids are non-Newtonian fluids such as soap and oil, and water and rubber.

The *viscosity* of a fluid is the ratio of the applied shear stress to the rate of angular deformation, given by Newton's *law of viscosity*,

$$\tau = \mu \frac{du}{dy}$$

where τ is the shear stress, μ is the viscosity, and du/dy is the rate of change of velocity normal to the velocity, which is the angular deformation velocity.

Dimension System

In general any consistent dimensional system may be used with the corresponding systematic units (see Sec. 3). In this section the FLT system is used with the units pound force, foot, and second. The unit of mass is the slug (32.174 pounds) and the unit of viscosity is pound (force)-second per foot squared.

Table of Symbols and Abbreviations

a	acceleration, constant, complex constant	i	$\sqrt{-1}$
b	water surface width, constant	k	adiabatic constant, roughness height, per-
c	celerity of wave propagation, constant		meability coefficient, constant
cc	cubic centimeter	l	mixing length, length, direction cosine
cfs	cubic feet per second	ln	natural logarithm
d	differential coefficient, constant	m	mass, constant, direction cosine
e	error	n	integer, Manning roughness factor, direc-
f	friction factor (Darcy-Weisbach), function		tion cosine
g	acceleration due to gravity	p	pressure intensity
gpm	gallons per minute	psi	pounds per square inch
h	vertical distance, liquid head, head loss	q	velocity, discharge per unit width

r	radius, coordinate	V	uniform velocity, average velocity, volume
s	specific gravity	W	weight
t	wall thickness, temperature, time	X	extraneous force component per unit mass,
u	velocity, velocity component in x-direction		x-direction
v	velocity, velocity component in y-direction	Y	extraneous force component per unit mass,
v'	velocity fluctuation		y-direction
w	velocity component in z-direction, complex	Z	extraneous force component per unit mass,
	potential		z-direction
x	coordinate axis		
y	coordinate axis	α	angle, constant, kinetic energy correction
z	coordinate axis		factor
		β	momentum correction factor
A	cross-sectional area, point designation	γ	specific weight
B	point designation	δ	layer thickness
C	point designation, coefficient, constant	ϵ	kinematic eddy viscosity, diffusion coeffi-
D	point designation, drag force, diameter		cient
E	modulus of elasticity, specific energy	ζ	complex variable
F	Fahrenheit, function, force	η	eddy viscosity, real variable
F_B	buoyant force	θ	angle
G	center of gravity	κ	circulation
H	head	μ	viscosity, strength, constant
I	moment of inertia	ν	kinematic viscosity
J	point designation	$\dot{\nu}$	normal velocity
K	constant, loss coefficient	ξ	real variable
K_m	bulk modulus of elasticity	$\bar{\omega}$	coordinate
L	length	π	constant 3.14159
M	metacenter, mass	ρ	mass density
O	point designation	σ	surface tension
P	point designation, force, perimeter	τ	shear stress
Q	discharge	ϕ	velocity potential
R	gas constant, gage difference	ψ	stream function
$\mathbf{R}$	Reynolds' number	ω	angular velocity
S	tensile stress, surface area, slope	Δ	increment
T	absolute temperature, tension in pipe wall,	Σ	summation
	kinetic energy, time, torque	Ω	extraneous force potential
U	uniform velocity	∇	del, the Laplacian operator

2. FLUID PROPERTIES

The fluid properties which are of use in the mechanics of incompressible flow are given in the form of either tables or charts. Since great accuracy in fluid flow calculations is generally impossible, the data should provide information of sufficient accuracy for practical applications.

Density and Compressibility

Density is defined as the mass per unit volume, and it depends upon the temperature and pressure intensity. The density of pure water is given in Fig. 1. Since the density variation in liquids is small for most engineering applications, densities of some of the other

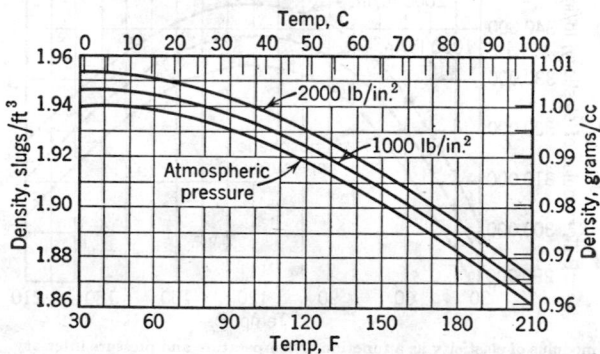

FIG. 1. Density p of pure water as a function of temperature and pressure intensity. By permission from *Fluid Mechanics for Hydraulic Engineers*, by Hunter Rouse, copyright 1938, McGraw-Hill Book Company, Inc.

common liquids are given in Table 1 for standard atmospheric pressure and 60 F. The *specific weight* γ is also given in the table.

Table 1. Density Characteristics of Common Liquids under Atmospheric Pressure at 60 F

Liquid	Density ρ slug/ft^3	Specific Weight γ lb/ft^3	Liquid	Density ρ slug/ft^3	Specific Weight γ lb/ft^3
Alcohol, ethyl.............	1.53	49.3	Mercury..................	26.3	847
Benzene..................	1.71	54.9	Oil		
Brine (20% NaCl)........	2.23	71.6	lubricating..............	1.65–1.70	53–55
Carbon tetrachloride.......	3.09	99.5	crude..................	1.65–1.80	53–58
Gasoline..................	1.28–1.34	41–43	fuel....................	1.80–1.90	58–61
Glycerin..................	2.45	78.8	Water		
Kerosene.................	1.51–1.59	49–51	fresh...................	1.94	62.4
			salt....................	1.99	64.0

The density of a gas varies widely with temperature and pressure, following closely the ideal-gas equation

$$\rho = \frac{pM}{RT}$$

where ρ is the mass density (slug/ft^3 or lb/ft^3), p is the absolute pressure (lbf/ft^2), T is the absolute temperature ($459.7 + F^\circ$), M is the molecular weight of the gas (see Sec. 14), and R is the gas constant (49,700 with slug/ft^3, 1,545 with lb/ft^3).

Table 2.   Units of Absolute Viscosity

Viscosity Unit	Force Unit	Mass Unit	g_c	Conversion Factor*
Poise	dyne	gram	1	0.01
(Kilogram-force)(sec)/m^2	kilogram	kilogram	9.807	1.02×10^{-4}
(Pound-force)(sec)/ft^2	pound	slug	1	2.09×10^{-5}
(Poundal)(sec)/ft^2	poundal	pound	1	6.72×10^{-4}
(Pound-mass)/(ft)(sec)	pound	pound	32.17	6.72×10^{-4}

* Multiply centipoises by factor to obtain other units.

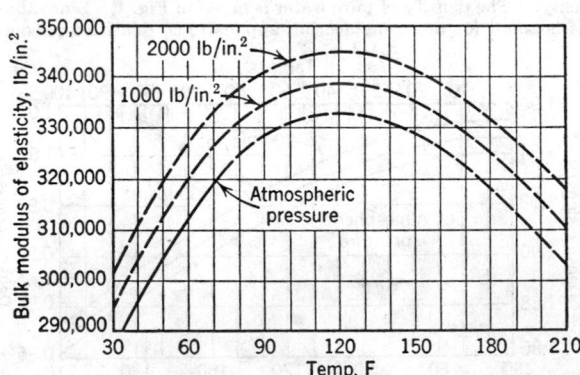

FIG. 2.   Bulk modulus of elasticity as a function of temperature and pressure intensity.   By permission from *Fluid Mechanics for Hydraulic Engineers*, by Hunter Rouse, copyright 1938, McGraw-Hill Book Company, Inc.

The compressibility of a liquid is given by its *bulk modulus of elasticity* K_m, defined by

$$K_m = - \frac{\Delta p}{\Delta V / V}$$

where Δp is the increment of pressure intensity applied to the volume of liquid V to cause a decrease in volume $-\Delta V$. K_m has the dimensions of the pressure intensity and depends upon both the pressure intensity and the temperature. The bulk modulus of elasticity for water is given in Fig. 2 as a function of temperature with pressure intensity as a parameter. Figure 3 shows K_m for a wide range of pressure intensities for three temperatures.

Example. The specific weight of water is desired at 68 F and 80,000 psi.
Rewriting the equation of bulk modulus of elasticity and integrating between the limits shown

$$\int_{14.7}^{80,000} \frac{dp}{K_m} = - \int_{V_0}^{V} \frac{dV}{V} = \ln \frac{V_0}{V}$$

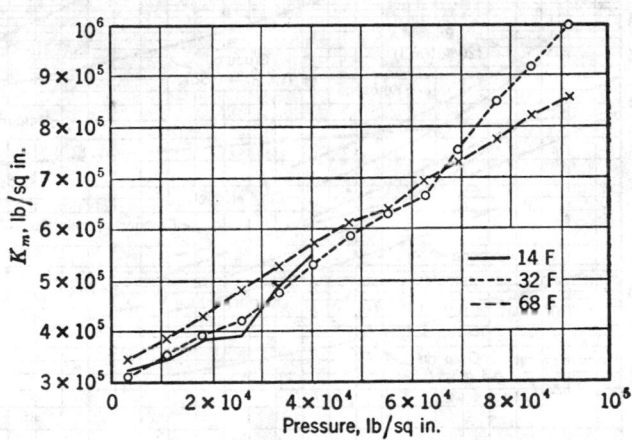

FIG. 3. Bulk modulus of elasticity for water.

The right-hand integral is evaluated as shown; the left-hand integral is evaluated by numerical integration. Plotting $1/K_m$ as ordinate against p as abscissa, the area under the curve between 14.7 and 80,000 psi is the value of the integral. This is 0.151. Hence

$$\ln \frac{V_0}{V} = 0.151, \qquad \frac{V_0}{V} = 1.163$$

and $V = 0.86 \, V_0$. The specific weight is $(62.4/0.86) = 72.5$ lb per cu ft.
Tables 3 and 4 contain compressibility information for various liquids.

Table 3. Bulk Modulus of Elasticity (K_m) of Various Liquids at 68 F

Fluid	Pressure Range, psi	K_m, psi	Fluid	Pressure Range, psi	K_m, psi
Methyl alcohol	15– 7,350	1.76×10^5	Carbon tetrachloride	14–1,435	1.63×10^5
	7,350–14,700	2.28×10^5		1,450–2,900	1.68×10^5
	14,700–22,050	2.94×10^5		2,900–4,350	1.79×10^5
	22,050–29,400	3.54×10^5		4,350–5,800	1.98×10^5
	29,400–36,750	4.20×10^5		5,800–7,250	2.13×10^5
Ethyl alcohol	15– 7,350	1.75×10^5			
	7,350–14,700	2.36×10^5			
	14,700–22,050	3.00×10^5			
	22,050–29,400	3.52×10^5			
	29,400–36,750	4.16×10^5			

Table 4.   Compressibility Formulas for Kerosene and Lubricating Oils

$$K_m = (a + bp - cp^2 + dp^3)10^6$$
K_m, lb per sq in.; p, lb per sq in. abs

Kind of Oil	Temp, °F	a	$10^6 b$	$10^{12} c$	$10^{17} d$	Range, psi
Kerosene...............	68	0.191	4.21	9.24	1.71	0–176,000
Paraffin oil.............	93.2	.176	4.63	19.09	11.25	0– 63,000
"Mobil A".............	104	.256	4.30	0	0	0– 20,800
"Bayonne".............	104	.251	4.30	0	0	0– 22,000

Viscosity

Viscosity is that property by which a fluid offers resistance to shear stress.   It is independent of pressure (except for extreme pressures); it increases with temperature for a gas; and it decreases with temperature for a liquid.   The resistance to shear stress depends

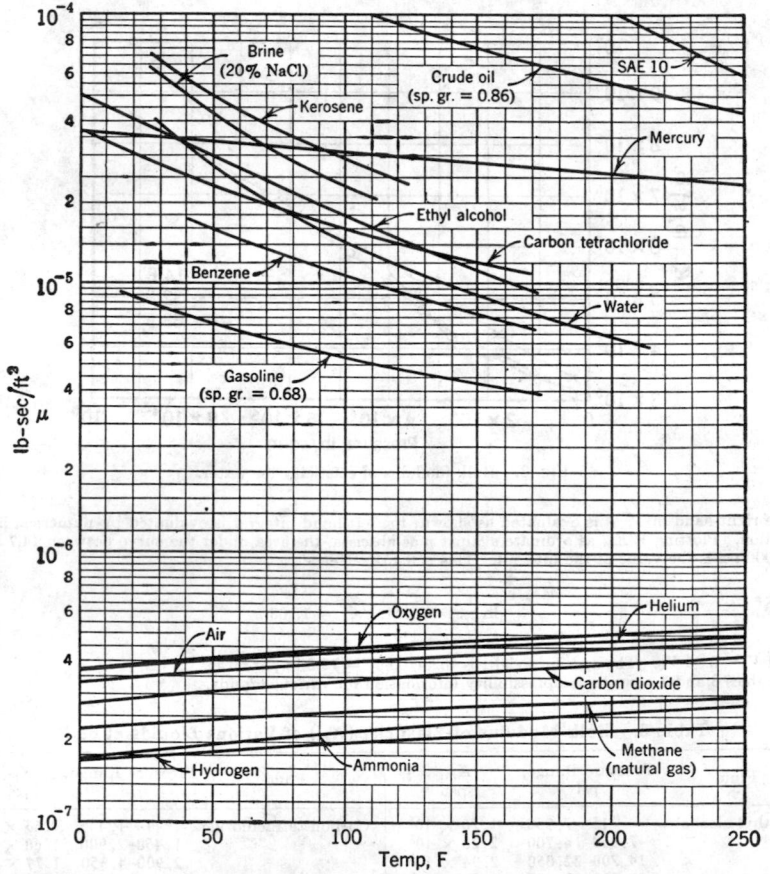

Fig. 4.   Dynamic viscosity versus temperature for common gases and liquids.   Courtesy of Prof. Hunter Rouse, State University of Iowa, Iowa City.

upon cohesion of the molecules and upon the molecular transfer of momentum from one layer to another in the fluid.   Cohesion predominates in liquids, and, since cohesion decreases with temperature, the viscosity does likewise.   Cohesion is relatively unimportant in gases, hence the increased molecular activity with temperature causes an increase in molecular momentum transfer with a corresponding increase in viscosity.

The dimensions of viscosity are determined from Newton's law of viscosity

$$\tau = \mu \frac{du}{dy}$$

where τ is the shear stress, μ is the viscosity, and du/dy is the rate of change of velocity normal to the flow. The dimensions are $ML^{-}T^{-1}$ or $FL^{-2}T$. In English units, which have no name, it is in slugs per foot second, or pound seconds per square foot. In the cgs system, the unit is the *poise*, which is 1 gram (mass) per centimeter second, or 1 dyne second per square centimeter. The viscosity as outlined above is referred to as the

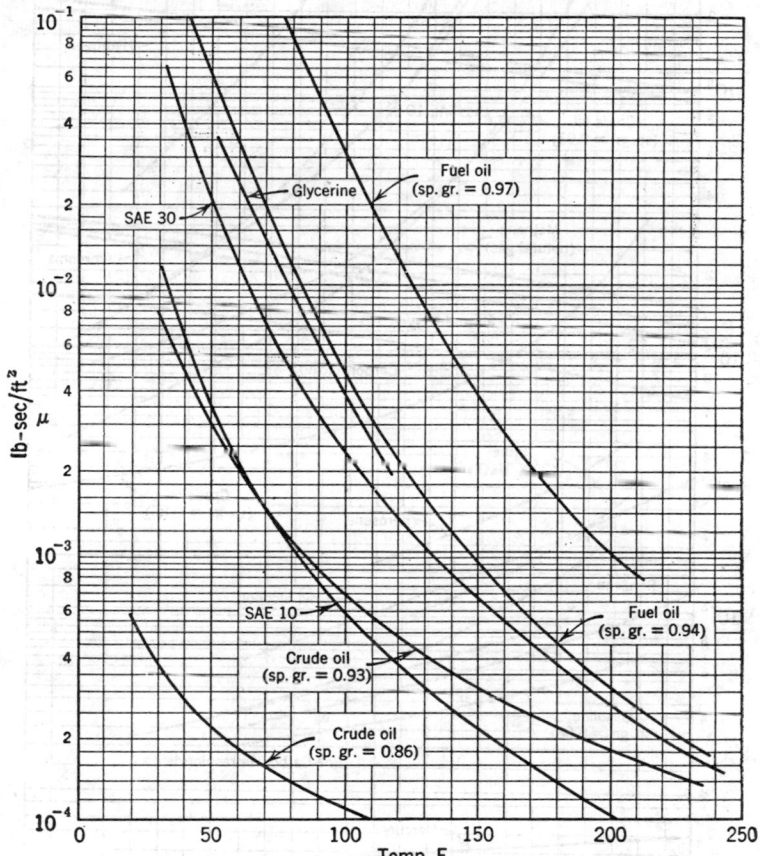

FIG. 5. Dynamic viscosity versus temperature for typical grades of oil. Courtesy of Prof. Hunter Rouse, State University of Iowa, Iowa City.

absolute, or *dynamic*, viscosity to distinguish it from the arbitrary viscosity units of commercial viscometers.

The ratio of viscosity to density is called the *kinematic* viscosity and has the dimensions L^2T^{-1}. The English unit is *feet squared per second;* the cgs unit is the *stoke*, in centimeters squared per second.

The viscosities of common gases and liquids are given in Figs. 4 and 5. The kinematic viscosity in English units is given in Fig. 6.

Viscosity determinations are made by several types of commercial viscometers. These viscometers generally yield the kinematic viscosity in some arbitrary measure, such as the time for a fixed volume to flow through a capillary tube. Owing to entrance effects and the varying head on the tube, an analytical conversion to absolute units is impracticable. An empirical conversion chart is given by Fig. 7 showing Saybolt seconds (United States),

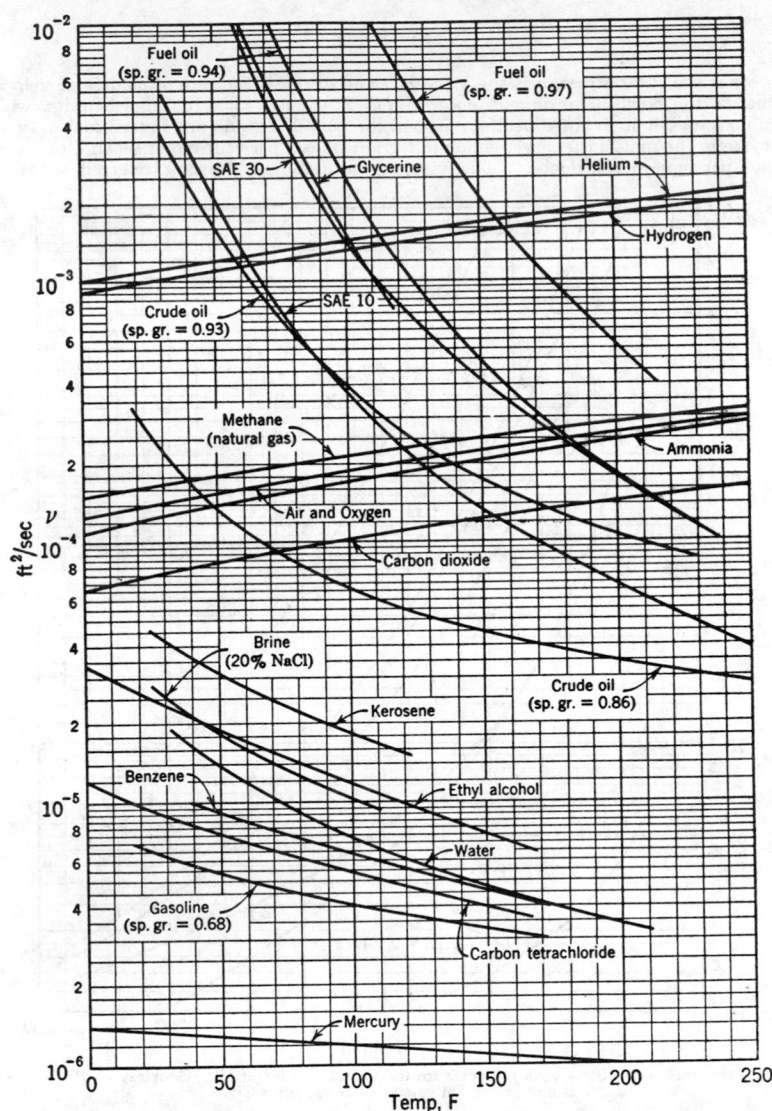

FIG. 6.   Kinematic viscosity versus temperature for common fluids.   Courtesy of Prof. Hunter Rouse, State University of Iowa, Iowa City.

Barbey seconds (France), Redwood seconds (England), and Engler degrees (German) in terms of the English and cgs units.

The English unit of viscosity is much larger than the poise, the conversion factor being 478. Similarly the English unit of kinematic viscosity is much larger than the stoke, the conversion factor being 929. Hence 1 poise is $1/478$ slug per ft sec or $1/478$ lb sec per sq ft. One stoke is $1/929$ ft squared per sec. (See Table 2.)

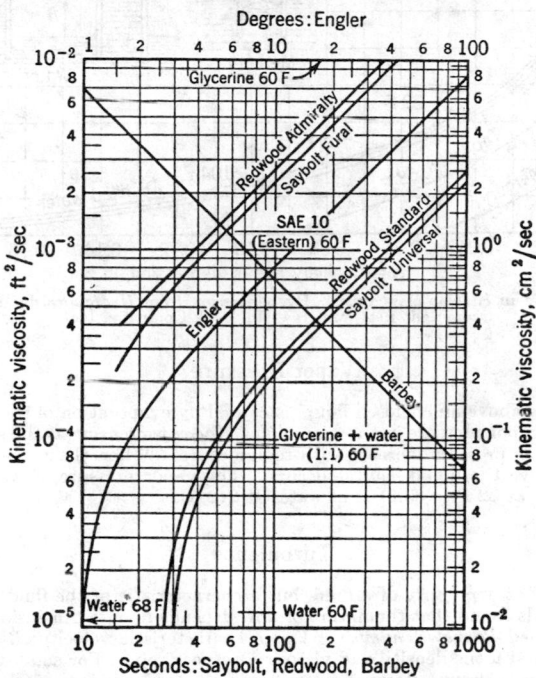

Fig. 7. Viscometer characteristics. By permission from *Fluid Mechanics for Hydraulic Engineers*, by Hunter Rouse, copyright 1938, McGraw-Hill Book Company, Inc.

Surface Tension, Capillarity

The cohesion of a fluid in contact with another fluid causes the formation of a free surface between them which requires surface energy. For all practical purposes this energy may be considered as a tension in a surface film called *surface tension*. Surface tension has the dimensions FL^{-1}. It affects the flow of small jets or thin layers of liquid and causes waves, known as *capillary* waves. The surface tension of several liquids in contact with air are given in Table 5. It varies with temperature; for example, pure water varies from 0.0051 lb per ft at 47 F to 0.0041 lb per ft at 203 F.

Table 5. Surface Tension of Common Liquids in Contact with Air at 68 F

Liquid	Surface Tension σ lb/ft	Liquid	Surface Tension σ lb/ft
Alcohol, ethyl.	0.00153	Mercury, in air.	0.0352
Benzene.	.00198	in water.	.0269
Carbon tetrachloride.	.00183	in vacuum.	.0333
Kerosene.	0.0016–0.0022	Oil, lubricating.	0.0024–0.0026
Water.	.00498	crude.	0.0016–0.0026

Surface tension causes *capillary rise*, which depends upon the two fluids, the solid, and the temperature. When the liquid does not *wet* the solid, as with mercury and glass, the capillary rise is negative.   Figure 8 gives the capillary rise for glass tubes with water and mercury.

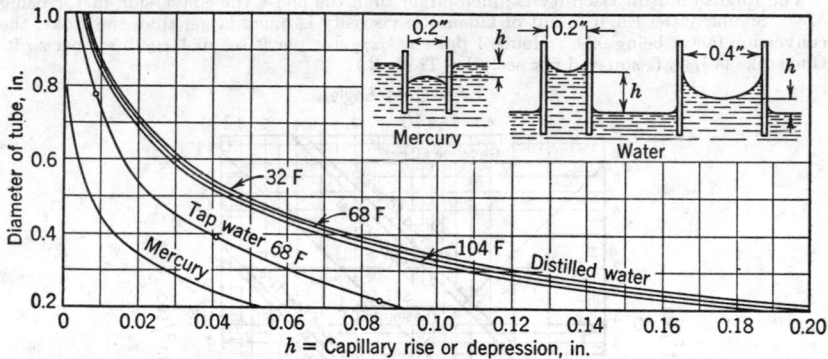

FIG. 8.   Capillarity in circular glass tubes.   By permission from *Hydraulics*, by R. L. Daugherty, copyright 1944, McGraw-Hill Book Company, Inc.

Vapor Pressure

Vapor pressure above an enclosed liquid is caused by evaporation of the liquid.   In the saturated state vapor pressure in the space equals the *vapor tension* at the surface.   Vapor tension and vapor pressure increase with temperature.   When the vapor tension equals the pressure above the liquid, *boiling* occurs.   The vapor pressure of water varies from 0.203 ft of water at 32 F to 33.91 ft of water at 212 F.

Turbulence

Turbulence is not a *property* of a fluid, but a characteristic of the fluid and its motion. Osborne Reynolds found that the nature of flow depends upon a dimensionless parameter $VL\rho/\mu$, now called *Reynolds' number*. V is a characteristic velocity, L a characteristic length, with ρ and μ the density and viscosity, respectively.   For small values of Reynolds' number the viscous effects predominate, all tendencies toward turbulence are damped, and laminar motion occurs.   For large values of Reynolds' number the laminar motion is unstable, and a very complex motion occurs, known as *turbulent flow*.   Eddies are generated at the boundaries and are diffused throughout the flow, causing an erratic fluctuation in velocity and pressure.   In spite of the erratic motion, temporal mean velocities and pressures remain constant for over-all steady conditions.

A measure of the turbulence can be obtained from an expression similar to Newton's flow of viscosity,

$$\bar{\tau} = \eta \frac{d\bar{v}}{dy}$$

where $\bar{\tau}$ is the mean shear stress, dv/dy is the mean velocity gradient, and η is the *eddy viscosity* which is analogous to μ, the dynamic viscosity. η depends upon the mean size of eddies and upon some characteristics of the velocity fluctuation, such as $\sqrt{\overline{v'^2}}$ the root-mean-square of the velocity fluctuation at a point, as well as the fluid density.   A characteristic of the motion alone is obtained by dividing η by ρ,

$$\epsilon = \frac{\eta}{\rho}$$

a factor having dimensions L^2T^{-1}, which is called the kinematic eddy viscosity.   Since ϵ depends upon the mean size of eddies l and upon the velocity fluctuations $\epsilon = l\sqrt{\overline{v'^2}}$, it is frequently called the diffusion coefficient.

Using a mean value of ϵ over a section, and dividing by the kinematic viscosity, ν, results in a Reynolds' number of the type VL/ν,

$$\frac{\epsilon_m}{\nu} \sim \frac{VL}{\nu}$$

indicating, in general, that the mixing or diffusion process increases with Reynolds' number.

3. FLUID STATICS

A fluid at rest, or moving so that every particle has the same (constant) velocity, must satisfy the equations for statics. That is, the summation of all forces acting on any free body of fluid must be zero, and the summation of all moments about any axis must be zero.

General

From the definition of a fluid any shear stress sets it in motion; hence a fluid at rest cannot be acted upon by shear stresses. With the absence of shear stresses on any free body of fluid, the only surface stresses must be normal stresses. Pressure intensity on a small area at a point in a static fluid is independent of the particular orientation of the area element.

The law of static variation of pressure intensity in an incompressible fluid is easily obtained from Fig. 9. Taking as free body a right circular cylinder with axis vertical and the upper end in the free surface, the atmospheric pressure pushes downward on the top with a force $p_0 A$, where p_0 is the local atmospheric intensity and A is the cross-sectional area. The weight of cylinder is $\gamma h A$, where γ is the specific weight of fluid and h the distance of base from the free surface. The force pushing upward at the base is pA. Writing the equation of equilibrium for the vertical direction yields, after dividing by the area,

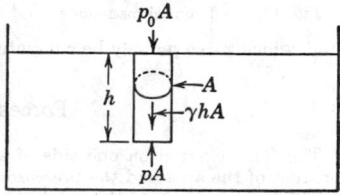

FIG. 9. Free body of fluid in form of vertical right circular cylinder.

$$p = p_0 + \gamma h$$

showing that the pressure intensity varies directly as the distance from the free surface and directly as the specific weight. The relation between difference in pressure intensity for difference in elevation is

$$\Delta p = \gamma \, \Delta h$$

With γ in lb per cu ft and Δh in ft, Δp is in lb per sq ft. A liquid at rest always has a horizontal free surface. All points at the same elevation in the same continuous mass of fluid at rest have the same pressure intensity.

Manometers and Gages *

A *Bourdon gage*, Fig. 10, measures the difference in pressure intensity inside and outside the curved tube. Since the outside is exposed to local atmospheric pressure, the gage

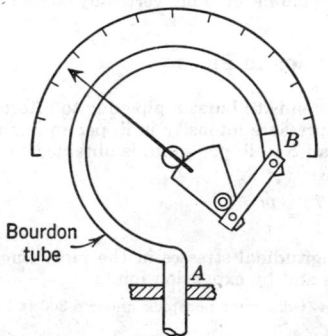

FIG. 10. Essential features of a Bourdon gage.

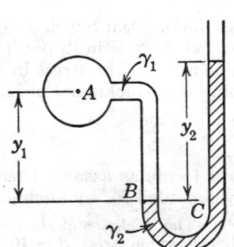

FIG. 11. Simple manometer.

* Figures 10 to 12, courtesy of Prof. R. C. Binder, Purdue University, Lafayette, Ind.

measures pressure intensities above or below it as datum; hence gage pressure zero is local atmospheric pressure. *Absolute* pressures are obtained by adding local atmospheric pressure intensity to gage pressure intensity.

Manometers are of various types, the principal ones being: *simple, differential, inclined,* and *compound.* They work on the general principle of balancing the pressure intensity to be measured against one or more liquid columns. Figure 11 illustrates a simple manometer, and Fig. 12 shows a differential manometer. A manometer reading is converted into a pressure intensity by writing an equation from one end of the manometer to the other, expressing the pressure at each meniscus in terms of the pressure at the preceding meniscus. For example, in Fig. 12,

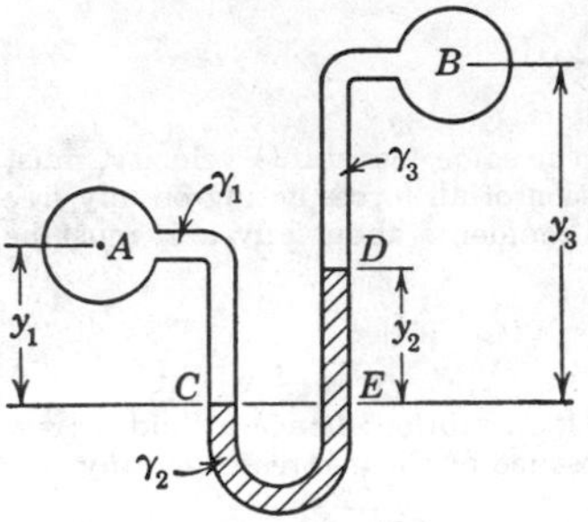

Fig. 12.   Differential manometer.

$$p_A + \gamma_1 y_1 - \gamma_2 y_2 - \gamma_3(y_3 - y_2) = p_B$$

from which $p_a - p_B$ may be computed. The same procedure is used for all manometers.

Forces on Plane Surfaces

The force exerted on one side of a plane surface submerged in a fluid is equal to the product of the area and the pressure intensity at the centroid of the area.

The line of action of the resultant force exerted on the surface is perpendicular to the area and acts at a point, called the *center of pressure,* that is below the centroid a distance $I_G/(\bar{y}A)$ measured in the plane of the area. I_G is the moment of inertia of the area about a horizontal centroidal axis in the area; A is the area; and $\bar{y}$ is the distance from the centroid to the free surface, measured in the plane of the area.

Forced Components on Curved Surfaces

The resultant force on a curved surface is best determined from the vertical component and two horizontal components at right angles. These components are determined from the following rules:

The horizontal component of pressure on a curved surface is equal to the pressure force on a vertical projection of the curved surface. The plane of the vertical projection is normal to the horizontal direction. The line of action of the horizontal component passes through the pressure center for the projected area.

The vertical component of pressure on a curved surface is equal to the weight of fluid, real or imaginary, which extends vertically above the curved surface up to the real or imaginary free surface. In case there is no actual free surface, an imaginary one can be taken at such a height that the same pressure intensity occurs over the surface. When the fluid is confined below the curved surface, the calculation can be made as if the fluid were above the curved surface up to the real or imaginary free surface. The line of action of the vertical component is through the centroid of the volume of fluid vertically above the curved surface.

Bursting Tension in Pipes

The circumferential tensile force T which tends to burst a pipe due to internal pressure is $T = pr$, where T is in lb per in., p is the pressure intensity in lb per sq in., and r is the pipe radius in in. The stress in the pipe wall S in lb per sq in. is obtained by dividing T by the pipe wall thickness t in in., hence

$$S = \frac{T}{t} = \frac{pr}{t}$$

The above formulas assume there is no longitudinal stresses in the pipe, since these are normally provided for by anchors at elbows and by expansion joints.

Example. The thickness of steel plate for a 24-ft-diameter penstock under a 300-ft head of water is desired. Allowable stress, $S = 10,000$ psi.

$$t = \frac{pr}{S} = \frac{300 \times 62.4}{144} \times \frac{12 \times 12}{10,000} = 1.87 \text{ in.}$$

Buoyancy

An object floating or submerged in a static fluid is acted upon by a vertical resultant force due to the fluid. Since the vertical projection of opposite sides of an object are always identical, there can be no resultant horizontal force. Using the rule for vertical force components, the force is found to be equal to the weight of fluid displaced, *Archimedes' law*. The line of action is through the centroid of the displaced volume.

Stability of Floating or Submerged Bodies *

An object floats with vertical stability, since a slight upward displacement decreases the buoyant force and hence results in a net downward force, and a slight downward displacement increases the buoyant force with a resultant net upward force. Bodies sub-

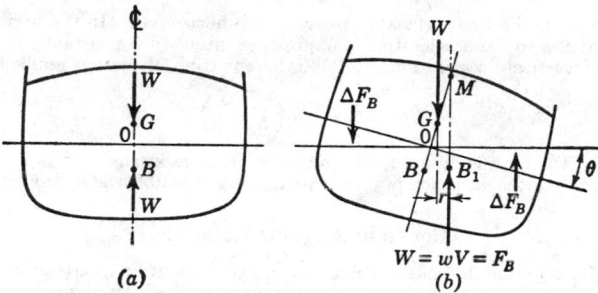

$$W = wV = F_B$$

(a)   (b)

FIG. 13.   Metacenter and metacentric height.

merged in a liquid have practically no vertical stability because of the very slight changes in specific weight.

Stability with respect to overturning is generally of much more importance. The buoyant force may be considered as acting at the center of buoyancy and the weight as acting through the center of gravity. When the center of gravity is below the center of buoyancy, the object is always stable. When the center of buoyancy is below the center of gravity, it is stable only when the center of buoyancy shifts sufficiently to set up a righting moment.

Figure 13b illustrates the shift in center of buoyancy due to the left-hand wedge coming out of the liquid and the right-hand wedge entering the liquid. The intersection of the vertical line through the new center of buoyancy with the center line *BG* extended at *M* is called the *metacenter*. When *M* is above *G* the object is in stable equilibrium. The *metacentric* height is *MG* and is a direct measure of the stability.

For prismatic bodies, the new center of buoyancy can be computed from the submerged cross section for any assumed rotation. For non-prismatic bodies, the metacentric height for very small angles of rotation is given by

$$\overline{MG} = \frac{I}{V} \mp \overline{GB}$$

where the minus sign is used when *G* is above *B*, the plus sign when *G* is below *B*. *I* is the moment of inertia of the horizontal section of object at the liquid surface about the axis through *O* (normal to the cross section). *V* is the volume of liquid displaced.

Submerged bodies have stability with respect to rotation only when the center of gravity is below the center of buoyancy.

4. ACCELERATED LIQUIDS IN RELATIVE EQUILIBRIUM

When a fluid is accelerated in such a manner that it moves as a solid, i.e., with no relative motion of one layer of fluid with respect to an adjacent layer, shear forces do not occur in the fluid, and the analysis of pressure distribution is relatively simple.

* Figure 13, courtesy of Prof. R. A. Dodge, University of Michigan. Ann Arbor.

Uniform Linear Acceleration

When a vessel containing a liquid is accelerated uniformly in a horizontal direction the liquid undergoes an adjustment in position and pressure distribution, and then moves as a solid. Since there is no vertical acceleration, the variation of pressure along a vertical line is hydrostatic; i.e., the pressure intensity h feet below the free surface is γh, where γ is the specific weight. Along a horizontal line in the fluid in the direction of the acceleration the rate of change of pressure intensity $\Delta p / \Delta l$ is $\gamma a / g$, where a is the acceleration. The rate of change of head h along this line is the slope of free surface, and is

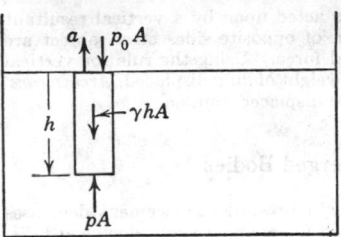

Fig. 14.   Uniform vertical acceleration of a liquid.

$$\frac{\Delta h}{\Delta l} = \frac{a}{g} = \tan \theta$$

where θ is the angle the free surface makes with the horizontal. In the horizontal direction at right angles to the acceleration, the pressure intensity is constant.

For uniform vertical acceleration, Fig. 14, the equation of motion applied to the free body yields

$$p - p_0 = \gamma h \left(1 + \frac{a}{g} \right)$$

When $a = -g$, the tank is falling freely, and the pressure intensity is p_0 throughout the fluid. When $a = g$, the variation in pressure intensity is double the hydrostatic variation.

Constant Angular Rotation

Rotation of a container at constant angular velocity about any vertical axis causes the fluid therein to assume a characteristic surface profile (if a liquid) and a characteristic pressure distribution. The only acceleration is a linear one directed radially inward toward the axis of rotation. Since there is no component of the acceleration in the vertical direction, the hydrostatic-pressure variation holds for any vertical line.

Writing the equation of motion in the radial direction for an element of fluid, and integrating, results in

$$p = p_0 + \frac{\gamma \omega^2 r^2}{2g}$$

where p is the pressure intensity at the distance r from the axis, p_0 is the pressure intensity at the axis of rotation *in the same horizontal plane*, γ is the specific weight, and ω the angular velocity (radians per second).

When a free surface exists and the equation is applied to the plane through the free surface at the axis, p_0 becomes zero, and

$$\frac{p}{\gamma} = h = \frac{\omega^2 r^2}{2g}$$

where h is the height above the horizontal plane and extends to the free surface. Hence the free surface takes the shape of a paraboloid of revolution. If the rotation occurs about the axis of a circular cylinder, the depression of level at the center equals the rise in liquid at the periphery.

5. DYNAMICS OF A NON-VISCOUS, INCOMPRESSIBLE FLUID *

The ideal-fluid theory leads to problem solutions that are of value in real fluids of low viscosity when large expanses of fluid are involved. The resulting velocity and pressure distributions are valid except in a narrow region near the boundaries and in the wake.

Fundamental Relationships

The flow of a non-viscous, incompressible fluid must satisfy (a) the equations of motion, (b) the equation of continuity, (c) the prescribed boundary conditions, and the (d) conditions as to rotation or lack of rotation of fluid elements.

* Illustrations and material used by permission from *Fluid Dynamics* by V. L. Streeter, copyright 1948, McGraw-Hill Book Company, Inc.

Continuum. Although fluids are composed of molecules between which are spaces much larger than the molecules themselves, the assumption of a continuous medium must be made in order to approach the motion from a mathematical viewpoint. This assumption, however, is not tenable when the causes of viscosity and diffusion are under consideration. The quantities density, pressure intensity, velocity, and acceleration are assumed to vary continuously throughout the fluid, except for special points, lines, surfaces, or volumes of discontinuity.

Euler's Equations of Motion. Consideration of forces on a fluid element leads to the Euler equations:

$$X - \frac{1}{\rho}\frac{\partial p}{\partial x} = u\frac{\partial u}{\partial x} + v\frac{\partial u}{\partial y} + w\frac{\partial u}{\partial z} + \frac{\partial u}{\partial t}$$

$$Y - \frac{1}{\rho}\frac{\partial p}{\partial y} = u\frac{\partial v}{\partial x} + v\frac{\partial v}{\partial y} + w\frac{\partial v}{\partial z} + \frac{\partial v}{\partial t}$$

$$Z - \frac{1}{\rho}\frac{\partial p}{\partial z} = u\frac{\partial w}{\partial x} + v\frac{\partial w}{\partial y} + w\frac{\partial w}{\partial z} + \frac{\partial w}{\partial t}$$

where X, Y, Z, are the components of the body (extraneous) forces per unit mass in the xyz-directions, respectively; ρ is the mass density; p is the pressure intensity at a point (independent of direction); u, v, w are velocity components in the xyz-directions at any point x, y, z. The terms on the right-hand side of the equations are acceleration components, the first three of which are known as the *convective* acceleration and the fourth as the *local* acceleration. The Euler equations must hold for every point in the flow with the exception of singular points.

Continuity Equation

The continuity equation for an ideal fluid states that the net flow rate into any small volume must be zero, except when the volume contains a singular point. In equation form

$$\frac{\partial u}{\partial x} + \frac{\partial v}{\partial y} + \frac{\partial w}{\partial z} = 0$$

or in vector notation

$$\nabla \cdot \mathbf{q} = 0$$

i.e., the divergence of the velocity vector $\mathbf{q}$ is everywhere zero except at singular points (see Sec. 2 for vector analysis).

Boundary Conditions

A kinematic boundary condition must be satisfied at every solid boundary, namely, that the component of velocity normal to the boundary must be equal to the velocity component of the boundary normal to itself. For a stationary boundary

$$lu + mv + nw = 0$$

where l, m, n, are direction cosines of the normal to the boundary. For a moving boundary

$$lu + mv + nw = \dot{\nu}$$

where $\dot{\nu}$ is the velocity of the boundary normal to itself. If $F(x, y, z, t)$ is the equation of the boundary surface, the boundary condition may be expressed

$$u\frac{\partial F}{\partial x} + v\frac{\partial F}{\partial y} + w\frac{\partial F}{\partial z} + \frac{\partial F}{\partial t} = 0$$

Dynamic boundary conditions arise when two fluids are in contact. It is necessary that the pressure be continuous across the interface.

Irrotational Flow—Velocity Potential

Rotation of a fluid element may be represented by a vector that has a length proportional to the magnitude of the rotation (radians per second) and a direction parallel to the instantaneous axis of rotation. The right-handed rule is adopted; i.e., the positive direction of the vector is the direction a right-handed screw would progress when rotating in the same sense as the element. In vector notation the *curl* of the velocity vector is twice the rotation vector

$$\nabla \times q = 2\omega$$

Scalar components of the rotation vector w_x, w_y, w_z in the directions of the xyz-axes may be used in place of the vector itself. Defining a rotation component of an element about an axis as the average angular velocity of two infinitesimal line segments through the point, mutually perpendicular to themselves and the axis,

$$\omega_x = \frac{1}{2}\left(\frac{\partial w}{\partial y} - \frac{\partial v}{\partial z}\right), \quad \omega_y = \frac{1}{2}\left(\frac{\partial u}{\partial z} - \frac{\partial w}{\partial x}\right), \quad \omega_z = \frac{1}{2}\left(\frac{\partial v}{\partial x} - \frac{\partial u}{\partial y}\right)$$

Irrotational flow may be defined as that flow where there is an absence of rotation at every point except singular points, hence the following equations must be satisfied

$$\frac{\partial v}{\partial x} = \frac{\partial u}{\partial y}, \quad \frac{\partial w}{\partial y} = \frac{\partial v}{\partial z}, \quad \frac{\partial u}{\partial z} = \frac{\partial w}{\partial x}$$

A visual concept of irrotational flow may be obtained by considering as a free body a small element of fluid in the form of a sphere. As the fluid is frictionless, no tangential stresses or forces may be applied to its surface. The pressure forces act normal to its surface, and hence through its center. Extraneous, or body, forces act through its mass center, which is also its geometric center for constant density. Hence, it is evident that no torque may be applied about any diameter of the sphere. The angular acceleration of the sphere must always be zero. If the sphere is initially at rest, it cannot be set in rotation by any means whatsoever; if it is initially in rotation, there is no means of changing its rotation. As this applies to every point in the fluid, one may visualize the fluid elements as being pushed around by boundary movements but not being rotated if initially at rest. Rotation or lack of rotation of the fluid particles is a property of the fluid itself and not its position in space.

The velocity potential ϕ is a scalar function of space such that its negative rate of change with respect to any direction is the velocity component in that direction. In vector notation

$$q = -\ \text{grad}\ \phi$$

or in terms of cartesian coordinates

$$u = -\frac{\partial \phi}{\partial x}, \quad v = -\frac{\partial \phi}{\partial y}, \quad w = -\frac{\partial \phi}{\partial z}$$

The assumption of irrotational flow is equivalent to the assumption of a velocity potential.

Bernoulli's Equation

Assuming that the flow is irrotational and that the extraneous force components are derivable from a force potential

$$X = -\frac{\partial \Omega}{\partial x}, \quad Y = -\frac{\partial \Omega}{\partial y}, \quad Z = -\frac{\partial \Omega}{\partial z}$$

The Euler equations may be integrated, resulting in the Bernoulli equation

$$\frac{1}{2}q^2 - \frac{\partial \phi}{\partial t} + \Omega + \frac{p}{\rho} = F(t)$$

where q is the speed at any point. $q^2 = u^2 + v^2 + w^2$, and $F(t)$ is an arbitrary function of the time.

For steady flow, i.e., flow such that conditions at any point do not change with the time, $\partial \phi/\partial t = 0$, and $F(t)$ become a constant, hence

$$\frac{1}{2}q^2 + \Omega + \frac{p}{\rho} = C$$

where C is to be determined from known conditions at some point. With gravity the only extraneous force, and taking the z-axis as positive upward,

$$\Omega = gz$$

Considering the pressure term as composed of two parts, one due to static conditions p_s, the pressure intensity which would exist if there were no motion and that due to dynamic conditions p_d, Bernoulli's equation for steady flow may be written

$$\frac{1}{2} q^2 + \frac{p_d}{\rho} = C - \left(gz + \frac{p_s}{\rho} \right) = \text{a constant}$$

since $gz + p_s/\rho$ = a constant is the law of hydrostatic variation of pressure. The subscript on the dynamic pressure term is usually dropped, resulting in a simple form of the equation

$$p - p_0 = \frac{\rho}{2} (q_0{}^2 - q^2)$$

where p_0, q_0 have values determined for some point in the flow.

Laplace Equation

The *Laplace equation* results when the continuity equation is written in terms of the velocity potential

$$\frac{\partial^2 \phi}{\partial x^2} + \frac{\partial^2 \phi}{\partial y^2} + \frac{\partial^2 \phi}{\partial z^2} = 0$$

This is usually written $\nabla^2 \phi = 0$, where, in this shortened form, the equation may be in terms of any orthogonal coordinate system. For plane polar coordinates (r, θ)

$$\nabla^2 \phi = \frac{\partial \phi}{\partial r} + r \frac{\partial^2 \phi}{\partial r^2} + \frac{1}{r} \frac{\partial^2 \phi}{\partial \theta^2} = 0$$

For cylindrical coordinates $(x, \bar{\omega}, \omega)$ where $\bar{\omega}$ is the distance from the x-axis and ω is the angle a plane through the point and the x-axis makes with an initial plane,

$$\nabla^2 \phi = \frac{\partial^2 \phi}{\partial x^2} + \frac{\partial^2 \phi}{\partial \bar{\omega}^2} + \frac{1}{\bar{\omega}} \frac{\partial \phi}{\partial \bar{\omega}} + \frac{1}{\bar{\omega}^2} \frac{\partial^2 \phi}{\partial \omega^2} = 0$$

For spherical polar coordinates (r, θ, ω), where r is the distance from the origin, θ is the polar angle, and ω is the meridian angle,

$$\nabla^2 \phi = \frac{\partial}{\partial r} \left(r^2 \frac{\partial \phi}{\partial r} \right) + \frac{1}{\sin \theta} \frac{\partial}{\partial \theta} \left(\sin \theta \frac{\partial \phi}{\partial \theta} \right) + \frac{1}{\sin^2 \theta} \frac{\partial^2 \phi}{\partial \omega^2} = 0$$

Any ϕ which satisfies the Laplace equation is a possible flow case. In steady flow any streamline may be taken as a solid boundary, because it satisfies the boundary condition. The pressure distribution may then be found by use of the Bernoulli equation.

Two-dimensional Flow

In two-dimensional flow all lines of motion are parallel to a fixed plane, say the xy-plane, and the flow patterns (networks of equipotential lines and streamlines) in all planes parallel to this plane are identical. In Cartesian coordinates

$$\nabla^2 \phi = \frac{\partial^2 \phi}{\partial x^2} + \frac{\partial^2 \phi}{\partial y^2} = 0$$

Stream Function. A streamline is a continuous line drawn through the fluid in such a way that at every point it has the direction of the velocity vector. The stream function is a scalar function of space such that the rate of flow across any line connecting two points is given by the difference in values of the stream function at these points. Let A and P represent two points in one of the planes, Fig. 15, and consider that the flow has unit thickness; i.e., the flow is between two planes, say $z = 0$ and $z = 1$. The rate of flow across any two lines ACP, ABP must be the same if the density is constant (and if no fluid is created or destroyed within the region) as a consequence of the continuity equation. Considering A as a fixed point and P as a variable point, the flow rate across any line connecting the two points is a function of the position of P and of the time. Let this function be ψ, and take as sign convention that it denotes the flow rate from right to left across any line

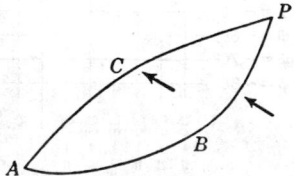

Fig. 15. Fluid region showing the positive flow direction used in the definition of a stream function.

when the observer is at A looking along the line toward P. Thus

$$\psi = \psi(x, y, t)$$

is defined as the stream function. Since there is no flow across a streamline, the value of ψ is constant along a streamline. Velocity components, u and v in the x and y directions, may be expressed in terms of ψ

$$u = -\frac{\partial \psi}{\partial y}, \quad v = \frac{\partial \psi}{\partial x}$$

From the definition of stream function, it is subject to addition of an arbitrary constant. Substituting the above values of u, v into the continuity equation

$$\frac{\partial u}{\partial x} + \frac{\partial v}{\partial y} = 0$$

yields

$$\frac{\partial^2 \psi}{\partial x^2} + \frac{\partial^2 \psi}{\partial y^2} = 0$$

or, since $\nabla^2 \psi = 0$, ψ may also be taken as velocity potential for some case of irrotational flow. Lines in the flow having constant value of the velocity potential are called *equipotential lines*. They are everywhere orthogonal to streamlines, except at singular points, since there is no component of the velocity vector tangent to an equipotential surface. Relations between ϕ and ψ are given by

$$\frac{\partial \phi}{\partial x} = \frac{\partial \psi}{\partial y}, \quad \frac{\partial \phi}{\partial y} = -\frac{\partial \psi}{\partial x}$$

Application of Complex Variables to Irrotational Flow. Let $z = x + iy$ be a complex variable where x and y are real and $i = \sqrt{-1}$. Any function of z

$$w = f(z) = (x + iy) = \phi + i\psi$$

that is defined throughout a region and that has a derivative throughout the region gives rise to two possible irrotational flow cases, since it may be shown that

$$\nabla^2 w = \nabla^2 \phi + i \nabla^2 \psi = 0$$

and hence $\nabla^2 \phi = 0$ and $\nabla^2 \psi = 0$. w is called the complex potential, ϕ is the real part of w, and ψ is the pure imaginary part of w. The complex velocity

$$\frac{dw}{dz} = -u + iv$$

provides a simple method of finding velocity components from the complex potential. Stagnation points occur at those points in the flow where the velocity is zero, i.e.,

$$\frac{dw}{dz} = 0$$

Conformal Mapping. Each of the two complex variables w and z introduced in the preceding paragraphs may be represented by plotting on a graph. The w-plane is a graph having ϕ as abscissa and ψ as ordinate, showing values of $\phi = \pm nc$, $\psi = \pm nc$, where n is every integer and c is a constant. Figure 16 is a flow net of the simplest form showing equi-

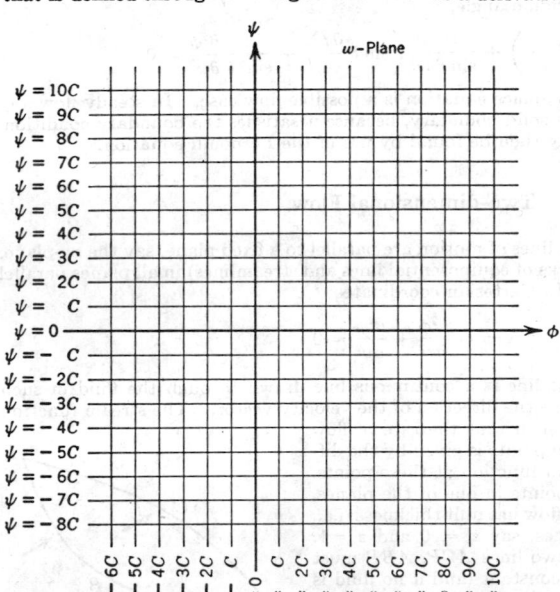

FIG. 16.   Flow net in the w-plane.

potential and streamlines as series of parallel straight lines. The z-plane is a plot of the equipotential and streamlines of the w-plane, with x as abscissa and y as ordinate. The particular flow net in the z-plane depends entirely upon the functional relation between w and z, i.e., $w = f(z)$. Since

$$w = f(z) = \phi + i\psi, \quad z = x + iy$$

where ϕ, ψ, x, y, are real

$$\phi = \phi(x, y), \quad \psi = \psi(x, y)$$

The values of ϕ = constant and ψ = constant can be plotted on the z-plane from the functional relations of ϕ and ψ with x, y.

Examples of Two-dimensional Flow. Since the Laplace equation is linear in ϕ, the sum of two solutions is also a solution, and the product of a solution by a constant is a solution. Several of the important solutions are given.

Rectilinear Flow. Uniform straight-line flow is given by

$$w = -Uz + iVz, \quad \phi = -Ux - Vy, \quad \psi = Vx - Uy$$

where the character of flow is easily seen from the complex velocity

$$\frac{dw}{dz} = -u + iv = -U + iV, \quad u = U, \quad v = V$$

and U, V are the $+ x$, y-components of the velocity, respectively.

Source or Sink. A source in two-dimensional flow is a straight line (parallel to the z-axis) from which fluid flows outward uniformly in all directions normal to the line. A sink is a negative source; i.e., the flow is into the line.

$$w = -\mu(z - a), \quad \phi = -\mu \ln r, \quad \psi = -\theta, \quad \frac{dw}{dz} = -\frac{\mu}{z - a}$$

where $2\pi\mu$ is the outward flow from the line per unit length, known as the *strength*, a is the position of the source in the z plane, Fig. 17, with r_1, θ_1, shown in the figure. Stream-

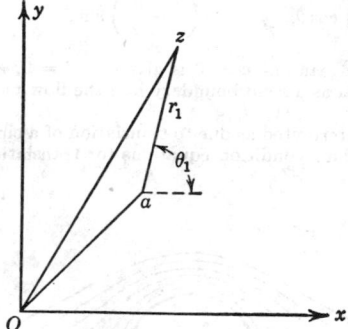

FIG. 17. Special coodinates for source at $z = a$.   FIG. 18. Flow net for source or vortex.

lines are the radial lines of Fig. 18, and equipotential lines are the concentric circles. When μ is negative the equations are for a sink.

Vortex. The equations for a source, multiplied by $-i$, yield relations for the vortex

$$w = i\mu \ln (z - a) = i\mu \ln r_1 - \mu\theta_1, \quad \phi = -\mu\theta_1, \quad \psi = \mu \ln r_1$$

Figure 18 is the flow net for a vortex with the radial lines now equipotential lines and the circles streamlines. The vortex causes circulation about itself. Circulation about any closed curve is defined as the line integral of the velocity around the curve. The circulation about the vortex is $\kappa = 2\pi\mu$, considered positive in the counterclockwise direction around the vortex.

Doublet. The doublet is defined as the limiting case of a source and a sink of equal strength which approach each other such that the product of the strength by the distance between them remains a constant.   The constant, μ, is called the strength of the doublet.

$$w = \frac{\mu}{z}, \quad \phi = \frac{\mu x}{x^2 + y^2}, \quad \psi = -\frac{\mu y}{x^2 + y^2}, \quad \frac{dw}{dz} = -\frac{\mu}{z^2}$$

Figure 19 shows the flow net.   The axis of the doublet is in the direction from sink to source and is parallel to the $+x$-axis as given here.   The equipotential lines are circles having their centers on the x-axis; the streamlines are circles having centers on the y-axis.

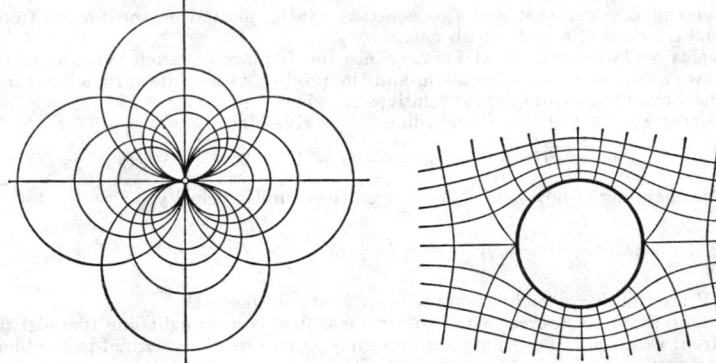

FIG. 19.   Flow net for doublet, $w = \mu/z$.

FIG. 20.   Flow pattern for uniform flow around a circular cylinder without circulation.

Flow around a Circular Cylinder.   The superposition of a uniform flow in the $-x$-direction on a doublet with axis in the $+x$-direction results in flow around a circular cylinder. Adding the two flows, taking $\mu = Ua^2$,

$$w = \frac{Ua^2}{z} + Uz, \quad \phi = U\left(r + \frac{a^2}{r}\right)\cos\theta, \quad \psi = U\left(r - \frac{a^2}{r}\right)\sin\theta$$

where polar coordinates are employed.   The streamline $\psi = 0$ is given by $\theta = 0$, π and by $r = a$; hence the circle, $r = a$, may be taken as a solid boundary and the flow pattern of Fig. 20 is obtained.

The equation for the doublet alone may be interpreted as due to translation of a circular cylinder through an infinite fluid.   The boundary condition equations for translation of

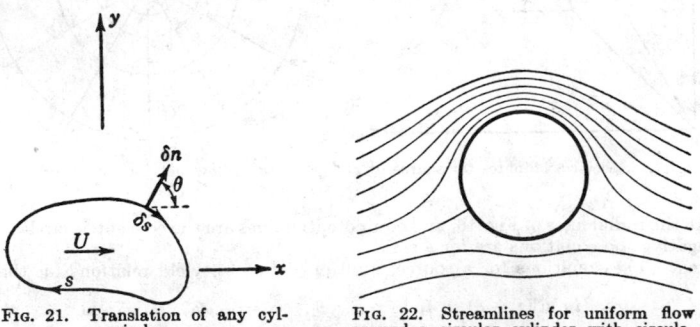

FIG. 21.   Translation of any cylinder.

FIG. 22.   Streamlines for uniform flow around a circular cylinder with circulation.

any cylinder through an infinite fluid may be simply expressed in terms of the stream function.   Let the cylinder, Fig. 21, have the velocity U in the $+x$-direction.   At the surface of the cylinder

$$\psi = -Uy + \text{constant} = -Ur\sin\theta + \text{constant}$$

where the constant is arbitrary. Substituting the stream function for a doublet into the boundary condition

$$\psi = -Ur \sin \theta + \text{constant} = -\frac{Ua^2}{r} \sin \theta$$

showing that the boundary condition is satisfied by taking the constant as zero, and $r = a$.

When, in addition to the doublet at the origin and a uniform flow, a vortex is superposed at the origin, the case of flow around a circular cylinder with circulation is obtained, Fig. 22. The addition of circulation to flow around any cylinder causes a force to be exerted on the cylinder at right angles to the uniform flow. This force is called *lift* and has the magnitude $\rho U \kappa$, where κ is the value of the circulation, ρ is the mass density of the fluid, and U is the approach velocity. There is no resultant force on the cylinder in

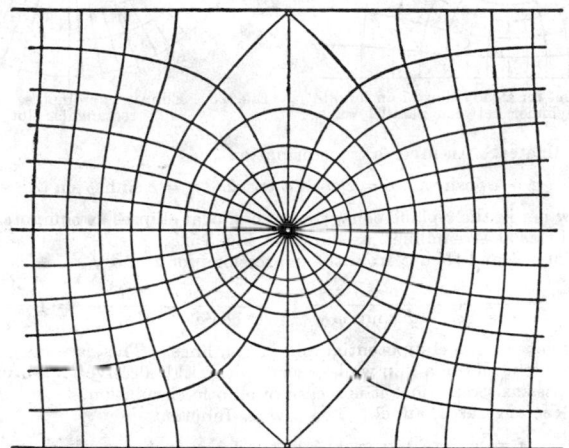

Fig. 23. Flow net for one of a series of equal and equidistant sources.

the direction of the approach velocity; i.e., no *drag* forces result from irrotational flow about a body.

Series of Equal Sources. An infinite series of equal sources, uniformly spaced, say along the y-axis at $y = 0, \pm a, \pm 2a, \pm 3a \cdots$ is given by

$$w = C \ln \sinh \frac{\pi z}{a}, \quad \phi = \frac{1}{2} C \ln \frac{1}{2} \left(\cosh \frac{2\pi x}{a} - \cos \frac{2\pi y}{a} \right), \quad \psi = C \tan^{-1} \frac{\tan \pi y/a}{\tanh \pi x/a}$$

the flow net for the strip $-a/2 \leq y \leq +a/2$ is given in Fig. 23. Owing to symmetry the flow may be interpreted as due to a well midway between parallel impervious walls, or due to the pumping of a series of wells.

Flow through a Grating of Cylindrical Bars. Equal doublets with axes in the $+x$-direction, spaced uniformly a distance a apart along the y-axis and with a uniform flow in the $-x$-direction, result in flow around a series of cylinders. When the cylinders take up no more than half the spacing between doublets, they are, for all practical purposes, circular. The equations are

$$w = U_z + C \coth \frac{\pi z}{a}$$

$$\phi = U_x + C \frac{\sinh 2\pi x/a}{\cosh 2\pi x/a - \cos 2\pi y/a}$$

$$x = U_y - C \frac{\sin 2\pi y/a}{\cosh 2\pi x/a - \cos 2\pi y/a}$$

Figure 24 illustrates a typical flow net.

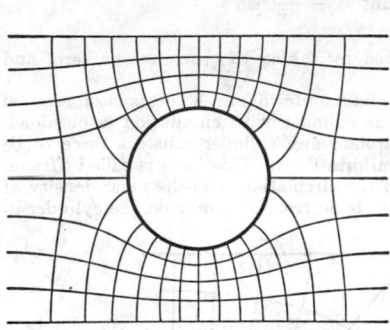

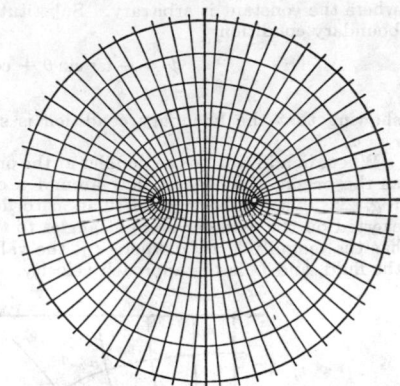

FIG. 24.   Flow net for steady flow of an infinite fluid around a cylinder between parallel walls.

FIG. 25.   Elliptic coordinates.   Flow through a rectangular slot.

Elliptic Coordinates.   An inverse transformation

$$z = c \cosh w, \quad x = c \cosh \phi \cos \psi, \quad y = c \sinh \phi \sin \psi$$

results in a flow net in the z-plane consisting of confocal ellipses as equipotential lines and confocal hyperbolas as streamlines, as shown in Fig. 25.

Eliminating first ψ and then ϕ from the expressions for x and y,

$$\frac{x^2}{c^2 \cosh^2 \phi} + \frac{y^2}{c^2 \sinh^2 \phi} = 1, \quad \frac{x^2}{c^2 \cos^2 \psi} - \frac{y^2}{c^2 \sin^2 \psi} = 1$$

yields the equations of the equipotential and streamlines.   This case may be interpreted as flow through a long slot in a thin wall, or as flow through a converging-diverging section. Taking the ellipses as streamlines it is a case of elliptic circulation.

Flow into a Rectangular Channel.   The inverse function

$$z = w + e^w, \quad x = \phi + e^\phi \cos \psi, \quad y = \psi + e^\phi \sin \psi$$

results in flow into a rectangular channel, as in Fig. 26.   The velocity becomes infinite at the ends of the channel walls.

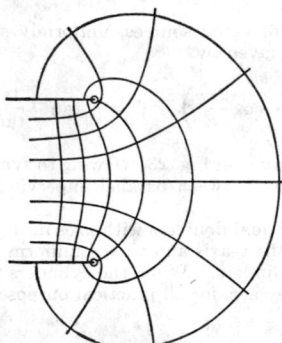

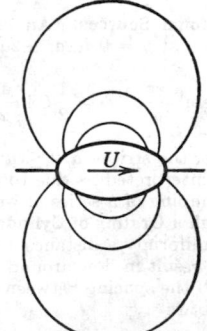

FIG. 26.   Flow into a rectangular channel.

FIG. 27.   Unsteady streamlines for translation of elliptic cylinder parallel to major axis.

Translation of an Elliptic Cylinder.   The boundary conditions for translation of any cylinder, developed in connection with the circular cylinder, for motion in the x-direction, is

$$\psi = -Uy + \text{constant}$$

the relations

$$w = Ce^{-\zeta}, \quad z = c \cosh \zeta, \quad \zeta = \xi + i\eta$$

which result in

$$\phi = Ce^{-\xi}\cos\eta, \quad \psi = -Ce^{-\xi}\sin\eta$$

satisfy the boundary condition if the arbitrary constant is zero for one value of ξ, say ξ_0, determined by

$$Ce^{-\xi_0} = Uc\sinh\xi_0$$

$\xi = \xi_0$ is the equation of an elliptic cylinder with semi-major and semi-minor axes, a, b, respectively

$$a = c\cosh\xi_0, \quad b = c\sinh\xi_0, \quad c = \sqrt{a^2 - b^2}$$

Eliminating C, the stream function becomes

$$\psi = -Ub\sqrt{\frac{a+b}{a-b}}\,e^{-\xi}\sin\eta$$

with the streamlines shown in Fig. 27.

For translation in the direction of the minor axis, say the $+y$-direction, the boundary condition is

$$\psi = Vx + \text{constant}$$

and the resulting stream function

$$\psi = Va\sqrt{\frac{a+b}{a-b}}\,e^{-\xi}\cos\eta$$

The streamlines are shown in Fig. 28. The two ψ's may be added to give translation of an elliptic cylinder in any direction through an infinite fluid. The kinetic energy of the fluid T is given by

$$T = -\frac{\rho}{2}\int\phi\frac{\partial d}{\partial n}\,dS = -\frac{\rho}{2}\int\phi\,d\psi$$

where the latter expression is for two-dimensional flow only. The integrals are taken over the boundaries, with n the normal to the boundary. The kinetic energy for translation in the y-direction is

$$T = \frac{\pi\rho}{2}V^2 a^2$$

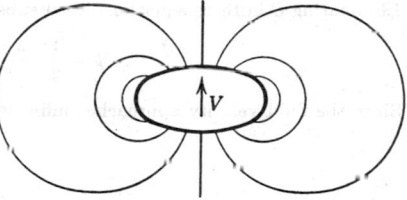

FIG. 28. Unsteady streamlines for translation of elliptic cylinder parallel to minor axis.

For steady flow around an elliptic cylinder, a uniform flow is superposed on the system such that the cylinder is brought to rest. Streamlines are shown in Fig. 29 for one case.

Making $b = 0$ reduces the case to translation of a plane lamina, or steady flow around a lamina.

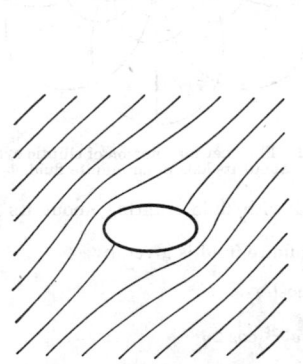

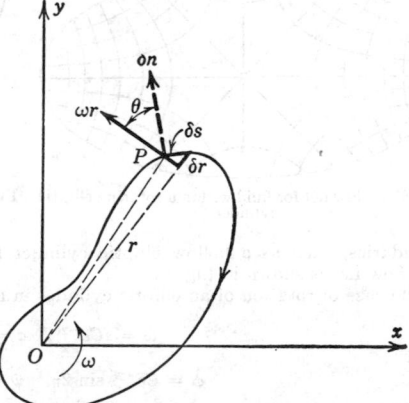

FIG. 29. Steady flow around an elliptic cylinder $U = V$.

FIG. 30. Rotation of any cylinder about origin.

Rotating Elliptic Cylinder. Referring to Fig. 30, the boundary condition that the component of the velocity normal to the boundary must equal the velocity of the boundary normal to itself may be written

$$\frac{\partial \psi}{\partial s} = \omega r \cos \theta = \omega r \frac{dr}{ds}$$

which holds for external or internal boundaries. Integrating

$$\psi = \frac{1}{2} \omega r^2 + \text{constant}$$

For fluid within a rotating elliptic cylinder,

$$\omega = iAz^2, \quad \phi = -2Axy, \quad \psi = A(x^2 - y^2)$$

may be made to satisfy the boundary condition

$$\psi = A(x^2 - y^2) = \frac{1}{2} \omega(x^2 + y^2) - C$$

where Cartesian coordinates are used in the boundary condition. Rearranging

$$\frac{x^2}{C/(\frac{1}{2}\omega - A)} + \frac{y^2}{C/(\frac{1}{2}\omega + A)} = 1$$

which is an ellipse when $A < \frac{1}{2} \omega$. Denoting the semi-major and semi-minor axes by a, b respectively

$$a^2 = \frac{C}{\frac{1}{2}\omega - A}, \quad b^2 = \frac{C}{\frac{1}{2}\omega + A}$$

Eliminating C in these equations, then substituting the value of A in the stream function

$$\psi = \frac{1}{2} \omega \frac{a^2 - b^2}{a^2 + b^2} (x^2 - y^2)$$

Since the fluid velocity approaches infinity at infinity, this must be a case with external

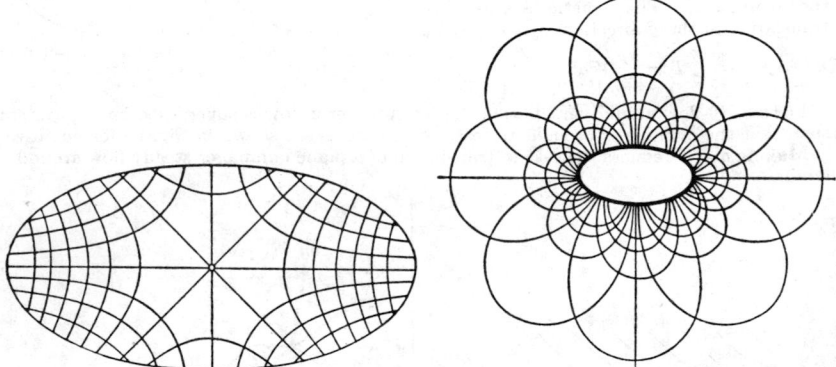

FIG. 31. Flow net for fluid within a rotating elliptic   FIG. 32. Flow net for rotation of elliptic cylinder
cylinder.   about its axis in an infinite fluid.

boundaries, such as a hollow elliptic cylinder filled with fluid rotating about its axis. The flow net is shown in Fig. 31.

The case of rotation of an elliptic cylinder in an infinite fluid is given by

$$\omega = iCe^{-2\zeta}, \quad z = c \cosh \zeta$$

with

$$\phi = Ce^{-2\xi} \sin 2\eta, \quad \psi = Ce^{-2\xi} \cos 2\eta$$

Substitution into the boundary condition gives

$$C = \frac{1}{4} \omega(a + b)^2$$

The flow net is given in Fig. 32.

Three-dimensional Flow

Three-dimensional flows in general are more difficult to handle than those in two dimensions, primarily owing to lack of methods comparable to the use of complex variables and conformal mapping. A velocity potential must be found which satisfies the Laplace equation and the boundary conditions. When the velocity potential is *single valued*, it may be shown that the solution is unique. For a body moving through an infinite fluid, otherwise at rest, a necessary condition is that the solution be such that the fluid at infinity remain at rest. Flow cases are usually found by investigating solutions of the Laplace equation to determine the particular boundary conditions which they satisfy.

Stokes' Stream Function. The Stokes' stream function is defined only for those three-dimensional flow cases which have axial symmetry, i.e., where the flow is in a series of planes passing through a given line and where the flow pattern is identical in each of these planes. The intersection of these planes is the axis of symmetry.

In any one of these planes through the axis of symmetry select two points A, P, such that A is fixed and P is variable. Draw a line connecting AP. The flow through the surface generated by rotating AP about the axis of symmetry is a function of the position of P. Let this function be $2\pi\psi$, and let the axis of symmetry be the x-axis. Then ψ, which is Stokes' stream function, is a function of x and ω, where

$$\omega = \sqrt{y^2 + z^2}$$

is the distance from P to the x-axis. The surfaces $\psi =$ constant are stream surfaces. Since A is an arbitrary point, the stream function is always subject to the addition of an arbitrary constant.

Velocity components may be determined in terms of ψ,

$$u = -\frac{1}{\omega}\frac{\partial\psi}{\partial\bar\omega}, \quad v' = \frac{1}{\bar\omega}\frac{\partial\psi}{\partial x}$$

where u, v' are in the x, $\bar\omega$-directions, respectively. The relations between ϕ and ψ are obtained by equating expressions for velocity components

$$\frac{\partial\phi}{\partial x} = \frac{1}{\bar\omega}\frac{\partial\psi}{\partial\bar\omega}, \quad \frac{\partial\phi}{\partial\bar\omega} = -\frac{1}{\bar\omega}\frac{\partial\psi}{\partial x}$$

Stokes' stream function has the dimensions *volume per unit time*. To express ψ in spherical polar coordinates, let r be the distance from the origin and θ be the angle the radius vector makes with the x-axis; the meridian angle is not needed because of axial symmetry. Then

$$v_r = -\frac{1}{r^2\sin\theta}\frac{\partial\psi}{\partial\theta}, \quad v_\theta = \frac{1}{r\sin\theta}\frac{\partial\psi}{\partial r}$$

and

$$\frac{1}{\sin\theta}\frac{\partial\psi}{\partial\theta} = r^2\frac{\partial\phi}{\partial r}, \quad \frac{\partial\psi}{\partial r} = -\sin\theta\frac{\partial\phi}{\partial\theta}$$

These equations are useful in dealing with flow about spheres, ellipsoids, and discs, and through apertures.

Examples of Three-dimensional Flow. Several examples of three-dimensional flow with axial symmetry follow.

Source in a Uniform Stream. A point source is a point from which fluid issues at a uniform rate in all directions. Its strength m is the flow rate from the point, and the velocity potential and stream function for a source at the origin are

$$\phi = \frac{m}{4\pi r}, \quad \psi = \frac{Ur^2}{2}\sin^2\theta$$

The flow net is shown in Fig. 33. Superposing a uniform flow on a point source results in a *half body*, with the flow equations,

$$\phi = \frac{m}{4\pi r} + Ur\cos\theta, \quad \psi = \frac{m}{4\pi}\cos\theta + \frac{Ur^2}{2}\sin^2\theta$$

The resulting flow net is shown in Fig. 34. The body extends to infinity in the downstream direction and has the asymptotic cylinder $\bar\omega = \sqrt{m/(\pi U)}$. The equation of the half body is

$$r = \frac{1}{2}\sqrt{\frac{m}{\pi U}}\sec\frac{\theta}{2}$$

The pressure intensity on the half body is given by

$$p = \frac{\rho}{2}U^2\left(\frac{3m^2}{16\pi^2 r^4 U^2} - \frac{m}{2\pi r^2 U}\right)$$

showing that the dynamic pressure drops to zero at great distances downstream along the body.

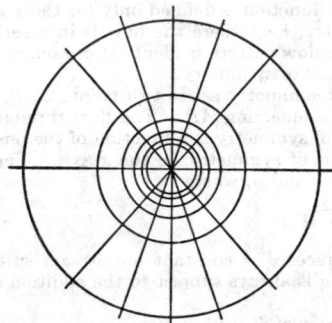

FIG. 33. Streamlines and equipotential lines for a source.

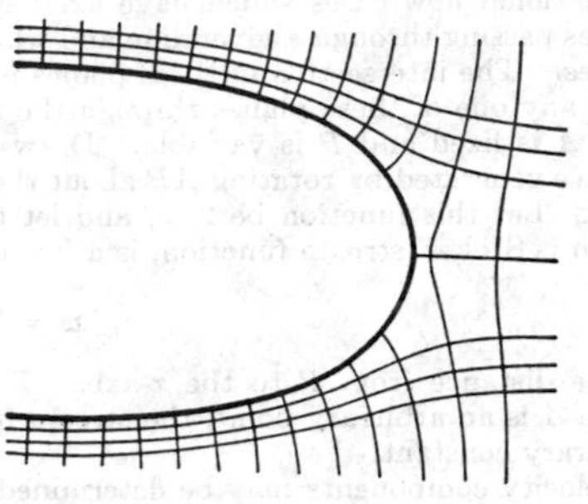

FIG. 34. Streamlines and equipotential lines for a half body.

Flow around a Sphere.   A doublet is defined as the limiting case as a source and sink approach each other, such that the product of their strength by the distance between them remains a constant.   The doublet has directional properties.   Its axis is positive from the sink towards the source.   For a doublet at the origin, with axis in the $+x$-direction

$$\phi = \frac{\mu}{r^2}\cos\theta, \quad \psi = -\frac{\mu}{r}\sin^2\theta$$

where μ is the strength of the doublet.
Superposing a uniform flow, $u = -U$, on the doublet results in flow around a sphere,

$$\phi = \frac{Ua^3}{2r^2}\cos\theta + Ur\cos\theta, \quad \psi = -\frac{Ua^3}{2r}\sin^2\theta + \frac{Ur^2}{2}\sin^2\theta$$

where $\mu = Ua^3/2$.   This is shown to be the case, since the streamline $\psi = 0$ is satisfied by

$$\theta = 0, \pi; \quad r = a$$

The flow net is shown in Fig. 35.

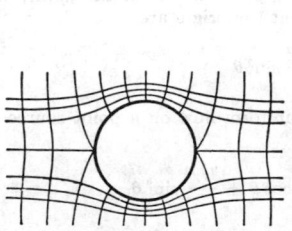

FIG. 35. Streamlines and equipotential lines for uniform flow about a sphere at rest.

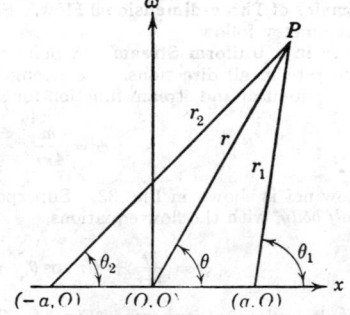

FIG. 36. Auxiliary coordinate systems used for Rankine body.

Rankine Body. An oval body of arbitrary breadth and length (such that the length is always greater than the breadth), with an arbitrary uniform flow U, can be obtained by using a sink of strength m at $(0, -a)$, a source of the same strength at $(0, a)$, and a uniform flow $u = -U$. Using the notation of Fig. 36,

$$\phi = Ux + \frac{m}{4\pi}\left(\frac{1}{r_1} - \frac{1}{r_2}\right) = Ux + \frac{m}{4\pi}\left[\frac{1}{\sqrt{(x-a)^2 + \bar{\omega}^2}} - \frac{1}{\sqrt{(x+a)^2 + \bar{\omega}^2}}\right]$$

$$\psi = \frac{1}{2}Ur^2\sin^2\theta + \frac{m}{4\pi}(\cos\theta_1 - \cos\theta_2)$$

The equation of the body is given by $\psi = 0$. Its half breadth h is obtained by trial from

$$h^2 = \frac{m}{\pi U}\frac{a}{\sqrt{h^2 + a^2}}$$

Similarly the half length x_0 is given by trial from

$$U - \frac{max_0}{\pi(x_0^2 - a^2)^2} = 0$$

Eliminating m/U in the two expressions

$$\frac{m}{U} = \frac{\pi}{a}\frac{(x_0^2 - a^2)^2}{x_0} = \frac{\pi}{a}h^2\sqrt{h^2 + a^2}$$

the value of a may be computed for given x_0, h. Then the ratio m/U is determined. The flow net for a typical example is given in Fig. 37. These oval bodies are called *Rankine bodies*.

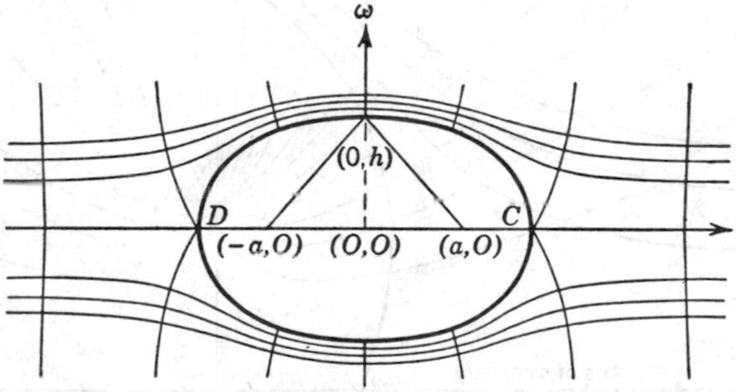

Line Source. A line source is a line segment over which point sources are uniformly distributed. Its strength is defined as the flow rate from the line per unit length of line. A flow net for a line source is given in Fig. 38. Combinations of point sources and sinks with line sources and sinks plus a uniform flow permit great

FIG. 37.   Rankine body.

variation in the shape characteristics of the resulting body. It is essential that the net flow out of all sources and sinks be zero within a closed body.

A method using a sequence of equal-length line sources of unknown strength has been developed by von Kármán * for determining the velocity potential and stream function for a given body. As many points on the body are specified as the number of source (or sink) segments on the axis of symmetry. A set of simultaneous linear equations is obtained which may be solved for the unknown strengths of the line source segments.

Source Distributions Not on the Axis of Symmetry. The various bodies discussed above have been produced by sources on the axis of symmetry. Distributions of sources over lines or areas, or through volumes, symmetrically located with respect to the axis result in bodies which may be blunt-nosed, or in certain cases the body may be annular.

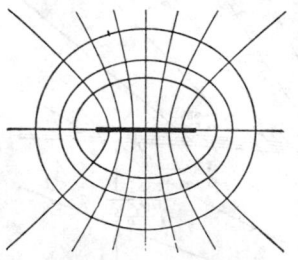

FIG. 38.   Streamlines and equipotential lines for finite line source.

Streamlines are shown in Fig. 39 for a ring source. The ring has its center on the axis of symmetry and is in a plane normal to the axis. Superposition of a relatively low uniform flow results in a half body with a concave nose, Fig. 40. A higher uniform flow results in an annular half body, Fig. 41. The equations require the use of incomplete elliptic

---

* Th. von Kármán, "Berechnung der Druckverteilung an Luftschiffkörpern," *Abh. a. d. Aero. Institut an der Tech. Hochschule Aachen*, No. 6, 1927, pp. 3–7, Translation NACA Tech. Memo., No. 574, 1930.

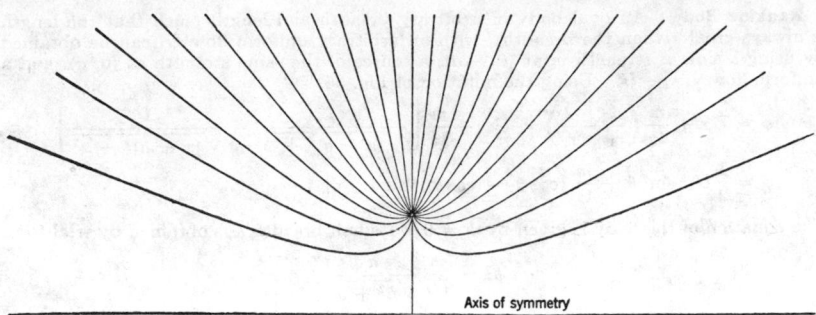

FIG. 39.   Streamlines for ring source.

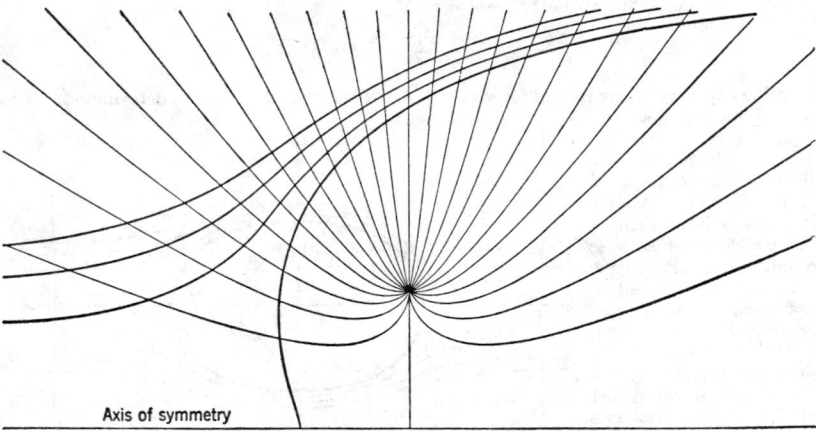

FIG. 40.   Superposition of uniform flow (low velocity) upon a ring source.

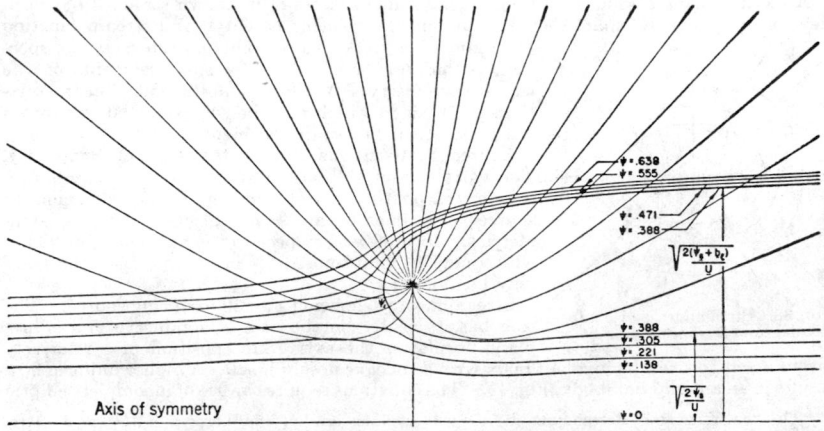

FIG. 41.   Superposition of uniform flow (high velocity) upon a ring source.

integrals of the first and second kinds. After the ring-source streamlines of Fig. 39 are obtained, subsequent combined flows are obtained graphically by superposing the other patterns and adding. Figure 42 shows a closed annular body in the flow.

Streamlines for uniform distribution of source strength over a disc are shown in Fig. 43. By changing the scale and keeping the same strength per unit area, one disc source may be subtracted from a larger disc source, resulting in an annular disc source as shown in Fig. 44. Two combinations are shown in Figs. 45 and 46. The equations again contain elliptic functions, and combined flows are obtained by graphical superposition of elemental flows.

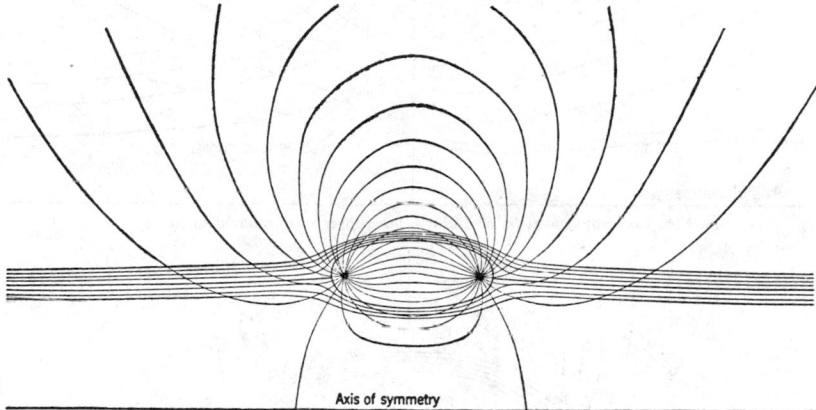

Fig. 42. Superposition of ring source, ring sink with uniform flow.

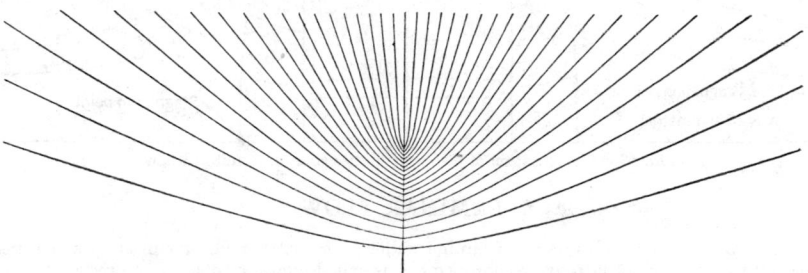

Fig. 43. Streamlines for uniform disc source.

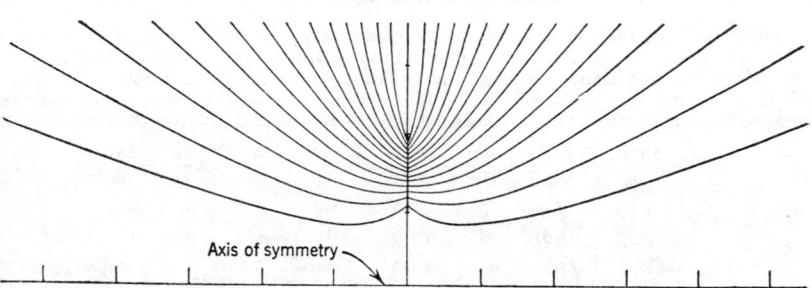

Fig. 44. Streamlines for annular disc source.

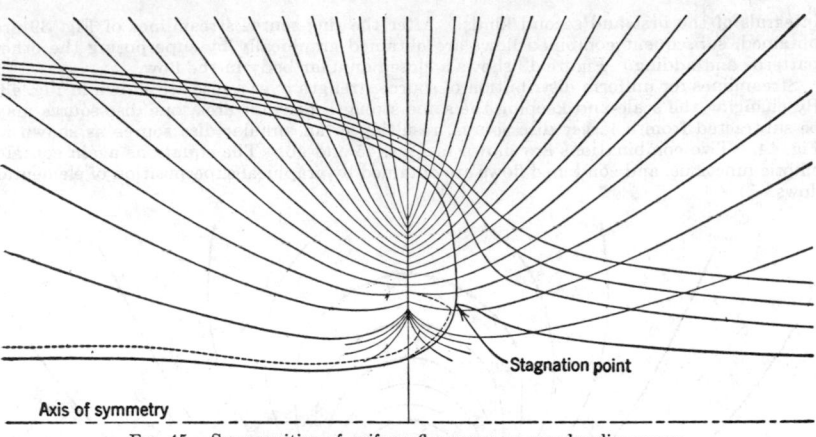

Fig. 45.   Superposition of uniform flow upon an annular disc source.

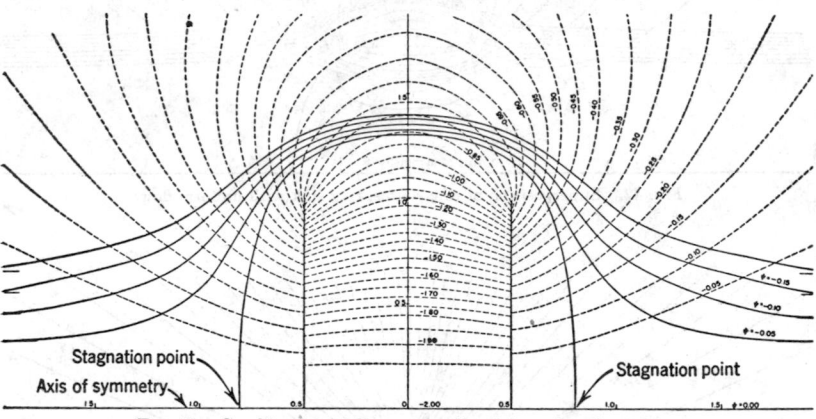

Fig. 46.   Combination of disc source, disc sink, and uniform flow.

6. LAMINAR FLOW *

Fluid motion in which one fluid lamina slides over another lamina in such a manner that all turbulent fluctuations are damped out is called *laminar flow*. The predominating forces acting are due to viscous shear in the fluid.

Fundamental Relationships

A real fluid must obey certain relations and conditions, stated in the following paragraphs, which in some cases permit unique solutions of flow problems to be obtained.
Navier-Stokes Equations.   Writing the equation of motion for a particle of fluid in each of the coordinate directions, taking into account the shear stresses which may be expressed in terms of viscosity and velocity gradients, results in the *Navier-Stokes equations:*

$$X - \frac{1}{\rho}\frac{\partial p}{\partial x} + \nu\left(\frac{\partial^2 u}{\partial x^2} + \frac{\partial^2 u}{\partial y^2} + \frac{\partial^2 u}{\partial z^2}\right) = u\frac{\partial u}{\partial x} + v\frac{\partial u}{\partial y} + w\frac{\partial u}{\partial z} + \frac{\partial u}{\partial t}$$

$$Y - \frac{1}{\rho}\frac{\partial p}{\partial y} + \nu\left(\frac{\partial^2 v}{\partial x^2} + \frac{\partial^2 v}{\partial y^2} + \frac{\partial^2 v}{\partial z^2}\right) = u\frac{\partial v}{\partial x} + v\frac{\partial v}{\partial y} + w\frac{\partial v}{\partial z} + \frac{\partial v}{\partial t}$$

$$Z - \frac{1}{\rho}\frac{\partial p}{\partial z} + \nu\left(\frac{\partial^2 w}{\partial x^2} + \frac{\partial^2 w}{\partial y^2} + \frac{\partial^2 w}{\partial z^2}\right) = u\frac{\partial w}{\partial x} + v\frac{\partial w}{\partial y} + w\frac{\partial w}{\partial z} + \frac{\partial w}{\partial t}$$

* Illustrations and material used by permission from *Fluid Dynamics* by V. L. Streeter, copyright 1948, McGraw-Hill Book Company, Inc.

X, Y, Z are extraneous force components per unit mass, in the xyz-directions, respectively; ρ is the mass density, considered constant; p is the average pressure at a point; ν is the kinematic viscosity; u, v, w are the velocity components in the xyz-directions.

These simultaneous, non-linear, differential equations cannot be integrated except for extremely simple flow cases where many of the terms are neglected. They contain the basic assumption that the stresses on a particle may be expressed as the most general linear function of the velocity gradients.

Boundary Conditions. A real fluid in contact with a solid boundary must have a velocity exactly equal to the velocity of the boundary; this is much more restrictive than for a non-viscous fluid, where no restrictions are placed on tangential velocity components at a boundary.

When two fluids are flowing, a dynamical boundary condition arises at the interface. Applying the equation of motion to a thin layer of fluid enclosing a small portion of the interface shows that the terms containing mass are of higher order of smallness than the surface stress intensities, and hence that the stresses must be continuous through the surface.

In general, the boundary conditions at a solid surface give rise to rotational flow. Although the Navier-Stokes equations are satisfied by a velocity potential, since the viscous terms drop out, the boundary conditions cannot be satisfied.

Continuity. The continuity equation must hold, as in the case of non-viscous flow. It is

$$\frac{\partial u}{\partial x} + \frac{\partial v}{\partial y} + \frac{\partial w}{\partial z} = 0$$

Similitude Relationships. Reynolds' Number. Due to the complexity of the viscous flow equations, it has been necessary to resort to experimental means for the solutions of many fluid problems. Even though the Navier-Stokes equations have not been solved, much can be learned from them through considerations of similarity.

For two flow cases to be dynamically similar the following criteria must be met:

(a) The geometrical boundaries must be similar.

(b) The boundary conditions must be the same.

(c) The streamlines, or flow patterns, must be geometrically similar, or the dynamic pressures at corresponding points must bear a fixed ratio to each other.

In applying the Navier-Stokes equations to two flow cases, the alteration of units of length, time, and pressure should transform one equation into the other for complete similarity. The condition for this complete similarity comes out to be

$$\frac{\rho_1 u_1 l_1}{\mu_1} = \frac{\rho_2 u_2 l_2}{\mu_2} = R$$

where subscript 1 refers to one case and subscript 2 to the other. The dimensionless quantity $\rho u l / \mu$ is called *Reynolds' number*. For two viscous flow cases to be dynamically similar, each case must have the same Reynolds' number.

The drag on a body can be considered as made up from a pressure difference times an area. From the dimensional reasoning the drag D may be written in two forms:

$$D = k_1 l^2 \rho u^2$$

$$D = k_2 \mu l u$$

where the ratios k_2 / k_1 is

$$\frac{k_2}{k_1} = \frac{l \rho u}{\mu} = R$$

for dynamically similar flows. The dimensionless quantities, k_1, k_2, are constant for dynamic similarity (1 Reynolds' number) and, in general, vary with Reynolds' number,

$$D = f_1(R) \rho l^2 u^2$$

$$D = f_2(R) \mu l u$$

where f_1, f_2 are unknown functions which must in general be determined by experiment.

When Reynolds' number is very small, the viscosity terms predominate; k_2 must not depend on ρ and hence must be constant. Similarly, for large Reynolds' numbers, the inertial terms predominate, and viscosity should have no effect; hence k_1 must be a constant.

These two relations have been observed to hold experimentally. For very large Reynolds' numbers, solutions of the Euler equations apply closely, except for a narrow region along the boundaries and possibly in the wake. For very small Reynolds' numbers, the initial terms (those containing density) may be omitted from the Navier-Stokes equations permitting special solutions to be effected. For the broad range of Reynolds' numbers between the two extremes, theory contributes little and recourse must be had to experimental methods.

Examples of Viscous Flow

Several examples of flow at low Reynolds' numbers are given. It is assumed that in each case any turbulent fluctuations are completely damped out by viscous action.

Flow between Parallel Boundaries. For steady flow between fixed parallel boundaries at low Reynolds' numbers, the Navier-Stokes equations can be greatly reduced. Taking the coordinates as shown in Fig. 47, the differential equations reduce to

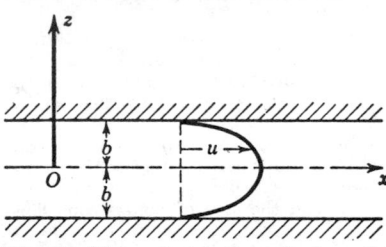

$$\frac{\partial}{\partial x}(p + \gamma h) = \mu \frac{\partial^2 u}{\partial z^2}$$

$$\frac{\partial}{\partial y}(p + \gamma h) = \mu \frac{\partial^2 v}{\partial z^2}$$

$$\frac{\partial}{\partial z}(p + \gamma h) = 0$$

Fig. 47. Viscous flow between fixed parallel boundaries.

where h is measured vertically upward and γ is the specific weight of the fluid. Integrating and introducing the boundary conditions $u = v = O$ for $z = \pm b$,

$$u = \frac{z^2 - b^2}{2\mu}\frac{\partial}{\partial x}(p + \gamma h) = \frac{\partial}{\partial x}\left[(p + \gamma h)\left(\frac{z^2 - b^2}{2\mu}\right)\right] = -\frac{\partial \phi}{\partial x}$$

$$v = \frac{z^2 - b^2}{2\mu}\frac{\partial}{\partial y}(p + \gamma h) = \frac{\partial}{\partial y}\left[(p + \gamma h)\left(\frac{z^2 - b^2}{2\mu}\right)\right] = -\frac{\partial \phi}{\partial y}$$

where $-\phi$ is the quantity in brackets. For this special viscous flow case a velocity potential exists, given by ϕ.

Using this case as an analogy to potential flow, Hele-Shaw * constructed an apparatus consisting of two closely spaced glass plates. A transparent fluid is caused to flow between the plates, and dye is continuously injected into the fluid at regular intervals along the upstream edge of the plates. An object placed between the plates causes the fluid to deviate in flowing around it in such a way that the dyed portions of the fluid trace out streamlines for two-dimensional flow. The results are confirmed by potential theory and by other experimental means.

For motion of the upper plate in the x-direction with velocity U, the boundary conditions become

$$u = v = 0 \quad \text{for} \quad z = -b, \quad u = U, \quad v = 0 \quad \text{for} \quad z = +b$$

The velocity components are

$$u = \frac{U}{2}\left(1 + \frac{z}{b}\right) + \frac{z^2 - b^2}{2\mu}\frac{\partial}{\partial x}(p + \gamma h)$$

$$v = \frac{z^2 - b^2}{2\mu}\frac{\partial}{\partial x}(p + \gamma h)$$

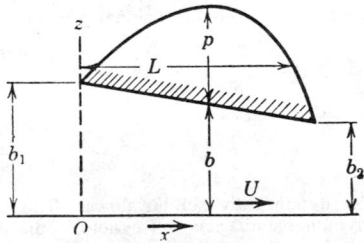

Fig. 48. Sliding bearing.

The maxium velocity has been displaced from the middle plane. When $(p + \gamma h)$ is constant, the gradient is zero and flow results due to motion of the upper plate only.

Theory of Lubrication. The equations for two-dimensional viscous flow are applicable to the case of a slider bearing and can be applied to journal bearings. The simple case of a bearing of unit width is developed here, under the assumption that there is no flow out of the sides of the block, i.e., normal to the plane of Fig. 48, where the clearance b is shown to a greatly exaggerated scale. The motion of a bearing block sliding over a plane surface, inclined slightly so that fluid is crowded between the two surfaces, develops large sup-

---

* H. J. S. Hele-Shaw, "Investigation of the Nature of the Surface Resistance of Water and of Streamline Motion under Certain Experimental Conditions," *Trans. Inst. Naval Architects*, Vol. 40, 1898.

porting forces normal to the surfaces. The angle of inclination is very small, therefore the differential equations of the preceding article apply. Since elevation changes also are very small and flow is in the x-direction only, the equations reduce to

$$\frac{\partial p}{\partial x} = \mu \frac{\partial^2 u}{\partial z^2}, \quad \frac{\partial p}{\partial z} = 0$$

Considering the inclined block stationary and the plane surface in motion, and taking the pressure at the two ends of the block as zero, the boundary conditions become

$$x = 0, \quad x = L, \quad p = 0; \quad u = U, \quad z = 0; \quad u = 0, \quad z = b$$

Integrating the equations, and considering unit width normal to the figure, the discharge Q and pressure distribution are determined,

$$Q = \frac{Ub_1b_2}{b_1 + b_2}, \quad p = \frac{6\mu Ux(b - b_2)}{b^2(b_1 + b_2)}$$

The last relation shows that b must be greater than b_2 for positive pressure build-up in the bearing. The point of maximum pressure intensity and its value are

$$x\Big|_{p_{max}} = \frac{b_1 L}{b_1 + b_2}, \quad p_{max} = \frac{3}{2}\frac{\mu UL}{b_1 b_2}\frac{b_1 - b_2}{b_1 + b_2}$$

The force P, which the bearing will sustain, is

$$P = \int_0^L p\,dx = \frac{6\mu UL^2}{b_2^2(k - 1)^2}\left[\ln k - \frac{2(k - 1)}{k + 1}\right]$$

The maximum bearing load is obtained for $k = 2.2$, yielding

$$P = 0.16\mu \frac{UL^2}{b_2^2}, \quad D = 0.75\mu \frac{UL}{b_2}$$

the ratio

$$\frac{P}{D} = 0.21\frac{L}{b_2}$$

can be made very large, since b_2 is small. The pressure distribution is shown for one case in Fig. 48. For $k = 2.2$, the line of action of the bearing load is at $x = 0.58L$. In general the line of action is given by

$$\bar{x} = \frac{L}{2}\left[\frac{2k}{k - 1} - \frac{k^2 - 1 - 2k \ln k}{(k^2 - 1)\ln k - 2(k - 1)^2}\right]$$

Journal bearings are computed in an analogous manner. In general the clearances are so small compared with the radius of curvature of bearing surface that the equations for plane motion can be applied.

Flow through Circular Tubes. For steady flow through circular tubes with values of Reynolds' number $vD\rho/\mu$ less than 2000, the motion is laminar. Equilibrium conditions on a cylindrical element concentric with the tube axis show that the shear stress varies linearly from zero at the pipe axis to a maximum $\Delta p/\Delta L\, r_0/2$ at the wall, where $\Delta p/\Delta L$ is the drop in pressure intensity per unit length of tube. In one-dimensional laminar flow the relation between shear stress and velocity gradient is given by Newton's law of viscosity,

$$\tau = -\mu \frac{du}{dr}$$

Using this relation for value of shear stress, the velocity distribution is found to be

$$u = \frac{\Delta p}{\Delta L}\frac{r_0^2 - r^2}{4\mu}$$

where r is the distance from the tube axis. The maximum velocity is at the axis.

$$u_{max} = \frac{\Delta p r_0^2}{\Delta L 4\mu}$$

The average velocity is half the maximum velocity

$$V = \left(\frac{\Delta p}{\Delta L}\right)\left(\frac{r_0^2}{8\mu}\right)$$

The discharge, obtained by multiplying average velocity by cross-sectional area, is

$$Q = \left(\frac{\Delta p}{\Delta L}\right)\left(\frac{\pi D^4}{128\mu}\right)$$

which is the *Hagen-Poisseuille law*. The above equations are independent of the surface condition in the tube, and they therefore hold for either rough or smooth tubes.

Flow through Porous Media. In the flow of fluid through pervious materials, such as flow of water through sand, the velocities and the flow passages are usually very small. Although it is impossible to write equations for an individual particle, the mass-flow relationship is obtained by neglecting such terms as $u \, \partial u/\partial x$, $v \, \partial u/\partial y$, $w \, \partial u/\partial z$, and assuming steady flow. The equations become

$$\nabla^2(p + \gamma h) = 0$$

where h is the vertical distance to the point from an arbitrary datum, and γ is the specific weight. A velocity potential exists, with the velocity components given by

$$u = -\frac{k}{\gamma}\frac{\partial}{\partial x}(p + \gamma h)$$

$$v = -\frac{k}{\gamma}\frac{\partial}{\partial y}(p + \gamma h)$$

$$w = -\frac{k}{\gamma}\frac{\partial}{\partial z}(p + \gamma h)$$

where k is the permeability coefficient which has the dimensions of velocity. The permeability coefficient takes into account the size, shape, and spacing of the solid particles as well as the physical properties of the fluid. For two-dimensional flow cases, such as flow under a long dam, the flow net may be constructed graphically or by any other means available to ideal fluid theory. When the porous media is stratified, k takes on directional properties and is no longer a constant.

Viscous Flow around a Sphere. Stokes' Law. The flow of an infinite viscous fluid around a sphere at very low Reynolds' numbers has been solved by Stokes.* The Navier-Stokes equations with the acceleration terms omitted must be satisfied, as well as continuity and the boundary condition that the velocity vanish at the surface of the sphere. Stokes' solution is

$$u = U\left[\frac{3}{4}\frac{ax^2}{r^3}\left(\frac{a^2}{r^2} - 1\right) + 1 - \frac{1}{4}\frac{a}{r}\left(3 + \frac{a^2}{r^2}\right)\right]$$

$$v = U\frac{3}{4}\frac{axy}{r^3}\left(\frac{a^2}{r^2} - 1\right)$$

$$w = U\frac{3}{4}\frac{axz}{r^3}\left(\frac{a^2}{r^2} - 1\right)$$

$$p = -\frac{3}{2}\frac{\mu Uax}{r^3}$$

The radius of a sphere is a; the undisturbed velocity is in the x-direction, $u = U$; and p is the average dynamic pressure intensity at a point. The drag on a sphere is made up from the pressure difference over the surface and from the shear stress. The viscous drag due to shear stress is twice as great as that due to the pressure difference. The total drag is

$$D = 6\pi a\mu U$$

This is known as *Stokes' law*.

The settling velocity of small spheres may be obtained by writing the equation for drag, weight of particle, and buoyant force. Solving for settling velocity

$$U = \frac{2}{9}\frac{a^2}{\mu}(\gamma_s - \gamma)$$

where γ_s is the specific weight of the solid particle. Stokes' law has been found by experiment to hold for Reynolds' numbers below 1, i.e.,

$$\frac{2a\rho U}{\mu} < 1$$

---

* G. Stokes, *Trans. Cambridge Phil. Soc.*, Vol. 8, 1845, and Vol. 9, 1851.

7. BASIC RELATIONS IN MECHANICS OF REAL FLUIDS

Hydraulics is the practical application of the mathematical development of fluid mechanics, Articles 5 and 6, and of the body of experimental information which is available. It is not an exact science and much depends upon judgment and experience. The recent trends are toward rational developments with use of experimental information where needed.

General

Steady flow is defined as flow in which conditions (velocity, pressure, density, etc.) at any point do not change with the time. In turbulent flow, the definition is generally extended to include flows such that the temporal mean conditions at any point do not change with the time. **Unsteady** flow refers to changing conditions at any point with respect to time.

Uniform flow is defined as flow in which the velocity vector is everywhere the same at any instant. For open or closed conduit flow this definition is generalized to include prismatic flow with the average velocity the same at all cross sections at any instant. **Non-uniform** flow refers to those cases where the velocity (or average velocity at a section) changes with position at any instant.

A **streamline** is a continuous line through the fluid that has the direction of the velocity vector at every point. A *stream tube* is the tube of flow contained within the streamlines which pass through any small closed curve.

In steady flow a streamline is fixed in space and the path of a particle coincides with the streamline. Similarly, a stream tube is fixed in space in steady flow, and, in addition, the mass per unit time passing any cross section is the same. This relation, called the *continuity equation*, may be written

$$\rho_1 v_1 \delta A_1 = \rho_2 v_2 \delta A_2$$

where the subscripts refer to any two cross sections, v is the velocity normal to the elemental cross section δA, and ρ is the mass density. For incompressible flow, ρ is constant, and the expression reduces to an equality of volume rate of flow. Summing up the flow through all stream tubes passing through a cross section, the equation becomes

$$V_1 A_1 = V_2 A_2$$

where V_1, V_2 are the average velocities normal to the cross sections 1 and 2.

Energy Equation

Writing the equation of motion for a fluid particle, in the direction of the streamline through the point, and assuming steady flow of a non-viscous, incompressible fluid, results in the *Bernoulli equation*

$$\frac{v^2}{2g} + \frac{p}{\gamma} + z = \text{constant}$$

which states that the mechanical energy of the fluid particle per unit weight is constant along the streamline. The kinetic energy of unit weight of the fluid having velocity v is $v^2/2g$; the energy of position is the work required to raise the fluid vertically from an arbitrary datum to its position, which is z for unit weight, z being measured vertically upward from the datum. The capacity of the fluid to do work by means of its *sustained* pressure intensity $pa\delta l$ for the volume $a\delta l$, obvious when a piston of area a is allowed to move in a cylinder a distance δl under pressure p. For unit weight the work is divided by the weight of the volume of fluid $\gamma a\delta l$, resulting in p/γ. Due to the lack of viscosity, no mechanism is available by which mechanical energy can be converted into heat. Hence, the sum of the three mechanical energy terms must remain constant along the streamline. The Bernoulli equation may be interpreted as applying to the total flow in a conduit, when the flow is essentially one-dimensional. The average velocity V for the cross section then replaces the velocity v on a particular streamline. Kinetic-energy correction factors are discussed in Article 9. In applying the equation to two sections, it is generally written

$$\frac{V_1^2}{2g} + \frac{p_1}{\gamma} + z_1 = \frac{V_2^2}{2g} + \frac{p_2}{\gamma} + z_2$$

When a real (viscous) fluid is considered and the flow is such that one layer moves relative to an adjacent layer, mechanical energy is converted into heat. The amount of the loss may be expressed by insertion of a loss term h_l on the downstream side of the equation,

$$\frac{V_1{}^2}{2g} + \frac{p_1}{\gamma} + z_1 = \frac{V_2{}^2}{2g} + \frac{p_2}{\gamma} + z_2 + h_l$$

where section 2 is downstream from section 1. Due to the complex mechanism of turbulent flow, it is generally necessary to determine h_l experimentally. Discussion of specific cases appear in the following articles.

Variation of Pressure Intensity across Streamlines. Neglecting hydrostatic pressure changes, the changes in pressure intensity along lines drawn orthogonal to the streamlines is given by

$$p = \rho \int \frac{v^2}{r}\,(dr) + C$$

where v is the velocity, and r is the radius of curvature of streamline at each point along the path. The constant of integration C may be found from known conditions at any point on the orthogonal line. In actual cases it may be necessary to perform the integration by numerical or graphical methods.

Linear Momentum. Newton's second law of motion may be applied to a fluid element and integrated in such a manner that the *linear momentum* equation is obtained,

$$\Sigma \mathbf{F} = Q\rho(\mathbf{V}_2 - \mathbf{V}_1)$$

This is a vector equation: $\Sigma \mathbf{F}$ is the resultant force acting on a free body of fluid; $Q\rho$ is the mass of fluid per unit time having its velocity changed from $\mathbf{V}_1$ to $\mathbf{V}_2$, the initial and final velocities, respectively. As a scalar equation for any direction, such as the x-direction, it becomes

$$\Sigma F_x = Q\rho(V_{x2} - V_{x1})$$

These equations are restricted to steady flow in the simple forms given above.

The momentum equation is useful in those situations where the initial and final velocities are known, since it yields the resultant force acting. As an example, the force exerted on a vane by an open jet is easily determined. In Fig. 49 the fluid is deflected through the

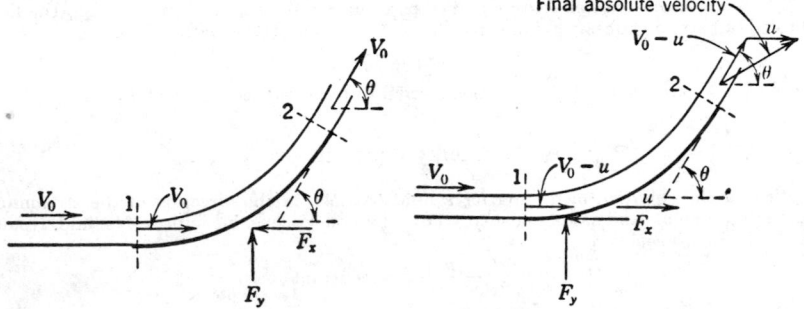

FIG. 49.   Flow over a fixed vane.   FIG. 50.   Deflection of jet by moving vane.

angle θ by the fixed vane. To find the force components F_x, F_y exerted on the fluid by the vane, the momentum equation is applied to the free body of fluid between sections 1 and 2, assuming the frictional resistance between fluid and vane is negligible. The only force in the x-direction is $-F_x$, hence

$$-F_x = Q\rho(V_0 \cos \theta - V_0)$$

or

$$F_x = Q\rho V_0(1 - \cos \theta)$$

In the y-direction

$$F_y = Q\rho V_0 \sin \theta$$

For a vane moving in the x-direction with velocity u, Fig. 50, the final velocity is shown as the sum of the final velocity relative to the vane and the velocity of the vane.

If a series of vanes are mounted on a wheel such that a vane always intercepts the jet approximately as shown in Fig. 50, the force components on the fluid free body are

$$F_x = Q\rho(V_0 - u)(1 - \cos \theta)$$

$$F_y = Q\rho(V_0 - u) \sin \theta$$

The force acting on the vanes *in the direction of motion* is equal and opposite to F_x, and yields the power given up to the wheel when multiplied by u:

$$\text{Power} = Q\rho(V_0 - u)u(1 - \cos\theta)$$

For maximum power $\theta = 180°$ and $u = V_0/2$. The impulse turbine is designed along these lines to convert kinetic energy in a fluid jet into mechanical energy of rotation of a shaft.

Moment of Momentum

The *moment of momentum*, or *angular momentum*, is useful in considering the flow relations in steady rotating channels, such as in a centrifugal pump or a reaction turbine. Since momentum is a vector quantity, its moment about an axis may be determined in a manner analogous to the moment of a force about the axis. The component of the momentum in the radial direction contributes nothing to the moment of momentum. The equation is

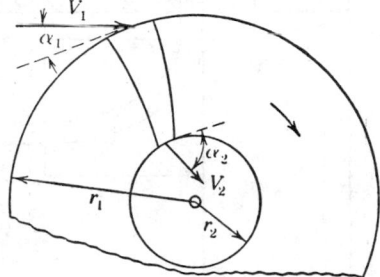

$$T = Q\rho(V_{t_2}r_2 - V_{t_1}r_1)$$

where T is the resultant torque on the fluid, $Q\rho$ is the mass per second having its angular momentum changed, V_t is the tangential component of velocity, subscript 2 refers to the final condition, and subscript 1 to the initial condition.

FIG. 51. Rotating channel, as in a turbine runner.*

Referring to Fig. 51, and expressing the tangential velocities in terms of the absolute velocities at entrance and exit,

$$T = Q\rho(V_2 r_2 \cos\alpha_2 - V_1 r_1 \cos\alpha_1)$$

Hence if the velocities and discharge are known, the torque may be determined. Multiplication of torque by angular velocity yields power obtainable from the fluid.

8. MEASUREMENT OF FLOW *

The measurement of fluid flow may be effected by many methods, the principal ones being gravimetric or volumetric devices, displacement meters, rate meters, and velocity measuring devices.

Displacement Meters. Displacement meters record the number of distinct and successive characteristic volumes of the fluid which pass through the device. They may consist of lobular rotating units, rotating cups, successive piston oscillations, rotations of a nutating disc, or the movements of a traveling screen or partition. The average rate of flow can be determined by recording the number of oscillations in a given line.

Venturi Meters. These meters are used to determine rate of flow in a closed conduit. In Fig. 52 flow is from left to right. The increase in velocity at the throat, section 2, causes a decrease in pressure intensity there. Writing Bernoulli's equation between sections 1 and 2 results in an equation relating discharge to drop in hydraulic grade line

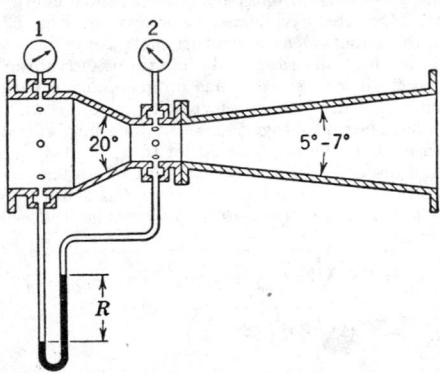

FIG. 52. Venturi meter.

$$Q = C_v A_2 \sqrt{2g \frac{(p_1 + z_1 - p_2 - z_2)}{1 - (A_2/A_1)^2}}$$

where Q is the discharge, A_2 is the area at section 2, z_1, z_2 are elevations, p_1, p_2 are pressure intensities, and C_v is an experimentally determined coefficient,† usually between 0.95 and

* Figures 51, 55, and 56, courtesy of Prof. R. C. Binder, Purdue University, Lafayette, Ind.
† For additional information on flow measurement consult *Fluid Meters and Their Application*, A.S.M.E.

0.99.   Using a differential manometer between sections 1 and 2, the equation becomes

$$Q = C_v A_2 \sqrt{\frac{2gR\left(\frac{s_0}{s_1} - 1\right)}{1 - (A_2/A)^2}}$$

where R is the gage difference (difference in elevation of menisci), and s_0, s_1, are the specific

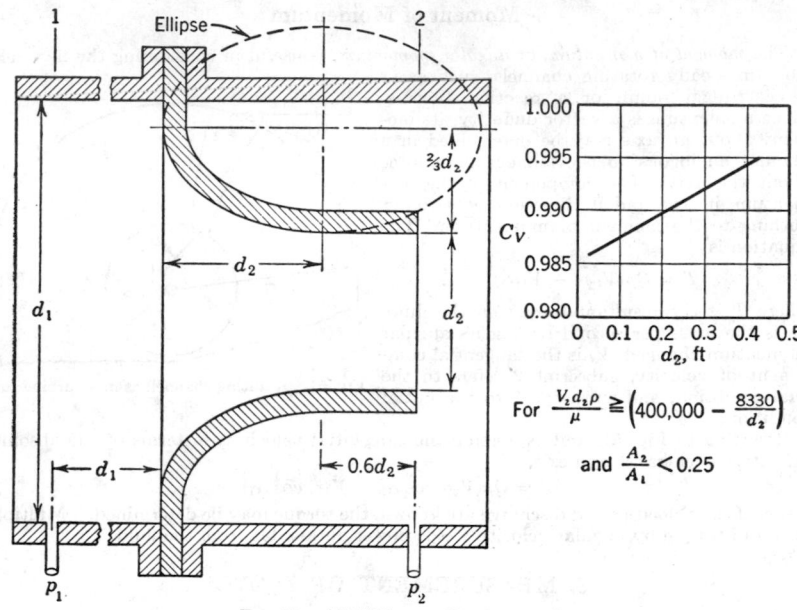

For $\frac{V_2 d_2 \rho}{\mu} \geqq \left(400,000 - \frac{8330}{d_2}\right)$

and $\frac{A_2}{A_1} < 0.25$

FIG. 53.   ASME long-radius flow nozzle.

gravities of gage fluid and flowing fluid, respectively. In this form it is evident that R is independent of orientation of the meter since z_1, z_2 do not appear in the expression. The gradually diverging section aids in converting the high kinetic energy at the throat back into pressure energy.

Nozzles. Orifices. A *nozzle*, as shown in Fig. 53, has the same equation as a venturi meter, and its coefficient is in the same range. It has the disadvantage of larger over-all loss, since it has no gradually diverging section to aid in recovery of velocity head.

The *orifice*, shown in Fig. 54, is usually sharp edged; it causes a contraction of the fluid jet to a value A_2 less than A, the area of orifice. The amount of contraction is expressed by a contraction coefficient C_c, defined by $C_c = A_2/A$. The equation may be arranged in the forms

$$Q = CA \sqrt{2g\left(\frac{p_1}{w} + z_1 - \frac{p_2}{w} - z_2\right)}$$

$$= CA \sqrt{2gR\left(\frac{s_0}{s_1} - 1\right)}$$

where

$$C = \frac{C_v C_c}{\sqrt{1 - C_c^2(A/A_1)^2}}$$

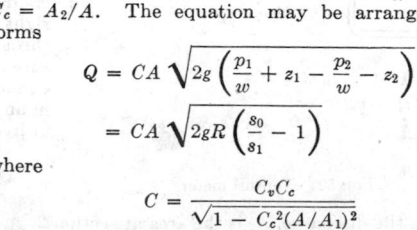

FIG. 54.   Sharp-edged orifice.

Values of C vary from about 0.60 to 1.0, depending upon A/A_1 and upon the location of the pressure taps.

Current Meters, Anemometers, Rotometers, and Hot-wire Anemometers. *Current meters* and *anemometers* are used to measure the velocity of flow in liquids and gases, respectively. They are of the cup type or vane (propeller) type; there is a relation between the number of revolutions per unit time and velocity, which must be determined by calibration.

The *rotometer* determines the flow through a gradually diverging transparent vertical tube by the position of a suspended body of a plumb-bob shape which has slots cut into it causing it to rotate. Larger discharges cause the suspended body to move upward, providing larger annular clearances for the flow. Graduations are marked on the tube to indicate the flow.

The *hot-wire anemometer*, most successfully used to measure gas velocities, measures the amount of cooling caused by the gas stream flowing over a short segment of fine platinum wire heated by an electric current. The electric circuits are of the constant-current or constant-voltage type, and they must be calibrated. The instrument has been valuable for taking turbulence measurements in air streams. Section 8 contains additional information on hot-wire anemometers.

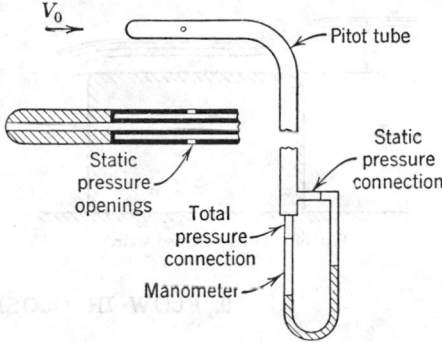

FIG. 55. Combined pitot-static tube.

Pitot Tubes. A *pitot tube* is a velocity measuring device which provides a means of measuring the *dynamic* head $v^2/2g$ of a fluid stream. It has an opening directed upstream, Fig. 55, which measures the dynamic pressure plus the *static* pressure in the fluid, while the static (or undisturbed) pressure alone is measured by openings in the walls of the cylindrical tube. Bringing these two pressures to opposite legs of a differential manometer provides a means of reading the difference, which is the dynamic pressure. The equation for velocity v becomes

$$v = C \sqrt{2gR \left(\frac{s_0}{s_1} - 1 \right)}$$

where R is the gage difference; s_0, s are the specific gravities of the manometer fluid and flowing fluid; and C is an experimentally determined coefficient, usually near unity.

In place of a slender tube directed upstream, a pitot tube can be made from a cylinder

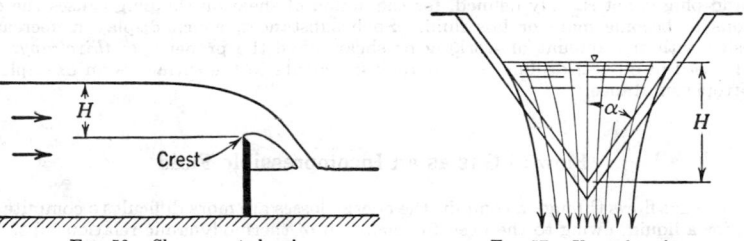

FIG. 56. Sharp-crested weir. FIG. 57. V-notch weir.

placed normal to the flow, or it can be made from a sphere, by proper location of the openings.

Weirs are used for determining flow rate in open channels or rivers. A weir is an obstruction placed in the channel to cause the liquid to rise upstream from it and flow over or through it. By suitably measuring the height of liquid surface above the weir crest H, the flow may be expressed as a function of H. A *sharp-crested* weir is shown in Fig. 56. Theory indicates the discharge from such a weir varies as $H^{3/2}$. The equation may be written in the form

$$Q = CLH^{3/2}$$

where L is the length of weir. The coefficient C is determined by calibration, unless the exact geometry of previous tests is used, in which case these coefficients also apply.

For small discharges a *V-notch* weir is frequently employed; it consists of a triangular notch cut in a flat plate, as shown in Fig. 57. For this type weir the discharge varies approximately as $H^{5\!/_2}$. For a 90° notch the formula has been found to be $Q = 2.48$, $H^{2.48}$ for polished brass plates, and $Q = 2.52H^{2.47}$ for rough steel plate.

Other common weir shapes are the *broad-crested* weir, Fig. 58, and the *Ogee* weir, Fig. 59.*

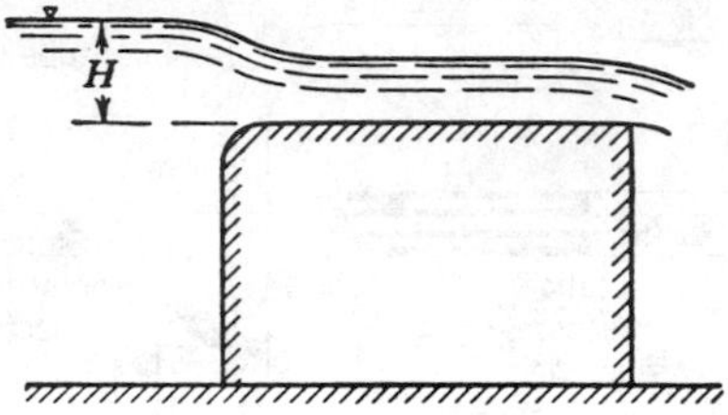

FIG. 58.   Broad-crested weir.

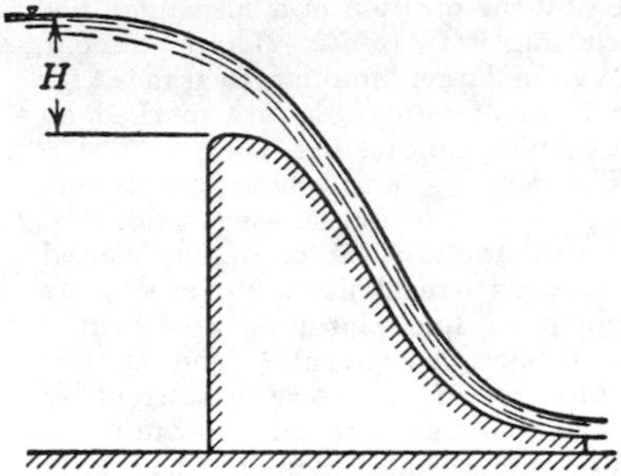

FIG. 59.   Ogee weir.

9. FLOW IN CLOSED CONDUITS

Flow through a conduit may be steady or unsteady, uniform or non-uniform, and laminar or turbulent.   Steady flow refers to flow at constant rate, uniform flow to prismatic sections of conduit, and laminar flow to those cases where viscous forces predominate and the losses are a linear function of the velocity; whereas for turbulent flow the losses vary as the velocity to some power (1.7 to 2.0), depending in part upon Reynolds' number.

General Considerations

Many substances pumped through conduits are not Newtonian fluids.   These substances vary from fats in solid form, heavy oils, and paper stock to sand-water mixtures.   Works on rheology treat flow of plastics and semi-solids.   Plug flow is a special case of plastic flow in which the central core of substance moves through the conduit as a solid mass, lubricated by a fluid film at the conduit walls; in general, the line of separation between film and plug is not sharply defined, for the action of shear on the plug causes the outer portions to become more or less fluid.   Such substances, which display a decrease in viscosity with the amount of working or shear, have the property of *thixotropy*.   The "mud" used in oil-well drilling to float rock fragments to the surface is an example of a thixotropic substance.

Flow of Gas as an Incompressible Fluid

When a gas flows through a conduit, the energy losses are more difficult to compute than those for a liquid, owing to the need for inclusion of thermodynamic relationships.   The error resulting from the assumption of constant density depends primarily upon the actual variation in density which takes place.   Equations for error are presented for two cases: isothermal flow and adiabatic flow in a horizontal pipe.   Error for adiabatic flow is not strictly due to flow in an insulated pipe, because the heat generated by friction is neglected, and the resistance coefficient is assumed to be constant.   The error e in per cent is defined as

$$e = 100 \frac{\Delta p_c - \Delta p_i}{\Delta p_c}$$

where Δp_c is the pressure drop obtained from the compressible-flow formula for given flow conditions, and Δp_i is the pressure drop for incompressible flow through the same length of pipe at a density equal to that at the upstream section in the compressible case.

Isothermal Flow.   For steady flow the Reynolds' number is constant; hence, for a

* For more complete information on weirs, consult King's *Handbook of Hydraulics*, 4th Edition, New York, McGraw-Hill Book Co., 1954.

given pipe, the friction factor is also constant along the pipe. The percentage of error is

$$e = 100 \left[1 - \frac{1 + \dfrac{p_2}{p_1}}{2 \left(1 + \dfrac{2D}{fL} \ln \dfrac{p_1}{p_2} \right)} \right]$$

where p_1, p_2 are the absolute pressures at upstream and downstream sections of the pipe, a distance L apart; D is the inside diameter of the pipe; and f is the Darcy-Weisbach fric-

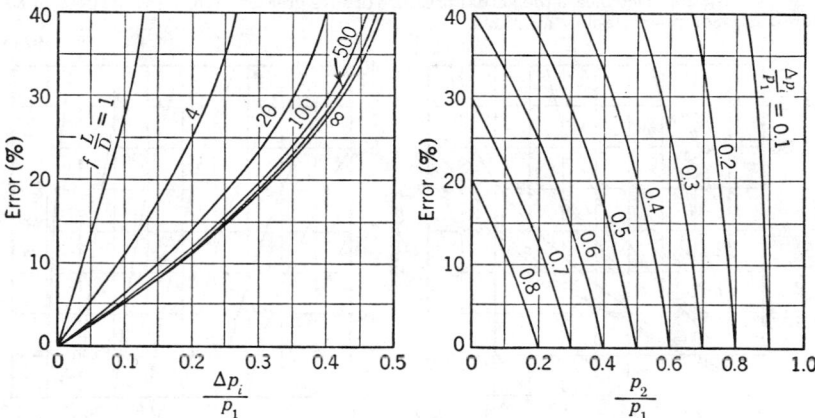

FIG. 60.   Pressure-error chart for isothermal flow of a compressible fluid.

tion factor, as given in Fig. 68. The formulas for solution of the isothermal flow problems are presented in graphical form in Fig. 60.

Example. Air flows under isothermal conditions at the rate of 0.5 lb/sec through 2000 ft of 4-in. ID clean brass pipe. The upstream pressure is 20 psi abs and the temperature is 70 F. Find the pressure drop in psi.

Solving first as an incompressible flow problem to find Δp_i, using the methods of Fig. 68, for upstream conditions,

$$0.5 \text{ lb/sec} = VA\gamma, \quad A = \frac{\pi}{36} \text{ sq ft}, \quad \gamma = \frac{p}{RT} = \frac{20 \times 144}{53.3 \times 529.4} = 0.102 \text{ lb/ft}^3$$

$$V = \frac{0.5 \times 36}{0.102\pi} = 56.2 \text{ ft/sec}, \quad \mu = 3.9 \times 10^{-7} \text{ lb/sec/ft}^2$$

$$\mathbf{R} = \frac{VD\rho}{\mu} = 56.2 \times \frac{1}{3} \times \frac{0.102}{32.2 \times 3.9 \times 10^{-7}} = 151{,}800, \quad f = 0.016$$

$$\Delta p_i = f\rho \frac{L}{D} \frac{V^2}{2} = 0.016 \times \frac{0.102}{32.2} \times \frac{2000}{1/3} \times \frac{\overline{56.2}^2}{2} = 600 \text{ lb/ft}^2$$

using Fig. 60:

$$\frac{fL}{D} = \frac{0.016 \times 2000}{1/3} = 96, \quad \frac{\Delta p_i}{p_1} = \frac{600}{20 \times 144} = 0.208$$

the error in per cent is found in the left-hand chart $e = 12.4\%$. From the right-hand chart $p_2/p_1 = 0.76$, $p_2 = 15.2$ psi abs, and the pressure drop is 4.8 psi. The pressure drop is always greater than for incompressible flow.

Adiabatic Flow. Using the adiabatic relationship

$$\frac{p}{\rho^k} = \text{constant}$$

and assuming the friction factor constant, as determined from upstream conditions, the percentage of error is

$$e = 100 \left[1 - \frac{k}{k+1} \frac{1 - \left(\dfrac{p_2}{p_1} \right)^{\frac{k+1}{k}}}{\left(1 - \dfrac{p_2}{p_1} \right) \left(1 + \dfrac{2D}{kfL} \ln \dfrac{p_1}{p_2} \right)} \right]$$

k is the ratio of specific heat at constant pressure to specific heat at constant volume. Figure 61 is a chart for determining error in using incompressible flow formulas, for $k = 1.40$ (air); in Fig. 62 $k = 1.32$ (steam). Solving for Δp_c in terms of error

$$\Delta p_c = \frac{\Delta p_i}{1 - \dfrac{e}{100}}$$

The right-hand charts in Figs. 61 and 62 may also be used to determine Δp_c, since $p_c = p_1 - p_2$. Again, in each case the actual pressure drop is greater than the value determined from the incompressible flow formulas.

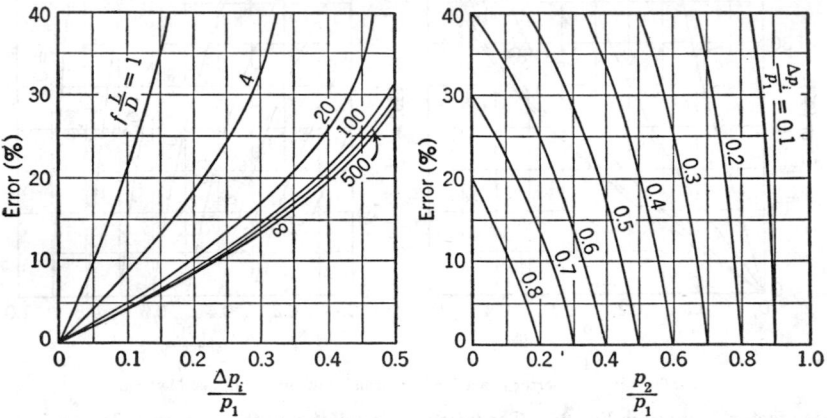

FIG. 61.   Pressure-error chart for adiabatic flow of air ($k = 1.40$).

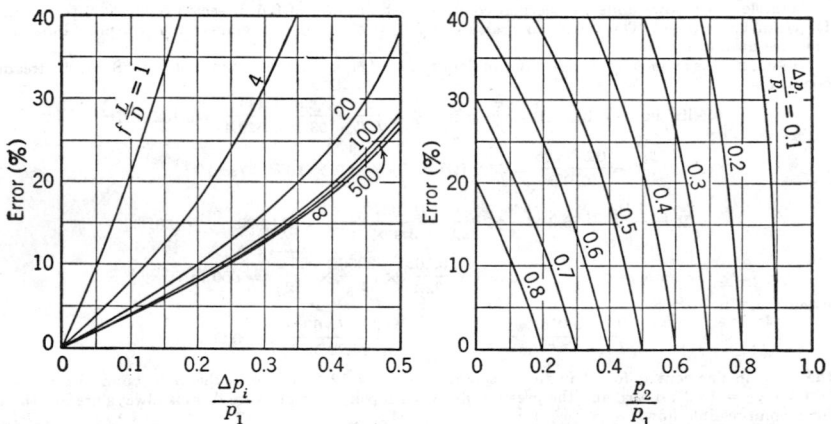

FIG. 62.   Pressure-error chart for adiabatic flow of steam ($k = 1.32$).

General Methods of Solution

The classical methods of hydrodynamics applying to an ideal fluid are of little value in solving flow problems in conduits, although they are extremely useful in connection with flow around immersed bodies. The nature of turbulent flow, on the other hand, is not sufficiently well understood to permit computation of the energy losses for given boundary conditions and rates of flow, and hence recourse must generally be taken to experimentation.

Much has been accomplished in the last three decades in the application of similitude and dimensionless presentation; the energy-loss terms, or their coefficients, have been found by experiment to be dependent upon the conduit geometry, the Reynolds' number, and the surface roughness. The energy equation (Bernoulli's equation) is of the first importance in solving flow problems. Momentum relationships are of use in certain cases in which the forces acting on the fluid are known, or are desired. In special situations where both the energy and momentum equations are applicable, the energy loss may be computed without recourse to experimentation. The continuity equation in steady flow usually states that the flow past every cross section is the same. When the fluid is compressible this is a statement of mass or weight flow, but for liquids it is sufficient to deal with volume rates only. The three types of equations (energy, momentum, and continuity), together with the experimentally determined loss relationships, provide the general framework for solving closed-conduit problems.

Velocity Distribution

The velocity distribution for established laminar flow through round tubes and between parallel plates has been discussed under laminar fluid motion, Article 6.

Establishment of Flow. The velocity distribution for fully established uniform flow in a closed conduit is determined by the relationship between the radial velocity gradient and the shear stress. In turbulent flow the velocity distribution cannot be derived exactly, although much has been accomplished in recent years in the analytical approach to rational velocity-distribution equations.

Downstream from any change in cross section or direction, there is a length over which the velocity distribution regains its characteristic form, depending upon the shape of cross section, the wall roughness, and the Reynolds' number. For example, when the flow passes from a reservoir through a rounded entrance into a conduit, the velocity is practically constant over the section at the upstream end of the conduit. Such flow during the initial stages is therefore practically irrotational, since the boundary layer is very thin. The effect of boundary resistance, however, is to retard the fluid in the wall vicinity, resulting through lateral transmission of shear, in a continuous growth of the boundary layer with distance from the inlet. Since the mean velocity must nevertheless remain constant, the central portion of the fluid is simultaneously accelerated, until the forces of shear and pressure gradient reach equilibrium as the velocity distribution of uniform flow becomes fully established some distance downstream.

For laminar flow experiments by Nikuradse give the nominal length L for establishment of flow as

$$\frac{L}{D} = 0.06R$$

where D is the pipe diameter and Reynolds' number is based on average velocity and diameter.

In turbulent flow, the transition to the established velocity distribution is effected in a much shorter reach, because of the pronounced mixing action which then prevails. Nikuradse's experiments indicate that a distance of 25 to 40 diameters is sufficient, and that the length is not so dependent upon Reynolds' number.

Rational Formulas for Established Turbulent Flow. In turbulent flow, the ratio of shear stress to velocity gradient depends not only upon physical properties of the fluid, but also upon characteristics of the flow.

Stanton first stated that the turbulent velocity distribution in the central portions of a conduit has a form which is independent of the wall roughness and viscous effects, provided that the wall shear remains the same. In equation form

$$\frac{v_{max} - v}{\sqrt{\tau_0/\rho}} = F\left(\frac{r}{r_0}\right)$$

where v_{max} is the velocity at the pipe axis, v is the velocity at the distance r from the axis, r_0 is the pipe radius, τ_0 is the wall shear, ρ is the mass density of fluid, and F is an unknown function. The proof is evident by an inspection of the Nikuradse data on smooth- and sand-roughened pipes, given in Fig. 63; k is the diameter of sand grains cemented to the pipe walls.

Based on the above, von Kármán [*] obtained the formula for smooth pipes,

* Kármán, Th. von, "Turbulence and Skin Friction," *J. Aeronautical Sci.*, Vol. 1, No. 1, p. 1, 1934.

$$\frac{v}{\sqrt{\tau_0/\rho}} = C_1 + \frac{1}{\kappa} \ln \left(\sqrt{\frac{\tau_0}{\rho}} \frac{y}{\nu} \right)$$

where κ is a universal constant, having the value 0.40, and ν is the kinematic viscosity. Figure 64, based on Nikuradse's tests, shows the value of C_1 to be 5.5 for best agreement with the data.   In the immediate vicinity of the pipe wall, through a film called the

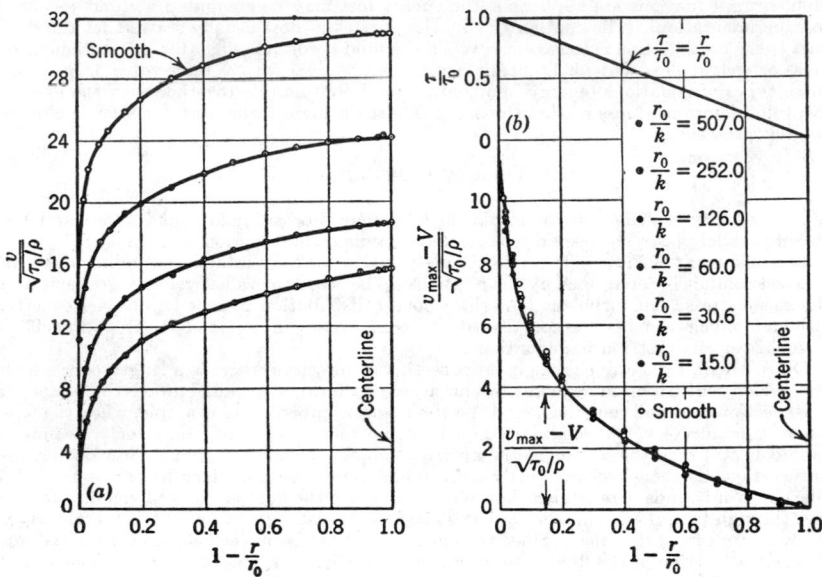

FIG. 63.   Generalized plot of velocity distribution for smooth and rough pipes.

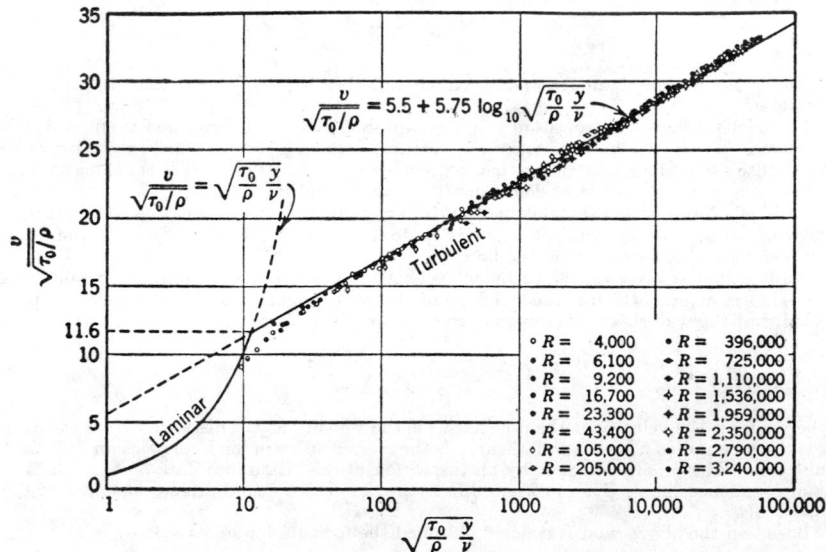

FIG. 64.   Universal velocity distribution for smooth pipes.

laminar sublayer, the velocity is given closely by $v = y\tau_0/\mu$. This may be written

$$\frac{v}{\sqrt{\tau_0/\rho}} = \sqrt{\frac{\tau_0}{\rho}\frac{y}{\nu}}$$

plotted in Fig. 64. The intersection of the two curves may be taken arbitrarily as the border between the two types of flow, although actually there is a transition from the laminar to the turbulent zone. From the figure, the laminar film has the thickness

$$\delta = \frac{11.6\nu}{\sqrt{\tau_0/\rho}}$$

For rough pipes von Kármán obtained the formula

$$\frac{v}{\sqrt{\tau_0/\rho}} = C_2 + \frac{1}{\kappa}\ln\left(\frac{y}{\kappa}\right)$$

Figure 65 shows the Nikuradse sand-roughened pipe tests. From these data, $C_2 = 8.5$. The two logarithmic equations do not give a zero slope of the velocity distribution curve at the center line. This is a defect in the formulas which, nevertheless, has little signifi-

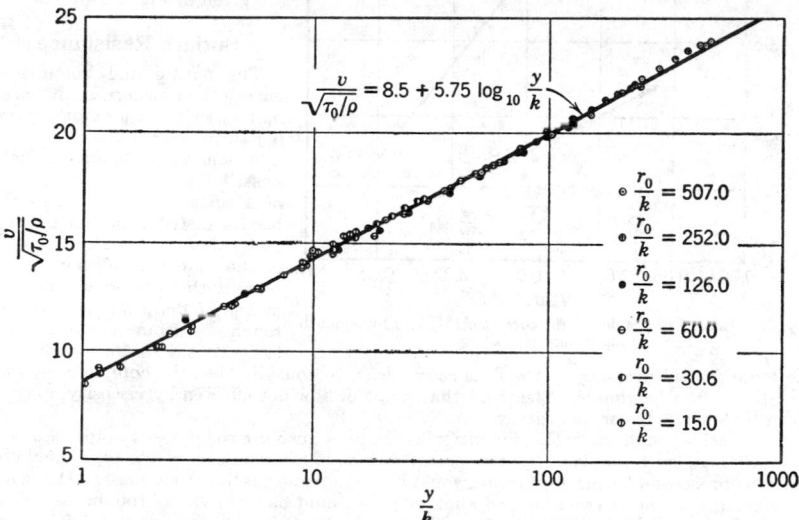

FIG. 65. Universal velocity distribution for rough pipes.

cance from a practical viewpoint. The equations actually portray the true velocity distribution in the central region of the flow very well, although they were derived for the region near the wall.

Energy and Momentum Correction Factors. In writing the Bernoulli equation between two cross sections of a conduit, it is usually satisfactory to express the mean kinetic energy per unit weight simply as $V^2/2g$, where V is the average velocity at a section. This is strictly true, however, only when the velocity is constant over both cross sections. For laminar flow, the correction α by which $V^2/2g$ must be multiplied to give the true mean value is 2.0, and for sections where there is back flow the factor may be even larger.

The mean kinetic energy per unit weight is best obtained by determining the kinetic energy passing a cross section in unit time and dividing by the weight of fluid passing the section in the same time. In equation form

$$\frac{1}{\gamma V A}\int_A \frac{\rho v^2}{2}v\,dA = \alpha\frac{V^2}{2g}$$

or

$$\alpha = \frac{1}{A}\int_A \left(\frac{v}{V}\right)^3 dA$$

The quantity α is called the *kinetic energy correction factor* and is always equal to or greater than unity.

Similarly, in applying the momentum equation to conduit flow, the true momentum per second passing a given section is not given by the mass passing per second times the average velocity, but by the integral

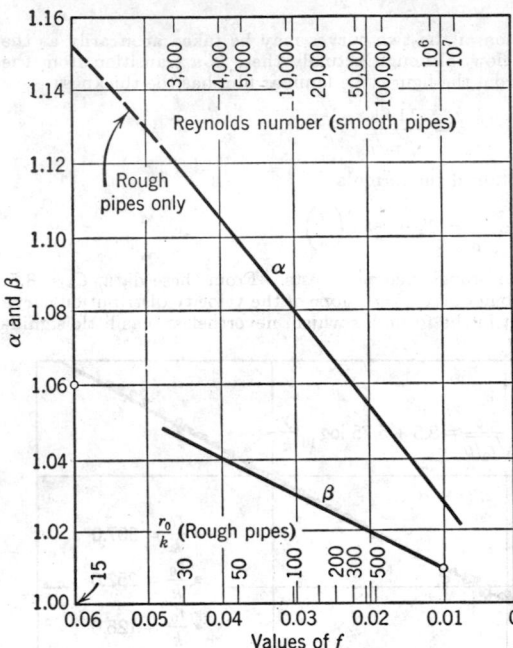

$$\beta Q \rho V = \rho \int_A v^2 dA$$

in which

$$\beta = \frac{1}{A} \int_A \left(\frac{v}{V}\right)^2 dA$$

The quantity β is called the *momentum correction factor* and is equal to or greater than unity. Values of α and β based on the Prandtl-Kármán turbulent velocity distribution equations are given in Fig. 66.

FIG. 66. Energy and momentum correction factors for smooth and rough pipes.

Surface Resistance

The nature and amount of energy loss occurring in prismatic sections are discussed under surface resistance.

Absolute and Relative Roughness. The absolute roughness of a surface depends upon the height, distribution, angularity, and other geometrical aspects of the roughness elements. At present there is no satisfactory means of determining absolute roughness from a sample section of the surface. It has therefore proved necessary, as well as convenient, to consider that the complete description of absolute roughness, at least for that range of flow not affected by viscosity, may be given by a single linear quantity k.

The relative roughness is the dimensionless ratio of absolute roughness to pipe radius or diameter. As an example of the two classifications of roughness, concrete pipes made by the same process and using steel forms would have the same absolute roughness. However, one having a diameter twice that of another pipe would have a relative roughness half as great. Irregularities in the bituminous lining of a 10-foot pipe, consisting of folds and wrinkles 1/4-inch deep, would give it a relative roughness equal to that of similarly shaped roughness only 0.004-inch deep in a 2-inch pipe.

The concept of relative roughness as the surface factor which influences the energy loss has been confirmed for sand-coated pipes by Nikuradse. He utilized three sizes of pipes and cemented to their inner surfaces sand grains of such average diameters k that the same relative roughness was obtained for the three pipes. He then found that for a given Reynolds' number the resistance coefficient was the same for the three pipes. This was repeated so that five relative roughnesses in all were used, covering the range from fairly smooth to very rough.

Formulas for Energy Loss. Rational expressions for the resistance coefficient f in the Darcy-Weisbach equation

$$h_f = f \frac{L}{D} \frac{V^2}{2g}$$

have been derived from the universal velocity distribution equations by Prandtl and von Kármán. h_f is the energy loss per unit weight due to surface resistances in the length L, D is the inside diameter of the pipe, and V is the average velocity. The coefficient f has been found to be a function of the Reynolds' number and the relative roughness.

For smooth pipes, using the constants which best conform to the Nikuradse data,

$$\frac{1}{\sqrt{f}} = 2 \log_{10} \mathbf{R}\sqrt{f} - 0.8$$

For rough pipes, again using constants which agree best with the sand-roughened pipe data

$$\frac{1}{\sqrt{f}} = 2 \log_{10} \frac{r_0}{k} + 1.74$$

This latter equation does not contain Reynolds' number and is valid only for fully developed boundary turbulence.

In the transition region for rough pipes, Colebrook and White * obtained the semi-empirical equation

$$\frac{1}{\sqrt{f}} = 1.74 - 2 \log_{10} \left(\frac{k}{r_0} + \frac{18.7}{R\sqrt{f}} \right)$$

in which k represents the effective roughness of the surface in question; that is, that value which yields the same magnitude of f as the Nikuradse sand roughness for the limiting

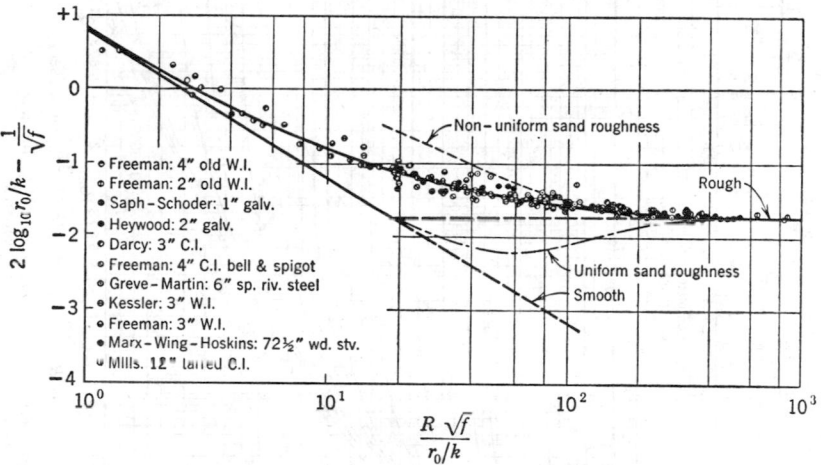

FIG. 67. Transition function for commercial pipe materials.

conditions of the rough-pipe formula. The transition function evidently reduces to the smooth-pipe equation for very small values of k/r_0. Figure 67 shows the agreement of this transition equation with tests on commercial pipes.

When the thickness of the laminar film is large compared with the height of the roughness projections, the loss is the same as if the pipe were smooth; conversely, when the height of the roughness projections is large compared with the thickness of the laminar film, the resulting boundary flow is that of fully developed turbulence.

Resistance Diagram. Based on the Colebrook and White equation, a Stanton diagram has been prepared by L. F. Moody,† Fig. 68, which shows the relation between friction-factor and Reynolds' number with the relative roughness as a parameter.

Suggested values of absolute roughness k for clean materials are listed in the lower left-hand corner of the diagram. Since k varies so widely, it is not necessary to determine it to more than one significant figure. The average value listed for k should be taken, unless there is some reason to believe that the particular pipe is very smooth or very rough for its class. Procedures for old pipes are discussed in the latter part of this article.

Three types of problems are encountered in the computation of simple pipe flow; they may be outlined as follows:

	Given	Required
(a)	D, L, ρ, μ, k and Q or V	h_f
(b)	h_f, D, L, ρ, μ, k	Q or V
(c)	h_f, Q, L, ρ, μ, k	D

* Colebrook, C. F., "Turbulent Flow in Pipes, with Particular Reference to the Transition Region between the Smooth and Rough Pipe Laws," *J. Instit. Civil Engineers*, Vol. 11, pp. 133-156, London, 1938-39.

† Moody, L. F., "Friction Factor for Pipe Flow," *Trans. ASME*, November, 1944.

For the first type, where the head loss is required, the Reynolds' number may be computed at once; with k and D known, k/D determines the parametric curve, and f is read from Fig. 68. Direct substitution into the Darcy-Weisbach equation yields the head loss h_f in energy per unit weight.

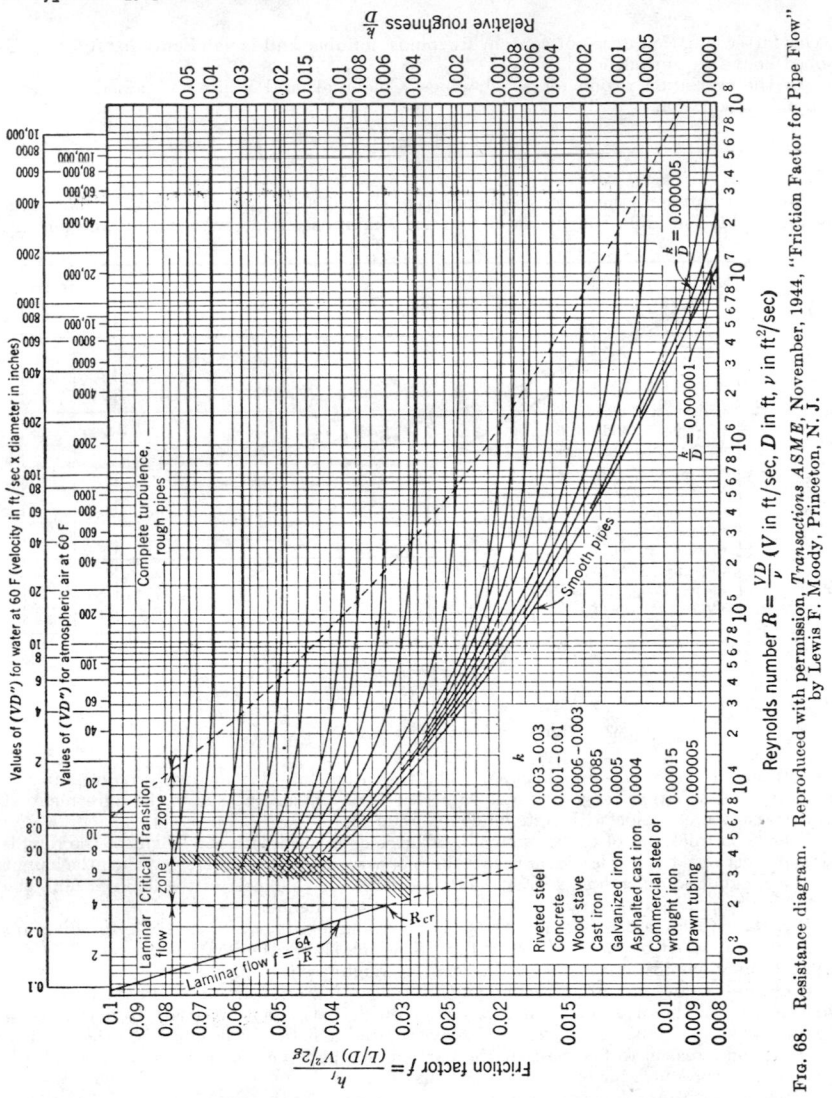

FIG. 68.   Resistance diagram.   Reproduced with permission, *Transactions ASME*, November, 1944, "Friction Factor for Pipe Flow" by Lewis F. Moody, Princeton, N. J.

For the second type, substitution into the Darcy-Weisbach equation and into the Reynolds'-number equation yields

$$h_f = f \frac{L}{D} \frac{V^2}{2g}, \quad \text{or} \quad fV^2 = C_1; \quad \frac{VD\rho}{\mu} = R, \quad \text{or} \quad R = C_2 V$$

where C_1, C_2 are known, and k/D is also known.   A trial procedure is now used: assume an f, compute V from $fV^2 = C_1$; substitute into $R_1 = C_2 V$ to find a trial R_1, then look up a new f for this R_1 and k/D.   If it is different from the assumed f, use the new f and repeat the process. This procedure rapidly converges, and usually the second f is sufficiently close.

For the third type, where the diameter is required, a trial-and-error solution is also needed. Through continuity, V and D are related

$$V = \frac{4Q}{\pi D^2}$$

when substituted into Darcy-Weisbach and Reynolds' number

$$\frac{D^5}{f} = \frac{8LQ^2}{h_f \pi^2 g} = C_3, \quad RD = \frac{4Q}{\pi \nu} = C_4$$

where C_3 and C_4 are known for the specific problem. Assume a value of f; solve for D, R, and k/D; then look up f in Fig. 68. Using this f, a new D can be obtained, etc. The procedure converges very rapidly. Agreement of the last two values of f yields the correct value for D.

Example 1. What is the pressure drop in psi per 1000 ft for flow of 3000 gpm.of crude oil (specific gravity 0.855) at 60 F through a steel pipe line of 18-in. diameter?

$$V = \frac{3000}{7.48 \times 60\pi \dfrac{2.25}{4}} = 3.78 \text{ ft/sec}$$

From Fig. 6,

$$\nu = 9.1 \times 10^{-5} \text{ ft}^2/\text{sec}; \quad R = \frac{VD}{\nu} = \frac{3.78 \times 1.5}{9.1 \times 10^{-5}} = 62{,}500$$

$$\frac{K}{D} = \frac{0.00015}{1.5} = 0.0001$$

From Fig. 68, $f = 0.020$. The head loss is

$$h_f = f\frac{L}{D}\frac{V^2}{2g} = 0.020 \times \frac{1000}{1.5} \times \frac{\overline{3.78}^2}{64.4} = 2.96 \text{ ft}$$

Hence the pressure drop is

$$2.96 \times 0.855 \times \frac{62.4}{144} = 1.09 \text{ psi}$$

Example 2. The rate of flow of water at 100 F through 4-in. cast iron pipe is to be determined when the head loss is 0.50 psi per 100 ft of pipe.

$$\frac{h_f}{L} = \frac{0.50 \times 144}{62.4 \times 100} = 0.0115$$

From Fig. 6, $\nu = 7.7 \times 10^{-6}$, hence

$$\frac{K}{D} = \frac{0.00085}{1/3} = 0.0026, \quad fV^2 = \frac{h_f}{L} 2gD = 0.0115 \times 64.4 \times \frac{1}{3} = 0.2475$$

$$R = \frac{VD}{\nu} = \frac{V}{3 \times 7.7 \times 10^{-6}} = 43{,}400V$$

Assume $f = 0.020$, then

$$V = \sqrt{\frac{0.2475}{0.02}} = 3.52 \text{ ft/sec}, \quad R = 43{,}400 \times 3.52 = 153{,}500$$

Looking up f for the R and k/D, $f = 0.026$. Repeating

$$V = \sqrt{\frac{0.2475}{0.026}} = 3.08, \quad R = 43{,}400 \times 3.08 = 133{,}800, \quad f = 0.026$$

Hence, the rate of flow is

$$3.08 \times \frac{\pi}{36} \times 60 \times 7.48 = 120 \text{ gpm}$$

Example 3. It is desired to find the size of commercial steel pipe which will carry 100 gpm of kerosene at 50 F a distance of 2000 ft with a head loss of 10 ft.

$$Q = \frac{100}{7.48 \times 60} = 0.223 \text{ cfs}, \quad \nu = 3.2 \times 10^{-5} \text{ ft}^2/\text{sec}, \quad k = 0.00015 \text{ ft}$$

Then

$$\frac{D^5}{f} = \frac{8LQ^2}{h_f \pi^2 g} = \frac{8 \times 2000 \times \overline{0.223}^2}{10\pi^2 32.2} = 0.25, \quad RD = \frac{4Q}{\pi\nu} = \frac{4 \times 0.223}{\pi 3.2 \times 10^{-5}} = 8860$$

Assume $f = 0.020$, then $D = 0.346$, $R = 25,600$, $k/D = 0.000433$; from Fig. 68, $f = 0.026$. Repeating, using $f = 0.026$, $D = 0.366$, $R = 24,200$, $k/D = 0.00041$, $f = 0.026$, hence

$$D = 0.366 \times 12 = 4.4 \text{ in.}$$

Use a 5-in. pipe.

Effects of Aging. The values of k given for the various pipe materials of Fig. 68 are for new, clean pipes. In general, pipes become increasingly rough with age, owing to deposition or corrosion. Colebrook and White have determined an approximately linear increase in absolute roughness with time, which may be expressed as

$$k = k_0 + \alpha t$$

where k_0 is the absolute roughness of the new material, α is a constant, and k is the absolute roughness at time t.

Example 4. After 10 years of service, a 10-in. cast-iron pipe line in water service has a drop of 3.13 psi per 1000 ft for a flow of 1000 gpm. What is the estimated pressure drop for 1200 gpm after 20 years of service?

$$V = \frac{1000}{7.48 \times 60 \frac{\pi}{4}\left(\frac{5}{6}\right)^2} = 4.1 \text{ ft/sec}, \quad \frac{V^2}{2g} = 0.262 \text{ ft}$$

$$h_f = \frac{3.13 \times 144}{62.4} = f \frac{1000}{5/6} 0.262, \quad f = 0.023$$

$$R = \frac{4.1 \times 5}{6 \times 1.2 \times 10^{-5}} = 285,000 \text{ (taking } \nu = 1.2 \times 10^{-5} \text{ ft}^2/\text{sec)}$$

From Fig. 68 for the above values of f and R, $k/D = 0.0017$ and $k = 0.00142$ ft. For new cast iron take $k_0 = 0.001$ ft, hence for 10 years.

$$0.00142 = 0.001 + 10\alpha$$

$$\alpha = 0.000042 \text{ ft/yr}$$

and for 20 years

$$k = 0.001 + 20 \times 0.000042 = 0.00184 \text{ ft}$$

$$\frac{k}{D} = 0.0022, \quad V = \frac{1200}{1000} \times 4.1 = 4.92, \quad R = \frac{4.92}{4.1} \times 285,000 = 342,000$$

From Fig. 68, $f = 0.025$; then

$$h_f = 0.025 \times \frac{1000}{5/6} \frac{\overline{4.92}^2}{64.4} = 11.3 \text{ ft}$$

$$\Delta p = \frac{11.3 \times 62.4}{144} = 4.9 \text{ psi}$$

Conduits of Non-circular Cross Section. The Darcy-Weisbach equation may also be applied to non-circular conduits if the diameter D is replaced by some equivalent linear measure of the cross section. The hydraulic radius R, widely used in open channel equations, can be related to D for the circular cross section; this relationship is usually assumed to be a valid replacement of D in the pipe formula. The hydraulic radius is defined as the ratio of the cross-sectional area to the wetted perimeter. For a circular cross section

$$R = \frac{\pi D^2/4}{\pi D} = \frac{D}{4}$$

Hence the diameter may be replaced by four times the hydraulic radius in the Reynolds' number, the relative roughness, and the resistance equation; the resistance equation becomes

$$h_f = f \frac{L}{4R} \frac{V^2}{2g}$$

Although satisfactory for conduits which are reasonably comparable to pipes in cross-sectional form, this equation cannot be expected to give accurate results for cross sections which are at great variance therefrom.

Example. Find the head loss per 1000 ft for a flow of 200 gpm of water at 150 F through a clean cast-iron conduit of rectangular cross section 3 in. by 6 in.

From Fig. 6, $\nu = 4.8 \times 10^{-6}$ ft^2/sec

$$R = \frac{3 \times 6}{12 + 6} = 1 \text{ in.} = \frac{1}{12} \text{ ft}, \quad Q = \frac{200}{7.48 \times 60} = 0.446 \text{ cfs}$$

$$V = \frac{0.446}{18} \times 144 = 3.57 \text{ ft/sec}, \quad R = \frac{V4R}{\nu} = 248{,}000$$

$$\frac{k}{D} = \frac{k}{4R} = \frac{0.00085}{4/12} = 0.00255$$

From Fig. 68, $f = 0.025$; hence

$$h_f = f \frac{L}{4R} \frac{V^2}{2g} = \frac{0.025 \times 1000}{4 \times 1/12} \frac{\overline{3.57}^2}{64.4} = 14.88 \text{ ft}$$

Form Resistance

In the preceding paragraphs, methods were presented for solving simple pipe problems in which only surface resistance was considered. In addition to surface effects, there are also losses due to changes in cross section and in flow direction, which depend upon the geometrical form of the conduit. Any such change in cross section or flow direction disturbs the normal velocity distribution in the conduit. Owing to these disturbances, additional mechanical energy is converted into heat through the action of turbulence.

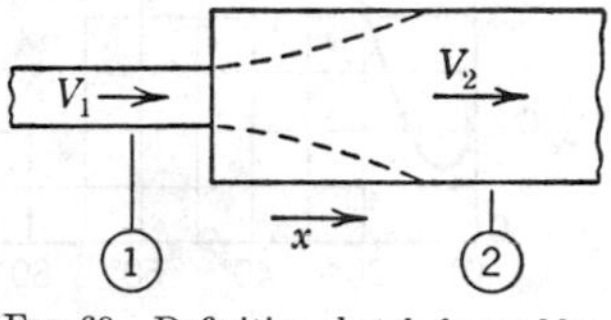

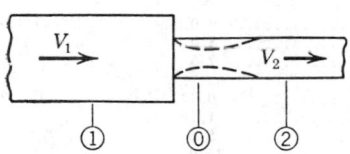

FIG. 69. Definition sketch for sudden expansions.

Sudden Changes in Section. A sudden expansion, as in Fig. 69, results in a loss in mechanical energy due to the inefficient conversion of velocity head into pressure head as the jet expands to fill the larger section. The momentum equation may be applied to the free body of liquid between sections 1 and 2. This equation together with Bernoulli's equation may be solved for the loss in energy per unit weight,

$$h_l = \frac{(V_1 - V_2)^2}{2g} = \frac{V_1^2}{2g}\left[1 - \left(\frac{A_1}{A_2}\right)\right]^2 = K \frac{V_1^2}{2g}$$

in which K, the loss coefficient for sudden expansion, depends only upon the area ratio. This expression agrees well with the loss as determined from experiment.

Values of K as a function of area ratio are presented in Table 6.

Table 6. Loss Coefficients for Sudden Expansions

A_1/A_2	0	0.1	0.2	0.3	0.4	0.5	0.6	0.7	0.8	0.9	1.0
K	1.00	.81	.64	.49	.36	.25	.16	.09	.04	0.1	0

For the case of infinite expansion, say from a pipe into a reservoir, A_2 is infinite and the loss is equal to the velocity head in the pipe, i.e., $K = 1$.

FIG. 70. Definition sketch for sudden contractions.

Sudden contractions, illustrated by Fig. 70, display characteristics similar to sudden expansions, in that the flow first contracts and then expands to fill the smaller pipe. Since the loss due to accelerated flow is very small compared with that in decelerated flow, the loss between sections 1 and 0 may be disregarded, and that between sections 0 and 2 taken as the total loss due to the change in section. Information is needed as to the amount of contraction which occurs at 0 to apply the loss term for a sudden expansion. The only experimental information is on the contraction of free jets, determined by Weisbach. The loss equation is

$$h_l = \frac{V_2^2}{2g}\left(\frac{1}{C_c} - 1\right)^2 = K \frac{V_2^2}{2g}$$

Table 7 lists values of K as a function of the area ratio.

Table 7. Loss Coefficients for Sudden Contractions

A_1/A_2	0	0.1	0.2	0.3	0.4	0.5	0.6	0.7	0.8	0.9	1.0
K	.38	.36	.34	.31	.27	.22	.16	.10	.05	.01	0

For a reservoir the computed loss is usually taken as $K = 1/2$. If the inlet is re-entrant (i.e., when the pipe extends into the reservoir), the loss coefficient may be taken as $K = 1.0$. For rounded entrances the loss is very small: K varies from perhaps 0.01 to 0.05, and therefore may usually be neglected.

Transitions are sections of conduit which connect one prismatic portion to another by a gradual change in cross section. Since, owing to the inherent stability of accelerated flow, losses are small in gradual contractions, transition design is usually determined by factors other than energy loss. For example, it is often important that the pressure decrease continuously to that of the reduced section, so that the sections will be both *separation-proof* and *cavitation-proof*.

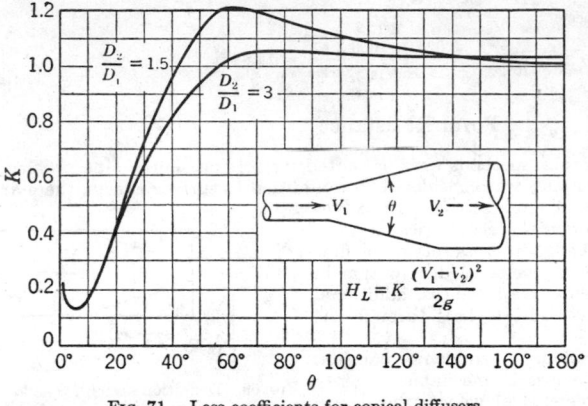

In expanding transitions or diffusers, wherein kinetic energy is converted into potential energy, it is even more essential that separation be avoided. The slowly moving fluid near the wall, which is retarded by surface resistance, is also retarded by the adverse pressure gradient

FIG. 71.   Loss coefficients for conical diffusers.

due to the flow expansion. If the adverse pressure gradient acts over a sufficient length, it is certain to result in boundary-layer separation. Once separation occurs, with the backflow and eddies which accompany it, the losses become high. A series of experiments was conducted by Gibson * on conical diffusers, the results of which are shown in Fig. 71.

Valves, Bends, and Elbows. Losses due to valves, bends, and elbows in turbulent flow are closely proportional to the square of the velocity. The loss is generally given by a loss coefficient K, as for sudden expansions and contractions, in the form

$$h_l = K \frac{V^2}{2g}$$

Table 8 lists representative values of K for various types of valves, the first seven items being taken from a tabulation published by the Crane Company † and the last fifteen from investigations by the Armour Research Foundation.

Table 8.   Loss Coefficients for Valves

Type	K
Globe valve, fully open	10.0
Angle valve, fully open	5.0
Swing check valve, fully open	2.5
Gate valve	
Fully open	0.19
3/4 open	1.15
1/2 open	5.6
1/4 open	24.0
Plug globe, or stop valve (600 psi)	
Fully open	4.0
3/4 open	4.6
1/2 open	6.4
1/4 open	780.0
Diaphragm valve	
Fully open	2.3
3/4 open	2.6
1/2 open	4.3
1/4 open	21.0
Plug valve (1/4 turn from closed to full open)	
Fully open	0.77
99% open	0.86
98% open	0.95
95% open	1.45
90% open	2.86
80% open	9.6
70% open	28.0

* Gibson, A. H., "Hydraulics and Its Applications," *London Constable*, 1912.
† "Flow of Fluids," *Tech. Paper 409*. Crane Company, Chicago, May 1942.

Flow through a bend causes an increase in pressure around the outside (and a corresponding decrease in velocity) and a decrease in pressure (with a corresponding increase in velocity) around the inside, as in Fig. 72. The pressure changes induce a secondary flow in the form of a double vortex, also shown in Fig. 72. The loss in energy is due to the separation at the inside as well as to the effects of the secondary flow in disarranging the velocity distribution. Increasing the width and decreasing the depth reduces losses by lessening the secondary flow.

For plumbing fittings it is common practice to assume K values of 0.9, 0.75, and 0.6, respectively, for standard, medium-sweep, and long-sweep elbows of 90°, and to modify these values in direct proportion for equal-radius elbows of larger or smaller angles.

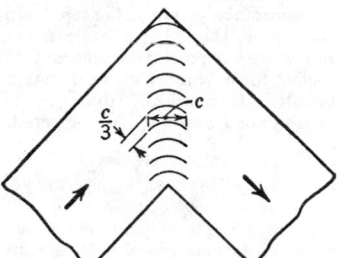

FIG. 73. Simple vane installation for a miter bend.

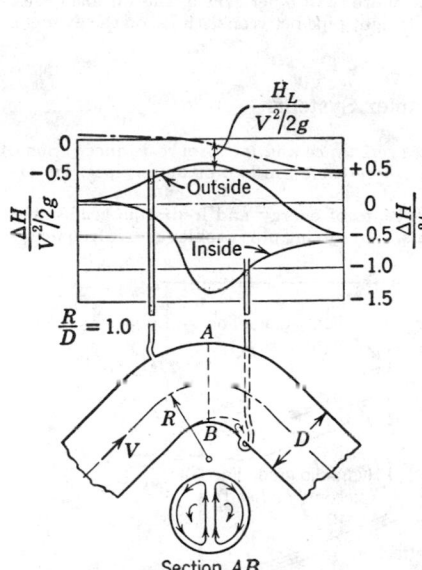

Section AB
FIG. 72. Variation in head at a 90° bend.

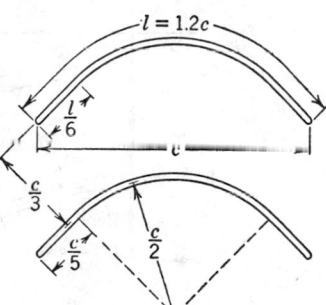

FIG. 74. Dimensions of turning vanes for miter bends.

The introduction of a series of deflecting vanes in a simple miter joint, Fig. 73, may reduce the loss in a 90° bend to $K = 0.15$ with very carefully designed variable thickness vanes. A value of $K = 0.20$ can be obtained with vanes of sheet metal and of circular-arc shape, or with end tangents as shown in Fig. 74, following investigations of Wirt, [*] and of Klein, Tupper, and Green.[†]

Nozzles. The loss through a nozzle at the end of a conduit is traditionally in terms of a velocity coefficient C_v, which is defined as the ratio of actual efflux velocity to that which would occur for the same pressure at the base of the nozzle if no loss occurred. This is related to K by

$$K = \left(\frac{1}{C_v{}^2} - 1 \right) \left[1 - C_c \left(\frac{D_2}{D_1} \right)^4 \right]$$

where C_c is the ratio of jet area to outlet area; D_1, D_2 are base and outlet diameters, respectively. Nozzles with cylindrical tips have practically zero contraction (i.e., $C_c = 1$), and at high Reynolds' numbers C_v is about 0.96–0.99.

Principle of Equivalent Lengths. By equating the Darcy-Weisbach formula to the form-loss expression, one can determine the equivalent length of straight pipe L_e which

[*] Wirt, L., "New Data for the Design of Elbows in Duct Systems," *Gen. Elec. Rev.*, Vol. 30, pp. 236–296, 1927.

[†] Klein, G. J., Tupper, K. F., and Green, J. J., "The Design of Corners in Fluid Channels," *Canadian J. Research*, Vol. 3, pp. 272–825, 1930.

would result in the same energy loss as the fitting, thus

$$h_l = K \frac{V^2}{2g} = f \frac{L_e}{D} \frac{V^2}{2g}$$

from which

$$L_e = \frac{KD}{f}$$

In instances where f is known or may be approximated, the form losses can be expressed as equivalent lengths and added to the actual length of conduit to simplify calculations.

Similarly, two pipes may be said to be equivalent if the same over-all head loss occurs for the same discharge.

Importance of Form Losses. Due to uncertainties in determining the value of f for a given pipe, the friction loss cannot generally be predicted closer than within 5 per cent; and with old pipes the accuracy is undoubtedly much less. Therefore it is feasible to neglect form (minor) losses if they do not total more than 5 per cent of the surface losses. Usually if there are 1000 diameters or more of straight pipe between fittings, on the average, such minor losses may be neglected.

Analysis of Complex Systems

Losses have been examined above for surface resistance and for form resistance. Some of the situations where both are involved, or where there are two or more pipes, are examined in the remainder of this article.

Energy and Hydraulic Gradients. The concepts of energy and hydraulic grade lines are of value in the analysis of many flow systems. In solving pipe problems, it is frequently

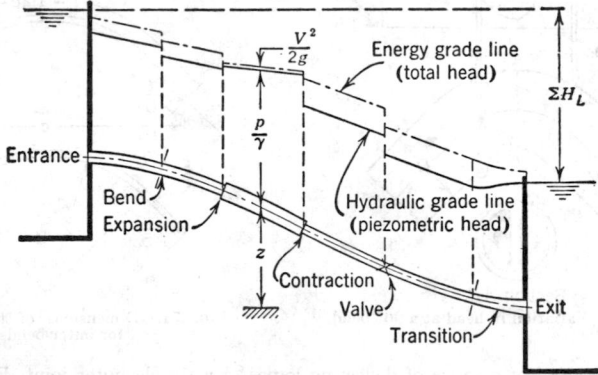

FIG. 75.   Schematic representation of energy and hydraulic grade lines.

advantageous to write Bernoulli's equation from an upstream point to a downstream point, including all the energy loss terms between the two points, i.e.,

$$\frac{V_1^2}{2g} + \frac{p_1}{\gamma} + z_1 = \frac{V_2^2}{2g} + \frac{p_2}{\gamma} + z_2 + \Sigma h_l$$

where subscript 1 refers to the upstream point, and subscript 2 to the downstream point. These points are usually selected where the most information is available, to reduce the number of unknowns in the equation. Since each term has the units length, the equation may be given a graphical interpretation by plotting the terms as ordinates against distance along the pipe as abscissa. The energy grade line is a plot of the total head $V^2/2g + p/\gamma + z$ for each point along the pipe. This line always decreases in the downstream direction unless a pump or other mechanism is available to add energy to the fluid.

The hydraulic grade line is a plot of $p/\gamma + z$ for each point along the pipe; it is also known as the piezometric head, since piezometer tubes would permit liquid to rise in them to the hydraulic grade line. The hydraulic grade line is always $V^2/2g$ below the energy grade line. An illustration of the two grade lines is given by Fig. 75. The hydraulic grade line slopes upward in those reaches where kinetic energy is being converted into pressure or potential energy at a rate greater than the energy loss, as in a diffuser. The two grade lines coincide at the surface of reservoirs.

Siphons. A siphon is a conduit which conveys liquid from a reservoir to a lower elevation, after raising it to a higher elevation at an intermediate section. The pressure is necessarily reduced at the summit, a condition which limits its performance. As long as the minimum pressure is well above the vapor pressure of the liquid, the flow may be determined by use of Bernoulli's equation. When vapor pressure is reached in the pipe, the liquid boils, and flow occurs in a two-phase system. Bernoulli's equation is no longer applicable when applied to points on either side of the discontinuity, but, since the pressure is known at the critical points, it may be applied from the upstream reservoir to this point as a means of determining flow. Practically, the amount of suction must be limited because of the gases coming out of solution and collecting at the summit.

Changes of Head at Pumps and Turbines. A pump in a pipeline adds energy to the flow, the power gained being $Q\gamma H$. If the suction and discharge lines are of different sizes, the change in elevation of hydraulic head will differ from H by an amount depending upon the change in velocity head. The term H is included on the upstream side of Bernoulli's equation.

A turbine withdraws energy from a conduit flow, and hence lowers the energy and hydraulic grade lines. It may be treated as an energy loss H which will occur on the downstream side of the equation.

Series Pipes. When two or more pipes are connected in series, as in Fig. 75, the same discharge prevails in each pipe and the total head loss for the system is the sum of the head losses for the various pipes. Two types of problem are encountered: one in which the discharge is given and the total head loss required, and one in which the discharge is required for a given total head loss.

The first type of problem, with the velocity head immediately obtainable from the discharge, simply requires the addition of all the losses due to the various pipes and fittings. The second type is best solved by writing the Bernoulli equation between upstream sections, including all loss terms. Then, from the continuity equation, all velocity heads may be expressed in terms of the velocity head in one of the pipes. This leaves an equation with several unknown values of f (depending upon the number of distinct pipes) and one unknown velocity head. By assuming arbitrary values of each f (they may all be assumed the same for convenience), the approximate velocity may be computed, and therefrom the approximate Reynolds' number for each pipe. Using the appropriate relative roughness, the corresponding values of f are taken from Fig. 68. With these new values, the velocities are again computed. Usually the resistance coefficients resulting from the first trial are found to be satisfactory.

Example. Two water reservoirs are connected by 2000 ft of 24-in. pipe, followed by 1000 ft of 18-in. pipe. There are three elbows and two gate valves in each pipe with a square-edged entrance from the first reservoir, a sudden contraction between the two pipes, and a submerged exit into the second reservoir. The pipes are new wrought iron. What discharge will result from a difference in elevation of 12 ft?

Writing the Bernoulli equation from one reservoir surface to the other and including all losses,

$$12 = 0.5\,\frac{V_1^2}{2g} + 2 \times 0.19\,\frac{V_1^2}{2g} + 3 \times 0.9\,\frac{V_1^2}{2g} + 1000\,f_1\,\frac{V_1^2}{2g} +$$
$$0.18\,\frac{V_2^2}{2g} + 2 \times 0.19\,\frac{V_2^2}{2g} + 3 \times 0.9\,\frac{V_2^2}{2g} + \frac{V_2^2}{2g} + 667\,f_2\,\frac{V_2^2}{2g}$$

where subscript 1 refers to the 24-in. pipe. Then

$$\frac{V_1^2}{2g} = \left(\frac{3}{4}\right)^4 \frac{V_2^2}{2g}$$

the introduction of which reduces the equation to

$$12 = \frac{V_2^2}{2g}\,(5.39 + 316\,f_1 + 667\,f_2)$$

assuming $f_1 = f_2 = 0.020$; then $V_2 = 5.55$ fps, and $V_1 = 3.12$ fps. Using the VD'' scale along the top of Fig. 68 with $k_1/D_1 = 0.000075$, $k_2/D_2 = 0.00001$, $f_1 = 0.014$, $f_2 = 0.014$. For these values $V_2 = 6.25$, and

$$Q = 6.25 \times \frac{\pi}{4}\,2.25 = 11.05\ \text{cfs}$$

Series problems may also be solved by using the principle of equivalent lengths. The form losses, including entrance and exit, are expressed as equivalent lengths of the one pipe.

Parallel Pipes. It is common practice to supplement a pipeline with one or more additional lines, connected in parallel, to increase the flow capacity. Two types of problems are encountered: that of determining the capacity of a system when the elevation of hydraulic grade line is known at the junction points; and that of determining the division of a given flow among the pipes and the corresponding drop in grade line.

In parallel-pipe problems, Fig. 76, the drop in hydraulic grade line is the same for each pipe, since they all have common junction points. Moreover, the discharge through the parallel pipes must equal the discharge in the entrance and exit lines, since flow into and out of a junction must be the same.

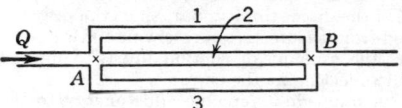

The first type of problem is simply that of solving for the discharge in each pipe corresponding to the given drop in hydraulic grade line and then adding the discharges.

FIG. 76. Parallel-pipe system.

The second type is best handled by a special procedure, the successive steps of which are as follows:

(a) Assume a discharge in one pipe, say Q'_1, and compute the corresponding drop in hydraulic grade line, h'_f.

(b) Using this drop, compute the discharges Q'_2, Q'_3, $\cdots$ in the other pipes.

(c) Apportion the given total discharge Q among the pipes in the same ratio as the computed trial discharges; that is,

$$Q_1 = \frac{Q'_1}{Q'_1 + Q'_2 + Q'_3, \cdots} \times Q, \quad Q_2 = \frac{Q'_2}{Q'_1 + Q'_2 + Q'_3, \cdots} \times Q, \quad \cdots$$

(d) Using the values, Q_1, Q_2, Q_3, $\cdots$ obtained in this fashion, compute the head loss along each pipe. If these values are sufficiently close, the problem is solved; if not, the procedure is repeated, using one of the new discharges.

Example. In Fig. 76, water at 60 F flows at the rate of 14 cfs through three parallel pipes of new asphalted cast iron. $L_1 = 5000$ ft, $D_1 = 16$ in., $L_2 = 3000$ ft, $D_2 = 12$ in., $L_3 = 6000$ ft, $D_3 = 12$ in. Determine the distribution of flow and the head loss from A to B.

$$\frac{k_1}{D_1} = 0.0003, \quad \frac{k_2}{D_2} = 0.0004, \quad \frac{k_3}{D_3} = 0.0004$$

Assume $Q'_1 = 6.0$ cfs, then

$$V'_1 = 4.3 \text{ ft/sec}, \quad h'_f = 0.016 \times \frac{5000}{1.333} \frac{\overline{4.3}^2}{64.4} = 17.3 \text{ ft}$$

$$3000 f'_2 \frac{V'_2{}^2}{2g} = 17.3, \quad f'_2 = 0.017, \quad V'_2 = 4.66 \text{ ft/sec}, \quad Q'_2 = 3.66 \text{ cfs}$$

$$6000 f'_3 \frac{V'_3{}^2}{2g} = 17.3, \quad f'_3 = 0.018, \quad V'_3 = 3.21 \text{ ft/sec}, \quad Q'_3 = 2.52 \text{ cfs}$$

Hence, $Q'_1 + Q'_2 + Q'_3 = 12.18$,

$$Q_1 = \frac{6}{12.18} \times 14 = 6.90 \text{ cfs}, \quad Q_2 = \frac{3.66}{12.18} \times 14 = 4.21 \text{ cfs}, \quad Q_3 = \frac{2.52}{12.18} \times 14 = 2.89 \text{ cfs}$$

Check on solution:

$$(h_f)_1 = 0.016 \times \frac{5000}{1.333} \frac{\overline{4.87}^2}{64.4} = 22.2 \text{ ft}$$

$$(h_f)_2 = 0.017 \times \frac{3000}{1.0} \frac{\overline{5.36}^2}{64.4} = 22.8 \text{ ft}$$

$$(h_f)_3 = 0.018 \times 6000 \frac{\overline{3.68}^2}{64.4} = 22.7 \text{ ft}$$

The head losses are within 3 per cent of each other, which is generally considered satisfactorily close for this type of problem.

Branching Pipes. Several reservoirs may be connected by pipe lines in various complex arrangements. A simple case, illustrated by Fig. 77, is discussed as a means of demonstrating the general approach to these problems. A system of branching pipes connects three reservoirs. The problem is to determine the flow through each line for particular elevations of the reservoirs when pipe characteristics and fluid properties are known. The hydraulic grade lines coincide with the water surfaces in each reservoir and have a common elevation at the junction point. The flow through each pipe is determined by the slope of each hydraulic grade line, which depends upon the elevation of hydraulic grade line at the junction J. The continuity equation takes one of the following forms,

$$Q_1 = Q_2 + Q_3, \quad Q_1 + Q_2 = Q_3$$

depending upon whether the elevation of hydraulic grade line at J is above or below reservoir B. The computation procedure is as follows:

(a) Assume an elevation of hydraulic grade line at J.

(b) Compute Q_1, Q_2, Q_3 for this assumed condition.

(c) If continuity is satisfied the problem is solved.

(d) If the flow into the junction does not equal the flow out of the junction, assume a new elevation of hydraulic grade line at J, *raising* the hydraulic grade line there for more flow in than out, and *lowering* it for less flow in than out.

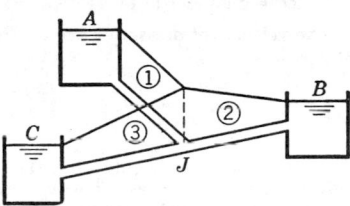

FIG. 77. Branching pipe system with single junction.

(e) Repeat the procedure until the continuity equation is satisfied.

Pipe Networks. In networks of pipes there are three conditions which must be satisfied: (a) the algebraic sum of successive pressure drops around any circuit must be zero; (b) the flow into each junction must equal the flow out of the junction; and (c) the Darcy-Weisbach relation between head loss and velocity must be satisfied for each length.

It is not practical to solve complex networks analytically, since the equations are not linear except for laminar flow, and the large number of non-linear simultaneous equations precludes a direct algebraic solution. Methods are available, however, which successively improve an assumed distribution of flows or heads. The Hardy Cross * method is one in which a correction to the assumed flow or head is computed for each circuit and applied to bring the circuits successively into closer balance.

Economics of Design. Although the power required to maintain a given flow in a pipe line may be reduced to any desired magnitude by increasing the pipe diameter, the larger size requires a larger capital investment. The most economical system is the one which has a minimum combined annual cost of operation and of capital investment, which may be determined as follows: For a given pipe line, the initial cost of the system may be estimated for each of several diameters. By reducing this to an annual cost (i.e., interest on capital plus depreciation and maintenance) and adding to the annual cost of power for operation, a plot may be made of total annual cost against diameter, from which the diameter having the lowest total annual cost may be selected.

Water Hammer

As an example of unsteady flow in a closed conduit, the phenomenon known as *water hammer*, due to sudden stoppage of flow, is briefly considered. When a valve is rapidly closed at the downstream end of a long pipe carrying a liquid, the liquid comes to rest progressively due to a pressure wave in the system. The pressure wave moves upstream from the valve with the velocity of sound. Bringing the moving liquid to rest requires a pressure increase at the valve. The pressure increase causes a compression of the liquid and an expansion of the pipe walls, storing additional energy in both. When the pressure wave reaches the upstream end of the pipe, all liquid is at rest, but the pressure intensity is everywhere increased. Immediately after the pressure wave reaches the upstream end, a negative pressure wave starts downstream and the fluid flows upstream slightly, resulting in a lower-than-normal pressure at the valve. The liquid then starts moving downstream again, and this cycle is repeated with diminishing pressure changes until the effects of fluid friction and imperfect elasticity of pipe walls have absorbed the original kinetic energy of the flowing liquid.

The head rise due to sudden closure is obtained by use of the energy equation and is

$$h = \cfrac{V}{\sqrt{\gamma g \left(\dfrac{1}{K_m} + \dfrac{D}{Et} \right)}}$$

---

* "Analysis of Flow in Networks of Conduits or Conductors," Univ. of Illinois Bulletin 286, Nov. 1946.

where h = head rise due to water hammer
  V = mean velocity of flow before closure
  γ = specific weight of liquid
  K_m = bulk modulus of elasticity of liquid
  E = modulus of elasticity of pipe walls
  t = thickness of pipe walls
  D = diameter of pipe

The velocity of pressure wave v_L, from the momentum equation and the head equation above, is

$$v_L = \frac{1}{\sqrt{\dfrac{\gamma}{g}\left(\dfrac{1}{K_m} + \dfrac{D}{Et}\right)}}$$

These equations apply only when the closure is so rapid that the valve is completely closed before the negative pressure wave returns to the valve; i.e., the closure time is less than $2L/v_L$ where L is the length of pipe. Consistent units must be employed throughout in the above equations.

10. FLOW IN OPEN CHANNELS

Open channel flow introduces complications not encountered in closed channel flow in that the free surface, which has constant pressure, is usually unknown.

General Considerations

Laminar flow in an open channel occurs only for low Reynolds' numbers, which means extremely viscous liquids, or very small velocities, or very shallow flow, such as in a film. For cases of very simple boundary geometry, application of Newton's law of viscosity may be applied to determine the velocity distribution.

Turbulent flow has much greater engineering significance than laminar flow, and hence the following treatment refers exclusively to it.

Open channel flow may be *steady* or *unsteady*, *uniform* or *non-uniform*. *Steady uniform* flow is well understood. Empirical equations are used which predict the flow within the usual accuracy required. *Steady non-uniform* flow occurs at entrances and exits to channels, at transitions and controls, and throughout most canals which are not extremely long. For gradual changes in the flow, several methods of computation are available based on the uniform flow equations. *Unsteady uniform* flow rarely occurs in open channels. *Unsteady non-uniform* flow is most difficult, and solutions have been obtained for only a few simple situations.

Specific Energy—Critical Depth

The mechanical energy per unit weight, with elevation datum taken as the bottom of the channel, is called *specific energy*. It is simply the sum of depth of flow and velocity head. In steady uniform flow, when all cross sections are identical, the specific energy is constant along the channel.

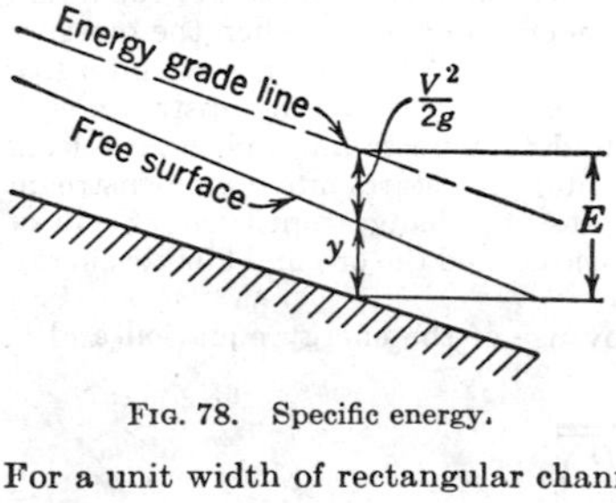

Referring to Fig. 78, the specific energy is

$$E = y + \frac{V^2}{2g}$$

assuming uniform distribution of velocity over the cross section. For a given discharge Q, the specific energy varies with the depth of flow. Substituting $V = Q/A$, where A is the cross-sectional area and a function of y,

$$E = y + \frac{Q^2}{2gA^2}$$

FIG. 78. Specific energy.

For a unit width of rectangular channel, with q the discharge per unit width,

$$E = y + \frac{q^2}{2gy^2}$$

A plot of specific energy against depth, Fig. 79, for a constant q, reveals that a certain minimum specific energy is required for the flow, found by setting $dE/dy = 0$. Calling this depth y_c the *critical depth*, we have

$$y_c = \left(\frac{q^2}{g}\right)^{\frac{1}{3}}$$

In terms of the velocity, $V_c = \sqrt{gy_c}$. Hence critical depth is the depth at which the velocity of flow V_c is just equal to the velocity of an elementary wave $\sqrt{gy}$ in still liquid.

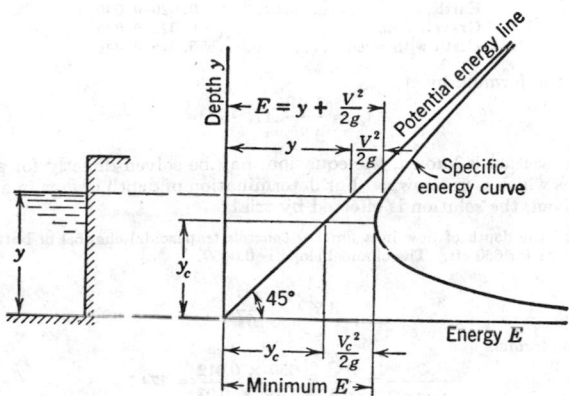

FIG. 79. Specific energy diagram.*

Greater specific energy is required for both greater and lesser depths of flow. It is obvious from Fig. 79 that there are two depths at which the flow has the same specific energy. For non-rectangular channels the critical depth occurs

$$\frac{Q^2}{g} = \frac{A^3}{b}$$

where b is the top width of the cross section at the liquid surface.

Steady, Uniform Flow

In steady uniform flow, the slope of channel bottom, free surface (hydraulic grade line), and energy grade line are the same ($\tan \theta$, Fig. 80). For very wide channels, the shear stress varies linearly with distance from the free surface y, given by $\tau = y\gamma \sin \theta$. For other channels, the average shear stress τ_0 at the solid boundary is $\tau_0 = \gamma R \sin \theta$; where R is the hydraulic radius, which is defined as the ratio of area of cross section A to wetted perimeter P. The liquid velocity at the solid boundary is zero; it increases generally with distance from a boundary. The maximum velocity is usually below the free surface.

The *Manning formula* is the most commonly used open-channel formula,

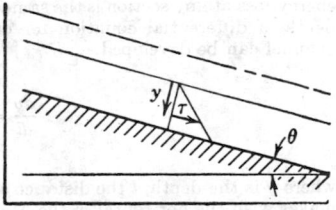

FIG. 80. Uniform flow.

$$V = \frac{1.49}{n} R^{\frac{2}{3}} S^{\frac{1}{2}}$$

where V is the average velocity; R is the hydraulic radius; S is the sin θ, Fig. 80; and n is an absolute roughness factor, having the dimensions $L^{\frac{1}{6}}$, whose values for different surfaces are determined experimentally. Table 9 lists many of these values. Since the constant in the Manning formula is not dimensionless, it is necessary to use the foot-pound-second system of units.

* Courtesy of Prof. R. A. Dodge, University of Michigan, Ann Arbor.

Table 9.   Values of the Manning Roughness Factor for Various Boundary Materials

Boundary Surface	Manning n (feet$^{1/6}$)
Planed wood	0.010–0.014
Unplaned wood	0.011–0.015
Finished concrete	0.011–0.013
Unfinished concrete	0.013–0.016
Cast iron	0.013–0.017
Riveted steel	0.017–0.020
Brick	0.012–0.020
Rubble	0.020–0.030
Earth	0.020–0.030
Gravel	0.022–0.035
Earth with weeds	0.025–0.040

Multiplying the formula by A,

$$Q = \frac{1.49}{n} A R^{2/3} S^{1/2}$$

When the cross section is known, the equation may be solved directly for any one of the other quantities which is unknown. For determination of depth of flow in a given section, with Q, n, S given, the solution is effected by trial.

Example. Find the depth of flow in a *finished concrete* trapezoidal channel of bottom width 10 ft and side slopes 1 : 1 for 650 cfs. The channel slope is 0.0009.
Writing

$$AR^{2/3} = \frac{A^{5/3}}{P^{2/3}}$$

from the Manning formula

$$\frac{Qn}{1.49 S^{1/2}} = \frac{A^{5/3}}{P^{2/3}} = \frac{650 \times 0.012}{1.49 \times 0.03} = 174.7$$

$A = 10D + D^2$, $P = 10 + 2\sqrt{2}D$, hence

$$f(D) = \frac{(100 + D^2)^{5/3}}{(10 + 2\sqrt{2}D)^{2/3}} = 174.7$$

Trying $D = 5$, $f(D) = 160$; hence D must be larger. Trying $D = 5.5$, $f(D) = 191$. By straight line interpolation $D = 5.24$, $f(D) = 174$, which is a satisfactory check. Hence $D = 5.24$ ft is the answer sought.

The cross section having the least perimeter for given conditions is called the *most efficient* cross section. The semi-circular section is most efficient of all cross sections, since it has the least perimeter for a given area. The most efficient *rectangular* channel has a bottom width twice the depth. The most efficient *trapezoidal* channel is half of a hexagon.

Steady, Non-uniform Flow

Gradually varied channel flow is steady flow in which changes in depth, section, slope, and roughness with respect to length along the channel are small. By assuming that the energy loss at any section is the same as in uniform flow at the same discharge and the same depth, a differential equation for change in depth as a function of distance along the channel can be developed:

$$\frac{dy}{dl} = \frac{S_0 - \dfrac{n^2 Q^2}{(1.49)^2 A^2 R^{4/3}}}{1 - \dfrac{Q^2 b}{g A^3}}$$

where y is the depth, l the distance along the channel, S_0 the sine of the angle the bottom makes with the horizontal, n the Manning roughness factor, Q the discharge, A the cross-sectional area, R the hydraulic radius, and b the width of cross section at the liquid surface. Solving for l

$$l = \int \frac{1 - \dfrac{Q^2 b}{g A^3}}{S_0 - \dfrac{n^2 Q^2}{(1.49)^2 A^2 R^{4/3}}} \, dy$$

For constant S_0 and n, the integrand is a function of y only, and l may be determined as a function of y, usually by numerical integration. When the integrand is zero, $Q^2 b / g A^3 = 1$, which is the condition for critical depth. Hence for a change in depth there is no change in l; i.e., neglecting the effects of the curvature of streamlines and the non-hydrostatic pressure distribution, the liquid surface is vertical as the flow goes through critical. When the denominator is zero, uniform flow occurs, and there is no change in depth along the channel.

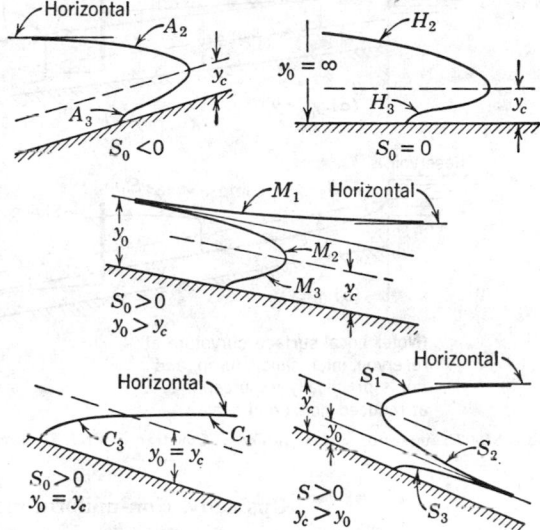

The various possible free-surface profiles given by the above equation are shown in Fig. 81. In each case the flow is from left to right. y_0 is the normal depth, i.e., the depth given by the Manning uniform flow equation. y_c is the critical depth. When the normal depth is greater than the critical depth, the slope of channel is *mild*; when normal depth equals critical depth, the slope is *critical*; when normal depth is less than critical depth the slope is *steep*. The two other cases are *horizontal* and *adverse*.

The determination of surface profiles from the equation are effected by starting the numerical integration at

Fig. 81. Surface profiles on adverse, horizontal, mild, critical, and steep slopes. Courtesy of Prof. Hunter Rouse, State University of Iowa, Iowa City.

a *control* section. When the flow is above critical depth, the control is always downstream, and the depth is evaluated first for the control section, and then use is made of the gradually varied flow equation. Writing the equation in the form

$$l = \int F(y) \, dy$$

a plot of $F(y)$ as ordinate against y as abscissa is made, starting with the control depth and varying y in the direction indicated by the characteristic curves. This plot, Fig. 82, gives

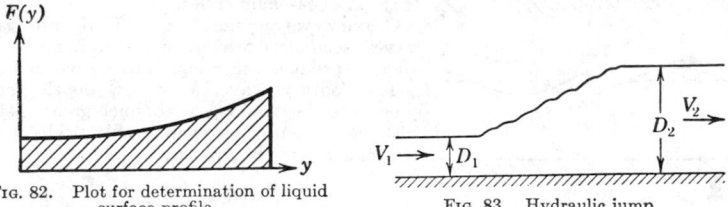

Fig. 82. Plot for determination of liquid surface profile.

Fig. 83. Hydraulic jump.

the value of l from the control section to the new depth y as the area under the curve between the values of y. In this manner the whole profile may be worked out.

When the depth of flow is less than critical, the control section is upstream (i.e., flow is out from under a gate) and the integration is handled in a similar fashion for determination of profile downstream from the control.

A phenomenon known as the *hydraulic jump* occurs under certain conditions in channel flow. The flow prior to the jump must always be below critical depth, and when the downstream depth is such that the momentum equation is satisfied for the liquid contained in the jump, the hydraulic jump will occur. The momentum equation applied to the liquid between sections 1 and 2 of Fig. 83, for a rectangular channel, yields the relation between depths

$$D_2 = -\frac{D_1}{2} + \sqrt{\frac{2 D_1 V_1^2}{g} + \frac{D_1^2}{4}}$$

Examples of occurrence of the surface profiles, including situations where the jump results, are given in Fig. 84.

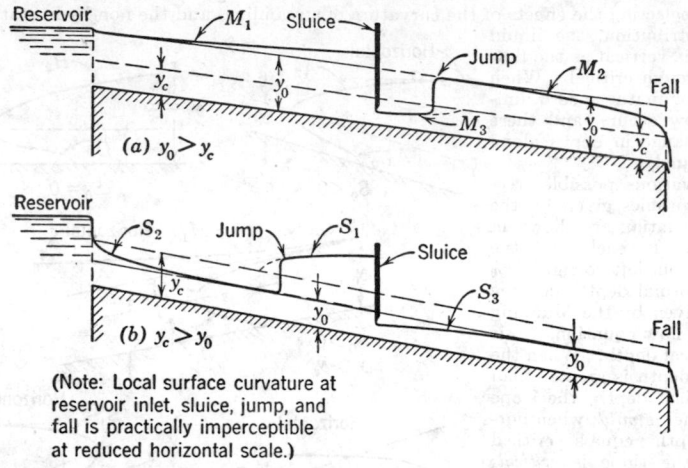

Reservoir
M_1
Sluice
Jump
M_2
Fall
y_c
y_0
M_3
(a) $y_0 > y_c$
Reservoir
S_2
Jump
S_1
Sluice
S_3
Fall
(b) $y_c > y_0$
(Note: Local surface curvature at reservoir inlet, sluice, jump, and fall is practically imperceptible at reduced horizontal scale.)

Fig. 84. Examples of surface profiles. Courtesy of Prof. Hunter Rouse, State University of Iowa, Iowa City.

Unsteady, Non-uniform Flow

Any change in discharge *in an open channel* generally results in an unsteady non-uniform flow until steady conditions again prevail. The changes in regime are usually effected by means of surface waves which travel upstream or downstream and thereby alter the flow.

Waves are of three classes: *capillary* waves, dependent upon surface tension; *elastic* waves, dependent upon compressibility; and *gravity* waves, dependent upon fluid weight. Capillary waves are of little interest in the usual open-channel problem. Elastic waves are of principal interest in closed-channel flow, as in water hammer. Gravity waves are of the upmost importance in open-channel flow.

(a)
(b)
(c)

Fig. 85. Generation of (*a*) solitary waves, (*b*) oscillatory waves, and (*c*) surges.

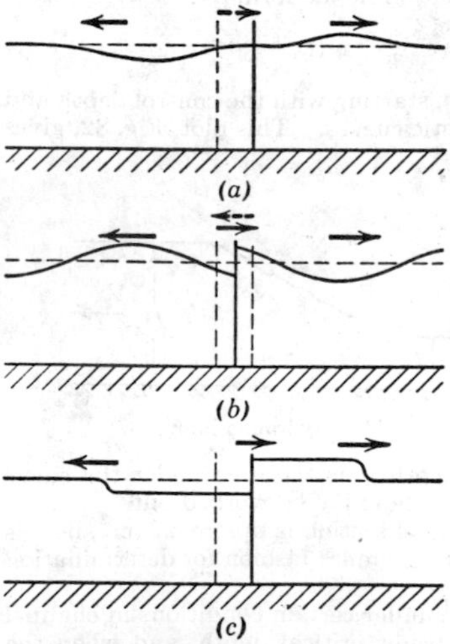

Gravity waves may be divided into *solitary* waves, *oscillatory* waves, and *surges*. all of which may be produced in a canal, as shown in Fig. 85.

The solitary wave, a shallow depth phenomenon, translates along a channel great distances with very small change in profile. The velocity of propagation c is given by

$$c = \sqrt{gy}\left(1 + \frac{a}{y}\right)^{1/2}$$

where y is the undisturbed depth, and a is the height of wave. A *negative* solitary wave travels with the speed c, as given above, with a negative. The negative wave is inherently unstable and cannot be transmitted long distances.

Oscillatory waves are propagated in channels. Two approximate solutions have been effected: one is for relatively short waves known as *Stokian waves*; the other is for long waves where the effects of the channel bottom are important, known as *cnoidal waves*. The limiting case for cnoidal waves, as the height increases, is the solitary wave.

The propagation velocity of a discontinuous surge in a rectangular channel may be computed by application of the momentum and continuity equations to the fluid, as shown in Fig. 86. It is

$$c = \sqrt{gy_1} \left[\frac{1}{2} \frac{y_2}{y_1} \left(\frac{y_2}{y_1} + 1 \right) \right]^{\frac{1}{2}}$$

where c is the surge velocity relative to the undisturbed fluid of velocity V_1.
Intermittent surges occur on steep slopes although the inflow is essentially steady.
These waves, known as *roll waves* or as *slug flow*, arise from the instability of the flow.
They have been studied analytically and experimentally by H. A. Thomas.* The individual waves are stable in form, but their number, height, and velocity are affected by the frequency of initial disturbances and the length of the channel.
The waves will generally form if the channel slope is more than four times the critical slope,

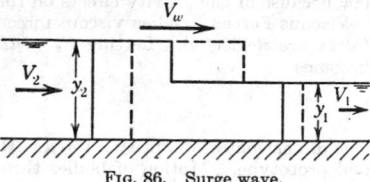

FIG. 86. Surge wave.

or if the velocity is more than twice the celerity of an elementary wave, i.e., more than $2\sqrt{gy}$ at the design depth y.

11. HYDRAULIC MODELS

Many practical flow situations are so complex that a complete analytical solution is ruled out. Recourse may be taken to hydraulic model studies, which have been successful in a great variety of problems. Generally speaking, it is impossible to obtain complete similarity in all details; hence, the important factors are considered with possible corrections for minor items.

Similitude Requirements

For complete similitude the model must be *geometrically, kinematically,* and *dynamically similar* to the prototype.

Geometric similitude occurs when all corresponding linear dimensions of prototype and model bear a constant fixed ratio to each other. This should also include the surface finish.

Kinematic similitude requires complete similarity of corresponding streamlines. The ratios of velocity components at homologous points must be equal.

Dynamic similitude requires that the ratio of forces at homologous points be the same.
With geometric similitude, complete similitude relations may be obtained by use of Newton's second law of motion:

$$m\mathbf{a} = \mathbf{F}_g + \mathbf{F}_v + \mathbf{F}_t + \mathbf{F}_e$$

where $m\mathbf{a}$, the mass of a fluid element times its resultant acceleration, is taken as the inertial force; $\mathbf{F}_g$ is the force due to gravity; $\mathbf{F}_v$ is the viscous force; $\mathbf{F}_t$ is the surface tension force; and $\mathbf{F}_e$ is the elastic compression force. For complete similarity the following equations must be satisfied:

$$\frac{m_m a_m}{m_p a_p} = \frac{(\mathbf{F}_g + \mathbf{F}_v + \mathbf{F}_t + \mathbf{F}_e)_m}{(\mathbf{F}_g + \mathbf{F}_v + \mathbf{F}_t + \mathbf{F}_e)_p} = \frac{(\mathbf{F}_g)_m}{(\mathbf{F}_g)_p} = \frac{(\mathbf{F}_v)_m}{(\mathbf{F}_v)_p} = \frac{(\mathbf{F}_t)_m}{(\mathbf{F}_t)_p} = \frac{(\mathbf{F}_e)_m}{(\mathbf{F}_e)_p}$$

The subscripts m and p refer to model and prototype, respectively.

Gravity Forces. When gravity predominates, the other forces may be disregarded.
Writing $m\mathbf{a} = \rho L^2 V^2$ and $F_g = \gamma L^3$

$$\frac{V_m}{\sqrt{g_m L_m}} = \frac{V_p}{\sqrt{g_p L_p}}$$

The dimensionless number $V/\sqrt{gL}$ is called the *Froude number* and must have the same value for both model and prototype. Rewriting the above equation

$$\frac{V_r}{\sqrt{g_r L_r}} = 1$$

* H. A. Thomas, "The Propagation of Waves in Steep Prismatic Conduits," *Proc. of Hydraulics Conference*, Univ. of Iowa Studies in Engineering Bulletin 20, 1940, p. 214.

where the *subscript* r is the model-to-prototype ratio. Since g_r is usually unity, the velocities vary as the square root of the length ratio. The discharge ratio Q_r is found by multiplying the velocity ratio by the area ratio,

$$Q_r = A_r V_r = (L_r)^{5/2}$$

Hydraulic structure models, as well as river and harbor models, depend on the Froude law because of the gravity effects on the free surfaces.

Viscous Forces. When viscous forces predominate, as in closed conduit flow, the other forces are neglected. Letting $F_v = \mu L V$, where μ is the dynamic viscosity, the ratio becomes

$$\frac{L_r V_r}{(\mu/\rho)_r} = 1$$

The dimensionless quantity $LV\rho/\mu$ is Reynolds' number and must be the same for model and prototype. Motion of bodies through large expanses of fluid or *fluid flow in closed conduits* require Reynolds' number equality in model studies.

Surface Forces. Neglecting other forces, and expressing the surface force as $F_t = \sigma L$, where σ is the surface tension per unit length, the ratio of inertia forces to surface forces is

$$\frac{V_r}{\sqrt{\sigma_r/\rho_r L_r}} = 1$$

The dimensionless parameter $V/\sqrt{\sigma/(\rho L)}$ is the *Weber number*. For situations where surface tension forces predominate, models should have the same Weber number as the prototype.

Elastic Forces. Elastic forces result from compression of the fluid and are expressed in terms of the bulk modulus of elasticity K_m of the fluid, $F_e = K_m L^2$. Neglecting other forces

$$\frac{V_r}{\sqrt{(K_m)_r/\rho_r}} = 1$$

The dimensionless number $V/\sqrt{K_m/\rho}$ is the *Cauchy number*, or *Mach number*, and must be the same in model and prototype.

12. FLOW AROUND IMMERSED AND FLOATING BODIES

The flow around circular and elliptic cylinders, as well as spheres and Rankine bodies, has been discussed from the viewpoint of an idealized fluid in Article 5. In 1904 Prandtl developed the concept of the boundary layer, which provided the link between hydrodynamics and hydraulics. For fluids of relatively small viscosity, the effect of internal friction is confined to a thin layer of fluid adjacent to the boundaries. Since this layer does not appreciably affect the boundary form, the flow may be determined outside the boundary layer by use of potential theory. Due to the thinness of the layer and lack of acceleration normal to it, the boundary layer takes the pressure distribution as determined by the potential flow. The application of the momentum equation to the boundary layer permits the shear stress at the surface of the body to be determined.

General Considerations

When motion of a fluid having very small viscosity is started from rest, the flow is essentially irrotational in the first instants. Since the fluid at the boundaries must have zero velocity relative to the boundaries, there is a sharp velocity gradient from the velocity given by the potential flow to the boundary, which sets up shear forces in a real fluid. The fluid layer which has had its velocity affected by the boundary shear is called the *boundary layer*. Considering flow around a streamlined body, the boundary layer is very thin at the upstream end, and becomes thicker in the downstream direction due to the retarding action of the boundary shear.

The component of the boundary shear in the direction of the relative velocity of approach is known as *skin friction*. For streamlined bodies with practically complete closure of the streamlines behind the body, skin friction is the predominate resistance. When the streamlines adjacent to the boundary become detached from the body, the turbulent region inside these streamlines and downstream from the body is known as the *wake*. The pressure in the wake is less than the pressure given by potential theory with no separation; this results in a resistance known as *form drag*. A force exerted by the fluid on a body at right angles to the relative approach velocity is known as the *lift* and is treated extensively in Section 8.

The phenomenon of the detachment of streamlines from the boundary, known as *separation*, results from an adverse pressure gradient which slows the boundary layer and brings it to rest. The flow near the boundary must then be deflected away from the boundary, resulting in a very thick boundary layer or wake. The potential flow equations do not yield the pressure in the wake or downstream from the separation point.

The Navier-Stokes equations may be simplified when applied to two-dimensional flow in the boundary layer. Using the notation of Article 6,

$$\mu \frac{\partial^2 u}{\partial y^2} = \frac{\partial p}{\partial x} + \rho \left(u \frac{\partial u}{\partial x} + v \frac{\partial u}{\partial y} + \frac{\partial u}{\partial t} \right)$$

which in connection with the continuity equation

$$\frac{\partial u}{\partial x} + \frac{\partial v}{\partial y} = 0$$

permits solutions for *thin* boundary layers.

For steady flow of an infinite fluid along a flat plate, von Kármán * has applied the momentum equation to the boundary layer, yielding

$$\tau_0 = \frac{\partial}{\partial x} \int_0^\delta \rho(U - u)u \, dy$$

where U is the undisturbed flow in the x-direction, δ is the boundary layer thickness, and τ_0 is the boundary shear stress, $\mu \dfrac{\partial u}{\partial y}\Big|_{y=0}$.

When floating objects are in motion relative to a fluid, they suffer *wave resistance* in addition to form drag and skin friction. Waves are set up by the boundary shape and motion, and these waves affect the fluid pressure distribution. Since the waves have energy, this energy must come from the moving object, which results in increased resistance.

Lift and drag on specific objects is treated in Section 8.

BIBLIOGRAPHY

ADDISON, HERBERT, *Hydraulic Measurements*, John Wiley, 1946.
ALLEN, J., *Scale Models in Hydraulic Engineering*, Longmans, Green, London, 1947.
BAKHMETEFF, B. A., *The Mechanics of Turbulent Flow*, Princeton University Press, 1941.
GOLDSTEIN, S., *Modern Developments in Fluid Dynamics*, Vols. I and II, Oxford, 1938.
HUNSAKER, J. C., and RIGHTMIRE, B. G., *Engineering Applications of Fluid Mechanics*, McGraw-Hill, 1947.
Hydraulic Models, A.S.C.E. Manual of Engineering Practice, No. 25, 1945.
LAMB, H., *Hydrodynamics*, Dover, 1945.
MILNE THOMPSON, L. M., *Theoretical Hydrodynamics*, Macmillan, 1938.
ROUSE, HUNTER, *Elementary Mechanics of Fluids*, John Wiley, 1946.
ROUSE, HUNTER, Ed. in Chief, *Engineering Hydraulics*, John Wiley, 1950.
ROUSE, HUNTER, *Fluid Mechanics for Hydraulic Engineers*, McGraw-Hill, 1938.
STREETER, V. L., *Fluid Dynamics*, McGraw-Hill, 3rd ed., 1962.
VENNARD, JOHN K., *Elementary Fluid Mechanics*, John Wiley, 4th ed., 1961.
WOODWARD, S. M., and POSEY, C. J., *Hydraulics of Steady Flow in Open Channels*, John Wiley, 1941.

* Kármán, Th. von, "Turbulence and Skin Friction," *J. Aeronautical Sci.*, Vol. 1, No. 1, p. 1, 1934.

$$\nu\left(\frac{\partial^2 u}{\partial x^2}+\frac{\partial^2 u}{\partial y^2}\right) = \frac{\partial u}{\partial t}+u\frac{\partial u}{\partial x}+v\frac{\partial u}{\partial y}+\frac{1}{\rho}\frac{\partial p}{\partial x}$$

$$\frac{\partial u}{\partial x}+\frac{\partial v}{\partial y}=0$$

BIBLIOGRAPHY

SECTION 7

AERONAUTICS AND ASTRONAUTICS

BY

STAFF OF THE UNITED STATES AIR FORCE ACADEMY

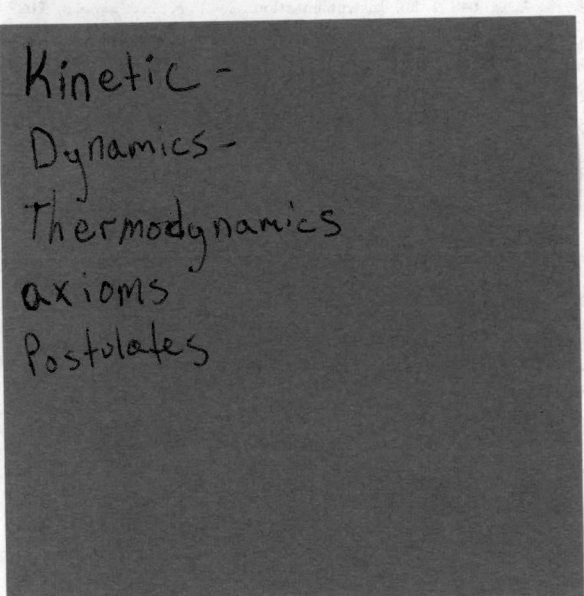

AERONAUTICS AND ASTRONAUTICS

This section presents the most important phases of aeronautics and astronautics. The discussions in most cases are brief and to the point, emphazing the assumptions, highlights of the theory, and conclusions. Working formulas are presented wherever possible. The reader is referred to the references at the end of each article for details.

KINETIC THEORY OF MATTER

By C. A. Forbich, Jr. and L. M. Nicolai

The kinetic theory of matter is based on the fact that all matter is made up of microscopic particles, namely molecules or atoms. In developing kinetic theory, it is assumed that classical Newtonian dynamics govern the motion of each molecule. The methods of kinetic theory are more complex than those used in classical macroscopic thermodynamics, but the results give reasonable foundations to many of the axioms and postulates used to develop classical thermodynamics. This is done in kinetic theory: *first*, by determining the energy distribution among the molecules and determining how this affects macroscopic variables; *second*, by making use of the energy distribution among the particles to predict transport phenomena such as viscosity and diffusion coefficients; and *finally*, by using the energy distribution to develop the conservation equations of gas motion by using the Boltzmann equation.

1. KINETIC THEORY OF AN IDEAL GAS

The kinetic theory for an ideal gas is based on the very simple model of a molecular gas.[2,3] The gas molecules are assumed to be hard spheres which do not interact except through collisions. Intermolecular collisions or wall collisions are assumed to be perfectly elastic, with particle speed ranging from zero to infinity, moving in all directions with equal probability.

Using this model, it is possible to compute the time rate of change of the momentum of the molecules colliding with the wall which gives the average force F per unit area dA (assumed to be the macroscopic pressure p) as

$$\frac{F}{dA} = p = nmv_z^2 \tag{1}$$

with n the number of molecules per unit volume, m the molecular mass, and v_z the molecular velocity normal to the wall. (See Fig. 1.)

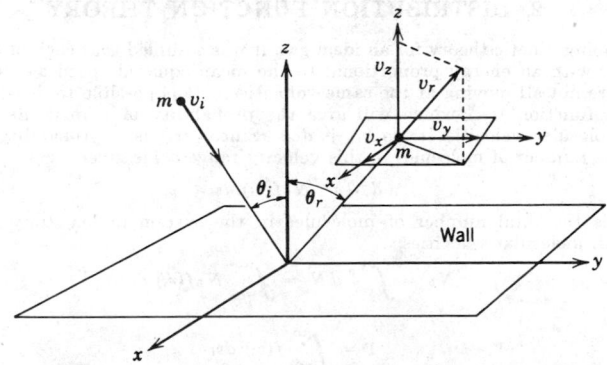

FIG. 1. Coordinates for molecular-wall collision.

Making use of the assumption that all molecules are moving with equal probability in all directions, the mean square speed $\overline{v^2}$ is given by

$$\overline{v^2} = \overline{v_x{}^2} + \overline{v_y{}^2} + \overline{v_z{}^2}$$

with

$$\overline{v_x{}^2} = \overline{v_y{}^2} = \overline{v_z{}^2}$$

so that

$$\overline{v^2} = 3\overline{v_z{}^2}$$

Using this result, eq. 1 can be rewritten as

$$p = \frac{1}{3} nm\overline{v^2} \tag{2}$$

Since p is assumed to be identical to the macroscopic pressure, it is also given by classical thermodynamics as

$$p = nmRT \tag{3}$$

In comparing the pressure given by eqs. 2 and 3, for equivalence it is necessary that[2,3]

$$\overline{v^2} = 3RT$$

When this is rewritten in terms of the mean molecular energy $\overline{E}$,

$$\overline{E} = \frac{1}{2} m\overline{v^2} = \frac{3}{2} mRT$$

or

$$\overline{E} = \frac{1}{2} m\overline{v^2} = \frac{3}{2} kT$$

where k is Boltzmann's constant. It is clearly seen here that the macroscopic temperature T is proportional to the mean translational (kinetic) energy of the molecules in the system. The value of Boltzmann's constant depends on the system of units of the problem, for example

$$1.3803 \times 10^{-23} \frac{\text{joules}}{\text{molecule } °K} \text{ [MKS system]}$$

$$1.3803 \times 10^{-16} \frac{\text{ergs}}{\text{molecule } °K} \text{ [cgs system]}$$

$$7.23 \times 10^{-27} \frac{\text{Btu}}{\text{molecule } °R} \text{ [Engineering]}$$

2. DISTRIBUTION FUNCTION THEORY

In developing kinetic theory for an ideal gas, it was assumed that each of the molecules was moving with an energy proportional to the mean squared speed $\overline{v^2}$. However, the molecules are not all moving at the same velocities. It is possible to develop a velocity distribution function $f(v_i)$ which will give the probability of a molecule being in the range of molecular velocities v_i to $v_i + dv_i$. Since $f(v_i)$ is a probability distribution function, the number of molecules in this velocity range dv_i is given by

$$dN = N_0 f(v_i) \, dv_i \tag{4}$$

where N_0 is the total number of molecules in the system under study. Integrating eq. 4 over all molecular velocities,

$$N_0 = \int_0^{N_0} dN = \int_{-\infty}^{\infty} N_0 f(v_i) \, dv_i$$

or

$$1 = \int_{-\infty}^{\infty} f(v_i) \, dv_i \tag{5}$$

This equation describes the mathematical characteristics of the distribution function.

With this definition, any property of the system which is a function of the molecular velocities can be calculated; for example, the mean or average velocity $\bar{v}_i$ is

$$\bar{v}_i = \frac{\int_{-\infty}^{\infty} v_i N_o f(v_i)\, dv_i}{\int_{-\infty}^{\infty} N_o f(v_i)\, dv_i} = \int_{-\infty}^{\infty} v_i f(v_i)\, dv_i = 0$$

and the mean square of the molecular velocity v_i is

$$\overline{v_i^2} = \int_{-\infty}^{\infty} v_i^2 f(v_i)\, dv_i$$

and the mean molecular energy in the ith direction is

$$\frac{1}{2}\overline{mv_i^2} = \int_{-\infty}^{\infty} \frac{1}{2} mv_i^2 f(v_i)\, dv_i$$

The usual form of the velocity distribution function for a gas in equilibrium is given by[1,2,3]

$$f(v_i) = \left(\frac{m}{2\pi kT}\right)^{3/2} \exp\left(-\frac{mv_i^2}{2kT}\right) \tag{6}$$

and is referred to as the *Maxwell-Boltzmann velocity distribution function*. Other molecular distribution functions can be obtained from the velocity distribution function by mathematical transformations; for example, the distribution function for the molecular speed v is

$$f(v) = \frac{4v^2}{\sqrt{\pi}}\left(\frac{m}{2kT}\right)^{3/2} \exp\left(-\frac{mv^2}{2kT}\right) = \begin{array}{l}\text{Maxwell-Boltzmann}\\ \text{speed distribution function}\end{array} \tag{7}$$

Also, the molecular energies are distributed as

$$f(E) = 2\left(\frac{E}{\pi}\right)^{1/2}\left(\frac{1}{kT}\right)^{3/2} \exp\left(-\frac{E}{kT}\right) = \begin{array}{l}\text{Maxwell-Boltzmann}\\ \text{= energy distribution}\\ \text{function}\end{array} \tag{8}$$

Using eq. 7, the mean or average speed of a particle is

$$\bar{v} = \int_{0}^{\infty} vf(v)\, dv = \sqrt{\frac{8kT}{\pi m}}$$

and the mean square speed is

$$\overline{v^2} = \int_{0}^{\infty} v^2 f(v)\, dv = \frac{3kT}{m}$$

3. MOLECULAR COLLISIONS

It has been assumed thus far that the molecules interact only through collisions. For gases at low pressures and temperatures (below 1 atm of pressure and below about $2000°R$) this is the most probable type of interaction!

Molecules are assumed to interact in a system only through collisions. The distance traveled between molecular collisions, the molecular free path, is not a constant distance. However, it is useful to know the mean free path λ, which is the mean distance traveled by a molecule between collisions. Introducing the concept of a collision cross section σ, for a uniform gas of molecules having a radius r,

$$\sigma = \text{collision cross section} = 4\pi r^2$$

and the collision frequency ν is given by

$$\nu = \text{collision frequency} = \sigma n\bar{v}$$

From the definitions, it is easy to show that in a time t, the mean distance between molecular collisions is

$$\lambda = \frac{\bar{v}t}{n\sigma\bar{v}t} = \frac{1}{n\sigma} \tag{9}$$

For air, the mean speed $\bar{v}$, the collision frequency ν, and the mean free path λ which exist in the atmosphere as a function of altitude are illustrated in Figs. 4, 5, and 6 of Article 11.

4. TRANSPORT PHENOMENA

Transport phenomena are those properties of matter which exist as a result of non-equilibrium. That is, whenever there is a property gradient in matter, it will tend to equilibrate at a rate proportional to the transport property of the matter. The transport of mass, momentum, and energy due to a nonequilibrium situation can be treated by kinetic theory methods. It is important to realize that this is not possible in classical thermodynamics except by empirical methods.

Mass Transport. A system having a nonequilibrium concentration of molecules will tend to equilibrium. This tendency is expressed empirically in classical thermodynamics by Ficks law:

$$J = -D \frac{dn}{dx} \tag{10}$$

where D is the coefficient of self-diffusion and n is the number density of particles. Calculating the net rate of molecular transport from kinetic theory, it is found that[1,2]

$$J = -\frac{1}{3} \nu\lambda^2 \frac{dn}{dx}$$

$$= -\frac{1}{3} \bar{v}\lambda \frac{dn}{dx} \tag{11}$$

Comparing Eqs. 10 and 11, the coefficient of self-diffusion D can be determined in terms of molecular collision properties as

$$D = \frac{1}{3} \bar{v}\lambda \tag{12}$$

Momentum Transport. In a system undergoing a shear, there is a nonequilibrium distribution of momentum. In classical thermodynamics this nonequilibrium state is described empirically by the shear force equation

$$\tau = \mu \frac{du}{dy} \tag{13}$$

where μ is the coefficient of viscosity and u is a macroscopic velocity component of the system. Using the concepts of kinetic theory, it is possible to show that the shearing force comes about from molecular collisions, and in terms of collisional properties the shear force is given by[1,2]

$$\tau = \frac{1}{3} nm\bar{v}\lambda \frac{du}{dy}$$

$$= \frac{1}{3} \frac{m\bar{v}}{\sigma} \frac{du}{dy} \tag{14}$$

Thus the coefficient of viscosity is given by

$$\mu = \frac{1}{3} \frac{m\bar{v}}{\sigma} = \frac{1}{3} nm\bar{v}\lambda \tag{15}$$

depending on the particular gas properties known upon comparing Eqs. 13 and 14.

Energy Transport. When there is a temperature gradient in a system, there will be a transport of energy as the system tends to equilibriom. This phenomenon is described classically by Fouriers law of heat transfer

$$q = -\kappa \frac{dT}{dy} \tag{16}$$

Again the transport of energy comes about as a result of molecular collision. Thus in terms of collisional parameters, it can be shown that[1,2]

$$q = -\frac{1}{6} n\bar{v}fk\lambda \frac{dT}{dy} \tag{17}$$

where f is the number of degrees of freedom that the molecules possess; for example,

$$f = \begin{cases} 3 \text{ for monatomic gases} \\ 7 \text{ for diatomic gases} \end{cases}$$

Again by comparing Eqs. 16 and 17, the coefficient of heat transfer κ is given by

$$\kappa = \frac{1}{6} n\bar{v}vk\lambda$$

$$= \frac{1}{6} \frac{\bar{v}fk}{\sigma} \tag{18}$$

It should be recognized that an analytical method for calculating transport phenomena is more desirable than the experimental methods of classical thermodynamics. This is one of the most significant contributions of kinetic theory to thermodynamics.

5. THE BOLTZMANN EQUATION

Using the concept of a distribution function described earlier, it is possible to determine the velocity and position of each molecule in a system as a function of time. Consider the distribution function $f(x_i, c_i, t)$ where

x_i = position vector
c_i = total velocity of particle
  $= u_i + v_i$
u_i = macroscopic velocity of system
v_i = thermal or molecular velocity of particle

The quantity $f(x_i, c_i, t) \, dc_i \, dx_i$ is the probability of a particle having velocity in the range c_i to $c_i + dc_i$ and position in the range x_i to $x_i + dx_i$ at time t. Thus, if $f(x_i, c_i, t)$ is known, the system of particles is completely described.

The rate of change of $f(x_i, c_i, t)$ with time is given by the Boltzmann equation[1,4,5]

$$\frac{\partial(nf)}{\partial t} + c_i \frac{\partial(nf)}{\partial x_i} + a_i \frac{\partial(nf)}{\partial c_i} = \left[\frac{\partial(nf)}{\partial t}\right]_{collision} \tag{19}$$

where a_i = acceleration vector due to an external force and $[\partial(nf)/\partial t]_{collision}$ is the change in the distribution function due to collisions. Equation 19 is a nonlinear integro-differential equation (due to the collision term) and is very difficult to solve. The Boltzmann equation has been solved for a few restricted cases such as equilibrium flow or by approximating the collision term with simple expressions (see Chap. 10, Ref. 1). It must be emphasized that $f(x_i, c_i, t)$ is the quantity being sought and the Boltzmann equation is the governing equation for f.

6. THE KNUDSEN NUMBER

It is useful to nondimensionalize the Boltzmann equation and examine its character. The appropriate characteristic quantities are a macroscopic length R, a microscopic length λ, and the macroscopic free stream velocity. The nondimensional Boltzmann equation (starred quantities are nondimensional) is

$$\frac{\partial f^*}{\partial t^*} + c_i^* \frac{\partial f^*}{\partial x_i^*} + a_i^* \frac{\partial f^*}{\partial c_i^*} = \frac{1}{\sqrt{2}\pi} \frac{1}{K_n} \left[\frac{\partial f^*}{\partial t^*}\right]_{\text{collision}} \tag{20}$$

where $K_n = \lambda/R$, the Knudsen number, is the ratio of mean free path to a macroscopic characteristic length.

The character of the Boltzmann equation can be categorized according to the magnitude of K_n.[5] These categories are shown in Table 1. For small K_n (less than 0.01) the fluid is collision-dominated and is the continuum regime. For large K_n (greater than about 100) the collisions between particles are insignificant compared to the incident particle impact on a surface and the flow regime is called free molecular or rarefied gas (see Art. 11).

Table 1

K_n	Flow Regime	Influence of Collisions
<1	Continuum	Very significant, no slip condition valid
<1	Slip flow	Significant, slip at the surface
1	Transition	Same order of magnitude as convection
>1	Free molecular	Not significant

In the continuum flow regime the governing similarity parameters are Reynolds number R_e and Mach number M. These parameters are related to the Knudsen number by

$$K_n = 1.88\sqrt{\gamma}\,\frac{M}{R_e}$$

7. CONTINUUM FLOW CONSERVATION EQUATIONS

The Boltzmann equation is multiplied by a quantity $Q(c_i)$ and then integrated over all c_i (i.e., integrated over all velocity space). The quantity $Q(c_i)$ is chosen to be equal to the mass, momentum, and energy which are conserved during collisions. When the collision term (right-hand side of Eq. 19) is multiplied by $Q(c_i)$ and integrated over c_i, the result is zero. The left-hand side is simplified by using the fact that $\int_{-\infty}^{\infty} Q(c_i) f(x_i,c_i,t)\,dc_i = Q\,(x_i,t)$ where the bar indicates a quantity averaged over velocity space. Thus $\int_{-\infty}^{\infty} c_i\, f(x_i,c_i,t)\,dc_i = u_i \int_{-\infty}^{\infty} f(x_i,c_i,t)\,dc_i + \int_{-\infty}^{\infty} v_i\, f(x_i,c_i,t)\,dc_i = u_i + 0 = u_i$ and the general result is the following:[1,5]

(1) For $Q(c_i) = m$ ..., the continuity equation

$$\frac{\partial \rho}{\partial t} + \frac{\partial}{\partial x_i}(\rho u_i) = 0 \tag{21}$$

(2) For $Q(c_j) = mc_j = m(u_j + v_j)$ ..., the momentum equation

$$\frac{\partial(\rho u_j)}{\partial t} + \frac{\partial}{\partial x_i}(\rho u_j u_i) = \rho a_j - \frac{\partial}{\partial x_i}(\overline{\rho v_i v_j}) \tag{22}$$

The term $\overline{\rho v_i v_j}$ is a tensor quantity representing the transport of momentum by molecular motion. The diagonal terms are identified, using Eq. 2, as the pressure p. The off-diagonal terms are identified by the Chapman-Enskog solution[1] to be the shear stress tensor τ_{ij} (see Art. 10).

(3) For $Q(c_i) = (1/2)mc_ic_i$ ..., the energy equation

$$\frac{\partial}{\partial t}\left[\frac{1}{2}\rho u^2 + \frac{1}{2}\overline{\rho v^2}\right] + \frac{\partial}{\partial x_i}\left[u_i\left(\frac{1}{2}\rho u^2 + \frac{1}{2}\overline{\rho v^2}\right)\right]$$

$$= \rho a_i u_i - \frac{\partial}{\partial x_i}\left(\frac{1}{2}\overline{\rho v_i v^2}\right) - \frac{\partial}{\partial x_i}(\overline{\rho v_i v_k u_k}) \tag{23}$$

The term $1/2\ \rho u^2$ is the macroscopic kinetic energy of the system and the term $1/2\ \overline{\rho v^2} = 3/2\ nkT$ is the thermal or mean molecular energy of the particles in the system.

The term $1/2 \, \rho \overline{v_i v}^2$ is the transport by molecular motion of the mean molecular energy. This term is identified as heat transfer by conduction at the macroscopic level. The last term, $\rho \overline{v_i v_i} u_k$, is the rate of work done by the shear stress and the pressure.[1]

Equations 21, 22, and 23 are called the continuum conservation equations and form a most useful set of equations. However, it must be remembered that this set is an approximation to a more accurate system[6]; i.e., the Boltzmann equation. The process of integrating over all velocity space is an averaging process and removes the velocity as an independent variable.

BIBLIOGRAPHY

1. VINCENTI, W. G., and KRUGER, C. H. JR., *Introduction to Physical Gas Dynamics*, New York, Wiley, 1965.
2. LEE, J. F., SEARS, F. W., and TURCOTTE, D. L., *Statistical Thermodynamics*, Reading, Massachuetts, Addison-Wesley, 1963.
3. PRESENT, R. D., *Kinetic Theory of Gases*, New York, McGraw-Hill, 1958.
4. CHAPMAN, S., and COWLING, T. G., *The Mathematical Theory of Non-Uniform Gases*, London, Cambridge, Univ. Press, 1952.
5. EMMONS, H. W. (ed.), *High Speed Aerodynamics and Jet Propulsion, Volume III, Fundamentals of Gas Dynamics*, Princeton, New Jersey, Princeton Univ. Press, 1958.
6. MOORE, F. K., *High Speed Aerodynamics and Jet Propulsion, Volume IV, Theory of Laminar Flows*, Princeton, New Jersey, Princeton Univ. Press, 1964.

CONSERVATION EQUATIONS OF CONTINUUM FLOW

By L. M. Nicolai

8. THE CONTINUUM APPROACH

In this article the conservation equations in both integral and differential form that describe the continuum fluid flow phenomena are presented. We consider the fluid as a continuum since in many engineering applications our primary interest lies not in motions of molecules but rather in the gross behavior of the fluid thought of as a continuous material. This means that the fluid region under study must contain enough molecules so that statistically averaged properties are meaningful.

The system of equations can be derived from two general approaches.[5] One is by the method of kinetic theory (Art. 1) where the behavior of the molecules is of interest, and the velocity and position of the molecules are given by a distribution function. By assuming a simplified model for the molecular interactions an equation can be formulated[1,2,3] that describes the rate of change, with respect to position and time, of the distribution function. This equation is called the Boltzmann equation[1,2,3] (see Eq. 19, Art. 5). If velocity moments of the Boltzmann equation are integrated over the velocity space, the resulting equations are the general form of the continuum conservation equation[1,2] (see Eqs. 21, 22, and 23, Art. 7).

The second approach is to replace the detailed molecular structure that is actually present in any gas or liquid by a continuous model of matter with statistically averaged properties at every point.

9. INTEGRAL EQUATIONS

The second, or continuum, approach is the one from which the conservation equations will be derived. A control region of volume V, surface area S, and surface normal n_i is considered fixed with respect to the coordinate system. We then examine the fluid property fluxes through the control region, the influences of body and surface forces, and the heat and work interactions at the surface.

The conservation of mass equation follows then from the statement that the time rate of change of mass within V plus the flux of mass across S is equal to zero;

$$\int_V \frac{\partial \rho}{\partial t} \, dV + \int_S \rho(u_i n_i) \, dS = 0 \tag{1}$$

where u_i = mean macroscopic motion of fluid.

The conservation of momentum equation follows from Newton's laws of motion for a system of fixed mass. The system boundaries are assumed coincident with the control region boundaries[4] so that the time rate of change of the momentum within the system boundaries is equal to the external forces on the system, which is equal to the local time rate of change of momentum and the momentum flux through the control region V. This is expressed as:

$$\int_V \frac{\partial(\rho u_i)}{\partial t}\, dV + \int_S \rho u_i(u_j n_j)\, dS = \int_V \rho f_i\, dV + \int_S \sigma_{ij} n_j\, dS \tag{2}$$

where f_i represents the body forces per unit mass (gravity, Lorentz forces, etc.) and σ_{ij} is a symmetric tensor representing the surface stresses.

The conservation of energy equation follows from the First Law of Thermodynamics for a system. Here again the system boundaries coincide with the boundaries of the control region[4] so that the rate of heat and work interaction at the surface S is equal to the local rate of change of total energy and the flux of total energy through the volume V. The total energy is the sum of the internal, kinetic, and potential energies. This conservation of energy is expressed in the integral form as:

$$\int_V \frac{\partial}{\partial t}\left[\rho\left(e + \frac{u_i u_i}{2} + gz\right)\right] dV + \int_S \rho\left(e + \frac{u_i u_i}{2} + gz\right) u_j n_j\, dS$$

$$= -\int_S q_i n_i\, dS + \int_V \rho w_x\, dV + \int_V \dot{Q}\, dV + \int_V \rho u_i f_i\, dV + \int_S (\sigma_{ij} u_i n_j)\, dS \tag{3}$$

where q_i represents the rate of energy transfer into the system by conduction, w_x represents the rate of work done per unit mass on the system by a work reservoir of some other system, $\dot{Q}$ is the rate at which energy is generated internal to the system (such as by Joule heating), and the last term is the rate of work done on the system by pressure and shearing forces at the surface. Equation 3 neglects the energy transfer by radiation.

10. DIFFERENTIAL EQUATIONS

The differential form of the conservation equations follows directly from the integral form by assuming that the parameters have sufficiently many derivatives and using the divergence theorem of calculus to transform the surface integrals into volume integrals. The conservation of mass equation becomes

$$\frac{\partial \rho}{\partial t} + \frac{\partial}{\partial x_i}(\rho u_i) = 0 \tag{4}$$

Before writing the momentum equation in the form usually called the Navier-Stokes equation, we must express the surface stress term σ_{ij} in a more useful form. We assume:

(1) Newtonian fluid—the rate of strain of a fluid element is relatively small such that a linear relationship exists between the rate of strain and the rate of stress.

(2) The normal stress in the absence of deformation is equal to the negative of the thermodynamic pressure P.

(3) The properties of the fluid are isotropic.

The stress tensor σ_{ij} is expressed as[3]

$$\sigma_{ij} = -P\delta_{ij} + \tau_{ij} = -P\delta_{ij} + \mu\left(\frac{\partial u_i}{\partial x_j} + \frac{\partial u_i}{\partial x_i}\right) + \left(\mu' - \frac{2}{3}\mu\right)\frac{\partial u_k}{\partial x_k}\delta_{ij} \tag{5}$$

where μ is the ordinary viscosity coefficient, μ' is the dilational coefficient,[3,5] and δ_{ij} is the kronecker delta function.

The momentum equation or Navier-Stokes equation is finally expressed as

$$\rho\frac{\partial u_i}{\partial t} + \rho u_j\frac{\partial u_i}{\partial x_j} = -\frac{\partial P}{\partial x_i} + \rho f_i + \frac{\partial \tau_{ij}}{\partial x_j} \tag{6}$$

The differential form for the conservation of energy equation is (omitting the potential energy term):

$$\frac{\partial}{\partial t}\left[\rho\left(e + \frac{1}{2}u_i u_i\right)\right] + \frac{\partial}{\partial x_j}\left[\rho u_j\left(e + \frac{1}{2}u_i u_i\right)\right] = -\frac{\partial q_i}{\partial x_i}$$

$$+ \rho u_i f_i - \frac{\partial(Pu_i)}{\partial x_i} + \frac{\partial(u_j \tau_{ij})}{\partial x_i} + \rho w_x + \dot{Q} \tag{7}$$

The heat conduction term is usually expressed as $q_i = -k\ (\partial T/\partial x_i)$ according to Fourier's heat conduction laws. Also, the pressure work term can be expressed as $-\ (D/Dt)\ (P/\rho) + (1/\rho)\ (\partial P/\partial t)$ using the continuity equation.

Finally, the energy equation is rewritten as

$$\rho \frac{D}{Dt}\left(h + \frac{1}{2}u_i u_i\right) = \frac{\partial}{\partial x_i}\left(k\frac{\partial T}{\partial x_i}\right) + \rho u_i f_i + \frac{\partial P}{\partial t} + \frac{\partial}{\partial x_i}(u_j \tau_{ij}) + \rho w_x + \dot{Q} \quad (8)$$

where $h = e + (P/\rho)$, the static enthalpy per unit mass.

BIBLIOGRAPHY

1. VINCENTI, W. G., and KRUGER, C. H., JR., *Introduction to Physical Gas Dynamics*, New York, Wiley, 1965.
2. UMAN, M. A., *Introduction to Plasma Physics*, New York, McGraw-Hill, 1964.
3. EMMONS, H. W., *High Speed Aerodynamics and Jet Propulsion, Volume III, Fundamentals of Gas Dynamics*, Princeton, New Jersey, Princeton Univ. Press, 1958.
4. PAO, R. H. F., *Fluid Dynamics*, Columbus, Ohio, Merrill, 1967.
5. MOORE, F. K., *High Speed Aerodynamics and Jet Propulsion, Volume IV, Theory of Laminar Flows*, Princeton, New Jersey, Princeton Univ. Press, 1964.

ATMOSPHERIC STRUCTURE

By W. C. Bauman

11. PROPERTIES OF THE ATMOSPHERE

The first modern standard atmosphere was developed in the United States and Europe in the 1920s to satisfy a need for standardizing aircraft instruments and performance. Subsequently, several groups have met to obtain international adoption of past work. With the advent of reliable instrumentation in rockets and satellites it has been possible to extend the standard atmosphere to altitudes near 1000 km. The material presented here has been drawn from the results of the United States Committee on Extension of the Standard Atmosphere (COESA) which represented twenty-nine U.S. scientific and engineering organizations.

The atmosphere enclosing the earth is composed of at least seventeen different gases. The two principal gases are nitrogen and oxygen, 78.084 and 20.947% by volume, respectively. In describing the atmosphere it is convenient to classify the various layers by their thermal structure. The names of the regions and boundaries, up to 100 km, are shown in Fig. 1 and are those adopted by the World Meterological Organization.

The *U.S. Standard Atmosphere, 1962*, is defined in geopotential height up to 90 km (300,000 ft) and in geometric height above 90 km.

When using atmospheric properties, two basic equations are used to relate pressure, temperature, and density. The hydrostatic equation is

$$dp = -\rho g\ dz = -\rho g_0\ dh$$

where p is pressure, ρ is density, g is the acceleration of gravity at height z, g_0 is standard acceleration of gravity at sea level, z is the geometric height, and h is the geopotential height, defined by the equation $dh = g/g_0\ dz$.

The equation of state for air is

$$p = \rho R T_m/M_0 = \rho R_a T$$

where ρ is the density, R is the universal gas constant, R_a is the gas constant for air, T is the temperature (kinetic temperature), and M_0 is the average molecular weight of air at sea level. The average molecular weight from sea level up to approximately 90 km (30 × 10⁵ ft), is constant and equal to 82.964. It then decreases to 22.52 at 10⁶ ft and 16.77 at 2 × 10⁶ ft. Above 90 km the molecular weight begins to decrease because of molecular dissociation and this decrease is approximately linear with altitude. The molecular temperature T_m is defined as

$$T_m = M_0/M T$$

where M is the average molecular weight of air at the specified altitude, M_0 as defined above.

The composition of the atmosphere is virtually constant up to 90 km, above which molecular dissociation increases with altitude (see Chapter 6, Ref. 2). The pressure and density decrease approximately exponentially with altitude (see Fig. 2). Water vapor decreases irregularly with altitude except for a slight increase between 35,000 and 115,000 ft.

The coefficient of viscosity is defined as the coefficient of internal friction developed where gas regions move adjacent to one another at different velocities. Two of the more widely used formulas for determining the coefficient of viscosity (μ) are Sutherland's law and the power law. Sutherland's equation gives reasonably good results for the temperature range 180 to 3400°R (at atmospheric pressure) and was used for the U.S. Standard Atmosphere, 1962. Sutherland's equation is

$$\mu = \frac{\beta T^{3/2}}{T + S}$$

where T is absolute temperature, S is Sutherland's constant, and β is a constant; both values are listed in Table 1 for the English and Metric system of units. The power law is somewhat easier to use and yields good results in the temperature range of 300 to 900°R.

$$\mu = \mu_0 \left(\frac{T}{T_0} \right)^{0.76}$$

where μ_0 and T_0 are reference values for viscosity and absolute temperature, $\mu_0 = 3.02 \times 10^{-7}$ lb-sec/ft^2 and $T_0 = 392$°R.

The properties of air at standard sea level pressure (14.696 psia) are given in Table 2. The generally accepted method of computing viscosity (μ) is by use of Sutherland's equation. Beyond the upper temperature limit of 3400°R the properties are based on molecular dissociation having taken place.

An abbreviated listing of the various properties of the atmosphere (U.S. Standard Atmosphere, 1962[1]) up to 2,300,000 ft is given in Table 3. Above 300,000 feet (90 km) the speed of sound (c), viscosity (μ), and thermal conductivity (k) are omitted as a result

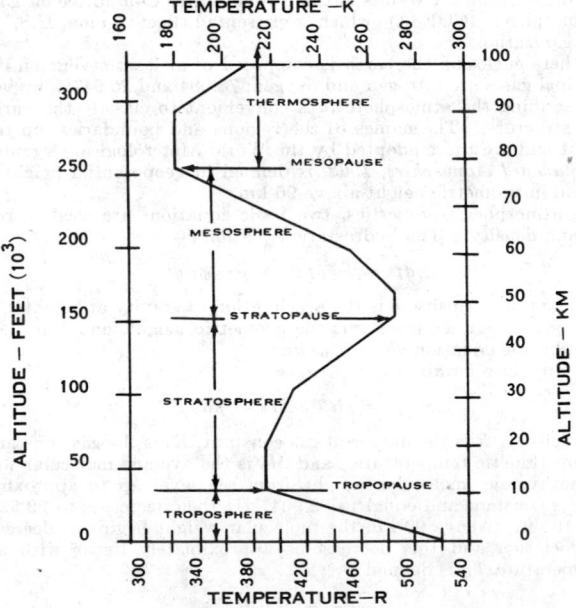

Fig. 1. Variation of temperature in 1962 standard atmosphere.

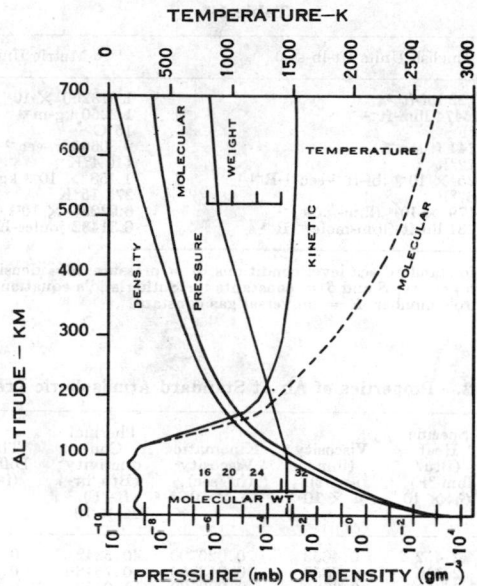

Fig. 2. Variation in properties in 1962 standard atmosphere.

of the increased mean free path of the air molecules and the lack of meaning of these conventional properties.

The *U.S. Standard Atmosphere, 1962*, represents an average of the (spring and fall) properties at approximately 45° north latitude. Atmospheric properties for various north latitudes for the different seasons of the year are contained in Ref. 4.

Those individuals and agencies requiring computer oriented functions for computation of a standard atmospheric pressure and density are referred to Ref. 4, Part 4.

Figure 3 presents the ratio γ of the specific heat at constant pressure to that at constant volume for air, obtained from measurements by Heck[5].

At lower altitudes the air is dense enough so that it may be considered a continuum under practically all flight conditions. However, at higher altitudes the air may be so rare that it may no longer be considered a continuum as in ordinary gas dynamics but rather an aggregate of rapidly moving molecules. When the mean free path λ of the

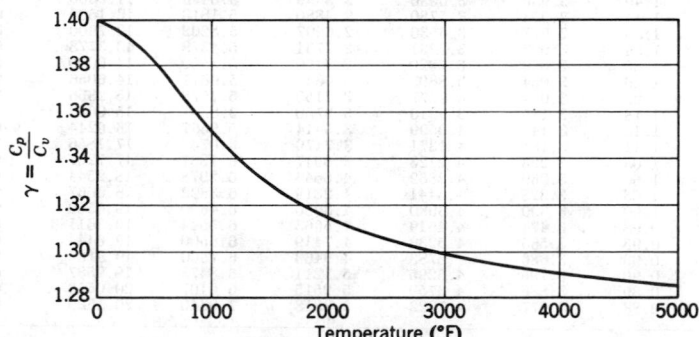

Fig. 3. Specific heat ratio of pure dry air.

Table 1[1]*

Symbol	English Units (ft-lb-sec)	Metric Units (m-k-s)
P_0	2116.22 lbf-ft^{-2}	1.013250×10^5 Newtons-M^{-2}
ρ_0	0.076474 lbm-ft^{-3}	1.2250 kg-m^{-3}
t_0	59.0°F	15°C
g_0	32.1741 ft-sec^{-2}	9.80665 m-sec^{-2}
S	198.72°R	110.4°K
β	7.3025×10^{-7} lbf-ft^{-1}-sec^{-1}-R$^{-1/2}$	1.458×10^{-6} kg-sec^{-1}-m^{-1}-k$^{-1/2}$
T_i	491.67°R	273.15°K
N	2.73179×10^{26} (lbm-mole)$^{-1}$	6.02257×10^{26} (kg-mole)$^{-1}$
R	1545.31 lbf-ft-(lbm-mole)$^{-1}$R^{-1}	8.31432 joules-K^{-1}-mole^{-1}

* Subscript 0 refers to standard sea level conditions. p = pressure, ρ = density, t = temperature, g = acceleration due to gravity, S and β = constants for Sutherland's equation, T_i = ice point temperature, N = Avogadro's number, R = universal gas constant.

Table 2.   Properties of Air at Standard Atmospheric Pressure[3]*

Temp (°F), t	Density (lbm/cu ft), $\rho \times 10^2$	Specific Heat (Btu/lbm-°F), $C_p \times 10$	Viscosity (lbm/sec-ft), $\mu \times 10^5$	Kinematic Viscosity (ft^2/sec), $\nu \times 10^3$	Thermal Conductivity (Btu/hr-ft-°F), $K \times 10^2$	Thermal Diffusivity (ft^2/hr), a	Prandtl Number, P_r
−280	22.48	2.452	0.4653	0.020700	0.5342	0.09691	0.770
−200	15.64	2.416	0.6659	0.043856	0.7648	0.20863	0.755
−100	11.04	2.403	0.8930	0.080620	1.0450	0.39390	0.739
0	8.66	2.401	1.0926	0.10960	1.3124	0.54874	0.720
100	7.10	2.404	1.2750	0.18102	1.5647	0.92477	0.706
200	5.99	2.414	1.4413	0.24213	1.8047	1.25633	0.694
300	5.23	2.429	1.5951	0.29293	2.0320	1.53656	0.686
400	4.62	2.450	1.7390	0.36471	2.2481	1.92489	0.681
500	4.14	2.474	1.8743	0.45420	2.4570	2.40600	0.680
600	3.75	2.512	2.0027	0.53587	2.6536	2.82656	0.680
700	3.42	2.538	2.1231	0.62122	2.8431	3.27811	0.682
800	3.14	2.568	2.2390	0.71310	3.0220	3.74800	0.684
900	2.92	2.596	2.3498	0.80562	3.2003	4.22567	0.686
1000	2.71	2.628	2.4569	0.90602	3.3710	4.72967	0.690
1100	2.54	2.659	2.5600	1.0075	3.5313	5.22667	0.693
1200	2.39	2.690	2.6569	1.1117	3.6911	5.74333	0.697
1300	2.25	2.717	2.7513	1.2222	3.8428	6.28333	0.701
1400	2.13	2.742	2.8450	1.3380	3.9933	6.84667	0.703
1500	2.02	2.767	2.9367	1.4547	4.1472	7.42111	0.706
1600	1.93	2.791	3.0234	1.5722	4.2811	7.97067	0.710
1700	1.84	2.815	3.1090	1.6900	4.4100	8.51400	0.714
1800	1.76	2.840	3.1918	1.8222	4.5383	9.11844	0.718
1900	1.68	2.864	3.2736	1.9542	4.6676	9.72622	0.722
2000	1.61	2.887	3.3513	2.0853	4.8003	10.34633	0.725
2100	1.55	2.909	3.4280	2.2171	4.9284	10.9600	0.728
2200	1.49	2.930	3.5030	2.3499	5.0496	11.5600	0.732
2300	1.44	2.953	3.5780	2.4880	5.1813	12.1600	0.736
2400	1.39	2.977	3.6530	2.6297	5.3202	12.7600	0.740
2500	1.34	3.002	3.7231	2.7731	5.4378	13.3778	0.745
2600	1.30	3.028	3.7920	2.9170	5.5500	14.0000	0.749
2700	1.26	3.054	3.8603	3.0648	5.6611	14.6056	0.755
2800	1.22	3.081	3.9277	3.2150	5.7678	15.2356	0.760
2900	1.18	3.110	3.9910	3.3750	5.8567	15.9633	0.764
3000	1.15	3.143	4.0599	3.5414	5.9567	16.6244	0.771
3100	1.11	3.181	4.1371	3.7176	6.0733	17.1856	0.779
3200	1.08	3.223	4.2123	3.8917	6.1867	17.7267	0.790
3300	1.06	3.269	4.2862	4.0644	6.2978	18.2544	0.801
3400	1.03	3.328	4.3441	4.2319	6.3822	18.6667	0.815
3500	1.00	3.390	4.3980	4.3980	6.4600	19.0500	0.831
3600	0.98	3.474	4.4619	4.5663	6.5544	19.3611	0.849
3700	0.95	3.565	4.5239	4.7419	6.6500	19.6444	0.869
3800	0.93	3.686	4.5783	4.9463	6.7500	19.8167	0.898
3900	0.90	3.795	4.6295	5.1211	6.8473	19.9587	0.923
4000	0.89	3.886	4.6759	5.2515	6.9407	20.0553	0.942
4100	0.88	3.977	4.7222	5.3818	7.0340	20.1520	0.961

* Symbols: t = temperature; ρ = density; C_p = specific heat at constant pressure; μ = dynamic viscosity; ν = kinematic viscosity; K = thermal conductivity; a = thermal diffusivity ($a = K/\rho C_p$); P_r = Prandtl number ($P_r = \nu/a$).   Note: To determine property value, use equation at top of column and set equal to tabulated value.   Example: Viscosity at 1000°F, $\mu \times 10^5$ = 2.4569.   Therefore μ = 2.4569 × 10^{-5} lbm/sec-ft.

ATMOSPHERIC STRUCTURE 651

Table 3. Properties of the U.S. Standard Atmosphere, 1962*

Geometric Altitude (z) ft	Temperature (t) °F	Temperature (t) °C	Speed of Sound (ft/sec)	Pressure (P) lbf/ft²	Density (ρ) lbm/ft³	Coefficient of Viscosity (μ) lbm/ft-sec	Kinematic Viscosity (ν) ft²/sec	Thermal Conductivity (K) Btu/sec-ft-°F	Acceleration Due to Gravity (g) ft/sec²
0	59.0	15.0	1116.5	2.112 +3	7.647 −2	1.202 −5	1.572 −4	4.067 −6	32.174
1,000	55.4	13.0	1112.6	2.041 +3	7.426 −2	1.196 −5	1.611 −4	4.042 −6	32.171
2,000	51.9	11.0	1108.8	1.968 +3	7.209 −2	1.190 −5	1.650 −4	4.017 −6	32.168
3,000	48.3	9.1	1104.9	1.897 +3	6.998 −2	1.183 −5	1.691 −4	3.992 −6	32.165
4,000	44.7	7.1	1101.0	1.828 +3	6.792 −2	1.177 −5	1.732 −4	3.967 −6	32.162
5,000	41.2	5.1	1097.1	1.761 +3	6.590 −2	1.170 −5	1.776 −4	3.942 −6	32.159
6,000	37.6	3.1	1093.2	1.696 +3	6.393 −2	1.164 −5	1.820 −4	3.917 −6	32.156
7,000	34.0	1.1	1089.3	1.633 +3	6.200 −2	1.157 −5	1.866 −4	3.891 −6	32.152
8,000	30.5	−0.8	1085.3	1.572 +3	6.012 −2	1.150 −5	1.914 −4	3.866 −6	32.149
9,000	26.9	−2.8	1081.4	1.513 +3	5.828 −2	1.144 −5	1.963 −4	3.840 −6	32.146
10,000	23.4	−4.8	1077.4	1.456 +3	5.648 −2	1.137 −5	2.013 −4	3.815 −6	32.143
11,000	19.8	−6.8	1073.4	1.400 +3	5.473 −2	1.131 −5	2.066 −4	3.789 −6	32.140
12,000	16.2	−8.8	1069.4	1.346 +3	5.302 −2	1.124 −5	2.120 −4	3.764 −6	32.137
13,000	12.7	−10.7	1065.4	1.294 +3	5.135 −2	1.117 −5	2.175 −4	3.738 −6	32.134
14,000	9.1	−12.7	1061.4	1.244 +3	4.973 −2	1.110 −5	2.233 −4	3.713 −6	32.131
15,000	5.5	−14.7	1057.4	1.195 +3	4.814 −2	1.104 −5	2.293 −4	3.687 −6	32.128
16,000	2.0	−16.7	1053.3	1.148 +3	4.659 −2	1.097 −5	2.354 −4	3.661 −6	32.125
17,000	−1.6	−18.7	1049.2	1.102 +3	4.508 −2	1.090 −5	2.418 −4	3.636 −6	32.122
18,000	−5.1	−20.6	1045.2	1.058 +3	4.361 −2	1.083 −5	2.484 −4	3.610 −6	32.119
19,000	−8.7	−22.6	1041.1	1.015 +3	4.217 −2	1.076 −5	2.553 −4	3.584 −6	32.115
20,000	−12.3	−24.6	1036.9	9.733 +2	4.077 −2	1.070 −5	2.623 −4	3.558 −6	32.112
21,000	−15.8	−26.6	1032.8	9.333 +2	3.941 −2	1.063 −5	2.697 −4	3.532 −6	32.109
22,000	−19.4	−28.5	1028.7	8.946 +2	3.808 −2	1.056 −5	2.772 −4	3.506 −6	32.106
23,000	−22.9	−30.5	1024.5	8.573 +2	3.679 −2	1.049 −5	2.851 −4	3.480 −6	32.103
24,000	−26.5	−32.5	1020.3	8.212 +2	3.553 −2	1.042 −5	2.932 −4	3.454 −6	32.100
25,000	−30.0	−34.5	1016.1	7.863 +2	3.431 −2	1.035 −5	3.017 −4	3.428 −6	32.097
26,000	−33.6	−36.5	1011.9	7.527 +2	3.311 −2	1.028 −5	3.104 −4	3.401 −6	32.094
27,000	−37.2	−38.4	1007.7	7.203 +2	3.195 −2	1.021 −5	3.195 −4	3.375 −6	32.091
28,000	−40.7	−40.4	1003.4	6.890 +2	3.082 −2	1.014 −5	3.289 −4	3.349 −6	32.088
29,000	−44.3	−42.4	999.1	6.588 +2	2.974 −2	1.007 −5	3.387 −4	3.322 −6	32.085
30,000	−47.8	−44.3	994.9	6.297 +2	2.866 −2	9.996 −6	3.488 −4	3.296 −6	32.082
31,000	−51.4	−46.3	990.6	6.016 +2	2.762 −2	9.925 −6	3.594 −4	3.270 −6	32.079
32,000	−54.9	−48.3	986.2	5.746 +2	2.661 −2	9.853 −6	3.703 −4	3.243 −6	32.076
33,000	−58.5	−50.3	981.9	5.486 +2	2.563 −2	9.781 −6	3.817 −4	3.217 −6	32.072
34,000	−62.1	−52.3	977.5	5.235 +2	2.468 −2	9.709 −6	3.935 −4	3.190 −6	32.069
35,000	−65.6	−54.2	973.1	4.994 +2	2.375 −2	9.637 −6	4.057 −4	3.163 −6	32.066
36,000	−69.7	−56.2	968.7	4.761 +2	2.285 −2	9.564 −6	4.185 −4	3.137	32.063
37,000	−69.7	−56.5	968.1	4.539 +2	2.181 −2	9.553 −6	4.379 −4	3.133	32.060
38,000	−69.7	−56.5	968.1	4.326 +2	2.079 −2	9.553 −6	4.550 −4	3.133	32.057
39,000	−69.7	−56.5	968.1	4.124 +2	1.982 −2	9.553 −6	4.819 −4	3.133	32.054
40,000	−69.7	−56.5	968.1	3.931 +2	1.889 −2	9.553 −6	5.056 −4	3.133	32.051
41,000	−69.7	−56.5	968.1	3.748 +2	1.801 −2	9.553 −6	5.304 −4	3.133	32.048
42,000	−69.7	−56.5	968.1	3.572 +2	1.717 −2	9.553 −6	5.564 −4	3.133	32.045
43,000	−69.7	−56.5	968.1	3.405 +2	1.637 −2	9.553 −6	5.837 −4	3.133	32.042

Table 3.   Properties of the U.S. Standard Atmosphere, 1962 (continued)

Geometric Altitude (z) ft	Temperature (t) °F	Temperature (t) °C	Speed of Sound (ft/sec)	Pressure (P) lbf/ft²	Density (ρ) lbm/ft³	Coefficient of Viscosity (μ) lbm/ft-sec	Kinematic Viscosity (ν) ft²/sec	Thermal Conductivity (K) Btu/sec-ft-°F	Acceleration Due to Gravity (g) ft/sec²
44,000	−69.7	−56.5	968.1	3.246	1.560	9.553	6.123	3.133	32.039
45,000	−69.7	−56.5	968.1	3.095	1.487	9.553	6.423	3.133	32.036
46,000	−69.7	−56.5	968.1	2.950	1.418	9.553	6.738	3.133	32.033
47,000	−69.7	−56.5	968.1	2.812	1.352	9.553	7.068	3.133	32.030
48,000	−69.7	−56.5	968.1	2.681	1.288	9.553	7.414	3.133	32.026
49,000	−69.7	−56.5	968.1	2.555	1.228	9.553	7.778	3.133	32.023
50,000	−69.7	−56.5	968.1	2.436	1.171 −2	9.553	8.159 −4	3.133	32.020
55,000	−69.7	−56.5	968.1	1.918	9.219 −3	9.553	1.036 −3	3.133	32.005
60,000	−69.7	−56.5	968.1	1.510	7.259	9.553	1.316	3.133	31.990
65,000	−69.7	−56.5	968.1	1.189 +2	5.716 −3	9.553 −6	1.671 −3	3.133 −6	31.974
70,000	−67.4	−55.2	970.9	9.373 +1	4.479 −3	9.599 −6	2.143 −3	3.150	31.959
75,000	−64.7	−53.7	974.2	7.399	3.511	9.655	2.750	3.170	31.944
80,000	−62.0	−52.2	977.6	5.851	2.758	9.711	3.521	3.191	31.929
85,000	−59.3	−50.7	980.9	4.635	2.170	9.766	4.501	3.211	31.913
90,000	−56.5	−49.2	984.3	3.678	1.710	9.821	5.743	3.231	31.898
95,000	−53.8	−47.7	987.6	2.923	1.350	9.876	7.316	3.251	31.883
100,000	−51.1	−46.2	990.9	2.327	1.068 −3	9.931 −6	9.302 −3	3.272	31.868
110,000	−41.3	−40.7	1002.7	1.484 +1	6.647 −4	1.013 −5	1.524 −2	3.345	31.837
120,000	−26.1	−32.3	1020.8	9.601 +0	4.151	1.043	2.512	3.457	31.807
130,000	−10.9	−23.8	1038.5	6.310	2.653	1.072	4.069	3.568	31.777
140,000	4.3	−15.4	1055.9	4.206	1.699	1.101	6.480 −2	3.678	31.746
150,000	19.4	−7.0	1073.0	2.842	1.112	1.130	1.016 −1	3.787	31.716
160,000	27.5	−2.5	1082.0	1.942	7.471 −5	1.145	1.532	3.845	31.686
170,000	27.5	−2.5	1082.0	1.330 +0	5.116 −5	1.145	2.238	3.845	31.656
180,000	18.9	−7.3	1072.4	9.105 −1	3.558	1.129	3.173	3.783	31.626
190,000	8.1	−13.3	1060.3	6.155	2.466	1.109	4.495	3.705	31.596
200,000	−2.7	−19.3	1048.0	4.135	1.696	1.088	6.416	3.628	31.566
210,000	−22.0	−30.0	1025.6	2.744	1.175 −5	1.051	8.944 −1	3.487	31.536
220,000	−43.5	−41.9	1000.1	1.784	8.036 −6	1.008 −5	1.255 0	3.328	31.506
230,000	−64.9	−53.8	974.0	1.135 −1	5.389	9.650 −6	1.791	3.168	31.476
240,000	−86.4	−65.8	947.1	7.041 −2	3.535	9.207	2.604	3.007	31.446
250,000	−107.8	−77.7	919.5	4.248	2.263	8.753	3.868	2.843	31.42
260,000	−129.3	−89.6	891.1	2.484	1.409 −6	8.289	5.881 0	2.678	31.39
270,000	−134.5	−92.5	884.0	1.418 −2	8.172 −7	8.173	1.000 +1	2.638	31.36
280,000	−134.5	−92.5	884.0	8.086 −3	4.661	8.173	1.754	2.638	31.33
290,000	−134.5	−92.5	884.0	4.615	2.660	8.173	3.073	2.638	31.30
300,000	−126.8	−88.2	894.5	2.372 −3	1.488 −7	8.343 −6	5.608 −1	2.638 −6	31.27
350,000	−24.5	−31.4		4.459 −4	1.012 −8				31.12
400,000	233.9	112.2		1.757	1.164 −9				30.97
450,000	734.1	390.1		9.760 −5	2.600 −10				30.83
500,000	1203.8	651.0		6.217	1.020 −10				30.68
550,000	1491.7	810.9		4.185 −6	5.453 −10				30.54
600,000	1647.2	897.3		4.185	3.350				30.40

650,000	1754.9	957.2	2.909		2.177 −11	30.26
700,000	1835.7	1002.1	2.067		.467	30.12
750,000	1912.5	1044.8	1.496 −6		.009 −12	29.98
800,000	1964.3	1073.5	1.100 −7		.135 −12	29.84
850,000	2010.6	1099.2	8.197		.121	29.70
900,000	2053.4	1123.0	6.180		.724	29.57
950,000	2092.9	1144.9	4.710		.742	29.43
1,000,000	2124.6	1162.5	3.627		.046	29.30
1,100,000	2160.3	1182.4	2.205		.179 −12	29.03
1,200,000	2189.3	1198.5	1.380 −7		.020 −13	28.77
1,300,000	2214.6	1212.5	8.874 −8		.301	28.51
1,400,000	2217.2	1214.0	5.834		.724	28.25
1,500,000	2221.2	1212.2	3.907		.761	28.00
1,600,000	2232.1	1222.3	2.661		.160 −13	27.75
1,700,000	2233.7	1223.1	1.840 −8		.814 −14	27.50
1,800,000	2232.9	1222.7	1.286 −9		.351	27.26
1,900,000	2241.4	1227.4	9.089		.704 −14	27.02
2,000,000	2250.8	1232.7	6.486		.596	26.79
2,100,000	2251.7	1233.2	4.664		.843	26.55
2,200,000	2253.9	1234.4	3.378 −9		.318 −14	26.32
2,300,000	2254.0	1234.4	2.463		.491 −15	26.10

* The digit preceded by plus or minus indicates the power of 10 by which the tabulated values should be multiplied.

molecules is small but not negligible compared to the dimensions of a body moving through the air, the phenomenon of slip flow appears; i.e., the gas no longer adheres to the body surface but actually slips over the surface with a finite velocity.   Furthermore, if the mean free path is much larger than the body dimensions, i.e., $K_N > 10$, the flow becomes free-molecule, in which event the body is sprayed, as it were, with pellets which do not interfere with one another after they strike the body.   Figures 4, 5, and 6 present the mean particle speed $\bar{v}$, collision frequency, and mean free path λ for air.

BIBLIOGRAPHY

1. *U. S. Standard Atmosphere, 1962*, Washington 25, D.C., Bureau of Documents, Government Printing Office.
2. *Handbook of Geophysics and Space Environment, 1965*, Washington 25, D.C., Bureau of Documents, Government Printing Office.
3. *Tables of Thermodynamic and Transport Properties of Air, Argon, Carbon Dioxide, Carbon Monoxide, Hydrogen, Nitrogen, Oxygen and Steam*, originally published as National Bureau of Standard's Circular 564.
4. *U.S. Standard Atmosphere Supplements, 1966*, Washington 25, D.C., Bureau of Documents, Government Printing Office.
5. HECK, R. C. H., "The New Specific Heats," *Mech. Eng.*, Vol. 63, pp. 126–135, 1941.

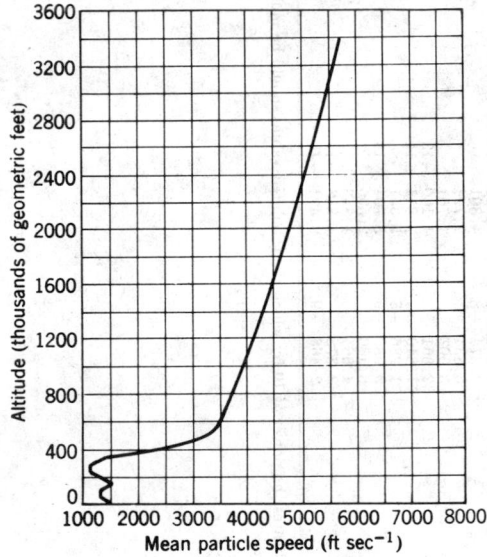

FIG. 4.   Particle speed vs. geopotential altitude.

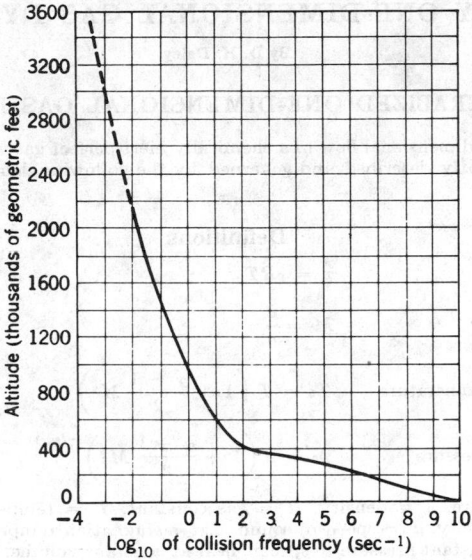

FIG. 5.   Collision frequency vs. geopotential altitude.

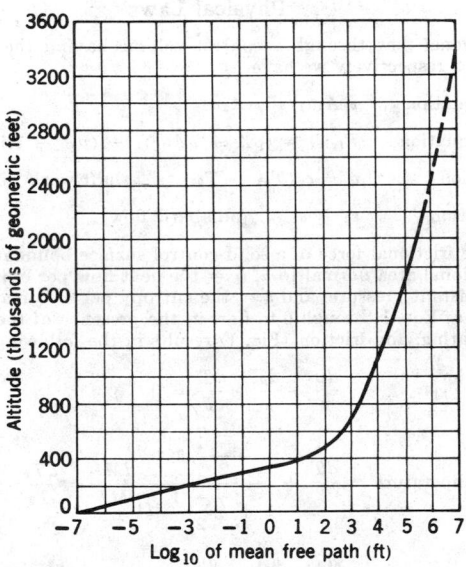

FIG. 6.   Mean free path vs. geopotential altitude.

STEADY ONE-DIMENSIONAL GAS DYNAMICS

By D. H. Daley

12. GENERALIZED ONE-DIMENSIONAL GAS DYNAMICS

The steady one-dimensional flow of a chemically inert perfect gas with constant specific heats is conveniently described and governed by the following definitions and physical laws.[6]

Definitions

Perfect gas

$$p = \rho RT \tag{1}$$

Mach number

$$M = \frac{u}{a} \tag{2}$$

Stagnation temperature

$$T_0 = T\left(1 + \frac{\gamma - 1}{2} M^2\right) \tag{3}$$

Stagnation pressure

$$p_0 = p\left(1 + \frac{\gamma - 1}{2} M^2\right)^{(\gamma/\gamma-1)} \tag{4}$$

where p = pressure, ρ = density, R = gas constant, T = temperature, M = Mach number, u = velocity, a = speed of sound, T_0 = stagnation temperature, γ = ratio of specific heat at constant pressure to specific heat at constant volume, and p_0 = stagnation pressure. It is conventional to denote those stream properties at the point in the flow where $M = 1$ by p^*, u^*, etc.

Physical Laws

For one-dimensional flow through a control volume having the single inlet and exit flow section 1 and 2 respectively we have

Continuity equation

$$\rho_1 A_1 u_1 = \rho_2 A_2 u_2 \tag{5}$$

Momentum equation

$$F_{\text{frict}} = (pA + \rho Au^2)_1 - (pA + \rho Au^2)_2 \tag{6}$$

Energy equation

$$q = c_p(T_{02} - T_{01}) \qquad (\text{shaft work} = 0) \tag{7}$$

Entropy equation

$$s_2 \geq s_1 \qquad (\text{adiabatic flow}) \tag{8}$$

where F_{frict} = the frictional force of a solid control surface boundary on the flowing gas, the flow cross-sectional area normal to u, q = the heat flow per unit mass flow, c_p = the specific heat at constant pressure, and s = the entropy per unit mass flow.

The application of Eqs. 1 through 6 to flow in the presence of the simultaneous effects of area-change, heating, and friction (Fig. 1) results in the following set of equations

Perfect gas

$$\frac{dp}{p} + \frac{d\rho}{\rho} + \frac{dT}{T} = 0 \tag{9}$$

Stagnation temperature

$$\frac{dT}{T} + \frac{\dfrac{\gamma - 1}{2} M^2}{1 + \dfrac{\gamma - 1}{2} M^2} \frac{dM^2}{M^2} = \frac{dT_0}{T_0} \tag{10}$$

Continuity

$$\frac{d\rho}{\rho} + \frac{dA}{A} + \frac{du}{u} = 0 \tag{11}$$

Stagnation pressure

$$\frac{dp}{p} + \frac{\dfrac{\gamma M^2}{2}}{1 + \dfrac{\gamma - 1}{2} M^2} \frac{dM^2}{M^2} = \frac{dp_0}{p_0} \tag{12}$$

Momentum
$$\frac{dp}{p} + \gamma M^2 \frac{du}{u} + \frac{\gamma M^2}{2}\frac{4f\,dx}{D} = 0 \tag{13}$$

Mach number
$$2\frac{du}{u} - \frac{dT}{T} = \frac{dM^2}{M^2} \tag{14}$$

In these equations heat effects are measured in terms of the stagnation temperature change according to Eq. 7. The entropy condition of Eq. 8 is also applicable if $dT_0 = 0$. If $dT_0 \neq 0$, then the entropy requirement is $ds \geq (dq/T)$.

The six dependent variables M^2, u, p, ρ, T, and p_0 in the above set of six linear algebraic equations may be expressed in terms of the three independent variables, A, T_0, and $4fx/D$. The solution of this set of equations is given in Table 1.

General conclusions can be made relative to the variation of the stream properties of the flow with each of the independent variables by the relations of Table 1.[6] An example of the relation given for (du/u) at the bottom of the table indicates that, in a constant area adiabatic flow, friction will increase the stream velocity in subsonic flow and will decrease the velocity in supersonic flow. Similar reasoning may be applied to determine the manner in which any dependent property varies with a single independent variable.

Table 1.   Influence Coefficients for Steady One-Dimensional Flow

Dependent	Independent		
	$\dfrac{dA}{A}$	$\dfrac{dT_0}{T_0}$	$\dfrac{4f\,dx}{D}$
$\dfrac{dM^2}{M^2}$	$-\dfrac{2\left(1 + \dfrac{\gamma-1}{2}M^2\right)}{1 - M^2}$	$\dfrac{(1+\gamma M^2)\left(1 + \dfrac{\gamma-1}{2}M^2\right)}{1 - M^2}$	$\dfrac{\gamma M^2\left(1 + \dfrac{\gamma-1}{2}M^2\right)}{1 - M^2}$
$\dfrac{du}{u}$	$-\dfrac{1}{1 - M^2}$	$\dfrac{1 + \dfrac{\gamma-1}{2}M^2}{1 - M^2}$	$\dfrac{\gamma M^2}{2(1 - M^2)}$
$\dfrac{dp}{p}$	$\dfrac{\gamma M^2}{1 - M^2}$	$\dfrac{-\gamma M^2\left(1 + \dfrac{\gamma-1}{2}M^2\right)}{1 - M^2}$	$\dfrac{-\gamma M^2[1 + (\gamma-1)M^2]}{2(1 - M^2)}$
$\dfrac{d\rho}{\rho}$	$\dfrac{M^2}{1 - M^2}$	$-\dfrac{\left(1 + \dfrac{\gamma-1}{2}M^2\right)}{1 - M^2}$	$\dfrac{-\gamma M^2}{2(1 - M^2)}$
$\dfrac{dT}{T}$	$\dfrac{(\gamma-1)M^2}{1 - M^2}$	$\dfrac{(1 - \gamma M^2)\left(1 + \dfrac{\gamma-1}{2}M^2\right)}{1 - M^2}$	$\dfrac{-\gamma(\gamma-1)M^4}{2(1 - M^2)}$
$\dfrac{dp_0}{p_0}$	0	$-\dfrac{\gamma M^2}{2}$	$-\dfrac{\gamma M^2}{2}$

Table is read:
$$\frac{du}{u} = \left[-\frac{1}{1 - M^2}\right]\frac{dA}{A} + \left[\frac{1 + \dfrac{\gamma-1}{2}M^2}{1 - M^2}\right]\frac{dT_0}{T_0} + \left[\frac{\gamma M^2}{2(1 - M^2)}\right]\frac{4f\,dx}{D}$$

13. SIMPLE FLOWS

A simple flow is defined as one in which all but one of the independent variables in Table 1 are zero. Three types of simple flows are summarized in Table 2 by presenting for each simple flow (i) the independent effects present, (ii) a schematic of the flow situation, (iii) the locus on a temperature–entropy diagram of the possible states attained for

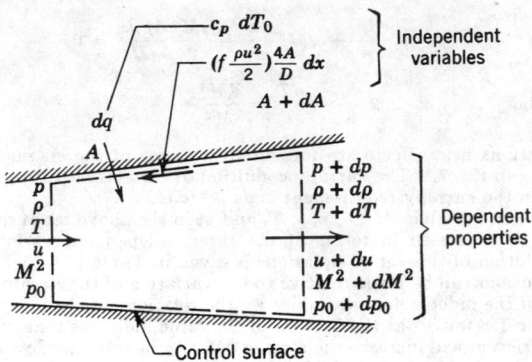

FIG. 1.   Independent and dependent variables for generalized one-dimensional flow, where f = friction coefficient and D = hydraulic diameter.

each flow, and (iv) useful functions obtained by integration of the relations of Table 1 or the basic Eqs. 1 thru 6.   These useful functions are tabulated in Ref. 1.

In the temperature–entropy diagrams of Table 2 the path lines of states corresponding to simple area flow, simple heating flow, and simple friction flow, respectively, are shown. These path lines are called the isentrope, Rayleigh, and Fanno lines respectively.

To proceed downward along the *isentrope line* from point a of the diagram, the flow area is decreased until the sonic point at b is reached.   The area must be increased after point b in order to continue down to point c.   By proper adjustments in the flow area and boundary pressure the flow may be made to proceed through point b in either direction along the isentrope line.   Point a represents the isentropic stagnation condition for all points on the isentrope line ac.

The *Rayleigh line* shows the series of possible states in a steady, frictionless, constant area flow.   Motion along the Rayleigh line is caused by changes in the stagnation temperature produced by heating effects which, in turn, produce entropy changes in the manner indicated on the line.   Heating in an initially subsonic (point f) flow causes the flow Mach number to approach one (point e).   Neither heating nor cooling alone can continuously alter the flow from subsonic to supersonic speeds, or from supersonic to subsonic speeds.

The *Fanno line* represents the possible series of states in a steady, constant area, constant stagnation temperature flow.   Frictional effects alone produce motion along the Fanno line.   Consequently the flow progression along the line must always be one of increasing entropy toward the sonic point h.   The flow is subsonic on the Fanno line above h and supersonic below.   Since the entropy decreases along the Fanno line from point h, it is impossible in simple friction flow to proceed by continuous changes through sonic conditions at point h.

14. NOZZLE OPERATING CHARACTERISTICS

The operating characteristics of a nozzle are governed by the ratio of its exit area to throat area (A_e/A_t) and by the ratio of its reservoir chamber pressure to exhaust region ambient pressure (p_c/p_a).   The operating regimes of a nozzle for isentropic (except for shock waves), steady one-dimensional flow of the perfect gas air are depicted in Fig. 2 on a diagram with nozzle pressure ratio and nozzle area ratio as the coordinate axes.

The curve in Fig. 2 with the two branches labeled design-expansion and sonic-limit obtained from the simple area isentropic functions of Table 2 by plotting the reciprocal of p/p_0 versus A/A^*.   The design-expansion line corresponds to sets of values of nozzle pressure ratio and nozzle area ratio for which the flow is shock free with the nozzle exit section Mach number supersonic and with the exit pressure p_e equal to the exhaust region ambient pressure p_a.   For any given area ratio the sonic-limit line locates the nozzle pressure ration which will produce sonic conditions at the throat and subsonic flow elsewhere.   Alternatively, the sonic-limit branch corresponds to the minimum nozzle pressure

Table 2. Simple Flows

		SIMPLE AREA	SIMPLE HEATING	SIMPLE FRICTION
EFFECTS	AREA	PRESENT	O	O
	HEATING	O	PRESENT	O
	FRICTION	O	O	PRESENT
SCHEMATIC OF FLOW SITUATION				

Table 2. Simple Flows—*Continued*

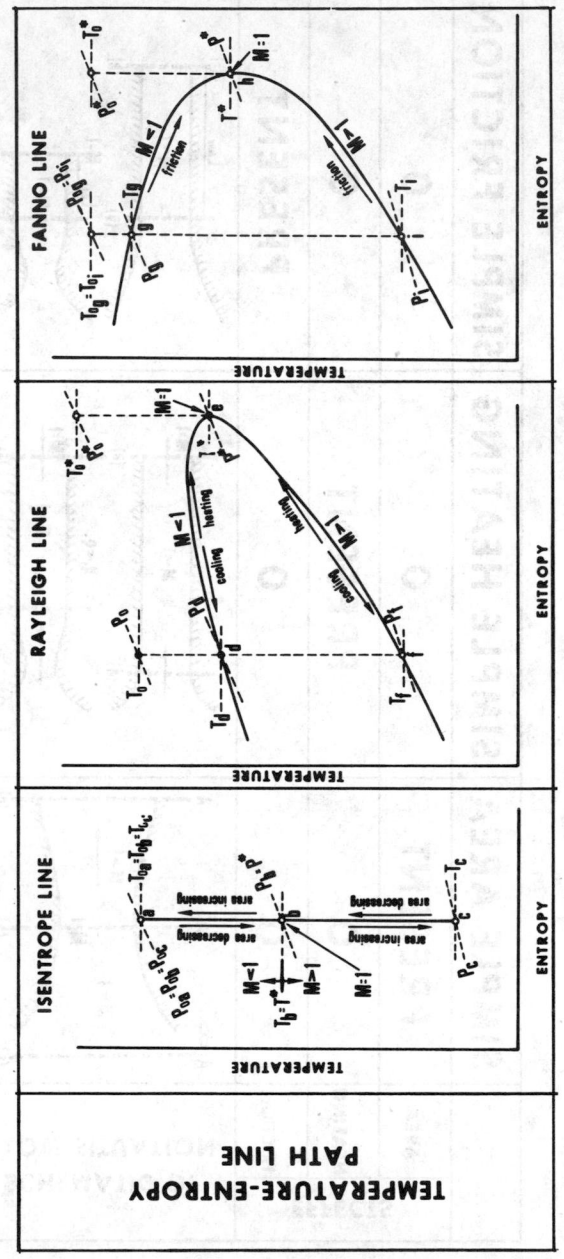

Table 2.   Simple Flows—Continued

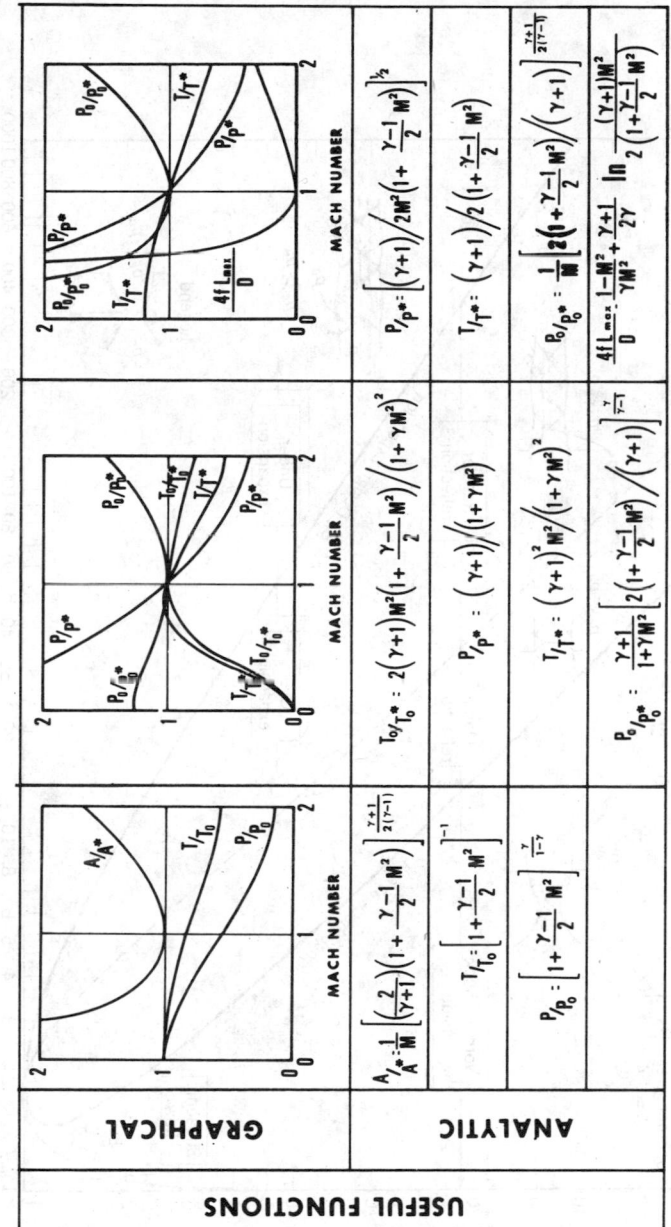

USEFUL FUNCTIONS

GRAPHICAL

ANALYTIC

$$\frac{A}{A^*} = \frac{1}{M}\left[\left(\frac{2}{\gamma+1}\right)\left(1+\frac{\gamma-1}{2}M^2\right)\right]^{\frac{\gamma+1}{2(\gamma-1)}}$$

$$\frac{T}{T_0} = \left[1+\frac{\gamma-1}{2}M^2\right]^{-1}$$

$$\frac{p}{p_0} = \left[1+\frac{\gamma-1}{2}M^2\right]^{\frac{\gamma}{1-\gamma}}$$

$$\frac{T_0}{T_0^*} = 2(\gamma+1)M^2\left(1+\frac{\gamma-1}{2}M^2\right)/(1+\gamma M^2)^2$$

$$\frac{p}{p^*} = (\gamma+1)/(1+\gamma M^2)$$

$$\frac{T}{T^*} = (\gamma+1)^2 M^2/(1+\gamma M^2)^2$$

$$\frac{p_0}{p_0^*} = \frac{\gamma+1}{1+\gamma M^2}\left[\frac{2\left(1+\frac{\gamma-1}{2}M^2\right)}{(\gamma+1)}\right]^{\frac{\gamma}{\gamma-1}}$$

$$\frac{p}{p^*} = \frac{1}{M}\left[(\gamma+1)/2M^2\left(1+\frac{\gamma-1}{2}M^2\right)\right]^{1/2}$$

$$\frac{T}{T^*} = (\gamma+1)/2\left(1+\frac{\gamma-1}{2}M^2\right)$$

$$\frac{p_0}{p_0^*} = \frac{1}{M}\left[2\left(1+\frac{\gamma-1}{2}M^2\right)/(\gamma+1)\right]^{\frac{\gamma+1}{2(\gamma-1)}}$$

$$\frac{4fL_{max}}{D} = \frac{1-M^2}{\gamma M^2}+\frac{\gamma+1}{2\gamma}\ln\frac{(\gamma+1)M^2}{2\left(1+\frac{\gamma-1}{2}M^2\right)}$$

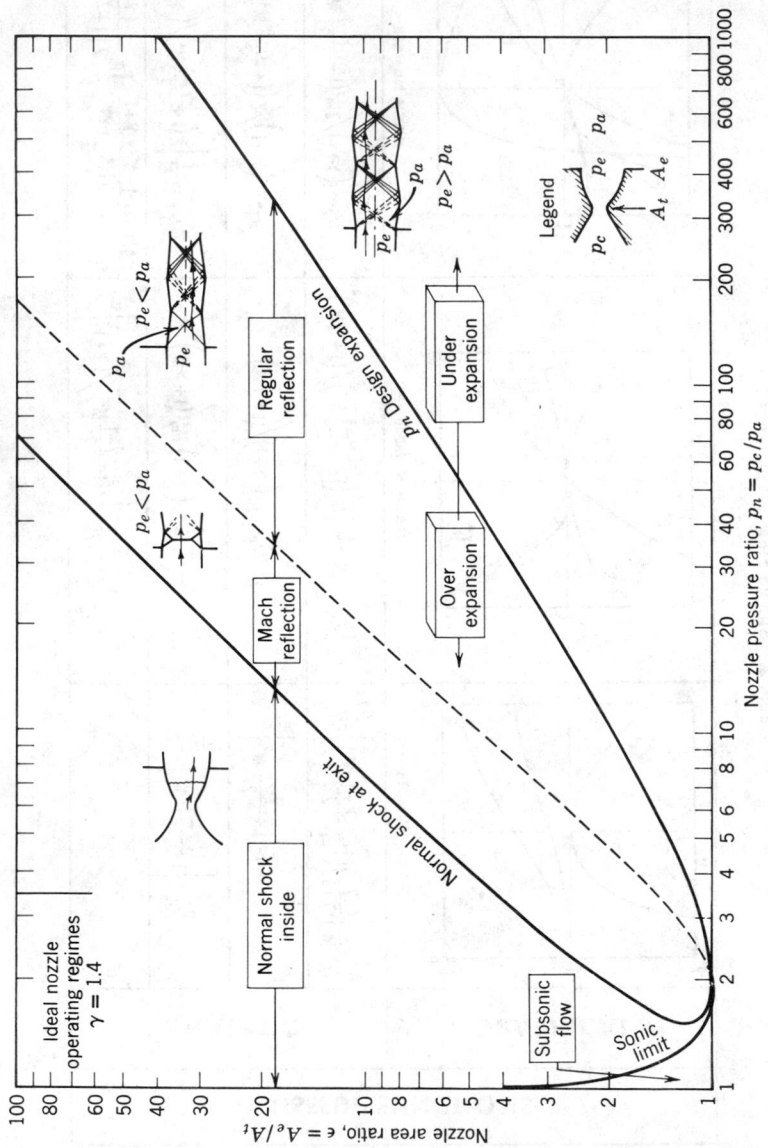

FIG. 2. Nozzle operating diagram. (From Ref. 7.)

ratio that will give the maximum flow through a given area ratio nozzle with specified throat area and reservoir conditions. The flow is subsonic in a nozzle which has its operating point located below the sonic-limit line. In the region above the design-expansion line $p_e > p_a$, the nozzle is underexpanded, and the flow expands to ambient conditions in the exhaust jet.

The area between the sonic-limit and the design-expansion curves is divided into two regions wherein shock waves occur inside the nozzle or in the jet exhaust. The common boundary of these two regions, labeled shock-exit, is the locus of pressure ratios which will produce a normal shock at the nozzle exit. Between the shock-exit line and the sonic-limit line, normal shock waves occur inside the nozzle. Between the shock-exit and the design-expansion lines oblique shocks, as shown, occur in the jet exhaust.

15. NORMAL SHOCK WAVES

A discontinuous change in stream properties can be sustained in a one-dimensional flow. When this phenomenon occurs normal to the flow direction, it is called a normal shock wave (Fig. 3). The process in this case is irreversible adiabatic and, therefore, with increasing entropy. The velocity of the flow entering the shock is supersonic, and that leaving is subsonic.

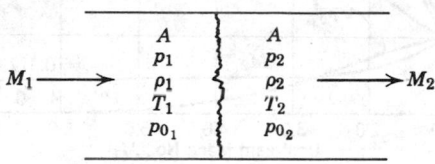

FIG. 3.   Normal shock.

Various important relations are:

$$\frac{p_2}{p_1} = \left(\frac{2\gamma}{\gamma + 1}\right) M_1{}^2 - \left(\frac{\gamma - 1}{\gamma + 1}\right) \tag{15}$$

$$\frac{T_2}{T_1} = \frac{\left(\frac{2\gamma}{\gamma - 1} M_1{}^2 - 1\right)\left(1 + \frac{\gamma - 1}{2} M_1{}^2\right)}{\frac{(\gamma + 1)^2}{2(\gamma - 1)} M_1{}^2} \tag{16}$$

$$\frac{\rho_2}{\rho_1} = \frac{u_1}{u_2} = \left[\frac{\gamma - 1}{\gamma + 1} + \frac{2}{(\gamma + 1)M_1{}^2}\right]^{-1} \tag{17}$$

$$\frac{p_{0_2}}{p_{0_1}} = \frac{A_1{}^*}{A_2{}^*} = \left[\frac{\frac{\gamma + 1}{2} M_1{}^2}{1 + \frac{\gamma - 1}{2} M_1{}^2}\right]^{(\gamma/\gamma - 1)} \left[\frac{2\gamma}{\gamma + 1} M_1{}^2 - \frac{\gamma - 1}{\gamma + 1}\right]^{(1/1 - \gamma)} \tag{18}$$

$$M_2 = \left[\frac{M_1{}^2 + \frac{2}{\gamma - 1}}{\frac{2\gamma}{\gamma - 1} M_1{}^2 - 1}\right]^{1/2} \tag{19}$$

and the Rayleigh pitot formula,

$$\frac{p_{0_2}}{p_1} = \left[\frac{\gamma + 1}{2} M_1{}^2\right]^{(\gamma/\gamma - 1)} \left[\frac{2\gamma}{\gamma + 1} M_1{}^2 - \frac{\gamma - 1}{\gamma + 1}\right]^{(1/1 - \gamma)} \tag{20}$$

Numerical values for Eqs. (15)–(20) are given in Refs. 1, 2, and 3. Equations (15)–(19) are plotted in Fig. 4 and Eq. 20 is shown on Fig. 5.

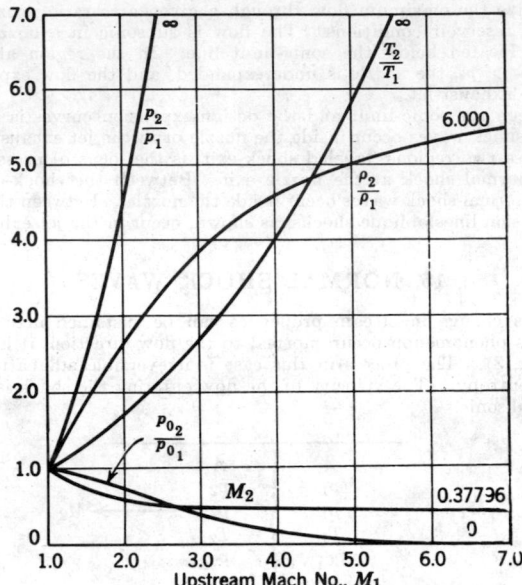

Fig. 4.   Property variations across a normal shock.

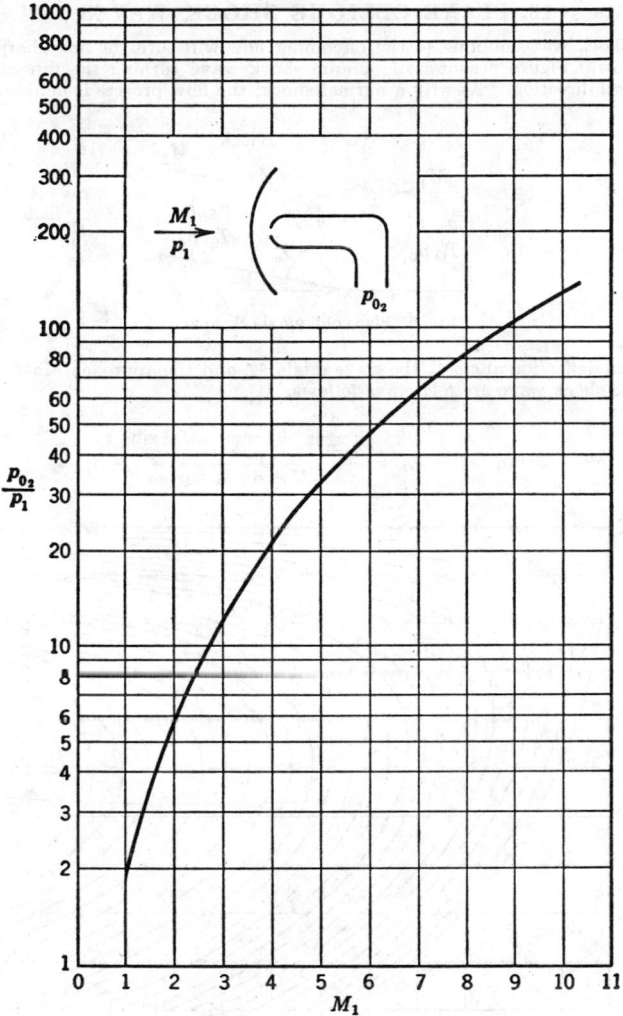

Fig. 5.   Pitot-static pressure ration for supersonic flow of air ($\gamma = 1.4$).

16. PLANE OBLIQUE SHOCK WAVES

A plane shock wave oblique to the oncoming flow will turn the flow sharply into the oncoming flow. Figure 6 shows an oblique shock wave turning the flow at a sudden change in wall direction. As with a normal shock, the flow process is nonisentropic

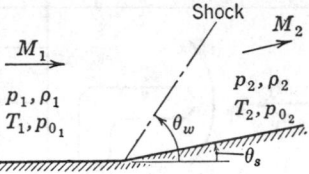

FIG. 6 Plane oblique shock wave.

The stream deflection angle θ_s, the wave angle θ_w, and the approach Mach number M_1 of an oblique shock wave are related as follows

$$\tan (\theta_w - \theta_s) = \frac{\dfrac{2}{\gamma + 1} + \dfrac{\gamma - 1}{\gamma + 1} M_1{}^2 \sin^2 \theta_w}{M_1{}^2 \sin \theta_w \cos \theta_w} \qquad (21)$$

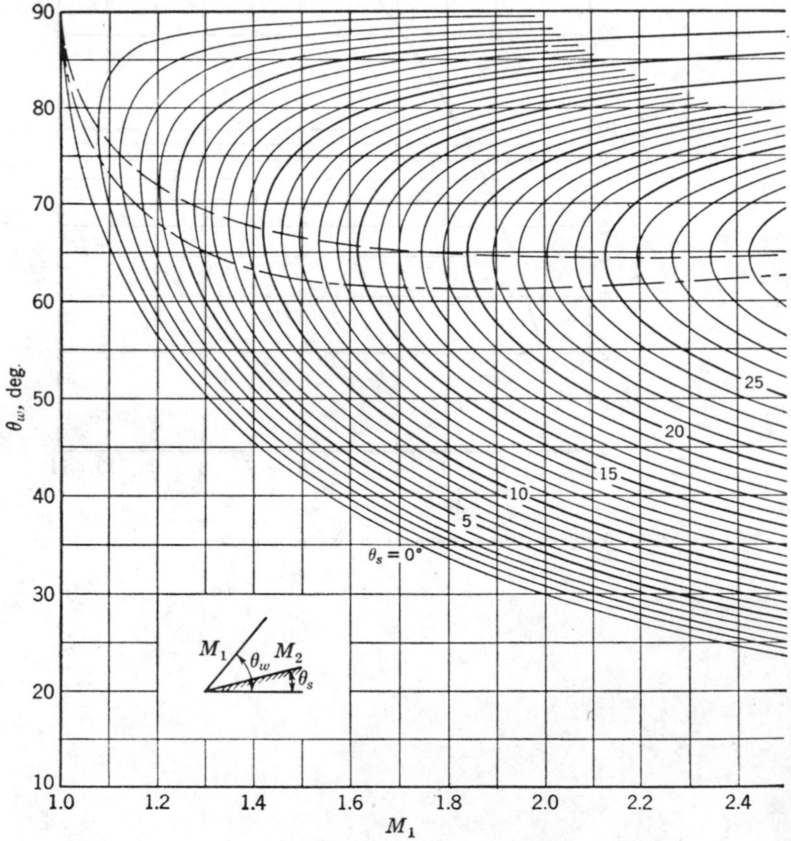

FIG. 7.   Wave angle for a plane

This relation is plotted in Fig. 7.   Observe from the figure that two waves are possible. However, when the shock is attached to the corner (Fig. 6), only the wave with the smaller wave angle, the so-called weak wave, occurs in nature.   In detached shocks, such as the curved shock in front of a blunt wedge, both solutions, strong and weak, occur.   The line of minimum M_1 for constant θ_s is the locus of shock-wave detachment.

Relations that relate the pressures, temperatures. densities, stagnation pressures, and Mach numbers before and after plane oblique shocks can be obtained in terms of M_1 and θ_w.   In fact, when the products $M_1 \sin \theta_w$ and $M_2 \sin (\theta_w - \theta_s)$ are substituted for M_1 and M_2, respectively, in the normal shock eqs. 15 through 19, they become directly applicable to flow through plane oblique shocks.   Thus by entering the graph of Fig. 4 with $M_1 \sin \theta_w$ (where θ_w may be obtained from Fig. 7 for known values of M_1 and θ_s), the values of the property ratios plotted correspond to those for a plane oblique shock wave.   Similarly, the curve labeled M_2 in Fig. 4 gives $M_2 \sin (\theta_w - \theta_s)$ for a plane oblique shock wave.   The variation of pressure ratio, downstream Mach number, and total pressure ratio with flow deflection and upstream Mach number is plotted in Figs. 8 and 9.

References 1, 2, and 8 present the oblique shock functions in tabular or graphical form.

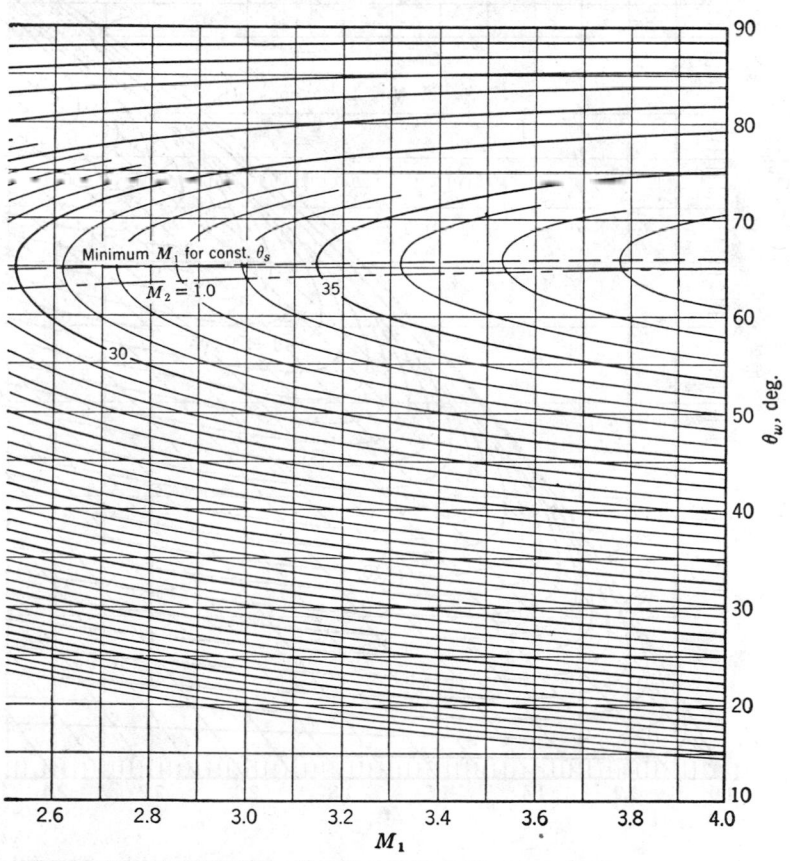

oblique shock wave ($\gamma = 1.4$).

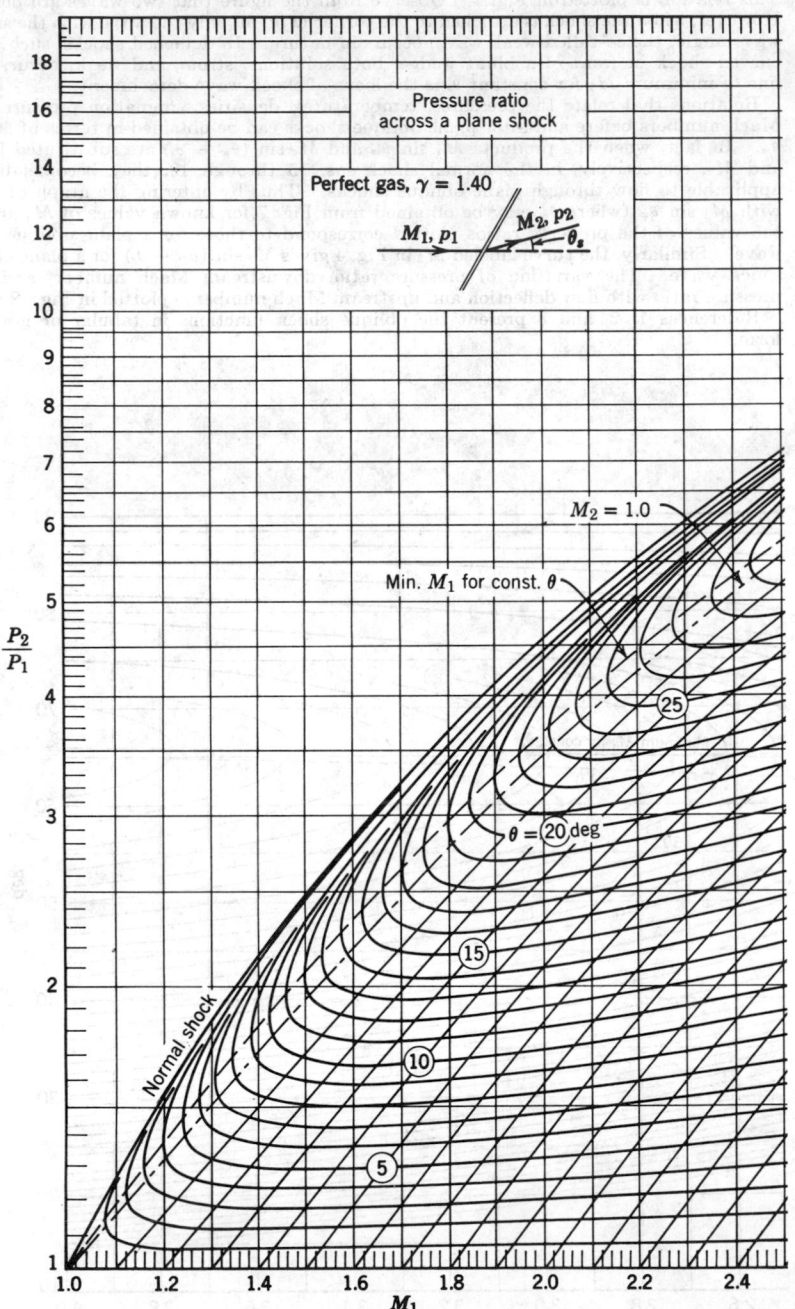

FIG. 8.   Variation of pressure ratio and downstream Mach number with

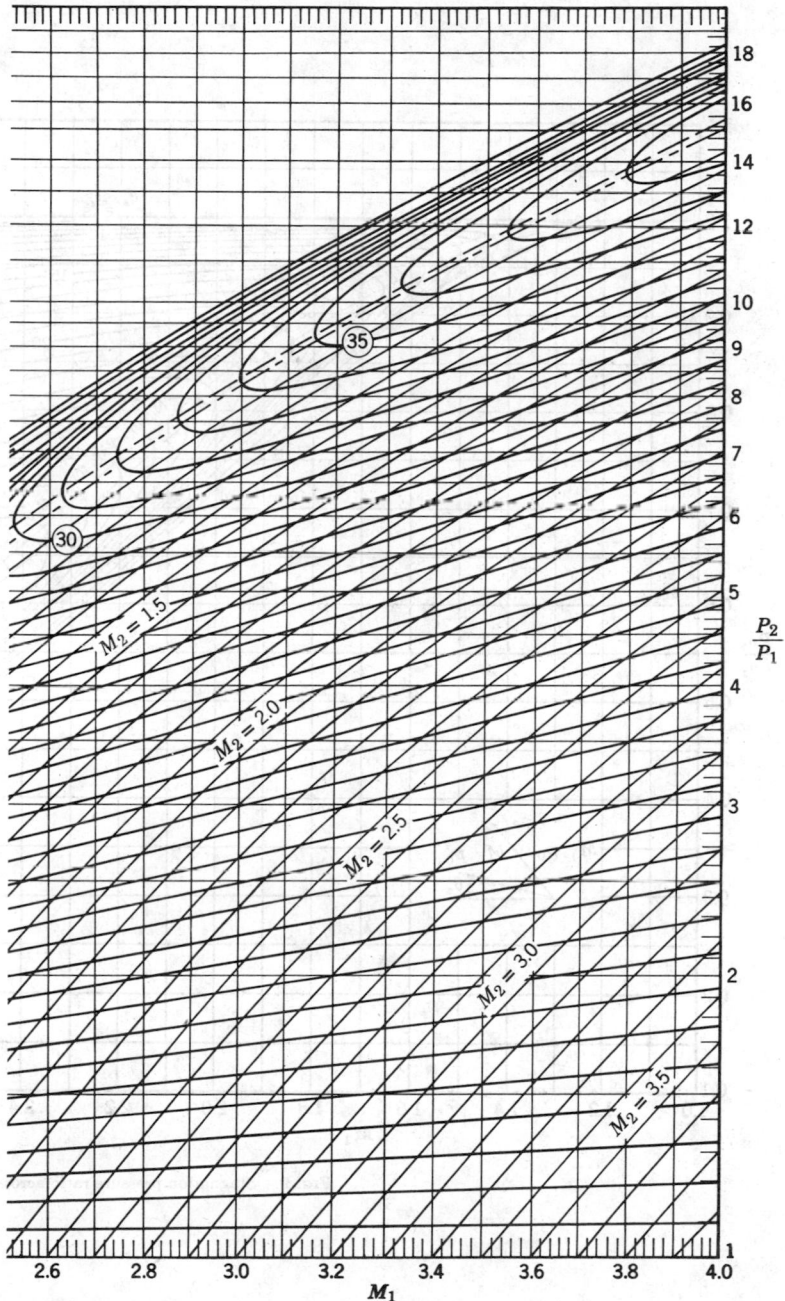

flow deflection angle θ_s and upstream Mach number M_1. (From Ref. 3.)

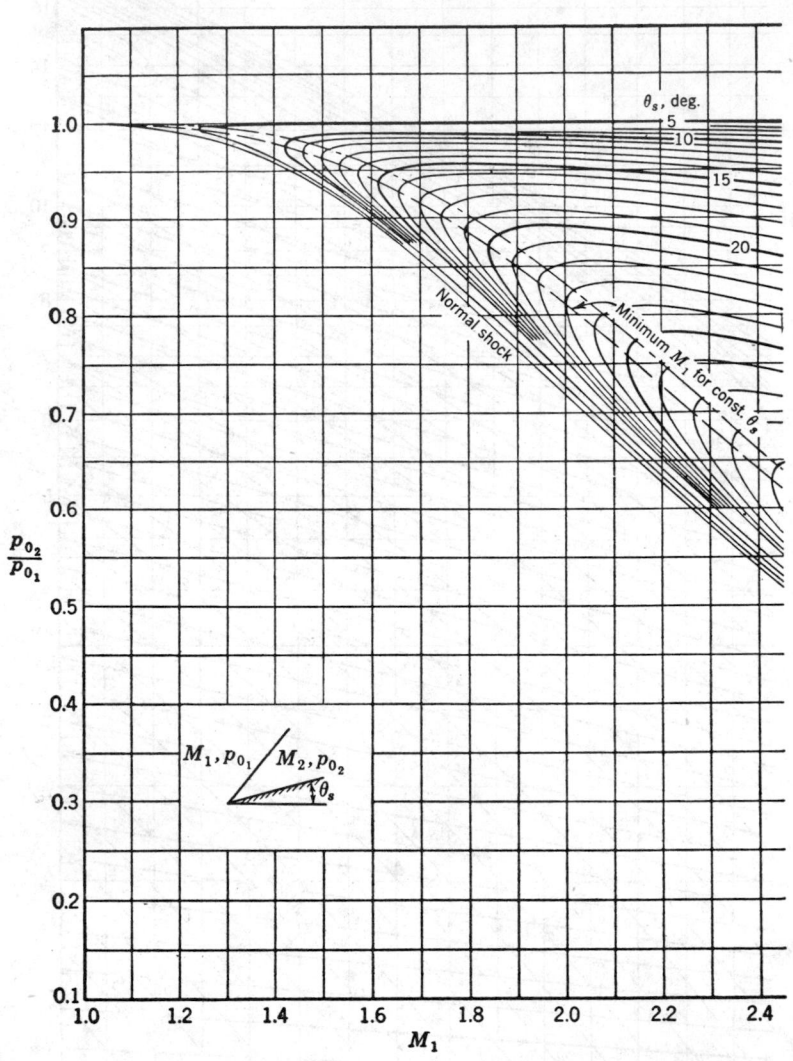

$$\frac{p_{0_2}}{p_{0_1}}$$

FIG. 9.   Stagnation pressure ratio across

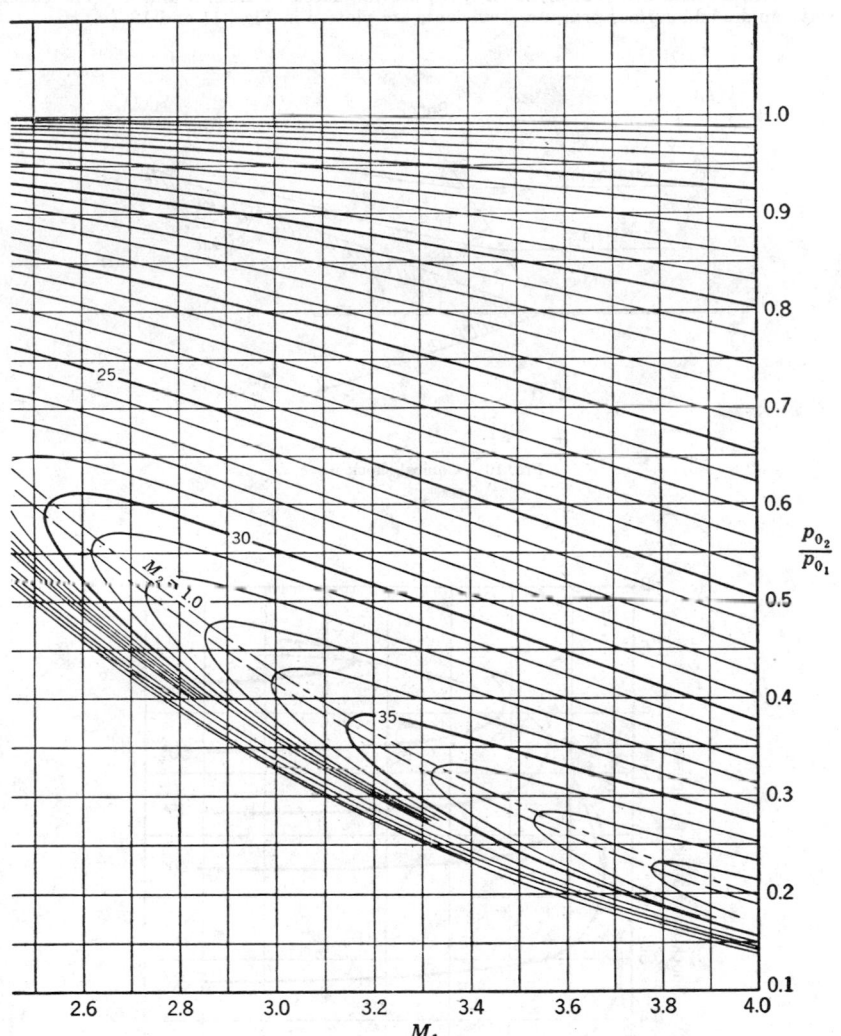

a plane oblique shock wave ($\gamma = 1.4$).

17. CONICAL SHOCKWAVES

Conditions on the surface of a cone, Fig. 10, with attached shock wave has been calculated[4] on the basis of the Taylor-Maccoll theory.[5] All fluid properties are constant along radial lines through the nose of the cone. Data for the surface stream properties on a cone with an attached conical shock wave are tabulated in Refs. 2 and 4. The shock wave angles and surface pressure coefficients are plotted in Figs. 11 and 12.

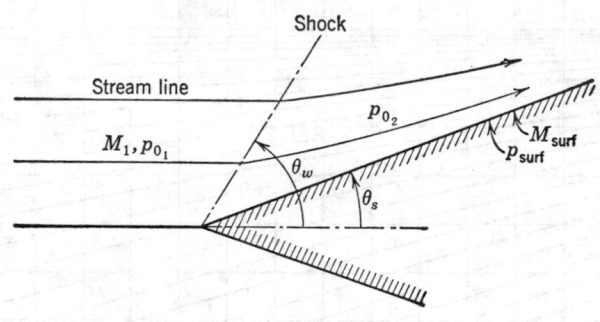

Fig. 10.   Conical shock wave.

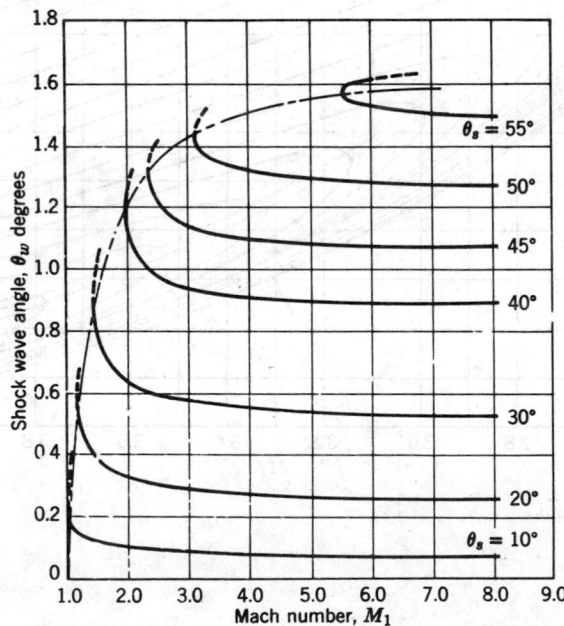

Fig. 11.   Conical shock wave angle θ_w vs. Mach number for various cone angles.

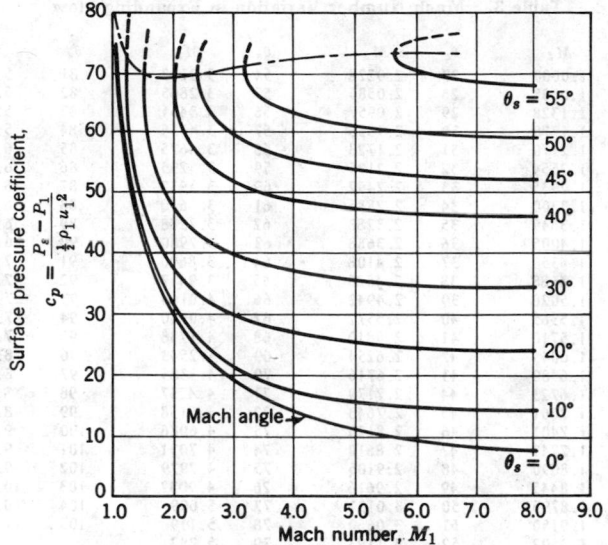

FIG. 12. Surface pressure coefficient bs. Mach number for various cone angles.

18. PRANDTL-MEYER EXPANSION

Supersonic flow turning away from the oncoming flow results in expansion of a gas to a higher Mach number. Figure 13 shows such an expansion around a corner starting with an initial Mach number of unity. The flow is here assumed two-dimensional and the process is downward from $M = 1$ along the isentrope line of Table 2.

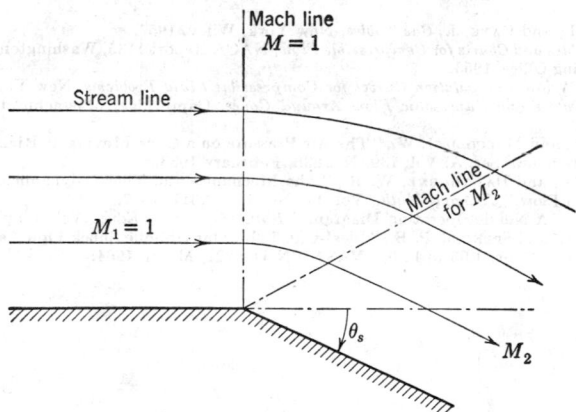

FIG. 13. Prandtl-Meyer expansion at a corner.

The angle through which a stream must turn in order to expand from $M_1 = 1$ to M_2 is

$$\theta_s = \sqrt{\frac{\gamma + 1}{\gamma - 1}} \tan^{-1} \sqrt{\frac{\gamma - 1}{\gamma + 1}(M_2{}^2 - 1)} - \tan^{-1}\sqrt{M_2{}^2 - 1} \qquad (22)$$

Numerical values of Eq. 22 are given in Refs. 1 and 2 and Table 3 for air ($\gamma = 1.4$).

Table 3.   Mach Number Variation in Expanding Flow

θ_s	M_2	θ_s	M_2	θ_s	M_2	θ_s	M_2
0	1.0000	27	2.0222	54	3.2293	81	5.470
1	1.0808	28	2.0585	55	3.2865	82	5.595
2	1.1328	29	2.0957	56	3.3451	83	5.724
3	1.1770	30	2.1336	57	3.4055	84	5.867
4	1.2170	31	2.1723	58	3.4675	85	6.008
5	1.2554	32	2.2105	59	3.5295	86	6.155
6	1.2935	33	2.2492	60	3.5937	87	6.311
7	1.3300	34	2.2885	61	3.6610	88	6.472
8	1.3649	35	2.3288	62	3.7288	89	6.643
9	1.4005	36	2.3688	63	3.7980	90	6.820
10	1.4350	37	2.4108	64	3.8690	91	7.008
11	1.4688	38	2.4525	65	3.9417	92	7.202
12	1.5028	39	2.4942	66	4.0164	93	7.407
13	1.5365	40	2.5372	67	4.0940	94	7.623
14	1.5710	41	2.5810	68	4.1738	95	7.852
15	1.6045	42	2.6254	69	4.2543	96	8.093
16	1.6380	43	2.6716	70	4.3385	97	8.350
17	1.6723	44	2.7179	71	4.4257	98	8.622
18	1.7061	45	2.7643	72	4.5158	99	8.907
19	1.7401	46	2.8120	73	4.6086	100	9.210
20	1.7743	47	2.8610	74	4.7031	101	9.539
21	1.8090	48	2.9105	75	4.7979	102	9.887
22	1.8445	49	2.9616	76	4.9032	103	10.260
23	1.8795	50	3.0131	77	5.009	104	10.658
24	1.9150	51	3.0660	78	5.119	105	11.081
25	1.9502	52	3.1193	79	5.232		
26	1.9861	53	3.1737	80	5.349		

Values given in Table 3 can be applied to isentropic flow along any convex curved surface.   The downstream Mach number M_2 after expanding M_1 through an angle $\Delta\theta$ can be found by adding $\Delta\theta$ to the θ_s corresponding to M_1 and then finding the M_2 corresponding to $\theta_s + \Delta\theta$ in Table 3.

BIBLIOGRAPHY

1. KEENAN, J. H., and KAYE, J., *Gas Tables*, New York, Wiley, 1957.
2. *Equations, Tables and Charts for Compressible Flow*, NACA Report 1135, Washington, D.C., Government Printing Office, 1953.
3. DAILEY, and WOOD, *Computation Curves for Compressible Fluid Problems*, New York, Wiley, 1949.
4. KOPAL, Z., *Tables and Supersonic Flow Around Cones*, Cambridge, Massachusetts, MIT Press, 1947.
5. TAYLOR, G. I., and MACCOLL, J. W., "The Air Pressure on a Cone Moving at High Speeds," *Proc. Roy. Soc.* (London), Ser. A, Vol. 139, No. 838, February 1933.
6. SHAPIRO, A. H., and HAWTHORNE, W. R., "The Mechanics and Thermodynamics of Steady One-Dimensional Flow," *J. Appl. Math.*, Vol. 14, No. 4, p. A317, 1947.
7. DALEY, D. H., "A Nozzle Operating Diagram," *Bull. Mech. Eng. Educ.*, Vol. 6, pp. 293–300, 1967.
8. DENNARD, J. S., and SPENCER, P. B., "Ideal-Gas Tables for Oblique Shock Flow Parameters in Air Mach Numbers From 1.05 to 12.0," NASA TN D-2221, March 1964.

PROPULSION

By R. B. Oliver and E. L. Larson

19. GENERAL

Several types of thermal-air propulsion systems have been developed in addition to the conventional reciprocating aircraft engine. Included in this category are the ram-jet, pulse-jet, turbo-jet, and turbo-prop. Compounding of reciprocating engines, which consists of gearing the torque of an exhaust gas turbine back into the drive shaft, is also being developed. The rocket, although a self-contained system not of the thermal-air type, is included herein for comparative purposes. The above engines, or variations of them, singly or in combination, provide power for practically all present-day aircraft, either piloted or pilotless.

The several types are compared briefly [1] in Fig. 1. The individual parameters are described in detail later in the text.

The rocket is a self-contained power plant which does not depend on atmospheric air for operation. Fuel for a rocket is either liquid or solid, depending on the type. An oxidizer is carried, along with the fuel, and an ignition system is provided for starting combustion if such is necessary. Most rockets burn fuel and oxidizer at a constant rate and hence produce nearly a constant thrust. The burning gases expand out of a nozzle and produce mechanical energy. The static pressure at the nozzle exit influences the thrust; since expansion to low pressure aids in the conversion of thermal energy into mechanical energy, thrust is increased with altitude.

The ram-jet is perhaps the simplest of engines in principle. Air is compressed through a diffuser into a combustion chamber by relative motion of the engine and the air. Fuel is introduced and burned in the combustion chamber. The hot products of combustion expand out of the nozzle, producing an exit velocity in excess of the free-stream velocity of the engine. The resulting increase in momentum of the combustion products determines the thrust produced by the unit. Since the combustion-chamber pressure must be greater than ambient, the ram-jet cannot produce thrust without relative velocity between it and the air. However, a modification of the ram-jet, the pulse-jet, is capable of producing static thrust, because intake valves are provided which alternately open and close as a function of the intake pressure and back pressure from combustion, thus making possible a working cycle. Since the pulse-jet vibrates badly and operates at a rather high noise level, its use will probably be limited to pilotless aircraft. The specific fuel consumption of the pulse-jet is somewhat higher than that of the turbo-jet for subsonic flight speeds. This tends to limit the usefulness of the pulse-jet, inasmuch as at supersonic flight speeds the ram-jet appears more promising than the pulse-jet.

The turbo-jet engine differs from the ram-jet principally in only one major aspect: the incoming ram air is further compressed by a compressor before fuel is introduced and burned in the combustion chamber. A turbine placed behind the combustion chamber, and integrally connected with the compressor, is driven by the expanding products of combustion. This in turn drives the compressor. Inasmuch as the turbo-jet can produce pressure above ambient in the combustion chamber even when the engine is at rest, a static thrust can be produced. This desirable characteristic is obtained without the vibration associated with the pulse-jet, mentioned previously. Because the combustion pressure is higher than for a ram-jet at the same speed, the efficiency is greater, up to Mach numbers between 2 and 3, where allowable temperatures for turbine and compressor blades limit the present-day turbo-jet.

The turbo-prop is essentially a turbo-jet with a propeller attached to the same shaft as the compressor and turbine. Of course, reduction gearing is necessary between the actual shaft and the propeller, which produces the majority of the thrust. The rest of the thrust is obtained from jet action. The high efficiency of this engine at relatively low speeds, along with its small frontal area, makes it desirable for long-range patrol planes and the like.

The reciprocating engine is discussed in detail elsewhere. Little need be said here of its characteristics, except that for aircraft application it is doubtful that this engine will be used for air speeds much in excess of 500 mph, frontal area, weight, cooling drag, and propeller tip losses being limiting factors. However, reciprocating aircraft engines will undoubtedly be widely used for many years to come for a large number of airplanes.

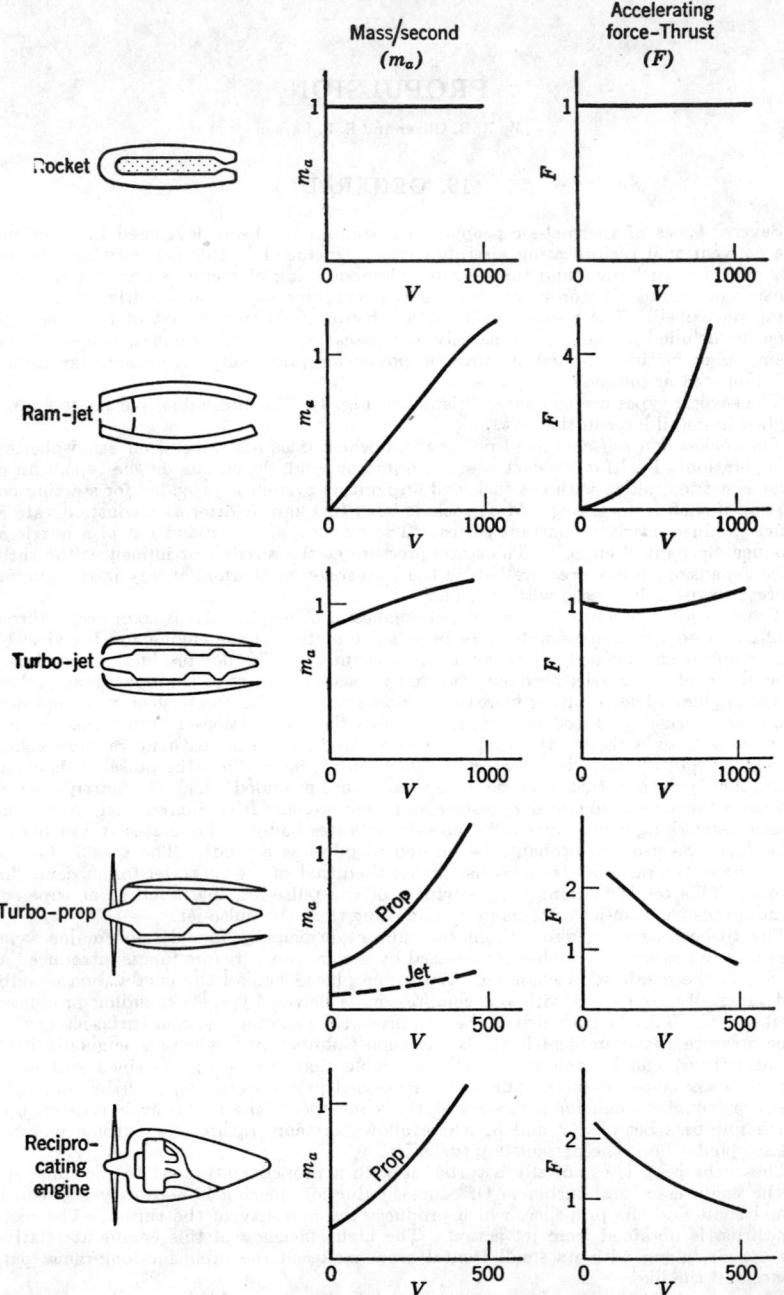

Fig. 1. Power plant characteristics. V = remote velocity relative to the engine. (Method of
and Hage,

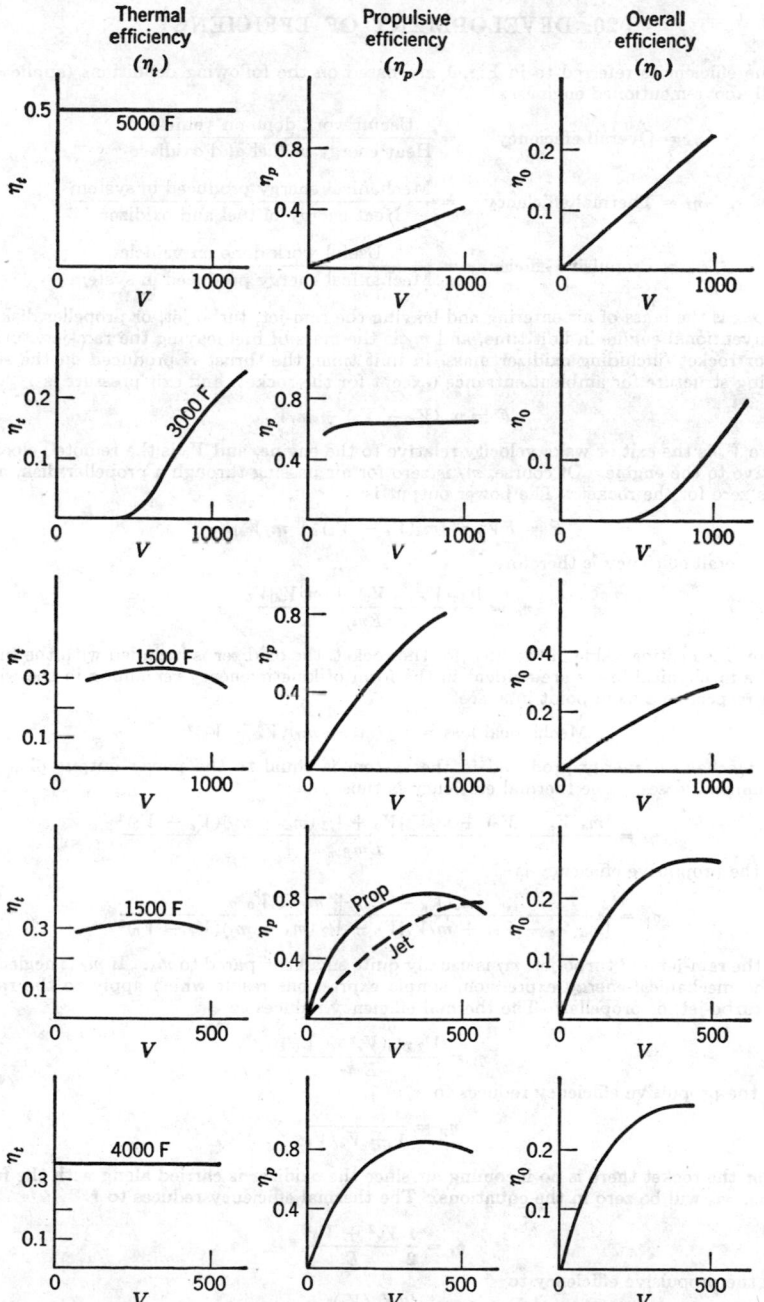

presentation suggested by Westinghouse "Jet Propulsion" brochure, B3834, 5M-2-47, see Perkins Ref. 1.)

20. DEVELOPMENT OF EFFICIENCY

The efficiencies referred to in Fig. 1 are based on the following definitions (applicable to all above-mentioned engines):

$$\eta_o = \text{Overall efficiency} \quad = \frac{\text{Useful work done on vehicle}}{\text{Heat energy of fuel and oxidizer}}$$

$$\eta_t = \text{Thermal efficiency} \quad = \frac{\text{Mechanical energy produced in system}}{\text{Heat energy of fuel and oxidizer}}$$

$$\eta_p = \text{Propulsive efficiency} = \frac{\text{Useful work done on vehicle}}{\text{Mechanical energy produced in system}}$$

If m_a is the mass of air entering and leaving the ram-jet, turbo-jet, or propeller disk of a conventional engine in unit time, and m_f is the mass of fuel leaving the ram-jet, turbo-jet, or rocket (including oxidizer mass) in unit time, the thrust F produced on the supporting structure for ambient entrance (except for the rocket) and exit pressure is

$$F = m_a(V_e - V_0) + m_f V_e \tag{1}$$

where V_e is the exit or wake velocity relative to the engine, and V_0 is the remote velocity relative to the engine. Of course, m_f is zero for air passing through a propeller disk, and m_a is zero for the rocket. The power output is

$$P = FV_0 = [m_a(V_e - V_0) + m_f V_e]V_0 \tag{2}$$

The overall efficiency is therefore

$$\eta_o = \frac{[m_a(V_e - V_0) + m_f V_e]V_0}{Em_f} \tag{3}$$

where E is heating value of the fuel (for the rocket, the oxidizer is included with the fuel).

The mechanical losses are evident in the form of kinetic energy remaining in the wake with respect to a fixed point in space.

$$\text{Mechanical loss} = \tfrac{1}{2}(m_a + m_f)(V_e - V_0)^2$$

The mechanical energy produced in the system is equal to the power output plus the mechanical losses. The thermal efficiency is thus

$$\eta_t = \frac{[m_a(V_e - V_0) + m_f V_e]V_0 + \tfrac{1}{2}(m_a + m_f)(V_e - V_0)^2}{Em_f} \tag{4}$$

and the propulsive efficiency is

$$\eta_p = \frac{[m_a(V_e - V_0) + m_f V_e]V_0}{[m_a(V_e - V_0) + m_f V_e]V_0 + \tfrac{1}{2}(m_a + m_f)(V_e - V_0)^2} \tag{5}$$

For the ram-jet and turbo-jet, m_f is usually quite small compared to m_a. If m_f is neglected in the mechanical-energy expression, simple expressions result which apply to the ram-jet, turbo-jet, or propeller. The thermal efficiency reduces to

$$\eta_t = \frac{\tfrac{1}{2} m_a(V_e^2 - V_0^2)}{Em_f} \tag{6}$$

and the propulsive efficiency reduces to

$$\eta_p = \frac{2}{1 + V_e/V_0} \tag{7}$$

For the rocket there is no incoming air since the oxidizer is carried along with the fuel. Hence m_a will be zero in the equations. The thermal efficiency reduces to *

$$\eta_t = \frac{1}{2}\frac{V_e^2 + V_0^2}{E} \tag{8}$$

and the propulsive efficiency to

$$\eta_p = \frac{2(V_e/V_0)}{1 + (V_e/V_0)^2} \tag{9}$$

---

* The velocity V_0 in equation 8, representing the energy of motion of the fuel before combustion, is omitted by some authors. (See Fig. 1, in which η_t is sketched independent of V_0.)

21. ROCKET MOTOR

The rocket (Fig. 2) consists of a fuel and oxidizer injection system (if liquid fuel is used), a combustion chamber, and a Laval-type exit nozzle. A Laval-type nozzle is a nozzle with a throat and expanding exit so that if a sufficient ratio of internal pressure to external pressure exists the gases may expand beyond the throat, thus yielding supersonic speeds.

Calculation of the thrust produced by a rocket can be made if the combustion pressure p_{T_c}, combustion temperature T_{T_c}, and the mass flow of fuel and oxidizer m_f are known.

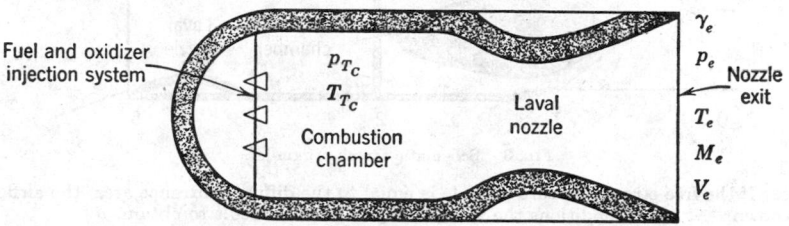

FIG. 2. Schematic diagram of rocket motor.

Denoting the exit pressure, temperature, Mach number, and nozzle pressure recovery factor by p_e, T_e, M_e, and ζ_e respectively, the following relations hold where full expansion of the gases to ambient pressure, p_0, is assumed (subscript T = total conditions):

$$p_e = p_0$$

$$p_{T_e} = \zeta_e p_{T_c} = \zeta_e p_0 \left(1 + \frac{\gamma - 1}{2} M_e^2\right)^{\gamma/(\gamma - 1)}$$

$$T_{T_c} = T_{T_e} = T_e \left(1 + \frac{\gamma - 1}{2} M_e^2\right)$$

For proper use of these expressions, γ should be a mean value. The value of the mean γ depends on the static and total temperatures at the exit. If a value of γ is assumed, and p_{T_c}, p_0, and ζ_e are known, M_e can be calculated. Use of the known T_{T_c}, the assumed mean γ, and the calculated M_e will allow determination of T_e. The mean γ associated with any T_e and T_{T_e} can be calculated for each combination of rocket propellants under consideration. A check of the mean γ for the T_e and T_{T_e} obtained above will indicate how closely the assumed γ was matched. If too much discrepancy exists, an iteration should be carried out. A mean γ near 1.2 is often obtained.

When T_e and the value of γ associated with it are known, along with the gas constant R_e, for the particular chemical combination used, the local exit acoustic velocity, a_e, can be obtained,

$$a_e = \sqrt{\gamma_e R_e T_e}$$

It follows that

$$V_e = M_e a_e$$

and

$$F = m_f V_e$$

where V_e and F are the exit velocity and thrust respectively. The specific impulse, I (sometimes called specific thrust), defined as the thrust F divided by the weight flow of fuel w_f, is

$$I = \frac{F}{w_f} = \frac{m_f V_e}{m_f g} = \frac{V_e}{g}$$

where g is the acceleration due to gravity.

22. RAM-JET ENGINE *

A ram-jet engine (Fig. 3) consists essentially of an inlet diffuser (1-2), a flame holder and fuel injectors (2), a combustion chamber (2-3), and a Laval-type exit nozzle (3-5). Performance of a ram-jet is usually expressed in terms of a thrust coefficient C_F, and a specific impulse I. The thrust coefficient is defined as the thrust F divided by the product of

* Figures 4-11 and the approximate equations of the turbo-jet section were adapted from methods developed by the Aerophysics Laboratory, North American Aviation, Inc.

the incompressible dynamic pressure $q = \frac{1}{2}\rho V^2$ and a representative area A. The reference area most frequently used as a basis of thrust coefficients is the ram-jet combustion-chamber area, although a great many data are based on the diffuser-entrance area. The calculation of ram-jet performance can be broken into three steps: (a) diffusion and airflow, (b) combustion-chamber analysis, and (c) velocity and pressure calculation at the nozzle

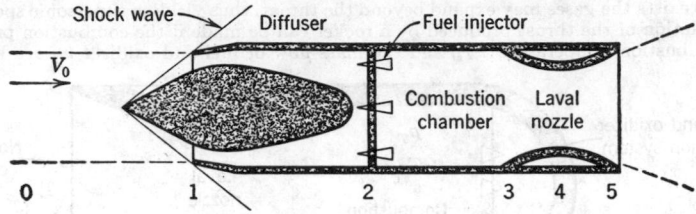

Shock wave   Diffuser   Fuel injector

V_0

Combustion Laval
chamber   nozzle

0   1   2   3   4   5

FIG. 3.   Schematic sketch of ram-jet.

exit. If the free-stream capture area A_0 is equal to the diffuser-entrance area, the airflow is known. At other conditions the airflow may be more difficult to obtain.

Starting with the diffuser, the following relation is obtained from continuity:

$$A_0\theta_0 = A_2\theta_2\zeta_d \tag{10}$$

where (see Article 3)

$$\theta = \frac{\rho V}{\rho_T a_T} = \frac{M}{\left(1 + \dfrac{\gamma - 1}{2}M^2\right)^{(\gamma+1)/[2(\gamma-1)]}}$$

and

$$\zeta_d = \text{total pressure ratio} = \frac{p_{T_2}}{p_{T_0}}$$

Symbols:

A = cross-sectional flow area perpendicular to body center line.
γ = specific heat ratio, c_p/c_v.
p = pressure.
M = section Mach number.
V = section velocity.
a = speed of sound.
ρ = fluid density.

Subscripts:

T = total (stagnation) conditions.
0 = remote conditions (A_0 is capture area).
2 = entrance to combustion chamber.

Across the combustion chamber, the following expression obtained from the energy equation and the perfect gas law is conventionally used:

$$\left[(1+f)\left(\frac{R_3 T_{T_3}}{R_2 T_{T_2}}\right)^{\frac{1}{2}}\right]\left[\frac{\gamma_2^{\frac{1}{2}}M_2\left(1 + \dfrac{\gamma_2 - 1}{2}M_2^2\right)^{\frac{1}{2}}}{1 + (1 - N/2)\gamma_2 M_2^2}\right] = \left[\frac{\gamma_3^{\frac{1}{2}}M_3\left(1 + \dfrac{\gamma_3 - 1}{2}M_3^2\right)^{\frac{1}{2}}}{1 + \gamma_3 M_3^2}\right] \tag{11}$$

Symbols:

f = fuel-air ratio.
R = gas constant.
$N = \Delta p/q_2$ = ratio of pressure lost due to flame holder and friction to combustion-chamber-entrance dynamic pressure ($\frac{1}{2}\rho_2 V_2^2$).
T = absolute temperature.
M = Mach number.
γ = specific heat ratio.

Subscripts:

T = total (stagnation) conditions.
2 = entrance to combustion chamber.
3 = exit of combustion chamber.

The total pressure ratio ζ_c between sections 2 and 3 is expressed as

$$\zeta_c = \frac{pT_3}{pT_2} = \frac{\left(1 + \dfrac{\gamma_3 - 1}{2} M_3{}^2\right)^{\gamma_3/(\gamma_3 - 1)}}{1 + \gamma_3 M_3{}^2} \div \frac{\left(1 + \dfrac{\gamma_2 - 1}{2} M_2{}^2\right)^{\gamma_2/(\gamma_2 - 1)}}{1 + \gamma_2(1 - N/2)M_2{}^2} \tag{12}$$

Solution of equations 11 and 12 can be facilitated by means of charts similar to those shown in Figs. 4,[2] 5, and 6. A chart relating temperatures and fuel-air ratio is also necessary for the solution of the equations; such a chart [2] is shown in Fig. 7.

Conditions through the exit nozzle follow the equation

$$A_3\theta_3 = A_5\theta_5\zeta_N \tag{13}$$

in which ζ_N is the total pressure ratio across the nozzle.

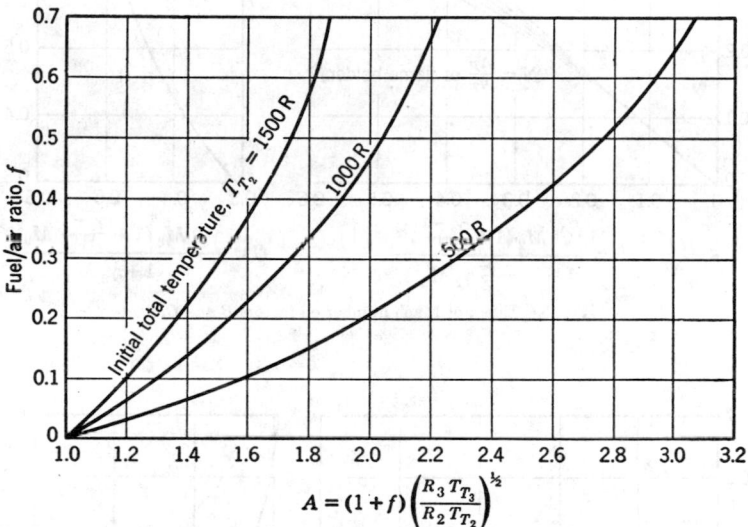

$$A = (1 + f)\left(\frac{R_3 T_{T_3}}{R_2 T_{T_2}}\right)^{\frac{1}{2}}$$

Fig. 4. Effect of fuel-air ratio on factor A. Data for kerosene.

In practice, due account must be made of the change in γ caused by large temperature change. Changes in γ with temperature and fuel-air ratio can be seen [2] in Fig. 8; changes in the gas constant with temperature (due to dissociation) and fuel-air ratio can be seen [2] in Fig. 9.

The total pressure at the exit, pT_5, will be

$$pT_5 = pT_0\left(\frac{pT_2}{pT_0}\right)\left(\frac{pT_3}{pT_2}\right)\left(\frac{pT_5}{pT_3}\right) = pT_0\zeta_d\zeta_c\zeta_N \tag{14}$$

where ζ_d is obtained from theory or experiment.
ζ_c is obtained from Fig. 6.
ζ_N is obtained from theory or experiment (approximately 0.90–0.95).

If a full expansion nozzle is provided ($p_5 = p_0$), one must assume an effective mean γ_5 and solve for M_5 directly as explained above for a rocket motor. If either an over-expanded or underexpanded nozzle is provided, equation 13 must be used also, and both M_5 and p_5 obtained.

Assuming that the temperature T_{T_3} is known, then $T_{T_5} = T_{T_3}$ if no heat transfer through the walls is considered. As for the rocket, the assumed mean γ_5 must be checked with the mean γ_5 associated with T_5 and T_{T_5}. Iteration may be necessary again. The exit acoustic velocity is then

$$a_5 = \sqrt{\gamma_5 R_5 T_5}$$

and

$$V_5 = M_5 a_5$$

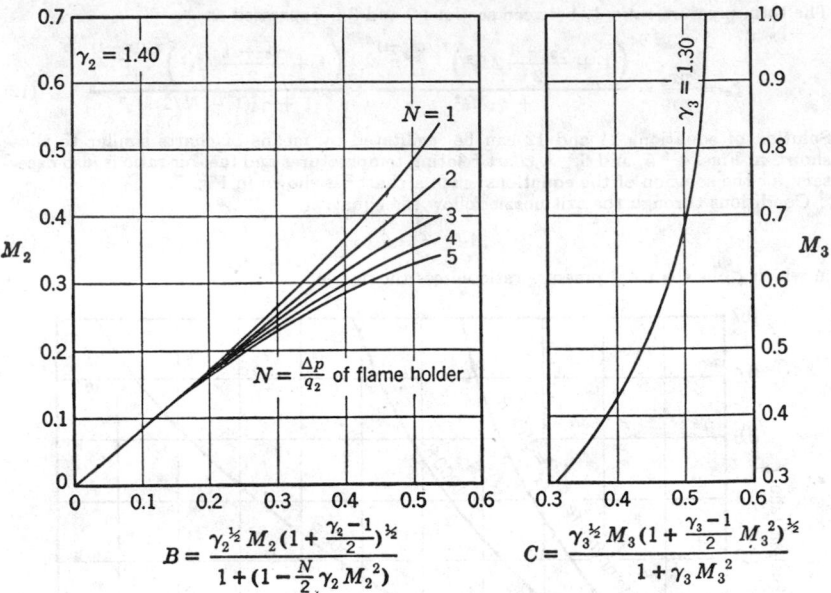

$$B = \frac{\gamma_2^{\frac{1}{2}} M_2 (1 + \frac{\gamma_2 - 1}{2})^{\frac{1}{2}}}{1 + (1 - \frac{N}{2} \gamma_2 M_2^2)}$$

$$C = \frac{\gamma_3^{\frac{1}{2}} M_3 (1 + \frac{\gamma_3 - 1}{2} M_3^2)^{\frac{1}{2}}}{1 + \gamma_3 M_3^2}$$

FIG. 5.   Effect of Mach number on factors B and C.

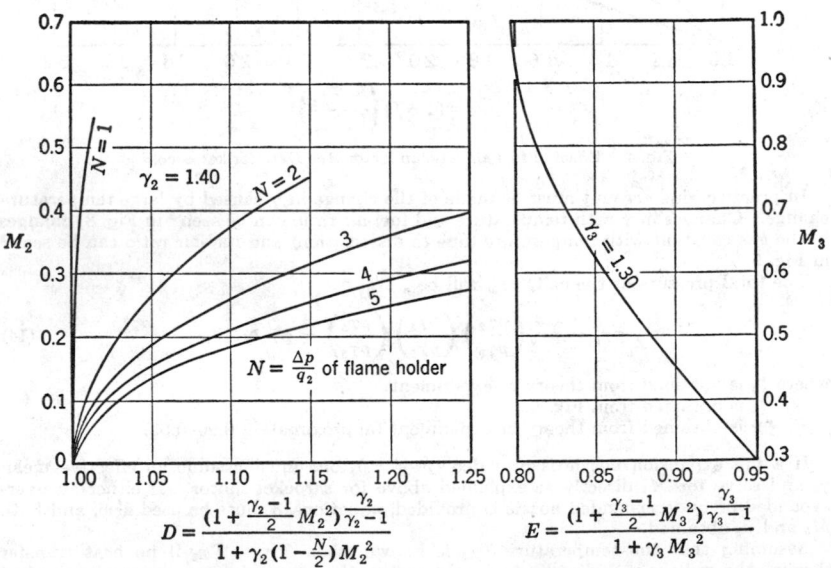

$$D = \frac{(1 + \frac{\gamma_2 - 1}{2} M_2^2)^{\frac{\gamma_2}{\gamma_2 - 1}}}{1 + \gamma_2 (1 - \frac{N}{2}) M_2^2}$$

$$E = \frac{(1 + \frac{\gamma_3 - 1}{2} M_3^2)^{\frac{\gamma_3}{\gamma_3 - 1}}}{1 + \gamma_3 M_3^2}$$

FIG. 6.   Effect of Mach number on factors D and E.

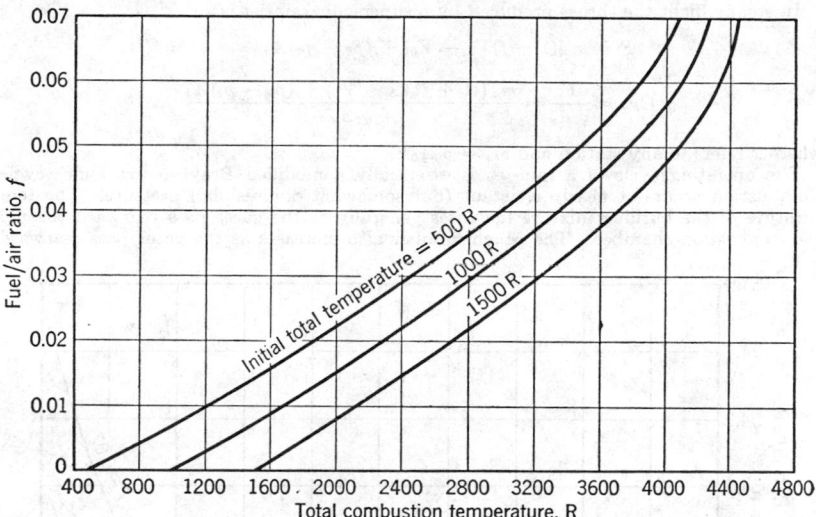

FIG. 7. Effect of fuel-air ratio on combustion temperature. Data for kerosene.

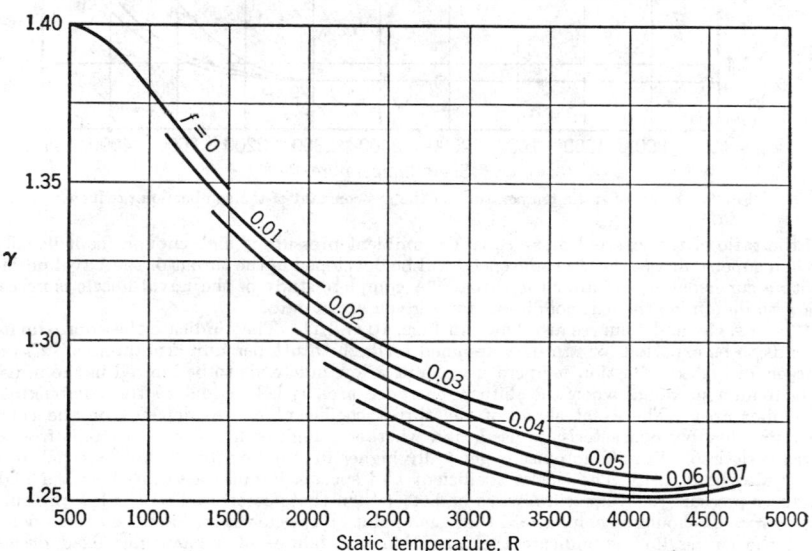

FIG. 8. Effect of static temperature on the ratio of specific heats of the combustion products.

In steady flight the thrust produced by a supersonic ram-jet is

$$F = m_a[(1 + f)V_5 - V_0] + (p_5 - p_0)A_5$$

or

$$C_{F_x} = \frac{F}{q_0 A_x} = \frac{m_a [(1 + f)V_5 - V_0] + (p_5 - p_0)A_5}{q_0 A_x}$$

where x refers to any station and $m_a = \rho_0 A_0 V_0$.

The operating cycle of a ram-jet is essentially a modified Brayton (or Joule) cycle. Combustion occurs at nearly constant (but somewhat diminishing) pressure. The temperature of the burning mixture increases generally as the gases pass rearward through the combustion chamber. The Mach number also increases as the gases pass rearward.

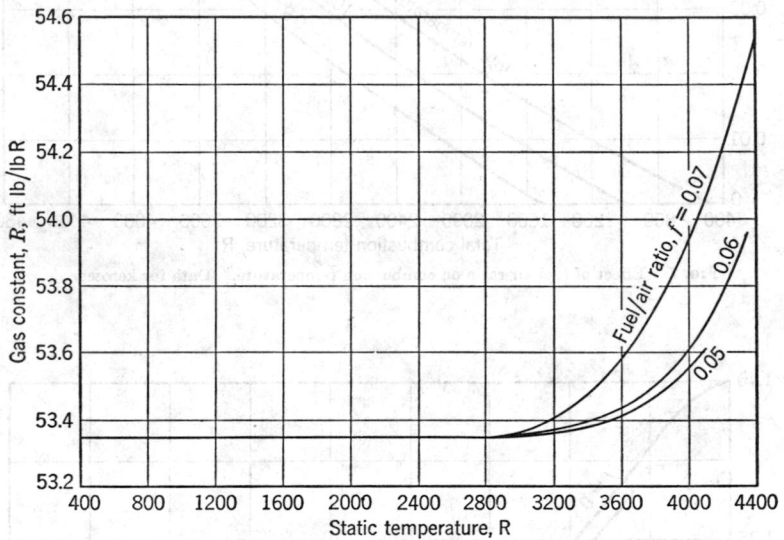

FIG. 9.   Effect of static temperature on the gas constant of the combustion products.

If the ratio of the internal pressure to the ambient pressure is high enough (as it usually is in a supersonic ram-jet), a sonic speed will be developed in the throat of the Laval nozzle with a supersonic speed after the throat. A complete study of the Laval nozzle is necessary to determine the exit conditions for a given set of data.

Typical C_{F_2} and I curves are shown in Figs. 10 and 11. They indicate the approximate trends to be expected for ram-jets designed for high thrust per unit frontal area. C_{F_2} is larger for high combustion temperature than for low but tends to be limited in the upper Mach number range when the diffuser-entrance area is held equal to the combustion-chamber area. The exact shape of the thrust coefficient curves depends on the combustion-chamber characteristics used, and whether a high or low thrust per unit frontal area is desired. Specific impulse is generally higher for the low thrust, but external drag may also be higher. The thrust coefficient and specific impulse associated with a high diffuser pressure recovery will always be better than those associated with a low pressure recovery. It should be emphasized that each point on the curves represents a design point and the curves do not indicate the operation possibilities of a particular fixed design operating at other than its design conditions.

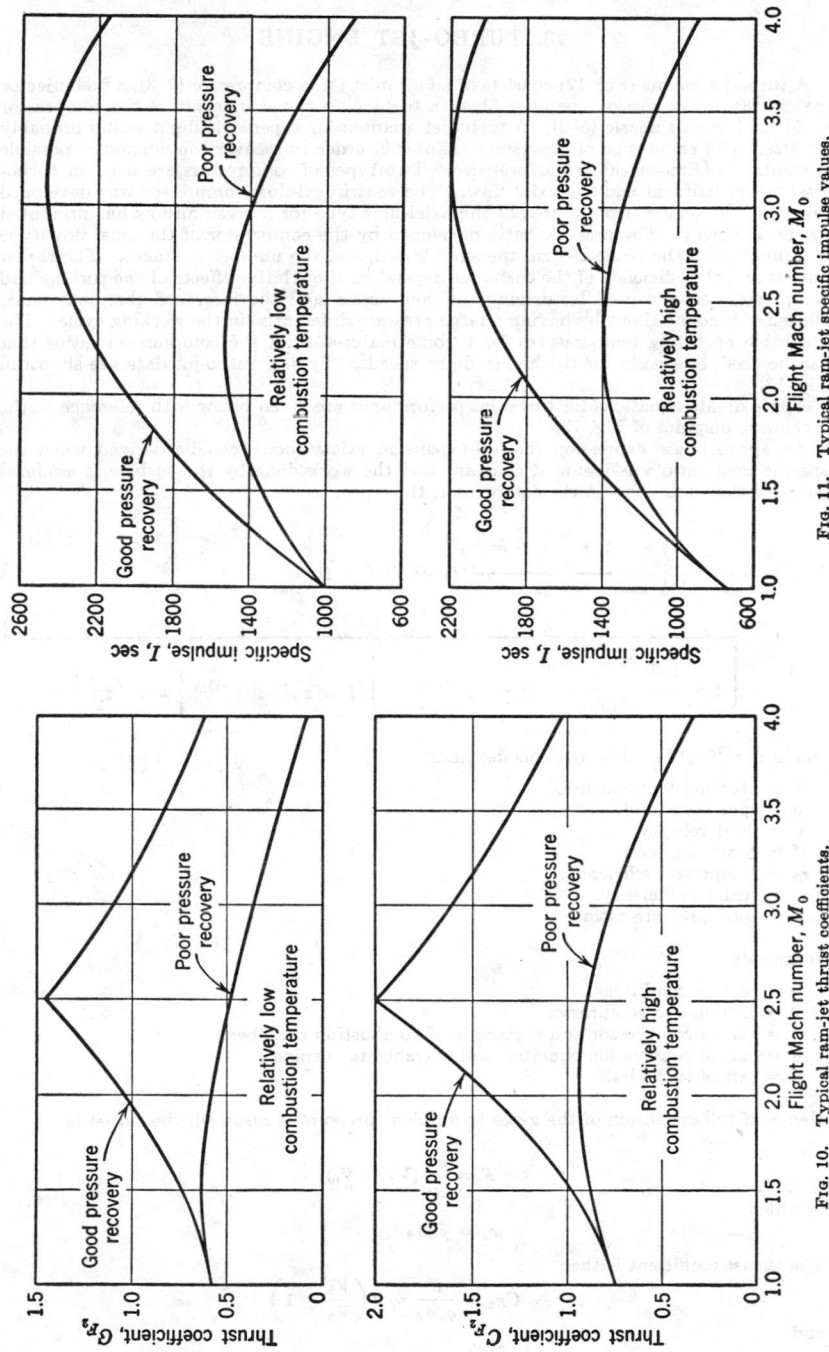

Fig. 11. Typical ram-jet specific impulse values.

Fig. 10. Typical ram-jet thrust coefficients.

23. TURBO-JET ENGINE

A turbo-jet engine (Fig. 12) consists of an air inlet (1), a compressor (2–3), a fuel injector system (3), a combustion chamber (3–4), a turbine on the same shaft as the compressor (4–5), and an exit nozzle (5–6). A turbo-jet adapted for supersonic flight would probably be fitted with some type of supersonic diffuser in order to recover the maximum possible percentage of free-stream total pressure. Two types of compressors are used in turbo-jets: the centrifugal and the axial flow. The centrifugal-flow compressor was developed first, but the smaller frontal area of the axial-flow type has prompted its development. The pressure ratio developed by the compressor of the axial-flow type is a function of the blade design, the shaft speed, and the number of stages. The thrust output and the efficiency of the turbo-jet depend on the relative effects of the turbine and compressor losses caused by driving the compressor and the increased thermodynamic efficiency brought about by having a large pressure differential in the working cycle. The allowable operating temperatures for turbine material limit the compression ratios that can be used, especially for the higher flight speeds. Typical turbo-jet data are shown in Fig. 13.

Approximate equations for turbo-jet performance are given below with reference to the schematic diagram of Fig. 12.

An approximate expression for full-expansion exit velocity can be derived when the specific heat ratio γ is assumed constant and the work done by the turbine is assumed equal to the work done on the compressor, thus:

$$\left(\frac{V_6}{a_0}\right)^2 = \frac{2}{\gamma - 1}\left\{\tau - \frac{1 + \frac{\gamma - 1}{2}M_0^2}{\eta_c}[\zeta_{2,3}^{(\gamma-1)/\gamma} - 1]\right\}$$

$$\left\{1 - \frac{1}{\left[1 + \frac{1 + \frac{\gamma-1}{2}M_0^2}{\tau\eta_c\eta_t}(1 - \zeta_{2,3}^{(\gamma-1)/\gamma})\right][\zeta_{3,4}\zeta_{2,3}\zeta_{0,2}]^{(\gamma-1)/\gamma}\left[1 + \frac{\gamma-1}{2}M_0^2\right]}\right\}$$

(15)

where $\tau = T_{T_4}/T_0$. The symbols used are:

T = absolute temperature.
a = speed of sound.
V = fluid velocity.
M = Mach number.
η_c = compressor efficiency.
η_t = turbine efficiency.
ζ = total pressure ratio.

Subscripts:

0 = remote conditions.
2 = entrance to compressor.
3 = exit of compressor and beginning of combustion chamber.
4 = exit of combustion chamber and entrance to turbine.
6 = exit of turbo-jet.

Hence, if full expansion of the gases to ambient pressure is assumed, the thrust is

$$F = \frac{w_a}{g}(V_6 - V_0)$$

in which

$$w_a = \rho_0 A_0 V_0 g$$

The thrust coefficient is then

$$C_{F_0} = \frac{F}{q_0 A_0} = 2\left(\frac{V_6}{V_0} - 1\right)$$

and

$$C_{F_{A\max}} = C_{F_0}\frac{A_0}{A_{\max}}$$

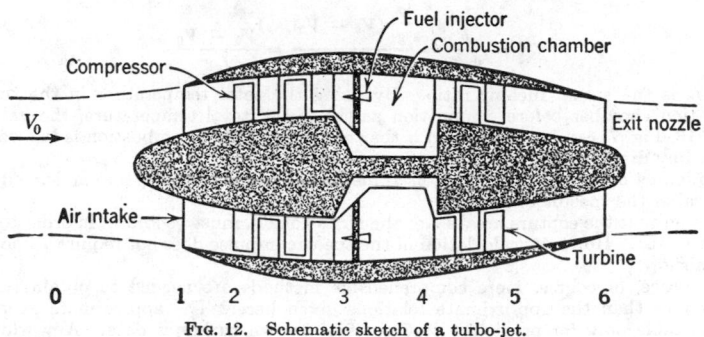

FIG. 12. Schematic sketch of a turbo-jet.

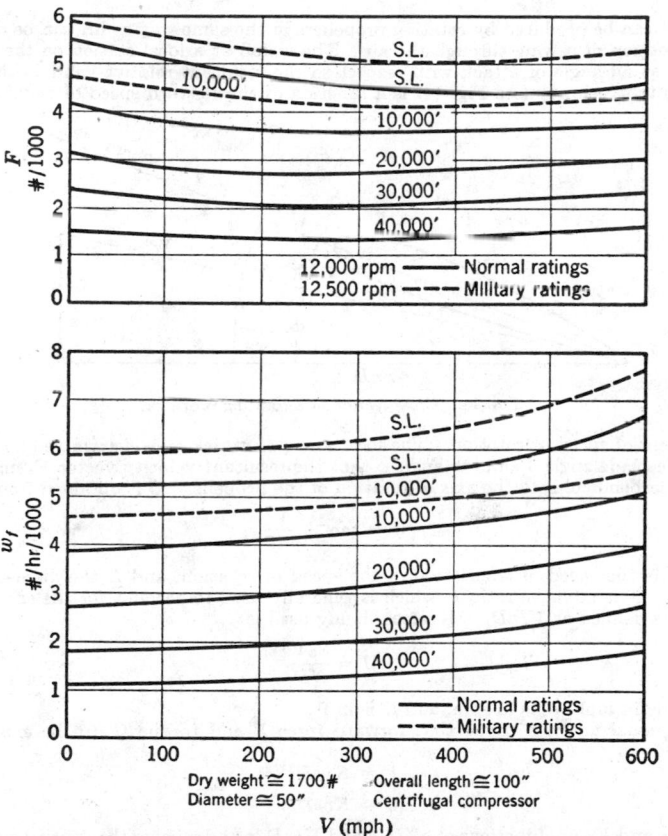

Dry weight ≅ 1700#   Overall length ≅ 100"
Diameter ≅ 50"   Centrifugal compressor

V (mph)

FIG. 13. Estimated "Nene" performance. (Estimated from Rolls-Royce Turbo-jet Power-plant brochure published by Taylor Turbine Corp., August, 1946, see Perkins and Hage, ref. 1.)

The specific impulse I (thrust divided by weight flow of fuel) is

$$I = \frac{F}{w_a f_a} = \frac{\frac{w_a}{g}(V_6 - V_0)}{w_a f_a} = \frac{V_6 - V_0}{f_a g} \tag{16}$$

where f_a is the actual fuel-air ratio. By using the total temperature in the turbo-jet combustion chamber before combustion as the initial total temperature, the valve of f in Art. 13, Fig. 7, can be determined if the temperature after combustion is known. f_a is obtained by dividing f by the combustion efficiency, η_b.

Specific fuel consumption is often used instead of specific impulse and is defined as the reciprocal of the specific impulse.

It is seen that the capture area A_0, or the airflow itself, must be known in order to calculate the thrust. However, calculation of the specific impulse does not require a knowledge of the airflow.

In practice, of course, more comprehensive methods are needed to obtain turbo-jet performance than the approximate relations given here. The approximate expressions are applicable only for predicting variations of known turbo-jet data. A γ which will give good correlation with known data should be used. A value of γ near $4/3$ often fulfills this requirement.

24. PROPELLERS

Thrust can be produced by rotating propellers in the same way as lift can be developed by the motion of a wing through the air. The airfoil at a local station on the propeller must be at an angle of attack with respect to the resultant relative wind at that point. The resultant velocity (see Fig. 14) is a function of the forward speed of translation and

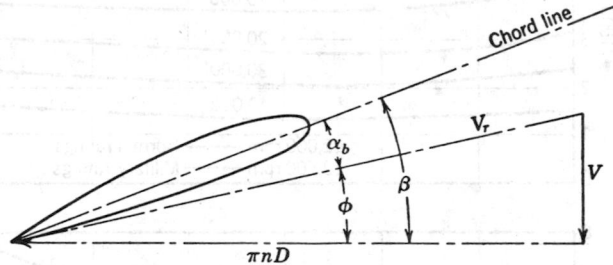

FIG. 14.   Cross section of a rotating propeller.

the tangential speed of rotation. The local angle of attack α_b is determined by the local geometric blade angle β and the angle ϕ that the resultant velocity vector V_r makes with a plane perpendicular to the axis of rotation of the propeller. ϕ is obtained from

$$\phi = \tan^{-1}\frac{V}{\pi n D}$$

where V is the speed of translation, n the speed of rotation, and D the diameter of the propeller. The advance ratio J, which is generally used to account for the effects of the angle ϕ, is defined as V/nD. More commonly used is

$$J = \frac{88V}{ND}$$

where V is in mph, N is in rpm, and D is in ft.

Dimensional analysis shows that forward force F and torque Q can be expressed as follows:

$$F = K_F \rho D^2 V^2$$

$$Q = K_Q \rho D^3 V^2$$

where K_F and K_Q are functions of $\rho V D/\mu$ and V/nD, ρ and μ being the density and viscosity of the fluid, respectively. Alternative expressions are

$$F = C_F \rho n^2 D^4$$

$$Q = C_Q \rho n^2 D^5$$

The power output is

$$P = 2\pi nQ = C_P \rho n^3 D^5$$

Propeller propulsive efficiency can be defined as

$$\eta = \frac{FV}{P} = \frac{C_F \rho n^2 D^4 V}{C_P \rho n^3 D^5} = \frac{C_F}{C_P} \frac{V}{nD} = \frac{C_F J}{C_P}$$

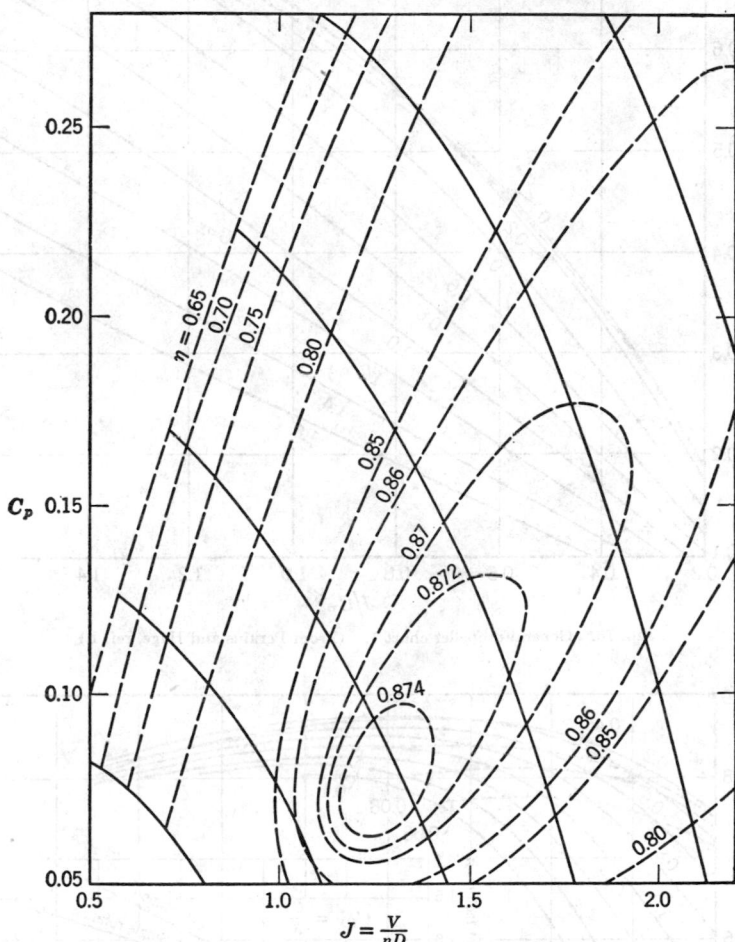

$$J = \frac{V}{nD}$$

FIG. 15. Typical propeller characteristics chart. Ref.: NACA ARR 3125.

If engineering units are used,

$$C_F = 0.1518 \frac{F/1000}{\sigma(N/1000)^2(D/10)^4}$$

$$C_P = 0.5 \frac{BHP/1000}{\sigma(N/1000)^3(D/10)^5}$$

where σ is the ratio of the air density at altitude to the air density at sea level, F is in lb, N in rpm, and D in ft.

Propeller data[3] obtained from tests are generally plotted as in Fig. 15.

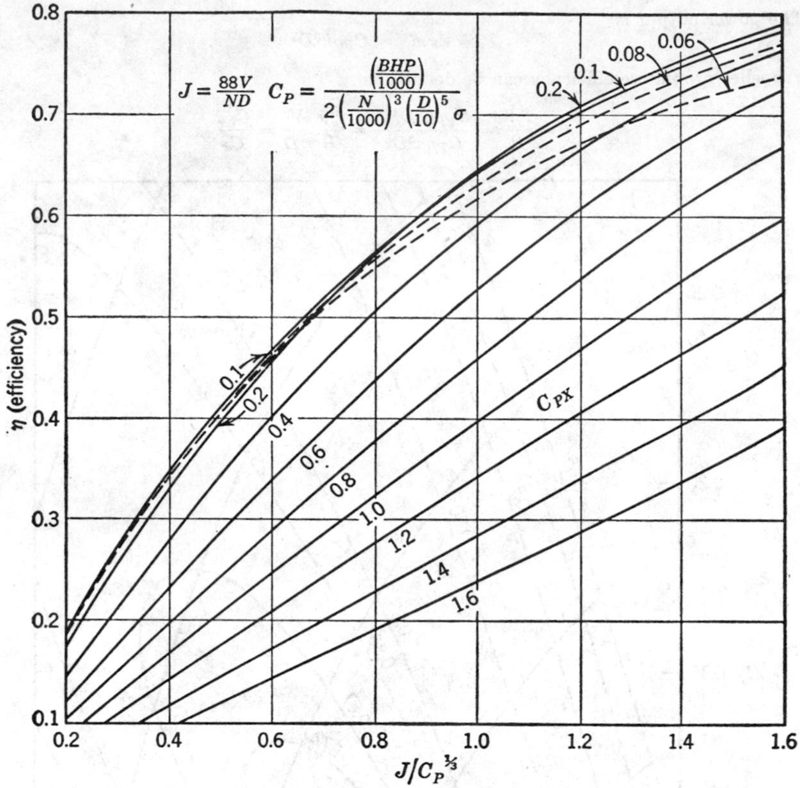

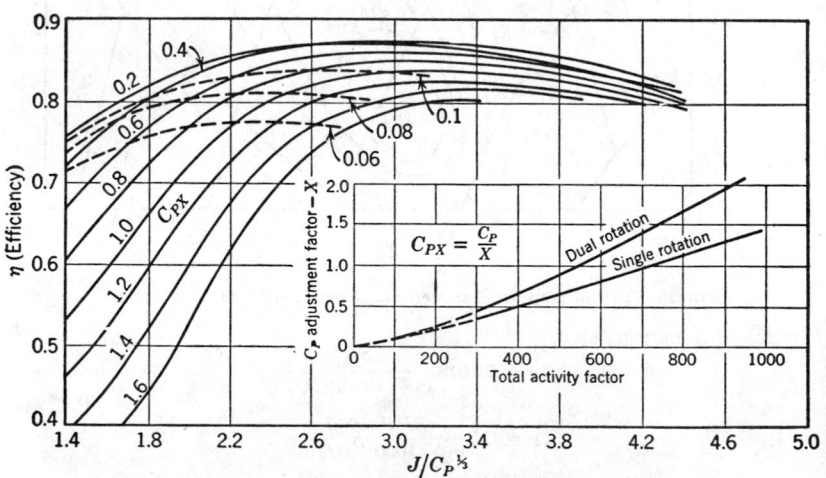

FIG. 16.   General propeller chart 1.   (From Perkins and Hage, ref. 1.)

FIG. 17.   General propeller chart 2.   (From Perkins and Hage, ref. 1.)

$$J = \frac{88V}{ND} \quad C_P = \frac{\left(\frac{BHP}{1000}\right)}{2\left(\frac{N}{1000}\right)^3\left(\frac{D}{10}\right)^5 \sigma}$$

$$C_{PX} = \frac{C_P}{X}$$

The capacity of a propeller blade to absorb power is indicated by the "activity factor," which is defined as

$$\text{Activity factor} = \text{No. blades} \times \frac{100,000}{16} \int_{0.2}^{1.0} \left(\frac{r}{R}\right) \left(\frac{b}{D}\right) \left(d\,\frac{r}{R}\right)$$

where r = distance measured radially outward from axis of rotation.
b = blade width at a given r.
D = propeller diameter.
R = propeller radius.

A convenient method- of incorporating the effect of the activity factor into the propulsive efficiency is given in Figs. 16 and 17.

25. ELECTRIC PROPULSION

Electrical rocket propulsion has application in space vehicles which can use low thrust devices where the massive power conversion and conditioning equipment currently necessary can be tolerated. The thrust levels range from micropounds to pounds, but have operation times of long duration. Electrical rocket propulsion systems break down into three broad classifications;[5] electrostatic, electrothermal, and electromagnetic.

In electrostatic propulsion systems, ions or colloid particles are accelerated to high velocity in an electrostatic field. The typical ion rocket uses an ion production chamber into which an uncharged propellant is pumped. The propellant is ionized and ions and electrons are withdrawn in separate streams. The ions are then accelerated by electrostatic field to produce thrust. The field is normally established by maintaining the ionization chamber at a positive potential and the final electrode at "ground." After acceleration the beam is neutralized by electrons expelled from the rocket. Ion engines can be further classified by ion source; either bombardment ionization or contact ionization. The principle of operation of colloidal rockets is similar. However, attempts to form uniform charged colloidal particles (1 to 0.01 μ in diameter) have been unsuccessful. In addition to the thrust producing mechanism, electrical rockets require a propellant storage and supply system and an electrical power source.

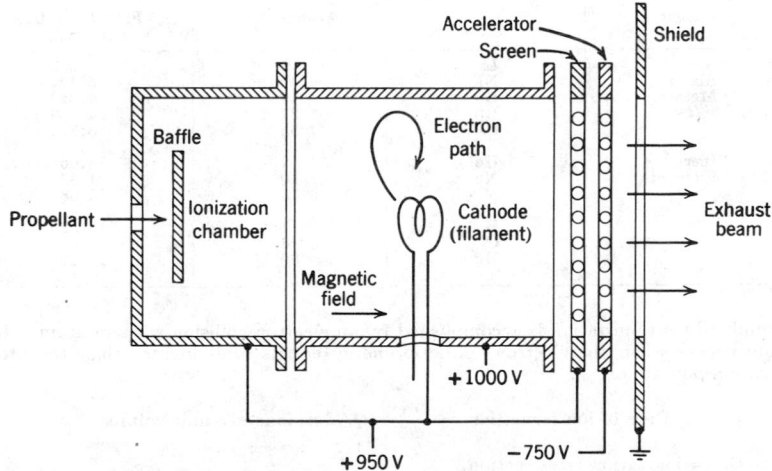

FIG.19.   Schematic diagram of a low-density ion engine.   (From Ref. 6.)

Figure 19 is a schematic of a typical ion rocket showing the ion engine, power source, and propellant system. Some important relations for ion engines are:

$$\text{Specific impulse} = I_{sp} = \frac{1}{g_e} \sqrt{2 \frac{q}{m} V_a}$$

where q = charge of particle.
   m = particle mass.
   V_a = potential difference particle passes through.
   g_e = acceleration due to gravity at surface of earth.

$$\text{Beam power} = P_b = I V_a = \frac{\dot{m} u_e^2}{2}$$

where I = total beam current = $\left(\dfrac{q}{m}\right) \dot{m}$.

   $\dot{m}$ = particle mass flow rate.
   u_e = exhaust velocity.

$$\text{Thrust} = \tau = 2P_b/g_e I_{sp} = \dot{m} u_e$$

$$\text{Power efficiency } \eta = \frac{\text{beam power}}{\text{input electrical power}}$$

$$\eta \approx \frac{\dfrac{1}{2} m g_e^2 I_{sp}^2}{\dfrac{1}{2} m g_e^2 I_{sp}^2 + e_l}$$

where $1/2 m g_e^2 I_{sp}^2$ = exhaust kinetic energy per particle.
   e_l = energy loss/ion $\sim$ charging energy.

The ability to ionize an atom depends on the electronic structure of the atom. The alkali metal elements lithium, sodium, potassium, rubidium, and cesium have single electrons in unfilled outer shells and are easiest to ionize. Table 1 indicates the first ionization potentials for these elements and others for reference.

Table 1. Ionization Potentials

Element		Atomic No.	First Ionization Potential, c_s
Alkali Metals	Li	3	5.363
	Na	11	5.12
	K	19	4.318
	Rb	37	4.159
	Cs	55	3.87
Inert Elements	He	2	24.46
	Ne	10	21.47
	A	18	15.68
	H	1	13.527
	C	6	11.217
	Hg	80	10.39

Bombardment ionization is accomplished by an electron collision with an atom. Ionization occurs when the electron's kinetic energy equals or is greater than the atom's ionization potential.

$$\text{Rate of ion formation} = \frac{dn_i}{dt} = Q_i N_e n_a \text{ ions/sec-unit volume}$$

where Q_i = ionization cross section.
   N_e = electron flux per second per unit volume.
   n_a = atom density.

The ion engine shown schematically in Fig. 1 is a low current density device using an electron-bombardment ion source. The propellant flows axially through the ionization chamber where it is bombarded by electrons generated from the cathode. As the electrons flow from the cathode to the anode (case) they are diverted into curved paths by the magnetic field and held in the chamber until at least one collision occurs. The approximate strength of the magnetic field is given by[5,6]

$$B = \frac{\sqrt{8 \left(\frac{m_e}{e}\right) \Delta V_a}}{r_a \left[1 - \frac{r_c^2}{r_a^2}\right]}$$

where B = magnetic field strength.
m_e = electron mass.
e = electron charge.
r_a = anode radius.
r_c = cathode radius.
ΔV_a = anode potential relative to cathode.

Engines similar to the one shown in Fig. 1 have produced ions with an energy consumption of the order of 450 V per ion and propellant efficiencies (ratio of ion mass flow to total propellant flow) greater than 90%. Beam interception by the accelerating electrode has been less than 5% and current densities of the order of 1.5 to 2.0 mA/cm² have been achieved.

Contact ionization will occur if the work function of the contact surface is greater than the ionization potential of the propellant. Therefore combinations of low work function alkali metals with high work function surfaces are utilized. The most frequently used combination is cesium and tungsten. The cesium contact ion engine has produced current densities of 20 to 80 mA/cm² at reasonable power efficiency and with very low beam interception by the accleration system.

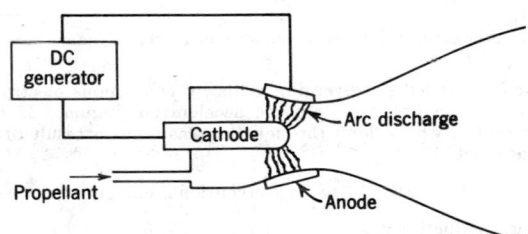

Fig. 20.   Schematic diagram of an arcjet.   (From Ref. 6.)

Electrothermal propulsion uses an arcjet or resistance jet. Figure 20 is a schematic of an arcjet. The propellant is heated to a high temperature in an electric arc and then expanded in a conventional nozzle.

The high-current-density arc discharge is maintained by a large voltage difference between cathode and anode. Problem areas include heat transfer to the electrodes, which is especially severe at the cathode, anode erosion due to electron impact, and cathode erosion due to ion impact.

The efficiency of an arcjet is the ratio of kinetic energy flux in the exhaust to electrical power supplied. Efficiencies in the region of 50% for specific impulses of 1200 sec have been attained.

The resistancejet, or resistojet, operates in much the same way as the arcjet, except the propellant is heated by an electrical heating element instead of an arc. Resistojets have been used functionally on several military satellites.

The fraction f of the total energy supplied to the propellant which appears as thermal energy is

$$f = \frac{e_k}{e_k + \alpha_d e_d + \alpha_I e_I + \alpha_{II} e_{II}}$$

where e_k = internal energy added to the propellant per unit mass.
 e_d = energy of dissociation.
 e_I = energy of first ionization.
 e_{II} = energy of second ionization.
 α_d = dissociation fraction.
 α_I = first ionization fraction.
 α_{II} = second ionization fraction.

Helium, hydrogen, and ammonia are possible propellants in an arcjet. Helium is superior to hydrogen for specific impulses up to 2000 sec. It has no dissociation losses, is chemically inert, has a high ionization potential, produces a high arc column resistance at high temperature giving good efficiency, and the electrode potential drops are low. Ammonia is heavy, storable, and is a liquid at normal operating temperatures. It dissociates when heated but is the preferred propellant for use in resistojets.

An example of an electromagnetic propulsion device is the steady crossed-field accelerator shown in Fig. 21.

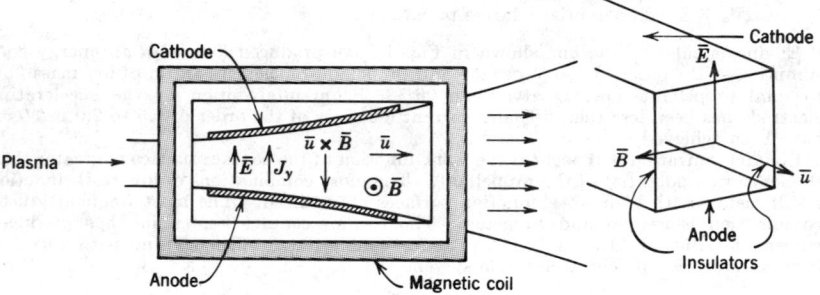

Fig. 21.   Schematic of a typical steady flow magnetogasdynamic accelerator.

The accelerator is bounded on diverging walls by continuous electrodes. A plasma, from a source such as an arcjet, enters the acceleration region. If Hall effects* are negligible, the current j_y which flows through the plasma as a result of the applied and induced electrostatic field is

$$j_y = \sigma(E - uB) \qquad (J_y \text{ is conventional current } + \text{ to } -)$$

where σ = plasma conductivity.
 B = magnetic field intensity.
 E = electrostatic field intensity.
 u = plasma velocity.

In order to achieve useful conductivities (1 mho/cm or more) it may be necessary to seed the propellant gas with an alkali metal. One combination which can be used is argon seeded with potassium. Acceleration of the plasma occurs primarily as a result of the Lorentz force per unit volume which acts on the moving electrons in the current flow j_y where

$$\overline{F} = \overline{j} \times \overline{B}$$

The electrons collide with molecules in the plasma, thus transmitting kinetic energy to the plasma. Pressure forces may contribute some momentum increase.

The thrust developed by the accelerator can be determined from

$$\tau = BhI_t$$

---

* A resultant current or electron drift component to the right in the schematic is due to the Lorentz force and is called the Hall current.[5]   This current produces a desired momentum exchange, but reduces the apparent conductivity, changes the direction of the net force on the gas, and can adversely affect the thrust. It can be significantly reduced by appropriate arrangement of the electrodes. In some thrusters the Hall effect is used to advantage in accelerating the plasma.

where B = magnetic field intensity.
     h = channel height parallel to j.
     I_t = total current.

The efficiency of the accelerator is

$$\eta = \frac{\tau^2/2m + \tau u_1}{p}$$

where p = total electrical power.
     u_1 = entrance velocity of plasma.
     m = plasma mass flow rate.

The pulsed-plasma accelerator is another form of electromagnetic propulsion device. This accelerator does not require an electromagnet since the plasma current is used to generate the magnetic field which yields the Lorentz force. These devices produce high exhaust velocities and relatively low electrode erosion.

The treveling wave accelerator is a third type of plasma accelerator which requires neither external magnets nor electrodes. It relies on currents being induced in the plasma by a traveling magnetic wave. The magnetic wave is established by a number of sequentially energized external conductor loops along the duct.

BIBLIOGRAPHY

1. PERKINS, COURTLAND D., and HAGE, ROBERT E., *Airplane Performance, Stability, and Control*, Chapter 3, New York, John Wiley, 1949.
2. TOPPS, J. E. C., *The Thermal Properties of the Combustion Products of Kerosene Air Mixtures*, National Gas Turbine Establishment, Report No. R14, April, Z947.
3. GRAY, W. H., and MASTROCOLA, NICHOLAS, *Representative Operating Charts of Propellers Tested in the NACA 20 ft. Propeller Research Tunnel*, NACA, ARR 3125, September, 1943.
4. SUTTON, G. P., *Rocket Propulsion Elements*, New York, John Wiley, 3rd ed., 1963.
5. JAHN, R. G., *Physics of Electric Propulsion*, New York, McGraw-Hill, 1968.
6. HILL, P. G., and PETERSON, C. R., *Mechanics and Thermodynamics of Propulsion*, Reading, Massachusetts, Adison-Wesley, 1965.

EXPERIMENTAL METHODS

By R. D. Marker and R. W. Milling

26. SUBSONIC AND SUPERSONIC WIND TUNNELS

Wind tunnels are devices for producing a controlled flow of gas for the study of aerodynamic phenomena. They are generally classed by their speed range and test section size. Most of the subsonic tunnels are similar to the closed circuit unit shown in Fig. 1. Heat exchangers are seldom required in subsonic wind tunnels to remove the energy due to the heat of compression; however, for large high speed facilities they must be employed.

One of the largest supersonic wind tunnels is the 16 ft × 16 ft Propulsion Wind Tunnel located at the USAF Arnold Engineering Development center in Tennessee. This facility consists of two adjacent continuous-flow circuits which share a 216,000 hp drive system to operate separate axial-flow compressors. The two 16 ft × 16 ft test sections in the transonic and supersonic circuits develop Mach numbers up to 4.5.

The cost of both constructing and operating such a large wind tunnel is generally prohibitive for most users. Therefore, a blowdown wind tunnel is a common type of facility used to achieve subsonic, transonic, and supersonic (trisonic) Mach numbers. Figure 2 is a schematic of a 1 ft × 1 ft trisonic wind tunnel. Compressed air is stored in a reservoir and then allowed to expand through a two-dimensional fixed-contour nozzle into the test section. Interchangeable nozzle blocks and a transonic test section are used to achieve a variety of Mach numbers. The test model is mounted on a mechanism which provides variable model attitudes. Electrical transducers and recorders are used to collect aerodynamic data during the relatively short test runs.

Precision instruments are used to calibrate the test section flow and to measure the various forces, pressures, and temperatures associated with a test. In a subsonic or

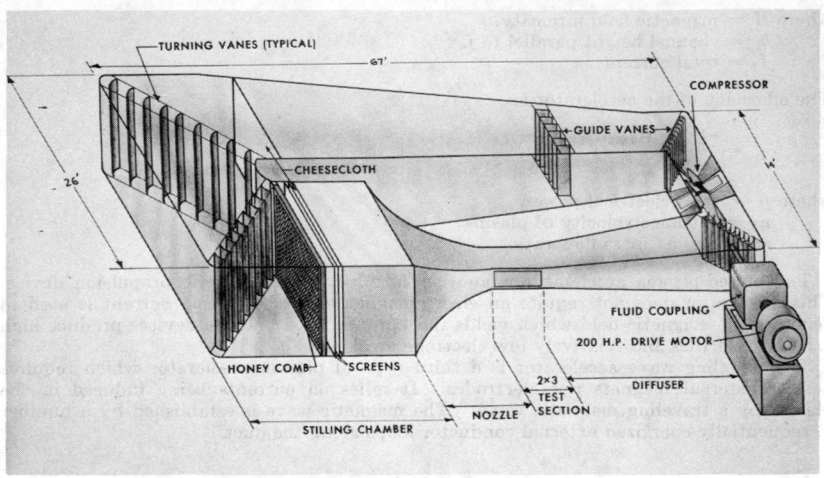

FIG. 1.   USARA 2 × 3 Subsonic Wind Tunnel.

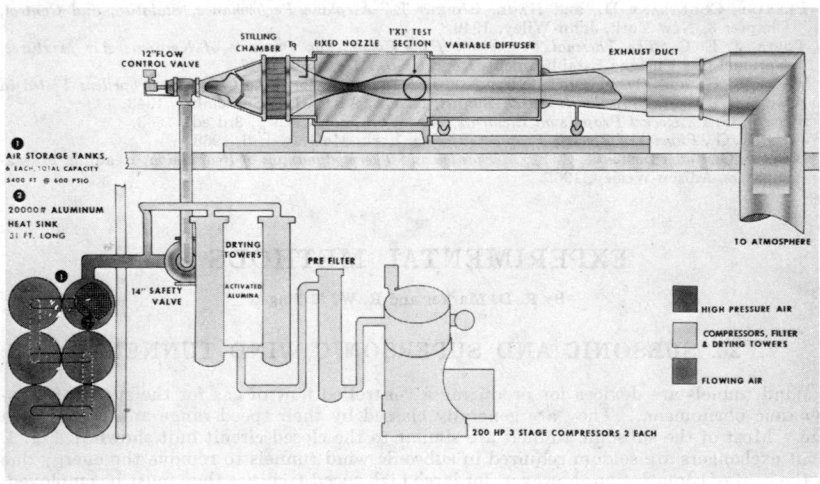

FIG. 2.   Schematic of USAFA Trisonic Wind Tunnel.

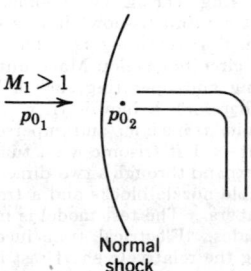

FIG. 3.   Schematic of a total pressure probe in a supersonic flow.

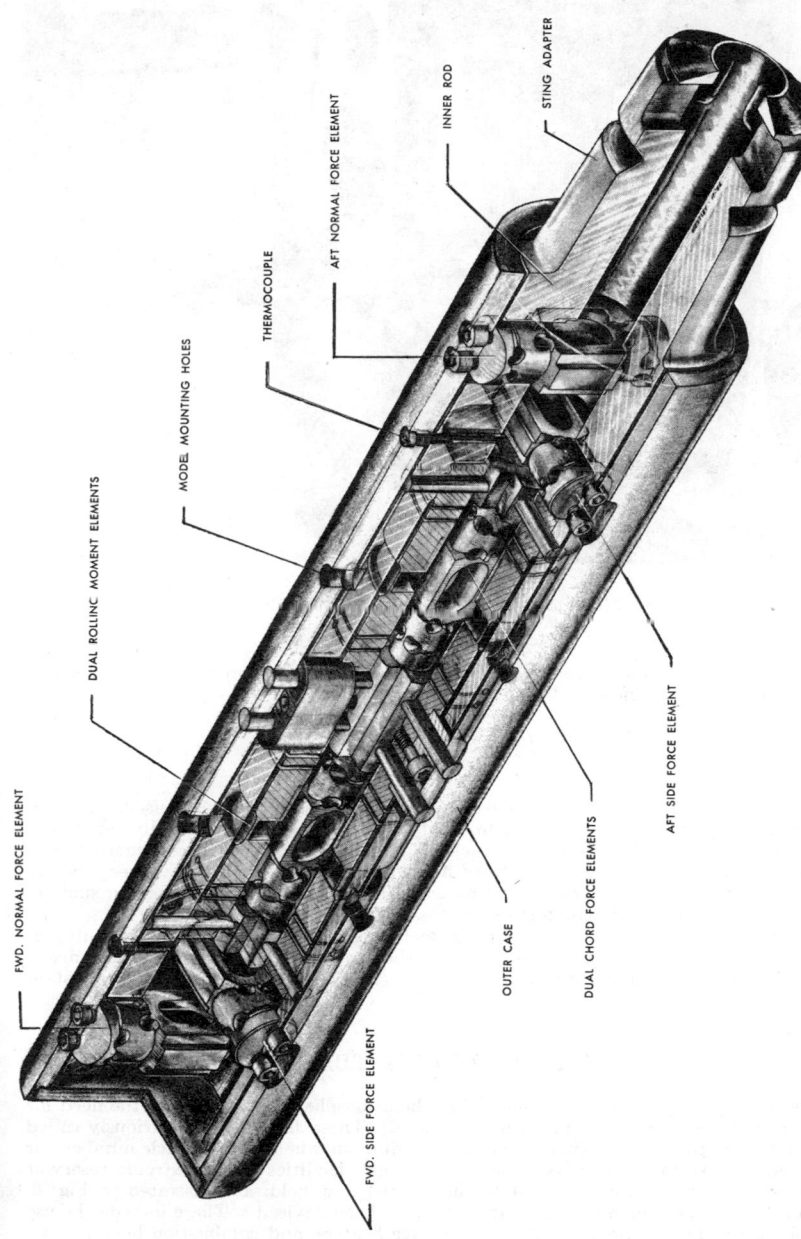

FWD. NORMAL FORCE ELEMENT

DUAL ROLLING MOMENT ELEMENTS

MODEL MOUNTING HOLES

THERMOCOUPLE

AFT NORMAL FORCE ELEMENT

INNER ROD

STING ADAPTER

FWD. SIDE FORCE ELEMENT

OUTER CASE

DUAL CHORD FORCE ELEMENTS

AFT SIDE FORCE ELEMENT

FIG. 4. Six-component internal strain-gage balance (courtesy of the Task Corp., Anaheim, California).

FIG. 5.   Model on sting in the USAFA Transonic Wind Tunnel.

transonic tunnel, the Mach number is determined from Eq. 1 where the total and static pressures, p_0 and p, are measured with a pitot-static probe:

$$\frac{p}{p_0} = \left(1 + \frac{\gamma - 1}{2} M^2 \right)^{-\gamma/(\gamma-1)} \tag{1}$$

At supersonic speeds, a single total pressure probe as shown in Fig. 3 is used.   Since a detached normal shock wave forms in front of the probe, the probe actually senses the total pressure downstream of a normal shock (p_{02}), which can be used in conjunction with the free stream total pressure in Eq. 18 of Art. 15 to obtain the Mach number.   Static pressures on the surface of a model are measured through small orifices in the surface, connected to multiple-tube manometers or pressure transducers.

Aerodynamic forces and moments on the model are generally measured with a multiple-component internal strain-gage balance such as the one shown in Fig. 4.   The balance is installed inside the model and supported from the rear by a movable sting support system as in Fig. 5.

27. HYPERVELOCITY TUNNELS

As the velocity of vehicles traveling within the atmosphere has increased, the need for higher velocity test facilities has become evident.   These facilities are variously called hypersonic or hypervelocity wind tunnels, depending on whether the Mach numbers or the velocity is extreme.   The basic problem of such facilities is the extreme reservoir conditions required to properly similate the desired flow field, as illustrated in Fig. 6. Many methods of heating air at high pressure have been devised.   These include the use of pebble bed storage heaters, electrical resistance heaters, and combustion heaters.   Of the more common methods currently in use, three will be discussed here: (1) the shock tunnel, (2) the "hot shot" tunnel, and (3) the arc heated plasma tunnel.

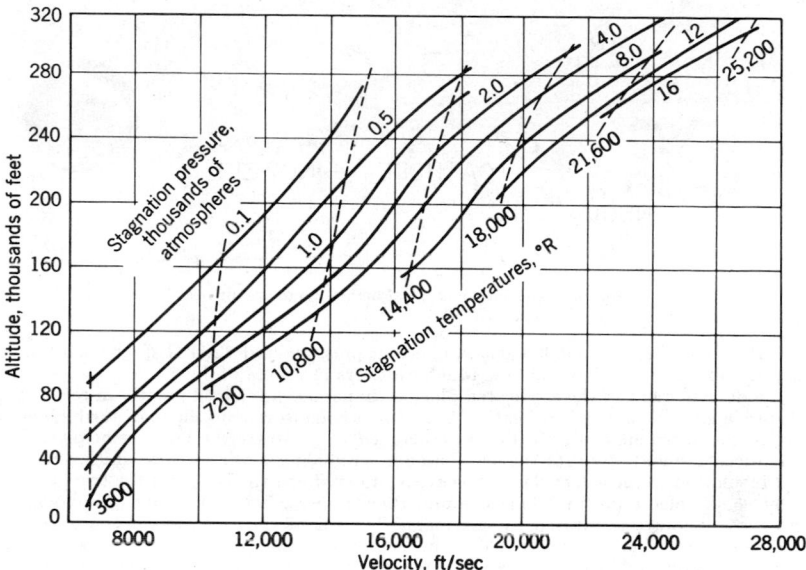

FIG. 6. The stagnation pressure and temperatures required to duplicate flight conditions over a wide range of speeds and altitudes. (From Ref. 2.)

The shock tunnel is an obvious extension of the shock tube which results from the installation of a convergent-divergent nozzle at the driven end of a shock tube. The test gas which has been compressed by both the incident and reflected shock waves provides a reservoir of high pressure, high temperature gas of constant properties of a short duration. Stagnation conditions of 2000 atm pressure at 14,000°R, for test periods of 6 or 7 msec have been obtained in this type facility.

The "hot shot" tunnel is illustrated in Fig. 7. The arc chamber is initially filled with air (or nitrogen) at pressures up to 10,000 psia while the remainder of the circuit is exhausted to a hard vacuum. To initiate the flow, electrical energy from a capacitance or inductance storage system is discharged into the arc chamber. This addition of energy, causes the pressure and temperature within the arc chamber to increase until the diaphragm ruptures, thus starting the flow. The high velocity flow in the test section may last for as long as 100 msec, but the flow properties vary continuously due to the decay of the pressure and temperature in the arc chamber with time. Even so, relatively constant flow properties are obtained for periods of 10 to 20 msec, which warrants their use.

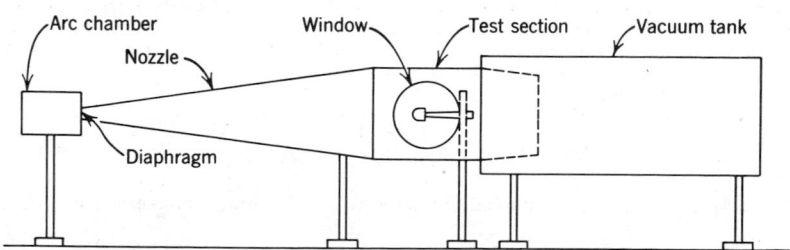

FIG. 7. Hot shot hypervelocity tunnel.

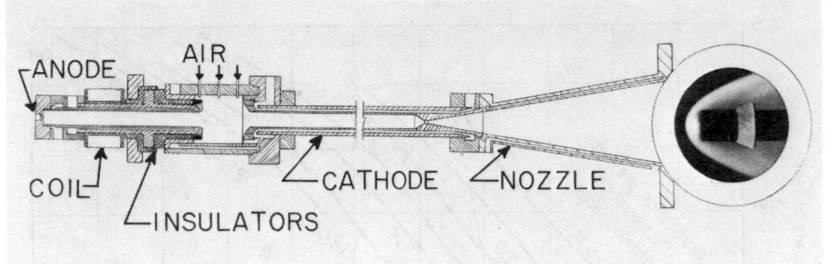

FIG. 8.　Schematic of typical high-voltage arc air heater.

Operating conditions of 30,000 psia stagnation pressure and 7200°R stagnation temperature are obtained, thus giving flow Mach numbers in excess of 20.

The development of the arc heated plasma tunnel has advanced greatly in recent years. There are many designs of arc heaters, both direct and alternating current; the high voltage design shown schematically in Fig. 8 is included to illustrate the basic principles.　Air is introduced tangentially into the arc chamber, producing a vortex flow which offers some heat insulation to the electrodes.　The arc is drawn down the center of the two cylindrical water-cooled electrodes in a long column, thus giving efficient transfer of energy to the test gas.　The arc is rotated by the magnetic field of the coil around the anode, thus reducing contamination of the flow and increasing the life of the electrodes.　After passing through the heater, the air is expanded through an axisymmetrical nozzle which is usually conical.

Arc-heated plasma tunnels may be operated essentially continuously and are rated in thousands or even millions of watts.　One of the largest facilities of this type uses a 50-million watt arc heater and produces a stagnation enthalpy of up to 10,000 Btu/lb, stagnation pressures up to 2000 psia, and core flow diameters of up to 5 ft.　The stream velocity in the plasma tunnel is generally very high (20,000 ft/sec); however, the Mach number is usually less than 3.0 since the speed of sound is also high.　The primary advantages of the plasma tunnel are continuous operation and the ability to develop very high velocities and high heating rates (300 Btu/ft²-sec).　The disadvantages are the existence of nozzle throat erosion at high power levels and contamination of the flow by electrode material.

28. SHOCK TUBE

The shock tube provides a simple means of producing rapid changes in the state of a gas in order to observe chemical reactions such as relaxation effects.　In its simplest form the shock tube is a straight tube divided by a diaphragm into two sections, the driver and the driven tubes.　The driver side is pressurized until the diaphragm ruptures and produces a shock wave which propagates through the test gas in the driven tube, thus raising its temperature and pressure.　The nomenclature and flow regions are shown in Fig. 9.　Region (1) is the undisturbed test gas; region (2), the shock heated test gas; region (3), the expanding driver gas; and region (4) is the undisturbed driver gas.　The test time is generally determined by the time between the shock wave passage and the arrival of the contact surface; however, near the end of the tube the testing time may be terminated by the arrival of the reflected shock wave.

The shock velocity is given by

$$c_s = M_1 a_1 = a_1 \left(\frac{\gamma - 1}{2\gamma} + \frac{\gamma + 1}{2\gamma} \frac{p_2}{p_1} \right)^{1/2} \qquad (2)$$

The shock strength, p_2/p_1, is implicitly related to the diaphragm pressure ratio, p_4/p_1, by

$$\frac{p_4}{p_1} = \frac{p_2}{p_1} \left[1 - \frac{(\gamma_4 - 1)(a_1/a_4)(p_2/p_1 - 1)}{\sqrt{2\gamma_1} \sqrt{2\gamma_1 + (\gamma_1 + 1)(p_2/p_1 - 1)}} \right]^{-2\gamma_4/(\gamma_4 - 1)} \qquad (3)$$

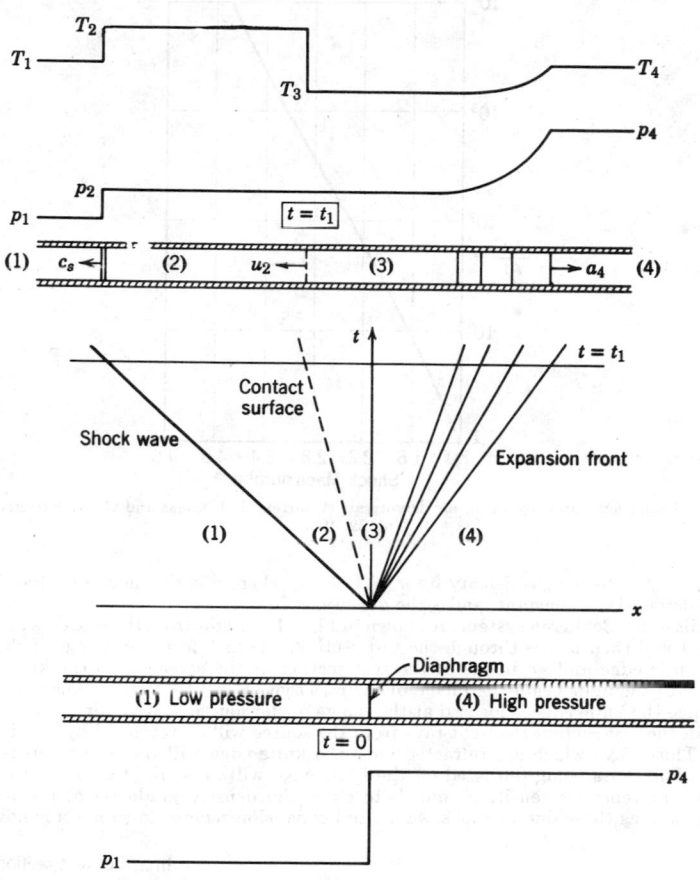

Fig. 9. Motion in a shock tube. (From Ref. 5.)

Other conditions across the shock may be found from the Rankine-Hugoniot relations of Art. 15, equations 15, 16, and 17.

The maximum shock velocity obtainable is limited by the physical requirements dictated by the need for very high diaphragm pressure ratios, as illustrated in Fig. 10 for the case of air driving air. Since the required diaphragm pressure ratio depends on the ratio of the speeds of sound of the driver and driven gases, the use of light driver gases (such as helium or hydrogen) and driver preheating are common.

29. FLOW VISUALIZATION

The visualization of the flow about a model in subsonic flow can be achieved with short strings (tufts) attached to the model or held in the vicinity of the model on a grid network. Oil films can also be painted on the model to detect flow patterns and separation. Smoke tunnels are also used to visualize low speed flow.

In transonic and supersonic wind tunnels shadowgraph and Schlieren systems are used extensively. Both systems make use of the fact that the index of refraction of a gas is

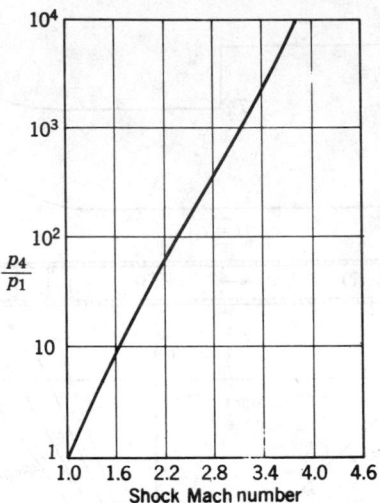

Fig. 10.   Diaphragm pressure ratio, air driving air (Courtesy I. I. Glass and G. N. Patterson, *JAS*, Vol. 22, 1955).

directly related to the gas density by $n = 1 + k\rho$, where n is the index of refraction, k is the Gladstone-Dale constant, and ρ the density.

Details of the Schlieren system are shown in Fig. 11.   Light from the source is collimated by lens 1 and then passes through the test section.   Lens 2 forms an image of the source at the knife edge and an image of the test section at the screen.   As the knife edge is moved into the light beam the image of the test section on the screen is uniformly darkened since the knife edge is located at the image of the source.   With air flowing over the model in the test section the light rays from the source will be refracted by density gradients.   Those rays which are refracted onto the knife edge will cause dark areas on the screen while those being refracted off the knife edge will cause light areas on the screen.   Schlieren systems are sensitive enough to show the density gradients of the boundary layer as well as those due to shock waves and expansion regions in supersonic flow.

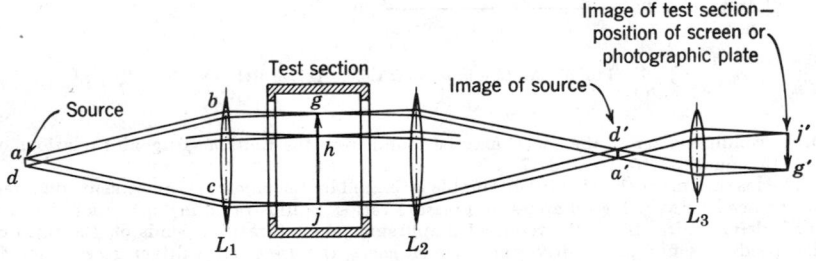

Fig. 11.   Schlieren system.

30. INSTRUMENTATION

The hot wire anemometer is an instrument designed to study fluid velocities and turbulence by heat conduction from a very thin wire several thousandths of an inch in diameter.   The sensing element and probe are extremely small and cause a minimum of disturbance to the fluid.   The wire, normally tungsten, is supported between two needle-like supports and heated by an electrical current passing through it.   The wire is then cooled through forced convection by the flowing medium under test.   Since the electrical properties of the wire change with temperature, and the heat transfer is a function of velocity, either the voltage or current supplied to the wire may be used to determine the velocity. Hot wire anemometers are generally classed as being a constant temperature or constant current system. In the constant temperature system, the wire is maintained at a constant temperature under all velocity conditions by controlling the direct current supplied to the wire. From a freestream calibration of current vs. velocity, unknown velocities can be determined. This type of system is very valuable for determining the transition point, velocity profile, and thickness of a boundary layer.

The constant current anemometer is used primarily for the measurement of turbulence. The sensor provides a means for measuring the fluctuating component of velocity which is superimposed on the mean velocity. In this case an a-c amplifier is used to amplify the voltage signal produced by the turbulence. Although the heat capacity of the wire will cause the temperature and voltage to lag the rapid fluctuations, under ideal conditions a frequency response of 320 kc can be achieved through the use of an electrical compensating circuit.

Pressure sensors must be carefully selected, based upon the range, accuracy, time response, and the physical size of the sensor required. (See Table 1.)

Table 1. Pressure and Vacuum Measuring Devices

Type	Usable Pressure Range	Accuracy	Time Response
Manometer	4 atm–0.01 Torr	Excellent	Long
Bourdon Tube	Down to 1mm	Good	Long
Null Balance Transducer	±500 psia down to ±1 psia	0.5% of full scale	Long
Strain Gage Transducer	5000 psia down to 1 psia	1.0% of full scale	Short
Piezoelectric Transducer	Above 5000 psia to less than 1 psia	1.0% of full scale	Very short (1 kc)
McLeod Gage	5 to 10^{-5} Torr	See text	Long
Pirani Gage	2 to 10^{-3} Torr	See text	Short
Thermocouple Gage	1 to 10^{-3} Torr	See text	Short
Ionization Gage	10^{-3} to 10^{-10} Torr	See text	Short

One of the most common pressure measuring devices is a liquid manometer in which the height of a column of liquid and the density of the liquid is used to determine an unknown pressure by

$$p = (\Delta h)(62.4)(\text{sp gr})$$

where p = pressure, lb/ft² gage
  Δh = column height, ft
  sp gr = specific gravity of the liquid compared to water

The range and sensitivity of manometers can be changed through the use of liquids with different specific gravities. Some of the commonly used liquids and their specific gravities are: alcohol 0.82, water 1.0, TBE (tetrabromoethane) 2.95, and mercury 13.56. Several types of manometers are available including U-tube, cistern, multiple cistern, inclined, and micromanometers. The last two are the most sensitive, with the micromanometer capable of measuring to any accuracy of ±0.001 in. of water pressure.

Sophisticated vacuum gages which are used solely for vacuum measurements differ primarily in range and accuracy. The accuracy of the various vacuum gages is dependent upon the range in which the instrument is operated, foreign gases and contaminants, and may vary from 3 to 100% of the reading. The specifications of the particular instruments must be reviewed carefully to select a gage for a particular requirement.

The McLeod gage is an extremely accurate mercury manometer and is considered a standard for the calibration of high vacuum gages. It covers a wide range of vacuum with an accuracy of 1.5 to 2.0%. The pirani and thermocouple gages are general purpose thermal conductivity type, with the Pirani being the more accurate. The ionization gage covers a wide range, extending to extremely low presssures.

BIBLIOGRAPHY

1. POPE, A., and HARPER, J. J., Low Speed Wind Tunnel Testing, New York, Wiley, 1966.
2. POPE, A., and GOIN, K. L., High Speed Wind Tunnel Testing, New York, Wiley, 1965.
3. GOETHERT, B. H., Transonic Wind Tunnel Testing, New York, Pergamon Press, 1961.

4. *Handbook of Supersonic Aerodynamics*, NAVORD Report 1488, Washington, D.C, Bureau of Ordnance, U.S. Navy Department, 1950.
5. LIEPMANN, H. W., and ROSHKO, A., *Elements of Gasdynamics*, New York, Wiley, 1957.
6. SHAPIRO, A. H., *The Dynamics and Thermodynamics of Compressible Fluid Flow*, New York, Ronald Press, 1953.
7. HOLMAN, J. P., *Experimental Methods for Engineers*, New York, McGraw-Hill, 1966.
8. BECKWITH, T. G., and BUCK, N. L., *Mechanical Measurements*, Reading, Massachusetts, Addison-Wesley, 1961.
9. CERNI, R. H., and FOSTER, L. E., *Instrumentation for Engineering Measurement*, New York, 1962.

AERODYNAMICS OF WINGS

By L. M. Nicolai

31. INTRODUCTION

The fundamental problem of aerodynamics is to determine the aerodynamic force and moment on a body (i.e., airfoil, wing, body, or combination) immersed in a moving fluid.* The aerodynamic force is resolved into a force normal to the free stream velocity V_∞ called the lift and a force parallel to V_∞ called the drag (see Fig. 1a). The integration of the stress distribution (pressure and shearing stress) over the body will yield the lift, drag, and moment. This stress distribution is the solution of the governing fluid equations and the boundary conditions describing the body surface and far field property values. These governing equations, discussed in Art. 10, are a complicated set of nonlinear equations.

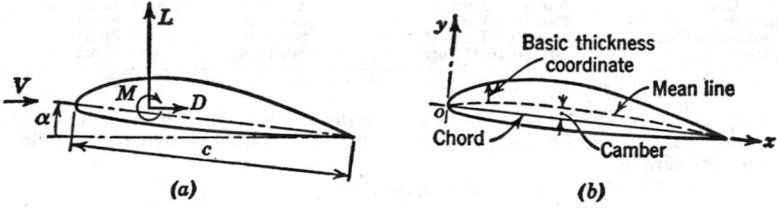

(a)   (b)

FIG. 1.   Airfoil notation and geometry.

The governing equations are simplified somewhat by ignoring the viscous and heat conduction terms and considering what is called an inviscid flow situation. This seemingly drastic action is justified as follows. It can be shown[1] that the lift and moment for a thin body at a small to moderate angle of attack α (up to about 10°) depends primarily on the pressure distribution over the body. The tangential shearing stress contributing a negligible amount. In 1904 Prandtl hypothesized that for large Reynolds number R_e (see Sec. 6) the effects of viscosity are confined to a thin region next to the surface of the body called a boundary layer. It can be shown[1] that for slender bodies at moderate α and large R_e the static pressure at the surface is the same as the pressure at the edge of the boundary layer; i.e., the pressure is constant across the thin boundary layer. Thus if we seek only the pressure distribution we can consider the flow at the edge of the boundary layer which is the inviscid flow region. The Reynolds number of aircraft is at all times large.

The above arguments justify the basic approach for estimating the aerodynamic forces and moments over wings and bodies. The flow is broken into an inviscid region (outside the boundary layer) and a viscous region (boundary layer and wake). The inviscid flow theory will give us the pressure distribution from which we can determine the lift, moment, and inviscid drag. The viscous flow theory will give us the skin friction drag, drag due to separation, stall phenomena, and aerodynamic heating.

The following Arts. 32 to 37 will discuss the inviscid flow theory. Articles 45 to 49 will present the viscous flow theory.

---

* It is assumed that a Gallelian transformation may always be applied so that the body of interest is stationary while the fluid streams past it.

32. GENERAL INVISCID FLOW THEORY

The flow is assumed to be steady, adiabatic, and irrotational* (see Sec. 6) with a free stream velocity V_∞. A slender solid body is placed in this uniform stream and disturbs the basic motion as shown in Fig. 2.

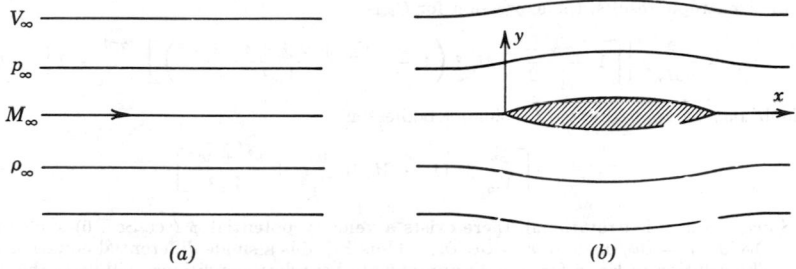

$$V_\infty$$

$$p_\infty$$

$$M_\infty$$

$$\rho_\infty$$

(a)   (b)

FIG. 2.   Perturbation of a uniform flow by a slender body: (a) uniform flow; and (b) perturbed flow.

The total velocity $\overline{V}$ in the perturbed flow has components $V_\infty + u$, v, and w in the x, y, and z directions, respectively, where u, v, and w are called perturbation velocity components.

The governing equations for steady, adiabatic, and irrotational flow are (from Art. 10)

$$\text{Continuity} \quad \frac{\partial}{\partial x_i}(\rho u_i) = 0 \tag{1}$$

$$\text{Momentum} \quad \rho u_j \frac{\partial u_i}{\partial x_j} = -\frac{\partial p}{\partial x_i} \tag{2}$$

$$\text{Energy} \quad \frac{u_i u_i}{2} + h = h_0 \tag{3}$$

Since an adiabatic irrotational flow is also isentropic we can use the useful isentropic relation

$$\frac{p}{p_0} = \left(\frac{\rho}{\rho_0}\right)^\gamma \tag{4}$$

and the definition of the acoustic velocity (speed of sound)

$$a^2 = \left(\frac{\partial p}{\partial \rho}\right)_s \tag{5}$$

Equations 1–5 are combined to give the single governing differential equation[2] (SGDE)

$$(1 - M_\infty{}^2)\frac{\partial u}{\partial x} + \frac{\partial v}{\partial y} + \frac{\partial w}{\partial z} = M_\infty{}^2\left[(\gamma + 1)\frac{u}{V_\infty} + \frac{\gamma + 1}{2}\frac{u^2}{V_\infty{}^2} + \frac{\gamma - 1}{2}\frac{v^2 + w^2}{V_\infty{}^2}\right]\frac{\partial u}{\partial x}$$

$$+ M_\infty{}^2\left[(\gamma - 1)\frac{u}{V_\infty} + \frac{\gamma + 1}{2}\frac{v^2}{V_\infty{}^2} + \frac{\gamma - 1}{2}\frac{w^2 + u^2}{V_\infty{}^2}\right]\frac{\partial v}{\partial y}$$

$$+ M_\infty{}^2\left[(\gamma - 1)\frac{u}{V_\infty} + \frac{\gamma + 1}{2}\frac{w^2}{V_\infty{}^2} + \frac{\gamma - 1}{2}\frac{u^2 + v^2}{V_\infty{}^2}\right]\frac{\partial w}{\partial z}$$

$$+ M_\infty{}^2\left[\frac{v}{V_\infty}\left(1 + \frac{u}{V_\infty}\right)\left(\frac{\partial u}{\partial y} + \frac{\partial v}{\partial x}\right) + \frac{w}{V_\infty}\left(1 + \frac{u}{V_\infty}\right)\left(\frac{\partial u}{\partial z} + \frac{\partial w}{\partial x}\right)\right.$$

$$\left. + \frac{vw}{V_\infty{}^2}\left(\frac{\partial w}{\partial y} + \frac{\partial v}{\partial z}\right)\right] \tag{6}$$

where $M_\infty = V_\infty/a_\infty$ is the free stream Mach number.

* Since the field of flow about an airfoil is uncontaminated by viscosity (except in the thin boundary layer), the fluid motion can be considered irrotational. This follows from the fact that irrotational motion in the approaching stream cannot become rotational owing to pressure forces in the field alone.

The pressure coefficient is defined by

$$C_p = \frac{p - p_\infty}{\frac{1}{2}\,\rho_\infty V_\infty{}^2} = \frac{2}{\gamma M_\infty{}^2}\left[\frac{p - p_\infty}{p_\infty}\right]$$

Using isentropic results, the expression for C_p is

$$C_p = \frac{2}{\gamma M_\infty{}^2}\left\{\left[1 + \frac{\gamma - 1}{2}\,M_\infty{}^2\left(1 - \frac{(V_\infty + u)^2 + v^2 + w^2}{V_\infty{}^2}\right)\right]^{\gamma/\gamma-1} - 1\right\} \quad (7)$$

which may be approximated for slender bodies[2] as

$$C_p = -\left[\frac{2u}{V_\infty} + (1 - M_\infty{}^2)\,\frac{u^2}{V_\infty{}^2} + \frac{v^2 + w^2}{V_\infty{}^2}\right] \quad (8)$$

Since the flow is irrotational, there exists a velocity potential ϕ (see Sec. 6) such that $u = \partial\phi/\partial x$, $v = \partial\phi/\partial y$, and $w = \partial\phi/\partial z$. Thus Eq. 6 is a single differential equation for ϕ. The solution of Eq. 6 for ϕ, with appropriate boundary conditions, will yield the distribution of perturbation velocities. Equation 8 would then be used to determine the pressure coefficient distribution over the body.

Equation 6 cannot be solved in practice because of its nonlinearity. Thus the flow is broken down into flow regimes, each having its own set of conditions and assumptions. The SGDE is then solved according to simplifications appropriate to the particular flow regime. These flow regimes are:

1. Incompressible $0 \leqslant M_\infty \leqslant 0.2$
2. Subsonic $0.2 < M_\infty < 1$
3. Transonic $M_\infty \approx 1$
4. Supersonic $1 < M_\infty < M_{\text{hypersonic}}$
5. Hypersonic $M_{\text{hypersonic}} < M_\infty$

The Mach number for the hypersonic flow regime is not clearly defined and depends upon the thickness of the body in question. Generally, $M_{\text{hypersonic}}$ is greater than 5 (see Art. 42).

33. INCOMPRESSIBLE FLOW THEORY

When $M_\infty \approx 0$, the right-hand side of Eq. 6 vanishes and the governing equation becomes Laplace's equation

$$\nabla^2\phi = 0 \quad (9)$$

Thus the flow is governed by a linear differential equation. Physically what has happened is that the flow speed is so low that the changes in pressure are small which results in infinitesimal changes in density, i.e., $\rho = \text{constant}$. Also the kinetic energy of the fluid is very small compared to the internal energy so that the fluid temperature is constant in spite of velocity perturbations.

Two-Dimension Flow

The equation of continuity for the two-dimensional motion of an incompressible fluid is

$$\frac{\partial u}{\partial x} + \frac{\partial v}{\partial y} = 0 \quad (10)$$

As a consequence of continuity in steady motion, a stream function ψ exists (see Sec. 6) such that

$$u = \frac{\partial\psi}{\partial x} \quad \text{and} \quad v = -\frac{\partial\psi}{\partial x} \quad (11)$$

The condition for irrotationality is $\nabla \times V = 0$ which for two-dimensional flow is

$$\frac{\partial v}{\partial x} - \frac{\partial u}{\partial y} = 0 \tag{12}$$

Putting Eq. 11 into 12 yields

$$\frac{\partial^2 \psi}{\partial x^2} + \frac{\partial^2 \psi}{\partial y^2} = 0 \tag{13}$$

which is linear. It is seen that the potential lines (constant ϕ) and streamlines (constant ψ) are everywhere perpendicular to each other. This forms the basis of the conformal mapping discussed in Sec. 6.

Circulation

A concept called *circulation* is very useful in incompressible flow theory. It is defined as the line integral of tangential velocity components along a closed path, thus (Fig. 3a)

$$\Gamma = \oint V \cos \theta ds \tag{14}$$

taken positive in the counterclockwise direction. It is seen from Fig. 3b that

$$\Gamma = \Sigma \Delta \Gamma \tag{15}$$

since internal line integrals cancel one another.

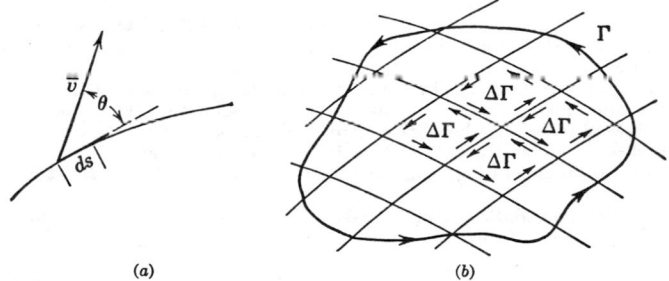

(a)   *(b)*

Fig. 3. Circulation.

Performing the line integration around the fluid element in Fig. 4 one obtains

$$d\Gamma = \left(\frac{\partial v}{\partial x} - \frac{\partial u}{\partial y}\right) dx\, dy \tag{16}$$

whence

$$\frac{d\Gamma}{dx\, dy} = \frac{\partial v}{\partial x} - \frac{\partial u}{\partial y} = 2\omega \tag{17}$$

which states that the circulation per unit area is equal to twice the angular velocity at that point. From Eqs. 15 and 16

$$\Gamma = \iint \left(\frac{\partial v}{\partial x} - \frac{\partial u}{\partial y}\right) dx\, dy \tag{18}$$

from which it can be concluded that the circulation around a circuit enclosing fluid in irrotational motion is zero.

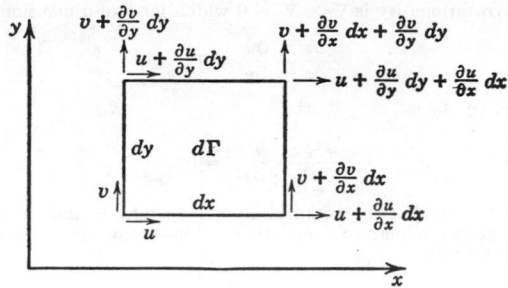

FIG. 4.   Velocity components on a fluid element.

Thomson's Theorem

Thomson's theorem states that the circulation along a closed line composed of fluid particles and lying wholly in a frictionless region remains constant with time; i.e., $d\Gamma/dt = 0$. This is to be expected because pressure and body forces are not tangential forces and therefore cannot produce a net change in the angular momentum of the enclosed system.

Forces on a Body in a Frictionless Fluid

Any irrotational physically possible* two-dimensional flow has a stream function and velocity potential that satisfy Laplace's equation (see Sec. 6). Conversely, any solution of Laplace's equation in two-dimensions represents the stream function or velocity potential of an irrotational physically possible flow. Because Laplace's equation is linear, the sum of any number of solutions is also a solution. Thus the solution to complicated flow patterns can be obtained by adding together the flow patterns of simple potential flows. For example, Sec. 6 shows how the solution for the flow about a rotating cylinder can be obtained by adding together the solution for a uniform stream, a doublet, and a free vortex flow.

The pressure distribution over a body of arbitrary shape can be found by using Bernoulli's equation (see Sec. 6)

$$p + \frac{1}{2}\rho V^2 = p_\infty + \frac{1}{2}\rho V_\infty^2 = \text{const} \tag{19}$$

and the velocity potential or stream function obtained by superposing appropriate potential solutions. This pressure distribution, when integrated over the surface of the body, yields[1]

$$L' = \rho V_\infty \Gamma, \text{ lift per span}$$

$$D' = 0, \qquad \text{drag per span} \tag{20}$$

where Γ is the circulation of the body and is positive clockwise. This very important result is called the Kutta-Joukowski theorem.

Joukowski Airfoils

The Joukowski airfoils represent the simplest class of theoretically determinable airfoils. They are obtained by means of the mathematical method called conformal transformation (see Sec. 6). By this method the flow about a cylinder is transformed into the flow about a body having the shape of an airfoil (Fig. 5). The transformation is $\zeta = \xi + i\eta = z + a^2/z$, where a is a constant and $z = x + iy$.

---

* Any flow that satisfies continuity is termed physically possible.

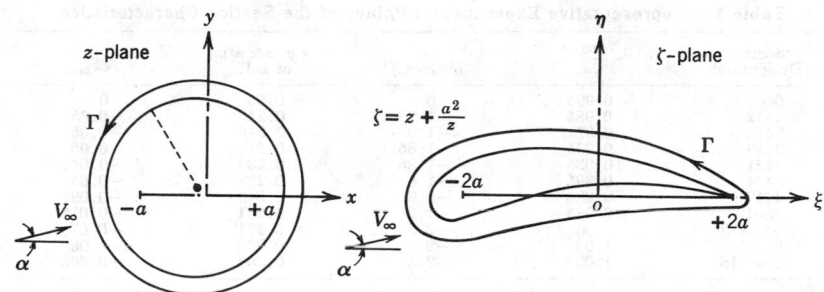

FIG. 5. Joukowski transformation.

The Joukowski profiles, however, are not the best for practical purposes. Structurally, the nose is too massive and the tail cusped; aerodynamically, the pressure distribution is too asymmetric, which precludes high-speed and/or low-drag airfoils. Furthermore, the Joukowski family of airfoils is restricted in shape to the cambered type and hence does not include such shapes as the reflexed type generally necessary for the flying wing. Although Joukowski airfoils are not used in practice, their theoretical analysis predicts such general characteristics as (a) an aerodynamic center exists (see Properties of Airfoils); (b) the aerodynamic center is near the quarter-chord point for thin airfoils; (c) the moment about the aerodynamic center is proportional to camber for this airfoils; (d) lift $= 2\pi\alpha$ for thin airfoils and small angles of attack α.

Generation of Lift

The theoretical calculation of the lift on a Joukowski airfoil, or any airfoil with sharp trailing edge, requires the adjustment of the circulation about the airfoil such that the rear stagnation point coincides with the trailing edge. This will insure that the fluid will flow smoothly off the trailing edge of the airfoil. The above requirement is called the *Kutta condition.*

The Kutta condition is satsified physically through the action of viscosity. Consider the course of events upon sudden starting of an airfoil. (See Fig. 6 and note that in airfoil practice the circulation Γ is positive clockwise.)

At the first instant of starting, the flow past the airfoil is such that the circulation immediately around the airfoil is zero: i.e., no lift occurs on the airfoil and the rear stagnation point is found on the upper surface (Fig. 6a); also the circulation along a circuit far from the airfoil is zero. However, as time progresses, the viscosity of the fluid takes effect in that, at the trailing edge, the fluid separates from the edge (Fig. 6b) and curls up, forming a vortex which is swept downstream with the flow (Fig. 6c). Yet the circulation along the fluid line which is far from the body and unaffected by viscous action remains zero by Thomson's theorem. Hence, circulation must exist along a circuit immediately around but outside of the boundary layer of the airfoil, and its value must be equal and opposite to that around the shed vortex, the so-called starting vortex. Lift equal to $\rho V \Gamma$ will now act on the airfoil.

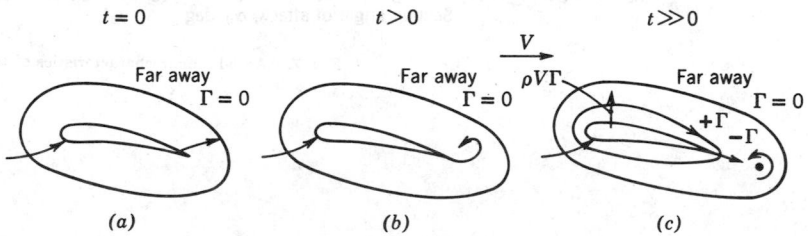

FIG. 6. Development of circulation.

Table 1. Representative Experimental Values of the Section Characteristics

Section Designation	$\frac{m_0}{2\pi}$	α_{OL} (degrees)	a.c. x/c aft of L.E.	$c_{m_{ac}}$
0009	0.995	0	0.25	0
2412	0.985	−1.9	0.243	−0.05
2415	0.97	−1.9	0.246	−0.05
2418	0.935	−1.85	0.242	−0.05
2421	0.925	−1.85	0.239	−0.045
2424	0.895	−1.8	0.228	−0.04
4412	0.985	−3.9	0.246	−0.095
23012	0.985	−1.2	0.241	−0.015
64₁–418	1.06	−2.9	0.271	−0.07
65₁–418	1.03	−2.5	0.266	−0.06
66₁–418	1.00	−2.5	0.264	−0.065

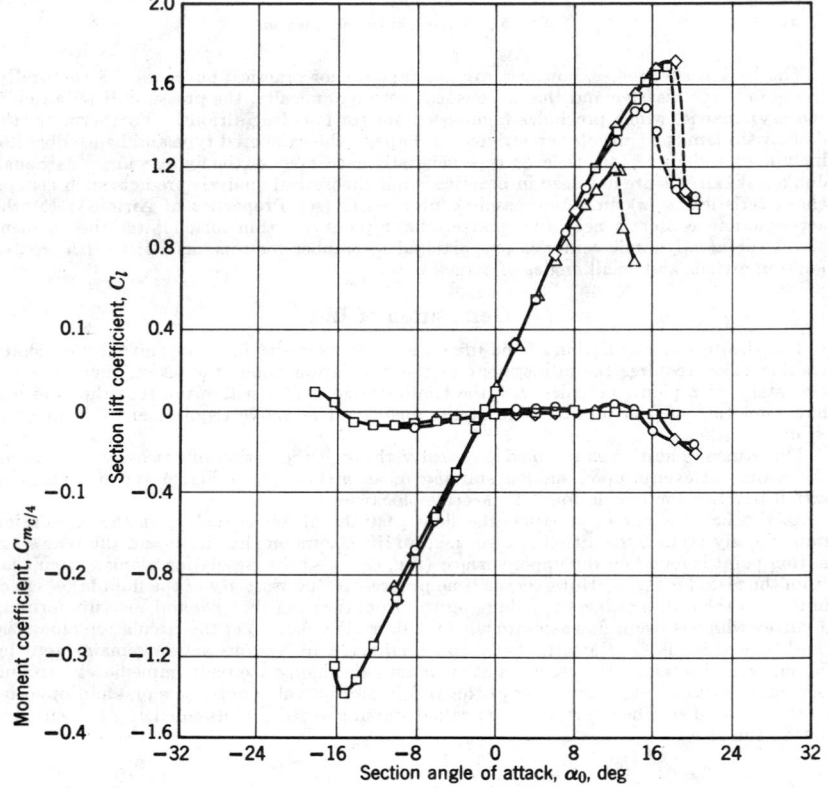

Fig. 7. Aerodynamic characteristics of the

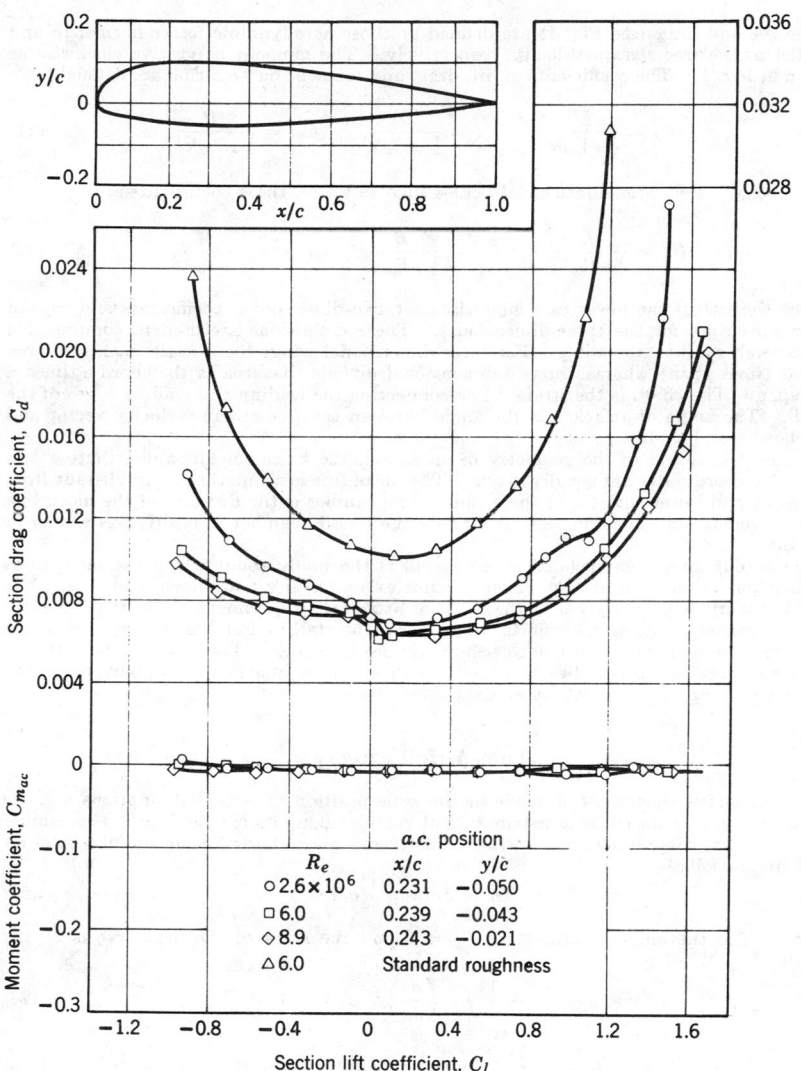

NACA 23015 airfoil section, 24-in. chord.

Properties of Airfoils

The lift and drag (see Fig. 1) are defined as those aerodynamic forces normal to and parallel to the free stream velocity, respectively. The moment is positive clockwise as shown in Fig. 1. The coefficients of lift, drag, and moment on an airfoil are defined by

$$C_l = \frac{L''}{\frac{1}{2}\rho V_\infty^2 c}, \qquad C_d = \frac{D'}{\frac{1}{2}\rho V_\infty^2 c}, \qquad C_m = \frac{M'}{\frac{1}{2}\rho V_\infty^2 c^2} \qquad (21)$$

respectively. For three-dimensional bodies such as wings, these coefficients are

$$C_L = \frac{L}{\frac{1}{2}\rho V_\infty^2 S}, \qquad C_D = \frac{D}{\frac{1}{2}\rho V_\infty^2 S}, \qquad C_M = \frac{M}{\frac{1}{2}\rho V_\infty^2 S c} \qquad (22)$$

Notice the use of the lower case subscripts for two-dimensional coefficients and capital letter subscripts for the three-dimensional. These definitions are used in compressible flow as well we incompressible. For three-dimensional wings the S is the planform area (chord times span) whereas in two-dimensional airfoils the area is the chord c times a unit span. The chord is the straight line connecting the leading and trailing edges of the airfoil. The angle of attack α is the angle between the free-stream velocity vector and the chord (see Fig. 1a).

In the description of the geometry of an airfoil, the basic thickness distribution and mean-line coordinates are usually given. The mean line is a line that is equidistant from the upper and lower surfaces of the airfoil. The camber is the distance of the mean line from the chord and is a function of x along the chord (camber is positive as shown in Fig. 1a).

The aerodynamic center (a.c.) of an airfoil is the point about which the moment is independent of angle of attack. Such a point exists for any two-dimensional airfoil,[1] as long as the lift is proportional to the angle of attack. (Experiment shows that an aerodynamic center exists also for finite wings below the stall.) For thin airfoils, this point is located one-quarter chord length behind the leading edge. For practical airfoils, the aerodynamic center usually lies in the neighborhood of the quarter-chord point; hence the moment data are usually given about the quarter-chord point.

Thin Airfoil Theory

In thin airfoil theory use is made of the superposition of potential solutions and the airfoil section is replaced by a distribution of vortices along its camber line. The camber line is given as some function $z = f(x)$. The theory gives the following useful results for cambered airfoils:[1]

$$C_l = 2\pi(\alpha - \alpha_{OL}) \qquad (23)$$

where α_{OL} is the angle of attack for $C_l = 0$ and the slope of the lift curve is 2π per radian.

$$\alpha_{OL} = -\frac{1}{\pi}\int_0^\pi \left(\frac{dz}{dx}\right)(\cos\theta - 1)\, d\theta \qquad (24)$$

$$C_{mac} = \frac{1}{2}\int_0^\pi \left(\frac{dz}{dx}\right)(\cos 2\theta - \cos\theta)\, d\theta \qquad (25)$$

where the dz/dx is the local slope of the camber line line. The θ appears through the change of variable $x = c/2(1 - \cos\theta)$ where x is the distance from leading edge. The aerodynamic center is located at $0.25c$ behind the leading edge.

From Eqs. 24 and 25 it can be seen that α_{OL} and C_{mac} will always be a negative number for an airfoil whose mean camber line lies above the chord line ($z > 0$, positive camber). The shape of the camber in the vicinity of the trailing edge (T.E.) has a very powerful effect on α_{OL} and C_{mac} whereas the camber line near the leading edge (L.E.) has very little influence on these properties. It is for this reason that flaps are put on the T.E.

The results of thin airfoil theory agree very well with experimental results on two-dimensional airfoils. Some experimental results are shown in Figure 7[17] and Table 1.

Most of the NACA (now NASA) airfoils are classified among the three types, the four-digit, the five-digit, and the series 6 series. The meanings of these designations are illustrated by the examples below.

NACA 4415
  4—The maximum camber of the mean line is 0.04c.
  4—The position of the maximum camber is at 0.4c.
  15—The maximum thickness is 0.15c.

NACA 23012
  2—The maximum camber of the mean line is approximately 0.02c.
    The design lift coefficient is 0.15 times the first digit for this series.
  30—The position of the maximum camber is at 0.30/2 = 0.15c.
  12—The maximum thickness is 0.12c.

NACA 65₃-421
  6—Series designation.
  5—The minimum pressure is at 0.5c.
  3—The drag coefficient is near its minimum value over a range of lift coefficients of 0.3 above and below the design lift coefficient.
  4—The design lift coefficient is 0.4.
  21—The maximum thickness is 0.21c.

Finite Wing Theory

In the Prandtl wing theory the wing is replaced by a bundle of vortex segments at its aerodynamic center. Since the circulation varies along the span, the lengths of the vortex segments must be such that the sum of the strengths (circulations) at each spanwise point of the wing is equal to the circulation around the wing at that section. Furthermore, since a vortex filament cannot end in a fluid (according to Helmholz), it is assumed that "trailing" vortices extend downstream indefinitely from the ends of the vortices "bound" to the wing line. The strengths of the trailing vortices are equal to the strengths of the bound vortices. The pattern of each vortex filament is that of a horseshoe extending to infinity (see Fig. 8).

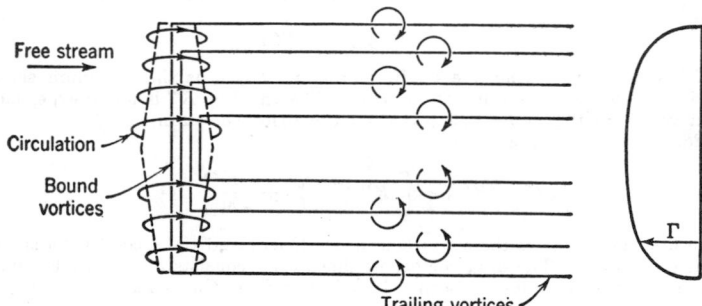

Free stream

Circulation

Bound vortices

Γ

Trailing vortices

FIG. 8. Vortex system for lifting wing.

Induced Drag

The trailing vortices induce downward velocities (downwash) along the lifting line and behind the wing. At each wing section the downward velocity w, combined with the free-stream velocity V_∞ yields a resultant velocity V_{res}, inclined downwards at an angle α_i. Therefore, the local resultant pressure force vector L_0 (per unit length of span), being perpendicular to the local resultant velocity vector V_{res}, is inclined downstream, thus giving rise to a local induced drag D_i (Fig. 9). Since the induced angle of attack α_i is small in practice, the local lifting force L is approximately equal to the local resultant pressure force L_0; also $D_i \approx \alpha_i L$. Summation of the local conditions (strip theory) over

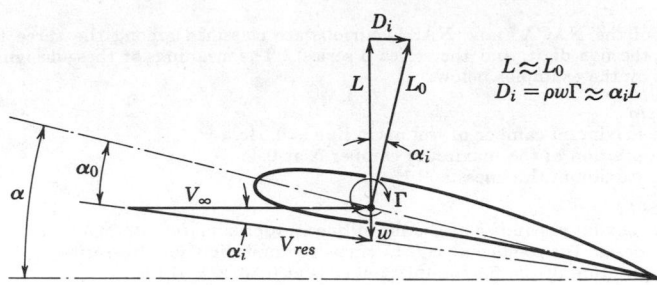

FIG. 9.   Induced drag vector diagram.

the span then yields an effective angle of attack and a resultant induced drag for the entire wing. Symbols for the aerodynamics of the whole wing remain the same as for the local sections.

Prandtl found that the planform yielding minimum resultant induced drag (pressure drag due angle of attack) is elliptic and that the corresponding spanwise lift (circulation) distribution is also elliptic. The formula for resultant induced drag is

$$C_{Di} = \frac{C_L{}^2}{\pi \mathit{R}} \tag{26}$$

where R is the aspect ratio defined as the ratio of the span squared to the wing area, and C_L is the resultant lift coefficient. The total drag coefficient is then given by

$$C_D = C_{Dp} + \frac{C_L{}^2}{\pi \mathit{R}} \tag{27}$$

in which C_D is the airfoil-section drag coefficient due to separation; i.e., C_{Dp} is the drag coefficient of a two-dimensional (infinite aspect ratio) wing with the same airfoil section. For low and moderate values of C_l the airfoil section drag coefficient can be approximated very well by

$$C_d = C_{dp\min} + K''C_l{}^2 \tag{28}$$

where K'' is the section drag due to lift factor and $C_{dp\min}$ is the minimum separation drag coefficient (usually the value at $C_l = 0$). The value of K'' is found experimentally and has a value of 0.003 at $R_e = 9 \times 10^6$ for the airfoil shown in Fig. 7.

Equation 27 is often expressed as

$$C_D = C_{D0} + \left[K'' \left(\frac{m_0}{m} \right)^2 + \frac{1}{\pi \mathit{R}} \right] C_L{}^2 \tag{29}$$

where the first term, C_{D0}, is the zero lift drag coefficient and the second term is the drag coefficient due to lift. The drag coefficient due to lift consists of a viscous separation term plus and inviscid (induced) term. In subsonic flow, C_{D0} is made up of $C_{dp\min}$ and skin friction.

The formula for angle of attack is

$$\alpha = \alpha_0 + \alpha_i = \alpha_0 + \frac{C_L}{\pi \mathit{R}} \tag{30}$$

where α_0 is the effective angle of attack of the entire wing, i.e., the angle of attack for infinite aspect ratio at the same lift coefficient.

Equation 30 can be rewritten in terms of a lift-curve slope, m, thus

$$m = \frac{dC_L}{d\alpha} = \frac{m_0}{1 + m_0/\pi \mathit{R}} \tag{31}$$

where m_0 is the two-dimensional wing or section value. The value of m_0 is close to the theoretical value of 2π for most airfoils (see Table 1).

Thus far it has been assumed that the wings have an elliptical lift distribution, i.e., a constant downwash along the aerodynamic center. A wing having an elliptical planform has this elliptical lift distribution. For wings not having such a distribution the downwash and induced angle of attack will vary along the span as will the section lift coefficient C_l. The effect of nonelliptical distribution may be accounted for by the insertion of appropriate correction factors into the equations for the elliptical wing. Letting τ and δ represent the correction factors for induced angle of attack and induced drag, respectively, we get:

$$\alpha = \alpha_0 + \frac{C_L(1 + \tau)}{\pi R} \tag{32}$$

$$C_{D_i} = \frac{C_L{}^2(1 + \delta)}{\pi R} \tag{33}$$

$$m = \frac{dC_L}{d\alpha} = \frac{m_0}{1 + \frac{m_0(1 + \tau)}{\pi R}} \tag{34}$$

Values of τ and δ are given in Fig. 10 as functions of taper ratio and aspect ratio. The fact that τ and δ are always greater than or equal to zero illustrates that minimum induced drag is obtained for an elliptical wing since, then, $\tau = \delta = 0$. Note that taper ratios in the range of 0.2 to 0.5 give results very close to the elliptic wing.

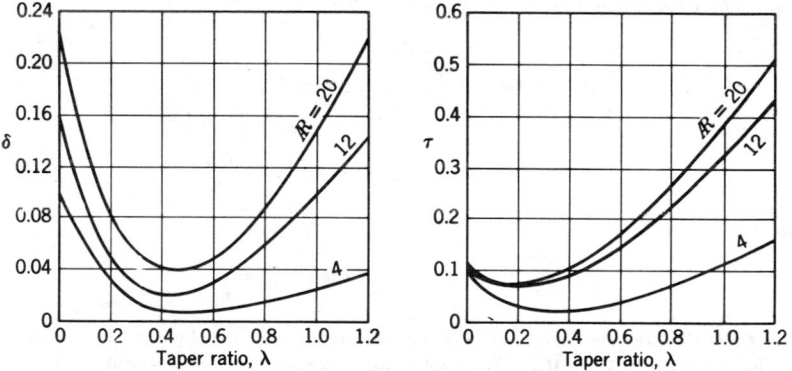

FIG. 10. Values of τ and δ for various aspect and taper ratios.

Helmbold[5] suggested a first-order correction to Eq. 31 to give better agreement with experimental results on wings of moderate thickness ratio. The modified expression for finite wing lift curve slope is

$$\frac{dC_L}{d\alpha} = \frac{m_0}{\frac{m_0}{\pi R} + \sqrt{1 + \left(\frac{m_0}{\pi R}\right)^2}} \tag{35}$$

Equation 35 predicts the lift curve slope for moderate and high aspect ratio wings very well up to the region of stall. For low aspect ratio wings, i.e., aspect ratio 2 or less, the normal force due to the separated flow around the wing tip is significant compared to the wing lift. This normal force is called the nonlinear lift and at large α (i.e., greater than about 10°) causes the lift curve slope to be nonlinear. In this case the expression for wing C_L is given by

$$C_L = \left(\frac{dC_L}{d\alpha}\right)\alpha + \text{(nonlinear term)}$$

This nonlinear term for low aspect ratio wings is discussed in Ref. 16.

34. SUBSONIC FLOW THEORY

Here the flow is considered compressible but $M_\infty < 1$. Generally aerodynamic wings and bodies are slender in shape such that the perturbation velocities u, v, and w are all very small compared to V_∞. The assumption of small perturbations, i.e., $u/V_\infty \ll 1$, $v/V_\infty \ll 1$, and $w/V_\infty \ll 1$, permits the neglect of the entire right-hand side of Eq. 6 as long as M_∞ is not close to 1 (i.e., the flow is not transonic). The SGDE reduces to the linear governing equation:

$$(1 - M_\infty{}^2)\frac{\partial^2\phi}{\partial x^2} + \frac{\partial^2\phi}{\partial y^2} + \frac{\partial^2\phi}{\partial z^2} = 0 \tag{36}$$

Equation 36 holds for both subsonic and supersonic $(M_\infty > 1)$ flow.

Two-Dimensional Subsonic Flow

Equation 36 in two dimensions is an elliptic-type partial differential equation and can be solved (i.e., separation of variables) for given boundary conditions. Ackeret's solution for a wavy wall is a classical and very useful example.[2]

By a simple change of variables, Eq. 36 in two dimensions can be transformed into Laplace's equation. Thus an incompressible flow past an airfoil can be related to the compressible flow past a similar shape and subsonic Mach number by the use of similarity rules. Three useful rules are:[3]

Rule 1:

$$\text{If } \tau_c = \beta\tau_i \text{ and } \alpha)_c = \beta\alpha)_i, \text{ then } C_{pc} = C_{pi}$$

Rule 2:

$$\text{If } \tau_c = \tau_i \text{ and } \alpha)_c = \alpha)_i, \text{ then } C_{pc} = \frac{C_{pi}}{\beta}$$

Rule 3:

$$\text{If } \tau_c = \frac{\tau_i}{\beta} \text{ and } \alpha)_c = \frac{\alpha)_i}{\beta}, \text{ then } C_{pc} = \frac{C_{pi}}{\beta^2}$$

where C_p is the pressure coefficient, τ is the thickness ratio, α is the angle of attack, and β is the quantity $\sqrt{1 - M_\infty{}^2}$. The subscript c denotes the compressible case at free stream Mach number M_∞ and the subscript i denotes the incompressible $(M \to 0)$ case. The bodies can have different τ's and angles of attack but must be of similar shape (i.e., affinely related). The rules are valid only for thin airfoils at small angles of attack. They are not valid near stagnation points. Furthermore, they apply only to subcritical speeds, i.e., only to those flight speeds which produce local velocities less than those of sound. The same rules apply to the lift coefficient and lift curve slope.

Rule 2, i.e., thickness ratios and angles of attack the same, is called the Prandtl-Glauret rule and expresses directly the effect of compressibility on the airfoil section characteristics, i.e.,

$$C_{pc} = \frac{C_{pi}}{\sqrt{1 - M_\infty{}^2}} \tag{37}$$

$$C_{lc} = \frac{C_{li}}{\sqrt{1 - M_\infty{}^2}} \tag{38}$$

$$\left(\frac{dC_l}{d\alpha}\right)_c = \frac{(dC_l/d\alpha)_i}{\sqrt{1 - M_\infty{}^2}} \tag{39}$$

The comparison of Eqs. 38 and 39 with experimental data is shown in Figs. 11 and 12.

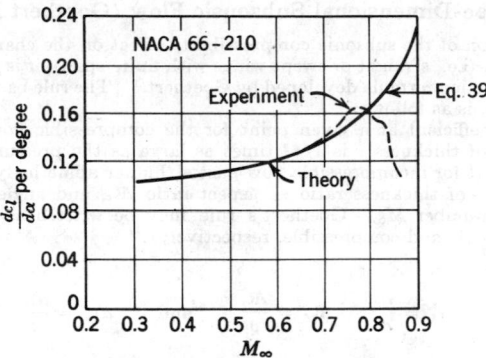

FIG. 11. Comparison of theoretical and experimental values of $dC_l/d\alpha$ for a NACA series 6 airfoil of 10% thickness ratio.

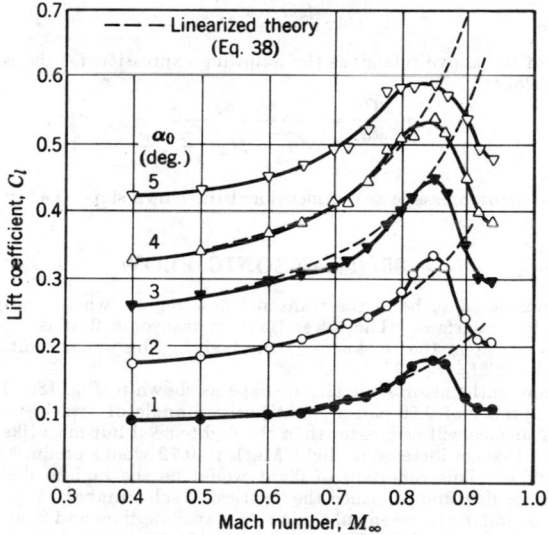

FIG. 12. Effect of compressibility on the lift of the NACA 0006-34 airfoil.

Laitone's Modification of Prandtl-Glauret Rule

The Laitone compressibility relation for two-dimensional flow is[4]

$$C_{pc} = \frac{C_{pi}}{\sqrt{1 - M_\infty^2} + \dfrac{M_\infty^2\left(1 + \dfrac{\gamma - 1}{2} M_\infty^2\right)}{2\sqrt{1 - M_\infty^2}} C_{pi}} \tag{40}$$

This relation reduces to the Prandtl-Glauret rule for low Mach numbers but gives higher values than the Prandtl-Glauret rule at higher Mach numbers. Equation 39 agrees with experimental C_p data very well.

Three-Dimensional Subsonic Flow (Goethert Rule)

In the calculation of the subsonic compressibility effect on the characteristics of three-dimensional wings (i.e., straight or swept wings with finite span), it is necessary to use the more general correction formula developed by Goethert.[7]   The rule based upon a similarity solution of Eq. 36, is as follows:

The pressure coefficient at a given point for the compressible flow at Mach number M_∞ past a body of thickness τ is $1/\beta^2$ times as large as the pressure coefficient at the corresponding point for incompressible flow past a thinner affine body of thickness $\beta\tau$.

Consider a wing of thickness ratio τ_c, aspect ratio $\mathcal{R}_c$, and angle of attack $\alpha)_c$ in a stream of Mach number M_∞.   Goethert's rule may be written as (subscripts i and c denote incompressible and compressible, respectively):

If

$$\tau_c = \frac{\tau_i}{\beta}, \qquad \mathcal{R}_c = \frac{\mathcal{R}_i}{\beta}, \qquad \text{and} \qquad \alpha)_c = \frac{\alpha)_i}{\beta}$$

then

$$C_{pc} = \frac{C_{pi}}{\beta^2}$$

and

$$C_{Lc} = \frac{C_{Li}}{\beta^2}$$

Application of the above rule gives the following expression for the compressible finite span lift curve slope:[3]

$$\frac{dC_L}{d\alpha} = \frac{m_0}{\sqrt{1 - M_\infty^2} + \dfrac{m_0}{\pi \mathcal{R}}} \tag{41}$$

where m_0 is the incompressible two-dimensional lift curve slope (i.e., generally $m_0 = 2\pi$ per radian).

35. TRANSONIC FLOW

A body is considered to be in the transonic flow regime when locally sonic flow first occurs on the body surface.   The lower limit of transonic flow is some M_∞ less than unity and depends upon the thickness of the body.   The upper limit is generally considered to be about $M_\infty = 1.3$.

Consider a conventional subsonic airfoil shape as shown in Fig. 13.   If this airfoil is at a flight Mach number of 0.50 and a slight positive angle of attack, the maximum local velocity on the surface will be greater than the flight speed but most likely less than sonic speed.   Assume that an increase in flight Mach to 0.72 would produce the first evidence of local sonic flow.   This condition of flight would be the highest flight speed possible without supersonic flow and is termed the "critical Mach number."   Thus critical Mach number is the boundary between subsonic and transonic flow and is an important point of reference for all compressible effects encountered in transonic flight.

As critical Mach number is exceeded, an area of supersonic flow is created on the wing surface.   The acceleration of the airflow from subsonic to supersonic is smooth and unaccompanied by any shock waves.   However, the transition from supersonic to subsonic occurs through a shock wave and since there is no change in direction of the flow, the wave formed is a normal shock wave.

One of the principal effects of the normal shock wave is to produce a large increase in the static pressure of the airstream behind the wave.   If the shock wave is strong, the boundary layer may not have sufficient kinetic energy to withstand the large adverse pressure gradient and separation will occur.   At speeds only slightly beyond the critical Mach number the shock wave formed is not strong enough to cause separation or any noticeable change in the aerodynamic force coefficients.   However, an increase in speed sufficiently above the critical Mach number to cause a strong shock wave will produce separation and yield a sudden change in the force coefficients.   Such a flow condition is shown in Fig. 13 by the flow pattern for $M = 0.77$.   Notice that a further increase in Mach number to 0.82 can enlarge the supersonic area on the upper surface and form an additional area of supersonic flow and a normal shock wave on the lower surface.

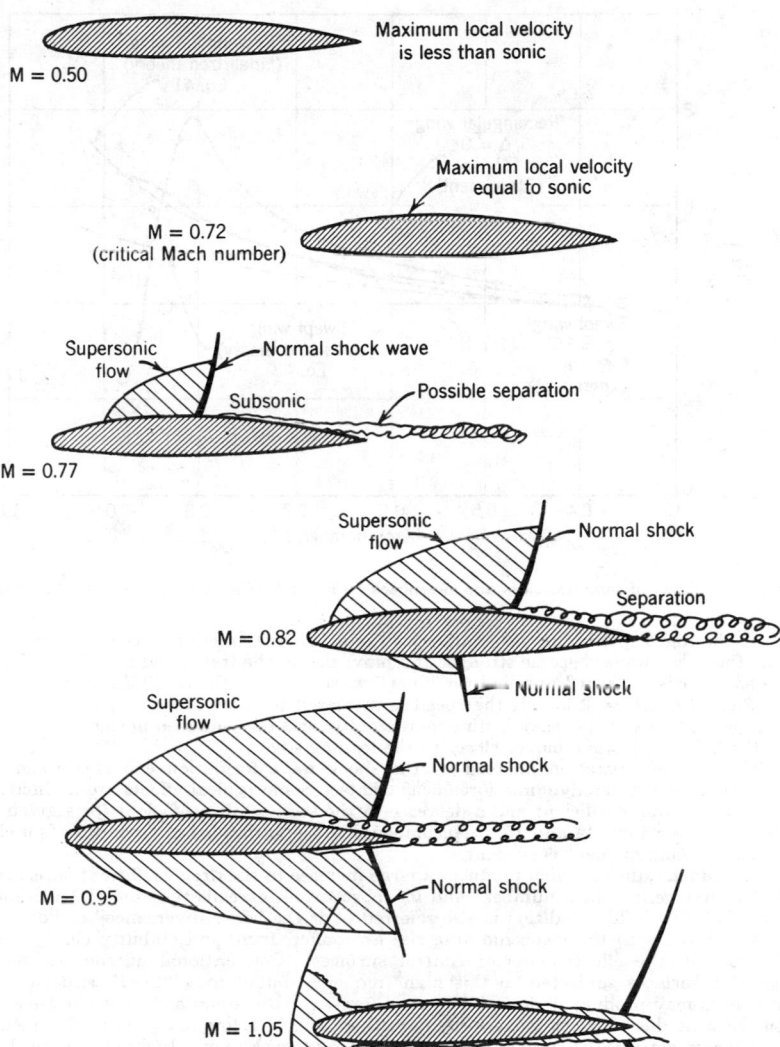

M = 0.50

Maximum local velocity
is less than sonic

M = 0.72
(critical Mach number)

Maximum local velocity
equal to sonic

Supersonic
flow

Normal shock wave

Subsonic

Possible separation

M = 0.77

Supersonic
flow

Normal shock

Separation

M = 0.82

Normal shock

Supersonic
flow

Normal shock

M = 0.95

Normal shock

M = 1.05

Subsonic
airflow

"Bow wave"

Fig. 13.   Transonic flow patterns (Courtesy NASA).

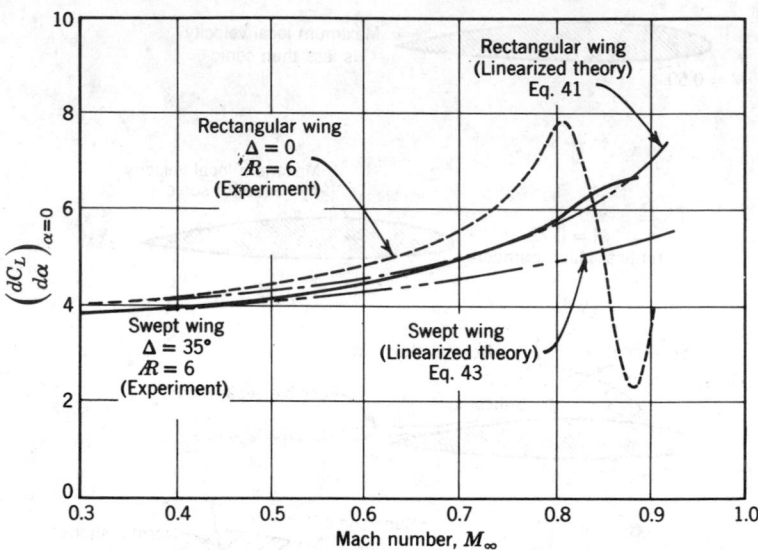

Fig. 14.   Effect of compressibility and sweepback on lift of NACA 00012 airfoil with aspect ratio 6.

As the flight speed approaches the speed of sound the areas of supersonic flow englarge and the shock waves become stronger and move nearer the trailing edge. When the flight speed exceeds the speed of sound the "bow" wave forms at the leading edge as illustrated in Fig. 13 at $M = 1.05$. If the speed is increased to some higher supersonic value all oblique portions of the wave incline more greatly and the detached normal shock portion of the bow shock wave moves closer to the leading edge.

The airflow separation induced by the shock wave formation can create significant variations in the aerodynamic force coefficients. Some typical effects are an increase in the section drag coefficient and a decrease in the section lift coefficient for a given angle of attack (see Figs. 14 and 15). Accompanying the variations in C_l and C_d is a change in the pitching moment coefficient.

The Mach number which produces a large increase in the drag coefficient is termed the "force divergence Mach number" and for most airfoils exceeds the critical Mach number by 5 to 10%. This condition is also referred to as the "drag divergence" or "drag rise."

Associated with the transonic drag rise are buffet, trim, and stability changes, and a decrease in the effectiveness of control surfaces. Conventional aileron, rudder, and elevator surfaces subjected to this high frequency buffet may "buzz" and changes in moments may produce undesirable control forces. Also, when airflow separation occurs on the wing due to shock wave formation, there will be a loss of lift and subsequent loss of downwash aft of the affected area. If the wings shock unevenly due to physical shape differences or sideslip, a rolling moment may be created and can contribute to control difficulty. If the shock induced separation occurs symmetrically near the wing root, the resulting decrease in downwash on the horizontal tail will create a diving moment and the aircraft will "tuck under."

Since most of the difficulties of transonic flight are associated with shock wave induced flow separation, any means of delaying or alleviating this separation will improve the aerodynamic characteristics of an aircraft.

Wing Sweep

One of the most effective means of delaying and reducing the effects of shock wave induced flow separation is the use of sweep. Generally the effect of wing sweep will apply either to sweep back or sweep forward. While the swept forward wing has been used in rare instances, sweepback has been found to be more practical for ordinary application.

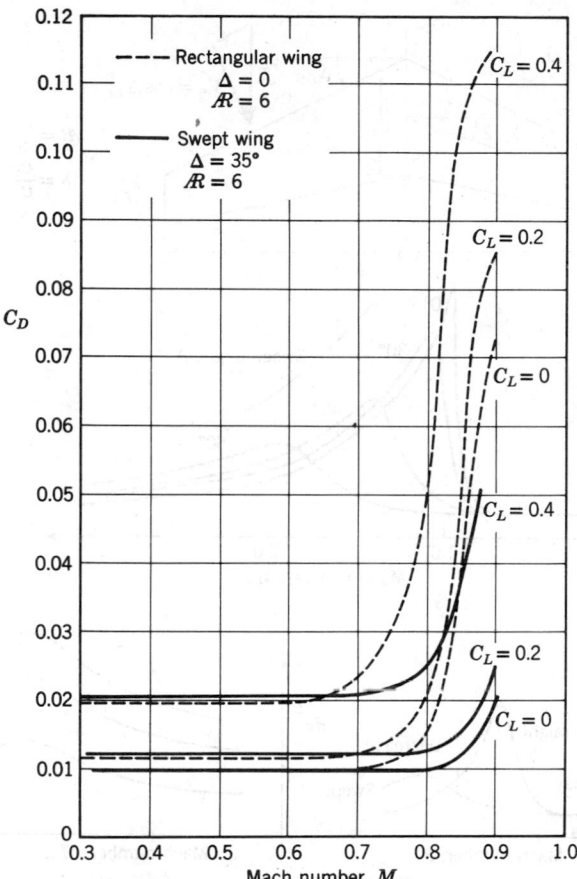

FIG. 15.   Effect of compressibility and sweepback on drag of NACA 00012 airfoil with aspect ratio 6.

A method of visualizing the effect of sweepback is shown in Fig. 16.   The swept wing shown has the streamwise velocity vector resolved into components perpendicular and parallel to the leading edge.   The component parallel to the leading edge could be visualized as moving across constant sections and, in doing so, does not contribute to the pressure distribution on the wing.   The component perpendicular to the leading edge ($V_\infty \cos \Delta$) is less than free stream velocity, and it is this component which determines the magnitude of the pressure distribution and the aerodynamic force coefficients.

Hence, sweep of a surface in high-speed flight produces a beneficial effect, since higher flight speeds may be obtained before components of velocity perpendicular to the leading edge produce critical conditions on the wing.   Thus sweepback will increase the critical Mach number, force divergence Mach number, and increase the Mach number at which the drag rise will peak.   In other words, sweep will delay the onset of compressibility effects.   The critical Mach number M_{cr} is increased by $(M_{cr})_{\Delta=0}/\cos \Delta$.

In addition to the delay of the onset of compressibility effects, sweepback will reduce the magnitude of the changes in force coefficients due to compressibility.   Since the component of velocity perpendicular to the leading edge is less than free stream velocity, the magnitude of all pressure forces on the wing will be reduced (approximately by the

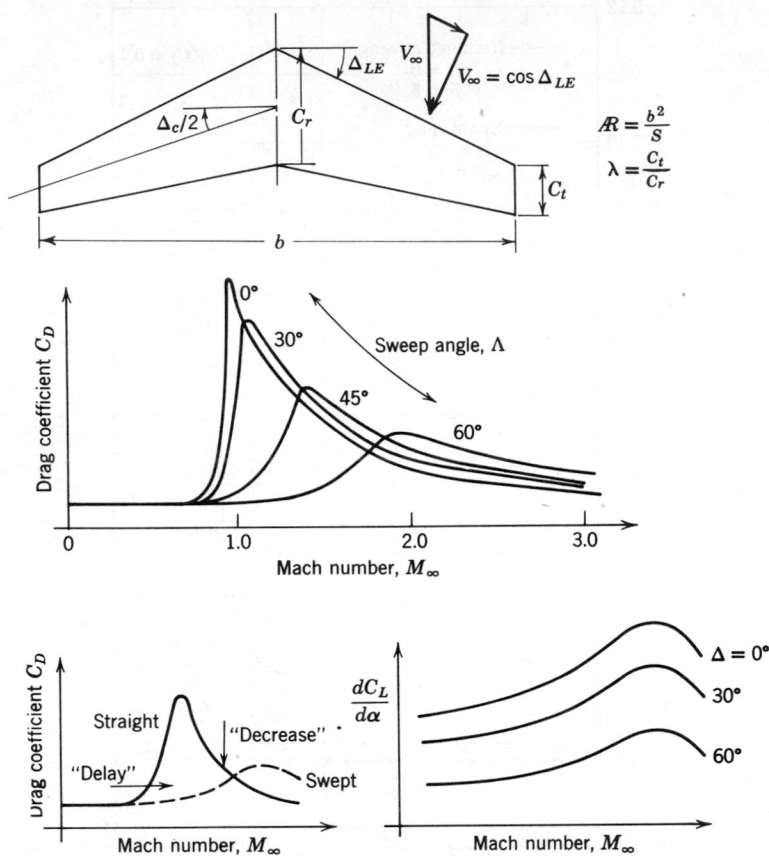

FIG. 16.   General effects of sweepback.

square of the sweep angle).   Since compressibility force divergence occurs due to change
in pressure distribution, the use of sweepback will "soften" the force divergence.   This
effect is illustrated by the graph of Fig. 16 which shows the typical variation of drag
coefficient with Mach number for various sweepback angles.   The straight wing shown
begins drag rise at about $M = 0.70$, reaches a peak near $M = 1.0$, and begins a continual
drop past $M = 1.0$.   Note that use of sweepback then delays the drag rise to some higher
Mach number and reduces the magnitude of the rise in drag coefficient.   It is evident
from the figure that small angles of sweep provide very little benefit.   If sweep is to be
used at all, at least 35 to 45° should be used.

A disadvantage of wing sweep is the decrease in wing lift curve slope.   Reference 6
shows this effect to be approximately

$$\left(\frac{dC_L}{d\alpha}\right)_\Delta = \left(\frac{dC_L}{d\alpha}\right)_{\Delta=0} \cos \Delta \qquad (42)$$

This means that a swept wing aircraft will have to land and take-off at higher angles of
attack than a straight wing aircraft.

If we combine the corrections for sweep (Eq. 42) and Mach number (Eq. 39) with Eq. 35, we get the following very useful expression for the finite wing lift curve slope:

$$\frac{dC_L}{d\alpha} = \frac{2\pi \mathcal{R}}{2 + \sqrt{4 + \mathcal{R}^2\beta^2 \left(1 + \frac{\tan^2 \Delta_{c/2}}{\beta^2}\right)}} \tag{43}$$

where $\beta = \sqrt{1 - M_\infty^2}$ and $\Delta_{c/2}$ is the sweep of the half chord line. Equation 43 predicts the linear portion of the lift curve slope quite well for a wide range of planform shapes and $M_\infty < 1$. (See Fig. 14.)

Other disadvantages to wing sweep are a reduction in $C_{L\max}$ and tip stall. The early flow separation at the tip is due to the spanwise flow causing a thickening of the boundary layer near the tips and hastening flow separation.

Supercritical Wing

Another way of delaying the drag rise due to shock wave induced separation is by using an airfoil shape called a supercritical section. A typical supercritical airfoil shape is shown in Fig. 17a. The section is shaped such that the normal shocks occur at locations where the surface pressure gradient is decreasing[1] (favorable pressure gradient) or zero. Thus the boundary layer is better able to cope with the pressure jump across the normal shock and resist the tendency to separate. For a given thickness ratio the critical Mach number stays the same but the divergence Mach number can be delayed by a supercritical shape as shown in Fig. 17b.

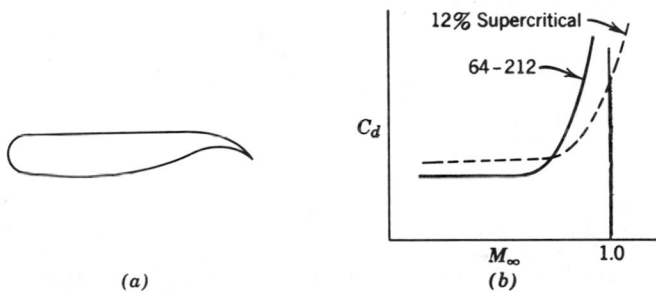

(a)   (b)

FIG. 17. Supercritical wing shape and effect of divergence Mach number: (a) 12% thick supercritical airfoil; and (b) comparison of two 12% airfoils.

Transonic Similarity Solutions

Thus far the discussion of transonic flow effects has largely been based upon experimental observations. This is because the transonic flow theory is very meager. Equation 6 cannot be linearized by assumptions of small perturbation for flow near $M_\infty = 1$. The governing flow equation is very nonlinear and general solutions or approximations are not possible.

Equation 6 can be simplified somewhat by assuming small perturbations and discarding terms of small orders of magnitude. The resulting approximate equation for two-dimensional flow is:

$$(1 - M_\infty^2)\frac{\partial^2\phi}{\partial x^2} + \frac{\partial^2\phi}{\partial y^2} = \frac{(\gamma + 1)M_\infty^2}{V_\infty}\frac{\partial\phi}{\partial x}\frac{\partial^2\phi}{\partial x^2} \tag{44}$$

A similarity solution of Eq. 44 yields the following similarity rule for similar shaped bodies in transonic flow.

Consider two different transonic flow situations denoted by subscripts 1 and 2. If two similar shaped bodies of different thickness ratios τ_1 and τ_2 have their Mach numbers and gases (denoted by ratio of specific heats γ) related such that the following condition is satisfied:

$$\frac{\tau_1 M_1{}^2(\gamma_1 + 1)}{(1 - M_1{}^2)^{3/2}} = \frac{\tau_2 M_2{}^2(\gamma_2 + 1)}{(1 - M_2{}^2)^{3/2}} \tag{45}$$

then the pressure coefficients on the two bodies are related as

$$C_{p2} = C_{p1} \left(\frac{\gamma_1 + 1}{\gamma_2 + 1}\right)\left(\frac{M_1}{M_2}\right)^2\left(\frac{1 - M_2{}^2}{1 - M_1{}^2}\right)$$

Here, as before, a known flow solution is extended to another flow solution for affinely related shapes. However, notice that the transonic similarity rule is much more restrictive than the subsonic rules. For example, the same body ($\tau_1 = \tau_2$) cannot be compared at different Mach numbers unless the gas is changed (i.e., γ_1 and γ_2 are different values).

36. SUPERSONIC FLOW THEORY

Supersonic Thin Airfoil Theory (Two-Dimensional)

In supersonic flow Eq. 6 can again be linearized if the airfoil is thin and always at small angles of attack. Not only must M_∞ be greater than unity but also it must be less than some $M_{\text{hypersonic}}$. Clearly, if M_∞ were large (i.e., $M_\infty > 5$), there are some terms on the right-hand side of Eq. 6 that could not be considered negligible compared to particular terms on the left-hand side. $M_{\text{hypersonic}}$ will be better defined in Art. 42 but generally $M_{\text{hypersonic}} \approx 5$.

Thus, for thin airfoils,* small angles of attack, and $1 < M_\infty < M_{\text{hypersonic}}$, Eq. 6 can be approximated in two dimensions by

$$(1 - M_\infty{}^2)\frac{\partial^2\phi}{\partial x^2} + \frac{\partial^2\phi}{\partial y^2} = 0 \tag{47}$$

This equation is a hyperbolic partial differential equation and its solution is of the wave type. Ackeret's solution for a wavy wall is a classical and very useful example.[2]

The solution of Eq. 47 for the pressure coefficient is

$$C_p = \frac{2\theta}{\sqrt{M_\infty{}^2 - 1}} \tag{48}$$

where θ is the local flow deflection angle.

Since the governing equation is linear, the pressure distributions from different shapes can be added to give the pressure distribution over a particular shape. Thus the pressure distribution over an arbitrary airfoil shape can be resolved into contributions due to angle of attack, camber, and thickness. This is shown in Fig. 18.

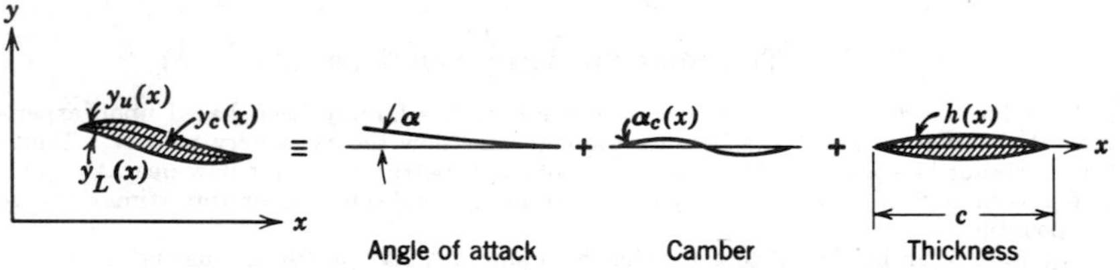

FIG. 18.   Linear resolution of arbitrary airfoil into angle of attack, camber, and thickness.

The lift coefficient is given by

$$C_l = \frac{L'}{qc} = \int_0^c \frac{P_L - P_u}{qc}\, dx = \frac{1}{c}\int_0^c (C_{PL} - C_{Pu})\, dx$$

where the subscripts L and u denote lower and upper surface, respectively.

---

* The airfoils also have sharp leading edges such that the shock waves are attached oblique shocks. Round leading edges would result in detached shock (normal) waves and the pertubation velocities would not be small at the stagnation point.

For the angle of attack contribution the integral is

$$C_l = \frac{4\alpha}{\sqrt{M_\infty{}^2 - 1}} \tag{49}$$

There is no contribution from the mean camber line because the integral of the slope is zero. Also, the thickness envelope is nonlifting. Therefore Eq. 49 represents the entire lift coefficient for any thin airfoil.

The drag coefficient is given by (using small angle approximations)

$$C_d = \frac{D'}{qc} = \frac{1}{c} \int_0^c \left[C_{PL}\theta_L - C_{Pu}\theta_u \right] dx$$

All three components shown in Fig. 18 contribute to the drag and the result is[2]

$$C_d = \frac{4}{\sqrt{M_\infty{}^2 - 1}} \left[\alpha^2 + \overline{\alpha_c{}^2(x)} + \overline{\left(\frac{dh}{dx}\right)^2} \right] \tag{50}$$

where

$$\overline{\alpha_c{}^2(x)} = \frac{1}{c} \int_0^c \alpha_c{}^2(x)\ dx \tag{51}$$

and

$$\overline{\left(\frac{dh}{dx}\right)^2} = \frac{1}{c} \int_0^c \left(\frac{dh}{dx}\right)^2 dx \tag{52}$$

The drag expressed by Eq. 50 is an inviscid drag and is called the wave drag. It is caused by the shock wave system over a body immersed in a supersnoic flow. It is made up of a drag due to lift, a drag due to camber, and a drag due to thickness. For the supersonic shapes shown in Fig. 19 we have:

Shape	$\overline{\left(\dfrac{dh}{dx}\right)^2}$
Flat plate	0
Double wedge	$\dfrac{t}{c}$
Biconvex	$\dfrac{4}{3}\dfrac{t}{c}$

If the airfoil shape is symmetric (i.e., no camber) such as those in Fig. 19, $\overline{\alpha_c{}^2(x)} = 0$.

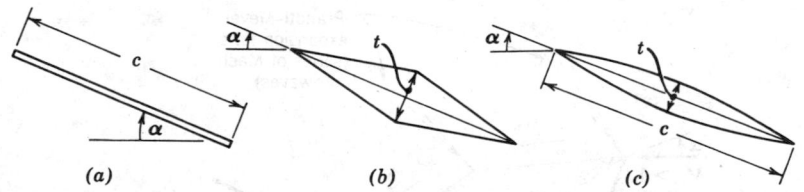

$$(a) \qquad\qquad (b) \qquad\qquad (c)$$

FIG. 19. Typical supersonic airfoil sections: (a) flat plate; (b) double wedge; and (c) biconvex.

Equation 50 is often rewritten in terms of a zero lift part and a drag due to lift part (see Eq. 28) as

$$C_d = C_{d0} + C_{dL} = C_{d0} + KC_l{}^2 \tag{53}$$

where

$$C_{d0} = C_f + C_{dw}$$

$$C_f = \text{skin friction (see Art. 48)}$$

$$C_{dw} = \text{wave drag} = \frac{4}{\sqrt{M_\infty^2 - 1}}\left[\overline{\alpha_c^2(x) + \left(\frac{dh}{dx}\right)^2}\right]$$

$$C_{dL} = \text{drag due to lift} = \frac{4\alpha^2}{\sqrt{M_\infty^2 - 1}} = \frac{\sqrt{M_\infty^2 - 1}}{4}C_l^2 \qquad (54)$$

The moment about the leading edge is

$$C_{m\text{L.E.}} = \frac{M'_{\text{L.E.}}}{qc^2} = -\frac{1}{c^2}\int_0^c (C_{PL} - C_{Pu})x\,dx$$

The moment contribution due to thickness is zero. The result is

$$C_{m\text{L.E.}} = -\frac{2\alpha}{\sqrt{M_\infty^2 - 1}} + \frac{4}{\sqrt{M_\infty^2 - 1}}\overline{(\alpha_c x)} \qquad (55)$$

where

$$\overline{\alpha_c x} = \frac{1}{c^2}\int_0^c \alpha_c(x)x\,dx \qquad (56)$$

and $\overline{\alpha_c x}$ is zero for symmetric airfoils. Since the mean camber line contributes no lift (in this linear theory), its moment contribution is a pure couple. Thus Eq. 55 indicates that the midchord is the point about which C_m is independent of angle of attack. Therefore the aerodynamic center is located at the midchord for all thin airfoils in supersonic flow.

The thin airfoil theory is a solution to the linearized equation and as such is an approximate method. The results agree very well with exact solutions for local flow deflections up to about 10°; up to θ's of 15° the thin airfoil theory results are still satisfactory. An exact solution is the useful shock-expansion method which is discussed next.

Shock-Expansion

The shock-expansion method makes use of the exact oblique shock solutions and Prandtl-Meyer expansion discussed in Arts. 16 and 18, respectively. The supersonic flow over a two-dimensional body is broken into regions of compression and expansion as shown in Fig. 20.

For the airfoil in Fig. 20, regions 1 and 6 are compression regions and Fig. 8 of Art 16 could be used to find the pressures and Mach numbers in these regions. Regions 2, 3, 4, and 5 are expansion regions, and Table 3 of Art. 18 could be used to find the Mach numbers in these regions. The pressures in the expansion regions could then be determined using the isentropic results of Art. 13. A slip surface is a surface across which the

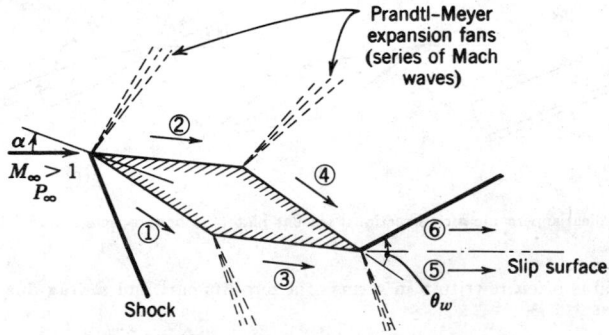

FIG. 20.   Double wedge airfoil in supersonic flow.

pressures and flow directions are the same but the temperatures, densities, and entropy levels may be different. Such a surface is shown in Fig. 20 and exists behind supersonic shapes developing lift.

Finite Wings

Rectangular Wings. The flow field about a thin rectangular wing moving at supersonic speed in a direction normal to its leading edge is made up of two parts: (1) the regions within the Mach cones (infinitesminal wave fronts) emanating from the wing tips; and (2) the region between the Mach cones where the flow is two-dimensional (see Fig. 21).

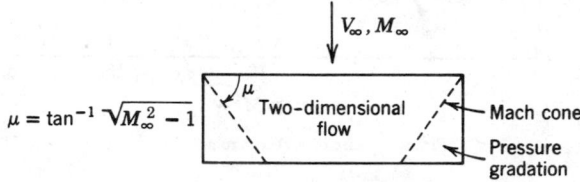

$$\mu = \tan^{-1} \sqrt{M_\infty^2 - 1}$$

FIG. 21.   Rectangular wing in supersonic flow.

The effect of the wing tips on the wave drag of a rectangular wing depends[8,9] upon the product $\cancel{R}\sqrt{M_\infty^2 - 1}$. If $\cancel{R}\sqrt{M_\infty^2 - 1} \geq 1$, i.e., the Mach cones from the leading edge corners fall within the trailing edge corners, there is no effect of wing tip regardless of wing section. When $\cancel{R}\sqrt{M_\infty^2 - 1} < 1$, i.e., the Mach cones fall outside of the trailing edge corners, the wave drag is less than that of a two-dimensional wing, the variation depending upon the wing section. For a double-wedge wing (symmetric fore and aft) the wave drag is shown in Fig. 22.

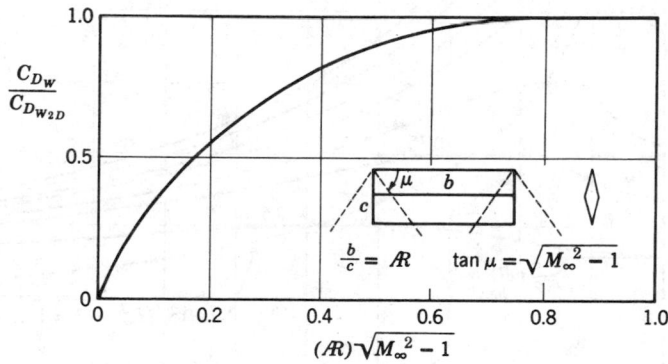

FIG. 22.   Wave drag of double-wedge rectangular airfoil.

In the case of lift, if $\cancel{R}\sqrt{M_\infty^2 - 1} \geq 1$, the slope of the lift curve is given by[10]

$$\frac{dC_L}{d\alpha} = \frac{4}{\sqrt{M_\infty^2 - 1}} \left(1 - \frac{1}{2\cancel{R}\sqrt{M_\infty^2 - 1}} \right) \tag{57}$$

and plotted in Figs. 23 and 24. The lift is independent of the wing section. As $\cancel{R}\sqrt{M_\infty^2 - 1}$ decreases below unity, the theory becomes increasingly more complicated[11,12] and difficult to evaluate numerically. As $\cancel{R}\sqrt{M^2 - 1} \rightarrow 0$, presumably[13] $dC_L/d\alpha \rightarrow (\pi/2)\,\cancel{R}$, which is indicated in Figs. 23 and 24.

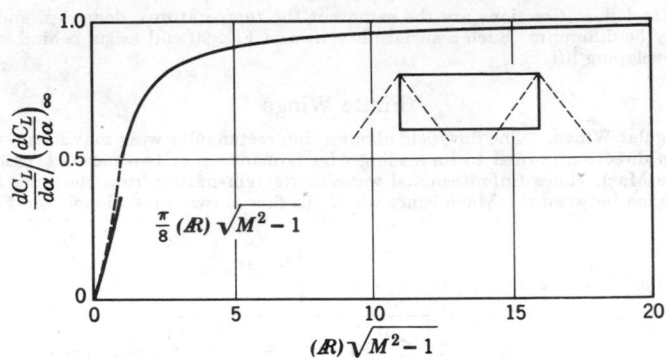

FIG. 23. Lift of a rectangular wing.

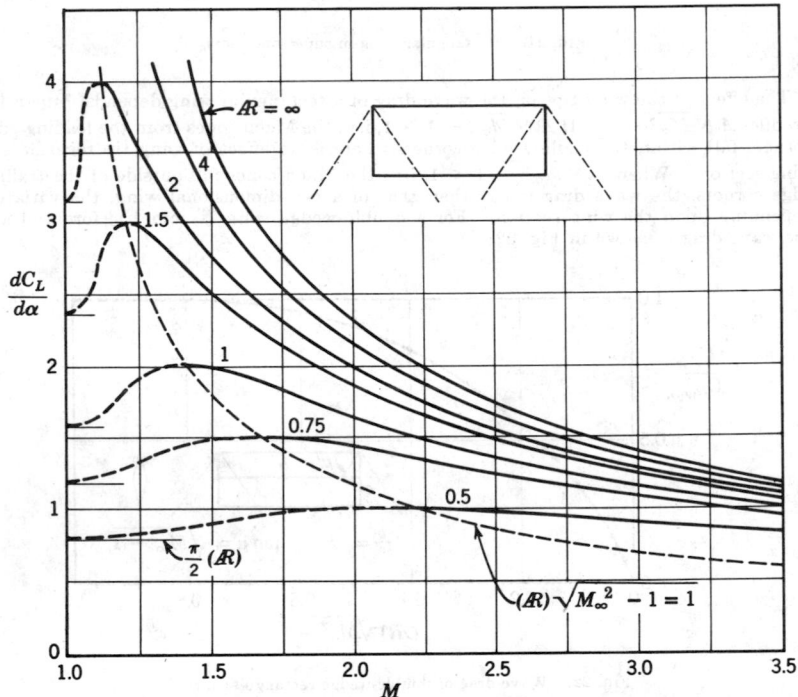

FIG. 24. Lift of a rectangular wing.

For rectangular wings the drag due to lift is obtained by multiplying the lift by the angle of attack, thus $C_{DL} = \alpha C_L$.

The center of pressure of the rectangular wing is located at

$$\frac{cp}{c} = \frac{\mathcal{R}\sqrt{M_\infty^2 - 1} - 2/3}{2\mathcal{R}\sqrt{M_\infty^2 - 1} - 1} \tag{58}$$

from the leading edge. (Since in the linearized theory the c.p. does not change with angle of attack, it has the same position as the aerodynamic center.) According to Eq. 58, the c.p. travels forward from the midchord point at $\mathcal{R} = \infty$ to the third-chord point at $\mathcal{R}\sqrt{M_\infty^2 - 1} = 1$. On the other hand, for subsonic flight, the a.c. moves forward very slowly from the quarter-chord point as aspect ratio decreases. Hence, a transonic stability problem arises which can be alleviated by use of low-aspect-ratio wings.

Delta Wings. Low-aspect-ratio wings are important in supersonic missile design for both aerodynamic and structural reasons. The delta wing offers some definite advantages over the low-aspect-ratio straight wing, and therefore its properties[14] are presented here.

Linearized theory shows that for the double-wedge delta wing (Fig. 25) the wave drag coefficient C_{DW} is generally related to the Mach number and wing geometry by

$$\frac{C_{DW}\sqrt{M_\infty^2 - 1}}{\tau^2} = f\left(\frac{\sqrt{M_\infty^2 - 1}}{\tan \Lambda},\, a,\, b\right) \tag{59}$$

in which the symbols are defined in Fig. 25. The ratio $\sqrt{M_\infty^2 - 1}/\tan \Lambda$, which is equal to $\tan \mu/\tan \Lambda$, indicates the relative position of the leading edge with respect to the Mach cone. If $\sqrt{M_\infty^2 - 1}/\tan \Lambda$ is greater, equal to, or less than unity, the leading edge is designated supersonic, sonic, or subsonic, respectively. The function (Eq. 59) is plotted

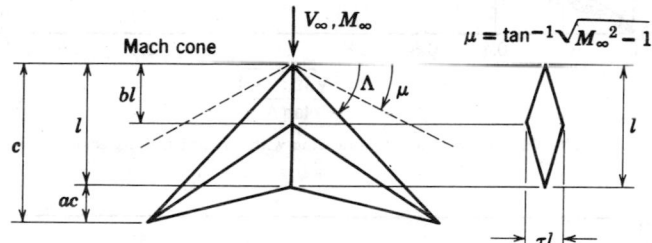

Fig. 25. Delta wing.

in Figs. 26 and 27. The peaks of the curves indicate when the Mach cone coincides successively with the trailing edge, maximum thickness line, and leading edge as Mach number increases. In Fig. 28 are plotted wave drag coefficients for wings with 60° and 75° sweepback, each for $a = 0$ and 0.5. (The value $b = 0.2$ was chosen for Fig. 28 because, at about this value of b, the drag is found to be the minimum as long as the line of maximum thickness is subsonic.) It is seen from these figures that the drag is appreciably less than the two-dimensional-wing value only if the line of maximum thickness is subsonic. However, the maximum wave-drag peak is far below that of the two-dimensional wing (infinite when $M_\infty = 1$) and considerably delayed, this being the general effect of sweepback.

The slope of the lift curve is plotted in Figs. 29 and 30. Because of the difficulty of obtaining solutions to the lift problem when the trailing edge is subsonic, i.e., $\sqrt{M_\infty^2 - 1}/\tan \Lambda \geq |a|$, the curves be in at the point where the trailing edge is sonic. It is noted that, for a wide range of Mach number, the delta wing with a sweptback trailing edge has a higher lift-curve slope than the two-dimensional wing. In the limit when $a = 0$ and $\tan \Lambda \to \pi/2$, it will be found[15] that $dC_L/d\alpha \to (\pi/2)\,\mathcal{R}$.

The travel of the c.p. (measured from the nose) with Mach number and wing geometry is shown in Fig. 31. The graph shows that the effect of Mach number is generally not great and the c.p. (or the a.c.) is located at the c.g. of the planform.

When the leading edge of the wing is supersonic, the drag due to lift is obtained by merely multiplying the lift by the angle of attack as is done for rectangular wings in supersonic flight. However, when the leading edge is subsonic, a suction occurs at the leading edge (as for subsonic flight of a two-dimensional airfoil), resulting in a leading edge thrust or decrease in drag below that given by $C_{DL} = \alpha C_L$. The corrected drag is plotted in Fig. 32 for the range $|a| \leq \sqrt{M_\infty^2 - 1}/\tan \Lambda \leq 1$.

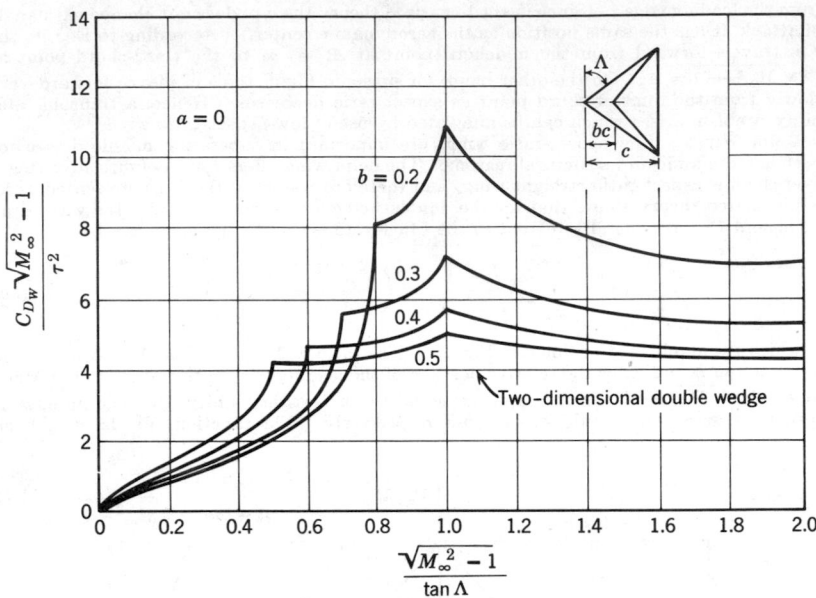

FIG. 26.   Wave drag of delta wing with straight trailing edge.

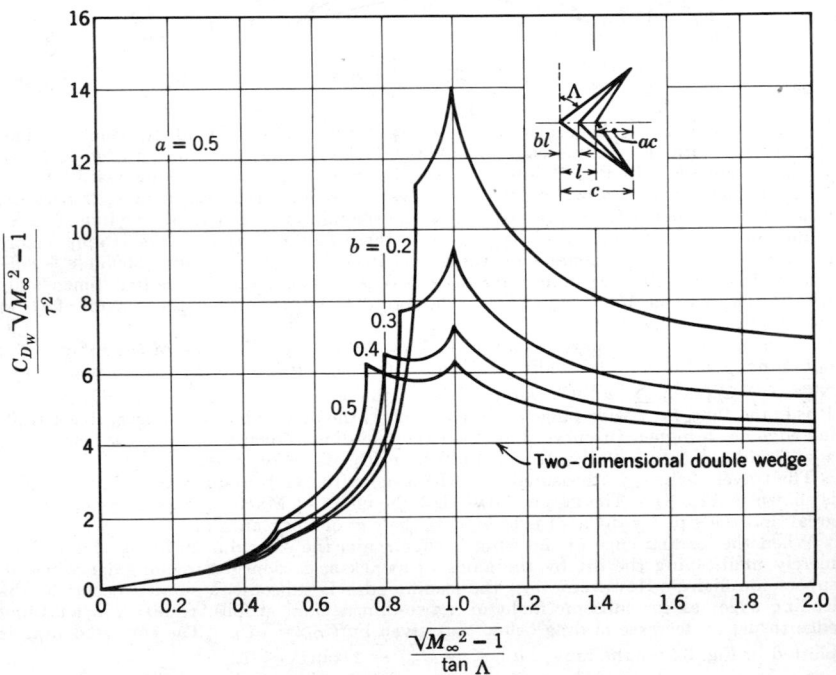

FIG. 27.   Wave drag of delta wing with swept trailing edge.

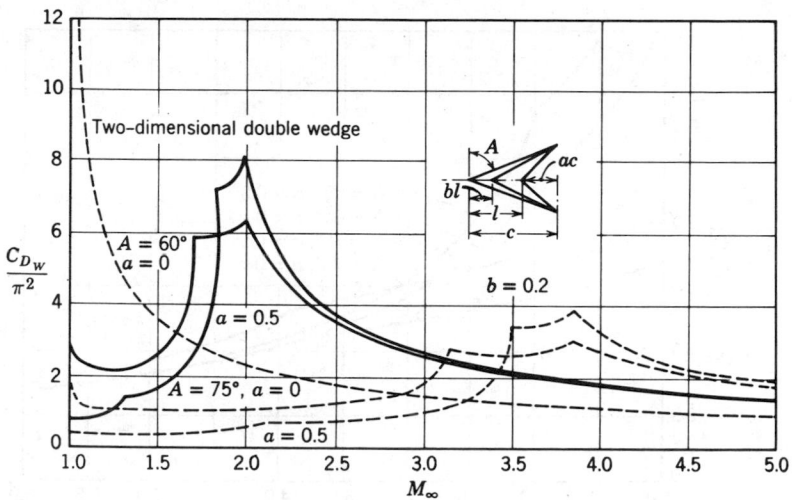

FIG. 28. Wave drag of delta wing.

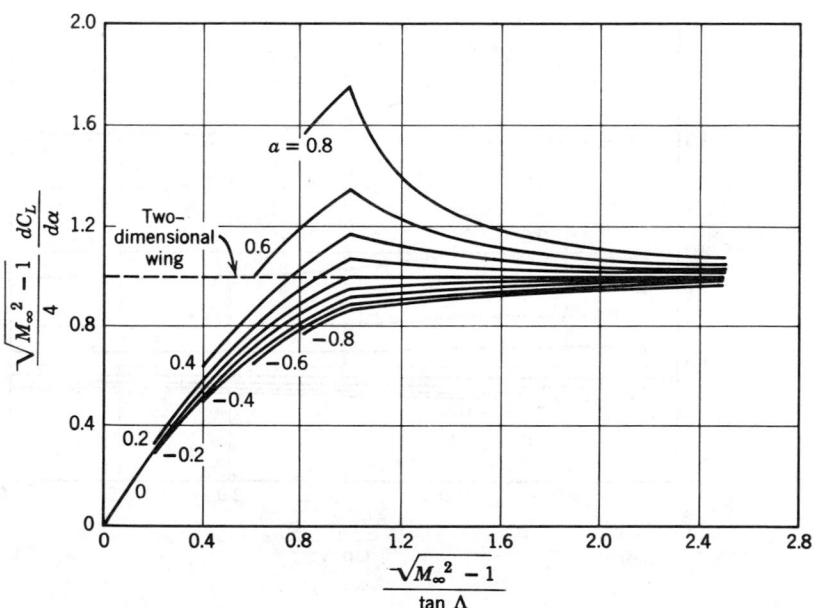

FIG. 29. Lift of delta wing.

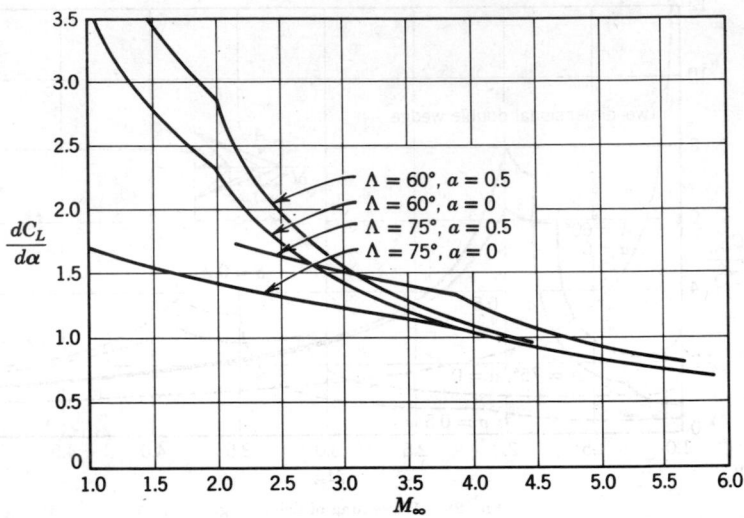

FIG. 30.   Lift of delta wing.

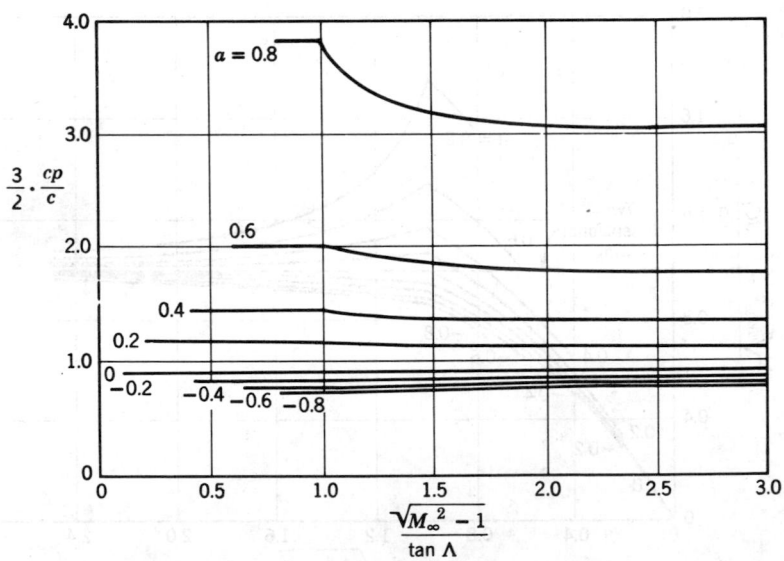

FIG. 31.   Center of pressure location for delta wing.

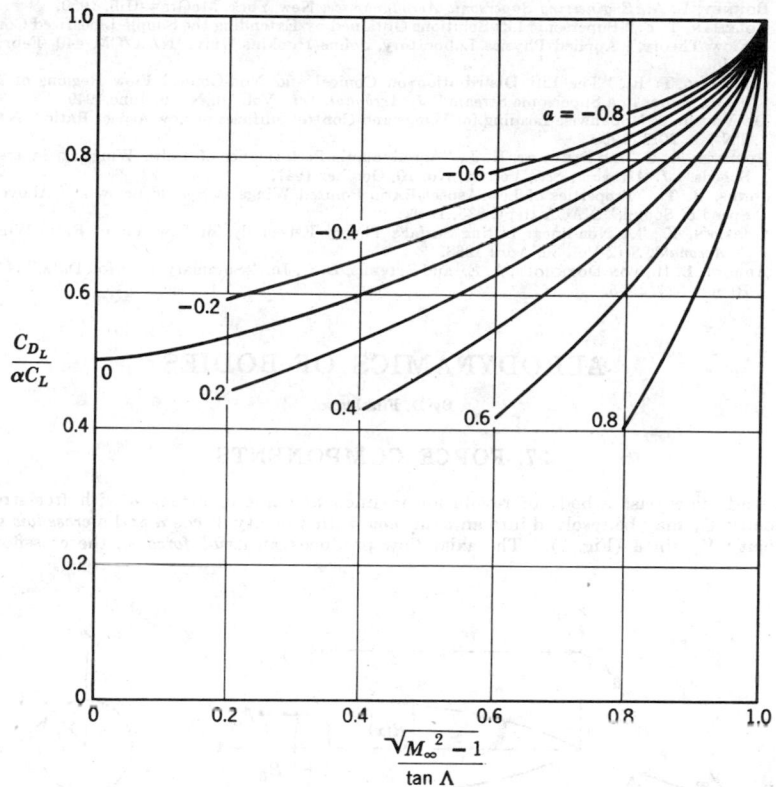

FIG. 32. Drag due to lift for delta wing.

Comparison of low-aspect-ratio delta and straight wings will show that: (1) delta wings have considerably less drag than straight wings in transonic and moderate supersonic speeds; (2) the lift-drag ratio of delta wings is better than that of straight wings at moderate supersonic speeds but worse at subsonic and high supersonic speeds; and (3) the stability and control changes of delta wings are smaller than those of straight wings.

BIBLIOGRAPHY

1. KUETHE, A. M., and SCHETZER, J. D., *Foundations of Aerodynamics*, New York, Wiley, 2nd ed., 1959.
2. LIEPMANN, H. W., and ROSHKO, A., *Elements of Gasdynamics*, New York, Wiley, 1957.
3. SHAPIRO, A. H., *The Dynamics and Thermodynamics of Compressible Fluid Flow*, Vol. I, New York, Ronald, 1954.
4. LAITONE, E. V., "New Compressibility Correction for Two-Dimensional Subsonic Flow," *J. Aeronaut. Sci.*, Vol. 18, No. 5, p. 350, 1951.
5. HELMBOLD, H. B., "Der unverwundene Ellipsenflügelals tragende Flache," in *Jahrbuch 1942 ser Deutshcen Luftfahrtforschung*, Munich, R. Oldenbuourg, pp. 111–113.
6. LOWRY, J. G., and POLHAMUS, E., NACA TN 3911, 1957.
7. GOETHERT, B., "Plane and Three-Dimensional Flow at High Subsonic Speeds," NACA TM No. 1105, 1946. (Traeslated from *Lilienthal Gesellshaft*, Vol. 127, 1940.
8. Anon "Handbook of Supersonic Aerodynamics for Three-Dimensional Airfoils," NAVORD Report 1488, Vol. 3, Section 7, Bureau of Ordnance Publication, 1958.
9. VON KARMAN, TH., "Supersonic Aerodynamics-Principles and Applications," J. Aeronautics Sciences, Vol. 14, No. 7, July 1947.

10. BONNEY, E. A., *Engineering Supersonic Aerodynamics*, New York, McGraw-Hill, 1950.
11. COLEMAN, T. F., "Supersonic Lift Solutions Obtained by Extending the Simple Linearized Conical Flow Theory," Applied Physics Laboratory, Johns Hopkins Univ., NAA/CM 440, February 1948.
12. GOODMAN, T. R., "The Lift Distribution on Conical and Non-Conical Flow Regions of Thin Finite Wings in a Supersonic Stream," *J. Aeronaut. Sci.*, Vol. 16, No. 6, June 1949.
13. DE YOUNG, J., "Spanwise Loading for Wings and Control Surfaces of Low Aspect Ratio," NACA TN No. 2011, 1950.
14. PUCKETT, A. E., and STEWART, H. J., "Aerodynamic Performance of Delta Wings at Supersonic Speeds," *J. Aeronaut. Sci.*, Vol. 14, No. 10, October 1947.
15. JONES, R. T., "Properties of Low-Aspect-Ratio Pointed Wings at Speeds Below and Above the Speed of Sound," NACA Rept. 835, 1946.
16. GERSTEN, K., "A Non-linear Lifting Surface Theory Especially for Low Aspect Ratio Wings," *J. Aeronaut. Sci.*, Vol. 30, April 1963.
17. ABBOTT, L. H., VON DOENHOFF, A. E., and STIVERS, L. S., JR., "Summary of Airfoil Data," NACA Report 824, 1945.

AERODYNAMICS OF BODIES

By D. Finkleman

37. FORCE COMPONENTS

Steady flow past a body of revolution inclined at angle of attack α with free-stream velocity V_∞ may be resolved into an *axial flow* with velocity $V \cos \alpha$ and a *crossflow* with velocity $V_\infty \sin \alpha$ (Fig. 1). The axial flow produces an *axial force* X, the crossflow a

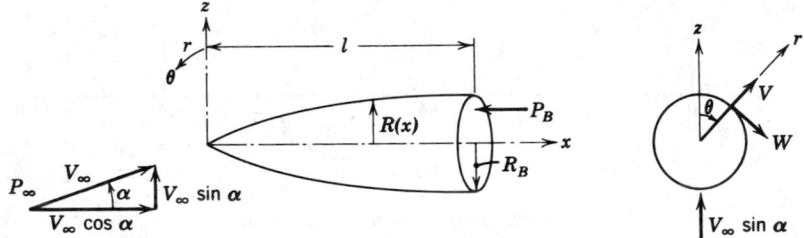

FIG. 1.   Inclined body of revolution.

normal force N and *pitching moment* M_0 about the nose (Fig. 2). Often the force is resolved into *drag* D and *lift* L, which are related by

$$D = X \cos \alpha + N \sin \alpha$$
$$L = N \cos \alpha - X \sin \alpha$$
$$X = D \cos \alpha - L \sin \alpha$$
$$N = L \cos \alpha + D \sin \alpha$$

Corresponding relations connect the *force coefficients*, which are usually referred to base area S_B:

$$C_x = \frac{X}{\frac{1}{2}\rho V_\infty^2 S_B}, \qquad C_N = \frac{N}{\frac{1}{2}\rho V_\infty^2 S_B}, \qquad C_D = \frac{D}{\frac{1}{2}\rho V_\infty^2 S_B}, \qquad C_L = \frac{L}{\frac{1}{2}\rho V_\infty^2 S_B}$$

The pitching moment coefficient is referred to the body length l:

$$C_{M0} = \frac{M_0}{\frac{1}{2}\rho V_\infty^2 S_B l}$$

where $M_0 = -\,dN$.

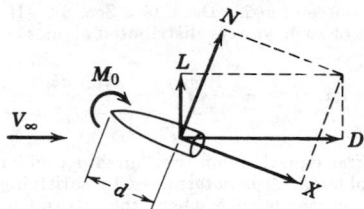

Fig. 2. Force system for body of revolution.

Since the governing equation for flow past a slender body at a small angle of attack can be approximated by a linear equation, the solution to the problem can be expressed as the sum of an axial flow solution and a two-dimensional crossflow solution. In terms of perturbation velocity potentials, the solution is

$$\phi(x, r, \theta) = \phi_a(x,r) + \phi_c(x, r, \theta) \tag{1}$$

where

$$u = \frac{\partial \phi}{\partial x}, \quad v = \frac{\partial \phi}{\partial r}, \quad \text{and} \quad w = \frac{1}{r}\frac{\partial \phi}{\partial \theta} \tag{2}$$

The local resultant velocity $\overline{V}$ is expressed as (see Fig. 1),

$$\overline{V}^2 = (V_\infty \cos \alpha + u)^2 + (V \sin \alpha \cos \theta + v)^2 + (V_\infty \sin \alpha \sin \theta - w)^2 \tag{3}$$

and the local pressure coefficient is

$$C_p = \frac{2}{\gamma M_\infty{}^2}\left[\left(1 + \frac{\gamma - 1}{2}\right) M_\infty{}^2 \left(1 - \frac{\overline{V}^2}{V_\infty{}^2}\right)^{\gamma/\gamma-1} - 1\right] \tag{4}$$

Pressure coefficients on bodies of revolution can be obtained theoretically for inviscid fluids by using the method of sources and sinks[1,2] for the axial flow component of the free-stream velocity, and doublets[3,4] for the crossflow component. The first and most difficult part of the procedure is the determination of the source-sink and doublet distributions necessary to form separately the given body. The axial flow solutions will be discussed in Arts. 38 and 39, and the crossflow case in Art. 40.

38. SLENDER BODY THEORY, AXIAL FLOW CASE

Here the body is considered to be very slender, i.e., the ratio of its maximum thickness R_B to length l is very much less than unity. The governing equation (see Eq. 6, Art. 32) is again linearized by assuming small perturbations. For the axisymmetric case the linearized equation[10] is

$$(1 - M_\infty{}^2)\frac{\partial^2\phi}{\partial x^2} + \frac{\partial^2\phi}{\partial r^2} + \frac{1}{r}\frac{\partial \phi}{\partial r} = 0 \tag{5}$$

which is valid as long as M_∞ is not close to unity or too large. The pressure coefficient, correct to first order, is

$$C_p = -\frac{2\partial\phi/\partial x}{V_\infty} - \left(\frac{\partial\phi/\partial y}{V_\infty}\right)^2 \tag{6}$$

Incompressible Flow

For incompressible flow, $M_\infty \to 0$, Eq. 5 is Laplace's equation which has the basic solution

$$\phi_i = \frac{-A}{\sqrt{x^2 + r^2}} \tag{7}$$

Equation 7 represents a source of strength A (see Sec. 6). If $f(\xi)$ is the source strength per unit length, the effect of such sources distributed along the x-axis is

$$\phi_i(x,\, r) \;=\; -\,\frac{1}{4\pi}\int_0^l \frac{f(\xi)\; d\xi}{\sqrt{(x-\xi)^2 + r^2}} \tag{8}$$

Equation 8 is an integral equation for the "airship problem" of incompressible flow theory.[1,3] In a given problem, $f(\xi)$ is determined by satisfying the boundary conditions. The solution is usually a numerical one where the integral is approximated by a finite sum of sources and sinks such as Eq. 7.

Using the boundary condition that the flow at the surface must be tangent to the surface, it can be shown that the solution for the source strength is

$$f(x) \;=\; V_\infty R\,\frac{dR}{dx} \;=\; \frac{V_\infty}{2\pi}\frac{dS}{dx} \tag{9}$$

where $R(x)$ is the r distance to the body surface and $S(x) = \pi R^2$ is the cross-sectional area of the body at x. Equation 9 follows also from the reasoning that the volumetric outflow per unit length should be proportional to the streamwise rate of change of cross-sectional area.

Laitone[5] derived the following approximate formula for the pressure coefficient at the surface of a body of revolution:

$$C_p \;=\; \frac{1}{\pi}\left[\frac{1 - 2x/l}{1 - x/l}\frac{S'}{2x} - \left(1 + \ln\frac{R}{2l\,\sqrt{x(1 - x/l)/l}}\,S''\right)\right.$$

$$\left.+\; \frac{l}{4}\left(1 - \frac{2x}{l}\right)S''' + \frac{l^2}{24}\left(1 - \frac{2x}{l} + \frac{2x^2}{l^2}\right)S'''' + \ldots\right] \tag{10}$$

where l in the length of the body, x is the distance aft of the nose, $S' = dS/dx$, $S'' = d^2S/dx^2$, etc.

Figures 3 and 4 show axial flow results for ellipsoids of revolution.

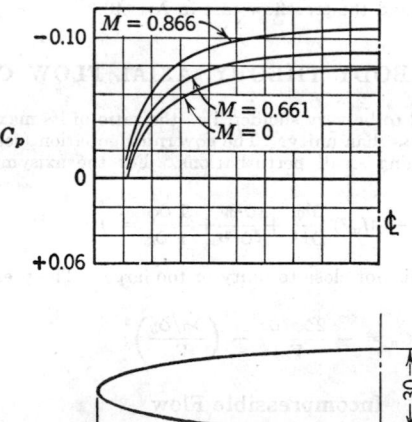

FIG. 3. Pressure coefficient along the meridian of an ellipsoid of revolution of fineness ratio 3.3 according to linearized theory.

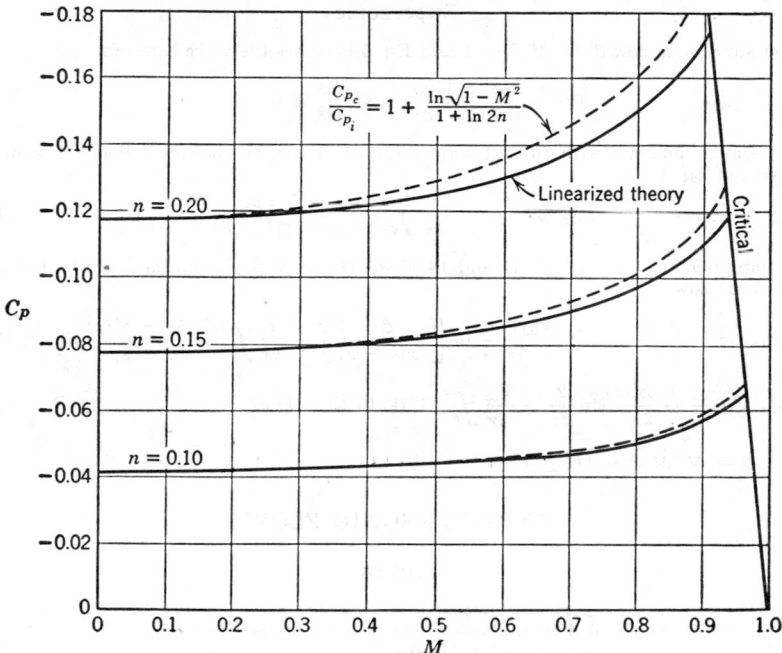

$$\frac{C_{p_c}}{C_{p_i}} = 1 + \frac{\ln \sqrt{1 - M^2}}{1 + \ln 2n}$$

FIG. 4. Pressure coefficient at midsection of ellipsoids of revolution.

High Subsonic Flow

For compressible subsonic flow, the coordinate transformation $r' = \beta r$, where $\beta^2 = 1 - M_\infty^2 < 0$, transforms Eq. 5 into Laplace's equation. Thus the general solution for the subsonic potential is

$$\phi_{\text{sub}}(x,r) = -\frac{1}{4\pi} \int_0^l \frac{f(\xi)\, d\xi}{\sqrt{(x - \xi)^2 + \beta^2 r^2}} \tag{11}$$

The method of solving this integral equation for a given body shape is the same as discussed for Eq. 8.

Another method for determining the subsonic compressible pressure coefficient is to extend an affinely related incompressible result by means of the Goethert rule[6] (generalized Prandtl-Glauert rule). The rule states that the linearized subsonic compressible flow pressure coefficient on a given slender three-dimensional body at small angle of attack is equal to $1/(1 - M_\infty^2)$ times the linearized incompressible-flow pressure coefficient on another body whose thickness ratio and angle of attack are $(1 - M_\infty^2)^{1/2}$ times as great as those of the given body. M_∞ is the free-stream Mach number. For further discussion and experimental investigation, see Ref. 7. The following formula [8,9] expresses the effect of compressibility on the maximum pressure coefficient for ellipsoids of revolution at zero angle of attack:

$$\frac{C_{pc}}{C_{pi}} = 1 + \frac{\ln (1 - M_\infty^2)^{1/2}}{1 + \ln 2n} \tag{12}$$

where n is the thickness ratio (maximum diameter to length) and the subscripts c and i refer to compressible and incompressible flow, respectively. This formula, which is plotted in Fig. 4, can be used to give an indication of the first-order effect of compressibility on pressure at the central portion of very slender bodies of revolution in subsonic motion.

Supersonic

For supersonic flow $\beta^2 = M_\infty^2 - 1$ and Eq. 5 becomes the wave equation

$$\frac{\partial^2 \phi}{\partial r^2} + \frac{1}{r}\frac{\partial \phi}{\partial r} - \beta^2 \frac{\partial^2 \phi}{\partial x^2} = 0 \tag{13}$$

By a similar analogy with the subsonic solution, it can be shown[10,11] that the general solution of Eq. 13 is

$$\phi_{\text{sup}}(x,r) = -\frac{1}{4\pi}\int_0^{x-\beta r} \frac{f(\xi)\,d\xi}{(x-\xi)^2 - \beta^2 r^2} \tag{14}$$

Combining Eq. 9 with Eqs. 11 and 14 yields the axial flow solution for subsonic and supersonic flow,

$$\phi_{\text{sub}} = \frac{V_\infty S'(x)}{2\pi}\ln r + \frac{V_\infty S'(x)}{4\pi}\ln\left[\frac{\beta^2}{4x(l-x)}\right] - \frac{1}{4\pi}\int_0^l \frac{S'(\xi) - S'(x)}{|x-\xi|}\,d\xi \tag{15}$$

$$\phi_{\text{sup}} = -\frac{V_\infty S'(x)}{2\pi}\ln\frac{2}{\beta r} - \frac{V_\infty}{2\pi}\int_0^x S''(\xi)\ln(x-\xi)\,d\xi \tag{16}$$

where $\beta = \sqrt{|M_\infty^2 - 1|}$.

39. SUPERSONIC FLOW

Cones

Exact solutions for supersonic flow past inclined cones have been calculated numerically. Kopal[12,13,14] gives detailed tabulations of the flow variables expanded in powers of α up to α^2. Typical results are plotted in Figures 5 and 6.

Method of Characteristics

Supersonic flow past slender bodies can be calculated numerically by the method of characteristics. The reader is referred to Refs. 15, 16, and 17 for a detailed discussion of this very useful method.

Slender Body Theory

Using Eq. 16, the pressure coefficient over an axisymmetric body is[10,11]

$$C_p = \frac{S''(x)}{\pi}\log\frac{2}{\beta R} + \frac{1}{\pi}\frac{d}{dx}\int_0^x S''(x)\log(x-\xi)\,d\xi - \left(\frac{dR}{dx}\right)^2 \tag{17}$$

The zero lift drag coefficient, referenced to the maximum cross-sectional area $S_{\max}$, is given by

$$C_{Do} = \frac{D}{q_\infty S_{\max}} = \frac{1}{S_{\max}}\int_0^l C_p \frac{dS}{dx}\,dx + C_{DB} + C_F \tag{18}$$

where the first term is called the wave drag C_{DW}, C_{DB} is the drag due to the base pressure P_b, and C_F is the skin friction drag. The C_{DB} is determined by the mechanics of the viscous wake and is discussed further in Art. 41. If the slender body is closed at the base, C_{DB} is approximately zero. The C_F is discussed in Art. 47.

Combining Eqs. 17 and 18 and performing the integration for a body with a pointed nose and either a pointed base or zero slope at the base yields

$$C_{DW} = -\frac{1}{2\pi S_{\max}}\int_0^l \int_0^l S''(\xi)S''(x)\log|x-\xi|\,d\xi\,dx \tag{19}$$

Equation 19 indicates that the wave drag coefficient of a slender body in supersonic flow is independent of Mach number.

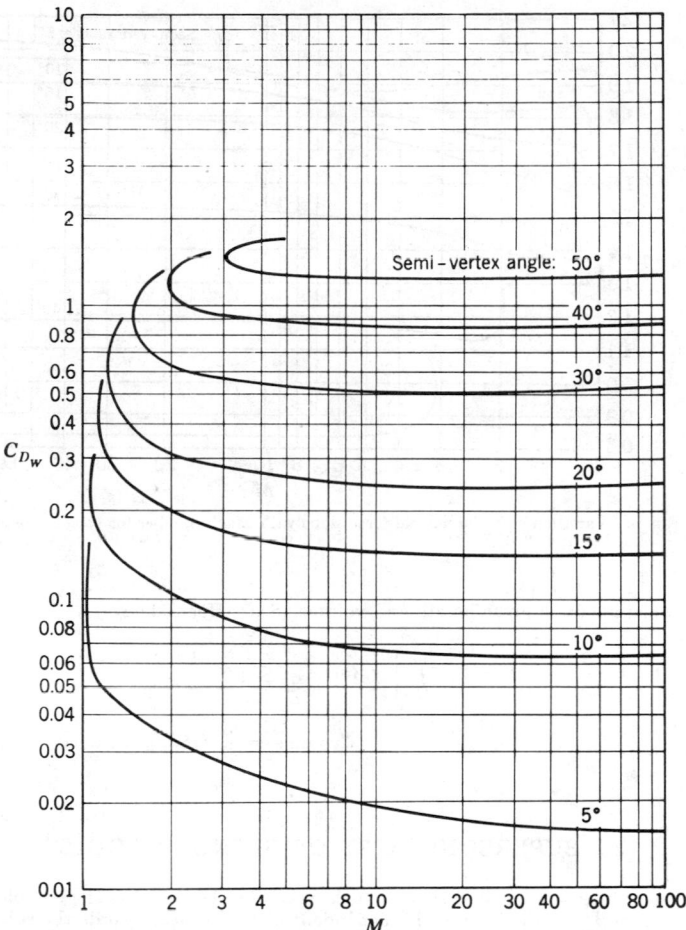

FIG. 5. Variation of drag coefficient with Mach number for various cones at zero angle of attack.

Minimum Wave Drag Bodies

It is of interest to find area distributions of slender bodies which, for given length, volume, or some other constraint, give the lowest possible wave drag. Standard variational methods may be applied to Eq. 19 with appropriate constraint equations. Some results are:[11,18]

1. Half body of given length and diameter (von Karman ogive):

$$\left(\frac{R}{R_{max}}\right)^2 = \frac{S}{S_{max}} = \frac{1}{\pi}\left[\frac{2x}{l}\sqrt{1 - \left(\frac{2x}{l}\right)^2} + \cos^{-1}\left(\frac{-2x}{l}\right)\right], \text{ for}$$

$$-\frac{l}{2} \le x < \frac{l}{2}, \qquad \text{volume} = \tfrac{1}{2}lS_{max}, \qquad C_{DW} = \frac{D_W}{\tfrac{1}{2}\rho V_\infty^2 S_{max}} = \frac{4S_{max}}{\pi l^2}$$

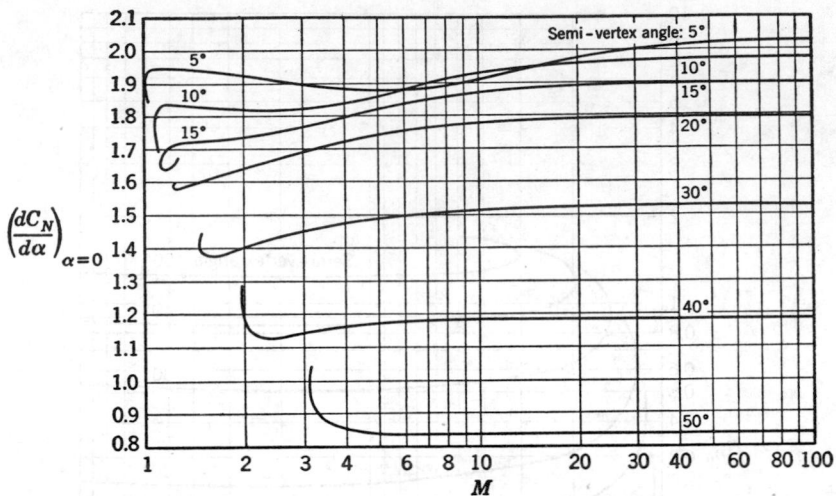

FIG. 6.   Variation of initial normal-force slope with Mach number for various cones.

2. Complete body of given length and volume (Sears-Haack body):

$$\left(\frac{R}{R_{max}}\right)^2 = \frac{S}{S_{max}} = \left[1 - \left(\frac{2x}{l}\right)^2\right]^{3/2}, \text{ for } -\frac{l}{2} \leq x \leq \frac{l}{2}$$

$$\text{volume} = \frac{3}{16}\pi l S_{max}, \qquad C_{DW} = \frac{9}{2}\frac{\pi}{l^2} S_{max}$$

40. SUPERSONIC LIFT OF SLENDER BODIES

For slender bodies at small angle of attack, it can be shown[11] that the crossflow case is independent of β near the body and independent of x. In other words the subsonic and supersonic crossflow can be considered incompressible and two dimensional. For a circular cross-section the solution is the potential flow about a cylinder (see Sec. 6), i.e.,

$$\phi_c(r,\theta) = V_\infty \alpha R^2 \frac{\cos \theta}{r} \tag{20}$$

The crossflow pressure coefficient is[10]

$$C_{PC} = -4\alpha \frac{dR}{dx}\cos\theta + (1 - 4\sin^2\theta)\alpha^2 \tag{21}$$

and the formal force coefficient, reference to S_{max}, is

$$C_N = 2\alpha \tag{22}$$

The crossflow contributes a term to the axial force which is similar to the induced drag of a subsonic finite wing. The total supersonic drag coefficient, referenced to S_{max}, is

$$C_D = C_{D0} + \alpha^2 \tag{23}$$

41. DRAG OF VARIOUS BODIES

The drag on a body immersed in a moving fluid is expressed as

$$D = C_D \text{ (shape, orientation, } R_e, M_\infty) \tfrac{1}{2}\rho V_\infty^2 S$$

where C_D is the drag coefficient, R_e the Reynolds number based upon a characteristic dimension of the body, M_∞ the free-stream Mach number, and S the reference area of the body. Figures 7 and 8 show the variation of drag coefficient with Reynolds number for spheres and cylinders, respectively, when the compressibility effect is negligible. The sudden drop in C_D between R_e of 10^5 to 10^6 is due to the boundary layer transitioning from laminar to turbulent and the associated delay in the flow separation on the surface of the bodies.

Figure 9 shows the effect of compressibility on the drag coefficient for various bodies.

At the rear of a body with a blunt base in supersonic flow, the flow tries to expand 90°. Inviscid flow theory would predict that the base pressure P_B would be zero. However,

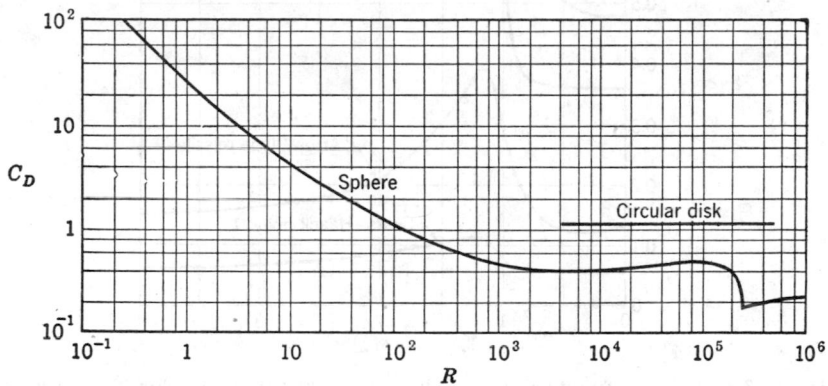

Fig. 7. Drag of a sphere in incompressible flow.

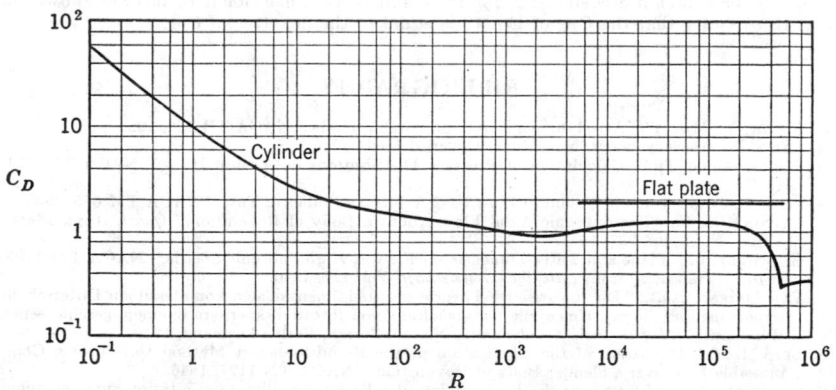

Fig. 8. Drag of a cylinder in incompressible flow.

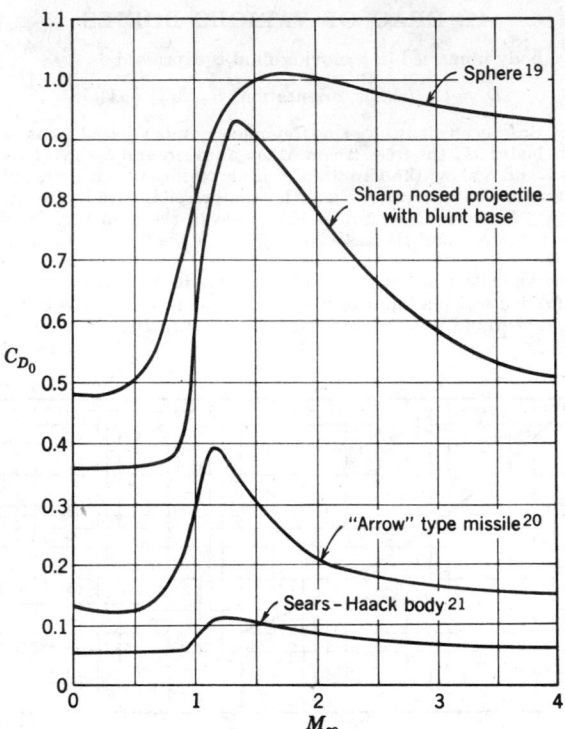

FIG. 9.   Zero lift drag coefficient for various bodies showing effect of comressibility (C_{D0} referenced to S_{max}).

in a viscous fluid the base pressure is not zero but is some value less than ambient pressure P_∞. This is due to the boundary layer bleeding into the separated flow region at the base giving a turbulent or "dead water" region and a $0 < P_B < P_\infty$. Experimental values of base pressure coefficients for two- and three-dimensional bodies are shown on Fig. 10. Notice that the C_{DB} of Eq. 18 is equal to the negative of C_{PB}.

BIBLIOGRAPHY

1. MUNK, M. M., "The Aerodynamic Forces on Airship Hulls," NACA TR 184, 1924.
2. LAMB, H., Hydrodynamics, New York, Dover, 6th ed., 1945.
3. VON KARMAN, TH., "Calculation of Pressure Distributions on Airship Hulls," NACA TM 574, 1930.
4. LOTZ, I., "Calculation of Potential Flow Past Arbitrary Bodies in Yaw," NACA TM 675, 1932.
5. LAITONE, E. V., "The Subsonic Axial Flow About a Body of Revolution," Quart. Appl. Math., Vol. 5, No. 2, pp. 227, 1947.
6. GOETHERT, B., "Plane and Three-Dimensional Flow at High Subsonic Speeds," NACA TM 1105, 1946. (Translated from Lilienthal Gesselschaft, Vol. 127, 1940.
7. VAN DRIEST, E. R., "Die linearisierte Theorie der dreidimensionalen komptessiblen Unterschall-stroemung und die experimentelle Untersuchung von Rotationskoerpern in einem geschlossenen Windkanel," Mitteil. Inst. Aerodynamik, No. 16, Zurich, Verlag Leemann, 1949.
8. LEES, L., "A Discussion of the Application of the Prandtl-Glauert Method to Subsonic Compressible Flow over a Slender Body of Revolution," NACA TN 1127, 1946.
9. SCHMIEDEN C., and KAWALKI, K. H., "Einfluss der Kompressibilitaet bei rotationssymmetrischer Umstroemung eines Ellipsoids," Forschungsbericht No. 1633, Duesche Luftfahrtforschung, 1942. (See also NACA TM 1233, 1949.)
10. LIEPMANN, H. W., and ROSHKO A., Elements of Gasdynamics, New York, Wiley, 1957.

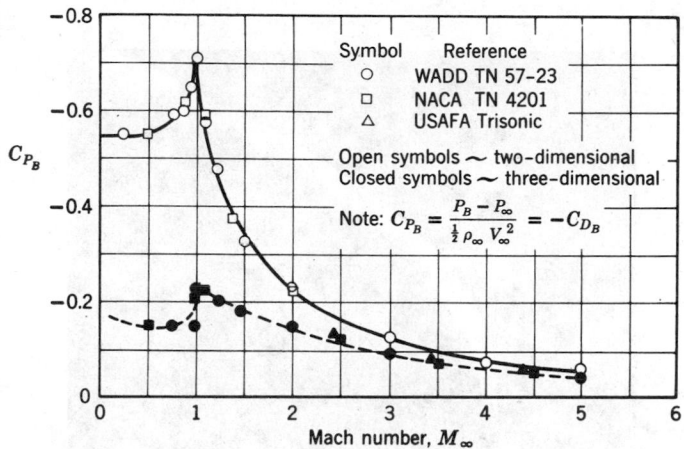

Symbol | Reference
○ | WADD TN 57-23
□ | NACA TN 4201
△ | USAFA Trisonic

Open symbols ∼ two-dimensional
Closed symbols ∼ three-dimensional

Note: $C_{P_B} = \dfrac{P_B - P_\infty}{\frac{1}{2}\rho_\infty V_\infty^2} = -C_{D_B}$

Mach number, M_∞

Fig. 10. Experimental values of base pressure coefficient for two- and three-dimensional bodies.

11. Ashley, H., and Landahl, M., *Aerodynamics of Wings and Bodies*, Reading, Massachusetts, Addison-Wesley, 1965.
12. Kopal, Z., "Supersonic Flow of Air Aound Cones," MIT Tech. Rept. 1, 1947.
13. Kopal, Z., "Supersonic Flow Around Yawing Cones," MIT Tech. Rept. 3, 1947.
14. Kopal, Z., "Supersonic Flow Around Cones of Large Yaw," MIT Tech. Rept. 5, 1949.
15. Ferri, A., *Elements of Aerodynamics of Supersonic Flows*, New York, Macmillan, 1949.
16. Shapiro, A. H., *The Dynamics and Thermodynamics of Compressible Fluid Flow*, New York, Ronald, 1953.
17. Courant, R., and Friedrichs, K. O., *Supersonic Flow and Shook Waves*, New York, Interscience, 1948.
18. Sears, W. R., "On Projectiles of Minimum Wave Drag," *Quart. Appl. Math.*, Vol. 6, pp. 361-366, 1947.
19. Charters, A. C., and Thomas, R. N., "The Aerodynamic Performance of Small Spheres From Subsonic to High Supersonic Velocities," *J. Aeronaut. Sci.*, Vol. 12, No. 10, October 1945.
20. Seifert, H. S., Mills, M. W., and Summerfield, M., "Physics of Rockets—Dynamics of Long Range Rockets," *Amer. J. Phys.*, May–June 1947.
21. Hall, C. F., "Lift, Drag and Pitching Moment of Low Aspect Ratio Wings at Subsonic and Supersonic Speeds," NACA RM A53A30, January 1958.

HYPERSONIC FLOW

By J. T. Clay

42. GENERAL FEATURES OF HYPERSONIC FLOW

Hypersonic flow can be defined as existing when the nonlinearity of the equation (i.e., Eq. 6, Art. 32) describing the supersonic flow over an object becomes an essential feature due to the largeness of $M_\infty^{1,2}$. Thus the breakdown of supersonic linear theory (Art. 36) serves almost as a definition for hypersonic flow. Supersonic linear theory breaks down when the hypersonic similarity parameter $M_\infty \tau \geq 1$, where M_∞ is the free-stream Mach number and τ is proportional to the thickness ratio or flow deflection angle of the body.

A certain Mach number, M_∞, might be hypersonic for one object immersed in the flow while it is supersonic for another object. For very slender bodies, hypersonic flow is usually present for $M_\infty \geq 5$. On the other hand, a blunt object such as a sphere is in a hypersonic flow situation at $M_\infty = 3^4$. Thus we admit to a certain arbitrariness in the term hypersonic; however, hypersonic flow is generally assumed to be $M_\infty > 5$.

FIG. 1.   Schlieren photograph of flow past a 2-in. hemisphere at $M_\infty = 4.38$ (USAFA Trisonic Tunnel).

The manner of analyzing the flow situation in the hypersonic flow regime is very much different than for the other flow regimes.   Potential flow theory used in treating slender bodies and thin airfoils in supersonic flow is not applicable at hypersonic speed.   In hypersonic flow the shock waves lie close to the body (see Fig. 1) and are curved, thus there are large lateral entropy gradients and the flow field is highly rotational.   In addition, the high temperatures which are generated behind the strong shock waves cause real gas effects such as dissociation and ionization and must be accounted for.[3,4]

The effect of compressibility is to thicken the boundary layer.   Thus at hypersonic speeds the boundary layer may thicken to the extent that it cannot be considered thin and there may be a shock wave–boundary layer interaction.

The remainder of the discussion will consider the fluid as an inviscid perfect gas; however, the real gas features mentioned above should be kept in mind.

43. EXACT SOLUTIONS

For shapes composed of wedges, cones, and straight sections the pressure coefficient on the body surface can be determined using the plane oblique shock, conical shock, and Prandtl-Meyer expansion solutions discussed earlier in Arts. 16, 17, and 18.   These solutions are exact[6,7] and the method is the same as for supersonic flow.

The method of characteristics, combined with a method for defining the flow field between the shock wave and the body at some station downstream of the nose, can be used to solve the problem of flow around more complex bodies in hypersonic flight.   Reference 4 discusses the manner in which gas property deviations from a perfect gas can be incorporated.

44. APPROXIMATE SOLUTIONS

Newtonian Impact Theory

In this theory it is assumed that the momentum normal to the surface is entirely converted into pressure so that[5]

$$C_P = \frac{P_S - P_\infty}{\frac{1}{2}\rho_\infty V_\infty^2} = 2 \sin^2 \theta \tag{1}$$

where θ is the local flow deflection angle. The theory also assumes that the pressure coefficient in any expansion region is zero.

A more accurate pressure distribution may be found by replacing the 2 in Eq. 1 by the value of the stagnation pressure coefficient on the body. The modified expression is called the Modified Newtonian and is expressed as

$$C_P = C_{P_{stag}} \sin^2 \theta \tag{2}$$

where $C_{P_{stag}}$ is given by the Rayleigh pitot formula (see Art. 15)

$$C_{P_{stag}} = \frac{2}{\gamma M_\infty^2} \left[\left(\frac{\gamma + 1}{2} M_\infty^2 \right)^{\gamma/\gamma-1} \left(\frac{\gamma + 1}{2\gamma M_\infty^2 - \gamma + 1} \right)^{1/1-\gamma} - 1 \right] \tag{3}$$

The Modified Newtonian approximation gives quite good results when the shock wave is close to the body, which is the condition necessary for the assumption that the momentum of the gas normal to the surface is lost. Normally the Newtonian approximation is applied to predicting the pressure distribution on the nose of blunt bodies and gives reliable estimates down to $M_\infty = 3.0$.[4] Figure 2 shows the measured and calculated (using Eq. 2) values of pressure coefficient around a hemisphere at $M_\infty = 4.38$.

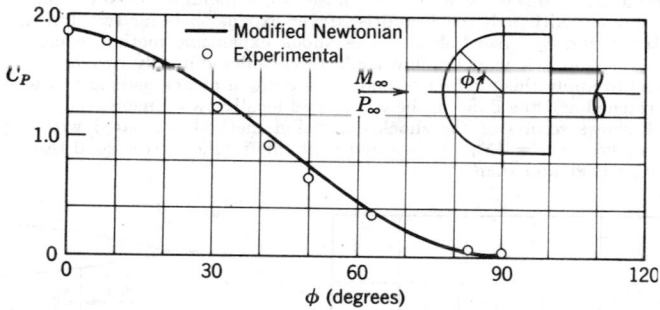

FIG. 2. Measured and calculated (modified Newtonian) pressure coefficient distribution for hemisphere in Fig. 1 (Data from USAFA Trisonic Tunnel).

Newtonian flow theory can be applied to flat surfaces to provide an approximation for the aerodynamic coefficients. For a flat plate at an angle of attack α, the coefficients are

$$C_L = 2 \sin^2 \alpha \cos \alpha \tag{4}$$

$$C_{DL} = \text{drag coefficient due to lift} = 2 \sin^3 \alpha \tag{5}$$

$$C_D = C_{D0} + 2 \sin^3 \alpha \tag{6}$$

Tangent-Wedge and Tangent-Cone Approximations

The tangent-wedge and tangent-cone approximations assume that the pressure coefficient at any point on the surface of a body is the same as that on a wedge or cone which has the same surface inclination to the free stream as the point of interest. The equation for the surface pressure coefficient for a wedge in hypersonic flow can be approximated by[2,5]

$$C_P \approx \theta^2_W \left[\frac{\gamma + 1}{2} + \sqrt{\left(\frac{\gamma + 1}{2}\right)^2 + \frac{4}{(M_\infty \theta_W)^2}} \right] \tag{7}$$

This equation is applicable to a two-dimensional body which has small surface inclination θ_W.

Lees[8] developed an approximate equation for the surface pressure on a cone in hypersonic flow which is applicable to determining the surface pressure on bodies of revolution. This expression is

$$C_P = \theta^2_c \left[\frac{4}{\gamma + 1} \left(\frac{K_s^2 - 1}{K_c^2} \right) + \frac{2(K_s - K_c)^2}{K_c^2} \left(\frac{\gamma + 1}{\gamma - 1 + 2/K_c^2} \right) \right] \tag{8}$$

where θ_c = semivertex angle

$K_c = M_\infty \theta_c$, hypersonic similarity parameter

$$K_s = \frac{\gamma + 1}{\gamma + 3} K_c + \sqrt{\left(\frac{\gamma + 1}{\gamma + 3}\right)^2 K_c^2 + \frac{2}{\gamma + 3}}$$

This formula applies when the shock wave lies close to the body and when $M_\infty \theta_c \geq 1$.

Shock Expansion Method

The shock expansion method is applicable to slender, pointed, two-dimensional, or axisymmetric bodies with attached shock waves. In this method the expansion waves generated by a curved body in a supersonic flow are assumed to be absorbed by the shock wave and will not reflect to interact with the flow field or body downstream of the original wave. Thus, if the body shape is known, then the flow conditions at the nose immediately behind the shock wave can be determined from known solutions for a wedge[6] or cone[6,7] in supersonic flow. The Prandtl-Meyer expansion is then used to determine the flow characteristics downstream of the nose by expanding through the appropriate expansion angle. Although the Prandtl-Meyer expansion is for two-dimensional flows, Eggers, Savin, and Syverson[9] showed the shock expansion method could be used for bodies of revolution as long as the flow could be considered locally two dimensional. It is also possible to apply this method to bodies at angle of attack as long as the shock wave remains attached and the flow can be considered locally two dimensional.

Figure 3 shows results of the shock expansion method compared with experimental values for ogives at $\alpha = 15°$. It is noted that this method gives good results when the value of $M_\infty \tau$ is greater than 1.

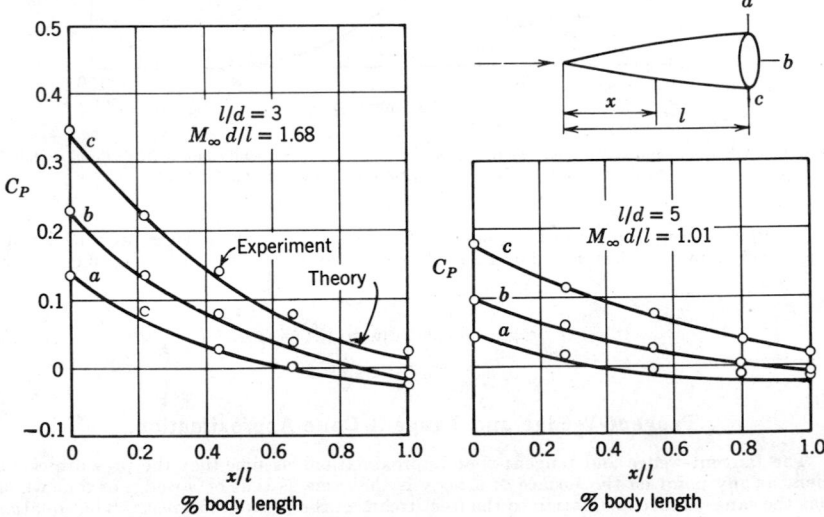

FIG. 3.   Pressure coefficient distributions over two ogives at $M_\infty = 5.05$ and $\alpha = 15°$.

Small Disturbance Theory

Small disturbance theory encompasses a variety of hypersonic flow theory appropriate for slender bodies. Even though the bodies are slender and the perturbation velocities are small compared to the free-stream velocity, they are not small compared to the free-stream speed of sound. Thus the small disturbance theory is essentially a nonlinear one. If the substitution $x = V_\infty t$ is made in the small perturbation equations and boundary conditions,[5] then they become identical with the equations for unsteady flow in one less space dimension. This relationship is the "equivalance principle."[10] The equivalance principle is important because it means that a number of existing methods and solutions for unsteady flow problems become available for steady hypersonic flows, and an important extension of the principle has resulted in methods of treating the effects of nose blunting on the flow past a slender body.[4,5]

The small disturbance theory leads to the important conclusion that in the hypersonic small perturbation flow equations the Mach number and the body-shape parameter τ appear only in the combination $M_\infty\tau$, the "hypersonic similarity parameter." This means that for flow past affinely related slender bodies that have the same value of $M_\infty\tau$, the values of the nondimensional variables should be similar. This hypersonic similarity rule is discussed next.

Hypersonic Similarity

The hypersonic similarity parameter $M_\infty\tau$ makes it possible to extend the experimental data of a slender body at a particular Mach number to other bodies of similar shape but at different Mach numbers. Two bodies are said to be similar (geometrically) if their nondimensional thickness distributions $h(\xi)$ are the same. The surface pressure coefficient is given as[2,4,5]

$$C_P = \tau^2 f\left(\gamma, M_\infty\tau, h, \frac{\alpha}{\tau}\right)$$

where γ is the ratio of specific heats and α is the angle of attack. As before, τ is the thickness ratio of the slender body.

If two similar bodies A and B (i.e., $h_A = h_B$) have their M_∞ and α adjusted such that $(M_\infty\tau)_A = (M_\infty\tau)_B$ and $(\alpha/\tau)_A = (\alpha/\tau)_B$, then $C_P/\tau^2)_A = C_P/\tau^2)_B$ and the lift and drag coefficients are related as follows:[5]

1. Two dimensional bodies (coefficients based upon planform area)

$$\left(\frac{C_L}{\tau^2}\right)_A = \left(\frac{C_L}{\tau^2}\right)_B, \qquad \left(\frac{C_D}{\tau^3}\right)_A = \left(\frac{C_D}{\tau^3}\right)_B$$

2. Three-dimensional bodies (coefficients based upon cross-sectional area)

$$\left(\frac{C_L}{\tau}\right)_A = \left(\frac{C_L}{\tau}\right)_B, \qquad \left(\frac{C_D}{\tau^2}\right)_A = \left(\frac{C_D}{\tau^2}\right)_B$$

Nose Bluntness Effects

Nose blunting has a significant effect on the flow field about a body in hypersonic flow because of the strong pressure gradients which are generated in the nose region. This effect can be significant for hundreds of nose diameters downstream. The blast wave analogy of Lees[11] for treating flow past blunt nosed cylinders and flat plates and the methods of Chernyi[2] and Cheng[13] for treating flow about blunt nosed cones and wedges are significant because of the data correlation parameters which resulted from these theories. Important aerodynamic quantities such as surface pressures, heat transfer coefficient, and shock wave shape can be correlated against parameters involving the geometry of the aerodynamic configuration. Table 1 shows a partial listing of the correlation parameters resulting from the theories.

Experimental evidence of the validity of these correlation parameters is shown in Fig. 4 in which $\dfrac{c_p}{2\theta_c{}^2}$ is plotted vs. $\dfrac{\theta_c{}^2}{\sqrt{\epsilon C_{dn}}}\left(\dfrac{x}{d_n}\right)$ for blunt cones with several bluntness ratios, cone angles, and free stream Mach numbers. Note that the pressure data correlate

Table 1.   Correlation Parameters for Nose Bluntness effects *

Geometrical Shape	Aerodynamic Quantity	Correlation
Blunted Cylinder	Surface Pressure	$\dfrac{p}{p_\infty}$ vs. $\dfrac{1}{M_\infty{}^2\sqrt{C_{dn}}}\left(\dfrac{x}{d_n}\right)$
	Shock Wave Shape	$\dfrac{1}{M_\infty\sqrt{C_{dn}}}\left(\dfrac{r_s}{d_n}\right)$ vs. $\dfrac{1}{M_\infty\sqrt{C_{dn}}}\left(\dfrac{x}{d_n}\right)$
Blunted Flat Plate	Surface Pressure	$\dfrac{p}{p_\infty}$ vs. $\left\{\dfrac{1}{M_\infty{}^3 C_{dn}}\left(\dfrac{x}{d_n}\right)\right\}^{2/3}$
	Shock Wave Shape	$\dfrac{1}{M_\infty{}^2 C_{dn}}\left(\dfrac{y_s}{d_n}\right)$ vs. $\dfrac{1}{M_\infty{}^3 C_{dn}}\left(\dfrac{x}{d_n}\right)$
Blunted Cone	Surface Pressure	$\dfrac{C_p}{2\theta_c{}^2}$ vs. $\dfrac{\theta_c{}^2}{\sqrt{\epsilon C_{dn}}}\dfrac{x}{d_n}$
	Heat Transfer	$\dfrac{(\epsilon C_{dn})^{1/4}\,C_H\,\sqrt{Re_{\infty,d}}}{M_\infty\theta_c{}^2\,\sqrt{C^*}}$ vs. $\dfrac{\theta_c{}^2}{\sqrt{\epsilon C_{dn}}}\dfrac{x}{d_n}$
Blunted Wedge	Surface Pressure	$\dfrac{C_p}{2\theta_w{}^2}$ vs. $\theta_w{}^2\left(\dfrac{x}{C_{dn}\epsilon d_n}\right)^{2/3}$

C_{dn} = Nose drag coefficient.
C_H = Coefficient of heat transfer.
C_p = Pressure coefficient.
d_n = Nose diameter.
M_∞ = Free stream Mach number.
p = Surface pressure.
p_∞ = Free stream static pressure.
r_s = Shock wave radius.
$Re_{\infty,d}$ = Reynolds number based upon nose diameter.
x = Axial distance.
y_s = Distance of shock wave from the centerline.
ϵ = $(\gamma - 1)/(\gamma + 1)$.
γ = Ratio of specific heats.
θ_c = Cone half-angle.
θ_w = Wedge half-angle.
ψ = Nose bluntness ratio.

quite well and are relatively insensitive to Mach number.   These data may be used to obtain estimates of the pressure distributions over cones which have cone angles and bluntness ratios different from those tested.   Similar agreement has been found for the other dynamic parameters and for other geometries.

BIBLIOGRAPHY

1. HAYES, W. D., and PROBSTEIN, R. F., Hypersonic Flow Theory, New York, Academic, 1959.
2. CHERNY, G. G., Introduction to Hypersonic Flow, New York, Academic, 1961.
3. DORRANCE, W. H., Viscous Hypersonic Flow, New York, McGraw-Hill, 1962.
4. HAYES, W. D., and PROBSTEIN, R. F., Hypersonic Flow Theory, Volume I, Inviscid Flow, New York, Academic, 1966.
5. COX, R. N., and CRABTREE, L. F., Elements of Hypersonic Aerodynamics, London, English Univ. Press, 1965.
6. Ames Research Staff, "Equations Tables and Charts for Compressible Flow," NACA Report 1135, 1953.
7. KOPAL, Z., "Tables of Supersonic Flow Around Coves," MIT Tech. Rept. No. 1, 1947.
8. LEES, L., "Note on the Hypersonic Similarity Law for an Unyawed Cone," J. Aeronaut. Sci., Vol. 18, pp. 700–702 1951.

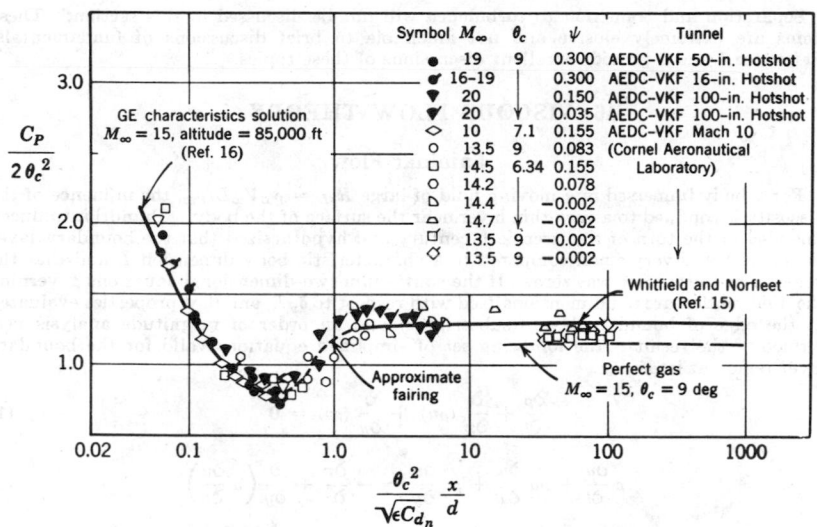

FIG. 4. Correlation of zero-lift cone pressures, $M_\infty = 10$ to 20 (Griffith and Lewis[14]).

9. EGGERS, A. J., SAVIN, R. C., and SYVERTSON, C. A., "The Generalised Shock Expansion Method and its Application to Bodies Traveling at High Supersonic Airspeeds," *J. Aeronaut. Sci.*, Vol. 22, pp. 231–238, 248, 1955.
10. HAYES, W. D., "On Hypersonic Similitude," *Quar.. Appl. Math.*, Vol. 5, pp. 105–106, 1947.
11. LEES, L., "Inviscid hypersonic flow over blunt noses slender bodies," GALCIT Hypersonic Research Project, Memo No. 31, 1956.
12. WAGNER, R. D., and WATSON, R., "Induced pressures and Shock Shapes on Blunt Cones in Hypersonic Flow," NASA Technical Note, TN–D–2182, 1964.
13. CHENG, H. K., "Hypersonic Flow With Combined Leading-Edge Bluntness and Boundary-Layer Displacement Effects," Cornell Aeronautical Laboratory Report No. AF–1285–A–4, 1960.
14. GRIFFITH, B. J., and LEWIS, C. H., "A Study of Laminar Heat Transfer to Sherically Blunted Cones and Hemisphere-Cylinders at Hypersonic Conditions," Arnold Engineering Development Center Technical Documentary Report No. AEDC–TDR–63–102, 1963.
15. WHITFIELD, J. D., and NORFLEET, G. D., "Source Flow Effects in Conical Hypervelocity Nozzles," Arnold Engineering Development Center Technical Documentary Report No. AEDC–TDR–62–116, 1962.
16. LEWIS, C. H., "Pressure Distribution and Shock Shape over Blunt Slender Cones at Mach Numbers from 16 to 19," Arnold Engineering Development Center Technical Note No. AEDC–TN–61–81, 1961.

VISCOUS FLOWS

By L. M. Nicolai and J. J. Beoddy

45. INTRODUCTION

The purpose of this article is to present the fundamental ideas of viscous continuum fluid flow over aircraft and hypervelocity vehicles. The discussions will consider the theory of viscous flows and methods for predicting skin friction and aerodynamic heating. For the discussions the flow may be assumed to be laminar if the local Reynolds number is

$$R_e = \frac{\rho_e u_e x}{\mu_e} < 5 \times 10^5$$

where x is the distance from the leading edge to the point of interest and the subscript e denotes conditions at the outer edge of the boundary layer.

Separation and transition to turbulence will not be discussed in this section. These items are extremely elusive and not amenable to brief discussions of fundamentals. Reference 1 and 9 provide excellent discussions of these topics.

46. VISCOUS FLOW THEORY

Laminar Flow

For a body immersed in a moving fluid of large $R_{eL} = \rho_\infty V_\infty L/\mu_\infty$, the influence of the viscosity is confined to a very thin layer near the surface of the body. Prandtl introduced this idea at the turn of the twentieth century and hypothesized that the boundary layer thickness δ was very small compared to a characteristic body dimension L and that the velocity at the surface was zero. If the continuum two-dimensional equations governing the fluid motion are nondimensionalized with respect to δ, L, and flow properties evaluated at the edge of boundary layer (subscript e) and an order of magnitude analysis performed,[1,9] the result is the following set of simplified equations valid for the boundary layer region at large R_e:

$$\frac{\partial \rho}{\partial t} + \frac{\partial}{\partial x}(\rho u) + \frac{\partial}{\partial y}(\rho v) = 0 \tag{1}$$

$$\rho \frac{\partial u}{\partial t} + \rho u \frac{\partial u}{\partial x} + \rho v \frac{\partial u}{\partial y} = -\frac{\partial p}{\partial x} + \frac{\partial}{\partial y}\left(\mu \frac{\partial u}{\partial y}\right) \tag{2}$$

$$\frac{\partial p}{\partial y} = 0 \tag{3}$$

$$\rho \frac{\partial H}{\partial t} + \rho u \frac{\partial H}{\partial x} + \rho v \frac{\partial H}{\partial y} = -\frac{\partial}{\partial y}\left(k \frac{\partial T}{\partial y}\right) + \frac{\partial}{\partial y}\left(\mu u \frac{\partial u}{\partial y}\right) \tag{4}$$

where $H = h + (u^2 + v^2)/2$, total enthalpy, and $h = \int_0^T C_p \, dT$, static enthalpy. The boundary conditions are

$$\begin{aligned} u = 0 \text{ (no slip condition) and } v = 0 \text{ at } y = 0 \\ u = u_e \qquad \text{at } y \to \infty \end{aligned} \tag{5}$$

Equation 3 is the y-momentum equation and expresses the important result that the pressure along the body surface is equal to the pressure p_e at the edge of the boundary layer. Since the flow outside the boundary layer can be considered inviscid, the pressure gradient term is given by

$$\rho \frac{\partial u_e}{\partial t} + \rho u_e \frac{\partial u_e}{\partial x} = -\frac{\partial p}{\partial x} \tag{6}$$

Incompressible. *Zero Pressure Gradient, $dp/dx = 0$.* The solution of the velocity field for the incompressible flat plate at zero angle of attack was determined by H. Blasius in 1908. The system of equations under consideration has no preferred length, thus it is reasonable to suppose that the velocity profiles at varying distances from the leading edge are similar to each other. An appropriate choice for the reference velocity and distance are $u_e(x)$ and $\delta(x)$, respectively. Thus we seek a function g such that

$$\frac{u}{u_e} = g(\eta) \tag{7}$$

where $\eta = y/\delta$. It can be shown[1] that $\delta \sim \sqrt{\dfrac{\mu x}{\rho u_e}}$ so that we define $\eta = y \sqrt{\dfrac{\rho u_e}{\mu x}}$.

A nondimensional stream function $f(\eta)$ is defined as

$$\psi = \sqrt{\frac{\mu x u_e}{\rho}} f(\eta) \tag{8}$$

The velocity components become

$$u = \frac{\partial \psi}{\partial y} = \frac{\partial \psi}{\partial \eta}\frac{\partial \eta}{\partial y} = u_e f'(\eta) \tag{9}$$

$$v = \frac{1}{2}\sqrt{\frac{\mu u_e}{\rho x}}\,(\eta f' - f) \tag{10}$$

and our function g in Eq. 7 is $f'(\eta)$.

Inserting our new variables into the steady incompressible flat plate continuity and momentum equations gives us

$$2f''' + f''f = 0 \tag{11}$$

with $f = f' = 0$ at $\eta = 0$ and $f' = 1$ at $\eta \to \infty$. The details of the solution of Eq. 11 are in Ref. 1 and are presented in tabular form in Table 1.

Table 1. The function f, f' and f'' for laminar boundary layer with dp/dx = 0 (Blasius solution)

η	$f(\eta)$	$f'(\eta)$	$f''(\eta)$
0	0	0	0.3321
0.2	0.0066	0.0664	0.3320
0.4	0.0266	0.1328	0.3315
0.6	0.0597	0.1989	0.3301
0.8	0.1061	0.2647	0.3274
1.0	0.1656	0.3298	0.3230
1.5	0.3716	0.4867	0.3023
2.0	0.6500	0.6298	0.2668
2.5	0.9953	0.7518	0.2173
3.0	1.3968	0.8460	0.1614
3.5	1.8282	0.9128	0.1080
4.0	2.3058	0.9555	0.0642
4.5	2.7852	0.9794	0.0343
5.0	3.2833	0.9916	0.0159
5.5	3.7806	0.9969	0.0068
6.0	4.2796	0.9990	0.0024
6.5	4.7793	0.9997	0.0008
7.0	5.2793	0.9999	0.0002

If we define our boundary layer thickness δ as that distance from the wall where $u = -0.99u_e$, we see from Table 1 that $f'(\eta) = 0.99$ at $\eta = 5.0$. Thus

$$\delta = 5\sqrt{\frac{\mu x}{\rho_e \mu_e}} = \frac{5x}{\sqrt{R_{ex}}} \tag{12}$$

The shearing stress at the wall is

$$\tau_w = \left(\mu\frac{\partial u}{\partial y}\right)_w = \mu u_e\sqrt{\frac{\rho u_e}{\mu x}}\,f''(0)$$

and from Table 1, $f''(0) = 0.332$, such that the local skin friction drag coefficient for one side of a flat plate is

$$C_f = \frac{\tau_w}{\frac{1}{2}\rho_e\mu_e^2} = \frac{0.664}{\sqrt{R_{ex}}} \tag{13}$$

and the average skin friction coefficient for one side of a flat plate is

$$C_F = \frac{1}{L}\int_0^L C_f\,dx = \frac{1.328}{\sqrt{R_{eL}}} \tag{14}$$

Nonzero Pressure Gradient. Let us consider that the velocity u_e at the outer edge of the boundary layer is given by[2]

$$u_e(x) = cx^m \tag{15}$$

where c is a constant and x is measured from the stagnation point. This $u_e(x)$ describes the velocity along a wedge of included angle $\pi\beta$ $(0 < \beta < 2, m > 0)$ where $\beta = 2m/(m + 1)$. Following the idea of the last section we transform the independent variable y into

$$\eta = y \sqrt{\frac{m + 1}{2} \frac{\rho u_e}{\mu x}} = y \sqrt{\frac{m + 1}{2} \frac{c\rho x^{m-1}}{\mu}} \tag{16}$$

The continuity equation is satisfied by the introduction of a stream function

$$\psi(x,y) = \sqrt{\frac{2}{m + 1} \frac{\mu c}{\rho}} \, x^{\frac{m-1}{2}} f(\eta)$$

The momentum equation is transformed into

$$f''' + ff'' + \beta(1 - f'^2) = 0 \tag{17}$$

with boundary condition $f = f' = 0$ at $\eta = 0$ and $f' = 1$ at $\eta \to \infty$. Equation 17 was first introduced by Falkner and Skan.[2] We observe that when $\beta = 0$ the equation reduces to the Blasius expression for a flat plate (when Eq. 16 is put in the Blasius form for η). The $\beta = 1$, $m = 1$ case is the situation for two-dimensional stagnation flow.

The numerical solution of Eq. 17 is presented in Ref. 1. The value of the local skin friction on a wedge can be determined from

$$C_f = \frac{\tau_w}{\frac{1}{2}\rho u_e^2} = \frac{\mu c f''(0)}{\frac{1}{2}\rho u_e^2}$$

where $f''(0)$ is the slope of the velocity distribution $u/u_e = f'(\eta)$ at $\eta = 0$.

Compressible. Prandtl number $P_r = \dfrac{\mu C_p}{k} = 1$ and $dp/dx = 0$. The right-hand side of Eq. 4 can be expressed in terms of total enthalpy H giving

$$\rho \left(u \frac{\partial H}{\partial x} + v \frac{\partial H}{\partial y} \right) = \frac{\partial}{\partial y} \left(\frac{\mu}{P_r} \frac{\partial H}{\partial y} \right) + \frac{\partial}{\partial y} \left[\mu u \left(1 - \frac{1}{P_r} \right) \frac{\partial u}{\partial y} \right] \tag{18}$$

For $P_r = \mu C_p/k = 1$ and $dp/dx = 0$ we compare the momentum and energy equations:

$$\rho \left(u \frac{\partial u}{\partial x} + v \frac{\partial u}{\partial y} \right) = \frac{\partial}{\partial y} \left(\mu \frac{\partial u}{\partial y} \right)$$

$$\rho \left(u \frac{\partial H}{\partial x} + v \frac{\partial H}{\partial y} \right) = \frac{\partial}{\partial y} \left(\mu \frac{\partial H}{\partial y} \right) \tag{19}$$

with boundary conditions

$$u = 0 \quad \text{and} \quad H = h_w \quad \text{at } y = 0$$

$$u = u_e \quad \text{and} \quad H = H_e \quad \text{at } y = \delta$$

The similarity of the two equations (both diffusion type equations) indicates a solution for the energy equation of

$$\frac{H - h_w}{H_e - h_w} = \frac{u}{u_e} \tag{20}$$

The solution in terms of static enthalpy is

$$h = h_w + (H_e - h_w) \frac{u}{u_e} - \frac{u^2}{2} \tag{21}$$

and we note that since $P_r = 1$, H_e is equal to the recovery enthalpy or adiabatic wall enthalpy h_{aw}. The heat transfer per unit area is

$$-q = k \frac{\partial T}{\partial y} = \frac{k}{c_p} \frac{\partial h}{\partial y} = \frac{k}{c_p} \left[\frac{H_e - h_w}{u_e} \frac{\partial u}{\partial y} - u \frac{\partial u}{\partial y} \right]$$

and at the wall $y = 0$, $u = 0$ such that

$$- q_w = \frac{k}{c_p} \frac{(H_e - h_w)}{u_e} \left(\frac{\partial u}{\partial y} \right)_w = \frac{k}{c_p} \frac{(h_{aw} - h_w)}{u_e} \left(\frac{\partial u}{\partial y} \right)_w \qquad (22)$$

In terms of the Stanton number $St = \dfrac{-q}{\rho_e u_e (h_{aw} - h_w)}$ and $C_f = \dfrac{\tau_w}{\frac{1}{2} \rho_e u_e^2}$ we have Reynolds Analogy

$$St = \tfrac{1}{2} C_f \qquad (23)$$

which postulates that energy and momentum are transferred by the same mechanism.

As a result of $P_r = 1$ and $dp/dx = 0$ we determined the interesting results (due to Crocco) $St = \frac{1}{2} C_f$ and that $H(y)$ is a linear function of $u(y)$, but we have not solved for the velocity field yet. This will come in the next section.

$P_r \neq 1$ and $dp/dx = 0$. The steady momentum and energy Eqs. 2 and 18 are transformed using a linear viscosity law $\mu = BT$ and the transformation (credit to Dorodnitzen or Howarth[9])

$$S = x, \qquad n = \int_0^y \frac{\rho}{\rho_e} \, dy \qquad (24)$$

which gives an equivalent incompressible set of equations[9] (i.e., momentum equation is independent of the temperature field). Then we introduce the Blasius nondimensional variables

$$\eta = \frac{n}{\delta} = n \sqrt{\frac{\rho_e u_e}{\mu_e S}} \qquad (25)$$

and

$$f(\eta) = \frac{\psi}{\sqrt{\mu_e u_e S / \rho_e}} \qquad (26)$$

The resulting set of equations is

$$f''' + 2 f f'' = 0 \qquad (27)$$

$$H'' + \tfrac{1}{2} P_r f H' = u_e^2 (P_r - 1)(f''^2 + f' f''') \qquad (28)$$

with boundary conditions

$$f = f' = 0 \text{ at } \eta = 0, \quad f' = 1 \text{ at } \eta \to \infty$$

$$H = h_w \text{ at } \eta = 0, \quad H = H_e \text{ at } \eta \to \infty$$

The Dorodnitzen-Howarth transformation and the linear viscosity law have uncoupled the momentum and energy equation. Equations 27 and 28 put the solution for the velocity field in the Blasius form. Thus the solution for u/u_e is tabulated in Table 1 in terms of the transformed variable n. Equation 28 is solved numerically using the Blasius solution for $f(\eta)$. The details of the solution are discussed in Ref. 9. The solution for the local skin friction at the wall is

$$C_f = 0.664 \frac{1}{\sqrt{\rho_e u_e x}} \sqrt{\frac{\mu_w T_e}{\mu_e T_w}} = \frac{0.664}{\sqrt{R_{ex}}} \sqrt{\frac{\mu_w T_e}{\mu_e T_w}} \qquad (29)$$

Clearly the linear viscosity law is an approximation to the more correct Sutherland's equation (see Art. 11). The law $\mu = BT = C(\mu_e/T_e)T$, where C is a constant, can satisfactorily describe the variation in viscosity over reasonable temperature intervals if $C < 1$. From Eq. 29

$$C_{f_e} = \frac{0.664}{\sqrt{R_{ex}}} \sqrt{C} = C_{f_i} \sqrt{C}$$

where the subscripts c and i denote compressible and incompressible, respectively. Thus the effect of compressibility, i.e. dissipation and heat transfer, is a decrease in the local skin friction. This conclusion agrees with experimental results and is indicated on Fig. 1.

Another conclusion which is imbedded in the transformed y coordinate is that compressibility thickens the boundary layer. This result is shown on Fig. 2.

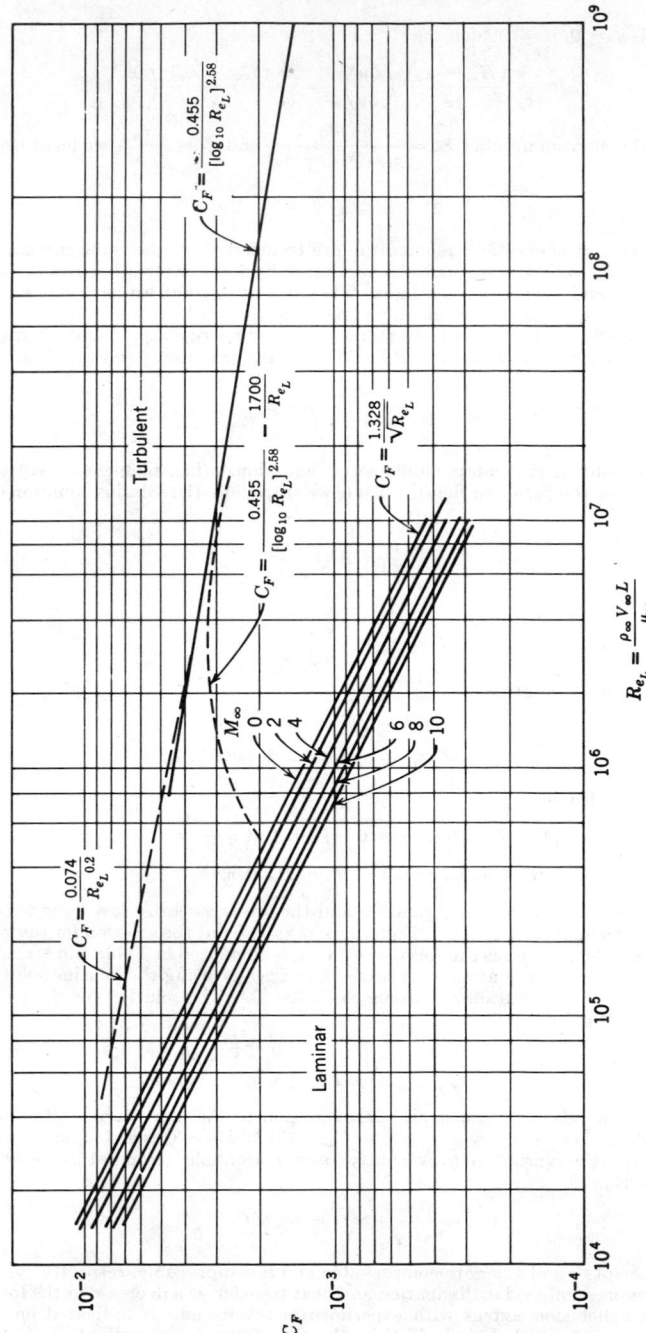

FIG. 1.   Averaged skin friction coefficient vs. Reynolds number for flow over an insulated flat plate.

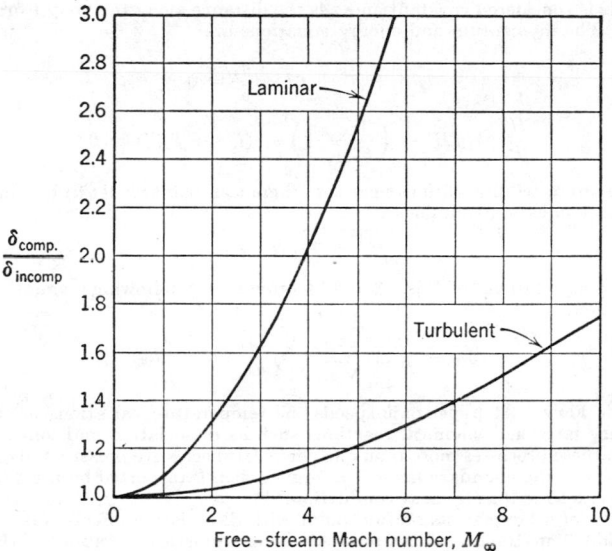

FIG. 2.   Effect of compressibility on the thickness of the boundary
layer on an insulated flat plate.

The solution for the heat transfer at the wall is

$$St = \frac{0.332}{P_r^{2/3}} \frac{1}{\sqrt{\rho_e u_e x / \mu_e}} \sqrt{\frac{\mu_w T_e}{\mu_e T_w}} \tag{30}$$

Combining C_f and St we obtain an extended Reynolds' analogy for general P_r,

$$\frac{St}{C_f} = \frac{P_r^{-2/3}}{2} \tag{31}$$

Cohen and Reshotko[8] considered a compressible laminar boundary layer with a nonzero pressure gradient.   Their analysis parallels the discussion above except that their x and y coordinates are transformed according to

$$X = \int_0^x B \frac{p_e a_e}{p_0 a_0} dx, \qquad Y = \frac{a_e}{a_0} \int_0^y \frac{\rho}{\rho_0} dy$$

where a is the speed of sound, B is the constant in the linear viscosity law, and the subscript 0 denotes free stream stagnation values.   The power law velocity distribution of Falkner and Skan, Eq. 14 is used so that $p_e(x)$ is known.   The details of their analysis and solution are presented in Ref. 8.   One important result from Cohen and Reshotko's work is the variation of Reynolds' analogy with the pressure gradient parameter β. For $\beta \neq 0$ (i.e., a nonzero pressure gradient) Eq. 31 is not valid.

Stagnation Point Heating.   The steady continuity, momentum, and energy (in the form of Eq. 18) equations are transformed as before using a linear viscosity law, the Dorodnitzen, Howarth transformation and Eqs. 25 and 26.   The pressure gradient term is determined by the steady form of Eq. 6.   Near the forward stagnation point the velocity u_e just outside the boundary layer may be expressed as

$$u_e = \frac{du_e}{dx} x$$

where du_e/dx is considered constant and x is the distance along the body from the stagnation point. The momentum and energy equations are[9]

$$f''' + ff'' + f'^2 + \frac{\rho_e}{\rho} = 0 \tag{32}$$

$$\frac{H''}{P_r} + fH' - \left(\frac{1}{P_r} - 1\right) u_e^2 (f''^2 + f'f''') = 0 \tag{33}$$

Since we are interested in solutions near the stagnation point we let u_e become small such that the energy equation simplifies to

$$H'' + P_r f H' = 0 \tag{34}$$

The numerical solution [9] of Eqs. 32 and 34 provides the following stagnation point heat transfer relation:

$$q_{sp} = \frac{0.5}{P_r^{2/3}} \frac{\mu_w \rho_w}{\sqrt{\mu_e \rho_e}} \sqrt{\frac{du_e}{dx}} (H_e - h_w) \tag{35}$$

Hypersonic Flow. At hypersonic speeds the temperature variations in the boundary layer are very large and chemical reactions such as dissociation and ionization may be present. In these processes more than one chemical species are present and concentration gradients occur in the boundary layer. Energy is then transported by the diffusion of the chemical species as well as by heat conduction.[9,10]

The heating of a laminar stagnation point with dissociation effects was considered by Fay and Riddell in Ref. 8. They carried out a numerical solution of the governing equations for a stagnation point boundary layer using tabulated values of the properties of high temperature air in dissociation equilibrium. Their results are given by the expression

$$q_{sp} = \frac{0.54 \sqrt{k+1}}{P_r^{0.6}} (H_e - H_w) (\rho_w \mu_w)^{0.1} (\rho_e \mu_e)^{0.4} \sqrt{\frac{du_e}{dx}} \left[1 + (L_e^m - 1) \frac{h_D}{H_e}\right] \tag{36}$$

where h_D is the enthalpy of dissociation, L_e is the Lewis number (relative rate of mass diffusion to heat conduction), $k = 0$ for a two-dimensional body, $k = 1$ for an axisymmetrical body, $m = 0.52$ for equilibrium flow, and $m = 0.63$ for frozen flow.

Turbulent Flow

Exact solutions to the governing differential equations of the boundary layer are not readily obtained for the turbulent flow case. With turbulence the particles of fluid experience highly irregular fluctuations which lead to high frequency variations in both pressure and velocity at a fixed point in the flow. Since the flow is a fluctuating or random process, we work in statistical mechanics, seeking not an explicit account of the motion of every particle of fluid, but only a statistical description in terms of various averaged functions.[13]

The fluid particles in the turbulent boundary layer undergo large scale eddy motion. The transfer mechanisms of energy and momentum are microscopic molecular transport (as in the laminar boundary layer) and macroscopic eddy motion; the macroscopic transfer being much larger than the molecular transport. Since the macroscopic transfer mechanism of momentum is the same as that for energy, i.e., the large scale motion of lumps of fluid,[13] we would expect a turbulent Reynolds' analogy to be valid in the fully turbulent region away from the wall.

A very thin region near the body surface has a laminar character and is called the laminar sublayer. The presence of this sublayer is essential since the large scale eddy motion cannot extend all the way to the wall. In this sublayer the viscous action and heat transfer take place under circumstances like those in a laminar boundary layer.

Experimentally it is observed that the analogy between heat transfer and skin friction expressed by Eq. 31 retains its validity everywhere in the turbulent boundary layer. Also, the analogy remains approximately true in cases where the pressure gradient is different from zero.

Because of the macroscopic fluid motion in the fully turbulent region, the velocity distribution is fairly uniform (compared to the laminar boundary layer distribution) away from the wall. This results in a steep velocity gradient at the wall (through the laminar sublayer) and large skin friction. The velocity distribution in the turbulent boundary layer can be approximated by

$$\frac{u}{u_e} = \left(\frac{y}{\delta}\right)^{1/n}$$

where $n = 7$ at $R_e = 5 \times 10^5$ and increases slightly with increasing R_e.[1] The turbulent velocity profile is compared with the laminar profile on Fig. 3.

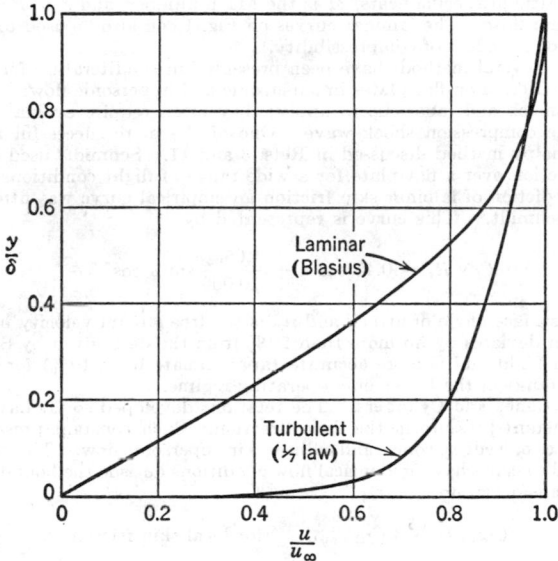

FIG. 3. Velocity profiles for laminar and turbulent boundary layers (incompressible flow).

Using the empirical 1/7 power law, Blasius developed the following very useful expressions for turbulent boundary layer thickness δ_T and local skin friction coefficient:[1]

$$\delta_T = \frac{0.37x}{R_{ex}^{0.2}} \tag{37}$$

$$C_f = \frac{0.059}{R_{ex}^{0.2}} \tag{38}$$

47. SKIN FRICTION METHODS

Laminar

Incompressible. For smooth surfaces with zero pressure gradient the average, integrated skin friction drag coefficient for *one* side of a flat plate is given by the Blasius solution,

$$C_F = \frac{1.328}{\sqrt{R_{eL}}} = 2C_{fx=L} \tag{14}$$

where $R_{eL} = \rho_e u_e L/\mu_e$ is evaluated for a characteristic dimension L and the reference area is the area wetted by the flow (on one side).

Compressible. The most useful method for estimating the compressible laminar skin friction is the Reference Temperature method.[3,11] Essentially this method relies on the empirical observation that the incompressible formulas for heat transfer and skin friction (for laminar and turbulent flows) may be used provided that all physical properties are evaluated at a temperature (or enthalpy for large variations in temperature) intermediate between the values at the surface (T_w) and outer edge of the boundary layer (T_e). Equation 14 is used with all of the properties evaluated at a reference temperature T^* given by

$$T^* = 0.5(T_e + T_w) + 0.22r \left(\frac{\gamma - 1}{2} \right) M_e{}^2 T_e$$

where γ is the ratio of specific heats, M is the Mach number, and $r = P_r{}^{1/2}$ the recovery factor for laminar flow.   The laminar curves on Fig. 1 can also be used to obtain a first-order estimate of the effect of compressibility.

Hypersonic.   Several methods have been presented in the literature for the prediction of laminar skin friction on flat plates in supersonic and hypersonic flow.   These methods are usually complex and laborious to apply; they often require evaluation of the flow field behind the compression shock wave.   One of the more successful methods is the Reference Enthalpy method discussed in Refs. 3 and 11.   Schmidt[5] used this method to predict skin friction over a flat plate for a wide range of flight conditions.   To simplify further, the prediction of laminar skin friction an empirical curve was fitted to the data presented by Schmidt.[4]   This curve is represented by

$$C_f \sqrt{R_e} = 0.45 \cos \alpha + \frac{4.65 u_\infty}{10,000} \sin \alpha \cos^{2.2} \alpha \tag{39}$$

where α is the surface angle of attack and u_∞ is the free stream velocity in ft/sec.

This equation deviates by no more than 20% from the data given by Schmidt for low altitude, high α flight and is more accurate (approximate by $\pm 10\%$) for the rest of the altitude and α range in the hypersonic operating regime.

Extension to Cones and Cylinders.   The relations developed so far have been for flat plates with constant pressure in the flow direction.   Such constant pressures also arise along the surfaces of wedges, cones, and cylinders in supersonic flow.   The flat plate results can be extended to cones having identical flow conditions outisde the boundary layer using the Mangler transformation

$$C_{f\,\text{cone}} = \sqrt{3}\, C_{f\,\text{flat plate}} \qquad \text{for local skin friction}$$

$$C_{F\,\text{cone}} = \frac{2}{\sqrt{3}}\, C_{F\,\text{flat plate}} \qquad \text{for total skin friction}$$

Skin friction coefficients on the surface of a cylinder whose axis is aligned with the free stream are identical with the ones for a flat plate, providing the radius of the cylinder is large compared with the boundary layer thickness.

Turbulent

Incompressible.   For turbulent flow the local and averaged skin friction coefficients can be estimated from the Blasius and Prandtl-Schlichting empirical relationships.[1]   These expressions are

$$C_f = \frac{0.059}{R_{ex}{}^{0.2}} \tag{38}$$

$$C_F = \frac{0.455}{(\log_{10} R_{eL})^{2.58}} \tag{40}$$

The Prandtl-Schlichting expression for C_F agrees with measured results very well up to $R_e = 10^9$ and is shown on Fig. 1.

The R_{eL} at which transition occurs is variable, depending on the turbulence level in the free stream and the roughness of the surface.   Assuming transition at $R_e = 5 \times 10^5$, the following expression can be used to estimate the averaged skin friction coefficient in the transition region (see Fig. 1).

$$C_F = \frac{0.455}{(\log_{10} R_{eL})^{2.58}} - \frac{1700}{R_e}$$

Compressible. Similar to the laminar compressible flow, the turbulent compressible skin friction can be determined using the incompressible expression (Prandtl-Schlichting) but evaluating the fluid properties at a reference temperature T^*:

$$T^* = 0.5(T_e + T_w) + 0.22r \left(\frac{\gamma - 1}{2}\right) M_e{}^2 T_e$$

where $r = P_r{}^{1/3}$, the recovery factor for turbulent flow. Experimental results show that the effect of compressibility is to decrease the turbulent skin friction. Hayes and Probstein[15] found that an excellent fit for the experimental results is given by

$$\frac{C_{Fc}}{C_{Fi}} = (1 + 0.144 M_\infty{}^2)^{-0.65}$$

Hypersonic. Schmidt[6] used the reference enthalpy method to construct a comprehensive series of graphs for the prediction of turbulent skin friction on a flat plate for a wide range of flight parameters. Hankey and Neumann[4] determined a curve fit to Schmidt's analysis which results in the very useful expression for the local turbulent skin friction coefficient

$$C_f(R_e)^{0.2} = 0.48 \sin (4.5\alpha) + 0.7 \frac{u_\infty}{10,000} \cos^{2.25} \alpha \sin^{1.5} \alpha \qquad (41)$$

where

$$R_e = \frac{\rho_e u_\infty L}{\mu_e}$$

and α = surface angle-of-attack.

The average integrated skin friction coefficient is then

$$C_F = 1.39 \left(\frac{S_w}{S}\right) S_\alpha \qquad (42)$$

where the characteristic length is the distance to the trailing edge. The factor 1.39 is used for swept flat plates or delta wings. Also

S = reference wing area.

S_w = compression surface wetted area.

$S_\alpha = 1 + \dfrac{1}{|\alpha| + 1.15^{|\alpha|}}$ where α is in degrees.

Extension to Cones. Van Driest[7] developed what is referred to as the "cone rule" for turbulent flow by extending the momentum integral method to the case of the cone. This rule can be expressed as

$$C_{F_{cone}} = 1.15 C_{F_{flat\ plate}}$$

This rule can be used for other axisymmetric bodies near $\alpha = 0°$.

48. HIGH SPEED AERODYNAMIC HEATING

Flat Plate—Laminar and Turbulent

The method for estimating the heat transfer at the surface of a body (with zero pressure gradient) is the Reference Temperature (or enthalpy) method and the extended Reynolds' analogy. The discussion here is equally valid for both laminar and turbulent boundary layers in high speed ($M_\infty > 2$) flow.

The local Stanton number is again defined as

$$St = \frac{-q}{\rho_e u_e (h_{aw} - h_w)}$$

and the adiabatic wall enthalpy or recovery enthalpy is

$$h_{aw} = h_e + r \frac{u_e{}^2}{2} \qquad (43)$$

where

$$r = \begin{cases} P_r^{1/2} \text{ for laminar flow} \\ P_r^{1/3} \text{ for turbulent flow} \end{cases}$$

The heat transfer per unit area per unit time at the wall is

$$q_w = -k \left(\frac{\partial T}{\partial y}\right)_w = h'_c(h_{aw} - h_w) \tag{44}$$

where h'_c is the forced convection heat transfer coefficient. If the temperature variation in the boundary layer is not large such that C_p is fairly constant, then Eq. 44 is expressed as

$$q_w = h_c(T_{aw} - T_w) \tag{45}$$

where

$$h_c = C_p h'_c$$

The laminar and turbulent extended Reynolds' analogy is [1],[11]

$$\frac{St}{C_f} = \frac{P_r^{-2/3}}{2} \tag{31}$$

which is valid for $dp/dx \approx 0$.

The reference temperature or enthalpy is determined from

$$T^* = 0.5(T_w + T_e) + 0.22r \left(\frac{\gamma - 1}{2}\right) M_e^2 T_e$$

$$h^* = 0.5(h_w + h_e) + 0.22r \left(\frac{\gamma - 1}{2}\right) M_e^2 h_e$$

Next the local skin friction coefficient C_f is determined using the methods presented in the last section. Then, using the extended Reynolds' analogy, Eq. 31, the local Stanton number is obtained and finally the local heat transfer q_w.

The above procedure is in widespread use today and is referred to as the Reference Enthalpy method. Its accuracy has been checked by comparison with the results obtained by boundary layer solutions[4],[5],[6]. Agreement within $\pm 4\%$ has been found for $400 < T_\infty < 800°R$, $400 < T_w < 3000°R$, and Mach numbers up to 16.

For surfaces having laminar and turbulent regions, the average heat transfer for the surface is obtained by integrating the local values over the surface. We consider an abrupt transition to turbulence at a local Reynolds number of 5×10^5. The expression for the average Stanton number over a surface is

$$St\, P_r^{2/3} = \frac{\int_0^L \frac{1}{2}C_f \, dx}{\int_0^L dx}$$

$$= \frac{1}{L}\left[\int_0^{x_c} \frac{0.332}{R_e} \, dx + \int_{x_c}^L \frac{0.228}{(\log_{10} R_e)^{2.58}} \, dx\right]$$

where $R_e = \rho u_\infty x/\mu$ and x_c corresponds to $R_e = 5 \times 10^5$.

If $dp/dx \neq 0$ for the laminar case, the results of Cohen and Reshotko[8] should be used to determine the proper expression for Reynolds' analogy.

Stagnation Point

The heat transfer rate per unit area at the stagnation point q_{sp} can be determined using Eq. 35. An estimate of the velocity gradient du_e/dx can be obtained using Modified Newtonian (see Art. 49)

$$\frac{p_e(x)}{p_{sp}} = \cos^2 \theta$$

where θ is measured from the stagnation point to the point x on the blunt surface. Using $u_e{}^2 = \dfrac{2}{\rho_e} (p_{sp} - p_e)$ we obtain

$$\frac{du_e}{dx} \approx \frac{u_\infty}{R} \sqrt{\frac{\gamma - 1}{\gamma}}$$

where R is the radius of the blunt surface.

Detra and Hidalgo[14] used empirical data for the air properties and obtained the following approximate expression for Eq. 35:

$$q_{sp} = \frac{865}{\sqrt{R}} \left(\frac{u_\infty}{10^4}\right)^{3.15} \sqrt{\frac{\rho_\infty}{\rho_{sL}}} \tag{46}$$

where ρ_{sL} is the density at sea level, R is the nose radius in feet, u_∞ is the free stream velocity in ft/sec, and q_{sp} is the heating rate in Btu/ft²-sec. Equation 46 agrees $\pm 10\%$ with experimental results for $6{,}000 \leq u_\infty \leq 26{,}000$ ft/sec and sea level to 400,000 feet.

The distribution of heating rate about a blunt surface behaves similar to the pressure distribution. Thus, using Modified Newtonian

$$\frac{q}{q_{sp}} \approx \cos \theta$$

The importance of a blunt surface with a large radius R in order to decrease the stagnation point heating is readily apparent from Eq. 46.

The expression developed by Fay and Riddel, Eq. 36, should be used for estimating the stagnation point heating rate when dissociation effects are present.

BIBLIOGRAPHY

1. SCHLICHTING, H., *Boundary Layer Theory*, New York, McGraw-Hill, 1960.
2. FALKNER, V. M., and SKAN, S. W., "Some Approximate Solutions of the Boundary Layer Equations," *Phil. Mag.*, Vol. 12, pp. 865, 1931.
3. ECKERT, E. R. G., "Survey of Heat Transfer at High Speeds," WADC TR 54–70, 1964.
4. HANKEY, W. L., and NEUMANN, R. D., "Design Procedures for Computing Aerodynamic Heating at Hypersonic Speeds," WADC TR 59–610, 1960.
5. SCHMIDT, H., "Laminar Skin Friction and Heat Transfer Parameters for a Flat Plate at Hypersonic Speeds in Terms of Free-Stream Flow Properties," NASA TN D–8, September, 1959.
6. SCHMIDT, H., "Turbulent Skin Friction and Heat Transfer Coefficients for an Inclined Plate at Hypersonic Speeds in Terms of Free-Stream Flow Properties," NASA TN D–869, May, 1961.
7. VAN DRIEST, E. R., "Turbulent Boundary Layer on a Cone in Supersonic Flow at Zero Angle of Attack," *J. Aeronaut. Sci.*, Vol. 19, No. 1, pp. 55, 1952.
8. HARTNETT, J. P. (ed.), *Recent Advances in Heat and Mass Transfer*, New York, McGraw-Hill, 1961.
9. MOORE, F. K. (ed.), *Theory of Laminar Flows, Vol. IV, High Speed Aerodynamics and Jet Propulsion*, Princeton, New Jersey, Princeton Univ. Press, 1964.
10. COX, R. N., and CRABTREE, L. F., *Elements of Hypersonic Aerodynamics*, New York, Academic, 1965.
11. ECKERT, E. R. G., "Survey of Boundary Layer Heat Transfer at High Velocities and High Temperatures," WADC TR 59–624, 1960.
12. FRANCIS, W. L., MALVESTUTO, F. S., and STUART, J. W., "Study to Determine Skin-Friction Drag in Hypersonic Low Density Flow," ASD–TR 61–433, Vol. I, 1962.
13. HINZE, J. O., *Turbulence*, New York, McGraw-Hill, 1959.
14. DETRA, R. W., and HIDALGO, H., *ARS JOURNAL*, March 1961.
15. HAYES, W. D., and PROBSTEIN, R. F., *Hypersonic Flow Theory*, New York, Academic, 1959.

AERODYNAMICS OF WING-BODY COMBINATIONS

By L. M. Nicolai

49. WING-BODY LIFT

The combined wing-body lift characteristics are expressed as

$$C_L = (C_{L\alpha})_{WB}\alpha + C_1\alpha^2 \tag{1}$$

where $(C_{L\alpha})_{WB}$ is the wing-body linear lift curve slope and C_1 is the nonlinear lift factor. For wings with sharp leading edges and aspect ratios less than 2, the nonlinear lift is quite pronounced. Values of C_1 are given in Refs. 1 and 2.

The lift characteristics of a wing and a body do not add directly to give the wing-body lift. Rather, there are interference effects of one component on the other.[3] A method that gives good results for the wing-body linear lift curve slope is

$$(C_{L\alpha})_{WB} = F(C_{L\alpha})_W \tag{2}$$

where $(C_{L\alpha})_W$ is the linear lift curve slope (based upon the exposed wing area) of the wing alone and F is a wing-body lift interference factor given on Fig. 1. The $(C_{L\alpha})_W$ is determined using Eq. 43 of Art. 35 for subsonic flow and the methods of Art. 36 for supersonic flow. The curve for F on Fig. 1 was determined using the method of Ref. 11. The $(C_{L\alpha})_{WB}$ is referenced to the exposed wing area S_e.

50. WING-BODY DRAG

The combined wing-body drag characteristics can be expressed as (see Eq. 29 of Art. 33)

$$C_D = C_{D0} + KC_L^2 \tag{3}$$

where C_{D0} = zero lift drag coefficient = $C_{Dp} + C_F + C_{DW}$.

C_{Dp} = pressure drag coefficient due to flow separation at $C_L = 0$. Base drag C_{DB} is included here.

C_F = skin friction drag coefficient.

C_{DW} = wave drag coefficient, due to thickness and camber.

K = drag due to lift factor.

Subsonic Drag

The subsonic C_{D0} for slender wing-body combinations is primarily due to skin friction ($C_{DW} = 0$ for subsonic flow). The C_F's for the wing alone and body alone are determined using the methods of Art. 52 and then added together as

$$C_F = 1.05 \left[(C_F)_{body} \frac{S_{BW}}{S_{Ref}} + (C_F)_{wing} \frac{S_{WW}}{S_{Ref}} \right] \tag{4}$$

where S_{BW} is the body wetted area, S_{WW} is the wing wetted area (usually $2S_e$), and S_{Ref} is the reference area in the expression

$$\text{Drag} = C_D \tfrac{1}{2} \rho_\infty V_\infty^2 S_{Ref} \tag{5}$$

The coefficient of 1.05 in Eq. 4 accounts for a drag interference of 5%.

The drag due to lift factor is discussed in Article 33 and shown in Fig. 2 for wing-body combinations having delta planforms.[5,10] A very useful empirical expression for the subsonic drag due to lift factor K for $\textit{Æ} > 2$ is given by[6]

$$K_{WB} = \frac{0.95}{(C_{L\alpha})_{WB}} \tag{6}$$

Transonic and Supersonic Drag

Wave drag interference effects in the transonic and supersonic range are greater than those in the subsonic region because of the higher local velocities of the individual components and the greater propagation of these perturbations from this source. The most successful and by far the most systematic method for predicting the transonic and supersonic drag is the area-rule concept.

The area-rule method is based upon the supersonic slender body theory discussed in Art. 39. Reference 7 shows that this supersonic theory can be extended down to a Mach number of 1 as a limiting case. It can be assumed that at large distances from the body the disturbances are independent of the arrangement of the components and only a function of the cross-section area distribution.[8] This means that the drag of a wing-body combination can be calculated as though the combination were a body of revolution with equivalent-area cross sections. This is shown in Fig. 3 for a Mach number of 1.

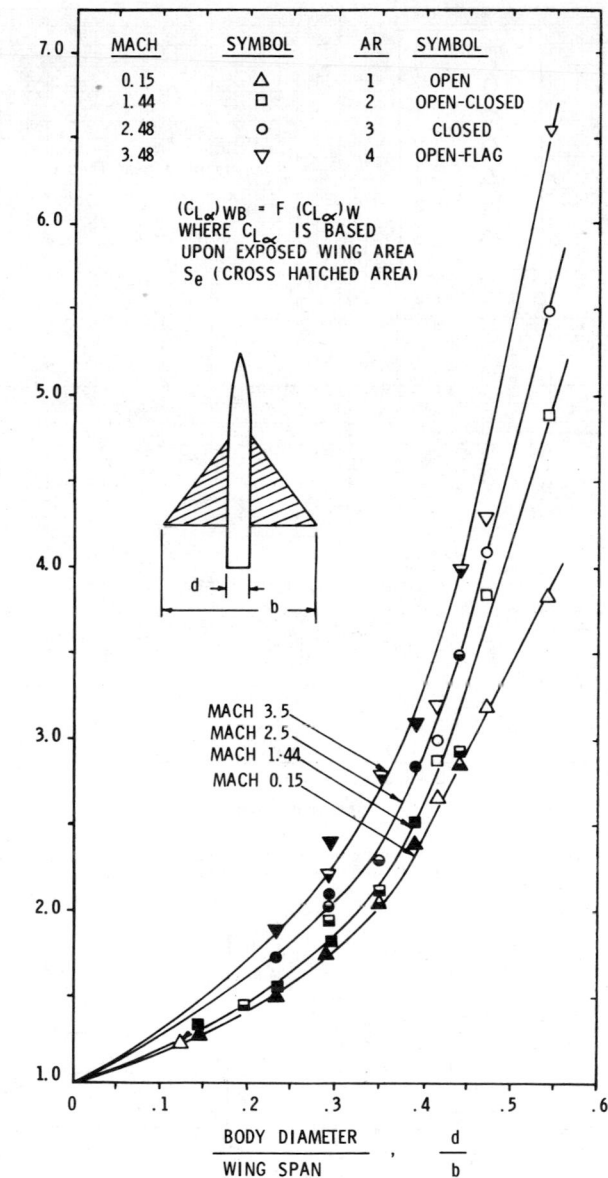

MACH	SYMBOL	AR	SYMBOL
0.15	△	1	OPEN
1.44	□	2	OPEN-CLOSED
2.48	○	3	CLOSED
3.48	▽	4	OPEN-FLAG

$$(C_{L\alpha})_{WB} = F\,(C_{L\alpha})_W$$
WHERE $C_{L\alpha}$ IS BASED
UPON EXPOSED WING AREA
S_e (CROSS HATCHED AREA)

MACH 3.5
MACH 2.5
MACH 1.44
MACH 0.15

$$\frac{BODY\ DIAMETER}{WING\ SPAN}\ ,\ \frac{d}{b}$$

FIG. 1.   Wing-body lift interference factor.

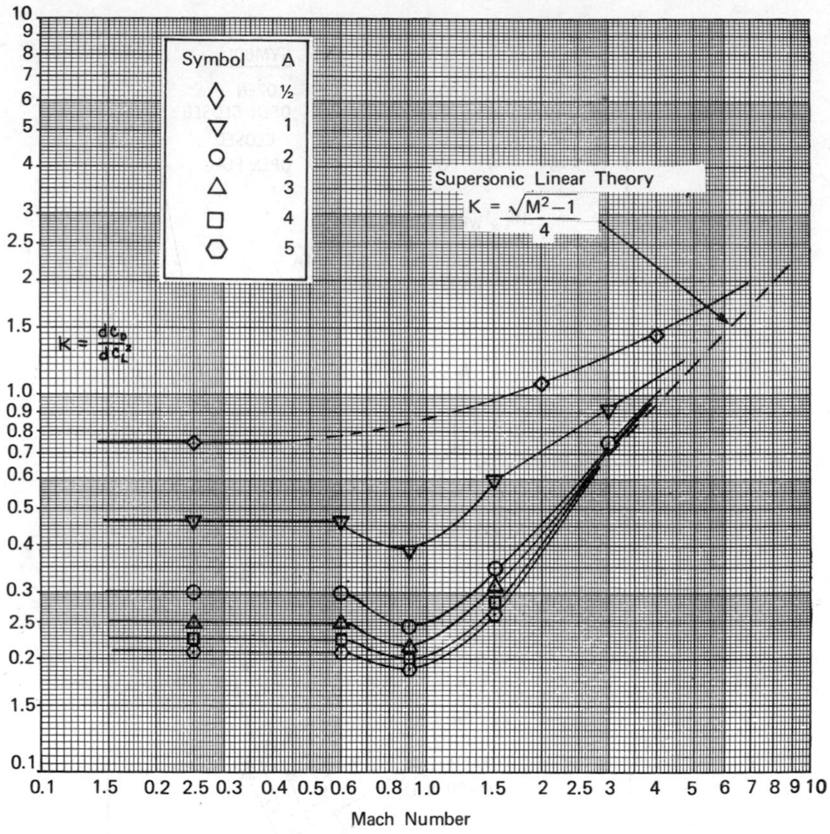

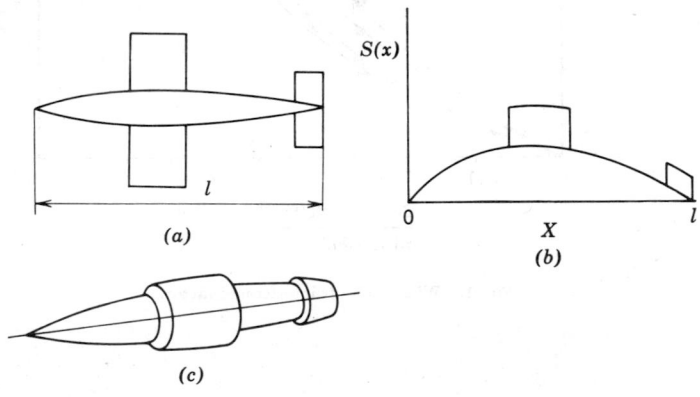

FIG. 2.   Drag due to lift factor for wing-body combinations with delta planforms.   (Data from USAFA Trisonic Tunnel and Refs. 5 and 10.)

FIG. 3.   Equivalent body for a wing-body-tain combination at $M_\infty = 1$: (a) wing-body; (b) cross-section area distribution; and (c) equivalent body.

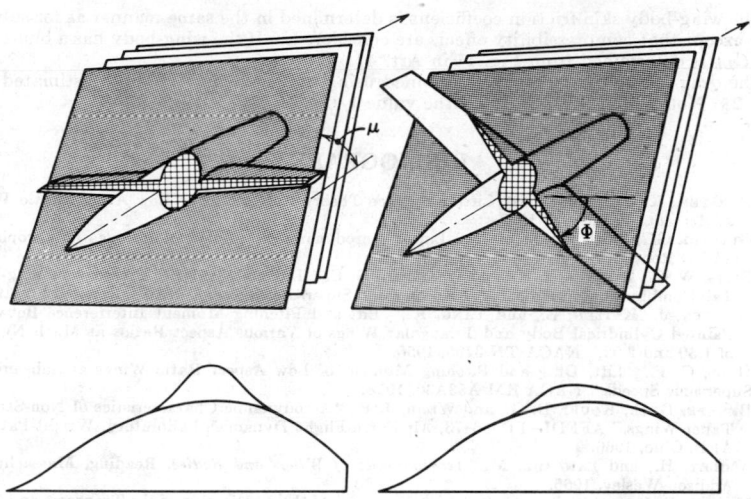

FIG. 4.   Area distribution given by intersection of Mach planes for $M_\infty > 1$.

At Mach numbers greater than 1, the cross-section areas $S(x)$ are along planes inclined at the angle $\mu = \arc\sin 1/M_\infty$ to the x-axis.   There is a different $S(x)$ for each roll angle Φ. This is shown in Fig. 4.

Once the area distribution $S(x)$ is determined [one $S(x)$ for each Φ angle for $M_\infty > 1$], the wave drag is calculated using Eq. 19 in Art. 39.   For $M_\infty > 1$ the C_{DW} is determined for each roll angle Φ and then averaged.   Application of the area-rule method usually requires automatic computing equipment.[9]

The area-rule method indicates the most desirable way to arrange the vehicle components for minimum wave drag at a particular M_∞.   The most common example of this is to indent or "coke bottle" the fuselage enough to permit the wing to be added without a sharp discontinuity appearing on the $S(x)$ distribution.   If the cross-section area distribution of a wing-body combination at a particular M_∞ is the same as a Sears-Haack distribution (see Art. 39), the configuration produces minimum wave drag at that M_∞.   Thus a wing-body can be configured to give minimum wave drag at one Mach number but will usually aggravate the wave drag at other Mach numbers.   Figure 5 demonstrates this.

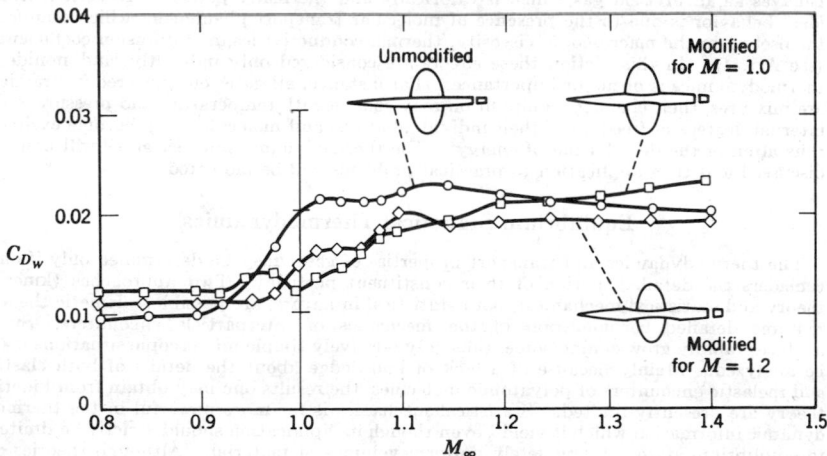

FIG. 5.   Wave drag of bodies with elliptic wings.

The wing-body skin friction coefficient is determined in the same manner as for subsonic flow except that compressibility effects are considered. If the wing-body has a blunt base, the C_{DB} is determined from Fig. 11 in Art. 41. The drag due to lift factor is determined using the charts in Art. 36 or estimated from Fig. 2. Notice that K approaches the value $1/C_{L\alpha}$ at $M_\infty > 1$.

BIBLIOGRAPHY

1. GERSTEN, K., "A Non-Linear Lifting Surface Theory Especially for Low Aspect Ratio Wing," *J. Aeronaut. Sci.*, April 1963.
2. GERSTEN, K., "Calculation of Non-Linear Aerodynamic Stability Derivatives of Aeroplanes," AGARD Rept. 342, 1961.
3. PITTS, W. C., NIELSEN, J. N., and KAATTARI, G. E., "Lift and Center of Pressure of Wing-Body-Tail Combinations at Subsonic, Transonic and Supersonic Speeds," NACA Report 1307, 1959.
4. NIELSEN, J., KATZEN, E., and TANG, K., "Lift and Pitching Moment Interference Between a Painted Cylindrical Body and Triangular Wings of Various Aspect Ratios at Mach Numbers of 1.50 and 2.02," NACA TN 3795, 1956.
5. HALL, C. F., "Lift, Drag and Pitching Moment of Low Aspect Ratio Wings at Subsonic and Supersonic Speeds," NACA RM A53A30, 1958.
6. BENEPE, D. B., KOURI, B. G., and WEBB, J. B., "Aerodynamic Characteristics of Non-Straight-Taper Wings," AFFDL–TR–66–73, Air Force Flight Dynamics Laboratory, Wright-Patterson AFB, Ohio, 1966.
7. ASHLEY, H., and LANDAHL, M., *Aerodynamics of Wings and Bodies*, Reading, Massachusetts, Addison-Wesley, 1965.
8. NELSON, R. L., and WELSH, C. J., "Some Examples of the Application of the Transonic and Supersonic Area Rules to the Prediction of Wave Drag," NASA TN D–446, September 1960.
9. CAHN, M. S., and OLSTAD, W. B., "A Numerical Method for Evaluating Wave Drag," NACA TN 4258, June 1958.
10. OSBORNE, R. S., and KELLY, T. C., "A Note on the Drag Due to Lift of Delta Wings at Mach Numbers Up to 2.0," NASA TN D–545, November 1960.
11. NICOLAI, L. M., and SANCHEZ, F., "Correlation of Wing-Body Combination Lift Data," *AIAA Journal of Aircraft*, February 1973.

REAL GASES

· By D. Finkleman

51. THERMODYNAMICS OF REAL GASES

Most classical analyses in aerodynamics assume the medium to behave ideally. Although the results of such investigations have been quite accurate in the past, no fluid behaves as an inviscid gas which is calorically and thermally perfect. Deviation from ideal behavior is due to the presence of molecular transport phenomena which manifest themselves in the macroscopic viscosity, thermal conductivities, and diffusion coefficients (see Art. 4). In this section these effects are considered only indirectly, and nonideal thermodynamics is of major importance. For instance, all gases encountered in practice are mixtures, their chemical composition may change with temperature and pressure, and internal degrees of freedom of their individual atoms and molecules may become excited, thus altering the distribution of energy. The thermodynamics of real gases will first be discussed and then application to practical problems will be indicated.

Equilibrium Statistical Thermodynamics

The thermodynamics and transport properties of gases may be determined only if one considers the detailed motion of their constituent particles. Two approaches (kinetic theory and statistical mechanics), both statistical in nature, are possible. Kinetic theory requires detailed consideration of the mechanics of interparticle encounters. Such analyses rapidly grow cumbersome, thus only relatively simple microscopic situations may be analyzed. Mainly because of a lack of knowledge about the details of both elastic and inelastic encounters of polyatomic molecules, the results one may obtain from kinetic theory are presently limited. Statistical mechanics is far more powerful in the thermodynamic information which it yields, even though its application should strictly be limited to equilibrium states of completely uniform volumes of material. Although theories of nonequilibrium statistical thermodynamics exist, most rely upon the postulate that the thermodynamic properties of a system subject to nonequilibrium processes are, at each point the same functions of local state variables that they would be if the gas were in equilibrium at those conditions.[1]

At the heart of all thermodynamics is the concept of entropy. For most purposes entropy may be regarded as a means of keeping track of the processes which may occur within a system. Statistically, this is reflected by enumerating each possible energy state in which each of the particles may reside under given macroscopic conditions with the Boltzmann relation

$$S = k \ln \Omega$$

The evaluation of Ω depends upon the manner in which the particles behave. If no two particles of the same type may reside in the same energy level, so-called Fermi-Dirac statistics apply. Thus if the number of particles of energy E is N_j and there are g_j possible states with that energy, the total number of ways of assigning the particles to the states is

$$W_j = \prod_j \frac{(g_j)!}{(g_j - N_j)! N_j!}$$

On the other hand, if any number of particles may be crowded into a single energy level, Bose-Einstein statistics apply and

$$W_j = \prod_j \frac{(N_j + g_j - 1)!}{(g_j - 1)! N_j!}$$

A particular choice of the N_j's for a total number of particles, N, is called a distribution. When all possible distributions are considered, the total number of microstates of a system is

$$\Omega = \sum_{\substack{\text{all possible} \\ \text{sets of } N_j \\ \text{such that}}} W_j \quad \begin{cases} N = \sum_j N_j \\ E = \sum_j N_j \epsilon_j \end{cases}$$

It can be shown that only the largest term in the summation contributes significantly.[2] For large numbers of particles and an even larger number of possible states (so that the energy levels are sparsely populated) it follows that if all microstates are equally probable in equilibrium (when S is a maximum), the most probable distribution of particles among the energy states is:

$$\frac{N_j^*}{N} = \frac{g_j e^{-(\epsilon_j/kT)}}{\sum_j g_j e^{-(\epsilon_j/kT)}} \tag{1}$$

independent of the type of statistics applied. The denominator in the exponent, kT, and the identification of k as Boltzmann's constant are a consequence of comparison of the statistical entropy expression and resulting reciprocity relations with the first law of thermodynamics. The quantity in the demominator of Eq. 1 is the sum-over-states (Zustandsumme) or partition function and is often denoted by Q. We see that the various terms in the summation are proportional to the probabilities of the states to which they refer. The partition functions may be found only if the energies ϵ_j and degeneracies g_j are known, and they are determined using quantum mechanics.

Partition Functions

The partition function is defined as

$$Q = \sum_j g_j e^{-(\epsilon_j/kT)}$$

It is assumed that the energies partitioned in the various degrees of freedom are all independent. Therefore the total partition function can be represented as

$$Q = Q_{\text{trans}} Q_{\text{vib}} Q_{\text{rot}} Q_{\text{elect}} \cdots$$

Thus we can look at the individual energy degrees of freedom and determine the Q contribution of each energy mode independently.

In a gas of structureless particles (particles with neither internal degrees of freedom nor interparticle potentials) contained in a rectangular "potential box" of dimensions a_1, a_2, a_3, the allowed energies are:[2]

$$\epsilon_{n_1,n_2,n_3} = \frac{h^2}{8m}\left(\frac{n_1^2}{a_1^2} + \frac{n_2^2}{a_2^2} + \frac{n_3^2}{a_3^2}\right)$$

where all integral n's are allowed and the levels are nondegenerate. The translational partition function is

$$Q_{\text{trans}} = \sum_{n_1=1}^{\infty}\sum_{n_2=1}^{\infty}\sum_{n_3=1}^{\infty} e^{-\epsilon_{n_1,n_2,n_3}/kT} = V\left(\frac{2\pi mkT}{h^2}\right)^{3/2} \quad (2)$$

where h is Planck's constant. In monatomic gas the only other mode of excitation is electronic.

Electronic excitation requires large amounts of energy, and electrons which deviate large distances from their nuclei may fall under the influence of other particles. This depends upon the density of the gas, which provides a density dependent cutoff to the electronic partition function. This is important only at high temperatures since at moderate temperatures the exponential terms in the electronic partition function

$$Q_{\text{elect}} = \sum_{l=j}^{\infty} g_j e^{-(\epsilon_j/kT)} \quad (3)$$

rapidly become small with increasing j. Values for g_j and ϵ_j are given in Ref. 5.

Diatomic gases may possess rotational and vibrational energy as well as translational and electronic energy. The Schroedinger equation appropriate to a rigid rotator requires that the molecule exist in energy levels

$$\epsilon_l = \frac{h^2}{8\pi^2 I} l(l+1)$$

whose degeneracy is $g_l = (2l+1)$. The degeneracy is the number of orientations of a quantized angular momentum vector of length l. Thus the expression for the rotational partition function is

$$Q_{\text{rot}} = \sum_{l=0}^{\infty} (2l+1)\exp\left(\frac{-l(l+1)h^2}{8\pi^2 IkT}\right) \quad (4)$$

The energies of the nondegenerate vibrational levels are

$$\epsilon_{\text{vib}} = (n + \tfrac{1}{2})h\nu$$

For a gas with a finite number of equally spaced vibrational levels below the dissociation limit, it can be shown that

$$Q_{\text{vib}} = \frac{e^{-h\nu/kT}(1 - e^{-D/kT})}{1 - \exp(-h\nu/kT)} \quad (5)$$

In the previous expressions I is the moment of inertia about a centroidal axis, ν is the vibrational frequency, m is the mass of the particle, D the dissociation energy, and all of the electronic energies are properties of the particle in question.[5] They may be obtained from spectroscopic data or from observation of other quantities which depend upon the internuclear or interparticle potentials. Table 1 presents some of these quantities.

Table 1.   Molecular Vibrational and Rotational Data*

Molecule	$1/\lambda$, cm^{-1}	B, cm^{-1}
O_2	1560	1.45
H_2	4400	60.8
CO	2170	1.93
NO	1900	1.70
N_2	2360	1.99
CN	2070	1.90

* $1/\lambda$ = wave number = ν/c. B = band spectral constant = $h/8\pi^2 Ic$.

Thermodynamics in Terms of Partition Functions

Once partition functions are available, all of the thermodynamic properties of a substance may be found. For a gas all of whose modes of excitation are characterized by a single temperature, the appropriate form of the First Law of Thermodynamics is[2]

$$T \, dS = dE + P \, dV - \mu \, dN \tag{6}$$

where E and μ are the internal energy and chemical potential (Gibbs free energy) per particle. The statistical entropy relation leads to

$$S = Nk \left\{ \ln (Q/N) + 1 + T \frac{\partial \ln Q}{\partial T} \right\} \tag{7a}$$

$$E = NkT^2 \frac{\partial \ln Q}{\partial T} \tag{7b}$$

$$P = NkT \frac{\partial \ln Q}{\partial V} \tag{7c}$$

$$\mu = -kT \ln (Q/N) \tag{7d}$$

Note that since only the translational partition function depends upon the volume, V, only translational motion can contribute to hydrostatic pressure. Use of Eq. 2 in Eq. 7c yields the thermally perfect equation of state. Again it is stressed that the treatment outlined here applies only to weakly interacting particles. If the particles interact strongly, the quantum states of the system cannot be described in terms of those of the individual particles.

The principles applied to a system with only one constituent may be used in the study of gas mixtures as well (see Ref. 3).

Nonequilibrium Thermodynamics

Thus far only the thermodynamics of gases in equilibrium has been discussed. Certainly equilibrium exists in many instances, but it must be realized that when physical conditions change abruptly, equilibrium is approached at a finite rate. On the microscopic scale there is a certain number of collisions required among the individual particles for the energies of the various modes of excitation to be equibrated. From previous comments on the activation energies of the various degrees of freedom, one may deduce that for particles whose masses do not differ greatly, translational degrees of freedom are equilibrated first and electronic ones last. Flow with translational nonequilibrium, i.e., with nonuniformity in the distribution of translational energy, is most appropriately dealt with when one considers the Boltzmann equation and approximations thereto. These are discussed in Art. 5. A discussion of rate processes would draw upon these concepts.

Surveys of relaxation processes in gases, i.e., their approach to equilibrium, may be found in Ref. 2 on an elementary level and in Ref. 3 in a more advanced form.

52. REAL GAS DYNAMICS

Governing Equations

The control volume derivation of the phenomenological relations which govern all gas dynamics may be found in Art. 8. The extended thermodynamic and state relations previously mentioned assure the applicability of these equations to completely general situations. The conservation of mass, momentum, and energy are, in the absence of transport phenomena,

$$\frac{D\rho}{Dt} + \rho \operatorname{div} \mathbf{V} = 0 \tag{8}$$

$$\frac{D\mathbf{V}}{Dt} + \frac{1}{\rho} \operatorname{grad} p = 0 \tag{9}$$

$$\rho \frac{Dh}{Dt} - \frac{Dp}{Dt} = - \operatorname{div} \mathbf{q} \qquad (10)$$

where, in the absence of molecular transport phenomena, the heat flux is due only to radiation. For a general system one may allow for N individual species, each with a number of nonequilibrium internal processes denoted by α_{si} for the ith nonequilibrium variable associated with species s. By examining the net rate of flow into a control volume and the rate of progress of each reaction therein, all but translational nonequilibrium phenomena are governed by an equation of the form

$$\frac{D\alpha_{sj}}{Dt} = \frac{\chi_{sj}}{\tau_{sj}} \qquad (11)$$

where χ and τ may depend on all properties of the fluid. The rate of production of species "s," is governed by an equation of this form also. However, since the overall density is $\rho = \sum_{s=1}^{N} \rho_s$ only $(N-1)$ of the species "rate" equations are independent if Eq. 8 is employed. If the molecular masses of the various species differ widely, it must be expected that their translational energies will not be equilibrated with each other. If one assigns the translational temperature T_s to each species, the appropriate equation must be derived from consideration of the work done on the boundaries of the control volume due to normal momentum transfer. The relaxation of translational temperatures is governed by:

$$\rho_s C_{strans} \frac{DT_s}{Dt} + p_s \operatorname{div} \mathbf{V} = \sum_i \frac{\chi_{si}}{\tau_{si}} + \left(\frac{V^2}{2} - e_s\right) \frac{D\rho_s}{Dt} + Q_s \qquad (12)$$

where Q_s is the energy lost from species "s" due to collisional and radiative processes. Note the additional "rate of work" term in Eq. 12 which does not appear in Eq. 11. Since the enthalpy depends upon all of the internal and translational contributions to each of the e_s, one of the equations of the type will not be independent of the others if the overall energy equation is employed. Translational nonequilibrium of this type is important mostly in ionized gases.

Equations 8 through 12, in addition to $(N-1)$ chemical "rate" equations, complete the relations which govern the motion of real gases. If radiation is included in the formulation, then the radiative transfer equation (see Art. 53) is required as well. The boundary conditions for the fluid mechanics are those normally employed. If the nonequilibrium parameters are specified in the undisturbed medium, it can be proved that data cannot be prescribed for them if diffusion is not allowed. The fluid mechanics and nonequilibrium chemistry are, in most cases, intimately bound together.

Throughout this discussion, when a choice of two variables to specify the state of a gas is to be made, we shall choose pressure and temperature. Experience dictates that density is an inconvenient quantity to deal with. One reason is that the density hardly varies behind a strong shock wave, hence if density and temperature are employed it is difficult to obtain accurate pressure distributions. Furthermore, temperature must be retained if radiative processes are included.

At this point the stage is set for the study of interesting real gas problems such as the structure of shock waves of the nonequilibrium flow in nozzles. These and other real gas problems are discussed in detail in Ref. 2.

BIBLIOGRAPHY

1. PRIGOGINE, I., *Introduction to Thermodynamics of Irreversible Processes*, New York, Wiley, 2nd ed., 1962.
2. VINCENTI, W. G., and KRUGER, C. H., *Introduction to Physical Gas Dynamics*, New York, Wiley, 1965.
3. CLARKE, J. F., and McCHESNEY, M., *The Dynamics of Real Gases*, London, Butterworths, 1964.
4. BOND, J. W., WATSON, K. M., and WELCH, J. A., *Atomic Theory of Gas Dynamics*, Reading, Massachusetts, Addison-Wesley, 1965.
5. LEE, J. F., SEARS, F. W., and TURCOTTE, D. L., *Statistical Thermodynamics*, Reading, Massachusetts, Addison-Wesley, 1963.

RADIATION GASDYNAMICS

By C. A. Forbrich, Jr.

53. GENERAL FEATURES OF RADIATION GASDYNAMICS

High temperature gases emit energy in the form of electromagnetic radiation resulting from rotational, vibrational, and electronic energy level transitions in the gas atoms. When radiative energy transfer occurs, the motion of the radiating gas may be affected, i.e., if the gas is at rest initially it may have some velocity after the energy transfer. Radiation gasdynamics is the study of this interaction between the radiative transfer occurring and the motion of the gas.

In the analysis of radiative gasdynamics, it is assumed that the radiation field is composed of photon particles having energy $h\nu$ moving in various directions all at a speed c, the speed of light. The radiation field interacts with the molecular field (the gas) through the emission and absorption of radiation, and the motion of the gas. When the energy absorbed by a gas is just equal to the energy emitted at every frequency, then the gas is in radiative equilibrium. When the energy absorbed is not equal to that emitted, the gas is in radiative nonequilibrium. Non equilibrium effects can be significant and must be considered in many cases.

As a first step in studying the interaction between the gas motion and the radiative energy transfer, it is necessary to write a conservation equation for the photons so that the rate of energy transfer due to photon fluxes can be determined. Such an equation is given by [4]

$$\frac{1}{c}\frac{\partial I_\nu}{\partial t} + l_j \frac{\partial I_\nu}{\partial x_j} = \rho(j_\nu - K_\nu I_\nu) \tag{1}$$

which is the equation of radiative transfer, with I_ν the specific radiation intensity, l_j the direction cosine, ρ the mass density, j_ν the mass emission coefficient, and K_ν the mass absorption coefficient. Thus the specific radiation intensity I_ν is governed by a first-order partial differential equation which in principle can be solved to give $I_\nu(r_i, l_i, t)$ explicitly.

Emission and Absorption of Radiation

Radiative transfer is essentially a microscopic phenomena. It is possible to consider microscopic processes alone to write an explicit expression for the right-hand side of Eq. 1 in terms of the actual atomic transitions which occur.

There are three types of transitions to be considered related to atomic or molecular structure of the gas. The more complicated molecular transitions are excluded here for simplicity. The first of these possible transitions is the *bound-bound* transition which occurs between two bound energy states of the atom. These transitions are characterized by discrete spectral line emission or absorption. A *bound-free* transition occurs when an atom having an electron in a bound state is so highly excited that the atom is ionized and the electron is free. A transition of this type is characterized by continuous spectral radiation emission (or absorption). *Free-free* transitions occur as a result of interactions between two free elecrtons in the gas or by the interaction of a free electron and an onized particle and are charcterized by continuous radiation.

To describe the term on the right-hand side of Eq. 1 in terms of microscopic processes, only *bound-bound* transition will be considered. In this case *Einstein transition probability coefficients* can be introduced to characterize the radiation field. There are three Einstein transition probability coefficients: The Einstein coefficient for spontaneous radiation emission A_{mn}, defined to be the probability per unit time for a spontaneous transition from state n to state m to occur for a single atom; B_{mn}, the probability per unit time for an absorption to occur for a single atom resulting in transition from state m to state n; and B_{nm}, defined to be the probability per unit time for induced emission to occur for a single atom resulting in a transition from state n to state m.

The Einstein coefficients are not independent and can be shown to be related at equilibrium in the following manner [4]

$$\frac{B_{mn}}{B_{nm}} = 1 \qquad \frac{A_{nm}}{B_{nm}} = \frac{2h\nu^3}{c^2} \tag{2}$$

where h is Planck's constant and ν is the radiation frequency. Letting n equal the number of atoms in a particular state, the mass emission term ρj_ν on the right-hand side of Eq. 1 is equivalent to $n_n(A_{nm} + B_{nm}I_\nu)h\nu$, which gives the rate at which radiative energy is emitted, and the mass absorption term $\rho K_\nu I_\nu$ is equivalent to $n_m N_{mn}I_\nu h\nu$, which gives the rate at which energy is absorbed. The assumption of local thermodynamic equilibrium (LTE) can be used to further simplify the equation of radiative transfer. LTE is characterized by the gas atoms having local rotational equilibrium (thermodynamic temperature) not necessarily in equilibrium with the radiation field. This can occur because the population of the atomic states locally is governed by atomic collisions resulting in an equilibrium distribution of states corresponding to a local temperature which may vary from point to point in the gas. With the assumption of LTE there is a Boltzmann distribution of equilibrium states corresponding to the local temperature and the equation of radiative transfer becomes

$$\frac{1}{c}\frac{\partial I_\nu}{\partial t} + l_j\frac{\partial I_\nu}{\partial x_j} = \rho K_\nu \left[1 - \exp\left(-\frac{h\nu}{kT}\right)\right][B_\nu - I_\nu]$$

where B_ν is defined as the Planck function representing the equilibrium specific radiation intensity given vy

$$B_\nu \equiv \text{Planck function} \equiv \frac{2h\nu^3/c^2}{\exp\{h\nu/kT\} - 1}$$

Formal Solution of the Equation of Radiative Transfer

The equation of radiative transfer can be rewritten in terms of the volume absorption coefficient α_ν as

$$\frac{1}{c}\frac{\partial I_\nu}{\partial t} + l_j\frac{\partial I_\nu}{\partial x_j} = -\alpha_\nu(I_\nu - B_\nu) \tag{3}$$

where

$$\alpha_\nu \equiv \rho k_\nu \left[1 - \exp\left(-\frac{h\nu}{kT}\right)\right]$$

To solve Eq. 3 the time derivative term is neglected. This can be done because characteristic velocities (not necessarily the flow velocity) in radiative gasdynamics problems are small compared to the speed of light c.

If a coordinate r is chosen such that it is measured in a direction opposite to the direction of radiation propagation as shown in Fig. 1, then

$$l_j\frac{\partial I_\nu}{\partial x_j} = -\frac{\partial I_\nu}{\partial r}$$

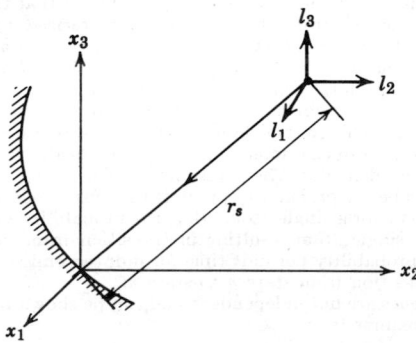

FIG. 1.   Radiation field coordinate geometry.

Under these conditions the equation of radiative transfer becomes

$$\frac{\partial I_\nu}{\partial r} = \alpha_\nu (I_\nu - B_\nu)$$

This equation can be formally integrated to obtain

$$I_\nu(r) = I_\nu(r_s) \exp\{ - \tau_\nu(r_s)\} + \int_0^{r_s} \alpha_\nu B_\nu(r) \exp\{ - \tau_\nu(r)\}\, dr \qquad (4)$$

where $\tau_\nu(r) \equiv \int_0^r \alpha_\nu\, dr$ is defined as the optical depth and $r = 0$ is the point in question. In this formal solution the first term on the right-hand side is due to the contribution of the boundary attenuated by the net absorption from the boundary to the point in question, while the integral term accounts for self-emission, absorption, and attenuation within the gas.

It remains to evaluate the radiation heat-addition term which will appear in the energy equation of a radiative gas dynamics problem. The heat addition is given by the negative divergence of the radiative heat flux vector $q_j{}^R$. The heat-addition term is given by

$$-\frac{\partial q_j{}^R}{\partial x_j} = \int_0^\infty \alpha_\nu \left(\int_0^{4\pi} I_\nu\, d\Omega - 4\pi B_\nu \right) d\nu \qquad (5)$$

In certain cases approximations to this formal solution are possible to obtain simplified results. Some of these approximations and simplifications will now be discussed with the radiation heat-addition terms evaluated.

Emission-Dominated Approximation. Many physical situations exist in which the specific intensity I_ν is much less than its equilibrium value B_ν. This is the case when the energy being emitted from the gas is much larger than that being absorbed, so the gas is far from radiative equilibrium. When $I_\nu \ll B_\nu$, the equation of radiative transfer is

$$\frac{\partial I_\nu}{\partial r} = -\alpha_\nu B_\nu$$

which can be readily integrated to give

$$I_\nu(r) = \int_0^{r_s} \alpha_\nu B_\nu\, dr$$

in which the radiation contribution resulting from radiating boundaries $I_\nu(r_s)$ is neglected. The radiation heat-addition term $\partial q_j{}^R/\partial x_j$ is given by

$$-\frac{\partial q_j{}^R}{\partial x_j} = -4\pi \int_0^\infty \alpha_\nu B_\nu\, d\nu = -4\alpha_P \sigma T^4$$

where α_P is the Planck mean absorption coefficient defined by

$$\alpha_P(p, T) \equiv \frac{\int_0^\infty \alpha_\nu B_\nu\, d\nu}{\int_0^\infty B_\nu\, d\nu} = \frac{\pi}{\sigma T^4} \int_0^\infty \alpha_\nu B_\nu\, d\nu$$

Optically Thin Gas Approximation. The optically thin gas is characterized by the condition that the optical depth τ_ν is small compared to unity, i.e., $\tau_\nu \ll 1$. In this case the general solution to the radiation equation becomes

$$I_\nu(r) = I_\nu(r_s) + \int_0^{r_s} \alpha_\nu B_\nu(r)\, d\nu$$

This equation states that the specific intensity at the point r is given by the specific intensity $I_\nu(r_s)$ at the radiating boundary plus the unattenuated sum of all emission along the direction of propagation. The relation between the optically thin gas the the emission-dominated gas can be given by the further condition that

$$I_\nu(r) = I_\nu(r_s) + \int_0^{r_s} \alpha_\nu B_\nu(r)\, dr \ll B_\nu$$

If $B_\nu(r) \cong B_\nu$ at $r = 0$, then this relationship reduces to a condition on $I_\nu(r_s)$ compared to B_ν since the optically thin assumption requires the integral term to be small, leading again to the condition $I_\nu(r_s) \ll B_\nu$. So it is seen that the assumption of a gas being optically thin is a necessary but not sufficient condition for the radiation to be emission dominated. In the case of an optically thin, emission-dominated gas, the radiation heat-addition term is the same as that for the emission-dominated gas.

Optically Thick Gas Approximation. An optically thick gas is one in which the optical depth τ_ν is large compared to unity, i.e., $\tau_\nu \gg 1$, for all frequencies. When the point in the gas under consideration is far enough from any radiating boundaries, the radiation contribution due to the boundaries $I_\nu(r_s)$ is neglected and the formal solution to the equation of radiative transfer at $r = 0$ is

$$I_\nu(0) = \int_0^{r_s} \alpha_\nu(r) B_\nu(r) \exp\{-\tau_\nu(r)\}\, dr$$

Expanding $B_\nu(r)$ in terms of a Taylor series expansion about $r = 0$, the integral can be evaluated yielding for the solution

$$I_\nu(0) = B_\nu(0) \left[1 + \frac{1}{\alpha_\nu(0) B_\nu(0)} \left(\frac{dB_\nu}{dt}\right)_{r=0} \left(\frac{\partial T}{\partial r}\right)_{r=0} + \cdots \right]$$

This result indicates that $I_\nu(0)$ is made up of the equilibrium radiation at the point of interest $B_\nu(0)$ plus a small additional amount due to emission from the gas in the immediate neighborhood of the point of interest. The optically thick condition exists when I_ν differs only a slight amount from B_ν, thus the departure from radiative equilibrium is small as opposed to the cases of emission-dominated and optically thin gases. It should be mentioned that the optically thick gas is somewhat unrealistic because a gas cannot be thick at all frequencies if it radiates primarily in lines or bands.

The radiation heat addition term for the case of the optically thick gas can be shown (see Ref. 4, page 469) to be given by

$$-\frac{\partial q_j{}^R}{\partial x_j} = \frac{16\sigma T^3}{3\alpha_R} \frac{\partial^2 T}{\partial x_j \partial x_j}$$

where α_R is the Rosseland mean absorption coefficient defined as

$$\alpha_R \equiv \frac{\displaystyle\int_0^\infty \frac{\partial B_\nu}{\partial T}\, d\nu}{\displaystyle\int_0^\infty \frac{1}{\alpha_\nu} \frac{\partial B_\nu}{\partial T}\, d\nu} = \frac{4\sigma T^3}{\pi \displaystyle\int_0^\infty \frac{1}{\alpha_\nu} \frac{\partial B_\nu}{\partial T}\, d\nu}$$

Grey-Gas Approximation. The grey-gas approximation is based on the ad hoc assumption that there exists some unspecified, frequency-averaged absorption coefficient $\alpha(p, T)$ which remains a function of the gas pressure p and temperature T. This assumption does not come from a logical theoretical limit as do the previous assumptions; however, it provides a means of studying radiative gasdynamics from the pedagogical viewpoint and eliminates in the most part extensive use of computers for solutions. In this case the formal solution to the equation of radiative transfer is

$$I(r) = I(r_s) \exp\{-\tau_\nu(r_s)\} + \frac{\sigma}{\pi} \int_0^{r_s} \alpha T^4 \exp\{-\tau_\nu(r)\}\, dr$$

where $I \equiv \int_0^\infty I_\nu\, d\nu$. The radiation heat-addition term is given by

$$-\frac{\partial q_j{}^R}{\partial x_j} = \int_0^{4\pi} \frac{\partial I}{\partial r}\, d\Omega = \alpha \left[\int_0^{4\pi} I\, d\Omega - 4\sigma T^4 \right]$$

where σ is the Stephan-Boltzmann constant and $d\Omega$ is the differential solid angle subtended.

Other Approximations. There exist other approximate solutions to the equation of radiative transfer. They will be only mentioned here; however, in many cases they are the most important approximations in the field since they can more accurately describe a realistic physical situation. Among these approximations are:

(a) *The one-dimensional approximation*, in which the specific intensity I_ν is a function of one space coordinate only. In many cases the integral expression in the radiation heat-addition term can be evaluated and the full problem solved explicitly.

(b) The *exponential approximation*, which arises in the case of a grey gas in the one-dimensional approximation. In this case the expression for the radiation heat-addition term reduces to exponential integrals which can be evaluated by approximate exponential functions. This approximation has been used to advantage many times in one-dimensional radiative gasdynamics and astrophysics.[6]

(c) The *differential approximation*, which is a generally applicable method used in three-dimensional problems including the case of nonisotropic radiation. This approximation is concerned with satisfying certain moments of the equation of radiative transfer and appropriately truncating the assumed series solution usually in terms of spherical hormonics.[3] This method produces a general result which yields the other approximations, i.e., optically thin, optically thick, etc., with an appropriate choice of parameters arising in the solution as discussed in Vincenti and Kruger.[4]

54. BASIC NONLINEAR EQUATIONS OF RADIATIVE GASDYNAMICS

The conservation equations of radiative gasdynamics will now be presented. In doing this the radiation energy density and pressure will be neglected, being insignificant except in cases of astrophysical interest and nuclear explosions. In this case the governing equations in a radiative gasdynamics problem are:

$$\frac{D\rho}{Dt} + \rho \frac{\partial u_j}{\partial x_j} = 0 \tag{6a}$$

$$\rho \frac{Du_i}{Dt} + \frac{\partial p}{\partial x_i} = 0 \tag{6b}$$

$$\rho \frac{Dh}{Dt} - \frac{Dp}{Dt} = -\frac{\partial q_j{}^R}{\partial x_j} \tag{6c}$$

where h is the enthalpy and p is the pressure and $D(\ )/Dt$ is the substantial derivative defined by $D(\ )/Dt = \partial(\ )/\partial t + u_j[\partial(\ )/\partial x_j]$. Assuming that the gas is perfect, an equation of state can be written in the form

$$h = \frac{\gamma}{\gamma - 1} \frac{p}{\rho} \tag{6d}$$

where p is given by the ideal gas law $p = \rho RT$, with R the ordinary gas constant and γ the ratio of specific heats assumed constant. The radiation heat-addition term on the right-hand side of the energy equation is given by Eq. 5 which has been evaluated in several asymptotic limits and for certain simplified cases earlier.

These equations constitute a set of eleven scalar equations with eleven unknowns ρ, u_i, p, h, $q_j{}^R$, T, and I_ν. They are a set of coupled nonlinear integro-differential equations to which no closed form solution exists.

55. LINEARIZED EQUATIONS OF RADIATIVE GASDYNAMICS

Linearization of the complete set of conservation equations governing a problem in radiation gasdynamics is made relative to a uniform gas at rest. Thus it is possible to write $u = u_i'$, $p = p_0 + p'$, $T = T_0 + T'$, $q_i{}^R = q_{i0}{}^R + q_i{}^{R'}$ etc., where the subscript 0 represents uniform conditions in an undisturbed gas and primed quantities denote small perturbations of the particular quantity of interest. Substituting into the conversion equations, a set of linearized equations is obtained of the form

$$\frac{\partial \rho'}{\partial t} + \rho_0 \frac{\partial u_j'}{\partial x_j} = 0 \tag{7a}$$

$$\rho_0 \frac{\partial u_i'}{\partial t} + \frac{\partial p'}{\partial x_i} = 0 \tag{7b}$$

$$\rho_0 \frac{\partial h'}{\partial t} - \frac{\partial p'}{\partial t} = - \frac{\partial q_j^{R'}}{\partial x_j} \tag{7c}$$

$$- \frac{\partial q_j^{R'}}{\partial x_j} = \alpha_0 [I_A' - 16\sigma T_0{}^3 T'] \tag{7d}$$

$$\frac{\partial I'_A}{\partial x_i} = - 3\alpha_0 q_i^{R'} \tag{7e}$$

Introducing the velocity potential ϕ, the set of equations can be combined into one fifth-order partial differential equation

$$\frac{\partial^3 A_s}{\partial t \partial x_k \partial x_k} + \frac{16}{B_0} \alpha_0 a_{s0} \frac{\partial^2 A_T}{\partial x_k \partial x_k} - 3\alpha_0 \frac{\partial A_s}{\partial t} = 0 \tag{8}$$

where

$$A_s = \frac{1}{a_{s0}} \frac{\partial^2 \phi}{\partial t^2} - \frac{\partial^2 \phi}{\partial x_j d x_j}$$

$$A_T = \frac{1}{a_{T0}} \frac{\partial^2 \phi}{\partial t^2} - \frac{\partial^2 \phi}{\partial x_j \partial x_j}$$

$$a_{s0} = \sqrt{\gamma R T} = \text{isentropic speed of sound}$$

$$a_{T0} = \sqrt{R T} = \text{isothermal speed of sound}$$

and

$$B_0 = \frac{\gamma R \rho_0 a_{s0}}{(\gamma - 1)\sigma T_0{}^3} = \text{Boltzmann number}$$

The Boltzmann number B_0 is the most important dimensionless parameter in radiation gasdynamics, providing a measure of the relative importance of convective and radiative process in the energy flux.

The two sets of Eqs. 6 and 7, which govern respectively a generalized gasdynamic problem and a linearized radiation gasdynamics problem, can now be solved with the appropriate boundary conditions for a large range of problems. Particular solutions to these equations will not be presented here because any such treatment would by its brevity do an injustice to the solutions which are readily available in the literature and textbooks. A comprehensive list of references with solutions to some of the classical problems of fluid mechanics, including radiation effects, can be found in Refs. 1, 2, 4, and 5.

BIBLIOGRAPHY

1. BIBERMAN, L. M., IAKUBOU, I. T., NORMAN, G. E., and VOROBYOV, V. S., "Radiation Hearing under Hypersonic Flow," *Astronaut. Acta*, Vol. 10, p. 238, 1964.
2. CESS, R. D., and SPARROW, E. M., *Radiation Heat Transfer*, Belmont, California, Brooks/Cole, 1966.
3. CHENG, P., "Study of the Flow of a Radiating Gas by a Differential Approximation," Ph.D. Dissertation, Stanford University, 1965.
4. VINCENTI, W. G., and KRUGER, C. H., JR., *Introduction to Physical Gas Dynamics*, New York, Wiley, 1965.
5. ZEL'DOVICH, YA., B., and RAIZER, YU. P., *Physics of Shock Waves and High Temperature Hydrodynamic Phenomena*, New York, Academic, 1966.
6. VINCENTI, W. G., and BALDWIN, B. S., "Effect of Thermal Radiation On the Propogation of Plane Acoustic Waves," *J. Fluid, Mech.*, Vol. 12, Part 3, p. 449, 1962.

MAGNETOGASDYNAMICS

By C. A. Forbrich, Jr.

56. CHARGED PARTICLE MOTION IN A PLASMA

Magnetogasdynamics deals with a unique type of matter called a plasma. A plasma is a fluid in which there are charge carriers but which as a whole is electrically neutral because the numbers of positive and negative charges are equal. The presence of the charge carriers renders the plasma electrically conducting. Generally a gaseous plasma is created by the ionization of some or all of the gaseous molecules by collisional processes.

To study the interaction of a plasma with electric or magnetic fields it is necessary to consider certain aspects of the microscopic motion of the particles composing the plasma in order to understand the macroscopic phenomena observed. On a microscopic scale, certain regions of a plasma may have large deviations from electrical neutrality as a result of applied electrical or magnetic fields. If the mean kinetic energy of a particle due to its thermal motion is compared to the potential energy of the field produced by the ionized particles, a characteristic length is revealed, giving the relationship[5,7]

$$h \equiv \sqrt{\frac{kT}{4\pi n e^2}} \equiv \text{Debye length}$$

where k is the Boltzmann constant, n the number density of electrons, e the electron charge, and T the absolute temperature. The Debye length h is a measure of the extent of deviation from electrical neutrality. Assuming the microscopic region of gas being studied can be enclosed in a sphere of radius L, then if $h \ll L$, the gas can be considered as electrically neutral and the region of gas can be studied using a macroscopic description. The condition $h \ll L$ defines an ionized gas as a plasma. If $h \geq L$, then the gas cannot be considered electrically neutral and can be studied only from a microscopic point of view. For gas dynamic considerations it is most convenient to study the coupling action between a magnetic field and a plasma using the macroscopic approach (the condition $h \ll L$ being implied) so the conventional equations of macroscopic gasdynamics can be used when appropriately modified.

If a magnetic field $\mathbf{B}$ is impressed on a plasma, the charged particles will spiral around the magnetic lines of force. The radius of gyration r of the charged particles about the magnetic force lines is given by

$$r = \frac{3kTm}{e|\mathbf{B}|}$$

where m is the charged mass, ion, or electron, all other terms having been previously defined. The electron obviously has a smaller radius of gyration r_e than the more massive ion radius of gyration r_i. The ion and electron rotate about the magnetic force lines in opposite directions, which can be deduced from the relation above for the radius of gyration, since the electron charge is $-e$ and that of an ion is $+e$ for singly ionized particles.

If, in addition to the external magnetic field $\mathbf{B}$, an external electric field $\mathbf{E}$ is applied to a plasma, the charged particles drift sideways in the direction $\mathbf{E} \times \mathbf{B}$ rather than following the lines of force. This effect is called $\mathbf{E} \times \mathbf{B}$ drift. The electrical conductivity of the plasma will vary, with and without $\mathbf{E} \times \mathbf{B}$ drift, having in general a tensor character as a result of the various drifts which occur.[3,5]

57. MAGNETOGASDYNAMIC EQUATIONS OF MOTION

Having established the condition that the dimensions of the sphere of radius L under study are large compared to the Debye length h, a macroscopic description of a plasma is justified. The electromagnetic character of a plasma is governed by Maxwell's equations although displacement currents are usually neglected in laboratory plasma physics studies. The fluid motion is governed by the usual conservation equations of gasdynamics however, with additional terms in the momentum and energy equations. The equation for the conservation of mass remains unchanged. In the momentum equation, the external forces must include the Lorentz force density $\mathbf{j} \times \mathbf{B}$, where $\mathbf{j}$ is the electrical current in the field. In place of this term the Maxwell stress tensor may also be used.[3]

The Maxwell stress tensor used in the momentum equation leads directly to a Bernoulli equation for magnetogasdynamics contributing a new term known as the magnetic pressure,[7] given by $B^2/8\pi$. The energy equation must include additional energy terms accounting for the electromagnetic work and Joule heating occurring in the system. These considerations lead to a complex set of coupled differential equations given by:

Maxwells Equations

$$\operatorname{curl} \mathbf{H} = \mathbf{j} + \epsilon \frac{\partial \mathbf{D}}{\partial t}$$

$$\operatorname{div} \mathbf{B} = 0$$

$$\operatorname{curl} \mathbf{E} = -\frac{\partial \mathbf{B}}{\partial t}$$

$$\operatorname{div} \mathbf{D} = \omega$$

Phenomenological Equations

$$\mathbf{D} = \epsilon \mathbf{E}$$

$$\mathbf{B} = \mu \mathbf{H}$$

$$\mathbf{j} = \sigma(\mathbf{E} + \mathbf{V} \times \mathbf{B})$$

$$\sigma = \sigma(T,p)$$

Fluidmechanical Equations

Mass: $\quad \dfrac{\partial \rho}{\partial t} + \operatorname{div} (\rho \mathbf{V}) = 0$

Momentum: $\rho \dfrac{D\mathbf{V}}{Dt} = -\operatorname{grad} p + \operatorname{div} (\tau_{ij} + T_{ij})$

Energy: $\quad \rho \dfrac{D[h + (V^2/2)]}{Dt} = \dfrac{\partial p}{\partial t} + \mathbf{j} \cdot \mathbf{E} + \operatorname{div} \mathbf{V} \tau_{ij} - \operatorname{div} \mathbf{q}$

State: $\quad p = (1 + \alpha) NkT = \rho(1 + \alpha)RT; \ \alpha = \alpha(p,T)$

Heat: $\quad \mathbf{q} = -K\boldsymbol{\nabla}T$

where α is the degree of ionization, K the heat conductivity, T_{ij} the Maxwell stress tensor, and τ_{ij} the viscous stress tensor, with all other terms in the fluidmechanical equations having their usual meaning. In the electromagnetic and phenomenological equations, $\mathbf{H}$ is the magnetic field strength, $\mathbf{j}$ the electric current, ϵ the dielectric constant, $\mathbf{D}$ the displacement current, $\mathbf{B}$ the magnetic induction, $\mathbf{E}$ the electric field strength, ω the charge density, μ the magnetic permeability, and σ the electrical conductivity. Transforming these equations by eliminating $\mathbf{j}$ and $\mathbf{E}$ in favor of ρ, $\mathbf{V}$, p, and $\mathbf{B}$ as variables, the governing equations of magnetogasdynamics become

Mass: $\quad \dfrac{D\rho}{Dt} + \rho \operatorname{div} \mathbf{V} = 0$

Momentum: $\rho \dfrac{D\mathbf{V}}{Dt} = -\operatorname{grad} p - \operatorname{grad} \left(\dfrac{B^2}{8\pi\mu}\right) + \dfrac{1}{4\pi\mu} (\mathbf{B} \cdot \operatorname{grad}) \mathbf{B}$

Induction: $\dfrac{D\mathbf{B}}{Dt} + \mathbf{B} \operatorname{div} \mathbf{V} - (\mathbf{B} \cdot \operatorname{grad}) \mathbf{V} = \dfrac{c^2}{4\pi\mu\sigma} \operatorname{grad} \operatorname{div} \mathbf{B}$

State: $\quad \dfrac{D}{Dt}\left(\dfrac{p}{\rho^\gamma}\right) = 0$

with the condition that $\operatorname{div} \mathbf{B} = 0$ and where $D(\ )/Dt = \partial(\ )/\partial t + \mathbf{V} \cdot \operatorname{grad} (\ )$. Thus this set of differential equations describes the plasma gas motion when appropriate boundary conditions are applied.

Applications

A classic example of a magnetofluidynamic flow situation is the Hartmann problem,[5] where an incompressible conducting fluid in a channel is subjected to external magnetic and electric fields. This problem points out many of the fundamental interaction phenomena between a plasma and external fields. The results show that the velocity profile becomes flatter and the velocity gradient near the channel wall becomes steeper as the magnetic field strength is increased.

A plasma can be accelerated to very high velocities by the Lorentz force resulting from crossed electric and magnetic fields.[5]   Such devices, sometimes called Hall accelerators, are used as propulsion devices (see Art. 25) and high-speed wind tunnels (see Art. 27).

It is possible to "push" the ionized bow shock away from the nose of a re-entry vehicle[4] and to balance an electric arc in a supersonic flow[8] by the proper orientation and magnitude of external magnetic fields.   These and other interesting examples are found in the references.

58. ALFVÉN WAVES

The wave phenomena in magnetogasdynamics is a very complicated study except in the simplest of cases.   The classical wave analysis is that of Alfvén who studied plasma waves in an incompressible plasma assuming an applied magnetic field **B** perturbing the plasma slightly in one direction, say z.   The assumption of incompressibility reduces the continuity equation to div **V** = 0.   Further assuming the fluid is inviscid having infinite electrical conductivity σ, the governing equations of magnetogasdynamics can be linearized to obtain

$$Mass: \quad \frac{\partial v_i}{\partial x_i} = 0$$

$$Momentum: \quad \frac{\partial^2}{\partial x_i \partial x_i}\left[\frac{p}{\rho} + \frac{B^2}{8\pi\rho}\right] = 0$$

$$Energy: \quad \frac{\partial v_i}{\partial t} = \frac{B_0}{4\pi\rho}\frac{\partial b_i}{\partial z} - \frac{\partial}{\partial x_i}\left[\frac{p}{\rho} + \frac{B^2}{8\pi\rho}\right]$$

$$Induction: \quad \frac{\partial b_i}{\partial t} = B_0 \frac{\partial v_i}{\partial z}$$

Taking the gradient of the energy equation and combining it with the induction equation results in an equation in terms of one unknown function b_i or v_i:

$$\frac{\partial^2 v_i}{\partial t^2} = V_a^2 \frac{\partial^2 v_i}{\partial z^2}$$

$$\frac{\partial^2 b_i}{\partial t^2} = V_a^2 \frac{\partial^2 b_i}{\partial t^2}$$

where $V_a^2 = B_0^2/4\pi\rho = $ (Alfvén speed)^2.   These are typical wave equations describing transverse waves traveling at a speed $+ V_a$ in the direction of the applied magnetic field **B**$_0$.   These waves are called Alfvén waves and V_a is the Alfvén wave speed. It is emphasized that this is the simplest type of wave motion possible under rather restrictive assumptions. It can be shown by a slightly more complicated analysis that Alfvén waves are damped by considering a viscous fluid having a finite electrical conductivity.

If the fluid is compressible but nonviscous having an infinite electrical conductivity σ, the magnetogasdynamic equations reveal that there is a combination of transverse Alfvén waves and the standard longitudinal sound waves of gasdynamics.   If the analysis is carried out in full, the wave speed at some angle θ with respect to the applied magnetic field **B**$_0$ is given by Thompson[7] as

$$V_\pm{}^2 = \tfrac{1}{2}\left[(a_s{}^2 + V_a{}^2) \pm \sqrt{(a_s{}^2 + V_a{}^2)^2 - 4V_a{}^2 a_s{}^2 \cos^2\theta}\right]$$

where $V_\pm$ are fast (+) or slow (−) waves, a_s is the wave speed of sound, and V_a is the Alfvén wave speed.   This equation can be plotted to show the magnitude of the fast and slow waves as given in Fig. 1.

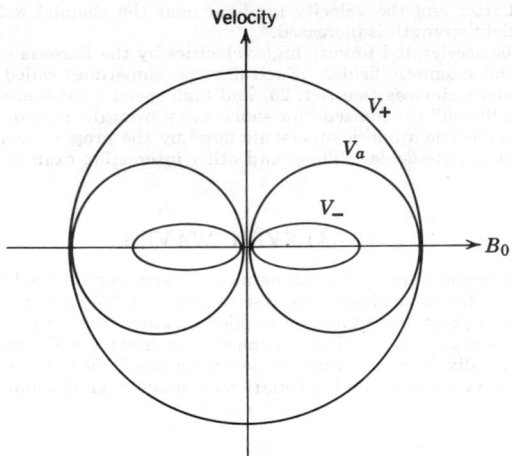

FIG. 1. Magnetofluidmechanical waves.

BIBLIOGRAPHY

1. ALFVÉN, H., and FALTHAMMAR, G. G., *Cosmical Electrodynamics*, New Jersey, Oxford Univ. Press, 1963.
2. CAMBEL, A., *Plasma Physics and Magnetofluidmechanics*, New York, McGraw-Hill, 1963.
3. LONGMIRE, C., *Elementary Plasma Physics*, New York, Wiley, 1963.
4. PARKER, E. N., *Interplanetary Dynamical Processes*, New York, Wiley, 1963.
5. SUTTON, G. W., and SHERMAN, A., *Engineering Magnetohydrodynamics*, New York, McGraw-Hill, 1965.
6. STIX, T., *The Theory of Plasma Waves*, New York, McGraw-Hill, 1962.
7. THOMPSON, W. B., *An Introduction to Plasma Physics*, Reading, Massachusetts, Addison-Wesley, 1962.
8. NICOLAI, L. M., "An Experimental and Theoretical Analysis of the Convected Balanced Arc," NASA CR-1267, February 1969.

MODERN CONTROL THEORY

By R. E. Willes

59. MODERN CONTROL THEORY

Modern control theory addresses the problems of multi-input-output, nonlinear, time varying, stochastic systems particularly common to aerospace vehicles.

Optimal Control Theory

It is natural to seek controllers for systems that satisfy objectives in an optimal way. The analytical and numerical definitions of such controls is the subject of optimal control theory.[1,5]

A large number of engineering systems in which stochastic considerations are not of dominant importance can be described by an ordinary differential equation of the form

$$\dot{x} = f(x,u)$$

where x is a vector variable describing the state of the system, u is a vector of control variables, and $f(x,u)$ is a possibly nonlinear vector function of x and u. In terms of such a system, most optimal control problems may then be formulated as follows: find the admissible control that drives the system from some initial $x_{(t_0)}$ to a particular target while minimizing the first component of the state vector, $x_{0(t_f)}$ at the final time, t_f. The necessary conditions that such a control must satisfy are obtained from the Minimum principle of Pontryagin.[3]

From the set of all admissible controls, U, the optimal control, $u(t)$, must minimize the Hamiltonian, $H(x,u,\lambda)$, defined as

$$H \equiv \lambda^T_{(t)} f(x_{(t)}, u_{(t)})$$

for all t between t_0 and t_f. The adjoint, or costate vector, $\lambda(t)$, is required to be normal to the target surface at t_f and is defined by the differential equation

$$\dot{\lambda}^T = -\frac{\partial H}{\partial x}$$

Further, H is constant and if t_f is not specified, then $H = 0$. In some cases the optimal control can be identified analytically, see Fig. 1, but numerical techniques are usually required.

Numerical Methods

Methods of obtaining numerical solutions for the optimal control basically involve initially satisfying some of the conditions of the Pontryagin minimum principle and iterating numerically until the others are satisfied. A straightforward method is to guess initial values of the costate variables and integrate the state and costate equations forward in time, using the control, u, that minimizes $H(x,\lambda,u)$. The target surface and the normal condition on the costate vector will be missed, but in principle can eventually be satisfied by observing how variations in the initial costate change final values of the state and costate.[4]

A less intuitive but more widely used numerical method is the Method of Steepest Descent.[5] Starting with a control that causes the system to nearly hit the target, the adjoint equations may be integrated backwards using the final condition

$$\lambda_{(t_f)} = \delta x_{(t_f) \text{ desired}}$$

where $\delta x_{(t_f) \text{ desired}}$ is the required change in the final state to be produced on the next iteration. The change in the control for the next iteration is then

$$\delta u_{(t)} = K \left(\frac{\partial f}{\partial u}\right)^T \lambda_{(t)}$$

where K is a constant "step size" that must be judiciously chosen between the limits of 0 and 1. This iteration procedure is continued until the boundary conditions on the problem are satisfied and an acceptable decrease in the cost component of the state vector has been obtained.

Dynamic Programming

To allow for physical implementation of the control in a "feedback" system, it is desirable to generate the optimal control as a function of the current state of the system, viz.,

$$u_{(t)} = u(x_{(t)})$$

in the manner illustrated in Fig. 1. Dynamic programming is an optimal control technique that generates this map in a straightforward, but computationally difficult, manner. It seeks solution to the Hamilton-Jacobi-Bellman[7] partial differential equation

$$\frac{\partial x_0}{\partial t} + \frac{\min}{u} \frac{\partial x_0}{\partial x} f(x,u) = 0$$

for the minimum cost remaining, $x_0(x,t)$ as a function of the current state and time. This subsequently generates the optimal control as a function of current state. This equation is generated by asking what is the optimal way to get to the target from a position in the near vicinity of the target, and then to repeat this question from states farther and farther from the target. Solution of this partial differential equation directly is always computationally expensive for complex systems. The standard Method of Characteristics

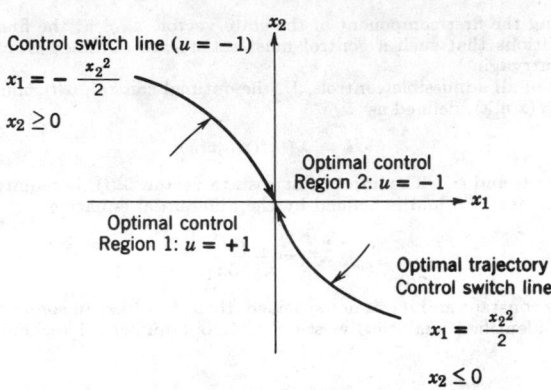

FIG. 1.   Phase plane representation of the optimal trajectories and the optimal control, $u(x_1,x_2)$, for for the system $x_0 = 1$, $\dot{x}_1 = x_2$, $\dot{x}_2 = u$, $|u| - 1$, and target $x_1 = 0$, $x_2 = 0$.

solution to this equation yields some of the conditions of the Pontryagin Minimum Principle previously given.

Stochastic Controllers[8,9]

Unknown random inputs as well as errors in measuring outputs exist in all systems. Usually this "noise" is in some sense small, so that it is meaningful to linearize the system equations about some nominal path, possibly the solution to the deterministic optimal control problem

$$\delta\dot{x} = F\delta x + G\delta u + w$$

where δx and δu are a small deviation in the state and control, respectively, F and G are the partial of $f(x,u)$ with respect to x and u, respectively, both evaluated along the nominal trajectory. w is a random input with covariance W. The control δu that minimizes the expected value of the "quadratic" cost

$$J = \delta x^T_{(t_f)}S_f\delta x_{(t_f)} + \int_{t_0}^{t_f} (\delta x^T A\ \delta x + \delta u^T B\ \delta u)\ dt$$

is

$$\delta u = -C\delta x$$

where

$$C = -B^{-1}G^TSx$$

and S satisfies the Riccati equation

$$\dot{S} = -SF - F^TS + SGB^{-1}G^TS - A$$

with the boundary condition

$$S_{(t_f)} = S_f$$

The estimate of $\delta\hat{x}$, δx, is computed from the differential equation

$$\delta\dot{\hat{x}} = F\ \delta x + G\delta u + K(\delta z - H\ \delta x)$$

and the deviation in measurement of x, δz is related to δx by

$$\delta z = H\ \delta x + v$$

H is the partial derivative of the measurement, z, with respect to state, x, evaluated along the nominal trajectory and v is the error in the measurement, with zero mean and covariance V.

The weighting, K, applied to the difference between the expected and actual measurement, is computed from

$$K = P H^T V^{-1}$$

where P, the covariance of the error in the estimate of the state, is determined from

$$\dot{P} = FP + FP^T + W - P H^T V^{-1} HP$$

The initial value, $P(t_0)$, is known for a biased estimate of δx, or set arbitrarily large for an unbiased estimate.

An a priori estimate of the performance (the "expected" performance) of this controller can be obtained in the form of a covariance of the deviation of the state, X

$$X \equiv \overline{\delta \hat{x} \, \delta \hat{x}^T}$$

$$= P + Y$$

where the covariance of the estimate of the state, Y,

$$Y \equiv \overline{\delta x \, \delta x^T}$$

is determined from

$$\dot{Y} = (F - GC)Y + Y(F - GC)^T + P H^T R^{-1} HP$$

where

$$Y_{(t_0)} = 0$$

An estimate of the control effort required is

$$\overline{\delta u \, \delta u^T} = C Y C^T$$

An estimate of the cost required to control the system between t_0 and t_f starting at the initial state $\delta x_{(t_0)}$ is

$$J = \delta x_{(t_0)} S_{(t_0)} \delta x^T_{(t_0)}$$

which is also the system Liapunov Function,[2] so that all such controllers are asymptotically stable.

These relations are the basis of a powerful tool for developing and evaluating the expected performance of linear time varying controllers that keep a system in the vicinity of some prescribed trajectory.

Controllability and Observability[5]

For any linear control synthesis technique to be successful (including the one presented in the previous section), the linearized system must be controllable and observable over all intervals of time in t_0 to t_f. The problem that otherwise occurs can be seen by transforming the system

$$\delta \dot{x} = F \, \delta x + G \, \delta u$$

into

$$\delta \dot{y} = PFP^{-1} \, \delta y + PG \, \delta u$$

where

$$\delta y = P \, \delta x$$

P is a matrix, such that the matrix PFP^{-1} becomes the Jordan Canonical Form, for all practical purposes a diagonal matrix. P is composed of a row of column vectors

$$P_x = (p^1 \mid \cdots \mid p^n) \tag{1}$$

where the column vectors, p^i are the n-eigenvector solutions to the equation

$$Pp = \lambda p \tag{2}$$

The eigen values, λ_i, are the diagonal values in the diagonal matrix PFP^{-1}, specifically,

$$PFP^{-1} = \begin{matrix} \lambda & \cdots & 0 \\ & \cdots & \\ & \cdots & \\ & \cdots & \\ 0 & \cdots & \lambda_n \end{matrix} \qquad (3)$$

(See Ref. 11 for Control Theory.)
This transformation effectively uncouples separate "modes," (components of the vector δy) of the system. If the vector $PG\,\delta u$, over some interval of time, does not have a component associated with each mode of the system, then this mode and hence the system is uncontrollable during this interval of time. Similarly if the vector

$$\delta z = H\,\delta x = HP^{-1}\,\delta y$$

does not have every component of δy represented over every interval of time, then that mode and hence the system is not observable in δz during that interval of time. It is important to realize that real systems are never strictly uncontrollable or unobservable but difficulties can result when systems are nearly (asymptotically) so.

Stability[2]

Stability has a number of broad definitions for a nonlinear system. Generally they all require a system to remain in the vicinity of a normal trajectory when the initial conditions are perturbed and the same control is applied over a finite interval of time. When the linearized system of equations has constant coefficients, stability is easily defined. For the linear system

$$\delta x = F_0\,\delta x$$

where F_0 is a constant coefficient matrix, the solution for an arbitrary initial condition δx_0 is

$$\delta x_{(t)} = \mathcal{L}^{-1}\,[sI - F_0]^{-1}\,\delta x_0$$

where $\mathcal{L}^{-1}$ is the inverse Laplace transform and s is the Laplace variable, $\delta x_{(t)}$ approaches zero for any δx_0, and hence the system is stable if and only if the real parts of the roots of the equation

$$\text{Det } [sI - F_0] = 0$$

are negative.

BIBLIOGRAPHY

1. BRYSON, A. E., JR., and HO, Y., *Applied Optimal Control, Optimization, Estimation and Control*, Waltham, Massachusetts, Blaisdel, 1969.
2. LA SALLE, J., and SEFSCHETZ, S., *Stability by Liapunov's Direct Method*, New York, Academic 1961.
3. PONTRYAGIN, L. S., BOLTYANSKII, V. G., GAMKRELIDZE, R. V., and MISHCHENKO, E. F., *The Mathematical Theory of Optimal Processes*, New York, Wiley, 1962.
4. LIETMANN, G. (ed.), *Optimization Techniques*, New York, Academic, 1962.
5. ATHANS, M., and FALB, P. L., *Optimal Control*, New York, McGraw-Hill, 1966.
6. BRYSON, A. E., JR., and DENHAM, W. F., "A Steepest-Ascent Method for Solving Optimum Programming Problems," *J. Appl. Mech.*, June 1962.
7. BELMANN, R., *Dynamic Programming*, Princeton, New Jersey, Princeton Univ. Press, 1957.
8. KALMAN, R. E., and BUCY, R., "New Results in Linear Filtering and Prediction," *Trans. ASME*, Vol. 83D, p. 95, 1961.
9. BRYSON, A. E., "Applications of Optimal Control Theory in Aerospace Engineering," *J. Spacecraft Rockets*, Vol. 4, p. 545, 1967.
10. BREAKWELL, J. V., SPEYER, J. L., and BRYSON, A. E., Jr., "Optimization and Control of Nonlinear Systems Using the Second Variation," *SIAM Journal*, Vol. 1, p. 193, 1963.
11. ZADEH, L., and DESOER, C. A., *Linear System Theory*, New York, McGraw-Hill, 1963.

FLIGHT DYNAMICS

By R. E. Willes, T. J. Forster, and G. E. Thompson

The motion of most flight vehicles can be divided into two specific types; motion of rapid nature about the vehicle's center of mass and a somewhat slower motion of the vehicle's center of mass itself. The motion of the center of mass is the subject of performance analysis[4,5] while the motion about the center of mass is the topic of stability and control analysis.[1,6]

60. AIRCRAFT PERFORMANCE

The central problem of aircraft performance analysis is to determine the values of particular performance parameters (rate of climb, speed, range, endurance, turn rate, turn radius, etc.) as a function of other parameters that control the motion of the aircraft's center of mass (engine speed, Mach number, altitude, and load factor). Because of the largely numerical format of data concerning both the propulsive and the aerodynamic forces on an aircraft, it is convenient to perform these calculations numerically. The usual method of displaying these calculations is to plot the performance indices as a function of the aircraft's Mach number M, its altitude h, and its load factor n for some engine speed N. A number of typical performance parameter plots are shown in the accompanying figures.

An aircraft can exchange kinetic for potential energy much more rapidly than it can change the combined total energy by excess thrust. The combined kinetic and potential energy expressed in units of energy altitude h_e where

$$h_e = h + \frac{V^2}{2g}$$

is a convenient parameter measuring the joint altitude and velocity capability of an aircraft at any particular time. The rate at which energy altitude can be changed (the specific energy rate), P_s, is

$$P_s = \frac{F - D}{W} V$$

where F is the engine thrust, D is the aircraft total drag, W is the aircraft weight, and V is the aircraft velocity. In Fig. 1 contours of constant energy altitude and specific energy rate are plotted as functions of Mach number and altitude, at a load factor of one and at maximum rotational engine speed (rpm) in full afterburner. The maximum Mach number normally is limited at high altitudes by temperature and at low altitudes by dynamic pressure. The maximum energy climb is obtained by selecting the altitude-Mach number combination that results in the largest P_s value for the existing energy level, h_e.

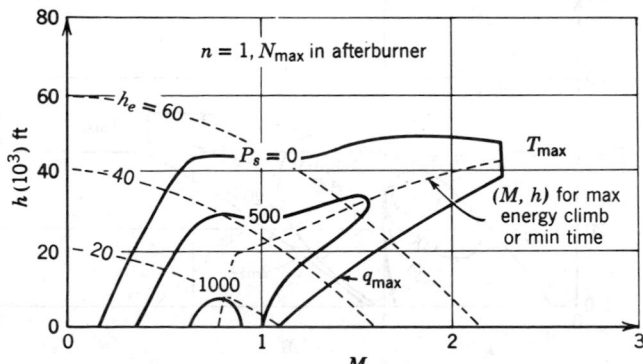

FIG. 1. Specific energy rate, power (ft/sec), and energy altitude (ft).

The energy gain per unit fuel expended, specific energy efficiency E_s, is

$$E_s \equiv \frac{dh_e}{dW_f} = \frac{(F - D)}{W} \frac{V}{Fc}$$

where W_f is the weight of fuel and c is the thrust specific fuel consumption. Plots of energy efficiency with the same constraints of load factor and engine speed are shown in Fig. 2. A minimum fuel climb (maximum energy efficiency) is determined by selecting the altitude-Mach number pair that results in the highest E_s value for the existing energy level h_e.

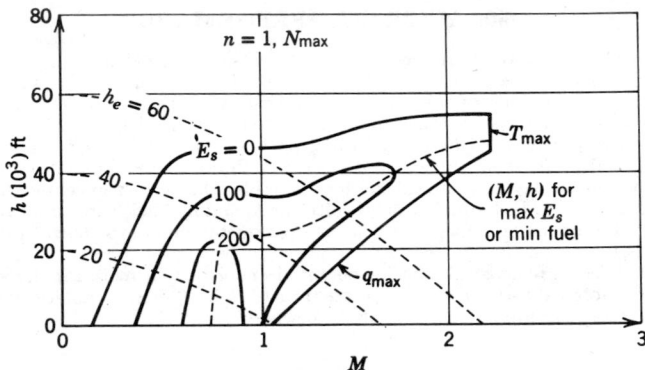

FIG. 2.   Specific energy efficiency (ft/lb fuel).

The range per unit fuel expended, the specific range R_s is

$$R_s \equiv \frac{dR}{dW_f} = \frac{V}{cF}$$

Figure 3 shows contours of specific range plotted as a function of Mach number and altitude for a load factor of one and the engine thrust set equal to drag. The maximum range occurs near the maximum altitude and just below Mach 1 for subsonic flight; minimum afterburner and near maximum altitude will result in maximizing R_s in supersonic flight.

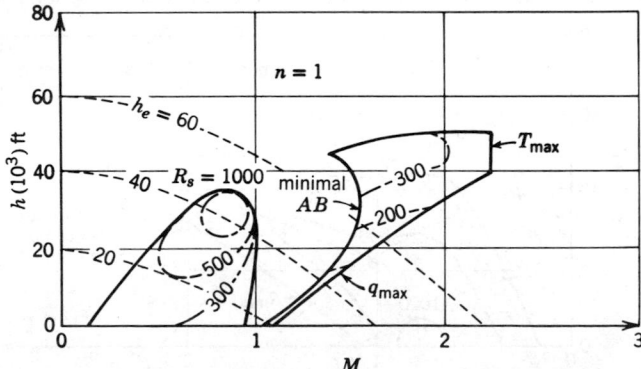

FIG. 3.   Specific range (nautical miles/10,000 lb fuel).

Specific endurance is defined as a time per unit fuel used and is given by

$$E \equiv \frac{dt}{dW_f} = \frac{1}{cF}$$

Endurance is plotted as a function of Mach number and altitude (with load factor and thrust as before) in Fig. 4. Maximum endurance occurs near the tropopause and near Mach 1 subsonically, and at minimum operating afterburner limits supersonically.

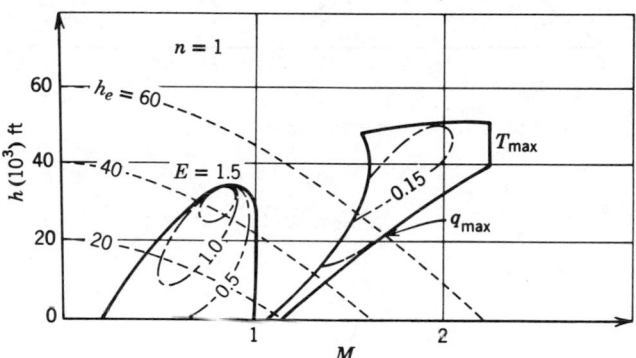

FIG. 4. Specific endurance (hr/10,000 lb fuel).

Turn rate ω and radius r are approximately given by

$$\omega \cong \frac{ng}{v} \quad , \quad r \cong \frac{v^2}{ng}$$

Contours of constant turn rate and constant radius are shown in Figs. 5 and 6 along with typical lift coefficient and load factor limits. Observe that the maximum turn rate and

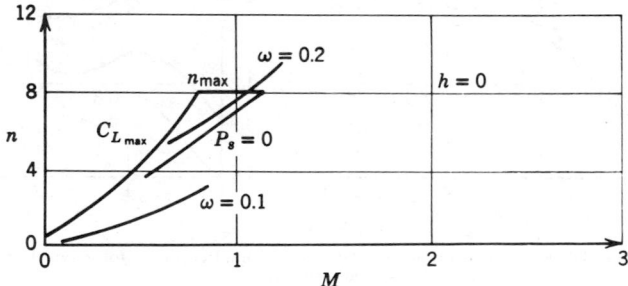

FIG. 5. Turn rate (radians/sec).

the minimum turn radius occur at the maximum lift coefficient-load factor intersection. The data for these plots is obtained in the following manner. Numerical data for engine thrust F,

$$F = \frac{p}{p_0} F_0(M, N_c)$$

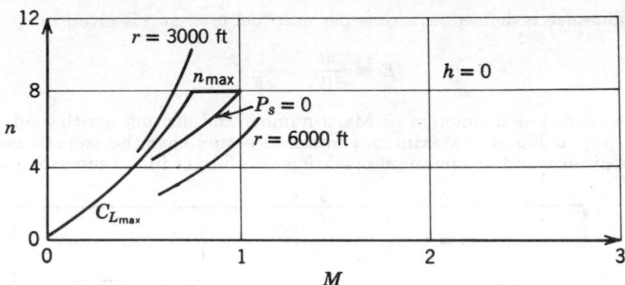

FIG. 6.   Turn radius (ft).

where

$$N_c = N \sqrt{\frac{T_0}{T}}$$

and specific consumption c,

$$c = \sqrt{\frac{T_0}{T}} \, c_0(M, N_c)$$

are tabulated as a function of Mach number M, corrected engine speed N_c, pressure p, and temperature T.   The subscript 0 indicates the value of the quantity at the planetary surface.   Typical dependences may be seen in Fig. 7.

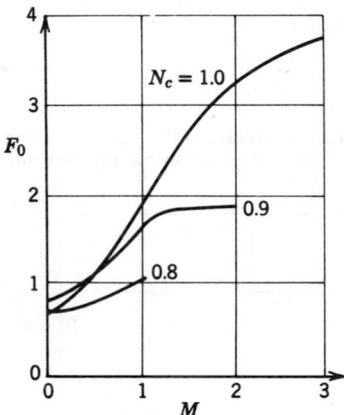
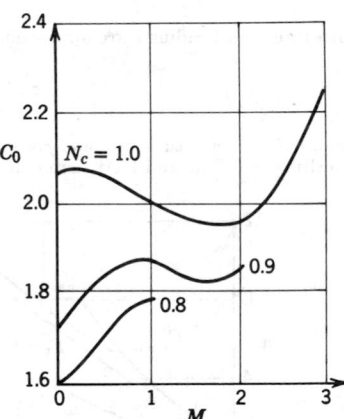

FIG. 7.   Thrust (10,000 lb) and thrust specific fuel consumption (hr^{-1}) for a typical turbojet in full afterburner.

Atmospheric pressure p and temperature T are functions of altitude.   As a first approximation the temperature is constant.   The pressure then approximately satisfies the relation

$$p = p_0 e^{-\beta h}$$

where h is the altitude, and β is the inverse atmospheric scale height.

$$\beta = \frac{g}{RT}$$

The drag ocefficient, C_D, is a function of Mach number M and lift coefficient C_L. See Figs. 8 and 9. The lift coefficient C_L is specified by requiring that the lift L equal the normal load nW and is computed from

$$C_L = \frac{nW}{(\gamma/2)pM^2S}$$

where γ is the ratio of specific heats and S is the wing area. The drag by definition is

$$D = C_D \frac{\gamma}{2} pM^2S$$

Aircraft velocity V is related to the speed of sound a and Mach number M by

$$V = Ma$$

where a is the speed of sound

$$a = \sqrt{\gamma RT}$$

and R is the gas constant.

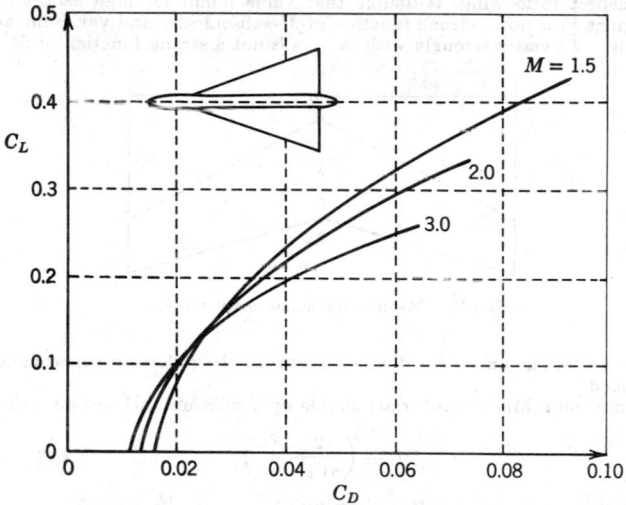

FIG. 8. Lift coefficient vs. drag coefficient and Mach number of a typical supersonic aircraft.

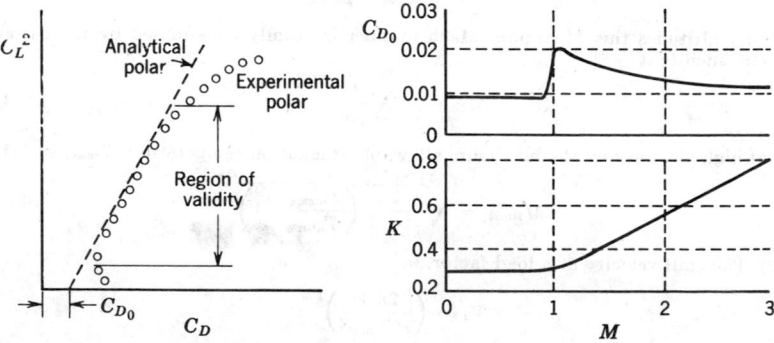

FIG. 9. Approximate quadratic drag polar for a typical supersonic aircraft.

It is desirable to have analytical relations that express performance indices as a function of aircraft configuration. These analytical relations are based on approximating the drag polar with the quadratic form

$$C_D = C_{D0} + K C_L^2$$

where C_{D0} and K are judiciously picked to conform to the actual drag polar over a range of M and C_L of interest. See Fig. 9. K is sensitive to angle-of-attack subsonically after separation occurs and may double before maximum C_L is reached. For high aspect ratio wings, it is approximately

$$K \cong \frac{1}{\pi \mathcal{R}}$$

where aspect ratio $\mathcal{R}$ is defined as

$$\mathcal{R} = \frac{b^2}{S} = \frac{b}{\bar{c}}$$

where b is the span, and $\bar{c}$ is the mean aerodynamic chord. See Fig. 10. The value of K for low aspect ratio wings is double that value found for high aspect ratios. For turbojet engines F_0 is not a strong function of M subsonically and varies linearly with M supersonically. F_0 varies strongly with N. c_0 is not a strong function of M and N.

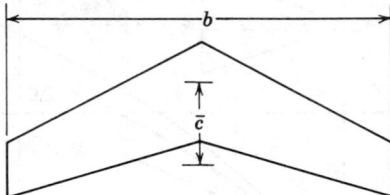

FIG. 10. Mean aerodynamic chord and span.

Using these relations and discarding small terms, the following performance relations can be obtained:

(1) The maximum Mach number attainable by a subsonic turbojet aircraft is given by

$$M \cong \left(\frac{2}{\gamma C_{D0}} \frac{F_0}{p_0 S} \right)^{1/2}$$

while that for a supersonic aircraft is given by

$$M \cong \frac{2}{\gamma C_{D0}} \frac{dF_0/dM}{p_0 S}$$

At lower altitudes this Maximum Mach number is usually constrained by the dynamic pressure such that

$$M_{limit} = \sqrt{\frac{2}{\gamma p} q_{max}}$$

and at high altitudes by the maximum allowable stagnation temperature, T_{max}, such that

$$M_{limit} = \sqrt{\frac{2}{\gamma - 1} \left(\frac{T_{max}}{T_{atm}} - 1 \right)}$$

(2) The stall velocity at a load factor, n,

$$V_s = \left(\frac{2nW}{\rho C_{Lmax} S} \right)^{1/2}$$

is determined by the largest lift coefficient C_{Lmax}, that can be developed by the aircraft.

This stall velocity is usually less than the minimum sustainable speed (due to excessive drag at high values of C_L), so that

$$M_{\min} = \left(\frac{2 K n^2 W^2 p_0}{\gamma p^2 F_0 S}\right)^{1/2}$$

(3) The subsonic maximum rate of climb, or equivalently the maximum energy rate of climb for high performance aircraft occurs at the transonic drag rise. The maximum angle of climb occurs at the velocity for L/D a maximum, V_R, where

$$V_R = \sqrt{\frac{2W}{\rho S}} \sqrt[4]{\frac{K}{C_{D_0}}}$$

The corresponding rate of climb is

$$\frac{dh}{dt} = \frac{dh_e}{dt} = \left(\frac{p}{p_0}\frac{F_0}{W} - 2\sqrt{KC_{D_0}}\right) V_R$$

The maximum rate of climb is given by

$$\frac{dh}{dt} = \frac{dh_e}{dt} = \frac{pMa}{W}\left(\frac{F_0}{p_0} - \frac{1}{2} C_{D_0}\gamma M^2 S\right)$$

(4) The ceiling (maximum altitude) occurs at the altitude

$$h = -\frac{1}{\beta} \ln \left[2\sqrt{KC_{D_0}} \frac{W}{F_0}\right]$$

which corresponds to the ambient pressure

$$p = 2\sqrt{KC_{D_0}} \frac{W}{F_0} p_0$$

(5) The maximum rate of climb trajectory also gives minimum fuel for climb, and the fuel consumed per unit time is

$$\frac{dW_f}{dt} = \frac{cp}{p_0} F_0$$

(6) In level flight, maximum range R is

$$R = \frac{1.14 V_{Ri}}{c\sqrt{KC_{D_0}}}\left(1 - \sqrt{\frac{W}{W_i}}\right)$$

when flown at the following velocity

$$V = 1.316 V_R$$

Maximum endurance E is flown at L/D a maximum and is

$$E = \frac{1}{2c} \frac{\ln\left[W_i/(W_i - W)\right]}{\sqrt{KC_{D_0}}}$$

and is flown at

$$V = V_R.$$

(7) Somewhat longer range and endurance are usually obtained if the aircraft is slowly "cruise climbed" at constant engine speed. The profile

$$V = 1.189 V_R$$

results in maximum range

$$R = \frac{0.384 V_{Ri}}{c} (KC_{D_0})^{-3/4} \left(\frac{F}{W_i}\right)^{1/2} \ln\left(\frac{W_i}{W_i - W}\right)$$

For maximum endurance,

$$V = V_R$$

results in

$$E = \frac{1}{2c} (KC_{D0})^{-1/2} \ln \left(\frac{W_i}{W_i - W} \right)$$

The altitude increase from start to end of the cruise climb is

$$h_f - h_i = \frac{a^2}{Kg} \ln \left(\frac{W_i}{W_i - W} \right)$$

(8) The rate of turn ω is asymptotically a constant for high performance jet aircraft turning at large load factors with fixed thrust and no loss of energy. The rate of turn ω for this case is approximately

$$\omega \cong \frac{ng}{V} = \frac{g}{W} \left(\frac{F \rho S}{2K} \right)^{1/2}$$

The turn radius decreases for increasing C_L and reaches a minimum at the maximum lift coefficient boundary. The minimum turn radius r is given by

$$r \cong \frac{V^2}{ng} = \frac{2}{g \rho S} \left(\frac{KnW^3}{F} \right)^{1/2}$$

It can be seen from these relations that turn rate and turn radius improve dramatically with increased wing area, increased thrust, and lower altitude.

61. ATMOSPHERIC EXIT AND RE-ENTRY

To obtain orbital velocity from the surface of a planet or to land on the surface of a planet from orbital velocity, the planetary atmosphere must be traversed. This transition is the topic of atmospheric exit and re-entry.

Launch Trajectories

Aerodynamic forces may be ignored when exiting from planetary atmospheres if the vehicle is kept at zero angle of attack. This is usually necessary when passing through maximum dynamic pressure to assure that the vehicle is not destroyed by lateral loads. Once free of the planetary atmosphere, it can be shown that a desired velocity is obtained with minimum fuel if the thrust vector remains at a constant orientation angle. The launch trajectory is, therefore, usually divided into three parts, a vertical ascent, a gravity turn at zero angle of attack, and a constant thrust angle phase.

Vertical Ascent. During the vertical ascent, the rocket's nondimensional mass m, velocity v, and altitude h, are described by the following integrals of the dynamic equations

$$m = 1 - \dot{m}t$$

$$v = - \ln m - t$$

$$h = + \frac{m}{\dot{m}} \ln m + t - \frac{t^2}{2}$$

where the nondimensional mass m, mass flow $\dot{m}$ (a positive constant), time t, velocity v, and altitude h are defined as

$$m = \frac{m'}{m'_0}, \qquad \dot{m} = \frac{\dot{m}'v_e}{mg}, \qquad t = \frac{t'g}{v_e}, \qquad v = \frac{v'}{v_e}, \qquad h = \frac{h'g}{v_e^2}$$

where m' is the dimensional mass, t' is dimensional time, m'_0 is the initial rocket mass, v_a is the rocket exhaust velocity, and g is the acceleration of gravity. Notice that the magnitude of m should be as large as possible to gain maximum v or h with fixed expenditure of m.

Gravity Turn. Aerodynamic forces reach their largest value during the gravity turn, but both aerodynamic and gravity accelerations along the flight path are small in comparison to thrust accelerations. Approximate integrals of the dynamic equations are obtainable, but not as simple functions, though they may be straightforwardly expressed as a series. They are

$$m = 1 - \dot{m}t$$

$$v = -\ln m - t$$

$$\sin \gamma = \frac{z - 1}{z + 1}, \quad z = -\int_1^m \frac{d\tau}{\tau(-\ln \tau + 1)}$$

$$h = -\int_1^m -\frac{\ln \tau + 1}{\tau}\left(1 - \left(\frac{z - 1}{z + 1}\right)^2\right)^{1/2} d\tau$$

where γ is the angle between the velocity vector and the horizontal, the nondimensional down range distance x is

$$x = \frac{x'g}{v^2_c}$$

where v_c is the circular orbit velocity and all other variables are as previously defined. Maximum dynamic pressure, "max q," occurs where $\rho v^2/2$ is a maximum or where

$$m(1 - \ln m)^2 = \frac{2\dot{m}}{\beta}$$

The nondimensional atmospheric scale height, $1/\beta$, is defined as

$$\beta = \beta' v^2_c / g$$

where β' is dimensional inverse atmospheric scale height.

Constant Thrust Inclination. The minimum fuel path to a given velocity is flown with the thrust at some constant inclination to the horizontal. If such a path is flown after the vehicle escapes the sensible atmosphere in a gravity turn, the integrals of the equation of motion that describe this path are:

$$m = 1 - \dot{m}t$$

$$v_x = \cos \gamma_T \ln m + v_{x0}$$

$$v_y = -t - \sin \gamma_T \ln m + v_{y0}$$

$$x = v_{x0}t - \cos \gamma_T \left(\frac{m \ln m}{\dot{m}} - t\right)$$

$$h = v_{y0}t - \tfrac{1}{2}t^2 - \sin \gamma_T \left(\frac{m \ln m}{\dot{m}} - t\right)$$

where v_x and v_y are the components of the velocity vector in the horizontal and vertical direction, respectively, γ_T is the inclination of the thrust vector, and all other variables are as previously defined.

After the rocket has obtained the required velocity for orbit or escape, the engine is cut off, and from this point forward the principles of orbital mechanics apply.

Multistage Rockets

Single stage chemical rockets have a fundamental limit to the maximum velocity they can obtain which is associated with the rocket's exhaust velocity and the portion of the rocket that must be allocated to structure. If velocities greater than this velocity are desired, the rocket must be multistaged.

Neglecting the usually small drag and gravity acceleration along the flight path, the equation describing the final velocity obtained by the kth stage of a chemical rocket is

$$v_{fk} = v_{0k} + v_{ek} \ln \frac{m_{fk}}{m_{0k}}$$

The total velocity acquired by n stages is

$$v = \sum_{k=1}^{n} v_{ek} \ln \frac{m_{0k}}{m_{fk}}$$

Defining the structural efficiency factor, ϵ, and the π payload ratio for each stage as

$$\epsilon = \frac{m_s}{m_s + m_p}$$

$$\pi = \frac{m_l}{m_0}$$

where m_s is the mass of the structure, m_p is the mass of the propellant, m_l is the mass of the payload, m_0 is the initial mass, and m_f if the final mass; it is then possible to rewrite the equation for v as

$$v = \sum_{k=1}^{n} v_{ek} \ln \frac{1}{\epsilon_k + (1 - \epsilon_k)\pi_k}$$

It is easily seen that, for maximum v, v_e should be as large as possible and ϵ_k as small as possible for each stage.   For the case where v_e and ϵ are equal for all stages, it can be shown that v is a maximum if all π are equal so that

$$\frac{m_f}{m_{01}} = \pi^n$$

$$\frac{m_f}{m_{01}} = \left[\frac{\exp\left[-(v/nv_e) \right] - \epsilon}{1 - \epsilon} \right]^n$$

which is the ratio of payload to initial mass that can be taken to a velocity v by n stages with a structural efficiency of ϵ.   This has an upper level as $n \to \infty$ of

$$\frac{m_f}{m_{01}} = \exp\left(-\frac{v/v_e}{1 - \epsilon} \right)$$

which gives the upper limit of the payload ratio that can be gained by staging.
   In the limit of no payload,

$$v = nv_e \ln \frac{1}{\epsilon}$$

which gives the maximum velocity that can be obtained with n stages of a rocket with structural efficiency ϵ.

Atmospheric Entry

The objectives of atmospheric entry are to decelerate a space vehicle from orbital or superorbital velocities with acceptable aerodynamic and heating loads and land on some specified portion of the planetary surface or to maneuver inside the planetary atmosphere to obtain a change of orbital elements.   This type of flight poses some unusual problems.
   All planetary atmospheres are thin in comparison to the planetary radius.   A measure of this thinness is ϵ, the ratio of the atmosphere scale height $1/\beta$ to the planetary radius r_0:

$$\epsilon = \frac{1}{\beta r_0}$$

See Table 1.
   The time required for a vehicle at orbital velocity to pass through one atmospheric scale height at moderate flight path angles is on the order of ϵ times the orbital period. A vehicle can quickly pass through many scale heights if the flight path angle is not kept extremely small, and thus encounter severe heating and deceleration problems.

Table 1. Planetary Flight Data

Planet	Equatorial Radius	Equatorial Gravitational Acceleration	Inverse Atmospheric Scale Height	Equatorial Radius/ Atmospheric Scale Height	Orbital Velocity Mach No.	Deceleration loads when	
						$p = 0(\epsilon)$	$p = 0(1)$
	r_0 (ft)	g_0 (ft/sec²)	β_0 (ft⁻¹)	$\frac{1}{\epsilon} = \beta_0 r_0$	M_0	$g_0/g_0\otimes$	$g_0/g_0\otimes$
Venus	2.03×10^7	28.3	4.9×10^{-5}	1006	26	.88	1000
Earth	2.09×10^7	32.2	4.3×10^{-5}	900	25	1.0	900
Mars	1.11×10^8	12.2	1.0×10^{-5}	132	9.5	.39	5
Jupiter	2.27×10^8	83.7	1.7×10^{-5}	3600	49	2.61	9100
Saturn	1.87×10^8	36.6	1.6×10^{-5}	3000	45	1.12	3360

Aerodynamic accelerations are large for vehicles having velocities of the order of orbital velocity when deep within any planetary atmosphere. Thus entry trajectories must be kept shallow at entry altitudes where the atmosphere is sparce; however, the danger of skipping out of the sensible atmosphere also exists at these shallow entry trajectories.

At orbital velocities the entry Mach number M_0 is approximately

$$M_0 \cong \sqrt{\frac{1}{\epsilon}}$$

and is therefore large. This implies that aerodynamic heating, convective and/or radiative, is severe during entry.

Equilibrium Glide. The earliest analytical solutions for flight at near orbital velocity were produced by Sänger.[3] They describe deceleration at small flight path angles where the difference between gravity and centrifugal forces is balanced by lift. The appropriate analytical solutions are

$$v^2 = \frac{1}{(C_L/2)\epsilon p + 1}$$

$$\gamma = -2\epsilon \frac{C_D}{C_L}$$

$$\theta = \theta_0 + \frac{1}{2}\frac{C_L}{C_D} \ln\left(\frac{1 - v^2}{1 - v_0^2}\right)$$

where velocity V and pressure p have been nondimensionalized as

$$v = \frac{v'}{\sqrt{r_0 g_0}}$$

$$p = \frac{p'}{m g_0/A}$$

$$p' = p_0^{\exp(-\beta h')}$$

where g_0 is the acceleration of gravity at the planet's surface, p_0 is the value of pressure at the planet's surface, h is height, m is the vehicle's mass, and A is the vehicle's reference area. Other symbols are defined in Fig. 2.

Load factor n, due to total aerodynamic force, measured in multiples of earth gravity acceleration, is

$$n = \frac{1}{2\epsilon}(C_L^2 + C_D^2)^{1/2}pv^2 \frac{g_0}{g_{0 \text{ earth}}}$$

The load factor stays low for equilibrium glide but slowly increases as the trajectory descends.

Heating loads per unit time are relatively small with small rediative heating loads dominating at high altitudes and velocities, and convective loads dominating at lower altitudes. The total energy absorbed can be quite large if attempts are not made to reradiate the energy to free space. The heating loads can be evaluated using empirical relations of the form

$$\dot{q} = C_Q \rho^i v^{2j}/r^k$$

where for stagnation point heating (laminar flow)

$$C_Q = 1.55 \times 10^{-5}, \quad i = 0.5, j = 1.5, k = 0.5$$

and for sonic point heating (turbulent flow)

$$C_Q = 7.45 \times 10^{-4}, \quad i = 0.8, j = 1.5, k = 0.2$$

and for stagnation point radiant heating

$$C_Q = 1.23 \times 10^{-25}, \quad i = 1.6, j = 4.2, k = -1$$

where C_Q is a heat transfer coefficient and r is the vehicle nose radius. For equilibrium glide the maximum heating rate occurs deep in the atmosphere.

Ballistic and Skip Entry. The Ballistic and Skip solution, associated with the names of Allen and Eggers,[7] is valid when the aerodynamic drag dominates the flight path component of gravity and when the difference between centrifugal and gravity accelerations are small, or when all normal forces are dominated by lift.

For ballistic entry where there is no lift, the solution is

$$v^2 = v^2{}_0 \exp\left(\frac{C_D(p - p_0)}{\sin \gamma_0}\right)$$

$$\cos \gamma = \cos \gamma_0$$

$$\theta = \theta_0 + \cot \gamma_0 (h - h_0)$$

where all symbols are as previously defined. The maximum load factor due to deceleration occurs at the altitude where

$$p = \frac{\sin \gamma_0}{C_D}$$

and has a value

$$n = \frac{\sin \gamma_0}{\epsilon} \frac{v_0{}^2}{e}$$

The maximum heating rate occurs at the altitude where

$$p = -\frac{i}{j} \frac{\sin \gamma_0}{C_D}$$

and has a value

$$\dot{q} = C_Q p v_0{}^{2j} \exp\left(-jC_D \frac{p}{\sin \gamma_0}\right)$$

When lift is present (skip entry) the solution is

$$v^2 = v_0{}^2 \exp\left[\frac{2C_D}{C_L}(\gamma - \gamma_0)\right]$$

$$\cos \gamma = \cos \gamma_0 + \frac{C_L}{2}(p - p_0)$$

Maximum heating rate and deceleration occur near the bottom of the skip where $\gamma = 0$ at the altitude corresponding to

$$p = p_0 + \frac{2}{C_L} (1 - \cos \gamma_0)$$

The load factor due to total aerodynamic force on the vehicle at this point is given by

$$n = \frac{1}{2_\epsilon} (C_L{}^2 + C_D{}^2)^{1/2} \, p v_0{}^2 \exp \left\{ -2\gamma_0 \frac{C_D}{C_L} \right\} \frac{g_0}{g_0 \text{ earth}}$$

and the heating rate is obtained from the expression given above for ballistic entry.

Conclusion

A number of other relations are available for both two- and three-dimensional flight. One important use of the relations presented is in the calculations of "entry corridors" from superorbital speed. The corridor is defined by the class of Keplerian orbits which are bounded on the bottom by a skip at maximum lift-up that will not exceed heating or acceleration limits and on the top by an equilibrium glide trajectory with maximum lift-down.

BIBLIOGRAPHY

1. ETKIN, B., *Dynamics of Flight*, London, Wiley, 1959.
2. HALFMAN, R. L., *Dynamics: Particles, Rigid Bodies, and Systems*, Reading, Massachusetts, Addison-Wesley, 1962.
3. LOH, W. H., *Dynamics and Thermodynamics of Planetary Entry*, Englewood Cliffs, New Jersey, Prentice-Hall, 1963.
4. MIELE, A., *Flight Mechanics*, Reading, Massachusetts, Addison-Wesley, 1962.
5. PERKINS, C. D., and HAGE, R. E., *Airplane Performance Stability and Control*, London, Wiley, 1949.
6. SECKEL, E., *Stability and Control of Airplanes and Helicopters*, New York, Academic, 1964.
7. WILLES, R. E., FRANCISCO, M. C., REID, J. G., LUN, W. K., "An Application of Matched Asymptotic Expansion to Hypervelocity Flight," AIAA Paper No. 67-598, August 67.

FOUNDATION OF ORBITAL MECHANICS

By A. E. Preyss, R. R. Bate, and J. D. Hines

Newton's laws of motion and his law of universal gravitation form the foundation of orbital mechanics. With these laws the motion of a body orbiting in a gravitational field can be described by means of a vector differential equation of second order. One way of stating Newton's law of universal gravitation is as follows:

One particle attracts another with a force which is directed along a line connecting them and is proportional to the product of their masses and inversely proportional to the square of the distance separating them.

Thus, for a system comprised of n particles wherein only gravitational forces are acting, the motion of the ith particle with respect to an inertial frame of reference is governed by the equation

$$\left(m_i \frac{d^2 r_i}{dt^2} \right) = \sum_{\substack{j=1 \\ j \neq i}}^{n} G \frac{m_i m_j}{r_{ij}^3} (r_j - r_i) \tag{1}$$

where the proportionality factor G is the universal gravitation constant, m_i is the mass of the ith particle, r_i is the position vector of the ith particle, and r_{ij} is the distance between the ith and jth particles, $r_{ij} = |r_j - r_i|$. Since the motion of the ith particle depends on the relative positions of all n particles, it is necessary, in general, to write down a set of n equations, each having the same form as Eq. 1, and to solve them simultaneously in order to describe the behavior of the system of particles or of any one particle. Only in the case of two particles is it possible to analytically obtain a complete solution to this problem. For the case of three particles a useful first integral of the motion, the Jacobi integral, has been derived.[3,6] For the general case of n particles certain first integrals are also available and are found in Art. 63. Some physical and orbital parameters for bodies in the solar system are listed in Tables 2, 3, and 4.

62. GRAVITATIONAL POTENTIALS

Although the bodies of the solar system can usually be treated as mass particles, there are important situations, such as a satellite orbiting in close proximity to a planet, in which the finite size of one body has a significant influence on the motion of another. When the finite dimensions of a body must be considered, it is useful to introduce the concept of a potential function, the gradient of which gives the attractive force. For example, the gravitational potential produced by the distribution of the earth's mass can be expressed in terms of Legendre polynomials as follows:

$$V(r,\phi) = \frac{Gm_E}{r} \left[1 - \sum_{k=2}^{\infty} J_k \left(\frac{r_E}{r} \right)^k P_k(\cos \phi) \right] \tag{2}$$

where the spherical coordinates r and ϕ are the radius and latitude, respectively, of the point at which the potential is to be computed, r_E is the earth's radius, m_E is the mass of the earth, and the J_k are coefficients of the series expansion which must be determined experimentally. Typical values are tabulated in Table 1.

Table 1. J_k, Coefficients of the Potential Function ($\times 10^6$)

$J_2 = 1,082.28 \pm 0.03$
$J_3 = -2.3 \pm 0.2$
$J_4 = -2.12 \pm 0.05$
$J_5 = -0.2 \pm 0.1$
$J_6 = 1.0 \pm 0.8$

Table 2. Basic Characteristics of the Solar System[10]

Body	Symbol	Semi-Major Axis to Sun (AU)*	Period Earth-Years ($\oplus = 1$)	Mean Diameter ($\oplus = 1$)	Mass ($\oplus = 1$)	No. of Natural Satellites	Surface Escape Velocity ($\oplus = 1$)
Sun	☉	–	–	109.0	3×10^5	–	55.0
Mercury	☿	0.387	0.241	0.38	0.054	0	0.371
Venus	♀	0.723	0.616	0.97	0.815	0	0.915
Earth	⊕	1.00	1.00	1.00	1.00	1	1.00
Mars	♂	1.52	1.88	0.52	0.108	2	0.449
Jupiter	♃	5.20	11.9	11.0	318.0	12	5.38
Saturn	♄	9.54	29.5	9.03	95.1	9	3.26
Uranus	♅	19.2	84.0	3.72	14.5	5	1.97
Neptune	♆	30.1	165.0	3.38	17.0	2	2.24
Pluto	♇	39.5	248.0	1.02?	0.8	0	0.85?
Earth's Moon	☾	–	0.075	0.27	0.012	0	0.212

* 1 AU = 92,959,670 miles.

Rodrigues' formula can be used for the explicit determination of the Legendre polynomials,

$$P_k(x) = \frac{1}{2^k k!} \frac{d^k (x^2 - 1)^k}{dx^k} \tag{3}$$

Equation 2 is based on the assumption of symmetry about the polar axis. Contributions to the potential from terms in the summation are small and decay rapidly with increasing distance from the earth. Therefore, when $r \gg r_E$ the sum makes a negligible contribution to the potential and the earth then attracts essentially like a point mass.

If a completely general representation of a gravitational potential is required, then the assumption of axial symmetry must be removed and Eq. 2 replaced by

$$V(r,\phi,\theta) = \frac{Gm}{r} + \sum_{k=1}^{\infty} \frac{A_k}{r^{k+1}} P_k(\cos\phi) + \sum_{k=1}^{\infty} \sum_{j=1}^{k} \frac{B_k{}^j}{r^{k+1}} P_k{}^j(\cos\phi)\cos j\theta$$

$$+ \sum_{k=1}^{\infty} \sum_{j=1}^{k} \frac{C_k{}^j}{r^{k+1}} P_k{}^j(\cos\phi)\sin j\theta \tag{4}$$

where the constant coefficients can be evaluated from the expressions

$$A_k = G \iiint \rho^{k+2} D(\rho,\beta,\lambda) P_k(\cos\phi)\sin\beta \, d\rho \, d\beta \, d\lambda \tag{5}$$

$$B_k{}^j = 2G \frac{(k-j)!}{(k+j)!} \iiint \rho^{k+2}(\rho,\beta,\lambda) P_k{}^j(\cos\beta)\cos j\lambda \sin\beta \, d\rho \, d\beta \, d\lambda \tag{6}$$

$$C_k{}^j = 2G \frac{(k-j)!}{(k+j)!} \iiint \rho^{k+2} D(\rho,\beta,\lambda) P_k{}^j(\cos\beta)\sin j\lambda \sin\beta \, d\rho \, d\beta \, d\lambda \tag{7}$$

$$m = \iiint D(\rho,\beta,\lambda)\rho^2 \sin\beta \, d\rho \, d\beta \, d\lambda \tag{8}$$

where $D(\rho,\beta,\lambda)$ is the density distribution of matter, and r, ϕ, and θ are spherical coordinates. Due to current emphasis on near earth satellites for communications, weather reconnaissance, etc., generalized representations such as Eq. 4 have received much attention.[5,6,8] Considerable effort has been expended on the experimental determination of the coefficients in the leading terms of the expansion and on the determination of suitable approximate models of gravitational potentials for use in analytic studies.

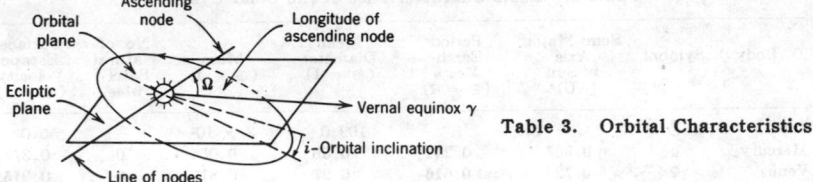

Table 3.   Orbital Characteristics

Symbol Planet	☿ Mercury	♀ Venus	⊕ Earth	♂ Mars
Semi-Major Axis a, AU	0.387099	0.723332	1.000000	1.523691
Perihelion Distance, AU $= a(1 - e)$	0.307501	0.718422	0.983277	1.381416
Aphelion Distance, AU $= a(1 + e)$	0.466697	0.728242	1.016723	1.665966
Oribital Eccentricity e	0.205628	0.006788	0.016723	0.093375
Mean Orbital Velocity ($\oplus = 1$)	1.607271	1.175794	1.00	0.806855
km/sec	47.90	35.05	29.77	24.02
NM/sec	25.87	18.92	16.80	12.97
ft/sec	157,186.0	114,958.0	97,702.1	78,805.7
Sidereal Mean Daily Motion, sec	14,732.4	5767.670	3548.193	1886.519
Period of Revolution ($\oplus = 1$)	0.2411	0.6156	1.00	1.8822
Oribital Inclination i to Ecliptic, deg	7.00412	3.39431	0	1.84989
Inclination of Equatorial Plane to Orbit, deg		<10	23.443597	23.99
Mean Longitude of Ascending Node Ω, deg	47.94364	76.38541		49.30530
True Longitude of Perihelion $\bar{\omega}$, deg (Epoch 1967 April 20.0)	76.94664	131.11097	102.3781	335.45705
Mean Anomaly M, deg	253.906	353.764	105.02313	240.222
Axial Rotational Period	$59^d \pm 2^d$	$242.6^d \pm 0.6^d$ (retrograde)	$23^h56.07^m$	$24^h37.38^m$
Escape Velocity $\left(\dfrac{2\mu}{R}\right)^{1/2}$, ft/sec	13,600	33.500	36,675	16,500

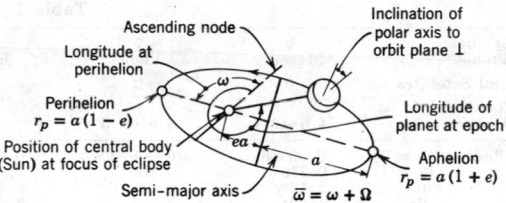

of the Planets[10]

♃ Jupiter	♄ Saturn	⛢ Uranus	Ψ Neptune	♇ Pluto
5.202694	9.538836	19.25290	30.04427	39.64092
4.951857	9.012979	18.285114	29.765219	29.719670
5.453531	10.064693	20.220686	30.323321	49.562170
0.048213	0.055128	0.050267	0.009288	0.250278
0.438411	0.323782	0.2283249	0.1823988	0.1590757
13.05	9.64	6.797	5.43	4.73
7.047	5.20	3.6705	2.93	2.55
42,817.6	31,595.2	22,302.0	17,802.7	15,560
299.128	120.455	42.235	21.532	14.283
11.86	29.46	84.0	164.8	247.7
1.30601	2.48875	0.77250	1.77310	17.12631
3.067	26.733	97.884	28.8	
100.1176	113.4342	73.9208	131.3902	109.7729
13.5550	92.3756	169.1552	56.2011	222.7818
108.0054	277.3342	3.3525	176.5732	328.0351
9^h50^m	10^h14^m	10^h49^m	15^h48^m	6^d
197,500	119,500	72,500	82,400	31,300

Table 4. Physical Characteristics

Symbol Planet	☿ Mercury	♀ Venus	⊕ Earth	♂ Mars
No. of Natural Satellites	0	0	1	2
Apparent Equatorial Angular Diameter, sec	4.6–12.7	9.9–64.5		3.5–25.1
Equatorial Radius ($\odot = 1$)	0.00346	0.00873	0.00915	0.00488
($\oplus = 1$)	0.379	0.956	1.00	0.535
km	2.42^3	6.05^3	6.37816^3	3.41^3
miles	1.50^3	3.78^3	3.96320^3	2.12^3
NM	1.31^3	3.26^3	3.44393^3	1.84^3
Oblateness f	0	0	1/298.3	1/192
J_2	0	0	1.082^{-3}	1.92^{-3}
Volume ($\oplus = 1$)	0.054	0.87	1.00	0.153
Mean Density ($\odot = 1$)	4.06	3.68	3.94	2.78
($\oplus = 1$)	1.03	0.952	1.00	0.705
g/cm³	5.7	5.25	5.52	3.89
lb/ft³	355	329	344.6	243
Mass ($\odot = 1$)	1.64^{-7}	2.4477^{-6}	3.04039^{-6}	3.236^{-7}
(Including ($\oplus = 1$) Satellites)	0.0546	0.81485	1.01230	0.1077
$GM = \mu = g_0R^2$				
km³/sec²	2.18^4	3.2485^5	3.98604^5	4.293^4
$\mu = \dfrac{aV_e^2}{2} = \begin{cases}$ mi³/sec²	5.23^3	7.7936^4	9.56302^4	1.0299^4
ft³/sec²	7.70^{14}	1.1472^{16}	1.40766^{16}	1.516^{15}
AU³/day²	4.85^{-11}	7.2430^{-10}	8.88757^{-10}	9.576^{-11}
Equatorial Surface ($\oplus = 1$)	0.380	0.893	1.00	0.377
Gravity g_e (cm/sec²)	372	873	978.031	369
Albedo	0.06	0.76	0.36	0.15
Maximum Surface Temperature, °F	750	210	140	90

* Superscripts denote exponents of 10; e.g., $4.6^{-8} = 4.6 \times 10^{-8}$. $G = 6.6695 \times 10^{-8}$ (cgs units) = universal gravitational constant.

63. INTEGRALS FOR THE N-BODY PROBLEM

It is a relatively simple matter to show that in the absence of external forces acting on a system of n bodies, the total linear momentum is conserved

$$\sum_{i=1}^{n} m_i \mathbf{r}_i = \mathbf{c}_1 t + \mathbf{c}_2 \tag{9}$$

This result is obtained by summing Eq. 1 over i and integrating twice. By taking the cross product of Eq. 1 with $\mathbf{r}_i$ and again summing on i, it is possible to show after a straightforward integration that the total angular momentum of the system is conserved,

$$\sum_{i=1}^{n} \mathbf{r}_i \times m_i \frac{d\mathbf{r}_i}{dt} = \mathbf{c}_3 \tag{10}$$

A plane, commonly referred to as the invariable plane, can be defined by the total angular momentum vector $\mathbf{c}_3$ which is normal to it. In this plane the center of mass of the system

$$\mathbf{r}_{cm} = \frac{\displaystyle\sum_{i=1}^{n} m_i \mathbf{r}_i}{\displaystyle\sum_{i=1}^{n} m_i} \tag{11}$$

of the Planets[10] *

♃ Jupiter	♄ Saturn	♅ Uranus	♆ Neptune	♇ Pluto
1	9	5	2	
30.8–50.0	14.9–20.6	3.4–4.2	2.2–2.4	0.4–0.6
0.102	0.0865	0.0337	0.0320	0.010
11.14	9.47	3.69	3.50	1.1
7.14^4	6.04^4	2.35^4	2.23^4	$7.^3$
4.43^4	3.75^4	1.46^4	1.39^4	$4.^3$
3.85^4	3.26^4	1.27^4	1.21^4	$4.^3$
1/16.1	1/10.4	1/16	1/50	0
1.47^{-2}	0.67^{-2}	1.5^{-2}	$5.^{-3}$	0
1400	850	50.	43	1.3
0.89	0.44	1.14	1.6	2.4
0.227	0.112	0.290	0.40	0.6
1.25	0.62	1.60	2.2	3.3
78	39	100	138	200
9.5475^{-4}	2.857^{-4}	4.360^{-5}	$5.^{-5}$	$(2.5 \pm 0.3)^{-6}$
317.89	95.12	14.52	17	0.8 ± 0.1
1.2671^8	3.792^7	5.788^6	6.8^6	3.2^5
3.0399	9.098^6	1.388^6	1.6^6	7.7^4
4.4747^{18}	1.339^{18}	2.044^{17}	2.4^{17}	1.1^{16}
2.8252^{-7}	8.454^{-8}	1.290^{-8}	1.5^{-8}	7.4^{-10}
2.54	1.06	1.07	1.4	0.7
2490	1040	1050	1400	700
0.51	0.50	0.66	0.62	0.16
−200	−240	−270	−330	−370

moves in a straight line with constant velocity. Finally, if Eq. 1 is written in the form

$$m_i \frac{d^2 \mathbf{r}_i}{dt^2} = \nabla_i U \tag{12}$$

where ∇_i is the gradient operator with respect to the components of $\mathbf{r}_i$, and U is the force function, taking the dot product of Eq. 12 with $d\mathbf{r}_i/dt$, summing on i, and integrating shows that the total energy of the system is conserved,

$$T - U = c \tag{13}$$

where c is a constant, $-U$ is the potential energy, and T is the kinetic energy. By definition

$$T = \frac{1}{2} \sum_{i=1}^{n} m_i \frac{d\mathbf{r}_i}{dt} \cdot \frac{d\mathbf{r}_i}{dt} \tag{14}$$

and

$$U = \frac{G}{2} \sum_{i=1}^{n} \sum_{\substack{j=1 \\ j \neq 1}}^{n} \frac{m_i m_j}{r_{ij}} \tag{15}$$

which is the total work done by the gravitational forces in bringing the n bodies from infinity to some initial configuration.

These integrals provide ten constants c_1, c_2, c_3, and c. To completely describe the motion of the system, $6n$ integrals are required, but for the n-body problem it is known that all integrals are functions of these ten.[9]

64. TWO-BODY PROBLEM

When, as is often the case, only the relative motion of a two-particle system is of interest, a general analytical solution can be derived. Since such a solution requires six integrations, having ten known integrals makes this possible. By writing down Eq. 1 for each particle and letting $r = r_2 - r_1$, an equation of relative motion is obtained,

$$\frac{d^2 r}{dt^2} + \frac{\mu}{r^3} r = 0 \tag{16}$$

where $\mu = G(m_1 + m_2)$. A first integral of the motion results from crossing r with Eq. 16 and integrating to find that

$$h = r \times \frac{dr}{dt} = \text{constant} \tag{17}$$

where h is angular momentum per unit mass and, since it is constant, the motion is planar. Another integral is obtained by taking the cross product of h with Eq. 16 and integrating,

$$\frac{dr}{dt} \times h = \frac{\mu}{r} (r + re) \tag{18}$$

where e is the eccentricity vector and is a constant. When the scalar product between r and Eq. 18 is taken the result is

$$r = \frac{h^2/\mu}{1 + e \cos \nu} \tag{19}$$

where ν is the angle between e and r and is called the true anomaly. This is the equation of a conic whose semilatus rectum or parameter p is given by h^2/μ and whose classification depends on the magnitude of e:

Circle: $e = 0$
Ellipse: $0 < e < 1$
Parabola: $e = 1$
Hyperbola: $e > 1$

From Eq. 19 is it apparent that r is a minimum when $\nu = 0$ and, therefore, the eccentricity vector points in the direction of periapsis, that is, where the two bodies come closest to each other during their orbits. It also can be shown that each of the two bodies describes a conical path with respect to a fixed origin as well as describing a conical path with respect to each other.

Since $h \cdot e = 0$, h and e provide only five independent constants of integration, and these specify the size and shape of the conic, its orientation in the plane of motion, and the orientation of the plane. A sixth constant of integration is obtained when determining the time dependency of the relative motion. Time and position can be related by recognizing that the angular momentum per unit mass can be expressed in the form

$$h = r^2 \frac{d\nu}{dt} \tag{20}$$

and thus that r can be eliminated between Eq. 19 and 20 to give

$$\sqrt{\frac{\mu}{p^3}} \, dt = \frac{d\nu}{(1 + e \cos \nu)^2} \tag{21}$$

Integrating this expression yields the final constant which is often taken to be the time when $\nu = 0$ and the bodies are nearest each other.

65. ORBITAL ELEMENTS

It is common practice to refer to the six constants of integration of two-body motion as orbital elements. Since two-body motion is governed by Eq. 16, these six constants are $r(t_0)$ and $dr(t_0)/dt$, where t_0 is some reference time or epoch. Historically, however, different sets of constants related to r_0 and v_0 and frequently having some simple geometric interpretation are selected as orbital elements.

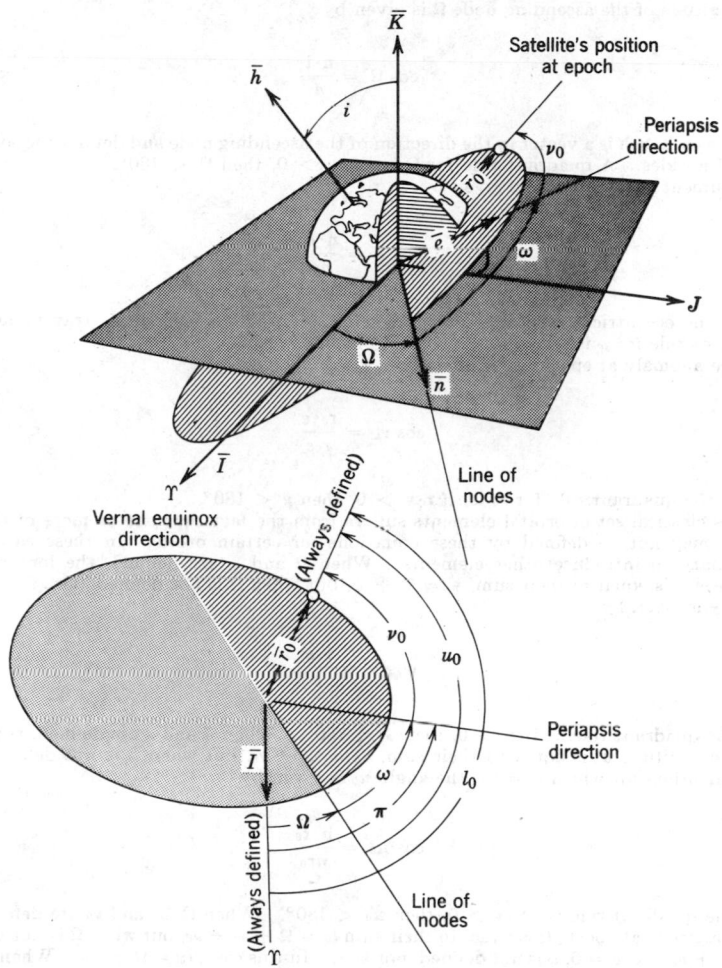

FIG. 1.   Classical orbital elements.

A classical choice for such a set of orbital elements is illustrated in Fig. 1.   Three Euler angles, the inclination i, the longitude of the ascending node Ω, and the argument of periapsis ω, are used to specify the spatial orientation of the orbit.   Size and shape of the orbit are fixed by the semimajor axis a and either the semilatus rectum p or the eccentricity e.   Position of the body in the orbit at epoch is established by ν_0, the true anomaly at the reference time t_0.   The relationship between this set of elements and the constants $\mathbf{r}_0$ and $\mathbf{v}_0$ is readily obtained from the geometry of Fig. 1.   The four angles i, Ω, ω, and ν_0 may be defined by taking the scalar product between the appropriate pair of vectors.

Inclination, i, is given by

$$\cos i = \frac{\mathbf{h} \cdot \mathbf{k}}{h} \tag{22}$$

where $\mathbf{h} = \mathbf{r}_0 \times \mathbf{v}_0$ and the inclination is always less than $180°$.

Longitude of the ascending node Ω is given by

$$\cos \Omega = \frac{\mathbf{n} \cdot \mathbf{i}}{n} \tag{23}$$

where $n = \mathbf{k} \times \mathbf{h}$ is a vector in the direction of the ascending node and defines the so-called line of apsides. A quadrant rule for Ω is: if $n_j > 0$, then $\Omega < 180°$.
Argument of periapsis ω is given by

$$\cos \omega = \frac{\mathbf{n} \cdot \mathbf{e}}{ne} \tag{24}$$

where the eccentricity vector is given by $\mathbf{e} = (1/\mu)\{[v^2 - (\mu/r)]\mathbf{r} - (\mathbf{r} \cdot \mathbf{v})\mathbf{v}\}$ and the quadrant rule for ω is: if $e_k > 0$, ghen $\omega < 180°$.
True anomaly at epoch, ν_0, is given by

$$\cos \nu_0 = \frac{\mathbf{r}_0 \cdot \mathbf{e}}{r_0 e} \tag{25}$$

where the quadrant rule for ν_0 is: if $\mathbf{r} \cdot \mathbf{v} > 0$, then $\nu < 180°$.
This classical set of orbital elements suffers from the fact that one or more of the elements may not be defined by these equations for certain orbits. In these cases it is customary to introduce other elements. When Ω and ω are defined, the longitude of periapsis π is equal to their sum, $\pi = \Omega + \omega$, but when Ω is not defined (i.e., $i = 0$) the angle π is given by

$$\cos \pi = \frac{\mathbf{i} \cdot \mathbf{e}}{e} \tag{26}$$

and the quadrant rule is: if $e_j > 0$, then $\pi < 180°$. When ω and ν are defined, the argument of latitude u_0 is equal to their sum, $u_0 = \omega + \nu_0$, but when ω is not defined (i.e., circular orbits for which $e = 0$), the angle u_0 is given by

$$\cos u_0 = \frac{\mathbf{n} \cdot \mathbf{r}_0}{n r_0} \tag{27}$$

and the quadrant rule is: if $r_k > 0$, then $u_0 < 180°$. When Ω, ω, and ν_0 are defined the true longitude at epoch, l_0 is equal to their sum $l_0 = \Omega + \omega + \nu_0$, but when Ω is not defined $l_0 = \pi + \nu_0$. If $e = 0$, ω is not defined, nor is π. In this case, $l_0 = \Omega + u_0$. When $\mathbf{n}$ and $\mathbf{e}$ are both zero, the true longitude at epoch is given by

$$\cos l_0 = \frac{\mathbf{r}_0 \cdot \mathbf{i}}{r_0} \tag{28}$$

where the quadrant rule is: if $r_j > 0$, then $l_0 < 180°$. A summary of the four possible cases and the definitions which apply in each situation is given in Fig. 2.
From Eq. 19 it is already known that the parameter p is given by

$$p = h^2/\mu \tag{29}$$

and therefore, since $\mathbf{h} = \mathbf{r}_0 \times \mathbf{v}_0$, the connection between this orbital element and the constants of integration $\mathbf{r}_0$ and $\mathbf{v}_0$ is clear. Equation 29 may be interpreted as saying that the angular momentum determines the shape of the orbit.

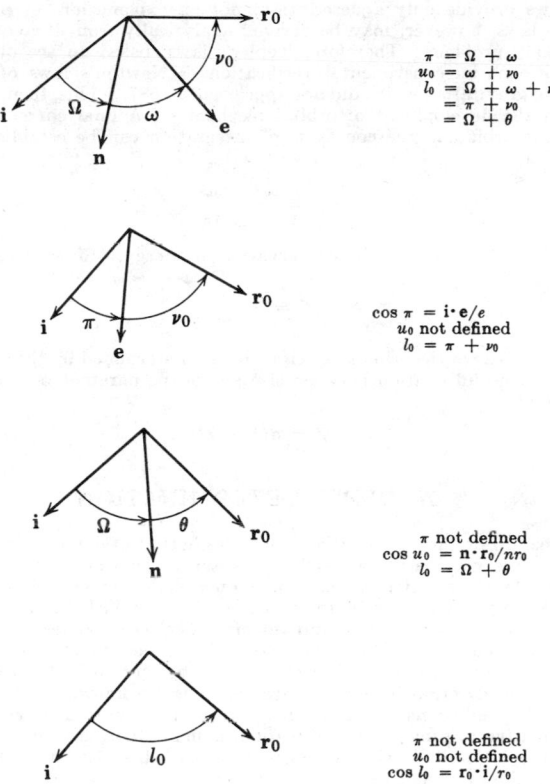

$$\pi = \Omega + \omega$$
$$u_0 = \omega + \nu_0$$
$$l_0 = \Omega + \omega + \nu_0$$
$$= \pi + \nu_0$$
$$= \Omega + \theta$$

$$\cos \pi = \mathbf{i} \cdot \mathbf{e}/e$$
$$u_0 \text{ not defined}$$
$$l_0 = \pi + \nu_0$$

$$\pi \text{ not defined}$$
$$\cos u_0 = \mathbf{n} \cdot \mathbf{r}_0/nr_0$$
$$l_0 = \Omega + \theta$$

$$\pi \text{ not defined}$$
$$u_0 \text{ not defined}$$
$$\cos l_0 = \mathbf{r}_0 \cdot \mathbf{i}/r_0$$

Fig. 2.   Classical orbital elements—special cases.

66. KEPLER'S LAWS

Since the time of Aristotle, who taught that circular motion was the only perfect and natural motion and that the heavenly bodies, therefore, necessarily moved in circles, the planets were assumed to revolve in circular paths or combinations of smaller circles moving on larger ones (epicycles).   However, early in the seventeenth century Johann Kepler, working with the accurate observations of Tycho Brahe, found immense difficulty in reconciling any such theory with the actual data on celestial motion.   From 1601 until 1609 he tried fitting various geometrical curves to Brahe's measurements on the motion of Mars.   Finally, after struggling for almost a year to remove a discrepancy of only 8 minutes of arc, Kepler hit upon the ellipse as a possible solution. It fit.   The orbit was found and in 1609 Kepler published his first laws of planetary motion.   The third law followed in 1619.   These laws, which mark an epoch in the history of mathematical science, are as follows:

Kepler's Laws

First Law.   The orbit of each planet is an ellipse, with the sun as a focus.

Second Law.   The line joining the planet to the sun sweeps out equal areas in equal times.

Third Law.   The square of the period of a planet is proportional to the cube of its mean distance from the sun.

Kepler's laws provide only a description, not an explanation, of planetary motion. Each of these laws, however, may be derived analytically from the equation of motion for the two-body problem. Therefore, Kepler's laws, based on the observational data of Brahe, represent an experimental verification of Newton's laws of motion and his universal law of gravitation which did not appear until 1687. This, then, is their historical significance in the development of orbital mechanics. A final connection between the geometry of the orbit and the constants of integration can be established through the vis viva integral

$$\zeta = \frac{v_0{}^2}{2} - \frac{\mu}{r_0} \tag{30}$$

which is simply a statement of the conservation of energy. Since it can be shown that

$$\zeta = -\frac{\mu}{2a} \tag{31}$$

it is clear that the size of the orbit as specified by a is determined by the energy as specified by r_0 and v_0. A useful relation between size and shape parameters, easily derived from Eq. 19, is given by

$$p = a(1 - e^2) \tag{32}$$

67. ORBIT DETERMINATION

One fundamental problem is celestial mechanics in that of determining the character of an orbit from observational data on the trajectory. Depending on the nature of the observational data, the solution to this problem may follow one or more approaches, some of which have their origin back in the times of Galileo. With the advent of radar and the ability to measure both position and velocity of celestial bodies, the orbit determination problem has been considerably simplified. Since a measurement of $\mathbf{r}$ and $\mathbf{v}$ at a given time provides the six constants of integration for the motion in the two-body case, the orbit is completely determined once these quantities are known. Without this type of information, orbit determination from such observational data as angular displacements of one satellite relative to some "fixed" direction, is much more difficult.[5,6,8]
When initial position and velocity can be measured directly, this data can be used to predict future values of $\mathbf{r}$ and $\mathbf{v}$.
The simpler problem of expressing $\mathbf{r}$ and $\mathbf{v}$ as functions of the true anomaly is treated now, and in the next section the more complex problem of relating $\mathbf{r}$ and $\mathbf{v}$ to time will be examined.
Given $\mathbf{r}(t_0)$ and $\mathbf{v}(t_0)$, a set of unit vectors which determine the orientation of the orbital plane is readily found. Expressing $\mathbf{r}$ and $\mathbf{v}$ in this frame at times t and t_0 permits the elimination of the unit vectors from the equations with the result,

$$\mathbf{r} = \left\{ 1 - \frac{r}{p} [1 - \cos (\nu - \nu_0)] \right\} \mathbf{r}_0 + \frac{rr_0}{\sqrt{\mu p}} \sin (\nu - \nu_0) \mathbf{v}_0 \tag{33}$$

$$\mathbf{v} = \left\{ \frac{\mathbf{r}_0 \cdot \mathbf{v}_0}{pr_0} [1 - \cos (\nu - \nu_0)] - \frac{1}{r_0} \sqrt{\frac{\mu}{p}} \sin (\nu - \nu_0) \right\} \mathbf{r}_0$$

$$+ \left\{ 1 - \frac{r_0}{p} [1 - \cos (\nu - \nu_0)] \right\} \mathbf{v}_0 \tag{34}$$

Thus given the angular difference $\nu - \nu_0$, $\mathbf{r}$ and $\mathbf{v}$ are known once r and p are determined. Equation 29 gives p and Eq. 10 gives r; therefore the solution is complete.

Position and Velocity in Two-Body Orbits

Classically, the starting point for establishing the time dependence of position and velocity in a two-body orbit is Eq. 21. Unfortunately, integrating this equation leads to certain undesirable results. First, the integration requires the introduction of a dummy variable which, although it may be interpreted geometrically, is different for each type of conic, thus necessitating a special treatment for each case. Second, as the eccentricity approaches the value of 1 (the parabolic case), most of the known methods of solution blow-up because of singular behavior. Third, the integration provides a time dependence on position instead of the more favorable inverse. Integrating Eq. 21 leads to Barker's formula for the parabolic case, Kepler's equation for the elliptic case, and an analogous form of Kepler's equation for the hyperbolic case. Finding inverse expressions (i.e., position dependence on time) for Kepler's equation is impossible in closed form because of its transcendental nature.[6,8] Attempts to find solutions to this problem have led to the development of new branches of mathematics. In this section a unified approach to time of flight problems is described. This approach is a re-formulation of Battin's[4] work on universal formulas for conic orbits and is designed to eliminate certain indeterminacies which would otherwise arise.[11]

Instead of beginning with Eq. 21, this unified approach starts with the vis-viva integral and the resolution of the velocity vector into its radial component r and its transverse component $r\dot{v}$. If an auxiliary variable x, applicable to all conics, is introduced according to

$$\dot{x} = \sqrt{\mu/r} \tag{35}$$

then an expression for the time of flight along any conic can be derived

$$t = \frac{\mathbf{r}_0 \cdot \mathbf{v}_0}{\mu} x^2 C\left(\frac{x^2}{a}\right) + \frac{x^3}{\sqrt{\mu}}\left[1 - \frac{r_0}{a}\right] S\left(\frac{x^2}{a}\right) + \frac{r_0 x}{\sqrt{\mu}} \tag{36}$$

where the two transendental functions C and S, which are introduced to avoid problems of indeterminacy, can be defined by the power series expansions,

$$C(z) = \frac{1}{2!} - \frac{z}{4!} + \frac{z^2}{6!} - \frac{z^3}{8!} + \cdots \tag{37}$$

$$S(z) = \frac{1}{3!} - \frac{z}{5!} + \frac{z^2}{7!} - \frac{z^3}{9!} + \cdots \tag{38}$$

An alternate way of expressing these functions is

$$C(z) = \begin{cases} \dfrac{1 - \cos \sqrt{z}}{z} & z > 0 \text{ (ellipse)} \\[2ex] 1/2 & z = 0 \text{ (parabola)} \\[2ex] \dfrac{\cos h \sqrt{-z} - 1}{-z} & z < 0 \text{ (hyperbola)} \end{cases} \tag{39}$$

$$S(z) = \begin{cases} \dfrac{\sqrt{z} - \sin \sqrt{z}}{z^3} & z > 0 \text{ (ellipse)} \\[2ex] 1/6 & z = 0 \text{ (parabola)} \\[2ex] \dfrac{\sin h \sqrt{-z} - \sqrt{-z}}{(-z)^{3/2}} & z < 0 \text{ (hyperbola)} \end{cases} \tag{40}$$

In the derivation of Eq. 36 it has been assumed that $x = 0$ when $\mathbf{r} = \mathbf{r}_0$ and $\mathbf{v} = \mathbf{v}_0$. To solve for x when t is given is a simple matter, since Newton's method of iteration may be applied successfully. As the time is a monotonically increasing function of increasing x, if x_n is an approximation to the solution, then a better approximation is obtained from

$$x_{n+1} = x_n + \frac{t - t_n}{(dt/dx)_{x=x_n}} \tag{41}$$

where the slope of t vs. x is given by

$$\frac{dt}{dx} = \frac{1}{\sqrt{\mu}}\left\{ x^2 C\left(\frac{x^2}{a}\right) + \frac{\mathbf{r}_0 \cdot \mathbf{v}_0}{\sqrt{\mu}} x\left[1 - \frac{x^2}{a} S\left(\frac{x^2}{a}\right)\right] + r_0\left[1 - \frac{x^2}{a} C\left(\frac{x^2}{a}\right)\right]\right\} \tag{42}$$

With these results it is a simple matter to determine position in an orbit when time is given.

Using the iteration just described to find x and the identify

$$r = a(1 - e \cos E) \tag{43}$$

it follows directly that for an elliptical orbit

$$x = \sqrt{a} \ (E - E_0) \tag{44}$$

and for a hyperbolic orbit

$$x = \sqrt{-a} \ (H - H_0) \tag{45}$$

These relations provide an interpretation of x in terms of the eccentric anomaly E and hyperbolic anomaly H, respectively. For a parabolic orbit, the relation

$$x = \sqrt{p} \ \left(\tan \frac{\nu}{2} - \tan \frac{\nu_0}{2} \right) \tag{46}$$

can be obtained from Barker's formula. Finally, using the identity

$$\tan \frac{\nu}{2} = \sqrt{\frac{1 + e}{1 - e}} \tan \frac{E}{2} \tag{47}$$

for an ellipse and the identity

$$\tan \frac{\nu}{2} = \sqrt{\frac{e + 1}{e - 1}} \tanh \frac{H}{2} \tag{48}$$

for a hyperbola, it is possible to relate x to the true anomaly ν for each type of conic section.

When time is given and x is found, the equations above provide a means to determine ν. Equations 33 and 34 may then be used to obtain the position and velocity vectors $\mathbf{r}$ and $\mathbf{v}$. However, $\mathbf{r}$ and $\mathbf{v}$ can be expressed directly in terms of x, avoiding the use of anomalies entirely. This may be accomplished by using Eqs. 33 and 34 and the relations between x and the anomalies. The derivation proceeds in a straightforward manner and results in

$$\mathbf{r}(t) = \left[1 - \frac{x^2}{r_0} C \left(\frac{x^2}{a} \right) \right] \mathbf{r}_0 + \left[t - \frac{x^3}{\sqrt{\mu}} S \left(\frac{x^2}{a} \right) \right] \mathbf{v}_0 \tag{49}$$

$$\mathbf{v}(t) = \frac{\sqrt{\mu}}{r r_0} \left[\frac{x^3}{a} S \left(\frac{x^2}{a} \right) - x \right] \mathbf{r}_0 + \left[1 - \frac{x^2}{r} C \left(\frac{x^2}{a} \right) \right] \mathbf{v}_0 \tag{50}$$

A computational algorithm written in Algol, which permits the evaluation of $\mathbf{r}$ and $\mathbf{v}$ given $\mathbf{r}_0$, $\mathbf{v}_0$ and the time t, follows. Comments have been added to the program which make the logic easy to follow. Use of this algorithm provides a complete solution of the Kepler problem quickly and accurately by means of digital computation.

```
procedure Kepler (r1, v1, dt, r2, v2); value dt;
array r1, v1, r2, v2; real dt;
comment r1 and v1 are the initial position and velocity of an object moving in a Keplerian orbit.
    Kepler computes the final position and velocity, r2 and v2, after a time, dt, has elapsed;
begin
    integer i;
    real ta, tb, tc, td, f, xp, x, s, c, z, te, g, tf;
    mag (r1); mag (r2); comment: magnitude of vectors r1 and r2;
    ta: = dot (r1, v1);
    tb: = 2/r1[0] − (v1[1] ↑ 2 + v1[2] ↑ 2 + v1[3] ↑ 2);
    xp: = dt/r1[0];
    if tb × xp ↑ 2 < −10 then
    begin tc: = sqrt(−tb) × sign(dt);
        xp: = ln(abs(2 × dt × tb/(ta + (1 − tb × r1[0])/tc)))/tc;
    end;
    for i: = 0 step 1 until 50 do
    begin
        while xp × dt < 0 ∨ tb × xp ↑ 2 > 9.49₁₀23 ∨
```

```
tb X xp ↑ 2 < −1.12₁₀4 do xp: = .5 X (xp + x);
x: =xp; z: = tb X xp ↑ 2;
Battinf(z, s, c); comment: compute s(2) and c(2);
tc: = x X (1 − z X s);
te: = c X x ↑ 2;
td: = ta X te + r1[0] X tc + x + 3 X s;
if abs((td − dt)/dt) < ₁₀−9 then go to LB;
tf: = ta X tc + r1[0] X (1 − re X tb) + te;
xp: = x − (td − dt)/tf;
end;
error ("Kepler failed to converge in 50 iterations");
LB: f: = 1 − re/r1 [0];
g: = dt − x ↑ 3 X s;
for i: = 1 step 1 until 3 do r2[i]: = f X r1[i] + g X v1[i];
mag(r2);
f: = − tc/(r1[0] X r2[0];
g: = 1 − te/r2[0]
for i: = 1 step 1 until 3 do v2[i]: = f X r1[i] + g X v 1[i];
mag(v2);
end Kepler;
```

Orbit Determination from Two Positions and Time

Another classical problem of orbital mechanics is that of determining a two-body orbit when it is known that the trajectory passes through positions r_1 at time t_1 and r_2 at time t_2. A two-body orbit, that is, a conic path, is the lowest order solution to the n-body problem. In other words, it is a good approximation to the actual path of, say, a spacecraft when n-bodies are present but when one of them is the dominant attractive body during a given time period. For this reason, two-body orbits are of interest in such operations as planning for interplanetary missions. Since tasks like mission planning may require the solution of Lambert's problem[3,8] for a large number of initial and final conditions, it is desirable to have available a convenient technique to solve this class of orbital boundary value problems. Again, by introducing the dummy variable x, it is possible to derive a unified method of approach which is applicable to all conic sections.[11]

The geometry of the problem is illustrated in Fig. 3. If the radial distances r_1 and r_2 are expressed in terms of the dummy variable x, that is,

$$r_1 = a \left(1 + e \sin \frac{c_0}{\sqrt{a}} \right) \tag{51}$$

$$r_2 = a \left(1 + e \sin \frac{x + c_0}{\sqrt{a}} \right) \tag{52}$$

where $c_0 = \sqrt{a} \sin^{-1}\{ (1/e)[(r_0/a) − 1]\}$, then writing an expression for the chord length l and doing some algebraic and trigonometric manipulation leads to the quadratic equation

$$2y^2 − 4(r_1 + r_2)y + (r_1 + r_2)^2 zC(z) + l^2(2 − zC(z)) = 0 \tag{53}$$

where $y \equiv x^2 C(z)$ and $z \equiv x^2/a$. Doing the algebraic and trigonometric manipulation necessary to obtain this result is equivalent to the geometrical process of constructing the location of the vacant focus in the case of an ellipse or a hyperbola and of the directerix in the case of a parabola. Solving Eq. 53 provides two answers.

$$y_{1,2} = r_1 + r_2 \pm \beta \sqrt{2 − zC(z)} \tag{54}$$

where $\beta \equiv \sqrt{r_1 r_2 + r_1 \cdot r_2}$. To obtain a unique solution, the proper choice of sign in Eq. 54 must be determined.

Making the proper choice is not difficult. The fact that the transit between P_1 and P_2 can be accomplished in either of two directions can be used to resolve the ambiguity. Equation 35 is used to derive an expression for the time of flight which permits an explicit identification of the possibilities. This expression takes the form

$$\sqrt{\mu}\, t = x^3 S \pm \sigma(\sqrt{z}) \beta \sqrt{y} \tag{55}$$

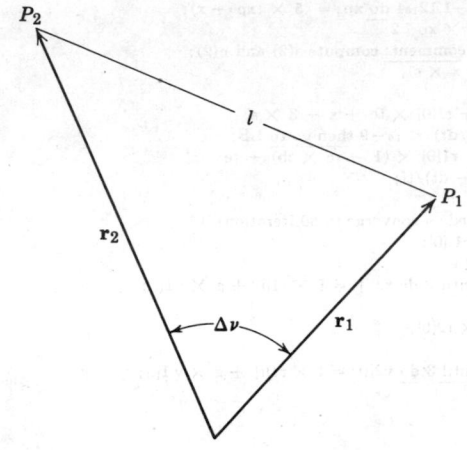

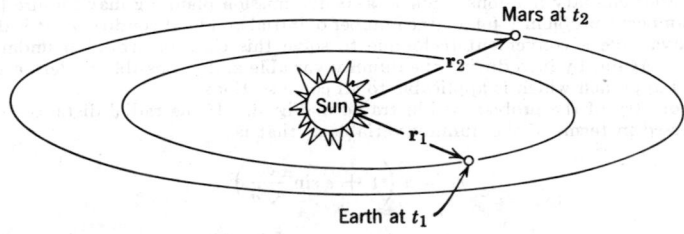

FIG. 3.   Geometry of the orbit determination problem.

where by definition

$$\sigma(\omega) = \left\{ \begin{array}{l} 1, \ \omega \ \text{imaginary} \\ (-1)^n, \ n\pi < \omega < (n + 1)\pi \end{array} \right\}$$

Writing the time of this form permits a useful generalization; namely, it allows transfers from P_1 to P_2 which pass through P_2 any number of times before terminating at P_2. That is, n circuits of the same transfer orbit are allowed.   The ambiguity with respect to the direction of motion is resolved by the fact that $\mathbf{v}_1$ has a component along the positive direction of $\mathbf{r}_2$ whenever, for some integer $n \ \Delta\nu - 2n\pi < \pi$. This allows Eqs. 54 and 55 to be rewritten as

$$y = r_1 + r_2 - \frac{A(1 - zS)}{\sqrt{C}} \qquad (56)$$

and

$$\sqrt{\mu} \, t = x^3 S + A\sqrt{y} \qquad (57)$$

respectively, where $A = \mathrm{e}\,(\Delta\nu)\beta$ and is the parameter dependent on the direction of motion.

Note that t has a monotonic dependence on z and, in general, there may be two times associated with two different values of z but associated with the same value of the semi-major axis.

A solution of the problem of determining the orbit for a transfer between r_1 and r_2 in time t follows from these results in a straightforward manner. An initial guess for z is made and time of flight computed is compared with the given time of flight. A Newton iteration based on the derivative

$$\mu \frac{dt}{dz} = x^3 S' - \frac{3C'S}{2C} + \left(\frac{3S \sqrt{y}}{C} + \frac{A}{x} \right) \tag{58}$$

is used to improve initial and subsequent estimates of z until convergence on t occurs. A computational algorithm for solving Lambert's problem follows.

```
procedure Deyst (r1, v1, tf, dir, br, r2, v2, poss);
value tf, dir, br; array r1, v1, r2, v2;
real tf; integer dir, br; boolean poss;
comment Deyst method (ref. R. H. Battin, Astronautical Guidance, McGraw-Hill, 1964, pp 85-87).
  Given position vectors, r1 and r2, the time of flight, tf, the direction of flight, dir, and the branch,
br, find velocity vectors, v1 and v2.
begin
  own integer i;
  own real rs, a, ze, z, t, y, x, s, c, ds, dc, csr, ysr, dtdz, zl, zu;
  procedure comp;
  begin i: = i + 1;
    if i < 100 then
      begin Battinder (z, s, c, ds, dc);
        csr: = sqrt (c);
        y: = rs − a × (1 − z × s)/crs;
        if y > 0 then
        begin ysr: = sqrt(y);
          x: = ysr/csr;
          t: = x ↑ 3 × s + a × ysr;
          dtdz: = x ↑ 3 × (ds − 1.5 × x ↑ 2 × s × dc/y) + a/8 × (3 × s × x ↑ 2/ysr + a/x);
        end
      end else
        error ("Deyst failed to converge in 100 iterations")
  end comp;
  i: = 0;
  mag (r1); mag (r2);
  rs: = r1[0] + r2[0];
  a: = dir × sqrt (r1[0] × r2[0] + dot (r1, r2));
  if br = 0 then
  begin z: = zl: = 0;
    zu: = pi2 ↑ 2;
    LG: comp;
    if y ≤ 0 then
    begin z: = .5 × (z + zl);
      go to LG.
    end;
    if abs (t − tf) > 10 −9 × abs (tf) then
    begin zl: = z;
      ze: = z + (tf − t)/dtdz;
      z: = if ze ≥ zu then .5 × (z + zu) else ze;
      go to LG
    end
  end else
  begin zl: = ( (br + 1) ÷ 2) × pi2;
    zu: = (zl + pi2) ↑ 2;
    zl: = zl ↑ 2;
    if odd (br) then
    begin z: = zu;
      LC : z : = .5 × (zl + z);
```

```
comp;
if dtdz ≥ 0 ∨ t ≤ tf then go to LC;
LD : if abs (t − tf) >₁₀−9 × abs (tf) then
begin z := z + ιtf − t)/dtdz;
    comp;
    if dtdz ≥ 0 ∨ z ≥ zu then go to LB;
        go to LD
    end
end else
begin z := zl;
    LE : z := .5 × (zu + z);
    comp;
    if dtdz ≤ 0 ∨ t ≤ tf then go to LE;
    LF : if abs (t − tf) >₁₀−9 × abs (tf) then
    begin z := z + (tf − t)/dtdz;
        comp;
        if dtdz ≤ 0 ∨ z ≤ zl then
        begin LB : poss = false;
            go to LA
        end;
            go to LF
        end
    end
end;
poss := true;
ds := 1 − y/r1[0];
dc := a × ysr;
for i := 1 step 1 until 3 do v1[i] := (r2[i] − ds × r1[i])/dc;
mag (v1);
ds := 1 − y/r2 [0];
for i := 1 step 1 until 3 do v2 [i] := (ds × r2[i] −
r1 [i])/dc;
mag (v2);
LA :
end Deyst;
```

After the value of z, y, and x which yield the desired t are determined by iteration, v_1 and v_2 may be found from

$$v_1 = \frac{1}{g} (r_2 - fr_1) \qquad (59)$$

$$v_2 = \frac{1}{g} (\dot{g}r_2 - r_1) \qquad (60)$$

where

$$f = 1 - (y/r_1) \qquad (61)$$

$$g = A\sqrt{y/\mu} \qquad (62)$$

$$\dot{g} = 1 - (y/r_2) \qquad (63)$$

Using the methods described in Sec. 5, the classical orbital elements may be found.

68. MISSION PLANNING

Introduction

In the previous section a solution was developed to the problem of determining the elements of an orbit passing through a position r_1 at a time t_1, and through r_2 at t_2. The orbit determination problem is usually part of an even larger problem, mission planning. Mission planning is partly concerned with the determination of suitable departure and arrival times for given extraterrestrial missions. In this section there is presented a simple technique for finding "launch windows," that is, time intervals suitable for departing point P_1. Although there are usually many other constraints which must be satisfied in addition to the basic requirements of any mission, the sole criterion used herein to assess the suitability of a launch time is the so-called "characteristic velocity." By assumption space maneuvering will be considered to be a sequence of impulsive velocity changes, and by definition, the characteristic velocity is the scalar sum of all the velocity changes required to complete a given mission. To illustrate this idea, Fig. 4 shows a transfer orbit

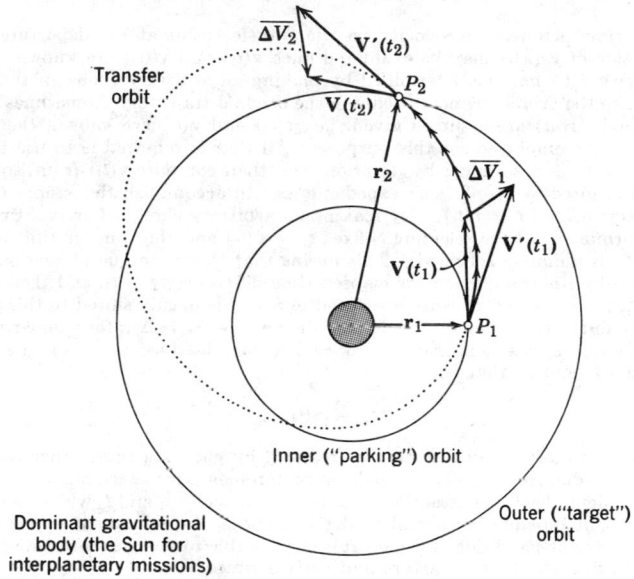

FIG. 4. A coplanar transfer between circular orbits.

between two coplanar circular orbits. If the object of the mission were to maneuver from the inner orbit to the outer orbit and it is elected to make the change as shown, two impulsive velocity changes would be required, ΔV_1 and ΔV_2. In this particular case, the characteristic velocity v_c, is equal to the algebraic sum of $|\Delta V_1|$ and $|\Delta V_2|$. Clearly, there are many ways the transfer could be executed. In each of these cases the characteristic velocity can be computed, and for a given mission its magnitude will depend on such variables as the time of departure, the number of impulsive velocity changes, and the time at which they are made. It is beyond the scope of this text and, in fact, it is sometimes beyond the reach of the most sophisticated analysis, to determine the optimum sequencing of velocity changes to give the minimum value of the characteristic velocity (note—the cost of a mission increases with increasing v_c and this is why the characteristic velocity has such significance in mission planning). Therefore, this section is limited to a search for the better ways to accomplish those missions which can be performed using one, or at most two, impulsive velocity changes. Launch windows for the mission will be identified as a result of this search, thereby satisfying the main objective of this brief treatment of mission planning.

Single-Impulse Transfers

There are several missions of current interest which require at least one impulsive velocity change to accomplish. Among them are certain types of intercepts, probes of deep space by instrumented craft, and nonstop flyby reconnaissance of celestial bodies. Typically, these missions do not impose stringent boundary conditions on the spacecraft's velocity at P_2; usually there is considerable latitude in the specification of the velocity at arrival. As a result, once the transfer orbit is established, there is generally no further need to correct the spacecraft's velocity, except for minor navigational adjustments enroute.

To establish the launch windows for this class of missions, the magnitude of the velocity change necessary to establish the transfer orbit is computed first. With the reference to Fig. 4 it is seen that the change amounts to

$$\Delta v_1 = |\Delta \mathbf{v}_1| = |\mathbf{v}'(t_1) - \mathbf{v}(t_1)| \tag{64}$$

where the prime denotes the velocity on the transfer orbit at the departure P_1. The right-hand side of Eq. 64 may be evaluated once $\mathbf{v}(t_1)$ and $\mathbf{v}'(t_1)$ are known.

The velocity $\mathbf{v}(t_1)$ may be determined by making an arbitrary choice of the departure time t_1. Since the orbital elements defining the original trajectory (sometimes referred to as a "parking" orbit) are assumed given, i.e., $\mathbf{r}(t_0)$ and $\mathbf{v}(t_0)$ are known, the method of Section 8 may be employed for this purpose. All that is required is to use the elapsed time difference $t_1 - t_0$ to find z by iteration, and then compute $\mathbf{v}(t_1)$ from Eq. 50. The algorithm presented in Sec. 8 is an expedient way to accomplish this step. Choosing a departure time t_1 fixes $\mathbf{r} \equiv \mathbf{r}(t_1)$. By making an arbitrary choice of arrival time t_2, $\mathbf{v}'(t_1)$ may be determined. Since selecting t_2 fixes $\mathbf{r}_2 = \mathbf{r}(t_2)$ and the transfer time $t_2 - t_1$, the transfer orbit is uniquely specified. This means that the technique of Section 9 may be used to find z by iteration, using the elapsed time difference $t_2 - t_1$, and then computing $\mathbf{v}'(t_1)$ from Eq. 59. The algorithm presented in Sec. 9 is ideally suited to this purpose.

With $\mathbf{v}(t_1)$ and $\mathbf{v}'(t_1)$ evaluated, the cost of the mission, $v_c = \Delta v_1$, for a departure at time t_1 and an arrival at time t_2 is known. Since the cost clearly depends on the arbitrarily selected times t_1 and t_2, that is,

$$v_c = v_c(t_1, t_2) \tag{65}$$

it is reasonable to ask if the cost may be reduced by choosing some other pair of times (t_1, t_2). In fact, due to the severe propulsion requirements for space missions, it is not at all unreasonable, indeed it is essential, to ask for a choice of t_1 and t_2 which minimizes the cost. As it is not possible in general to derive a closed form analytic expression of the function on the right-hand side of Eq. 65, it is not possible to use conventional optimization techniques to find the best departure and arrival times.

A brute force procedure for finding the best times is to program a high-speed digital computer to do a double iteration on t_1 and t_2, to have the computed v_c for each pair (t_1, t_2) printed out, and to examine the results for the minimum v_c. Figure 5 illustrates what the hypothetical result of executing such a program might look like. In the figure those pairs $(t_1, t_2 - t_1)$ having the same v_c requirements are connected, and the contours so produced have been fictitiously labeled to show how the characteristic velocity might vary. Time of flight, $t_F = t_2 - t_1$, is used for the ordinate of the plot instead of t_2 to make the contours easier to interpret. These contours need not necessarily be closed and simply connected nor will there always be a unique minimum for v_c. A suitable time for departure can be chosen simply by stipulating that the v_c requirement be within the performance capability of the rocket used to make the velocity change. Thus, if the rocket can produce velocity changes of up to say $1.0 Du/Tu$ (approximately 26,000 fps), then the launch window is the time span indicated on Fig. 5. The range of flight times corresponding to this window is also shown. One should remember that with each pair of times there is uniquely associated a transfer trajectory whose orbital elements have been found by the method of Sec. 5.

As mentioned earlier, there may be other constraints on the mission which would interfere with a launch within the window. For example, the transfer trajectory selected might intersect the path of some other celestial body at a moment when there is a possibility of a collision, or perhaps the time of flight is too long for the endurance of the on-board life support system. Factors such as these must normally be considered in mission planning.

Although there are special cases of single-impulse maneuvers of some interest, which can be treated analytically, they will not be examined herein. The next topic is the general case of a two-impulse maneuver.

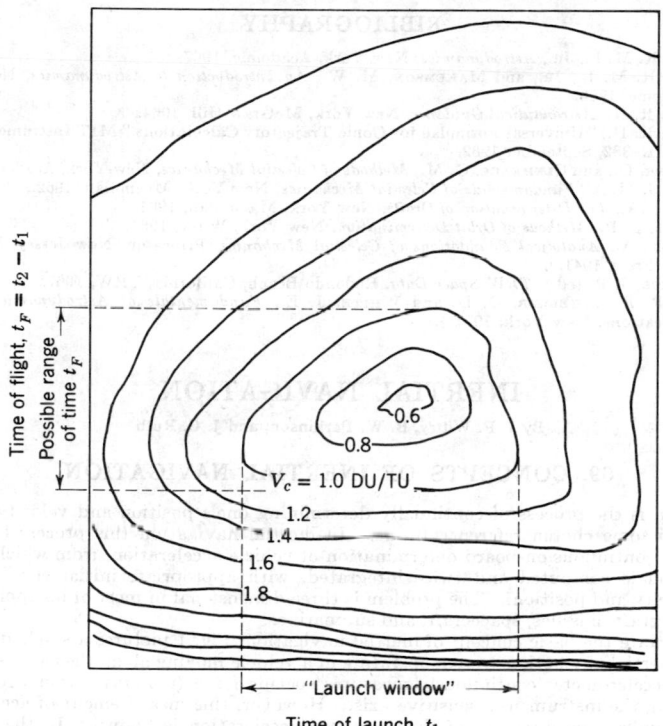

FIG. 5.   Single-impulse transfer.

Two-Impulse Transfers

For a large class of space missions at least two velocity changes are required.   Generally speaking, the first change is necessary to establish the vehicle on a transfer orbit to the destination.   The second one is needed at the destination in order to satisfy some mission-imposed boundary conditions.   For example, a spaceship may be required to orbit the destination planet so that a landing site may be selected.   In this case a velocity change must be made to switch from the transfer orbit to the planetocentric orbit.   Other missions, exemplary of this type, are spacecraft rendezvous and soft landings.

Referring again to Fig. 4, it is easy to see that the procedure described in the previous section for computing Δv_1 at P_1 is also applicable to the computation of Δv_2 at P_2.   Therefore, if the velocity at P_2 on the transfer orbit is denoted by $\mathbf{v}'(t_2)$ and the velocity desired in the terminal orbit at P_2 is denoted by $\mathbf{v}(t_2)$, then the velocity change required is

$$\Delta v_2 = |\Delta \mathbf{v}_2| = |\mathbf{v}(t_2) - \mathbf{v}'(t_2)| \tag{66}$$

and the characteristic velocity for the entire mission is

$$v_c = \Delta v_1 + \Delta v_2 \tag{67}$$

Again, the launch windows are determined by computing the characteristic velocity, this time by Eq. 67, for pairs of launch and arrival times and identifying the time intervals for which the required v_c is within the performance capability of the rocket.

BIBLIOGRAPHY

1. BAKER, R. M. L., JR., *Astrodynamics*, New York, Academic, 1967.
2. BAKER, R. M. L., JR., and MAKEMSON, M. W., *An Introduction to Astrodynamics*, New York, Academic, 1960.
3. BATTIN, R. H., *Astronautical Guidance*, New York, McGraw-Hill, 1964.
4. BATTIN, R. H., "Universal Formulae for Conic Trajectory Calculations," MIT Instrumental Lab. Rept. R–382, September 1962.
5. BROUWER, D., and CLEMENCE, G. M., *Methods of Celestial Mechanics*, New York, Academic, 1961.
6. DANBY, J. M. A., *Fundamentals of Celestial Mechanics*, New York, Macmillan, 1962.
7. DUBIAGO, A., *The Determination of Orbits*, New York, Macmillan, 1961.
8. ESCOBAL, P. R., *Methods of Orbit Determination*, New York, Wiley, 1965.
9. WINTNER, A., *Analytical Foundations of Celestial Mechanics*, Princeton, New Jersey, Princeton Univ. Press, 1941.
10. KENDRICK, J. B. (ed.), *TRW Space Data*, Redondo Beach, California, TRW, 1967.
11. BATE, R. R., MUELLER, D. D. and WHITE, J. E., *Fundamentals of Astrodynamics*, Dover Publications, New York, 1971.

INERTIAL NAVIGATION

By J. P. Wittry, B. W. Parkinson, and J. C. Ruth

69. CONCEPTS OF INERTIAL NAVIGATION

Navigation is the process of continually determining one's position and velocity vectors relative to some chosen reference frame. In inertial navigation this process is accomplished by continuous on-board determination of vehicle acceleration, from which inertial acceleration is computed and twice integrated, with appropriate initial conditions, to yield velocity and position. The problem is three dimensional in most of its applications; namely aircraft, missiles, spacecraft, and submarines.

To illustrate the basic concept of inertial navigation, Fig. 1 pictures a single-degree of-freedom inertial naviagtion system operating in a vehicle moving along the earth's surface. A linear accelerometer continuously measures specific force (nongravitational force/unit mass) along the instrument's sensitive axis. However, this measurement of acceleration is useless unless the direction of the measured acceleration is known. In the example shown, the gyro is used to provide a stable accelerometer platform on the vehicle. The gyro senses any platform rotation due to irregular vehicle rotation by sensing relative displacement between the gyro spin axis and the gyro case. This relative displacement is sensed as an electrical error signal, amplified, and used to drive a gimbal torque motor which rotates the platform relative to the vehicle so as to null the error signal from the gyro. This continuous feedback action from the gyro to the platform maintains the platform, and hence the sensitive axis of the accelerometer, in its initial angular orientation with respect to an inertial reference frame. Thus the direction of the measured acceleration is always known, and the vehicle velocity and position can be computed.

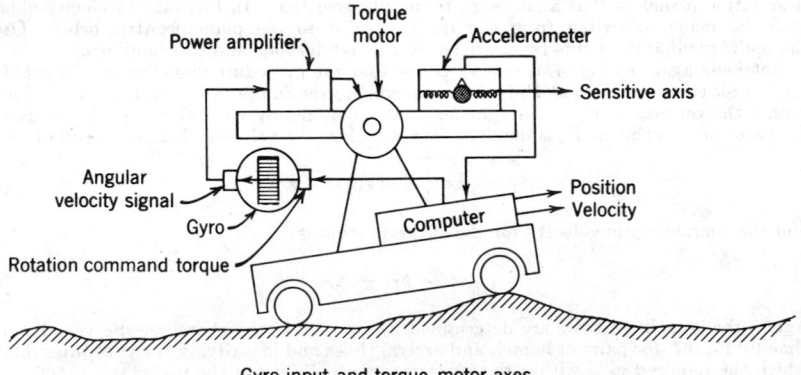

FIG. 1.   Simple single-degree-of-freedom inertial navigational system (small scale).

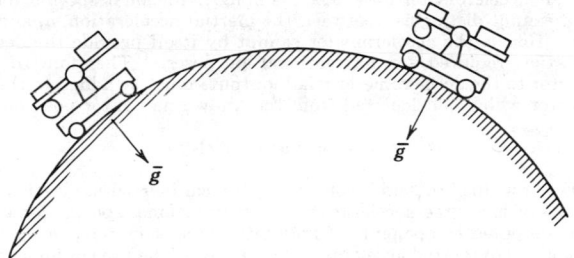

Fig. 2. Simple single-degree-of-freedom inertial navigation system (large scale).

Figure 2 shows the same problem in larger perspective to further illustrate the concept. As the vehicle moves along the surface of the curved earth, it is desired that the sensitive axis of the accelerometer be kept oriented not in a fixed direction with respect to an inertial frame but rather fixed along the local horizontal so that velocity and position along the earth's surface is computed. To accomplish this, the platform is caused to rotate as a function of changing position as determined by the navigation computer. The desired platform orientation change is effected by the same gyro-platform servo loop previously described. The appropriate platform alignment signal generated by the computer causes a current to be fed to a torque generator in the gyro (Fig. 1), causing the gyro spin axis to precess and thus generate an "artificial" error signal which causes the desired platform rotation. This same system provides for initial alignment of the platform to the desired orientation.

The concept outlined above for a single-degree-of-freedom inertial navigation system can readily be implemented in three degrees of freedom by providing a multigimballed platform carrying three accelerometers with their sensitive axes arranged in an orthogonal triad, and with three gyros mounted on the platform to sense platform rotation about three axes. In practice, two two-degree-of-freedom gyros are sometimes used in place of three single-degree-of-freedom gyros; in systems which must navigate for more than 10 or 15 minutes, only horizontal accelerometers may be used, with navigation in the vertical accomplished by another means. It should be noted that the process of platform stabilization described here is continuous and functions on the basis of nearly infinitesimal error signals from the gyro, so that precise platform orientation is continuously maintained even in the presence of rapid and large changes of vehicle attitude.

The concept described above is based on the mechanization of a stabilized accelerometer platform. An alternative method is the "strapdown" system in which the accelerometers are fastened to the vehicle so that the measured acceleration components are in a vehicle-fixed reference frame. Gyros are also fastened to the vehicle to measure the attitude changes of the vehicle, hence of the vehicle-fixed reference frame. The gyro-derived attitude information is input to the navigation computer along with the measured accelerations, and the computer calculates coordinate transformations so that the measured accelerometer readings can be properly interpreted.

In either the stabilized platform or strapdown mechanizations, the inertial navigation system is completely self-contained and independent of external signals or information after the initial alignment.

Accelerometers

The external forces acting on the vehicle are of two basic types; (1) that due to gravitational attraction, and (2) nongravitational forces such as thrust and aerodynamic lift and drag. That is,

$$\bar{f} + \bar{g} = \bar{a} \qquad (1)$$

where $\bar{f}$ = summation of nongravitational specific forces acting.

$\bar{g}$ = gravitational specific force.

$\bar{a}$ = inertial acceleration.

Unfortunately, an accelerometer measures the nongravitational specific force, output = $\bar{f} = \bar{a} - \bar{g}$, and cannot distinguish between the inertial acceleration, $\bar{a}$, and gravitational acceleration, $\bar{g}$. Hence, the accelerometer cannot by itself provide the determination of inertial acceleration required for navigational purposes. Therefore to get the total acceleration vector to the accelerometer triad, outputs must be added to the gravitational acceleration vector which is calculated from the known navigator position and attitude:

$$\bar{a} = \overline{\text{output}} + \overline{g(\bar{r})} \qquad (2)$$

Figure 3 shows how this problem is solved in a typical inertial navigation system. The three components of measured acceleration (nongravitational specific force), f_x, f_y, f_z, are combined with computed components of gravitational specific force, g_x, g_y, g_z, to provide the three components of inertial acceleration, a_x, a_y, a_z. The navigation computer doubly integrates each of these components of inertial acceleration to determine velocity and position and also continuously computes new values of the components of the gravitational specific force according to a stored algebraic function relating this specific force to the position.

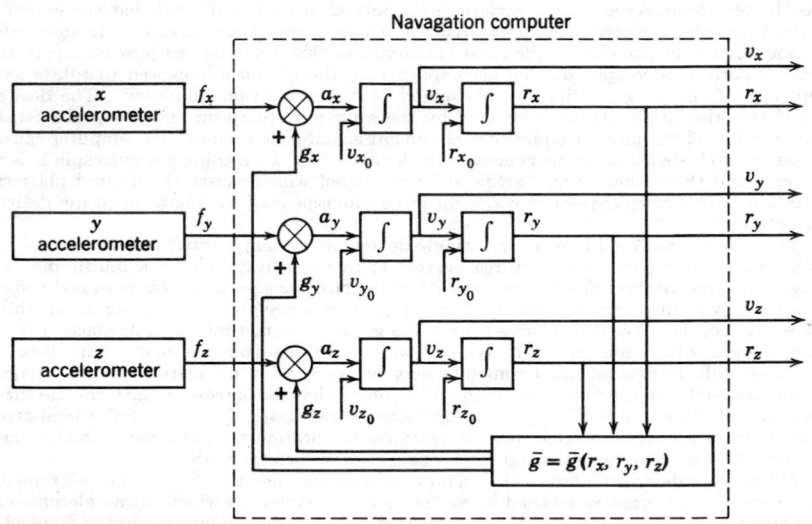

FIG. 3. Typical inertial navigation problem.

The position and velocity outputs shown in Fig. 3 relate to some defined xyz inertial coordinate frame. If it is desired to present this navigation information in terms of some other coordinate frame (e.g., to an aircraft pilot in terms of latitude, longitude, altitude coordinates), the navigation computer performs the additional mathematical task of coordinate transformation based upon the known relationship between the desired display coordinate frame and the inertial frame.

Gyroscopes

A gyroscope is a spinning rotor mounted so as to allow the spin axis to rotate freely about either one or two axes orthogonal to the spin axis. The dynamic motion of the rotor is described by Euler's Equation, viz.

$$\overline{T} = \frac{d}{dt}(\overline{H}) \qquad (3)$$

where $\overline{T}$ = summation of external torques acting on the gyroscope.

$\overline{H}$ = angular momentum of the gyroscope = $I\overline{\Omega}$ where I is the inertia tensor, and $\overline{\Omega}$ is the angular velocity of the gyroscope.

$\dfrac{d}{dt}$ ( ) = time derivative with respect to an inertial reference frame.

In a practical gyro the angular momentum of the rotor is very high, accounting for virtually all of the total gyro angular momentum, and $\overline{H}$ can be taken to be the angular momentum of the rotor itself.

In accordance with basic vector kinematics (the "law" of Coriolis):

$$\frac{d}{dt}\,(\overline{H})_{\substack{\text{inertial}\\ \text{frame}}} = \frac{d}{dt}\,(\overline{H})_{\substack{\text{Gimbal}\\ \text{frame}}} + \overline{\omega} \times \overline{H} \tag{4}$$

where $d/dt\,(\overline{H})_{\substack{\text{Gimbal}\\ \text{frame}}}$ is the time derivative of the rotor angular momentum with respect to the gimbal containing the rotor, and $\overline{\omega}$ is the angular velocity (precessional velocity) of this gimbal with respect to an inertial reference frame. In practice, $d/dt\,(\overline{H})_{\substack{\text{Gimbal}\\ \text{frame}}} \cong$ 0 by careful control of the rotor spin velocity so that the practical equation describing the gyroscope's behavior is

$$\overline{T} = \overline{\omega} \times \overline{H} \tag{5}$$

Figure 4 schematically represents a single-degree-of-freedom gyro mounted on a platform and showing the relationship between the rotor angular momentum vector, applied torque vector (due to platform rotation, for example), and the precession vector. Since the precession is uniquely related to the applied torque, measurement of the precession by the signal generator provides a self-contained measurement of platform rotation about one axis.

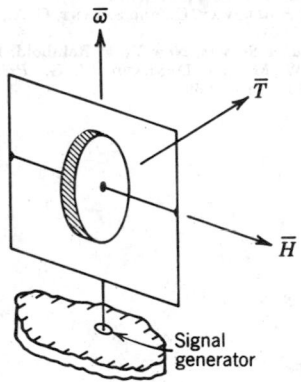

FIG. 4. Schematic of a single-degree-of-freedom gyro.

In practice, this precession angle remains very small, either by nulling with platform torquers (gimballed system) or by nulling using a gyro torquer (strapdown system). Other torques due to various mechanical imperfections in a real gyro also cause precession, referred to as gyro drift because these unwanted torques cause an undesired drift of the spin axis away from its assumed orientation in space. In practice, other intentional torques are introduced by spring restraint and viscous damping on the precession axis to modify the gyro's dynamic behavior.

Incorporating an additional gimbal allows the gyro to receive inputs and experience precession about two axes. Of course, this two-degree-of-freedom gyro accurately senses rotations along two orthogonal axes.

Vertical Reference

In many mechanizations of inertial navigation systems it is convenient to provide a vertical reference, vertical being the direction of the local gravity vector.   A pendulum is a good vertical indicator when mounted on a stationary base, but on an accelerating base the pendulum moves as a result of the gravitational and inertial forces.   If it were possible to construct a pendulum which would rotate (when accelerated) at just the rotation rate of local gravity vector (as it moves over the earth, it would always indicate the vertical in spite of the base acceleration.   This problem is present in the process of flying over a spherical earth and was first presented, along with a theoretical solution, by Dr. M. Schuler in 1923.   (See Appendix A of Ref. 8 for an English translation of Dr. Schuler's paper.)   For flight over the earth's surface the pendulum must be constructed so as to have an undamped period of 84.4 min.   It is impossible to achieve a physical free pendulum with this characteristic, but it is practical to construct a feedback control system using gyroscopes and accelerometers mounted on the stable platform of the inertial navigation system so as to achieve an equivalent to the physical Schuler pendulum.   Such a system is referred to as "Schuler tuned" and will provide a true vertical reference in spite of the base acceleration.

BIBLIOGRAPHY

1. BROXMEYER, C., *Inertial Navigation Systems*, New York, McGraw-Hill, 1964.
2. DRAPER, C., WRIGLEY, W., and HOVORKA, J., *Inertial Guidance*, New York, Pergamon Press, 1960.
3. FERNANDEZ, M., and MACOMBER, G., *Inertial Guidance Engineering*, Englewood Cliffs, New Jersey, Prentice-Hall, 1962.
4. LEONDES, C. T., *Guidance and Control of Aerospace Vehicles* (University of California Engineering and Science Extension Series), New York, McGraw-Hill, 1963.
5. MARKEY, W., and HOVORKA, J., *The Mechanics of Inertial Position and Heading Indication*, New York, Wiley, 1961.
6. O'DONNELL, C. G., *Inertial Navigation Analysis and Design*, New York, McGraw-Hill, 1964.
7. PARVIN, R. H., *Inertial Navigation*, New York, Van Nostrand Co., 1962.
8. PITMAN, G. R., *Inertial Guidance*, New York, Wiley, 1962.
9. SAVANT, C., JR., HOWARD, R., SOLLOWAY, C., and SAVANT, C. A., *Principles of Inertial Navigation*, New York, McGraw-Hill, 1961.
10. SLATER, J. M., *Inertial Guidance Sensors*, New York, Reinhold, 1964.
11. WRIGLEY, W., HOLLISTER, W. M., and DENHARD, W. G., *Principles of Instrumentation*, Cambridge, Massachusetts, MIT Press, 1969.

SECTION 8

ENGINEERING THERMODYNAMICS

BY

MILTON C. STUART AND PAUL J. KIEFER*

* This section is a revision by the authors, in collaboration with G. A. Hawkins, of Section 7 of the first edition.

ENGINEERING THERMODYNAMICS

PRINCIPLES OF THERMODYNAMICS

The science of engineering thermodynamics develops means for determining the nature and amounts of the energy transformations occurring in engineering processes. Articles 1 to 7 present those principles which are common to all processes, regardless of the particular kind of working fluid used and the particular form of the machine or device in which the process is carried out.

1. ENERGY FORMS

The energies encountered in engineering thermodynamics are of several well-defined forms, named and described as follows:

(a) **Mechanical Potential Energy** is the form in which energy is stored by virtue of the relative vertical distance of a body above a horizontal reference plane.

(b) **Mechanical Kinetic Energy** is the form in which energy is stored by virtue of the relative motions existing between parts of a system.

(c) **Internal Energy** is the form in which energy is stored within a body, such as a quantity of gas, liquid, or solid, by virtue of the relative motions of, and the forces between, the molecules or atoms composing the body. It is evidenced by some properties of the body, usually temperature, but also by pressure and specific volume, and by physical phase, i.e., whether solid, liquid, or gaseous. Energy stored in substances and released in chemical reactions also comes under this classification of internal energy. In this case, the energy is ascribed to the arrangement of the atoms within the molecules. Considerable needless confusion has been introduced in many works by referring to internal energy in an ad lib. fashion as " heat." See discussion of the proper use of the term *heat* below.

The energy forms (a), (b), and (c) just described have the common characteristic that they are forms in which energy may be stored away for possible future use. The two forms (d) and (e) about to be described have quite different characteristics in that they represent the forms or means by which energy may be transferred or transformed.

(d) **Work** is a transient form of mechanical energy by means of which certain transformations of other forms of energy are brought about through the agency of a force acting through a distance. Several examples of energy transformation by means of work may be given. Work done by lifting a body stores mechanical potential energy in the system consisting of the body and the earth. Work done on a body to set it in motion stores mechanical kinetic energy in the system of which the body is one part. When a gas is compressed, work is transformed into internal energy which then exists as energy stored in the gas.

(e) **Heat**, like work, is a transient form of energy. Heat is defined as the energy in transition or transfer from one body to another by virtue of a temperature difference existing between the bodies. Conduction and radiation are the two methods through which heat transfer occurs.

Conduction is energy transfer from one body to another body at a lower temperature by tangible contact. It may be thought of as the transfer of some of the internal molecular kinetic energy of the hotter body to the more slowly moving molecules of the colder body.

Radiation is energy transmission through space from a hotter body to a colder body. In most processes involving heat flow it is impossible to distinguish the exact amounts transmitted by radiation and by conduction, but this distinction need seldom be made in practice as the category heat includes all transfer of energy due to temperature differences existing between two systems.

In identifying the various forms of energy, care must be exercised to preserve two distinctions. First there must be recognized the distinction between the group consisting of the *three stored forms of energy*, mechanical potential, mechanical kinetic, and internal energy; and the group comprising the *two transient forms*, work and heat. A second important distinction is that which exists between *internal energy and heat*. Energy stored in any molecular or atomic system such as energy in air, steam, or fuel comes under the category of internal energy. The name heat is used only to identify energy actually in transition or flow between systems at two different temperatures.

Convection is not a form of energy or energy transformation. When a body with its associated energy is moved from one position to another without state changes or energy transformations, the process is designated by the term *convection*. A typical example of convection is the movement of heated air from one part of a room to another. The internal energy has moved from one location to another, but no energy transformation has occurred. Another example of convection is the movement through a pipe of the kinetic energy or the potential energy associated with the fluid flowing in the pipe, without any energy transformations taking place.

2. UNITS OF ENERGY

Units of Energy. Any form of energy may be expressed quantitatively in any unit of energy. Fundamental units of energy are the *erg* in the metric system and the *foot-pound* in the English system. These units are derived from the standards of mass, length, and time. From the erg or foot-pound are derived numerous other energy units such as the joule, meter-kilogram, kilowatthour, horsepower-hour, etc.

From the earliest days of the science of thermodynamics there has existed an independent set of energy units, the *calorie* and the *British thermal unit* (Btu), related in no way to the standards of force or distance, but independently defined in terms of the thermal properties of water. For example, a widespread definition of the Btu is $1/180$ of the quantity of energy required (heat required) to change 1 lb of water from the ice-point to the steam-point at standard atmospheric pressure. With respect to these units there have evolved no accepted standards of definition of the units, methods of experimental procedure, or numerical results.

In order to recognize a common basis for all energy units and at the same time retain the calorie and Btu, the International Steam Table Conference, meeting in London in 1929, recommended that the international calorie be defined as $1/860$ of an International watthour. By arithmetic, using the defined relations between the English and metric systems, the Btu can be thus established, by definition, as 778.26 ft-lb.*

In honor of Dr. Joule who, in his classical experiment of raising the temperature of water by stirring it, demonstrated the conversion of mechanical potential energy into internal energy by the agency of work, the factor 778.26 is called *Joule's equivalent* and is represented by the symbol *J*. For all engineering purposes *J* may be taken as equal to 778. From the viewpoint of this treatment, *J* is not the "mechanical equivalent of heat," but merely a conversion factor between two energy units.

Since *power* is the time rate of expenditure of energy, we have a series of units of power which correspond to the energy units. Here, also, any power unit may be used to measure the expenditure or transition of energy by any method; either by work, or heat, or electrical energy. The watt (1 joule per sec) and the horsepower (550 ft-lb per sec) are the commonest forms of power units.

Factors for the ready conversion of the various energy and power units are found in Section 1, Tables 47 and 48. Below are given some of the definitions and conversion factors for the energy and power units most frequently encountered in engineering practice.

Conversion Factors for Energy Units.

1 Btu = 778 ft-lb (by definition) = 1055 joules = 252 calories.
1 erg = 1 dyne-cm (by definition).
1 joule = 10^7 ergs (by definition) = 0.73756 ft-lb.
1 horsepower-hour = 2545 Btu = 0.7457 kwhr.
1 kilowatthour = 860 international calories (by definition of calorie) = 1.341 **hp-hr.**

Conversion Factors for Power Units.

1 horsepower = 550 ft-lb per sec (by definition) = 33,000 ft-lb per min = 2545 Btu per hr = 0.7457 kw.
1 watt = 1 joule per sec (by definition).
1 kilowatt = 1000 watts (by definition) = 1.341 hp = 3413 Btu per hr.

---

* See Eric Thorkelson, "The Mechanical Equivalent of Heat, Its Rise and Fall," *Mechanical Engineering*, p. 347, June, 1934. Also E. F. Mueller, "The Passing of the Mechanical Equivalent of Heat," *Mechanical Engineering*, February, 1930.

3. ENERGY TRANSFORMATIONS

Processes. In a broad sense, a *process* is any event in nature in which a redistribution or transformation of energy occurs. Engineering thermodynamics considers chiefly those processes in which energy transformation occurs by means of changes in physical state of fluids.

The processes of engineering thermodynamics may be conveniently classified into well-defined groups in several ways. One classification is into the groups of reversible and irreversible processes. Another classification which is of much value is into the divisions called flow and non-flow processes.

Reversible Processes are those which meet two criteria as follows:

(*a*) The process, after completion, may be caused to occur in an exactly reverse order whereby the immediate system and any other systems associated with it may be returned from their last to their initial state.

(*b*) All the energy which was transformed during the process may be returned from its final to its original form, location, and amount.

Irreversible Processes are those which do not meet the above criteria. Among the conditions which contribute to the irreversibility of a process are the following:

(*a*) Heat flow from a higher to a lower temperature, (*b*) mixing of fluids at different temperatures, (*c*) fluid turbulence, (*d*) fluid or solid friction, (*e*) inelastic deformation.

Some examples of reversible processes are:

(*a*) A frictionless pendulum swinging in a vacuum, (*b*) a gas expanding slowly without friction or turbulence in a heat-insulated system.

Non-flow Processes are those occurring in a container or a space in such a way that the fluid does not flow in or out of the container or space during the process. An example is the expansion of steam in a cylinder during the period when the valves are closed.

Steady-flow Processes are those in which the fluid passes continuously through a region under conditions of steady flow. The conditions for the ideal steady-flow process are:

(*a*) At entrance to the region or apparatus considered, all properties which fix the state of the fluid maintain fixed values, that is, they do not vary with respect to time. Velocity of the fluid at entrance does not vary.

(*b*) Likewise, at exit from the region, fluid properties and velocity do not vary. But note that the exit properties may be and usually are quite different from the entrance properties.

(*c*) The mass flow at exit must equal the mass flow at entrance.

(*d*) At points between entrance and exit, not only may properties change in value, but they may be quite variable and unsteady. It is only at entrance and exit that conditions must be steady.

The steady-flow process or a process which closely approximates steady flow exists in most of the devices and machines employed in engineering practice. Examples are the steam engine, turbine, condenser, pump, boiler, nozzle, valve, and most heat-exchange appliances.

Other common types of processes are defined as follows:

Constant-pressure Process, in which the pressure of the fluid is constant throughout the process.

Constant-volume Process, in which the volume of the fluid is constant throughout the process.

Isothermal Process, in which the temperature of the fluid is constant throughout the process.

Adiabatic Process, in which no heat is added to or removed from the fluid during the process. (Caution: Refer to definition of heat in the previous section.)

Isentropic or Constant-entropy Process, in which the entropy is constant throughout the process.

Each of the five preceding processes may, in general, be either flow or non-flow, reversible or irreversible.

4. ENERGY EQUATIONS

Principal Symbols of Arts. 4, 5 and 6

E = internal energy, Btu per pound.

H = enthalpy, Btu per pound = $E + PV/J$.

J = Joule's equivalent, 778 ft-lb per Btu.

M = mass, pounds (numerically, the same as weight).

P = absolute pressure, pounds per square foot.

PV = flow work (displacement energy), foot-pounds per pound.

Q = energy supplied to or departing from a system or region as heat, Btu per pound.

$d'Q$ = a small quantity of heat (the inexact differential of Q).

Q_R = the unavailable energy of a process, Btu per pound.

S = entropy per pound.

T = absolute temperature, degrees fahrenheit.

T_R = absolute temperature of a receiver into which the unavailable energy of a process is rejected.

t = ordinary temperature, degrees fahrenheit.

U = velocity, feet per second.

$\dfrac{U^2}{64.34}$ = mechanical kinetic energy of a fluid stream or jet, foot-pounds per pound ($64.34 = 2 \times 32.17$).

V = specific volume, cubic feet per pound.

W = a general symbol for work, including all the mechanical energies or effects, namely: flow work, shaft work, and mechanical kinetic and potential energies.

W_s = work supplied to or departing from a system by means of a shaft, or its equivalent, foot-pounds per pound of fluid.

Z = elevation, ft.

$(\int)$ = symbol for the integral of a function or quantity which completes a cycle.

Subscript 1 designates entrance to a system, or start of a process.

Subscript 2 designates exit from a system, or end of a process.

The First Law of Thermodynamics is the *Principle of the Conservation of Energy*. A statement of this principle most useful in this work is: In any process the net amount of energy added to a system equals the net change in energy within the system. This principle is expressed in a general energy equation as follows:

$$W + JQ = J(E_2 - E_1) \qquad (1)$$

in which W stands for work and all other forms of mechanical energy; Q stands for heat, and $(E_2 - E_1)$ represents the change in internal energy within the system.

Equation 1 as written is for work and heat both entering the system. If work or heat leaves the system during the process, use minus signs for W or Q, or transpose W or Q to the right-hand side of the equation.

For mechanically reversible processes

$$W = -\int P\,dV \qquad (2)$$

The minus sign is required because W is taken as positive when work is added to the system, and for this condition dV is negative.

For all reversible processes,

$$Q = E_2 - E_1 + \int P\,dV/J \qquad (3)$$

and in the differential form

$$d'Q = dE + P\,dV/J \qquad (4)$$

Note that equations 2, 3, and 4 apply for reversible processes only, but 1 applies for irreversible processes also.

Application of the Conservation of Energy Principle to the Steady-flow Process is made by reference to a diagrammatical sketch (Fig. 1) representing any device through which *steady flow* exists.

Energy may enter the system in the following forms and places.

(a) Mechanical potential energy (Z_1) entering with the fluid at section 1.

(b) Mechanical kinetic

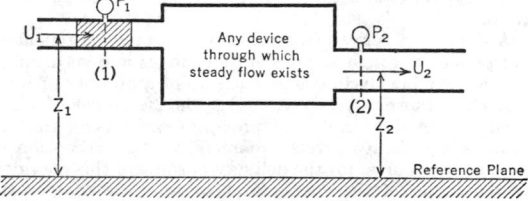

FIG. 1. Steady Flow Device

energy $\dfrac{U_1^2}{64.34}$ entering with the fluid at section 1.

(c) Internal energy (E_1) entering with the fluid at section 1.

(d) Work required to cause flow into the system against the pressure existing at the entrance section 1 in amount equal to the product of pressure and specific volume, viz., P_1V_1 foot-pounds per pound. This energy is termed flow work or displacement energy.

(e) Heat (Q) may be supplied to the system at any point between the entrance section 1 and the exit section 2, for example, as to the water in a boiler.

(f) Shaft work (W_s) may be added to the system at any point between the entrance and exit by means of a shaft, for example, as in a compressor.

Energy may leave the system in six forms analogous to those entering: potential energy, kinetic energy, internal energy, and flow work, at the point at which the fluid leaves the system, and heat or work delivered from the system at some point between the entrance and exit. The total amount of energy within the system remains constant.

The steady flow energy equation then becomes:

$$Z_1 + \frac{U_1{}^2}{64.34} + JE_1 + P_1V_1 + W_{\text{in}} + JQ_{\text{in}} = Z_2 + \frac{U_2{}^2}{64.34} + JE_2 + P_2V_2 + W_{\text{out}} + JQ_{\text{out}}$$

Enthalpy. The combination $JE + PV$ occurs here and in other places with such frequency that there has been established a universal custom of combining them into a single term. Since E, P, and V are each properties of the fluid, the sum $E + PV/J$ is a property which is designated by the symbol H, so that, by definition, $H = E + PV/J$. Modern usage gives to this property or function the name *enthalpy* (pronounced ĕn-thăl'py, with accent on the second syllable). Other names in wide usage are heat content and total heat.

The mechanical potential energy term Z is of importance in hydraulic applications, but is usually of negligible value in thermodynamics.

The simplified energy equation for the steady-flow process is:

$$\frac{U_1{}^2}{64.34} + JH_1 + W_s + JQ = \frac{U_2{}^2}{64.34} + JH_2 \tag{5}$$

where W_s and JQ are *net* values.

In *non-flow* processes energy may enter or leave the system as shaft work, W_s, or heat Q, and changes of energy within the system are limited to changes in internal energy, E. The non-flow energy equation is therefore similar to equation 1.

$$W_s + JQ = J(E_2 - E_1) \tag{6}$$

For reversible non-flow processes this becomes

$$JQ = J(E_2 - E_1) + \int P\,dV \tag{7}$$

5. ENTROPY AND THE AVAILABILITY OF ENERGY

The energy equations, though perfectly true, do not contain in themselves the means for evaluating them for all processes. For example, in a certain ideal (reversible) adiabatic process, the energy equation merely states that the work obtained from the process is equal to the change in the internal energy of the fluid, but no information whatever is given as to the amount of work which will be obtained or the magnitude of the change in the internal energy. The first law of thermodynamics is embodied in the energy equations. The second law of thermodynamics, as expressed in the entropy function, is required to make possible the computation of the maximum possible amount of work which may be obtained in certain ideal processes, particularly the reversible adiabatic process, and the reversible cycles upon which ideal heat engines operate.

In order to trace briefly how the entropy concept is developed from the second law of thermodynamics and the Carnot principle, a number of concepts and definitions must be stated.

A **Cycle** is a series of processes which may be repeated in a given order, the working fluid passing through various state changes and returning periodically to its initial state.

A **Reversible Cycle** is one made up of a number of reversible processes.

A **Heat Power Cycle** is a fluid cycle the object of which is to obtain work from heat. Regardless of the number of processes comprising the heat power cycle, three essential elements are always present, namely, (a) the reception of energy as heat from a high-temperature source, (b) the delivery of some of this energy as work, and (c) the rejection of the remainder of the energy as heat to a low-temperature receiver. The reception of energy may consist alternatively of a combustion process occurring with the working fluid, as in the internal-combustion engine.

The **Cycle Efficiency** is the ratio of the work delivered to the heat supplied.

Let Q = heat supplied from the source.

Q_R = heat rejected to the receiver.

W/J = work delivered by the cycle. Then for all cycles, reversible or irreversible,

$$Q = Q_R + W/J$$

(since in a cycle $\Delta E = 0$)

$$\text{Efficiency} = \frac{W/J}{Q} = \frac{Q - Q_R}{Q} = 1 - \frac{Q_R}{Q} \tag{8}$$

The Carnot Principle. In 1824 Sadi Carnot, in one of the most important contributions ever made to science, established the factors which affect the efficiency of cycles, and stated the necessary conditions for maximum efficiency. The full import of *the Carnot principle* may be given by stating it in the form of the three propositions which follow. These propositions are the embodiment of the *second law of thermodynamics*.

Proposition I. No cycle which continuously delivers work by the reception or conversion of energy at a high temperature (with rejection of any residue of energy at a lower temperature) can be more efficient than a reversible cycle operating between a source and receiver at these temperatures.

Proposition II. The efficiency of all reversible cycles working between the same temperatures is the same, irrespective of the differences in the character of the cycle or of the working fluid.

Proposition III. The efficiency of a reversible cycle depends only on the temperatures of the source and of the receiver.

Working with the third proposition of the Carnot principle, Kelvin established a thermodynamic temperature scale which made temperature a function of the amount of heat rejected and the amount of heat received in a reversible cycle. The Carnot principle states that Q_R/Q is a function of the temperature of the receiver and of the source. Kelvin made this temperature function as simple as possible by establishing the ratio of any two temperatures as being equal to the ratio of the amount of heat rejected and the amount of heat received by a reversible cycle operating between these temperatures. By computing the efficiency of a particular reversible cycle (the Carnot cycle) which used a perfect gas as the working substance, and then using the perfect gas as a thermometer, numerical values of the thermodynamic temperature scale were obtained.

The Kelvin Temperature Scale is summarized in the following relation, which may be appropriately called the *Carnot–Kelvin relation*

$$\frac{T_R}{T} = \frac{Q_R}{Q}$$

where T_R = absolute temperature of a receiver.

T = absolute temperature of a source.

Q_R = heat rejected from the reversible cycle to a receiver at a constant temperature T_R.

Q = heat added to a reversible cycle from a source at a constant temperature T.

Since the efficiency of a cycle is $1 - \dfrac{Q_R}{Q}$, the efficiency of any *reversible* cycle operating between the constant temperatures of T and T_R is

$$1 - \frac{T_R}{T} \quad \text{or} \quad \frac{T - T_R}{T} \tag{9}$$

The Carnot Cycle. Carnot further described a particular reversible cycle of much theoretical interest. Contrary to impressions sometimes received, the Carnot cycle need not be operated with a perfect gas as a working fluid. Any fluid may be used, since the Carnot principle states that the efficiency of the reversible cycle depends only on the temperatures of the source and the receiver and not on the nature of the working fluid or on any other consideration. The *Carnot cycle* is described essentially and completely by describing four processes of which it is composed, as follows:

1. A reversible isothermal expansion in which heat in the amount Q is received from a source at a constant temperature T.

2. A reversible adiabatic expansion in which the fluid passes from the source temperature T to the receiver temperature T_R.

3. A reversible isothermal compression in which heat in amount Q_R is rejected to a receiver at a constant temperature T_R.

4. A reversible adiabatic compression in which the fluid is returned from the receiver temperature T_R to the source temperature T, and its original state.

During processes 1 and 2, work is delivered to an external system. During processes 3 and 4, work must be supplied from an external system.

For this cycle, as for all cycles, the net work output $= Q - Q_R$ and the efficiency is $\dfrac{Q - Q_R}{Q}$ or $1 - \dfrac{Q_R}{Q}$. Since, from the Kelvin temperature scale, $\dfrac{Q_R}{Q} = \dfrac{T_R}{T}$, the efficiency of the Carnot cycle becomes $1 - T_R/T$. This is also the efficiency of all reversible cycles operating between temperatures of source and receiver of T and T_R respectively.

Availability of Energy. The idea of availability of energy, like the idea of reversibility, is of essential importance in thermodynamics. The availability of energy refers to the maximum amount of energy of a given process that may be transformed into work. Energy which is in the mechanical stored forms, kinetic or potential mechanical energy, may, by a reversible process, be converted wholly into mechanical work. A part of the internal energy of a gas under pressure may be converted into work. By the process of combustion, the internal chemical energy of a fuel may be converted into heat and then a portion of this heat converted into work by means of the heat power cycle. Any quantity of energy which is already in the form of mechanical energy or which may be converted into mechanical energy or work is called available energy. In all reversible processes there is no change in the availability of the energy evolved in the process. An irreversible process always decreases the available portion of energy involved and thus increases the unavailable portion. When available energy becomes unavailable through the agency of an irreversible process, the energy is said to be degraded.

The availability of the energy for a reversible process is determined by the efficiency of a reversible cycle (any reversible cycle, the Carnot cycle, for convenience) or more simply by the mere statement of the Kelvin temperature scale. If, in a process, a quantity of heat Q is supplied to a system at a temperature T, the amount of energy Q_R which cannot under any circumstances be converted into work, and which must therefore become unavailable, depends upon the realizable receiver temperature T_R and the temperature T at which Q is added. This statement is in accordance with the fundamental relation which has been referred to as the Carnot–Kelvin relation.

$$\frac{Q_R}{Q} = \frac{T_R}{T}; \quad \text{or} \quad \frac{Q_R}{T_R} = \frac{Q}{T}; \quad \text{or} \quad Q_R = T_R \frac{Q}{T}$$

If the temperature T of the source varies from T_1 to T_2 during the addition of the heat Q the expressions above must be given in the integral form

$$\frac{Q_R}{T_R} = \int_1^2 \frac{d'Q}{T} \quad \text{or} \quad Q_R = T_R \int_1^2 \frac{d'Q}{T} \tag{10}$$

It is only for reversible processes that the above expressions are true.

Recalling that the reversible process gives the maximum amount of available energy and therefore the minimum amount of unavailable energy, it follows that for an irreversible process the amount of unavailable energy which must be discarded will be greater than the amount of unavailable energy of the reversible cycle. The important conclusion follows that, for the *irreversible* process,

$$Q_R > T_R \int \frac{d'Q}{T} \quad \text{and} \quad \frac{Q_R}{T_R} > \int \frac{d'Q}{T} \tag{10a}$$

Entropy. The above expression is known as the *inequality of Clausius*. Because of its essential utility in evaluating the availability of energy of a *reversible process* as given in equation 10, the function $\int \frac{d'Q}{T}$ was given by Clausius the name *entropy*, applicable, however, only to reversible processes.

For $d'Q$ in the above entropy expression we may substitute its value from the general energy equation for reversible process, $d'Q = dE + PdV/J$, and obtain

$$S = \int \frac{dE + PdV/J}{T}$$

This function, S, may be shown to have a very important property, namely, that its value for any given process in which a fluid passes from state 1 to state 2 is also its value for every process by which the fluid may pass from state 1 to state 2. This property is described by saying that *entropy is a point function*, meaning that the change in its value for processes depends only on the end points of the process and not at all upon the particular path taken by the process between the end points. For various reversible processes between states 1 and 2, Q may have various values, but $\int_1^2 \frac{d'Q}{T}$ and $\int_1^2 \frac{dE + PdV/J}{T}$ may have but one value. Also, it may be shown that the ratio Q_R/T_R for all processes between two states will have a single value regardless of the nature of the process and whether reversible or irreversible. For this reason entropy is sometimes defined as Q_R/T_R. An alternate definition of entropy change, true, however, only for reversible processes, is:

$$S_2 - S_1 = \int_1^2 \frac{dE + PdV/J}{T} \tag{11}$$

If the process is reversible and the path is known, the entropy change may be computed directly from equation 11. If the process is irreversible the fluid is not in a state of equilibrium at any point. The path may not be stated functionally and may not be depicted graphically. We may, however, substitute for any irreversible process any reversible process or series of reversible processes passing between the same end points. The computed entropy change for the reversible processes thus substituted is the entropy change for the original irreversible process. But *note carefully* that for the irreversible process

$$S_2 - S_1 \neq \int_1^2 \frac{d'Q}{T} \tag{12}$$

The entropy change $(S_2 - S_1)$ for any process being known, the amount of available energy Q_a for the process may be determined thus:

$$\begin{aligned} Q_R &= T_R (S_2 - S_1) \\ Q_a &= Q - Q_R \end{aligned} \tag{13}$$

In a reversible adiabatic process, $Q =$ zero, $S_2 - S_1 = \int_1^2 \frac{d'Q}{T} = 0$, $Q_R =$ zero, and there is no change in availability. The principal use of entropy in engineering computations is to determine the final state in the reversible adiabatic for which $S_2 = S_1$.

Entropy as a Coordinate. Since entropy is a property or state function of a fluid it may be employed as a coordinate, in connection with various of the other properties, in the graphical representation of state changes. The more common diagrams in which entropy is so employed, are: (a) the temperature-entropy (T, S) diagram in which absolute temperature is the ordinate and entropy per pound of fluid is the abscissa, and (b) the "Mollier" diagram (named after its inventor) in which enthalpy is the ordinate and entropy again the abscissa. The latter is of more particular convenience in connection with the portrayal of processes in which steady flow occurs.

The temperature-entropy diagram acquires a particular utility from the relationships between the properties T and S as they enter into the definition of the change of entropy of a fluid as follows:

For any process, reversible or irreversible,

$$Q_R/T_R = \Delta S, \quad \text{or} \quad Q_R = T_R \Delta S \tag{14}$$

For reversible processes,

$$\left. \begin{aligned} Q_R/T_R &= Q/T = \Delta S \\ Q &= T \Delta S \\ Q &= \int_a^b T dS \end{aligned} \right\} \tag{15}$$

To illustrate the use of these expressions in connection with the T, S diagram let the line ab in Fig. 2 be a graphic record of the simultaneous values of the absolute temperature and the entropy of a fluid during any assumed process and state change, and let the horizontal line mn be a contour line of constant temperature at absolute temperature T_R. From equation 14 the change in the unavailable energy chargeable to the fluid as a result of the process is represented by the cross-hatched area below the line mn (the area $mnb'a'm$) since that area is the product of T_R and ΔS. Furthermore, from equation 15, if the process were one without fluid friction or other internal irreversibility and were accomplished solely by the reception of energy by the fluid (as heat), then the area between the line ab and the S axis (the area $abb'a'a$) would represent the amount of heat received by the fluid during the process, since that area represents the summation of all infinitesimal products of T and dS between the limits of a and b, or equals $\int_a^b T dS$.

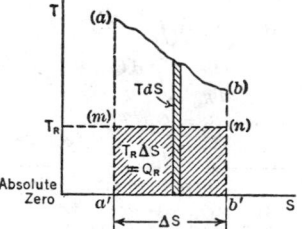

FIG. 2. *T–S* Relation in Process

A constant temperature (isothermal) state change is represented by a straight line parallel to the S axis. Also a reversible adiabatic, being a constant entropy or isentropic process, is a straight line parallel to the T axis. For a fluid undergoing the sequence of reversible processes of the Carnot cycle the state changes are represented by the sequence of lines forming the rectangular figure $abmna$ of Fig. 3. In that figure the area $a'abb'a'$ again represents and is a measure of the "heat" energy received by the fluid during the isothermal expansion at T, the area $a'mnb'a'$ measures the unavailable energy rejected during the isothermal compression at T_R and also, since the work output of the cycle is the difference between the energy supplied and that rejected, the area $mabnm$ measures the available portion of the energy supplied. The thermal efficiency of the cycle is thus the ratio $\dfrac{mabnm}{a'abb'a'}$, which equals the ratio $\dfrac{T - T_R}{T}$. The advantageous effect of increased T and of decreased T_R is thus portrayed graphically.

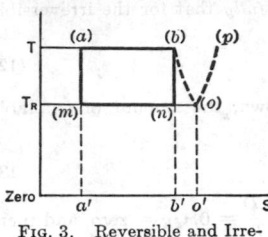

FIG. 3. Reversible and Irreversible Cycles

For the irreversible adiabatic, due to the increase of unavailable energy (Q_R) which is inherent in any such process, the essential and invariable characteristic is an increase in the entropy of the fluid as a result of the process, this irrespective of whether the temperature may tend to decrease (as in an expansion) or to increase (as in a compression). Thus an irreversible adiabatic expansion from T to T_R starts at point b and ends at some point such as o, with increase of entropy, and a like irreversible adiabatic compression may start at o and end at some point p with an increase of entropy. The exact paths for irreversible processes cannot be shown graphically. For the expansion the increase of unavailable energy with respect to a receiver at T_R would be represented by the area $b'noo'b'$.

6. SUMMARY OF ENERGY EQUATIONS AND ENTROPY EXPRESSIONS

Table 1. Distinctions to be Made between Reversible and Irreversible Processes

For All Reversible Processes		For All Irreversible Processes *
$W + JQ = J(E_2 - E_1)$	(1)	Same as for reversible
$W = -\int PdV$	(2)	$W < \int PdV *$
$Q = (E_2 - E_1) + \int PdV/J$	(3)	$Q < (E_2 - E_1) + \int PdV/J$
$d'Q = dE + PdV/J$	(4)	$d'Q < dE + PdV/J *$
$dS = \dfrac{dE + PdV/J}{T}$	(11)	For an irreversible process, substitute any reversible process between the same end points, and compute entropy change from (11)
$dS = \dfrac{d'Q}{T}$		$dS > \dfrac{d'Q}{T}$
$S_2 - S_1 = \int_1^2 \dfrac{d'Q}{T}$	(12)	$S_2 - S_1 > \int_1^2 \dfrac{d'Q}{T}$
$\dfrac{Q_R}{T_R} = \int \dfrac{d'Q}{T}$	(10)	$\dfrac{Q_R}{T_R} > \int \dfrac{d'Q}{T}$ (10a)
$S_2 - S_1 = Q_R/T_R$	(13)	Same as for reversible
$Q_R = T_R (S_2 - S_1)$		Same as for reversible
For Reversible Adiabatic		**For Irreversible Adiabatic**
$Q = 0$		Same as for reversible
$\int \dfrac{d'Q}{T} = 0$		Same as for reversible
$S_2 - S_1 = 0$		$S_2 > S_1$
$Q_R = 0$		$Q_R > 0$
$\dfrac{Q_R}{T_R} = 0 = S_2 - S_1$		$\dfrac{Q_R}{T_R} = (S_2 - S_1) > 0$
$dE = -PdV/J$		$dE \neq PdV/J *$
$W = J \int dE = -\int PdV$		$\int dE = W/J \neq \int PdV/J *$

* For irreversible processes, the path cannot be definitely stated. Such expressions as $\int PdV$ must be considered as pertaining to a substitute reversible path between the same end points.

Table 1—*Continued*

For All Reversible Cycles	For All Irreversible Cycles
$(\oint)\, dS = 0$	Same as for reversible
$(\Sigma)Q = (\Sigma)W/J$	Same as for reversible
$(\oint)\, dE = 0$	Same as for reversible
$(\Sigma)\dfrac{Q_R}{T_R} = (\oint)\dfrac{d'Q}{T} = 0$	$(\Sigma)\dfrac{Q_R}{T_R} > (\oint)\dfrac{d'Q}{T}$
Efficiency $= \dfrac{Q - Q_R}{Q} = \dfrac{W}{JQ}$ (8)	Same as for reversible
Efficiency $1 - T_R/T$ (9) (with constant source temperature)	Eff. $< (1 - T_R/T)$

Steady-flow Energy Equation for Reversible and Irreversible Processes

$$\frac{U_1{}^2}{64.34} + JH_1 + W_s + JQ = \frac{U_2{}^2}{64.34} + JH_2 \qquad (5)$$

Adiabatic Flow through Orifices, Nozzles, etc.

Reversible	Irreversible
$\dfrac{U_2{}^2 - U_1{}^2}{64.34} = J(H_1 - H_2)_{S-C}$ $= \displaystyle\int_1^2 V dP$	$\dfrac{U_2{}^2 - U_1{}^2}{64.34} = J(H_1 - H_2)_{S_2 > S_1}$ and $\quad (H_1 - H_2) < (H_1 - H_2)_{S-C}$

Adiabatic Engines, Turbines, etc. $\quad (W_s = $ shaft work$)$

Reversible	Irreversible
$W_s = J(H_1 - H_2)_{S-C}$ $= \displaystyle\int_1^2 V dP$	$W_s = J(H_1 - H_2)_{S_2 > S_1}$ and $\quad (H_1 - H_2) \neq (H_1 - H_2)_{S-C}$

For All Adiabatic Flow Processes

Reversible	Irreversible
$J(H_1 - H_2)_{S-C} = \displaystyle\int_1^2 V dP$	$J(H_1 - H_2) \neq \displaystyle\int_1^2 V dP$

Throttling Flow (Irreversible)

$W_s = 0; \quad Q = 0; \quad \Delta\dfrac{V^2}{64.43} = 0$	$P_2 < P_1$ $H_1 = H_2 \quad$ and $\quad S_2 > S_1$

Summary of Energy Equations for Non-Flow Processes

For all non-flow processes,

$$W_s + JQ = J(E_2 - E_1) \qquad (6)$$

in which $\qquad\qquad W_s = $ shaft work, or equivalent.

For all reversible non-flow processes,

$$JQ = J(E_2 - E_1) + \int P dV \qquad (7)$$

For all reversible processes, $\quad W_s = -\displaystyle\int P dV$

For irreversible processes, $\quad W_s \neq \displaystyle\int P dV.$

For constant volume processes, $W_s = 0 \qquad Q = E_2 - E_1.$

For constant pressure reversible processes, $W_s = P(V_1 - V_2), \quad Q = (H_2 - H_1)$

For adiabatic reversible processes, $Q = 0, \quad S_2 = S_1;$

$$W_s = J(E_2 - E_1)_{S-C} = -\int P dV.$$

7. GENERAL THERMODYNAMIC RELATIONS

Equations expressing general relations which exist between and among the various thermodynamic properties of any substance are called general thermodynamic relations. These equations are of value in calculating various properties of fluids from experimentally determined values of other properties. They are also of use in the development and application of the general theory of thermodynamics in the fields of physics and chemistry. The general thermodynamic relations are derived by the application of certain mathematical methods to the general energy equations presented in Art. 4 and 5.

The following symbols are used in this article. No units are attached to the quantity, and the Joule's equivalent J is not used, since it is desired to leave the equations in a general form, applicable in any consistent system of units.

c_P = specific heat at constant pressure.
c_V = specific heat at constant volume.
d = "differential of."
∂ = "partial differential of."
E = internal energy per unit mass.
H = $E + PV$, or enthalpy per unit mass.
l_P = latent heat of pressure change.
l_V = latent heat of expansion.
P = pressure, force per unit area.
Ψ = $E - TS$, or "psi" function, per unit mass.
Q = energy in transition by conduction or radiation, per unit mass.
R = the gas constant (1545/mol wt, in ft-lb-°F system, and 82.9×10^6/mol wt, in cm-gram-°C system).
S = entropy per unit mass.
T = absolute temperature.
V = volume per unit mass.
W = mechanical effects, per unit mass of fluid.
Z = $E + PV - TS$, or "zeta" function, per unit mass.

Inserting in the general energy equation,

$$W + Q = E_2 - E_1 \tag{1}$$

the expression

$$W = -\int P \, dV \tag{2}$$

and the entropy expression,

$$Q = \int T \, dS \tag{15}$$

there is obtained an equation which contains only property functions, and which is the basis for the derivation of the general thermodynamic relations. This basic equation is, for reversible processes,

$$-\int P \, dV + \int T \, dS = E_2 - E_1$$

or in differential form:

$$T \, dS - P \, dV = dE \tag{16}$$

In addition to the five properties or functions T, S, P, V, and E of the foregoing equations, there are defined three additional property functions as follows:

$H = E + PV$,   enthalpy
$\Psi = E - TS$,   psi function (after Gibbs)
$Z = E + PV - TS$,   zeta function (after Gibbs)

A very great number of relations may be written connecting these eight state functions.* In this work only a few are presented that have been found to be particularly useful in connection with the engineering aspects of thermodynamics. These formulas provide significant relations between the property functions of a fluid and are relations that must be true for any fluid whatever. Relations between partial derivatives of the state functions are:

---

* Bridgman in his *Condensed Collection of Thermodynamic Formulas*, Harvard University Press, 1925, presents an ingenious collection of tables by which any of the general thermodynamic relations may be readily obtained. He notes that, including relations involving W and Q, 11×10^6 relations between first derivatives are possible and 9.5×10^{21} between second derivatives.

$$\left(\frac{\partial T}{\partial V}\right)_S = -\left(\frac{\partial P}{\partial S}\right)_V \quad (17)$$

$$\left(\frac{\partial E}{\partial V}\right)_S = \left(\frac{\partial \Psi}{\partial V}\right)_T = -P \quad (21)$$

$$\left(\frac{\partial T}{\partial P}\right)_S = \left(\frac{\partial V}{\partial S}\right)_P \quad (18)$$

$$\left(\frac{\partial E}{\partial S}\right)_V = \left(\frac{\partial H}{\partial S}\right)_P = T \quad (22)$$

$$\left(\frac{\partial S}{\partial V}\right)_T = \left(\frac{\partial P}{\partial T}\right)_V \quad (19)$$

$$\left(\frac{\partial H}{\partial P}\right)_S = \left(\frac{\partial Z}{\partial P}\right)_T = V \quad (23)$$

$$\left(\frac{\partial S}{\partial P}\right)_T = -\left(\frac{\partial V}{\partial T}\right)_P \quad (20)$$

$$\left(\frac{\partial \Psi}{\partial T}\right)_V = \left(\frac{\partial Z}{\partial T}\right)_P = -S \quad (24)$$

A very useful characteristic of a fluid is one which shows the energy required as heat to effect a unit change in the magnitude of the pressure, the temperature, or the specific volume of a unit mass of the fluid when one of these properties is maintained constant. Such a characteristic of the fluid is known as a *thermal capacity*. The familiar specific heat at constant pressure c_P is thus one of the several thermal capacities, being the amount of heat energy required to change the temperature of unit mass of the substance one degree while the pressure is maintained constant. Four thermal capacities are of particular practical utility. These, together with their names and conventional symbols, are:

$$\left(\frac{\partial' Q}{\partial T}\right)_P = c_P; \quad \text{specific heat at constant pressure} \quad (25)$$

$$\left(\frac{\partial' Q}{\partial T}\right)_V = c_V; \quad \text{specific heat at constant volume} \quad (26)$$

$$\left(\frac{\partial' Q}{\partial P}\right)_T = l_P; \quad \text{latent heat of pressure change} \quad (27)$$

$$\left(\frac{\partial' Q}{\partial V}\right)_T = l_V; \quad \text{latent heat of expansion} \quad (28)$$

The notation $d'Q$ and $\partial'Q$ which appears in these relations is a reminder of the important consideration that Q is not a point function. Relations among these coefficients and the other properties are as follows:

$$dS = \frac{c_P}{T}\,dT + \frac{l_P}{T}\,dP \quad (29)$$

$$dS = \frac{c_V}{T}\,dT + \frac{l_V}{T}\,dV \quad (30)$$

$$l_P = T\left(\frac{\partial S}{\partial P}\right)_T = -T\left(\frac{\partial V}{\partial T}\right)_P \quad (31)$$

$$l_V = T\left(\frac{\partial S}{\partial V}\right)_T = T\left(\frac{\partial P}{\partial T}\right)_V \quad (32)$$

$$c_P = -l_P\left(\frac{\partial P}{\partial T}\right)_S = T\left(\frac{\partial V}{\partial T}\right)_P\left(\frac{\partial P}{\partial T}\right)_S \quad (33)$$

$$c_V = -l_V\left(\frac{\partial V}{\partial T}\right)_S = -T\left(\frac{\partial P}{\partial T}\right)_V\left(\frac{\partial V}{\partial T}\right)_S \quad (34)$$

$$c_P - c_V = l_V\frac{dV}{dT} - l_P\frac{dP}{dT}$$

$$= l_V\left(\frac{\partial V}{\partial T}\right)_P\left[\text{or} \quad = -l_P\left(\frac{\partial P}{\partial T}\right)_V\right]$$

$$= T\left(\frac{\partial P}{\partial T}\right)_V\left(\frac{\partial V}{\partial T}\right)_P \quad (35a)$$

or

$$= -T\left(\frac{\partial P}{\partial T}\right)_V^2\left(\frac{\partial V}{\partial P}\right)_T \quad (35b)$$

or

$$= -T\left(\frac{\partial V}{\partial T}\right)_P^2\left(\frac{\partial P}{\partial V}\right)_T \quad (35c)$$

$$c_P/c_V = k = -\frac{\left(\frac{\partial V}{\partial T}\right)_P \left(\frac{\partial P}{\partial T}\right)_S}{\left(\frac{\partial P}{\partial T}\right)_V \left(\frac{\partial V}{\partial T}\right)_S} = -\left(\frac{\partial V}{\partial T}\right)_P \left(\frac{\partial T}{\partial P}\right)_V \left(\frac{\partial P}{\partial V}\right)_S$$

$$= \left(\frac{\partial V}{\partial P}\right)_T \left(\frac{\partial P}{\partial V}\right)_S \qquad (36)$$

$$\left(\frac{\partial c_P}{\partial P}\right)_T = -T\left(\frac{\partial^2 V}{\partial T^2}\right)_P \qquad (37)$$

$$\left(\frac{\partial c_V}{\partial V}\right)_T = T\left(\frac{\partial^2 P}{\partial T^2}\right)_V \qquad (38)$$

The utility of equations 31 to 38, correlating as they do the thermal capacities and the property functions P, V, T, and S, is quite varied. They afford an opportunity for cross-checking the consistency of the more difficultly and generally less accurately measurable calorimetric data on thermal capacities by use of the more accurately determinable relations among the property functions of a fluid.

Equation 32, which is in effect the historical Clapeyron's equation, has the particular utility that it enables the computation of the latent heat of vaporization of a fluid solely from data on the change of volume during vaporization and the relation between the pressure and temperature of saturated vapor. Thus, interpreting the latent heat of expansion in terms of the heat supplied during vaporization at constant temperature,

$$l_V = \frac{Q_{\text{vaporization}}}{\Delta V_{\text{vaporization}}} = \frac{H_{fg}}{V_{fg}}$$

(see Art. 16), whence, from equation 32,

$$H_{fg} = V_{fg}\, T\left(\frac{dP}{dT}\right)$$

In this expression, since the ratio dP/dT for a saturated vapor is the same for any process whether at constant volume or otherwise, it is written as the total derivative instead of the partial derivative. Physically it is the rate of change of the saturation pressure with temperature (the slope of the P–V curve) at the temperature T.

Several useful expressions which correlate the internal energy and enthalpy with the thermal capacities and the property functions follow.

$$dE = c_V\, dT + \left[T\left(\frac{\partial P}{\partial T}\right)_V - P\right]dV \qquad (39)$$

$$\left(\frac{\partial E}{\partial V}\right)_T = T\left(\frac{\partial P}{\partial T}\right)_V - P \qquad (40)$$

$$\left(\frac{\partial E}{\partial T}\right)_V = c_V \qquad (41)$$

$$dH = c_P\, dT + \left[-T\left(\frac{\partial V}{\partial T}\right)_P + V\right]dP \qquad (42)$$

$$\left(\frac{\partial H}{\partial P}\right)_T = V - T\left(\frac{\partial V}{\partial T}\right)_P \qquad (43)$$

$$\left(\frac{\partial H}{\partial T}\right)_P = c_P \qquad (44)$$

The Joule–Thomson Experiment. A type of physical experiment which is of much value in assisting to ascertain and check the characteristics of a fluid is the *Joule–Thomson* experiment. In this experiment the fluid is caused to flow to, throttle through, and pass from a porous plug, without energy reception or departure as heat and without appreciable velocity or velocity change in any phase of the process. The pressures and temperatures are measured on each side of the plug.

The energy equation for this pure throttling and wholly irreversible process is $H_1 = H_2$. Alternatively, its characteristic is that $dH = 0$. Introducing this consideration in equation 42, the relation for the process becomes

$$c_P\, (\partial T)_H = \left[T\left(\frac{\partial V}{\partial T}\right)_P - V\right](\partial P)_H$$

or

$$\left(\frac{\partial T}{\partial P}\right)_H = \frac{1}{c_P}\left[T\left(\frac{\partial V}{\partial T}\right)_P - V\right] \qquad (45)$$

The coefficient $(\partial T/\partial P)_H$ in this expression is called the *Joule–Thomson coefficient* (μ).

It is evidently to be interpreted as the change of temperature of the fluid per unit fall of pressure under the above irreversible adiabatic throttling conditions. It may be shown that for a perfect gas, which has the state equation $PV = RT$, there would be no temperature change. With most gases, excepting hydrogen, there is a temperature drop or *cooling effect* at normal temperature levels, but a temperature *rise* at temperature levels above the so-called *inversion temperature*. For most gases the inversion temperature is high but for hydrogen it is at about -80 C.

This experiment is further useful for reducing the readings of a gas thermometer to the energy or thermodynamic scale of temperature and for determining the location of the absolute thermodynamic zero of temperature.

The cooling effect which is characteristic of the actual gases is used regeneratively for the liquefaction of gases by the *Linde* process.

GASES

Symbols and Notations

A = area, square feet.
a = a coefficient in specific heat formulations.
b = a coefficient in specific heat formulations.
c_p = specific heat at constant pressure, Btu per pound.
c_v = specific heat at constant volume, Btu per pound.
E = molecular (internal) energy, Btu per pound.
f = a coefficient in specific heat formulations; also "a function of."
H = enthalpy $(E + PV/J)$, Btu per pound.
J = Joule's equivalent (= 778 ft-lb per Btu).
k = ratio of c_p/c_v (a variable, but approaching constancy).
M' = mass rate of flow, pounds per second.
m = molecular weight.
n = an exponential constant (as in the property relation PV^n = a constant).
P = absolute pressure, pounds per square foot.
P_r = relative pressure.
Q = energy transferred by heat (radiation or conduction), Btu per pound.
R = the gas constant for any given gas.
S = entropy, per pound.
T = absolute temperature, degrees Rankine (= degrees fahrenheit + 460).
t = temperature, degrees fahrenheit.
U = velocity, feet per second.
u = internal energy per unit mass.
V = specific volume, cubic feet per pound.
v_r = relative volume.
W = shaft-work, foot-pounds per pound.

$$\phi = \int_{T_0}^{T} \frac{c_p}{T}\, dT,\ \text{Btu/lb F abs.}$$

8. GAS LAWS

The Perfect Gas

P–V–T Relation for the Perfect Gas. The relation between the pressure, specific volume, and temperature of a perfect gas in any given condition or state is given by the *characteristic equation for gases*, which is:

$$PV = RT \tag{1}$$

in which R is a constant for any gas, and is called the *gas constant*.

The above relation embodies Boyle's and Charles' laws.

Boyle's Law states that, when the temperature of a given mass of gas is held constant, the volume and pressure vary inversely, or PV = a constant.

Charles' Law states that, when the volume of a given mass of gas is held constant, the change in the pressure of the gas is proportional to the change in temperature, or $\dfrac{\Delta P}{\Delta T}$ = a constant.

Gas Constant. Actual gases at moderate pressure and also very low-pressure vapors approach this perfect gas criterion that R is a constant. However, for many practical purposes it may be sufficiently accurate to neglect this variation, and it is therefore customary to assign to the various gases values of R which correspond to some standard state. In Table 1 are presented, for the gases of more usual interest, the values of the gas constant as determined by physical measurement of the specific volume (V_0) at the standard atmospheric pressure (14.7 lb per sq in. abs) and 32 F. There are also tabulated the molecular weights of the fluids (m, referred to oxygen as 32), the product of the molecular weight and the gas constant ($= mR$), and the product mV_0. For a more complete table of the properties of gases see Section 16.

Table 1. Characteristic Properties of Gases

Gas	V_0, at 14.7 lb and 32° F	R	m	mR	mV_0
Sulfur dioxide (SO_2).............	5.47	23.6	64.0	1512	350
Carbon dioxide (CO_2)............	8.10	34.9	44.0	1536	356
Oxygen (O_2).....................	11.2	48.3	32.0	1546	358
Atmospheric air..................	12.4	53.3	(29.0)	(1545)	(358)
Nitrogen (N_2) (atmos.)*.........	12.74	54.9	(28.1)	(1545)	(358)
Nitrogen (N_2) (chem.)...........	12.8	55.1	28.0	1543	358
Carbon monoxide (CO)...........	12.8	55.1	28.0	1545	358
Ammonia (NH_3).................	20.8	89.5	17.0	1516	354
Helium (He)....................	89.7	386	4.0	1544	359
Hydrogen (H_2).................	178	767	2.0	1546	358.5

Representative value of $mR = 1545$; of $mV_0 = 358$.

* By "atmospheric nitrogen" is meant the residue of atmospheric air after abstraction of the oxygen but retaining the argon, carbon dioxide, etc., which exist in traces in the atmosphere.

It is significant that in this table the values of mR and of mV_0 are found to show relative uniformity, particularly for the permanent gases of less than two atoms per molecule. For convenience it is customary to select as representative the values of 1545 and 358 for these respective quantities, designating them as the "universal gas constant" (mR) and the "standard mol volume" (mV_0).* It follows from this approximate constancy of the product mV_0 that the specific volumes of the permanent gases at a given temperature and pressure are inversely proportional to their molecular weights, or that the densities are directly proportional.†

Real Gases

The deviation of actual, or real, gases under certain conditions from the ideal or perfect gas law is considerable. This fact is to be expected, since real gases do not follow the basic assumptions set forth for an ideal gas. In general, the real gas deviates from an ideal gas in that the molecules of the real gas occupy a definite volume that may not be negligible in comparison with the total volume of the gas, and attractive forces exist between the molecules.

Compressibility Factors. If a term known as the *compressibility factor* is introduced, the ideal, or perfect, gas equation may be used for real gases. The ideal gas equation modified for use for real gases is:

$$pv = CRT$$

In this relation, C is the compressibility factor; it depends upon the pressure, temperature, and the gas.

Compressibility factors have been plotted against pressures and temperatures and presented in the technical literature. The main objection to these charts is that a separate diagram is required for each gas. One diagram may be constructed for several gases if the law of corresponding states is used as a foundation. According to this law, all gases behave alike whenever any two of the variables *pressure, volume,* or *temperature* bear a given ratio to the variables at the critical condition. In other words, if gases A and B at different state conditions have equal values for the ratio of the stated pressure to the critical pressure, and equal values for the ratio of the stated temperature to the critical temperature, the gases A and B will have approximately the same thermodynamic prop-

---

* Since a mass of m pounds of a fluid is commonly known as a "mol" the volume of m pounds is known as the "mol volume."

† This relationship is one of the bases for Avogadro's "law," which states that for all gases at a given pressure and temperature the number of molecules per unit volume is the same. If that were strictly true the densities of various gases would necessarily be proportional to their molecular weights and their specific volumes would be inversely proportional. The above relative uniformity of mV_0 is an approximate verification of the hypothesis.

erties. Under this condition one point on a graph of compressibility factor versus ratio of pressure to critical pressure for a given ratio of temperature to critical temperature will represent both gases A and B. The ratios of the variables pressure, volume, and temperature to the corresponding critical values are termed the *reduced coordinates* or *variables*. The reduced pressure (p_r) is defined as the ratio of the pressure of the gas to the critical pressure. The reduced temperature and reduced volume are defined in a similar manner. Therefore, according to the law of corresponding states, the functional relationship which holds approximately for all gases is:

$$v_r = f(p_r, T_r)$$

A general reduced-pressure compressibility chart for various reduced temperatures is shown in Figs. 1 and 2. The critical temperatures and pressures for several gases are given in Table 2 for use in connection with Figs. 1 and 2.

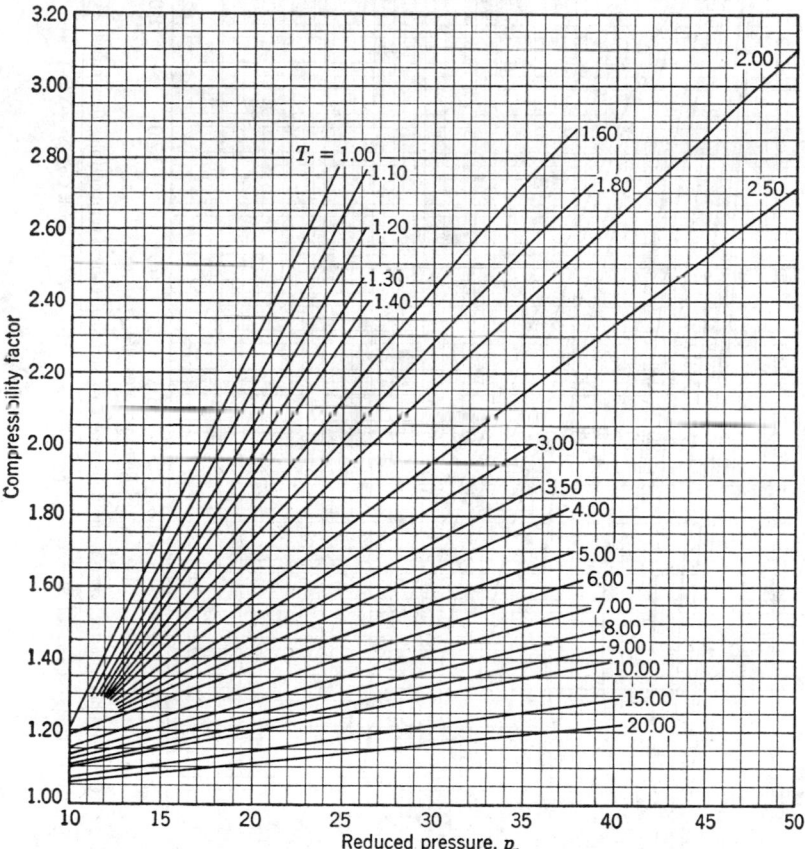

FIG. 1.* Compressibility factor *versus* reduced pressure for a series of reduced temperatures (high-pressure range)

An example will be used to illustrate the use of the material presented.

Example. Compute the specific volume of air at a temperature and pressure of 495 F and 10,940 psia. The reduced temperature and pressure values are:

$$T_r = \frac{955}{238.8} = 4$$

$$p_r = \frac{10,940}{547.0} = 20$$

---

* Reproduced by permission from B. F. Dodge, *Chemical Engineering Thermodynamics*, McGraw-Hill Book Co., New York, 1944.

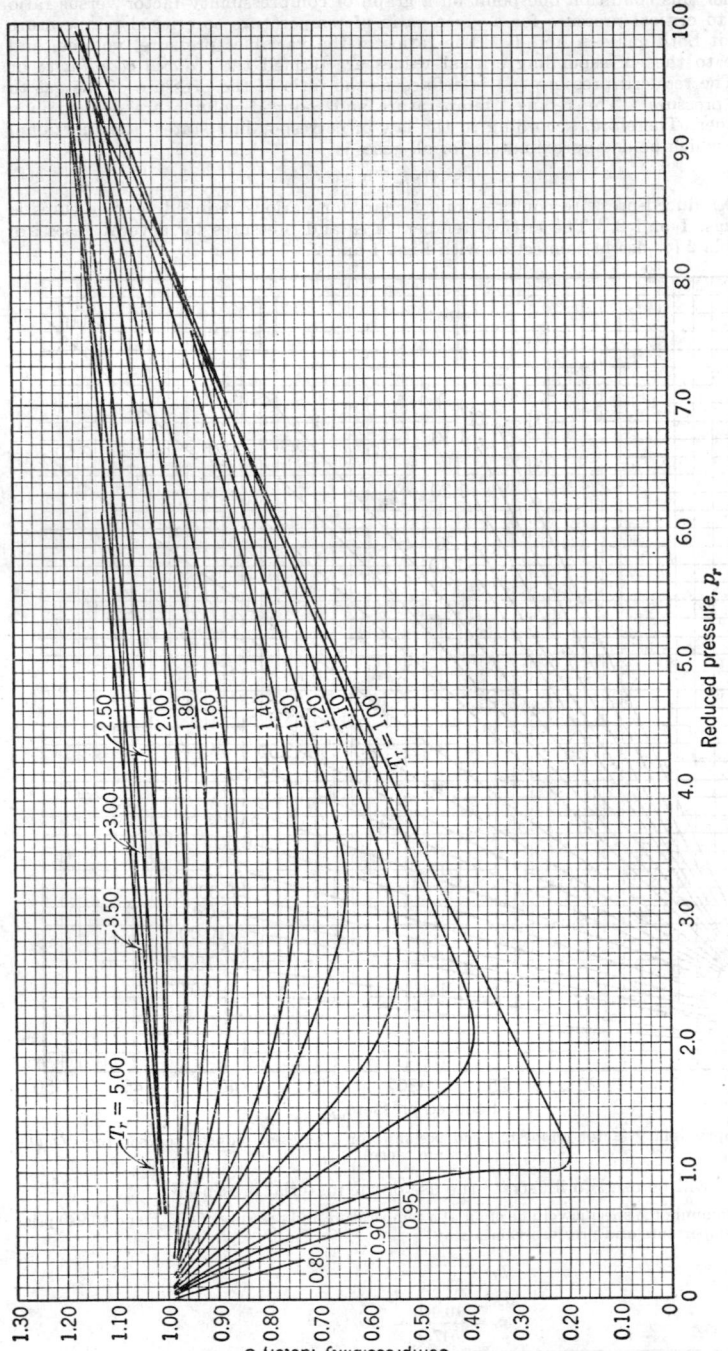

Reduced pressure, p_r

Compressibility factor, C

FIG. 2.* Compressibility factor *versus* reduced pressure for a series of reduced temperatures (low-pressure range)

* Reproduced by permission from B. F. Dodge, *Chemical Engineering Thermodynamics*, McGraw-Hill Book Co., New York, 1944.

The compressibility factor obtained from Fig. 1 equals 1.42.   Hence the specific volume equals

$$v = \frac{CRT}{p} = \frac{1.42(1544)955}{10,940(144)29} = 0.0458 \text{ ft}^3/\text{lb}$$

Table 2.   Approximate Critical Constants For Several Gases *

Gas	Critical Temperature		Critical Pressure	
	K	R	atm	psi
1. Air	132.4	238.8	37.2	547.0
2. Ammonia	405.5	730.1	111.5	1639.0
3. Argon	151.0	272.2	48.0	705.0
4. Carbon dioxide	304.1	547.8	72.9	1071.0
5. Carbon monoxide	134.4	242.2	34.6	508.2
6. Chlorine	417.0	751.0	76.0	1116.0
7. Ethane	305.2	549.8	48.8	717.0
8. Ethylene	282.8	509.5	50.7	745.0
9. Helium	5.2	10.0	2.3	33.8
10. Hydrogen	33.2	60.5	12.8	188.0
11. Isopentane	460.9	830.0	32.9	483.6
12. Mercury	1172.0	2109.0	180.0	2646.0
13. Methyl chloride	416.2	749.7	65.8	967.0
14. Neon	44.4	79.0	25.9	377.8
15. Nitrogen	126.0	227.2	33.5	492.5
16. Oxygen	154.3	278.1	49.7	730.9
17. Propane	369.9	666.3	42.0	617.0
18. Sulfur dioxide	430.3	775.0	77.7	1141.0
19. Water vapor	647.3	1165.4	218.2	3206.0

* Reproduced by permission from G. A. Hawkins, *Thermodynamics*, John Wiley & Sons, Inc., New York, 1947.

9. SPECIFIC HEAT OF GASES

Specific Heat.   When energy is delivered to or withdrawn from a substance *as heat* (by conduction or radiation) and during the process the temperature of the substance changes, the ratio of the quantity of heat interchange per unit mass of the substance to the change in temperature is called the *specific heat, c,* of the substance for the particular process. Symbolically,

$$c = \frac{Q}{\Delta t} \tag{2}$$

where c = specific heat of the substance for the process in question.

Q = energy interchange as heat during the process, per unit mass of substance, regarded as positive if incoming and negative if outgoing.

Δt = change of temperature of the substance, regarded as positive if increasing and negative if decreasing.

In the case of solids and liquids there is only one characteristic process which is of usual concern, namely, the heat interchange while the substance is at constant pressure. Therefore when the specific heat of such a substance is quoted it may be taken implicitly to be the specific heat *at constant pressure*.   With gases, however, there is in practice an infinite variety of processes or state changes of the gas during which energy may be transferred as heat, frequently under circumstances where energy transition as work occurs simultaneously and affects correspondingly the temperature change.   Thus the specific heat of a given gas may have an infinite range of values.   For definiteness, however, there is commonly quoted for any particular gas its specific heat for a *constant volume* state change and that for a *constant pressure* state change, the two being designated respectively as the *specific heat at constant volume* (c_v) and the *specific heat at constant pressure* (c_p).

Not only is the specific heat of a substance dependent upon the type of process but also, for a given process, it may vary with other conditions, such as the instantaneous temperature at which the process is operating.   In that case, to determine the specific heat at a particular temperature we must pass to differential changes of temperature, dt, in place of Δt and write

$$c = \frac{d'Q \text{ *}}{dt} \tag{2a}$$

---

* See p. 827 for meaning of symbol $d'Q$.

Transposing, $d'Q = c\,dt$, and for a finite rise from temperature t_1, to temperature t_2,

$$Q = \int_{t_1}^{t_2} c\,dt \tag{3}$$

This expression shows the use to which information concerning the specific heat is frequently put, namely, that of computing the amount of energy transferred *as heat* during the performance of a specified process. 'If the specific heat varies with the temperature (as it usually does) a functional relation between c and t must be known before the expression may be integrated. A functional relation which is frequently found to be suitable is of the form,

$$c = a + bt + ft^2$$

where a, b, and f are coefficients which are determined by experiment.

Specific Heat of Actual Gases. For the perfect gas, c_v and c_p must either be constants or functions only of the temperature. Whether they are constants or variables their difference must be constant ($c_p - c_v = R/J$).

In all the real gases these specific heats are found actually to vary rather materially with the temperature and to a slight degree with the pressure,* and also there is evidence that their difference is not exactly constant nor does it exactly equal R/J, where R is the gas constant as ascertained from specific volume data (Table 3). However, a degree of

Table 3.   Specific Heats of Gases

(At Standard Atmospheric Pressure)
According to (a) Partington and Shilling, (b) Goodenough and Felbeck.

Expressions for c_v at the Absolute Temperature T (deg R)	Mean Values, 32° to 400° F			
	$c_p - c_v$ $(= R/J)$	c_v	c_p	k
Carbon dioxide (CO_2)				
(a) $0.1260 + 0.0_457\,T - 0.0_872\,T^2$	.0457	0.162	0.208	1.28
(b) $0.1037 + 0.0_3115\,T - 0.0_7284\,T^2 + 0.0_{11}247\,T^3$	.0451	0.171	0.216	1.26
Oxygen (O_2)				
(a) $0.1539 + 0.0_530\,T + 0.0_830\,T^2$	.0622	0.156	0.218	1.40
(b) $0.1545 + 0.0_8375\,T^2$	.0621	0.156	0.218	1.40
Air				
(a) $0.1699 + 0.0_533\,T + 0.0_833\,T^2$	.0687	0.173	0.241	1.40
Nitrogen, atmospheric				
(a) $0.1751 + 0.0_545\,T + 0.0_834\,T^2$	.0708	0.177	0.248	1.40
Nitrogen (chem.) and carbon monoxide				
(a) $0.1760 + 0.0_534\,T + 0.0_834\,T^2$	.0711	0.178	0.249	1.40
(b) $0.1766 + 0.0_8428\,T^2$	.0710	0.179	0.250	1.40
Hydrogen (H_2)				
(a) $2.314 + 0.0_3193\,T$	.985	2.44	3.43	1.40
(b) $1.990 + 0.0_3331\,T$	.986	2.21	3.20	1.44

(Note that, for example, $0.0_53\,T$ is the symbol for $0.000003\,T$ or $0.3\,T \times 10^{-5}$.)

accuracy which is sufficient for most practical purposes as well as a great gain in convenience is attained if these specific heats are regarded as effectively constant when the temperature range is moderate and as a function only of the temperature for greater ranges, and also if their differences may be regarded as virtually constant.

In Table 3 are presented, for certain gases of more usual engineering interest, expressions for the relations between the specific heat at constant volume (c_v) and the absolute temperature (T) and also corresponding values of the mean ($c_p - c_v$) ($= R/J$), and finally of the mean values of c_p and c_v and of the mean value of the ratio c_p/c_v ($= k$) for the temperature range between 32 and 400 F. Unless materially greater temperature ranges are involved, these mean values may be employed as generally representative figures. In the table two values are presented for certain of the gases, those designated by the

* For air at 140 F the specific heat has been shown to vary with the pressure according to the relation $c_p = 0.2414 - 0.0_2014\,P - 0.0_92478\,P^2 - 0.0_{13}795\,P^3$. For ranges less than 1500 lb per sq in. the relation $c_p = 0.2414 - 0.0_42\,p$ is quite adequately accurate.

bracketed a being ones proposed by Partington and Shilling * as the result of exhaustive analyses of all available data and those designated by the bracketed b being ones proposed by Goodenough and Felbeck † as the result of like analyses. The two are observed to check closely except for H_2 and CO_2. The differences in these at least reflect the difficulties in the way of accurate experimental determinations.

All of the relations are observed to be of the form

$$c = a + bT + fT^2 + \cdots$$

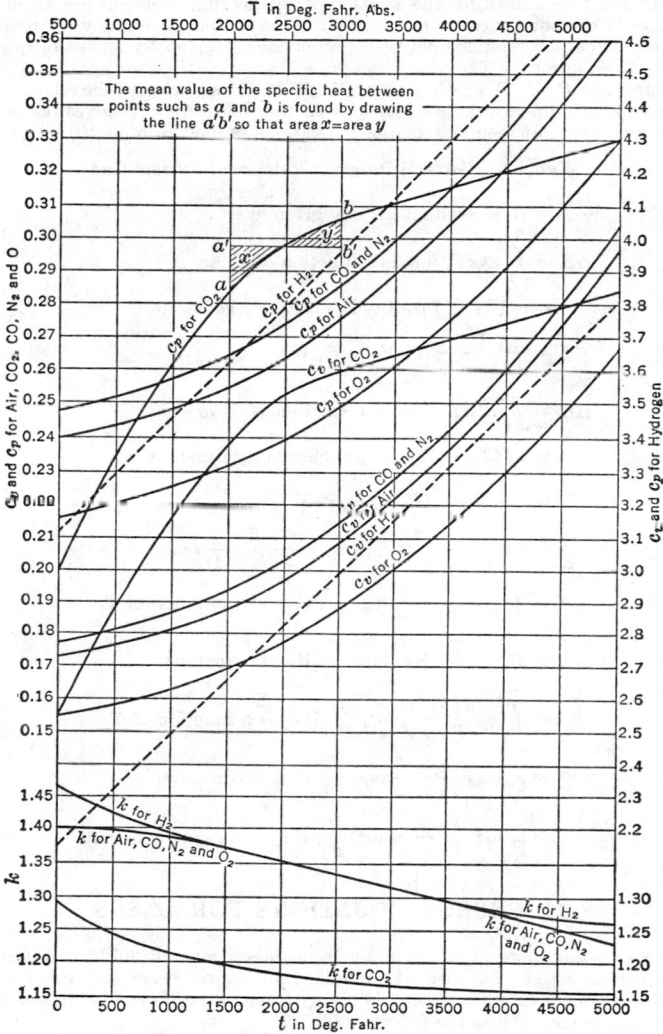

FIG. 3. Chart of c_p, c_v, and k for Several Gases ‡

where a, b, and f are numerical coefficients which are established from the experimental evidence.

---

* Partington and Shilling, *The Specific Heat of Gases*, Benn Ltd., London, 1924.
† Goodenough and Felbeck, *Univ. of Illinois Bull.* 139, 1923.
‡ Reproduced by permission from Barnard, Ellenwood, and Hirshfeld, *Heat-Power Engineering*, Part I, John Wiley & Sons, 1926.

Figure 3 shows the values graphically of the specific heat at constant pressure (c_p), specific heat at constant volume (c_v), and their ratio ($k = c_p/c_v$) for air, CO_2, N_2, O_2, CO, and H_2 in the temperature range 0 F to 5000 F.

10. PROPERTY RELATIONS OF GASES

For vapors such as steam, ammonia, etc., the relations which exist between the properties which must be known for the solution of engineering problems are given in tables of properties of the substances in question. For gases that adequately approach perfect gas characteristics, the relations between properties are given by means of quite simple mathematical equations. The properties and characteristic constants of gases which are of interest are P, V, T, E, H, S, c_v, c_p, k, and R. In addition to the relations between specific heats and temperatures given on p. 842 the following table gives the more useful relations which exist between and among the various properties of gases.

Table 4. General Property Relations: Perfect Gas

$$\frac{PV}{T} = R, \text{ a constant for any given gas.} \tag{4}$$

$$\Delta E = \int_1^2 c_v \, dT, \text{ if } c_v \text{ is a function of } T, \text{ or}$$

$$= c_v \,(T_2 - T_1), \text{ if } c_v \text{ is effectively constant, or} \tag{5}$$

$$= \frac{1}{J\,(k-1)}\,(P_2 V_2 - P_1 V_1). \tag{5a}$$

$$\Delta H = \int_1^2 c_p \, dT, \text{ if } c_p \text{ is a function of } T, \text{ or}$$

$$= c_p \,(T_2 - T_1), \text{ if } c_p \text{ is effectively constant, or} \tag{6}$$

$$= \frac{k}{J\,(k-1)}\,(P_2 V_2 - P_1 V). \tag{6a}$$

$$\frac{c_p}{c_v} = k; \quad c_p - c_v = \frac{R}{J}; \quad c_p = \frac{Rk}{J\,(k-1)} \tag{7}$$

$$\Delta S = \int_1^2 c_v \frac{dT}{T} + \frac{R}{J} \log_e \frac{V_2}{V_1}, \text{ if } c_v \text{ is a function of } T, \tag{8}$$

$$= c_v \log_e \frac{T_2}{T_1} + \frac{R}{J} \log_e \frac{V_2}{V_1}, \text{ if } c_v \text{ is constant;} \tag{8a}$$

$$= \int_1^2 c_p \frac{dT}{T} - \frac{R}{J} \log_e \frac{P_2}{P_1}, \text{ if } c_p \text{ is a function of } T, \tag{8b}$$

$$= c_p \log_e \frac{T_2}{T_1} - \frac{R}{J} \log_e \frac{P_2}{P_1}, \text{ if } c_p \text{ is constant;} \tag{8c}$$

$$= c_p \log_e \frac{V_2}{V_1} + c_v \log_e \frac{P_2}{P_1}, \text{ if } c_p \text{ and } c_v \text{ are constant.} \tag{8d}$$

11. ENERGY EQUATIONS FOR GASES

All the energy equations of Art. 4 apply to processes with all fluids, therefore to processes with perfect gases. The general method of solving processes with gases is to set up the energy equation in a form in which the energy transformation for the process is equated to a property change, such as the following:

For a non-flow reversible constant pressure process $Q = H_2 - H_1$. The property H must then be expressed in terms of some directly measurable properties such as P, V, and T, and a characteristic constant such as R or specific heat. In the above example,

$$H_2 - H_1 = \Delta H = c_p(T_2 - T_1) \quad \text{or} \quad = \frac{k}{J(k-1)}\,(V_2 - V_1)P$$

if c_p is taken as constant.

Table 5 below gives a summary of the special property and energy relations required for the usual processes encountered with gases. Note that the term *mechanical effects*

Table 5. Special Property and Energy Relations for State Changes of Perfect Gas

Constant (specific) volume:

$$\frac{P}{T} = \frac{R}{V} = \text{a constant during the state change} \tag{9}$$

$$\Delta E = \int_1^2 c_v \, dT = \frac{1}{J(k-1)}(P_2 - P_1)V$$

$$\Delta H = \int_1^2 c_p \, dT = \frac{k}{J(k-1)}(P_2 - P_1)V$$

$$\Delta S = \int_1^2 c_v \frac{dT}{T}, \quad \text{or} \quad = c_v \log_e \frac{P_2}{P_1} \text{ if } c_v \text{ is constant}$$

Mechanical effects, if reversible = zero; $Q = \Delta E$.

Constant pressure:

$$\frac{V}{T} = \frac{R}{P} = \text{a constant during the state change} \tag{10}$$

$$\Delta E = \int_1^2 c_v \, dT = \frac{1}{J(k-1)}(V_2 - V_1)P$$

$$\Delta H = \int_1^2 c_p \, dT = \frac{k}{J(k-1)}(V_2 - V_1)P$$

$$\Delta S = \int_1^2 c_p \frac{dT}{T}, \quad \text{or} \quad = c_p \log_e \frac{V_2}{V_1} \text{ if } c_p \text{ is constant}$$

Mechanical effects, if reversible = $P(V_1 - V_2)$; $Q = \Delta H$.

Isothermal (constant temperature):

$$PV = RT = \text{a constant during the state change} \tag{11}$$

$$\Delta E = \text{zero}; \quad \Delta H = \text{zero}; \quad \Delta S = \frac{R}{J}\log_e\frac{P_1}{P_2} \quad \text{or} \quad = \frac{R}{J}\log_e\frac{V_2}{V_1}$$

Mechanical effects, if reversible $= -RT \log_e (P_1/P_2)$ or
$$-RT \log_e (V_2/V_1) \tag{12}$$

$$Q = \frac{RT}{J}\log_e\frac{P_1}{P_2}, \quad \text{or} \quad \frac{RT}{J}\log_e\frac{V_2}{V_1} \tag{13}$$

Reversible adiabatic (isentropic):

$PV^k = \text{a constant during the change}$; $TV^{k-1} = \text{a constant}$;

$$\frac{T}{P^{(k-1)/k}} = \text{a constant} \tag{14, 15, 16}$$

$$\Delta E = \int_1^2 c_v \, dT \quad \text{or} \quad = \frac{R}{J(k-1)} T_1\left[\left(\frac{P_2}{P_1}\right)^{(k-1)/k} - 1\right] \text{ if } k \text{ is constant} \tag{17}$$

$$\Delta H = \int_1^2 c_p \, dT \quad \text{or} \quad = \frac{kR}{J(k-1)} T_1\left[\left(\frac{P_2}{P_1}\right)^{(k-1)/k} - 1\right] \text{ if } k \text{ is constant} \tag{18}$$

$\Delta S = \text{zero}$; $Q = \text{zero}$; mechanical effects $= J\Delta E$.

Irreversible adiabatics may frequently be characterized approximately by the property relations:

$$PV^n = \text{a constant}; \quad TV^{n-1} = \text{a constant}; \quad \frac{T}{P^{(n-1)/n}} = \text{a constant} \tag{19, 20, 21}$$

$$\Delta E = \int_1^2 c_v \, dT \quad \text{or} \quad = \frac{R}{J(k-1)} T_1\left[\left(\frac{P_2}{P_1}\right)^{(n-1)/n} - 1\right] \tag{22}$$

$$\Delta H = \int_1^2 c_p \, dT \quad \text{or} \quad = \frac{kR}{J(k-1)} T_1\left[\left(\frac{P_2}{P_1}\right)^{(n-1)/n} - 1\right] \tag{23}$$

$$\Delta S = c_v \frac{n-k}{n}\log_e\frac{P_2}{P_1}, \quad \text{where} \quad \begin{cases} n > k \text{ in a compression.} \\ n < k \text{ in an expansion.} \end{cases}$$

Mechanical effects $= J\Delta E$ but $\neq -\int_1^2 P \, dV$; $Q = \text{zero}$.

Table 5—*Continued*

Throttling process is primarily characterized by the relation:

$$H_2 = H_1 \tag{24}$$

Polytropics, which virtually are internally reversible but are not adiabatic, may frequently be characterized by the property relations:

$$PV^n = \text{a constant}; \quad TV^{n-1} = \text{a constant}; \quad \frac{T}{P^{(n-1)/n}} = \text{a constant} \tag{19, 20, 21}$$

$$\Delta E = \int_1^2 c_v \, dT = \frac{R}{J(k-1)} T_1 \left[\left(\frac{P_2}{P_1} \right)^{(n-1)/n} - 1 \right] \tag{22}$$

$$\Delta H = \int_1^2 c_p \, dT = \frac{kR}{J(k-1)} T_1 \left[\left(\frac{P_2}{P_1} \right)^{(n-1)/n} - 1 \right] \tag{23}$$

$$S = c_v \frac{n-k}{n} \log_e \frac{P_2}{P_1}, \quad \text{where} \quad \begin{cases} n < k \text{ in a compression with heat emission.} \\ n > k \text{ in an expansion with heat emission.} \end{cases}$$

$$\text{Mechanical effects} = \frac{R}{n-1} T_1 \left[\left(\frac{P_2}{P_1} \right)^{(n-1)/n} - 1 \right] \tag{25}$$

$$Q = \frac{R}{J} \frac{(n-k)}{(k-1)(n-1)} T_1 \left[\left(\frac{P_2}{P_1} \right)^{(n-1)/n} - 1 \right] \tag{26}$$

includes shaft work, flow work, and changes in mechanical, kinetic, or potential energy. Mechanical effects $= W_s + (P_1V_1 - P_2V_2) + \dfrac{U_1^2 - U_2^2}{64.34} + (Z_1 - Z_2)$. For non-flow processes, mechanical effects consist only of shaft work.

12. GRAPHICAL REPRESENTATION OF GAS STATE CHANGES

The property relations and state changes of gases may with advantage be portrayed graphically by curves drawn to coordinate axes of selected properties. The properties which are so employed with most frequency are the absolute pressure P as ordinate and specific volume V as abscissa. To these coordinates may be drawn contour lines of constant volume, constant pressure, or constant entropy (reversible adiabatic) state changes

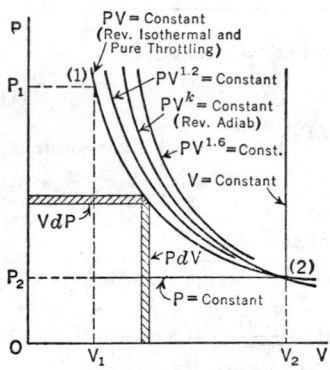

Fig. 4.   *P–V* Curves for a Gas

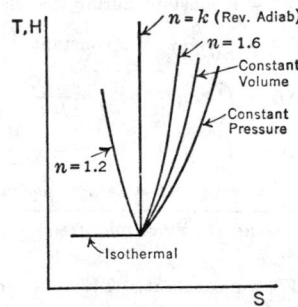

Fig. 5.   *T–S* or *H–S* Curves for a Gas

as well as any other sort of state changes. The typical forms of curves representing the commonly occurring state changes, as they would appear on *P–V* coordinates, are shown in Fig. 4. One evident feature of the family of curves is their increasing "steepness" with increase in the magnitude of n.

Aside from the general utility of the *P–V* curves they offer the usual particular utility that $-\int_1^2 P \, dV$ is represented by the area below the state change curve and that this area is thus a graphic measure of the work in a mechanically reversible non-flow process, as well as the sum of all the *mechanical effects* in a mechanically reversible process, but *only* for a reversible one. In this connection note similarly that $\int_1^2 V\,dP$ is the area "back of" the state change curve and that, referring to the figure,

$$P_1V_1 + \int_1^2 P\,dV - P_2V_2 = -\int_1^2 V\,dP$$

However, for a reversible steady-flow process,

$$W + (U_1^2 - U_2^2)/64.34 + P_1V_1 - P_2V_2 = -\int_1^2 P\,dV$$

whence

$$W + (U_1^2 - U_2^2)/64.34 = -P_1V_1 - \int_1^2 P\,dV + P_2V_2 = \int_1^2 V\,dP$$

FIG. 6.   *T–S* and *H–S* Diagrams for Diatomic Gases

This is to say that for a steady-flow reversible process the summation of the shaft work and the change of kinetic energy, or either individually if the other is negligible or not germane, is measured graphically by the area "back of" the state change curve.

Curves T-S and H-S, though not commonly provided for gases, may for some purposes be used to advantage, particularly with T and H plotted jointly against S, Fig. 5. A T, H-S diagram for air, N_2, O_2, and CO appears in Fig. 6. For the plotting of the properties of these four diatomic gases, advantage is taken of the unique phenomenon that for all, the specific heat per mol-mass (per m pounds, where m is the molecular weight) is the same and therefore that their enthalpy and entropy per mol-mass are identical. The relation for the specific heat per mol-mass, or the "molar" specific heat, of these gases, is

$$mc_v = 4.924 + 0.0_495\ T + 0.0_796\ T^2; \quad m(c_p - c_v) = 1.99 \text{ (Partington and Shilling)}$$

$$mc_v = 4.945 + 0.0_612\ T^2; \quad m(c_p - c_v) = 1.98 \text{ (Goodenough and Felbeck)}$$

The relation of Partington and Shilling has been used for the plotting of the curves, Fig. 6.

The values of mH and mS used in the figure are ones relative to a standard reference state at 32 F (492 R) and 14.7 lb per sq in. abs, at which state the entropy and the internal energy are regarded as zero. The values of T and mH have been plotted to logarithmic scales in order to make the pressure contours virtually straight lines (departing from those only by reason of specific heat variation with temperature). On this figure any polytropic or adiabatic state change conforming to the relation "PV^n = constant" would be representable by a straight line of the proper slope. The slope corresponding to various values of n may be found by passing a line through the reference state point, at 32 F and 14.7 lb, and the proper one of the various indices appearing in left and upper margins. For a state change passing through any other state, pass a line of proper slope through the desired state point.

13. DRY AIR TABLES

The solution of thermodynamic problems involving gases frequently requires laborious calculations because of the involved relationship between specific heat and temperature. The calculations can be made less time-consuming if suitable tables of the thermodynamic properties of the gas are available. Such tables have been prepared by Keenan and Kaye, and the table for dry air is presented in part as Table 6.

For many engineering problems, the ideal-gas relation may be used as a suitable equation of state for air at low-pressure conditions. At 32 F air deviates from an ideal gas by about 1 per cent at 300 psia and 0.1 per cent at atmospheric pressure. At low pressures and high temperatures the departure of air from an ideal gas is negligible for engineering calculations.

For an ideal gas the internal energy and enthalpy are functions of temperature alone. This is shown by the following relations:

$$u - u_0 = \int_{T_0}^{T} c_v\ dT \tag{27}$$

and

$$h - h_0 = \int_{T_0}^{T} c_p\ dT \tag{28}$$

In the above equations the lower limits of the integrals refer to an arbitrary standard state which is to be used as the base of the tables. For convenience, Keenan and Kaye assigned to h_0 the value $h_0 = 0$, when the absolute temperature is $T_0 = 0$ R. With the foregoing value for enthalpy as a base, and by integrating equations 27 and 28, between $T_0 = 0$ and T, and expressing c_p and c_v as functions of temperature, the values of h and u as functions of temperature are obtained. Other thermodynamic properties can then be calculated from thermodynamic relationship such as $h = u + pv = u + RT$. Table 6, taken from the tables of Keenan and Kaye, presents values for the internal energy, u, and enthalpy, h, for air.

For an ideal gas which experiences a change of state between state 1 and state 2, the changes in its internal energy and enthalpy may be expressed as follows:

$$u_2 - u_1 = (u_2 - u_0) - (u_1 - u_0) = \int_{T_0}^{T_2} c_v\ dT - \int_{T_0}^{T_1} c_v\ dT \tag{29}$$

and

$$h_2 - h_1 = (h_2 - h_0) - (h_1 - h_0) = \int_{T_0}^{T_2} c_p\ dT - \int_{T_0}^{T_1} c_p\ dT \tag{30}$$

These relations are useful in the solution of many types of problems.

Table 6. Air at Low Pressures (for One Pound) *

T	t	h	p_r	u	v_r	ϕ	T	t	h	p_r	u	v_r	ϕ
400	−59.7	95.53	0.4858	68.11	305.0	0.52890	700	240.3	167.56	3.446	119.58	75.25	0.66321
405	−54.7	96.73	0.5073	68.97	295.7	0.53187	705	245.3	168.77	3.533	120.45	73.91	0.66493
410	−49.7	97.93	0.5295	69.82	286.8	0.53481	710	250.3	169.98	3.623	121.32	72.60	0.66664
415	−44.7	99.13	0.5524	70.67	278.3	0.53771	715	255.3	171.18	3.714	122.18	71.32	0.66834
420	−39.7	100.32	0.5760	71.52	270.1	0.54058	720	260.3	172.39	3.806	123.04	70.07	0.67002
425	−34.7	101.52	0.6003	72.38	262.3	0.54341	725	265.3	173.60	3.900	123.91	68.86	0.67169
430	−29.7	102.71	0.6253	73.23	254.7	0.54621	730	270.3	174.82	3.996	124.78	67.67	0.67335
435	−24.7	103.91	0.6510	74.09	247.5	0.54898	735	275.3	176.03	4.093	125.65	66.52	0.67501
440	−19.7	105.11	0.6776	74.93	240.6	0.55172	740	280.3	177.23	4.193	126.51	65.38	0.67665
445	−14.7	106.30	0.7048	75.79	233.9	0.55442	745	285.3	178.44	4.294	127.38	64.28	0.67828
450	−9.7	107.50	0.7329	76.65	227.45	0.55710	750	290.3	179.66	4.396	128.25	63.20	0.67991
455	−4.7	108.70	0.7617	77.50	221.27	0.55975	755	295.3	180.87	4.501	129.12	62.14	0.68152
460	0.3	109.90	0.7913	78.36	215.33	0.56235	760	300.3	182.08	4.607	129.99	61.10	0.68312
465	5.3	111.10	0.8218	79.22	209.60	0.56495	765	305.3	183.30	4.716	130.86	60.09	0.68470
470	10.3	112.30	0.8531	80.07	204.08	0.56751	770	310.3	184.51	4.826	131.73	59.11	0.68629
475	15.3	113.50	0.8852	80.93	198.77	0.57004	775	315.3	185.73	4.937	132.60	58.14	0.68786
480	20.3	114.69	0.9182	81.77	193.65	0.57255	780	320.3	186.94	5.051	133.47	57.20	0.68942
485	25.3	115.89	0.9521	82.63	188.70	0.57503	785	325.3	188.16	5.167	134.34	56.28	0.69098
490	30.3	117.08	0.9868	83.49	183.94	0.57749	790	330.3	189.38	5.285	135.22	55.38	0.69251
495	35.3	118.28	1.0224	84.34	179.34	0.57992	795	335.3	190.59	5.404	136.09	54.49	0.69405
500	40.3	119.48	1.0590	85.20	174.90	0.58233	800	340.3	191.81	5.526	136.97	53.63	0.69558
505	45.3	120.68	1.0965	86.06	170.60	0.58472	805	345.3	193.03	5.650	137.84	52.78	0.69709
510	50.3	121.87	1.1349	86.92	166.46	0.58707	810	350.3	194.25	5.775	138.72	51.96	0.69860
515	55.3	123.07	1.1743	87.76	162.46	0.58942	815	355.3	195.46	5.903	139.60	51.15	0.70010
520	60.3	124.27	1.2147	88.62	158.58	0.59173	820	360.3	196.69	6.033	140.47	50.35	0.70160
525	65.3	125.47	1.2560	89.48	154.84	0.59403	825	365.3	197.90	6.165	141.34	49.57	0.70308
530	70.3	126.66	1.2983	90.34	151.22	0.59630	830	370.3	199.12	6.299	142.22	48.81	0.70455
535	75.3	127.86	1.3416	91.19	147.72	0.59855	835	375.3	200.34	6.435	143.10	48.07	0.70601
540	80.3	129.06	1.3860	92.04	144.32	0.60078	840	380.3	201.56	6.573	143.98	47.34	0.70747
545	85.3	130.26	1.4315	92.90	141.03	0.60299	845	385.3	202.79	6.714	144.86	46.62	0.70892
550	90.3	131.46	1.4779	93.76	137.85	0.60518	850	390.3	204.01	6.856	145.74	45.92	0.71037
555	95.3	132.66	1.5255	94.61	134.77	0.60736	855	395.3	205.23	7.001	146.62	45.24	0.71180
560	100.3	133.86	1.5742	95.47	131.78	0.60950	860	400.3	206.46	7.149	147.50	44.57	0.71323
565	105.3	135.06	1.6240	96.33	128.88	0.61164	865	405.3	207.67	7.298	148.38	43.91	0.71465
570	110.3	136.26	1.6748	97.19	126.08	0.61376	870	410.3	208.90	7.450	149.27	43.26	0.71606
575	115.3	137.47	1.7269	98.05	123.34	0.61586	875	415.3	210.12	7.604	150.15	42.62	0.71747
580	120.3	138.66	1.7800	98.90	120.70	0.61793	880	420.3	211.35	7.761	151.02	42.01	0.71886
585	125.3	139.86	1.8345	99.76	118.13	0.61999	885	425.3	212.58	7.920	151.91	41.39	0.72025
590	130.3	141.06	1.8899	100.62	115.65	0.62204	890	430.3	213.80	8.081	152.80	40.80	0.72163
595	135.3	142.26	1.9467	101.48	113.22	0.62406	395	435.3	215.03	8.245	153.68	40.21	0.72301
600	140.3	143.47	2.005	102.34	110.88	0.62607	900	440.3	216.26	8.411	154.57	39.64	0.72438
605	145.3	144.67	2.064	103.20	108.59	0.62807	905	445.3	217.49	8.580	155.46	39.07	0.72574
610	150.3	145.88	2.124	104.06	106.38	0.63005	910	450.3	218.72	8.752	156.34	38.52	0.72710
615	155.3	147.08	2.186	104.91	104.22	0.63200	915	455.3	219.95	8.925	157.23	37.97	0.72845
620	160.3	148.28	2.249	105.78	102.12	0.63395	920	460.3	221.18	9.102	158.12	37.44	0.72979
625	165.3	149.49	2.313	106.64	100.08	0.63589	925	465.3	222.41	9.281	159.01	36.92	0.73113
630	170.3	150.68	2.379	107.50	98.11	0.63781	930	470.3	223.64	9.463	159.89	36.41	0.73245
635	175.3	151.89	2.446	108.36	96.17	0.63971	935	475.3	224.87	9.647	160.79	35.90	0.73377
640	180.3	153.09	2.514	109.21	94.30	0.64159	940	480.3	226.11	9.834	161.68	35.41	0.73509
645	185.3	154.30	2.584	110.08	92.47	0.64347	945	485.3	227.35	10.024	162.57	34.92	0.73640
650	190.3	155.50	2.655	110.94	90.69	0.64533	950	490.3	228.58	10.216	163.46	34.45	0.73771
655	195.3	156.71	2.727	111.80	88.96	0.64718	955	495.3	229.82	10.412	164.36	33.97	0.73900
660	200.3	157.92	2.801	112.67	87.27	0.64902	960	500.3	231.06	10.610	165.26	33.52	0.74030
665	205.3	159.12	2.877	113.53	85.63	0.65083	965	505.3	232.30	10.811	166.15	33.07	0.74158
670	210.3	160.33	2.953	114.40	84.03	0.65263	970	510.3	233.53	11.014	167.05	32.63	0.74287
675	215.3	161.54	3.032	115.26	82.47	0.65443	975	515.3	234.77	11.221	167.94	32.19	0.74413
680	220.3	162.73	3.111	116.12	80.96	0.65621	980	520.3	236.02	11.430	168.83	31.76	0.74540
685	225.3	163.94	3.193	116.99	79.47	0.65798	985	525.3	237.26	11.643	169.73	31.34	0.74666
690	230.3	165.15	3.276	117.85	78.03	0.65973	990	530.3	238.50	11.858	170.63	30.92	0.74792
895	235.3	166.36	3.360	118.72	76.62	0.66147	995	535.3	239.74	12.077	171.53	30.52	0.74917

* Reproduced by permission from J. H. Keenan and J. Kaye, *Gas Tables*, John Wiley & Sons, New York, 1948.

Table 6. Air at Low Pressures (for One Pound)—*Continued*

T	t	h	p_r	u	v_r	φ	T	t	h	p_r	u	v_r	φ
1000	540.3	240.98	12.298	172.43	30.12	0.75042	1300	840.3	316.94	32.39	227.83	14.868	0.81680
1005	545.3	242.23	12.523	173.33	29.73	0.75166	1305	845.3	318.24	32.86	228.78	14.711	0.81779
1010	550.3	243.48	12.751	174.24	29.34	0.75290	1310	850.3	319.53	33.34	229.73	14.557	0.81878
1015	555.3	244.72	12.981	175.14	28.97	0.75413	1315	855.3	320.82	33.82	230.68	14.404	0.81976
1020	560.3	245.97	13.215	176.04	28.59	0.75536	1320	860.3	322.11	34.31	231.63	14.253	0.82075
1025	565.3	247.21	13.452	176.95	28.22	0.75657	1325	865.3	323.41	34.80	232.58	14.104	0.82172
1030	570.3	248.45	13.692	177.84	27.87	0.75778	1330	870.3	324.69	35.30	233.52	13.958	0.82270
1035	575.3	249.70	13.935	178.75	27.51	0.75899	1335	875.3	325.99	35.80	234.47	13.813	0.82367
1040	580.3	250.95	14.182	179.66	27.17	0.76019	1340	880.3	327.29	36.31	235.43	13.670	0.82464
1045	585.3	252.20	14.432	180.56	26.82	0.76139	1345	885.3	328.59	36.83	236.38	13.529	0.82561
1050	590.3	253.45	14.686	181.47	26.48	0.76259	1350	890.3	329.88	37.35	237.34	13.391	0.82658
1055	595.3	254.71	14.943	182.38	26.15	0.76377	1355	895.3	331.18	37.87	238.29	13.253	0.82753
1060	600.3	255.96	15.203	183.29	25.82	0.76496	1360	900.3	332.48	38.41	239.25	13.118	0.82848
1065	605.3	257.21	15.467	184.19	25.51	0.76614	1365	905.3	333.79	38.94	240.22	12.983	0.82944
1070	610.3	258.47	15.734	185.10	25.19	0.76732	1370	910.3	335.09	39.49	241.17	12.851	0.83039
1075	615.3	259.72	16.005	186.01	24.88	0.76848	1375	915.3	336.39	40.04	242.13	12.721	0.83134
1080	620.3	260.97	16.278	186.93	24.58	0.76964	1380	920.3	337.68	40.59	243.08	12.593	0.83229
1085	625.3	262.22	16.557	187.84	24.27	0.77080	1385	925.3	338.99	41.16	244.05	12.465	0.83323
1090	630.3	263.48	16.838	188.75	23.98	0.77196	1390	930.3	340.29	41.73	245.00	12.340	0.83417
1095	635.3	264.74	17.124	189.66	23.69	0.77311	1395	935.3	341.60	42.30	245.97	12.217	0.83511
1100	640.3	265.99	17.413	190.58	23.40	0.77426	1400	940.3	342.90	42.88	246.93	12.095	0.83604
1105	645.3	267.26	17.704	191.50	23.12	0.77540	1405	945.3	344.21	43.47	247.90	11.974	0.83697
1110	650.3	268.52	18.000	192.41	22.84	0.77654	1410	950.3	345.52	44.06	248.86	11.855	0.83790
1115	655.3	269.77	18.300	193.33	22.57	0.77768	1415	955.3	346.83	44.66	249.83	11.738	0.83883
1120	660.3	271.03	18.604	194.25	22.30	0.77880	1420	960.3	348.14	45.26	250.79	11.622	0.83975
1125	665.3	272.29	18.912	195.17	22.04	0.77993	1425	965.3	349.45	45.87	251.75	11.507	0.84067
1130	670.3	273.56	19.223	196.09	21.78	0.78104	1430	970.3	350.75	46.49	252.72	11.394	0.84158
1135	675.3	274.82	19.538	197.01	21.52	0.78215	1435	975.3	352.06	47.12	253.69	11.282	0.84250
1140	680.3	276.08	19.858	197.94	21.27	0.78326	1440	980.3	353.37	47.75	254.66	11.172	0.84341
1145	685.3	277.34	20.181	198.86	21.02	0.78437	1445	985.3	354.69	48.39	255.63	11.062	0.84432
1150	690.3	278.61	20.51	199.78	20.771	0.78548	1450	990.3	356.00	49.03	256.60	10.954	0.84523
1155	695.3	279.87	20.84	200.71	20.530	0.78658	1455	995.3	357.32	49.68	257.57	10.848	0.84614
1160	700.3	281.14	21.18	201.63	20.293	0.78767	1460	1000.3	358.63	50.34	258.54	10.743	0.84704
1165	705.3	282.41	21.51	202.56	20.059	0.78876	1465	1005.3	359.95	51.01	259.51	10.639	0.84794
1170	710.3	283.68	21.86	203.49	19.828	0.78985	1470	1010.3	361.27	51.68	260.49	10.537	0.84884
1175	715.3	284.95	22.21	204.41	19.601	0.79093	1475	1015.3	362.58	52.36	261.46	10.436	0.84973
1180	720.3	286.21	22.56	205.33	19.377	0.79201	1480	1020.3	363.89	53.04	262.44	10.336	0.85062
1185	725.3	287.49	22.91	206.26	19.157	0.79308	1485	1025.3	365.21	53.73	263.40	10.237	0.85151
1190	730.3	288.76	23.28	207.19	18.940	0.79415	1490	1030.3	366.53	54.43	264.38	10.140	0.85239
1195	735.3	290.03	23.64	208.12	18.725	0.79522	1495	1035.3	367.85	55.14	265.36	10.043	0.85328
1200	740.3	291.30	24.01	209.05	18.514	0.79628	1500	1040.3	369.17	55.86	266.34	9.948	0.85416
1205	745.3	292.58	24.38	209.99	18.307	0.79734	1505	1045.3	370.50	56.58	267.32	9.854	0.85504
1210	750.3	293.86	24.76	210.92	18.102	0.79840	1510	1050.3	371.82	57.30	268.30	9.761	0.85592
1215	755.3	295.14	25.14	211.86	17.900	0.79945	1515	1055.3	373.15	58.04	269.28	9.669	0.85679
1220	760.3	296.41	25.53	212.78	17.700	0.80050	1520	1060.3	374.47	58.78	270.26	9.578	0.85767
1225	765.3	297.69	25.92	213.71	17.504	0.80155	1525	1065.3	375.79	59.53	271.24	9.489	0.85853
1230	770.3	298.96	26.32	214.65	17.311	0.80258	1530	1070.3	377.11	60.29	272.23	9.400	0.85940
1235	775.3	300.24	26.72	215.59	17.120	0.80362	1535	1075.3	378.44	61.06	273.21	9.313	0.86026
1240	780.3	301.52	27.13	216.53	16.932	0.80466	1540	1080.3	379.77	61.83	274.20	9.226	0.86113
1245	785.3	302.80	27.54	217.46	16.747	0.80569	1545	1085.3	381.09	62.61	275.18	9.141	0.86199
1250	790.3	304.08	27.96	218.40	16.563	0.80672	1550	1090.3	382.42	63.40	276.17	9.056	0.86285
1255	795.3	305.37	28.38	219.34	16.383	0.80774	1555	1095.3	383.75	64.20	277.15	8.973	0.86370
1260	800.3	306.65	28.80	220.28	16.205	0.80876	1560	1100.3	385.08	65.00	278.13	8.890	0.86456
1265	805.3	307.94	29.23	221.22	16.030	0.80978	1565	1105.3	386.42	65.81	279.13	8.808	0.86541
1270	810.3	309.22	29.67	222.16	15.857	0.81079	1570	1110.3	387.74	66.63	280.11	8.728	0.86626
1275	815.3	310.50	30.11	223.10	15.687	0.81180	1575	1115.3	389.08	67.46	281.10	8.648	0.86710
1280	820.3	311.79	30.55	224.05	15.518	0.81280	1580	1120.3	390.40	68.30	282.09	8.569	0.86794
1285	825.3	313.08	31.01	224.99	15.352	0.81381	1585	1125.3	391.74	69.15	283.09	8.491	0.86878
1290	830.3	314.36	31.46	225.93	15.189	0.81481	1590	1130.3	393.07	70.00	284.08	8.414	0.86962
1295	835.3	315.65	31.92	226.88	15.027	0.81580	1595	1135.3	394.41	70.86	285.07	8.338	0.87046

Table 6. Air at Low Pressures (for One Pound)—*Continued*

T	t	h	p_r	u	v_r	ϕ	T	t	h	p_r	u	v_r	ϕ
1600	1140.3	395.74	71.73	286.06	8.263	0.87130	1900	1440.3	477.09	141.51	346.85	4.974	0.91788
1605	1145.3	397.08	72.61	287.06	8.189	0.87213	1905	1445.3	478.47	143.01	347.87	4.934	0.91860
1610	1150.3	398.42	73.49	288.05	8.115	0.87297	1910	1450.3	479.85	144.53	348.91	4.896	0.91932
1615	1155.3	399.76	74.39	289.05	8.042	0.87380	1915	1455.3	481.23	146.06	349.95	4.857	0.92004
1620	1160.3	401.09	75.29	290.04	7.971	0.87462	1920	1460.3	482.60	147.59	350.98	4.819	0.92076
1625	1165.3	402.43	76.20	291.03	7.900	0.87545	1925	1465.3	483.97	149.14	352.01	4.781	0.92148
1630	1170.3	403.77	77.12	292.03	7.829	0.87627	1930	1470.3	485.36	150.70	353.05	4.744	0.92220
1635	1175.3	405.11	78.05	293.03	7.760	0.87709	1935	1475.3	486.74	152.27	354.09	4.707	0.92291
1640	1180.3	406.45	78.99	294.03	7.691	0.87791	1940	1480.3	488.12	153.87	355.12	4.670	0.92362
1645	1185.3	407.79	79.94	295.03	7.623	0.87873	1945	1485.3	489.50	155.48	356.16	4.634	0.92433
1650	1190.3	409.13	80.89	296.03	7.556	0.87954	1950	1490.3	490.88	157.10	357.20	4.598	0.92504
1655	1195.3	410.47	81.85	297.02	7.490	0.88035	1955	1495.3	492.25	158.73	358.24	4.562	0.92574
1660	1200.3	411.82	82.83	298.02	7.424	0.88116	1960	1500.3	493.64	160.37	359.28	4.527	0.92645
1665	1205.3	413.16	83.81	299.02	7.359	0.88197	1965	1505.3	495.02	162.02	360.32	4.493	0.92715
1670	1210.3	414.51	84.80	300.03	7.295	0.88278	1970	1510.3	496.40	163.69	361.36	4.458	0.92786
1675	1215.3	415.85	85.80	301.03	7.231	0.88358	1975	1515.3	497.79	165.37	362.40	4.424	0.92856
1680	1220.3	417.20	86.82	302.04	7.168	0.88439	1980	1520.3	499.17	167.07	363.43	4.390	0.92926
1685	1225.3	418.54	87.84	303.03	7.106	0.88519	1985	1525.3	500.55	168.77	364.48	4.357	0.92996
1690	1230.3	419.89	88.87	304.04	7.045	0.88599	1990	1530.3	501.94	170.50	365.53	4.323	0.93066
1695	1235.3	421.24	89.91	305.05	6.984	0.88678	1995	1535.3	503.32	172.25	366.57	4.290	0.93135
1700	1240.3	422.59	90.95	306.06	6.924	0.88758	2000	1540.3	504.71	174.00	367.61	4.258	0.93205
1705	1245.3	423.94	92.01	307.06	6.864	0.88837	2005	1545.3	506.10	175.77	368.66	4.226	0.93274
1710	1250.3	425.29	93.08	308.07	6.805	0.88916	2010	1550.3	507.49	177.55	369.71	4.194	0.93343
1715	1255.3	426.64	94.16	309.08	6.747	0.88995	2015	1555.3	508.88	179.35	370.75	4.162	0.93412
1720	1260.3	428.00	95.24	310.09	6.690	0.89074	2020	1560.3	510.26	181.16	371.79	4.130	0.93481
1725	1265.3	429.35	96.34	311.10	6.633	0.89152	2025	1565.3	511.65	182.99	372.84	4.099	0.93550
1730	1270.3	430.69	97.45	312.10	6.576	0.89230	2030	1570.3	513.04	184.81	373.88	4.069	0.93618
1735	1275.3	432.05	98.57	313.12	6.521	0.89308	2035	1575.3	514.43	186.67	374.93	4.038	0.93687
1740	1280.3	433.41	99.69	314.13	6.465	0.89387	2040	1580.3	515.82	188.54	375.98	4.008	0.93756
1745	1285.3	434.76	100.83	315.14	6.411	0.89464	2045	1585.3	517.22	190.42	377.03	3.978	0.93823
1750	1290.3	436.12	101.98	316.16	6.357	0.89542	2050	1590.3	518.61	192.31	378.08	3.949	0.93891
1755	1295.3	437.48	103.13	317.17	6.304	0.89619	2055	1595.3	519.99	194.23	379.12	3.919	0.93958
1760	1300.3	438.83	104.30	318.18	6.251	0.89697	2060	1600.3	521.39	196.16	380.18	3.890	0.94026
1765	1305.3	440.19	105.48	319.20	6.198	0.89774	2065	1605.3	522.78	198.11	381.23	3.861	0.94094
1770	1310.3	441.55	106.67	320.22	6.147	0.89850	2070	1610.3	524.18	200.06	382.28	3.833	0.94161
1775	1315.3	442.91	107.87	321.23	6.095	0.89927	2075	1615.3	525.57	202.03	383.33	3.805	0.94229
1780	1320.3	444.26	109.08	322.24	6.045	0.90003	2080	1620.3	526.97	204.02	384.39	3.777	0.94296
1785	1325.3	445.63	110.30	323.27	5.995	0.90079	2085	1625.3	528.35	206.03	385.43	3.749	0.94363
1790	1330.3	446.99	111.54	324.29	5.945	0.90155	2090	1630.3	529.75	208.06	386.48	3.721	0.94430
1795	1335.3	448.35	112.78	325.30	5.896	0.90231	2095	1635.3	531.15	210.10	387.54	3.694	0.94497
1800	1340.3	449.71	114.03	326.32	5.847	0.90308	2100	1640.3	532.55	212.1	388.60	3.667	0.94564
1805	1345.3	451.08	115.30	327.35	5.799	0.90383	2105	1645.3	533.95	214.2	389.65	3.640	0.94630
1810	1350.3	452.44	116.57	328.37	5.752	0.90458	2110	1650.3	535.35	216.3	390.71	3.614	0.94696
1815	1355.3	453.81	117.86	329.39	5.705	0.90534	2115	1655.3	536.75	218.4	391.77	3.588	0.94762
1820	1360.3	455.17	119.16	330.40	5.658	0.90609	2120	1660.3	538.15	220.5	392.83	3.561	0.94829
1825	1365.3	456.54	120.47	331.43	5.612	0.90684	2125	1665.3	539.55	222.6	393.88	3.536	0.94895
1830	1370.3	457.90	121.79	332.45	5.566	0.90759	2130	1670.3	540.94	224.8	394.93	3.510	0.94960
1835	1375.3	459.27	123.12	333.48	5.521	0.90833	2135	1675.3	542.34	226.9	395.99	3.485	0.95026
1840	1380.3	460.63	124.47	334.50	5.476	0.90908	2140	1680.3	543.74	229.1	397.05	3.460	0.95092
1845	1385.3	462.00	125.82	335.53	5.432	0.90982	2145	1685.3	545.14	231.3	398.10	3.435	0.95157
1850	1390.3	463.37	127.18	336.55	5.388	0.91056	2150	1690.3	546.54	233.5	399.17	3.410	0.95222
1855	1395.3	464.74	128.56	337.57	5.345	0.91129	2155	1695.3	547.95	235.8	400.23	3.386	0.95287
1860	1400.3	466.12	129.95	338.61	5.302	0.91203	2160	1700.3	549.35	238.0	401.29	3.362	0.95352
1865	1405.3	467.49	131.35	339.63	5.259	0.91277	2165	1705.3	550.76	240.3	402.35	3.338	0.95417
1870	1410.3	468.86	132.77	340.66	5.217	0.91350	2170	1710.3	552.16	242.6	403.41	3.314	0.95482
1875	1415.3	470.23	134.20	341.69	5.175	0.91424	2175	1715.3	553.57	244.9	404.48	3.290	0.95546
1880	1420.3	471.60	135.64	342.73	5.134	0.91497	2180	1720.3	554.97	247.2	405.53	3.267	0.95611
1885	1425.3	472.97	137.10	343.75	5.093	0.91570	2185	1725.3	556.37	249.5	406.59	3.244	0.95675
1890	1430.3	474.35	138.55	344.78	5.053	0.91643	2190	1730.3	557.78	251.9	407.66	3.221	0.95740
1895	1435.3	475.72	140.03	345.81	5.013	0.91715	2195	1735.3	559.19	254.2	408.72	3.198	0.95804

Table 6. Air at Low Pressures (for One Pound)—*Continued*

T	t	h	p_r	u	v_r	φ	T	t	h	p_r	u	v_r	φ
2200	1740.3	560.59	256.6	409.78	3.176	0.95868	2500	2040.3	645.78	435.7	474.40	2.125	0.99497
2205	1745.3	562.00	259.0	410.85	3.154	0.95932	2505	2045.3	647.22	439.4	475.50	2.112	0.99554
2210	1750.3	563.41	261.4	411.92	3.131	0.95996	2510	2050.3	648.65	443.0	476.58	2.099	0.99611
2215	1755.3	564.82	263.9	412.99	3.109	0.96059	2515	2055.3	650.08	446.7	477.67	2.085	0.99668
2220	1760.3	566.23	266.3	414.05	3.088	0.96123	2520	2060.3	651.51	450.5	478.77	2.072	0.99725
2225	1765.3	567.64	268.8	415.12	3.066	0.96186	2525	2065.3	652.94	454.2	479.86	2.059	0.99782
2230	1770.3	569.04	271.3	416.18	3.045	0.96250	2530	2070.3	654.38	458.0	480.94	2.046	0.99838
2235	1775.3	570.45	273.8	417.24	3.024	0.96313	2535	2075.3	655.81	461.8	482.03	2.034	0.99895
2240	1780.3	571.86	276.3	418.31	3.003	0.96376	2540	2080.3	657.25	465.6	483.13	2.021	0.99952
2245	1785.3	573.27	278.9	419.38	2.982	0.96439	2545	2085.3	658.69	469.5	484.22	2.008	1.00008
2250	1790.3	574.69	281.4	420.46	2.961	0.96501	2550	2090.3	660.12	473.3	485.31	1.9956	1.00064
2255	1795.3	576.10	284.0	421.52	2.941	0.96564	2555	2095.3	661.55	477.2	486.41	1.9832	1.00120
2260	1800.3	577.51	286.6	422.59	2.921	0.96626	2560	2100.3	662.99	481.1	487.51	1.9709	1.00176
2265	1805.3	578.93	289.3	423.67	2.901	0.96689	2565	2105.3	664.42	485.1	488.60	1.9587	1.00232
2270	1810.3	580.34	291.9	424.74	2.881	0.96751	2570	2110.3	665.86	489.1	489.69	1.9465	1.00288
2275	1815.3	581.75	294.6	425.80	2.861	0.96813	2575	2115.3	667.30	493.1	490.79	1.9345	1.00344
2280	1820.3	583.16	297.2	426.87	2.841	0.96876	2580	2120.3	668.74·	497.1	491.88	1.9225	1.00400
2285	1825.3	584.58	299.9	427.95	2.822	0.96937	2585	2125.3	670.17	501.2	492.97	1.9107	1.00456
2290	1830.3	585.99	302.7	429.01	2.803	0.96999	2590	2130.3	671.61	505.3	494.07	1.8989	1.00511
2295	1835.3	587.41	305.4	430.09	2.784	0.97061	2595	2135.3	673.05	509.4	495.17	1.8872	1.00567
2300	1840.3	588.82	308.1	431.16	2.765	0.97123	2600	2140.3	674.49	513.5	496.26	1.8756	1.00623
2305	1845.3	590.24	310.9	432.24	2.746	0.97184	2605	2145.3	675.93	517.6	497.36	1.8641	1.00678
2310	1850.3	591.66	313.7	433.31	2.728	0.97246	2610	2150.3	677.37	521.8	498.46	1.8527	1.00733
2315	1855.3	593.08	316.5	434.39	2.709	0.97307	2615	2155.3	678.81	526.1	499.56	1.8414	1.00788
2320	1860.3	594.49	319.4	435.46	2.691	0.97369	2620	2160.3	680.25	530.3	500.65	1.8302	1.00843
2325	1865.3	595.90	322.2	436.53	2.673	0.97429	2625	2165.3	681.69	534.6	501.75	1.8190	1.00898
2330	1870.3	597.32	325.1	437.60	2.655	0.97489	2630	2170.3	683.13	538.9	502.85	1.8079	1.00953
2335	1875.3	598.74	328.0	438.68	2.637	0.97550	2635	2175.3	684.57	543.2	503.95	1.7970	1.01008
2340	1880.3	600.16	330.9	439.76	2.619	0.97611	2640	2180.3	686.01	547.5	505.05	1.7861	1.01063
2345	1885.3	601.58	333.9	440.83	2.602	0.97672	2645	2185.3	687.46	551.9	506.15	1.7753	1.01117
2350	1890.3	603.00	336.8	441.91	2.585	0.97732	2650	2190.3	688.90	556.3	507.25	1.7646	1.01172
2355	1895.3	604.42	339.8	442.99	2.567	0.97793	2655	2195.3	690.35	560.7	508.34	1.7539	1.01226
2360	1900.3	605.84	342.8	444.07	2.550	0.97853	2660	2200.3	691.79	565.2	509.44	1.7434	1.01281
2365	1905.3	607.26	345.8	445.14	2.533	0.97913	2665	2205.3	693.24	569.7	510.55	1.7329	1.01335
2370	1910.3	608.68	348.9	446.22	2.517	0.97973	2670	2210.3	694.68	574.2	511.65	1.7225	1.01389
2375	1915.3	610.11	351.9	447.31	2.500	0.98033	2675	2215.3	696.12	578.8	512.75	1.7121	1.01443
2380	1920.3	611.53	355.0	448.38	2.483	0.98092	2680	2220.3	697.56	583.3	513.85	1.7019	1.01497
2385	1925.3	612.94	358.1	449.45	2.467	0.98152	2685	2225.3	699.01	587.9	514.95	1.6917	1.01551
2390	1930.3	614.37	361.3	450.54	2.451	0.98212	2690	2230.3	700.45	592.5	516.05	1.6817	1.01605
2395	1935.3	615.79	364.4	451.62	2.435	0.98271	2695	2235.3	701.90	597.2	517.15	1.6716	1.01659
2400	1940.3	617.22	367.6	452.70	2.419	0.98331	2700	2240.3	703.35	601.9	518.26	1.6617	1.01712
2405	1945.3	618.64	370.7	453.78	2.403	0.98390	2705	2245.3	704.80	606.6	519.37	1.6518	1.01766
2410	1950.3	620.07	374.0	454.87	2.387	0.98449	2710	2250.3	706.24	611.3	520.47	1.6420	1.01819
2415	1955.3	621.50	377.2	455.96	2.372	0.98508	2715	2255.3	707.69	616.2	521.57	1.6323	1.01873
2420	1960.3	622.92	380.5	457.02	2.356	0.98567	2720	2260.3	709.13	620.9	522.68	1.6226	1.01926
2425	1965.3	624.34	383.8	458.11	2.341	0.98626	2725	2265.3	710.58	625.8	523.78	1.6130	1.01979
2430	1970.3	625.77	387.0	459.20	2.326	0.98685	2730	2270.3	712.03	630.7	524.88	1.6035	1.02032
2435	1975.3	627.20	390.4	460.28	2.311	0.98744	2735	2275.3	713.48	635.6	525.99	1.5941	1.02085
2440	1980.3	628.62	393.7	461.36	2.296	0.98802	2740	2280.3	714.93	640.5	527.10	1.5847	1.02138
2445	1985.3	630.05	397.1	462.45	2.281	0.98861	2745	2285.3	716.38	645.4	528.21	1.5754	1.02191
2450	1990.3	631.48	400.5	463.54	2.266	0.98919	2750	2290.3	717.83	650.4	529.31	1.5662	1.02244
2455	1995.3	632.91	403.9	464.61	2.252	0.98976	2755	2295.3	719.27	655.4	530.42	1.5570	1.02296
2460	2000.3	634.34	407.3	465.70	2.237	0.99035	2760	2300.3	720.72	660.5	531.53	1.5480	1.02348
2465	2005.3	635.77	410.8	466.79	2.223	0.99093	2765	2305.3	722.17	665.6	532.64	1.5389	1.02401
2470	2010.3	637.20	414.3	467.88	2.209	0.99151	2770	2310.3	723.62	670.7	533.74	1.5299	1.02453
2475	2015.3	638.63	417.8	468.96	2.194	0.99208	2775	2315.3	725.07	675.8	534.85	1.5210	1.02506
2480	2020.3	640.05	421.3	470.05	2.180	0.99266	2780	2320.3	726.53	681.0	535.96	1.5122	1.02558
2485	2025.3	641.49	424.9	471.14	2.166	0.99324	2785	2325.3	727.97	686.2	537.06	1.5035	1.02610
2490	2030.3	642.91	428.5	472.22	2.153	0.99381	2790	2330.3	729.42	691.4	538.17	1.4948	1.02662
2495	2035.3	644.35	432.1	473.31	2.139	0.99439	2795	2335.3	730.88	696.7	539.29	1.4861	1.02714

Table 6. Air at Low Pressures (for One Pound)—*Continued*

T	t	h	p_r	u	v_r	ϕ	T	t	h	p_r	u	v_r	ϕ
2800	2340.3	732.33	702.0	540.40	1.4775	1.02767	2900	2440.3	761.45	814.8	562.66	1.3184	1.03788
2805	2345.3	733.79	707.3	541.51	1.4690	1.02818	2905	2445.3	762.91	820.8	563.78	1.3110	1.03838
2810	2350.3	735.24	712.7	542.62	1.4606	1.02870	2910	2450.3	764.37	826.8	564.90	1.3037	1.03889
2815	2355.3	736.70	718.1	543.74	1.4522	1.02922	2915	2455.3	765.84	832.9	566.02	1.2964	1.03939
2820	2360.3	738.15	723.5	544.85	1.4439	1.02974	2920	2460.3	767.29	839.0	567.13	1.2892	1.03989
2825	2365.3	739.60	728.9	545.95	1.4356	1.03025	2925	2465.3	768.75	845.2	568.25	1.2820	1.04039
2830	2370.3	741.05	734.4	547.06	1.4274	1.03076	2930	2470.3	770.21	851.3	569.37	1.2749	1.04089
2835	2375.3	742.51	740.0	548.18	1.4192	1.03128	2935	2475.3	771.67	857.5	570.48	1.2679	1.04139
2840	2380.3	743.96	745.5	549.29	1.4112	1.03179	2940	2480.3	773.13	863.8	571.60	1.2608	1.04188
2845	2385.3	745.42	751.1	550.40	1.4031	1.03230	2945	2485.3	774.59	870.0	572.72	1.2539	1.04238
2850	2390.3	746.88	756.7	551.52	1.3951	1.03282	2950	2490.3	776.05	876.4	573.84	1.2469	1.04288
2855	2395.3	748.33	762.4	552.63	1.3872	1.03332	2955	2495.3	777.51	882.7	574.95	1.2401	1.04336
2860	2400.3	749.79	768.1	553.74	1.3794	1.03383	2960	2500.3	778.97	889.1	576.07	1.2332	1.04386
2865	2405.3	751.25	773.8	554.86	1.3715	1.03434	2965	2505.3	780.44	895.5	577.20	1.2265	1.04435
2870	2410.3	752.71	779.6	555.98	1.3638	1.03484	2970	2510.3	781.90	902.0	578.32	1.2197	1.04484
2875	2415.3	754.16	785.3	557.09	1.3561	1.03535	2975	2515.3	783.37	908.5	579.44	1.2130	1.04534
2880	2420.3	755.61	791.2	558.19	1.3485	1.03586	2980	2520.3	784.83	915.0	580.56	1.2064	1.04583
2885	2425.3	757.07	797.0	559.31	1.3408	1.03637	2985	2525.3	786.29	921.6	581.68	1.1998	1.04631
2890	2430.3	758.53	802.9	560.43	1.3333	1.03687	2990	2530.3	787.75	928.2	582.79	1.1932	1.04682
2895	2435.3	759.99	808.8	561.55	1.3258	1.03738	2995	2535.3	789.22	934.9	583.92	1.1867	1.04730
3000	2540.3	790.68	941.4	585.04	1.1803	1.04779	3400	2940.3	908.66	1613.2	675.60	0.7807	1.08470
3010		793.61	955.0	587.29	1.1675	1.04877	3410		911.64	1633.9	677.89	0.7732	1.08558
3020		796.54	968.7	589.53	1.1549	1.04974	3420		914.61	1654.8	680.17	0.7657	1.08645
3030		799.47	982.5	591.78	1.1425	1.05071	3430		917.58	1675.9	682.46	0.7582	1.08732
3040		802.41	996.4	594.03	1.1302	1.05168	3440		920.55	1697.2	684.75	0.7508	1.08818
3050	2590.3	805.34	1010.5	596.28	1.1181	1.05264	3450	2990.3	923.52	1718.7	687.04	0.7436	1.08904
3060		808.28	1024.8	598.52	1.1061	1.05359	3460		926.50	1740.4	689.32	0.7365	1.08990
3070		811.22	1039.2	600.77	1.0943	1.05455	3470		929.48	1762.3	691.61	0.7294	1.09076
3080		814.15	1053.8	603.02	1.0827	1.05551	3480		932.45	1784.5	693.90	0.7224	1.09162
3090		817.09	1068.5	605.27	1.0713	1.05646	3490		935.42	1806.8	696.19	0.7155	1.09247
3100	2640.3	820.03	1083.4	607.53	1.0600	1.05741	3500	3040.3	938.40	1829.3	698.48	0.7087	1.09332
3110		822.97	1098.5	609.79	1.0488	1.05836	3510		941.38	1852.1	700.78	0.7020	1.09417
3120		825.91	1113.7	612.05	1.0378	1.05930	3520		944.36	1875.2	703.07	0.6954	1.09502
3130		828.86	1129.1	614.30	1.0269	1.06025	3530		947.34	1898.6	705.36	0.6888	1.09587
3140		831.80	1144.7	616.56	1.0162	1.06119	3540		950.32	1922.1	707.65	0.6823	1.09671
3150	2690.3	834.75	1160.5	618.82	1.0056	1.06212	3550	3090.3	953.30	1945.8	709.95	0.6759	1.09755
3160		837.69	1176.4	621.08	0.9951	1.06305	3560		956.28	1969.8	712.24	0.6695	1.09838
3170		840.64	1192.5	623.35	0.9848	1.06398	3570		959.26	1993.9	714.54	0.6632	1.09922
3180		843.59	1208.7	625.60	0.9746	1.06491	3580		962.25	2018.3	716.84	0.6571	1.10005
3190		846.53	1225.1	627.86	0.9646	1.06584	3590		965.23	2043.0	719.14	0.6510	1.10089
3200	2740.3	849.48	1241.7	630.12	0.9545	1.06676	3600	3140.3	968.21	2067.9	721.44	0.6449	1.10172
3210		852.43	1258.5	632.39	0.9448	1.06768	3610		971.20	2093.0	723.74	0.6389	1.10255
3220		855.38	1275.5	634.65	0.9352	1.06860	3620		974.18	2118.4	726.04	0.6330	1.10337
3230		858.33	1292.7	636.92	0.9256	1.06952	3630		977.17	2144.0	728.34	0.6272	1.10420
3240		861.28	1310.0	639.19	0.9162	1.07043	3640		980.16	2169.9	730.64	0.6214	1.10502
3250	2790.3	864.24	1327.5	641.46	0.9069	1.07134	3650	3190.3	983.15	2196.0	732.95	0.6157	1.10584
3260		867.19	1345.2	643.73	0.8977	1.07224	3660		986.14	2222.4	735.26	0.6101	1.10665
3270		870.15	1363.1	646.00	0.8886	1.07315	3670		989.13	2249.0	737.57	0.6045	1.10747
3280		873.11	1381.2	648.27	0.8797	1.07405	3680		992.12	2275.8	739.87	0.5990	1.10828
3290		876.06	1399.5	650.54	0.8708	1.07495	3690		995.11	2302.9	742.17	0.5936	1.10910
3300	2840.3	879.02	1418.0	652.81	0.8621	1.07585	3700	3240.3	998.11	2330.3	744.48	0.5882	1.10991
3310		881.98	1436.6	655.09	0.8535	1.07675	3710		1001.11	2358.0	746.79	0.5829	1.11071
3320		884.94	1455.4	657.37	0.8450	1.07764	3720		1004.10	2385.9	749.10	0.5776	1.11152
3330		887.90	1474.5	659.64	0.8366	1.07853	3730		1007.10	2414.0	751.41	0.5724	1.11233
3340		890.86	1493.7	661.92	0.8283	1.07942	3740		1010.09	2442.4	753.73	0.5672	1.11313
3350	2890.3	893.83	1513.0	664.20	0.8202	1.08031	3750	3290.3	1013.09	2471.1	756.04	0.5621	1.11393
3360		896.80	1532.6	666.48	0.8121	1.08119	3760		1016.09	2500.0	758.35	0.5571	1.11473
3370		899.77	1552.5	668.76	0.8041	1.08207	3770		1019.09	2529.2	760.66	0.5522	1.11553
3380		902.73	1572.6	671.04	0.7962	1.08295	3780		1022.09	2558.7	762.98	0.5473	1.11633
3390		905.69	1592.8	673.32	0.7884	1.08383	3790		1025.09	2588.4	765.29	0.5424	1.11712

Table 6. Air at Low Pressures (for One Pound)—Continued

T	t	h	p_r	u	v_r	ϕ	T	t	h	p_r	u	v_r	ϕ
3800	3340.3	1028.09	2618.4	767.60	0.5376	1.11791	3900	3440.3	1058.14	2934.4	790.80	0.4923	1.12571
3810		1031.09	2648.9	769.92	0.5328	1.11870	3910		1061.15	2967.6	793.12	0.4881	1.12648
3820		1034.09	2679.5	772.23	0.5281	1.11948	3920		1064.16	3001.1	795.44	0.4839	1.12725
3830		1037.10	2710.3	774.55	0.5235	1.12027	3930		1067.17	3034.9	797.77	0.4797	1.12802
3840		1040.10	2741.5	776.87	0.5189	1.12105	3940		1070.18	3069.0	800.10	0.4756	1.12879
3850	3390.3	1043.11	2772.9	779.19	0.5143	1.12183	3950	3490.3	1073.19	3103.4	802.43	0.4715	1.12955
3860		1046.11	2804.6	781.51	0.5098	1.12261	3960		1076.20	3138.1	804.75	0.4675	1.13031
3870		1049.12	2836.6	783.83	0.5054	1.12339	3970		1079.22	3173.0	807.08	0.4635	1.13107
3880		1052.13	2869.0	786.16	0.5010	1.12416	3980		1082.23	3208.3	809.41	0.4595	1.13183
3890		1055.13	2901.6	788.48	0.4966	1.12494	3990		1085.24	3243.8	811.73	0.4556	1.13259

For a reversible process one can write:

$$dQ = T\,ds = du + \frac{p\,dv}{J}$$

The internal energy term du may be eliminated from this equation by differentiating the expression for enthalpy, solving for the internal energy term, and substituting. Thus

$$T\,ds = dh - \frac{v\,dp}{J} \tag{31}$$

For a perfect gas, however, $dh = c_p\,dT$. Substituting the last expression into equation 29 gives:

$$ds = c_p\frac{dT}{T} - \frac{v\,dp}{J} = c_p\frac{dT}{T} - \frac{R}{J}\frac{dp}{p} \tag{32}$$

Integrating equation 32 between the base temperature $T_0 = 0$ and the base pressure p_0 gives:

$$s - s_0 = \int_{T_0}^{T} c_p\frac{dT}{T} - \frac{R}{J}\ln\frac{p}{p_0} \tag{33}$$

Let $\phi = \int_{T_0}^{T} c_p\frac{dT}{T}$. Then equation 33 reduces to:

$$s - s_0 = \phi - \phi_0 - \frac{R}{J}\ln\frac{p}{p_0} \tag{34}$$

The parameter ϕ is a function of temperature alone, and the values of ϕ for various temperatures have been calculated and tabulated in Table 6. By means of equation 34 the entropy change for an ideal gas which passes from state 1 to state 2, may be found. Thus

$$(s_2 - s_0) - (s_1 - s_0) = (\phi_2 - \phi_0) - (\phi_1 - \phi_0) - \frac{R}{J}\ln\frac{p_2}{p_0} + \frac{R}{J}\ln\frac{p_1}{p_0}$$

or

$$s_2 - s_1 = \phi_2 - \phi_1 - \frac{R}{J}\ln\frac{p_2/p_0}{p_1/p_0} = \phi_2 - \phi_1 - \frac{R}{J}\ln\frac{p_{r2}}{p_{r1}} = \phi_2 - \phi_1 - \frac{R}{J}\ln\frac{p_2}{p_1} \tag{35}$$

For an isentropic process for an ideal gas equation 33 may be written as follows:

$$\ln\frac{p}{p_0} = \frac{J}{R}\int_{T_0}^{T} c_p\frac{dT}{T} = \frac{J}{R}\phi \tag{36}$$

or

$$\frac{p}{p_0} = e^{J\phi/R} = p_r \tag{37}$$

The symbol p_r is introduced to simplify the expression. The ratio of pressures will be termed the relative pressure, p_r. If the base temperature, T_0, is fixed, then the relative pressure, p_r, is a function only of temperature. Values of p_r are tabulated in Table 6 for a base temperature of absolute zero on the Fahrenheit scale.

For an ideal process which involves the change of state of a gas between states 1 and 2, the following relations may be written:

$$\frac{p_1}{p_0} = p_{r1} \quad \text{and} \quad \frac{p_2}{p_0} = p_{r2}$$

Eliminating p from these two equations and solving for p_2 gives:

$$p_2 = p_1 \cdot \frac{p_{r2}}{p_{r1}} \qquad (38)$$

This relation is useful in the solution of various types of problems.

A volume ratio for a perfect gas may be established by dividing the ideal gas relations for the base condition by the relation for any other state condition. Hence

$$\frac{v}{v_0} = \frac{T}{T_0} \cdot \frac{p_0}{p} \qquad (39)$$

This relation is perfectly general. However, if the pressure ratio from equation 37 is substituted, then it is restricted to isentropic changes. That is

$$\frac{v}{v_0} = \frac{T}{T_0} e^{J\phi/R} = v_r \qquad (40)$$

The symbol v_r is introduced to simplify the expression, and the ratio of the volumes is termed the relative volume. If the base temperature, T_0, is fixed, then the relative volume, v_r, is a function of temperature alone. Using the base temperature of absolute zero on the Fahrenheit scale, values have been computed for v_r and tabulated in Table 6.

For an ideal process which involves the change of state of a gas between states 1 to 2 the following relations may be written:

$$\frac{v_1}{v_0} = v_r \quad \text{and} \quad \frac{v_2}{v_0} = v_{r2}$$

Eliminating v_0 from these two equations and solving for v_2 gives

$$v_2 = v_1 \cdot \frac{v_{r2}}{v_{r1}} \qquad (41)$$

This equation will be found useful in the solution of many types of problems.

Several examples will be presented to demonstrate the use of the gas tables and the equations developed above. Other useful tables are presented in the original references, and for additional information the reader is referred to the text cited.

Example 1. One pound of dry air undergoes a reversible isothermal change at 1000 R. Compute the increase in entropy if the ratio of the initial and final pressures is 6.

The gas constant for R used in the preparation of the table values is 53.34 ft-lb/lb R or 0.06855 Btu/lb R. By use of equation 35, the entropy change equals:

$$s_2 - s_1 = 0.75042 - 0.75042 - 0.06855 \ln \tfrac{1}{6}$$
$$= -0.06855 \ln \tfrac{1}{6} = 0.06855 \ln 6$$
$$= 0.06855(1.7918) = 0.123 \text{ Btu/lb R}$$

Example 2. One pound of dry air is heated under reversible constant pressure conditions from an initial temperature of 500 R to a final temperature of 2000 R. Compute (a) the enthalpy change, and (b) the change in entropy.

(a) The enthalpy change may be found by subtracting the enthalpy values found in the table. Hence

$$h_2 - h_1 = 504.71 - 119.48 = 385.23 \text{ Btu/lb}$$

(b) The entropy change may be found by substituting suitable table values for ϕ in equation 35.

$$s_2 - s_1 = 0.93205 - 0.58233 = 0.34972 \text{ Btu/lb R}$$

Example 3. One pound of air undergoes an isentropic process, during which the temperature increases from 500 R to 800 R. If the initial pressure is 14.7 psia, compute (a) the initial specific volume, (b) the final specific volume, and (c) the final pressure.

(a) The initial specific volume from the perfect gas law equals:

$$v = \frac{RT}{p} = \frac{53.34(500)}{14.7(144)} = 12.59 \text{ ft}^3/\text{lb}$$

(c) Since the temperatures are known, values for v_{r1} and v_{r2} may be found from the table. The final volume may be found from equation 41.

$$v_2 = v_1 \cdot \frac{v_{r2}}{v_{r1}} = 12.59 \cdot \frac{53.63}{174.90} = 3.86 \text{ ft}^3/\text{lb}$$

(d) The final pressure may be found by use of equation 38. Values for p_{r1} and p_{r2} may be found in the table.

$$p_2 = p_1 \cdot \frac{p_{r2}}{p_{r1}} = 14.7 \cdot \frac{5.526}{1.059} = 76.7 \text{ psia}$$

Example 4. One pound of dry air expands isentropically through a nozzle from an initial temperature and pressure of 1600 R and 6 atm to a final pressure of 1 atm. Compute (a) the final temperature, and (b) the enthalpy change.

(a) Using the pressure ratio value from the table for a temperature of 1600 R and equation 38 gives the final pressure ratio.

$$p_{r2} = p_{r1} \cdot \frac{p_2}{p_1} = 71.73 \cdot \frac{1}{6} = 11.95$$

The temperature corresponding to this value from the table is 992 R.

(b) The enthalpy change may be found by use of the enthalpy values obtained from the table for the temperatures of 1600 R and 922 R.   Hence

$$h_2 - h_1 = 395.74 - 239.00 = 156.74 \text{ Btu/lb}$$

Example 5. A tank having a volume of 5000 cu ft contains dry air at a pressure and temperature of 60 psia and 1000 R. The air is removed from the tank by passage through a properly designed nozzle. If the process is assumed isentropic, determine the temperature and pressure of the air in the tank after one half of the original weight of air has been removed.

If W represents the original weight of air in the tank, then the following values represent the initial and final specific volumes:

$$v_1 = \frac{5000}{W} \quad \text{and} \quad v_2 = \frac{5000}{W/2} = \frac{10,000}{W}$$

The value for v_{r1} at 1000 R from the table is 30.12.   The term v_{r2} may be found by use of equation 41.

$$v_{r2} = v_{r1} \cdot \frac{v_2}{v_1} = \frac{30.12(10,000)W}{5000W} = 60.24$$

The temperature corresponding to this value of v_{r2} from the table is 764 R.   By use of equation 38 and the values of p_r at 1000 R and 764 R, the final pressure may be found.   Hence

$$p_2 = p_1 \cdot \frac{p_{r2}}{p_{r1}} = \frac{60(4.694)}{12.298} = 22.9 \text{ psia}$$

Example 6. A turbine operating on air expands the gas from an initial pressure and temperature of 5 atm and 1660 R to a final pressure of 1 atm. If the efficiency of the turbine is 95 per cent, determine the horsepower developed per pound of air flow per second assuming isentropic expansion.

The following values for the initial condition from the table refer to the temperature of 1660 R:

$$p_{r1} = 82.83 \quad h_1 = 411.82$$

The value for p_{r2} may be found by use of equation 38.   Hence

$$p_{r2} = p_{r1} \cdot \frac{p_2}{p_1} = \frac{82.83(1)}{5} = 16.56$$

The temperature and enthalpy for this value of p_{r2} as found in the table are 625.3 R and 262.22.   For a flow rate of one pound per second the theoretical work equals:

$$W = (h_1 - h_2)778 = (411.82 - 262.22)778 \text{ ft-lb/sec}$$

The actual horsepower rating equals:

$$hp = \frac{(411.82 - 262.22)778(0.95)}{550} = 201$$

14. FLOW OF GASES THROUGH PIPES, NOZZLES, AND ORIFICES

Flow of Gases through Pipes or Ducts

For flow through pipes or ducts, if the energy transitions as heat and velocity changes are negligible, the energy equation becomes $H_1 = H_2$ ($dH = 0$).   A further condition is actual increase of entropy in the amount of $-\frac{1}{J}\int \frac{V}{T} dP$ (if $dH = 0$), or of $\frac{R}{J}\log_e \frac{P_1}{P_2}$ for a perfect gas.   The pressure *drop* that actually occurs in passage from (1) to (2) is thus an indicator of the entropy increase and corresponding energy degradation.

The usual practical concern in such flow is this pressure drop, and its relation to the features that influence it.   Those features are (1) the general geometrical form and relative interior roughness of the flow-channel and (2) its length if a channel of extensive character; (3) the diameter and area of a circular channel, the equivalent diameter ($= \sqrt{\text{Area}/0.7854}$) and area of an extensive non-circular channel, or any size dimension of a non-extensive channel such as a valve or elbow; (4) the mass-rate of flow of the gas through the channel; and (5) the temperature, pressure, gas constant, and absolute viscosity of the gas.

For conditions of relatively small pressure drop along an extensive channel, or of greater pressure drop if the gas temperature remains effectively constant, an expression suitably relating these factors is

$$\frac{\Delta P}{L} = \frac{f'}{64.34}\frac{RT}{P}\frac{(M'/A)^2}{D}, \quad \text{or} \quad = \frac{f'}{40}\frac{RT}{P}\frac{(M')^2}{(D)^5} \tag{42}$$

where A = cross-sectional area of channel, square feet.

D = diameter of circular channel, feet, or equivalent diameter of non-circular channel ($= \sqrt{\text{area}/0.7854}$).

f' = "friction factor" (see below).

L = length of channel, feet.

M' = mass-rate of flow, pounds per second.

P = *mean* pressure in line, pounds per square foot, or pressure at any point if pressure drop is relatively small.

ΔP = pressure drop in length L, pounds per square foot.

R = gas constant, foot-pounds/(pounds × degrees Rankine).

T = mean absolute temperature of gas, degrees Rankine ($= °F + 460$).

The value of the friction factor, f', depends on the interior form and relative surface roughness of the channel through which the flow occurs. For a given character of channel it bears further a relatively definite relation to certain criteria that serve indirectly as indices of the general character of the flow (that is, as regards whether turbulent or stream line in character and, if turbulent, the order of turbulence). The basic one of such criteria is the ratio

$$\frac{DU}{\mu V}, \quad \text{or} \quad \frac{DUP}{\mu RT} \text{ for a gas}$$

where U = velocity, feet per second.

V = specific volume, cubic feet per pound.

μ = absolute viscosity of the fluid, pounds per foot-second ($=$ absolute viscosity in poises × 0.0672).

This criterion will be recognized as the Reynolds' index (Section 6, Art. 6).

For purposes of practical computation it is advantageous to recognize that $M' = AU/V = (\pi/4)D^2U/V$, whereby the above index may be modified to a more convenient form, $\frac{M'}{\mu D}\left(= \frac{\pi}{4} \times \text{Reynolds' index}\right)$. Further convenient criteria are combinations of f' and the above, namely, $\frac{\Delta P}{L}\frac{P}{RT}\frac{(M')^3}{(\mu)^5}$ and $\frac{\Delta P}{L}\frac{P}{RT}\frac{(D)^3}{(\mu)^2}$. $\left(\text{These are respectively } \frac{f'(RI)^5}{133} \text{ and } \frac{f'(RI)^2}{64.34}\right)$.

Table 7 gives reasonable values of f' as found by experiment for the flow through *circular*

Table 7. Turbulent Flow through Circular Channels

(Stream-line Flow Would Rarely be Encountered with Gas)

Reynolds' Index $\frac{DU}{\mu V}$	"A" $\frac{M'}{\mu D}$	"B" $\frac{\Delta P}{L}\frac{P}{RT}\frac{M'^3}{\mu^5}$	"C" $\frac{\Delta P}{L}\frac{P}{RT}\frac{D^3}{\mu^2}$	Friction Factor, f'
		Criteria		
	Interior Surface, Relatively Fairly Rough			
3×10^3	2.36×10^3	9.4×10^{13}	7.2×10^3	0.052
10^4	7.85×10^3	3.3×10^{16}	6.8×10^4	0.044
3×10^4	2.36×10^4	2.0×10^{18}	5.3×10^5	0.038
10^5	7.85×10^4	2.6×10^{21}	5.4×10^6	0.035
3×10^5	2.36×10^5	6.0×10^{23}	4.6×10^7	0.033
10^6 and up	7.85×10^5 and up	2.4×10^{26} and up	5.0×10^8 and up	0.032
	Interior Surface, Relatively Quite Smooth			
3×10^3	2.36×10^3	9.1×10^{13}	7.0×10^3	0.050
10^4	7.85×10^3	2.6×10^{16}	5.4×10^4	0.035
3×10^4	2.36×10^4	5.3×10^{18}	4.0×10^5	0.029
10^5	7.85×10^4	1.7×10^{21}	3.6×10^6	0.023
3×10^5	2.36×10^5	3.7×10^{23}	2.8×10^7	0.020
10^6	7.85×10^5	1.3×10^{26}	2.7×10^8	0.017
3×10^6	2.36×10^6	2.7×10^{28}	2.1×10^9	0.015
10^7	7.85×10^6	1.0×10^{31}	2.0×10^{10}	0.013

ducts of various relative interior-surface roughnesses at the various characters of flow indicated by the quoted values of the above criteria. For other than circular ducts equation 42 still applies, but further specific information would be required as regards the proper values of f' and of its relation to the Reynolds' index.

Units for the items of the several criteria are as above.

In using the table to ascertain f' in équation 42:

(a) If M', μ and D are specified, plus all except f' and one other term of equation 42, as for example ΔP, use column "A."

(b) If P, R, T, μ, M', ΔP, and L are specified and the required diameter (D) is to be estimated, use column "B."

(c) If P, R, T, μ, ΔP, D, and L are specified and the attainable M' is to be estimated, use column "C."

For conditions of relatively small pressure drop through non-extensive channels and obstructions, such as pipe fittings or valves, an expression suitably relating the pressure drop and the controlling variables is

$$\Delta P = \frac{r'}{40} \frac{RT}{P} \frac{(M')^2}{(D)^4} \tag{43}$$

where all symbols and units are as in equation 42, excepting that D = any representative linear dimension of the obstructing device, such as the interior diameter of a pipe fitting; and r' = a "resistance factor" the magnitude of which depends on the geometrical form of the flow channel through the device, and also depends in principle on $M'/\mu D$, the modified form of the Reynolds' index, Table 7, but in practice is nearly constant *for a particular character of flow channel*, irrespective of the value of the criterion. For example, the value of r' for a short-radius, 90-deg circular elbow is about 1.0.

Flow through Orifices, Nozzles, etc., with Large Velocity Changes

For such flow, which is almost invariably adiabatic ($Q = 0$), the energy equation becomes

$$\frac{U_2{}^2 - U_1{}^2}{64.34} = J(H_1 - H_2)$$

with an ideal constancy of the entropy if ideal flow, without fluid friction or turbulence, were securable. Because of friction, there is always some entropy increase. A necessary adjunct in the usual analysis of such flow is the continuity equation

$$M' = A_1 U_1 / V_1 = A_2 U_2 / V_2$$

The usual objective in the analyses of orifice or nozzle flow is an estimate either of the velocity U_2 at any section of area A_2, or of the mass-rate of flow M'. As it is convenient to evolve such estimates by considering ideal and thus isentropic flow, it is advantageous to combine the energy and continuity equations to give the relations

$$U_{2,\text{ ideal}} = 8.02 \sqrt{\frac{J(H_1 - H_2)s}{1 - (A_2/A_1)^2(V_1/V_2)s^2}}$$

and

$$\frac{M'}{A_2}\text{ (ideal)} = \frac{U_{2,\text{ ideal}}}{V_{2,\,S}} = \frac{8.02}{V_{2,\,S}} \sqrt{\frac{J(H_1 - H_2)s}{1 - (A_2/A_1)^2(V_1/V_2)s^2}},$$

where the subscript S indicates the constancy of the entropy for ideal expansion.

For the flow of gases (and of supersaturated vapors, p. 880), introduction of their isentropic relations (p. 845) enables the transpositions of the above equations to more useful ones involving the initial temperature T_1, the initial and terminal pressures, P_1 and P_2, and pertinent characteristics of the gas. The resultant equations are

$$U_{2,\text{ ideal}} = \frac{8.02 \left\{ \begin{array}{l} \sqrt{RT_1}, \text{ or} \\ \sqrt{P_1 V_1} \end{array} \right\} \sqrt{\dfrac{k}{k-1} \left[1 - \left(\dfrac{P_2}{P_1} \right)^{(k-1)/k} \right]}}{\sqrt{1 - \left(\dfrac{A_2}{A_1} \right)^2 \left(\dfrac{P_2}{P_1} \right)^{2/k}}} \tag{44}$$

$$\frac{M'}{A_2}\text{ (ideal)} = \frac{8.02 \left\{ \begin{array}{l} P_1/\sqrt{RT_1}, \\ \text{or } \sqrt{P_1/V_1} \end{array} \right\} \sqrt{\dfrac{k}{k-1} \left[\left(\dfrac{P_2}{P_1} \right)^{2/k} - \left(\dfrac{P_2}{P_1} \right)^{(k+1)/k} \right]}}{\sqrt{1 - \left(\dfrac{A_2}{A_1} \right)^2 \left(\dfrac{P_2}{P_1} \right)^{2/k}}} \tag{45}$$

or

$$= \frac{8.02 \left\{ \begin{matrix} P_2/\sqrt{RT_1}, \text{ or} \\ P_2/\sqrt{P_1V_1} \end{matrix} \right\} \sqrt{\frac{k}{k-1} \left(\frac{P_1}{P_2}\right)^{(k-1)/k} \left[\left(\frac{P_1}{P_2}\right)^{(k-1)/k} - 1 \right]}}{\sqrt{1 - \left(\frac{A_2}{A_1}\right)^2 \left(\frac{P_2}{P_1}\right)^{2/k}}} \quad (46)$$

For air at moderate temperature, with $R = 53.3$ and $k = 1.4$, these become

$$U_{2,\text{ ideal}} = \frac{109.7 \sqrt{T_1} \sqrt{1 - (P_2/P_1)^{0.286}}}{\sqrt{1 - (A_2/A_1)^2 (P_2/P_1)^{1.429}}}$$

$$\frac{M'}{A_2} (\text{ideal}) = \frac{2.055 (P_1/\sqrt{T_1}) \sqrt{(P_2/P_1)^{1.429} - (P_2/P_1)^{1.714}}}{\sqrt{1 - (A_2/A_1)^2 (P_2/P_1)^{1.429}}}$$

or

$$= \frac{2.055 (P_2/\sqrt{T_1}) \sqrt{(P_1/P_2)^{0.286} [(P_1/P_2)^{0.286} - 1]}}{\sqrt{1 - (A_2/A_1)^2 (P_2/P_1)^{1.429}}}$$

In these equations the denominator originated in effecting an elimination of the term U_1 from the original energy equation. Consequently its reciprocal is frequently regarded and designated as a "correction factor for the initial velocity." It may evidently be omitted when U_1 is of relatively negligible magnitude, or in general when (A_2/A_1) is less than 0.10.

It is evident from equations 45 or 46 that, if facilities are to be provided for ideal or isentropic flow at a particular rate and with a given initial area, a definite relation exists between the transverse areas A_2 at successive sections in a flow channel and the pressure P_2 at those sections. Solution of the equations for A_2 at successive values of P_2 would further disclose that, again if facilities for isentropic flow are to be provided, the fluid stream, (a) must first be convergent (that is, with decreasing area) until the pressure attains a certain proportion of the initial pressure, and (b) must thereafter diverge.

Except under rather abnormal conditions as regards U_1, at the consequent point of minimum area or the *throat* of the stream,

$$\left(\frac{P_2}{P_1}\right)_{\text{throat, ideal flow}} = \left(\frac{2}{k+1}\right)^{k/(k-1)}.$$

This particular ratio is conventionally known as the *critical pressure ratio*. By introduction of the ratio into equation 45 it develops that

$$\frac{M'}{A_{\text{throat}}} (\text{ideal}) = 8.02 \left\{ \begin{matrix} P_1/\sqrt{RT_1} \\ \text{or } \sqrt{P_1/V_1} \end{matrix} \right\} \sqrt{\frac{k}{k+1} \left(\frac{2}{k+1}\right)^{2/(k-1)}} \quad (47)$$

$$= 0.53 \, P_1/\sqrt{T_1} \text{ for air at moderate temperature.}$$

Values of various of the functions of P_2, P_1, and k that appear in the above equations are tabulated for convenience in Table 8 for values of P_2/P_1 from 0.9 to 0.1 and at values of k of 1.3, 1.4, and 1.67, these being fair average values of k at moderate temperature for, respectively, triatomic, diatomic, and monatomic gases.

For values of P_2/P_1 between 1.0 and 0.9 (that is, for values of $\Delta P/P_1$ from zero to 0.1), accurate evaluation of these functions is bothersome. Convenient equivalents that give results accurate within 0.5 per cent are tabulated herewith.

Nozzles. Regarding the term *nozzle* as inclusive of any simple flow channel that aims to provide ideally suitable facilities for an isentropic expansion and acceleration of a flowing gas, the foregoing relations provide all that is necessary for estimates of the velocity or the mass-rate of flow ideally securable with specified P_1 and T_1 and upon expansion to pressure P_2 at a section of area A_2. Recall that, from above thermodynamic considerations, a convergent channel is adequate for an expansion to a region that is at any pressure greater than or equal to the critical pressure (= 0.528 P_1 for air or other diatomic gases, see Table 8), whereas convergence to a throat and a subsequent carefully controlled divergence are required to enable an approach to isentropic expansion to a region having a pressure less than the critical.

Table 8. Pertinent Functions of P_2, P_1, and k

$\dfrac{P_2}{P_1}$	$\left(\dfrac{P_2}{P_1}\right)^{2/k}$			$\sqrt{\dfrac{k}{k-1}\left[1-\left(\dfrac{P_2}{P_1}\right)^{(k-1)/k}\right]}$			$\sqrt{\dfrac{k}{k-1}\left[\left(\dfrac{P_2}{P_1}\right)^{2/k}-\left(\dfrac{P_2}{P_1}\right)^{(k+1)/k}\right]}$			$\sqrt{\dfrac{k}{k-1}\left(\dfrac{P_1}{P_2}\right)^{(k-1)/k}\left[\left(\dfrac{P_1}{P_2}\right)^{(k-1)/k}-1\right]}$		
$k=$	1.3	1.4	1.67	1.3	1.4	1.67	1.3	1.4	1.67	1.3	1.4	1.67
0.9	0.850	0.860	0.881	0.323	0.322	0.321	0.297	0.299	0.302	0.330	0.332	0.335
0.8	.709	.727	.765	.466	.465	.462	.393	.396	.404	.491	.496	.505
0.7	.578	.601	.652	.585	.582	.577	.445	.451	.465	.635	.645	.665
0.6	.456	.482	.542	.694	.689	.680	.469	.479	.500	.781	.798	.834
0.5	.344	.371	.435	.800	.793	.778	.469	.484	.513	.939	.967	1.027
0.4	.244	.270	.333	.909	.898	.875	.449	.467	.505	1.123	1.166	1.264
0.3	.157	.179	.236	1.026	1.009	.978	.406	.427	.475	1.354	1.424	1.584
0.2	.084	.100	.145	1.159	1.136	1.089	.336	.360	.415	1.681	1.799	2.074
0.1	.029	.037	.063	1.337	1.299	1.227	.227	.250	.308	2.274	2.508	3.081
							At P_2/P_1 = critical pressure ratio					
0.546	0.394			0.752			0.472			(0.865)		
.528		0.402			0.764			0.484			(0.917)	
.487			0.422			0.791			0.513			(1.054)

Table 9. Equivalents of Pertinent Function, for $\Delta P/P_1$ from 0.0 to 0.10

Function	Equivalent when $\Delta P/P_1$ is not greater than—	
	0.10	0.01
$(P_2/P_1)^{2/k}$	$1 - \dfrac{2}{k}\left(\dfrac{\Delta P}{P_1}\right)$	$\dfrac{P_2}{P_1}$
$\sqrt{\dfrac{k}{k-1}\left[1-\left(\dfrac{P_2}{P_1}\right)^{(k-1)/k}\right]}$	$\sqrt{\dfrac{\Delta P}{P_1}\left[1+\dfrac{0.5}{k}\dfrac{\Delta P}{P_1}\right]}$	$\sqrt{\dfrac{\Delta P}{P_1}}$
$\sqrt{\dfrac{k}{k-1}\left[\left(\dfrac{P_2}{P_1}\right)^{2/k}-\left(\dfrac{P_2}{P_1}\right)^{(k+1)/k}\right]}$	$\sqrt{\dfrac{\Delta P}{P_1}\left[1-\dfrac{1.5}{k}\dfrac{\Delta P}{P_1}\right]}$	$\sqrt{\dfrac{\Delta P}{P_1}}$
$\sqrt{\dfrac{k}{k-1}\left(\dfrac{P_1}{P_2}\right)^{(k-1)/k}\left[\left(\dfrac{P_1}{P_2}\right)^{(k-1)/k}-1\right]}$	$\sqrt{\dfrac{\Delta P}{P_2}\left[1-\dfrac{(1.5-k)}{k}\dfrac{\Delta P}{P_2}\right]}$	$\sqrt{\dfrac{\Delta P}{P_2}}$

Actual flow conditions are conventionally accounted by employing *coefficients* of *velocity* and *discharge*, also a *nozzle efficiency*, where

Coefficient of velocity (C_U) $= U_{2,\text{ actual}}/U_{2,\text{ ideal}}$

Coefficient of discharge (C_D) $= M'_{\text{actual}}/M'_{\text{ideal}}$

Nozzle efficiency $= \dfrac{(U_2{}^2 - U_1{}^2)_{\text{actual}}}{(U_2{}^2 - U_1{}^2)_{\text{ideal}}}$

$$= \dfrac{U_2{}^2\ (\text{actual})}{U_2{}^2\ (\text{ideal})} \text{ if } U_1 \text{ is negligible}$$

Thus

$$U_{2,\text{ actual}} = C_U \times U_{2,\text{ ideal}}, \quad M'_{\text{actual}} = C_D \times M'_{\text{ideal}}$$

$$\dfrac{U_2{}^2}{64.34}\ (\text{actual}) = \text{Eff.}_{\text{nozzle}} \times \dfrac{U_2{}^2}{64.34}\ (\text{ideal}) \ (\text{if } U_1 \text{ is negligible})$$

The discharge coefficient for the flow of a gas through well-formed convergent nozzles or through the convergent portion of convergent-divergent nozzles attains a value of 0.98 and over at conventional rates of flow, but so decreases at very low rates (sensible decrease beginning at a Reynolds' number of about 2×10^5) as in general to require calibration when nozzles are employed at such rates. The efficiency of advantageously formed convergent nozzles or of the convergent section of convergent-divergent nozzles attains values of 0.96 and over.

The inertia of the rapidly flowing fluid entering the divergent section of convergent-divergent nozzles does not permit so efficient an expansion in that section, reducing the efficiencies of such nozzles to 0.92 or less. Improper design or unsuitable operation, in the feature of an $\dfrac{\text{Exit area, } A_2}{\text{Throat area, } A_{\text{th}}}$ ratio greater than that suitably corresponding to the actual $\dfrac{\text{Discharge region pressure, } P_2}{\text{Throat pressure, } P_{\text{crit}}}$ ratio (that is, *over-expansion*), or the reverse condition (*under-expansion*), will act to further and materially decrease the efficiency.

Orifices. The control or the metering of gases is frequently effected (without concern as to efficient jet formation) by directing flow through circular apertures located in plates or diaphragms. These apertures, or orifices, may be simply straight-sided holes through a thin plate (or its equivalent), or a converging profile may be provided on the upstream side of an orifice in a thicker plate. The orifice plates may be located at the entrance end or the exit end of a duct or as a constrictive passage in the run of a duct or pipe, the location being in general one of convenience or expediency. Provisions are necessary for measurement of at least the supply pressure (P_1) and temperature (T_1) and the pressure differential (ΔP) across the orifice.

Several unique characteristics pertain to orifice flow. One pertains more directly to a smoothly convergent orifice, and relates to the circumstance that by reason of its convergence such an orifice provides for and effects a full and efficient expansion within its confines from the initial pressure down to any pressure that exceeds or equals the critical pressure. By the same token, however, it will do no more, with the consequences that:

(a) If the pressure in the discharge region *exceeds* (or equals) the critical pressure the gas will have expanded *to the discharge-region pressure* when it has reached the orifice throat, so that the ideal mass-rate of flow is directly computable by equations 45 or 46 (see also Table 8 and/or 9), with P_2 regarded as the discharge-region pressure and A_2 as the throat-area of the orifice. Rounded orifices may advantageously be employed that provide facilities for direct measurement of P_2 at the throat.

(b) If the pressure in the discharge region *is less than* the critical pressure the gas will have expanded at the orifice throat *only to the critical pressure*, so that the ideal mass-rate of flow is to be computed either by equations 45 or 46 (see also Table 8), with A_2 regarded as the throat area and P_2 as the critical pressure, or preferably it is directly computed by equation 47. The persistence of the critical pressure at the orifice throat, irrespective of the amount by which the discharge-region pressure may be less than the critical, is well substantiated by experiment. Subsequent expansion from the critical to the discharge-region pressure occurs after passing the throat, quite as in a convergent-divergent nozzle, but now in a highly turbulent and irreversible manner because of the absence of a suitably divergent guide channel. Flow through orifices to regions at a pressure less than the critical is frequently known as *unaffected* flow, as the flow rate is not affected by the discharge-region pressure.

With a thin-plate orifice somewhat parallel conditions exist. They differ, however, in the features that:

(c) Even in the absence of a convergent channel, the fluid will itself establish and issue as a convergent stream and will consequently have expanded efficiently to any pressure exceeding or equal to the critical pressure when it has reached the stream-section of minimum area, or the *vena contracta;* but

(d) The location and area of this *vena contracta* may be quite variable. Its location may range from a distance that is downstream from the orifice by perhaps the diameter of the orifice when the pressure differential (ΔP) is small and the flow rate low, to a location virtually in the plane of the orifice at maximum flow rates, thus making uncertain the suitable location at which to measure P_2. Its area may correspondingly range from some 60 per cent of the orifice area when the flow rate is low to effectively 100 per cent of the orifice area when the flow rate is maximum and the *vena contracta* is in the orifice plane.

Actual flow rates through orifices are estimated by multiplying the ideal rate, computed in the manner and by the equations indicated above, by a suitable coefficient of discharge. This coefficient may be as high as 0.99 for rounded-entrance orifices and also for thin-plate orifices at maximum flow rates, but for the latter at lower rates it may be as low as about 0.6 if in the use of the equations the actual orifice area is used as A_2, rather than the (usually not measurable) stream area at the *vena contracta*. At quite low rates of flow ($\Delta P/P_1$ not much greater than 0.01) some degree of coordination of coefficients may be secured for geometrically similar orifices, if installed in geometrically similar environs and with geometrically similar locations of the pressure taps, by correlating the coefficient with the Reynolds' index.

When metering orifices are employed in circumstances where the velocity of approach of the stream to the orifice is not inherently negligible and a correction for initial velocity is necessary, that correction may be made by routine inclusion of the denominator of equations 45 or 46, or it may alternatively be made by measuring P_1 by means of a suitably placed impact tube in the upstream, when the denominator may be taken as unity.

Diffuser and Ejector. When a fluid is supplied at a lower pressure but higher and sufficient velocity the corresponding mechanical kinetic energy of the stream may be caused to effect its delivery against a higher pressure. The process is a decelerative compression that is the reverse of the accelerative expansion occurring in a nozzle; its energy relation is identical with that of a nozzle excepting as regards a reverse significance of subscript 1 as pertaining to a supply region now of *lower pressure and higher velocity* and of subscript 2 as pertaining to a discharge region of *lower velocity and higher pressure*. The required channel form is effectively the reverse of that of a nozzle, or is alternatively that of a nozzle through which the direction of flow is reversed, with need simply for a divergent

(vs. convergent) channel if the $\dfrac{\text{Lower (supply) pressure}}{\text{Higher (discharge) pressure}}$ ratio exceeds the critical ratio

of the fluid, and a convergent-divergent channel otherwise. Given the pressure, temperature, and velocity of the entering stream, the relation expressing the discharge pressure against which it might be caused to deliver if deceleration were continued to a negligible velocity is:

$$P_{2, \text{ideal}} = P_1 \left[\frac{k-1}{kRT_1} \frac{U_1^2}{64.34} + 1 \right]^{k/(k-1)}$$

and

$$P_{2, \text{actual}} = P_1 \left[\frac{k-1}{kRT_1} \frac{U_1^2}{64.34} + 1 \right]^{n/(n-1)}$$

(48)

where n = the effective P–V exponent of the more or less irreversible adiabatic compression process, the value of which exponent will exceed the k of the gas in a degree that corresponds to the degree of turbulence and irreversibility associated with the flow.

The principle of diffuser action is employed in the after-section of the Venturi meter, in the impact and Pitot tube (see below), in the delivery sections of centrifugal blowers, and in ejectors. In this last a lower-pressure and high-velocity jet acts to entrain mechanically a second fluid, thereby to withdraw it from some region of like pressure and then to deliver the mixture by diffuser action to a region of higher pressure, with, however, the encountering of high turbulence and rather low efficiency.

Impact and Pitot Tube. These devices employ primarily a small tube the open end of which is squarely across the path of a flowing stream and the other end of which is extended to a point outside of the stream and is closed by some suitable pressure-measuring device. The action at the stream-end of the tube is a decelerating compression of the thread of fluid as it approaches along the (extended) axis of the open tube, essentially equivalent to that action as it occurs in a diffuser. The velocity of that stream thread is directly computable by equation 44 without denominator or its equivalent (Table 9), P_1 being the *impact* pressure measured as indicated above and P_2 the *static* pressure of the region in which the fluid is flowing as measured wholly without impact effect. The Pitot tube provides facilities for both measurements by a single instrument consisting of two concentric tubes, with impact pressure imposed on the open end of the center one and the static pressure transmitted to the interior of the outer tube by small flush openings in its walls. The value of T_1 is that taken with a thermometer which likewise is exposed to impact effect.

The impact or Pitot tube determines only a point velocity, whereas the mass or volumetric rate of flow in a stream depends on an average velocity. For ascertaining an average, a conventional procedure is to take an arithmetical mean of a series of velocity determinations made at a number of points so located that the velocity at each individual one is reasonably representative of the mean velocity in one of the same number of equal subdivisions of the aggregate cross-sectional area of the stream. For a circular stream or duct a schedule that reasonably accomplishes this specifies locations taken at points along one or more diameters, and along each at radial distances on each side of the center line of the stream as follows:

$$r_1 = R\sqrt{1/n}, \quad r_2 = R\sqrt{3/n}, \quad r_3 = R\sqrt{5/n}, \quad r_{n/2} = R\sqrt{(n-1)/n}$$

where R = radius of stream, and n = total number of locations to be employed in traversing a diameter, with $n/2$ locations on each side of the center line. An alternative procedure permits tube locations at any desired points along a diameter and ascertains the mean velocity by finding the mean ordinate of a curve plotted with individual velocities as ordinates and with the square of the corresponding radial distances of the tube from the center line as abscissas, the curve extending to the square of the stream or duct radius.

15. COMPRESSION OF GASES IN COMPRESSORS

A compressor is actually or in effect a steady-flow device. From the steady-flow energy equation,

$$W_{\text{in}}, \text{ft-lb per lb of fluid} = \frac{U_2^2 - U_1^2}{64.34} + Jc_p(T_2 - T_1) + JQ_{\text{out}}$$

or

$$= \frac{U_2^2 - U_1^2}{64.34} + \frac{k}{k-1}(P_2V_2 - P_1V_1) + JQ_{\text{out}}$$

where the subscripts 1 and 2 refer to conditions respectively at the intake and the discharge regions of the compressor. Under most conditions of compressor operation any energy transition by conduction or radiation (as heat) is an outward one; therefore, the term Q_{out} is introduced in these relations. Work energy is always put into the cycle, so the term W_{in} is also introduced.

Isothermal Compression. Using the characteristic equation of state of a gas, $PV = RT$, and considering the process as reversible isothermal,

$$W_{in}, \text{ reversible isothermal compression, ft-lb per lb} = \frac{U_2{}^2 - U_1{}^2}{64.34} + RT_1 \log_e \frac{P_2}{P_1}$$

or

$$= \frac{U_2{}^2 - U_1{}^2}{64.34} + P_1V_1 \log_e \frac{P_2}{P_1}$$

For cases where the kinetic energy term can be omitted the equations become,

$$W_{in}, \text{ reversible isothermal compression, ft-lb per lb} = P_1V_1 \log_e \frac{P_2}{P_1}$$

$$= RT_1 \log_e \frac{P_2}{P_1}$$

If $\dfrac{P_2 - P_1}{P_1}$ does not exceed 0.01,

$$W_{in} \text{ (ideal)} = \frac{U_2{}^2 - U_1{}^2}{64.34} + (P_2 - P_1)V_1$$

which is the usual equation for low-pressure fan performance.

The power requirement for isothermal compression is

$$Hp \text{ input, reversible isothermal compression} = \frac{M'P_1V_1 \log_e P_2/P_1}{550} = \frac{P_1V'_1 \log_e P_2/P_1}{550}$$

or

$$= \frac{M'RT_1 \log_e P_2/P_1}{550}$$

When $\dfrac{P_2 - P_1}{P_1}$ does not exceed 0.01,

$$Hp \text{ input} = \frac{M'[(P_2 - P_1)V_1 + (U_2{}^2 - U_1{}^2)/64.34]}{550}$$

In these relations,

M' = pounds of gas delivered per second.

$M'V_1 = V'_1$ = volume of gas, in cubic feet per second, at the state under which it exists in the region from which the compressor suction is taken.

Adiabatic Compression, Single Stage. The work required for single-stage adiabatic compression,

$$W_{in}, \text{ reversible adiabatic compression, ft-lb per lb} = Jc_pT_1[(P_2/P_1)^{(k-1)/k} - 1]$$

or

$$= \frac{k}{k-1} P_1V_1[(P_2/P_1)^{(k-1)/k} - 1]$$

The power required:

$$Hp \text{ input, reversible adiabatic compression} = \frac{M'Jc_pT_1[(P_2/P_1)^{(k-1)/k} - 1]}{550}$$

or

$$= \frac{\dfrac{k}{k-1} P_1V'_1[(P_2/P_1)^{(k-1)/k} - 1]}{550}$$

When $\dfrac{P_2 - P_1}{P_1}$ does not exceed 0.10,

$$W_{in}, \text{ reversible adiabatic compression, ft-lb per lb} = RT_1 \frac{P_2 - P_1}{P_1}\left(1 - \frac{1}{2k}\frac{P_2 - P_1}{P_1}\right)$$

or

$$= V_1(P_2 - P_1)\left(1 - \frac{1}{2k}\frac{P_2 - P_1}{P_1}\right)$$

Also,

$$Hp \text{ input, reversible adiabatic compression} = \frac{M'RT_1\dfrac{P_2 - P_1}{P_1}\left(1 - \dfrac{1}{2k}\dfrac{P_2 - P_1}{P_1}\right)}{550}$$

or

$$= \frac{V'_1(P_2 - P_1)\left(1 - \dfrac{1}{2k}\dfrac{P_2 - P_1}{P_1}\right)}{550}$$

When $\dfrac{P_2 - P_1}{P_1}$ does not exceed 0.01, the last term in the expression $\left(1 - \dfrac{1}{2k}\dfrac{P_2 - P_1}{P_1}\right)$ becomes negligible.

Adiabatic Compression, Multistage with Complete Recooling between Stages.

W_{in}, ideal two-stage isentropic compression $= 2Jc_pT_1[(P_2/P_1)^{(k-1)/2k} - 1]$

or

$$= \frac{2k}{k-1} P_1V_1[(P_2/P_1)^{(k-1)/2k} - 1]$$

Where, with N stages of compression, the intermediate pressures follow the schedule:

$$P' = [(P_1)^{(N-1)}(P_2)]^{1/N}; \qquad P'' = [(P_1)^{(N-2)}(P_2)^2]^{1/N};$$

$$P''' = [(P_1)^{(N-3)}(P_2)^3]^{1/N}; \quad \cdots; \quad P^{N-1} = [(P_1)(P_2)^{(N-1)}]^{1/N}$$

the corresponding minimum work requirement for the entire compressor is:

$$W_{\text{in}} = NJc_pT_1[(P_2/P_1)^{(k-1)/Nk} - 1]$$

or

$$= \frac{Nk}{k-1} P_1V_1[(P_2/P_1)^{(k-1)/Nk} - 1]$$

LIQUIDS AND VAPORS

Symbols

f = a subscript attached to any property of a saturated liquid.

g = a subscript attached to any property of a (dry) saturated vapor.

fg = a subscript employed to designate the change of any property between the saturated liquid and saturated vapor states, at constant pressure.

E = molecular (internal) energy, Btu per pound.

H = enthalpy ($E + PV/J$), Btu per pound.

J = Joule's equivalent ($= 778$ ft-lb per Btu).

p = pressure, pounds per square inch (absolute).

P = pressure, pounds per square foot (absolute).

Q = energy transferred as heat (by conduction or radiation), Btu per pound.

S = entropy, per pound.

t = temperature, degrees fahrenheit.

T = temperature, degrees Rankine (fahrenheit absolute).

U = velocity, feet per second.

V = specific volume, cubic feet per pound mass.

W = shaft work, foot-pounds per pound mass.

x = quality of a vapor-liquid mixture; also a subscript to any property to characterize it as that of a vapor-liquid mixture.

16. PHYSICAL CONDITIONS

A liquid and its vapor may exist in several conditions defined as follows:

(a) **Saturated Liquid.** If heat is added to a liquid which is maintained at a certain constant pressure, the liquid will not boil until a certain temperature is reached. This temperature is called the saturation temperature corresponding to the pressure on the liquid, and the liquid which is at the saturated temperature is called saturated liquid.

(b) **Compressed Liquid** is liquid at a temperature less than the saturation temperature corresponding to the pressure, or conversely it is liquid having an imposed pressure greater than the saturation pressure corresponding to the temperature of the liquid.

(c) **Saturated Vapor** is the dry vapor at the saturation temperature corresponding to the pressure. It is the vapor at the temperature and pressure of the saturated liquid from which it is formed. The saturated state represents an equilibrium condition between a liquid and its vapor.

(d) **Wet Vapor** is a physical mixture of saturated vapor and saturated liquid; the ratio between the mass of vapor and the mass of the mixture is known as the quality of the mixture, and is represented by the symbol x.

(e) **Superheated Vapor** is the state in which vapor is not in contact with the liquid and its temperature is above the saturated temperature corresponding to the pressure, the excess of temperature being known as the degrees of superheat.

Table 1.   Dry Saturated Steam: Pressure Table *

Abs Press., lb sq in. p	Temp., F t	Specific Volume		Enthalpy			Entropy			Internal Energy		Abs Press., lb sq in. p
		Sat. Liquid v_f	Sat. Vapor v_g	Sat. Liquid h_f	Evap. h_{fg}	Sat. Vapor h_g	Sat. Liquid s_f	Evap. s_{fg}	Sat. Vapor s_g	Sat. Liquid u_f	Sat. Vapor u_g	
1.0	101.74	0.01614	333.6	69.70	1036.3	1106.0	0.1326	1.8456	1.9782	69.70	1044.3	1.0
2.0	126.08	0.01623	173.73	93.99	1022.2	1116.2	0.1749	1.7451	1.9200	93.98	1051.9	2.0
3.0	141.48	0.01630	118.71	109.37	1013.2	1122.6	0.2008	1.6855	1.8863	109.36	1056.7	3.0
4.0	152.97	0.01636	90.63	120.86	1006.4	1127.3	0.2198	1.6427	1.8625	120.85	1060.2	4.0
5.0	162.24	0.01640	73.52	130.13	1001.0	1131.1	0.2347	1.6094	1.8441	130.12	1063.1	5.0
6.0	170.06	0.01645	61.98	137.96	996.2	1134.2	0.2472	1.5820	1.8292	137.94	1065.4	6.0
7.0	176.85	0.01649	53.64	144.76	992.1	1136.9	0.2581	1.5586	1.8167	144.74	1067.4	7.0
8.0	182.86	0.01653	47.34	150.79	988.5	1139.3	0.2674	1.5383	1.8057	150.77	1069.2	8.0
9.0	188.28	0.01656	42.40	156.22	985.2	1141.4	0.2759	1.5203	1.7962	156.19	1070.8	9.0
10	193.21	0.01659	38.42	161.17	982.1	1143.3	0.2835	1.5041	1.7876	161.14	1072.2	10
14.696	212.00	0.01672	26.80	180.07	970.3	1150.4	0.3120	1.4446	1.7566	180.02	1077.5	14.696
15	213.03	0.01672	26.29	181.11	969.7	1150.8	0.3135	1.4415	1.7549	181.06	1077.8	15
20	227.96	0.01683	20.089	196.16	960.1	1156.3	0.3356	1.3962	1.7319	196.10	1081.9	20
25	240.07	0.01692	16.303	208.42	952.1	1160.6	0.3533	1.3606	1.7139	208.34	1085.1	25
30	250.33	0.01701	13.746	218.82	945.3	1164.1	0.3680	1.3313	1.6993	218.73	1087.8	30
35	259.28	0.01708	11.898	227.91	939.2	1167.1	0.3807	1.3063	1.6870	227.80	1090.1	35
40	267.25	0.01715	10.498	236.03	933.7	1169.7	0.3919	1.2844	1.6763	235.90	1092.0	40
45	274.44	0.01721	9.401	243.36	928.6	1172.0	0.4019	1.2650	1.6669	243.22	1093.7	45
50	281.01	0.01727	8.515	250.09	924.0	1174.1	0.4110	1.2474	1.6585	249.93	1095.3	50
55	287.07	0.01732	7.787	256.30	919.6	1175.9	0.4193	1.2316	1.6509	256.12	1096.7	55
60	292.71	0.01738	7.175	262.09	915.5	1177.6	0.4270	1.2168	1.6438	261.90	1097.9	60
65	297.97	0.01743	6.655	267.50	911.6	1179.1	0.4342	1.2032	1.6374	267.29	1099.1	65
70	302.92	0.01748	6.206	272.61	907.9	1180.6	0.4409	1.1906	1.6315	272.38	1100.2	70
75	307.60	0.01753	5.816	277.43	904.5	1181.9	0.4472	1.1787	1.6259	277.19	1101.2	75
80	312.03	0.01757	5.472	282.02	901.1	1183.1	0.4531	1.1676	1.6207	281.76	1102.1	80
85	316.25	0.01761	5.168	286.39	897.8	1184.2	0.4587	1.1571	1.6158	286.11	1102.9	85
90	320.27	0.01766	4.896	290.56	894.7	1185.3	0.4641	1.1471	1.6112	290.27	1103.7	90
95	324.12	0.01770	4.652	294.56	891.7	1186.2	0.4692	1.1376	1.6068	294.25	1104.5	95
100	327.81	0.01774	4.432	298.40	888.8	1187.2	0.4740	1.1286	1.6026	298.08	1105.2	100
110	334.77	0.01782	4.049	305.66	883.2	1188.9	0.4832	1.1117	1.5948	305.30	1106.5	110
120	341.25	0.01789	3.728	312.44	877.9	1190.4	0.4916	1.0962	1.5878	312.05	1107.6	120
130	347.32	0.01796	3.455	318.81	872.9	1191.7	0.4995	1.0817	1.5812	318.38	1108.6	130
140	353.02	0.01802	3.220	324.82	868.2	1193.0	0.5069	1.0682	1.5751	324.35	1109.6	140
150	358.42	0.01809	3.015	330.51	863.6	1194.1	0.5138	1.0556	1.5694	330.01	1110.5	150
160	363.53	0.01815	2.834	335.93	859.2	1195.1	0.5204	1.0436	1.5640	335.39	1111.2	160
170	368.41	0.01822	2.675	341.09	854.9	1196.0	0.5266	1.0324	1.5590	340.52	1111.9	170
180	373.06	0.01827	2.532	346.03	850.8	1196.9	0.5325	1.0217	1.5542	345.42	1112.5	180
190	377.51	0.01833	2.404	350.79	846.8	1197.6	0.5381	1.0116	1.5497	350.15	1113.1	190
200	381.79	0.01839	2.288	355.36	843.0	1198.4	0.5435	1.0018	1.5453	354.68	1113.7	200
250	400.95	0.01865	1.8438	376.00	825.1	1201.1	0.5675	0.9588	1.5263	375.14	1115.8	250
300	417.33	0.01890	1.5433	393.84	809.0	1202.8	0.5879	0.9225	1.5104	392.79	1117.1	300
350	431.72	0.01913	1.3260	409.69	794.2	1203.9	0.6056	0.8910	1.4966	408.45	1118.0	350
400	444.59	0.0193	1.1613	424.0	780.5	1204.5	0.6214	0.8630	1.4844	422.6	1118.5	400
450	456.28	0.0195	1.0320	437.2	767.4	1204.6	0.6356	0.8378	1.4734	435.5	1118.7	450
500	467.01	0.0197	0.9278	449.4	755.0	1204.4	0.6487	0.8147	1.4634	447.6	1118.6	500
550	476.94	0.0199	0.8424	460.8	743.1	1203.9	0.6608	0.7934	1.4542	458.8	1118.2	550
600	486.21	0.0201	0.7698	471.6	731.6	1203.2	0.6720	0.7734	1.4454	469.4	1117.7	600
650	494.90	0.0203	0.7083	481.8	720.5	1202.3	0.6826	0.7548	1.4374	479.4	1117.1	650
700	503.10	0.0205	0.6554	491.5	709.7	1201.2	0.6925	0.7371	1.4296	488.8	1116.3	700
750	510.86	0.0207	0.6092	500.8	699.2	1200.0	0.7019	0.7204	1.4223	598.0	1115.4	750
800	518.23	0.0209	0.5687	509.7	688.9	1198.6	0.7108	0.7045	1.4153	506.6	1114.4	800
850	525.26	0.0210	0.5327	518.3	678.8	1197.1	0.7194	0.6891	1.4085	515.0	1113.3	850
900	531.98	0.0212	0.5006	526.6	668.8	1195.4	0.7275	0.6744	1.4020	523.1	1112.1	900
950	538.43	0.0214	0.4717	534.6	659.1	1193.7	0.7355	0.6602	1.3957	530.9	1110.8	950
1000	544.61	0.0216	0.4456	542.4	649.4	1191.8	0.7430	0.6467	1.3897	538.4	1109.4	1000
1100	556.31	0.0220	0.4001	557.4	630.4	1187.8	0.7575	0.6205	1.3780	552.9	1106.4	1100
1200	567.22	0.0223	0.3619	571.7	611.7	1183.4	0.7711	0.5956	1.3667	566.7	1103.0	1200
1300	577.46	0.0227	0.3293	585.4	593.2	1178.6	0.7840	0.5719	1.3559	580.0	1099.4	1300
1400	587.10	0.0231	0.3012	598.7	574.7	1173.4	0.7963	0.5491	1.3454	592.7	1095.4	1400
1500	596.23	0.0235	0.2765	611.6	556.3	1167.9	0.8082	0.5269	1.3351	605.1	1091.2	1500
2000	635.82	0.0257	0.1878	671.7	463.4	1135.1	0.8619	0.4230	1.2849	662.2	1065.6	2000
2500	668.13	0.0287	0.1307	730.6	360.5	1091.1	0.9126	0.3197	1.2322	717.3	1030.6	2500
3000	695.36	0.0346	0.0858	802.5	217.8	1020.3	0.9731	0.1885	1.1615	783.4	972.7	3000
3206.2	705.40	0.0503	0.0503	902.7	0	902.7	1.0580	0	1.0580	872.9	872.9	3206.2

* Abridged by permission from Joseph H. Keenan and Frederick G. Keyes, *Thermodynamic Properties of Steam*, John Wiley & Sons, Inc., New York, 1937.

Table 2.   Dry Saturated Steam: Temperature Table *

Temp., F, t	Abs Press., lb sq in. p	Specific Volume			Enthalpy			Entropy			Temp., F, t
		Sat. Liquid v_f	Evap. v_{fg}	Sat. Vapor v_g	Sat. Liquid h_f	Evap. h_{fg}	Sat. Vapor h_g	Sat. Liquid s_f	Evap. s_{fg}	Sat. Vapor s_g	
32	0.08854	0.01602	3306	3306	0.00	1075.8	1075.8	0.0000	2.1877	2.1877	32
35	0.09995	0.01602	2947	2947	3.02	1074.1	1077.1	0.0061	2.1709	2.1770	35
40	0.12170	0.01602	2444	2444	8.05	1071.3	1079.3	0.0162	2.1435	2.1597	40
45	0.14752	0.01602	2036.4	2036.4	13.06	1068.4	1081.5	0.0262	2.1167	2.1429	45
50	0.17811	0.01603	1703.2	1703.2	18.07	1065.6	1083.7	0.0361	2.0903	2.1264	50
60	0.2563	0.01604	1206.6	1206.7	28.06	1059.9	1088.0	0.0555	2.0393	2.0948	60
70	0.3631	0.01606	867.8	867.9	38.04	1054.3	1092.3	0.0745	1.9902	2.0647	70
80	0.5069	0.01608	633.1	633.1	48.02	1048.6	1096.6	0.0932	1.9428	2.0360	80
90	0.6982	0.01610	468.0	468.0	57.99	1042.9	1100.9	0.1115	1.8972	2.0087	90
100	0.9492	0.01613	350.3	350.4	67.97	1037.2	1105.2	0.1295	1.8531	1.9826	100
110	1.2748	0.01617	265.3	265.4	77.94	1031.6	1109.5	0.1471	1.8106	1.9577	110
120	1.6924	0.01620	203.25	203.27	87.92	1025.8	1113.7	0.1645	1.7694	1.9339	120
130	2.2225	0.01625	157.32	157.34	97.90	1020.0	1117.9	0.1816	1.7296	1.9112	130
140	2.8886	0.01629	122.99	123.01	107.89	1014.1	1122.0	0.1984	1.6910	1.8894	140
150	3.718	0.01634	97.06	97.07	117.89	1008.2	1126.1	0.2149	1.6537	1.8685	150
160	4.741	0.01639	77.27	77.29	127.89	1002.3	1130.2	0.2311	1.6174	1.8485	160
170	5.992	0.01645	62.04	62.06	137.90	996.3	1134.2	0.2472	1.5822	1.8293	170
180	7.510	0.01651	50.21	50.23	147.92	990.2	1138.1	0.2630	1.5480	1.8109	180
190	9.339	0.01657	40.94	40.96	157.95	984.1	1142.0	0.2785	1.5147	1.7932	190
200	11.526	0.01663	33.62	33.64	167.99	977.9	1145.9	0.2938	1.4824	1.7762	200
210	14.123	0.01670	27.80	27.82	178.05	971.6	1149.7	0.3090	1.4508	1.7598	210
212	14.696	0.01672	26.78	26.80	180.07	970.3	1150.4	0.3120	1.4446	1.7566	212
220	17.186	0.01677	23.13	23.15	188.13	965.2	1153.4	0.3239	1.4201	1.7440	220
230	20.780	0.01684	19.365	19.382	198.23	958.8	1157.0	0.3387	1.3901	1.7288	230
240	24.969	0.01692	16.306	16.323	208.34	952.2	1160.5	0.3531	1.3609	1.7140	240
250	29.825	0.01700	13.804	13.821	216.48	945.5	1164.0	0.3675	1.3323	1.6998	250
260	35.429	0.01709	11.746	11.763	228.64	938.7	1167.3	0.3817	1.3043	1.6860	260
270	41.858	0.01717	10.044	10.061	238.84	931.8	1170.6	0.3958	1.2769	1.6727	270
280	49.203	0.01726	8.628	8.645	249.06	924.7	1173.8	0.4096	1.2501	1.6597	280
290	57.556	0.01735	7.444	7.461	259.31	917.5	1176.8	0.4234	1.2238	1.6472	290
300	67.013	0.01745	6.449	6.466	269.59	910.1	1179.7	0.4369	1.1980	1.6350	300
310	77.68	0.01755	5.609	5.626	279.92	902.6	1182.5	0.4504	1.1727	1.6231	310
320	89.66	0.01765	4.896	4.914	290.28	894.9	1185.2	0.4637	1.1478	1.6115	320
330	103.06	0.01776	4.289	4.307	300.68	887.0	1187.7	0.4769	1.1233	1.6002	330
340	118.01	0.01787	3.770	3.788	311.13	879.0	1190.1	0.4900	1.0992	1.5891	340
350	134.63	0.01799	3.324	3.342	321.63	870.7	1192.3	0.5029	1.0754	1.5783	350
360	153.04	0.01811	2.939	2.957	332.18	862.2	1194.4	0.5158	1.0519	1.5677	360
370	173.37	0.01823	2.606	2.625	342.79	853.5	1196.3	0.5286	1.0287	1.5573	370
380	195.77	0.01836	2.317	2.335	353.45	844.6	1198.1	0.5413	1.0059	1.5471	380
390	220.37	0.01850	2.0651	2.0836	364.17	835.4	1199.6	0.5539	0.9832	1.5371	390
400	247.31	0.01864	1.8447	1.8633	374.97	826.0	1201.0	0.5664	0.9608	1.5272	400
410	276.75	0.01878	1.6512	1.6700	385.83	816.3	1202.1	0.5788	0.9386	1.5174	410
420	308.83	0.01894	1.4811	1.5000	396.77	806.3	1203.1	0.5912	0.9166	1.5078	420
430	343.72	0.01910	1.3308	1.3499	407.79	796.0	1203.8	0.6035	0.8947	1.4982	430
440	381.59	0.01926	1.1979	1.2171	418.90	785.4	1204.3	0.6158	0.8730	1.4887	440
450	422.6	0.0194	1.0799	1.0993	430.1	774.5	1204.6	0.6280	0.8513	1.4793	450
460	466.9	0.0196	0.9748	0.9944	441.4	763.2	1204.6	0.6402	0.8298	1.4700	460
470	514.7	0.0198	0.8811	0.9009	452.8	751.5	1204.3	0.6523	0.8083	1.4606	470
480	566.1	0.0200	0.7972	0.8172	464.4	739.4	1203.7	0.6645	0.7868	1.4513	480
490	621.4	0.0202	0.7221	0.7423	476.0	726.8	1202.8	0.6766	0.7653	1.4419	490
500	680.8	0.0204	0.6545	0.6749	487.8	713.9	1201.7	0.6887	0.7438	1.4325	500
520	812.4	0.0209	0.5385	0.5594	511.9	686.4	1198.2	0.7130	0.7006	1.4136	520
540	962.5	0.0215	0.4434	0.4649	536.6	656.6	1193.2	0.7374	0.6568	1.3942	540
560	1133.1	0.0221	0.3647	0.3868	562.2	624.2	1186.4	0.7621	0.6121	1.3742	560
580	1325.8	0.0228	0.2989	0.3217	588.9	588.4	1177.3	0.7872	0.5659	1.3532	580
600	1542.9	0.0236	0.2432	0.2668	617.0	548.5	1165.5	0.8131	0.5176	1.3307	600
620	1786.6	0.0247	0.1955	0.2201	646.7	503.6	1150.3	0.8398	0.4664	1.3062	620
640	2059.7	0.0260	0.1538	0.1798	678.6	452.0	1130.5	0.8679	0.4110	1.2789	640
660	2365.4	0.0278	0.1165	0.1442	714.2	390.2	1104.4	0.8987	0.3485	1.2472	660
680	2708.1	0.0305	0.0810	0.1115	757.3	309.9	1067.2	0.9351	0.2719	1.2071	680
700	3093.7	0.0369	0.0392	0.0761	823.3	172.1	995.4	0.9905	0.1484	1.1389	700
705.4	3206.2	0.0503	0	0.0503	902.7	0	902.7	1.0580	0	1.0580	705.4

* Abridged by permission from Joseph H. Keenan and Frederick G. Keyes, *Thermodynamic Properties of Steam*, John Wiley & Sons, Inc., New York, 1937.

Table 3. Properties of Superheated Steam *

Temperature—Degrees Fahrenheit

Abs Press, lb sq in. (Sat. Temp)		200	300	400	500	600	700	800	900	1000	1100	1200	1400	1600
1 (101.74)	v	392.6	452.3	512.0	571.6	631.2	690.8	750.4	809.9	869.5	929.1	988.7	1107.8	1227.0
	h	1150.4	1195.8	1241.7	1288.3	1335.7	1383.8	1432.8	1482.7	1533.5	1585.2	1637.7	1745.7	1857.5
	s	2.0512	2.1153	2.1720	2.2233	2.2702	2.3137	2.3542	2.3923	2.4283	2.4625	2.4952	2.5566	2.6137
5 (162.24)	v	78.16	90.25	102.26	114.22	126.16	138.10	150.03	161.95	173.87	185.79	197.71	221.6	245.4
	h	1148.8	1195.0	1241.2	1288.0	1335.4	1383.6	1432.7	1482.6	1533.4	1585.1	1637.7	1745.7	1857.4
	s	1.8718	1.9370	1.9942	2.0456	2.0927	2.1361	2.1767	2.2148	2.2509	2.2851	2.3178	2.3792	2.4363
10 (193.21)	v	38.85	45.00	51.04	57.05	63.03	69.01	74.98	80.95	86.92	92.88	98.84	110.77	122.69
	h	1146.6	1193.9	1240.6	1287.5	1335.1	1383.4	1432.5	1482.4	1533.2	1585.0	1637.6	1745.6	1857.3
	s	1.7927	1.8595	1.9172	1.9689	2.0160	2.0596	2.1002	2.1383	2.1744	2.2086	2.2413	2.3028	2.3598
14.696 (212.00)	v	......	30.53	34.68	38.78	42.86	46.94	51.00	55.07	59.13	63.19	67.25	75.37	83.48
	h	......	1192.8	1239.9	1287.1	1334.8	1383.2	1432.3	1482.3	1533.1	1584.8	1637.5	1745.5	1857.3
	s	......	1.8160	1.8743	1.9261	1.9734	2.0170	2.0576	2.0958	2.1319	2.1662	2.1989	2.2603	2.3174
20 (227.96)	v	......	22.36	25.43	28.46	31.47	34.47	37.46	40.45	43.44	46.42	49.41	55.37	61.34
	h	......	1191.6	1239.2	1286.6	1334.4	1382.9	1432.1	1482.1	1533.0	1584.7	1637.4	1745.4	1857.2
	s	......	1.7808	1.8396	1.8918	1.9392	1.9829	2.0235	2.0618	2.0978	2.1321	2.1648	2.2263	2.2834
40 (267.25)	v	......	11.040	12.628	14.168	15.688	17.198	18.702	20.20	21.70	23.20	24.69	27.68	30.66
	h	......	1186.8	1236.5	1284.8	1333.1	1381.9	1431.3	1481.4	1532.4	1584.3	1637.0	1745.1	1857.0
	s	......	1.6994	1.7608	1.8140	1.8619	1.9058	1.9467	1.9850	2.0212	2.0555	2.0883	2.1498	2.2069
60 (292.71)	v	......	7.259	8.357	9.403	10.427	11.441	12.449	13.452	14.454	15.453	16.451	18.446	20.44
	h	......	1181.6	1233.6	1283.0	1331.8	1380.9	1430.5	1480.8	1531.9	1583.8	1636.6	1744.8	1856.7
	s	......	1.6492	1.7135	1.7678	1.8162	1.8605	1.9015	1.9400	1.9762	2.0106	2.0434	2.1049	2.1621
80 (312.03)	v	......	......	6.220	7.020	7.797	8.562	9.322	10.077	10.830	11.582	12.332	13.830	15.325
	h	......	......	1230.7	1281.1	1330.5	1379.9	1429.7	1480.1	1531.3	1583.4	1636.2	1744.5	1856.5
	s	......	......	1.6791	1.7346	1.7836	1.8281	1.8694	1.9079	1.9442	1.9787	2.0115	2.0731	2.1303
100 (327.81)	v	......	......	4.937	5.589	6.218	6.835	7.446	8.052	8.656	9.259	9.860	11.060	12.258
	h	......	......	1227.6	1279.1	1329.1	1378.9	1428.9	1479.5	1530.8	1582.9	1635.7	1744.2	1856.2
	s	......	......	1.6518	1.7085	1.7581	1.8029	1.8443	1.8829	1.9193	1.9538	1.9867	2.0484	2.1056
120 (341.25)	v	......	......	4.081	4.636	5.165	5.683	6.195	6.702	7.207	7.710	8.212	9.214	10.213
	h	......	......	1224.4	1277.2	1327.7	1377.8	1428.1	1478.8	1530.2	1582.4	1635.3	1743.9	1856.0
	s	......	......	1.6287	1.6869	1.7370	1.7822	1.8237	1.8625	1.8990	1.9335	1.9664	2.0281	2.0854

Abs. Press. (Sat. Temp)		C1	C2	C3	C4	C5	C6	C7	C8	C9	C10	C11	C12	C13
140 (353.02)	v	......	......	3.468	3.954	4.413	4.861	5.301	5.738	6.172	6.604	7.035	7.895	8.752
	h	......	......	1221.1	1275.2	1326.4	1376.8	1427.3	1478.2	1529.7	1581.9	1634.9	1743.5	1855.7
	s	......	......	1.6087	1.6683	1.7190	1.7645	1.8063	1.8451	1.8817	1.9163	1.9493	2.0110	2.0683
160 (363.53)	v	......	......	3.008	3.443	3.849	4.244	4.631	5.015	5.396	5.775	6.152	6.906	7.656
	h	......	......	1217.6	1273.1	1325.0	1375.7	1426.4	1477.5	1529.1	1581.4	1634.5	1743.2	1855.5
	s	......	......	1.5908	1.6519	1.7033	1.7491	1.7911	1.8301	1.8667	1.9014	1.9344	1.9962	2.0535
180 (373.06)	v	......	......	2.649	3.044	3.411	3.764	4.110	4.452	4.792	5.129	5.466	6.136	6.804
	h	......	......	1214.0	1271.0	1323.5	1374.7	1425.6	1476.8	1528.6	1581.0	1634.1	1742.9	1855.2
	s	......	......	1.5745	1.6373	1.6894	1.7355	1.7776	1.8167	1.8534	1.8882	1.9212	1.9831	2.0404
200 (381.79)	v	......	......	2.361	2.726	3.060	3.380	3.693	4.002	4.309	4.613	4.917	5.521	6.123
	h	......	......	1210.3	1268.9	1322.1	1373.6	1424.8	1476.2	1528.0	1580.5	1633.7	1742.6	1855.0
	s	......	......	1.5594	1.6240	1.6767	1.7232	1.7655	1.8048	1.8415	1.8763	1.9094	1.9713	2.0287
220 (389.86)	v	......	......	2.125	2.465	2.772	3.066	3.352	3.634	3.913	4.191	4.467	5.017	5.565
	h	......	......	1206.5	1266.7	1320.7	1372.6	1424.0	1475.5	1527.5	1580.0	1633.3	1742.3	1854.7
	s	......	......	1.5453	1.6117	1.6652	1.7120	1.7545	1.7939	1.8308	1.8656	1.8987	1.9607	2.0181
240 (397.37)	v	......	......	1.9276	2.247	2.533	2.804	3.068	3.327	3.584	3.839	4.093	4.597	5.100
	h	......	......	1202.5	1264.5	1319.2	1371.5	1423.2	1474.8	1526.9	1579.6	1632.9	1742.0	1854.5
	s	......	......	1.5319	1.6003	1.6546	1.7017	1.7444	1.7839	1.8209	1.8558	1.8889	1.9510	2.0084
260 (404.42)	v	......	......	......	2.063	2.330	2.582	2.827	3.067	3.305	3.541	3.776	4.242	4.707
	h	......	......	......	1262.3	1317.7	1370.4	1422.3	1474.2	1526.3	1579.1	1632.5	1741.7	1854.2
	s	......	......	......	1.5897	1.6447	1.6922	1.7352	1.7748	1.8118	1.8467	1.8799	1.9420	1.9995
280 (411.05)	v	......	......	......	1.9047	2.156	2.392	2.621	2.845	3.066	3.286	3.504	3.938	4.370
	h	......	......	......	1260.0	1316.2	1369.4	1421.5	1473.5	1525.8	1578.6	1632.1	1741.4	1854.0
	s	......	......	......	1.5796	1.6354	1.6834	1.7265	1.7662	1.8033	1.8383	1.8716	1.9337	1.9912
300 (417.33)	v	......	......	......	1.7675	2.005	2.227	2.442	2.652	2.859	3.065	3.269	3.674	4.078
	h	......	......	......	1257.6	1314.7	1368.3	1420.6	1472.8	1525.2	1578.1	1631.7	1741.0	1853.7
	s	......	......	......	1.5701	1.6268	1.6751	1.7184	1.7582	1.7954	1.8305	1.8638	1.9260	1.9835
350 (431.72)	v	......	......	......	1.4923	1.7036	1.8980	2.084	2.266	2.445	2.622	2.798	3.147	3.493
	h	......	......	......	1251.5	1310.9	1365.5	1418.5	1471.1	1523.8	1577.0	1630.7	1740.3	1853.1
	s	......	......	......	1.5481	1.6070	1.6563	1.7002	1.7403	1.7777	1.8130	1.8463	1.9086	1.9663
400 (444.59)	v	......	......	......	1.2851	1.4770	1.6508	1.8161	1.9767	2.134	2.290	2.445	2.751	3.055
	h	......	......	......	1245.1	1306.9	1362.7	1416.4	1469.4	1522.4	1575.8	1629.6	1739.5	1852.5
	s	......	......	......	1.5281	1.5894	1.6398	1.6842	1.7247	1.7623	1.7977	1.8311	1.8936	1.9513

* Abridged by permission from Joseph H. Keenan and Frederick G. Keyes, *Thermodynamic Properties of Steam*, John Wiley & Sons, Inc., New York, 1937.

Table 3. Properties of Superheated Steam *—*Continued*

Temperature—Degrees Fahrenheit

Abs Press., lb sq in. (Sat. Temp)		500	550	600	620	640	660	680	700	800	900	1000	1200	1400	1600
450 (456.28)	v	1.1231	1.2155	1.3005	1.3332	1.3652	1.3967	1.4278	1.4584	1.6074	1.7516	.8928	2.170	2.443	2.714
	h	1238.4	1272.0	1302.8	1314.6	1326.2	1337.5	1348.8	1359.9	1414.3	1467.7	1521.0	1628.6	1738.7	1851.9
	s	1.5095	1.5437	1.5735	1.5845	1.5951	1.6054	1.6153	1.6250	1.6699	1.7108	1.7486	1.8177	1.8803	1.9381
500 (467.01)	v	0.9927	1.0800	1.1591	1.1893	1.2188	1.2478	1.2763	1.3044	1.4405	1.5715	1.6996	1.9504	2.197	2.442
	h	1231.3	1266.8	1298.6	1310.7	1322.6	1334.2	1345.7	1357.0	1412.1	1466.0	1519.6	1627.6	1737.9	1851.3
	s	1.4919	1.5280	1.5588	1.5701	1.5810	1.5915	1.6016	1.6115	1.6571	1.6982	1.7363	1.8056	1.8683	1.9262
550 (476.94)	v	0.8852	0.9686	1.0431	1.0714	1.0989	1.1259	1.1523	1.1783	1.3038	1.4241	1.5414	1.7706	1.9957	2.219
	h	1223.7	1261.2	1294.3	1306.8	1318.9	1330.8	1342.5	1354.0	1409.9	1464.3	1518.2	1626.6	1737.1	1850.6
	s	1.4751	1.5131	1.5451	1.5568	1.5680	1.5787	1.5890	1.5991	1.6452	1.6868	1.7250	1.7946	1.8575	1.9155
600 (486.21)	v	0.7947	0.8753	0.9463	0.9729	0.9988	1.0241	1.0489	1.0732	1.1899	1.3013	1.4096	1.6208	1.8279	2.033
	h	1215.7	1255.5	1289.9	1302.7	1315.2	1327.4	1339.3	1351.1	1407.7	1462.5	1516.7	1625.5	1736.3	1850.0
	s	1.4586	1.4990	1.5323	1.5443	1.5558	1.5667	1.5773	1.5875	1.6343	1.6762	1.7147	1.7846	1.8476	1.9056
700 (503.10)	v	......	0.7277	0.7934	0.8177	0.8411	0.8639	0.8860	0.9077	1.0108	1.1082	1.2024	1.3853	1.5641	1.7405
	h	......	1243.2	1280.6	1294.3	1307.5	1320.3	1332.8	1345.0	1403.2	1459.0	1513.9	1623.5	1734.8	1848.8
	s	......	1.4722	1.5084	1.5212	1.5333	1.5449	1.5559	1.5665	1.6147	1.6573	1.6963	1.7666	1.8299	1.8881
800 (518.23)	v	......	0.6154	0.6779	0.7006	0.7223	0.7433	0.7635	0.7833	0.8763	0.9633	1.0470	1.2088	1.3662	1.5214
	h	......	1229.8	1270.7	1285.4	1299.4	1312.9	1325.9	1338.6	1398.6	1455.4	1511.0	1621.4	1733.2	1847.5
	s	......	1.4467	1.4863	1.5000	1.5129	1.5250	1.5366	1.5476	1.5972	1.6407	1.6801	1.7510	1.8146	1.8729
900 (531.98)	v	......	0.5264	0.5873	0.6089	0.6294	0.6491	0.6680	0.6863	0.7716	0.8506	0.9262	1.0714	1.2124	1.3509
	h	......	1215.0	1260.1	1275.9	1290.9	1305.1	1318.8	1332.1	1393.9	1451.8	1508.1	1619.3	1731.6	1846.3
	s	......	1.4216	1.4653	1.4800	1.4938	1.5066	1.5187	1.5303	1.5814	1.6257	1.6656	1.7371	1.8009	1.8595
1000 (544.61)	v	......	0.4533	0.5140	0.5350	0.5546	0.5733	0.5912	0.6084	0.6878	0.7604	0.8294	0.9615	1.0893	1.2146
	h	......	1198.3	1248.8	1265.9	1281.9	1297.0	1311.4	1325.3	1389.2	1448.2	1505.1	1617.3	1730.0	1845.0
	s	......	1.3961	1.4450	1.4610	1.4757	1.4893	1.5021	1.5141	1.5670	1.6121	1.6525	1.7245	1.7886	1.8474
1100 (556.31)	v	......	......	0.4532	0.4738	0.4929	0.5110	0.5281	0.5445	0.6191	0.6866	0.7503	0.8716	0.9885	1.1031
	h	......	......	1236.7	1255.3	1272.4	1288.5	1303.7	1318.3	1384.3	1444.5	1502.2	1615.2	1728.4	1843.8
	s	......	......	1.4251	1.4425	1.4583	1.4728	1.4862	1.4989	1.5535	1.5995	1.6405	1.7130	1.7775	1.8363
1200 (567.22)	v	......	......	0.4016	0.4222	0.4410	0.4586	0.4752	0.4909	0.5617	0.6250	0.6843	0.7967	0.9046	1.0101
	h	......	......	1223.5	1243.9	1262.4	1279.6	1295.7	1311.0	1379.3	1440.7	1499.2	1613.1	1726.9	1842.5
	s	......	......	1.4052	1.4243	1.4413	1.4568	1.4710	1.4843	1.5409	1.5879	1.6293	1.7025	1.7672	1.8263
1400 (587.10)	v	......	......	0.3174	0.3390	0.3580	0.3753	0.3912	0.4062	0.4714	0.5281	0.5805	0.6789	0.7727	0.8640
	h	......	......	1193.0	1218.4	1240.4	1260.3	1278.5	1295.5	1369.1	1433.1	1493.2	1608.9	1723.7	1840.0
	s	......	......	1.3639	1.3877	1.4079	1.4258	1.4419	1.4567	1.5177	1.5666	1.6093	1.6836	1.7489	1.8083

Abs. Press. lb/sq in. (Sat. Temp)													
1600 (604.90)	v	0.7545	0.6738	0.5906	0.5027	0.4553	0.4034	0.3417	0.3271	0.3112	0.2936	0.2733	
	h	1837.5	1720.5	1604.6	1487.0	1425.3	1358.4	1278.7	1259.6	1238.7	1215.2	1187.8	
	s	1.7926	1.7328	1.6669	1.5914	1.5476	1.4964	1.4303	1.4137	1.3952	1.3741	1.3489	
1800 (621.03)	v	0.6693	0.5968	0.5218	0.4421	0.3986	0.3502	0.2907	0.2760	0.2597	0.2407	......	
	h	1835.0	1717.3	1600.4	1480.8	1417.4	1347.2	1260.3	1238.5	1214.0	1185.1	......	
	s	1.7786	1.7185	1.6520	1.5752	1.5301	1.4765	1.4044	1.3855	1.3638	1.3377	......	
2000 (635.82)	v	0.6011	0.5352	0.4668	0.3935	0.3532	0.3074	0.2489	0.2337	0.2161	0.1936	......	
	h	1832.5	1714.1	1596.1	1474.5	1409.2	1335.5	1240.0	1214.8	1184.9	1145.6	......	
	s	1.7660	1.7055	1.6384	1.5603	1.5139	1.4576	1.3783	1.3564	1.3300	1.2945	......	
2500 (668.13)	v	0.4784	0.4244	0.3678	0.3061	0.2710	0.2294	0.1686	0.1484	......	......	......	
	h	1826.2	1706.1	1585.3	1458.4	1387.8	1303.6	1176.8	1132.3	......	......	......	
	s	1.7389	1.6775	1.6088	1.5273	1.4772	1.4127	1.3073	1.2687	......	......	......	
3000 (695.36)	v	0.3966	0.3505	0.3018	0.2476	0.2159	0.1760	0.0984	......	......	......	......	
	h	1819.9	1698.0	1574.3	1441.8	1365.0	1267.2	1060.7	......	......	......	......	
	s	1.7163	1.6540	1.5837	1.4984	1.4439	1.3690	1.1966	......	......	......	......	
3206.2 (705.40)	v	0.3703	0.3267	0.2806	0.2288	0.1981	0.1583	......	......	......	......	......	
	h	1817.2	1694.6	1569.8	1434.7	1355.2	1250.5	......	......	......	......	......	
	s	1.7080	1.6452	1.5742	1.4874	1.4309	1.3508	......	......	......	......	......	
3500	v	0.3381	0.2977	0.2546	0.2058	0.1762	0.1364	0.0306	......	......	......	......	
	h	1813.6	1689.8	1563.3	1424.5	1340.7	1224.9	780.5	......	......	......	......	
	s	1.6968	1.6336	1.5615	1.4723	1.4127	1.3241	0.9515	......	......	......	......	
4000	v	0.2943	0.2581	0.2192	0.1743	0.1462	0.1052	0.0287	......	......	......	......	
	h	1807.2	1681.7	1552.1	1406.8	1314.4	1174.8	763.8	......	......	......	......	
	s	1.6795	1.6154	1.5417	1.4482	1.3827	1.2757	0.9347	......	......	......	......	
4500	v	0.2602	0.2273	0.1917	0.1500	0.1226	0.0798	0.0277	......	......	......	......	
	h	1800.9	1673.5	1540.8	1388.4	1286.5	1113.9	753.5	......	......	......	......	
	s	1.6640	1.5990	1.5235	1.4253	1.3529	1.2204	0.9235	......	......	......	......	
5000	v	0.2329	0.2027	0.1696	0.1303	0.1036	0.0593	0.0263	......	......	......	......	
	h	1794.5	1665.3	1529.5	1369.5	1256.5	1047.1	746.4	......	......	......	......	
	s	1.6499	1.5839	1.5066	1.4034	1.3231	1.1622	0.9152	......	......	......	......	
5500	v	0.2106	0.1825	0.1516	0.1143	0.0880	0.0463	0.0262	......	......	......	......	
	h	1788.1	1657.0	1518.2	1349.3	1224.1	985.0	741.3	......	......	......	......	
	s	1.6369	1.5699	1.4908	1.3821	1.2930	1.1093	0.9099	......	......	......	......	

* Abridged by permission from Joseph H. Keenan and Frederick G. Keyes, *Thermodynamic Properties of Steam*, John Wiley & Sons, Inc., New York, 1937.

17. PROPERTIES OF LIQUIDS AND VAPORS

The properties of liquids and vapors which are of use in engineering computations are pressure, temperature, specific volume, entropy, enthalpy, and internal energy. It becomes necessary to be able to obtain the numerical values of these properties for fluids in any of the conditions described above. For gases the relations among the several properties are given by relatively simple mathematical formulations known as characteristic equations, so that tables of properties are not required. For liquids and vapors, however, the characteristic equations are found to be entirely too complex in form for direct use in practical work. As a consequence it is the custom of the engineer to depend upon the physicist for the primary experimental determinations of the functional relations and to use for practical computations tabular statements of the properties of the liquids and vapors at a great variety of states.

Among the engineering fluids for which adequately complete and reliable tables of properties are available are steam and the refrigerants ammonia (NH_3) and carbon dioxide (CO_2); less complete tables exist for a number of other vapors. These tables are usually drawn up in sections which give separately the properties of saturated liquid and vapor and of superheated vapor, these generally being arranged variously with pressure or temperature or perhaps entropy as the major argument. An abridgment of the latest American tables of the properties of steam appears on pp. 8-44 to 8-49. The symbols are those which at this writing are accepted as standard and which are used throughout this section. It will be observed that the subscripts f and g are used to denote properties at the saturated liquid state and at the saturated vapor state respectively, and that the subscript fg denotes the change of a property between the same two states.

Properties of Superheated Vapor. Steam tables also provide extensive tabular data on the properties of superheated steam. In such tabulations, because a vapor which is not contiguous with its liquid may be superheated to any temperature above the saturation temperature corresponding to the pressure, the properties are given for a wide range of temperatures or of degrees of superheat at each pressure. An abridged edition of the superheated steam section of the Keenan and Keyes table appears in Table 3, pp. 868 to 871.

The properties listed are seen to be the specific volume, enthalpy, and entropy. For convenience there is repeated directly below the pressure item the corresponding saturation temperature.

Properties of Compressed Liquid and Wet Vapor. The tables described have provided means for the direct determination of all the necessary properties of saturated liquid, saturated vapor, and superheated vapor. The two additional conditions of fluids for which data are frequently required are those of compressed liquid and wet vapor. It will be found that the table for the saturated condition supplies the requisite primary data for determination of the properties at these latter conditions, but some computations are necessary. The following describes the procedures to be followed.

Compressed Liquid is defined as liquid at a temperature less than the saturation temperature corresponding to the pressure under which it may be placed. An equivalent but alternative and frequently useful definition is that of a liquid under an externally imposed pressure which exceeds the saturated vapor pressure corresponding to the temperature. Simple illustrative conditions would be that of water in an open stream under atmospheric pressure but at 70 F instead of at the saturation temperature of 212 F which corresponds to the pressure, or that of boiler feedwater at 328 F but under a pump pressure of 400 lb per sq in. instead of the saturation pressure of 100 lb per sq in. which corresponds to the temperature.

As regards the specific volume, internal energy, and entropy of compressed water, those items are predominantly dependent on and influenced by the temperature but are unquestionably influenced slightly by superpressure. Therefore, for accurate work, especially at high liquid processes, it becomes necessary to have data showing the effect of superpressure in the values of these properties. However, for most ordinary work the values of V, E, and S for compressed water are regarded as the same as those of saturated water at the specified temperature (not pressure) of the compressed water.

As regards the enthalpy of compressed water the situation differs, for the reason that the enthalpy property ($E + PV/J$) contains directly the pressure property as an inherent component and so must be definitely influenced by any material superpressure. At a given temperature, it may be estimated from saturation data with good accuracy by adding to the value of the enthalpy of saturated water (H_f), as quoted in the saturated vapor table at the specified temperature (not pressure), the excess PV/J occasioned by the superpressure. More specifically,

$$H_{\text{compressed liquid at } t \text{ and } P} = H_{f, \text{ at } t} + (P - P_f)V_f/J \qquad (1)$$

In many practical instances in which P does not greatly exceed P_f the item $(P - P_f)V_f/J$ may be found to be relatively minor in magnitude and with adequate accuracy the enthalpy of the compressed liquid may be taken as that of saturated liquid at the temperature (not pressure) to which it is compressed.

Wet Vapor is a physical mixture of saturated vapor with droplets or fog of saturated but unvaporized liquid which is mechanically entrained with the vapor. The ratio of mass of vapor to the mass of the mixture of liquid and vapor is called the *quality* and is designated by the symbol x. For a wet vapor the temperature is the same as the saturation temperature corresponding to the pressure. The specific volume, enthalpy, and entropy may be computed as follows:

$$V_x = V_f + xV_{fg} \qquad (2)$$

$$H_x = H_f + xH_{fg} \qquad (3)$$

$$S_x = S_f + xS_{fg} \qquad (4)$$

18. GRAPHICAL REPRESENTATION OF VAPOR PROPERTIES

One may choose as coordinates any pair of properties of a fluid and depict graphically the manner of their joint variation as the fluid progresses through any characteristic sequence of states. Of the various diagrams which might be so devised those which have been found to be most generally useful are the $P-V$ diagram, the $T-S$ diagram and the $H-S$ or Mollier diagram. These are described and discussed in the following paragraphs.

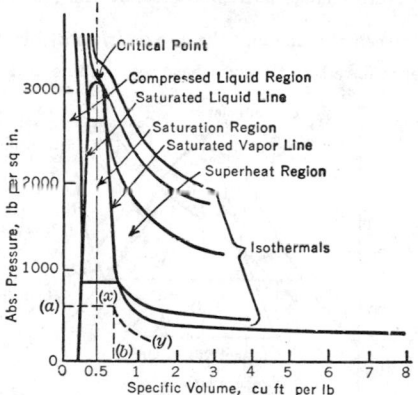

FIG. 1. $P-V$ Diagram, Water and Water Vapor

P–V Diagram. In the $P-V$ diagram, as represented for a typical fluid in Fig. 1, the absolute pressure is employed as the ordinate and the specific volume as the abscissa. In the diagram the slightly oblique and curved line on the left which is labeled the *saturated liquid line* shows the manner of variation of the specific volume of saturated liquid as the pressure (and temperature) is progressively increased. The state of a compressed liquid at any given pressure (but a temperature less than saturation) would be represented by a point to the left of the saturated liquid line, this by reason of the lesser specific volume at the lesser temperature. Consequently the general region which lies to the left of the saturated liquid line is known as the *compressed liquid region*.

The sloping line labeled the *saturated vapor line* similarly shows the relation between the pressure and specific volume of (dry) saturated vapor. The length of any horizontal line intercepted by the liquid and vapor lines thus represents the increase of specific volume during vaporization at constant pressure (V_{fg}). Therefore, since by the equation $V_x = V_f + xV_{fg}$, the state of a wet vapor mixture of quality x and at any given pressure would be represented by a point on that pressure line and lying at a distance xV_{fg} to the right of the liquid line. The region between the two saturation lines is known as the *saturation region*.

Since the superheating of a vapor at a given pressure effects an increase of the specific volume the state of a superheated vapor would be represented by a point to the right of the saturated vapor line, whence the region to the right of that line is known as the *superheat region*.

The saturated liquid and saturated vapor lines are observed to join at a *critical pressure*, which for water is at 3206.2 lb per sq in., and at which pressure the saturation temperature of water is 705.4 F. The saturation temperature at the critical pressure is similarly known as the *critical temperature*. The diagram indicates that a liquid under that pressure would on warming pass imperceptibly into a dense superheated vapor state without the appearance of the characteristic phase of vaporization at constant temperature.

The *isothermal lines* which appear in the figure are contour lines of constant temperature and indicate the specific volume of a fluid at any given temperature and pressure or, alternatively, the pressure corresponding to a given state as designated by the temperature and specific volume. Since the temperature is constant during vaporization at constant pressure the isothermals become straight horizontal lines within the saturation region. In the compressed liquid region the isothermals are very nearly vertical, showing the very slight change (decrease) of volume of a liquid as it is compressed at constant temperature. In the superheat region the isothermals are approximately hyperbolic curves, approaching that curve more exactly at very low pressures or considerable superheats.

For temperatures which are above the critical a vapor may not be condensed at any finite pressure. Some writers consider the critical temperature isothermal to provide an arbitrary line of demarcation between the *gas phase* (above the line) and the *vapor phase* (below the line) of a substance.

In addition to the isothermal lines one might draw contour lines of *constant quality*, *constant superheat*, *constant internal energy*, *constant enthalpy*, or *constant entropy*, which lines might be useful for various particular purposes. Also any sort of state change whatsoever may be represented by a suitable curve or line on the diagram.

In connection with the P–V diagram it should be remarked that the product PV corresponding to any state of a fluid is represented by the rectangular area bounded by the coordinates, by a constant pressure line through the point, and by a constant volume line through the point, that is, by an area such as *oaxb* for state x. Also for any manner of state change as represented by the line xy, the $\int_x^y P\,dV$ is represented by the area between the line and the V-axis (that is, by the area *below* the line), and the $\int_x^y V\,dP$ is represented by the area between the line and the P-axis (the area *back of* the line).

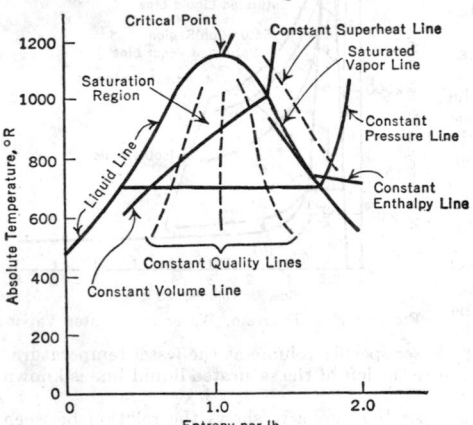

T–S Diagram. In the T–S diagram (Fig. 2) *absolute* temperature is the ordinate, and entropy per pound mass (relative to liquid at 32 F) is the abscissa. The saturated liquid line, the critical point, and the saturated vapor line have the same significance as in the P–V diagram.

Except at quite high pressures, constant pressure contour lines nearly coincide with the saturated liquid line up to the saturation temperature corresponding to a particular pressure, from which point that pressure line proceeds across the saturation region at the saturation temperature of vaporization. The compressed liquid region is a thin band along the saturated liquid line.

At any temperature (or pressure) the horizontal intercept between the liquid and saturated vapor lines measures the increase of entropy S_{fg}

FIG. 2. T–S Diagram, Water and Water Vapor

during vaporization, and, since for a wet vapor mixture $S_x = S_f + x S_{fg}$, the state of a wet vapor of quality x is represented by a point on the pressure line at a distance $x S_{fg}$ to the right of the liquid line. Contour lines of constant quality will appear as in the diagram.

Typical contour lines of constant specific volume, constant enthalpy, and constant superheat also appear in the diagram.

The principal utility of the T–S diagram will be recalled to arise from the facts that for any process

$$Q_R = T_R\,\Delta S$$

and for any reversible process

$$Q = \int T\,dS$$

Thus for any state change of a fluid, as plotted on T–S coordinates, the change of the unavailable energy (Q_R) with reference to a receiver temperature T_R and chargeable to a pound mass of the fluid is represented by the area bounded by the S-axis below, by the T_R

line above, and on the sides by the initial and final entropy lines. Likewise the heat energy reception or departure which accompanies a *reversible* state change is represented by the area between the state change curve and the S-axis (*below* the curve). A reversible adiabatic process is an isentropic (constant entropy) one and is represented by a straight vertical line.

H–S ("Mollier") Diagram. The coordinates of the Mollier diagram are the enthalpy and entropy of the fluid, each per pound mass and each with reference to zero relative values of those properties for a liquid at 32 F. By reason of the intimate utility of the enthalpy property for all circumstances in which steady-flow processes occur, a portion of the diagram is invariably incorporated with the formal tables of vapor properties.

Referring to Fig. 3, the saturated liquid line, the critical point, and the saturated vapor line have the same significance as in the P–V and T–S diagrams. The compressed liquid region is practically restricted to a rather narrow band immediately above the saturated liquid line.

Constant pressure lines in this compressed liquid band approach the saturated liquid line as the excess of their pressure over the saturation pressure approaches zero. Within the saturation region the constant pressure lines are straight lines and are also constant temperature contours. The vertical distance between the inter-

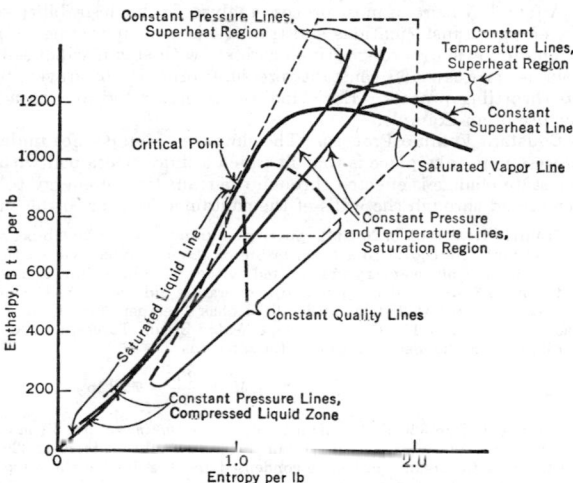

FIG. 3. *H–S* Diagram, Water and Water Vapor

sections of a constant pressure line with the saturated liquid and saturated vapor lines measures the increase of enthalpy during vaporization, and the horizontal distance measures the entropy of vaporization.

On entering the superheat region the constant pressure lines veer upward, and the constant temperature contours approach constant *H* lines with increase of superheat.

Contour lines of constant quality and of constant superheat are also drawn in Fig. 3, and other contour lines might be shown. However, the two noted are usually the ones which appear. It may be remarked that the common Mollier chart, as used for many practical purposes, shows to larger scale only a small region around the saturated vapor line, such as that enclosed by dotted lines in the upper portion of Fig. 3.

19. STATE CHANGES AND PROCESSES

In the consideration of the various actual processes which are encountered in engineering thermodynamics certain classes of state changes of fluids occur with marked frequency. Several of these will be examined in detail.

For the analysis of any process it is the first requisite that the initial state of the fluid be adequately specified. For the fluids and the states which are encountered in engineering thermodynamics information concerning any two properties is, in general, sufficient for designation of the state. In practice it is commonly found that one or more of the specified properties will be those which we are best equipped to measure readily, that is, the pressure or temperature or perhaps the specific volume.

A second requisite is the specification of the character of the state change. In this connection it will be found that many of the actual processes are characterized by the actual or effective constancy of some property of the fluid during the state change. To illustrate, a state change taking place in a fluid which is retained within a closed and non-extensible vessel is one in which the specific volume remains constant or is a constant volume process; one taking place during the passage of a fluid through a device such as a boiler or superheater may closely approach constant pressure conditions and be taken as effectively such; the ideal state change taking place during the expansion of a fluid in an engine or turbine is the reversible adiabatic, which is characterized by a constant entropy of the fluid.

As a third feature it will be observed that, as the result of such constancy of one property of the fluid during a change of its state and thus a change of its other properties, this particular property is known at the cessation of the process. Therefore, to designate the final state of the fluid and thus to complete the description of the process, it will in general be necessary to specify the magnitude of one additional property at the final state. With sufficient information thus available for fixing that state, all the other properties may then be determined.

A fourth feature of most process analyses is the desirability or necessity for determining the energy transformations and transitions accompanying the processes.

The following representative processes with steam vapor will be analyzed: (a) constant volume process, (b) constant pressure process, (c) reversible adiabatic process, and (d) throttling process. Each analysis is illustrated and to a large extent developed by the use of an example.

Constant Volume Process. The simplest circumstance under which a constant volume process may take place is that in which a fluid is retained within a closed container while the state change is effected through the transition of energy to or from the system as heat conducted through the walls of the container from or to the surrounding region.

Example 1. A tank of 20 cu ft capacity contains steam at 200 lb per sq in. abs and 500 F. The tank and contents are cooled to a temperature of 326 F. What will be the final pressure and quality of the steam, how much energy was emitted by the steam in cooling, and what was the change of enthalpy of the steam? At what pressure and temperature did the steam become (dry) saturated?

Solution. Saturated steam at 200 lb abs has a temperature of 382 F, whence at state (1) the steam had a superheat of 118 F. From Superheated Steam Tables, at 200 lb and 500 F, $V_1 = 2.73$ cu ft per lb, whence the mass of steam in the tank was

$$M = \frac{20}{2.73} = 7.33 \text{ lb}$$

After cooling to 326 F at constant volume *the specific volume* V_2 *must be the same.* From the Saturated Steam Tables the specific volume of (dry) saturated steam at 326 F is 4.538 cu ft per lb, whence a portion of the steam must have condensed and the fluid in the tank become a mixture of saturated liquid and saturated vapor (wet steam). The pressure of this saturated mixture at 326 F must be 97.5 lb per sq in. abs (from Saturated Steam Tables).

To find the quality x_2

$$x_2 = \frac{V - V_{f,2}}{V_{fg,2}} = \frac{2.73 - 0.0177}{4.521} = 0.60$$

The energy emitted as heat in cooling must have come wholly from the internal energy of the steam since no energy emission, such as shaft work, flow work, or kinetic energy, was possible.

Therefore, for each pound of fluid,

$$Q_{out} = E_1 - E_2$$

$$= \left[H_1 - \frac{P_1 V_1}{J} \right] - \left[(H_{f,2} + x_2 H_{fg,2}) - \frac{P_2 V_{x,2}}{J} \right]$$

$$= \left[1268.9 - \frac{200 \times 144 \times 2.73}{778} \right]$$

$$- \left[(296.5 + 0.60 \times 890.2) - \frac{97.5 \times 2.73}{5.4} \right]$$

$$= 1167.9 - 781.3 = 386.6 \text{ Btu per lb}$$

Total energy emitted, as heat, $= 7.33 \times 386.6 = 2834$ Btu.

Change of enthalpy $= H_2 - H_1$

$$= (H_{f,2} + x_2 H_{fg,2}) - H_1$$

$$= 830.6 - 1268.9 = -438.3 \text{ Btu per lb}$$

To find the pressure and temperature at which the steam became (dry) saturated it is necessary to search the Saturated Steam Tables for the state at which $V_g = 2.73$. This is found at a pressure of 166.4 lb per sq in. abs and a temperature of 366.7 F.

Note that in this non-flow, constant volume circumstance only the change of internal energy $(E_2 - E_1)$ measures or denotes an energy quantity, the change of PV becoming simply a change in that product of properties, *without an energy significance.*

Constant Pressure Process. A constant pressure process may readily be conceived to occur (a) without flow, as in a cylinder with piston, or (b) with steady flow, as in any vessel such as a steadily steaming boiler, a steadily operating condenser, etc., the state change being effected in either case by energy transition to or from the fluid, this usually as heat.

Example 2. A pound of water *in a cylinder* (*with piston*) is initially at 200 F and 235.3 lb per sq in. gage, and by energy reception as heat is transformed to steam of 98 per cent quality at the same pressure, the constancy of pressure being attained by permitting the piston to retreat as the fluid expands.

Compute the amount of energy required to effect the process and also that portion of this energy which is stored internally in the fluid.

Solution.

$$JQ = J(E_2 - E_1) - W$$

$$= J(E_2 - E_1) + W_{\text{out}}$$

In this device the work output at the piston face equals $\int_1^2 P \, dV = P(V_2 - V_1)$, since the pressure P is constant. Therefore

$$Q_{\text{in}} = (E_2 - E_1) + P(V_2 - V_1)/J$$

$$= (E_2 + PV_2/J) - (E_1 + PV_1/J)$$

$$= H_2 - H_1, \text{ for a non-flow constant-pressure process}$$

For the specific data of the problem:

$$H_1 = 167.9 + \frac{(250 - 11.5) \times 144 \times 0.0166}{778}$$

$$= 168.7 \text{ (exactly)} = 168.0 \text{ (approx.)}$$

$$H_2 = 376.0 + 0.98 \times 825.1 = 1184.6$$

$$Q = \Delta H = 1184.6 - 168.7 = 1015.3 \text{ Btu per lb}$$

$$E_1 = 167.9 - \frac{11.5 \times 144 \times 0.0166}{778} = 167.9$$

$$E_2 = 1184.6 - \frac{250 \times 0.98 \times 1.844}{5.4} = 1101.1$$

$$\Delta E = 1101.1 - 167.9 = 933.2 \text{ Btu per lb}$$

The difference between the energy supplied (ΔH) and that stored as molecular energy in the steam (ΔE) is transferred directly via the fluid to the piston face and is there delivered as work [$P(V_2 - V_1)$, = 1015.3 − 932.6 = 82.7].

The foregoing example has had to do with a non-flow constant pressure process. For the circumstance of steady flow at constant pressure through a device such as a boiler, since there is no shaft work,

$$Q_{\text{in}} = (H_2 - H_1) + (U_2{}^2 - U_1{}^2)/64.34J$$

As the change of kinetic energy in the usual device of this character would very rarely indeed exceed 1 Btu per lb and would usually be much less, this in comparison with a change of enthalpy which commonly would be some thousand Btu in magnitude, the kinetic energy term becomes effectively negligible and may be omitted. Thus for both a non-flow and a steady-flow constant pressure process the energy received by the fluid is measured by its change of enthalpy.

Reversible Adiabatic Process. Recalling that, by definition, *any process whatsoever in which no energy enters or departs from the system as heat is designated as an adiabatic process,* we may immediately write the following general energy equations for an adiabatic non-flow and an adiabatic steady-flow process:

$$W = J(E_2 - E_1), \text{ for a non-flow adiabatic process}$$

$$W + (U_1{}^2 - U_2{}^2)/64.34 = J(H_2 - H_1), \text{ for a steady-flow adiabatic process}$$

These equations are universally applicable whether the adiabatic process may be ideal and reversible or actual and thus irreversible. For the further condition of a *reversible* adiabatic process there is introduced the additional feature that the entropy remains constant during the process, or the process is isentropic. It may well be remarked that the outstanding practical utility of the entropy property lies in this very feature of its constancy during a reversible adiabatic process. To denote such constancy we employ the letter S as a subscript attached to any change of properties of a fluid during an isentropic process. Thus, rewriting the above equations but now in the special form which is applicable to the reversible adiabatic,

$W = J(E_2 - E_1)_S$, for a non-flow reversible adiabatic

$W + (U_1{}^2 - U_2{}^2)/64.34 = J(H_2 - H_1)_S$, for a steady-flow reversible adiabatic

For an easy and accurate determination of the fluid properties after isentropic expansion from a given initial state to a specified final pressure the H–S (Mollier) chart obviously is useful. Also the final quality after expansion to a given pressure or temperature *within* the saturation region may be computed directly by the equation

$$x_2 = \frac{S - S_{f,2}}{S_{fg,2}} = 1 - \frac{S_{g,2} - S}{S_{fg,2}} \tag{5}$$

where S is the initial and constant entropy. The final quality being known, the other properties may be computed as required.

Example. Steam at 200 lb per sq in. abs and a quality of 0.995 is delivered to an ideal steam turbine, expands adiabatically and reversibly therein to a pressure of 14.7 lb per sq in. abs, and is discharged at that pressure.
Compute the entropy, quality, and enthalpy of the steam leaving the turbine and the shaft work which would ideally be obtainable per pound of steam used, assuming negligible kinetic energies entering and leaving the turbine.
Solution.

$$S_1 = S_{g,1} - (1 - x_1)S_{fg,1} = 1.5453 - 0.005 \times 1.0018 = 1.5403$$

$$S_2 = S_1 = 1.5403$$

$$x_2 = \frac{1.5403 - 0.3120}{1.4446} = 0.850$$

$$H_2 = 1150.4 - 0.15 \times 970.3 = 1004.9$$

$$H_1 = 1198.4 - 0.005 \times 842.4 = 1194.2$$

$$W = J(H_1 - H_2)_S = J(1194.2 - 1004.9)$$

$$= 778 \times 189.3 = 147{,}100 \text{ ft-lb}$$

Throttling Process. Technically, a pure throttling process is one in which a fluid expands adiabatically but thoroughly irreversibly through a labyrinthal or porous obstruction in a line, the arrangement being such that the velocities of the stream approaching, through, and departing from the obstruction shall be very small and their differences wholly negligible. No shaft work is performed.

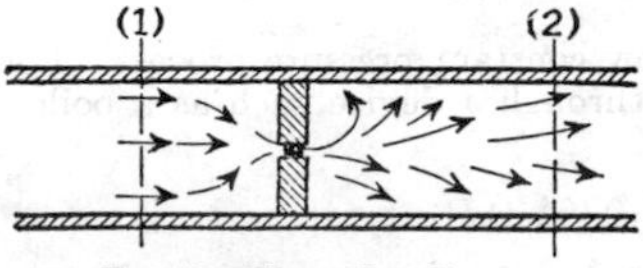

(1)   (2)

This process is approached closely in practice in the "wire-drawing" of a vapor (or gas) through a constriction in a line such as a partially closed valve or small orifice. In such devices there is unquestionably an accretion of velocity and kinetic energy directly in the constriction, but this momentary unidirectional velocity is immediately dissipated by reason of the high degree of turbulence existing in the stream directly after the constriction, whereby the kinetic energy in the stream finally departing from the region $(U_2{}^2/64.34)$ is negligible or differs negligibly from that in the stream approaching the region $(U_1{}^2/64.34)$. (See Fig. 4.)

Fig. 4. Throttling Process

The process is also approached, but less exactly (owing to energy dissipation as heat), in the flow of a vapor through a long pipe line in which there is a distinct pressure drop due to friction.

To discern the working scheme for analyzing the throttling process, note that in fitting the general steady-flow energy equation (equation 5, Art. 4) to the conditions of the process the Q and W terms disappear and the kinetic energy terms are either negligible or effectively canceled, whence

$$H_1 = H_2$$

This equality of the initial and final enthalpy is the outstanding characteristic of the throttling process.

Example. Steam at 250 lb abs and 99 per cent quality is throttled to a pressure of 200 lb abs. What is the state of the steam after throttling, and what is the loss of available energy, assuming that the steam were being supplied to a reversible engine operating with a 70 F exhaust?
Solution.

$$H_1 = 1192.8 = H_2$$

To ascertain the second state, since H_g at 200 lb abs is 1198.4 the fluid must still be a slightly wet mixture. To ascertain the quality,

$$H_2 = 1192.3 = H_{f,2} + x_2 H_{fg,2} = 355.4 + x_2\ 843.0,\ x_2 = 0.9935$$

To ascertain the increase of unavailable energy:

$$S_1 \text{ (at 250 lb and 99\%)} = 1.5167;\quad S_2 \text{ (at 200 lb and 99.35\%)} = 1.5438$$

$$\Delta S = 0.0271;\quad \Delta Q_R = T_2\ \Delta S = (70 + 460) \times 0.0271 = 14.36 \text{ Btu per lb.}$$

The Throttling Calorimeter. Both by reference to the above example and to a Mollier diagram (see Fig. 5) it becomes apparent that if a wet vapor passes through a throttling process the requisite equality of the initial and final enthalpies will in general produce an increase in the quality of the vapor.* Also if the initial quality of the vapor is sufficiently high and the pressure drop sufficiently great, some degree of superheat may well exist in the exit or lower pressure vapor. By reason of the latter circumstance and the additional

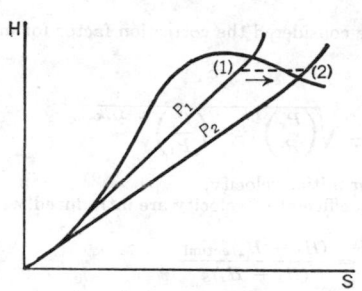

FIG. 5. H–S Diagram for Throttling Process   FIG. 6.   Throttling Calorimeter

facts (a) that *in the superheat region the pressure and temperature will suffice to determine the state and enthalpy of the fluid* and (b) that these P and T properties are both directly measurable ones, the throttling phenomenon may be and is regularly employed in a device which is known as the *throttling calorimeter* and is used for ascertaining the enthalpy and quality of steam which contains a limited amount of moisture.

Referring to Fig. 6, a simple arrangement of such a device is there indicated. In the figure, A is a steam line through which is passing the steam the enthalpy and quality of which are desired. Pipe B is a *sampling tube* through which it is endeavored to draw a true sample of the steam in A. Gage C or thermometer D determines the saturation pressure or temperature of the steam supply. These obviously should "check" if the steam is saturated. Between the flanges at E is located a plate with a small ($1/16$ in. to $1/8$ in. diameter) orifice through which the sample is throttled to the discharge pressure, which is usually about atmospheric. Thermometer F, and mercury manometer G, together with data on the barometric pressure, provide the requisite data on the temperature and pressure of the throttled fluid by which its state may be ascertained from the Superheated Steam Tables. Thus the enthalpy of the leaving steam and thereby that of the entering steam is determinable.

The quality of the steam sample is found by solving for x in the expression for the enthalpy of wet steam:

$$H_2 = H_1 = H_{f,1} + x_1 H_{fg,1};\quad \text{or}\quad x_1 = (H_2 - H_{f,1})/H_{fg,1}$$

An alternative method is to enter the Mollier chart at the intersection of the P_2 and t_2 lines, pass across at constant H to P_1, and read the quality directly.

This device if properly constructed and operated will give the enthalpy and quality of the sample with good accuracy, although a slight correction for radiation may be desirable. A serious practical difficulty lies in the securing of a truly representative sample of the steam in line A. Also its use is limited to fairly dry and fairly high-pressure steam since an excess of moisture or low steam pressure will prevent the superheating, and *its whole utility depends on the possibility of fixing the state of superheated steam from knowledge of its pressure and temperature.*

---

* This statement must be made with some qualification. To illustrate, inspection of the Mollier diagram of steam (Keenan tables) shows that with initial pressures in excess of about 500 lb per sq in. a throttling through a moderate pressure range may in fact act to increase the moisture content.

20. FLOW OF VAPORS THROUGH NOZZLES AND ORIFICES

The general energy equations for a nozzle are invariably applicable whether the fluid flowing is a gas or a vapor, whence we may refer immediately to the basic energy equations. Those equations are repeated here for convenience:

$$\frac{U_x^2 - U_1^2}{64.34} = J(H_1 - H_x) \tag{6}$$

$$M' = \frac{U_1 A_1}{V_1} = \frac{U_x A_x}{V_x} \text{ (continuity equation)} \tag{7}$$

$$U_x = 223.7 \sqrt{\frac{H_1 - H_x}{1 - \left(\frac{A_x}{A_1}\right)^2 \left(\frac{V_1}{V_x}\right)^2}} \tag{8}$$

where the factor $\sqrt{\dfrac{1}{1 - \left(\frac{A_x}{A_1}\right)^2 \left(\frac{V_1}{V_x}\right)^2}}$ is to be considered the correction factor for initial velocity.

$$\frac{M'}{A_x} \text{ (ideal)} = 8.02 \sqrt{\frac{k}{k-1}} \frac{P_1}{\sqrt{RT_1}} \sqrt{\left(\frac{P_x}{P_1}\right)^{2/k} - \left(\frac{P_x}{P_1}\right)^{(k+1)/k}} \tag{9}$$

which does not include the correction factor for initial velocity.

In actual flow the nozzle efficiency and the coefficient of velocity are introduced, e_n and C_v, respectively:

$$e_n = \frac{(U_x^2 - U_1^2)_{\text{actual}}}{(U_x^2 - U_1^2)_{\text{ideal}}} = \frac{(H_1 - H_x)_{\text{actual}}}{(H_1 - H_x)_S} \tag{10}$$

$$C_v = \frac{U_{x,\text{ actual}}}{U_{x,\text{ ideal}}} = \sqrt{e_n} \tag{11}$$

Quite as it was with the flow of a gas, given the state of the fluid as supplied, the problem of applying these equations in developing either the ideal or the actual nozzle form is that of ascertaining the simultaneous values of H_x and V_x as expansion proceeds to successively lower pressures. However, as may be anticipated, the detailed procedure in ascertaining these values may differ in the cases of the gas and of the vapor, particularly if the vapor is one for which tables of properties are available. Such tables being available for the vapors of major engineering importance, it is the practice (at least among engineers of the United States) to utilize them, although it will appear subsequently that in considerations of actual flow certain difficulties are encountered in that procedure.

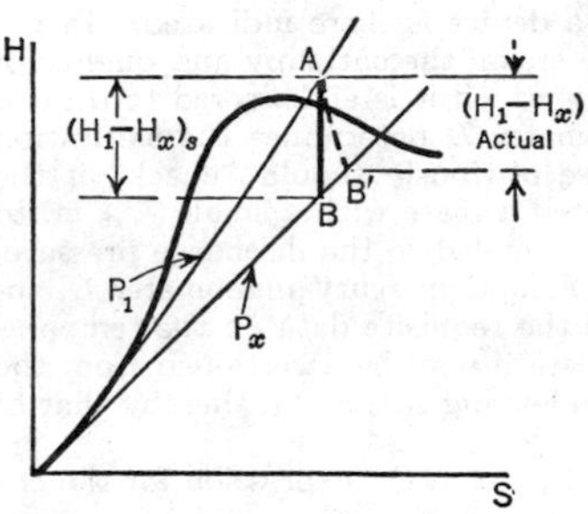

For the circumstance of the ideal expansion the values of the desired properties are easily obtained by recognizing the isentropic character of the expansion and then following the methods of the example on p. 878, using either the vapor tables alone or those jointly with a Mollier (H–S) chart. The appearance of an ideal state change as represented on H–S coordinates would be that of a line such as AB in Fig. 7. The essential characteristic of the line is the ideal constancy of the entropy.

FIG. 7. H–S Diagram of Nozzle Flow Process

The procedure in investigating the nozzle form for ideal expansion is perhaps shown to best advantage by an illustrative example.

Example. Determine the characteristic form and dimensions of a steam nozzle which shall provide for the flow and reversible expansion of 2 lb of steam per second from an initial state of 200 lb per sq in. abs and 420 F to a discharge region at a pressure of 60 lb abs. Do this by computing the areas required upon expansion to successive pressures of 160, 120, 115, 109, 85, and 60 lb. Assume the ideal case of reversible adiabatic expansion and also assume negligible entrance velocity.
Solution for P_x of 160 lb per sq in.

$$H_1 = 1222.9; \quad V_1 = 2.438; \quad S_1 = 1.5738$$

$$S_x = S_1 = 1.5738; \quad t_x \text{ (from Superheat table)} = 376.4 \text{ F}$$

$$H_x = 1202.7; \quad \Delta H = 19.6; \quad U_x = 223.7 \sqrt{19.6} = 990 \text{ ft per sec}$$

$$V_x = 2.894; \quad M'/A_x = 990/2.894 = 342; \quad A_x = 0.00585 \text{ sq ft}$$

Solution for P_x of 120 lb per sq in.

$$S_x = S_1 = 1.5735; \quad x_x = 1 - \frac{1.5874 - 1.5735}{1.0956} = 1 - 0.0127 = 0.987$$

$$H_x = 1189.8 - 0.0127 \times 877.4 = 1178.7; \quad \Delta H = 43.6; \quad U_x = 1480$$

$$V_x = 3.725 - 0.0127 \times 3.707 = 3.678$$

$$M'/A_x = 1480/3.678 = 401.5; \quad A_x = 0.00498$$

The major results at these and the other specified pressures are tabulated below.

Tabular Results

p_x (lb per sq in.)	t, or x	H_x	ΔH	U_x	V_x	M'/A_x	A_x (sq ft)
200	420	1222.9	....	....	2.438	....	....
160	376	1202.7	19.6	990	2.894	342	0.00585
120	0.987	1178.7	43.6	1480	3.678	402	0.00498
115	0.984	1175.3	47.0	1530	3.817	402	0.00498
109	0.981	1171.1	51.2	1600	4.001	400	0.00500
85	0.964	1151.1	71.2	1890	4.978	379	0.00528
60	0.943	1124.4	97.9	2210	6.7605	327	0.00612

Scrutiny of the above tabular results indicates that:

(a) For a given initial state a particular enthalpy, specific volume and area (per unit mass-rate of flow) are again associated with each successive pressure.

(b) The critical pressure phenomenon is again evidenced by the minimum area, or throat, which occurs in the above example upon expansion to about 57 per cent (115/200) of the supply pressure.

(c) For expansion to any pressure equal to or less than the critical a convergent passage is proper.

(d) A subsequent divergent passage is required in order to provide properly for a controlled and directed expansion to a pressure which is less than the critical.

(e) For expansion to a discharge region pressure which is greater than the critical the required exit area for a given mass-rate of flow will depend on both the supply state and the discharge region pressure or, conversely, the mass-rate of flow at given supply and discharge region pressures will be proportional to the exit area.

(f) On the contrary, for expansion to a discharge region which is at a pressure equal to or less than the critical, the invariable existence of the critical pressure at the throat and a fixed value of the critical ratio will act to cause the required throat area for a given mass-rate of flow to depend only on the state of the fluid supply or, conversely, the mass-rate of flow obtained with a given supply state depends only on the throat area and is independent of the discharge region pressure (although the proper final exit area for suitable accommodation of the entire expansion must depend also on the discharge region pressure).

One feature in which vapor flow and gas flow will differ slightly is that the critical ratio for a vapor does not appear to be strictly a constant. Thus, in the example, in which the steam supply was moderately superheated, the critical ratio appears to have the value of about 0.57. By similar investigation it would be found that the critical ratio for steam would apparently range from about 0.55 for highly superheated steam supply to about 0.58 for saturated steam supply. However, this variation does not require much concern, owing to the very slight change of the cross-sectional area of the nozzle at pressures between about 55 and 60 per cent of the supply pressure.

An important characteristic of this ideal expansion, which is clearly indicated in the above tabulation and was evident in Fig. 7, is the progressive condensation of a portion of the vapor with progressive expansion. Quoting specific figures, the reversible adiabatic expansion of the initially superheated supply would require the actual condensation of about 2 per cent of the vapor prior to reaching the throat and about 6 per cent prior to reaching the 60 lb exit pressure, and the example above would indicate that 13 per cent should condense if the expansion were continued to atmospheric pressure. In making these quality computations it was inherently assumed that a stable thermal equilibrium was maintained between the liquid and the vapor as the expansion proceeded, whence the expansion would be designated as being of the equilibrium type. A further consideration will be given to the possibility of an actual expansion of this type.

Actual Expansion of a Vapor. Any actual expansion through a nozzle is necessarily irreversible by reason of the unescapable friction and turbulence accompanying the high-velocity flow. The influence of such irreversibility and the general manner of accounting for it in air-nozzle design has been considered above. It will similarly be considered for steam-nozzle design in the following paragraphs, but first it needs be noted that without question additional causes for irreversibility exist in the case of vapor flow, or more particularly in the case of a vapor which is supplied in a saturated or only slightly superheated state. This additional irreversibility is occasioned by a temporary unstable condition which has been found to exist in rapidly expanding vapor. The condition is that known as supersaturation and is characterized by a retarding or deferring of the partial condensation which we have seen would accompany the equilibrium type of expansion. Experiments indicate that complete supersaturation and action like a superheated vapor persist until a pressure is reached which is about one-half of that at which condensation would normally begin. This is approximately equivalent to stating that for initially dry and saturated steam complete supersaturation persists down to and somewhat beyond the critical pressure. Assuming superheated and supersaturated steam with frictionless adiabatic expansion having the basic property relation

$$PV^{1.3} = \text{a constant} \tag{12}$$

the following relations are derived:

$$\frac{M'}{A_x} \text{ (ideal)} = 16.7 \sqrt{\frac{P_1}{V_1}\left[\left(\frac{P_x}{P_1}\right)^{1.54} - \left(\frac{P_x}{P_1}\right)^{1.77}\right]} \tag{9a}$$

or, if $(P_1 - P_x)/P_1$ does not exceed about 0.33,

$$\frac{M'}{A_x} \text{ (ideal)} = 8.02 \frac{1}{\sqrt{P_1V_1}} \sqrt{P_x(P_1 - P_x) - 0.15(P_1 - P_x)^2} \tag{9b}$$

or, if $(P_1 - P_x)/P_1$ does not exceed about 0.1,

$$\frac{M'}{A_x} \text{ (ideal)} = 8.02 \sqrt{\frac{P_x(P_1 - P_x)}{P_1V_1}} \tag{9c}$$

or, if $(P_1 - P_x)/P_1$ does not exceed about 0.01,

$$\frac{M'}{A_x} \text{ (ideal)} = 8.02 \sqrt{\frac{P_1 - P_x}{V_1}} \tag{9d}$$

Also for the special consideration of a value of P_x equal to the critical pressure, in which circumstance the flow rate per unit area is recalled to be the maximum,

$$\frac{M'}{A_{\text{throat}}} \text{ (ideal)} = 3.78 \sqrt{P_1/V_1} \tag{9x}$$

The critical ratio for the flow of superheated and supersaturated steam becomes

$$\frac{P_{\text{critical}}}{P_1} = \left(\frac{2}{1.3 + 1}\right)^{1.3/0.3} = 0.545 \tag{13}$$

Diffuser and Ejector. The diffuser is in principle the reverse of a nozzle and is a device by which a lower pressure, high-velocity jet may deliver itself at a low velocity at and against a higher pressure. Referring to the Mollier diagram of Fig. 8, a fluid passing reversibly through a nozzle from pressure P_1 to the lower pressure P_2 would change state along the path ab, at constant entropy, decreasing enthalpy, increasing specific volume, and increasing velocity. A diffuser in which the action was wholly reversible would take the high-velocity fluid under the conditions existing at the nozzle outlet, b, and return it along state path ba to its original state and to a region of the higher original pressure P_1. All the state changes which occurred in the ideal nozzle would occur in the ideal diffuser in the reverse order.

The contour of the channel required by the diffuser is determined from exactly the same considerations which determine the contour of a nozzle, namely, consideration of the steady-flow energy equation and the continuity of flow principle. Parallel features of the two devices would be that, just as expansion in a nozzle to a lower pressure which exceeds about one-half of the higher calls for a converging section only, so a compression in a diffuser from a lower pressure in excess of one-half of the higher would require only a diverging section. Similarly, if the stream velocity to the diffuser were sufficient to enable discharge to a region the pressure of which exceeded twice that of the lower-pressure supply region, the diffuser channel would be initially convergent and then divergent from a throat. The general form of a diffuser for operation under the latter condition is shown in Fig. 9. Below the diffuser are depicted the decreasing velocity and specific volume and the increasing pressure and enthalpy as the fluid flows from the lower-pressure to the higher-pressure region.

In Fig. 8 there is also shown the character of the *actual* processes in a serial arrangement of nozzle and diffuser. The line *ac* represents the actual expansion in the nozzle from

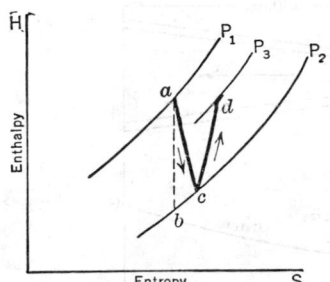

FIG. 8.   *H–S* Diagram of Diffuser
Process

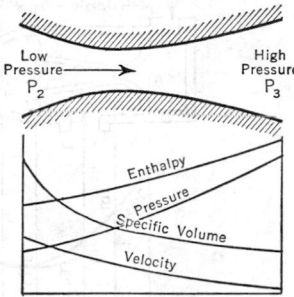

FIG. 9.   Form of Diffuser and Characteristic Curves

pressure P_1 to pressure P_2, and the line *cd* represents the recompression in the diffuser. The paths are both of increasing entropy and so, if in general the recompression may be one to a final enthalpy H_3 equal to the original enthalpy H_1, the final pressure P_3 must be less than the original pressure P_1 by an amount which depends on the degree of irreversibility of the two processes.

For the compression of a gas in a diffuser, if the stream were carried to a negligible exit velocity U_3, the pressure to and against which it might be delivered would depend on the pressure, temperature, and velocity of the supply stream. Thus, by a development of the energy equation,

$$\frac{U_2{}^2}{64.34} = \frac{k}{k-1} RT_2 \left[\left(\frac{P_3}{P_2} \right)^{(n-1)/n} - 1 \right]$$

or

$$P_{3,\,\text{max}} = P_2 \left[\frac{k-1}{kRT_2} \frac{U_2{}^2}{64.34} + 1 \right]^{n/(n-1)} \tag{14}$$

in which $P_{3,\,\text{max}}$ = maximum discharge pressure against which the stream could be delivered from diffuser after deceleration to a negligible velocity.

P_2 = pressure of the region of the supply stream.

U_2 = velocity of the supply stream, feet per second.

T_2 = supply region temperature, degrees Rankine.

$k = c_p/c_v = 1.40$ for air.

n = the effective P-V exponent of the more or less irreversible adiabatic compression process, the value of which exponent would equal k for ideal reversible compression, would exceed k for irreversible compression, and would be an index of the character of the diffuser performance.

Common applications of the diffuser are found in the downstream section of the Venturi meter and in centrifugal pumps and compressors. In the latter, high-velocity streams are created at the periphery of a rotating element and stationary diffusers are employed to utilize the kinetic energy of the streams for delivery to discharge regions of higher pressure.

The diffuser forms an essential component of the *ejector*, which is a device in which the kinetic energy of one fluid is used to pump another fluid from a region of lower to one of higher pressure. In the steam air ejector, Fig. 10, a high-velocity jet of steam is produced by the steam nozzle AC. Air drawn into the ejector from a region of low pressure D mixes with the jet, and the mixture is compressed in the diffuser EG and discharged into a region of higher pressure beyond G. The changes in pressure, velocity, enthalpy, and entropy occurring throughout the ejector are shown graphically in Fig. 11. The steam is supplied to the nozzle at a pressure P_1 considerably higher than the pressure P_3 at which the air is to be discharged. Along AC occurs the increase of velocity in the nozzle. Along DC' is a slight increase in velocity as the air is drawn into the ejector. Between C and some indefinite point E the steam and air mix at nearly constant pressure and with an increase in air velocity and decrease of steam velocity. This mixing is considered to occur by a wholly irreversible process described as inelastic impact, but with conservation of momentum of the streams of fluid. From E to G the mixture is compressed with decreasing velocity and increasing entropy and enthalpy. The diffuser requires a minimum section

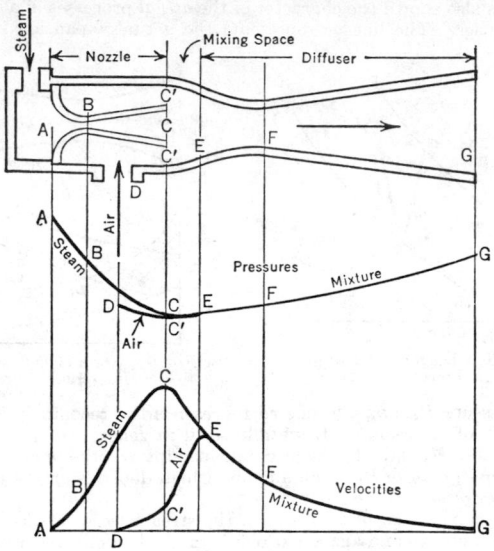

Fig. 10.   Ejector and Characteristics

at F if the absolute pressure at E is less than about half the pressure at G. Because of lack of knowledge of just what occurs during the mixing process and also of the efficiency of the compression process in the diffuser, no satisfactorily exact analysis of the complete ejector has been made. Design is largely by empirical methods. However, it is not difficult to compute the efficiency of an ejector from performance data.

Let

M'_s = mass of steam used in unit time.

M'_a = mass of air handled in unit time.

$M'_s(H_1 - H_3)_S$ = adiabatic decrease in enthalpy of steam from nozzle supply pressure P_1 to air discharge (not suction) pressure P_3.

M'_aW_a = adiabatic work of compression of air from air suction pressure P_2 to air discharge pressure P_3.

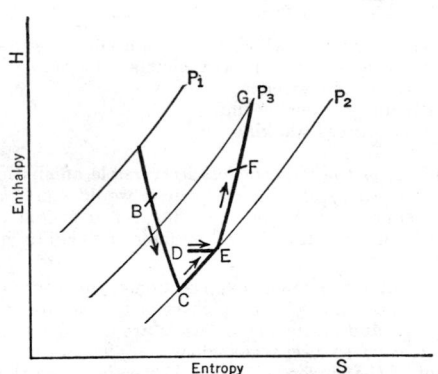

Fig. 11. H–S Diagram for Ejector Process

Then efficiency of ejector = work required to compress air adiabatically, divided by available energy supplied by steam above discharge pressure, or

$$\text{Ejector eff.} = \frac{M'_aW_a}{JM'_s(H_1 - H_3)_S} \tag{15}$$

The efficiency computed in this manner corresponds to the efficiency which should be used if the air were compressed in a steam-driven piston or centrifugal air pump directly by the use of mechanical work. The additional kinetic energy obtained in the steam jet by expansion from P_3 to P_2 is not charged to the ejector, since exactly that amount of energy is ideally required to compress the same steam back from P_2 to P_3 in the diffuser.

Example. A steam-air ejector is operated with dry saturated steam at a gage pressure of 100 lb per sq in. Air is discharged into the atmosphere from a vessel under a vacuum of 24.4 in Hg and at a temperature of 70 F. The barometer is 30.5 in. Hg. Three pounds of steam are used per pound of air discharged. Compute:

(a) Ideal velocity of discharge from steam nozzle.
(b) Ideal velocity of mixture of steam and air at entrance to diffuser.
(c) Efficiency of ejector.

Solution. (a) Atmospheric pressure, $P_3 = 15.0$ lb per sq in. abs.
Steam pressure, $P_1 = 115$ lb per sq. in. abs.
Air pressure in vessel, $P_2 = 3$ lb per sq in. abs.
Ideal enthalpy drop in entire nozzle, 115 to 3 lb $= 1189 - 945 = 244$ Btu.
Corresponding jet velocity $= 223.7\sqrt{244} = 3485$ ft per sec.

(b) Assume velocity of air at point c, just before mixing, to be 200 ft per sec. Applying principle of equal momenta before and after mixing $(3485 \times 3) + 200 =$ velocity of mixture $\times 4$. Ideal velocity of mixture $= 2664$ ft per sec.

If the mixing occurred without loss of kinetic energy the velocity of mixture would be 3030 ft per sec. As stated, little is known as to what actually occurs at this point.

(c) The work required to compress the air is computed on the basis of reversible adiabatic compression from a pressure of 3 lb per sq in. abs to a pressure of 15 lb per sq in., which gives 57,000 ft-lb per lb of air. The adiabatic enthalpy drop in the steam nozzle from 115 lb to 15 lb is 148.8 Btu per lb. This is the energy available to produce the compression of the *air* in the diffuser. The overall efficiency of the ejector is then:

$$\frac{\text{Ideal energy required for air compression}}{\text{Ideal energy available in steam nozzle}} = \frac{57,000}{148.8 \times 778 \times 3} = 0.164 = 16.4 \text{ per cent}$$

In steam condenser applications, the air withdrawn from the condenser is saturated with water vapor at the temperature of the air-vapor mixture. In computing the work of compression it is this mixture which must be considered. For high vacuum in condensers it becomes necessary to use two or three stages of ejectors in series, condensing the steam between stages to reduce the amount of steam which otherwise would need to be compressed in the diffusers.

Flow of a Vapor through an Orifice. An orifice will perform the function of only the convergent portion of a nozzle. Therefore, a vapor will expand in the actual zone of the orifice to any discharge region pressure which is greater than or equal to the critical pressure, but not to any lower pressure. Any further expansion must take place beyond the orifice.

It follows that we again encounter the two general conditions of:

(a) Flow with the discharge region pressure greater than the critical, when computation of the ideal rate of discharge may be based upon expansion in the orifice to the lower pressure, and

(b) Flow with the discharge region pressure equal to or less than the critical, when computation of the ideal rate of discharge must be based upon expansion in the orifice only to the critical pressure.

THE STEAM POWER PLANT

21. THE RANKINE CYCLE

Symbols

e = efficiency.
f = a subscript designating the saturated liquid state.
g = a subscript designating the dry saturated vapor state.
H = enthalpy, Btu per pound of fluid.
J = Joule's equivalent, 778 (ft-lb per Btu).
M = proportional mass of vapor passing to a particular heater in a regenerative cycle.
P = pressure (absolute), pound per square foot.
p = absolute pressure, pounds per square inch.
Q = energy transferred as heat, Btu per pound of fluid.
S = entropy per pound mass, and also a subscript designating constancy of entropy, during a process.
T = absolute temperature, degrees Rankine (= degrees Fahrenheit absolute).

t = temperature, degrees Fahrenheit.
U = linear velocity, feet per second.
V = specific volume, cubic feet per pound.
W = energy in transition as shaft work, foot-pounds per pound of fluid.

Two general methods are at present available for the production of power from the energy stored in the fuels. These are the internal-combustion engine and the type of composite device familiarly exemplified by the *steam power plant*. The thermodynamic principles which underlie the steam power plant cycles are presented here.

The Carnot Cycle is the classical example of a wholly reversible temperature-engine cycle in which energy that is being supplied at an elevated temperature is reduced to a lower temperature, in which process the maximum possible portion of that energy is con-

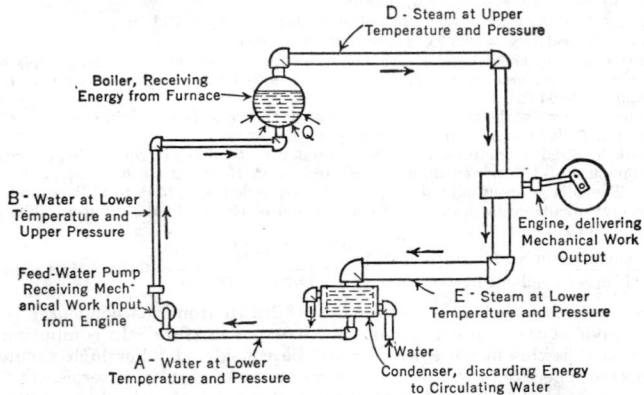

Fig. 1.   Elementary (Rankine Cycle) Steam Power Plant

verted into work. Also this cycle, notwithstanding its impracticability and unattainability, is in certain features the prototype of the various steam power plant cycles and in all cases it acts as an invaluable guide in the selection and development of those cycles. The Carnot cycle is described on p. 829.

The Rankine Cycle, of the actual steam plant cycles, is of concern not only because of its simplicity, practicability, and historical importance but also because the other, more complex cycles may in a sense be regarded as refinements of the Rankine. The ideal Rankine cycle consists in general of (a) the warming, constant temperature vaporization and, perhaps, further superheating of the working fluid of the cycle at a constant upper

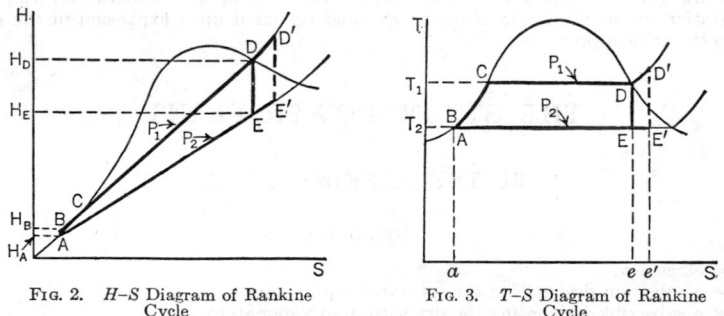

Fig. 2. H–S Diagram of Rankine
Cycle

Fig. 3. T–S Diagram of Rankine
Cycle

pressure, this through the addition of energy from the source whereby the fluid is brought from the lower to the upper temperature limit of the cycle; (b) the isentropic expansion of the fluid to the lower temperature and pressure limits of the cycle; (c) the complete condensation of the fluid by the emission of energy at the lower temperature and pressure, this energy departure constituting the rejection of the unavailable energy to the receiver; and (d) the return of the condensed fluid to the upper pressure by reversible adiabatic compression, whereupon the fluid cycle may be repeated. These processes take place successively in the boiler and superheater, the engine, the condenser, and the feed pump.

To facilitate the description and analysis of the cycle, reference will be made to Fig. 1, which is a diagrammatic representation of the arrangement of the essential pieces of apparatus in which the processes are carried out, and to Figs. 2, 3, 4, in which the state changes which occur progressively in the cycle are depicted respectively on H–S, T–S, and P–V coordinates.

There is selected arbitrarily as a starting point in the cycle the point and state A at which the water exists as saturated liquid at the lowest pressure and temperature of the cycle. This state occurs in the suction line of the *pump* which acts to raise the pressure of the water from this lowest pressure to the highest pressure of the cycle.

From this initial state, the fluid passes progressively through the following series of processes:

(1) The pressure of the water is raised by a reversible adiabatic compression in the *pump* from the lower pressure P_2 to the upper pressure P_1, causing a change from the saturated liquid state A to a subcooled liquid state B. The compression is not accompanied by any appreciable change of temperature, internal energy, specific volume, or entropy.

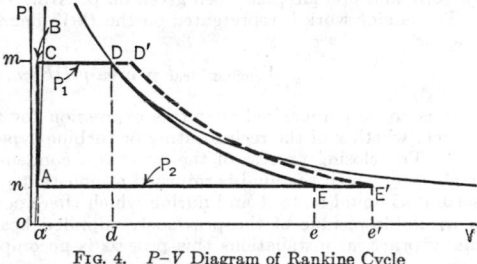

Fig. 4. P–V Diagram of Rankine Cycle

The energy required for the pumping process is nearly:

$$W_{\text{pump}} = (P_1 - P_2)V_{f,2}$$

where $V_{f,2}$ = specific volume of saturated liquid of temperature T_2. In Fig. 4 the product $(P_1 - P_2)V_{f,2}$ is represented by the area $mBAnm$, which area thus represents the pump work.

(2) The second distinctive process of the Rankine cycle is the warming, vaporizing, and perhaps superheating of the fluid, all occurring under a virtually constant pressure. The energy supply for accomplishing these several phases of the cycle is provided by heat conduction through the heating surfaces of the boiler and the superheater (if the latter is supplied) and is furnished directly or indirectly from the primary high-temperature energy source, namely, the hot products of combustion of the fuel in the furnace.

In the feedwater heating portion of the process, the water is warmed at P_1, from temperature T_2 up to the saturation temperature which corresponds to that pressure, whence the state of the liquid after completion of the heating is represented by the saturated liquid state point C. If the boiler is assumed to accomplish a complete vaporization of the water, the state of the fluid leaving the boiler is represented by the dry saturated vapor state point D. If in addition a superheater is supplied the fluid leaves that device at the superheated state point D', which point must lie on the P_1 line in the superheat region.

The heat required for this complete process of energy supply as heat is

$$Q_{\text{supplied}} = H_{D \text{ (or } D')} - H_B, \text{ Btu per lb}$$

Referring to the figures, the total energy supply as heat is measured directly on the H–S diagram by the vertical distance of point D (or D') above B and on the T–S diagram by the area $aBCDea$ (or $aBCDD'e'a$). The inherently unavailable portion of the heat energy supply is represented by the area $aAEea$ (or $aAE'e'a$) on the T–S diagram.

(3) The next state change in the ideal Rankine cycle is the reversible adiabatic expansion of the fluid from the upper temperature and pressure to the lower. To accomplish this the fluid is delivered to the *engine* at constant pressure and expands therein, after which it departs from the engine at a constant lower pressure. The term engine is here used in a generic sense to designate the main prime mover of the plant, whether that be actually a reciprocating engine or turbine engine. It is in this composite delivery, expansion, and rejection process which takes place in the engine that the available portion of the previously supplied heat energy is transformed to shaft work.

Because the property characteristic of a reversible adiabatic state change is the constancy of the entropy, the expansion through the ideal engine is represented on the H–S and T–S diagram by the vertical isentropic line DE (or $D'E'$). The line DE in the P–V diagram portrays the typical appearance of this reversible adiabatic state change on those coordinates.

The work delivered by the engine is

$$W_{engine} = J(H_D - H_E)[\text{or } J(H_{D'} - H_{E'})] \text{ ft-lb per lb}$$
$$= J(H_1 - H_2)_S$$

where the subscript S indicates constant entropy.

The work output of the ideal engine is measured on the H–S diagram by the vertical distance of point D above E. The method of computing the enthalpy at E, given the entropy and pressure, has been given on p. 873.

The engine work is represented on the P–V diagram by the area back of the expansion line, or

$$W_{engine, \, ideal} = \text{area } mDEnm, \text{ or } = +\int_1^2 V dP$$

It is to be emphasized that this expression for the engine work applies to any ideal engine, whether of the reciprocating or turbine type.

(4) The closing process of the cycle is a condensation of the exhaust steam from the engine, by which the fluid is returned at constant pressure from the state E to the initial saturated liquid state A and during which there occurs the discarding of the inescapably unavailable residue of the previously supplied heat energy. In the ideal cycle and in many practical installations this process is accomplished during passage through a *condenser* in which the energy departure and the condensation are effected by energy transfer as heat conducted from the steam through the condenser structure to circulating water. The same state change and energy rejection may be accomplished, however, in various other practical manners, as by condensation in the atmosphere, in a heating system, or in any other process in which the low temperature energy of the exhaust steam may be employed to advantage.

The heat rejected in the condensation process is

$$Q_{condenser} = H_E - H_A \text{ (or } H_{E'} - H_A), \text{ Btu per lb}$$

On the H–S diagram the vertical distance between state points E and A measures the energy rejected in the condensation.

On the T–S diagram this energy discard as heat $= T_2(S_A - S_E)$ is represented by the area $aAEea$ (or $aAE'e'a$). Therefore, as the area $aBCDea$ measured the amount of energy supplied from the source as heat, the difference between these two areas, or the area $ABCDEA$, measures the net amount of work obtainable from the operation of the cycle, that is, the net amount after debiting the engine output with the portion of its work which must be employed to drive the pump.

The cycle is thus completed.

22. MEASURES OF EFFICIENCY AND PERFORMANCE

The Thermal Efficiency of a Heat Engine Cycle, or of a heat engine alone, is defined as the ratio between the amount of work output and the amount of heat energy which must be supplied in order to accomplish that output, or

$$\text{Thermal efficiency } (e_t) = \frac{\text{Work output from system}}{\text{Heat supplied to system}}$$

The thermal efficiency of the Rankine cycle $= \dfrac{(H_1 - H_2) - (P_1 - P_2)V_{f,2}/J}{(H_1 - H_{f,2}) - (P_1 - P_2)V_{f,2}/J}.$ This is a very nearly exact value. An approximation is

$$\frac{(H_1 - H_2)_S}{H_1 - H_{f,2}}$$

where	$H_1 = $ enthalpy of the steam as supplied at P_1.
$(H_1 - H_2)_S = $ decrease of enthalpy by isentropic expansion to P_2.
$H_{f,2} = $ enthalpy of saturated water at P_2.
$V_{f,2} = $ specific volume of saturated water at P_2.

The error arising from use of the approximation is usually regarded as negligible except when the modern higher boiler pressures (500 lb per sq in. and up) are employed.

In order to measure the effectiveness with which a cycle might ideally utilize the temperature range encountered in the cycle and thus to designate the degree to which the ideal cycle efficiency approaches the Carnot efficiency for that temperature range there is employed a figure of merit which is known as the *type efficiency* of the cycle. This is defined as the ratio between the ideal thermal efficiency of a cycle and the thermal efficiency of a Carnot cycle operating between the extremes of temperature found in the cycle under consideration, or

$$\text{Type efficiency} = \frac{\text{Ideal thermal efficiency of cycle}}{(T_1 - T_2)/T_1}$$

No actual engine can attain a performance equal to that of an ideal engine operating on the same cycle, because of the numerous irreversible actions which will inescapably occur in any real process. Heat will be radiated or otherwise dissipated from higher to lower temperature levels without full realization of its available energy; and mechanical irreversibility will occur owing to fluid friction, throttling and turbulence, and to the friction of moving parts of the engine mechanism, etc. The result must be less actual work output per pound of fluid than that ideally obtainable in the ideal cycle upon which the engine operation is based.

In order to evaluate the relative effectiveness of an actual engine and the degree to which it approaches ideal performance there is employed a measure of performance which is known as the *engine efficiency* * and which is defined as the ratio between the output actually obtained per pound of steam supplied to the engine and the output ideally obtainable upon isentropic expansion of that steam, or

$$\text{Engine efficiency} = \frac{\text{Actual output per pound of steam supplied to engine}}{\text{Work ideally obtainable by isentropic expansion}}$$

$$= \frac{\text{Actual efficiency}}{\text{Ideal efficiency}}$$

The actual output of an engine is necessarily determined by test. Such a test requires a determination of the actual power output of the engine and the mass of steam supplied per unit time.

The power delivered to the piston face of a reciprocating engine may be measured by the engine indicator, the shaft horsepower output of any engine may likewise be measured by a suitable dynamometer, or for a direct-connected engine-generator set the electrical power output may be determined by proper electrical instruments. The mass-rate of steam supply may be determined by condensing and weighing the steam or by various types of steam flow-meters or calibrated orifices. The state of the steam is determined by its absolute pressure and its quality (ascertained by means of a calorimeter) or its temperature, if superheated. The various pressures are found by gages or manometers.

By reason of the variety of manners and positions which may be selected in designating the output of an engine a like variety of engine efficiencies may be assigned to a given machine. Each output mentioned above is successively less than the preceding one by reason of the progressive energy dissipation through mechanical and electrical losses. It is obvious that due attention must be given to this condition in quoting values of engine efficiency or in interpreting quoted values.

If adequate test data are available on the actual steam requirements and the power output of an engine, these data may be combined with the factors 2545 (Btu per horsepower-hour) or 3413 (Btu per kilowatthour), and the numerator of the efficiency equation may thereby be evaluated. Thus

Actual output per pound of steam supplied, Btu

$$= \frac{\text{Power output, hp (or kw)} \times 2545 \text{ (or 3413)}}{\text{Steam supplied, pounds per hour}}$$

The denominator of the efficiency equation, that is, the output ideally obtainable per pound of steam supplied, may be computed for an engine according to the conditions of the particular type of cycle upon which the system operates. For the Rankine cycle the work ideally obtainable per pound of steam is $(H_1 - H_2)_S$. This is called the *adiabatic enthalpy drop*.

The ratio between the steam supplied per hour and the power output is known as the *steam rate* of an engine and is commonly expressed in terms of the pounds of steam supplied per horsepower-hour (or kilowatthour) or

$$\text{Actual steam rate} = \frac{\text{Pounds of steam supplied per hour}}{\text{Output in hp (or kw)}}$$

---

* Unfortunately various terms have been used to designate this efficiency: "efficiency ratio," "relative efficiency," "Rankine efficiency," and "turbine efficiency." *Engine efficiency* is the term employed in the Power Test Codes of the American Society of Mechanical Engineers.

The engine efficiency may be expressed concretely in the form:

$$\text{Engine efficiency} = \frac{2545/\text{Actual steam rate, lb per hp-hr}}{\text{Output ideally obtainable, Btu per lb}}$$

$$= \frac{3413/\text{Actual steam rate, lb per kwhr}}{\text{Output ideally obtainable, Btu per lb}}$$

The Actual Thermal Efficiency is a figure which in its basic significance is quite parallel to the ideal thermal efficiency of a cycle or an engine, which is broadly defined as the fraction $\dfrac{\text{Work output from system}}{\text{Heat supplied to system}}$. However, in evaluating this actual efficiency for an operating engine or engine-generator unit or for a complete operating plant the useful output may in practice be variously regarded as the indicated power output of a reciprocating engine, or as the shaft power output of any engine, or as the electrical power output of an engine-generator unit, or as the net electrical output at the bus-bars of a complete plant. Also in ascribing a value to the energy supply, that quantity may be taken variously as the energy supplied as heat (at the boiler), that supplied as chemical energy in the fuel, etc. It is thus apparent that a considerable variety of interpretations may be applied to the term without need for confusion if due care is employed in making the interpretation. In any individual case, and irrespective of its particular interpretation, the thermal efficiency may invariably be regarded as a ratio between the useful output delivered and the energy supplied for the procuring of that output, or

$$\text{Actual thermal efficiency} = \frac{\text{Useful output from system}}{\text{Energy supplied to system}}$$

In utilization of actual test data, it is commonly more convenient to express it in terms of a specific output or supply, that is, an output (or supply) per unit time, per unit mass of steam supplied to an engine or per unit mass of fuel supplied to a plant. For example, the specific output per pound of steam to an engine is expressed by the relation

$$\text{Actual output, Btu per lb of steam to engine} = \frac{2545 \text{ (or 3413)}}{\text{Actual steam rate, lb per hp-hr (or kwhr)}}$$

A parallel relation for the complete plant, in terms of the fuel rate, is

$$\text{Actual output, Btu per lb of fuel to plant} = \frac{2545 \text{ (or 3413)}}{\text{Actual fuel rate, lb per hp-hr (or kwhr)}}$$

Either of these relations may be employed directly as the numerator of the efficiency fraction of the above equation.

Several distinctive practices are followed in interpreting the denominator of the efficiency fraction in the various circumstances in which efficiency is to be determined. For the simple Rankine cycle it is the established practice to regard the energy supply which is assignable to each pound of steam and is chargeable against the engine as the enthalpy difference between that of saturated liquid at the temperature of the engine exhaust and that of the steam as supplied to the engine. It is to be observed that for this case no concern is given to the actual temperature at which the condensate may be leaving the condenser; this is in order to avoid penalizing the engine for any subcooling of the condensate which may occur by reason of unfavorable characteristics of the condenser. For the engine of the Rankine cycle, the efficiency thus becomes

$$\text{Actual thermal eff. of engine} = \frac{2545 \text{ (or 3413)}/\text{Steam rate}}{H_1 - H_{f,2}}$$

For the entire power plant it is the established practice to regard the chemical energy released by a complete combustion of the fuel as the energy supply. That energy when evaluated per pound of fuel is known as the calorific value or heating value of the fuel. It is determined by calorimetric measurement made upon representative samples of the fuel. Introducing this quantity in the efficiency fraction,

$$\text{Actual thermal eff., complete plant} = \frac{2545 \text{ (or 3413)}/\text{Fuel rate}}{\text{Heating value of fuel, Btu per lb}}$$

The *heat rate of an engine or of a plant* is defined as the amount of energy actually to be supplied per horsepower-hour (or per kilowatthour) of useful output. This may readily be expressed in terms of the steam rate, or the fuel rate, and the energy supply per pound of steam or of fuel, or

Actual heat rate, Btu per hp-hr (or kwhr)

$$= \text{Steam rate of engine} \times \text{Energy supplied per lb of steam}$$

or

$$= \text{Fuel rate of plant} \times \text{Heating value, Btu per lb of fuel}$$

Example 1. A Rankine cycle type of steam plant is operating with a boiler pressure of 300 lb per sq in. abs and a 20 lb per sq in. abs back pressure. There is no superheater, and the steam delivered by the boiler has a 99 per cent quality. Compute the properties of the steam or water at each significant point of the ideal cycle, the ideal feed pump work (in foot-pounds and in Btu per pound of steam), the requisite heat supply, and the ideal minimum energy discard per pound of steam.

How many pounds of steam would be required in the ideal Rankine plant per net horsepower-hour output? Compute the Rankine cycle thermal efficiency and the type efficiency.

Solution.

Water leaving condenser:

$$P_A = P_2 = 20 \times 144; \quad t = 227.96\,F; \quad V_A = V_{f,2} = 0.01683$$

$$H_A = H_{f,2} = 196.16\,\text{Btu/lb}; \quad S_A = S_{f,2} = 0.3356$$

Water leaving pumps:

$$P_B = P_1 = 300 \times 144; \quad V_B = V_{f,2} = 0.01683; \quad t = 227.96\,F$$

$$H_B = 196.16 + \frac{(300 - 20)144}{778} \times 0.01683 = 196.97; \quad S_B = S_A = 0.3356$$

Steam leaving boiler:

$$P_D = 300 \times 144; \quad V_D = 0.01890 + 0.99(1.5433) = 1.5470; \quad t_D = 417.33\,F$$

$$H_D = 393.84 + 0.99(809.0) - 1193.8$$

$$S_D = 0.5879 + 0.99(0.9225) = 1.5019$$

Steam leaving engine:

$$P_E = 20 \times 144; \quad t_D = 227.96; \quad X_E = \frac{1.5010 - 0.3356}{1.3962} = 0.835; \quad S_E = S_D = 1.5019$$

$$H_E = 196.16 + 0.835(960.1) = 998.16$$

$$V_E = 0.01683 + 0.835(20.072) = 16.767$$

Pump work:

$$W_P = 144(0.01683)(300 - 20) = 680\,\text{ft-lb} = 0.874\,\text{Btu/lb}$$

Engine work:

$$W_E = 778[-393.84 + 0.99(809.0) + 196.16 + 0.835(960.1)] = -152,200\,\text{ft-lb} = -195.7\,\text{Btu/lb}$$

Net work output:

$$W_E - W_P = -152,200 + 680 = -151,520\,\text{ft-lb}; \quad -194.8\,\text{Btu/lb}$$

Energy supplied as heat:

$$[393.84 + 0.99(809.0)] - \left[196.1 + \left(\frac{300 - 20}{778}\right)144 \times 0.01683\right] = 996.87\,\text{Btu/lb}$$

Energy discard:

$$-[196.16 + 0.835(960.1)] + 196.1 = -802.06\,\text{Btu/lb}$$

Steam ideally required per horsepower-hour of plant output:

$$\frac{33,000 \times 60}{W_{\text{net}}} = \frac{1,980,000}{151,520} = 13.1 \quad \text{or} \quad \frac{2545}{\dfrac{W_{\text{net}}}{778}} = 13.1\,\frac{\text{lb}}{\text{hp-hr}}$$

Rankine cycle efficiency:

$$\eta_R = \frac{-194.8}{996.87 - 0.874} = 0.196; \ 19.6\%$$

Carnot cycle thermal efficiency:

$$\eta_C = \frac{417.33 - 227.96}{(417.33 + 460)} = 0.216; \ 21.6\%$$

Type efficiency:

$$\eta_T = \frac{0.196}{0.216} = 0.908; \ 90.8\%$$

Example 2. The engine of the cycle of Example 1 develops 300 hp and uses 6000 lb of steam per hour. Compute actual thermal efficiency, heat rate, and engine efficiency.

Solution.

Steam rate:

$$6000/30 = 20\,\text{lb/hp-hr}$$

Actual thermal efficiency:

$$\eta T_a = \frac{2545}{20(996.87)} = 0.128; 12.8\%$$

Heat rate:

$$20(996.87) = 19,950 \text{ Btu/hp-hr}$$

Engine efficiency:

$$\eta E = \frac{0.128}{0.196} = 0.653; 65.3\%$$

$$= \frac{2545}{20(194.8)} = 0.653; 65.3\%$$

23. THE ENERGY ("HEAT") BALANCE

In the thermodynamic design of a steam power plant and in the analysis of its actual operation attention is very commonly given to what is known as the *"heat" balance* of the plant. By this is meant, more exactly, a comprehensive energy analysis and energy accounting for the purpose of ascertaining the ultimate destination of all portions of the original energy supply in the fuel and also for the more important purpose of correcting in so far as may be practicable any wastage of available energy.

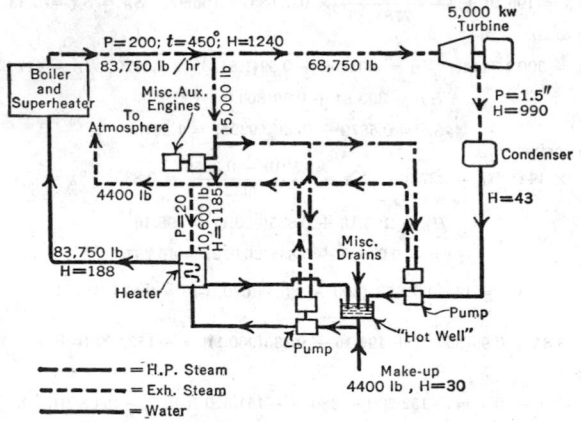

FIG. 5.   Diagram for First Law Energy Balance

Two viewpoints are available in the development of such an energy analysis. On the simpler but less valuable basis concern is given only to the distribution of the energy with regard to its physical location and destination, ascribing losses only as energy may obviously be dissipated by radiation to the atmosphere from hot pipes or surfaces, in the hot stack-gases from the furnace, by conduction to the circulating water, or by mechanical friction in the moving parts of machines. Because the development of such an accounting is dependent only on the conservation of energy principle it may be designated as a *first law energy balance.*

The evolution of a first law balance must proceed from specific information concerning the type, arrangement, and interconnections of all steam-using equipment which is installed in the plant and from concrete data or estimates concerning operating conditions and operating efficiencies obtaining in the numerous individual units which comprise the plant. From such information a detailed bookkeeping account may be developed which shall record specifically the amount and character of all energy entering or departing from each device and shall permit a summarization portraying the direction, magnitude, and destination of all component parts of the energy stream en route through the plant.

Such a summarization of the energy account may, as usual, be presented to advantage graphically. Either one of two general types of procedure is commonly employed in the graphic representation. One is illustrated in Fig. 5, in which are depicted diagrammatically the primary components of a simple steam power plant, the interconnecting pipes joining the several devices, the direction of flow between them, the amount of fluid flowing per hour, and pertinent properties of the fluid en route. In the figure there are intentionally presented a plant character and performance which are *not* representative of better modern practice, in order that certain unfortunate features of the plant may the better be recognized. It is observed that the plant is so devised that the various *auxiliary* engines exhaust at about atmospheric pressure and that the exhaust steam from those engines is condensed as far as possible in a heater in which some degree of feedwater heating is thus accomplished.

The second method of presentation of the first law balance appears in Fig. 6. In that figure the character of the interconnections between parts or the fluid properties at specific points are not emphasized but instead there is shown graphically the relative magnitude of the energy quantities at each significant stage in the energy stream. The energy magnitudes are in general expressed in terms of the total enthalpies of the mass of fluid passing a given section in a unit time. The enthalpies are those with respect to a relative zero of enthalpy for water at 32 F, and in that feature the diagram may be to an extent mis-

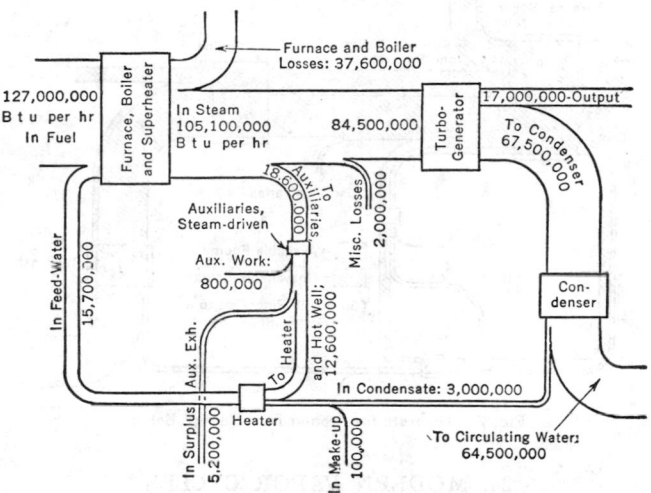

FIG. 6. Diagram for First Law Energy Balance

leading as the fluid does not and practically could not attain that temperature at any point in the cycle. This procedure is followed because it is the conventional and more convenient one. The figure portrays with maximum clarity the general directions of the energy stream through the plant, the manner of distribution of the energy en route, and the unfortunately small proportion of the initial energy from the fuel which actually is made available as work output.

It is to be observed that in Fig. 5 the combined steam exhaust from the various auxiliary engines is shown to be in excess of that which can practicably be condensed in the feedwater heater, whereby there exists a surplus of auxiliary exhaust which must be rejected to the atmosphere. This condition would obviously be an inefficient one. To avoid its existence, as well as to avoid the opposite circumstance of an exhaust steam supply which is insufficient for adequate feedwater heating, is one frequent objective of the development of an energy balance in the design of a plant. Also, in recognition of the necessity of assuring a proper balance between the amount of **a**uxiliary exhaust steam and the amount which can profitably be employed for the purpose of feedwater heating, the notion of energy balancing and the terms *heat-balance design, heat-balance arrangement, heat-balance diagram*, etc., are frequently employed in a somewhat limited sense referring only to the arrangement and character of the auxiliary machinery as that influences the proper balance between the exhaust steam supply and the possibility for its utilization.

The maintenance of such a suitable balance under all normal operating conditions in the large and complex power plant becomes a very nice problem indeed and one offering many varieties of solution. It becomes still more important that the designer adequately recognize the requirement of avoiding to the maximum practicable limit the *dissipation of available energy to unavailable* by reason of any serious degree of irreversibility in any of the numerous energy transition processes which take place in the plant. A valuable device for the detection and avoidance of such irreversibilities is what is known as a *second law energy balance*.

The general procedure in the development of a second law balance is first that of determining the energy quantities throughout the energy stream, quite as in the first law balance, and then the separation of the energy passing each significant point in the stream into its inherently unavailable and ideally available portions. The results of such a subdivision of the energy stream of Fig. 6 are presented in Fig. 7.

The features which are perhaps of more outstanding significance in the figure are (a) the pronounced dissipation of available energy which occurs in the energy transition from the high-temperature furnace region to the relatively moderate-temperature steam and (b) the quite appreciable dissipation of available energy and consequent loss of overall plant efficiency which result from inefficient auxiliary machinery and thermally irreversible heat transmission occurring in the feedwater heating equipment.

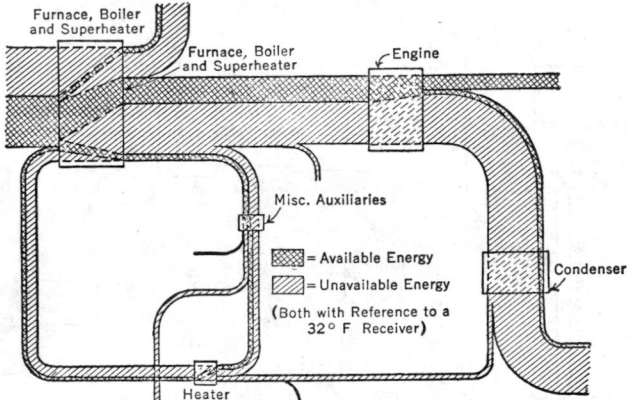

FIG. 7.   Diagram for Second Law Energy Balance

24. MODERN VAPOR CYCLES *

From consideration of the Carnot cycle and its efficiency fraction, $(T_1 - T_2)/T_1$, it is evident that the efficiency of a temperature-engine cycle may be increased by increase of the temperature at which the fluid receives its energy from the source, or by decrease of the temperature to which the fluid expands reversibly in the engine and at which therefore it discards the unavailable residue of energy to the receiver. In a vapor cycle such as the Rankine the temperature of energy rejection is established primarily and practically by that of the atmosphere or of the natural supplies of water in the streams, lakes, or seas. The attainment of suitable cyclic efficiency requires that the fluid shall expand as closely as practicable to this temperature. With water as the working fluid, this requires that a very low pressure (high vacuum) be maintained in the condenser. In order to gain efficiency by increase of the temperature of energy reception in the Rankine vapor cycle the temperature of vaporization may be raised by increase of vapor pressure or the average temperature of energy reception may be raised moderately by superheating after vaporization is completed.

For a like temperature range in each, the Rankine cycle may not attain the thermal efficiency of a wholly reversible cycle such as the Carnot. For a cycle operating without superheat this is in one sense due to the lower average temperature of the liquid during warming from the condensing to the vaporizing temperature. From an alternative and perhaps more fundamental viewpoint the reduced efficiency is due to thermal irreversibility associated with the liquid-warming phase of the cycle. This irreversibility may be reduced by the use of *stage* or *regenerative* feed-heating, and the cyclic efficiency may thereby be sensibly bettered. The improvement of efficiency resulting from superheating may be doubly realized by the expedient of *resuperheating* the vapor after partial expansion through the engine. The benefits of high vaporization temperatures may be obtained without the disadvantage of high attendant pressures by the use of a fluid of lower vapor pressure than water. All these expedients are employed in modern power plant practice.

---

* For a more complete discussion of this subject see Barnard, Ellenwood and Hirshfeld, *Heat Power Engineering*, Part I, Chap. XVIII, John Wiley & Sons, 1926.

The **Regenerative Feed-heating Cycle** employs a method of progressive liquid warming in a series of heaters which are activated by steam extracted from the engine after more or less complete expansion therein. Thermal irreversibility is thus reduced by accomplishing the first portion of the warming with lower-temperature steam the available energy of which has been well realized in the engine, by similarly accomplishing the next step in the warming of the liquid by medium-temperature steam, and so on. This particular method of efficiency betterment by reduction of thermal irreversibility in the feed-heating is simply illustrative of the gains invariably to be secured by the reduction of any irreversible features. The cycle is best analyzed in detail by writing and solving the energy equation for each successive heater in order from the highest-pressure heater down. Thereby the relative amount of steam to be extracted for each heater may be computed, whereupon the available energy output obtainable in each successive portion of the engine may be determined.

The **Reheating Cycle** may effect an ideal efficiency betterment only if the resuperheating of the partially expanded steam is accomplished by energy supplied directly from the

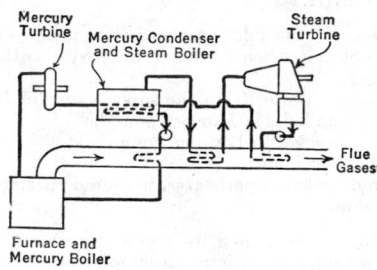

Mercury Turbine Mercury Condenser and Steam Boiler Steam Turbine

Flue Gases'

Furnace and Mercury Boiler

FIG. 8. Binary-fluid (Mercury-water) Cycle

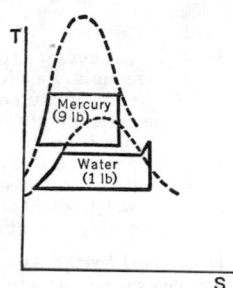

T

Mercury (9 lb)

Water (1 lb)

S

FIG. 9. T–S Diagram for Mercury-water Cycle

primary source (the hot furnace gases) and if the resuperheating serves to raise the average temperature of energy reception by the fluid from that source. The analysis of the cycle is quite like that of the Rankine cycle except in the necessity for recognizing the two-stage character of the expansion in the engine and for properly accounting the additional energy supplied to the fluid in the resuperheater. The cycle lends itself well to the addition of regenerative feed-heating.

·Binary-fluid Cycles are employed for the purpose of permitting high average temperatures of energy reception from the source without the handicap of difficultly high fluid pressures in the higher-temperature portion of the cycle. This is accomplished by employing in that portion of the cycle a fluid which has favorable vapor pressure-temperature characteristics. Such advantageous pressure-temperature characteristics may be utilized by employing a fluid of relatively lower vapor pressure in a higher-temperature section of the composite cycle and a fluid of higher vapor pressure in a lower-temperature section. A compound cycle of this character which employs the two fluids mercury and water is in successful use at this writing. The general scheme of arrangement of the apparatus of this binary-fluid cycle is shown in Fig. 8. Observe that the energy rejected from the condensing mercury forms the energy source for the vaporizing water. In Fig. 9 there is shown a T–S chart for 9 lb of mercury and for 1 lb of water. These are about the requisite proportions between the rates of mercury and water circulation in the cycle, because the relative latent heats are such that the heat delivered in condensing 1 lb of mercury after expansion through the mercury turbine is about one-ninth of the heat required to evaporate 1 lb of water.

25. THE STEAM GENERATING UNIT *

Steam Generating Unit is the modern name used to describe the combined equipment required in the steam power cycle for the addition of heat to the working fluid, water. This consists not only of a furnace and boiler but, in the more modern plants, may include some or all of other heat-absorbing elements such as a superheater, economizer, air preheater, steam reheater, and air- or water-cooled furnace walls.

---

* This section is reprinted in a modified form by permission from Barnard, Ellenwood, and Hirshfeld, Heat Power Engineering, Part II, John Wiley & Sons, 1935.

The Steam-generating Capacity of the unit may be expressed either in terms of

(1) The maximum rate of heat absorption by the unit, expressed either in *kilo Btu* (kB) *per hour*, or in *mega Btu* (mB) *per hour* (kB = 1000 Btu, mB = 1000 kB = 1,000,000 Btu), or

(2) The maximum rate of steam formation, in *pounds per hour*, or multiples thereof, together with a statement of the pressure and temperature of the steam and of the feedwater.

The Efficiency of a Steam-generating Unit is the fraction that shows what portion of the heat supplied by the fuel used in a definite time has been absorbed by the water and steam passing through the unit in the same time. When a true measure of the performance of a unit is required, the energy consumption of its auxiliaries must be deducted from the gross output. The ratio of the resulting net output of the unit to the energy supplied in the form of fuel is the true efficiency of the steam generating unit.

$$\text{Eff.} = \frac{w_n(h_2 - h_1)}{w_f(\text{H.V.})}$$

in which h_1 = the enthalpy of the feedwater at the inlet to the first heating element of the unit (e.g., to the economizer when installed, otherwise to the boiler itself), in Btu per pound.

$\quad h_2$ = the enthalpy of the steam at the outlet from the superheater (or from the boiler if no superheater is installed), in Btu per pound.

$\quad$ H.V. = the heating value of the fuel, *as fired*, in Btu per pound.

$\quad w_f$ = weight of fuel fired, in pounds per hour.

$\quad w_n$ = net weight of steam delivered by the superheater, in pounds per hour.

$\qquad$ = gross weight − auxiliary steam.

The Energy Used by the Auxiliaries may be consumed by steam jets, steam engines, turbines, or electric motors needed to operate the various pieces of equipment which comprise the generating unit. The electric energy consumed should be converted into its steam equivalent. The total steam equivalent of all auxiliaries should then be deducted from the total weight of steam produced in order to determine the net energy output of the generating unit. The weight of auxiliary steam equivalent to the electrical energy consumption of the auxiliaries, in pounds per hour, is

$$w_s = (F_n)(1 - E_a) \begin{pmatrix} \text{Rate of} \\ \text{firing,} \\ \text{tons} \\ \text{per} \\ \text{hour} \end{pmatrix} \begin{pmatrix} \text{Auxiliary energy} \\ \text{of the steam-} \\ \text{generating unit,} \\ \text{kilowatthours} \\ \text{per ton of fuel} \end{pmatrix} \begin{pmatrix} \text{Actual} \\ \text{evaporation,} \\ \text{pounds of} \\ \text{steam per} \\ \text{pound of fuel} \end{pmatrix}$$

or

$$w_s = (F_n)(1 - E_a) \begin{pmatrix} \text{Auxiliary power of} \\ \text{the steam-generating} \\ \text{unit, in kilowatts} \end{pmatrix} \begin{pmatrix} \text{Actual} \\ \text{evaporation,} \\ \text{pounds of} \\ \text{steam per} \\ \text{pound of fuel} \end{pmatrix},$$

in which F_n = net fuel rate of the station, in pounds of fuel fired per kilowatthours sent out from the station.

$\quad E_a$ = fraction of the main generator gross output required to operate all the station auxiliaries, as determined by test.

The Losses of Energy that Occur in a Steam-generating Unit and thereby affect its efficiency are the following:

(1) *Loss from the incomplete combustion of the fuel* as evidenced by the presence of:
$\quad$ (a) combustible in the refuse collected in the ash pit;
$\quad$ (b) combustible, such as solid carbon, carbon monoxide, hydrogen, and hydrocarbons, in the exit gas.

(2) *Loss to the stack not caused by incomplete combustion but due to:*
$\quad$ (a) the dry exit gases leaving the last heating surface at a higher temperature than that of the atmosphere;
$\quad$ (b) the water vapor, which accompanies the dry gas, leaving the last heating surface at a higher temperature than that of the air and fuel from which it came.

(3) *Loss by heat transfer from the entire unit to the surrounding air.*

The magnitudes of these losses are variables which depend upon the type and quality of fuel used, size and type of equipment, load on the unit, and the manner in which the plant is operated.

The Boiler Energy Balance

A study of the magnitudes of the losses is of importance to the operating engineer. It serves as a basis for comparison of the performance of different steam-generating units. Any tabulation prepared with the object of evaluating the various energy items of a plant constitutes an energy balance. Such a tabulation applied to a steam-generating unit is usually called a boiler energy balance. The items generally found in such a statement are:

1. Energy absorbed by steam in the boiler.
2. Energy carried away by the dry flue gas.
3. Energy loss due to unburned combustible in refuse.
4. Energy loss due to incomplete combustion of C to CO instead of to CO_2.
5. Energy carried away by surface moisture in the fuel.
6. Energy carried away by moisture resulting from combustion of hydrogen.
7. Energy carried away by moisture in the air used for combustion.
8. Radiation and unaccounted-for losses.

Convenient methods for determining each of these items will be outlined briefly in the succeeding paragraphs.*

1. Energy Absorbed by Steam in Boiler.

$$Q = (H_2 - H_1)W \tag{1}$$

where H_2 = enthalpy of steam at exit, Btu per pound.
H_1 = enthalpy of water at entrance, Btu per pound.
Q = heat absorbed by steam, Btu per pound of fuel.
W = pounds of steam per pound of fuel as fired.

2. Energy Carried Away by Dry Flue Gas. The weight of dry flue gas per pound of fuel may be found in the following manner: The gases found in the analysis of a sample of the dry products of combustion of a fuel are CO_2, CO, O_2 and N_2. These constituents are determined as volumetric percentages, or the analysis gives the volume of each constituent per 100 volumes of mixture.

Therefore since a molecular weight in pounds of any gas occupies approximately 358.83 cu ft at 32 F and atmospheric pressure, the relative weight of the percentage volume of each gas in the mixture will be its percentage volume multiplied by *molecular weight in pounds*.

If the carbon balance is used as a basis of calculation, it is assumed that all the carbon appearing in the flue gas in the form of CO_2 and CO has come from the carbon in the fuel *that was burned*.

Hence

$$\frac{\text{Lb of dry flue gas}}{\text{Lb } (CO_2 \text{ and } CO)} = \frac{44/358\ CO_2 + 28/358\ CO + 32/358\ O_2 + 28/358\ N_2}{44/358\ CO_2 + 28/358\ CO}$$

$$= \frac{44CO_2 + 28CO + 32O_2 + 28N_2}{44CO_2 + 28CO}$$

But CO_2 is, by weight, $12/44$ carbon and $32/44$ oxygen, and CO is, by weight, $12/28$ carbon and $16/28$ oxygen; therefore

$$\frac{\text{Lb dry flue gas}}{\text{Lb carbon in fuel}} = \frac{44CO_2 + 28CO + 32O_2 + 28N_2}{12/44 \times 44CO_2 + 12/28 \times 28CO}$$

$$= \frac{44CO_2 + 28CO + 32O_2 + 28N_2}{12(CO_2 + CO)}$$

Since a pound of fuel does not contain a pound of carbon, and since, in any event, all the combustible is seldom burned, the above expression can be multiplied by the percentage of carbon in the fuel *actually burned*, C_g (equal to C in fuel − C in refuse), to get the weight of flue gas per pound of fuel. Thus

$$W_{fg} = \frac{\text{Lb dry flue gas}}{\text{Lb coal}} = C_g \times \frac{44CO_2 + 28CO + 32O_2 + 28N_2}{12(CO_2 + CO)} \tag{2}$$

* For a complete treatment, refer to the *Power Test Codes*, 1946, The American Society of Mechanical Engineers, Stationary Steam-Generating Units.

Sulfur dioxide has been purposely omitted from the above expression since the quantity of this gas present in a sample of dry flue gas is generally quite small, and also since its presence cannot be detected with the ordinary apparatus available for flue-gas analysis. The weight and volume of SO_2 can be computed easily if desired from the elementary combustion equation $S + O_2 = SO_2$.

Assuming specific heat and pressure each to be essentially constant, the energy carried away by the dry flue gas is

$$h_2 = W_{fg} \times C_p \times (t_c - t_1) \tag{3}$$

where t_c = gas temperature, at boiler outlet, degrees Fahrenheit.

t_1 = air temperature at entrance to furnace, degrees Fahrenheit.

C_p = mean specific heat of dry flue gas between t_1 and t_c, Btu per pound per degree Fahrenheit.

C_p = 0.240–0.250 average value.

3. Energy Loss Due to Unburned Combustible in Refuse. The unburned combustible in the refuse consists of small amounts of unburned or incompletely burned fuel, and may be assumed to be essentially carbon. The energy loss involved is

$$h_3 = C_r \times 14,600 \tag{4}$$

where C_r = weight of combustible in refuse per pound of fuel.

14,600 = approximate heating value of this combustible per pound.

4. Energy Loss Due to Incomplete Combustion of C to CO. When insufficient air is supplied for combustion or when air and fuel are not intimately and thoroughly mixed, part of the carbon may burn to CO instead of CO_2. A pound of C burning to CO_2 releases 14,540 Btu per lb of carbon, whereas when a pound of carbon is burned to CO, only 4350 Btu are released per pound of carbon. The loss here, then, is $14,540 - 4350 = 10,190$ Btu per lb C burning to CO.

The weight of C burning to CO can be determined from the flue-gas analysis by a method similar to that employed in paragraph 2.

$$\frac{\text{Weight of carbon monoxide}}{\text{Weight of carbon monoxide and dioxide}} = \frac{{}^{28}/_{358}\ CO}{{}^{44}/_{358}\ CO_2 + {}^{28}/_{358} \times CO}$$

$$= \frac{28CO}{44CO_2 + 28CO}$$

where CO and CO_2 are percentages of carbon monoxide and carbon dioxide by volume.

Since all the carbon in the flue gas (present as CO_2 and CO) must have come from that carbon in the fuel *that was actually burned*, and since, by weight, carbon monoxide is $^{12}/_{28}$ carbon, and carbon dioxide is $^{12}/_{44}$ carbon, then the weight of carbon, W_c, present in the CO content of the flue gases is

$$\frac{{}^{12}/_{28} \times 28CO}{{}^{12}/_{44}\ CO_2 + {}^{12}/_{28}\ CO} \times C_g = \frac{CO}{CO_2 + CO} \times C_g$$

where C_g, CO_2, and CO are as in paragraph 2.

The energy loss resulting from the CO is

$$h_4 = W_c \times 10,160 = 10,160 \times \frac{CO}{CO_2 + CO} \times C_g \tag{5}$$

5. Energy Carried Away by Surface Moisture in Fuel. Most of the surface moisture in a fuel probably evaporates between temperatures of 70 and 120 F. To accomplish this, energy must be supplied by the fuel in sufficient amount (a) to raise the moisture to the vaporization temperature, (b) to evaporate it at this temperature, and (c) to raise its temperature to that of the flue gas at exit from the boiler. Steam tables would be required for a scientific calculation of this item. In lieu of this rather elaborate method the following simplification can be used:

$$h_5 = W_{mf}(1089 - t + 0.46t_c)\ \text{if}\ t_c < 575\ \text{F} \tag{6a}$$

$$h_5 = W_{mf}(1066 - t + 0.5t_c)\ \text{if}\ t_c > 575\ \text{F} \tag{6b}$$

where W_{mf} = pounds of surface water per pound of fuel.

t = temperature of fuel, degrees Fahrenheit.

t_c = temperature of gases at boiler outlet, degrees Fahrenheit.

6. Energy Carried Away by Moisture Resulting from Combustion of H_2. At precisely what temperature the formation of water vapor takes place is unknown. It is generally assumed, however, that it occurs at the outset of combustion. Whence, the parenthetical quantity in the preceding equation can be used to determine the loss per pound of moisture, or since 9 lb of moisture result from the combination of 1 lb of H_2

$$h_6 = 8.936H(1066 - t + 0.5t_c) \text{ if } t_c > 575 \text{ F} \qquad (7a)$$

$$h_6 = 8.936H(1089 - t + 0.46t_c) \text{ if } t_c < 575 \text{ F} \qquad (7b)$$

where H = percentage of H_2 in fuel by weight; other symbols as in equation (6).

7. Energy Carried Away by Moisture in Air. The moisture in the air used for combustion has already been raised in temperature to the vaporization point and has been completely vaporized. Hence the only energy absorbed by it within the furnace is that required to raise the temperature of this vapor to that of the flue gases at boiler outlet, or

$$h_7 = W_{ma}0.46(t_c - t_1) \qquad (8)$$

where t_c = flue gas temperature at boiler outlet, degrees Fahrenheit.

t_1 = air temperature at entrance to furnace.

W_{ma} = weight of moisture in air per pound of fuel.

8. Radiation and Unaccounted-for Losses. It is generally assumed that the preceding seven items constitute the principal manners in which the heating value of the fuel is distributed. Hence all other items are included under the head of radiation and unaccounted-for, or

$$h_8 = \text{H.H.V. of fuel} - \text{items 1-7 inclusive} \qquad (9)$$

Example. The following typical problem will illustrate the determination of the various items of the boiler heat balance.

Given: Coal analysis, as fired, C, 65%; O, 8%; H, 5%; N, 1%; ash, 13%; H_2O, 8%; heating value, 11,850 Btu per lb. Pounds of steam per pound of coal = 8.5.

Flue gas analysis (by volume) CO_2, 12.8%; CO, 0.6%; O_2, 5.4%; N_2, 81.2%.

Air and fuel enter furnace at 70 F; gas temperature at boiler outlet = 470 F; temperature of steam in boiler = 400 F; combustible in refuse = 20%; moisture in air = 0.5% of weight of dry air. Boiler pressure = 210 lb per sq in. abs. Feedwater temperature = 180 F.

Solution.

Refuse = Ash + carbon.

Weight of refuse = $\dfrac{0.13}{0.80} \times 0.1625$ lb per lb coal, as fired.

Weight of combustible in refuse = $\dfrac{0.13}{0.80} \times 0.20 = 0.0325$ lb per lb coal, as fired.

Hence, carbon burned = $C_g = 0.65 - 0.0325 = 0.6175$ lb per lb coal, as fired. Weight of dry flue gas, by equation 2 = 11.62 lb per lb coal, as fired.

Weight of dry air supplied

lb fuel = lb coal, as fired

$$\frac{\text{lb } DA + \text{lb fuel}}{\text{lb fuel}} = \frac{\text{lb } DFG + \text{lb } H_2O + \text{lb refuse}}{\text{lb fuel}}$$

$$\frac{\text{lb } DA}{\text{lb fuel}} = 11.62 + 0.163 + 0.45 - 1 = 11.23$$

Boiler Energy Balance

Eq.	Name	Calculation	Btu	Per Cent
1	Absorbed by boiler	8.5(1208.8 − 147.92)	9,025	76.13
2	Absorbed by DFG	11.62 × 0.24(470 − 70)0.6175 × 18.84	1,118	9.44
3	Loss due to C in refuse	0.0325(14,600)	475	4.01
4	Loss due CO	$0.6175(10,160)\left(\dfrac{0.006}{0.128 + 0.006}\right)$	281	2.38
5	Absorbed by H_2O in fuel	0.08[(1090 − 70 + 0.46(470)]	99	0.84
6	Absorbed by H_2O from H_2	9(0.05)[1090 − 70 + 0.46(470)]	556	4.70
7	Absorbed by H_2O in air	11.23(0.005)[0.46(470 − 70)]	10.3	0.087
8	Radiation, etc.	By difference	285.7	2.413
	Total		11,850	100.000

Overall boiler efficiency = 9025 ÷ 11,850 = 76.13%

Interpretation of Boiler Balance. The energy balance as calculated gives the distribution of the *actual* energy quantities. Some of the losses can be reduced or eliminated entirely; others are always present and are *inherent* losses. An energy balance setting forth the inherent losses will enable the reader to tell at a glance where improvement may be made, and where no further gain will be possible. For instance, air supplied for combustion may be reduced, thus decreasing the excess air, and hence energy carried away by dry flue gas; but the air cannot be reduced below the theoretical requirements. Similarly, by suitable baffling and maintenance of optimum flue-gas velocity and rate of fuel firing, the temperature of the flue gas may be reduced, but it can never be reduced below the temperature of the steam leaving the boiler (without an economizer). The following tabulations will further clarify the point. All items are per pound of fuel as fired.

Distribution of Inherent Losses

Eq.	Name	Calculation	Btu	Per Cent
2	Absorbed by DFG	$9.22 * \times 0.24(400 - 70)$	730.0	6.16
3	Loss due C in refuse		0.0	
4	Loss due CO		0.0	
5	Absorbed by H_2O in fuel	$0.08[1090 - 70 + 0.46(400)]$	96.3	0.813
6	Absorbed by H_2O from H_2	$9(0.05)[1090 - 70 + 0.46(400)]$	542.0	4.57
7	Absorbed by H_2O in air	$8.87*(0.005)[0.46(400 - 70)]$	6.7	0.057
8	Radiation, etc.		0.0	
1	Absorbed by ideal boiler	By difference	10,475.0	88.4
	Total		11,850	100.000

* Theoretical, for complete combustion.

Calculations

$$\frac{\text{lb of } DFG}{\text{lb of fuel}} = [0.65(2.67) + 0.05(7.94) + 0.08 - 0.08]4.32 + 0.01 = 9.22$$

$$\frac{\text{lb of air required}}{\text{lb of fuel}} = [0.65(2.67) + 0.05(7.94) - 0.08]4.32 = 8.82$$

Comparison of Actual and Inherent Losses

Item	Actual, %	Inherent, %
Absorbed by boiler	76.13	88.4
Absorbed by dry flue gas	9.44	6.16
Loss due C in refuse	4.01	
Loss due CO	2.38	
Absorbed by H_2O in fuel	0.84	0.813
Absorbed by H_2O from H_2	4.70	4.57
Absorbed by H_2O in air	0.087	0.057
Radiation, etc.	2.413	
Total	100.000	100.000

26. STEAM ENGINES AND TURBINES

The engine of the steam cycle treated in Arts 21–24 may be either of two general types, the reciprocating engine, commonly referred to as the *steam engine*, or the turbine engine, commonly referred to as the *steam turbine*. Given the same initial conditions of the steam and the same exhaust conditions the ideal thermal efficiencies of the steam engine and the turbine are exactly the same. The use of one or the other for a specific purpose is determined by practical or economic considerations. The measures of efficiency and performance developed in Art. 22 are applicable to both the steam engine and the steam turbine.

A **Simple Steam Engine** is one in which the expansion of steam is completed in one cylinder.

A **Compound Steam Engine** is one in which the steam progressively expands in two or more cylinders. The term *compound*, without qualification, refers to the two-cylinder arrangement. A *triple-expansion* engine means that the expansion is completed in three stages or cylinders, and a *quadruple-expansion* engine means that the expansion is completed in four stages or cylinders. Compound engines may have the cylinders arranged in various ways such as: tandem-compound, cylinders placed one behind the other in line; angle-compound, cylinder axes placed at right angles to one another; cross-compound, cylinders arranged side by side; and vertical triple expansion, three cylinders placed with axes vertical.

An engine may operate *non-condensing*, i.e., exhausting to the atmosphere, or *condensing*, i.e., exhausting into a condenser in which a vacuum is maintained.

A **Single-acting Engine** is one in which steam is admitted only to one side of the piston. A *double-acting* engine is one in which steam is admitted alternately to each side of the piston.

Valves. A *slide valve* is a valve which controls the inlet and exhaust of steam by sliding across the ports. A *poppet valve* is a disk, fitting a port, which is raised or lowered for the control of inlet and exhaust. A *Corliss valve* has a cylindrical surface which oscillates through a small angle to open or close the port. Each event is separately controlled. The main feature of this valve is the quick closure of the steam port, obtained by a trip-gear on the inlet valve. The more recent high-speed Corliss engine has positive control of the valves at all times.

The exhaust valve of the *uniflow engine* is the piston itself, which functions with respect to the exhaust ports located at the middle of the cylinder. The inlet valves, located at the ends of the cylinder, may be of any type.

Steam Engine Speed is governed in two ways: by throttling the steam supply, or by varying the cut-off.

Indicator Diagram. An indicator diagram is a diagram (see Fig. 10), showing the steam pressure in the engine cylinder at each point of the stroke. Such a diagram may be actually obtained by means of a steam-engine indicator, which is an instrument which causes a pencil to record on paper the pressure in the cylinder at every point of the stroke. The diagram drawn by the pencil shows whether the valves are properly adjusted, and it is also used in figuring the power developed in the cylinder, and the approximate steam consumption. A diagram of a non-condensing engine in which the steam is cut off at about one-quarter of the stroke is shown in Fig. 10.

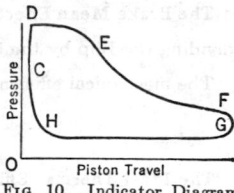

FIG. 10.   Indicator Diagram of a Simple Steam Engine

The lines and points have the following significance.

Point of Admission, C, is the point at which the steam valve opens.

Admission Line, CD, shows the rise of pressure due to the admission of steam to the cylinder by opening the steam valve.

Steam Line, DE, is drawn when the steam valve is open and steam is being admitted to the cylinder

Point of Cut-off, E, is the point where the admission of steam is stopped by the closing of the valve. It is often difficult to determine the exact point at which the cut-off takes place. It is usually located where the outline of the diagram changes its curvature from convex to concave.

Expansion Curve, EF, shows the fall in pressure as the steam in the cylinder expands doing work.

Point of Release, F, shows when the exhaust valve opens.

Exhaust Line, FG, represents the change in pressure that takes place when the exhaust valve opens.

Back-pressure Line, GH, shows the pressure against which the piston acts during its return stroke.

Point of Exhaust Closure, H, is the point where the exhaust valve closes. It cannot be located definitely, as the change in pressure is at first due to the gradual closing of the valve.

Compression Curve, HC, shows the rise in pressure due to the compression of the steam remaining in the cylinder after the exhaust valve has closed.

Initial Pressure is the pressure acting on the piston at the beginning of the stroke.

Terminal Pressure is the pressure above the line of perfect vacuum that would exist at the end of the stroke if the steam had not been released earlier. It is found by continuing the expansion curve to the end of the diagram.

The Indicated Mean Effective Pressure, found by dividing the diagram area by its length and multiplying the quotient by the pressure scale of the diagram, is in a broader sense that equivalent pressure which, if acting unopposed upon the piston face during the power stroke of a cycle, would produce the net amount of work output actually delivered per cycle.

Indicated Work per Cycle, ft-lb,

$$= \text{Imep} \times A \times L$$

$$= P_m \times A \times L$$

where P_m = abbreviation for the mean effective pressure, pounds per square foot.
　A = piston area, upon which the pressure acts, square feet.
　$P_m A$ = mean effective force acting on piston face, pounds.
　L = length of piston stroke, feet.

The indicated horsepower is

$$\text{Ihp} = \frac{P_m A L N}{33,000}$$

where N = the number of power strokes per minute (2 per revolution for simple double-acting engine).

The Brake Horsepower of an engine is calculated by the use of

$$\text{Bhp} = \frac{2\pi r w n}{33,000}$$

where r = radius of brake arm, feet.
　w = force acting at end of brake arm, pounds.
　n = number of revolutions per minute.

The Brake Mean Effective Pressure, Bmep, useful for comparison purposes, is found by dividing the Bhp by the term $\dfrac{L A N}{33,000}$.

The mechanical efficiency of the engine may be stated,

$$\text{Mechanical efficiency} = \frac{\text{Bmep}}{\text{Imep}} = \frac{\text{Bhp}}{\text{Ihp}}$$

The Ideal Thermal Efficiency of the steam engine is the same as that for the Rankine cycle, Art. 22. The actual thermal efficiency of the steam engine is the same as that for the steam cycle, but may be based on either indicated horsepower or brake horsepower. Likewise the engine efficiency, heat rate, and steam rates of the steam engine are the same as for the steam cycle, and each may be based on either indicated horsepower or brake horsepower.

Tests. For detailed methods of tests of steam engines, as well as of all other power machinery, reference should be made to the Power Test Codes published by the American Society of Mechanical Engineers, New York.

The following summary represents briefly the generally accepted ideas regarding the cylinder action and the losses in the reciprocating steam engine.

A. The losses of available energy in the steam engine are due to the following causes, arranged in the approximate order of magnitude of loss:

1. Initial or surface condensation.
2. Incomplete expansion and early release.
3. Throttling of steam flowing through partly open valves.
4. Leakage of steam past piston and closed valves.
5. Friction of moving parts.
6. Heat loss by radiation and conduction.
7. Compression and early admission.

B. Initial or surface condensation is increased by the following conditions:

1. Early cut-off and large ratio of expansion.
2. Large pressure and temperature range in a single cylinder.
3. Small cylinder diameters, because the exposed surface is large in comparison with the volume.
4. Steam supply saturated or wet.
5. Long valve passages and the use of the same passage for admission and exhaust steam.

C. Conditions favorable to reduction of initial condensation are:

1. Late cut-off and small expansion ratio (but this increases loss due to incomplete expansion in simple engines).
2. Superheat.
3. Separate valves for admission and exhaust, and short valve passages.
4. High compression.
5. Compounding.
6. Uniflow principle.

Types of Turbines

In the Turbine Engine or steam turbine the energy conversion to work is accomplished by means of the conversion of the available portion of the internal energy of the steam into mechanical kinetic energy of a jet in stationary or moving nozzles, and then the conversion of this mechanical kinetic energy of the jet into work on a shaft by allowing the jet to change its direction of motion while passing over moving vanes or blades.

In the Impulse Turbine the steam pressure drop and consequent development of kinetic energy take place solely in the stationary nozzles, and the work is obtained by the conversion of this kinetic energy into work on moving blades. In the *reaction* turbine only a part (about half) of the kinetic energy conversion occurs in the stationary nozzle, the remainder of the kinetic energy conversion being accomplished by a pressure drop in the steam as it passes through the moving blades. The efficiencies and other performance of

Table 1 *

Classification	Type	Characteristics	
		Physical	Operating
Impulse	De Laval	One nozzle or set of nozzles. Single disk with one row of blades. One passage of steam across blades.	High steam velocities. High wheel velocities. Large pressure drop in nozzle.
	Simple Curtis	One nozzle or set of nozzles. Single disk with two or more rows of blades. Intermediate reversing blades. One passage of steam across each blade row.	Moderate wheel speed. Large pressure drop in nozzle. Same pressure throughout stage.
	Re-entry (a) Axial flow (b) Tangential type	One nozzle or set of nozzles. Single row of blades or buckets. Reversing chambers to redirect steam on blades one or more times.	Usually moderate wheel speed. High wheel speeds in geared sets. Large pressure drop in nozzle.
	Rateau	Series of De Laval wheels with intermediate diaphragms carrying orifices. Usually, large number of stages.	Small pressure drop per stage. Most efficient ratio of wheel speed to steam speed can be secured.
	Multistage Curtis	Series of simple Curtis wheels separated by diaphragms carrying orifices. Usually relatively few stages.	Relatively large pressure drop per stage.
Reaction	Axial flow, Parsons	Series of alternate rows of converging fixed and moving blades. Moving blades mounted usually on several drums. Steam flows axially.	Small pressure drop per row. Large number of rows. Long spindles. Most efficient ratio of wheel speed to steam speed possible.
	Radial flow, Ljungstrom	Series of radial rings of converging reaction blades. Alternate rings revolve in opposite directions.	High ratio of blade speed to steam speed. Elaborate steam packing devices.
Combination Types	Curtis-Rateau	One Curtis stage followed by several simple-impulse stages.	Only moderate temperatures and pressures in casing due to large pressure drop in first nozzle.
	Curtis-Parsons	One Curtis stage followed by a series of Parsons stages. Usually disk and drum construction.	Temperatures in casing low, since steam is expanded in nozzle.

* Abstracted by permission from *Kent's Mechanical Engineers' Handbook*, John Wiley & Sons, New York, 1950.

the steam turbine are the same as for the reciprocating engine except that for the turbine the items of indicated horsepower and mechanical efficiency are missing.

De Laval Turbine. The characteristic features of the *De Laval turbine* are the diverging nozzles which expand the steam to the back pressure in a single stage, and a single steel disk, mounted on a slender flexible shaft, carrying the blades on its periphery.

The Rateau Turbine consists of a number of De Laval elements in series. The steam expands in several pressure stages until completely expanded to the back pressure.

The Parsons Turbine is a reaction turbine in which there are a large number of rows of blades mounted on a rotor or revolving drum. Between each of these rows of blades is a row of stationary blades attached to the casing. The steam expands to a lower pressure in both sets of blades. A set of stationary blades and the following set of moving blades constitute what is known as a *stage*.

Several combinations of these different types are in use. A summary of the principal features of construction and the operating characteristics of several widely used types of turbines is given in Table 1.

27. STEAM CONDENSERS AND OTHER HEAT TRANSFER APPARATUS

Steam Condensers are either of the *jet* type, in which the condensate mixes with the cooling water and from which both are discharged together, or of the *surface* type, in which the two fluids are kept separate by thin metallic walls and are handled independently.

The Weight of Water necessary to condense a given weight of steam may be found for any type of condenser from the equation that expresses the energy balance for the condenser. Neglecting the effect of the presence of air, changes in velocity, and heat lost to the atmosphere,

$$w_s(H_s - H_c) = w_w(H_2 - H_1)$$

where w_s and w_w = the respective weights, in pounds, of steam and condensing water flowing through the condenser in a unit of time, usually the hour.

H_1 and H_2 = the respective enthalpies of the condensing water entering and leaving the condenser, in Btu per pound.

H_s = the enthalpy of the steam entering the condenser, in Btu per pound.

H_c = the enthalpy of the condensate leaving the condenser, in Btu per pound.

Then, the weight of condensing water per unit weight of steam becomes

$$\frac{w_w}{w_s} = \frac{H_s - H_c}{H_2 - H_1}$$

or, very closely,

$$\frac{w_w}{w_s} = \frac{H_s - (t_c - 32)}{t_2 - t_1} = \frac{H_s - (t_c - 32)}{t_s - (t_1 + \theta_b)}$$

where t_1 and t_2 = the respective temperatures of the condensing water entering and leaving the condenser, in degrees Fahrenheit.

t_c = the temperature of the condensate leaving the condenser, in degrees Fahrenheit.

t_s = the temperature of the exhaust steam (assuming no superheat), in degrees Fahrenheit.

$\theta_b = t_s - t_2$ = terminal temperature difference.

With jet condensers, $t_c = t_2 = t_{\text{mix.}} = t_s - \theta_b$. Therefore,

$$\frac{w_w}{w_s} = \frac{H_s - t_{\text{mix.}} + 32}{t_{\text{mix.}} - t_1}$$

$$= \frac{H_s - t_s + \theta_b + 32}{t_s - (t_1 + \theta_b)}$$

Most problems of heat transmission in surface condensers can be solved by the use of the following:

$$w_s(H_s - H_c) = \frac{A U(t_2 - t_1)}{\log_e \left(\dfrac{t_s - t_1}{t_s - t_2} \right)} = w_w c_w(t_2 - t_1)$$

where A = the amount of condenser heating surface, in square feet.
c_w = the specific heat of the condensing water.
U = the thermal transmittance, in Btu per hour per degree Fahrenheit per square foot of condenser heating surface.
e = Napierian base of logarithms = 2.71828.

The mean temperature difference

$$\frac{t_2 - t_1}{\log_e \left(\dfrac{t_s - t_1}{t_s - t_2} \right)}$$

is commonly called the *logarithmic mean temperature difference*.

Assuming c_w as unity, permissible for fresh water where extreme accuracy is not essential,

$$\log_e \left(\frac{t_s - t_1}{t_s - t_2} \right) = \frac{U}{(w_w/A)}$$

in which (w_w/A) is the number of pounds of circulating water used per hour per square foot of condensing surface.

$$t_s = t_1 + (t_2 - t_1) + \theta_b = t_1 + \frac{H_s - H_c}{(w_w/A)} + \theta_b$$

$$= t_1 + \frac{H_s - H_c}{(w_w/w_s)} \cdot \left[\frac{\theta_b}{t_2 - t_1} + 1 \right]$$

$$= \frac{t_1 + (t_2 - t_1)}{1 - (e)^{-U/(w_w/A)}}$$

These equations are useful in showing how the steam temperature, and the corresponding absolute pressure, are influenced by the variations in each of the individual factors and ratios appearing in the right-hand members.

Feedwater heaters are either of the *open* type, in which the two fluids mix and from which both are discharged together, or of the *closed* type, in which the two fluids are kept separate and are handled separately.

Neglecting heat loss through the shell

$$w_s(H_a - H_b) = w_f(H_2 - H_1)$$

in which w_s and w_f = the quantities of steam and feedwater flowing through the heater, respectively.
H_a and H_b = the respective enthalpies of the hot fluid entering and leaving the heater, in Btu per pound.
H_1 and H_2 = the respective enthalpies of the feedwater entering and leaving the heater, in Btu per pound.

Since the condensate formed in the open-type heater mixes with the feedwater, H_b and H_2 are equal in this type of heater. For moderate feedwater temperatures c_p of water is near unity; the term $(t_2 - t_1)$ may be substituted for the term $(H_2 - H_1)$, where t_2 and t_1 are the respective temperatures of the feedwater leaving and entering the heater. The problems of heat transmission in the closed feedwater heater are treated the same as those of the surface condenser.

An economizer is an apparatus by which some of the heat from gases leaving the boiler is recovered and used principally to heat feedwater. For heat transmission in economizers,

$$\text{Mean temperature difference} = \frac{\theta_a - \theta_b}{\log_e \dfrac{\theta_a}{\theta_b}}$$

where θ_a = the initial temperature difference of the two fluids, degrees Fahrenheit.
θ_b = the final temperature difference of the two fluids, degrees Fahrenheit.
And

$$w_g c_g(t_a - t_b) = \frac{A U(\theta_a - \theta_b)}{\log_e \dfrac{\theta_a}{\theta_b}} = w_f c_f(t_2 - t_1)$$

where w_g = weight, in pounds, of flue gas through the economizer per unit time.
w_f = weight, in pounds, of feedwater through the economizer per unit time.
c_g and c_f = the respective specific heats of the flue gas and feedwater.

Air preheaters, used also to recover some of the heat from gases leaving the boiler, may be of two general types, recuperative and regenerative. For the recuperative type in which the heat is transferred from the flue gases to the air through thin metallic walls the principles of heat transmission are the same as for the economizer. In the regenerative type the heat-transmitting surfaces are exposed alternately to the heat-surrendering gases and to the heat-absorbing air.

THE INTERNAL-COMBUSTION ENGINE

28. GENERAL CHARACTERISTICS

The **Internal-combustion Engine** constitutes one of the two general types of devices for realizing as shaft work some portion of the available energy associated with the high-temperature products of the combustion of a fuel, the other device being represented by the composite arrangement of furnace, steam boiler, and steam engine or turbine treated in Arts. 21-27. It will be recalled that in the latter arrangement the stored chemical energy of the fuel is released by a process of combustion carried out in a furnace at high temperature levels but at constant atmospheric pressure, that the energy so released is then in part delivered by conduction to moderate-temperature steam which is formed in the boiler under high pressure, and that this steam acts as a carrier through the medium of which energy is delivered to the engine, in which there is finally accomplished a partial transformation of the energy to shaft work and from which the unavailable residue of energy is discarded at about atmospheric temperature. In distinction to this serial procedure in the steam plant the internal-combustion engine effects a more direct utilization of the chemical energy released by the combustion by so confining the burning fuel mixture that a *high pressure is produced* (or *maintained*) *in the hot combustion products themselves,* whereby through a suitable mechanism these products may act directly upon a piston to motivate a shaft against whatever resisting force may be imposed by the external load.

The **Characteristic Conditions** for which the internal-combustion engine must be designed may be summarized as follows:

(A) A major portion of the material which must be introduced into the engine cylinder is the atmospheric air which supplies the oxygen necessary for the combustion. The induction of air from the atmosphere to the engine cylinder may be accomplished in various ways. A very common method is to employ one complete piston stroke for this purpose. An engine which does so is said to operate upon a *four-stroke* cycle. That cycle is described in the following. The alternative or *two-stroke* cycle is considered briefly later.

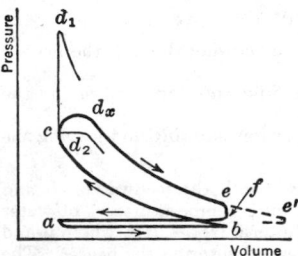

FIG. 1. Indicator Diagram of Internal Combustion Engines

In a continuous graphic record of (a) the pressure existing within the space enclosed by the cylinder and piston at any point of the piston travel and (b) the total volume of that space at the same piston position, such as is furnished by the engine *indicator,* this induction or *suction* stroke would be represented by a line such as the line *ab* in Fig. 1. During this entry of the fluid through the *intake* valve the pressure within the cylinder will be less than that at the entrance to the induction system by the amount of the pressure drop necessary to accelerate the fluid and to overcome fluid friction in the intake passages.

(B) For both practical and theoretical reasons it is necessary that prior to the combustion of the fuel the fuel mixture or at least the air portion of this mixture shall have been brought to a pressure and temperature materially above those of the atmosphere. This is accomplished by a compression which is caused to take place by a return or *compression* stroke of the piston with all valves closed. For this stroke the simultaneous fluid pressures and volumes would be represented by points on the line *bc.*

(C) For motivation of the engine it is necessary that during the next or *power* stroke of the piston the average fluid pressure shall be materially greater than that during the other strokes. This pressure increase is effected by a pronounced temperature increase of the fluid during and resulting from combustion while confined within the cylinder. This combustion is initiated at or about a point in the piston travel which would be represented by point *c* in the figure.

The rate of combustion (and thus the rates of energy release and of temperature and pressure rise as a result of combustion) is a characteristic which is in effect the major basis for classification of modern engine types. Suitable control of the conditions under which the combustion occurs might permit or cause it to be so rapid and explosive in character as to take place at virtually constant volume and produce a more or less "vertical" rise of temperature and pressure, such as indicated by the line cd_1 in Fig. 1. Alternatively the combustion might be caused to proceed with such lesser rapidity that it would continue through a portion of the power stroke, with the production of a somewhat lesser temperature and pressure rise than would occur under the first conditions. Lines cd_x or cd_2 would be representative of the latter condition.

The exact rate of energy release under this second schedule might be caused to be virtually anything from the one producing almost constant volume combustion to one so relatively slow as simply to maintain a constant temperature and thus to permit a gradual pressure drop following the initiation of the combustion. The line cd_x may be taken as a general one representing any sort of combustion conditions which might obtain. However, the prototype of one of the standard cycles which are employed in present-day internal-combustion engine practice, that is, the *Diesel cycle*, is one in which it is assumed that the pressure shall be maintained constant during and by the combustion, producing a combustion line as represented by cd_2 in the figure. The constant volume combustion of line cd_1 is a basic characteristic of the *Otto cycle*.

To permit the constant volume combustion of the Otto cycle it is necessary that at point c in the cycle the engine cylinder shall contain the fuel and air in intimate mixture and thus in effective condition for ignition (by electric spark) and *explosive* combustion. This condition is obtainable by (1) the use of either gaseous fuels or relatively volatile liquid fuels and (2) the atomization and intimate mixing of the fuel with the air in some portion of the cycle prior to the completion of the compression. This mixing is ordinarily done outside of the engine cylinder in a mixing valve or *carburetor* and during the induction of the air on stroke ab. It is to be noted, however, that this very presence of a combustible mixture in the cylinder during the compression stroke imposes certain definite limits on the extent of the compression which may be employed in the Otto cycle engine, because of the temperature rise produced by the compression and the resultant tendency toward spontaneous pre-ignition of the mixture, and also by reason of the tendency of numerous fuels toward an abnormally rapid combustion and the production of abnormally and seriously high combustion temperature and pressure if compression is carried beyond certain limits. This latter phenomenon is known as *detonation*.

The progressive, constant pressure combustion of the conventional *Diesel cycle* is securable (1) by withholding the fuel from the engine cylinder until about the completion of the compression stroke, which implies that only air shall have entered during the suction stroke, and (2) by a subsequent progressive injection of an atomized jet of the (liquid) fuel into the cylinder during the combustion period *at a properly controlled rate*. Except in various smaller engines operating on a so-called *semi-Diesel* cycle, the ignition of the fuel is accomplished by compression of the air to so high a pressure P_c that at the end of compression it shall have attained a temperature high enough to induce spontaneous ignition upon the fuel injection. Even under this condition, however, to accomplish suitable ignition and combustion requires that the fuel must be injected in a very finely atomized state.

(D) Returning to a consideration of the events of the cycle—except as there is some incompleteness of combustion at point d by reason of chemical equilibrium and thus a persistence of combustion, or *after-burning*, beyond that point, the *remainder* of the power or *expansion stroke* provides solely for an expansion of the high-temperature products of combustion. This continues until the end of the stroke is reached, at or somewhat before which point the exhaust valve is opened. Line de is representative of this expansion phase. It will be observed that an expansion to the same total volume as that at the beginning of compression (V_b) predicates a release of the combustion products at a pressure which is materially higher than the intake or atmospheric pressure, and also their rejection or exhaust at a temperature very considerably above atmospheric temperature. The available energy loss associated with this temperature and pressure excess might be reduced by arranging for a continuation of the expansion to some greater volume such as the volume $V_{e'}$ of the figure. Such cycles have been and are frequently proposed but have not been developed commercially.

(E) Directly upon opening of the exhaust valve there occurs a rapid drop of pressure within the cylinder due to the escape of a portion of the combustion products through the exhaust passages, and upon the following or *exhaust* stroke in a four-stroke cycle the remainder of the products are ejected by the returning piston, except as a portion is left in the *clearance space* between the piston and the head end of the cylinder. These two exhaust processes are represented by lines ef and fa in Fig. 1. During the exhaust stroke the pressure within the cylinder would exceed that of the exterior by the amount necessary to accelerate the fluid and to overcome fluid friction in the exhaust system. The particular mass of products which remains in the clearance space at the end of any given exhaust stroke is replaced by an equivalent mass at the end of the next cycle; this clearance residue may therefore be regarded as continuously entrapped within the cylinder.

The performance of a complete cycle has thus been. described. Observe that for its accomplishment four strokes of the piston have been required, and therefore, as noted above, the cycle would properly be designated as a *four-stroke* cycle. An engine operating on this four-stroke schedule would be described as a *four-cycle* engine whether it might conform to the Otto cycle or the Diesel cycle or any modification of either as regards the degree of compression or the character of combustion.

In lieu of the four-stroke schedule it is possible and wholly practicable to accomplish the introduction of the air or the mixture into the cylinder, its compression, the combustion and expansion, and a sufficient clearing of the combustion products from the cylinder, in two piston strokes.

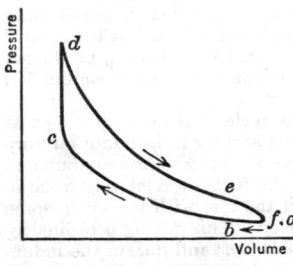

FIG. 2.   Indicator Diagram of a
2-cycle Engine

A cycle so carried out is designated, as remarked above, as a *two-stroke* cycle, and an engine operating on the cycle is a *two-cycle* engine. The arrangements by which the cycle is accomplished differ in detail in various engines, but the general procedure is in effect (a) to cause the exhaust valves (or ports) to open rather early in the *expansion* stroke, thus facilitating as far as possible the escape of the combustion products without the provision of a separate exhaust stroke, and (b) to force the air (or the mixture) into the cylinder and jointly to sweep out the residual combustion products by the use of a low-pressure *scavenger* blower which is caused to discharge to the cylinder during the early portion of the compression stroke, thus eliminating a separate intake stroke. The general appearance of the pressure-volume diagram of a two-stroke cycle is shown in Fig. 2, in which equivalent phases in the two- and the four-stroke cycles are denoted by the same letters as were used in Fig. 1.

The idealized conditions of operation of an internal-combustion engine may be taken to involve the following general features:

(a) Zero resistance and thus zero pressure drop through the intake system, whereby the cylinder pressure during admission might be that of the atmosphere, P_a.

(b) A reversible adiabatic compression of the air or mixture, the adiabatic feature being in distinction to the actual condition of energy transfer as heat between the fluid and the water-cooled or air-cooled cylinder of the real engine.

(c) A reversible adiabatic combustion phase, which again is in distinction to the actual condition of energy escape as heat from the burning mixture to the cylinder during combustion.

(d) A reversible adiabatic expansion of the combustion products, with, however, the possibility of some persistence of combustion during and after expansion if conditions of chemical equilibrium were encountered during combustion.

(e) An instantaneous pressure drop upon opening of the exhaust valve *at the end* of the expansion stroke, and zero resistance and thus zero pressure drop through the exhaust system, whereby the cylinder pressure during exhaust might be that of the atmosphere, P_a.

29. THE OTTO CYCLE

A pressure-volume diagram for an Otto cycle conforming to the above-mentioned ideal features would be like that of Fig. 2. The extent of the compression and expansion employed is a feature which affects the final products' temperature and thus the efficiency of the cycle. The outstanding influence of these features may be indicated approximately by quite artificial and unreal analyses which are known as the *air standard* analyses. In these analyses it is presumed (1) that air alone is caused to pass through the typical state changes of the cycles, (2) that the temperature and pressure rises which in the actual cycles are produced by the chemical energy release during combustion shall in this hypothetical cycle be produced by a like supply of energy to the air as heat from some external source, (3) that the energy discarded by the cycle equals the energy emission during a constant-volume cooling of the air from the temperature T_e to the original temperature T_1, and (4) that the specific heat of the air may be regarded as constant and as that of air at normal atmospheric temperature.

A significant item associated with the *internal-combustion engine* cycle is the ratio between (1) the total volume in the cylinder at the end of intake and beginning of compression and (2) the clearance or residual volume in the cylinder when the piston is at the end of the compression stroke. This ratio is known as the *ratio of compression* and will be denoted as r_c. Referring to Fig. 3, which is representative of the events in the ideal Otto cycle,

$$\text{Compression ratio, } r_c, = \frac{V_b}{V_c}$$

A second generally significant ratio is that between the volume at the end of expansion and the volume at the beginning of expansion. This ratio is known as the *ratio of expansion* and will be denoted as r_e. Referring again to Fig. 3,

$$\text{Expansion ratio, } r_e, = \frac{V_e}{V_d}$$

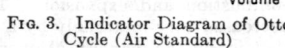

FIG. 3. Indicator Diagram of Otto Cycle (Air Standard)

It is observed that for the ideal Otto cycle the expansion ratio and the compression ratio are equal.

An air standard analysis of the Otto cycle indicates that its thermal efficiency depends only upon the compression ratio of the cycle and increases with increase in compression ratio in accordance with the relation

$$\text{Thermal efficiency, air standard Otto cycle} = 1 - \frac{1}{(r_c)^{0.4}} \tag{1}$$

Although more exact estimates of ideal performance as well as actual engine tests show that efficiency is likewise influenced by the proportions of the fuel-air mixture, with improved efficiency at mixtures with some excess of air, the evidence of the air standard analysis is well corroborated with respect to the effect of the compression ratio.

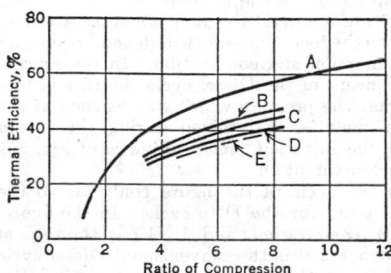

FIG. 4. Thermal Efficiencies of Otto Cycles

The evidence offered by equation (1) that, for the air standard Otto cycle, the thermal efficiency is determined solely by the compression or expansion ratio, is indicated graphically by the curve A of air standard thermal efficiency vs. compression ratio in Fig. 4. In the figure there are also represented, in curves B, C, and D, the results of computations made by Goodenough and Baker * for the ideal (indicated) thermal efficiency of an Otto cycle employing real mixtures of gasoline and air. In their computations careful attention was given to the effect of specific heat variation and of chemical equilibrium during combustion, employing the then (1926) best available data. Curve E shows the actual thermal efficiencies (based on indicated horsepower) attained on test of a very high-grade, single-cylinder Otto engine in which the compression ratio could be varied at will, the results being those reported by Ricardo † when using a non-detonating fuel and a fuel mixture containing air 15 per cent in excess of the amount theoretically required for complete combustion of the fuel.

The results of the ideal efficiency computations (curves B, C, and D) and the actual efficiency results (curve E) distinctly corroborate the evidence offered by the air standard cycle computations in so far as those predict the material influence of compression and expansion ratio on efficiency. In addition the ideal efficiency curves indicate a material influence of the character of the fuel mixture on the ideal efficiency. Test results on actual engines operating with mixture strengths which are readily explosive further corroborate the evidence of the ideal curves, to wit, that decrease of the mixture strength acts to increase efficiency quite distinctly.

Progress in the use of distinctly weak mixtures is handicapped by the difficulty in ignition and rapid burning of the mixture and by the material reduction in the power

---

* Goodenough and Baker, *A Thermodynamic Analysis of Internal Combustion Engine Cycles*, Univ. of Illinois Engineering Experiment Station Bull. 160, 1927.

† See Ricardo, *Engines of High Output*, Macdonald and Evans, London, 1926.

capacity of a given size engine with reduction in mixture strength, maximum power being secured both in the ideal cycle and in the actual engine with mixtures in which the air supply is some 5 to 10 per cent less than the theoretical requirement. Progress in the direction of the higher compression ratio is handicapped by the detonation tendency of many fuels but similarly is facilitated by the development and utilization of fuels which are less detonative in character (such as gasoline-benzol blends, gasoline treated with various catalysts, alcohol, etc.). The tendency of the efficiency curves toward a gradual flattening with increase of compression ratio and a very pronounced increase of the peak pressure of the cycle without commensurate increase of the engine power will undoubtedly act to limit the ratio for which it is economically worth while to strive. With ordinary gasolines, values of the ratio in excess of about 6 cannot be used to advantage owing to detonation.

The departure of actual engine efficiency from the ideal is attributed to a number of factors, such as (1) a failure to burn portions of the fuel which are in intimate contact with the cylinder walls, (2) untimely burning of portions of the fuel during the compression stroke, this occasioned by localized contacts with hot portions of the cylinder or exhaust valve, and (3) energy conduction through the cylinder walls as heat during the combustion and expansion. The efficiency of multiple-cylinder engines is frequently decreased very materially by a poor distribution of a properly proportioned fuel mixture, that is, by the delivery of over-weak mixture to certain cylinders and over-rich mixture to others. The efficiency of two-cycle Otto engines is commonly rather less than that of a four-cycle engine by reason of a direct wastage of fuel mixture through the exhaust during the dual exhaust-induction phase of that cycle.

30. THE DIESEL CYCLE

The distinctive characteristics of the *Diesel cycle* are the induction and compression of air only to a temperature well above the ignition temperature of the fuel, whereby upon injection of an atomized fuel oil jet at the end of the compression stroke ignition will automatically occur and combustion will continue so long as fuel injection is continued

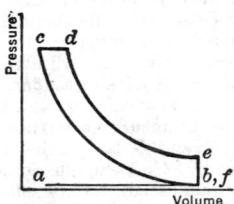

and the oxygen or air supply in the cylinder does not become depleted. The resulting temperatures and pressures attained by the products during the fuel admission will depend primarily on the rate of fuel injection and combustion. In the *conventional* present-day concept of the Diesel cycle the rate is presumed to be such that the pressure which was reached at the end of compression shall be maintained during the piston travel to the end of the injection phase. Line *cd* of Fig. 5 is representative of such a condition.

Fig. 5. Indicator Diagram of Diesel Cycle (Air Standard)

The remainder of the cycle of the figure conforms to the idealized conditions noted for the Otto cycle. In the figure, V_b/V_c is the *ratio of compression*, and V_e/V_d is the *ratio of expansion*. It is obvious that in the conventional Diesel cycle the expansion ratio is less than the compression ratio. For any given compression ratio the duration of the energy supply phase, *cd*, would be increased and the expansion ratio would be correspondingly decreased with increase in the *amount* of energy supplied to the engine per cycle. Interpreted in association with the actual engine, the amount of energy supplied per cycle is measured by the amount of fuel supplied per cycle and thus by the rate and duration of the fuel injection. Note in this connection that the ideal limiting amount of fuel which might properly be injected per cycle is established by the amount of oxygen and thus of air which had been inducted into the engine cylinder on the intake stroke of the cycle. Decreased fuel injection per cycle implies increased expansion ratio and increased air excess.

For the Conventional Diesel Cycle with constant pressure combustion the air standard analysis indicates that the cycle efficiency is influenced jointly by the compression ratio and the expansion ratio in accordance with the relation

$$\text{Thermal efficiency, air standard Diesel cycle} = 1 - \frac{1}{(r_c)^{0.4}} \frac{(r_c/r_e)^{1.4} - 1}{1.4(r_c/r_e - 1)} \tag{2}$$

For a given compression ratio the Diesel efficiency is less than that of the Otto and departs progressively from the Otto with decreasing values of the expansion ratio r_e. This last fact is due directly to the increase of the temperature T_e (at the end of the expansion and the beginning of the energy rejection phases of the cycle) with decrease of the expansion ratio. This temperature increase is in turn due jointly to the increasing temperature T_d

resulting from greater energy supply to the fluid during the longer injection phase and the concurrently lesser opportunity for temperature drop during the expansion phase. In spite of this apparent handicap the actual Diesel engine is in general able to attain thermal efficiencies higher than those obtained by the Otto, owing to the compensating advantage that the detonating character of the fuel used with the Otto engine and the very high maximum pressures encountered in that cycle limit its practicable compression ratio to about 6 to 8, whereas the Diesel engine may and commonly does employ compression ratios of 12 or over.

In the Diesel engine a very thorough atomization of the fuel oil jet is necessary as well as a jet momentum sufficient to cause the fuel particles to penetrate the mass of highly compressed air, and also there needs be a considerable turbulence in that air, all in order that the fuel and oxygen molecules may quickly become intimately associated and rapid combustion may thereby be facilitated. It has been the usual practice to assist the atomization of the fuel and the mixing by introducing with the fuel jet a jet of very highly compressed air which was furnished by an accessory compressor. With more recent advances in the art, fuel nozzles are being developed which provide adequate pulverization and dissemination of the fuel without the aid of the air jet. The two methods of introducing the fuel are known respectively as *air injection* and *airless injection* or, as the latter is frequently although rather inaptly termed, *solid injection*.

For Methods of Testing and Reporting the Performance of internal-combustion engines, reference should be made to the Test Code for Internal Combustion Engines published by the American Society of Mechanical Engineers, 345 E. 47th St., New York City.

It may be stated that the items brake horsepower, indicated horsepower, mechanical efficiency, brake mean effective pressure, indicated mean effective pressure, etc., have the same significance and are determined in quite the same manner as described for the steam engine, Art. 26.

For the single-acting four-cycle engine $Ihp = \dfrac{PALN}{2 \times 33,000}$ per cylinder, and for the two-cycle single-acting engine the Ihp per cylinder is $\dfrac{PALN}{33,000}$.

In American practice the thermal efficiency of the internal-combustion engine is based upon the higher heating value of the fuel.

REFRIGERATION

Symbols and Abbreviations

C.P. = coefficient of performance.
c_p = specific heat at constant pressure.
H = enthalpy, Btu per pound of fluid.
J = Joule's equivalent (778 ft-lb per Btu).
k = specific heat ratio, c_p/c_v.
P = pressure, pounds per square foot, absolute.
P_1 = pressure of refrigerant in condenser or cooler.
P_2 = pressure of refrigerant in evaporator or refrigerator.
Q = energy transferred as heat, Btu per pound of fluid.
S = entropy per pound mass of fluid.
T = temperature, degrees Rankine (degrees Fahrenheit absolute).
V = specific volume, cubic feet per pound.

The refrigerating machine is the practical example of the reversed heat engine, in that it acts to abstract energy as heat from a region of lower temperature and to deliver heat to a region of high temperature and does this by the supplying of energy to the machine usually as shaft work but also possibly as heat from some still higher temperature level.

The objective of the heat engine is the transformation to work of a maximum portion of the energy which is supplied from a high-temperature source, but a necessary accompaniment of the engine is the rejection of an unavailable residue of energy to that universal energy receiver, the atmosphere. The heat engine thus operates through a temperature range which extends from atmospheric temperature upward. In distinction, the refrigerating machine has for its practical objective only the securing and maintenance of a temperature which is below that of the atmosphere, but again the energy which it discards must go to the same general energy receiver, the atmosphere. Consequently the refrigerating machine operates through a temperature range which extends downward from the temperature of the atmosphere.

31. THE CARNOT REFRIGERATION CYCLE

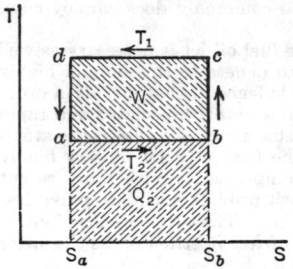

FIG. 1. Carnot Refrigeration Cycle

The Reversed Carnot Cycle may be described as one abstracting energy from a lower-temperature region and returning it to a region of higher temperature by means of the work supplied. Examining this reversed Carnot cycle by the aid of the $T\text{--}S$ diagram of Fig. 1 it is seen that, by operating the cycle in the direction indicated, heat Q_2 may be abstracted from a region at the lower temperature T_2 in the amount $T_2(S_b - S_a)$, and heat Q_1 may be delivered to a region at the higher temperature T_1 in the amount $T_1(S_c - S_d)$ or $T_1(S_b - S_a)$. For the operation of this complete cycle work will be required in the amount $Q_1 - Q_2$, or alternatively, $W/J = Q_1 - Q_2 = (T_1 - T_2)(S_b - S_a)$.

32. PERFORMANCE AND CAPACITY

As a natural objective in refrigeration is the removal of energy from the lower temperature region with a minimum expenditure of work, a suitable index of the performance of a refrigeration machine is the ratio between the amount of heat removed, or the refrigerating effect, and the work required. This ratio is called the *coefficient of performance* of the machine, or

$$\text{Coefficient of performance (C.P.)} = \frac{\text{Heat removed } (Q_2)}{\text{Work required } (W/J)}$$

For the reversed Carnot cycle the coefficient thus becomes

$$\text{C.P.} = \frac{Q_2}{W/J} = \frac{T_2(S_b - S_a)}{(T_1 - T_2)(S_b - S_a)} = \frac{T_2}{T_1 - T_2}$$

This Carnot or reversible cycle coefficient is the maximum coefficient of performance conceivably obtainable with any refrigerating machine operating in any manner with any fluid between a given (constant) lower temperature T_2 and a given (constant) upper temperature T_1.

The Carnot refrigeration cycle may ideally be realized with any vapor. It is instructive to see how a fluid such as water might be used. Referring to the $T\text{--}S$ diagram of a vapor, as presented in Fig. 2, and starting with the substance at state d in a saturated liquid state at the upper temperature (and pressure) of the cycle, let the liquid be expanded adiabatically and reversibly in a cylinder to state a. As it expands, the temperature drops, part of the liquid vaporizes, and some work output is obtained. Between states a and b additional liquid is caused to vaporize at the constant lower temperature and pressure of the cycle by the absorption of energy as heat from the cold region. Here the refrigeration is accomplished. After the partial vaporization to some suitable state b, work energy is supplied and the vapor and liquid mixture is compressed isentropically to a state c which in the figure is the saturated vapor state at the upper pressure and temperature. The vapor is now condensed at the constant upper temperature and pressure to the original state at d, condensation being effected by heat energy rejection in an amount equal to that absorbed from the lower-temperature region plus the *net* work of the cycle. The net work is that required for the compression less that regained by the expansion.

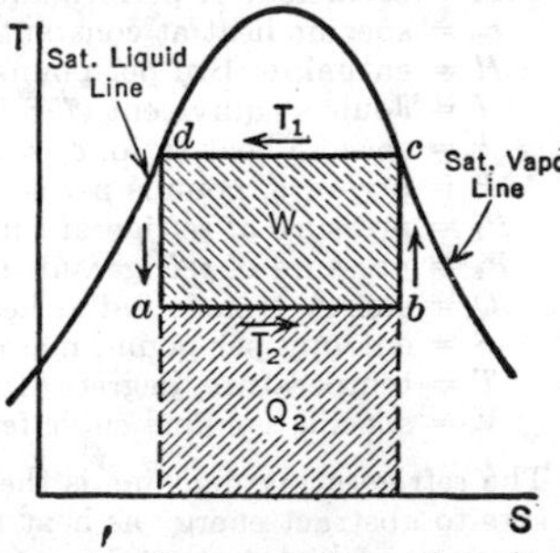

FIG. 2. Carnot Refrigeration Cycle with a Vapor

Example. A Carnot refrigeration cycle is operated between the temperatures of 50 and 100 F. What vapor pressures would be required if the fluid were steam, and what would be the coefficient of performance? What refrigeration would be accomplished per pound of steam?

Solution. From the saturated steam tables the necessary pressure would be 0.178 and 0.95 lb per sq in. abs, respectively. The coefficient of performance would equal $(50 + 460)/(100 - 50)$ or 10.2. The heat rejected per pound of steam to the upper temperature region at 100 F would be the enthalpy or "latent heat" of evaporation at that temperature or, from the steam tables, 1036.3 Btu per lb. From the definition of coefficient of performance and the necessary energy relations of the cycle,

$$10.2 = \frac{\text{Refrigeration}}{\text{Work}} = \frac{1036.3 - \text{Work per lb}}{\text{Work per lb}}$$

Solving this equation for the work term, the net work required per pound of steam circulating equals 92.2 Btu. Finally therefore the refrigeration accomplished per pound of steam $= Q_1 - W/J = 1036.3 - 92.2 = 944.1$ Btu.

For evaluating the rate of available energy delivery from an engine or power plant the horsepower (or the kilowatt) is employed as the conventional unit of capacity. The analogous unit which is used for expressing the rate of heat energy intake from the cold region of a refrigerating machine, or its *refrigerating capacity*, is the *tons of refrigeration per 24 hours*. This term is frequently abbreviated, and the rate of refrigerating effect is stated simply in *tons*.

The standard commercial ton is arbitrarily defined as a removal of energy as heat from the cold region at the rate of 288,000 Btu per 24 hours or $\left(\dfrac{288,000}{24 \times 60}\right) = 200$ Btu per minute. The unit originates from the fact that the "latent heat of fusion" of ice is approximately 144 Btu per lb or $(144 \times 2000 =)$ 288,000 Btu per short ton, whence a refrigerating machine which is operating at a capacity of one ton is absorbing energy at a rate equal to that which would exist if one ton of ice were melting in the refrigerated region each 24 hours.

An index of refrigerating-machine performance which is employed in practice rather more than the coefficient of performance is a derived one which is associated with the ton unit of capacity, namely, *horsepower required per ton* of refrigeration per 24 hours. The relation between the performance so expressed and the coefficient of performance is readily obtained by recalling that

$$\text{Refrigerating effect, Btu per minute} = \text{tons (per 24 hours)} \times 200$$

and

$$\text{Work required, Btu per minute} = \text{Hp} \times \frac{2545}{60} = \text{Hp} \times 42.4$$

whence

$$\text{Coefficient of performance} = \frac{\text{Tons} \times 200}{\text{Hp} \times 42.4} = \frac{4.71}{\text{Hp per ton}}$$

It is practically advantageous to modify the reversed Carnot cycle, or Carnot refrigeration cycle, in some particulars and to carry out the resulting fluid cycles in a succession of apparatus through which the fluid is caused to flow progressively and in each of which one phase of the cycle is accomplished. These practical adaptations fall in general into three classifications:

(*a*) *Vapor cycles*, in which the transfer of the vapor from the lower temperature and pressure of the cycle to the upper is accomplished through the agency of work supply to a *compressor*.

(*b*) *Gas (air) cycles*, in which the transfer of the fluid from the lower temperature and pressure to the upper is likewise effected by a *compressor*.

(*c*) *Vapor cycles* in which the transfer of the vapor from the lower pressure and temperature to the upper is accomplished in an *absorption system*, primarily through the agency of heat supplied at a still higher temperature.

These several practical cycles are considered in some detail in the following articles.

33. THE VAPOR COMPRESSION SYSTEM

The practical vapor compression system of refrigeration differs in principle from the Carnot cycle in two features. The one is the segregation of the several phases of the cycle into various devices to and through which the vapor is caused to flow progressively. The other is that, owing to the failure to develop as yet a practicable engine or turbine in which to accomplish the expansion of the condensed liquid from the upper temperature and pressure of the cycle to the lower, this expansion is in practice done in an irreversible throttling process which takes place in an *expansion valve*.

The resulting cycle as it would appear with these modifications, but still idealized in some respects, is represented in the T–S and H–S diagrams of Figs. 3 and 4, respectively. A diagrammatic representation of the arrangement of the apparatus appears in Fig. 5.

Starting at point a in each figure the fluid, which is called the *refrigerant*, leaves a storage tank in which it exists as a liquid at about atmospheric temperature and at its corresponding saturation pressure P_1. Between a and b the fluid is permitted to escape under control through the expansion valve into the low-pressure region of the cycle. Owing to the throttling character of the process the enthalpy of the fluid at state b must equal that at state a, with the result that a sensible portion of the liquid must vaporize, but owing to the lower pressure of the vapor its temperature must have fallen. Specifically

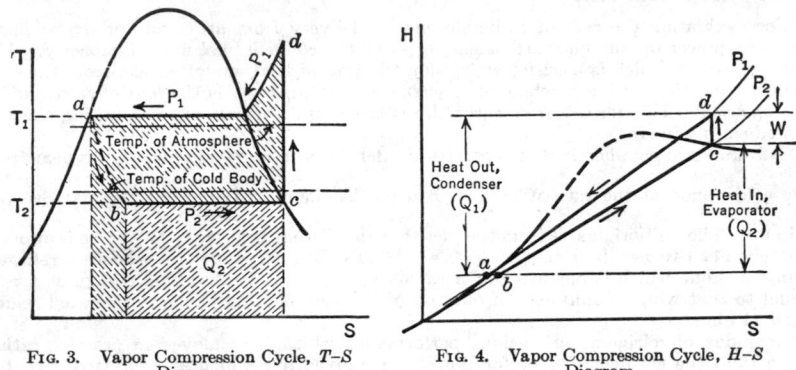

FIG. 3. Vapor Compression Cycle, T–S
Diagram

FIG. 4. Vapor Compression Cycle, H–S
Diagram

it must have dropped to the saturation temperature corresponding to the lower pressure.

After leaving the expansion valve the low-temperature and low-quality vapor mixture enters the *evaporator*. Here the provision is made for the absorption of energy as heat by the fluid from the region or substance which it is desired to refrigerate. The pipe coils which constitute the evaporator may be placed directly in the region that is to be cooled, in which event the system would be said to be one with *direct expansion*, or they may be placed in and act to cool some liquid with low freezing point, such as calcium chloride brine, which is circulated to the region that is to be cooled and which thus acts simply as a convective energy carrier. The latter system would be known as one with *indirect expansion*. In either event the reception of energy by the refrigerant will act to cause it to evaporate. The amount of energy received by each pound of the refrigerant and thus the degree of completeness of its evaporation will depend on the rapidity of its circulation, on the temperature difference between the fluid in the evaporator and its environs, and on other operating conditions. In the figures the fluid is shown as leaving the evaporator at state c, which is the saturated vapor state. It may in fact leave as a wet vapor mixture or even moderately superheated. In connection with the process of evaporation in the evaporator it is to be observed that the temperature therein must be *below* the temperature of the cold region in order that heat transition may occur in the desired direction. The pressure maintained therein must be the saturation pressure corresponding to the temperature, which pressure we shall designate as P_2.

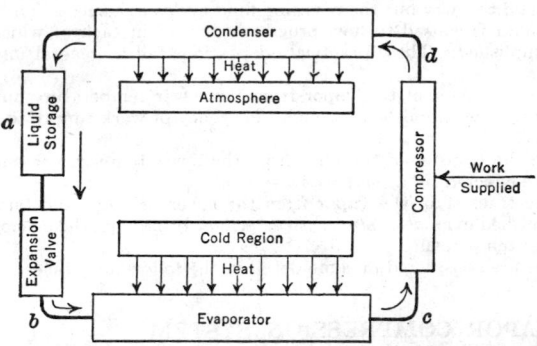

FIG. 5.   Elements of Vapor-compression Refrigeration Cycle

From the evaporator the refrigerant passes to the *compressor*. The function of the compressor is twofold. The one function is that of withdrawing the fluid from the evaporator at a rate sufficient to maintain the necessary reduced pressure and temperature in the evaporator. The other is that of compressing and delivering the fluid at a temperature which is adequately *above* that of the atmosphere or of the region or substance to which the fluid must next discard its "load" of energy. The increase of temperature of the vapor must be accompanied by increase of pressure. In this compression process, a reversible adiabatic or isentropic compression must be regarded as the best that might be obtained. The figures show such a compression. The state change represented by line c–d is one that will dry and superheat the refrigerant.

After leaving the compressor at state d the vapor enters the *condenser* in which it must first be de-superheated and then condensed. The departing energy passes as heat to *circulating water* which may be expected to be at or about atmospheric temperature. The temperature during the condensation must moderately exceed that of the water in order that the heat transition may proceed in the desired direction. The refrigerant pressure which must be delivered by the compressor is the saturation pressure P_1 corresponding to the condensing temperature. The state change of the refrigerant within the condenser is represented in the figures by line d–a.

A variety of fluids have been used as the refrigerant in vapor compression systems. For larger commercial installations ammonia is used almost universally at the present time. For the small self-contained household installations sulfur dioxide vapor is very popular. In marine installations on combatant ships neither of the foregoing is considered desirable on account of their active toxicity, and carbon dioxide is preferred. Various hydrocarbons such as ethyl chloride and propane are also used. Ammonia offers the advantages of moderate condenser pressures and moderate specific volume at the evaporator pressures. Sulfur dioxide offers the further advantage of even more moderate pressures but requires considerably more compressor displacement on account of fairly high specific volumes at the evaporator pressure. Carbon dioxide has the disadvantage of requiring very high pressures in both the condenser and the evaporator.*

Probably the most effective means of improving the performance of a refrigerating process, but at the same time one that is too frequently overlooked, is that of providing adequate, well-designed and well-maintained heat-transfer surfaces in the evaporator and in the condenser and an ample amount of cool circulating water to the condenser. The effect is to reduce as far as practicable the total temperature range, $T_1 - T_2$, through which the refrigerant must pass by enabling its lower temperature T_2 to be a minimum amount below the requisite cold region temperature and its upper temperature T_1 to be a minimum amount above the circulating water supply temperature.

The energy relations for the several phases of the cycle, both for actual operation and for the ideal cycle, are (regarding all kinetic energy terms as negligible), per pound of vapor circulated,

for the expansion valve,
$$H_a = H_b$$

for the evaporator,
$$H_b + Q_{in,\,b-c} = H_c, \quad \text{or}$$

for the compressor,
$$Q_{in,\,b-c} = H_c - H_b = H_c - H_a$$

$$JH_c + W_{in,\,c-d} = JH_d + JQ_{out,\,c-d}, \quad \text{or}$$

$$W_{in,\,c-d} = J(H_d - H_c) = JQ_{out,\,c-d}\ \dagger$$

$$= J(H_d - H_c)s \text{ for isentropic compression}$$

for the condenser,
$$H_d = H_a + Q_{out,\,d-a}, \quad \text{or}$$

$$Q_{out,\,d-a} = H_d - H_a$$

for the complete cycle,

$$W_{in} = J[(Q_{out,\,d-a} - Q_{in,\,b-c}) + Q_{out,\,c-d}] = J(H_d - H_c + Q_{out,\,c-d})\ \dagger$$

---

* For a complete description of the various refrigerants, a discussion of their adaptability under various conditions, and tables and charts of their thermodynamic properties see Macintire, *Handbook of Mechanical Refrigeration*, revised printing, John Wiley & Sons, 1940.

† As with any compressor, energy departure as heat *which might act to reduce the enthalpy increase* acts to reduce instead of increase the work requirement. In distinction, any heat emission occasioned by mechanical friction in moving parts obviously acts to increase the work.

$$\text{Coefficient of performance} = \frac{H_c - H_a}{H_d - H_c} = Q_{\text{out, } c-d}$$

$$= \frac{H_c - H_a}{(H_d - H_c)_S} \text{, for isentropic compression}$$

In these relations, the several subscripts are to be interpreted as referring to locations with respect to the various apparatus of the plant rather than to particular states indicated by the letters on the foregoing H–S or T–S diagrams.

Example. For an indirect-expansion ice-making installation operating with ammonia assume that the brine leaves the cooler (evaporator) at 14 F and that to obtain this temperature the necessary vapor temperature in the evaporator is 5 F. Also assume that the cooling water is supplied at 70 and leaves at 80 F, enabling the ammonia to condense at 86 F.

Assuming dry saturated vapor at the compressor suction, isentropic compression, and no sub-cooling of the liquid from the condenser, ascertain the pressures in the evaporator and the condenser, the temperature after compression, and the ideal coefficient of performance and horsepower per ton of refrigeration. Compare with the Carnot cycle performance between the vapor temperatures in the evaporator and condenser and with a Carnot cycle performance between the lowest brine temperature and the cooling water supply temperature. Indicate the reasons for the lower performance of the vapor cycle as compared with the Carnot between the same temperature limits.

Solution. From tables of the properties of ammonia (Bureau of Standards):

Evaporator (suction) pressure = saturated pressure corresponding to 5 F = 34.3 lb per sq in. abs (= 19.6 lb gage).

Condenser (discharge) pressure = saturated pressure at 86 F = 169.2 lb per sq in. abs (= 154.5 lb gage).

For saturated vapor at 5 F, $H = 613.3$ (= H_c), and $S = 1.3253$ (= S_c).

After compression at a constant entropy of 1.3253 to 169.2 lb abs, t (from charts) = 210 F and $H = 713$ (= H_d).

For saturated liquid at 86 F, $H = 138.9$ (= H_a and H_b).

$$\text{Coefficient of performance} = \frac{613.3 - 138.9}{713 - 613.3} = \frac{474.4}{99.7} = 4.76.$$

$$\text{Horsepower per ton of refrigerating effect per 24 hours} = \frac{4.71}{4.76} = 0.99.$$

$$\text{Carnot performance between } (460 + 5 =) \text{ 465 R and } (460 + 86 =) \text{ 546 R} = \frac{465}{546 - 465} = 5.74.$$

$$\text{Carnot performance between } (460 + 14 =) \text{ 474 R and } (460 - 70 =) \text{ 530 R} = \frac{474}{530 - 474} = 8.47.$$

The ammonia cycle coefficient is lower than the Carnot (4.76 vs. 5.74) because of (a) the irreversible throttling in the expansion valve and (b) the higher temperature at which the heat is rejected during the de-superheating of the ammonia vapor leaving the compressor.

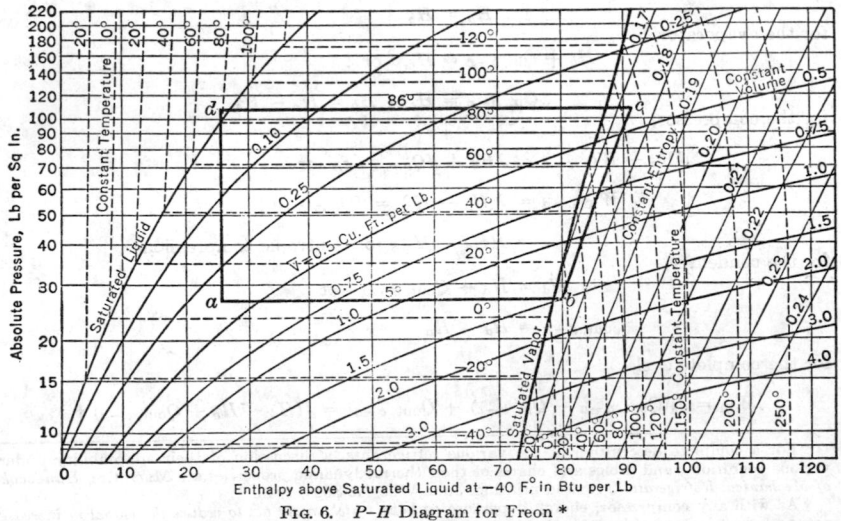

FIG. 6. P–H Diagram for Freon *

* Redrawn by permission from Barnard, Ellenwood, and Hirshfeld, *Heat Power Engineering*, Part III, John Wiley & Sons, 1933.

Charts Having Absolute Pressure Plotted against Enthalpy are found most useful to show the refrigeration process. Figure 6 shows such a chart for the refrigerant freon, upon which is drawn a characteristic cycle diagram. The throttling process is represented by the line da; the absorption of energy as heat in the evaporator is represented by the line ab; the adiabatic compression process is represented by the line bc; and the discarding of energy as heat to the atmosphere is represented by the line cd.

Table 1 shows the physical properties of the refrigerants in most common use.

Table 1. Some Properties of Refrigerants

Refrigerant	Symbol	Molecular Weight	Boiling Point at 1 Atmos., °F	Critical Temperature, °F	Critical Pressure, lb/sq in. abs	Freezing Point, °F	Liquid Density at Boiling Point, lb/cu ft	Gas Density,* lb/cu ft	Ratio of Specific Heats, $k = c_p/c_v$
Dichloromethane	CH_2Cl_2	84.906	105.0	421.0	670±	−142.0	186.9	0.225	1.14
Hexane	C_6H_{14}	86.112	155.9	454.6	435	−137.7	38.29	0.205	1.0
Pentane	C_5H_{12}	72.906	97.1	387.0	470	−203.4	38.09	0.185	1.0
Ethyl chloride	C_2H_5Cl	64.497	54.5	369.0	764	−217.7	56.50	0.177	1.13
Butane ‡	C_4H_{10}	58.080	30.9	307.8	510	−211.0	37.51	0.1549	1.11
Sulfur dioxide	SO_2	64.064	14.0	315.0	1127	− 97.6	91.15	0.1694	1.26
Isobutane	C_4H_{10}	58.080	+ 10.0	272.7	522	−229.0	37.20	0.1543	1.11
Methyl chloride	CH_3Cl	50.481	− 11.4	289.6	954	−153.2	62.30	0.1336	1.20
Dichlorodifluoromethane §	CCl_2F_2	120.916	− 22±			−247±		0.338±‡	1.12
Ammonia	NH_3	17.032	− 28.0	270.3	1636	−107.9	42.56	0.0446	1.32
Propane	C_3H_8	44.064	− 44.0	204.1	647	−309.8	36.35	0.1158	1.15
Propylene	C_3H_6	42.048	− 53.7	196.1	645	−301.4	38.12	0.1087	
Carbon dioxide	CO_2	44.00	−109.3†	88.0	1058	− 69.9		0.1144	1.30
Ethane	C_2H_6	30.048	−127.5	90.1	694	−277.6	34.13	0.0782	1.22
Ethylene	C_2H_4	28.032	−154.7	− 48.9	717	−272.9	35.49	0.0729	
Methane	CH_4	16.032	−258.9	−115.7	657	−297	19.20	0.0415	1.32

* At 30 in. Hg and 70 F.   † Solid.   ‡ At 00 F.   § Also called "F-12" and "freon."

34. THE AIR COMPRESSION SYSTEM

The apparatus required for an air compression system is in principle the equivalent of that required for a vapor system, *except that* for obtaining the low air temperature it is necessary that the compressed air shall do work in its expansion, instead of being permitted simply to throttle through an expansion valve. An air engine is therefore employed for effecting the pressure and temperature drop which must occur as the air passes to the cooling coils.

The arrangement of the elemental components of an air compression system of refrigeration is shown diagrammatically in Fig. 7 and the ideal state changes of the cycle are represented in the T, H–S, and P–V diagrams of Figs. 8 and 9, respectively.

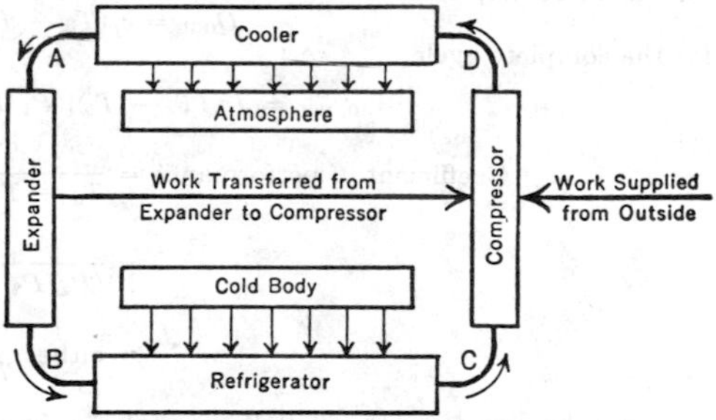

Referring to the several locations and state points indicated on the figures, state a represents air at slightly above atmospheric tem-

FIG. 7. Elements of Air-compression Refrigeration Cycle

perature but under a pressure considerably above atmospheric. The air at this state enters the *expander cylinder* or engine through which it would ideally expand isentropically to a state b, at which state the pressure commonly would still exceed considerably the atmospheric pressure but the temperature would be 50 F or more below the temperature of the region which it is desired to refrigerate. The work output from the expander engine is available for and would actually be used for assisting in driving the compressor of the system.

Between states b and c the air passes at ideally constant pressure through the refrigerating coils and there absorbs energy by heat transition from the cold region, departing from the refrigerating coils at a temperature still moderately below that of the cold region.

After leaving the coils at state c the air enters the compressor cylinder and is compressed to its original pressure at a but to a temperature considerably above that of the atmosphere. An isentropic state change, $c - d$, is represented in the figures, but as usual heat emission from the air during its compression would act to reduce the work required for

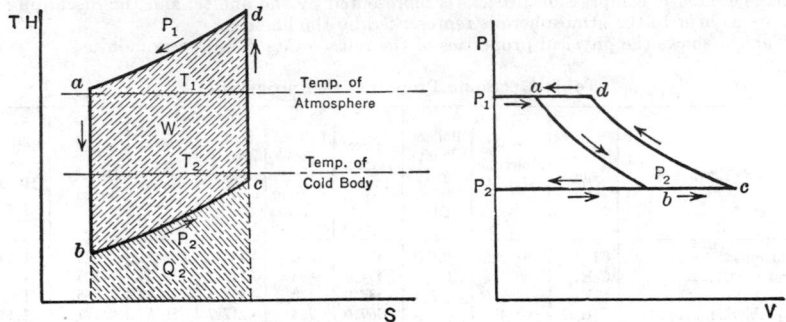

FIG. 8. T–S Diagram for Air-compression Cycle

FIG. 9. P–V Diagram for Air-compression Cycle

the compression. The work necessary for driving the compressor comes partly from an external source and partly from the expander engine.

Upon leaving the compressor at state d the hot air passes at ideally constant pressure through a cooler where it emits energy as heat to the circulating water and is thus recooled to the original state a.

Presuming the ideal conditions of constant pressure cooling and warming and isentropic expansion and compression the energy relations for the various phases of the cycle are:

for the expander cylinder, per pound of air,

$$W_{\text{out}} = Jc_p(T_a - T_b)$$
$$= Jc_pT_b[(P_1/P_2)^{(k-1)/k} - 1]$$

for the refrigerating coils,

$$Q_{\text{in}} = c_p(T_c - T_b)$$

for the compressor,

$$W_{\text{in}} = Jc_p(T_d - T_c)$$
$$= Jc_pT_c[(P_1/P_2)^{(k-1)/k} - 1]$$

for the air cooler,

$$Q_{\text{out}} = c_p(T_d - T_a)$$

for the complete cycle,

$$W_{\text{in, net}} = Jc_p(T_c - T_b)[(P_1/P_2)^{(k-1)/k} - 1]$$

$$\text{Coefficient of performance} = \frac{T_c - T_b}{T_d - T_a - T_c + T_b}, \quad \text{or}$$

$$= \frac{1}{(P_1/P_2)^{(k-1)/k} - 1}, \quad \text{or}$$

$$= \text{either } \frac{T_b}{T_a - T_b} \quad \text{or} \quad \frac{T_c}{T_d - T_c}$$

It may be seen from the T–S diagram of Fig. 8 or from the last expressions above that for a useful temperature range $T_a - T_c$ the air cycle coefficient of performance must be materially lower than that of the Carnot cycle, owing to the necessary temperature depression between b and c and the temperature elevation between d and a. At a given capacity the performance may be improved by a more rapid mass-rate of circulation of air, due to the consequent possibility of reducing the temperature range in the refrigerator $(T_c - T_b)$ and in the air cooler $(T_d - T_a)$.

The air compression system once held sway in marine installations, as the *dense air* system, but now it is practically abandoned partly because of its relatively poorer coefficient of performance and partly because of operating difficulties arising from the freezing of any moisture in the air and the consequent stoppage of the expander valves, etc.

35. THE ABSORPTION SYSTEM—REFRIGERATION THROUGH THE SUPPLYING OF ENERGY AS HEAT

In the refrigerating systems hitherto considered the energy required for passing the refrigerant from the lower temperature and pressure of the cycle to the upper temperature and pressure was supplied as work at the compressor. Instead of effecting this by work supply through a power-driven compressor, arrangements may also be made whereby it is accomplished through a direct supply of heat energy from a high-temperature source. Several typical forms and arrangements of apparatus have been devised and are in use for this purpose, all of which may be designated, however, as *absorption systems*. Those in practical use employ ammonia as the refrigerant and act by the alternate absorption of the cold ammonia vapor by some *adsorbent* and subsequent elimination of the ammonia at a higher temperature by the application of heat energy. The system in most common use employs water as the ammonia adsorbent.

The elements of an ammonia absorption system are shown diagrammatically in Fig. 10. The usual condenser, expansion valve, and evaporator are evident. Through these occurs

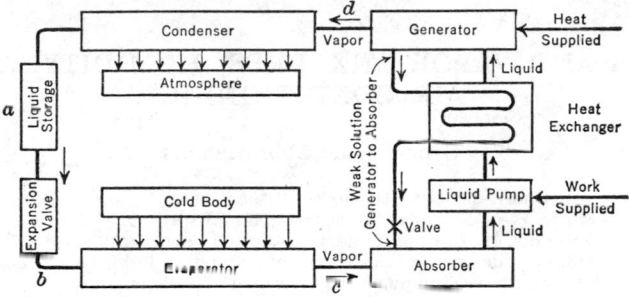

FIG. 10. Elements of Vapor-absorption Refrigeration Plant

the customary progressive flow of the refrigerant. The distinctive items of the system are the *absorber*, *liquid pump*, *heat exchanger*, and *generator*. These jointly replace the conventional compressor of the compression system, and through them there is a closed circuit of an ammonia-water solution. The general actions in this circuit are as follows.

The cold vapor leaving the evaporator passes to the absorber and is there avidly absorbed by a weak solution of water and ammonia known as the *weak aqua*. This absorption process has the characteristic of emitting energy, whence it is necessary that cooling water be circulated through coils in the absorber in order to maintain the temperature of the liquid sufficiently low and thus permit the building up of an adequate ammonia concentration. The weak aqua in the absorber thus becomes eventually a *strong aqua* and is then delivered by the liquid pump through the heat interchanger to the generator. In the generator heat energy is supplied, usually by a low-pressure steam coil, and the temperature of the strong aqua is thereby raised sufficiently that the ammonia which had been taken up in the absorber is eliminated from the strong aqua and is thus delivered as ammonia vapor to the condenser. The resulting weak aqua then completes the aqua circuit by returning through the heat exchanger to the absorber. The action in the exchanger is the mutual one of warming the strong aqua which is en route from the absorber to the generator by the cooling of the weak aqua which is returning from the generator to the absorber.

In certain smaller absorption systems the liquid pump, which is observed to be the only moving part required in the system, is dispensed with by employing the same element alternately as the absorber and as the generator, with a consequent intermittent action of the system. Also there is an ingenious system in use in household units in which the pump is dispensed with by introducing hydrogen gas in the evaporator and absorber, with the result that the system is under virtually the same *total* pressure throughout. Circulation is brought about by differences in the densities of the liquids in various parts of the system. The total pressure in the evaporator is the sum of the partial pressures of the hydrogen gas and the ammonia vapor, but the temperature is established only by the partial pressure of the saturated ammonia vapor.

Various other adaptations of the absorption system employ other absorbents, such as anhydrous ammonium nitrate and silica gel.

The energy relations for the condenser, expansion valve, and evaporator of an absorption system are in principle the same as those for a compression system, but in practical application they may become more complicated because there may be a carry-over of some of the adsorbent (such as water vapor in the aqua system) from the generator to and through the condenser and evaporator. For the absorber, pump, exchanger, and generator, together with the various accessory apparatus which are found in the actual plant, the energy relations become still more involved by reason of the semi-chemical character of the processes of vapor absorption by and elimination from the adsorbent.

It will have been observed that the elemental requisites for refrigeration are simply a supply of a liquid refrigerant under some pressure, an expansion valve, and the coils in which the expanded refrigerant may evaporate at the lower pressure and temperature. If an unlimited supply of the liquid refrigerant were available and if its saturation pressure at the desired refrigeration temperature should perchance be above the pressure of the atmosphere, then continued refrigeration might be secured simply by letting the refrigerant escape. It would thus appear that the greater portion of the equipment which is employed for effecting a practical, economical, and continuously operating refrigeration system is required only in order that the refrigerant may be recovered and be reliquefied at atmospheric temperature.

GAS AND VAPOR MIXTURES, HUMIDITY, AND AIR CONDITIONING

Symbols and Abbreviations

a = subscript designating the air constituent of a mixture.

f = subscript designating the saturated liquid state.

g = subscript designating the (dry) saturated vapor state.

fg = subscript designating the change of a property during change of a fluid from the saturated liquid to the saturated vapor state.

l = subscript designating a liquid constituent of a mixture.

m = subscript designating a mixture.

P = subscript designating constancy of pressure.

V = subscript designating constancy of specific volume.

v = subscript designating a vapor constituent of a mixture.

w = subscript designating the water (vapor, liquid, or both) constituent in a mixture.

x, y, z = subscripts designating the several constituents x, y, and z of a (gas) mixture.

ϕ = relative humidity.

c = specific heat, of a constituent or a mixture.

E = molecular energy, Btu per pound of a constituent or of a mixture.

H = enthalpy, Btu per pound of a constituent or of a mixture.

k = ratio of specific heats (c_p/c_v).

m = molecular weight of a constituent (or of mixture).

M = mass of a constituent (or of mixture).

P = partial pressure of a constituent (or total pressure of mixture).

Q = energy transferred by radiation or conduction (as heat).

R = gas constant of a constituent (or of a mixture).

S = entropy per pound of a constituent (or of mixture).

t = temperature, degrees Fahrenheit.

T = temperature, degrees Rankine (degrees Fahrenheit absolute).

$V_{x, y, \text{ or } z}$ = volume (not specific volume) of a constituent x, y, or z in a mixture if the constituent is segregated and put under the temperature and total pressure of the mixture, as in volumetric analysis.

V_g = specific volume of (dry) saturated vapor.

x = quality (saturated-vapor fraction) of a mixture of saturated liquid and vapor.

Various of the fluids which are encountered in engineering processes are mixtures either of gases or of a gas and vapor. Thus dry air is itself a mixture primarily of oxygen and nitrogen, and atmospheric air is a further mixture of dry air with superheated water vapor or with saturated vapor and perhaps some unvaporized liquid, as in clouds or during fog and rainfall. Several of the simpler of such types of mixtures are presented in the following.

36. GAS MIXTURES

The **Actual Gas Mixture** consists of a haphazard aggregation of molecules of each constituent, the molecules of any single constituent being distributed uniformly throughout the entire space and also acting or moving quite as if they occupied the space alone. As a consequence the total pressure exerted by the mixture against the walls of its container by reason of the molecular bombardment of those walls may be distributed between several *partial* pressures, each of which is individually attributable to the molecules of a corresponding individual constituent and each of which would be the pressure exerted by that constituent if it alone should fill the space or volume occupied by the mixture (V_m) at the common temperature (T_m) of it and of all the components of the mixture. The values of the partial pressures of the gases in a mixture are:

$$P_x = M_x R_x \frac{T_m}{V_m} = 1545 \frac{M_x}{m_x} \frac{T_m}{V_m}$$

$$P_y = M_y R_y \frac{T_m}{V_m} = 1545 \frac{M_y}{m_y} \frac{T_m}{V_m}$$

$$P_z = M_z R_z \frac{T_m}{V_m} = 1545 \frac{M_z}{m_z} \frac{T_m}{V_m}$$

where P_x, P_y, and P_z = the partial pressures attributable individually to constituents x, y, and z.

$\quad M_x$, M_y, and M_z = the relative masses of those constituents.

$\quad R_x$, R_y, and R_z = their individual gas constants.

$\quad m_x$, m_y, and m_z = their molecular weights.

$\quad\quad T_m$ = the mixture temperature.

$\quad\quad V_m$ = the volume occupied by the mixture.

The **Relation between the Partial Pressure** of any one constituent (as x) and that of any other (as y) is

$$\frac{P_x}{P_y} = \frac{M_x}{M_y} \frac{R_x}{R_y} = \frac{M_x/m_x}{M_y/m_y} \qquad (1)$$

Also,

$$\frac{P_x}{P_m} = \frac{P_x}{P_x + P_y + P_z} = \frac{M_x R_x}{M_x R_x + M_y R_y + M_z R_z}$$

$$= \frac{(M_x/M_m) R_x}{(M_x/M_m) R_x + (M_y/M_m) R_y + (M_z/M_m) R_z}$$

$$= \frac{\dfrac{M_x/M_m}{m_x}}{\dfrac{M_x/M_m}{m_x} + \dfrac{M_y/M_m}{m_y} + \dfrac{M_z/M_m}{m_z}}$$

where M_m = mass of mixture, $M_x + M_y + M_z$.

$\quad P_m$ = total pressure, $P_x + P_y + P_z$.

The several ratios M_x/M_m, M_y/M_m, and M_z/M_m in the last relation evidently represent the mass proportions of each constituent with respect to the mass of the whole, as would be determined by or would determine the gravimetric analysis of the mixture. It frequently happens that, as in the practical analysis of furnace gases or of an internal-combustion engine exhaust, the several constituents of a gas mixture are in effect segregated progressively and their relative volumes are determined *when maintained under the original temperature but brought (individually) to the original total pressure of the mixture.* An expression of the proportional composition of a mixture in terms of the ratios of the individual volume of the constituent, when so segregated and at the temperature and total pressure of the mixture, to the actual mixture volume is known as its *volumetric* analysis.

The **Volume of Each Component** under such temperature and pressure conditions is

$$V_x = M_x R_x \frac{T_m}{P_m} = 1545 \frac{M_x T_m}{m_x P_m}, \text{ etc.,}$$

and the ratio between those volumes for any pair of the constituents is

$$\frac{V_x}{V_y} = \frac{M_x R_x}{M_y R_y} = \frac{M_x/m_x}{M_y/m_y} \left(\text{and} = \frac{P_x}{P_y}, \text{ by equation 1} \right) \tag{2}$$

Also relations are readily obtainable by which to convert from gravimetric to volumetric analyses, and *vice versa.* Thus

$$\frac{V_x}{V_m} = \frac{V_x}{V_x + V_y + V_z} = \frac{M_x/m_x}{M_x/m_x + M_y/m_y + M_z/m_z}$$

$$= \frac{\dfrac{M_x/M_m}{m_x}}{\dfrac{M_x/M_m}{m_x} + \dfrac{M_y/M_m}{m_y} + \dfrac{M_z/M_m}{m_z}} \tag{3}$$

and

$$\frac{M_x}{M_m} = \frac{M_x}{M_x + M_y + M_z} = \frac{V_x m_x}{V_x m_x + V_y m_y + V_z m_z}$$

$$= \frac{(V_x/V_m)m_x}{(V_x/V_m)m_x + (V_y/V_m)m_y + (V_z/V_m)m_z} \tag{3a}$$

Relations are thus available by which to ascertain the partial pressure of any constituent of a gas mixture and by which to convert from gravimetric to volumetric analysis, or the reverse. It remains to be observed that any given mixture has its own effective or equivalent gas constant, molecular weight, and specific heats, which may be used in analyses of mixture processes quite as for any single gas.

R_m, **the Equivalent Gas Constant for the Mixture,** is obtained from

$$R_m = (M_x/M_m)R_x + (M_y/M_m)R_y + (M_z/M_m)R_z \tag{4}$$

Since, for the gases, $mR = 1545$, an equivalent molecular weight may be computed for a mixture by the relation

$$m_m = \frac{1545}{R_m} \tag{4a}$$

c_m, **an Equivalent Specific Heat for the Mixture,** is obtained from

$$c_m = \frac{M_x}{M_m} c_x + \frac{M_y}{M_m} c_y + \frac{M_z}{M_m} c_z \tag{5}$$

With the foregoing relations at hand it is in general possible to proceed with analyses of engineering processes with a gas mixture quite as though the mixture were a single gas, *so long as the mass proportions of the mixture do not change* during the process. In this connection it will be recalled that the practice of regarding a gas mixture as effectively equivalent to a single gas is exemplified by the treatment of the properties and energy relations of air as though it were a single gas.

37. GAS AND VAPOR MIXTURES

There are Several Distinctive Conditions of Gas-vapor Mixtures which are of general concern and which are classifiable as:

(a) Mixtures of gas and (low-pressure) superheated vapor, as exemplified by the air and superheated water vapor which constitutes normal atmospheric air.

(b) Mixtures of gas and (low-pressure) saturated vapor, as exemplified by the atmosphere at the incipiency of fog or rainfall.

(c) Mixtures of gas, saturated vapor, and saturated liquid, as exemplified by the atmosphere during fog or rainfall.

Considering these in order:

(a) The low-pressure vapors, when even very moderately superheated, and in fact even when saturated if at a sufficiently low vapor pressure, are found to conform closely to the gas characteristics. As a consequence, engineering computations involving gas-superheated vapor mixtures are commonly made as for gas mixtures, employing the foregoing gas mixture and gas relations. However, with respect to a given mixture, before one may proceed with confidence upon that basis it is necessary first to verify the fact of, or to ascertain the degree of, superheat of the vaporous constituent. To do so it is necessary to possess data concerning the temperature and total pressure of the mixture and the mass proportion of the vapor. It is further necessary to recall that the vapor pressure (or temperature) of a *saturated* vapor is established solely and unchangeably by its temperature (or pressure), as is also the density or any other property of the saturated vapor. The following example will illustrate the principles involved perhaps better than a collection of formulas, since recourse must be had to the steam tables in any case.

Example 1. A parcel of atmospheric air at a (total) pressure of 14.7 lb per sq in. (abs) and a temperature of 70 F is found (by hygrometric determination) to contain 0.009 lb of water vapor per lb of air (not mixture). Is the vapor superheated and, if so, by how many degrees? What is the partial pressure of the air? Compute the density of each constituent, the density of the mixture, the mass of vapor per pound of mixture, and the density of moisture-free air at the above temperature and pressure. What is the equivalent gas constant and molecular weight of the mixture? Compute the enthalpy per pound of mixture, relative to air and water at 32 F.

Solution. By equation 1 *et seq.*:

$$P_v, \text{ if superheated } = 14.7 \ \frac{\dfrac{0.009/1.009}{18}}{\dfrac{0.009/1.009}{18} + \dfrac{1.0/1.009}{29}} = 0.210 \text{ lb per sq in.}$$

Temperature of saturated vapor at 0.21 lb per sq in. = 54.4 F (from steam tables).
Vapor is therefore superheated by (70 − 54.4 =) 15.6 F.
P_{air} = 14.7 − 0.21 = 14.49 lb per sq in.

$$\text{Density, vapor} \quad = \frac{0.21 \times 144}{86 \times (70 + 460)} = 0.00066 \ (86 = R \text{ for superheated vapor}).$$

$$\text{Density, air} \quad = \frac{14.49 \times 144}{53.3 \times (70 + 460)} = 0.07393$$

$$\text{Density, mixture} \quad\quad = 0.07459 \text{ lb per cu ft.}$$

$$\text{Vapor per pound of mixture} = \frac{0.00066}{0.07459} \left(\text{or} \ \frac{0.009}{1.009} \right) = 0.00892 \text{ lb.}$$

$$\text{Density, moisture-free air} = \frac{14.7 \times 144}{53.3(70 + 460)} = 0.07493 \text{ lb per cu ft.}$$

R_m = 0.00892 × 86 + 0.99108 × 53.3 = 53.6; m_m = 1545/53.6 = 28.8
H_{air}, per lb of mixture = 0.99108 × 0.241 × (70 − 32) = 9.07 Btu.
H_{vapor}, per lb of mixture = $(M_v/M_m)H_{g.70°}$ (since for low-pressure vapor H depends solely on temperature) = 0.00892 × 1090.8 = 9.73 Btu.
$H_{mixture}$, per lb = 9.07 + 9.73 = 18.80 Btu.

(b) In the consideration of mixtures consisting of a gas and a saturated vapor, or so-called *saturated* mixtures, the items of outstanding significance are that, although the strictly gaseous constituents continue to conform to the typical gas characteristics, the *saturated* vapor constituent now acts independently, that is, in conformity with the typical vapor characteristics. The application of these considerations, as well as certain additional features and properties of a saturated mixture, are developed in the following examples.

Example 2. An air-water vapor mixture at 14.7 lb per sq in. total pressure is known to be saturated (i.e., the vapor constituent is saturated vapor) at 70 F. Ascertain the partial pressures and densities of the vapor and air constituents and the density of the mixture. Compute the mass of vapor per pound of air and per pound of mixture, and the relative enthalpy per pound of mixture.

Solution.
P, saturated vapor at 70 F, by steam tables = 0.363 lb per sq in.
P of air = 14.7 − 0.363 = 14.337 lb per sq in.

$$\text{Density of vapor} = \frac{1}{869}, \text{ by tables, or } = \frac{0.363 \times 144}{86 \times (460 + 70)} = 0.00115$$

$$\text{Density of air} \quad = \frac{14.337 \times 144}{53.3 \times (460 + 70)} = 0.07301$$

$$\text{Density of mixture} \qu\quad = 0.07416$$

$$\text{Vapor per pound of air} = 0.00115/0.07301, \text{ or (equation 1) } \frac{0.363}{14.337} \times \frac{18}{29} = 0.0157.$$

Vapor per pound of mixture = 0.00115/0.07416, or 0.0157/1.0157 = 0.0155.
R_m = 0.0155 × 86 + 0.9845 × 53.3 = 53.8.
H_{vapor}, per lb of mixture = 0.0155 × 1090.8 = 16.91 Btu
H_{air}, per lb of mixture = (1.000 − 0.0155) × 0.241 × (70 − 32) = 9.02 Btu

$H_{mixture}$, per lb = 25.93 Btu

Example 3. Referring to the mixture of example 1 (0.009 lb of vapor per lb of air, not mixture), ascertain the temperature at which the mixture would become saturated upon cooling at constant (total) pressure, and compute the amount of energy removal as heat necessary to have effected the specified cooling, Btu per pound of mixture.

Solution. As the low-pressure vapor conforms to gas characteristics down to saturation, and as the same mass proportions between the vapor and air will persist so long as the cooling is not sufficient to have produced any condensation of the vapor, the vapor pressure is the same as in example 1 or 0.210 lb per sq in. The temperature at which the vapor (and mixture) will become saturated is therefore its saturation temperature corresponding to that pressure, or 54.4 F.

$H_{air, per lb of mixture}$ = 0.99108 × 0.241 × (54.4 − 32) =   5.35 Btu
$H_{vapor, per lb of mixture}$ = 0.00892 × 1083.7   =   9.67 Btu

$H_{mixture, per lb}$ = 15.02 Btu

Energy abstracted for constant pressure cooling = ΔH = 18.80 − 15.02 = 3.78 Btu per lb of mixture.

R is the same as in example 1.

In connection with the last example it is to be noted that the temperature at which a (unsaturated) mixture becomes saturated upon cooling at constant pressure is known as the *dew point* of the original mixture.

(c) Mixtures of a gas, saturated vapor, and saturated liquid may be encountered when a saturated mixture is cooled or compressed, or when a volatile liquid is introduced into a space containing a gas but in a greater amount than can exist in the space as saturated vapor at the existing temperature. Again the gas constituent conforms to the gas characteristics, but the properties of the saturated liquid and vapor are independently interrelated in accordance with the saturation characteristics of the vapor. The following example illustrated one phase of these considerations.

Example 4. The mixture of examples 1 and 3 is further cooled at a total pressure of 14.7 lb per sq in. to a temperature of 40 F. Ascertain the partial pressures and densities of the remaining vapor and of the air; the masses of saturated vapor and of the fog or rain which must have condensed out, pound per pound of air and per pound of mixture; the quality of the vapor-fog mixture; the enthalpy of the mixture; and the heat-energy removal necessary to have cooled from the dew point to 40 F.

Solution.

P, saturated vapor at 40 F, by steam tables, = 0.1217 per sq in.
P of air = 14.7 − 0.1217 = 14.578 lb per sq in.
Density of vapor (by tables) = 1/2445 = 0.00041.

$$\text{Density of air} = \frac{14.578 \times 144}{53.34 \times (460 + 40)} = 0.07878.$$

Vapor per lb of air = 0.00041/0.07878 = 0.0052 lb.
Fog per lb of air = 0.009 − 0.0052 = 0.0038 lb.
Fog per lb of mixture = 0.0038/1.009 = 0.00376 lb.
Vapor per lb of mixture = 0.0052/1.009 = 0.00515 lb.
Quality of vapor-fog component = 0.0052/0.009 = 0.577 = 57.7%.
$H_{vapor, per lb of mixture}$ = (0.0052/1.009) × 1077.1   = 5.55 Btu
$H_{fog, per lb of mixture}$ = (0.0038/1.009) × 8.05   = 0.03 Btu
$H_{air, per lb of mixture}$ = (1.0/1.009) × 0.241 × (40 − 32) = 1.91 Btu

H, per lb of mixture   = 7.49 Btu

Energy abstracted, = ΔH = 15.02 − 7.49 = 7.53 Btu per lb of mixture

38. HUMIDITY AND HYGROMETRY

In many circumstances, as in scientific ventilation and air conditioning, it becomes most essential that air of closely prescribed temperature and moisture content shall be provided. This moisture content may be designated in several ways, but the index commonly employed by the engineer is that of the *relative humidity* of the mixture. This is defined as the *ratio between the density of the actual (superheated) vapor in the mixture and the density of saturated vapor at the temperature of the mixture.* As for both circumstances the density of the (low-pressure) vapor is satisfactorily computable by the gas relation, density = P/RT, it follows that the relative humidity is also expressed by the ratio between the actual partial pressure of the vapor in the mixture and its saturation pressure corresponding to the mixture temperature. Thus, denoting relative humidity by the symbol ϕ

$$\phi = \frac{\text{Actual vapor density}}{\text{Density, sat. vapor at mixture temperature}}$$

or

$$= \frac{\text{Actual vapor pressure}}{\text{Sat. pressure at mixture temperature}}$$

To illustrate, the mixture of example 1 would be said to have a relative humidity of (0.00066/0.00115 or 0.210/0.363 =) 0.575, or 57.5 per cent. In the same parlance, the saturated mixture of example 2 would be said to have a relative humidity of 100 per cent. Likewise, in accordance with example 3, the mixture of example 1 would attain a 100 per cent relative humidity upon cooling to its dew point (at 54.4 F). Its humidity might alternatively have been brought to 100 per cent if supplied with (0.0155 − 0.00892) 0.0066 lb of moisture (per pound of mixture) while maintained at 70 F and atmospheric pressure.

Devices Which Determine the Humidity of gas-vapor mixtures are known as *hygrometers* or *psychrometers*. The type which is most used by the engineer consists simply of two thermometers, with the bulb of one *dry* and the bulb of the other covered by a wick which is kept *wet* with the same liquid as exists in the vapor phase in the mixture (as the water in atmospheric air). An active current of the mixture under test is caused to pass across the wet bulb, effecting a continuous vaporization from the bulb and consequent cooling and temperature depression, due to the energy absorption required for producing the vaporization.

The immediate environs of the wet bulb are thus a region to which there pass steadily the air and superheated vapor in the air mixture under test, where the water which passes up the wick is evaporated, and from which there departs a mixture of air and *saturated* vapor. By writing the steady-flow energy equation for this region it is possible to develop an expression for direct evaluation of the mass of the vapor per pound of air (not mixture) in the air mixture as supplied. From these data the relative humidity is readily computable by the methods of example 1 and the above.

In developing the equation the following symbols are employed:

t_d = dry bulb (actual mixture) temperature, degrees Fahrenheit.

t_w = temperature registered by wet bulb, degrees Fahrenheit.

$M_{v,1}$ = mass of (superheated) vapor per pound of air (not mixture) in the approaching mixture at t_d.

$M_{v,2}$ = mass of (saturated) vapor per pound of air (not mixture) in the *mixture* leaving the immediate surface of the wet-bulb wick, at t_w.

$M_{v,2} - M_{v,1}$ = mass of liquid passing up wick, at t_w and vaporizing, pound per pound of air (not mixture).

$H_{f,1}$ = enthalpy of saturated liquid at t_d, Btu per pound.

$H_{g,1}$ = enthalpy of (superheated) vapor in the approaching mixture at t_d, Btu per pound. At the very low vapor pressures encountered this effectively equals the enthalpy of saturated vapor (H_g) *at the same temperature,* t_d. (See example 1.)

$H_{fg,1}$ = enthalpy of vaporization at t_d.

$H_{f,2}$ = enthalpy of liquid passing up wick to wet bulb, at t_w, Btu per pound.

$H_{g,2}$ = enthalpy of saturated vapor leaving wet bulb, at t_w.

$H_{fg,2}$ = enthalpy of vaporization at t_w.

$H_{a,1}$ = enthalpy of air (not mixture) approaching at t_d Btu per pound.

$H_{a,2}$ = enthalpy of air (not mixture) leaving at t_w Btu per pound.

c_p = specific heat of air at constant pressure = 0.241.

Q_r = energy to wet bulb as heat by radiation, etc., from its warmer environs, Btu per pound of air (not mixture).

Writing the steady-flow energy equation for the wet bulb by equating the sum of the energies coming to the bulb via the water which passes up the wick, via the approaching mixture and by radiation, to the sum of the energies leaving via the departing air and saturated vapor, all energies being in Btu per pound of air (not mixture),

$$[(M_{v,2} - M_{v,1}) \times H_{f,2}] + [1.0 \times H_{a,1} + M_{v,1} \times H_{g,1}] + Q_r = [1.0 \times H_{a,2} + M_{v,2} \times H_{g,2}]$$

or, collecting terms in $M_{v,1}$,

$$M_{v,1}(H_{g,1} - H_{f,2}) = M_{v,2}(H_{g,2} - H_{f,2}) - (H_{a,1} - H_{a,2}) - Q_r$$

But

$$H_{g,1} - H_{f,2} = H_{f,1} + H_{fg,1} - H_{f,2} = H_{fg,1} + (t_d - t_w)$$

$$H_{g,2} - H_{f,2} = H_{fg,2}$$

and

$$H_{a,1} - H_{a,2} = c_p(t_d - t_w) = 0.241(t_d - t_w)$$

whence,

$$M_{v,1} = \frac{M_{v,2}H_{fg,2} - 0.241(t_d - t_w) - Q_r}{H_{fg,1} + (t_d - t_w)} \qquad (6)$$

In this relation the value of $M_{v,2}$ is directly computable from equation 1 (Art. 36), that is, $M_{v,2}$ = (for air and water vapor) 18 $P_{v,2}/29$ $(P_{\text{mix.}} - P_{v,2})$, where $P_{v,2}$ is the saturation pressure of the vapor at t_w. (See also example 2.)

Measurement of Q_r is quite impracticable in actual use of the wet and dry bulb hygrometer, but tests indicate that with a conventional wick-covered mercury thermometer the radiation correction may be compensated with reasonable accuracy if the observed wet-bulb temperature be reduced by about 0.016 (1.6 per cent) of the observed wet-bulb depression $(t_d - t_w)$.

The application of equation 6 is shown by the following example, which contains data such as would be secured in the use of a wet and dry bulb hygrometer with the mixture of example 1.

Example 5. Observed psychrometric data taken with a wet and dry bulb hygrometer in air at a barometric pressure of 14.7 lb per sq in. were

$$t_d = 70 \text{ F}; \quad t_w = 60.65 \text{ F}$$

Compute the mass of vapor per pound of air (not mixture) and the relative humidity of the mixture as supplied.

Solution.

Corrected wet-bulb temperature = 60.65 − 0.016(70 − 60.65) = 60.50 F.

Saturated vapor pressure at 60.50 F = 0.2608 lb per sq in.

$M_{v,2}$ = 18 × 0.2608/29 × (14.7 − 0.2608) = 0.01121.

$H_{fg,1}$ (at 70 F) = 1052.7; $H_{fg,2}$ (at 60.50 F) = 1057.9.

$$M_{v,1} = \frac{0.01121 \times 1057.9 - (0.241 \times 9.50)}{1052.7 + 9.50} = \frac{9.58}{1062.2} = 0.0090.$$

$$\frac{P_{v,1}}{14.7 - P_{v,1}} = \text{(by equation 1)} \frac{0.009/18}{1/29} = \frac{1}{69}; \quad P_{v,1} = 14.7/70 = 0.210.$$

Relative humidity = 0.210/0.363 = 57.8 per cent.

Psychrometric Charts for Air-water Vapor Mixtures at or about a total pressure of 14.7 lb per sq in., as well as tables giving the mass of saturated vapor per pound of air at various temperatures but at the standard atmospheric pressure, are provided in the various mechanical engineering handbooks and give in graphical and tabular form the results of computations such as the foregoing. Their utility is in general limited, however, to the above total pressure. A psychrometric chart in common use is that devised by H. W. Carrier and shown on p. 8-105 as Fig. 1.

The **Ferrel Psychrometric Formula,*** with empirical constants that hold for temperatures between −40 and 140 F, is:

$$\frac{p' - p}{t_d - t_w} = 0.000367 p_m \left(\frac{t_w + 1539}{1571}\right)$$

where p = the actual partial pressure of the water vapor in the mixture, in inches of mercury absolute.

p' = the saturation pressure at the wet-bulb temperature, in inches of mercury absolute.

t_d = the dry-bulb temperature, in degrees Fahrenheit.

t_w = the wet-bulb temperature, in degrees Fahrenheit.

p_m = the total pressure of the mixture (commonly the barometric pressure), in inches of mercury absolute.

This relation is the wet-bulb energy equation with certain constants supplied. With the dry-bulb and wet-bulb temperatures and the total mixture pressure experimentally determined, the actual partial pressure of the water vapor in the mixture may be found from this formula.

The saturation pressure of water vapor for temperatures at 32 F and above may be found from steam tables or charts. For convenience, Table 1 is included, in which the

---

* U. S. Dept. of Agriculture, Weather Bureau Bull. 235.

saturation pressures of water vapor in contact with ice at temperatures from −20 to 31 F are given. The values in this table are from the Weather Bureau Bulletin 235.

With the partial pressure, p, of the water vapor in the mixture known, the *dew point* can then be found, since the dew point of the mixture is the saturation temperature cor-

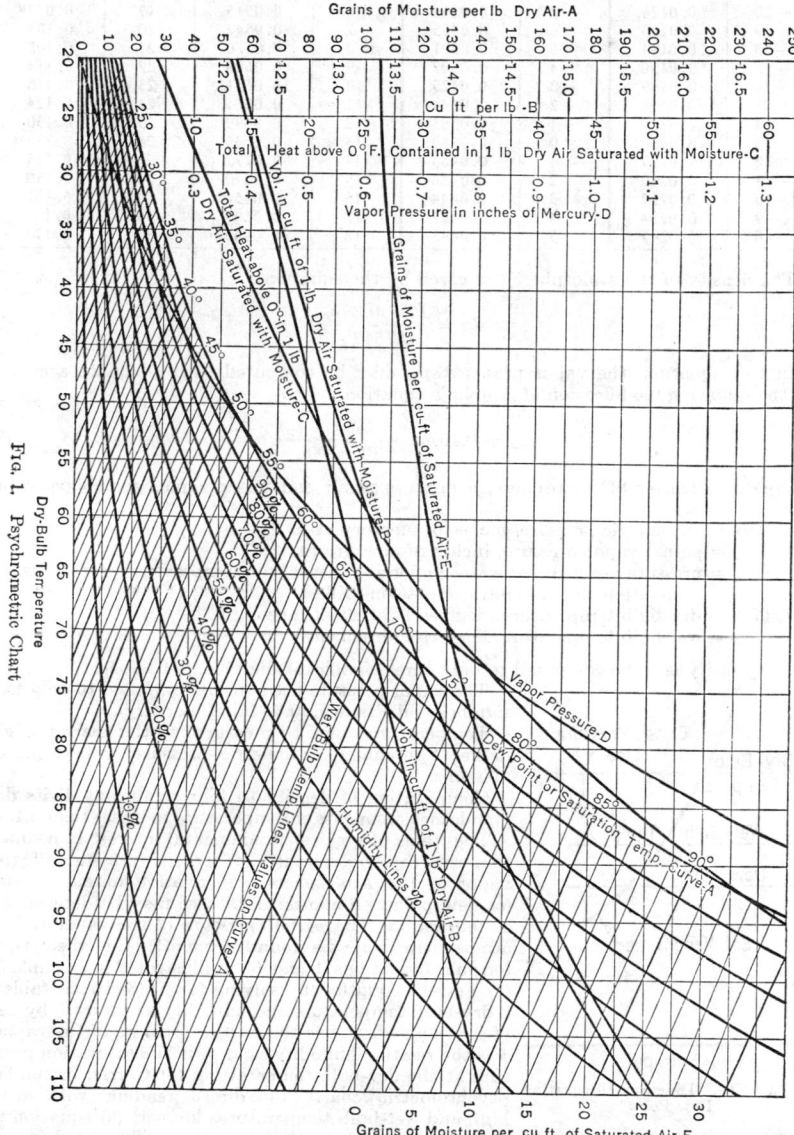

Fig. 1.   Psychrometric Chart

responding to this vapor pressure. If the mixture were cooled at constant pressure, the dew point would be that temperature at which condensation of the water vapor would start. The dew point is shown as t_3 in Fig. 2. The relationship between dry-bulb temperature, wet-bulb temperature, and dew point is also shown in the same figure for the mixture in which the water vapor is superheated.

Table 1. Pressures of Saturated Water Vapor in Contact with Ice

t °F	p in. hg	t °F	p in. hg	t °F	p in. hg	t °F	p in. hg
−20	0.0126	−7	0.0260	6	0.0515	19	0.0979
−19	0.0133	−6	0.0275	7	0.0542	20	0.103
−18	0.0141	−5	0.0291	8	0.0570	21	0.108
−17	0.0150	−4	0.0307	9	0.0600	22	0.113
−16	0.0159	−3	0.0325	10	0.0631	23	0.118
−15	0.0168	−2	0.0344	11	0.0665	24	0.124
−14	0.0178	−1	0.0363	12	0.0699	25	0.130
−13	0.0188	0	0.0383	13	0.0735	26	0.136
−12	0.0199	+1	0.0403	14	0.0772	27	0.143
−11	0.0210	2	0.0423	15	0.0810	28	0.150
−10	0.0222	3	0.0444	16	0.0850	29	0.157
−9	0.0234	4	0.0467	17	0.0891	30	0.164
−8	0.0247	5	0.0491	18	0.0933	31	0.172

The density of the atmosphere * is given by the equation:

$$d_m = \frac{B - 0.38 p_v}{0.754 T_d} \tag{7}$$

In this equation, the vapor pressure, p_v, may be computed with adequate accuracy by the following modification of Apjohn's equation:

$$p_v = 0.49 p_w - \frac{B}{30} \frac{(t_d - t_w)}{90} \tag{8}$$

where d_m = density of atmosphere, a mixture of air and water vapor, pounds per cubic foot.

B = barometric pressure, inches of mercury at 32 F.

p_v = actual vapor pressure, inches of mercury at 32 F.

p_w = pressure of saturated water vapor at the wet bulb temperature, in pounds per square inch (as found from steam tables).

t_d (T_d) = dry-bulb temperature, degrees Fahrenheit (absolute).

t_w = wet-bulb temperature, degrees Fahrenheit.

The density is to be computed to only three significant digits (e.g., 0.0764).

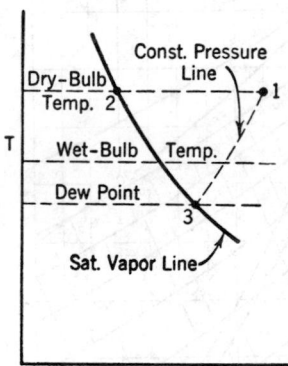

Const. Pressure Line

Dry-Bulb Temp. 2

Wet-Bulb Temp.

Dew Point

Sat. Vapor Line

Entropy

FIG. 2.   T–S Diagram Illustrating Dew Point

The vapor pressure needs to be computed to only two significant digits (e.g., 0.63 in. of mercury).

Temperatures must be measured to the nearest single degree Fahrenheit; barometric pressure to 0.01 in. of mercury.

Computation of density to four significant digits demands measurements of temperature to closer than 0.1 F.

The items of greatest fundamental interest in connection with hygrometry are (a) partial pressure of the vapor in the atmosphere, (b) relative humidity, and (c) dew point. Summarizing, these may be determined as follows: (a) The partial pressure of the vapor in the atmosphere may be found from (1) the theoretical formula (equations 1 and 6) as illustrated in example 5; (2) Ferrel's equation by solving for p; (3) steam tables, if dry-bulb temperature and humidity are known, by use of $p = \phi p_g$, where p is the partial pressure of the vapor, ϕ is the relative humidity, and p_g is the saturation pressure at the dry-bulb temperature; (4) Carrier or similar psychrometric charts by direct reading with dry-bulb and wet-bulb temperatures known; (5) equation 8.

(b) Relative humidity may be found from (1) the theoretical formula as illustrated in example 5; (2) Ferrel's equation, in which ϕ = p/saturation pressure at dry-bulb temperature; (3) Carrier or similar psychrometric charts by direct reading, with dry-bulb and wet-bulb temperatures known; (4) tables giving values of humidity for various dry-

* Based upon considerations discussed in three articles by Professor C. H. Berry of Harvard University, published in *Combustion*, Vol. 6, August, 1934, p. 15; September, 1934, p. 24; October, 1934, p. 21.

bulb and wet-bulb temperatures. (c) The *dew point* may be found from (1) Carrier or similar psychrometric chart by direct reading with dry-bulb and wet-bulb temperatures known; (2) steam tables, the temperature of saturation at the partial pressure of the vapor; (3) dew-point instrument, which measures dew-point temperature directly.

39. AIR CONDITIONING

Air Conditioning consists chiefly of the simultaneous control of temperature, humidity, motion, and purity of air to meet the requirements of human comfort. Some of the principal factors that affect human comfort and welfare as influenced by air environment are the (1) dry-bulb temperature, (2) humidity, (3) motion, (4) distribution, (5) dust content, (6) bacteria content, and (7) odors. Other factors which may influence comfort, but the effects of which are not so well established at the present time, are (8) light, (9) ozone content, (10) ionic content, and (11) pressure.

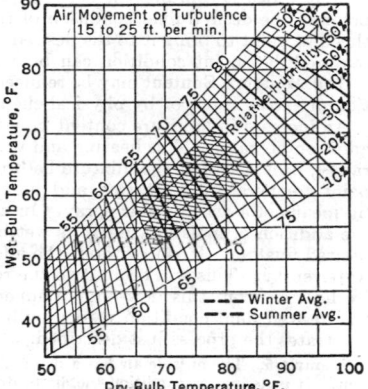

FIG. 3. Effective Temperature and Comfort Zones

Human Occupancy of a confined space produces at least six important alterations in the properties of the air: (1) the oxygen content is decreased slightly; (2) the carbon dioxide content is increased slightly; (3) products of decomposition, usually accompanied by odors, are given off; (4) the air temperature is raised; (5) the humidity is increased by evaporation of moisture from the skin and lungs; and (6) the number of positive and negative ions in a unit volume of the air is decreased.

Effective Temperature may be defined as an arbitrary index of the degree of warmth or cold felt by the human body in response to temperature, humidity, and air movement. The numerical value of the effective temperature index for any given air condition is fixed by the temperature of saturated air which, at a velocity or turbulence of 15 to 25 ft per min, induces a sensation of warmth or cold like that of the given condition. Figure 3 shows the effective temperature-humidity relation at various values of humidity and dry-bulb temperature when the air movement is between 15 and 25 ft per min. The intersection of the dry-bulb and the wet-bulb temperature lines will give the relative humidity and also the effective temperature. This figure shows the comfort zone as established by experiment, upon which are shown the average comfort lines for winter and summer. The fact that the summer average is greater than the winter average is probably due partly to adaptation to seasonal weather and partly to differences in clothing worn in the two seasons.

It is now recognized that the importance of air motion in air conditioning ranks second only to temperature. In an occupied space, air having all the other essential qualities but lacking motion feels stagnant, stuffy, and depressing, because the vitiated air next to the body is not replaced by air possessing satisfactory qualities. The air in an occupied space should be in constant motion sufficient to give uniformity in temperature and humidity without objectionable draft.

Air Purity is a vital problem principally in manufacturing establishments, crowded rooms, or tenement buildings. The air should be free from all toxic, unhealthful, or disagreeable gases and fumes and relatively free from odors and dust.

In an occupied space, at least 10 cu ft of fresh air per minute per person should be provided from an outside source to remove body heat, body odors, and products of respiration. In rooms where the occupants smoke, an additional 6 to 7 cu ft of fresh air per minute per person should be provided. This volume of air change is inadequate to provide proper air motion. It is customary to provide the minimum quantity of fresh air required for the removal of heat and odors and to recirculate the additional volume required for proper air motion.

Control of Relative Humidity is very important if comfortable conditions are to be maintained. Cold, outside air in winter usually will have a very low relative humidity after it has been heated. Such air should be humidified. On the other hand, air in the summer months may have a too great relative humidity. The standard established by the American Society of Heating and Ventilating Engineers states: The relative humidity shall be not less than 30 per cent, nor more than 60 per cent, in any case. The effective temperature shall range between 64 and 69 F when heating or humidification is required and between 69 and 73 F when cooling or dehumidification is required.

Air-conditioning Equipment has evolved from the air washer which has been used in large buildings for many years. In the air washer, air is forced through a chamber containing a water spray and then over baffle plates which eliminate free moisture and wet particles of solid matter. Although the air washer was initially used for the removal of dust from the air, its principal use in air conditioning is to change the water-vapor content of the air. Under typical winter conditions, merely heating the air will result in a water-vapor content too low to satisfy comfort requirements. Under usual summer conditions, mere cooling to the desired final temperature will result in too high a water-vapor content. The most satisfactory method which has been devised for effecting a desired water-vapor content at a desired dry-bulb temperature, and one which is used in the large majority of cases, is that which is known as the dew-point method. In the dew-point method of temperature and humidity control, air is brought to a saturated condition at the temperature at which it contains the desired quantity of moisture. This temperature is the dew point of the desired final condition of the air. The air may then be merely heated from this dew point to bring it to the desired dry-bulb temperature and relative humidity. The required dew-point condition can be reached only by the removal or addition of water.

The **Moisture Content** may be reduced by sufficiently cooling the air to cause condensation of excess vapor or by physical-chemical means.

An increase in moisture content may be accomplished by a combination of heating and addition of water. This heating and water addition may be done simultaneously or separately, provided that the mixture be brought to a saturated condition at the desired dewpoint temperature. Heating and humidification may be accomplished simultaneously by means of a warm water spray. In another process, the air is first preheated, without the addition of moisture, until its wet-bulb temperature is equal to the dew point of the desired final air. Then, without the addition or removal of heat, the air is *saturated* in a water spray. This will bring it to the required, saturated, dew-point condition. Finally, by heating from this dew-point temperature, the air-vapor mixture will be brought to the desired final condition of dry-bulb temperature and relative humidity. Example 6 illustrates the process just described.

Example 6. The outside air has a dry-bulb temperature of 40 F and a relative humidity of 65 per cent. It is desired to condition the air to dry-bulb temperature of 74 F and relative humidity of 45 per cent, which will give an effective temperature of about 69 F.

The air is (1) heated before it enters the washer until the wet-bulb temperature is equal to the dew point of the final air, then (2) saturated in the spray to the temperature equal to the dew point of the final condition, then (3) heated to the dry-bulb temperature required. This brings it to the final desired relative humidity.

A tabulation of the air condition in the various stages will illustrate the process.

Table 2

	Initial Air Condition	After Heater	After Spray	Final Air Condition
Dry-bulb temperature, °F........	40.0	72.0	51.3	74.0
Relative humidity..............	0.65	0.205	1.00	0.45
Wet-bulb temperature, °F........	35.6	51.3	51.3	60.5
Dew point, °F.................	29.6	29.6	51.3	51.3
Vapor pressure, lb per sq in.......	0.0791	0.0791	0.1869	0.1869

The preheat temperature and the temperature of the water spray need not be definitely as described as long as the air from the spray is saturated and at the temperature of 51.3 F. If the water is warmer than 51.3 F, not so much preheating is required. Example 7 illustrates the dehumidification of air under summer conditions.

Example 7. The outside air has a dry-bulb temperature of 95 F and a relative humidity of 65 per cent. It is desired that the final condition of example 6 be attained.

Table 3

	Initial Condition	After Cooler	Final Condition
Dry-bulb temperature, °F....................	95.0	51.3	74.0
Relative humidity........................	0.65	1.00	0.45
Wet-bulb temperature, °F....................	84.0	51.3	60.5
Dew point, °F............................	81.4	51.3	51.3
Vapor pressure, lb per sq in....................	0.5297	0.1869	0.1869

In this case the cooler must decrease the temperature of the saturated mixture to the desired dew point of 51.3 F.

BIBLIOGRAPHY

1. ADAMS, A. S., *Fundamentals of Thermodynamics*, New York and London, Harper, 1945.
2. ALLEN, J. R., and BURSLEY, J. A., *Heat Engines*, New York, McGraw-Hill, 5th ed., 1941.
3. American Society of Heating and Ventilating Engineers' *Guide*.
4. American Society of Mechanical Engineers, *Fluid Meters Report*.
5. American Society of Refrigerating Engineers' *Data Book*.
6. BADGER, W. L., *Heat Transfer and Evaporation*, Chemical Catalog Co., 1926.
7. BAILEY, N. P., *Heat Engineering*, New York, John Wiley, 1942.
8. BARNARD, W. N., ELLENWOOD, F. O., and HIRSHFELD, C. F., *Elements of Heat Power Engineering*, New York, John Wiley, 3rd ed., 1933.
9. BEHAR, M. F., *Manual of Instrumentation*, Instruments Publishing, 1932.
10. BINDER, R. C., *Fluid Mechanics*, Prentice-Hall, 1943.
11. BIRTWISTLE, G., *Principles of Thermodynamics*, New York, Macmillan, 1925.
12. BOYER, G. C., *Diesel and Gas Engine Power Plants*, New York, McGraw-Hill, 1943.
13. BRIDGMAN, P. W., *Nature of Thermodynamics*, Harvard University Press, 1941.
14. BROWN, S. P., *Air Conditioning and Elements of Refrigeration*, New York, McGraw-Hill, 1947.
15. BROWN, A. I., and MARCO, S. M., *Introduction to Heat Transfer*, New York, McGraw-Hill, 1942.
16. BUTTERFIELD, T. E., JENNINGS, B. H., and LUCE, A. W., *Steam and Gas Engineering*, New York, Van Nostrand, 4th ed., 1947.
17. CARSLAW, H. S., *Mathematical Theory of the Conduction of Heat in Solids*, Dover Publications, 2nd ed., 1945.
18. CHURCH, E. F., JR., *Steam Turbines*, New York, McGraw-Hill, 1928.
19. CORK, J. M., *Heat*, New York, John Wiley, 1942.
20. COX, G. N., and GERMANO, F. J., *Fluid Mechanics*, New York, Van Nostrand, 1941.
21. CRAIG, D. P., and ANDERSON, H. J., *Steam Power and Internal Combustion Engines*, New York, McGraw-Hill, 2nd ed., 1937.
22. CROFT, H. O., *Thermodynamics, Fluid Flow and Heat Transmission*, New York, McGraw-Hill, 1938.
23. DEGLER, H. E., *Internal Combustion Engines*, New York, John Wiley, 1938.
24. DIEDRICHS, H., and ANDRAE, W. C., *Experimental Mechanical Engineering*, New York, John Wiley, 1930.
25. DODGE, B. F., *Chemical Engineering Thermodynamics*, New York, McGraw-Hill, 1944.
26. DOOLITTLE, J. S., and ZERBAN, A. H., *Engineering Thermodynamics*, Scranton, International Textbook, 1949.
27. ELLENWOOD, F. O., and MACKEY, C. O., *Vapor Charts*, New York, John Wiley, 2nd ed., 1944.
28. ELLIOT, B. G., and CONSOLIVER, E. L., *The Gasoline Automobile*, New York, McGraw-Hill, 4th ed., 1932.
29. EMSWILER, J. E., and SCHWARTZ, F. L., *Thermodynamics*, New York, McGraw-Hill, 5th ed., 1939.
30. EVERETT, H. A., *Thermodynamics*, New York, Van Nostrand, 2nd ed., 1941.
31. EWING, J. A., *Thermodynamics for Engineers*, New York, Macmillan, 1920.
32. FAIRES, V. M., *Applied Thermodynamics*, New York, Macmillan, 1938.
33. FAIRES, V. M., *Elementary Thermodynamics*, New York, Macmillan, 1938.
34. FISHENDEN, M., and SAUNDERS, H. M., *Calculation of Heat Transmission*, London Stationery Office, 1932.
35. FRAAS, A. P., *Combustion Engines*, New York, McGraw-Hill, 1948.
36. FRASER, E. S., and JONES, R. B., *Motor Vehicles and Their Engines*, New York, Van Nostrand, 4th ed. (rev.), 1930.
37. GAFFERT, G. A., *Steam Power Stations*, New York, McGraw-Hill, 3rd ed., 1946.
38. GEBHARDT, G. F., *Steam Power Plant Engineering*, New York, John Wiley, 6th ed., 1925.
39. GLASSTONE, S., *Thermodynamics for Chemists*, New York, Van Nostrand, 1947.
40. GREENE, A. M., JR., *Principles of Heating, Ventilating and Air Conditioning*, New York, John Wiley, 1936.
41. HARDING, L. A., *Steam Power Plant Engineering*, New York, John Wiley, 1932.
42. HARDING, L. A., and WILLARD, A. C., *Heating, Ventilating and Air Conditioning*, New York, John Wiley, 2nd ed., 1921.
43. HASLAM, R. T., and RUSSELL, R. P., *Fuels amd Their Combustion*, New York, McGraw-Hill, 1926.
44. HATSOPOULOS, G. N., and KUNAN, J. H., *Principles of General Thermodynamics*, New York, John Wiley, 1966.
45. HOARE, F. E., *Textbook of Thermodynamics*, London, E. Arnold & Co., 1938.
46. Hoffman, J. D., *Handbook for Heating and Ventilating Engineers*, New York, McGraw-Hill, 2nd ed., 1926.
47. HOTTEL, H. C., WILLIAMS, G. C., SATTERFIELD, C. N., *Thermodynamic Charts for Combustion Processes*, New York, John Wiley, 1949.
48. HOUGEN, O. A., WATSON, K. M., *Chemical Process Principles, Part I (Material and Energy Balances)*, New York, John Wiley, 1947.
49. HOUGEN, O. A., WATSON, K. M., *Chemical Process Principles, Part II (Thermodynamics)*, New York, John Wiley, 1947.
50. HOUGEN, O. A., WATSON, K. M., *Industrial Chemical Calculations*, New York, John Wiley, 2nd ed., 1936.
51. INGERSOLL, L. R., ZOBEL, O. J., INGERSOLL, A. C., *Heat Conduction*, New York, McGraw-Hill, 1948.
52. JAKOB, M., and HAWKINS, G. A., *Elements of Heat Transfer*, New York, John Wiley, 3rd ed., 1957.
53. JENNINGS, B. H., and LEWIS, S. R., *Air Conditioning and Refrigeration*, Scranton, International Textbook, 3rd ed., 1957.
54. JORDAN, R. C., PRIESTER, G. B., *Refrigeration and Air Conditioning*, New York, Prentice-Hall, 1948.
55. JOST, W., *Explosion and Combustion Processes in Gases* (trans. by H. O. Croft), New York, McGraw-Hill, 1946.
56. KEENAN, J. H., *Thermodynamics*, New York, John Wiley, 1941.
57. KEENAN, J. H., and KAYE, J., *Gas Tables—Thermodynamic Properties of Air, Products of Combustion and Component Gases—Compressible Flow Functions*, New York, John Wiley, 1948.
58. KEENAN, J. H., KEYES, F. G., HILL, P. G., and MOORE, J. G., *Steam Tables: Thermodynamic Properties of Water Including Vapor, Liquid, and Solid Phases*, New York, John Wiley, 1969.

59. KENT, R. T., *Mechanical Engineer's Handbook*, New York, John Wiley, 1950.
60. KIEFER, P. J., and Stuart, M. C., *Principles of Engineering Thermodynamics*, New York, John Wiley, 2nd ed., 1954.
61. LICHTY, L. C., *Internal Combustion Engines*, New York, McGraw-Hill, 5th ed., 1939.
62. LICHTY, L. C., *Thermodynamics*, New York, McGraw-Hill, 1947.
63. MACINTIRE, H. J., *Refrigeration Engineering*, New York, John Wiley, revised printing, 1947.
64. MACKEY, C. O., *Air Conditioning Principles*, Scranton, International Textbook, 1947.
65. MACNAUGHTON, E., *Elementary Steam Power Engineering*, New York, John Wiley, 3rd ed., 1948.
66. MALEEV, V. L., *Internal Combustion Engines—Theory and Design*, New York, McGraw-Hill, 2nd ed., 1945.
67. MCADAMS, W. H., *Heat Transmission*, New York, McGraw-Hill, 2nd ed., 1942.
68. MORRISON, L. H., *American Diesel Engines*, New York, McGraw-Hill, 2nd ed., 1939.
69. MORSE, F. T., *Power Plant Engineering and Design*, New York, Van Nostrand, 1932.
70. MOYER, J. A., *Power Plant Testing*, New York, McGraw-Hill, 3rd ed., 1926.
71. MOYER, J. A., and FITTZ, R. V., *Air Conditioning*, New York, McGraw-Hill, 2nd ed., 1938.
72. MOYER, J. A., CALDERWOOD, J. P., POTTER, A. A., *Elements of Engineering Thermodynamics*, New York, John Wiley, 6th ed., 1941.
73. NORRIS, E. B., THERKELSEN, E., *Heat Power*, New York, McGraw-Hill, 2nd ed., 1939.
74. OBERT, E. F., *Internal Combustion Engines*, Scranton, Pa., International Textbook, 1950.
75. OBERT, E. F., *Thermodynamics*, New York, McGraw-Hill, 1948.
76. PERRY, J. H., *Chemical Engineers' Handbook*, New York, McGraw-Hill, 1941.
77. POLSON, J. A., *Internal Combustion Engines*, New York, John Wiley, 2nd ed., 1942.
78. POTTER, A. A., CALDERWOOD, J. P., *Elements of Steam and Gas Power Engineering*, McGraw-Hill, 4th ed., 1938.
79. PYE, D. R., *Internal Combustion Engines*, Vols. I and II, Oxford, England, Clarendon Press, 1931.
80. RABER, B. F., and HUTCHINSON, F. W., *Refrigeration and Air Conditioning Engineering*, New York, John Wiley, 1945.
81. ROYDS, R., *Heat Transmission in Boilers, Condensers, and Evaporators*, New York, Van Nostrand, 1921.
82. RUMMEL, A. J., and VOGELSONG, L. O., *Practical Air Conditioning*, New York, John Wiley, 1941.
83. SAWYER, R. T., *The Modern Gas Turbine*, New York, Prentice-Hall, 1945.
84. SCHACK, A., SMITH, H. W., and PARTRIDGE, E. P., *Industrial Heat Transfer*, New York, John Wiley, 1965.
85. SEVERNS, W. H., *Heating, Ventilating and Air Conditioning*, New York, John Wiley, 2nd ed., 1949.
86. SEVERNS, W. H., and DEGLER, H. E., *Steam, Air, and Gas Power*, New York, John Wiley, 5th ed., 1954.
87. SHARPE, N., *Refrigeration Principles and Practices*, New York, McGraw-Hill, 1949.
88. SHOOP, C. F., and TUVE, G. L., *Mechanical Engineering Laboratory Practice*, New York, McGraw-Hill, 3rd ed., 1941.
89. SOLBERG, H. J., CROMER, O. C., and SPALDING, A. R., *Elementary Heat Power*, New York, John Wiley, 1946.
90. SPARKS, N. R., *Theory of Mechanical Refrigeration*, New York, McGraw-Hill, 1948.
91. SPORN, P., AMBROSE, E. R., BAUMEISTER, T., *Heat Pumps*, New York, John Wiley, 1947.
92. STEINER, L. E., *Introduction to Chemical Thermodynamics*, New York, McGraw-Hill, 1941.
93. STOEVER, H. J., *Applied Heat Transmission*, New York, McGraw-Hill, 1941.
94. SUTTON, G. P., *Rocket Propulsion Elements*, New York, John Wiley, 1949.
95. TAYLOR, C. F., and TAYLOR, E. S., *The Internal Combustion Engine*, Scranton, Pa., International Textbook, rev. ed., 1938.
96. TIETJENS, O. G., *Applied Hydro- and Aero-Mechanics*, New York, McGraw-Hill, 1934.
97. TIETJENS, O. G., *Fundamentals of Hydro- and Aero-Mechanics*, New York, McGraw-Hill, 1934.
98. TRINKS, W., *Industrial Furnaces*, New York, John Wiley, Vol. I, 4th ed., 1967; Vol. II, 1961.
99. VENNARD, J. K., *Elementary Fluid Mechanics*, New York, John Wiley, 4th ed., 1961.
100. WEBER, H. C., *Thermodynamics for Chemical Engineers*, New York, John Wiley, 1939.
101. WENNER, R. R., *Thermochemical Calculations*, New York, McGraw-Hill, 1941.
102. WRANGHAM, D. A., *The Theory and Practice of Heat Engines*, Cambridge University Press, 1942.
103. YOUNG, V. W., and YOUNG, G. A., *Elementary Engineering Thermodynamics*, New York, McGraw-Hill, 3rd ed., 1947.
104. ZEMANSLEY, M. W., *Heat and Thermodynamics*, New York, McGraw-Hill, 1943.
105. ZUCROW, M. J., *Principles of Jet Propulsion and Gas Turbines*, New York, John Wiley, 1948.

SECTION 9

ELECTROMAGNETICS AND CIRCUITS

REVISED BY

JOHN A. M. LYON *

* This section is a revision of Section 9 of the second edition. The original material was prepared under the authorship of Charles Weyl, Irven Travis, and Carl C. Chambers, of the Moore School of Electrical Engineering of the University of Pennsylvania.

ELECTROMAGNETICS AND CIRCUITS

Scope

This section presents the principles of electricity and magnetism together with formulas basic to the solution of electric and magnetic circuits.

1. SYSTEMS OF UNITS

The rationalized system of meter-kilogram-second units, often referred to as the Giorgi system, is used throughout this section unless otherwise indicated. The so-called mks system may be considered as formed through the use of practical units for the electromagnetic quantities in addition to the meter as a unit of length, the kilogram as a unit of mass, and the second as a unit of time, in order to convert the practical system into an "absolute" system. Historically, several systems of units preceded the mks system. The electrostatic and electromagnetic systems were both "absolute" systems based on the centimeter as a unit of length, the gram as a unit of mass, and the second as a unit of time. The electrostatic unit of charge, the *statcoulomb*, is defined on the basis of Coulomb's law for the force which exists between two charges with a given distance between them and in a homogeneous medium. On the basis of this law, the electrostatic unit of charge is related to the fundamental units of length, mass, and time. The electromagnetic unit of charge follows directly from the electromagnetic unit of current, which is defined by Ampère's law for the force between two current elements with a given distance of separation. On this basis the electromagnetic unit of charge, the *abcoulomb*, was a much larger unit of charge than the statcoulomb, the corresponding unit quantity of charge in the electrostatic system. For practical usage, certain units which constitute the practical system were found convenient. These units are: the unit of resistance, the *ohm;* the unit of potential, the *volt;* the unit of current, the *ampere;* the unit of quantity of charge, the *coulomb;* the unit of capacitance, the *farad;* the unit of inductance, the *henry;* the unit of power, the *watt;* and the unit of energy, the *watt-second* or *joule.* Working backwards from these practical units of length, mass, and time associated with the mks system, and substituting all of these in Ampère's law, it was found possible to define the permeability of free space μ_0. Substituting these units, or whichever ones are appropriate, into Coulomb's law, it was found possible to define the permittivity of free space ϵ_0. In this way, the practical system has been extended into an "absolute" system, thus giving rise to the mks system of units. Since the mks system as here used includes the coulomb as a unit charge, it is sometimes designated the mksc system. By occasional reference to Section 3 on Physical Units and Standards, it is possible for the reader to transfer from one system of units to another. Standard symbols for quantities are given in Section 1, Table 4. When vector relations are involved, the symbol is in bold-face type.

ELECTRON THEORY

2. ELECTRICITY AND MATTER

Electrical Entities. Contemporary physics identifies electricity with matter. For example, an atom of the light isotope of the element hydrogen is at present thought to consist of a single nuclear quantity with a positive charge of electricity called a *proton*, about which rotates a negative charge of electricity called an *electron*. These two charges are equal in magnitude and opposite in sign. Each represents what is believed to be the smallest quantity of electricity that can exist discretely. The simplest atom, the light isotope of hydrogen, consists of one orbital electron moving about a nucleus composed of one proton.

An element may have a number of *isotopes*. Isotopes are alternative forms of atoms, each atom having the same number of protons but differing from each other with respect to the number of neutrons. The nucleus of the light hydrogen isotope is the simplest nucleus known; it consists of one proton. The nucleus of the heavier hydrogen isotope, called *deuterium*, is next to the simplest in degree of complexity; it consists of one proton and one neutron. Deuterium occurs as 1 part in 5000 in ordinary hydrogen. The heaviest hydrogen isotope, *tritium*, has one proton and two neutrons in the nucleus. Tritium occurs rarely in nature and must be artificially produced.

All atoms have a diameter of about 10^{-10} m. Coulomb's law, discussed in a later paragraph in this section, has been authenticated by experiment for distances considerably greater than this atomic distance.

Although there are many particles of theoretical physics, those of primary interest in electrical matters are: the *atom*, the smallest unit of matter, which remains unchanged in chemical reactions; the *proton*, a basic particle of atomic nuclei which has a positive charge equal to that of the negative *electron*, and a mass of 1.67×10^{-27} kilogram.

The *neutron* when not associated with another particle does not enter into the passage of electric current since the neutron itself has no charge. Electrically speaking, the neutron is of interest primarily by its contribution of mass in the nucleus of an atom.

The *positron* was discovered shortly after the neutron. The positron has the same mass as an electron but a charge of opposite sign. It has no place in the atom model described above. The positron has apparently a very short life, which added to the difficulty of discovery.

An additional electrical concept is the *photon*. It can be defined as a quantum of electromagnetic radiation having a definite frequency and a definite energy content. It has been related to the energy radiated or absorbed when an electron passes from one energy level to another within an atom.

The study of electricity has to do with the properties of these entities, in particular the proton and negative electron.

Mass, Energy, and Frequency Relations. Modern quantum physics identifies mass with energy. Thus the law of conservation of energy has been revised to require that the total mass plus energy of the universe be conserved. Quantum physics also to a certain degree identifies electric particles with electromagnetic waves. Under some circumstances entities which have properties normally ascribed only to particles exhibit characteristics normally observed only in waves. In such cases the energy of the particle and the frequency of the wave are proportional.

The above principles are stated in the following experimental laws:

(a) Whenever energy is transformed into mass or mass is transformed into energy

$$E = mc^2 \tag{1}$$

in which E is the energy transformed, m is the mass transformed, and c is the velocity of light in vacuo. ($c = 3 \times 10^8$ meters per second, approximately.)

(b) A photon of electromagnetic radiation always has associated with it an energy proportional to its frequency.

$$E = h\nu \tag{2}$$

in which E is the energy of the photon, ν is its frequency, and h is a constant factor known as Planck's constant. ($h = 6.624 \times 10^{-34}$ joule-second.)

In nuclear disintegration, in emission of particles or waves by bombardment, and in transmutation of elements the above principles find application. For example, if a photon of radiation strikes a nucleus it may cause the emission of a positron and an electron. In this case

$$\frac{h\nu}{c^2} = \text{Mass of positron} + \text{Mass of electron}$$

$$+ \frac{1}{c^2} \text{(Energy of positron} + \text{Energy of electron)} \tag{3}$$

3. PROPERTIES OF ELECTRICAL ENTITIES

The Negative Electron. The results of experiment justify the following assumptions regarding the electron:

(a) No charge of electricity can be produced which is not an integral multiple of the charge carried by a simple electron. The value of this charge is

$$e = 1.602 \times 10^{-19} \text{ coulomb} \tag{4}$$

(b) The mass of an electron may be defined as thé force required to give it unit acceleration.* Experiments indicate that the mass of an electron, as thus defined, is

$$m = 9.1066 \times 10^{-31} \text{ kilogram for } v \ll e \tag{5}$$

The "effective" mass of the electron increases with increase in velocity (see eq. 8).

Unless otherwise stated, wherever the expression "mass of an electron" is used in this article, it is to be understood that this effective mass is meant.

(c) An electron having a charge e and moving with velocity v produces the same magnetic field as an elementary length ds of a conduction current of strength i, where $ev = ids$ (see Art. 11).

(d) Every neutral atom of matter contains at least one electron which is held in position by forces analogous to elastic forces; that is, an electron may oscillate within the atom, or may be displaced by an impressed electrostatic field. The electrons are always of the same nature irrespective of the substance in which they exist.

(e) An electron may be forced from the atom by the influence (mutual repulsion) of a "free" electron moving at a high velocity ín its immediate vicinity. This action is usually spoken of as a bombardment, or collision, although it is not necessary to assume that the free electron hits the atom.

(f) When a neutral atom loses an electron, this atom manifests the properties of a positively charged body and is called a positive ion.

(g) When a neutral atom gains an electron, this atom manifests the properties of a negatively charged body and is called a negative ion.

(h) In every substance there exists, in addition to the electrons within the atoms, a certain number of "free" electrons, i.e., electrons which can move freely in the interatomic spaces; these free electrons may also pass from one substance to another. Under certain conditions, e.g., in a gas at ordinary pressures, a free electron may attach to itself one or more atoms or molecules.

The Positive Particles. Experiment also justifies the following assumptions regarding the positively charged particles:

(a) The smallest positive charge that it has been found possible to produce is numerically equal to that of an electron, viz., 1.602×10^{-19} coulomb.

(b) This smallest possible positive charge when associated with a particle having a mass of the same order of magnitude as that of a hydrogen atom is called a *proton*. The mass of a proton is 1.6734×10^{-27} kilogram.

(c) This smallest possible positive charge when associated with a particle having a mass of the same order of magnitude as that of the negative electron is called a positive electron, or *positron*.

(d) Other positively charged particles may consist of atoms from which one or more electrons have been expelled. Their masses therefore depend upon the number of protons in the nuclei of the atoms. A "free" positively charged particle may also attach to itself one or more neutral atoms or molecules.

(e) The term "ion" is usually reserved to designate any charged body having a mass of the order of magnitude of that of a molecule or atom. In this sense a positive or negative electron is not an ion, but the proton is an ion. If, however, the electron becomes attached to an atom or molecule, then this combination forms an ion. In the discharge of electricity through gases at low pressure, the negative electron, although it is not attached to an atom or molecule, is sometimes called an ion.

The Photon. Experiment justifies the following assumptions regarding the photon:

(a) The photon is a discrete quantity or *quantum* of electromagnetic radiation (see p. 935).

(b) The photon is electrically neutral, not being attracted or repelled by electric charges.

(c) Photons are capable of producing ionization. A photon may be absorbed by an atom, an electron being emitted in the process. It is thought that occasionally a photon may vanish and give rise to a pair of oppositely charged electrons. When this happens the charge of the universe remains unchanged and the energy of the photon is converted into the mass of the two electrons.

(d) Deflection of photons by magnetic fields has not been observed.

(e) The energy and frequency of a photon always satisfy the relation $E = h\nu$.

The Neutron. Experiment justifies the following assumptions regarding the neutron:

(a) The neutron is electrically neutral, not being deflected from its path by electric charges or magnetic fields.

(b) The mass of the neutron is nearly the same as that of the proton. The best measured value of the mass of the neutron is 1.6747×10^{-27} kilogram.

---

* That is, $f = ma$, where f is the force, a is the acceleration, and $m = f/a$ is the mass, computed at low electron velocities (less than one-tenth the velocity of light).

(c) Transmutation can be effected by neutrons. For example, a neutron plus a nitrogen nucleus yields a boron nucleus plus a helium nucleus.

(d) Neutrons are capable of producing ionization through the intermediary of recoiling nuclei.

Determination of the Constants. A few fundamental principles properly combined yield a powerful set of working tools for the investigation of the properties and the constants of the entities described above. These principles are:

(a) If a particle with charge Q falls through a difference of potential V, it acquires a kinetic energy equal to the loss of potential energy.

$$^1/_2 \, mv^2 = QV \tag{6}$$

(b) If a particle with charge Q, mass m, and velocity v is acted upon by a magnetic field B which is perpendicular to v, the path is a circle, the radius of curvature of which is

$$\rho = \frac{mv}{BQ} \tag{7}$$

(c) Supersaturated water vapor condenses quickly on charged particles formed in the presence of the supersaturated vapor.

(d) Matter inserted in the path of a moving particle may absorb energy and thus reduce the velocity of the particle. When a photon encounters a barrier of matter it also may lose energy; this changes the frequency of the photon but does not change its velocity.

The cloud chamber of C. T. R. Wilson, which makes use of principle (c) above, enables the observer to photograph the paths of particles or waves through a gas. The character of the paths under various conditions allows many of the fundamental constants to be calculated.

Mass of the Electron. Experiment shows that the mass as defined by $m = f/a$ (see footnote, p. 9-03) varies with the velocity of the electron, changing appreciably as the velocity of the electron becomes greater than about one-tenth that of light. The formula which describes this experimental fact is

$$m = \frac{m_0}{\sqrt{1 - (v/c)^2}} \tag{8}$$

where m_0 (= 9.1066×10^{-31} kilogram) is the "rest" or minimum mass of the electron, v is the velocity of the electron, and c is the velocity of light (3×10^8 meters per sec). There are two theories as to the constitution of the mass m.

1. The earlier theory assumes that the mass is made up of two parts: one a purely mechanical or Newtonian mass invariant with velocity; and a second part, electromagnetic in origin, which is associated with the charge of the moving electron and which represents the energy of the magnetic field caused by the moving electron. This second part is a function of the velocity of the electron.

2. A second theory dispenses entirely with the concept of mechanical mass and attributes the inertial properties of the electron wholly to the energy associated with the electron.

Relativity, on theoretical grounds, requires that the mass of the electron ($m = f/a$) should change as indicated in eq. 8.

4. EMISSION PHENOMENA

Thermionic Emission of Electrons. When a metal is heated to a high temperature in a high vacuum it gives off electrons freely. The amount of the electron emission is subject to the temperature of the metal or cathode and to the electric field at its surface, as due, for example, to a neighboring anode.

The theory assumes that in accordance with the kinetic theory of matter the electrons within the metal are in a constant state of motion, and that of those nearest the surface a certain proportion escape, a few at high, the greater number at lower, velocities in accordance with Maxwell's distribution law. When an electron leaves the surface the metal becomes positively charged, thus putting a retarding force on the electron and tending to make it return. The distance traversed by the electron is greater the greater its initial velocity, and in the presence of a neighboring metal charged surface it may not return at all. The higher the temperature of the heated metal or cathode the greater the velocities of the electron leaving it, and the greater the volume of electron discharge.

An electron can escape from a metal when, and only when, its kinetic energy is greater than the work which must be done in escaping; that is, when

$$^1/_2 \, mv^2 = w_0 \tag{9}$$

where m is the mass of the electron, v is the component of its velocity normal to the emitting surface, and w_0 is the electron affinity or internal work of electron evaporation of the emitting substance.

With a given cathode temperature there is a maximum electron current beyond which an increase in anode voltage is ineffective. Under these conditions all the electrons emitted from the cathode are drawn to the anode. This condition is known as *voltage saturation* or *temperature limited*.

This electron current follows the law

$$i = Ne = A T^\lambda \epsilon^{-(W_0/kT)} \tag{10}$$

where W_0 is the work function in electron volts, i is the saturation current per square centimeter of emitting surface, N is the number of electrons emitted per second per square centimeter, e is the charge per electron, A a constant of the material, T the absolute temperature in degrees Kelvin, λ a number not much different from unity, ϵ the base of the Napierian system of logarithms, and k the Boltzmann constant, 8.620×10^{-5} electron volts per degree Kelvin. As far as the individual electron is concerned, this is a statistical law. The value of the current given in eq. 10 is independent of the potential of the anode.

With a given anode voltage there is a maximum electron current beyond which an increase in cathode temperature is ineffective. When the cathode is first heated, electrons are emitted with varying velocities. If the anode voltage is zero the first few of these electrons experience no external force except a slight attraction toward the cathode. These first few electrons ultimately land on the anode or on the walls of the container. The succeeding electrons experience a repulsion due to the electrons, previously emitted but still near the cathode, in the space between the anode and cathode; and if the energies of the newly emitted electrons are small, their motion will be retarded or even reversed. There is thus built up an electric field of force tending to decelerate an electron emerging from the cathode and to make it return to the cathode. If there is no removal of electrons from the space surrounding the cathode this electric field increases in strength until a condition of equilibrium is reached, that is, until as many electrons are entering the cathode as are leaving it. The charge due to the presence of these electrons constitutes the *space charge*. From the above it is evident that for a given anode voltage an increase in cathode temperature beyond a certain value will serve only to increase the space charge without increasing the current to the anode. This condition is known as *temperature saturation* or *space charge limited*.

Electron Tubes. A glass or metal bulb, either evacuated or gas filled, and containing two or more electrodes, is called an *electron tube*. Electrons (and positive ions) are produced in such tubes by thermionic emission, by photoelectric emission (q.v.), or by bombardment, and are caused to move by the action of an electric field. This electron flow is made to serve various purposes depending upon its magnitude, velocity, and method of control. Two-electrode tubes can be used for rectification since current will pass more readily in one direction than the others. Tubes with three or more electrodes are important since the voltage signal at one or more electrodes can exert great influence to obtain a much higher level of signal at another electrode. See Sec. 10, Electronics.

Semiconductor diodes and three-electrode transistors are based upon the production of a single crystal of material with distinct regions, each containing a specified amount and distribution of an impurity. The single crystal is in turn a fragment or chip of a much larger single crystal. Germanium and silicon are the most commonly used crystals. The diode contains one junction of two regions, whereas the transistor has two junctions corresponding to three regions. The terminals of the diode are simply called the anode and cathode. The semiconductor diode serves as a rectifying device since it passes current more readily in one direction than in the other. The terminals of the transistor are called emitter, base, and collector, with three regions similarly designated. The transistor is characterized by providing control of current to one electrode by a much smaller signal current to another electrode. In the most usual connection the application of a small current change to the base electrode will change the current in the collector electrode circuit by a larger amount. The distinct regions in these junction devices are associated with different impurities deliberately introduced into the regions in carefully controlled amounts. The selected species of impurity atoms actually take the place of germanium or silicon atoms in the crystal structure as the case may be. Only a minute amount of impurity is used. See Sec. 10, Electronics.

Cathode Rays and Canal Rays. At the occurrence of an electric discharge in an electron tube and at a certain value of the pressure of the gas contained in the tube, it is noticed that a greenish fluorescence occurs on the walls of the tube. By placing solid bodies in the tube it appears that the fluorescence is due to something proceeding in straight lines perpendicularly from the surface of the cathode. The name *cathode rays* has been given to this agent which produces the fluorescence. These rays are deflected by both electric and magnetic fields and convey a negative electric charge to an insulated conductor. They consist of a stream of rapidly moving electrons. The amount of the deflection of the rays, under the action of magnetic and electrostatic fields, can be calculated or observed.

Cathode-ray tubes with hot cathodes (incandescent filament) have been constructed in which potential differences of a million or more volts are impressed between anode and cathode. A window of extremely thin aluminum foil permits a substantial proportion of the electron stream to emerge into the open air. The behavior of chemical substances and of living tissue under the bombardment of these rays is being studied. Cathode-ray tubes are used as oscillographs and as receiving tubes for television.

If the cathode of a tube producing cathode rays is perforated, faint luminous streaks are seen to proceed through these perforations in a direction opposite to that of the cathode rays. The name *kanalstrahlen* or *canal rays* has been given to this phenomenon. These rays are also deflected by both electric and magnetic fields but to a lesser extent and in relatively the opposite direction from the deflection of cathode rays, and they communicate a positive charge to an insulated conductor.

In terms of the electron theory these rays are positively charged ions. Calculations from actual measurements, making this assumption, show that the charge carried by each of these ions is numerically equal but opposite to that of an electron, and that the mass of each ion is never less than the mass of a hydrogen atom, but may be greater, depending upon the nature of the gas and the pressure in the tube. That is, the cathode particle is an electron traveling in one direction, and the canal-ray particle is what is left of the atom which has lost an electron, traveling in the opposite direction.

Electromagnetic Waves. Electromagnetic disturbances which are propagated through space can be described by electric and magnetic fields associated with the points in space. Such fields can be formulated mathematically as

$$H = F_1(x - vt) \tag{11}$$

$$E = F_2(x - vt) \tag{11a}$$

in which E and H are the electric and magnetic fields (q.v.), x is the distance from some datum to the point in question along the direction of propagation, v is the velocity of propagation, and t is the time. A disturbance which propagates through space is called a wave. An electromagnetic wave is often described in terms of time and space variations of electric and magnetic fields as shown in eqs. 11 and 11a.

The physical concept of an electromagnetic wave is aided if one visualizes a medium (this medium is called the ether) in which the wave exists as a transverse vibration. Many of the experimentally observed properties of electromagnetic waves can be explained on the basis of the assumption of an ether through which the waves are propagated. An electromagnetic wave which is a simple harmonic vibration is termed a monochromatic wave. A monochromatic wave has a definite *frequency* which is the reciprocal of the number of periods per second. The character of a given wave is dependent upon its frequency. The electromagnetic spectrum is often divided arbitrarily into bands and sub-bands according to usage. Some of the schemes of designation overlap. In order of increasing frequency one division into bands would be:

(a) Commercial power.
(b) Induction heating.
(c) Radio.
(d) Television.
(e) Radar.
(f) Microwave spectroscopy.
(g) Infrared rays.
(h) Visible light.
(i) Ultraviolet rays.
(j) X-rays.
(k) Gamma rays.
(l) Cosmic rays.

The mechanism of producing the radiations or waves in groups (a) to (f) from about 10^4 to about 3×10^{10} cycles per second is that of accelerating electrons in electric circuits. The mechanism of producing the radiations in groups (g) to (l) from about 10^{13} to more than 10^{22} cycles per second is that associated with the increase of temperature of a substance, the sudden stopping of high-velocity electrons, or the forced or spontaneous disruption of atoms or atomic nuclei. It is with these extremely high-frequency radiations that the quantum theory is associated.

Photoelectric Effect. Experiment shows that when light of sufficient energy falls upon a substance electrons are liberated from the surface. The number of electrons released per unit time is directly proportional to the intensity of the incident light. The maximum energy of electrons released is independent of the intensity of the incident light but directly proportional to the frequency of the light. The energy equation representing this effect is

$$h\nu = 1/2\ mv^2 + \phi \tag{12}$$

in which h is Planck's constant, ν is the frequency of the incident radiation, $1/2\ mv^2$ is the kinetic energy of the liberated electron, and ϕ is the energy required to free the electron from the parent body. ϕ is called the *work function* of the given surface in joules. If eq. 12 is written

$$1/2\ mv^2 = h(\nu - \nu_0) \tag{13}$$

then ν_0 is the threshold frequency of the substance.

The work function for the alkali metals is lower than that for other substances. Special films having extremely low work functions have been produced by the use of potassium, rubidium, and other elements. It is possible to construct photoelectric surfaces having selective frequency characteristics such that color sensitivity is produced.

Photoelectric cells consisting of a photoelectric surface and an anode in an evacuated glass tube are used commercially for automatic control, product testing, television, etc.

X-rays. When cathode rays strike a target of metal, x-rays are radiated from the metal surface. The wavelength spectra of such x-rays depend primarily on the velocity of the impinging electrons and the material of the target. In general, the higher the voltage between anode and cathode, and the denser the metal of the target, the shorter will be the wavelength of the x-radiation.

The x-ray output of a tube operating at constant potential difference between anode and cathode consists of a continuous spectrum superposed with a number of characteristic lines.

The cathode stream has been shown to consist of electrons traveling at high velocities. Thomson and Stokes have shown, by means of the electromagnetic theory of Maxwell, that the deceleration of these electrons at the anode may produce a pulse of electromagnetic radiation. They have shown further that the radiated energy traversing a unit area at a point P, distant from the region of deceleration of one of the electrons, is for electron velocities small compared to the velocity of light

$$S_p = \frac{ae^2\beta}{4\pi r^2 c^2} \sin^2 \theta \tag{14}$$

in which a is the electron deceleration.

e is the charge of the electron.

β is ratio of velocity of electron before deceleration to velocity of light in vacuo.

r is distance from P (at which energy flux is S_p) to the source of radiation.

c is velocity of light in vacuo.

θ is angle measured from line of travel of electron before deceleration to line joining source of radiation to the point P.

A further development of this electromagnetic pulse theory, by expressing the pulse as a Fourier integral, predicts the spectrum of the radiation to be of the form:

$$I_\lambda\ d\lambda = K \sin^2 (\pi l/\lambda)\ d\lambda \tag{15}$$

in which I_λ is the x-ray intensity between the wavelengths λ and $(\lambda + d\lambda)$.

l is the pulse thickness, i.e., the product of its velocity c and the time for deceleration of the electron producing the pulse.

K is a constant.

This expression indicates a continuous radiation with all frequencies from zero to infinity. Duane found that the continuous x-ray spectrum was limited by a maximum frequency ν_{max} in such a way that

$$h\nu_{max} = eV \tag{16}$$

in which e is the electronic charge and V is the maximum x-ray tube voltage—which agrees with the quantum theory. The electromagnetic theory offers no suggestion concerning the cause of the characteristic line spectra mentioned above.

X-rays are of immense value in diagnosis and therapeutics in medicine, in the examination and inspection of engineering materials for structure and flaws, and in the study of the constitution of matter. The molecules of matter act as diffraction gratings of an x-ray. The resulting interference patterns yield important evidence with regard to the arrangement and spacing of the atoms composing the substance.

Absorption, Scattering, Reflection, and Refraction. The following principles observed in the study of x-rays apply with slight modifications to the other electromagnetic radiations.

When monochromatic x-rays traverse matter, the intensity (I) of the beam after traversing a distance x in a given material is associated with the initial intensity (I_0) by:

$$I_x = I_0 \epsilon^{-\mu x} \qquad (17)$$

where ϵ is the base of the natural system of logarithms and μ is the linear absorption coefficient, defined by the relation above shown. Intensity is a measure of energy per unit area per unit time.

Other absorption coefficients defined for convenience are:

The mass absorption coefficient $\mu_m = \mu/\rho$, where ρ is the density of the absorbing material.

The atomic absorption coefficient $\mu_a = \mu/n$, where n is the number of atoms per cubic meter of the volume occupied by the absorbing material.

If the heterogeneous radiation from an x-ray tube traverses matter the absorption coefficient varies with the thickness of the absorbing material. Its value is found to depend chiefly upon the x-ray tube voltage and upon the atomic number of the absorbing material, assumed to be an element.

Two effects are noted when x-rays traverse matter. Electrons are ejected from the absorbing material at high velocities (see Photoelectric Effect). The radiation is scattered, reducing the intensity along the lines of incidence of the primary beam. Thus μ may be conveniently divided into two parts, τ and σ. The coefficient τ represents the transformation of the energy of the original beam of x-rays into high-speed electrons; σ is the coefficient describing the scattered radiation. Experiment shows that not all the scattered radiation has the same wavelength as the primary beam. The number of photoelectrons ejected by x-rays, at such speeds that the energy of the photoelectron is nearly equal to the energy $h\nu$ of an incident photon, has been shown to be equal, within the limits of experimental error, to the number of photons ("truly absorbed") represented by accurate measurements of τ.

Measurement of x-ray wavelengths can be made by taking advantage of the regularity of atomic structure in crystals. W. L. and W. H. Bragg showed that x-rays of wavelength λ incident upon the face of a crystal at a glancing angle θ would, if the crystal planes were separated by a distance d, be reinforced upon reflection, if:

$$n\lambda = 2 d \sin \theta \quad (n = \text{integer}) \qquad (18)$$

This is known as Bragg's law.

By calculations based on an assumed crystal structure, experimental data of density, Avogadro's number, and the molecular weight, the value of d may be obtained for some more simply constructed crystals. Then Bragg's law may be used to measure wavelengths of an x-ray beam.

Ionization of Gases. Under ordinary circumstances at atmospheric pressure, gases are very poor conductors of electricity. By irradiation with x-rays or radioactive substances, by brush discharge, corona, or contact with hot flame, and by other known means, it is possible to make a gas a comparatively good conductor. The process is known as *ionization*. It appears that agents such as those mentioned have the power of breaking down a portion of the gas molecules into free electrons and positively charged particles.

According to the quantum theory, radiation is absorbed by an atom in discrete amounts. The absorption of these quanta makes an electron jump from its then orbit to one of greater radius. If sufficient energy is absorbed the electron leaves the atom altogether and becomes, at least temporarily, a free electron. This leaves the remainder of the atom positively charged. The positively charged particle is known as a positive ion. The electrons and positive ions thus released respond to the force exerted by an electric field and in moving give rise to an electric current. The gas thus becomes conducting. In the absence of such an electric field the electrons and positive ions tend to recombine, and, when the radiation producing ionization is removed, this recombination eventually renders the gas non-conducting. At sufficiently high gas pressures the free electrons and positive ions tend to attract neutral molecules to which they attach themselves. At low pressures, however, the mean free paths of the molecules are so great that the charged particle rarely comes within attracting distance of the molecule.

It is assumed that all substances contain at all times some free ions. In a gas under the action of an electric field these ions may attain sufficient velocity to detach electrons from neutral molecules by "collision." The electric field in this instance becomes an ionizing agent. As the ionization increases the gas becomes more and more conducting until the rate of recombination of ions balances the rate of ionization.

Theory of Corona. When the voltage between two conductors becomes sufficiently great, the electric field, which is greatest at the surface, ionizes the air at the surfaces of the conductor. The increase in the conductivity of the air or other gas surrounding the conductor is equivalent to an increase in the effective diameter of the conductor to the value at which the decreasing electric field is balanced by the dielectric strength of the air. The phenomenon is known as *corona*. When the corona is formed, there is visible a faint violet light near the conductor. If the voltage is raised, a *brush discharge* occurs in which bluish streaks like the bristles of a brush are visible near the surface of the conductor. If the voltage is still further increased, a disruptive spark discharge takes place between conductors.

In transmission lines, corona is accompanied by power losses, frequently of serious proportions. It can be controlled by the use of larger-diameter conductors and by operating with voltages low enough to prevent its formation.

DIRECT-CURRENT CIRCUITS

5. ELECTRIC CHARGE, CURRENT, AND ELECTROMOTIVE FORCE

Electric Charge and Electric Current. The unit of electric charge (or quantity of electricity) is called a *coulomb* in the practical system of units. In terms of the charge carried by an electron it requires 6.242×10^{18} electrons to equal a coulomb. An electric charge in motion constitutes an electric current. The rate at which a quantity of electricity flows past a given point in a circuit determines the value of the current. In the rationalized mks system, and also in the practical system of units, the unit of current, called an *ampere*, is the flow of 1 coulomb per second. (See Section 3 for the relationships between various units and for international standards for units.)

Electromotive Force. The agent which tends to produce or to maintain an electric current in a circuit is called an *electromotive force* (often abbreviated emf). The names in common use are *difference of potential* and *voltage*, the latter being derived from the practical unit of electromotive force, the *volt* (see Section 3). Electromotive force is not a force in the mechanical sense. Quantitatively it is the electrical energy developed per unit charge

$$E = W/Q \qquad (1)$$

where E is the electromotive force in volts, Q is the charge in coulombs, and W is energy or work in joules. (One joule = 1 watt-second = newton-meter = 10^7 ergs.) The energy W is that required to move the charge Q between two points the difference of potential of which is E. For a more fundamental definition of emf see Section 3.

Sources of Emf. The primary sources of emf are:

1. *Electromagnetic Induction* (see p. 960, Art. 12).
2. *Contact of Two Dissimilar Bodies.* Under this head, apart from primary cells and storage batteries, may be included the following sources of emf:

(a) *Volta effect* (contact emf). When two dissimilar substances are brought in contact an emf is developed between them. This is explained by the electron theory on the basis of the fact that certain substances have less tendency to give up electrons than other substances. Zinc releases electrons more easily than copper; therefore, if zinc is brought into contact with copper, some of the electrons leave the zinc and enter the copper, making the copper negative and the zinc positive. The emf's thus produced are usually very small. According to the electron theory frictional electricity is due to the same cause. However, where the conductivities of the substances are low, friction is required to bring the surfaces into sufficiently intimate contact for the transference of electrons.

(b) *Thermoelectric effects.* If a closed circuit is made of two different metals and the two junctions are held at the same temperature, no current will flow because although emf's are developed at both junctions (see Volta effect, above) they are equal and opposite. If, however, one junction is held at a higher temperature than the other a current will flow as long as the difference of temperature is maintained. This is known as the *Seebeck effect.* According to the electron theory the relative release of electrons is a function of the temperature of the metal. Such junctions are used in thermometry and also for current measurements. The *Peltier effect* is the inverse of the Seebeck effect. If a current is sent through the junction of two dissimilar metals, heat is absorbed by the junction when the current flows in one direction and is emitted by the junction when the current is reversed. The sign of the Peltier effect is connected with the direction of the thermoelectric current which would be produced at a given junction. A current which cools a junction (Peltier effect) flows in the same direction as the thermoelectric current when that junction is heated, and conversely. The *Thomson effect* is apparently an extension of the Peltier effect. In certain metals heat is carried in the direction of the current flow when the current flows from hot to cold regions. When the current flows from cold to hot regions, the hotter parts are cooled. In other metals these effects are reversed.

(c) *Pyroelectric effect.* When certain crystals such as quartz and tourmaline are heated an emf is developed between faces. If the crystal is broken into fragments or even ground to powder, each particle will exhibit this same electrical effect.

(d) *Piezoelectric effect.* When certain crystals such as quartz and Rochelle salt are subjected to mechanical pressure, electric charges proportional to the pressure appear on certain parts of the crystal. If a crystal of Rochelle salt (cut so that two of its faces are perpendicular to its optical axis) is compressed between two electrically connected metal plates applied to the two opposite crystal faces, and a metal band is placed around the middle of the crystal between the end plates, an emf of considerable magnitude will be developed between the band and end plates, upon small changes of pressure between end plates. Compression produces a positive charge on one plate and a negative charge on the other plate. Tension causes the charges to reverse. The pressure due to sound waves may under proper conditions produce emf's of several volts. The piezoelectric effect is reversible. A variable emf applied between the plates and band will cause the crystal to expand and contract mechanically in synchronism with the varying applied emf. This effect has been experimentally applied in sound reproducers. Certain crystals, notably quartz, vibrate at natural periods, dependent upon the thickness of the crystal and its elastic constants, if an emf is applied which has the same frequency as the natural frequency of the crystal. The energy supplied by the electrical oscillator may cause very violent vibrations of the crystal. The constancy of frequency of properly constructed quartz vibrators is so great that they are universally employed to control master oscillators for radio. The natural frequency of vibrations of such crystals lies between about ten thousand and several million vibrations per second.

Direction of an Emf. The direction of the emf in any portion of a circuit is taken as the direction in which a positive charge would be forced around a circuit containing only this one source of emf. A closed circuit may contain several sources of emf; in this case the resultant emf acting around the circuit is the algebraic sum of all these emf's, those acting around the circuit in one direction being taken as positive and those acting around the circuit in the opposite direction being taken as negative. Those which act in the opposite direction to the resultant emf are called "back" or "counter" emf's.

A convenient symbol for a source of constant emf is shown in Fig. 1; the long light line represents the positive terminal and the short heavy line the negative terminal. Figure 2 shows two emf's acting around a circuit in the same direction, and Fig. 3 shows two acting around the circuit in opposite directions. In the first instance the resultant emf is $E_1 + E_2$ and the two emf's are said to be "in series aiding," whereas in the second the resultant emf acting in the clockwise direction around the circuit is $E_1 - E_2$ and the two emf's are said to be "in series opposing." When E_2 is less than E_1 then E_2 is a back or counter emf.

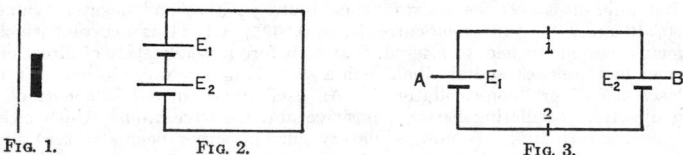

FIG. 1.   FIG. 2.   FIG. 3.

Difference of Electric Potential. Consider a portion of a path A between any two points 1 and 2 in an electric circuit (Fig. 3). The total work done on a unit charge, by the emf's external to this portion of the path when a unit charge moves from 1 to 2, is called the "drop of electric potential" from 1 to 2. The term "difference" of electric potential is also commonly used to designate this quantity; the term "drop" is preferable since it signifies the direction of the difference. Electric potential drop is of the same nature as emf and is expressed in the same units. It is frequently abbreviated "p.d." Potential drop may be due either (1) to a back emf, analogous to the back pressure of a pump, or (2) to an opposing force analogous to that due to the resistance of a pipe to the flow of water. The symbol for potential drop is V when it is desired to distinguish it from electromotive force.

Positive and Negative Terminals. That terminal of a device which is at the higher potential is called the positive terminal, the other being called the negative terminal. The drop of potential is always from the positive to the negative terminal, irrespective of the direction of flow of electricity through the device.

Conductors and Insulators. All substances may for practical purposes be divided into three classes: conductors, semiconductors, and insulators. With conductors a continuously applied emf is accompanied by a continuous flow of electric current. With dielectrics upon first applying an emf there is a transient flow of current until the back emf produced by the establishment of a positive charge at one terminal and a negative charge at the other terminal exactly balances the applied emf, after which current ceases to flow. Strictly speaking, every substance conducts electricity to some extent; that is, a constantly applied emf is always accompanied by a constant flow of current. In those substances which are classed as dielectrics this constant flow is negligibly small. A perfect dielectric may be defined as a substance through which it is impossible to maintain a continuous flow of electricity; no such conductor exists, but very poor conductors are approximately such. The electron theory of conduction assumes that any substance contains, in addition to the electrons and protons which form its atoms and molecules, a certain number of free electrons which like the molecules are in violent agitation. These free electrons or ultimate particles of negative electricity tend to drift towards the point of highest potential when an emf is applied to the substance. This drift constitutes the electric current. According to the theory, good conductors have, great numbers of these free electrons, and poor conductors and dielectrics have comparatively very few. In dielectrics the momentary flow of electricity is believed to be due to the displacement of *bound electrons* in the atoms without their actual removal from the atom to which they are bound. The greater the applied emf the greater the displacement of the electron within the atom until a point is reached at which the electron is torn from the atom and the dielectric "breaks down," becoming a partial conductor. This breakdown tends to destroy the solid dielectrics.

Metals are the best conductors and are almost universally employed for electric circuits. For the properties of metals as conductors see Section 12. Carbon and most moist substances are fair conductors; dry non-metallic bodies such as air and other gases (at normal pressure), porcelain, glass, rubber, and dry paper are very good insulators.

A wire covered with an insulating substance or supported on insulating substances is said to be "insulated," even though its ends are connected to a source of emf. For the properties of insulating materials see Section 13.

If decomposition takes place when a current of electricity is passed through liquids which are chemical compounds, the liquid is called an *electrolyte*. Certain solid chemical compounds and fused salts behave as electrolytes. For a discussion of electrolysis and electrolytes see Section 11.

Continuous, Direct, Pulsating, Alternating, and Transient Current. A *continuous* electric current is defined as a current which does not vary with time. A *direct* current is a current which is always in the same direction but may vary or pulsate in value. The term *direct current* is ordinarily used to designate either a continuous current or a current which varies or pulsates only by an inappreciable amount, such as the current from a battery or direct-current generator (q.v.). A *pulsating* current is a direct current which pulsates by an appreciable amount, such as the current from a rectifier (q.v.). An *alternating* current is a current which reverses in direction, being first positive and then negative, but alternates between constant maximum positive and negative values (see p. 978, Arts. 19 and 20). A *transient* current (see p. 995, Art. 24) is a current which flows when a circuit is closed, opened, or altered, that is, before a steady state of direct or alternating current has been reached or when such a state is in any way altered. Transients may be "oscillatory" or "non-oscillatory." An oscillatory current is a current which reverses in direction, oscillating between positive and negative values which either decrease or increase with time. A non-oscillatory current either begins at zero and rises to a steady value, or begins at some finite value and decreases to zero, in either event without oscillation.

For continuous currents

$$I = Q/T \qquad (2)$$

where I is the current in amperes, Q is the charge of electricity in coulombs passing a given cross section of the circuit, and T is the time in seconds. When the flow of current is not continuous its value at any instant is given by

$$i = dq/dt \qquad (3)$$

where i is the *instantaneous* value of the current in amperes and dq/dt is the instantaneous rate of flow of charge through a given cross section of circuit in coulombs per second.

6. OHM'S LAW, RESISTANCE AND CONDUCTANCE

Ohm's Law. If a steady difference of potential V (in volts) is impressed across a conductor which (a) is held at constant temperature and in which (b) there is no internal emf,

$$V = rI \qquad (4)$$

where I is the steady current in amperes which will flow through the conductor and r is the factor of proportionality called the *resistance* of the conductor. The drop in potential V is therefore equivalent to the drop in potential rI, this latter being called the *resistance drop*. Equation 4 may also be written

$$I = gV \qquad (5)$$

where I and V are the same as before and g is termed the *conductance* of the conductor. Obviously *under the conditions stated*

$$g = 1/r \qquad (6)$$

The practical unit of resistance is the *ohm*, and the practical unit of conductance is the *mho*. For standards and conversion factors see Section 3 and tables in Section 1.

Equations 4 and 5 are forms of what is called Ohm's law. If there is one or more counter emf's between the terminals across which V is impressed the relationship is given by

$$V = rI + E \qquad (7)$$

where V, $\dot{r}$, and I are the same as in eq. 4 and E is the algebraic sum of *counter emf's* within the circuit. In this computation all counter emf's have positive signs and all emf's in the direction of the current have negative signs; hence such an emf is a negative counter emf. Equation 7 is sometimes referred to as the *modified* Ohm's law.

Terminal and Impressed Voltage. The application of this equation is shown by considering a simple circuit containing a generator and a motor or other receiving device, Fig. 4. Let I be the current in this circuit, r_g the internal resistance of the generator, r_m the internal resistance of the motor, r_l the total resistance of the "line" or connecting wires, E_g the emf developed by

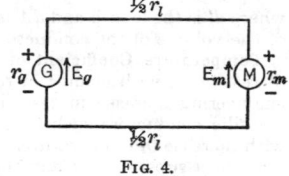

FIG. 4.

the generator, E_m the back emf developed by the motor, V_g the terminal voltage of the generator, and V_m the terminal voltage of the motor. Through the generator the current flows from the − to the + terminal, the emf is from the − to the + terminal, and the drop of potential is from the + to the − terminal. Hence

$$V_g = E_g - r_g I \qquad (7a)$$

i.e., the terminal voltage is less than the generator emf by an amount equal to the resistance drop in the armature circuit. Through the motor the current flows from the + to the − terminal, the emf is from the − to the + terminal, and the drop of potential is from the + to the − terminal. Hence

$$V_m = E_m + r_m I \qquad (7b)$$

i.e., the terminal voltage is greater than the back emf of the motor by an amount equal to the resistance drop through the motor.

The terminal voltage of the motor is less than that of the generator by an amount equal to the resistance drop in the line, i.e.,

$$V_m = V_g - r_l I \qquad (7c)$$

The expression *impressed* emf is also used to designate the rise of potential from the negative to the positive terminal of any receiving device, for example, a motor or a bank of lamps, regardless of whether or not the device contains a source of emf.

Resistance and Conductance. It should be carefully noted that the definitions of resistance and conductance expressed by eqs. 4 and 5 hold only when there is no emf in the portion of the circuit under consideration; this condition is realized only when the current remains constant in value (i.e., a continuous current) and the conductor is of uniform material and at constant temperature throughout. Also, the definition is meaningless unless the same current flows through each cross section of the conductor and the drop of potential is the same between all points in the two end surfaces; i.e., the end surfaces must be equipotential surfaces. For cylindrical conductors whose ends are equipotential surfaces and for any homogeneous conductor or wire of constant cross section in which the diameter of cross section is small in comparison with the length of the conductor, and through which the current is continuous and uniformly distributed,

$$r = \rho \frac{l}{A} \quad \text{and} \quad g = \sigma \frac{A}{l} \qquad (8)$$

where r is the resistance in ohms, l is the length of the conductor in meters, A is the cross-sectional area in square meters, and ρ is called the *resistivity* in ohms per meter cube or, more briefly, ohm meters. Similarly g is the conductance in mhos and σ is the *conductivity* in mhos per meter cube or mho meters. Note that the word "cube" used in the resistivity-conductivity units refers to a cube of unit dimensions and is *not* an exponent. The ρ may be defined as the resistance in ohms of a cube of the material under consideration having a length of 1 meter and a cross section of 1 sq meter. Similarly σ may be defined as the conductance in mhos of a cube of the material under consideration having a length of 1 meter and a cross section of 1 sq meter. It is evident from eqs. 6 and 8 that

$$\rho = 1/\sigma \tag{9}$$

ρ and σ depend upon the material and temperature of the conductor and upon the choice of units (see next paragraph), but are independent of the shape and size of the conductor. Hence ρ is sometimes called the *specific resistance* and σ the *specific conductance* of a substance.

Equation 8 cannot be directly applied to irregularly shaped conductors, but the following equation may be applied to elementary volumes of irregular conductors and the total resistance found by processes of summation or integration:

$$r = \rho \, \frac{dl}{dA} \quad \text{and} \quad g = \sigma \, \frac{dA}{dl} \tag{10}$$

where dl is the length and dA is the cross-sectional area of an elementary cylinder or prism of the volume of the conductor.

Temperature Coefficient of Resistance. Practically all substances show a variation of resistance with change in temperature. All metals and most alloys used in electrical engineering increase in resistance with increase in temperature; the resistance of non-metallic conductors such as carbon and electrolytes, also of most dielectrics, decreases with increase in temperature. In general, the relation between the resistance r_t of a given mass of a substance at any temperature t may be expressed in terms of its resistance r_o at zero degrees by the following power series:

$$r_t = r_o(1 + at + bt^2 + ct^3 + \text{etc.}) \tag{11}$$

where r_t and r_o are in ohms, t is in degrees Centigrade, and a, b, c, etc., are constants. With metallic conductors such as copper and aluminum the following expression is sufficiently accurate for most practical purposes:

$$r_t = r_o(1 + \alpha_o t) \tag{12}$$

where r_t and r_o are as above and α_o is called the *zero degree temperature coefficient of resistance*. α_o is the change in ohms per ohm per degree in the neighborhood of zero Centigrade. For dielectrics the linear relationship expressed by eq. 12 is not sufficiently accurate. When resistance increases with temperature, α is positive; when resistance decreases with temperature, α is negative. (For properties of resistance materials see Section 12.)

7. DIRECT-CURRENT NETWORKS

Resistances and Conductances in Series. When several conductors are connected end to end so that the same current flows through each of them (Fig. 5), they are said to be connected *in series*. Let I_{12} be the current in each conductor in the direction from 1 to 2; let r', r'', r''', etc., be the resistances of the various conductors and E'_{12}, E''_{12}, E'''_{12}, etc., the emf's in the circuit between 1 and 2 in the direction from 1 to 2. Then the potential drop from 1 to 2 is

$$V_{12} = r'I_{12} - E'_{12} + r''I_{12} - E''_{12} + r'''I_{12} - E'''_{12} + \text{etc.}$$

Therefore, the resistances between the points 1 and 2 are equivalent to a single resistance

$$r = r' + r'' + r''' + \text{etc.} \tag{13}$$

and the emf's between the points 1 and 2 are equivalent to a single emf:

$$E_{12} = E'_{12} + E''_{12} + E'''_{12} + \text{etc.} \tag{13a}$$

The equivalent conductance of several conductances g', g'', g''', etc., in series when there are no emf's in the path, is g, where

$$\frac{1}{g} = \frac{1}{g'} + \frac{1}{g''} + \frac{1}{g'''} + \text{etc.} \tag{13b}$$

Resistances and Conductances in Parallel. When several conductors are connected to two common junction points so that the same potential drop is established through each

(Fig. 6), they are said to be *in parallel*. Let the currents, emf's, and resistances be as designated in the figure. Then, since $\Sigma I = 0$ (Kirchhoff's law),

$$I_{12} = I'_{12} + I''_{12} + I'''_{12} + \text{etc.}$$

and from equation 7a

$$V_{12} = r'I'_{12} - E'_{12} = r''I''_{12} - E''_{12} = r'''I'''_{12} - E'''_{12} = \text{etc.}$$

from which relations the currents in the individual branches may be calculated.

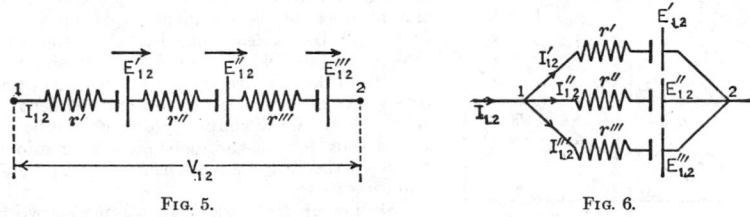

FIG. 5.   FIG. 6.

When there are no emf's in the various branches, the combined resistance of the several branches from 1 to 2 is r, where

$$\frac{1}{r} = \frac{1}{r'} + \frac{1}{r''} + \frac{1}{r'''} + \text{etc.} \tag{14}$$

and the combined conductance is

$$g = g' + g'' + g''' + \text{etc.} \tag{14a}$$

where g', g'', g''', etc., are the individual conductances.

In the special case of two conductors in parallel, and no emf in either, the combined resistance is

$$r = \frac{r_1 r_2}{r_1 + r_2} \tag{14b}$$

Series-parallel Circuits. When a circuit is made up of several conductors some of which are in series and some in parallel, it is called a *series-parallel circuit*. The total resistance or conductance of such a circuit can be calculated from the constants of the several branches by applying successively the formulas for series and for parallel circuits.

Kirchhoff's Network Laws. The relations given above for conductors in series and in parallel are special cases of two general laws, namely:

1. The algebraic sum of the currents coming to any junction in a network of conductors is always zero.

2. The algebraic sum of the potential drops around any closed loop in a network of conductors is always zero.

These two statements are known as Kirchhoff's laws. By making use of them one can always predetermine (a) the current in each branch of a network when the resistance of each branch and the emf in each branch are known, or (b) the emf in each branch when the current in each branch and the resistance of each branch are known.

It should be carefully borne in mind in applying these laws that a current leaving a point is equivalent to a negative current entering that point, and that an emf in any chosen direction is equivalent to a rise of potential in that direction. In working out any problem concerning a network of circuits it is convenient to make a diagram of the network and to place on each branch in this diagram a number or symbol to represent the value of the current (the directions of the current are arbitrarily assumed), and wherever there is an emf to place a number or symbol to represent its value and an arrow or subscripts to indicate its direction. Then at any junction point those currents represented by arrows pointing toward the point are to be considered positive and those represented by arrows pointing away from the point are to be considered negative; and for any closed mesh those currents and emf's represented by arrows pointing around the mesh in the clockwise direction are to be considered positive and those pointing around the mesh in the counterclockwise direction are to be considered negative. With this understanding, these laws may be written

$$\Sigma I = 0 \text{ at every point} \tag{15}$$

$$\Sigma E - \Sigma r I = 0 \text{ for every closed mesh} \tag{16}$$

where I, r, and E represent the current, the resistance, and the emf respectively in each branch of the mesh, and the symbol Σ indicates the algebraic sum of the quantities following it.

These equations enable one to write down a set of simultaneous equations for the given network, but it will be found, for networks having more than one mesh, that at least one of the current equations may be derived directly from the other current equations, and that at least one of the potential equations may be derived from the other potential equations. That is, the number of independent equations of each form will be at least one less than the number which it is possible to write down. It should also be noted that it is frequently unnecessary to write down formally all the possible independent equations; many of the simpler problems can be solved by writing down two independent expressions for the potential drop between each pair at points and equating them.

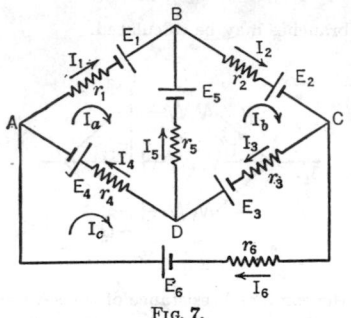

FIG. 7.

Solution of Networks. In solving networks of conductors there are several alternative methods all of which are equivalent as far as results are concerned. Directions for solution by one method are here given. In Fig. 7 let the resistances and emf's be given. It is desired to find the currents in the five branches. There are four junction points, A, B, C, and D. Since the currents and their directions are unknown some assumption of direction must be made. Assume directions of currents in all branches and indicate by arrows and symbols I_1, I_2, I_3, I_4, I_5, and I_6. This is shown in the figure. Apply the first law to junctions A, B, and C as follows, remembering that a current approaching a junction is positive and a current leaving a junction is negative. Since $\Sigma I = 0$

At junction A $\qquad\qquad I_4 + I_6 - I_1 = 0$ $\qquad\qquad\qquad\qquad$ (17a)

At junction B $\qquad\qquad I_1 + I_5 - I_2 = 0$ $\qquad\qquad\qquad\qquad$ (17b)

At junction C $\qquad\qquad I_2 - I_3 - I_6 = 0$ $\qquad\qquad\qquad\qquad$ (17c)

If the first law is applied to the currents at the junction D, there will result an equation derivable from eqs. 17. The second law is next applied to the closed loops formed by branches. The potential drops will be summed *clockwise* about the loops. The following conventions will be adopted:

1. All potential drops will be positive, and all potential rises will be negative.

2. In summing potential drops around a mesh, a given emf will be considered positive when we proceed from its positive to its negative pole, *regardless of the direction of the current through it.*

3. In summing potential drops around a mesh, a given emf will be considered negative when we proceed from its negative to its positive pole, *regardless of the direction of the current through it.*

4. In summing potential drops around a mesh, the difference of potential across a given resistance will be considered positive if the assumed direction of the current is clockwise with respect to that mesh.

5. In summing potential drops around a mesh, the difference of potential across a given resistance will be considered negative if the assumed direction of the current is counterclockwise with respect to that mesh.

Applying the above conventions to Fig. 7 the equations representing the second law may be written

Mesh $ABDA$ $\qquad r_1I_1 - E_1 + E_5 - r_5I_5 + r_4I_4 + E_4 = 0$ $\qquad\qquad$ (18a)

Mesh $BCDB$ $\qquad r_2I_2 + E_2 + r_3I_3 - E_3 + r_5I_5 - E_5 = 0$ $\qquad\qquad$ (18b)

Mesh $ADCA$ $\qquad -E_4 - r_4I_4 + E_3 - r_3I_3 + r_6I_6 - E_6 = 0$ $\qquad\qquad$ (18c)

Any further mesh equations are derivable from eqs. 18. The six equations 17 and 18 permit the solution for currents in all branches. When these equations are solved a positive sign in front of a current indicates that the assumed direction of the current was correct. When a negative sign appears in front of a current, that current actually flows in a direction *opposite* to the direction assumed.

It is frequently convenient to use fictitious currents called *mesh currents* since the first law is automatically taken care of when writing the second-law equations by means of mesh currents. Referring again to Fig. 7, the circular arrows represent the mesh currents. The second-law equations may then be written

Mesh a $\qquad r_1 I_a + r_5 (I_a - I_b) + r_4 (I_a - I_c) = E_1 - E_5 - E_4$ (19a)

Mesh b $\qquad r_2 I_b + r_3 (I_b - I_c) + r_5 (I_b - I_a) = E_3 + E_5 - E_2$ (19b)

Mesh c $\qquad r_6 I_c + r_4 (I_c - I_a) + r_3 (I_c - I_b) = E_6 + E_4 - E_3$ (19c)

These three equations may be solved for I_a, I_b, and I_c. The branch currents can then be found from

$$
\left.
\begin{aligned}
I_1 &= I_a \\
I_2 &= I_b \\
I_3 &= I_b - I_c \\
I_4 &= I_a - I_c \\
I_5 &= I_b - I_a \\
I_6 &= I_c
\end{aligned}
\right\}
\tag{20}
$$

The law of the *superposition of currents and voltages* simplifies the calculation of certain types of distributing networks. If electric energy is being supplied over a network to a number of individual loads: (1) the current at any point in the network is equal to the algebraic sum of the currents which would flow if the individual load currents were considered in succession instead of simultaneously; and (2) the voltage drop from the source to any point in the network is equal to the algebraic sum of the drops to that point, each drop being calculated on the basis of individual load currents, taken successively instead of simultaneously.

8. ENERGY AND POWER

Electric Energy and Power. From the definition of emf and potential drop it follows that the total work done by the external agents in forcing Q units of electricity from any point 1 in a circuit to any point 2 is $W = VQ$. (See Eq. 1.) From the definition of electric current the quantity of electricity carried from 1 to 2 when the current I is established from 1 to 2 is $Q = It$, where t is the time during which the current exists. Hence when a current I is established through any device for a time t by an impressed voltage V, the energy input to this device is

$$W = VIt \tag{21}$$

and the power input, i.e., energy input per unit time, is

$$P = VI \tag{22}$$

When V and I are expressed in volts and amperes respectively, the power input is in watts; if t is in seconds, the energy input is in joules or watt-seconds. Another frequently used energy unit is the electron-volt; this is the energy acquired by an electron falling through a potential difference of one volt. An electron-volt is equivalent to 1.602×10^{-19} joule.

Applying the above relations to the simple circuit shown in Art. 6, Fig. 4, containing a generator and a motor (armature circuit only is considered), the power input to the motor armature is

$$P_i = E_m I + r_m I^2 \tag{22a}$$

the power output of the generator armature is

$$P_o = E_g I - r_g I^2 \tag{22b}$$

and the power lost in the line is

$$P_l = r_l I^2$$

The term $r_g I^2$ represents the power lost in heating the armature circuit of the generator due to its resistance, and the term $r_m I^2$ represents the power dissipated as heat in the armature circuit of the motor. The net electric input to the generator armature is $E_g I$, and the gross mechanical output of the motor armature is $E_m I$. The gross mechanical input to the generator is greater than $E_g I$ and the net mechanical output of the motor is less than $E_m I$ by an amount equal to the friction and "iron loss" in the respective machines.

Joule's Law. That portion of the power input to any device which is equal to the product of the resistance of the conductors forming the winding of the device and the square of the current through this winding is always converted into heat. That is, when a current I flows through a resistance r, heat is always "dissipated" in this resistance, and the rate of dissipation is

$$P_h = rI^2 \qquad (23a)$$

This experimental fact is known as *Joule's law*. This law applies directly only to continuous or non-varying currents. The relation $P_h = rI^2$ is, however, used as the basis for defining the "effective" resistance of a conductor to an alternating current (q.v.).

Effective Resistance and Conductance. In Art. 6, eq. 4, resistance was defined as the factor of proportionality in Ohm's law. On this basis it was shown that resistance is a variable factor dependent upon material, temperature, size, and shape of conductor and upon current distribution. Furthermore, eq. 4 applies only where there is no internal emf in the circuit. The most general definition of the resistance r of a substance between any two equipotential surfaces intersecting the path of a current I is

$$r = P_h/I^2 \qquad (23b)$$

where P_h is the power dissipated as heat between the two equipotential surfaces and I is the effective value of the total current from one surface to the other. With a varying current this dissipation of heat may occur in four different ways, viz.: (1) as heat due to the conduction current through the substance, (2) as heat due to dielectric hysteresis accompanying the displacement current through the substance (when the substance is an insulator), (3) as heat due to magnetic hysteresis accompanying the varying magnetic flux produced by the current, and (4) as heat due to eddy currents induced in neighboring conductors.

In a continuous current the last three effects do not occur, and the heat is that due to the conduction current only. The resistance offered by a substance to a continuous current is called the "true," "ohmic," "continuous-current," or "direct-current" resistance, as distinguished from the "effective" resistance offered by the substance to a varying current. The effective resistance, even when there are no losses due to dielectric or magnetic hysteresis, is in general greater than the ohmic resistance, owing to the skin effect.

Similarly, the general definition of conductance is

$$g = P_h/V^2 \qquad (23c)$$

where P_h has the same meaning as above and V is the effective value of the potential difference between the two equipotential surfaces. When the voltage is non-varying the conductance is called the "true," "ohmic," "continuous-current," or "direct-current" conductance as distinguished from the "effective" conductance to a varying current.

It should be noted that the ohmic resistance and conductance are reciprocals of each other, but that this is not true for the effective resistance and conductance.

ELECTROKINETICS AND THE MAGNETIC CIRCUIT

9. MAGNETS AND MAGNETISM

Magnets and Magnetic Substances. A magnet may be defined as any body which possesses the property of attracting pieces of iron or steel * and which when freely suspended takes up a definite position with respect to the geographical meridian. A magnetic substance is any body which acquires this property when it is placed near a magnet or near a conductor carrying an electric current. A body which is given this property is said to be "magnetized." A magnetic needle is a magnetized needle of iron or steel; the north-seeking end of such a needle is called its north pole, and the south-seeking end its south pole. When such a needle is freely † suspended near a magnet or a conductor carrying an electric current a couple is bound to be exerted upon it which causes it to take up a definite direction. The needle is said to "point" in the direction of a line drawn through it from its south to its north pole.

See Section 12 for the magnetic properties of substances used in electrical engineering.

Magnetic Effects. The following classification due to S. R. Williams covers the various types of phenomena associated with magnetism.

1. Magneto-magnetics is covered by the paragraph immediately following in this section.

---

* With a force in excess of the gravitational force, which is extremely small.

† A needle is said to be freely suspended when no controlling force is exerted upon it through its suspension tending to make it take up any definite position.

2. Magneto-mechanics includes mechanical strains due to magnetic stresses and magnetic strains due to mechanical stresses.

3. Magneto-acoustics covers the production of sound by magnetization and the influence of mechanical vibrations on magnetism.

4. Magneto-electrics involves the relationship between magnetic and electric circuits, the principal phenomena of which are covered in following paragraphs.

5. Magneto-thermics covers the influence of heat on magnetism, and vice versa.

6. Magneto-optics covers the influence of magnetism on light, and vice versa.

7. Cosmical magnetism.

Magnetic Field of Force. Any region in which a magnetic substance (e.g., a piece of soft iron), when placed therein, becomes magnetized is said to be a *magnetic field*. A magnetic field exists in and around every magnetized substance and around every electric current. The direction of the magnetic field at any point P is arbitrarily taken as the direction in which a small magnetic needle point would point when placed at P without disturbing appreciably the existing conditions.

Ferromagnetism. Much of the contemporary theory of ferromagnetism is based on spectrum analysis and interpreted by the Bohr-Sommerfeld atomic model and its modifications. Specifically the elementary magnetic particle is the so-called "spinning" electron. The model of the atom used to account for the elementary magnetic effect requires that the electron spin about an axis passing through its center, as distinguished from the rotation in circular or elliptical orbits around the atomic nucleus. In this manner each orbital electron has a magnetic momentum due to its moving electrical charge and an angular momentum due to its moving mass.

Uncompensated Spins. It is assumed that all such orbital electrons spin, but that in general, at any particular energy level or shell within the atom, all electrons may be divided into two equal groups—those that spin in one direction and those that spin in the opposite direction, thus producing a *null* magnetic effect. This effect is known as compensated electron spins. In certain elements, however, it is consistent with the theory to believe that uncompensated electron spins occur in one or more shells. In other words, there are, for example, more electrons spinning in one direction than in the other, in a given shell. These excess or uncompensated electron spins are an important factor in the phenomena of ferromagnetism.

Exchange. Fundamentally, ferromagnetism consists of the reorientation of the magnetic moments of the uncompensated electron spins due to the magnetizing force of an externally applied magnetic field. Although this accounts for the major ferromagnetic properties of the ferromagnetic elements such as iron, cobalt, and nickel, it fails to account for the absence of such ferromagnetic properties in other elements also known to have uncompensated electron spins. This apparent discrepancy is removed when the so-called exchange forces are considered. For an element to exhibit ferromagnetic properties it is required that, in addition to the existence of uncompensated electron spins, these spins must be parallel in contiguous atoms. Were this not the case the magnetic moments of individual atoms would be random in direction and hence the resultant magnetic moment of an appreciable region within the substance would be nil. It has been found that a certain ratio must exist between the diameter of an atom and the diameter of an electron shell that has uncompensated spins in order to permit the alignment of magnetic moments in contiguous atoms. This ratio is necessary because the electron spins and charges influence each other, dependent upon the distance between them. This influence, which is known as the *exchange*, must have a proper value in order that the uncompensated spins can be aligned to produce ferromagnetic effects. These forces of exchange tend to keep the spins parallel in neighboring atoms while the forces of thermal agitation tend to destroy this alignment. When the temperature of the substance becomes great enough the forces of exchange are completely overcome and the substance loses its ferromagnetic properties. The temperature at which this occurs is the well-known Curie point. According to the theory, magnetic saturation depends both on the uncompensated electron spins and the exchange. Rough approximation makes the saturation point a function of the product of the number of uncompensated electron spins and the exchange.

Domain. In ferromagnetic substances the forces of exchange are sufficiently large so that the uncompensated electron spins in neighboring atoms are more stable when their magnetic moments are parallel than under any other orientation, even when no external magnetic field is applied. However, this situation holds true only over very small regions in the given specimen of the substance. These regions, which are called "domains," have been found experimentally to have the volume equivalent to a cube approximately $1/1000$ in. on an edge. Ferromagnetic substances are completely divided into such domains, each domain being magnetized to saturation in a definite direction. Any specimen of a ferromagnetic substance is said to be unmagnetized when the directions of magnetization of the individually magnetically saturated domains are oriented at random with respect to each other. Thus, the application of an external magnetic field tends to reorient the individually magnetically saturated domains in the direction of this applied field.

Crystal Structure. X-ray analysis has shown that most materials are of crystalline structure. In the case of the ferromagnetic substances these crystals are too small to be seen individually. However, their properties have been studied by means of spectrum analysis and photomicrography. Owing to this crystalline structure there is in general more than one axis of stable magnetic saturation. In the cubic crystal characteristic of iron there are six equally stable axes of magnetic saturation.

Magnetization Curves. On the basis of uncompensated electron spins, forces of exchange, the existence of domains and the multiple axes of equally stable magnetization, it is now possible to interpret the major ferromagnetic phenomena. When a small external field (say less than 1000 amperes per meter) is applied to an unmagnetized specimen of a ferromagnetic substance, the saturation magnetization within each crystal changes its direction from its axis of stable magnetization to that axis of stable magnetization which is most nearly in line with the applied field. This process takes place as sudden individual reorientations within the individual crystals, and the strength of the field required to produce these effects is dependent on the original deviation of these orientations from that of the applied field. These sudden jumps constitute the explanation of the well-known Barkhausen effect. The portion of the magnetization curve represented by the foregoing process is that up to but not including the knee of the curve. At the knee it is assumed that all individual crystals have directions of stable magnetization which lie nearest to the direction of the applied field. When stronger fields than this are applied (order of magnitude 1000 to 100,000 amperes per meter), the domains themselves are rotated as a whole until the axes of magnetization of the crystal are coincident with the direction of the applied field. Needless to say, beyond this point an increased applied field produces no further increase in flux density due to the ferromagnetic phenomena.

On the basis of the foregoing theory it is possible to account for the shape of the magnetization curve and actually to calculate many of the known quantities which are used in ferromagnetic practice.

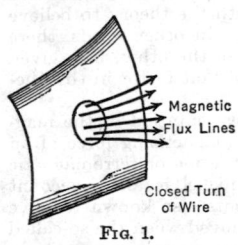

Magnetic Flux Lines

Closed Turn of Wire

FIG. 1.

Magnetic Flux. Consider a small closed turn of wire, Fig. 1, placed in a magnetic field with its plane perpendicular to the direction of the field. Experience shows that when such a turn of wire is removed from the field in any manner whatever (the coil remaining short-circuited on itself or forming part of a closed circuit), or when the magnetic field is caused to disappear in any manner whatever, a momentary emf is set up or *induced* in this coil, which in turn causes a momentary electric current to flow through the coil. This emf exists only while the coil is moving across the field or while the field through the coil is varying.

The time integral of the induced emf when the coil is removed entirely from the magnetic field is taken as the measure of the magnetic flux existing through the coil when in its original position. That is, calling e the emf induced in the coil at any instant by its motion through the field, and t the time during which the emf exists in the coil, then the magnetic flux Φ through the coil when in its original position is

$$\Phi = \int_0^t e \, dt \qquad (1)$$

When e is in volts and t is in seconds, Φ is in webers. See Section 3. This quantity Φ is readily measured by means of a ballistic galvanometer.

Magnetic Flux Density. Experience shows that the magnetic flux through any closed loop, such as the turn of wire described above, depends upon the area inclosed by this loop. The magnetic flux per unit area through any surface perpendicular to the direction of the field is defined as the *magnitude* of the *magnetic flux density* at this surface and is usually represented by the symbol B. By the magnitude of the flux density at any point is meant the magnitude of the flux density at any infinitely small surface drawn perpendicular to the field at this point. The *direction* of the *magnetic flux density* at any point is the same as that in which a magnetic needle would point if placed at this point; i.e., the direction of the flux density and the direction of the magnetic field are the same. The vector having the above-defined magnitude and direction is called the flux density and is usually designated by the symbol B. When the flux density has the same value B at every point of a surface of area A and is perpendicular to this surface, then the total flux through this surface is

$$\Phi = BA \quad \text{webers} \qquad (2)$$

The total magnetic flux across any surface S may in general be expressed mathematically by the surface integral

$$\Phi = \int (B \cos \alpha)\, ds \quad \text{webers} \tag{3}$$

where ds represents any elementary area of this surface and $(B \cos \alpha)$ the component of the flux density perpendicular to ds. Magnetic flux density in the mks system is expressed in webers per square meter.

Magnetic Flux Lines. Magnetic flux can be represented by lines so drawn in the field that their direction coincides at each point with the direction of the field at that point, and of such a number that their density at each point (number per unit area perpendicular to their direction) is equal to the magnetic flux density at that point. Such lines are called "magnetic flux lines." Experience shows that lines thus drawn in a magnetic field always form closed loops; i.e., a magnetic flux line has no ends. As a consequence of this fact the total magnetic flux coming up to any surface in a magnetic field is always equal to the total flux leaving that surface.

10. ELECTROMAGNETISM

Magnetic Fields Due to Electric Currents. Experience shows that every filament or stream line of electric current is always accompanied by a magnetic field the flux lines of which link the stream line of current. That is, the flux lines thread the loops formed by the stream lines and the stream lines thread the loops formed by the flux lines; see Fig. 2.

Right-handed Screw Law. The direction of the current flowing around any electric circuit and the direction in which the flux lines due to that current thread this circuit are related to each other in the same manner as the direction of motion of a point on the edge of the head of a right-handed screw placed at the center of the circuit and the direction of advance of the screw. Or, if one faces the electric circuit looking in the direction of the flux lines threading it, the current producing these lines is in the clockwise direction around the circuit. The relative direction of the current and its magnetic flux may be briefly described by saying that the current is in the right-handed screw direction with respect to the flux which it produces.

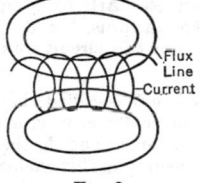

Fig. 2.

Induced Emf. The measure of magnetic flux is based on the experimental fact that, whenever the magnetic field threading an electric circuit changes, an emf is induced in that circuit. When the circuit is formed by a single turn of wire this induced emf is, from the definition above, equal to the rate of change of this flux with respect to time; that is, $e = d\Phi/dt$. When the circuit is in the form of a coil each turn of which links the flux, the emf induced in each turn is equal to $d\Phi/dt$, where Φ is the flux which links that particular turn. When each turn links the same number of flux lines, then the total induced emf in a coil of N turns is

$$e = N \frac{d\Phi}{dt} \quad \text{volts} \tag{4}$$

where Φ is in webers.

When the change in flux is due to a motion of a circuit or part of a circuit through a magnetic field the induced emf in any conductor is also equal to the negative of the number of flux lines which cut across this conductor per unit time.

Magnetic Linkages. The condition that each turn of a coil be linked by the same flux Φ seldom exists; some of the flux lines usually link only part of the turns. In general, the total emf is

$$e = -\frac{d}{dt}(n_1 \Phi_1 + n_2 \Phi_2 + \cdots + n_n \Phi_n) \tag{4a}$$

where Φ_1, Φ_2, etc., represent the fluxes linking the various numbers of turns, n_1, n_2, etc., respectively. The sum $(\lambda_1 + \lambda_2 + \cdots + \lambda_n)$ may be called the total number of *magnetic linkages* or *flux linkages*, and may be conveniently represented by the symbol λ, viz.,

$$\lambda = \lambda_1 + \lambda_2 + \cdots + \lambda_n \quad \text{weber-turns} \tag{4b}$$

and the total induced emf may then be written

$$e = d\lambda/dt \quad \text{volts} \tag{4c}$$

When all the N turns link the same flux, Φ, then $\lambda = N\Phi$.

Direction of Induced Emf. The direction of the induced emf around a circuit is found to be in the left-handed screw direction with respect to the increase of flux; viz., if one faces the circuit looking in the direction of the increase of flux, the induced emf is in the counterclockwise direction. The current which would be set up by this emf, however, would produce a flux linking the circuit in the right-handed screw direction. Hence a change in the magnetic flux through an electric current always sets up an emf which tends to produce a current around this circuit in such a direction as to set up an opposing flux. This fact may be expressed mathematically by writing a minus sign before $d\Phi/dt$ as in eq. 4, i.e., by putting

$$e = -N\frac{d\Phi}{dt} \tag{4d}$$

The value of $\left(-N\dfrac{d\Phi}{dt}\right)$ is then the emf induced in the circuit in the right-handed screw direction with respect to the increase of flux. Or stated in other words $\left(-N\dfrac{d\Phi}{dt}\right)$ represents the rise of electric potential and $N(d\Phi/dt)$ represents the drop of potential around the circuit in the right-handed screw direction with respect to the increase of flux.

11. MAGNETIZING FORCE AND THE MAGNETIC CIRCUIT

Simple Magnetic Circuit. In certain simple magnetic circuits the treatment is mathematically analogous to that of simple electric circuits. An expression similar to Ohm's law (see Art. 6) is employed. Consider a closed magnetic circuit such as a uniformly wound torus, Fig. 3.

For this circuit we may write

$$\mathfrak{F} = \mathfrak{R}\Phi \tag{5}$$

where $\mathfrak{F}$ is called the *magnetomotive force* (abbreviated mmf) and is analogous to electromotive force; $\mathfrak{R}$ is a factor of proportionality called the *reluctance* and is analogous to resistance; and Φ, the flux, is analogous to electric current. It must be remembered in considering this analogy that an unvarying flux is thought of as a static condition, whereas electric current is defined as the motion of electric charge. Therefore, there is no true *physical* analogy. Equation 5 is sometimes called Ohm's law for magnetic circuits.

Magnetomotive Force. The mmf $\mathfrak{F}$, of eq. 5, may also be expressed

$$\mathfrak{F} = NI \tag{6}$$

where $\mathfrak{F}$ is in *amperes*, N is the total number of turns of the coil, and I is the current in amperes flowing through the coil. The current-turns, NI, are frequently spoken of as the *ampere-turns* when I is given in amperes. Obviously $\mathfrak{F}$ is in direct proportion to the ampere-turns and dimensionally is the same as the ampere. In general, mmf may be expressed

$$\mathfrak{F} = \Sigma NI \tag{6a}$$

where ΣNI represents the algebraic summation of all current-turns linking the magnetic circuit. The mmf is taken as positive when the current links the flux lines in the right-

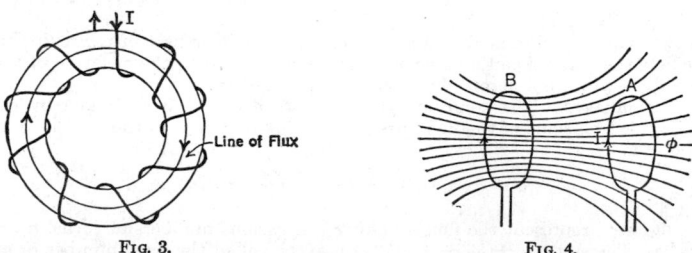

FIG. 3.   FIG. 4.

handed screw direction, and negative when the current links the flux lines in the left-handed screw direction.

Work Done by a Varying Magnetic Flux. Consider a coil A (Fig. 4) of N turns of wire, and let each of these N turns be linked by a flux Φ due to some external agent, e.g., another coil B in which an electric current is flowing. Let the flux Φ through A due to B be increasing at any instant at the rate $d\Phi/dt$ in the left-handed screw direction with respect to the current I in A at this instant. Then there is induced in A at this instant an emf in the direction of I equal to $e = N(d\Phi/dt)$, and therefore the electric power developed in A at this instant is $ei = NI(d\Phi/dt)$. This power is transmitted to the coil A as a result of the varying flux through it; hence the power

$$p = NI\frac{d\Phi}{dt} \tag{7}$$

may be looked upon as the magnetic power input, this power being converted within the coil into electric power. In this example the negative is omitted because of the assumed direction in coil A.

Magnetizing Force or Magnetic Field Intensity. Experience shows that the magnetic flux density produced at any point by a given mmf depends (a) upon the position of the point with respect to the source of the mmf and (b) upon the nature of the substances through which this mmf produces the magnetic flux. These facts lead to the conception of the flux density at any point in a magnetic field as being due to a "magnetizing force", H at that point, this magnetizing force H depending solely upon the mmf producing the field and the distribution of the flux lines, as distinguished from the flux density B which depends not only upon these two items but also upon the nature of the medium at the point in question.

The magnetizing force (also called the "magnetic field intensity") at successive points along any closed path in a magnetic field may be defined by the relation that its line integral around such a path is equal to the total mmf acting around this path, viz.,

$$\Sigma NI = \int (H \cos \theta)\, dl \tag{8}$$

where dl represents any elementary length of this path (see Fig. 5); ($H \cos \theta$) the value of the component of the magnetizing force at dl in the direction of dl; and ΣNI the total number of current turns linked by the path. Experience shows that such a definition leads to a simple means of expressing in a quantitative manner the interrelations of a number of experimental facts. Magnetizing force may also be expressed as the force in newtons which would act on a "unit positive magnetic pole."

When the path coincides in direction with the magnetizing force at each point, eq. 8 may be written

$$\Sigma NI = \int H\, dl \tag{8a}$$

Direction of the Magnetizing Force. Experience shows that *except* for points inside a permanent magnet the magnetizing force H and the flux density B are always in the same direction. For points inside a permanent magnet the direction of the magnetic field in-

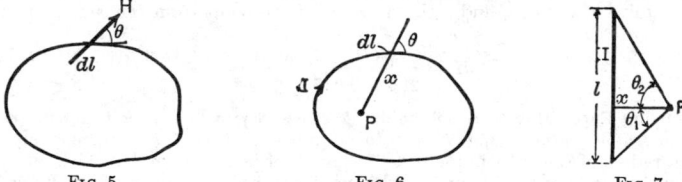

FIG. 5.   FIG. 6.   FIG. 7.

tensity H, due solely to the magnet itself, is opposite to the direction of the flux lines; i.e., a permanent magnet produces a "demagnetizing force" on itself.

Lines of Magnetizing Force. The magnetizing force at any point in a magnetic field may be represented by lines drawn in the field in a direction such as to coincide with the direction of the magnetizing force at each point, and of such a number per unit area perpendicular to their direction that their density at each point gives the value of the magnetizing force at that point. Such lines are called *lines of magnetizing force* or *lines of magnetic field intensity*. In general the lines of magnetizing force and the magnetic flux lines coincide in direction (except within the substance of permanent magnets), but their densities are different. In non-magnetic substances the flux lines and lines of magnetizing force measured in cgs units coincide in both number and direction. The simple expression "lines of force" is frequently used to designate either the flux lines or the lines of magnetic intensity, but it is evident that such loose use of this term is likely to lead to confusion when speaking of the magnetic field within a magnetic substance.

Magnetizing force is of the nature of mmf per unit length. The mks unit of magnetizing force is called an ampere per meter.

When the medium surrounding the stream lines of an electric current is of a uniform magnetic nature throughout, the magnetizing force at any point may be calculated from the shape and distribution of the stream lines of the current, irrespective of whether the medium is non-magnetic or highly magnetic.

Magnetizing Force at Any Point Due to an Element of a Current-stream Line (Fig. 6). Consider any closed stream line of electric current and let the surrounding medium be uniform in its magnetic properties throughout the region in which the magnetic field produced by this stream line exists. It can be shown that each elementary length dl of this stream line may be considered as contributing to the magnetizing force H at any point P in this region an amount

$$dH = \frac{(I \sin \theta)\, dl}{4\pi x^2} \quad \frac{\text{amperes}}{\text{meter}} \tag{9}$$

where I is the current flowing along this stream line, x the distance from P to dl, and θ the angle between x and dl. The direction of dH is perpendicular to the plane determined by x and dl. The total magnetizing force at P is then the vector sum or vector integral of dH for all the elementary lengths into which the stream line is divided.

Magnetizing Force Due to a Straight Wire (Fig. 7). Applying eq. 9 to the case of a straight wire of circular cross section carrying a current I, the magnetizing force at any point P due to a length l of this wire is

$$H = \frac{I}{4\pi x}(\sin \theta_1 + \sin \theta_2) \quad \frac{\text{amperes}}{\text{meter}} \tag{9a}$$

where x is the perpendicular distance from P to the wire, and θ_1 and θ_2 the angles designated in Fig. 7.

If the wire is very long compared with x, this becomes

$$H = \frac{I}{2\pi x} \quad \frac{\text{amperes}}{\text{meter}} \tag{9b}$$

This formula also holds approximately for any point outside a wire of any shaped cross section, provided x is large compared with the maximum diameter of this section. For a point inside a long wire of circular cross section of radius a the magnetizing force is also given by eq. 9b when I is taken to represent that part of the current inside the circle through P concentric with the axis of the wire. When the current density is uniform over the cross section, as is usual, the magnetizing force inside the wire is

$$H_i = \frac{xI}{2\pi a^2} \quad \frac{\text{amperes}}{\text{meter}} \tag{9c}$$

Magnetizing Force on the Axis of a Circular Coil of N Turns. Let I be the current, r the mean radius of the coil, and x the distance of the point from the center of the circle; then

$$H = \frac{NIr^2}{2(r^2 + x^2)^{3/2}} \quad \frac{\text{amperes}}{\text{meter}} \tag{9d}$$

Magnetizing Force Due to a Solenoid. A solenoid is a helical coil of wire, each turn having the same radius. Let N = total number of turns, I = current in amperes, r = mean radius of the helix in meters, l = length of helix in meters. Then at any point on the axis of the helix (inside or outside), at a distance of x meters from its center, the magnetizing force is

$$H = \frac{NI}{2l}\left[\frac{0.5l + x}{\sqrt{r^2 + (0.5l + x)^2}} + \frac{0.5l - x}{\sqrt{r^2 + (0.5l - x)^2}}\right] \quad \frac{\text{amperes}}{\text{meter}} \tag{9e}$$

This formula holds only when the thickness of the winding is small compared with the mean radius r. When l is large compared with r this reduces to

$$H = \frac{NI}{l} \quad \frac{\text{amperes}}{\text{meter}} \tag{9f}$$

For all points inside the solenoid (whether on the axis or not) at a distance from the ends large compared with r, that is, inside the central portion of a long solenoid, the field is uniform over the cross section of the solenoid and its value is given by eq. 9f. An exact formula for field intensity at any point inside a solenoid is given by O. Billieux in *Rev. gén. élec.*, 6, p. 827, Dec. 13, 1919.

Magnetizing Force Inside a Torus (Fig. 8). When a torus is uniformly wound with an insulated wire so that the turns of the wire are close together and cover the entire surface of the torus, the magnetic field is confined entirely within the space inclosed by these turns, and therefore, when the core on which the wire is wound is of uniform magnetic material throughout, both the lines of magnetizing force and the flux lines must be concentric circles, as shown in the figure. The magnetizing force will have the same value at every point on the circumference of any one of these circles, and, therefore, from eq. 8a the value of H at any point P within the core is

$$H = \frac{NI}{l} \quad \frac{\text{amperes}}{\text{meter}} \tag{9g}$$

where N is the total number of turns on the core, I the current in each turn, and l the length of the circumference through P. Unless the ring has a large radius compared with the radius of the cross section of the core, H will not be uniform over this section, since l for the various points in the cross section will differ considerably.

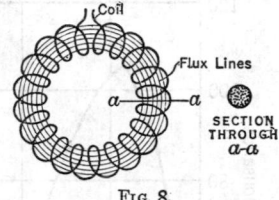

FIG. 8.

It should be noted that the value of H is independent of the material of the core provided only that the core is of uniform material throughout. That is, eq. 9g applies to an iron core as well as to an air or wood core, provided the iron is uniform throughout and there is no air gap across the path of the flux lines. Even a mechanically perfect contact between two pieces of iron of the same kind, however, is sufficient to make the above formula useless.

Magnetic Permeability. In free space, or vacuo, magnetic flux density is related to magnetic field intensity by the defining formula

$$B = \mu_0 H \tag{10}$$

where μ_0 the permeability of free space, has the value $4\pi \times 10^{-7}$ henry per meter. In other materials the magnetic flux density at a point is related to the magnetic intensity at the same point by

$$B = \mu H = \mu_0 \mu_r H \tag{11}$$

where

$$\mu = \mu_0 \mu_r \tag{11a}$$

The relative permeability μ_r is a numeric comparing a given material with free space. The magnetic permeability μ for a medium is the product of the relative permeability of that medium and the permeability of free space.

The relative permeability of most substances differs inappreciably from unity. Aside from iron, steel, nickel, cobalt, and synthetic magnetic alloys, all the materials used in

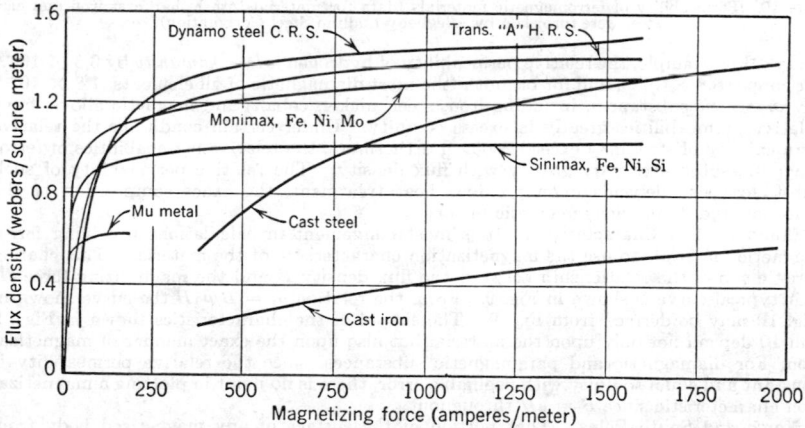

FIG. 9. Magnetization curves of ferromagnetic materials (data for materials other than cast iron and cast steel were furnished by Allegheny-Ludlum Steel Corporation).

electrical engineering have a relative permeability which may be considered equal to that of air. The relative permeability of non-magnetic substances is for all practical purposes a constant irrespective of the flux density. A physical picture of relative permeability may be gained from the following. A coil carrying an unvarying current produces a magnetic field in the air surrounding it. If a substance is placed, for example, inside this coil without changing the amount or distribution of the flux inside the coil, this substance has a relative permeability of unity. If the flux inside the coil is altered in value or distribution, the relative permeability of the introduced substance differs from unity.

Paramagnetic, Diamagnetic, and Ferromagnetic Substances. Considering empty space as being truly non-magnetic, all substances may be divided into three classes. *Paramagnetic* substances have a relative permeability slightly greater than unity. For example, the relative permeability of air is *greater than unity* by 3.8×10^{-7} and for aluminum by 2.3×10^{-5}. *Diamagnetic* substances have a relative permeability slightly less than

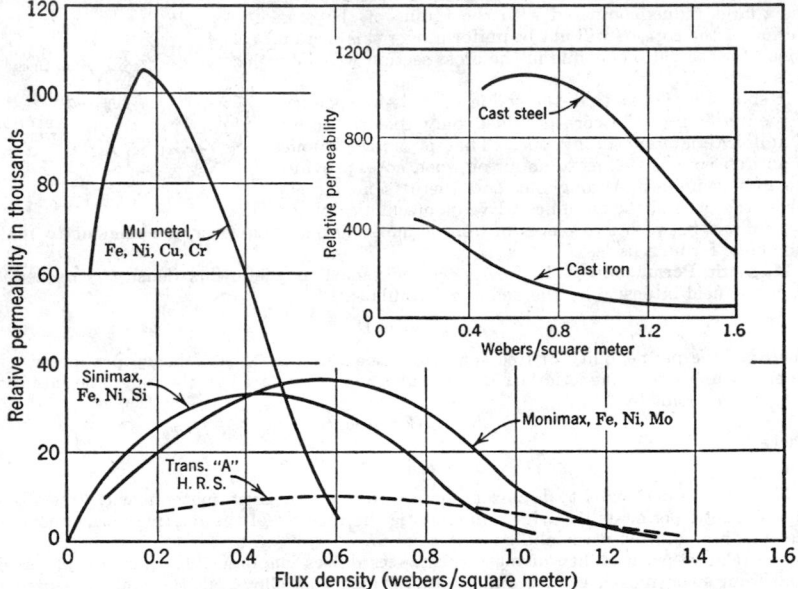

FIG. 10.   Permeability of ferromagnetic materials (data for materials other than cast iron and cast steel were furnished by Allegheny-Ludlum Steel Corporation).

unity. For example, the relative permeability of hydrogen is *less than unity* by 6.3×10^{-8}, for copper 8.8×10^{-6}, and for bismuth (the most diamagnetic of all elements) 1.8×10^{-4}. *Ferromagnetic* substances, including iron, steel, nickel, cobalt, and magnetic alloys, have relative permeabilities greatly in excess of unity. Under certain conditions the relative permeability of steel may exceed 2000. Furthermore, the relative permeabilities of ferromagnetic substances vary greatly with flux density. The relative permeability of such substances also depends upon previous heat treatment, the exact composition of the material, and its previous magnetic history.

Magnetization Characteristic. It is most convenient in calculations involving ferromagnetic materials to use the magnetization characteristic of the material. This characteristic shows the relationship between the flux density B and the magnetizing force H.

A typical curve is shown in Fig. 9. From the relation $\mu_r = B/\mu_0 H$ the curve shown in Fig. 10 may be derived from Fig. 9. The shapes of the characteristics shown in Figs. 9 and 10 depend not only upon the material but also upon the exact manner of magnetization. For diamagnetic and paramagnetic substances, since the relative permeability is constant and equal to unity with negligible error, there is no point in plotting a magnetization characteristic since $B = \mu H$ throughout.

North and South Poles. That portion of the surface of any magnetized body from which the flux lines pass out into the air (or any substance of lower relative permeability) is said to be a north magnetic pole, and that portion of the surface at which the flux lines enter the body is said to be a south magnetic pole. A *unit north pole* is a pole from which one line of magnetic flux or one weber emerges into the surrounding air. When a magnetic needle is placed near the surface of a magnetized body, its north-seeking end points away from the surface when this surface is a north pole and toward the surface when this surface is a south pole.

Difference of Magnetic Potential. Consider any two points 1 and 2 in a magnetic field (Fig. 11), and let the path between them from 1 to 2 pass through an electric circuit producing a mmf in the direction from 1 to 2; then the expression

$$U_{12} = \int_1^2 (H \cos \theta) \, dl - \mathfrak{F}_{12} \quad \text{amperes} \qquad (12)$$

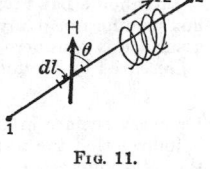

Fig. 11.

is called the *drop of magnetic potential* from 1 to 2. From the definition of magnetizing force, eq. 8, it follows that around any closed circuit the drop of magnetic potential is always zero. A magnetomotive force $\mathfrak{F}_{12}$ is, therefore, equivalent to a rise of magnetic potential from 1 to 2.

When there is no source of mmf between 1 and 2 and the path coincides with a line of magnetizing force, the drop of magnetic potential is

$$U_{12} = \int_1^2 H \, dl \quad \text{amperes} \qquad (12a)$$

Magnetic potential difference is of the same nature as mmf and may, therefore, be expressed in the same units, viz., amperes.

Magnetic Equipotential Surfaces. A surface, drawn in a magnetic field in such a manner that this surface is perpendicular at each point to the magnetizing force at this point (i.e., to the line of magnetizing force through this point) is called a "magnetic equipotential surface."

Magnetic Reluctance. To establish a magnetic flux Φ through a given portion of a substance which is not itself linked by a source of mmf, a difference of magnetic potential must always be established between the end surfaces of this substance. Let U be the magnetic potential drop established from one surface to the other; then the quotient

$$\mathfrak{R} = \frac{U}{\Phi} \quad \frac{\text{amperes}}{\text{weber}} \qquad (13)$$

is defined as the magnetic reluctance of the given portion of the substance. It should be noted that the above definition is meaningless except when applied to a portion of a substance of which the end surfaces are magnetic equipotential surfaces and through every cross section of which the same flux passes.

Factors upon Which Reluctance Depends. The magnetic reluctance of a given portion of a substance included between two equipotential surfaces and bounded laterally by a surface which no flux line passes depends upon (a) the magnetic permeability of the substance, (b) the dimensions of this portion of the substance, and (c) the distribution of the flux lines over each cross section perpendicular to them. The relations are identical with those which determine the electrical resistance of a conductor, the magnetic permeability taking the place of the electric conductivity. For example, for a straight bar of constant cross section A and length l, through which the flux lines are straight, parallel, and uniformly distributed, the reluctance is

$$\mathfrak{R} = \frac{l}{\mu A} = \frac{l}{\mu_0 \mu_r A} \quad \frac{\text{amperes}}{\text{weber}} \qquad (13a)$$

Note particularly that this formula for reluctance is applicable only under the special conditions just stated. Formulas for other cases are much more complex; see J. F. H. Douglas, "Reluctance of Irregular Magnetic Fields," *Proc., A.I.E.E.*, 34, May, 1915.

Magnetic reluctance is not a constant quantity even for a given material and given flux distribution, unless this material is non-magnetic. For all highly magnetic materials μ_0 depends upon the magnetizing force and therefore also upon the flux density. It should also be noted that the magnetic reluctance does not represent a "resistance" in the sense of something which causes a dissipation of energy.

Magnetic Permeance. The reciprocal of magnetic reluctance is called *magnetic permeance*. The permeance of a straight bar under the conditions specified above is

$$\mathcal{P} = \frac{\mu A}{l} = \frac{\mu_0 \mu_r A}{l} \quad \frac{\text{webers}}{\text{ampere}} \qquad (13b)$$

The permeability of a substance is, therefore, equal to the permeance of a unit cube of this substance when the flux through the cube is parallel to four edges of the cube and is uniformly distributed over the section at right angles to these four edges.

Magnetic permeance is analogous to electric conductance, except that it is not a factor which affects the dissipation of energy in a substance. It does, however, enter into the expression for the energy stored in a magnetic field in the same way that the electrostatic capacity of a dielectric is a determining factor in the expression for the energy stored in the electric field (see p. 977 , Art. 18).

Kirchhoff's Laws for the Magnetic Circuit. As already noted (eq. 3) the total magnetic flux coming up to any surface in a magnetic field is always zero, provided a flux leaving a surface is considered as a negative flux coming up to that surface. This fact may be represented by the formula

$$\Sigma\Phi = 0 \tag{14}$$

for every surface in the field. Similarly, from the definition of magnetic potential drop, it follows that the total magnetic potential drop around any closed circuit is zero, or that the total mmf acting around any closed circuit is equal to the sum of the reluctance drops around that circuit, which may be represented by the formula

$$\Sigma\mathfrak{F} = \Sigma\mathfrak{R}\Phi \tag{15}$$

These two equations are identical in form with those representing Kirchhoff's laws for the electric circuit. They are, however, not so easy to use for practical calculations, for the magnetic flux is not confined to approximately geometrical lines like the currents in a network of insulated wires, but in general fills all space surrounding the coils which establish the mmf's; also, when there is iron or other magnetic material in the circuit, the permeability depends on the flux density and the previous history of the iron. (The distribution of magnetic flux in and around an iron circuit is analogous to the distribution of current in and around an uninsulated mass of copper of the same shape as the iron circuit immersed in a liquid having a conductivity about equal to that of carbon.) Only in the special case of a uniformly wound circular ring or toroid are the lines of induction confined entirely to an iron circuit; in general a certain number also exist in the air and in whatever other substances are in the vicinity of the iron circuit.

12. ELECTROMAGNETIC INDUCTION

Induction. When the current in a given electric circuit varies with time the magnetic flux accompanying this current also varies with time, and, since this flux links the current which produces it, an emf is induced in each turn of the circuit equal to the rate at which the flux through this turn is varying, and in such a direction as to oppose the change in the current. That is, an increasing electric current is always accompanied by a back emf due to the increase in the magnetic flux which accompanies this increase in current. Again, when an electric current decreases, its accompanying flux decreases and an emf is set up in the circuit tending to oppose this decrease, i.e., tending to keep the current from decreasing. This is analogous to the effect of inertia in ordinary matter. These relations can be summarized in the formula, $e = -N(d\Phi/dt)$, known as *Faraday's law*.

If the current in a given circuit is constant, but there is a variation with time of the permeances of the flux paths or of the number of turns linking a given flux path, an emf is produced in the circuit in a direction such as to tend to produce a current which would oppose the variation. The preceding statement is often designated as *Lenz's law*.

These phenomena are known as *induction*.

Coefficient of Self-induction or Inductance. In general, the coefficient L by which the rate of change of the current (di/dt) in any circuit must be multiplied to give the self-induced emf e is called the *coefficient of self-induction* or simply the *inductance* of the electric circuit. In general, then, the self-induced emf in an electric circuit is

$$e = -L\frac{di}{dt} \text{ volts} \tag{16}$$

where L is the inductance of the electric circuit in henrys, and di/dt represents the change in the current in amperes per unit of time in seconds. Since $e = -d\lambda/dt$ (see eq. 4c), where λ is the number of magnetic linkages between the electric circuit and the flux established by the current i, the inductance may also be defined by the relation

$$L = \frac{\partial\lambda}{\partial i} \frac{\text{weber-turns}}{\text{ampere}} \text{ or henrys} \tag{16a}$$

That is, the inductance is equal to the increase in the number of linkages per unit increase in the current; in the mks system 1 henry equals 1 weber-turn per ampere.

It should be noted that e in eq. 16 is not the *total* induced emf, even of an isolated circuit, this being given by

$$e = -\frac{\partial \lambda}{\partial i}\frac{di}{dt} - \frac{\partial \lambda}{\partial t} \quad \text{volts} \tag{17}$$

in which the first term is the self-induced emf, and the second term is an emf induced by motion.

When the permeability of the magnetic circuit is independent of the current, the inductance is also independent of the current. For example, the expression for the inductance of a coil which is not in the vicinity of any ferromagnetic substance may be written

$$L = \lambda/i \quad \text{henrys} \tag{18}$$

where λ is the total flux linkages corresponding to the current i in the coil. Since in this case λ is directly proportional to i, L is at most a function of time and is independent of the value of i. When every flux line is linked by every stream line of electric current, and the permeability of the entire magnetic circuit is independent of current,

$$L = N^2/\mathfrak{R} \quad \text{henrys} \tag{18a}$$

where N is the number of turns forming the electric circuit and $\mathfrak{R}$ is the reluctance of the complete magnetic circuit.

When the eq. 18 holds the *total* induced emf may be written

$$e = -(d/dt)(Li) \quad \text{volts} \tag{19}$$

The mks unit of inductance is called the henry; for the relation between the henry and the abhenry and millihenry see Section 3.

Coefficient of Mutual Induction or Inductance. In general, the coefficient M_{ab} by which the negative of the rate of change of the current (di_b/dt) in a circuit B must be multiplied to give the electromotive force e induced by this current in another circuit A is called the *coefficient of mutual induction* or simply the *mutual inductance* between A and B. See Fig. 4, Art. 11. In general, then, the emf induced in any circuit A by a varying current i_b in any other circuit B is

$$e_a = -M_{ab}(di_b/dt) \quad \text{volts} \tag{20}$$

where M_{ab} is the mutual inductance between A and B. The first letter of the double subscript indicates the circuit in which the flux linkages are considered; the second, the circuit in which the current is considered.

It can be shown from the principle of the conservation of energy that the mutual inductance of a circuit A with respect to a second circuit B must be equal to the mutual inductance of B with respect to A; that is, $M_{ab} = M_{ba}$. Whence, the emf induced in B when the current in A increases by an amount di_a is

$$e_b = -M_{ba}(di_a/dt) \quad \text{volts}$$

Since $e_a = -(d\lambda_{ab}/dt)$ where λ_{ab} is the number of linkages between the circuit A and the flux through A due to the current i_b, the mutual inductance may also be defined by the relation

$$M_{ab} = \frac{\partial \lambda_{ab}}{\partial i_b} \quad \frac{\text{weber-turns}}{\text{ampere}} \quad \text{or henrys} \tag{21}$$

That is, the mutual inductance between two circuits A and B is equal to the increase in the number of magnetic linkages of the circuit A per unit increase of the current in B, and vice versa. When the permeability of the magnetic circuit is independent of the current, the mutual inductance is also independent of the current and equal to the linkages of A per unit current in B, and vice versa. When every flux line linking both A and B is linked by every turn in A and every turn in B, then

$$M_{ab} = N_a N_b/\mathfrak{R}_{ab} \quad \text{henrys} \tag{22}$$

where N_a and N_b are the number of turns forming the circuits A and B, respectively, and $\mathfrak{R}_{ab}$ is the reluctance of that part of the magnetic circuit through which the flux from A to B passes when there is current in one coil only.

The units of mutual inductance are the same as those of self-inductance.

Instantaneous Potential Drop through a Coil. Consider a coil of wire which has a resistance r and an inductance L. Then, when this coil contains no other source of emf than its own self-induced emf, the expression for the instantaneous potential drop through the coil is

$$v = ri + L\frac{di}{dt} \tag{23}$$

where i is the instantaneous value of the current and di/dt the increase in this current per unit of time.

When there is another coil near the first coil, and the two coils have resistances r_1 and r_2 and self-inductances L_1 and L_2 respectively, and a mutual inductance M, then the potential drops through them are respectively

$$\left. \begin{aligned} v_1 &= r_1 i_1 + L_1 \frac{di_1}{dt} + M \frac{di_2}{dt} \\[2mm] v_2 &= r_2 i_2 + L_2 \frac{di_2}{dt} + M \frac{di_1}{dt} \end{aligned} \right\} \tag{24}$$

where i_1 and i_2 are the currents in the two coils in the same direction with respect to the magnetic circuit. These differential equations are perfectly general and can be applied to the solution of any circuit containing constant r's, L's, and M's regardless of the nature of the variation of any applied emf in the circuit. The solution of transient as well as steady-state conditions are obtainable from eqs. 24 as will be indicated later (see Arts. 24 to 26).

Skin Effect. A conductor of finite cross section may be looked upon as made up of separate filaments each carrying its portion of the total current. When the same potential gradient is established through all the filaments of the conductor, the exterior filaments are linked by fewer flux lines than the interior filaments. If the emf producing the potential gradient through the wire is an alternating one, the induced back emf in the interior filaments will be greater than in those nearer the surface. Since the potential drop is the same across all the filaments the resistance drops in the internal filaments is less than in the external ones. This can be brought about only by the current distributing itself over the cross section of the conductor in such a manner that the current density in the interior of the wire will be less than at the surface, i.e., the current is forced toward the surface filaments or "skin" of the wire; hence the term *skin effect* is applied to this phenomenon.

The self-induced emf depends not only upon the amount of flux set up but also upon the rapidity of its variation; hence the skin effect becomes more pronounced the greater the frequency of the impressed emf. It is also greater the larger the cross section of the conductor, the greater its conductivity, and the greater its magnetic permeability. The skin effect also depends slightly upon the temperature since the conductivity changes with temperature.

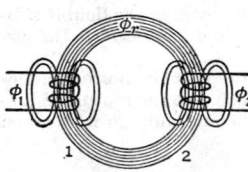

FIG. 12.

Leakage Inductance. In discussing the action of a transformer which consists of two electric circuits linking the same iron core, it is necessary to deal with the resultant flux due to the currents in both electric circuits or windings; the fluxes due to the two windings separately have no meaning. Referring to Fig. 12, let Φ_r represent that portion of the resultant flux due to the currents i_1 and i_2 in the two windings 1 and 2, and let Φ_1 represent that part of the total flux which links 1 only and Φ_2 that part of the total flux which links 2 only. Let λ_1 be the linkages between Φ_1 and circuit 1, and λ_2 the linkages between Φ_2 and circuit 2. Then

$$L'_1 = \lambda_1/i_1 \quad \text{and} \quad L'_2 = \lambda_2/i_2 \tag{25}$$

are called the *leakage inductances* of the two windings respectively; the reluctances of the paths of Φ_1 and Φ_2 are practically constant since the air portion of these paths forms a greater part of the reluctance in each.

Let i_1 be the current in winding 1 and i_2 the current in winding 2 linking the flux path in the opposite direction to i_1 (this is the actual relation in a transformer during most of the time), and let e_1 be the impressed emf across the terminals of the first or primary winding and e_2 the terminal emf at the terminals of the second or secondary winding when the current i_2 is flowing. Then

$$\left. \begin{aligned} e_1 &= r_1 i_1 + L'_1 \frac{di_1}{dt} + N_1 \frac{d\Phi_r}{dt} \\[2mm] e_2 &= N_2 \frac{d\Phi_r}{dt} - r_2 i_2 - L'_2 \frac{di_2}{dt} \end{aligned} \right\} \tag{26}$$

where r_1 and r_2 are the resistances of the two windings and N_1 and N_2 are the numbers of turns in the two windings respectively.

Comparing eq. 26 with 24 and noting that $e_1 = v_1$ and $e_2 = -v_2$, and i_2 in eq. 26 is taken in the opposite direction from i_2 in 24, it may be shown that

$$\left.\begin{array}{l} L'_1 = L_1 - \dfrac{N_1}{N_2} M \\[2mm] L'_2 = L_2 - \dfrac{N_2}{N_1} M \end{array}\right\} \tag{27}$$

Whence the leakage inductance of each winding is very much less than the total self-inductance of that winding.

Energy of the Magnetic Field. Energy is required to establish a flow of electricity just as energy is required to set a column of water in motion, this "energy of motion" of electricity being analogous to the kinetic energy of a moving body. This energy of motion is most conveniently expressed in terms of the magnetic field which accompanies the flow of electricity or electric current; the mathematical expression for it may be put into various forms.

Magnetic Energy of a Single Electric Circuit, Permeability Constant. For example, consider a single electric circuit and the magnetic field which is established around this circuit when the current in it increases from zero to a value of I. When the current increases by di, the linkages increase by an amount $d\lambda$ in the right-handed screw direction with respect to the current. The electric energy output of the circuit during this change is, from eq. 4c, $i\,d\lambda$, which is also equal to the magnetic energy input into the magnetic circuit which it links. Hence the total energy input into the magnetic circuit or magnetic field is

$$W = \int_0^{LI} i\, d\lambda = \int_0^I L i\, di \tag{28}$$

since by definition $d\lambda = L\,di$; see eqs. 4c and 16. When the permeability is constant L is also constant, whence for constant permeability

$$W = \tfrac{1}{2} L I^2 \quad \text{joules} \tag{29}$$

This equation may also be written

$$W = \tfrac{1}{2}\,\lambda I = \mathfrak{F}^2/2\mathfrak{R} = \tfrac{1}{2}\,\mathfrak{R}\Phi^2 \quad \text{joules} \tag{29a}$$

where $\mathfrak{F}$ is the mmf ($= NI$ when the coil has N turns in a concentrated winding), $\mathfrak{R}$ is the reluctance of the magnetic circuit, and Φ the total magnetic flux.

Since the impressed mmf per unit length of a magnetic flux line is equal to the magnetizing force H, and the flux per unit area perpendicular to this flux line is equal to the flux density B, the energy per unit volume of the magnetic field is

$$w = \frac{HB}{2} = \frac{\mu_r \mu_v H^2}{2} = \frac{B^2}{2\mu_r \mu_0} \quad \text{joules} \tag{30}$$

These various formulas should be compared with the corresponding formulas (p. 977) for electrostatic energy.

Magnetic Energy of Two or More Electric Circuits, Permeability Constant. It can also be readily shown that the total energy required to establish currents I_1, I_2, etc., in several electric circuits linking one or more magnetic circuits of constant reluctance is

$$W = \tfrac{1}{2} L_1 I_1^2 + \tfrac{1}{2} L_2 I_2^2 + \tfrac{1}{2} L_3 I_3^2 + \cdots + M_{12} I_1 I_2 + M_{13} I_1 I_3 + M_{23} I_2 I_3 + \cdots \tag{31}$$

where the L's and M's represent the self- and mutual inductances, respectively. This may also be written

$$W = \frac{1}{2} \sum_{n=1}^{N} \Phi_n I_n \tag{31a}$$

where the summation is an algebraic one and includes every complete turn of the electric circuit, I_n being the current in the nth turn and Φ_n the magnetic flux linking this turn in the right-handed screw direction with respect to the current.

The energy per unit volume at any point in the magnetic field due to any number of electric circuits is represented by eq. 30, where H and B are taken as the resultant magnetizing force and flux density respectively at this point.

When the permeability is not constant the energy transferred to unit volume of the magnetic field due to any number of currents is

$$w = \int_0^B H\, dB \quad \frac{\text{joules}}{\text{cubic meter}} \tag{32}$$

provided the magnetizing force H and the flux density B are either in the same or directly opposite directions, as, for example, in a uniformly magnetized iron torus. To integrate this expression requires a knowledge of the relation between B and H. Note also that, owing to the phenomenon of magnetic hysteresis, part of the energy required to establish a magnetic field in iron or other magnetic substance is dissipated as heat and is not recoverable when the field disappears; therefore, only part of the energy represented by this formula is "stored" in the field in a recoverable form.

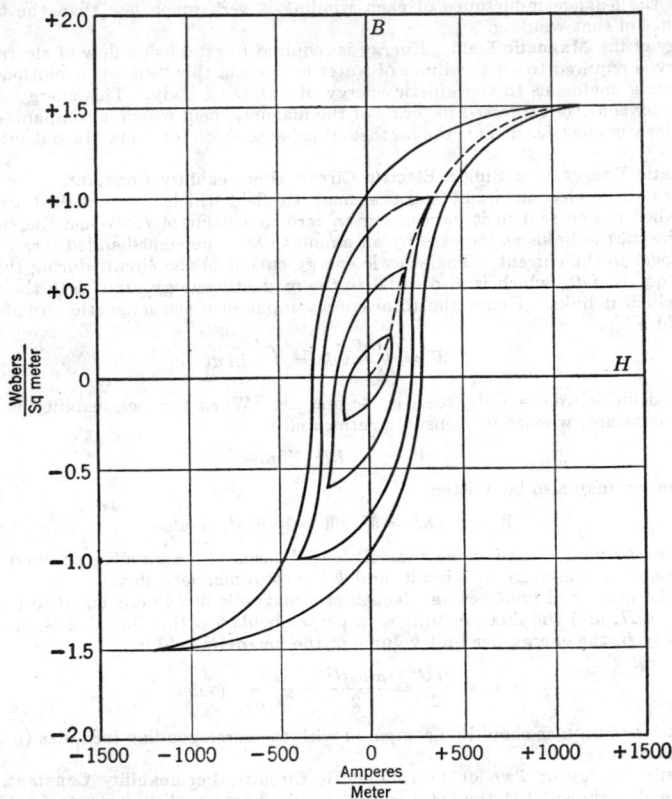

FIG. 13. Hysteresis loops of a magnetic material.

Hysteresis. When any completely demagnetized ferromagnetic substance is subjected to an increasing magnetizing force H, a curve such as Fig. 9 results. The dotted curve in Fig. 13 is also a representation of the phenomenon. Such a curve is variously termed a *rising characteristic*, a *virgin curve*, or a *neutral curve*; all these names are reserved for the special case of a magnetization curve of a previously *demagnetized* body.

If, from any positive value of H, the magnetizing force is decreased, the flux density will also decrease, but not along the same curve as the original rising characteristic. For decreasing values of H, B is greater than for the corresponding value on the rising characteristic. The magnetized body retains a part of its magnetization. It will be noticed that, when the magnetizing force is reduced to zero, there is still a positive value of B, which is dependent upon the maximum positive value of H on the rising characteristic, before H is decreased. The flux density which remains when H has been reduced to zero is called the *residual magnetism* and is in the nature of permanent or semi-permanent magnetism. If now the magnetizing force is reversed (e.g., by reversing the direction of the current through the magnetizing coil) a value of H will be found which will reduce B to zero. This value of H is called the *coercive force*. By proceeding through a cycle of values for the magnetizing force—from zero to a positive maximum, to zero, to a negative maximum, and back to zero—a closed curve will be obtained. A loop of this type is called a *hysteresis loop*, and each substance is capable of having an infinity of different hysteresis loops depending on the conditions of magnetization. During part of the magnetizing cycle the electric circuit is transferring energy to the magnetic circuit, and during the remainder of the cycle energy is retransferred from the magnetic to the electric circuit. When any ferromagnetic substance is present in the magnetic circuit more energy is transferred to the magnetic circuit than is retransferred to the electric circuit. The difference between these two amounts of energy is a loss and is dissipated as heat. Part of this loss is called the *hysteresis loss* and is presumably due to friction accompanying the change in orientation of the elementary magnets within the ferromagnetic substance. It may be shown that the hysteresis loss per cycle per cubic meter in any magnetic substance due to a complete cycle of changes of the flux density and the magnetizing force is equal to the area of the corresponding hysteresis loop provided this loop is determined under such conditions that there is no mechanical motion and no change in the relative distribution of the lines of force. This relation is based on the further assumption that unit distance on one scale represents 1 weber per square meter and that unit distance on the other scale represents 1 ampere per meter. The following empirical formula is useful in determining hysteresis loss.

$$w = \eta B_m^{1.6} \quad \text{watts per cubic meter per cycle} \tag{33}$$

where η is a coefficient depending upon the substance, and B_m is the maximum flux density reached during a cycle.

The power loss per unit weight of magnetic material is given by

$$P_h = kfB_m^{1.6} \quad \text{watts per pound} \tag{33a}$$

where f is the frequency in cycles per second and k is a constant.

Eddy Currents. Since a varying magnetic field induces an emf in every path which links the flux, such an emf will in general cause a flow of current in the magnetic materials composing the magnetic circuit. Such currents, called *eddy currents* or *Foucault currents*, cause rI^2 losses. Eddy currents can be distinguished from ordinary induced currents by the fact that they are due solely to the lines of force which pass *through* the space occupied by the conducting mass. Ordinary induced currents are produced by lines of force which *link* the conductor but do not pass through the space occupied by the conducting mass. It is customary wherever possible to laminate magnetic cores in order to increase the resistance of the path of induced eddy currents. The eddy-current loss in metal sheets is given by

$$P_e = (\pi^2/8\rho)(afB_m)^2 \quad \text{watts per cubic meter} \tag{34}$$

where ρ is the resistivity of the material in ohm-meters, a is the thickness of each sheet, f is the frequency of the eddy currents, and B_m is the maximum value of the flux density during a cycle. Equation 34 holds provided (*a*) that the lines of force are parallel to the planes of the laminations, (*b*) that the laminations are thoroughly insulated from each other, and (*c*) that the thickness of each lamination is small in comparison with its other dimensions. The combination of hysteresis loss and eddy-current loss is called *total core loss*.

13. MECHANICAL FORCES OF ELECTROMAGNETIC ORIGIN

Mechanical Forces on Conductors Carrying Current. A. M. Ampère observed forces between wires carrying electrical currents and conceived the law relating the forces and currents. Ampère's fundamental law can be formulated thus:

$$F = \frac{\mu_0}{2\pi} \frac{i_1 i_2}{r^2} \quad \text{newtons per meter} \tag{35}$$

where F, the force, is expressed in newtons per meter of conductor length, i_1 and i_2 are in amperes, r is the distance between conductors in meters, and μ_0 is $4\pi \times 10^{-7}$. The form of the law as given here is limited to two parallel conductors far apart compared to their diameters, and the force observed is in the plane of the two conductors and perpendicular to each conductor. If the two currents are in the same direction, the force on each of the two conductors is such as to tend to move the conductors together.

For conductors of relatively large cross section and small spacing between conductors, eq. 35 may be in serious error. The necessary formula is then much more complex (see H. B. Dwight, "Repulsion between Strap Conductors," *Elec. World*, 70, p. 522, Sept. 15, 1917).

Other forms of Ampère's law incorporating differential vector elements are of greater generality than the simple form given above. Ampère's law is an important experimental one which is the basis of electromagnetic force effects. It also may be used as the basis of evaluating the permeability of free space, μ_0, from known currents in amperes, a given separation in meters, and an observed force in newtons. Actually the basic experiments may be repeated somewhat more easily with loops of wire than with straight wires.

Mechanical Forces in the Magnetic Field. Experience shows that all bodies in which an electric current exists, and all bodies in which a magnetic flux exists, exert in general mutual mechanical forces upon one another tending to produce a relative motion of these bodies such as will increase the energy of the magnetic field. Let f be the component of the force tending to move any body in the field in a given direction and let dW be the increase in the energy of the field due to displacing the body a distance dx in this direction; then

$$f = dW/dx \tag{36}$$

provided this displacement does not alter the existing mmf's in the field.

Similarly calling T the component of the torque tending to turn any body in the field about a given axis, and dW the increase in the magnetic energy due to the turning of the body through an angle $d\alpha$ radians about this axis,

$$T = dW/d\alpha \tag{36a}$$

provided this displacement does not alter the existing mmf's in the field.

Equations 36 and 36a also give the actual force and torque respectively during a change in position which does cause a change in the mmf's in the field, provided dW is taken to represent the net increase in the energy of the field. This force and torque may differ greatly from the steady-state force and torque.

Force Produced by a Magnetic Field on a Coil Carrying a Current. When the magnetic flux, threading an electric circuit in the right-handed screw direction with respect to the current, increases by an amount $d\Phi$, the energy output of the circuit is $dw = NI\,d\Phi$, where N is the number of turns linked by this increase in flux and I is the current in each turn. This is the energy input into the magnetic field. Whence if an increase in flux $d\Phi$ is produced through the coil when it moves a distance dx, the force acting on the coil in the direction of dx is

$$f = NI(d\Phi/dx) \tag{37}$$

$d\Phi$ represents the increase in the flux linking the coil in the right-handed screw direction with respect to the current in it or the number of flux lines which cut the coil as the result of its motion. Similarly, when the coil is so mounted that it can move only about a fixed axis, then the value of the torque tending to turn it about this axis is

$$T = NI(d\Phi/d\alpha) \tag{37a}$$

where $d\Phi$ represents the increase in the flux linking the coil in the right-handed screw direction with respect to the current in it when the coil turns through an angle α (in radians).

From these relations it follows that a coil carrying an electric current, when in a magnetic field due to any other agent (current or permanent magnet), always tends to take up that position in which it will embrace the maximum possible flux linking the coil in the right-handed screw direction with respect to the current in it. This accounts for the attraction of two parallel coils when they carry currents in the same direction, and the repulsion of two parallel coils when they carry currents in opposite directions. This principle is useful in determining the direction of motion of the moving element in such devices as the electric motor, galvanometer, current balance, and electrodynamometer.

Torque on the Coil When Its Plane Is Parallel to the Magnetic Field. When the coil is placed with its plane parallel to the flux lines due to some other agent (e.g., a permanent magnet or another coil carrying a current), the flux linking the coil due to this agent is zero; see Fig. 14. Let the two circles represent sections of the two sides of the coil, its plane being perpendicular to the page; and let the dot in the left-hand circle indicate that the current is up through this side of the coil and the cross in the other circle that it is down through the other

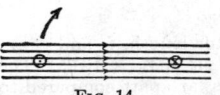

FIG. 14.

side. Let B be the flux density of the field, constant for each point along the flux line since the flux lines are parallel; let A be the area of the coil; and let $\Phi = BA$, that is, Φ represents the total flux which would be produced through the coil by a uniform field of flux density B at right angles to it. Then the torque on the coil when its plane is parallel to the field is

$$T = N\Phi I \tag{38}$$

This relation is useful in calculating the torque on the moving element of a galvanometer, ammeter, electrodynamometer, wattmeter, etc.

Average Torque on a Coil Rotating in a Magnetic Field. Consider a coil which is rotating with an angular velocity ω about a fixed axis in a magnetic field due to some other agent (e.g., an armature coil rotating in the magnetic field produced by the current in the field coils). Let the current in this coil be constant and in the same direction with respect to the coil while the coil turns from the position in which it embraces the maximum flux Φ in the left-handed screw direction to the position (a half revolution to a 2-pole machine) when it embraces this same maximum flux Φ in the right-handed screw direction. The total change in the flux while the coil turns through this angle, π radians in a 2-pole machine, is 2Φ, whence the average torque turning the coil through this half revolution is

$$T = (2N/\pi)I\Phi \tag{39}$$

That is, the average torque is proportional to the product of the current and the total flux per pole. When a commutator is provided to change the direction of the current every half turn, the torque is in the same direction for a complete turn.

Force on a Wire in a Magnetic Field. Consider a wire of length l forming part of a closed circuit, Fig. 15. Let B be the value of the flux density at the wire and I the cur-

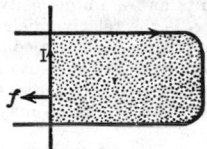

FIG. 15.

rent in the direction indicated, and let the lines representing the flux be perpendicular to the wire in the direction from the eye to the page. When this wire moves a distance dx to the left the flux threading the closed loop formed by the circuit is increased by an amount $d\Phi = Bl\,dx$, whence the force acting on the wire is (from eq. 37):

$$f = BlI \tag{40}$$

Equation 40 may also be considered an alternative statement of Ampère's law (35).

Left-hand Rule. The relative directions of this force, the flux density B, and the current I may be conveniently determined by pointing the forefinger of the left hand in the direction of the flux and the middle finger in the direction of the current: then if the thumb is held perpendicular to these two fingers it will point in the direction in which the force tends to move the wire. Compare this with the right-handed rule for emf.

Forces on Magnetic Bodies in a Magnetic Field. In general, the reluctance of a magnetic field to the flux set up by a given mmf depends upon the relative positions of the various magnetic bodies in the field with respect to one another and with respect to the electric circuit producing this mmf. When any magnetic body in the field is displaced the total reluctance will, in general, be changed owing chiefly to the change in the dimensions of the air portion of the circuit. From eqs. 29a and 36 it can be shown that the force acting on any magnetic body in the field is in the direction of the flux lines threading it and has the value

$$f = -\,{}^{1}/_{2}\,\Phi^2(d\Re/dx) \tag{41}$$

(provided the mmf remains constant) where Φ represents the total flux threading the body and $d\Re$ represents the increase in the reluctance of the magnetic circuit corresponding to a displacement dx of the body in the direction of the flux lines. The minus sign in this formula indicates that the force is always in the direction in which a motion of the body would decrease the reluctance of the circuit. In deducing this expression it is assumed that the permeability of each body in the field is constant. It can also be shown that to a close approximation the same formula holds for actual magnetic bodies, for which the permeability is not a constant.

The above relation accounts for the attraction of one magnet for another when their unlike poles are nearer each other than their like poles, and the repulsion of two magnets when their like poles are nearer than their unlike poles. It also accounts for the attraction of iron or other paramagnetic substance by either pole of a magnet or by either "face" of an electric circuit, and the repulsion of a diamagnetic substance by either pole of a magnet or either face of an electric circuit.

ELECTROSTATICS AND THE DIELECTRIC CIRCUIT

14. ELECTROSTATIC FIELDS, POTENTIALS, AND CURRENTS

Coulomb's Law

Coulomb observed that an electric charge in the vicinity of another electric charge had a force exerted on it. If the charges were like charges, the force on each charge was such as to move the charges apart; for unlike charges the forces were such as to move the charges together. Coulomb's formulation of the results of his experiments produced the following basic relation known as Coulomb's law

$$F = \frac{Q_1 Q_2}{4\pi\epsilon_r\epsilon_0 r^2} \quad \text{newtons} \tag{1}$$

In this law F is the magnitude of the force in newtons; Q_1 and Q_2 are charges in coulombs separated a distance r meters; ϵ_0 is the dielectric constant of free (evacuated) space in the rationalized mks system of units and is numerically equal to 8.85×10^{-12} farad per meter or coulomb per meter-volt; ϵ_r is the relative dielectric constant of the medium and is a dimensionless number. The relative dielectric constant is 1 for vacuum; it is very nearly 1 for all gases. The product $\epsilon_r\epsilon_0$ is the absolute dielectric constant of the medium and is symbolized by ϵ without a subscript. The forces on the two charges have the same magnitude but opposite directions. Either charge may be considered as in the electric field of the other.

Electric Fields of Force. In any portion of a substance in which the electricity is acted upon by a force tending to move it, there is said to be an *electric field of force*, or briefly an *electric field*. An electric field is also said to exist in any region of free space where a charge, if placed there, would have a force exerted upon it tending to move it.

Intensity of an Electric Field. The *intensity of an electric field* E at any point is defined as the force exerted on a unit positive charge at this point by the agent or agents producing the field, i.e., by the agent or agents tending to move the charge. The direction of the field intensity, or the direction of the field, is defined as the direction of the force acting on a positive charge at this point. A positive charge then moves or tends to move in the direction of the field, and a negative charge moves or tends to move in the opposite direction.

The intensity of electric field can be obtained from Coulomb's law by setting one charge, say Q_2, equal to 1 coulomb, − unit charge, then:

$$E = \frac{Q}{4\pi\epsilon_r\epsilon_0 r^2} \quad \text{newtons per coulomb or volts per meter} \tag{2}$$

The unit of electric field intensity has not been given any special name, but, since it is of the same nature as emf per unit distance, the intensity at any point may be conveniently expressed as so many volts per meter or per inch. Alternatively, it may be expressed as a force per unit charge, such as so many newtons per coulomb. See Section 3.

Lines of Electric Force. Lines of Electric Intensity. A line drawn in an electric field in such a manner that its direction at each point coincides with the direction of the field at that point is called a *line of electric force*. A line of force is usually a curved line, though in certain special cases it may be straight. Any number of such lines may be drawn in an electric field, but no two of these lines can intersect. The density of these lines, i.e., the number drawn through unit area perpendicular to their direction, may be chosen arbitrarily to represent the value of the field intensity at this area, and when so drawn are preferably called *lines of electric intensity*, as distinguished from flux and stream lines defined below. The term *lines of force*, however, is frequently used to designate any one of these three sets of lines, and this loose use of the term is likely at times to lead to much confusion. The term *line of electric force* will be used in this article to designate merely the direction of the field at any point; in any statement involving the density of these lines the proper one of the other terms will be employed.

Electric Equipotential Surfaces. A surface drawn in an electric field in such a manner that it is perpendicular at each point to the line of force through that point is called an *electric equipotential surface*. The electric intensity has no component along such a surface, and therefore no work is required to move a charge from one point to another over any path wholly on such a surface.

Electric Potential Difference. The electric potential difference between two points 2 and 1 (see Fig. 1) is the work per unit positive charge in moving the unit positive charge from point 1 to point 2 against the electric field intensity, viz.:

$$V_2 - V_1 = \int_1^2 -(E \cos \theta)\, dl \quad \frac{\text{newton-meters}}{\text{coulomb}} \text{ or } \frac{\text{joules}}{\text{coulomb}} \text{ or volts} \qquad (3)$$

where dl represents an elementary length of the path, and $-E \cos \theta$ the component of force to overcome the field intensity E along dl.

If $V_2 - V_1$ is positive, then point 2 is said to be at a higher electric potential than point 1. If $V_2 - V_1$ is negative, point 2 is at a lower potential than point 1.

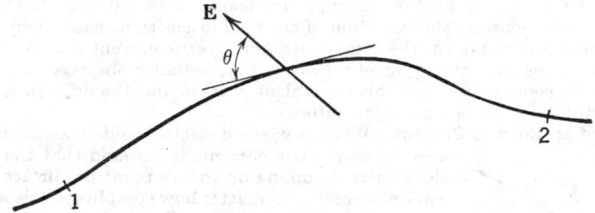

FIG. 1.

Electric Potential. Electric potential is based on the definition of difference of potential. A point may be designated as zero electric potential for reference purposes; this designation may be entirely arbitrary, although in some physical cases it is often both logical and convenient to consider a special point as being at zero potential. The electric potential at a point is the difference of potential between that point and a reference point which has been designated as being at zero electric potential.

Voltage or Potential Gradient. Equation 3 can be rewritten:

$$V_2 - V_1 = \int_1^2 dV = \int_1^2 -E \cos \theta\, dl = \int_1^2 -E\, dn \qquad (3a)$$

where dn represents the differential element taken perpendicular to electric equipotential surfaces. Differentiating eq. 3a with respect to dn results in:

$$E = -dV/dn \qquad (3b)$$

The bold-face type indicates that vector quantities are involved. Now dV/dn is called the potential or voltage gradient; it is the maximum value of the directional derivative of voltage with respect to distance. Symbolically ∇V represents gradient of V, thus:

$$E = -\text{grad } V = -\nabla V \qquad (3c)$$

Flow of Electricity. Whenever an electric field is set up in a substance by any means whatever, a displacement of the electricity in that substance always takes place, the nature of the displacement depending upon the nature of the substance. The positive electricity within the substance is displaced or orientated in the direction of the field intensity and the negative electricity in the opposite direction, until an opposing force is set up which just balances the forces due to the impressed field. In metallic conduction, the flow of electrons in a direction opposite to the field constitutes the electric current. In electrolytes there is a migration of positive ions in the direction of the field, and of electrons opposite to the direction of the field. It is believed that, in good insulators, actual migration is negligible but that molecules of the substance are deformed and reorientated in such a manner as to produce a momentary motion of positive electricity in the direction of the field, and of negative electricity opposite to the direction of the field. (See Arts. 27 and 28.)

The displacement of the electricity within a substance cannot be measured directly, but only in terms of some effect produced thereby. Two effects which always result when electricity is displaced are: (1) a magnetic field is established around the path along which the displacement takes place (but disappears when the electricity comes to rest); and (2) heat is developed in the path of the displacement. The magnetic field produced by a displacement or flow of electricity is usually taken as the measure of the rate of flow, i.e., of the quantity of electricity displaced per unit time through a surface perpendicular to the direction of the displacement. This rate of flow is called the *intensity* of the electric current, or simply the electric current.

A flow of positive electricity in one direction is equivalent magnetically to a flow of an equal amount of negative electricity in the opposite direction; hence the total flow along the given path is the sum of the positive electricity displaced per unit time in one direction past a point in this path plus the negative electricity displaced per unit time past this point in the opposite direction. The *direction* of the electric current is taken as the direction in which the positive electricity is displaced, and it is, therefore, the same as the direction of the field intensity.

Current Due to Varying Electric Field. When the electric field in any substance is varying, the total magnetic effect produced is found to depend not only upon the rate of displacement of the electricity within the substance but also upon the rate of change of the electric field. In fact, a magnetic field is produced around a path in free space along which the intensity of the electric field is varying. In dealing with varying electric fields it is found convenient to consider the variation of the field intensity as equivalent to an actual flow of electricity, and to take as the total equivalent electric current the flow of electricity which would produce the same magnetic field as that actually observed; the actual flow of electricity is in general less than this equivalent current, but the difference is negligible except in substances which are good insulators.

Continuity of an Electric Current. When a varying electric field is considered as equivalent to an electric current, it is found that the total equivalent current coming up to any point or surface in any network of circuits, no matter how complicated, is always equal to the total current leaving that point or surface, irrespective of the nature of the substances through which the currents are constant or are varying. For example, in Fig. 2,

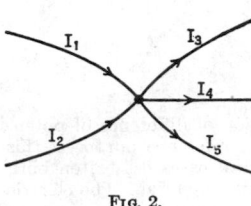

$$i_1 + i_2 = i_3 + i_4 + i_5 \qquad (4)$$

Or calling the currents coming up to any point positive, and the currents leaving that point negative, the algebraic sum of all the currents at any point in a network of circuits is always zero. This fundamental principle is conveniently expressed by the formula

$$\Sigma i = 0 \qquad (4a)$$

which holds at every junction point both for variable and for continuous currents. This is a generalized form of Kirchhoff's first law. See Art. 7, eq. 15.

Stream Lines of Electric Current. Current Density. As a consequence of the continuity of an electric current, the total current in any substance of any size or shape may be looked upon as made up of a number of small streams of electricity flowing side by side, the strength (quantity of electricity per second) of each stream being constant throughout its length. If the cross section of each stream at any point is so chosen that each stream represents unit current (unit quantity per second), then the number of these streams crossing unit area of a surface perpendicular to their direction will be equal to the current per unit area of this surface, or to the current density at this surface. Each such stream may be represented graphically by a line coinciding with its axis; such lines are called *stream lines*. When the stream lines are drawn as described, their direction at any point gives the direction of the current at this point, and the number of these lines per unit area perpendicular to their direction is equal to the current density at this point.

For an insulated wire the stream lines are parallel to the axis of the wire, except in the immediate vicinity of its ends; in a long wire this non-uniformity at the ends is negligible. All the stream lines in an ordinary wire may also, in practice, be considered as coinciding, and the wire may, therefore, be treated as a geometrical line as regards external effects. However, for a short rod or strip (such as an ammeter shunt) connected in the circuit by wires attached to its ends, the stream lines are not in general parallel but diverge from one terminal and converge toward the other.

Conduction Current and Displacement Current. Experience shows that, when an electric field is established in any substance, the total equivalent electric current set up depends (1) upon the value of the field intensity, (2) upon the rate of change of the field intensity, and (3) upon the nature of the substance in which the field is established. The current density J at any point at any instant may in general be expressed by the relation:

$$J = \sigma E + (d/dt)(\epsilon_0 \epsilon_r E) \quad \text{amperes per square meter} \qquad (5)$$

where E is the field intensity, σ and ϵ_r are coefficients depending upon the chemical nature and physical condition of the substances at the point in question, and d/dt means the rate of change with respect to time. ϵ_0 is the dielectric constant of free space. σ is the conductivity of a substance in mho-meter. (See Conductance in Art. 6.) The total current density may then be considered as the sum of the two components having respectively the densities

$$J_c = \sigma E \quad \text{and} \quad J_d = (d/dt)(\epsilon_0 \epsilon_r E) \tag{6}$$

The first of these components, J_c, is called the conduction current density; the second, J_d, is commonly called the displacement current density. The term displacement current density, however, is something of a misnomer, for both components of the total current density are probably due, in part at least, to a displacement of electricity. The conduction current density is the only appreciable component in substances usually classed as conductors, and the displacement current density is appreciable only in substances ordinarily classed as dielectrics. The conduction current density in a dielectric is usually small, though measurable; it is frequently called the *leakage current density*. When the electric field in a dielectric is rapidly varying, the displacement current density may be many times greater than the conduction or leakage current density through the dielectric.

Conductivity and Resistivity. The quotient of the density, J_c, of the conduction current by the field intensity E, i.e., the coefficient σ in the expression $J_c = \sigma E$, is called the conductivity or specific conductance of the substance at the point in question. Since in an ordinary conductor the displacement current density is inappreciable, the conductivity of an ordinary conductor is also equal to the density of the total current divided by the field intensity; i.e., for a conductor

$$J = \sigma E \quad \text{amperes per square meter} \tag{7}$$

where J represents the density of the total current. Experience shows that for a given conductor at constant temperature (and also at constant pressure, in a gas) this coefficient, σ, is a constant irrespective of the strength, distribution, or time variation of the current. The value of σ for a dielectric, however, is not in general a constant but depends upon the time variation of the field intensity.

The above relation between J and E may also be written

$$E = \rho J \tag{7a}$$

where ρ is the reciprocal of the conductivity σ. The constant ρ is called the *resistivity* or specific resistance of the substance. Values of σ and ρ for various conductors and insulating materials are given in Section 12. For the units of conductivity and resistivity, see Section 3.

15. DIELECTRIC FLUX

Dielectric Flux and Dielectric Flux Density. As noted above, the displacement current density through a dielectric at any point depends upon the rate of change of the electric field intensity and upon the nature of the dielectric. The density of this displacement current at any point may be expressed by the relation

$$J_d = (d/dt)(\epsilon_0 \epsilon_r E) \tag{8}$$

where E is the field intensity, ϵ_0 is the dielectric constant of free space, and ϵ_r is the relative dielectric constant of the material. Another name for dielectric constant is *specific inductive capacitance*.

The quantity $\epsilon_0 \epsilon_r E$, whose rate of change is equal to the density of the displacement current, is called the *dielectric flux density* and may be represented by the symbol D. Then

$$D = \epsilon_0 \epsilon_r E \tag{9}$$

The direction of the dielectric flux density is arbitrarily chosen to be the same as that of the electric field intensity E. Through any surface of area A at each point of which the dielectric flux density has a constant value D and is perpendicular to that surface, there is said to exist a *dielectric flux* equal to DA. The total dielectric flux through a surface may be represented by the symbol ψ.

In general, the total dielectric flux through any surface is

$$\psi = \int D \cos \alpha \, ds \tag{10}$$

where ds represents any elementary area of this surface, $D \cos \alpha$ the component of flux density normal to the surfaces at ds, and $\int$ the sum of all the products $D \cos \alpha \, ds$ for that surface. The total displacement current through this surface is then:

$$i_d = d\psi/dt \tag{11}$$

Lines of Dielectric Flux. The electric flux through any surface may be represented by lines drawn in the same direction as the lines of electric intensity, but of such a density that their number per unit area perpendicular to their direction at any point is equal to the dielectric flux density at this point. The number of these lines cutting any surface is then equal to the total dielectric flux through this surface. The ratio of the number of flux lines through any surface to the number of lines of electric intensity through that surface is equal to the dielectric coefficient of the substance in which the field exists.

Electric Charge and Dielectric Flux. Within any substance of uniform structure throughout, the dielectric flux lines are continuous lines; i.e., the number of these lines coming up to one side of a surface within such a substance is equal to the number of these lines leaving the other side of that surface. Experience shows that it is impossible to produce an appreciable dielectric flux in those substances ordinarily classed as conductors; hence dielectric flux lines cannot pass through a good conductor, but terminate at its surface. Every dielectric is a conductor to at least a slight extent, and on account of this fact not all the dielectric flux lines coming up through one dielectric to the surface of contact between this dielectric and another pass through the second dielectric, but some of them terminate at this surface.

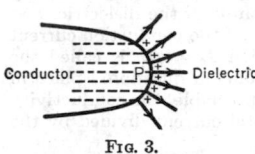

Conductor —— P —— Dielectric

FIG. 3.

Experience shows that, to establish an electric field in the dielectric around a conductor, electricity must be conducted through the conductor to the surface of contact between the conductor and the dielectric. For example, consider a good conductor in contact with a perfect dielectric (Fig. 3); a momentary conduction current must flow through the conductor along the stream lines of the conduction current, represented by the dotted lines. While the field is being established (and therefore varying), a displacement current is set up in the dielectric requiring an equal conduction current in the conductor, and consequently the rate of change of the dielectric flux (ψ) established in the dielectric must be equal to the conduction current (i) flowing up to this surface through the conductor; i.e.,

$$d\psi/dt = i$$

or

$$\psi = \int i\,dt = Q \quad \text{coulombs} \tag{12}$$

where Q is the quantity of electricity conducted through the conductor to this surface.

This relation is a general one, viz., the total dielectric flux from any area A in the surface of a conductor is equal to the total charge on this area. Hence every flux line originates at a positively charged conducting surface and terminates at a negatively charged conducting surface, a line connecting each unit positive to each unit negative charge.

The quantity of electricity conducted through a conductor when a momentary current is established through it can be measured readily by means of a ballistic galvanometer, and consequently the dielectric flux (equal to Q) may readily be determined.

Dielectric flux may be expressed in the same units as electric charge, viz., coulombs, statcoulombs, or abcoulombs; see Section 3.

Surface Density of Charge. When there is no current in a conductor there can be no electric field within it (see eq. 7); and therefore the surface of a conductor in which no current is flowing is always an equipotential surface. Hence the lines of electrostatic intensity, in the surrounding dielectric, and therefore the dielectric flux lines also, must leave or enter this surface in a direction perpendicular to it. The dielectric flux density in the dielectric just outside a conducting surface in which there is no electric current is perpendicular to this surface and has the magnitude

$$D = \sigma_s \tag{13}$$

where σ_s is the charge per unit area of the surface at this point, or the *surface density* of the charge.

Dielectric Flux Density Due to a Number of Charged Conductors. It can be shown that, when any number of charged conductors are surrounded by a uniform dielectric, the dielectric flux density at any point in the field may be expressed by considering each elementary surface having a charge q as producing at any point P at a distance r from q a flux density equal to $q/4\pi r^2$, in the direction of the line from q to P when q is positive and in the direction of the line from P to q when q is negative. The total flux density at P due to all the charges is then the vector summation

$$D = \frac{1}{4\pi} \sum \frac{q}{r^2}\, \hat{r} \tag{14}$$

in which $\hat{r}$ is a unit vector in the direction from q to P.

16. ELECTROSTATIC CAPACITANCE, CONDENSERS, AND INDUCTION

Electrostatic Capacitance and Condensers. To establish a given dielectric flux ψ through a given dielectric a certain difference of electric potential is always required. Consider any portion of an electric field (Fig. 4) between the two equipotential surfaces S and S_1 bounded laterally by a surface tangent at each point to the flux line through that point. Let V be the drop of potential from S to S_1, and let ψ be the dielectric flux through this region. When there is no source of emf between S and S_1, the quotient

$$C = \psi/V \qquad (15)$$

is defined as the electrostatic capacitance of this portion of the field.

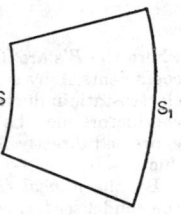

FIG. 4.

When the equipotential surfaces S and S_1 are the surfaces of two conductors, the two conductors and the dielectric between them are said to form a capacitor or an *electric condenser*. When all the flux lines from one conductor end on the second conductor (e.g., when they are given equal and opposite charges by connecting them respectively to the two terminals of a source of emf), then the flux from one to the other is equal to Q, where Q is the numerical value of the total charge on either conductor. The capacitance of the condenser may then be written

$$C = Q/V \qquad (15a)$$

When there are several charged conductors in the field, the total flux from one conductor does not in general end on another single conductor, but some of the flux lines from conductor 1 may run to 2, some to 3, etc. Let ψ_{12} be that portion of the flux from any conductor 1 which ends on any other conductor 2, and let V_{12} be the drop of potential from 1 to 2; then the capacitance between conductor 1 and conductor 2 is

$$C_{12} = \psi_{12}/V_{12} \qquad (15b)$$

Or, calling Q_{12} that portion of the charge on 1 which is balanced by an equal and opposite charge on 2, the capacitance between 1 and 2 is

$$C_{12} = Q_{12}/V_{12} \qquad (15c)$$

The unit of capacitance in the mks and practical systems of units is the farad, but as this is a very large unit, a unit equal to one-millionth of a farad, called the microfarad, is usually employed. The cgs electrostatic unit is called the statfarad, and the cgs electromagnetic unit the abfarad. See Section 3.

Potential Coefficients, Electrostatic Induction Coefficients. Consider any number of conductors 0, 1, 2, 3, etc., either (1) at a great distance from all other conductors or (2) completely surrounded by a hollow conducting shell, the inside surface of which is to be considered as one of the conductors, say 0, of the system. The electrostatic condition of such a system of conductors is uninfluenced by any electrostatic effects produced outside the system; it may therefore be called an *electrostatically independent system*.

Any conductor of an electrostatically independent system may be chosen as a conductor of reference; let this reference conductor be designated as conductor 0. Let v_{10}, v_{20}, v_{30}, etc., represent the potential drops from 1 to 0, from 2 to 0, from 3 to 0, etc., and let q_0, q_1, q_2, q_3, etc., represent the charges on 0, 1, 2, 3, etc. Then, if the relative positions of the various conductors and insulators in the field remain unaltered and the specific inductance capacitances (dielectric constants) of the various insulating materials between the conductors are constant (not necessarily the same for each insulating material, however), the following relations hold for all values of the charges on and potential drops between conductors irrespective of how the conductors may be connected (provided the connecting wires are of small cross section compared with the dimensions of the conductors):

$$
\left.
\begin{aligned}
v_{10} &= A_{11}q_1 + A_{12}q_2 + A_{13}q_3 + \text{etc.} \\
v_{20} &= A_{12}q_1 + A_{22}q_2 + A_{23}q_3 + \text{etc.} \\
v_{30} &= A_{13}q_1 + A_{23}q_2 + A_{33}q_3 + \text{etc.} \\
q_0 &= -(q_1 + q_2 + q_3 + \text{etc.})
\end{aligned}
\right\} \qquad (15d)
$$

where all the A's are constants depending upon the distances apart of the conductors and

the nature of the insulating medium between them. The coefficients A in these equations may be called the *potential coefficients* of the system of conductors.

The above equations may also be written:

$$\left.\begin{array}{l} q_1 = B_{11}v_{10} + B_{12}v_{20} + B_{13}v_{30} + \text{etc.} \\ q_2 = B_{12}v_{10} + B_{22}v_{20} + B_{23}v_{30} + \text{etc.} \\ q_3 = B_{13}v_{10} + B_{23}v_{20} + B_{33}v_{30} + \text{etc.} \\ q_0 = -(q_1 + q_2 + q_3 + \text{etc.}) \end{array}\right\} \tag{15e}$$

where the B's are also constants and may be expressed directly in terms of the potential coefficients A by solving the equations for q_1, q_2, q_3, etc. The constants B are called the electrostatic induction coefficients, and like the constants A are independent of how the conductors may be charged and of how they may be interconnected. The B's may be expressed directly in terms of the normal and grounded capacitances of the various conductors.

By the *normal capacitance* between any two conductors is meant the capacitance of the condenser formed by these two conductors when all the other conductors are connected to one another and to the conductor of reference, the two conductors of course being insulated therefrom. The normal capacitance between any two conductors of a system, say 1 and 2, is then, from eq. 15e,

$$C_{12} = \frac{B_{11}B_{12} - B_{12}^2}{B_{11} + B_{22} + 2B_{12}} \tag{15f}$$

When the arrangement of the conductors is perfectly symmetrical (as in a three-conductor cable), $B_{11} = B_{22}$ and the normal capacitance between 1 and 2 is

$$C_{12} = {}^1/_2 (B_{11} - B_{12}) \tag{15g}$$

By the *grounded capacitance* of any conductor of a system is meant the capacitance of the condenser formed by this conductor as one plate, and all the other conductors, including the conductor of reference, connected together as the other plate. The grounded capacitance of conductor 1, say, is then, from eq. 15g,

$$C_{1g} = B_{11} \tag{15h}$$

That is, the electrostatic coefficient of self-induction of any given conductor is the same as the grounded capacitance of this conductor.

Electrical Images. The distribution of charge upon conductors may often be found most simply by a method devised by Lord Kelvin. Let it be supposed that a charge $+q$ is placed in the neighborhood of an infinite conducting plane. There will be produced on the plane of charge density σ_s which together with the charge $+q$ will produce a given distribution of electric field intensity throughout the region. This field intensity can be shown to be equal to the field intensity which would be produced, in the absence of the conducting plane, by the charge $+q$ and a second charge $-q$ placed at a position such that $+q$ and $-q$ are symmetrical about the plane. The charge $-q$ is called the *electrical image* of the charge $+q$, the name being suggested by the optical analogy.

The capacitance of an actual condenser formed by a wire and the earth (assuming the earth to be an infinite conducting plane parallel to the wire) is twice the capacitance of the condenser formed by the conductor and its image.

Factors upon Which Capacitance Depends. The capacitance of a given portion of a dielectric depends upon (a) the dielectric coefficient or specific inductive capacitance of free space ϵ_0; (b) the relative dielectric constant ϵ_r for the insulating medium; (c) the length of the dielectric flux lines through the dielectric; (d) the cross section of the dielectric at right angles to the flux lines; and (e) the distribution of the flux lines over this cross section (compared with electric conductance). The product $\epsilon_r\epsilon_0$ is often designated by ϵ. In general, the capacitance of any portion of a dielectric bounded laterally by flux lines and at the ends by equipotential surfaces (Fig. 4) can be expressed by the formula

$$C = \frac{\epsilon_0\epsilon_r\psi}{\int D\, dl} \tag{16a}$$

where ϵ_0 and ϵ_r are as just described, ψ is the total dielectric flux through the given portion of dielectric, dl any elementary length along one of the flux lines, and D the component of the dielectric flux density in the direction of dl at this point, the integral being taken along the flux line from one end surface to the other. When the end surfaces are conductors charged with $+Q$ and $-Q$ units respectively, then

$$C = \frac{\epsilon Q}{\int D \, dl} = \frac{\epsilon_0 \epsilon_r Q}{\int D \, dl} \tag{16b}$$

By the application of this formula the capacitance of various practical forms of condensers may be calculated. It should be noted that the capacitance of a condenser depends upon the distribution of the dielectric flux ($\epsilon_0 \epsilon_r$ or ϵ being assumed constant), but not upon the absolute value of the flux; i.e., for a given dielectric and given distribution of flux the capacitance is a constant. In general, when any conductor or dielectric of a different specific inductive capacitance is placed in the electric field set up by the charged plates of a condenser, the distribution of the flux, and therefore the capacitance of the condenser, are altered.

Relation between Conductance and Capacitance. Comparing $r = \dfrac{\rho \int J \, dl}{I}$ and eq. 16a, it is apparent that, when the dielectric flux lines and the current stream lines have the same distribution in any given region, the ratio of the conductance of this region to the capacitance of this region is σ/ϵ, where ϵ is the dielectric coefficient and σ the conductivity of the material in this region. Hence the formulas for the capacitance and conductance of the dielectric between the plates of any shape or size of condenser, differ only by a constant coefficient. That is, if C is the capacitance of any condenser, then

$$g = (\sigma/\epsilon)C \tag{17}$$

is the total conductance between the plates where $\epsilon = \epsilon_0 \epsilon_r$, and σ is the conductance of the dielectric.

Charge and Discharge of a Condenser. To charge a condenser a difference of electric potential must be established between its plates. This may be done, as noted above, by

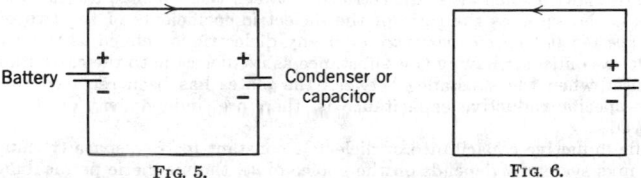

Battery	Condenser or capacitor	

| FIG. 5. | FIG. 6. |

connecting the two plates of the condenser respectively to the two terminals of any source of emf; see Fig. 5. If the dielectric has a very high resistance and the source of emf has a constant value E, the current set up in this circuit will continue only until a difference of potential equal to E has been established across the two plates of the condenser, or until a charge equal to CE has been transferred from the *negative* to the *positive* plate of the condenser. The establishment of the electric flux through the dielectric of the condenser may be looked upon as setting up in the dielectric itself an opposing force analogous to the opposing force set up in a spring when it is compressed. When the opposing force just balances the impressed force a steady state is attained, just as the compressing of a spring ceases when the force producing the compression is just balanced by the opposing force due to the elasticity of the spring.

When a condenser has thus been charged, the wires connecting it to the source of emf may be removed and the condenser remains charged for a length of time depending upon the resistance of the dielectric separating the plates; the higher this resistance, the longer the time that the condenser remains charged. If the plates are moved apart they still retain their charges, one plate a positive charge and the other a negative charge, but the distribution of these charges on the plates will in general become altered. Experience shows that a mechanical force is required to separate the charge plates irrespective of whether or not they are connected to the source of emf.

When the two charged plates are short-circuited by a wire, as shown in Fig. 6, a momentary current is established through the wire; the electric field between the plates as well as the charges disappear. The quantity of electricity discharged through the wires is equal to the quantity of charge originally on either plate. A charged condenser, therefore, acts like a source of emf, the direction of this emf around the circuit containing the condenser being in the direction through the condenser from its negative to its positive plate. A condenser when it is being charged may also be looked upon as producing a back emf, that is, an emf opposing the emf which charges it. When a condenser is considered from this point of view only the conducting portion of the circuit is to be considered in applying Kirchhoff's laws. When the condenser has an appreciable leakage its resistance must be considered to be in parallel with its emf.

Charging Current and Leakage Current. The displacement current through the dielectric of a condenser is frequently called the *charging* current. The conduction current through the dielectric is called the *leakage* current. Let C be the capacitance of the condenser, g the conductance of the dielectric, and v the voltage across the condenser; then the total current through the condenser. is

$$i = gv + C(dv/dt) \qquad (18)$$

where dv/dt represents the rate of change of v with time. The component gv of this current is the leakage current, and the component $C(dv/dt)$ is the charging current.

Capacitances in Series. When several capacitances * are connected end to end so that the same dielectric flux passes through each of them, they are said to be *in series*. The total capacitance of any number of individual capacitances C_1, C_2, C_3, etc., connected in series is C, where

$$\frac{1}{C} = \frac{1}{C_1} + \frac{1}{C_2} + \frac{1}{C_3} + \cdots \qquad (19)$$

Capacitances in Parallel. When several capacitances are connected between the same pair of equipotential surfaces so that the same potential drop is established through each, they are said to be *in parallel*. When there are no emf's in any of the circuits between the two equipotential surfaces the total equivalent capacitance of any number of capacitances C_1, C_2, C_3, etc., connected in parallel is

$$C = C_1 + C_2 + C_3 + \cdots \qquad (20)$$

Specific Inductive Capacitance and Dielectric Coefficient. From eq. 16a it is evident that, when the capacitance of a given condenser is measured (1) with a dielectric A between the plates and (2) with some other dielectric B between these plates, the ratio of the two capacitances is the same as the ratio of the dielectric coefficients of the two dielectrics. The *relative specific inductive capacitance* ϵ_r of any dielectric is defined as the ratio of the capacitance of a condenser having this substance as its dielectric to the capacitance of the same condenser when the substance between the plates has been replaced by vacuum. The relative specific inductive capacitance is, therefore, independent of the system of units employed.

The specific inductive capacitance or dielectric constant for free space (vacuum) in the rationalized mks system ϵ_0 depends on the choice of μ_0, the magnetic permeability of free space. The value of μ_0 has been taken as $4\pi/10^7$ henrys per meter in the rationalized mks system. Wave theory relates the velocity of light or electromagnetic waves in free space c with μ_0 and ϵ_0 thus:

$$\frac{1}{\sqrt{\mu_0 \epsilon_0}} = c \text{ meters per second} \qquad (21)$$

The velocity of light c has been measured accurately as 2.998×10^8 meters per second. This results in (21) yielding ϵ_0 as 8.854×10^{-12} farad per meter. If 3×10^8 meters per second is used as c for a convenient approximation, then the corresponding value of ϵ_0 is $1/(36\pi 10^9)$ farad per meter.

Values of the relative specific inductive capacitance or dielectric constant ϵ_r of various insulating materials are given in Section 13.

17. DIELECTRIC LOSSES

Electric Absorption and Residual Charge. The value of the relative dielectric coefficient ϵ_r of a given dielectric is not strictly a constant unless the dielectric is perfectly homogeneous. For such non-homogeneous substances as glass, mica, rubber, paper, and cloth, the dielectric coefficient is found to depend upon the time of electrification, i.e., upon the length of time that the voltage is applied, its value increasing with the time of electrification. This phenomenon is sometimes described as *electric absorption*, the idea being that the charge from the plates of the condenser soaks into the dielectric, for an increase in the dielectric coefficient for a given impressed voltage means a greater quantity of electricity conducted to the plates. This idea is also in accord with the experimental fact that when such a condenser is discharged by short-circuiting it with a wire, Fig. 6, the wire then being removed, a *residual* charge appears on the plates after a lapse of a few seconds.

---

* Either condensers or dielectrics of different kinds, sizes, or shapes in contact along equipotential surfaces.

Dielectric Hysteresis. A phenomenon closely associated with electric absorption is the fact that when the electric field in a heterogeneous dielectric is caused to vary rapidly an amount of heat is dissipated in the dielectric greatly in excess of that which can be accounted for in terms of its leakage resistance as determined by continuous-current measurements. This may be due in part to an actual increase in the resistance of the dielectric with the speed of variation of the field, or it may be due to a phenomenon analogous to magnetic hysteresis, i.e., to a lag of the dielectric flux density behind the electric field intensity. Whatever may be the cause of this extra loss of power for rapidly varying fields, it is generally described as the loss due to "dielectric hysteresis." The heat developed is in many cases quite appreciable.

Dielectric Strength. When an electric field is established by a charge on a conductor, and the intensity of this field is increased, corona forms in the dielectric at the surface of conductors, and if this field is further increased a point is reached at which the dielectric breaks down. Under some conditions (corona) the breakdown is not permanent but results in the acquisition of a much higher conductivity by the dielectric only while the voltage gradient is maintained above the critical value, the dielectric regaining its insulating property when the field is reduced below this critical value.

18. ELECTROSTATIC ENERGY AND MECHANICAL FORCES

Electrostatic Energy. From the general relation expressed by eq. 18, it is evident that when the potential difference between the plates of a condenser is increased from 0 to V the energy input is

$$W = \int_0^t gv^2 \, dt + \int_0^V Cv \, dv \quad \text{joules or watt-seconds} \tag{22}$$

where g is in mhos, v in volts, C in farads, and t in seconds. The energy represented by the first term on the right-hand side of this equation is dissipated as heat in the dielectric, but the energy represented by the second term, which, when C is constant, may be written

$$W = \tfrac{1}{2} CV^2 \quad \text{joules or watt-seconds} \tag{22a}$$

does not represent a dissipation of heat; this is a fact of experience. Moreover, when the condenser is discharged, by short-circuiting its plates with a wire, this same amount of energy $\tfrac{1}{2} CV^2$ is transferred to the wire. Hence, the energy represented by $\tfrac{1}{2} CV^2$ is said to be stored in the condenser, or preferably in the dielectric of the condenser, for the electric field E exists only in the dielectric. This stored energy is called the *electrostatic energy*. It is analogous to the energy stored in a spring when the spring is compressed or stretched. Equation 22 may also be written

$$W = \frac{1}{2} \frac{Q^2}{C} = \frac{1}{2} QV = \frac{1}{2} \psi V \quad \text{joules or watt-seconds} \tag{22b}$$

The electrostatic energy per unit volume of an electric field may be written

$$w = \frac{DE}{2} = \frac{\epsilon_0 \epsilon_r E^2}{2}$$

$$= \frac{D^2}{2\epsilon_0 \epsilon_r} = \frac{D^2}{2} \quad \text{joules per cubic meter} \tag{22c}$$

where ϵ is $\epsilon_0 \epsilon_r$, ϵ_0 the dielectric constant for free space, ϵ_r the relative dielectric constant, E the magnitude of the electric field intensity or potential drop per meter, and D the magnitude of the dielectric flux density in coulombs per square meter.

It should be noted that eqs. 22a and 22b are based on the assumption that ϵ_r is a constant, independent of the value of E. When this condition does not hold, the energy required to establish the field is $\int_0^V Cv \, dv$, the evaluation of which depends upon the relation between C and v.

Mechanical Forces in an Electric Field. Experience shows that all bodies (conductors or insulators) in an electric field exert, in general, mutual mechanical forces upon one another tending to produce such a relative motion as will decrease the energy of the field. Let F be the component of the force tending to move any body in the field in a given direction, and let dW be the increase in the energy of the field due to displacing the body a distance dx in this direction; then

$$F = -dW/dx \tag{23}$$

provided this displacement does not cause a change in the existing electric charges in the field. As a consequence of this general relation it can be shown that every charged surface exerts a force of repulsion on every other surface charged with electricity of the same sign and a force of attraction on every surface charged with electricity of the opposite sign.

In the special case of the two conductors forming a condenser the force of attraction exerted by one conductor on the other is

$$F = -\frac{V^2}{2}\frac{dC}{dx} \text{ newtons} \tag{23a}$$

where V is the potential difference across the condenser in volts, C is the capacitance of the condenser in farads, and dC represents the increase in the capacitance of the condenser when one conductor moves a differential distance dx away from the other. The distance x is measured in meters. This relation results from the substitution of eq. 22b in 23. For example, the capacitance of a parallel-plate condenser is approximately

$$C = \frac{\epsilon_0 \epsilon_r A}{x} \text{ farads}$$

where ϵ_0 is the dielectric constant of vacuum, ϵ_r the relative dielectric constant, A the area of the smaller plate in square meters, and x the distance between the plates in meters. Hence the force of attraction is

$$F = \frac{V^2 \epsilon_0 \epsilon_r A}{2x^2} \tag{24}$$

ALTERNATING-CURRENT CIRCUITS

19. GENERAL DEFINITIONS

The definitions given below are applicable to currents, electromotive forces, potential differences, or any other function of time. They are the definitions recommended by the sectional committee on electrical definitions of the I.E.E.E.

Periodic Quantities

An **Oscillating Quantity** is a quantity which as a function of some independent variable (such as time) alternately increases and decreases in value, always remaining within finite limits.

A **Periodic Quantity** is an oscillating quantity the values of which recur for equal increments of the independent variable.

The **Period** of a periodic quantity is the smallest value of the increment of the independent variable which separates recurring values of the quantity.

A **Cycle** is the complete series of values of a periodic quantity which occur during a period. A cycle per second is designated by the unit Hertz or Hz.

The **Frequency** of a periodic quantity, in which time is the independent variable, is the reciprocal of the period. Frequency is expressed as a number of Hz.

The **Angular Velocity** of a periodic quantity is the frequency multiplied by 2π.

An **Alternating Quantity** is a periodic quantity which has alternately positive and negative values.

As **Examples** and to avoid repetition the following statements will be given in terms of electric currents.

If a given current is represented by the equation

$$i = f(t) \tag{1}$$

and if the function $f(t)$ has the property that

$$f(t) = f(t + T) \tag{2}$$

in which T is a constant, then the current is said to be periodic in time and T is the period. If time is measured in seconds then $1/T$ represents the number of periods per second and is usually denoted by f. The frequency f may also be said to be the number of cycles per second. If, as often happens, the current is expressible more simply as

$$i = f(\omega t) \tag{3}$$

in which ω is defined by

$$f(\omega t) = f(\omega t + 2\pi) \tag{4}$$

then the constant ω, being equal to 2π divided by the period or 2π times the frequency, is the angular velocity as defined above. That is

$$\omega = 2\pi f = 2\pi / T \tag{5}$$

Maximum, Average, and Rms or Effective Values

The **Instantaneous Value** of an alternating current is the value of the current at any instant. Instantaneous values of current, potential difference, and emf will be designated by lower-case letters throughout this article, viz., i, v, and e.

The **Maximum Value** of an alternating current is the numerical value of its maximum instantaneous value. Maximum values will be designated by capital letters with the subscript m.

The **Half-period Average Value** of a symmetrical alternating current * is the absolute value of the algebraic average of the values of the current taken throughout a half period, beginning with a zero value. If the current has more than two zeros during a cycle, that zero shall be taken which gives the largest half-period average value. The expression for the half-period average of a symmetrical alternating current $i(t)$ having a period T is

$$I_{av} = \frac{2}{T} \int_{t_0}^{t_0 + \frac{1}{2}T} i(t) \, dt \tag{6}$$

where t_0 is chosen such that $i(t) = 0$ at t_0 and I_{av} is the largest value obtainable.

The **Rms or Effective Value.** The square root of the means of the squares of the instantaneous values of an alternating current over a complete period is called the *rms* or the *effective* value of the alternating current. In specifying the value of an alternating current as so many amperes, this rms value is always meant unless specifically stated otherwise. In the same manner the square root of the mean of the squares of the instantaneous values of an alternating potential difference over a complete period is called the rms value of the alternating potential difference. When the value of an alternating potential difference is specified as so many volts, this rms value is always meant unless specifically stated otherwise.

The reason for selecting the above particular function of the instantaneous values of an alternating current or a potential difference as a measure of the current or the potential difference is that the average power dissipated as heat in a resistance, r, when an alternating current of rms value I flows through it, is rI^2. A similar argument exists for the use of rms values of potential.

Rms values will be designated throughout this article by capital letters without subscripts. The general expression for the rms value of an alternating current is

$$I = \sqrt{\frac{1}{T} \int_0^T i^2 \, dt} \tag{7}$$

and similarly for an alternating potential difference.

Form Factor, Crest or Peak Factor, Deformation Factor

The **Form Factor** of a symmetrical alternating current is the ratio of the effective value of the current to its half-period average value.

The **Peak or Crest Factor** of an alternating current is the ratio of the maximum value of the current to its effective value.

The **Equivalent Sinusoidal Current** of a given alternating current is a sinusoid having the same period and the same effective value as the given alternating current.

The **Deformation Factor** of an alternating current is the ratio of the maximum value of the equivalent sinusoidal current to the maximum difference between the corresponding values of the current considered and the equivalent sinusoid, when the two are superimposed in such a way as to make this difference a minimum.

---

* A symmetrical alternating quantity is one of which all values separated by a half period have the same magnitude but opposite sign. The term half-period average has no meaning for alternating currents which are not symmetrical.

Power, Power Factor, Volt-amperes, Reactive Power

Power. Let v be the value at any instant of the potential drop from any point 1 to any other point 2, and let i be the instantaneous value of the current from 1 to 2 at this same instant; then the *power input* at this instant is

$$p = vi \tag{8}$$

When v and i are both positive (i.e., in the direction from 1 to 2, say) or when they are both negative, the power input is positive; but when v is positive and i negative, or vice versa, the power input is negative, i.e., there is an actual power output.

The average value of the product vi over a complete period for both v and i (or over any whole number of periods) is the *average power* input or output, usually called simply the power input or output (input when the average of vi is positive, output when the average of vi is negative), the word average being understood. That is, the average power input is

$$p = \frac{1}{T} \int_0^T p \, dt = \frac{1}{T} \int_0^T vi \, dt \tag{9}$$

T being the time for a complete period.

Power Factor. Only in certain special cases (see below) is the average power input P equal to the product of the rms value V of the potential difference by the rms value I of the current; it can never be greater and as a rule is less. The ratio of the average power P to the product of the rms value V of the potential difference by the rms value I of the current is called the *power factor* of the circuit between the terminals considered; i.e.,

$$\text{Power factor} = P/VI \tag{10}$$

When V is expressed in volts and I in amperes, then P must be in watts; when V is expressed in kilovolts and I in amperes, P must be in kilowatts.

Apparent Power. The product of the rms volts across the terminals of a circuit and the rms amperes through it is called the *volt-amperes* or *apparent power* taken by the circuit; this product divided by 1000 is called the kilovolt-ampere input. Or, when V is in volts and I in amperes,

$$\text{Volt-amperes} = VI \tag{11}$$

$$\text{Kilovolt-amperes} = VI/1000 \tag{11a}$$

Reactive Power has no accepted definition when either the current or emf is non-sinusoidal. For reactive power with sinusoidal currents and emf's see below.

20. SINUSOIDAL CURRENTS AND VOLTAGES

A Simple Sinusoidal Current (simple harmonic current) is an alternating current the instantaneous values of which are equal to the product of a constant and the sine of an angle having values varying linearly with time. Thus

$$i = I_m \sin(\omega t + \theta) \tag{12}$$

where t represents time in seconds, measured from any arbitrarily chosen instant; I_m the maximum value of the current; $\omega = 2\pi f = 2\pi/T$, where f is the frequency in cycles per second and T the period as a fraction of a second; and θ a constant which depends upon the instant chosen as the zero of time.

The quantity I_m is often called the *amplitude* of the sinusoidal current.

The Phase of a periodic current for a particular value of the independent variable is the fractional part of a period through which the independent variable has advanced, measured from an arbitrary origin. For a simple sinusoidal current, the origin is usually taken as the last previous passage through zero from the negative to positive direction. The phase angle is the angle obtained by multiplying the phase by 2π if the angle is to be expressed in radians, or by $360°$ if the angle is to be expressed in degrees.

When a sine-wave emf is impressed on a circuit having constant parameters, the resulting current is likewise a sine function of time (transient state ignored) having the same frequency, but the emf and current do not reach their maximum values simultaneously. Let the current be represented by eq. 12 and the voltage be given by

$$v = V_m \sin \omega t \tag{13}$$

where t is the time measured from the instant when $v = 0$, and is increasing in the positive direction. The voltage reaches its maximum value when $t = \pi/2\omega$; the current reaches its maximum value when $t = (\pi/2\omega) - (\theta/\omega)$. Hence when θ is positive the voltage or potential drop reaches its maximum value θ/ω seconds after the current reaches its maximum, or the current reaches its maximum value θ/ω seconds before the potential drop reaches its maximum; when θ is negative the current reaches its maximum value θ/ω seconds after the potential drop reaches its maximum. In the first case, the current is said to "lead" the potential drop; and in the second, the current is said to "lag" the potential drop. The angle θ is called the angular phase difference between the current and potential drop.

When the phase difference is zero the current and potential drop are said to be "in phase"; when the phase difference is $\pi/2$ radians or 90° the current and potential drop are

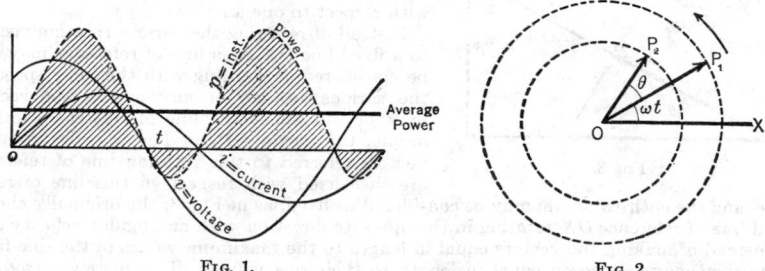

Fig. 1.   Fig. 2.

said to be "in quadrature"; when the phase difference is π radians or 180° the current and potential drop are said to be "in opposition."

Power and Power Factor. Let the voltage drop from terminal 1 to terminal 2 through any piece of apparatus be $v = \sqrt{2}V \sin(\omega t + \theta_v)$, and the current from terminal 1 to terminal 2 be $i = \sqrt{2}I \sin(\omega t + \theta_i)$, where V and I are the rms values and $\sqrt{2}V$ and $\sqrt{2}I$ are the maximum values. Then the instantaneous power input is

$$p = vi = VI\left[\cos(\theta_v - \theta_i) - \cos(2\omega t + \theta_v + \theta_i)\right] \tag{14}$$

A study of Fig. 1 will show the physical meaning of this expression. The average power input is

$$P = VI \cos(\theta_v - \theta_i) \tag{14a}$$

where $(\theta_v - \theta_i)$ is the angular phase difference between the voltage and current. Putting θ for this difference in phase, viz., $\theta = \theta_v - \theta_i$, eq. 14a may be written:

$$P = VI \cos\theta \tag{14b}$$

Whence the power factor of the load supplied to the apparatus is (from eq. 10):

$$\cos\theta = P/VI \tag{14c}$$

Since with sine-wave currents and voltages the power factor is equal to the cosine of the angle which expresses the difference in phase between them, this difference in phase is frequently called the "power-factor angle." If the wave shape is not a pure sine curve, the power factor cannot be interpreted as the cosine of the phase difference, for phase difference has no definite meaning except in reference to sine waves; see definitions above. A non-sinusoidal voltage and current may both reach their zero values at the same instant, and in a sense may be said to be "in phase," but the power factor as defined by eq. 10 may be far from unity.

The Reactive Power in a circuit in which a sinusoidal current is flowing is equal to the effective emf times the effective current times the sine of the phase difference between them. When the emf and current are in volts and amperes respectively the reactive power is in *vars*.

Vector Representation of Sinusoids. Consider any sine function

$$i = I_m \sin\omega t$$

The value of i at any instant may be represented graphically, see Fig. 2, by the vertical projection (i.e., the vertical distance from P_1 to OX) of a point P_1 at the end of a radius $OP_1 = I_m$ which revolves * at a constant angular velocity ω about a fixed point O, the

---

* Counterclockwise rotation was adopted (1911) as standard by the International Electrotechnical Commission.

angle ωt being measured from the horizontal fixed line OX. Similarly, any other sine function

$$v = V_m \sin (\omega t + \theta)$$

may be represented by the vertical projection of the point P_2 at the end of a radius $OP_2 = V_m$ also revolving about O with a constant angular velocity ω, the angle between OP_1 and OP_2, when the frequency of both i and v is the same, remaining fixed in value and equal to the difference in phase θ between v and i. That is, v and i may be represented by rotating vectors. When v and i are of the same frequency, the relative position of the two vectors remains fixed. Similarly any number of currents and voltages of the same frequency may be represented by rotating vectors which remain fixed with respect to one another.

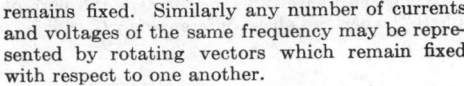

Fig. 3.

Instead of referring the various rotating vectors to a fixed line OX, this line of reference may also be considered as rotating with the same speed as the various vectors, or any one of the vectors may be chosen as the line of reference, for example, the vector OP_1 in Fig. 2. The rotating vectors referred to this rotating line of reference are then fixed with respect to this line of reference, and the entire diagram may be considered as fixed, as in Fig. 3, the originally chosen fixed line of reference OX rotating in the opposite direction with an angular velocity ω.

Instead of making the vectors equal in length to the maximum values of the sine functions they may be chosen equal in length to their rms values. This merely introduces a factor $\sqrt{2}$ so that the instantaneous value of the quantity which any rotating vector is considered to represent is equal to $\sqrt{2}$ times the perpendicular distance from the end of the vector to the fixed line of reference.

Since the rms values and phase relations of sine-wave currents and voltages may be represented by vectors, sine-wave currents may be added in exactly the same manner as vectors are added, and similarly for sine-wave voltages. To add any two sine-wave currents or voltages not only their effective values but also their phase relation must be known; the resultant of two alternating voltages of rms values V_1 and V_2 is never the arithmetical sum of V_1 and V_2, except when the two voltages are exactly in phase, and similarly for alternating currents.

In Fig. 3, considering OI as equal to the rms value I of the current and OV as representing the rms value V of the voltage, the vector voltage V may be considered as made up of two components, viz.:

$$V_1 = V \cos \theta \text{ in phase with the vector } I$$

$$V_2 = V \sin \theta \text{ in quadrature with the vector } I$$

The average power corresponding to the component $V_1 = V \cos \theta$ is, from eq. 14b, $V_1 I = VI \cos \theta$, and is equal to the total power corresponding to V and I. The average power corresponding to the component $V_2 = V \sin \theta$, since the angle between the current and this component of the voltage is 90°, is equal to zero. The voltage component $V_1 = V \cos \theta$ is therefore frequently called the "power" component of the voltage, and the component $V_2 = V \sin \theta$ is frequently called the "wattless" component of the voltage. These terms, however, are not recommended. It is preferable to refer to these two components as the in-phase and quadrature components respectively. The terms active and reactive components are also used.

Similarly, the vector current I may be considered as made up of two components, viz.:

$$I_1 = I \cos \theta \text{ in phase with } V$$

$$I_2 = I \sin \theta \text{ in quadrature with } V$$

The first component is called the in-phase component of the current and the second the quadrature component of the current.

21. NON-SINUSOIDAL PERIODIC CURRENTS AND VOLTAGES

Fourier Series, Harmonics. It can be shown that any single-valued, periodic function (which fulfills certain mathematical conditions always fulfilled by electric currents) may be represented by a sum of sinusoids of which the frequencies form an arithmetical progression. Hence any current $i(t)$ can be written

$$i(t) = A_1 \sin 2\pi \frac{t}{T} + A_2 \sin 2\pi \frac{2t}{T} + \cdots + A_n \sin 2\pi \frac{nt}{T} + \cdots$$

$$+ \frac{B_0}{2} + B_1 \cos 2\pi \frac{t}{T} + B_2 \cos 2\pi \frac{2t}{T} + \cdots + B_n \cos 2\pi \frac{nt}{T} + \cdots \quad (15)$$

where the values of the coefficients A and B are given by the integrals:

$$A_n = \frac{2}{T} \int_0^T i(t) \sin 2\pi \frac{nt}{T} dt \qquad n = 1, 2, 3, \text{ etc.} \quad (15a)$$

$$B_n = \frac{2}{T} \int_0^T i(t) \cos 2\pi \frac{nt}{T} dt \qquad n = 0, 1, 2, 3, \text{ etc} \quad (15b)$$

The series (15) is called a *Fourier series*. If the wave is not symmetrical with respect to the chosen axis a component $B_0/2$ will exist; this is often called the direct component in speaking of voltage and current wave forms.

If an analytic expression for $i(t)$ is available and if the integrations (15a) and (15b) can be performed, the terms of the Fourier series (15) may be written out. In practice the function $i(t)$ will usually be obtainable only in the form of a plotted curve and the evaluation of the A's and B's must be carried out by graphical means. This process is called *harmonic analysis*. Various means of harmonic analysis have been devised, of which the Fisher-Hinnen method is perhaps the most convenient in ordinary cases. See Pender's *Electrical Engineers' Handbook*.

If the terms involving the same frequencies are combined, and if $2\pi/T = \omega$, eq. 15 may be written

$$i(t) = I_0 I_1 \sin (\omega t - \theta_1) + I_2 \sin (2\omega t - \theta_2) + \cdots + I_n \sin (n\omega t - \theta_n) + \cdots \quad (16)$$

I_0 designates a component not varying with time—a direct component. The term "$I_1 \sin (\omega t - \theta_1)$" is called the *fundamental*, and the terms "$I_n \sin (n\omega t - \theta_n)$" are called *harmonics*. The *order* of a harmonic is the ratio of the harmonic frequency to the frequency of the fundamental. Each harmonic is characterized by two constants, its amplitude I_n and the angle θ_n. It should be noted that this angle may be given in either of two ways:

$$I_n \sin (n\omega t - \theta_n) \quad (17a)$$

$$I_n \sin n (\omega t - \theta'_n) \quad (17b)$$

In the first notation the angle θ_n is taken relative to the period of the harmonic, whereas in the second notation the angle θ'_n is taken relative to the period of the fundamental. It follows that

$$\theta_n = n\theta'_n \quad (18)$$

Rms Value of Non-sinusoidal Currents. The rms value of a non-sinusoidal current can be obtained directly from eq. 7, Art. 19, or if I_0 and the rms values I_1, I_2, I_3, etc., of the fundamental and harmonics are known the rms value of the resultant current is

$$I = \sqrt{I_0^2 + I_1^2 + I_2^2 + I_3^2 + \cdots} \quad (19)$$

A like relation holds for the rms value of a non-sinusoidal voltage. Similarly, if for example a 25-Hertz emf, say E_{25}, and a direct emf, say E_0, are acting in series in the same circuit, the rms value of the resultant emf of the combination is

$$E = \sqrt{E_{25}^2 + E_0^2}$$

The Power Corresponding to Non-sinusoidal Currents and Voltages may be computed as follows:

Let I_0 be the value of the direct component of current, and V_0 the value of the direct component of voltage; let I_1 be the rms value of the fundamental of the current, V_1 the rms value of the fundamental of the voltage, and θ_1 the difference in phase between these two fundamentals, both being of the same frequency; let I_2, V_2, and θ_2 be the corresponding quantities for the second harmonic; let I_3, V_3, and θ_3 be the corresponding values for the third harmonic, etc. Then the average power is

$$p = V_0 I_0 + V_1 I_1 \cos \theta_1 + V_2 I_2 \cos \theta_2 + V_3 I_3 \cos \theta_3 + \text{etc.} \quad (20)$$

That is, each harmonic contributes an amount to the total power equal to the power it would develop were the other harmonics not present. If, for example, the third harmonic is not present in the current wave, then this harmonic contributes nothing to the average power even though there may be a large third harmonic in the voltage wave. Again, when a 25-Hertz alternating emf E_{25} and a direct emf E_0 are acting in series on the same circuit, the power developed is the sum of the powers which each would develop if they acted separately, but the resultant emf of the combination is not $E_{25} + E_0$, but, as noted above, $\sqrt{E_{25}{}^2 + E_0{}^2}$.

22. CURRENT AND VOLTAGE RELATIONSHIPS

Electric Circuit Parameters. The following are restatements of principles given in foregoing articles:

(a) The flow of a current i through a resistance R is always accompanied by a drop of electric potential, or voltage drop, in the direction of this current, equal at each instant to the product of the resistance and the current.

$$v_r = Ri \tag{21}$$

(b) The flow of a current i through a self-inductance L is always accompanied by a voltage drop, in the direction of this current, equal at each instant to the product of the inductance and the rate of increase of the current.

$$v_L = L(di/dt) \tag{22}$$

(c) The flow of a current i_b through a conductor B which has, with respect to another conductor A, a mutual inductance M, is always accompanied by a voltage drop in the conductor A equal at each instant to the product of the mutual inductance and the rate of increase of the current in the conductor B.

$$v_M = M(di_B/dt) \tag{23}$$

The direction of the mutual inductive drop in the conductor A is such as to link *in the right-hand screw direction* the magnetic flux produced by the current in the conductor B.

(d) The flow of a (displacement) current i through a capacitance C is always accompanied by a time rate of change of voltage drop equal at each instant to the current divided by the capacitance.

$$dv_C/dt = i/C \tag{24}$$

The voltage v_C is therefore equal to the integral of i/C from the instant at which v_C was zero up to the instant under consideration

$$v_C = \int_{t_0}^{t} i\,\frac{dt}{C} \tag{24a}$$

Since the value of this integral is always zero at the lower limit, it is customary to omit the limits of integration and write simply

$$v_C = \frac{1}{C} \int i\,dt \tag{24b}$$

The quantities R, L, C, and M are called *electric circuit parameters*.

Operative Impedance. Any portion of an electric circuit which contains a resistance, an inductance, or a capacitance, or which contains two or more of these quantities, is said to offer an impedance to the flow of an electric current.

The drop of electric potential through any portion of an electric circuit due solely to its resistance, inductance, and capacitance is called the *impedance drop* in this portion of the circuit.

The impedance drop through an impedance formed by a resistance R, inductance L, and capacitance C in series, due to the flow of a current i through it, is

$$v = Ri + L\frac{di}{dt} + \frac{1}{C}\int i\,dt \tag{25}$$

The drop of potential

$$v_A = M(di_B/dt)$$

in any conductor A due to the mutual inductance between A and any other conductor B is called the *mutual impedance drop* in A due to the current i_B in B.

The group of operations represented by the three terms in the right-hand member of 25 is often represented by the single symbol $\mathbf{Z}$. The function $\mathbf{Z}$ operates on i_0

$$\mathbf{Z} = R + L\frac{d}{dt} + \frac{1}{C}\int dt \tag{25a}$$

The operator Z is referred to as the *operative impedance* of the portion of the circuit under consideration.

The term *operative mutual impedance* is used when the mutual impedance drop is written $Z_m i$, where

$$Z_m = M(d/dt) \tag{25b}$$

It is customary in electric circuit theory to use the notation $p = d/dt$. Thus Z defined above is often written $Z(p)$. It is assumed that the energy storage is zero initially.

Emf. Electromotive force has been defined as that property of a device which tends to produce an electric current in a circuit. It is convenient in electric circuit theory to narrow this definition to include only those sources of potential difference not defined as impedance drops. Making use of this restricted definition of emf, Kirchhoff's second law is stated: "The sum of the emf's is, at any instant, equal to the sum of the impedance drops," whereas making use of the more general definition this law is stated: "The sum of the emf's is at any instant equal to the sum of the resistance drops."

Impedance Equations for a Network. Kirchhoff's second law applied to mesh 1 of Fig. 4 gives

$$Z_{12}i_{12} + Z_{13}i_{13} + Z_{14}i_{14} + Z_{15}i_{15}$$
$$= e_{12} + e_{13} + e_{14} + e_{15} \tag{26}$$

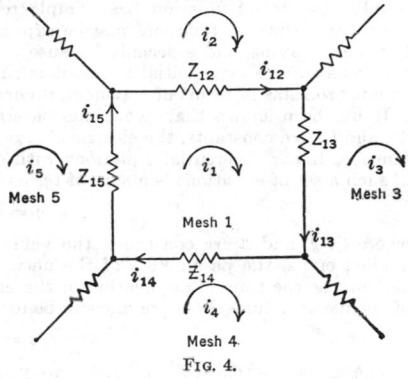

Fig. 4.

where the Z's are the operative impedances of the several branches which make up this mesh, the i's are the *branch currents*, and the e's are the emf's, if any, in the respective branches of the mesh, the positive sense of each emf being taken as clockwise around the mesh.

In terms of the *mesh currents* defined in Art. 7, the equation may also be written

$$(Z_{12} + Z_{13} + Z_{14} + Z_{15})i_1 - Z_{12}i_2 - Z_{13}i_3 - Z_{14}i_4 - Z_{15}i_5 = e_{12} + e_{13} + e_{14} + e_{15} \tag{27}$$

This relation may be simplified by putting

$$Z_{11} = Z_{12} + Z_{13} + Z_{14} + Z_{15} \tag{27a}$$

and

$$e_1 = e_{12} + e_{13} + e_{14} + e_{15}$$

Equation 27 then becomes

$$Z_{11}i_1 - Z_{12}i_2 - Z_{13}i_3 - Z_{14}i_4 - Z_{15}i_5 = e_1 \tag{28}$$

In this equation the operator Z_{11} is the sum of the operative impedances of the several branches of mesh 1, and e_1 is the algebraic sum, in the clockwise direction, of all emf's in this mesh.

If as many independent mesh equations, of the form 28, as there are mesh currents in a given network, are written down, the set of equations can be solved for the various currents. Such equations are called the *impedance equations* of the network.

Differential Equations in Terms of Quantity of Electricity. When capacitances exist in any of the meshes of the network the impedance equations contain some derivatives and some integrals. Equations of this type are more difficult to handle than true differential equations. It is often more convenient therefore to express the relation shown by eq. 28 in terms of the quantity of electricity q. If we write

$$i_r = dq_r/dt \tag{29}$$

then eq. 28 becomes

$$Z'_{11}q_1 - Z'_{12}q_2 - Z'_{13}q_3 - Z'_{14}q_4 - Z'_{15}q_5 = e_1 \tag{30}$$

in which

$$Z'_{rs} = \left(Z_{rs}\frac{d}{dt}\right) = \left(\frac{1}{C_{rs}} + R_{rs}\frac{d}{dt} + L_{rs}\frac{d^2}{dt^2}\right) \tag{31}$$

Equation 30 is a true differential equation.

Superposition. *Linear differential equations* are differential equations, such as the above, having coefficients which are constants, or which are functions of the independent variable only. They have the property that the sum of the solutions is also a solution.

That is, if x_1 and x_2 are two solutions of a given linear differential equation then $x_1 + x_2$ is also a solution of the given equation.

Suppose the equation to be non-homogeneous, and let the non-homogeneous term be A_1. Call the particular solution of the equation corresponding to this condition x_1. Now let the non-homogeneous term be given the value A_2, the remainder of the equation remaining the same, and let the particular solution corresponding to this condition be x_2. Then it can be shown that $x_1 + x_2$ is a solution of the equation when the non-homogeneous term is given the value $A_1 + A_2$. A_1 and A_2 can each be considered a source.

The above principles jointly are called the *principle of superposition*. Electrical networks in which this principle holds are called *linear networks*.

Exponential Emf, Charge, Current, and Impedance. Derivatives and integrals of the simple exponential function bear simple relationships to the function itself. Furthermore, any arbitrary function may be expressed in terms of a sum of exponential terms. For these reasons, and especially because the simple sinusoidal function is simply expressible as a sum of exponential terms, it is customary to formulate the solution of electric circuit problems in terms of exponential currents.

It has been shown that, when the parameters of a given network (the factors R, L, M, and C) are constants, the electric charge equations for this network are a set of simultaneous, linear differential equations with constant coefficients. The complete solution of such a set of equations is a sum of terms of the form

$$q = Qt^g \epsilon^{\beta t} \tag{32}$$

where Q, g, and β are constants, the values of which can be expressed in terms of the applied emf's, the parameters of the network, and the conditions which existed in the network at the time of application of the emf's. The solutions of the current equations of the network may therefore likewise be expressed as a sum of the terms of the form

$$i = It^g \epsilon^{\beta t} \tag{33}$$

The general forms (32) and (33) rarely occur, and so the usual case in which $g = 0$ will be assumed here. In the exponential current

$$i = I\epsilon^{\beta t}$$

the quantities I and β are often called the *initial value* and the *periodicity* of the current respectively.

From eq. 25a the impedance drop corresponding to the current

$$i = I\epsilon^{\beta t}$$

is

$$v = Zi = \left(R + L\beta + \frac{1}{\beta C} \right) I\epsilon^{\beta t} \tag{34}$$

The factor in the parenthesis is a *constant* for any given value of β. That is, for a current of the form $I\epsilon^{\beta t}$ the operative impedance

$$Z = R + L\frac{d}{dt} + \frac{1}{C}\int dt \tag{35}$$

reduces to the simple algebraic quantity

$$Z(\beta) = R + \beta L + \frac{1}{\beta C} \tag{35a}$$

This quantity will be referred to as the *exponential impedance*.

From the above it is evident that an exponential voltage drop is always associated with an exponential charge and current.

23. STEADY-STATE SOLUTION

Definition of a Steady State. The complete solution of a non-homogeneous linear differential equation having constant coefficients consists of a particular function which satisfies the given equation plus the most general solution of the homogeneous equation (see Section 2).

In stable * electrical systems the charges and currents represented by the solutions of the homogeneous equations decrease rapidly with time. After a brief interval following the application of emf's to such a system the currents and charges are represented by the

---

* For a definition of stability refer to Art. 25.

particular solutions of the non-homogeneous equations only. When this condition has been reached the system is said to have attained a *steady state*. During the interval before this condition is reached the system is said to be in a *transient state*.

In linear electrical systems the steady-state currents and usually the charges have the same wave form (i.e., are represented by time functions of the same type) as the impressed voltages.

Sinusoidal Emf's. A sinusoidal emf given by

$$e = \sqrt{2}E \cos (\omega t + \phi) \tag{36}$$

may be written as the sum of the two exponential components

$$e' = E' \epsilon^{j\omega t} \tag{37}$$

$$e'' = E'' \epsilon^{-j\omega t} \tag{37a}$$

in which $j = \sqrt{-1}$ and in which

$$E' = \frac{1}{\sqrt{2}} E \epsilon^{j\phi} \tag{37b}$$

$$E'' = \frac{1}{\sqrt{2}} E \epsilon^{-j\phi} \tag{37c}$$

In eq. 36 the constant E is the rms value of the emf, and ω is its angular velocity.

In electric-circuit theory the phase differences between various quantities having the same frequency are of importance. Since these differences in phase are independent of the instant at which they are evaluated, it is customary to refer to the angle ϕ (which is the phase angle when $t = 0$) as the phase angle of the emf. The two constants E' and E'' in eqs. 37 are the initial values of the two exponential components of e, and $j\omega$ and $(-j\omega)$ are the periodicities of these two components. Note that the two initial values E' and E'' are conjugate complex numbers,* and that the two periodicities $j\omega$ and $(-j\omega)$ are conjugate imaginaries.

Complex Number Solution. It can be proved that during steady state an exponential current having a given periodicity exists in a stable linear network when and only when there is impressed upon the network an exponential emf having the given periodicity.

Each current which flows in a given network due to the application of a sinusoidal emf is therefore composed of two exponential components, each due to one of the exponential components of the sinusoidal emf. The superposition theorem states that these two components may be computed separately.

Let the sinusoidal emf, given by eq. 36, be applied to a mesh (say mesh 1) of a network, and consider the currents flowing due to the component e', eq. 37. Each of these currents will be of the form

$$i'_k = I'_k \epsilon^{j\omega t} \tag{38}$$

and each self-impedance drop (see 35a) will have the form

$$Z_{kk} i'_k = \left(r_k + j\omega L_k + \frac{1}{j\omega C_k} \right) I'_k \epsilon^{j\omega t} \tag{39}$$

Mutual impedance drops will be of a similar form. Hence any one of the impedance equations such as 28 will contain the factor $\epsilon^{j\omega t}$ in every term, and this factor can be divided out. This results in a set of equations containing as unknowns the complex numbers representing the initial values of the exponential currents and as coefficients the complex numbers obtained by putting $p = j\omega$ in the various operative impedances. The initial value of the exponential current i'_k flowing in the kth mesh may be written

$$I'_k = A_k \epsilon^{j\theta_k} \tag{40}$$

The currents which flow in the network due to the other component e'' of the sinusoidal emf will be of the form

$$i''_k = I''_k \epsilon^{-j\omega t} \tag{41}$$

The initial values of these currents will be obtained from a set of complex number equations in which every term is the conjugate of the corresponding term in the set of equations for I'_k. The initial value of any one of these currents, say I''_k, can be shown to be the conjugate of the initial value, I'_k, of the corresponding current, viz.,

$$I''_k = A_k \epsilon^{-j\theta_k} \tag{42}$$

The actual steady-state sinusoidal current flowing in the kth mesh due to the appli-

---

* Complex numbers will be printed in bold-face type throughout this article.

cation of the sinusoidal emf to the first mesh is the sum of the two exponential currents i'_k and i''_k,

$$i_k = i'_k + i''_k = 2A_k \cos{(\omega t + \theta_k)} \tag{43}$$

If, in solving the mesh equations, the initial values 37b and 37c were replaced by the values $E\epsilon^{j\phi}$ and $E\epsilon^{-j\phi}$, then a new value of A_k equal to $\sqrt{2}$ times the value in eq. 43 would be obtained. Let this value be called I_k; i.e.,

$$I_k = \sqrt{2}A_k \tag{44}$$

In terms of the quantity I_k the current i_k is

$$i_k = \sqrt{2}I_k \cos{(\omega t + \theta_k)} \tag{45}$$

From the above we have the following simple procedure for finding the steady-state currents produced in a network by a group of sinusoidal emf's of a given frequency:

(a) Write the operative impedance eqs. 27 and 28 described in Art. 22.

(b) Replace each emf in the network by a complex number having a modulus equal to the rms value of the given emf, and a phase angle equal to the phase angle of the given emf (when expressed in cosine form).

(c) Replace the self- and mutual operative impedances of the several meshes of the network by complex numbers obtained by putting $p = (j\omega)$, where ω is the angular velocity of the given sinusoidal electromotive forces.

(d) Replace each current in the network by an unknown complex number, and solve for these complex numbers.

(e) The rms values of the various currents are the moduli of the corresponding complex numbers, and the phase angles of the currents (when expressed in cosine form) are the phase angles of these complex numbers.

This method is known as the *complex number method* for alternating currents.

The complex numbers obtained from the operative impedances are usually referred to as *vector impedances* or often simply as *impedances*. Although the term impedance is used for both the differential operator and the complex number, the term used without qualification usually means the complex number. A still more general definition of impedance is given below. The algebra of complex numbers is treated in the section on mathematics.

Impedance, Admittance, Reactance, Susceptance. In previous paragraphs the terms operative impedance and vector impedance were defined. Although the term *impedance* is loosely used to designate either of these, the accepted definition of impedance is: The impedance of a portion of an electric circuit, to a completely specified periodic current and potential difference, is the ratio of the effective value of the potential difference between the terminals to the effective value of the current, there being no source of emf in the portion under consideration.

Impedance defined in this way is a real number which in a sinusoidal current and voltage reduces to the modulus value of the vector impedance defined above.

Admittance is defined as the reciprocal of the impedance. The *vector admittance* is the reciprocal of the complex number representing the vector impedance.

Impedance and admittance, as thus defined, both depend upon the frequency and wave shape of the current. Impedance is expressed in the same units as resistance (e.g., ohms), and admittance in the same units as conductance (e.g., mhos).

The *reactance* of a portion of a circuit for a sinusoidal current, and hence for any one of the frequencies of a periodic current, is the ratio of the quadrature component of the potential difference for a particular frequency to the value of the current for that frequency, there being no source of emf in the portion of the circuit under consideration.

The *susceptance* of a portion of a circuit for a sinusoidal potential difference, and hence for any one of the frequencies of a periodic potential difference, is the ratio of the quadrature component of the current for a particular frequency to the value of the potential difference for that frequency, there being no source of emf in the portion of the circuit under consideration.

From the definitions given above it may be shown that for sine-wave currents and voltages of a given frequency the following relations hold for any portion of a circuit:

$$
\left.
\begin{array}{ll}
Z = r + jx & Y = g - jb \\
Z = \sqrt{r^2 + x^2} & Y = \sqrt{g^2 + b^2} \\
Z = 1/Y & Y = 1/Z \\
r = g/Y^2 & g = r/Z^2 \\
x = b/Y^2 & b = x/Z^2
\end{array}
\right\} \tag{46}
$$

where r = effective resistance, x = reactance (taken positive when inductive and negative when capacitive), g = effective conductance, b = susceptance (taken positive when inductive and negative when capacitive), Z = impedance, Y = admittance, $\mathbf{Z}$ = vector impedance, and $\mathbf{Y}$ = vector admittance, all for the given portion of circuit.

Vector Impedances in Simple Circuits. The vector impedances of the various circuit elements to a sinusoidal current having an angular velocity ω are:

In a resistor $\qquad \mathbf{Z}_r = R$

In an inductor $\qquad \mathbf{Z}_L = j\omega L$

In a capacitor $\qquad \mathbf{Z}_C = 1/j\omega C = -j(1/\omega C)$

The mutual impedance in a device having mutual inductance is

$$\mathbf{Z}_M = j\omega M$$

The product (ωL) or (ωM) corresponding to a self-inductance L, or to a mutual inductance M, is called the *reactance* of this inductance at the angular velocity ω. Similarly the expression $-1/\omega C$ corresponding to a capacitance C is called the *reactance* of this capacitance at the angular velocity ω. Note that the reactance of a capacitance is always negative. These statements are in agreement with the previous definition of reactance.

In general, then, the vector impedance of any mesh, or the mutual vector impedance of any two meshes, may be written in the form

$$\mathbf{Z} = r + jx \tag{47}$$

where r is the sum of the effective resistances and x is the algebraic sum of the reactances of the several portions of the network which make up this impedance.

The vector impedance $(r + jx)$ may also be expressed in the form

$$\mathbf{Z} = Z \angle \theta \tag{48}$$

where

$$Z = \sqrt{r^2 + x^2} \tag{48a}$$

is the modulus of the complex number $(r + jx)$ and

$$\theta = \tan^{-1}(x/r) \tag{48b}$$

is the phase angle of this complex number.

In eq. 48 the symbol $\angle$ is read "at an angle of" and has the same mathematical significance as e^{j}. Thus $Z \angle \theta$ is identical with $Z e^{j\theta}$.

When two or more vector impedances $\mathbf{Z}_1$, $\mathbf{Z}_2$, etc., are connected in series, so that the same current I flows through each, the total impedance drop through the group is

$$\mathbf{V} = \mathbf{Z}_1 I + \mathbf{Z}_2 I + \cdots \tag{49}$$

$$= (\mathbf{Z}_1 + \mathbf{Z}_2 + \mathbf{Z}_3 \cdots) I$$

hence two or more vector impedances in series are equivalent to a single vector impedance equal to the sum of the several impedances.

When two or more vector impedances $\mathbf{Z}_1$, $\mathbf{Z}_2$, etc., are connected in parallel, so that the voltage drop is the same for each, the total current through the group is

$$I = \frac{V}{\mathbf{Z}_1} + \frac{V}{\mathbf{Z}_2} + \cdots$$

$$= \left(\frac{1}{\mathbf{Z}_1} + \frac{1}{\mathbf{Z}_2} + \cdots \right) V$$

$$= (\mathbf{Y}_1 + \mathbf{Y}_2 + \cdots) V \tag{50}$$

It is customary to express this law in terms of the vector admittances $\mathbf{Y}_1$, $\mathbf{Y}_2$, $\cdots$ as shown in the last expression of eq. 50. Two or more vector admittances in parallel are equivalent to a single vector admittance equal to the sum of the several admittances.

Calculation of Power. The average power in a circuit in which the current is i and the voltage drop is v is

$$P = \frac{1}{T} \int_0^T vi \, dt$$

With a sinusoidal current and voltage

$$v = \sqrt{2}V \cos(\omega t + \theta_v) \quad \text{and} \quad i = \sqrt{2}I \cos(\omega t + \theta_i)$$

this power is equal to

$$P = VI \cos(\theta_v - \theta_i)$$

The vector voltage and vector current in this case are

$$V = V_\epsilon{}^{j\theta_v}$$

$$I = I_\epsilon{}^{j\theta_i}$$

of which the product *is not* the average power. The average power is given by the *real part* of the product of the vector voltage by the *conjugate* of the vector current.

$$P = \text{real part of } (V_\epsilon{}^{j\theta_v}) (I_\epsilon{}^{-j\theta_i}) \tag{51}$$

$$= \text{real part of } V\hat{I} \tag{51a}$$

This may be expressed in terms of the components of the vector current and voltage as follows: Let

$$V = V_1 + jV_2$$

$$I = I_1 + jI_2$$

then
$$\hat{I} = I_1 - jI_2$$

$$P = V_1I_1 + V_2I_2 \tag{51b}$$

Often P_q is used as the symbol for reactive power to distinguish from P or P_r, the real power as in eq. 51. The reactive power is given by the imaginary part of the product of the vector voltage by the conjugate of the vector current,

$$P_q = \text{imaginary part of } (V_\epsilon{}^{j\theta_v})(I_\epsilon{}^{-j\theta_i})$$

$$= \text{imaginary part of } V\hat{I}$$

Using components of voltage and current,

$$V = V_1 + jV_2$$

$$I = I_1 + jI_2$$

$$\hat{I} = I_1 - jI_2$$

$$P_q = jV_2I_1 - jV_1I_2 \tag{52}$$

Using the conjugate of current results in lagging reactive power being designated by a positive sign followed by the j symbol.

If both real and reactive power are computed by the product of the conjugate of voltage by the current, the real power will be obtained as in eq. 51, but lagging reactive power will now be designated with a negative sign followed by the j symbol.

Complicated Circuits. The vector impedance equations described give the complete steady-state solution of any complicated electrical network. If there are n meshes in the network, these equations are

$$Z_{11}I_1 - Z_{12}I_2 - Z_{13}I_3 \cdots - Z_{1n}I_n = E_1$$

$$- Z_{12}I_1 + Z_{22}I_2 - Z_{23}I_3 \cdots - Z_{2n}I_n = E_2$$

$$- Z_{13}I_1 - Z_{23}I_2 + Z_{33}I_3 \cdots - Z_{3n}I_n = E_3 \tag{53}$$

$$- Z_{1n}I_1 - Z_{2n}I_2 - Z_{3n}I_3 \cdots + Z_{nn}I_n = E_n$$

which can be solved either by successive substitutions or by the method of determinants. If the number of equations is small, the solution by substitution usually results in less labor. If there are many meshes, the method of determinants often results in a saving of time. Methods for solving algebraic equations in complex numbers are described in the section on mathematics.

General Network Theorems

A linear impedance is an impedance that is independent of the amount of current passing through it. A bilateral impedance is an impedance that allows current in either direction to flow equally well. Resistors, capacitors, and air-core inductances can often be considered linear and bilateral elements. An iron-core inductance must often be considered non-linear, but it is still bilateral. Electron tubes and semiconductor rectifiers are neither bilateral nor linear, although there may be a narrow range over which reasonable linearity may be assumed. A resistive element such as an incandescent lamp sometimes must be considered non-linear. All the network theorems listed in this section assume bilateral linear circuit elements.

Theorem 1. Superposition Theorem. If a network has two or more sources of emf in it, the current through any branch of the network is equal to the sum of the currents obtained by considering each source of emf separately while each of the other sources is replaced by its internal impedance.

Theorem 2. Thévenin's Theorem. If a load impedance is connected to two terminals of a network which contains sources of emf, the current through the impedance so connected will be equal to the voltage between the terminals when the load impedance is removed divided by the sum of the load impedance and the impedance of the network between the two terminals with the load resistance removed and all sources of emf in the network replaced by their internal impedances.

Theorem 3. Reciprocity Theorem. If a source of emf in a given branch of a network causes a current to flow in another branch of the same network and if the source of emf is inserted in the branch where the current was observed, it will now be found that the same value of current will now flow in the conductor connecting the points in the branch where the source of emf was first utilized.

Theorem 4. Maximum Power Transfer Theorem. If a power source is to supply at its terminals maximum power to a load impedance, then (a) the load impedance must be the conjugate of the impedance looking back into the terminals of the source or (b), if the phase angle of the load impedance cannot be changed, then the absolute values of the load impedance and the impedance looking back into the terminals of the source must be equal.

Coupled Circuits. Two circuits are coupled if energy can pass from either one to the other. Figure 5 shows simple examples of resistive, capacitive, and inductive coupling.

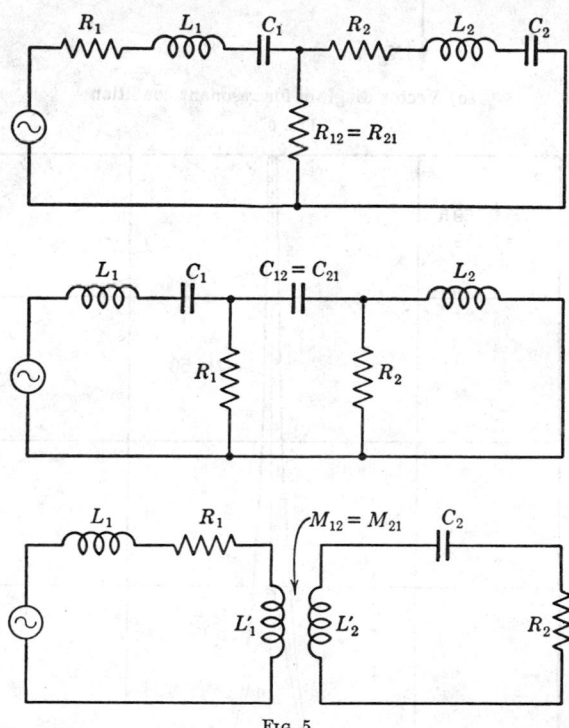

Fig. 5.

Quality Factor. The quality factor of a circuit is the ratio at resonant frequency of the stored energy per cycle to the dissipated energy per cycle. See the following paragraphs for illustrative examples.

Series Resonance. Series resonance in a circuit containing resistance, inductance, and capacitance occurs when the voltage drop across the inductance is equal to the voltage drop across the capacitance; series resonance is also characterized by unity power factor. Figure 6 illustrates a series circuit.

For practical circuit elements some resistance must be associated with the inductor and likewise some resistance with the capacitance; in each case the value of the resistance is such as to account for the power loss in the inductor or capacitor. The resistance of the

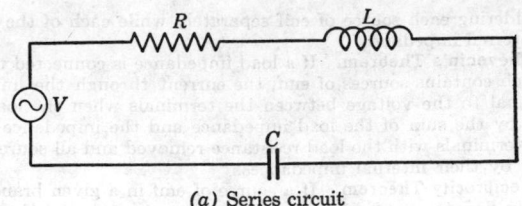

(a) Series circuit

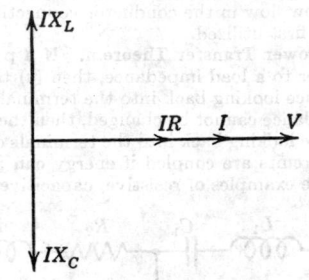

(b) Vector diagram for resonant condition

Fig. 6.

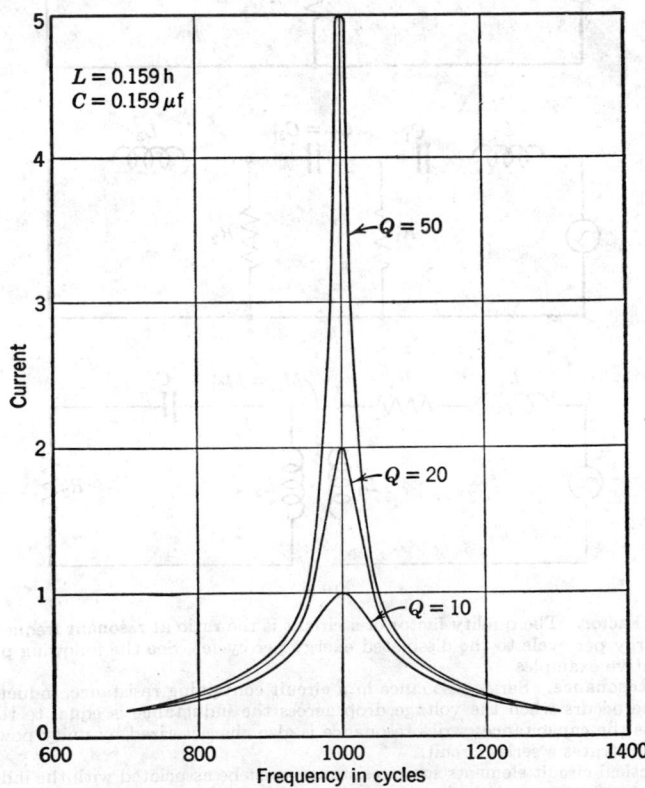

$L = 0.159$ h
$C = 0.159$ μf

$Q = 50$

$Q = 20$

$Q = 10$

Current

Frequency in cycles

Fig. 7. Resonance curves—current *vs.* cycles per sec.

capacitor can often be neglected. Figure 6 also shows the vector diagram depicting the voltage and current relationships for series resonance. The resonant frequency is obtained by equating the inductive reactance to the capacitive reactance, yielding:

$$f_0 = \frac{1}{2\pi\sqrt{LC}} \quad \text{cycles per second} \tag{54}$$

where L is the self-inductance in henrys and C is the capacitance in farads. The variation in the current of a series circuit for variation of the frequency of the applied voltage of fixed magnitude is shown in Fig. 7.

Curves have been drawn for the circuit with three different values of resistance corresponding to the indicated values of quality factor Q. High-Q series circuits are characterized by steep resonance curves.

The Q factor of the series circuit of Fig. 6 can be expressed by:

$$Q = \frac{\omega_0 L}{R} = \frac{1}{\omega_0 CR} = \frac{1}{R}\sqrt{\frac{L}{C}} \tag{55}$$

where L is in henrys, C is in farads, and ω_0 is 2π times the resonant frequency f_0.

Parallel resonance occurs when the reactive component of current in the inductive branch equals the reactive component of current in the capacitive branch. The parallel

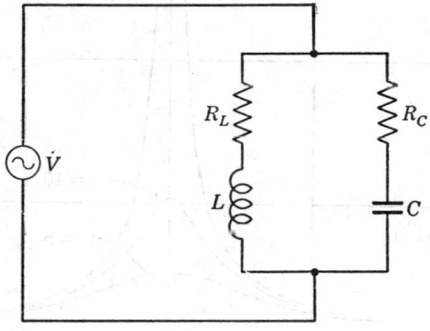

(a) Parallel circuit

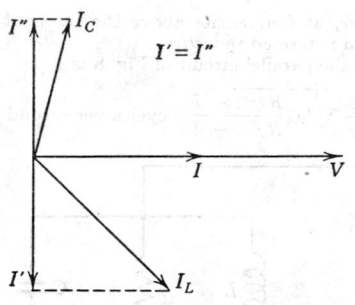

(b) Vector diagram of
resonant conditions

FIG. 8.

circuit may have resistance in the inductive branch as well as resistance in the capacitive branch. Figure 8 shows a typical circuit together with the vector diagram for the resonant condition.

At resonance the impedance of the parallel circuit is very high and is a pure resistance. The variation in the impedance of a parallel circuit with the frequency of the applied voltage of fixed amplitude is shown in Fig. 9. The maximum of impedance of a parallel circuit does not occur, in general, exactly at the resonant frequency. For circuits of high Q the frequency at which maximum impedance is obtained will be very near the frequency of resonance. At frequencies below the resonant frequency the circuit has a leading power

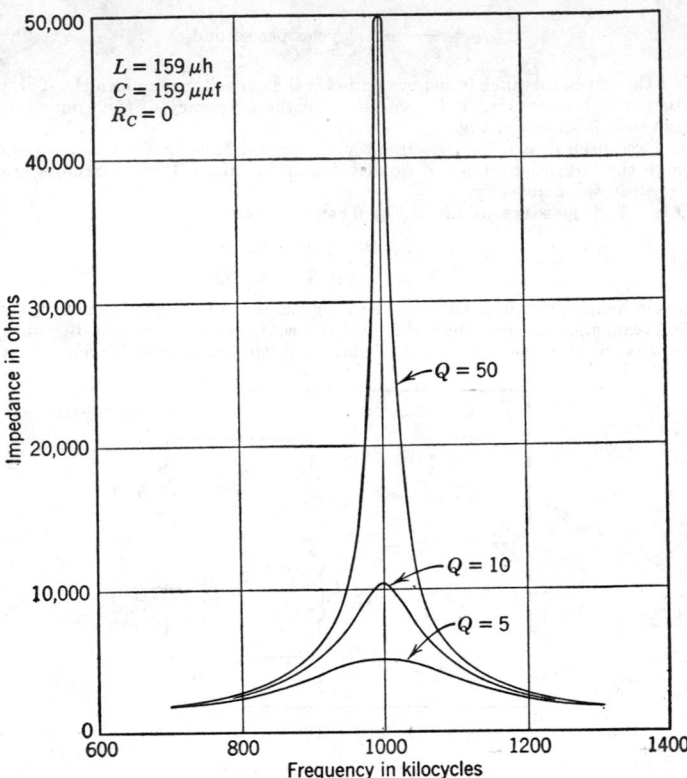

FIG. 9.   Resonance curves—impedance $vs.$ kilocycles per sec.

factor and is termed *capacitive;* at frequencies above the resonant frequency the circuit has a lagging power factor and is termed *inductive.*

The resonant frequency for the parallel circuit of Fig. 8 is

$$f_0 = \frac{1}{2\pi} \sqrt{LC \frac{R_L{}^2 C - L}{R_C{}^2 C - L}} \quad \text{cycles per second} \tag{56a}$$

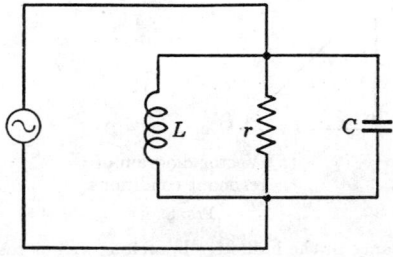

FIG. 10.

where R is in ohms, L in henrys, and C in farads. If the term under the radical is negative there is no real frequency for which resonance will occur. When R_L and R_C are equal the expression simplifies to

$$f_0 = \frac{1}{2\pi\sqrt{LC}} \quad \text{cycles per second} \tag{56b}$$

also, if $R_L = R_C = 0$ or can be considered to be so as an approximation, then the simplified form of eq. 56b may be used.

The quality factor Q_r of the series loop of the parallel circuit of Fig. 8 is given by

$$Q_r = \frac{\omega_0 L}{R_C + R_L} \tag{57}$$

ω_0 is 2π times the resonance frequency f_0.

For a parallel circuit as in Fig. 10, the resonant frequency is given by eq. 56b. The quality factor of such a circuit designated Q_p is given by

$$Q_p = r/\omega_0 L = r\sqrt{C/L} \tag{58}$$

TRANSIENTS

24. GENERAL EQUATIONS

Definition of a Transient State. The currents in an electric circuit are said to be in a transient state in the interval of time between a change in the emf or impedance conditions in the circuit and the establishment of a steady state (q.v.). During this interval, currents flow which usually decrease rapidly with time. Some transient currents persist for a comparatively long period of time.

Indicial Admittance. The current response in a mesh of an electric network initially in equilibrium to a unit emf applied to a pair of terminals in the given mesh is called the *input indicial admittance* of the network at those terminals. The current response in another mesh of the network under the same conditions is called the *transfer indicial admittance* from the mesh in which the emf is inserted to the mesh in which the current is considered. The indicial admittance consists of a sustained or steady-state term plus a transient term, the boundary conditions for the transient term being those of equilibrium. The indicial admittance has the important property that an expression for the total current (including both sustained and transient currents) which flows in a network due to the application of an emf of any type whatever can be found when the indicial admittance is known. This expression, known as the Duhamel integral, is

$$i(t) = e(0)A(t) + \int_0^t A(t - \lambda)e'(\lambda)\, d\lambda \tag{1}$$

or the alternative forms

$$i(t) = \frac{d}{dt}\int_0^t A(t - \lambda)e(\lambda)\, d\lambda$$

or

$$i(t) = \frac{d}{dt}\int_0^t A(\lambda)e(t - \lambda)\, d\lambda$$

in which $i(t) =$ the total current in the mesh; $e(t) =$ the emf impressed in the same mesh or any other mesh; $A(t) =$ indicial admittance or transfer indicial admittance as the case may be; $\lambda =$ a variable of integration.

Unit Function. In electric circuit problems it is often convenient to use a unit function designated $u(t)$. It is a time function that is zero for all times previous to $t = 0$, suddenly rises to the value of 1, and remains 1 for all values of time greater than 1. This function is depicted in Fig. 1.

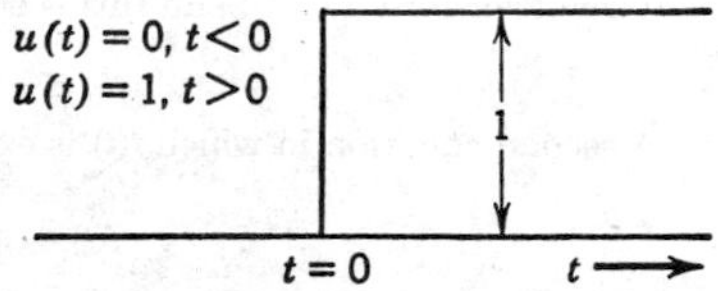

$u(t) = 0, t < 0$
$u(t) = 1, t > 0$

FIG. 1.

Unit function is often called *unit step*. It may be a unit step of voltage or current; in the case of non-electrical circuits unit steps of force or torque are often employed.

Formulation of the Transient Problem. Loop or node analysis may be used in the study of transients in a given electrical network. For example, Fig. 2 shows a simple two-mesh network. Branch currents will be considered in the loop analysis of this circuit.

Writing Kirchhoff's second circuital law

$$e(t) = R_1 i_1(t) + L_1 \frac{di_1(t)}{dt} + M \frac{di_2(t)}{dt} \tag{2}$$

$$0 = R_2 i_2(t) + \frac{q_2(t)}{C_2} + L_2 \frac{di_2(t)}{dt} + M \frac{di_1(t)}{dt}$$

but

$$q_2 = \int_0^t i_2(t) \, dt + q'_2$$

where q'_2 is the initial valve of q_2, and, therefore,

$$0 = R_2 i_2(t) + \frac{1}{C_2} \int_0^t i_2(t) \, dt + L_2 \frac{di_2(t)}{dt} + M \frac{di_1(t)}{dt} \tag{3}$$

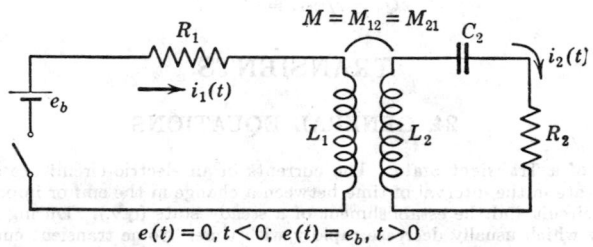

$$e(t) = 0, t < 0; \; e(t) = e_b, \, t > 0$$

FIG. 2.

Equations 2 and 3, considered as simultaneous differential equations, can be solved according to the methods of classical differential equations or by the Laplace transformation method.

In a similar way equations for a circuit of any number of meshes may be written. This is true whether steady-state or transient conditions are being considered. Equations 53 of Art. 23 illustrate the method for steady-state conditions using vector impedances corresponding to an impressed frequency.

Loop analysis may be made using either mesh or branch currents as described in Art. 22; in either case the network equations are expressions of Kirchhoff's second law.

A node analysis may also be made. Potentials of various nodes with respect to a reference node are used. Current generators are used in place of voltage generators. Admittances are used in place of impedances. The network equations resulting are then expressions of Kirchhoff's first law involving the summation of currents at each node.

Laplace Transformation Method. Transients in electric circuits often can be analyzed by means of a transformation based on the integral equation

$$F(s) = \int_0^\infty e^{-st} f(t) \, dt \tag{4}$$

where t = a real variable; $f(t)$ = a real function of t; $f(t) = 0$, $t < 0$; $F(s)$ = a function of s; s = a complex variable; and e = base of the natural logarithms. The reader is referred to Art. 50 of Section 2 for a treatment of the Laplace transformation. $F(s)$ is called the Laplace transform of $f(t)$, and this is expressed

$$\mathcal{L}[f(t)] = F(s) \tag{5}$$

A second equation in which $f(t)$ is expressed in explicit fashion is:

$$f(t) = \frac{1}{2\pi j} \int_{c-j\infty}^{c+j\infty} e^{st} F(s) \, ds \tag{6}$$

where the terminology is the same as for eq. 4 with the addition that c is chosen to the right of any singularity of $F(s)$.

Equation 4 defines uniquely a function $F(s)$ for any time function $f(t)$. Conversely, eq. 6 defines uniquely a time function, $f(t)$, for any function $F(s)$. There are certain restrictions on these statements, and the use of eqs. 4 and 6 is beyond the scope of this discussion. Table 1 lists important theorems and transform pairs. The recognition of these forms is helpful in the solution of differential equations.

Table 1. Laplace Transformation Theorems and Corresponding Functions

t function	s function
1. $u(t) = $ unit function	$1/s$
2. $af_1(t) + bf_2(t)$	$aF_1(s) + bF_2(s)$
3. $f'(t)$	$sF(s) - f(+0)$ †
4. $f^n(t)$	$s^n F(s) - s^{n-1}f(+0) - s^{n-2}f'(+0) - \cdots f^{n-1}(+0)$
5. $\int_0^t f(t)\,dt$	$\dfrac{1}{s}F(s)$
6. $e^{-at}f(t)$	$F(s+a)$
7. $f(t-a) = f_a(t)$ ‡	$e^{-as}F(s)$
8. $\dfrac{t^{n-1}}{(n-1)!}$	$\dfrac{1}{s^n}$
9. e^{-at}	$\dfrac{1}{s+a}$
10. $\sin at$	$\dfrac{a}{s^2+a^2}$
11. $\cos at$	$\dfrac{s}{s^2+a^2}$
12. $\int_0^t f_1(t-\lambda)f_2(\lambda)\,d\lambda = f_1 * f_2$ §	$F_1(s)F_2(s)$
13. $\displaystyle\sum_{n=1}^{n=k} \dfrac{p(a_n)}{q'(a_n)}e^{a_n t}$	$\dfrac{p(s)}{q(s)}$, where $q(s) = (s-a_1)(s-a_2)\cdots(s-a_k)$
14. $e^{at}\displaystyle\sum_{n=1}^{n=r} \dfrac{\gamma^{(r-n)}(a)}{(r-n)!}\dfrac{t^{n-1}}{(n-1)!} + \cdots$	$\dfrac{p(s)}{q(s)} = \dfrac{\gamma(s)}{(s-a)^r}$

† $f(|0)$ is the value of $f(t)$ as t approaches 0 through positive values.
‡ $f(t-a) = f_a(t)$ is a function which is zero for all times prior to $t = a$.
§ $f_1 * f_2$ is read the convolution of f_1 and f_2.

25. CIRCUIT STABILITY

Stable and Unstable Systems. Let a given dynamical system be supposed to be moving under the action of applied forces in some known manner described by a set of differential equations. If any small disturbing influences be applied to the system it may deviate only slightly from the previous condition of motion or it may depart from it further and further. If the deviation is slight the system is said to be *dynamically stable*, otherwise the system is *dynamically unstable*. It is clear that a given motion might be stable for one type of disturbance and unstable for another type. The disturbance will be supposed to be general in nature so that the motion will be considered stable only if it is stable for all kinds of disturbances.

An electric network in which currents flow under the action of applied emf's may be stable or unstable in the sense defined above depending upon the character of the parameters of the network.

Two overcompounded direct-current generators operating in parallel without an equalizer connection have the property that a disturbing influence which increases the emf of one machine slightly results ultimately in a considerable increase in emf and an unbalance of load which renders the system inoperative. Such a system is said to be unstable. Other examples of electric systems which may be unstable are electric arcs, vacuum-tube amplifiers or oscillators, alternators in parallel, and discharges in gas-filled tubes.

In electric power systems, protective equipment which is designed to maintain continuous operation is always provided. When such equipment is actuated as the result of some abnormal condition, the system is electrically different before and after such a device has performed its function. The differential equations are therefore changed, and a system which might have been unstable is often transformed into one in which no current or emf departs further and further from its normal value. The term stable is usually applied to a system which *with its protective equipment* will not be rendered inoperative by any small disturbance. For other special meanings ascribed to stability and not consistent with *dynamical stability* see Electric Circuits and Electric Lines in Pender's *Electrical Engineers' Handbook*.

Method of Small Oscillations. Let the variables in the system of differential equations describing a dynamical system be q_1, q_2, q_3, $\cdots$, q_n, and let the disturbances applied to the system be such that these variables are replaced by

$$(q_1 + S_1), (q_2 + S_2), \cdots, (q_n + S_n)$$

The quantities S_1, S_2, $\cdots$, S_n are initially small since the disturbance was assumed to be small. A quantity is said to be small when it is in absolute value less than some quantity λ which is such that its square may be neglected.

If, after the disturbance, S_1, S_2, $\cdots$, S_n always remain small, the motion is stable, otherwise unstable. The investigation of the stability of a given motion by a study of the behavior of the S's has proved extremely fruitful. The method is known as the *method of small oscillations.*

Criteria for Stability. An inspection of the time functions representing the transient currents in an electrical system indicates that the given system will be unstable if the characteristic determinant has:

(a) Real roots which are positive.
(b) Complex roots having positive real parts.
(c) Pure imaginary roots which are repeated.

Certain criteria for the existence of any of the above were deduced by Routh and later developed independently and in a somewhat different form by Hurwitz. These criteria are somewhat difficult to apply in practice and are applicable only to systems described by differential equations having constant coefficients. They are useful, however, in electrical systems which are approximately linear and are valuable in various other dynamical systems, such as the dynamical system of airplanes.

Test Functions of Routh and Hurwitz. Let the characteristic determinant be written in the form

$$D(\beta) = a_0\beta^m + a_1\beta^{m-1} + a_2\beta^{m-2} + \cdots + a_{m-1}\beta + a_m$$

then the conditions that there be no complex root having a positive real part or a real root which is positive are:

1. A necessary but not sufficient condition is that all the a's have the same sign.
2. A necessary and sufficient condition is that the following test functions all are positive when the equation $D(\beta) = 0$ is put in such form that a_0 is positive.

$$T_1 = a_1$$

$$T_2 = \begin{vmatrix} a_1 & a_0 \\ a_3 & a_2 \end{vmatrix} \quad T_3 = \begin{vmatrix} a_1 & a_0 & 0 \\ a_3 & a_2 & a_1 \\ a_5 & a_4 & a_3 \end{vmatrix} \quad T_4 = \begin{vmatrix} a_1 & a_0 & 0 & 0 \\ a_3 & a_2 & a_1 & a_0 \\ a_5 & a_4 & a_3 & a_2 \\ a_7 & a_6 & a_5 & a_4 \end{vmatrix} \quad T_5 = \begin{vmatrix} a_1 & a_0 & 0 & 0 & 0 \\ a_3 & a_2 & a_1 & a_0 & 0 \\ a_5 & a_4 & a_3 & a_2 & a_1 \\ a_7 & a_6 & a_5 & a_4 & a_3 \\ a_9 & a_8 & a_7 & a_6 & a_5 \end{vmatrix}$$

and so on, until a determinant which is identically zero is obtained. The total number of test functions required is in general equal to the degree of $D(\beta)$.

Sufficient Conditions. A set of conditions which are often encountered in electric circuits can be shown to be sufficient but not necessary for the system to be stable. These conditions are:

(a) All resistances, inductances, and capacitances are positive.
(b) The total resistance in any mesh is greater than the sum of the mutual resistances of the given mesh with respect to the various other meshes.
(c) The total capacitance in any mesh is greater than the sum of the mutual capacitances of the given mesh with respect to the various other meshes.
(d) The product of the self-inductances in any two meshes is greater than the square of the mutual inductance between the two meshes.

26. EXAMPLES OF SIMPLE TRANSIENTS

Transient with Resistance and Inductance. The establishment of a direct current in a coil having resistance R in ohms, and inductance L in henrys, due to the action of a constant emf E in volts, is of practical importance. Let the current in the coil at the instant of impressing the emf be I_0, and let Fig. 3a represent the circuit. By Kirchhoff's second law the following results:

$$E = L\frac{di}{dt} + iR \tag{7}$$

Separating variables gives

$$dt = \frac{L\,di}{E - iR}$$

Integrating both sides yields

$$t = \frac{L}{R} \log_e (E - iR) + K \tag{8}$$

where K is a constant of integration. The above equation can be readily changed to an equation of exponential form

$$K'e^{-(R/L)t} = E - iR \tag{8a}$$

At this point it is convenient to introduce the boundary condition $i(t) = I_0$ when $t = 0$ and evaluate the constant K'.

$$K' = E - I_0R$$

Hence, there results upon substituting in eq. 8a

$$i(t) = \frac{E}{R} - \left(\frac{E}{R} - I_0\right) e^{-(R/L)t} \tag{8b}$$

Equation 8b gives the current t seconds after closing the switch; the term E/R is the sustained current, and $(I_0 - E/R)$ is the initial transient current.

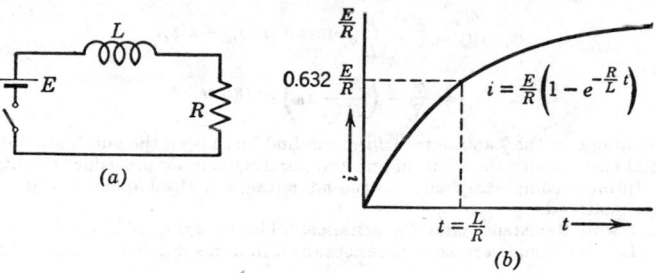

Fig. 3.

For the special case where the initial current is zero, the value of current t seconds after closing the switch is:

$$i(t) = \frac{E}{R} (1 - e^{-(R/L)t}) \tag{8c}$$

Figure 3b shows the growth of current according to the solution 8c. After an infinite time the current i would have a value E/R. In many practical circuits it can be assumed that the final value E/R is obtained in a finite time, often a matter of seconds or a fraction of a second.

The time constant of a circuit such as in Fig. 3a is the time for the current to reach $1/e$ times final value, assuming the initial current to be zero; for a simple R and L circuit the time constant T is L/R.

If E is put equal to unity, eq. 8c becomes the indicial admittance.

$$A(t) = \frac{1}{R} (1 - \epsilon^{-(R/L)t}) \tag{8d}$$

The short-circuit current of a coil which has resistance and inductance only and in which a current I_0 is flowing is given by eq. 8e when E is put equal to zero:

$$i = I_0\epsilon^{-(R/L)t} \tag{8e}$$

The result of eq. 8b may be obtained by the Laplace transformation method. Applying the battery voltage E of Fig. 3a by means of the switch at time $t = 0$ is equivalent to saying that the forcing function of the circuit is E times unit function. Equation 7 can be rewritten as

$$Eu(t) = L\frac{di}{dt} + iR \tag{7}$$

Each term of this equation in order from left to right is transformed using items 7 and 2 of Table 1 and eq. 4; there results:

$$\frac{E}{s} = sLI(s) - LI_0 + RI(s) \tag{9}$$

Solve this for $I(s)$.

$$I(s) = \frac{E/s + LI_0}{sL + R} \tag{9a}$$

Obtain the inverses using Table 1.

$$\mathcal{L}^{-1}I(s) = i(t) \tag{9b}$$

Using item 13 of Table 1:

$$\mathcal{L}^{-1}\frac{E}{s(sL+R)} = \mathcal{L}^{-1}\frac{E/L}{s\left(s+\dfrac{R}{L}\right)} = \left|\frac{E/L}{s+R/L}\right|_{s=0}e^{0t} + \left|\frac{E/L}{s}\right|_{s=-R/L}e^{-(R/L)t}$$

$$= \frac{E}{R} - \frac{E}{R}e^{-(R/L)t} \tag{9c}$$

Then using item 9 of Table 1:

$$\mathcal{L}^{-1}\frac{LI_0}{sL+R} = \mathcal{L}^{-1}\frac{I_0}{s+R/L} = I_0 e^{-(R/L)t} \tag{9d}$$

Finally, combining these individual steps yields:

$$i(t) = \frac{E}{R} - \frac{E}{R}e^{-(R/L)t} + I_0 e^{-(R/L)t}$$

$$= \frac{E}{R} - \left(\frac{E}{R} - I_0\right)e^{-(R/L)t} \tag{8b}$$

The advantages of the Laplace transform method have been the automatic introduction of the initial current into the solution and the purely algebraic procedure in obtaining the solution. In more complicated circuits the advantages of the Laplace transform method are more pronounced.

Transient with Resistance and Capacitance. The charging of a condenser of capacitance C in farads through a resistor of resistance R in ohms due to the action of a constant

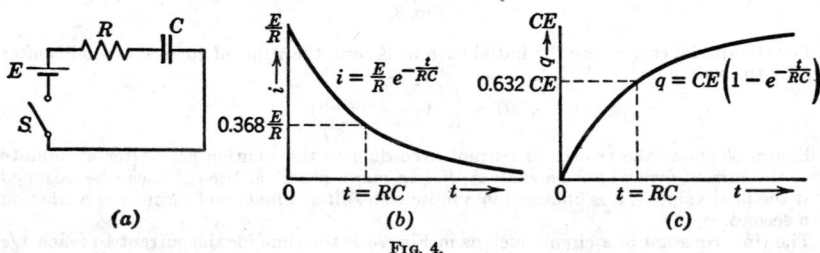

FIG. 4.

emf E in volts will be considered here. It is assumed that the conductance of the condenser and the inductance of the resistor are zero. The condenser is assumed to have a charge q_0 at the instant of closing the switch. Figure 4a describes the circuit.

Writing Kirchhoff's second law yields:

$$E = iR + \frac{q}{C} \quad \text{volts} \tag{10}$$

But $i = dq/dt$; hence

$$E = R\frac{dq}{dt} + \frac{q}{C} \quad \text{volts} \tag{10a}$$

Separating variables gives

$$\frac{dt}{CR} = \frac{dq}{CE - q} \tag{10b}$$

Integrating,

$$\frac{t}{CR} + K = -\log_e (CE - q) \tag{11a}$$

where K is a constant of integration.

Equation 11a may be written in exponential form.

$$e^{-t/RC}e^{-K} = CE - q \tag{11b}$$

Let
$$K' = e^{-K}$$
$$K'e^{-t/RC} = CE - q \tag{11c}$$
At this point the initial condition $q = q_0$ at $t = 0$ is introduced and K' is evaluated.
$$K' = CE - q_0$$
Inserting this value of K' in eq. 11c and rearranging results in
$$q(t) = CE - (CE - q_0)e^{-t/RC} \quad \text{coulombs} \tag{11d}$$
For the special case where the initial charge is zero the charge t seconds after closing the switch is given by
$$q(t) = CE(1 - e^{-t/RC}) \quad \text{coulombs} \tag{11e}$$
By differentiating eq. 11d an expression for the current results
$$i(t) = \frac{E}{R} e^{-t/RC} - \frac{q_0}{RC} e^{-t/RC} \quad \text{amperes} \tag{12a}$$
The current for zero initial charge is
$$i(t) = \frac{E}{R} e^{-t/RC} \quad \text{amperes} \tag{12b}$$

The time for the current of eq. 12b to decrease to $1/e$ times the initial value E/R is RC, the value of the time constant T for the circuit.

If E is put equal to unity, eq. 12b becomes the indicial admittance.
$$A(t) = \frac{1}{r} \epsilon^{-(1/RC)t} \tag{12c}$$

The discharge of a condenser, initially charged to a potential $V_0 = q_0/C$, through a resistor is given by eq. 12b when E is put equal to zero. The current is
$$i = -\frac{V_0}{R} \epsilon^{-(1/RC)t} \tag{12d}$$

In Fig. 4 are shown representations of the variations defined by eqs. 11e and 12b. The Laplace transform solution for Fig. 4 is as follows:
$$E = iR + \frac{q}{C} \quad \text{volts} \tag{10}$$
But $q = \int_0^t i\, dt + q_0$, and the battery voltage is suddenly applied at $t = 0$; therefore
$$Eu(t) = iR + \int_0^t i\, dt + \frac{q_0}{C} \tag{13}$$
Taking transforms of each term,
$$\frac{E}{s} = RI(s) + \frac{1}{Cs} I(s) + \frac{q_0}{Cs} \tag{13a}$$
Solving for $I(s)$,
$$I(s) = \frac{(E/s - q_0/Cs)Cs}{RCs + 1} \tag{13b}$$
$$= \frac{E/R}{s + 1/RC} - \frac{q_0/RC}{s + 1/RL}$$
Inverting yields
$$i(t) = \frac{E}{R} e^{-t/RC} - \frac{q_0}{RC} e^{-t/RC} \quad \text{amperes} \tag{12a}$$

Using eq. 10a and transforming, it would be possible to obtain a solution for $q(t)$.

Transient with Resistance, Inductance, and Capacitance in Series. If a single series circuit containing parameters R in ohms, L in henrys, and C in farads is closed upon an emf E in volts at the instant $t = 0$, and if the charge on the condenser is q_0 and the current through the circuit is I_0 at $t = 0$, the current t seconds after the switch is closed is
$$i = I_0 \epsilon^{-\alpha t} \cos \omega t - \left[\frac{\alpha}{\omega} I_0 + \frac{\alpha^2 + \omega^2}{\omega} (q_0 - CE) \right] \epsilon^{-\alpha t} \sin \omega t \tag{14}$$

and the charge on the condenser is

$$q = CE + \frac{I_0 + \alpha(q_0 - CE)}{\omega} \epsilon^{-\alpha t} \sin \omega t + (q_0 - CE) \epsilon^{-\alpha t} \cos \omega t \qquad (14a)$$

in which

$$\alpha = \frac{R}{2L} \qquad (14b)$$

$$\omega = \sqrt{\frac{1}{LC} - \frac{R^2}{4L^2}} \qquad (14c)$$

If the initial conditions ($I_0 = 0$, $q_0 = 0$) are imposed and E is put equal to unity in eq. 14 the indicial admittance is obtained

$$A(t) = \frac{1}{\omega L} \epsilon^{-\alpha t} \sin \omega t \qquad (15)$$

The term $\epsilon^{-\alpha t}$ appearing in eqs. 14 and 15 produces a decrease with time of the function which it multiplies; it is called the *damping factor* of the current or charge in which it appears. The constant ω is the natural angular velocity of the network. This constant can have a real or imaginary value depending upon the relative values of $1/LC$ and $R^2/4L^2$. When the resistance is extremely small α may be negligible in comparison with $1/LC$ and ω may be taken as $1/\sqrt{LC}$. This value of ω is called the undamped angular velocity of the circuit. If the resistance is increased, the amount of damping may be increased to such an extent that ω is zero. This condition is known as *critical* damping and is the borderline between oscillatory currents and non-oscillatory currents. If the resistance is increased further, overdamping results, ω becomes imaginary, and the current and charge are represented by hyperbolic functions instead of by trigonometric functions of time.

BIBLIOGRAPHY

1. A.I.E.E., *Elements of Nucleonics for Engineers*, a series of 11 articles, March, 1949.
2. A.I.E.E., *Report on Proposed American Standard Definitions of Electrical Terms*, 1932.
3. ATTWOOD, STEPHEN S., *Electric and Magnetic Fields*, New York, John Wiley, 1949.
4. BOHR, NIELS, "On the Constitution of Atoms and Molecules," *Philosophical Magazine*, Vol. 26, p. 476, 1913.
5. BOZORTH, R. M., "Present Status of Ferromagnetic Theory," *Elec. Eng.*, Vol. 54, p. 1251, November, 1935.
6. BRAGG, W. H., *An Introduction to Crystal Analysis*, New York, Van Nostrand, 1928.
7. BUSH, V., *Operational Circuit Analysis*, New York, John Wiley, 1929.
8. CAMPBELL, N. R., *Modern Electrical Theory*, New York, Macmillan, 1922.
9. CARSON, J. R., *Electric Circuit Theory and the Operational Calculus*, New York, McGraw-Hill, 1927.
10. CHURCHILL, RUEL V., *Modern Operational Mathematics in Engineering*, New York, McGraw-Hill, 1944.
11. COMPTON, A. H., *X-rays and Electrons*, New York, Van Nostrand, 1926.
12. COMPTON, K. T., and I. LANGMUIR, "Electrical Discharges in Gases, Part I," *Rev. Modern Phys.*, Vol. 3, p. 191, 1931.
13. DAHL, O. G. C., *Electric Circuits, Theory and Applications*, New York, McGraw-Hill, 1928.
14. DARROW, K. K., "Statistical Theories of Matter, Radiation and Electricity," *Rev. Modern Phys.*, Vol. 1, p. 90, 1929.
15. FARADAY, M., *Experimental Researches in Electricity*, 3 vols., London, R. and J. E. Taylor, 1839-1855.
16. FOWLER, R. H., "Thermionic Emission Constant A," *Proc. Roy. Soc.*, Vol. A122, p. 36, 1929.
17. FRENKEL, J., *Elementary Theory of Wave Mechanics*, New York, Oxford University Press, 1932.
18. GARDNER, M. F., and J. L. BARNES, *Transients in Linear Systems*, Vol. 1, New York, John Wiley, 1942.
19. GOLDMAN, STANFORD, *Transformation Calculus and Electrical Transients*, New York, Prentice-Hall, 1949.
20. HAAS, A., *Wave Mechanics and the New Quantum Theory*, London, Constable, 1928.
21. HAGUE, B., *Electromagnetic Problems in Electrical Engineering*, London, Oxford University Press, 1929.
22. HEAVISIDE, O., *Electromagnetic Theory*, 5 vols., London, Benn Bros., 1925.
23. HEAVISIDE, O., *Electrical Papers*, 2 vols., Boston, Copley, 1925.
24. HUGHES, A. L., and L. A. DU BRIDGE, *Photoelectric Phenomena*, New York, McGraw-Hill, 1932.
25. JEANS, J. H., *Mathematical Theory of Electricity and Magnetism*, Cambridge, England, University Press, 1908.
26. KRONIG, R. DE L., "Theory of Supraconductivity, Part II," *Zeits. Physik*, Vol. 80, p. 203, 1933.
27. LANGMUIR, I., and K. T. COMPTON, "Electrical Discharges in Gases, Part II," *Rev. Modern Phys.*, Vol. 3, p. 191, 1931.
28. LINFORD L. B., "Recent Developments in the Study of the External Photoelectric Effect," *Rev. Modern Phys.*, Vol. 5, p. 34, 1933.
29. LORENTZ, H. A., *The Theory of Electrons*, Leipzig, B. G. Teubner, 1909.
30. MAXWELL, J. C., *A Treatise on Electricity and Magnetism*, 2 vols., London, Oxford University Press, 1904.
31. MILLIKAN, R. A., *The Electron*, University of Chicago Press, 1924.
32. PENDER, H., and S. R. WARREN, JR., *Electric Circuits and Fields*, New York, McGraw-Hill, 1943.
33. ROENTGEN, WILHELM CONRAD, "On a New Form of Radiation," *Electrician*, Vol. 36, p. 415, January 24; p. 850, April 24, 1896.
34. RUSSELL, A., *The Theory of Alternating Currents*, Cambridge, England, University Press, 1914.
35. RUTHERFORD, E., *Radio-activity*, Cambridge, England, University Press, 1904.
36. SOMMERFELD, A., *Atomic Structure and Spectral Lines*, London, Methuen, 1934.
37. STARLING, S. G., *Electricity and Magnetism for Advanced Students*, London, Longmans, Green, 1924.
38. STEINMETZ, C. P., *Transient Electric Phenomena and Oscillations*, New York, McGraw-Hill, 1920.
39. TAYLOR, H. S., *A Treatise on Physical Chemistry*, Chapter XI by J. R. Partington, New York, Van Nostrand, 1942.
40. THOMPSON, S. P., *Elementary Lessons in Electricity and Magnetism*, New York, Macmillan, 1915.
41. THOMSON, J. J., *Elements of Electricity and Magnetism*, New York, Macmillan, 1921.
42. THOMSON, J. J., and G. P., *Conduction of Electricity through Gases*, London, MacMillan, 1933.
43. VAN VLECK, J. H., *The Theory of Electric and Magnetic Susceptibilities*, London, Oxford University Press, 1932.
44. VON LAUE, M., "Phénomenes d'interference des rayon de Roentgen," *Le Radium*, Vol. 10, p. 47, 1913.
45. WEBSTER, A. G., *Theory of Electricity and Magnetism*, London, Macmillan, 1897.
46. WILSON, A. H., "Theory of Electronic Semi-Conductors, Part II," *Proc. Roy. Soc.*, Vol. A134, p. 277, 1931.
47. ZWORYKIN, V. K., and E. G. RAMBERG, *Photoelectricity and Its Application*, New York, John Wiley, 1949.

SECTION 10

ELECTRONICS

BY

JOSEPH TUSINSKI

ELECTRONICS

1. DEFINITION

An outgrowth in the evolution of electrical engineering, electronics had been limited to electrical conduction by charged particles in circuitry where Ohm's law does not apply. This concept is no longer accepted in modern technology, and a specific boundary does not confine the application. The field of electronics engineering overlaps into many specialized areas so vast that it would require several pages to give only a cursory treatment. The fundamental aspects of electronics, however, are common to many of the specialized fields and can be defined simply as *the control of the fundamental charge* with the definition extended to include the lack of the electronic charge as in semiconductor devices.

The basic physics of electronics is presented under Electron Theory in Section 9, Electricity and Magnetism.

2. ELECTRON TUBES

Emission phenomena (see Art. 4, Sec. 9) form the basis of electron tubes in general. The various tubes are classified by the number of electrodes, purposes, or electronic characteristics utilized.

The term *vacuum tube* encompasses glass or metal bulbs that have been evacuated or filled with an inert gas to alter the conduction process. These latter tubes are classed as gas-filled types. The "soft" tube must be differentiated from a gas tube. "Soft" refers to a high-vacuum that has become contaminated by a foreign element, such as air. Softness generally manifests itself as a purplish glow inside the vacuum tube. At times some types of glass will fluoresce when bombarded by electrons; however, this effect should not be construed to indicate a "soft" tube.

Diodes. The diode (hot cathode, cold anode) represents the simplest form of an electron tube.

The source of the electrons is the cathode that liberates a copious number of electrons when excited by some form of energy. Heat, light, and a high electric field are the usual forms of energy which may be applied to the cathode's surface. The electrons, being negative, have a mutual repulsion, hence a definite number will be capable of occupying the space in the vicinity of the cathode. A state of equilibrium will develop to form a dense area near the cathode that is called the *space-charge* region.

The ultimate objective is to collect or control the number of electrons through an external circuit. The second electrode in the diode collects the electrons and is called an anode or plate. A positive voltage applied to the plate with respect to the cathode will attract the electrons that become the circuit current flow (i_p). However, by virtue of the space-charge, all of the electrons are not collected. The approximate amount of current (electron) flow is influenced by the magnitude of the plate voltage (and cathode tempei a-ture) and follows a three-halves power law, at times called Child's law.* Plate current can be approximated by the equation

$$i_p = K e_p^{3/2} \tag{1}$$

The constant K depends on the geometry of the elements within the tube, such as cathode area and spacing.

When a negative potential is applied to the plate with respect to the cathode, the field will repel the electrons and circuit current flow ceases. This natural phenomena is the basis of rectifiers (a-c to d-c converters) and other unidirectional circuit types. Some attributes of the diode are depicted in Fig. 1.

It should be noted that the curve near the dotted section has square-law characteristics. This factor is utilized at times in devices or circuits were the objective is to measure some function that follows a square-law. Second, it is shown that when e_p is zero, a small current can exist. That is, the initial energy imparted to the electron, plus the repulsion by electrons in the negative space charge region, force electrons to the plate. If the plate circuit is returned to the cathode, a measurable amount of current will flow. Normally this current can be considered insignificant for most applications.

Triodes are tubes wherein a third electrode (grid) is placed in the envelope between the cathode and plate. The grid is composed of fine wires somewhat like a coil spring

---

* C. B. Child, *Phys. Rev.*, Vol. 32, p. 492, May 1911.

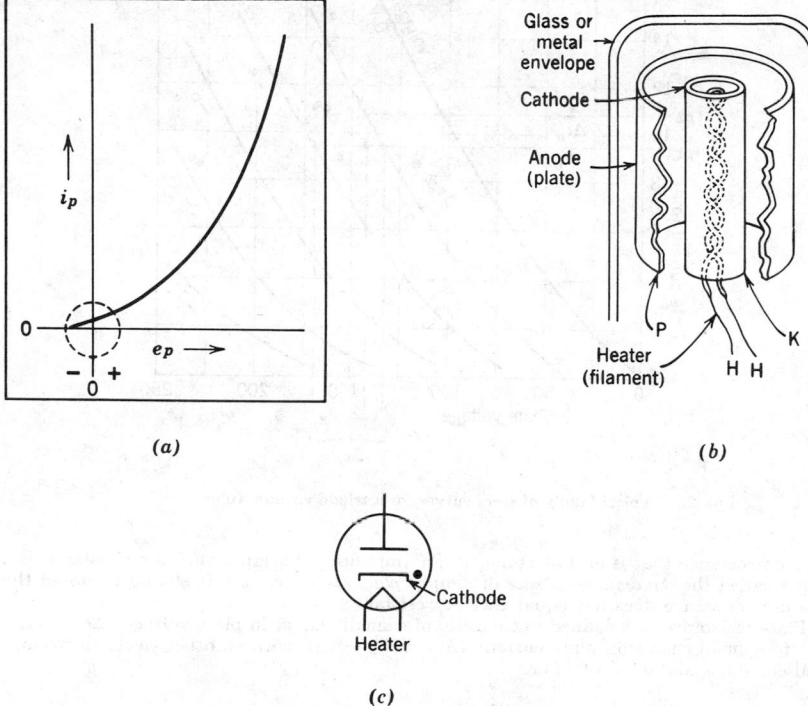

(a)

(b)

(c)

Fig. 1. (a) Typical vacuum-diode i_p–e_p characteristic curve. The dotted section has been exaggerated to show that a current will flow when e_p is zero (see text). (b) H-H heater connections. K is the cathode and P is the plate connection. (c) Schematic diagram of a diode. The black dot represents a gas tube.

encircling the cathode; however, other physical configurations may be employed. The prime factor is that the grid structure occupies the space between the cathode and plate. By applying voltages to the grid, the electrostatic field influences the amount of plate current. With fixed discrete changes of grid voltages, a family of curves is obtained. A typical set of these curves is shown in Fig. 2.

Operating Coefficients. There are three coefficients, similar to Ohm's law ($E = IR$), that are common to most tubes. These coefficients may be obtained from the tube's characteristic e_p–i_p curves in the vicinity of the desired operating point.

The *amplification factor* μ is defined as the ratio of a small change in output voltage ($\Delta e_p = e_{p3} - e_{p1}$) to a small change of input voltage ($\Delta e_g = e_{c2} - e_{c1}$), with all other electrode voltages and currents maintained constant. For the triode curves shown in Fig. 2, this may be expressed by the equation,

$$\mu = \frac{\partial i_p/\partial e_g}{\partial i_p/\partial e_p} \cong \frac{\Delta e_p}{\Delta e_g}\bigg| \Delta i_p = \text{zero (dimensionless)} \qquad (2)$$

From this, μ may be thought of as the tube's voltage characteristic.

The tube's *transconductance* g_m is defined as the ratio of a small change of plate current $\Delta i_p = (i_{p2} - i_{p1})$ to a small change of grid voltage ($\Delta e_{g2} = e_{c2} - e_{c1}$), with all other electrode voltages and currents maintained constant. Thus g_m may be thought of as the current characteristic of the tube in a broad sense and may be expressed by

$$g_m = \frac{\Delta i_p}{\Delta e_{g2}}\bigg| \Delta e_p = \text{zero (mhos)}$$

The resistive property of the tube has a dual nature. That is, without a signal, d-c components of E_p and I_p will yield a d-c *static-resistance*. However, it is the variational

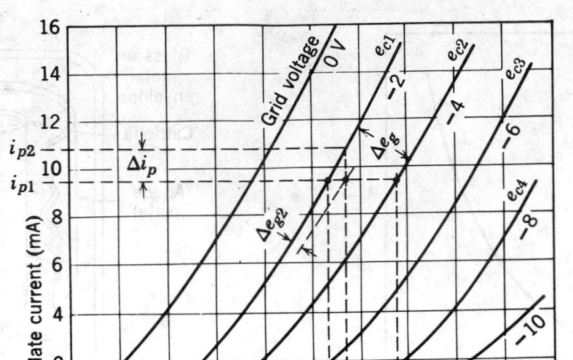

FIG. 2.   Typical family of i_p–e_p curves for a triode vacuum tube.

or a-c resistance that is of importance in an amplifier.   Variational or a-c resistance is at times called the *dynamic resistance* or simply *plate resistance* r_p.   It should be noted that the d-c resistance does not equal the a-c resistance.

Plate resistance r_p is defined as the ratio of a small change in plate voltage ($\Delta e_p = e_{p2} - e_{p1}$) to a small change in plate current ($\Delta i_p = i_{p2} - i_{p1}$), with all other electrode voltages and current maintained constant:

$$r_p = \frac{\Delta e_p}{\Delta i_p}\bigg|\; \Delta e_g = \text{zero (ohms)} \tag{4}$$

From these three basic equations the interrelationship is expressed as

$$\mu = g_m r_p \tag{5}$$

Normally the control grid is operated with an average negative potential; however, in some cases it can be operated with positive potentials with respect to the cathode.   With negative potentials on the grid, the plate current may be expressed approximately as

$$i_p \cong K(e_p + \mu e_g)^{3/2}$$

where i_p = plate current.
   K = physical geometric constant.
   e_p = plate voltage.
   e_g = grid voltage.

The coefficients μ, g_m, and r_p vary considerably, and it should be emphasized that the derived values hold only in the operating region from which they were obtained.   Figure 3a shows how the dynamic plate resistance is graphically determined.   Figure 3b depicts a typical variation of the coefficients of a triode.

Grid Bias.   In order to maintain an average negative potential on the grid when a signal is applied, a d-c voltage is inserted in series with the signal voltage.   This voltage E_c is called bias and it determines the operating region for a given value of plate supply voltage E_b.   From Eq. 2 it can be seen that the grid voltage is μ times as effective in controlling plate current as is the plate voltage.   In order to reduce the plate current to zero (cut-off), the approximate grid bias may be computed from

$$-E_c = \frac{E_p}{\mu} \quad \text{(plate current cut-off)} \tag{6}$$

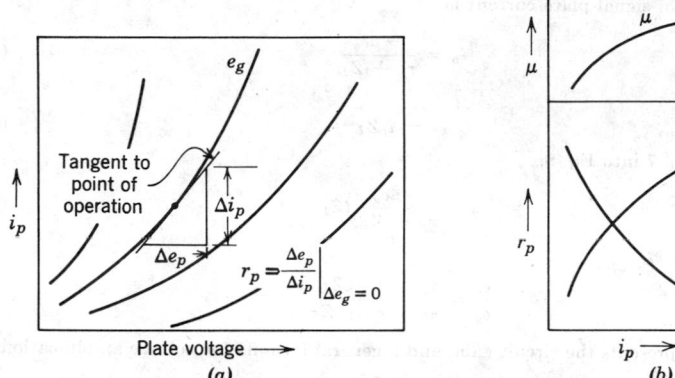

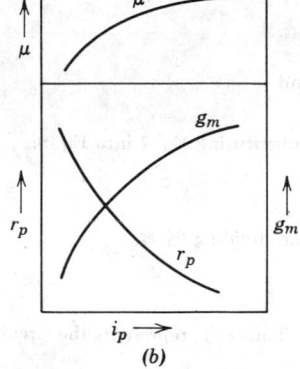

Plate voltage ⟶

(a)

i_p ⟶

(b)

FIG. 3. (a) Graphically determining r_p from a set of characteristic curves. (b) Variations of a triode's coefficients as a function of the plate current.

Amplifier Characteristics. The plate current flows through an impedance Z_L in the output circuit, and for a circuit where the cathode is common to the input (grid) and the output (plate), the configuration is called conventional or a *common-cathode amplifier*. A schematic diagram of a common-cathode amplifier is shown in Fig. 4a.

If the schematic diagram of Fig. 4a is Thévenized, the circuit of Fig. 4b evolves. Note that all d-c operating voltages do not appear in the equivalent circuit and that e_g is equal to e_i and e_p is equal to e_0. The minus sign of μe_g does not have an algebraic connotation; however, it does mean that a *phase-inversion* of the output voltage e_0 takes place.

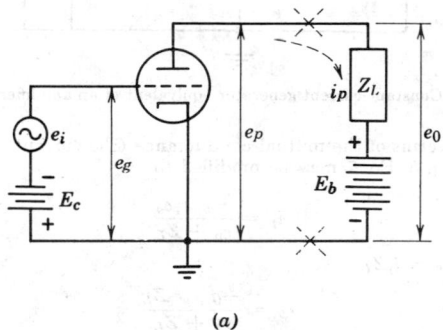

(a)

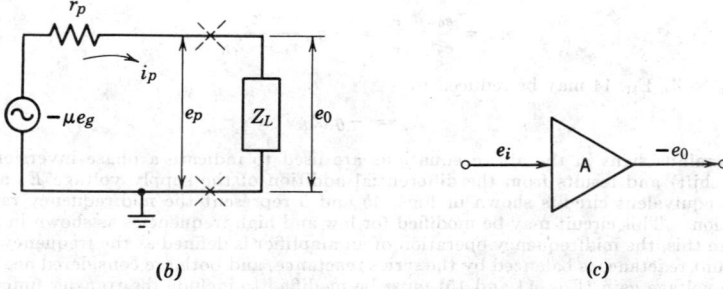

(b)

(c)

FIG. 4. (a) Schematic diagram of a typical common-cathode amplifier. (b) Amplifier's constant-voltage generator equivalent circuit. (c) Symbolic representation of an amplifier.

The amount of signal plate current is

$$i_p = \frac{-\mu e_g}{r_p + Z_L} \tag{7}$$

and

$$e_0 = i_p Z_L \tag{8}$$

Substituting Eq. 7 into Eq. 8,

$$e_0 = \frac{-\mu e_g}{r_p + Z_L} (Z_L) \tag{9}$$

and dividing by e_g,

$$\frac{e_0}{e_g} = \frac{e_0}{e_i} = \frac{-\mu Z_L}{r_p + Z_L} \tag{10}$$

Thus e_0/e_i represents the circuit gain, and a general formula for voltage amplification is

$$A_v = \frac{-\mu Z_L}{r_p + Z_L} \tag{11}$$

The same circuit may be represented by a constant-current form (Norton's theorem) as shown in Fig. 5.

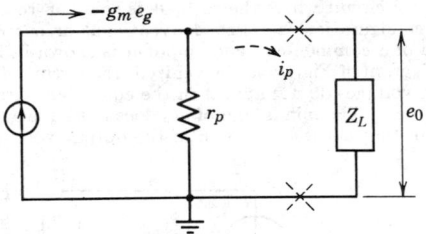

FIG. 5.   Constant-current generator equivalent of an amplifier.

A gain formula in terms of the mutual conductance (the current characteristic) may be developed. From Eq. 5, Eq. 7 may be modified to

$$i_p = \frac{-g_m r_p e_g}{r_p + Z_L} \tag{12}$$

and substituting in $e_0 = i_p Z_L$

$$e_0 = \frac{-g_m r_p e_g Z_L}{r_p + Z_L} \tag{13}$$

Dividing by e_g,

$$A_v = \frac{e_0}{e_g} = \frac{e_0}{e_i} = -g_m \frac{r_p Z_L}{r_p + Z_L} \tag{14}$$

If $r_p \gg Z_L$ Eq. 14 may be reduced to

$$A_v = -g_m Z_L \tag{15}$$

The minus signs in the above equations are used to indicate a phase inversion (not phase shift) and results from the differential addition of the supply voltage E_b and e_L.

The equivalent circuits shown in Figs. 4b and 5 represent the midfrequency range of operation.   This circuit may be modified for low and high frequencies as shown in Fig. 6.

From this, the midfrequency operation of an amplifier is defined as the frequency where the shunt reactance is balanced by the series reactance, and both are considered negligible.

The voltage gain (Eqs. 11 and 15) must be modified to include the transfer function of the significant reactive elements. However, the low-frequency response is normally

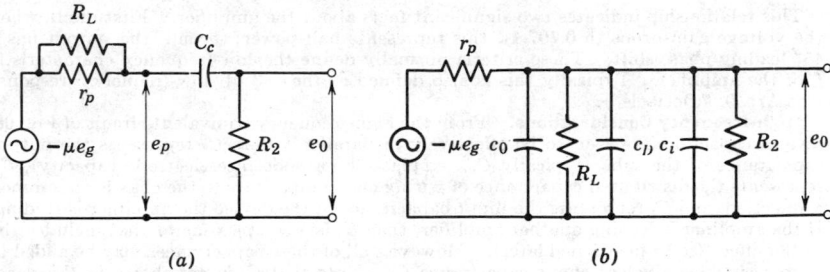

FIG. 6. (a) Low-frequency equivalent of an R-C coupled amplifier. (b) High-frequency equivalent of an R-C coupled amplifier.

defined as the frequency at which the series reactance is equal to the equivalent series resistance. If $R_2 \gg r_p R_L/(r_p + R_L)$, then the decrease in output voltage is determined primarily by C_c and R_2. Specifically, e_p may be treated as a source for the high-pass network C_c and R_2. The circuit is capacitive, hence a leading current causes e_0 to lead the applied volage e_p. Also, e_p is an inverted form of $e_i = e_g$. Graphically, the phasors would be represented as shown in Fig. 7.

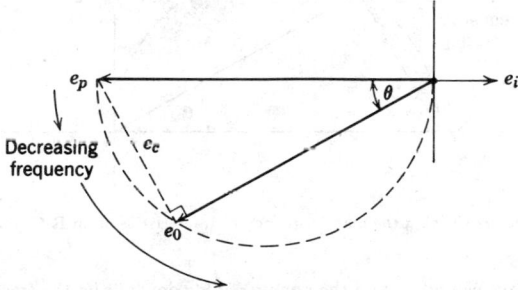

FIG. 7. Vector representation of the low-frequency characteristics of an R-C coupled amplifier (see Fig. 6).

The phase-angle θ (Fig. 7) represents the amount of leading phase-shift suffered by e_p. The normal 180° inversion of e_p is considered as a zero phase shift.

In general the low frequency transfer function K_{lf} may be expressed in several forms,

$$K_{lf} = \frac{e_0}{e_p} = \frac{1}{1 - j/\omega R_2 C_c} = \frac{R_2}{\sqrt{R_2{}^2 + X_c{}^2}} = \frac{1}{\sqrt{1 + (X_c/R_2)^2}}$$

where phase-shift $\theta = \tan^{-1}(X_c/R_2)$
$$= \tan^{-1} 1/(\omega C_c R_2)$$

and $\omega = 2\pi f$ and the product $R_2 C_c$ is expressed as the circuit time-constant.

The low-frequency gain of an R–C coupled amplifier may then be expressed as

$$A_{v-lf} = K_{lf} A_v \tag{16}$$

A special case exists when $R_2 = 1/\omega C_c$, assuming as before that $R_2 \gg r_p \parallel R_L$. The transfer function is

$$K_{fl} = \frac{1}{\sqrt{2} \underline{/-45°}}$$

and stage gain or amplification is

$$A_{fl} = A_v 0.707 \underline{/45°}$$

This relationship indicates two significant facts about the amplifier. First of all, when the voltage gain drops to $0.707A_v$, this represents half-power; second, the output has a 45° leading phase-shift. These criteria normally define the low-frequency characteristic f_1 of the amplifier. Typically this ls also defined as the -3 db low-frequency response (see Art. 9, "Decibels").

High-frequency Considerations. From the high-frequency equivalent circuit of Fig. 6b, several capacitors are seen to be effectively in parallel. Here C_0 represents the output capacitance of the tube (typically C_{pk} = plate to cathode interelectrode capacity), C_D represents the distributed capacitance of wiring and components to the chassis or common connection, and C_i represents the input capacitance of the device the amplifier is feeding. If the amplifier is feeding another amplifier, then C_i is a complex factor that includes the Miller effect (to be mentioned later). However, all of these capacitances may be added to form a single equivalent shunt capacitance C_t. Note that C_c is not shown in this schematic because its reactance is too low to be significant.

A similar method is used to signify the -3 db high-frequency response of an R-C coupled amplifier. That is, f_2 represents the frequency where the shunt reactance is equal to the equivalent shunt resistance of the circuit. The voltage gain is 0.707 times A_v with a 45° lagging phase-shift. From these definitions the total bandwidth (BW) of the amplifier is defined as $f_2 - f_1$. A vector representation of the amplifier operating at a high frequency is shown in Fig. 8.

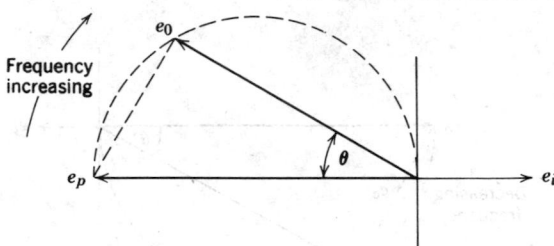

FIG. 8.   Vectors showing the high-frequency characteristics of an R-C coupled amplifier.

Note that e_0 lags behind e_p, and the amount of lag depends on the frequency of operation and the total shunt capcitance and equivalent shunt resistance.

Miller Effect (After John M. Miller, 1918). When the input of an amplifier is viewed from the generator, the input current i_i divides into two components. With reference to Fig. 9, i_1 is the current flowing into the capacitance formed by the grid-cathode C_{gk}. Second, i_2 is the current flowing into the plate-grid capacitance C_{pg}. The total signal current is then

$$i_i = i_1 + i_2 \tag{17}$$

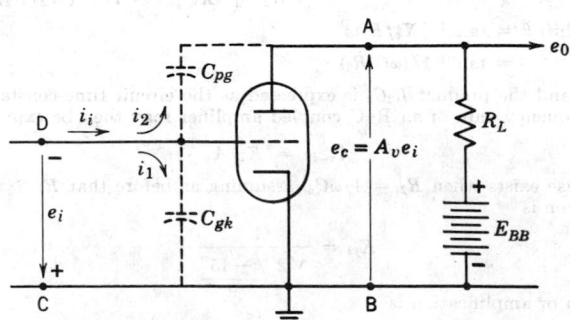

FIG. 9.   Distribution of currents in the input capacitance of a triode.

The arrows of e_i and e_0 are shown displaced by 180°, as normally represented by the phase-inversion of the given conventional amplifier. Thus if the voltages are summed from A-B → C-D, the capacitor C_{pg} sees

$$e_{C_{pg}} = A_v e_i + e_i$$

or

$$e_{C_{pg}} = e_i(1 + A_v) \tag{18}$$

If admittance is used in lieu of impedance, $(|Y| = 1/|Z|)$, the currents may be represented as,

$$i_1 = e_i Y_1 \quad \text{and} \quad e_{C_{pg}} Y_2 = e_i(1 + A_v) Y_2$$

Substituting in Eq. 17,

$$i_i = i_1 + i_2 = e_i Y_1 + e_i(1 + A_v) Y_2$$

or

$$i_i = e_i[Y_1 + (1 + A_v) Y_2]$$

Also,

$$Y_i = i_i/e_i \quad \text{and} \quad i_i = Y_i e_i$$

Substituting,

$$Y_i e_i = e_i Y_1 + (1 + A_v) Y_2$$

Dividing by e_i,

$$Y_i = Y_1 + (1 + A_v) Y_2 \tag{19}$$

Considering only the susceptive component of Y,

$$j\omega C_i = j\omega C_{gk} + j\omega C_{pg}(1 + A_v)$$

Dividing by $j\omega$,

$$C_i = C_{gk} + C_{pg}(1 + A_v) \tag{20}$$

Thus with a high amplification the input capacitance may be quite high even though C_{pg} may be small.

Normally in pentodes the plate-to-grid capacitance is quite small and the input capacitance may be simply C_{gk}. At high frequencies even this small C_{pg} must be considered.

The Miller effect is used to advantage in some devices to obtain an extremely large value of capacitance by bridging a relatively small capacitor from the output to the input of an amplifier (Miller integrator, timing circuits, operational amplifiers).

Tetrodes. By adding another grid, commonly called a *screen-grid*, some of the inherent qualities of the triode are improved. The screen grid occupies the space between the control grid and the plate. A positive voltage is applied to the screen; however, it is normally by-passed to the cathode (common) or B-minus, so that essentially the signal impedance of the screen circuit is zero. The intervening space (pitch) between wires is generally greater than the space between the control grid wires. Thus the screen grid does afford a certain degree of electrostatic shielding of the grid from the plate and reduces the plate to grid capacitance.

Second, the space charge is reduced in the normal operating region (see B, C of Fig. 10a) so that essentially the plate current becomes independent of the plate voltage. Hence higher amplification factors and plate resistances are feasible. These factors may be realized by applying the data given for triodes to the curves of Fig. 10.

A negative resistance area exists between points A and B as shown in Fig. 10a of the tetrode curves. In fact, it is possible for the plate current to reverse itself, a seemingly impossible effect. This effect may be explained by remembering that the screen grid is operated at a fixed value of positive voltage. Thus when E_p is zero, the screen current is high. As the plate voltage is increased the plate current increases gradually; however, further plate voltage increases cause higher electron velocities which cause secondary electrons to be emitted from the plate. The secondary electrons do not return to the plate but are collected by the screen. Thus the net flow from the plate by secondary emission exceeds the impinging primary electrons. Further increases of plate voltage causes the plate current to increase to a value that is determined by the magnitude of the control grid voltage. From B to C of the figure, the plate current increases very little for a large change in plate voltage.

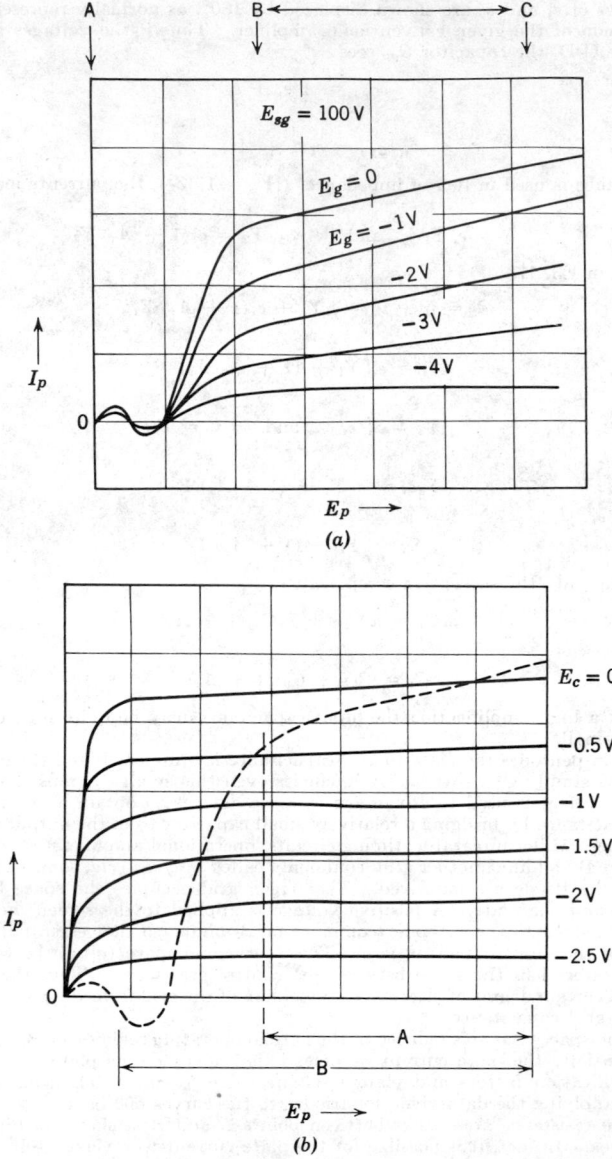

Fɪɢ. 10. (a) Family of tetrode characteristic curves. (b) Typical pentode characteristics with a single tetrode curve shown by the dashed line. Compare the operating range A of the tetrode with B of the pentode.

Pentodes. By adding a third grid (suppressor grid) between the screen and plate, the effect of the secondary emission phenomenon may be eliminated. The suppressor is operated at a potential near or at the cathode potential. Hence the suppressor field repels the secondary electrons back to the plate. A typical family of characteristic curves is shown in Fig. 10b.

The pentode is characterized by high amplification factors and plate resistance equal to or greater than 1 megohm.

By altering the pitch of the control grid wires, some operating characteristics are changed. One such change uses a grid that has a nonuniform pitch of the grid wires; that is, the wires are close together in one area and further apart in another. This type of arrangement results in a device called a remote cut-off, variable-mu tube, or a super-control tube. This type of structure is adaptable to both triode or pentodes.

3. SEMICONDUCTORS

In an intrinsic material of germanium (Ge) or silicon (SI), all of the valence electrons of an atom are shared with an adjacent atom. This covalent bonding exists until sufficient energy is imparted to an electron to cause it to leave its parent atom. This can be brought about by heat. The generation of ions increases exponentially with temperature increase. In the parlance of semiconductor physics, the removal of an electron leaves a void which is called a *hole*. The number of electrons and holes created depends on the so-called energy gap of the material (Fig. 1).

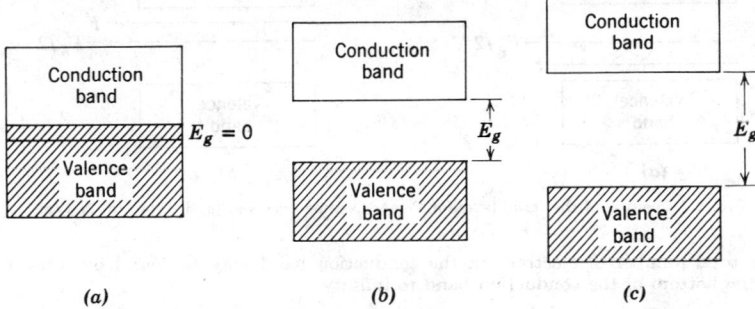

(a)   (b)   (c)

FIG. 1.   Relative energy gaps for various materials: (a) metals; (b) semiconductors; and (c) insulators.

When an electric field is applied, an electron will leave an atom and fill the previously created hole, and thus create a hole. In essence it can be said that the negative charge carriers are moving in one direction while the position charge carriers (holes) move or drift in the opposite direction. Electrons, however, have a greater mobility as shown in Table 1. The drift velocity of electrons or holes when an electric field is applied may be expressed as

$$V_d = \mu E$$

where μ = carrier mobility of electrons or holes (cm^2/V-sec).
  E = electric field (V/cm).
  V_d = drift velocity (cm/sec).

The energy gap for semiconductors is between that of metals and insulators.

Mobility. The electron or hole mobility is a measure of freedom that a charge has in moving under various influencing factors. Thermal states, crystal dislocations, and grain boundaries are a few of the determining factors. Perhaps the least understood fact is the difference between hole mobility and electron mobility. This is probably the result of thinking in terms of electron-hole pair generation in intrinsic materials. However, it must be remembered that holes have less energy than free electrons and their mean free

Table 1

	E_g (eV)	Hole mobility (cm^2/V-sec)	Electron mobility (cm^2/V-sec)
Silicon	1.1	500	1400
Germanium	0.7	1900	3900

paths as well as their translational velocities are lower. That is, holes are still attached to atoms and thus are not as mobile. However all considerations given to an electron must be given the hole as to a general drift in one direction when an electric field is imposed.

Note that Ge and Si are both in Group IV of the periodic chart. If a Group III material (Al, Ga, In) is added to a crystal of Ge or Si, an electron is captured from the Group IV crystal to create a hole. Thus the Group III material "accepted" an electron from the intrinsic crystal and a "P"-type crystal is formed. Similarly if a Group V (P, As) material is added to a Group IV crystal, a free electron is created after the octet of valence electrons has been satisfied. Thus the Group V material donated (donor) an electron and an "N"-type crystal is formed.

Fermi Level. Simply stated, the Fermi level (F) may be defined as the energy level at which there is a 50% probability of an electron occupancy. Thus in an intrinsic material, the Fermi level is equal to $E_g/2$. It follows then that a greater probility of finding an electron in an N-type crystal would be near the conduction band. Conversely the Fermi level would be closer to the valence band for a P material (Fig. 2).

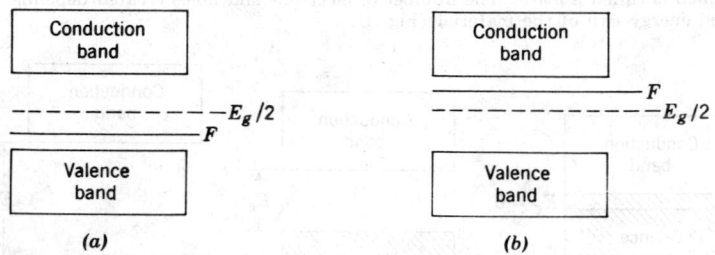

(a)   (b)

FIG. 2. Position of the fermi level for P- and N-type crystals: (a) P-type; (b) N-type.

The total number of electrons in the conduction band may be found by integrating from the bottom of the conduction band to infinity:

$$\#e = \int_{Ec}^{\infty} n(E)\epsilon^{-(E-F)/kT}\,dE \tag{1}$$

where $n(E)$ = the number of energy states.
  E = energy state.
  F = Fermi level.
  k = Boltzmann's constant.
  T = absolute temperature ($273 + °C$).
  dE = energy interval.
  $\#e$ = total number of electrons.

The number of electrons created in the conduction band will influence the conductivity ($1/\rho$), and an expression for conductivity is

$$\sigma = e(P\mu_p + N\mu_n) \tag{2}$$

where σ = conductivity (mhos/cm).
  $e = 1.6 \times 10^{-19}$ coulomb.
  μ_p = hole mobility.
  P = hole density.
  N = electron density.
  μ_n = electron mobility.
  ρ = resistivity.

For example, pure (intrinsic) germanium at room temperature exhibits a conductivity of approximately 0.020 mho/cm, whereas silicon at room temperature has a conductivity of 5.2×10^{-6} mho/cm.

Minority and Majority Carriers. When an intrinsic semiconductor is at a temperature greater than absolute zero, electron-hole pairs are created. By doping the intrinsic crystal with an impurity atom, it has been shown that N- and P-type materials are created. In an N-type crystal, electrons are the majority carriers; however, holes are present by

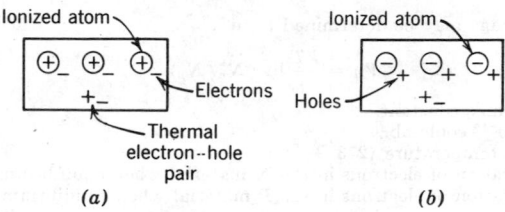

Fig. 3. (a) N-type; (b) P-type.

virtue of the inherent ionization process. The holes in this case would be termed the minority carriers. Conversely in a P material, holes are the majority carriers and electrons are the minority carriers. On a very small scale, the distribution of charges may be graphically depicted as shown in Fig. 3.

The effect of applying an external potential will cause the appropriate free charges to move. The arrows attached to the signs show the direction of charge drift (Fig. 4). Note that the majority of charges moving in the N-type are the electrons and holes in the P-doped material.

P-N Junction. When a *single* crystal is doped N on one end and P on the other end, the intersection of the two doped regions is called a junction. It must be emphasized that the material represents a single crystal, not two joined crystals. Using the methods of Figs. 3 and 4, when a PN junction is formed a diffusion of charges at the junction takes place as shown in Fig. 5. The free electrons in the N side diffuse into the P side, neutralizing an existing hole. This results in a positively charged ionized atom in the N section. Conversely a hole can neutralize an electron in the N side, leaving a negative charge in the P section. Further migration of electrons into the P material is prevented by the negative charge at the junction of the P material. That is, an electron must surmount this potential barrier or hill; however, it will require additional energy to do so.

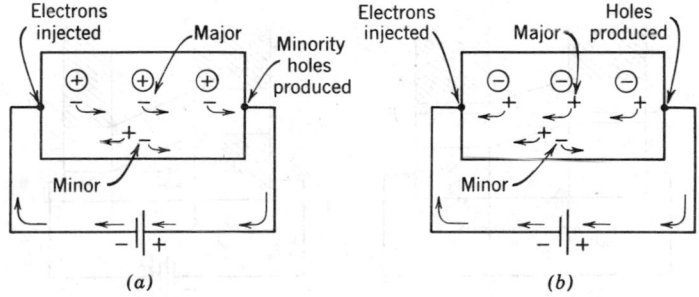

Fig. 4. Majority and minority charge movement with an applied field: (a) N-type; (b) P-type.

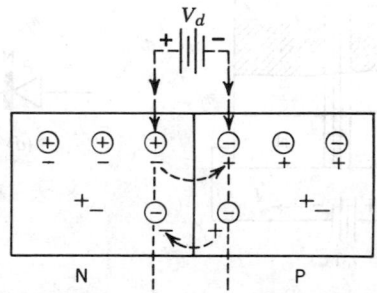

Fig. 5. PN junction showing the diffusion of an electron into the P-region and a hole into the N-region.

The diffusion voltage may be determined from

$$V_d = \frac{kT}{q} \ln (N_n / N_p) \tag{3}$$

where k = Boltzmann's constant.
  $q = 1.6 \times 10^{-19}$ coulomb.
  T = absolute temperature (273 + °C).
  N_n = concentration of electrons in the N material when equilibrium is reached.
  N_p = concentration of electrons in the P material when equilibrium is reached.

Minority carriers may also contribute, on a small scale, to the diffusion process. The area between the two ionic barriers is referred to as the *depletion* region, *space-charge* region, or *transition* region.

This void region can be likened to the dielectric of a capacitor and the ionized regions as the two plates of a capacitor. By applying a negative voltage to the P section with respect to the N section, the negatively charged ions move away from the junction. The positively charged ions also move away from the junction, and the effective capacity is reduced. Thus the capacity may be varied by controlling the reverse bias on the junction. This is the basis of some parametric amplifiers that employ a voltage-controlled capacitor, or at times it is called a silicon-capacitor (see below under Variable Capacitance Diode).

At equilibrium and without an external potential, the Fermi levels of the P and N regions will equalize. Conventionally this is graphically shown as a potential hill that is a result of the diffusion of charges. Three states are shown in Fig. 6.

The total current for the diode may be expressed by

$$I = I_s(\epsilon^{eV/kT} - 1) \tag{4}$$

where I = junction current.
  I_s = saturated reverse (leakage) current.
  $e = 1.6 \times 10^{-19}$ coulomb.
  T = absolute temperature (273 + °C).
  $k = 1.3 \times 10^{-23}$ joule/°K.
  V = applied potential.

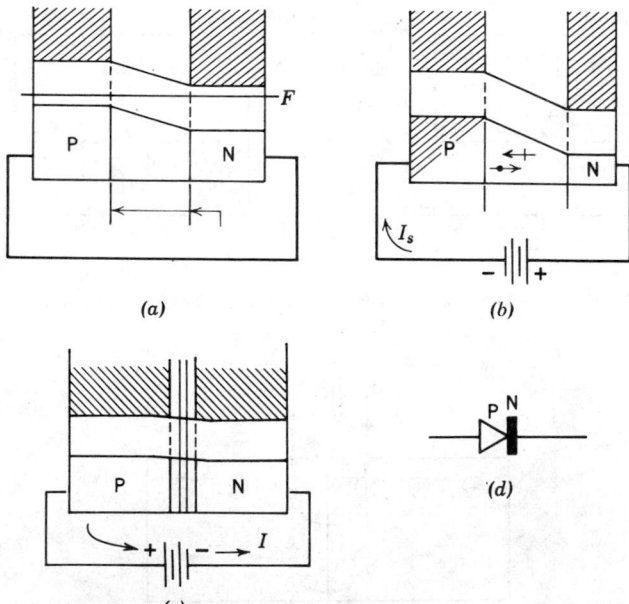

(a)   (b)

(c)   (d)

FIG. 6. (a) PN junction, zero bais; (b) reverse bias, I_s = saturation current (minority carriers); (c) forward bias, I = majority carriers plus minority carriers; (d) schematic symbol of a PN junction or diode. Note that the arrow always points toward the N-doped material.

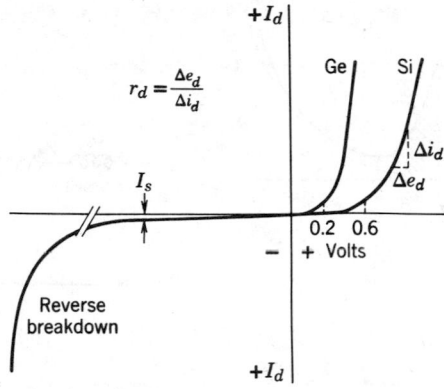

FIG. 7. Typical diode characteristic curve. Scale is compressed in the third quadrant.

With reference to Fig. 7, it can be seen that neither Ge nor Si diodes conduct appreciably until the forward bias voltage reaches a certain value. This is approximately 0.15 to 0.2 V for Ge, and 0.6 to 0.7 V for Si.

PIN Diodes. PIN diodes are made with very intrinsic silicon, and doped P on one end and N on the other end. However, a relatively large intrinsic area is left between the two doped regions. Hence the name, P-I-N. In order to control the constancy of the intrinsic area, it is lightly P-doped.

In a normal forward biased diode, recombinations take place at a rapid rate, and when it is reverse biased the stored charges are swept out of the narrow depletion region in a relatively short time. This factor primarily determines how fast a diode can switch off. By injecting carriers into a relatively long intrinsic area, the recombination process takes a longer time. Thus the diode remains conducting with reverse bias. It might seem obvious that ultimately current will cease, as it will; however, if an extremely high frequency is used as the switching agent, the diode appears to be *on* at all times. By virtue of the frequency, the trapped carriers in the intrinsic region suffer several reversals, hence dissipate some of their energy in the resistance of this region.

The prime use of PIN diodes is in microwave attenuators, modulators, and power levelers. A pictorial sketch of the device is shown in Fig. 8.

Tunnel Diode. The "Esaki" diode is more popularly known as the tunnel diode. Tunneling is a process where the electron does not follow conventional concepts of surmounting a potential hill. That is, the concepts must be explained using quantum mechanics. The significant point is that an electron having a discrete energy level appears on the other side of the thin potential barrier without an increase in energy level. The process is based on the electromagnetic property (wave theory) propagating through the barrier at the speed of light (3×10^8 m/sec) and displacing an electron from the boundary of the barrier. Thus the total effect is one where conduction appears as though the electron has penetrated the potential hill by tunneling through.

These properties are obtained by heavily doping the N and P regions of the crystal. This causes the Fermi levels to actually overlap into the conduction and valence bands of the N and P areas of the crystal.

A typical characteristic curve of a tunnel diode is shown in Fig. 9. By examining the characteristic curve it is seen that the tunnel diode does not exhibit rectifying properties.

Variable Capacitance Diode (Varactors). A capacitor is formed by the charge carriers on either side of the depleted region of a reverse-biased junction. Increasing or decreasing the reverse bias will cause an inverse change of capacitance. Maximum capacitance occurs with zero bias and minimum capacitance occurs prior to reverse breakdown. Low-

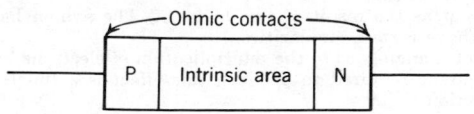

FIG. 8. Pictorial sketch of a pin diode.

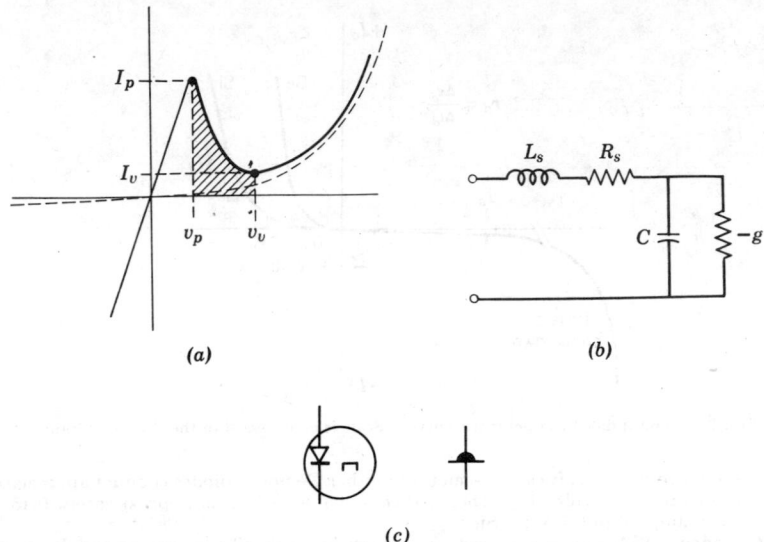

FIG. 9. (a) Tunnel diode characteristics. I_p = peak point of current; I_v = valley current; V_v = valley voltage; V_p = peak point of voltage. Dashed line shows the characteristics of a conventional diode. The shaded area represents the region of negative conductance. (b) An equivalent circuit of a tunnel diode: L_s = total series inductance; R_s = sereis resistance; C = total capacitance; $-g$ = negative conductance in the region shown in a. (c) Schematic symbols.

and high-frequency equivalents are depicted in Fig. 10, as well as popular schematic symbols and a simple application diagram.

Rate of capacitance variation is influenced by the contact potential of the material used (0.6 V for Si) and how the junction is graded. An abrupt junction has a value of $n = 0.45$, as expressed in

$$C = \frac{1}{[1 + (V_R/K)^n]} (C_0) \tag{5}$$

where C_0 = zero bias capacitance.
C = capacitance with bias V_R.
K = contact potential, dependent on the material.
n = varies with the junction gradient.

The figure of merit or Q of the diode may be ascertained from

$$Q = 1/(2\pi f R_s C) \tag{6}$$

where f = frequency in Hz.
R_s = series resistance in ohms.
C = capacitance in farads.
Q = dimensionless figure of merit.

Applications include harmonic generation and modulators as well as tuning circuitry.

Zener Diodes. Referring to Fig. 7, it can be seen that by increasing the reverse bias of a diode, a phenomenon of reverse breakdown occurs. That is, after the applied voltage exceeds a certain value, the reverse current increases very rapidly and is limited only by the circuit resistance. Conventional diode rectifiers are operated at peak voltage values less than the breakdown voltage (V_B).

Breakdown can occur as the result of two factors. The avalanche and Zener effects are independent but have certain similarities.

The avalanche effect is analogous to the multiplication of electrons by collision in a gas. The magnitude of voltage required to produce this effect is a function of the physical attributes of the junction.

The Zener effect may be analogous to cold-cathode emission or simply field-emission. That is, a concentrated electric field literally rips out a charge carrier to manifest itself as

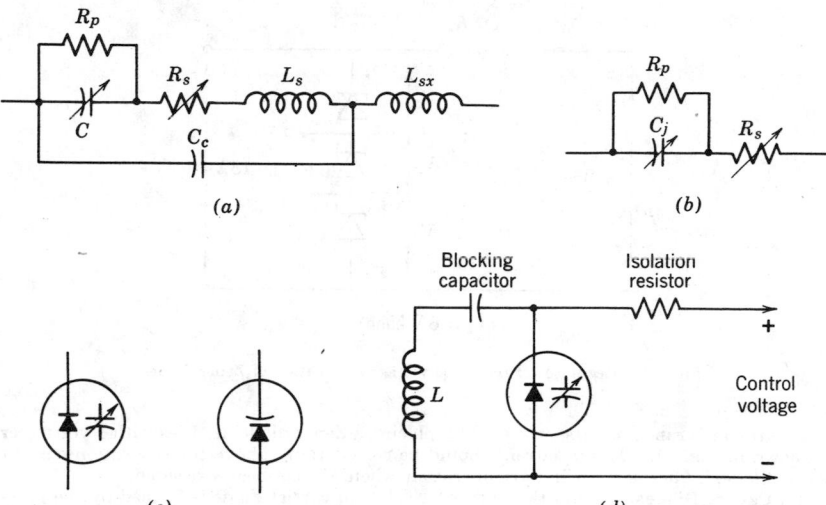

Fig. 10. Variable capacitator diode. (a) High-fequency equivalent circuit: R_p = parallel resistance, R_s = series resistance, L_s = series inductance, L_{sx} = extenal inductance, C_c = package capacitance, and C_j = junction capacitance. (b) Low-frequency equivalent circuit. (c) Schematic symbols. (d) Simple application.

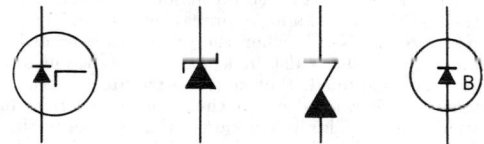

Fig. 11. Zener diode schematic symbols.

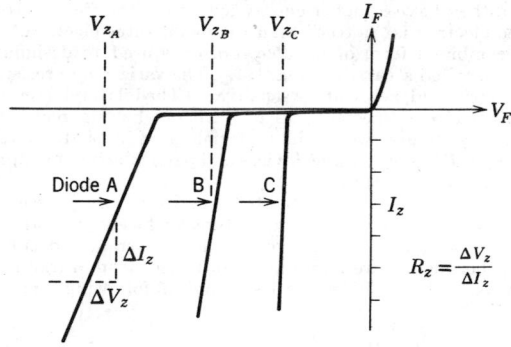

Fig. 12. Zener diode characteristics. Curves for three diodes are shown.

a current. It follows that if the resistivity of a material (doping) is controlled, the Zener breakdown voltage may be controlled. For silicon diodes the Zener voltage V_z may be expressed by

$$V_z = 39\rho_n + 8\rho_p \tag{7}$$

where ρ_n and ρ_p represent the bulk resistivity of the PN mtaerial (ohm-cm).

Zener diodes may be used as regulators, reference voltage sources, protective devices, and even meter range extenders. The diodes may be cascaded to extend their individual voltage ratings, as shown in Fig. 13.

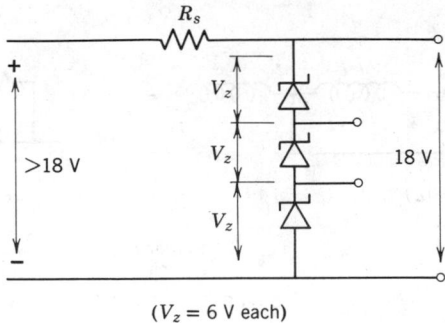

$(V_z = 6 \text{ V each})$

Fɪɢ. 13.   Regulator or reference voltage application of Zener diodes.

A voltmeter's range may be extended by placing a Zener diode in series with a voltmeter as shown in Fig. 14. Meter current should be considerably above the saturation current of the diode. The meter will start operation when V_z has been exceeded.

Hot Carrier Diodes. (Schottky barrier). The hot carrier diode is formed by the junction of a metal and a semiconductor as opposed to a junction formed with a P- and N-doped semiconductor. By choosing metals with proper work-functions, a rectifying junction is formed with the semiconducting material. The semiconductor is normally N-type. In this material the electron (carrier) has a higher mobility, hence it may contribute to faster operation.

Operation of the hot carrier diode is based on majority carriers contributing to current flow. The PN junction, on the other hand, requires electrons to be injected into the P-side and holes to be injected into the N-side when the junction is forward biased. When bias is removed, these minority carriers exist in a foreign environment for a short time; however, when reverse bias is applied, they may contribute to a continuation of current flow. Current continues to flow until all of the minority carriers have recombined or return to their proper domain. This is the factor that prevents the PN junction from acting as a high-speed device. The hot carrier diode's operation may be ascertained by referring to the energy level diagram of Fig. 15.

The term "hot" comes from the fact that the carriers, in this case electrons, are injected into the metal when they have a higher energy level than the free electrons of the metal. That is, the injected electron is "hotter" than the metal's free electron.

Varistors. The combined form of *variable resistor* is used to designate a form of bi-lateral resistive device called a varistor (Fig. 16). The variation of resistance is caused by a variation of the electric field, i.e., voltage sensitive. The bilateral device is at times called symmetrical as opposed to a special class of varistors having rectifier characteristics. Silicon carbide and Thyrite are two of the materials used in bilateral varistors.

Varistors are widely used as protective devices and particularly for suppressing transients in both a-c and d-c switching circuitry.

Thermistors. Temperature-sensitive bilateral resistors are classed as thermistors. Generally the devices exhibit a negative temperature coefficient. That is, high resistances will prevail at low temperatures and, conversely, low resistances occur when the device is hot. Physically, thermistors are manufactured in sizes from a fraction of a millimeter to massive proportions. The small devices are capable of following temperature variations

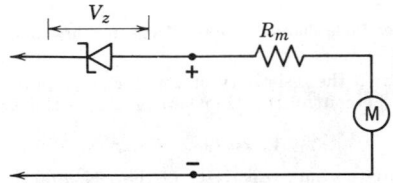

Fɪɢ. 14.   Voltmeter range extender.

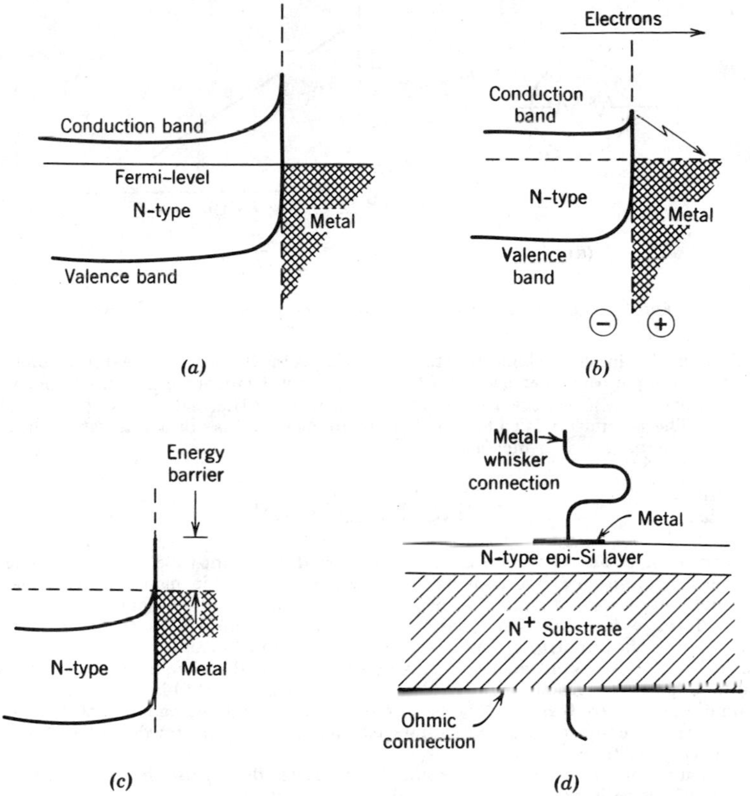

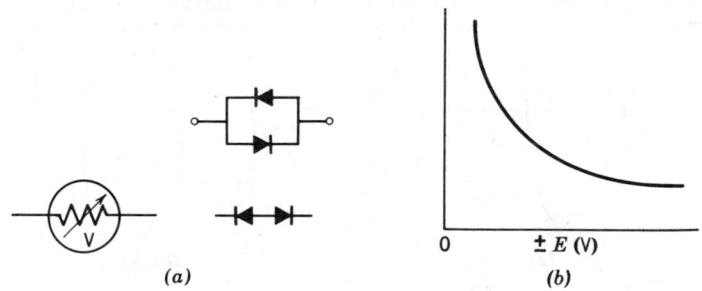

Conduction band

Fermi–level

N–type

Valence band

Metal

(a)

Electrons

Conduction band

N–type

Valence band

Metal

$\ominus$ $\oplus$

(b)

Energy barrier

N–type

Metal

(c)

Metal→ whisker connection

Metal

N–type epi-Si layer

N^+ Substrate

Ohmic connection

(d)

Fig. 15. Hot carrier diode: (a) no bias, (b) forward bias, (c) reverse bias, and (d) construction.

$\pm E$ (V)

(a) (b)

Fig. 16. (a) Symbols used to designate varistors. (b) E–R characteristics of a virastor.

of hundreds of times per second. Thus they are capable of being used as RF carrier envelope detectors with certain limits.

The curve shown in Fig. 17 depicts a typical thermistor operated above a specific temperature. Thermistors are used as power detectors, temperature-sensing transducers, and in circuit protection circuits.

One use of the thermistor would be in series with a tungsten filamentary device. When cold, tungsten exhibits a very low resistance, hence a high surge of current may be sup-

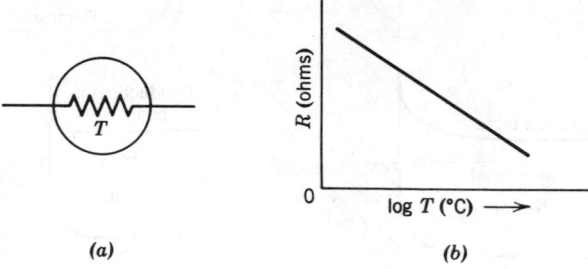

(a)　　　　　　　　　　　　　(b)

FIG. 17. (a) Thermistor symbol. (b) Thermistor characteristic curve.

pressed with the inherent high resistance of the cold thermistor. After a time, the thermistor's temperature increases ($I^2 Rt$) and the circuit current comes up to normal.

Thermistors are also widely used as protection against thermal runaway in transistor circuitry. The thermistor is so arranged as to reduce the bias of a transistor whose environmental temperature is increasing.

4. TRANSISTORS

The transistor is the solid-state analog of the triode vacuum tube. Its small size and low power consumption are responsible for its replacing tubes in many applications.

Two ends of a *single* crystal of either germanium or silicon are doped with impurity atoms to form a dual junction transistor. The middle, known as the base, contains relatively few of the impurity atoms. A graphic representation and schematic symbols are shown in Fig. 1. The emitter-base forms one junction and the collector-base forms the second junction. In normal operation the emitter-base is forward biased and the collector base junction is reverse biased. The base may be considered analogous to the grid of the vacuum tube, the emitter is likened to the cathode, and the collector does the duty of the plate of a vacuum tube.

Even though some of the newer forms of the transistor do not use the name "junction," junctions are normally part of a great number of these devices.

The d-c and a-c resistance of a forward-biased junction is quite low; however, a reverse-biased junction has a characteristic high resistance. By referring to the graphic representation of Fig. 2, it would seem that output (collector) current would be zero regardless of the input (emitter-base) current. However, the base region is very narrow, and the

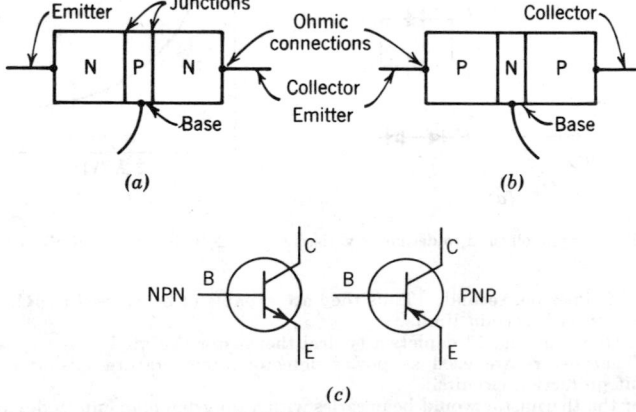

(a)　　　　　　　　　　　　　(b)

(c)

FIG. 1. (a) NPN junction transistor. (b) PNP junction transistor. (c) Schematic symbols: E = emitter, B = base, C = collector.

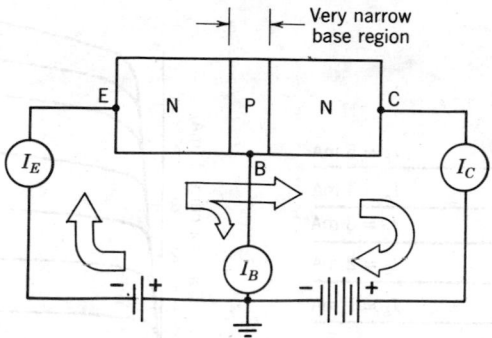

FIG. 2.   Distribution of current in an NPN transistor circuit.

electrons injected from the emitter region generally pass into the collector region; however, a few do combine with holes in the P-region and show up as base current.   If the transistor is a PNP type, the voltage polarities and current directions would be reversed.

Alpha.   In an amplifier configuration that maintains the base as a common element, the current gain is called alpha (α), and is expressed by

$$\alpha = \frac{\Delta I_c}{\Delta I_E} \cong 0.99 \text{ or less than unity} \tag{1}$$

From Fig. 2 it should be noted that I_E represents the largest current and is equal to I_B plus I_C.

$$I_E = I_B + I_C \tag{2}$$

Amplification.   From the definition of Eq. 1, alpha is seen to be less than unity.   It would seem that the so-called amplifier does not really have a gain; however, it is possible to have both a voltage and power gain when the current gain is less than one.   This is possible by virtue of the inherent resistances of the input (emitter-base) and of the output (collector-base).   Remember the emitter-base is forward biased, hence it will possess a low resistance, and for purposes of discussion let $R_i = 100$ ohms.

Conversely, the output is reverse biased, hence it can be represented by a high resistance, and for purposes of discussion let $R_0 = 500$ kohms.

Referring to Fig. 3, suppose that $i_e = 1$ mA $= 1000$ μA, and if alpha is 0.99, $i_c = 1000$ μA $\times 0.99 = 990$ μA.   Then if R_L is arbitrarily chosen to be 10 kohms, the voltage appearing across R_L would be

$$e_0 = I_c \times R_L = 990 \text{ } \mu\text{A} \times 10\text{K} = 9.9 \text{ V}$$

and

$$e_i = i_E \times R_i = 1 \text{ mA} \times 100 = 0.1 \text{ V}$$

The voltage gain would be

$$A_v = \frac{e_0}{e_i} = \frac{9.9}{0.1} = 99$$

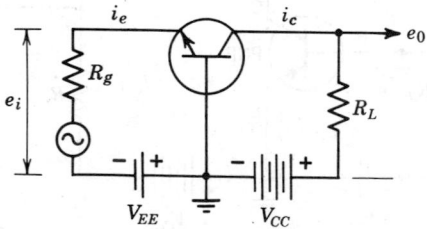

FIG. 3.   Common-base amplifier.

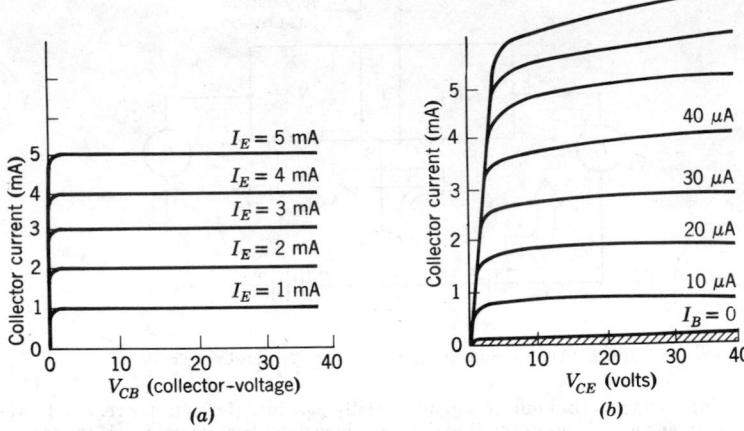

FIG. 4.   Typical family of characteristic curves.   (a) Common base.   (b) Common emitter.

Power gain for this circuit would be,

$$A_p = \frac{P_0}{P_i} = \frac{I^2 R_L}{I^2 R_i} = \frac{I_c^2 R_L}{I_E^2 R_i} = \frac{(990 \times 10^{-6})^2 \times 10^4 \text{ W}}{(10^{-3})^2 \times 10^2 \text{ W}} = 98$$

In the vacuum tube it was noted that voltages controlled the operation whereas the transistor is a current-operated device.

Beta.   By rearranging the common element of a transistor amplifier, variations in operational characteristics are noted.   Figure 5 depicts a common-emitter amplifier configuration using a PNP transistor.   By using this configuration, a base to collector amplification factor called beta is defined as

$$\text{Beta} = \beta = \frac{\Delta I_C}{\Delta I_B}\bigg|_{\Delta VCE=0} \tag{3}$$

This operating parameter may be obtained graphically from a family of operating characteristic curves Fig. 6.   Note that beta is much greater than unity, and typically will be approximately 100.

Common-Emitter Amplifier (CE).   The common-emitter amplifier is very much like the common-cathode amplifier, and at times it is called a conventional amplifier.   The word "conventional" is not as popular as it was during the transistor's infancy.   A great similarity exists between the operation of the CE amplifier and the common-cathode or grounded-cathode amplifier.   This terminology is also extended to the other forms: (1) grounded-base, (2) grounded-emitter, and (3) grounded-collector (covered later).

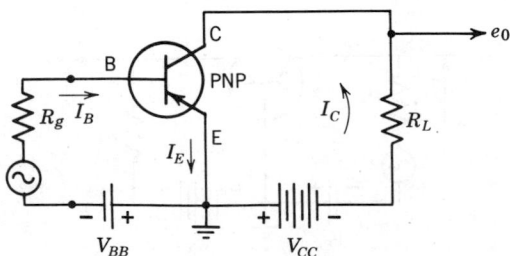

FIG. 5.   Common-emitter amplifier.

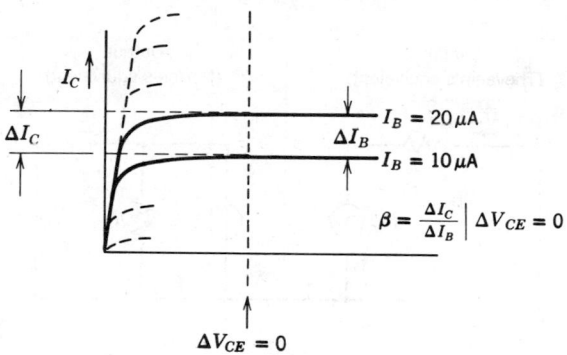

$$\beta = \frac{\Delta I_C}{\Delta I_B} \bigg| \Delta V_{CE} = 0$$

$$\Delta V_{CE} = 0$$

Fig. 6. Graphically determining beta.

Beta (Eq. 3) may be expressed in terms of alpha (Eq.1). $\beta = \Delta I_C/\Delta I_B$ and $\alpha = \Delta I_C/\Delta I_E$, $I_C = \alpha I_E$, then $\beta = \alpha I_E/I_B$, and from Eq. 2, $I_B = I_E - I_C$, then $\beta = \alpha I_E/(I_E - I_C)$ and $\alpha I_E/(I_E - \alpha I_E)$. Factoring the denominator,

$$\beta = \frac{\alpha}{1 - \alpha} \tag{4}$$

Similarly, alpha may be expressed in terms of beta,

$$\alpha = \frac{\beta}{1 + \beta} \tag{5}$$

From Eq. 5 it is seen that alpha will always be less than unity.

Common-Collector Amplifier (CC). The common-collector amplifier is analogous to the common-plate amplifier; however, the vacuum tube counterpart is more popularly known as a cathode follower. The name "common-collector" similarly is not as popular as the name "emitter follower." A schematic diagram of an emitter follower is shown in Fig. 7.

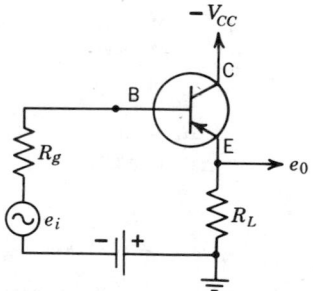

Fig. 7. Emitter-follower amplified (common-collector amplifier).

Hybrid Parameters. Complex circuit problems may be resolved by reducing a circuit to a simplified *equivalent circuit*. Transistor amplifier solutions lend themselves to a mixture (hybrid) of equivalent circuit types, i.e., Thevenin's and Norton's equivalents. A general model for hybrid (h) parameter solutions is shown in Fig. 8.

Writing loop equations for the input and output circuits will yield the following equations:

(input) $$V_i = h_i i_i + h_r V_0 \tag{6}$$

(output) $$i_0 = h_f i_i + h_0 V_0 \tag{7}$$

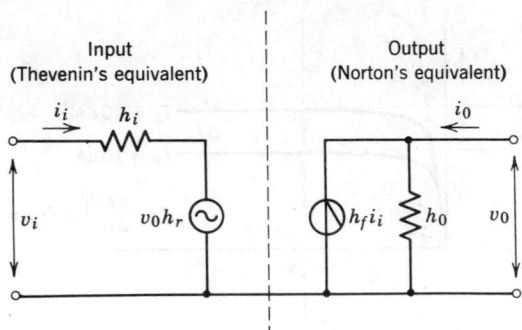

$$\text{Fig. 8.} \quad h\text{-parameter model.}$$

The subscripts e, c, or b are used to denote a common-emitter, common-collector, or common-base respectively. For example, the input impedance of a common-emitter amplifier would be stated as h_{ie}. Futher definitions are as follows:

h_i = input impedance with output a-c shorted (ohms).
h_f = forward transfer current ratio with output a-c shorted (dimensionless).
h_r = reverse transfer voltage ratio with input a-c open-circuited (dimensionless).
h_0 = output admittance with input a-c open-circuited (mhos).

The forward transfer current ratio is commonly referred to as alpha or beta depending on the configuration. From Eq. 7,

$$h_{fe} = \frac{\Delta i_0}{\Delta i_i}\bigg|_{\Delta V_0 = 0}$$

For a common-emitter amplifier $i_0 = i_c$, $i_i = i_B$, and $V_0 = V_{CE}$, then,

$$\beta = h_{fe} = \frac{\Delta i_c}{\Delta i_B}\bigg|_{\Delta V_{CE} = 0}$$

Then for a common-base amplifier,

$$h_{fb} = \frac{\Delta i_0}{\Delta i_i}\bigg|_{\Delta V_0 = 0}$$

and

$$\Delta i_0 = \Delta i_c$$
$$\Delta i_i = \Delta i_E$$
$$\Delta V_0 = \Delta V_{CB}$$

Then,

$$\alpha = h_{fb} = \frac{\Delta i_c}{\Delta i_e}\bigg|_{\Delta V_{CB} = 0}$$

See Table 1.

Z and Y Parameters. The Z parameters are also referred to as the open-circuit impedance measurements, and the Y parameters are called the short-circuit admittance measurements. The equivalent circuit representations are either voltage or current generators, but not both as in the hybrid configurations (Fig. 9).

Input (1) and output (2) circuits will yield two equations:

$$v_1 = z_{11}i_1 + z_{12}i_2 \tag{8}$$

$$v_2 = z_{21}i_1 + z_{22}i_2 \tag{9}$$

From Eqs. 8 and 9,

Input impedance $\qquad\qquad z_{11} = \dfrac{\Delta v_1}{\Delta i_1}\bigg|_{\Delta i_2 = 0}$ $\qquad\qquad\qquad$ (10)

Table 1. h-Parameters

Amplifier configuration (see Table 4.2)	CB	CE	CC
Input impedance	$h_{ib} = \dfrac{\Delta v_{eb}}{\Delta i_e}\bigg|_{\Delta v_{cb}=0}$	$h_{ie} = \dfrac{\Delta v_{be}}{\Delta i_b}\bigg|_{\Delta v_{ce}=0}$	$h_{ic} = \dfrac{\Delta v_{bc}}{\Delta i_b}\bigg|_{\Delta v_{ec}=0}$
Output admittance	$h_{ob} = \dfrac{\Delta i_c}{\Delta v_{cb}}\bigg|_{\Delta i_e=0}$	$h_{oe} = \dfrac{\Delta i_c}{\Delta v_{ce}}\bigg|_{\Delta i_b=0}$	$h_{oc} = \dfrac{\Delta i_e}{\Delta v_{ec}}\bigg|_{\Delta i_b=0}$
Forward transfer current ratio	$h_{fb} = \alpha = \dfrac{\Delta i_c}{\Delta i_e}\bigg|_{\Delta v_{cb}=0}$	$h_{fe} = \beta = \dfrac{\Delta i_c}{\Delta i_b}\bigg|_{\Delta v_{ce}=0}$	$h_{cr} = \dfrac{\Delta i_e}{\Delta i_b}\bigg|_{\Delta v_{ec}=0}$
Reverse transfer voltage ratio	$h_{rb} = \dfrac{\Delta v_{eb}}{\Delta v_{cb}}\bigg|_{\Delta i_e=0}$	$h_{re} = \dfrac{\Delta v_{be}}{\Delta c_{ce}}\bigg|_{\Delta i_b=0}$	$h_{rc} = \dfrac{\Delta v_{bc}}{\Delta v_{ec}}\bigg|_{\Delta i_b=0}$

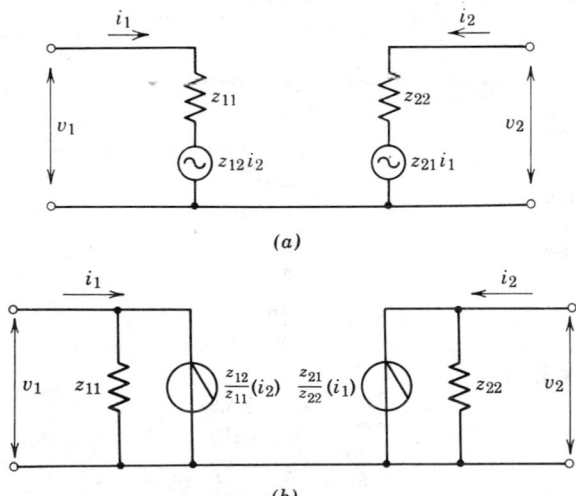

FIG. 9. General Z-parameter forms. (a) Voltage form. (b) current form.

Reverse $(2\rightarrow1)$ transfer impedance	$z_{12} = \dfrac{\Delta v_1}{\Delta i_2}\bigg|_{\Delta i_1=0}$	(11)
Forward $(1\rightarrow2)$ transfer impedance	$z_{21} = \dfrac{\Delta v_2}{\Delta i_1}\bigg|_{\Delta i_2=0}$	(12)
Output impedance	$z_{22} = \dfrac{\Delta v_2}{\Delta i_2}\bigg|_{\Delta i_1=0}$	(13)

Equations 10, 11, 12, and 13 may be adapted to the three forms of amplifier configurations listed in Table 2.

The Y forms are derived from the input and output equations.

(input) $\qquad\qquad i_1 = y_{11}v_1 + y_{12}v_2$ $\qquad\qquad$ (14)

(output) $\qquad\qquad i_2 = y_{21}v_1 + y_{22}v_2$ $\qquad\qquad$ (15)

Table 2. Z-Parameters

	Common–base (CB)	Common–emitter (CE)	Common–collector (CC)
Simplified amplifier configurations	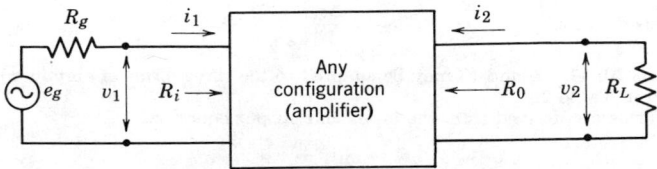		
Input impedance	$z_{ib} = \dfrac{\Delta v_{eb}}{\Delta i_e}\Big\|_{\Delta i_c=0}$	$z_{ie} = \dfrac{\Delta v_{be}}{\Delta i_b}\Big\|_{\Delta i_c=0}$	$z_{ic} = \dfrac{\Delta v_{bc}}{\Delta i_b}\Big\|_{\Delta i_e=0}$
Output impedance	$z_{ob} = \dfrac{\Delta v_{cb}}{\Delta i_c}\Big\|_{\Delta i_e=0}$	$z_{oe} = \dfrac{\Delta v_{ce}}{\Delta i_c}\Big\|_{\Delta i_b=0}$	$z_{oc} = \dfrac{\Delta v_{ec}}{\Delta i_e}\Big\|_{\Delta i_b=0}$
Forward transfer impedance	$z_{fb} = \dfrac{\Delta v_{cb}}{\Delta i_e}\Big\|_{\Delta i_c=0}$	$z_{fe} = \dfrac{\Delta v_{ce}}{\Delta i_b}\Big\|_{\Delta i_c=0}$	$z_{fc} = \dfrac{\Delta v_{ec}}{\Delta i_b}\Big\|_{\Delta i_e=0}$
Reverse transfer impedance	$z_{rb} = \dfrac{\Delta v_{eb}}{\Delta i_c}\Big\|_{\Delta i_e=0}$	$z_{re} = \dfrac{\Delta v_{be}}{\Delta i_c}\Big\|_{\Delta i_b=0}$	$z_{rc} = \dfrac{\Delta v_{bc}}{\Delta i_e}\Big\|_{\Delta i_b=0}$

From these equations, four parameters are extracted:

Input admittance
$$y_{11} = \frac{\Delta i_1}{\Delta v_1}\Big|_{\Delta v_2=0} \tag{16}$$

Output admittance
$$y_{22} = \frac{\Delta i_2}{\Delta v_2}\Big|_{\Delta v_1=0} \tag{17}$$

Forward transfer admittance
$$y_{21} = \frac{\Delta i_2}{\Delta v_1}\Big|_{\Delta v_2=0} \tag{18}$$

Reverse transfer admittance
$$y_{12} = \frac{\Delta i_1}{\Delta v_2}\Big|_{\Delta v_1=0} \tag{19}$$

Amplifiers (Fig. 10).

$$R_i = \frac{h_{11} + \Delta^h R_L}{1 + h_{22}R_L} \tag{20}$$

$$R_0 = \frac{h_{11} + R_g}{\Delta^h + h_{22}R_g} \tag{21}$$

$$A_i = \frac{i_2}{i_1} = \frac{h_{21}}{1 + h_{22}R_L} \tag{22}$$

$$A_v = \frac{h_{21}R_L}{h_{11} + \Delta^h R_L} \tag{23}$$

$$PG = A_i A_v = \frac{(A_i)^2 R_L}{R_i} = \frac{(h_{21})^2 R_L}{(1 + h_{22}R_L)(h_{11} + \Delta^h R_L)} \tag{24}$$

Fig. 10. General form used to compute operation characteristics.

Where

$$\Delta^h = h_{11}h_{22} - h_{21}h_{12} \qquad (25)$$

and R_i = input resistance.
R_0 = Amplifiers output resistance.
A_i = current gain.
A_v = voltage gain.
PG = power gain.
Δ^h = hybrid determinant.
R_g = generator or source resistance.
R_L = load resistance.
e_g = generator open-circuit voltage.
i_i, v_1 = input current and voltage.
i_2, v_2 = output current and voltage.

Biasing the Transistor. Comparing a family of pentode vacuum tube characteristic curves with those of a common-emitter transistor characteristic curves, a physical similarity exists. By close examination, however, it is seen that zero bias creates opposite results. A great deal of current flows in the tube: however, only a small leakage current flows in the transistor. Thus, when a signal is applied, it would be desirable for the output current to duplicate the input current in most cases.

It can thus be seen that in a vacuum tube the plate current is reduced when the grid is made more negative. Note that in most cases it is not desirable to allow the control grid to go positive. In the transistor, collector current would increase with an increase in base current; however, the base-emitter junction current cannot be made to reverse under normal operation. Thus it can be seen that signals of only one polarity would produce an output. If a sinusoidal signal were to be used, only one-half of the sine wave would be reproduced in the output of the amplifier. Some method must then be established so that the amplifier is capable of reproducing a signal faithfully, i.e., with fidelity. This is accomplished by fixing the no-signal operating current somewhere between a maximum and minimum value. This procedure is called biasing.

Load Line. In order to establish limits of operation, the output load (R_L) must be utilized. Graphically, this can be done simply and yields good results for large signal emplifiers. Note that Eq.22 or 23 does not mention bias but assumes that bias is present to establish an operating value of h_{21}. Several approaches may be used to draw a typical load line. However, the method presented here will be adequate for the engineer not versed in electronics. The procedure for drawing a load line will be presented in steps.

1. Establish maximum operating potentials and the power rating of the transistor. These facts will be found in the specification sheets. If the family of curves does not have a maximum collector dissipation curve imprinted, one should be incorporated. This is done by determining the maximum power dissipation from the specification sheet, and from $P = EI$ the curve is drawn. Thus if the rating is given as 60 MW, then 66 MW/2 V $= I_{max}$, 60 MW/4 V $= I_{max}$, 60 MW/6 V $= I_{max}$, etc., and connecting all of the I_{max} points on the characteristic curves results in a P_{max} curve as shown in Fig. 11. Continuous operation must be below this curve.

2. Establish one point at $I_C = 0$ and V_{CC}, see ① in Fig. 11. V_{CC} is the collector supply voltage (see Fig. 12).

3. Point ② is established by V_{CC}/R_L and $V_{CE} = 0$. An alternate method, if R_L is not known, is to draw a suitable line under P_{max} and establish R_L from the slope of the load line:

$$R_L = -1/\text{slope} \qquad (26)$$

The load line shown is for 2 kohms. A 4-kohm load line is shown as a dashed line. A 1-kohm load is shown as a dotted line and crosses the P_{max} curve, hence it is not desirable.

The next step is to establish the bias current for this amplifier. A rule-of-thumb is to choose a value of one-half of V_{CC} and from the abscissa of Fig. 11 draw a vertical line up to the load line, point ③. The intersection of these two lines simultaneously intersect a value of I_B. This value of I_B is the required bias and in this case will establish a quiescent or no-signal collector current of 4 mA and a collector voltage of 8 V.

With the application of a signal source that is capable of changing the base current from 14 μA to zero and from 14 to approximately 30 μA, the output current will vary and thus the collector voltage will vary in unison.

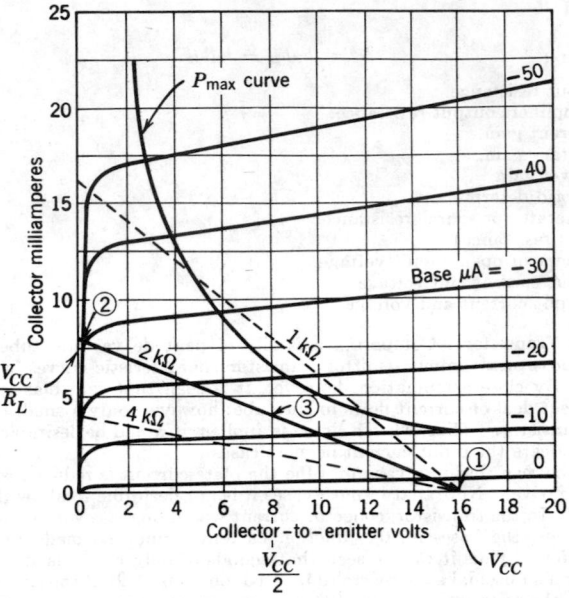

Fig. 11.   Construction of a load-line on a typical family of curves.

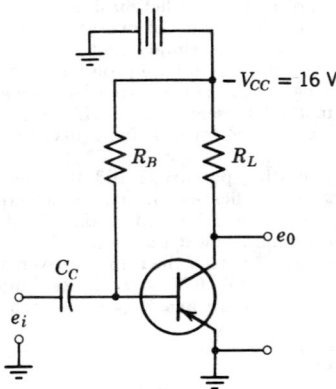

Fig. 12.   Simple CE amplifier.

The next step is to determine the value of R_B in Fig. 12.   Note that the base emitter must be forward biased, and as such this junction represents a low value of resistance (r_b). Now, in order to establish a bias current of 14 μA with a 16-V source (V_{CC}), the total circuit resistance must be

$$R_T + \frac{E}{I} = \frac{14}{16 \times 10^{-6}}$$

use 910 kohms.   The total resistance is

$$R_B + r_b = R_T \quad \text{or} \quad R_T \cong R_B$$

where r_b will generally be less than 100 ohms and can be discounted.

The method of biasing shown in Fig. 12 has a disadvantage of changes with temperature.   Some of the temperature effects may actually damage the transistor.   A desirable method of biasing a common-emitter amplifier is shown in Fig. 13a.

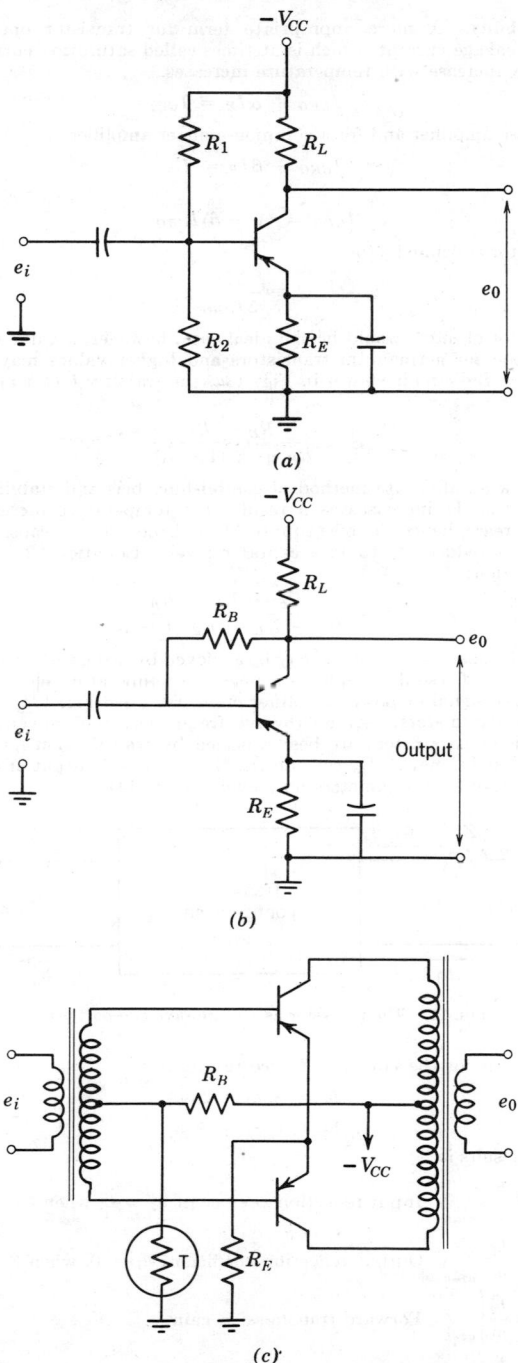

FIG 13.   (a) Preferred method of biasing a CE amplifier.   (b) A dual form of d-c feedback to achieve circuit stability.   (c) A thermistor used to enhance circuit stabilization.

Operating Stability. A more appropriate term for transistor operation would be "instability". Leakage current, which is at times called saturation currents (I_{CBO}), and d-c beta normally increase with temperature increases.

$$I_{CBO} + \alpha I_E = I_C \tag{27}$$

for a common-base amplifier and for a common-emitter amplifier,

$$I_{CEO} + \beta I_B = I_C \tag{28}$$

Note that

$$I_{CEO} = (1 + \beta)I_{CBO} \tag{29}$$

A stability factor is defined as

$$S \equiv \frac{\Delta I_C}{\Delta I_{CBO}} \tag{30}$$

A stability factor of unity would be the ideal case; however, a value of less than 5 can be considered good for germanium transistors and higher values may be tolerated for silicon units. For the circuit shown in Fig. 13a, the stability factor may be determined from

$$S = \frac{R_E + R_2}{R_E + R_2(1 - \alpha)} \tag{31}$$

Figure 13b show an alternate method of establishing bias and stability by using feedback. Note that as I_C increases as a result of a temperature increase, the collector potential will decrease, hence the bias supply (V_C) also decreases, causing I_B to decrease. The decrease of I_B reduces I_C to an acceptable level. Equation 32 is used to establish the stability criterion:

$$S = \frac{R_E + R_L + R_B}{R_E + R_L + R_B (1 - \alpha)} \tag{32}$$

Stabilization of transistor amplifiers may be achieved by using thermistors (temperature-sensitive resistors). These devices have a negative temperature characteristic and lend themselves to large signal or power amplifier circuitry. See Fig. 13c.

Scattering (S) Parameters. At microwave frequencies the conventional methods of describing transistor parameters are best replaced by transmission line thoughts of the incident and reflected signals. By considering the input and output of the transistor as a two-port device, a set of S-parameters may evolve (Fig. 14).

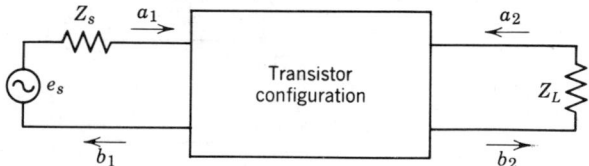

FIG. 14.   The transistor as a microwave two-port device.

Writing equations for the two-port device results in

$$b_1 = S_{11}a_1 + S_{12}a_2 \tag{33}$$

$$b_2 = S_{21}a_1 + S_{22}a_2 \tag{34}$$

Solving for S results in

$$S_{11} = \left.\frac{b_1}{a_1}\right|_{a_2=0} \quad \text{Input reflection coefficient } a_2 = 0, \text{ when } Z_L = Z_0. \tag{35}$$

$$S_{22} = \left.\frac{b_2}{a_1}\right|_{a_2=0} \quad \text{Output reflection coefficient } a_1 = 0, \text{ when } Z_s = Z_0. \tag{36}$$

$$S_{21} = \left.\frac{b_2}{a_1}\right|_{a_2=0} \quad \text{Forward transmission gain.} \tag{37}$$

$$S_{12} = \left.\frac{b_1}{a_2}\right|_{a_1=0} \quad \text{Reverse transmission gain.} \tag{38}$$

The input and output reflection coefficients may be plotted on a Smith chart and converted to an equivalent impedance.

Z_s, Z_L, and Z_0 signify the sending generator impedance, the load impedance and the characteristic impedance of the system used to mount the transistor (usually a microstrip).

In the above equations, a_1, a_2, b_1, and b_2 are defined as

$$a_1 = \frac{V_1 + i_1 Z_0}{2Z_0} = \frac{\text{Voltage incident on port 1}}{Z_0} = \frac{V_{i1}}{Z_0}$$

$$a_2 = \frac{V_2 + i_2 Z_0}{2Z_0} = \frac{\text{Voltage incident on port 2}}{Z_0} = \frac{V_{i2}}{Z_0}$$

$$b_1 = \frac{V_1 - i_1 Z_0}{2Z_0} = \frac{\text{Voltage reflected from port 1}}{Z_0} = \frac{V_{r1}}{Z_0}$$

$$b_2 = \frac{V_2 - i_2 Z_0}{2Z_0} = \frac{\text{Voltage reflected from port 2}}{Z_0} = \frac{V_{r2}}{Z_0}$$

5. UNIJUNCTION TRANSISTOR

The unijunction transistor is used most frequently in timing and triggering applications. The physical construction may vary; however, the cube and bar structure are perhaps the most popular. The two methods of construction are shown in Fig. 1. A schematic symbol and equivalent circuit for the device are shown in Fig. 2.

The interbase resistance R_{BB} is equal to $R_{B1} + R_{B2}$ and may vary from about 5 to 12 kohms. R_{BB} is thus a function of the dopant concentration. Normally V_{BB} is positive at B_2 with respect to B_1; however, a few devices are present that utilize opposite polarities.

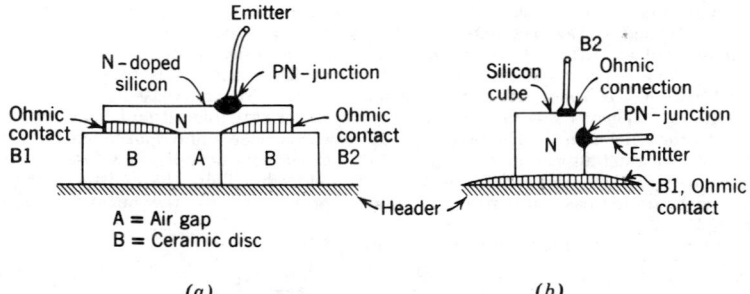

(a)   (b)

Fig. 1. (a) Bar construction. (b) Cube construction.

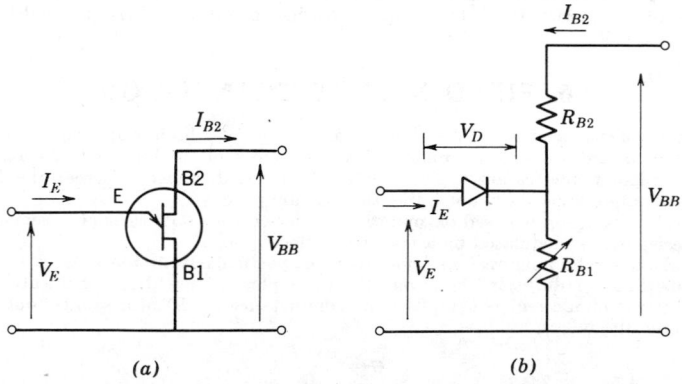

(a)   (b)

Fig. 2. (a) Unijunction schematic. (b) Unijunction equivalent circuit.

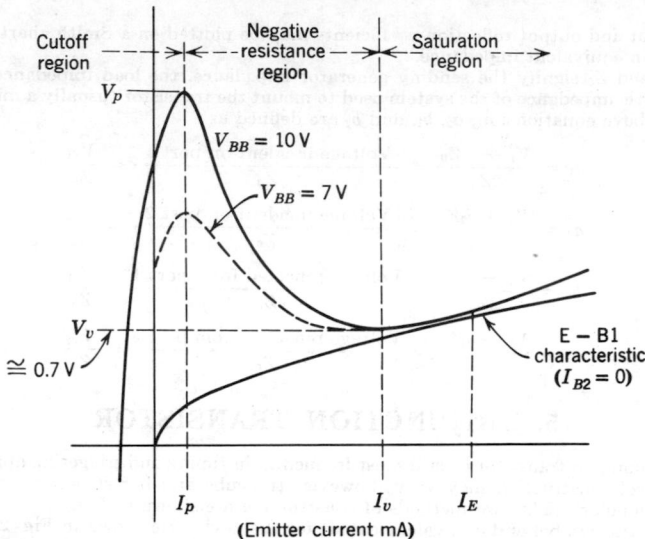

FIG. 3.　Unijunction emitter characteristics. The curve shown is for the bar-structure, a cube-structure brings I_v closer to I_p.

The operation of the device may be explained with the aid of its characteristic curve Fig. 3. Note that voltage is plotted as a function of current.

Without emitter current flowing, the bar acts as a resistive divider where a certain fraction of V_{BB} appears at the emitter junction. This fraction is called the intrinsic stand-off ratio η. If $V_E < \eta V_{BB}$, the emitter junction is reverse biased and only leakage current flows. When $V_E > \eta V_{BB}$, the junction is forward biased and holes will be injected into the silicon bar. The holes migrate towards Base 1 and consequently an equal number of electrons appear in the R_{B1} region. Since conductivity is a function of the population of carriers, the resistance of R_{B1} is reduced. Thus the emitter voltage decreases with an increase of current. This is represented by the negative resistance region of Fig. 3.

The intrinsic stand-off ratio η may be expressed by

$$\eta = \frac{V_p}{V_{BB} + V_D} \quad \text{or} \quad \frac{R_{B1}}{R_{B1} + R_{B2}}$$

The emitter saturation voltage $V_{E(\text{sat})}$, is normally measured with an emitter current of 50 mA and an interbase voltage (V_{BB}) of 10 V.

The emitter junction is located closer to B_2, hence typical values of η will be slightly greater than 0.50.

6. FIELD EFFECT TRANSISTOR

The field effect transistor (FET) has the advantage of a high input impedance. Thus it can serve in areas where the inherent low impedance of the bipolar transistor is detrimental. Bipolar devices are typically current operated devices whereas the FET is a voltage-operated device very much like the vacuum tube.

An FET's operation is based on majority carrier current flow through a semiconductor whose resistance is modulated by a transverse electric field.

Several physical configurations have evolved, particularly in the area of the popular planar methods of diffusion; however, the principles discussed are applicable to most types. In the ohmic region (Fig. 3a), the conductance $(1/R)$ of a semiconductor is defined classically as

$$G = \frac{\sigma W T}{L}$$

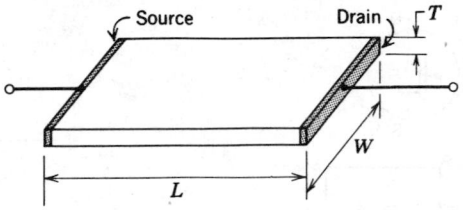

FIG. 1

where $q = 1.602 \times 10^{-19}$ coulomb.

μ (n or p) = mobility of the majority carriers.

σ = conductivity = $q(N\mu_n + P\mu_p)$.

n = electron density.

P = hole density.

W, T, L = defined in Fig. 1.

The source in analogous to a cathode of a vacuum tube. The gate corresponds to a grid and the drain corresponds to the plate.

The subscript may appear as the first or second letter, and used in this manner it refers to the source as the common or reference point. However, when it is used as the third subscript, the undesignated terminal is shorted to the source. I_{GSS} would mean the gate-source current with the drain shorted to the source. A third subscript O would mean that the undesignated terminal is open-circuited.

FET Operation. The junction FET is a depletion operated device. Conductivity from source to drain is controlled by the reverse-biased junction whose area of depletion (space charge) extends into the conducting channel. The depletion region is influenced by the gate to source voltage as well as the gradient voltage field from drain to source.

A significant term is the drain current (I_{DSS}) with the gate shorted to the source ($V_{GS} = 0$). Increasing the drain-source voltage (V_{DS}) from zero causes an increase of current that is a function of the channel conductivity. However, a value of V_{DS} is reached where further increases do not materially influence the drain current I_D. This value of current is I_{DSS} and the value of V_{DS} is called the pinch-off voltage V_P. Further increase of V_{DS} results in saturation current of the reverse-biased junction or drain diode. Thus the depletion region distorts toward the drain and pinches-off channel conduction. Operation of the FET should normally be in the pinch-off region. The pinch-off region is at times called the constant-current or saturation region. Applying some reverse bias to the gate will produce a similar curve with a lower value of I_D. A typical set of characteristic curves for a junction FET is shown in Fig. 3.

Note that the gate junction is always reverse biased and the drain voltage has a polarity that induces majority carrier drift toward the drain. The normal magnitude of V_{GS} ranges from zero to V_P.

The transconductance of an FET is defined as

$$g_m = \frac{\Delta I_D}{\Delta V_{GS}}\bigg|_{\Delta V_{DS}=0}$$

The gain of a common source amplifier may then be expressed by

$$A_v = g_m R_L \parallel R_{DS}$$

Normally $R_{DS} \gg R_L$ and the equation reduces to

$$A_v = g_m R_L$$

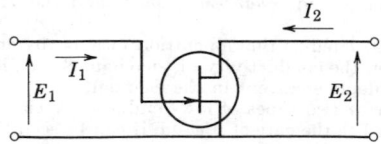

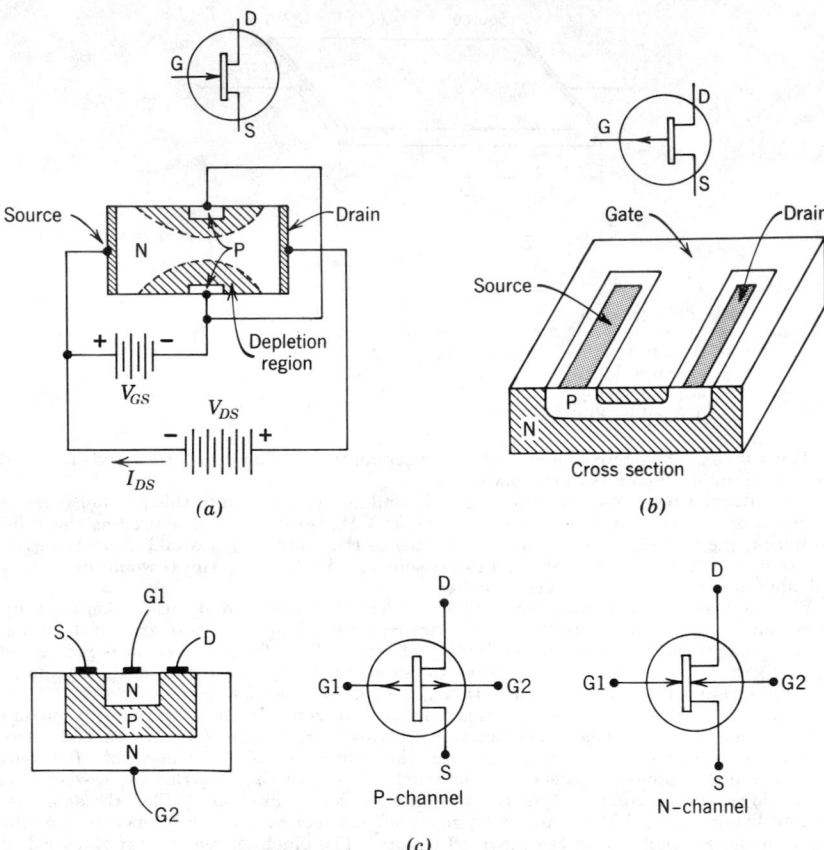

FIG 2.   (a) N-channel bar construction.   (b) P-channel planar construction.   (c) G1 and G2 may be internally connected or brought ou as separate leads.

where $I_1 = Y_{is}E_1 + Y_{rs}E_2$.
$I_2 = Y_{fs}E_1 + Y_{0s}E_2$.   ($Y_s$ parameters.)
Y_{is} = input admittance with the output shorted.
Y_{rs} = reverse transfer admittance with the input shorted.
Y_{fs} = forward transfer admittance with the output shorted.
Y_{0s} = output admittance with the input shorted.

Insulated Gate FET (IGFET).   The insulated gate FET is also known as a metal oxide semiconductor FET (MOSFET).   Although many similarities exist, there are significant differences between IGFETs and junction FETs (JFET).   The main difference lies in the manner in which conductivity is controlled.

JFETs, for example, are considered "ON" devices as compared to a bipolar transistor. That is, JFETs are biased to depletion to affect a reduction in channel current whereas an "OFF" device like the bipolar transistor is biased so that conduction is enchanced. This type of operation is called *enhancement mode* and the JFETs operation is called, *depletion mode.*

IGFETs may be constructed so that operation may be by enhancement or depletion. Basically this depends on the conductivity of the channel, i.e., low-conductivity channels are biased to reduce or deplete current in the channel.

A cross-sectional view of two types of low-conductivity channels are shown in Fig. 4. With zero voltage applied to the gate of Fig. 4a, it can be seen that a reverse-biased junc-

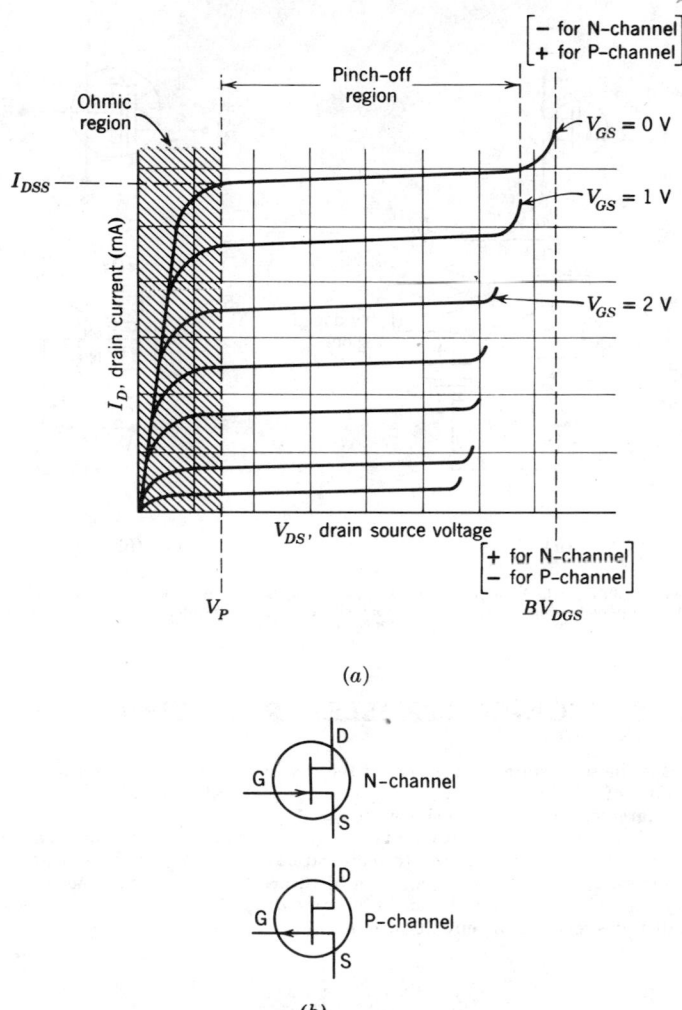

Fig. 3.   (a) JFET characteristics.   (b) JFET schematic symbols.

tion is formed and the conductance would be very low.   By applying a positive voltage to the gate plate (of the capacitor formed by the gate and channel), electrons are pulled from the highly doped N-material to the P-section opposite the dielectric, hence conductivity is enhanced from source to drain.   A signal applied in series with the gate bias supply would modulate the area of the induced channel, thus controlling drain current.

The SiO dielectric has good insulating properties and leakage resistance night be on the order of 10^{15} ohms.   Thus the total *circuit* leakage resistance might be influenced by the resistance of the mounting or even a printed circuit board, so care must be exercised in this region; however, a more significant factor is the fragility of the dielectric.   The dielectric may be punctured by picking up the device and discharging static body charges through the dielectric.   Precautions must also be taken about static charges on soldering devices.   Thus it is considered good practice to leave all of the leads shorted on the device until it is installed.   Frequently breakdown diodes are placed in the gate circuitry to afford some protection to the device.   All precautions and installation procedures recommended by the manufacturer should be followed.

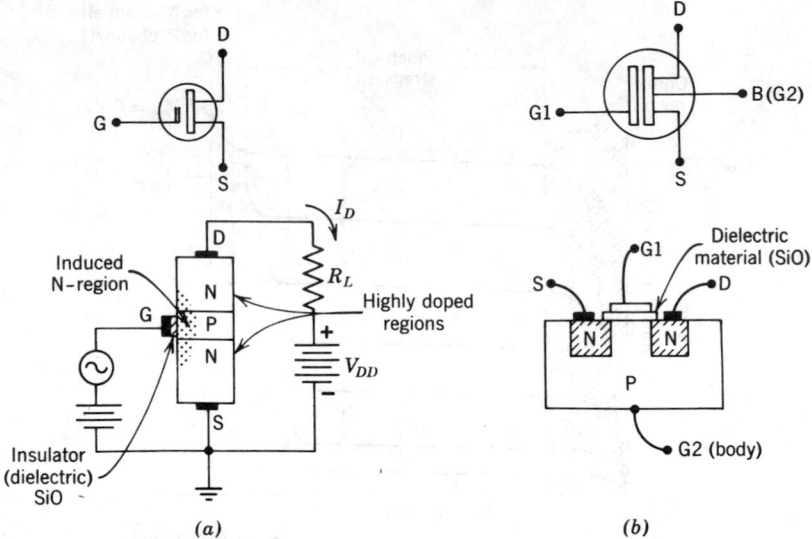

FIG. 4. Two possible methods of producing an N-channel (induced) IGFET. P-channels would be represented by opposite doping and voltage polarities. (a) N-channel MOSFET. (b) N-channel MOSFET.

7. SILICON-CONTROLLED RECTIFIER (SCR)

The SCR is the solid-state counterpart of the thyratron. The SCR is a member of the Thyrite family of PNPN or four-layer devices. A schematic diagram, symbolic sketch, an typical characteristic curve are shown in Fig. 1.

As a controlled rectifier, the SCR may be controlled (fired) by a pulse or sinusoidal voltage. Its use, however, is not limited to rectification. After firing, the circuit current is limited by external circuit resistance and can be turned off by removing the anode voltage or reducing the voltage so that the anode current drops below I_H, the holding current.

For a half-wave rectifier the circuit current is

$$i = \frac{E \sin \omega t}{R_L} \bigg|_{0 < \omega t < \pi} \tag{1}$$

From Fig. 2b, if current is allowed to flow from θ to π, the average or d-c current may be determined by integrating over the interval 0 to 2π, remembering the conditions set forth in Eq. 1.

$$I_{d\text{-}c} = \frac{1}{2\pi} \int_0^{2\pi} i \, d\omega t = \frac{1}{2\pi} \int_\theta^\pi \frac{E \sin \omega t}{R_L} \, d\omega t = \frac{E}{2\pi R_L} (1 + \cos \theta) \tag{2}$$

E is the peak value of the impressed sinusoidal voltage and ω is the angular velocity of the a-c component $2\pi f$. Equation 2 also assumes R_L to be resistive.

For more precise control of circuit power, SCR's may be connected in a back-to-back arrangement. This circuit configuration is also called an inverse-parallel connection and is shown in Fig. 2c.

Silicon-Controlled Switches (SCS). An SCS is a four-layer device having leads brought out from all four layers. A schematic representation is shown in Fig. 3. This device has a wider latitude than the SCR, for it may be turned on and off with pulses.

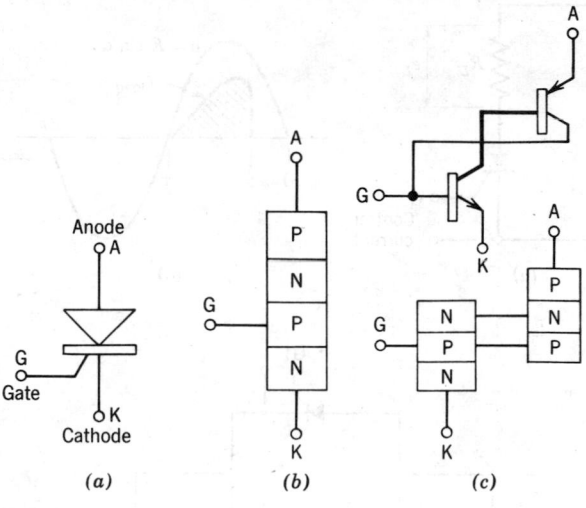

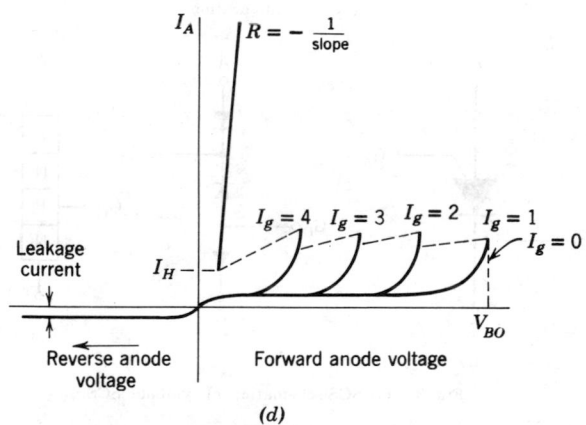

Fig. 1.   Silicon-controlled rectifier.   (a) Schematic.   (b) Symbolic figure.   (c) Two-transistor analogy.
(d) Typical characteristic curves.

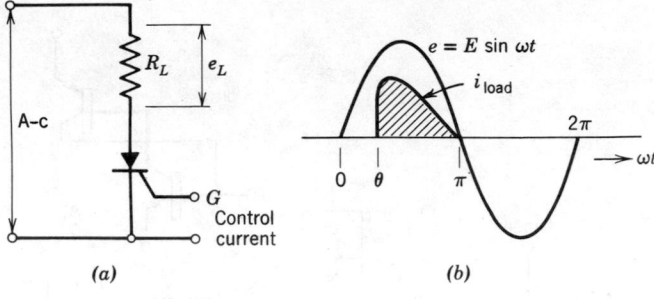

(a)   *(b)*

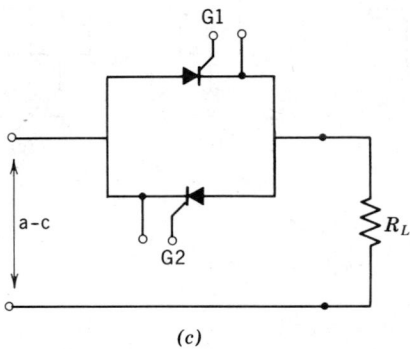

(c)

Fig. 2.   (a) SCR circuit.   (b) Controlled current wave compared to the input voltage wave.   (c) Back-to-back configuration.

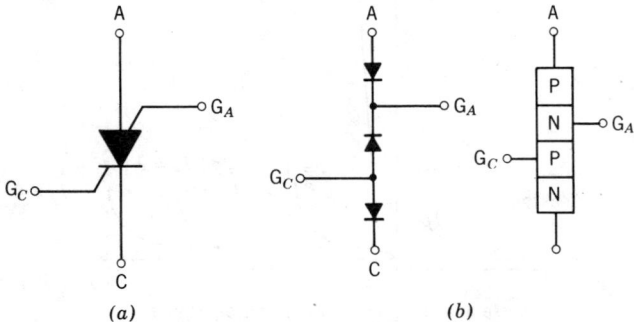

(a)   *(b)*

Fig. 3.   (a) SCS schematic; (b) symbolic SCS.

8. OPERATIONAL AMPLIFIERS

Basically *any* stable amplifier may be called an operational amplifier. This classification, however, was originally intended for a special class of high-gain amplifier used for analog computation. These amplifiers incorporate a great deal of feedback, hence their gain and stability are a function of the feedback elements. A symbolic representation of an operational amplifier (op-amp) is shown in Fig. 1a.

Z_i and Z_f in Fig 1 can assume any configuration of R, C, or L.

In the field of computation, op-amps are used to add (sum), substract, average, differentiate, and integrate. This is accomplished by choosing proper components for Z_i and Z_f.

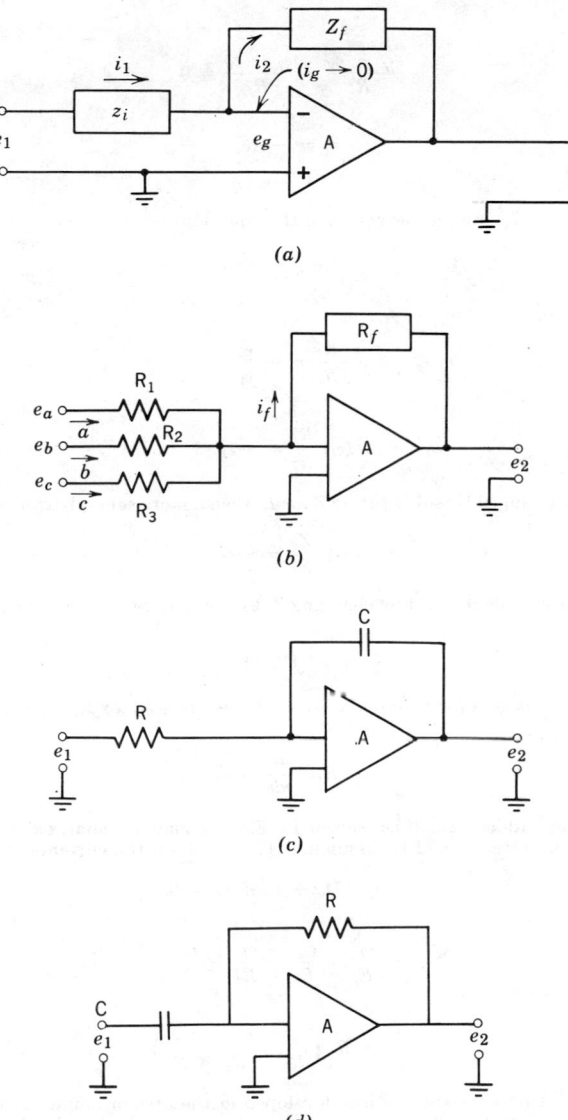

FIG. 1. (a) Symbolic operational amplifier. (b) S summing amplifier. (c) An integrating amplifier.
(d) A differentiating amplifier.

In the field of instrumentation, the op-amp is called upon to perform a myriad of tasks.
It may be used as a comparator, converter, oscillator, or other signal processing functions.
Some of the mathematical functions are shown symbolically in Fig. 1.

If Z_i and Z_f are both resistive, and the open-loop gain of the amplifier is considered
ideal ($A = \infty$), then the currents added at the amplifier's input are

$$i_1 + i_2 = i_g = 0 \tag{1}$$

and

$$\frac{e_i - e_g}{R_i} + \frac{e_2 - e_g}{R_f} = 0 \tag{2}$$

also from

$$e_2 = -Ae_g$$

$$e_g = -e_2/A \tag{3}$$

Thus if $A \equiv \infty$, e_g approaches zero and the equation becomes,

$$\frac{e_1}{R_i} + \frac{e_2}{R_f} = 0 \tag{4}$$

Rearranging terms,

$$\frac{R_f}{R_i} = -\frac{e_2}{e_1} \tag{5}$$

or

$$(e_1)\frac{R_f}{R_i} = -e_2 \tag{5a}$$

If Eq. 5a is rearranged by substituting Z for R, then a more general expression is realized:

$$(e_1)\frac{Z_f}{Z_i} = -e_2 \tag{6}$$

An integrator may be realized by changing Z_f to a pure capacitance. Then the output is

$$-\frac{1}{R_iC_f}\int e_i \, dt = e_2 \tag{7}$$

If a capacitance is used for R_i and a resistive element is used for R_f, then a differentiating circuit evolves:

$$-R_fC_i\frac{de_1}{dt} = e_2 \tag{8}$$

The summing (adder) amplifier shown in Fig. 1b may be analyzed as follows: By Kirchoff's law of currents, and by assuming $A = \infty$, then the currents are

$$-i_f + i_a + i_b + i_c = 0$$

and

$$-\frac{e_2}{R_f} = \frac{e_a}{R_1} + \frac{e_b}{R_2} + \frac{e_c}{R_3}$$

Then

$$e_2 = -R_f\left(\frac{e_a}{R_1} + \frac{e_b}{R_2} + \frac{e_c}{R_3}\right) \tag{9}$$

Definitions. A new vocabulary has developed in the area of amplifiers used for operational amplifiers. Such factors as slew-rate and common-mode rejection ratio are not easily recognizable. Slew-rate may best be described with the aid of a diagram (see Fig. 2). A typical value might be represented as 2 V/us. Common-mode rejection ratio is the ratio of the voltage which must be applied to both inputs of a differential amplifier (most op-amps are differential types) together to produce a given output voltage to the differential input voltage required to produce the same output:

$$CMR = \frac{\text{gain of the difference signal}}{\text{gain of the common-mode signal}} \quad \text{(expressed in db)}$$

Normally, a differential op-amp requires a dual power supply, i.e., plus and minus voltage with respect to a common terminal. With this type of arrangement the output is capable of a plus and/or minus swing with respect to a common point.

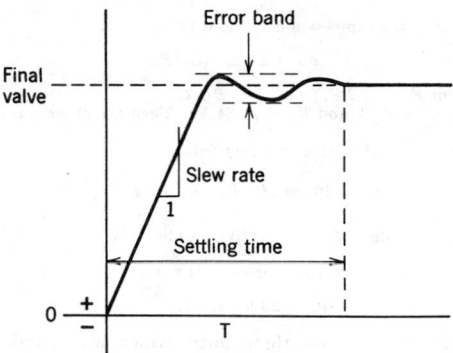

FIG. 2. Operational amplifier terms as applied to a step function.

9. DECIBELS

The magnitude of sound levels, amplification (voltage or power), and even system losses can be mathematically unwieldy. The decibel can be used to reduce these numbers to practical values. A second advantage is that being a logarithmic quantity, many calculations can be reduced to addition and subtraction.

Most significant is the fact that the decibel represents a *ratio* of powers. Thus it cannot be used by *itself* to represent a magnitude unless a reference quantity is specified.

Voltage and current can be related to power and are used frequently in lieu of power. Their use, however, is restricted to a mathematical consideration that is covered later in the text.

The bel is defined in terms of two powers as

$$\log \ (P_o/P_i)$$

and the submultiple in decimal form is

$$\text{decibel (db)} \equiv 10 \log \ (P_o/P_i)* \tag{1}$$

Example. Determine the decibel gain of the given system

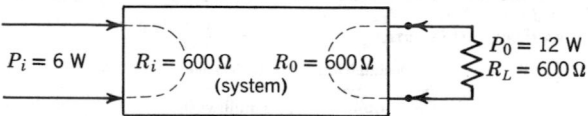

The db gain (+) is 10 log (12/6) = 10 log 2 = 10 × 0.301 = +3.01. This is normally rounded to 3 db.

If voltages are to be used in calculations, they must be derived from the expression for power.

$$db = 10 \log \frac{P_o}{P_i} = 10 \log \frac{E_o^2/R_o}{E_i^2/R_i}$$

where R_i = system input resistance.
R_o = system output resistance.
P_i = input power.
P_o = power delivered to the load (R_L).

Then

$$db = 10 \log \frac{E_o^2 \ R_i}{E_i^2 \ R_o} = 10 \log \left(\frac{E_o}{E_i} \right)^2 + 10 \log \left(\frac{R_i}{R_o} \right)$$

or

$$db = 20 \log \frac{E_o}{E_i} + 10 \log \frac{R_i}{R_o} \tag{2}$$

---

* Log x is used to denote the use of base ten, and must be used for decibel calculations. Logarithms using the base ϵ are used to connote the system involving the neper.

If the ratio R_i/R_o is unity, the expression is reduced to

$$db = 20 \log (E_o/E_i) \tag{3}$$

The voltage across R_i or R_o is $\sqrt{PR}$, from $P = E^2/R$.
In the above example, $E_i = 60$ V and $E_o = 84.84$ V. Then the db gain is 20 log (84.84/60) = 20 log 1.414 = 20 × 0.15 = 3.
Similar calculations are made when currents are used.

$$db = 10 \log (P_o/P_i) = 10 \log \frac{I_o^2 R_o}{I_i^2 R_i}$$

$$db = 20 \log (I_o/I_i) + 10 \log ((R_o/R_i))$$

Similarly, if the ratio R_o/R_i is unity, the expression is reduced to

$$db = 20 \log (I_o/I_i)$$

Frequently the ratio P_o/P_i is less than unity, which would indicate a system has a power loss. To avoid the use of a minus sign in the logarithm, the ratio may be inverted and a minus sign affixed to the decibel value.

Example. Determine the decibel loss of the given network.

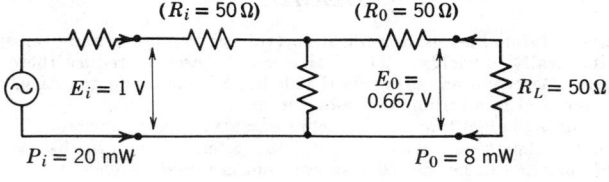

$$
\begin{aligned}
db &= 10 \log (8 \text{ mW}/20 \text{ mW}) \\
&= 10 \log 0.4 \\
&= 10 \times 9.602 - 10 \\
&= 96.02 - 100 \\
&= -3.98
\end{aligned}
$$

By inverting the ratio and affixing a minus sign, db = 10 log (20/8) = 10 log 2.5 = 10 × 0.398 = −3.98.

On occasion the practicing engineer encounters such terms as dbm, dbw, etc. The third letter desngnates the reference quantity, and a tangible quantity may be derived from this dbx value. Normally the reference value will replace the denominator in the ratios involving P, E, or I.
The more popular designations are

Designation	Reference
dbm	1 milliwatt
dbw	1 watt
dbv	1 volt

At times other reference values may be used to subscript the db, and that specific value is used in calculations.

Example. Determine the magnitude of voltage on a 600-ohm transmission line when the level is given as +8 dbm.

Solution

$$
\begin{aligned}
8 \text{ dbm} &= 10 \log (P_x/1 \text{ mW}) \\
0.8 &= \log P_x \\
\log^{-1} 0.8 &= P_x \\
P_x &= 6.31 \text{ mW} \\
\text{Then from } E &= \sqrt{PR}, \\
&= \sqrt{6.31 \times 10^{-3} \times 600} \\
&= \sqrt{3.786} \\
&= 1.94 \text{ V}
\end{aligned}
$$

Use of Tables. The decibel table will yield ratios in terms of power and voltage or current. Care must be exercised when the voltage or current ratios are used for these values are correct only for identical input and output resistances.   Table 1 lists db equivalents in tenths of a decibel up to 10 db and in 10 db steps to 100 db.

Example. Using the tables, find the voltage and power loss of a device that is rated, -36.5 db.

(a) *Power ratio.* The power ratio of -30 db (the highest multiple or submultiple of 10) is 10^{-3}. Then the decibel ratio of -6.5 [from $-36.5 - (-30)$] is 0.2239.   The two ratios are multiplied, 0.2239×10^{-3} or 2.239×10^{-4}.   This means that P_o is 2.239×10^{-4} as great as P_i. For example, if P_i is equal to 4 W, the output of the device is $4 \times 2.239 \times 10^{-4}$ W or 0.8956 mW.

One unique characteristic of the decibel is that zero db does not represent an absence of power.   It only means that the ratio of voltage or power is unity and that the value represented in the numerator is equal to the value in the denominator (the reference).

(b) *Voltage ratio.* The nearest tens multiple of 36.5 is 30, thus in the loss column of voltage, -30 db $= 0.0316$.   Also from the table, -6.5 db is equal to a voltage ratio of 0.4732.   Combining these two ratios by multiplying, $3.16 \times 10^{-2} \times 4.732 \times 10^{-1} = 1.495 \times 10^{-2}$, or E_o equals 0.01495 times the input voltage.

Use in Complex Systems.

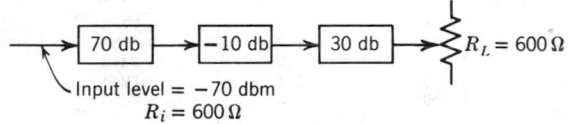

Input level $= -70$ dbm
$R_i = 600\,\Omega$

Example. Determine the voltage across R_L of the given system.
The overall db gain of the system would be the sum of all the individual db gains or losses:

$$70 - 10 + 30 = 90 \text{ db (total gain)}$$

From the given data, voltages at the input and output can be computed.   Note that the input level is given as -70 dbm.   The voltage across the 600-ohm input resistance with 1 mW delivered is

$$E = \sqrt{PR} = \sqrt{10^{-3} \times 6 \times 10^{-2}} = 0.774 \text{ V}$$

From the tables, -70 db is found to correspond to a voltage ratio of 3×10^{-4} and the actual input voltage would then be $0.774 \times 3 \times 10^{-4} = 232\ \mu$V.   Then the actual voltage appearing across the output load resistor (R_L) is, from the tables, 90 db $= 3.162 \times 10^4$, then $E_o = 232 \times 10^{-6} \times 3.162 \times 10^4 = 7.34$ V.

Ratios Involving Complex Impedances in the form, $Z = R + jX$.   When complex impedances are involved, special consideration must be allowed for the circuit power factor (pf).   However if Z_o and Z_i are equal, Eqs. (1) and (3) will be valid.   When Z_o is not equal to Z_i, then the db levels are computed from

$$\text{db} = 20 \log \frac{E_o}{E_i} + 10 \log \frac{|Z_i|}{|Z_o|} + 10 \log \frac{pf_o}{pf_i} \qquad (4)$$

where pf_o = power factor of $Z_o = R_o/|Z_o| = \cos \theta_o$
pf_i = power factor of $Z_i = R_i/|Z_i| = \cos \theta_i$
θ = phase angle between current and voltage
$\quad = \tan^{-1}(X/R)$
X = reactive component of Z

(See Sec. 9.)

Volume Units (VU).   The broadcast industry has long realized that a considerable amount of confusion can result when different reference levels are used.   For example, 6 mW, 50 mW, and 1 W are a few of the standard reference powers in use.   The standard level chosen depends on its suitability to a particular field.   Thus it was decided that a separate unit called a VU be used and that 1 mW into 600 ohms be the recommended reference power.

Voltmeters can thus be calibrated in VU or dbm units; however, meters used in the broadcast field must have specific damping and ballistic characteristics.

db Sound Level.   Sound and noise (unwanted sound) levels are measured in terms of the decibel.   The reference level is the threshold of hearing produced by a certain sound pressure change or an equivalent electrical power.   (See Sec. 11.)

Sound level meters are calibrated in terms of db and can be weighted to correspond to the characteristics of the human ear.   For example, dbA would be a sound level weighted (corrected) to the response of the ear at certain intensities.

Table 1.   Decibel Table
in terms of power and voltage or current

| System gain, +db | | | System loss, −db | |
Voltage or current ratio	Power ratio	db	Power ratio	Voltage or current ratio
1.012	1.023	0.1	0.9772	0.9886
1.023	1.047	0.2	0.9550	0.9772
1.035	1.072	0.3	0.9333	0.9661
1.047	1.097	0.4	0.9120	0.9550
1.059	1.122	0.5	0.8913	0.9441
1.072	1.148	0.6	0.8710	0.9333
1.084	1.175	0.7	0.8511	0.9226
1.097	1.202	0.8	0.8318	0.9120
1.109	1.230	0.9	0.8128	0.9016
1.122	1.259	1.0	0.7943	0.8913
1.135	1.288	1.1	0.7763	0.8811
1.148	1.318	1.2	0.7586	0.8710
1.162	1.349	1.3	0.7413	0.8610
1.175	1.380	1.4	0.7244	0.8511
1.189	1.413	1.5	0.7080	0.8414
1.202	1.445	1.6	0.6918	0.8318
1.216	1.479	1.7	0.6761	0.8222
1.230	1.514	1.8	0.6607	0.8128
1.245	1.549	1.9	0.6457	0.8035
1.259	1.585	2.0	0.6310	0.7943
1.274	1.622	2.1	0.6166	0.7852
1.288	1.660	2.2	0.6026	0.7763
1.303	1.698	2.3	0.5888	0.7674
1.318	1.738	2.4	0.5754	0.7586
1.334	1.778	2.5	0.5623	0.7499
1.349	1.820	2.6	0.5495	0.7413
1.365	1.862	2.7	0.5370	0.7328
1.380	1.905	2.8	0.5248	0.7244
1.396	1.950	2.9	0.5129	0.7161
1.413	1.995	3.0*	0.5012	0.7080
1.429	2.042	3.1	0.4898	0.6998
1.445	2.089	3.2	0.4786	0.6918
1.462	2.138	3.3	0.4677	0.6839
1.479	2.188	3.4	0.4771	0.6761
1.496	2.239	3.5	0.4467	0.6683
1.514	2.291	3.6	0.4365	0.6607
1.531	2.344	3.7	0.4266	0.6531
1.549	2.399	3.8	0.4169	0.6457
1.567	2.455	3.9	0.4074	0.6383
1.585	2.512	4.0	0.3981	0.6310
1.603	2.570	4.1	0.3891	0.6237
1.622	2.630	4.2	0.3802	0.6166
1.641	2.692	4.3	0.3715	0.6095
1.660	2.754	4.4	0.3631	0.6026
1.679	2.818	4.5	0.3548	0.5957
1.698	2.884	4.6	0.3467	0.5888
1.718	2.951	4.7	0.3389	0.5821
1.738	3.020	4.8	0.3311	0.5754
1.758	3.090	4.9	0.3236	0.5689
1.778	3.162	5.0	0.3162	0.5623
1.799	3.236	5.1	0.3090	0.5559
1.820	3.311	5.2	0.3020	0.5495
1.841	3.388	5.3	0.2951	0.5433
1.862	3.467	5.4	0.2884	0.5370
1.884	3.548	5.5	0.2818	0.5309
1.905	3.631	5.6	0.2754	0.5248
1.928	3.715	5.7	0.2692	0.5188
1.950	3.802	5.8	0.2630	0.5129
1.973	3.891	5.9	0.2570	0.5070
1.995	3.981	6.0	0.2512	0.5012

Table 1.   Decibel Table—*Continued*

System gain, +db		db	System loss, −db	
Voltage or current ratio	Power ratio	db	Power ratio	Voltage or current ratio
2.018	4.074	6.1	0.3455	0.4958
2.042	4.169	6.2	0.2399	0.4898
2.065	4.266	6.3	0.2344	0.4842
2.089	4.365	6.4	0.2291	0.4786
2.114	4.467	6.5	0.2239	0.4732
2.138	4.571	6.6	0.2188	0.4677
2.163	4.677	6.7	0.1238	0.4624
2.188	4.786	6.8	0.2089	0.4571
2.213	4.898	6.9	0.2042	0.4519
2.239	5.012	7.0	0.1995	0.4467
2.265	5.129	7.1	0.1950	0.4416
2.291	5.248	7.2	0.1906	0.4365
2.317	5.370	7.3	0.1862	0.4315
2.344	5.495	7.4	0.1820	0.4266
2.371	5.623	7.5	0.1778	0.4217
2.399	5.754	7.6	0.1738	0.4169
2.427	5.888	7.7	0.1698	0.4121
2.455	6.026	7.8	0.1660	0.4074
2.483	6.166	7.9	0.1622	0.4027
2.512	6.310	8.0	0.1585	0.3981
2.541	6.457	8.1	0.1549	0.3936
2.570	6.607	8.2	0.1514	0.3891
2.600	6.761	8.3	0.1479	0.3846
2.630	6.918	8.4	0.1445	0.3802
2.661	7.079	8.5	0.1413	0.3758
2.696	7.244	8.6	0.1380	0.3715
2.723	7.413	8.7	0.1349	0.3673
2.754	7.586	8.8	0.1318	0.3631
2.786	7.762	8.9	0.1288	0.3589
2.818	7.943	9.0	0.1259	0.3548
2.851	8.128	9.1	0.1230	0.3508
2.884	8.318	9.2	0.1202	0.3467
2.917	8.511	9.3	0.1175	0.3428
2.951	8.710	9.4	0.1148	0.3389
2.985	8.913	9.5	0.1122	0.3350
3.020	9.120	9.6	0.1097	0.3311
3.055	9.333	9.7	0.1072	0.3273
3.090	9.550	9.8	0.1047	0.3236
3.126	9.772	9.9	0.1023	0.3199
3.162	10.000	10.0	0.1000	0.3162
10.000	10^2	20.0	10^{-2}	0.1000
31.62	10^3	30.0	10^{-3}	0.0316
10^2	10^4	40.0	10^{-4}	10^{-2}
316.2	10^5	50.0	10^{-5}	0.0032
10^3	10^6	60.0	10^{-6}	10^{-3}
3162.0	10^7	70.0	10^{-7}	0.0003
10^4	10^8	80.0	10^{-8}	10^{-4}
3.162×10^4	10^9	90.0	10^{-9}	3×10^{-5}
10^5	10^{10}	100.0	10^{-10}	10^{-5}

* 3 db is generally rounded to indicate a doubling of power and a 1.4 increase of voltage or current.
−3 db is generally rounded to indicate half-power and 0.7 of the voltage or current.

Neper. In the napierian system the unit that is similar to the decibel is called the neper, and at times it is called the transmission unit (TU):

$$\text{Neper} = \frac{1}{2} \ln (P_o/P_i)$$

To convert a neper to decibels, multiply by 8.686 and, conversely, to convert decibels to nepers, multiply by 0.1151.

BIBLIOGRAPHY

1. ANDERSON, R. W., *S-Parameter Techniques for Faster More Accurate Network Design*, *Hewlett Packard Journal*, Vol. 18, No. 6, February 1967.
2. BLACKWELL L. and KOTZEBUE, K., *Semiconductor Diode Parametric Amplifier*, Englewood Cliffs, New Jersey, Prentice-Hall, 1961.
3. BORDEEN, J., and BRATTAIN, W. H., *The Transistor, A Semiconductor Triode*, *Phys. Rev.*, Vol. 74 (July 1948).
4. CLEARY, J. F. (Editor), et al., *Transistor Manual*, Syracuse, New York, General Electric Co., 1964.
5. CUTTER P., *Electronic Circuit Analysis*, Vol. 1, New York, McGraw-Hill, 1960.
6. CUTTER, P., *Semiconductor Circuit Analysis*, McGraw-Hill, 1964.
7. FITCHEN, F. C., *Transistor Circuit Analysis and Design*, Princeton, New Jersey, Van Nostrand, 1962.
8. HAYT, W. JR., and KEMMERLY, J., *Engineering Circuit Analysis*, New York, McGraw-Hill, 1962.
9. INTERNATIONAL TELEPHONE AND TELEGRAPH CORP., *Reference Data for Radio Engineers*, 4th Ed., New York, 1957.
10. KUROKAWA, K., "Power Waves and the Scattering Matrix," *IEEE Transactions on Microwave Theory and Techniques*, Vol. MIT-13, No. 2, March 1965.
11. LANDEE, R., DAVIS, D., and ALBRECHT, A., *Electronic Designer's Handbook*, New York, McGraw-Hill, 1957.
12. LEVINE, S. N., *Principles of Solid-State Microelectronics*, New York, Holt, Rinehart, and Winston, 1963.
13. MIDDLEBROOK, R. D., *An Introduction to Junction Transistor Theory*, New York, Wiley, 1957.
14. MILLMAN and SEELY, *Electronics*, New York, McGraw-Hill, 1941.
15. MILLMAN and TAUB, *Pulse, Digital and Switching Waveforms*, New York, McGraw-Hill, 1965.
16. PENDER, H., and DEL MAR, W., (Editors), *Electrical Engineer's Handbook* (Wiley Engineer Handbook Series), New York, Wiley 4th ed., 7th Printing, 1965.
17. RISTENBATT and RIDDLE, *Transistor Physics and Circuits*, Englewood Cliffs, New Jersey, Prentice-Hall, 2nd ed., 1966.
18. ROMANOWITZ, H. A., and PICKETT, R. F., *Introduction to Electronics*, New York, Wiley, 1968.
19. RYDER, D., *Electronic Fundamentals and Application*, Englewood Cliffs, New Jersey, Prentice-Hall, 3rd ed., 1964.
20. RYDER, J. D., *Engineering Electronics*, New York, Mc-Graw Hill, 1957.
21. SHEA, R. F., *Transistor Applications*, New York, Wiley, 1964.
22. Shockley, W., *Electronics and Holes in Semiconductors*, New York, Van Nostrand, 1950.
23. SIEDMAN, A. H., and MARSHALL, S. L., *Semiconductor Fundamentals: Devices and Circuits*, New York, Wiley, 1963.
24. SMITH, P. A., *Electronic Applications of the Smith Chart*, New York, McGraw-Hill, 1969.
25. SMITH, R. J., *Circuits, Devices and Systems*, New York, Wiley, 1966.
26. SORENSEN, H. O., "Hot Carrier Diodes," *Hewlett Packard Journal*, Vol. 17, No. 4, December 1965.
27. SOULERS, M., *The Engineers' Companion*, New York, Wiley, 1966.
28. STANLEY, W. D., *Transform Circuit Analysis for Engineering and Technology*, Englewood Cliffs, New Jersey, Prentice-Hall, 1968.
29. SURINA, T., and HERRICK, C., *Semiconductor Electronics*, New York, Hot, Reinhart, and Winston, 1964.
30. VEATCH, H. C., *Transistor Circuit Action*, New York, McGraw-Hill, 1968.
31. ZENER, C., "A Theory of Electrical Breakdown of Solid Dielectrics," *Proc. Roy. Soc.* (London), Vol. 145A, p. 523 (1934).

SECTION 11

RADIATION, LIGHT, AND ACOUSTICS

BY

ERNST WEBER*

ACOUSTICS

BY

R. B. LINDSAY

* This section is a revision of Section 10 of the second edition.

RADIATION, LIGHT, AND ACOUSTICS

THEORY OF RADIATION

By Ernst Weber

1. FUNDAMENTALS

Radiation is the transportation of energy through space; a detectable medium of transmission has not been discovered as yet. The assumption of an all-pervading ether as the carrier of radiated energy has proved unsatisfactory; it is, however, not impossible that ether or some other medium, not yet known, does exist.
The Nature of Radiation. At one time, all radiations were believed to be electromagnetic waves of various wavelengths (see Art. 3). At present the corpuscular nature of radiation must be conceded, as many phenomena can be explained only if the energy of radiation is associated with particles (see Art. 4). The wave and corpuscular points of view are now being adopted alternatively as required by the occasion. The common basis of both aspects, the true nature of radiation, has not been established.

Radiation travels with a velocity v which depends on the medium through which it is propagated. Since the frequency ν of radiation is dependent only on the source, it does not vary with the medium. The ratio $v/\nu = \lambda$ is called the wavelength of radiation and depends like v on the medium.

Fundamental Definitions

Total Emissive Power is the time rate of total radiant energy emitted per unit area of the radiating body. It is measured in ergs per second per square centimeter or in watts per square centimeter.

Spectral Distribution of Emissive Power. The total emissive power of a body may be composed of radiations of many wavelengths. The graph of the emissive power as a function of wavelength is called the spectral distribution of emission. This spectrum is continuous if emissive power is a smooth continuous function of wavelength, or discontinuous if emission occurs only over small ranges of wavelengths. The discontinuous spectral distribution can have the character of lines, if a sharp selective radiation at certain wavelengths is evident, or of bands, if many lines are so close that the discontinuity is apparent only on a very large scale of wavelength. The total emissive power is the integral of the spectral emissive power over all wavelengths from zero to infinity, and is, of course, identical with the area under the spectral distribution curve.

Monochromatic Emissive Power is the time rate of radiant energy emitted per unit area of a radiating body, at a particular wavelength. It can be determined most conveniently from the spectral distribution of emissive power as the ordinate at this particular wavelength.

Intensity of Radiation is the amount of power transmitted through unit area perpendicular to the direction of propagation of radiation.

Density of Radiation is the total radiant energy contained in unit volume. The radiation of the sun has an energy density at the earth's surface of approximately 4.3×10^{-5} erg per cu cm.

Absorption. Radiation of incident intensity I_0 passing through matter of thickness δ loses intensity according to the exponential law

$$I = I_0\, \varepsilon^{-\mu\delta} \tag{1}$$

where μ is called the absorption coefficient and has the dimension $[\mu] = [L]^{-1}$. The extinction coefficient, or index of absorption, is defined as the numerical parameter

$$\kappa = \frac{\lambda}{2\pi}\,\mu \tag{2}$$

where λ is the wavelength of radiation within the absorbing medium. Since the wavelength in vacuum is $\lambda_0 = n\lambda$, where n is the index of refraction, another coefficient can be defined as

$$k = n\kappa = \frac{\lambda_0}{2\pi}\,\mu \qquad (2a)$$

The use of the various definitions is not standardized, and many confusing statements have resulted.

Absorptivity, A, is the ratio of absorbed energy to incident energy of radiation and is a pure numeric always less than unity. A perfectly absorbing body, $A = 1$, is usually referred to as a *black body*.

Scattering is a particular kind of absorption whereby the radiation is diffusely reflected in all directions by the interaction with the inner structure of the substance and thus reduced in intensity in the direction of propagation.

Spectral Distribution of Absorptivity. The absorptivity varies greatly with wavelength. The graph of absorptivity against wavelength gives the spectral distribution of the absorption characteristics of a particular substance, and can be continuous (gradually changing) or discontinuous (selective). In case of discontinuous spectral distribution it can have the character of lines, if a sharp selective absorption at certain wavelengths is evident, or of bands, if many lines are so close that the discontinuity is apparent only on a very large scale of wavelength.

Mass Absorption Coefficient μ_d is the absorption coefficient μ divided by the density d of the absorbing substance, $\mu_d = \mu/d$. Its unit is cm^2 per gram.

Reflection. If radiation is incident upon a surface of a substance, part of it will enter the substance, part of it will be turned back, reflected. If the surface is rough, reflection will occur in all directions, and the radiation is said to be *diffusely* reflected. If the irregularities in the surface are small compared with the wavelength, the radiation will be *regularly* reflected, i.e., the reflected radiation will propagate in a definite direction determined by the equality of the angle of reflection and the angle of incidence. The angle of reflection is the angle which the outward normal to the surface makes with the direction of propagation of the reflected radiation; the angle of incidence similarly is the angle which the inward normal to the surface makes with the direction of propagation of the incident radiation.

Reflectivity or the Coefficient of Reflection, R, is the ratio of reflected to incident energy of radiation and is a pure numeric always less than unity. A perfectly reflecting body, $R = 1$, is usually referred to as a *perfect mirror*. The reflectivity depends upon the angle of incidence (see Reflection) as well as upon the wavelength. If for a particular wavelength $R = 1$ the reflecting body is said to act like a perfect mirror at this wavelength.

Total Reflection. If radiation comes from a medium of higher refractive index and enters a medium of lower refractive index, total reflection may occur, i.e., for angles of incidence larger than a certain critical value θ_c all the radiation is reflected; the critical angle is designated as the angle of total reflection.

Lambert's Law. If radiation falls perpendicularly upon a surface which reflects diffusely, the intensity of the reflected radiation varies approximately with the cosine of the angle made with the normal to the surface; for larger angles with the normal, the approximation becomes poor.

Transmissivity, T, of a body is the ratio of the intensity of radiation leaving the body to the incident intensity of radiation; it is a numerical value, and varies greatly with the thickness of the body and wavelength. If, for any particular wavelength, $T = 0$, i.e., no radiation is transmitted, the body is said to be *totally opaque* for this wavelength; if on the other hand $T = 1$, i.e., all the radiation is transmitted without loss of intensity, the body is said to be *totally transparent* for this wavelength. Obviously, the sum of the three characteristic numerics must always be equal to unity

$$A + R + T = 1 \qquad (3)$$

Refraction. Radiation changes its direction and velocity of propagation with the medium in which it travels. The ratio of the velocity in vacuum to that in a medium is called the *refractive index*, or the *index of refraction* for that medium. It is a numeric always larger than unity and varies greatly with the wavelength.

Refractivity is defined as the difference between refractive index and unity, $n - 1$. Like the refractive index, it varies with the wavelength. *Specific refractivity* is the quotient of refractivity of a medium and its density.

Snell's Law of Refraction. If radiation in one medium is incident upon the surface of another medium at an angle α, it will proceed into this medium with an angle β between the normal to the surface and the direction of propagation. β is called the angle of refraction, and the ratio of the sines of the two angles is the relative index of refraction of the second medium against the first

$$n_{21} = \frac{\sin \alpha}{\sin \beta} = \frac{n_2}{n_1} \tag{4}$$

The *relative index of refraction* is equal to the quotient of the absolute indices of refraction of each medium.

Dispersion. The fact that the refractive index varies with wavelength is called dispersion of radiation. Polychromatic radiation (containing radiation of many wavelengths) can be separated into its monochromatic components by utilizing the fact of dispersion (see Spectroscopy). Historically, normal and anomalous dispersion are differentiated. The former covers the cases of increasing index of refraction with decreasing wavelength; the latter refers to the converse dependents.

Diffraction. The normal laws of refraction and reflection cease to be valid if the objects in the path of radiation are of a size comparable with the wavelength of the radiation. The peculiar disturbance caused by such an object is called diffraction. A regular array of diffracting objects is usually referred to as a *diffracting grating*, and is used to determine the wavelength of the incident radiation.

2. TYPES OF RADIATION

The various types of radiation that are known can be grouped either according to their characteristics or according to their origin. The more important kinds of radiation, grouped according to their most evident nature, are briefly defined below.

Wave Radiations

Thermal Radiation is that radiation emitted by solids or liquids which depends only on the temperature of the substance. The spectral distribution of emissive power is continuous (see Art. 12) and changes smoothly with temperature. Thermal radiation is thought to be produced entirely by the thermal agitation of atoms or molecules.

Characteristic Line Radiation is emitted by gases and vapors if properly "excited." Excitation may occur in the flame (by thermal agitation), in electric arcs or sparks, by bombardment with electrons or atoms, and by absorption of radiation of suitable wavelength. The spectral distribution of emissive power in this case is discontinuous and either of distinct line character or in the form of bands (see Art. 10).

X-rays. The term x-rays is applied to the secondary radiation of a substance upon which high-speed electrons are impinging. The energy distribution has the same character for all elements and consists of a continuous energy spectrum upon which is superimposed a line spectrum which in its details is characteristic of the element serving as the secondary emitter (see Art. 14).The very hardest x-rays produced in the course of natural disintegration of radioactive substances are called γ-rays and show again a characteristic line spectral distribution in energy.

Fluorescence and Phosphorescence are the secondary emission of characteristic visible radiation if a substance is excited by radiation of suitably shorter wavelength even without perceptible rise of the temperature of the emitter. In the case of fluorescence the secondary radiation ceases with the primary radiation; in the case of phosphorescence the secondary radiation persists for an appreciable time after the primary radiation has been stopped.

Corpuscular Radiations

Electron Radiation is the emission of corpuscular electrons from the atoms or molecules of substances. As primary radiation, it may be caused by thermal agitation (thermionic emission, especially in electron tubes), by very high electric fields (cold emission, corona discharge, and cathode rays), and by the natural decay of radioactive substances (as β-rays). As secondary radiation, it can be produced by ultraviolet rays and x-rays (photoelectricity) or by bombardment with high-speed ionized atoms or molecules.

Positive Radiation is the emission of corpuscular positively charged particles from atoms or molecules of substances. If these particles have a small mass approximately corresponding to that of electrons, they are called *positrons*; as secondary radiation they are produced by bombardment of heavy metals with high-energy γ-rays or high-speed α-particles. *Proton radiation* is the emission of positively charged hydrogen nuclei with a mass approximately corresponding to that of the hydrogen atom; it occurs as secondary radiation if substances are bombarded with high-speed α-rays. α-*Rays* are positively charged particles identified as helium nuclei with approximately the mass of helium atoms. As primary radiation they may be produced by the natural decay of the heaviest elements (radioactivity), by intense electric fields in the very low-pressure discharge tube (canal rays). As secondary radiation, they are the result of artificial atomic disintegration by bombardment with high-speed protons.

Neutron Radiation is the secondary emission of corpuscular electrically neutral particles from atoms of substances if bombarded with very high-speed α-rays so as to produce artificial disintegration. The mass of the neutron corresponds approximately to that of the hydrogen atom.

Cosmic Rays are apparently extremely high-speed protons from interstellar space. In penetrating through the atmosphere they produce short-lived *mesons*, particles of either positive or negative charges and with masses reported from 20 to 1000 times the mass of the electron.

3. WAVE ASPECTS OF RADIATION

Definition of Wave. The simplest type of a wave is the sinusoidal wave which is presented by any phenomenon the intensity of which varies sinusoidally with time at any point in space, as well as with distance at any particular instant of time (see Fig. 1). The mathematical formulation for a one-dimensional (or plane) wave is given by

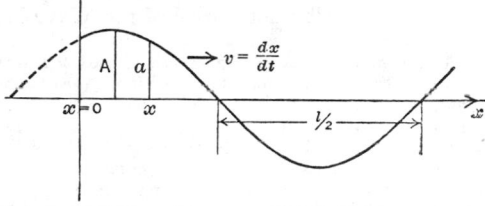

$$a = A \sin \frac{2\pi}{\lambda} (x \pm vt) \qquad (5)$$

This form is symmetrical in x and vt so that the variation in space and time is sinusoidal. A is the amplitude or crest value of the wave, λ is called the wavelength;

Fig. 1. Definition of a wave. Distribution in space at a particular instant of time.

if $(x \pm vt) = n\lambda + \phi$, where n is any integer and ϕ any arbitrary quantity, the variable a takes on the same values, thus demonstrating its periodicity. The velocity with which the wave propagates in space is v, the upper sign denoting progression in the direction of the negative x-axis, the lower sign in the direction of the positive x-axis.

$$\nu = v/\lambda \qquad (6)$$

is the frequency of the local variation in time.

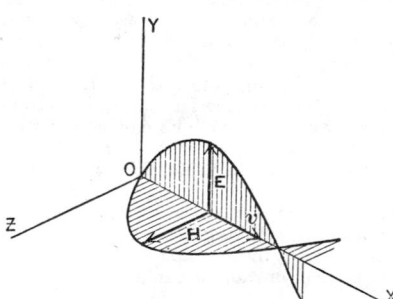

Fig. 2. The plane electromagnetic wave.

More complex, yet periodic, wave forms can be obtained by an algebraic superposition of several waves of different parameters. Instead of plane waves, cylindrical or spherical waves can arise if the origin of the waves is a linear infinitely long emitter or a point source, respectively.

Definition of Wave Packet. In dispersive media (see Dispersion, Art. 1), the velocity of propagation of radiation is a function of the frequency, and is related to the velocity in vacuum by $v = c/n$, if n is the refractive index. The superposition of a number of waves with infinitesimally different frequencies will, at a certain time t, give an absolute maximum for a point x. It can be shown that the progressing individual waves will superimpose at any other instant to give a maximum which appears to travel with a velocity

$$U = \frac{c}{\dfrac{d}{d\nu}(n\nu)},$$ which is called the group velocity of the wave train.

Longitudinal and Transverse Waves. The oscillation constituting the radiation can be either in the direction of propagation or perpendicular to it. In material substances waves of both types are possible; in vacuum only transverse waves are known. Radiation proper, which does not need any specifically known medium for its propagation, is composed of transverse waves.

The Electromagnetic Wave. The simplest case is the plane electromagnetic wave which is progressing in one direction and entirely homogeneous in all directions perpendicular to the direction of propagation. It is characterized by the electric field vector E, the magnetic field vector H, and the velocity of propagation vector v, which form a right-handed coordinate system (see Fig. 2). The wave is transversal, which means that both field vectors are oscillating in time perpendicularly to their direction of propagation. The velocity of propagation in any medium is given by

$$v = \frac{c}{\sqrt{\epsilon\mu}} \tag{7}$$

where c is the velocity of electromagnetic waves in vacuum (identical with the velocity of light); ϵ and μ are the relative dielectric constant and relative magnetic permeability of the medium, respectively.

The superposition of various plane electromagnetic waves leads to more complex forms of electromagnetic vibrations, and the theory can easily be extended to cylindrical and spherical electromagnetic waves.

Characteristics of the Wave Nature of Radiation

Certain phenomena can be explained only if a wave nature of radiation is assumed.

Interference is the property by which waves of the same frequency but traveling in different directions superimpose in such a manner that they give alternate regions of stationary high and low (or zero) intensities. In the case of light phenomena, dark and light spots or bands will alternate on a screen. Interference can be produced by diffraction through narrow slits (Young, 1801), reflection from mirrors under different angles (Fresnel, 1811), diffraction from ruled gratings (Rowland, 1885), and reflection from mirrors at different distances (Michelson, 1895). The interference patterns can serve to determine the wavelength of the radiation, as was first done for light by Young in 1801. The most recent applications were the confirmation of the wave nature of electrons by Davisson and Germer (1927), and of protons and helium atoms by O. Stern (1930).

Polarization. If radiation is a transverse wave phenomenon the oscillations may occur in any direction whatsoever. In the case of perfect symmetry about the direction of propagation, the radiation is said to be unpolarized or natural. If the oscillations constituting the radiation occur in one plane only, the radiation is said to be *plane polarized*, and the plane of oscillation is called plane of polarization. If the oscillations are uniformly rotating without change in amplitude, the radiation is said to be *circularly polarized*, either left-handed or right-handed according to the direction of rotation as seen by an observer. If the oscillations rotate and change in amplitude with the same period as they rotate, the radiation is said to be *elliptically polarized*, either left-handed or right-handed according to the direction of rotation as seen by an observer. Circularly polarized radiation can be decomposed into two plane-polarized radiations at right angles to each other, of equal amplitude, and in quadrature time phases. Elliptically polarized radiation can be decomposed into two plane-polarized radiations in the direction of two conjugate diameters with proper time phases. Plane-polarized radiation can be decomposed into two oppositely rotating circularly polarized radiations. Any experiment establishing polarization of a radiation suggests transverse waves.

Velocity of Propagation in Different Media. From the laws of refraction (see Art. 1) it follows that the velocity of propagation of waves in media should decrease with increasing refracting index. This was first established experimentally by Foucault in 1850 for light radiation and therefore serves to support the wave aspect of radiation in general.

The Electromagnetic Spectrum of Radiation

Electromagnetic Theory of Light. Maxwell was the first to assume the identity of light and electromagnetic waves, and Hertz demonstrated it by showing the phenomena of reflection and interference for short electromagnetic waves. Further experiments have shown that the state of polarization is determined by the properties of the field vector H which with the direction of propagation determines the plane of polarization. If the vector H is oscillating without changing direction, the electromagnetic wave is said to be plane polarized; if it is rotating, the wave can be circularly or elliptically polarized (see Polarization, above). The photographic action of light is due to the electric field vector E. The index of refraction is identified with $n = \sqrt{\epsilon\mu}$, where, of course, ϵ and μ, the dielectric constant and the magnetic permeability, have to be chosen as variable with the wavelength in order to account for the fact of dispersion.

Atomic Spectra. To account for the origin of atomic spectra (characteristic line radiation), Bohr assumed a model for the atom consisting of a positively charged nucleus at the center, which contained most of the mass of the atom, surrounded by moving electrons, which he postulated could exist only in certain orbits. This model is now known to be essentially correct. The nucleus is composed of protons and neutrons, particles whose masses are about 1836 times that of the electron, the proton carrying a positive charge equal in magnitude to that of the electron and the neutron carrying no net charge. Thus, in the neutral atom the number of electrons is equal to the number of protons in the nucleus (this is the atomic number, Z). The fact that only certain electronic orbits are allowed is accounted for by the stationary solutions of Schroedinger's wave equation or the matrix quantum rules of Heisenberg and Born. These allowed orbits tend to fall into groups and give rise to the orbital electronic shell structure of atoms. Electronic transitions between allowed states are associated with the emission or absorption of an electromagnetic wave (a photon) of frequency ν determined by

$$h\nu = E_2 - E_1 \tag{8}$$

where $E_2 > E_1$, emission occurring if the electron is initially in the state of higher energy, E_2; absorption if the transition is from the state of lower energy, E_1. h is Planck's quantum of action. These orbital electronic transitions give rise not only to visible radiation (light), but to ultra-violet and x-radiation as well, as shown in Table 1.

Table 1. The Electromagnetic Spectrum *

Wavelength Range	Source	Name	Name of Discoverer and Means of Detection and Study
∞ to 100 cm	Movements of electricity in large systems with capacity and self-inductance	Radio waves	Predicted by Maxwell, discovered by Hertz
100 cm to 0.022 cm	Electrical oscillations of minute systems, metal filings, magnetrons, klystrons, and traveling wave tubes	Microwaves	Nichols and Tear. Large gratings, interferometers, electrical detectors. Also Glagoliewa-Arkadiewa
0.022 cm to 7 $\times$ 10^{-5} cm	Oscillations or vibrations of charged atomic or molecular systems, ions in crystals or gas molecules; rotation of dipoles	Infrared or heat waves	Rubens and Bayer. Paschen. Gratings and residual rays
7 $\times$ 10^{-5} cm to 4 $\times$ 10^{-5} cm	Loosely bound outer valence electrons	Visible light	Prisms, gratings, interferometers
4 $\times$ 10^{-5} cm to 1.6 $\times$ 10^{-5} cm	Outer electrons more tightly bound	Ultraviolet	Schumann and Lyman gratings
1.6 $\times$ 10^{-5} cm to 1.2 $\times$ 10^{-6} cm	Inner electrons of light atoms, or electrons in stripped light atoms or shells next to valence shells of heavy atoms	Extreme ultraviolet or soft x-rays	Millikan and Bowen. Vacuum spectrograph and gratings
1.2 $\times$ 10^{-6} cm to 1.6 $\times$ 10^{-8} cm	Interior electrons of elements	Soft x-rays	Thibaud reflected from glass gratings at grazing incidence
1.6 $\times$ 10^{-8} cm to 1.25 $\times$ 10^{-9} cm	Innermost electrons of atoms, shorter waves apply to heaviest elements	X-rays hard and soft, K and L series	Crystal gratings. Method of Laue, Bragg and also from $Ve = 1/2 \; mv^2 = h\nu ; \nu = c\lambda$
1.25 $\times$ 10^{-9} cm to 5.56 $\times$ 10^{-11} cm	Nuclear electrons. Latter are hardest ones from RaC	γ-rays	Robinson, deBroglie, Ellis, Meitner. From $1/2 \; mv^2 = h\nu$, ν from magnetic fields
? $\times$ 10^{-12} cm to ? $\times$ 10^{-13} cm	Possibly creation of nuclei of complex atoms in space	Cosmic rays	Absorption coefficients in water and air. Kohlhörster, Hess, Millikan. Estimated from absorption. May be neutrons or particles

*From L. B. Loeb and A. S. Adams, *The Development of Physical Thought*, John Wiley & Sons, New York, 1933.

Molecular Spectra. The number of electrons in the outermost electronic shell determines the valence of an atom. The chemical binding of atoms into molecules is due to interactions between the valence electrons of the participating atoms. The electrons of a molecule give rise to spectra similar to those of the orbital electrons in atoms. In addition to molecular electronic spectra, vibrations of nuclei within molecules and molecular rotations give rise to radiations. These are similarly due to transitions between allowed energy states but are in the infrared and microwave regions of the spectrum. The line character of infrared spectra is often unresolved and hence they are called band spectra. Thermal energies at ordinary temperatures are not great enough to excite electronic transitions, but fall in the region of vibrational energies, hence the association of the infrared with thermal radiation.

Wave Nature of Matter. The attempts to explain the interaction of matter and radiation led to the conclusion that any particle of mass m moving with a speed v can be thought to be associated with a wave the frequency of which is given by the interrelation of energy:

$$h\nu = W = \frac{mc^2}{\sqrt{1 - (v/c)^2}} \tag{9}$$

where h is the quantum of action of Planck and c the velocity of light in vacuum. The velocity of propagation of the wave is defined as

$$V = c^2/v > c \tag{10}$$

in order to satisfy the wavelength relation for radiation

$$\lambda\nu = V \tag{11}$$

but no physical meaning can be attached to a velocity larger than the velocity of light in vacuum. The associated wave for electrons has a wavelength similar to those of very hard x-rays, and this has been verified by interference experiments on crystals by G. P. Thomson.

Electromagnetic Spectrum of Radiation. Table 1 shows the complete range of electromagnetic radiations from longest to shortest wavelengths with the supposed sources of radiation.

4. CORPUSCULAR ASPECTS OF RADIATION

A Corpuscle is defined as that which possesses a momentum of amount $p = mv$ and a kinetic energy $W = \frac{1}{2} mv^2$ according to Newtonian mechanics, or $p = \frac{mv}{\sqrt{1 - (v/c)^2}}$ and $W = \frac{mc^2}{\sqrt{1 - (v/c)^2}}$ according to relativistic mechanics. Its presence becomes obvious by collisions with other particles which can be made visible in certain devices (Wilson's cloud chamber).

The Photon. Electromagnetic radiation transports energy in amounts which are integer multiples of fundamental quanta ($h\nu$). Einstein suggested that a quantum of radiation, $h\nu$, be considered as associated with the characteristics of a corpuscle of velocity c, the velocity of light in vacuum. From the relativistic expression of energy (equation 9) it is seen that the mass of a photon must approach zero as its velocity approaches c. The momentum of a photon is defined as $p = h\nu/c$.

The Photon Gas. Radiation as a corpuscular phenomenon can be treated with statistical methods similar to the kinetic theory of gases with the fundamental difference that the distribution in frequency ν takes the place of the distribution in energy of gas molecules; each photon is assumed to have the same velocity c in vacuum. For statistical methods radiation is called the photon gas.

Characteristics of the Corpuscular Nature of Radiation

Rectilinear Propagation of Radiation. The most striking corpuscular characteristic of radiation is its propagation along straight lines so that definite shadows are cast. This led Newton to his corpuscular theory and initiated the whole field of geometrical optics (see Art. 5).

Collision of Radiation and Electrons. If monochromatic radiation of quantum $h\nu_1$ falls upon a scattering substance, the secondary radiation from the substance is of longer wavelength than the incident radiation, $\nu_2 < \nu_1$. This can be explained only if the photon colliding with an electron, or atom, loses some of its energy (its speed c cannot change) to the electron or atom which will be accelerated and proceed in a direction determined by the law of the conservation of momentum. The experimental evidence is due to A. H. Compton, and the phenomenon is consequently named *Compton effect*.

If monochromatic radiation of quantum $h\nu$ falls upon metals, electrons will be liberated with a kinetic energy given by the Einstein relation

$$^1/_2 \, mv^2 = h\nu - \omega_0 \tag{12}$$

where ω_0 is the amount of energy necessary to free the electron from its surroundings within the metal, and is independent of the frequency of the incident light. This *photoelectric effect* is explainable only by assuming corpuscular collision of radiation and electrons according to the laws of conservation of energy and momentum.

Penetrating Power of Radiation. Einstein's hypothesis attributes a momentum $p = h\nu/c$, proportional to frequency ν, to the photon, so that the shorter its wavelength, the more penetrating the radiation is from the corpuscular point of view. This agrees well with experimental facts. It is customary to measure the penetrating power by the absorption coefficient (see Art. 1) in a standard material.

The Corpuscular Spectrum of Radiation. The proportionality of momentum (corpuscular aspect) and frequency (wave aspect) of the photon leads to the same scheme of identification as Table 1 for the electromagnetic spectrum.

BIBLIOGRAPHY ON THEORY OF RADIATION

1. Bohr, N., *The Theory of Spectra and Atomic Constitution*, London, Cambridge University Press, 1922.
2. Bragg, W., *The Universe of Light*, New York, Macmillan, 1934.
3. Condon, E. U., and Shortley, G. H., *The Theory of Atomic Spectra*, London, Cambridge University Press, 1935.
4. Herzberg, G., *Atomic Spectra and Atomic Structure*, New York, Dover, 1944.
5. Jackson, J. D., *Classical Electrodynamics*, New York, John Wiley, 1962.
6. Jordan, E. C., *Electromagnetic Waves and Radiating Systems*, Englewood Cliffs, New Jersey, Prentice-Hall, 1950.
7. Kronig, R. de L., *Band Spectra and Molecular Structure,* London, Cambridge University Press, 1930.
8. Pauling, L., and Goudsmit, S., *The Structure of Line Spectra*, New York, McGraw-Hill, 1930.
9. Pringscheim, P., *Fluorescence and Phosphorescence*, New York, Interscience Publ., 1949.
10. Ramsey, N. F., "The Hydrogen Maser As a Frequency and Time Standard", *Metrologia*, p. 13 (1965).
11. Schiff, L. I., *Quantum Mechanics*, New York, McGraw-Hill, 1949.
12. Sommerfeld, A., *Atomic Structure and Spectral Lines*, London, Methuen, 1930.

GEOMETRY OF RADIATION

By Ernst Weber

Geometry of radiation deals with "rays" which are identical with the directions of propagation of the electromagnetic waves in the wave picture and with the paths of the photons in the corpuscular picture of radiation. It is applicable only if the interaction of radiation and matter involves objects whose dimensions are large compared with the wavelengths of the radiation considered. Thus, any problem of electromagnetic radiation from antennas can be treated by the geometrical method if the objects can be considered as large compared with the wavelength. Historically the first, and still the most important, application of the geometric method lies in the field of optics, although with proper interpretation the results can be applied to any other type of radiation.

5. PRINCIPLES OF GEOMETRICAL OPTICS

The ray from the light source is called the *incident* ray. If light in air is incident upon a polished surface of a transparent medium part of it will be regularly reflected, the other part will enter the medium. Referring to Fig. 1, the point of incident P is the point where an incident ray strikes the surface of the medium. The plane determined by the outside normal n to the surface at this point, and the incident ray, is called the plane of incidence, and the angle α which these two directions make is the angle of incidence.

Laws of Reflection and Refraction. Both the reflected and refracted rays lie in the plane of incidence, on opposite sides of the surface of the medium. The angle γ of the reflected ray with the outside normal, the angle of reflection, is equal to α, $\gamma = \alpha$; the angle β of the refracted ray with the inside normal, the angle of refraction, is determined from the law of Snell

$$\frac{\sin \alpha}{\sin \beta} = n \tag{1}$$

where n is the relative index of refraction of the medium against air. $n > 1$ for all transparent media. For normal incidence and regular optical behavior $\alpha = \beta = \gamma = 0$, no deviation of the ray occurs. If a plane surface is totally reflecting it is called a plane mirror.

Real and Virtual Image. Referring to Fig. 2, if MN indicates the surface of a plane mirror, the rays issuing from the source P and striking the mirror at the points A and B

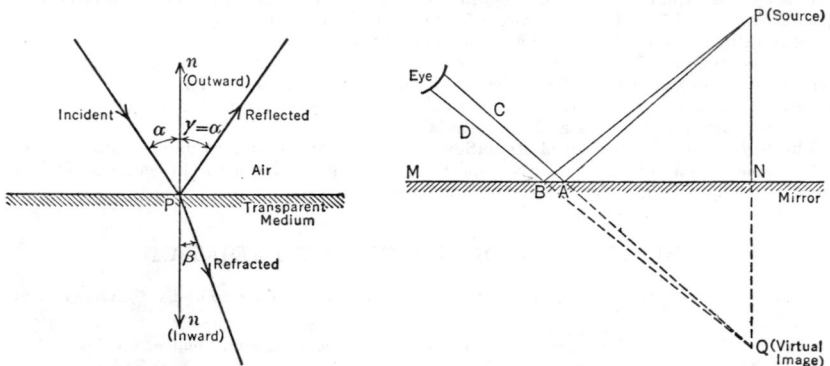

FIG. 1.   Reflection and refraction of a ray.   FIG. 2.   Virtual image from a plane mirror.

will be reflected and reach the eye as the rays C and D. To the eye these rays will appear as coming from the point Q behind the mirror. Q is called the *image* of P and more specifically a *virtual image* because the rays C and D do not actually pass through Q and therefore cannot be received there on a screen and be made visible. A *real image* is, then, one that can be received on a screen and be made visible.

Total Reflection. If light passes from a medium of large refractive index into one of small refractive index, the relative index of refraction according to equation 1 becomes less than unity. The angle α_c for which $\sin \beta = 1$ so that $\sin \alpha_c = n$ is called the critical angle because no light enters the second medium. For angles $\alpha > \alpha_c$, the light is said to be totally reflected as there is no refracted ray.

Optical Distance of two points in a medium of refractive index n is the actual length of the path l covered by the ray between the two points multiplied by the value of the refractive index. If a ray passes through different media, the optical distance between any two points is the sum of the optical paths within each medium, $\sum_i l_i n_i = \sum_i o_i$. A surface is called *aplanatic* if for each of its points the sum of the optical paths to two fixed points has a constant value.

Fermat's Principle. Using the definition of the optical distance, the laws of refraction and reflection can be expressed, according to Fermat, in a single principle: rays of light (and for that matter, of any radiation) travel along such lines that the optical distance between any two points of the rays is a minimum.

6. SPHERICAL MIRRORS AND LENSES

Concave Spherical Mirrors (Fig. 3). A spherical mirror is concave if its center of curvature C is on the side from which the light is incident. By the general law of reflection, a point source P will form a real image at a point Q so that PCQ lie on a straight line, the optical axis, and that the angle of incidence is equal to the angle of reflection. Rays through the center C are reflected in themselves; rays parallel to the optical axis converge at a point F midway between C and M, the principal focus, or simply *focus*, of the mirror. If the distances are designated as $MF = f$, $MP = p$, and $MQ = q$, the relation exists

$$1/p + 1/q = 1/f \tag{2}$$

which holds for any position of P if p, q, and f are counted positive on the right-hand side and negative on the left-hand side of the mirror center M. For negative values of q the images become virtual.

Convex Spherical Mirrors (Fig. 4). A spherical mirror is convex if the center of curvature C is on the side opposite to that from which the light is incident. The same relations

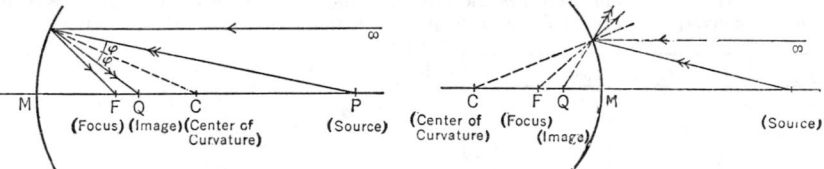

FIG. 3. Reflection from a concave spherical mirror.   FIG. 4. Reflection from a convex spherical mirror.

hold as for the concave mirror, except that the focal distance f has negative values in accordance with the proper interpretation of equation 2.

Images Formed by Spherical Mirrors (Figs. 5 and 6). The position of the image Q' of any point P' near but not in the optical axis of a mirror is found most expediently by the intersection of two principal rays, the ray through the center of curvature C and the

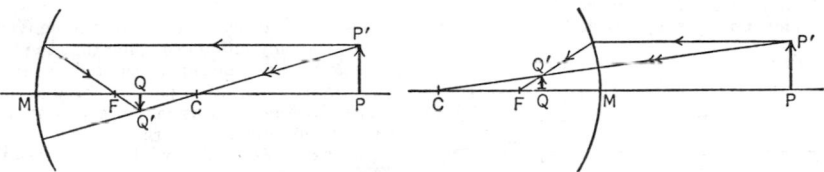

FIG. 5. Image formed by a concave spherical mirror.   FIG. 6. Image formed by a convex spherical mirror.

reflected ray through the focus F. Any object PP' perpendicular to the optical axis will produce an image QQ' again vertical to the axis and of definite characteristics; Table 1 shows all the possibilities of position and character of images formed by spherical mirrors.

Table 1.   Images Formed by Spherical Mirrors

(Figs. 3 and 4)

Mirror	Position of Object (PP') P	Position of Image (QQ') Q	Character of Image
Concave	At ∞	At F	Real
	Between ∞ and C	Between F and C	Real, inverted, diminished
	At C	At C	Real, inverted, same size
	Between C and F	Between C and ∞	Real, inverted, magnified
	At F	At ∞	
	Between F and M	From ∞ behind M to M	Virtual, erect, magnified
	At M	At M	erect, same size
Convex	At ∞	At F	Virtual
	Between ∞ and M	Between F to M	Virtual, erect, diminished
	At M	At M	erect, same size

Thin Spherical Lenses. A lens is a portion of a refracting medium which is bounded by two spherical surfaces or by one spherical and one plane surface. The straight line through the center of curvature of the two surfaces is called the axis of the lens. If one of the surfaces is a plane, then the axis is a straight line perpendicular to this plane and going through the center of the spherical surface. Any plane through the axis is said to be a principal section of the lens; Fig. 7 shows the principal sections of the common types of lenses.

Thin Convex Spherical Lenses (Fig. 8). By the general laws of refraction, a ray coming from the point source P in the axis of the lens will be refracted twice as it passes through the lens and form an image at the point Q. For thin lenses (large radii of curvature) relation 2 will hold true if the distances p and q are measured from the nearest surface of the lens and if both are taken positive when on opposite sides of the lens. f is the focal length and is defined by

$$1/f = (n - 1)(1/r_1 - 1/r_2) \qquad (3)$$

where r_1 and r_2 are the radii of curvature of the two surfaces, both taken positive if on the same side of the lens, and n is the refractive index of the lens. All rays parallel to the axis converge at a point F the principal focus, or simply focus, at a distance f from

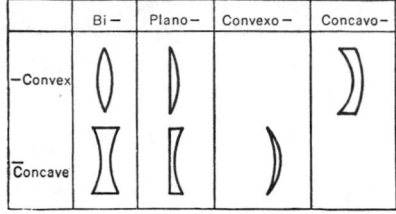

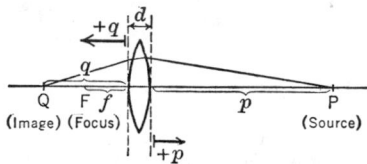

FIG. 7.   Principal sections of spherical lenses.   FIG. 8.   Refraction by thin convex lens.

the nearest surface.   Rays passing through the optical center will not be deviated.   For thin lenses the optical center can be approximately identified with the two points where the axis meets the surfaces of the lens.

Thin Concave Spherical Lenses (Fig. 9).   The same relations hold as for the convex lens, the only difference being that the focus is on the same side as the light source so that the rays, after passing through the lens, diverge; for this reason the concave lens is often called diverging lens.

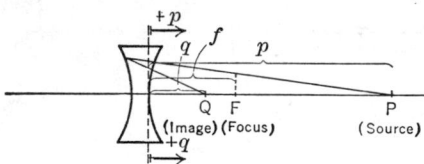

FIG. 9. ᶦRefraction by thin concave lens.

Images Formed by Thin Spherical Lenses (Figs. 10 and 11).   The form of the image Q' of any point P' near but not in the axis of the lens is found approximately but most expediently by the intersection of two principal rays, the ray through the optical center and the ray parallel to the axis, refracted through the focus. For the actual construction it suffices to consider the plane through the optical center and perpendicular to the axis as the refracting plane.   Any object PP' perpendicular to the axis will produce an image QQ' again perpendicular to the axis and of definite characteristics;   Table 2 shows the possible positions and characteristics of images formed by convex and concave lenses.

Table 2.   Images Formed by Spherical Lenses

(Figs. 8 and 9)

Lens	Position of Object	Position of Image	Character of Image
Convex	$u = +\infty$	$v = f$	Real
	$\|2f\| < u < +\infty$	$f < v < 2f$	Real, inverted, diminished
	$u = \|2f\|$	$v = 2f$	Real, inverted, same size
	$\|f\| < u < \|2f\|$	$2f < v < \infty$	Real, inverted, magnified
	$u = \|f\|$	$v = \infty$	
	$0 < u < \|f\|$	$-d < v < -\infty$	Virtual, erect, magnified
Concave	$u = +\infty$	$v = f$	Virtual
	$0 < u < +\infty$	$0 < v < f$	Virtual, erect, diminished

Spherical Aberration.   The rays of light incident, for example, upon a spherical mirror, from a light source P (Fig. 3), form an image at point Q; this is approximately true if the angle between light ray and optical axis is small.   For larger angles the image points occupy all the points on the axis between Q and M instead of meeting at a single point; an image line will be formed which constitutes the longitudinal,spherical aberration.   If a screen is placed at point Q the reflected rays will, for the same reason, illuminate a circular area instead of a sharp point image; this is referred to as lateral spherical aberration. The same holds for all types of spherical mirrors and lenses.

Chromatic Aberration.   Since the refractive index varies with wavelength, the image produced by white light from a convex lens, for example, will be, in fact, a superposition of many monochromatic images at slightly different foci. The yellowish green image is the brightest, and the one upon which the screen will be focused, so that the other images superimpose slightly out of focus and thus blur the resultant white image. This effect is

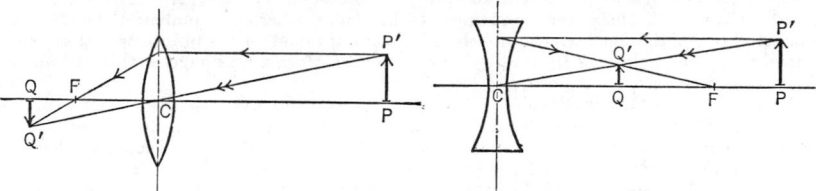

FIG. 10.   Image formed by thin convex lens.   FIG. 11.   Image formed by thin concave lens.

known as chromatic aberration, and can be minimized by using several lenses instead of a single one. An arrangement which tends to refocus the monochromatic images into a single sharp image is called achromatic. Chromatic aberration is produced with any type of single lens.

The Focal Length of a convex lens can most readily be determined on an optical bench. In front of a light source fine cross-wires are placed. The light, after passing through the lens, forms an image on a screen which is clearly focused. If p is the distance of the cross-wires, q that of the screen, from the center plane of the lens, then by equation 2 the focal length can be computed. With a concave lens it is necessary to use a convex lens also. From the resultant focal length F and the known focal length f of the convex lens that of the concave lens, f', can be determined by

$$1/f' = 1/F - 1/f$$

7. OPTICAL INSTRUMENTS *

Magnification, increasing visibility of detail, is secured by bringing the image of an object nearer to the eye, or by any other means of increasing the visual angle which the object subtends. Vision with the unaided normal eye is, however, most distinct at a distance of 25 to 30 cm because the accommodating mechanism of the eye is unable to focus sharply on the retina points nearer than this. A magnifying optical system produces in effect the required increase in the visual angle while forming an image (real or virtual) farther from the eye than the least distance of distinct vision. Often the arrangement is such that the eye views a virtual image at an infinite distance, so that the muscles of accommodation may be completely relaxed.

The Simple Microscope (Fig. 12). A single converging lens if placed closer to an object than the principal focal length produces an enlarged virtual image, which is seen on looking through the lens. The magnification produced is $1 + d/f$ for an eye whose least distance of distinct vision is d. A simple plano-convex lens, with the plane side toward the eye, gives good images for magnification less than eight diameters, that is, with focal lengths greater than about 3 cm. The image may be much improved,

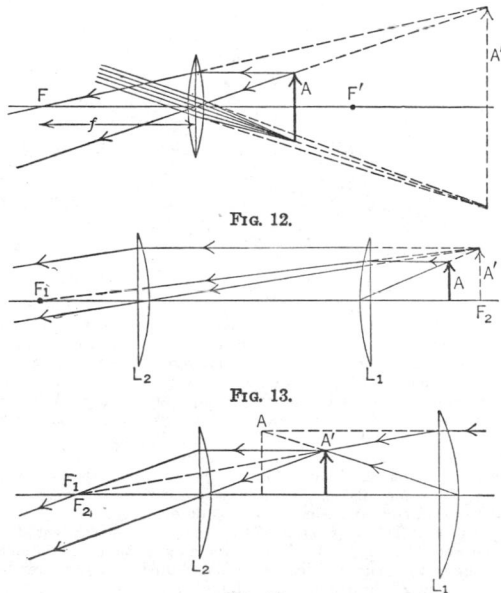

FIG. 12.

FIG. 13.

FIG. 14.

* From Merriman and Wiggin. *American Civil Engineers' Handbook*, 5th Edition, 1930.

especially where the magnification is considerable, by the use of special combinations of lenses designed to reduce spherical and chromatic aberration so as to give a fairly large field of view approximately free from distortion and color.

The **Ramsden** or **Positive Eyepiece** (Fig. 13) consists of two converging lenses, usually plano-convex, with their convex surfaces facing each other, of equal focal length, and separated by $2/3$ the focal length of either. A virtual image of the object or real image A is formed by the field-lens L_1 at A'. The eye-lens L_2 forms an image of this at infinity. This eyepiece is fairly, but not quite, achromatic.

Huyghens or **Negative Eyepiece** (Fig. 14). Two converging lenses, usually plano-convex, with the plane surfaces toward the eye, are so arranged as to divide equally between them the deviation produced on incident light parallel to and close to the axis. The field-lens L_1 has three times the focal length of the eye-lens L_2, and the two are separated by the difference in their focal lengths. Light which if unhindered would converge at A is deviated by L_1 to form an image at A', of which L_2 forms an image at infinity. This eyepiece is highly achromatic and free from disturbing spherical aberration.

For Measuring Microscopes and Telescopes in which the eyepiece is fitted with cross-hairs, the positive eyepiece is far more suitable than the negative because the image of the hairs being formed by both lenses is corrected for both chromatic and spherical aberration, and because the cross-hairs can be easily adjusted to suit different eyes by altering their distance from the eyepiece.

FIG. 15.

The **Compound Microscope** (Fig. 15) in its simplest form consists of two converging lenses. The objective L_1 forms within the tube a real, inverted, magnified image A' of the object A. This image is viewed through the eyepiece L_2 and further magnified. A microscope is usually fitted with either a Huyghens or a Ramsden eyepiece, according to the purpose for which it is to be used. The objective is also generally a combination of several lenses to overcome spherical and chromatic aberration while admitting as much light as possible. In microscopes of the highest power a drop of oil of cedar is placed between the slide and the objective; this is known as "immersion."

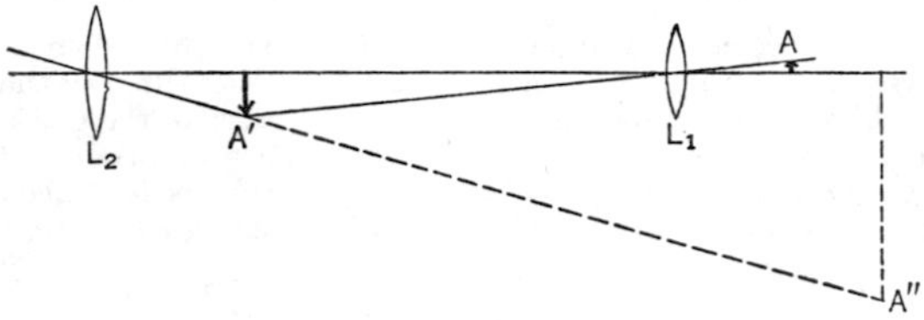

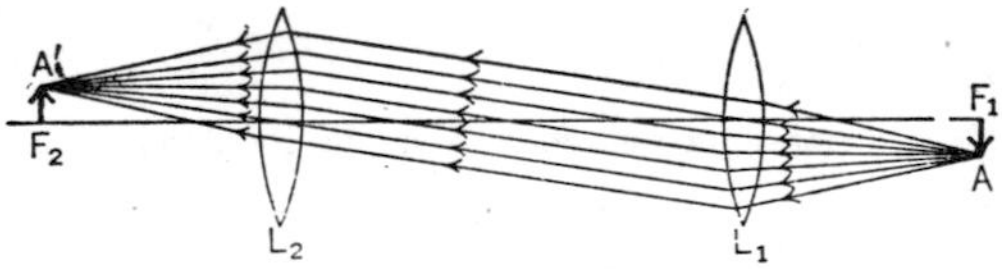

FIG. 16.

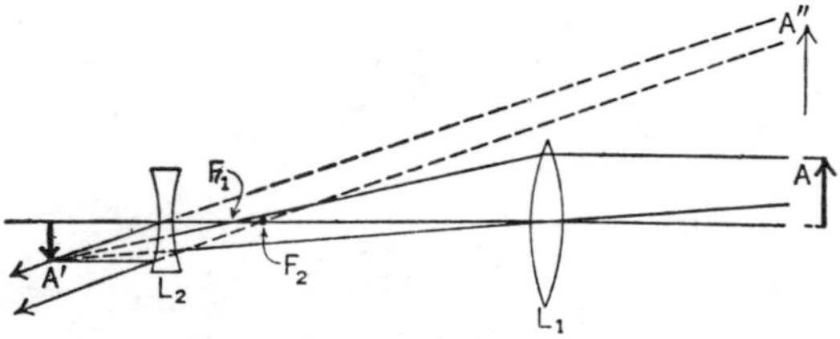

FIG. 17.

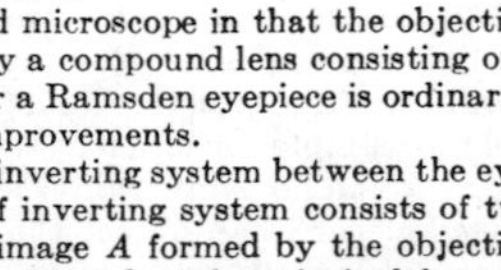

FIG. 18.

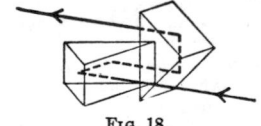

The smallest interval that can be optically resolved is about 0.00005 mm, and the limit of resolution of the microscope is attained when the total magnification is about 1200.

The **Astronomical Refracting Telescope** differs from the compound microscope in that the objective forms a reduced image of a distant object. The objective is generally a compound lens consisting of a convex lens of crown and a concave lens of flint glass. A Huyghens or a Ramsden eyepiece is ordinarily used; but the best instruments employ eyepieces embodying later improvements.

The **Terrestrial Telescope** (Fig. 16) produces an erect image by an inverting system between the eyepiece and the inverted image formed by the objective. One form of inverting system consists of two converging lenses of equal focal length so placed that the inverted image A formed by the objective is at the principal focus of the first lens. An erect image A' is then formed at the principal focus of the second lens, and is magnified by an eyepiece.

Galileo's Telescope (Fig. 17) consists of a convex lens L_1 for objective and a concave lens L_2 for eyepiece. The light from L_1, converging so that if unhindered it would form at A' a real, inverted image of the distant object A, is intercepted by L_2 and rendered parallel or slightly divergent as if it came from A'', which is a virtual, erect image of A. The use of the diverging eye-lens limits considerably the angular field of view. Ordinary field-glasses and opera-glasses are Galilean telescopes.

The **Prism Binocular** (Fig. 18) secures the wide field of view that accompanies the use of a converging eyepiece, at the same time avoiding the inconvenient length of the ordinary terrestrial telescope. This is accomplished by employing four total reflections within two right-angled prisms to invert the image formed by the objective. Otherwise the construction is the same as that of the astronomical telescope. This prism construction permits a considerable shortening of the telescope by separating the prisms, since the light traverses the distance between them three times. In addition the stereoscopic effect due to binocular vision may be greatly increased by placing the centers of the objectives much farther apart than the pupils of the eye. The increased field of view is obtained at a sacrifice of illumination.

In **Reflecting Telescopes** the object lens is replaced by a concave mirror with 'parabolic surface. The mirror is of glass and coated with a thin film of either silver or aluminum. The parabolic surface of the mirror renders it completely achromatic and, therefore, of special value in astronomic photography. The first reflecting telescope was constructed by Newton, in 1670, and is known as the Newton Reflector. The telescope in use at the Mount Wilson Observatory in California is 100 inches in diameter and 43 feet in focal length. A new reflecting mirror of 200 inches in diameter has just been completed and is installed at the Mount Palomar Observatory.

BIBLIOGRAPHY ON GEOMETRY OF RADIATION

1. CONRADY, A. E., *Applied Optics and Optical Design*, Oxford University Press, England, 1929.
2. DITCHBURN, R. W., *Light*, New York, John Wiley, 2nd ed., 1963.
3. HARDY, A. C., PERRIN, F. H., *The Principles of Optics*, New York, McGraw-Hill.
4. HOUSTOUN, R. A., *A Treatise on Light*, London, Longmans, Green, 1933.
5. KÖNIG, A., *Geometrische Optik*, Akademische Verlags-Gesellschaft, Leipzig, 1929.
6. MARTIN, L. C., *Optical Measuring Instruments*, London, Blackie & Sons, Limited.
7. STEINEIL, A., VOIT, E., *Applied Optics* (Trans. from the German by J. W. FRENCH), London, Blackie & Sons, Limited.

PHYSICS OF RADIATION
By Ernst Weber

For electromagnetic radiation from resonating circuits and antennas, and thermionic emission, see the volume *Electrical Communication* of this handbook series.

8. PROPAGATION OF LIGHT

The Velocity of Light * in vacuum is approximately 186,000 miles per second, and was first determined in 1675 by O. Roemer, a Danish astronomer. According to recent measurements by Michelson (1926), the velocity of light in vacuum is $2.99796 \pm 0.00004 \times 10^{10}$ cm per sec, but a slight variation is suspected to take place.

The Color of Light is determined by its spectral distribution. The range of visible light extends from about $\lambda = 380\ m\mu = 3800\ A$ (extreme violet), to about $\lambda = 780\ m\mu = 7800\ A$ (extreme red).

The Refractive Index of light varies considerably with wavelength, normally increasing in value with decreasing wavelength (normal dispersion). It is customary to give refractive indices for yellow light of $\lambda = 589.3\ m\mu$ wavelength (bright radiation of sodium vapor) in order to facilitate comparison. For *gases* the refractive index is not much different from unity and varies only slightly with wavelength. Table 1 gives some representative

Table 1. Index of Refraction of Gases

(From *Smithsonian Tables;* reduced to 0 C and 760 mm Hg, at wavelength $\lambda = 589.3\ m\mu$, sodium light)

Substance	Symbol	Index of Refraction	Substance	Symbol	Index of Refraction
Hydrogen.......	H_2	1.000 132	Carbon monoxide	CO	1.000 346
Water vapor.....	H_2O	1.000 249 to 1.000 259	Carbon dioxide..	CO_2	1.000 448 to 1.000 454
Oxygen.........	O_2	1.000 27J	Sulfur dioxide...	SO_2	1.000 686
Air.............	.....	1.000 2926	Benzene........	C_6H_6	1.001 700 to 1.001 823
Nitrogen........	N_2	1.000 296 to1.000 298			

values. For *liquids* the refractive index lies between 1 and 2 and shows variation with temperature as well as with wavelength. Table 2 gives some representative values.

Transparent Solids, if optically *isotropic*, refract regularly. The most important representative is glass, which has very marked dispersive properties in the visible range; this

* The expression "light" is commonly used to designate monochromatic visible radiation (scientific usage) or total visible radiation from a light source.

fact is utilized in prism spectroscopy. Table 3 shows the variation of the refractive index with wavelength for various standard kinds of glass in general use for optical purposes.

Table 2. Index of Refraction for Liquids

(At 20 C and at wavelength λ = 589.3 mμ, sodium light; various sources)

Substance	Symbol	Index of Refraction	Substance	Symbol	Index of Refraction
Water............	H_2O	1.332 99	Benzene........	C_6H_6	1.501
Ethyl ether......	$C_2H_5OC_2H_5$	1.351	Phenol..........	C_6H_5OH	1.550
Ethyl alcohol....	C_2H_5OH	1.361	Bromine........	Br	1.654
Gylcerin.........	$C_3H_8O_3$	1.474			

Table 3. Index of Refraction of Glass

(From *Handbook of Chemistry and Physics*, 1949)

Variety of Glass	Wavelength in Millimicrons							
	361	434	486	589	656	768	1200	2000
Zinc crown................	1.539	1.528	1.523	1.517	1.514	1.511	1.505	1.497
Higher dispersion crown.....	1.546	1.533	1.527	1.520	1.517	1.514	1.507	1.497
Light flint................	1.614	1.594	1.585	1.575	1.571	1.567	1.559	1.549
Heavy flint................	1.705	1.675	1.664	1.650	1.644	1.638	1.628	1.617
Heaviest flint................	....	1.945	1.919	1.890	1.879	1.867	1.848	1.832

Transparent Crystals, if optically anisotropic, have, in general, different propagation properties in all three space directions and therefore must be characterized by three principal refractive indices. They exhibit certain symmetries with respect to two definite crystallographic directions, called the optical axes; the crystals themselves are called *optically biaxial*.

Certain *transparent crystals* exhibit rotational symmetry with respect to a single crystallographic direction, called the optical axis; the crystals themselves are called *optically uniaxial* and must be characterized by two principal refractive indices. The most widely known representatives are quartz and Iceland spar. Very complete tables of the indices of refraction can be found in the International Critical Tables.

Metals absorb as well as reflect light; in general, they are entirely opaque, except when used in very thin films. The optical behavior of metals is characterized by refractive index n, index of absorption K, and reflectivity R. The optical indices of metals are not directly measurable; they must be computed from the observation of changes of polarization (see Art. 9) in the reflected light. In general, the metals show an apparent irregular and large variation of their optical indices with the wavelength of the incident light; this has led to a theory of resonance of the conduction electrons. Table 4 gives the optical

Table 4. Optical Constants of Metals

(At a wavelength λ = 589.3 mμ; various sources)

Metal	Symbol	Index of Refraction n	Index of Absorption K	Reflectivity R	Metal	Symbol	Index of Refraction n	Index of Absorption K	Reflectivity R
Aluminum...	Al	1.44	5.32	0.83	Nickel......	Ni	1.79	1.86	0.62
Bismuth.....	Bi	1.90	1.93	0.65	Silver......	Ag	0.18	20.2	0.95
Copper.....	Cu	0.64	4.08	0.73	Steel.......	......	2.7	1.28	0.57
Gold........	Au	0.36	7.70	0.85	Tungsten...	W	3.46	0.94	0.54
Mercury.....	Hg	1.73	2.87	0.78					

indices for a few metals at a wavelength λ = 589.3 mμ (sodium light) as determined by Drude.* All metals show low reflectivity for ultraviolet, increasing values through the visible spectrum, and act as almost perfect mirrors for infrared and longer wavelengths.

Most common materials reflect diffusely. Table 5 gives the diffuse reflective power for the more important engineering materials. Rather complete tables of spectral reflec-

* P. Drude, *Wied. Annalen*, 1890, Vol. 36, p. 885; Vol. 39, p. 481.

Table 5. Diffuse Reflecting Power

(For visible spectrum; various sources)

Material	Reflecting Power in Per cent	Material	Reflecting Power in Per cent
Cloth		Paper—*Continued*	
White linen, dull finish	81	Olive green	15
Red cotton (diamine fast red)	44	Ultramarine blue	3.5
Blue woolen (lanacyl)	25	Black	5
Blue woolen (salacine)	15	Pigment	
Blue flannel	17.5	Chromium oxide, Cr_2O_3	27
Blue linen (navy blue)	17	Cobalt oxide, Co_2O_3	3
Black cotton (diamine)	33	Lead oxide, PbO	52
Black cotton (columbia)	29	Lead carbonate, $PbCO_3$	90
Black cotton (sulfur)	2.4	Magnesium oxide, MgO	86
Black felt	14	Red iron oxide, Fe_2O_3	52
Black velvet	1.8	Zinc oxide, ZnO	82
Black woolen (salacine)	12	Stone and minerals	
Green leaf (tulip tree)	22	Asphalt (pavement)	15
Lampblack (paint)	3.2	Macadam road	12
Paints		Granolith (pavement)	17
White lead	75	Bluestone (sandstone)	18
Zinc lead	69	Limestone (Indiana)	43
Paper		Brick, light buff	48
White blotting	82	darker	40
Cheap white	70	red	30
Light gray	73	darker and glazed	23
Medium gray	45	Feldspar	39
Dark gray	20	Quartz (powder)	81
Pink	60	Marble white, unpolished	53.5
Buff	60	Slate (dark clay)	6.7
Chocolate brown	20		

tivity of pigments and transmissivity of dyes can be found in M. Luckiesh, "The Basis of Color Technology," *Journal of the Franklin Institute*, Vol. 184, p. 73, 1917; and M. Luckiesh, "The Measurement of the Transmission Factor," *Journal of the Franklin Institute*, Vol. 186, p. 111, 1918.

9. POLARIZATION OF LIGHT

Double Refraction and Polarization. The fact that optically uniaxial crystals propagate light selectively with respect to their optical axis can best be explained by the assumption of the polarization of light. As defined in Art. 3, a light wave is called plane polarized if the oscillations of each of the electromagnetic field vectors occur in a fixed plane. In the case of optically uniaxial crystals (in relation to which the concept of polarization was originally introduced), double refraction takes place; that is, an incident pencil of light, propagating in a direction other than the optical axis, is broken up into two pencils: the ordinary and the extraordinary "rays," both of which travel with different velocities within the crystal. The two resulting rays, on emerging, are plane polarized in two perpendicular planes. It is customary to call "ordinary ray" that ray which is plane polarized *in* the principal plane (a plane parallel to the optical axis and the incident ray); this ray is uniformly refracted in all directions and thus behaves like ordinary light. It is customary to call "extraordinary ray" that ray which is plane polarized perpendicular to the principal plane; this ray is non-uniformly refracted, the refractive index varying symmetrically about the optical axis. The extraordinary ray thus behaves differently from ordinary light. Upon leaving the crystal, the two plane-polarized waves have, on account of their different velocities within the crystal, a phase difference

$$\delta = \frac{2\pi d}{\lambda}(n_1 - n_2) \tag{1}$$

where d is the length of the path within the crystal, λ the wavelength in air, and n_1 and n_2 the refractive indices for the two rays in the directions of propagation. The resulting light from a uniaxial crystal is, therefore, in general, elliptically polarized.

Production of Plane-polarized Light. Double refraction provides a means of producing plane-polarized light from natural light by suppressing the ordinary ray, as for example, in a Nicol prism. In the case of tourmaline the high value of absorptivity for the ordinary ray naturally suppresses it so that the plane-polarized extraordinary ray remains. If light falls upon transparent substances at such an angle of incidence that reflected and refracted rays are perpendicular to each other, the reflected light will be almost completely plane polarized (Brewster's law).

Photoelasticity. Isotropic transparent substances, if subjected to strain, become doubly refracting. Plane-polarized light, in passing through the substance, is then decomposed into two rays, polarized, respectively, parallel and perpendicular to the direction of stress. On account of their different velocities of propagation they suffer a relative phase difference, so that the light leaving the medium is, in general, elliptically polarized (see Art. 3). The amount of the phase difference is proportional to the stress and the thickness of the medium, so that a stress analysis can be performed. For practical purposes, models of the structure to be investigated are cut out of xylonite,* and subjected to stresses proportional to those acting on the original.

Optical Rotation. In many substances plane-polarized light experiences a rotation of the plane of polarization proportional to the length of the path within the substance. According to Fresnel, it is thought that the plane-polarized ray decomposes into two oppositely rotating circularly polarized rays which propagate with different velocities, so that they suffer a relative phase difference; on leaving the substance, they combine into a plane-polarized ray, the plane of polarization of which appears to be rotated with respect to that of the incident light. The specific rotation is measured in angular degrees per decimeter for liquids and solutions, and in angular degrees per millimeter in solids; it is called negative if the rotation is left-handed, and positive if it is right-handed, as seen by an observer. Substances exhibiting optical rotation are called *optically active.* Applications are found in commerce and medicine in the test for sugar. Instruments used to measure the optical rotation are called saccharimeters or polarimeters. Table 6 shows the values of specific rotation of a few common substances at 20° C, for sodium light. Quartz shows a very marked increase of the specific rotation with decreasing wavelength. Rather complete tables can be found in the International Critical Tables, Volume 7.

Table 6. Specific Optical Rotation

(At 20 C and λ = 589.3 mμ, sodium light; various sources)

Substance	Specific Rotation	Unit
Solids:		
Quartz...............................	+ 21.68	Angular degrees per millimeter
Sodium bromate.......................	+ 2.8	" " " "
Liquids:		
Amyl alcohol..........................	− 5.7	" " " decimeter
Nicotine..............................	−162	" " " "
Turpentine............................	− 37	" " " "
Solutions:*		
Albumine..............................	− 25 to 38	" " " "
Dextrose..............................	+ 52.25	" " " "
Lactose...............................	+ 52.4	" " " "
Levulose (fruit sugar).................	− 87.1	" " " "
Maltose...............................	+138.3	" " " "
Sucrose (cane sugar)..................	+ 66.3	" " " "

* Solvent; water (1 g in 100 g).

Magnetic Rotation. Optically isotropic substances, when exposed to a strong magnetic field, rotate the plane of polarization of plane-polarized light. This is called the *Faraday effect* after its discoverer (1845), and depends on the magnetic field intensity H. It is termed positive if the rotation occurs in the direction of the current which produces the magnetic field. The angle of rotation in angular minutes is given by

$$\theta = rlH \cos \alpha$$

where r is Verdet's constant (see Table 7), l (centimeters) the length of the path of light within the magnetic field of intensity H (oersteds), and α the angle which H makes with the ray of light.

* Tuzi, "A New Material for the Study of Photoelasticity," *Sci. Papers Ins. Phys. Chem. Res.*, Tokio, 7 (1927), p. 79.

Table 7. Magneto-optic Rotation

(At 20 C and wavelength λ = 589.3 mμ, sodium light; various sources)

Substance	Verdet's Constant	Remarks
Solids:		
Flint glass (medium)................	+ 0.0420	
Quartz ($\perp$ to optical axis)..........	+ 0.0172	
Liquids and solutions:		
Benzene........................	+ 0.0297	
Ethyl alcohol.....................	+ 0.0107	
Ferric chloride...................	− 0.2026	Aqueous solution, 1.4331 g per cm^3
Ferrous chloride.................	+ 0.0025	" " 1.6933 "
Sodium carbonate.................	+ 0.0140	" " 1.1006 "
Water...........................	+ 0.0130	
Gases (at atmospheric pressure):		
Air.............................	+ 6.83 × 10^{-6}	
Carbon dioxide...................	+13.00 "	
Nitrogen........................	+ 6.92 "	
Oxygen.........................	+ 6.28 "	

Kerr Effect. Transparent substances, when subjected to a strong electrostatic field, become doubly refracting. When plane-polarized light passes through a transparent substance, it is decomposed into two rays, one of which is polarized in the direction of the electric field and the other of which is polarized perpendicular to the electric field. On account of their different velocities of propagation, the rays suffer a relative phase difference, so that the light leaving the medium is in general elliptically polarized. A high-voltage condenser with liquid dielectric (preferably nitrobenzene) employing this electro-optical effect is called a *Kerr cell*, and is of use in some phases of television.

10. LIGHT SPECTROSCOPY

The Visible Spectrum * is the graphic arrangement of the visible radiant energy against wavelength or frequency. The wavelength is usually measured in millimicrons, 1 mμ = 10^{-7} cm, or in angstroms, 1A = 10^{-8} cm (practically). Three different kinds of spectra are commonly distinguished: the emission spectrum, representing the spectral distribution of emissive power of a light source; the reflection spectrum, representing the spectral distribution of reflectivity of a regularly reflecting surface; and the absorption spectrum, representing the spectral distribution of the absorbing power of transparent substances. The best-known representative of the last type is the solar absorption spectrum, or the Fraunhofer lines, discovered in 1817.

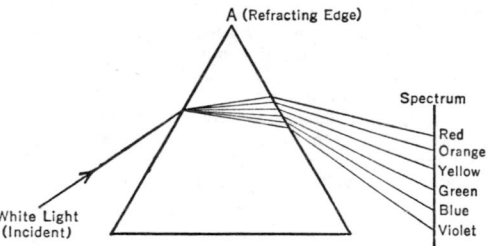

Fig. 1. Dispersion of white light by a glass prism.

Spectroscopy. The spectral distribution of radiant energy composing light can be directly observed in a spectroscope. The most common type of spectroscope, first used by Newton (1666), utilizes the dispersion of light passing through a prism of glass (see Fig. 1). Later types of spectroscopes use the diffraction of light by fine gratings (Rowland, 1882), and the interference patterns resulting from reflections by mirrors (interferometer by Michelson). A very high-power spectroscope, using an echelon grating (Michelson, 1898), has been particularly successful in finding the fine structure of spectral lines. Monochromatic radiations are located by comparison with standard radiations. Table 8 gives a few of the wavelengths useful for the calibration of spectroscopes and their respective sources. The primary standard for absolute wavelength measurements is the wavelength of the red cadmium line determined by Benoist, Fagry, and Perot (1907), as 6438.4696 angstroms, and adopted internationally.†

---

* The term visible spectrum is also used to designate the band of colors constituting white light.

† The adoption of the wavelength standard as exactly 6438.4696 angstroms represents, in fact, an absolute definition of the unit angstrom which is slightly different from the practical definition 1A = 10^{-8} cm as derived unit. It is, therefore, generally preferred to give wavelengths in millimicrons.

Table 8.   Wavelengths for Calibration of the Spectroscope

(Selected from R. A. Houstoun, *A Treatise on Light*)

Source	Wavelength mμ	Color	Corresponding Fraunhofer Absorption Line
Potassium nitrate on Pt wire in Bunsen flame...	{ 770. 2 { 766. 8	Red	
Lithium sulfate on Pt wire in Bunsen flame	670.78	Red	
Hydrogen vacuum tube (α line)...............	656. 28	Red	C
Sodium bicarbonate on Pt wire in Bunsen flame..	{ 589.00 { 589.59	Orange	D_2 D_1
Mercury arc...............................	546.07	Green	
Thallium chloride on Pt wire in Bunsen flame....	535.07	Green	
Hydrogen vacuum tube (β line)..............	486.14	Greenish blue	F
Hydrogen vacuum tube (γ line).............	434.04	Indigo	f
Cadmium spark in air......................	396.84	Violet	H

The Performance of a Spectroscope is judged by the brightness of the spectrum, which depends in the first place on the aperture of the instrument, i.e., the ratio of diameter to focal length of the collimator lens which is the convex lens between light source and prism or grating; and by the *resolving power*, which is the ratio of any given wavelength λ to the smallest increment $d\lambda$ that can be distinctly observed, i.e., $\lambda/d\lambda$. The resolving power varies with the wavelength and spectroscope; it is largest for diffraction and echelon gratings, where it can reach 250,000.

Origin of Line Spectra. The most important part of spectroscopy is the one dealing with the line emissions or characteristic spectra of the various chemical elements. The grouping of lines into series was first established empirically for the hydrogen spectrum, and a formula was given by Balmer (1885), later generalized by Rydberg (1890) as

$$n = N \left(\frac{1}{(m_1 + \beta_1)^2} - \frac{1}{(m + \beta)^2} \right) \qquad (2)$$

where N is a universal constant, the Rydberg constant, and n are the wave numbers (number of wavelengths per centimeter) of the various lines of any one series obtained by letting m take all integer values from unity on. The constants β and β_1 are characteristic for each individual series, and m_1 is a definite integer which determines the type of a series; thus $m_1 = 1$ gives the Lyman (ultraviolet), $m_1 = 2$ the Balmer (visible), $m_1 = 3$ the Paschen, and $m_1 = 4$ the Brackett (the last two in the infrared), series of the hydrogen atom. In general each spectral line can be either a singlet, if even with largest resolving power only a single line can be detected, or a doublet, triplet, up to octet. The whole series is then referred to as a multiplet series, its order being equal to that of the highest order line that occurs in the series.

Energy Level Diagrams. For all elements the wave numbers of their characteristic line radiations can be brought into a form similar to equation 2 which was interpreted by Bohr (see Art. 3) as the difference of two definite energy terms. If multiplied by hc (h = Planck's constant, c = velocity of light in vacuum), equation (2) becomes

$$hcn = h\nu_{rs} = W_r - W_s \qquad (3)$$

where ν_{rs} is the frequency of the spectral line. The energy terms can be computed from the known wavelengths of the radiations, and if arranged systematically, ascending from lower to higher absolute values of energy, the whole system of terms for an element is called its *energy level diagram*, which can be represented graphically. In the diagram each energy level is a *term* and belongs to a *sequence* which is designated by a capital letter as S (sharp), P (principal), D (diffuse), F (fundamental). The main terms of each sequence are differentiated by integer order numbers (the values m or m_1 in equation 2) in front of the letter symbols; a superscript indicates the multiplicity of the sequence (determined by the highest order multiple in the sequence); and a subscript designates the component of the multiple within the main term. For example, 3^4P_2 is the second component of the third main term of the principal sequence which is a quadruplet. The spectral lines then are given by transitions like $3^4P_2 \rightarrow 2^3D_1$, the difference in the corresponding energies determining uniquely the frequency of radiation emitted according to equation (3). Empirical selection rules prohibit certain lines so that transitions are possible only between neighboring sequences, and only if the changes of the subscripts are ±1 or 0.

Excitation of Line Spectra. Emission of the characteristic radiation occurs when electrons "fall" from higher to lower energy levels. For this to happen, atoms must first absorb energy and be "raised" to higher energy levels, which can be done either by bombardment with high-speed electrons (arcs, sparks), irradiation with x-rays, resonance radiation (incident radiation of same frequency as emitted radiation), or by collisions with atoms of the same or other elements.

Doppler Effect. A light source moving towards an observer with a relative radial velocity s appears to emit radiations of wavelength λ'

$$\lambda' = \frac{v - s}{v} \cdot \lambda$$

where v is the velocity of light and λ the absolute wavelength emitted by the source (Doppler, 1843). Spectroscopic identification of stars and planets reveals their relative velocities, s, when the spectral lines are compared with earthly sources. At very low gas pressures the thermal velocity of molecules can also produce a Doppler effect.

Zeeman Effect. Magnetic fields split the spectral lines of elements into a number of weaker lines, whose observed polarizations depend on the direction of observation relative to the magnetic field direction. The amount of splitting depends on the magnetic field strength and on the strength of the magnetic moment of the interacting atomic system. Zeeman (1896) first discovered this effect and propounded a simple theory. The effect is explained in detail by quantum mechanics. A similar effect occurs in absorption spectra. This is called the "inverse Zeeman effect," and occurs, for example, in the sun spots, indicating vast magnetic storms.

Stark Effect. Stark (1913) discovered that strong electric fields split spectral lines of the elements into a number of weaker lines. The effect is similar to the Zeeman effect, and again the observed polarization of the components depends upon the relative direction of observation with respect to the applied electric field direction. The effect is explained by quantum mechanics, and the applied electric field is generally treated as a perturbation on the atomic system.

Raman Effect. Transparent substances, if illuminated with strictly monochromatic light, can exhibit in their spectrum in addition to the strong line of the incident frequency (the only one to be expected according to the classical theory of scattering light) fainter lines of lower and higher frequencies (Raman, 1928). The presence of the new frequencies is explained by assuming that light is composed of photons which either lose energy to molecules or gain energy from excited molecules, as they pass through the substance.

11. ULTRAVIOLET RADIATION

The ultraviolet spectrum extends from approximately 380 mμ wavelength down to the softest x-rays. The longer ultraviolet rays, of vital importance for organisms, are often called *actinic rays*.

Absorption. Most optically transparent substances absorb ultraviolet radiation. Thus ordinary window glass does not transmit wavelengths shorter than approximately 350 mμ, whereas uviol glass, 1 cm thick, stops transmitting at approximately 300 mμ as do ordinary crown and flint glass. Quartz 1 mm thick is transparent for ultraviolet radiation down to a wavelength $\lambda = 170$ mμ, fluorite even to 100 mμ. One millimeter of air (at a pressure of 760 mm Hg) absorbs all wavelengths below $\lambda = 170$ mμ; the solar spectrum as measured on the surface of the earth stops at 250 mμ on account of the absorption by the air.

Refraction and Reflection. The refractive indices of the ultraviolet-transparent glasses and crystals show normal dispersion. Metals are highly absorbing and have low reflectivity in the ultraviolet region.

Spectroscopy of Ultraviolet. Since all glass absorbs ultraviolet radiation to a considerable extent, prisms and lenses for ultraviolet spectroscopy must be made of quartz or fluorite. In using quartz, correction must be made for the effects of double refraction and optical rotation. For spectrophotographic investigations of ultraviolet radiation of the shorter wavelengths, a vacuum spectrograph must be used on account of the absorption characteristics of air. Convenient sources of ultraviolet radiation are electric sparks between metals; each metal has a definite and characteristic spectral line distribution, one of which can be used as a standard for wavelength measurements. The most efficient source is the mercury-vapor arc.

The ultraviolet spectrum can be made visible by fluorescent screens or by photography.

Fluorescence. Many organic substances in solution emit characteristic monochromatic visible radiation, that is fluorescent, when illuminated by white light or ultraviolet radiation. These fluorescent substances strongly absorb radiations in the near and extreme ultraviolet region. Hence, to explain fluorescence, quantum theory assumes that this range will excite the electrons and raise them to high energy levels from which they may fall back in one or in several energy steps; in the latter case this gives rise to characteristic visible radiations. Table 9 gives a few organic substances exhibiting strong fluorescence (from *Handbook of Chemistry and Physics*, 1933). Inorganic gases and vapors, as well as crystals, show fluorescence under certain conditions. Much valuable information can be found in R. W. Wood's *Physical Optics*. Practical use of fluorescence is made in luminous paints and in the analysis of paints, oils, and rubber, whereby impurities in composition can be detected by differences in fluorescent properties.

Table 9. Fluorescence of Organic Substances in Solution

(From *Handbook of Chemistry and Physics*)

Substance	Solvent	Wavelength (mμ)	Color
Anthracene	Alcohol	400, 430, 436	Violet
Quinine sulfate	Water	437	Violet
Esculine	Alcohol	460	Blue
Fluorescin	Water (alkaline)	542	Green
Rhodamin	Water	554	Yellow
Eosine	Alcohol or water	589	Yellow
Naphthalin, red	Alcohol	632	Orange
Resorcin, blue	Water	650	Red

Phosphorescence. A number of crystals show persistent fluorescence, even after the exciting source has been removed. Usually any prolonged luminescence of crystals is called phosphorescence; however, a substance is truly phosphorescent only if small metallic impurities cause the storage of radiation energy.

Photography. Light-sensitive emulsions of silver bromide in gelatin on glass are capable of forming pictures by chemical action. Certain very small sulfur-containing organic bodies present in the gelatin react with the incident light and upon "development" form centers for the reduction of silver bromide to metallic silver. Ordinary dry plates are light sensitive in the wavelength range of 220 to 500 mμ with a maximum of sensitiveness in the violet. Below 250 mμ the gelatin begins to absorb strongly so that special plates are necessary. By utilizing fluorescence of organic substances in the various ranges of the visible spectrum, emulsions can be formed which are sensitive up to 720 mμ (known as verichromatic or panchromatic plates or films).

12. THERMAL (INFRARED) RADIATION

Thermal Radiation is assumed to be produced by the agitation of the molecules or atoms of a substance and extends continuously over a wavelength range from a characteristic minimum frequency up to wavelengths in the far ultraviolet. It represents a continuous energy distribution depending only upon, and being characteristic of, the temperature of the body. The term *infrared radiation* refers to that part of the spectrum which extends from approximately 780 mμ up to the very shortest electromagnetic oscillations produced in electric circuits.

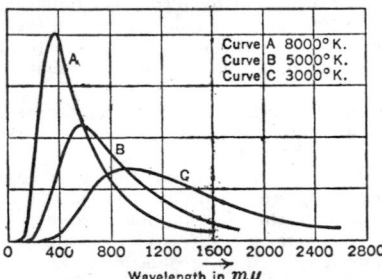

Curve A 8000° K.
Curve B 5000° K.
Curve C 3000° K.

Wavelength in $m\mu$

Fig. 2. Spectral energy distribution of a black body. The ordinates of the curves *B* and *C* must be divided by 6 and 50, respectively, in order to obtain the proper relative values.

Kirchhoff's Law. For any body in thermal equilibrium at a certain temperature T the ratio of monochromatic emissive power E_λ to absorptivity A_λ is the same function of wavelength, or in other words, every substance emits as much heat radiation as it receives for any wavelength and temperature. This law was deduced analytically by Kirchhoff (1860) and has been verified experimentally.

Black-body Radiation. A black body is defined as being perfectly absorbing for all wavelengths, $A_\lambda = 1$. In this case Kirchhoff's law leads to $(E_\lambda/A_\lambda)_b = E_{\lambda b}$, or the characteristic function of Kirchhoff is identical with the spectral distribution of emissive power of a black body. For practical purposes, Lummer and Pringsheim showed that the characteristics of the black body are closely approximated by a heated hollow sphere with blackened inner surface. Through a small hole in the surface of the sphere the internal state of radiation can be observed with a bolometer. The experimental curve of the spectral distribution of black-body radiation is given in Fig. 2 for several absolute temperatures.

Wien's Displacement Law. The wavelength of the black-body spectrum at which maximum radiation occurs for any particular temperature is connected with this temperature by the relation

$$\lambda_m T = 2900\mu^0 K \tag{4}$$

formulated by Wien. With this law, Wien also showed that, given a spectral distribution of radiation energy for one temperature T_1, the distribution curve for any other temperature T_2 can be constructed.

Spectral Distribution of Thermal Radiation. Introducing the concept of the "quantum" $h\nu$ as a definite fundamental unit of energy associated with the frequency ν, Planck (1900) formulated a theory of black-body radiation in which the various molecules are considered as oscillators having the same frequency but different and integer multiples of the energy unit. The monochromatic emissive power obtained as the statistical average over the contributions of all the molecules is then given by the exponential form

$$E_{\lambda b} = \frac{c_1}{\lambda^5} \cdot \frac{1}{e^{c_2/\lambda T} - 1} \tag{5}$$

where c_1 and c_2 are universal constants. Equation 5 gives very accurately the experimental curves obtained by Lummer and Pringsheim. For very small values of (λT) this leads to Wien's approximate expression for the spectral emissive power

$$E_{\lambda b} \approx c_1 \lambda^{-5} e^{-c_2/\lambda T} \tag{5a}$$

which is much simpler to use and gives good approximation even for the maxima of the spectral curves. For very large values of (λT) equation 5 leads to Rayleigh's approximate form for the spectral emissive power

$$E_{\lambda b} \approx \frac{c_1 T}{c_2 \lambda^4} \tag{5b}$$

which, again, is simpler to use than Planck's form.

Stefan-Boltzmann Law. Stefan showed empirically that for thermal radiation the total emissive power of a body varies with the fourth power of the absolute temperature. If two bodies of different temperatures exchange heat radiation, the rate of flow of energy per unit area from the higher temperature T_1 to the lower T_2 is given by $[\sigma \cdot (T_1{}^4 - T_2{}^4)]$, where σ is the Stefan constant of radiation, $\sigma = 5.735 \times 10^{-5}$ erg per sec per sq cm per deg Kelvin to the fourth power. This law was tested experimentally by Lummer and Pringsheim over a range from 100 to 1300 C, and found to hold accurately.

Optical Pyrometry. The temperature of glowing solids and vapors can be determined by using the radiation characteristics of a black body. In many cases the emitter can be considered approximately as a black body, so that Wien's displacement law, equation 4, can be applied directly if λ_m is determined; or Stefan's law can be applied if a thermocouple is used to measure the total radiation. In the visible range, a photometer calibrated in terms of known temperatures may be employed. For emitters deviating appreciably from the black-body characteristics, corrections have to be made according to the method used. The temperature of the sun was found to be 6000 C; that of the carbon-arc crater, between 3500 and 4000 C.

Measurement of Heat Radiation. Small energies, radiated for example by glowing bodies, can best be measured by a thermocouple which by the heat absorbed in the junction generates a small thermal electromotive force indicated by a galvanometer. To increase the deflection, a *thermopile*, a series arrangement of a number of thermocouples, can be used. The *bolometer* utilizes the increase of resistance of a thin blackened wire with the absorption of the heat rays; the wire forms one arm in a Wheatstone bridge. The *radiometer* measures the torque exerted when the heat rays fall upon small mica vanes in a low-pressure tube, the torque being produced by the unequal pressures set up on the two sides of the vanes.

Transmission of Infrared Radiation. Most kinds of glass do not transmit wavelengths longer than about 2.5 μ; quartz will transmit up to 4 μ and fluorite up to 11 μ. The best material for use in infrared spectroscopy is rock salt and sylvin, which transmit wavelengths up to 18 μ. Metals show, in general, high reflectivity over the infrared spectrum, and act as almost perfect mirrors.

Spectroscopy in the Infrared. The spectral distribution of emissive power in the infrared spectrum is best determined by bolometric or thermocouple measurements. The refracting system must use fluorite, or rock salt, in order to avoid absorption of the infrared rays. The calibration of the infrared range is best accomplished by reference to the known absorption bands of water which extend beyond 6μ. Tungsten lamps or straight Nernst lamps serve as sources of infrared spectra. For photographic spectroscopy specially prepared plates are required.

Residual Rays. Quartz, fluorite, and rock salt reflect certain wavelengths in the far infrared almost perfectly. If, therefore, heat radiation is reflected several times from the surfaces of such crystals, all other radiations will be weakened considerably and almost pure monochromatic radiation remains, which is called *residual rays*.

13. PHOTOELECTRICITY

Photoelectric Effect. When a metallic surface is illuminated by light (preferably ultraviolet), x-rays, or γ-rays, electrons are emitted from the surface. Their number is strictly proportional to the intensity; their velocity is proportional to the frequency of the incident radiation. This phenomenon was discovered first by Hertz, and later by Hallwachs; it is often referred to as *Hallwachs' effect*.

Einstein's Relation. If the photoelectric discharge is influenced by a supporting or counteracting potential, a definite negative stopping potential is found for each frequency at which no emission occurs however strong the intensity of the incident light may be. This indicates that the frequency of the incident light is an essential factor in the liberation of the electrons. Einstein (1905) propounded a theory of light quanta in analogy to Planck's energy quanta of thermal radiation and proposed the relation

$$1/2 \, mv^2 = h\nu - p \tag{6}$$

for the kinetic energy and thus the velocity of an emitted electron. The kinetic energy, therefore, appears as the difference between the energy $h\nu$ of the incident light quantum, $h = 6.55 \times 10^{-27}$ erg sec (Planck's quantum of action), and the work function p, i.e., the amount of energy needed to just force an electron from the metal; obviously p determines the stopping potential. The work p can be divided into one part ω needed to detach an electron from the atom and, therefore, a function of the position of the electron within the atom; and another part p' constituting the work to liberate the electron from the metal as a whole and, therefore, in close relation to Richardson's work function for thermionic emission (see Thermonics). Equation 6 has been found true even for incident x-radiation, the electrons liberated coming then from the inner shells of the atoms; consequently ω may now assume the most prominent part.

Photoconductivity. When a semi- or non-conductor with a voltage applied to its ends is exposed to light, either transverse or parallel to the direction of the electric field, its resistance is instantaneously lowered and an appreciable current will flow. The primary effect is a true photoelectric effect, following the relation of equation 6, but many superimposed secondary effects mask it. In general, photoconductivity refers to any phenomenon wherein light influences the conductivity of the substance. Selenium and zincblende are the most striking materials exhibiting photoconductive properties.

Photoionization. Many gases and vapors when illuminated by light of short wavelength, preferably from the ultraviolet, show marked ionization phenomena as a result of the instantaneous absorption of photons. The vapors of the alkali metals being readily ionizable amplify the photoelectric effect normally present.

Photoelectric Cells. The industrial utilization of the photoelectric and photoconductive effects has resulted in various devices called photoelectric cells. In general, these cells are evacuated or gas filled and shielded from light except for a glass or quartz window through which the incident radiation enters and falls upon a target substance; the resulting photoelectric current is then indicated by a meter. According to the application of the cell, the window material can be used to correct the light sensitiveness of the target substance either to approximate that of the human eye, or to restrict it to some definite portion of the spectrum. Such cells are used as intensity meters in illumination design, as protective or active relays in power circuits, as regulating or governing devices, and in the transmission of pictures (television) and sound. Complete information is furnished by the catalogs of the manufacturing companies.

Spectrophotometry. Photoelectric cells are especially well adapted for comparative measurements of the spectral energy distribution of radiations, particularly in transmission and reflection spectra. Individual observation can be replaced by automatic recorders; color analyses as well as spectral line analyses are thus greatly facilitated.

14. X-RAYS

The **X-ray Region** in the electromagnetic spectrum extends from the shortest ultraviolet wavelengths at about 0.8 mμ down to 0.008 mμ. X-rays are essentially a secondary radiation produced when very high-speed electrons impinge upon a metallic target. They were discovered accidentally by Roentgen (1895).

Absorption of X-rays. X-rays are absorbed by all substances to a varying degree, depending upon wavelength and absorber. Usually a specimen of aluminum is taken as the reference standard for absorption measurements, and the mass absorption index μ_d of this specimen is used as a measure of the hardness of x-rays; the smaller μ_d, the harder the x-rays. A related measure is the penetration power giving the thickness of aluminum required to reduce the intensity of x-rays to a definite fraction of the incident intensity. The mass absorption coefficient shows very marked peak values for certain wavelengths which depend on the atomic number of the element and indicate a relation to the structure of the inner, more closely bound, electrons of the K, L, and M levels (see X-ray Spectroscopy).

Refraction of X-rays. The refractive index of most substances for x-rays is very slightly less than unity and shows in general normal dispersion. The measurement of the refractive index is extremely difficult and is mainly based upon the critical angle for total reflection. In the region of the wavelengths corresponding to the characteristic line radiation of substances, x-rays show very distinct anomalous dispersion which can be used for examining the structure of matter.

Ionization by X-rays. Quantitative measurements of x-ray intensities are in general based upon the ionization effects in air or gases. For this purpose special ionization chambers are used which give an indication of the rate of ionization produced by, and proportional to the intensity of, the x-rays.

Photographic Action of X-rays. X-rays produce in emulsions used for photographic purposes the same effects as ultraviolet light. On account of their relatively low intensity, the time of exposure is long, but can be shortened considerably by using "intensifying" screens, usually coverings of fluorescent character. Since x-rays cast shadows, and by their penetrating power render most substances transparent in varying degrees, photographs can be made which reveal the inner structural outlines of organisms. This application of x-ray photography is especially valuable in the medical sciences as a fundamental diagnostic help.

Diffraction of X-rays. An x-ray passing through a thin crystal plate produces on a photographic plate a bright central spot in the direction of the incident ray surrounded by a regular pattern of spots of distinctly different intensities. This pattern changes with the relative orientation of x-ray and crystal plate and obviously is caused by the diffractive interaction of the crystal structure and the x-ray photons. It is usually referred to as a *Laue pattern* and forms the basis of x-ray crystallography since its arrangement gives an indication of the internal structure of the crystals.

Bragg's Law. In order to explain the regularity of the pattern, Bragg assumed that each plane of atoms within the crystal individually reflects the x-rays, and he arrived at the relation

$$n\lambda = 2d \sin \theta \qquad (7)$$

where n is a variable integer, λ the wavelength of the incident x-rays, d the spacing between parallel atomic planes, and θ the angle between an atomic plane and the incident x-rays. The angle θ can be observed or computed from the distances between spots in the Laue pattern, so that equation 7 gives the relation between wavelength λ and grating space d of the atomic planes.

Table 10. Grating Spacings of Various Crystals

(From A. H. Compton and S. K. Allison, *X-rays in Theory and Experiment*)

Crystal	Parallel to Plane	Grating Spacing in Angstroms	Linear Expansion Coefficient (perpendicular to crystal plane)
Calcite	Cleavage	3.029	1.02×10^{-5} per °C
Gypsum	Cleavage	7.579	3.78×10^{-5} " "
Mica	Cleavage	9.928	1.53×10^{-6} " "
Quartz	Prism	4.245	1.04×10^{-5} " "
Rock salt	(1, 0, 0)	2.814	4.0×10^{-5} " "
Sugar	(1, 0, 0)	10.57	

X-ray Crystallography. Bragg's law provides the means of exploring the atomic structure of crystals by observation of the Laue pattern. If the wavelength of the x-rays is known and the reflection angle θ observed, equation 7 leads to values for d. From Laue patterns with the x-rays incident in normal directions to the principal crystallographical planes, the distances of all possible atomic planes can be computed and finally the arrangement of the chemical atoms can be constructed. In locating the atomic planes the *Miller indices* are used; these constitute a triplet of numbers each designating the reciprocal of the intersects the plane with a cartesian coordinate system. The triplet (2, 1, 0), for example, defines all planes parallel to one which intersects the x-axis at one half, the y-axis at one arbitrary unit, and is parallel to the z-axis. Table 10 gives for various crystals the spacings between atomic planes parallel to the plane indicated, and also how the expansion coefficient which takes into account the change of the spacings with temperature. The value for rock salt (a cubic crystal) is defined for the face parallel planes as 0.281400 mμ and is used as a primary standard for crystallographic measurements.

X-ray Analysis of Materials. The Laue pattern and its variation as the Debye-Scherrer pattern for pulverized crystals are used in industry for testing and checking of materials including the study of temperature influences, elasticity, hardening, melting points, and so on. *Radiography* refers to the industrial applications of x-ray photography and fluroscopy for detecting flaws, inhomogeneities, yielding points in structural parts; to the investigation of materials under operating conditions; and to other uses. In art, the use of x-ray photographs permits the identification of true antiques, old paintings, and the like.

X-ray Spectroscopy. If x-rays fall upon a solid substance, a characteristic secondary radiation is observed which has for each chemical element a definite energy spectrum, although being similar in form for all elements. The spectral distribution of the secondary radiation can be investigated by utilizing Bragg's law (equation 7) if a crystal of known grating d is employed as analyzer. The intensity of the radiation can be measured either with the ionization chamber or by photography. Two types of spectra are discernible, a characteristic line spectrum and a continuous spectrum. The line spectrum is due to electrons in the inner orbits which by collision with the incident x-ray photons are raised to outer orbits and in returning emit radiation of a particular frequency. The continuous spectrum is due to the absorption by atoms of x-radiation of insufficient amounts to raise their electrons to higher orbits. There is for any hardness of incident x-rays a definite critical shortest wavelength at which the continuous spectrum abruptly ends. This wavelength limit constitutes the converse to the photoelectric effect, and the continuous spectrum is often referred to as the *inverse photoelectric effect*. From the line spectra, *energy level diagrams* can be deduced in the same way as in light spectroscopy (see Art. 10). The results of x-ray spectroscopy were instrumental in building up the nuclear models of atoms.

BIBLIOGRAPHY ON PHYSICS OF RADIATION

1. ALLEN, H. S., *Photo-Electricity*, London, Longmans, Green, 1913.
2. BALY, E. C. C., *Spectroscopy*, New York, Longmans, Green, 3rd ed., 1924.
3. BULLOCK, E. R., *Chemical Reactions of the Photographic Latent Image*, New York, Van Nostrand, 1927.
4. CAMPBELL, N. R., and RITCHIE, D., *Photoelectric Cells*, London, Pitman & Sons, 1934.
5. CLARK, G. L., *Applied X-rays*, New York, McGraw-Hill, 3rd ed., 1940.
6. COBLENTZ, W. W., and EMERSON, W. B., *Luminous Radiation from a Black Body*, Bureau of Standards, Sci. Paper 305, 1917.
7. COMPTON, A. H., and ALLISON, S. K., *X-rays in Theory and Experiment*, New York, Van Nostrand, 1935.
8. FORSYTHE, W. E., *Measurement of Radiant Energy*, New York, McGraw-Hill, 1937.
9. FOWLER, A., *Report on Series in Line Spectra*, London, Fleetway Press, 1922.
10. *Handbuch der Physik*, Vols. 19, 20 and 21, Berlin, J. Springer, 1928–1929.
11. HERZBERG, G., *Infrared and Raman Spectra*, New York, Van Nostrand, 1945.
12. HICKS, W. M., *Treatise on the Analysis of Spectra*, Macmillan, 1922.
13. HOUSTOUN, R. A., *A Treatise on Light*, London, Longmans, Green, 1933.
14. HUGHES, A. L., and DURBRIDGE, L. A., *Photoelectric Phenomena*, New York, McGraw-Hill, 1932.
15. *Industrial Electronics Reference Book*, Westinghouse Electric, New York, John Wiley, 1948.
16. LANGE, B., *Photoelements and Their Applications*, Reinhold, 1938.
17. LEWIS, S. J., *Spectroscopy in Science and Industry*, London, Blackie & Sons, Limited, 1945.
18. LOWRY, T. M., *Optical Rotatory Power*, London, Longmans, Green, 1935.
19. LUCKIESH, M., *Ultra-Violet Radiation*, New York, Van Nostrand, 1922.
20. LYMAN, T., *The Spectroscopy of the Extreme Ultraviolet*, London, Longmans, Green, 2nd ed., 1928.
21. MEES, C. E. K., "Artificial Illuminants for Use in Practical Photography," *Trans. Illum. Engg. Soc.*, Vol. 10, p. 947, 1915.
22. *Photography as a Scientific Implement*, London, Blackie & Son, Limited, 1923.
23. *Pyrometry*, American Institute of Mining and Metallurgical Engineers, New York, 1920.
24. RAWLING, S. O., *Infrared Photography*, Blackie & Son, Limited, London, 1933.
25. RICHTMYER, F. K., and KENNARD, E. H., *Introduction to Modern Physics*, New York, McGraw-Hill, 4th ed., 1947.

26. ROBERTSON, J. K., *Introduction to Physical Optics*, New York, Van Nostrand, 1941.
27. ST. JOHN, A., and ISENBURGER, H. R., *Industrial Radiography*, New York, John Wiley, 1943.
28. SIEGBAHN, M., *Spectroscopy of X-rays* (trans. by G. Lindsay), Oxford University Press, England, 1925.
29. SUTHERLAND, G. B. B. M., *Infrared and Raman Spectra*, Methuen, 1935.
30. TERRILL, H. M., and ULREY, C. T., *X-ray Technology*, New York, Van Nostrand, 1930.
31. VALASEK, J., *Elements of Optics*, New York, McGraw-Hill, 1928.
32. VON HEVESY, G., *Chemical Analysis by X-rays and Its Applications*, New York, McGraw-Hill, 1932.
33. WAGNER, A. F., *Experimental Optics*, New York, John Wiley, 1929.
34. WOOD, R. W., *Physical Optics*, Macmillan, 2nd ed., 1934.
35. WYCKOFF, R. W. G., *Structure of Crystals*, New York, Chemical Catalog Co., 1931.
36. ZWORYKIN, V. K., and RAMBERG, E. G., *Photoelectricity and Its Application*, New York, John Wiley, 1949.

Tables

1. *Handbook of Chemistry and Physics*, Cleveland, Ohio, Chemical Rubber Publishing Co.
2. *International Critical Tables*, New York, McGraw-Hill, 1926.
3. LANDHOLT-BOERNSTEIN, *Physikalische-Chemische Tabellen*. Berlin, J. Springer, 1923–1931.
4. *The Smithsonian Tables*, Washington, D. C., 1916.

PHYSIOLOGY OF RADIATION

By Ernst Weber

15. VISION

The Eye. Radiation enters the eye through the cornea or outer coating; penetrates the anterior chamber (filled with aqueous humour), the lens, the posterior chamber (filled with a gelatinous, vitreous humour); and is received on the retina, a network of fine fibers which transmit in some way the sensory impressions through the optic nerve to the brain where they are transformed into perceptions. The range of radiation leading to visual perception is from approximately 380 to 780 mμ. The most sensitive point of the retina is called the yellow spot, and has a depression, the fovea centralis, where vision is most distinct. To bring objects into such relative position to the eye that they will be received at the fovea centralis requires a delicate adjustment apparatus which is provided by the muscle fibers holding the eye. Being a multirefracting medium, the eye suffers from both spherical and chromatic aberration (see Art. 5).

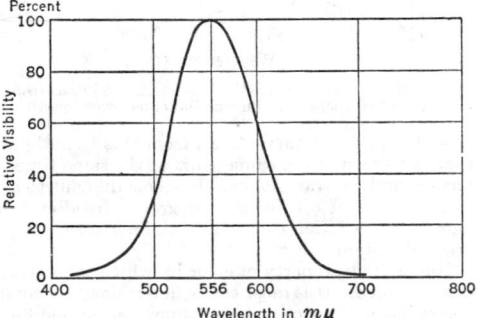

FIG. 1. Spectral distribution of relative human visibility.

Fechner's Law. The sensation of light as produced by the eye varies logarithmically with the intensity of the stimulus. This law, found by Fechner to hold true for most sensory perceptions, is an excellent illustration of natural economy.

Spectral Visibility. The visual sensation of the eye varies greatly with the wavelength of light, reaching a maximum at about 556 mμ (yellow). If white light is used as a standard to be compared with the spectral monochromatic radiations, the relative intensity of white light causing the same sensation as the spectral light is called the *relative visibility*. The spectral distribution of relative visibility is shown in Fig. 1, assuming the visibility of yellow-green light, λ = 556 mμ, as 100 per cent. The visibility falls off very sharply from this maximum, which indicates that essentially only the wavelengths between 480 and 640 mμ constitute the easily visible spectrum.

Color Vision. The retina is composed of nervous fibers known as rods and cones, of which only the cones seem to be responsive to colors; the cones are, at the same time, fairly insensitive so that, at low intensities of light, colors cannot be distinguished (twilight vision). The rods are more sensitive; they contain a light-sensitive substance, the visual purple, but apparently they cannot transmit color sensation. Perception of color is not instantaneous; of all the spectral colors the blue sensations are most active, whereas the green sensations are most sluggish.

Purkinje Effect. At low intensities, before twilight vision is reached, there is a distinct shift of the maximum in the visibility curve (Fig. 1) towards a shorter wavelength; this is known as the Purkinje effect. This shift of the spectral visibility curve tends to make color sensations variable quantities, and color photometry very difficult.

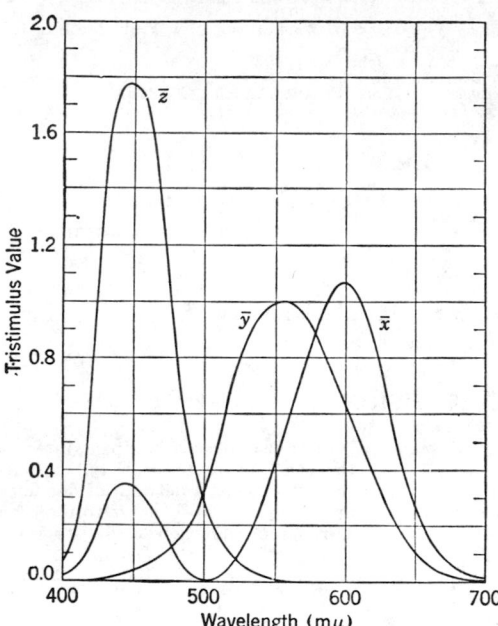

Theories of Color Vision. Two types of theories have been advanced to explain the facts of normal and abnormal color sensitiveness of the eye. The first type of theory is due to Young and Helmholtz, and proposes the existence of three distinct modes of response of the eye corresponding to the three fundamental color sensations: red, green, and blue. Each of these partial responses represents a definite spectral distribution, the perception of color being the superposition of the three partial responses; Fig. 2 indicates the amounts of the three primaries required to match a unit amount of energy at each wavelength. The relative stimulation is measured in terms of this tristimulus value, where $\bar{x}$, $\bar{y}$, and $\bar{z}$ represent the red, green, and blue sensations, respectively.[*] This theory is capable of accounting very readily for the specific types of color blindness known as red and green color blindness. The second type of theory initiated by Hering is based upon antagonistic colors and associates chemical building-up and breaking-down processes with the sensation of complementary colors; from this hypothesis it follows that complementary colors cannot be seen at the same point at the same time. This theory readily accounts for afterimages and contrast colors. It seems difficult to accept any one theory exclusive.

FIG. 2. The amounts of tristimulus energy required to match a unit of energy having the indicated wavelength.

Saturation of a color is its degree of freedom from admixture with white light. Monochromatic radiation can be said to have a saturation of 100 per cent, and white light is of zero saturation.

Hue is that property of color by which the various spectral regions are characteristically distinguished. It is most easily determined by comparison with the spectral colors as blue, yellowish, and so on. Equal hues can still differ in saturation. Complementary hues produce white light when mixed. White can be said to have no hue.

Brightness of a color is closely related to its diffuse reflection factor, and can be classified accordingly. It is customary for artistic purposes to distinguish the values of colors by their relative brightnesses. Table 1 gives the values of colors as suggested by Luckiesh.

Tints are unsaturated colors of definitely recognizable hue.

Table 1.   Scale of Color Values

(From M. Luckiesh, *Light and Shade and Their Applications*)

Artist's Scale		Suggested Scale Reflection factors in percentage	Artist's Scale		Suggested Scale Reflection factors in percentage
Symbols	Values		Symbols	Values	
B	Black	0–10	LL	Low light	50–60
LD	Low dark	10–20	L	Light	60–70
D	Dark	20–30	HL	High light	70–80
HD	High dark	30–40	W	White	80–90
M	Medium	40–50			

[*] See references 9, 10, 23, and 32.

The Color Triangle (Fig. 3). The laws of color mixtures and relative intensities can readily be visualized in the so-called color triangle. Any point within the triangle indicates by its relative distance from the three fundamental colors in the apices the relative amounts of these fundamental colors. The resultant color, or hue, can be determined by drawing a line from the center (white) through the color point; where it intersects the triangle, the equivalent monochromatic wavelength can be read. The relative distance of the point on this line from the center gives the saturation in percentage.

The ICI Chromaticity Diagram, Fig. 4, is a recommended method of plotting the chromaticity of colors, specified by dominant wavelength and purity. Considering color as a psychophysical stimulus as recommended by the Optical Society of America, the International Commission on Illumination adopted standards for international use. The charts in Figs. 1, 2, and 4 result from these standards. If X, Y, and Z represent the three primary stimuli, red, green and blue, respectively, one defines as chromaticity coordinates $x = X/(X + Y + Z)$, and analogously y, and z. The plot of any two chromaticity coordinates against each other as in Fig. 4 gives the locus of visible spectral radiation. The straight-line portion connecting the extremes represents the magentas and purples. Thus, spectrum colors will lie on the horseshoe-shaped locus, and mixtures thereof will always lie within the locus. Points within the area may represent the location of all actual colors.

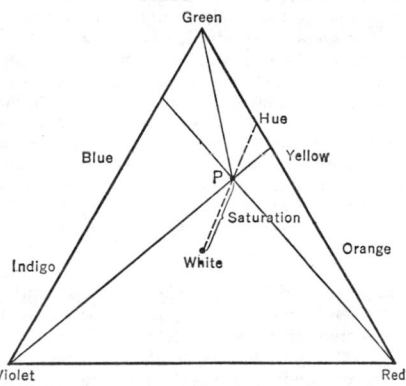

Fig. 3. The color triangle; point P has a ratio of the fundamental sensations $R : G : V = 9.7 : 6 : 11.5$, a saturation 58%, and yellow hue.

In Fig. 4 the points A, B, and C represent standard ICI illuminants. Point C is of particular importance and represents white or neutral gray. It is called the point of zero saturation. Any point, such as A or B, on a straight line joining C to a spectrum color on the curve, represents a color which can be obtained by mixing the spectrum color with white or neutral gray in amounts inversely proportional to the ratio of the distances separating the point from the spectrum color and the point C.*

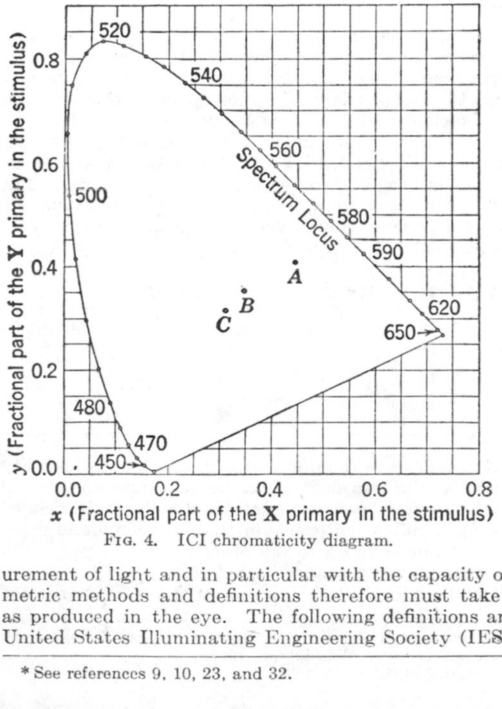

Analysis of Colors. An objective means to determine the spectral energy distribution of light is provided by the spectral photometer (Art. 16). Automatic methods have been developed which are particularly useful in color matching and mixing, and in determining the characteristics of illuminants and reflecting and transmitting media.

Fig. 4. ICI chromaticity diagram.

y (Fractional part of the Y primary in the stimulus)
x (Fractional part of the X primary in the stimulus)

Spectrum Locus

520 540 560 500 580 590 620 650 480 470 450

A B C

16. PHOTOMETRY

Definitions and Units

Photometry deals with the measurement of light and in particular with the capacity of light to affect the eye. All photometric methods and definitions therefore must take into consideration light sensations as produced in the eye. The following definitions are based upon the standards of the United States Illuminating Engineering Society (IES).

* See references 9, 10, 23, and 32.

Luminous Flux, Φ, is the rate of flow of radiant energy evaluated with reference to visual sensation. Although defined as a rate of flow of energy, it may be regarded for photometric purposes as an entity. The *lumen* is the unit of luminous flux and is defined as the luminous energy radiated per second into unit solid angle from a point source of uniform luminous intensity 1 candela (see below). (A point source having a uniform luminous intensity of 1 candela would therefore emit a total of 4π lumens.)

Luminous Intensity, I, of a source of light in a given direction is the solid angular intensity of the luminous flux emitted by the source in the direction considered, when the flux involved acts, as far as computations and measurements are concerned, as if it came from a point; mathematically it can be defined as $I = dF/d\omega$, where $d\omega$ is the elemental solid angle. The unit of luminous intensity is the *candela*, which has been internationally adopted as the standard primary source (January 1948) and which has a luminous intensity approximately equal to the previous standard international candle. (1 candela = 98.1% of the luminous intensity of one international candle.) On January 1, 1948, the National Bureau of Standards announced that the following new definition for luminous intensity had been agreed upon by national and international organizations:*

The new system of photometric units takes the definition of the candela to be such that the luminance of a full radiator at the temperature of freezing platinum is 60 candela per square centimeter. The spectral distribution of such a full radiator is similar to that of a tungsten filament lamp operating at a color temperature of 2042°K (1769°C). Other units are derived from the candela, with the provision that when differences of color are involved the evaluation shall be made by means of standard spectral luminosity factors that have been adopted by the International Commission on Illumination and the International Committee on Weights and Measures. For the types of lamps now in common use the ratings under this new system will be practically the same as those now in effect.

Mean Horizontal Candlepower of a source is the average intensity in a horizontal plane normal to the axis of the source. The *mean spherical candlepower* of a source is the average intensity over all directions; if the luminous flux is known, the mean spherical candlepower can be computed by dividing the flux by 4π.

Mechanical Equivalent of Light.† The luminous equivalent of radiation at maximum visibility (at $\lambda = 556$ mμ) is defined by

$$1 \text{ lumen} = 0.00147 \text{ watt}$$

$$1 \text{ watt} = 680 \text{ lumens}$$

Illumination at a given point of a surface is the luminous flux incident on the surface at this point; mathematically it can be defined as $E = dF/dS$, where dS is the surface element receiving the flux element dF. A point source, giving a uniform luminous intensity I in a certain direction, sets up through the small surface dS perpendicular to this direction and at a distance R from the source the elemental flux $dF = I \cdot d\omega$; since the surface element is $dS = R^2 \cdot d\omega$, it follows that $E = I/R^2$ (inverse square law). The units of illumination depend on the reference area chosen and are called *lux* for 1 lumen per square meter, *foot-candle* for 1 lumen per square foot, and *phot* for 1 lumen per square centimeter. The term "foot-candle" is still widely used for the illumination unit in the fps system. It must be noted that on our new standard primary-source definition, a more precise term would be the "foot-candela." 1 foot-candela = 0.98 foot-candle.

Table 2. Interrelations between the Units of Illumination

1 Unit is equal to	Lux	Foot-candela	Phot	Milliphot
1 lux..................................	1	0.0929	10^{-4}	10^{-1}
1 foot-candela...........................	10.76	1	0.001076	1.076
1 phot.................................	10^4	929	1	10^3
1 milliphot..............................	10	0.929	10^{-3}	1

Brightness of a surface element dS is defined as the quotient of the luminous intensity produced by the surface element divided by the area as projected into the direction normal to the line from the observer to the surface element; mathematically it is defined as $b = dI(\theta)/dS \cdot \cos\theta$, where $dI(\theta)$ is the luminous intensity in the direction θ and θ the angle between the normal to dS and the line from dS to the observer.

There are two related systems of units for luminance in terms of:
1. The luminous flux per unit solid angle radiated from unit projected area of the surface.

---

* Reproduced from "Announcement of Changes in Electrical and Photometric Units," *Circular of the National Bureau of Standards C459,* U. S. Printing Office, Washington, 1947.
† H. E. Ives, *Journal Optical Soc. of America,* Vol. 12, p. 75, 1926.

Luminance Units (Direct)

System	cgs	mks	fps
Photometric unit	Stilb (Sb)	Nit (nt)	cd/ft^2
System unit	cd/cm^2	cd/m^2	cd/ft^2

2. A comparison, such that the illumination of the specimen surface (as defined in 1 above) is compared directly with the illumination of a standard uniform diffuser which radiates a total luminous flux of 1 lumen per unit area of surface.

Luminance Units (by comparison)

System	cgs	mks	fps
Illumination of standard diffuser	1 lm/cm^2	1 lm/m^2	1 lm/ft^2
Equivalent luminance of surface	1 lambert	1 apostilb	1 foot lambert
Symbol	L	asb	fL

The relation between the units of 1 and 2 above: A uniform diffuser of luminance (L) radiates a total luminous flux of πL lumens/unit area surface.
One lumen/unit area is equivalent to a luminance of $1/\pi$ cd/unit area.

$$1 \text{ lm/cm}^2 = 1 \text{ lambert} = 1/\pi \text{ cd/cm}^2$$
$$1 \text{ lm/m}^2 = 1 \text{ apositilb} = 1/\pi \text{ cd/m}^2$$
$$1 \text{ lm/ft}^2 = 1 \text{ ft lambert} = 1/\pi \text{ cd/ft}^2$$

Table 3. Interrelations between the Units of Luminance

1 Unit is equal to	Candela per sq cm	Lumen per sq cm per steradian	Candela per sq in.	Lambert	Milli-lambert
1 candela per sq cm............	1	1	6.4516	3.1416	3141.6
1 lumen per sq cm per steradian..	1	1	6.4516	3.1416	3141.6
1 candela per sq in.............	0.1550	0.1550	1	0.4869	486.9
1 lambert....................	0.3183	0.3183	2.0538	1	1000
1 millilambert...............	0.3183 × 10^{-3}	0.3183 × 10^{-3}	0.0020538	0.001	1
1 foot lambert..............	0.00034	0.00034	0.00221	0.00108	1.076

Lambert's Laws. A perfectly diffusing surface receives an illumination or emits a luminous intensity (by reflection or true emission) which varies with the cosine of the angle between the normal to the surface and the direction of the incident or emitted (reflected) ray; for perfectly diffusing surfaces the illumination and brightness are therefore constant under all angles of observation. These cosine laws of incidence and emission are only approximately true for the practical substances so that in measurements of illumination or brightness the angle of observation always should be stated.

Instruments and Methods

Equality and Contrast Methods. Most photometric instruments compare an unknown light source with a known light source either simultaneously (comparison) or, by means of a third light source, alternatively (substitution method). To observe the equivalence of two sources either an equality method can be employed in which the two sources illuminate different parts of the same surface, the adjustment being made for equal brightness, or a *contrast method* can be used in which four fields are observed, two receiving light from one source and the other two from the other source, an absorbing medium decreasing the flux received on one field in each instance, and adjustment being made for equal contrast. The special type of table upon which most photometric measurements are made is called a photometer bench.

Bunsen Photometer. A piece of white paper with a grease or wax spot in its center is held between the two sources of light being compared. The relative distances between paper and sources are adjusted until the paper presents uniform brightness. If the luminous intensities of the two light sources are I_1 (known) and I_2 (unknown), and the distance for observation from one side of the paper r_1 and r_2, from the other side of the paper r'_1 and r'_2, then the unknown intensity follows as

$$I_2 = I_1 \frac{r_2 r'_2}{r_1 r'_1} \qquad (1)$$

The advantages are simplicity and cheapness; the error may, however, be as great as 5 per cent or more.

Lummer and Brodhun Photometer. The most essential part is the cube, a combination of two rectangular prisms so fitted together that the light reflected by mirrors from the two sources can be either compared for equality or adjusted for equal contrast. Calibrated filters may be used to adjust the light intensities. The accuracy is very high, about one-half of one per cent, and is considered the highest obtainable accuracy in photometric work.

Illuminometers and Brightness Meters. These are small portable photometers, usually employing some form of Lummer and Brodhun cube and having a small built-in electric lamp with either adjustable distance, or filters, or both. On a test screen the light from the known source is compared with either the light incident on a surface (illumination) or the light emitted by a surface (brightness). American designs are the Macbeth illuminometer and Luckiesh-Taylor brightness meter. Care must be taken to shield properly and to avoid errors on account of deviations from the cosine laws.

Ulbricht Sphere. To measure the mean spherical candlepower, the Ulbricht sphere has come into predominant use. It consists of a large hollow sphere, coated with highly reflecting paint on its inner surface and having a window of transparent glass. The brightness of this window, shielded from the direct light of the source at the center of the sphere, is proportional to the mean spherical candlepower of the source and can be determined by an illuminometer.

Color Photometry. Methods have been devised which try to eliminate the influence of color in photometric work. The flicker photometer of Rood is perhaps the most useful means of utilizing the fact that two colored surfaces, if presented in quick alternation, appear colorless and thus permit a comparison of illumination. The most objective method is suggested by spectrophotometry, which, however, does not yield results immediately useful in photometry because the relative visibility of the human eye is not taken into account.

Colorimeters are instruments for the measurement of the hue and the saturation of colors. *Trichromatic* colorimeters mix the three fundamental sensations by means of glass filters in order to match the color to be determined; the hue and saturation can then be computed from the color triangle (see Art. 15). *Monochromatic* colorimeters first determine by comparison with the spectral colors the dominant hue of the unknown color and then mix it with white light until the proper saturation is obtained.

17. ELEMENTS OF ILLUMINATION *

Artificial Light Sources

Artificial Light Sources. At the present time the important sources of artificial light are incandescent filament lamps, gas or vapor discharge lamps, and fluorescent lamps. Important commercial examples of gas discharge lamps are various mercury-vapor lamps, neon gas lamps, and sodium-vapor lamps. Each type has its specific advantages from the viewpoint of efficiency, color characteristics, life, first cost including necessary associated fixtures, ballast, etc. Each source has its particular field of application and limitations.

Luminescence, Incandescence, Fluorescence, and Phosphorescence. Radiant energy within the visible range 380 to 760 mμ may be produced (1) by heating a body to incan-

---

* Revised by C. C. Whipple.

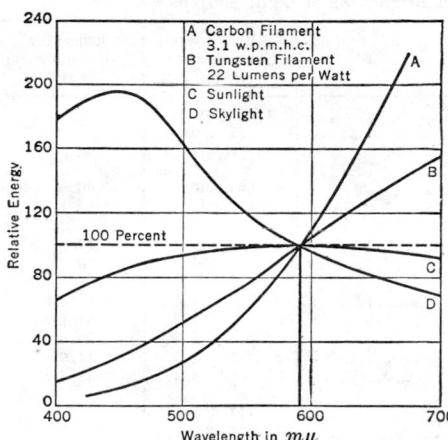

FIG. 5. Spectral distribution of radiant energy in some common illuminants.

descence, for example in the incandescent lamp; (2) by luminescence, involving molecular excitation, usually by electric action in a gas or vapor as in the mercury-vapor lamp; (3) by fluorescence, in which invisible radiant energy, usually from a mercury-arc discharge, impinges upon and is absorbed by certain materials called phosphors which reradiate the energy at another wavelength within the visible range, as for example in the present-day fluorescent lamp; and (4) by phosphorescence, a process similar to fluorescence except that the phosphor continues to radiate as a glow after the removal of the exciting energy.

In the practical fluorescent lamp the inside surface of a hollow cylindrical glass tube is coated with a phosphor. A mercury-arc discharge between end electrodes provides radiant energy mainly at a wavelength of 253.7 mμ which excites the phosphor to radiate energy within the visual range. Neon gas tubes are used mainly for signs. The neon may be mixed with other gases to provide various color effects.

Luminosity Curve. Figure 5 gives the spectral-energy-distribution curves over the visible range for several sources of radiation. From such spectral energy curves the luminous energy curves may be obtained if the relative energy at each wavelength is multiplied by the corresponding point on the luminosity curve from Fig. 1. The resulting curve is a very important and useful characteristic as it shows graphically the distribution of the visible energy from the source. The area under such a curve is a measure of the total light energy output of the source. It is observed that a large amount of the visible radiation from the sun is near the point of maximum visibility at $\lambda = 556$ mμ. Since the energy distribution of most artificial sources of light is often very different from that of the sun, or the sky, these sources will usually produce a different color response; i.e. the color of an object illuminated by them will not appear "natural."

Luminous Efficiency. A comparison of the luminous energy output curve of a light source with its radiation energy spectrum reveals the visual ineffectiveness of much of the radiation. It is customary to designate the ratio of the total luminous output in lumens to the total power radiated in watts as the luminous efficiency. The overall efficiency of a source of light, such as an incandescent lamp, is usually expressed as ratio of total lumen output to the watts input to the lamp, and is thus stated in lumens per watt. Table 4 gives the overall efficiency of several light sources. Since the mechanical equivalent of luminous radiation (see Art. 16) is 680 lumens per watt, it is seen that the luminous efficiency of all common light sources, including the sun, is relatively low.

Brightness of Light Sources. Table 4 also gives the brightness of various light sources, including that of the sun and of a black body.

Characteristics of Electric Incandescent Lamps. The characteristic curves of incandescent lamps give the variation, with voltage, of the light output in lumens, the efficiency in lumens per watt, the power input in watts, the current in amperes, and the resistance in ohms, taking all the values as 100 per cent at rated voltage, which also is considered as 100 per cent. These characteristics can, with a reasonable degree of approximation, be brought into the exponential form

$$Q_2 = Q_1 (V_2/V_1)^\alpha \qquad (2)$$

where Q_1 and Q_2 are values of any of the characteristic quantities and V_1 and V_2 the corresponding voltage values.

The greatest change occurs in the life and lumen output. The exponent α for gas-filled filament lamps is 3.38 when Q is in lumens, $\alpha = 1/13.1$ when Q is in hours of life.

Lamp Life. An important economic aspect of illumination is the expected life of the artificial source. At present incandescent lamps for commercial use are rated at 750 (below 300 watt) and 1000 (above 300 watt) hours, hot fluorescent lamps are now rated at 7500 hours, cold cathode at 15,000 or 25,000 hours, and mercury-vapor lamps 4000 to 6000 hours. The life of all mercury-vapor and hot-cathode fluorescent lamps is very much affected by the number of hours on for each switching.

Table 4.   Efficiency and Brightness of Light Sources *

Light Source	Brightness (candles per sq in.)	Efficiency (lumens per watt)
Sun, as observed from earth's surface at meridian	1,066,000	
at horizon	3,870	
Moon as seen from earth's surface	1.64	
Clear sky (average)	5.14	
Black body, at 6500 K	1,895,000	90
at 4000 K	157,000	52.2
Tungsten filament		
500 watt, gas filled	7,800	20
25 watt, vacuum lamp	1,335	10
Fluorescent lamps		
standard cool white	3.8	46.5 †
standard warm white	4.2	51.2 †
de luxe cool white	2.4	29.0 †
de luxe warm white	2.5	31.0 †
daylight	3.4	41.3 †
Electric arcs		
plain carbon-arc crater	113,000	
high-intensity searchlight arc	575,000	
high-pressure mercury-arc type H4	5,750	26.8 †

* From *IES Lighting Handbook*, 1952, and GE Lamp Bull. No. LD-1.
† Includes ballast loss.

Principles of Good Lighting

Requirements of Good Lighting. A good interior lighting installation will have the following characteristics: (1) the illumination is sufficient to enable one to accomplish the seeing task satisfactorily; (2) the distribution of the illumination is uniform throughout the seeing area; (3) the light is properly directed and diffused by the fixtures and by proper painting of the surroundings; (4) the fixtures shield the light source so that a brightness near the horizontal (to at least 30° below) is low ($1/2$ candle per square inch for large fixtures to 2 for small ones); (5) shadows, although important in providing form and depth to objects, are soft; (6) the color of the light source and that of the wall paint is acceptable to the type of service and the preference of the individuals involved; (7) glare is entirely eliminated; (8) contrasts in brightness are not too great, as they may be the cause of glare and visual discomfort.

Glare is the discomforting effect of ill-directed light. It can be caused by excessive luminous flux entering the eye from a large light source at a relatively small distance, by excessive brightness of a light source in the field of view as unshaded electric lamps or flames; or by excessive contrast in brightness between light source and surroundings as reflections from metals and polished surfaces or dark ceilings and walls. Glare is not only uncomfortable but also injurious to the eye if exposure is prolonged. It is especially important to avoid glare under working conditions. The best means to avoid glare is obviously to eliminate its causes by using indirect lighting, direct lighting from large distances, or properly designed diffusers, and by painting the walls and ceilings in light colors, with a matt surface.

Brightness Ratios. In order to provide comfortable seeing conditions in schoolrooms and offices the IES recommends the limits of brightness ratios given in Table 5. In order to achieve these ratios the reflection factors of walls, ceiling, furniture, and floors should be about as follows: ceilings, 80–85%; walls, 50–60%; desk and table tops, 35–50%; furniture, 35%; floors, 30%. To avoid reflected glare, all surfaces should have a matt finish.

Table 5.   Recommendations for Limits of Brightness Ratios in Schoolrooms

(Rooms used by faculty or pupils for educational purposes)

Reference Areas	Ratio
Between the "central visual field" (the seeing task) and immediately adjacent surfaces, such as between task and desk top, with the task the brighter surface *	3 to 1
Between the "central visual field" (task) and the more remote darker surfaces in the "surrounding visual field," such as between task and floor	10 to 1
Between the "central visual field" (task) and the more remote brighter surfaces in the "surrounding visual field," such as between task and ceiling †	1 to 10
Between luminaires or windows and surfaces adjacent to them in the visual fields	20 to 1

* Chalkboard and some art and shop tasks are illustrations of cases where the reverse ratio of 1 to 3

may apply.

† These ratios apply for areas of appreciable visual size as measured by the solid visual angle subtended at the eye. Luminous areas on luminaires are generally small in size in this respect. For brightness limitations of luminaires see last entry.

Source: "American Standard Practice for School Lighting," sponsored by IES, approved by ASA, 1948.

Reflection Factors. Table 6 shows reflection factors for illumination by incandescent lamps.

Table 6. Reflection Factors from Colored Surfaces

Reflection Factor in Per Cent	Colors of Surfaces					
80 to 75	Ivory white					
75 to 70	Ivory					
70 to 65	Bright yellow	Light gray				
65 to 60		Lichen gray	Buff	Latin green		
60 to 55		Silver gray	Tan (Ivory)		Pale azure + white	
55 to 50		Medium gray	Buff stone			Shell pink
50 to 45		French gray		Light green		Pink
45 to 40		Darker gray		Bright sage	Pale azure	
40 to 35			Light brown		Light blue	
35 to 30					Sky blue	
30 to 25			Tan			
25 to 20		Dark gray	Brown	Olive green		
20 to 15		Dark gray	Cocoanut brown	Forest green		Cardinal red

Recommended Illumination Levels for Various Interiors. Table 7 gives some typical recommended values of illumination taken from the *IES Lighting Handbook*.

Table 7. Recommended Values of Illumination

(*IES Lighting Handbook*, Second Edition, 1952)

A. Commercial and Public Areas

Type of Use	Maintained Foot-Candle
Auditoriums (assembly only)	10
Banks, in offices and cages	50
Barber shops, beauty parlors	50
Cars, day coach, Pullman	30
Churches, auditoriums	5–10
Sunday school rooms	20
Drafting rooms	50
Gymnasiums, exhibition and matches	30
general exercise	20
lockers and shower room	10
Hallways and corridors	5
Hotels, lobby	20
dining room	10
writing room, using supplementary illumination	30
Libraries, general reading room	30
reading from detail	50
stacks, open to public	30
closed to public	10
Offices, bookkeeping, typing, accounting and other difficult seeing tasks	50
ordinary seeing tasks, general office work	30
Schools, class and study rooms	30
corridors, stairway	5
drawing and drafting rooms	50
laboratory, general work	30
close work from detail	50
Theatres, auditoriums during intermissions	5
lobby	20

B. Industrial Interiors

Auto manufacturing	30–200
Electrical equipment manufacturing	30–100
Machine shops	20–200
Woodworking shops	30–100

18. THERAPY OF RADIATION

All radiations of the electromagnetic spectrum have been found useful for therapeutic purposes. The physical properties of the radiations are to be found under Physics of Radiation, page 10 6 5.

Infrared Radiation

Physiological Action. Infrared radiation produces heat (Art. 12) when absorbed by organic tissues, which, in general, show more or less selective absorptivity. In the visible range most of the radiation is absorbed by the skin and the superficial layers of flesh; for deep therapy, wavelengths in the range from 700 to 1500 mμ (near infrared) prove most penetrating, the skin being apparently more transparent for these wavelengths. Beyond 2μ, the longer the wavelength the less penetrating will be the radiation   The heat produced by infrared radiation has decided therapeutic effects on arthritis, fractures, and internal injuries and acts stimulatingly on the muscles and tendons.

Sources of Infrared Radiation. The natural and most efficient source for infrared therapy is the sun. Next come the gas-filled tungsten lamps glowing at about 3000 K, which give off very high radiation energies in the near infrared. Carbon lamps at lower temperatures and other glowing sources and flames have characteristic radiations with band-spectral energy distributions of varying widths and locations; they are in general less efficient than tungsten lamps.

Ultraviolet Radiation

Physiological Action. Ultraviolet radiation (Art. 11) is highly stimulating and seems to be of utmost importance for organic life; the wavelength range from 300 to 400 mμ constituting the shortest-wavelength part of the daylight spectrum, and often referred to as "actinic rays," is of particular practical value. Shorter wavelengths also have decided therapeutic effects on human beings; they are, however, easily absorbed by air so that proper care has to be taken to use non-absorbing materials in the path of the rays. Wavelengths at about 280 mμ are supposed to be injurious to organisms and should be avoided in therapeutic uses. In the range from 200 to 300 mμ ultraviolet radiation is highly germicidal and is used practically for the sterilization of food and water supplies. Ultraviolet therapy is entirely superficial, is absorbed by the skin, and produces the tan or sunburn. It is important to time the exposure to ultraviolet radiation and to increase the exposure gradually from very short intervals to not more than about one-half hour, depending on the strength of the source and its spectrum.

Photochemical Actions. Industrially the strongly actinic effects of ultraviolet radiation are utilized in chemical processes of bleaching, vulcanizing, and so on. The more recently found food vitamins which are essential for health and the proper performance of biological functions are influenced by ultraviolet radiation, and their production can be stimulated by short-time irradiations. In particular, vitamin D, known to possess antirachitic qualities, is produced industrially in this way.

Sources of Ultraviolet Radiation. Again the most important source is natural sunlight. Artificial sources are provided by the pure carbon arcs; the blue and white burning arcs have slightly different characteristic line radiations in the ultraviolet region. The most important source for therapeutic uses is, however, the high-pressure mercury-arc lamp which produces very strong line radiations in the near ultraviolet—the most active range; high-power mercury-arc lamps use water cooling to reduce the heating effects.

X-radiation

Physiological Action. X-radiation is injurious to most organic tissues but can be utilized, if properly directed, to destroy malignant internal growths as well as to cure skin diseases. It is the most widely used weapon in non-surgical medical treatment. The penetrating power of x-rays depends on the voltage of the x-ray tubes; their destroying action, on the voltage and the time of exposure (dosage). Over-exposure causes very severe burns lasting for a disproportionately long time; great care has to be taken to avoid even slight burns.

19. HIGH-ENERGY RADIATION

Production. The production of high-energy radiation will be recalled from reference to Table 1 of the Electromagnetic Spectrum (p. 10 57) where the region of interest is beyond the ultraviolet into the shorter wavelength regions which are comprised of photons of increasing photon energy as the wavelength of the emitted electromagnetic radiation decreases. For tabulation convenience these regions are divided into the ultraviolet, x-ray, γ-ray, and cosmic-ray regions which overlap to a certain extent but which are essentially radiated electromagnetic photons (see also Art. 14, X-rays). The diagram indicates how matter behaves generally to high-energy radiation.

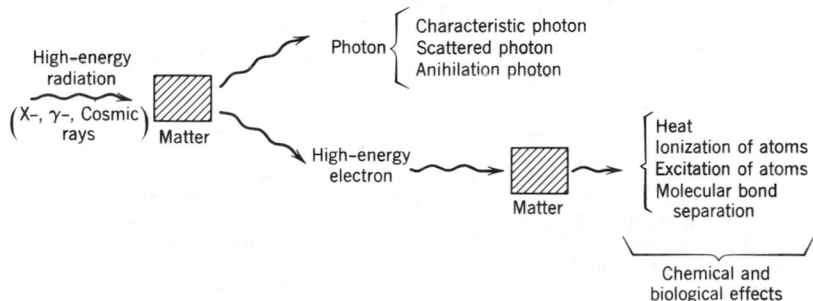

Exposure to Radiation. Exposure to high-energy radiation causes changes in the atomic structure of matter which in turn causes changes in the chemical and biological properties. These effects manifest themselves as changes in the electrical, mechanical, and thermal properties of the irradiated matter. The exposure of living tissue to high-energy irradiation results in local burns together with further delayed biological consequences. The irradiation of solids has become an extremely important technique in solid-state research.

Health Physics. Whatever the high-energy radiation, extreme care must be exercised by the operator to prevent personal exposure with subsequent consequences. To this end a branch of physics (Health Physics) has been evolved to reduce exposure hazards to a minimum. This branch of physics utilizes the knowledge gained by research experience in all high-radiation energy sciences to provide adequate and reasonable limits for (a) radiation exposure, (b) measurements of radiation exposure, and (c) radiation shielding.

The International Commission or Radiological units and Measurements (1956) has established the following dose units: (1) the roentgen as a unit of exposure dose, and (2) the rad as the unit of absorption.

Roentgen (R). This exposure unit has been defined as the quantity of x- or γ-radiation which produces an ionization change of 1 esu in 0.001 291 g of dry air, this being equivalent to 1 cm³ of dry air at 0°C and 760 mm Hg.

Rad (rad). This absorption unit corresponds to an absorption of 100 erg/g in any medium from an incident X- or γ-radiation. The roentgen and the rad are designed to be very nearly equal in magnitude, but note that the roentgen is specifically the absorption in air whereas the rad concerns the absorption in any other medium.

Radiation Absorption in Living Tissue

Roentgen Equivalent Physical (rep). This absorption unit is defined as that amount of incident radiation which results in the absorption of 83.9 erg/g at a localized spot in soft living tissue (or water).

Relative Biological Effectiveness (rbe). This unit is defined as that amount of incident radiation of any type which produces a specific biological damage in living tissue equal to that caused by a standard incident x- or γ-radiation which would transfer 3 keV of energy linearly into water when delivered at the rate of 10 rad/min.

Roentgen Equivalent Man (rem). This unit is designed to relate the effect of radiation (other than x- or γ-radiation) to the rad. 1 rem is defined as that amount of ionizing radiation of any type which imparts the same energy to 1 g of living tissue as 1 rad of (200 to 250) kV x-radiation.

Maximum Exposure Dose Rate. The recognized safety level of maximum permissible dose rate is 0.1 rem/wk (8 hr day) or 2.5_m rem/hr (8 hr day). Personal exposure to x- or γ-radiation should never exceed 300 to 600 mr/wk (8 hr day). The relative biological effectiveness of ionizing radiation can be obtained from the following table.

Maximum Permissible Flux

| | | Dose rate 0.1 rem per week | | |
Type of radiation	rbe	Exposure, rad/week	Energy level	Flux Units/sec-cm²
Gamma- and x-rays	1	0.100	Various	1400/MeV
Electrons	1	0.100	1 MeV	15
Thermal neutrons	2.5	0.040	0.025 eV	667
Fast neutrons	10	0.010	2 MeV	13
Alpha particles	10	0.010	5 MeV	0.005
Protons	10	0.010	5 MeV	0.06
Heavy ions	20	0.005	5 MeV	0.0002

Measurement of X- or γ-ray Quantity—Dosimetry. There are several recognized methods dependent on the magnitude of the dose rate to be measured, the accuracy of measurement required, and the portability of the measuring device.

For large dose rates. Considerable attention has been paid to the oxidation of ferrous sulfate to ferric sulfate in dilute solution and to the change in the UV absorption spectrum resulting from oxidizing radiation. This method has been utilized as the basis of an accurate dosemeter known as the Fricke dosemeter.

Standardization method. The internationally established method of incident high-energy radiation is by the direct measurement of the absorbed energy by calorimetric methods. Since the increase in temperature involved is extremely small, accurate sensitive thermometers are essential.

Ionization of gases. The ability of x- or γ-radiation to ionize air and make it conducting make this method the most acceptable method for the standard measurement of x- or γ-ray quantity. Air has been chosen for the following reasons: (1) ready availability, (2) its average atomic number is almost identical with that of soft tissue, and, (3) the energy absorbed per gram for air and soft tissue is approximately the same for all ranges of incident radiation energy.

Measurement of Radiation Exposure (Bragg-Gray principle). This method is used as the basis of many measurements of ionizing radiation and can be briefly summarized by the following formula: The ionizing radiation energy loss per unit mass of an absorber (b) is given by:

$$\frac{dE}{dm_b} = P_b J_g W_g$$

where P_b = relative mass stopping power of absorber (b) with respect to the ionizing radiation absorbed in a small gas filled cavity (g) placed in medium (b).

W_g = average energy required to produce an ion pair in gas (g).

J_g = the number of ion pairs per unit mass of gas (g) actually produced by the ionizing radiation.

Radiation Shielding

Ionizing radiation may be attenuated by means of several materials ranging from lead to concrete depending on the area to be shielded and the cost of the shielding materials chosen.

X-ray Shielding. Every precaution should be taken to shield operators from the primary beam of such equipment generating x-rays. Such protection is usually by one of more of the following methods:

(1) Protective screen of lead sheet, not less than 0.5 mm thick which may also be molded around such parts as the exit part of an x-ray machine. A 1.5-mm thickness of lead is sufficient to completely absorb 100 kV x-rays. An additional 0.01 mm/kV can be estimated for higher ranges up to 150 kV.

(2) Fluorescent screens used in direct observation methods must be backed with lead glass (up to 20 mm thickness of glass may be required).

(3) Protective lead-rubber waistcoats and other clothing are available for certain experimental methods. (This is equivalent to a thickness of 0.25 mm of lead.)

(4) Water windows containing high-density dissolved salts such as lead nitrate may be used between transparent partitions in certain circumstances as direct observation shields.

In the case of x-ray units it is always advisable to check the x-ray housing, the back of the ionization chamber detector or fluorescent screen, and the rectifying unit with a portable radiation dosemeter in order to be properly aware of any stray primary or secondary scattered x-radiation.

γ-ray Shielding Attenuation. The attenuation of a primary γ-ray beam with distance can be stated as follows:

(1) For a plane source the ratio of the intensities of the beam at distance x (I_x) within the material to that of the primary beam (I_0) in air is given by:

$$\frac{I_x}{I_0} = b_e{}^{-\mu x} \qquad (I \text{ in MeV/cm}^2\text{-sec})$$

where b = "build up" factor for shields of appreciable thickness.
μ = attenuation coefficient.

(2) For a point source surrounded by a sphere of absorbing material, the current density J_x in MeV/cm^2-sec at radius x is given by

$$J_x = \frac{b I_0 e^{-\mu x}}{4\pi x^2}$$

Note. b is determined experimentally and is found to depend upon μx and I_0 as well as the shielding material.

Shielding Properties of Materials

	Water	Iron	Lead	Portland Concrete	Barite Concrete
Density, g/cm³	1.00	7.78	11.3	2.37	3.49
γ at 0.5 MeV					
μ, cm⁻¹	0.096	0.653	1.72	0.20	0.30
Build-up b					
At $\mu x = 1$	2.46	2.80	1.51		
At $\mu x = 10$	71.5	34.2	3.01		
γ at 6 MeV					
μ, cm⁻¹	0.027	0.237	0.503	0.063	0.103
Build-up b					
At $\mu x = 1$	1.46	1.30	1.14		
At $\mu x = 10$	5.18	7.10	4.20		

BIBLIOGRAPHY ON PHYSIOLOGY OF RADIATION

1. COBB, P. W., "Some Experiments on the Speed of Vision," *Trans, IES.* Vol. 19, p. 150, 1924.
2. ELLIS, C., and WELLS, A. A., *The Chemical Action of Ultra-violet Rays* (revised and enlarged by F. F. Heyrother), New York, Reinhold, 1941.
3. HELMHOLTZ, H. V., *Handbuch der physiologischen Optik* (English trans. by J. C. Southall), 3 vols., Bata, 1896. HOUSTOUN, R. A., *Vision and Colour Vision*, London, Longmans, Green, 1932.
4. HOWELL, W. H., *Textbook of Physiology*, Philadelphia, Saunders, 11th ed., 1931.
5. LAURENS, H., *Physiological Effects of Radiant Energy*, New York, Chemical Catalog Co., 1933.
6. LUCKIESH, M., *Application of Germicidal, Erythemal and Infrared Energy*, New York, Van Nostrand, 1946.
7. LUCKIESH, M., *Artificial Sunlight*, New York, Van Nostrand, 1930.
8. LUCKIESH, M., *Ultra-Violet Radiation*, New York, Van Nostrand, 1923.
9. POHLE, E. A., and CHAMBERLAIN, W. E., *Theoretical Principles of Roentgen Therapy*, Lea and Febiger, 1938.
10. ROBERTSON, J. K., *Radiology Physics*, New York, Van Nostrand, 1941.
11. SPOEHY, H. A., *Photosynthesis*, New York, Chemical Catalog Co., 1926.

Illumination

1. AMICK, C. L., *Fluorescent Lighting Manual*, New York, McGraw-Hill, 1947.
2. BARROWS, W. E., *Light, Photometry and Illuminating Engineering*, New York, McGraw-Hill, 3rd ed., 1951.
3. BARBROW, L. F., and WILSON, S. W., "Vertical Distribution of Light from Gas-Filled Candlepower Standards," *Illum. Eng.*, Vol. 53, p. 645, 1958.
4. BENFORD, F. E., "Visible Radiation from the Low Pressure Mercury Arc," *Soc. M.P.E.*, Vol. 30, pp. 365, 390, 1927.
5. COTTON, H., *Principles of Illumination*, New York, Wiley, 1961.
6. DOBSON, G. M. B., GRIFFITH, I. O., and HARRISON, D. N., *Photographic Photometry*, England, Oxford University Press, 1926.
7. EVANS, RALPH M., *An Introduction to Color*, New York, Wiley, 1948.

9. HARDY, A. C., *Handbook of Colorimetry*, Massachusetts Institute of Technology, Cambridge, Massachusetts, 1936.
9. HURST, G. M., *Handbook of the Theory of Color*, New York, Van Nostrand, 1916.
10. I.E.S., *Lighting Handbook*, New York, Illum. Eng. Soc.,
11. KRAEHENBUEHL, J. O., *Electrical Illumination*, New York, Wiley, 1951.
12. LUCKIESH, M., TAYLOR, A. H., and SINDEN, R. H., "Data Pertaining to Visual Discrimination and Desired Illumination Intensities," *J. Franklin Inst.*, Vol. 192, p. 757, 1921.
13. MILLER, C. W., *Principles of Photographic Reproduction*, New York, Macmillan, 1942.
14. MILLER, S. C., *Neon Signs*, New York, McGraw-Hill, 1952.
15. MOON, P., *The Scientific Basis of Illuminating Engineering*, New York, McGraw-Hill, 1936.
16. WALSH, J. W. T., *Photometry*, New York, Dover, 3rd ed. 1958.
17. WINCH, G. T., "Photometry and Colorimetry of Fluorescent Lamps," *Trans. Illum. Eng. Soc.*, London, Vol. 11, p. 107, 1946.
18. WINTRINGHAM, W. T., "Color Television and Colorimetry," *Proc. I.R.E.*, Vol. 39, p. 1135, October 1951.
19. WYSZICK, "The Measurement of Brightness and Color," *Metrologia* p. 111, 1966.

High-Energy Radiation

1. BAERG, A. P., "Measurement of Radioactive Disintegration Rate by the Coincidence Method", *Metrologia*, p. 23, 1966.
2. "Choice and Operation of Detecters," *Metrologia*, p. 26, 1966.
3. ETTER, *The Science of Ionizing Radiation*, Thomas, 1965.
4. GRODSTEIN, G. W., *X Ray Attenuation Coefficients*, Natl. Bur. Standards Circ. 583, Washington, D.C., 1957.
5. International Commission on Radiological Units and Measurements, (ICRU), Handbook No. 84 (1962) and No. 88 (1963), Natl. Bur. Standards, Washington, D.C.
6. *Permissible Dose from External Sources of Ionizing Radiation*, NBS Handbook, U.S. Government Printing Office, Washington, D.C.
7. SKOLDBOM, H., *Acta Radiology*, Suppl. 187, 1959.

ACOUSTICS

By R. B. Lindsay

20. NATURE OF SOUND

Sound, from the physical point of view, consists of compressional disturbances produced and propagated in solid or fluid media. This includes not only those disturbances which are ultimately (by transmission through the air) audible to the individual but also a large field of disturbances (ultrasonic, etc.) which are outside the range of audibility and are rendered evident to the senses in other ways, e.g., by touch, etc. The whole field is usually known by the name *acoustics*.

Consider a metal bar which is struck with a hammer. Originally all parts of the bar are at rest with respect to one another; equilibrium exists. The impact of the hammer upsets the equilibrium and squeezes a portion of the bar by producing a relative displacement. (Note that if the bar were rigid it would have to move as a whole—but no material medium is rigid.) The motion of the "squeeze" along the rod constitutes what is called a *compressional wave*. If the rod is in air some of the motion is communicated to the air in its neighborhood and a compressional wave in air results.

The study of acoustics is concerned with the production of such disturbances, their propagation, and their absorption. It must be remembered that the displacements from equilibrium involved here are usually very small. In particular in the case of harmonic or sinusoidal sound waves in air the maximum fractional change in density due to the passage of the waves having the intensity of conversational speech is of the order of 10^{-7} and the maximum corresponding change in pressure (the so-called excess pressure) is about 1 dyne per sq cm. A compressional wave of the same intensity in a metal like platinum at ordinary temperature will correspond to a maximum change in stress of about 370 dynes per sq cm but the corresponding fractional change in density is only of the order of 10^{-10}. The solid is of course much less compressible than the gas. The displacement velocities involved in such displacements are also ordinarily very small, varying from 10^{-2} cm per sec for plane waves in gases to about 10^{-5} cm per sec for similar waves in solids. This velocity, however, must not be confused with the velocity of propagation of the sound wave itself, which may be very large (see Art. 21).

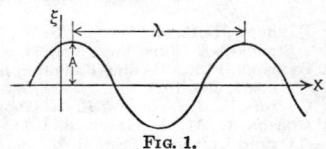

FIG. 1.

General Properties of Sound Waves. For practical purposes the most important type of wave is the simple harmonic or sinusoidal. A snap-shot picture of a plane wave of this character progressing in the x direction is symbolized in Fig. 1 where a part of such a wave is shown. The ordinate measures the displacement ξ which is being propagated. The maximum ordinate, A, is called the *amplitude* of the wave. A positive maximum is said to correspond to a *crest*, a negative one to a *hollow*. The distance between two successive maxima is the *wavelength*, designated by λ. The wave propagation corresponds to the motion of the whole figure to the right along the x axis with a velocity c. Mathematically such a situation can be symbolized by

$$\xi = A \sin \frac{2\pi}{\lambda} (ct - x) \tag{1}$$

It may be noted that either the sine or the cosine may be used. The number of complete wavelengths which pass a given point per second is the *frequency* of the wave, usually denoted by ν and referred to in cycles or cycles per second. The relation

$$\lambda = \frac{c}{\nu} \tag{2}$$

is fundamental in the theory of harmonic waves. Written in terms of frequency

$$\xi = A \sin (\omega t - kx) \tag{3}$$

in which $\omega = 2\pi\nu$ and $k = \omega/c = 2\pi/\lambda$. This form is often preferred because of its greater simplicity. The quantity $\frac{2\pi}{\lambda} (ct - x) = \omega t - kx$ is called the *phase* of the wave.

Types of Waves. The wave symbolized by equation (1) is called a plane wave because the displacement at a given time and value of x is the same over the whole plane through this point perpendicular to the x axis, or, as is sometimes said, all points in this plane are in the same phase. Using still another common phraseology the *wave front* of the wave is a plane. Another type of wave of great value in acoustics is the *spherical* wave in which the wave fronts are spheres. Such a wave results when a disturbance takes place at a point in a homogeneous medium and spreads in all directions from this point as center with the same velocity. Far away from the source a small portion of the wave front may be considered a plane.

Although sound waves have been considered as compressional waves in a material medium, it is quite possible for other types of elastic waves in solids to give rise to acoustic waves in air. Thus the transverse waves in strings, membranes, plates, and rods are of importance in sound. These will be mentioned later in Art. 27.

The progress of a sound wave in air can be made visible by a method due to Foley. (*Physical Review*, Vol. 35, 373, 1912.) The wave sent out by a spark discharge is photographed in the light of another suitably timed spark, the excess density produced by the passage of the wave providing enough change in the refractive index of the air to register on the photographic plate. This method is valuable in the picturization of the reflection, refraction, and diffraction of sound. Since 1936 great progress has been made in the use of optical methods for the study of sound propagation, particularly at high frequency. (Cf. bibliography, reference 45.)

21. VELOCITY OF SOUND

The general expression for the velocity of a compressional wave in a deformable medium is

$$c = \sqrt{\frac{dp}{d\rho}} \tag{4}$$

where dp is the change in *pressure* accompanying the change $d\rho$ in *density*. Since the compressions and rarefactions accompanying the propagation of sound through a gas are adiabatic, the relation between pressure p and volume V is the well-known one $pV^\gamma = $ constant (γ being the ratio (c_p/c_v) of specific heat at constant pressure to that at constant volume), and equation (4) becomes for a gas

$$c = \sqrt{\frac{\gamma p}{\rho}} \tag{5}$$

In the case of air, for example, for $p = 1$ atmosphere $= 1.01 \times 10^6$ dynes per sq cm and

$\rho = 0.001205$ gram per cu cm at 20 C and 76 cm of Hg and $\gamma = 1.403$, $c = 344$ meters per sec in good agreement with the experimental value.

The Velocity of Sound in a Perfect Gas is independent of the pressure. This is practically true of all real gases, although a slight increase is observed with pressure of the order of 50 to 100 atmospheres (see International Critical Tables, Vol. VI). The temperature dependence is obtained from equation 5 by the use of the general gas equation, which for a perfect gas is $pV = RT$, T being the absolute temperature and R the gas constant. Thus the velocity at any temperature t C in terms of the velocity at 0 C is given by the formula

$$c_t = c_0 \sqrt{1 + t/273} \qquad (6)$$

For air in the neighborhood of 0 C and room temperature the increase in c is about 0.61 meter per sec per degree Centigrade. The velocity of sound for gases is always lowered slightly by confinement in a tube.

The Velocity of Sound in Air depends slightly on the presence of water vapor. Thus for saturated air at 20 C and standard pressure, c is about 0.40 per cent greater than for dry air at the same temperature and pressure.

For velocity of sound in a *liquid* equation (4) takes the form

$$c = \sqrt{\frac{E}{\rho}} = \sqrt{\frac{1}{K\rho}} \qquad (7)$$

where E is the bulk modulus or coefficient of volume elasticity, and in the alternative form K is the compressibility. Both E and ρ depend on the temperature and pressure, but since the equation of state of liquids is not known in general form the precise temperature and pressure dependence of c in this case can be described only by empirical formulas relating to specific substances. The velocity of sound in liquids is almost uniformly greater by a factor of several times than the velocity in gases, the one notable exception being hydrogen where c at 0 C is 1270 meters per sec. In general, aqueous solutions have higher c than water.

The Velocity of Sound in Solids is given by the formula

$$c = \sqrt{\frac{E + 4\mu/3}{\rho}} \qquad (8)$$

in which μ is the *shear* modulus or *rigidity* of the solid. For the case of compressional waves in long narrow rods equation (8) must be replaced by

$$c = \sqrt{\frac{Y}{\rho}} \qquad (9)$$

Y being Young's modulus. For a large number of solids the former velocity is greater than the latter by roughly 10 per cent. It is interesting to note that the velocity of transverse waves (i.e., flexural or torsional) in a solid is given by $c = \sqrt{\mu/\rho}$, which is less than the value from equation 8 for any given solid under the same conditions. In general the velocity in metallic solids exceeds that in non-metallic solids, although glass, marble, and some varieties of wood are decided exceptions. In the latter case the velocity naturally depends considerably on the relation of the direction of the sound propagation with respect to the grain of the wood. Some data are available on the direct measurement of the velocity of sound in crystals, metallic and otherwise. Sound-velocity measurements are very valuable in the precise determination of elastic constants, both for polycrystalline and crystalline materials. (Cf. bibliography, references 32, 45.)

Change of Velocity with Frequency. Equation (4) would indicate that the velocity of sound is independent of the frequency. This is found to be approximately true for sound in gases. Pierce * found, however, that the velocity of sound in air at 0 C varies from 332.45 m per sec at 41,009 cycles per sec to 331.64 m per sec at 1,479,900 cycles per sec. This is for entirely unconfined air. The velocity of sound in liquids is slightly dependent on the frequency but the change is small, that in water being a decrease of only 5 per cent as the frequency ranges from 40,000 cycles to 600,000 cycles. The variation with frequency of sound velocity in solids has not yet been studied thoroughly enough to state conclusive results, though *configurational* dispersion does exist in solid rods, i.e., change of longitudinal wave velocity with frequency in the range where the wavelength is of the order of magnitude of the transverse dimensions of the rod. (Cf. bibliography, reference 45.)

---

* *Proc. Am. Acad. Arts and Sci.*, Vol. 60, p. 271, 1925.

When gases and liquids are confined in tubes, the decrease in velocity due to viscosity may be considerable. This is also a function of the frequency, the change here growing smaller as the frequency increases (see Art. 25).

Since a sound wave is propagated through a material medium, the sound velocity depends on the large-scale motion of the medium. Hence in the free atmosphere the velocity is considerably affected by wind currents.

Wavelength of Sound Waves. A harmonic wave with frequency 256 cycles per sec has a wavelength in air at 20 C of 1.34 meters. In water at the same temperature the wavelength is 5.70 meters, and in a brass *rod*, 13.7 meters. The influence of the change in velocity with the nature of the medium is evident.

Doppler Effect. When a source of sound moves relatively to an observer, a phenomenon occurs which is known as the *Doppler effect.* Consider a stationary observer at B, Fig. 2

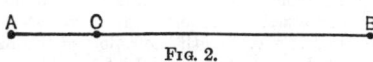

FIG. 2.

(i.e., stationary with respect to the air), with the sound source at A emitting waves of frequency ν moving towards him with velocity v which is assumed less than c, the velocity of sound in a stationary medium. Since the frequency of the source is ν, this is the number of waves sent out in 1 sec while the source is moving from A to C. These waves will strike the ear of the observer within $1 - v/c$ seconds. Consequently the frequency of the sound as he will observe it is not ν, but

$$\nu' = \frac{\nu}{1 - v/c} \qquad (10)$$

Otherwise expressed, there is an apparent increase in the pitch. This is easily observed when a train with whistle blowing approaches one, and is even more commonly noted with approaching automobiles.

When the source recedes from the observer, fewer waves are received per second and the apparent frequency, ν'', is lower than the true frequency. Then

$$\nu'' = \frac{\nu}{1 + v/c} \qquad (11)$$

The situation is similar though not identical when a moving observer approaches or recedes from a stationary source. Here the observer, so to speak, goes to meet or runs away from the waves, respectively. Analysis shows the relations between apparent and true frequencies, for approach and recession, respectively, to be:

$$\nu' = \nu(1 + v/c) \qquad (12)$$

$$\nu'' = \nu(1 - v/c) \qquad (13)$$

The velocity v may be larger than c, resulting in the "bow" wave, characteristic, for example, of the flight of a rifle bullet through the air.

22. REFLECTION AND REFRACTION

The concept of an acoustic wave traveling forward steadily in one direction or spreading outward from a single source is a highly idealized one. Actually all acoustic waves move in bounded media, and wherever boundaries are encountered the phenomena of reflection and refraction occur, resulting in more or less complicated wave patterns.

Reflection. As a simple illustration, suppose that a plane wave of sound progressing to the right meets a rigid wall parallel to the wave front. Reflection takes place, i.e., the space to the left of the wall is now occupied by a wave moving toward the left as well as the original wave moving toward the right. Symbolically the disturbance may be represented by

$$\xi = A \sin (\omega t - kx) + B \sin (\omega t + kx + \varepsilon) \qquad (14)$$

where the term with the plus sign before the kx represents the reflected wave, A and B are the amplitudes of the two waves respectively, and ε is the phase change that accompanies reflection. ε can be shown to be zero when a plane wave of sound is reflected in going from an acoustically rare to an acoustically dense medium, from air to a rigid wall, for example. When the sound is reflected in going from a dense to a rare medium, e.g., water to air, the change of phase is π. It is to be understood in this result that the positive direction of particle displacement is always in the direction in which the wave is being propagated.

The Acoustic Density is measured by the product of the ordinary density and the velocity of sound in the medium. This quantity is called in technical acoustics the *specific acoustic resistance* * for a plane wave (see Arts. 24 and 34).

Echo. When sound in air is reflected from a large wall or barrier the result is known as an *echo*. This will be recognized as a distinct sound if the reflecting surface is sufficiently far away from the observer so that the original sound has died down by the time the reflected wave reaches him.

Reverberation. If a multitude of echoes follow on each other at very short time intervals the result is known as *reverberation*. This is observed 'n large closed spaces and if the walls and other objects in the room are not sufficiently absorbing can constitute a serious impediment to hearing in halls (see Art. 33).

Reflecting Surfaces and Media. If the surface dimensions are large compared with the wavelength of the sound, the incident plane wave beam will be reflected approximately as such, but if the dimensions are small compared with the wavelength the reflected sound will be scattered in all directions. In the former case the sound that might be scattered outside the beam is nullified by the interference of the reflected waves (i.e., the cancellation of the crest of one wave by the hollow of another) from various parts of the reflector. This interference is no longer materially operative when the reflector is small. It thus turns out that plane reflectors exercise a selective action.

Sound waves can also be reflected in other types of mirrors, e.g., parabolic, in which if a point source of sound is placed at the focus the reflected sound will consist largely of plane waves with wave front normal to the axis of the mirror. Owing to the relatively large wavelength of audible sound in air the focusing of sound in such a mirror is not very precise. The situation is even worse in liquids.

Reflection occurs not only when a sound wave encounters the boundary between two media but also when sound passing down a tube meets an abrupt change in area. The fractional sound energy (see Art. 28) reflected compared with the incident energy can be readily computed † and turns out to be

$$\left(\frac{S_1 - S_2}{S_1 + S_2}\right)^2 \tag{15}$$

where S_1 and S_2 are the two areas of cross section in question. It is usually assumed for elementary purposes (Art. 27) that when a plane wave of sound passing through a tube is incident on the open end, the reflection is complete. This of course is not entirely true since some radiation takes place.

Oblique Reflection. The discussion of reflection so far has been confined to the case where the incident sound wave front is parallel to the reflecting surface. In unconfined regions, of course, oblique incidence is more common. Reflection then follows the optical law: *the angle of reflection equals the angle of incidence.* The ratio of the sound energy per square centimeter per second reflected to that incident is *

$$\left(\frac{\rho_2/\rho_1 - \cot\theta_2/\cot\theta_1}{\rho_2/\rho_1 + \cot\theta_2/\cot\theta_1}\right)^2 \tag{16}$$

where ρ_1 and ρ_2 are the mean densities of the two media respectively at the boundary of which the reflection takes place, and θ_1 and θ_2 are the angles of incidence and refraction respectively.

Refraction. With oblique incidence there also occurs transmission into the second medium accompanied by bending of the wave front, i.e., *refraction.* The law of refraction is the same as that of Snell in optics, namely,

$$\frac{\sin\theta_1}{\sin\theta_2} = \frac{c_1}{c_2} \tag{17}$$

Very good illustrations of the refraction of sound are provided by transmission through the atmosphere in which a temperature gradient and winds exist. A wave traveling from a region of high temperature to one of low temperature has its front bent toward the normal. Actually there are usually no sharp boundaries and the bending is a more or less continuous process. The increased range of sound in still air when the temperature gradient is negative, i.e., cold air near the ground and warmer air higher up, is to a considerable extent accounted for by this refraction. Acoustic mirages have the same source.

---

* See bibliography, references 3, 7, and 11.

† See bibliography, reference 11.

Refraction also takes place when sound passes through a region where the wind velocity is changing from place to place. This is called convective refraction, and it is well to note that the wave front is then no longer normal to the *ray* as it is in a stationary medium. Reflection and refraction play a considerable rôle in the scattering of sound in the atmosphere, which is found in general to be greater for high frequencies than for low.

23. DIFFRACTION

Effect of Rigid Obstacles. When sound in passing through a médium encounters a more or less rigid obstacle in addition to reflection and refraction there results *diffraction*, i.e., the *bending* of the sound waves around the obstacle. Here the situation is precisely analogous to the behavior of light waves in passing the edge of an obstacle. Acoustical diffraction acounts for our ability to hear sound around a corner and at the same time for the difficulty of producing sharp acoustic shadows with ordinary audible sounds; the sound waves with their greater length are able to bend more than the very much shorter light waves. The reception of sound by the individual auditor is largely conditioned by the diffraction caused by the human head. A careful study has been made of this, for the effect both on speech and on hearing. It is found,* for example, that for a point source on a rigid sphere there are two directions of maximum intensity, one directly in front of the source and the other directly behind the sphere, the latter intensity being naturally the smaller of the two. The intensity falls to a minimum in a direction approximately at right angles to that of maximum intensity. The angle increases with increasing frequency of sound and the magnitude of the minimum intensity decreases as the frequency increases, i.e., short-wavelength sounds do not so readily bend around the spherical obstacle. For a given frequency the variation in intensity around the obstacle is greater near than far from the sphere.

Sound of wavelength very short compared with the dimensions of an obstacle in its path is more or less completely shadowed by the obstacle, and we can treat the sound propagation in terms of "rays," analogous to the light rays of geometrical optics. Accoustics then becomes "geometrical acoustics."

A similar discussion can be given for a sphere having two sound receivers on the opposite ends of a diameter (e.g., the head with the two ears). It turns out that the intensity of an outside source of sound at the sphere is greatest when the line joining the two receivers is directed toward the source and least when the line is perpendicular to the source direction. Here again for given source distance the difference is more marked for short waves than for long and for a given wavelength more marked at short distances than long.

These results have an immediate application in respect to the human head and to sound-receiving instruments like microphones for mapping the distribution of sound intensity in a room. The diffraction of sound due to the presence of such instruments can thus be compensated for theoretically.

Scattering. The special phenomenon of diffraction by an obstacle of dimensions much smaller than the wavelength is known as *Rayleigh scattering*. Here the sound is scattered in all directions and its intensity varies inversely as the fourth power of the wavelength, analogously to light scattering. "Harmonic" echoes are due to this type of scattering. When a compound musical note is sounded near a group of obstacles like a grove of trees the intensity of the octave is raised above that of the fundamental by the scattering.

When sound emerges from an opening, diffraction produces in general a scattering in all directions. If the frequency is very high this effect is much reduced and a more or less sharp beam of sound results. This suggests the use of ultrasonic waves in signaling (see Art. 30).

Diffraction is also important in the use of baffle plates in some loud speakers. The actual calculation of the diffracting effect of such plates is extremely complicated.

24. INTENSITY OF SOUND

Definition. The propagation of sound waves involves the transfer of energy, and it is this idea which is expressed in the concept of acoustic intensity, which is the *average* rate of flow of energy per unit area normal to the direction of propagation. Care must be taken to distinguish between intensity and loudness. The former is a purely physical quantity, whereas the latter is a physico-psychological one. We shall discuss it further

---

* See bibliography, reference 11.

below. At the moment consider merely the intensity. It can be shown that the intensity I of a plane or spherical wave (the two predominant types) is given most simply by the formula

$$I = \frac{1}{2} \frac{p^2_{max}}{\rho_0 c} \qquad (18)$$

where p_{max} = maximum excess pressure, ρ_0 = average density, and c = velocity of the wave. The quantity $\rho_0 c$ is the *specific acoustic resistance* of a plane wave. For a given p_{max}, a large acoustic resistance means a small intensity, and vice versa. This has considerable bearing on all acoustic transmission problems, e.g., sound proofing and reduction of noise (see Art. 34).

For a sound wave in air at 20 C and 76 cm of Hg the intensity of ordinary conversational speech (with $p_{max} = 1$ dyne per cm^2) is approximately 1.21 $\times$ 10^{-9} watt per cm^2 = 1.21 $\times$ 10^{-3} microwatt per cm^2. Interestingly enough it develops that the ear is sensitive to sounds of intensity as low as 4 $\times$ 10^{-10} microwatt per cm^2. It should be noted that the frequency does not enter (18) explicitly. On the other hand, if the intensity is expressed in terms of displacement amplitude instead of pressure, the frequency ν is necessarily involved, equation (18) being replaced by

$$I = \tfrac{1}{2} \rho_0 c \omega^2 \xi^2_{max} \qquad (19)$$

with ξ_{max} = maximum displacement and $\omega = 2\pi\nu$. For constant amplitude the intensity thus varies as the square of the frequency. Hence for ultrasonic waves of reasonable amplitude the intensity can become relatively very large; e.g., for $\nu = 500{,}000$ cycles per sec in air and $\xi_{max} = 10^{-4}$ cm, $I = 2.07 \times 10^5$ microwatts per cm^2. For acoustic instruments which record pressure changes, equation (18) is the more useful one. For others, equation (19) is employed.

The Measurement of Intensity in absolute units is now largely replaced by the logarithmic unit, the *bel*, which is of particular advantage in denoting the relative levels of intensity of several sounds. The difference in intensity of two sounds whose absolute intensities are I_1 and I_2 respectively is α bels, where

$$\alpha = \log_{10} I_2/I_1 \qquad (20)$$

More commonly the *decibel* (*db*), which is one-tenth of a bel, is used. To illustrate, if one calls an average excess pressure of 1 dyne per cm^2 (strictly this is the root-mean-square value which = 0.707 p_{max}) the normal level for ordinary speech, the threshold of audibility has a minimum lying about 70 db below this level, and the threshold of feeling (i.e., where hearing is painful) has a maximum lying about 70 db above this level. We may think of audible sound therefore as lying in a maximum range of some 140 db.*

25. INFLUENCE OF THE MEDIUM ON THE PROPAGATION OF SOUND

Spherical Waves. From a point source, sound spreads in a perfectly homogeneous medium in the form of a spherical wave, the mathematical expression for the displacement in a harmonic wave of this kind being

$$\xi = \frac{A}{r} \sin (\omega t - kr) \qquad (21)$$

From equation (19) it is seen that the intensity of such a wave is given by the expression

$$I_s = \tfrac{1}{2} \rho_0 c \omega^2 A^2 / r^2 \qquad (22)$$

i.e., the intensity falls off inversely as the square of the distance from the source. This is in contrast to a plane wave (see equation 3), for which the intensity is independent of distance. The diminution of intensity of a spherical wave is a purely geometrical phenomenon—as the wave spreads in all directions, the same amount of energy at any instant flows across spheres of larger radii, whose surfaces increase as the square of the radii. In addition to this geometrical decrease in intensity with distance all material media act to dissipate acoustic energy and so decrease intensity by absorption.

Causes of Sound Absorption in Fluid Media

Viscosity acts to retard the relative motion of adjacent layers of the medium, and so

---

adds a damping force to the elastic restoring force normally present. The effect is to produce an exponential decrease in displacement amplitude with distance; i.e., the amplitude A must be multiplied by $e^{-\alpha r}$ where e is the Naperian base ($= 2.71828 \cdots$) and $\alpha = {}^2/_3 \, \mu \omega^2/\rho_0 c^3$. μ, the coefficient of viscosity for air at 20 C, is approximately 18×10^{-5} gram per cm per sec and for water at 20 C is 10^{-2} gram per cm per sec. It will be observed that the damping coefficient increases with the square of the frequency. For $\omega = 1000$ in air, $\alpha = 0.24 \times 10^{-8}$; i.e., a wave in air could proceed a distance of about 4000 km without having its amplitude reduced to more than $1/e$ as far as viscosity is concerned. This is a very slight effect and hardly observable except at high frequencies. The effect is much more marked when the medium is confined as in a tube, for here the fluid tends to stick to the walls. For a tube whose diameter is large compared with $\sqrt{8\pi^2\mu/\rho_0\omega}$ the theory *

again indicates exponential damping with $\alpha = \dfrac{1}{ac}\sqrt{\dfrac{\omega\mu}{2\rho_0}}$, where $a = $ radius of the tube.

For a tube of radius 1 cm and $\omega = 1000$, $\alpha = 2.5 \times 10^{-4}$ for air at 20 C. Associated with the absorption is a decrease in the velocity of sound which in the instance just cited is theoretically 0.86 per cent. Actually both the absorption and the decrease in velocity depend on the material of the confining tube and are greater in practice than the theory predicts.

For capillary tubes, i.e., where $a \lll \sqrt{8\pi^2\mu/\rho_0\omega}$, the viscous absorption is naturally

very much greater, the value of α being $\dfrac{2}{a}\sqrt{\dfrac{\omega\mu}{p_0}}$, where p_0 is the average value of the fluid pressure in the tube. Here the waves are very rapidly damped out, with α having, for moderate frequencies, values in the neighborhood of 10 to 100. This accounts for the large absorption of sound in carpets and other materials with small interstices.†

The effect of viscosity is enhanced by the conduction by the vibrating gas or liquid of the heat produced during the compressions accompanying the passage of the wave. The order of magnitude here is the same as that involved in viscosity damping.

The absorption of sound in extended liquid media (e.g., the open sea or lakes) is very much larger than can be accounted for by viscosity or heat conduction alone. Recent research indicates that it is due to some kind of thermal or structural relaxation process. (Cf. bibliography, reference 45.)

Changes in Physical Properties. In the actual propagation of sound through such media as the atmosphere and the sea the numerous changes in the physical properties of the medium due to temperature, etc., produce by reflection, refraction, and diffraction a damping of a directed beam which is much more pronounced than that due to any of the causes just mentioned.

Recently it has been found by Knudsen ‡ that the absorption of sound in air depends on temperature and humidity in a characteristic manner. Thus the absorption coefficient per centimeter (α) for air at 20 C and for a frequency of 1500 cycles per sec has a maximum value of 10^{-4} at a relative humidity of 10 per cent. For a frequency of 10,000 cycles per sec the maximum is about 6.5×10^{-4} and occurs at a relative humidity of about 18 per cent. These figures are of course much larger than those theoretically predicted on the basis of viscosity absorption. It has been established that this absorption is due almost entirely to the energy transfers between the oxygen molecules of the air. This is a relaxation process which opens up a new field for the study of sound absorption.

26. VIBRATING SYSTEMS AS SOURCES OF SOUND

Free Oscillations. Most vibrating systems do not oscillate as a whole but are the seat of more or less complicated wave patterns. For the sake of simplicity, however, it is often convenient to think of such bodies as if they moved as a whole. Thus a membrane in which the amplitude at the center is greater than that elsewhere may often conveniently be assumed to oscillate like a piston moving to and fro. The same is true of the air in the opening of a Helmholtz resonator.

The vibration of a body conceived from this point of view is governed by what may be called its "elements," viz., its *inertia* (measured by *mass*), *stiffness* (elastic restoring factor), and *dissipative resistance* (damping factor). The *stiffness* is defined analytically as the restoring force per unit displacement; the *resistance* is the damping force per unit velocity. These are the analogs of inductance, capacitance (strictly reciprocal of capacitance), and

---

* See bibliography, references 5 and 11.

† See bibliography, references 5 and 11.

‡ *J. Acous. Soc. Am.*, Vol. 6, p. 199, 1935; *Science*, Vol. 81, p. 578, 1935.

resistance, respectively, in an oscillating electric circuit. A knowledge of these elements is sufficient to describe completely the behavior of the system if disturbed from equilibrium and thereafter left to itself. Under such circumstances the system is said to be *free*.

Let m denote the mass, f the stiffness, and R the damping factor of the system. It turns out that a disturbance from equilibrium will lead to oscillations if and only if §

$$R^2/4m^2 < f/m \qquad (23)$$

and the frequency of the oscillation is

$$\nu_f = \frac{1}{2\pi} \sqrt{\frac{f}{m} - \frac{R^2}{4m^2}} \qquad (24)$$

If the damping factor is small enough so that the left side of the inequality (23) is very much smaller than the right, the free or natural frequency is approximately

$$\frac{1}{2\pi} \sqrt{\frac{f}{m}} \qquad (25)$$

As an illustration consider a membrane (e.g., a telephone diaphragm) which when replaced by an equivalent piston has

$$m = 0.50 \text{ gram}, \quad f = 2.5 \times 10^7 \text{ dynes per cm}$$

$$R = 250 \text{ dyne sec per cm}$$

Then $R^2/4m^2 = 6.25 \times 10^4 \text{ sec}^{-2}$, while $f/m = 5 \times 10^7 \text{ sec}^{-2}$, so that in equation (24) $\nu_f = 1124.6$ vibrations per sec whereas equation 25 gives 1125.4 vibrations per sec, the difference being negligible for most practical purposes.

Although the effect of the damping on the frequency is very small, the amplitude of the harmonic oscillation dies away exponentially; i.e., the displacement ξ may be written

$$\xi = Ae^{-(R/2m) \cdot t} \cos (2\pi\nu_f t + B) \qquad (26)$$

where A = initial amplitude and B = initial phase. The amplitude decreases to $A/e = 0.37 A$ in the time $t' = 2m/R$, which in the above illustration is $1/250$ sec. All free physical oscillations are damped, the damping arising from a variety of factors, such as internal viscosity of the vibrator, viscosity of the medium, or energy communicated to the medium by the vibrator and carried away in the form of radiation.

Forced Oscillations. More interesting and important than free oscillations are those imposed on the system by an external harmonic force, e.g., the fluctuating magnetic field in the telephone diaphragm. If the frequency of the force is $\nu = \omega/2\pi$ and its amplitude is F_0, the displacement is *

$$\xi = \frac{F_0}{\omega Z} \cos (\omega t - \alpha) \qquad (27)$$

where $Z = \sqrt{(m\omega - f/\omega)^2 + R^2}$, and is defined as the *mechanical impedance* of the system (cf. analogy to electrical impedance). The quantity α is the phase difference between the force and displacement and equals $\tan^{-1} \dfrac{R}{m\omega - f/\omega}$. The displacement velocity is usually more important. Denoting this by $\dot{\xi}$ we have

$$\dot{\xi} = \frac{F_0}{Z} \cos (\omega t - \beta) \qquad (28)$$

where $\beta = \tan^{-1} \dfrac{m\omega - f/\omega}{R}$ = phase difference between force and velocity ($\alpha + \beta = 90°$).

It is seen that the magnitude of both ξ and $\dot{\xi}$ depends on the impedance; a large impedance, other things being equal, corresponds to a small displacement, and vice versa (cf. the alternating-current analogy: current = emf ÷ impedance).

The impedance is a minimum for $\nu = \dfrac{\omega}{2\pi} = \dfrac{1}{2\pi} \sqrt{\dfrac{f}{m}}$. This is called the *resonance frequency*. For systems with small damping it coincides effectively with the free oscillation frequency. Physically this means that a system will vibrate most vigorously if subjected to a harmonic force with frequency equal to its own natural frequency. At resonance the force contributes energy to the system at maximum average rate and the system

---

§ See bibliography, references 2, 5, and 9.

* See bibliography, references 2, 5, 9, and 11.

itself dissipates energy at an equal rate: if a vibrating tuning-fork is held over the open end of an air column in resonance with the fork, the air in the tube vibrates vigorously and the note is much reinforced; at the same time the fork more rapidly uses up its original store of energy and comes to rest sooner than it otherwise would. Examples of resonance are at hand throughout acoustical phenomena, as well as in vibration phenomena in general.

The damping exercises an important influence on resonance; if the damping factor is small the response of the system at resonance is great but the tuning is sharp; that is, the response drops rapidly as the frequency of the force is altered from the resonance frequency. On the other hand, large damping leads to a smaller resonance response but a broader peak and more diffuse tuning. The same effect is well known in electrical oscillating circuits.

A good illustration of such a vibrating system as we have been considering is a Helmholtz resonator which consists of an enclosure usually but not necessarily spherical in shape communicating with the outside by means of a small orifice. The mass of the system is the mass of the air in the opening which vibrates under external influence against the air cushion provided by the air in the enclosure or resonator chamber, the latter supplying the stiffness of the system. The damping comes mainly from the radiation of sound into surrounding medium. The approximate resonance frequency is *

$$\nu = \frac{1}{2\pi} c \sqrt{\frac{c_0}{V}} \qquad (29)$$

in which c = velocity of sound in air, V = volume of the resonator chamber, and c_0 = a quantity with dimensions of length called the conductivity of the opening. It is rather difficult to compute but is of the order of magnitude of the diameter of the orifice.

Resonators have many practical uses, among them being the construction of sensitive sound detectors and instruments for the measurement of sound intensity (see Art. 35).

27. SPECIAL VIBRATING SYSTEMS

Vibration of Strings. If a string with line density (mass per unit length) ρ and stretched with tension τ is pulled aside from its equilibrium position and let go, a *transverse* wave travels along the string with velocity †

$$v = \sqrt{\frac{\tau}{\rho}} \qquad (30)$$

If the string is of finite length l and is fastened at both ends so that no motion is possible there, it turns out that the string is the seat of waves in both directions which combine to form standing waves of only certain frequencies, viz.,

$$\nu = \frac{n}{2l} \sqrt{\frac{\tau}{\rho}} \qquad (31)$$

where n is *any* integer. These allowed values of ν are the *characteristic* or *natural* frequencies of the stretched string. $n = 1$ corresponds to the fundamental, and $n = 2, 3, 4 \cdots$ to the successive harmonics or overtones. By the imposition of a harmonic force the string can, of course, be forced to vibrate with any frequency, but *resonance* (Art. 26) ensues when the frequency of the force coincides with one of the set (equation

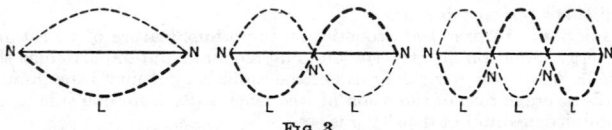

Fig. 3.

31). When a string vibrates with one of its characteristic frequencies, it is said to be in a characteristic mode of oscillation. Each mode is also characterized by the presence of certain points where no motion ever takes place: these are the so-called nodal points or *nodes* in distinction to the places of maximum disturbance or *loops*. Figure 3 indicates the

---

* See bibliography, references 2 and 11.
† See bibliography, references 5 and 9.

first three characteristic modes of vibration of a stretched string. The full line in each sketch represents the undisturbed position of the string; the dotted and dashed lines represent the extreme positions assumed by the string. Nodes and loops are indicated by N and L respectively. Of course the ends are, strictly speaking, nodes by virtue of the fastening. Obviously the distance between successive nodes is one-half wavelength (see equation 2 in connection with equation 31).

String Instruments. The most important acoustical application of the vibration of strings occurs in musical instruments. Here strings may be vibrated by plucking (harp), by bowing (violin), or by striking with a hammer (piano). If a string is plucked at one of its possible nodes, the corresponding harmonic or harmonics will be absent. This is known as Young's law: thus plucking at the center removes all harmonics of even order. In any event the higher harmonics are relatively weak in the plucked string. The theory of the struck string has not been completely worked out. Much depends on the precise manner and duration of the blow. But Young's law is again true and is utilized to minimize the harmonics above the sixth. The same is true of the bowing of the violin string. No matter how a string is excited, much of the effect depends on the presence of resonating surfaces and air spaces.

The Vibrations of Air Columns (e.g., organ pipes) are mathematically quite analogous to those of strings. Here, to be sure, the characteristic frequencies depend on whether both ends of the tube are open or only one. If both ends are open the natural frequencies * are $\frac{nc}{2l}$ (c = velocity of sound in air); if only one end is open they are $\frac{(2n-1)c}{4l}$. The difference comes from the fact that the wave pattern must have a node at the closed end of the tube while it has a loop at the open end. The values cited are not quite correct, for radiation from the open end has been neglected. To take account of this the length, l, in the formula must be slightly increased.

Quality. It is the combination of harmonics in the vibration which gives the sound radiation from such a system its characteristic *quality* (see Art. 32 on the analysis of musical sounds); it is this which distinguishes sounds of essentially the same frequency or pitch produced by different systems (e.g., violin and flute).

Membranes and Diaphragms. For the practical production of sound, circular membranes and diaphragms are of the utmost importance. A membrane is supposed to be perfectly flexible; i.e., to possess no rigidity, and many of the diaphragms used in sound generators more nearly approximate plates. The mathematical details in either instance are too difficult to give here. A circular membrane of radius a clamped around the periphery has a very complicated set of characteristic frequencies. The simplest of these correspond to vibrations symmetrical about the center and have frequencies given approximately by

$$0.77\,\frac{c}{2a},\quad 1.76\,\frac{c}{2a},\quad 2.76\,\frac{c}{2a} \tag{32}$$

where c = velocity of flexural waves in the membrane = $\sqrt{\tau/\rho}$, in which τ = superficial tension or force per unit length in the membrane and ρ = mass per unit area. It will be noted that the harmonics bear no integral multiple relationship to one another, differing thus from the harmonics of the string and air column. Corresponding to the nodal points of these systems, the membrane has a set of nodal lines which in symmetrical vibrations are concentric circles. In the higher modes of vibration there also exist nodal diameters. The situation for vibrating plates is even more involved and the wave patterns more complicated, although the characteristic frequencies have been computed in special instances. As has been intimated in Art. 26, the tendency in practical work is to operate with the equivalent piston vibrator.

Energy Radiation. In practical acoustics an important feature of vibrating systems is the rate of energy radiation into the surrounding medium, and in particular the efficiency of the system as a radiator, viz., the ratio of the radiation output to the input. It can be shown that the average rate of radiation of frequency $\omega/2\pi$ from one side of a diaphragm into a semi-infinite medium of density ρ is

$$\frac{1}{4}\,\pi\rho\,\frac{\omega^2}{c}\,a^4\dot{\xi}_0{}^2 \tag{33}$$

where a is the radius of the *equivalent piston* and $\dot{\xi}_0$ is the maximum displacement velocity. Here c is the velocity of sound in the medium. The efficiency of telephone diaphragms is rather low except at the resonance frequencies. The Fessenden oscillator † for under-

---

* See bibliography, references 5 and 9.
† See bibliography, reference 11.

water sound signaling has, however, an efficiency as high as 50 per cent at an output of 500 watts. The output of all sources is very largely conditioned by the type of acoustic coupling used: a diaphragm placed in the throat of a horn will have a larger output for the characteristic frequencies of the horn, since the resonance principle comes into play.

28. TRANSMISSION THROUGH TUBES

Many important acoustic problems involve the transmission of sound through tubes; for example, speaking tubes, horns for the production and amplification of sound as well as for its reception, stethoscopes, special types of acoustic filters, and apparatus for the measurement of acoustical intensity and impedance.

Effect of Length. If a plane sound wave produced by a diaphragm or other source is led into a cylindrical tube, the wave form will depend in the first place on the length of the tube. If the tube is short, the reflected wave from the end of the tube away from the source will combine with the direct wave to form a more or less complicated pattern; if the tube is very long the reflected wave may be largely damped out and a wave traveling effectively in one direction results.

Change in Cross-section. Consider a wave traveling in one direction and note what happens to it when the size of the tube is altered.* At A, in Fig. 4a, the wave coming from the left encounters an abrupt increase in cross section. In Art. 22 it was noted that

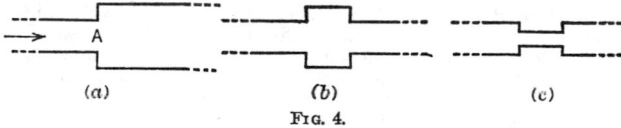

(a)   (b)   (c)

FIG. 4.

reflection ensues so that to the left of A there are waves in both directions. The wave that goes across A is therefore weakened in intensity and indeed by an amount which depends on the ratio of the cross sections (formula 15). The ratio of the total energy flow (watts) in the transmitted wave to the right of the boundary to that in the incident wave may be called the power transmission ratio and will be denoted by P_r. There is no change in phase of the transmitted wave. Incidentally P_r is the same in either direction. The loss in intensity can be very much mitigated by the use of a connector joining the two tubes so as to make the change in dimensions less abrupt; the longer the connector the greater is P_r.

In Fig. 4b, again, a plane wave coming from the left meets an expansion of finite length which joins the main line. In Fig. 4c the analogous case of a constriction is illustrated.

Calculation and experiment indicate that P_r in both examples is a function of the frequency, being fairly close to unity for low frequencies but becoming very small for a considerable range of high frequencies, then rising and falling again in a series of alternate peaks and hollows, as in Fig. 5. Such a structure is said to be *selective*. Change in phase of the transmitted wave is also involved. This has important application to the pinching of a sound tube in order to control intensity.

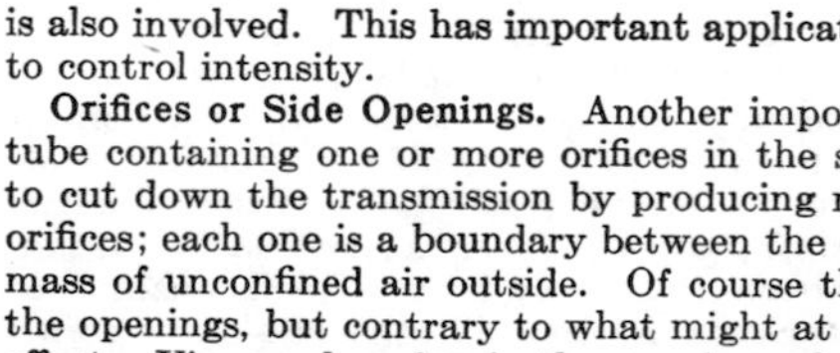

FIG. 5.

Orifices or Side Openings. Another important case of transmission is that through a tube containing one or more orifices in the side. The physical effect of the openings is to cut down the transmission by producing reflection due to the inertia of the air in the orifices; each one is a boundary between the confined air in the tube and the much larger mass of unconfined air outside. Of course there is some actual radiation of sound from the openings, but contrary to what might at first be supposed this is less than the inertia effect. Viscous damping in the openings also plays an insignificant rôle unless they are very narrow. Calculation and experiment indicate that P_r is small for low frequencies

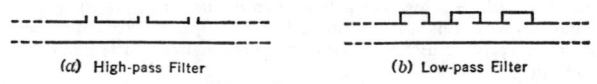

(a) High-pass Filter   (b) Low-pass Filter

FIG. 6.

and rises as the frequency increases. If more than one hole is used, as their number (they are assumed to be evenly spaced) increases, P_r for low frequency decreases until finally

---

* See bibliography, reference 11.

no sound passes through the tube up to a certain frequency: it acts as a high-frequency-pass acoustic filter (see the schematic diagram in Fig. 6a).'‡

It is seen that all the transmission problems discussed so far in this section involve acoustic filtration to a certain extent. The same is true of a Helmholtz resonator placed over an orifice in the side of a tube. Here, as might be expected, P_r is reduced materially only in the neighborhood of the resonance frequency of the resonator. If, however, one attaches a whole series of resonators to evenly spaced openings (Fig. 6b), the resulting structure is found to act as a low-pass filter; i.e., P_r is different from zero up to a certain frequency dependent on the dimensions of the various parts; it then falls to zero and stays so for a considerable frequency interval. It then rises again and transmission and attenuation bands succeed each other in a way which can be predicted theoretically. The use of more elaborate side branch attachments permits the construction of filters having transmission bands at any desired frequency and of any desired frequency range.

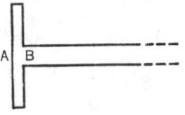

FIG. 7.

The Stethoscope. Another important instrument which is essentially an acoustic filter is one form of the stethoscope (see Fig. 7) consisting of a thin air chamber with a broad base of hard rubber or other solid substance.* Sound is incident on A from the left; some of it passes through the base (which may vibrate as a whole if it is thin, or act as a genuine transmitting medium if it is thick) and then through the air chamber to B where more reflection takes place and a final transmitted wave progresses up the narrow tube. The theory is too complicated for presentation here. An illustration will suffice. Consider the stethoscope immersed in water. Assume that the ratio of the area of cross section at A to that at $B = 225$ and that the thickness of the chamber is 0.3 mm; then at $\nu = 750$ cycles per sec $P_r = 0.13$. This contrasts with $P_r = 0.0012$ for the ordinary power-transmission ratio going directly from water to air without change in tube cross section. It must be emphasized that this type of stethoscope is highly selective. For medical purposes, small conical horns are now being used extensively (see Art. 29).

29. HORNS

The horn occupies a significant place in the discussion of acoustic transmission.†
Strictly speaking, any tube of finite length may be a horn. However, it is usual to restrict the term to tubes of varying cross section.

The Principal Types are shown diagrammatically in Fig. 8, where (a) represents a *conical* horn in which the area of cross section $S = S_0 x^2$ (x denoting the coordinate distance along the horn axis from some chosen origin and S_0 the cross-sectional area for unit x); (b) represents an *exponential* horn with $S = S_0 e^{mx}$, where m is a parameter governing the flare (this horn has the property that the area S is the geometric mean of the areas at the same distance on either side of S); (c) represents a *parabolic* horn in which $S = S_0 x$. There are

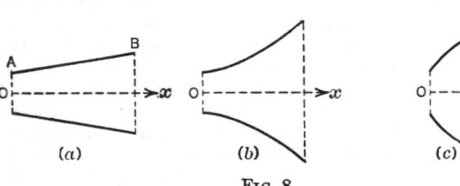

(a)　　　(b)　　　(c)

FIG. 8.

other types of flaring horns but their properties are very similar to the exponential variety.

The Horn as a Receiver of Sound. If the horn were not there, little difference in the intensity of the sound would be noted at the points A and B. However, with the horn in place, and sound directed toward the large end, there is considerable increase in the intensity at the throat (i.e., the small end) over the intensity at B with no horn present. The ratio of these intensities is the *intensity amplification* (I.A.) of the horn and a measure of the horn's utility as a receiver. The mathematical theory by which this quantity is computed is complicated, and it must be made clear that the qualitative idea of the horn's "concentrating" the sound by reflection from the walls, etc., is quite inadequate except at high frequencies. The action of a horn like that of any air column is mainly dependent on *resonance* characteristics. This is clearly shown in the fact that when I.A. for a conical horn, for example, is plotted as a function of frequency, it shows a number of peaks and hollows, the peaks corresponding to the resonance frequencies of the horn considered as a vibrating air column.

This selectivity is most marked in the conical horn. The parabolic horn gives a much

---

‡ See bibliography, references 3, 7, and 11.

* See bibliography, reference 11.

† See bibliography, references 3, 7, and 11.

more uniform response and is being widely used in connection with sound-motion pictures. The exponential horn, other things being equal, gives the largest I.A. and is often used in loud-speaker units for public-address systems.

The Horn as a Transmitter. A good receiving horn is a good transmitting horn. Naturally the behavior always depends on the acoustical attachments. In general, however, the resistance component of the impedance (Arts. 24 and 26) of the horn at the throat is the best criterion of the horn's action. This is greater for flaring horns than for conical horns of similar dimensions.

The use of horns in musical instruments is well known. Recently large exponential horns have proved valuable in the acoustic detection of aircraft.

30. ULTRASONICS

Considerable interest has been displayed during the past two decades in the use of high-frequency, inaudible sound in under-water signaling, for ultrasonic waves are not so easily diffracted as audible waves and hence can be directed more or less in a beam. Moreover, high-frequency sound oscillators have a much higher radiation efficiency than low-frequency ones; i.e., much more of the energy of the oscillator actually goes into radiation. For signaling, frequencies of 50,000 to 100,000 have been employed. The effective damping increases with the frequency and hence precludes the use of much higher frequencies.

Sources. A widely used source of ultrasonic radiation is the **quartz** oscillator (first used for signaling purposes by Langevin in France during World War I). Quartz like some other asymmetric crystals (notably Rochelle salt) when subjected to stress becomes electrically polarized. Thus if a slab of quartz cut parallel to the optic axis is inserted between two metal plates connected to an a-c circuit, the slab will oscillate with the impressed alternating emf, the oscillations being particularly vigorous at resonance.

Another source of ultrasonic waves is the *magnetostriction oscillator* developed by Pierce.[*] When a rod of magnetic material is magnetized it changes its length; when placed in an oscillating magnetic field mechanical vibrations are thus set up with the frequency of the field.

Other Uses. In addition to their use for signaling, ultrasonic waves have been studied for their physical, chemical, and biological effects. Wood and Loomis [†] have produced sound waves of frequency about 300,000 cycles and with very high intensity, in some cases reaching a radiation pressure of 3000 dynes per cm[2]. When the vibrations are communicated to liquids, stable emulsions are often formed, e.g., mercury in water. These waves have also been used in the measurement of the velocity of sound in liquids (see Art. 21). It is now perfectly possible to produce ultrasonic radiation of frequency 500 million cycles, and the transmission of such high-frequency radiation through solid liquid and gaseous media has been studied extensively. (Cf. bibliography, reference 45.) It is becoming a very useful tool in the analysis of the properties of matter, even in the molecular domain.

31. PHYSIOLOGICAL ACOUSTICS

Physiological acoustics comprises all the acoustical phenomena associated with the reception of sound by the ear and the production of speech sounds.

The Human Ear consists of the auricles or pinnae, the canal with the eardrum at the end of it, the middle ear with the three little bones or ossicles, and the inner ear with the liquid-filled cochlea containing the basilar membrane which is presumably the seat of the auditory process. Sound passing through the canal is communicated to the inner ear via the drum and ossicles. The actual mechanism of audition is still a subject of much investigation, and there are several theories of hearing, the chief of which is still that of Helmholtz which assigns resonance characteristics to the fibers in the basilar membrane.[‡]

Intensity and Loudness. As explained in Art. 24, the ear is sensitive to sounds of intensity as low as 4×10^{-10} microwatt per cm[2] and the threshold of feeling as well as that of audibility are both functions of frequency or pitch (the physiological interpretation of frequency). For most normal persons the frequency range of audibility runs from about 20 to 16,000 cycles per sec. A few people can hear up to 20,000. The work of Stevens and Davis [§] should be consulted for details. The connection between intensity and loudness is a matter of some interest: the former is a purely physical quantity (Art. 24); the

[*] *Proc. Am. Acad. Arts and Sci.*, Vol. 63, p. 1, 1928.
[†] *Phil. Mag.*, Vol. 4, p. 417, 1927.
[‡] See bibliography, references 18, 23, 43, and 50.
[§] See bibliography, reference 40.

latter is physiological. The general law connecting the two is a special case of the so-called Weber-Fechner psychophysical law that equal increments of sensation correspond to equal increments in the *logarithm* of the stimulus. This makes it particularly convenient to use the decibel notation (Art. 24) for loudness. Thus the difference in the loudness level of two sounds of physical intensity I_1 and $I_2 (I_2 > I_1)$ respectively is

$$L = 10 \log I_2/I_1 \tag{34}$$

in decibels. The relation is complicated by differences associated with pitch. We cannot go into the discussion of the loudness of complex tones.

Frequency. To every frequency corresponds a minimum perceptible intensity difference; e.g., for a sound or sensation level of 50 db above the audible threshold, $\Delta I/I = 0.10$. There exists a similar minimum perceptible pitch difference, which varies from 0.5 cycle per sec at 50 cycles to 9 cycles per sec at 3000 cycles per sec. For further details, the work of Knudsen should be consulted.

Beats. The presentation simultaneously to the ear of two sounds of definite frequencies ν_1 and ν_2 leads, when the two sounds are of approximately the same intensity and ν_1 and ν_2 do not differ too much, to the well-known phenomenon of beats. Even when the sounds differ widely, a "difference" tone of frequency $\nu_2 - \nu_1$ is heard, and what at first seems more remarkable, a summation tone of frequency $\nu_1 + \nu_2$ together with various linear combinations of these frequencies. The explanation has been sought in the fact that the eardrum is an asymmetrical vibrator, i.e., that the displacement for given excess pressure is not the same in the positive as in the negative direction.

Sound Direction. One of the most striking aspects of hearing is the localization of the direction of a source of sound. The listening individual tends to turn so that this direction is the perpendicular bisector of the line joining the two ears. Presumably two dis tinct effects are involved: the *binaural intensity* and the *binaural phase* effects respectively.* If the phase is the same at both ears a variation in intensity about the head will produce an angular displacement of the apparent source of sound. However, this is very uncertain and for many observers the effect fails to exist at all in frequency regions where localizability still persists. On the other hand, if the intensity is maintained constant and the phase is varied, the apparent shift about the head persists for all frequencies for which tests have been made, although observers differ considerably on the upper limit (about 1300 for most people).

The Origin of Speech is found in the currents of air which are forced by the lungs to pass through the vocal passages exciting to vibration the so-called vocal cords in the larynx and then passing through the cavities of nose and throat. The functions of tongue and lips in the formation of sounds must not be forgotten, of course. The whole mechanism is too involved for description here, and reference should be made to Fletcher's very complete discussion.

The Flow of Energy in Speech varies greatly from one individual to another and in a given individual from one sound to another. Thus for a single speaker the *average* intensity involved in a certain syllable may be of the order of 100 microwatts per cm^2 while the *peak* intensity in the same syllable may be as high as 2000 microwatts per cm^2. Naturally, these figures depend on the point in front of the mouth where measurements are made, since the intensity falls off with distance. It has been found that the average intensity in normal speech is about 7 microwatts per cm^2 at a point 9 cm from the mouth. Among speech sounds the vowels rank highest in energy content, the semi vowels next, and con sonants lowest.

The work of D. C. Miller has made it clear that the peculiar quality of vowel sounds depends on certain frequency regions regardless of the fundamental pitch used. Thus the quality of the vowel *u* (the vowel sound in "pool") depends essentially on a low-frequency component in the neighborhood of 400 cycles per sec and a high-frequency around 800 cycles per sec. Corresponding figures for long *e* (the vowel sound in "team") are 375 and 2400. Complete tables are given in Fletcher's book.

Articulation. By the use of electric and acoustic filters, certain frequency ranges can be eliminated from speech and the effect on articulation studied. By articulation is under stood the percentage of intelligible syllables in a selected list. The results may be sum marized as follows: (*a*) frequencies above 1550 cycles per sec are just as important for articulation as those below; specifically by using only frequencies above or only fre quencies below this value an articulation of 65 per cent results; (*b*) the elimination of all frequencies below 1000 cycles per sec results in an articulation of 86 per cent; (*c*) the elimi nation of all above 1000 cycles per sec, an articulation of 40 per cent; (*d*) the elimination of all frequencies below 1000 yields the same articulation as the elimination of all fre-

---

* See bibliography, references 11, 18, 23, and 43.

quencies above 3000. With music the elimination of the low frequencies renders it thin and metallic, and cutting out the high frequencies makes it fuller and, so to speak, thicker (H. Fletcher, *Speech and Hearing*).†

32. MUSICAL SOUNDS

A **Musical Sound** may be defined as a single sustained sound of definite pitch or a composite of such sounds having frequencies related to each other by simple integral ratios. The frequencies used cover practically the whole audible range (Art. 31). The intensity of musical sounds is in general much greater than that involved in speech. For principal sources of musical sounds, viz., strings, organ pipes, membranes, and horns, see Art. 26.

Characteristics Which Distinguish Music from Other Sound. Musical sounds are seldom simple; they usually are composites of several sounds, viz., the fundamental or note of lowest frequency and the overtones or harmonics with frequencies which are integral multiples of the fundamental. This implies that any complex tone can be analyzed into a set of harmonic components (often called partials): *Ohm's acoustical law*. The *quality* of a complex tone depends on the number and relative strength of the various harmonics or partials and does not depend on their differences in phase. The early work on the analysis of complex sounds was due to Miller * using the phonodeik and mechanical analyzers. More detailed studies have recently been made at the Bell Telephone Laboratories, and complete acoustic spectra of many instruments have been obtained using electrical analyzers and filters.†

The **Basis of Musical Scales** is found in the fact that the ear recognizes as pleasing combinations of simple tones whose frequencies are in the ratio of *small* whole numbers. These are the so-called consonant intervals ‡ (a musical interval is the ratio of the two frequencies concerned), and the principal ones are: octave, 1 : 2; fifth, 2 : 3; fourth, 3 : 4; major third, 4 : 5; minor third, 5 : 6; major sixth, 3 : 5; minor sixth, 5 : 8. As the numbers become larger, dissonance results; the interval 7 : 8 is recognized as dissonant. It is probable, however, that the range of accepted consonance can be increased by custom and education.

The **Ideal Musical Scale** would consist of a set of simple notes in approximately equal steps and including only consonant intervals. This has never been achieved exactly in practice. The two most important scales, the consonant diatonic and the equally tempered scales, are indicated in the accompanying table.

Table 1.   Musical Scales (A = 440)

Note	C	D	E	F	G	A	B	C'
Frequency on diatonic scale............	264	297	330	352	396	440	495	528
Interval with C......................	1	9 : 8	5 : 4	4 : 3	3 : 2	5 : 3	15 : 8	2 : 1
Frequency on equally tempered scale . . .	261.7	293.7	329.7	349.2	392	440	493.9	523.3

Most of the notes on the scale form consonant intervals with each other, but the steps are decidedly unequal, so that in using it one cannot play in any key at will; i.e., the frequencies bearing the same relation to D as those in the scale bear to C are not all found in the list, and hence to play in the key of D would involve introducing extra frequencies. This complicates matters considerably, hence in the piano the equally tempered scale is used. This has 12 intervals in the octave: the individual notes are C, C♯, D, D♯, E, F, F♯, G, G♯, A, A♯, B, C. (Not all these are listed in the table.) The interval between each successive pair of notes is $2^{1/12} = 1.05946$, and therefore music can be played in any key with this scale. There are *no* exactly consonant intervals, but apparently the defects are too slight to be observed by most people.

Musical Pitch Specification has been very variable, and various standards have been used within the history of modern music. The classical pitch of Handel and Mozart assigned a frequency of 422 cycles per sec to A whereas the modern "American concert pitch" uses 440 for the same note. The modern "international pitch" assigns 435 cycles per sec to A. In the equally tempered scale this gives middle C a frequency of 258.65 cycles per sec.

---

* See bibliography, reference 34.
† See bibliography, reference 23.
‡ See bibliography, reference 12.

33. ARCHITECTURAL ACOUSTICS

The first important experiments on the acoustical properties of rooms were carried out by W. C. Sabine in Cambridge, Massachusetts, in the last decade of the nineteenth century. He was apparently the first to realize the importance of reverberation in its relation to the amount of absorbing material present.

Sound Absorption and Reverberation. When a sound is produced in a closed space, if the walls were perfect reflectors there would be no diminution in the intensity and if the production of sound continued the total energy would therefore increase indefinitely. Actually, however, the sound is absorbed not only by the air but also and primarily by the walls and other furnishings, so that a given amount of energy is dissipated in a certain time. Technically the reverberation time is that which elapses after the stopping of a sustained note having one million times the minimum audible intensity to the instant when the resulting reverberant sound has dropped to the audible threshold. The relation between this time, T, in seconds and the dimensions of the room is given approximately for moderate-sized and not too greatly absorbing halls by the formula *

$$T = 0.049 \ V/a \qquad (35)$$

in which V = volume of the room in cubic feet, and a = total absorbing power of the walls and materials in the room. This latter quantity represents the average fraction of the incident sound energy per unit area which is not thrown back into the room (i.e., average absorption coefficient) multiplied by the total surface area exposed. It is calculated by multiplying the absorption coefficient for each particular article by its area and summing over the whole area. Thus, denoting by α_i the absorption coefficient for a certain group of similar articles with area S_i, the total absorbing power is

$$a = \Sigma \alpha_i S_i = \bar{\alpha} S$$

where the sum is taken over all the absorbing material. The quantity α is called the average absorption coefficient, and S is the total surface.

Sabine's experimental measurements of T for most of the auditoriums which he tested were in rather good agreement with equation (35). In particular the precise shape of the room (barring such things as deep recesses, etc.) does not seem to affect T. Sabine's experimental method was simply to start and stop an organ pipe and measure the time for the decaying sound to become inaudible.

Recent work on reverberation indicates that Sabine's formula (equation 35) is unsatisfactory for very large halls or rooms where the absorption is very great, i.e., where $\bar{\alpha}$ is greater than 0.5. Eyring † has recently derived a more general formula, viz.,

$$T = -\frac{0.05V}{S \log (1 - \bar{\alpha})} \qquad (36)$$

This agrees with Sabine's quite closely for $\bar{\alpha}$ small but departs from it considerably when $\bar{\alpha}$ is larger than 0.5. Eyring's formula checks rather well with experimental determinations for "dead" rooms. Other methods of averaging lead to still different though analogous formulas. The whole question is at present somewhat unsettled. For example, the above formula does not take into account the absorption of sound in the air of the room, which further reduces T.

It is clear that, if sufficient intensity of sound is secured, the most important single factor in the satisfactory acoustical use of a room is the reverberation time.‡ If it is too great, any given sound like a spoken syllable or musical note will take so long to build up and decay that speech becomes inarticulate and music blurred. A short reverberation time greatly increases articulation. Of course too short a time gives the impression of acoustic "deadness," particularly with music. In general a larger value of T can be tolerated for music than for speech; hence it is easier to set an optimum value in the latter case. Knudsen has made the most complete tests of this matter. He has found, for example, that the optimum T for rooms with volume about 300,000 cu ft (without audience) is about 2.75 sec. In general, for a given T, the articulation is greater for vowels than for consonants. The figures depend on the size of the room. For smaller halls of volume up to 40,000 cu ft to be used for both speech and music, a period of about 1 sec is indicated.

Absorption Coefficients. For a room of given size the control of T rests with the absorption. The unit commonly used is 1 sq ft of open window, which is effectively a per-

---

* See bibliography, reference 39.
† *J. Acous. Soc. Am.*, Vol. 1, p. 217, 1930.
‡ See bibliography, references 27, 39, and 42.

fect absorber. The absorption coefficient α_i for any particular material is then defined as the ratio of the absorbing power of 1 sq ft of the substance to that of 1 sq ft of open windows. The measurement of absorption has been carried out by a number of methods. Table 2 gives some typical values.* It is of interest to note that some substances show highly selective absorption, in particular the high absorbers.

Sound Proofing has become a significant branch of architectural acoustics. When sound strikes a wall, part is reflected and part is absorbed in the true sense of being ultimately dissipated into heat. Another part passes through the wall in the form of compressional waves. Finally some of the incident sound energy causes the wall to vibrate as a whole, and it is indeed in this way that most of the sound energy gets across the wall into the room space on the other side. This suggests that to prevent sound transmission from air to air through a wall should be made as rigid as possible. On the other hand, rigidity facilitates wave transmission through the walls and flooring of sounds originating there, so that the problem of complete sound proofing is somewhat complicated. The use of double walls with a fairly wide air space is of considerable help, as is also the use of inner floors floating on insulating material resting in turn on the main structural floor.

Recent work (Cf. bibliography, references 51, 55) has shown that the reverberation time is not always a sufficient measure of the acoustical qualities of a room. The relation between the sound that reaches the auditor directly from the source and that which reaches him via reflections of the walls is also important. This involves the concept of "liveness." The normal modes of vibration of the room are also important. To accentuate one mode relative to the rest may distort hearing. Recent research has also shown that the absorption coefficients of the walls may often be less important than the specific acoustic impedance (Cf. Art. 21) of the wall surfaces.

Table 2. Sound-absorption Coefficients

Material	Sound-Absorption Coefficients Frequencies 128 512 2048	Weight	Composition	Thicknesses	Stock Sizes
Acousti - Celotex, Single B, 5/8 in.	0.11 0.45 0.68	13 oz per sq ft	Cane fiber tile perforated with 441 holes per sq ft	5/8 in. 13/16 in. 1 1/4 in.	6 in. by 12 in. 12 in. by 12 in. 12 in. by 24 in.
Acoustic Flexfelt	0.27 0.56 0.68		Rock wool felted between metal netting and stucco lath	Required thickness	4 ft by 4 ft 4 ft by 8 ft
Acoustone 1/2 in.	0.48 0.59	1 1/4 lb. per sq ft	Artificial stone filaments bonded together in tile form in a large variety of shapes and colors	1/2 in. 3/4 in. 1 in.	6 in. by 6 in. 6 in. by 12 in. 12 in. by 12 in. 9 in. by 18 in. Special sizes up to 24 in. by 36 in. available
Akoustolith Plaster 1/2 in.	0.21 0.29 0.37		Light-weight plastic material	Applied 1/2 in. in thickness over usual ground coats	
Akoustolith A, Tile, 1 in.	0.14 0.48 0.83	About 4 lb per sq ft	Artificial stone	1 in. 1 1/2 in. 2 in.	3 in. by 16 in. 4 in. by 8 in. 5 in. by 10 in. 12 in. by 12 in. 8 in. by 16 in. for 1-in. thickness
Balsam wool 1 in.	0.15 0.52 0.66	320 lb per 1000 sq ft	Balsam wool mat covered one side with a Kraft paper liner, other side with fireproof cloth mesh	1 in.	Packed in rolls 34 in. wide, containing 124 sq ft

* Selected values from Knudsen's *Architectural Acoustics* (John Wiley and Sons, 1932). Used by special permission. For more complete list of data, reference should be made to the original table, probably the most complete in existence.

Table 2. Sound-absorption Coefficients—*Continued*

Material	Sound-Absorption Coefficients Frequencies 128 512 2048	Weight	Composition	Thicknesses	Stock Sizes
Fir-Tex, 1 in.	0.32 0.39 0.41	1.2 lb per sq ft	Made from Douglas fir. Contains about 10 per cent of bark; 0.1 per cent resin for fiber water-proofing set upon fiber by 0.1 per cent alum solution giving considerable fireproofing	1 in. 1 1/2 in.	12 in. by 12 in.
Insulite Acoustile, Type 37	0.21 0.38 0.46	750 lb. per 1000 sq ft	Wood fiber fabricated into tiles	3/4 in.	Sizes range from 6 in. by 6 in. up to 24 in. by 24 in.
Kalite No. 102, 3/4 in., on metal lath	0.37 0.40 0.53	About one-half of ordinary plaster	Made of graded sizes of a special pumice mixed with a gypsum binder. The inherent porosity of the pumice together with the method of mixing gives a plaster with many communicatting channels. Calcined gypsum $(CaSO_4 \cdot 1/2\ H_2O)$, 34.85 per cent; pumice 60.33 per cent	Can be applied in thicknesses of 1/2, 3/4, 1 in.	
Macoustic plaster, stippled to depth of 1/2 in.	0.12 0.31 0.58	2 lb. per sq ft	A fibrous plaster	Applied 1/2 in.	
Masonite	0.18 0.32 0.33	700 lb per 1000 sq	Made chiefly of long-leaf pine and southern gum	7/16 in.	4 ft by 12 ft
Rockoustile 1 in.	0.18 0.57 0.72	1.5 lb per sq ft	Rock wool product	1 in.	6 in. by 12 in. 12 in. by 12 in.
Sabinite	0.34 0.49	2 to 3 lb. per sq ft	Gypsum base plaster. No stippling or special art in application required. No. 38 is a special humidity - resisting material, utilizing a hydraulic binder, for the acoustical treatment of natatoria and similar humid rooms	Applied in two coats, about 1/2 in.	
Silent-Ceal	0.29 0.68 0.75	3 lb. per sq ft	Rock wool fill; special metal furring; No. 20 gage perforated metal primary membrane; fabric secondary membrane	Determined by requirements of job	
Transite tile, 1 in.	0.19 0.81 0.72	3 lb. per sq ft	1 in. sound-absorbing block faced with perforated Transite 3/16 in. asbestos paper	1 3/16 in.	6 in. by 6 in. 12 in. by 12 in.

34. NOISE AND ITS PREVENTION

Noise. Previously the scientific definition ˜of noise was any sound or set of sounds having no definite pitch or pitch components. At the present time it would be perhaps fairer to consider as noise any *disturbing* sound, regular or otherwise. For some people, a large amount of radio "music" is quite definitely noise. The endeavor is now under way to study very thoroughly the effect of noise in all forms on the individual, particularly with respect to fatigue and the lowering of mental and motor efficiency. No safe generalizations are yet at hand, though it is widely believed that city noise is ultimately harmful to man.*

Noise Measurement. In order to study noise it is necessary to measure its intensity (e.g., by means of a microphone coupled with some kind of amplifier system), its wave form (by oscillographic observation), and the distribution of energy among the various component frequencies. One difficulty encountered is the fact that physical intensity must be translated into physiological loudness; i.e., one must use a noise meter which closely simulates the response of the ear.

In the use of noise meters the usual plan is to vary the intensity of a standard note until it is judged subjectively to be of the same loudness as the noise being measured. The level of the sound in decibels (db) above the audible threshold is then taken as a measure of the noise. The results in all cases depend more or less on the frequency. Thus the average street noise may run from 10 db at 8000 cycles per sec to 35 db at 1000 cycles per sec. At a particularly noisy corner of a busy city, the level may reach 50 db (Fletcher). A boiler factory in operation runs from 95 to 100 db (Davis). In a quiet residence the figure should not be above 15. These last values are averages over a considerable range of frequency. Of course there is fluctuation with time.

For the important effect of noise on speech articulation the work of Fletcher should be consulted.

In the Reduction of Noise from Machinery, which is probably the largest single source of indoor noise and much outdoor noise as well, the general principle is the prevention of the transfer of vibrational energy to surfaces large enough to communicate this energy to the air in the form of acoustical radiation. In general, the amount of sound given to the air directly by the moving parts is negligible compared with that due to the vibration of the supports and foundation. It is therefore desirable as far as possible to insert between moving parts and supports material of very different specific acoustical resistance (see Art. 24) from either. The material should also have as large a damping coefficient as possible.

For a very complete discussion of the possibility of noise prevention in large cities, reference should be made to *City Noise*, published by the Noise Abatement Commission, Department of Health, New York, 1930.

35. CERTAIN ACOUSTICAL INSTRUMENTS AND MEASUREMENTS

The measurements of leading importance in acoustical work are those of the intensity of sound and acoustic impedance.

The Rayleigh Disk. The classical method for measuring intensity is that employing the Rayleigh disk.† This is a small thin disk of glass about $1/2$ cm in diameter suspended by a quartz fiber so that its plane faces are vertical. When such a disk is immersed in a sound beam it tends to set itself so that its plane is normal to the direction of particle displacement. The sound wave thus exerts a torque on the supporting fiber which can be measured by the angular deflection of the disk. Theory indicates that this torque depends on the mean square of the particle velocity in the sound wave. It is therefore a measure of the intensity (Art. 24).

Webster's Phonometer is another intensity-measuring instrument consisting of a tunable cylindrical resonator with a diaphragm mounted in its mouth.† This diaphragm is tuned to resonance with the sound whose intensity is to be measured by changing the tension in the wires supporting it. The diaphragm vibrates under the action of the air vibrations produced by the sound in the mouth of the resonator, and the motion of the diaphragm causes rotation of a small concave mirror from which light is reflected. The displacement amplitude of the diaphragm is proportional to the pressure amplitude of the sound at the diaphragm, and hence the intensity can be computed from equation (18).

---

* See bibliography, reference 3.
† See bibliography, references 3 and 11

The Hot-wire Microphone. A fine platinum wire heated to red heat and exposed to a fluctuating current of air in the neck of a resonator suffers a decrease in resistance which is proportional to the mean square particle velocity and hence to the intensity of the sound. This arrangement is called the hot-wire microphone.*

It is of interest to note the important rôle played by the resonator in the last two instruments mentioned.

Of the Electroacoustical Methods of measuring intensity the condénser microphone and the ribbon microphone are the two principal devices.† The condenser microphone concists of a thin stretched circular diaphragm fastened about the periphery and attracted toward a metal plate by means of a static electric charge. When sound waves impinge on the diaphragm, they alter the capacity of the system and hence produce a small alternating current in the circuit of which the diaphragm and plate are a part. This current is amplified and used as a measure of intensity on proper calibration. The great advantage of the instrument is its remarkably uniform sensitivity over a wide range of frequencies.

The ribbon microphone consists of a light metallic ribbon suspended in a magnetic field and exposed to air vibrations on both sides. The vibrations of the ribbon produced by an incident sound wave lead to the induction of an emf corresponding to the oscillations in the wave. This instrument has the advantage of possessing directional characteristics; a property not possessed by the condenser microphone and other pressure-operated instruments.

Piezoelectric crystal microphones (quartz, Rochelle salt, ammonium dihydrogen phosphate, and more recently ceramic barium tibanate) are receiving increasing use in intensity measurements for sounds of all frequencies.‡

The Measurement of Acoustic Impedance is important as a means of estimating the performance of acoustic apparatus such as horns (Art. 29) and also in architectural acoustics. Reference should be made to the literature.

BIBLIOGRAPHY

British and American

General

1. BRAGG, SIR WILLIAM, *The World of Sound*, London, Bell. Latest printing, 1933.
2. CRANDALL, J. B., *Theory of Vibrating Systems and Sound*, New York, Van Nostrand, 1926.
3. DAVIS, A. H., *Modern Acoustics*, London, Bell, 1934.
4. FLEMING, J. A., *Waves and Ripples*, New York, Macmillan, 4th issue, 1923.
5. LAMB, SIR HORACE, *Dynamical Theory of Sound*, London, Arnold, 2nd ed., 1925.
6. MORSE, P. M., *Vibration and Sound*, New York, McGraw-Hill, 2nd ed., 1948.
7. OLSON, H. F., and MASSA, F., *Applied Acoustics*, Philadelphia, Blakiston, 1934.
8. OLSON, H. F., *Elements of Acoustical Engineering*, New York, Van Nostrand, 1940.
9. RAYLEIGH, LORD, *The Theory of Sound*, London, Macmillan, 2 vols., 2nd rev. and enl. ed., 1929.
10. RICHARDSON, E. G., *Sound: A Physical Text-book*, London, Arnold, 1927.
11. STEWART, G. W., and LINDSAY, R. B., *Acoustics*, New York, Van Nostrand, 1930.
12. STEWART, G. W., *Introductory Acoustics*, New York, Van Nostrand, 1932.
13. TYNDALL, JOHN, *Lectures on Sound*, New York, Appleton, 1867.
14. WATSON, F. R., *Sound*, New York, John Wiley, 1935.
15. WOOD, A. B., *A Text-book of Sound*, London, Bell, 1930.
16. WOOD, A. B., *Acoustics*, New York, Interscience Publishers, 1941.

Special Subjects

17. BAGENAL, H., and WOOD, ALEXANDER, *Planning for Good Acoustics*, London, Methuen, 1931.
18. BEATTY, R. T., *Hearing in Man and Animals*, London, Bell, 1932.
19. BERANEK, L. L., *Acoustic Measurements*, New York, John Wiley, 1949.
20. DAVIS, A. H., and KAYE, G. W. C., *The Acoustics of Buildings*, London, Bell, 1927.
21. DOUGLAS, D., *The Science of Voice*, New York, Carl Fischer, 1929.
22. DRYSDALE, C. V., and others, *Mechanical Properties of Fluids*, London, Blackie, 1923.
23. FLETCHER, H., *Speech and Hearing*, New York, Van Nostrand, 1929.
24. GLOVER, C. W., *Practical Acoustics for the Constructor*, Cleveland, Ohio, 1934.
25. HART, M. D., and SMITH, W. W., *The Principles of Sound Signalling*, London, Constable, 1925.
26. HUMPHREYS, W. J., *Physics of the Air*, Philadelphia, Lippincott, 2nd ed., 1928.
27. KNUDSEN, V. O., *Architectural Acoustics*, New York, John Wiley, 1950.
28. KNUDSEN, V. O., and HARRIS, C. M., *Acoustical Designing in Architecture*, New York, Van Nostrand, 1942.
29. LAMB, SIR HORACE, *Hydrodynamics*, Cambridge, University Press, 6th ed., 1932.

---

* See bibliography, references 3 and 11.
† See bibliography, references 3, 7, and 11.
‡ See bibliography, references 19, 32 and 45.

30. LOVE, A. E. H., *Mathematical Theory of Elasticity*, Cambridge, University Press, 3rd ed., 1920.
31. MASON, W. P., *Electromechanical Transducers and Wave Filters*, New York, Van Nostrand, 1942.
32. MASON, W. P., *Piezoelectric Crystals and Ultrasonics*, New York, Van Nostrand, 1950.
33. McLACHLAN, N. W., *Loud Speakers*, Oxford, University Press, 1934.
34. MILLER, D. C., *The Science of Musical Sounds*, New York, Macmillan, 1916.
35. MORSE, P. M., and BOLT, R. H., "Sound Waves in Rooms," *Reviews of Modern Physics*, Vol. 16, 69, 1944.
36. NATIONAL RESEARCH COUNCIL, COMMITTEE ON ACOUSTICS, *Certain Problems in Acoustics*, Washington, 1922.
37. RICHARDSON, E. G., *The Acoustics of Orchestral Instruments and the Organ*, London, Arnold, 1929.
38. SABINE, P. E., *Acoustics and Architecture*, New York, McGraw-Hill, 1932.
39. SABINE, W. C., *Collected Papers on Acoustics*, Harvard University Press, Cambridge, Mass., 1922.
40. STEVENS, S. S., and DAVIS, H., *Hearing*, New York, John Wiley, 1938.
41. TIMOSHENKO, D., *Vibratory Problems in Engineering*, New York, Van Nostrand, 1928.
42. WATSON, F. R., *Acoustics of Building*, New York, John Wiley, 3rd ed., 1941.
43. WILKINSON, G., and GRAY, A. A., *The Mechanism of the Cochlea*, London, Macmillan, 1924

German

44. AIGNER, F., *Unterwasserschalltechnik*, Berlin, Krayn, 1922.
45. BERGMANN, L., *Der Ultraschall*, Zürich, Hirzel Verlag, 1949.
46. FISCHER, F., and LICHTE, H., *Tonfilm*, Leipzig, Hirzel, 1931.
47. *Handbuch der Physik*, Vol. VIII, Akustik, Berlin, Springer, 1927.
48. *Handbuch der experimental Physik*, Vol. XVII (2 and 3), Technische Akustik, Leipzig, Akad. Verlag., 1934.
49. *Handbuch der experimental Physik*, Vol. XVII (1), Ultra Akustik (all so far published). Leipzig, Akad. Verlag., 1934.
50. HELMHOLTZ, H. VON, *Die Lehre von den Tonempfindungen*, Braunschweig, Vieweg, 6th ed., 1913. English translation, *On the Sensations of Tone as a Physiological Basis for the Theory of Music* (from the 3rd German ed. by A. J. Ellis). London, Longmans, Green, 1875.
51. MÜLLER-POUILLET, *Lehrbuch der Physik*, Vol. 1, part 3, Akustik. Braunschweig, Vieweg, 1929.
52. PETERS, I., *Die Mathematischen und physikalischen Grundlagen der Musik*, Leipzig and Berlin, Teubner, 1924.
53. POHL, R. W., *Einführung in die Mechanik und Akustik*, Berlin, Springer, 1930. English translation by W. M. Deans. London, Blackie, 1932.
54. TRENDELENBURG, F., *Fortschritte der physikalischen und technischen Akustik*, Leipzig, Akad. Verlag., 2nd ed., with bibliography, 1934.
55. WAGNER, K. W., *Einführung in die Lehre von den Schwingungen und Wellen*, Wiesbaden, Dieterich'sche Verlagsbuchhandlung, 2nd ed., 1947.

French

56. DAVID, P., *L'électro-acoustique*, Paris, Hermann, 1930.
57. FOCH, A., *Acoustique*, Paris, Libraire Armand Colin, 1934
58. GALBRUN, H., *Propagation d'un onde sonore dans l'atmosphere*, Paris, Gauthier-Villars, 1931.

SECTION 12

HEAT TRANSFER

BY

WARREN M. ROHSENOW

HEAT TRANSFER

1. INTRODUCTION

There are in general two kinds of heat transfer: conduction and radiation. Conduction is the transmission of heat by molecular vibrations from one part of a body to another part of the same body or from one body to another body in physical contact with it, without appreciable displacement of the particles of the body.

Radiation is the transmission of heat in the form of electromagnetic waves. All bodies give off heat in the form of radiant energy. When this radiant energy falls on another body, part of it is absorbed by the body, and part of it is reflected. The difference between the amount of radiant energy which a surface gives off and the amount it absorbs from other surrounding surfaces which it "sees" is the net heat transferred by radiation.

Convection refers to the motion of a fluid. Fluid circulated by the action of gravity due to differences of density of the hotter and colder portions of the fluid is said to circulate by "natural convection." If a mechanical device causes the circulation, the flow process is called "forced" convection. In any of these convection heat processes, heat transferred between a surface and the fluid takes place by pure conduction at the wall.

In most cases the actual heat transfer is accomplished by both of these types of heat transfer; however, in some cases the major part of the heat transfer is accomplished by one of the types, and the heat transferred by the other may be neglected. For example, in the furnace cavity most of the heat transferred to the first row of boiler tubes is accomplished by radiation; in later tubes both conduction and radiation heat transfer must be considered; in the economizer ane condenser, conduction heat transfer is most important.

Most heat-transfer problems applied to a power plant are of a steady-state type. The operating conditions, such as temperature and pressure, do not vary appreciably with time, and the rate of heat transfer from or to a particular surface is constant with time. Consideration will be given to only this type of problem—one of steady-state operation.

2. CONDUCTION

Flat Solid Wall. The rate of heat transfer through a flat solid wall of a single material is calculated from the equation

$$q = -kA \frac{\partial T}{\partial x} = kA \frac{T_1 - T_2}{x} \tag{1}$$

where k is called the thermal conductivity of the material and is a property of the kind of material and the average temperature of the material. It is the amount of heat transferred per hour through 1 sq ft area of a wall 1 ft thick with a 1°F temperature difference across the wall.

Cylindrical Solid Wall. The rate of heat transfer through a cylindrical solid wall of length L is calculated from the equation

$$q = 2\pi Lk \frac{T_1 - T_2}{\ln (r_2/r_1)} \tag{2}$$

where r_2 and r_1 are outside and inside radii in feet or inches, and ln is logarithm with base e.

Spherical Solid Wall. Similarily

$$q = \frac{4\pi k r_1 r_2 (T_1 - T_2)}{(r_2 - r_1)} \tag{3}$$

Surface Coefficient of Heat Transfer

When a fluid flows parallel to a wall, the thin layer of fluid which is touching the surface clings to the wall and does not move. The fluid just next to this stagnant layer moves in sheets parallel to the wall; this is called the laminar layer. At the outer edges of the laminar layer the fluid in the transition zone tends to break away from its sheetlike formation to enter the turbulent region where it has extremely random motion (Fig. 1). Heat transferred between the surface and the main stream must be transferred through these various layers. Heat is transferred by conduction at the surface.

The rate of heat transfer between the surface at temperature T_s and the main body of the fluid at temperature T_f is calculated from

$$q = hA(T_s - T_f) = -k_f A \left(\frac{\partial T}{\partial y}\right)_{\text{surface}} \tag{4}$$

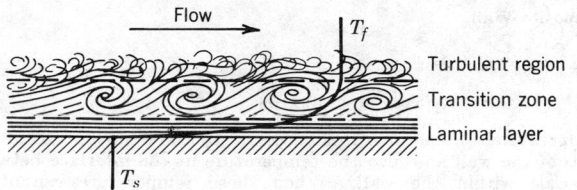

which defines the film coefficient of heat transfer h having the units Btu per hour per deg F temperature difference per square foot of heat-transfer surface.   In bounded convection such as in ducts, T_f is taken as the enthalpy mixed mean temperature.   In unbounded convection, as in natural convection around a hot plate in a large atmosphere of fluid, T_f is the uniform temperature in the fluid far away from the plate.

Composite Walls

An important case of heat transfer is that from a hot fluid on one side of a solid wall to a cooler fluid on the other side.   The wall may be a cylindrical or a flat wall of a single material or composite wall of layers of different materials.   The rate of heat transfer is calculated from

$$q = UA(T_h - T_c) \tag{5}$$

where T_h and T_c are the temperatures of the hot and cold fluids, respectively.

Flat Composite Wall.   Equation 5 is used to calculate the rate of heat transfer from a hot fluid successively through the hot-side film, layers of solid material of the wall, and cold-side film to the cold fluid (Fig. 2).   For this case, U of Eq. 5 is

$$U = \frac{1}{(1/h_h) + (x_A/k_A) + (x_B/k_B) + \ldots + (1/h_c)} \tag{6}$$

where h_h and h_c are the hot- and cold-side film coefficients of heat transfer, and the other terms are as defined in Eq. 4.

Cylindrical Composite Wall.   Equation 5 is also used to calculate the rate of heat transfer from a hot to a cold ffluid through a composite cylindrical pipe wall (Fig. 3). For this case the product UA of Eq. 5 is

$$UA = \frac{1}{\dfrac{1}{2\pi r_1 L h_h} + \dfrac{\ln(r_2/r_1)}{2\pi L k_A} + \dfrac{\ln(r_3/r_2)}{2\pi L k_B} + \ldots + \dfrac{1}{2\pi r_n L h_c}} \tag{7}$$

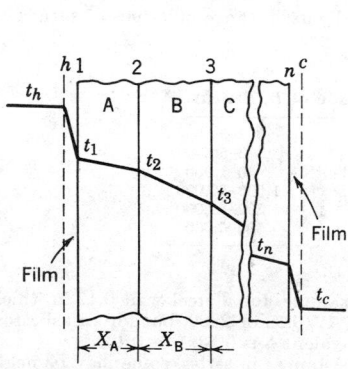

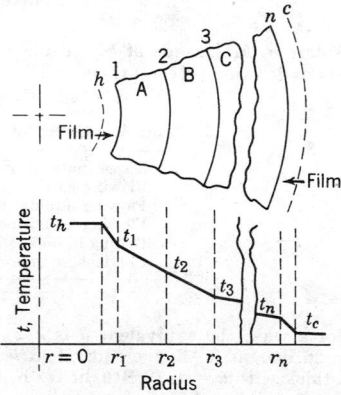

Spherical Composite Wall.

$$UA = \cfrac{1}{\cfrac{1}{4\pi r_1^2 h_h} + \cfrac{r_2 - r_1}{4\pi k r_1 r_2} + \ldots + \cfrac{1}{4\pi r_n^2 h_c}} \qquad (8)$$

Surface Temperature. Sometimes it is necessary to know the temperature of the inner and outer faces of the wall and also the temperature at the interface between layers of different materials within the wall. When these temperatures cannot be measured conveniently, they may be estimated by calculation. Consider the flat composite wall shown in Fig. 2. The rate of heat transfer q through the wall is calculated from Eqs. 5 and 8. In the case of steady-state heat transfer the rates of heat transfer through the films and through each layer of material of the wall are all equal and given by this value of q. The rate of heat transfer through the films is given by Eq. 4 and through each layer of the composite wall by Eq. 1. The temperature of the hot and cold fluids may be measured; and values of film coefficients h_h and h_c may be estimated by various correlations. Then the surface temperatures T_1 and T_n may be calculated from Eq. 4

$$T_h - T_1 = q/h_h A$$

$$T_n - T_c = q/h_c A$$

and the interface temperatures may be calculated by successively using Eq. 1. Since T_1 is now known, calculated T_2 from

$$T_1 - T_2 = (q x_A)/k_A A$$

Since T_2 is now known, calculate T_3 from

$$T_2 - T_3 = (q x_B)/k_B A$$

This process may be continued until all interface temperatures are known.

Critical Radius of Insulation. A tube or sphere with a single layer of insulation and with finite h_c on the outside and essentially infinite h_h on the inside has a maximum heat flux at a finite thickness of insulation. For such a cylinder

$$q = 2\pi L(T_1 - T_c) \frac{1}{(1/h_c r_0) + (1/k_{INS}) \ln (r_0/r_i)}$$

set $dq/dr_0 = 0$; this is a maximum magnitude of q at

$$\text{cylinder } (r_0)_{critical} = \frac{k}{h_c} \qquad (9)$$

similarly

$$\text{sphere } (r_0)_{critical} = \frac{2k}{h_c} \qquad (10)$$

Orders of Magnitude of h. Table 1 shows the usual range of values of surface coefficients under various conditions.

Table 1. Order of Magnitude of h, Btu/hr ft^2 °F

Gases (natural convection)	0.9–5
Flowing gases	2–50
Flowing liquids (nonmetallic)	30–1,000
Flowing liquid metals	1,000–50,000
Boiling liquids	200–50,000
Condensing vapors	500–5,000

For a wall, the equivalent h is k/x. For example, for a steel wall 0.12 in. thick and $k = 26$ Btu/hr ft °F, the equivalent h is $26 \times 12/0.12$ or 2600, but for an asbestos wall 1 ft thick with $k = 0.13$ Btu/hr ft °F, the equivalent h is 0.13/1 or 0.13.

For certain combinations of these various resistances in series, some may be negligible compared with others.

General Conduction Equation.   The differential equations for temperature distributions are

$$\frac{\partial^2 T}{\partial x^2} + \frac{\partial^2 T}{\partial y^2} + \frac{\partial^2 T}{\partial z^2} + \frac{W_i}{k} = \frac{1}{\alpha}\frac{\partial T}{\partial t} \tag{11}$$

or

$$\nabla^2 T + \frac{W_i}{k} = \frac{1}{\alpha}\frac{\partial T}{\partial t}$$

Cylinder:

$$\left[\frac{1}{r}\frac{\partial}{\partial r}\left(r\frac{\partial T}{\partial r}\right) + \frac{1}{r^2}\frac{\partial^2 T}{\partial \theta^2} + \frac{\partial^2 T}{\partial z^2}\right] + \frac{W_i}{k} = \frac{1}{\alpha}\frac{\partial T}{\partial t} \tag{12}$$

Sphere, where θ is the meridianal angle and ϕ is the azimuthal angle:

$$\frac{1}{r^2}\left[\frac{\partial}{\partial r}\left(r^2\frac{\partial T}{\partial r}\right) + \frac{1}{\sin\theta}\frac{\partial}{\partial \theta}\left(\sin\theta\frac{\partial T}{\partial \theta}\right) + \frac{1}{\sin^2\theta}\frac{\partial^2 T}{\partial \phi^2}\right] + \frac{W_i}{k} = \frac{1}{\alpha}\frac{\partial T}{\partial t} \tag{13}$$

Here W_i is a quantity of energy per unit volume and time resulting from such actions as electric current ($I^2 R$), chemical reactions, or nuclear reactions; these are known as heat sources.

Heat Sources, One-Dimensional.   In rectangular coordinates

$$\frac{d^2 T}{dx^2} = -\frac{W_i}{k} \tag{14}$$

or

$$T = -\frac{W_i}{2k}x^2 + C_1 x + C_2 \tag{15}$$

With symmetry around a centerline at $x = 0$ ($dT/dx = 0$) and a known temperature, T_0, on both faces at $x = +x_0$, the temperature distribution is

$$T - T_0 = \frac{W_i}{2k}(x_0{}^2 - x^2) \tag{16}$$

In cylindrical coordinates

$$\frac{d}{dr}\left(r\frac{dT}{dr}\right) = -\frac{W_i r}{k} \tag{17}$$

$$T = -\frac{W_i}{4k}r^2 + C_1 \ln r + C_2 \tag{18}$$

For a solid cylinder with T_0 at r_0, Eq. 18 becomes

$$T - T_0 = \frac{W_i}{4k}(r_0{}^2 - r^2) \tag{19}$$

For these two special cases the heat flux at the outer surface is

$$\frac{q}{A} = x_0 W_i \tag{20}$$

$$\frac{q}{A} = \frac{r_0}{2}W_i \tag{21}$$

Fins.   Figure 4 represents a fin of uniform area transferring heat from a wall at T_b to air at T_a.   Consider a steady state with k and h uniform.   The fin is assumed to be so thin that the temperature distribution normal to the fin is essentially uniform.   Then the differential equation is

$$\frac{d^2(T - T_a)}{dx^2} - \frac{hP}{kS}(T - T_a) = 0 \tag{22}$$

For $T = T_b$ at $x = 0$ and $\partial T/\partial x = 0$ at $x = L$

$$\frac{(T - T_a)}{T_b - T_a} = \frac{\cosh B(L - x)}{\cosh BL} \tag{23}$$

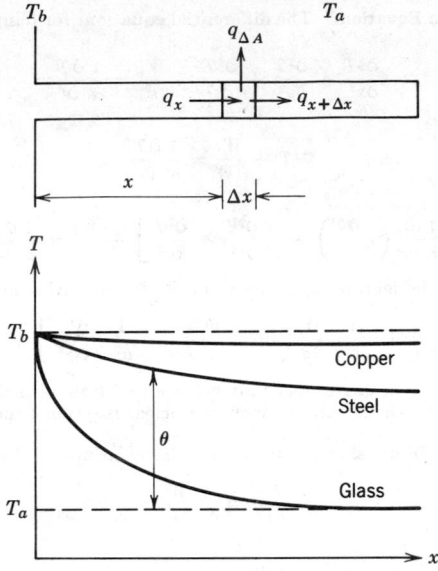

FIG. 4.   Steady temperature distribution in a thin rod.

where

$$B \equiv \sqrt{\frac{hP}{kS}} \tag{24}$$

and q at $x = 0$ is

$$q = \sqrt{hPkS} \; (T_b - T_a) \tanh BL \tag{25}$$

A performance factor η called fin efficiency is the ratio of the actual heat transferred through the fin to that which would be transferred if the entire fin were at T_b:

$$\eta = \frac{\sqrt{hPkS} \tanh BL \; (T_b - T_a)}{hPL(T_b - T_a)} = \frac{\tanh BL}{BL} \tag{26}$$

From this definition

$$q = h(T_b - T_a)\eta A_f \tag{27}$$

where A_f is the exposed area of the fin.

Similar results are available for a variety of fin shapes (1).

Multidimensional Problems.   Various special cases of one-, two-, and three-dimensional conduction problems have been solved and the results presented graphically in ref. 2.

Unsteady Conduction.   When a solid such as a large flat plate with an initially uniform temperature is heated on both sides by a hot fluid at T_f, the temperature distribution varies with time as sketched in Fig. 5.   If the plate is thin (Fig. 5b) or if the conductivity is high, the temperature gradients within the body may be negligible and a single value of temperature T may be used to describe the thermal state of the plate at any instant of time. A similar description is valid for bodies of any shape.

Bodies with Negligible Temperature Gradients.   In time interval dt, the energy balance for a body of arbitrary shape with negligible temperature gradients may be wtitten,

$$dQ = Ah(T_f - T) \, dt = V\rho c \, dT \tag{28}$$

$$t = \frac{V\rho c}{Ah} \ln \frac{T_f - T_1}{T_f - T_2} \tag{29}$$

$$\frac{T_f - T_2}{T_f - T_1} = \exp\left(-\frac{Ah}{V\rho c} t\right) \tag{30}$$

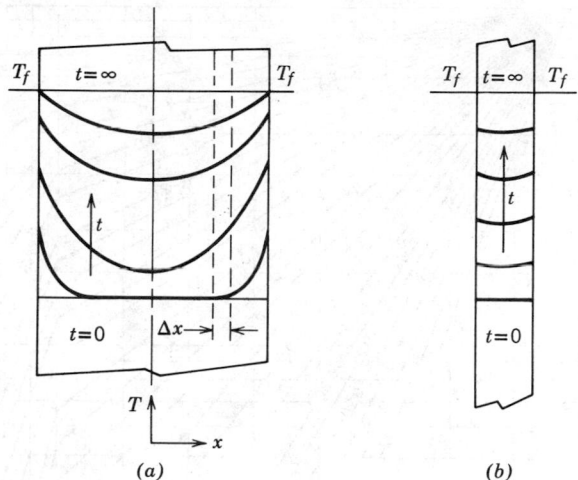

FIG. 5. Transient temperature distribution. (a) Large rectangular plate. (b) Thin plate.

Bodies With Internal Temperature Gradients. For the flat plate of Fig. 5a, the applicable differential equation from Eq. 11 is

$$\frac{\partial T}{\partial t} = \alpha \frac{\partial^2 T}{\partial x^2} \tag{31}$$

The solution is presented in graphical form in Figs. 6, 7, and 8 for the following boundary condition:

$$T(x,0) = T_i \text{ uniform}$$

$$\frac{\partial T(0,t)}{\partial x} = 0 \text{ and } \frac{\partial T(r_0,t)}{\partial x} = -\frac{h}{k}[T(r_0,t) - T_f] \tag{32}$$

Similar solutions exist for cylinders and spheres. (2, 4).

Semi-Infinite Body. A useful special case of transient conduction is a semi-infinite body initially at a uniform temperature T_i and suddenly exposed to a fluid at T_f and a constant h. The results are shown graphically in Fig. 9.

3. RADIATION

Solid bodies, liquids, and some gases emit radiant energy which may be in many forms such as X-rays resulting from high-speed electron bombardment of a metal plate, γ-rays emitted from radioactive material, ultraviolet light emitted from an electric discharge in a gas, and thermal radiation resulting from the thermal excitation of the molecules. Modern theory describes the nature of this radiation in terms of electromagnetic waves which all travel at the velocity of light. The various forms of radiation differ only in wavelength (Fig. 10). In this section discussion will be confined to thermal radiation. The emission of this energy depends upon the temperature and character of the emitting body.

Emission and Absorption Characteristics. The emission and absorption characteristics of a body may be treated independent of each other but are related under certain circumstances.

For a surface exposed to a quantity of radiant energy e, Btu/hr ft², the absorptivity α, the transmissivity τ, and the reflectivity ρ are, respectively, the fractions of e shich are absorbed, transmitted, and reflected. Then

$$\alpha + \tau + \rho = 1$$

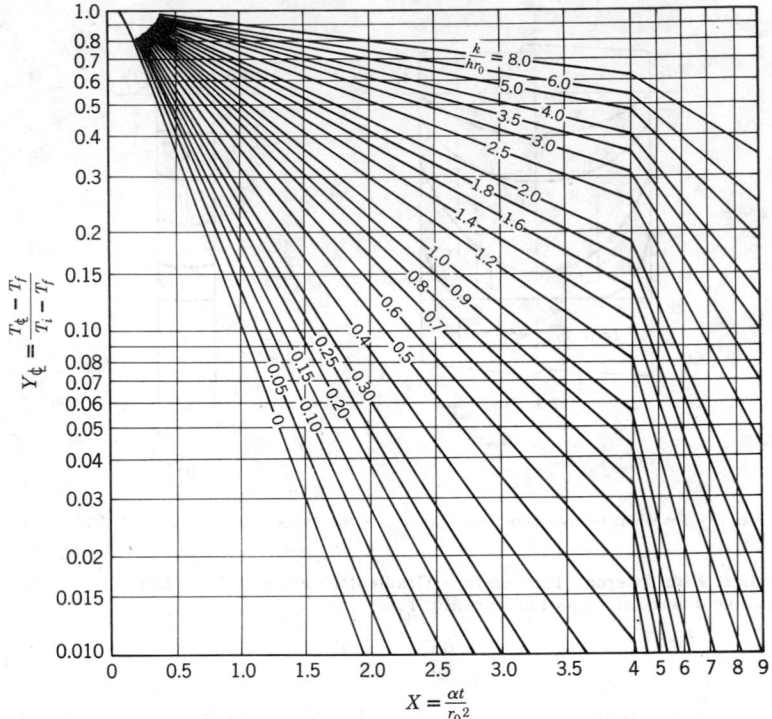

$$X = \frac{\alpha t}{r_0{}^2}$$

FIG. 6.   Temperature history at the center of an infinite slab [Hottel (3)].

The part of the radiant energy which is absorbed or emitted by materials which are electric conductors is all completely absorbed or emitted in a layer of approximately 0.00005 in. below the surface.   In electric nonconductors this layer is of the order of 0.05 in. thick. Because this region of influence is so small, the emissive and absorptive properties at a surface are not seriuosly influenced by strong temperature gradients at and normal to the surface.   For most industrial applications, liquids and solid bodies are almost always thick enough to be opaque to radiation ($\tau = 0$), so $\rho = 1 - \alpha$.

The reflection from a highly polished surface may be specular (Fig. 11a); from a rough surface it is usually diffuse (Fig. 11b).

In general, unoxidized surfaces of good electric conductors have low values of α, and surfaces of poor conductors and oxidized surfaces have high values.   A black body is defined as one which absorbs all oncoming (or incident) radiation, $\alpha = 1$, $\rho = 0$.   The concept of a black body is useful because the laws governing its radiation are simple and many real bodies may be idealized as black bodies.   Although no bodies in nature are black bodies, many industrial materials have very high values of α, and hence approach black-body conditions.   The term black is used because if a surface actually does absorb all radiant energy falling on it, the surface will appear black to the eye; this, of course, ignores the emitted radiation.   Some surfaces absorb nearly all incident radiation yet do not appear black to the eye because they do not absorb all visible light rays.   Indeed, freshly fallen snow, soot, and whitewashed walls all have absorptivities greater than 0.95 for thermal radiation.

Black bodies also emit radiation at all wavelengths, and the total intensity e_b (Bru/hr ft²) depends only on the surface temperature.   Then even without reference to any solid surface, black-body radiation of intensity e_b has an associated temperature.   Further, since the radiation travels at finite velocity, an enclosed space may be imagined to be "filled" with radiation.

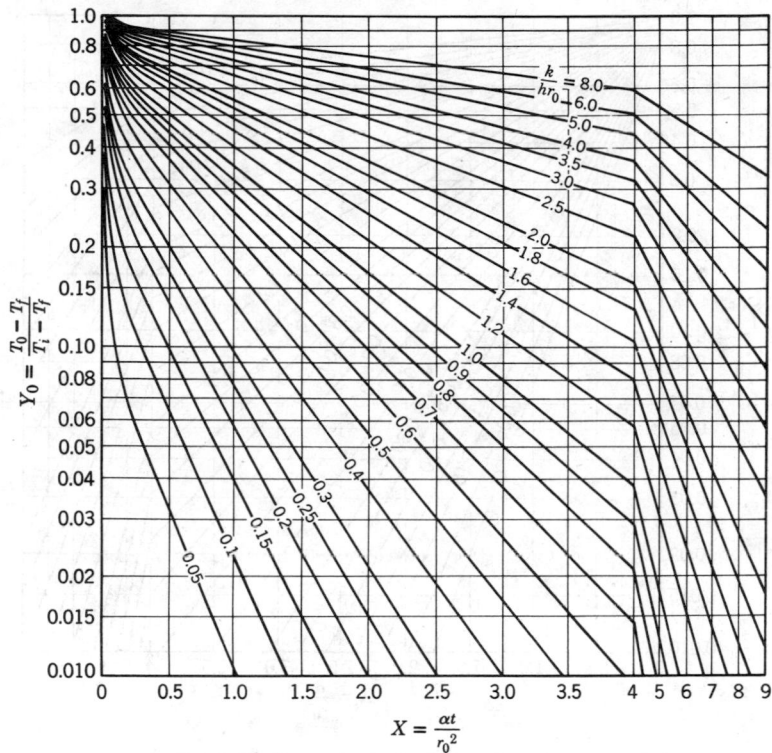

$$X = \frac{\alpha t}{r_0^2}$$

The emissive power of black bodies is

$$e_b = \sigma T^4$$

where $\sigma = 0.1713 \times 10^{-8}$ Btu/ft^2 hr R^4.

Radiation between Surfaces Separated by Nonabsorbing Media. The net rate of heat transferred by radiation from one surface to another separated by a vacuum or a gas which is transparent to the radiation is calculated from the equation

$$q = A_1 \mathcal{F}_{12} \sigma (T_1^4 - T_2^4) \tag{34}$$

where A_1 is the area of surface 1. The factor $\mathcal{F}_{12}$ is calculated from the equation

$$\mathcal{F}_{12} = \frac{1}{(1/F_{12}) + [(1/\epsilon_1) - 1] + (A_1/A_2)[(1/\epsilon_2) - 1]} \tag{35}$$

where ϵ_1 and ϵ_2 are emissivities of the two surfaces, and A_1 and A_2 are areas of the two surfaces. The emissivity ϵ of a surface depends upon the character of the surface and temperature of the surface. The angle factor F_{12} depends upon the size, shape, and arrangement of the two surfaces exchanging heat by radiation. The value of F_{12} is not known for all surface arrangements and shapes. It has been evaluated for only a few of the more common ones.

Angle Factor, F_{12}. Surfaces for which $F_{12} = 1.00$:

1. A small flat surface radiating to a wall which forms a large hemisphere.
2. Concentric spheres or long concentric cylinders such as long pipes.
3. Large, closely spaced rectangular or circular plates, where ratio of side or diameter to distance between planes is large (greater than 15).

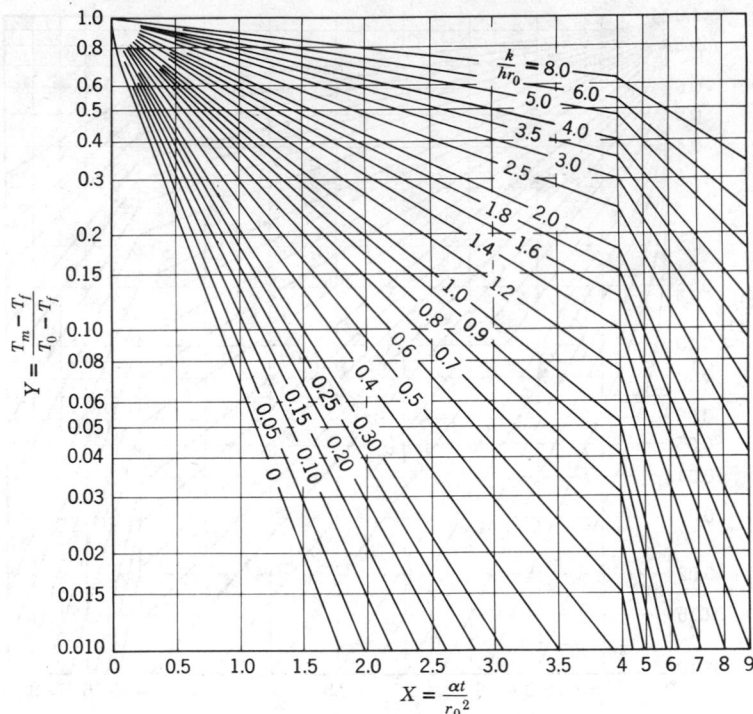

$$Y = \frac{T_m - T_f}{T_0 - T_f}$$

$$X = \frac{\alpha t}{r_0^2}$$

FIG. 8.   Space-mean-temperature history in an infinite slab [Hottel (3)].

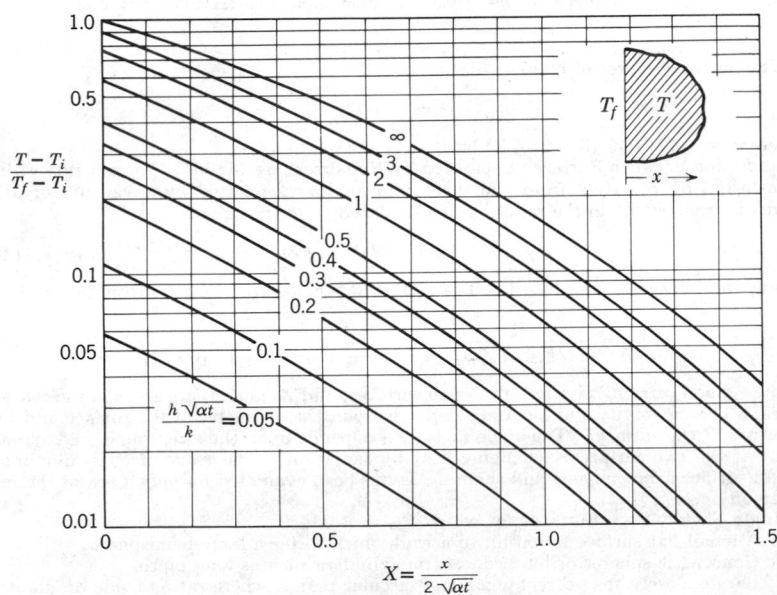

$$\frac{T - T_i}{T_f - T_i}$$

$$X = \frac{x}{2\sqrt{\alpha t}}$$

FIG. 9.   Temperature history in a semi-infinite solid initially at a uniform temperature and suddenly exposed at its surface to a fluid at constant temperature [Schneider (2)].

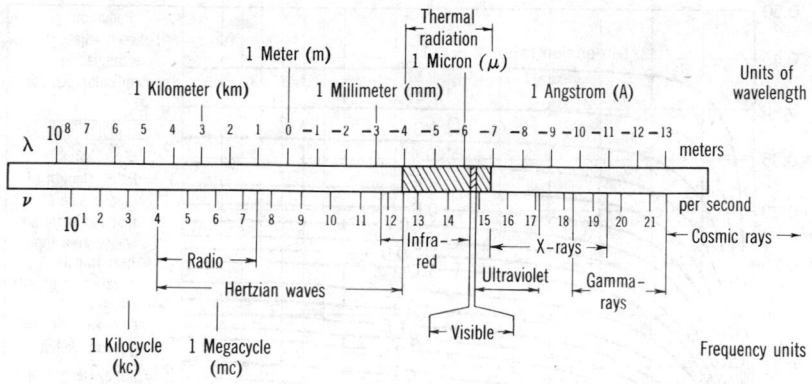

Fig. 10. Electromagnetic wave spectrum.

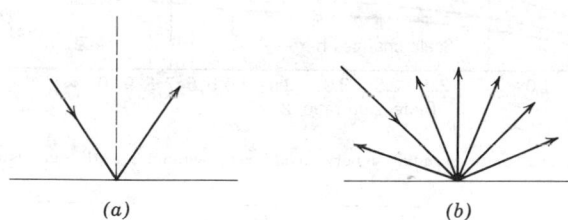

(a)   (b)

Fig. 11. Reflecting surfaces. (a) Specular. (b) Diffuse.

Surfaces for which $F_{12} \neq 1.00$:

1. Figure 12 is for perpendicular rectangular planes.
2. Figure 13 shows the angle factor for parallel planes (disks, squares, and rectangles) directly opposite each other. Lines 5, 6, 7, and 8 are for planes separated by refractory (nonconducting but reradiating) walls.
3. Figure 14 is for radiation between a plane and one or two rows of tubes parallel to the plane.

Problems of furnace design involve radiation of gases, flames, and clouds. A detailed study of these problems is found in McAdams (5).

If the surfaces are separated by adiabatic ($q = 0$) refractory surfaces (Fig. 15), replace F_{12} by $\overline{F}_{12}$ which is given in Fig. 13 for special cases.

Additional cases are discussed in Refs. 4 and 5.

Convection and Radiation

Heat transfer by convection and radiation occurs at the same time from a given surface. Usually one has the greater effect, and the other may be neglected. In some cases, as in heat transfer from bare and insulated pipes, the convection and radiation are of the same order of magnitude. Then it is convenient to use a combined convection and radiation coefficient $(h_v + h_r)$. The rate of heat transfer becomes

$$q = (h_v + h_r) A (T_s - T_f) \tag{36}$$

where h_v is defined by Eq. 4 and is determined from experimental data given under the heading Film Coefficient, and h_r is determined by q_r calculated from Eqs. 34 and 35 from

$$h_r = \frac{q_r}{A(T_s - T_f)} \approx 4 \, \mathfrak{F}_{12} \sigma \, T_{\text{mean}}^3 \tag{37}$$

where T_s and T_f are temperatures of the surface and surrounding fluid and walls, respectively.

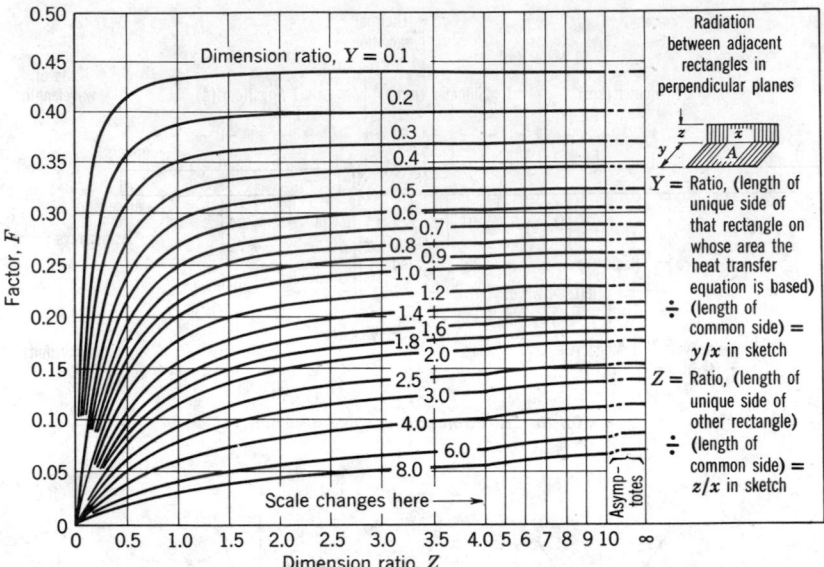

FIG. 12.   View factor F for direct radiation between adjacent rectangles in adjacent planes [Hottel (3)].

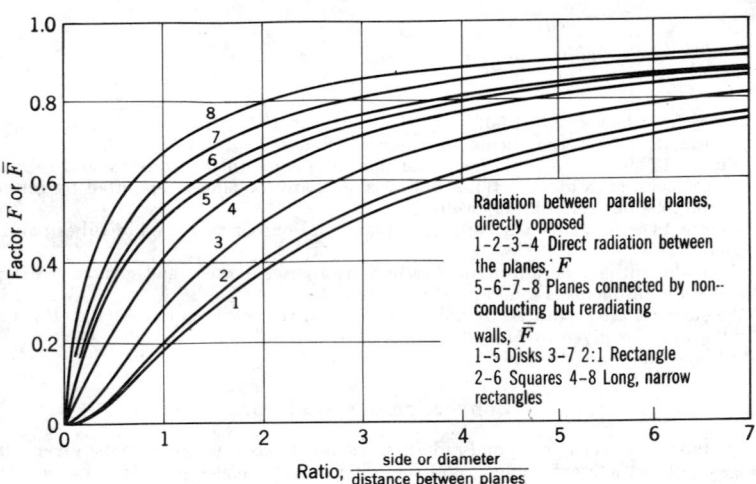

FIG. 13.   View factor F and interchange factor F̄ for parallel planes directly opposed (5).

For certain special cases, values of the combined coefficient $(h_v + h_r)$ are given also under Magnitudes of Surface Coefficients, pages 25 and 26.

4. HEAT EXCHANGERS

The equation for calculating the rate of heat transfer in a heat exchanger is

$$q = Y U A \frac{\theta_1 - \theta_2}{\ln (\theta_1/\theta_2)} \tag{38}$$

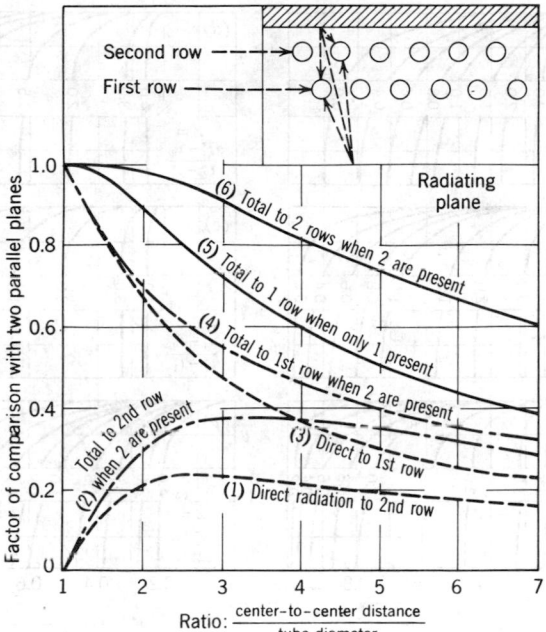

FIG. 14.　View factor F and interchange factor F from a plane to one or two rows of tubes parallel to the plane (5).

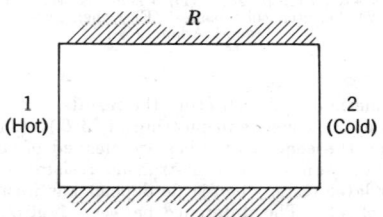

FIG. 15.　Enclosure of black surfaces with a refractory (no net flux) surface.

This equation assumes constant U, constant specific heats of the two fluids, constant rates of flow, no changes in phase, and negligible heat loss to the atmosphere. The product UA is evaluated as in Eq. 6 or 7.

Factor Y.　For a pure counterflow heat exchanger, $Y = 1$. For various multipass and crossflow heat exchangers, Y is determined from Figs. 16 and 17.

Steam Condenser.　In a steam condenser one fluid is a condensing vapor and the other is a liquid that does not change phase.　The rate of heat transfer is calculated from

$$q = UA \frac{T_{w2} - T_{w1}}{\ln \left[(T_s - T_{w1})/(T_s - T_{w2}) \right]} \tag{39}$$

For condenser tubes with cooling water inside the tubes, the value for UA in Eq. 39 is

$$UA = \frac{1}{\dfrac{1}{2\pi r_1 L h_c} + \dfrac{1}{2\pi r_1 L h_s} + \dfrac{\ln (r_2/r_1)}{2\pi L k_a} + \dfrac{1}{2\pi r_2 L h_h}} \tag{40}$$

where h_s is the coefficient of heat transfer for the scale deposit in the tubes.

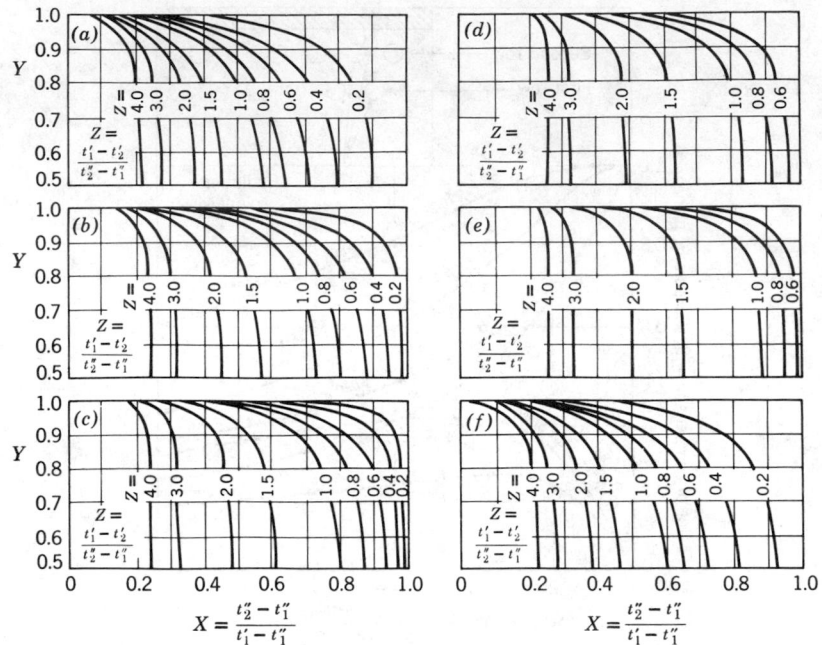

FIG. 16. Factor Y in reversed-current heat exchangers (shell side well mixed at a given cross section). (A) 1 shell pass and 2, 4, 6, etc., tube passes. (B) 2 shell passes and 4, 8, 12, etc., tube passes. (C) 3 shell passes, and 6, 12, 18, etc., tube passes. (D) 4 shell passes, and 8, 16, 24, etc., tube passes. (E) 6 shell passes, and 12, 24, 36, etc., tube passes. (F) 1 shell pass, and 3, 6, 9, etc., tube passes. [See McAdams (5).]

A valuable graphical method of interpreting the results of tests on surface condensers, proposed by E. E. Wilson (2), consists in plotting $(1/A U)$ vs. $(1/V)^{0.8}$ and drawing the best straight line. When the condenser tubes are cleaned of scale deposit, the value of $1/A U$ at $1/V^{0.8} = 0$ is the sum of the heat-transfer resistance of the condensing vapor and the tube wall. For later tests, when scale deposit has formed, the straight line will be higher on the curve sheet. The difference between $1/A U$ for the "clean" and the "dirty" condition at $1/V^{0.8} = 0$ measures the amount of scale formed and may be used as an indicator for the necessity of cleaning the condenser tubes.

5. MAGNITUDES OF SURFACE COEFFICIENTS OF HEAT TRANSFER

The value of the film coefficients depends upon two kinds of properties—physical properties of the materials, such as thermal conductivity, viscosity, specific heat, and density, and geometrical properties, such as shape and arrangement of heating surface and velocity of flow across the surface. The problem of prediction of film coefficients divides itself into three velocity ranges. In the high-velocity range, turbulent flow exists as described in Fig. 1. As the velocity is decreased, the laminar layer increases until the entire tube stream flows in a laminar flow. A further reduction in velocity to near zero causes natural convection currents to govern the prediction of film coefficients.

Turbulent Flow in Ducts. Most applications met in practice are forced flow of gases or liquids in turbulent flow. An empirical equation correlating the heating of gases and liquids inside horizontal pipes when Re $> 10,000$ is

$$\mathrm{Nu} = 0.023(\mathrm{Re})^{0.8}(\mathrm{Pr})^{0.4} \qquad (41)$$

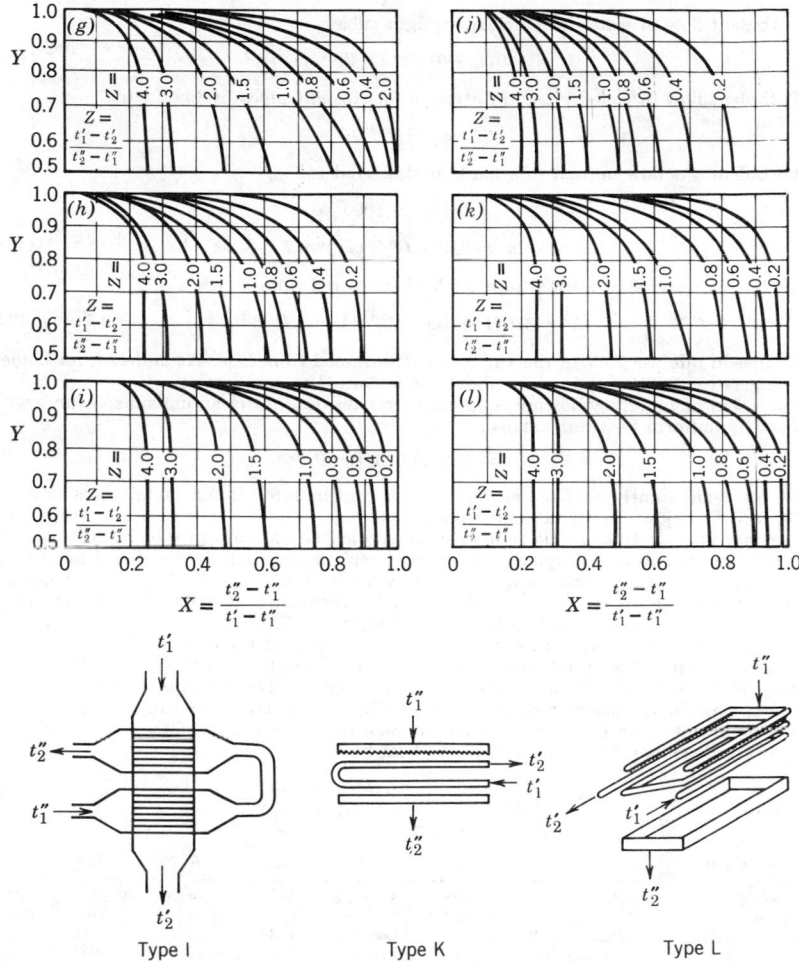

Fig. 17.   Factor Y in cross-flow heat exchangers. (G) Cross flow, both fluids unmixed, 1 tube pass. (H) Cross flow, shell fluid mixed, 1 tube pass. (I) Cross flow, shell fluid mixed, 2 tube passes—shell fluid flows across second and first passes in series. (J) Cross flow, shell fluid mixed, 2 tube passes—shell fluid flows over first and second passes in series. (K) Cross flow (drip type), 2 horizontal passes with U-bend connection (trombone type). (L) Cross flow (drip type), helical coils with 2 turns. [See McAdams(5).]

where the dimensionless numbers are

$$\text{Nu} = \text{Nusselt number} = hD/k$$

$$\text{Re} = \text{Reynolds number} = DG/\mu$$

$$\text{Pr} = \text{Prandtl number} = c_p\mu/k$$

The following simplified equations for specific cases are deduced from Eq. (41): Turbulent flow of gases inside clean straight tubes:

$$h = 0.024[c_p G^{0.8}/(D')^{0.2}] \qquad (42a)$$

Turbulent flow of water inside clean straight tubes:

$$h = 160(1 + 0.012\,T)[V^{0.8}/(D')]^{0.2} \tag{42b}$$

Turbulent flow in coiled pipes: Multiply h for straight pipes by bhe factor

$$1 + 3.5(D/D_c)$$

Turbulent gas flow normal to a bank of staggered tubes:

$$DG/\mu > 1000$$

$$h = 0.36[c_p G^{0.6}/(D')^{0.4}] \tag{42c}$$

Turbulent water flow normal to a bank of staggered tubes:

$$h = 370(1 + 0.0067\,T_f)[V^{0.6}/(D')^{0.4}] \tag{42d}$$

For tubes in line use 75% of the value of h determined from Eqs. 42c and 42d; for baffled exchangers use 60% of the values above for unbaffled exchangers.

For liquid metals the following equation correlates data for clean unoxidized tubes with essentially uniform heat flux axially:

$$\text{Nu} = 7 + 0.025(\text{Re} \cdot \text{Pr})^{0.8} \tag{43}$$

Data for even slightly oxidized surfaces or for contiminated liquid metals falls as much as 40 to 80% below the calculations from Eq. 43.

Laminar Flow In Ducts. If the fluid enters a duct with a uniform velocity distribution, this velocity distribution changes along the length, becoming fully developed after some length which in round tubes is approximately $L/D = 0.02\text{Re}_D$. If heating or cooling begins at the entrance, both the velocity and temperature distribution changes along the length. If heating or cooling begins after the fully developed velocity distribution is established, only the temperature distribution changes along the length. In this entrance region the heat transfer coefficient decreases with length. Ultimately, both a fully developed velocity and temperature distribution exists; beyond this asymptotic point the heat transfer coefficient remains constant—beyond $L/D_e = 0.02\text{Re}_{D_e} \cdot \text{Pr}$.

The following is an empirical relation representing h in the entrance region both for uniform wall temperature and for uniform heat flux (6).

$$\text{Nu} = \text{Nu}_\infty + \frac{K_1[(D/x)\text{Re}_D\text{Pr}]}{1 + K_2[(D/x)\text{Re}_D\text{Pr}]^n} \tag{44}$$

Wall condition	Velocity	Pr	Ny	Nu_∞	K_1	K_2	n
Uniform T_0	Parabolic V	Any	Average	3.66	0.0668	0.04	2/3
Uniform T_0	Developing V	0.7	Average	3.66	0.104	0.016	0.8
Uniform $(q/A)_0$	Parabolic V	Any	Local	4.36	0.023	0.0012	1.0
Uniform $(q/A)_0$	Developing V	0.7	Local	4.36	0.036	0.0011	1.0

In the fully developed asymptotic region the constant heat transfer coefficient or Nu is given in the following table.

Geometry	Velocity distribution	Condition at wall	$\dfrac{hD_e}{k}$
Circular tube	Parabolic	$(q/A)_0$ Uniform	4.36
Circular tube	Parabolic	T_0 Uniform	3.66
Circular tube	Slug	$(q/A)_0$ Uniform	8.00
Circular tube	Slug	T_0 Uniform	5.75
Parallel plates	Parabolic	$(q/A)_0$ Uniform	8.23
Parallel plates	Parabolic	T_0 Uniform	7.60
Triangular duct	Parabolic (Ref. 2)	$(q/A)_0$ Uniform	3.00
Triangular duct	Parabolic (Ref. 2)	T_0 Uniform	2.35

If the temperature difference between the wall and fluid is large, natural convection may be superimposed on the laminar flow. This increases the h; therefore the above magnitudes represent conservative values. Correction for natural convection are found in Ref. 7.

Flow over Flat Plates. Flow over flat plates results in the following relations for predicting heat transfer coefficients.

1. Uniform wall temperature $Re_x < 3 \times 10^5$ laminar

$$Pr > 0.6 \qquad Nu_x = 0.332 Pr^{1/3} Re_x^{1/2} \tag{45}$$

$$0.006 \leq Pr \leq 0.03 \qquad Nu_x = \frac{0.564}{Pr^{-1/2} + 0.90} Re_x^{1/2} \tag{46}$$

2. Uniform heat flux $Re_x < 3 \times 10^5$ laminar

$$Pr > 0.6 \qquad Nu_x = 0.458 Pr^{1/3} Re_x^{1/2} \tag{47}$$

$$0.006 \leq Pr \leq 0.03 \qquad Nu_x = \frac{0.880}{Pr^{-1/2} + 1.317} Re_x^{1/2} \tag{48}$$

3. Turbulent flow $Re_L > 10^6$

$$Pr > 0.6 \qquad Nu_x = 0.0296 \, Pr^{1/3} Re_x^{0.8} \tag{49}$$

Flow Over Submerged Bodies. For water and hydrocarbon oils, data (5) suggest the following relation for liquid flowing across single cylinders:

$$1 < \frac{D_0 G}{\mu_f} < 1000 \qquad \frac{h_m D_0}{k_f} (Pr_f)^{-0.3} = 0.35 + 0.56 \left(\frac{D_0 G}{\mu_f} \right)^{0.52} \tag{50}$$

For air flowing across single cylinders:

$$1000 < \frac{D_0 G}{\mu_f} < 50,000 \qquad \frac{h_m D_0}{k_f} (Pr_f)^{-0.3} = 0.26 \left(\frac{D_0 G}{\mu_f} \right)^{0.60} \tag{51}$$

For flow of water and oil $(7.3 < Pr < 380)$ past single spheres $(0.279$ in. $< D_0 < 0.496$ in.) data of Kramers was correlated by McAdams in the range $1 < D_0 G/\mu_f < 1000$ as follows:

$$\frac{h_m D_0}{k_f} (Pr_f)^{-0.3} = 0.97 + 0.68 \left(\frac{D_0 G}{\mu_f} \right)^{0.50} \tag{52}$$

As $D_0 G/\mu_f$ gets smaller and approaches unity, the magnitude of Nu should approach 2, which is the calculated result for pure conduction of a shpere at uniform temperature in an infinite stagnant medium.

For flow across banks of tubes that are 10 or more rows deep, the data of Colburn (8) for $D_0 G_{max}/\mu_f > 6000$ can be correlated by

$$\frac{h_m D_0}{k_f} = C \left(\frac{D_0 G_{max}}{\mu_f} \right)^{0.6} Pr_f^{1/3} \tag{53}$$

where $C = 0.33$ for staggered tubes and 0.26 for in-line tubes.

Using their test data, Kays and Lo (9) suggest the magnitudes in the following table for banks of tubes with less than 10 rows.

Ratio h_m for N Rows to h_m for 10 Rows Deep

N	1	2	4	6	8	10
Staggered tubes	0.68	0.75	0.90	0.95	0.98	1.0
In-line tubes	0.64	0.80	0.89	0.94	0.98	1.0

Data (10) for mercury $(Pr = 0.022)$ flowing over a 10-row-deep bank of staggered 1/2 in. tubes, 1.375 pitch-to-diameter ratio, was correlated as

$$Nu = 4.03 + 0.228 (Re_{max} Pr)_f^{0.67} \tag{54}$$

for the range $20,000 < Re_{max} < 80,000$.

For flow through packed beds of spheres, Eckert (11) suggests the following equation based on data of Denton:

$$\frac{h D_p}{k} = 0.80 \left(\frac{D_p G_0}{\mu} \right)^{0.7} Pr^{1/3} \tag{55}$$

Correlations of others are found in McAdams (5).

Natural Convection. Coefficients of heat transfer from surfaces to an infinite atmosphere of fluid are correlated with the following dimensionless groups:

$$\text{Grashof number } N_{Gr} \equiv g\beta L^3 \rho^2 \Delta T/\mu^2$$

$$\text{Prandtl number } N_{Pr} \equiv C_p\mu/k$$

The correlation equations for fluids with $N_{Pr} > 0.5$ are

$$10^4 \leq N_{Gr}N_{Pr} \leq 10^9 \qquad h = C_L k[(g\beta\rho^2/\mu^2)\,N_{Pr}]^{1/4}(\Delta T/L)^{1/4} \tag{56a}$$

$$10^9 \leq N_{Gr}\dot{N}_{Pr} \leq 10^{12} \qquad h = C_T k[(g\beta\rho^2/\mu^2)\,N_{Pr}]^{1/3}(\Delta T)^{1/3} \tag{56b}$$

Values for C_L, C_T, and L are listed below:

Surface	L	C_L	C_T
Vertical plates	b	0.59	0.23
Horizontal cylinders	D	0.48	0.11
Long vertical cylinders	b	0.50	0.12
Spheres	D/2	0.60	0.15

These equations may be simplified when air at temperatures in the neighborhood of 100°F is the infinitely large fluid.

Vertical Plates:

$$10^{-2} < b^3\Delta T < 10^3 \qquad h = 0.29(\Delta T/b)^{1/4} \tag{56c}$$

$$10^3 < b^3\Delta T < 10^5 \qquad h = 0.21(\Delta T)^{1/3} \tag{56d}$$

Horizontal Pipe:

$$10^{-2} < D^3\Delta T < 10^3 \qquad h = 0.25(\Delta T/D)^{1/4} \tag{56e}$$

$$10^3 < D^3\Delta T < 10^6 \qquad h = 0.18(\Delta T)^{1/3} \tag{56f}$$

Horizontal Hot Plate Facing Up (Cold Plate Facing Down):

$$0.1 < b^3\Delta T < 20 \qquad h = 0.27(\Delta T/b)^{1/4} \tag{56g}$$

$$20 < b^3\Delta T < 30,000 \qquad h = 0.22(\Delta T)^{1/3} \tag{56h}$$

Horizontal Hot Plate Facing Down (Cold Plate Facing Up):

$$0.3 < b^3\Delta T < 30,000 \qquad h = 0.12(\Delta T/b)^{1/4} \tag{56i}$$

Condensing Vapors. When a pure vapor, superheated or saturated, contacts a wall whose temperature is below saturation temperature, condensate forms and wets the tube, resulting in film-type condensation. If $4\Gamma/\mu$ is less than around 2100 and the vapor velocity is small, the condensate flow is laminar and the following equations may be used:

Horizontal Tubes:

$$h = 0.728\left[1 + 0.2\,\frac{c_l\Delta T}{h_{fg}}\,(n-1)\right]\sqrt[4]{\frac{g\rho(\rho - \rho_v)k^3 h'_{fg}}{nD\mu\Delta T}} \tag{57}$$

Vertical Plates and Tubes (Outside):

$$h = 0.943\sqrt[4]{\frac{g\rho(\rho - \rho_v)k^3 h'_{fg}}{L\mu\Delta T}} \tag{58}$$

where $h'_{fg} = h_{fg} + 0.68c_l\Delta T + c_v(T_v - T_{sat})$

$\Delta T = T_{sat} - T_{surf}$

n = number of horizontal tubes in a vertical bank

$$\Gamma = \frac{A}{\text{per}}\,\frac{h\Delta T}{h_{fg} + (3/8)c_l\Delta T + c_v(T_v - T_{sat})}$$

horizontal tubes: $\dfrac{A}{\text{per}} = L$

vertical plate and tube: $\dfrac{A}{\text{per}} = \text{width or } \pi D$

For condensing steam at 1 atm pressure, Eq. 58, reduces to

$$h = 4000/L^{1/4}\Delta T^{1/4} \tag{59}$$

Liquid Metals. For condensing liquid metals at pressures below 1 atm an additional resistance at the liquid vapor interface becomes important. This resistance in terms of an interface heat transfer coefficient is (12)

$$h_i = \sqrt{\frac{1}{2\pi R}}\frac{h_{f_g}^2}{v_{f_g}T_v^{3/2}} = \frac{q/A}{T_v - T_i} \tag{60}$$

Then in Eq. 57 or 58 with $\Delta T = T_i - T_{surf}$ and

$$h = \frac{q/A}{T_i - T_{surf}} \tag{61}$$

Solve Eqs. 60 and 61 with h given either by Eq. 57 or 58 for T_i. Then q/A can be determined from Eq. 61. This additional resistance represented by Eq. 60 may be neglected when the pressure is above 0.2 atm for liquid metals. It is negligible for all nonmetals.

Inside of Tubes. Condensation of Freon-12 and -22 in forced convection was studied by Bae, Maulbetsch and Rohsenow [13] and Traviss, Baron and Rohsenow [14] who modified the Dukler film analysis [15] to include the effect of total pressure drop. The friction pressure drop was evaluated using the Lockhart–Martinelli two-phase flow pressure drop correlation.

The local heat-transfer coefficient is correlated [126] with $\pm$ 15% by

$$0 < F(X_{tt}) < 1 \qquad \frac{NuF_2}{Pr_l Re_l} = F(X_{tt}),$$

$$\tag{62}$$

$$1 < F(X_{tt}) < 15 \qquad \frac{NuF_2}{Pr_l Re_l} = [F(X_{tt})]^{1.75}$$

where

$$F(X_{tt}) \equiv 0.15\,[X_{tt}^{-1} + 2.85\,X_{tt}^{-0.476}],$$

$$X_{tt} \equiv \left(\frac{\mu_l}{\mu_v}\right)^{0.1}\left(\frac{1-x}{x}\right)^{0.9}\left(\frac{\rho_v}{\rho_l}\right)^{0.5},$$

$$Re_l < 50 \qquad F_2 = 0.707\,Pr_l Re_l^{0.5}, \tag{63a}$$

$$50 < Re_l < 1125 \qquad F_2 = 5Pr_l + 5\ln\,[1 + Pr_l(0.09636Re_l^{0.585} - 1)], \tag{63b}$$

$$Re_l > 1125 \qquad F_2 = 5Pr_l + 5\ln\,(1 + 5Pr_l) + 2.5\ln\,(0.00313Re_l^{0.812}), \tag{63c}$$

$$Re_l = \frac{G(1-x)D}{\mu_l}. \tag{64}$$

The length of tube Δz required to change the quality Δx from an energy balance is

$$\Delta z = \frac{Gh_{f_g}D\Delta x}{4h_z\Delta T}. \tag{65}$$

To predict the length of tube to change the quality from x_i at inlet to x_e at exit determine Δz for each Δx, say 0.05, along the tube. Given a particular G, D, x_i and x_e, divide the calculations into steps of $\Delta x = 0.05$ or 0.10. Then Pr_l is known and Re_l is calculated at that zone by eqn. (64). Evaluate F_2 by the appropriate eqn. (63). Calculate X_{tt}, $F(X_{tt})$ and Nu or h_z from eqn. (62). The Δz for the assumed Δx is calculated from eqn. (65). Repeat this calculation at each Δx step along the tube.

At each of the Δx positions the pressure gradient may be calculated by

$$\left(\frac{dp}{dz}\right) = \left(\frac{dp}{dz}\right)_f + \left(\frac{dp}{dz}\right)_m + \left(\frac{dp}{dz}\right)_g \tag{66}$$

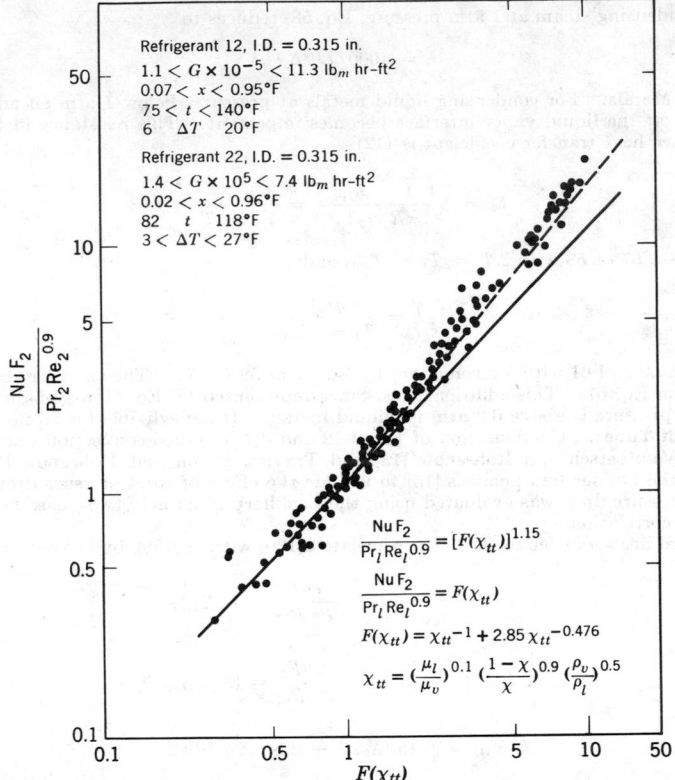

$$\frac{Nu\,F_2}{Pr_2\,Re_2^{0.9}}$$

Refrigerant 12, I.D. = 0.315 in.
$1.1 < G \times 10^{-5} < 11.3$ lb$_m$ hr–ft^2
$0.07 < x < 0.95\,°F$
$75 < t < 140\,°F$
$6 \quad \Delta T \quad 20\,°F$

Refrigerant 22, I.D. = 0.315 in.
$1.4 < G \times 10^5 < 7.4$ lb$_m$ hr–ft^2
$0.02 < x < 0.96\,°F$
$82 \quad t \quad 118\,°F$
$3 < \Delta T < 27\,°F$

$$\frac{Nu\,F_2}{Pr_l\,Re_l^{0.9}} = [F(\chi_{tt})]^{1.15}$$

$$\frac{Nu\,F_2}{Pr_l\,Re_l^{0.9}} = F(\chi_{tt})$$

$$F(\chi_{tt}) = \chi_{tt}^{-1} + 2.85\,\chi_{tt}^{-0.476}$$

$$\chi_{tt} = \left(\frac{\mu_l}{\mu_v}\right)^{0.1}\left(\frac{1-x}{x}\right)^{0.9}\left(\frac{\rho_v}{\rho_l}\right)^{0.5}$$

$F(\chi_{tt})$

Fig. 18. Comparison of forced convection condensation data and analysis (14).

where for friction

$$\left(\frac{dp}{dz}\right)_f \frac{g_0 D \rho_v}{x^2 G^2} = -\,0.09\left(\frac{\mu_v}{GD}\right)^{0.2}[1 + 2.85 X_{tt}^{0.523}]^2 \qquad (66a)$$

When noncondensible gases are present, heat transfer coefficients are reduced greatly by small amounts of gas since the vapor must diffuse through the gas which tends to accumulate at the cold surfaces. This problem can be treated by a method suggested by Votta (16).

Boiling Liquids. Liquids are often evaporated by heat transfer from submerged tubes heated internally either by condensing vapor or by electricity. The rate of heat transfer rises to a maximum value as the tube surface temperature is increased. At this point the bubbles formed at the surface become so numerous that they coalesce to form a vapor film. Figure 19 shows a characteristic curve for boiling liquids. If electricity causes the heating, when q/A reaches point (a), Fig. 19, further increase in electric current results in a rapid change from (a) to (b). Thus point (a), the maximum heat flux point, is often called the "burnout" point.

The position of the curve in Fig. 19 depends on the kind of liquid and the kind and character of the surface in addition to the pressure, the geometry of the heating surface, and the agitation of the liquid.

Data in the nucleate boiling region [to the left of point (a)] have been correlated (Ref. 18) by

$$\frac{C_1(T_{surf} - T_{sat})}{h_{fg}} = C_{sf}\left(\frac{q/A}{\mu_1 h_{fg}}\sqrt{\frac{g_0 \sigma}{g(\rho_1 - \rho_v)}}\right)^{0.33}\left(\frac{C\mu}{k}\right)_1^{1.0} \qquad (67)$$

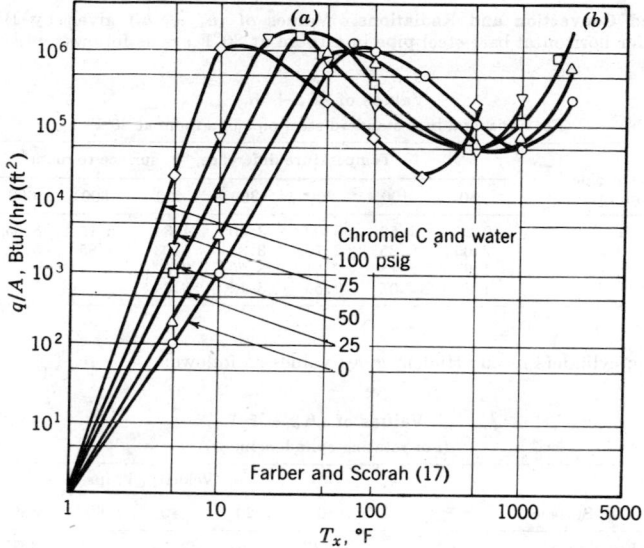

FIG. 19. Characteristic curve for boiling liquids.

where the following values of C_{sf} have been obtained:

Surface–liquid	C_{sf}
Water–platinum	0.013
Benzene–chromium	0.010
Ethyl alcohol–chromium	0.003
n-Pentane–chromium	0.015
Water–brass	0.006

For water boiling at atmospheric pressure on a brass surface, the following simplified expression expresses the data:

$$q/A \cong 10(T_{surf} - T_{sat})^3 \tag{68}$$

The values for maximum heat flux [point (a), Fig. 19] have been correlated for water, n-propane, n-pentane, n-heptane, benzene, and ethanol by

$$\frac{(q/A)_{max}}{\rho_v h_{fg}} = 143 \left(\frac{\rho_l - \rho_v}{\rho_v}\right)^{0.6} \left(\frac{g}{g_0}\right)^{1/4} \text{ ft per hr}$$

with a spread of data of $\pm 11\%$.

When water boils in a tube with forced convection flow, the total heat transfer rate is found to be a superposition of the heat transfer rate associated with "pool boiling," Eq. 67, and that assciated with nonboiling forced convection flow, Eq. 41. For a particular geometry and liquid mass, the rate of flow is

$$(q/A)_{nonboil} = h(T_{surf} - T_{liq}) \tag{69}$$

where h is calculated from Eq. 16. Then as a good approximation

$$q/A = (q/A)_{nonboil} + (q/A)_{boil} \tag{70}$$

There exists a maximum heat flux for boiling with forced convention flow. In these systems the magnitude of the maximum heat flux is a function of the type of pump characteristics, the temperature of the entering liquid, and the length-to-diameter ratio of the heated section. For a more comprehensive review of boiling heat transfer, refer to Tong (20).

Combined Convection and Radiations. Values of $(h_v + h_r)$ given by Bailey and Lyell (21) for horizontal bare steel pipe in a room at 80°F are as follows.

Values of $(h_v + h_r)$

(For horizontal bare standard steel pipe in a room at 80°F)

Nominal pipe diam, in.	Temperature difference, °F, surface to room							
	50	100	200	300	400	500	600	700
1	2.26	2.50	3.00	3.60	4.34	5.16	6.05	6.98
3	2.05	2.25	2.73	3.31	4.03	4.85	5.71	6.66
5	1.95	2.15	2.61	3.20	3.90			
10	1.87	2.07	2.54	3.12	3.84			

For rotating cylinders in air, Hinlein gives values as follows:

Values of $(h_v + h_r)$

(For rotating cylinders in air)

Surface	Velocity, V, fps					
	0	20	40	60	80	100
Smooth polished copper	0.47	1.97	2.79	3.14	3.30	3.38
Rough, dull, black-varnished	0.47	2.49	3.32	3.95	4.50	5.10

For flat surfaces contacting air, Lichty (Univ. Illinois Bull., 1915) gives values in the left table below. For external surfaces multiply $(h_v + h_r)$ by factors shown in the right table below.

Surface in still air	$h_v + h_r$
Cement plaster finish	0.9
Concrete or corkboard	1.3
Brickwork, sheet asbestos, or wood	1.4
Plate glass or magnesia board	1.5

	5	10	15	20
Air velocity, mph	5	10	15	20
Factor for brickwork	2.4	3.0	3.8	4.2
Factor for wood	1.2	2.0	2.9	3.0

Scale Deposits. Heat-transfer coefficients h_s for scale deposits from water for use in Eqs. 39 and 40 are as follows.

Heat-transfer Coefficients h_s for Scale Deposits from Water *

Temperature of heating medium	Up to 240°F		240–400°F	
Temperature of water	125°F		Above 125°F	
Water velocity, fps	3 and less	Over 3	3 and less	Over 3
Distilled water	2000	2000	2000	2000
Sea water	2000	2000	1000	1000
Treated boiler feed water	1000	2000	500	1000
Treated make-up for cooling tower	1000	1000	500	500
City, well, Great Lakes	1000	1000	500	500
Brackish, clean river water	500	1000	330	500
River water, muddy, silty†	330	500	250	330
Hard (over 15 grains/gal)	330	330	200	200
Chicago Sanitary Canal	130	170	100	130

* From "Standard of Tubular Exchanger Manufacturers Association," 1941.
† Delaware, East River (New. York), Mississippi, Schuylkill, and New York Bay.

Nomenclature

A	area.	q/A, q''	heat flux.
b	width.	r	radius.
c_p	specific heat at constant pressure.	S	cross-sectional area.
D, d	diameter, D' in inches.	T	temperature.
D_c	coil diameter	t	time.
D_e	equivalent diameter.	U	overall heat transfer coefficient.
e_b	black body emissive power.	V	volume.
F_{12}, $\bar{F}_{12}$	geometric shape factor between black bodies.	V	velocity.
$\mathfrak{F}$	geometric shape-emissivity factor between grey surfaces.	W_i	heat source or rate of heat generation per unit volume.
G	mass flow rate, $(G = w/S = \rho V)$.	x, y, z	coordinate axes.
g	gravitational acceleration.	α	thermal diffusivity, $k/\rho c_p$.
g_0	constant, 32.17 $lb_m ft/lb_f sec^2$, 4.17 $\times$ 10^8 lb_m ft/lb_f hr^2.	α	absorptivity.
		β	thermal coefficient of volume expansion.
h	heat transfer coefficient.		
h_x	local heat transfer coefficient.	Γ	mass rate of flow of condensate per unit width of wetted wall.
h_{fg}	latent heat of condensation or evaporation.	η	efficiency.
k	thermal conductivity.	ρ	mass density.
L, l	length.	ρ	reflectivity.
N, n	number in general.	σ	Stefan-Boltzmann constant = 0.1713 $\times$ 10^{-8} Btu/Ft^2hr R^4.
P	wetted perimeter.	τ	transmissivity.
q	rate of heat transfer.	τ	shear stress.

BIBLIOGRAPHY

1. KRAUS, ALLAN D., *Extend Surfaces*, Baltimore Spartan Books, 1964; also K. A. GARDNER, *Trans. ASME*, Vol. 68, No. 8, 621–31, November 1945.
2. SCHNEIDER, P. J., *Temperature Response Charts*, New York, Wiley, 1963.
3. HOTTEL, H. C., Personal Communication.
4. ROHSENOW, W. M., and CHOI, H. Y., *Heat, Mass and Momentum Transfer*, Englewood Cliffs New Jersey, Prentice-Hall, 1961.
5. MCADAMS, W. H., *Heat Transmission*, New York, McGraw-Hill, 3rd ed., 1954.
6. KAYS, W. M., *Trans. ASME*, Vol. 77, p. 1265, 1955.
7. MARTINELLI, R. C., and BOELTER, L. M. K., University of California Publication 5:2, pp. 23–58, 1942.
8. COLBURN, A. P., *Trans. Amer. Inst. Chem. Eng.*, Vol. 29, pp. 174–210, 1933.
9. KAYS, W. M., and LO, R. K., TR-15, NR-035-104, Mech. Engrg. Dept., Stanford University, August 1952.
10a. HOE, R. J., DROPKIN, D., and DWYER, O. E., *Trans. ASME*, Vol. 79, p. 899, 1957.
10b. RICHARDS, C. L., DWYER, O. E., and DROPKIN, D., *ASME Paper* 57-HT-11, 1957.
11. Eckert, E. R. G., *Introduction to Heat and Mass Transfer*, New York, McGraw-Hill, 1950.
12. WILCOX, S. J., and ROHSENOW, W. M., "Film Condensation of Potassium Using Copper Condensing Block for Precise Wall Temperature Measurement," ASME Paper No. 69-WA/HT-29.
13. BAE, S., MAULBETSCH, J. S., and ROHSENOW, W. M., "Refrigerant Forced-Convection Condensation Inside Tubes," Rept. No. 79760–59, Heat Transfer Laboratory, M.I.T., Nov. 1968.
14. TRAVIS, D. P., BARON, A. B., and ROHSENOW, W. M., "Forced Correction Condensation In Serlias" ASHRAE Trans. *79*, I, 1973.
15. DUKLER, A. E., "Fluid Mechanics and Heat Transfer for Vertical Falling Films," 3rd Nat. Ht. Trans. Conf., Storrs, Conn., May 1959. Preprint No. 10.
16. VOTTA, F., JR., "Condensing from Vapor-Gas Mixtures," *Chem. Eng.*, pp. 223–228, June 1964.
17. FARBER, E. A., and SCORAH, R. L., *ASME Trans.*, Vol. 70, p. 369, 1948.
18. ROHSENOW, W. M., "Heat Transfer with Evaporation," University of Michigan Heat Transfer Symposium, 1952, University of Michigan Press.
19. ROHSENOW, W. M., and GRIFFITH, P. J., "Correlation of Maximum Heat Flux Data for Boiling Liquids," Paper No. 9, AICE-ASME Joint Heat Transfer Symposium, Louisville, Kentucky, March 21, 1955.
20. TONG, L. S., *Boiling Heat Transfer and Two-Phase Flow*, New York, Wiley, 1965.
21. BAILEY, A., and LYELL, N. C., *Engineering*, Vol. 147, pp. 60–62, 1939.

AUTOMATIC CONTROL

BY

C. H. BARKELEW

AUTOMATIC CONTROL

1. INTRODUCTION

In the extraordinary advance of science and technology that has taken place during the last quarter century, one of the most conspicuous elements has been the increasing use of machines in homes, offices, laboratories, and factories. New concepts and theories of control, major advances in electronic science, new types of control equipment, new computing devices which can be used for control, and new areas of application have been appearing almost continuously.

Much of the incentive for this rapid growth of what is now being called automation has come from the great research and development programs sponsored by the government of the United States. Strong pressure for development has come from business and industry, whose increasing complexity makes them more and more difficult to operate with human hands. Pressure for improved measuring tecnhiques comes from scientists whose experiments are becoming more sophisticated and are more difficult to interpret.

The role of the control engineer in circumstances such as these is a difficult one. He must be constantly aware of developments that might be of use to him; he must be able to discriminate between those ideas that are truly significant to him and those which may not be useful for one reason or another. He must be of course thoroughly competent in the standard practices of his field, because a technology can remain useful only through continued self-examination by its own creative experts.

The primary job of a control engineer usually requires him to specify and design control systems wherever they may be needed. It is to this man, the working engineer, that this handbook section is directed. In it he will find assembled in one place enough of the basic principles of his field to enable him to start the solution of most of his design problems. Perhaps more important, it will tell him where to turn if he should need more than he finds here.

Finally, the section is intended to make him realize that his field is a growing one, and that all of its problems have not been solved in principle. By observing what others have been able to accomplish, and also what they have not accomplished, he may see how he himself can participate in the growth of this important part of science, and create new knowledge for the use of the many who will follow him.

Discussion of a purely mechanical nature has been kept to a minimum; thus for description of equipment such as transmitters, valves, motors, rectifiers, and the like, he should look into handbooks where these subjects are covered in full detail.

Symbols

The symbols to be used in the mathematical expressions are defined when they are first used. The usage is very nearly the same as that commonly found in the literature on control, and should lead to no uncertainty. The single expeption is a preference for "i" to represent $\sqrt{-1}$, rather than "j" as commonly used by electrical engineers.

A list of symbols used repeatedly in the text follows. Units are not specified, since no particular set has been recommended. In a number of cases a single symbol has been used to mean more than one quantity. The meaning is always clear from the context, and no confusion should result.

A	amplitude of a disturbance.		k	gain of controller.
α	angular variable.		L	Laplace transform operator.
C	Capacitance.		λ	root of stability determinant.
D	derivative time constant.		M	mass, modulus of complex function.
$D(A)$	describing function.		$O(t)$	output function.
$\delta(t)$	impulse function.		P	pressure.
$E(X)$	expected value of X.		p	pole of transfer function.
ϵ	error.		R	resistance, modulus of complex function.
F	flow, function.		ρ	density.
f	forces, function.		s	complex Laplace transformation variable, with dimension of frequency.
$G(s)$	transfer function of a process.			
$G_{xx}(\omega)$	power spectrum.		S	entropy.
g	gain.		σ	standard deviation.
H	Hamiltonian function.		T	time constant, temperature.
$H(s)$	transfer function of controller.		t	time.
h	height.		τ	time, dead time.
i	current, $\sqrt{-1}$.		$\mu(t)$	control function.
I	integral time constant.		V	voltage, volume, Liapunov function.
J	functional to be minimized.			

$W(t)$	time response function.		ϕ, θ	phase angle of complex variable.
x	state variable.		$\phi_{xx}(\tau)$	autocorrelation function.
y	state variable.		$\phi_{xy}(\tau)$	cross correlation function.
z	transformation variable.		ψ	adjoint variable.
ζ	damping factor.		ω	frequency.

Definitions

A number of words and phrases are used repeatedly in the text with meanings that are in common use by those familiar with control theory but which may lead to puzzlement in some who do not read extensively in the field. Therefore, a short list of definitions of terms is included here.

Adaptive Control. A type of control in which the controller continuously adjusts itself for changing process characteristics, in particular for changing dynamics or gain.

Advisory Control. A computer installation in which measurement signals are sent directly to the computer, which makes calculations and prints results. Action based on these results is accomplished manually.

Analog Computer. An assembly of amplifiers and other electronic elements which can be interconnected to construct a network which is mathematically analogous to some other physical system.

Asymptotic Stability. A system is asymptotically stable if it eventually returns to its equilibrium state after a disturbance which is sufficiently small.

Autocorrelation. A measure of the persistence of a random variable. Autocorrelation is defined as

$$\phi_{xx}(\tau) = \lim_{T \to \infty} \frac{1}{2T} \int_{-T}^{T} x(t)x(t + \tau)\, dt$$

Automation. A concept rather than a precisely defined engineering term. The use of machines to perform operations which are traditionally manual.

Bode Diagram. A set of two curves, one of which represents the logarithm of the amplitude of the open-loop frequency response of a system, the other the phase angle. The abscissa in both cases is the logarithm of the frequency.

Cascade Control. A closed-loop control system in which one controller, the master, is used to provide the set-point of another, the slave.

Closed-loop Control. A control system in which a controller acts on a process in such a way as to correct an error which it detects by means of a direct measurement.

Critical Damping. A particular nonoscillatory response of a system to a disturbance, characterized by coalescence of the two dominant poles of its transfer function. The damping factor is unity.

Cross Correlations. A defined measure of similarity between two random functions.

$$\phi_{xy}(\tau) = \lim_{T \to \infty} \frac{1}{2T} \int_{-T}^{T} x(t)y(t + \tau)\, dt$$

Damping Factor. In an exponentially dissipative system, the damping factor is the coefficient of time in the dominant (smallest) exponent.

Dead Time. A definite delay between two related actions. Synonymous with **Transport Lag.**

Decibel. The unit of power or intensity ratio. Two amplitudes which differ by one decibel are in a ratio of 1 to 1.122, which is $10^{1/20}$.

Describing Function. If a nonlinear system is excited by a sinusoid, the amplitude of the fundamental in the response is a function of the amplitude of the exciting signal. Their ratio is the describing function.

Deviation Ratio. The frequency response function of a closed-loop system to periodic disturbances in the controlled variable.

Digital Computer. A computer in which quantities are represented as combinations of discrete states. In modern usage the term is restricted to an electronic system, with its program stored internally.

Direct Digital Control. Use of a digital computer in a controlled system in such a way that more conventional types of controllers are not required. Generally the computer generates the signals which are sent directly to the final control elements.

Distribution Function. In statistics the distribution function of a random variable, with argument x, is the probability that the random variable will be equal to or less than x.

Dynamic Programming. A sophisticated mathematical concept used in searching for an optimal policy, characterized by continuing reexamination and search in successive stages.

Error Signal. In controlled systems a quantity which is proportional to the difference between the measured value of a variable and its desired value.

Feedback Control. A type of control in which a measurement is used to make an adjustment, which in turn affects the measured quantity.

Feed-forward Control. A type of control in which the action is taken by a controller before it affects the subsequent behavior of the system.

Floating Control. A control made in which the rate of change of signal to the actuator depends on the error.

Fourier Transforms. A class of integral transformations of a function. The prototype is

$$f(\tau) = \int_{-\infty}^{\infty} \phi(\omega) \, e^{i\omega\tau} \, d\omega$$

Frequency Domain. In linear dynamic systems, analysis of time functions is equivalent to analysis of their Fourier or Laplace transforms, in which the variable of transformation is frequency. In using the latter procedure one is said to be operating in the frequency domain.

Frequency Function. In statistics the derivative of the Distribution Function.

Gain. The ratio of the amplitude of the response of a linear device to that of a sinusoid applied to its input. Commonly used to characterize controllers and amplifiers.

Gain Crossover. That frequency for which the open-loop gain of a control system is unity.

Gain Margin. The gain of an open-loop control system at the frequency where the phase lag is 180°.

Hardware. In computer jargon the electronic and mechanical parts of a digital computer.

Hierarchy. In computer jargon an organization of computer control systems in which the proper modes of operation of the primary controllers are computed periodically by a central separate computer, which in turn may be one of several dictated to by a master.

Hill Climber. In control jargon a feedback controller which automatically adjusts a system for optimal operation.

Histogram. In statistics a chart showing the frequency of occurrence of each of several values taken by a random variable, as a function of the value.

Impulse Function. Dirac's "δ-function," it may be defined as

$$\lim_{\epsilon \to 0} \frac{1}{\sqrt{2\pi}\,\epsilon} \, e^{-x^2/2\epsilon^2}$$

Interaction Coefficient. In a multiply-variate control system a particular ratio of transfer functions which quantitatively describes the degree of interaction between loops.

Interface. In computer jargon that part of a combined analog and digital computing system which translates information passing between the two types of computers.

Laplace Transform.

$$L(s) = \int_{0}^{\infty} f(t) \, e^{-st} \, dt$$

Limit Cycle. A stable oscillatory state of a nonlinear dynamic system.

Master Controller. In cascade control the master receives measurement signals from the variable to be controlled and transmits a set point to the slave.

Maximum Principle. A mathematical technique for determining optimal trajectories. It was developed by L. S. Pontryagin, and is closely related to Hamilton's Principle in classical mechanics and to Dynamic Programming.

Mean. In statistics the average value of a random variable.

Negative Feedback. A procedure for closing a control loop, in which an error causes compensating action to be taken.

Nichols Diagram. A Cartesian graph in which the ordinates of the two curves of a Bode diagram are plotted against each other. The diagram also includes contours of constant closed-loop response.

Noise. An undesired and unpredicted disturbance.

Nyquist Diagram. A polar plot of amplitude versus phase response of an open loop.

On-Off Control. A type of control system which has only two output states.

Open Loop. A feedback control system which has been broken at some point in the signal path. A loop may be opened for test purposes, for emergencies, or for observations.

Operational Amplifier. A high gain electronic amplifier with high input and low output impedances. An essential component of an analog computer.

Optimal Control. Control in such a manner as to make some quantity take its most desirable value.

PD Control. A controller which used proportional plus rate or derivative action.

Phase. An adjective which has two distinct meanings in the terminology of automatic control: (1) when modifying "angle," "lag," "lead," or "margin" it refers to the relative angular displacement of two sinusoids; (2) when modifying "plane" it refers to a two-dimensional Cartesian system where the coordinates are a variable and its derivative.

Phase Crossover. The frequency at which an open loop has a phase lag of 180°.

Phase Margin. 180° minus the phase angle at which an open loop has a gain of unity.

PID Control. Proportional plus reset and rate action in a controller.

Positive Feedback. In a closed loop, if the effect of a disturbance acts in such a way as to reinforce that disturbance, the loop is said to have positive feedback.

Process Identification. A term used in applied control technology, referring to the determination, by some method, of the important dynamic characteristics of a system which is to be controlled.

Proportional Band. 100 divided by the gain of a controller. The unit is percent.

Proportional Control. A controller action in which the signal to the actuator is directly proportional to the difference between the measured variable and its set point.

Random Event. In statistics an event whose output cannot be predicted with certainty.

Random Process. A set of random variables whose elements are functions of time.

Random Variable. A function obtained by assigning a numerical value to each of a set of random events.

Rate Action. In a controller the additive part of the output signal which is directly proportional to the rate of change of the measured variable. This is sometimes called derivative control.

Rate Gain. The high frequency limit of the gain of a controller which uses rate action.

Reset Action. In a controller the additive part of the output signal which is directly proportional to the time integral of the error signal.

Reset Gain. The low frequency limit of the gain of a controller which uses reset action.

Reset Windup. Jargon for an undesirable effect in controllers using reset action, in which the accumulated integral term, which may arise when the loop has been open for an extended time, may cause saturation.

Root Locus. The path in the complex frequency plane which the poles of the closed-loop transfer function follow as the gain of the controller is varied.

Sampled-data System. A system involving intermittent, rather than continuous, transfer of information about a measured variable to a control system.

Set Point. The desired value at which a controller is to hold a system variable.

Simulation. The development and use of models for the study of ideas, systems, and situations.

Slave Control. In a cascade system the inner loop which takes its set-point signal from the master.

Software. In computer jargon the assemblage of special-purpose programs furnished with the computer.

Standard Deviation. In statistics the square root of the mean value of the square of the difference between a random variable and its average.

State Space. A non-Euclidian space in which the coordinates are identified with the variables of a system which characterize its state. The two-dimensional phase plane is an example of a simple state space. The state variables might be position and velocity of a moving object, for example.

Subsidence Rate. In an oscillating system the subsidence rate is the ratio of successive peak amplitudes of the oscillating variable.

Supervisory Control. A type of digital computer control in which the computer adjusts the set points of conventional control instruments, usually at intervals after optimization calculations.

Time Domain. In control jargon a study of the dynamics of a system made through the differential equations rather than through their transfer functions is said to be made in the time domain.

Transfer Function. The transfer function of a linear element is the ratio of Laplace transforms of output to input. It is a function of the complex frequency, s, and contains all possible information about the dynamic character of the element.

White Noise. A random signal which is characterized by a uniform distribution of frequencies.

z-Transform. The z-transform of a sampled signal is the Laplace transform expressed as a function of $z \equiv e^{Ts}$, where T is the sampling interval.

FUNDAMENTALS OF CONTROL THEORY

The essential idea of an automatic control system is that measurements are made on the device or process to be controlled, that these measurements are analyzed by the controller, and that action is taken according to the results of the analysis. The simplest kind of control system is called a *feedback* or closed-loop control system. In it the measurement is compared with the desired value or set point. The difference, sometimes called the error signal, can then be used to generate corrective action. In a manual control system the comparison and adjustments are made by a human being. In an automatic control system they are made by a hydraulic, mechanical, or electrical device. Another somewhat more complicated type of control is finding increasing use. It is sometimes called *feedforward* control. In it the effects of disturbances are predicted rather than measured and control action is taken on the basis of the predictions. We will be concerned primarily with feedback control systems.

The simplest way to visualize the operation of a control system is through a block diagram such as is shown in Fig. 1. A block diagram of this sort is a conventional repre-

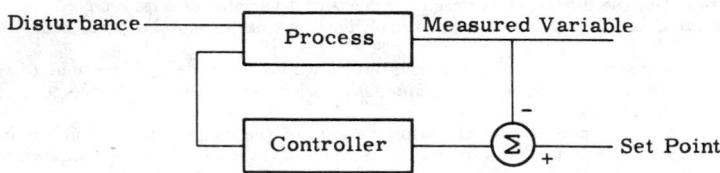

FIG. 1.   Block diagram.

sentation of a differential equation which describes the process or device plus the control system. The conventions for the construction of these block diagrams are the following:

Lines represent the flow of signals, with arrowheads indicating the direction.

Rectangles are used to represent those elements of the system in which the relationship of input and output is a dynamic one, dependent on time in some way or another. Often some indication of the dynamic action of the element is written within the rectangle.

Circles are used to represent algebraic operations which are independent of time. Usually they mean addition or subtraction, but sometimes also multiplication, amplification, or attenuation of a signal.

A typical system which might be represented by the above block diagram is a tank into which a fluid is flowing at an uncontrolled rate, and from which the fluid is being pumped (Fig. 2). The controller can be used to keep the liquid surface at a desired level by measuring the level, comparing the generated signal with another signal which represents the desired level, and using the difference to generate a third signal which operates a control valve on a line leading out from the pump. The relationship between liquid level and flow disturbance is a dynamic one, in that the level is related to the time integral of the difference between inflow and outflow.

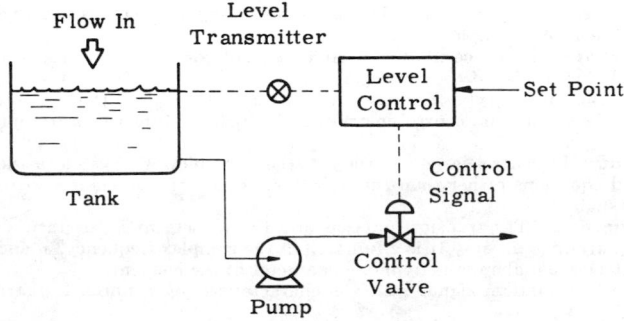

FIG. 2.   A simple control system.

2. MATHEMATICS OF LINEAR CONTROL THEORY

Naturally enough the kinds of systems that have been most fully described mathematically are those which are linear, and in particular those whose dynamic characteristics can be described by linear differential equations with constant coefficients. It is characteristic of systems of this sort that they can be broken into component parts, the dynamic characteristics of the component parts analyzed, and then the individual analyses assembled into a description of the whole. Mathematically this means the differential equation which describes the behavior of a component is not altered by the fact that the component is connected to some other component, except by the addition of terms which describe the interconnections. This characteristic is related to the property of superposition of linear systems, which states that in a homogeneous linear differential equation the sum of two solutions is also a solution. The response of the system to several disturbances simultaneously can be obtained by analyzing the response of the system to individual disturbances and then combining the results. A group of typical linear components with the differential equations which describe their behavior is given in Table 1. Any system which is composed of these or similar components can be analyzed by a linear method. A number of simple linear systems are shown in Fig. 3.

There are many physical situations in which a simple linear description is not good enough, and for these cases special techniques must be used. For example a linear system might be distributed in space, in which case its mathematical description might be a set of partial differential equations rather than ordinary differential equations. A linear system might be such that some of the coefficients in its ordinary differential equation are time-dependent. There are many physical situations which are essentially nonlinear and which cannot be handled by the linear techniques to be described; for example, any electrical network which includes diodes, air resistance in a supersonic aircraft or missile,

Table 1. Typical Linear Components

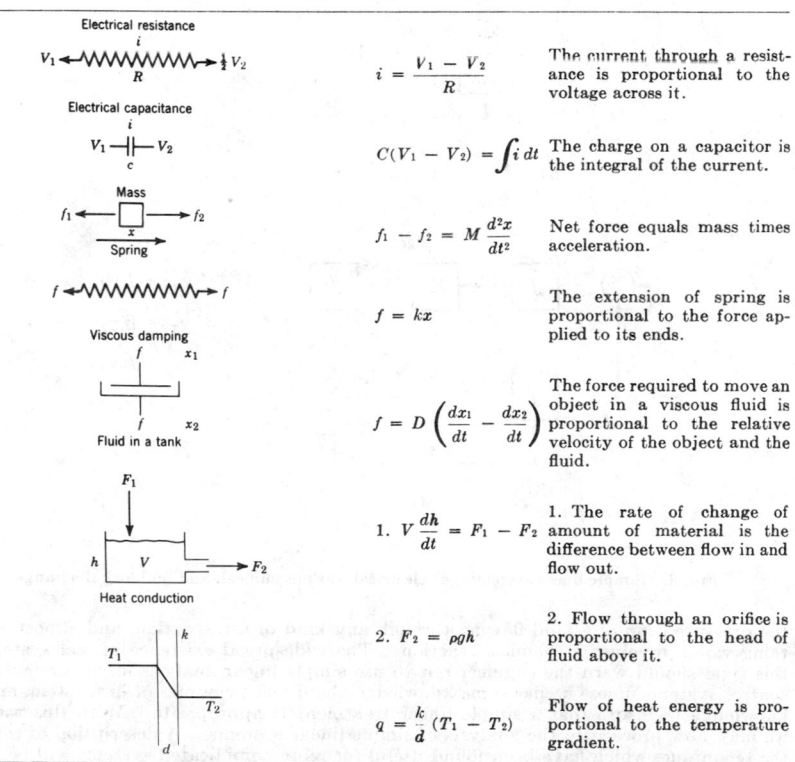

Electrical resistance $V_1 \xleftarrow{} \text{\^{}} \xrightarrow{} \frac{1}{2} V_2$ R	$i = \dfrac{V_1 - V_2}{R}$	The current through a resistance is proportional to the voltage across it.
Electrical capacitance $V_1 \dashv\mathrel{\|}\vdash V_2$ c	$C(V_1 - V_2) = \displaystyle\int i\, dt$	The charge on a capacitor is the integral of the current.
Mass $f_1 \xleftarrow{} \boxed{} \xrightarrow{} f_2$ x Spring	$f_1 - f_2 = M\dfrac{d^2x}{dt^2}$	Net force equals mass times acceleration.
$f \xleftarrow{} \text{\^{}} \xrightarrow{} f$	$f = kx$	The extension of spring is proportional to the force applied to its ends.
Viscous damping $f \qquad x_1$ $f \qquad x_2$ Fluid in a tank	$f = D\left(\dfrac{dx_1}{dt} - \dfrac{dx_2}{dt}\right)$	The force required to move an object in a viscous fluid is proportional to the relative velocity of the object and the fluid.
F_1 $h \quad V \xrightarrow{} F_2$	1. $V\dfrac{dh}{dt} = F_1 - F_2$	1. The rate of change of amount of material is the difference between flow in and flow out.
Heat conduction $T_1 \quad k$ T_2 d	2. $F_2 = \rho g h$ $q = \dfrac{k}{d}(T_1 - T_2)$	2. Flow through an orifice is proportional to the head of fluid above it. Flow of heat energy is proportional to the temperature gradient.

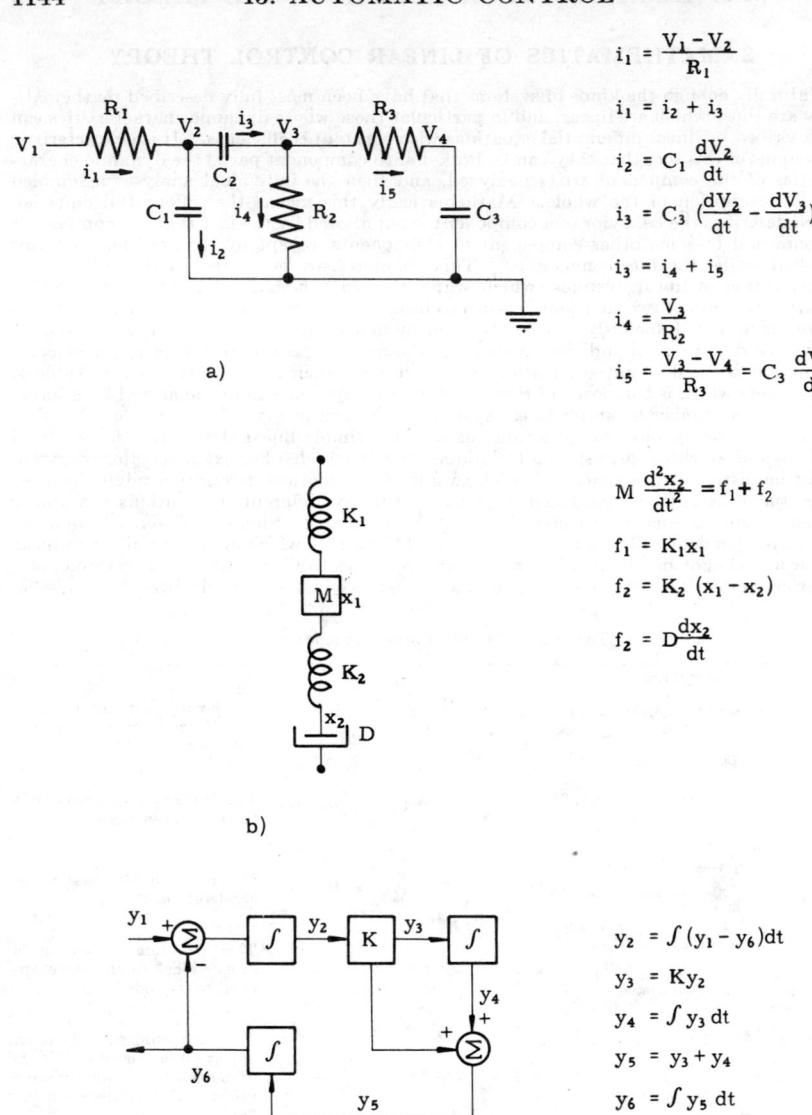

$$i_1 = \frac{V_1 - V_2}{R_1}$$

$$i_1 = i_2 + i_3$$

$$i_2 = C_1 \frac{dV_2}{dt}$$

$$i_3 = C_3 \left(\frac{dV_2}{dt} - \frac{dV_3}{dt}\right)$$

$$i_3 = i_4 + i_5$$

$$i_4 = \frac{V_3}{R_2}$$

$$i_5 = \frac{V_3 - V_4}{R_3} = C_3 \frac{dV_4}{dt}$$

a)

$$M \frac{d^2 x_2}{dt^2} = f_1 + f_2$$

$$f_1 = K_1 x_1$$

$$f_2 = K_2 (x_1 - x_2)$$

$$f_2 = D \frac{dx_2}{dt}$$

b)

$$y_2 = \int (y_1 - y_6) dt$$

$$y_3 = K y_2$$

$$y_4 = \int y_3 \, dt$$

$$y_5 = y_3 + y_4$$

$$y_6 = \int y_5 \, dt$$

c)

FIG. 3.   Simple linear systems: (a) electrical, (b) mechanical, and (c) block diagram.

the pressure drop of a fluid flowing through any kind of a restriction, and almost every-thing which involves a chemical reaction.   The widespread existence of real systems of this type should warn the engineer not to use simple linear analysis in the design of his control systems unless he has some knowledge about the dynamics of his system, enough knowledge to insure that a simple linear treatment is appropriate.   With this warning we may now proceed to the analysis of simple linear systems.   A description of some of the techniques which have been found useful for more complicated systems will be found later in this section.

The Laplace Transformation

The methods of the operational calculus, first proposed by Oliver Heaviside and later made rigorous by Bromwich, offer a convenient and easily understood way to analyze the dynamics of a complicated, interconnected linear system.

The Laplace transformation is useful in control work because of the simply way in which derivatives and integrals transform. It is defined by the integral:

$$F(s) \equiv \int_0^\infty f(t) \; e^{-st} \, dt \tag{1}$$

where $F(s)$ is the transform.

$f(t)$ is the function to be transformed.

t is a real variable, usually time in control theory.

s is a complex variable.

In a common shorthand notation, $F(s) = \mathcal{L}(f)$ and $f(t) = \mathcal{L}^{-1}[F(s)]$.

The following properties of the transformation are useful. They will not be derived here, since a more complete treatment will be found elsewhere in this handbook, and also in many other places.

$$f(t) = \frac{1}{2\pi i} \int_{c-\infty i}^{c+\infty i} F(s) \; e^{st} \, ds \tag{2}$$

where c is a real constant, greater than the real part of all the singular points of $F(s)$.

$$\mathcal{L}(kf) = k\mathcal{L}(f) \tag{3}$$

where k is any constant.

$$\mathcal{L}(f_1 + f_2) = \mathcal{L}(f_1) + \mathcal{L}(f_2) \tag{4}$$

$$\mathcal{L}\left(\frac{df}{dt}\right) = s\mathcal{L}(f) - f(0) \tag{5}$$

where $f(0)$ is the initial value of f.

$$\mathcal{L}\left[\int_0^t f(\tau) \, d\tau \right] = \frac{1}{s} \mathcal{L}(f) \tag{6}$$

if $\mathcal{L}^{-1}F_1(s) \equiv f_1(\tau)$, and $\mathcal{L}^{-1}F_2(s) \equiv f_2(\tau)$, then

$$\mathcal{L}^{-1}[F_1(s) \cdot F_2(s)] = \int_0^t f_1(t - \tau) \, f_2(\tau) \, d\tau \tag{7}$$

This is called the convolution integral.

$$\mathcal{L} f(t - \tau) = e^{-\tau s} \mathcal{L}(f) \tag{8}$$

$$\lim_{t \to \infty, \, (0)} f(t) = \lim_{s \to 0, \, (\infty)} sf(s) \tag{9}$$

These are called the initial and final value theorems.

Other useful properties may be found in the detailed treatises. Table 2 shows a few useful transforms.

Table 2

Function	Transform
Step:	
$f(t) = 0$ for $t < T$	
$f(t) = 1$ for $t > T$	$F(s) = \dfrac{e^{-sT}}{s}$
Ramp:	
$f(t) = kt$	$F(s) = \dfrac{k}{s^2}$
Sine wave:	
$f(t) = \sin(\omega t + \phi)$	$F(s) = \dfrac{\omega \cos \phi + s \sin \phi}{s^2 + \omega^2}$

Using the above properties, the engineer can derive the transformed system of equations which describe the behavior of his process or device. If the resulting equation can be solved for the transformed variable, then the inversion integral, or a table of transforms, can be used to find the solution of the system of equations.

More often than not in control work, it is more convenient to use the transformed system of equations than to use original equations or their solutions. This fact can be illustrated in the following way. Suppose we have a system described by the equation

$$\frac{dy}{dt} + ky = z(t) \tag{10}$$

for example, a tank of liquid, or a simple RC network. Then the Laplace transformation leads to:

$$s\,Y(s) + k\,Y(s) = Z(s) \tag{11}$$

and

$$Y(s) = \frac{Z(s)}{k+s} \tag{12}$$

if the initial condition is disregarded. In block diagram form, we can illustrate the original equation by

the transformed equation by

or equivalently by

In the latter form, $1/(s+k)$ is called the *transfer function* of the block. It is the Laplace transform of the output divided by the Laplace transform of the input. The transfer function of two elements in series is the product of the transfer functions of the individual elements.

A short list of useful transfer functions is given in Table 3.

Table 3.   Useful Transfer Functions *

Operation	Equation	Transfer Functions
Amplification	$y = kz$	k
Integration	$y = \int_0^1 z(k)\,dk$	$\dfrac{1}{s}$
First-order lag	$T\dfrac{dy}{dt} + y = z$	$\dfrac{1}{1+Ts}$
Simple delay	$y(t) = z(t-\Delta)$	$e^{-\Delta s}$
Second-order, damped	$\dfrac{d^2y}{dt^2} + \zeta\dfrac{dy}{dt} + \omega^2 y = z$	$\dfrac{1}{s^2 + \zeta s + \omega^2}$

* T is often referred to as the "time constant," Δ as the "dead time," and ζ as the damping factor.

Detailed treatments of the Laplace transformation and extensive tables can be found in references (3), (13), (22), (34), (42) and (43).

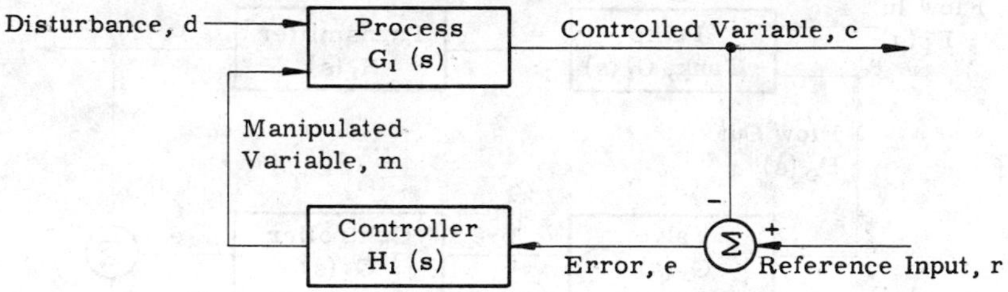

Fig. 4. General control loop.

Transfer Functions

Figure 4 shows a generalized diagram of a control loop. This is the same diagram used in Fig. 1, except that the transfer functions, $G_1(s)$ for the process and $H_1(s)$ for the controller, have been added and the Laplace transforms of the variables have been identified by symbols.

The transfer function equations are

$$c = G_1(s)[m + d]$$
$$e = (r - c)$$
$$m = H_1(s)(e)$$

from which

$$c = \frac{G_1(s)\,H_1(s)r + G_1(s)d}{1 + G_1(s)\,H_1(s)} \tag{13}$$

The overall transfer function from the reference input or set point to the controlled variable is thus

$$\frac{c}{r} = \frac{G_1(s)\,H_1(s)}{1 + G_1(s)\,H_1(s)} \tag{14}$$

The transfer function from the disturbance, to the controlled variable is

$$\frac{c}{d} = \frac{G_1(s)}{1 + G_1(s)\,H_1(s)} \tag{15}$$

Equation 14 contain all information necessary to determine the dynamic response of the controlled system to a change in the set point of the controlled variable. Similarly, Eq. 15 can be used to obtain the dynamic response of the controlled system to a load disturbance. The behavior of the system will clearly be different for the two kinds of changes, since the transfer functions have different numerators.

It is sometimes of interest to compare the transfer function of a *closed-loop* system with an *open-loop* transfer function of the same system in which the signal transfer has been inhibited at some point. For example, in Fig. 4 the measuring device could be disconnected so that the signal "c" cannot reach the summing point. Then the open-loop transfer function from r to c would be

$$\frac{c}{r} = H_1(s)G_1(s) \tag{16}$$

The relationship between open-loop and closed-loop transfer function is of considerable importance, as will be seen.

This sort of analysis can be readily extended to obtain the transfer function of systems which are far more complicated. The block diagram which goes with Fig. 2, for example, might be as shown in Fig. 5. The Laplace transforms of the time variables have been identified with symbols, as before, and the time-dependent behavior of the valve, level transmitter, and set-point element have been added.

The transfer function from reference to level is

$$\frac{c(s)}{r(s)} = \frac{G_1 G_2 G_3 H_2}{1 + G_1 G_2 G_3 H_1} \tag{17}$$

as can be readily verified.

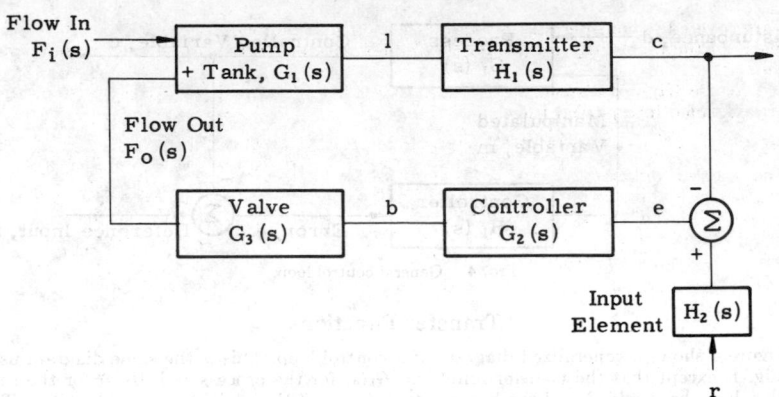

FIG. 5.   Block diagram for a level control.

The transfer function of a system may be used for many different purposes, as will be apparent in later parts of this chapter.   The first use to be illustrated will be the determination of the transient response of a system to step disturbances in set point and load. This can be done best by going back to the two-block system of Eqs. 14 and 15, thus avoiding algebraic complexities not essential to the illustration.

A typical process might react dynamically by a first-order lag whose transfer function is

$$G_1(s) = \frac{1}{1 + \tau s}$$

where τ is the time constant of the lag.   A typical controller might be described by the transfer function

$$H_1(s) = g\left(1 + \frac{I}{s}\right)$$

where g and I are constants, commonly called the "gain" and "reset rate."   The response of such a system to a set point change would be

$$\frac{c(s)}{r(s)} = \frac{[1/(1 + \tau s)]g[1 + (I/s)]}{1 + [1/(1 + \tau s)]g[1 + (I/s)]} = \frac{g(s + I)}{gI + s(1 + g) + \tau s^2} \tag{18}$$

If the set point should be changed suddenly by an amount Δr, then the transfer function of the change in c would be

$$\Delta c(s) = \frac{g(s + I)}{gI + s(1 + g) + \tau s^2} \frac{\Delta r}{s} \tag{19}$$

since the transfer function of the step is $\Delta r/s$.

Δc is an analytical function of the complex variable s.   It has a zero at $s = -I$, and three poles, one at $s = 0$, the others at the negative roots of the quadratic expression in the denominator.   The residues at these poles may be found by expansion into partial fractions:

$$\Delta c(s) = \frac{R_1}{s} + \frac{R_2}{s + p_1} + \frac{R_3}{s + p_2} \tag{20}$$

where R_1, R_2, R_3 are the residues and $-p_1$ and $-p_2$ are the roots of the quadratic.   The poles are at $-p_1$ and $-p_2$.   If real, p_1 and p_2 are the reciprocals of the time constants of the system.

Thus the engineer who wants to determine the transient response of a complicated system from its transfer function must first be able to locate the roots of a polynomial of any degree, since linear component transfer functions are almost always ratios of polynomials.   Second, he must be able to decompose a ratio of polynomials into partial fractions.   These two operations of elimentary algebra will not be explained here, since they are covered in considerable detail in Sec. 2 of this handbook.   A review of the essential parts of the theory of functions of a complex variable will also be found in that same section.

The inverse Laplace transformation is most readily performed when the transfer function is expanded into partial fractions.   Thus

$$C(t) \equiv \mathcal{L}^{-1}\Delta c(s) = R_1 + R_2 e^{-p_1 t} + R_3 e^{-p_2 t} \tag{21}$$

A number of pertinent observations may be made from this form of $C(t)$.   First, if either p_1 or p_2 is negative or has a negative real part, that is if the poles are in the positive half-plane of s, then the system will be exponentially unstable, and its response to a step disturbance will grow without limit.   In the simple system of the illustration, this can happen, for example, if either g or I should be negative, since

$$p_1, \ p_2 = \frac{(1 + g) \pm \sqrt{(1 + g)^2 - 4gI\tau}}{2\tau} \tag{22}$$

For this reason controllers should always be put into a loop in such a way that the corrective action opposes the disturbance.   This situation is called *negative feedback*.   *Positive feedback*, in which some component of a control signal tends to reinforce a disturbance, can lead to instability.   In higher order systems with more than two poles, instabilities can arise in other ways.   This circumstance will be described more fully later.

A second observation which can be made on $C(t)$ is that if p_1 and p_2 are complex, there may be oscillations, since $e^{\alpha + \beta i} = e^{\alpha}(\cos \beta + i \sin \beta)$.   From the form of p_1 and p_2, α above must be negative.   Complex roots will occur if $(4gI\tau) > (1 + g)^2$, in which case p_1 and p_2 are complex conjugates.   A damped oscillatory response to a disturbance will occur if gI is sufficiently large.   As before, in higher order systems the criterion is more complicated and will be described later.

A third significant observation concerning $C(t)$ is that if p_1 and p_2 are both real and positive, the system can overshoot but not oscillate. with C jumping past its ultimate value and returning more slowly.   If this happens, there is some value of t for which $dC/dt = 0$.

$$\frac{dC}{dt} = -p_1 R_2 e^{-p_1 t} - p_2 R_3 e^{-p_2 t} = 0 \text{ if } p_1 R_2 e^{-p_1 t} = -p_2 R_3 e^{-p_2 t} \tag{23}$$

and hence if

$$\frac{e^{-p_1 t}}{e^{-p_2 t}} = -\frac{p_2 R_3}{p_1 R_2} \tag{24}$$

thus

$$t_{max} = \frac{1}{-p_1 + p_2} \log \rho \left(-\frac{p_2 R_3}{p_1 R_2} \right)$$

For t_{max} to be real and positive, we must have $p_2 R_3 / p_1 R_2 < -1$ if p_1 is the smaller root.   If this ratio is positive, the logarithm is not real; if it is negative but greater than -1, t_{max} will be real and negative and hence of no physical interest.

The existence of overshoot in a feedback system is related to the location of the zeros of the transfer function, as will be shown.   In the partial fraction expansion of $c(s)$,

$$R_2 = \frac{-p_1 + I}{-p_1(p_2 - p_1)} \tag{25}$$

and

$$R_3 = -\frac{-p_2 + I}{-p_2(p_1 - p_2)} \tag{26}$$

as can be readily verified.   The condition for overshoot is thus

$$\frac{p_2 R_3}{p_1 R_2} = -\frac{(-p_2 + I)}{(-p_1 + I)} < -1 \tag{27}$$

thus

$$\frac{-p_2 + I}{-p_1 + I} > 1$$

$$-p_2 + I > -p_1 + I, \text{ if } -p_1 + I \text{ is positive}$$

$$-p_2 + I < -p_1 + I, \text{ if } -p_1 + I \text{ is negative}$$

The first is not possible, since p_2 has been taken greater than p_1, but the second condition can occur. Therefore, there will be overshoot if $I < p_1$. Since positive I is necessary for inherent stability, this means that overshoot will take place if there are zeros sufficiently close to the origin of s. The analysis of a more general system for overshoot is quite complicated and will not be attempted in this section. A complete treatment involving graphical procedures can be found in the book by Truxal, for example.

A fourth observation on $C(t)$ is that if $I = 0$, the transfer function reduces to

$$c'(s) = \frac{g}{s(1 + g + \tau s)} = \frac{g/\tau}{s\left(s + \dfrac{1 + g}{\tau}\right)} = \frac{g}{1 + g}\frac{1}{s} - \frac{g}{1 + g}\frac{1}{\left(s + \dfrac{1 + g}{\tau}\right)} \tag{28}$$

$$C'(t) = \frac{g}{1 + g}\left[1 - e^{-(1+g)(t/\gamma)}\right] \tag{29}$$

C no longer approaches its set point as time continues, but rather is offset from it by a ratio of $1/(1 + g)$. This phenomenon is referred to as *proportional offset*; it always occurs in a feedback control system whose feedback element does not integrate the error. $H(s)$ must include a term in $1/s$ (integration) for the offset to be reduced to zero.

If $G_1(s)$ is characterized by a simple first-order time constant, such as in the example, a feedback controller without integration can never cause overshoot or oscillation, since there is then only a single exponential term. The proportional offset can be reduced to as small as desired by making g large enough.

A fifth phenomenon of interest is the special case where

$$4gI\tau = (1 + g)^2$$

In this case the two poles coalesce ($p_1 = p_2$), and the system is said to be *critically damped*. The partial fraction expansion then takes the form

$$C(s) = \frac{R_1}{s} + \frac{R_2}{s + p} + \frac{R_3}{(s + p)^2} \tag{30}$$

where $-p$ is the value of s at the double pole, $-(1 + g)/2\tau$. p is necessarily real and positive. The critically damped state is intermediate between the exponential behavior characteristic of poles on the negative real axis and the damped oscillation which occurs when the poles are at complex values of s. With more complicated systems the behavior as control parameters change is described in the section on *Root-Locus Methods*. Figure 6 shows some typical response curves.

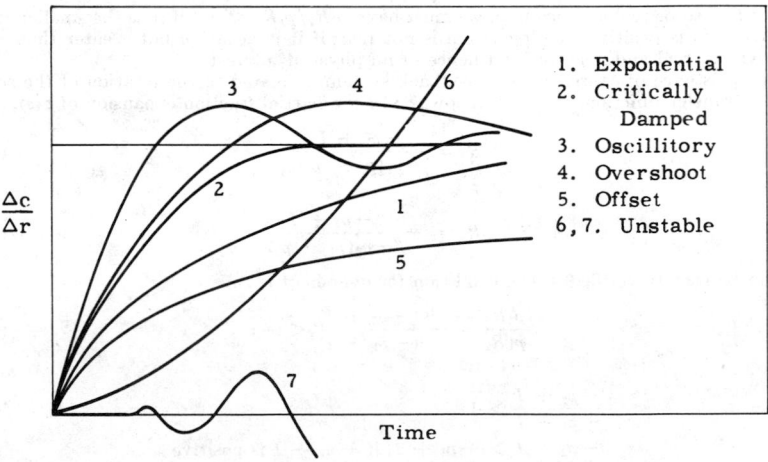

1. Exponential
2. Critically Damped
3. Oscillitory
4. Overshoot
5. Offset
6,7. Unstable

$\dfrac{\Delta c}{\Delta r}$

Time

FIG. 6.   Transient response curves.

In the same way, the analysis of response to a load change can be made. Comparison of Eqs. 14 and 15 shows that the transfer functions for load disturbance and set-point change have the same poles but different zeros. Instead of the expression 18, the transfer function c/d is given by

$$\frac{c}{d} = \frac{1/(1 + \tau s)}{1 + [1/(1 + \tau s)]g[1 + (I/s)]} = \frac{s}{s(1 + \tau s) + g(s + I)} \tag{31}$$

The response of c to a step change in d would be described by

$$\Delta c(s) = \frac{\Delta d}{s(1 + \tau s) + g(s + I)} = \frac{R_3}{s + p_1} + \frac{R_4}{s + p_2} \tag{32}$$

One of the poles has disappeared, that at $s = 0$. This is a general feature of load disturbances, and has a number of consequences.

$C(t) \equiv \mathcal{L}^{-1}\Delta c(s)$ takes the form $R_3 e^{-p_1 t} + R_4 e^{-p_2 t}$, where the p's are the same as for set-point changes but the R's are different. This function goes to zero as time increases, which means that the effect of a load change will be eventually removed by a controller whose transfer function is of the form

$$H_1(s) = g[1 + (I/s)] \tag{33}$$

so long as p_1 and p_2 are positive. The response will be oscillatory if the p's are complex, as before.

If I should be zero, then one of the poles moves to $s = 0$.

$$\Delta c(s) = \frac{R_3}{s} + \frac{R_4}{s + p} \tag{34}$$

and

$$C(t) = R_3 + R_4 e^{-pt} \tag{35}$$

As was the case with set-point changes, there is an offset in the controlled variable when the controller transfer function contains no term in $1/s$.

Figure 7 shows typical response curves for load disturbances.

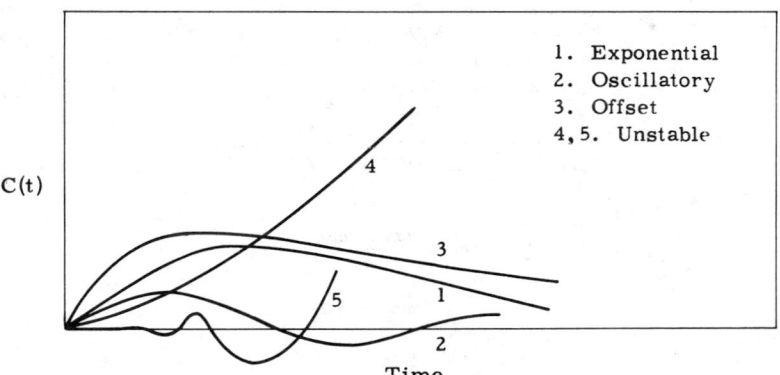

1. Exponential
2. Oscillatory
3. Offset
4, 5. Unstable

C(t)

Time

FIG. 7.   Response to load disturbance.

We have seen how the transfer function of a linear system is related to its transient response to two kinds of disturbances. A more complete analysis of the transient response of controlled systems will be deferred until later in this section. First, some of the other important properties of transfer functions will be described.

A stable system can be loosely defined as one in which all transients ultimately decay to zero. An unstable system is one in which this is not the case. Instability can be of two kinds, one in which the response increases without bound for a bounded disturbance, the other in which the response remains bounded but oscillatory. The latter kind is characteristic of some nonlinear systems.

We have seen that if a pole of the closed-loop transfer function is located in the right half-plane of the complex variable s, then the system is exponentially unstable. If there are poles on the imaginary axis, excepting the origin, then the system will oscillate periodically. The engineer who designs a control system must be sure that all poles of his system are in the left half-plane.

Descriptions of uses of transfer functions in control system design can be found in references (1), (3), (4), (9), (10), (12), (14), (18), (20), (22), (35), (42) and (43).

The Routh and Routh-Hurwitz Criteria for Stability

It is not always convenient to determine the location of all poles of a system since the algebra for locating roots of equations can become tedious if a number of combinations of control parameters is to be investigated. Therefore, stability can best be analyzed by one of the special methods developed for this purpose. One of the oldest is the Routh criterion. It was devised as a method for determining if a polynominal has roots with positive real parts, and if so, the number of such roots. It does not give the location of these roots, it just indicates their existence. The absence of such roots in the denominator of the transfer function is a necessary condition for stability in a linear system.

The method is applied in the following way.

Step 1.   Write the denominator of the transfer function in the form:

$$D = As^m(a_0s^n + a_1s^{n-1} + a_2s^{n-2} + \ldots + a_n) \tag{36}$$

where A is a constant.   Normally a_0 will be unity, but it has been left in here for generality. It should always be made positive by choice of A.   The a's may be numbers or literal quantities.

Step 2.   Examine the signs of the a's.   If any are zero or negative, the system will be unstable.   A necessary but not sufficient condition for stability is that all coefficients in the polynominal denominator be of the same sign.

Step 3.   Form the following array:

Index		Terms			
n	a_0	a_2	a_4	a_6	$\ldots$
$n-1$	a_1	a_3	a_5	a_7	
$n-2$	b_1	b_2	b_3	$\ldots$	
$n-3$	c_1	c_2	c_3	$\ldots$	
$n-4$	d_1	d_2			
—					
—					
—					

where

$$b_1 = \frac{a_1a_2 - a_0a_3}{a_1}$$

$$b_2 = \frac{a_1a_4 - a_0a_5}{a_1}$$

$$b_3 = \frac{a_1a_6 - a_0a_7}{a_1}$$

$$c_1 = \frac{b_1a_3 - a_1b_2}{b_1}$$

$$c_2 = \frac{b_1a_5 - a_1b_3}{b_1}$$

and so forth.   A number of special cases may arise in performing this part of the calculation.

Case 1. The first term in a row may be zero. In this event a small positive literal number, δ, should be temporarily substituted. δ will appear as a divisor in the following row, but will disappear in the formation of the row after that. Since any row may be multiplied by any positive constant, the row in which δ appears as a divisor may be multiplied by δ, and then δ replaces by zero. For example, the array

$$
\begin{array}{cccc}
3 & a_0 & a_2 & a_4 \\
2 & 0 & a_3 & a_5
\end{array}
$$

becomes

$$
\begin{array}{cccc}
3 & a_0 & a_4 a_2 & a_4 \\
2 & \delta & a_3 & a_5 \\
1 & \dfrac{\delta a_2 - a\delta a_3}{\delta} & \dfrac{\delta a_4 - a\delta a_5}{\delta} & \\
& \dfrac{\dfrac{\delta a_2 a_3 - a_0 a_3{}^2}{\delta} - \dfrac{\delta^2 a_4 - \delta a_0 a_5}{\delta}}{\dfrac{\delta a_2 - a_0 a_3}{\delta}} & &
\end{array}
$$

In the row of index 0, $1/\delta$ is a common factor of numerator and denominator. As δ approaches zero, the dominant term in this row becomes a_3. In the row of index 1 the terms become $-a_0 a_3$ and $-a_0 a_5$. Notice that if this should happen in the second row, as above, then the polynominal is missing its second term and the system is unstable through the condition of step 2.

Case 2. If all terms in a row are zero, a pair of roots of equal magnitude but opposite sign is indicated. Such a system is necessarily unstable since either one of the roots is real and positive or else the two are conjugate imaginaries. To continue the process in this event, perform the following operations, a procedure which is not proved here.

(a). Form a polynominal whose coefficients are the terms in the preceding row, whose highest power of s is the index of the row, and whose successive powers decrease by two.

(b). Differentiate this polynominal. The coefficients in the resulting polynominal are to be put in place of the row of zeros, and the calculation proceeds.

Step 4. After the complete array has been formed, examine the first column of terms. The number of changes of sign between successive terms in the columns is equal to the number of roots with positive real parts. This is the Routh criterion of instability.

Step 5. Replace s by $-s$ in the polynominal denominator of the transfer function, and repeat steps 1 through 3 above. This replaces all roots by their negatives. Examine the signs in the first columns of terms. The numbers of sign changes in this column is equal to the number of roots of the original denominator with negative real parts. If this number plus the number found in *Step* 4 is not equal to the order of the polynominal, then roots are present on the imaginary axis, and the system is unstable. In designing a control system the engineer can guarantee stability by choosing his control parameters in such a way that the terms in the column of *Step* 5 alternate regularly in sign.

A related procedure was developed by Hurwitz, and has become to be known as the Routh-Hurwitz criterion. The test is applied by constructing a series of determinants of increasing order, up to the order of the demonimator of the transfer function.

$$
\Delta_e =
\begin{vmatrix}
a_1 & a_3 & \cdots & a_{2l-1} \\
a_0 & a_2 & a_4 & a_{2l-3} \\
0 & a_1 & a_3 & - \\
0 & a_0 & a_2 & - \; - \\
0 & 0 & a_1 & - \; - \\
- & - & - & - \; a_l
\end{vmatrix}
\tag{37}
$$

The rule for forming the determinant of order l is that the term in the lower right-hand corner is a_l, the terms above it increase by one index per row, a_l, a_{l+1} a_{l+2}, $\ldots$ until a_n is reached, then they are all zeros. The terms moving to the left in any row decreases in index by 2 until a_0 or a_1 is reached, then they are all zeros. The index of terms in any column increase by one moving upward from the bottom. The successive Routh-Hurwitz determinants for a 5th degree polynominal would be

$$\Delta_1 \quad |a_1|$$

$$\Delta_2 = \begin{vmatrix} a_1 & a_3 \\ a_0 & a_2 \end{vmatrix}$$

$$\Delta_3 = \begin{vmatrix} a_1 & a_3 & a_5 \\ a_0 & a_2 & a_4 \\ 0 & a_1 & a_3 \end{vmatrix}$$

$$\Delta_4 = \begin{vmatrix} a_1 & a_3 & a_5 & 0 \\ a_0 & a_2 & a_4 & 0 \\ 0 & a_1 & a_3 & a_5 \\ 0 & a_0 & a_2 & a_4 \end{vmatrix}$$

$$\Delta_5 = \begin{vmatrix} a_1 & a_3 & a_5 & 0 & 0 \\ a_0 & a_2 & a_4 & 0 & 0 \\ 0 & a_1 & a_3 & a_5 & 0 \\ 0 & a_0 & a_2 & a_4 & 0 \\ 0 & 0 & a_1 & a_3 & a_5 \end{vmatrix} \qquad (38)$$

$$\Delta_6 \ldots = 0$$

The Routh-Hurwitz criterion for stability is that Δ_l should be positive for all $l \leq n$, if a_0 is chosen to be positive.

These two tests for system stability have the advantage of being relatively easy to apply to the transfer function. They avoid the troublesome factorization of polynomials; however, they do not indicate the performance of the system nor give a simple indication of how an unstable system may be stabilized by design changes. These important factors are provided by the powerful stability criterion to be described next.

The Nyquist Criterion for Stability

It has been shown that if the open-loop response of a controlled variable to a change in its set point is given by a transfer function $G(s)$, then the closed-loop response of the system is described by

$$\frac{c(s)}{r(s)} = \frac{G(s)}{1 + G(s)} \qquad (39)$$

In Eq. 14, for example, $G(s)$ was labeled as $G_1(s) H_1(s)$ because there was more than one component in the loop. This distinction will be of no significance in what is to follow.

The stability of the system of Eq. 39 is determined by the location of the roots of the function $1 + G(s)$, or equivalently by the values of s for which $G(s) = -1$. Poles of $G(s)$ do not contribute to instability when the loop is closed; however, they do have an effect on the Nyquist criterion for stability. The criterion will be explained first for the situation where $G(s)$ is the transfer function of a stable system.

It will be assumed that $G(s)$ does not contain branch points or essential singularities. Consider Fig. 8a in which the location of a number of poles of a hypothetical *closed-loop* transfer function have been marked. If there are poles to the right of the imaginary axis, the corresponding system is unstable, as we have seen. This situation is equivalent to there being poles within the contour of Fig. 8b. A simple pole at the origin does not cause instability, hence the closed contour need not enclose it. Poles on the imaginary axis of s do contribute to instability, particularly if they are of order higher than one.

The basis of the Nyquist criterion is the conformal mapping of this closed contour onto the plane of $G(s)$. If poles are within the contour, they remain within it after the mapping. The particular usefulness of the Nyquist criterion comes from the fact that the contour on G can be rather easily generated by frequently response methods, as will be shown later. For the present discussion, it will be assumed that a contour can be drawn.

There are four distinctly different parts of this contour, as marked on Fig. 8b. The mapping of each of these parts onto the G plane will be considered separately.

Path 1 is a large semicircle, beyond all poles of the transfer function. It will be recalled from the theory of the Laplace transform that a condition for existence of a unique inversion integral is that the transformed function approach zero as the variable of transformation becomes indefinitely large. Physically this means that if there is a unique

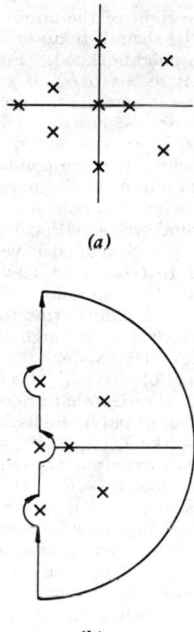

(a)

(b)

FIG. 8. s-Plane poles: (a) poles of $\dfrac{G(s)}{1 + G(s)}$, (b) integration contour.

transient open-loop response to a disturbance, then the open-loop transfer function will go to zero as s becomes large.   Therefore, the large semicircle in s maps onto the origin of $G(s)$.   This is indicated on Fig. 9.

Path 2 on Fig. 8 is a small semicircle passing around a singularity on the imaginary s axis.   If the diameter is made small, $G(s)$ gets very close to -1, and in the limit approaches it.   These points thus map onto the -1 point of G, as indicated on Fig. 9.

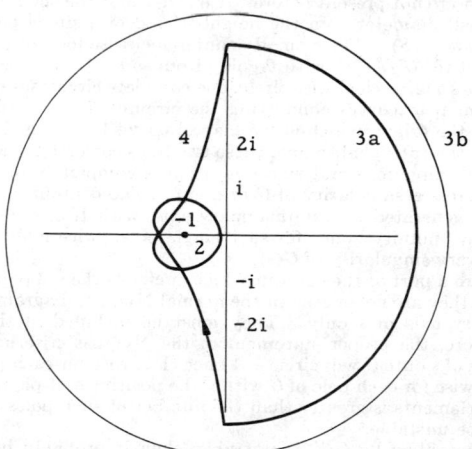

FIG. 9. $G(s)$ plane.

Path 3 is a small semicircle to the right of the origin of s. If the origin is an ordinary point of G, then this semicircle can be shrunk to an ordinary point on G. If the origin is a simple pole, then $G \approx G^1/s$ in its neighborhood. For $s = i\epsilon$, where ϵ is a positive infinitesimal, G increases without limit as $-i(G^1/\epsilon)$, if $s = -i\epsilon$, g behaves as $i(G^1/\epsilon)$. If $s = \epsilon e^{i\alpha}$, G becomes $(G^1/\epsilon)e^{-i\alpha}$. Thus the small semicircle in s maps into a large semicircle in G, preserving the sense of the arrows as showing in Fig. 9, curve 3a.

If the origin is a double pole of G, then $s = \epsilon e^{i\alpha}$ maps into $(G^1/\epsilon)e^{-2i\alpha}$, a large circle, starting on the negative real axis, where it corresponds to $s = -i\alpha$ encircling the plane in the counterclockwise sense, and ending on the negative real axis. This is shown in Fig. 9, curve 3b. In general, if the origin is a pole of G of order n, the map on the G plane of contour 3 will be a set of n large semicircles, with counterclockwise movement in s corresponding to clockwise movement in G, with real positive s corresponding to real positive G, and with $s = \pm i\epsilon$, corresponding to $G/\epsilon = G^1/(\pm i)^n\epsilon$. These latter points are the significant ones because they indicate the end points of path 4.

Path 4 is the imaginary axis of s. It transforms to a curve on the G plane which is symmetrical about the real axis, which goes through the origin, and which may or may not pass through $G = -1$, depending on the existence or absence of poles for pure imaginary s. It terminates on the "end points" of curve 3. Its direction of approach to the origin may be inferred from the form of G. The G's that are of interest in the design of feedback control systems are rational fractions of polynomials in s. For $s = \pm iA$, where A is a large positive number, $G(s)$ behaves like $1/(\pm i)^m A^m$, where m is the excess of poles over zeros of G. If this excess is unity, for example, the G-path will approach the origin from the negative imaginary direction as s goes to $+\infty$. If it is two, the approach will be from the negative real direction. A typical path 4 is shown in Fig. 9.

The Nyquist criterion for stability may now be stated. All poles of $G/(1 + G)$ map to $G = -1$. If any poles are within the closed contour in the s plane, the system is unstable. This contour maps into a closed contour in the G plane. If the -1 point is "within" the closed contour, therefore, the system is unstable. These Nyquist plots are in general rather complicated since a single point in the G plane can correspond to several points in the s plane. The contour can loop around the -1 point several times, in which event the number of times is equal to the number of singularities with positive real parts. A few typical plots are shown in Fig. 10.

The foregoing analysis was based on the assumption that $G(s)$ itself had no poles in the right half-plane. This restriction comes from the necessary requirement that a mapping function must be analytic within the region to be conformally transformed. An unstable $G(s)$, therefore, does not qualify as a proper mapping function. Nevertheless, the Nyquist criterion can be stated for systems which are unstable when the loop is open. Consider Fig. 11 in which the X's indicate poles of a function $G(s)$. The s-plane has been cut from each singularity to "infinity." The region within the indicated closed contour can certainly be mapped onto $G(s)$ because it now contains no singularities. The map will differ from the simple map considered previously in that it contains pieces of the contour, labeled 5, 6, and 7, which were not present before. Consider first the small loop 5, and allow it to shrink to very small diameter. In the neighborhood of a single pole, at S, $G(s)$ can be represented by $G^1/(s - S)$. For a small counterclockwise loop of radius ϵ, this representation is equivalent to $(G^1/\epsilon)e^{-i\theta}$, with θ going from some θ_0 to $2\pi + \theta_0$. Therefore, $G(s)$ on the loop describes a large clockwise circle, one complete circuit for each order of the pole.

Lines 5 and 7 map to curves connecting the circumference of this large circle to the small loop around the origin, which is the map of curve 1. In the limit as the contour is expanded to cover the entire half-plane, these two lines coalesce toward a single line which is traversed in both directions and whose location is completely immaterial so long as it does not directly cross a singularity of G or a cut. The contour of Fig. 11 thus maps to whatever curve is generated by the imaginary axis, with 1, 5, 6, and 7, mapping into a clockwise circuit at "infinity" once for each single pole, twice more for each double pole, and so forth for every singularity of $G(s)$.

These circuits are a part of the contour which encircles the -1 point for each singularity of $G/(1 + G)$, but they are never seen in the normal Nyquist diagram which is constructed from the imaginary axis of s only. They must be included in the stability criterion, however. Therefore, the proper statement of the Nyquist criterion is that the map of the imaginary axis of s onto G will circle -1 once clockwise for each pole of $G/(1 + G)$ and once counterclockwise for each pole of G within the positive half-plane. If the net number of clockwise encirclements is greater than the number of such poles of G, then the closed-loop control will be unstable.

If the system described by $G(s)$ is unstable, then it probably has feedback elements included within itself. Therefore, it can be analyzed in exactly the same way to determine

Diagram
(Positive Imaginary s Only)

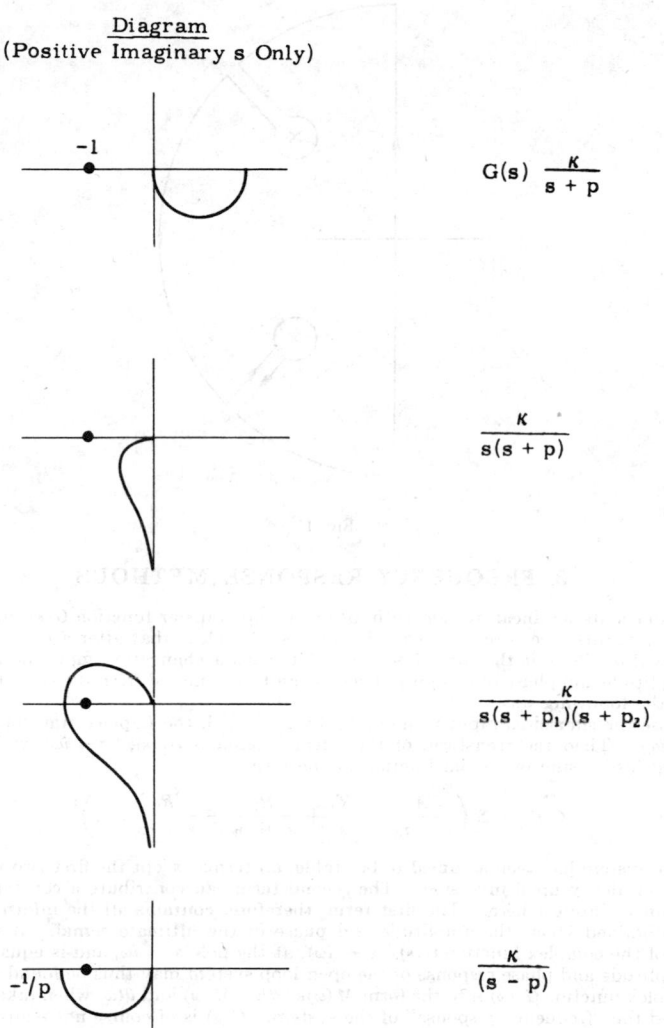

$$G(s) \quad \frac{K}{s + p}$$

$$\frac{K}{s(s + p)}$$

$$\frac{K}{s(s + p_1)(s + p_2)}$$

$$\frac{K}{(s - p)}$$

FIG. 10.   Typical Nyquist diagrams.

the number of poles which cause its instability.   Full application of the Nyquist criterion for stability thus can involve several nested simple plot constructions.

If poles of G are located on the imaginary axis of s, they should be excluded from the closed contour by making a small semicircle to the right, as was done for the origin.   The behavior of G in the neighborhood of such points must be considered in the analysis of the diagram.   This situation can cause considerable complexity.   Fortunately, it is uncommon in real systems.

In the foregoing, the full power of the Nyquist method has not been completely developed.   We will return to it after a description of frequency response methods, which it complements very nicely.   Details of its use can be found in references (9), (10), (12), (20), (21), (22), (42) and (43).

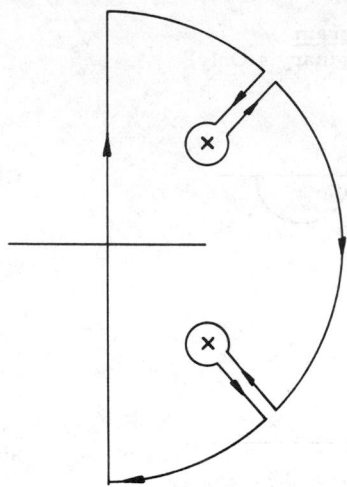

FIG. 11.

3. FREQUENCY RESPONSE METHODS

Consider a stable linear system with an open-loop transfer function $G(s)$ into which a sinusoidal disturbance is introduced. It is physically clear that after a time the system output will oscillate at the same frequency, but with a change in amplitude and phase. The amplitude and phase of this output signal can be calculated from the transfer function $G(s)$ in the following way.

Suppose the sinusoidal input is described by $\epsilon e^{i\omega t}$, with the Laplace transform equal to $\epsilon/(s - i\omega)$. Then the transform of the output signal is $\epsilon G(s)/(s - i\omega)$, which can be expanded into a sum of partial fraction of the form

$$C(s) = \Sigma \left(\frac{A}{s - i\omega} + \frac{R_0}{s} + \frac{R_1}{s + p_1} + \frac{R_2}{s + p_2} \cdots \right) \qquad (40)$$

Since the system has been assumed to be stable, all terms except the first two will lead to exponential decay upon inversion. The second term can contribute a constant at most and is not of interest here. The first term, therefore, contains all the information that can be obtained about the amplitude and phase of the ultimate signal. A is just the residue of the complex function $G(s)/(s - i\omega)$, at the pole $s = i\omega$, and is equal to $G(i\omega)$. The amplitude and phase response of the open-loop system may thus be found by putting the complex function $G(i\omega)$ into the form $M(\omega)e^{i\theta(\omega)}$. $M(\omega)$ and $\theta(\omega)$ when taken together are called the "frequency response" of the system. $G(s)$ is of course not really restricted to open loops. It can also be the transfer function of a system under feedback control.

Two points of interest emerge from the above brief description. $G(i\omega)$ is just the ordinary Fourier transform of the response to a pulse at $t = 0$. It is exactly the function which is plotted on a Nyquist diagram. This latter point establishes a useful connection between frequency response and the Nyquist diagram, with no need for consideration of the Laplace integral or the transfer function. The entire theory of linear control systems could be developed by starting with frequency analysis and proceeding through a description of Nyquist diagrams and ending with transfer function and root locus considerations. This is the way control theory is taught in many schools. Either that approach to the subject or the one chosen here are equally acceptable; the important part is the ultimate unification of all different approaches.

The function $M(\omega)$ is usually called the *gain*, or *amplitude ratio*. $\theta(\omega)$ is called the *phase angle*. If θ is negative, it is usually referred to as a *lag*; if positive, as a *lead*. Plots of $\log_{10} M$ and θ against $\log_{10} \omega$ taken together form the important *Bode diagram* of the system.

A simple test for closed-loop stability can be made from the frequency response of an open loop. Figure 12 shows what might be a number of typical curves of excitation and

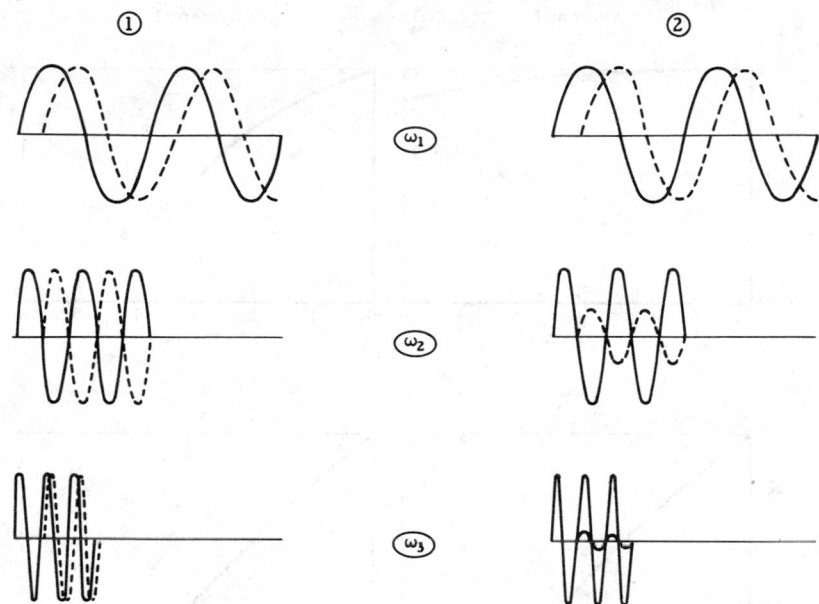

FIG. 12. Frequency response.

response of two uncontrolled systems. In both systems, as the frequency is increased, the phase lag also increases and passes 180° or π radians. In system 1 the gain at the point where the lag is exactly 180° is greater than 1, in system 2 it is less than 1. If the loop should be closed, sending the output back to the input with a sign reversal, system 1 will clearly amplify itself. System 2, on the other hand, will damp itself out. This is the basis of the simple test for stability that can be derived from a Bode plot. It may be clearly stated in the following way. "If the gain at the frequency corresponding to 180° phase shift is greater than 1 with the control loop opened and if the phase shift is 180° or more for all higher frequencies, closing the loop will result in instability."

Bode plots are often made with coordinate units that are a bit unusual. Abscissas are usually labeled in decades, corresponding to units in the common logarithm of the frequency. Occasionally the *octave* is used for a unit of the abscissa. This means the length corresponding to a factor of 2. The units of frequency normally used are cycles or radians per unit of time. Occasionally one sees degrees per unit time. A commonly used scale of gain is again the decade. Since this may be too coarse for many purposes, it is often subdivided into *decibels* (db), one decibel being defined as 1/20th of a decade. 0 db corresponds to a gain of unity, 6 db to a gain of approximately 2, 20 db to a gain of 10, and so forth. The common unit of phase angle is degrees.

There are at least two procedures for constructing the Bode diagram for a real system. The first is to make direct measurements on the system itself; for example, by disturbing the system with a sinusoid and plotting it and a signal corresponding to the output on a two-pen recorder. Amplitude and phase (mod 360°) can be read directly from the recording. If a sufficient number of frequencies are used, the plot can be made easily.

The second method of constructing a Bode plot starts from a block diagram. It will be recalled that transfer functions of a number of blocks in series can be multiplied to give the overall transfer function. Any transfer function can be put into phase-amplitude form, from which it can be seen that the amplitude of a product is the product of the amplitudes, and the phase of a product is the sum of the phases:

$$R_1 e^{i\theta_1} \cdot R_2 e^{i\theta_2} \cdot R_3 e^{i\theta_3} = R_1 R_2 R_3 e^{i(\theta_1 + \theta_2 + \theta_3)} \tag{41}$$

The Bode method exploits this relationship in a neat way, since the ordinate of amplitude is logarithmic and that of phase is linear. This makes it possible to construct easily by graphical methods the plot for a system from the individual plots for components.

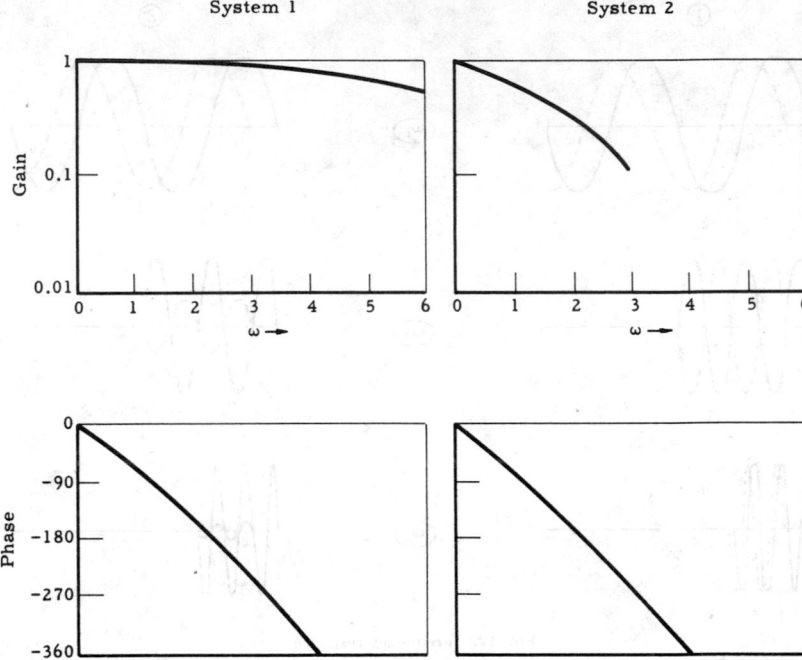

Fig. 13.   Bode diagrams.

Bode plots for a few simple components are shown in Figs. 14 through 18.

The Bode plot of a product of transfer functions can be easily constructed graphically from the plots of the factor functions.   However, it is also necessary to construct the plots for sums of transfer functions, at summing points of feedback control systems for example.   This operation is considerably more complicated than the construction of a product, and cannot be handled conveniently graphically unless the amplitudes are greatly different for each frequency of interest.

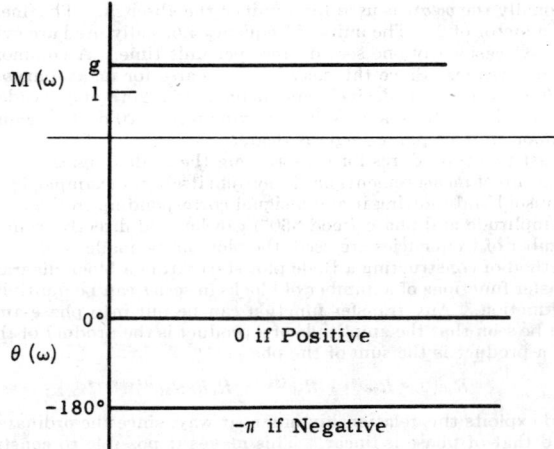

Fig. 14.   Bode plot of pure gain.

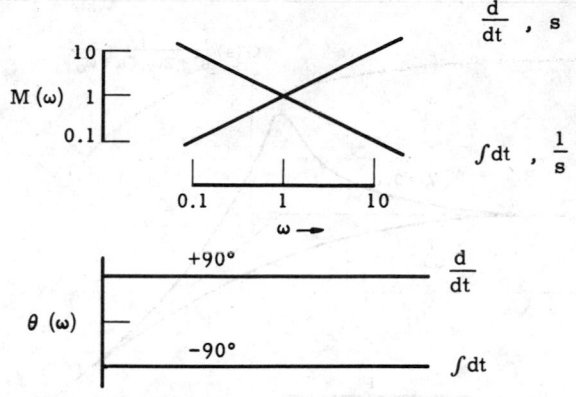

FIG. 15. Bode plots of integration and differentiation.

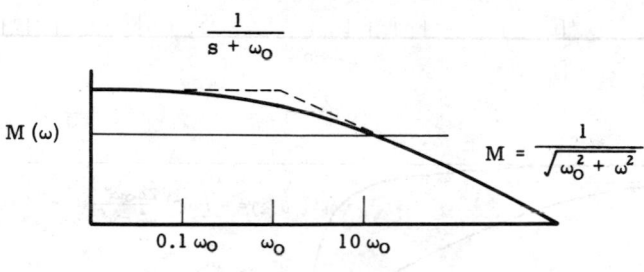

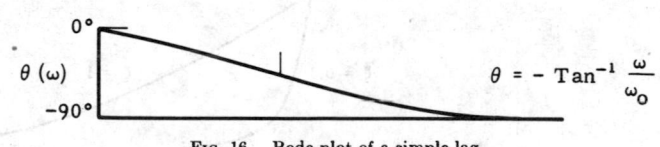

FIG. 16. Bode plot of a simple lag.

The sum of two complex quantities may be calculated from the ordinary algebra of complex numbers

$$M_\Sigma e^{i\theta_2} = M_1 e^{i\theta_1} + M_2 e^{i\theta_2}$$
$$= M_1(\cos\theta_1 + i\sin\theta_1) + M_2(\cos\theta_2 + i\sin\theta_2) \qquad (42)$$
$$= (M_1\cos\theta_1 + M_2\cos\theta_2) + i(M_1\sin\theta_1 + M_2\sin\theta_2)$$

$$M_\Sigma = \sqrt{M_1{}^2 + M_2{}^2 + 2M_1 M_2 \cos(\theta_1 - \theta_2)} \qquad (43)$$

$$\theta_\Sigma = \tan^{-1}\frac{M_1\sin\theta_1 + M_2\sin\theta_2}{M_1\cos\theta_1 + M_2\cos\theta_2} \qquad (44)$$

If $M_1 = M_2$,

$$M_\Sigma = 2M_1\cos\tfrac{1}{2}(\theta_1 - \theta_2) \qquad (45)$$

$$\theta_\Sigma = \tfrac{1}{2}(\theta_1 + \theta_2) \qquad (46)$$

If $\theta_1 = \theta_2$,

$$M_\Sigma = M_1 + M_2 \qquad (47)$$

$$\theta_\Sigma = \theta_1 \qquad (48)$$

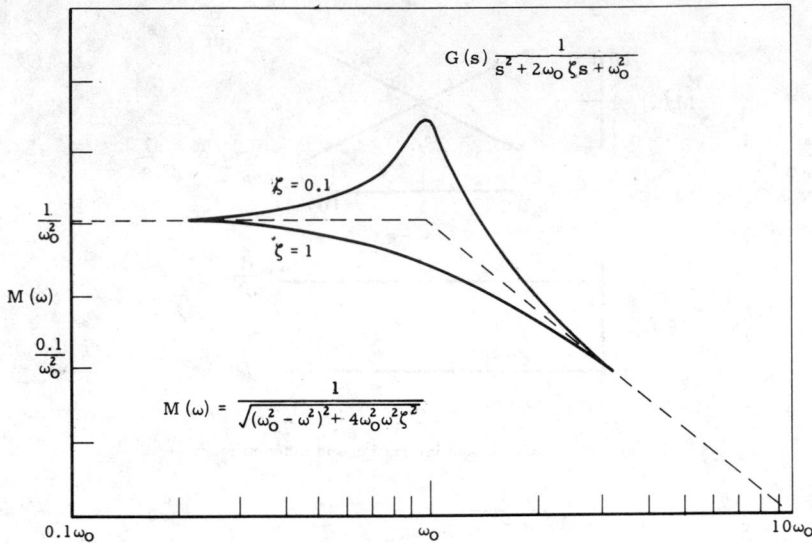

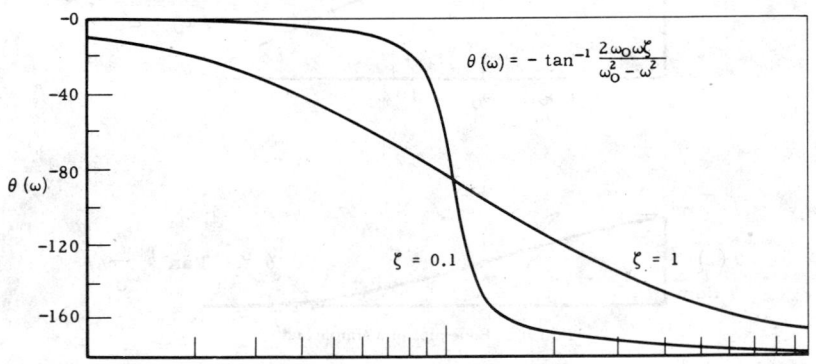

FIG. 17.   Bode plot of a quadratic lag.

If $\theta_1 = \theta_2 + \pi$,

$$M_\Sigma = M_1 - M_2 \tag{49}$$

$$\theta_\Sigma = \theta_1 \tag{50}$$

A simple way to construct the plot of a sum is to draw first regions where the amplitudes are greatly different, where the amplitude and phase of the sum may be set equal to the amplitude and phase of the larger component.   The next step is to put in the three above points.   Often both curves may be sketched in with sufficient accuracy from these points alone.   Additional points may be calculated from the formulas 43 and 44 if necessary.

There are many ways in which the control engineer can use a Bode plot.   Determination of stability has already been mentioned: if the open-loop gain is greater than 1 at the point where the plase is $-180°$, then the closed loop will be unstable.   A number of terms

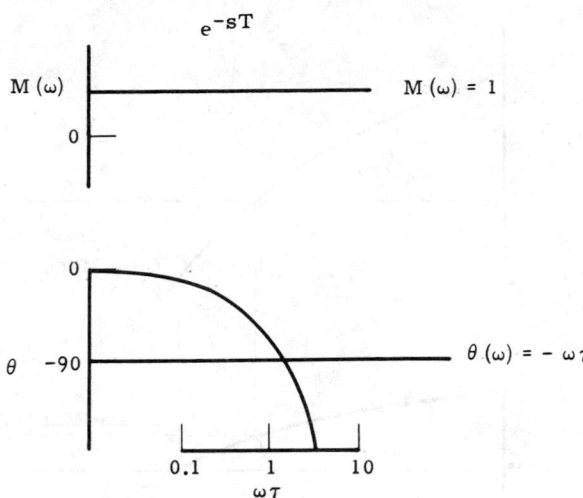

FIG. 18. Bode plot of a time delay.

commonly used to characterize qualitatively the closeness of a system to instability include the following:

(a). *Gain cross-over frequency.* The frequency at which the open-loop gain is unity.

(b). *Phase cross-over frequency.* The frequency at which the open-loop phase lag is 180°.

(c). *Gain margin.* The gain at the phase cross-over frequency.

(d). *Phase margin.* The difference between 180° and the phase lag at the gain cross-over frequency.

Sensitivity, or near instability of a system, in indicated if the phase cross-over frequency is only slightly greater than the gain cross-over frequency, if the gain margin is slightly less than one, or if the phase margin is small. Often a large number of possible control systems can be scanned quickly by making approximate Bode plots and examining them for phase and gain margins. The more promising ones can be investigated in greater detail.

A Bode diagram often resembles a series of straight-line segments with rounded corners. This commonly occurs when zeros and poles of the transfer function are real and well separated; if there are complex poles their imaginary parts should be relatively small for this to happen. Under these conditions the Bode plot of a system will have a corner that is concave downward for each zero, and a corner that is concave upward for each pole. The change of slope at the corner as frequency increase is $+1$ for a zero, -1 for a pole, -2 for a conjugate pair of complex poles, $+n$ for an n-fold zero, $-n$ for an n-fold pole. The frequency at the corner is the reciprocal of the time constant at the associated zero or pole.

The slope of the amplitude ratio plot at the gain cross-over point is a qualitative increase of relative stability. A system with low slope in this region will be less sensitive than one with a sharply dropping amplitude ratio.

Because the Bode and Nyquist diagrams are intimately related to each other, the ability to construct the former in a simply way can be transferred to the latter, which is generally very useful for design purposes. The significant part of a Nyquist diagram, it will be recalled, is the map of the imaginary axis ($s = i\omega$) onto the complex plane of the open-loop transfer function, $G(s)$. $G(i\omega)$ can be expressed as $M(\omega)e^{i\theta(\omega)}$, where M and θ are the Bode functions, and also the polar coordinates of a point which moves along the Nyquist locus as frequency varies. The Nyquist diagram is just a transformation of the Bode plot to polar coordinates, with frequency as a parameter. The use of this fact allows the engineer to avoid the complicated algebra which is necessary when the construction is made by conformal mapping. Gain and plase margin appear in a simple way on the Nyquist diagram, as shown in Fig. 19. The phase at the point where the line crosses the unit circle is the phase margin, the amplitude at the point where the line crosses the negative real axis is the gain margin.

References (10), (11), (12), (14), (22) and (42) contain descriptions of frequency response methods.

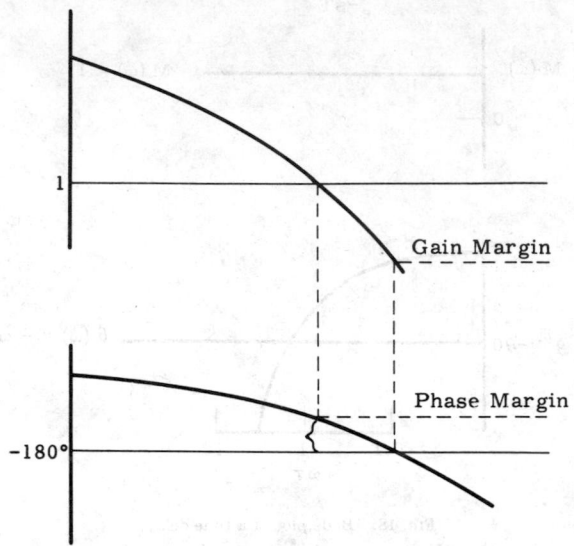

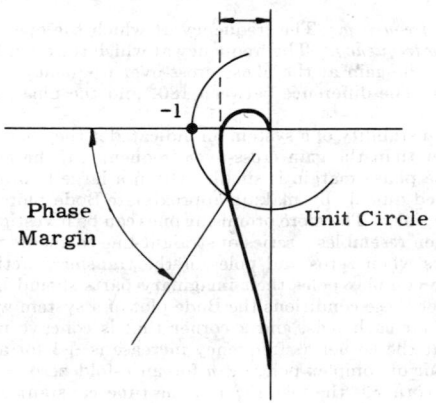

Fig. 19. Phase and gain margins.

Closed-Loop Response

It will always be necessary to investigate the response of the control system with the feedback loop closed. It has been shown how the response function can be constructed with the loop open, and indeed the same procedure can be used to determine the closed-loop characteristics, although it is somewhat more difficult to apply in this case. A preferable procedure is to generate the closed-loop functions from those of the open-loop system.

If the open-loop transfer function is $M_0 e^{i\theta}$, then the closed-loop function is

$$M e^{i\alpha} = \frac{M_0 e^{i\theta}}{1 + M_0 e^{i\theta}} \tag{51}$$

This may be written

$$\frac{1}{M} e^{-i\alpha} = \frac{1}{M_0} e^{-i\theta} + 1$$

from which   (52)

$$\frac{1}{M} = \frac{1}{M_0^2} + 1 + \frac{2}{M_0} \cos \theta$$

and

$$\alpha = \tan^{-1} \frac{(1/M_0) \sin \theta}{(1/M_0) \cos \theta + 1}$$ (53)

These formulas may be used to find the closed-loop response curves from the Bode or Nyquist plots.

A simple graphical method equivalent to these formulas may be derived by super-imposing curves of constant M and α directly onto the Nyquist plot. The shape of the curves can be derived easily from Eqs. 52 and 53 by transforming to Cartesian coordinates, using $x^2 + y^2 = M_0^2$ and $x = M_0 \cos \theta$, $y = M_0 \sin \theta$.

This transformation gives

$$\left(x - \frac{M^2}{1 - M^2}\right)^2 + y^2 = \left(\frac{M}{1 - M^2}\right)^2$$ (54)

and

$$(x + \tfrac{1}{2})^2 + \left(y - \frac{1}{2 \tan \alpha}\right)^2 = \left(\frac{1}{2 \sin \alpha}\right)^2$$ (55)

Curves of constant M are circles with center at $x = M^2/(1 - M^2)$, $y = 0$, and radius $M/(1 - M^2)$. Curves with constant α are circles with center at $x = \tfrac{1}{2}$, $y = 1/(2 \tan \alpha)$, and radius $1/(2 \sin \alpha)$. The singular case where $M = 1$ transforms to $x = -1/2$. These two families of circles are shown in Figs. 20 and 21. The closed-loop response corresponding to the superimposed Nyquist diagram is shown in Fig. 22.

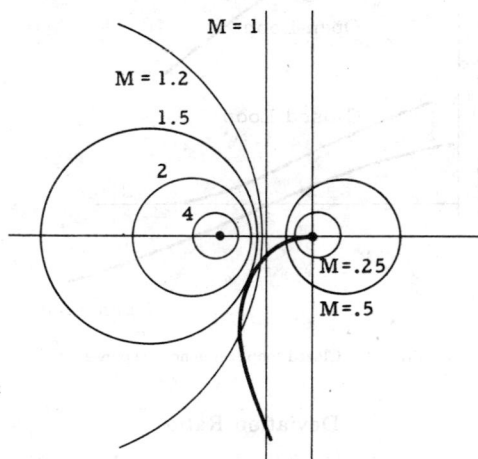

FIG. 20.   Contours of constant M.

The Nyquist diagram is a polar plot of the Bode diagram with frequency as a parameter. A Cartesian plot of the Bode diagram is called a "Nichols chart" if lines of constant M and α are superimposed upon it. A simplified Nichols chart is shown in Fig. 23. It is, of course, a periodic function of phase angle, and the chart repeats indefinitely by mirror reflections right and left.

A particularly useful feature of the Nichols chart is that the maximum closed-loop gain can be determined directly by finding the M-curve to which the G-curve is just tangent.

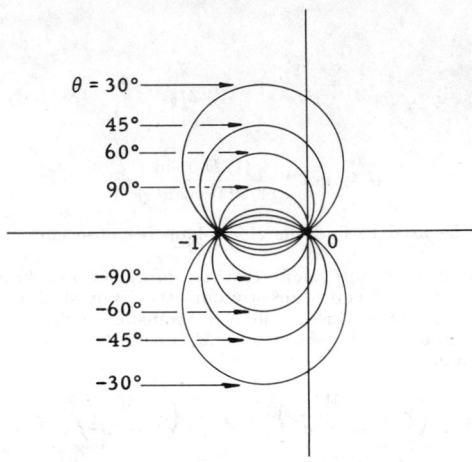

FIG. 21.   Contours of constant phase.

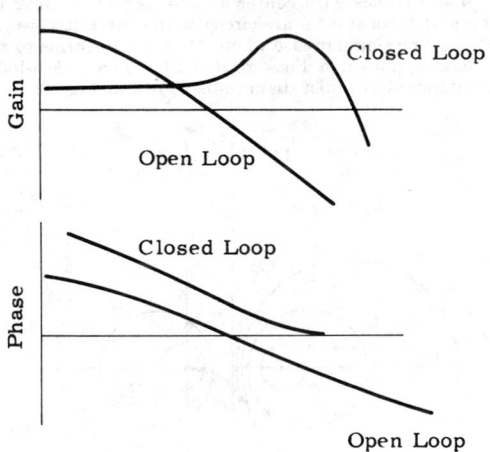

FIG. 22.   Closed-loop frequency response.

Deviation Ratio

A dynamic function of considerable utility in design is the *deviation ratio*, which is the frequency response of a closed-loop system to disturbances in loading, as contrasted with the Bode plot, which is the response of the open loop to set-point changes.   The relation of deviation ratio to transfer function depends on where in the loop disturbances enter and is particularly simple when the entry is beyond the dynamic components as shown in Fig. 24.

The amplitude response is the most useful part of the deviation ratio.   It is simply related geometrically to the Nyquist plot, as shown in Fig. 25.   The length of the vector from -1 to a point on the curve is the reciprocal of the amplitude of the deviation ratio at the frequency of the point on the curve.   Figure 26 shows a typical deviation ratio.

Control systems should be designed so that the deviation ratio is as small as possible in the regions where disturbances occur.   The deviation ratio usually has at least one

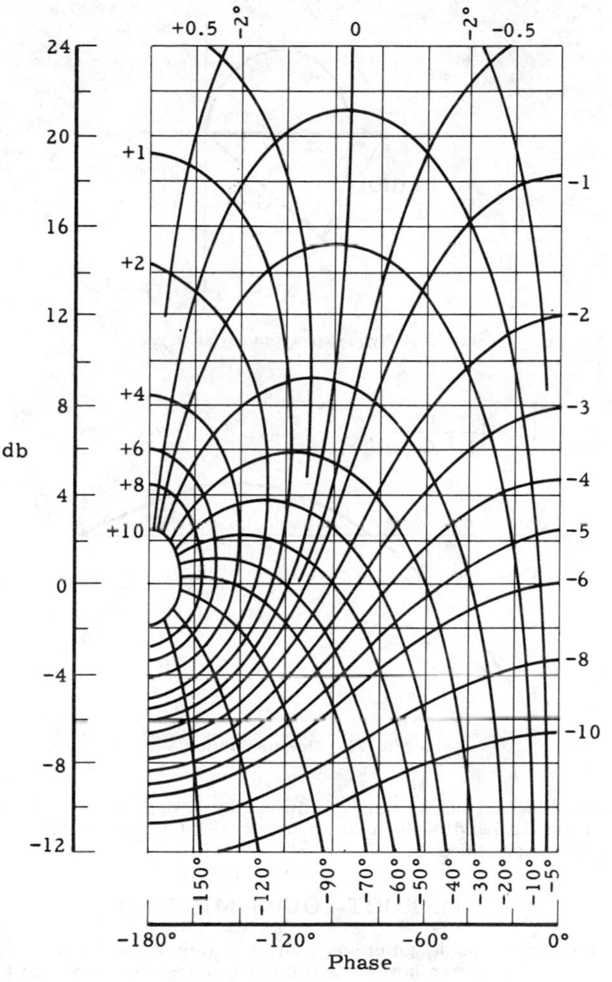

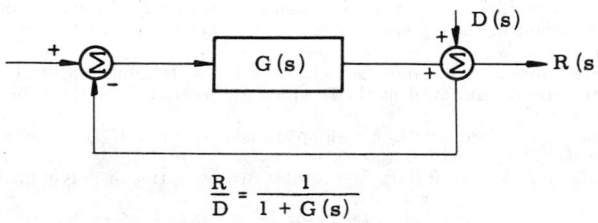

FIG. 23. Nichols chart.

FIG. 24. Deviation ratio.

$$\frac{R}{D} = \frac{1}{1 + G(s)}$$

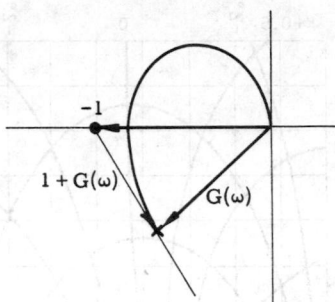

FIG. 25.   Nyquist plot and deviation ratio.

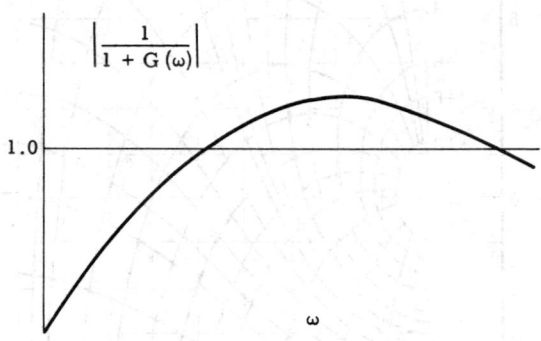

FIG. 26.   Deviation ratio.

peak whose position and height can be adjusted by design of the system.   This peak should be outside the band of disturbance frequencies for good design, otherwise some disturbances can be amplified.

4. ROOT-LOCUS METHOD

A central problem in the design of any control system is the choice of the parameters of that system.   We have seen how the performance of a system is related to the location of the poles and zeros of its transfer function.   If the system has adjustable parameters, and if the engineer knows the effect that parameter values have on the location of poles and zeros, he can then choose those values in such a way as to give the best possible system behavior.   The *root-locus* method, developed by Evans, is specifically a procedure for graphically studying the movement of closed-loop poles as a single parameter (a multiplying constant) of the open-loop transfer function is varied.   In this article, the main emphasis will be on stability considerations.   Specific procedures will be discussed later.

The root-locus method starts with an open-loop transfer function of the form $kG(s)$ and ends with a plot in the s-plane of the positions of all roots of $1 + kG(s)$ as k varies from 0 to ∞.

Algebraic description of the root-locus method is not feasible—rather its great utility depends on the simple graphical methods which are available for sketching the essential character of the loci.

The root locus of a system with open-loop transfer function $G(s)$ is the set of all points for which $1 + kG(s) = 0$, with $0 \leqslant k \leqslant \infty$, or $G(s) = -(1/k) = (1/k)e^{\pi i}$.

Since transfer fractions of interest are almost always ratios of polynomials,

$$G(s) = \frac{(s + Z_1)\ (s + Z_2)\ \dots}{(s + P_1)\ (s + P_2)\ \dots} \tag{56}$$

If each factor in Eq. 46 is written in polar form, we have

$$G(s) = \frac{A_1 e^{i\phi_1} \cdot A_2 e^{i\phi_2} \cdots}{B_1 e^{i\theta_1} \cdot B_2 e^{i\theta_2} \cdots} \tag{57}$$

$$= \frac{A_1 A_2 \cdots}{B_1 B_2 \cdots} e^{i(\phi_1 + \phi_2 \cdots, -\theta_1 - \theta_2 \cdots)} \tag{58}$$

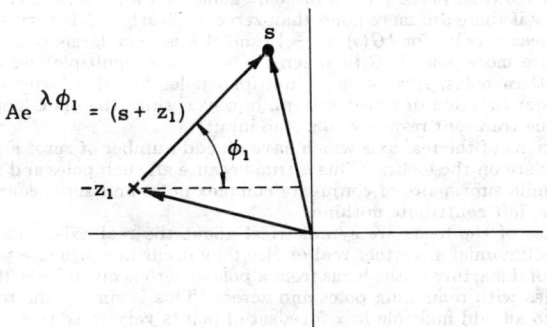

FIG. 27.   Construction of root locus.

Each ϕ is the angle that a straight line, drawn from a zero of $G(s)$ to a test point, s, makes with a line parallel to the real axis of s.   For the test point to lie on the root locus, the algebraic sum of all such angles must add to an odd multiple of π, the angles of vectors from zeros of G being counted as positive, the angles of vectors from poles of G counted as negative.   The geometric construction of a root locus consists of repeatedly testing points for satisfaction of this angle summation condition.   When a point is found, the appropriate k may be obtained by measuring the vectors from roots and poles to give $A_1, A_2 \ldots B_1, B_2 \ldots$ from which

$$k = \frac{A_1 \cdot A_2 \cdots}{B_1 \cdot B_2 \cdots} \tag{59}$$

A typical root locus is shown in Fig. 28.

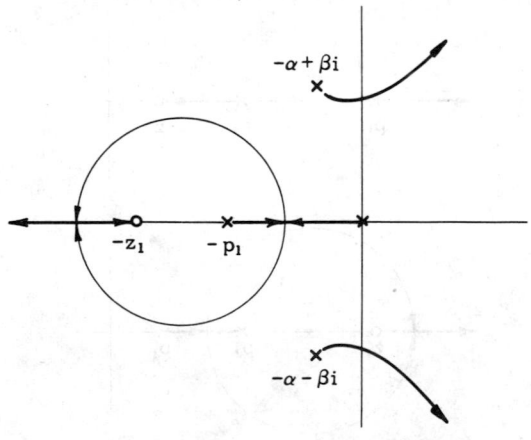

FIG. 28.   Root locus of $G(s) = \dfrac{s + z_1}{s(s + p_1)\,[s^2 + 2\alpha s + (\alpha^2 + \beta^2)]}$.

Clearly, if this trial procedure were the only method for finding the locus, the root-locus method for control-system design could not be particularly useful because the search of the entire complex plane would be too tedious.   Fourtunately, there are systematic simplifications that can be made.   The most useful ones are the following:

1. The number of continuous branches is equal to the number of closed-loop poles, or zeros of $1 + kG(s)$.   Here a branch is the path traced by one pole as k increases from zero to infinity.

2. Each branch "starts," for $k = 0$, either at an open-loop pole or else at infinity if there are more zeros than poles.   Each branch "ends," for $k = \infty$, either at an open-loop zero or at infinity if there are more poles than zeros.   Clearly if k is very small, $G(s)$ must be very large (near a pole) for $kG(s) = -1$, and if k is very large, $G(s)$ must be near a zero.   If there are more poles in G than zeros, $s = \infty$ is a (multiple) zero of G.   If there are more zeros than poles, $s = \infty$ is a (multiple) pole.   In this latter case $G(s)$ would not be the transfer function of a real system, however, since the open-loop system would not have a unique transient response to a step input.

3. Those portions of the real axis which have an odd number of zeros and poles to their right on the axis are on the locus.   This is true because all such poles and zeros contribute π each to the angle sum, pairs of conjugate complex poles and zeros contribute nothing, and points to the left contribute nothing.

4. All branches of the locus are symmetrical about the real axis.   This is so because roots of a real polynomial are either real or else they occur in conjugate pairs.

5. The angle of departure of the locus from a pole or zero is given by $\pm(2n + 1)\pi$ minus the sum of angles with remaining poles and zeros.   This is simply the requirement that the angles add to an odd multiple fo π for a set of points very close to a pole or zero.

6. Asymptotes may be drawn easily. As s increases, $G(s)$ behaves like s^{-n}, where n is the excess of poles over zeros.   For ks^{-n} to equal -1 for large positive k, s^{-n} must be small and negative real; in other words, s must move in the direction of the nth root of -1.   These asymptotes do not, in general, go through the origin, but rather through the centroid of the group of zeros and poles of G.

7. The region between two real poles of G is on the locus if there are an odd number of zeros and poles to the right, by 3 above.   Therefore, as k increases, the two branches must move toward each other from the poles, meet at some finite k, and break away to head toward infinity or the appropriate zeros.   The location of this "break-point" can often be estimated by noting that it is a double root of $1 + kG(s)$, and therefore it must be a root of its derivative, dG/ds.

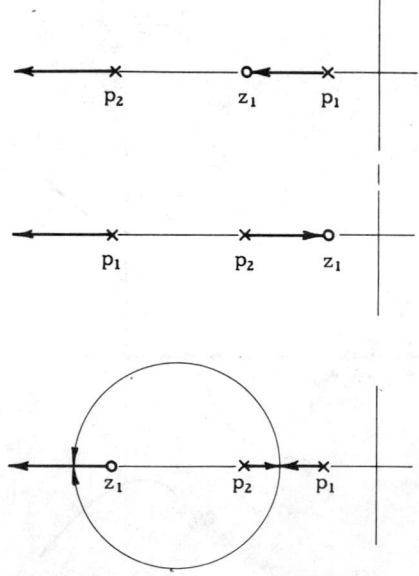

FIG. 29.   Root locus for $G(s) = \dfrac{s + z_1}{(s + p_1)(s + p_2)}$.

8. The intersections of the individual branches with the imaginary axis may be determined by a variation of the Routh test for positive zeros. If the rows in the Routh array which have two elements only are written, then either $\alpha\omega^2 + \beta = 0$ or $\gamma\omega^2 + \delta = 0$ gives the value of ω at which the locus crosses the imaginary axis. Setting $\alpha\delta - \gamma\beta = 0$ enables the value of k to be determined. Some common root-locus plots are shown in Figs. 29 through 31.

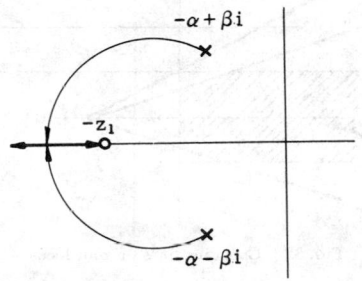

FIG. 30. Root locus for $G(s) = \dfrac{s + z_1}{s^2 + 2\alpha s + (\alpha^2 + \beta^2)}$.

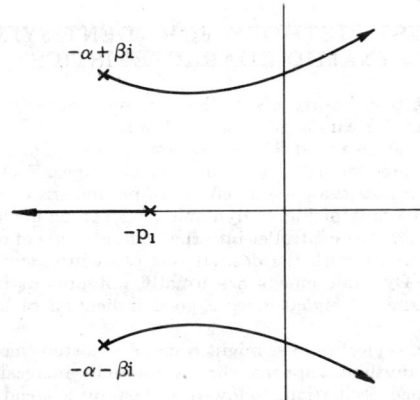

FIG. 31. Root locus for $G(s) = \dfrac{1}{(s + p_1)\,[s^2 + 2\alpha s + (\alpha^2 + \beta^2)]}$.

The usefulness of the root locus in design is based primarily on the concept in Fig. 32 which shows lines on which various characteristics of the transient response are constant. Lines parallel to the real axis are lines of constant frequency. A pole within a conjugate pair of such lines at $\pm i\omega_d$ contributes a term in the inversion integral equal to $e^{-\alpha t}(\cos \omega t + i \sin \omega t)$, where $\omega < \omega_d$.

Lines parallel to the imaginary axis are lines of constant decay time. A pole to the left of $-\alpha_d$ results in the term $e^{-\alpha t}(\cos \omega t + i \sin \omega t)$, with $\alpha > \alpha_d$.

Radial lines through the origin are lines of constant quadratic damping factor, ζ. If a term like $1/(s^2 + 2\zeta\omega_0 s + \omega_0^2)$ is in the transfer function, ζ represents the degree of damping of a sinusoid. If $\zeta = 1$, the system is critically damped.

An upper frequency limit, a minimum damping factor, and a minimum decay time thus determine a region of the s-plane in which all poles of the closed-loop transfer function must lie. The designer can usually select k from the root locus in such a way that this occurs.

By moving the poles and zeros which are characteristic of the controls and making corresponding root-locus plots, the designer can see the effects that control parameters have on the behavior of the total system.

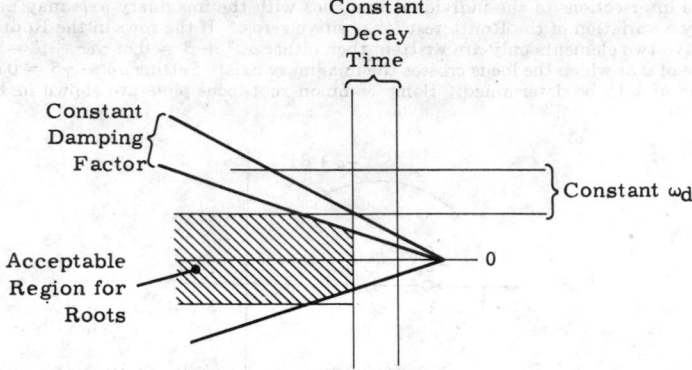

FIG. 32.   Constant lines on root locus.

Full treatment of the root locus method will be found in references (21), (22), (42), and (43).

5. TEST METHODS FOR IDENTIFYING DYNAMIC CHARACTERISTICS

The information that is necessary about the dynamic characteristics of a system to design an effective controller cannot always be obtained by analysis or theory.   Sometimes control systems must be added to existing equipment.   Then it is necessary to test the equipment itself.   Three methods for doing this with linear systems will be described here, with a few hints on how the possible effects of nonlinearities may be assessed (22).

In a system such as the tank of Fig. 5, dynamic tests can be made by first opening the control loop, say by putting the controller into the manual mode of operation, varying the manual setting in accordance with the desired test procedure, and observing the signal from the transmitter.   Dynamic effects are usually not present in manually operated controllers, so the transmitted signal gives a good indication of how the process itself responds.

Figure 33 shows how a typical process might respond to a step change.   There might be a *dead time*, T_d, before anything happens, then a gradually increasing rise, with perhaps an overshoot and damped oscillation, followed at last by a steady output with super-

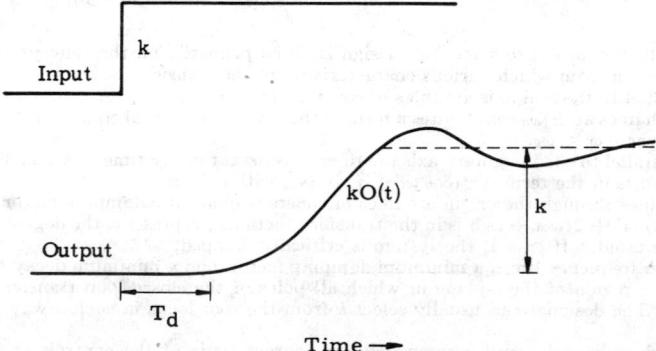

FIG. 33.   Step response.

imposed noise. k/h is the gain. The frequency response of the system may be calculated from the step response, in principle at least, by using the definition of transfer function:

$$G(s) = \frac{k \int_0^\infty 0(t)e^{-st}dt}{h/s} \tag{60}$$

$$\frac{G(s)}{(s)} = \frac{k}{h}\int_0^\infty 0(t)e^{-st}dt \tag{61}$$

or

$$\frac{g(i\omega)}{i\omega} = \frac{k}{h}\int_0^\infty 0(t)e^{-i\omega t}dt \tag{62}$$

The integrals may be evaluated numerically for several values of ω to give a rough idea of the shape of the Bode plot; however, noise tends to obscure the behavior at high frequency.

If the dynamic response can be represented satisfactorily by two real-time lags in series, then a special graphical procedure attributed to Oldenbourg and Sartorius may be conveniently used to analyze the response to a step change.

Figures 34 and 35 show the procedure. The ratio T_C/T_A, determined by a straight line through the inflection point of the curve, is used as the intercept of the straight line on each axis of Fig. 35. The ratios T_1/T_A and T_2/T_A may be read directly from the intersections with the curve. One such line is drawn in on Fig. 35 for $T_C/T_A = 0.8$. The corresponding values of T_1/T_A and T_2/T_A are 0.663 and 0.137, respectively.

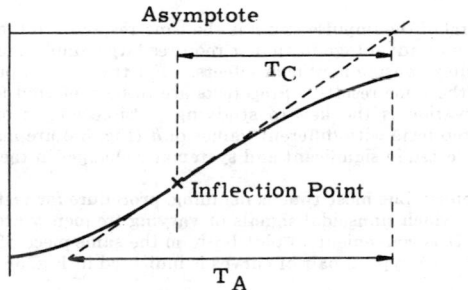

FIG. 34.   Second-order step response.

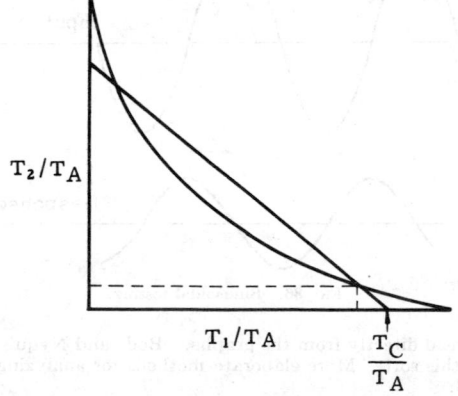

FIG. 35.   Oldenbourg-Sartorius method.

The coordinates of the Oldenbourg-Sartorius curve of Fig. 35 satisfy

$$-\frac{T_1}{T_A} \log \frac{T_1}{T_A} = -\frac{T_2}{T_A} \log \frac{T_2}{T_A} \tag{63}$$

The curve is symmetrical about the main diagonal, and passes through the point $T_1/T_A = T_2/T_A = 1/e$, where its slope is -1. At the point $(0,1)$, the curve is vertical; at $(1,0)$ it is horizontal. The short table of values below, plus the above considerations, should enable the user to reconstruct a curve of sufficient accuracy.

$\dfrac{T_1}{T_A}$	$\dfrac{T_2}{T_A}$
0.00	1.0000
0.02	0.9183
0.05	0.8364
0.10	0.7292
0.20	0.5666
0.30	0.4402
0.3679	0.3679

If T_C/T_A is less than $2e^{-1} \cong 0.7358$, the system cannot be represented by two simple time lags.

Impulse testing may also be used in the same way. An impulse is an idealized derivative of a step change, hence

$$G(i\omega) = \frac{\int_0^\infty h(t)e^{-i\omega t}dt}{I} \tag{64}$$

where I is the integral of the impulse and h is the time response to the impulse. Impulse testing suffers from a disadvantage in that it requires large amplitude and short duration of disturbance, possibly causing nonlinear effects. For this reason, pulse and step testing do not always give the same results. Step tests are to be preferred for this reason.

This latter observation is the key to studying possible effects of nonlinearities. If several successive step tests with different values of h (Fig. 33) are made, significant nonlinearities will be reflected in significant and systematic changes in the calculated dynamic parameters.

The most satisfactory, but most time consuming, procedure for testing is the frequency response method, in which sinusoidal signals of varying frequency are introduced and responses observed. It is convenient to plot both on the same piece of paper, using a two-pen recording device. A typical pair of curves is indicated in Fig. 36. Phase and ampli-

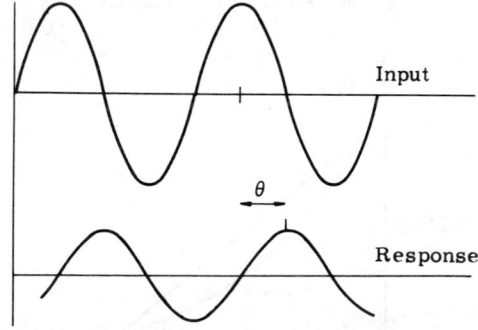

Fig. 36.   Sinusoidal testing.

tude may often be read directly from the graphs. Bode and Nyquist plots may be easily made from data of this sort. More elaborate methods for analyzing frequency test data have been described.

The use of natural noise in a system for determining its dynamic characteristics will be described under "Adaptive Control."

6. THEORY OF SAMPLED-DATA SYSTEMS

In the preceding description of the construction and use of system transfer functions, the need for continuity of signals was not involved.   As a matter of fact, certain types of discontinuities can be included in the analysis of linear control systems.   The introduction of one particular type of discontinuity forms the subject of the next article of this section.

Two fairly recent developments in control equipment involve the generation of step-wise approximation to a continuous signal.   These are: (1) Certain types of analytical instruments require time for the performance of the analysis.   A gas chromatograph is an example.   A small sample is taken at time t, then results are generated over an interval δt, a new sample is taken at time $t + \Delta t$ with $\Delta > \delta$, and the process repeats.   Results are known only at the instants of time $t, t + \Delta t, t + 2\Delta t$, etc.   (2) Serial computing equipment must sample all inputs intermittently, usually periodically.   Some signals can be sampled more frequently than others, but none can be sampled continuously.

It is clear that sampling must degrade the information in some way, since parts of it are not being used.   The nature of this degradation is the subject of sampled-data theory. The theory can be used by design engineers to relate sample intervals to design parameters and to quality requirements of controlled systems.

The linear theory of sampled-data systems has much in common with the theory of communications, from which much of the following is derived.

A sampling device consists of two major parts; an input device for generating a sequence of signals representing the values of the input signal at selected time intervals, and a filter which acts in some way on the sequence to generate usable input signals for the component which follows in the system.

The transfer function of the first part, or sampler, may be derived rather simply.

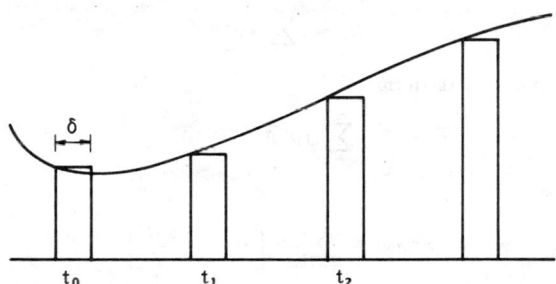

FIG. 37.   Analysis of data sampling.

For a sequence of signals such as is shown in Fig. 37, the Laplace transform is given by

$$\mathcal{L}(f^*) \equiv \int_0^{\infty} e^{-st} \, [\text{sampled } f(t) \, [dt$$

$$\equiv \int_{t_0}^{t_0+\delta} e^{-st} f(t) \, dt + \int_{t_0}^{t_1+\delta} e^{-st} f(t) \, dt \ldots \tag{65}$$

where the asterisk is used to indicate that the function is sampled.

$$\mathcal{L}(f^*) = \frac{f(t_0 + k_0)}{s} \, [e^{-st_0} - e^{-s(t_0+\delta)}]$$

$$+ \frac{f(t_1 + k_1)}{s} \, [e^{-st_1} - e^{-s(t_1+\delta)}] \ldots \tag{66}$$

where $0 \leq$ (each k) $\leq \delta$.

For sufficiently small δ, this expression is approximately equal to

$$\mathcal{L}(f^*) \approx e^{-st_0} f(t_0) \left[\frac{1 - e^{-s\delta}}{s} \right] + e^{-st_1} f(t_1) \left[\frac{1 - e^{-s\delta}}{s} \right] \tag{67}$$

$$= \left[\frac{1 - e^{-s\delta}}{s} \right] [e^{-st_0} f(t_0) + e^{-st_1} f(t_1) \ldots] \tag{68}$$

If $t_1 - t_0 = t_2 - t_1 \ldots = T$, where T may be called the sampling interval, we have finally

$$\mathcal{L}(f^*) \approx (e^{-st_0}) \left[\frac{1 - e^{-s\delta}}{s} \right] [f(t_0) + e^{-sT} f(t_1) + e^{-2sT} f(t_2) \ldots] \tag{69}$$

Without loss in generality, t_0 may be taken as zero, whence

$$\mathcal{L}(f^*) = \left[\frac{1 - e^{-s\delta}}{s} \right] \left[\sum_{n=0}^{\infty} f(nt) e^{-nsT} \right] \tag{70}$$

The first factor is the transform of a single unit pulse of width δ, the second factor is the transform of a function which is sampled at intervals of T by impulses.

The two factors may individually be put into frequency response form by substituting $i\omega$ for s. The first factor is

$$\frac{1 - e^{-i\omega\delta}}{i\omega}$$

which is equal to

$$\left(\frac{2}{\omega} \sin \frac{\omega\delta}{2} \right) e^{-i\omega\delta/2} \tag{71}$$

This is the "spectrum" of the pulse of width δ at $t = 0$.

The spectrum of the second factor can be related to the spectrum of $f(t)$ by noting that the factor is the transform of $f(t) \cdot p(t)$, where $p(t)$ is a train of impulses at interval T, which can be expanded to

$$p(t) = \sum_{k=-\infty}^{\infty} e^{2\pi i k t/T} \tag{72}$$

The sample function is of the form

$$\sum_{k=-\infty}^{\infty} f(t) e^{2\pi i k t/T} \tag{73}$$

which transforms to

$$F^*(i\omega) \equiv \sum_{k=-\infty}^{\infty} F \left[i(\omega) - \frac{2\pi k}{T} \right] \tag{74}$$

where F is the spectrum of f.

The spectrum of the sampled function is thus the original spectrum, but repeated at intervals of $2\pi/T$. This is shown if Fig. 38, where the amplitude only is shown and

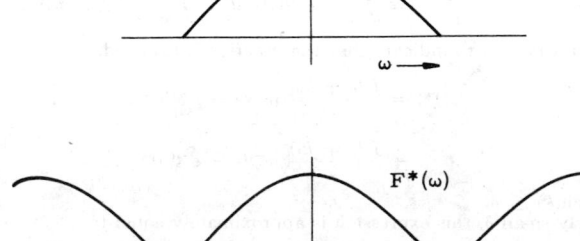

Fig. 38.   Spectrum of sampled function.

$f(2\pi/T)$ is assumed to be small compared to $F(0)$. This is the particular effect that sampling has on the function to be sampled. It introduces extraneous frequencies into the signal, which must ultimately be filtered out.

The input signal into a sampling device is very often of limited bandwidth; that is, it contains no frequencies greater than some upper limit ω_L. An important theorem states that such signals are completely described by sampling at intervals of π/ω_L; that is, twice per cycle of the highest frequency in the system Conversely, if a signal is sampled at intervals of T, it contains no useful information about frequencies greater than π/T, or $1/2$ the sampling frequency

Filter Characteristics

A perfect filter which can reconstruct a signal from a periodic sampling is obviously impossible, since discarded information between samples cannot be anticipated by a purely mechanical device.

The simplest practical filter is just a square-wave generator, which holds the most recent sampled value until a new one appears. A filter of this type is commonly included in the sampling equipment. It is sometimes called a *zero-order data hold*.

Figure 39 shows how a signal might look after passing through a sampler and zero-order data hold.

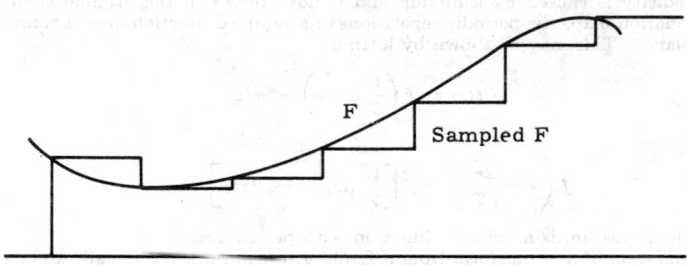

$$\text{FIG. 39. Zero-order hold.}$$

The transfer function of the zero-order hold is

$$G_{0k}(s) = \frac{1}{s}(1 - e^{-Ts}) \tag{75}$$

Its frequency response is of course the same as that for the single pulse derived before:

$$G_{0k}(i\omega) = \frac{2}{\omega}\sin\frac{\omega T}{2}\,e^{-i\omega/2} \tag{76}$$

The first-order filter is equivalent to linear extrapolation in time. Its output is given by

$$f_1(t) = f^*(nT) + \frac{t - nT}{T}[f^*(nT) - f^*(n-1)T] \tag{77}$$

where f^* is the output of the sampler and $nT \le t \le (n+1)T$. Its transfer function is

$$G_k(s) = T(1 + sT)\left(\frac{1 - e^{-sT}}{sT}\right)^2 \tag{78}$$

Its frequency response is

$$G_{1k}(i\omega) = T\sqrt{1 + \omega^2 T^2}\left(\frac{\sin(\omega T/2)}{\omega T/2} \times e^{-i\omega T} + i\tan^{-1}\omega T\right) \tag{79}$$

A filter can be very much more complicated. For example, it can include a digital computer performing an extensive calculation on the pulse train coming from the sampler. Needless to say, the transfer function of such components can be derived in principle if everything is linear. However, as we shall see, the inclusion of nonlinearities can be a distinct advantage when sophisticated computing equipment is used, so in practice the only kinds of filters to which a linear analysis is applied are the simple data-holding devices described above.

z-Transform Analyses

Transfer functions of the sort described above can be combined in the usual way with transfer functions of other components to make the previously described methods of system analysis usable when sampling occurs. Nyquist, Bode, and root-locus plots can be made as before. Such plots can become complicated under certain circumstances, and it has been found useful to proceed by a slightly different approach, commonly called *z-transform* analyses.

The complex variable z is defined as e^{Ts}, where T is the sampling period and s is the ordinary Laplace variable. The transfer function of the *pulse-sampled* variable of Eq. 70 can thus be written

$$F^*(z) = \sum_{n=0}^{\infty} f(nT)z^{-n} \tag{80}$$

Equation 74 shows that $F^*(s)$ is periodic in the imaginary direction, with period $2\pi i/T$. This periodicity is caused by sampling and is not present in the original signal. The z-transformation maps the periodic repetitions of a sampled function onto a common part of the z plane. This may be shown by letting

$$f(s) = f\left(\frac{1}{T} \log z\right) \equiv \phi(z)$$

Then

$$f\left(s + \frac{2\pi i}{T}\right) = f\left[\frac{1}{T}(\log z + 2\pi i)\right] = \phi(z) \tag{81}$$

because the logarithm is a periodic function with period $2\pi i$.

The generation of a z-transform from a Laplace transform is not usually to be done by replacing s with $(1/T) \log z$. Although this would appear to be mathematically correct, it does not correspond to the physical situation of interest which is to find the z-transform of the *sampled* variable which corresponds to the Laplace transform of the *continuous* variable. In this sense, the z-transform corresponding to a Laplace transform whose expansion

$$F(s) = \sum_i \frac{R_i}{s + p_i}$$

is

$$F^*(z) = \sum_i \frac{R_i}{1 + z^{-1} e^{p_i T}} \tag{82}$$

This can be proved by using the convolution theorem for transforming a function of the form $f(t) \cdot g(t)$, where $g(t)$ is a pulse train. The proof will not be repeated here, since it can be found elsewhere.

The inversion integral for z-transforms is simply

$$f(nT) = \frac{i}{2\pi i} \int_\Gamma z^{n-1} F^*(z) \, dz \tag{83}$$

where the closed contour Γ encloses all singularities of F^*. In systems of interest, where all poles are within the unit circle, it can be taken as just that circle.

A short list of z-transforms is given in Table 4. More detailed lists can be found in the text by Ragazzini and Franklin.

The z-transformation has no effect on the Nyquist and Bode plots, since these are properties of a system and its transfer function, and not the form of its independent variable. Of course, *sampling* can effect these curves because it alters the transfer function.

The root-locus plot, on the other hand, is a construction in the complex plane of the transformation, and thus depends explicity on the form of the variable. Graphical

Table 4. z-Transforms

Time Function	Laplace Transform	z-Transforms of Pulsed Functions
$\delta - (t - nT)$	e^{-snT}	z^{-n}
Step at $(t = nT)$	$\dfrac{1}{s} e^{-snT}$	$\dfrac{z}{z - 1} z^{-n}$
nT	$\dfrac{1}{s^2}$	$\dfrac{Tz^{-1}}{(1 - z^{-1})^2}$
e^{-anT}	$\dfrac{1}{s + a}$	$\dfrac{1}{1 - e^{-aT} z^{-1}}$

construction of a root-locus in the z-plane is identical to that in the s-plane. Poles and zeros of the open-loop transfer function, $G^*(z)$, are located, and angles summed as before.

Stability of a system requires that roots of $1 + kG(z)$ be within the unit circle. If a branch of the locus goes outside, then the gain of the controller must be limited as before.

Lines of constant maximum frequency, which are parallel to the real axis in the s-plane, become radial lines in the z-plane. If frequencies in the closed loop are to be limited, then the roots must lie within the appropriate segment of the unit circle. Lines of constant transient response rate which were parallel to the imaginary axis in s, become concentric circles around the origin in z. If the response rate must be greater than some minimum, then the roots must lie within the appropriate circle in z. Lines of constant damping factor, which were radial in s, become logarithmic spirals in z. The roots must lie between two conjugate spirals for the controller to be damped to some desired degree. These regions are indicated in Fig. 40.

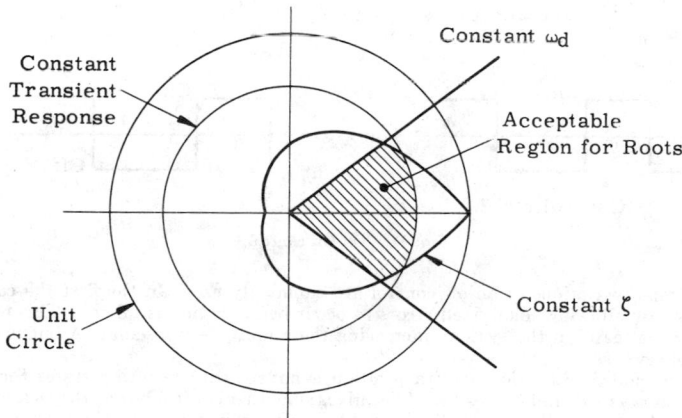

FIG. 40. z-Plane characteristics.

Sampled data systems and z-transforms are treated in detail in references (8), (26), (38), and (41).

7. MODES OF CONTROL

As described to this point, the theory of automatic control has been concerned with the controlled system as a whole, i.e., the process or machine plus the control elements. The process part of a controlled system is not always easily adaptable to the requirements for good control; accordingly, adjustment of the control elements is often the only way a system can be made to operate most effectively and smoothly.

In Fig. 5 a typical set of control elements was shown in block diagram, including a measuring device, a transmitting device for the measured variable signal, a controller which acts upon the signal, a transmitter for the control signal, and an actuator, which might be a valve or a motor. Each of these components has an effect on the dynamic response of the controlled system. Those which are perfectly linear may have only dynamic lags or time delays of the sort already considered. Others may introduce extraneous noise or nonlinearities in addition to those inherent in the process. A discussion of these latter phenomena will follow the theory of a simple linear controller, the third in the above list of components.

On-Off Control

Strictly speaking, the simplest type of control is not linear and cannot be analyzed by the theoretical methods already considered. However, because of its simplicity and extensive use, and because its mathematical description is straightforward, it is considered here first. An on-off controller acts by applying full corrective action whenever an error is detected. It has two states only, one of which corrects positive deviation of the system, the other, negative.

A system controlled in this way is always in a transient state. With no load or set-point disturbances, oscillation between limits will always occur. The controller compensates for disturbances by varying the relative time in the two control states. Figure 41 shows a typical time response of a system under on-off control. A common example is the furnace control in a home heating system.

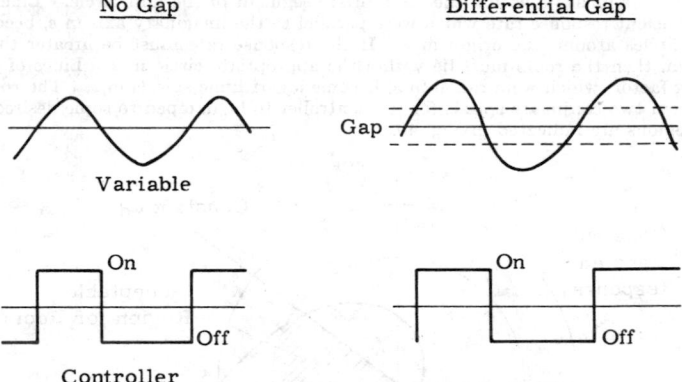

FIG. 41. On-Off control.

Two simple variations of on-off control are commonly used. In the first the controller is modified by allowing small fixed errors to occur before action is taken. This results in a "differential gap" in the system, increasing the amplitude of oscillation but decreasing the frequency.

In the second modification a third position is added, with resulting states that might be called *on-plus*, *off*, and *on-minus*. This gives smoother control but at the cost of a more complicated system. The controlled variable is always in a transient condition, as with the two-position control.

Floating On-Off Control

A *floating* controller is a simple variation of the two-position controller in which the final control element is moved continuously or intermittently at an average rate which depends on the sign of the deviation. It is, therefore, nothing more than a two or three-position *on-off* system to which an additional dynamic element, a pure integrator, has been added. Oscillations are always present, as before. Figure 42 shows the action of a typical floating control.

Proportional Control

The simplest linear controller which can be analyzed using frequency response or transform methods is *proportional* or *throttling* control, in which the output of the controller is a linear algebraic function of its input. A change in the measured variable instantly

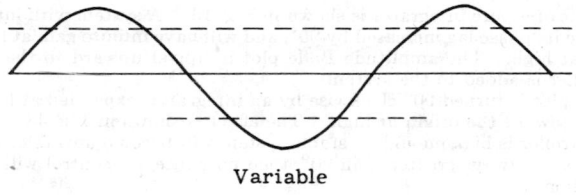

Variable

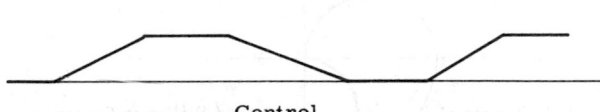

Control

FIG. 42. Three-position floating control with gap.

produces a proportionate change in the control signal. The ratio of control signal change to measured variable change is called the *gain* of the controller. One hundred times the reciprocal of the gain is often called the *propotional band* of the controller. The terminology comes from the concept that a change of some fraction of the total range of the input variable, measured in percentage units, is required to move the control signal over one hundred percent of its range. Thus a gain of 10 is equivalent to "10% proportional band," a gain of 0.1 is "1000% proportional band," etc. Low gain is equivalent to a wide proportional band, high gain to a narrow proportional band.

In analyzing the effects of this and the control modes to follow, the system diagrammed in Fig. 4 will be used. For a proportional control, $H_1(s) = k$. The open-loop transfer function to load change is $G_1(s)$, that of the closed loop is $G_1/(1 + kG_1)$. The transform of the transient response to a step change in load is

$$\frac{1}{s} \frac{G_1(s)}{1 + kG_1(s)}$$

In the limit of large time, by Eq. 9, the time function goes to

$$\frac{G_1(0)}{1 + kG_1(0)}$$

Since this ratio is unlikely to be zero, proportional control is almost always characterized by an error, or *proportional offset*. For large k, the offset is inversely proportional to k.

The effect of the gain of a proportional controller on the Bode, Nyquist, and root-locus diagrams is minimal, since there is no phase change associated with it. Changing k moves the amplitude plot up and down, radially expands the Nyquist diagram, and has no effect on the root-locus. If the open-loop phase lag exceeds 180° at any frequency, increasing the gain will result in instability. Although systems with ultimate phase lag less than 180° can be imagined in theory, they are rare in practice because there are almost always short time delays or nonlinear effects in equipment that is usually assumed to be perfectly linear. The controller itself might have a small time lag. Therefore, a system in which the gain can be increased indefinitely without instability is unusual.

Integral Control

The effect of proportional offset can be eliminated by using a control system in which the measured error signal is continuously integrated with respect to time, with $C = kI \int \epsilon \, dt$. This mode of control is often called *reset action* because it resets the control system in such a way as to eliminate error or offset.

The transfer function of an integrator is kI/s. Therefore, the transfer function of an integrally controlled system is

$$G(s)/s \left[1 + \frac{kI}{s} G(s) \right]$$

By the final value theorem, the ultimate error is zero after a step-change.

The Bode plot of a pure integrator is shown in Fig. 16. A system with integral control will always have its phase lag increased by 90°, and will have infinite gain at low frequency and zero gain at high. The amplitude Bode plot is tipped upward to the left when an integral controller is added to the system.

The Nyquist plot is turned 90° clockwise by an integrator, expanded at low frequency, and contracted toward the origin at high. The effect is shown in Fig. 43. Clearly if the gain of the controller is high enough, a stable system will become unstable. If the phase lag of the process is always greater than 90°, then pure integral control will always cause unstable operation.

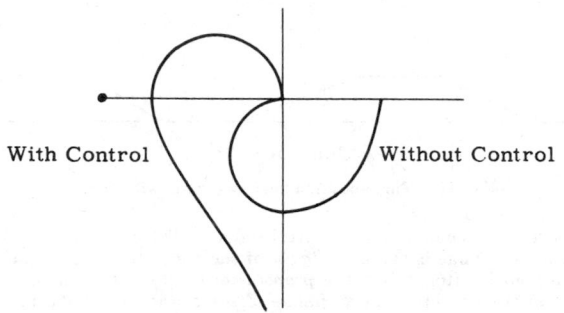

With Control   Without Control

FIG. 43.   Effect of integrating control.

Integral control adds a pole at the origin in the root-locus plot.

Commercially available controllers with integral action are usually not perfect integrators in the sense that for a very small input signal the output might behave like a lag with a very long time constant, that is, as though its transfer function were $1/(s + p)$ with a very small real p, instead of like $1/s$. The gain at very low frequency is then $1/p$, commonly called the "reset gain" of the controller. Too low a value for this figure is not desirable because it can lead to offset.

Proportional Plus-Reset Control

Proportional control can be fast-acting because it does not attenuate moderately high frequencies. However, it does give offset. Integral control, on the other hand, has no offset but is sluggish because of its attenuation of high frequencies. A mode of control which can be used to give the best features of both is equivalent to the sum, that is, with a transfer function of form $k[1 + (I/s)]$, or a time-behavior of $C = k(\epsilon + \int \epsilon \, dt)$. k is called the gain as before; I is called the reset rate, whose common unit is "repeats per unit time." If the transfer function is written $(k/s)(s + I)$, it can be seen that this control action adds a pole at the origin of the root locus, and adds a zero at $s = -I$.

A typical time response of a proportional + integral (PI) controller is shown in Fig. 44, which also indicates how the pure-action controllers would behave under the same stimulus.

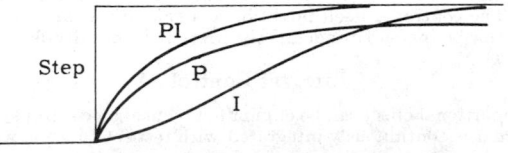

PI

Step   P

I

FIG. 44.   Set-point responses.

A complete Bode plot of a PI controller, with the imperfections that are always present, is shown in Fig. 45. At low frequency the reset gain limits the response of the system. The region of slope -1 is where integral action predominates, that with zero slope is where proportional action is most important. At very high frequency the gain drops because of the imperfections in the proportional part. The phase lag in this region may

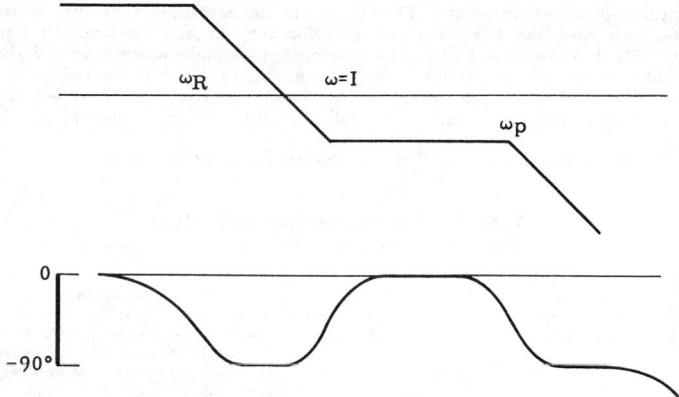

Fig. 45. Bode diagram of PI controller.

drop well below the 90° characteristic of a pure first-order lag. Proper design of the control equipment requires that the critical frequencies ω_R and ω_P be far apart. Proper adjustment of the controller in a loop requires that the parameters k and I be properly "tuned" according to the dynamics of the process.

PI control is probably the most commonly used mode of control. Its principal disadvantage is that extended open loop operation can lead to comparatively large overshoot in the integral part of the action through the effect known as "reset windup." Large uncontrolled errors cause rapid integration and a large integral term in the response. For steady operation this integral term must just balance the proportional offset; however, the effect of extended error may cause the integral term to be much larger than its steady-state limit, and it cannot decrease after the loop is closed until the error changes sign. Some manufacturers have designed special controllers to minimize this effect.

Derivative or Rate Action

A third linear control mode is occasionally used in which the output of the controller is proportional to the rate of change of the error signal. Thus if an error begins to develop, a controller with such "rate action" can anticipate and begin to take action before the error becomes large. The transient response is $c = kD(d\epsilon/dt)$. The transfer function is kDs. If the action were perfect, the gain would increase steadily with frequency and the phase lead would always be 90°. This sort of control action would be highly undesirable, since high frequency noise would be excessively amplified. Derivative controllers must therefore be modified to limit their action at high frequency. The modified transfer function is usually of the form $kD[s/(T_Ds + 1)]$ with T_D sufficiently small.

A pure derivative controller would not be a controller at all, since there is no unique relationship between error and control action. Derivative action is, therefore, used only to augment proportional or proportional-plus-integral controllers to improve their high-frequency response.

The constant D is called the "rate time," with units of time. A large value of D is equivalent to very rapid response. D/T_D is called the "rate gain." It is an important characteristic of controllers with rate action, since it determines the filtering action of the system against unwanted noise.

The effect of derivative action on the root locus is the addition of an open-loop zero at $s = 1/D$ for a PD controller. There is a more complicated effect with PID control. This effect of derivative action is always stabilizing if the system is perfectly linear with no time delay. Since this is rarely the case in practice, the use of rate response is infrequent, although there are cases where it can be effectively used to improve speed of control action. Tuning for proper action is critical.

Inverse Derivative Control

A particular form of imperfect PI control, called "inverse derivative," is sometimes used. This designation applies when the point ω_R' of Fig. 45 is near the upper end of the frequency range of process interest.   Then the controller acts like P for ordinary frequency, like PI for high, and like P with a low gain for very high.   The transfer function is $(kTs + 1)/(Ts + 1)$, with $k < 1$.   The frequency plot looks something like the mirror image of that of the PD controller, hence the name.   Its use is limited to pneumatic systems, in series with a second control, where the combination might be less susceptible to mechanical limitations than either controller would be alone.   It acts as a high frequency filter.

Table 5 summarizes the characteristics of the various control modes.

Table 5.   Summary of Control Modes

Mode	Time Response (ϵ = error)	Transfer Function	Comments
On-off	$k \times$ sign (ϵ)	—	May be modified by adding a differential gap or third position. Use restricted to application where oscillation can be tolerated.
Floating	$k \int (\text{sign } \epsilon)$	—	Smoother and slower action than on-off.
P	$k\epsilon$	k	Use where offset can be tolerated, moderately high speed of action is required.
I	$kI \int \epsilon$	kI/s	Eliminates offset but reduces speed of response. Can cause instability if gain is too high.
PI	$k \left(\epsilon + I \int \epsilon \right)$	$k \left(1 + \dfrac{I}{s} \right)$	Most commonly used mode. If properly adjusted has good speed of response and no offset. Can lead to overshoot on large changes of set point.
PD	$k \left(\epsilon + D \dfrac{d\epsilon}{dt} \right)$	$k (1 + Ds)$	Improved speed of response over purely proportional control. If properly tuned, gain can be increased. Should not be used in systems with time delay or dead time.
PID	$k \left(\epsilon + I \int \epsilon + D \dfrac{d\epsilon}{dt} \right)$	$k \left(1 + \dfrac{I}{s} + Ds \right)$	Combines best action of all of above Should not be used in systems with dead time.

Figure 46 shows the time response of the linear controllers to a step change in the error.

8. ADJUSTMENT OF CONTROLLERS

The mode of control that is used in a system depends on the dynamic characteristics of the uncontrolled system and/or the type of control action that is needed.   Having chosen the control mode, the engineer must make certain that it operates in the intended manner.   If the system could be described exactly by the linear methods in the preceding articles of this section, then the proper adjustments could be made by calculation.   Two features of the system and its control must be specified to do this.   (1) The disturbances which might come into the system must be characterized in some way.   If they are random, what are their statistical characteristics both with respect to frequency and amplitude?   If they are changes in load or method of operation, how rapidly do they change and how large are the changes?   If there are to be changes in operating points or set points, how often do they occur and how large are they?   When this has been determined, then (2), the desired response of the sytem to these disturbances must be specified.

Figure 47 indicates a typical response of a controlled system to a disturbance.   If controller settings are to be calculated, the desired character of the response must be known.   It can be that the maximum excursion of the error is to be made as small as possible for the worst possible disturbance, or for the most probable disturbance.   It can be that the subsidence rate, h_2/h_4, is to be made as small as possible.   The total area under the error curve should perhaps be minimized.   Minimum value of the integral of the square of the error is a common measure of control quality.

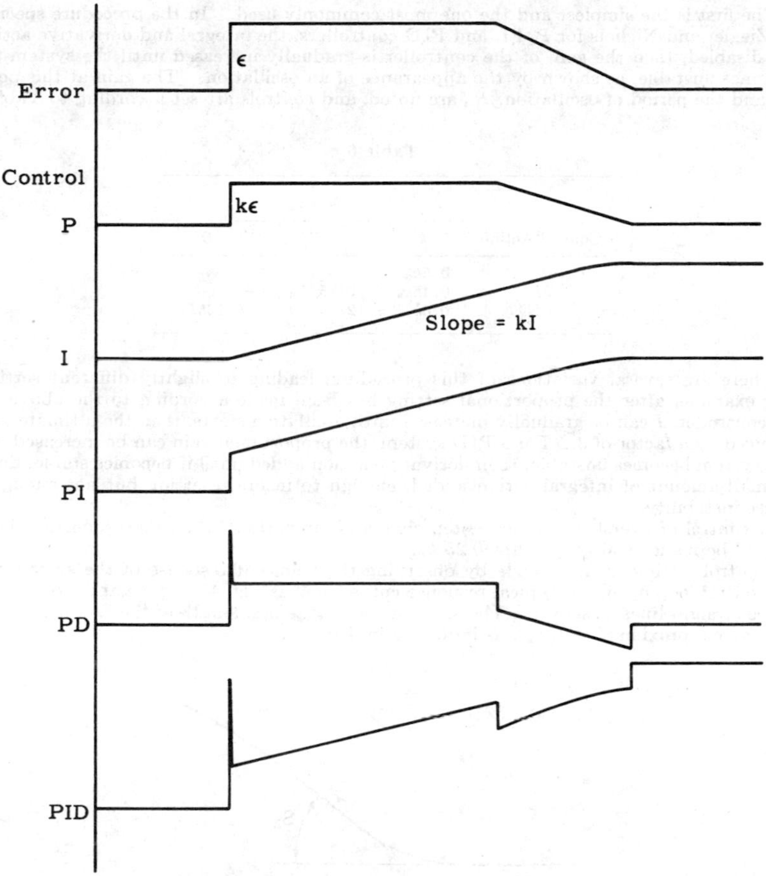

Fig. 46. Time response to error.

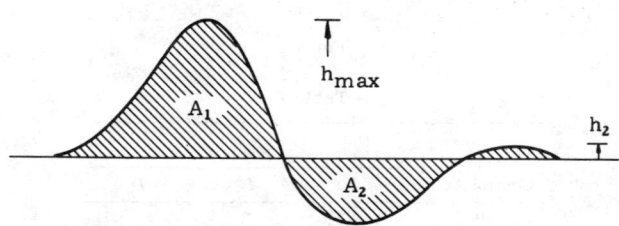

Fig. 47. Time response of a controlled system.

Once these characteristics are specified, then the best controller settings can be calculated for the perfectly linear system. However, this is not always possible, particularly in the process industries, since nonlinear effects may be important and the mathematical description of the process and controllers may be imperfect. In these circumstances it is necessary to adjust the control system after it is installed and the system is in operation.

There are three principal methods for doing this in practice: using the stability limit, using transient response to a step change, and using the frequency response.

The first is the simplest and the one most commonly used.   In the procedure specified by Ziegler and Nichols for P, PI, and PID controllers, the integral and derivative actions are disabled, then the gain of the controller is gradually increased until the system first becomes unstable, as shown by the appearance of an oscillation.   The gain at this point, k_u, and the period of oscillation, P_u are noted, and controls are set according to Table 6.

Table 6

Control Action	Settings		
	k	I	D
P	$0.5k_u$		
PI	$0.45k_u$	$0.25/P_u$	
PID	$0.6k_u$	$2/P_u$	$0.125P_u$

There are several variations of this procedure, leading to slightly different settings. For example, after the proportional setting has been made according to the above in a PI controller, I can be gradually increased until oscillation starts, then the ultimate value reduced by a factor of 3.   For a PID system, the proportional gain can be increased until the system becomes unstable, then derivative action added until it becomes stable, finally a small amount of integral action added, enough to insure resetting but not enough to cause instability.

If control of overshoot on large step changes is important, then the proportional gain should be made smaller, perhaps 0.25 k_u.

Control settings can be made by observing the transient response of the system with its control loop open.   A typical response curve, such as Fig. 48, can be approximated by three straight lines as shown.   The slope of the second and length of the first can be used to make approximate settings, as indicated in Table 7.

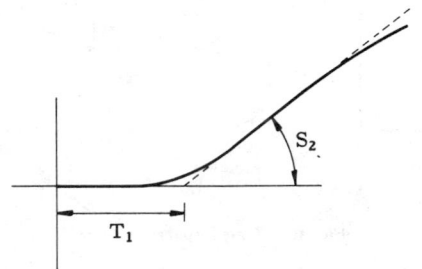

Fig. 48.

Table 7

Control Action	Settings		
	k	I	D
P	$1/S_2T_1$		
PI	$0.9/S_2T_1$	$0.3/T_1$	
PID	$1.2/S_2T_1$	$0.5/T_1$	$0.5/T_1$

This empirical method of controller adjustment is quick and easily implemented; however, the transient response may be obscured by disturbances unless the upsets are large.

If frequency response information about the process is available, controller settings can be found by considering the Nyquist or Bode plots.   A gain margin of about 0.3, or a phase margin of about 45°, is a satisfactory setting.   The control parameters can be set to give values near these, although there is considerable latitude, as can be imagined.

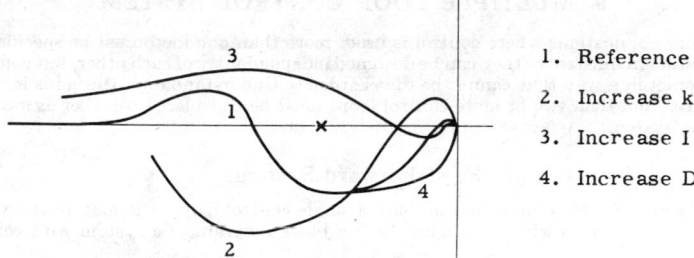

1. Reference

2. Increase k

3. Increase I

4. Increase D

FIG. 49.   Effects of control action.

Typically, the effect of a PID controller on the Nyquist plot of a process might be as shown in Fig. 49.   Integral action always increases the lag by 90° at low frequency, derivative action decreases it by 90° at high frequency.   Proper choice of k, I, and D indents the Nyquist contour in its most sensitive region, near $G(s) = 1$.   The system shown is conditionally stable, since it can be made unstable either by increasing or decreasing the gain of the controller.

Figures 50 and 51 show the sort of stability boundaries that might be expected with PI and PD controllers.   Control settings within the shaded area are generally satisfactory.   Optimal settings, of course, depend on the criterion of optimality, but they always tend to be close to the boarder of stability, near the maximum gain setting.

Further information on control modes will be found in references (9), (10), (11), (12), (14), (17), (18), (20), (22), (42), and (48).

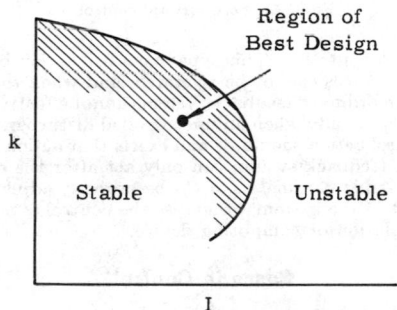

Region of
Best Design

k

Stable   Unstable

I

FIG. 50.   Gain-reset plot.

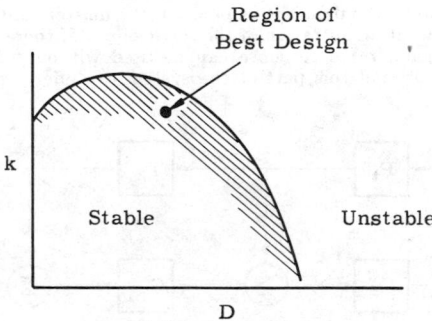

Region of
Best Design

k

Stable   Unstable

D

FIG. 51.   Gain-rate plot.

9. MULTIPLE LOOP CONTROL SYSTEM

In many applications where control is used, more than one loop must be specified. In the simplest circumstances they can be designed independently of each other, but commonly they interact in a way that cannot be disregarded. Understandably, this adds increasing complexity, and then two or more control loops must be considered together as a common multiloop system. A few examples follow (22), (42).

Feed-Forward Systems

A *feed-forward* system may contain only a single control loop, or it may have two loops controlling the same variable. Figure 52 is a block diagram of a system with combined

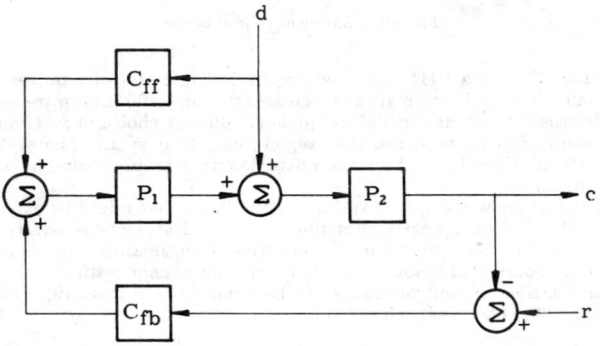

FIG. 52. Feed-forward control.

feed-forward and feedback controls. This type of control action becomes necessary when the effects of large disturbances cannot be tolerated, and when the dynamic character of the process is such that an ordinary feedback system cannot effectively control them. The feed-forward system is active only when an uncontrolled disturbance enters, at which time it predicts what the control action should be and exerts that action before the effects of the disturbance are felt. A feedback system can only act after the effects. It is necessary that the control and dynamic parameters of C_{ff} be properly adjusted to match the static and dynamics of the rest of the system, otherwise the control is not effective. Typically C_{ff} is some kind of simulation or computing device.

Cascade Control

A double-loop system in which one controller, the *master*, provides the set point for a second, the *slave*, is called a *cascade control system*. Figure 53 is a typical block diagram. C_m is the master controller, C_s is the slave. There are two principal uses for cascade control systems: (1) to reduce the effects of load disturbances and nonlinearities of a slave loop, and (2) to improve the dynamic response of the master control. Both uses depend on stable, fast, and accurate operation of the slave loop. If there is a part of a controlled system where high-gain, rapid response can be used without causing instability, then closing a slave loop around this part of the system can effectively improve the overall control.

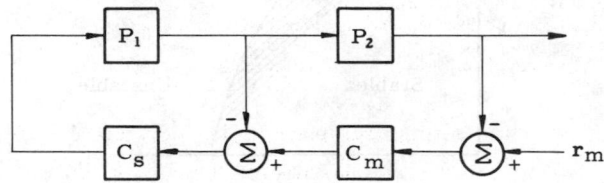

FIG. 53. Cascade control.

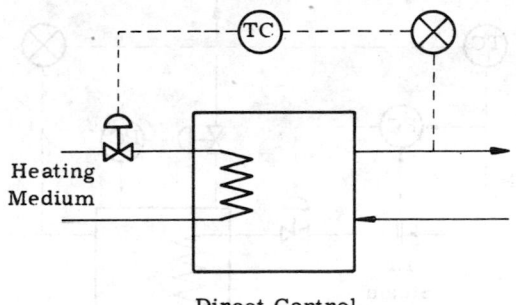

Direct Control

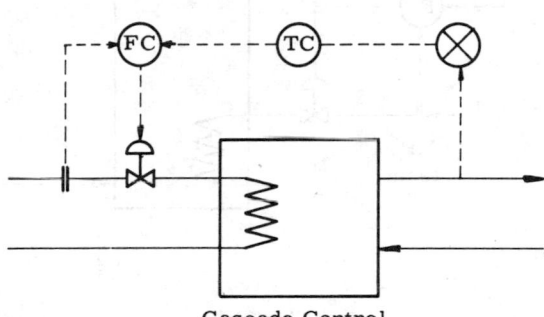

Cascade Control

Fig. 54.   Direct and cascade control of temperature.

An effective use of a cascade system is shown in Fig. 54.   Flow control is very fast acting, hence the effects of variation in temperature of the heating medium can be quickly compensated by the slave loop.

Split-Range Operation

Operation of two different final control elements from a single controller is possible, and sometimes of advantage.   In the example of Fig. 54, there might have been two sources of steam, one at low pressure, and inexpensive, the other at high pressure, and costly. The arrangement of Fig. 55 shows how the low-pressure steam can be used alone, as long as there is enough, with high-pressure steam coming on only when the low-pressure valve is fully open.

Override Control

There are often considerations, usually concerned with safety, when it is necessary to limit some particular variable.   Under normal operating conditions, a controller for this limited quantity may be totally inoperative, but when the quantity reaches its limit, the control must act in such a way that the limit is not exceeded, without regard for what the other loops are doing.   A pressure control which could be used in this way is shown in Fig. 55.   If the pressure should reach a preset limit, the valve opens and the flow is diverted to a safe place.

Interacting Multiple Variable Controls

A multiple variable system is here distinguished from the cascade system described previously by the number of independent set points.   In the cascade system there was one, in a true multiple system there are more than one.

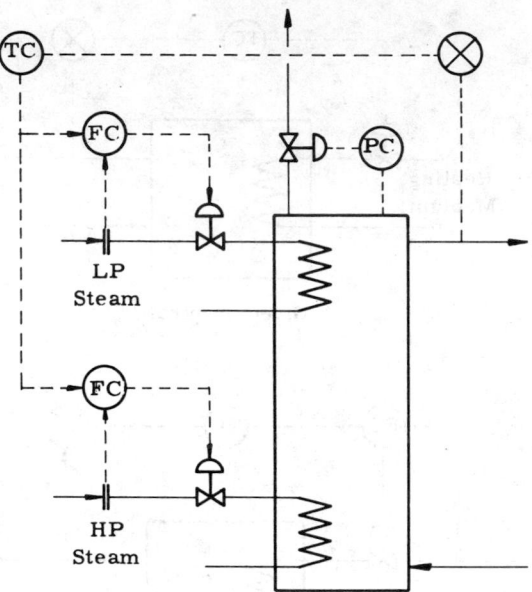

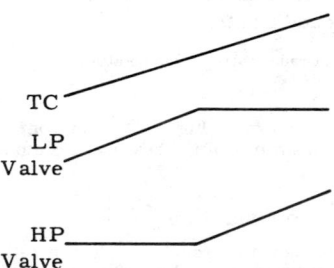

TC

LP
Valve

HP
Valve

Fɪɢ. 55.   Split-range control.

Figure 56 is a block diagram of doubly controlled system.   Its transient behavior might
be described by the system of ordinary differential equations:

$$D_1(x_1) = f_1(x_1, x_2, c_1, c_2) \Big\}$$
$$D_2(x_2) = f_2(x_1, x_2, c_1, c_2) \Big\}$$

(84)

where D_1 and D_2 are dynamical operators and the relationship between x_1 and c_1, and x_2
and c_2 may be assumed to be those of ordinary controllers.
    It is perfectly clear that if these equations are linear then, in principle, the system can
be put into the form of a single equation with constant coefficients.   An ordinary transfer
function can be derived, the stability can be analyzed, and the effect of control parameters
determined exactly as before.   Often the algebra of stability analysis can be considerably
simplified by the use of matrix methods.   Thus if Eq. 84 were linear, it might be written as

$$x_1(s) = G_{11}H_1(-x_1 + r_1) + G_{21}H_2(-x_2 + r_2) \Big\}$$
$$x_2(s) = G_{12}H_1(-x_1 + r_1) + G_{22}H_2(-x_2 + r_2) \Big\}$$

(85)

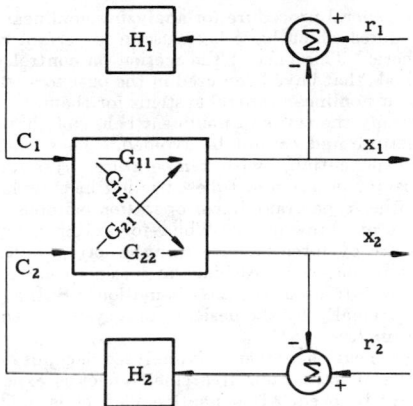

FIG. 56. Interacting loops.

In matrix form,

$$\left|\begin{matrix} r_1 \\ r_2 \end{matrix}\right| = \left\{\begin{matrix} (G_{11}H_1 + 1) & G_{21}H_2 \\ G_{12}H_1 & (G_{22}H_2 + 1) \end{matrix}\right\} \left|\begin{matrix} x_1 \\ x_2 \end{matrix}\right| \tag{86}$$

or

$$|r| = (M)|x| \tag{87}$$

and

$$|x| = (M^{-1})|r| \tag{88}$$

(M^{-1}) is a matrix form of the transfer function, which is a function of the transform variable s. System stability requires that those values of s for which the determinant of (M) vanishes have only negative real parts. The determinant may be written out as

$$(1 + G_{11}H_1)(1 + G_{22}H_2) - (G_{12}H_1)(G_{21}H_2) = 0$$

or

$$1 - E\,\frac{G_{11}H_1}{1 + G_{11}H_1}\,\frac{G_{22}H_2}{1 + G_{22}H_2} = 0 \tag{89}$$

where $E(s) \equiv G_{12}G_{21}/G_{11}G_{22}$ is an *interaction coefficient*. With no interaction terms the condition for stability is identical with the condition for stable operation of the independent loops, as it should be. For practical operation the individual loops should always be stable independently, since in many cases a process may operate with some of its control inoperative or manually adjusted, and the system should remain stable when this is done.

In general, interaction between loops is always likely to narrow the limits of stable operation of a system. Therefore, the parameter adjustments should be somewhat farther from the stability limits of the individual loops than would normally be the case with no interaction.

In strongly interacting systems, adjustment is very difficult, and existence of strong interaction is one of the major justifications for using sophisticated control systems, such as digital computers.

10. NONLINEAR SYSTEMS

In what has been said thus far, the possible effects of nonlinear behavior have been mentioned as complicating the analysis, and perhaps degrading the performance of a control system under some conditions. This should not be taken to mean that nonlinearities in control systems should be avoided, because this is definitely not the case. Observation of the behavior of a nonlinear process can give valuable clues to the design of effective controls, and the use of the right kind of nonlinear control systems can make possible great improvement in performance.

There is, of course, no general procedure for analyzing nonlinear behavior of controlled systems. Specific procedures which have been useful in certain cases cannot always be successfully used elsewhere. This part of the section on control will therefore take the form of a review of methods that have been used in the past to analyze nonlinear systems, and to synthesize linear or nonlinear control systems for them.

The first step is to classify the types of nonlinear behavior that might be encountered. Some are imposed by nature and cannot be avoided. These include rates of chemical reactions, which vary exponentially with temperature, hysteresis in magnetic fields, current–voltage relationships in vacuum tubes, friction in moving parts, and, of course, many many more. Another type of nonlinear operation is imposed into a control system by an unavoidable choice of components. Thus, for example, temperature is measured by thermocouples, pressure by force balance using a spring, and flow by pressure drop across an orifice. Electric motors or turbines are used to move things, flows are controlled by valves, and gears are used to transmit motion. Still a third class of nonlinear behavior is imposed intentionally by the designer of a system. On-off control limits and saturation effects are examples.

The way in which a nonlinear phenomenon is analyzed depends on its dynamic character and the magnitude of its effect. Thus if a small effect is expected, *linearization* and *perturbation* methods may be used. If a nonlinear effect is independent of time, then perhaps the static and dynamic effects may be separated and studied separately, resulting in effective simplification.

There are three essential characteristics of nonlinear systems that are of importance in automatic control theory. The first is that the behavior of the system depends on the amplitude of the disturbance. The response to large changes may be entirely different in a qualitative way from what it is for small. The second essential characteristic is that a nonlinear system may oscillate with fixed period and amplitude in a limit cycle. The third is that the response to a sinusoidal disturbance will contain frequencies different from that of the disturbance, usually harmonics, occasionally subharmonics, often anharmonic. The occurrence of one or more of these types of behavior is almost universal in all physical systems, and should be regarded as the rule rather than the exception. The proper question to ask regarding nonlinearity in a system is not "does it exist?", but rather "can the effects be ignored or must they be considered in detail?" In what has preceded, the analytical treatment has been strictly applicable to the cases where nonlinearities can be ignored. In what is to follow, procedures are given for studying systems whose nonlinearities have important effects.

Describing Function Analysis

Naturally enough, the first attempts to work out detailed description of nonlinear controlled systems involved the fitting of the nonlinear mathematical behavior into the frame of linear analysis. The most useful consequences of this approach to the problem has been called the *Describing Function* method. Its successful use depends on two major assumptions about the system to be studied.

1. For mathematical purposes all nonlinear phenomena can be grouped into a single element whose behavior is independent of frequency. Although frequency-dependent describing functions can be defined and constructed, their use required so much computation that it is usually more useful to proceed in other ways.

2. In a closed loop, harmonics or other frequencies generated when a nonlinear element is excited sinusoidally are effectively filtered by the linear elements in the loop. These assumptions make it possible to analyze nonlinear phenomena by frequency response methods.

The describing function of a nonlinear element is defined as the gain and phase of the fundamental component of the output signal as a function of the amplitude of an input sinusoid.

In the block diagram of Fig. 57, the nonlinear block has a describing function $D(A)$, the linear block a gain of $G(i\omega)$, and the system has an open-loop transfer function of DG. If DG should equal -1 for any combination of amplitude and frequency, then the system is unstable, and a limit-cycle will occur, with a frequency equal to that at which $DG = -1$.

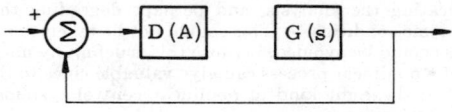

FIG. 57.

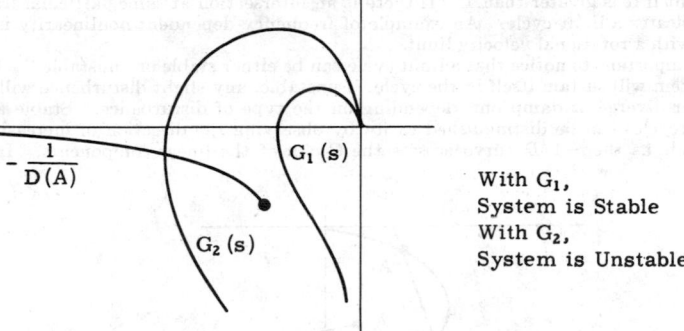

With G_1,
System is Stable
With G_2,
System is Unstable

FIG. 58.   Graphical describing function analysis.

The Nyquist diagram is a most convenient method of using the describing function. Figure 58 shows how it can be done. $G(i\omega)$ is plotted in the usual way. $-1/D(A)$ is plotted in polar coordinates of gain versus phase. If the two curves intersect, then a limit cycle is indicated, with amplitude given by distance from the origin and frequency given by the position on the G-locus. Notice that the curve $-1/D(A)$ replaces the -1 point of the linear Nyquist stability criterion.

The extension of this procedure to describing functions which depend on the frequency is straightforward but tedious. $-1/D(A)$ is plotted for each of a number of frequencies of interest, as indicated in Fig. 59. The plot of Fig. 60 can be generated from the inter-

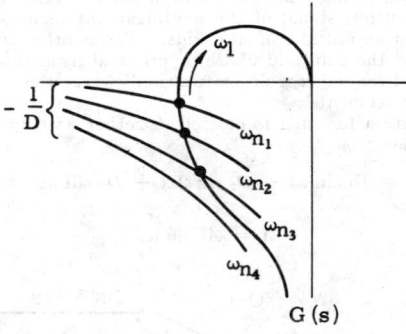

FIG. 59.   Frequency-dependent describing function.

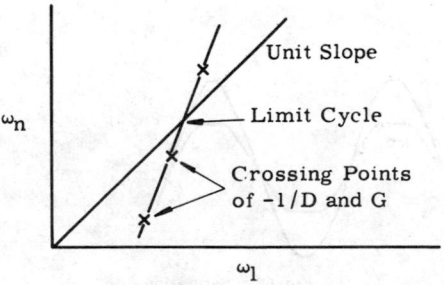

FIG. 60.   Approximate frequency of limit cycle.

sections.   If there is no intersection, the system is stable if DG is always less than 1 and divergent if it is greater than 1.   If there is an intersection at some particular frequency, this indicates a limit cycle.   An example of frequency-dependent nonlinearity is a servo motor with a rotational velocity limit.

It is important to notice that a limit cycle can be either stable or "unstable."   If stable, the system will sustain itself in the cycle; if unstable, any slight disturbance will cause it to either diverge or damp out, depending on the type of disturbance.   Stable and "unstable" cycles can be distinguished easily by observing the direction of increasing input amplitude as the $-1/D$ curve crosses the G-plot of the linear components.   In Fig. 61

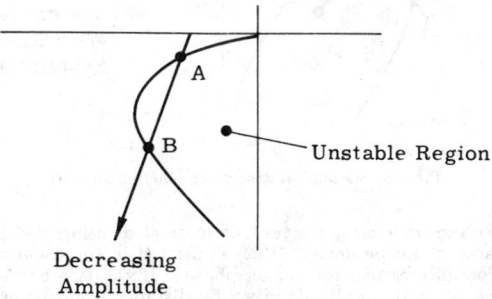

Fig. 61.   Stable and unstable limit cycles.

point B is an unstable point, A indicates a stable limit cycle.   At B, a slight increase in amplitude moves the system into a state where it is unstable and the amplitude grows further.   At A the reverse is the case, and slight disturbances are counteracted by the system.

The calculation of a describing function is complicated, in general.   The procedure is to expand the calculated output signal of the nonlinear device into a Fourier series for a number of selected input amplitudes of sinusoids.   Terms other than those of the fundamental are rejected, and the gain and phase determined from those retained.   The procedure will be illustrated for the case of "saturation" where the nonlinearity is described by a set of straight-line segments.

In Fig. 62 the saturation function is plotted, together with typical input and output time functions.   Assume

$$D = D_1 \sin \omega t + D_2 \sin 2\omega t + D_3 \sin 3\omega t \ldots \tag{90}$$

and

$$A = A_0 \sin \omega t \tag{91}$$

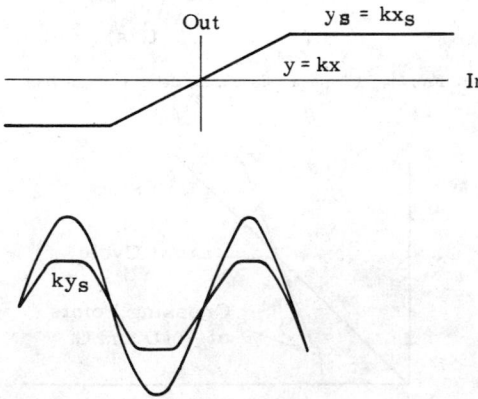

Fig. 62.   Saturation.

D_1/A_0 is the describing function for $A_0 < A_s$

$$D_1 = \frac{4\omega}{\pi} \int_0^{\pi/2\omega} A_0 \sin^2 \omega t \tag{92}$$

$$= A_0, \text{ hence } D_1/A_0 = 1 \tag{93}$$

The integral can be taken over 1/4 cycle because of symmetry. For $A_0 > A_s$,

$$D_1 = \frac{4\omega}{\pi} \int_0^{(1/\omega) \sin^{-1} (A_s/A_0)} A_0 \sin^2 \omega t \, dt + \frac{4\omega}{\pi} \int_{(1/\omega) \sin^{-1} (A_s/A_0)}^{\pi/2\omega} A_s \sin \omega t \, dt \tag{94}$$

$$= \frac{4A_s}{\pi} \int_0^{\sin^{-1} (A_s/A_0)} \sin^2 \mu \, d\mu + \frac{4A_s}{\pi} \int_{\sin^{-1} (A_s/A_0)}^{\pi/2} \sin \mu \, d\mu \tag{95}$$

$$= \frac{4A_s}{\pi} \left[\cos \left(\sin^{-1} \frac{A_s}{A_0} \right) \right] \tag{96}$$

$$\frac{D_1}{A_0} = \frac{2}{\pi} \left[\sin^{-1} \frac{A_s}{A_0} - \sin \left[\left(\sin^{-1} \frac{A_s}{A_0} \right) \right] \cos \sin^{-1} \frac{A_s}{A_0} \right] + 2 \frac{A_s}{A_0} \cos \sin^{-1} \frac{A_s}{A_0} \tag{97}$$

$$= \frac{2}{\pi} \left[\sin^{-1} \frac{A_s}{A_0} + \frac{A_s}{A_0} \sqrt{1 - \frac{A_s^2}{A_0^2}} \right] \tag{98}$$

93 and 98 together form the describing function for saturation. In this case there is no phase effect.

A collection of describing functions for simple nonlinearities is given in Table 8.

It must be emphasized that the use of the describing function in this way is at best a rough approximation to reality, and may be seriously in error. The ignoring of harmonics is not a trivial matter. Some nonlinear elements have no fundamental component whatsoever in the output signal—a squaring device for example. There is no sure way to assess errors except to examine the magnitude of the harmonic terms. If they are of significant magnitude compared with the fundamental, over any part of the amplitude range of interest, then the describing function method will give a poor description of the actual performance of a system.

Nevertheless, the method offers an easy way to investigate the qualitative behavior of some of the ordinary nonlinearities that can occur. If their effects are small enough so that they do not dominate the behavior of the complete system, then the use of method is justifiable.

Phase-Plane Methods

In the preceding, the use of frequency domain methods in the analysis of nonlinear systems was reviewed. Naturally enough, the methods are not strictly applicable, and approximate procedures of doubtful validity are all that can be used in many circumstances. A somewhat more rigorous approach, albeit somewhat less convenient to use, has been called the *time-domain* method. This really means nothing more than investigating the differential equations that describe the system, rather than their Laplace or Fourier transforms.

The natural procedure for investigating a complex system is to generate a sufficient number of solutions of the equations that describe it. Unfortunately this almost always requires a large digital computer or some simulation device such as an analog computer. Simulation will be discussed in its place, but first a few of the analytical procedures will be described. The first of these is graphical stability analysis in the *phase plane*.

It should first be mentioned that the word "phase" is used here with a different meaning than heretofore. The user must be careful not to confuse the two. The meaning will always be clear from the context.

Phase-plane methods are useful for a special class of systems only, those described by ordinary differential equations of second order, in which the independent variable (time) does not appear explicitly.

An equation of this sort can usually be written

$$\frac{d^2y}{dt^2} = f\left(\frac{dy}{dt}, y\right) \tag{99}$$

Table 8. Describing Functions

A = Amplitude. Unit slope assumed.

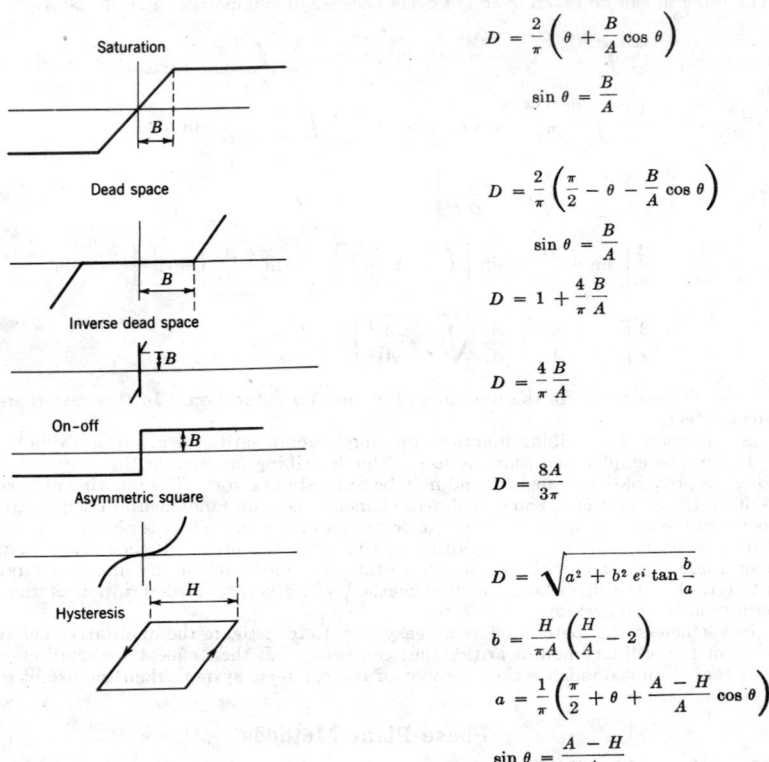

	Describing Function
Saturation	$D = \dfrac{2}{\pi}\left(\theta + \dfrac{B}{A}\cos\theta\right)$
	$\sin\theta = \dfrac{B}{A}$
Dead space	$D = \dfrac{2}{\pi}\left(\dfrac{\pi}{2} - \theta - \dfrac{B}{A}\cos\theta\right)$
	$\sin\theta = \dfrac{B}{A}$
Inverse dead space	$D = 1 + \dfrac{4}{\pi}\dfrac{B}{A}$
On-off	$D = \dfrac{4}{\pi}\dfrac{B}{A}$
Asymmetric square	$D = \dfrac{8A}{3\pi}$
Hysteresis	$D = \sqrt{a^2 + b^2}\, e^{i\tan\frac{b}{a}}$
	$b = \dfrac{H}{\pi A}\left(\dfrac{H}{A} - 2\right)$
	$a = \dfrac{1}{\pi}\left(\dfrac{\pi}{2} + \theta + \dfrac{A - H}{A}\cos\theta\right)$
	$\sin\theta = \dfrac{A - H}{A}$

which is equivalent to

$$\frac{dp}{dt} = f(p, y)$$

$$\frac{dy}{dt} = p \tag{100}$$

and

$$\frac{dp}{dy} = \frac{f(p, y)}{p} \tag{101}$$

In an alternative formulation of the equations of a system, we might have something like

$$\frac{dy}{dt} = f(y, \zeta) \tag{102}$$

$$\frac{d\zeta}{dt} = g(y, \zeta)$$

whence

$$\frac{dy}{d\zeta} = \frac{f(y, \zeta)}{g(y, \zeta)} \tag{103}$$

Here ζ might be a control parameter rather than the generalized velocity of (100). In either case, a Cartesian coordinate system in which the ordinate and abscissa are y and ζ, or y and p, or y and dy/dt, is called the phase plane of the system.

From Eq. 103, at almost every point in the phase plane there will be a unique derivative $dy/d\zeta$, and through each such point there will be a unique curve which describes the instantaneous relationship between y and ζ. This curve is called a "phase portrait." A typical one is shown in Fig. 63. Notice that direction of motion in the time cannot be determined from the phase portrait alone. This must be done by considering one of the original equations.

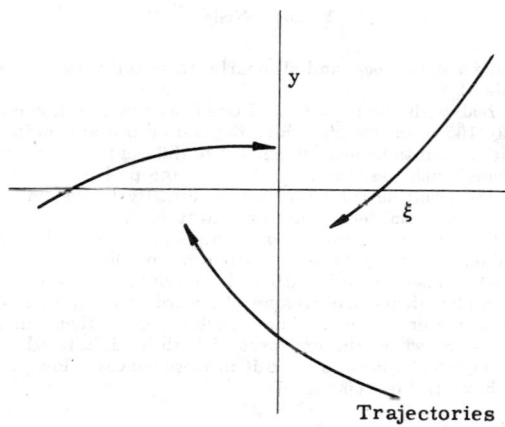

FIG. 63. Phase portrait.

The phase plane may have singular points, where f and g in Eq. 103 are simultaneously zero and thus $dy/d\zeta$ is indeterminate. The behavior of the system in the neighborhood of a singularity may be discovered by expanding f and g in Taylor's series and dropping terms past the first order. The linear system which results can be analyzed by ordinary linear methods. Thus, if y, ζ, g_0, and f_0 characterize a singular point.

$$\left.\begin{aligned}\frac{dt}{dt} - \left(\frac{dy}{dt}\right)_0 &= (f - f_0) = a(y - y_0) + b(\zeta - \zeta_0)\\[2mm]\frac{d\zeta}{dt} - \left(\frac{d\zeta}{dt}\right)_0 &= (g - g_0) = c(y - y_0) + d(\zeta - \zeta_0)\end{aligned}\right\} \tag{104}$$

The solution of Eq. 104 is

$$\left.\begin{aligned}y - y_0 &= k_1 e^{\lambda_1 t} + k_2 e^{\lambda_2 t}\\[2mm](\zeta - \zeta_0) &= k_3 e^{\lambda_1 t} + k_4 e^{\lambda_2 t}\end{aligned}\right\} \tag{105}$$

where λ_1 and λ_2 are the eigenvalues of the matrix $\begin{Bmatrix} a & b \\ c & d \end{Bmatrix}$, that is, the values of λ for which

$$\begin{vmatrix} a - \lambda & b \\ c & d - \lambda \end{vmatrix} = 0 \tag{106}$$

We having the following definitions of types of singular points.

1. If λ_1 and λ_2 are both real and positive, all trajectories in the immediate vicinity move away from the singularity. It is called an *unstable node*. Similarly, if λ_1 and λ_2 are real

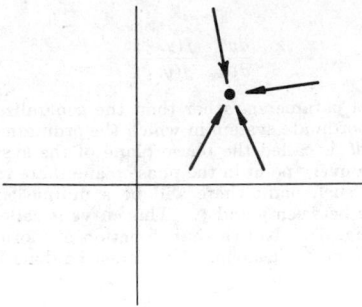

FIG. 64.   Node.

and negative, we have a *stable node*, and all nearby trajectories move inward.   Figure 64 shows a typical node.
2. If both λ's are real, with one positive and one negative, the singular point is called a *saddle point*.   In Eq. 105 there are k's, which depend on initial condition of the system. Different sets of initial conditions lead in general to different trajectories.   If the sets of conditions can be found such that the coefficients of the positive exponential in Eq. 105 are both zero, then the resulting trajectory moves directly to the singular point.   Similarly, if the negative exponential term is not present, then the curve moves away from the singular point.   Such a pair of trajectories can always be found in the vicinity of a saddle point.   They are called the "separatrices."   Other nearby phase portraits have the separatrices as asymptotes.   They move inward and then outward, as in Fig. 65.
3. If the λ's are complex, then the curves spiral inward or outward according to whether the real parts are negative or positive.   The singular point is then called a stable or unstable *focus*.   In the limit where the real part of both λ's is zero, the singular point is called a *center*, and trajectories move around it in close curves.   Portraits in the vicinity of these points are shown in Figs. 66 and 67.

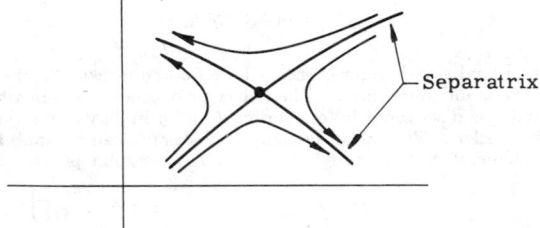

FIG. 65.   Saddle point.

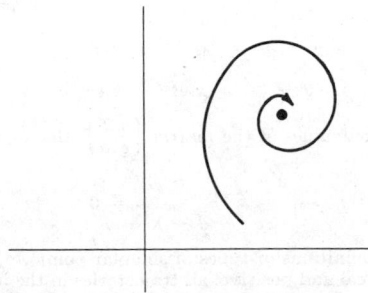

FIG. 66.   Focus.

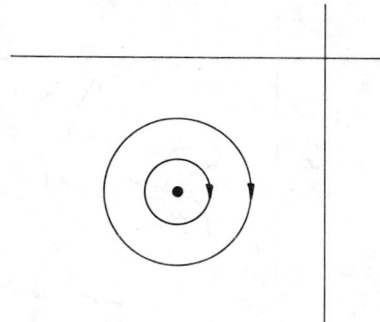

Fig. 67. Center.

Some readers will have noticed that the λ's associated with a singular point are precisely the roots of the transfer function of the linearized system. This establishes a curious connection between frequency methods and phase-plane methods. However, this same reader should remember that a phase plane may have many singular points if the system deviates from linearity in a complicated way. The linearization will be different near different points, hence so will the λ's.

Examination of the location and character of the singular points of a nonlinear system can tell much about its transient behavior. Thus trajectories (except separatrices) cannot cross each other. Any one trajectory must either be closed, or else both ends must terminate at a singular point or infinity.

A closed trajectory is called a *limit cycle* if other trajectories are asymptotic to it. It is stable if they approach it, unstable if they diverge. Detection of a limit cycle is not always easy in phase-plane methods, although it is one of the more important questions to be asked about a system.

A sometimes useful consideration in this regard is the *index* of a singular point. If the point be enclosed by a continuous curve, and to each point on this curve a vector be assigned with components f and g from Eq. 103, then a clockwise circuit around this closed curve will be associated with a rotating vector, which in one cycle will move through a clockwise angle of $2\pi\eta$. η is called the index of the singular point. The index of a node, focus, or center is $+1$, that of a saddle is -1, that of an ordinary point is zero. The principal useful property of indices is that the index around any closed curve is the sum of the indices of all singular points within it. By sketching a grid of directed slope vectors and drawing in a number of closed curves, one can use indices to find the approximate locations and character of singularities without finding the roots $f = g = 0$. He can then sketch the profiles crudely, after which he can usually infer the existence of limit cycles.

The Bendixon theorem of non-existence of limit cycles says that if $(\partial f/\partial y) + (\partial g/\partial \zeta)$ is everywhere different from zero in a region of y, ζ, then no limit cycles can exist in that region. Therefore, a limit cycle must enclose a point where this sum is zero. If the points where $(\partial f/\partial y) + (\partial g/\partial \zeta) = 0$ can be located, then a limit cycle can often be detected by drawing a small circuit and a large circuit around the point in question, as indicated in Fig. 68. Vectors with components dy/dt and $d\zeta/dt$ are put in at points around the circumference. If all these vectors point inward into the annular space, then a stable limit cycle exists. If they all point outward, then an "unstable" limit cycle exists. "Unstable limit cycle" is a misnomer. Existence of one means that a system will be stable for small disturbances within it, divergent or unstable for large.

In summary, phase-place methods can give useful qualitative information for second-order, nonlinear systems. Consideration of systems of higher order can be made in much the same way; however, the simple geometric interpretations are no longer available.

Stability of Nonlinear Control Systems

The modern algebraic theory of the stability of systems of nonlinear differential equations has descended from pioneering work of Liapunov late in the last century. It is applicable under broad conditions to analysis of nonlinear control systems, and it is in this context that it will be described here.

A precise definition of stability is the first requirement in the description. A system is said to be stable about the origin with respect to a disturbance if given an $\epsilon > 0$ there

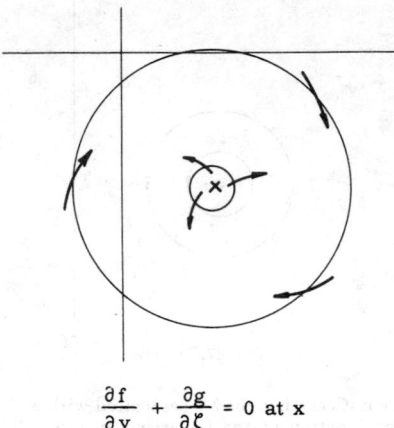

$$\frac{\partial f}{\partial y} + \frac{\partial g}{\partial \zeta} = 0 \text{ at x}$$

FIG. 68. Bendixson theorem.

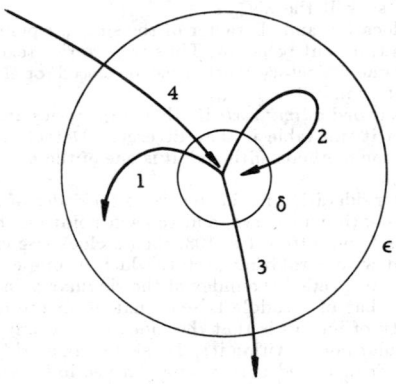

FIG. 69. Stability in two dimensions.

exists a δ such that when $|x_k(t_0)| \leq \delta$, $|x_k(t)| < \epsilon$ for all $t > t_0$. The $|x_k$'s| are components in n-dimensional phase space, or *state space* as is sometimes called, where n is the order of the differential equation that describes the system. Figure 69 indicates several different types of stability under the above definition for $n = 2$. Curve 1 is a *stable* trajectory. After a sufficiently small disturbance, the system stays within ϵ of $|x| = 0$. Curve 2 is said to be *asymptotically stable*, and the system eventually reaches $|x| = 0$. Curve 3 is unstable. Curve 4 is a trajectory of a system which is *asympotically stable in the large*. All trajectories end at $x = 0$. The theory of the Liapunov function distinguishes among these stability types.

The physical interpretation of the Liapunov theory can best be understood by using a simple, linear example. A damped sinusoidal oscillator is described by the equation

$$\ddot{x} + a\dot{x} + bx = 0 \tag{107}$$

with a and b positive. Multiplication by $\dot{x}$ gives

$$\ddot{x}\dot{x} + a\dot{x}^2 + bx\dot{x} = 0 \tag{108}$$

Integration gives

$$\frac{\dot{x}^2}{2} + a \int_0^t (\dot{x})^2 \, dt + \frac{bx^2}{2} = C \tag{109}$$

where C is a positive constant equal to

$$\frac{\dot{x}^2(0)}{2} + \frac{bx^2(0)}{2}$$

This can be written

$$V \equiv \dot{x}^2 + bx^2 = C - a \int_0^t (\dot{x})^2 \, dt \tag{110}$$

with

$$\dot{V} = 2\dot{x}\ddot{x} + 2bx\dot{x} = -a(\dot{x})^2 \tag{111}$$

From Eq. 110, V is always positive (in this simple problem it is the available energy). From Eq. 111, V always decreases with time. The function V, which is called a Liapunov function of the system, was constructed from the differential equation without the need for generating a solution.

In using the Liapunov stability criterion, one determines, if possible, a function like the energy that is always positive and always decreases with time. In precise form, the criterion states that if a function $V(x_n)$ can be found from the set of differential equations which describe a system, which is zero for $(x_n) = 0$, and which is positive everywhere else, then the system is stable in a region of state space if $dV/dt \leq 0$ over that region and asymptotically stable over that region if $dV/dt < 0$ everywhere except $(x_n) = 0$.

The theorems should not, of course, be interpreted to mean that if such a function cannot be found, then the system is not stable. The condition is sufficient but not necessary. The use of the Liapunov technique to define regions of stability requires a good bit of ingenuity on the part of the user. No general method for constructing good Liapunov functions, one that is easily usable in every case, can be recommended. Functions over the state space which are always positive except at the origin can, of course, be easily found. A weighted sum of squares of the variables is such a function. It is just one of a set of quadratic forms whose general expression is

$$V = (x_n^*)(B)(x_n) \tag{112}$$

where B is a square positive definite matrix, (x_n) is the vector whose components are the coordinates of a general point in state space, and (x_n^*) is its transpose. This general quadratic form includes sums of squares of the form $(x_1 - \alpha x_2)^2 + (x_1 - \beta x_4 - \gamma x_7)^2$, . . ., etc., and hence gives a wider variety of possible V's that can be tried.

An even broader class of Liapunov function is often appropriate in control studies because of the special form nonlinear control equations take. If the nonlinearity is entirely in the control section of a system, then a typical set of control equations might be

$$(\dot{x}_n) = (A)(x_n) + bf$$
$$f = \phi(\epsilon) \tag{113}$$

Here A is a matrix of order $n \times n$, f is a scalar control parameter, and ϕ is a function of the error, ϵ, which is not necessarily linear but which is always positive for positive ϵ, always negative for negative ϵ, and for which

$$\int_0^\epsilon \phi(\epsilon) \rightarrow \infty \text{ as } \epsilon \rightarrow \infty$$

If such a function ϕ occurs in a control system, and this is quite common, then

$$V = (x^*)\beta(x) + \int_0^\epsilon \phi(\epsilon) \tag{114}$$

is a usable Liapunov function for the system, with $(x^*)\beta(x)$ defined as before.

Given a positive definite form in the state space, and a system of equation describing the transients of the state variables, such as Eqs. 111 and 113, one can find the time derivative of the quadratic form by direct substitution. This derivative will, in some areas, be positive in some areas, negative in others. Those regions where it is negative are regions in which the control system is guaranteed to be asymptotically stable. By using a variety of B-matrices, one can often increase the area of known stability, since these areas add to each

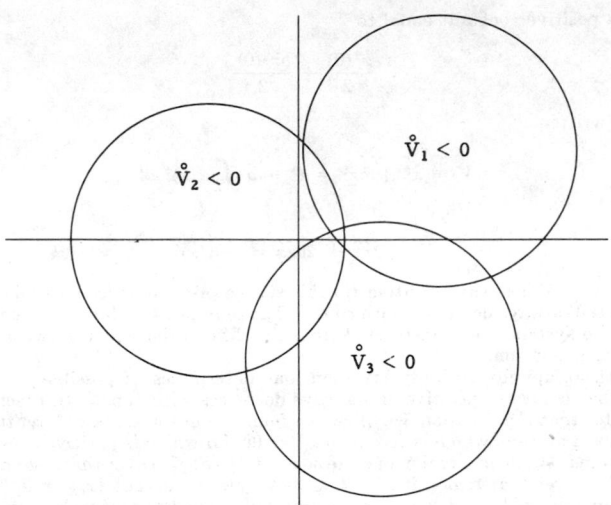

FIG. 70.   Stability by the Liapunov method.

other.   The procedure is illustrated in Fig. 70, where the stability regions in a two-dimensional space for three different Liapunov functions are indicated.   Stability for a disturbance within the total shaded area is assured.

A procedure which gives the exact boundary of asymptotic stability, but which is far less convenient to use, is due to Zubov.   He showed that for a system of equations of the form,

$$\frac{dx_1}{dt} = f_1(x_1, x_2, x_3, \ldots)$$

$$\frac{dx_2}{dt} = f_2(x_1, x_2, x_3, \ldots)$$

(115)

a Liapunov function is given by the solution of the partial differential equation

$$\frac{\partial V}{\partial x_1} f_1 + \frac{\partial V}{\partial x_2} f_2 + \ldots = \phi(x_1, x_2, \ldots)(1 + V)\sqrt{1 + f_1^2 + f_2^2}$$

(116)

where ϕ must always be positive but can otherwise be chosen in a way to make the solution of the system possible.   Furthermore, the Liapunov function constructed in this way gives the actual boundary of asymptotic stability.

Nonlinear methods are treated more fully in references (8), (11), (19), (22), (23), (28), (30), (31), and (37).

11. RANDOM PROCESSES IN CONTROL

There are two fundamentally different kinds of operation that are meant by the words "automatic control."   On the one hand there are those control operations which are intended to cause a machine or process to proceed correctly through a planned series of steps, where external influences are of minor significance.   On the other hand, there are situations in which the control is intended to compensate for disturbances which by their nature cannot be predicted.   These disturbances are commonly called *noise*.   The effects of noise on the behavior of controlled systems forms the next subject to be considered.

A *random event* may be defined as an event whose consequences cannot be precisely predicted, that is there is always some uncertainty in its outcome.   If in a sequence or collection of random events a numerical value is assigned to each consequence of the various random events, a numerical function called a "random variable" is generated.   A simple example is in coin-tossing.   A random variable, x, can be assigned of either zero or one, depending on the outcome of the toss, "heads" or "tails." The instantaneous value of the temperature at some selected spot in a room has many characteristics of a random variable defined over a continuous "space" of samples.

The concept of probability is essentially intuitive. In a set of random events, some are more likely to occur than others. A number assigned to each event such that its value is a measure of its likelihood of occurrence is called the *probability* of that event if the sum of the numbers over all events in the set is 1. The term probability will not be more precisely defined here. The probability of an event can often be estimated by observing the frequency with which the particular event occurs in a large sample.

The *distribution function* $F(x)$, of a random variable X, is defined as the probability that X is less than or equal to x. This is conveniently written

$$F(x) = \Pr(X < x)$$

$F(x)$ is a monotonic increasing function of x, which is 0 for $x = -\infty$ and $+1$ for $x = +\infty$. The *probability density function* or *frequency function*, $f(x)$, is defined as the derivative of $F(x)$. If X takes on only discrete values, $X_1, X_2, \ldots, X_i, \ldots$, then $f(x_i)$ may be defined as $F(x_i) - F(x_{i-1})$. In the following discussion it will be assumed that X is continuous. $\int_{-\infty}^{\infty} f(x) \, dx$ is unity by definition.

Some writers use the term "distribution function" to mean what is called "frequency funtion" above. Here the terms will always be used as defined

The *average value* or mean of a random variable over a sequence of events is given by

$$\overline{X} \equiv \frac{n_1 X_1 + n_2 X_2 + \ldots}{n_1 + n_2 + \ldots} \tag{117}$$

where n_i is the number of times the event whose value is X_i occurred. If the quantity $n_i/\Sigma n_i$ is replaced by $p_i = f(x_i)$, the frequency with which the event can be expected to occur, then the sum is called the *expected value*, or *mean of the distribution*

$$E(X) \equiv \Sigma f(x_i)x_i = \overline{X} \tag{118}$$

or in the continuous case

$$E(X) \equiv \int_{-\infty}^{\infty} x f(x) \, dx \tag{119}$$

The *mean square* of a variable or the *second moment* of a distribution function is

$$E(X^2) = \int_{-\infty}^{\infty} x^2 f(x) \, dx \tag{120}$$

Similarly, the third and higher moments may be defined as

$$E(X^n) \equiv \int_{-\infty}^{\infty} x^n f(x) \, dx \tag{121}$$

The mean square of the difference between the variable and the mean is called the *variance* of the distribution, or the *second central moment*.

$$\sigma^2(X) = E(X - \overline{X})^2 = \int_{-\infty}^{\infty} (x - \overline{X})^2 f(x) \, dx \tag{122}$$

$$= \int_{-\infty}^{\infty} x^2 f(x) \, dx - 2\overline{X} \int_{-\infty}^{\infty} x f(x) \, dx + \overline{X}^2 \int_{-\infty}^{\infty} f(x) \, dx \tag{123}$$

$$= E(X^2) - [E(X)]^2 \tag{124}$$

The square root of the variance is called the *standard deviation*, σ.

$$\sigma(X) \equiv \sqrt{E(X^2) - [E(X)]^2} \tag{125}$$

The notion of probability can be readily extended to more than one random variable. The *joint distribution function* of several variables $x, y, z, \ldots$ is the probability that $X \leqslant x$, $Y \leqslant y, Z \leqslant z$, etc. The "$ij$th" moment of a distribution of two variables is given by

$$\alpha_{ij} \equiv E[(X)^i \cdot (Y)^j]$$

$$= \int_{-\infty}^{\infty} x^i \, dx \int_{-\infty}^{\infty} y^j f(x, y) \, dy \tag{126}$$

Thus, the mean of X is

$$\alpha_{10} = \int_{-\infty}^{\infty} x \, dx \int_{-\infty}^{\infty} f(x_1 y_1) \, dy \tag{127}$$

and the variance of X is

$$\mu_{20} \equiv \sigma^2(X) = \alpha_{20} - (\alpha_{10})^2 \tag{128}$$

The expected value of the quantity $(X - a_{10})(Y - a_{01})$, which is designated μ_{11}, is an important statistical parameter called the *covariance*. The normalized covariance is called the *correlation coefficient*, σ_{12}.

$$\sigma_{12} \equiv \frac{\mu_{11}}{\sqrt{\mu_{20}\mu_{02}}} \tag{129}$$

There are many distribution functions which have been found useful in control applications. The commonest is the *Gaussian* or *normal* distribution.

$$F(x) = \frac{1}{\sigma\sqrt{2\pi}} \int_{-\infty}^{(x-\mu)/\sigma} e^{-t^2/2\sigma^2} \, dt$$

σ and μ are the two parameters needed to specify the function completely.

The *uniform distribution* function is a straight line. The distance between values of x where $F(x) = 0$ and $F(x) = 1$ is called the *range*.

The physical significance of a frequency function is easily grasped when it is in graphical form. If observed quantities are plotted in the form of a frequency function, the resultant diagram is called a *histogram*. Figure 71 shows a normal frequency function with a typical histogram superimposed.

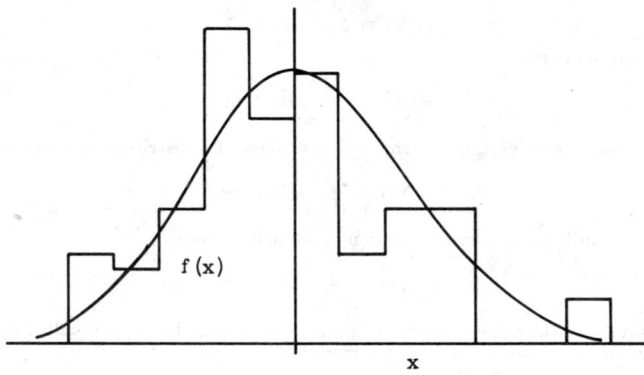

FIG. 71. Histogram.

A *random process* may be defined as a set of random variables whose elements are functions of time. The notion of quantities which vary in a random fashion with time is a central one in control theory. Those random functions which are considered to be outside the precise area of design of a control system have been called noise.

If the statistical parameters of a random process, like μ and σ, do not change with time, the process is called *stationary*. We will here be concerned primarily with stationary processes.

The *autocorrelation function* of a stationary random variable, $x(t)$, is given by:

$$\phi_{xx}(\tau) = \lim_{T \to \infty} \frac{1}{2T} \int_{-T}^{T} x(t)x(t + \tau) \, dt \tag{130}$$

Important properties of ϕ_{xx} are:

$$\phi_{xx}(\tau) \leqslant \phi_{xx}(0) \equiv \sigma^2(x) \tag{131}$$

$$\phi_{xx}(\tau) = \phi_{xx}(-\tau) \tag{132}$$

The *cross-correlation* function between two random variables, x and y, is given by:

$$\phi_{xy}(\tau) = \lim_{T \to \infty} \frac{1}{2T} \int_{-T}^{T} x(t)y(t + \tau) \, dt \tag{133}$$

In control theory it has become traditional to call the square of the value of a random variable at a particular time the *instantaneous power* associated with that variable. This quantity is proportional to the thermodynamic function called "power" if the random variable is a voltage or current. Here $x^2(t)$ is called "power" although the dimensionality may be inconsistent with the conventional definition. The reader should be careful to avoid confusion. The "energy" associated with the random variable is defined by analogy as

$$\int_{t_1}^{t_2} [x(t)]^2 \, dt$$

The *"power spectrum"* associated with a random variable is defined as

$$G(\omega, x) = \lim_{T \to \infty} \frac{1}{T} \left(\frac{1}{2\pi} \right) \left| \int_{-T}^{+T} x(t) e^{-i\omega t} \, dt \right|^2 \tag{134}$$

whenever the limit exists. The power spectrum function is identically equal to the Fourier transform of the autocorrelation function for stationary processes.

$$G(\omega, x) = \frac{1}{\pi} \int_{-\infty}^{\infty} \phi_{(xx)}(\tau) e^{-i\omega \tau} \, d\tau \tag{135}$$

$$\phi_{xx}(0) \text{ is equal to } \int_{-\infty}^{\infty} G(\omega, x) \, d\omega \tag{136}$$

These concepts form a background to the consideration of the reaction of a controlled system to unpredictable disturbances. The statistical parameters of random processes defined above are just those that are called for by the analysis of the dynamical properties of a system. Consider the time behavior of a linear system, such as the one of Fig. 4, for example. In an otherwise steady system, the response of the controlled variable to an impulse at time τ will be a function of time, t, $W(t, \tau)$. If the system equations are linear with constant coefficients, that is, if the system has a transfer function of the form considered before, then $W = W(t - \tau)$. W is in fact the inverse Laplace transform of the transfer function The transient response, $y(t)$, to a time varying input, $x(t)$, is just the convolution integral

$$y(t) = \int_{0}^{\infty} W(\tau) x(t - \tau) \, d\tau \tag{137}$$

The product of the y's at two different times is

$$y(t_1)y(t_2) = \int_{0}^{\infty} W(\tau_1) x(t_1 - \tau_1) \, d\tau_1 \int_{0}^{\infty} W(\tau_2) x(t_2 - \tau_2) \, d\tau_2 \tag{138}$$

$$= \int_{0}^{\infty} W(\tau_1) \, d\tau_1 \int_{0}^{\infty} W(\tau_2) x(t_2 - t_1) x(t_2 - \tau_2) \, d\tau_2 \tag{139}$$

Then

$$\phi_{yy}(\tau) = \int_{0}^{\infty} W(\tau_1) \, d\tau_1 \int_{0}^{\infty} W(\tau_2) \, \phi_{xx}(\tau - \tau_2 + \tau_1) \, d\tau_2 \tag{140}$$

and

$$\phi_{yy}(0) = \int_{0}^{\infty} W(\tau_1) \, d\tau_1 \int_{0}^{\infty} W(\tau_2) \phi_{xx}(\tau_1 - \tau_2) \, d\tau_2 \tag{141}$$

This is an expression for the mean square value of y, showing how it depends on the time response of the system and the autocorrelation of the disturbances. The mean square error in the output of the system is a commonly used measure of performance of a control system. Varying parameters in the controls will lead to variation in $W(t)$ and hence to variation in mean square output.

The power spectrum of y is similarly related to that of x by making the Fourier transformation of Eq. 139. After some algebra there results

$$G_{yy}(\omega) = \frac{1}{\pi} \int_{\infty}^{\infty} \phi_{yy}(\tau)e^{-i\omega\tau}\,d\tau$$

$$= G(i\omega)G(-i\omega)G_{xx}(\omega) \qquad (142)$$

where $G(i\omega)$ is just the ordinary transfer function for the process. $G(-i\omega)$ is the square of the deviation ratio. From Eqs. 142 and 136 we have

$$G_{yy}(0) = \bar{y^2} = \int_{-\infty}^{\infty} G(i\omega)G(-i\omega)G_{xx}(\omega)\,d\omega \qquad (143)$$

Disturbances pass through a controlled process in those frequency bands where the power spectrum of the disturbances and the deviation ratio are both large. If one or the other is small, then disturbances will not propagate. Therefore, an effective design procedure is to choose control parameters in such a way that peaks in the deviation ratio do not overlap peaks in the noise spectrum.

If minimum mean-square error is chosen as the criterion by which an optimum control system is to be chosen, then Eq. 143 can be used as the basis of a design procedure. The analytical treatment has been developed by Wiener. A description of its advantages and disadvantages will be found under *Optimal Control*.

For a more thorough treatment of random process, see references (6), (7), (8), (27), (29), (31), (36), (45), and (46).

12. OPTIMAL CONTROL

The preceding discussion of random processes has led to the consideration of one type of *optimal control*. This subject will be continued in what is to follow, but first it will be pointed out that there are at least two other kinds of automatic control processes which have been called "optimal." Here all three will be described in order. The first is that based on the theory of random processes, in which an optimal control system is one which makes the mean-square error, or some other related statistical function, a minimum. The second type of optimal control is that which takes a system from one state to another in minimum time, or with minimum expenditure of energy, or with maximum of some kind of value function. This type of control has been called *dynamic programming* by Bellman (2) and has been extensively studied by Pontryagin and his followers (37). The third type of optimal control is one in which a control system is used to continually assess some economic quantity, to determine if it can be improved, and if so, to take the necessary action. This last type of control is now being used to some extent in the process industries, with large digital computing systems included in the controls.

We consider first that part of optimal control theory which is concerned with minimizing the effects of random disturbances on the controlled variable. Here it is convenient to regard the controlled system as a generalized filter into which noisy signals are sent and from which comparatively noise-free outputs are to be generated. If this is done, then the detailed studies of filtering techniques which have been worked out for communication systems become appropriate for control systems.

There are three implicit assumptions which must be made to apply conventional filter theory to control system. The system must be linear, the statistical properties of the disturbances must be stationary, and minimum mean-square error must be an acceptable criterion of performance. The impulse response corresponding to optimal control in this sense is given by the solution of the Wiener-Hopf equation:

$$\phi_{yx}(t) = \int_0^t W(\tau)\phi_{xx}(t - \tau)\,d\tau \qquad (144)$$

which can be derived from Eq. 137 by variational methods. Here ϕ_{yx} is the cross-correlation function between input and desired output, ϕ_{xx} is the autocorrelation of the disturbance. The solution of Eq. 144 gives W, the response function of the optimally controlled system. Knowing this, it is then possible in principle to synthesize a control mechanism by the methods already described. For example, since the integral in Eq. 144 is a convolution, the Laplace transform of W, which is the optimal, closed-loop system transfer function, may be found in terms of the Laplace transforms of the correlation functions. From Eq. 15 one may solve for the transfer function of the optimal controller. If this corresponds to a physically realizable control system, then the technique gives a useful result. More often than not, however, the controller indicated by the transfer function found in this way cannot be implemented.

It is possible, in principle, to relax the assumption concerning the stationary character of the statistical properties and still derive and solve the corresponding Wiener-Hopf equation. The interested reader should refer to the literature for details.

A more directly applicable procedure for finding a locally optimal control, that is, a control system which is "tuned" to give minimum mean-square error, starts with the expression for mean-square error, Eq. 143. If G, the transfer function for the system contains one or more adjustable parameters, then best values of these parameters can often be found by differentiating Eq. 143 with respect to the adjustable parameters, setting the derivatives equal to zero, and solving for the locally optimal values. This gives, in general, a larger mean-square error than the Wiener-Hopf equation, but it gives the advantage of starting with a physically realizable controller.

One point must always be considered when a control system is used which is optimal in the sense of minimum mean-square error. The closed-loop system may be near the border of stability. This can lead to undesirably large overshoot if set-point changes are made, it can lead to possible instability if process parameters change, or if the analysis has been applied to a linearized version of a nonlinear system. Accordingly, the designer should not assume that a controller which minimizes mean-square error is optimal in every sense.

Dynamic Programming and the Maximum Principle

The second part of the theory of optimal control may be called the "determination of an optimal trajectory."

Given a system of differential equations that describe the dynamic behavior of a system,

$$\dot{x} = f(x, \mu, t)$$

with $x = x(t_0)$ at $t = t_0$, find $u(t)$ such that

$$J \equiv \phi[x(T), T] + \int_{t_0}^{T} G(x, u, t)\, dt \tag{145}$$

takes a minimum value.

In practical situations, J is usually an economic quantity and $u(t)$ is the output of some sort of control system. If there are no restrictions on $u(t)$ and if J can be differentiated, then the solution for the minimum K may be obtained by the classical methods of the calculus of variations, either through the use of the Euler-Lagrange equations or the Hamilton-Jacobi partial differential equation. The use of these procedures will not be described here because there are almost always restrictions on the control function u. It is for these cases that the elegant methods of Bellman and Pontryagin are particularly applicable.

Dynamic programming is the name which Bellman has given to the procedure he originated. It applies to a wide variety of minimization problems of the type of Eq. 145, particularly including discrete systems where Eq. 145 is replaced by a finite difference equation.

Bellman's principle of optimality states that "an optimal policy has the property that whatever the initial state (x) and initial decisions (u) are, the remaining decisions must constitute an optimal policy with respect to the state resulting from the first decision." From this principle, the equation

$$-\frac{\partial J^0(x, t)}{\partial t} = \min_{(u)} \left[G(x, u, t) + \frac{\partial J^0(x, t)}{\partial x} \frac{dx}{dt} \right] \tag{146}$$

can be derived. Here J^0 is the value function which results from optimal control. This reduces to the Hamilton-Jacobi equation

$$-\frac{\partial J^0}{\partial t} = G(x, u, t) + \frac{\partial J^0}{\partial x} \frac{dx}{dt} \tag{147}$$

when G can be differentiated with respect to u, and when u is continuous.

Bellman's methods are widely applicable and in principle can be used to find optimal control conditions where no other method will work. Unfortunately, the amount of computation required in any but the simplest cases makes the procedure very difficult to apply practically.

The equivalent to Eq. 146 for a stagewise process is

$$J_N = \min_{(u)} [G_N(u_N) + J_{N-1}]$$

where J_N is the optional value function at the Nth stage, G_N is the added contribution in moving from the $N-1$th to the Nth. This form is usually more convenient to use in dynamic programming calculations because of possible difficulties in solving the partial differential equation. Its use is illustrated in Fig. 72, where the problem is to move from

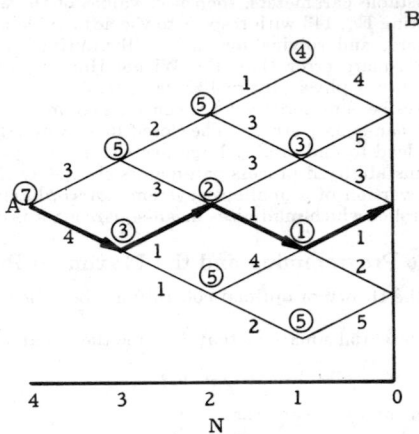

FIG. 72.   A simple dynamic program.

point A to line B jumping from one intersection to the next, to line B, with a minimum sum of values associated with the path segments. These path numbers correspond to the values of A in Eq. 147, and u_N corresponds to a choice of direction at a point. Values of J_N are circled below each node.

The solution proceeds from the right. At each node corresponding to $N = 1$ the minimum path value is determined and entered on the diagram. At $N = 2$, the value of J_2 is the lowest sum of (path $+J_1$), and so forth. The optimal path is indicated by arrows.

Pontryagin has called his method of finding an optimal control policy the *maximum principle*. It is a necessary but not sufficient condition for optimality.

Given the dynamic state equations for n variables,

$$\frac{dx^i}{dt} = f^i(x, n) \qquad i = 1, 2, 3, \ldots, n \tag{148}$$

find the control function, which causes the functional

$$x^0 = \int_{t_0}^{t} G(x, u)\, dt \tag{149}$$

to take its minimum possible value. Pontryagin starts by forming the linear homogeneous system

$$\frac{d\psi_i}{dt} = -\sum_{0}^{n} \frac{\partial f^\alpha}{\partial x^i} \psi_\alpha \tag{150}$$

The ψ_i are called the "adjoint" variables; Eq. 150 the adjoint equations. Notice that f^0, or G, the integrand of Eq. 149, is included in Eq. 150.

He next forms the function $H(x, \psi, u)$, defined by

$$H = \sum_{a=0}^{n} \psi_a f^\alpha(x, u) \tag{151}$$

Notice that

$$\frac{d\psi_i}{dt} = -\frac{\partial H}{\partial x^i} \tag{152}$$

and

$$\frac{dx^i}{dt} = \frac{\partial H}{\partial \psi_i} \tag{153}$$

The maximum principle states that the optimal control function u, that is, the $u(t)$ which causes x_0 to take its minimum value is that u which causes H to take its maximum possible value at every t. There is no requirement for continuity of u; indeed, it is quite common for the optimal u to jump back and forth between constraints, resulting in *bang-bang* control.

A simple illustration of how this important method can be used follows: Consider the system described by

$$\frac{dx^1}{dt} = x^2 \tag{154}$$

$$\frac{dx^2}{dt} = u \tag{155}$$

with

$$|u| \leqslant 1$$

Minimize the time for the system to reach $x^1 = x^2 = 0$, starting from any given initial state, $x^1(0)$, $x^2(0)$. Let

$$x^0 \equiv \int_0^t dt \tag{156}$$

Then minimum x^0 is the desired condition for optimality. f^0 of Eq. 149 is 1. Since f^0, f^1, and f^2 do not depend on x^0 in this problem, ψ_0 may be disregarded. Form

$$\frac{d\psi_1}{dt} \equiv \frac{\partial f^1}{\partial x^1} \psi_1 - \frac{\partial f^2}{\partial x^1} \psi_2 = 0 \tag{157}$$

$$\frac{d\psi_2}{dt} = -\frac{\partial f^1}{\partial x^2} \psi_1 - \frac{\partial f^2}{\partial x^1} \psi_2 = -\psi_1 \tag{158}$$

The solution of Eqs. 157 and 158 is

$$\psi_1 = C_1 \tag{159}$$

a constant, and

$$\psi_2 = C_2 - C_1 t \tag{160}$$

Then

$$H = C_1 x^2 + (C_2 - C_1 t) u(t) \tag{161}$$

H takes its maximum possible value for $u = +1$ when ψ_2 is positive, and for $u = -1$ when ψ_2 is negative. This is a case of "bang-bang" control. The optimal u will start at (1) $\times$ (sign of C_2), and will change its sign once only.

The determination of integration constants corresponding to C_1 and C_2 is the most difficult part of using the maximum principle. In the simple case above, they can be found from the requirement that x^1 and x^2 must start at given initial values and end at (0, 0). In the general case where the equations are not linear, the problem of finding integration constants or the equivalent problem of finding a suitable set of initial conditions for Eq. 150 is not trivial. It is likely to require extensive machine computation, just like dynamic programming does, and makes widespread use of the method unattractive except in the simplest cases.

A comparison between the results of an analysis by dynamic programming and a solution of the Pontryagin equation is of interest. If a system has a single well-defined optimal policy, then the two give identical results, as they should. If several locally optimal policies exist, then dynamic programming will find the true optimum, the maximum principle will locate one of the local optima. A continued search over the set of possible initial conditions may detect the presence of other local optima, however. In the presence of disturbances the two methods differ slightly in the way they are applied. An optimal control policy determined by the maximum principle results in a single trajectory in state space, that in which the expected value is a maximum or minimum. Dynamic programming results in a family of trajectories with the particular one chosen in an individual case depending on the actual disturbance pattern that has occurred. Therefore the time-dependent behavior of the controls in the two cases will not be the same, with dynamic programming giving the true optimal policy in the sense of an extreme in the expected value. The difference is that the maximum principle attempts to form a complete policy from the start, whereas dynamic programming allows for continuing changes in policy which are made necessary by unforeseen disturbances.

This description of both of these extremely effective procedures in control implementation has been necessarily brief. There is an extensive literature on both subjects which should be consulted for a thorough explanation. See references (2), (6), (7), (8), (31), (32), (33), (36), (37), and (40).

COMPUTERS IN CONTROL

We now turn to that part of the subject of automatic control where the use of conventional theory and conventional control systems is not possible. This area is still in the state of active development, hence descriptions here must necessarily be less quantitative than those already given. The common characteristic of the *modern control theory* appears to be that machine computation of some sort is required.

13. OPTIMAL CONTROL

The third type of control action which can be called *optimizing* operates in such a way as to keep the controlled system in or near that steady state which is of greatest value. This kind of control can be used best when the quantity whose maximum is sought varies with time only very slowly, such that the associated time scale is long compared with those of the other dynamic processes. It is, of course, possible to use the same techniques when the time scales are not so conveniently separated, but then care must be taken to account properly for the interactions which inevitably result. There must be some unpredictable time dependence of the value function, of course, since otherwise the control problem would be of no significance, and the optimal control procedures would have to be determined only once and used unchanged thereafter.

In steady-state optimal control the controller is a feedback device which takes signals from the process under control and uses them to determine if the current state of operation can be improved. If so, it takes appropriate action. A common control system for this sort of operation is a digital computer into which are continually read operating conditions and other pertinent information. The optimal steady-state conditions are determined by techniques used in operations research. The result of these calculations is a set of optimal control points such as temperatures, flows, and speeds.

The operating conditions in most systems are almost always constrained in some way, which means that they cannot be obtained by the methods of differential calculus. This circumstance is illustrated in Fig. 73 which shows a constrained function of two variables. Contours of the value function indicate an optimum at point 1, but constraints limit the usable region in such a way that 2 is the highest value that can be obtained. After a disturbance the value function might change in such a way as to change the character of the optimum. The contours of Fig. 73 might change to those of Fig. 74, for example.

Because of the existence of constraints, it is often necessary to use highly sophisticated mathematical methods to find the optimal controls. A few of these will be described briefly here—for more detail the reader should refer to the literature, which is voluminous. See, for example, references (2), (5), (7), (8), (22), (31), (32), (33), (40) and (41).

Linear programming is the simplest technique. Its objective is to find the set of x's which maximize a linear function

$$F = \sum_j C_j x_j \tag{162}$$

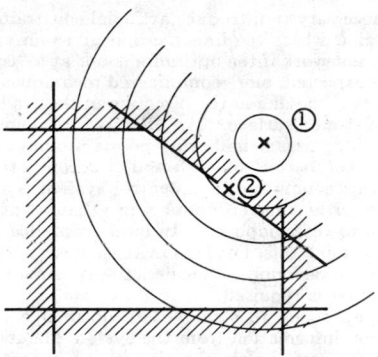

FIG. 73. Constrained optimum.

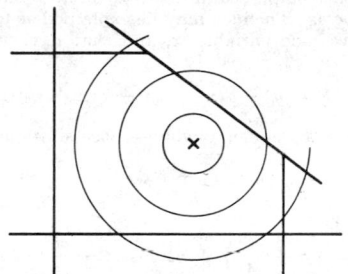

FIG. 74. Effect of a disturbance on the optimum.

subject to a set of linear constraints

$$\sum_j a_{ij}x_j \leq b_i \tag{163}$$

If one regards the set of x's as the coordinates of an n-dimensional space, then the constraints, Eq. 163, form "hyperplanes" in that space which bound a "hyperpolyhedron." Thus in Fig. 73 the constraints are lines in a two-dimensional space, the region in which the system is constrained is a polygon. It is evident that since Eq. 162 is linear, the curved contours of Fig. 73 cannot exist and the maximum of F will occur at a corner. This fact is used in the *simplex method* of solving linear programming problems. A corner is located as a starting point, then each edge through that corner is investigated to determine if F increases along it. If not, then the corner is the optimum point. If so, then the corner at the other end of an appropriate edge is investigated in the same way. If several favorable edges are found, the one along which F increases most is chosen. Ultimately the optimum will be reached.

As has been the case in many other situations already, this simplest linear method does not usually work in control applications. The reason is that systems of real interest usually have nonlinear objective functions and constraints.

The degree of modification necessary to make a prodecure of this sort work in a system which is not linear depends, of course, on the complexity of the system. The simplest approach is to break the space into regions, in each of which the nonlinear functions can be approximated linearly. The problems of crossing from one region to another make the use of the technique considerably more complicated than the simplex method.

A different approach, which requires somewhat more computation than the above, is to linearize the equations about the point under test—the current corner point in the simplex method—using the first two terms of the Taylor's series. This linearized model is then used with the simplex method to generate a new point to test. The linearization is made again about this point, and the whole process repeated until the objective function changes no further. It is often necessary to introduce artificial constraints at each step, outlining the boundaries of the region where the linearization gives an acceptable approximation. These modifications do not work if the optimum is not at a "corner," such as in Fig. 74. If this type of behavior is expected, more complicated techniques must be used. Methods have been proposed involving moving in the direction of the gradient of the value function or its projection on a constraint surface. An effective "probing" method has been suggested in which the computer tries neighboring points around a current test point systematically or at random. If a "better" one is found, it becomes the next center for probing.

The real point of all this computing in a control system is to determine if there is a way to operate that is better in some economic sense than what was in effect at the time measurements were given to the computer. In large, complicated systems much computation is necessary to determine this, but it is in these large systems where incentives for doing this sort of thing are to be found. Needless to say, an accurate and detailed mathematical model of the system is required.

A somewhat different approach to steady-state optimal control avoids the use of a model and instead relies on information from the system about its state of "optimality." This approach may involve direct experimentation on the system, disturbing it at intervals by small changes in each of the significant inputs, or continuously probing it with periodic test signals.

The latter method has been employed in a variety of automatic self-optimizing systems, sometimes called *hill-climbers*. The idea may be explained as follows. Suppose the value function, F, depends on a single variable, x, and that F varies quadratically about the optimum:

$$F = a + b(x - x_0) + c(x - x_0)^2 \qquad (164)$$

Let x vary sinusoidally at a frequency where phase shifts are small:

$$x = x_0 + \delta \cos \omega t \qquad (165)$$

Then

$$F = \left(a + \frac{c\delta^2}{4}\right) + b\delta \cos \omega t + \frac{c\delta^2}{4} \cos 2\omega t \qquad (166)$$

There is a sinusoidal component at the test frequency whose amplitude is proportional to b. If b is positive, the signal is in phase with the input; if b is negative, the component is 180° out of phase. This is enough information to enable the control system to seek a new approximation to the optimum performance, where $b = 0$. A new value of x_0 would be used.

There are practical difficulties with this approach to optimization, of course, but most of them appear to be surmountable. If noise is present containing a significant power density at ω, then it interferes with the operation. If ω is too small, the response is slow. If ω is too large, phase effects may become significant and a more complicated controller would be necessary. The treatment of constrained optima is not clear. If more than one independent variable is present, a different test frequency for each of the variable would have to be used and, therefore, possible interference by harmonics and sum or difference frequencies would have to be considered.

This method is simple enough to implement computationally, and it requires a minimum of model building, only that required to determine noise characteristics and phase shifts at a few frequencies. The procedure will not give a true dynamic optimum in the sense of Bellman, but it does appear to be capable of reaching nearly an optimal steady state.

14. ADAPTIVE CONTROL SYSTEMS

An adaptive controller may be defined as a control system which continually and automatically readjusts itself for proper operation in the presence of changing system dynamics or noise characteristics by using measurements or signals from the system.

Figure 75 shows the two parts of an adaptive controller, which have to be considered separately. The first part is where signals from the system are analyzed to yield information about its dynamics, the second is where this information is used to readjust the controls.

The first accomplishes what has loosely been called *process identification*. It is the critical part of the adaptive procedure, and invariably involves finding some part of the cross-correlation function between input and output signals of the system. We have seen that

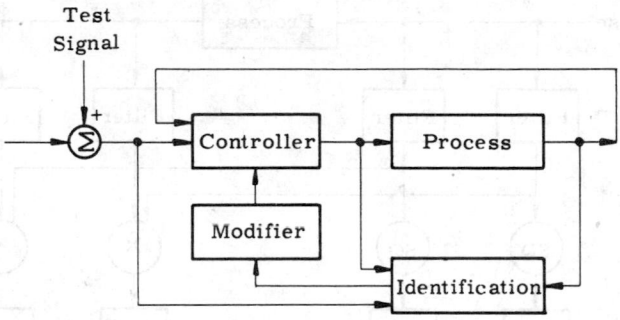

FIG. 75. Adaptive control system.

this cross-correlation is related to the autocorrelation of the input signal and the transfer function of the elements in between by

$$G_{xy}(s) = G(s)G_{xx}(s) \tag{167}$$

or its equivalent in frequency terms obtained by replacing s by $i\omega$. $G_{xy}(i\omega)$ and $G_{xx}(i\omega)$ are then the ordinary power spectra. An equivalent relationship is, of course,

$$\phi_{xy}(t) = \int_0^\infty W(\tau)\phi_{xx}(t - \tau) \, d\tau \tag{168}$$

where the ϕ's are the correlation function and W is the impulse response of the system. Automatic "process identifiers" use one or the other of these relationships.

For example, "white noise" might be used as an input signal. It can be deliberately added continuously or else that which is naturally present can be used. The former method is preferred if the system can tolerate the disturbance, because if the controls are properly designed there will be little overlap between the noise spectrum and the frequency range of interest in the system. "White noise" here means a random signal with a broad flat power spectrum. Its autocorrelation is a peak centered about $\tau = 0$. From Eq. 167, the amplitude frequency response of the system is just the power spectrum corresponding to the cross correlation between input and output when noise is introduced. From Eq. 168, since $\phi_{xx}(t - \tau) \approx \delta(t - \tau)$

$$\phi_{xy}(t) = W(t) \tag{169}$$

The time-response of the system to an impulse is equal to the cross-correlation function.

Adaptive methods rarely require the use of the entire spectrum or time response, therefore, implementation in a practical situation need not be complicated. Figure 76 shows

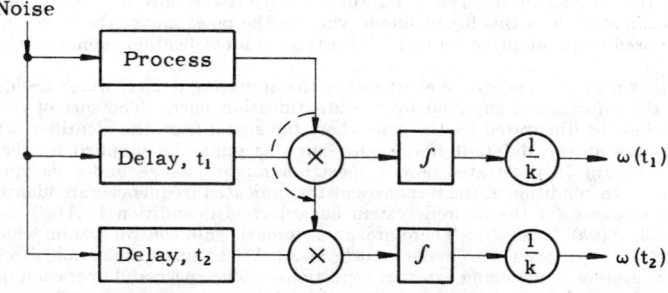

FIG. 76. Time-response identification.

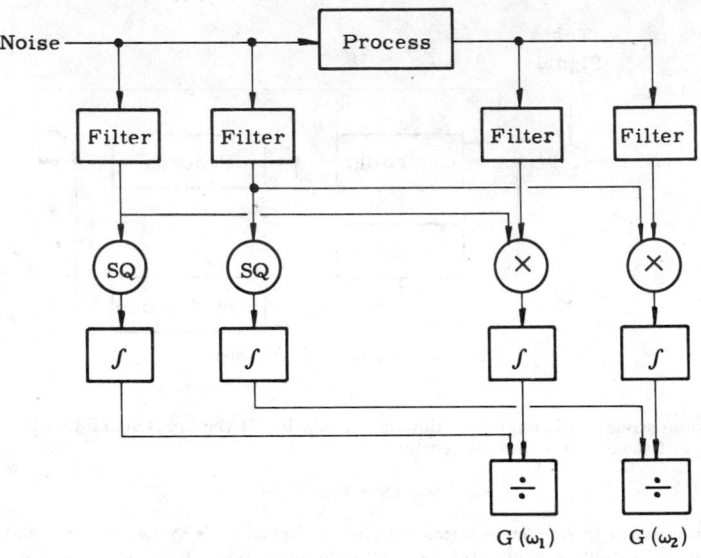

FIG. 77.   Frequency identification.

how the value of $W(t)$ for two values of t may be found.   Figure 77 shows how the amplitude of $G(i\omega)$ may be obtained for two frequencies.   In the first case the integration must be over a fixed interval, long enough to establish an accurate correlation coefficient, but not related to the time delay.   This procedure is not always handy to use because accurate time delay is difficult and requires a high-speed memory.   The second procedure is often more useful because suitable filters can easily be provided.

If the noise should be ideally "white," its power spectrum would be constant, and hence neither the two filters on the input nor the division would be necessary.   If the noise is just that which is naturally present, for example, if external disturbances are undesirable, then all the filters must be in the system.

A useful variation of this scheme is to inject as a disturbance only those frequencies which are of interest in the transfer function.   Input filtering is then completely unnecessary and output filtering becomes simpler because unnecessary signals have not been used.

A different type of identification has been proposed for linear systems where a single pair of poles in the root locus dominate the transfer function.   The aim of an adaptive controller might be to keep the location of these poles fixed as external effects tend to move them.   Movement toward the axis tends to make the system less stable, as has been shown.   It also changes the effective damping rate.   Therefore, if the damping rate can be detected, it can be used to automatically adjust the system.   This is done in one type of adaptive controller, used in an Autopilot, by periodically disturbing the system with impulses, then counting the number of changes in sign of the output response in a fixed time interval.   Analysis can show how this figure should vary as the poles move; therefore, in principle, it can be used in an adaptive control.   This type of identification cannot work in a noisy system.

The other part of an adaptive controller is the actuating device, which decides what to do with the information supplied by the identification part.   The sort of thing that is done can best be illustrated by the case where the signal from the identifier is the amplitude response at two different frequencies, such as might be supplied by the device of Fig. 77.   Figure 78 illustrates how a spectrum might change under varying external conditions.   In condition 2, the responses at the indicated frequencies are identical.   This might be the case for the desired system behavior.   In condition 1, $A(\omega l) > A(\omega_n)$, in condition 3, $A(\omega h) > A(\omega l)$.   Therefore, an automatic gain control system which changes the controller settings so as to make $|A(\omega l) - A(\omega h)|$ as small as possible can be used to adapt this system to changing external conditions.   The successful operation of this kind of adaptive control depends on the ability of the designer to find a pair of frequencies like ωh and ωl which are sensitive indicators of the dynamic characteristics of the system.

Conditions

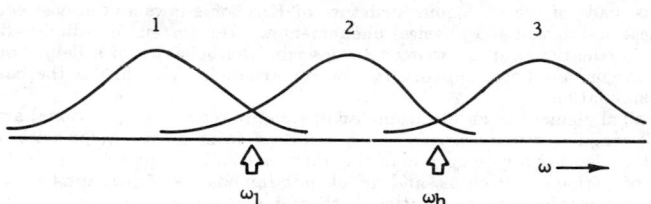

Fig. 78.   Frequency spectrum in adaptive control.

It will be clear to the reader that if a large computer is to be part of an adaptive control system, then far more sophisticated computations than the one just described can be used. A variety of adaptive methods have been proposed, and the interested user must consult the literature for details.

15. SIMULATION

The impact of "the computer revolution" which started in the 1950s has not been stronger anywhere than in the field of automatic control.   The first of the two basic ways modern computing equipment can be used in control engineering is to perform some of the algebra, trial and error estimates, graphical procedures, and calculations which have been described in the preceding pages.   This is simulation.   The second way is, of course, to use the computer directly as a part of the control system.

Simulation has been defined as "the development and use of models for the study of ideas, systems, and situations."   Therefore, a simulation need not involve a computer. However, here we will be concerned especially with the situation where it does.

In approaching a problem where simulation is indicated, the engineer or scientist generally proceeds through a series of steps about as shown in Fig. 79.   He must correctly formulate the problem, correctly proceed to a solution, and above all he must assure himself that the answers he develops are the correct ones.   He must develop internal checks where necessary, and he must be prepared to verify and provide a rational explanation for unexpected results.   A modern computer can generate answers so quickly and easily that there is a tendency to forget that it cannot think.

The science of real-time simulation using digital computers has not yet reached the stage where it is widely and generally usable, as analog computer simulation has been for some years.   A number of digital simulation languages, most of which are quite easy to use, have been made available by manufacturers.   However, even with the highest speed computers they tend to be slow and cannot be made to run synchronously except with further slowing.   The acronymic titles of the digital simulation languages that were proposed as of the end of 1966 include DAS, MIMIC, PACTOLUS, DSL-90, DES-1, DYANA, ECAP, and PACER.   The most important parts of these simulation programs are compilers which connect programs prepared from simple block diagrams to machine language, integration procedures, printout and display programs, and error diagnostics. There is much hope that this type of simulation program, with its increasing use of integrated circuits, will make parallel digital computation fast enough and cheap enough so that analog computation, with its high-speed but limited accuracy, will become less attractive.   However, analog computers and their descendent hybrids, are now the only real-time tools for satisfactory simulation, whereas digital computers can be used effectively for their memory, logic, peripheral equipment, and high accuracy in slow computation.

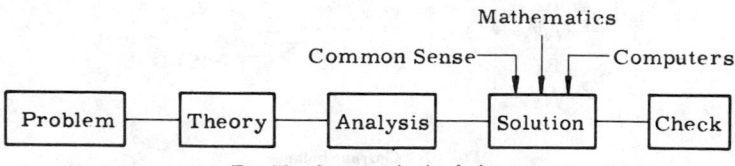

Fig. 79.   Sequence in simulation.

A modern analog computer is in essence a large programmable electrical network, in which use is made of the analogous structure of Kirchoff's laws and almost every other mathematical description of a physical phenomenon.   If a system of ordinary differential and algebraic equations can be written to describe the behavior of a device or process, then an analogous electrical network can be constructed.   This fact is the basis of the science of simulation.

The principal element of an electronic analog computer is the operational amplifier, a carefully designed electronic circuit which is linear from d-c to frequencies of several hundred kilocycles, has a voltage gain of the order of a million (always negative), an input impedance of perhaps 10^{12} ohms, and an output impedance of perhaps 0.01 ohm.   Its commonly used symbolic representation is that of Fig. 80.

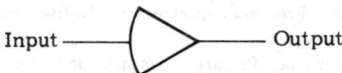

Input ————⟩———— Output

Fig. 80.   High gain operational amplifier.

If the voltage at the input to an amplifier, commonly referred to as the *summing junction*, exceeds a few microvolts, then the amplifier becomes *saturated* or *overloaded* because the power supply cannot force the output voltage high enough to give the correct gain.   Therefore, if an amplifier is to be operated correctly, its summing junction must be very close to ground potential, and is usually treated as such in programming.   In Fig. 81 the current balance around this summing point is

$$\frac{V_1}{R_1} + \frac{V_2}{R_2} + \frac{V_3}{R_3} + \frac{V_4}{R_4} = i_{sj} \tag{170}$$

or

$$V_4 \cong - \left(\frac{R_4}{R_1} V_1 + \frac{R_4}{R_2} V_2 + \frac{R_4}{R_3} V_3 \right) \tag{171}$$

if i_{sj} is neglected.   Typical V/R values are of the order of 10^{-4} A, thus since i_{sj} is typically 10^{-12} A (Eq. 171), it is sufficiently accurate, particularly since this degree of permanence in resistor values is questionable.   With suitable choice of resistor values, an operational amplifier can generate a voltage which is equal to the negative of a weighted sum of voltages fed into it.   If there is a single input, and the *input* and *feedback* resistors are equal, then the amplifier inverts the voltage.   The sum of one voltage and a second voltage which has been inverted is equivalent to subtraction.   If input and feedback resistors are different sizes, the amplifier multiplies the input voltage by a negative constant.   Figure 82 shows a simpler symbol that is sometimes used for a summer-inverter.

A second device which can be used for multiplying a voltage by a fixed constant between zero and $+1$ is the potentiometer.   This is commonly a helical resistance wire around which a movable slider can be positioned.   Two symbols commonly used are shown in Fig. 83.

Since electrical circuits are very nearly linear, the generation of even the simplest non-linearity, like a product, requires special consideration.   One way this can be done is to

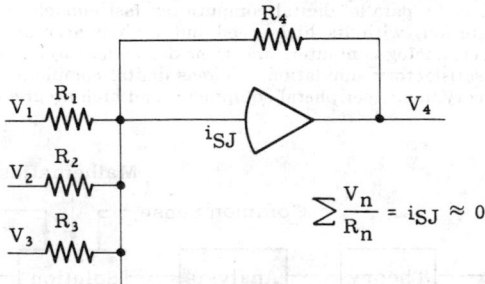

Fig. 81.   Current balance.

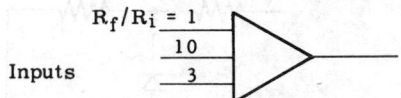

FIG. 82. Summer-inverter.

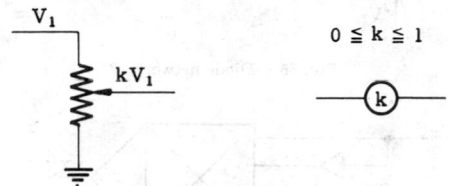

FIG. 83. Potentiometer.

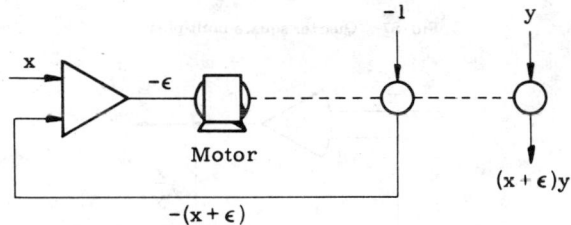

FIG. 84. Servomultiplier.

use the feedback control concept as shown in Fig. 84. Two or more potentiometers are mounted rigidly with a common shaft turning the sliders exactly together. A negative unit voltage, which commonly is 10 to 100 V, is applied across one of the potentiometers. The slider voltage is added to the voltage representing one of the factors. If they differ, then the motor turns the shaft in the direction which eliminates the error. The second factor voltage is applied across the other potentiometer. Its slider then reads the product. This device is called a *servo-multiplier*. It is quite slow in operation because of the motor. Faster multiplication requires the use of special nonlinear networks using diodes. A diode allows current to flow in one direction only if polarities are in the indicated directions in the symbolic representations of Fig. 85. Combinations of diodes and resistors can be made such that the current-voltage curve is a series of straight-line segments. The circuit of Fig. 86 gives the current indicated there. Almost any shape of curve can be approximated by straight-line segments in this way, in particular a network can be made in which the current into the summing junction of an amplifier is approximately proportional to the square of the voltage applied.

The *quarter square* multiplier uses two of these circuits, based on the identity

$$xy = \frac{1}{4}\left[(x + y)^2 - (x - y)^2\right] \tag{172}$$

Its symbol is shown in Fig. 87. In most modern computers both polarities of input voltage must be provided. This has been indicated by two amplifiers on the inputs.

Solid State   Vacuum Tube

FIG. 85. Diode.

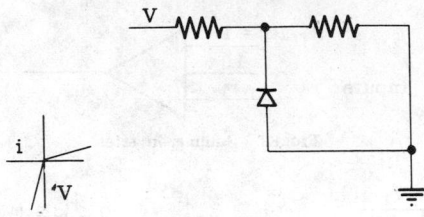

FIG. 86.   Diode network.

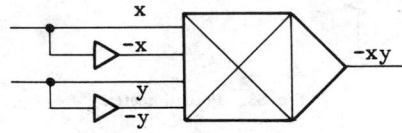

FIG. 87.   Quarter square multiplier.

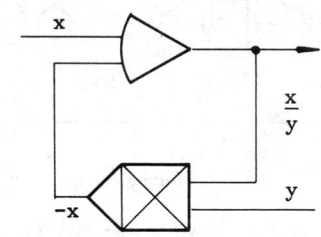

FIG. 88.   Division.

Division of voltages is accomplished by using a multiplier in the feedback path of an amplifier, as indicated in Fig. 88.

With the quarter-square multiplier, multiplication and divisions may be performed accurately at high frequencies. These complete the description of how the ordinary algebraic operations are performed on an analog.

Next, we turn to the operations of the calculus, which enable differential equations to be solved on the analog. Consider the circuit of Fig. 89. The charge on a capacitor is the integral of the current, therefore the derivative of the charge is the current. The charge is the potential times the capacitance, which is fixed, thus current balance around the summing junction gives

$$\frac{V_1}{R} + C\dot{V}_2 = 0 \tag{173}$$

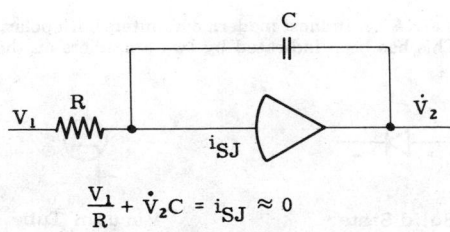

$$\frac{V_1}{R} + \dot{V}_2 C = i_{SJ} \approx 0$$

FIG. 89.   Current balance.

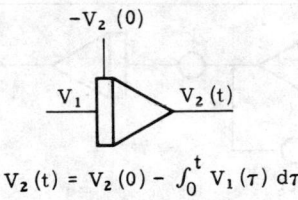

$$V_2(t) = V_2(0) - \int_0^t V_1(\tau)\, d\tau$$

Fig. 90. Integrator.

or

$$V_2(t) = V_2(0) - \frac{1}{RC}\int_0^t V_1(\tau)\, d\tau \tag{174}$$

Thus this circuit is an accurate integrator with a proportionality factor or *time constant* of RC. A common symbol is shown in Fig. 90. The initial condition can be set in with a special network furnished with most computers.

Differentiation can be performed with an input capacitor and a feedback resistor; however, it is not generally recommended because input signals are often contaminated by high-frequency noise, and the differentiator circuit amplifies this noise in direct proportion to its frequency. Since differential equations can always be written in terms of integrals, this causes no inconvenience.

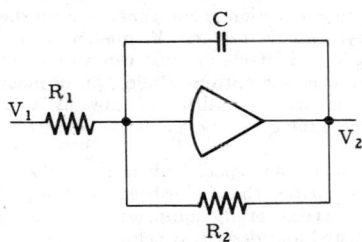

Fig. 91. Time lag.

The circuit of Fig. 91 generates a simple time lag, as can be derived from its current balance:

$$\frac{V_1}{R_1} + \dot{V}_2 C + \frac{V_2}{R_2} = 0 \tag{175}$$

or

$$\dot{V}_2 + \frac{1}{R_2 C} V_2 = \frac{V_1}{R_1 C} \tag{176}$$

The transfer function of this circuit is

$$\frac{1}{R_1 C}\frac{1}{s + (1/R_2 C)}$$

Its time response to a pulse input is a simple exponential decay, with its time constant equal to $R_2 C$.

The circuit corresponding to a damped sinusoid is shown in Fig. 92. Its differential equation is

$$\ddot{y} + 2k\zeta\dot{y} + k^2 y = f(t) \tag{177}$$

The transfer function is

$$G(s) = \frac{1}{s^2 + 2k\zeta s + k^2} \tag{178}$$

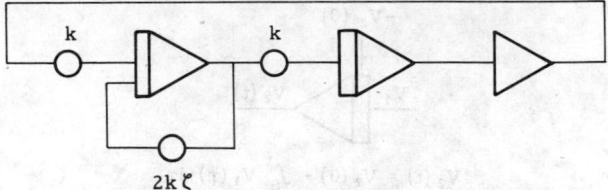

FIG. 92. Sinusoid.

A particular precaution that must be taken with analog computers is in scaling voltages to represent other physical variables. This must be done consistently for all variables, and must be done in such a way that no amplifier is ever overloaded. Since amplifier outputs can never be fully predicted, some problems may require rescaling because of a bad initial guess about the range of a dependent variable. This is particularly troublesome if nonlinear equipment is involved in the simulation. The necessity for proper prior scaling is one of the major disadvantages of analog computation.

With these basic circuit elements, the control engineer can proceed directly from a block diagram of a physical system to a wired program board without writing down differential equations, because the circuits corresponding to the common transfer functions are combinations of those shown in Figs. 90, 91, and 92. Special nonlinearities can be simulated by special networks. Simple ones such as multiplication, squaring, exponentiation, and generation of logarithms are often prewired directly to the program board.

Therefore, by suitable interconnection of components, even the most complicated system can be simulated in full dynamic character. Frequency response, stability limits, sensitivity to parameter changes, and effects of noise can all be tested and studied. Adjustments of the simulated controls for optimum rejection of noise can be done. Different types of control can be tried experimentally. If care has been taken in formulating the system dynamics, then the analog can be made to behave precisely like the device it simulates.

Figures 93 through 97 show some special diode networks for simulating some of the special discontinuous nonlinearities that have been discussed previously.

Modern analog computers are generally equipped with additional types of special equipment, some of which are listed and described below.

A. Comparator: equivalent to the on-off control of Fig. 94.

B. Resolver: a circuit for rapidly and continuously converting from polar to Cartesian coordinates or the reverse.

C. Relays and switches, operated by voltages generated in the simulation.

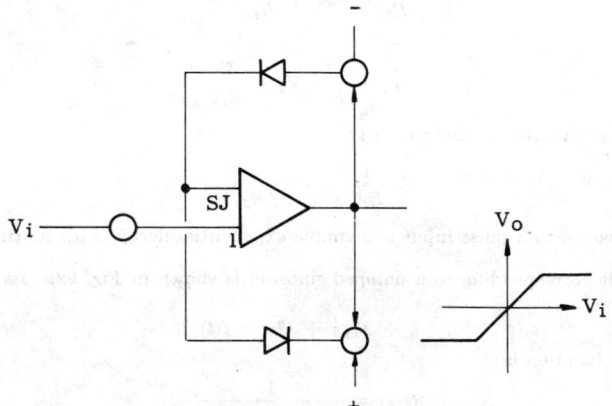

FIG. 93. Saturation.

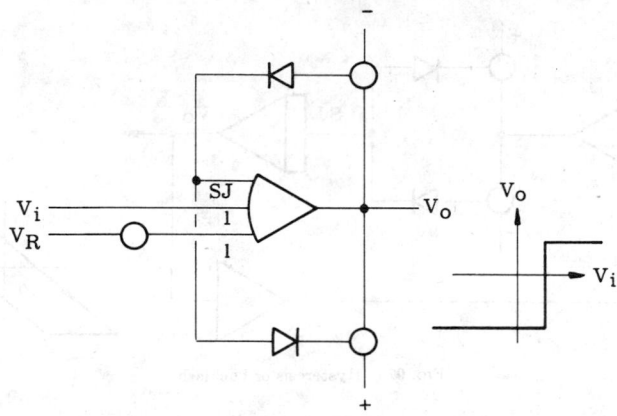

FIG. 94. On-off controller.

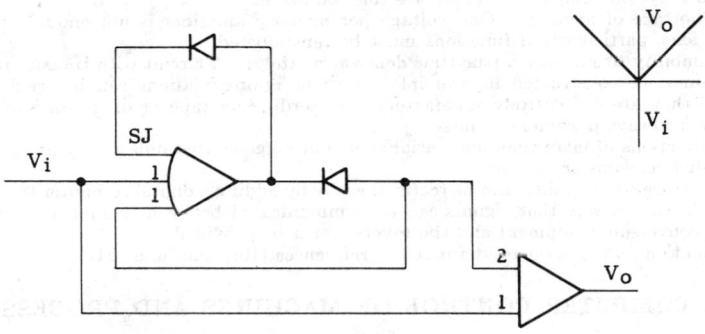

FIG. 95. Rectifier.

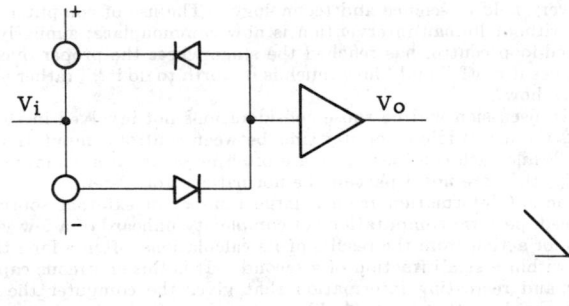

FIG. 96. Dead zone.

D. Memory amplifiers: An integrator into which an initial condition is set, but to which no integrating input is patched, will "remember" the initial condition.

E. Repetitive operation at high speed: solutions can be generated to repeat hundreds of times per second.

F. Outputs can be displayed on an oscilloscope, plotted on coordinate paper, or on continuous recording equipment.

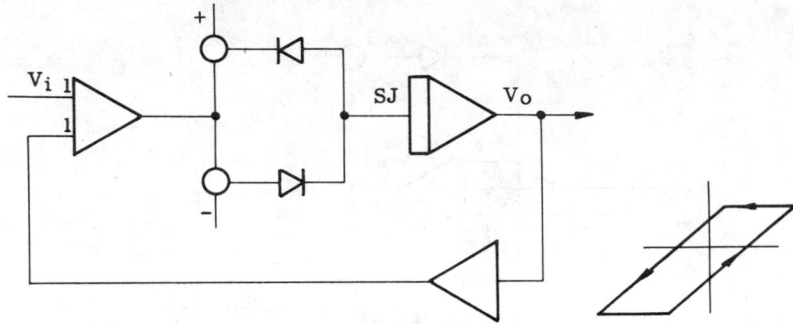

FIG. 97. Hysteresis or backlash.

All these features make an analog a highly useful tool for design of control systems. However, there are shortcomings which can only be overcome by adding digital equipment to form a hybrid computer. These are the vollowing.

A. Shortage of memory. One voltage per memory amplifier is not enough in many simulations, particularly if functions must be remembered.

B. Inability to simulate a true time delay accurately. A circuit with transfer function e^{-Ts} cannot be constructed in principle. Various approximations can be tried, but in general they are not entirely satisfactory. Recording on tape or drum for subsequent reading is always possible, of course.

C. Functions of more than one variable cannot be generated unless they are of special form, such as sums or products.

All of these shortcomings can be rectified easily by adding a digital computer that is connected in such a way that signals can be communicated between them—i.e., analog-to-digital conversion equipment and the reverse must be provided.

Simulation and its uses are discussed in references (10), (22), and (24).

16. COMPUTER CONTROL OF MACHINES AND PROCESSES

It will be clear from what has been said than an analog computer can be used directly as a process controller with considerable sophistication built into it. We now turn to the use of the modern digital computer, the device whose development has already had enormous influence on every field of science and technology. The use of computers for processing information without human intervention is now commonplace; similarly the use of computers in closed-loop control has reached the stage where the proper questions to ask are "how much does it cost?," and "how much is it worth to do it?," rather than "can it be done?" or "if so, how?."

Excluded from this discussion will be those considerations not involved in the control system itself, such as the use of idle computer time between control computations for accounting purposes. While such considerations are often necessary if a computer is to be justified economically, they are not a part of the normal control system.

A computer can accept information from a large number of external sources at an incredible rate of speed, perform computations of complexity unheard of a few years ago, and generate signals for action from the results of its calculations, often a long time after the fact, sometimes within a small fraction of a second. It is this enormous capacity for accepting, digesting, and recreating information that gives the computer the valuable position in modern technology that it now holds.

There are a great many different types of technological applications for digital computers in control. There are almost as many characteristics about the way they are used; so many that only an outline can be given here.

One of the largest applications has been in the area of electrical utilities Computers have been used for logging, monitoring, and alarming functions in steam-power plants. This includes periodic scanning of all possible points where dangerous conditions may arise, calculation of heat and material balances and efficiencies, and determination by linear programming of optimum operating conditions as loads change. Starting and stopping sequences can be rather complicated because of the many requirements for safety and the many different conditions that can occur when a shutdown commences. The kind of logic required is well suited to fast computer control. At each step in the procedure the program can check to make sure that all prerequisite conditions have been met.

The dispatching of power from several sources to scattered load points is another part of the utilities system where computers can be used effectively. Here mathematical programming techniques can determine the most economical distribution pattern, taking into account costs at each of the power stations, transmitting losses, and insuring adequate reserve in case of failure at any point.

The steel, aluminum, cement, glass, paper, and textiles industries involve large, complex operation, where continuous logging, calculation of efficiencies, and optimum performance conditions can create strong incentives for the use of computers in control.

The manufacturing industries have found many applications for computers for testing and production control. While this is not control in the sense that has been discussed, it is made possible by the ability of the computer to absorb enormous quantities of information, to check against standards, and to direct the performance of fixed procedures. Thus diodes and transistors can be tested individually for each of the twenty or thirty performance specifications each should meet. Bearings, engines, valves, and so forth can be put through programmed endurance or wear tests. In all of these applications the computer is programmed to proceed through a fixed sequence of operations, to collect measurements, to perhaps perform calculations against standards, and to report results. The petroleum, gas, and chemical industries have made extensive use of computer control in recent years. In petroleum, usage has extended from the oil field through control of dispatching in pipelines, to control of individual refining units, and to control of the distribution of finished products. In the chemical industry several ethylene, vinyl chloride, acrylonitrile, and ammonia processes have been reported as having been operated under computer control.

This brief review has been for the purpose of illustrating how broadly the concept of computer control has been used and how diversified are the types of use. From this point on, the subject will turn to specific ways a computer can be used, with minimum reference to any particular industry or process. The great mass of recent literature should be consulted for details about specific circumstances under which computer control has been applied or proposed.

Figure 98 is a diagram of a generalized computer-controlled system. Here the used "process" can mean anything which is under computer control. Signals move both ways along the indicated paths. Several different types of computer usage have been given special names, of which the following are the most widely used:

Off-line advisory control; Paths 1, 2, and 3 are used. Information from the process is given to the computer by hand, and instructions from the computer are put back into the process instruments by one or more human operators. This type of computer usage is sometimes called open-loop operation. In the initial stages of installation and operation of a computer control system, advisory control is used for a period before any loops are closed.

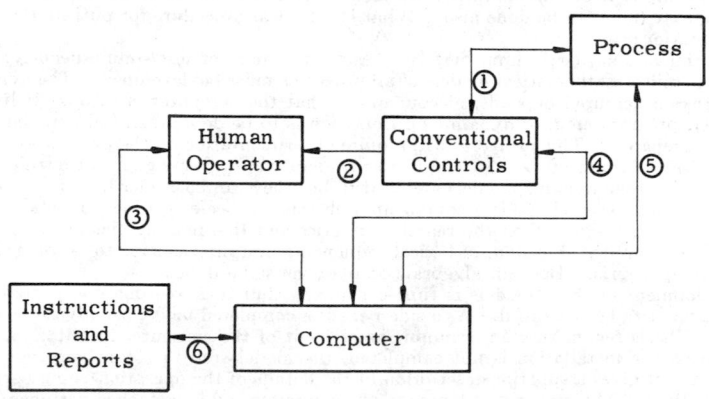

FIG. 98. Computer control system.

On-line supervisory control; Paths 1 and 4 are the paths for data movement in this most widely used form of computer control. The computer ideally can be used for direct optimization, for adaptive control, and for programmed sequence operation. This sort of computer usage is true closed-loop control in which all the dynamic factors discussed thus far must be brought into consideration if effective stable operation is to be insured.

Direct digital control or DDC; here path 5 is the only form of communication between the computer and its controlled system. Conventional control instruments have been removed and their functions taken over by the computer. In this ultimate and most sophisticated form of computer control it is easily possible to take advantage of many advanced control methods not available in ordinary control systems, such as favorable nonlinear feedback elements, feed-forward methods, special adaptive procedures, and on-line optimization, both static and dynamic.

The Computer Hierarchy

In Fig. 98 an external source of instruction is indicated via channel 6. The device that furnishes this instruction can itself be a computer; for example, several DDC computers may each be exercising dynamic control continuously over some process in an oil refinery. The scheduling of interrelationships between these units may be done less frequently by a separate computer, say once a day, so that the entire refinery can be operated in something approaching optimal fashion. It should be noted that if a petroleum company has several refineries and oil fields, there may be a higher computer in the hierarchy which dictates weekly or monthly schedules to the individual sites.

A computer-control installation must, of course, pay for itself promptly in improved operation or reduced expenses. A rational first step in deciding for or against a large computer in a control system is to make a realistic study of just what the computer can do for the process, and what it is worth if it succeeds. In many instances this is a perfectly straightforward economic evaluation, since the uses and effects of a control system can be precisely defined. This is the situation in many manufacturing processes where the value of good and safe sequence control, proper automatic testing procedures, and strict monitoring of every step in a complicated assembly process is easily assessed. In many other circumstances, however, this accurate economic evaluation is very difficult or even impossible. This is the case in many petroleum processes, for example, where mathematical models are very difficult and expensive to derive, where external disturbances may be frequent and severe, and where accuracy of the primary measurements is questionable. In these latter cases good judgment is an essential part of the "yes? of no?" decision. If "yes," then all concerned must recognize that they are gambling. The stakes are high because computers are costly and many man-years of program and system development may be needed to reach a successfully operating installation. The rewards are correspondingly great, however.

If the decision is "yes," then the details of the design and installation must be worked out exactly. For illustration it will be supposed that the computer is to do an extremely difficult job—a combination of direct digital and closed-loop supervisory control, with feed-forward, optimizing, and adaptation, and with uncertain mathematical models. Modern computers can divide their time between two or more users, and therefore we may suppose that this is to be done also. What is a typical procedure for putting the system into operation?

Two things must begin immediately, presumably more or less simultaneously. Sufficiently detailed mathematical models of all processes must be developed. The amount of detail that is required depends, of course, on what the computer is to do. If linear or nonlinear programming, or dynamic optimization is to be done, then full detailed models must be prepared. This probably will require experimentation on the processes, and very much off-line computation If feed-forward, adaptive, or test-signal optimizing control is to be done, then dynamic models are needed, but they can be cruder than those required for "operations research." If a complicated chemical process is to be controlled in some sort of optimal fashion, then the reaction kinetics and thermodynamics must be worked out. If a complicated system of logical sequencing and interlocks is to be programmed, then the appropriate Boolean algebraic analysis must be done.

Development of these models is time-consuming, but it is extremely important if the installation is to be successful. As a side-benefit a completed model usually gives a much-improved basis for making an economic assessment of the computer installation. If for any reason the installation is not completed, the model study is often very valuable in itself, since it gives a superior description of the details of the operation being performed.

While the model is under development, the computer and its attached peripherals must be specified. The number of input and output signals, their types, the number of separate operators consoles, and the kind of peripheral equipment desired (such as typewriters, printers, card readers and punches, tape equipment, visual display devices, and automatic plotters) must all be given consideration.   In the computer itself the size of memory, the type of bulk storage, and the division of computation between *hardware* and *software*, i.e., between (1) completely electronic implementation of all possible arithmetic and logical operations, and (2) specially-prepared programs for performing the more complicated operations using a machine with a simple basic command structure, must be specified.

It must be determined how the computer is to control the process itself.   The question of what to do in the event of computer failure is recurrent.   Decision must be made among adequate manual "backup" in a parallel control system, a spare computer, or complete process shutdown when the computer is inoperative.   The second alternative is certainly feasible in a computer hierarchy, since periodic reassessment to insure optial operation can be suspended in a case of emergency.   If a computer has been assigned that function, it can be made available for other uses when necessary.   The third alternative in case of failure is to stop the process until the computer is operative.   This may be an entirely reasonable alternative even if a shut-down is expensive, because modern computers are becoming highly reliable, with mean-time between failures specified in terms of months and even years.   If this trend continues, an occasional shutdown may be less costly than a parallel system of emergency equipment that is only rarely used.   When the system is specified, it might look conceptually like the diagram of Fig. 99.   Some of the features shown there require special comment.   Signals from a process or machine are of a variety of types.   They may be indications of "on" or "off," i.e., bits of information.   They may involve multivalued logic like "low," "middle," and "high," or some other indications of a variety of logic levels.   They may be pulse trains or frequency-modulated or amplitude-modulated a-c carriers.   They may be register contents.   All these except those involving the a-c carriers are commonly referred to as "digital" inputs, although the last-named is the only type that is strictly digital in nature.   Continuous inputs, such as voltages, currents, shaft rotational position, other position indications, and pressure levels must be converted to digital form before the computer can operate on them.   Signals of this type are commonly called "analog" because analog computers operate with continuous variables.   The analog-to-digital (A/D) conversion equipment is usually quite expensive and often must be shared by several input devices.   Hence a multiplexer is a common part of a computer-process *interface.*

The device marked *interrupt* is an essential part of a control and time-shared computer system.   Every operation performed by the computer must be assigned a priority level according to its urgency.   Thus operations which are strictly synchronous, like reading an exterior clock into a memory cell to keep track of the time some event occurred, must have

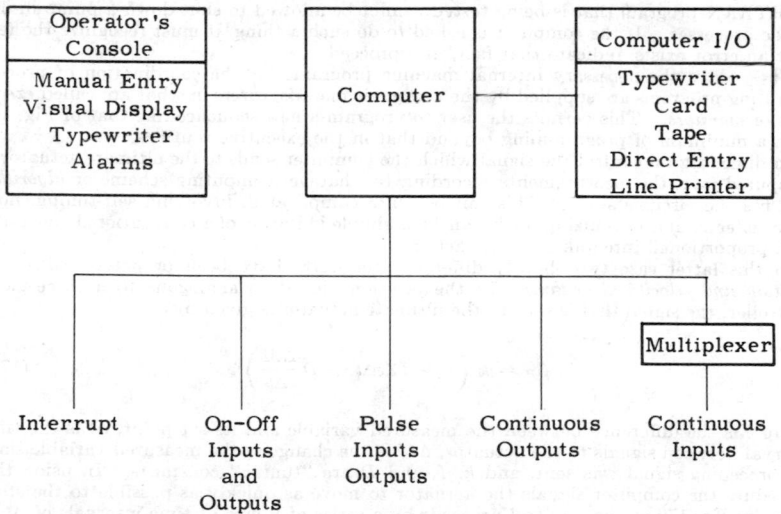

FIG. 99.   Computer control system.

a high priority. Operations connected with safety must have very high priority. Operations connected with control functions other than safety can have a somewhat lower priority, periodic heat and material balances still lower, and optimizations and adaptations lower yet. The priority signal instructs the computer to stop if it is doing something of lower priority and serve the demands of the signal being sent to it. The new program can, of course, be interrupted by another of higher priority, and it must wait if a more urgent task is taking place when it enters. Modern control computers are organized to keep track of priority levéls, to store all necessary information in an interrupted program, and to form a waiting line or *queue* in strict order of priority if several things that are to be done enter at nearly the same time.

In most modern control computers, entry of a signal into the memory is direct; that is, the data do not pass through the arithmetic units of the central processor. This feature saves considerable computer time and is absolutely essential in control systems where large quantities of data are handled. As a new piece of data enters, it may require pretreatment, such as a check for validity by comparison with known limits. If this sort of thing can be done in a satellite console, considerable computer time may be saved.

The emphasis on time saving in the programming of control computers is important because synchronous or *real-time* operation is required. If an operation must be done every ten milliseconds, the user must first be assured that it can be done once in 10 msec, and second that the amount of time left in each 10-msec interval is enough to process the tasks that have to be done every 100 msec, including the time lost by shifting information around at each interrupt signal. Detailed consideration of the execution times of every program step in the computer muṣt be made, for if times overlap, the system cannot operqte properly.

In any real operating computer system there will be continuing program changes as the process changes, as the mathematical model is improved, as control duties become more advanced, and as new ideas are tried. The entry of program changes without disturbing the normal control function of the system requires an interruptable *compiler* or *assembler* if the program is written in any of the modern programming languages such as FORTRAN. If program entry and "debugging" requires disconnection from the controlled process, it is not likely to be a useful system. On the other hand, on-line compilers require large computer memories and therefore make the installation expensive.

Operators must have the ability to ask the system for information if they need it. This function is called the demand log and is initiated by an operator interrupt.

A flow chart of a typical control program is shown in Fig. 100. The program cycles continuously around the left loop until an interrupt signal is received.

A very important feature of a computer system that is shared between users and which permits on-line compiling is program protect. This is a part of the hardware which prevents one user from interfering with the program of another. For example, a "bug" in a FORTRAN program that is being tested cannot be allowed to shut down a pump in one of the processes. If the computer is asked to do such a thing, it must recognize the fact that an error exists, indicate that fact, and proceed.

These days the necessary internal machine programs for the coordination of process operating programs are supplied by the computer manufacturers in what are called *executives* or *monitors*. This permits the user to program a new sequence like that of Fig. 100 with a minimum of programming beyond that in the executive routines.

In direct digital control the signal which the computer sends to the ultimate actuator is computed from the measurements according to whatever computing scheme or *algorithm* best fits the circumstances. This can be quite complicated, involving self-tuning, nonlinear effects, and optimizing, or it can be a simple imitation of a conventional controller with proportional, integral, and rate action.

In this latter case two slightly different procedures have been proposed, called the *position and velocity algorithms*. In the position algorithm analogous to a three-mode controller, the signal that is sent to the ultimate actuator is given by

$$C_P = k \left(e + I\Sigma e\Delta t + D \frac{\Delta M}{\Delta t} \right) \tag{179}$$

where e is the difference between the measured variable and its set point, Δt is the time interval between signals to the actuator, ΔM is the change in the measured variable since the preceding signal was sent, and k, I, and D are "tuning" constants. In using this procedure the computer signals the actuator to move as quickly as possible to the state given by Eq. 179 so that control proceeds by a series of jumps at time intervals of Δt.

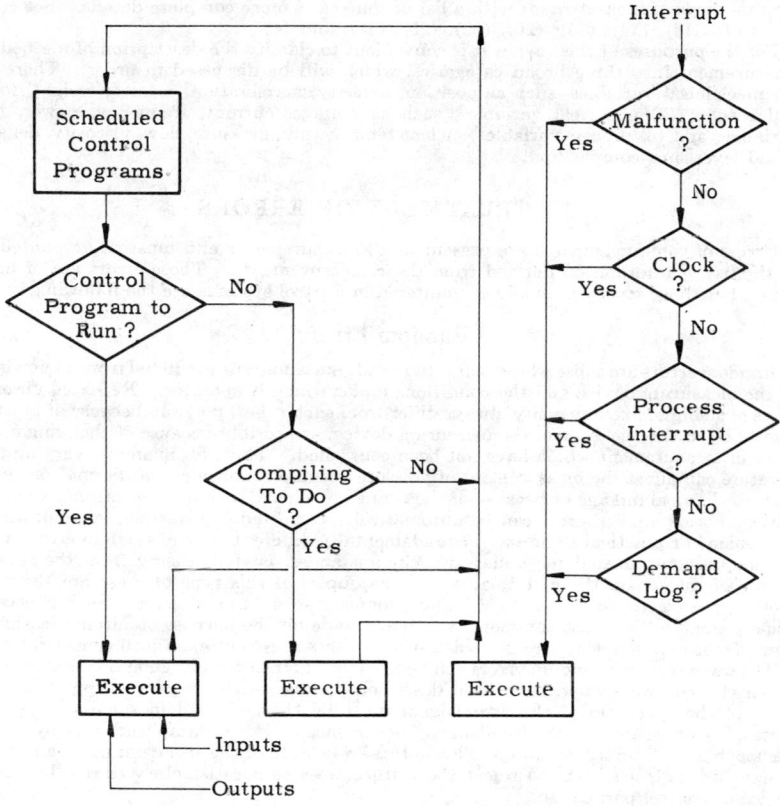

FIG. 100.   Typical computer control program.

In the velocity algorithm the signal is sent to an integrator, perhaps external, perhaps part of the digital hardware.   The ultimate signal to the actuator is given by

$$C_v(t + \Delta t) = C(t) \int_{t}^{t+\Delta t} k \left(\frac{\Delta M}{\Delta t} + Ie + D \frac{\Delta^2 M}{(\Delta t)^2} \right) dt \qquad (180)$$

The actuator is told to move at the indicated rate for a time Δt.   A choice between these two output methods in any given situation can be made by noting that the first is the best in principle, provided that the control calculations are exactly right.   However, if the control computation gives only an approximate dynamic correction to an upset, as is usually the case, then the velocity method gives smoother response.

The particular advantage of DDC is that the algorithm can be modified easily by the state of the system.   Thus "reset windup" can be easily eliminated by disabling the integral action in either the velocity or position algorithm when the error exceeds a certain limit.   Making the gain a function of the error or of some process state is quite straightforward.   This can give greatly improved control over the simple three-mode method if it is properly designed by consideration of process dynamics.

METHODS OF MEASUREMENT

The art and science of measurement of variables for use in automatic control systems is extensive and very well documented.   It is feasible to include here only a review of common methods of measurement with a list of sources of more complete details.   See references (4), (10), (14), (15), (20), (22), (25), (35), and (43).

For the purposes of this review it is convenient to classify the description of methods of measurement into three broad categories, which will be discussed in order.   There are (1) mechanical variables, such as position, velocity, acceleration, mass or weight, force, and torque; (2) electrical variables, such as voltage, current, frequency, power, field variables; and (3) process variables, such as temperature, pressure, flow, viscosity, density, liquid level, and composition.

17. TREATMENT OF ERRORS

Errors of some magnitude are present in all measurements and must be accounted for in the use of information derived from those measurements.   These errors are of many types, but those most commonly encountered in control systems are the following.

Random Errors

Random errors are those whose magnitude and sign cannot be predicted from a knowledge of the measuring device and the conditions under which it operates.   Repeated observations of a single, fixed quantity always differ from each other, possibly because of inherent limitation in the accuracy of the measuring device, or possibly because of changing conditions in measurement which have not been controlled.   Thus, for example, varying temperature can affect the measurement of pressure because of the effect of thermal expansion on a mechanical linkage or because of the change of capacitance in an electronic amplifier. Although this kind of error can be automatically corrected in principle, it is not always convenient or practical to do so.   A fundamentally different type of random error, which cannot be compensated for under any circumstances, is that arising from the random nature of matter on the molecular scale.   Examples of this type of noise are Brownian motion, the electronic "shot effect," and "Johnson noise" in a resistor.   Such effects are usually small in those measurements which are made for the purpose of automatic control; however, if extreme accuracy is needed, their effects must be taken into consideration.

The ways in which random errors can be accounted for in the design and operation of an automatic control system have been described, in the article on random processes.   In essence, the spectrum of the error signal must be characterized in some way, say by measuring or estimating the frequency of occurrence of various amplitudes, or by finding the mean and standard deviation.   From this knowledge of the spectrum the engineer can design his control system to reject the disturbances to a satisfactory degree by proper choice of control parameters.

Systematic Errors of Measurement

Systematic errors are those that are consistently reproduced with a repetition of a measurement.   This type of error is insidious in a control system because process operators have no method for detecting it.   Bearings in flowmeters become worn, resistance thermometers change characteristics because of strains, and mechanical linkages develop frictional resistance as dust accumulates.   Errors of this sort cannot be analyzed by statistical methods, and the control engineer cannot automatically correct for them. In those cases where systematic errors must be avoided, the only available tool is calibration of the measuring devices against known standards at sufficient frequency.   The necessary accuracy and interval between calibrations depend entirely on the application.

Dynamic Errors

It must never be assumed that a measuring device responds instantly to changes in the quantity being measured.   It always has its own inherent time delays which must always be included in the transfer function of the control system.   In certain types of signal transmission, such as the penumatic systems in common use in the process industries, time lags must always be accounted for.

18. MEASUREMENT OF MECHANICAL VARIABLES

Acceleration

Acceleration is the rate of change of velocity, either linear or angular, and by Newton's second law of motion it is equal to the net force on an object divided by its mass.   A useful

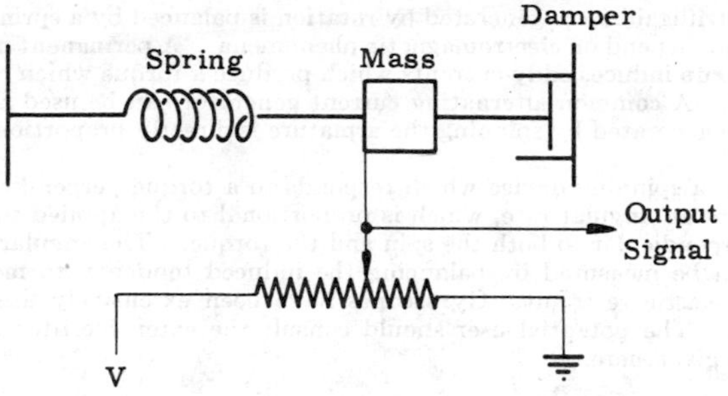

FIG. 101. Accelerometer.

qualitative concept in measurement of acceleration is *inertia*, which may be defined as the resistance of an object to a change in its velocity. Accelerometers are commonly called inertial devices because they depend on this resistance for their operation.

The construction of a typical accelerometer is shown in Fig. 101. A mass is supported in an equilibrium position by one or more springs with a damping device of some sort rigidly attached. Usually the mass is constrained to move in one direction only. The differential equation which ideally describes the motion of the mass is

$$m\ddot{x} + c\dot{x} + kx = -m\ddot{z} \qquad (181)$$

where m is the mass.

x is the displacement from equilibrium.
c is the damping constant.
k is the spring constant.
$\ddot{z}$ is the acceleration of the frame to which the accelerometer is attached.

Equation 181 is a second-order differential equation whose solution, as we have seen, can have its own characteristic frequency of oscillation. This characteristic must be accounted for in designing a device for measuring acceleration. The "natural frequency" of a spring-mass system is defined as

$$f_n = \frac{1}{2\pi} \sqrt{\frac{k}{m}} \qquad (182)$$

The frequency response of a system described by Eq. 181 has been displayed in Fig. 17 as that of a quadratic lag. There is a critical damping coefficient

$$c_c = 2\sqrt{km} \qquad (183)$$

the lowest damping factor for which a damped oscillation will not occur, and a resonant frequency for $c < c_c$ which is given by

$$f_r = f_n \sqrt{1 - \frac{c^2}{c_c^2}} \qquad (184)$$

Clearly an accelerometer will not give easily usable information in the neighborhood of such a resonant peak. A common design practice is to make $c \approx 0.7c_c$, for which the frequency response is nearly flat out to about half the natural frequency.

Velocity or Speed

Since acceleration is the derivative of velocity, a measure of velocity may be obtained by integrating the output of an accelerometer.

The two other common methods for measuring velocity involve tachometers and gyroscopes. This first is a measuring device for the speed of rotation. If it is to be used for linear velocity, some angular rotation rate proportional to the velocity must be generated. The speedometer of an automobile is a simple example.

A tachometer can be a mechanical counting device which displays the number of revolutions in a fixed time, it can be an arrangement of weights known as *flyballs* in which the outward or "centrifugal" force generated by rotation is balanced by a spring or pneumatic force, or else it can depend on electromagnetic phenomena. A permanent magnet rotating in an aluminum cup induces eddy currents which produce a torque which can be balanced against a spring. A common alternating current generator can be used as a tachometer since the voltage generated by spinning the armature is directly proportional to the speed of rotation.

A gyroscope is a spinning device which responds to a torque perpendicular to its spin axis by moving at an angular rate, which is proportional to the applied torque, about an axis which is perpendicular to both the spin and the torque. The angular rate of motion of the frame can be measured by balancing the induced tendency to motion against a spring or other elastic restraint. Gyroscopes have been extensively used in aerospace instrumentation. The potential user should consult the extensive literature for details which cannot be given here.

Position or Displacement

Measurement of position for control purposes is, of course, quite simple. It can be transmitted by mechanical linkages, by conversion to a voltage through a slide wire, or it can be obtained by integrating velocity.

Mass or Weight

Mass and weight are not equivalent concepts; however, except for aerospace applications the systems that the control engineer deals with are in a fixed gravitational field and he need not distinguish between them.

The simplest method of measuring weight is exemplified by the common laboratory balance in which the measurement is made by comparison with known weights. This type of measurement is not convenient in automatic control systems although it is in common use in many industries where manual operation is involved.

Measurement of strain is quite commonly used to measure weight or mass. Extension of a spring is one of the earliest. Hooke's law states that the elongation is directly proportional to the weight up to the elastic limit of the spring. Recently the use of strain gages has come into prominence. A strain gage is an electrical resistive element whose resistance is changed by deformation. In the most common use, one or more gages are bonded directly to a piece of metal which is deformed by the weight or applies stress. A "Wheatstone bridge" arrangement is often used, with four gages mounted in such a way that lateral stresses cancel each other. Strain gages are sensitive to temperature and must be compensated if accurate measurement is needed. They may also be equipped with cooling fins.

A most important method for weight measurement is the pneumatic or hydraulic load cell. In the pneumatic type the pressure required to bring a loaded platform to a fixed position is supplied. In the hydraulic type the weight is supported by a liquid whose pressure can be measured by techniques to be described later.

19. MEASUREMENT OF ELECTRICAL VARIABLES

Electrical variables are often used in process control to transmit signals related to measurement, hence the measurement of these quantities is not of primary interest. If and when they occur in such a way as to be the primary quantities to be controlled, then their measurement is of importance. This is, of course, the situation in the power industry. The prime variables of interest are direct and alternating current and voltage, power, power factor, and frequency.

We have seen that d-c voltage may be converted to a form usable in a control system with some kind of operational amplifier, such as commonly found in all electronic controllers. Thus any electrical quantity which can be converted to a related d-c voltage can be measured for control purposes. Direct current is measured by observing the voltage drop across a standard resistance.

For alternating currents, it is often adequate to rectify and filter to obtain an equivalent d-c measurement. Other types of a-c measuring devices use the reaction between induced fields in coils or vanes as a measure of alternating current. Commonly the output signal originates in a shaft rotation, and because of the nature of the fields it can be quite nonlinear.

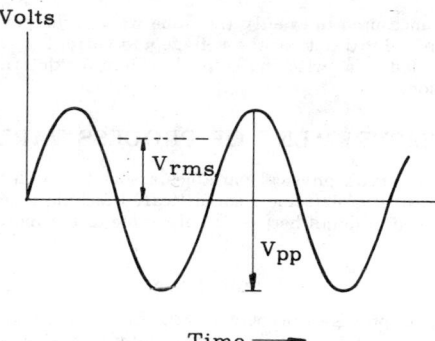

FIG. 102.   A-c voltage relationships.

It is necessary to distinguish between several different measures of alternating voltage or current.   In a periodic waveform, Fig. 102, peak-to-peak voltage is the indicated potential difference between extremes.   The quantity usually measured is

$$V_{rms} = \sqrt{\frac{1}{t} \int_0^t V^2\, dt}$$ (185)

rms here means root mean square.   If the voltage varies sinusoidally,

$$V_{pp} = V_{rms} \times 2\sqrt{2}$$ (186)

Instantaneous power is the rate at which electrical energy is dissipated, and is equal to the product of voltage and current.   The average power in a periodic state is just the integral of this product over a single period multiplied by the frequency.   If voltage and current are sinusoidal, then

$$P = \frac{V_{max} I_{max}}{t} \int_0^t \sin(wt)\, \sin(wt - \phi)\, dt$$ (187)

where V_{max} and I_{max} are the peak values of voltage and current, and ϕ is the phase lag of the current behind the voltage.   This integral is equal to

$$P = V_{rms} I_{rms} \cos \phi$$ (188)

Power is measured commonly by the reaction between two magnetic fields, one generated by the instantaneous current and the other by a current proportional to the instantaneous voltage.   With suitable filtering the instrument can be made to read average power.

A common integrating watt-hour meter, which measures total dissipated energy over an elapsed time, uses eddy currents generated in an aluminum disk to rotate the disk on a shaft coupled to a counter.

The measurement of power factor is somewhat more complicated.   In one type of instrument a shaft position depends on the mutual interaction of three coils, one of which has its field in-phase with the current, a second is in-phase with the voltage, and the third 90° out-of-phase with the voltage.   The instrument is designed in such a way that the angular position taken by the coils is a unique function of the power factor.   The principle can be used with three-phase power systems equally well, although the construction is understandably somewhat more complicated.

Frequency may be measured by any of the counting procedures previously described for angular velocity.   In essence, the number of oscillations in a given time can be counted electronically or mechanically.   Use of resonant circuits is possible in which the dependence of impedance on frequency can be used to position a rotating shaft as a unique function of frequency, so long as frequency does not deviate greatly from a standard.

A common problem in interconnection of power systems is synchronization.   When two generators are to be connected together, their voltage, frequency, and phase must be matched exactly.   We have seen how voltage and frequency can be measured, and the relative phase can be measured in exactly the same way as the power factor, except that two voltages are compared rather than the voltage and current in the same circuit. This kind of information is fed to a switching control system for determination of the proper time for interconnection.

20. MEASUREMENT OF PROCESS VARIABLES

The measurement of certain physical variables associated with the control of modern processes requires a number of extremely sophisticated techniques. For obvious reasons, these cannot be described in detail here, and only some of the more common devices will be reviewed.

Liquid Level

Most chemical and oil processes involve storage of many liquids, some in large tanks and some in small accumulators between process subunits. Knowledge of the quantity of liquid in each of these vessels can be important for inventory purposes, for insuring that pumps always have enough head to operate, for blending purposes, for monitoring the product as it passes from stage to stage in a continuous system, and for filtering out flow disturbances. Hence the need for level measurement and control. There are two essentially different uses of liquid level measurement which must use different procedures for measuring the level. In the simplest, the measuring device indicates only that the level is above or below some fixed point in the vessel. Optical devices, X-rays, gamma-rays, acoustic signals, electrical conductivity, capacitance between fixed plates, and simple floats have all been used for level detection in this way. They can be used to operate a switch, relay, or logic system to indicate to an external controller the state of the level. In many systems this kind of level control is enough.

In other cases it is necessary to know where the level is. Here one can simply measure the pressure at or near the bottom of the vessel, follow the position of a float, or measure the buoyant force on a partially immersed weight. Each of these methods for level measurement is in common use in the process industries; however, since the first is really a pressure measurement, it will be described more fully under that heading. Most process instruments these days use a balancing method in which a force exerted on a float or weight is balanced by air or hydraulic pressure or by an electromagnetic field. More will be said later about how this is done. Here it will be assumed that it is done so that an adequate output signal is available from the level gage.

A simple float gage is shown in Fig. 103. Rotation of the lever produced by a change in level translates motion to the force balance device which sends a signal to the control or indicating system. A buoyancy gage is shown in Fig. 104.

Pressure Measurement

In the middle of the last century Bourdon discovered that fluid pressure inside a bent tube would tend to straighten the tube. This discovery led to one of the most commonly used pressure-sensing elements in automatic process control. The simplest type of Bourdon tube is a flat double ribbon, bent into an arc of about 270°, which a lever mechanism for indication or a flapper-nozzle arrangement for force balance (see Pneumatic Controls in Art. 23). Figure 105 shows the sort of arrangement that is commonly used. In typical use the range of movement of the tip is about 1/4 in., with the pressure range corresponding to this movement depending on the stiffness of the tube. The gage can be used for vacuum measurement equally well.

The Bourdon tube takes two other forms, a spiral and a helix, where increased sensitivity is necessary. In the sprial form the tube makes several revolutions instead of

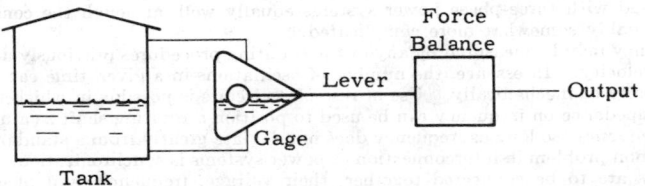

FIG. 103.   Float gage.

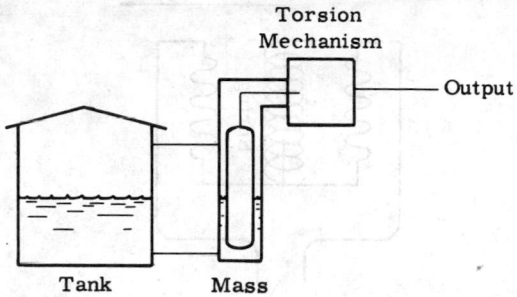

Fig. 104. Displacement gage.

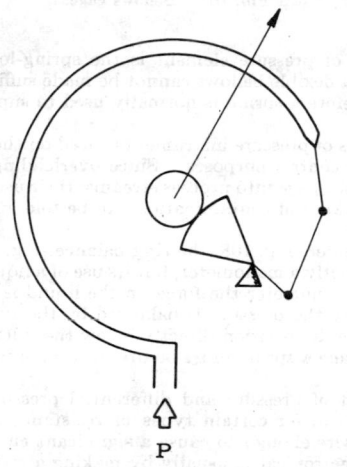

Fig. 105. Bourdon tube.

stopping at 270°. The deflection of the tip is proportional to arc length and pressure.
In the helical form the tube is wound in such a way as to resemble the surface of a cylinder.
In both types the mechanical linkage which amplifies the motion of the tip of the tube
can be greatly simplified.

A second type of sensing element in common use is the diaphragm gage, shown in Fig.
106.   The pressure is sensed by movement of flexible metallic diaphragms, usually formed
into flat, disk-like capsules, often with several such capsules stacked together.   As in the
Bourdon gage, the force due to pressure within the element is usually balanced by the
elasticity of the diaphragm, hence use of this gage is restricted to lower pressure ranges.

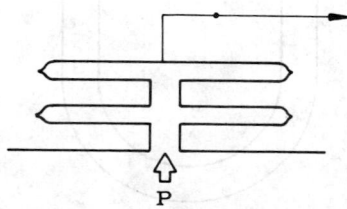

Fig. 106. Diaphragm gage.

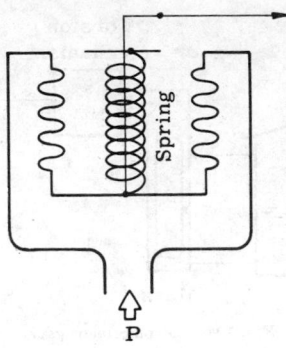

FIG. 107.   Bellows gage.

A third common type of pressure element is the spring-loaded bellows, diagrammed in Fig. 107.   In general, a flexible bellows cannot be made sufficiently elastic to accommodate high pressures, therefore a spring is normally used to supply the necessary restoring force.

A large number of types of pressure instruments based on the manometer principle have been used in the past for control purposes.   Since overloading can cause the manometer liquid to be ejected from the gage into process streams, their use must be restricted to those circumstances where this kind of contamination can be tolerated.   Their operating range is normally very small.

Typical are the manometer, Fig. 108, the ring balance, Fig. 109, and the bell gage, Fig. 110.   The latter is not strictly a manometer, but its use of a liquid seal allows it to be classified with them.   In the manometer the force on the liquid is balanced by its differential head.   In the ring balance the pressure is balanced by the torque exerted by the weight which is moved away from its position directly below the fulcrum when a pressure difference exists.   In the bell gage a spring must be provided to supply the necessary balancing force.

Electrical measurement of pressure and differential pressure by strain gages is quite common.   As we have seen, for certain types of resistance wire an applied stress can stretch or compress the wire enough to cause a significant change in its resistance, which can them be measured electronically, usually by making a resistance bridge.   Figure 111 shows a bonded strain gage in which the resistance wire is cemented directly to the surface

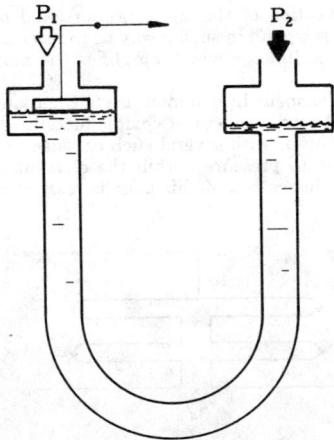

FIG. 108.   Manometer.

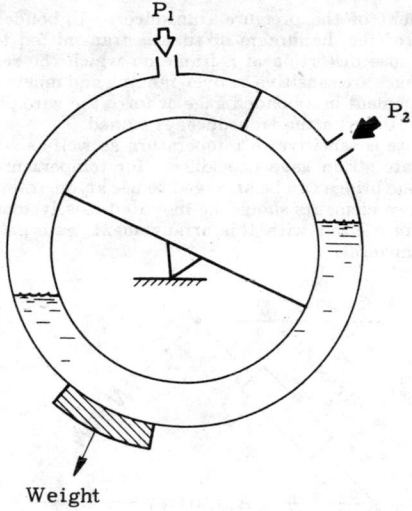

P₁

P₂

Weight

FIG. 109.   Ring balance.

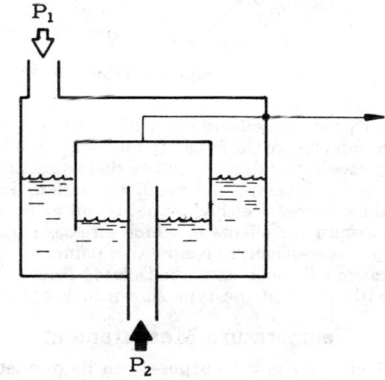

P₁

P₂

FIG. 110.   Bell gage.

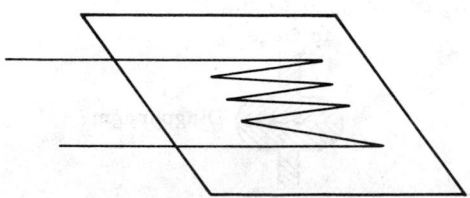

Gage Bonded to Stressed Surface

FIG. 111.   Strain gage.

of the elastic component of the pressure transducer. Unbonded strain gages are also used in which motion of the diaphragm or tube is transmitted through levers and bars in such a way as to cause distortion of a frame on which the resistance wire is wound. Both types of strain gages are sensitive to over-ranging and must be protected. Excessive motion can break the cement in a bonded gage or force the wire past its elastic limit in an unbonded gage. In both cases the transducer is ruined.

The resistance of wire is sensitive to temperature as well as strain, and it is therefore necessary to compensate strain gage transducers for temperature changes. Figure 112 shows how a Wheatstone bridge can be arranged to use an unstrained gage for temperature compensation. The two elements should be mounted closely together so that they have a common temperature. Even with this arrangement, external compensation is often required for exact compensation.

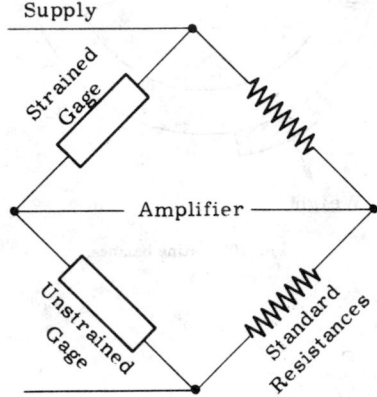

FIG. 112.   Strain gage bridge.

Other types of electrical pressure sensing elements include :(1) capacitors, where distortion by an applied force can change the capacity and hence the frequency of an oscillator in which it is a part, (2) piezoelectric devices, where distortion of certain types of crystals causes potential differences to develop, (3) magnetic action, in which movement of a magnetic element can induce currents either by induction or by changing reluctance in a magnetic circuit, (4) semiconductor effects in which stresses can cause dramatic changes in conductivity, and (5) pressure-sensitive conductive paint.

In many process applications the gage must be isolated from a corrosive or fouling fluid. This is commonly done with a seal of the type shown in Fig. 113.

Temperature Measurement

Temperature measurement for control purposes can be divided into five broad classes of measuring device: (1) filled bulbs, (2) thermocouples, (3) resistance thermometers, (4) radiation pyrometers, and (5) bimetallic devices. Their characteristics will be described in order.

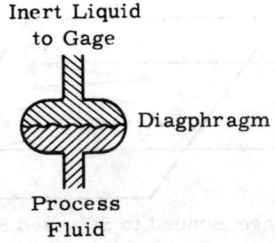

FIG. 113.   Seal for pressure transducer.

Filled Bulb Systems. The common household thermometer depends on expansion of a liquid for determination of temperature, likewise control thermometers depend on the fact that the pressure or volume of a closed system will change in a predictable way as the temperature changes. A thermometer used in a closed-loop control system must have some transmittable signal generated by the expansion. In most of the commercially available instruments today this is just a pressure transmitter of one of the types already described.

The bulb may be filled with a liquid, like mercury or alcohol, it may be filled with an inert gas, like nitrogen, in which case the variation of pressure with temperature is given by the gas law, or it may be filled with a vaporizable liquid, in which case the relationship of pressure to temperature is given by the Clapeyron equation

$$\frac{dp}{dT} = \frac{\Delta S}{\Delta V} \tag{189}$$

where ΔS and ΔV are, respectively, the entropy change and volume change with vaporization. Pressure is very nearly an exponential function of temperature under these conditions.

It is clear that the entire closed volume, including the Bourdon tube or diaphragm cells, must be at the same constant temperature if the use of an equation of state is to be used with full accuracy. Since this is obviously impossible under most circumstances, compensation for external temperature variation should always be provided with a filled-bulb device. For example, if a long capillary passing through air connects the bulb and the Bourdon tube, then a parallel capillary and duplicate Bourdon tube can be used to determine and cancel out the effect of the capillary. Figure 114 shows a filled-bulb thermometer as it might be used in a process control system. The compensation system usually includes a mechanical linkage between the two gages.

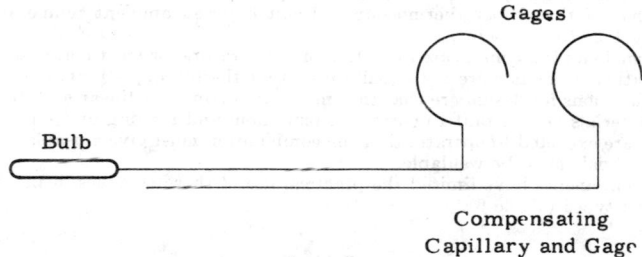

FIG. 114. Filled-bulb thermometer.

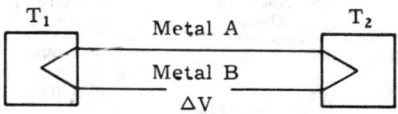

FIG. 115. Thermocouple.

Thermocouples. This most common and useful method of measuring temperature depends on the Seebeck effect for its operation. In the circuit of Fig. 115 a small potential difference develops at the indicated spot when the two junctions of the dissimilar metals are at different temperatures. A measuring device which depends on the use of this potential difference is called a thermocouple. The voltage is commonly of the order of a few millivolts in commercial devices and is referred to as the thermal emf. One junction is placed in the spot whose temperature is to be measured, the other at some standard or reference temperature. The thermal emf developed is not affected by intermediate conductors of different metals, provided that corresponding junctions are at the same temperature. If the junctions with these intermediate conductors are not at a common temperature, then, of course, the resulting thermal emf's affect the overall voltage developed. No thermal emf can develop along a conductor of a single metal, therefore thermocouple leads can pass through a temperature gradient without affecting its emf, except posssibly that thermal conduction along the wires may affect the actual temperature at the junction. If in any circuit the temperature is uniform between two points, then the thermal emf between these two points is entirely independent of what is in the circuit between them, and is the same as it would be if they were put into contact.   Thus, for example, a single junction can be connected to an amplifier as shown in Fig. 116.   The thermal emf developed across the amplifier is exactly the same as it would be if the amplifier were replaced by a junction at the same temperature and the circuit broken somewhere else.   The temperature must be uniform in the amplifier, of course.

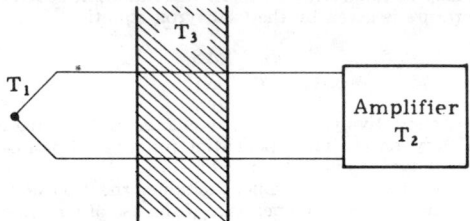

FIG. 116.   Thermocouple amplifier.

Since the temperature of the amplifier of Fig. 116 may vary with time, so will the thermal emf, even though the temperature at the measured point is constant.   This means that for accurate measurement, compensation must be made for ambient temperature changes.   This is normally done by incorporating a temperature-sensitive element, like a thermistor, into the amplifier itself in such a way that a counter-emf is developed exactly equal to that of the proper thermocouple circuit between ambient temperature and a standard such as the ice point.

The thermal emf is a complicated function of temperature for most materials; therefore, all combinations of metals are not equally useful for thermocouple junctions.   The common considerations for design are that the emf be approximately linear with temperature; that the materials be resistant to corrosion, oxidation, and melting under the conditions where they are expected to operate; that the combination must give a usable voltage; and that the materials must be weldable.

These requirements have limited the practical use of thermocouples in process control to five or six types (Table 9).

Table 9

ISA Type	Positive Element	Negative Element	Approx Temp Range, °C	Join by
S	90% Pt, 10% Rh	Pt	$0 \to 1600$	Electric arc or oxyacetylene
R	87% Pt, 13% Rh	Pt	$0 \to 1600$	Electric arc or oxyacetylene
K	Chromel (90% Ni, 10% Cu)	Alumel (94% Ni, 27% Al, 4% Si)	$-200 \to 1200$	Oxyacetylene
J	Fe	Constantan (60% Cu, 40% Ni)	$-200 \to 800$	Oxyacetylene
T	Cu	Constantan	$-200 \to 300$	Silver solder
	Chromel	Constantan	$-100 \to 1000$	

In joining wires, high speed of response may be best obtained by butt-welding.   Excessive twisting, bending, beating, and hammering should in general be avoided, as these can lead to stresses which can change the emf or cause the couple to crack.

Because of these effects and because of variation in composition, thermocouples must be calibrated before use if accurate measurements are required.

A number of thermocouples placed in series form what is known as a thermopile, in which the emf's or the individual junction pairs are added.   If enough junctions are used, extreme sensitivity can be obtained.

Resistance Thermometers. Resistance thermometry depends on the change in electrical resistivity of metals or semiconductors with temperature. The resistance bulb is commonly placed in one arm of a bridge. Common resistance elements are copper, nickel, platinum, and certain semiconductor materials called "thermistors" which have a strong negative temperature coefficient of resistance.

Resistance windings within a bulb should be double helical to avoid inductive effects. They should be in good thermal contact with the bulb surface but electrically insulated, two requirements that are not always consistent. The connection to a remote bridge circuit should be made in such a way that lead resistances do not contribute errors, for example, as shown in Fig. 117, where three separate leads are used. The leads to the bridge resistors should be matched in resistance and routing.

Several characteristics of the most widely used resistance elements are shown in Table 10.

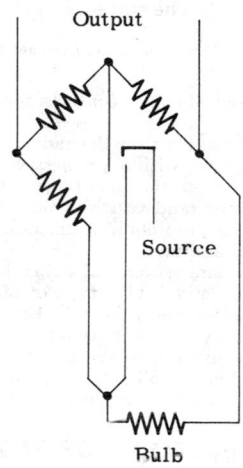

Output

Source

Bulb

Fig. 117.   Resistance thermometer.

Table 10

	Resistivity at 0°C, microhm-cm	% Change per 100°C	Practical Range, °C
Copper	1.56	43.1	−200 to 120
Nickel	6.38	66.0 (not linear)	−150 to 300
Platinum	9.83	39.2	−250 to 1000

A thermistor is a semiconductive device, usually made of mixtures of metal oxides, fused, sintered, or compressed, into a ceramic-like solid. Excitation of conducting holes or electrons into the band of energy levels where they can drift in an electric field is responsible for the temperature effect. Since this is a rate process which involves an activation energy, the resistance-temperature relationship is exponential-like in character, being derived from the Boltzmann distribution of population of energy levels.

A thermistor can be used in a bridge circuit just like a resistance bulb; however, care must be taken not to allow the power dissipation in the thermistor to change its temperature significantly. Excitation currents must generally be small.

Pyrometry. A radiating black body by definition emits a spectrum of radiation that depends only on its temperature. The total radiant flux from a hot black surface of unit area is thus a function of temperature alone. If a portion of this flux is focused upon a thermopile or other temperature-measuring device in fixed surroundings, then an optical temperature-measuring device known as a radiation pyrometer results. The signal from the thermopile can be used in a closed loop to control the temperature of the surface being

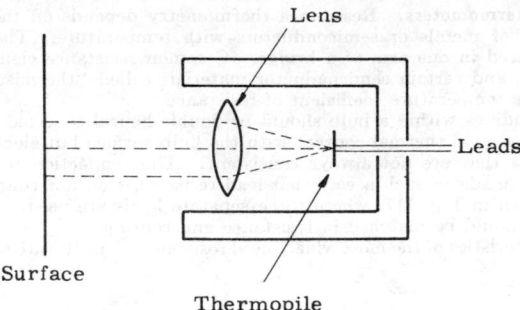

Lens

Leads

Surface

Thermopile

FIG. 118.   Radiation pyrometer.

observed.   Such devices have use in furnace controls, for example.   A typical configuration is shown in Fig. 118.

The Stefan-Boltzman equation states that the radiant flux is directly proportional to the fourth power of the temperature, with a predictable proportionality constant if the surface is black.   Nonblackness, where the emissivity of the surface is less than one, and thermal conduction at the detector tend to make the temperature–voltage relationship somewhat different from that of a pure equilibrium radiator.   Accordingly, pyrometers must be calibrated if accuracy is required.

Bimetallic Elements.   When a composite metal strip is heated, the differential expansion of the metals causes the strip to bend.   This type of temperature transducer has been used in on-off home thermostats for many years.   For small temperature changes the deflection of the end of the element, which is usually wound in spiral or helical form, is proportional to the temperature difference, the square of the length of the strip, and inversely proportional to its thickness.   The device is inherently nonlinear, and it therefore finds its greatest use in on-off controls and in indicators.

21. MEASUREMENT OF FLOW RATE

Process flow can be measured in five different ways: (1) by measuring the pressure drop across some standard restriction, such as an orifice, nozzle, or Venturi tube; (2) by varying an area available to flow, such as with a rotameter; (3) by measuring the volume flowing past a fixed point by emptying and filling of chambers of known volume, (4) by using the impact of the moving fluid to rotate a turbine of deflect a vane, and (5) by observing the voltage generated as a conducting fluid moves through a magnetic flied.

Metering by Pressure Drop

Devices based on this principle are commonly known as "head flowmeters."   Figures 119, 120, and 121 show diagramatically the three major types of restrictions that are used to restrict the flow and cause its pressure to drop.

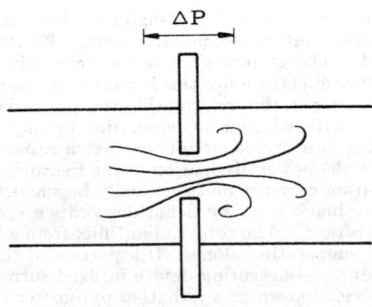

ΔP

FIG. 119.   Orifice flow meter.

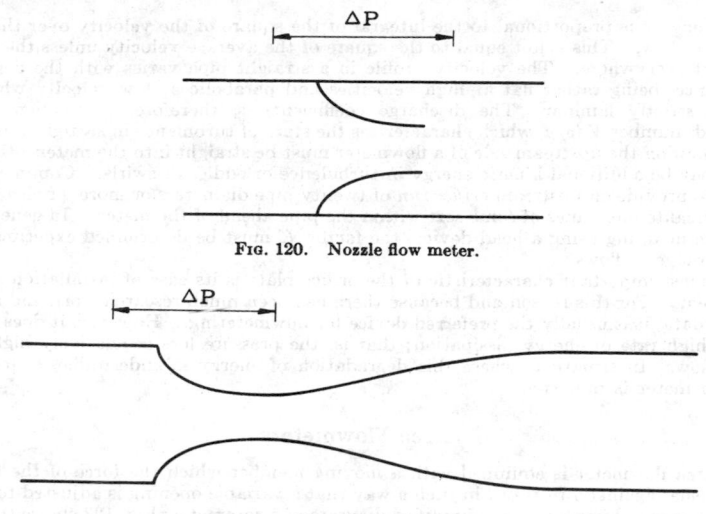

FIG. 120. Nozzle flow meter.

FIG. 121. Venturi meter.

All these devices are mathematically described by a particular form of the law of conservation of energy known as Bernoulli's equation. In its most general form it is:

$$E_1 + \frac{P_1}{\rho_1} + \frac{V_1^2}{2} + gh_1 + \Delta = E_2 + \frac{P_2}{\rho_2} + \frac{V_2^2}{2} + gh_2 \tag{190}$$

where E is the internal energy per unit mass of the fluid.
P is the pressure.
ρ is the density.
V is the velocity.
h is the height above some datum.
Δ is added energy per unit mass.
subscripts 1 and 2 refer to two different positions; for example, the two locations indicated by arrows in Figs. 119 through 121.

In most metering situations the terms in elevation and internal energy can be ignored, leading to the conclusion that

$$\frac{P}{\rho} + \frac{V^2}{2} = \text{constant} \tag{191}$$

except for the term Δ which accounts for interchange of energy with the surroundings. This term Δ cannot be accounted for exactly, its main component is heat loss to the structure, with viscous dissipation being the source of heat. It must not be ignored, and it is usually taken into account in head meter formulas by use of an empirical *discharge coefficient*.

If the fluid is incompressible, then the velocity must be inversely proportional to the area through which the fluid flows. Therefore, if it is assumed that the downstream pressure tap is at the point where the area is equal to the minimum cross section in the meter, we have

$$\text{Flow} = V_1 A = K_1 A \sqrt{\frac{\Delta P}{[(A/a)^2 - 1]\rho}} \tag{192}$$

where K includes the empirical correction for dissipation and allows for the fact that the downstream pressure tap cannot be located exactly at the right spot. As will be apparent, it may depend on the flow rate.

Use of a formula like Eq. 190 in any of these metering devices requires the assumption that the kinetic energy of a flowing fluid is proportional to the square of its mean velocity. In actual practice this is in error for two reasons. If the flow is rectilinear, the total kinetic energy is proportional to the integral of the square of the velocity over the cross section of flow. This is not equal to the square of the average velocity unless the flow is constant everywhere. The velocity profile in a straight pipe varies with the degree of turbulence, being rather flat at high velocities and parabolic at low velocity when the flow is strictly laminar. The discharge coefficient, is, therefore, a function of the Reynolds number $Vd\rho/\mu$, which characterizes the state of turbulence in straight pipe flow.

The flow on the upstream side of a flowmeter must be straight into the meter, otherwise there may be additional kinetic energy in turbulence or eddies or swirls. Common practice is to provide an upstream *orifice run* of twenty pipe diameters or more, or else to provide straightening vanes of some sort within the pipe ahead of the meter. In general, for accurate metering using a head device, the factor K must be determined experimentally over a range of flows.

The most important characteristic of the orifice plate is its ease of installation and replacement. For this reason and because there has been much research work and a great deal of data, it is usually the preferred device for flowmetering. However, it does have a rather high rate of energy dissipation; that is, the pressure loss is relatively high for a given flow. In situations where this degradation of energy is undesirable, a nozzle or Venturi meter is preferred.

Area Flowmeters

An area flowmeter is equipped with a moving member which the force of the flowing fluid pushes against in such a way that a variable opening is adjusted to give a force balance. Figure 122 is a simplified diagram of a rotameter; Fig. 123 shows the principle of an area meter. In the rotameter the float is pushed upward against gravity in a tapered tube by the fluid until the annular area is just large enough for the pressure drop across the float to support its weight. The position is usually sensed by a rod attached to the float. The name "rotameter" originated from one particulr type in which the float was made with helical slots and rotated continuously.

In the second type of area meter the fluid moves a weighted or spring-loaded piston past an orifice, again until pressure force just balances the load on the piston. If a gas is being metered in a piston device, damping must sometimes be provided to eliminate resonant oscillation.

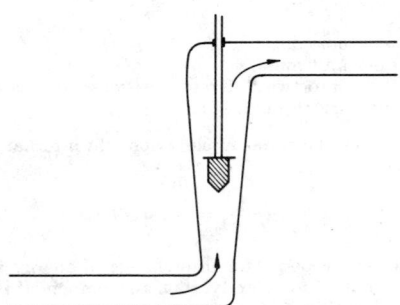

Fig. 122. Rotameter.

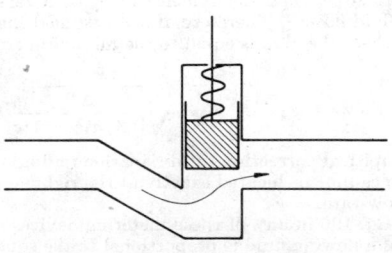

Fig. 123. Area meter.

Positive Displacement Meters

The two types of meters just described depend on energy and force balances for their operation. Because of dissipation and other effects, they cannot be relied on for better accuracy than a few percent. If more precise measurement of flow is needed for control purposes, in blending of gasoline for example, then some other more accurate metering device is necessary. The various types of positive displacement meters fill this need. In them, a precisely known volume is emptied and filled, with the number of cycles per unit time giving an accurate indication of flow to within a fraction of 1%. The principal sources of error are friction and wear, which cause leakage. Typical types of meters are those in which the known volume is created (1) by a slotted nutating disk in an appropriately shaped chamber, (2) by rotating oval gears in a cylindrical chamber, (3) by a rotor with sliding vanes, and (4) by movable pistons. These devices take the energy for their motion from the fluid, and they commonly rotate a shaft and gears for transmitting an output signal to a controller.

Rotating Turbines

Another highly accurate flow-measuring device that is commonly available now is the turbine meter. Fluid causes rotation of a specially shaped turbine wheel about an axis parallel to the flow. The blades are commonly made of magnetic material so that simple electrical pickup devices can be used to generate transmittible pulse trains for control purposes. This type of meter is accurate to within a few tenths of a percent if calibrated for the particular fluid being metered, but it is subject to wear in bearings and blades, and to fouling. Both of these can affect its accuracy.

Magnetic Flowmeters

Movement of a conductive fluid through a magnetic field generates a perpendicular electromotive force, proportional to its velocity, just as movement of a wire would do. This application of Faraday's law of induction can be used effectively in metering liquids whose properties prohibit the use of ordinary meters, i.e., fluids which are corrosive, highly viscous or non-Newtonian, slurries, and emulsions. A magnetic field is generated by saddle-shaped coils on opposite sides of the pipe through which the fluid moves. They are shaped and located in such a way that the field is perpendicular to the flow. In the third perpendicular axis small electrodes are mounted, usually flush with the wall. Some advantages are linearity, bidirectionality, and insensitivity to viscosity, density, and conductivity (above $\sim$200 μ). A particular disadvantage is that the tube must be nonconductive.

22. OTHER PROCESS MEASUREMENTS

Space does not permit description of the many types of process measurements that can be used in automatic control. Here they will be simply listed.

Continuous devices are available for measuring density, pH, refractive index, dielectric constant,, electrical and thermal conductivity, oxygen partial pressure, boiling and freezing points, vapor pressure, dew points, infrared, ultraviolet, and visible spectral characteristics, thickness by nuclear radiation or X-rays, viscosity, and humidity.

Intermittent sampling devices are coming into increasing use in process control, the most widely used now being the gas chromatograph. A small sample of volatile fluid whose chemical composition is desired is injected into one end of a packed tube through which it is carried by a flow of inert gas. Selective absorption equilibria on the packing, which is usually wetted with a nonvolatile absorbent, causes differing net velocities of the various components. Emerging from the far end of the "column" are a series of nonoverlapping "peaks," each one being a pure component of the sample if the column has been properly sized. These concentrations of materials in the carrier gas are detected by changes in thermal conductivity, by changing ionic concentration of ions in a flame, or by other means. The gas chromatograph is one of the most versatile and widely used analytical tools in use today.

23. CONTROLLERS AND TRANSMITTERS

In a feedback control a measuring device generates a signal which often must be transformed into an easily transmitted signal, like pressure, voltage, or current. This signal proceeds, often for some distance, to a control device which decides if and how a corrective action should be supplied. The controller must send this information to an actuator, which must act.

If the control is not of the feedback type, as in a timing device for example., then the measurement and signal transmission devices are not present, but the control and actuation still take place. These parts of control systems have much in common and will be described together here as "controllers."

There are three major subdivisions of controllers, according to the method by which they are supplied with energy. They are (1) pneumatic, (2) hydraulic, and (3) electric.

Pneumatic Controls

The heart of a pneumatic device is the amplifying relay where motion generated by the primary measuring device can be used to generate a pressure which balances the force which caused the motion, and which itself is transmitted or used to determine a control action. There are many types of force-balancing relays, and the components of a typical system are shown in Figs. 124 and 125. In Fig. 124 motion caused by the sensor is transmitted by levers or gears to the error-detecting vane or "flapper." If it moves away from the nozzle, the pressure inside the diaphragm capsule drops, the diaphragm moves, and the pilot valve opens. In Fig. 125 the nozzle-pilot valve arrangement is indicated by a block. Changing pressure in the output causes movement of the assembly, which brings the nozzle back into near contact with the flapper by expanding or contracting the bellows. In a similar way the output pressure can be used to generate a nulling force on the primary measuring device, such as by generating an opposing torque in a displacement level control, or an

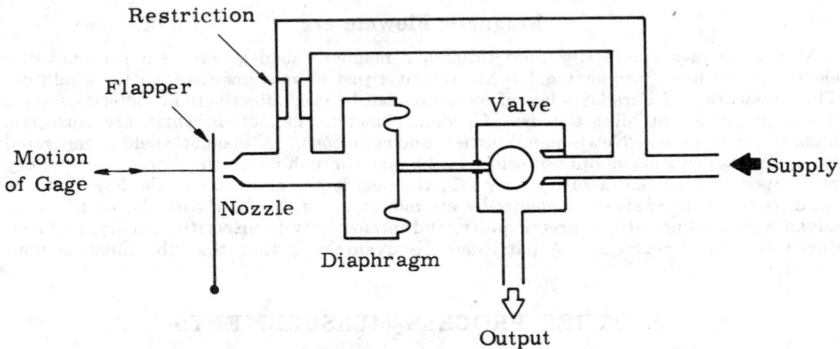

Fig. 124. Pilot valve.

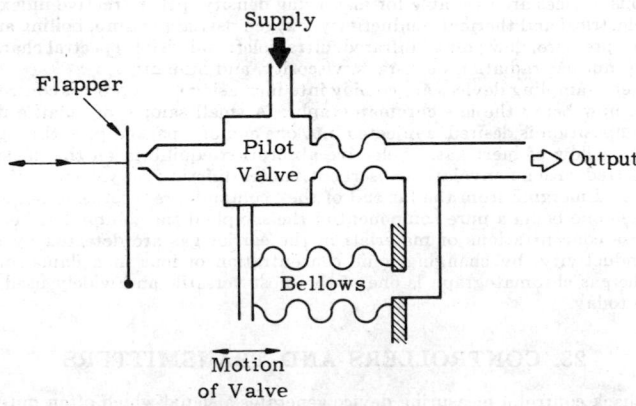

Fig. 125. Pneumatic relay.

opposing pressure in a pressure gage or expansion temperature bulb. All pneumatic transmitters and controllers depend in one way or another on this principle of operation.

The great advantage of pneumatic control systems is their inherent safety in hazardous environments. They tend to be rugged, highly reliable, and easy to maintain, and are used almost exclusively in oil refineries and chemical plants.

Implementation of three-mode control with pneumatics is illustrated in Fig. 126. Ultimate balance will be attained only when the force from the signal and set point bellows are equal; however, unbalance will result in changing output pressure. A continuously changing signal will result in a high output pressure because immediate force balance is delayed by the restriction R_d. Any permanent offset in signal from set point will cause a changing output as the R bellows tries to bring the flapper back to the nozzle but is inhibited by the resistance R_i.

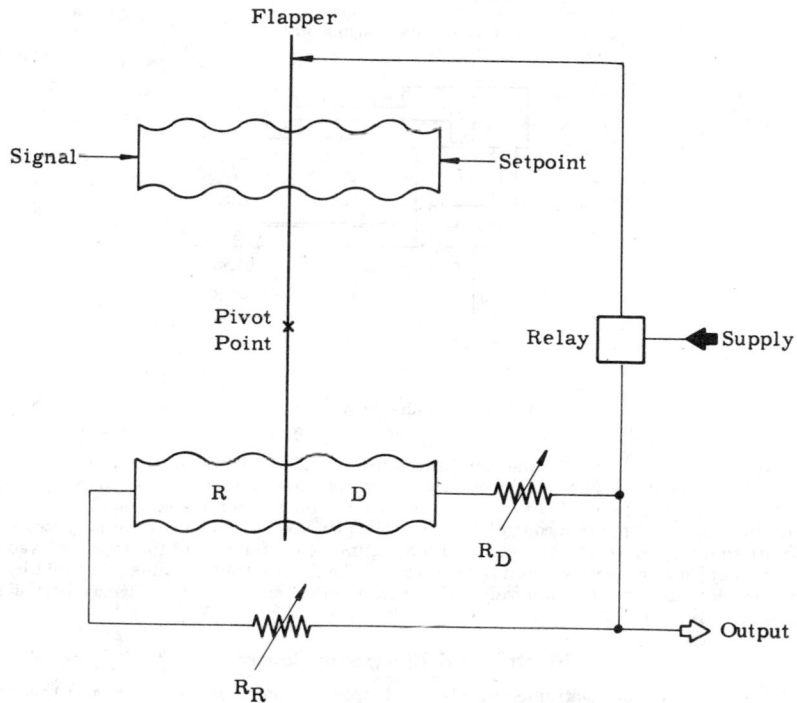

FIG. 126. Three-mode pneumatic controller.

Pneumatic valves are extensively used in processes. The action is extremely simple—pressure from the controller is changed to motion by a diaphragm—spring combination called a "motor." Figure 127 shows a typical arrangement.

Signal pressures for pneumatic controllers are normally in the range 3 to 15 psig. Supply pressures commonly range from 25 to 100 psig.

Hydraulic Controls

It will be clear that hydraulic systems can be used in a way that is exactly analogous to pneumatics. There are certain advantages in some circumstances. since incompressible fluids offer higher speed of response. Because signals are transmitted by volume changes, simple pumps can be used for supply rather than the large and expensive compressors that pneumatic systems call for. However, the hydraulic circuitry is more complicated because whereas air can be bled off at the point of ultimate delivery, hydraulic fluid must be returned to the system.

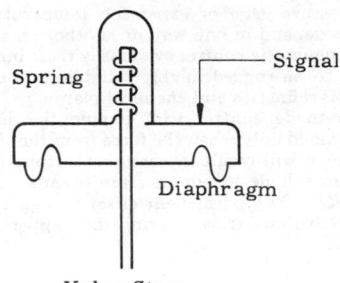

Valve Stem

FIG. 127.   Diaphragm motor.

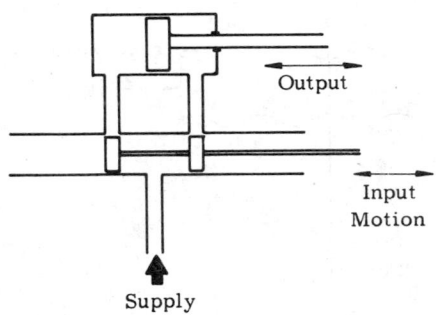

FIG. 128.   Hydraulic pilot valve.

A typical hydraulic pilot value is indicated in Fig. 128.   Motion of the shaft connected to the signal opens one valve or the other to the supply, and fluid is forced into the cylinder, moving the output shaft.   This motion in its turn can operate a force-balancing mechanism and can be transmitted to a control device.   The principle on which hydraulic controls act is quite similar to that already described for pneumatics.   Balance of forces is achieved by a system of linkages, bellows, and restrictions.   The final output signal is generated by an arrangement just like the pilot valve of Fig. 128, where the output shaft may be, for example, a valve stem.

Electric and Electronic Controls

The availability of electronic and electric amplifiers makes it natural to use this type of energy source in control.   We have seen how measuring devices like thermocouples, resistance thermometers, strain gages, and signals from detectors in chromatographs, for example, generate signals which are inherently electric in nature and which can easily be amplified to give power outputs.   It is also readily apparent that motion can be converted to an electrical signal through motion of a potentiometer slide wire, or even through the operation of on-off switches.   Forces and motion can be transduced to electrical signals by strain gages.

A *servo-mechanism* contains a reversible motor which is driven by a properly amplified signal dependent on the differences between a measured value and a reference, the motion of the motor being such as to generate some sort of corrective action.   This motor can drive, for example, the shaft on a potentiometer, and simultaneously a control valve. Proportional derivative and reset action can be designed into the servo-amplifier.

Another method of implementing electrical control systems has grown rapidly in recent years.   In essence, a control action is treated like a computation, and programmed with operational amplifiers and circuit elements just like an analog computer.   Most of the modern process-type control systems in use today use exactly this principle.   Because of the advanced state of the art in analog programming, control system designers find it quite easy to incorporate desirable features like elimination of *reset windup*, to put limiting values into the controller, to add alarms, and to accept pulsed inputs. Interfacing with computers is easy and natural. Interfacing with pneumatic and hydraulic actuators is simple because special convertors have been developed for just this purpose.

Transmission of signals from electronic controllers has not been standardized among the various manufacturers. The most common method is through current, with a *live zero*. Typically, full range controller output signals will be 1 to 5, 4 to 20, and 10 to 50 mA. If two or more devices are to receive signals from the same source, they must be put in series.

Electronic controls have many advantages. Their principal disadvantage at the time of this writing is their cost.

Program Controllers

There are automatic control systems in which feedback of information is not involved but which operate a system in some fixed, timed sequence, or in a sequence which varies from one implementation to the next through manual interpretation of feedback in the form of logic signals.

The design of controls incorporating logic has much in common with design of computing equipment and cannot be covered in detail here. This section will conclude with a review of timing devices and their use in controls.

If a synchronized sequence of signals is to be generated, some sort of clock must be included in the system. For high accuracy and high speed, electronic oscillators using crystals or tuning forks are available. These can generate pulse trains which can be counted electronically in toggles and "flip-flops" to give highly accurate and reproducible time intervals.

Next in order of accuracy, and perhaps first in convenience, is the 60 cycle synchronous motor or *electric clock*. Because the frequency of electric power generators is very carefully controlled, this form of timing is always accurate to a few hundredths of a second in the absence of gear backlash and slippage.

Escapement-pendulum timers (*alarm clocks*) find occasional use where electrical operation is not desirable, although they are obviously less convenient and less reliable than the other types.

A timing motor usually is attached through a gear train to a shaft on which are mounted the necessary cam-operated relays. The latter can be electric, pneumatic, or hydraulic. If the timer depends on an electronic oscillator, the counting device can, of course, operate the relays directly without cams.

BIBLIOGRAPHY

1. ALEXANDER, J. E. and BAILEY, J. M., *Systems Engineering Mathematics*, Englewood Cliffs, New Jersey, Prentice-Hall, 1962.
2. BALAKRISHNAN, A. V., and Neustadt, L. W., *Computing Methods in Optimization Problems*, New York, Academic Press, 1964.
3. BARNES, J. L., *Transients in Linear Systems*, New York, Wiley, 1942.
4. BAUMEISTER, T. (ed.), *Mark's Mechanical Engineers Handbook*, New York, McGraw-Hill, 6th ed., 1958.
5. BELLMAN, R. E., *Adaptive Control Processes*, Princeton, New Jersey, Princeton Univ. Press, 1961.
6. BELLMAN, R. E., and DREYFUS, S. E., *Applied Dynamic Programming*, Princeton, New Jersey, Princeton Univ. Press, 1962.
7. BOLLINGER, L. E. (ed.), *Proceeding of the First International Symposium on Optimizing and Adaption Control, Rome 1962*, Pittsburgh ISA, 1962.
8. BROIDA, V. (ed.), *Automatic and Remote Control* (Second Congress of IFAC, Basle, 1963), Vol. 1, Theory: Vol. 2.-Applications, London, Butterworths, 1964. See also First Congress, Moscow, 1964.
9. BROWN, G. S., and Campbell, D. P., *Principles of Seromcechanisms*, New York, Wiley, 1948.
10. BUCKLEY, P. S., *Techniques of Process Control*, New York, Wiley, 1964.
11. CALDWELL, W. I., COON, G. A., and ZOSS, L. M., *Frequency Response for Process Control*, New York, McGraw-Hill, 1959.
12. CHESTNUT, H., and MAYER, R. W., *Servomechanisms and Regulating System Design*, New York, Wiley, 1951.
13. CHURCHILL, R. V., *Modern Operational Mathematics in Engineering*, New York, McGraw-Hill, 1949.
14. CONSIDINE, D. M. (ed.), *Process Instruments and Controls Handbook*, New York, McGraw-Hill, 1957.
15. CONSIDINE, D. M., and ROSS, S. D., *Handbook of Applied Instrumentation*, New York, McGraw-Hill, 1964.
16. COSGRIFF, R. L., *Nonlinear Control Systems*, New York, McGraw-Hill, 1958.

17. COUGHANOWR, D. R., and KOPPEL, L. R., *Process Systems Analysis and Control*, New York, McGraw-Hill, 1965.
18. DEL TORO, V., and PARKER, S. R., *Principles of Control Systems Engineering*, New York, McGraw-Hill, 1960.
19. DORF, R. C., *Time-Domain Analysis and Design of Control Systems*, Reading, Massachusetts, Addison-Wesley, 1965.
20. ECKMAN, D. P., *Automatic Process Control*, New York, Wiley, 1958.
21. EVANS, W. R., *Control Systems Dynamics*, New York, McGraw-Hill, 1954.
22. GRABBE, E. M., RAMO, S., and WOOLDRIDGE, D. E., *Handbook of Automation, Computation and Control*, Vol. 1, Control Fundamentals; Vol. 2, Computers and Data Processing, and Vol. 3, Systems and Components, New York, Wiley, 1958.
23. HAHN, W., *Theory and Applications of Liapunov's Direct Method*, Englewood Cliffs, New Jersey, Prentice-Hall, 1963.
24. JACKSON, A. S., *Analog Computation*, New York, McGraw-Hill, 1960.
25. KNOWLETON, A. E. (ed.), *Standard Handbook for Electrical Engineers*, New York, McGraw-Hill, 1949.
26. KUO, B. C., *Analysis and Synthesis of Sampled Data Control Systems*, Englewood Cliffs, New Jersey, Prentice-Hall, 1964.
27. LANING, J. H., and BATTIN, R. H., *Random Processes in Automatic Control*, New York, McGraw-Hill, 1956.
28. LA SALLE, J., and LEFSCHETZ, S., *Stability by Liapunov's Direct Method*, New York, Academic Press, 1961.
29. LEE, R. C. K., *Optimal Estimation, Identification, and Control*, Cambridge, Massachusetts, MIT Press, 1964.
30. LEFSCHETZ, S., *Stability of Nonlinear Control Systems*, New York, Academic Press, 1965.
31. LEONDES, C. T. (ed.), *Advances in Control Systems*, Vol. 1, 2, 3 and 4, New York, Academic Press, 1963–1966.
32. MILLER, W. E. (ed.), *Digital Computer Applications to Process Control* (Stockholm, 1964), New York, Plenum Press, 1965.
33. MISHKIN, E. and BRAUN, L., *Adaptive Control Systems*, New York, McGraw-Hill, 1961.
34. NIXON, F. E., *Handbook of Laplace Transformations*, Englewood Cliffs, New Jersey, Prentice-Hall, 1961.
35. PERRY, R. H., CHILTON, C. H., and KIRKPATRICK, S. D., *Perry's Chemical Engineers Handbook*, New York, Mc-Graw-Hill, 1963.
36. PETERSON, E. L., *Statistical Analysis and Optimization of Systems*, New York, Wiley, 1961.
37. PONTRYAGIN, L. S., et al, *The Mathematical Theory of Optimal Processes*, New York, Wiley-Interscience, 1962.
38. RAGAZZINI, R., JR., and FRANKLIN, G. F., *Sampled-Data Control Systems*, New York, McGraw-Hill, 1958.
39. ROBERTS, S. M., *Dynamic Programming in Chemical Engineering and Process Control*, New York, Academic Press, 1964.
40. SAVAS, E. S., *Computer Control of Industrial Processes*, New York, McGraw-Hill, 1965.
41. TOU, J. T., *Digital and Sampled-Data Control Systems*, New York, McGraw-Hill, 1959.
42. TRUXAL, J. G., *Automatic Feedback Control System Synthesis*, New York, McGraw-Hill, 1955.
43. TRUXAL, J. G. (ed.), *Control Engineers Handbook*, New York, McGraw-Hill, 1958.
44. TSIEN, H. S., *Engineering Cybernetics*, New York, McGraw-Hill, 1954.
45. WIENER, N., *Cybernetics—Control and Communication in the Arrival of the Machine*, New York, Wiley, 1948.
46. WIENER, N., *The Extrapolation, Interpolation, and Smoothing of Stationary Time Series with Engineering Applications*, New York, Wiley, 1949.
47. WILLIAMS, T. J., "Annual Review of Computers and Process Control," *Ind. Eng. Chem.*, Vol. 57, No. 12, p. 33, 1965; Vol. 58, No. 12, p. 55, 1966.
48. ZIEGLER, J. G., and NICHOLS, N. B., "Optimum Settings for Automatic Controllers," *Trans. ASME 64*, p. 759, 1942; *65*, p. 433, 1943.

SECTION 14

CHEMISTRY

BY

C. DEAN NEWNAN

CHEMISTRY

Symbols

A	Helmholtz work function	V	volume
a	activity	v	molal volume
C	heat capacity, concentration	x	mole fraction in liquid phase
E	potential, volts	y	mole fraction in vapor phase
E_a	activation energy	Z	compressibility factor
F	Faraday constant		
f	fugacity	β	isothermal compressibility
G	Gibbs free energy	γ	activity or fugacity coefficient
H	enthalpy	Δ	change in, final-initial
J	energy conversion, joules/cal	μ	chemical potential
K	equilibrium constant	ν	stoichiometric coefficient
k	reaction rate constant	ρ	density
M	molarity		
m	molality	*superscripts*	
N	number of moles	-	partial molal quantity
n	number of equivalents	$\circ$	standard condition, defined
P	total pressure		
p	partial pressure	*subscripts*	
R	gas constant	c	critical
S	entropy	i, j	refer to components
T	temperature	pc	pseudocritical
t	time	R	reduced variable

1. ATOMIC STRUCTURE AND THE PERIODIC TABLE

Atoms. Atoms contain a dense, positively charged nucleus surrounded by a cloud of negatively charged orbital electrons that occupy discrete energy levels and orbital configurations. The nucleus consists of Z positively charged protons and $A - Z$ uncharged neutrons. Atomic number Z determines the name of the element and the number of orbital electrons in a neutral atom. Mass number A equals the number of protons plus neutrons in the nucleus. Gain or loss of electrons results in ionized atoms with a net charge. Chemical reactions are concerned only with electronic structure changes, while nuclear reactions involve changes in the constitution of the nucleus. Isotopes are atoms with the same atomic number, the same number of electrons, and the same chemical reactions, but differing in mass number because the nucleus contains a different $(A - Z)$ number of neutrons.

Atomic Weights of Elements. The mass of an element's naturally occurring distribution of isotopes is its atomic weight. This may be expressed in atomic mass units (amu) or in grams for an Avogadro number (6.023×10^{23}) of atoms. Atomic weights are relative to 12.000 for the $^{12}_6C$ isotope of carbon, an atom whose Z is 6 and A is 12. Examples: Naturally occurring chlorine contains 75.53 atom % of $^{35}_{17}Cl$ and 24.47% of $^{37}_{17}Cl$, it is assigned an atomic weight of 35.453. Naturally occurring tin consists of a presumably geographically invarient distribution of ten stable isotopes ranging from mass number 112 to 124. These yield a weighted average atomic weight of 118.69.

Atomic Radii. Radii of atoms, though nebulous and difficult to estimate, range from 0.4 to 2.5 Angstrom units ($Å = 10^{-8}$ cm). For an element of low atomic number, radius may be deduced from quantum mechanical calculations to agree with experimentally determined crystal ionic radii. Ionic radius differs from atomic radius due to a greater or lesser occupancy of the outer electron orbitals.

Periodicity of the Elements. The periodic chart is an arrangement of elements in rows and columns in order of increasing atomic number according to similarity of chemical behavior. Differences in electron energy levels, deduced from transitions that absorb or emit spectral energy, yield a similar and consistent pattern of electron buildup in numbers and energy levels as shown in Fig. 1, (see front end papers).

Electronic Orbitals. Rationalization of spectroscopic data via quantum mechanics leads to four types of quantum numbers for describing the building up of shells of electron orbitals with discretely increasing energy level for each orbital. This buildup, with electrons, individaully occupying the lowest vacant energy level, results in calculated spectra that closely match the experimental. These four types of quantum numbers are:

(a). *Principal quantum number*, n, represents a major grouping of energy levels. n ranges from 1 to 7, with 1 representing the lowest grouping of levels.

(b). *Azimuthal quantum number, l,* denotes different shapes of orbitals within the major grouping. These shapes are labeled s, p, d, and f for $l = 0$, 1, 2, and 3 respectively. Although l may range from 0 to $n - 1$, there is no occupancy of orbitals at the lowest or ground energy level beyond $l = 3$; this corresponds to 32 electrons within a single principal quantum number, n.

(c). *Magnetic quantum number, m,* may be $+1$, -1, or zero. Each value of m defines a magnetic property of the electron resulting from its orbital motion.

(d). *Spin magnetic quantum number, s,* may have a value of $+\frac{1}{2}$ or $-\frac{1}{2}$ depending on the direction of electronic spin.

Azimuthal Characteristics of s, p, d, and f Electronic Orbitals. Directionality of the most probable electronic position about the nucleus is provided by sets of mathematically symmetrical and orthogonal orbitals shown in Fig. 2. Each orbital can contain two electrons of opposite spin. An electron has the highest probability of being located on the surfaces of revolution shown. The size of these surfaces (distance from the nucleus) increases with the principal quantum number.

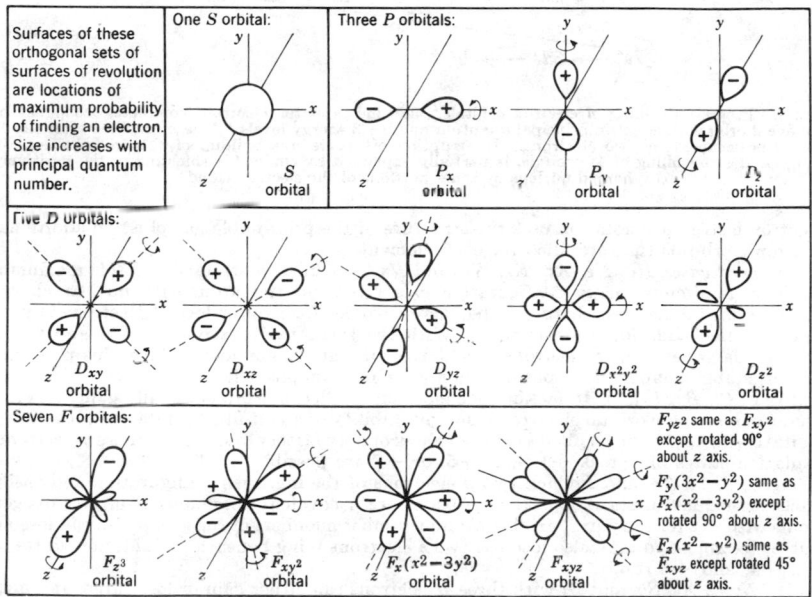

FIG. 2. Azimuthal characteristics of electronic orbitals.

Electronic Structure of Elements in the Ground State. As atomic number increases through the periodic chart, and the buildup of electronic structure follows an interrupted, through regular, pattern. One s, three p, five d and seven f orbitals can contain 2, 6, 10, and 14 electrons (with different magnetic and spin quantum numbers) respectively. Electron energy levels associated with quantum number designations are sequentially filled, beginning with the lowest energy level available. Figure 3 shows the pattern of electron occupancy of orbitals for elements from hydrogen ($Z = 1$) to lawrencium ($Z = 103$) at their lowest or ground energy state.

Oxidation States of the Elements. The oxidation state of the elements is the atomic charge attainable by chemical methods. This is correlated with the systematic buildup of electrons in orbitals. The addition or loss of electrons is subject to potential barriers which become essentially unsurmountable with filled shells exemplified by the noble gas structures. In the discussion of groups of elements in each vertical column of the periodic chart, Fig. 1, the known oxidation states are cited and related to electronic structure. This does not mean that all can exist as stable states in contact with oxygen, environmental moisture, or aqueous solution, for many of the oxidation states are particularly reactive.

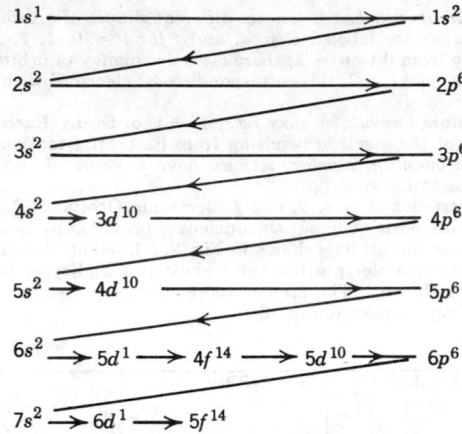

FIG. 3. Progressive filling of electron orbitals. $3d^{10}$ indicates ten electron (complete) occupancy of the five d orbitals existent in principal quantum number 3 energy levels. These are filled after the $4s$ orbital is occupied by two electrons. Interruption of progressive buildup of the $5d$ sequence, by fourteen electron filling of $4f$ orbitals, is partially explained by coulombic shielding of the positively charged nucleus by inner portions of the electron cloud.

Electronic structure notation uses the structure of the prior noble gas plus the additional electronic orbitals that are filled for each element.

(a). *Inert gases He, Ne, Ar, Kr, Xe, and Rn* have filled p orbitals, and this maximum stability, minimum energy configuration exists as a neutral atom with nuclear charge equaling orbital electron charge. The energy barrier for gain or loss of electrons is very high and the oxidation state is zero. With the exception of xenon (XeF_6) and radon compounds, where inner electrons shield the nuclear charge and reduce this potential barrier, stable compound formation has not been accomplished.

(b). *F, Cl, Br, I, and At*, lacking one electron of filled p orbitals, readily gain one electron; can lose one p orbital electron, reaching stability of a half-filled orbital; can lose five p orbital electrons, leaving a filled s orbital below; or may farther lose both of those s electrons. Oxidation states of -1, 0, $+1$, $+3$, $+5$, or $+7$ are possible.

(c). *O, S, Se, Te, and Po* lack two p electrons of the inert gas configuration and easily reach the oxidation state of -2. Peroxides are a covalent combination of two oxygen atoms with -2 for the pair. Higher atomic number members of the group also can lose all four p electrons and may also lose the two s electrons lying beneath. Oxidation states of -2, $+4$, and $+6$ result.

(d). *N, P, As, Sb, and Bi* with three p electrons can either gain or lose three, and may also lose two s electrons beneath (exception: NO or NO_2 gases). Oxidation states of ± 3 and $+5$ exist for the group.

(e). *C, Si, Ge, Sn, and Pb* with two p electrons can lose these, plus two s electrons beneath, resulting in $+2$ and $+4$ oxidation states for the group. Carbon forms numerous covalent bonds with an oxidation state of -4.

(f). *B, Al, Ga, In, and Tl* lose their single p electron plus the two s electrons below for an oxidation state of $+3$. Thallium has an additional stable $+1$ oxidation state.

(g). *Zn, Cd, and Hg*, with 10 electrons filling the d orbitals in stable configurations, lose two s electrons forming $+2$. Exception: mercurous dimer Hg_2^{2+}.

(h). *Cu, Ag, and Au*, with filled d orbitals, can lose their one s electron, forming $+1$ oxidation state. Exceptions for the column are Cu^{+2} and Au^{+3} due to removal of additional d electrons over low energy barriers.

(i). *Ni, Pd, and Pt*, with some overlap of d and s orbital energies, exhibit oxidation states of $+2$, $+4$, and $+6$.

(j). *Co, Rh, Ir* and the *Fe, Ru, Os* transition groups tend to have two electrons in the s orbital at the expense of vacant d orbitals. Oxidation states of $+2$, $+3$, $+4$, and $+6$ occur, with the $+8$ oxidation state reached in Ru and Os.

(k). *Mn, Tc, and Re*, with five electron, half-filled set of d orbitals and filled s orbital, tend to lose the s orbital electrons for $+2$ and any in the d orbitals for a maximum oxidation state of $+7$. Exception: Re also forms a -1 ion; it is the only transition element to do so.

(l). *Cr, Mo,* and *W,* with four electrons in the *d* orbitals and two in the *s* orbital, can have oxidation states ranging from $+2$ to $+6$.

(m). *V, Nb,* and *Ta* group, with three or four electrons in *d* orbitals and one or two in the *s* orbital, can have oxidation states ranging from $+2$ to $+5$.

(n). *Ti, Zr,* and *Hf* group, with two *d* and two *s* electrons, tends to have an oxidation state of $+4$.   Exception: Ti^{+3} also occurs.

(o). *Sc, Y, La,* and *Ac* group, with one *d* and two *s* electrons, can lose these for an oxidation state of $+3$.

(p). *The fourteen rare earths (lanthanides),* following lanthanum and beginning with cerium, have the lanthanum $+3$ oxidation state and structure with approximately progressive filling of the seven $5f$ orbitals.   Ce, Nd, Pr, and Nd form $+4$ ions while Sm, Eu, Tm, and Yb also form $+2$ ions.

(q). *The fourteen actinides,* beginning with thorium, paralleling cerium above, approximately repeat the structure of the rare earths with 14 electron filling of $5f$ orbitals.   Some exceptions occur, forming the half-filled orbital situation, and the $+3$ oxidation state can be increased to $+6$ due to much electronic shielding of the nuclear positive charge.

(r). *Beryllium and the alkaline earth* group have 2 electrons filling the *s* orbital.   Loss of these electrons results in a $+2$ oxidation state.

(s). *Hydrogen and the alkalai metals* have a single electron in the *s* orbital.   Its loss results in a $+1$ oxidation state.   Exception: gain of an electron by hydrogen to fill the *s* orbital results in the hydride, H^{-1}.

Excited Electronic States and Ionization Energy.   Excited electronic states are produced through energy absorption by an electron of a gaseous element; this raises the electron's energy level to that of an unfilled orbital.   Emission of that particular energy increment as a photon occurs within about 10^{-8} sec.   Energy absorption of magnitude greater than the highest available energy level is synonomous with having achieved an electronic escape velocity and ionization occurs.   This ionization energy ranges from 3.9 eV (90 kcal/g atom) for cesium to 24.5 eV for helium.

Electron Affinity.   The counterpart to ionization potential, electron affinity is energy released when an isolated monatomic gaseous atom gains an electron and becomes an anion. For halogens this energy is of the order of 3 to 4 eV; experimental problems have precluded measurement for many of the other elements that form negative ions.

Electronegativity.   The electron-attracting power of the elements has been estimated from chemical bond energies and from average electron densities.   The result is a consistent series of relative numbers based on zero for inertness of the noble gases, and ranging from a maximum of 4.0 for fluorine (which readily forms the anion F^-) to 2.5 for carbon (that forms equally shared covalent bonds) to a minimum of 0.7 for cesium (that forms the cation Cs^+).   Large differences in electronegativities of two chemically bonded atoms indicate an ionic bond with the more electronegative atom monopolizing the electron pair.   Small differences in electronegativity indicate a covalent bond with approximately equal sharing of the electron pair.

2. MOLECULAR STRUCTURE AND CHEMICAL BONDING

Molecules.   Molecules consist of groups of atoms held together in geometric arrangement by chemical bonds.   Chemical bonds are based on electronic configuration of the individual atoms and electron occupancy of vacancies existent in the atomic orbitals. Mutual occupancy of orbitals of two atoms by an electron pair constitutes a chemical bond between atoms and results in joint configuration of greater stability, or lower energy, than possessed by the individual atoms.   The total charge on a molecule is zero, with the number of orbital electrons matching the summation of positive nuclear charges.

Chemical Bonds.   Mutual occupancy of orbitals, the shared electron pair, leads to two general classifications of chemical bonds.   These are ionic and covalent.   An ionic bond exists where the shared electron pair is monopolized by one atom, giving that portion of the molecule a net negative charge and leaving an electron deficiency or net positive charge on the other.   A covalent bond results when both atoms have approximately the same electronegativity or electron affinity; the electron pair is shared without monopoly by either atom.   One shared electron pair constitutes one chemical bond, commonly designated by a dash between symbols of the elements.   The spatial geometry of molecules is related to orientation of the atomic orbitals involved in the bond.

Bond Energy.   Energy liberated on bond formation, or energy required to rupture a bond, is the bond energy.   This varies with the parent atoms and the particular molecular configuration involved, see Table 1 in Art. 5.   The summation of all bond energies is related to the heat of formation of a compound from its elements.

Chemical Compounds and Ions. Compounds can be classified into categories according to their ions in aqueous solution. Ionic compounds are those with ionic bonds between major atomic groupings, and these ionize in aqueous solution to form positive cations and negative anions. The naturally ocucrring combinations are used for inorganic chemical nomenclature. Metals, having low ionization potential, readily form positive cations in agreement with their electronic configuration. Examples: Na^+, Ca^{2+}, Al^{3+}, Fe^{2+} and Fe^{3+}, Zn^{2+}, Ag^+, etc. Halogens, having high electron affinity, form halide ions: F^-, Cl^-, Br^-, and I^-. Oxygen forms oxides, and readily combines to other atoms to form MnO_4^-, $Cr_2O_7^{2-}$, SO_4^{2-}, SO_3^{2-}, NO_2^-, NO_3^-, PO_4^{-3}, CO_3^{-2}, SiO_3^{-2}, and similar ions.

Electrical Neutrality of Compounds results from summation of component oxidation states and permits assignment of an oxidation state to component atoms in the molecule.

Example: Determine oxidation state of Cr in $NaCrO_4$.

Sodium's electronic structure, revealed by its position on the periodic chart, definitely assigns its ion as Na^+. Oxygen, except in peroxides, assumes -2 oxidation state. Chromium's usual oxidation states are $+2$, $+3$, and $+6$, as shown on the periodic chart. Therefore, the oxidation state of Cr is $+6$, as required for compound electrical neutrality.

Inorganic Chemical Nomenclature. Systematic and unambiguous nomenclature for inorganic and organic compounds and ions is necessitated by proliferation of known species. Traditional (often trivial) names of common compounds are paralleled by more cumbersome names following rules of the International Union of Pure and Applied Chemistry (IUPAC). Traditional nomenclature for inorganic compounds, Table 1, uses simple prefix and suffix terminology.

Table 1.   Prefix and Suffix Terminology

Cations (+ ions)	Anions (− ions)
-ic. Higher usual oxidation state of cation. Examples: Fe^{+3} is ferric, and Cr^{3+} is chromic.	*-ide.* Usual oxidation state of anion in binary compounds. Examples: oxide O^{2-}, chloride Cl^-, and hydride H^-.
-ous. Lower usual oxidation state of cation. Examples: Au^+ is aurous and Cr^{2+} is chromous.	*-ate.* Higher usual positive oxidation state of major element in oxygenated acid anions. Examples: ZnO_2^{-2} is zincate, SO_4^{-2} is sulfate, and ClO_3^- is chlorate.
-yl. Oxygenated cation containing the -ic or higher usual oxidation state of the major element. Examples: UO_2^{2+} is uranyl, TiO^{2+} is titanyl.	*-ite.* Lower usual positive oxidation state of major element in oxygenated acid anions. Examples: NO_2^- is nitrite, SO_3^{-2} is sulfite, and ClO_2^- is chlorite.
No suffix is used for cations that form only one ion. Examples: Zn^{+2} is zinc and Na^+ is sodium.	*hypo- -ite* or *-ous.* Lower oxidation state than -ite or -ous implies. Example: ClO^- is hypochlorite.
meta- and *ortho-.* Infer smaller and larger numbers of oxygen atoms in the -ate and -ite acid anions. Examples: SiO_3^{-2} is meta silicate and SiO_4^{-4} is ortho silicate.	*per- -ate* or *-ic.* Higher oxidation state than -ate or -ic implies. Example: ClO_4^- is perchlorate and MnO_4^- is permanganate.

pyro-. Infers a dimeric structure (a di- -ate) of the oxygenated acid anion. Example: $P_2O_7^{4-}$ is pyrophosphate.

thio-. Means single surfur substitution for oxygen in oxygenated acid anions. Example: $S_2O_3^{-2}$ is thiosulfate (SO_4^{-2} with S replacing one O).

Acids and Bases. Three acid-base definitions are recognized:

(a). The traditional Arrhenius definition, that acids yield H^+ and bases yield OH^- suffices for inorganic reactions in aqueous solution.

(b). Brönsted introduced the broader definition that acids are proton (H^+) donors and bases are proton acceptors. Since H^+ in aqueous solution exists hydrated as the hydronium ion, H_3O^+, water is by definition a base because it can accept a proton. In nonaqueous media this definition recognizes hydrides, alcoholates, alkyl lithium compounds, and ammonia as the powerful bases they are. Brönsted broadens the traditional definition because OH^- does accept a proton.

(c). Lewis defined acids as molecules or ions capable of accepting electron pairs, and bases as molecules or ions capable of donating electron pairs. Thus $AlCl_3$, BF_3, $FeCl_3$,

$ZnCl_2$, and $SnCl_4$ are Lewis acids because they lack pairs of electrons from filled orbitals. These strong acids are important in nonaqueous organic reactions. Lewis bases have an uncoordinated electron pair available; for example, on the nitrogen atom of ammonia, $:NH_3$.

The Brönsted definition of proton donors and acceptors, broadened by the Lewis definition of electron pair donors and acceptors, is necessary to explain reactions in nonaqueous media. The remaining sections on inorganic chemistry utilize the traditional simplicity of acid H^+ and base OH^- because these ions do result when any of the more broadly defined acids and bases react with water.

Strong and Weak Acids and Bases. The strong or weak designation refers to extent of ionization; this does not refer to concentration in solution. Strong acids and bases are completely dissociated into their ions, in water, while weak ones are only slightly dissociated. A 1-M solution of a strong acid like HCl yields a hydrogen ion concentration, H^+, of 1 mole/liter, while a 1-M solution of a weak acid such as acetic, CH_3COOH, yields a H^+ of only 4×10^{-3} mole/liter. A similar classification of bases is made, based on the extent of ionization in the solvent employed.

Complex Ions and Coordination Compounds. Unfilled electron orbitals of transition metal elements, those occurring in the center of the periodic chart, permit forming complex ions and coordination compounds that appear inconsistent with the usual oxidation state of the metallic element. But complex ions are consistent with the metal's electronic structure; its unfilled orbitals are filled by sharing of electron pairs to reach a stable configuration. Ligands (see Table 2) are groups attached to the central metal atom.

Table 2. Coordination Compound Ligands

F^-	NH_3, PH_3	$S_2O_3{}^{2-}$ thiosulfate
Cl^- Br^-	NH_2—CO—NH_2 (urea)	$C_2O_4{}^{2-}$ oxalate
CN^-	H_2O	$S_x{}^{2-}$ polysulfide
OH^-	CO, NO	
SCN^-	Various unsaturated hydrocarbons, including acetylide ion $(HC\equiv C)^-$ and cyclopentadienyl ion $(C_5H_5)^-$.	

Each ligand contributes an electron pair to be shared with and occupy vacant orbitals. This presumably results in a minimum energy configuration. Coordination number is the number of ligands surrounding the central metal atom. Solvation in water (hydration) or in liquid ammonia (ammoniation), formation of hydrated and double salts, and the numerous complex ions of inorganic and metal organic chemistry are examples of complex formation. The geometry of the complexes is spatially related to the hybridized orbitals involved. In the following examples, carbon monoxide, ammonia, and cyanide ion are ligands:

$Ni(CO)_4$	nickel carbonyl
$Cr(NH_3)_6{}^{3+}$	hexamminechromium III ion
$Fe(CN)_6{}^{3-}$	ferricyanide, or hexacyanoiron III ion

In nickel carbonyl, nickel has an elemental or zero oxidation state with one $4s$ and three $4p$ orbitals unfilled before reaching the next noble gas structure, Kr. An oxygen from each of the four CO molecules provides an electron pair which effectively occupies one of the four equal SP^3 hybrid orbitals to generate a stable compound.

The hexamminechromium III ion consists of Cr^{3+} symmetrically surrounded by six ammonia molecules, each providing an electron pair from its nitrogen atom. The resultant complex ion has gained stability by having all the $4d$ orbitals filled and the $4p$ orbitals half-filled. Further electron sharing with three negative ions will yield a hexamminechromium III salt; this effectively gives the chromium atom the stable electronic structure of Kr.

The ferricyanide ion consists of Fe^{3+} surrounded by six CN^- ions, each N of which shares an electron pair to occupy the six d^2sp^3 hybrid orbitals. Three of these same cyanide ions further share an electron to fill the remaining three vacancies and give the iron atom an effective Kr electronic structure. Charge on the remaining three CN^- ions gives the very stable ferricyanide ion its $3+$ charge.

A similar rationale can be applied to other coordination compounds composed of a central metal atom or ion and mixtures of ligands.

Chelate Compounds. Chelating (clawlike) ligands have a stereospecific configuration that permits multiple attachment to the metallic ion to form complexes. Examples are EDTA (ethylenediamine tetraacetate ion), acetylacetonate ion, and ethylenediamine. Multiple attachment can consist of a chemical bond plus coordination with oxygen or nitrogen atoms of the ligands.

Dissociation of Complex Ions. Variable stability of complexes is represented by equilibrium constants for their dissociation. Many complexes are very stable, while others can exist only in specific environments.

3. CHEMICAL REACTIONS AND STOICHIOMETRY

Chemical Reactions. Chemical reactions fit into two broad categories: equilibrium-controlled conversions of reactants to products without changes in oxidation state, and equilibrium control of oxidation-reduction reactions where changes in oxidation state occur.

Conversion of reactants to products can proceed only to an equilibrium condition because no driving force exists to go beyond equilibrium. Reactions can be optimized by continuously removing product from the reaction system so conversion can proceed without reaching equilibrium and stopping.

Le Chatelier's principle is basic to the understanding of all chemical equilibria. This universally applicable principle states that "systems adjust to reduce applied stress." For example, equilibrium of the reaction $aA + bB \rightleftarrows cC + dD$ can be shifted to the right when excess concentration of reactants A or B are present, or when products C or D are removed from the reaction mixture. Conversely, equilibria may be shifted to the left by adjustment of reactant or product concentrations.

Equilibrium Controlled Conversions. Many chemical operations use processes which consist of mixtures of chemical species that proceed toward equilibrium, and whose conversion to a desired product is governed by product removal from the reaction system. Product removal from the reaction system, as shown in Table 1, occurs through removal of ions: (a) by forming slightly ionized products, (b) by forming slightly soluble products, (c) by complex ion formation, and (d) by gas evolution.

Table 1. Equilibrium-controlled Conversions

Product Removal Method	Example	Net Reaction of Example	Equilibrium Governing Conversion Completeness
Slightly ionized product	Acid base neutralization, formation of weak acid or base	$H^+ + OH^- \rightleftarrows H_2O$ $H^+ + CH_3COO^- \rightleftarrows CH_3COOH$	Dissociation of water Dissociation of weak acid or base
Slightly soluble products (least soluble product of several alternatives will prevail)	Precipitation reactions leaving uninvolved ions remaining in solution	$Ca^{2+} + SO_4^{2-} \rightleftarrows CaSO_4$	Solubility product of precipitate
Complex ion formation (most stable complex of several alternatives will prevail)	Removal of ions from solution (convert them to soluble species that do not participate in the reaction)	$Ag^+ + 2CN^- \rightleftarrows Ag(CN)_2^-$	Dissociation of complex ion
Volatile product formation	Loss of CO_2 from aqueous solution	$H_2CO_3 \rightleftarrows H_2O + CO_2$	Solubility of gas at the reaction temperature

Oxidation-Reduction (redox) Reactions. Most chemical reactions have changes of element oxidation states as molecular configurations and chemical bondings are rearranged. Redox terminology is given in Table 2. Electron transfer takes place from an oxidation source to a reduction sink. Electrons are not lost, only transferred. The number of electrons transferred determines the relative stoichiometry between their oxidation half-reaction source and their reduction half-reaction sink. Redox reactions are recognized by finding changes in oxidation state. Except in disproportionation, oxidation occurs for

Table 2. Redox Terminology

Oxidation	Process producing electrons; electron source; occurs at anode in electrochemistry; electron loss produces an increase in oxidation state.
Oxidizing agent	Species that is reduced, and is electron sink, consumes electrons produced on oxidation of other things; presumed able to cause oxidation of other things.
Oxidation half-reaction	A charge balanced equation for an electron source (reversing the equation makes it a reduction half-reaction).
Oxidation potential	Energetics of the oxidation half-reaction, expressed thermally as free energy change or electrically as voltage, under standard conditions.
Disproportionation	Self-redox; reaction occurring when a metastable oxidation state simultaneously oxidizes and reduces the atom to more stable higher and lower oxidation states.
Reduction	Process consuming electrons; electron sink; occurs at cathode in electrochemistry; electron gain causes a reduction in oxidation state.
Reducing agent	Species that is oxidized and is an electron source, producing electrons for reduction of other things; presumed able to cause reduction of other things.
Reduction half-reaction	A charge balanced equation for an electron sink.
Reduction potential	Energetics of a reduction half-reaction. Positive sign means the reaction can proceed spontaneously as written. Reversal of reduction half-reaction changes it to an oxidation half-reaction and changes algebraic sign of the potential.

one atom and reduction occurs for another. Changes in oxidation state are determined by considering the following:

(a). Elemental state of the elements; i.e., S, O_2, H_2, Fe, etc.: oxidation state is 0.

(b). Oxygen in other molecules: Oxidation state is -2 (exception, when covalently bonded in peroxide, -1).

(c). Hydrogen in other molecules: oxidation state is $+1$ (exception in hydride, -1).

(d). Other elements according to position on periodic chart and list of usual oxidation states (Fig. 1, Art. 1).

(e). Summation of oxidation states in any molecule = 0, and in any ion = net charge on the ion.

(f). Consult table of reduction potentials (Table 1, Art. 9) for typical half-reactions of the elements in acidic or neutral and basic aqueous media.

Stoichiometric Equations. Stoichiometric equations can be written for reactions relating initial reactants and final products without regard to details such as equilibria, steps in the reaction mechanism, kinetics, percent conversion in a closed system, intermediate products, excess reactants required in the process, process variables, solvents, catalysts, etc. The balanced stoichiometric equation relates mass and moles of each product species to mass and moles of each of the reactants, assuming that the reaction goes to completion as written. The stoichiometric equation is the balanced chemical equation.

Balanced chemical equations involve three basic principles: (1) *a mass balance*, based on conservation of mass, which requires that total mass in = total mass out; (2) *an atom balance*, based on constant mass of each atomic species involved in ordinary nonnuclear reactions, which requires that there is no change in the numbers of each atomic species; and (3) *an electron balance*, based on conservation of the number of electrons involved in oxidation-reduction reactions.

Example. (1) *Reaction not involving change in oxidation states* is a mixture of ions that may proceed to complete conversion because a product is removed from the reaction system. The stoichiometric equation: $CaCl_2 + Na_2SO_4 \rightarrow CaSO_4 + 2NaCl$ can be written as a net ionic reaction: $Ca^{2+} + SO_4{}^{2-} \rightarrow CaSO_4$ for the formation of slightly soluble $CaSO_4$, whose solubility is quantified by its solubility product (see Chemical Equilibrium). Atom balance and mass balance apply; but electron balance, although valid, is not required, since no oxidation-reduction occurs.

(2) *Reaction involving redox* requires atom, mass, and electron balances. A gross reaction for dissolving copper in concentrated nitric acid (with liberation of brown N_2O_4 gas resulting from air oxidation of NO, the reaction product) can be written as two half-reactions:

$$\text{(a)} \quad NO_3{}^- + 4H^+ + 3e^- \rightarrow NO + 2H_2O$$

$$\text{(b)} \quad Cu \rightarrow Cu^{2+} + 2e^-$$

Change balance requires electron balance so twice (a) provides $6e^-$ that match the requirement of three times (b) for the net result: $2NO_3{}^- + 8H^+ + 3Cu \rightarrow 2NO + 4H_2O + 3Cu^{2+}$. It is apparent that $8H^+$ must have come from $8HNO_3$, and the product $3Cu^{2+}$ must be accompanied by $6NO_3^-$ for charge balance.   The final stoichiometric equation is $3Cu + 8HNO_3 \rightarrow 2NO + 4H_2O + 3Cu(NO_3)_2$. Equations (a) and (b) above can be found among the equations in the table of reduction potentials in the electrochemistry section (Table 1, Art.9).

(3) *Balance an unbalanced redox equation given*: $Ag_2S + CN^- + O_2 \rightarrow S + Ag(CN)_2^- + OH^-$ Atom, mass, and electron balances are required.   Water can participate in the reaction, but H^+ cannot exist in alkaline solution.   Determine which one element is oxidized and between which oxidation states.   Do the same for the element reduced.   (a) Reduction of O_2 from zero to -2 oxidation state in basic solution.   (b) Oxidation of S from -2 to zero oxidation state in basic solution. Write net reactions for (a) and (b), or find those reactions in the table of reduction potentials.

(a) $O_2 + 2H_2O + 4e^- \rightarrow 4OH^-$
Note the need for H_2O inclusion to balance this half-reaction.

(b) $S^{2-} \rightarrow 2 + 2e^-$ or $Ag_2S \rightarrow S + 2Ag^+ + 2e^-$
Equation (a) supplies $4e^-$, meeting the electron needs of twice equation (b) for the net result:

$$2Ag_2S + O_2 + 2H_2O \rightarrow 4OH^- + 2S + 4Ag^+$$

This matches that given, except it needs the CN^- added to give the $Ag(CN)_2^-$ complex.   The final stoichiometric equation is:

$$2Ag_2S + O_2 + 2H_2O + 8CN^- \rightarrow 4OH^- + 2S + 4Ag(CN)_2^-$$

A redox equation cannot be correctly balanced unless the electron balance requirement is first met. Trial and error on the basis of an atom balance usually leads to erroneous results.

(4) *Balance the disproportionation*: $ClO^- \rightarrow ClO_3^- + Cl^-$.   The Oxidation state of chlorine changes from $+1$ to $+5$ and -1.   The simplified oxidation half-reaction is $Cl^+ \rightarrow Cl^{5+} + 4e^-$.   The simplified reduction half-reaction is $Cl^+ + 2e^- \rightarrow Cl^-$.   Electron balance equates the oxidation half-reaction with twice the reduction half-reaction.   Addition produces a net skeleton reaction of $3Cl^+ \rightarrow Cl^{5+} + 2Cl^-$. The final redox equation is $3ClO^- \rightarrow ClO_3^- + 2Cl^-$ after oxygens are appropriately inserted to obtain required mass and atom balance for the equation.

Molar Volume of Gases.   At standard conditions of $0°C$ and 1 atm ($32°F$, 14.7 psia), the molar volume of ideal gases is approximately 22.4 l/g mole or 359 ft^3/lb mole.   Application of the ideal gas law ($Pv = RT$) using ratios of absolute pressure and absolute temperature permits estimation of molar volumes at other temperatures and pressures. These molar volumes of gases can usefully elaborate the mass and molar quantities of gases directly available from the stoichiometric equations.

Material Balances.   Material balances for processes can be written on many bases; i.e. per ton mole of product, per hour of operation, per batch, etc.   All are based on a stoichiometric equation for the process and on a definition of the system such that mass or molar balances may be made at the steady state, using "out less in equals no accumulation" as the guiding principle.   Selection of the element(s) used in the material balance depends upon the system; for example, the nitrogen of air in combustion processes passes through unchanged, and quantities of other gases may be related to it via gas analyses.

Example.   (1).   Determine the amount of CaO required to neutralize a short ton of 37% wt H_2SO_4 waste.

(a) Establish a balanced stoichiometric equation.   Add the molecular weights of the species involved.

$$CaO + H_2SO_4 \rightarrow CaSO_4 + H_2O$$
$$56 \qquad 98 \qquad 136 \qquad 18$$

Note that a material balance exists.

(b) Determine the weight of 100% H_2SO_4 available.

$$2000 (0.37) = 740 \text{ lb.}$$

(c) Ratio reactants and products according to the stoichiometric equation.

$$\frac{\text{wt CaO}}{56} = \frac{740}{98} = \frac{\text{wt CaSO}_4}{136} = \frac{\text{wt H}_2\text{O}}{18}$$

(2).   Per 1000 standard cubic feet (scf) of methane, calculate: 1) maximum carbon black available from partial combustion, 2) scf dry air required for that combustion, and, 3) composition of wet flue gas.

(a) Establish balanced stoichiometric equation.   Add atomic and molecular weights of all reactants and products.

$$CH_4 + O_2 \rightarrow C + 2H_2O$$
$$16 \qquad 32 \qquad 12 \quad 2(18)$$

Note that 1 lb mole of CH_4 can produce 1 lb atom of carbon black.

(b) Convert 1000 scf CH_4 to lb moles and lb.   359 scf gas = 1 lb mole; therefore,

$$\frac{1000}{359} = 2.79 \text{ lb moles or } 44.7 \text{ lb of CH}_4$$

(c) Ratio reactants and products according to the stoichiometric equation.

$$\frac{44.7}{16} = \frac{lb\ O_2}{32} = \frac{lb\ moles\ O_2}{1} = \frac{lb\ C}{12} = \frac{lb\ moles\ H_2O}{2}$$

Obtain 2.79 lb moles O_2, 2.79 lb moles or 33.5 lb C, and 5.58 lb moles of H_2O.

(d) Dry air composition is 21% volume = 21 mole % oxygen, and 79 mole % nitrogen and inerts.

$$Dry\ air\ required\ =\ 2.79\ \frac{(100)}{(21)}\ =\ 13.3\ lb\ moles\ or\ 13.3\ (359)\ =\ 4770\ scf$$

(e) Assume complete electrostatic precipitation of carbon black and no water condensation, then the flue gases consist of nitrogen, inerts, and H_2O vapor.   Nitrogen + inerts in = nitrogen + inerts out.

$$13.3\ (0.79)\ =\ 10.5\ \ lb\ moles\ N\ +\ inerts$$
$$\underline{\qquad\qquad\quad 5.58\ lb\ moles\ H_2O}$$
$$total:\ 16.08\ lb\ moles$$

$$N_2\ =\ \frac{10.5}{16.08}\ =\ 65.3\%\ by\ volume$$

$$H_2O\ =\ \frac{5.58}{16.08}\ =\ 34.7\%\ by\ volume$$

(3).   Products of hydrocarbon combustion, analyzed on a dry basis are: 10.34 vol % CO_2, 0.80 vol % CO, 5.16 vol % O_2, and the remainder N_2.   Water is undetermined.   Calculate: 1) carbon/hydrogen wt ratio of the hydrocarbon fuel, assuming all hydrogen burns to water before any carbon is converted to other products, and 2) percent excess air entering the combustion zone.

(a) Establish 100 lb moles dry products as basis of calculation.   On a molar basis products are:

$$10.34\ lb\ moles\ CO_2$$
$$0.80\ lb\ moles\ CO$$
$$5.16\ lb\ moles\ O_2$$
$$\underline{83.70\ lb\ moles\ N_2}$$
$$total:\ 100.00\ lb\ moles\ dry\ products$$

(b) Use a nitrogen balance to determine O_2 entering the combustion zone.

$$83.70\ lb\ moles\ N_2\ in\ =\ 83.70\ lb\ moles\ N_2\ out$$

$$O_2\ of\ air\ entering\ =\ 83.70\ \frac{(21)}{(79)}\ =\ 22.25\ lb\ moles$$

(c) Determine lb moles oxygen accounted for in known products

$$in\ CO_2:\ 10.34\ lb\ moles$$
$$in\ CO\ :\ \frac{0.80}{2}\ =\ 0.40\ lb\ moles$$
$$\underline{as\ O_2\ \ :\ \ 5.16\ lb\ moles}$$
$$total:\ 15.90\ lb\ moles$$

(d) Remainder of oxygen must exist as undetermined water.

$$22.25\ -\ 15.90\ =\ 6.35\ lb\ moles\ O_2\ in\ water$$

therefore

$$2(6.35)\ =\ 12.70\ lb\ moles\ H_2O\ are\ in\ products$$

(e) Hydrogen in products = 12.70(1) = 12.70 lb moles = 25.4 lb.

(f) Carbon in products = 10.34(1) + 0.80(1) = 11.14 lb atoms = 133.5 lb; therefore:

$$C/H\ wt\ ratio\ =\ \frac{(133.5)}{(25.4)}\ =\ 5.26$$

(g) Percent excess air:

$$(100)\ \frac{(total\ O_2\ entering)\ -\ (O_2\ required\ for\ complete\ combustion)}{O_2\ required\ for\ complete\ combustion}$$

(h) Alternate method for calculating percent excess air.   Free oxygen in combustion products = 5.16 lb moles, oxygen required to burn CO to CO_2 = 0.80/2 = 0.40 lb moles; therefore excess O_2 beyond that required for complete combustion = 4.76 lb moles.

$$\%\ excess\ O_2\ =\ \%\ excess\ air\ =\ 100\ \frac{(4.76)}{(17.49)}\ =\ 27.2\%$$

Mixture and Dilution Calculations. Mixtures with properties linearly dependent on mass, volume or mole fractions of components in the mixture can be handled by a simple summation:

$$P = \sum_i X_i P_i \tag{1}$$

where X_i is mass, volume, or mole fraction, as appropriate, of component i. P_i is the property P of pure i.

$$P \text{ of a binary mixture} = X_i P_i + (1 - X_i) P_j \tag{2}$$

Examples. (1) Mixture of density 3.6 is desired from components having densities of 2.6 and 7.5:

$$1 (3.6) = (x) 2.6 + (1 - x) 7.5, x = \frac{3.9}{4.9} = 0.796 \text{ and } 1 - x = 0.204$$

Since density is weight/volume, this is a weight summation when x is the volume fraction.
(2) Dilute 37% wt solution with 3% wt solution to make 20% wt solution.

$$1 (20) = x (37) + (1 - x) 3, x = 0.50 \text{ and } (1 - x) = 0.50$$

Since (wt) $\times$ (% wt.) = wt, this is also a weight summation when x is mass or weight fraction. Densities and assumption of volume additivity of ideal solutions are not required unless volume fractions are sought.

4. CHEMICAL THERMODYNAMICS

Chemical thermodynamics deals with work-energy relationships and the driving forces of chemical reactions. It is based on a wider variety of energy terms than the expansion-compression work, thermal, flow, kinetic, and potential energy terms of engineering thermodynamics. Basic relationships presented in the engineering thermodynamics section are augmented with other forms of work and energy that are applicable to the systems considered.

Energy Terms of Thermodynamics. Energy terms are the product of an intensive property (independent of quantity) and an extensive property (varies with amount) as shown in Table 1. When any of these terms are involved in a chemical thermodynamic system, they are included in the basic equations of engineering thermodynamics.

Chemical Potential, μ. Chemical potential at constant temperature and pressure is Gibbs free energy per mole, expressed in Btu/lb mole or cal/g mole. For a multicomponent system, μ is the summation of potentials (partial molal free energies) of all components present:

$$\mu = \sum_i \frac{G_i}{N_i} \tag{1}$$

Table 1.   Energy Terms of Thermodynamics

Kind of Energy	Intensive Property	Extensive Property	Energy Term and Its Units
Expansion-contraction work	P, pressure	dV, volume	$P\,dV$; cc atm, ft lb, cal, Btu, etc.
Thermal	T, temperature	dS, entropy	$T\,dS$; cal, Btu, etc.
Mechanical	F, force	dX, displacement	$F\,dX$; ft lb., etc.
Electrical	E, volts	dQ, charge	$E\,dQ$ or $EF\,dn$, joules
Magnetic	H, magnetic field intensity	dM, magnetization	$H\,dM$, work of magnetization
Surface	γ, surface tension	dA, area	$\gamma\,dA$; work of increasing area
Chemical	μ, chemical potential (partial molal free energy)	dN, no. of moles	$\mu \cdot dN$; energy due to undergoing reaction or crossing phase boundaries

where G_i is Gibbs free energy of component i, and N_i is moles of i present. At constant temperature and volume, chemical potential is the Helmholtz work function per mole.

Chemical Energy. Like other forms of energy, all products of an intensive and an extensive property, chemical energy in calories added to a system is $\mu_i \, dN_i$ for one component, and $\sum \mu_i \, dN_i$ for all components of the system.

Chemical Thermodynamic Equations. Basic differential equations of engineering thermodynamics are expanded by adding the chemical energy and other appropriate energy terms. The thermal energy sign convention is used for chemical energy, positive terms are energy added, and negative terms are energy removed from the system.

Relation of chemical potential to internal energy:

$$dU = T \, dS - P \, dV + \sum_i \mu_i \, dN_i \tag{2}$$

Relation to enthalpy:

$$dH = T \, dS + V \, dP + \sum_i \mu_i \, dN_i \tag{3}$$

Relation to Gibbs free energy:

$$dG = -S \, dT + V \, dP + \sum_i \mu_i \, dN_i \tag{4}$$

Relation to Helmholtz work function:

$$dA = -S \, dT - P \, dV + \sum_i \mu_i \, dN_i \tag{5}$$

For constant temperature ($dT = 0$) and constant pressure ($dP = 0$) reaction systems, Eq. 4 reduces to the most widely used relationship:

$$dG = \sum_i \mu_i \, dN_i \tag{6}$$

Standard Thermodynamic Data. Augmenting usual engineering sources of thermodynamic data (steam tables, gas tables, and special compilations for working fluids), useful tabulations of thermochemical data are available at 25°C (298°K).

ΔH_f° = heat of formation from elements; kcal/g mole (ΔH_f° of all elements in their standard state is arbitrarily taken as zero)

ΔG_f° = Gibbs free energy of formation from elements; kcal/g mole (Some references use ΔF° than than ΔG°)

S° = entropy, cal/g mole °K

C_p° = heat capacity, cal/g mole °K

The superscript $^\circ$ designates standard state at 25°C and unit activity or 1 atm fugacity. For a pure compound with more than one allotropic form, this refers to the stable modification. Aqueous solution data are given at 1 molal activity. The standard state is independent of pressure or concentration variables because activity (as a thermodynamically corrected concentration) and fugacity (as a corrected pressure to give ideal gas behavior) have been introduced as substitute variables.

Activity, a. Activity is a thermodynamically effective concentration used in lieu of actual concentrations to compensate for deviations of gases from ideality, incomplete ionization of strong electrolytes in solution, and other discrepancies between calculated and experimental behavior.

Fugacity, f. Fugacity (escaping tendency) of a gas equals its activity. Fugacity is an effective partial pressure, expressed in atmospheres. For ideal gases fugacity, f, exactly equals partial pressure. Since real gases in the standard state (in their usual phase at 1 atm pressure and 25°C) deviate slightly from ideality, activity of real gases is defined as the ratio of fugacity, f, to fugacity, f°, in a standard state where $f^\circ = 1$ atm. For all practical purposes, f° is 1 atmosphere pressure. Nonidealities of real gases at other temperatures and pressures are handled by an activity coefficient, γ.

Activity Coefficient, γ. Activity coefficient is the ratio of fugacity to partial pressure, or the ratio of activity to molal concentration.

For ideal gases: $\gamma = 1$.

For real gases: γ varies with T,P and is the ratio f/p, where f is fugacity of the gas and p is its partial pressure.

For ions in solution: $\gamma = a/m$, where a is activity of the ion and m is its molality in solution.

(a) *Activity of gases in mixtures.* *Activity of any gas in a mixture* (Lewis and Randall rule):

$$a_i = p_i \, \gamma_i \tag{7}$$

where p_i is partial pressure in atmospheres and γ_i is the fugacity coefficient of pure i at the temperature and total pressure of the mixture.

Calculation of ΔG° at Elevated Temperatures. Using Eq. 4 at constant pressure and constant number of moles, $dP = 0$ and $dN_i = 0$, we note that $dG = -S\,dT$. The calculation of $\Delta G_T{}^\circ$ for solids, liquids, and real gases at any temperature is dependent only on an expression for S as $f(T)$. From the thermodynamic definitions that $G = H - TS$ and $dH = C_p\,dT$, a general expression for ΔG at any temperature T can be derived using a power series expression for C_p as $f(T)$:

$$\Delta G_T{}^\circ = H_0 - \alpha T \ln T - \frac{\beta T^2}{2} - \frac{\gamma T^3}{6} + (\text{constant})\ T \tag{8}$$

H_0 is a constant derived from the integrated enthalpy term at some known temperature:

$$\Delta H_T{}^\circ = H_0 + \int_0^T \Delta C_p \, dT \tag{9}$$

The other constant is evaluated by substituting the known value of $\Delta G_T{}^\circ$ at that same known temperature. α, β, and γ are the constants in a usual power series expression for C_p: $C_p = \alpha + \beta T + \gamma T^2$. An example using this important general equation is:

Calculate $\Delta G^0{}_{298}$ for the reaction: $CO + \frac{1}{2} O_2 \rightleftarrows CO_2$ at unit fugacity and activity of all products and reactants at (a) 298°K and (b) 800°K.

(a) At standard conditions 25°C (298°K) and 1 atm:

$\Delta G_f{}^0$ for CO $= -32.81$ kcal/g mole

$\Delta G_f{}^0$ for $O_2 = 0$

$\Delta G_f{}^0$ for $CO_2 = -94.26$ kcal/g mole;

$\Delta G^0{}_{298}$ for the reaction $= G^0{}_{298}$ for products $- G^0{}_{298}$ for reactants

(1) $(-94.26) - (1)\ (-32.81) - (\frac{1}{2})\ (0) = -61.45$ kcal

(b) At 800°K and A atm, use general Eq. 8:

$$\Delta G_T{}^0 = H_0 - \alpha T \ln T - \frac{\beta}{2} T^2 - \frac{\gamma T^3}{6} + H_0 - (\text{constant } T)$$

for CO: $C_p = 6.3424 + 1.8363 \times 10^{-3}T - 0.2801 + 10^{-6}T^2$, cal/g mole °K,

for O_2: $G_p = 6.0954 + 3.2533 \times 10^{-3}T - 1.0171 \times 10^{-6}T^2$

for CO_2: $C_p = 6.393 + 10.100 \times 10^{-3}T - 3.405 \times 10^{-6}T^2$

C_p for the reaction $= C_p$ for products $- C_p$ for the reactants

$C_p = 1\ (C_p{}_{CO_2}) - \frac{1}{2}\ (C_p{}_{O_2}) - (C_p{}_{CO}) = -2.997 + 6.637 \times 10^{-3}\ T - 2.716 \times 10^{-6}T^2$

$= \alpha + \beta T + \gamma T^2$

$\Delta H_f{}^0$ for CO $= -26.42$ kcal/g mole

$\Delta H_f{}^0$ for $O_2 = 0$

$\Delta H_f{}^0$ for $CO_2 = -94.05$ kcal/g mole

$\Delta H^0{}_{298}$ for the reaction $= H^0{}_{298}$ for products $- H^0{}_{298}$ for reactants

(1) $(-94.05) - (1)\ (-26.42) - (\frac{1}{2})\ (0) = -67.63$ kcal $= -67,630$ cal

$$\Delta H^0{}_{298} = -2.997\ (298) + \frac{6.637 \times 10^{-3}\ (298)^2}{2}$$

$$- \frac{2.716 \times 10^{-6}\ (298)^2}{3} + H_0 = -67,630 \text{ cal}$$

Therefore, $H_0 = 67,010$ cal.

$$\Delta G^0{}_{298} = +2.997\ (298)\ (\ln 298) - \frac{6.63 \times 10^{-3}\ (298)^2}{2}$$

$$+ \frac{2.716 \times 10^{-6}\ (298)^3}{6} - 67,010 + (\text{constant})\ (298)$$

$$= -61,450 \text{ cal}$$

Therefore, the constant $= +2.56$.

With constants evaluated, ΔG_T^0 may be calculated at any temperature represented by the heat capacity data.

$$\Delta G_{800}^0 = +2.997\ (800)\ (\ln 800) - \frac{6.637 \times 10^{-3}}{2}\ (800)^2$$

$$+\frac{2.716 \times 10^{-6}}{6}\ (800)^3 - 67{,}010 + 2.56\ (800)$$

$$= -50{,}820\ \text{cal} = -50.82\ \text{kcal}$$

for reaction at 1 atm fugacity for gases, and unit activity for solids, liquids, and soluble species.

Calculation of ΔG at Elevated Pressure.* At constant temperature and constant number of moles $dT = 0$ and $dN_i = 0$, Eq. 4 reduces to $dG = +V\ dP$. The calculation of ΔG for ideal gases, real gases, solids, and liquids at any pressure is dependent only on an expression for V as $f(P)$.

For ideal gases $PV = RT$, and $dG = (RT/P)\ dP = RT\ d(\ln P)$. Integrating between limits gives:

$$\Delta G = RT \ln \frac{P_2}{P_1} \tag{10}$$

For real gases more complicated equations of state relating V as $f(T,P)$ can be substituted for ideal gas law and the integration performed. It is more expedient to use fugacity, f, instead of pressure, P, to correct for deviations from ideality. Ratios of f/P are obtained for gases by entering Fig. 4, Art. 7, with reduced temperatures and pressures; then

$$\Delta G = RT \ln \frac{f_2}{f_1} \tag{11}$$

For incompressible liquids and solids, V is constant, so

$$\Delta G = V(P_2 - P_1) \tag{12}$$

using appropriate units. The magnitude of this change for all condensed phases is very small.

For compressible liquids and solids, V is dependent on isothermal compressibility, β. $V = V_0\ (1 - \beta P)$

$$dG = V_0\ (1 - \beta P)\ dP = V_0\ dP - \beta V_0 P\ dP \tag{13}$$

or

$$\Delta G = V_0\ (P_2 - P_1) - \frac{\beta V_0}{2}\ (P_2^2 - P_1^2)$$

Standard Gibbs Free Energy Change and Reactivity. ΔG_T° for a reaction is

$$\Delta G_T^\circ = \sum_i \nu_i G_{Ti}^\circ \tag{14}$$

where ν_i is the stoichiometric coefficient of species i, taken as positive for products and negative for reactants. ΔG_T° is standard Gibbs free energy change for a reaction at temperature T when all reactants and products are at unit activity (unit fugacity for gases). ΔG_T° is independent of pressure; its magnitude as a driving force for predicting reactivity is very important:

If ΔG_T° is negative:	Reaction can occur as written; a forward driving force or chemical potential exists.	(15)
If ΔG_T° is zero:	No driving force exists; system is at equilibrium.	
If ΔG_T° is positive:	Reverse reaction can occur; a reverse driving force or chemical potential exists.	

Rate of reaction is dependent on temperature and activation energy; this calculation, considering only the reactant and product states, cannot tell anything about rate of reaction.

---

* Note: ΔG is not the same as pressure independent ΔG° introduced above.

Examples.   (1) Reaction solid $1CaO$ + gaseous $1CO_2$ = solid $1CaCO_3$

$$\Delta G_f^0 = \qquad \Delta G_f^0 = \qquad \Delta G_f^0 =$$
$$-144.4 \qquad -94.26 \qquad -269.78$$
$$\text{kcal/gmole} \qquad \text{kcal/gmole} \qquad \text{kcal/gmole}$$

ΔG^0 for the reaction as written = $(1)\,(-269.78) - (1)\,(-144.4) - (1)\,(-94.26) = -31.1$ kcal for the reaction as written at $298°K$.

(2) Phase change

$$C_{\text{graphite}} = C_{\text{diamond}}$$
$$\Delta G_f^0 = 0 \qquad \Delta G_f^0 = +0.68$$

$\Delta G^0{}_{298} = 1(+0.68) - 1\,(0) = +0.68$ kcal for the reaction as written. (The positive value of ΔG^0 indicates that the reaction as written cannot proceed at the standard condition, but energy is existent for the reverse reaction of diamond transition to graphite.)

Thermodynamics of Equilibrium.   At equilibrium the capability of any system for doing work is zero because there is no net energy available as driving force to modify the status quo.   In chemical systems at constant temperature and pressure, the change in the Gibbs free energy, ΔG, is zero at equilibrium between reactants and products; it is negative when a thermodynamic driving force for forward reaction is existent.   (For equilibrium at constant temperature and constant volume, the change in the Helmholtz work function, ΔA, is zero.)   Since most significant chemical equilibria are at constant temperature and pressure, this section contains only the relationships involving Gibbs free energy.   The conditions for chemical equilibrium are:

$$dG = 0, \text{ at constant } T,P \qquad dA = 0, \text{ at constant } T,V \qquad (16)$$

$$dG = \sum_i \mu_i \, dN_i = 0 \qquad dA = \sum_i \mu_i \, dN_i = 0$$

Standard Gibbs Free Energy Change and the Equilibrium Constant.   The magnitude of the driving force, ΔG_T°, at unit activity or unit fugacity of reactants and products, available to cause the reaction to proceed, was calculated for temperature T.   The standard free energy, ΔG_T°, was considered pressure independent because unit activities or fugacities were involved.

At equilibrium ΔG_T° is related to equilibrium constant K, when K is expressed using activities (or fugacities of gases).   This particular K is also considered pressure independent because the product of any set of equilibrium fugacities, each to the power of its stoichiometric coefficient, is a constant.

$$\Delta G_T^\circ = -RT \ln K \qquad (17)$$

where K is expressed using activities (or fugacities of gases).

Temperature Dependence of the Equilibrium Constant.   The temperature dependence of ΔG_T° directly affects K.   Combining Eqs. 8 and 17 yields:

$$\Delta G_T^\circ = -RT \ln K = H_0 - \alpha T \ln T - \frac{\beta T^2}{2} - \frac{\gamma T^3}{6} + \text{(constant)} \, T \qquad (18)$$

then

$$\ln K = \frac{-H_0}{RT} + \frac{\alpha \ln T}{R} + \frac{\beta T}{2R} + \frac{\gamma T^2}{6R} - \frac{\text{constant}}{R}$$

where α, β, γ are net coefficients of the power series for heat capacities of products less reactants, and H_0 is a constant which is derived from the standard heat of reaction ΔH_{298}° as shown by the previous example.   An integrated form of the Van't Hoff equation

$$\ln K = \frac{\Delta H_{298}^\circ}{RT} + \text{constant} \qquad (19)$$

approximates the above over a limited temperature range if ΔH remains constant and if heat capacities of reactants and products are equal.

Pressure Dependence of the Equilibrium Constant.   For an equilibrium constant expressed in activities and fugacity, the effect of pressure is introduced through change of activity and fugacity terms.

For gases: Fugacity equals partial pressure of ideal gases, and corrects for nonlinearities in partial pressure of real gases.

For liquids or solids: Activities of solids and liquids are essentially invarient. The change in ΔG with pressure for these condensed phases, calculated from Eq. 12 [$\Delta G = V_0 (P_2 - P_1)$], is of small magnitude.

For solutes: The changes of activity coefficient and equilibrium constant with pressure are given by

$$\frac{d(\ln \gamma_i)}{dP} = \frac{\bar{v}_i}{RT} \tag{20}$$

where $\bar{v}_i$ is partial molal volume of component i. The magnitude of pressure dependence of the equilibrium constant is extremely small.

$$\frac{d(\ln K)}{dP} = \frac{\sum_i \nu_i \bar{v}_i}{RT} \tag{21}$$

where $\sum_i \nu_i \bar{v}_i$ is the difference in partial molal volumes of products and reactants.

Thermodynamics and Phase Equilibria. Phase equilibrium exists, as with other equilibria, when change in Gibbs free energy, $\Delta G = 0$. Since chemical potential of component i, μ_i is defined as $(dG/dN_i)_{T,P,N_1,N_2}$ μ_i is partial molal free energy G_i at constant temperature, pressure, and moles of other species present. At constant temperature and pressure, $\Delta G = 0 = \sum_i \mu_i \, dN_i$. Phase equilibrium exists (with dN_i moles of component i transferred between phases) when partial molal free energy $\bar{G}_i$ or chemical potential μ_i of each component is the same in all phases:

$$\mu_i(\text{phase 1}) = \mu_i(\text{phase 2}) \tag{22}$$

Since $\bar{G}_i = \bar{G}_i{}^\circ + RT \ln a_i$, or $\mu_i = \mu_i{}^\circ + RT \ln a_i$, it follows that activities in each phase can be used with standard free energies to determine chemical potentials. For liquid phases, activities can be related to concentrations, and for vapor phases, activities are related to partial pressures. These relationships allow calculation for the effects of changes in temperature, pressure, and concentrations.

Partial Molal Quantities. Mixture nonidealities are often empirically treated by summing partial molal quantities contributed by each component. Partial molal quantities, such as molal volume or molal free energy, are of interest in phase equilibria:

$$dV = \sum_i \bar{v}_i \, dN_i \tag{23}$$

$$dG = \sum_i \bar{G}_i \, dN_i$$

where the bar indicates partial molal quantities.

5. THERMOCHEMISTRY

ΔH° **Standard Heats of Reaction and Formation.** Energy is transferred in all reactions and phase changes. Enthalpy added during an endothermic reaction is positive, in agreement with the basic sign convention of thermodynamics. Conversely, enthalpy evolved in an exothermic process is negative.

Data at a standard condition of 25°C (298°K) and 1 atm (or at unit fugacity of gases and unit activity of solutes) are designated by the $^\circ$ superscript. Standard heats of formation of compounds from their elements are available, all based on a reference state of $\Delta H^\circ = 0$ for elements in their stable phase at the standard condition. These data are given in kcal/g mole.

Example:

$$H_2(g) + \tfrac{1}{2} O_2 (g) \rightarrow H_2O (g) \qquad \Delta H_f{}^0 = -57.80 \text{ kcal/gmole of water (g)}$$
$$H_2 (g) + \tfrac{1}{2} O_2 (g) \rightarrow H_2O (1) \qquad \Delta H_f{}^0 = -68.32 \text{ kcal/gmole of water (l)}$$

Parentheses enclose the phase of each species involved. Reversal of direction for any reaction changes the sign of $\Delta H_f{}^0$.

Additivity of Extensive State Functions. Extensive properties that are state functions, such as H, G, and S, are additive over successive reactions carried out at constant temperature and pressure, according to the principle known as Hess's law:

$$\Delta X = \sum_i (\nu_i X_i)_1 + \sum_i (\nu_i X_i)_2 + \ldots \tag{1}$$

where ν_i is the stoichiometric coefficient of species i. Thus ΔX for a reaction may be obtained from the summation of a series of simpler or better-known reactions that give the same overall chemical balance. Thus the heat of a reaction may be obtained from the heats of combusion or formation of the compounds involved in the reaction. If X is any extensive property of the system, and X_i is the corresponding molal property of pure i for the reaction:

$$\Delta X = \sum_i \nu_i X_i \qquad (2)$$

In simultaneous or successive reactions, where the numerical subscripts refer to the different reactions:

$$dN_i = (dN_i)_1 + (dN_i)_2 + \cdots \qquad (3)$$

and

$$N_i = N_i^\circ + (\nu_i \Delta N)_1 + (\nu_i \Delta N)_2 + \cdots \qquad (4)$$

Estimating Heat of Formation and Reaction from Bond Energies. In the absence of heat of formation data for organic compounds, the standard heat of formation can be estimated by summing average nonpolar energies for each of the bonds of the compound. These energies are for gaseous molecules and are to be used with caution since bond strengths are influenced by molecular configuration, being reduced in resonance structures and increased in polar compounds. Heats of formation can be expected with in $\pm 10\%$ of the experimentally determined values using the values of Table 1.

Table 1. Bond Energies, kcal/g mole at 25°C *

H—H	104.2	F—F	36.6	H—F	134.6
O=O	119.1	Cl—Cl	58.0	H—Cl	103.2
N≡N	225.8	Br—Br	46.1	H—B	87.5
C=O (carbon		I—I	36.1	H—I	71.4
monoxide type)	255.8	C—C	82.6	C—F	116
C—H	98.7	C=C	145.8	C—Cl	81
N—H	93.4	C≡C	199.6	C—Br	68
O—H	110.6	C—N	72.8	C—I	51
S—H	83	C=N	147	C—S	65
P—H	76	C≡N	212.6	N—F	65
N—N	39	C—O	85.5	N—Cl	46
N=N	100	C=O (carbon		O—F	45
O—O	35	dioxide type)	192	O—Cl	52
S—S	54	C=O (aldehyde		O—Br	48
N—O	53	type)	176		
N=O	145	C=O (ketone			
		type)	179		

* These bond energies apply to energies of formation of gaseous molecules from gaseous atoms. For graphitic carbon to gaseous carbon atoms the heat of sublimation and atomization is 172 kcal/gmole. Heat of vaporization of water is 10.5 kcal/gmole.

Example: Heat of formation of acetaldehyde H—C—C=O:
(with H, H above and H below)

$$2C(s) + 2H_2 + \tfrac{1}{2}O_2 \rightarrow CH_3CHO$$

Energy evolved to make products bonds:
4 C—H bonds at 98.7 = −394.8
1 C—C bond = −82.6
1 C=O aldehyde type bond = −176
−653.4 kcal/mole

Energy added to break reactant bonds:
2 H—H bonds at 104.2 = +208.4
½ O=O bond at 119.1 = +59.6
2 (heat of sublimation and atomization of graphite at 172) = +344
+612.0 kcal/mole

net bond energy change for the reaction = −653.4 + 612.0
= −41.4 kcal/mole

Compare with −39.8 actually evolved.

Estimating Heats of Combustion. Standard heats of reaction in the vapor phase can similarly be extimated by considering only the energy added to break specific bonds of reactants and energy released in forming the new product bonds. Where reactants and products are liquid or solid, enthalpies of vaporization or sublimation must be used to get reactants and products into and out of the vapor phase where the calculated bond energies are valid.

ΔH **for Changes of Phase.** Thermal energy changes during phase transitions are significant factors in the overall energetics of chemical reaction systems. The numerical value of enthalpy of vaporization, sublimation, fusion, and other phase transitions is a $f(T,P)$.

Trouton's Rule for Estimating ΔH_v. Based on an estimated average entropy change of 21 cal/°K g mole occurring at the normal boiling point at constant pressure with no change in chemical potential between phases:

$$dH = T\,dS + V\,dP + \sum_i \mu_i d N_i \tag{5}$$

reduces to $\Delta H_v = 21(T°K)$. This generality applies for a number of nonpolar materials but is grossly inaccurate for others where $\Delta S \neq 21$.

Clausius-Clapyron Equation for Phase Transitions.

$$\frac{dP}{dT} = \frac{\Delta H}{T\Delta V} = \frac{\Delta S}{\Delta V} \tag{6}$$

where ΔH and ΔS are enthalpy and entropy changes of phase transitions at temperature T, and ΔV is the associated molal volume change. An integrated form, assuming ideal gas behavior of the vapor phase and negligible volume for condensed phases, is applicable to sublimation and vaporization:

$$\frac{d \ln P}{dT} = \frac{\Delta H}{R T^2} \quad \text{or} \quad \frac{d(\ln P)}{d(1/T)} = -\frac{\Delta H}{R} \tag{7}$$

Enthalpies of Solution, Dilution, Solvation, Adsorption, Crystallization and Mixing. Very limited published data usually require direct experimental measurement in all but the most widely used industrial systems. Energies involved in solvation, association, and hydration are small relative to those normally associated with chemical bond rupture and formation.

Heat Capacity. The basis relationship that $\Delta H = \int C_p\,dT$ for temperature changes of a system, plus enthalpy of phase transitions, provides the basis for thermal calculations at other than the standard temperature. Further thermodynamic use of heat capacity data is in calculation of entropy changes via $\Delta S = \int (C_p/T)\,dT$ and calculation of Gibbs free energy changes using the definitive relationship $G = H - TS$.

Heat Capacity Equations. Power functions of temperature best fit molar C_p data over a wide temperature range. These are of the form $C_p = \alpha + \beta T + \gamma T^2$. Although C_p is expressed as either cal/g mole °K or Btu/lb mole °R, care must be used in either inserting T in °K in the heat capacity equation or else using °R/1.8 to obtain the proper numerical value of C_p. Typical heat capacity equations are given in table 3.

Mean Molar Heat Capacity. Integration of heat capacity equations between 25°C (298°K, 77°F, 537°R) and temperature T yields a mean molar heat capacity over the range 25°C to T that grossly simplifies thermal calculations. Graphs of these results for the usual gases are given by Smith and Van Ness. Mean molar heat capacities between any pair of temperatures can be obtained by difference.

Heat Capacities of Ideal Gases. C_p molar heat capacity at constant pressure is independent of pressure. C_v molar heat capacity at constant volume is independent of volume. $C_p - C_v = R$ where $R = 1.985$ cal/g mole °K or Btu/lb mole °R. The oversimplifications that $C_v = \frac{3}{2}R$ for monatomic, $\frac{5}{2}R$ for diatomic, and $\frac{7}{2}R$ for triatomic ideal gases at ambient temperature are inadequate for most engineering purposes. These values are based on the kinetic theory of gases.

Zero Pressure Heat Capacity of Real Gases $C_p°$. Zero pressure heat capacities for an ideal gas state can be calculated from spectroscopic data. Since C_p for ideal gases is independent of pressure and is a function of temperature only, it can be used for real gases under T,P conditions where the real gas does not grossly deviate from ideal gas behavior. For conditions near critical and near the vapor-liquid phase change, recourse is made to more detailed compilations of specific thermal data for that system (steam tables, gas tables, etc.).

C_p **of Liquids and Solids.** The relative incompressibility of liquids and solids usually reduces the difference between C_p and C_v to less than experimental error in their measurement. The specific heat of liquids and solids increases with temperature but to a lesser extent than for gases. Experimental data are reported on both molar and unit mass bases at one temperature, average over a temperature range, or as power functions of T.

Kopp's Rule. Molar $C_p{}^\circ$ of solid molecules at 25°C is approximated by summation of contributory atomic $C_p{}^\circ$ values for constituents. Average values are given in Table 2.

Table 2. Contributory Atomic $C_p{}^\circ$ Values for Kopp's Rule

C	1.8	O	4.0	S	5.4
H	2.3	F	5.0	Mg	5.7
B	2.7	N	5.0	Al	5.8
Si	3.5	P	5.4	for all elements of higher molecular weight use an average value of 6.4 cal/g atom °K (Dulong and Petit)

Example: $Ag_2CO_3 = 2(6.4) + 1.8 + 3(4.0) = 26.6$ cal/g mole °K, (actual 26.8)

C_p **of Solutions and Mixtures.** In lieu of experimental data, heat capacity of a solution or mixture can be estimated by:

$$C_{p \text{ mixture}} = \sum_i y_i C_{pi} \tag{8}$$

where y_i is mole fraction and C_{pi} is molar heat capacity of component i.

Effect of Temperature on Heat of Reaction. Consider the reactants as one state and the products as another, and that thermodynamically one can proceed from one state function to another regardless of path. From reactants at temperature T, one can change temperature (using heat capacities and the enthalpies of any phase changes) to 25°C, then utilize the standard heat of reaction at 25°C and 1 atm, then change products back to temperature T.

Example: Calculate the heat of the ammonia synthesis reaction, $N_2 + 3H_2 \rightarrow 2NH_3$, as a function of temperature T. Given the standard heat of reaction $\Delta H^0 = -22.08$ kcal for the equation as written at 25°C (298°K) and 1 atm.

Heat capacity equations for reactants and products:

$$H_2: \quad 6.62 + 0.00081T \text{ (°K) cal/g mole °K or}$$

$$6.62 + \frac{0.00081}{1.8} T \text{ (°R)} \frac{\text{Btu}}{\text{lb mole °R}}$$

$$N_2: \quad 6.50 + 0.00100T \text{ °K}$$

$$NH_3: 6.70 + 0.00630T \text{ °K}$$

(1) Cool reactants from T down to 298°K:

$$\Delta H_1 = \sum \int_T^{298} C_p \, dt = 1 \left[(6.50)(298 - T) + 0.00100 \frac{(298^2 - T^2)}{2} \right]$$

$$+ 3 \left[(6.62)(298 - T) + 0.00081 \frac{(298^2 - T^2)}{2} \right]$$

$$= 26.36(298 - T) + 0.00343 \left(\frac{298^2 - T^2}{2} \right)$$

$$= -26.36(T - 298) - 0.00343 \left(\frac{T^2 - 298^2}{2} \right)$$

(2) Standard heat of reaction $\Delta H^0 = -22.08$ kcal/gmole quantities for $N_2 + 3H_2 = 2NH_3$ at 298°K. Given as $\Delta H^0 = -11.04$ kcal for $\frac{1}{2}N_2 + 3/2H_2 = NH_3$ in tables. Doubling stoichiometric quantities doubles heat of reaction. Multiply by 1000 to convert to Btu /lb mole quantities.

(3) Heat products from 298°K up to T°K

$$\Delta H_2 = \int_{298}^T C_p \, dt = 2 \left[6.70 \, T - 298) + 0.00630 \left(\frac{T^2 - 298^2}{2} \right) \right]$$

$$= 13.40(T - 298) + 0.01260 \left(\frac{T^2 - 298^2}{2} \right)$$

Table 3. Molar Heat Capacities of Gases in Ideal Gaseous State *

$C_p^0 = a + bT + cT^2 + dT^3$, where T is in $°K$

Compound	Formula	a	$b \times 10^3$	$c \times 10^6$	$d \times 10^9$	Range, $°K$	Accuracy, %
Bromine	Br_2	8.4228	0.9379	−0.3555		300–1500	0.5
Chlorine	Cl_2	7.5755	2.4244	−0.9650		300–1500	0.5
Fluorine	F_2	6.115	5.864	−4.186	0.9797	273–2000	0.5
Hydrogen	H_2	6.9469	−0.1999	0.4808		300–1500	0.3
Iodine	I_2	8.504	1.3135	−1.0684	0.3125	273–1773	0.1
Nitrogen	N_2	6.4492	1.4125	−0.0807		300–1500	1
Oxygen	O_2	6.0954	3.2533	−1.0171		300–1500	0.5
Sulfur	S_2	6.499	5.298	−3.888	0.9520	273–1773	0.5
Air		6.557	1.477	−0.2148		273–3773	1
Ammonia	NH_3	6.189	7.887	−0.728		273–1000	0.5
Carbon dioxide	CO_2	6.393	10.100	−3.405		300–1500	1
Carbon monoxide	CO	6.3424	1.8363	−0.2801		300–1500	0.5
Cyanogen	$(CN)_2$	9.892	14.484	−6.207		291–1000	0.5
Hydrogen bromide	HBr	5.5776	0.9549	0.1581		300–1500	0.5
Hydrogen chloride	HCl	6.732	0.4325	0.3697		300–1500	0.5
Hydrogen cyanide	HCN	5.974	10.208	−4.317		300–1000	0.5
Hydrogen iodide	HI	6.702	0.4546	1.216	−0.4813	273–1873	0.5
Hydrogen sulfide	H_2S	6.385	5.704	−1.210		298–1500	1
Nitric oxide	NO	7.020	−0.370	2.546	−1.087	298–1500	0.5
Nitrous oxide	N_2O	6.529	10.515	−3.571		298–1500	1
Phosgene, carbonyl chloride	$COCl_2$	10.35	1.653	−8.408		273–973	0.5
Phosphine	PH_3	4.496	14.372	−4.072		298–1500	0.5
Phosphorus pentachloride	PCl_5	4.739	107.329	−119.2		298–500	2
Sulfur dioxide	SO_2	6.147	13.844	−9.103	2.057	273–1773	0.5
Sulfur trioxide	SO_3	6.077	23.537	−0.687		298–1200	1.5
Stannic chloride	$SnCl_4$	21.72	6.33			273–573	1
Water	H_2O	7.219	2.374	0.267		298–1500	0.5
Methane	CH_4	3.381	18.044	−4.300		298–1500	1
Ethane	C_2H_6	2.247	38.201	−11.049		298–1500	0.5
Propane	C_3H_8	2.410	57.195	−17.533		298–1500	1.5
n-Butane	C_4H_{10}	4.453	72.270	−22.214		298–1500	1.5
2-Methyl propane	C_4H_{10}	3.332	75.214	−23.734		298–1500	1.5
n-Pentane	C_5H_{12}	5.910	88.449	−27.388		298–1500	1.5
n-Hexane	C_6H_{14}	7.477	104.422	−32.471		298–1500	1.5
2,2-Dimethyl butane	C_6H_{14}	0.593	133.001	−52.878		298–1000	0.5
n-Heptane	C_7H_{16}	9.055	120.352	−37.528		298–1500	1.5
n-Octane	C_8H_{18}	10.626	136.398	−42.592		298–1500	1.5
2,2,4-Trimethyl pentane	C_8H_{18}	−3.2	152.5			400–500	2
Ethene	C_2H_4	2.830	28.601	−8.726		298–1500	1.3
Propene	C_3H_6	2.253	45.116	−13.740		298–1500	1
1-Butene	C_4H_8	5.132	61.760	−19.322		298–1500	1.5
cis-2-Butene	C_4H_8	1.625	64.836	−20.047		298–1500	1.3
trans-2-Butene	C_4H_8	4.967	59.961	−18.147		298–1500	0.8
Ethyne	C_2H_2	7.331	12.622	−3.889		298–1500	1.5
Propyne	C_3H_4	6.334	30.990	−9.457		298–1500	1
2-Butyne	C_4H_6	5.700	48.207	−14.479		298–1500	0.5
Benzene	C_6H_6	−0.409	77.621	−26.429		298–1500	3
Toluene	$C_6H_5CH_3$	0.576	93.493	−31.227		298–1500	2.5
Cyclopropane	C_3H_6	−6.481	82.06	−55.77	15.61	273–973	0.5
Cyclopentane	C_5H_{10}	−5.763	97.377	−31.328		298–1500	2.4
Cyclohexane	C_6H_{12}	−7.701	125.675	−41.584		298–1500	2
Methyl cyclohexane	$C_6H_{11}CH_3$	−4.624	140.877	−46.698		298–1500	2
Formaldehyde	$HCHO$	4.498	13.953	−3.730		291–1500	0.5
Acetaldehyde	CH_3CHO	7.422	29.029	−8.742		298–1500	1.5
Acetone	$(CH_3)_2CO$	5.371	49.227	−15.182		298–1500	1.5
Methanol	CH_3OH	4.398	24.274	−6.855		273–1000	2
Ethanol	C_2H_5OH	6.990	39.741	−11.926		298–1500	0.8
2-Propanol	$(CH_3)_2CHOH$	0.7936	85.02	−50.16	11.56	273–1473	0.2
n-Propanol	$C_2H_5CH_2OH$	−1.307	92.35	−58.00	14.14	273–1473	0.5
Ethyl ether	$(C_2H_5)_2O$	−24.83	338.7	−593		300–400	2
Ethylene oxide	$(CH_2)_2O$	−1.12	4.925	−23.89	3.149	273–973	0.3

(Continued)

Table 3. **Molar Heat Capacities of Gases in Ideal Gaseous State *—**Continued*

Compound	Formula	a	$b \times 10^3$	$c \times 10^6$	$d \times 10^9$	Range, °K	Accuracy, %
Bromomethane	CH_3Br	4.184	22.445	−7.496		300–1200	1
Chloromethane	CH_3Cl	3.563	22.998	7.571		273–773	0.5
Fluoromethane	CH_3F	3.616	18.239	−2.035		298–600	0.5
Iodomethane	CH_3I	4.105	24.487	−9.733		300–600	0.5
Dichloromethane	CH_2Cl_2	4.309	31.67	−16.35		250–600	0.5
Fluorochloromethane	CH_2FCl	4.292	27.025	−10.605		250–600	0.5
Tribromomethane, bromoform	$CHBr_3$	9.356	32.319	−21.272		300–600	0.1
Trichloromethane, chloroform	$CHCl_3$	7.052	35.598	−21.686		273–773	0.5
Methyl cyanide	CH_3CN	5.018	27.935	−9.302		291–1200	0.5

*Reprinted with permission from Wilson and Ries, *Principles of Chemical Engineering Thermodynamics*, New York, McGraw-Hill, 1956.

(4) Algebraically sum the three ΔH values to get ΔH_T for the reaction at temperature T °K

$$\Delta H_1 = \qquad -26.36\,(T - 298) - 0.00343 \left(\frac{T^2 - 298^2}{2} \right)$$

$$\Delta H^0 = -22.08$$

$$\Delta H_2 = \qquad +13.40\,(T - 298) + 0.01260 \left(\frac{T^2 - 298^2}{2} \right)$$

$$\overline{\Delta H_T = -22.08 - 12.96\,(T - 298) + 0.00917 \left(\frac{T^2 - 298^2}{2} \right)}$$

kcal/gmole quantities for equation as written.

(5) Can convert to Btu and lb mole units for T °R, since

$$298°K = 537\ °R$$

$$\Delta H_T = -22.08\,(1000) - \frac{12.96}{1.8}\,(T - 537) + \frac{0.00917}{1.8} \left(\frac{T^2 - 537^2}{2} \right)$$

Btu/lb mole quantities for equation as written.

Combustion Calculations. (a) *The Composition of dry air* for purposes of combustion calculations is assumed to have a molecular weight of 29 and the following composition:

oxygen 21% volume = 21 mole %
nitrogen* 79% volume = 79 mole %

This composition allows combustion calculation on the basis of 100 lb moles (2900 lb) of dry air entering a reaction zone with 21 lb moles oxygen available for reaction and 79 lb moles nitrogen and inerts that pass through unreacted.

(b) *The moisture in ambient air* entering a combustion zone can often be ignored in approximate combustion calculations; however, it cannot be ignored in the exhaust gases. Gases, saturated with water vapor, contain a partial pressure of water vapor equal to the vapor pressure of water at that temperature. Saturation is 100% relative humidity; and the dew point, being the temperature of condensation initiation, is then the ambient temperature. The vapor pressure of water is available from steam tables.

Example. At a dew point of 77° F, vapor pressure of water from steam tables (saturated steam, temperature table) is 0.459 psia. Applying Dalton's law of partial pressures, the mole fraction of water vapor in gases at 1 atm is (0.459 psia)/(14.7 psia) = 0.031. At a dew point of 160°F, the vapor pressure is 4.739 psia and the mole fraction of water vapor in gases at 28 in. Hg pressure is

$$\frac{4.739\ \text{psia}}{\dfrac{28}{30}\,(14.7\ \text{psia})} = 0.347$$

(c) *Psychrometric charts*, useful for air conditioning and humidification calculations, express moisture content in grains of water per pound of dry air (7000 grains = 1 lb) and

* Includes 1% argon and traces of carbon dioxide.

utilize wet bulb (essentially dew point) and dry bulb (ambient) temperatures as variables for determining percent relative humidity.

(d) *Percent excess air.* Stoichiometric air is the amount of air theoretically necessary to burn all hydrogen in fuels to water, and to burn all carbon to carbon dioxide. With insufficient air, hydrogen is preferentially oxidized to water and carbon may be released as elemental carbon (soot), carbon monoxide, or mixtures of carbon monoxide and dioxide. Significant amounts of excess air and residence time in the combustion zone are required for complete combustion of fuel-contained carbon to carbon dioxide.

(e) *Basis for calculation.* All stoichiometric calculations should be preceded by a basis of calculation in order to readily permit scale-up or -down of the result. Examples of usual bases are: per 100 lb of fuel, per 100 lb moles of a reactant, and per ton of a product.

(f) *Heat balances.* Thermal calculations for reaction systems are most easily made for steady-state conditions involving an accounting calculation which equates all forms of energy in plus heat of reaction to all forms of energy out plus losses. Forms of energy involved are thermal, chemical, electrical, and mechanical in the form of shaft work, kinetic, potential, and flow energies, plus enthalpies of all phase transitions involved. Heat balances are necessarily predicated on valid material balances, and vice versa. Calculation can be based on mass flow of some component, a unit of operating time, or any other useful basis chosen to simplify calculation and utilize available thermal data.

6. CHEMICAL EQUILIBRIUM

Equilibrium Constant. Equilibrium condition for any reaction $a\text{A} + b\text{B} + \ldots = c\text{C} + d\text{D} + \ldots$ is defined by the expression:

$$K = \frac{[\text{C}]^c[\text{D}]^d \ldots}{[\text{A}]^a[\text{B}]^b \ldots} \tag{1}$$

Exponents are stoichiometric coefficients of the equation as written, and bracketed terms are activities of each species existent at equilibrium. The wide numerical range of K encountered has promoted usage of $pK = -\log_{10} K$ for tabulated data. The equilibrium constant is an expression relating activities of reactants and products such that the net chemical potential is zero at the temperature and pressure of reaction. The standard free energy of a reaction is based on unit activity of all reactants and products. K varies with temperature, and has a pressure dependence introduced via fugacity of gaseous species, being derived from the relationship $\Delta G_T^\circ = -RT \ln K$ (see Eq. 17, Art. 4).

For reactions where equilibrium constant data are unavailable, combining K's from other known reactions will yield the one desired. This is a consequence of the additivity of reactions and their thermodynamic state functions.

Example:

$$CO_2 = CO + \tfrac{1}{2}O_2$$

$$H_2O = H_2 + \tfrac{1}{2}O_2 \qquad K = \frac{[H_2][O_2]^{1/2}}{[H_2O]} = 7.25 \times 10^{-1} \text{ at } 1000°K$$

$$CO_2 + H_2 = CO + H_2O \qquad K = \frac{[CO][H_2O]}{[CO_2][H_2]} = 8.70 \times 10^{-11} \text{ at } 1000°K$$

$$\therefore CO_2 = CO + \tfrac{1}{2}O_2 \qquad K = \frac{[CO][O_2]^{1/2}}{[CO_2]} = (7.25 \times 10^{-1})(8.70 \times 10^{-11})$$

$$= 6.3 \times 10^{-11} \text{ at } 1000°K$$

Activities of solids and solvents are substantially invarient, each in its own phase; they are assigned numerical values of 1. The activity of gases is their fugacity in atmospheres. Fugacity is related to partial pressure via a fugacity coefficinet $\gamma = f/p$ which varies with pressure and temperature. The fugacity coefficient of ideal gases is 1, but the deviations of real gases necessitate its consideration. For real gases it is expendient to divide K, based on activities, into two parts:

$$K = K_\gamma K_p \tag{2}$$

where K_γ is an activity coefficient product and K_p is the pressure product.

Equilibrium Composition of Reaction Mixtures. The effects of pressure and temperature on equilibrium composition, qualitatively predicted by LeChatelier's principle (Art.3), are increased conversion with pressure rise if moles of reactants exceeds moles of products, and decreased conversion with temperature rise in an exothermic reaction. Quantitatively, equilibrium composition may be calculated from the equilibrium constant. The equilibrium composition is completely separate from kinetics of the reaction, for the time to reach this equilibrium composition under the the given T,P conditions may be infinite due to a large activation energy barrier.

Effect of Inerts in Gaseous Reactions. Inert gases reduce the partial pressures of reactant and product species. As with pressure, where there is a difference in number of moles between reactants and products, the mole fractions at equilibrium are shifted.

Sample Calculation of Equilibrium Composition for Gaseous Reactions.

$$\tfrac{3}{2}H_2(g) + \tfrac{1}{2}N_2(g) = NH_3(g) \qquad K = \frac{[NH_3]^1}{[H_2]^{3/2}[N_2]^{1/2}} = 0.0431 \text{ atm}^{-1} \text{ at } 600°K$$

Note that the equilibrium constant is dimensional, and its units are useful in reconstructing the equilibrium constant expression and the reaction equation on which it is based.

Example:

$$K = \frac{(f_{NH_3})^1}{(f_{H_2})^{3/2}(f_{N_2})^{1/2}} = 0.0431 \text{ atm}^{-1}$$

applies for NH_3 synthesis. For a stoichiometric ratio of hydrogen and nitrogen, without inerts, calculate mole fraction of ammonia in the equilibrium mixture at (a) 1 atm and (b) 100 atm.

At 1 atm total pressure, where the assumption of ideal gas behavior for this system may be valid, the activity coefficient γ is assumed and confirmed to be 1 for all species. At 100 atm γ should be estimated for each species by entering the $\gamma = f/p$ chart (see Art. 7,) using reduced temperature and pressure based on the reaction temperature and total pressure. Significant deviation of K_γ from 1 signals the necessity of its inclusion in the calculation.

	P_C, atm	T_C, °K	T_R	P_R at $P = 100$ atm	$\gamma = f/p$ at $P = 100$ atm	P_R at $P = 1$ atm	$\gamma = f/p$ at $P = 1$ atm
NH_3	111.3	405.5	0.90	1.48	ca 0.92	0.009	~1
H_2	12.8	33.3	7.8	18.	ca 1.1	0.078	~1
N_2	33.5	126.2	2.98	4.76	ca 1.0	0.030	~1

$$K_\gamma = \frac{(0.92)^1}{(1.1)^{3/2}(1.0)^{1/2}} = \frac{0.92}{(1.16)(1.0)} = 0.79 \text{ at } 100 \text{ atm}$$
$$= 1.0 \text{ at } 1 \text{ atm}$$

Calculation of mole fraction NH_3.

At total $P = 1$ atm, $K_\gamma = 1$

Let
$$X = \text{partial pressure of } N_2 \text{ in the reaction mixture}$$
$$3X = \text{partial pressure of } H_2$$
$$1 - 4X = \text{partial pressure of } NH_3$$
$$K = K_\gamma K_p = 0.0431 \text{ atm}^{-1} = \frac{(1 - 4X)}{(3X)^{3/2}(X)^{1/2}}$$
$$1 - 4X = 0.0431 (5.2X^2)$$

This is solved by quadratic formula or successive approximation to give $X \cong 0.246$

mole fraction, $y_{NH_3} = \dfrac{p_{NH_3}}{p_{total}} = \dfrac{1 - 4X}{1} = 0.016$

At total $P = 100$ atm, $K = 0.79$. Let $X = p_{N_2}$
$$3X = p_{H_2}$$

$$K = 0.0431 = \frac{0.79(100 - 4X)}{(3X)^{3/2}(X)^{1/2}} \qquad 100 - 4X = p_{NH_3}$$

$$100 - 4X = \frac{0.0431}{0.79}(5.2X^2) \qquad X = 13.0$$

mole fraction $y_{NH_3} = \dfrac{p_{NH_3}}{p_{total}} = \dfrac{100 - 4X}{100} = 0.48$

Applications to Solution Chemistry. Equilibrium constants have important quantitative applications in solution chemistry, and specific K's are given appropriate subscripts:

(1) K_A = ionization constant for weak acids.
(2) K_B = ionization constant for weak bases.
(3) K_W = dissociation constant for water.
(4) K_D = dissociation constant for complex ions.
(5) K_{sp} = solubility product for slightly soluble species.

Solutions and Solution Concentrations. Solutes are smaller amounts of any phase, dissolved in larger amounts of liquid or solid solvents, resulting in single phase solutions.

Solution concentrations are expressed in three ways:

(1) *Molar* concentration, M, is gram moles solute/liter of solution. Small temperature dependence exists due to density changes.

(2) *Molal* concentration, m, is gram moles solute/kilogram of solvent. Molal concentrations are not temperature dependent. Since volume of solution = [(grams solute + grams solvent)/(solution density)], molality, m, approximately equals molarity, M, only in dilute aqueous solutions.

(3) *Mole fraction* x or y expresses solution concentration. x symbolizes mole fraction in liquid phases, and y applies to vapor phases. Mole fraction is the ratio of moles of component i to total moles of all species present in that phase.

$$x_i \text{ or } y_i = \frac{N_i}{N_{\text{total}}} \tag{3}$$

where N = number of moles.

Activity-Concentration Relationship for Solutes. Activity coefficient γ is ratio of activity a to molality m. At infinite dilution $\gamma = 1$ for all solutes, but γ varies with temperature, concentration, and the species involved (see Art. 9). Activity is correctly used for equilibrium calculations, although molar concentrations are widely used as approximations.

Hydrogen Ion Concentration, pH. pH is an exponential notation using the operator p meaning "$-\log_{10}$ of" to express a wide range of $[H^+]$. At 10^{-7} M $[H^+]$, resulting from dissociation of water at ambient temperature, $pH = 7$.

7. PHASE EQUILIBRIA

Temperature-Pressure Phase Diagrams. Pure chemical species, in the absence of thermal decomposition, have temperature-pressure phase diagrams of the same general shape exemplified by Fig. 1. In this figure, changes of state occur along transition lines,

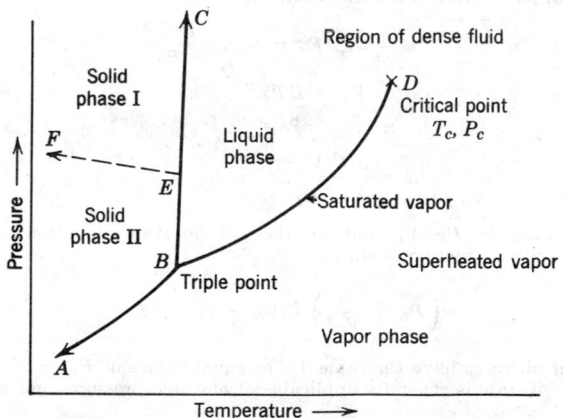

FIG. 1. Generalized temperature-pressure phase diagram.

i.e., sublimation along AB, fusion along BC, and vaporization along BD. Each phase transition is accompanied by volume, enthalpy, and entropy changes. Phase equilibrium exists at any T,P condition on a transition line. If multiple solid phases exist, transition occurs at T,P along lines such as EF.

Excepting material (H_2O, Bi) which expand on solidification where line BC slopes to the left, BC slopes only slightly to the right. For different species the coordinates of triple and critical points differ, but the diagram retains the same general form. Logarithmic compression of scales is often helpful. The normal boiling point occurs along transition line BD at $P = 1$ atm. The normal freezing point occurs approximately at the temperature of the triple point because of the almost vertical slope of line BC.

Equations of State. The PVT relations for gases are most simply expressed by the ideal gas law $Pv = RT$, where v is molar volume, applicable with small error at pressures well below the critical pressure and at temperatures well above the condensation temperature. More precise equations of state are necessary for gases near their condensation conditions where atomic size becomes important and interaction must be considered. Several modifications are used, and all contain additional terms to improve precision near the liquid state. One of the several historical and more precise equations of state is van der Waals:

$$\left(P - \frac{a}{v^2}\right)(v - b) = RT \tag{1}$$

where a and b are dimensional and specific Van der Waal's constants applicable to individual gases. Constant a is a measure of attractive force between molecules, and constant b is due to incompressibility and finite molecular volume.

The Redlich-Kwong equation is another example:

$$P = \frac{RT}{v - b} - \frac{a}{T^{1/2}v(v - b)} \tag{2}$$

where a and b are constants calculated from P_c and T_c available in Table 1.

Virial coefficients are used in more easily computed power series expansions used as equations of state:

$$Pv = RT + \frac{\beta}{v} + \frac{\gamma}{v^2} + \frac{\delta}{v^3} \quad \text{and} \quad Pv = RT + BP + CP^2 + DP^3 + \ldots \tag{3}$$

where β, γ, δ, and B, C, D are 2nd, 3rd, and 4th virial coefficients.

Reduced Equations of State. Coefficients of Van der Waal's equation of state, a and b, vary for individual gases and also vary to some extent with temperature. Coefficients a and b and the gas constant R can be generalized and expressed in terms of the critical constants for all gases using consistent units:

$$a = 3V_c^2 P_c = (3/4)^3 \frac{R^2 T_c^2}{P_c} \tag{4}$$

$$b = \frac{V_c}{3} = \frac{RT_c}{8P_c} \tag{5}$$

$$R = \frac{8}{3} \frac{P_c V_c}{T_c} \tag{6}$$

Using reduced variables P_R, V_R, and T_R, the van der Waals equation of state yields a dimensionless reduced equation of state:

$$\left(P_R + \frac{3}{V_R^2}\right)(3V_R - 1) = 8T_R \tag{7}$$

This infers that all gases have the same V_R at equal values of P_R and of T_R. This reduced equation of state is generally applicable at elevated pressures and near the critical point.

Corresponding States. The reduced equation of state, similar shapes of the basic T,P phase diagrams, maximum ΔH_v and ΔS_v at the triple point, and disappearance of ΔH_v and ΔS_v at the critical point have led to use of reduced pressure, volume, and temperature

Table 1. Critical Constants; Latent Heat of Vaporization at Normal Boiling Point *

Compound	Formula	T_B, °K	P_c, atm	ρ_c, g/cm³	T_c, °K	ΔH_v at T_B, cal/g mole
Bromine	Br₂		102	1.18	584	
Chlorine	Cl₂	239.1	76.1	0.573	417	4878
Hydrogen	H₂	20.39	12.797	0.0310	33.24	216
Nitrogen	N₂	77.36	33.5	0.311	126.0	1333
Oxygen	O₂	90.19	50.14	0.430	154.78	1630
Sulfur	S₂	717.8	120	0.4	1313	2500
Air			37.2	0.35	132.5	
Ammonia	NH₃	239.8	111.5	0.235	405.6	5581
Carbon dioxide	CO₂	194.7	72.9	0.459	304.1	6100
Carbon monoxide	CO	81.7	34.53	0.301	133.0	1444
Carbon disulfide	CS₂	319.4	76	0.4	546.2	6400
Hydrogen bromide	HBr	206.4	84		363	4210
Hydrogen chloride	HCl	188.1	81.6	0.421	324.6	3860
Hydrogen cyanide	HCN	298.86	50.0	0.20	456.7	6027
Hydrogen sulfide	H₂S	212.8	88.9	0.349	373.6	4463
Nitric oxide	NO	121.4	65	0.52	179	3292
Nitrous oxide	N₂O	184.7	71.7	0.45	309.7	3956
Phosgene	COCl₂	280.7	56	0.52	455.0	5832
Phosphine	PH₃	185.4	64.5	0.30	324.5	3490
Sulfur dioxide	SO₂	263.1	77.7	0.518	430.4	5950
Sulfur trioxide	SO₃	316.5	83.8	0.633	491.5	9990
Stannic chloride	SnCl₄	380	37	0.74	522	8300
Water	H₂O	373.2	218.4	0.323	647.3	9717
Methane	CH₄	111.67	45.8	0.162	191.0	1955
Ethane	C₂H₆	184.5	48.2	0.203	305.5	3517
Propane	C₃H₈	231.1	42.0	0.022	370.0	4487
n-butane	C₄H₁₀	272.7	37.5	0.228	425.2	5350
2-Methyl propane	C₄H₁₀	261.4	36.0	0.221	408.15	5008
n-Pentane	C₅H₁₂	309.2	33.3	0.232	470.1	6160
n-Hexane	C₆H₁₄	341.9	29.9	0.234	507.9	6900
2,2-Dimethyl butane	C₆H₁₄	322.9	30.7	0.242	489.4	6290
n-Heptane	C₇H₁₆	371.6	27.0	0.235	540.2	7580
n-Octane	C₈H₁₈	398.8	24.6	0.233	569.4	8215
2,2,4-Trimethyl pentane	C₈H₁₈	372.4	25.4	0.243	544	7410
Ethene	C₂H₄	169.5	50.0	0.227	282.5	3237
Propene	C₃H₆	225.5	45.6	0.233	364.9	4405
1-Butene	C₄H₈	266.9	39.6	0.232	419.7	5240
cis-2-Butene	C₄H₈	276.9	40.8	0.239	428.2	5580
trans-2-Butene	C₄H₈	274.05	40.8	0.239	428.2	5440
Ethane	C₂H₂	184.7	61.6	0.231	308.7	4270
Propyne	C₂H₄		52.8		401	
2-Butyne	C₄H₆		60		489	
Benzene	C₆H₆	353.3	48.3	0.304	562.1	7350
Toluene	C₇H₈	383.8	41.6	0.291	593.8	8000
Cyclopropane	C₃H₆	240.2	54		398	
Cyclopentane	C₅H₁₀	322.4	44.6	0.270	511.8	6525
Cyclohexane	C₆H₁₂	353.9	40	0.272	553.7	7190
Methyl cyclohexane	C₇H₁₄	374.1	34.32	0.285	572.3	7580
Acetaldehyde	CH₃CHO	293.3	44	0.26	461	6500
Acetone	(CH₃)₂CO	329.35	46.6	0.273	508.7	7100
Methanol	CH₃OH	337.9	78.5	0.272	513.2	8430
Ethanol	C₂H₅OH	351.7	63.0	0.2755	516	9220
i-Propanol	(CH₃)₂CHOH	355.36	53	0.27	509	9650
n-Propanol	C₃H₇OH	370.5	50.2	0.273	537	9890
Methyl ether	(CH₃)₂O	248.3	53	0.271	400.1	5141
Ethyl ether	(C₂H₅)₂O	307.8	35.6	0.263	467.0	6220
Ethylene oxide	(CH₂)₂O	283.7	71	0.31	469	6101
Chloromethane	CH₃Cl	248.9	65.9	0.35	416.3	5150
Fluoromethane	CH₃F	195.1	58.0	0.300	317.8	4230
Trichloromethane, chloroform	CHCl₃	334.4	55	0.516	536	7020
Methyl cyanide, acetonitrile	CH₃CN	354.7	47.7	0.240	547.9	7830

* Data of Kobe and Lynn, *Chem. Rev.*, Vol. 52, p. 117, 1953.

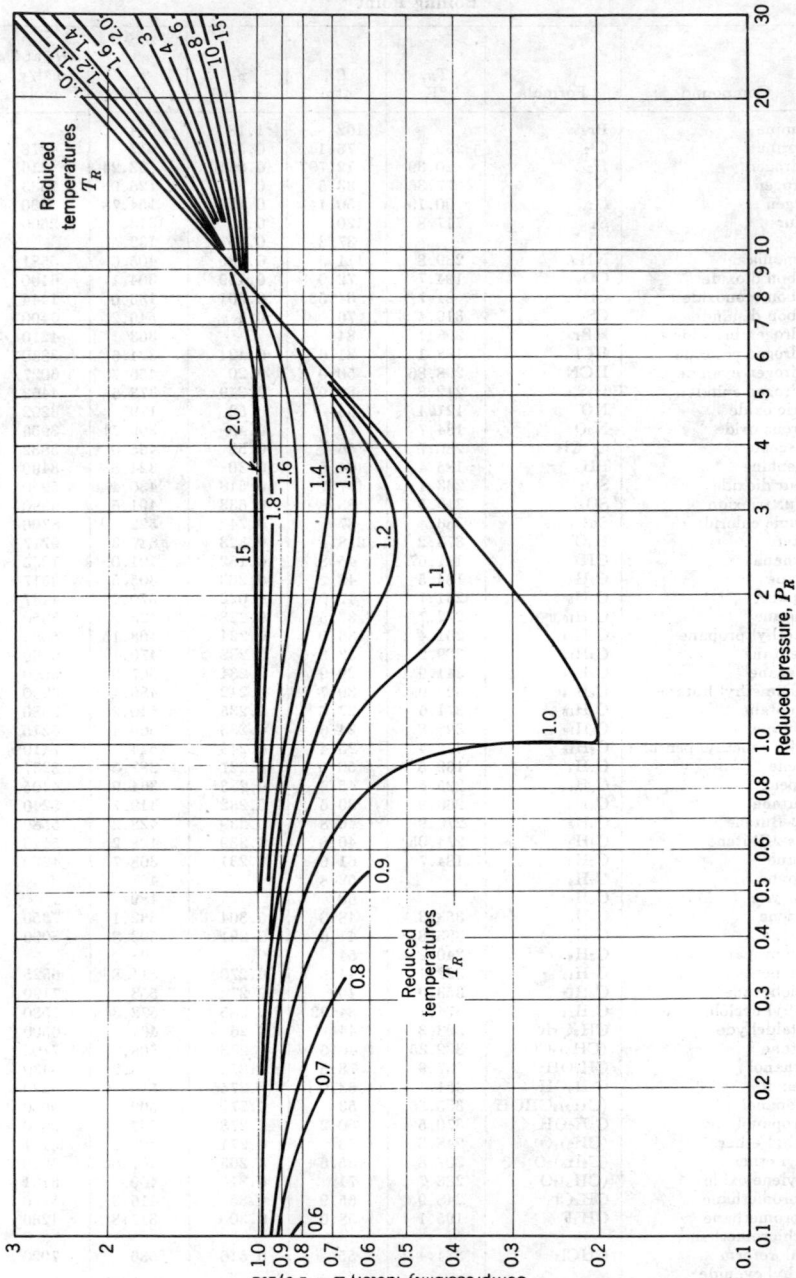

FIG. 2. Compressibility factor. (Reprinted by permission from Souders, *Engineers Companion*, New York, Wiley, 1966.)

for correlation of other properties. These reduced variables P_R, V_R, and T_R are defined as ratios of the existent to the critical variable:

$$P_R = \frac{P}{P_c}, \qquad V_R = \frac{V}{V_c}, \qquad \text{and} \qquad T_R = \frac{T}{T_c} \qquad (8)$$

Two gases are in corresponding states when they have the same values for two of the three reduced variables. Corresponding states permit property estimation by relating gases or liquids to comparable species where more complete PVT data are available.

Compressibility. The PVT relations of any gas may be expressed as:

$$Pv = ZRT \qquad (9)$$

where Z is the compressibility factor or ratio of the volume of a real gas to that of an ideal gas. If the parameters P and T are replaced by their ratios to the corresponding critical values, called reduced properties, all gases are in corresponding states and can be approximately represented by a general chart, Fig. 2, of Z versus P_R on lines of constant T_R.

Most individual gases deviate from such a general graph by 2 to 7%. Hydrogen and helium, however, deviate so much that the reduced properties for them are computed by adding 8 atm to the critical pressure and 8°K to the critical temperature.

Generalized Fugacity Chart. A generalized fugacity chart can be constructed from the compressibility chart by integration, along each isotherm, of

$$\ln \frac{f}{p} = \int_0^{P_R} (Z - 1) \, d(\ln P_R) \qquad (10)$$

The result of these integrations is Fig. 3.

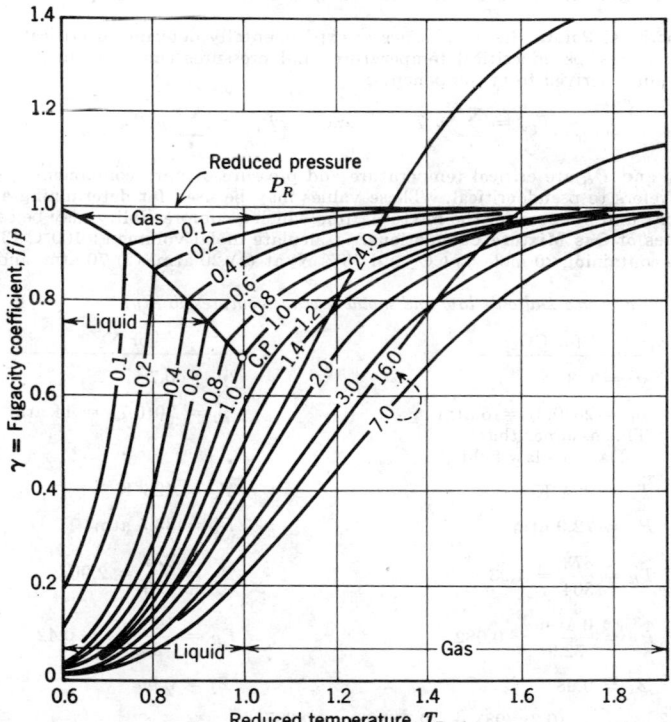

Fig. 3. Fugacity of gases and liquids. (Reprinted by permission from Hougen, Watson, and Ragatz, *Chemical Process Principles*, New York, Wiley, 1959.)

Gas Mixtures. Mixtures of ideal gases have a compressibility $Z = 1$ at low pressures; however, mixtures of real gases deviate from ideality as pressure is increased. A compressibility factor for the mixture can be determined to simplify calculation. At pressures below 50 atm Dalton's law of additive partial pressures can be used for calculating mixture pressure, while above 300 atm Amagat's law of additive molal volumes allows better calculation of mixture volume. The pseudocritical point method may be used to estimate mixture compressibility between 50 and 300 atm.

Dalton's Law of Additive Partial Pressures. For real gases below 50 atm, where the actual volume of the molecules is small relative to the volume occupied and where molecular interaction does not have a large effect, Dalton's law is valid:

$$P = \sum_i p_i \tag{11}$$

where $p_i = Py_i$ and P is total pressure, p_i is partial pressure of species i, and y_i is mole fraction of species i.

Amagat's Law of Additive Molal Volumes. Amagat's law is valid at pressures above 300 atm:

$$V = \sum_i v_i \tag{12}$$

where $v_i = Vy_i$ and v_i is molal volume of species i.

Molal Average Compressibility.

$$Z = \sum_i y_i Z_i \tag{13}$$

where Z_i is determined using T_R and P_R for species i at the temperature of the mixture, and at p_i for low total pressures (50 atm max.) or at P for high total pressures (300 atm min).

Pseudocritical Point. In the absence of experimentally determined critical conditions for gas mixtures, pseudocritical temperatures and pressures can be estimated as molal average values derived from components:

$$T_{pc} = \sum_i y_i T_{ci} \quad \text{and} \quad P_{pc} = \sum_i y_i P_{ci} \tag{14}$$

where T_{ci} and P_{ci} are critical temperature and pressure of pure component i, and subscript pc refers to pseudocritical. These values may be used for determining a reduced temperature and pressure for the mixture, from which a compressiblity may be estimated.

Examples of Gas Mixture Calculations. Calculate molal volume at 100°C (373°K) of a mixture containing 30 mole % CO_2 and 70% N_2 at (1) 20 atm, (2) 70 atm, and (3) 400 atm.

(1) *At 20 atm use Dalton's law and molal average compressibility:*

for CO_2	for N_2
$y_i = 0.3$	$y_j = 0.7$
$p_i = 20(0.3) = 6$ atm	$p_j = 20(0.7) = 14$ atm
This assumes that Dalton's law holds.	
$T_c = 304°K$	$T_c = 126°K$
$P_c = 72.9$ atm	$P_c = 33.5$ atm
$T_R = \dfrac{373}{304} = 1.23$	$T_R = \dfrac{373}{126} = 2.96$
$P_R = \dfrac{6\text{ atm}}{72.9} = 0.082$	$P_R = \dfrac{14\text{ atm}}{33.5} = 0.42$
$Z_i = 0.98$	$Z_j = 0.99$

$$Z_{\text{mixture}} = (0.3)(98) + (0.7)(99) = 0.987$$

$$v = \frac{ZRT}{P} = \frac{(0.987)(0.0821)}{(20)\text{ atm}} \frac{\text{liter atm}}{\text{g mole °K}} \frac{(373)°K}{} = 1.5 \frac{\text{liters}}{\text{mole of mixture}}$$

(2) *At 70 atm use pseudocritical point to determine mixture compressibility:*

mixture $T_{pc} = 0.3(304) + 0.7(126) = 179°K$

mixture $P_{pc} = 0.3(72.9) + 0.7(33.5) = 45.3$ atm

$$T_R = \frac{373}{179} = 2.08$$

$$P_R = \frac{70}{45.3} = 1.54$$

$$Z_{mixture} = 0.980$$

$$v = \frac{ZRT}{P} = \frac{(0.980)\,(0.0821)\,(373)}{70} = 0.43 \frac{liter}{mole\ of\ mixture}$$

(3) *At 400 atm use Amagat's law and molal average compressibility:*

for CO_2	for N_2
$T_R = \dfrac{373}{304} = 1.23$	$T_R = \dfrac{373}{126} = 2.96$
$P_R = \dfrac{400}{72.9} = 5.49$	$P_R = \dfrac{400}{33.5} = 8.38$
$Zi = 0.76$	$Z_j = 1.08$

$$Z_{mixture} = (0.3)\,(0.76) + (0.7)\,(1.08) = 0.984$$

$$v = \frac{ZRT}{P} = \frac{(0.984)\,(0.0821)\,(373)}{400} = 0.075 \frac{liter}{mole\ of\ mixture}$$

Ideal Solutions. The activity of each constituent of ideal liquid solutions is equal to its mole fraction under all conditions of temperature, pressure, and concentration. The solution volume exactly equals the summation of component volumes. The enthalpy of mixing of components is zero. The total vapor pressure is the summation of the contribution of individual components following Raoult's law: individual component vapor pressure contribution is the product of its mole fraction and the vapor pressure of the pure component. This also applies to the vapor pressure of solutions containing nonvolatile components. The freezing point of solvent in ideal solutions occurs at that temperature where the vaopr pressure of the solution equals the vapor pressure of the solid solvent.

Real Solutions. Actual liquid solutions are seldom ideal, showing deviations from the above conditions of ideality. Most significant are positive or negative vapor pressure deviations from direct summation of component contributions; these affect behavior on distillation for separating the components. Deviations from ideality increase with solute concentration; i.e., dilute solutions behave reasonably ideally.

Henry's Law. Solubility, and hence activity, of gaseous solute in liquid solvent is directly proportional to partial pressure (fugacity) of the solute vapor phase in equilibrium with the solution. If the solution and vapor phases behave ideally, mole fraction of gas solute in solution at low concentration equals solute activity ($X_i = a_i$). Henry's law is:

$$Py_i = h_T x_i \qquad (15)$$

where y_i and x_i are mole fractions of i in vapor and liquid phases, P is total pressure, and h_T is Henry's law coefficient for i at temperature T. Solubility of gases in liquids is inversely proportional to temperature, and frequently chemical similarity between solute and solvent leads to a higher solubility. Henry's law applies to the same molecular species of solute in the solution and in the gas phases. For example, for NH_3, CO_2, SO_2, H_2S, Cl_2, HCl, etc. in water it applies to the unhydrated, undissociated species in equilibrium.

For nonreactive, unhydrated, and undissociated gases such as O_2, N_2, H_2, CO, and CH_4 in water, h_T is of the order of 10^{-5} atm^{-1} between 0 and 80°C. Similar data of different magnitudes are existent for nonaqueous solvents, but a wider range of this type of data is presented in vapor-liquid equilibrium diagrams.

Raoult's Law. For ideal vapors and ideal liquid solutions at any temperature, the equilibrium vapor phase partial pressure of a liquid solution component equals the product of the vapor pressure of that pure component and its mole fraction in solution:

$$Py_i = p_i{}^\circ x_i = p_i \tag{16}$$

where y_i and x_i are mole fractions of i in vapor and liquid phases, P is total pressure, $p_i{}^\circ$ is vapor pressure of pure i, and p_i is the actual partial pressure of i, all at temperature T. For ideal solutions Raoult's law holds for all mole fractions of a component, while for real solutions it can be assumed valid only for mole fractions near unity. Interaction between dissimilar species in the liquid phase leads to large deviations from ideal solution behavior, and hence divergence from conformity with Raoult's law. Raoult's law can be considered a special case of Henry's law where Henry's law coefficient h_T is equal to the vapor pressure of pure component i, $p_i{}^\circ$ at that temperature.

Nernst Distribution Law. Equilibrium distribution of solute between two immiscible liquid phases at temperature T is constant regardless of concentration, and equal to the ratio of solute activities in the two phases:

$$K = \frac{a_1}{a_2} \tag{17}$$

where a_1 is activity in phase 1 and a_2 is activity in phase 2. In dilute solutions molar concentrations can be used to replace activities.

Gibbs' Phase Rule. The phase rule gives the relationship to define heterogeneous phase equilibria:

$$P + F = C + 2 \tag{18}$$

P is the number of separate phases involved, C is the number of components (minimum number of chemical species necessary to define composition of all phases), and F is the number of degrees of freedom or number of variables, such as temperature, pressure, or concentration, of each component that can be varied independently without changing the number of phases.

Example: (1) *Water at the triple point.* Solid, liquid, and vapor phases are in equilibrium. A pure compound is a single component, and its concentration in each phase is determined to be 100 mole %. Therefore, $P = 3$, $C = 1$, and F is to be determined by substituting these values in Eq. 18.

$$3 + F = 1 + 2$$

Hence $F = 0$, the system is invariant; temperature and pressure are fixed, and a shift in either will reduce the number of phases in equilibrium.

(2) *For a 3-component system, defined by T,P but not composition,* determine the number of phases present. $F = 2$, $C = 3$, and P is to be determined.

$$P + 2 = 3 + 2$$

Therefore $P = 3$ phases. Further defining concentration of one of the three components raises F to 3 and reduces the number of phases to 2. Defining the concentrations of two components also establishes the concentration of the third, raises F to 4, and reduces number of phases to one.

(3) *Two component system at liquid-vapor equilibrium* (2 phases). Concentrations in the two phases are not independent. $P = 2$, $C = 2$, and F is to be determined.

$$2 + F = 2 + 2$$

Two of the three independent variables (T and P, or concentration of only one component) are required to define the system; the third is fixed.

Colligative Properties of Solutions. Colligative properties depend on the number, rather than the nature, of particles in solution: nonelectrolytes, being undissociated, yield 1 particle per molecule unless they are dimerized. Strong electrolytes, ionized to an extent indicated by their activity, yield several ions. Colligative properties are: (1) boiling point elevation, as a direct consequence of vapor pressure reduction, (2) freezing point depression, also as a direct consequence of vapor pressure reduction, and (3) osmotic pressure. Since ideal solutions are expected at low concentrations, changes in these colligative properties are predictable.

Binary Solution Vapor–Liquid Equilibria. In the vapor phase:

$$P_{\text{total}} = p_i + p_j \tag{19}$$

$$N_{\text{total}} = N_i + N_j$$

$$y_i = \frac{p_i}{P_{\text{total}}}$$

$$y_i + y_j = 1$$

where p_i is the partial pressure, y_i is the mole fraction, and N_i is the number of moles of i.

In the liquid phase:

$$x_i = \frac{N_i}{N_{total}} \qquad (20)$$

$$x_i + x_j = 1$$

$$N_{total} = N_i + N_j$$

where x_i is the mole fraction of i.

(a) *Ideal solutions.* Each component of an ideal solution obeys Raoult's law, Eq. 16, relating concentrations in vapor and liquid phases.

(b) *Real solutions.* Numerical distillation calculations often use a vapor-liquid equilibrium ratio k for each component:

$$k_i = \frac{y_i}{x_i} = \frac{f \text{ of pure } i \text{ at its vapor pressure at } T \text{ of the system}}{f \text{ of pure } i \text{ at } T,P \text{ of the system}} \qquad (21)$$

The volatility ratio between components of binary solutions is:

$$\alpha = \frac{k_i}{k_j} = \frac{y_i x_j}{x_i y_y} \qquad (22)$$

where α is the volatility ratio. For ideal binary solutions, where α is constant, manipulation of Eqs. 21 and 22 relates mole fraction in the vapor phase y_i to volatility ratio α and mole fraction in theliquid phase x_i:

$$y_i = \frac{\alpha x_i}{1 + x_i(\alpha - 1)} \qquad (23)$$

Binary Solution Vapor Pressure–Composition Diagrams. (1) Ideal solutions follow Raoult's law, Eq. 16. As shown in Fig. 4(a), the vapor pressure of each component is linear and proportional to the mole fraction, and the vapor pressure of the mixture is the

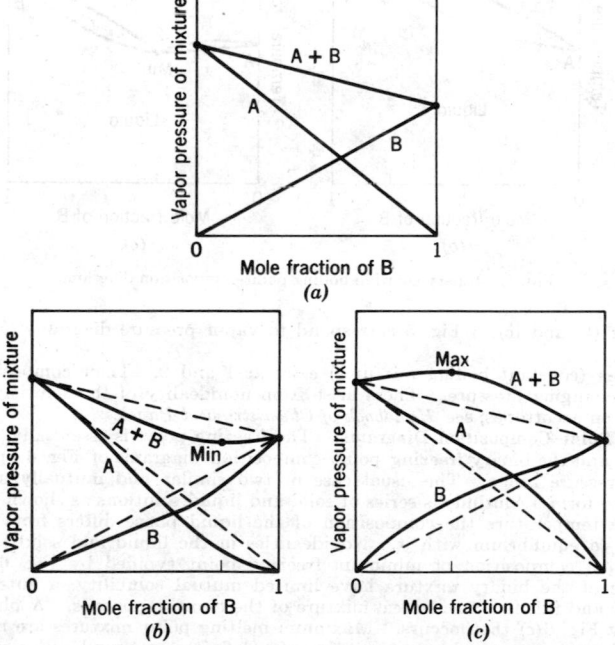

Fig. 4. Binary solution vapor pressure-composition diagrams.

simple sum of the component vapor pressures. The diagrams shown represent one fixed temperature

(2) Real solutions show significant deviations from linearity of individual vapor pressures with the mole fraction. At low mole fractions in the liquid phase, Henry's law (Eq. 15) is followed, while at high mole fractions, Raoult's law tends to be followed. These deviations from linearity are shown in Figs. 4(b) and 4(c).

Boiling Point–Composition Diagrams. Figure 5 shows three types of boiling point–composition diagrams. These are drawn for a single pressure. Diagram 5(a) is the usual case existent for ideal solutions; it corresponds with vapor pressure diagram 4(a). At any given temperature, vapor composition y is in equilibrium with liquid composition x. This ideal type of boiling point diagram exists for many combinations of chemically similar materials.

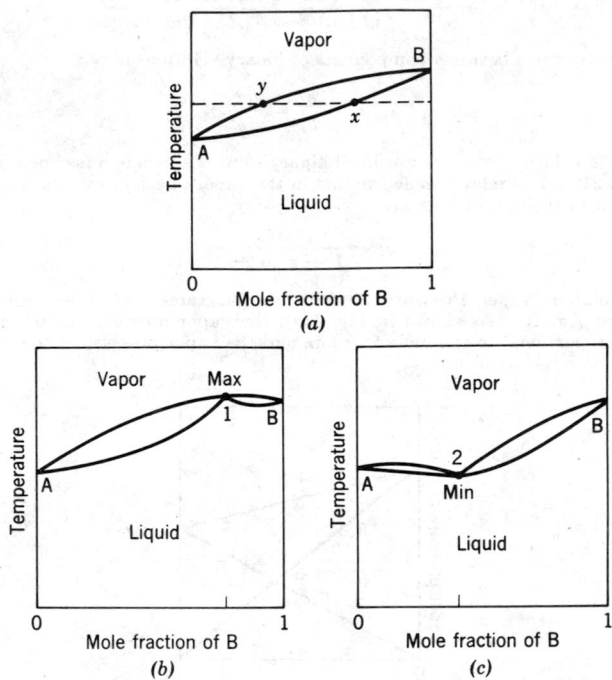

FIG. 5. Binary solution boiling point-composition diagrams.

Diagrams (b) and (c) of Fig. 5 correspond to vapor pressure diagrams (b) and (c) of Fig. 4.

Azeotropes (constant boiling mixtures) exist at 1 and 2. Their composition can be altered by changing pressure. These arise from nonideality of the solutions. For collected data on azeotropes, see *Handbook of Chemistry and Physics*.

Freezing Point–Composition Diagrams. The freezing point is essentially independent of pressure, and the binary freezing point–composition diagrams of Fig. 6 apply over an extended pressure range. The usual case of two similar and mutually soluble components is to form a continuous series of solid and liquid solutions as shown in Fig. 6(a). At any one temperature the composition of the liquid phase differs from that of the solid phase in equilibrium with it. Nonidealities in the liquid and solid solutions can give rise to a composition of minimum freezing point typified by Fig. 6(b). Where components of the binary mixture have limited mutual solubility, a eutectic composition exists and this is a mechanical mixture of the two components. A phase diagram idealized by Fig. 6(c) then occurs. Maximum melting point mixtures are rare, and the existence of a maximum is due to formation of a definite compound. Figure 6(d) shows

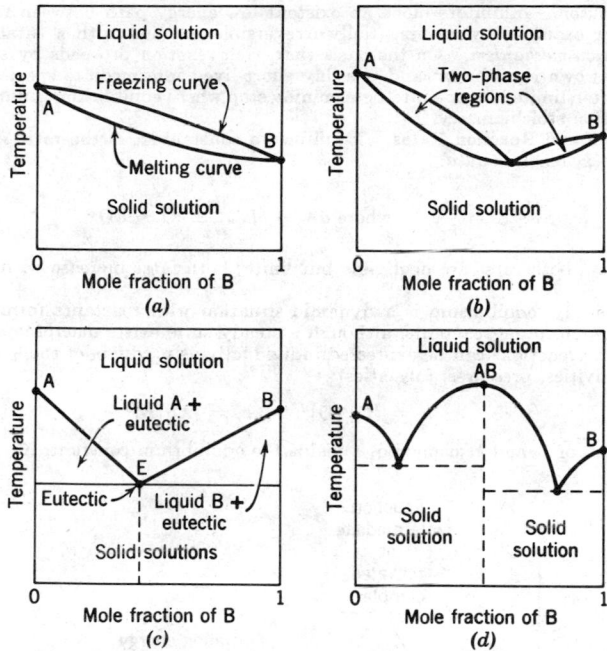

FIG. 6. Binary solution freezing point-composition diagrams.

the existence of an equimolar compound A_1B_1. The resultant phase diagram then shows two eutectic diagrams placed side by side. One of these eutectics exists for component A and compound A_1B_1; the other exists for compound A_1B_1 and component B.

Numerous proliferations of these basic types of phase diagram exist. All may be interpreted on the basis of the temperature and composition range for mutual solubility, the existence of nonideal solutions, and the formation of compounds stable over an often limited temperature range.

8. CHEMICAL REACTION RATES

Process Kinetics and Conversion. The kinetics of chemical reactions vary widely, ranging from a rapid approach to equilibrium for ionic reactions in aqueous media (primarily limited by diffusion and mixing) to a much slower approach to equilibrium for organic reactions involving covalent bonds. The net reaction rate at equilibrium is zero, and further conversion of reactants to products depends upon the rate of product removal from the reaction system so the reaction can again proceed at a finite rate toward equilibrium. Only temperature can shift the equilibrium constant of the reaction system; however, physical control of reactant and product concentrations can allow a reaction to proceed to complete conversion of reactants to product, despite an equilibrium situation that may permit a maximum of a few percent of product in the system at any instant.

Kinetics of Reaction is concerned with the rate of approach to a temperature-dependent equilibrium condition. The rate varies with:

(a). *Displacement from equilibrium.* This is dependent on the reactant and product concentrations.

(b). *Temperature.* In addition to changing the numerical value of the equilibrium constant and altering the diffusional mobility, increased temperature raises the fraction of molecules with enough energy to react.

(c). *Catalysis.* This modifies the reaction mechanism and may provide a lower energy path from reactants to products and vice versa. Catalysis has no effect on equilibrium.

(d). *Inhibition.* Inhibitors block an existent low energy path between reactants and products; for example, by an essentially irreversible reaction with a catalytic surface.

(e). *Reaction mechanism.* On the basis that each reaction proceeds by several steps, each with its own equilibrium and possibly short-lived intermediate species, an overall reaction is often limited by one rate-determining step whose equilibrium is unrelated to the overall reaction stoichiometry.

Equilibrium and Reaction Rates. Equilibrium constant K is the ratio of forward k_1 to backward k_2 reaction rate:

$$K = \frac{k_2}{k_1}, \qquad \text{where } a\text{A} + b\text{B} \underset{k_2}{\overset{k_1}{\rightleftharpoons}} c\text{C} + d\text{D} \tag{1}$$

At equilibrium both rates are negligible but finite; both rates increase with system concentration.

Microscopically, equilibrium is a dynamic situation with reactants forming products and products reforming reactants, although a steady state exists macroscopically.

Rates of the reactions can be expressed individually as products of the species concentrations (activities, pressures, fugacities):

$$k_2 = [\text{C}]^c[\text{D}]^d, \qquad k_1 = [\text{A}]^a[\text{B}]^b \tag{2}$$

A combination of k_2 and k_1 using Eq. 1 yields the equilibrium constant K.

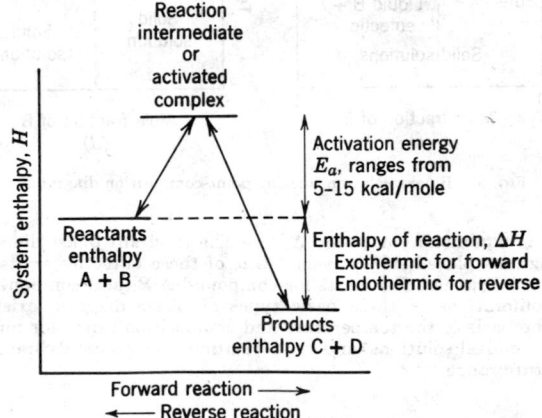

Fig. 1.   Enthalpy-reaction diagram.

Activation Energy. An enthalpy versus reaction diagram, Fig. 1, can be ideally drawn for any reaction. An activation energy barrier exists between reactants and products, and enthalpy of reaction ΔH separates the enthalpies of reactants and of products. Reactant molecules must receive a minimum activation energy E_a before they are energetic enough to react. They form some intermediate species which can spontaneously further react or decompose to products, liberating activation energy plus enthalpy of reaction ΔH. For the reverse reaction, product molecules must receive $E_a + \Delta H$ before they can react in the reverse direction. Catalysts modify the intermediate condition and result in a reduced activation energy by providing a lower energy path between reactants and products; by so doing, they effectively speed *both* forward and reverse reactions without affecting the equilibrium condition.

Collision Theory. Energy of activation results from the statistical probability that some molecules, after collision with others, conserving momentum, have energy enough for reaction to occur. An elevated temperature increases the Maxwell-Boltzmann energy distribution of molecules and raises the number of molecules having the energy required for the reaction as shown in Fig. 2.

Rate Dependence on Temperature. The population of high-energy molecules is directly correlated with the temperature dependence on the reaction rate using the

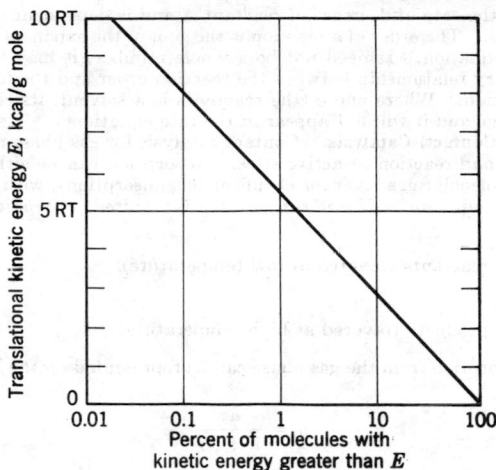

FIG. 2. Molecular energy distribution. The energy level increases with temperatures 10RT at 300°K = 6 kcal; at 400°K = 8 kcal; and at 600° K = 12 kcal. The population of molecules above some energy level increases exponentially with temperature.

Arrhenius equation (which has a form similar to dependence of equilibrium constant on temperature).

$$\frac{d \ln k}{dT} = \frac{E_a}{RT^2} \qquad (3)$$

where E_a is the activation energy, R is the gas constant, and T is the absolute temperature. To obtain constant E_a, Eq. 3 is integrated to $\ln k = -E_a/RT + \text{constant}$. A plot of $\ln k$ vs. $1/T$ yields a straight line, with slope equal to $-E_a/R$, from which activation energy can be obtained. It shows that a rate at 25°C will double for a 10°C temperature rise when the activation energy is 12.5 kcal/g mole. A lower activation energy gives a larger rate increase, and a higher energy gives a smaller rate increase. Caution: Changes of mechanism, via catalysis or otherwise, will alter the E_a of a reaction.

Reaction Mechanisms. The low statistical probability of trimolecular collisions essentially precludes the possibility that any reaction can proceed in one step according to the stoichiometric equation. More probable is a sequence of bimolecular reactions forming intermediate species, some of which may be too short-lived to isolate. Each of these stepwise reactions has its equilibrium constant and rates dependent on species concentrations. Usually some one step is limiting and becomes the rate-determining factor for the entire reaction. The intermediate species may be an activated complex that rearranges to produce the reaction products.

Different reaction mechanisms have been postulated and proved for many reactions; these include:

1. Simple bimolecular collision.
2. Chain reaction for polymerization, with initiation, propagation, and termination steps.
3. Free radical, photoinitiated reactions.
4. Acid- or base-catalyzed intermediate steps.
5. Activated complex on a heterogeneous catalyst surface, with adsorption-desorption contributing to the overall rate.
6. Solvolysis participation in the reaction sequences.

Many processes have been studied in detail, with reaction sequences, proven or postulated, consistent with experimentally determined kinetics.

Generalized Rate Equation. A reaction rate equation can be experimentally deduced and, from this, a reaction mechanism postulated. The reaction rate equation has the generalized form:

$$-\frac{dC_A}{dt} = kC_A{}^a C_B{}^b \ldots \qquad (4)$$

where $-dC_A/dt$ is the rate of decrease of reactant A and is dependent upon a product of concentration terms. The order of a reaction is the sum of the exponents of concentration terms in the rate equation. It need not be a whole number; it may be 0, $\frac{1}{2}$, 1, 3/2, 2. There is no necessary relationship between the reaction order and the form of the overall stoichiometric relation. Where one of the reactants is a solvent, its concentration does not effecively change and it will not appear in the rate equation.

Heterogeneous (Contact) Catalysis. Contact catalysis for gas phase reactions involves surface adsorption and reaction at active sites. Adsorption can be either physical, with a mono- or multimolecular gas layer, or chemical (chemisorption), with a monomolecular layer in some activated state. A large surface area is required. Steps involved in a catalyzed reaction are:

1. Adsorption of reactants (favored at low temperature).
2. Activation.
3. Reaction.
4. Desorption of products (favored at high temperature).

The amount of adsorption from the gas phase can be represented by the Langmuir adsorption isotherm:

$$y = \frac{ap}{1 + bp} \tag{5}$$

where $y =$ (mass adsorbed)/(adsorbant mass), p is pressure, and a and b are empirical constants applicable at one temperature only.

Adsorption from the liquid phase is represented by the Freundlich equation:

$$y = aC^{1/b} \tag{6}$$

where C is concentration of solute and a and b are empirical and applicable at one temperature only.

Catalyst poisoning involves blocking active sites on the catalyst surface by adsorbed material that is not readily desorbed.

Photochemical Reactions. Many reactions proceed via UV energy absorption. Photons from mercury vapor at 2536 Å $\cong$ 113 kcal/g mole are absorbed, one maximum per molecule, and have energy sufficient to disrupt chemical bonds (usual range of 80 to 100 kcal) and form reactive free radical species. Photons of IR wavelength 9000 Å $\cong$ 31.8 kcal/g mole do not have sufficient energy for bond disruption, although they do increase the overall thermal energy of a system.

An Einstein is an Avogadro number (6.02×10^{23}) of photons.

9. ELECTROCHEMISTRY

Chemical Energy of Redox Reactions. The chemical energy of redox reactions can be equated at constant T,P with Gibbs free energy using Eq. 4, Art. 4:

$$dG = -S\,dT + V\,dP + \sum_i \mu_i\,dN_i$$

Since dT and dP are both zero at constant T,P then $dG = \sum_i \mu_i\,dN_i$. Gibbs free energy then can be manifest as electrical energy by an electron flow through an external conductor provided there is electrical isolation of the oxidation reaction (electron source) and reduction reaction (electron sink). The external conductor transfers electrons from source to sink, and a liquid junction permits internal migration of ions for preservation of charge neutrality of the system as a whole.

Chemical energy at constant T,P can be considered convertible to electrical and/or thermal energy, for it can be shown by conservation of energy that $\sum_i \mu_i\,dN_i = -E\,dq$ $-\,T\,dS$. In the absence of $T\,dS$ thermal energy:

$$dG = -E\,dq \tag{1}$$

where E is volts, dq is coulombs/g mole, and dG is energy in J/g mole. Joules are convertible to calories by $J = 4.185$ J/cal.

Without source-sink electrical isolation, an internal short circuit results in energy conversion to thermal. Thus at constant T,P:

$$dG = -T \, dS \tag{2}$$

where T is °K, and dS is cal/g mole °K.

Faraday, F, is the 96,500 coulomb charge carried by 1 g equivalent or an Avogadro number of electrons. Hence:

$$dq = F \, dn \tag{3}$$

where dn is the number of gram equivalents.

Standard Potential, $E°$ Volts. The superscript ° refers to the standard conditions of 25°C (298°K), 1 atm fugacity (pressure), and 1 M activity of reactants and products, the same conditions as for $\Delta G°$, thus establishing dimensionally correct conversion of energy terms at the standard thermodynamic reaction condition:

$$dG° = -\frac{dnFE°}{J}, \quad \text{or} \quad \Delta G° = -\frac{nFE°}{J} \tag{4}$$

in integrated form. If the equilibrium constant for the reaction is known at 298°K, Eq. 17, Art. 4, $\Delta G_T° = -RT \ln K$, can be used to calculate $\Delta G°$ which can then be related to $E°$ via Eq. 4. $\Delta G°$ actually indicates the magnitude of displacement from equilibrium that exists at 298°K when all reactants and products are at unit activity:

$$\Delta G° = RT \ln (1) - RT \ln K = -RT \ln K = -\frac{nFE°}{J} \tag{5}$$

Potential $E_T°$ at Temperatures Other Than 298°K. $\Delta G_T°$ calculated for reactions at other temperatures (via Eq. 8, Art. 4) can be directly related to $E_T°$. Similarly, experimental values of the equilibrium constant K can be related to $\Delta G_T°$ (via Eq. 18, Art. 4) and then to $E_T°$ using:

$$\Delta G_T° = -RT \ln K = -\frac{nFE_T°}{J} \tag{6}$$

Potential E at Other Concentrations—Nernst Equation. No driving force exists at equilibrium where $\Delta G = 0$, and the driving force is directly related to displacement from equilibrium. Therefore, an E value can be corrected for reactant and product activities other than 1 at 298°K by using the general equation

$$G = RT \ln K' - RT \ln K = \frac{-nFE}{J} \tag{7}$$

where K' has the form and exponents of the reaction equilibrium constant, but uses actual activities of products and reactants in the system rather than those existent at equilibrium.

Combining Eqs. 5 and 7 yields the Nernst equation:

$$E = E_0 \frac{-JRT}{nF} \ln K' \tag{8}$$

At $T = 298°K$, this equation reduces to the usual equation:

$$E = E° - \frac{0.0592}{n} \log_{10} K' \tag{9}$$

Both equations reduce E to zero at equilibrium concentrations. At temperatures other than 298°K, $E_T°$ instead of $E°$ may be used in the general equation to obtain E actual.

Standard Reduction Potential, $E°$. $E°$, being related to both $\Delta G°$ and to equilibrium constant K at standard conditions, is a measure of the chemical energy available from redox reactions. Tabulated potentials for a number of half-reactions in aqueous solution can be combined to give the standard potential of all possible combinations of these half-reactions. Table 1 gives the standard potentials for reduction half-reactions in 1 M acid and neutral media, and in $1M$ base media because the reactions and their potentials change in many cases. The table is reversible because a reduction half-reaction (with

Table 1. Reduction Potentials

Acid Solution	E^0	Basic Solution	$E_B{}^0$
Aluminum			
$Al^{3+}+3e^-\rightarrow Al$	-1.66	$H_2AlO_3{}^-+H_2O+3e^-\rightarrow Al+4OH^-$	-2.35
Antimony			
$Sb+3H^++3e^-\rightarrow SbH_3$	-0.51	$Sb+3H_2O+3e^-\rightarrow 3bH_3+3OH^-$	~-1.3
$SbO^++2H^++3e^-\rightarrow Sb+3H_2O$	$+0.21$	$SbO_2{}^--2H_2O+3e^-\rightarrow Sb+4OH^-$	-0.66
$Sb_2O_5+6H^++4e^-\rightarrow 2SbO^++3H_2O$	$+0.58$	$H_3SbO_6{}^{4-}+H_2O+2e^-\rightarrow SbO_2{}^-+5OH^-$	~-0.4
Arsenic			
$As+3H^++3e^-\rightarrow AsH_3$	-0.60	$As+3H_2O+3e^-\rightarrow AsH_3+3OH^-$	-1.43
$HAsO_2+3H^++3e^-\rightarrow As+2H_2O$	$+0.25$	$AsO_2{}^-+2H_2O+3e^-\rightarrow As+4OH^-$	-0.68
$H_3AsO_4+2H^++2e^-\rightarrow HAsO_2+2H_2O$	$+0.56$	$AsO_4{}^{3-}+2H_2O+2e^-\rightarrow AsO_2{}^-+4OH^-$	-0.67
Barium			
$Ba^{2+}+2e^-\rightarrow Ba$	-2.90	$Ba(OH)_2\cdot 8H_2O+2e^-\rightarrow Ba+2OH^-+8H_2O$	-2.97
Beryllium			
$Be^{2+}+2e^-\rightarrow Be$	-1.85	$Be_2O_3{}^{2-}+3H_2O+4e^-\rightarrow 2Be+6OH^-$	-2.62
Bismuth			
$BiO^++2H^++3e^-\rightarrow Bi+H_2O$	$+0.32$	$Bi_2O_3+3H_2O+6e^-\rightarrow 2Bi+6OH^-$	-0.46
Boron			
$H_3BO_3+3H^++3e^-\rightarrow B+3H_2O$	-0.87	$H_2BO_3{}^-+3e^-\rightarrow B+4OH^-$	-1.79
Bromine			
$Br_2+2e^-\rightarrow 2Br^-$	$+1.07$	$Br_2+2e^-\rightarrow 2Br^-$	$+1.07$
$HBrO+H^++e^-\rightarrow \frac{1}{2}Br_2+H_2O$	$+1.59$	$BrO^-+H_2O+2e^-\rightarrow Br^-+2OH^-$	$+0.76$
$BrO_3{}^-+6H^++5e^-\rightarrow \frac{1}{2}Br_2+3H_2O$	$+1.52$	$BrO_3{}^-+3H_2O+6e^-\rightarrow Br^-+6OH^-$	$+0.61$
Cadmium			
$Cd^{2+}+2e^-\rightarrow Cd$	-0.40	$Cd(OH)_2+2e^-\rightarrow Cd+2OH^-$	-0.81
Calcium			
$Ca^{2+}+2e^-\rightarrow Ca$	-2.87	$Ca(OH)_2+2e^-\rightarrow Ca+2OH^-$	-3.03
Cerium			
$Ce^{4+}+e^-\rightarrow Ce^{3+}$	$+1.61$		
$Ce^{3+}+3e^-\rightarrow Ce$	-2.48 or -2.34		
Cesium			
$Cs^++e^-\rightarrow Cs$	-2.92	$Cs^++e^-\rightarrow Cs$	-2.92
Chlorine			
$Cl_2+2e^-\rightarrow 2Cl^-$	$+1.36$		
$HClO+H^++e^-\rightarrow \frac{1}{2}Cl_2+H_2O$	$+1.63$	$ClO^-+H_2O+e^-\rightarrow \frac{1}{2}Cl_2+2OH^-$	$+0.40$
$ClO_3{}^-+6H^++5e^-\rightarrow \frac{1}{2}Cl_2+3H_2O$	$+1.47$	$ClO_3{}^-+2H_2O+4e^-\rightarrow ClO^-+4OH^-$	$+0.50$
$ClO_4{}^-+2H^++2e^-\rightarrow ClO_3{}^-+H_2O$	$+1.20$	$ClO_4{}^-+H_2O+2e^-\rightarrow ClO_3{}^-+2OH^-$	$+0.36$
Chromium			
$Cr^{3+}+3e^-\rightarrow Cr$	-0.74	$CrO_4{}^{2-}+4H_2O+3e^-\rightarrow Cr(OH)_3+5OH^-$	-0.13
$Cr^{3+}+e^-\rightarrow Cr^{2+}$	-0.41	$CrO_2{}^-+H_2O+3e^-\rightarrow Cr+4OH^-$	-1.2
$Cr_2O_7{}^{2-}+14H^++6e^-\rightarrow 2Cr^{3+}+7H_2O$	$+1.33$	$Cr(OH)_3+3e^-\rightarrow Cr+3OH^-$	-1.3
Cobalt			
$Co^{2+}+2e^-\rightarrow Co$	-0.28	$Co(OH)_2+2e^-\rightarrow Co+2OH^-$	-0.73
$Co^{3+}+e^-\rightarrow Co^{++}$	$+1.82$	$Co(OH)_3+e^-\rightarrow Co(OH)_2+OH^-$	$+0.14$
Copper			
$Cu^{2+}+2e^-\rightarrow Cu$	$+0.34$	$Cu_2O+H_2O+2e^-\rightarrow 2Cu+2OH^-$	-0.36
$Cu^{2+}+e^-\rightarrow Cu^+$	$+0.15$	$2Cu(OH)_2+e^-\rightarrow Cu_2O+4OH^-$	-0.08
Fluorine			
$F_2+2e^-\rightarrow 2F^-$	$+2.85$		

Table 1. Reduction Potentials—*Continued*

Acid Solution	E^0	Basic Solution	$E_B{}^0$
Gallium			
$Ga^{3+}+3e^-\rightarrow Ga$	-0.53	$H_2GaO_3{}^-+H_2O+3e^-\rightarrow Ga+4OH^-$	-1.22
Germanium			
$GeO_2+4H^++4e^-\rightarrow Ge+2H_2O$	-0.15	$HGeO_3{}^-+2H_2O+4e^-\rightarrow Ge+5OH^-$	-1.0
Gold			
$Au^{3+}+3e^-\rightarrow Au$	$+1.50$	$H_2AuO_3{}^-+H_2O+3e^-\rightarrow Au+4OH^-$	$+0.7$
$Au^{3+}+2e^-\rightarrow Au^+$	$+1.41$		
Hafnium			
$Hf^{4+}+4e^-\rightarrow Hf$	-1.70	$HfO(OH)_2+H_2O+4e^-\rightarrow Hf+4OH^-$	-2.50
Hydrogen			
$H_2+2e^-\rightarrow 2H^-$	-2.23	$2H_2O+2e^-\rightarrow H_2+2OH^-$	-0.83
$2H^++2e^-\rightarrow H_2$	0 reference	$H_2+2e^-\rightarrow 2H^-$	-2.23
		$2H^+$ (at $pH=7$) $+\ 2e^-\rightarrow H_2$	-0.41
Indium			
$In^{3+}+3e^-\rightarrow In$	-0.34	$In(OH)_3+3e^-\rightarrow In+3OH^-$	-1.0
Iodine			
$I_2+2e^-\rightarrow 2I^-$	$+0.54$	$I_2+2e^-\rightarrow 2I^-$	$+0.54$
$HIO+H^++e^-\rightarrow \frac{1}{2}I_2+H_2O$	$+1.45$	$2IO^-+2H_2O+2e^-\rightarrow \frac{1}{2}I_2+4OH^-$	$+0.45$
$IO_3{}^-+5H^++4e^-\rightarrow HIO+2H_2O$	$+1.14$	$IO_3{}^-+2H_2O+4e^-\rightarrow IO^-+4OH^-$	$+0.14$
Iridium			
$Ir^{3+}+3e^-\rightarrow Ir$	$+1.15$	$Ir_2O_3+3H_2O+6e^-\rightarrow Ir+6OH^-$	$+0.1$
$IrO_2+4H^++e^-\rightarrow Ir^{3+}+H_2O$	$+0.7$	$2IrO_2+H_2O+2e^-\rightarrow Ir_2O_3+2OH^-$	$+0.1$
Iron			
$Fe^{2+}+2e^-\rightarrow Fe$	-0.44	$Fe(OH)_2+2e^-\rightarrow Fe+2OH^-$	-0.88
$Fe^{3+}+e^-\rightarrow Fe^{2+}$	$+0.78$	$Fe(OH)_3+e^-\rightarrow Fe(OH)_2+OH^-$	-0.56
Lanthanum			
$La^{3+}+3e^-\rightarrow La$	-2.52	$La(OH)_3+3e^-\rightarrow La+3OH^-$	-2.90
Lead			
$Pb^{2+}+2e^-\rightarrow Pb$	-0.13	$PbO_2+H_2O+2e^-\rightarrow PbO+2OH^-$	$+0.25$
$PbO_2+4H^++2e^-\rightarrow Pb^{2+}+2H_2O$	$+1.46$	$HPbO_2{}^-+H_2O+2e^-\rightarrow Pb+3OH^-$	-0.54
Lithium			
$Li^++e^-\rightarrow Li$	-3.05	$Li^++e^-\rightarrow Li$	-3.05
Magnesium			
$Mg^{2+}+2e^-\rightarrow Mg$	-2.37	$Mg(OH)_2+2e^-\rightarrow Mg+2OH^-$	-2.69
Manganese			
$Mn^{2+}+2e^-\rightarrow Mn$	-1.18	$Mn(OH)_2+2e^-\rightarrow Mn+2OH^-$	-1.55
$MnO_2+4H^++2e^-\rightarrow Mn^{2+}+2H_2O$	$+1.23$	$Mn(OH)_3+e^-\rightarrow Mn(OH)_2+OH^-$	-0.4
$MnO_4{}^-+4H^++3e^-\rightarrow MnO_2+2H_2O$	$+1.70$	$MnO_2+H_2O+2e^-\rightarrow Mn(OH)_2+2OH^-$	-0.05
		$MnO_4{}^-+2H_2O+3e^-\rightarrow MnO_2+4OH^-$	$+0.59$
Mercury			
$Hg^{2+}+2e^-\rightarrow Hg$	$+0.85$	$HgO+H_2O+2e^-\rightarrow Hg+2OH^-$	$+0.10$
$2Hg^{2+}+2e^-\rightarrow Hg_2{}^{2+}$	$+0.92$		
Molybdenum			
$Mo^{3+}+3e^-\rightarrow Mo$	-0.20	$MoO_2+2H_2O+4e^-\rightarrow Mo+4OH^-$	-0.87
$MoO_2{}^++4H^++2e^-\rightarrow Mo^{3+}+2H_2O$	0.0	$MoO_4{}^{2-}+2H_2O+2e^-\rightarrow MoO_2+4OH^-$	-1.40
$H_2MoO_4+2H^++e^-\rightarrow MoO_2{}^++2H_2O$	$+0.4$	$MoO_4{}^{2-}+4H_2O+6e^-\rightarrow Mo+8OH^-$	-1.05
Nickel			
$Ni^{2+}+2e^-\rightarrow Ni$	-0.25	$Ni(OH_2)+2e^-\rightarrow Ni+2OH^-$	-0.72
$NiO_2+4H^++2e^-\rightarrow Ni^{2+}+2H_2O$	$+1.78$	$NiO_2+2H_2O+2e^-\rightarrow Ni(OH)_2+2OH^-$	$+0.49$

Table 1.   Reduction Potentials—*Continued*

Acid Solution	E^0	Basic Solution	E_B^0
Niobium			
$Nb^{3+}+3e^-\rightarrow Nb$	-1.10		
$Nb_2O_5+10H^++10e^-\rightarrow 2Nb+5H_2O$	-0.65		
Nitrogen			
$2NO_3^-+4H^++2e^-\rightarrow N_2O_4+2H_2O$	$+.80$	$2NO_3^-+2H_2O+2e^-\rightarrow N_2O_4+4OH^-$	-0.86
$NO_2^-+4H^++3e^-\rightarrow NO+3H_2O$	$+.96$	$NO_3^-+H_2O+2e^-\rightarrow NO_2^-+2OH^-$	$+0.01$
$NO_3^-+2H^++e^-\rightarrow HNO_2+H_2O$	$+.94$		
Osmium			
$OsO_4+8H^++8e^-\rightarrow Os+4H_2O$	$+0.85$	$HOsO_5^-+4H_2O+8e^-\rightarrow Os+9OH^-$	$+0.02$
Oxygen			
$O_2+4H^++4e^-\rightarrow 2H_2O$	$+1.23$	$O_2+H_2O+2e^-\rightarrow O_2+2OH^-$	$+1.24$
$O_3+2H^++2e^-\rightarrow O_2+H_2O$	$+2.07$	$O_2+2H_2O+4e^-\rightarrow 4OH^-$	$+0.40$
		O_2+4H^+ (at pH $=7$) $+4e^-=2H_2O$	$+0.81$
Palladium			
$Pd^{2+}+2e^-\rightarrow Pd$	$+0.99$	$Pd(OH)_2+2e^-\rightarrow Pb+2OH^-$	$+0.07$
$Pd^{4+}+2e^-\rightarrow Pd^{2+}$	$+1.60$	$Pd(OH)_4+2e^-\rightarrow Pd(OH)_2+2OH^-$	$+0.73$
Phosphorus			
$P+3H^++3e^-\rightarrow PH_3$	-0.07	$P+3H_2O+3e^-\rightarrow PH_3+3OH^-$	-0.89
$H_3PO_2+H^++e^-\rightarrow P+2H_2O$	-0.51	$H_2PO_2^-+e^-\rightarrow P+2OH^-$	-2.05
$H_3PO_3+2H^++2e^-\rightarrow H_3PO_2+H_2O$	-0.51	$HPO_3^{2-}+2H_2O+2e^-\rightarrow H_2PO_2^-+3OH^-$	-1.57
$H_3PO_4+2H^++2e^-\rightarrow H_3PO_3+H_2O$	-0.28	$PO_4^{3-}+2H_2O+2e^-\rightarrow HPO_3^{2-}+3OH^-$	-1.12
Platinum			
$Pt^{2+}+2e\rightarrow Pt$	$+1.20$	$Pt(OH)_2+2e^-\rightarrow Pt+2OH^-$	$+0.15$
$Pt(OH)_2+2H^++2e^-\rightarrow Pt+2H_2O$	$+0.98$	$Pt(OH)_6^{2-}+2e^-+Pt(OH)_2+4OH^-$	$+0.20$
Potassium			
$K^++e^-\rightarrow K$	-2.93	$K^++e^-\rightarrow K^+$	-2.93
Rhenium			
$ReO_2+4H^++4e^-\rightarrow Re+2H_2O$	$+0.25$	$ReO_2+H_2O+4e^-\rightarrow Re+4OH^-$	-0.58
$ReO_4^-+4H^++3e^-\rightarrow ReO_2+2H_2O$	$+0.51$	$ReO_4^-+2H_2O+3e^-\rightarrow ReO_2+4OH^-$	-0.59
Rhodium			
$Rh^{3+}+3e^-\rightarrow Rh$	$+0.8$	$Rh_2O_3+3H_2O+6e^-\rightarrow 2Rh+6OH^-$	$+0.04$
Rubidium			
$Rb^++3e\rightarrow Rb$	-2.93	$Rb^++e^-\rightarrow Rb$	-2.93
Ruthenium			
$RuO_2+4H^++4e^-\rightarrow Ru+2H_2O$	$+0.79$	$RuO_2+2H_2O+4e^-\rightarrow Ru+4OH^-$	-0.04
		$RuO_4+H_2O+4e^-\rightarrow RuO_2+4OH^-$	$+0.58$
Scandium			
$Sc^{3+}+3e^-\rightarrow Sc$	-2.08	$Sc(OH)_3+3e^-\rightarrow Sc+3OH^-$	~-0.26
Selenium			
$Se+2H^++2e^-\rightarrow H_2Se$	-0.40	$Se+2e^-\rightarrow Se^{2-}$	-0.92
$H_2SeO_3+4H^++4e^-\rightarrow Se+3H_2O$	$+0.74$	$SeO_3^{2-}+3H_2O+4e^-\rightarrow Se+6OH^-$	-0.37
$SeO_4^{2-}+4H^++2e^-\rightarrow H_2SeO_3+H_2O$	$+1.15$	$SeO_4^{2-}+H_2O+2e^-\rightarrow SeO_3^{2-}+2OH^-$	$+0.05$
Silicon			
$H_2SiO_3+4H^++4e^-\rightarrow Si+3H_2O$	-0.87	$SiO_3^{2-}+3H_2O+4e^-\rightarrow Si+6OH^-$	-1.73
Silver			
$Ag^++e^-\rightarrow Ag$	$+0.80$	$Ag_2O+H_2O+2e^-\rightarrow 2Ag+2OH^-$	$+0.34$
Sodium			
$Na^++e^-\rightarrow Na$	-2.71	$Na^++e^-\rightarrow Na$	-2.71
Strontium			
$Sr^{2+}+2e^-\rightarrow Sr$	-2.89	$Sr(OH)_2\cdot 8H_2O+2e^-\rightarrow Sr+2OH^-$	
		$+8H_2O$	-2.99

Table 1. Reduction Potnetials—*Continued*

Acid Solution	E^0	Basic Solution	E_B^0
Sulfur			
$S+2H^++2e^-\rightarrow H_2S$	$+0.14$	$S+2e^-\rightarrow S^{2-}$	-0.51
$S_2O_3^{2-}+6H^++4e^-\rightarrow 2S+3H_2O$	$+0.50$	$S_2O_3^{2-}+3H_2O+4e^-\rightarrow 2S+6OH^-$	-0.74
$S_4O_6^{2-}+2e^-\rightarrow 2S_2O_3^{2-}$	$+0.08$	$S_4O_6^{2-}+2e^-\rightarrow 2S_2O_3^{2-}$	$+0.08$
$4H_2SO_3+4H^++6e^-\rightarrow S_4O_6^{2-}+6H_2O$	$+0.51$	$2SO_3^{2-}+3H_2O+4e^-\rightarrow S_2O_3^{2-}+6OH^-$	-0.58
$SO_4^{2-}+4H^++2e^-\rightarrow H_2SO_3+H_2O$	$+0.17$	$SO_4^{2-}+H_2O+2e^-\rightarrow SO_3^{2-}+2OH^-$	-0.93
Tantalum			
$Ta_2O_5+10H^++10e^-\rightarrow 2Ta+5H_2O$	-0.81		
Tellurium			
$Te+2H^++2e^-\rightarrow H_2Te$	-0.72	$Te+2e^-\rightarrow Te^{2-}$	-1.14
$TeO_2+4H^++4e^-\rightarrow Te+2H_2O$	$+0.53$	$TeO_3^{2-}+3H_2O+4e^-\rightarrow Te+6OH^-$	-0.57
$H_6TeO_6+2H^++2e^-\rightarrow TeO_2+4H_2O$	$+1.02$	$TeO_4^{2-}+H_2O+2e^-\rightarrow TeO_3^{2-}+2OH^-$	$+0.4$
Tallium			
$Tl^++e^-\rightarrow Tl$	-0.34	$Tl(OH)+e^-\rightarrow Tl+OH^-$	-0.34
$Tl^{3+}+2e^-\rightarrow Tl^+$	$+1.25$	$Tl(OH)_3+2e^-\rightarrow Tl(OH)+2OH^-$	-0.05
Thorium			
$Th^{4+}+4e^-\rightarrow Th$	-1.90	$Th(OH)_4+4e^-\rightarrow Th+4OH^-$	-2.48
Tin			
$Sn^{2+}+2e^-\rightarrow Sn$	-0.14	$HSnO_2^-+H_2O+2e^-\rightarrow Sn+3OH^-$	-0.93
$Sn^{4+}+2e^-\rightarrow Sn^{2+}$	$+0.15$	$Sn(OH)_6^{2-}+2e^-\rightarrow HSnO_2^-+H_2O$	
		$+3OH^-$	-0.90
Titanium			
$Ti^{2+}+2e^-\rightarrow Ti$	-1.63	$TiO_2(xH_2O)+4e^-\rightarrow Ti+4OH^-$	
$Ti^{3+}+e^-\rightarrow Ti^{2+}$	-0.37	$+(x-2)H_2O$	-1.69
$TiO^{2+}+2H^++e^-\rightarrow Ti^{3+}+H_2O$	$+0.10$		
Tungsten			
$WO_2+4H^++4e^-\rightarrow W+2H_2O$	-0.12	$WO_4^{2-}+4H_2O+6e^-\rightarrow W+8OH^-$	-1.05
$W_2O_5+2H^++2e^-\rightarrow 2WO_2+H_2O$	-0.04		
$2WO_3+2H^++2e^-\rightarrow W_2O_5+H_2O$	-0.03		
Uranium			
$U^{3+}+3e^-\rightarrow U$	-1.80	$U(OH)_3+3e^-\rightarrow U+3OH^-$	-2.17
$U^{4+}+e^-\rightarrow U^{3+}$	-0.61	$U(OH)_4+e^-\rightarrow U(OH)_3+OH^-$	-2.14
$UO_2^++4H^++e^-\rightarrow U^{4+}+2H_2O$	$+0.62$	$Na_2UO_4+4H_2O+2e^-\rightarrow U(OH)_4$	
$UO_2^{2+}+e^-\rightarrow UO_2^+$	$+0.05$	$+2Na^++4OH^-$	-1.61
Vanadium			
$V^{2+}+2e^-\rightarrow V$	-1.18	$VO_3^-+3H_2O+5e^-\rightarrow V+6OH^-$	-1.15
$V^{3+}+e^-\rightarrow V^{2+}$	-0.26		
$VO^{2+}+2H^++e^-\rightarrow V^{3+}+H_2O$	$+0.36$		
$VO_2^++2H^++e^-\rightarrow VO^{2+}+H_2O$	$+1.00$		
Ytrrium			
$Y^{3+}+3e^-\rightarrow Y$	-2.37	$Y(OH)_3+3e^-\rightarrow Y+3OH^-$	-2.8
Zinc			
$Zn^++2e^-\rightarrow Zn$	-0.76	$Zn(OH)_2+2e^-\rightarrow Zn+2OH^-$	-1.24
		$ZnO_2^{2-}+2H_2O+2e^-\rightarrow Zn+4OH^-$	-1.22
Zirconium			
$Zr^{4+}+4e^-\rightarrow Zr$	-1.53	$H_2ZrO_3+H_2O+4e^-\rightarrow Zr+4OH^-$	-2.36

its E°), on reversal, becomes an oxidation half-reaction (with the same numerical E° but of opposite sign).

Examples:

reduction: $Zn^{2+}+2e^-\rightarrow Zn$ $E^0 = -0.76$ V in neutral or acid
oxidation: $Zn\rightarrow Zn^{2+}+2e^-$ $E^0 = +0.76$ V in neutral or acid
reduction: $ZnO_2^{2-}+2H_2O+2e^-\rightarrow Zn+4OH^-$ $E_B^0 = -1.22$ V in base
oxidation: $Zn+4OH^-\rightarrow ZnO_2^{2-}+2H_2O+2e^-$ $E_B^0 = +1.22$ V in base

(a) *Sign convention.* Since E^0 for $2H^+ + 2e^- = H_2$ is taken as a zero reference potential, all positive E^0 values indicate the reaction can proceed as written, and all negative E^0 values indicate that the reverse reaction (with a positive E^0 value) can occur.

(b) *Oxidation potential.* Reversal of any reduction half-reaction and change of sign of the reduction potential generates an oxidation half-reaction with its oxidation potential.

Reduction Potentials for Other Than Tabulated Half-Reactions. These may be obtained by adding or subtracting known half-reactions, provided correction is made for the different numbers of electrons involved. This is done by summing volt equivalents nE° and dividing by n, the number of electrons involved in the final derived equation.

Example:

$$Fe^{2+} + 2e^- \rightarrow Fe \qquad E^0 = -0.44 \text{ V} \qquad nE^0 = -0.88$$
$$Fe^{3+} + e^- \rightarrow Fe^{2+} \qquad E^0 = +0.77 \text{ V} \qquad nE^0 = +0.77$$

$$Fe^{3+} + 3e^- \rightarrow Fe \qquad\qquad\qquad\qquad nE^0 = -0.11$$

Therefore, $E^0 = -0.11/3 = -0.04$ V.

Doubling stoichiometric coefficients of an equation does not change E^0, since E^0 is a molal quantity.

(a). *Relationship of tabulated reduction potentials to "electromotive series."* A compilation of standard reduction potentials, in acid, of metal-to-ion reduction half-reactions arranged in order of decreasing E° values comprises the usual abbreviated electromotive series tabulations.

Table of Standard Reduction Potentials in Aqueous Solution. The potentials listed are derived from Latimer with additions from the recent literature. E° values in volts, for the reduction equations as written, are based on a zero reference voltage for the hydrogen couple at 25°C. Not all reactions are pysically reversible, or achievable as written, under usual laboratory conditions. They are for 1 M activities for water-soluble species, for 1 atm fugacity for gases, and do not take into account overvoltages necessary for gas evolution at electrodes. Within the limitations imposed, these remain one of the most reliable sources of relative potential for reaction, although they infer nothing about reaction rate.

Significance of Reduction Half-Reactions and Potentials for Predicting Chemical Reactivity:

(1). Reaction with acids, water, and bases. Stability of solutions to air oxidation.

(2). Redox reactions. A source of balanced half-reactions that may be added to give balanced redox reactions.

(3). A systemmatic guide to descriptive inorganic chemistry and usual oxidation states.

(4). Compound disproportionation (self-oxidation and reduction).

Reactions with Acids, Bases, Water or Air add complexity to all chemical reactions and offer competing reactions. If a potential exists to oxidize or reduce water (consider the pH of the system), that oxidation or reduction can proceed. Atmospheric oxygen is easily reduced (is a moderately strong oxidizing agent) and its reactions in systems exposed to air may be significant. These important potentials, listed under hydrogen and oxygen in the reduction potential table, are pH-dependent and are summarized in Table 2.

Table 2. Summary of Air and Water Reduction Potentials

Acidic Solution, $pH = 0$	Neutral, $pH = 7$	Basic Solution, $pH = 14$
$O_2 + 4H^+ + 4e^- \rightarrow 2H_2O$ $E^0 = +1.23$ V $2H^+ + 2e^- \rightarrow H_2$ $E^0 = 0$ V	$O_2 + 4H^+ + 4e^- \rightarrow 2H_2O$ $E^0 = +0.81$ V $2H^+ + 2e^- \rightarrow H_2$ $E^0 = -0.41$ V	$O_2 + 2H_2O + 4e^- \rightarrow 4OH^-$ $E_B^0 = +0.40$ V $2H_2O + 2e^- \rightarrow H_2 + 2OH^-$ $E_B^0 = -0.83$ V

Examples: (1) Air can oxidize Sn^{2+} to Sn^{4+} in neutral solution:

$$O_2 + 4H^+ + 4e^- \rightarrow 2H_2O \qquad\qquad E^0 = +0.81 \text{ V}$$
$$2[Sn^{2+} \rightarrow Sn^{4+} + 2e^-] \qquad\qquad E^0 = -0.15 \text{ V}$$

$$2Sn^{2+} + O_2 + 4H^+ \rightarrow 2Sn^{4+} + 2H_2O \qquad E^0 = +0.66 \text{ V}$$

(2) Gold is dissolved by dilute cyanide solution with air oxidation in alkaline solution:

$$O_2 + 2H_2O + 4e^- \rightarrow 4OH^- \qquad\qquad E_B^0 = +0.40 \text{ V}$$
$$4[2CN^- + Au \rightarrow Au(CN)_2^- + e^-] \qquad\qquad E_B^0 = +0.60 \text{ V}$$

$$4Au + 8CN^- + O_2 + 4H^+ \rightarrow 2H_2O + 4Au(CN)_2^- \qquad E_B^0 = +1.00 \text{ V}$$

(3) Iron rusting in neutral solution:

(a) $O_2 + 4H^+$ (at pH 7) $+ 4e^- \rightarrow 2H_2O$ $E^0 = +0.81$

(b) $Fe + 2OH^- \rightarrow Fe(OH)_2 + 2e^-$ $E^0 = (+0.88)$ (this E^0 is reduced at pH 7 to $\sim +0.6$ V)

(c) $Fe(OH)_2 + OH^- \rightarrow Fe(OH)_3 + e^-$ $E_B{}^0 = +0.56$ V

The reduction of oxygen reaction (a) and the oxidation of iron reaction (b) are followed by further reactions (a) and (c). Alkaline solution reduces potential (a) to $+0.40$ and acid solution increases it to $+1.23$. Many reaction schemes are postulated for rusting; this one is plausible. Overall corrosion rate is increased at low pH and is dependent on oxygen diffusion rate.

(4) Aluminum dissolves in base to liberate hydrogen:

$$2 [Al + 4OH^- \rightarrow H_2AlO_3{}^- + H_2O + 3e^-] \qquad E^0{}_B = +2.35 \text{ V}$$
$$3 [2H_2O + 2e^- \rightarrow H_2 + 2OH^-] \qquad E_B{}^0 = -0.83 \text{ V}$$
$$\overline{2Al + 2OH^- + 4H_2O \rightarrow 2H_2AlO_3{}^- + 3H_2} \qquad E_B{}^0 = +1.52 \text{ V}$$

(5) Copper will not dissolve in a nonoxidizing acid:

$$Cu \rightarrow Cu^{2+} + 2e^- \qquad E^0 = -0.34 \text{ V}$$
$$2H^+ + 2e^- \rightarrow H_2 \qquad E^0 = 0 \text{ V}$$
$$\overline{Cu + 2H^+ \rightarrow Cu^{2+} + H_2} \qquad E^0 = -0.34 \text{ V and reaction cannot proceed spontaneously}$$

(6) Copper will dissolve in oxidizing acids (example: nitric):

$$3 [Cu \rightarrow Cu^{2+} + 2e^-] \qquad E^0 = -0.34 \text{ V}$$
$$2 [NO_3{}^- + 4H^+ + 3e^- \rightarrow 2H_2O + NO] \qquad E^0 = +0.96 \text{ V}$$
$$\overline{3Cu + 2NO_3{}^- + 8H^+ \rightarrow 3Cu^{2+} + 4H_2O + 2NO} \qquad E^0 = +0.62 \text{ V}$$

Electrolytic Cells. Electrolysis cells and the charging of batteries are comparable because an external applied voltage is used, depending on polarity, to augment the rate or to reverse the direction of spontaneous reaction. For example, charging a lead storage battery requires an opposing voltage greater than that of the spontaneous discharge reaction to reverse the direction of the oxidation and reduction reactions. Electrolysis cells can be operated using either molten salts or aqueous solutions, both of which are conductive.

(a). *Molten salt cells* are simple and straightforward in their operation.

Example: Electrolysis of fused $MgCl_2$ (with NaCl added to reduce melting point) liberates chlorine at the anode and magnesium metal at the cathode:

oxidation at anode: $2Cl^- \rightarrow Cl_2 + 3e^-$ $E^0 = -1.36$ V

reduction at cathode: $Mg^{2+} + 2e^- \rightarrow Mg$ $E^0 = -2.34$ V

$$E^0 = -3.70 \text{ V}$$

Possible competitive reactions do not occur because the minimum applied potential of 3.70 V (plus IR losses) reduces the magnesium ion more easily than the sodium ion:

$$Na^+ + e^- \rightarrow Na \qquad E^0 = -2.71 \text{ V}$$

Although E° is not directly applicable because of the nonaqueous media, elevated temperature, and high ionic concentrations, it can be helpful as a very crude approximation.

(b). *Aqueous solution* electrolysis cells are used for electroplating, electrolytic copper purification, and the conduct of many chemical reactions. Competing reactions such as electrolysis of water and overvoltages required for oxygen and hydrogen bubble formation and evolution are complications encountered.

Example of electrolysis in aqueous solution of electrolytic copper. Purification of impure copper requires its separation from Fe, Ag, Ni, Sb, and As metals. An electrolytic cell with an impure copper anode, $CuSO_4$ solution as electrolyte, and a pure copper cathode is used.

Reactions at anode		Reactions possible at cathode	
		(at low concentration)	
$Fe \rightarrow Fe^{2+} + 2e^-$	$E^0 = +0.44$ V	$Fe^{2+} + 2e^- \rightarrow Fe$	$E^0 = -0.44$ V
$Cu \rightarrow Cu^{2+} + 2e^-$	$E^0 = -0.34$ V	$Cu^{2+} + 2e^- \rightarrow Cu$	$E^0 = 0.34$ V
$Ag \rightarrow Ag^+ + e^-$	$E^0 = -0.80$ V	(at high concentrations)	

By application of a low voltage, Cu and the more easily oxidized metals Fe, Ni, Sb, and As are selectively dissolved at the anode, leaving Ag and more difficultly oxidized metals. Copper metal is deposited at the cathode and the more easily oxidized (more difficultly reduced) ions Fe^{2+}, Ni^{2+}, Sb^{3+}, and As^{3+} remain and accumulate in solution, replacing the Cu^{2+} of the electrolyte.

Table 3. Approximate Overvoltages for Gas Evolution

	At 0.01 A/cm²			At 0.1 A/cm²			At 1 A/cm²		
	O_2	H_2	Cl_2	O_2	H_2	Cl_2	O_2	H_2	Cl_2
Graphite	0.80	0.70	—	1.09	0.89	0.25	1.24	1.17	0.50
Pt black	0.40	0.03	0.02	0.64	0.04	0.03	0.79	0.05	0.08
Pt smooth	0.72	0.07	0.03	1.28	0.29	0.05	1.38	0.68	0.24
Ni	0.35	0.74	—	0.73	1.05	—	0.87	1.24	—
Cu	0.42	0.58	—	0.66	0.80	—	0.84	1.25	—
Ag	0.58	0.76	—	0.98	0.98	—	1.14	1.10	—
Au	0.67	0.39	—	1.24	0.59	—	1.68	0.80	—
Fe	—	0.56	—	—	0.82	—	—	1.29	—

Overvoltage for Gas Evolution. Gaseous products can be formed and react with an anode at voltages consistent with their $E°$ values; however, bubble formation and evolution from inert anodes require an overvoltage. Overvoltage is experimentally dependent upon anode surface and current density. The approximate values in Table 3 show the large potentials required, and the facts that hydrogen and halogen have very low overvoltages on platinum black, and that oxygen, though high, is minimum on that particular surface. Reaction with noninert electrodes occurs whenever possible. The effect of overvoltage on the selection of alternate reactions that may occur is shown in the following example:

Electrolysis of NaCl (basic aqueous solution) forms H_2 and Cl_2 on graphite electrodes:

(a)	$2\ [Na^+ + e^- \rightarrow Na]$	$E^0 = -2.71$ V
(b)	$2Cl^- \rightarrow Cl_2 + 2e^-$	$E^0 = -1.36$ V
(c)	$2\ [2H_2O + 2e^- \rightarrow H_2 + 2OH^-]$	$E_B{}^0 = -0.83$ V
(d)	$4OH^- \rightarrow O_2 + 2H_2O + 4e^-$	$E_B{}^0 = -0.40$ V

Reduction reactions (a) or (c) can occur; (c), the easiest reaction, occurs, and liberates hydrogen (despite the hydrogen evolution overvoltage). Oxidation reactions (b) or (d) can occur; they are dependent on the low concentration of OH^-, the high concentration of Cl^-, and the much lower overvoltage for Cl_2 than for O_2 evolution, and Cl_2 is preferentially evolved. The net result is that reactions (b) and (c) occur.

Definitions.

Anode   oxidation occurs at this electrode.
        site where reducing agent is oxidized.
        source of electrons to external circuit whether battery or electrolytic cell.
        electrode labeled negative terminal on batteries (physically labeled for condition of discharge).
        anode of a discharging battery becomes its cathode during charging (or acting as an electrolytic cell).
        name of electrode connected to + terminal of external voltage source for charging a battery or for electrolytic cell.
        anions ($-$ions) migrate toward it in an electrolytic cell.
Cathode reduction occurs at this electrode. Its characteristics are reversed from those of the anode.

Refer to redox terminology, Table 2, Art. 3.

Batteries. Batteries are redox cells that are physically arranged for external flow and internal ion mobility. When different electrolytes are used, gel structures or a porous membrane prevent mixing. The electrolyte(s) must permit ionization of the reactant species and be conductive. Mobility of ions in solution provides the mechanism for maintaining a charge balance and electrical neutrality within the electrolyte. By convention, batteries are labeled negative at the anode of the spontaneous discharge reaction.

(a). *Voltage.* Open circuit terminal voltage is concentration and temperature dependent due to changes in chemical potential, and is reduced by internal IR drop under current flow conditions. The current available depends on the reaction rate, which is a function of temperature and is not directly predictable from $E°$ values. $E°$ provides a reasonable voltage estimate for practical battery configurations.

(b). *Polarization.* Polarization is a localized accumulation of reaction products at an electrode until their concentration is reduced by diffusion, precipitation, formation of complexes, or further reaction to form new species not involved in the electrode reaction. Depolarization is the process of reducing the localized high concentration of reaction products; if depolarization is by a diffusional process, the cell may be reversible because the products retain their chemical identity and physical availability.

(c). *Irreversibility.* Although all redox reactions are ideally reversible near equilibrium, many cells operate far from equilibrium and with secondary reactions or physical configurations that render the cell partially, if not completely, irreversible for practical purposes. This situation results in having many cells that can convert their chemical energy to electrical, and a limited few that are reversible for use as practical and economic storage batteries.

(d). *Battery reactions* on discharge are listed below. Only a few are designed for reversibility.

(1) *Lead storage battery*, (H_2SO_4 electrolyte) (reversible). Oxidation reaction at lead plates, labeled electrically negative:

$$Pb + HSO_4^- \rightarrow PbSO_4 + 2e^- + H^+ \qquad E^0 = +0.36 \text{ V}$$

Reduction reaction at lead dioxide plates, labeled electrically positive:

$$PbO_2 + 3H^+ + HSO_4^- + 2e^- \rightarrow PbSO_4 + 2H_2O \qquad E^0 = +1.68 \text{ V}$$

(2) *Mercury cell* (KOH electrolyte saturated with ZnO). Oxidation ($-$ terminal):

$$Zn + 4OH^- \rightarrow ZnO_2^{2-} + 2H_2O + 2e^- \qquad E_B{}^0 = +1.22 \text{ V}$$

Reduction ($+$ terminal):

$$HgO + H_2O + 2e^- \rightarrow Hg + 2OH^- \qquad E_B{}^0 = +0.10 \text{ V}$$

(3) *Zinc-silver peroxide cell* (KOH electrolyte saturated with ZnO). Oxidation ($-$ terminal):

$$Zn + 4OH^- \rightarrow ZnO_2^{2-} + 2H_2O + 2e^- \qquad E_B{}^0 = +1.22 \text{ V}$$

Reduction ($+$ terminal):

$$Ag_2O + H_2O + 3e^- \rightarrow 2Ag + 2OH^- \qquad E_B{}^0 = +0.34 \text{ V}$$

(4) *LeClanche cell* (flashlight battery, NH_4Cl electrolyte): Oxidation ($-$ terminal):

$$Zn \rightarrow Zn^{2+} + 2e^- \qquad E^0 = +0.76 \text{ V}$$

Reduction ($+$ terminal):

$$2MnO_2 + 2NH_4^+ + 2e^- \rightarrow Mn_2O_3 + H_2O + 2NH_3 \text{ (for complexing } Zn^{2+}) \qquad E^0 = +0.74 \text{ V}$$

(5) *Alkaline-zinc-manganese dioxide* "alkaline flashlight battery" (KOH electrolyte). Oxidation ($-$ terminal):

$$Zn + 4OH^- \rightarrow ZnO_2^{2-} + 2H_2O + 2e^- \qquad E_B{}^0 = +1.22 \text{ V}$$

Reduction ($+$ terminal):

$$MnO_2 + 2H_2O + e^- \rightarrow Mn(OH)_3 + OH^- \qquad E_B{}^0 = +0.35 \text{ V}$$

(6) *Nickel-cadmium storage battery* (KOH electrolyte) (reversible) Oxidation ($-$ terminal):

$$Cd + 2OH^- \rightarrow Cd(OH)_2 + 2e^- \qquad E_B{}^0 = +0.81 \text{ V}$$

Reduction ($+$ terminal):

$$NiO_2 + 2H_2O + 2e^- \rightarrow Ni(OH)_2 \qquad E_B{}^0 = +0.49 \text{ V}$$

10. ORGANIC CHEMISTRY

Tetrahedral carbon. The electronic orbitals of carbon are consistent with the formation of four covalent bonds. Symmetry leads to a tetrahedral spatial orientation. The ability of carbon to bond to carbon leads to a multiplicity of organic compounds based on linear, branched chain, and ring sequences of carbon atoms.

Carbon-Carbon Bonds. Saturation refers to carbon-carbon single bonds, with the remaining three bonds to another carbon, to hydrogen, or to substituent groups. Double and triple bonds between carbon atoms are strained and quite reactive; compounds containing these are termed unsaturated because each double bond replaces two hydrogen atoms.

Classes of Compounds Based on Structure of the Carbon Chain

Aliphatic Compounds. Chains and branched chains of tetrahedral carbon atoms covalently bonded to each other give rise to a proliferation of aliphatic compounds. Bonds between carbon atoms forming the backbone chain may be single, double, or triple, and the remaining bonds available at each carbon atom are made to hydrogen or other substituent groups. Compounds have backbone and branch combinations ranging from C_1 to C_{40} or more. Since substituent groups other than hydrogen are possible at the carbon atoms, and can be placed in different sequences along the backbone, the numbers of organic compounds become very large—before considering the effect of spatial configuration than often permits one part of the molecule to combine with another reactive part of the same or a kindred molecule.

Cyclic Compounds. Ring structures of aliphatic compounds exist subject to bond strain and the possible three-dimensional configurations of the carbon atom chain. Cyclohexane, Fig. 10-1(a), C_6H_{12} (Fig. 1a), and cyclopentane, C_5H_{10} (Fig. 1b), most commonly exist as the backbone structures. These consist of tetrahedral carbon atoms connected by single bonds, and with two hydrogen atoms existent at each apex. Double bonds between carbon atoms can occur: cyclohexene, C_6H_{10} (Fig. 1c) and cyclopentadiene, C_5H_5 (Fig. 1d) are examples.

Note that these can be variously oriented on paper just as the molecules can be in space, so there is no correct or preferred orientation. The presence of a double bond limits the number of hydrogen atoms or substituent groups to one at the double-bonded carbon atoms. Exception: a double-bonded terminal carbon atom in a chain has two single bond positions available.

Aromatic Compounds. A particularly stable minimum energy configuration of six carbon atoms (Fig. 1e) consists of equal carbon atoms in planar configuration with single bonds to other things at each apex of the hexagon. The absence of other substituent groups infers the presence of hydrogen and the figure drawn signifies benzene, C_6H_6. Other groups may replace any or all hydrogen atoms subject to steric (space) considerations; two or three large groups or six small groups may each replace a hydrogen. Higher analogs of benzene (e.g., Figs. 1f, 1g, and 1h) are condensed planar ring structures con-

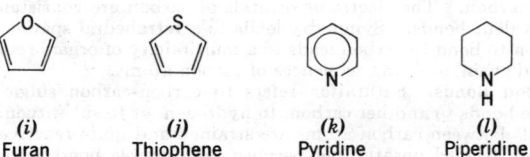

Cyclic Compounds

(a) Cyclohexane (b) Cyclopentane (c) Cyclohexene (d) Cyclopentadiene

Aromatic Compounds

(e) Benzene (f) Naphthalene (g) Phenanthrene (h) Anthracene

Heterocyclic Compounds

(i) Furan (j) Thiophene (k) Pyridine (l) Piperidine

FIG. 1. Cyclic, aromatic, and heterocyclic compounds.

taining H atoms at the protruding corners that are subject to similar space-limited substitutions.

Heterocyclic Compounds. Heterocyclic compounds are ring structures with O, S, or N substituted for one or more of the carbon atoms. Bonds at these substituent atoms are those expected from their electronic structure. Examples are furan, C_4H_4O (Fig. 1i) thiophene, C_4H_4S (Fig. 1j), pyridine, C_5H_5N (Fig. 1k), a nitrogen analog of the aromatic hydrocarbon benzene, and piperidine, $C_5H_{10}NH$ (Fig. 1l), an alicyclic amine.

Abbreviated Organic Chemical Nomenclature

Organic nomenclature is systematic and dependent upon an extensive set of rules of the IUPAC that discourages use of older trivial names. Use of the structural formula is encouraged because it is unambiguous. A few of the simpler rules are helpful:

Ring compounds are considered the basic unit and all appendages as secondary. If required, positions around ring compounds are numbered sequentially. Carbon atoms of linear and branched chain compounds are numbered along the longest string, starting at the end nearest a substituent group or branch. The positions of substituents and side chains are located by the number of the carbon atom where they are attached. Multiple bonds are located by the lowest numbered of the two numbered carbon atoms involved. Some of the prefix and suffix terms used in organic nomenclature are listed in Table 1.

Table 1. Organic Chemical Nomenclature

-ane	saturated hydrocarbon compound
-ene	unsaturated hydrocarbon compound with a double bond
-yne	unsaturated hydrocarbon compound with a triple bond
-yl	hydrocarbon group as substituent (see next section)
cis-	refers to stereochemistry
trans-	refers to stereochemistry
ortho-	1,2 positions on benzene ring (abbreviated o-)
meta-	1,3 positions on benzene ring (abbreviated m-)
para-	1,4 positions on benzene ring (abbreviated p-)
d-	dextro-rotatory optical isomer
l-	levo-rotatory optical isomer
r-	racemic, mixture of d- and l- isomers
n-	normal, or linear chain
iso-	branched chain; usually with an isopropyl group
t-	tertiary, three carbons attached to one
sec-	secondary, two carbons attached to one
neo-	four carbons attached to one
di-	two
tri-	three
cyclo-	ring structure

Three parts of a simple compound name are:

(1) A position number
(2) name of substituent group (see next section)
(3) name of parent structure

Examples:

CH₃ Br
CH₂=CH—CH—CH—CH₂—CH₃   3-methyl-4-bromo-hex-l-ene
        3   4   5   6

Br
⬡—Br   1,3-di-bromo-benzene

CH₃
⬡—C—CH₃   *t*-butyl-benzene
   CH₃

⬡   1,3-cyclo-hexa-di-ene

Common usage of trivial names that are unrelated to structure will continue.   The only recourse in a practical situation is to locate the compound and its physical constants in the *Merck Index*, or in the *Handbook of Chemistry and Physics.*

Classification of Compounds into Unreactive and Functional Groups

Although the number of known organic compounds is astronomic, the number of inexpensive and commercially available ones is drastically limited.   Their chemistry ranges from complete oxidation (during combustion) to conversion into specific compounds with academic or commercial application.

Reactivity of organic compounds occurs only at selected sites on the molecule.   Synthetic procedures attack sustituent groups, more reactive-than-normal hydrogen atoms, and multiple bonds in the carbon chains; the remainder of the molecule remains unchanged in the process.   This permits classification of compounds into a reactive portion of the molecule and one that is unaffected by the synthetic process.   The reactive portion is a substituent or functional group.   The unaffected portion is the remainder of the compound, and is called an aliphatic (alkyl), cyclic aromatic (aryl), or a heterocyclic R-group according to its structure; this R-group is the backbone that carries the reactive substituent group (Tables 2 and 3).

Table 2.   Alkyl and Aryl R-Groups

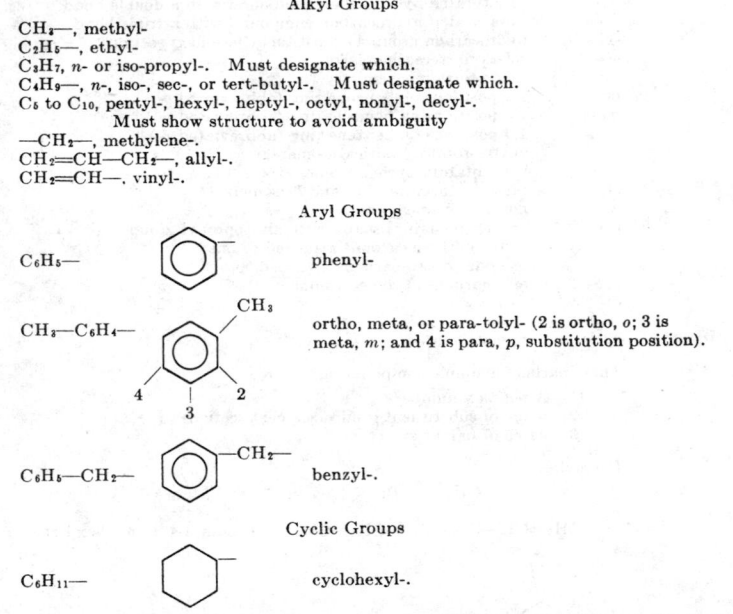

Alkyl Groups

CH_3—, methyl-
C_2H_5—, ethyl-
C_3H_7, n- or iso-propyl-.   Must designate which.
C_4H_9—, n-, iso-, sec-, or tert-butyl-.   Must designate which.
C_5 to C_{10}, pentyl-, hexyl-, heptyl-, octyl, nonyl-, decyl-.
        Must show structure to avoid ambiguity
—CH_2—, methylene-.
CH_2=CH—CH_2—, allyl-.
CH_2=CH—, vinyl-.

Aryl Groups

C_6H_5—   phenyl-

CH_3—C_6H_4—   ortho, meta, or para-tolyl- (2 is ortho, o; 3 is meta, m; and 4 is para, p, substitution position).

C_6H_5—CH_2—   benzyl-.

Cyclic Groups

C_6H_{11}—   cyclohexyl-.

Organic Reactions

Organic reactions make changes in the bonding to carbon atoms; their course is greatly influenced by reaction conditions that are experimentally established on the basis of reaction mechanism studies.   Complications arise because several different reactions may occur simultaneously, accompanied by cyclization, molecular rearrangements, and oxidation-reduction.   All occur in the direction of minimum energy and increased stability.   An encyclopedic literature is replete with tens of thousands of reactions, several hundred of which are generally useful in synthesis and are "name reactions" honoring their early investigators.   Studies of rate, catalysis, stepwise mechanism, and stereochemistry have

Table 3. Functional Groups

Group	Name	Group	Name
—OH	alcohol when attached to alkyl R, phenol when attached to aryl R	—SH	thioalcohol, mercaptan
—O—	ether	—S—	sulfide
—O—O—	peroxide	—S—S—	disulfide
—O—OH	hydroperoxide	$-\overset{O}{\underset{}{S}}-$	sulfoxide
—O—N=O	nitrite	$-\overset{O}{\underset{O}{S}}-$	sulfone
$-O-\overset{\displaystyle O}{\underset{\displaystyle O}{N}}$	nitrate	$-\overset{\displaystyle O}{\underset{\displaystyle OH}{S}}$	sulfinic acid
$-\overset{\displaystyle O}{\underset{\displaystyle O}{N}}$	nitro compound	$-\overset{\displaystyle O}{\underset{\displaystyle O}{S}}-OH$	sulfonic acid
—NH₂	primary amine	$-\overset{\displaystyle O}{C}-OH$	carboxylic acid
$-\overset{\displaystyle H}{N}-$	sec-amine	$-\overset{O}{C}-O-\overset{O}{C}-$	acid anhydride
$-N\big\backslash^{/}$	tert-amine	$-\overset{\displaystyle O}{C}-O-$	ester
=NH	imide	$-\overset{\displaystyle O}{C}-N\big\backslash^{/}$	amide
—C≡N	nitrile	$-\overset{\displaystyle O}{C}\big\backslash_{X}$	acid halide
—X	halide, where X = F, Cl, Br, or I	$-\overset{\displaystyle O}{\underset{\displaystyle \|}{C}}-$	ketone
—H	hydrogen	$-\overset{\displaystyle O}{C}\big\backslash_{H}$	aldehyde
$\overset{\backslash}{\underset{/}{C}}=\overset{/}{\underset{\backslash}{C}}$	double bond		
—C≡C—	triple bond		

coalesced much of the accumulated information into a few generalized categories of related mechanisms.

Substitution. In substitution reactions a group attached to a carbon atom is removed and another enters in its place. Reactions are designated S_{N1}, S_{N2}, S_{E1}, or S_{E2}, dependent on the nucleophilic or electrophilic nature of the reagent and the unimolecular or bimolecular dependence of the reaction rate. Nonpolar substitutions are dependent on a free radical mechanism. Solvolysis reactions, such as hydrolysis of esters or amides, and organic acid-base neutralizations are substitution reactions. Other reactions occur at both saturated and unsaturated carbon atoms. Intra- and intermolecular substitutions are the mechanism for rearrangements and cyclization (ring formation).

Addition. Addition of functional groups occurs at unsaturated carbon atoms, increasing the number of groups attached at those positions. The reagent species may be nucleophilic, electrophilic, or free radical.

Elimination. Reducing the number of functional groups bound to one carbon atom necessitates the formation of an unsaturated carbon bond to the adjacent carbon. Elimination thus involves removal of two functional groups from contiguous carbon atoms and is the reverse of addition reactions. Elimination reactions often occur simultaneously and competitively with substitution reactions at saturated carbon atoms.

Rearrangement. Many reactions are encountered where functional groups migrate within the molecule and result in products other than anticipated. These migrations tend to occur at adjacent carbon atoms; they may involve 5- and 6-membered heterocyclic ring intermediates or neighboring group participation in the reaction. The driving force for any rearrangement, as for any reaction, is increased stability of the resultant product species.

Oxidation–Reduction. Redox reactions are selectively applied to carbon-hydrogen bonds and carbon-carbon multiple bonds, but few oxidative procedures allow selective breaking of a specific carbon-carbon single bond without disrupting the entire molecule. The removal of hydrogen to form multiple carbon-carbon bonds or to make new bonds between carbon and oxygen, sulfur, nitrogen, or halogens is called oxidation. Primary alcohols are oxidized to carboxylic acids, and acids are reduced to alcohols.

Reagents of Organic Reactions

Further classification is made according to classes of reagent, the species which attacks a substrate to produce products. Regaents are collected into three classes based on their electronic structure:

(a). Nucleophilic. Lewis bases can donate unshared electron pairs, reducing agents.
(b). Electrophilic. Lewis acids can accept unshared electron pairs, oxidizing agents.
(c). Free radical. A reactive transient fragment that has one or more unpaired electrons. Examples: $Cl\cdot$, $\equiv C\cdot$, and $=C:$ are chlorine, carbon and carbene radicals, respectively.

The more common reagents are summarized in Table 4.

Catalysis

Catalysis is absolutely dependent on reaction mechanism; other than transition metal participation in intermediate species, acid and base catalysis are most important:

1. Acid catalysis facilitates production of electrophilic reagent species. Lewis acids (see Acids and Bases in Art.2) provide acid catalysis.

Table 4. Nucleophilic and Electrophilic Reagents

Nucleophilic Reagent, Bases and Reducing Agents, Donors of Unshared e^- Pair		Electrophilic Reagent, Acids and Oxidizing Agents, Acceptors of e^- Pair	
I^-	iodide ion	H_3O^+	hydronium ion
OH^-	hydroxyl ion	$\begin{cases} R_2C\!=\!\overset{+}{O}H \\ \overset{+}{R_2}C\!-\!OH \end{cases}$	these occur after protonation of ketones
RO^-	alcoholate ion		
RS^-	mercaptide ion	BF_3	boron trifluoride
CN^-	cyanide ion	$AlCl_3$	aluminum chloride
$H_2\ddot{O}\colon$	water	Cl_2	(their cleavage forms X^+ and
		Br_2	X^-) X^+ is actual electro-
$R\ddot{O}H$	alcohol	I_2	philic reagent
$\colon NH_3$	ammonia	NO_2^+	nitronium ion (formed after
Br^-	bromide ion		protonation of nitric acid)
Cl^-	chloride ion		
$\diagdown\!\!-C^-\colon\diagup$	carbanion	$\diagdown\!\!-C^+\diagup$	carbonium ion

2. Base catalysis facilitates production of nucleophilic reagent species. Lewis bases provide base catalysis.

3. Free radical initiation. Heat, light, or peroxide can initiate free radical formation, independent of both solvent and acid or base catalysis.

Solvents

The solvent for reactions may be one of the reactants or be introduced to facilitate mutual contact via solution. The choice is necessarily dependent on reaction conditions to avoid undesired solvent participation.

(a) Polar solvents are chosen to solubilize ionic species.
(b) Nonpolar solvents dissolve un-ionized molecules.

Table 5 summarizes the more widely used polar and nonpolar solvents.

Table 5. Solvents for Organic Reactions

Polar	Nonpolar
Water	Hydrocarbons
Formic acid	
Dimethylsulfoxide	Acetone—will dissolve some ionic species
Dimethylformamide	
Methanol	Ethers
Ethanol	Ethylene glycol dimethyl ether (diglyme)
Acetic acid	
Nitromethane	Tetrahydrofuran (THF)
Acetonitrile	
Liquid ammonia	Dioxane

11. NUCLEAR REACTIONS

The conventional notation for a nuclear reaction is

$$(_zC^A)_1 + (_zC^A)_2 \rightarrow (_zC^A)_3 + (_zC^A)_4 + Q \qquad (1)$$

in which z is the number of protons, A is the mass number, C is the chemical symbol for the atom, electron, or nucleon, and Q is the energy released.

The conservation equations are,

$$\sum Z_i = 0 \quad \text{and} \quad \sum A_i = 0 \qquad (2)$$

$$\text{Initial mass} - \text{final mass} = Q \qquad (3)$$

an example is the *fission reaction*:

$$_{92}U^{235} + _0n^1 \rightarrow _{92}U^{236} \rightarrow _{38}Sr^{94} + _{54}Xe^{140} + 2_0n^1 + Q$$

The strontium and xenon products are highly radioactive and decay further to other products. The final result is a spectrum of products.

Table 1 gives frequently encountered constants.

Table 1. Some Nuclear Constants

	amu	grams $\times 10^{-24}$	Rest Energy, MeV
Unit mass, m	1	1.65990	931.16
Electron, $_{-1}e^0$ or β^-	0.00054862	0.00091091	0.51083
Proton, p or $_1p^1$	1.007595	1.67247	938.17
Neutron, n or $_0n^1$	1.008983	1.67472	939.43
Hydrogen atom, $_1H^1$	1.00812	1.67338	938.68
Alpha particle, α or $(_2He^4)^{2+}$	4.00280	6.64424	3727.07

Charge on electron $(1.60206 + 0.00007) + 10^{-19}$ coulomb
Radius of electron $(2.81784 + 0.00010) + 10^{-13}$ cm
Radius of nucleus $(1.5 + 0.15) \sqrt{A} + 10^{-13}$ cm

Modes of Radioactive Decay

Negative beta e^- (electron) emission:

$$_0n^1 \rightarrow {}_{-1}e^0 + {}_1H^1 + \text{neutrino}$$

$$_{38}Sr^{94} \xrightarrow{\beta^-} {}_{39}Y^{94} \xrightarrow{\beta^-} {}_{40}Zr^{94}$$

$$_{54}Xe^{140} \rightarrow 4({}_{-1}e^0) + {}_{58}Ce^{140}$$

Positive beta β^+ (positron) emission:

$$_7N^{13} \rightarrow {}_{+1}e^0 + {}_6C^{13} + \text{neutrino}$$

$$_{+1}e^0 + {}_{-1}e^0 \rightarrow 2 \text{ gammas of } 0.51 \text{ MeV each}$$

Alpha emission:

$$_{94}Pu^{239} \rightarrow {}_2He^{2+} + {}_{92}U^{235}$$

Neutron emission:

$$_{53}I^{137} \xrightarrow{\beta^-} {}_{54}Xe^{137} \longrightarrow {}_0n^1 + {}_{54}Xe^{136}$$

Orbital electron (K) capture:

$$_{29}Cu^{64} + {}_{-1}e^0 \rightarrow {}_{28}Ni^{64}$$

Gamma emission by:

(a). Ejection of a gamma photon from an excited nucleus.
(b). Isomeric transition of a nucleus from one energy level to another.
(c). Annihilation of an electron following positive beta emission.

Nuclei with an excess of neutrons are usually electron emitters. Among nuclei with a deficiency of neutrons, the heavy ones usually decay by alpha emission, and the light ones by positron emission or orbital electron capture. Gamma emission often accompanies other types of decay.

Decay with Time

From statistical considerations the rate of decay (alpha emission) is proportional to the number N of radioactive nuclei present,

$$\frac{dN}{dt} = -\lambda N \qquad (4)$$

where the proportionality constant λ is called the *distintegration constant* or the *radioactive decay constant*. Integration of Eq. 4 gives

$$N = N_0 e^{-\lambda t} \qquad (5)$$

where N_0 is the initial number of radioactive nuclei. The *half-life* $t_{1/2}$ or the time required for one-half of the original atoms to decay, by substitution in Eq. 5 is

$$t_{1/2} = 0.693/\lambda \qquad (6)$$

When there is more than one radioisotope in the decay chain

$$P \xrightarrow{\lambda_1} Q \xrightarrow{\lambda_2} R \xrightarrow{\lambda_3}$$

for the second member or daughter Q

$$\frac{N_Q}{N_{P_1}} = \frac{\lambda_1}{\lambda_2 - \lambda_1} e^{-\lambda_1 t} + \frac{\lambda_1}{\lambda_1 - \lambda_2} e^{-\lambda_2 t} \qquad (7)$$

If the half-life of the parent P is longer than the half-life of the daughter $Q (\lambda_2 > \lambda_1)$, after a lapse of time $e^{-\lambda_2 t}$ becomes negligible and

$$\frac{N_Q}{N_P} = \frac{\lambda_1}{\lambda_2 - \lambda_1} e^{-\lambda_1 t} \qquad (8)$$

With native radioisotopes the half-life of the parent is very long compared to that of the daughter, so that Eq. 8 reduces to

$$\frac{N_Q}{N_P} = \frac{\lambda_1}{\lambda_2} e^{-\lambda_1 t} \qquad (9)$$

Halflife and emitted particle energy for selected radioisotopes are shown in Table 2.

Table 2. Selected Radioisotopes *

Atomic No.	Element	Mass No.	β, MeV	γ, MeV	Half-life
13	Aluminum	29	2.5, 1.4	1.28, 2.43	6.6 m
51	Antimony	122	1.40. 1.97	0.69, 0.56	2.8 d
		124	0.61, 2.37	0.60	60 d
		125	0.30, 0.16	0.43, 0.60	2.4 y
33	Arsenic	77	0.68	0.24	39 h
4	Beryllium	10	0.56	None	2.7 × 10⁶ y
83	Bismuth	210	1.16	None	5.0 d
35	Bromine	82	0.44	0.77, 0.55	36 h
48	Cadmium	115	0.59, 1.11	0.52, 0.49	53 h
20	Calcium	45	0.25	None	165 d
		49	2.0, 0.95	3.10, 4.05	8.8 m
6	Carbon	11	0.98 β⁺	0.51	20.4 m
		14	0.16	None	5.7 × 10³ y
58	Cerium	141	0.44, 0.58	0.14	33 d
17	Chlorine	36	0.17	None	3.2 × 10⁵ y
		38	4.81, 1.1, 2.77	2.15, 1.60	37 m
27	Cobalt	60	0.31	1.17, 1.33	5.3 y
29	Copper	61	1.22 β⁺	0.28, 0.66	3.3 h
63	Europium	154	0.15, 1.85	0.12, 0.72, 1.28	16 y
		155	0.15, 0.25	0.089	1.7 y
9	Fluorine	18	0.65 β⁺	0.51	1.87 h
31	Gallium	72	0.64, 0.96	0.84, 2.20	14.1 h
32	Germanium	71	e⁻ capture	None	11 d
		77	2.20, 1.38	0.22, 0.26	12 h
79	Gold	198	0.96	0.41	2.7 d
		199	0.30, 0.25, 0.46	0.21, 0.16	3.2 d
72	Hafnium	181	0.41	0.13, 0.48	45 d
1	Hydrogen	3	0.018	None	12.3 y
53	Iodine	131	0.61, 0.34	0.36, 0.080, 0.28, 0.64	8.0 d
77	Iridium	192	0.67, 0.53, 0.24	0.32, 0.47, 0.60	74 d
		194	2.24, 1.90	0.33, 0.65	19 h
26	Iron	59	0.46, 0.27	1.10, 1.29	45 d
36	Krypton	85	0.67, 0.15	—	10.5 y
57	Lanthanum	140	1.34, 1.10, 0.42	1.60, 0.49	40 h
12	Magnesium	27	1.75, 1.59	0.84, 1.02	9.5 m
80	Mercury	203	0.21	0.28	46 d
42	Molybdenum	99	1.23, 0.45	0.04, 0.74, 0.78	67 h
60	Neodymium	147	0.83, 0.38	0.09, 0.53	11 d
41	Niobium	95	0.16	0.77	35 d
7	Nitrogen	13	1.19 β⁺	0.51	10 m
76	Osmium	191	0.14	0.04, 0.13	15.0 d
		193	1.13, 1.06	1.58, 0.07, 0.14, 0.56	32 h
15	Phosphorus	32	1.71	None	14.3 d
78	Platinum	197	0.67, 0.48	0.08, 0.19	18 h
19	Potassium	40	1.32, e⁻ capture	1.42	1.3 × 10⁹ y
		42	3.54, 1.98	1.52	12.4 h
59	Praeseodymium	142	2.17, 0.59	1.60	19.3 h
		143	0.93	None	13.7 d
61	Promethium	147	0.23	Nil	2.5 y
75	Rhenium	186	1.07, 0.93	0.12, 0.14	91 h
45	Rhodium	105	0.57, 0.25	0.33	36 h
37	Rubidium	86	1.76, 0.68	1.08	18.7 d
44	Ruthenium	103	0.22, 0.14, 0.70	0.50	41 d
62	Samarium	153	0.70, 0.64, 0.81	0.103	47 h
21	Scandium	46	0.36	0.89, 1.12	85 d
14	Silicon	31	1.48	Nil	2.6 h
47	Silver	110	0.087, 0.53	0.66, 0.88	250 d
		111	1.05, 0.71	0.25, 0.34	7.5 d
11	Sodium	22	0.54 β⁺	1.28, 0.51	2.6 y
		24	1.39	1.37, 2.75	15.0 h
38	Strontium	89	1.46	None	50 d
		90	0.54	None	28 y
16	Sulfur	35	0.17	None	87 d
73	Tantalum	182	0.18, 0.36, 0.44	0.10, 1.12, 1.23	115 d

Table 2. Selected Radioisotopes *—*Continued*

Atomic No.	Element	Mass No.	β, MeV	γ, MeV	Half-life
43	Technetium	97	Isotope transition	0.09	91 d, > 3.6 × 10⁶ y
		99	0.30	0.67, 0.75	1.5 × 10⁶ y
52	Tellurium	127	Isotope transition, 0.70	0.089	105 d, 9.3 h
		129	IT, 1.45, 1.00	0.03, 0.10, 0.47	32 d, 72 m
		131	IT, 2.14, 1.69	0.15	30 h, 25 m
81	Thallium	204	0.77	None	3.9 y
22	Titanium	51	2.13, 1.52	0.32	5.8 m
74	Tungsten	185	0.43	Nil	75 d
		187	0.63, 1.32, 0.34	Many 0.13 to 0.87	24 h
54	Xenon	133	0.35	—	5.3 d
39	Yttrium	90	2.27	Nil	64 h

*Source: W. H. SULLIVAN, Trilinear Chart of Nuclides, Washington, D.C., U.S. Government Printing Office, 2nd ed., 1957, and subsequent revision sheets.

BIBLIOGRAPHY

1. CRAM, D. J., and HAMMOND, G. S., Organic Chemistry, New York, McGraw-Hill, 2nd ed., 1964.
2. DELAHAY, P., and TOBIAS, C. W., Advances in Electrochemistry and Electrochemical Engineering, Vols. 1–10, New York, Wiley, 1963–1974.
3. DENBIGH, K. G., Principles of Chemical Equilibrium, Cambridge, Cambridge Univ. Press, 3rd ed., 1971.
4. FERGUSON, F. D., and JONES T. K., The Phase Rule, London, Butterworths, 1966.
5. Handbook of Chemistry and Physics, WEAST, R. C., editor, Cleveland, CRS Press, 54th ed., 1973–1974.
6. HILDEBRAND, J. H., and POWELL, R. E., Principles of Chemistry, New York, Macmillan, 7th ed., 1964.
7. LANGE, N. A., Handbook of Chemistry, New York, McGraw-Hill, 10th ed., 1967.
8. LATIMER, W. M., Oxidation Potentials, New York, Prentice-Hall, 2nd ed., 1952.
9. MURPHY, G., Elements of Nuclear Engineering, New York, Wiley, 1966.
10. PERRY, R. H., and CHILTON, C. H., Chemical Engineers Handbook, New York, McGraw-Hill, 5th ed., 1974.
11. Selected Values of Chemical Thermodynamic Properties, NBS Circular 500, Washington, D.C., U.S. Government Printing Office, 1952.
12. SMITH, J. M., and VAN NESS, H. C., Introduction to Chemical Engineering Thermodynamics, New York, McGraw-Hill, 2nd ed., 1959.
13. SOUDERS, M., The Engineer's Companion, New York, Wiley, 1966.
14. STECHER, P. G., The Merck Index, Rahway, N.J., Merck, 8th ed., 1968.
15. SULLIVAN, W. H., Trilinear Chart of Nuclides, Washington, D.C., USAEC, U.S. Government Printing Office, 2nd ed., 1957 and continuing revision sheets.
16. WILSON E. D., and RIES, H. C., Principles of Chemical Engineering Thermodynamics, New York, McGraw-Hill, 1956.

SECTION 15

ENGINEERING ECONOMY

BY

MOTT SOUDERS

ENGINEERING ECONOMY

Principal Symbols

D	Net disbursement	P	present amount or worth
i	periodic interest rate	q	project interest rate
k	incremental interest rate	r	continuous interest rate
L	life of project	R	net receipt
n	number of interest periods	S	future amount or worth
p	minimum acceptable return		

1. TIME VALUE OF MONEY

Periodic Interest

By definition interest is the difference between the sum S received at the end and the deposit P made at the beginning of a period of time. The periodic rate of interest i is then $(S_1 - P)/P$ and

$$S_1/P = 1 + i \tag{1}$$

Amount at Compound Interest. The future amount S_n of an initial principal deposit P_0 accumulated over n periods at rate of interest i credited at the end of each period is given by

$$S_n/P_0 = (1 + i)^n \tag{2}$$

in which S_n/P_0 is known as the *compound amount factor* for a single payment. (See Table 2.)

Present Worth. The reciprocal of Eq. 2 gives the present worth P of an amount S_n payable at the end of n periods,

$$\frac{P}{S_n} = \frac{1}{(1 + i)^n} \tag{3}$$

in which P/S_n is known as the *present worth factor* for a single payment. (See Table 3.)

Continuous Interest

Interest which is applied at every instant or continuous interest r is often a mathematical convenience. Two different conventions are used with continuous interest. One of these assumes that receipts are concentrated at an instant of time such as at the end of a period. When continuous interest is used with this convention, the convenient tables of the exponential functions e^x and e^{-x} serve as the interest tables with $x = rn$. The second convention assumes that cash flow is uniformly distributed throughout the period, an assumption that simulates business experience. This convention eliminates much of the mathematical simplicity of continuous interest and may not be worth the extra labor, especially when future cash flows are uncertain.

Compound Amount Factor. If S is the compound amount at any time n, and r is the continuous rate of interest per period, usually one year,

$$dS = Sr \, dn \quad \text{and} \quad \int_{P_0}^{S_n} \frac{dS}{S} = \int_0^n r \, dn$$

or

$$\log_e \frac{S_n}{P_0} = rn \quad \text{and} \quad \frac{S_n}{P_0} = e^{rn} \tag{4}$$

in which S_n is the accumulated amount, with continuous interest r from a principal deposit P_0 at zero time.

If the receipt R_1 is distributed uniformly throughout any one period, the accumulated amount S_1 at the end of the period is given by

$$\frac{S_1}{R_1} = \frac{e^r - 1}{r} \tag{5}$$

Present Worth Factor. The reciprocal of Eq. 4 gives the present worth P of an amount S_n payable at time n with continuous interest at the rate r,

$$P/S_n = e^{-rn} \tag{6}$$

For a receipt or deposit R_c distributed uniformly throughout the period, the present worth P is given by

$$\frac{P}{R_c} = \frac{(e^r - 1)e^{-rn}}{r} \tag{7}$$

Table 8 gives solutions of Eq. 7 for a series of periods with various interest rates r.

Equivalent Values of i and r. Equations 2 and 4 do not yield the same numerical results since $(S_n)_r > (S_n)_i$ if $i = r$ and $i > r$ if $(S_n)_i = (S_n)_r$ except in the limit. The equivalent values of i and r per annum for $(S_n)_i = (S_n)_r$ are found by equating Eqs. 2 and 4 and substituting $n = 1$, from which

$$i = e^r - 1 \tag{8}$$

Uniform Series

For a uniform series of receipts R per n, there are four simple general formulas for solving interest problems. Table 1 shows these formulas for (1) periodic interest with R at the end of each period, (2) continuous interest with R at the end of each period, and (3) continuous interest with R uniformly distributed throughout each period. These three different conventions should be clearly distinguished when applying the formulas.

Table 1. Interest Formulas for Uniform Annual Series

	Periodic Interest i	Continuous Interest r	
	R end of year	R end of year	R continuous
S/R Compound amount factor Amount of unit annuity	$\dfrac{(1 + i)^n - 1}{i}$	$\dfrac{e^{rn} - 1}{e^r - 1}$	$\dfrac{e^{rn} - 1}{r}$
P/R Present worth factor Present worth of unit annuity	$\dfrac{1 - (1 + i)^{-n}}{i}$	$\dfrac{1 - e^{-rn}}{e^r - 1}$	$\dfrac{1 - e^{-rn}}{r}$
R/P Capital recovery factor Annuity with present worth of unity	$\dfrac{i}{1 - (1 + i)^{-n}}$	$\dfrac{e^r - 1}{1 - e^{-rn}}$	$\dfrac{r}{1 - e^{-rn}}$
R/S Sinking fund deposit factor Annuity that will amount to unity	$\dfrac{i}{(1 + i)^n - 1}$	$\dfrac{e^r - 1}{e^{rn} - 1}$	$\dfrac{r}{e^{rn} - 1}$

The Series Compound Amount Factor S/R or the *amount of unit annuity* is the accumulated amount of periodic deposits of unity with interest This amount is the summation of the series of $n - 1$ terms from Eq. 2 for $n = 1$ through $n = n - 1$ or the similar series of terms from Eqs. 4 and 5.

The Series Present Worth Factor P/R or the *present worth of unity annuity* is the discounted value of a series of periodic deposits of unity. This value is the summation of the series of n terms from Eqs. 3, 6, and 7 for $n = 1$ through $n = n$.

The Capital Recovery Factor R/P or the *annuity which has a present worth of unity* is the reciprocal of the series present worth factor. The capital recovery factor is also equal to the sinking fund factor plus the interest rate.

The Sinking Fund Deposit Factor R/S or the *periodic deposit that with interest will accumulate to unity* is the reciprocal of the series compound amount factor.

2. ECONOMIC CHOICE

Minimum Cost Point

Time Independent. When only immediate expenditures are relevant to the choice among alternatives, the time value of money is not involved. The problem then is to obtain the design with the minimum first cost. One such problem is the design of a power line where the variables are the pole height and the span length. Another one is the design of a multispan bridge where the variables are the costs of trusses and piers as functions of span length. These problems are solved graphically or analytically by expressing the costs as functions of the design variable (length of span), adding the separate costs, and locating the point of minimum total cost.

Economic Balance. When increase in annual capital charges results in decreased annual operating charges, both expressed in terms of a common design variable, the solution for minimum total annual cost is called economic balance. Choice of wire size for electrical transmission, insulation for a steam line, diameter of a pipe line, or lot size for a production run are all problems in economic balance. Although graphical or tabular constructions are generally applicable, many of these problems may be solved analytically. For example, if x is the common design variable a, b, c, and d are constants related to the specific problem, and

$$\text{annual capital cost} = ax + c$$
$$\text{annual operating cost} = b/x + d$$
$$\text{total annual cost} = ax + b/x + (c + d)$$

differentiating total cost and equating to zero gives the optimum solution as

$$x = (b/a)^{1/2} \tag{9}$$

When there are two common design variables x and y, and

$$\text{total annual cost} = ax + b/xy + cy + d$$

taking the partial derivatives with respect to x and y and equating them gives

$$x/y = c/a \tag{10}$$

from which a plot of total annual cost versus x at various values of y gives a cost curve with minimum point as the optimum. For limitations and other forms see Allen. *Mathematical Analysis for Economists.*

Measures of Profitability

In the choice among alternative investments the objective is to make the best possible use of a limited resource, capital. Decision ciriteria are thus required, among which the principal ones are described here using the convention of continuous interest r with periodic receipts and disbursements. To convert these equations to periodic interest substitute $(1 + i)^n$ for e^{rn}.

Periodic and continuous interest should not be intermixed in economic comparisons. All results should be converted to one form of interest by Eq. 8 before comparing profitabilities.

Minimum Acceptable Return. In the array of prospective investments, some more profitable than others, a minimum standard of comparison is needed. Minimum acceptable return p may be looked upon as the interest rate of the marginal investment that is generally available to the investor. For the state it is the cost of borrowed money. For a public utility it is the rate set by a regulatory agency, in theory the rate required to attract new capital. For a competitive enterprise it is commonly taken as the average rate of earnings on the total assets of the company, often called the *pool rate*. Rates of return are usually computed after income tax.

Internal Rate Method. This well-known method, recently referred to as *discounted cash flow* and *interest rate of return*, serves adequately for the great majority of investment and budgetary decisions. It may be applied to the incremental cash flow from an incremental investment in a project as well as to an array of unrelated projects. The internal rate method postulates that the algebraic sum of the compound amounts of all cash flows for a project is zero at some internal rate of return r found by trial-and-error solution of Eq. 11.

$$\sum_{n=1}^{n=L} R_n e^{r(L-n)} - \sum_{n=0}^{n=L} D_n e^{r(L-n)} = 0$$

or

$$\sum_{n=1}^{n=L} R_n e^{-rn} - \sum_{n=0}^{n=L} D_n e^{-rn} = 0$$

$$\sum_{n=1}^{n=L} R_n (1 + i)^{-n} - \sum_{n=0}^{n=L} D_n (1 + i)^{-n} = 0 \tag{11}$$

This method has two inherent defects: (1) it requires trial-and-error solution, and (2) the solution may be indeterminate (imaginary or multiple roots) when there is more than one reversal in the direction of annual net cash flow.

Proportional Gain Method. This method, attributed to Bernoulli, avoids trial-and-error solutions and is suitable for choosing between mutually exclusive alternates or for ranking an array of investment opportunities. This formulation postulates that the net receipts are accumulated in one account and the net investments in another account both at interest rate p. When the project is terminated at time L, the relative gain G is the ratio of the two accounts or the ratio of their present worths:

$$\frac{\sum_{n=1}^{n=L} R_n e^{p(L-n)}}{\sum_{n=0}^{n=L} D_n e^{p(L-n)}} = \frac{\sum_{n=1}^{n=L} R_n e^{-pn}}{\sum_{n=0}^{n=L} D_n e^{-pn}} = G$$

$$\frac{\sum_{n=1}^{n=L} R_n (1 + p)^{-n}}{\sum_{n=0}^{n=L} D_n (1 + p)^{-n}} = G \tag{12}$$

As this method is biased in favor of long-term investments, it is generally reliable only for comparing investments with nearly equal lives as originally proposed by Bernoulli.

Present Worth Method. This method, also referred to as venture worth and incremental present worth, is restricted to comparison of projects that have identical lives L or that cover the same total time span. If it can be assumed that the costs and returns of replacements will repeat those of the original asset, multiple cycles of a short-term project may be compared with a single long-term project covering the same time span, e.g., three 5-year lives considered equivalent to one 15-year life.

With this method present worth is the decision criterion, larger present worth being preferred. The present worth P of each project is computed, with r equal to the minimum acceptable return, as the algebraic sum of the present worths of the annual net cash flows with salvage value taken as a receipt at the end of the project life.

$$P = \sum_{n=1}^{n=L} R_n e^{-rn} - \sum_{n=0}^{n=L} D_n e^{-rn}$$

$$P = \sum_{n=1}^{n=L} R_n (1 + p)^{-n} - \sum_{n=0}^{n=L} D_n (1 + p)^{-n} \tag{13}$$

This method is suitable for projects that have no positive receipts, a situation in which the other methods are indeterminate.

Sinking Fund Method. This method assumes that the annual receipt R from a project is divided into two parts. The first part R_1 is the uniform annual amount sufficient to recover the original investment at the end of the project when deposited in a sinking fund earning at the minimum acceptable rate of return. The second part R_2, the remainder of the annual receipt, is the annual profit.

For a uniform series of receipts R and an initial investment P, the sinking fund method is known as *Hoskold's Formula*,

$$R = R_1 + R_2 = \frac{pP}{(1 + p)^n - 1} + kP$$

or

$$k = \frac{R}{P} - \frac{p}{(1 + p)^n - 1} \tag{14}$$

The sinking fund method may be applied to variable annual investments D_n and receipts R_n by accumulating the annual receipts in a sinking fund at rate p and subtracting the capital sinking fund P to obtain the total profit at the end of the project. The capital fund P is the present worth at rate p of all the investments D_n. Salvage value is a receipt at the end of the project. The ratio of profit to capital *each year* is the interest rate k. Hence k equals the ratio of total profit to the product of P and the series compound amount factor at rate p.

$$\frac{\sum_{n=1}^{n=L} R_n e^{p(L-n)} - P}{P \left(\frac{e^{pL} - 1}{e^p - 1} \right)} = k$$

or

$$\frac{\sum_{n=1}^{n=L} R_n (1 + p)^{(L-n)} - P}{P \dfrac{(1 + p)^L - 1}{p}} = k \tag{15}$$

Hoskold devised his formula in the 19th century for the valuation of exhausting resources such as mines and oil fields. Recently the method has been revived under names such as "return after amortization" and "venture return."

The sinking fund method may be used to rank projects having different lives L with some inherent biases. These biases are reduced as p approaches k. When p and k are equal, the sinking fund method is identical with the internal rate method.

Annual Cost Method. This method is used to choose among alternative means for obtaining a specified result when the differences between the alternatives are almost entirely differences in disbursements. To compare nonuniform disbursements and different equipment lives, it is necessary to obtain the *equivalent uniform annual cost* which is the annual cost of recovering the disbursements. This annual cost, at the minimum acceptable return p, is the product of the present worth of all disbursements D_n and the capital recovery factor. Salvage value is a negative disbursement at the time of salvage.

$$\sum_{n=0}^{n=L} D_n e^{-pn} \left[\frac{e^p - 1}{1 - e^{-pL}} \right] = \text{annual cost}$$

or

$$\sum_{n=0}^{n=L} D_n (1 + p)^{-n} \left[\frac{p}{1 - (1 + p)^{-L}} \right] = \text{annual cost} \tag{16}$$

3. INTEREST TABLES

Table 2 Amount at Compound Interest, $(1 + i)^n$

n	$1\frac{1}{2}\%$	2%	$2\frac{1}{2}\%$	3%	$3\frac{1}{2}\%$	4%	$4\frac{1}{2}\%$	5%
1	1.01500	1.02000	1.02500	1.03000	1.03500	1.04000	1.04500	1.05000
2	1.03023	1.04040	1.05062	1.06090	1.07122	1.08160	1.09203	1.10250
3	1.04568	1.06121	1.07689	1.09273	1.10872	1.12486	1.14117	1.15763
4	1.06136	1.08243	1.10381	1.12551	1.14752	1.16986	1.19252	1.21551
5	1.07728	1.10408	1.13141	1.15927	1.18769	1.21665	1.24618	1.27628
6	1.09344	1.12616	1.15969	1.19405	1.22926	1.26532	1.30226	1.34010
7	1.10984	1.14869	1.18869	1.22987	1.27228	1.31593	1.36086	1.40710
8	1.12649	1.17166	1.21840	1.26677	1.31681	1.36857	1.42210	1.47746
9	1.14339	1.19509	1.24886	1.30477	1.36290	1.42331	1.48610	1.55133
10	1.16054	1.21899	1.28008	1.34392	1.41060	1.48024	1.55297	1.62889
11	1.17795	1.24337	1.31209	1.38423	1.45997	1.53945	1.62285	1.71034
12	1.19362	1.26824	1.34489	1.42576	1.51107	1.60103	1.69588	1.79586
13	1.21355	1.29361	1.37851	1.46853	1.56396	1.66507	1.77220	1.88565
14	1.23176	1.31948	1.41297	1.51259	1.61869	1.73168	1.85194	1.97993
15	1.25023	1.34587	1.44830	1.55797	1.67535	1.80094	1.93528	2.07893
16	1.26899	1.37279	1.48451	1.60471	1.73399	1.87298	2.02237	2.18287
17	1.28802	1.40024	1.52162	1.65285	1.79468	1.94790	2.11338	2.29202
18	1.30734	1.42825	1.55966	1.70243	1.85749	2.02582	2.20848	2.40662
19	1.32695	1.45681	1.59865	1.75351	1.92250	2.10685	2.30786	2.52695
20	1.34685	1.48595	1.63862	1.80611	1.98979	2.19112	2.41171	2.65330
21	1.36706	1.51567	1.67958	1.86029	2.05943	2.27877	2.52024	2.78596
22	1.38756	1.54598	1.72157	1.91610	2.13151	2.36992	2.63365	2.92526
23	1.40838	1.57690	1.76461	1.97359	2.20611	2.46472	2.75217	3.07152
24	1.42950	1.60844	1.80873	2.03279	2.28333	2.56330	2.87601	3.22510
25	1.45095	1.64061	1.85394	2.09378	2.36324	2.66584	3.00543	3.38635
26	1.47271	1.67342	1.90029	2.15659	2.44596	2.77247	3.14068	3.55567
27	1.49480	1.70689	1.94780	2.22129	2.53157	2.88337	3.28201	3.73346
28	1.51722	1.74102	1.99650	2.28793	2.62017	2.99870	3.42970	3.92013
29	1.53998	1.77584	2.04640	2.35657	2.71188	3.11865	3.58404	4.11614
30	1.56308	1.81136	2.09757	2.42726	2.80679	3.24340	3.74532	4.32194
31	1.58653	1.84759	2.15001	2.50008	2.90503	3.37313	3.91386	4.53804
32	1.61032	1.88454	2.20376	2.57508	3.00671	3.50806	4.08998	4.76494
33	1.63448	1.92223	2.25885	2.65234	3.11194	3.64838	4.27403	5.00319
34	1.65900	1.96068	2.31532	2.73191	3.22086	3.79432	4.46636	5.25335
35	1.68388	1.99989	2.37321	2.81386	3.33359	3.94609	4.66735	5.51602
36	1.70914	2.03989	2.43254	2.89828	3.45027	4.10393	4.87738	5.79182
37	1.73478	2.08069	2.49335	2.98523	3.57103	4.26809	5.09686	6.08141
38	1.76080	2.12230	2.55568	3.07478	3.69601	4.43881	5.32622	6.38548
39	1.78721	2.16474	2.61957	3.16703	3.82537	4.61637	5.56590	6.70475
40	1.81402	2.20804	2.68506	3.26204	3.95926	4.80102	5.81636	7.03999
41	1.84123	2.25220	2.75219	3.35990	4.09783	4.99306	6.07810	7.39199
42	1.86885	2.29724	2.82100	3.46070	4.24126	5.19278	6.35162	7.76159
43	1.89688	2.34319	2.89152	3.56452	4.38970	5.40050	6.63744	8.14967
44	1.92533	2.39005	2.96381	3.67145	4.54334	5.61652	6.93612	8.55715
45	1.95241	2.43785	3.03790	3.78160	4.70236	5.84118	7.24825	8.98501
46	1.98353	2.48661	3.11385	3.89504	4.86694	6.07482	7.57442	9.43426
47	2.01328	2.53634	3.19170	4.01190	5.03728	6.31782	7.91527	9.90597
48	2.04348	2.58707	3.27149	4.13225	5.21359	6.57053	8.27146	10.4013
49	2.07413	2.63881	3.35328	4.25622	5.39606	6.83335	8.64367	10.9213
50	2.10524	2.69159	3.43711	4.38391	5.58493	7.10668	9.03264	11.4674

Table 2 (*continued*)

n	5½%	6%	7%	8%	10%	12%	15%	20%
1	1.055	1.060	1.070	1.080	1.100	1.120	1.150	1.200
2	1.113	1.124	1.145	1.166	1.210	1.254	1.322	1.440
3	1.174	1.191	1.225	1.260	1.331	1.405	1.521	1.728
4	1.239	1.262	1.311	1.360	1.464	1.574	1.749	2.074
5	1.307	1.338	1.403	1.469	1.611	1.762	2.011	2.488
6	1.379	1.419	1.501	1.587	1.772	1.974	2.313	2.986
7	1.455	1.504	1.606	1.714	1.949	2.211	2.660	3.583
8	1.535	1.594	1.718	1.851	2.144	2.476	3.059	4.300
9	1.619	1.689	1.838	1.999	2.358	2.773	3.518	5.160
10	1.708	1.791	1.967	2.159	2.594	3.106	4.046	6.192
11	1.802	1.898	2.105	2.332	2.853	3.479	4.652	7.430
12	1.901	2.012	2.252	2.518	3.138	3.896	5.350	8.916
13	2.006	2.133	2.410	2.720	3.452	4.363	6.153	10.699
14	2.116	2.261	2.579	2.937	3.797	4.887	7.076	12.839
15	2.232	2.397	2.759	3.172	4.177	5.474	8.137	15.407
16	2.355	2.540	2.952	3.426	4.595	6.130	9.358	18.488
17	2.485	2.693	3.159	3.700	5.054	6.866	10.761	22.186
18	2.621	2.854	3.380	3.996	5.560	7.690	12.375	26.623
19	2.766	3.026	3.617	4.316	6.116	8.613	14.232	31.948
20	2.918	3.207	3.870	4.661	6.727	9.646	16.367	38.338
21	3.078	3.400	4.141	5.034	7.400	10.804	18.821	46.005
22	3.248	3.604	4.430	5.437	8.140	12.100	21.645	55.206
23	3.426	3.820	4.741	5.871	8.954	13.552	24.891	66.247
24	3.615	4.049	5.072	6.341	9.850	15.179	28.625	79.497
25	3.813	4.292	5.427	6.848	10.835	17.000	32.919	95.396
26	4.023	4.549	5.807	7.396	11.918	19.040	37.857	114.475
27	4.244	4.822	6.214	7.988	13.110	21.325	43.535	137.370
28	4.478	5.112	6.649	8.627	14.421	23.884	50.065	164.845
29	4.724	5.418	7.114	9.317	15.863	26.750	57.575	197.813
30	4.984	5.743	7.612	10.063	17.449	29.960	66.212	237.376
31	5.258	6.088	8.145	10.868	19.194	33.555	76.143	284.851
32	5.547	6.453	8.715	11.737	21.114	37.582	87.565	341.822
33	5.852	6.841	9.325	12.676	23.225	42.091	100.700	410.186
34	6.174	7.251	9.978	13.690	25.548	47.142	115.805	492.223
35	6.514	7.686	10.677	14.785	28.102	52.799	133.175	590.668

Table 3 Present Worth, $(1 + i)^{-n}$

n	2%	3%	4%	5%	6%	7%	8%	10%
1	0.98039	0.97087	0.96154	0.95238	0.94340	0.9346	0.9259	0.9091
2	0.96117	0.94260	0.92456	0.90703	0.89000	0.8734	0.8573	0.8264
3	0.94232	0.91514	0.88900	0.86384	0.83962	0.8163	0.7938	0.7513
4	0.92385	0.88849	0.85480	0.82270	0.79209	0.7629	0.7350	0.6830
5	0.90573	0.86261	0.82193	0.78353	0.74726	0.7130	0.6806	0.6209
6	0.88797	0.83748	0.79031	0.74622	0.70496	0.6663	0.6302	0.5645
7	0.87056	0.81309	0.75992	0.71068	0.66506	0.6227	0.5835	0.5132
8	0.85349	0.78941	0.73069	0.67684	0.62741	0.5820	0.5403	0.4665
9	0.83676	0.76642	0.70259	0.64461	0.59190	0.5439	0.5002	0.4241
10	0.82035	0.74409	0.67556	0.61391	0.55839	0.5083	0.4632	0.3855
11	0.80426	0.72242	0.64958	0.58468	0.52679	0.4751	0.4289	0.3505
12	0.78849	0.70138	0.62460	0.55684	0.49697	0.4440	0.3971	0.3186
13	0.77303	0.68095	0.60057	0.53032	0.46884	0.4150	0.3677	0.2897
14	0.75788	0.66112	0.57748	0.50507	0.44230	0.3878	0.3405	0.2633
15	0.74301	0.64186	0.55526	0.48102	0.41727	0.3624	0.3152	0.2394
16	0.72845	0.62317	0.53391	0.45811	0.39365	0.3387	0.2919	0.2176
17	0.71416	0.60502	0.51337	0.43630	0.37136	0.3166	0.2703	0.1978
18	0.70016	0.58739	0.49363	0.41552	0.35034	0.2959	0.2502	0.1799
19	0.68643	0.57029	0.47464	0.39573	0.33051	0.2765	0.2317	0.1635
20	0.67297	0.55368	0.45639	0.37689	0.31180	0.2584	0.2145	0.1486
21	0.65978	0.53755	0.43883	0.35894	0.29416	0.2415	0.1987	0.1351
22	0.64684	0.52189	0.42196	0.34185	0.27751	0.2257	0.1839	0.1228
23	0.63416	0.50669	0.40573	0.32557	0.26180	0.2109	0.1703	0.1117
24	0.62172	0.49193	0.39012	0.31007	0.24698	0.1971	0.1577	0.1015
25	0.60953	0.47761	0.37512	0.29530	0.23300	0.1842	0.1460	0.0923
26	0.59758	0.46369	0.36069	0.28124	0.21981	0.1722	0.1352	0.0839
27	0.58586	0.45019	0.34682	0.26785	0.20737	0.1609	0.1252	0.0763
28	0.57437	0.43708	0.33348	0.25509	0.19563	0.1504	0.1159	0.0693
29	0.56311	0.42435	0.32065	0.24295	0.18456	0.1406	0.1073	0.0630
30	0.55207	0.41199	0.30832	0.23138	0.17411	0.1314	0.0994	0.0573
31	0.54125	0.39999	0.29646	0.22036	0.16425	0.1228	0.0920	0.0521
32	0.53063	0.38834	0.28506	0.20987	0.15496	0.1147	0.0852	0.0474
33	0.52023	0.37703	0.27409	0.19987	0.14619	0.1072	0.0789	0.0431
34	0.51003	0.36604	0.26355	0.19035	0.13791	0.1002	0.0730	0.0391
35	0.50003	0.35538	0.25342	0.18129	0.13011	0.0937	0.0676	0.0356
36	0.49022	0.34503	0.24367	0.17266	0.12274	0.0875	0.0626	
37	0.48061	0.33498	0.23430	0.16444	0.11579	0.0818	0.0580	
38	0.47119	0.32523	0.22529	0.15661	0.10924	0.0765	0.0537	
39	0.46195	0.31575	0.21662	0.14915	0.10306	0.0715	0.0497	
40	0.45289	0.30656	0.20829	0.14205	0.09722	0.0668	0.0460	0.0221
41	0.44401	0.29763	0.20028	0.13528	0.09172	0.0624	0.0426	
42	0.43530	0.28896	0.19257	0.12884	0.08653	0.0583	0.0395	
43	0.42677	0.28054	0.18517	0.12270	0.08163	0.0545	0.0365	
44	0.41840	0.27237	0.17805	0.11686	0.07701	0.0509	0.0338	
45	0.41020	0.26444	0.17120	0.11130	0.07265	0.0476	0.0313	0.0137
46	0.40215	0.25674	0.16461	0.10600	0.06854	0.0445	0.0290	
47	0.39427	0.24926	0.15828	0.10095	0.06466	0.0416	0.0269	
48	0.38654	0.24200	0.15219	0.09614	0.06100	0.0389	0.0249	
49	0.37896	0.23495	0.14634	0.09156	0.05755	0.0363	0.0230	
50	0.37153	0.22811	0.14071	0.08720	0.05429	0.0339	0.0213	0.0085

Table 3   (*Continued*)

n	12%	15%	20%	25%	30%	35%	40%	45%
1	0.8929	0.8696	0.8333	0.8000	0.7692	0.7407	0.7143	0.6897
2	0.7972	0.7561	0.6944	0.6400	0.5917	0.5487	0.5102	0.4756
3	0.7118	0.6575	0.5787	0.5120	0.4552	0.4064	0.3644	0.3280
4	0.6355	0.5718	0.4823	0.4096	0.3501	0.3011	0.2603	0.2262
5	0.5674	0.4972	0.4019	0.3277	0.2693	0.2230	0.1859	0.1560
6	0.5066	0.4323	0.3349	0.2621	0.2072	0.1652	0.1328	0.1076
7	0.4523	0.3759	0.2791	0.2097	0.1594	0.1224	0.0949	0.0742
8	0.4039	0.3269	0.2326	0.1678	0.1226	0.0906	0.0678	0.0512
9	0.3606	0.2843	0.1938	0.1342	0.0943	0.0671	0.0484	0.0353
10	0.3220	0.2472	0.1615	0.1074	0.0725	0.0497	0.0346	0.0243
11	0.2875	0.2149	0.1346	0.0859	0.0558	0.0368	0.0247	0.0168
12	0.2567	0.1869	0.1122	0.0687	0.0429	0.0273	0.0176	0.0116
13	0.2292	0.1625	0.0935	0.0550	0.0330	0.0202	0.0126	0.0080
14	0.2046	0.1413	0.0779	0.0440	0.0254	0.0150	0.0090	0.0055
15	0.1827	0.1229	0.0649	0.0352	0.0195	0.0111	0.0064	0.0038
16	0.1631	0.1069	0.0541	0.0281	0.0150	0.0082	0.0046	0.0026
17	0.1456	0.0929	0.0451	0.0225	0.0116	0.0061	0.0033	0.0018
18	0.1300	0.0808	0.0376	0.0180	0.0089	0.0045	0.0023	0.0012
19	0.1161	0.0703	0.0313	0.0144	0.0068	0.0033	0.0017	0.0009
20	0.1037	0.0611	0.0261	0.0115	0.0053	0.0025	0.0012	0.0006
21	0.0926	0.0531	0.0217	0.0092	0.0040	0.0018	0.0009	0.0004
22	0.0826	0.0462	0.0181	0.0074	0.0031	0.0014	0.0006	0.0003
23	0.0738	0.0402	0.0151	0.0059	0.0024	0.0010	0.0004	0.0002
24	0.0659	0.0349	0.0126	0.0047	0.0018	0.0007	0.0003	0.0001
25	0.0588	0.0304	0.0105	0.0038	0.0014	0.0006	0.0002	0.0001
26	0.0525	0.0264	0.0087	0.0030	0.0011	0.0004	0.0002	0.0001
27	0.0469	0.0230	0.0073	0.0024	0.0008	0.0003	0.0001	
28	0.0419	0.0200	0.0061	0.0019	0.0006	0.0002	0.0001	
29	0.0374	0.0174	0.0051	0.0015	0.0005	0.0002	0.0001	
30	0.0334	0.0151	0.0042	0.0012	0.0004	0.0001		
31	0.0298	0.0131	0.0035	0.0010	0.0003	0.0001		
32	0.0266	0.0114	0.0029	0.0008	0.0002	0.0001		
33	0.0238	0.0099	0.0024	0.0006	0.0002	0.0001		
34	0.0212	0.0086	0.0020	0.0005	0.0001			
35	0.0189	0.0075	0.0017	0.0004	0.0001			

Table 4 Amount of Annuity, $\dfrac{(1 + i)^n - 1}{i}$

n	$1\frac{1}{2}\%$	2%	$2\frac{1}{2}\%$	3%	$3\frac{1}{2}\%$	4%	$4\frac{1}{2}\%$	5%
1	1.00000	1.00000	1.00000	1.00000	1.00000	1.00000	1.00000	1.00000
2	2.01500	2.02000	2.02500	2.03000	2.03500	2.04000	2.04500	2.05000
3	3.04522	3.06040	3.07562	3.09090	3.10623	3.12160	3.13702	3.15250
4	4.09090	4.12161	4.15252	4.18363	4.21494	4.24646	4.27819	4.31013
5	5.15227	5.20404	5.25633	5.30914	5.36247	5.41632	5.47071	5.52563
6	6.22955	6.30812	6.38774	6.46841	6.55015	6.63298	6.71689	6.80191
7	7.32299	7.43428	7.54743	7.66246	7.77941	7.89829	8.01915	8.14201
8	8.43284	8.58297	8.73612	8.89234	9.05169	9.21423	9.38001	9.54911
9	9.55933	9.75463	9.95452	10.1591	10.3685	10.5828	10.8021	11.0266
10	10.70272	10.94972	11.2034	11.4639	11.7314	12.0061	12.2882	12.5779
11	11.86326	12.16872	12.4835	12.8078	13.1420	13.4864	13.8412	14.2068
12	13.04121	13.41209	13.7956	14.1920	14.6020	15.0258	15.4640	15.9171
13	14.23683	14.68033	15.1404	15.6178	16.1130	16.6268	17.1599	17.7130
14	15.45038	15.97394	16.5190	17.0863	17.6770	18.2919	18.9321	19.5986
15	16.68214	17.29342	17.9319	18.5989	19.2957	20.0236	20.7841	21.5786
16	17.93237	18.63929	19.3802	20.1569	20.9710	21.8245	22.7193	23.6575
17	19.20136	20.01207	20.8647	21.7616	22.7050	23.6975	24.7417	25.8404
18	20.48938	21.41231	22.3863	23.4144	24.4997	25.6454	26.8551	28.1324
19	21.79672	22.84056	23.9460	25.1169	26.3572	27.6712	29.0636	30.5390
20	23.12367	24.29737	25.5447	26.8704	28.2797	29.7781	31.3714	33.0660
21	24.47052	25.78332	27.1833	28.6765	30.2695	31.9692	33.7831	35.7193
22	25.83758	27.29899	28.8629	30.5368	32.3289	34.2480	36.3034	38.5052
23	27.22514	28.84496	30.5844	32.4529	34.4604	36.6179	38.9370	41.4305
24	28.63352	30.42186	32.3490	34.4265	36.6665	39.0826	41.6892	44.5020
25	30.06302	32.03030	34.1578	36.4593	38.9499	41.6459	44.5652	47.7271
26	31.51397	33.67091	36.0117	38.5530	41.3131	44.3117	47.5706	51.1135
27	32.98668	35.34432	37.9120	40.7096	43.7591	47.0842	50.7113	54.6691
28	34.48148	37.05121	39.8598	42.9309	46.2906	49.9676	53.9933	58.4026
29	35.99870	38.79223	41.8563	45.2189	48.9180	52.9663	57.4230	62.3227
30	37.53868	40.56808	43.9027	47.5754	51.6227	56.0849	61.0071	66.4388
31	39.10176	42.37944	46.0003	50.0027	54.4295	59.3283	64.7524	70.7608
32	40.68829	44.22703	48.1503	52.5028	57.3345	62.7015	68.6662	75.2988
33	42.29861	46.11157	50.3540	55.0778	60.3412	66.2095	72.7562	80.0638
34	43.93309	48.03380	52.6129	57.7302	63.4532	69.8579	77.0303	85.0670
35	45.59208	49.99448	54.9282	60.4621	66.6740	73.6522	81.4966	90.3203
36	47.27597	51.99437	57.3014	63.2759	70.0076	77.5983	86.1640	95.8363
37	48.98511	54.03425	59.7339	66.1742	73.4579	81.7022	91.0413	101.628
38	50.71989	56.11494	62.2273	69.1594	77.0289	85.9703	96.1382	107.710
39	52.48058	58.23724	64.7830	72.2342	80.7249	90.4091	101.464	114.095
40	54.26789	60.40198	67.4026	75.4013	84.5503	95.0255	107.030	120.800
41	56.08191	62.61002	70.0876	78.6633	88.5095	99.8265	112.847	127.840
42	57.92314	64.86222	72.8398	82.0232	92.6074	104.820	118.925	135.232
43	59.79199	67.15947	75.6608	85.4839	96.8486	110.012	125.276	142.993
44	61.68887	69.50266	78.5523	89.0484	101.238	115.413	131.914	151.143
45	63.61420	71.89271	81.5161	92.7199	105.782	121.029	138.850	159.700
46	65.56841	74.33056	84.5540	96.5015	110.484	126.871	146.098	168.685
47	67.55194	76.81718	87.6679	100.397	115.351	132.945	153.673	178.119
48	69.56522	79.35352	90.8596	104.408	120.388	139.263	161.588	188.025
49	71.60870	81.94059	94.1311	108.541	125.602	145.834	169.859	198.427
50	73.68283	84.57940	97.4843	112.797	130.998	152.667	178.503	209.348

Table 4   (*Continued*)

n	5½%	6%	7%	8%	10%	12%	15%	20%
1	1.00000	1.00000	1.000	1.000	1.000	1.000	1.000	1.000
2	2.05500	2.06000	2.070	2.080	2.100	2.120	2.150	2.200
3	3.16803	3.18360	3.215	3.246	3.310	3.374	3.472	3.640
4	4.34227	4.37462	4.440	4.506	4.641	4.779	4.993	5.368
5	5.58109	5.63709	5.751	5.867	6.105	6.353	6.742	7.442
6	6.88805	6.97532	7.153	7.336	7.716	8.115	8.754	9.930
7	8.26689	8.39384	8.654	8.923	9.487	10.089	11.067	12.916
8	9.72157	9.89747	10.260	10.637	11.436	12.300	13.727	16.499
9	11.2563	11.4913	11.978	12.488	13.579	14.776	16.786	20.799
10	12.8754	13.1808	13.816	14.487	15.937	17.549	20.304	25.959
11	14.5835	14.9716	15.784	16.645	18.531	20.655	24.349	32.150
12	16.3856	16.8699	17.888	18.977	21.384	24.133	29.002	39.580
13	18.2868	18.8821	20.141	21.495	24.523	28.029	34.352	48.497
14	20.2926	21.0151	22.550	24.215	27.975	32.393	40.505	59.196
15	22.4087	23.2760	25.129	27.152	31.772	37.280	47.580	72.035
16	24.6411	25.6725	27.888	30.324	35.950	42.753	55.717	87.442
17	26.9964	28.2129	30.840	33.750	40.545	48.884	65.075	105.931
18	29.4812	30.9057	33.999	37.450	45.599	55.750	75.836	128.117
19	32.1027	33.7600	37.379	41.446	51.159	63.440	88.212	154.740
20	34.8683	36.7856	40.995	45.762	57.275	72.052	102.443	186.688
21	37.7861	39.9927	44.865	50.423	64.002	81.699	118.810	225.025
22	40.8643	43.3923	49.006	55.457	71.403	92.502	137.631	271.031
23	44.1118	46.9958	53.436	60.893	79.543	104.603	159.276	326.237
24	47.5380	50.8156	58.177	66.765	88.497	118.155	184.167	392.484
25	51.1526	54.8645	63.249	73.106	98.347	133.334	212.793	471.981
26	54.9660	59.1564	68.676	79.954	109.182	150.334	245.711	567.377
27	58.9891	63.7058	74.484	87.351	121.100	169.374	283.568	681.852
28	63.2335	68.5281	80.698	95.339	134.210	190.699	327.103	819.223
29	67.7114	73.6398	87.347	103.966	148.631	214.582	377.169	984.067
30	72.4355	79.0582	94.461	113.283	164.494	241.332	434.744	1181.881
31	77.4194	84.8017	102.073	123.346	181.943	271.292	500.956	1419.257
32	82.6775	90.8898	110.218	134.214	201.138	304.847	577.099	1704.108
33	88.2248	97.3432	118.933	145.951	222.252	342.429	664.664	2045.930
34	94.0071	104.184	128.259	158.627	245.477	384.520	765.364	2456.116
35	100.251	111.435	138.237	172.317	271.024	431.663	881.168	2948.339

Table 5 Present Worth of Annuity, $\dfrac{1 - (1 + i)^{-n}}{i}$

n	2%	3%	4%	5%	6%	7%	8%	10%
1	0.98039	0.97087	0.96154	0.95238	0.94340	0.935	0.926	0.909
2	1.94156	1.91347	1.88609	1.85941	1.83339	1.808	1.783	1.736
3	2.88388	2.82861	2.77509	2.72325	2.67301	2.624	2.577	2.487
4	3.80773	3.71710	3.62990	3.54595	3.46511	3.387	3.312	3.170
5	4.71346	4.57971	4.45182	4.32948	4.21236	4.100	3.993	3.791
6	5.60143	5.41719	5.24214	5.07569	4.91732	4.767	4.623	4.355
7	6.47199	6.23028	6.00205	5.78637	5.58238	5.389	5.206	4.868
8	7.32548	7.01969	6.73274	6.46321	6.20979	5.971	5.747	5.335
9	8.16224	7.78611	7.43533	7.10782	6.80169	6.515	6.247	5.759
10	8.98258	8.53020	8.11090	7.72173	7.36009	7.024	6.710	6.144
11	9.78685	9.25262	8.76048	8.30641	7.88687	7.499	7.139	6.495
12	10.57534	9.95400	9.38507	8.86325	8.38384	7.943	7.536	6.814
13	11.34837	10.6350	9.98565	9.39357	8.85268	8.358	7.904	7.103
14	12.10625	11.2961	10.5631	9.89864	9.29498	8.745	8.244	7.367
15	12.84926	11.9379	11.1184	10.3797	9.71225	9.108	8.559	7.606
16	13.57771	12.5611	11.6523	10.8378	10.1059	9.447	8.851	7.824
17	14.29187	13.1661	12.1657	11.2741	10.4773	9.763	9.122	8.022
18	14.99203	13.7535	12.6593	11.6896	10.8276	10.059	9.372	8.201
19	15.67846	14.3238	13.1339	12.0853	11.1581	10.336	9.604	8.365
20	16.35143	14.8775	13.5903	12.4622	11.4699	10.594	9.818	8.514
21	17.01121	15.4150	14.0292	12.8212	11.7641	10.836	10.017	8.649
22	17.65805	15.9369	14.4511	13.1630	12.0416	11.061	10.201	8.772
23	18.29220	16.4436	14.8568	13.4886	12.3034	11.272	10.371	8.883
24	18.91393	16.9355	15.2470	13.7986	12.5504	11.469	10.529	8.985
25	19.52346	17.4131	15.6221	14.0939	12.7834	11.654	10.675	9.077
26	20.12104	17.8768	15.9828	14.3752	13.0032	11.826	10.810	9.161
27	20.70690	18.3270	16.3296	14.6430	13.2105	11.987	10.935	9.237
28	21.28127	18.7641	16.6631	14.8981	13.4062	12.137	11.051	9.307
29	21.84438	19.1885	16.9837	15.1411	13.5907	12.278	11.158	9.370
30	22.39646	19.6004	17.2920	15.3725	13.7648	12.409	11.258	9.427
31	22.93770	20.0004	17.5885	15.5928	13.9291	12.532	11.350	9.479
32	23.46833	20.3888	17.8736	15.8027	14.0840	12.647	11.435	9.526
33	23.98856	20.7658	18.1476	16.0025	14.2302	12.754	11.514	9.569
34	24.49859	21.1318	18.4112	16.1929	14.3681	12.854	11.587	9.609
35	24.99862	21.4872	18.6646	16.3742	14.4982	12.948	11.655	9.644
36	25.48884	21.8323	18.9083	16.5469	14.6210			
37	25.96945	22.1672	19.1426	16.7113	14.7368			
38	26.44064	22.4925	19.3679	16.8679	14.8460			
39	26.90259	22.8082	19.5845	17.0170	14.9491			
40	27.35548	23.1148	19.7928	17.1591	15.0463			
41	27.79949	23.4124	19.9931	17.2944	15.1380			
42	28.23479	23.7014	20.1856	17.4232	15.2245			
43	28.66156	23.9819	20.3708	17.5459	15.3062			
44	29.07996	24.2543	20.5488	17.6628	15.3832			
45	29.49016	24.5187	20.7200	17.7741	15.4558			
46	29.89231	24.7754	20.8847	17.8801	15.5244			
47	30.28658	25.0247	21.0429	17.9810	15.5890			
48	30.67312	25.2667	21.1951	18.0772	15.6500			
49	31.05208	25.5017	21.3415	18.1687	15.7076			
50	31.42361	25.7298	21.4822	18.2559	15.7619			

Table 5   (*Continued*)

n	12%	15%	20%	25%	30%	35%	40%	45%
1	0.893	0.870	0.833	0.800	0.769	0.741	0.714	0.690
2	1.690	1.626	1.528	1.440	1.361	1.289	1.224	1.165
3	2.402	2.283	2.106	1.952	1.816	1.696	1.589	1.493
4	3.037	2.855	2.589	2.362	2.166	1.997	1.849	1.720
5	3.605	3.352	2.991	2.689	2.436	2.220	2.035	1.876
6	4.111	3.784	3.326	2.951	2.643	2.385	2.168	1.983
7	4.564	4.160	3.605	3.161	2.802	2.507	2.263	2.057
8	4.968	4.487	3.837	3.329	2.925	2.598	2.331	2.109
9	5.328	4.772	4.031	3.463	3.019	2.665	2.379	2.144
10	5.650	5.019	4.192	3.571	3.092	2.715	2.414	2.168
11	5.938	5.234	4.327	3.656	3.147	2.752	2.438	2.185
12	6.194	5.421	4.439	3.725	3.190	2.779	2.456	2.196
13	6.424	5.583	4.533	3.780	3.223	2.799	2.469	2.204
14	6.628	5.724	4.611	3.824	3.249	2.814	2.478	2.210
15	6.811	5.847	4.675	3.859	3.268	2.825	2.484	2.214
16	6.974	5.945	4.730	3.887	3.283	2.834	2.489	2.216
17	7.120	6.047	4.775	3.910	3.295	2.840	2.492	2.218
18	7.250	6.128	4.812	3.928	3.304	2.844	2.494	2.219
19	7.366	6.198	4.844	3.942	3.311	2.848	2.496	2.220
20	7.469	6.259	4.870	3.954	3.316	2.850	2.497	2.221
21	7.562	6.312	4.891	3.963	3.320	2.852	2.498	2.221
22	7.645	6.359	4.909	3.970	3.323	2.853	2.498	2.222
23	7.718	6.399	4.925	3.976	3.325	2.854	2.499	2.222
24	7.784	6.434	4.937	3.981	3.327	2.855	2.499	2.222
25	7.843	6.464	4.948	3.985	3.329	2.856	2.499	2.222
26	7.896	6.491	4.956	3.988	3.330	2.856	2.500	2.222
27	7.943	6.514	4.964	3.990	3.331	2.856	2.500	2.222
28	7.984	6.534	4.970	3.992	3.331	2.857	2.500	2.222
29	8.022	6.551	4.975	3.994	3.332	2.857	2.500	2.222
30	8.055	6.566	4.979	3.995	3.332	2.875	2.500	2.222
31	8.085	6.579	4.982	3.996	3.332	2.857	2.500	2.222
32	8.112	6.591	4.985	3.997	3.333	2.857	2.500	2.222
33	8.135	6.600	4.988	3.997	3.333	2.857	2.500	2.222
34	8.157	6.609	4.990	3.998	3.333	2.857	2.500	2.222
35	8.176	6.617	4.992	3.998	3.333	2.857	2.500	2.222
∞	8.333	6.667	5.000	4.000	3.333	2.857	2.500	2.222

Table 6 Annuity with Present Worth of Unity, $\dfrac{i}{1-(1+i)^{-n}}$

n	2%	3%	4%	5%	6%	7%	8%	10%
1	1.020000	1.030000	1.040000	1.050000	1.060000	1.07000	1.08000	1.10000
2	0.515050	0.522611	0.530196	0.537805	0.545437	0.55309	0.56077	0.57619
3	0.346755	0.353530	0.360349	0.367209	0.374110	0.38105	0.38803	0.40211
4	0.262624	0.269027	0.275490	0.282012	0.288591	0.29523	0.30192	0.31547
5	0.212158	0.218355	0.224627	0.230975	0.237396	0.34389	0.25046	0.26380
6	0.178526	0.184598	0.190762	0.197017	0.203363	0.20980	0.21632	0.22961
7	0.154512	0.160506	0.166610	0.172820	0.179135	0.18555	0.19207	0.20541
8	0.136510	0.142456	0.148528	0.154722	0.161036	0.16747	0.17401	0.18744
9	0.122515	0.128434	0.134493	0.140690	0.147022	0.15349	0.16008	0.17364
10	0.111327	0.117231	0.123291	0.129505	0.135868	0.14238	0.14903	0.16275
11	0.102178	0.108077	0.114149	0.120389	0.126793	0.13336	0.14008	0.15396
12	0.094560	0.100462	0.106552	0.112825	0.119277	0.12590	0.13270	0.14676
13	0.088118	0.094030	0.100144	0.106456	0.112960	0.11965	0.12652	0.14078
14	0.082602	0.088526	0.094669	0.101024	0.107585	0.11434	0.12130	0.13575
15	0.077826	0.083767	0.089941	0.096342	0.102963	0.10979	0.11683	0.13147
16	0.073650	0.079611	0.085820	0.092270	0.098952	0.10586	0.11298	0.12782
17	0.069970	0.075953	0.082199	0.088699	0.095445	0.10243	0.10963	0.12466
18	0.066702	0.072709	0.078993	0.085546	0.092357	0.09941	0.10670	0.12193
19	0.063782	0.069814	0.076139	0.082745	0.089621	0.09675	0.10413	0.11955
20	0.061157	0.067216	0.073582	0.080243	0.087185	0.09439	0.10185	0.11746
21	0.058785	0.064872	0.071280	0.077996	0.085005	0.09229	0.09983	0.11562
22	0.056631	0.062747	0.069199	0.075971	0.083046	0.09041	0.09803	0.11401
23	0.054668	0.060814	0.067309	0.074137	0.081278	0.08871	0.09642	0.11257
24	0.052871	0.059047	0.065587	0.072471	0.079679	0.08719	0.09498	0.11130
25	0.051220	0.057428	0.064012	0.070952	0.078227	0.08581	0.09368	0.11017
26	0.049699	0.055938	0.062567	0.069564	0.076904	0.08456	0.09251	0.10916
27	0.048293	0.054564	0.061239	0.068292	0.075697	0.08343	0.09145	0.10826
28	0.046990	0.053293	0.060013	0.067123	0.074593	0.08239	0.09049	0.10745
29	0.045778	0.052115	0.058880	0.066046	0.073580	0.08145	0.08962	0.10673
30	0.044650	0.051019	0.057830	0.065051	0.072649	0.08059	0.08883	0.10608

Table 6 (Continued)

n	12%	15%	20%	25%	30%	35%	40%	45%
1	1.12000	1.15000	1.20000	1.25000	1.30000	1.35000	1.40000	1.45000
2	0.59170	0.61512	0.65455	0.69444	0.73478	0.77553	0.81667	0.85816
3	0.41635	0.43798	0.47473	0.51230	0.55063	0.58966	0.62936	0.66966
4	0.32923	0.35027	0.38629	0.42344	0.46163	0.50076	0.54077	0.58156
5	0.27741	0.29832	0.33438	0.37185	0.41058	0.45046	0.49136	0.53318
6	0.24323	0.26424	0.30071	0.33882	0.37839	0.41926	0.46126	0.50426
7	0.21912	0.24036	0.27742	0.31634	0.35687	0.39880	0.44192	0.48607
8	0.20130	0.22285	0.26061	0.30040	0.34192	0.38489	0.42907	0.47427
9	0.18768	0.20957	0.24808	0.28876	0.33124	0.37519	0.42034	0.46646
10	0.17698	0.19925	0.23852	0.28007	0.32346	0.36832	0.41432	0.46123
11	0.16842	0.19107	0.23110	0.27349	0.31773	0.36339	0.41013	0.45768
12	0.16144	0.18448	0.22526	0.26845	0.31345	0.35982	0.40718	0.45527
13	0.15568	0.17911	0.22062	0.26454	0.31024	0.35722	0.40510	0.45362
14	0.15087	0.17469	0.21689	0.26150	0.30782	0.35532	0.40363	0.45249
15	0.14682	0.17102	0.21388	0.25912	0.30598	0.35393	0.40259	0.45172
16	0.14339	0.16795	0.21144	0.25724	0.30458	0.35290	0.40185	0.45118
17	0.14046	0.16537	0.20944	0.25576	0.30351	0.35214	0.40132	0.45081
18	0.13794	0.16319	0.20781	0.25459	0.30269	0.35158	0.40094	0.45056
19	0.13576	0.16134	0.20646	0.25366	0.30207	0.35117	0.40067	0.45039
20	0.13388	0.15976	0.20536	0.25292	0.30159	0.35087	0.40048	0.45027

Table 7. Annuity Which Will Amount to 1 (Sinking Fund)

$$\frac{i}{(1+i)^n - 1}$$

n	1 1/2%	2%	2 1/2%	3%	3 1/2%	4%	4 1/2%	5%	5 1/2%	6%
1	1.000000	1.000000	1.000000	1.000000	1.000000	1.000000	1.000000	1.000000	1.000000	1.000000
2	0.496278	0.495049	0.493827	0.492611	0.491400	0.490196	0.488998	0.487805	0.486618	0.485437
3	0.328383	0.326755	0.325137	0.323530	0.321934	0.320349	0.318773	0.317209	0.315654	0.314110
4	0.244445	0.242624	0.240818	0.239027	0.237251	0.235490	0.233744	0.232012	0.230294	0.228591
5	0.194089	0.192158	0.190247	0.188355	0.186481	0.184627	0.182792	0.180975	0.179176	0.177396
6	0.160525	0.158526	0.156550	0.154598	0.152668	0.150762	0.148878	0.147017	0.145179	0.143363
7	0.136556	0.134512	0.132495	0.130506	0.128544	0.126610	0.124701	0.122820	0.120964	0.119135
8	0.118584	0.116510	0.114467	0.112456	0.110477	0.108528	0.106610	0.104722	0.102864	0.101036
9	0.104610	0.102515	0.100457	0.098434	0.096446	0.094493	0.092574	0.090690	0.088839	0.087022
10	0.093434	0.091327	0.089259	0.087231	0.085241	0.083291	0.081379	0.079505	0.077668	0.075868
11	0.084294	0.082178	0.080106	0.078077	0.076092	0.074149	0.072248	0.070389	0.068571	0.066793
12	0.076680	0.074560	0.072487	0.070462	0.068484	0.066552	0.064666	0.062825	0.061029	0.059277
13	0.070240	0.068118	0.066048	0.064030	0.062062	0.060144	0.058275	0.056456	0.054684	0.052960
14	0.064723	0.062602	0.060536	0.058526	0.056571	0.054669	0.052820	0.051024	0.049279	0.047585
15	0.059944	0.057825	0.055766	0.053767	0.051825	0.049941	0.048114	0.046342	0.044626	0.042963
16	0.055765	0.053650	0.051599	0.049611	0.047685	0.045820	0.044015	0.042270	0.040583	0.038952
17	0.052080	0.049970	0.047928	0.045953	0.044043	0.042199	0.040418	0.038699	0.037042	0.035445
18	0.048806	0.046702	0.044670	0.042709	0.040817	0.038993	0.037237	0.035546	0.033920	0.032357
19	0.045878	0.043782	0.041760	0.039814	0.037940	0.036139	0.034407	0.032745	0.031150	0.029621
20	0.043246	0.041157	0.039147	0.037216	0.035361	0.033582	0.031876	0.030243	0.028679	0.027185
21	0.040865	0.038785	0.036787	0.034872	0.033037	0.031280	0.029601	0.027996	0.026465	0.025005
22	0.038703	0.036631	0.034646	0.032747	0.030932	0.029199	0.027546	0.025971	0.024471	0.023046
23	0.036731	0.034668	0.032696	0.030814	0.029019	0.027309	0.025682	0.024137	0.022670	0.021278
24	0.034924	0.032871	0.030913	0.029047	0.027273	0.025587	0.023987	0.022471	0.021036	0.019679
25	0.033263	0.031220	0.029276	0.027428	0.025674	0.024012	0.022439	0.020952	0.019549	0.018227
26	0.031732	0.029699	0.027769	0.025938	0.024205	0.022567	0.021021	0.019564	0.018193	0.016904
27	0.030315	0.028293	0.026377	0.024564	0.022852	0.021239	0.019719	0.018292	0.016952	0.015697
28	0.029001	0.026990	0.025088	0.023293	0.021603	0.020013	0.018521	0.017123	0.015814	0.014593
29	0.027779	0.025778	0.023891	0.022115	0.020445	0.018880	0.017415	0.016046	0.014769	0.013580
30	0.026639	0.024650	0.022778	0.021019	0.019371	0.017830	0.016392	0.015051	0.013805	0.012649
31	0.025574	0.023596	0.021739	0.019999	0.018372	0.016855	0.015443	0.014132	0.012917	0.011792
32	0.024577	0.022611	0.020768	0.019047	0.017442	0.015949	0.014563	0.013280	0.012095	0.011002
33	0.023641	0.021687	0.019859	0.018156	0.016572	0.015104	0.013745	0.012490	0.011335	0.010273
34	0.022762	0.020819	0.019007	0.017322	0.015760	0.014315	0.012982	0.011755	0.010630	0.009598
35	0.021934	0.020002	0.018206	0.016539	0.014998	0.013577	0.012270	0.011072	0.009975	0.008974
36	0.021152	0.019233	0.017452	0.015804	0.014284	0.012887	0.011606	0.010434	0.009366	0.008395
37	0.020414	0.018507	0.016741	0.015112	0.013613	0.012240	0.010984	0.009840	0.008800	0.007857
38	0.019716	0.017821	0.016070	0.014459	0.012982	0.011632	0.010402	0.009284	0.008272	0.007358
39	0.019055	0.017171	0.015436	0.013844	0.012388	0.011061	0.009856	0.008765	0.007780	0.006894
40	0.018427	0.016556	0.014836	0.013262	0.011827	0.010523	0.009343	0.008278	0.007320	0.006462
41	0.017831	0.015972	0.014268	0.012712	0.011298	0.010017	0.008862	0.007822	0.006891	0.006059
42	0.017264	0.015417	0.013728	0.012192	0.010798	0.009540	0.008409	0.007395	0.006489	0.005683
43	0.016725	0.014890	0.013217	0.011698	0.010325	0.009090	0.007982	0.006993	0.006113	0.005333
44	0.016210	0.014388	0.012730	0.011230	0.009878	0.008665	0.007581	0.006616	0.005761	0.005006
45	0.015720	0.013910	0.012268	0.010785	0.009453	0.008262	0.007202	0.006262	0.005431	0.004701
46	0.015251	0.013453	0.011827	0.010363	0.009051	0.007882	0.006845	0.005928	0.005122	0.004415
47	0.014803	0.013018	0.011407	0.009961	0.008669	0.007522	0.006507	0.005614	0.004831	0.004148
48	0.014375	0.012602	0.011006	0.009578	0.008306	0.007181	0.006189	0.005318	0.004559	0.003898
49	0.013965	0.012204	0.010623	0.009213	0.007962	0.006857	0.005887	0.005040	0.004302	0.003664
50	0.013572	0.011823	0.010258	0.008865	0.007634	0.006550	0.005602	0.004777	0.004061	0.003444

For larger interest rates, obtain the sinking fund factor from Table 6 (Annuity with Present Worth of Unity) by subtracting the interest rate from the factor in Table 6.

Table 8 Present Worth of Unit R Uniformly Distributed
Over Period With Continuous Interest, $\dfrac{(e^r - 1)}{r} e^{-rn}$

Period	1%	5%	10%	15%	20%	25%	30%	35%	40%	50%
0–1	0.995	0.975	0.952	0.929	0.906	0.885	0.864	0.844	0.824	0.787
1–2	0.985	0.928	0.861	0.799	0.742	0.689	0.640	0.595	0.552	0.477
2–3	0.975	0.883	0.779	0.688	0.608	0.537	0.474	0.419	0.370	0.290
3–4	0.966	0.840	0.705	0.592	0.497	0.418	0.351	0.295	0.248	0.176
4–5	0.956	0.799	0.638	0.510	0.407	0.326	0.260	0.208	0.166	0.106
5–6	0.946	0.760	0.577	0.439	0.333	0.254	0.193	0.147	0.112	0.065
6–7	0.937	0.723	0.522	0.378	0.273	0.197	0.143	0.103	0.075	0.039
7–8	0.928	0.687	0.473	0.325	0.224	0.154	0.106	0.073	0.050	0.024
8–9	0.918	0.654	0.428	0.280	0.183	0.120	0.078	0.051	0.034	0.014
9–10	0.909	0.622	0.387	0.241	0.150	0.093	0.058	0.036	0.022	0.009

Table 9 Equivalent Values of Periodic and Continuous Interest

Periodic $i\%$	Nominal i or $r\%$	Continuous $r\%$	Periodic $i\%$	Nominal i or $r\%$	Continuous $r\%$
1.010	1	0.995	12.75	12	11.333
2.020	2	1.980	16.18	15	13.976
3.045	3	2.956	22.14	20	18.232
4.081	4	3.922	28.40	25	22.314
5.127	5	4.879	34.99	30	26.236
6.184	6	5.827	41.91	35	30.104
7.251	7	6.766	49.18	40	33.647
8.328	8	7.696	56.83	45	37.156
10.52	10	9.531	64.87	50	40.546

Table 10. Expectation of Life and Mortality Rate During Year of Age
(1970 U.S. Whole Population)*

Age	Years Life Expectation	Mortality Rate per 1000	Age	Years Life Expectation	Mortality Rate per 1000
0	70.5	20.4	45	29.9	4.8
5	67.3	0.6	50	25.7	7.5
10	62.4	0.3	55	21.8	11.6
15	57.6	0.8	60	18.1	17.1
20	52.9	1.3	65	14.8	25.6
25	48.2	1.3	70	11.8	38.7
30	43.5	1.5	75	9.2	55.8
35	38.9	2.1	80	8.4	81.5
40	34.3	3.1	85+	5.3	179.

* Source: U.S. Public Health Service, *Vital Statistics of the United States*.

Table 11. Cost Index Numbers, 1957–1959 = 100

Year	Consumers Prices*	Medical Care*	Wholesale Commodities*	Building Materials*	Building Labor*	Building Costs†	Construction Costs†
1940	49	50	43			41	36
1945	63	58	58	52	44	50	46
1950	84	73	87	83	68	71	67
1955	93	89	93	95	87	89	87
1960	103	108	101	100	109	106	108
1965	110	136	103	101	122	118	127
1970	135	165	117	127	180	158	181
1971	142		121	135	206	180	210

* U.S. Bureau of Labor Statistics, monthly and annual reports.
† U.S. Department of Commerce, *Construction Review*.

BIBLIOGRAPHY

1. ALLEN, R. G. C., *Mathematical Analysis for Economists*, London, Macmillan, 1947.
2. BARISH, N. N., *Economic Analysis for Engineering and Managerial Decision Making*, New York, McGraw-Hill, 1963.
3. *Cost Engineers Notebook*, American Association of Cost Engineers.
4. GRANT, E. L., and IRESON, W. G., *Principles of Engineering Economy*, New York, Ronald, 4th ed., 1960.
5. HOCKNEY, J. W., *Control and Management of Capital Projects*, New York, Wiley, 1965.
6. JELEN, E. C., *Cost and Optimization Engineering*, New York, McGraw-Hill, 1970.
7. RAUTENSTRAUCH, W., *The Economics of Business Enterprise*, New York, Wiley, 1939.
8. SCHWEYER, H. E., *Analytic Models for Managerial and Engineering Economics*, New York, Reinhold, 1964.
9. SOLOMON, E., *The Theory of Financial Management*, New York, Columbia Univ. Press, 1964.
10. *Wholesale Prices and Price Indexes*, Washington, D.C., Bureau of Labor Statistics, U.S. Department of Labor.

SECTION 16

PROPERTIES OF MATERIALS

BY

SIDNEY C. SINGER, JR.

Table 1. Physical Properties of Elemental Metals

Name	Atomic Weight	Density G per cu cm	Density Temp,* C	Melting Point, C	Latent Heat of Fusion, g-cal per g	Boiling Point, C	Latent Heat of Vaporization, g-cal per g	Specific Heat G-cal per g or Btu per lb	Specific Heat Temp, C	Thermal Coeff of Linear Expansion ×10^4 per °C	Thermal Expansion Temp, C	Thermal Conductivity G-cal per sec per sq cm per °C per cm	Thermal Conductivity Temp, C	Electrical Resistivity Microhm-cm	Electrical Resistivity Temp,* C	Temp Coeff of Resistivity Temp Coefficient	Temp Coeff Temp, C
Aluminum	26.97	2.70	20	657.1	93.0	2056	1950-2000	.226	0-100	.2545	20-300	.52	50	2.655	20	.00445	0-100
Antimony	121.76	6.618	20	630.5	38.26	1440	373	.0493	20-100	∥ .1129, ⊥ .080	20	.0444	0	39.1	0	.0036	20
Arsenic	74.91	5.73	14	814		615 Sublimes	74	.0822	18					35	0	.0042	20
Barium	137.36	3.5	20	850	345.5	1637	628	.068	-185 to +20	.0386	20			9.8	20	.0033	20
Beryllium	9.02	1.85	20	1285	318	2780		.425	0-100	.123	20			18.5	20		
Bismuth	209.00	9.781	20	271.3	12.46	1450	221	.029	20	∥ .1345, ⊥ .103	20	.0200	18	115.0	18	.004	20
Cadmium	112.41	8.648	20	320.9	13.17	766	227.5	.0547	27.9	∥ .54, ⊥ .20	0	.217	18	7.59	18	.0042	0-100
Calcium	40.08	1.55	20	851	78	1487		.157	0-100	.25	0-21			4.59	20	.00364	0-600
Cerium	140.13	6.90	20	775		1400		.0511	20-100					78	20		
Cesium	132.91	1.873	20	26.0	3.76	836.6	131.4	.0521	0-26	.97	0-26			20	0	.00478	-80 to +25
Chromium	52.01	7.14	25	1550	31.75	2482	14.71	.12	18-100	.081	20-100	.165		13.1	0		
Cobalt	58.94	8.9	21	1480	58.38	2900		.0989	20	.1208	20	.165		9.7	20	.00658	0-100
Copper	63.57	8.94	20	1083	50.6	2595	1756	.0918	18-100	.1642	20	.923	18	1.682	20	.00382	20
Gold	197.2	19.3	20	1063	16.11	2966	446	.0308	18	.144	17-100	.707	17	2.42	20	.0034	20
Iridium	193.1	22.42	17	2408.8	26.1	4900	340	.0323	18-100	.0641	20	.141	17	6.08	0	.00411	0-100
Iron (99.97%)	55.84	7.87	20	1535	65	2998	1110	.1075	20	.119	0-100	.19	0	9.8	20	.0065	0-100
Lead	207.21	11.342	20	327.4	6.26	1744	323	.0297	0	.295	20-100	.083	18	20.65	0	.00422	18
Lithium	6.94	0.534	20	186	32.81	1372		.79	50	.56	0-100			8.55	20	.0047	0
Magnesium	24.32	1.74	20	651	64.8	1107	1300-1500	.249	0-100	.257	0-178	.167	0	4.4611	0	.0040	20
Manganese	54.93	7.44	20	1242	64.8	2151	1044	.107	20-100	.228	20-300	.376	0-100	5.0	20		

Table 1. Physical Properties of Elemental Metals—*Continued*

Name	Atomic Weight	Density G per cu cm	Density Temp* C	Melting Point, C	Latent Heat of Fusion, g-cal per g	Boiling Point, C	Latent Heat of Vaporization, g-cal per g	Specific Heat G-cal per g or Btu per lb	Specific Heat Temp, C	Thermal Coefficient of Linear Expansion ×10^4 per °C	Lin. Exp. Temp, C	Thermal Conductivity G-cal per sec cm per °C per cm	Therm. Cond. Temp, C	Electrical Resistivity Microhm-cm	Elec. Res. Temp* C	Temp Coefficient of Resistivity	Temp Coeff. Temp, C
Mercury	200.61	13.546	20	−38.87	2.66	356.9	71	0.0332	17	†		0.020	0	95.783	20	0.0089	20
Molybdenum	95.95	10.2		2620		4803	176.8	.0647	0	.0549	25–100	.346	17	4.77	0	.0034	0–100
Nickel	58.69	8.85	20	1452	73.8	2900	1010	.112	0	.137	25–100	.14	0–100	6.9	20	.006	20
Osmium	190.82	22.48	20	2700		5493	350	.0311	19.98	.057	40			9.0	20		
Palladium	106.7	12.0	20	1555	34.2	3980	610	.0587	0	.1160	20	.161	18	10	20	.0033	20
Platinum	195.23	21.45	20	1773.5	27.1	4389	637	.0319	20–100	.088	20	.1664	18	9.83	0	.003	0
Potassium	39.10	0.87	20	63.5	14.5	773.9	513	.177	3.4	.83	0–50	.237	0	7.0	0	.0055	0
Rhodium	102.91	12.44	20	1966		4500	620	.0598	10–97	.089	6–21	.213	18	4.93	0	.0043	0–100
Silver	107.88	10.5	20	960.5	24.3	2001	551.6	.0558	0	.189	0–100	.974	0	1.629	18	.0038	20
Sodium	22.997	0.9712	20	97.5	27.53	892	1170	.295	0	.71	−190 to −17	.3225	0	4.6	0	.044	0
Strontium	87.63	2.60		771		1383	1045	.0735	15					22.76	20	.0031	20
Tantalum	180.88	16.6	20	2850		6093		.0356	58	.0655	0–100	.130	17	15.5	20		
Tellurium	127.61	6.24	20	452.2	7.305	1087	159	.0468	15–100	.168	20	.0143	45	200,000	19.6		
Thallium	204.39	11.85	20	303.3	7.185	1457	220	.0368	0–100	.280	40	.0915	0	18.1	0	.0040	0
Thorium	232.12	11.5	17	1845		5200		.0276	0–100					18	20	.0021	20–1800
Tin	118.70	7.30		231.89	14.4	2270	655	.548	25	(a) ∥ .224 / (a) ⊥ .464	20 / 20	.157	0	11.5	20	.0042	20
Titanium	47.90	4.5	18	1800		5100	1320	.142	0–100					3.0	20	.045	18
Tungsten	183.92	19.3		3370	44.0	5927	1183	.034	100	.0714	27	.476	17	5.48	20		
Uranium	238.07	18.7	13	1690		3500		.0276	0–98	.0444				60 ± 18	20		
Vanadium	50.95	5.68		1710		3000		.1153	0–100					26.0	0	.0037	20
Zinc	65.38	7.14		419.45	24.09	907.2	426.8	.0931	20–100	(b)	20–100	.268	18	(c)			
Zirconium	91.22	6.43		1700		5050		.063	0–100	(b)	20–100			141 ±		.00116	−80 to +30

* Where the temperature is not given, ordinary temperature is understood.

† $\frac{1}{v}\frac{dv}{dt} = 182.0 \times 10^{-6}$ at 20 C.

(a) Single crystal: 22.4 ∥ crystal axis; 46.4 ⊥ crystal axis.   (b) 32.5 pure, hot-rolled zinc with grain; 23.0 pure, hot-rolled zinc across rolling.   (c) Single crystal: 6.2 ∥ crystal axis; 5.8 ⊥ crystal axis.

Table 2. Chemical Composition of Some Common Metals*

Item No.	Metal	C	Mn	Fe	Cr	Ni	Cu	Si	Mo	Others
	Iron Base Alloys									
1	Steel, Low Carbon	0.10–0.30	0.50–0.80	Balance	—	—	—	0.10 min	—	
2	Cast Iron	2.75–3.50	0.50–0.80	Balance	—	—	—	1.90–2.25	—	
3	Ni Resist Type 1	3.0a	1.0 –1.5	Balance	1.75– 2.50	13.5–17.5	7.0	1.0 –2.5	—	
4a–h	Cr–Mo Steels	0.15a	1.0a	Balance	1.0 – 9.0	—	—	1.0a	0.5	
5	12 Cr Steel	0.15a	1.0a	Balance	11.50–13.0	—	—	0.5a	—	
6	Stainless—304	0.08a	2.0a	Balance	18.0 –20.0	8.0–11.0	—	1.0a	—	
7	Stainless—316	0.10a	2.0a	Balance	16.0 –18.0	10.0–14.0	—	1.0a	3.0	
8	Stainless—317	0.10a	2.0a	Balance	18.0 –20.0	11.0–14.0	—	—	4.0	
9	Worthite	0.07a	0.6	Balance	20.0	24.0	1.75	3.25	3.0	
10	Durimet 20	0.07a	0.65–0.85	Balance	19.0 –21.0	28.0–30.0	4.0	1.0	3.0	
11	25 Cr–12 Ni Steel	0.20a	2.0a	Balance	22.0 –24.0	12.0–15.0	—	1.0a	—	
12	25 Cr–20 Ni Steel	0.25a	2.0a	Balance	24.0 –26.0	19.0–22.0	—	1.5a	—	
13	Incoloy 800	0.10a	1.5a	Balance	19.0 –22.0	32.0–36.0	0.50	1.0a	—	Al 0.38; Ti 0.38
14	Silicon Iron	0.85a	0.50–0.65	Balance	—	—	—	14.5	—	
15	Durichlor	0.85a	0.65	Balance	—	—	—	14.5	3.0	
	Copper Base Alloys									
16	Copper	—	—		—	—	99.9	—	—	
17	Tin Bronze	—	—	—	—	—	90.0	—	—	Sn 10.0
18	Aluminum Bronze	—	—	—	—	—	95.0	—	—	Al 5.0
19	Ampco No. 18	—	—	3.5	—	—	Bal.	—	—	Al 11.0
20	Red Brass	—	—	—	—	—	85.0	—	—	Zn 15.0
21	Yellow Brass	—	—	—	—		65.0	—	↑	Zn 32.0; Sn 1.0; Pb 2.0
22	Muntz Metal	—	—	—	—		60.0	—	—	Zn 40.0
23	Admiralty	—	—	0.06a	—		70.0	—	—	Sn 1.0; Zn Balance
24	Silicon Bronze	—	—	—	—		97.0	3.0	—	
25	70 Cu–30 Ni	—	—	—	—	30.0	70.0	—	—	

No.	Alloy									Other
	Nickel Base Alloys									
26	Monel 400	0.15a	1.0	1.4	—	67.0	30.0	0.1a	—	Al 2.75; Ti 0.75
27	Monel K500	0.15a	0.75	0.9	—	66.0	29.0	0.5a	—	—
28	Nickel 200	0.05a	0.20	0.15	—	99.4	0.10	0.05	—	—
29	Inconel 600	0.08a	0.25	7.0	15.0	77.0	0.20	0.25	—	—
30	Hastelloy B	0.12a	1.0a	4.0–7.0	1.0a	60.0–65.0	—	1.0	28.0	Co 2.5a
31	Hastelloy C	0.15a	1.0a	4.0–7.0	13.0–16.0	55.0–60.0	—	1.0a	17.0	W 3.5–5.5; Co 2.5a
32	Hastelloy D	0.12a	0.80–1.25	1.0a	1.0a	Balance	4.0	8.5–10.0	—	—
33	Hastelloy F	0.05a	1.0–2.0	Balance	21.0–23.0	44.0–47.0	—	1.0a	6.5	W 1.0; Cb 2.0; Co 2.5a
34	Hastelloy X	0.15a	0.75a	Blanace	22.0	45.0	—	—	9.0	Co 1.5
	Super Alloys									
35	Nimonic 80	0.04	0.56	Balance	21.0	74.0	—	0.47	—	Ti 2.5; Al 0.6
36	Inconel X 750	0.08	0.53	6.0	13.0	70.0	—	0.40	—	Cb 1.0; Ti 2.5; Al 0.6
37	M-252	0.15	1.00	Balance	19.0	48.5	—	0.70	10.0	W 7.5; Co 10.0; Ti 2.5
38	Refractaloy 26	0.05	0.70	17.0	18.0	37.0	—	0.80	3.0	Co 20; Ti 2.8; Al 0.2
39	A-286	0.05	1.35	1.0	15.5	26.0	—	0.95	1.25	Ti 1.9; Al 0.2; Va 0.3
40	Discaloy	0.03	0.47	Balance	13.0	25.0	—	0.50	3.0	Ti 2.5; Al 0.2
41	N155 Alloy	0.10	1.50	Balance	20.0	20.0	—	0.50	3.0	W 2.0; Co 20.0; Cb 1.0
42	S-590	0.44	0.60	Balance	20.0	20.0	—	0.40	4.0	W 4.0; Co 20.0; Cb 4.0
43	S-816	0.38	1.30	2.95	20.0	20.0	—	0.59	4.0	W 4.0; Co 40.0; Cb 4.0
44	Haynes Alloy 31	0.50	1.0a	2.0a	25.0	10.0	—	1.0a	—	W 7.5; Co Balance
45	Haynes Alloy 25	0.08	1.50	3.0a	20.0	10.0	—	1.0a	—	W 15.0; Co Balance
46	Haynes Alloy 21	0.24	1.0a	2.0a	28.0	2.5	—	1.0a	6.0	B 0.007; Co Balance
47	16 Cr-25 Ni-6 Mo	0.15	1.14	Balance	16.0	25.0	—	0.57	6.0	N 0.13
48	19-9 DL	0.26	0.52	Balance	19.0	9.0	—	—	1.2	W 1.2; Cb 0.29; Ti 0.21
49	17-4 PH	0.05	—	—	16.5	4.0	4.0	—	—	Cb 0.25
	Other Metals									
50	Aluminum	—	—	—	—	—	—	—	—	Al 99.6
51	Lead	—	—	—	—	—	—	—	—	Pb 99.9
52	Titanium	0.20a	—	—	—	—	—	—	—	Ti 99.0 min
53	Zirconium	—	—	—	—	—	—	—	—	Zi 99.5
54	Tantalum	—	—	—	—	—	—	—	—	Ta 99.8 min

* Courtesy of Shell Development Company.
a Maximum.

Table 3. Physical Properties of Some Common Metals*

Item No.	Metal	Melting Point		Elect. Resist., Microhms/Cm³, 70°F	Density			Spec. Heat, Cal/G/°C or BTU/Lb/°F	Brinell Hardness (Annealed)	Mach. Rating, SAE 1112 = 100
		°F	°C		G/Cm³	Lb/Cu Ft	Lb/Cu In.			
	Iron Base Alloys									
1	Steel, Low Carbon	2,760	1,516	10	7.86	491	.284	0.11	130	70
2	Cast Iron	2,150	1,176	66	7.20	449	.260	0.13	180	50
3	Ni Resist—Type 1	2,250	1,231	140	7.30	456	.264	0.11	170	30
4a-h	Cr-Mo Steels	2,500	1,371	10	7.86	491	.284	0.11	180	70
5	12 Cr Steel	2,720	1,495	9	7.75	484	.280	0.11	150	70
6	Stainless 304	2,590	1,420	72	8.02	501	.290	0.12	160	50
7	Stainless 316	2,550	1,398	74	8.02	501	.290	0.12	165	50
8	Stainless 317	2,550	1,298	74	8.02	501	.290	0.12	165	50
9	Worthite Alloy 20	2,650	1,455	75	8.02	501		0.12	160	60
10	Durimet 20	2,650	1,454	75	8.02	501	.290	0.12	160	60
11	25 Cr-12 Ni Steel	2,650	1,454	78	8.02	501	.290	0.12	160	50
12	25 Cr-20 Ni Steel	2,650	1,454	78	8.02	501	.290	0.12	165	50
13	Incoloy 800	2,525	1,385	93	8.05	503	.291	0.12	184	45
14	Silicon Iron	2,300	1,260	63	7.00	437	.253	0.13	500	0
15	Durichlor	2,300	1,260	63	7.20	449	.260	0.13	500	0
	Copper Base Alloys									
16	Copper	1,980	1,080	2	8.91	556	.332	0.09	42	61
17	Tin Bronze	1,830	999	—	8.78	548	.317	0.09	60	130
18	Aluminum Bronze	1,900	1,038	19	7.78	486	.281	0.09	190	200
19	Ampco No. 18	1,900	1,038	19	7.60	474	.274	0.09	166	200
20	Red Brass	1,880	1,025	5	8.75	546	.316	0.09	50	100
21	Yellow Brass	1,710	931	7	8.47	529	.307	0.09	55	200
22	Muntz Metal	1,660	904	7	8.39	524	.303	0.09	80	200
23	Admiralty	1,720	937	7	8.53	533	.308	0.09	60	200
24	Silicon Bronze	1,870	1,019	27	8.53	533	.308	0.09	70	60
25	70 Cu-30 Ni	2,240	1,226	37	8.94	558	.323	0.09	70	120

No.	Material									
	Nickel Base Alloys									
26	Monel 400	2,460	1,350	48	8.83	551	.319	0.11	125	44
27	Monel K-500	2,460	1,348	48	8.47	529	.306	0.10	160	50
28	Nickel 200	2,640	1,449	7	8.91	556	.322	0.11	100	20
29	Inconel 600	2,600	1,426	103	8.43	526	.304	0.11	150	40
30	Hastelloy B	2,460	1,348	135	9.24	577	.334	0.09	230	23
31	Hastelloy C	2,380	1,303	133	8.94	558	.323	0.09	185	23
32	Hastelloy D	2,050	1,120	113	7.80	487	.282	0.11	400	0
33	Hastelloy F	2,350	1,286	112	8.20	512	.296	0.10	170	—
34	Hastelloy X	2,350	1,315	118	8.23	514	.297	0.11	183	30
	Super Alloys									
35	Nimonic 80	2,590	1,420	124	8.25	515	.298	0.10	185	—
36	Inconel X-750	2,600	1,425	122	8.25	515	.298	0.11	176	20
37	M-252	2,500	1,370	92	8.25	515	.298	0.10	250	
38	Refractaloy 26	2,450	1,344	91	8.21	513	.297	0.11	300	
39	A 286	2,550	1,400		7.94	496	.287	0.11		
40	Discaloy	2,520	1,380	98	7.37	498	.288	0.11	310	45
41	N-155 (carbon)	2,500	1,371	98	8.20	512	.296	0.10	185	—
42	S-590	2,400	1,315	—	8.36	519	.301	0.10		
43	S-816	2,400	1,315	93	8.36	541	.313	0.10		
44	Haynes Alloy 31	2,500	1,370	98	8.31	538	.311	0.10	340	
45	Haynes Alloy 25	2,570	1,409	88	9.15	571	.330	0.09	250	
46	Haynes Alloy 21	2,470	1,353	87	8.30	518	.300	0.11	237	
47	16 Cr–25 Ni–6 Mo	2,550	1,398	75	8.37	504	.292	0.11	165	
48	19-9 DL	2,600	1,427	78	7.34	496	.287	0.10	200	
49	17-4 PH	—	—	77	7.30	487	.282	0.13	400	—
	Other Metals									
50	Aluminum	1,220	660	3	2.72	170	.098	0.22	20	300–1500
51	Lead	621	327	21	11.55	709	.410	0.03	4	
52	Titanium	3,135	1,725	61	4.40	281	.163	0.13	150	500
53	Zirconium	3,355	1,845	40	6.45	403	.233	0.07	—	—
54	Tantalum	5,425	2,996	12	16.60	—	.600	0.04	100	—

* Courtesy of Shell Development Company.

Table 4. Tensile Strengths of Some Common Metals*

Tabular Values in 10^3 psi

Item No.	Metals	Temperature, °F													
		−325	−200	−100	70	200	400	600	800	1000	1200	1400	1600	1800	2000
	Iron Base Alloys														
1	Steel, Low Carbon	93	87	76	60	65	68	64	52	29	15	7	4	3	2
2	Cast Iron	—	—	—	—	—	—	—	—	—	—	—	—	—	—
3	Ni Resist—Type 1	—	—	—	—	—	—	—	—	—	—	—	—	—	—
4a	½ Mo	—	—	—	63	67	70	68	60	46	28	12	—	—	—
4b	1 Cr–½ Mo	—	—	—	65	67	68	66	62	52	32	11	—	—	—
4c	1¼ Cr–½ Mo	—	—	—	67	70	74	75	70	55	31	12	—	—	—
4d	2 Cr–½ Mo	—	—	—	65	—	—	65	60	46	29	11	—	—	—
4e	2¼ Cr–1 Mo	—	—	—	70	70	70	69	67	53	33	14	—	—	—
4f	5 Cr–½ Mo	—	—	—	71	67	62	58	54	44	25	13	—	—	—
4g	7 Cr–½ Mo	—	—	—	80	78	75	69	59	53	29	14	—	—	—
4h	9 Cr–1 Mo	—	—	—	84	82	77	72	64	53	30	13	—	—	—
5	12 Cr—Type 410	158	140	128	89	83	76	73	66	45	22	9	9	7	4
6	Stainless—304	232	200	168	85	79	74	73	69	58	45	30	16	9	6
7	Stainless—316	201	175	130	90	85	81	78	76	70	57	36	23	14	8
8	Stainless 317	—	—	—	—	—	—	—	—	—	—	—	—	—	—
9	Worthite Alloy 20	—	—	—	—	—	—	—	—	—	—	—	—	—	—
10	Durimet 20	—	—	—	—	—	—	—	—	—	—	—	—	—	—
11	25 Cr–12 Ni Steel	—	—	—	80	79	77	73	69	63	52	35	22	12	7
12	25 Cr–20 Ni Steel	—	—	—	88	86	83	81	77	70	58	41	27	17	10
13	Incoloy 800	—	—	—	82	—	77	76	75	72	54	32	—	—	—
14	Silicon Iron	—	—	—	—	—	—	—	—	—	—	—	—	—	—
15	Durichlor	—	—	—	—	—	—	—	—	—	—	—	—	—	—
	Copper Base Alloys														
16	Copper	50	44	37	32	31	23	20	17	9.0	4.5	—	—	—	—
17	Tin Bronze	—	—	—	—	—	—	—	—	—	—	—	—	—	—
18	Aluminum Bronze	102	93	85	68	63	57	38	20	11	—	—	—	—	—
19	Ampco No. 18	102	93	85	85	85	76	59	26	17	—	—	—	—	—
20	Red Brass	—	—	—	42	42	32	22	14	11	—	—	—	—	—

21	Yellow Brass	—	—	—	50	50	48	38	25	—	—	—	—	—	—
22	Muntz Metal	—	—	—	55	—	—	—	—	—	—	—	—	—	—
23	Admiralty	73	63	58	53	50	—	33	—	—	—	—	—	—	—
24	Silicon Bronze	92	182	172	58	55	50	40	27	17	—	—	—	—	—
25	70 Cu–30 Ni	—	—	—	55	—	47	—	—	—	—	—	—	—	—
	Nickel Base Alloys														
26	Monel 400	115	107	95	81	79	78	75	64	46	27	18	9	5	—
27	Monel K500	171	182	172	160	150	149	146	124	95	80	45	21	6.0	3.0
28	Nickel 200	110	95	82	67	67	67	66	44	32	22	14	—	—	—
29	Inconel 600	117	102	92	85	81	78	79	83	79	71	47	23	15	11
30	Hastelloy B	172	144	142	131	124	119	118	116	108	74	68	53	—	—
31	Hastelloy C	160	143	134	121	116	111	106	102	99	97	79	56	32	18
32	Hastelloy D	—	—	—	—	—	—	—	—	—	—	—	—	—	—
33	Hastelloy F	—	—	—	102	96	92	89	84	80	72	53	29	15	6.0
34	Hastelloy X	—	—	—	114	106	103	100	100	94	83	63	37	23	13
	Super Alloys														
35	Nimonic 80	—	—	—	155	—	134	128	—	—	113	89	40	12	—
36	Inconel X750	213	200	188	178	173	167	158	152	142	125	92	34	9.0	—
37	M-252	—	—	—	180	180	180	180	180	180	160	135	80	—	—
38	Refractaloy 26	—	—	—	154	153	151	148	145	143	136	105	48	—	—
39	A-286	205	185	172	145	141	138	138	138	132	104	62	—	—	—
40	Discaloy	—	—	—	145	145	142	137	129	125	104	75	—	—	—
41	N155	—	—	—	118	117	114	110	103	93	80	60	38	—	—
42	S-590	—	—	—	140	142	145	144	140	127	98	66	—	—	—
43	S-816	171	159	150	140	132	126	124	125	121	112	90	60	25	—
44	Haynes Alloy 31	—	—	—	172	163	149	135	121	105	91	74	—	—	—
45	Haynes Alloy 25	—	—	—	150	150	148	143	134	120	98	68	—	—	—
46	Haynes Alloy 21	—	—	—	154	153	150	143	133	120	100	78	—	—	—
47	16 Cr–25 Ni–6 Mo	—	—	—	107	105	101	98	93	86	72	50	—	—	—
48	19-9 DL	—	—	—	118	108	100	96	93	90	75	55	—	—	—
49	17-4 PH	260	240	220	200	185	175	160	150	95	40	—	—	—	—
	Other Metals														
50	Aluminum 5052-H34	50	45	30	28	28	18	7.5	—	—	—	—	—	—	—
51	Lead	—	—	—	2.5	—	—	—	—	—	—	—	—	—	—
52	Titanium A55	140	110	100	85	67	50	38	33	22	—	—	—	—	—
53	Zirconium	—	—	—	49	42	25	18	14	—	—	—	—	—	—
54	Tantalum	160	—	82	54	54	54	55	57	50	37	—	23	21	15

* Courtesy of Shell Development Company.

Table 5. Moduli of Elasticity of Some Common Metals*
Tabular Values in 10⁶ psi

Item No.	Metals	-325	-200	-100	70	200	400	600	800	1000	1200	1400	1600	1800	2000
	Iron Base Alloys														
1	Steel, Low Carbon	30.0	29.5	29.0	29.0	28.7	27.0	25.7	23.4	15.4	—	—	—	—	—
2	Cast Iron	—	—	—	13.4	13.2	12.6	11.7	10.2	—	—	—	—	—	—
3	Ni Resist—Type 1	—	—	—	—	—	—	—	—	—	—	—	—	—	—
4a	½ Mo	31.0	30.6	30.4	29.9	29.5	28.6	27.4	25.7	23.0	15.6	—	—	—	—
4b	1 Cr-½ Mo	31.0	30.6	30.4	29.9	29.5	28.6	27.4	25.7	23.0	15.6	—	—	—	—
4c	1¼ Cr-½ Mo	31.0	30.6	30.4	29.9	29.5	28.6	27.4	25.7	23.0	15.6	—	—	—	—
4d	2 Cr-½ Mo	31.0	30.6	30.4	29.9	29.5	29.6	27.4	25.7	23.0	15.6	—	—	—	—
4e	2¼ Cr-1 Mo	31.0	30.6	30.4	29.9	29.5	28.6	27.4	25.7	23.0	15.6	—	—	—	—
4f	5 Cr-½ Mo	29.4	28.5	28.1	27.4	27.0	26.4	25.4	24.2	22.8	20.8	18.1	—	—	—
4g	7 Cr-½ Mo	29.4	28.5	28.1	27.4	27.0	26.4	25.4	24.2	22.8	20.8	18.1	—	—	—
4h	9 Cr-1 Mo	29.4	28.5	28.1	27.4	27.0	26.4	25.4	24.2	22.8	20.8	18.1	—	—	—
5	12 Cr-Steel	30.8	30.3	29.8	29.2	28.7	27.7	26.0	23.1	18.6	12.2	—	—	—	—
6	Stainless—304	29.4	28.5	28.1	27.4	27.1	26.4	25.4	24.2	22.5	21.1	19.4	—	—	—
7	Stainless—316	29.4	28.5	28.1	28.1	28.1	26.9	25.4	24.2	22.8	21.5	20.0	—	—	—
8	Stainless—317	—	—	—	—	—	—	—	—	—	—	—	—	—	—
9	Worthite Alloy 20	—	—	—	—	—	—	—	—	—	—	—	—	—	—
10	Durimet 20	—	—	—	—	—	—	—	—	—	—	—	—	—	—
11	25 Cr-12 Ni Steel	—	—	28.6	28.2	28.2	26.8	25.5	23.1	22.6	21.8	20.5	19.2	—	—
12	25 Cr-20 Ni Steel	—	—	—	28.5	—	—	26.6	24.2	23.0	21.8	21.0	19.2	—	—
13	Incoloy 800	—	—	—	—	—	—	—	—	23.6	22.4	—	19.6	18.0	—
14	Silicon Iron	—	—	—	—	—	—	—	—	—	—	—	—	—	—
15	Durichlor	—	—	—	—	—	—	—	—	—	—	—	—	—	—
	Copper Base Alloys														
16	Copper	17.0	16.7	16.5	16.0	15.6	14.0	12.0	9.0	—	—	—	—	—	—
17	Tin Bronze	14.2	13.8	13.5	13.0	12.7	12.0	11.3	—	—	—	—	—	—	—
18	Aluminum Bronze	22.0	—	—	17.0	17.0	15.3	11.9	5.8	6.0	—	—	—	—	—
19	Ampco No. 18	22.0	—	—	17.0	17.0	15.3	11.9	7.5	—	—	—	—	—	—
20	Red Brass	—	—	—	15.0	—	—	—	—	—	—	—	—	—	—

21	Yellow Brass	15.0	14.7	14.5	14.0	13.7	13.0	12.2	—	—	—	—	—	—	—
22	Muntz Metal	15.0	14.7	14.5	14.0	13.7	10.0	—	—	—	—	—	—	—	—
23	Admiralty	—	—	—	15.0	15.0	13.5	7.5	—	—	—	—	—	—	—
24	Silicon Bronze	—	—	—	15.0	—	—	—	—	—	—	—	—	—	—
25	70 Cu–30 Ni	20.5	20.0	19.5	18.9	18.4	17.6	16.7	15.3	—	—	—	—	—	—
	Nickel Base Alloys														
26	Monel 400	26.8	26.6	26.4	26.0	26.8	25.6	25.6	24.8	23.7	22.6	21.3	18.3	—	—
27	Monel K500	—	—	—	26.0	—	—	—	—	—	—	—	—	—	—
28	Nickel 200	31.5	30.9	30.3	30.0	29.6	28.6	27.4	25.5	22.0	18.0	15.5	—	—	—
29	Inconel 600	32.6	31.9	31.5	31.0	31.0	31.0	29.5	28.0	25.0	20.0	17.0	14.0	—	—
30	Hastelloy B	—	—	—	31.0	—	—	—	—	—	—	—	—	—	—
31	Hastelloy C	—	—	—	29.8	29.8	29.8	29.8	29.8	24.8	24.5	22.7	19.5	15.0	10.5
32	Hastelloy D	—	—	—	—	—	—	—	—	—	—	—	—	—	—
33	Hastelloy F	—	—	—	29.0	29.0	29.0	29.0	27.7	24.8	20.8	19.5	18.9	—	—
34	Hastelloy X	—	—	—	28.6	28.6	23.8	25.5	21.4	24.3	22.5	20.1	18.7	9.1	6.1
	Super-Alloys														
35	Nimonic 80	—	—	—	27.0	—	—	—	—	27.0	—	—	14.0	6.0	—
36	Inconel X750	—	—	—	31.0	30.4	29.2	28.0	26.6	25.0	23.0	20.2	16.0	—	—
37	M-252	—	—	—	29.9	29.5	28.9	28.1	27.0	26.0	24.3	22.5	—	—	—
38	Refractaloy 26	—	—	—	30.6	—	—	—	—	26.3	25.0	23.0	19.5	—	—
39	A-286	29.6	—	28.4	29.1	28.4	27.2	26.2	24.8	23.5	22.2	20.6	18.9	—	—
40	Discaloy	—	—	—	28.0	27.0	25.0	24.0	23.0	22.0	21.0	—	—	—	—
41	N155 Alloy	—	—	—	29.3	28.8	27.6	26.3	25.2	24.2	—	—	—	—	—
42	S-590	—	—	—	31.1	—	—	—	—	—	24.6	—	—	—	—
43	S-816	—	—	—	35.2	34.5	33.3	32.2	31.0	29.9	28.4	27.1	25.1	—	—
44	Haynes Alloy 31	—	—	—	28.0	—	—	—	—	33.5	—	—	19.0	—	—
45	Haynes Alloy 25	—	—	—	32.6	32.1	31.0	29.6	28.5	27.1	25.0	23.5	22.0	21.0	—
46	Haynes Alloy 21	—	—	—	36.0	—	—	—	—	32.6	—	—	15.4	—	—
47	16 Cr–25Ni–6 Mo	—	—	—	32.5	—	—	—	—	—	17.9	—	—	—	—
48	19-9 DL	—	—	—	29.5	28.7	27.2	26.0	24.6	23.3	22.1	20.7	—	—	—
49	17-4 PH	29.8	29.0	28.7	28.5	27.5	26.0	24.8	23.3	—	—	—	—	—	—
	Other Metals														
50	Aluminum	11.3	11.1	10.9	10.6	10.4	9.5	—	—	—	—	—	—	—	—
51	Lead	—	—	—	2.0	—	—	—	—	—	—	—	—	—	—
52	Titanium A55	17.3	—	16.3	16.0	16.0	15.8	14.6	13.6	11.2	—	—	—	—	—
53	Zirconium	—	—	14.0	11.0	—	—	—	—	—	—	—	—	—	—
54	Tantalum	—	—	—	27.0	—	—	—	—	—	—	—	—	—	—

* Courtesy of Shell Development Company.

Table 6. Yield Strengths of Some Common Metals*

Tabular Values in 10³ psi

Item No.	Metals	Temperature, °F													
		-325	-200	-100	70	200	400	600	800	1000	1200	1400	1600	1800	2000
	Iron Base Alloys														
1	Steel, Low Carbon	—	—	—	40	38	35	29	23	17	10	3	—	—	—
2	Cast Iron	—	—	—	—	—	—	—	—	—	—	—	—	—	—
3	Ni Resist—Type 1	—	—	—	—	—	—	—	—	—	—	—	—	—	—
4a	½ Mo	—	—	—	40	38	36	33	29	23	15	5	—	—	—
4b	1 Cr–½ Mo	—	—	—	45	41	34	28	23	21	17	6	—	—	—
4c	1¼ Cr–½ Mo	—	—	—	48	48	46	42	34	26	17	7	—	—	—
4d	2 Cr–½ Mo	—	—	—	41	42	39	37	25	20	15	7	—	—	—
4e	2¼ Cr–1 Mo	—	—	—	43	28	26	24	33	28	21	8	—	—	—
4f	5 Cr–½ Mo	—	—	—	31	—	—	—	21	18	13	7	—	—	—
4g	7 Cr–½ Mo	—	—	—	56	—	—	—	—	37	20	6	—	—	—
4h	9 Cr–1 Mo	—	—	—	45	42	38	37	35	30	18	7.0	—	—	—
5	12 Cr—Type 405	148	115	94	47	46	44	41	35	24	13	5	2	—	—
6	Stainless—304	50	45	42	35	28	23	19	16	14	12	11	—	—	—
7	Stainless—316	108	63	55	42	39	35	31	27	24	21	18	—	—	—
8	Stainless 317	—	—	—	—	—	—	—	—	—	—	—	—	—	—
9	Worthite Alloy 20	—	—	—	—	—	—	—	—	—	—	—	—	—	—
10	Durimet 20	—	—	—	—	—	—	—	—	—	—	—	—	—	—
11	25 Cr–12 Ni Steel	119	88	63	57	55	50	46	41	36	32	26	—	—	—
12	25 Cr–20 Ni Steel	—	—	—	33	32	29	28	26	23	21	18	—	—	—
13	Incoloy 800	—	—	—	43	—	36	34	33	32	29	23	14	7	—
14	Silicon Iron	—	—	—	—	—	—	—	—	—	—	—	—	—	—
15	Duriehlor	—	—	—	—	—	—	—	—	—	—	—	—	—	—
	Copper Base Alloys														
16	Copper	12	11	10	10	6	6	—	—	—	—	—	—	—	—
17	Tin Bronze	—	—	—	20	—	—	—	—	—	—	—	—	—	—
18	Aluminum Bronze	—	—	—	40	33	33	33	20	10	—	—	—	—	—
19	Ampco No. 18	—	—	—	33	—	—	—	—	—	—	—	—	—	—
20	Red Brass	—	—	—	15	—	—	—	—	—	—	—	—	—	—

No.	Metal														
21	Yellow Brass	—	—	—	20	—	—	—	—	—	—	—	—	—	—
22	Muntz Metal	30	29	—	20	19	19	—	—	—	—	—	—	—	—
23	Admiralty	50	47	28	28	24	23	20	—	—	—	—	—	—	—
24	Silicon Bronze	—	—	44	42	40	—	—	—	—	—	—	—	—	—
25	70 Cu–30 Ni	—	—	—	22	—	—	—	—	—	—	—	—	—	—
	Nickel Base Alloys														
26	Monel 400	50	45	40	32	29	26	22	21	20	15	11	6.5	2.5	9.7
27	Monel K-500	153	143	134	111	108	103	105	105	92	80	30	—	—	—
28	Nickel 200	28	—	27	22	22	20	20	16	13	10	7	—	—	—
29	Inconel 600	—	—	40	36	32	28	27	28	22	22	19	39	—	—
30	Hastelloy B	83	61	59	56	52	46	42	42	42	42	40	38	18	8.0
31	Hastelloy C	96	79	65	58	55	51	47	45	44	43	41	—	—	—
32	Hastelloy D	—	—	—	—	—	—	—	—	—	—	—	—	—	—
33	Hastelloy F	—	—	—	52	—	—	—	—	—	—	—	26	16	—
34	Hastelloy X	—	—	—	52	51	49	43	44	42	40	38	—	—	—
	Super-Alloys														
35	Nimonic 80	131	—	—	87	—	81	—	—	—	77	68	30	8	10
36	Inconel X-750	—	125	120	122	116	112	110	108	107	105	92	74	55	14
37	M-252	—	—	—	122	120	118	117	115	111	108	104	70	—	13
38	Refractaloy 26	—	—	—	91	—	—	—	—	85	89	85	47	—	13
39	A-286	122	115	110	100	95	94	93	92	91	85	52	—	—	—
40	Discaloy	—	—	—	106	105	104	100	94	94	91	70	—	—	10
41	N155	—	—	—	57	56	55	53	51	48	43	35	—	—	—
42	S-590	—	—	—	75	78	82	83	82	78	70	58	—	—	—
43	S-816	110	95	—	67	58	46	44	44	44	44	40	—	—	—
44	Haynes Alloy 31	—	—	85	87	84	80	75	71	65	58	42	—	—	—
45	Haynes Alloy 25	—	—	—	70	67	64	60	56	52	47	41	34	15	—
46	Haynes Alloy 21	—	—	—	82	77	70	62	55	53	51	47	36	21	—
47	16 Cr–25 Ni–6 Mo	—	—	—	46	45	41	36	33	30	31	34	35	23	—
48	19-9 DL	—	—	—	69	65	56	50	45	40	37	33	38	26	—
49	17-4 PH	—	—	—	—	—	—	—	—	—	—	—	—	—	—
	Other Metals														
50	Aluminum 5052-H34	30	27	25	25	25	15	5	—	—	—	—	—	24	10
51	Lead	—	—	—	1.3	—	—	—	—	—	—	—	—	20	—
52	Titanium A55	—	—	74	63	47	33	25	20	14	—	—	—	13	—
53	Zirconium	—	—	—	28	24	14	8	7	—	—	—	—	—	—
54	Tantalum	134	—	—	38	—	—	23	21	17	15	13	—	—	—

* Courtesy of Shell Development Company.

Table 7. Thermal Conductivities of Some Common Metals*

Tabular Values in Btu/hr/sq ft/°F/ft

Item No.	Metals	-325	-200	-100	70	200	400	600	800	1000	1200	1400	1600	1800	2000
	Iron Base Alloys														
1	Steel, Low Carbon	—	25.8	—	30	27.6	26.8	25.5	24.5	23.2	22.2	21.1	—	—	—
2	Cast Iron	—	—	—	—	—	—	—	—	—	—	—	—	—	—
3	Ni Resist—Type 1	—	—	—	—	—	—	—	—	—	—	—	—	—	—
4a	½ Mo	—	—	—	28	25.8	24.6	23.1	21.6	20.3	—	—	—	—	—
4b	1 Cr–½ Mo	—	—	—	19.2	19.1	18.7	18.5	18.2	18.0	17.7	17.5	—	—	—
4c	1¼ Cr–½ Mo	—	—	—	18.8	18.3	17.9	17.5	17.0	16.9	—	—	—	—	—
4d	2 Cr–½ Mo	—	—	—	17.1	17.0	16.9	16.9	16.9	16.9	16.7	16.7	—	—	—
4e	2¼ Cr–1 Mo	—	—	—	15.0	15.1	15.1	15.2	15.4	15.5	15.9	—	—	—	—
4f	5 Cr–½ Mo	—	—	—	15.5	15.6	15.8	16.0	16.2	16.3	16.4	16.6	—	—	—
4g	7 Cr–½ Mo	—	—	—	—	—	—	—	—	—	—	—	—	—	—
4h	9 Cr–1 Mo	—	—	—	14.8	15.0	15.2	15.4	15.6	15.8	—	—	—	—	—
5	12 Cr—Type 410	5.0	6.7	8.1	13.0	14.4	14.7	15.4	15.9	13.7	14.6	15.5	—	—	—
6	Stainless—304	5.0	6.7	8.1	9.4	10.0	10.9	11.8	12.7	13.0	—	—	—	—	—
7	Stainless—316	—	—	—	9.4	—	—	—	—	—	—	—	—	—	—
8	Stainless 317	—	—	—	—	—	—	—	—	—	—	—	—	—	—
9	Worthite Alloy 20	—	—	—	—	—	—	—	—	—	—	—	—	—	—
10	Durimet 20	—	—	—	—	—	—	—	—	—	—	—	—	—	—
11	25 Cr-12 Ni Steel	—	—	—	8.0	8.50	9.50	10.7	11.7	13.0	14.1	15.3	—	—	—
12	25 Cr-20 Ni Steel	—	—	—	8.0	8.00	11.0	—	13.0	12.9	—	—	—	—	—
13	Incoloy 800	—	—	—	8.0	—	—	—	—	—	—	—	—	—	—
14	Silicon Iron	—	—	—	—	—	—	—	—	—	—	—	—	—	—
15	Durichlor	—	—	—	—	—	—	—	—	—	—	—	—	—	—
	Copper Base Alloys														
16	Copper	333	225	225	225	222	219	216	214	209	207	205	—	—	—
17	Tin Bronze	—	—	—	—	—	—	—	—	—	—	—	—	—	—
18	Aluminum Bronze	—	—	—	32.7	35.3	41.8	49.0	55.5	—	—	—	—	—	—
19	Ampco No. 18	—	—	—	92	—	—	—	—	—	—	—	—	—	—
20	Red Brass	—	—	—	—	—	—	—	—	—	—	—	—	—	—

Temperature, °F

#	Material														
21	Yellow Brass	35	47	56	69	—	—	—	—	—	—	—	—	—	—
22	Muntz Metal	—	48	55	73	—	—	—	—	—	—	—	—	—	—
23	Admiralty	—	—	—	64	—	—	—	—	—	—	—	—	—	—
24	Silicon Bronze	—	—	—	19	—	—	—	—	—	—	—	—	—	—
25	70 Cu–30 Ni	—	—	—	17	—	—	—	—	—	—	—	—	—	—
	Nickel Base Alloys														
26	Monel 400	9.4	10.8	11.6	12.6	13.8	16.0	18.0	20.0	22.0	24.6	25.8	27.6	—	—
27	Monel K500	—	—	—	10.1	11.3	13.0	14.8	16.6	18.3	20.1	21.8	16.7	—	—
28	Nickel 200	—	—	—	32.5	31.9	31.2	30.9	30.8	30.7	30.6	15.5	15.0	30.0	—
29	Inconel 600	—	—	—	8.6	9.1	10.1	11.1	12.1	13.7	14.3	13.7	13.7	—	—
30	Hastelloy B	—	—	—	6.5	—	7.1	7.6	8.1	9.0	9.7	—	—	—	—
31	Hastelloy C	—	—	—	7.3	5.60	6.55	7.45	8.40	9.30	10.2	—	—	—	—
32	Hasetloy D	—	—	—	—	—	—	—	—	—	—	—	—	—	—
33	Hastelloy F	—	—	—	9.4	—	—	—	—	—	—	—	—	—	—
34	Hastelloy X	—	—	—	5.25	6.33	7.30	8.45	9.80	11.2	12.4	13.7	15.0	16.3	—
	Super-Alloys														
35	Nimonic 80	—	—	—	7.0	7.0	8.0	9.0	10.0	10.9	11.9	12.8	13.7	14.0	—
36	Inconel X750	—	5.83	6.16	6.92	7.42	8.17	9.08	10.0	10.5	11.3	12.3	12.5	—	—
37	M-252	—	—	—	6.83	7.35	8.20	9.00	9.67	9.00	—	—	—	—	—
38	Refractaloy 26	—	—	—	—	7.80	7.9	8.0	8.0	8.2	14.0	—	—	—	—
39	A-286	—	—	—	—	—	7.3	—	—	—	—	—	—	—	—
40	Discaloy	—	—	—	7.75	8.20	9.20	9.95	10.9	12.0	13.1	14.2	—	—	—
41	N155	—	—	—	8.8	—	—	7.5	10.2	10.4	13.1	13.1	—	—	—
42	S-590	—	—	—	—	—	8.0	9.1	11.3	12.5	—	—	—	—	—
43	S-816	—	—	—	7.2	7.90	9.0	10.2	10.5	12.4	13.1	—	—	—	—
44	Haynes Alloy 31	—	—	—	8.6	—	8.5	10.1	—	11.5	—	—	—	—	—
45	Haynes Alloy 25	—	—	—	5.41	6.20	7.48	8.73	9.95	11.2	12.5	13.7	—	—	—
46	Haynes Alloy 21	—	—	—	—	—	8.40	9.30	10.0	11.4	12.3	—	—	—	—
47	16 Cr-25 Ni-6 Mo	—	—	—	9.0	—	9.10	9.90	10.7	11.5	—	—	—	—	—
48	19-9 DL	—	—	—	8.2	8.30	—	—	—	11.5	12.3	13.0	—	—	—
49	17-4 PH	—	—	—	10.2	—	—	—	—	10.7	—	12.1	—	—	—
	Other Metals														
50	Aluminum	—	124	—	131	133	137	141	—	—	—	—	—	—	—
51	Lead	—	—	—	20	10.9	10.4	10.5	—	—	—	—	—	—	—
52	Titanium A55	—	—	11.8	11.5	—	—	—	—	—	—	12.1	—	—	—
53	Zirconium	—	—	—	14.0	—	—	—	—	—	—	—	—	—	—
54	Tantalum	—	—	—	31.8	—	—	—	—	—	—	—	—	—	—

* Courtesy of Shell Development Company.

Table 8. Thermal Expansion of Some Common Metals*
Between 70°F and Indicated Temperature[a]

Item No.	Metals	Coefficient	-325	-150	-50	70	200	400	600	800	1000	1200	1400	1600	1800
	Iron Base Alloys														
1	Steel, Low Carbon	A	5.00	5.50	5.80	6.07	6.38	6.82	7.23	7.65	7.97	8.19	8.36	—	—
		B	-2.37	-1.45	-0.84	0	0.99	2.70	4.60	6.70	8.89	11.1	13.3	—	—
2	Cast Iron	A	—	—	—	—	5.75	6.10	6.47	6.83	7.19	—	—	—	—
		B	—	—	—	0	0.90	2.42	4.11	5.98	8.02	—	—	—	—
3	Ni Resist—Type 1	A	—	—	—	—	—	—	—	—	—	—	—	—	—
		B	—	—	—	—	—	—	—	—	—	—	—	—	—
4a	½ Mo	A	—	—	—	—	—	6.82	7.23	7.65	7.97	8.19	8.33	—	—
		B	—	—	—	—	—	2.70	4.60	6.70	8.90	11.1	13.3	—	—
4b	1 Cr-½ Mo	A	—	—	—	—	—	6.82	7.23	7.53	7.80	8.11	—	—	—
		B	—	—	—	—	—	2.70	4.60	6.60	8.70	11.0	—	—	—
4c	1¼ Cr-½ Mo	A	—	—	—	—	—	—	—	7.70	7.80	7.9	—	—	—
		B	—	—	—	—	—	—	—	6.75	8.70	10.7	—	—	—
4d	2 Cr-½ Mo	A	—	—	—	—	—	—	—	—	—	—	—	—	—
		B	—	—	—	—	—	—	—	—	—	—	—	—	—
4e	2¼ Cr-1 Mo	A	—	—	—	—	—	6.82	7.23	7.53	7.80	8.11	—	—	—
		B	—	—	—	—	—	2.70	4.60	6.60	8.70	11.0	—	—	—
4f	5 Cr-½ Mo	A	—	—	—	—	—	6.31	6.76	6.96	7.26	7.37	—	—	—
		B	—	—	—	—	—	2.50	4.30	6.10	8.10	10.0	—	—	—
4g	7 Cr-½ Mo	A	4.70	5.20	5.45	5.73	6.04	6.34	6.66	6.96	7.22	7.41	7.55	—	—
		B	-2.22	-1.37	-0.79	0	0.94	2.50	4.24	6.10	8.06	10.0	12.1	—	—
4h	9 Cr-1 Mo	A	—	—	—	—	—	6.31	6.76	6.96	7.17	7.30	—	—	—
		B	—	—	—	—	—	2.50	4.30	6.10	8.00	9.90	—	—	—

Temperature, °F

(Continued)

No.	Metal		C1	C2	C3	C4	C5	C6	C7	C8	C9	C10	C11	C12	C13
5	12 Cr—Type 410	A	—	—	6.90	6.78	6.63	6.39	6.13	5.81	5.50	5.24	5.00	4.70	4.30
		B	—	—	11.0	9.20	7.40	5.60	3.90	2.30	0.86	0	−0.72	−1.24	−2.04
6,7	Stainless—304, 316	A	—	—	10.6	10.5	10.3	10.1	9.82	9.59	9.34	9.11	8.90	8.60	8.15
		B	—	—	16.9	14.2	11.5	8.80	6.24	3.80	1.46	0	−1.28	−2.27	−3.85
8	Stainless 317	A	—	—	—	—	—	—	—	—	—	—	—	—	—
		B	—	—	—	—	—	—	—	—	—	—	—	—	—
9	Worthite Alloy 20	A	—	—	—	—	—	—	—	—	—	—	—	—	—
		B	—	—	—	—	—	—	—	—	—	—	—	—	—
10	Durimet 20	A	—	—	—	—	—	—	—	—	—	—	—	—	—
		B	—	—	—	—	—	—	—	—	—	—	—	—	—
11	25 Cr—12 Ni Steel	A	—	—	—	—	—	—	—	—	—	—	—	—	—
		B	—	—	—	—	—	—	—	—	—	—	—	—	—
12	25 Cr—20 Ni	A	10.0	9.7	9.18	9.08	8.92	8.68	8.38	8.08	7.76	7.48	7.20	6.85	6.35
		B	20.8	17.9	14.7	12.3	9.95	7.60	5.33	3.20	1.21	0	−1.04	−1.81	−3.00
13	Incoloy 800	A	—	10.2	10.0	9.60	9.40	9.20	9.00	8.70	8.00	—	—	—	—
		B	—	18.7	16.0	13.0	10.5	8.06	5.72	3.45	1.25	—	—	—	—
14	Silicon Iron	A	—	—	—	—	—	—	—	—	—	—	—	—	—
		B	—	—	—	—	—	—	—	—	—	—	—	—	—
15	Durichlor	A	—	—	—	—	—	—	—	—	—	—	—	—	—
		B	—	—	—	—	—	—	—	—	—	—	—	—	—
	Copper Base Alloys														
16	Copper	A	—	—	—	—	—	10.6	9.59	9.60	9.61	9.30	8.60	7.90	4.90
		B	—	—	—	—	—	9.30	6.10	3.80	1.30	0	−1.25	−2.09	−2.35
17	Tin Bronze	A	—	—	—	—	—	—	10.4	10.2	10.0	9.57	9.15	8.75	8.40
		B	—	—	—	—	—	—	6.64	4.05	1.56	0	−1.32	−2.31	−3.98
18	Aluminum Bronze	A	—	—	11.0	10.8	—	—	—	—	—	—	—	—	—
		B	—	—	14.9	12.1	—	—	—	—	—	—	—	—	—
19	Ampco No. 18	A	—	—	—	—	—	—	—	—	—	9.00	—	—	—
		B	—	—	—	—	—	—	—	—	—	0	—	—	—

Table 8. Thermal Expansion of Some Common Metals—*Continued*
Between 70°F and Indicated Temperature

Item No.	Metals	Coefficient	Temperature, °F												
			−325	−150	−50	70	200	400	600	800	1000	1200	1400	1600	1800
20	Red Brass	A	—	—	—	—	—	—	—	—	—	—	—	—	—
		B	—	—	—	—	—	—	—	—	—	—	—	—	—
21	Yellow Brass	A	6.2	8.50	9.5	10.4	9.76	10.2	10.7	11.2	11.6	12.1	—	—	—
		B	−2.96	−2.24	−1.35	0	1.52	4.05	6.80	9.78	13.0	16.4	—	—	—
22	Muntz Metal	A	—	—	—	—	—	—	—	—	—	—	—	—	—
		B	—	—	—	—	—	—	—	—	—	—	—	—	—
23	Admiralty	A	—	—	—	—	—	—	11.2	—	—	—	—	—	—
		B	—	—	—	—	—	—	7.12	—	—	—	—	—	—
24	Silicon Bronze	A	—	—	—	—	—	—	10.0	—	—	—	—	—	—
		B	—	—	—	—	—	—	6.35	—	—	—	—	—	—
25	70 Cu-30 Ni	A	6.65	7.40	7.80	8.16	8.54	8.90	9.0	—	—	—	—	—	—
		B	−3.15	−1.95	−1.13	0	1.33	3.52	5.73	—	—	—	—	—	—
	Nickel Base Alloys														
26	Monel 400	A	5.55	6.75	7.15	7.48	7.70	8.60	8.80	8.90	9.10	9.30	9.60	9.80	10.0
		B	−2.62	−1.79	−1.03	0	1.00	2.84	4.67	6.50	8.45	10.50	12.85	14.90	17.3
27	Monel K-500	A	5.35	6.45	6.80	7.12	7.48	7.90	8.30	8.70	9.10	9.50	9.89	—	—
		B	−2.53	−1.70	−0.98	0	1.17	3.13	5.28	7.62	10.2	12.9	15.8	—	—
28	Nickel 200	A	5.8	6.37	6.60	6.82	7.40	7.50	7.90	8.20	8.50	8.70	9.00	9.20	—
		B	−2.32	−1.68	−0.95	0	1.15	2.97	5.02	7.18	9.49	11.8	14.4	16.9	—
29	Inconel 600	A	5.6	6.05	6.35	6.70	7.00	7.50	7.90	8.25	8.50	8.60	8.90	9.10	—
		B	−2.24	−1.60	−0.91	0	1.09	2.97	5.02	7.23	9.49	11.7	14.2	16.7	—
30	Hastelloy B	A	—	—	—	—	5.60	—	6.41	6.57	6.66	6.73	6.88	7.78	—
		B	—	—	—	—	0.87	—	4.07	5.76	7.43	9.13	10.98	14.3	—

(Continued)

#	Metal		1	2	3	4	5	6	7	8	9	10	11	12	13
31	Hastelloy C	A	—	8.20	8.07	7.73	7.44	7.35	7.02	—	6.30	—	—	—	—
		B	—	15.1	12.9	10.5	8.30	6.44	4.46	—	0.98	—	—	—	—
32	Hastelloy D	A	—	—	—	—	—	—	—	—	—	—	—	—	—
		B	—	—	—	—	—	—	—	—	—	—	—	—	—
33	Hastelloy F	A	10.2	9.80	9.50	9.20	8.90	8.80	8.70	8.30	8.10	—	—	—	—
		B	21.2	18.0	15.2	12.5	9.93	7.71	5.53	3.29	1.26	—	—	—	—
34	Hastelloy X	A	9.20	9.00	8.80	8.56	8.39	8.15	7.90	7.82	7.70	—	—	—	—
		B	19.1	16.5	14.0	11.6	9.36	7.14	5.02	3.10	1.20	—	—	—	—
	Super Alloys														
35	Nimonic 80	A	—	—	—	—	—	—	—	—	—	—	—	—	—
		B	—	—	—	—	—	—	—	—	—	—	—	—	—
36	Inconel X-750	A	—	9.30	8.80	8.40	8.10	7.80	7.50	7.20	7.00	6.90	6.80	6.70	5.7
		B	—	17.1	13.9	11.4	9.04	6.83	4.77	2.85	1.09	0	-0.98	-1.77	-2.28
37	M-252	A	—	8.50	8.00	7.60	7.20	6.80	6.40	6.20	6.00	—	—	—	—
		B	—	15.6	12.8	10.3	8.04	5.96	4.07	2.46	0.94	—	—	—	—
38	Refractaloy 26	A	—	—	—	—	8.2	8.0	8.0	7.9	7.8	—	—	—	—
		B	—	—	—	—	9.15	7.05	5.10	3.14	1.24	—	—	—	—
39	A-286	A	—	—	—	9.88	9.78	9.64	9.47	9.35	9.17	—	—	—	—
		B	—	—	—	11.05	10.92	8.49	6.02	3.71	1.45	—	—	—	—
40	Discaloy	A	—	—	—	—	9.5	9.4	9.1	8.7	8.5	—	—	—	—
		B	—	—	—	—	10.60	8.29	5.80	3.45	1.34	—	—	—	—
41	N-155	A	10.0	9.90	9.70	9.40	9.10	8.70	—	8.50	7.90	—	—	—	—
		B	20.8	18.2	15.5	12.8	10.2	7.62	—	3.37	1.23	—	—	—	—
42	S-590	A	—	—	—	—	—	—	—	—	—	—	—	—	—
		B	—	—	—	—	—	—	—	—	—	—	—	—	—
43	S-816	A	9.20	9.00	8.70	8.40	8.03	7.89	7.65	7.45	7.38	—	—	—	—
		B	19.1	16.5	13.9	11.4	8.96	6.91	4.87	2.95	1.15	—	—	—	—
44	Haynes Alloy 31	A	—	9.19	9.05	8.75	8.38	8.08	7.84	—	—	—	—	—	—
		B	—	16.9	14.4	11.9	9.36	7.08	4.99	—	—	—	—	—	—

Table 8. Thermal Expansion of Some Common Metals—*Continued*

Between 70°F and Indicated Temperature

Item No.	Metals	Coefficient	Temperature, °F												
			-325	-150	-50	70	200	400	600	800	1000	1200	1400	1600	1800
45	Haynes Alloy 25	A	—	—	—	—	6.80	7.20	7.61	8.28	8.30	8.55	8.92	9.30	9.70
		B	—	—	—	0	1.06	2.85	4.84	7.25	9.26	11.6	14.2	17.1	20.1
46	Haynes Alloy 21	A	—	—	—	—	—	—	7.83	7.96	8.18	8.38	8.58	8.78	—
		B	—	—	—	0	—	—	4.98	6.97	9.13	11.4	13.7	16.1	—
47	16 Cr–25 Ni–6 Mo	A	—	—	—	—	8.40	8.70	9.00	9.40	9.60	10.2	10.8	11.7	—
		B	—	—	—	0	1.31	3.45	5.72	8.23	10.7	13.8	17.2	21.5	—
48	19-9 DL	A	—	—	—	8.40	8.5	9.10	9.3	9.60	9.8	9.90	—	10.1	—
		B	—	—	—	0	1.33	3.60	5.90	8.41	10.9	13.4	—	18.5	—
49	17-4 PH	A	—	—	—	—	—	—	—	—	—	—	—	—	—
		B	—	—	—	—	—	—	—	—	—	—	—	—	—
	Other Metals														
50	Aluminum	A	5.4	9.0	11.6	12.3	13.0	13.6	14.2	—	—	—	—	—	—
		B	-2.59	-2.3	-1.67	0	2.02	5.39	9.03	—	—	—	—	—	—
51	Lead	A	13.6	14.8	16.0	16.4	—	—	—	—	—	—	—	—	—
		B	-6.6	-3.91	-2.31	0	—	—	—	—	—	—	—	—	—
52	Titanium A55	A	—	—	—	—	4.80	5.20	5.30	5.40	5.50	5.60	5.70	5.80	5.90
		B	—	—	—	0	0.75	2.06	3.37	4.73	6.14	7.59	9.10	10.7	12.3
53	Zirconium	A	—	—	—	—	2.90	3.33	3.42	3.53	3.62	3.71	—	—	—
		B	—	—	—	0	0.45	1.31	2.18	3.09	4.04	5.03	—	—	—
54	Tantalum	A	—	—	—	—	—	—	—	—	—	—	—	—	—
		B	—	—	—	—	—	—	—	—	—	—	—	—	—

* Courtesy of Shell Development Company.
a Mean Coefficient of Thermal Expansion = $(A/10^6)$ (in./in./°F). Linear Thermal Expansion = B (in./100 ft).

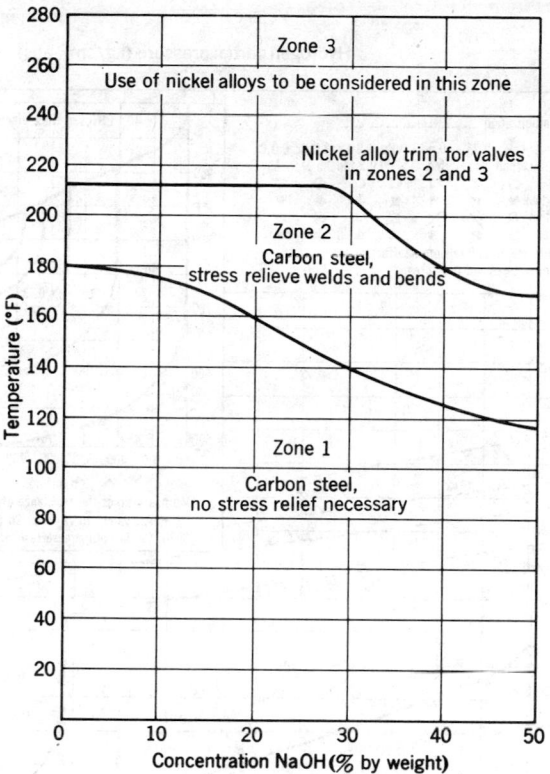

FIG. 1. Suitability of steels in caustic soda service.

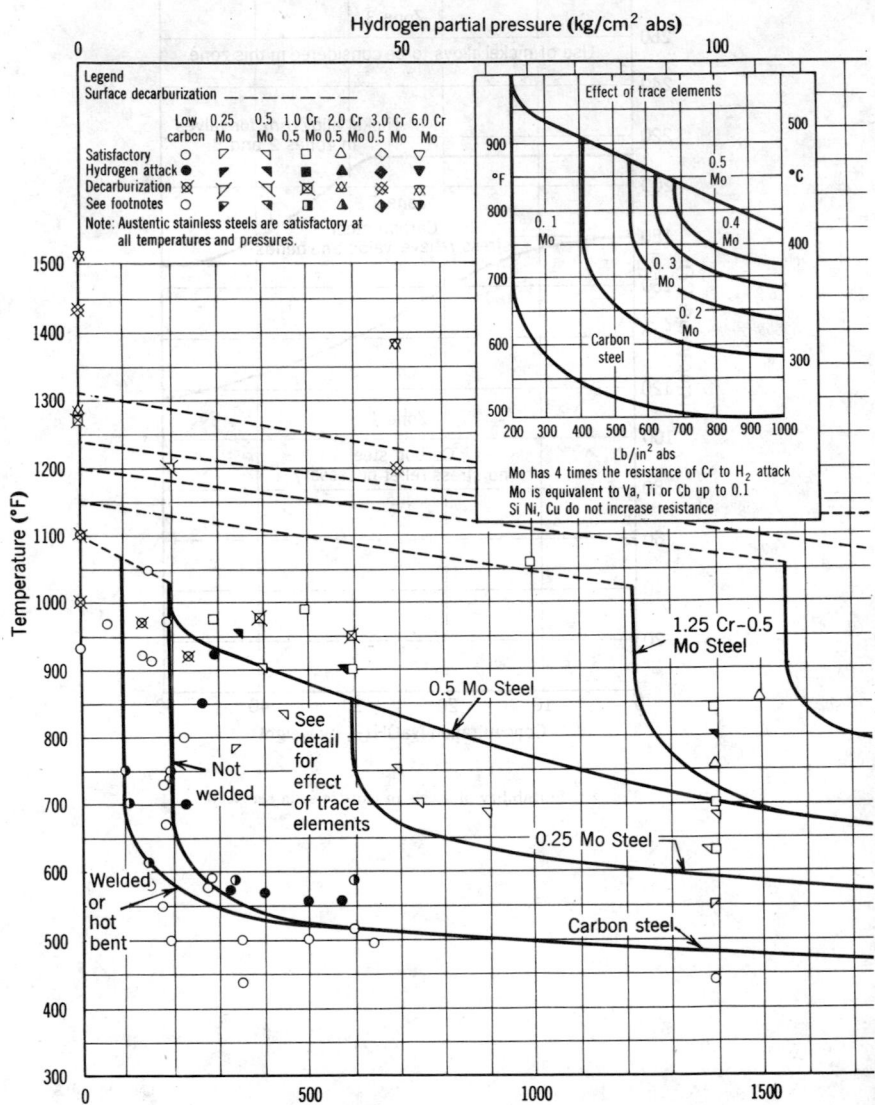

Fig. 2. Operating limits for steels in hydrogen service.

Courtesy of Shell Development Company.

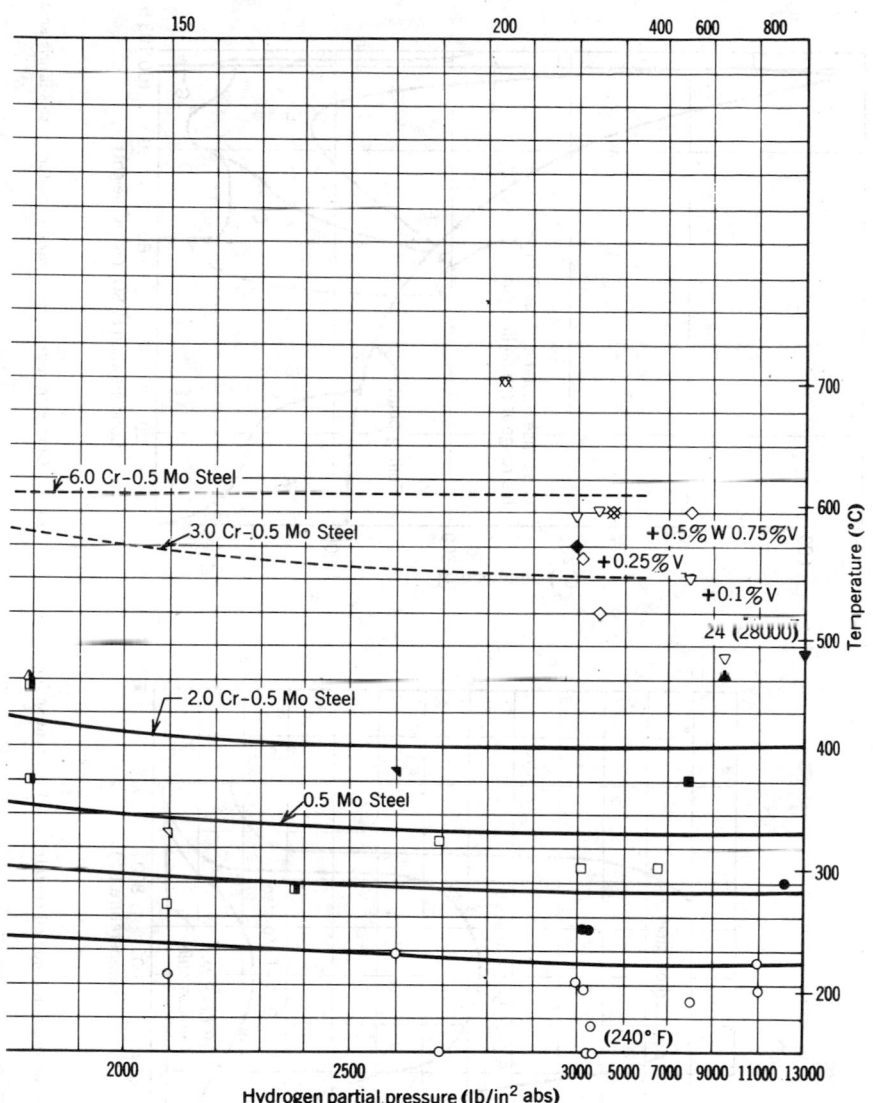

Hydrogen partial pressure (lb/in² abs)

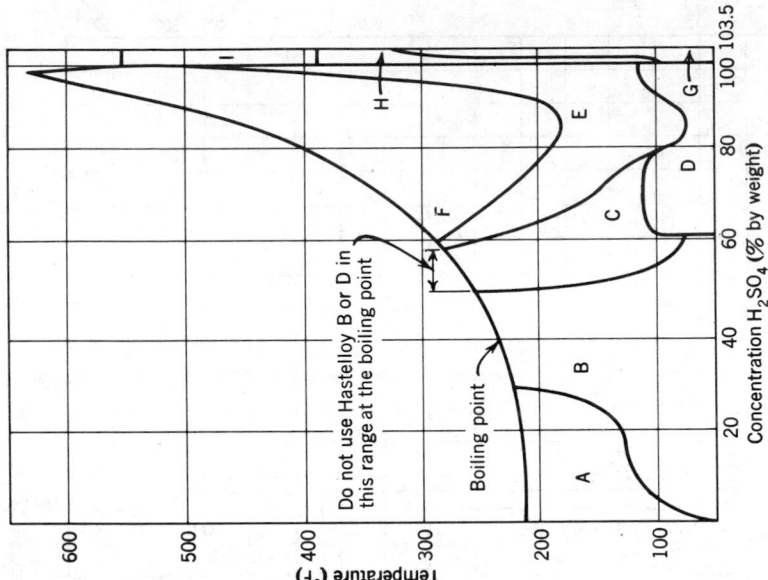

FIG. 4.   Corrosion resistance of various materials to sulfuric acid.

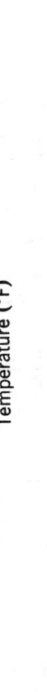

FIG. 3.   Corrosion rates for steels in crude oil containing 1.5 per cent sulfur.

Table 9.   Key For Sulfuric Acid Chart
Materials Having Reported Corrosion Rate Less Than 0.020″ per Year

AREA A

10% Aluminum Bronze
   (Air Free)
Illium G
Glass
Hastelloy B and D
Durimet 20
Worthite
Lead
Copper (Air Free)
Monel (Air Free)
Haveg 43
Rubber (up to 170°F)
Impervious Graphite
Tantalum
Gold
Platinum
Silver
Zirconium
Nionel
Tungsten
Molybdenum
Type 316 Stainless
   (up to 10% Aerated)

AREA B

Glass
Silicon Iron
Hastelloy B and D
Durimet 20 (up to 150°F)
Worthite (up to 150°F)
Lead
Copper (Air Free)
Monel (Air Free)
Haveg 43
Rubber (up to 170°F)
10% Aluminum Bronze
   (Air Free)
Ni Resist (up to 20% at 75°F)
Impervious Graphite
Tantalum
Gold
Platinum
Silver
Zirconium
Nionel
Tungsten
Molybdenum
Type 316 Stainless (up to 25%
   at 75°F) Aerated

AREA C

Glass
Silicon Iron
Hastelloy B and D
Durimet 20 (up to 150°F)
Worthite (up to 150°F)
Lead
Monel (Air Free)
Impervious Graphite
Tantalum
Gold
Platinum
Zirconium
Molybdenum

AREA D

Steel
Glass
Silicon Iron
Hastelloy B and D
Lead (up to 96% H_2SO_4)
Durimet 20
Worthite
Ni Resist
Type 316 Stainless
   (above 80%)
Impervious Graphite
   (up to 96% H_2SO_4)
Tantalum
Gold
Platinum
Zirconium

AREA E

Glass
Silicon Iron
Hastelloy B and D
Durimet 20 (up to 150°F)
Worthite (up to 150°F)
Lead (up to 175°F)
   and 96% H_2SO_4)
Impervious Graphite
   (up to 175°F and 96% H_2SO_4)
Tantalum
Gold
Platinum

AREA F

Glass
Silicon Iron
Tantalum
Gold
Platinum

AREA G

Glass
Steel
18 Cr-8 Ni
Durimet 20
Worthite
Hastelloy C
Gold
Platinum

AREA H

Glass
18 Cr-8 Ni
Durimet 20
Worthite
Gold
Platinum

AREA I

Glass
Gold
Platinum

Table 10.   Key For Hydrofluoric Acid Chart

Materials Having Reported Corrosion Rate Less Than 0.020" per Year

AREA A

Monel (Air Free)
Copper (Air Free)
70 Cu-30 Ni (Air Free)
Lead (Air Free)
Nickel (Air Free)
Alloy 20
Ni Resist
Hastelloy C
Platinum
Silver
Gold
Impervious Graphite
Haveg 43
Rubber
25 Cr-20 Ni Steel

AREA B

Monel (Air Free)
70 Cu-30 Ni (Air Free)
Copper (Air Free)
Lead (Air Free)
Nickel (Air Free)
Alloy 20
Hastelloy C
Platinum
Silver
Gold
Impervious Graphite
Rubber
Haveg 43

AREA C

Monel (Air Free)
70 Cu-30 Ni (Air Free)
Copper (Air Free)
Lead (Air Free)
Alloy 20
Hastelloy C
Platinum
Silver
Gold
Impervious Graphite
Haveg 43
Rubber

AREA D

Monel (Air Free)
70 Cu-30 Ni (Air Free)
Copper (Air Free)
Lead (Air Free)
Hastelloy C
Platinum
Silver
Gold
Impervious Graphite
Haveg 43

AREA E

Monel (Air Free)
70 Cu-30 Ni (Air Free)
Lead (Air Free)
Hastelloy C
Platinum
Silver
Gold
Impervious Graphite
Haveg 43

AREA F

Monel (Air Free)
Hastelloy C
Platinum
Silver
Gold
Haveg 43

AREA G

Carbon Steel
Monel (Air Free)
Hastelloy C
Platinum
Silver
Gold
Haveg 43

Table 11. Key For Hydrochloric Acid Chart
Materials Having Reported Corrosion Rate Less Than 0.020" per Year

AREA A

Chlorimet 2
Glass
Silver
Platinum
Tantalum
Hastelloy B
Durichlor (FeCl₃ Free)
Haveg
Saran
Rubber
Silicon Bronze (Air Free)
Copper (Air Free)
Nickel (Air Free)
Monel (Air Free)
Zirconium
Tungsten
Titanium—Up to 10% HCl at
  Room Temperature
Worthite—Up to 2% HCl at
  Room Temperature

AREA B

Chlorimet 2
Glass
Silver
Platinum
Tantalum
Hastelloy B
Durichlor (FeCl₃ Free)
Haveg
Saran
Rubber
Silicon Bronze (Air Free)
Zirconium
Molybdenum
Impervious Graphite

AREA C

Chlorimet 2
Glass
Silver
Platinum
Tantalum
Hastelloy B (Chlorine Free)
Durichlor (FeCl₃ Free)
Haveg
Saran
Rubber
Molybdenum
Zirconium
Impervious Graphite

AREA D

Chlorimet 2
Glass
Silver
Platinum
Tantalum
Hastelloy B (Chlorine Free)
Durichlor (FeCl₃ Free)
Monel (Air Free) (Up to 0.5% HCl)
Zirconium
Impervious Graphite
Tungsten

AREA E

Chlorimet 2
Glass
Silver
Platinum
Tantalum
Hastelloy B (Chlorine Free)
Zirconium
Impervious Graphite

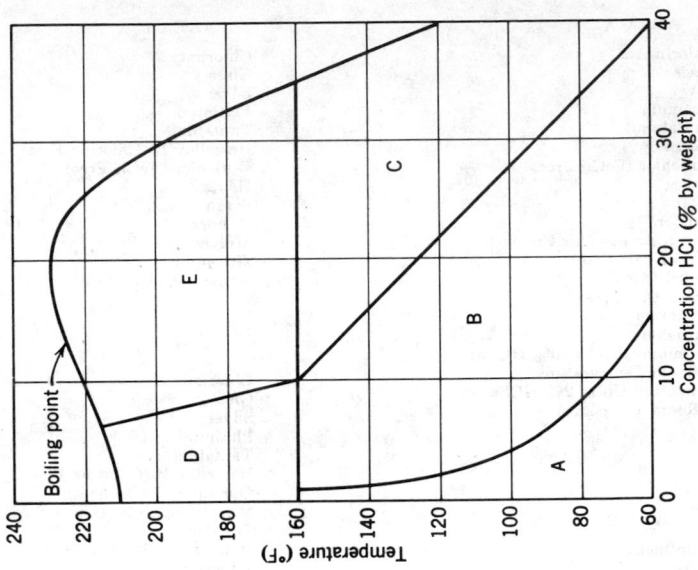

Fig. 6. Corrosion resistance of various materials to hydrochloric acid.

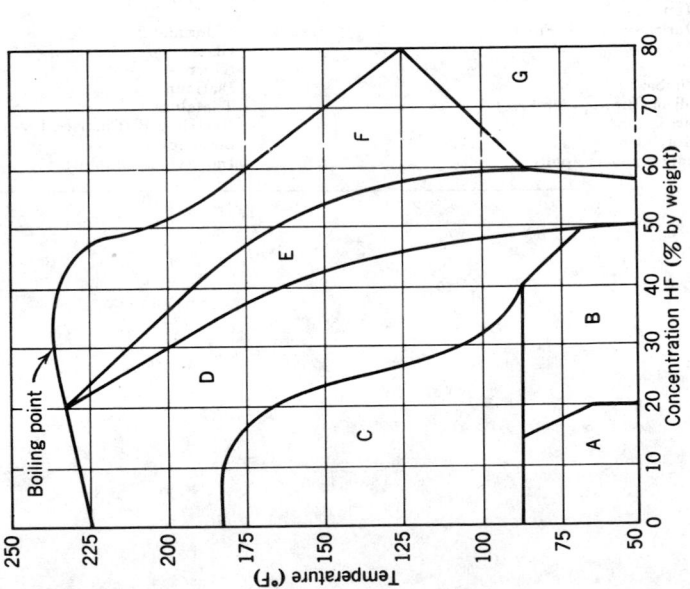

Fig. 5. Corrosion resistance of various materials to hydrofluoric acid.

Table 12. **Corrosion Ratings of Commercial Alloys of Iron, Nickel, and Chromium**

(Data by courtesy of the International Nickel Company)

Material	American Iron and Steel Institute Type Number	Outdoor Air			Specific Industrial Atmospheres (Wet)					Scaling Temp, °F, Note 1
		Rural	Urban	Marine	Ammonia	Hydrogen Sulfide	Hydrogen Chloride	Sulfur Dioxide	Chlorine	
Low-carbon steel	....	D$^+$	D	D	B	D	D	D	D	1050
Copper-bearing steel	....	D$^+$	D$^+$	D	B	D	D	D	D	1000
4–6% Cr steel	502	D$^+$	D$^+$	D$^+$	B	D$^+$	D	D	D	1200
12–14% Cr steel	420	B$^+$	B	C$^-$	A	C	D	D	D	1250
12–14% Cr 0.60% Mo	416	B$^+$	B	C$^-$	A	C	D	D	D	1250
16–18% Cr	430	A	B$^+$	C	A	B	D	D	D	1600
23–30% Cr	446	A	A	C$^+$	A	A	D	B	D	2100
7% Ni-17% Cr	301	A	A	B$^+$	A	A	D	C	C	Not used for oxidation res.
8% Ni-18% Cr	302 303 304	A	A	A$^-$	A	A	D	B	C	1650
8% Ni-18% Cr 1% Cb	347	A	A	A$^-$	A	A	D	B	C	1650
8% Ni-18% Cr 0.5% Ti	321	A	A	A$^-$	A	A	D	B	C	1650
14% Ni-23% Cr	309	A	A	A	A	A	D	B	C	2000
12% Ni-18% Cr 3% Mo	316	A	A$^+$	A$^+$	A	A	C	A	B	1650
20% Ni-25% Cr	310	A	A$^+$	A$^+$	A	A	D	B	C	2100
Nickel	....	A	A$^-$	A$^+$	C–E Note 5	B	C–B	C	B	Note 2, 1900 Note 3, 1000 Note 4, 700
30% Ni-70% Cu	....	A	A$^-$	A	C–E Note 5	C$^-$	C$^+$	B	C$^+$	....
Monel	....	A	A	A	C–E Note 5	B	B	C	B	Note 2, 1000 Note 3, 1000 Note 4, 650
80% Ni-20% Cr	....	A	A	A$^+$	A	A	B	B	B	2100
Inconel-80% Ni 7% Fe-13% Cr	....	A	A	A$^+$	A	A	B	B	B	Note 2, 2000 Note 3, 1500 Note 4, 1000

A. Practically complete resistance, or the alloy is the best of materials within its class.
B. Good resistance, as proved by being in common use. May replace materials given A rating to secure some other advantage.
C. Adequate resistance under favorable conditions which should be investigated beforehand.
D. Sufficient resistance if adequate precautions are taken to reduce effect of corrosive conditions, as by coatings, cathodic protection, re-design, etc., or where appearance is not important and appreciable corrosion may be provided for or tolerated.
E. Poor resistance; use only if no better material is available.
Plus and minus signs are used to permit a better differentiation between the corrosion-resistant qualities.

Notes

1. Values assume substantially constant-temperature operations; should be lowered for cyclic heating and cooling to an extent dependent upon the frequency and range of temperature fluctuations.
2. Scaling temperature in low-sulfur atmosphere.
3. Scaling temperature in high-sulfur oxidizing atmosphere.
4. Scaling temperature in high-sulfur reducing atmosphere.
5. Not recommended primarily to resist ammonia attack, but may be used where resistance to ammonia in low concentrations is an incidental requirement.

Table 13.   Analyses of Pig Irons or Cast Irons

Trade Name	Total C, per cent	Si, per cent	S, per cent	P, per cent	Mn, per cent
Low-phosphorus (Acid open-hearth)		0.50-3.00	0.035 max	0.035 max	0.75-1.25
Intermediate low-phosphorus	4.00-4.50	1.00-3.00	0.05 max	0.036-0.075	0.75-1.25
Bessemer	3.50-4.00	1.00-3.00	0.05 max	0.076-0.100	1.00-1.25
Malleable	3.75-4.50	0.75-3.50	0.05 max	0.101-0.300	0.50-1.25
Basic, northern	3.50-4.00	1.00-1.50	0.05 max	0.400 max	1.01-2.00
Basic, southern	3.50-4.00	1.00-1.50	0.05 max	0.700-0.900	0.40-0.75
Foundry, northern, low-phosphorus	4.00-4.50	1.75-3.50	0.05 max	0.300-0.500	0.50-1.25
Foundry, northern, high-phosphorus	4.00-4.50	1.75-3.50	0.05 max	0.501-0.700	0.50-1.25
Foundry, southern	3.50-4.00	1.75-3.50	0.05 max	0.700-0.900	0.40-1.00
Charcoal iron, southern		0.50-3.00	0.035 max	0.035 max	0.40-1.00
Silvery pig iron	0.75-1.00	5.00-17.00	0.05 max	0.300 max	1.00-2.00

Table 14.   Analyses of High-purity Iron

	Armco Ingot Iron %	Electrolytic Iron %
Carbon	0.012 *	0.006 *
Manganese	0.017 *	0.000 *
Phosphorus	0.005 *	0.005 *
Sulfur	0.025 *	0.004 *
Silicon	Trace *	0.005 *
Iron (about)	99.9	99.95

* National Metals Handbook, 1948 Edition, p. 424.

Table 15.   AISI Specification Numbering System

The American Iron and Steel Institute has developed a basic four-numeral series for identifying various steels. The first two digits of the series indicate the type of steel: e.g., 50xx, low-chromium. The last two digits indicate the approximate middle of the cargon content range: e.g., 32 represents a range of 0.30 to 0.35 per cent carbon content. In the case of some steels having average carbon contents greater than 0.99 per cent, the system gives rise to a five-numbered series: e.g., 50xxx. A prefix is used to indicate the steel-making process: e.g., B, acid Bessemer; E, electric furnace; either C or no prefix letter, basic open-hearth.

The various series of steel are as follows:

Series Designation	Types
10xx	Non-resulfurized basic open-hearth and acid Bessemer carbon-steel grades
11xx	Resulfurized basic open-hearth and acid Bessemer carbon-steel grades
13xx	Manganese, 1.75%
23xx	Nickel, 3.50%
25xx	Nickel, 5.00%
31xx	Nickel, 1.25%; chromium, 0.65 or 0.80%
33xx	Nickel, 3.50%; chromium, 1.55%
40xx	Molygdenum, 0.25%
41xx	Chromium, 0.95%; molybednum, 0.20%
43xx	Nickel, 1.80%;chromium, 0.50% or 0.80; molybdenum, 0.25%
46xx	Niclel, 1.80%; molybdenum, 0.25%
48xx	Nickel, 3.50%; molybdenum, 0.25%
50xx	Chromium, 0.30 or 0.60%
51xx	Chromium, 0.80, 0.95, or 1.05%
5xxxx	Carbon, 1.00%; chromium, 0.50, 1.00, or 1.45%
61xx	Chromium, 0.80 or 0.95%; vanadium, 0.10 or 0.15 min.
86xx	Nickel, 0.55%; chromium, 0.50%; molybdenum, 0.20%
87xx	Nickel, 0.55%; chromium, 0.50%; molybdenum, 0.25%
92xx	Manganese, 0.85%; silicon, 2.00%
93xx	Nickel, 3.25%; chromium, 1.20%; molybdenum, 0.12%
94xx	Manganese, 1.00%; nickel, 0.45%; chromium, 0.40%; molybdenum, 0.12%
97xx	Nickel, 0.55%; chromium, 0.17%; molybdenum, 0.20%
98xx	Nickel, 1.00%; chromium, 0.80%; molybdenum, 0.25%

Table 16. Influence of Alloying Elements upon the Properties of Steel
Elements Taken Individually in Considering Their Influence*

	Carbon C	Manganese Mn	Silicon Si	Aluminum Al	Nickel Ni	Chromium Cr	Molybdenum Mo	Vanadium V	Tungsten W	Cobalt Co	Copper Cu	Sulfur S	Phosphorus P	Titanium Ti	Tantalum Ta	Niobium (Cb) Nb
Yield strength	↗	↗	↗		↗	↗	↗	↗								
Tensile strength	⇗	⇗	↗		↗	↗	↗	↗	↗	↗	↗		↗	↗	↗	↗
Elongation	↙	↙	↙		↙	↙	↙		↙	↙	⇙		↙			
Tensile strength at elevated temperature	↗					⇗	⇗	↗	↗	↗			↗			
Creep strength	↗	↗	↗	↙	↗	↗	↗									
Fatigue strength	↗					↗	↗	⇗	↗	↗						
Ac1 point		⇙	⇗	↗	⇙	⇗	↗	↗	↗							
Ac3 point		↙	↗	↗	↙	↗	↗	↗	↗	↙	↙			↗	↗	↗
Austenite field		↗	↙	↙	↗	↙	↙	↙	↙	↗	↗			↙	↙	↙
Grain growth	↗	⇗	↗	↙	↙	↗		↙		↙			↗	↙		
Susceptibility to overheating		↗		↙		↙	↙		↙	↙	↙			↗		
Oxidation resistance			⇗	⇗		⇗				↗						
Red shortness												↗				
Critical cooling rate		↙	↙		↙	↙	↙	↙	↙	↗						
Hardenability	⇗	⇗	↗		⇗	↗	↗	↗	↗							
Hardness	↗	↗	↗		↗	↗	↗	↗	↗	↗	↗			↗	↗	↗
Tempering stability	↙		↗		↗	↗	⇗	⇗	↗	⇗					↗	↗
Carbide formation	⇗					↗	↗	⇗	↗						⇗	⇗

* Influence of element is ↗ increased, ⇗ greatly increased, ↙ decreased, ⇙ greatly decreased.

Table 17. Compositions, Characteristics, and Uses of Ferritic Stainless Steels*a

AISI Type Number	Chemical Composition, %							Magnetic	Characteristics and Typical Uses	SAE Type Number
	C max.	Mn max.	Si max.	P max.	S max.	Cr	Other elements			
405	0.08	1.00	1.00	0.040	0.030	11.50–13.50	Al 0.10–0.30	In all conditions	Non-hardenable by rapid cooling from high temperatures. Welded pressure vessels in petroleum industry.	51430
430	0.12	1.00	1.00	0.040	0.030	14.00–18.00	... .	In all conditions	Readily formed, superior corrosion resistance and mechanical properties. Automobile trim, chemical equipment.	51430
430F	0.12	1.25	1.00	.....	.....	14.00–18.00	b	In all conditions	Forming properties of type 430 partly sacrificed for machinability; remains bright under relatively severe conditions. Various screw machine items.	51430F
446	0.35	1.50	1.00	0.040	0.030	23.00–27.00	N 0.25 max.	In all conditions	Resists destructive scaling to 2000 F. Furnace parts, turbine blades, pump and valve parts.	51446

* From *Handbook of Engineering Materials*, Miner and Seastone, New York, Wiley, 1955.
a Not hardenable by thermal treatment.
b Phosphorus or sulfur or selenium, 0.07% min.; zirconium or modybdenum, 0.60% max.

Table 18. Compositions, Characteristics, and Uses of Martensitic Stainless Steels*ᵃ

AISI Type Number	Chemical Composition, %								Magnetic	Characteristics and Typical Uses	SAE Type Number
	C	Mn max.	Si max.	P max.	S max.	Cr	Ni	Other elements			
403	0.15 max.	1.00	0.50	0.040	0.030	11.50–13.00	....	....	In all conditions	Modified type 410. Steam turbine blades and other highly stressed parts.	51410
410	0.15 max.	1.00	1.00	0.040	0.030	11.50–13.50	....	....	In all conditions	Low-priced, general purpose, heat-treatable; good mechanical properties and general corrosion resistance.	
414	0.15 max.	1.00	1.00	0.040	0.030	11.50–13.50	1.25–2.50	....	In all conditions	Somewhat better mechanical properties than type 410 because of addition of nickel, but more difficult to fabricate. Tempered rules, springs, scraper knives.	51414
416	0.15 max.	1.25	1.00	.....	.....	12.00–14.00	....	ᵇ	In all conditions	Most machinable of the stainless steels; properties similar to type 410; galling and seizing resistance. Automatic screw machine products and all machined parts.	51416F
420	Over 0.15	1.00	1.00	0.040	0.030	12.00–14.00	....	....	In all conditions	High hardness attainable by heat treatment. Cutlery, dental and surgical instruments, ball bearings, valves.	51420
431	0.20 max.	1.00	1.00	0.040	0.030	15.00–17.00	1.25–2.50	....	In all conditions	High strength as annealed and cold-worked; best corrosion resistance of martensitic types. Marine applications, precision springs, parts for petroleum valves.	51431
440A	0.60–0.75	1.00	1.00	0.040	0.030	16.00–18.00	....	Mo 0.75 max	In all conditions	Heat-treated hardness increases and ductility decreases with increasing carbon content; corrosion-resistant only in the fully hardened condition. Carving knives, surgical and dental tools, springs, ball bearings, and all parts requiring excellent abrasion resistance.	51440A
440B	0.75–0.95	1.00	1.00	0.040	0.030	16.00–18.00	....	Mo 0.75 max	In all conditions		51440B
440C	0.95–1.20	1.00	1.00	0.040	0.030	16.00–18.00	....	Mo 0.75 max	In all conditions		51440C

* From *Handbook of Engineering Materials*, Miner and Seastone, New York, Wiley, 1955.
ᵃ Hardenable by thermal treatment.
ᵇ Phosphorus or sulfur or selenium, 0.07% min.; zirconium or molybdenum, 0.60% max.

Table 19. Compositions, Characteristics, and Uses of Austenitic Stainless Steels*ᵃ

AISI Type Number	Chemical Composition, %								Magnetic		Characteristics and Typical Uses	SAE Type Number
	C	Mn max.	Si max.	P max.	S max.	Cr	Ni	Other elements	When annealed	When cold-worked		
301	Over 0.08–0.20	2.00	1.00	0.040	0.030	16.00–18.00	6.00–8.00	....	No	Yes	Attains high strength by being cold-worked, while retaining good ductility. Frequently used in service in cold-worked condition; trim, springs, household and dairy equipment.	30301
302	Over 0.08–0.20	2.00	1.00	0.040	0.030	17.00–19.00	8.00–10.00	....	No	Slightly	Most widely used type; generally used in annealed condition; deep-drawing and severe forming applications; textile, paper, chemical, dairy, and food handling.	30302
302B	Over 0.08–0.20	2.00	2.00–3.00	0.040	0.030	17.00–19.00	8.00–10.00	....	No	Slightly	Scaling resistance superior to type 302 because of higher silicon. Used exclusively for high-temperature applications.	
303	0.15 max.	2.00	1.00	......	......	17.00–19.00	8.00–10.00	b	No	Slightly	Free machining; resists galling and seizing. Wood and cap screws, bushings, valve and pump parts.	30303F
304	0.08 max.	2.00	1.00	0.040	0.030	18.00–20.00	8.00–11.00	....	No	Slightly	Better welding characteristics and less susceptibility to intergranular corrosion than type 302. Welded and fabricated structures.	30304
304L	0.03 max.	2.00	1.00	0.040	0.030	18.00–20.00	8.00–11.00	....	No	Slightly	Low carbon prevents loss of corrosion resistant property during welding.	
305	0.12 max.	2.00	1.00	0.040	0.030	17.00–19.00	10.00–13.00	....	No	Very slightly	Minimum tendency to work-harden because of increased nickel content. Developed for spinning and other severe cold-working operations.	30305
308	0.08 max.	2.00	1.00	0.040	0.030	19.00–21.00	10.00–12.00	....	No	Very slightly	Better strength and oxidation resistance than type 304. Works well in drawing, forming, and welding; core wire for welding electrodes.	
309	0.20 max.	2.00	1.00	0.040	0.030	22.00–24.00	12.00–15.00	....	No	Very slightly	Scaling resistance to 2000 F; excellent high-temperature tensile and creep strengths. Welding electrodes and heat-resisting applications.	30309
310	0.25 max.	2.00	1.50	0.040	0.030	24.00–26.00	19.00–22.00	....	No	No	Similar to type 309, but properties somewhat better owing to higher chromium and nickel contents.	30310

314	0.25 max.	2.00	1.50–3.00	0.040	0.030	23.00–26.00	19.00–22.00	....	No	No	Highest heat-resisting properties of the austenitic stainless steels. Furnace parts, heat exchangers, valve parts.	
316	0.08 max.	2.00	1.00	0.040	0.030	16.00–18.00	10.00–14.00	Mo 2.00–3.00	No	Slightly	Superior corrosion resistance to sea water and many chemicals; excellent high-temperature tensile and creep properties. Chemical, petroleum, and paper industries.	30316
316L	0.03 max.	2.00	1.00	0.040	0.030	16.00–18.00	10.00–14.00	Mo 2.00–3.00	No	Slightly	Low carbon prevents loss of corrosion resistant property during welding.	
317	0.08 max.	2.00	1.00	0.040	0.030	18.00–20.00	11.00–14.00	Mo 3.00–4.00	No	Very slightly	Better properties than type 316 because of higher alloy content.	30317
321	0.08 max.	2.00	1.00	0.040	0.030	17.00–19.00	8.00–11.00	Ti 5 × C min.	No	Slightly	Stabilized, resists intergranular corrosion in unannealed welds and parts in service at 800–1650 F. Aircraft exhaust, jet engine, and gas turbine parts.	30321
347	0.08 max.	2.00	1.00	0.040	0.030	17.00–19.00	9.00–12.00	Cb 10 × C min.	No	Slightly	Similar to type 321, except type 347 also suitable for use in stabilized welding electrodes.	30347

* From *Handbook of Engineering Materials*, Miner and Seastone, New York, Wiley, 1955.
a Not hardenable by thermal treatment.
b Phosphorus or sulfur or selenium, 0.07% min.; zirconium or molybdenum, 0.60% max.

Table 20. Non-resulfurized Standard Wrought Plain Carbon Basic Open-hearth and Acid Bessemer Steels*

Chemical Ranges

(Basic open-hearth and acid bessemer carbon steels)

AISI No.	Chemical Composition Limits, %				Corresponding SAE No.
	C	Mn	P max.	S max.	
1008	0.10 max.	0.25–0.50	0.040	0.050	1008
1010	0.08–0.13	0.30–0.60	0.040	0.050	1010
1012	0.10–0.15	0.30–0.60	0.040	0.050	66
1015	0.13–0.18	0.30–0.60	0.040	0.050	1015
1016	0.13–0.18	0.60–0.90	0.040	0.050	1016
1017	0.15–0.20	0.30–0.60	0.040	0.050	1017
1018	0.15–0.20	0.60–0.90	0.040	0.050	1018
1019	0.15–0.20	0.70–1.00	0.040	0.050	1019
1020	0.18–0.23	0.30–0.60	0.040	0.050	1020
1021	0.18–0.23	0.60–0.90	0.040	0.050	....
1022	0.18–0.23	0.70–1.00	0.040	0.050	1022
1023	0.20–0.25	0.30–0.60	0.040	0.050	....
1024	0.19–0.25	1.35–1.65	0.040	0.050	1024
1025	0.22–0.28	0.30–0.60	0.040	0.050	1025
1026	0.22–0.28	0.60–0.90	0.040	0.050	....
1027	0.22–0.29	1.20–1.50	0.040	0.050	1027
1029	0.25–0.31	0.60–0.90	0.040	0.050	....
1030	0.28–0.34	0.60–0.90	0.040	0.050	1030
1033	0.30–0.36	0.70–1.00	0.040	0.050	1033
1035	0.32–0.38	0.60–0.90	0.040	0.050	1035
1036	0.30–0.37	1.20–1.50	0.040	0.050	1036
1037	0.35–0.42	0.40–0.70	0.040	0.050	....
1038	0.35–0.42	0.60–0.90	0.040	0.050	1038
1039	0.37–0.44	0.70–1.00	0.040	0.050	1039
1040	0.37–0.44	0.60–0.90	0.040	0.050	1040
1041	0.36–0.44	1.35–1.65	0.040	0.050	1041
1042	0.40–0.47	0.60–0.90	0.040	0.050	1042
1043	0.40–0.47	0.70–1.00	0.040	0.050	1043
1045	0.43–0.50	0.60–0.90	0.040	0.050	1045
1046	0.43–0.50	0.70–1.00	0.040	0.050	1046
1049	0.46–0.53	0.60–0.90	0.040	0.050	1049
1050	0.48–0.55	0.60–0.90	0.040	0.050	1050
1052	0.47–0.55	1.20–1.50	0.040	0.050	1052
1053	0.48–0.55	0.70–1.00	0.040	0.050	....
1055	0.50–0.60	0.60–0.90	0.040	0.050	1055
1060	0.55–0.65	0.60–0.90	0.040	0.050	1060
1065	0.60–0.70	0.60–0.90	0.040	0.050	1065
1069	0.65–0.75	0.40–0.70	0.040	0.050	....
1070	0.65–0.75	0.60–0.90	0.040	0.050	1070
1072	0.65–0.76	1.00–1.30	0.040	0.050	....
1075	0.70–0.80	0.40–0.70	0.040	0.050	....
1078	0.72–0.85	0.30–0.60	0.040	0.050	1078
1080	0.75–0.88	0.60–0.90	0.040	0.050	1080
1084	0.80–0.93	0.60–0.90	0.040	0.050	....
1085	0.80–0.93	0.70–1.00	0.040	0.050	1085
1086	0.82–0.95	0.30–0.50	0.040	0.050	1086
1090	0.85–0.98	0.60–0.90	0.040	0.050	1090
1095	0.90–1.03	0.30–0.50	0.040	0.050	1095
B1010	0.13 max.	0.30–0.60	0.07–0.12	0.060	....

Silicon. When silicon is required, the following ranges and limits are commonly used for basic open-hearth steel grades:

Standard Steel Designations	Silicon Ranges or Limits
Up to 1015 excl.	0.10 max.
1015 to 1025 incl.	0.10 max., 0.10–0.20, or 0.15–0.30
Over 1025	0.10–0.20, or 0.15–0.30

Copper. When required, copper is specified as an added element to a standard steel.

---

* From *Handbook of Engineering Materials*, Miner and Seastone, New York, Wiley, 1955.

Table 21. Resulfurized Standard Wrought Plain Carbon Basic Open-hearth Steels*

Chemical Ranges

AISI No.	Chemical Composition Limits, %				Corresponding SAE No.
	C	Mn	P max.	S max.	
1108	0.08–0.13	0.50–0.80	0.040	0.08–0.13	....
1109	0.08–0.13	0.60–0.90	0.040	0.08–0.13	1109
1110	0.08–0.13	0.30–0.60	0.040	0.08–0.13	....
1113	0.10–0.16	1.00–1.30	0.040	0.24–0.33	....
1115	0.13–0.18	0.60–0.90	0.040	0.08–0.13	1115
1116	0.14–0.20	1.10–1.40	0.040	0.16–0.23	1116
1117	0.14–0.20	1.00–1.30	0.040	0.08–0.13	1117
1118	0.14–0.20	1.30–1.60	0.040	0.08–0.13	1118
1119	0.14–0.20	1.00–1.30	0.040	0.24–0.33	1119
1120	0.18–0.23	0.70–1.00	0.040	0.08–0.13	1120
1125	0.22–0.28	0.60–0.90	0.040	0.08–0.13	....
1132	0.27–0.34	1.35–1.65	0.040	0.08–0.13	1132
1137	0.32–0.39	1.35–1.65	0.040	0.08–0.13	1137
1138	0.34–0.40	0.70–1.00	0.040	0.08–0.13	1138
1140	0.37–0.44	0.70–1.00	0.040	0.08–0.13	1140
1141	0.37–0.45	1.35–1.65	0.040	0.08–0.13	1141
1144	0.40–0.48	1.35–1.65	0.040	0.24–0.33	1144
1145	0.42–0.49	0.70–1.00	0.040	0.04–0.07	1145
1146	0.42–0.49	0.70–1.00	0.040	0.08–0.13	1146
1148	0.45–0.52	0.70–1.00	0.040	0.04–0.07	....
1151	0.48–0.55	0.70–1.00	0.040	0.08–0.13	1151

Silicon. When silicon is required, the following ranges and limits are commonly used for basic open-hearth steel grades:

Standard Steel Designation	Silicon Ranges or Limits
Up to C1113 excl.	0.10 max.
C1113 and over	0.10 max., 0.10–0.20 or 0.15–0.30

* From *Handbook of Engineering Materials*, Miner and Seastone, New York, Wiley, 1955.

Table 22. Resulfurized Standard Wrought Plain Carbon Acid Bessemer Steels*

Chemical Ranges

AISI No.	Chemical Composition Limits, %				Corresponding SAE No.
	C	Mn	P	S	
B1111	0.13 max.	0.60–0.90	0.07–0.12	0.08–0.15	1111
B1112	0.13 max.	0.70–1.00	0.07–0.12	0.16–0.23	1112
B1113	0.13 max.	0.70–1.00	0.07–0.12	0.24–0.33	1113

Silicon. Because of the technological nature of the process, acid bessemer steels are not furnished with specified silicon content.

* From *Handbook of Engineering Materials*, Miner and Seastone, New York, Wiley, 1955.

Table 23.   Compositions of Wrought Alloy Steels—Open-hearth and Electric Furnace, Bars, Billets, Blooms, and Slabs[a]

The ranges and limits in the table apply to steel not exceeding 200 sq in. in cross-sectional area, or 18 in. in width, or 10,000 lb weight per piece.

AISI No.	C	Mn	P	S	Si	Ni	Cr	Mo
1320	0.18-0.23	1.60-1.90	0.040	0.040	0.20-0.35	...	...	...
1321	0.17-0.22	1.80-2.10	0.050	0.050	0.20-0.35	...	...	...
1330	0.28-0.33	1.60-1.90	0.040	0.040	0.20-0.35	...	...	...
1335	0.33-0.38	1.60-1.90	0.040	0.040	0.20-0.35	...	...	...
1340	0.38-0.43	1.60-1.90	0.040	0.040	0.20-0.35	...	...	...
2317	0.15-0.20	0.40-0.60	0.040	0.040	0.20-0.35	3.25-3.75	...	...
2330	0.28-0.33	0.60-0.80	0.040	0.040	0.20-0.35	3.25-3.75	...	...
2335	0.33-0.38	0.60-0.80	0.040	0.040	0.20-0.35	3.25-3.75	...	...
2340	0.38-0.43	0.70-0.90	0.040	0.040	0.20-0.35	3.25-3.75	...	...
2345	0.43-0.48	0.70-0.90	0.040	0.040	0.20-0.35	3.25-3.75	...	...
E2512	0.09-0.14	0.45-0.60*	0.025	0.025	0.20-0.35	4.75-5.25	...	...
2515	0.12-0.17	0.40-0.60	0.040	0.040	0.20-0.35	4.75-5.25	...	...
E2517	0.15-0.20	0.45-0.60*	0.025	0.025	0.20-0.35	4.75-5.25	...	...
3115	0.13-0.18	0.40-0.60	0.040	0.040	0.20-0.35	1.10-1.40	0.55-0.75	...
3120	0.17-0.22	0.60-0.80	0.040	0.040	0.20-0.35	1.10-1.40	0.55-0.75	...
3130	0.28-0.33	0.60-0.80	0.040	0.040	0.20-0.35	1.10-1.40	0.55-0.75	...
3135	0.33-0.38	0.60-0.80	0.040	0.040	0.20-0.35	1.10-1.40	0.55-0.75	...
3140	0.38-0.43	0.70-0.90	0.040	0.040	0.20-0.35	1.10-1.40	0.55-0.75	...
3141	0.38-0.43	0.70-0.90	0.040	0.040	0.20-0.35	1.10-1.40	0.70-0.90	...
3145	0.43-0.48	0.70-0.90	0.040	0.040	0.20-0.35	1.10-1.40	0.70-0.90	...
3150	0.48-0.53	0.70-0.90	0.040	0.040	0.20-0.35	1.10-1.40	0.70-0.90	...
E3310	0.08-0.13	0.45-0.60*	0.025	0.025	0.20-0.35	3.25-3.75	1.40-1.75	...
E3316	0.14-0.19	0.45-0.60*	0.025	0.025	0.20-0.35	3.25-3.75	1.40-1.75	...
4017	0.15-0.20	0.70-0.90	0.040	0.040	0.20-0.35	...	...	0.20-0.30
4023	0.20-0.25	0.70-0.90	0.040	0.040	0.20-0.35	...	...	0.20-0.30
4024	0.20-0.25	0.70-0.90	0.040	0.035-0.050	0.20-0.35	...	...	0.20-0.30
4027	0.25-0.30	0.70-0.90	0.040	0.040	0.20-0.35	...	...	0.20-0.30
4028	0.25-0.35	0.70-0.90	0.040	0.035-0.050	0.20-0.35	...	...	0.20-0.30
4032	0.30-0.35	0.70-0.90	0.040	0.040	0.20-0.35	...	...	0.20-0.30
4037	0.35-0.40	0.70-0.90	0.040	0.040	0.20-0.35	...	...	0.20-0.30
4042	0.40-0.45	0.70-0.90	0.040	0.040	0.20-0.35	...	...	0.20-0.30
4047	0.45-0.50	0.70-0.90	0.040	0.040	0.20-0.35	...	...	0.20-0.30

SAE No.	C	Mn	P	S	Si	Ni	Cr	Mo
4053	0.50-0.56	0.75-1.00	0.040	0.040	0.20-0.35			0.20-0.30
4063	0.60-0.67	0.75-1.00	0.040	0.040	0.20-0.35			0.20-0.30
4068	0.63-0.70	0.75-1.00	0.040	0.040	0.20-0.35			0.20-0.30
4130	0.28-0.33	0.40-0.60	0.040	0.040	0.20-0.35		0.80-1.10	0.15-0.25
E4132	0.30-0.35	0.40-0.60	0.025	0.025	0.20-0.35		0.80-1.10	0.18-0.25 †
E4135	0.33-0.38	0.70-0.90	0.025	0.025	0.20-0.35		0.80-1.10	0.18-0.25 †
4137	0.35-0.40	0.70-0.90	0.040	0.040	0.20-0.35		0.80-1.10	0.15-0.25
E4137	0.35-0.40	0.70-0.90	0.025	0.025	0.20-0.35		0.80-1.10	0.18-0.25
4140	0.38-0.43	0.75-1.00	0.040	0.040	0.20-0.35		0.80-1.10	0.15-0.25
4142	0.40-0.45	0.75-1.00	0.040	0.040	0.20-0.35		0.80-1.10	0.15-0.25
4145	0.43-0.48	0.75-1.00	0.040	0.040	0.20-0.35		0.80-1.10	0.15-0.25
4147	0.45-0.50	0.75-1.00	0.040	0.040	0.20-0.35		0.80-1.10	0.15-0.25
4150	0.48-0.53	0.75-1.00	0.040	0.040	0.20-0.35		0.80-1.10	0.15-0.25
4317	0.15-0.20	0.45-0.65	0.040	0.040	0.20-0.35	1.65-2.00	0.40-0.60	0.20-0.30
4320	0.17-0.22	0.45-0.65	0.040	0.040	0.20-0.35	1.65-2.00	0.40-0.60	0.20-0.30
4337	0.35-0.40	0.60-0.80	0.040	0.040	0.20-0.35	1.65-2.00	0.70-0.90	0.20-0.30
4340	0.38-0.43	0.60-0.80	0.040	0.040	0.20-0.35	1.65-2.00	0.70-0.90	0.20-0.30
4608	0.06-0.11	0.25-0.45	0.040	0.040	0.25 max.	1.40-1.75		0.15-0.25
4615	0.13-0.18	0.45-0.65	0.040	0.040	0.20-0.35	1.65-2.00		0.20-0.30
E4617	0.15-0.20	0.45-0.65	0.025	0.025	0.20-0.35	1.65-2.00		0.20-0.27 ‡
4620	0.17-0.22	0.45-0.65	0.040	0.040	0.20-0.35	1.65-2.00		0.20-0.30
E4620	0.17-0.22	0.45-0.65	0.025	0.025	0.20-0.35	1.65-2.00		0.20-0.27
4621	0.18-0.23	0.70-0.90	0.040	0.040	0.20-0.35	1.65-2.00		0.20-0.30
4640	0.38-0.43	0.60-0.80	0.040	0.040	0.20-0.35	1.65-2.00		0.20-0.30
E4640	0.38-0.43	0.60-0.80	0.025	0.025	0.20-0.35	1.65-2.00		0.20-0.27
4812	0.10-0.15	0.40-0.60	0.040	0.040	0.20-0.35	3.25-3.75		0.20-0.30
4815	0.13-0.18	0.40-0.60	0.040	0.040	0.20-0.35	3.25-3.75		0.20-0.30
4817	0.15-0.20	0.40-0.60	0.040	0.040	0.20-0.35	3.25-3.75		0.20-0.30
4820	0.18-0.23	0.50-0.70	0.040	0.040	0.20-0.35	3.25-3.75		0.20-0.30
5045	0.43-0.48	0.70-0.90	0.040	0.040	0.20-0.35		0.55-0.75	
5046	0.43-0.50	0.75-1.00	0.040	0.040	0.20-0.35		0.20-0.35	
5120	0.17-0.22	0.70-0.90	0.040	0.040	0.20-0.35		0.70-0.90	
5130	0.28-0.33	0.70-0.90	0.040	0.040	0.20-0.35		0.80-1.10	
5132	0.30-0.35	0.60-0.80	0.040	0.040	0.20-0.35		0.75-1.00	
5135	0.33-0.38	0.60-0.80	0.040	0.040	0.20-0.35		0.80-1.05	
5140	0.38-0.43	0.70-0.90	0.040	0.040	0.20-0.35		0.70-0.90	
5145	0.43-0.48	0.70-0.90	0.040	0.040	0.20-0.35		0.70-0.90	
5147	0.45-0.52	0.70-0.95	0.040	0.040	0.20-0.35		0.85-1.15	
5150	0.48-0.53	0.70-0.90	0.040	0.040	0.20-0.35		0.70-0.90	
5152	0.48-0.55	0.70-0.90	0.040	0.040	0.20-0.35		0.90-1.20	
5160	0.55-0.65	0.75-1.00	0.040	0.040	0.20-0.35		0.70-0.90	

Table 23.   Compositions of Wrought Alloy Steels—Open-hearth and Electric Furnace, Bars, Billets, Blooms and Slabs—Continued

Chemical Composition Limits, %

AISI No.	C	Mn	P	S	Si	Ni	Cr	Mo
E50100	0.95–1.10	0.25–0.45	0.025	0.025	0.20–0.35	....	0.40–0.60	....
E51100	0.95–1.10	0.25–0.45	0.025	0.025	0.20–0.35	....	0.90–1.15	....
E52100	0.95–1.10	0.25–0.45	0.025	0.025	0.20–0.35	....	1.30–1.60	....
								V
6120	0.17–0.22	0.70–0.90	0.040	0.040	0.20–0.35	....	0.70–0.90	0.10 min.
6145	0.43–0.48	0.70–0.90	0.040	0.040	0.20–0.35	....	0.80–1.10	0.15 min.
6150	0.48–0.53	0.70–0.90	0.040	0.040	0.20–0.35	....	0.80–1.10	0.15 min.
6152	0.48–0.55	0.70–0.90	0.040	0.040	0.20–0.35	....	0.80–1.10	0.10 min.
								Mo
8615	0.13–0.18	0.70–0.90	0.040	0.040	0.20–0.35	0.40–0.70	0.40–0.60	0.15–0.25
8617	0.15–0.20	0.70–0.90	0.040	0.040	0.20–0.35	0.40–0.70	0.40–0.60	0.15–0.25
8620	0.18–0.23	0.70–0.90	0.040	0.040	0.20–0.35	0.40–0.70	0.40–0.60	0.15–0.25
8622	0.20–0.25	0.70–0.90	0.040	0.040	0.20–0.35	0.40–0.70	0.40–0.60	0.15–0.25
8625	0.23–0.28	0.70–0.90	0.040	0.040	0.20–0.35	0.40–0.70	0.40–0.60	0.15–0.25
8627	0.25–0.30	0.70–0.90	0.040	0.040	0.20–0.35	0.40–0.70	0.40–0.60	0.15–0.25
8630	0.28–0.33	0.70–0.90	0.040	0.040	0.20–0.35	0.40–0.70	0.40–0.60	0.15–0.25
8632	0.30–0.35	0.70–0.90	0.040	0.040	0.20–0.35	0.40–0.70	0.40–0.60	0.15–0.25
8635	0.33–0.38	0.75–1.00	0.040	0.040	0.20–0.35	0.40–0.70	0.40–0.60	0.15–0.25
8637	0.35–0.40	0.75–1.00	0.040	0.040	0.20–0.35	0.40–0.70	0.40–0.60	0.15–0.25
8640	0.38–0.43	0.75–1.00	0.040	0.040	0.20–0.35	0.40–0.70	0.40–0.60	0.15–0.25
8641	0.38–0.43	0.75–1.00	0.040	0.040–0.060	0.20–0.35	0.40–0.70	0.40–0.60	0.15–0.25
8642	0.40–0.45	0.75–1.00	0.040	0.040	0.20–0.35	0.40–0.70	0.40–0.60	0.15–0.25
8645	0.43–0.48	0.75–1.00	0.040	0.040	0.20–0.35	0.40–0.70	0.40–0.60	0.15–0.25
8647	0.45–0.50	0.75–1.00	0.040	0.040	0.20–0.35	0.40–0.70	0.40–0.60	0.15–0.25
8650	0.48–0.53	0.75–1.00	0.040	0.040	0.20–0.35	0.40–0.70	0.50–0.80	0.15–0.25
8653	0.50–0.56	0.75–1.00	0.040	0.040	0.20–0.35	0.40–0.70	0.40–0.60	0.15–0.25
8655	0.50–0.60	0.75–1.00	0.040	0.040	0.20–0.35	0.40–0.70	0.40–0.60	0.15–0.25
8660	0.55–0.65	0.75–1.00	0.040	0.040	0.20–0.35	0.40–0.70	0.40–0.60	0.15–0.25
8719	0.18–0.23	0.60–0.80	0.040	0.040	0.20–0.35	0.40–0.70	0.40–0.60	0.20–0.30
8720	0.18–0.23	0.70–0.90	0.040	0.040	0.20–0.35	0.40–0.70	0.40–0.60	0.20–0.30
8735	0.33–0.38	0.75–1.00	0.040	0.040	0.20–0.35	0.40–0.70	0.40–0.60	0.20–0.30
8740	0.38–0.43	0.75–1.00	0.040	0.040	0.20–0.35	0.40–0.70	0.40–0.60	0.20–0.30
8742	0.40–0.45	0.75–1.00	0.040	0.040	0.20–0.35	0.40–0.70	0.40–0.60	0.20–0.30
8745	0.43–0.48	0.75–1.00	0.040	0.040	0.20–0.35	0.40–0.70	0.40–0.60	0.20–0.30
8747	0.45–0.50	0.75–1.00	0.040	0.040	0.20–0.35	0.40–0.70	0.40–0.60	0.20–0.30
8750	0.48–0.53	0.75–1.00	0.040	0.040	0.20–0.35	0.40–0.70	0.40–0.60	0.20–0.30

Grade	C	Mn	P	S	Si	Ni	Cr	Mo
9255	0.50–0.60	0.70–0.95	0.040	0.040	1.80–2.20	……	……	……
9260	0.55–0.65	0.70–1.00	0.040	0.040	1.80–2.20	……	……	……
9261	0.55–0.65	0.75–1.00	0.040	0.040	1.80–2.20	……	0.10–0.25	……
9262	0.55–0.65	0.75–1.00	0.040	0.040	1.80–2.20	……	0.25–0.40	……
E9310	0.08–0.13	0.45–0.65	0.025	0.025	0.20–0.35	3.00–3.50	1.00–1.40	0.08–0.15
E9315	0.13–0.18	0.45–0.65	0.025	0.025	0.20–0.35	3.00–3.50	1.00–1.40	0.08–0.15
E9317	0.15–0.20	0.45–0.65	0.025	0.025	0.20–0.35	3.00–3.50	1.00–1.40	0.08–0.15
9437	0.35–0.40	0.90–1.20	0.040	0.040	0.20–0.35	0.30–0.60	0.30–0.50	0.08–0.15
9440	0.38–0.43	0.90–1.20	0.040	0.040	0.20–0.35	0.30–0.60	0.30–0.50	0.08–0.15
9442	0.40–0.45	1.00–1.30	0.040	0.040	0.20–0.35	0.30–0.60	0.30–0.50	0.08–0.15
9445	0.43–0.48	1.00–1.30	0.040	0.040	0.20–0.35	0.30–0.60	0.30–0.50	0.08–0.15
9747	0.45–0.50	0.50–0.80	0.040	0.040	0.20–0.35	0.40–0.70	0.10–0.25	0.15–0.25
9763	0.60–0.67	0.50–0.80	0.040	0.040	0.20–0.35	0.40–0.70	0.10–0.25	0.15–0.25
9840	0.38–0.43	0.70–0.90	0.040	0.040	0.20–0.35	0.85–1.15	0.70–0.90	0.20–0.30
9845	0.43–0.48	0.70–0.90	0.040	0.040	0.20–0.35	0.85–1.15	0.70–0.90	0.20–0.30
9850	0.48–0.53	0.70–0.90	0.040	0.040	0.20–0.35	0.85–1.15	0.70–0.90	0.20–0.30

a From *Handbook of Engineering Materials*, Miner and Seastone, New York, Wiley, 1955.

* For open-hearth steel the manganese is 0.40 to 0.60%.

† For open-hearth steel the molybdenum is 0.15 to 0.25%.

‡ For open-hearth steel the molybdenum is 0.20 to 0.30%.

Note 1. Grades shown in the list with the prefix letter E generally are manufactured by the basic electric furnace process. All others are normally manufactured by the basic open-hearth process but may be manufactured by the basic electric furnace process with adjustments in phosphorus and sulfur.

Note 2. The phosphorus and sulfur limitations for each process are:

Basic electric furnace	0.025% max.	Acid electric furnace	0.05% max.
Basic open hearth	0.04% max.	Acid open hearth	0.05% max.

Note 3. Small quantities of certain elements are present in alloy steels which are not specified or required. These elements are considered incidental and may be present to the following maximum amounts: copper, 0.35%; nickel, 0.25%; chromium, 0.20%; molybdenum, 0.06%.

Note 4. Minimum silicon limit for acid open-hearth or acid electric furnace alloy steel is 0.15%.

Note 5. Where minimum and maximum sulfur content are shown, they are indicative of resulfurized steels.

Table 24. Compositions of High-strength Steels[a]

Trade Name*	Producer	Analysis									
		C	Mn	P	S	Si	Cu	Ni	Cr	Mo	Other
Cor-Ten	U. S. Steel Corp.	0.12 max.	0.20–0.50	0.07–0.15	0.05 max.	0.25–0.75	0.25–0.55	0.65 max.	0.50–1.25	…	…
Man-Ten	Same	0.30 max.	1.10–1.60	0.045 max.	0.05 max.	0.30 max.	0.20 min.	…	…	…	…
Tri-Ten (formerly Mn-Ni-Cu)	Same	0.25 max.	1.30 max.	0.045 max.	0.05 max.	0.10–0.30	0.30–0.60	0.50–1.00	…	…	…
Aldecor	Republic Steel Corp.	0.12 max.	0.15–0.40	0.08–0.15	0.05 max.	0.35–0.75	0.35–0.60	…	…	0.16–0.28	…
Double Strength	Same	0.12 max.	0.50–1.00	0.04 max.	0.05 max.	…	0.50–1.00	0.50–1.10	…	0.10 min.	…
High Steel	Inland Steel Co.	0.12 max.	0.60–0.90	0.05–0.12	0.05 max.	0.15 max.	0.95–1.30	0.45–0.75	…	0.08–0.18	Al 0.12–0.27
Mayari-R	Bethlehem Steel Co.	0.12 max.	0.50–1.00	0.08–0.12	0.05 max.	0.10–0.50	0.50–0.70	0.25–0.75	0.40–1.00	…	Zr 0.05–0.15
NAX High Tensile	Great Lakes Steel Corp.	0.08–0.15	0.50–0.75	0.04 max.	0.05 max.	0.60–0.90	…	…	0.50–0.65	…	…
Yoloy	Youngstown Sheet and Tube Co.	0.15 max.	0.75 max.	0.05–0.10	0.05 max.	0.30 max.	0.75–1.25	1.50–2.00	…	…	…
Otiscoloy	Jones & Laughlin Steel Corp.	0.15 max.	0.90–1.40	0.08–0.13	0.05 max.	0.10 max.	0.30 min.	…	…	…	…
Dynalloy	Alan Wood Steel Co.	0.20 max.	1.25 max.	0.100 max.	0.05 max.	0.30 max.	0.60 max.	1.00 max.	…	0.10 min.	…

[a] From Handbook of Engineering Materials, Miner and Seastone, New York, Wiley, 1955.

* Some brands are produced by more than one company through license agreements.

Table 25. Typical Mechanical Properties, Characteristics, and Applications of Carbon Steel Hot-rolled Bars*

Type	AISI No.	Estimated, psi Tensile strength	Yield point	Cold Forming	Forging	Case-hardening	Flame and Induction Hardening	Welding	Brazing Furnace process	Torch method	Heat Treating (Quench and Temper)	Typical Applications
Dead soft	1008	42,000	21,000	Good	Good	Fair	...	Good	Good	Good	...	Low-strength assemblies requiring best forming and welding properties
	1010	45,000	22,500	Good	Good	Fair	...	Good	Good	Good	...	
Soft	1015	50,000	25,000	Fair	Good	Good	...	Good	Good	Good	...	"As rolled" for many low-strength parts made from forgings or bar stock not requiring much machining; shafting, rolled thread bolts and screws; casehardened and heat-treated for ratchets, pins, rollers, shifter forks, etc.; in numerous welded assemblies
	1016	56,000	28,000	Fair	Good	Good	...	Good	Good	Good	...	
	1020	56,000	28,000	Fair	Good	Good	...	Good	Good	Good	...	
	1022	56,000	31,500	Fair	Good	Good	...	Good	Good	Good	...	
	1025	63,000	30,000	Fair	Good	Good	...	Good	Good	Good	...	
	1030	68,000	34,000	Fair	Good	Good	...	Good	Good	Good	...	
Medium-carbon	1035	73,000	36,500	...	Good	...	...	Fair	Good	Good	Good	"As rolled" for shafting, bolts, and studs; "as forged" for many medium-strength parts; higher strength obtained by heat treatment
	1040	77,000	38,500	...	Good	...	...	Fair	Good	Good	Good	
Heat-treating	1045	84,000	42,000	...	Good	...	Good	...	Fair	Fair	Good	"As forged" or "as forged and annealed" for shafting and machine parts; widely used for heat-treated parts and flame and induction hardening applications
	1050	93,000	51,500	...	Good	...	Good	...	Fair	Fair	Good	
High-carbon	1055	96,000	53,000	...	Good	...	Good	...	...	...	Good	In heat-treated condition for battering tools, ground working tools, small springs, and wear-resisting applications
	1060	102,000	56,000	...	Good	...	Good	...	...	...	Good	
	1070	110,000	60,500	...	Good	...	Good	...	...	...	Good	
	1080	118,000	65,000	...	Good	...	...	...	...	...	Good	
Maximum hardness	1095	130,000	71,500	...	Good	...	...	...	...	...	Good	Edge tools, springs, parts requiring maximum hardness and wear resistance
Open-hearth screw steels	1115	57,000	28,500	...	Fair	Fair	...	...	Good	Good	...	Bar and forging applications where free machining is desirable; parts made from these steels may be casehardened; frequently used cold-drawn
	1117	62,000	31,000	...	Fair	Good	...	...	Good	Good	...	
	1118	64,000	32,000	...	Fair	Good	...	...	Good	Good	...	
	1120	62,000	31,000	...	Fair	Fair	...	...	Good	Good	...	
Medium-carbon free-cutting	1137	87,000	48,000	...	Fair	...	...	...	Good	Good	Good	Free machining bars or forgings which may be heat-treated
	1141	92,000	50,500	...	Fair	...	Good	...	Good	Good	Good	
	1151	93,000	51,000	...	Fair	...	Good	...	Good	Good	Good	
Soft bessemer	B1110	55,000	27,500	Fair	Fair	Fair	...	...	Good	Good	Good	
Bessemer screw steels	B1111	60,000	30,000	...	...	Fair	...	...	...	...	Good	Used where free machining and good finish on machined surfaces are of great importance; widely used for machine parts not subjected to high stresses; have low shock resistance and should not be used for applications involving shock loading; usually used in cold-finished conditions; sometimes casehardened
	B1112	60,000	30,000	...	...	Fair	...	...	...	...	...	
	B1113	60,000	30,000	...	...	Fair	...	...	...	...	...	

* From *ASTM Bull.*, p. 33, October 1947.

Table 26. **Typical Mechanical Properties of Steels Used for Seamless Tubing**[a]
(Based on information from National Tube Co., Pittsburgh)

Grade or Designation	Condition	Yield Point, psi	Ultimate Strength, psi	Elongation, % in 2 in.	Equivalent Rockwell	Equivalent Brinell
AISI 1015	Annealed *	30,000	48,000	40	....	...
	Hard-drawn	67,000	80,000	15	B-84	159
AISI 1019	Annealed	35,000	55,000	40	B-64	107
	Hard-drawn	70,000	85,000	10	B-87	170
AISI 1025	Annealed	35,000	55,000	35	B-64	107
	Hard-drawn	72,000	85,000	10	B-87	170
AISI 1035	Annealed	38,000	63,000	35	B-71	124
	Hard-drawn	80,000	90,000	10	B-89	179
AISI 1040	Annealed	40,000	70,000	30	B-77	137
	Hard-drawn	85,000	95,000	8	B-92	192
AISI 1045	Annealed	45,000	73,000	25	B-80	146
	Hard-drawn	90,000	100,000	7	B-95	207
AISI 1050	Annealed	60,000	78,000	20	B-83	156
	Hard-drawn	95,000	105,000	7	B-96	217
AISI 1115	Annealed	30,000	48,000	40	....	...
	Hard-drawn	60,000	75,000	10	B-81	149
AISI 2317	Annealed	40,000	65,000	35	B-73	128
	Hard-drawn	85,000	95,000	10	B-92	192
AISI 2330	Annealed	55,000	75,000	30	B-81	149
	Hard-drawn	100,000	125,000	10	C-25	255
AISI 2340	Annealed	60,000	80,000	25	B-84	159
	Hard-drawn	110,000	130,000	10	C-26	262
AISI 3115	Hot-rolled	45,000	70,000	25	B-77	137
	Hard-drawn	90,000	100,000	15	B-94	202
AISI 3140	Annealed	60,000	80,000	25	B-84	159
	Hard-drawn	110,000	130,000	10	C-28	269
AISI 4130	Annealed	50,000	75,000	30	B-81	149
	Hard-drawn	110,000	130,000	10	C-26	262
AISI 4140	Annealed	50,000	80,000	25	B-84	159
	Hard-drawn	120,000	140,000	10	C-30	285
AISI 4615	Annealed	35,000	70,000	40	B-77	137
	Hard-drawn	95,000	105,000	20	B-96	212
5% Cr-Mo	Annealed	25,000	60,000	30	B-87	163
	Hot-rolled	45,000	85,000	25	B-94	202
AISI 304	Annealed	35,000	80,000	50	B-80	146
(18 Cr-8 Ni)	Hard-drawn	50,000-150,000	100,000-200,000	...	B-100–C-35	241-331
AISI 347	Annealed	35,000	80,000	50	B-80	146
(18 Cr-8 Ni-Cb)	Hard-drawn	50,000-150,000	100,000-200,000	...	B-100–C-35	241-331
AISI 321	Annealed	35,000	80,000	50	B-80	146
(18 Cr-8 Ni-Ti)	Hard-drawn	50,000-150,000	100,000-200,000	...	B-100–C-35	241-331
USS Cor-Ten	Annealed	45,000	65,000	45	B-73	128
(High-strength Steel)	Hot-rolled	50,000	80,000	40	B-84	159

[a] From *Handbook of Engineering Materials*, Miner and Seastone, New York, Wiley, 1955.
* Soft-annealed; represents softest possible condition.

Table 27. Chemical Compositions and Characteristics of the Cast Corrosion-reistant Alloy Steels*

Alloy Casting Institute Designation	AISI Type	C	Mn, max.	P, max.	S, max.	Si, max.	Cr	Ni	Mo	Other Elements	Characteristics and Uses
CA-15 CA-40	310 420	0.15 max. 0.20–0.40	1.00 1.00	0.04 0.04	0.04 0.04	1.5 1.5	11.5–14.0 11.5–14.0	1.0 max. 1.0 max.	0.05 max. 0.05 max.	… …	Good wear and corrosion resistance. Hardenable by heat treatment. Used for valve trim.
CB-30	431	0.30 max.	1.00	0.04	0.04	1.0	18.0–21.0	2.0 max.		…	High corrosion resistance. Slightly hardenable by heat treatment. Valve bodies and trim. Nitric acid containers.
CC-50	446	0.50 max.	1.00	0.04	0.04	1.0	26.0–30.0	4.0 max.		…	Resistant to oxidizing corrodents.
CE-30		0.30 max.	1.50	0.04	0.04	2.0	26.0–30.0	8.0–11.0		…	Not hardenable by heat treatment. Valve and pump parts for corrosive liquors.
CF-8 CF-20	304	0.08 max. 0.20 max.	1.50 1.50	0.04 0.04	0.04 0.04	2.0 2.0	18.0–21.0 18.0–21.0	8.0–11.0 8.0–11.0		… …	The CF alloys are the most widely used. Not hardenable by heat treatment, and nonmagnetic. Corrosion resistance decreases, strength increases with carbon.
CF-3 CF-3M	304L 316L	0.30 max. 0.30 max.	1.50 1.50	0.04 0.04	0.04 0.04	2.0 2.0	18.0–21.0 18.0–21.0	8.0–11.0 9.0–12.0		… …	Low carbon prevents loss of corrosion resistant property upon welding.
CF-8C	347	0.08 max.	1.50	0.04	0.04	2.0	18.0–21.0	9.0–12.0		8 × C to 1.0 columbium	Resistant to intergranular attack.
CF-8M CF-12M CF-16F	316	0.08 max. 0.12 max. 0.16 max.	1.50 1.50 1.50	0.04 0.04 0.17	0.04 0.04 0.04	2.0 1.5 2.0	18.0–21.0 18.0–21.0 18.0–21.0	9.0–12.0 9.0–12.0 9.0–12.0	2.0–3.0 2.0–3.0 1.5 max.	… … 0.20–0.35 selenium	Resistant to dilute sulfuric acid. Resistant to dilute sulfuric acid. Free machining type.
CH-20 CK-20	309 310	0.20 max. 0.20 max.	1.50 1.50	0.04 0.04	0.04 0.04	2.0 2.0	22.0–26.0 23.0–27.0	12.0–15.0 19.0–22.0		… …	Chemical and paper industry.

* From *Handbook of Engineering Materials*, Miner and Seastone, New York, Wiley, 1955.

Table 28. Mechanical and Physical Properties of Cast Corrosion-resistant Alloy Steels*

Alloy Casting Institute Designation	AISI Type	Tensile Strength, psi	Yield Strength, psi	Elongation, % in 2 in.	Brinell Hardness	Mean Coefficient of Linear Thermal Expansion, per degree F	Modulus of Elasticity in Tension, psi
CA-15	410	120,000	110,000	18	241	70 to 1300—0.0000067	29,000,000
CA-40	420	135,000	120,000	10	269	70 to 1300—0.0000067	29,000,000
CB-30	431	75,000	50,000	10	170	{ 70 to 212—0.0000057 / 70 to 1300—0.0000067 }	29,000,000
CC-50	446	97,000	65,000	38	187	{ 32 to 212—0.0000059 / 32 to 1000—0.0000063 }	29,000,000
CE-30		97,000	67,000	18	170	120 to 1600—0.0000105	25,000,000
CF-8	304	78,000	38,000	55	140	{ 32 to 212—0.0000096 / 32 to 1100—0.0000104 }	28,000,000
CF-20		83,000	42,000	55	163	{ 32 to 212—0.0000096 / 32 to 1100—0.0000104 }	28,000,000
CF-8C	347	85,000	44,000	45	149	{ 32 to 212—0.0000093 / 32 to 950—0.0000103 }	28,000,000
CF-8M	316	88,000	45,000	50	156	{ 32 to 212—0.0000089 / 32 to 950—0.0000097 }	28,000,000
CF-12M		88,000	45,000	50	170	{ 32 to 212—0.0000089 / 32 to 950—0.0000097 }	28,000,000
CF-16F		80,000	42,000	45	...	{ 32 to 212—0.0000090 / 32 to 950—0.0000099 }	28,000,000
CH-20	309	88,000	50,000	38	184	{ 32 to 212—0.0000083 / 32 to 950—0.0000096 / 32 to 1850—0.0000115 }	28,000,000
CK-20	310	76,000	38,000	21	184	{ 32 to 212—0.0000080 / 32 to 950—0.0000092 / 32 to 1850—0.0000106 }	29,000,000

* From *Handbook of Engineering Materials*, Miner and Seastone, New York, Wiley, 1955.

Table 29. Chemical Compositions and Characteristics of Cast Heat-resistant Alloy Steels*

Alloy Casting Institute Designation	AISI Type	Composition, %									Characteristics and Uses
		C	Mn, max.	P, max.	S, max.	Si, max.	Cr	Ni	Mo	Other Elements	
HA		0.20 max.	0.35–0.65	0.04	0.04	1.0	8.0–10.0	....	0.90–1.20	....	Low strength but good heat resistance at high temperatures. Useful in high-sulfur atmospheres.
HC	446	0.50 max.	1.0	0.04	0.04	2.0	26.0–30.0	4.0 max.	0.50	....	Similar to type HC but better high-temperature strength.
HD	327	0.50 max.	1.5	0.04	0.04	2.0	26.0–30.0	4.0–7.0	0.50	....	Excellent scaling resistance to 2000 F.
HE	302B	0.20–0.50	2.0	0.04	0.04	2.0	26.0–30.0	8.0–11.0	0.50	....	Popular heat- and corrosion-resistant alloy for use to 1600 F.
HF		0.20–0.40	2.0	0.04	0.04	2.0	18.0–23.0	8.0–12.0	0.50	....	
HH	309	0.20–0.50	2.0	0.04	0.04	2.0	24.0–28.0	11.0–14.0	0.50	Nitrogen 0.20 max.	Widely used. High strength and corrosion resistance to 2000 F.
HI		0.20–0.50	2.0	0.04	0.04	2.0	26.0–30.0	14.0–18.0	0.50	....	Useful to 2150 F.
HK	310	0.20–0.60	2.0	0.04	0.04	2.0	24.0–28.0	18.0–22.0	0.50	....	Highly stable to phase changes. Good resistance to sulfur-bearing atmospheres.
HL		0.20–0.60	2.0	0.04	0.04	3.0	28.0–32.0	18.0–22.0	0.50	....	As above.
HT	330	0.35–0.75	2.0	0.04	0.04	2.5	13.0–17.0	33.0–37.0	0.50	....	Excellent resistance to oxidizing and reducing conditions at high temperatures.
HU		0.35–0.75	2.0	0.04	0.04	2.5	17.0–21.0	37.0–41.0	0.50	....	Heat-treating furnace parts.
HW		0.35–0.75	2.0	0.04	0.04	2.5	10.0–14.0	58.0–62.0	0.50	....	High resistance to oxidation, carburization, nitriding, and thermal shock. Furnace parts and quenching fixtures.
HX		0.35–0.75	2.0	0.04	0.04	2.5	15.0–19.0	64.0–68.0	0.50	....	

* From *Handbook of Engineering Materials*, Miner and Seastone, New York, Wiley, 1955.

Table 30. Mechanical and Physical Properties of Cast Heat-resistant Alloy Steels*

Alloy Casting Institute Designation	AISI Type	Condition	Tensile Strength, psi	Yield Strength, psi	Elongation, % in 2 in.	Brinell Hardness	Mean Coefficient of Linear Thermal Expansion, per degree F	Modulus of Elasticity in Tension, psi
HC	446	As cast	70,000	45,000	4	212	32 to 1000—0.0000063	29,000,000
		Aged	75,000	50,000	3	218	32 to 1850—0.0000076	
HE		As cast	98,000	50,000	10	200	120 to 1650—0.0000105	25,000,000
		Aged	102,000	55,000	7	215		
HF	302B	As cast	80,000	45,000	35	170	80 to 1600—0.0000104	28,000,000
		Aged	100,000	55,000	25	180		
HI		As cast	70,000	45,000	8	...	....	....
HK	310	As cast	75,000	47,000	17	170	32 to 950—0.0000092	29,000,000
		Aged	80,000	50,000	11	190	32 to 1850—0.0000106	
HL		As cast	82,000	52,000	19	192	....	29,000,000
HT	330	As cast	70,000	40,000	10	168	70 to 1200—0.0000089	24,000,000
		Aged	85,000	55,000	6	205	70 to 1850—0.0000097	
HU		As cast	70,000	35,000	9	170	32 to 1850—0.0000095	24,000,000
		Aged	75,000	41,000	7	200		
HW		As cast	68,000	34,000	6	179	32 to 1850—0.0000090	25,000,000
		Aged	83,000	55,000	5	190		
HX		As cast	75,000	37,000	9	200	32 to 1850—0.0000085	25,000,000
		Aged	88,000	44,000	8	205		

* From *Handbook of Engineering Materials*, Miner and Seastone, New York, Wiley, 1955.

Table 31.   Temper and Radiation Colors of Carbon Steel

°F	Approx. °C	Temper color
380–400	200	Pale yellow
420–440	220	Straw yellow
460–480	240	Yellowish brown
500–540	270	Bluish purple
540–560	285	Violet
560–580	300	Pale blue
600–640	325	Blue
		Radiation color
1000	540	Black
1100	590	Faint dark red
1200	650	Cherry red (dark)
1300	700	Cherry red (med.)
1400	760	Red
1500	815	Light red
1600	870	Reddish orange
1700	930	Orange
1800	980	Changes
1900	1040	to
2000	1090	Pale orange lemon
2100	1150	Lemon
2200	1205	Light lemon
2300	1260	Yellow
2400	1315	Light yellow
2500	1370	Yellowish gray: "white"

Table 32.   Maximum Service Temperature for Various Steels

Type of Steel	Maximum Temperature without Excessive Scaling (weight gain 10 mg per cm² in 1000 hours)	
	°F	Approx. °C
Carbon steel	1050	565
½ Mo steel	1050	565
1 Cr ½ Mo steel	1100	595
2¼ Cr 1 Mo steel	1175	635
5 Cr ½ Mo steel	1200	650
12 Cr AISI 410	1250	675
18 Cr 8 Ni Ti AISI 321	1600	870
18 Cr 8 Ni Nb AISI 347	1600	870
18 Cr 8 Ni Mo AISI 316	1600	870
25 Cr 20 Ni AISI 310	2100	1150

Table 33. Properties of Aluminum

		Sample Purity, per cent
Atomic weight....................................	26.97	....
Boiling point, °C.................................	2060	99.996
Crystal form.....................................	Face-centered cubic	....
Mean coefficient of expansion { 20 to 300 C.............	0.0000257	99.95 (cast)
{ 20 to 500 C.............	0.0000277	99.95 (cast)
Density at 20 C (68 F), g/cu cm....................	2.70	99.971 (wrought-annealed)
Density at melting point (solid), g/cu cm.............	2.55	....
Density at melting point (liquid), g/cu cm.............	2.38	....
Electric resistivity at 20 C:		
Microhm-cm..................................	2.688	99.968 (hard drawn)
Ohms (mil, ft) (A.I.E.E. Standard).................	17.01	Commercial
Temperature coefficient at 20 C (A.I.E.E. Standard)....	0.00403	Commercial
Electric conductivity at 20 C:		
Mass, per cent International Annealed Copper Standard	212.9	....
Volume, per cent International Annealed Copper Standard.....................................	64.9	99.996
Freezing point, °C................................	660.2	99.996
Heat of vaporization, g-cal/g......................	1950 to 2000	....
Latent heat of fusion { g-cal/g.......................	94.6	99.996
{ Btu/lb........................	170.27	....
Mechanical properties:		
Tensile strength, psi.............................	9000	Annealed
Yield strength, psi (set = 0.2 per cent)..............	3000	sheet
Elongation in 2 in., per cent......................	60	99.95
Brinell hardness, 10-mm ball, 500-kg load...........	17	99.996 (annealed)
Modulus of elasticity, psi..........................	10,000,000	....
Modulus of rigidity (torsion), psi...................	3,870,000	Commercial
Poisson's ratio....................................	0.33	....
Total reflectivity, per cent for white light..............	87	....
Mean specific heat, g-cal/g/°C (0-100 C)..............	0.226	....
Thermal conductivity, cgs units.....................	0.52	99.66
Watts/sq cm/°C/cm.............................	2.17	99.66
Btu/hr/sq ft/°F/in...............................	1509	....

Table 34. Four-Digit Designation System for Wrought Aluminum Alloys*

This diagram shows the significance of the numbers in The Aluminum Association standard designations for wrought alloys.

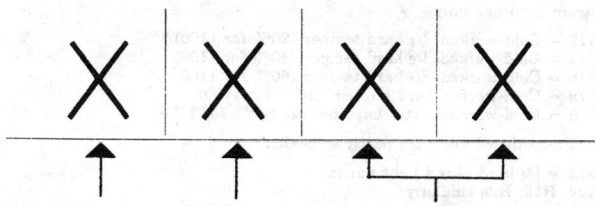

| This digit identifies alloy type | This digit identifies alloy modification. Modifications were formerly indicated by letters. In changing to the new system, the letter is replaced by the digit corresponding to its position in the alphabet. For example, A17S becomes 2117. Zero is the original alloy. | These two digits identify the aluminum purity or the specific aluminum alloy. For alloys in use prior to the adoption of the four-digit system, the digits are the same as the numbers in the old designation. For example, 24S becomes 2024. |

Type of Aluminum Alloy	Number	Group		
Aluminum—99.00% minimum and greater	1	x	x	x
Copper	2	x	x	x
Manganese	3	x	x	x
Silicon	4	x	x	x
Magnesium	5	x	x	x
Magnesium and silicon	6	x	x	x
Zinc	7	x	x	x
Other element	8	x	x	x
Unused series	9	x	x	x

* From Alcoa Aluminum Handbook, 1962 edition.

Table 35. Key to Temper Designations for Aluminum Alloys*

F = As fabricated (no treatment subsequent to forming)
O = Annealed (wrought products only)
H = Strain-hardened (cold-worked)

    H1 = Strain hardened only

        H12 = Cold-worked, $1/4$ hard temper (20% for 1100) [a]
        H14 = Cold-worked, $1/2$ hard temper (40% for 1100)
        H16 = Cold-worked, $3/4$ hard temper (60% for 1100)
        H18 = Cold-worked hard temper (80% for 1100)
        H19 = Cold-worked, extra hard temper (90% for 1100)

    H2 = Strain-hardened and then partly annealed

        H22 = $1/4$ hard plus a light anneal
        H24, H26, H28 similarly

    H3 = Strain-hardened and then stabilized

        H32 = $1/4$ hard plus a stabilizing heat treatment
        H34, H36, H38, H39 similarly

W = Solution-heat-treated and quenched (unstable temper)
T = Heat-treated to produce tempers other than F, O, or H [b]

    T2 = Annealed (cast products only)
    T3 = Solution-heat-treated and then cold-worked
    T4 = Solution-heat-treated and naturally aged to a stable condition
    T5 = Artificially aged only (no solution heat treatment—as in certain permanent-mold alloys and extrusions)
    T6 = Solution-heat-treated and then artificially aged (aged at an elevated temperature)
    T7 = Solution-heat-treated and then stabilized (somewhat overaged to provide dimensional stability)
    T8 = Solution-heat-treated, cold-worked, and then artificially aged
    T9 = Solution-heat-treated, artificially aged, and then cold-worked
    T10 = Artificially aged and then cold-worked.

* From *Handbook of Engineering Materials*, Miner and Seastone, New York, Wiley, 1955.
[a] Percentage reduction in cross-sectional area.
[b] For casting alloys the T designations may be followed by additional numbers designating a specific heat treatment; see *Metals Handbook*, under the specific casting alloy.

Table 36. Approximate Effect of Cold Working on Tensile Strength of Wrought Aluminum Alloys

Temper of Metal	Approximate Amount of Cold Working [a]	Approximate Relative Tensile Strength [b]
—O (soft)	0	1.0
—H12 (¼ hard)	20	1.18
—H14 (½ hard)	40	1.36
—H16 (¾ hard)	60	1.54
—H18 (hard temper)	80	1.72
—H19 (extra hard temper)	90	1.81

[a] Percentage reduction in cross-sectional area for Type 1100 alloy.
[b] From the relation: $T = T_0(1 + 0.9R)$; where T = tensile strength, psi; T_0 = tensile strength of the alloy in the fully annealed state, psi; and R = reduction in area expressed as a decimal.

Table 37. Nominal Composition of Wrought Aluminum Alloys*

Item No.	Alloy	Per Cent of Alloying Elements—Aluminum and Normal Impurities Constitute Remainder								
		Silicon	Copper	Manganese	Magnesium	Chromium	Nickel	Zinc	Lead	Bismuth
1	EC^a	99.60 per cent minimum aluminum								
2	1060	99.60 per cent minimum aluminum								
3	1100	99.00 per cent minimum aluminum								
4	2EC	0.40	...	...	0.6	...	...	...	...	...
5	2011	...	5.5	...	...	...	...	...	0.50	0.50
6	2014	0.8	4.4	0.8	0.40	...	...	...	...	...
7	2017	...	4.0	0.50	0.50	...	...	...	...	...
8	2018	...	4.0	...	0.6	...	2.0	...	...	...
9	2024	...	4.5	0.6	1.5	...	...	...	...	...
10	2025	0.8	4.5	0.8	...	...	...	...	...	...
11	2117	...	2.5	...	0.30	...	...	...	...	...
12	2218	...	4.0	...	1.5	...	2.0	...	...	...
13	2219^b	...	6.3	0.30	...	...	...	...	...	...
14	3003	...	...	1.2	...	...	...	...	...	...
15	3004	...	...	1.2	1.0	...	...	...	...	...
16	3105	...	...	0.9	0.50	...	...	...	...	...
17	4032	12.2	0.9	...	1.1	...	0.9	...	...	...
18	5005	...	...	...	0.8	...	...	...	...	...
19	5050	...	...	...	1.4	...	...	...	...	...
20	5052	...	...	...	2.5	0.25	...	...	...	...
21	5056	...	...	0.10	5.2	0.10	...	...	...	...
22	5083	...	...	0.8	4.45	0.10	...	...	...	...
23	5086	...	...	0.45	4.0	0.10	...	...	...	...
24	5154	...	...	...	3.5	0.25	...	...	...	...
25	5254	...	...	...	3.5	0.25	...	...	...	...
26	5356^c	...	...	0.10	5.0	0.10	...	...	...	...
27	5357	...	...	0.30	1.0	...	...	...	...	...
28	5454	...	...	0.8	2.75	0.10	...	...	...	...
29	5456	...	...	0.8	5.25	0.10	...	...	...	...
30	5457	...	...	0.30	1.0	...	...	...	...	...
31	5557	...	...	0.25	0.6	...	...	...	...	...
32	5652	...	...	...	2.5	0.25	...	...	...	...
33	6053	0.7	...	...	1.3	0.25	...	...	...	...
34	6061	0.6	0.25	...	1.0	0.25	...	...	...	...
35	6062	0.6	0.25	...	1.0	0.06	...	...	...	...
36	6063	0.40	...	...	0.7	...	...	...	...	...
37	6066	1.3	0.9	0.9	1.1	...	...	...	...	...
38	6151	1.0	...	...	0.6	0.25	...	...	...	...
39	6262	0.6	0.25	...	1.0	0.90	...	...	0.50	0.50
40	6463	0.40	...	...	0.7	...	...	...	...	...
41	6563	0.30	...	...	0.50	...	...	...	...	...
42	6951	0.30	0.25	...	0.6	...	...	...	...	...
43	7072	...	...	...	...	...	...	1.0	...	...
44	7075	...	1.6	...	2.5	0.30	...	5.6	...	...
45	7079	...	0.6	0.20	3.3	0.20	...	4.3	...	...
46	7178	...	2.0	...	2.7	0.30	...	6.8	...	...
47	7277	...	1.25	...	2.0	0.25	...	4.0	...	...

* From *Alcoa Aluminum Handbook*, 1962 edition.
^a Boron 0.02.
^b Titanium 0.06; vanadium 0.10; zirconium 0.18.
^c Nominal titanium content—0.10 per cent.

Table 38. Typical Physical Properties of Wrought Aluminum Alloys*

Item No.	Alloy and Temper	Specific Gravity	Weight, lbs/cu in.	Melting Range Approximate, °F	Electrical Conductivity at 20°C (68°F) Per cent of International Annealed Copper Standard	Thermal Conductivity at 25°C (77°F) CGS units[a]
1	EC-O	2.70	0.098	1195–1215	63	0.57
1a	EC-H19				62.5	0.56
2	1060-O	2.70	0.098	1195–1215	62	0.56
2a	1060-H18				61	0.55
3	1100-O	2.71	0.098	1190–1215	59[b]	0.53
3a	1100-H18					
4a	2EC-T6	2.70	0.098	1150–1210	57	0.52
4b	2EC-T61				59	0.53
4c	2EC-T62				58	0.52
4d	2EC-T64				60.5	0.54
5a	2011-T3	2.82	0.102	995–1190	39	0.36
5b	2011-T8				45	0.41
6	2014-O	2.80	0.101	950–1180	50	0.46
6a	2014-T4				34	0.32
6b	2014-T6				40	0.37
7	2017-O	2.79	0.101	955–1185	50	0.46
7a	2017-T4				34	0.32
8a	2018-T61	2.80	0.101	945–1180	40	0.37
9	2024-O	2.77	0.100	935–1180	50	0.46
9a–9f	2024-T3, -T36, -T4				30	0.29
	2024-T6, -T81, -T86				38	0.36
10a	2025-T6	2.79	0.101	970–1185	40	0.37
11a	2117-T4	2.74	0.099	950–1200	40	0.37
12a	2218-T72	2.80	0.101	940–1175	40	0.37

No.	Alloy/Temper					
13	2219-O	2.83	0.103	1010–1190	44	0.41
13a	2219-T31, -T37				28	0.27
13e	2219-T62, -T81, -T87				32	0.30
14	3003-O	2.73	0.099	1190–1210	46[b]	0.42
14a	3003-H18					
15	3004-O	2.72	0.098	1165–1210	42[b]	0.39
15a	3004-H38					
16a	3105-H25	2.73	0.099	1180–1215	:	.....
17	4032-O	2.69	0.097	990–1060	40	0.37
17a	4032-T6				35	0.33
18	5005-O	2.69	0.097	1170–1210	54[b,c]	0.49[c]
18a	5005-H38					
19	5050-O	2.69	0.097	1155–1205	50[b]	0.46
19a	5050-H38					
20	5052-O	2.68	0.097	1125–1200	35[b]	0.33
20a	5052-H38					
21	5056-O	2.64	0.095	1055–1180	29[b]	0.28
21a	5056-H38					
22	5083-O	2.66	0.096	1075–1185	29[b]	0.28
22a	5083-H343					
22b	5083-H113					
23	5086-O	2.66	0.096	1085–1185	32[b]	0.30
23a	5086-H34					
24	5154-O	2.66	0.096	1100–1190	32[b]	0.30
24a	5154-H38					
25	5254-O	2.66	0.096	1100–1190	32[b]	0.30
25a	5254-H38					
26	5356-O	2.64	0.095	1060–1180	29	0.28

(Continued)

Table 38. Typical Physical Properties of Wrought Aluminum Alloys—*Continued*

Item No.	Alloy and Temper	Specific Gravity	Weight, lbs/cu in.	Melting Range Approximate, °F	Electrical Conductivity at 20°C (68°F) Per cent of International Annealed Copper Standard	Thermal Conductivity at 25°C (77°F) CGS units[a]
27 27a	5357-O 5357-H38	2.70	0.098	1165–1210	43[b]	0.40
28 28a	5454-O 5454-H34	2.68	0.097	1115–1195	34[b]	0.32
29 29a	5456-O 5456-H321	2.66	0.096	1055–1180	29[b]	0.28
30 30a	5457-O 5457-H38	2.70	0.098	1165–1210	46[b]	0.42
31 31a	5557-O 5557-H38	2.70	0.098	1180–1215	49[b]	0.45
32 32a	5652-O 5652-H38	2.68	0.097	1125–1200	35[b]	0.33
33–33a 33b 33c	6053-O, -T5 6053-T4 6053-T6	2.69	0.097	1100–1205	45 40 42	0.41 0.37 0.39
34 34a 34b	6061-O 6061-T4 6061-T6	2.70	0.098	1100–1205	47 40 43	0.43 0.37 0.40
35 35a 35b	6062-O 6062-T4 6062-T6	2.70	0.098	1100–1205	51 44 47	0.47 0.41 0.43
36 36b 36c 36d–36e	6063-O 6063-T42 6063-T5 6063-T6, -T83	2.70	0.098	1140–1210	58 50 55 53	0.52 0.46 0.50 0.48

38	6151-O	2.70	0.098	1025–1200	54	0.49
38a	6151-T4				42	0.39
38b	6151-T6				45	0.41
39a	6262-T9	2.72	0.098	1100–1205	44	0.41
40	6463-O	2.70	0.098	1140–1210	58	0.52
40b	6463-T42				50	0.46
40c	6463-T5				55	0.50
40d	6463-T6				53	0.48
41a	6563-T4	2.70	0.098	1160–1210	50	0.46
41b	6563-T6				55	0.50
42	6951-O	2.70	0.098	1140–1210	...	....
42a	6951-T6				...	....
43	7072-O	2.72	0.098	1195–1215	59	0.53
44a	7075-T6	2.80	0.101	890–1175	33	0.31
45a	7079-T6	2.74	0.099	900–1180	32	0.30
46a	7178-T6	2.81	0.102	890–1165	31	0.30

* From *Alcoa Aluminum Handbook*, 1962 edition.

a CGS units = cal/cm/cm²/°C/sec.

b Average for range.

c When fabricated for use as electrical conductors, average values for electrical conductivity of the -O and -H19 tempers are 56 and 55 percent respectively, and the corresponding values for thermal conductivities are 0.51 and 0.50.

Table 39. Typical Mechanical Properties of Wrought Aluminum Alloys*[a]

Item No.	Alloy and Temper	Tension — Strength, psi Ultimate	Tension — Strength, psi Yield	Elongation in 2 inches, per cent 1/16-inch Thick Specimen	Elongation in 2 inches, per cent 1/2-inch Diameter Specimen	Hardness Brinell Number 500-kg load 10-mm ball	Shear Shearing Strength, psi	Fatigue Endurance[b] Limit, psi	Modulus — Modulus[c] of Elasticity, psi
1	EC-O[d]	10,000	4,000	..	..[e]	..	8,000	.....	10.0×10^6
1a	EC-H19	27,000	24,000	..	..[f]	..	15,000	7,000	10.0×10^6
2	1060-O	10,000	4,000	43	..	19	7,000	3,000	10.0×10^6
2a	1060-H18	19,000	18,000	6	..	35	11,000	6,500	10.0×10^6
3	1100-O	13,000	5,000	35	40	23	9,000	5,000	10.0×10^6
3a	1100-H18	24,000	22,000	5	15	44	13,000	9,000	10.0×10^6
4a	2EC-T6	32,000	28,000	..	19	70	22,000	9,000	10.0×10^6
4b	2EC-T61	25,000	20,000	..	22	55	17,000	9,000	10.0×10^6
4c	2EC-T62	30,000	25,000	..	20	65	21,000	9,000	10.0×10^6
4d	2EC-T64	17,000	9,000	..	24	35	13,000	7,000	10.0×10^6
5a	2011-T3	55,000[e]	43,000[g]	..	15	95	32,000	18,000	10.2×10^6
5b	2011-T8	59,000	45,000	..	12	100	35,000	18,000	10.2×10^6
6	2014-O	27,000	14,000	..	18	45	18,000	13,000	10.6×10^6
6a	2014-T4	62,000[m]	42,000[m]	..	20	105	38,000	20,000	10.6×10^6
6b	2014-T6	70,000[h]	60,000[h]	..	13	135	42,000	18,000	10.6×10^6
C 6	Alclad 2014-O	25,000	10,000	21	..	..	18,000	......	10.5×10^6
C 6a	Alclad 2014-T3	63,000[i]	40,000[i]	20	..	..	37,000	......	10.5×10^6
C 6b	Alclad 2014-T4	61,000[i]	37,000[i]	22	..	..	37,000	......	10.5×10^6
C 6c	Alclad 2014-T6	68,000[i]	60,000[i]	10	..	..	41,000	......	10.5×10^6
7	2017-O	26,000	10,000	..	22	45	18,000	13,000	10.5×10^6
7a	2017-T4	62,000	40,000	..	22	105	38,000	18,000	10.5×10^6
8a	2018-T61	61,000	46,000	..	12	120	39,000	17,000	10.6×10^6

Index	Alloy	$\times 10^6$							
9	2024-O	10.6×10^6	13,000	18,000	47	22	20	11,000	27,000
9a	2024-T3	10.6×10^6	20,000	41,000	120		18	50,000	70,000
9b	2024-T36	10.6×10^6	18,000	42,000	130		13	57,000	72,000
9c	2024-T4	10.6×10^6	20,000	41,000	120	19	20	47,000[A]	68,000[A]
9d	2024-T6	10.6×10^6	18,000	41,000	125	10		57,000	69,000
9e	2024-T81	10.6×10^6	18,000	43,000	128		6	65,000	70,000
9f	2024-T86	10.6×10^6		45,000	135		6	71,000	75,000
C 9	Alclad 2024-O	10.6×10^6		18,000			20	11,000	26,000
C 9a	Alclad 2024-T3	10.6×10^6		40,000			18	45,000[j]	65,000[j]
C 9b	Alclad 2024-T36	10.6×10^6		41,000			11	53,000[j]	67,000[j]
C 9c	Alclad 2024-T4	10.6×10^6		40,000			19	42,000[j]	64,000[j]
C 9e	Alclad 2024-T81	10.6×10^6		40,000			6	60,000[j]	65,000[j]
C 9f	Alclad 2024-T86	10.6×10^6		42,000			6	66,000[j]	70,000[j]
10a	2025-T6	10.4×10^6	18,000	35,000	110	19		37,000	58,000
11a	2117-T4	10.3×10^6	14,000	28,000	70	27		24,000	43,000
12a	2218-T72	10.8×10^6		30,000	95	11		37,000	48,000
13	2219-O	10.6×10^6					20	10,000	25,000
13a	2219-T31	10.6×10^6		33,000	100		17	37,000	54,000
13b	2219-T37	10.6×10^6		37,000	117		11	49,000	60,000
13c	2219-T62	10.6×10^6	15,000	37,000	115		11	42,000	61,000
13d	2219-T81	10.6×10^6	15,000	41,000	130		11	53,000	70,000
13e	2219-T87	10.6×10^6	15,000	41,000	130		10	58,000	70,000
14	3003-O	10.0×10^6	7,000	11,000	28	40	30	6,000	16,000
14a	3003-H18	10.0×10^6	10,000	16,000	55	10	4	27,000	29,000
C 14	Alclad 3003-O	10.0×10^6		11,000		40	30	6,000	16,000
C 14b	Alclad 3003-H12	10.0×10^6		12,000		20	10	18,000	19,000
C 14c	Alclad 3003-H14	10.0×10^6		14,000		16	8	21,000	22,000
C 14d	Alclad 3003-H16	10.0×10^6		15,000		14	5	25,000	26,000
C 14a	Alclad 3003-H18	10.0×10^6		16,000		10	4	27,000	29,000
15	3004-O	10.0×10^6	14,000	16,000	45	25	20	10,000	26,000
15a	3004-H38	10.0×10^6	18,000	21,000	77	6	5	36,000	41,000
C 15	Alclad 3004-O	10.0×10^6		16,000		25	20	10,000	26,000
C 15a	Alclad 3004-H32	10.0×10^6		17,000		17	10	25,000	31,000
C 15b	Alclad 3004-H34	10.0×10^6		18,000		12	9	29,000	35,000

(Continued)

Table 39. Typical Mechanical Properties of Wrought Aluminum Alloys—*Continued*

Item No.	Alloy and Temper	Tension Strength, psi Ultimate	Tension Strength, psi Yield	Elongation in 2 inches, per cent 1/16-inch Thick Specimen	Elongation in 2 inches, per cent 1/2-inch Diameter Specimen	Hardness Brinell Number 500-kg load 10-mm ball	Shear Shearing Strength, psi	Fatigue Endurance[b] Limit, psi	Modulus of Elasticity,[c] psi
C 15c	Alclad 3004-H36	38,000	33,000	5	9	:	20,000	......	10.0×10^6
C 15d	Alclad 3004-H38	41,000	36,000	5	6	:	21,000	......	10.0×10^6
16a	3105-H25	26,000	24,000	8	..	47	16,000	......	10.0×10^6
17a	4032-T6	55,000	46,000	..	9	120	38,000	16,000	11.4×10^6
18	5005-O	18,000	6,000	30	..	30	11,000	......	10.0×10^6
18a	5005-H38	29,000	27,000	5	..	51	16,000	......	10.0×10^6
19	5050-O	21,000	8,000	4	..	36	15,000	13,000	10.0×10^6
19a	5050-H38	32,000	29,000	6	..	63	20,000	18,000	10.0×10^6
20	5052-O	28,000	13,000	25	30	47	18,000	16,000	10.2×10^6
20a	5052-H38	42,000	37,000	7	14	77	24,000	20,000	10.2×10^6
21	5056-O	42,000	22,000	..	35	65	26,000	20,000	10.3×10^6
21a	5056-H38	60,000	50,000	..	15	100	32,000	22,000	10.3×10^6
22	5083-O	42,000	21,000	22	25	67	25,000	22,000	10.3×10^6
22a	5083-H343	52,000	41,000	8	..	92	30,000	......	10.3×10^6
23	5086-O	38,000	17,000	22	30	60	23,000	21,000	10.3×10^6
23a	5086-H34	47,000	37,000	10	14	82	28,000	23,000	10.3×10^6
24	5154-O	35,000	17,000	27	30	58	22,000	17,000	10.2×10^6
24a	5154-H38	48,000	39,000	10	..	87	28,000	22,000	10.2×10^6
25	5254-O	35,000	17,000	27	30	58	22,000	17,000	10.2×10^6
25a	5254-H38	48,000	39,000	10	..	87	28,000	22,000	10.2×10^6

No.	Alloy								E
27	5357-O	19,000	7,000	25	::	32	12,000	......	10.0×10^6
27a	5357-H38	32,000	30,000	6	::	55	18,000	......	10.0×10^6
28	5454-O	36,000	17,000	22	25	60	23,000	19,000	10.2×10^6
28a	5454-H34	44,000	35,000	10	16	81	26,000	21,000	10.2×10^6
29	5456-O	45,000	23,000	24	20	70	27,000	22,000	10.3×10^6
29a	5456-H321	51,000	37,000	16	16	90	30,000	23,000	10.3×10^6
30	5457-O	19,000	7,000	25	::	32	12,000	......	10.0×10^6
30a	5457-H38	32,000	30,000	7	::	55	18,000	......	10.0×10^6
31	5557-O	16,000	6,000	25	::	27	10,000	......	10.0×10^6
31a	5557-H38	30,000	28,000	8	::	55	18,000	......	10.0×10^6
32	5652-O	28,000	13,000	25	30	47	18,000	16,000	10.2×10^6
32a	5652-H38	42,000	37,000	7	8	77	24,000	20,000	10.2×10^6
33	6053-O	16,000	8,000	::	35	26	11,000	8,000	10.0×10^6
33c	6053-T6	37,000	32,000	::	13	80	23,000	13,000	10.0×10^6
34	6061-O	18,000	8,000	25	30	30	12,000	9,000	10.0×10^6
34a	6061-T4	35,000	21,000	22	25	65	24,000	13,000	10.0×10^6
34b	6061-T6	45,000 [l]	40,000 [l]	12	17	95	30,000	14,000	10.0×10^6
C 34	Alclad 6061-O	17,000	7,000	25	:::	:::	11,000	......	10.0×10^6 °
C 34a	Alclad 6061-T4	33,000	19,000	22	:::	:::	22,000	......	10.0×10^6 °
C 34b	Alclad 6061-T6	42,000	37,000	12	:::	:::	27,000	......	10.0×10^6 °
35	6062-O	18,000	8,000	:::	30	30	12,000	9,000	10.0×10^6
35a	6062-T4	35,000	21,000	22	25	65	24,000	14,000	10.0×10^6
35b	6062-T6	45,000	40,000	12	17	95	30,000	14,000	10.0×10^6
36	6063-O	13,000	7,000	:	:	25	10,000	8,000	10.0×10^6
36a	6063-T4	25,000	13,000	22	33	:	16,000	......	10.0×10^6
36b	6063-T42	22,000	13,000	20	22	48	14,000	10,000	10.0×10^6
36c	6063-T5	27,000	21,000	12	18	60	17,000	10,000	10.0×10^6
36d	6063-T6	35,000	31,000	12	:	74	22,000	10,000	10.0×10^6
36e	6063-T83	37,000	35,000	9	:	80	22,000	......	10.0×10^6
38b	6151-T6	48,000	43,000	:	17	100	32,000	12,000	10.2×10^6
39a	6262-T9	58,000 [p]	55,000 [p]	:	10	120	35,000	13,000	10.0×10^6

(Continued)

Table 39.　Typical Mechanical Properties of Wrought Aluminum Alloys—Continued

Item No.	Alloy and Temper	Tension — Strength, psi		Tension — Elongation in 2 inches, per cent		Hardness	Shear	Fatigue	Modulus
		Ultimate	Yield	1/16-inch Thick Specimen	1/2-inch Diameter Specimen	Brinell Number 500-kg load 10-mm ball	Shearing Strength, psi	Endurance Limit, psi	Modulus of Elasticity, psi
40	6463-O	13,000	7,000	: :	:	25	10,000	. . . . . .	10.0 × 10⁶
40a	6463-T4	25,000	13,000	22	:	. .	16,000	8,000	10.0 × 10⁶
40b	6463-T42	22,000	13,000	20	:	42	14,000	10,000	10.0 × 10⁶
40c	6463-T5	27,000	21,000	12	:	60	17,000	10,000	10.0 × 10⁶
40d	6463-T6	35,000	31,000	12	:	74	22,000	10,000	10.0 × 10⁶
41a	6563-T4	20,000	11,000	20	:	37	. . . . . .	. . . . . .	10.0 × 10⁶
41b	6563-T6	28,000	23,000	12	:	58	. . . . . .	. . . . . .	10.0 × 10⁶
42	6951-O	16,000	6,000	30	:	28	11,000	. . . . . .	10.0 × 10⁶
42a	6951-T6	39,000	33,000	13	:	82	26,000	. . . . . .	10.0 × 10⁶
44a	7075-T6	83,000 k,n	73,000 k,n	11	11	150	48,000	22,000	10.4 × 10⁶
C 44	Alclad 7075-O	32,000	14,000	17	:	. .	22,000	. . . . . .	10.4 × 10⁶ c
C 44a	Alclad 7075-T6	76,000	67,000	11	:	. .	46,000	. . . . . .	00.4 × 10⁶ c
45a	7079-T6	78,000	68,000	. .	14	145	45,000	22,000	10.4 × 10⁶
46a	7178-T6	88,000 k	78,000 k	10	11	160	52,000	22,000	10.4 × 10⁶
C 46	Alclad 7178-O	32,000	14,000	16	:	. .	22,000	. . . . . .	10.4 × 10⁶ c
C 46a	Alclad 7178-T6	81,000	71,000	10	:	. .	49,000	. . . . . .	10.4 × 10⁶ c

* From Alcoa Aluminum Handbook, 1962 edition.

a These typical properties are average for various forms, sizes and methods of manufacture, and may not exactly describe any one particular product or size.

b Based on 500,000,000 cycles of completely reversed stress using the R. R. Moore type of machine and specimen.

c Average of tension and compression moduli.　Compression modulus is about 2 per cent greater than tension modulus.

d Electrical conductor grade, 99.60 per cent minimum aluminum.

e EC-O wire will have an elongation of approximately 23 percent in 10 inches.
f EC-H19 wire will have an elongation of approximately 1½ per cent in 10 inches.
g Sizes greater than 1½ inches will have strengths slightly lower than these values.
h Extruded products more than ¾ inch thick will have strengths 15 to 20 per cent higher than these values.
i Sheet less than 0.040 inch thick will have strengths slightly lower than these values.
j Sheet more than 0.062 inch thick will have strengths slightly higher than these values.
k Extruded products will have strengths approximately 10 per cent higher than these values.
l Die forgings will have strengths approximately 5 per cent higher than these values.
m Die forgings will have a yield strength approximately 20 per cent lower than this value.
n Die forgings have strengths approximately 4 per cent lower than these values.
o Value shown is primary modulus. Secondary modulus is from 3 per cent to 10 per cent lower depending on thickness of cladding.
p Sizes greater than 2.000 inches will have strengths slightly lower than these values.

Table 40. Typical Tensile Properties of Wrought Aluminum Alloys At Various Temperatures*[a]

Item No.	Alloy and Temper	Temp., °F	Ultimate	Yield[b]	Elongation in 2 in., per cent
3	1100-O	-320	24,000	6,000	55
		-112	15,000	5,500	48
		-18	14,000	5,000	46
		75	13,000	5,000	45
		212	11,000	5,000	45
		300	8,500	4,500	55
		400	6,000	3,500	65
		500	4,000	2,000	75
		600	2,500	1,500	80
		700	2,000	1,000	85
3a	1100-H18	-320	35,000	26,000	30
		-112	26,000	23,000	16
		-18	24,000	22,000	15
		75	24,000	22,000	15
		212	22,000	19,000	15
		300	18,000	14,000	20
		400	6,000	3,500	65
		500	4,000	2,000	75
		600	2,500	1,500	80
		700	2,000	1,000	85
4a	2EC-T6	-320	43,000	34,000	24
		-112	36,000	31,000	20
		-18	34,000	30,000	19
		75	34,000	29,000	19
		212	32,000	26,000	20
		300	28,000	16,000	24
		400	19,000	7,000	40
		500	10,000	3,500	80
		600	5,000	2,500	100
		700	3,000	2,000	105
5a	2011-T3	75	55,000	43,000	15
		212	47,000	34,000	16
		300	28,000	19,000	25
		400	16,000	11,000	35
		500	6,500	4,000	45
		600	3,500	2,000	90
		700	2,500	1,500	125
6b	2014-T6	-320	84,000	69,000	14
		-112	74,000	63,000	14
		-18	72,000	61,000	13
		75	70,000	60,000	13
		212	63,000	56,000	14
		300	40,000	35,000	15
		400	16,000	13,000	35
		500	9,500	7,500	45
		600	6,500	5,000	65
		700	4,500	3,500	70
7a	2017-T4	75	62,000	40,000	22
		212	57,000	39,000	18
		300	40,000	35,000	15
		400	16,000	13,000	35
		500	9,000	7,500	45
		600	6,000	5,000	65
		700	4,500	3,500	70
8a	2018-T61	-320	72,000	51,000	12
		-112	64,000	47,000	12
		-18	63,000	46,000	12
		75	61,000	46,000	12
		212	56,000	43,000	12
		300	45,000	40,000	12
		400	19,000	13,000	25
		500	10,000	6,500	40
		600	5,500	3,000	60
		700	4,000	2,000	100
9a	2024-T3	-320	85,000	62,000	18
		-112	75,000	52,000	17
		-18	72,000	51,000	17
		75	70,000	50,000	17
		212	66,000	48,000	16
		300	55,000	50,000	11
		400	29,000	22,000	23
		500	12,000	9,000	55
		600	8,000	6,000	75
		700	5,500	4,000	100
9c	2024-T4	-320	81,000	57,000	19
		-112	71,000	49,000	19
		-18	70,000	48,000	19
		75	68,000	47,000	19
		212	64,000	45,000	19
		300	45,000	36,000	17
		400	27,000	20,000	27
		500	12,000	9,000	55
		600	8,000	6,000	75
		700	5,500	4,000	100
9d	2024-T6	-320	83,000	68,000	11
		-112	72,000	60,000	10
		-18	71,000	58,000	10
		75	69,000	57,000	10
		212	65,000	55,000	10
		300	45,000	36,000	17
		400	27,000	20,000	27
		500	12,000	9,000	55
		600	8,000	6,000	75
		700	5,500	4,000	100
9e	2024-T81	-320	85,000	77,000	8
		-112	75,000	68,000	7
		-18	72,000	67,000	7
		75	70,000	65,000	7
		212	66,000	61,000	8
		300	55,000	50,000	11
		400	29,000	22,000	23
		500	12,000	9,000	55
		600	8,000	6,000	75
		700	5,500	4,000	100

(Continued)

Table 40. Typical Tensile Properties of Wrought Aluminum Alloys At Various Temperatures—Continued

Item No.	Alloy and Temper	Temp., °F	Tensile Strength, psi Ultimate	Tensile Strength, psi Yield [b]	Elongation in 2 in., per cent
9f	2024-T86	-320	91,000	86,000	5
		-112	80,000	77,000	5
		-18	78,000	74,000	5
		75	75,000	71,000	5
		212	70,000	67,000	6
		300	55,000	51,000	11
		400	20,000	16,000	28
		500	12,000	9,000	55
		600	8,000	6,000	75
		700	5,500	4,000	100
11a	2117-T4	-320	56,000	33,000	30
		-112	45,000	25,000	29
		-18	44,000	24,000	28
		75	43,000	24,000	27
		212	36,000	21,000	16
		300	30,000	17,000	20
		400	16,000	12,000	35
		500	7,500	5,500	55
		600	4,500	3,500	80
		700	3,000	2,000	110
13c	2219-T62	-320	78,000	52,000	14
		-112	65,000	44,000	12
		-18	63,000	43,000	11
		75	61,000	42,000	11
		212	53,000	39,000	14
		300	44,000	30,000	15
		400	33,000	24,000	18
		500	27,000	20,000	18
		600	9,000	7,500	35
		700	4,500	3,500	100
13d	2219-T81	-320	86,000	65,000	13
		-112	73,000	56,000	12
		-85	72,000	54,000	11
		75	70,000	53,000	11
		212	63,000	50,000	13
		300	48,000	38,000	15
		400	34,000	29,000	16
		500	28,000	23,000	16
		600	7,000	6,500	40
		700	4,500	3,500	100
14	3003-O	-320	33,000	8,500	46
		-112	20,000	7,000	42
		-18	17,000	6,500	41
		75	16,000	6,000	40
		212	13,000	5,500	43
		300	11,000	5,000	47
		400	8,500	4,500	60
		500	6,000	3,500	65
		600	4,000	2,500	70
		700	3,000	2,000	70
14a	3003-H18	-320	41,000	33,000	23
		-112	32,000	29,000	11
		-18	30,000	28,000	10
		75	29,000	27,000	10
		212	26,000	21,000	10
		300	23,000	16,000	11
		400	14,000	9,000	18
		500	7,500	4,000	60
		600	4,000	2,500	70
		700	3,000	2,000	70
15	3004-O	-320	42,000	13,000	38
		-112	28,000	11,000	30
		-18	26,000	10,000	26
		75	26,000	10,000	25
		212	26,000	10,000	25
		300	22,000	10,000	35
		400	14,000	9,500	55
		500	10,000	7,500	70
		600	7,500	5,000	80
		700	5,000	3,000	90
15a	3004-H38	-320	58,000	43,000	20
		-112	44,000	38,000	10
		-18	42,000	36,000	7
		75	40,000	36,000	6
		212	40,000	36,000	7
		300	31,000	27,000	15
		400	22,000	15,000	30
		500	12,000	7,500	50
		600	7,500	5,000	80
		700	5,000	3,000	90
17a	4032-T6	-320	66,000	48,000	11
		-112	58,000	46,000	10
		-18	56,000	46,000	9
		75	55,000	46,000	9
		212	50,000	44,000	9
		300	37,000	33,000	9
		400	13,000	9,000	30
		500	8,000	5,500	50
		600	5,000	3,000	70
		700	3,500	2,000	90

(Continued)

Table 40. Typical Tensile Properties of Wrought Aluminum Alloys At Various Temperatures—*Continued*

Item No.	Alloy and Temper	Temp., °F	Tensile Strength, psi Ultimate	Tensile Strength, psi Yield [b]	Elongation in 2 in., per cent
19	5050-O	-320	36,000	10,000	.....
		-112	23,000	8,500	.....
		-18	22,000	8,000	.....
		75	21,000	8,000	.....
		212	21,000	8,000	.....
		300	19,000	8,000	.....
		400	14,000	7,500	.....
		500	9,000	6,000	.....
		600	6,000	4,000	.....
		700	4,000	3,000	.....
19a	5050-H38	-320	45,000	35,000	.....
		-112	34,000	30,000	.....
		-18	32,000	29,000	.....
		75	32,000	29,000	.....
		212	31,000	29,000	.....
		300	26,000	25,000	.....
		400	14,000	8,500	.....
		500	9,000	6,000	.....
		600	6,000	4,000	.....
		700	4,000	3,000	.....
20	5052-O	-320	44,000	16,000	45
		-112	30,000	13,000	36
		-18	28,000	13,000	32
		75	28,000	13,000	30
		212	28,000	13,000	37
		300	24,000	13,000	50
		400	18,000	11,000	60
		500	12,000	11,000	85
		600	7,500	8,000	110
		700	5,000	5,000	130
20a	5052-H38	-320	59,000	44,000	24
		-112	45,000	39,000	20
		-18	43,000	37,000	16
		75	42,000	37,000	14
		212	41,000	37,000	17
		300	34,000	29,000	25
		400	23,000	15,000	50
		500	12,000	8,000	75
		600	7,500	5,000	110
		700	5,000	3,000	130
22	5083-O	-320	60,000	24,000	36
		-112	43,000	21,000	30
		-18	42,000	21,000	27
		75	42,000	21,000	25
		212	41,000	21,000	37
		300	30,000	19,000	50
		400	22,000	17,000	60
		500	17,000	11,000	85
		600	11,000	7,500	110
		700	6,000	4,500	130
23	5086-O	-320	56,000	20,000	45
		-112	39,000	17,000	36
		-18	38,000	17,000	32
		75	38,000	17,000	30
		212	37,000	17,000	37
		300	29,000	17,000	60
		400	22,000	16,000	60
		500	17,000	11,000	85
		600	11,000	7,500	110
		700	6,000	4,500	130
24	5154-O	-320	53,000	20,000	45
		-112	36,000	17,000	36
		-18	35,000	17,000	32
		75	35,000	17,000	30
		212	35,000	17,000	37
		300	29,000	17,000	50
		400	22,000	16,000	60
		500	17,000	11,000	85
		600	11,000	7,500	110
		700	6,000	4,500	130
24a	5154-H38	-320	64,000	46,000	24
		-112	50,000	40,000	20
		-18	48,000	39,000	16
		75	48,000	39,000	14
		212	45,000	36,000	17
		300	39,000	33,000	25
		400	25,000	18,000	50
		500	18,000	11,000	75
		600	11,000	7,500	110
		700	6,000	4,500	130
28	5454-O	-320	54,000	20,000	39
		-112	37,000	17,000	30
		-18	36,000	17,000	27
		75	36,000	17,000	25
		212	36,000	17,000	30
		300	29,000	17,000	50
		400	22,000	16,000	60
		500	17,000	11,000	85
		600	11,000	7,500	110
		700	6,000	4,500	130

(*Continued*)

Table 40. Typical Tensile Properties of Wrought Aluminum Alloys At Various Temperatures—*Continued*

Item No.	Alloy and Temper	Temp., °F	Tensile Strength, psi Ultimate	Tensile Strength, psi Yield[b]	Elongation in 2 in., per cent
29	5456-O	-320	63,000	27,000	30
		-112	46,000	23,000	26
		-18	45,000	23,000	22
		75	45,000	23,000	20
		212	43,000	22,000	30
		300	30,000	20,000	50
		400	23,000	17,000	60
		500	17,000	11,000	85
		600	11,000	7,500	110
		700	6,000	4,500	130
33c	6053-T6	75	37,000	32,000	13
		212	32,000	28,000	13
		300	25,000	24,000	13
		400	13,000	12,000	25
		500	5,500	4,000	70
		600	4,000	2,500	80
		700	3,000	2,000	90
34b or 35b	6061-T6 or 6062-T6	-320	60,000	47,000	22
		-112	49,000	42,000	18
		-18	47,000	41,000	17
		75	45,000	40,000	17
		212	42,000	38,000	18
		300	34,000	31,000	20
		400	19,000	15,000	28
		500	7,500	5,000	60
		600	4,500	2,500	85
		700	3,000	2,000	95
36b	6063-T42	-320	34,000	16,000	44
		-112	26,000	15,000	36
		-18	24,000	14,000	34
		75	22,000	13,000	33
		212	22,000	14,000	18
		300	21,000	15,000	20
		400	9,000	6,500	40
		500	4,500	3,500	75
		600	3,000	2,500	80
		700	2,500	2,000	105
36c	6063-T5	-320	37,000	24,000	28
		-112	29,000	22,000	24
		-18	28,000	22,000	23
		75	27,000	21,000	22
		212	24,000	20,000	18
		300	20,000	18,000	20
		400	9,000	6,500	40
		500	4,500	3,500	75
		600	3,000	2,500	80
		700	2,500	2,000	105
36d	6062-T6	-320	47,000	36,000	24
		-112	38,000	33,000	20
		-18	36,000	32,000	19
		75	35,000	31,000	18
		212	31,000	28,000	15
		300	21,000	20,000	20
		400	9,000	6,500	40
		500	4,500	3,500	75
		600	3,000	2,500	80
		700	2,500	2,000	105
38b	6151-T6	75	48,000	43,000	17
		212	42,000	39,000	19
		300	27,000	25,000	22
		400	12,000	9,500	40
		500	6,500	5,500	50
		600	5,000	4,500	50
		700	4,000	3,500	50
44a	7075-T6	-320	102,000	92,000	9
		-112	90,000	79,000	11
		-18	86,000	75,000	11
		75	83,000	73,000	11
		212	70,000	65,000	14
		300	31,000	27,000	30
		400	16,000	13,000	55
		500	11,000	9,000	65
		600	8,000	6,500	70
		700	6,000	4,500	70
45a	7079-T6	-320	92,000	80,000	12
		-112	82,000	70,000	14
		-18	79,000	68,000	14
		75	78,000	68,000	14
		212	67,000	60,000	18
		300	33,000	28,000	37
		400	16,000	13,000	60
		500	11,000	8,500	100
		600	7,500	6,000	175
		700	5,500	4,500	175

* From *Alcoa Aluminum Handbook*, 1962 edition.

[a] Lowest strengths during 10,000 hours of heating at testing temperature under no load; stress applied at 5,000 psi/min to yield strength and then at strain rate of 0.05 in./in./min to failure. Under some conditions of temperature and time the application of heat will adversely affect certain other properties of some alloys. For specific information concerning the suitability of the various alloys for use at elevated temperatures, the nearest sales office of Aluminum Company of America should be consulted.

[b] Offset equals 0.2 per cent.

Table 41.   Thermal Expansion of Wrought and Casting Aluminum Alloys*

Equations of Linear Thermal Expansion for Aluminum Alloys:[a]

1. $L_{t(0 \text{ to } -320°F)} = L_0[1 + C(11.74t - 0.00125t^2 - 0.0000248t^3)10^{-6}]$

2. $L_{t(0 \text{ to } 1,000°F)} = L_0[1 + C(12.19t + 0.003115t^2)10^{-6}]$

L_0 = Length at 0° F

L_t = Length at temperature, t°F, within the range indicated

C = Alloy constant

Alloy	C	Alloy	C	Alloy	C	Alloy	C
			Constant C For Wrought Alloys [b]				
EC	1.000	2218	.950	5083	1.010	6062	.990
2EC	.995	2219	.955	5086	1.010	6063	.995
1060	1.000	2618	.945	5154	1.015	6151	.985
1100	1.000	3003	.985	5254	1.015	6262	.990
1345	1.000	3004	.985	5357	1.005	6463	.995
2011	.980	3105	.995	5454	1.005	6563	1.000
2014	.955	4032	.825	5456	1.015	6951	.990
2017	.970	4043	.940	5457	1.005	7072	1.000
2018	.950	5005	1.005	5557	1.000	7075	.990
2024	.970	5050	1.005	5652	1.010	7076	.980
2025	.965	5052	1.010	6053	.980	7079	1.005
2117	.990	5056	1.025	6061	.990	7178	.995
			Constant C For Casting Alloys [c]				
13	.880	138	.900	218	1.015	364	.895
43	.945	A140	.990	220	1.040	380	.885
108	.950	142	.955	319	.905	A380	.895
A108	.910	A142	.955	333	.880	384	.860
112	.965	195	.970	344	.920	A612	1.020
113	.940	B195	.950	355	.950	C612	1.000
C113	.920	212	.945	C355	.950	750	.990
122	.950	214	1.020	356	.910	A750	.965
A132	.845	A214	1.020	A356	.910	B750	.985
D132	.875	B214	.970	360	.890		
F132	.875	F214	1.005	A360	.900		

* From *Alcoa Aluminum Handbook*, 1962 edition.

[a] Empirical equations derived from expansion determinations on annealed high-purity aluminum within the temperature range −320° to 1,000°F, $C = 1.000$.

[b] Constants established from determinations made on Alcoa Alloys in annealed tempers, i.e., in states approaching dimensional stability through wide temperature ranges. With heat-treatable alloys, the application of equation 2 is restricted to temperatures below 600°F. With wrought alloys 7075, 7076, 7079 and 7178, application is restricted to temperatures below 400°F. The values of the constant when applied to alloys in their heat-treated tempers are approximately 0.015 greater than given above. With these tempers, application is further restricted to temperatures which do not appreciably exceed those used in the final aging treatments.

[c] Constants established from determinations made on Alcoa Alloys in annealed tempers; i.e., in states approaching dimensional stability through wide temperature ranges. With heat-treatable alloys, the application of Equation 2 is restricted to temperatures below 600°F. With casting alloys A612, C612, 750, A750, and B750, application is restricted to temperatures below 400°F. The values of the constant when applied to alloys in their heat-treated tempers are approximately 0.015 greater than given above. With these tempers, application is further restricted to temperatures which do not appreciably exceed those used in the final aging treatments.

Table 42. Relative Typical Characteristics of Wrought Aluminum Alloys*ᵃ

Item No.	Alloy and Temper	Resistance to Corrosion	Workability (Cold)	Machinability	Brazeability	Gas	Arc	Resistance Spot and Seam	Forgeability
1	EC-O	A	A	D	A	A	A	B	..
1a	-H19	A	C	C	A	A	A	A	..
2	1060-O	A	..	..	..	A	A	A	..
3	1100-O	A	A	D	A	A	A	B	A
3a	-H18	A	C	C	A	A	A	A	A
5a	2011-T3	D	C	A	D	D	D	D	..
5b	-T8	C	D	A	D	D	D	D	..
6a	2014-T4	C	C	B	D	D	B	B	C
6b	-T6	C	D	B	D	D	B	B	C
7a	2017-T4	C	C	B	D	D	B	B	..
8a	2018-T61	C	..	B	D	D	B	B	C
9a	2024-T3	C	C	B	D	D	B	B	..
9c	-T4	C	C	B	D	D	B	B	..
9b	-T36	C	D	B	D	D	B	B	..
9e	-T81	C	D	B	D	D	B	B	..
11a	2117-T4	C	B	C	D	D	B	B	..
12a	2218-T72	C	..	B	D	..	B	..	D
13a	2219-T31	C	C	B	D	A	A	A	C
13b	-T37	C	D	B	D	A	A	A	C
13d	-T81	C	D	B	D	A	A	A	C
13e	-T87	C	D	B	D	A	A	A	C
14	3003-O	A	A	D	A	A	A	B	A
14a	-H18	A	C	C	A	A	A	A	A
15	3004-O	A	A	D	B	B	A	B	..
15a	-H38	A	C	C	B	B	A	A	..
17a	4032-T6	C	..	..	D	D	B	C	..
18	5005-O	A	A	D	B	A	A	B	..
18a	-H38	A	C	C	B	A	A	A	..
19	5050-O	A	A	D	B	A	A	B	..
19a	-H38	A	C	C	B	A	A	A	..
20	5052-O	A	A	D	C	A	A	B	..
20a	-H38	A	C	C	C	A	A	A	..
21	5056-O	A	A	D	D	C	A	B	..
21a	-H38	A	C	C	D	C	A	A	..
22	5083-O	A	B	D	D	C	A	B	..
22b	-H113	A	C	D	D	C	A	A	..
22a	-H343	B	C	C	D	C	A	A	..
23	5086-O	A	A	D	D	C	A	A	..
23a	-H34	A	B	C	D	C	A	A	..
24	5154-O	A	A	D	D	C	A	B	..
24a	-H38	A	C	C	D	C	A	A	..

(Continued)

Table 42. Relative Typical Characteristics of Wrought Aluminum Alloys—*Continued*

Item No.	Alloy and Temper	Resistance to Corrosion	Workability (Cold)	Machinability	Brazeability	Weldability			Forgeability
						Gas	Arc	Resistance Spot and Seam	
27	5357-O	A	A	D	B	A	A	B	..
27a	-H38	A	C	C	B	A	A	A	..
28	5454-O	A	A	D	D	C	A	B	..
28a	-H34	B	B	C	D	C	A	A	..
29	5456-O	A	B	D	D	C	A	B	..
29a	-H321	A	C	D	D	C	A	A	..
30	5457-O	A	A	D	B	A	A	B	..
30a	-H38	A	C	C	B	A	A	A	..
31	5557-O	A	A	D	B	A	A	B	..
31a	-H38	A	C	C	B	A	A	A	..
34	6061-O	A	A	D	A	A	A	B	..
34a	-T4	A	B	C	A	A	A	A	..
34b	-T6	A	C	C	A	A	A	A	..
35	6062-O	A	A	D	A	A	A	B	..
35a	-T4	A	B	C	A	A	A	A	..
35b	-T6	A	C	C	A	A	A	A	..
36	6063-O	A	A	D	A	A	A	B	..
36c	-T5	A	B	C	A	A	A	A	..
36d	-T6	A	C	C	A	A	A	A	..
36b	-T42	A	A	C	A	A	A	A	..
36e	-T83	A	C	C	A	A	A	A	..
44a	7075-T6	C	D	B	D	D	D	B	D
45a	7079-T6	C	D	B	D	D	D	D	D
46a	7178-T6	C	D	B	D	D	C	B	D

* From *Alcoa Aluminum Handbook*, 1962 edition.

ᵃ Resistance to Corrosion, Workability (Cold), Machinability and Foregability ranges A, B, C, and D are relative ratings in decreasing order of merit.   Weldability and Brazeability ratings, A, B, C, and D are relative ratings defined as follows:

A. Generally weldable by all commercial procedures and methods.
B. Weldable with special technique or on specific applications which justify preliminary trials or testing to develop welding procedure and weld performance.
C. Limited weldability because of crack sensitivity or loss in resistance to corrosion, and all mechanical properties.
D. No commonly used welding methods have so far been developed.

Table 43. Nominal Composition of Casting Aluminum Alloys*

Alloy	Types of Castings			Per Cent of Alloying Elements—Aluminum and Normal Impurities Constitute Remainder						
	Sand	Permanent Mold	Die	Silicon	Copper	Manganese	Magnesium	Nickel	Zinc	Tin
13	..	..	√	12.0	...	...	...	...	...	...
43	√	√	√	5.0	...	...	...	...	...	...
108	√	..	..	3.0	4.0	...	...	...	...	...
A108	..	√	..	5.5	4.5	...	...	...	...	...
112	√	..	..	...	7.0	...	...	...	...	...
113	√	√	..	2.0	7.0	...	...	...	...	...
C113	..	√	..	3.5	7.0	...	...	...	...	...
122	√	√	..	...	10.0	...	0.20	...	...	...
A132	..	√	..	12.0	0.8	...	1.2	2.5	...	...
F132	..	√	..	9.5	3.0	...	1.0	...	...	...
138	..	√	..	4.0	10.0	...	0.30	...	...	...
A140	√	..	..	...	8.0	0.50	6.0	0.50	...	...
142	√	√	..	...	4.0	...	1.5	2.0	...	...
A142	√	..	..	...	4.1	...b	1.5	2.0	...	...
195	√	..	..	0.8	4.5	...	...	...	...	...
B195	..	√	..	2.5	4.5	...	...	...	...	...
212	√	..	..	1.2	8.0	...	...	...	...	...
214	√	..	..	...	...	...	3.8	...	...	...
A214	..	√	√	...	...	...	3.8	...	1.8	...
B214	√	..	..	1.8	...	...	3.8	...	...	...
F214	√	..	..	0.50	...	...	3.8	...	...	...
218	..	..	√	...	...	...	8.0	...	...	...
220	√	..	..	...	...	...	10.0	...	...	...
319	√	..	..	6.3	3.5	...	...	...	...	...
333	..	√	..	9.0	3.8	...	...	...	...	...
344	..	√	..	7.0	...	...	...	...	...	...
355	√	√	..	5.0	1.3	...	0.50	...	...	...
A335	√	..	..	5.0	1.4	0.8	0.50	0.8	...	...
C355	..	√	..	5.0	1.3	...	0.50	...	...	...
356	√	√	..	7.0	...	...	0.30	...	...	...
A356	..	√	..	7.0	...	...	0.30	...	...	...
360	..	..	√	9.5	...	...	0.50	...	...	...
364	..	..	√	8.5	...	...a	0.30	...	...	...
380	..	..	√	9.0	3.5	...	...	...	...	...
384	..	..	√	12.0	3.8	...	...	...	...	...
A612	√	..	..	...	0.50	...	0.7	...	6.5	...
C612	..	√	..	...	0.50	...	0.35	...	6.5	...
750	..	√	..	...	1.0	...	...	1.0	...	6.5
A750	√	..	..	2.5	1.0	...	...	0.50	...	6.5
B750	..	√	..	...	2.0	...	0.75	1.2	...	6.5

* From *Alcoa Aluminum Handbook*, 1962 edition. The alloy designations for aluminum castings contain three digits (except for alloys 43 and 13). The first digit of a casting designation identifies the alloy type; the other two digits are assigned arbitrarily.

a Also contains 0.35 per cent chromium and 0.03 per cent beryllium.

b Also contains 0.20 per cent chromium.

Table 44. Typical Physical Properties of Sand and Permanent-Mold Casting Aluminum Alloys*

Alloy and Temper	Specific Gravity [b]	Weight, lb per cu in. [b]	Approximate Melting Range, °F	Electrical Conductivity [e]	Thermal Conductivity at 25°C, CGS units [c]
43-F annealed [d]	} 2.69	0.097	1065–1170	{ 37 / 42	0.35 / 0.39
108-F	} 2.79	0.101	970–1160	{ 31 / 38	0.30 / 0.36
108 annealed [d]					
A-108-F [a]	2.79	0.101	970–1135	37	0.35
112-F	} 2.91	0.105	985–1165	{ 30 / 38	0.29 / 0.36
112 annealed [d]					
113-F [a]	} 2.92	0.106	965–1160	{ 30 / 38	0.29 / 0.36
113 annealed [a,d]					
C113-F [a]	2.92	0.106	970–1145	27	0.26
122-T61	2.95	0.107	965–1155	33	0.31
A132-T551 [a]	2.72	0.098	1000–1050	29	0.28
F132-T5 [a]	2.76	0.100	970–1080	26	0.25
138-F [a]	2.95	0.107	945–1110	25	0.25
A140-F	2.78	0.100	955–1120	..	....
142-T21				44	0.41
142-T571 [a]	} 2.81	0.102	990–1175	34	0.32
142-T61 [a]				33	0.31
142-T77				38	0.36
A142-T75	2.81	0.102	990–1175	..	....
195-T4	} 2.81	0.102	970–1190	{ 35 / 35	0.33 / 0.33
195-T62					
B195-T4 [a]	} 2.80	0.101	970–1170	{ 33 / 33	0.31 / 0.31
B195-T6 [a]					
212-F	2.89	0.104	965–1160	30	0.29
214-F	2.65	0.096	1110–1185	35	0.33
A214-F [a]	2.68	0.097	1075–1180	34	0.32
B214-F	2.65	0.096	1090–1170	38	0.36
F214-F	2.66	0.096	1090–1185	36	0.34
220-T4	2.57	0.093	840–1120	21	0.21
319-F	2.79	0.101	960–1120	27	0.26
333-F [a]				26	0.25
333-T5 [a]	} 2.77	0.100	960–1085	29	0.28
333-T6 [a]				29	0.28
333-T7 [a]				35	0.33
344-F	2.68	0.097	1065–1145	..	....
355-T51				43	0.40
355-T6				36	0.34
355-T6 [a]	} 2.71	0.098	1015–1150	39	0.36
355-T61				37	0.35
355-T7				42	0.39
A355-T51	2.74	0.099	1000–1145	32	0.30
C355-T61	2.71	0.098	1015–1150	39	0.36
356-T51				43	0.40
356-T6	} 2.68	0.097	1035–1135	39	0.36
356-T6 [a]				41	0.38
356-T7				40	0.37

(*Continued*)

Table 44.   Typical Physical Properties of Sand and Permanent-Mold
Casting Aluminum Alloys*—*Continued*

Alloy and Temper	Specific Gravity [b]	Weight, lb per cu in. [b]	Approximate Melting Range, °F	Electrical Conductivity [e]	Thermal Conductivity at 25°C, CGS units [c]
A356-T61	2.67	0.097	1035–1135	39	0.36
A612-F	2.81	0.102	1105–1195	35	0.33
C612-F [a]	2.84	0.103	1120–1190	40	0.37
750-T5 [a]	2.88	0.104	435–1200	47	0.43
A750-T5 [a]	2.83	0.103	440–1165	43	0.40
B750-T5 [a]	2.88	0.104	400–1175	45	0.41

* From *Alcoa Aluminum Handbook*, 1962 edition.
[a] Chill cast samples; all other samples cast in green sand molds.
[b] The specific gravity and weight data in this table assume solid (void-free) metal. Since some porosity cannot be avoided in commercial castings, their specific gravity or weight will be slightly less than the theoretical value.
[c] CGS units = calories per second per square centimeter per centimeter of thickness per degree Centigrade.
[d] While castings are not commonly annealed, similar effects on conductivities may result from the slower rate of cooling of thick sections as compared with thin ones and other variables in foundry practices. Comparison of the values for as-cast and annealed specimens will show the extent to which variations may be expected, depending upon differences in thermal conditions in the production of different types of castings.
[e] Per cent of International Annealed Copper Standard.

Table 45.   Typical Physical Properties of Die-Casting Aluminum Alloys*

Alloy	Specific Gravity	Weight, lb per cu in.	Approximate Melting Range, °F	Electrical Conductivity, per cent of International Annealed Copper Standard	Thermal Conductivity at 25°C, CGS units
13	2.65	0.096	1065–1080	31	0.30
43	2.69	0.097	1065–1170	37	0.35
A214	2.68	0.097	1075–1180	34	0.32
218	2.57	0.093	995–1150	24	0.24
360	2.64	0.095	1035–1105	28	0.27
A360	2.63	0.095	1035–1105	30	0.29
364	2.63	0.095	1035–1115	30	0.29
380	2.72	0.098	1000–1100	23	0.23
A380	2.71	0.098	1000–1100	25	0.25
384	2.70	0.098	960–1080	23	0.23

* From *Alcoa Aluminum Handbook*, 1962 edition.

Table 46. Typical Mechanical Properties of Sand-Casting Aluminum Alloys*

				Typical Values (not guaranteed)				
Alloy and Temper	Tensile Ultimate Strength,[b] psi	Tensile Yield Strength[b] (offset = 0.2%), psi	Elongation in 2 inches, per cent	Compressive Yield Strength[c] (offset = 0.2%), psi	Brinell Hardness,[b] 500-kg load, 10-mm ball	Shearing Strength, psi	Endurance Limited, R. R. Moore Type Specimen 500,000,000 Cycles, psi	Modulus[a] of Elasticity, psi
43-F	19,000	8,000	8.0	9,000	40	14,000	8,000	10.3×10^6
108-F	21,000	14,000	2.5	15,000	55	17,000	11,000	......
112-F	24,000	15,000	1.5	16,000	70	20,000	9,000	......
113-F	24,000	15,000	1.5	16,000	70	20,000	9,000	......
122-T61	41,000	40,000	[e]	43,000	115	32,000	8,500	10.7×10^6
A140-F	33,000	28,000	1.0	30,000	90	......	......	......
142-T21	27,000	18,000	1.0	18,000	70	21,000	8,000	10.3×10^6
142-T571	32,000	30,000	0.5	34,000	85	26,000	11,000	10.3×10^6
142-T77	30,000	23,000	2.0	24,000	75	24,000	10,500	10.3×10^6
A142-T75	31,000	......	2.0	......	75	......	......	......
195-T4[d]	32,000	16,000	8.5	17,000	60	26,000	7,000	10.0×10^6
195-T6	36,000	24,000	5.0	25,000	75	30,000	7,500	10.0×10^6
195-T62	41,000	32,000	2.0	34,000	90	33,000	8,000	10.0×10^f
212-F	23,000	14,000	2.0	14,000	65	20,000	9,000	......
214-F	25,000	12,000	9.0	12,000	50	20,000	7,000	10.3×10^6
B214-F	20,000	13,000	2.0	14,000	50	17,000	8,500	......
F214-F	21,000	12,000	3.0	13,000	50	17,000	8,000	......
220-T4	48,000	26,000	16.0	27,000	75	34,000	8,000	9.5×10^6
319-F	27,000	18,000	2.0	19,000	70	22,000	10,000	10.7×10^6
319-T5	30,000	26,000	1.5	27,000	80	24,000	11,000	10.7×10^6
319-T6	36,000	24,000	2.0	25,000	80	29,000	11,000	10.7×10^6
355-T51	28,000	23,000	1.5	24,000	65	22,000	8,000	10.2×10^6
355-T6	35,000	25,000	3.0	26,000	80	28,000	9,000	10.2×10^6
355-T7	38,000	36,000	0.5	38,000	85	28,000	10,000	10.2×10^6
355-T71	35,000	29,000	1.5	30,000	75	26,000	10,000	10.2×10^6
A355-T51	28,000	24,000	1.5	25,000	70	22,000	8,000	......
356-T51	25,000	20,000	2.0	21,000	60	20,000	8,000	10.5×10^6
356-T6	33,000	24,000	3.5	25,000	70	26,000	8,500	10.5×10^6
356-T7	34,000	30,000	2.0	31,000	75	24,000	9,000	10.5×10^6
356-T71	28,000	21,000	3.5	22,000	60	20,000	8,500	10.5×10^6
A612-F	35,000[f]	25,000[f]	5.0[f]	25,000	75[f]	26,000	8,000	9.7×10^6
750-T5	20,000	11,000	8.0	11,000	45	14,000	......	10.3×10^6
A750-T5	20,000	11,000	5.0	11,000	45	14,000	......	10.3×10^6
B750-T5	27,000	22,000	2.0	22,000	65	18,000	10,000	10.3×10^6

* From *Alcoa Aluminum Handbook*, 1962 edition.
 [a] Average of tension and compression moduli. Compression modulus is about 2 per cent greater than tension modulus.
 [b] Tension and hardness values determined from standard half-inch diameter tensile test specimens, individually cast in green-sand molds, and tested without machining off the surface.
 [c] Results of tests on specimens having an l/r ratio of 12.
 [d] On standing at room temperature for several weeks, the properties will approach those of the -T6 condition.
 [e] Less than 0.5 per cent.
 [f] From tests made approximately 30 days after casting.

Table 47.   Typical Mechanical Properties of Permanent-Mold Casting Aluminum Alloys*

Typical Values (not guaranteed)

Alloy and Temper	Tensile Ultimate Strength,b psi	Tensile Yield Strength b (offset = 0.2%), psi	Elongation in 2 inches, per cent	Compressive Yield Strength c (offset = 0.2%), psi	Brinell Hardness, b 500-kg load, 10-mm ball	Shearing Strength, psi	Endurance Limit, R. R. Moore Type Specimen 500,000,000 Cycles, psi	Modulus a of Elasticity, psi
43-F	23,000	9,000	10.0	9,000	45	16,000	8,000	10.3×10^6
A108-F	28,000	16,000	2.0	17,000	70	22,000	13,000	......
113-F	28,000	19,000	2.0	20,000	70	22,000	9,500	......
C113-F	30,000	24,000	1.5	25,000	85	24,000	9,500	......
122-T551	37,000	35,000	e	40,000	115	30,000	8,500	10.7×10^6
122-T65	48,000	36,000	e	36,000	140	36,000	9,000	10.7×10^6
A132-T551	36,000	28,000	0.5	28,000	105	28,000	13,500	......
A132-T65	47,000	43,000	0.5	43,000	125	36,000	......	......
F132-T5	36,000	28,000	1.0	......	105	......	......	11.2×10^6
138-F	30,000	24,000	1.5	30,000	100	24,000	......	......
142-T571	40,000	34,000	1.0	34,000	105	30,000	10,500	10.3×10^6
142-T61	47,000	42,000	0.5	44,000	110	35,000	9,500	10.3×10^6
B195-T4d	37,000	19,000	9.0	20,000	75	30,000	9,500	10.1×10^6
B195-T6	40,000	26,000	5.0	26,000	90	32,000	10,000	10.1×10^6
B195-T7	39,000	20,000	4.5	20,000	80	30,000	9,000	10.1×10^6
A214-F	27,000	16,000	7.0	17,000	00	22,000	......	......
333-F	34,000	19,000	2.0	19,000	90	27,000	14,500	......
333-T5	34,000	25,000	1.0	25,000	100	27,000	12,000	......
333-T6	42,000	30,000	1.5	30,000	105	33,000	15,000	......
333-T7	37,000	28,000	2.0	28,000	90	28,000	12,000	......
344-F	28,000	10,000	10.0	......	50	......	......	......
355-T51	30,000	24,000	2.0	24,000	75	24,000	......	10.2×10^6
355-T6	42,000	27,000	4.0	27,000	90	34,000	10,000	10.2×10^6
355-T62	45,000	40,000	1.5	40,000	105	36,000	10,000	10.2×10^6
355-T71	36,000	31,000	3.0	31,000	85	27,000	10,000	10.2×10^6
C355-T61	46,000	34,000	6.0	36,000	100	32,000	14,000	10.2×10^6
356-T6	38,000	27,000	5.0	27,000	80	30,000	13,000	10.5×10^6
356-T7	32,000	24,000	6.0	24,000	70	25,000	11,000	10.5×10^6
A356-T61	41,000	30,000	10.0	32,000	90	28,000	13,000	10.5×10^6
C612-F	35,000f	18,000f	8.0f	......	70f	......	11,000	......
750-T5	23,000	11,000	12.0	11,000	45	15,000	9,000	10.3×10^6
A750-T5	20,000	11,000	5.0	11,000	45	14,000	9,000	10.3×10^6
B750-T5	32,000	23,000	5.0	23,000	70	21,000	11,000	10.3×10^6

* From *Alcoa Aluminum Handbook*, 1962 edition.
a Average of tension and compression moduli.   Compression modulus is about 2 per cent greater than tension modulus.
b Tension and hardness values determined from standard half-inch diameter tensile test specimens, individually cast in a permanent mold, and tested without machining off the surface.
c Results of tests on specimens having an l/r ratio of 12.
d On standing at room temperature for several weeks, properties approach those of the -T6 condition.
e Less than 0.5 per cent.
f From tests made approximately 30 days after casting.

Table 48. Typical Mechanical Properties of Die-Casting Aluminum Alloys*

Alloy	Typical Mechanical Properties					
	Tensile Ultimate Strength,[b] psi	Tensile Yield Strength[b] (offset = 0.2%), psi	Elongation in 2 inches, per cent	Shearing Strength, psi	Endurance Limit, R. R. Moore Type Specimen 500,000,000 Cycles, psi	Modulus[a] of Elasticity, psi
13	43,000	21,000	2.5	28,000	19,000	10.3 × 10⁶
43	33,000	16,000	9.0	21,000	17,000	10.3 × 10⁶
A214	40,000	22,000	10.0	26,000	18,000	....
218	45,000	27,000	8.0	29,000	20,000	....
360	47,000	25,000	3.0	30,000	19,000	10.3 × 10⁶
A360[c]	46,000	24,000	5.0	29,000	18,000	....
364	43,000	23,000	7.5	26,000	18,000	....
380	48,000	24,000	3.0	31,000	21,000	10.3 × 10⁶
A380[c]	47,000	23,000	4.0	30,000	20,000	....
384	47,000	25,000	1.0	30,000	21,000	10.3 × 10⁶

* From *Alcoa Aluminum Handbook*, 1962 edition.

[a] Average of tension and compression moduli. Compression modulus is about 2 per cent greater than tension modulus.

[b] Tensile properties are average values obtained from ASTM standard round die-cast test specimen, ¼ inch in diameter, produced on a cold chamber (high-pressure) die-casting machine.

[c] The alloys whose numbers are prefixed by "A" differ from those without the prefix in that the impurities, notably iron, are controlled to lower limits.

Table 49. Typical Tensile Properties of Sand-Casting Aluminum Alloys at Elevated Temperatures*[a]

Alloy and Temper	Temperature, °F	Tensile Ultimate Strength, psi	Tensile Yield Strength,[b] psi	Elongation in 2 inches, per cent	Alloy and Temper	Temperature, °F	Tensile Ultimate Strength, psi	Tensile Yield Strength,[b] psi	Elongation in 2 inches, per cent
112-F	75	24,000	15,000	1.5	355-T51	75	28,000	23,000	1.5
	212	23,000	14,000	1.5		212	28,000	22,000	2.0
	300	22,000	13,000	1.5		300	24,000	19,000	3.0
	400	21,000	12,000	1.5		400	14,000	10,000	8.0
	500	14,000	10,000	3.5		500	9,500	5,000	16.0
	600	6,000	4,000	20.0		600	6,000	3,000	36.0
	700	4,500	2,500	40.0		700	3,500	2,000	50.0
122-T61	75	41,000	40,000	[c]	355-T6	75	35,000	25,000	3.0
	212	39,000	38,000	0.5		212	35,000	25,000	2.0
	300	36,000	35,000	1.0		300	33,000	25,000	1.5
	400	24,000	17,000	2.0		400	17,000	13,000	8.0
	500	17,000	11,000	6.0		500	9,500	5,000	16.0
	600	8,500	5,000	14.0		600	6,000	3,000	36.0
	700	5,000	2,500	30.0		700	3,500	2,000	50.0
142-T571	75	32,000	30,000	0.5	355-T71	75	35,000	29,000	1.5
	212	32,000	30,000	0.5		212	34,000	28,000	2.0
	300	30,000	28,000	0.5		300	30,000	26,000	3.0
	400	26,000	21,000	1.0		400	17,000	13,000	8.0
	500	13,000	8,000	8.0		500	9,500	5,000	16.0
	600	8,000	4,000	20.0		600	6,000	3,000	36.0
	700	5,000	3,000	40.0		700	3,500	2,000	50.0
142-T77	75	30,000	23,000	2.0	A355-T51	75	28,000	24,000	1.5
	212	30,000	23,000	2.0		212	27,000	23,000	1.5
	300	27,000	21,000	2.0		300	24,000	20,000	2.0
	400	20,000	15,000	3.0		400	17,000	12,000	4.0
	500	13,000	8,000	6.0		500	11,000	7,000	10.0
	600	8,000	4,000	20.0		600	7,500	5,000	10.0
	700	5,000	3,000	40.0		700	6,000	4,000	15.0
195-T6	75	36,000	24,000	5.0	356-T6	75	33,000	24,000	3.5
	212	34,000	23,000	5.0		212	32,000	24,000	4.0
	300	28,000	20,000	5.0		300	23,000	20,000	6.0
	400	15,000	9,000	15.0		400	12,000	8,500	18.0
	500	9,000	6,000	25.0		500	7,500	5,000	35.0
	600	4,000	3,000	75.0		600	4,000	3,000	60.0
	700	2,500	1,500	100.0		700	2,500	2,000	80.0
214-F	75	25,000	12,000	9.0	356-T7	75	34,000	30,000	2.0
	212	24,000	12,000	9.0		212	30,000	28,000	2.0
	300	22,000	12,000	7.0		300	23,000	20,000	6.0
	400	18,000	12,000	9.0		400	12,000	8,500	18.0
	500	13,000	8,000	12.0		500	7,500	5,000	35.0
	600	9,000	4,000	17.0		600	4,000	3,000	60.0
	700	5,000	2,000	35.0		700	2,500	2,000	80.0

* From *Alcoa Aluminum Handbook*, 1962 edition.
[a] Lowest strengths during 10,000 hours of heating at testing temperatures.
[b] Offset equals 0.2 per cent.
[c] Less than 0.5 per cent.

Table 50.   Typical Tensile Properties of Permanent-Mold Casting Aluminum Alloys at Elevated Temperatures*a

Alloy and Temper	Temperature, °F	Tensile Ultimate Strength, psi	Tensile Yield Strength,[b] psi	Elongation in 2 inches, per cent	Alloy and Temper	Temperature, °F	Tensile Ultimate Strength, psi	Tensile Yield Strength,[b] psi	Elongation in 2 inches, per cent
122-T551	75	37,000	35,000	[c]	355-T51	75	30,000	24,000	2.0
	212	34,000	33,000	[c]		212	28,000	24,000	3.0
	300	30,000	27,000	[c]		300	23,000	20,000	4.0
	400	25,000	20,000	1.0		400	15,000	10,000	19.0
	500	18,000	12,000	3.0		500	9,500	5,000	33.0
	600	8,500	5,000	10.0		600	6,000	3,000	38.0
	700	5,000	2,500	25.0		700	3,500	2,000	60.0
A132-T551	75	36,000	28,000	0.5	355-T6	75	42,000	27,000	4.0
	212	35,000	25,000	1.0		212	40,000	27,000	5.0
	300	31,000	22,000	1.0		300	32,000	25,000	10.0
	400	26,000	15,000	2.0		400	19,000	13,000	20.0
	500	18,000	10,000	5.0		500	9,500	5,000	40.0
	600	10,000	4,000	10.0		600	6,000	3,000	50.0
	700	5,000	3,000	45.0		700	3,500	2,000	60.0
F132-T5	75	36,000	28,000	1.0	355-T71	75	36,000	31,000	3.0
	212	33,000	27,000	1.0		212	33,000	29,000	4.0
	300	31,000	24,000	2.0		300	29,000	26,000	8.0
	400	25,000	16,000	3.0		400	19,000	13,000	20.0
	500	19,000	12,000	6.0		500	9,000	5,000	40.0
	600	12,000	8,000	15.0		600	6,000	3,000	50.0
	700	8,000	6,000	25.0		700	3,500	2,000	60.0
142-T571	75	40,000	34,000	1.0	356-T6	75	38,000	27,000	5.0
	212	40,000	34,000	1.0		212	30,000	25,000	6.0
	300	37,000	33,000	1.0		300	21,000	17,000	10.0
	400	28,000	22,000	2.0		400	12,000	8,500	30.0
	500	13,000	8,000	15.0		500	7,500	5,000	55.0
	600	8,000	4,000	35.0		600	4,000	3,000	70.0
	700	5,000	3,000	60.0		700	2,500	2,000	80.0
B195-T6	75	40,000	26,000	5.0	356-T7	75	32,000	24,000	6.0
	212	35,000	23,000	5.0		212	27,000	23,000	10.0
	300	29,000	23,000	5.0		300	21,000	17,000	20.0
	400	17,000	11,000	15.0		400	12,000	8,500	40.0
	500	7,000	4,000	25.0		500	7,000	5,000	55.0
	600	3,500	2,500	75.0		600	4,000	3,000	70.0
	700	2,500	1,500	100.0		700	2,500	2,000	80.0
C355-T61	75	46,000	34,000	6.0	A356-T61	75	41,000	30,000	10.0
	212	43,000	34,000	6.0		300	21,000	17,000	20.0
	300	38,000	34,000	10.0		400	12,000	8,500	40.0
	400	14,000	10,000	40.0		500	7,500	5,000	55.0
	500	6,500	5,500	60.0		600	4,000	3,000	70.0
	600	4,000	3,000	70.0		700	2,500	2,000	80.0
	700	3,500	2,500	90.0					

* From *Alcoa Aluminum Handbook*, 1962 edition.
a Lowest strengths during 10,000 hours of heating at testing temperatures.
b Offset equals 0.2 per cent.
c Less than 0.5 per cent.

Table 51. Typical Tensile Properties of Die-Casting Aluminum
Alloys at Elevated Temperatures*[a]

Alloy	Temperature, °F	Tensile Ultimate Strength, psi	Tensile Yield Strength,[b] psi	Elongation in 2 inches, per cent	Alloy	Temperature, °F	Tensile Ultimate Strength, psi	Tensile Yield Strength,[b] psi	Elongation in 2 inches, per cent
13	75	43,000	21,000	2.5	A360	75	46,000	24,000	5.0
	212	37,000	20,000	5.0		212	43,000	24,000	3.0
	300	32,000	19,000	8.0		300	34,000	23,000	5.0
	400	24,000	15,000	15.0		400	21,000	13,000	14.0
	500	13,000	9,000	30.0		500	11,000	6,500	30.0
	600	7,000	4,500	35.0		600	6,500	4,000	45.0
	700	4,500	2,500	40.0		700	4,000	2,500	45.0
43	75	33,000	16,000	9.0	380	75	48,000	24,000	3.0
	212	28,000	16,000	9.0		212	45,000	24,000	4.0
	300	22,000	15,000	10.0		300	34,000	22,000	5.0
	400	16,000	12,000	25.0		400	24,000	16,000	8.0
	500	9,000	6,000	30.0		500	13,000	8,000	20.0
	600	5,000	3,500	35.0		600	7,000	4,000	30.0
	700	3,500	2,500	35.0		700	4,000	2,500	35.0
218	75	45,000	27,000	8.0	A380	75	47,000	23,000	4.0
	212	40,000	25,000	8.0		212	44,000	23,000	5.0
	300	32,000	21,000	25.0		300	33,000	21,000	10.0
	400	21,000	15,000	40.0		400	23,000	15,000	15.0
	500	13,000	9,000	45.0		500	12,000	7,000	30.0
	600	8,500	4,500	45.0		600	6,000	4,000	45.0
	700	5,000	2,500	45.0		700	4,000	2,500	45.0
360	75	47,000	25,000	3.0	384	75	47,000	25,000	1.0
	212	44,000	25,000	2.0		212	46,000	25,000	1.0
	300	35,000	24,000	4.0		300	38,000	24,000	2.0
	400	22,000	14,000	8.0		400	26,000	18,000	6.0
	500	12,000	7,500	20.0		500	14,000	9,000	25.0
	600	7,000	4,500	35.0		600	7,000	4,000	45.0
	700	4,500	3,000	40.0		700	4,500	2,500	45.0

* From *Alcoa Aluminum Handbook*, 1962 edition.
[a] Lowest strengths during 10,000 hours of heating at testing temperatures.
[b] Offset equals 0.2 per cent.

Table 52. Characteristics and Typical Uses of Casting Aluminum Alloys*

Alloy	Usual Commercial Tempers	Types of Castings			Approximate Relative Ratings Based on Various Characteristics[a]								Typical Uses
		Sand	Permanent Mold	Die	Strength[b]	Castability	Resistance to Corrosion	Weldability (Torch and Arc)	Machinability	Electrical Conductivity	Pressure Tightness[a]	Hardness	
43	F	✓	✓	✓	D	A	B	A	D	C	A	D	General-purpose casting alloy, cooking utensils, architectural and marine applications and pipe fittings.
108	F	✓			C	B	D	B	B	C	B	C	General-purpose sand castings, manifold and valve bodies.
A108	F		✓		C	B	D	B	B	C	B	C	General-purpose permanent-mold castings and ornamental grills.
113	F	✓	✓		C	B	D	C	B	C	B	C	Housings, covers, hand wheels and washing machine agitators.
C113	F		✓		B	B	D	C	B	D	B	B	Washing machine agitators, automotive cylinder heads and timing gears.
122	T551, T61, T65	✓	✓		B, A	C	D	C	A	C	C	A	Bushings, bearing caps and meter parts.
A132	T551, T65		✓		B	C	C	B	C	D	B	B, A	Heavy-duty diesel pistons.
138	F		✓		B	C	D	C	B	D	C	B	Sole plates for electric hand irons.
142	T21, T571, T61, T77	✓	✓		C, A	C	D	C	B	B, C	C	C, B	Heavy-duty pistons and air-cooled cylinder heads.
195	T4, T6, T62	✓			B	C	C	C	B	C	C	C, B	Machinery and aircraft structural members.
B195	T4, T6, T7		✓		B	C	C	C	B	C	C	C, B	Aircraft fittings, gear housings and gun control parts.
212	F	✓			C	C	D	C	B	C		C	General-purpose sand castings.
214	F	✓			C	C	A	C	A	C	D	C	Food handling and chemical equipment, marine hardware and architectural applications.
A214	F		✓	✓	C	C	A	B	A	C	D	C	Cooking utensils and ornamental hardware.
B214	F	✓			C	C	A	B	B	C	C	C	Cooking utensils and pipe fittings.
F214	F	✓			C	C	A	B	A	C	D	C	Alumilite coated architectural parts and ornamental hardware.

Alloy	Temper											Typical uses
218	F	✓	⋮	A	D	A	…	A	D	D	…	Die castings requiring high strength, ductility and resistance to corrosion, such as marine fittings and hardware.
220	T4	✓	⋮	A	D	A	D	A	D	D	C	Sand castings requiring strength and shock resistance, such as aircraft structural members.
319	F, T6	✓	⋮	C, B	B	C	B	B	D	B	C, B	General-purpose sand castings, engine parts, automotive cylinder heads and piano plates.
333	F, T5, T6, T7	⋮	✓	B, A	B	C	C	B	D, C	B	B	General-purpose permanent-mold castings, engine parts, gas meter housings and regulator parts.
355	T51, T6, T61, T62, T7, T71	⋮	✓	C, B	A	B	B	A	C, B	A	C, B	Crankcases, accessory housings and aircraft fittings.
C355	T61	⋮	✓[c]	A	A	B	B	A	B	A	B	Aircraft, missile and other structural uses requiring high strength.
356	T51, T6, T7, T71	⋮	✓	C, B	A	B	B	A	C, B	A	C, B	Transmission cases, truck-axle housings and wheels, cylinder blocks, railway tank car fittings, marine hardware, valve bodies and bridge railing parts.
A356	T61	⋮	✓[c]	A	A	B	B	A	B	A	B	Aircraft, missile and other structural uses requiring high strength.
360	F	✓	⋮	A	A	B	…	C	D	A	…	General-purpose die castings, cover plates and instrument cases.
364	F, T5	✓	⋮	A	A	B	…	C	D	A	…	High ductility die castings where resistance to impact is important.
380	F	✓	⋮	A	B	C	C	B	D	B	…	General-purpose die castings.
A612	F	⋮	✓	B	C	B	C	C	C	D	C	General-purpose sand castings for brazing.
C612	F	✓	⋮	B	D	C	D	B	B	D	D	Torque converter blades and brazed parts.
750	T5	✓	⋮	D	D	C	D	A	A	D	D	Bearings
A750	T5	✓	⋮	D	D	C	D	A	A	D	D	Bearings
B750	T5	✓	⋮	B, C	D	C	D	A	A	D	B, C	Bearings

* From *Alcoa Aluminum Handbook*, 1962 edition.

[a] A, B, C and D are arbitrary relative ratings in decreasing order of merit. Where two letters shown, that on the left is for the softest temper listed and that on the right for the hardest temper listed.

[b] Refer to Tables 46, 47 and 48 for strength values.

[c] Also produced as premium strength castings.

Table 53.   Properties of Magnesium

(*Proc. A.S.T.M.*, Vol. 34, p. 299, 1934)

		Sample Purity, per cent
Atomic weight...	24.32	....
Boiling point, °C..	1097	....
Crystal form..	Close packed, hexagonal	....
Mean coefficient of expansion { 20 to 300 C...................	0.0000283	....
{ 20 to 500 C...................	0.0000299	....
Density at 20 C (68 F), g/cu cm.........................	1.74	....
Density at melting point (solid), g/cu cm................	1.64	....
Density at melting point (liquid), g/cu cm...............	1.57	....
Electrical resistivity at 20 C:		
Microhm/cu cm......................................	4.4611	....
Ohms (mil, ft) (A.I.E.E. Standard).....................	26.83	....
Temperature coefficient at 20 C (A.I.E.E. Standard)..........	0.0040	....
Electrical conductivity at 20 C:		
Mass, per cent International Annealed Copper Standard.....	197.7	....
Volume, per cent International Annealed Copper Standard...	38.6	....
Freezing point, °C......................................	651	....
Heat of vaporization, g-cal/g.............................	1300 to 1500	....
Latent heat of fusion { g-cal/g............................	70	....
{ Btu/lb............................	126	....
Mechanical properties:		
Tensile strength, psi.................................	27,000	Commercial
Yield strength, psi (set = 0.2 per cent)....................	10,000	annealed
Elongation in 2 in., per cent............................	15	sheet
Brinell hardness, 10-mm ball, 500-kg load.................	37	99.8 to 99.9
Modulus of elasticity, psi...............................	6,250,000	....
Mean reflectivity, per cent for white light..................	73	....
Mean specific heat, g-cal/g/°C (0–100 C)...................	0.249	....
Thermal conductivity, cgs units...........................	0.37	....
Watts/sq cm/°C/cm..................................	1.55	....
Btu/hr/sq ft/°F/in...................................	1102	....

Table 54. Physical Constants of Magnesium Alloys in Common Use*

ASTM Designation	Typical Composition, per cent					Specific Gravity	Density, lb/cu in.	Melting Point, °F	Thermal Conductivity, 68°F	Electric Resistivity, microohm-centimeters 20°C (68°F)	Uses
	Aluminum	Zinc	Manganese[a]	Other[b]	Magnesium						
AM100A	10.0	0.2[b]	0.13	0.55	Balance	1.81	0.066	1100	41	14.5	Sheet and plate
AZ31B	3.0	1.0	0.2	0.77	Balance	1.77	0.064	1160	56	10.0	Extrusions, sheets, and forgings
AZ61A	6.5	0.95	0.15	0.46	Balance	1.80	0.065	1145	46	12.5	Extrusions and forgings
AZ63A	6.0	3.0	0.15	0.71	Balance	1.83	0.067	1135	44	12.8	Sand and permanent mold castings
AZ80A	8.5	0.5	0.15	0.66	Balance	1.80	0.065	1130	44	14.5	Extrusions and forgings
AZ91B	9.0	0.7	0.13	1.11	Balance	1.81	0.066	875–1105	41	17.0	Die castings
AZ92A	9.0	2.0	0.10	0.41	Balance	1.82	0.066	830–1100	39	16.0	Sand and permanent mold castings
M1A	—	—	1.2	—	Balance	1.76	0.064	1200	73	5.0	Extrusions, forgings, sheets and special sand castings

* Modulus of elasticity, psi, 6,500,000. Modulus of rigidity, psi, 2,400,000. Coefficient of thermal expansion, in./in./°F (68–750°F), 0.000016. Poisson's ratio, 0.35.
[a] Minimum.
[b] Maximum.

Table 55. Typical Mechanical Properties of Magnesium Forgings

Alloy	Condition	Tension			Compression	Hardness	
		Tensile Strength, psi	Yield Strength,* psi	Elonga- tion, % in 2 in.	Yield Strength, psi	Brinell, 500-kg Load, 10-mm Ball	Rock- well E
		Typical	Typical	Typical			
AZ61A-F	As forged	45,000	33,000	16	19,000	60	72
M1A....	As forged	36,000	23,000	7	....	47	54
AZ80A-F.	As forged	46,000	31,000	8	25,000	69	80
AZ80A-T6	Heat treated and aged	52,000	34,000	5	27,000	72	82

* Yield strength is defined as the stress at which the stress-strain curve deviates 0.2 per cent from the modulus line.

Table 56. Typical Mechanical Properties of Magnesium Plate, Sheet and Strip

Alloy	Condition	Tension			Compression	Shear	Hardness	
		Tensile Strength, psi	Yield Strength,* psi	Elonga- tion, % in 2 in.	Yield Strength, psi	Ultimate Stress, psi	Brinell, 500-kg Load, 10-mm Ball	Rock- well E
		Typical	Typical	Typical				
AZ31B	Annealed	37,000	22,000	21	16,000	21,000	56	67
	Hard rolled	43,000	33,000	11	26,000	23,000	73	83
M1A ..	Annealed	33,000	18,000	16	12,000	18,000	48	55
	Hard rolled	37,000	28,000	7	20,000	17,000	56	67

* Yield strength is defined as the stress at which the stress-strain curve deviates 0.2 per cent from the modulus line.

Table 57. Typical Mechanical Properties of Magnesium Sand, Die, and Permanent Mold Casting Alloys

Alloy	Condition	Tension			Compression	Hardness		Shear	Im- pact
		Tensile Strength, psi	Yield Strength,* psi	Elonga- tion, % in 2 in.	Yield Strength, psi	Brinell, 500-Kg Load, 10-mm Ball	Rock- well E	Ultimate Stress, psi	Izod, ft
		Typical	Typical	Typical					
AZ92A-F	As cast	24,000	14,000	2	14,000	65	77	19,000	1
AZ92A-T4	Heat-treated	40,000	14,000	10	14,000	63	75	20,000	4
AZ92A-T6	Heat-treated and aged	40,000	23,000	2	23,000	84	90	21,000	1
AM100A-F	As cast	22,000	12,000	2	12,000	54	65	17,000	2
AM100A-T4	Heat-treated	40,000	13,000	6	13,000	52	62	19,000	4
AM100A-T6	Heat-treated and aged	40,000	16,000	4	16,000	69	80	21,000	2
AZ63A-F	As cast	29,000	14,000	6	14,000	50	59	18,000	3
AZ63A-T4	Heat-treated	40,000	14,000	12	14,000	55	66	19,000	5
AZ63A-T6	Heat-treated and aged	40,000	19,000	5	19,000	73	83	21,000	2
M1A-F	As cast	14,000	4,500	5	4,500	33	3	11,000	9
AZ91B	Die-cast	34,000	23,000	3	22,000	67	75	20,000	......

* Yield strength is defined as the stress at which the stress-strain curve deviates 0.2 per cent from the modulus line.

Table 58. Nominal Ranges of Mechanical Properties of Nickel*

Form and Condition	Tensile Properties			Hardness	
	Tensile Strength, 1000 psi	Yield Strength (0.20% offset), 1000 psi	Elongation in 2 in., %	Brinell, 3000 kg	Rockwell, scale as shown
Rod and bar					
Annealed....................	55–80	15–30	60–40	90–120	45–68B
Hot rolled, as rolled............	55–85	15–40	55–35	90–130	45–72B
Forged.......................	60–90	20–60	55–30	100–170	54–86B
Cold drawn, as drawn..........	80–115	60–90	35–15	130–200	72–93B
Wire, cold drawn					
Annealed....................	50–80	10–30	50–30		
No. 1 temper.................	90–105	75–90	20–5		
Regular temper...............	105–135	85–130	15–4		
Spring temper................	125–165	125–155	5–2		
Plate, hot rolled					
Annealed....................	55–80	15–30	50–35	90–120	45–68B
Hot rolled, as rolled............	60–100	20–75	45–30	100–200	54–92B
Sheet a					
Standard cold rolled					
Soft temper.................	55–80	15–30	55–35		55–70B
Hard temper...............	90–115	70–105	15–2		92B min
Special cold rolled					
Soft, spinning, and deep drawing......................	55–80	15–25	55–35		64B max
1/4 Hard temper............	60–80	30–50	45–30		71–79B
1/2 Hard temper............	65–85	50–65	40–25		80–85B
3/4 Hard temper............	70–90	55–70	25–15		86–91B
Strip a					
Standard cold rolled					
Soft temper.................	55–80	15–30	55–35		64B max
Spring temper..............	90–130	70–115	15–2		96B min
Special cold rolled					
Soft, spinning, and deep drawing......................	55–80	15–25	55–35		64B max
Skin hard temper...........	60–80	20–35	50–35		65–70B
1/4 Hard temper............	65–85	30–50	45–30		71–79B
1/2 Hard temper............	70–90	50–65	40–25		80–85B
3/4 Hard temper............	75–95	55–70	25–15		86–91B
Hard temper...............	90–115	70–105	15–2		92–95B
Tubing					
Soft........................	55–80	15–30	60–40		40–60B
As drawn....................	70–95	50–80	35–15		80–100B

* For properties of nickel and nickel alloys see tables 1–8 incl.

a Sheet and strip products are available on the basis of hardness for the various tempers shown.

Table 59.   Physical Properties of Zinc

(Compiled from various sources)

	Metric Units	English Units
Atomic weight..................................	65.38	65.38
Density (rolled), (see Note 1).................	7.14 g per cu cm	445.8 lb per cu ft
Melting point	419.45 C	787 F
Boiling point.................................	905 ± 2 C	1661 ± 4 F
Mean specific heat (20–100 C).................	0.0931 g-cal per g per °C	....
Latent heat of fusion.........................	26.6 g-cal per g	47.88 Btu per lb
Latent heat of vaporization...................	426.8 g-cal per g	768.2 Btu per lb
Coefficient of linear expansion		
Cast......................................	39.5 × 10⁻⁶ ⎫	21.9 × 10⁻⁶ ⎫
Rolled, with grain.........................	32 × 10⁻⁶ ⎬ Per °C	17.8 × 10⁻⁶ ⎬ Per °F
Rolled, across grain........................	23 × 10⁻⁶ ⎭	12.8 × 10⁻⁶ ⎭
Thermal conductivity........................	1.13 watts per sq cm per °C per cm 0.270 g-cal per sec per sq cm per °C per cm	
Electrical resistivity (see Note 2)..............	5.916 microhm-cm	2.33 microhm-in.
Temperature coefficient of resistance (0–100 C)...	0.00419 per °C	0.00233 per °F
Resonance....................................	See Note 3	

NOTES. 1. Solid zinc (rolled) is about 3 per cent heavier than liquid zinc at the melting point.
2. Based on soft rolled zinc strip 99.99 per cent pure. Resistivity of strip zinc may vary 3 per cent, depending on the rolling treatment, and in commercial grades about the same amount, depending on the purity.
3. Rolled zinc has extremely low resonance, which permits its use in various sound apparatus.

Table 60.   Commercial Grades of Zinc*

Grade	Composition, per cent			
	Lead, max	Iron, max	Cadmium, max	Zinc, min, by difference *a*
Special High Grade *b*....................	0.003	0.003	0.003	99.990
High Grade...........................	0.07	0.02	0.03	99.90
Intermediate.·.......................	0.20	0.03	0.40	99.5
Brass Special........................	0.6	0.03	0.50	99.0
Prime Western.......................	1.6	0.05	0.50	98.0

* From ASTM Designation B6-67.
  a Analysis need not regularly be made for copper, tin and aluminum in any grade.   Nevertheless, it is understood that the minimum per cent of zinc (by difference) takes into account the copper, tin, and aluminum contents if any, in addition to the impurities listed in the table.
  b Analysis need not regularly be made for tin in Special High Grade but, if found, shall not exceed 0.001 per cent.

Table 61. Composition and Mechanical Properties of Zinc Die Castings*

Designation		Alloy AG40A[a] (23)	Alloy AC41A[a] (25)
Composition[b] % by weight	Copper, per cent............................	0.25, max[c]	0.75 to 1.25
	Aluminum, per cent........................	3.5 to 4.3	3.5 to 4.3
	Magnesium, per cent.......................	0.020 to 0.05	0.03 to 0.08
	Iron, max, per cent........................	0.100	0.100
	Lead, max, per cent........................	0.005	0.005
	Cadmium, max, per cent...................	0.004	0.004
	Tin, max, per cent.........................	0.003	0.003
	Zinc, per cent.............................	remainder	remainder
Mechanical properties.	Charpy impact strength, ft-lb ¼ × ¼ in. bar as cast.............................	20	20
	Charpy impact strength, ft-lb after 8 yr aging indoors...............................	2	2
	Tensile strength, psi as cast.................	40,300	45,400
	Tensile strength, 8 yr indoor age.............	34,400	37,200
	Elongation % in 2 in. as cast................	5	3
	Elongation % in 2 in. 8 yr indoor age........	8	5
	Expansion, in./in. after 8 yr.................	0.0001	0.0001
Other properties and constants as cast	Brinell hardness...........................	74	79
	Compression strength, lb/in.^{2}................	60,500	87,300
	Electric conductivity, mhos/cm^3 at 20°C......	157,000	153,000
	Melting point °C..........................	380.9	380.6
	Modulus of rupture lb-in.^{2}..................	95,000	105,000
	Shearing strength, lb/in.^{2}...................	30,900	38,400
	Solidification point, °C....................	380.6	380.4
	Solidification shrinkage, in./ft..............	0.14	0.14
	Sp gr....................................	6.6	6.7
	Sp heat, cal/g/°C..........................	0.10	0.10
	Thermal conductivity, cal/sec/cm^3/°C.......	0.27	0.26
	Thermal expansion per °C..................	27.4 × 10^{-6}	27.4 × 10^{-6}
	Transverse deflection, in..................	0.27	0.16
	Wt, lb/cu in..............................	0.24	0.24

* From ASTM Designation B 86-64 and *Die Casting for Engineers*, The New Jersey Zinc Company.
  [a] Designations in parentheses are the former alloy designations.
  [b] Zinc alloy die castings may contain nickel, chromium, silicon, and manganese in amounts of 0.02, 0.02, 0.035, and 0.5 per cent respectively.   No harmful effects have ever been noted due to the presence of these elements in these concentrations and, therefore, analyses are not required for these elements.
  [c] For the majority of commercial applications, a copper content in the range of 0.25 to 0.75 per cent will not adversely affect the serviceability of die castings and should not serve as a basis for rejection.

Table 62. Chemical Composition of Titanium Strip, Sheet, and Plate*

Element	Grades 1 & 2	Grade 3	Grade 4	Grade 5	Grade 6	Grade 7
Nitrogen, max, per cent.....	0.05	0.07	0.07	0.07	0.07	0.07
Carbon, max, per cent......	0.10	0.10	0.15	0.10	0.10	0.15
Hydrogen, max, per cent....	0.015[a]	0.015	0.015	0.015	0.020	0.015
Iron, max, per cent........	0.20	0.30	0.50	0.40	0.50	0.50
Oxygen, max, per cent......	0.25	0.35	0.45	0.30	0.30	0.30
Manganese, per cent........	....	....	....	....	....	6.5 to 9.0
Aluminum, per cent........	....	....	....	5.5 to 6.75	4.0 to 6.0	....
Vanadium, per cent........	....	....	....	3.5 to 4.5		....
Tin, per cent.............	....	....	....	....	2.0 to 3.0	....
Titanium, per cent[b].......	remainder	remainder	remainder	remainder	remainder	remainder

* From ASTM Designation B265-58T.
  [a] Lower hydrogen may be obtained by negotiation with the manufacturer.
  [b] The percentage of titanium is determined by difference.

Table 63. Mechanical Properties of Titanium Strip, Sheet, and Plate*

Grade	Tensile Strength,[a] min, psi	Yield Strength[a] (0.2 per cent Offset)		Elongation in 2 in., min, per cent	Bend Test [c]	
		Minimum	Maximum		Under 0.070 in. in thickness	0.070 to 0.187 in., incl, in thickness
1.......	40 000	30 000	55 000	22	3T	4T [c]
2.......	50 000	40 000	65 000	20	3T	4T
3.......	60 000	50 000	75 000	18	4T	5T
4.......	80 000	70 000	95 000	15	5T	6T
5.......	130 000	120 000	....	10 [b]	9T	10T
6.......	115 000	110 000	....	10 [b]	8T	9T
7.......	120 000	110 000	....	10 [b]	6T	7T

* From ASTM Designation B265-58T.
[a] Minimum and maximum limits apply to tests taken both longitudinal and transverse to the direction of rolling. Mechanical properties for conditions other than annealed may be established by agreement between the manufacturer and the purchaser.
[b] For grades 5, 6, and 7, the elongation on material under 0.025 in. in thickness may be obtained only by negotiation.
[c] T equals the thickness of the bend test specimen.

Table 64. Typical Chemical Analysis of Commercial Copper

Type Copper	% Cu * Plus Ag	% O₂	% P	% Other
Electrolytic tough pitch........	99.945	0.045	.....	0.01
Oxygen-free...................	99.98	.003	.....	.015
Phosphorized.................	99.96	.....	0.025	.02
High-conductivity phosphorized.	99.97	.....	.006	.02
Silver-bearing tough-pitch †.....	99.945	.045	.....	.01
Silver-bearing oxygen-free †.....	99.98	.003	.....	.015
Tellurium, beryllium, cadmium, and other special coppers	The chemical composition of these varies widely to meet specific end uses.			

* These analyses represent the average values delivered by the producers and as such are above the minimum guaranteed by A.S.T.M. and other purchasing specifications.
† Ag 7 to 45 oz./ton as specified.

Table 65. Mechanical and Physical Properties of Electrolytic Copper (Tough Pitch)

Copper, 99.94%; oxygen, 0.04%; phosphorus, nil; silver, nil

Mechanical Properties

Rod

	Hard *	Soft †	Hot	Cold ‡	Cold §
				Forgings	
Tensile strength, psi (000 omitted).............	55	33	33–36	35–50	55
Apparent elastic limit, psi (000 omitted)........	40	4	4–8	8–35	44
Yield strength, 0.5% extension, psi (000 omitted)	48	8	8	22–46	50
Yield strength, 0.2% offset, psi (000 omitted)...	49	7	7	21–45	45
Yield strength, 0.1% offset, psi (000 omitted)...	43	6	6	10–40	44
Elongation, % in 2 in.......................	10	50	50–45	40–10	5
Reduction of area, %.......................	45	65	60–40	60–50	40
Endurance limit, psi (000 omitted).............	15	10	10	12–15	15
Rockwell hardness F, 1/16-in. ball, 60-kg load....	90	25	25–65	65–85	90
Rockwell hardness B, 1/16-in. ball, 100-kg load...	55	..	.....	15–50	55
Brinell hardness, 10-mm ball, 500-kg load.......	194	40	40–60	60–83	89
Modulus of elasticity, psi....................			16,000,000		

Physical Data

Melting point, °F (solidus)	1949
Coefficient of expansion, per °F from 68 to 572 F	0.0000098
Electrical conductivity, ‡ %, I.A.C.S. 68 F (vol. and wt. basis)	101
Thermal conductivity, ‡ Btu/sq ft/ft/hr/°F at 68 F	227
Density, lb per cu in.	0.322
Electrical resistivity, ohms (mil ft) at 68 F (annealed)	10.3
Specific heat, Btu/lb/°F at 68 F	0.092
Specific gravity	8.89–8.94
Forging range, °F	1250–1450
Forging quality	Good
Type structure	Single phase, alpha

* Refers to rod previously hard-drawn 50%; rod under 1 in. in dia, ready-to-finish, grain size 0.030 mm.
† Refers to 1100 F anneal for 1 hour.
‡ Material cold-forged from soft rod (5 to 40% reduction).
§ Material cold-forged from cold-worked condition (40%).

Table 66.   Designation and Classification of Coppers*

Designation	Type of Copper
CATH	Electrolytic cathode
Tough Pitch Coppers	
ETP	Electrolytic tough pitch [a]
FRHC	Fire-refined high conductivity tough pitch [a,b]
FRTP	Fire-refined tough pitch [c]
ATP	Arsenical, tough pitch [c]
STP	Silver bearing tough pitch [a,b]
SATP	Silver bearing arsenical, tough pitch [c]
CAST	Casting
Oxygen-Free Coppers	
OF	Oxygen-free without residual deoxidants [a]
OFP	Oxygen-free, phosphorus bearing
OFPTE	Oxygen-free, phosphorus and tellurium bearing
OFS	Oxygen-free, silver bearing [a]
OFTE	Oxygen-free, tellurium bearing
Deoxidized Coppers	
DHP	Phosphorized, high residual phosphorus [c]
DLP	Phosphorized, low residual phosphorus [d]
DPS	Phsophorized, silver bearing [c]
DPA	Phosphorized, arsenical [c]
DPTE	Phosphorized, tellurium bearing [c]

* From ASTM Designation B224-58.
[a] Types ETP, FRHC, STP, OF, and OFS are high-conductivity coppers.
[b] Low resistance lake copper is included under types FHRC and STP.
[c] High resistance lake copper is included under types ATP, SATP, DHP, DPS, DPA, DPTE, and FRTP.
[d] Type DLP can be furnished as a high-conductivity copper.

Table 67. Chemical and Physical Properties of Various Wrought Copper-base Alloys

(Variations must be expected in practice)

Item No.	Material	Form	Approximate Composition, per cent					Tensile Strength, psi		Elongation in 2 in., per cent		Yield Point, psi		Johnson's Elastic Limit, psi		Modulus of Elasticity, psi ×10⁻⁶
			Copper	Zinc	Lead	Tin	Nickel	Hard *	Soft	Hard *	Soft	Hard *	Soft	Hard *	Soft	Hard *
1	Commercial bronze 95% (gilding metal)	S	95.00	5.00				60,000	35,000	4	45	55,000	11,000			15.0
2	Commercial bronze 90%	S	90.00	10.00				67,000	37,000	3	40	53,000	11,000			15.0
3	Red brass 80%	S	80.00	20.00				85,000	43,000	4	50					15.0
4	Cartridge brass	S	70.00	30.00				90,000	45,000	5	50					14.0
5	Yellow brass	S	65.00	35.00				90,000	45,000	12	60					14.0
6		R	63.00	37.00				70,000	50,000	4	50					12.8
6a		S	63.00	37.00				84,000	50,000	2e	50					15.0
6b	Brass wire	W	63.00	37.00				125,000			50e			55,000	7,500	
7	Muntz metal	R	60.00	40.00				80,000	48,000	9.5	48		20,000			
8	Leaded commercial bronze	R	88.50	10.00	1.50			60,000	30,000	3	30	70,000	12,000	65,000	9,000	
9	Leaded brass	R	64.00	35.00	1.00			70,000	60,000	5	60	55,000	17,000	39,000	12,000	
10	Free-cutting brass	S	62.00	35.00	3.00			95,000	50,000	10	50	75,000	12,000	65,000	11,000	15.0
11	Admiralty	W	70.00	29.00		1.00		125,000	65,000	3e	65					
11a		T	70.00	29.00		1.00		100,000	45,000	2e	42e					
11b		R	70.00	29.00		1.00		85,000	57,000	4	65					
12	Naval brass (Tobin bronze)	S	60.00	39.25		0.75		100,000	55,000	15	50	60,000	25,000	72,500	16,000	15.0
12a		S	60.00	39.25		0.75		110,000	59,000	4	40	80,000	20,000	67,000	17,000	
13	Phosphor bronze A	S	95.00			5.00		85,000	58,000	3	50	87,000	20,000	75,000	20,000	15.0
14	Phosphor bronze B	R	92.00			8.00		90,000	50,000	5	60	90,000	24,000	69,000	16,000	14.0
15	Cupro-nickel	S	70.00				30.00		55,000	15	45	78,000	20,000	72,000	26,000	
16	18% nickel silver A	S	65.00	17.00			18.00	90,000	55,000	3	34	90,000	26,000			
17	18% nickel silver B	S	56.00	26.00			18.00	110,000	60,000	2	45	100,000	20,000	90,000	17,000	18.0

No.	Alloy																
18	5% aluminum bronze	S	95.00					Al 5.00	105,000	52,000	5	70					
18a		R	92.00					8.00	100,000	60,000	4	60					
19	10% aluminum bronze	R	90.00					10.00	125,000m	78,000	5m	36	67,000	41,000		22,000	15.0
20	Manganese bronze	R	59.00	39.00	Iron 1.00	1.00		Mn 0.30	90,000	65,000	10	35	65,000	25,000	65,000	22,000	
21	High-silicon bronze	R	96.00				Mn 1.00	Si 3.00	113,000	65,000	5	70			94,000	21,000	
22	Low-silicon bronze	R	98.25				0.25	1.50	100,000	40,000	6	60	65,000	12,000	55,000	7,000	
23	Extruded architectural bronze shapes		57.00	40.00	Lead 3.60	0.34		Iron 0.16	88,000	60,000	15	25	60,000	35,000	51,000	25,000	
24	Beryllium copper	S	97.40		Be 2.25		Ni 0.35		118,000	70,000	4.3	45	105,000	31,000	79,000	18,000	17.2
24a		S	97.40		2.25		0.35		192,000m	175,000p	2m	6.3p	138,000m	114,000p	130,000m	87,000n	18.4m

Table 67. Chemical and Physical Properties of Various Wrought Copper-base Alloys—*Continued*

Item No.	Material	Form	Shearing Strength, psi — Hard *	Shearing Strength, psi — Soft	Brinell Hardness No., 10-mm Ball, 500-kg Load — Hard *	Brinell — Soft	Rockwell Hardness No. "B," 1/16-in. Ball, 100 kg — Hard *	Rockwell — Soft	Melting Point, °C	Specific Gravity †	Density, lb per cu in.	Coefficient of Expansion $\times 10^7$ (j)	Resistivity ohms (mil, ft)	Conductivity per cent I.A.C.S.	Thermal Conductivity (u)	Use—Remarks
1	Commercial bronze 95% (gilding metal)	S			110	43	68	4	1065z	8.866d	0.320	181	18.98	54.6	0.576	For jewelry trade and manufacturing where soft, pliable metal is required.
2	Commercial bronze 90%	S		25,000	115	50	75	1	1045z	8.804d	0.318	182	25.36	40.90	0.446	For window screen wire and automobile radiators on account of resistance to corrosion and atmospheric action.
3	Red brass 80%	S	43,000R	27,000	150	53	86	11	1000z	8.667d	0.313	191	31.95‡	32.5‡	0.335	For its color and resistance to corrosion and atmospheric action.
4	Cartridge brass	S			157	53	87		955z	8.528d	0.308	199	37.61	27.58	0.290	For primers, shot shells, cartridges, seamless tubes, etc.
5	Yellow brass	S			153	52	85	30	930z	8.460d	0.306	202	38.68	26.8	0.285	For a large variety of articles, lamp fixtures, automobile radiators, and for ornamental purposes. Not good for exposure to weather.
6	Brass wire	R							920z	8.437d	0.305	205	39.97	25.95	0.285	For rivets, pins, screws, and other heading operations.
7	Muntz metal	S	35,000		155	80	87	42	905z	8.396	0.303	208	36.25	28.60	0.300	For bolts, nuts, sheathing, pin wire, etc.
8	Leaded commercial bronze	R			104		58			8.830d	0.319	183	25.61y	40.50y	0.432	Free cutting. For special shapes where high lead is detrimental to the bending or working of the stock.
9	Leaded brass	R								8.562d	0.309	200	37.65	27.55		
10	Free-cutting brass	R	36,000		120	54	77	16	885z	8.489d	0.307	204	41.46	25.0	0.258	For automatic machine work. Drills and turns easily.
11	Admiralty	S		28,000					935z	8.535b	0.308	202	42.07	24.65	0.263	Resists action of sea water. For condenser tubes.
12	Naval brass (Tobin bronze)	R	45,000	33,000	100	89	85	55	885z	8.404b	0.304	211	41.60	24.93	0.279	For piston rods, propeller shafts, nuts, bolts, plates, etc. Welding rod.
12a		S		33,000	165	90	93	55								
13	Phosphur bronze A	S			190	60	96	30	1050z	8.87b	0.320	178	56.46	18.37	0.195	For springs, electric switches, window weight chain. Bronze chain in general.
14	Phosphur bronze B	S	60,000		200	70	104	47	1025z	8.815b	0.318	182	79.8	13.00	0.150	For electric switches, contact fingers, diaphragms, radio parts, etc.
15	Cupro-nickel	R					81	37	1225z	8.950	0.323	162f	218.4	4.75	0.069	For condenser tubes.
16	18% nickel silver A	S			170		91	50	1110v	8.752b	0.316		175.0	5.91	0.080	For silver-plated forks, spoons, knives, hollow ware, etc.
17	18% nickel silver B	S	65,000		190	70	90	50	1055v	8.68b	0.314		186.5‡	5.56‡	0.071	Similar to 30% nickel silver but of lower resistance.

No.	Material	Form													Uses
18	5% aluminum bronze	S			176	67	93	20	1060t	8.176t	0.295		58.61	17.69	0.198	For diaphragms to withstand pressure; also for its color.
19	10% aluminum bronze	R			190	100	100	65	1040t	7.57b	0.273		76.80	13.5	0.157	For strength and resistance to ordinary corrosion and wear.
20	Manganese bronze	R					90			8.370b	0.302		42.15	24.6	0.241	For structural work due to strength and resistance to corrosion.
21	High-silicon bronze	R	60,000	33,000	200	70	95	40	1019b	8.539b	0.308	180	155.0	6.7	0.078	For strength and resistance to corrosion. Has strength of mild steel and corrosion resistance of copper. Welding rod. Subject to stress corrosion.
22	Low-silicon bronze	R					90	3	1055r	8.740b	0.316		86.4	12.0	0.129	For strength and resistance to corrosion, bolt stock, and sheet metal requiring high ductility.
23	Extruded architectural bronze shapes	S					84	63	884b	8.432b	0.305					For architectural shapes.
24	Beryllium copper	S					102	65–73	955b		0.297 ±0.01	170		17±p		For springs, diaphragms, low-duty bushings and bearings, and Bourdon tubes.
24a		S			m	p	114m	112.5p			0.297 ±0.01		18–25m	18–25m	0.25p 0.20m	High resistance to fatigue.

* For some alloys the figures given are for a temper slightly different from that commonly known as "hard."
† Compared to water at 4 C.
‡ Soft.

R. Rod.
S. Sheet.
T. Tube.
W. Wire.
b. Determination.
d. Scientific Paper 410, National Bureau of Standards.
e. Elongation in 10 in., per cent

f. Corning Glass Works.
j. Average linear coefficient per °C from 25 to 300 C. Tests on rod. Scientific Paper 410, National Bureau of Standards.
m. Cold-worked and heat-treated.
n. Guertler-Tammann constitution diagram.
p. Annealed, quenched, and heat-treated.
r. Smith constitution diagram.

t. Stockdale constitution diagram.
u. G-cal, per sec per sq cm per °C per cm at 20 C.
v. Tafel constitution diagram.
x. Bauer and Hansen constitution diagram.
y. Hard at 25 C.
z. Heyeock-Neville constitution diagram.

Table 68. Sand-cast Copper-base Alloys

Item No.	Name	Range of Composition, per cent						Tension 1000 psi				Brinell Hardness	Impact Strength Izod, ft-lb	Remarks and Uses
		Cu	Al	Zn	Si	Mn	Fe	Yield Strength	Ultimate Strength	Elongation in 2 in., per cent	Reduction in Area, per cent			
										CAST COPPER				
1 1a	Pure copper	99.6 99.9						8–10	20–25	20–25	75	40	....	Deoxidized with Si, B_2C_3, or carbon-free Mn. Electrical uses if pure.
						HIGH TENSILE BRONZES AND BRASSES								
2	Aluminum bronze	80–90	7–12			0–3	1–5	25–45	65–90	5–30	....	110–180	30–36	Aluminum bronzes. These alloys should not be slowly cooled from a high temperature if they are to be stressed at ordinary temperatures. Resistant to cold dilute H_2SO_4, cold weak HCl, sea water, etc. Pipe fittings, valves, and other equipment for chemical industry.
3	Silicon bronze	94–98			2–4	0–1		18–30	40–60	20–50	15–22	80–100	....	Resistant to H_2SO_4, HCl (in absence of air), sea water, caustic soda, and phenol.
4	Manganese bronze	56–65	0–6	30–44		0.25–5.0	0–5	25–80	60–110	8–30	5–40	80–200	7–40	Manganese-bronze. Propellers, engine frames, parts requiring strength and toughness. Resistant to sea water.

Name	Cu	Sn	Pb	Zn	Ni	Fe	Yield Strength	Ultimate Strength	Elongation in 2 in., per cent	Reduction in Area, per cent	Brinell Hardness	Impact Strength Izod, ft-lb	Remarks and Uses
5 Nickel silver	60–75	3–5	....	1–10	20–30	0–1	17–40	40–65	15–25	....	75–150	15–30	Nickel-bronze. Valve facing for severe conditions. Abrasion resistant. For hardness.

NOMINAL COMPOSITION

BRONZE AND RED BRASS FOR GENERAL ENGINEERING WORK

Name	Cu	Sn	Pb	Zn	Ni	Fe	Yield Strength	Ultimate Strength	Elongation in 2 in., per cent	Reduction in Area, per cent	Brinell Hardness	Impact Strength Izod, ft-lb	Remarks and Uses
6 Bronze	88	8	....	4	....	....	16–21	35–45	25–40	16–33	60–72	11–16	Bronze for water-tight castings, underwater fittings, machine parts. Non-corrosive.
7 Gear bronze	90	10	....	....	....	....	18–20	30–35	5–15	6–8	65–85	....	Gear bronze. Very resistant to abrasion. For heavy-duty gears and worm wheels.
8 Red brass	85	5	5	5	....	....	17–24	33–46	15–35	12–32	55–65	....	Red brass for pump bodies, valves, steam fittings, bearing backs, and metal patterns.
9 Semi-red brass	83	4	6	7	....	....	12–17	30–38	15–27	12–25	50–60	....	Semi-red brass. Very resistant to atmospheric corrosion. For overhead electrical fittings and for oil and water pumps.
10 Hard bronze	80–85	15–20	....	....	....	....	18–32	25–35	0–1	0–1	70–160	....	For hardness. Difficult to machine.

YELLOW BRASS

Name	Cu	Sn	Pb	Zn	Ni	Fe	Yield Strength	Ultimate Strength	Elongation in 2 in., per cent	Reduction in Area, per cent	Brinell Hardness	Impact Strength Izod, ft-lb	Remarks and Uses
11 Yellow brass	60–72	0.5–1.5	0.5–4.0	25–40	....	....	10–20	27–45	15–40	15–40	40–75	....	For common castings where cheapness and good machining properties are main considerations.

NOMINAL COMPOSITION

HIGH-CONDUCTIVITY ALLOYS, PRECIPITATION HARDENING

Name	Cu	Cr	Si	Be	Co	Yield Strength	Ultimate Strength	Elongation in 2 in., per cent	Brinell Hardness	% Cond.	Remarks and Uses
12	Remainder	0.8				35	52	18	100	85	High electrical conductivity combined with good mechanical properties.
12a	Remainder			0.45	2.5	70	85	7	160	45	

Table 69.   Chemical Requirements For Brass Alloy Die Castings*

	Alloy Z30A [b, c]	Alloy ZS331A [b, c]	Alloy ZS144A [b, c]
Copper, per cent................	57.0 min	63.0 to 67.0	80.0 to 83.0
Silicon, per cent................	0.25 max	0.75 to 1.25	3.75 to 4.25
Lead, max, per cent.............	1.50	0.25	0.15
Tin, max, per cent...............	1.50	0.25	0.25
Manganese, max, per cent........	0.25	0.15	0.15
Aluminum, max, per cent.........	0.25	0.15	0.15
Iron, max, per cent..............	0.50	0.15	0.15
Magnesium, max, per cent........	....	....	0.01
Other elements, [a] max, per cent.....	0.50	0.50	0.25
Zinc, per cent..................	30.0 min	remainder	remainder

* From ASTM Designation B176-62.

[a] Of the maximum of other elements specified, the amount of arsenic, antimony, and sulfur shall not exceed 0.05 per cent for each, and phosphorus shall not exceed 0.01 per cent.

[b] Prior to 1952 these alloys were designated as A, B, and C, respectively.

[c] These alloy designations were established in accordance with the Recommended Practice for Codification of Light Metals and Alloys, Cast and Wrought (ASTM Designation: B 275).

Table 70.   Composition and Typical Properties of Bronze Bearing Alloys

Alloy number..............	1	2	3	4	5	6	7
Chemical composition							
Copper, %................	80	84	88	80	83	78	70
Tin, %...................	20	16	10	10	7	7	5
Lead, %..................	....	......	......	10	7	15	25
Zinc, %..................	....	......	2	......	3	......	......
Mechanical properties							
Tensile strength, psi min....	31,000	28,000	36,000	30,000	34,000	30,000	21,000
Yield strength, psi min.....	21,000	20,000	19,000	19,000	17,000	16,000	
Elongation in 2 in., %......	0.5	1	15	12	20	15	10
BHN, 500-kg load........	130	90	65	65	60	55	48
Impact, Izod..............	1	1	7	5	4	5	5
Weight, lb/cu in...........	0.314	0.313	0.315	0.320	0.322	0.334	0.334
Compressive strength							
0.001 in. set.............	26,000	24,000	12,000	14,000	......	15,000	13,000
0.1 in. set...............	100,000	......	......	43,000	......	42,000	23,000

Table 71.   Composition of Copper-Base Sintered Metal Powder Bearing Alloys*

Element	Composition, per cent	
	Grade 1	Grade 2
Copper...	87.5–90.5	82.6–88.5
Iron, max......................................	1.0	1.0
Tin..	9.5–10.5	9.5–10.5
Lead..	—	2.0–4.0
Zinc, max.....................................	—	0.75
Nickel, max...................................	—	0.35
Antimony, max.................................	—	0.25
Carbon, max...................................	1.75 [a]	1.75 [a]
Total other elements, max......................	0.5	0.5

* From ASTM Designation B438-67.

[a] Commonly graphite. However a maximum of 1.5 per cent of another type of solid lubricant may be substituted.

Table 72. Composition and Physical Properties of White Metal Bearing Alloys*

Alloy Grade a	Specified Nominal Composition of Alloys				Specific Gravity c	Compositions of Alloys Tested				Yield Point, psi d		Johnson's Apparent Elastic Limit psi e		Ultimate Strength in Compression, psi f		Brinell Hardness g		Melting Point		Temperature of Complete Liquefaction		Proper Pouring Temperature	
	Tin, per cent	Antimony, per cent	Lead, per cent	Copper, per cent		Tin, per cent	Antimony, per cent	Lead, per cent	Copper, per cent	20°C	100°C	20°C	100°C	20°C	100°C	20°C	100°C	Deg Fahr	Deg Cent	Deg Fahr	Deg Cent	Deg Fahr	Deg Cent
No. 1	91.0	4.5	0.35	4.5	7.34	90.9	4.52	none	4.56	4400	2650	2450	1050	12850	6950	17.0	8.0	433	223	700	371	825	441
No. 2	89.0	7.5	0.35	3.5	7.39	89.2	7.4	0.03	3.1	6100	3000	3350	1100	14900	8700	24.5	12.0	466	241	669	354	795	424
No. 3	84.0	8.0	0.35	8.0	7.46	83.4	8.2	0.03	8.3	6600	3150	5350	1300	17600	9900	27.0	14.5	464	240	792	422	915	491
No. 7	10.0	15.0	75.0	0.50	9.73	10.0	14.5	75.0	0.11	3550	1600	2500	1350	15650	6150	22.5	10.5	464	240	514	268	640	338
No. 8	5.0	15.0	80.0	0.50	10.04	5.2	14.9	79.4	0.14	3400	1750	2650	1200	15600	6150	20.0	9.5	459	237	522	272	645	341
No. 13	6.0	10.0	remainder	0.50	.....	.....	.....	.....	.....	.....	.....	.....	.....	.....	.....	.....	.....	.....	.....	.....	.....	.....	.....
No. 15 b	1.0	16.0	remainder	0.60	10.05	.....	.....	.....	.....	.....	.....	.....	.....	.....	.....	21.0	13.0	479	248	538	281	662	350

* From ASTM Designation B23-66.

a Alloy grade No. 9 was discontinued in 1946 and grade Nos. 4, 5, 6, 10, 11, 12, 16, and 19 were discontinued in 1959.

b Also nominal arsenic 0.10 per cent.

c The specific gravity multiplied by 0.0361 equals the density in pounds per cubic inch.

d The values for yield point were taken from stress-strain curves at a deformation of 0.125 per cent reduction of gage length.

e Johnson's apparent elastic limit is taken as the unit stress at the point where the slope of the tangent to the curve is 2/3 times its slope at the origin. The compression test specimens were cylinders 1.5 in. in length and 0.5 in. in diameter, machined from chill castings 2 in. in length and 0.75 in. in diameter.

f The ultimate strength values were taken as the unit load necessary to produce a deformation of 25 per cent of the length of the specimen. The compression test specimens were cylinders 1.5 in. in length and 0.5 in. in diameter, machined from chill castings 2 in. in length and 0.75 in. in diameter.

g These values are the average Brinell number of three impressions on each alloy using a 10-mm ball and a 500-kg load applied for 30 sec. The tests were made on the bottom face of parallel machined specimens cast in a 2-in. diameter by 0.625-in. deep steel mold at room temperature.

Table 73.   Composition and Properties of Aluminum Bearing Alloys

Alloy*	750-T5	A750-T5	B750-T5
Form..	Permanent mold casting	Permanent mold casting	Permanent mold casting
Condition..................................	Heat-treated	Heat-treated	Heat-treated
Ultimate tensile strength, psi................	23,000	20,000	32,000
Yield strength (tension), psi†...............	11,000	11,000	23,000
Yield strength (compression), psi†...........	11,000	11,000	23,000
Elongation, %‡............................	12	5	5
BHN, 500-kg load..........................	45	45	70
Rockwell H hardness.......................	75	75	100
Shear strength, psi.........................	15,000	14,000	21,000
Endurance limit, psi§......................	9,000	9,000	11,000
Density....................................	2.88	2.83	2.88
Thermal conductivity‖.....................	0.47	0.43	0.45
Coefficient of thermal expansion¶...........	0.0000135	0.0000132	0.0000135
Composition			
Tin.....................................	6.5	6.5	6.5
Copper.................................	1.0	1.0	2.0
Nickel..................................	1.0	0.5	1.2
Silicon.................................	....	2.5	....
Magnesium.............................	....	....	0.8
Aluminum..............................	Bal.	Bal.	Bal.

* Alloy designations are those of the Aluminum Company of America.
† Offset, 0.2 per cent.
‡ Gage length, 4 × diameter of test bar.
§ Based on 500,000,000 cycles, R. R. Moore rotating beam test.
‖ EGS units, calculated from electrical conductivity.
¶ Per °F, temperature range 68–392°F.

Table 74.   Composition of Iron-Base Sintered Metal Powder Bearing Alloys*

Element	Composition, per cent			
	Grade 1	Grade 2	Grade 3	Grade 4
Copper.............................	...	...	7.0 to 11.0	10.8 to 22.0
Iron................................	96.25 min	95.9 min	remainder a	
Total other elements by difference, max..	3.0	3.0	3.0	3.0
Combined carbon b (on basis of iron only).	0.25 max	0.25 to 0.60	...	...
Silicon, max........................	0.3	0.3	...	...
Aluminum, max.....................	0.2	0.2	...	...

* From ASTM Designation B439-67.
a Total of iron plus copper shall be 97 per cent min.
b The combined carbon may be a metallographic estimate of the carbon in the iron.

Table 75. Chemical Composition of Lead-Tin Solders*[a,b]

ASTM Alloy Grade	Tin Desired, per cent	Lead, Nominal, per cent	Antimony, per cent			Silver, per cent			Bismuth, max, per cent	Copper, max, per cent	Iron, max, per cent	Aluminum, max, per cent	Zinc, max, per cent	Arsenic max, per cent
			Min	Desired	Max	Min	Desired	Max						
70A	70	30	...	...	0.12	...	...	...	0.25	0.08	0.02	0.005	0.005	0.03
70B	70	30	0.20	...	0.50	...	...	...	0.25	0.08	0.02	0.005	0.005	0.03
63A	63	37	...	...	0.12	...	...	...	0.25	0.08	0.02	0.005	0.005	0.03
63B	63	37	0.20	...	0.50	...	...	...	0.25	0.08	0.02	0.005	0.005	0.03
60A	60	40	...	...	0.12	...	...	...	0.25	0.08	0.02	0.005	0.005	0.03
60B	60	40	0.20	...	0.50	...	...	...	0.25	0.08	0.02	0.005	0.005	0.03
50A	50	50	...	...	0.12	...	...	...	0.25	0.08	0.02	0.005	0.005	0.025
50B	50	50	0.20	...	0.50	...	...	...	0.25	0.08	0.02	0.005	0.005	0.025
45A	45	55	...	...	0.12	...	...	...	0.25	0.08	0.02	0.005	0.005	0.025
45B	45	55	0.20	...	0.50	...	...	...	0.25	0.08	0.02	0.005	0.005	0.025
40A	40	60	...	...	0.12	...	...	...	0.25	0.08	0.02	0.005	0.005	0.02
40B	40	60	0.20	...	0.50	...	...	...	0.25	0.08	0.02	0.005	0.005	0.02
40C	40	58	1.8	2.0	2.4	...	...	...	0.25	0.08	0.02	0.005	0.005	0.02
35A	35	65	...	...	0.25	...	...	...	0.25	0.08	0.02	0.005	0.005	0.02
35B	35	65	0.20	...	0.50	...	...	...	0.25	0.08	0.02	0.005	0.005	0.02
35C	35	63.2	1.6	1.8	2.0	...	...	...	0.25	0.08	0.02	0.005	0.005	0.02
30A	30	70	...	...	0.25	...	...	...	0.25	0.08	0.02	0.005	0.005	0.02
30B	30	70	0.20	...	0.50	...	...	...	0.25	0.08	0.02	0.005	0.005	0.02
30C	30	68.4	1.4	1.6	1.8	...	...	...	0.25	0.08	0.02	0.005	0.005	0.02
25A	25	75	...	...	0.25	...	...	...	0.25	0.08	0.02	0.005	0.005	0.02
25B	25	75	0.20	...	0.50	...	...	...	0.25	0.08	0.02	0.005	0.005	0.02
25C	25	73.7	1.1	1.3	1.5	...	...	...	0.25	0.08	0.02	0.005	0.005	0.02
20B	20	80	0.20	...	0.50	...	...	...	0.25	0.08	0.02	0.005	0.005	0.02
20C	20	79	0.8	1.0	1.2	...	...	...	0.25	0.08	0.02	0.005	0.005	0.02
15B	15	85	0.20	...	0.50	...	...	...	0.25	0.08	0.02	0.005	0.005	0.02
10B	10	90	0.20	...	0.50	...	...	...	0.25	0.08	0.02	0.005	0.005	0.02
5A	5[c]	95	...	...	0.12	...	...	...	0.25	0.08	0.02	0.005	0.005	0.02
5B	5[c]	95	0.20	...	0.50	...	...	...	0.25	0.08	0.02	0.005	0.005	0.02
2A	2[d]	98	...	...	0.12	...	...	...	0.25	0.08	0.02	0.005	0.005	0.02
2B	2[d]	98	0.20	...	0.50	...	...	...	0.25	0.08	0.02	0.005	0.005	0.02
2.5S	0[e]	97.5	...	...	0.40	2.3	2.5	2.7	0.25	0.08	0.02	0.005	0.005	0.02
1.5S	1[f]	97.5	...	...	0.40	1.3	1.5	1.7	0.25	0.08	0.02	0.005	0.005	0.02
95TA	95	0.20 max	4.5	5.0	5.5	...	...	...	0.15	0.08	0.04	0.005	0.005	0.05
96.5TS	96.5	0.20 max	0.20	...	0.50	3.3	3.5	3.7	0.15	0.08	0.02	0.005	0.005	0.05

* From ASTM Designation B 32-66T. Analysis shall regularly be made only for the elements specificially mentioned in the above table. If, however, the presence of other elements is suspected, or indicated in the course of routine analysis, further analysis such as by qualitative spectroscopic analysis shall be made to determine that the total of these other elements is not in excess of 0.08 per cent.

[a] The chemical requirements of SAE specifications Nos. 1A, 2A, 2B, 3A, 3B, 4A, 4B, 5A, 5B, 6A, and E-07 conform substantially to the requirements for alloy grade Nos. 45B, 40B, 40C, 30B, 30C, 25B, 25C, 20B, 20C, 15B, and 2.5S, respectively.

[b] Federal specifications are similar to the above alloy grade Nos. 70B, 63B, 60B, 50B, 40B, 35C, 30C, 20C, 2.5S, and 95TA.

[c] Permissible tin range, 4.5 to 5.5 per cent.

[d] Permissible tin range, 1.5 to 2.5 per cent.

[e] Tin maximum, 0.25 per cent.

[f] Permissible tin range, 0.75 to 1.25 per cent.

Table 76.   Properties of Lead-Tin Solder Alloys*[a,b,c,d]

ASTM Alloy Grade	Nominal Composition, per cent			Specific Gravity	Melting Ranges				Uses
					Solidus		Liquidus		
	Tin	Lead	Antimony		deg Cent	deg Fahr	deg Cent	deg Fahr	
TIN-LEAD ALLOYS									
70A	70	30	...	8.32	183	361	192	378	For coating metals.
63A	63	37	...	8.40	183	361	183	361	As lowest melting (eutectic) solder for both by dip and by hand soldering methods.
60A	60	40	...	8.65	183	361	190	374	"Fine Solder." For general purposes, but particularly where the temperature requirements are critical.
50A	50	50	...	8.85	183	361	216	421	For general purposes. Most popular of all.
45A	45	55	...	8.97	183	361	227	441	For automobile radiator cores and roofing seams.
40A	40	60	...	9.30	183	361	238	460	Wiping solder for joining lead pipes and cable sheaths. For automobile radiator cores and heating units.
35A	35	65	...	9.50	183	361	247	477	General purpose and wiping solder.
30A	30	70	...	9.70	183	361	255	491	For machine and torch soldering.
25A	25	75	...	10.00	183	361	266	511	For machine and torch soldering.
20B	20	80	...	10.20	183	361	277	531	For coating and joining metals. For filling dents or seams in automobile bodies.
15B	15	85	...	10.50	227[f]	440[f]	288	550	For coating and joining metals.
10B	10	90	...	10.80	268[f]	514[f]	299	570	For coating and joining metals.
5A	5	95	...	11.30	270	518	312	594	For coating and joining metals.
TIN-LEAD-ANTIMONY ALLOYS									
40C	40	58	2	9.23	185	365	231	448	Same uses as (50-50) tin-lead but not recommended for use on galvanized iron.
35C	35	62.3	1.8	9.44	185	365	243	470	For wiping and all uses except on galvanized iron.
30C	30	68.4	1.6	9.65	185	364	250	482	For torch soldering or machine soldering, except on galvanized iron.
25C	25	73.7	1.3	9.96	184	364	263	504	For torch and machine soldering, except on galvanized iron.
20C	20	79	1	10.17	184	363	270	517	For machine soldering and coating of metals, tipping, and like uses, but not recommended for use on galvanized iron.
TIN-ANTIMONY ALLOYS									
95TA	95	...	5	7.25	234	452	240	464	For joints on copper: electrical plumbing, and heating.

SILVER-LEAD ALLOYS

	Tin	Lead	Silver						
2.5S	0	97.5	2.5	11.35	304	579	304	579	For use on copper, brass, and similar metals with torch heating. Not recommended in humid environments due to its known susceptibility to corrosion.
1.5S	1	97.5	1.5	11.28	309	588	309	588	For use on copper, brass, and similar metals with torch heating.

---

a Federal specifications similar to, and also those for, ASTM alloy grades 70B, 63B, 60B, 50B and 40B contain antimony in the range of 0.20 to 0.50 per cent in addition to a permissible limit of 0.25 per cent bismuth. Such formulation is intended to provide for reliability of soldered joints below 32°F, since there is a possibility of failure of the joints resulting from a phase change of the tin constituent of the solder. The change from beta tin to alpha tin is accompanied by a volume increase of 26 per cent which may be manifested in a powdery structural disintegration.

b It is recommended that the grade of solder metal be selected which contains the least amount of tin required to give suitable flowing and adhesive qualities for the work in hand. To provide for the reliability of soldered connections closely adjacent to other electrical conductors in compact equipment, the following precaution is suggested in order to reduce troubles developing in the nature of leaky or short circuits. The possibility of the growth of "tin whiskers" will be reduced when tin-lead solders containing 50 per cent tin, or less, are used.

c When soldering silver film surfaces (preferably 0.0005 in., min thickness) grade A solders containing 60 to 70 per cent tin with the addition of 1 to 3 per cent silver are suggested for use. The object is to avoid the detachment of the silver film by its solution in or migration into the solder.

d Valuable information on the use of solders and fluxes may be found in ASTM Special Technical Publication No. 189, "Symposium on Solder," June, 1956.

e Alloys are completely solid below the designated "solidus," and completely liquid only above the designated "liquidus." In the range of temperature between these two points the alloys are partly solid and partly liquid. In the 60 per cent tin, 40 per cent lead alloy the amount of solid portion is so small in the range given that it is practically unnoticeable. In the 40 per cent tin, 60 per cent lead alloy the proportion of solid and fluid metal in the range given makes this alloy suitable for use as wiping solder.

f For some engineering design purposes it is well to consider these alloys as having practically no mechanical strength at 183°C (361°F).

Table 77. Composition, Solidus and Liquidus For Brazing Filler Metals*

ALUMINUM-SILICON

AWS-ASTM Classification	Silicon, per cent a	Copper, per cent a	Iron, per cent a	Zinc, per cent a	Magnesium, per cent a	Manganese, per cent a	Chromium, per cent a	Titanium, per cent a	Aluminum, per cent	Other Elements, per cent a		Solidus, b deg Fahr	Liquidus, b deg Fahr	Brazing Temperature Range, deg Fahr
										Each	Total			
BAlSi-2	6.8 to 8.2	0.25	0.8	0.20	....	0.10	....	....	remainder	0.05	0.15	1070	1135	1110 to 1150
BAlSi-3	9.3 to 10.7	3.3 to 4.7	0.8	0.20	0.15	0.15	0.15	....	remainder	0.05	0.15	970	1085	1060 to 1120
BAlSi-4	11.0 to 13.0	0.30	0.8	0.20	0.10	0.15	....	....	remainder	0.05	0.15	1070	1080	1080 to 1120
BAlSi-5	9.0 to 11.0	0.30	0.8	0.10	0.05	0.05	....	0.20	remainder	0.05	0.15	1070	1095	1090 to 1120

COPPER-PHOSPHORUS

AWS-ASTM Classification	Phosphorus, per cent	Silver, per cent	Copper, per cent	Other Elements, Total, per cent a	Solidus, b deg Fahr	Liquidus, b deg Fahr	Brazing Temperature Range, deg Fahr
BCuP-1	4.75 to 5.25	...	remainder	0.15	1310	1650	1450 to 1700
BCuP-2	7.00 to 7.50	...	remainder	0.15	1310	1460	1350 to 1550
BCuP-3	5.75 to 6.25	4.75 to 5.25	remainder	0.15	1190	1485	1300 to 1500
BCuP-4	7.00 to 7.50	5.75 to 6.25	remainder	0.15	1190	1335	1300 to 1450
BCuP-5	4.75 to 5.25	14.50 to 15.50	remainder	0.15	1190	1475	1300 to 1500

SILVER

AWS-ASTM Classification	Silver, per cent	Copper, per cent	Zinc, per cent	Cadmium, per cent	Nickel, per cent	Tin, per cent	Lithium, per cent	Phosphorus, per cent [a]	Other Elements, Total, per cent [a]	Solidus, [b] deg Fahr	Liquidus, [b] deg Fahr	Brazing Temperature Range, deg Fahr
BAg-1	44 to 46	14 to 16	14 to 18	23 to 25					0.15	1125	1145	1145 to 1400
BAg-1a	49 to 51	14.5 to 16.5	14.5 to 18.5	17 to 19					0.15	1160	1175	1175 to 1400
BAg-2	34 to 36	25 to 27	19 to 23	17 to 19					0.15	1125	1295	1295 to 1550
BAg-3	49 to 51	14.5 to 16.5	13.5 to 17.5	15 to 17	2.5 to 3.5				0.15	1170	1270	1270 to 1500
BAg-4	39 to 41	29 to 31	26 to 30		1.5 to 2.5				0.15	1240	1435	1435 to 1650
BAg-5	44 to 46	29 to 31	23 to 27						0.15	1250	1370	1370 to 1550
BAg-6	49 to 51	33 to 35	14 to 18						0.15	1270	1425	1425 to 1600
BAg-7	55 to 57	21 to 23	15 to 19			4.5 to 5.5			0.15	1145	1205	1205 to 1400
BAg-8	71 to 73	remainder							0.15	1435	1435	1435 to 1650
BAg-8a	71 to 73	remainder					0.15 to 0.3		0.15	1410	1410	1410 to 1600
BAg-13	53 to 55	remainder	4.0 to 6.0		0.5 to 1.5				0.15	1325	1575	1575 to 1775
BAg-18	59 to 61	remainder				9.5 to 10.5		0.025	0.15	1115	1325	1325 to 1550
BAg-19	92 to 93	remainder					0.15 to 0.3		0.15	1435	1635	1610 to 1800

PRECIOUS METALS

AWS-ASTM Classification	Gold, per cent	Copper, per cent	Nickel, per cent	Other Elements, Total, per cent [a]	Solidus, [b] deg Fahr	Liquidus, [b] deg Fahr	Brazing Temperature Range, deg Fahr
BAu-1	37.0 +1 −0	remainder		0.15	1815	1860	1860 to 2000
BAu-2	79.5 +1 −0	remainder		0.15	1635	1635	1635 to 1850
BAu-3	34.5 +1 −0	remainder	2.5 to 3.5	0.15	1785	1885	1885 to 1995
BAu-4	81.5 +1 −0		remainder	0.15	1740	1740	1740 to 1840

Table 77. Composition, Solidus and Liquidus For Brazing Filler Metals—*Continued*

COPPER AND COPPER-ZINC

AWS-ASTM Classification	Copper, per cent	Zinc, per cent	Tin, per cent	Iron, per cent	Manganese, per cent	Nickel, per cent	Phosphorus, per cent[a]	Lead, per cent[a]	Aluminum, per cent[a]	Silicon, per cent[a]	Other Elements, Total, per cent[a]	Solidus,[b] deg Fahr	Liquidus,[b] deg Fahr	Brazing Temperature Range, deg Fahr
BCu-1	99.90 min						0.075	0.02	0.01		0.10	1980	1980	2000 to 2100
BCu-1a	99.0 min										0.30[c]	1980	1980	2000 to 2100
BCu-2[e]	86.5 min									[f]	0.50[f]	1980	1980	2000 to 2100
RBCuZn-A[g]	57 to 61	remainder	0.25 to 1.00					0.05[f]	0.01[f]		0.50[f]	1630	1650	1670 to 1750
RBCuZn-D[e]	46 to 50	remainder		[f]	[f]	9.0 to 11.0	0.25	0.05[f]	0.01[f]	0.04 to 0.25	0.50[f]	1690	1715	1720 to 1800

MAGNESIUM

AWS-ASTM Classification	Aluminum, per cent	Manganese, per cent	Zinc, per cent	Silicon, per cent[a]	Copper, per cent[a]	Nickel, per cent[a]	Iron, per cent[a]	Beryllium, per cent	Magnesium, per cent	Other Elements, Total, per cent[a]	Solidus,[b] deg Fahr	Liquidus,[b] deg Fahr	Brazing Temperature Range, deg Fahr
BMg-1	8.3 to 9.7	0.15 min	1.7 to 2.3	0.05	0.05	0.005	0.005	0.0002 to 0.0008	balance	0.30	830	1110	1120 to 1160
BMg-2	11.0 to 13.0		4.5 to 5.5						balance	0.30	770	1050	1080 to 1130
BMg-2a	11.0 to 13.0		4.5 to 5.5					0.0002 to 0.0008	balance	0.30	770	1050	1080 to 1130

NICKEL

AWS-ASTM Classification	Nickel, per cent	Chromium, per cent	Boron, per cent	Silicon, per cent	Iron, per cent a	Carbon, per cent a	Phosphorus, per cent	Other Elements, Total, per cent a	Solidus, b deg Fahr	Liquidus, b deg Fahr	Brazing Temperature Range, deg Fahr
BNi-1	remainder	13.0 to 15.0	2.75 to 4.00	3.0 to 5.0	4.0 to 5.0	0.6 to 0.9	...	0.50	1790	1900	1950 to 2200
BNi-2	remainder	6.0 to 8.0	2.75 to 3.5	4.0 to 5.0	2.0 to 4.0	.015	...	0.50	1780	1830	1850 to 2150
BNi-3	remainder	...	2.75 to 3.5	4.0 to 5.0	1.5	0.06	...	0.50	1800	1900	1850 to 2150
BNi-4	remainder	...	1.0 to 2.2	3.0 to 4.0	1.5	0.06	...	0.50	1800	1950	1850 to 2150
BNi-5	remainder	18.0 to 20.0	...	9.75 to 10.5	...	0.15	...	0.50	1975	2075	2100 to 2200
BNi-6	remainder	...	...	...	...	0.15	10.0 to 12.0	0.50	1610	1610	1700 to 1875
BNi-7	remainder	11.0 to 15.0	...	...	...	...	9.0 to 11.0	0.50	1630	1630	1700 to 1900

*From ASTM Designation B260-62T. (AWS Designation A5.8).

a Single values shown are maximum percentages, except where otherwise specified.

b Solidus and liquidus shown are for the nominal composition in each classification.

c Total other elements requirement pertains only to the metallic elements for this filler metal.

d These chemical requirements pertain only to the copper oxide and do not include requirements for the organic vehicle in which the copper oxide is suspended.

e Total other elements requirement pertains only to metallic elements for this filler metal. The following limitations are placed on the nonmetallic elements:

Constituent	per cent (max.)
Chlorides	0.4
Sulfates	0.1
Oxygen	remainder
Nitric acid insoluble	0.3
Acetone soluble matter	0.5

f Total other elements, including the elements marked with an f, shall not exceed the value specified

g This AWS-ASTM classification is intended to be identical with the same classification that appears in the Specification for Copper and Copper-Alloy Welding Rods (AWS Designation A5.7, ASTM Designation B 259).

h Cobalt 1.0 max, per cent (unless otherwise specified) if determined.

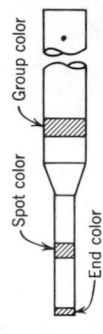

Spot color — Group color — End color

Table 78.   Standard Color Markings for Electrode Identification*

Mild- and Low-Alloy Steel Electrodes

AWS-ASTM classification	End color	Spot color	Group color
E6010	None	None	None
E6011	None	Blue	None
E6012	None	White	None
E6013	None	Brown	None
E6020	None	Green	None
E7010-A1	Blue	White	None
E7011-A1	Blue	Yellow	None
E7020-A1	Blue	Yellow	Silver
E10013	Green	Brown	Silver

Iron Powder Electrodes

AWS-ASTM classification	End color	Spot color	Group color
E6010	None	None	None
E6024	None	Yellow	None
E6014	None	None	None
E7016	Blue	Orange	Green
E8016-B2	White	Gray	Green
E9016-B3	Brown	Blue	Green

Low-Hydrogen Electrodes

AWS-ASTM classification	End color	Spot color	Group color
E6016	None	Orange	None
E7016	Blue	Orange	Green
E8016	White	Orange	Green
E8016-B1	White	Black	Green
E8016-C1	White	Blue	Green
E8016-B2	White	Gray	Green
E9016-B3	Brown	Blue	Green
E9016	Brown	Orange	Green
E10015	Green	Red	Green
E10016	Green	Orange	Green
E12016	Yellow	Orange	Green
E12015	Yellow	Red	Green
E15016	None	None	None

Electrodes for Welding Cast Iron

AWS-ASTM classification	End color	Spot color	Group color
EST	Orange	None	None
ENI	Orange	Blue	White

Stainless Steel Electrodes

AWS-ASTM classification	End color	Spot color	Group color
E308-15	Yellow	None	Black
E308-16	Yellow	None	Yellow
E308-ELC-15	Brown	None	Black
E308-ELC-16	Brown	None	Yellow
E347-15	Yellow	Blue	Black
E347-16	Yellow	Blue	Yellow
E316-15	Yellow	White	Black
E316-16	Yellow	White	Yellow
E316-ELC-15	Brown	White	Black
E316-ELC-16	Brown	White	Yellow
E317-15	Yellow	Brown	Black
E317-16	Yellow	Brown	Yellow
E309-15	Black	None	Black
E309-16	Black	None	Yellow
E310-15	Red	None	Black
E310-16	Red	None	Yellow
E307-15	None	Black	Black
E307-16	None	Black	Yellow
E308-MO-15	None	None	Black
E308-MO-16	None	None	Yellow
E502-16	Gray	Blue	Yellow

* By permission from *Canadian Metalworking/Machine Production*, May 1960.

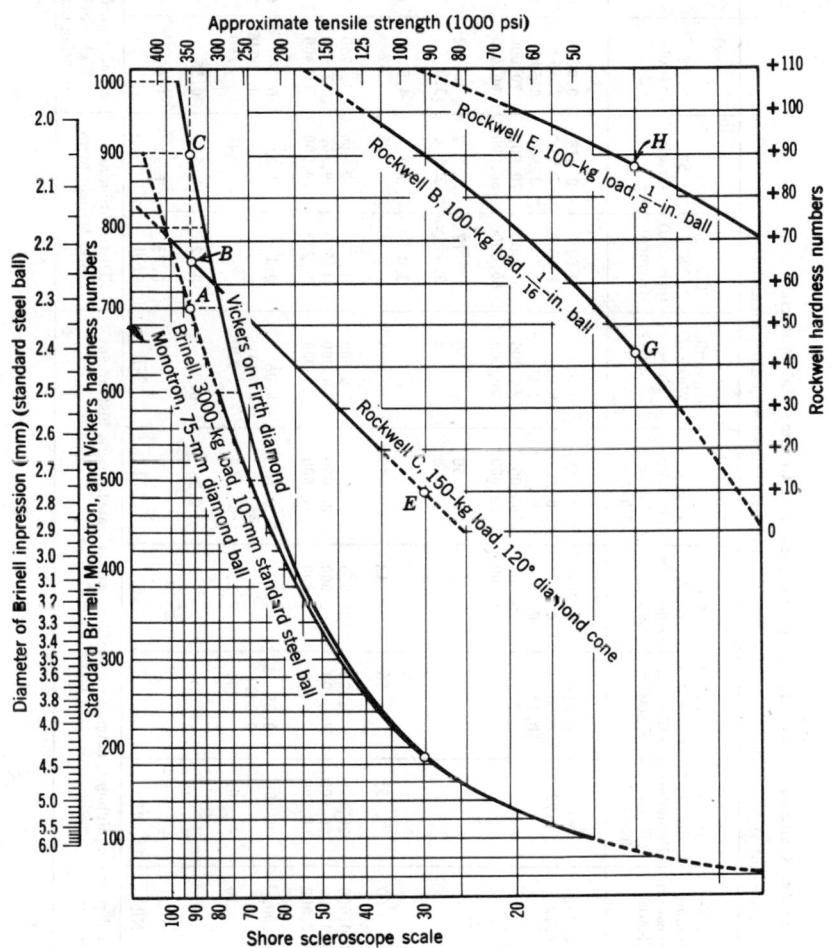

FIG. 7. Approximate relationship between hardnesses determined by various testing systems.

Verticals represent equivalent hardness.

For instance: Brinell or Monotran 700 (point A) equals Rockwell C-65 (point B), equals Vickers 930 (point C), equals Scleroscope 89 (bottom scale), and about 340,000 psi tensile (top scale).

Again: Rockwell B-91 (point F) equals Rockwell C-10 (point E), equals Brinell, Monotron or Vickers 187 and Scleroscope 29, and has about 92,000 psi tensile strength.

Finally, Rockwell B-43 (point G) equals Rockwell E-87 (point H). Chart based on data from Westinghouse Research Laboratories, Bureau of Standards, and the ASM Metals Handbook.

Courtesy of Shell Development Company.

Table 79. General Data on Magnetic Properties of Commercial Materials

Figures represent approximate average properties; individual samples may differ somewhat

	Supermalloy (Western Electric)	4-79 Molybdenum Permalloy (Western Electric)	78 Permalloy (Western Electric)	Mu metal (Allegheny Ludlum)	Hipernik V (oriented) (Westinghouse Electric)	Hipernik (Westinghouse Electric)	Trancor 3X (oriented) (Armco)	52 Grade	58 Grade	65 Grade
Physical										
Gage, in.	0.002		0.014	0.014	0.002	0.014	0.014	0.014	0.014	0.014
Density, gm/cm^3	8.77	8.72	8.60	8.58	8.25	8.25	7.65*	7.55*	7.55*	7.55*
Sheet weight, lb/sq ft			0.63	0.63		0.60	0.556	0.549	0.549	0.549
Ultimate tensile strength, lb/sq in.				70,000	65,000	65,000	67,000	70,000	70,000	69,000
Yield point, lb/sq in.				45,000	20,000	20,000	56,000	66,000	66,000	64,000
Elongation, per cent in 2-in. gage length				23	50	50	12	2	2	3
Ductility, Erichsen draw, mm.					8	8	5.5	2.5	3.0	3.5
Miscellaneous										
Approximate per cent silicon							3.0	5.0	4.5	4.0
Electrical resistivity, microhm-cm	60	55	16	42	45	45	47	65	62	59
Magnetic										
Initial permeability	100,000	20,000	9,000	20,000		6,000	1,500		750	700
Maximum permeability	800,000	100,000	100,000	100,000	80,000	90,000	51,500	9,800	7,600	6,700
Saturation induction (ferrio), gausses	7,900	8,700	10,700	6,500	16,000	16,000	20,200	19,200	19,400	19,600
Coercive force, oersteds (from 10,000 gauss tip) †	0.002	0.05	0.05	0.04	0.16	0.06	0.1	0.22	0.28	0.35
Steinmetz hysteresis coefficient			0.0001			0.00015			0.00046	0.00051
Hysteresis loss, ergs/cm^3/cycle ‡	10	100	200	200	500	220		640	790	1,020
Iron loss at 10,000 gausses, 60 cycles, watts/lb							0.32	0.52	0.58	0.65
Iron loss at 15,000 gausses, 60 cycles, watts/lb					0.38	0.25	0.73	1.33	1.46	1.62
Maximum aging, iron loss, per cent	Nil	Nil	Nil	Nil	Nil	Nil	Nil	3	3	3
Typical applications	Telephone equipment, magnetic amplifiers, audio transformers, instrument transformers						Distribution and power transformers		Small transformers, high-efficiency rotating machines	

Table 79. General Data on Magnetic Properties of Commercial Materials—*Continued*

	72 Grade	Dynamo or 82 Grade	Motor or 101 Grade	Electrical or 117 Grade	Armature or 130 Grade	Pure Iron Armco Iron or Equivalent	Low Carbon Steel Sheet or Shapes	Cast Steel (Annealed)	Cast Iron (Annealed)	Nodular Cast Iron (Annealed)
Physical										
Gage, in.	0.014	0.014	0.014	0.014	0.014					
Density, gm/cm³	7.55*	7.65*	7.65*	7.75*	7.75*	7.85	7.85	7.80	7.0	7.2
Sheet weight, lb/sq ft	0.549	0.556	0.556	0.563	0.563					
Ultimate tensile strength, lb/sq in.	68,000	67,000	63,000	50,000	43,000	40,000	45,000	60,000	25,000	70,000
Yield point, lb/sq in.	57,000	56,000	48,000	33,000	26,000	20,000	25,000	30,000		
Elongation, per cent in 2-in. gage length	11	12	18	26	29	25	18	20		15
Miscellaneous										
Ductility, Erichsen draw, mm.	4.5	5.5	6.0	6.5	7.0	10	9			
Approximate per cent silicon	3.6	3.25	2.5	1.25	0.5			0.4	2.0	
Electrical resistivity, microhm-cm	56	50	42	26	19	10.7	13	15	100	
Magnetic										
Initial permeability	600	500	400	350	325	275	250	175	125	175
Maximum permeability	5,800	5,700	5,800	5,600	5,200	4,500	2,500	1,500	500	1,550
Saturation induction (ferric), gausses	19,800	20,200	20,500	21,000	21,300	21,600	21,200	21,000	14,000	>15,000
Coercive force, oersteds (from 10,000 gauss tip) †	0.42	0.60	0.64	0.66	0.75	1.0	2.0	5.0	11.0	2.4
Steinmetz hysteresis coefficient	0.00056	0.00065	0.00081	0.00088	0.001	0.002	0.003	0.005	0.012	
Hysteresis loss, ergs/cm³/cycle ‡	1,250	1,750	1,940	2,040	2,150	3,200				
Iron loss at 10,000 gausses, 60 cycles, watts/lb	0.72	0.82	1.01	1.17	1.30					
Iron loss at 15,000 gausses, 60 cycles, watts/lb	1.78	2.00	2.46	3.16	3.65					
Maximum aging, iron loss, per cent	3	5	5	5	5					
Typical applications	Small transformers, high-efficiency rotating machines		Small motors, a-c magnets, starting transformers			Pole pieces, relays	Fields, frame of d-c and synchronous machines	Frames, solid poles		Frames

* ASTM assumed densities.
† Measured from tip induction of 10,000 gausses or from saturation induction if that is less than 10,000 gausses.
‡ Based on tip induction of 10,000 gausses or saturation, whichever is lower.

Table 80.   Normal Induction

Value of magnetizing force (oersteds)

Induction in Gausses	Supermalloy (Western Electric)	4-79 Molybdenum Permalloy (Western Electric)	78 Permalloy (Western Electric)	Mu Metal (Allegheny Ludlum)	Hipernik V (oriented) W	Hipernik W	Trancor 3X (Armco)	52 Grade	58 Grade	65 Grade
10			0.001					0.0087	0.0098	0.0125
16	0.00010		0.0017					0.0125	0.015	0.017
25	.00016	0.0012	0.0025					0.0175	0.020	0.0245
40	.00022	.0018	0.004					0.024	0.028	0.034
64	.00032	.0027	0.006			0.0038		0.033	0.040	0.048
100	.00044	.0044	0.008	0.003	0.030	0.0050		0.044	0.053	0.064
160	.0006	.007	0.01	0.004	0.047	0.0070		0.058	0.071	0.086
250	.0008	.01	0.016	0.006	0.065	0.0103		0.075	0.095	0.115
400	.00112	.015	0.022	0.009	0.095	0.0135		0.110	0.125	0.153
640	.00137	.021	0.028	0.013	0.112	0.0172		0.135	0.175	0.200
1,000	.00167	.027	0.032	0.018	0.125	0.0205		0.170	0.207	0.250
1,250	.00183	.03	0.034	0.021	0.127	0.0222		0.193	0.235	0.280
1,600	.00203	.034	0.035	0.029	0.130	0.0238		0.220	0.270	0.312
2,000	.00227	.036	0.036	0.045	0.132	0.026		0.252	0.311	0.362
3,000	.00328	.045	0.038	0.168	0.140	0.031		0.340	0.412	0.470
4,000	.0056	.053	0.041	0.55	0.145	0.036	0.105	0.440	0.53	0.55
5,000	.0117	.074	0.047	1.29	0.149	0.045	0.115	0.560	0.66	0.75
6,000	.0342	.115	0.058	5.0	0.152	0.060	0.127	0.710	0.82	0.93
7,000	.140	.218	0.0875		0.156	0.085	0.140	0.900	1.05	1.15
8,000		.666	0.167		0.159	0.132	0.157	1.140	1.32	1.44
9,000			0.474		0.160	0.235	0.175	1.450	1.70	1.81
10,000			2.0		0.161	0.448	0.198	1.900	2.30	2.32
11,000			200.0		0.162	1.13	0.230	2.620	3.07	3.15
12,000					0.163	2.63	0.276	4.400	4.85	4.7
13,000					0.167	5.10	0.350	10.7	8.50	8.0
14,000					0.203	9.25	0.470	26.5	18.75	17.5
15,000					0.500	40.0	0.720	53.0	44.0	40.0
16,000					10.0	400.0	1.65	91.0	100.0	90.0
17,000							5.75	142.0	210.0	195.0
18,000							19.20	220.0		
19,000								360.0		
20,000								720.0		
21,000										
22,000										

Data (Average Figures)

corresponding to induction

72 Grade	Dynamo or 82 Grade	Motor or 101 Grade	Electrical or 117 Grade	Armature or 130 Grade	Pure Iron Armco Iron or Equivalent (Annealed)	Low Carbon Steel Sheet or Shapes (Annealed)	Cast Steel (Annealed)	Cast Iron (Annealed)	Nodular Cast Iron (Annealed)	Induction, lines per sq in.
0.0155	0.016						0.05	0.08	0.06	64
0.0215	0.023						0.075	0.13	0.10	103
0.0295	0.0325						0.11	0.19	0.125	161
0.041	0.047						0.15	0.29	0.175	258
0.0508	0.065						0.22	0.44	0.26	423
0.077	0.089		0.15				0.30	0.65	0.37	645
0.104	0.116		0.185				0.42	0.91	0.52	1,030
0.155	0.185		0.230				0.57	1.2	0.72	1,610
0.195	0.245		0.295	0.33		0.47	0.79	1.6	1.00	2,580
0.245	0.310	0.370	0.38	0.42		0.58	1.03	2.0	1.25	4,230
0.305	0.362	0.450	0.455	0.50	0.16	0.72	1.23	2.5	1.63	6,450
0.345	0.390	0.500	0.495	0.55	0.17	0.80	1.38	2.8	1.82	10,300
0.390	0.425	0.56	0.545	0.60	0.21	0.89	1.57	3.3	2.02	12,900
0.435	0.470	0.61	0.59	0.66	0.24	0.96	1.8	4.0	2.18	16,100
0.555	0.585	0.73	0.69	0.78	0.30	1.14	2.3	7.0	2.53	19,400
0.695	0.720	0.84	0.785	0.88	0.37	1.35	2.8	14.0	2.91	25,800
0.860	0.880	0.96	0.880	1.00	0.43	1.57	3.3	25.0	3.39	32,300
1.06	1.06	1.14	0.972	1.14	0.50	1.83	3.9	45.0	3.90	38,700
1.32	1.30	1.36	1.13	1.30	0.585	2.16	4.5	70.0	4.50	45,200
1.63	1.62	1.63	1.34	1.50	0.690	2.63	5.2	81.0	5.35	51,600
2.02	2.02	2.03	1.62	1.76	0.825	3.32	6.2	160.0	6.72	58,100
2.54	2.59	2.58	2.00	2.15	1.000	4.35	7.5	220.0	10.0	64,500
3.28	3.80	3.37	2.72	2.75	1.25	5.72	9.5	310.0	13.3	71,000
4.62	4.75	4.90	4.30	3.85	1.58	7.65	12.5	410.0	18.1	77,400
7.85	7.50	7.75	7.40	6.00	2.06	10.3	16.0	620.0	32.3	83,900
16.6	14.0	13.5	13.5	10.00	2.85	16.4	22.0	1000.0	60.0	90,300
37.5	29.5	28.0	26.0	19.4	5.50	26.0	32.0		110.0	96,800
75.0	60.2	59.0	51.0	40.0	14.50	42.3	50.0			103,200
125.0	117.0	108.0	90.0	81.0	42.0	73.0	82.0			109,700
200.0	195.0	177.0	153.0	145.0	100.0	124.0	130.0			116,100
325.0	307.0	270.0	235.0	223.0	167.0	200.0	200.0			122,600
580.0	500.0	550.0	376.0	330.0	270.0		380.0			129,050
			660.0	660.0	470.0		800.0			135,400
					940.0		1800.0			141,900

Table 81. (Part 1). Approximate Frequency and Induction Core Loss Factors for 29 Gage (0.014 in.) Material

Induction, gausses	Frequency									Induction, lines per sq in.
	15 Cycles	25 Cycles	40 Cycles	50 Cycles	60 Cycles	80 Cycles	100 Cycles	160 Cycles	250 Cycles	
10	0.35×10^{-6}	0.60×10^{-6}	0.10×10^{-5}	0.13×10^{-5}	0.16×10^{-5}	0.22×10^{-5}	0.28×10^{-5}	0.47×10^{-5}	0.80×10^{-5}	64
16	0.10×10^{-5}	0.17×10^{-5}	0.30×10^{-5}	0.36×10^{-5}	0.46×10^{-5}	0.64×10^{-5}	0.80×10^{-5}	0.14×10^{-4}	0.24×10^{-4}	103
25	0.29×10^{-5}	0.50×10^{-5}	0.80×10^{-5}	0.10×10^{-4}	0.12×10^{-4}	0.18×10^{-4}	0.23×10^{-4}	0.40×10^{-4}	0.72×10^{-4}	161
40	0.80×10^{-5}	0.14×10^{-4}	0.23×10^{-4}	0.28×10^{-4}	0.35×10^{-4}	0.51×10^{-4}	0.64×10^{-4}	0.11×10^{-3}	0.21×10^{-3}	258
64	0.22×10^{-4}	0.41×10^{-4}	0.65×10^{-4}	0.82×10^{-4}	0.10×10^{-3}	0.15×10^{-3}	0.18×10^{-3}	0.32×10^{-3}	0.58×10^{-3}	423
100	0.58×10^{-4}	0.10×10^{-3}	0.18×10^{-3}	0.22×10^{-3}	0.26×10^{-3}	0.38×10^{-3}	0.47×10^{-3}	0.80×10^{-3}	0.15×10^{-2}	645
160	0.15×10^{-3}	0.28×10^{-3}	0.44×10^{-3}	0.56×10^{-3}	0.66×10^{-3}	0.92×10^{-3}	0.12×10^{-2}	0.21×10^{-2}	0.38×10^{-2}	1,030
250	0.36×10^{-3}	0.64×10^{-3}	0.10×10^{-2}	0.13×10^{-2}	0.16×10^{-2}	0.22×10^{-2}	0.30×10^{-2}	0.50×10^{-2}	0.90×10^{-2}	1,610
400	0.90×10^{-3}	0.15×10^{-2}	0.26×10^{-2}	0.30×10^{-2}	0.38×10^{-2}	0.56×10^{-2}	0.72×10^{-2}	0.12×10^{-1}	0.22×10^{-1}	2,580
640	0.20×10^{-2}	0.35×10^{-2}	0.62×10^{-2}	0.75×10^{-2}	0.93×10^{-2}	0.13×10^{-1}	0.17×10^{-1}	0.30×10^{-1}	0.53×10^{-1}	4,230
1,000	0.43×10^{-2}	0.74×10^{-2}	0.13×10^{-1}	0.16×10^{-1}	0.22×10^{-1}	0.28×10^{-1}	0.36×10^{-1}	0.61×10^{-1}	0.11	6,450
1,600	0.10×10^{-1}	0.16×10^{-1}	0.28×10^{-1}	0.36×10^{-1}	0.43×10^{-1}	0.60×10^{-1}	0.78×10^{-1}	0.14	0.26	10,300
2,000	0.14×10^{-1}	0.24×10^{-1}	0.40×10^{-1}	0.52×10^{-1}	0.62×10^{-1}	0.88×10^{-1}	0.11	0.20	0.38	12,900
2,500	0.20×10^{-1}	0.34×10^{-1}	0.60×10^{-1}	0.75×10^{-1}	0.92×10^{-1}	0.13	0.16	0.29	0.55	16,100
3,000	0.29×10^{-1}	0.46×10^{-1}	0.81×10^{-1}	0.10	0.12	0.18	0.22	0.37	0.74	19,400
4,000	0.45×10^{-1}	0.73×10^{-1}	0.13	0.17	0.20	0.28	0.36	0.64	1.25	25,800
5,000	0.66×10^{-1}	0.11	0.19	0.25	0.29	0.41	0.54	0.96	1.80	32,300
6,000	0.90×10^{-1}	0.15	0.26	0.34	0.40	0.56	0.72	1.30	2.50	38,700
7,000	0.12	0.20	0.35	0.44	0.52	0.92	0.96	1.70	3.30	45,200
8,000	0.15	0.25	0.44	0.55	0.66	1.14	1.20	2.10	4.1	51,600
9,000	0.19	0.31	0.54	0.68	0.82	1.40	1.48	2.60	5.1	58,100
10,000	0.23	0.38	0.64	0.82	1.00	1.70	1.80	3.3	6.2	64,500
11,000	0.28	0.46	0.76	0.98	1.20	2.05	2.25	4.0	7.7	71,000
12,000	0.33	0.55	0.91	1.15	1.45	2.5	2.8	4.9	9.5	77,400
13,000	0.39	0.64	1.07	1.37	1.70	2.90	3.3	5.9	11.3	83,900
14,000	0.45	0.74	1.25	1.60	2.00	2.90	3.9	7.1	13.0	90,300
15,000	0.52	0.86	1.47	1.90	2.35	3.40	4.6	8.5	15.0	96,800

NOTE. To find the approximate iron loss at any induction or frequency, multiply the loss, at 60 cycles, 10,000 gausses, by the factor shown in the table. Values are most accurate for about a 4 per cent silicon steel and are reasonably exact for lower silicon steels.

Table 81. (Part 2). Approximate Frequency and Induction Core Loss Factors for 29 Gage (0.014 in.) Material

Induction, gausses	Frequency									Induction, lines per sq in.
	400 Cycles	640 Cycles	1000 Cycles	1600 Cycles	2500 Cycles	4000 Cycles	6400 Cycles	10,000 Cycles	15,000 Cycles	
10	0.14×10^{-4}	0.30×10^{-4}	0.56×10^{-4}	0.13×10^{-3}	0.39×10^{-3}	0.79×10^{-3}	0.16×10^{-2}	0.38×10^{-2}	0.80×10^{-2}	64
16	0.47×10^{-4}	0.95×10^{-4}	0.20×10^{-3}	0.43×10^{-3}	0.10×10^{-2}	0.24×10^{-2}	0.56×10^{-2}	0.11×10^{-1}	0.30×10^{-1}	103
25	0.14×10^{-3}	0.28×10^{-3}	0.58×10^{-3}	0.14×10^{-2}	0.32×10^{-2}	0.80×10^{-2}	0.19×10^{-1}	0.44×10^{-1}	0.10	161
40	0.40×10^{-3}	0.85×10^{-3}	0.18×10^{-2}	0.44×10^{-2}	0.95×10^{-2}	0.24×10^{-1}	0.58×10^{-1}	0.14	0.31	258
64	0.11×10^{-2}	0.24×10^{-2}	0.51×10^{-2}	0.11×10^{-1}	0.27×10^{-1}	0.64×10^{-1}	0.15	0.38	0.84	423
100	0.28×10^{-2}	0.59×10^{-2}	0.13×10^{-1}	0.29×10^{-1}	0.67×10^{-1}	0.15	0.38	0.96	2.0	645
160	0.70×10^{-2}	0.15×10^{-1}	0.32×10^{-1}	0.70×10^{-1}	0.16	0.38	0.95	2.3	5.0	1,030
250	0.17×10^{-1}	0.36×10^{-1}	0.74×10^{-1}	0.17	0.38	0.93	2.2	5.3	11.0	1,610
400	0.41×10^{-1}	0.86×10^{-1}	0.18	0.39	0.88	2.2	5.2	13.0	27.0	2,580
640	0.95×10^{-1}	0.20	0.42	0.90	2.2	5.1	13.0	30.0	62.0	4,230
1,000	0.22	0.45	0.94	2.2	5.0	12.0	29.0	72.0		6,450
1,600	0.48	1.0	2.15	4.9	11.5	28.0	65.0			10,300
2,000	0.71	1.5	3.2	7.4	17.0	42.0				12,900
2,500	1.10	2.2	4.6	11.0	25.0	63.0				16,100
3,000	1.40	3.1	6.4	15.0	35.0					19,400
4,000	2.30	5.0	11.0	26.0	58.0					25,800
5,000	3.50	7.7	16.0	38.0	82.0					32,300
6,000	5.0	10.0	22.0	50.0						38,700
7,000	6.6	14.0	29.0	65.0						45,200
8,000	8.5	18.0	37.0							51,600
9,000	10.9	22.0	46.0							58,100
10,000	13.4	27.0	58.0							64,500
11,000	16.0	32.0								71,000
12,000	19.0	38.0								77,400
13,000	22.0									83,900
14,000	25.0									90,300
15,000	29.0									96,800

Note. To find the approximate iron loss at any induction or frequency, multiply the loss, at 60 cycles, 10,000 gausses, by the factor shown in the table. Values are most accurate for about a 4 per cent silicon steel and reasonably exact for lower silicon steels.

1435

Table 82. Core Losses in Watts per Pound at 60 Cycles for Commonly Used Commercial Grades of Electrical Sheets as a Function of Gage and Induction (Values Are Approximate Averages)

Induction in kilogausses

Grade	Gage	3	4	5	6	7	8	9	10	11	12	13	14	15	16	17
Tranor 3X (oriented)	29	.....	0.098	0.095	0.127	0.166	0.212	0.265	0.323	0.386	0.455	0.530	0.620	0.725	0.87	1.07
52	29	0.060	.120	.143	.200	0.26	0.33	0.41	0.50	0.61	0.75	0.91	1.10	1.28	1.41	1.55
58	29	.073	.130	.177	.240	0.310	0.385	0.470	0.56	0.68	0.82	0.98	1.18	1.35	1.52	1.69
65	29	.082		.196	.270	0.350	0.430	0.52	0.63	0.75	0.89	1.07	1.28	1.48	1.71	1.94
72	29	.088	.143	.216	.285	0.37	0.46	0.56	0.68	0.81	0.96	1.12	1.32	1.54	1.80	2.10
	26	.098	.159	.237	.320	0.42	0.52	0.64	0.77	0.93	1.10	1.29	1.53	1.80	2.10	2.48
	24	.110	.200	.300	.40	0.51	0.64	0.78	0.95	1.12	1.33	1.58	1.86	2.16	2.45	2.73
Dynamo	29	.100	.160	.242	.33	0.43	0.53	0.66	0.78	0.94	1.10	1.28	1.51	1.75	2.07	2.41
	26	.117	.187	.276	.36	0.49	0.60	0.74	0.89	1.05	1.24	1.44	1.70	1.95	2.29	2.65
	24	.132	.215	.325	.44	0.57	0.70	0.86	1.02	1.20	1.41	1.62	1.90	2.20	2.58	3.00
Motor	29	.125	.202	.300	.40	0.52	0.64	0.78	0.93	1.10	1.28	1.50	1.78	2.08	2.46	2.90
	26	.140	.227	.336	.46	0.59	0.73	0.89	1.07	1.25	1.47	1.70	2.00	2.35	2.67	3.28
	24	.157	.260	.383	.52	0.66	0.81	0.98	1.17	1.36	1.60	1.87	2.23	2.62	3.10	3.70
Electrical	29	.135	.220	.328	.45	0.58	0.73	0.89	1.06	1.27	1.51	1.80	2.17	2.62	3.20	3.97
	26	.157	.260	.390	.52	0.67	0.84	1.04	1.22	1.48	1.76	2.08	2.50	3.02	3.73	4.65
	24	.187	.313	.470	.64	0.82	1.03	1.25	1.50	1.79	2.13	2.55	3.10	3.78	4.65	5.85
	22	.240	.380	.575	.80	1.05	1.33	1.67	2.06	2.50	3.00	3.57	4.25	5.05	6.05	7.25
Armature	29	.157	.245	.358	.487	0.633	0.800	1.00	1.22	1.46	1.74	2.08	2.48	2.93	3.52	4.20
	26	.185	.290	.44	.58	0.76	0.96	1.19	1.41	1.68	1.96	2.28	2.72	3.20	3.95	4.80
	24	.240	.370	.57	.77	0.99	1.26	1.56	1.85	2.18	2.55	2.96	3.52	4.15	5.05	5.20
	22	.800	.450	.71	.97	1.27	1.57	1.93	2.30	2.70	3.18	3.72	4.50	5.30	6.60	8.25

Table 83. Apparent Incremental Permeability Standard E and I Laminations

Length of magnetic circuit = 5.6 in. Two lap joints. 29 gage material, 0.66 grade

A-C induction	D-C Magnetizing Force, in Oersteds							
	0	0.5	1.0	1.5	2	3	4	5
10 gausses	650	600	480	390	330	260	220	210
100 gausses	1230	1000	750	580	475	360	300	280
1000 gausses	2400	1450	1070	850	700	560	500	480

Data by American Rolling Mill Co., Middletown, Ohio. Data shown are for 60 cycles; 1000 cycles shows almost identical result.

Table 84. Incremental Permeability D-C Tests

29 gage material, 0.60 grade

	B = 10	B = 30	B = 100	B = 300	B = 1000	B = 3000
Steady magnetizing force = 0.0 oersted...	1000	1440	1970	2770	4460	7320
Steady magnetizing force = 0.1 oersted...	1000	1350	1910	2550	4030	6650
Steady magnetizing force = 0.3 oersted...	840	1090	1470	1985	3120	5200
Steady magnetizing force = 1.0 oersted...	578	740	934	1130	1570	2750
Steady magnetizing force = 3.0 oersted...	200	204	214	250	450	1000
Steady magnetizing force = 10.0 oersted..	62	63	65	70	100	310

Data from Westinghouse Electric tests. Ring samples, no air gaps. 60-cycle a-c tests with no air gaps have checked these figures closely.

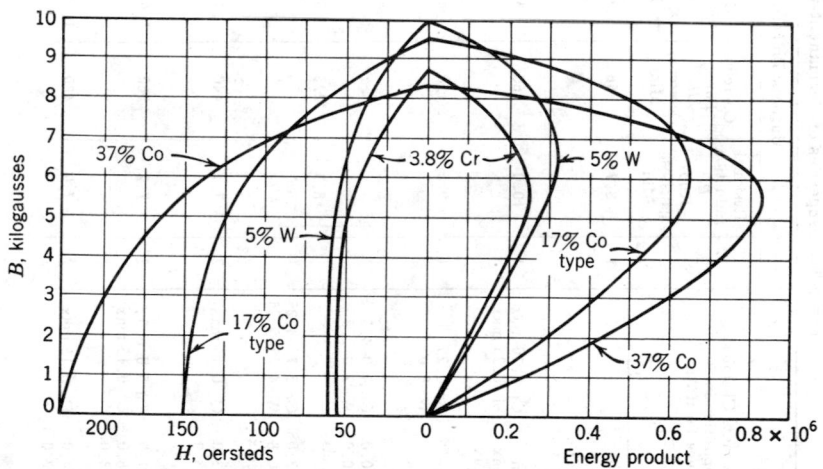

Fig. 8. Magnetic characteristics of cobalt, tungsten, and chromium types of retentive magnetic materials.

Where H = magnetic intensity or magnetizing force, in oersteds; B = magnetic induction, in kilogausses; and $B \times H$ = energy product, in kilogauss-oersteds.

Table 85. Properties of Permanent-magnet Materials, Approximate Values

Trade Name	Approximate Chemical Composition in %, Balance Fe If Any	Magnetic Properties						Description		Note
		Fig. No.	Residual Induction B_r, Kilogausses, Min	Coercive Force H_c Oersteds, Min	Energy Product $(B_d \times H_d)_m$, Min	"H_m" Oersteds, Min	Form	Mechanical Properties of Finished Magnets	Method of Fabrication	
Chromium Magnet	Cr 3.5–4.0, C 0.85–1.0, Mn 0.45–0.7, Si 0.25–0.4	1	8.75	56	250,000	200	Hot-rolled bars and sheets Rockwell C 30–40	Hard and strong	Hot forge, anneal for machining	
Tungsten	W 5.0 min, C 0.75 max, Mn 0.40 max, Cr 0.30 max, Si 0.25 max	1	10.0	60	320,000	200	Hot-rolled Rockwell C 38–42	Hard and strong	Hot forge, anneal for machining	
17% Cobalt	Co 17.0, C 0.75, Mn 2.5, Cr 0.25, W 8.0	1	9.5	150	650,000	1000	Rolled or cast	Hard and strong	Hot forge, anneal for machining	
High-cobalt bars	Co 37.0–40.0, C 0.87–0.95, Ni 0.50 max, Cr 4.0–4.5, W 2.0–2.5	1	8.3	225	830,000	1000	Hot-rolled bars	Hard and strong	Hot forge, anneal for machining	
High-cobalt cast	Co 35.0–37.0, C 0.65–0.8, Ni 0.30 max, Cr 3.75–4.25, W 4.75–5.25	1	8.3	225	830,000	1000	Cast	Hard and strong	Anneal for machining	
Alnico I	Al 11.0–13.0, Ni 19.0–23.0, Co 4.0–5.5, C 0.15 max	2	7.1	440	1,400,000	2000	Cast *	Hard and brittle	Grind	
Alnico II	Al 9.0–11.0, Ni 16.0–18.0, Co 12.0–13.0, Cu 5.0–6.5, C 0.15 max	2	7.2	560	1,640,000	2000	Cast *	Hard and brittle	Grind	
Alnico III	Al 11.0–14.0, Ni 25.0–27.0, C 0.15 max	2	7.0	480	1,350,000	2000	Cast *	Hard and brittle	Grind	

Material	Composition						Form	Mechanical properties	Fabrication	
Alnico IV	Al 12.0 Ni 28.0 Co 5.0	2	5.3	700	1,350,000	3000	Cast or sinter	Hard and brittle	Grind	
Alnico V	Al 7.0–9.0 Ni 13.0–15.0 Co 23.5–24.5 / Cu 2.75–3.25 C 0.15 max	2	12.5	575	4,500,000	3000	Cast *	Hard and brittle	Grind	†
Alnico VI	Al 8.0 Ni 14.0 Co 24.0 / Cu 3.0 Ti 1.0	2	10.2	775	3,600,000	3000	Cast or sinter	Hard and brittle	Grind	
Alnico XII	Al 6.0 Ni 18.0 Co 35.0 / Ti 8.0	2	6.0	950	1,850,000	3000	Cast or sinter	Hard and brittle	Grind	
Cunife I	Cu 60.0 Ni 20.0	3	5.8	600	1,900,000	2400	Wire or strip	Ductile and machinable	Cold work	‡
Cunife II	Cu 50.0 Ni 20.0 Co 2.5	3	7.3	260	830,000	2400	Wire or strip	Ductile and machinable	Cold work	‡
Cunico	Cu 50.0 Ni 21.0 Co 29.0	3	3.3	710	800,000	3200	Bar, wire strip or cast	Ductile and machinable	Machine	
Vectalite or Vectolite	Fe_2O_3 30.0 Fe_3O_4 44.0 Co_2O_3 26.0	3	1.6	900	500,000	4800	Sinter	Brittle and weak	Grind Machine	†
Vicalloy I	Co 52 V 10	3	9.0	300	1,100,000	1500	Wire or strip	Ductile and malleable	Machine Punch	‡
Vicalloy II	Co 52 V 10	3	10.6	470	3,330,000	1500	Wire or strip	Ductile and malleable	Machine Punch	‡ Aged
Remolloy or Conol	Co 12 Mo 17	3	10.4	240	1,100,000	1000	Forge or cast	Hard and brittle	Machine Grind	§

* May be obtained in sintered forms of approximately the same properties.
† Directional magnetic properties. Heat treated in magnetic field.
‡ Directional magnetic properties due to cold working.
§ May be formed and machined when hot.

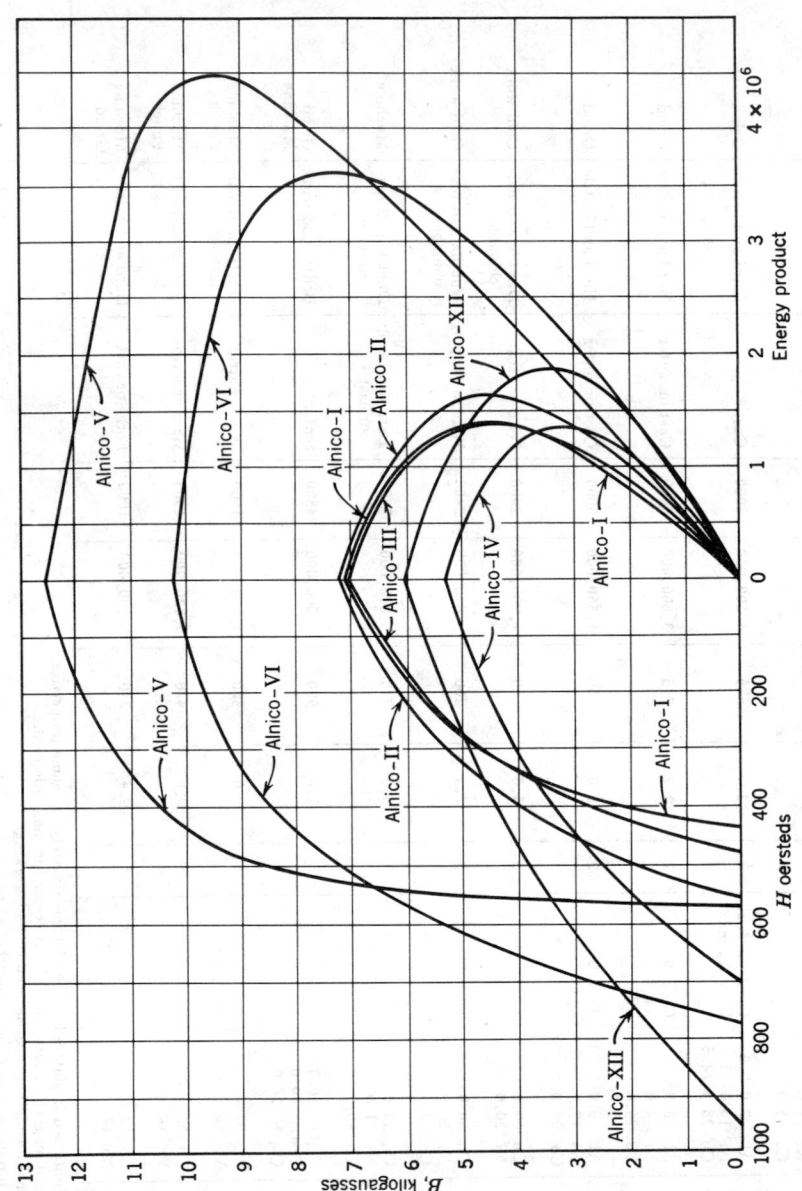

FIG. 9. Magnetic characteristics of Alnico types of retentive magnetic materials.

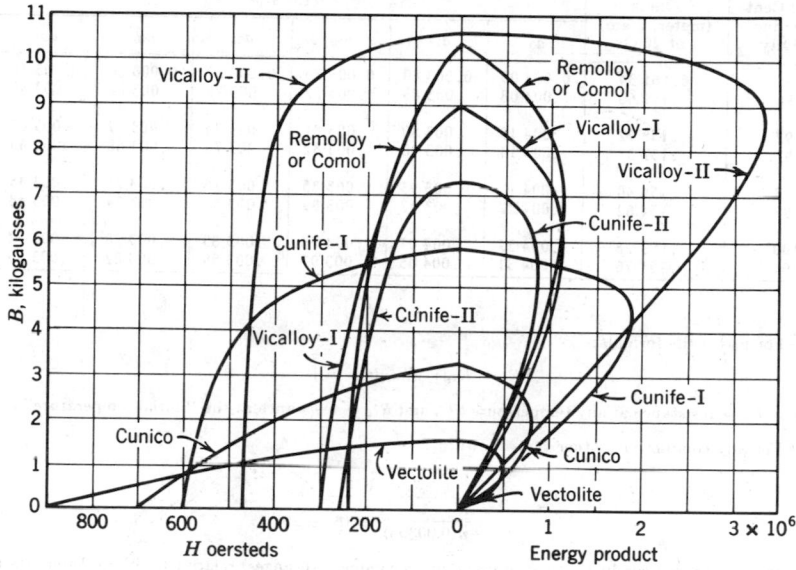

Fig. 10. Magnetic characteristics of some other commercial types of retentive magnetic materials.

where H = magnetic intensity or magnetizing force, in oersteds
B = magnetic induction, in kilogausses
B × H = energy product, in kilogauss-oersteds

Table 86.   International Annealed Copper Standard*

The International Electrotechnical Commission and The American Institute of Electrical Engineers have adopted the following normal values for Standard Annealed Copper:

1. The "volume" resistance, at 20°C, of a wire 1 m long and of uniform cross section of 1 sq. mm is 1/58 ohm = 0.017241 ohm.
2. The density, at 20°C, is 8.89 g/cc.
3. The constant "mass" temperature coefficient of resistance, at 20°C, measured between two potential points rigidly fixed to the wire, is 0.00393 = 1/254.45 per °C.
4. Hence the "mass" resistance, at 20°C, of a wire of uniform section, 1 m in length and weighing 1 g, is (1/58) × 8.89 = 0.15328 ohm.

Resistivities in various units (at 20°C) are as follows:

"Volume" Resistivity	"Mass" Resistivity
0.017241 ohm (meter, mm²)	0.015328 ohm (meter, gram)
1.7241 microhm-cm	875.20 ohms (mile, pound)
0.67879 microhm-in.	
10.371 ohms (mil, ft)	

* From *Handbook of Engineering Materials*, Miner and Seastone, New York, Wiley, 1955.

Table 87. Temperature Coefficients of Copper for Different Initial Temperatures (Centigrade) and Different Conductivities

Per Cent Conductivity [b]	Ohms (meter, gram) at 20 C	Factor a_{t_1} [a]					
		a_0	a_{15}	a_{20}	a_{25}	a_{30}	a_{50}
95	0.161 34	0.004 03	0.003 80	0.003 73	0.003 67	0.003 60	0.003 36
96	.159 66	.004 08	.003 85	.003 77	.003 70	.003 64	.003 39
97	.158 02	.004 13	.003 89	.003 81	.003 74	.003 67	.003 42
97.3	.157 53	.004 14	.003 90	.003 82	.003 75	.003 68	.003 43
98	.156 40	.004 17	.003 93	.003 85	.003 78	.003 71	.003 45
99	.154 82	.004 22	.003 97	.003 89	.003 82	.003 74	.003 48
100	.153 28	.004 27	.004 01	.003 93	.003 85	.003 78	.003 52
101	.151 76	.004 31	.004 05	.003 97	.003 89	.003 82	.003 55

[a] For use in the formula:

$$R_t = R_{t_1}[1 + a_{t_1}(t - t_1)]$$

where R_t = resistance at any temperature $t°C$, and R_{t_1} = resistance at the "initial temperature" $t_1°C$.

[b] For any conductivity other than shown:

$$a_{t_1} = \cfrac{1}{\cfrac{1}{n(0.00393)} + (t_1 - 20)}$$

where n = per cent conductivity of copper (within commercial ranges) relative to that of International Annealed Copper Standard at 20°C, expressed decimally; e.g., if per cent conductivity = 90, $n = 0.90$; and t_1 = any temperature in °C.

Table 88. Wire Table, Standard Annealed Copper
American Wire Gage (B. & S.). English Units

Gage No. A.W.G.	Diameter in Mils at 20 C	Cross Section at 20 C		Ohms per 1000 ft * at 20 C (= 68 F)	Pounds per 1000 ft	Feet per Pound	Feet per Ohm † at 20 C (= 68 F)	Ohms per Pound at 20 C (= 68 F)	Pounds per Ohm at 20 C (= 68 F)
		Circular Mils	Square Inches						
0000	460.0	211 600.	0.1662	0.049 01	640.5	1.561	20 400.	0.000 076 52	13 070.
000	409.6	167 800.	.1318	0.061 80	507.9	1.968	16 180.	0.000 1217	8219
00	364.8	133 100.	.1045	0.077 93	402.8	2.482	12 830.	0.000 1935	5169.
0	324.9	105 500.	.082 89	0.098 27	319.5	3.130	10 180.	0.000 3076	3251.
1	289.3	83 690.	.065 73	0.1239	253.3	3.947	8070.	0.000 4891	2044.
2	257.6	66 370.	.052 13	0.1563	200.9	4.977	6400.	0.000 7778	1286.
3	229.4	52 640.	.041 34	0.1970	159.3	6.276	5075.	0.001 237	808.6
4	204.3	41 740.	.032 78	0.2485	126.4	7.914	4025.	0.001 966	508.5
5	181.9	33 100.	.026 00	0.3133	100.2	9.980	3192.	0.003 127	319.8
6	162.0	26 250.	.020 62	0.3951	79.46	12.58	2531.	0.004 972	201.1
7	144.3	20 820.	.016 35	0.4982	63.02	15.87	2007.	0.007 905	126.5
8	128.5	16 510.	.012 97	0.6282	49.98	20.01	1592.	0.012 57	79.55
9	114.4	13 090.	.010 28	0.7921	39.63	25.23	1262.	0.019 99	50.03
10	101.9	10 380.	.008 155	0.9989	31.43	31.82	1001.	0.031 78	31.47
11	90.74	8234.	.006 467	1.260	24.92	40.12	794.0	0.050 53	19.79
12	80.81	6530.	.005 129	1.588	19.77	50.59	629.6	0.080 35	12.45
13	71.96	5178.	.004 067	2.003	15.68	63.80	499.3	0.1278	7.827
14	64.08	4107.	.003 225	2.525	12.43	80.44	396.0	0.2032	4.922
15	57.07	3257.	.002 558	3.184	9.858	101.4	314.0	0.3230	3.096
16	50.82	2583.	.002 028	4.016	7.818	127.9	249.0	0.5136	1.947
17	45.26	2048.	.001 609	5.064	6.200	161.3	197.5	0.8167	1.224
18	40.30	1624.	.001 276	6.385	4.917	203.4	156.6	1.299	0.7700
19	35.89	1288.	.001 012	8.051	3.899	256.5	124.2	2.065	0.4843
20	31.96	1022.	.000 802 3	10.15	3.092	323.4	98.50	3.283	0.3046
21	28.46	810.1	.000 636 3	12.80	2.452	407.8	78.11	5.221	0.1915
22	25.35	642.4	.000 504 6	16.14	1.945	514.2	61.95	8.301	0.1205
23	22.57	509.5	.000 400 2	20.36	1.542	648.4	49.13	13.20	0.075 76
24	20.10	404.0	.000 317 3	25.67	1.223	817.7	38.96	20.99	0.047 65
25	17.90	320.4	.000 251 7	32.37	0.9699	1031.	30.90	33.37	0.029 97
26	15.94	254.1	.000 199 6	40.81	0.7692	1300.	24.50	53.06	0.018 85
27	14.20	201.5	.000 158 3	51.47	0.6100	1639.	19.43	84.37	0.011 85
28	12.64	159.8	.000 125 5	64.90	0.4837	2067.	15.41	134.2	0.007 454
29	11.26	126.7	.000 099 53	81.83	0.3836	2607.	12.22	213.3	0.004 688
30	10.03	100.5	.000 078 94	103.2	0.3042	3287.	9.691	339.2	0.002 948
31	8.928	79.70	.000 062 60	130.1	0.2413	4145.	7.685	539.3	0.001 854
32	7.950	63.21	.000 049 64	164.1	0.1913	5227.	6.095	857.6	0.001 166
33	7.080	50.13	.000 039 37	206.9	0.1517	6591.	4.833	1364.	0.000 7333
34	6.305	39.75	.000 031 22	260.9	0.1203	8310.	3.833	2168.	0.000 4612
35	5.615	31.52	.000 024 76	329.0	0.095 42	10 480.	3.040	3448.	0.000 2901
36	5.000	25.00	.000 019 64	414.8	0.075 68	13 210.	2.411	5482.	0.000 1824
37	4.453	19.83	.000 015 57	523.1	0.060 01	16 660.	1.912	8717.	0.000 1147
38	3.965	15.72	.000 012 35	659.6	0.047 59	21 010.	1.516	13 860.	0.000 072 15
39	3.531	12.47	.000 009 793	831.8	0.037 74	26 500.	1.202	22 040.	0.000 045 38
40	3.145	9.888	.000 007 766	1049.	0.029 93	33 410.	0.9534	35 040.	0.000 028 54

* Resistance at the stated temperatures of a wire whose length is 1000 ft at 20 C.
† Length at 20 C of a wire whose resistance is 1 ohm at the stated temperatures.

Table 89. Specifications For Copper Wire

Diameter, in. [e]	Area at 20°C (68°F) cir mils	Area at 20°C (68°F) sq in.	Hard-Drawn [a] Tensile Strength, psi	Hard-Drawn [a] Elongation, per cent in 60 in.	Medium-Hard-Drawn [b] Tensile Strength, psi min	Medium-Hard-Drawn [b] Tensile Strength, psi max	Medium-Hard-Drawn [b] Elongation, min, per cent in 60 in.	Soft [c] Elongation, min, per cent in 10 in.	Tinned Soft [d] Elongation, min, per cent in 10 in.
0.4600	211,600	0.1662	49,000	3.75[f]	42,000	49,000	3.75[f]	35	30
0.4096	167,800	0.1318	51,000	3.25[f]	43,000	50,000	3.60[f]	35	30
0.3648	133,100	0.1045	52,000	2.80[f]	44,000	51,000	3.25[f]	35	30
0.3249	105,600	0.08291	54,500	2.40[f]	45,000	52,000	3.00[f]	35	30
0.2893	83,690	0.06573	56,100	2.17[f]	46,000	53,000	2.75[f]	30	25
0.2576	66,360	0.05212	57,600	1.98[f]	47,000	54,000	2.50[f]	30	25
0.2294	52,620	0.04133	59,000	1.79[f]	48,000	55,000	2.25[f]	30	25
0.2043	41,740	0.03278	60,100	1.24	48,330	55,330	1.25	30	25
0.1819	33,090	0.02599	61,200	1.18	48,660	55,660	1.20	30	25
*0.1650	27,220	0.02138	62,000	1.14	—	—	—	—	—
0.1620	26,240	0.02061	62,100	1.14	49,000	56,000	1.15	30	25
0.1443	20,820	0.01635	63,000	1.09	49,330	56,330	1.11	30	25
*0.1340	17,960	0.01410	63,400	1.07	—	—	—	—	—
0.1285	16,510	0.01297	63,700	1.06	49,600	56,660	1.08	30	25
0.1144	13,090	0.01028	64,300	1.02	50,000	57,000	1.06	30	25
*0.1040	10,820	0.008495	64,800	1.00	—	—	—	—	—
0.1019	10,380	0.008155	64,900	1.00	50,330	57,330	1.04	25	20
*0.0920	8,460	0.00665	65,400	0.97	—	—	—	—	—
0.0907	8,230	0.00646	65,400	0.97	50,660	57,660	1.02	25	20
0.0808	6,530	0.00513	65,700	0.95	51,000	58,000	1.00	25	20
*0.0800	6,400	0.00503	65,700	0.94	—	—	—	—	—
0.0720	5,180	0.00407	65,900	0.92	51,330	58,330	0.98	25	20
*0.0650	4,220	0.00332	66,200	0.91	—	—	—	—	—
0.0641	4,110	0.00323	66,200	0.90	51,660	58,660	0.96	25	20
0.0571	3,260	0.00256	66,400	0.89	52,000	59,000	0.94	25	20
0.0508	2,580	0.00203	66,600	0.87	52,330	59,330	0.92	25	20
0.0453	2,050	0.00161	66,800	0.86	52,660	59,660	0.90	25	20
0.0403	1,620	0.00128	67,000	0.85	53,000	60,000	0.88	25	20

0.0359	1.290	0.00101	—	—	—	—	—	—	25	20
0.0320	1.020	0.000804	—	—	—	—	—	—	25	20
0.0285	812	0.000638	—	—	—	—	—	—	25	20
0.0253	640	0.000503	—	—	—	—	—	—	25	20
0.0226	511	0.000401	—	—	—	—	—	—	25	20
0.0201	404	0.000317	—	—	—	—	—	—	20	15
0.0179	320	0.000252	—	—	—	—	—	—	20	15
0.0159	253	0.000199	—	—	—	—	—	—	20	15
0.0142	202	0.000158	—	—	—	—	—	—	20	15
0.0126	159	0.000125	—	—	—	—	—	—	20	15
0.0113	128	0.000100	—	—	—	—	—	—	20	15
0.0100	100	0.0000785	—	—	—	—	—	—	15	10
0.0089	79.2	0.0000622	—	—	—	—	—	—	15	10
0.0080	64.0	0.0000503	—	—	—	—	—	—	15	10
0.0071	50.4	0.0000396	—	—	—	—	—	—	15	10
0.0063	39.7	0.0000312	—	—	—	—	—	—	15	10
0.0056	31.4	0.0000246	—	—	—	—	—	—	15	10
0.0050	25.0	0.0000196	—	—	—	—	—	—	15	10
0.0045	20.2	0.0000159	—	—	—	—	—	—	15	10
0.0040	16.0	0.0000126	—	—	—	—	—	—	15	10
0.0035	12.2	0.00000962	—	—	—	—	—	—	15	10
0.0031	9.61	0.00000755	—	—	—	—	—	—	15	10

No requirement for tensile strength is specified.
No requirement for tensile strength is specified.

a From ASTM Designation B 1-56.
b From ASTM Designation B 2-52.
c From ASTM Designation B 3-63.
d From ASTM Designation B 33-63.
e The diameters marked by asterisks (*) are often specified for communication lines, but are not in the American Wire Gage (B & S Wire Gage) series, as are the other diameters listed.
f Per cent in 10 in.

Table 90.   Allowable Ampacities of Insulated Copper Conductors*
Not More than Three Conductors in Raceway or Cable or Direct Burial
[Based on Room Temperature of 86°F (30°C)]

Type of Insulated Conductor[a]	T	TW	RHW[b]	THW	THHN
Permissible Location	Dry	Dry or Wet	Dry or Wet		Dry
Temperature Rating °F	140		167		185
Temperature Rating °C	60		75		85–90
Size of Conductor, AWG or MCM					
14		15	15		25[c]
12		20	20		30[c]
10		30	30		40[c]
8		40	45		50
6		55	65		70
4		70	85		90
3		80	100		105
2		95	115		120
1		110	130		140
0		125	150		155
00		145	175		185
000		165	200		210
0000		195	230		235
250		215	255		270
300		240	285		300
350		260	310		325
400		280	335		360
500		320	380		405
600		355	420		455
700		385	460		490
750		400	475		500
800		410	490		515
900		435	520		555
1000		455	545		585
1250		495	590		645
1500		520	625		700
1750		545	650		735
2000		560	665		775

Correction Factors For Room Temperatures Over 86°F (30°C)

F.	C.			
104	40	.82	.88	.90
113	45	.71	.82	.85
122	50	.58	.75	.80
131	55	.41	.67	.74
140	60	...	.58	.67
158	70	...	.35	.52
167	75	...	...	.43
176	80	...	...	.30
194	90	...	...	...
212	100	...	...	...
248	120	...	...	...
284	140	...	...	...

* Adapted from the *National Electrical Code*, 1968 edition, copyrighted by the National Fire Protection Association, Boston, Mass.
  [a] Description of insulated conductors most commonly used in general service:
       T = Thermoplastic.
       TW = Moisture resistant thermoplastic.

RHW = Moisture and heat resistant rubber.
THW = Moisture and heat resistant thermoplastic.
THHN = Heat resistant thermoplastic.
For other types of insulated conductors, see *National Electrical Code*.
b For over 2000 V service, insulation shall be ozone-resistant.
c The ampacities for Type THHN conductors for sizes AWG 14, 12, and 10 shall be the same as designated for 167°F (75°C) conductors in this Table.

Table 91. Approximate Electric Conductivity of Aluminum

Metal	Conductivity, Per Cent of International Annealed Copper Standard
Aluminum (99.97 per cent)...............................	64.6
Aluminum conductors (hard drawn).........................	61
1100-O...	57
3003-O	50
3003-H18.	40
2017-T4.	30
6151-T6.	45

Table 92. Properties of Commercial Hard-Drawn Aluminum Conductors

Conductivity at 0°C, cgs electromagnetic units............................	38.36×10^{-6}
Mass resistivity, ohms (meter, gram) at 20°C............................	0.0764
Mass resistivity, ohms (mile, pound) at 20°C............................	436.0
Mass per cent conductivity relative to copper............................	200.7
Volume resistivity, microhm-cm at 20°C................................	2.828
Volume resistivity, microhm-in. at 20° C................................	1.113
Volume per cent conductivity relative to copper............................	61.0
Density, g/cu cm..	2.70
Density, lb/cu in..	0.0975

Table 93. Temperature Coefficient of Resistivity For Aluminum Wire

Temperature, °C................................	0	20
Coefficient of Resistivity, per °C *a*..................	0.00444	0.00403

a For 61 per cent conductivity annealed wire; coefficient is proportional to per cent conductivity.

Table 94. Comparison of Copper and Aluminum Wires For Equal Resistance per Unit Length

Property	Copper	Aluminum
Cross section..	1	1.61
Diameter..	1	1.27
Weight..	1	0.488
Breaking strength................................	1	0.64

Table 95. Wire Table, Aluminum

Hard-drawn aluminum wire at 20 C (68 F)
American Wire Gage (B. & S.), English units

Gage No.	Diameter, mils	Cross Section		Ohms per 1000 ft	Pounds per 1000 ft	Pounds per Ohm	Feet per Ohm
		Circular mils	Square Inches				
0000	460.	212 000.	0.166	0.0804	195.	2420.	12 400.
000	410.	168 000	.132	0.101	154.	1520.	9860
00	365.	133 000.	.105	0.128	122.	957.	7820.
0	325.	106 000.	.0829	0.161	97.0	602.	6200.
1	289.	83 700.	.0657	0.203	76.9	379.	4920.
2	258.	66 400.	.0521	0.256	61.0	238.	3900.
3	229.	52 600.	.0413	0.323	48.4	150.	3090.
4	204.	41 700.	.0328	0.408	38.4	94.2	2450.
5	182.	33 100.	.0260	0.514	30.4	59.2	1950.
6	162.	26 300.	.0206	0.648	24.1	37.2	1540.
7	144.	20 800.	.0164	0.817	19.1	23.4	1220.
8	128.	16 500.	.0130	1.03	15.2	14.7	970.
9	114.	13 100.	.0103	1.30	12.0	9.26	770.
10	102.	10 400.	.008 15	1.64	9.55	5.83	610.
11	91.	8230.	.006 47	2.07	7.57	3.66	484.
12	81.	6530.	.005 13	2.61	6.00	2.30	384.
13	72.	5180.	.004 07	3.29	4.76	1.45	304.
14	64.	4110.	.003 23	4.14	3.78	0.911	241.
15	57.	3260.	.002 56	5.22	2.99	0.573	191.
16	51.	2580.	.002 03	6.59	2.37	0.360	152.
17	45.	2050.	.001 61	8.31	1.88	0.227	120.
18	40.	1620.	.001 28	10.5	1.49	0.143	95.5
19	36.	1290.	.001 01	13.2	1.18	0.0897	75.7
20	32.	1020.	.000 802	16.7	0.939	0.0564	60.0
21	28.5	810.	.000 636	21.0	0.745	0.0355	47.6
22	25.3	642.	.000 505	26.5	0.591	0.0223	37.8
23	22.6	509.	.000 400	33.4	0.468	0.0140	29.9
24	20.1	404.	.000 317	42.1	0.371	0.008 82	23.7
25	17.9	320.	.000 252	53.1	0.295	0.005 55	18.8
26	15.9	254.	.000 200	67.0	0.234	0.003 49	14.9
27	14.2	202.	.000 158	84.4	0.185	0.002 19	11.8
28	12.6	160.	.000 126	106.	0.147	0.001 38	9.39
29	11.3	127.	.000 099 5	134.	0.117	0.000 868	7.45
30	10.0	101.	.000 078 9	169.	0.0924	0.000 546	5.91
31	8.9	79.7	.000 062 6	213.	0.0733	0.000 343	4.68
32	8.0	63.2	.000 049 6	269.	0.0581	0.000 216	3.72
33	7.1	50.1	.000 039 4	339.	0.0461	0.000 136	2.95
34	6.3	39.8	.000 031 2	428.	0.0365	0.000 085 4	2.34
35	5.6	31.5	.000 024 8	540.	0.0290	0.000 053 7	1.85
36	5.0	25.0	.000 019 6	681.	0.0230	0.000 033 8	1.47
37	4.5	19.8	.000 015 6	858.	0.0182	0.000 021 2	1.17
38	4.0	15.7	.000 012 3	1080.	0.0145	0.000 013 4	0.924
39	3.5	12.5	.000 009 79	1360.	0.0115	0.000 008 40	0.733
40	3.1	9.9	.000 007 77	1720.	0.0091	0.000 005 28	0.581

Table 96. Allowable Ampacities of Insulated Aluminum Conductors*

Not More than Three Conductors in Raceway or Cable or Direct Burial
(Based on Room Temperature of 86°F. 30°C.)

Type of Insulated Conductor [a]	T	TW	RHW	THW	THHN
Permissible Location	Dry	Dry or Wet	Dry or Wet		Dry
Temperature Rating, °F	140		167		185
Temperature Rating, °C	60		75		85–90
Size of Conductor, AWG or MCM					
12		15	15		25 [b]
10		25	25		30 [b]
8		30	40		40 [b]
6		40	50		55
4		55	65		70
3		65	75		80
2		75	90 [c]		95
1		85	100 [c]		110
0		100	120 [c]		125
00		115	135 [c]		145
000		130	155 [c]		165
0000		155	180 [c]		185
250		170	205		215
300		190	230		240
350		210	250		260
400		225	270		290
500		260	310		330
600		285	340		370
700		310	375		395
750		320	385		405
800		330	395		415
900		355	425		455
1000		375	445		480
1250		405	485		530
1500		435	520		580
1750		455	545		615
2000		470	560		650

Correction Factors For Room Temperatures Over 86°F (30°C)

F.	C.			
104	40	.82	.88	.90
113	45	.71	.82	.85
122	50	.58	.75	.80
131	55	.41	.67	.74
140	60	...	.58	.67
158	70	...	.35	.52
167	75	...	...	.43
176	80	...	...	.30
194	90	...	...	...
212	100	...	...	...
248	120	...	...	...
284	140	...	...	...

tection Association, Boston, Mass.

a Description of insulated conductors most commonly used in general service:

    T   = Thermoplastic.
    TW = Moisture resistant thermoplastic.
    RHW = Moisture and heat resistant rubber.
    THW = Moisture and heat resistant thermoplastic.
    THHN = Heat resistant thermoplastic.

For other types of insulated conductors, see *National Electrical Code.*

b The ampacities of Type THHN conductors for sizes AWG 12, 10, and 8 shall be the same as designated for 167°F (75°C) conductors in this table.

c For three wire single phase service and subservice circuits, the allowable ampacity of RHW aluminum conductors shall be for sizes #2 100 A, #1 110 A, #1/0 125 A, #2/0 150 A, #3/0 170 A, and #4/0 A.

Table 97. Electric Resistance Materials *

	Nominal Composition *a*							Resistivity, Ohms (mil, ft)	Temperature Coefficient of Resistance per °C	Coefficient of Linear Expansion per °C	Melting Point, °C
	Ni	Cr	Fe	Mn	Cu	Zn	Al				
Nickel chromium.......... Nichrome IV Chromel A Tophet A	80	20						650 (25 C)	0.0000937 (25 to 427 C)	0.000081 (20 to 500 C)	1320 (1100) *b*
Nickel chromium.......... Chromel C Nichrome Tophet C	60	15	25					675 (25 C)	0.000157 (25 to 427 C)	0.00015 (20 to 500 C)	1350 (1000) *b*
Manganin................	4			12	84			290 (20 C)	±0.00001		
Monel...................	69				28			268 (20 C)	0.00196 (20 to 100 C)	0.000015 (25 to 300 C)	1350 (500) *b*
Nickel silver............. Nickel silver German silver	18				65	17		175 (20 C)	0.00027 (20 to 100 C)		1110 (250) *b*
Copper nickel............. Advance, Cupron, Copel, Constantan, Ideal, etc.	45				55			294 (20 C)	±0.00001	0.0000149 (35 C)	1210 (500) *b*
Nickel-chromium-iron......	36	11	51	1.5				601 (20 C)		0.0000136 (0 to 100 C)	1450
Aluminum-chromium-iron (Ohmalloy)	.....	12.5	83				4.5	750 (20 C)	0.0003		(850) *b*
Nickel, pure.............	100–							58 to 64 (20 C)	0.00537 (20 to 100 C)	0.0000144 (20 to 300 C)	1440 (500) *b*
Iron, high-purity.........			99.97 †					9.8 (20 C)	0.0065 (0 to 100 C)	0.0000137 (0 to 400 C)	1535

* Properties may vary considerably with small variations in composition. Manufacturers' catalogs should be consulted for more specific information on any given material.

a Compositions given are approximate only; minor constituents and impurities have been omitted.

b Approximate values of maximum operating temperature in °C.

Table 98. Specifications for Carbon Brushes

Designation*	Grade A	Grade B	Grade D	Grade E
Specific resistance at 30 C, ohm-in......	0.00176 to 0.0021	0.0015 to 0.0019	0.0006 to 0.0012	0.000005 to 0.000025
Specific resistance at 250 C, ohm-in.....	0.0013 to 0.0017	0.0011 to 0.0015	0.0004 to 0.0010	....
Contact drop, volts at rated current....	1.5 min 3.2 max	1.8 min 3.0 max	1.6 min 3.0 max	Low
Coefficient of friction (no current)......	0.50 max	0.55 max	0.60 max	0.40 max
Coefficient of friction (carrying current)	0.35 max	0.40 max	0.45 max	....
Transverse strength, psi...............	2100 min	3000 min	1000 min	2500 min
Hardness, scleroscope................	43 to 58	50 to 65	8 to 16	8 to 20
Per cent ash........................	0.25 max	0.25 max	2.3 max	....
Density, lb/cu in....................	0.054 to 0.058	0.058 to 0.065	0.042 to 0.050	....
Per cent graphite...................	....	....	....	20 to 30

* From U.S. Navy.

Table 99. Approximate Current-carrying Capacities of Carbon Electrodes*

Nominal Diameter, in.	Area, sq in.	Current-carrying Capacity, amp	Current Density, amp/sq in.
ROUND ELECTRODES			
8	50	2,000–3,000	40–60
10	79	3,000–4,800	40–60
12	113	4,500–6,800	40–60
14	154	5,400–8,500	35–55
17	227	7,900–12,500	35–55
20	314	11,000–17,300	35–55
24	452	15,800–24,800	35–55
30	707	24,700–35,300	35–50
35	962	28,800–38,400	30–40
40	1257	37,700–50,200	30–40
SQUARE OR RECTANGULAR ELECTRODES			
Side, in.			
8	64	2,500–3,800	40–60
10	100	4,000–6,000	40–60
12	144	5,700–8,600	40–60
14	196	6,800–10,800	35–55
16	256	9,000–14,000	35–55
20	400	14,000–22,000	35–55
24	576	20,200–25,900	35–45
24 x 30	720	21,600–28,800	30–40

* From *Handbook of Engineering Materials*, Miner & Seastone, New York, Wiley, 1955.

Table 100.   Approximate Current-carrying Capacities of Graphite Electrodes*

Nominal Diameter, in.	Area, sq in.	Current-carrying Capacity, amp	Current Density, amp/sq in.
ROUND ELECTRODES			
2	3.1416	600–1,000	200–320
2 1/2	4.9087	800–1,500	160–310
3	7.0686	1,200–2,100	170–300
4	12.566	1,800–3,000	140–240
5 1/8	20.629	2,300–4,100	110–200
6	28.274	3,100–5,400	110–190
7	38.485	4,200–6,900	110–180
8	50.265	5,500–9,000	110–180
9	63.617	6,400–10,800	100–170
10	78.540	7,800–12,500	100–160
12	113.10	11,300–17,000	100–150
14	153.94	15,400–21,500	100–140
16	201.06	20,100–26,100	100–130
17	226.98	22,700–28,400	100–125
18	254.47	25,500–30,500	100–120
20	314.16	28,300–34,600	90–110
24	452	38,000–41,000	85–90
30	707	56,000–60,000	80–85
SQUARE ELECTRODES			
Side, in.			
2	4	650–1,000	160–250
4	16	1,900–3,000	120–180
6	36	3,600–4,900	100–140
8	64	6,400–8,300	100–130
10	100	9,000–13,000	90–130
12	144	13,000–17,300	90–120
16.35	267.32	21,400–26,700	80–100

* From *Handbook of Engineering Materials*, Miner and Seastone, New York, Wiley, 1955.

Table 101. Surface Conductances and Resistances for Air*

All conductance values expressed in Btu per (hr) (sq ft) (F deg temp diff)

SECTION A. Surface Conductances for Still Air^a

Position of Surface	Direction of Heat Flow	Surface Emissivity					
		Nonreflective ε = 0.90		Reflective ε = 0.20		Reflective ε = 0.05	
		C	R	C	R	C	R
Still air							
Horizontal..........	Upward	1.63	0.61	0.91	1.10	0.76	1.32
Sloping—45 deg......	Upward	1.60	0.62	0.88	1.14	0.73	1.37
Vertical............	Horizontal	1.46	0.68	0.74	1.35	0.59	1.70
Sloping—45 deg......	Downward	1.32	0.76	0.60	1.67	0.45	2.22
Horizontal..........	Downward	1.08	0.92	0.37	2.70	0.22	4.55
Moving air (any position)							
15 mph wind (for winter)...	Any	6.00	0.17				
7½ mph wind (for summer)..	Any	4.00	0.25				

SECTION B. Reflectivity and Emissivity Values of Various Surfaces and Effective Emissivities of Air Spaces

Surface	Reflectivity in Per cent	Average Emissivity ε	Effective Emissivity E of Air Space	
			With one surface having emissivity ε and other 0.90	With both surfaces of emissivity ε
Aluminum foil, bright........	92 to 97	0.05	0.05	0.03
Aluminum sheet..............	80 to 95	0.12	0.12	0.06
Aluminum coated paper, polished.........	75 to 84	0.20	0.20	0.11
Steel, galvanized, bright......	70 to 80	0.25	0.24	0.15
Aluminum paint.............	30 to 70	0.50	0.47	0.35
Building materials: wood, paper, glass, masonry, non-metallic paints.........	5 to 15	0.90	0.82	0.82

* Copyright by ASHRAE. Reprinted by permission from ASHRAE Handbook of Fundamentals, 1967.

^a Conductances are for surfaces of the stated emissivity facing virtual black-body surroundings at the same temperature as the ambient air. Values are based on a surface-air temperature difference of 10 deg and for surface temperature of 70° F.

Table 102. Thermal Conductances and Resistances of a Plane[a] Air Space*

¾ in. Thickness[b]

Position of Air Space	Direction of Heat Flow	Mean Temp,[c] °F	Temp Diff,[c] F deg	Thermal Conductance – C, Value of E[c,d]					Thermal Resistance[e] – R, Value of E[d]				
				0.03	0.05	0.2	0.5	0.82	0.03	0.05	0.2	0.5	0.82
Horiz.	Up	90	10	0.42	0.44	0.61	0.96	1.32	2.39	2.26	1.63	1.05	0.76
		50	30	0.58	0.60	0.73	1.0	1.24	1.72	1.67	1.37	1.0	0.78
		50	10	0.43	0.45	0.59	0.86	1.15	2.33	2.23	1.71	1.16	0.87
		0	20	0.55	0.56	0.66	0.86	1.07	1.83	1.79	1.52	1.16	0.93
		0	10	0.45	0.46	0.56	0.76	0.98	2.23	2.16	1.78	1.31	1.02
		-50	20	0.58	0.59	0.66	0.80	0.95	1.73	1.71	1.52	1.25	1.05
		-50	10	0.47	0.48	0.55	0.70	0.85	2.11	2.07	1.81	1.44	1.18
45° Slope	Up	90	10	0.33	0.36	0.53	0.87	1.24	3.01	2.81	1.90	1.15	0.81
		50	30	0.50	0.51	0.65	0.92	1.20	2.02	1.95	1.54	1.09	0.83
		50	10	0.34	0.36	0.50	0.77	1.06	2.94	2.78	2.02	1.30	0.94
		0	20	0.46	0.48	0.58	0.78	0.99	2.16	2.27	1.74	1.29	1.01
		0	10	0.36	0.37	0.47	0.67	0.88	2.81	2.71	2.13	1.49	1.13
		-50	20	0.50	0.51	0.58	0.72	0.88	1.99	1.96	1.72	1.38	1.14
		-50	10	0.38	0.39	0.46	0.60	0.75	2.66	2.59	2.19	1.67	1.33
Vertical	Horiz.	90	10	0.28	0.31	0.48	0.82	1.19	3.54	3.28	2.10	1.22	0.84
		50	30	0.34	0.36	0.49	0.76	1.04	2.95	2.80	2.04	1.32	0.96
		50	10	0.27	0.29	0.42	0.70	0.99	3.72	3.48	2.36	1.43	1.01
		0	20	0.31	0.32	0.37	0.62	0.84	3.23	3.10	2.73	1.60	1.19
		0	10	0.25	0.27	0.42	0.57	0.78	3.96	3.76	2.36	1.76	1.28
		-50	20	0.32	0.33	0.40	0.54	0.69	3.15	3.06	2.51	1.85	1.44
		-50	10	0.24	0.25	0.32	0.46	0.61	4.23	4.07	3.16	2.18	1.64
45° Slope	Down	90	10	0.29	0.31	0.48	0.82	1.19	3.51	3.24	2.09	1.21	0.84
		50	30	0.29	0.31	0.44	0.71	0.99	3.47	3.27	2.27	1.41	1.01
		50	10	0.26	0.28	0.42	0.69	0.98	3.82	3.57	2.40	1.45	1.02
		0	20	0.26	0.27	0.37	0.58	0.79	3.83	3.65	2.67	1.74	1.27
		0	10	0.23	0.25	0.35	0.55	0.76	4.28	4.04	2.88	1.82	1.31
		-50	20	0.25	0.26	0.33	0.47	0.62	4.05	3.90	3.05	2.13	1.61
		-50	10	0.21	0.22	0.29	0.43	0.58	4.84	4.63	3.48	2.33	1.72

	90	50	0	−50	90	50	0	−50	90	50	0	−50
	0.84	1.02	1.31	1.72	0.93	1.14	1.50	2.02	0.99	1.23	1.65	2.24
	1.21	1.44	1.82	2.33	1.42	1.71	2.21	2.90	1.56	1.91	2.54	3.40
	2.08	2.39	2.88	3.48	2.76	3.21	3.97	4.95	3.38	4.02	5.20	6.57
	3.25	3.55	4.04	4.63	5.24	5.74	6.59	7.63	8.08	8.94	10.9	12.3
	3.56	3.80	4.28	4.84	5.96	6.41	7.18	8.23	9.92	10.7	12.4	14.0
	1.19	0.98	0.76	0.58	1.07	0.88	0.67	0.50	1.01	0.81	0.61	0.45
	0.82	0.69	0.55	0.43	0.71	0.59	0.45	0.34	0.64	0.52	0.39	0.29
	0.48	0.42	0.35	0.29	0.36	0.31	0.25	0.20	0.30	0.25	0.19	0.15
	0.31	0.28	0.25	0.22	0.19	0.17	0.15	0.13	0.12	0.11	0.09	0.08
	0.28	0.26	0.23	0.21	0.17	0.16	0.14	0.12	0.10	0.09	0.08	0.07

1½ inch Thickness[b]

4 in. Thickness[b]

Down

Horiz.

Table 102. Thermal Conductances and Resistances of a Plane^a Air Space—Continued

¾ in. Thickness^b

Position of Air Space	Direction of Heat Flow	Mean Temp,^c °F	Temp Diff,^c F deg	Thermal Conductance — C — Value of E^{c,d}					Thermal Resistance^e — R — Value of E^d				
				0.03	0.05	0.2	0.5	0.82	0.03	0.05	0.2	0.5	0.82
Horiz.	Up	90	10	0.34	0.36	0.53	0.88	1.24	2.94	2.75	1.87	1.14	0.80
		50	30	0.47	0.48	0.62	0.89	1.17	2.14	2.06	1.62	1.13	0.85
		50	10	0.35	0.37	0.50	0.78	1.07	2.88	2.73	1.99	1.29	0.94
		0	20	0.44	0.45	0.55	0.75	0.98	2.28	2.22	1.81	1.33	1.03
		0	10	0.36	0.37	0.47	0.67	0.89	2.77	2.67	2.11	1.48	1.12
		-50	20	0.46	0.47	0.54	0.68	0.84	2.17	2.12	1.84	1.46	1.20
		-50	10	0.38	0.39	0.46	0.60	0.75	2.64	2.57	2.18	1.66	1.33
45° Slope	Up	90	10	0.31	0.33	0.51	0.85	1.21	3.22	3.00	1.98	1.18	0.82
		50	30	0.43	0.45	0.58	0.85	1.14	2.32	2.22	1.71	1.17	0.88
		50	10	0.31	0.33	0.47	0.74	1.04	3.18	3.00	2.13	1.34	0.96
		0	20	0.40	0.41	0.51	0.71	0.93	2.50	2.42	1.95	1.40	1.08
		0	10	0.32	0.34	0.44	0.64	0.85	3.09	2.97	2.49	1.57	1.17
		-50	20	0.42	0.43	0.50	0.64	0.79	2.39	2.33	2.00	1.56	1.26
		-50	10	0.34	0.35	0.42	0.56	0.71	2.97	2.89	2.40	1.79	1.41
Vertical	Horiz.	90	10	0.27	0.29	0.46	0.81	1.17	3.73	3.44	2.16	1.24	0.91
		50	30	0.36	0.38	0.52	0.78	1.07	2.74	2.62	1.94	1.28	0.94
		50	10	0.27	0.29	0.43	0.70	0.99	3.69	3.45	2.34	1.43	1.01
		0	20	0.34	0.35	0.45	0.65	0.87	2.98	2.86	2.22	1.53	1.16
		0	10	0.28	0.29	0.39	0.59	0.81	3.59	3.42	2.55	1.69	1.24
		-50	20	0.35	0.36	0.43	0.57	0.72	2.89	2.82	2.35	1.76	1.39
		-50	10	0.29	0.30	0.37	0.51	0.67	3.45	3.34	2.70	1.95	1.50
45° Slope	Down	90	10	0.21	0.23	0.40	0.74	1.11	4.84	4.36	2.50	1.34	0.90
		50	30	0.28	0.29	0.43	0.70	0.98	3.61	3.39	2.33	1.44	1.02
		50	10	0.21	0.23	0.36	0.64	0.93	4.79	4.41	2.75	1.57	1.08
		0	20	0.26	0.27	0.37	0.57	0.78	3.92	3.73	2.71	1.75	1.27
		0	10	0.21	0.23	0.33	0.53	0.74	4.67	4.39	3.05	1.89	1.35
		-50	20	0.26	0.27	0.34	0.48	0.64	3.82	3.68	2.92	2.06	1.57
		-50	10	0.22	0.23	0.30	0.45	0.60	4.47	4.29	3.29	2.24	1.67

Maximum Departures from Data (percent) of Interpolated Conductance Values Between ¾ and 4-inch Thick Air Spaces

Direction of Heat Flow	Mean Temp., °F	Temp. Diff., deg F	Value of E				
			0.03	0.05	0.2	0.5	0.82
45° Up	0	20	+ 5.7	—	+ 8.6	+ 6.3	+5.4
	−50	20	+10.0	+10.0			
Horiz.	90	10	+12.5	+11.5	+ 6.4		
	50	30	−10.4	− 8.3	+ 6.4		
	0	20	−12.7	−11.3	− 9.4	− 6.8	−5.2
	0	10	− 8.0	− 8.1	− 6.4		
	−50	20	−18.0	−17.9	−15.0	−11.3	−9.2
	−50	10	−20.3	−19.3	−16.6	−12.2	−9.5
45° Down	90	10	+35.5	+32.5	+18.0	+ 9.3	+6.4
	50	10	+27.0	+24.5	+14.1	+ 8.4	+5.7
	0	10	+15.8	+13.8	+ 8.8	+ 5.2	
	−50	20	−10.9	−11.0	− 8.4	− 5.9	
	−50	10	− 8.9	− 7.9	—	− 5.9	

* Copyright by ASHRAE. Reprinted by permission from *ASHRAE Handbook of Fundamentals*, 1967. Based on National Bureau of Standards data presented in Housing Research Paper No. 32, Housing and Home Finance Agency, 1954 (Superintendent of Documents, U. S. Government Printing Office, Washington 20025 D. C.).

a Spaces of uniform thickness bounded by moderately smooth surfaces.

b Interpolation between ¾ and 4-inch thick air spaces gives values of conductance that are within ±5.0 percent of the data, except for those values as indicated by the table of maximum departures. For more precise values, the reference should be consulted.

c Interpolation, and moderate extrapolation, of conductance values are permissible for other values of mean temperature, temperature difference, and effective emissivity E. Conductance values for 0 and −50°F mean temperatures were extrapolated from the experimental data using a dimensionless correlation.

d Effective emissivity of space, E, is given by $1/E = 1/\varepsilon_1 + 1/\varepsilon_2 - 1$, where ε_1 and ε_2 are the emissivities of the surfaces of the air space. (See Section B of Table 101.)

e The resistance values were determined from the relationship $R = 1/C$ prior to rounding off the conductance value to two decimal places for tabulation.

f The conductances of horizontal spaces with heat flow downward are substantially independent of temperature difference.

Table 103. Conductivities, Conductances, and Resistances of Building and Insulating Materials*ᵃ

These constants are expressed in Btu per (hour) (square foot) (Fahrenheit degree temperature difference). Conductivities (k) are per inch thickness, and conductances (C) are for thickness or construction stated, not per inch thickness

Material	Description	Density (Lb per Cu Ft)	Mean Temp °F	Conductivity (k)	Conductance (C)	Resistance ᵇ (R) Per inch thickness (1/k)	Resistance ᵇ (R) For thickness listed (1/C)
BUILDING BOARD ᶜ Boards, Panels, Flooring, Sheathing, Etc.	Asbestos-cement board	120	75	4.0	—	0.25	—
	Asbestos-cement board........⅛ in.	120	75	—	33.00	—	0.033
	Gypsum or plaster board........⅜ in.	50	75	—	3.10	—	0.32
	Gypsum or plaster board........½ in.	50	75	—	2.25	—	0.45
	Plywood	34	75	0.80	—	1.25	—
	Sheathing, wood fiber (impreg. or coated)	20	75	0.38	—	2.63	—
		22	75	0.41	—	2.44	—
	Wood fiber board, laminated or homogeneous	25	75	0.44	—	2.27	—
		26	75	0.42	—	2.38	—
		33	75	0.55	—	1.82	—
	Wood fiber, hardboard type........¼ in.	65	75	—	5.60	—	0.18
	Wood fiber, hardboard type	65	75	1.40	—	0.72	—
	Wood subfloor........²⁵⁄₃₂ in.	—	—	—	1.02	—	0.98
	Wood, hardwood finish........¾ in.	—	—	—	1.47	—	0.68
BUILDING PAPER	Vapor—permeable felt	—	75	—	16.70	—	0.06
	Vapor—seal, 2 layers of mopped 15 lb felt	—	75	—	8.35	—	0.12
	Vapor—seal, plastic film	—	75	—	—	—	Negl.
FINISH FLOORING MATERIALS	Carpet and fibrous pad	—	75	—	0.48	—	2.08
	Carpet and rubber pad	—	75	—	0.81	—	1.23
	Cork tile........⅛ in.	—	75	—	3.60	—	0.28
	Terrazzo........1 in.	—	75	—	12.50	—	0.08
	Tile—asphalt, linoleum, vinyl, rubber	—	75	—	20.00	—	0.05
INSULATING MATERIALS Blanket and Batt	Cotton fiber	0.8–2.0ᵈ	75	0.26	—	3.85	—
	Mineral wool, fibrous form processed from rock, slag, or glass	0.5ᵈ,ᵉ	75	0.32	—	3.12	—
		1.5–4.0ᵈ	75	0.27	—	3.70	—
	Wood fiber	3.2–3.6ᵈ	75	0.25	—	4.00	—

BOARD AND SLABS

Material	Density, lb/ft³	Mean temp, °F	k	C	1/k	1/C
Cellular glass	9	90	0.41	—	2.44	—
		60	0.39	—	2.56	—
		30	0.37	—	2.70	—
		0	0.35	—	2.86	—
		−30	0.33	—	3.00	—
		−60	0.32	—	3.12	—
Corkboard	6.5–8.0	90	0.28	—	3.57	—
		60	0.27	—	3.70	—
		30	0.26	—	3.85	—
		0	0.25	—	4.00	—
		−30	0.24	—	4.17	—
		−60	0.23	—	4.35	—
	12	90	0.31	—	3.22	—
		60	0.30	—	3.33	—
		30	0.29	—	3.45	—
		0	0.28	—	3.57	—
		−30	0.27	—	3.70	—
		−60	0.26	—	3.85	—
Glass fiber	4–9	90	0.26	—	3.85	—
		60	0.24	—	4.17	—
		30	0.22	—	4.55	—
		0	0.21	—	4.76	—
		−30	0.19	—	5.26	—
		−60	0.18	—	5.56	—
Expanded rubber –rigid	4.5	75	0.22	—	4.55	—
Expanded polyurethane (R-11 blown) (Thickness 1 in. and greater)	1.5–2.5	100	0.18	—	5.56	—
		75	0.17	—	5.88	—
		50	0.16	—	6.25	—
		25	0.17	—	5.88	—
		0	0.17	—	5.88	—
		−25	0.16	—	6.25	—
		−50	0.15	—	6.67	—
		−75	0.14	—	7.14	—
Expanded polystyrene, extruded	1.9	75	0.26	—	3.85	—
		60	0.25	—	4.00	—
		30	0.24	—	4.17	—
		0	0.22	—	4.55	—
		−60	0.19	—	5.26	—

Table 103. Conductivities, Conductances, and Resistances of Building and Insulating Materials—*Continued*

Material	Description	Density (Lb per Cu Ft)	Mean Temp °F	Conductivity (k)	Conductance (C)	Resistance[b] (R) Per inch thickness (1/k)	Resistance[b] (R) For thickness listed (1/C)
INSULATING MATERIALS BOARDS AND SLABS (*Continued*)	Expanded polystyrene, molded beads..............	1.0	75	0.28	—	3.57	—
			30	0.26	—	3.85	—
			0	0.24	—	4.17	—
			−60	0.21	—	4.76	—
	Mineral wool with resin binder..............	15	90	0.29	—	3.45	—
			60	0.28	—	3.57	—
			30	0.27	—	3.70	—
			0	0.25	—	4.00	—
			−30	0.24	—	4.17	—
			−60	0.23	—	4.35	—
	Mineral fiberboard, wet felted						
	Core or roof insulation..............	16–17	75	0.34	—	2.94	—
	Acoustical tile..............	18	75	0.35	—	2.86	—
	Acoustical tile..............	21	75	0.37	—	2.73	—
	Mineral fiberboard, wet molded						
	Acoustical tile[g]..............	23	75	0.42	—	2.38	—
	Wood or cane fiberboard						
	Acoustical tile[g]..............½ in.	—	75	—	0.84	—	1.19
	Acoustical tile[g]..............¾ in.	—	75	—	0.56	—	1.78
	Interior finish (plank, tile)..............	15	75	0.35	—	2.86	—
	Insulating roof deck						
	Approximately..............1½ in.	—	75	—	0.24	—	4.17
	Approximately..............2 in.	—	75	—	0.18	—	5.56
	Approximately..............3 in.	—	75	—	0.12	—	8.33
	Wood shredded (cemented in preformed slabs).....	22	75	0.60	—	1.67	—
LOOSE FILL	Macerated paper or pulp products..............	2.5–3.5	75	0.28	—	3.57	—
			90	0.30	—	3.33	—
	Mineral wool (glass, slag, or rock)..............	2.0–5.0	60	0.27	—	3.70	—
			30	0.25	—	4.00	—
			0	0.23	—	4.35	—
			−30	0.20	—	5.00	—
			−60	0.18	—	5.56	—

Material	Density	Temp.		
Perlite (expanded)	5.0–8.0	90	0.38	2.63
		60	0.36	2.78
		30	0.34	2.94
		0	0.32	3.12
		−30	0.30	3.33
		−60	0.28	3.57
		−90	0.26	3.85
		−120	0.24	4.17
		−115	0.30	3.33
		−115	0.26	3.84
		−115	0.23	4.35
		75	0.20	5.00
Sawdust or shavings	10.0	90	0.45	2.22
	7.5	60	0.17	5.88
	5.0	30	0.16	6.25
	3.0	0	0.15	6.67
Silica aerogel	0.8–15	−30	0.15	6.67
	7.6	−60	0.14	7.14
			0.13	7.69
Vermiculite (expanded)	7.0–8.2	90	0.48	2.08
		60	0.46	2.18
		30	0.44	2.27
		0	0.42	2.38
		−30	0.40	2.50
		−60	0.38	2.63
	4.0–6.0	90	0.45	2.22
		60	0.43	2.33
		30	0.40	2.50
		0	0.38	2.63
		−30	0.36	2.78
		−50	0.33	3.00
Wood fiber: redwood, hemlock, or fir	2.0–3.5	75	0.30	3.33
	3	90	0.31	3.22
Wood fiber: redwood bark	4	90	0.28	3.57
	4.5	60	0.26	3.85
		30	0.25	4.00
		0	0.23	4.35
		−30	0.21	4.76
		−60	0.20	5.00

Table 103. Conductivities, Conductances, and Resistances of Building and Insulating Material—*Continued*

Material	Description	Density (Lb per Cu Ft)	Mean Temp °F	Conductivity (k)	Conductance (C)	Resistance [b] (R) Per inch thickness (1/k)	Resistance [b] (R) For thickness listed (1/C)
ROOF INSULATION [h]	Preformed, for use above deck						
	Approximately..........½ in.	—	75	—	0.72	—	1.39
	Approximately..........1 in.	—	75	—	0.36	—	2.78
	Approximately..........1½ in.	—	75	—	0.24	—	4.17
	Approximately..........2 in.	—	75	—	0.19	—	5.26
	Approximately..........2½ in.	—	75	—	0.15	—	6.67
	Approximately..........3 in.	—	75	—	0.12	—	8.33
	Cellular glass	—	75	0.39	—	2.56	—
MASONRY MATERIALS CONCRETES	Cement mortar	116	—	5.0	—	0.20	—
	Gypsum-fiber concrete, 87½% gypsum, 12½% wood chips	51	—	1.66	—	0.60	—
	Lightweight aggregates including expanded shale, clay or slate; expanded slags; cinders; pumice; perlite; vermiculite; also cellular concretes	120	—	5.2	—	0.19	—
		100	—	3.6	—	0.28	—
		80	—	2.5	—	0.40	—
		60	—	1.7	—	0.59	—
		40	—	1.15	—	0.86	—
		30	—	0.90	—	1.11	—
		20	—	0.70	—	1.43	—
	Sand and gravel or stone aggregate (oven dried)	140	—	9.0	—	0.11	—
	Sand and gravel or stone aggregate (not dried)	140	—	12.0	—	0.08	—
	Stucco	116	—	5.0	—	0.20	—
MASONRY UNITS	Brick, common [i]	120	75	5.0	—	0.20	—
	Brick, face [i]	130	75	9.0	—	0.11	—
	Clay tile, hollow:						
	1 cell deep..........3 in.	—	75	—	1.25	—	0.80
	1 cell deep..........4 in.	—	75	—	0.90	—	1.11
	2 cells deep..........6 in.	—	75	—	0.66	—	1.52
	2 cells deep..........8 in.	—	75	—	0.54	—	1.85
	2 cells deep..........10 in.	—	75	—	0.45	—	2.22
	3 cells deep..........12 in.	—	75	—	0.40	—	2.50

Material						
Concrete blocks, three oval core:						
Sand and gravel aggregate................4 in.		75		1.40		0.71
........8 in.		75		0.90		1.11
........12 in.		75		0.78		1.28
Cinder aggregate........3 in.		75		1.16		0.86
........4 in.		75		0.90		1.11
........8 in.		75		0.58		1.72
........12 in.		75		0.53		1.89
Lightweight aggregate (expanded shale, clay, slate or slag: pumice)........3 in.		75		0.79		1.27
4 in.		75		0.67		1.50
8 in.		75		0.50		2.00
12 in.		75		0.44		2.27
Concrete blocks, rectangular core.[j]						
Sand and gravel aggregate						
2 core, 8 in. 36 lb.[k]		45		0.96		1.04
Same with filled cores[j]		45		0.52		1.93
Lightweight aggregate (expanded shale, clay, slate or slag, pumice):						
3 core, 6 in. 19 lb.[k]		45		0.61		1.65
Same with filled cores[l]		45		0.33		2.99
2 core, 8 in. 24 lb.[k]		45		0.46		2.18
Same with filled cores[l]		45		0.20		5.03
3 core, 12 in. 38 lb.[k]		45		0.40		2.48
Same with filled cores[l]		75		0.17	0.08	5.82
Stone, lime or sand			12.50			—
Gypsum partition tile:						
3 × 12 × 30 in. solid		75		0.79		1.26
3 × 12 × 30 in. 4 cell		75	,	0.74		1.35
4 × 12 × 30 in. 3-cell		75	—	0.60		1.67
PLASTERING MATERIALS						
Cement plaster, sand aggregate........½ in.	116	75	5.0	10.00	0.20	—
Sand aggregate	—	75		6.66		0.10
Sand aggregate........¾ in.		75				0.15
Gypsum plaster:						
Lightweight aggregate........½ in.	45	75		3.12		0.32
Lightweight aggregate........⅝ in.	45	75		2.67		0.39
Lightweight agg. on metal lath........¾ in.	—	75		2.13		0.47
Perlite aggregate	45	75	1.5		0.67	—
Sand aggregate	105	75	5.6		0.18	

Table 103. Conductivities, Conductances, and Resistances of Building and Insulating Materials—*Continued*

Material	Description	Density (Lb per Cu Ft)	Mean Temp °F	Conductivity (k)	Conductance (C)	Resistance[b] (R)	
						Per inch thickness ($1/k$)	For thickness listed ($1/C$)
PLASTERING MATERIALS (*Continued*)	Sand aggregate ½ in.	105	75	—	11.10	—	0.09
	Sand aggregate ⅝ in.	105	75	—	9.10	—	0.11
	Sand aggregate on metal lath ¾ in.	—	75	—	7.70	—	0.1
	Sand aggregate on wood lath	—	75	—	2.50	—	0.40
	Vermiculite aggregate	45	75	1.7	—	0.59	—
ROOFING	Asbestos-cement shingles	120	75	—	4.76	—	0.21
	Asphalt rool roofing	70	75	—	6.50	—	0.15
	Asphalt shingles	70	75	—	2.27	—	0.44
	Built-up roofing ⅜ in.	70	75	—	3.00	—	0.33
	Slate ½ in.	—	75	—	20.00	—	0.05
	Wood shingles	—	75	—	1.06	—	0.94
SIDING MATERIALS (On Flat Surface)	Shingles						
	Asbestos-cement	120	75	—	4.76	—	0.21
	Wood, 16 in., 7½ in. exposure	—	75	—	1.15	—	0.87
	Wood, double, 16-in., 12-in. exposure	—	75	—	0.84	—	1.19
	Wood, plus insul. backer board 5/16 in.	—	75	—	0.71	—	1.40
	Siding						
	Asbestos-cement, ¼ in., lapped	—	75	—	4.76	—	0.21
	Asphalt roll siding	—	75	—	6.50	—	0.15
	Asphalt insulating siding (½ in. bd.)	—	75	—	0.69	—	1.46
	Wood, drop, 1 × 8 in.	—	75	—	1.27	—	0.79
	Wood, bevel, ½ × 8 in., lapped	—	75	—	1.23	—	0.81
	Wood, bevel, ¾ × 10 in., lapped	—	75	—	0.95	—	1.05
	Wood, plywood, ⅜ in., lapped	—	75	—	1.59	—	0.59
	Architectural glass	—	75	—	10.00	—	0.10
WOODS	Maple, oak, and similar hardwoods	45	75	1.10	—	0.91	—
	Fir, pine, and similar softwoods	32	75	0.80	—	1.25	—
	Fir, pine, and similar softwoods 25/32 in.	32	75	—	1.02	—	0.98
	1⅝ in.	32	75	—	0.49	—	2.03
	2⅝ in.	32	75	—	0.30	—	3.28
	3⅝ in.	32	75	—	0.22	—	4.55

* Copyright by ASHRAE. Reprinted by permission from *ASHRAE Handbook of Fundamentals*, 1967.

a Representative values for dry materials are selected by the ASHRAE Technical Committee 2.4 on insulation. They are intended as design (not specification) values for materials of building construction in normal use. For conductivity of a particular product, the user may obtain the value supplied by the manufacturer or secure the results of unbiased tests.

b Resistance values are the reciprocals of C before rounding off C to two decimal places.

c See also Insulating Materials, Board.

d Includes paper backing and facing if any. In cases where the insulation forms a boundary (highly reflective or otherwise) of an air space, refer to Tables 101 and 102 to obtain the insulating value of the air space for the appropriate effective emissivity and temperature conditions of the space.

e Conductivity varies also with fiber diameter.

f These are values for aged board stock.

g Insulating values of acoustical tile vary depending on density of the board and on the type, size, and depth of the perforations. An average conductivity k value is 0.42.

h The U. S. Department of Commerce, *Simplified Practice Recommendation for Thermal Conductance Factors for Preferred Above Deck Roof Insulation*, No. R 257-55, recognizes the specification of roof insulation on the basis of the C values shown. Roof insulation is made in thicknesses to meet these values. Therefore, thickness supplied by different manufacturers may vary depending on the conductivity k value of the particular material.

i Face brick and common brick do not always have these specific densities. When the density is different from that shown, there will be a change in the thermal conductivity.

j Data on rectangular core concrete blocks differs from the above data on oval core blocks due to core configuration, different mean temperatures and possibly differences in unit weights. Weight data on the oval core blocks tested are not available.

k Weights of units is approximately 7⅜ in. high and 15⅝ in. long. These weights are given as a means of describing the blocks tested, but conductance values are all for one square foot of area.

l Vermiculite, perlite or mineral wool insulation. Where insulation is used, vapor barriers and or other precautions must be considered to keep insulation dry.

Table 104.   Dielectric Strength of Some Insulating Fabrics and Papers

	Thickness, in.	Breakdown Volts	
		1 Layer	Per mil
Treated Cloths			
Asbestos cloth, varnished................	0.047	3,780	80
Asbestos tape, varnished.................	.037	3,145	85
Cambric tape, varnished, black, bias........	.010	13,640	1364
Cambric tape, varnished, tan, bias.........	.010	11,515	1151
Cambric, rolls, varnished, black, straight....	.010	13,320	1332
Cambric, rolls, varnished, tan, straight......	.007	7,850	1121
Drilling, rolls, varnished, black, flexible.....	.020	9,250	462
Duck, rolls, varnished, black, flexible.......	.030	8,947	298
Friction, cloth tape, commercial............	.015	3,290	219
Friction, cloth tape, bias..................	.015	1,480	99
Friction cloth...........................	.024	1,815	76
Silk, oiled..............................	.004	4,450	1112
Surgical tape, varnished..................	.023	1,240	54
Treated Papers			
Asbestos paper, shellac 1 side..............	.010	2,120	212
Asbestos paper, shellac 1 side..............	.029	2,610	90
Express paper, paraffin 2 sides.............	.008	5,370	671
Fishpaper, shellac 1 side..................	.003	1,710	570
Fishpaper, paraffin 1 side.................	.010	8,840	884
Fullerboard, shellac 2 sides................	.013	3,390	260
Fullerboard, varnish 2 sides...............	.035	13,130	375
Japanese paper, shellac 2 sides.............	.003	1,136	379
Kraft paper, tan, shellac 1 side.............	.0045	1,270	282
Rope cement paper, varnish 2 sides.........	.008	9,746	1218
Rope cement paper, shellac 2 sides.........	.016	8,850	553

Cloths and papers tested flat between 2-in. circular electrodes.   Average of 10 breaks.   (60 cycles.)
Tapes tested by wrapping on 1-in. diameter rod in half-lapped layers.   Average of 10 breaks.
(60 cycles.)

Table 105.   Relative Dielectric Strengths of Gases (Air—1.00)

(Wolf)

Atmospheres Pressure	Carbon Dioxide	Nitrogen	Hydrogen
1	1.20	1.16	0.87
2	1.10	1.15	0.76
3	1.05	1.15	0.72
4	1.03	1.14	0.69
5	1.02	1.14	0.68

Table 106. Electrical Properties of Insulating Materials

Material	Dielectric Strength			Resistivity				Specific Inductive Capacity	
	Specimen thickness, mm	Kv per mm *	Authority	Volume, ohm-cm	Authority	Surface † Ohms, 30%	Surface † Ohms, 90%	Air unity	Authority
Asbestos paper	1.2 / 3.6	4.2	3	1.6×10^{11}	24				
Asphalt (Byerlyte)		14.0	4					2.7	25
Bakelite, wood molding mixture		17.7 to 21.6	5	1×10^{12}	5			4.5 to 5.5	5
Bakelite, asbestos molding mixture		up to 9.8	5	4×10^{11}	5				
Bakelite, Micarta-213		up to 31.4	6	5×10^{11}					
Cellophane	.022	51 to 66	7					5	6
Celluloid (clear)	0.25	12 to 28	1	2×10^{10}				8	6
Cellulose acetate	.019	48.0	6		26	8×10^{10}	2×10^{9}	5	6
Ceresin				over 5×10^{18}	26	8×10^{16}	8×10^{16}	5	6
Empire cloth, muslin	.38	48.0	9						
Fiber, vulcanized, including hard fiber, all colors	3.2	4.9 to 10.8	12	5 to 20×10^{9}	26	3×10^{10}	1×10^{7}	5	25
	6.4	3.9 to 8.9	12	5 to 20×10^{9}	26	3×10^{10}	1×10^{7}	5	25
Glass (ordinary)		8 to 9	13	9×10^{13}	26			5.5 to 9.1	28
Glass (plate)				2×10^{13}	27	3×10^{13}	2×10^{7}	5.5 to 9.1	28
Jute (impregnated)	.6	1.2	14						
Lava		3 to 10	15	2×10^{10}	26	6×10^{11}	1×10^{8}	3 to 4	29
Marble	.6	6.5	13	1 to 100×10^{9}	26	8×10^{10}	8×10^{10}	8.3	
Mica	1.6	21 to 28	13	.04 to 200×10^{15}	26	2×10^{13}	3×10^{9}	5 to 7	40
Micabond, plate	1.6	37.5	17						
Micabond, flexible	2.54	23.1	17						
Oil, insulating	0.13	10–16							
Paper		8.7	40	5×10^{4}	26			2.5	40
Paraffin (parowax)		11.5	13	1×10^{16}	26	1.5×10^{16}	5×10^{15}	2.6	31
Porcelain	.20	8.0	21	3×10^{14}	26	4×10^{13}	5×10^{8}	4.4	21
Pressboard (oiled)	1.58	29.2	40					5.0	32
Pressboard (varnished)	1.58	15.5	40					3	38
Rosin				5×10^{16}	26	8×10^{14}	2×10^{14}	2.0 to 3.5	30
Rubber (hard)	0.5	70	37	1×10^{18}	26	6×10^{15}	6×10^{9}	3.0 to 3.7	39
Shellac				1×10^{16}	26	2×10^{8}	1×10^{8}	6.6 to 7.4	35
Slate	10.3	1.3	22	1×10^{8}	26	2×10^{6}	1×10^{14}	2.9 to 3.2	39
Sulfur				1×10^{17}	26	1×10^{12}	2×10^{9}	4.1	32
Wood (maple), paraffined	15.2	4.6	32	3×10^{10}	26				

* To obtain volts per mil multiply kilovolts per millimeter by 25.4
† At 30 per cent and 90 per cent relative humidity.

Authorities for Table I: (1) Hobart and Turner. (2) Physicalische Technische Reichsanstalt. (3) Steinmetz. (4) Byerly & Sons, Manufacturers. (5) General Bakelite Co. (6) Westinghouse. (7) Continental Fiber Co. (8) Electrical Testing Laboratory. (9) Mica Insulator Co. (10) Bütlemann. (11) Formica Insulation Co. (12) William Eves, 3d. (13) Walter. (14) Baur. (15) American Lava Corporation. (16) Symons. (17) Chicago Mica Co. (18) Mica Insulation Co. (19) Minerallac Electric Co. (20) Canfield & Robinson. (21) The Locke Insulator Mfg. Co. (22) Massachusetts Institute of Technology. (23) Electrician. (24) Whittaker's Pocket Book. (25) Pirani. (26) Curtis. (27) Table of French Physical Society. (28) Coyne & Howe. (29) Schmidt. (30) Various. (31) Zietkowsky. (32) Hendrick. (33) Kinzbrunner. (34) Stadt Lab., Munich. (35) Schulze. (36) E. Miller. (37) C. C. Paterson, E. H. Raynor, A. Kinnes. (38) Boltzmann. (39) Williner. (40) Peek.

Table 107.   Types of Portland Cement

Type	Use
	Non-air-entraining [a]
I	For general concrete construction when special properties specified for the other types are not required. When no type is indicated, it assumed to be Type I.
II	For general concrete construction exposed to moderate sulfate action or where moderate heat of hydration is required.
III	For construction when high early strength is required.
IV	For construction when a low heat of hydration is required. This type is not generally carried in stock.
V	For construction when high sulfate resistance is required. This type is not generally carried in stock.
	Air-entraining [b]
IA IIA IIIA	Same as corresponding Types I, II, and III of non-air-entraining cement, but imparting to the concrete properties of greatly improved resistance to (1) severe weathering, and (2) the deleterious effect of applications of sodium or calcium salts to pavement surfaces for snow and ice removal.

[a] From ASTM Designation C 150-68.
[b] From ASTM Designation C 175-68.

Table 108.   Chemical Requirements For Portland Cements*

Non-air-entraining Air-entraining	Type I Type IA	Type II Type IIA	Type III Type IIIA	Type IV Type IVA	Type V Type VA
Silicon dioxide (SiO_2), min, per cent..........	...	21.0	...	...	...
Aluminum oxide (Al_2O_3), max, per cent.......	...	6.0	...	...	...
Ferric oxide (Fe_2O_3), max, per cent...........	...	6.0	...	6.5	...
Magnesium oxide (MgO), max, per cent......	5.0	5.0	5.0	5.0	5.0
Sulfur trioxide (SO_3), max, per cent:					
When $3CaO \cdot Al_2O_3$ is 8 per cent or less.....	2.5	2.5	3.0	2.3	2.3
When $3CaO \cdot Al_2O_3$ is more than 8 per cent..	3.0	...	4.0	...	...
Loss on ignition, max, per cent..............	3.0	3.0	3.0	2.5	3.0
Insoluble residue, max, per cent.............	0.75	0.75	0.75	0.75	0.75
Tricalcium silicate ($3CaO \cdot SiO_2$),[a] max, per cent	...	...	...	35	...
Dicalcium silicate ($2CaO \cdot SiO_2$),[a] min, per cent.	...	...	...	40	...
Tricalcium aluminate ($3CaO \cdot Al_2O_3$),[a] max, per cent............................	...	8	15[b]	7	5
Sum of tricalcium silicate and tricalcium aluminate, max, per cent...................	...	58[c]	...	...	...
Tetracalcium aluminoferrite plus twice the tricalcium aluminate[a] ($4\ CaO \cdot Al_2O_3 \cdot Fe_2O_3 + 2(3\ CaO \cdot Al_2O_3$), or solid solution ($4 CaO \cdot Al_2O_3 \cdot Fe_2O_3 + 2\ CaO \cdot Fe_2O_3$), as applicable, max, per cent..............	...	...	...	...	20.0

* From ASTM Designation C 150-68 (Portland Cement) and ASTM Designation C 175-68 (Air-entraining Portland Cement).
[a] The expressing of chemical limitations by means of calculated assumed compounds does not necessarily mean that the oxides are actually or entirely present as such compounds.
When the ratio of percentages of aluminum oxide to ferric oxide is 0.64 or more, the percentages of tricalcium silicate, dicalcium silicate, tricalcium aluminate and tetracalcium aluminoferrite shall

be calculated from the chemical analysis as follows:

Tricalcium silicate = (4.071 X per cent CaO) − (7.600 X per cent SiO_2) − (6.718 X per cent Al_2O_3) − (1.430 X per cent Fe_2O_3) − (2.852 X per cent SO_3)

Dicalcium silicate = (2.867 X per cent SiO_2) − (0.7544 X per cent C_3S)

Tricalcium aluminate = (2.650 X per cent Al_2O_3) − (1.692 X per cent Fe_2O_3)

Tetracalcium aluminoferrite = 3.043 X per cent Fe_2O_3

When the alumina-ferric oxide ratio is less than 0.64, a calcium aluminoferrite solid solution (expressed as ss (C_4AF + C_2F) is formed. Contents of this solid solution and of tricalcium silicate shall be calculated by the following formulas:

ss (C_4AF + C_2F) = (2.100 X per cent Al_2O_3) + (1.702 X per cent Fe_2O_3)

Tricalcium silicate = (4.071 X per cent CaO) − (7.600 X per cent SiO_2) − (4.479 X per cent Al_2O_3) − (2.859 X per cent Fe_2O_3) − (2.852 X per cent SO_3).

No tricalcium aluminate will be present in cements of this composition. Dicalcium silicate shall be calculated as previously shown.

In the calculations of C_3A, the values of Al_2O_3 and Fe_2O_3 determined to the nearest 0.01 per cent shall be used. In the calculation of other compounds the oxides determined to the nearest 0.1 per cent shall be used.

Values for C_3A and for the sum of C_4AF + $2C_3A$ shall be reported to the nearest 0.1 per cent. Values for other compounds shall be reported to the nearest 1 per cent.

[b] When moderate sulfate resistance is required for type III or Type IIIA cement, tricalcium aluminate shall be limited to 8 per cent. When high sulfate resistance is required, the tricalcium aluminate shall be limited to 5 per cent.

[c] This limit applies when moderate heat of hydration is required and tests for heat of dydration are not requested.

Table 109. Approximate Relative Strengths of Concrete as Affected by Type of Cement*

Type of Portland Cement	Compressive Strength—Per Cent of Strength of Normal Portland Cement Concrete[a]		
	3 days	28 days	3 months
1—Normal	100	100	100
2—Modified	80	85	100
3—High-early-strength	190	130	115
4—Low-heat	50	65	90
5—Sulfate-resistant	65	65	85

* From *Design and Control of Concrete Mixtures*, Portland Cement Association Bulletin T12-10, 10th ed., 1952.

[a] Values based on concrete moist-cured until tested.

Table 110. Maximum Sizes of Aggregate Recommended for Various Types of Construction*

Minimum Dimension of Section, in.	Maximum Size of Aggregate,[a] in.			
	Reinforced Walls, Beams, and Columns	Unreinforced Walls	Heavily Reinforced Slabs	Lightly Reinforced or Unreinforced Slabs
2½– 5	½– ¾	¾	¾–1	¾–1½
6–11	¾–1½	1½	1½	1½–3
12–29	1½–3	3	1½–3	1½–3
30 or more	1½–3	6	1½–3	3–6

* From *Recommended Practice for Selecting Proportions for Concrete*, American Concrete Institute Standard 613-54.

[a] Based on sieves with square openings.

Table 111.   Recommended Slumps for Various Types of Construction*

Types of Construction	Slump, in.[a]	
	Maximum	Minimum
Reinforced foundation walls and footings.................	5	2
Plain footings, caissons, and substructure walls.............	4	1
Slabs, beams, and reinforced walls.......................	6	3
Building columns..	6	3
Pavements...	3	2
Heavy mass construction................................	3	1

* From *Recommended Practice for Selecting Proportions for Concrete*, American Concrete Institute Standard 613-54.
[a] When high-frequency vibrators are used, the values given should be reduced about one-third.

Table 112.   Compressive Strength of Concrete for Various Water-Cement Ratios*

Water-Cement Ratio, gal. per bag [b] of cement	Probable Compressive Strength at 28 days,[a] psi	
	Non-air-entrained Concrete	Air-entrained Concrete
4	6000	4800
5	5000	4000
6	4000	3200
7	3200	2600
8	2500	2000
9	2000	1600

* From *Recommended Practice for Selecting Proportions for Concrete*, American Concrete Institute Standard 613-54.
[a] These average strengths are for concretes containing not more than the percentages of entrained and/or entrapped air shown in Table 113. For a constant water-cement ratio, the strength of the concrete is reduced as the air content is increased. For air contents higher than those listed in Table 113, the strengths will be proportionally less than those listed in this table.
Strengths are based on 6 × 12-in cylinders moist-cured under standard conditions for 28 days. See Method of Making and Curing Concrete Compression and Flexure Test Specimens in the Field (ASTM Designation C 31).
[b] 94 pounds, 1 cubic foot.

Table 113. **Approximate Mixing Water Requirements for Different Slumps and Maximum Sizes of Aggregates***

Slump, in.	Water,ᵃ gal. per cu yd of Concrete for Indicated Maximum Sizes of Aggregate							
	⅜ in.	½ in.	¾ in.	1 in.	1½ in.	2 in.	3 in.	6 in.
Non-air-entrained Concrete								
1 to 2	42	40	37	36	33	31	29	25
3 to 4	46	44	41	39	36	34	32	28
6 to 7	49	46	43	41	38	36	34	30
Approximate amount of entrapped air in non - air - entrained concrete, percent	3	2.5	2	1.5	1	0.5	0.3	0.2
Air-entrained Concrete								
1 to 2	37	36	33	31	29	27	25	22
3 to 4	41	39	36	34	32	30	28	24
6 to 7	43	41	38	36	34	32	30	26
Recommended average total air content, percent	8	7	6	5	4.5	4	3.5	3

* From *Recommended Practice for Selecting Proportions for Concrete*, American Concrete Institute Standard 613-54.

ᵃ These quantities of mixing water are for use in computing cement factors for trial batches. They are maxima for reasonably well-shaped angular coarse aggregates graded within limits of accepted specifications. If *more* water is required than shown, the cement factor, estimated from these quantities, *should* be increased to main desired water-cement ratio, except as otherwise indicated by laboratory tests for strength. If *less* water is required than shown, the cement factor, estimated from these quantities, *should not* be decreased except as indicated by laboratory tests for strength.

Table 114. Maximum Permissible Water-Cement Ratios (gal per bag) for Different Types of Structures and Degrees of Exposure*

Type of structures	Exposure conditions[a]					
	Severe wide range in temperature, or frequent alternations of freezing and thawing (air-entrained concrete only)			Mild temperature rarely below freezing, or rainy, or arid		
	In air	At the water line or within the range of fluctuating water level or spray		In air	At the water line or within the range of fluctuating water level or spray	
		In fresh water	In sea water or in contact with sulfates[b]		In fresh water	In sea water or in contact with sulfates[b]
A. Thin sections such as reinforced piles and pipe	5.5	5.0	4.5	6	5.5	4.5
B. Thin sections such as railings, curbs, sills, ledges, ornamental or architectural concrete, and all sections with less than 1-in. concrete cover over reinforcement	5.5	—	—	6	5.5	—
C. Moderate sections, such as retaining walls, abutments, piers, girders, beams	6.0	5.5	5.0	c	6.0	5.0
D. Exterior portions of heavy (mass) sections	6.5	5.5	5.0	c	6.0	5.0
E. Concrete deposited by tremie under water	—	5.0	5.0	—	5.0	5.0
F. Concrete slabs laid on the ground	6.0	—	—	c	—	—
G. Concrete protected from the weather, interiors of buildings, concrete below ground	c	—	—	c	—	—
H. Concrete which will later be protected by enclosure or backfill but which may be exposed to freezing and thawing for several years before such protection is offered	6.0	—	—	c	—	—

* From *Design of Concrete Mixtures*, Portland Cement Association Bulletin ST 100 (1963); adapted from *Recommended Practice for Selecting Proportions for Concrete*, American Concrete Institute Standard 603-54.

[a] Air-entrained concrete should be used under all conditions involving severe exposure and may be used under mild exposure conditions to improve workability of the mixture.

[b] Soil or ground water containing sulfate concentrations of more than 0.2 per cent. For moderate sulfate resistance, the tricalcium aluminate content of the cement should be limited to 8 per cent, and for high sulfate resistance to 5 per cent.

[c] Water-cement ratio should be selected on basis of strength and workability requirements, but should not exceed 9 gal per bag.

Table 115. Approximate Chemical Analyses of Refractories

Type of Brick	SiO_2	$Al_2O_3 + TiO_2$*	Fe_2O_3	FeO	CaO	MgO	$Cr_2O_3 + Al_2O_3$	Alkalies
Fireclay								
Superduty................	51.0	45.0	2.2	.....	0.3	0.4	....	1.0
High-heat-duty (aluminous).	55.0	41.0	2.2	.....	0.3	0.4	....	1.2
Intermediate-heat-duty.....	61.0	34.0	3.0	.....	0.3	0.4	....	1.5
High-heat-duty (siliceous)...	75.0	23.0	1.4	.....	0.2	0.2	....	0.2
High-alumina								
50% alumina class..........	43.0	53.0	1.8	.....	0.2	0.5	....	1.4
70% alumina class..........	23.0	74.0	1.7	.....	0.1	0.3	....	0.9
Silica......................	95.5	1.2	1.0	.....	2.1	0.2	....	0.2
Magnesite (fired).............	3	1	7	.....	3	85	....	....
Chrome (fired)...............	5	....	....	16	....	15.5	63 †	....

* The TiO_2 content will average about 5 per cent of the combined $Al_2O_3 + TiO_2$. Thus for a brick with 53 per cent $Al_2O_3 + TiO_2$, the TiO_2 content is about 2.7 per cent and the Al_2O_3 is about 50.3 per cent.

† In brick made from different raw materials, the Al_2O_3 content will vary from 12 to 30 per cent and the Cr_2O_3 will vary from 33 to 44 per cent.

Table 116. Specific Heats of Refractory Materials
(g-cal per g or Btu per lb)

Temperature, deg cent	0	100	200	300	400	500	600	700	800	900	1000	1100	1200	1300
Fireclay brick...........	.193	.199	.205	.211	.218	.225	.231	.238	.244	.249	.254	.258	.261	.265
Silica brick.............	.169	.189	.210	.230	.236	.245	.248	.253	.258	.262	.266	.269	.272	.275
Magnesite brick.........	.208	.220	.232	.241	.248	.254	.260	.265	.269	.274	.278	.282	.287	.292
Chrome brick...........	.170	.177	.182	.188	.193	.198	.202	.206	.210	.213	.217	.220	.222	.224

Table 117. Softening Point of Standard Pyrometric Cones Used in Connection with Refractories*

Cone No.	Softening Point[a] °C	Softening Point[a] °F	Cone No.	Softening Point[a] °C	Softening Point[a] °F	Cone No.	Softening Point[a] °C	Softening Point[a] °F
15	1435	2615	26	1595	2903	32 1/2	1725	3137
16	1465	2669	27	1605	2921	33	1745	3173
17	1475	2687	28	1615	2939	34	1760	3200
18	1490	2714	29	1640	2984	35	1785	3245
19	1520	2768	30	1650	3002	36	1810	3290
20	1530	2786	31	1680	3056	37	1820	3308
23	1580	2876	32	1700	3092	38	1835	3335

*Fairchild and Peters, "Characteristics of pyrometric cones," *Journal, Am. Ceramic Soc.*, Vol. 9, No. 11, p. 701, Nov., 1926.

[a] These temperatures, which were determined for a heating rate of 150 C per hr for cones 15 to 20, and or 100 C per hr for cones 23 to 38, other conditions being the same as specified, apply satisfactorily for all the conditions of this test method, but do not apply to conditions of the commercial firing and use of refractory materials.

Table 118. Pyrometric Cone Equivalents (P.C.E.) for Various Types of Refractory Brick

Type	P.C.E.	Temperature	
		°F	°C
Fireclay			
Superduty.....................	33	3173	1745
High-heat-duty (aluminous)	31–32	3056–3092	1680–1700
Intermediate-heat-duty.........	29	2984	1640
Low-heat-duty................	19	2759	1515
High-alumina			
50% alumina..................	34	3200	1760
60% alumina..................	35	3245	1785
70% alumina..................	36	3290	1810
80% alumina..................	38 (av.)	3335	1835
Silica...........................	31–32	3056	1680
Magnesite......................		3992	2200
Chrome........................		3542–3998	1950–2200
Forsterite.....................		3470	1910

Table 119. Porosity and True Specific Gravity for Various Types of Refractory Brick *

Material	Per Cent Porosity	True Specific Gravity
Fireclay		
Superduty............	12–16	2.6–2.7
High-heat-duty		
Dry press...........	18.6	2.6–2.7
Stiff mud...........	16.0	2.6–2.7
High-alumina...........	20–31	2.6–2.7
Silica..................	22–30	2.3–2.4
Magnesite..............	20–28	3.4–3.6
Chrome................	20–28	3.8–4.1
Forsterite..............	24–27	3.3–3.4

* From *Ceramic Data Book*, 1949–1950.

Table 120. Load-bearing Capacity of Special Refractories*

Type	Temperature (°C) for 10% Deformation	Test, psi
Commercial fused alumina	1716	25
Chrome	1404	25
Chrome-magnesite	1640	28
Kaolin (B&W K-80)	1593	25
Magnesite (sintered Grecian)	1400	25
Magnesite, fused	2240	10
Mullite, dense	1727	25
Silica, commercial	1646	25
Spinel	1510	25
Zirconia (clay-bonded baddeleyite)	1376	25
Zirconia, fused, lime stabilized	2200	25

* From F. H. Norton, *Refractories*, New York, McGraw-Hill, 3rd ed., 1949.

Table 121. Thermal Conductivity of Insulating Materials*

Btu/(hr)(sq ft)(°F/in.)

Type of Insulation	Linear Shrinkage, 5-hr Reheat, %, °F	Density, pcf	200	400	800	1200	1600	2000	2400
Insulating firebrick									
Kaolin base	1.0 at 2900	55	....	....	2.9	3.3	3.9	4.5	5.2
Kaolin base	1.0 at 2800	44	....	....	2.1	2.8	3.3	3.8	4.4
Fire clay base	3.1 at 2600	48	....	2.5	2.8	3.2	3.7	...	...
Kaolin base	0.1 at 2300	26	....	0.8	1.0	1.3	1.5	...	...
Insulating brick									
Raw diatomaceous earth base	1.0 at 1400	33	0.7	0.8	0.9	1.0		...	...
Calcined diatomaceous earth base	2.8 at 2300	43	....	....	1.4	1.6	2.0	...	...
High-temperature block insulation									
Diatomaceous earth base	3.1 at 1900	23	0.58	0.62	0.70	0.80	...	...	...
Vermiculite base	0.75 at 1600	18.5	0.55	0.67	0.80	0.95	...	...	...
Mineral wool base	1.69 at 1600	22	0.42	0.56	0.84	....	...	...	...
Kaolin base	...	19	0.58	0.67	0.91	1.21	...	...	...
Insulating blankets									
Mineral wool base	1000	8–12	0.38	0.52	0.78	....	...	...	...
Mineral wool base	800	8–12	0.35	0.51	....	....	...	...	...
Glass wool base	1000	3	0.31	0.36	0.57	....	...	...	...
Plastic insulating cements									
Diatomaceous base	1900	40	0.65	0.72	0.82	1.0	...	...	...
Mineral wool base	1200	24	0.67	0.75	0.93	....	...	...	...
Vermiculite base	1500	15	0.75	0.83	1.10	....	...	...	...
Asbestos base	1200	17	0.80	0.93	1.25	....	...	...	...
85% magnesia	600	15	0.49	0.55	....	....	...	...	...

* From F. H. Norton, *Refractories*, New York, McGraw-Hill, 3rd ed., 1949.

Table 122. Fire Clay Mortars*

Class	Pyrometric Cone Equivalent not Lower than
Super duty	No. 31
High heat duty	No. 28
Intermediate heat duty	No. 26
Low heat duty	No. 16

* From ASTM Designation C105–47 (1958).

Table 123. Spalling Resistance of Various Materials*

Material	Approximate Firing Temperature, °C [a]	Number of Quenches	Hardness
Fire clay			
No. 1	1200	26.5	Soft
No. 2	1430	23.0	Medium
No. 3	1316	15.6	Soft
No. 4	1316	12.2	Soft
No. 5	1200	8.4	Soft
No. 6	1200	5.2	Hard
No. 7	1200	2.7	Soft
Same as fire clay No. 2	1593	....	Hard
Sillimanite	1680	8.0/24.0	Hard
Silicon carbide	1540	....	Hard
Bauxite	1704	2.0/15.0	Very hard
Kaolin	1650	10.6	Hard
Spinel	1370	2.4	Hard
Silica	1480	0.3	Hard

* Adapted from F. H. Norton, *Refractories*, New York, McGraw-Hill, 3rd ed., 1949.
[a] In general, the spalling resistance decreases with increased firing temperature.

Table 124. Classifications for Degree of Spalling*

Conditions	Average Weight Loss, %
Average spalling	15
Moderately severe	10
Severe (malleable furnace bungs)	5–7.5
Extremely severe (super duty class)	4

* From F. H. Norton, *Refractories*, New York, McGraw-Hill, 3rd ed., 1949.

Table 125. Initial Slag Reaction Temperature of Some Commercial Refractories, °F*

Material	Zircon	Zirconia	Magnesite	Silicon Carbide	Chrome	Forsterite	Alumina	Kaolin
Zircon	....	3300	....	....	....	....	....	....
Magnesite	2800	....	....	2800	2800+	2800+	2800+	2500–2600
Zirconia	....	....	....	....	....	....	3300	....

* From F. H. Norton, *Refractories*, New York, McGraw-Hill, 3rd ed., 1949.

Table 126. Slag Reaction Temperature between Different Refractories, °C*

Type of Brick	Silica		Fire Clay		70% Al₂O₃		Chrome		Magnesite		Forsterite	
	A [a]	B [b]	A	B	A	B	A	B	A	B	A	B
Silica	....	....	1500	1600	1600	c	1650	....	1500	1600	1600	1700
Fire clay (high-duty)	1500	1600	....	....	....	....	1600	c	1400	1500	1500	1600
High alumina, 70% Al₂O₃	1600	c	....	....	....	....	1600	....	1500	1700	1650	1700
Chrome	1650	....	1600	c	1600	....	....	....	1700	....	1600	....
Magnesite	1500	1600	1400	1500	1500	1700	1700	....	....	....	1700	1700
Forsterite	1600	1700	1500	1600	1650	1700	1600	....	1700	1700	....	....

* Adapted from *Modern Refractory Practice*, Pittsburgh, Pa., Harbison-Walker Refractories Co., 1950.
[a] A = reaction first observed.
[b] B = reaction first becomes damaging.
[c] Not damaging.

Table 127. Thermal Expansion of Refractories*

Material	Firing Temperature, °C	Mean Coefficient of Expansion (0 F) and Shrinkage Temperature, × 10⁻⁶	Maximum Coefficient of Expansion between 300 and 700 C, × 10⁻⁶	Maximum Irreversible Contraction or Expansion up to 1700 C, % [a]	Temperature at which Shrinkage or Expansion Begins, °C
Silica	....	8.3	100.0	1	1550
Kaolin	1300	4.7	7.9	1	1050
Kaolin	1430	6.8	8.7	1	1380
Kaolin	1500	5.3	7.0	1	1580
Kaolin	1620	4.3	6.7	1	1610

Table 127.   Thermal Expansion of Refractories* —*Continued*

Material	Firing Temperature, °C	Mean Coefficient of Expansion (0 F) and Shrinkage Temperature, $\times 10^{-6}$	Maximum Coefficient of Expansion between 300 and 700 C, $\times 10^{-6}$	Maximum Irreversible Contraction or Expansion up to 1700 C, % [a]	Temperature at which Shrinkage or Expansion Begins, °C
Fire clay (Missouri)	....	5.4	8.0	Large	1300
Fire clay (Pennsylvania)	....	5.1	6.4	−5 [a]	1250
Fire clay (Colorado)	....	5.4	17.4	Large	1220
Fire clay (Maryland)	....	4.5	8.0	−3 [a]	1100
Silicon carbide	....	4.3	4.8	0	1700+
Zircon (white)	1650	6.4	9.2	Large	1510
Zircon (brown)	1590	4.2	4.8	2	1550
Zirconia	1675	5.9	8.7	1	1600
Mullite	1785	5.3	8.2	0	1700+
Magnesite (pure)	1680	14.2	15.1	0	1700+
Magnesite (commercial)	....	14.7	21.0	2	1440
Chrome (commercial)	....	10.4	12.4	2	1540
Spinel	1690	7.6	11.0	1	1600
Lime	1740	13.8	14.5	0	1700+
Alumina	1650	7.7	8.2	1	1580
Insulating	....	7.4	48.0	Large	1050
Zircon (electrically fused)	....	4.2	14.0	....	....
Magnesite (electrically fused)	....	10.2	12.0	....	....
Topaz	1675	4.3	15.1	....	....
Zircon (purified)	....	3.8	15.1	....	....

* From F. H. Norton, *Refractories*, New York, McGraw-Hill, 3rd ed., 1949.
[a] The minus sign indicates expansion.

Table 128.   Mean Thermal Heat Conductivity of Various Refractories*

Btu/(hr)(sq ft)(°F/in.)

Material	Apparent Porosity, %	Mean Temperatures, °F					
		400	800	1200	1600	2000	2400
Alumina (fused)	21.3	18.5	21.7	24.2	26.7	28.4	30.0
Carbon	....	....	....	....	18.5	22.0	....
Chrome	30.5	10.0	10.8	11.3	11.6	11.8	12.0
Fire clay	15.2	8.1	8.4	10.1	10.8	11.4	12.0
	18.4	7.3	8.7	9.8	10.6	11.6	11.8
	26.7	7.0	7.8	8.7	9.3	10.2	10.5
Super-duty fire clay	....	7.5	8.4	9.0	9.5	9.8	....
Graphite	....	730	500	400	310	220	....
Kaolin	10.8	13.5	15.0	16.3	18.0	18.8	19.7
	23.2	10.0	10.9	11.9	12.8	13.4	13.8
	49.1	3.0	4.4	5.2	6.1	6.7	7.1
Magnesite	29	25.8	22.0	18.2	15.7	13.7	12.6
	31.6	39.2	35.2	29.0	26.7	25.6	25.0
Silicon carbide-clay	28.3	26	26	29	34	38	50
Silicon carbide-clay, 5%	16.2	....	175	155	135	118	....
Silicon carbide-clay, 20%	21.0	....	70	61	58	57	....
Silicon carbide recrystallized	34.4	210	175	150	115	95	81
	35.3	....	143	126	105	93	80
Silica	28.0	6.5	8.5	10.5	12.4	14.0	15.7
	....	7.5	8.0	9.0	10.0	10.9	....
	30.4	8.1	10.0	11.8	13.0	14.2	15.3
Spinel	36.3	10.4	11.6	12.5	13.3	14.0	14.6
Alumina, fused 99% plus	21	....	27	20	18	17	....
	36	....	19	16	14	12	....
	47	....	13	11	9	8	....
	57	....	7	6	6	6	....
MgO, 96–99% fused	19	....	28	23	20	17	....
ZrO₂, 98–99% fused, stabilized with CaO	26	....	5	5	5	6	....
	44	....	3	3	4	5	....
Zironia, baddeleyite	29.5	10.4	11.3	12.2	13.0	13.7	14.3

* From F. H. Norton, *Refractories*, New York, McGraw-Hill, 3rd ed., 1949.

Table **129**.   Classification of Softwood Lumber*

Softwood lumber (this classification applies to rough or dressed lumber; sizes given are nominal)

- **Yard lumber** (lumber less than 5 in. thick, intended for general building purposes; grading based on use of the entire piece)
  - Finish (less than 3 in. thick and 12 in. and under in width)
    - A select
    - B select
    - C select
    - D select
  - Boards (less than 2 in. thick and 8 in. or over in width)
  - Strips (under 8 in. in width)
    - No. 1 boards
    - No. 2 boards
    - No. 3 boards
    - No. 4 boards
    - No. 5 boards
  - Dimension (2 in. and under 5 in. thick and of any width)
    - Planks (2 in. and under 4 in. thick and 8 in. and over wide)
      - No. 1 dimension
      - No. 2 dimension
      - No. 3 dimension
    - Scantling (2 in. and under 5 in. thick and under 8 in. wide)
      - No. 1 dimension
      - No. 2 dimension
      - No. 3 dimension
    - Heavy joists (4 in. thick and 8 in. or over wide)
      - No. 1 dimension
      - No. 2 dimension
      - No. 3 dimension
- **Structural material** (lumber 5 in. or over in thickness and width, except joist and plank; grading based on strength and on use of entire piece)
  - Joist and plank (2 to 4 in. thick and 4 in. and over wide)
  - Beams and stringers (5 in. and over thick and 8 in. and over wide)
  - Posts and timbers (6 x 6 in. and larger)
- **Factory and shop** (grading based on area of piece suitable for cuttings of certain size and quality)
  - Factory plank graded for door, sash, and other cuttings 1 to 4 in. thick and 5 in. and over wide
    - Factory clears upper grades
      - Nos. 1 and 2 clear factory
      - No. 3 clear factory
    - Shop clears lower grades
      - No. 1 shop
      - No. 2 shop
      - No. 3 shop
  - Shop lumber graded for general cut-up purposes
    - 1 in. thick (northern and western pine and Pacific Coast woods)
      - Select
      - Shop
    - All thicknesses (cypress, redwood, and North Carolina pine)
      - Tank and boat stock, firsts and seconds, selects
      - No. 1 shop
      - No. 2 shop, box

Grades

* From *Handbook of Engineering Materials*, Miner and Seastone, New York, Wiley, 1955.

Table 130. Summary of the Basic Grades for Yard Lumber*

Yard lumber (total products of a typical log arranged in series according to quality as determined by appearance and use)	Select (Lumber of good appearance and finishing qualities)	Suitable for natural finishes	Grade A—practically clear Grade B and better—Allows a few small defects or blemishes Grade C—Allows a limited number of small defects or blemishes that can be covered with paint
		Suitable for paint finishes	Grade D—Allows any number of defects or blemishes which do not detract from a finish appearance, especially when painted; intermediate between higher finishing grades and common grades, and partaking somewhat of the nature of both
	Common (Lumber containing defects or blemishes which detract from a finish appearance but which is suitable for general utility and construction purposes)†	Suitable for natural finishes	No. 1 Common—Sound and tight knotted stock; size of defects and blemishes limited; may be considered watertight lumber No 2 Common—Allows large and coarse defects; may be considered graintight lumber—less restricted in quality than No 1 but of the same general character No. 3 Common—Allows larger and coarser defects than No. 2 and occasional knotholes
		Suitable for paint finishes	No. 4 Common—Low-quality lumber admitting the coarsest defects such as rot and holes No. 5 Common—Must hold together under ordinary handling—lowest recognized grade

* Less than 5 in. thick and intended for general building purposes.

† Where members are designed to carry heavy loads, and where strength rather than stiffness is the controlling factor, as in heavy construction, the structural grades of joists and planks should be used in preference to the dimension grades; structural grades are more scientifically graded and definite working stresses can be assigned them.

Table 131. Strength Properties of Some Commercially Important Woods Grown in the United States*

(Results of tests on small,[1] clear specimens in the green and air-dry condition [2])

1	2	3	Static Bending			Work to—		Impact Bending		Compression Parallel to Grain		Compression Perpendicular to Grain	Shear Parallel to Grain	Hardness	
			4	5	6	7	8	9	10	11	12	13	14	15	16
Commercial and Botanical Name of Species	Moisture Content	Specific Gravity[3]	Fiber Stress at Proportional Limit	Modulus of Rupture	Modulus of Elasticity	Proportional Limit	Maximum Load	Fiber Stress at Proportional Limit	Height of Drop Causing Complete Failure (50-lb Hammer)	Fiber Stress at Proportional Limit	Maximum Crushing Strength	Fiber Stress at Proportional Limit	Maximum Shearing Strength	Load Required to Embed a 0.444-in. Ball to ½ Its Diameter — End	Side
	Per Cent		Lb per sq in.	Lb per sq in.	1000 lb per sq in.	In.-lb per cu in.	In.-lb per cu in.	Lb per sq in.	Inches	Lb per sq in.	Lb per sq in.	Lb per sq in.	Lb per sq in.	Lb	Lb
Alder, red (*Alnus rubra*)	98	0.37	3,800	6,500	1170	0.70	8.0	8,000	22	2620	2960	310	770	550	440
	12	.41	6,900	9,800	1380	1.85	8.4	11,600	20	4530	5820	540	1080	980	590
Ash, black (*Fraxinus nigra*)	85	.45	2,600	6,000	1040	.41	12.1	.....	33	1690	2300	430	860	590	.....
	12	.49	7,200	12,600	1600	1.57	14.9	.....		4520	5970	940	1570	1150	.....
Ash, commercial white [4] (*Fraxinus sp.*)	43	.54	5,300	9,500	1400	1.14	14.7	12,800	35	3360	4060	860	1350	1010	940
	12	.58	8,900	14,600	1680	2.68	15.6	17,000	37	5580	7280	1510	1920	1680	1260
Ash, Oregon (*Fraxinus oregona*)	48	.50	4,200	7,600	1130	.92	12.2	3,900	40	2760	3510	650	1190	850	790
	12	.55	7,000	12,700	1360	2.08	12.4	13,300	39	4100	6040	1540	1790	1430	1160
Aspen (*Populus tremuloides*)	94	.35	3,200	5,100	860	.69	6.4	7,000	33	1670	2140	220	660	280	300
	12	.38	5,600	8,400	1180	1.53	7.6	7,000	22	3040	4250	460	850	510	350
Basswood (*Tilia glabra*)	105	.32	2,700	5,000	1040	.40	5.3	6,300	21	1690	2220	210	600	290	250
	12	.37	5,900	8,700	1460	.85	7.2	9,800	16	3800	4730	450	990	520	410
Beech (*Fagus grandifolia*)	54	.56	4,300	8,600	1380	1.37	11.9	11,500	43	2550	3550	670	1290	970	850
	12	.64	8,700	14,900	1720	2.63	15.1	16,000	41	4880	7300	1250	2010	1590	1300
Birch [5] (*Betula sp.*)	62	.57	4,400	8,700	1560	.79	15.9	11,100	48	2640	3310	550	1160	910	970
	12	.63	10,100	16,700	2070	2.83	19.8	20,000	52	6200	8310	1250	2020	1660	1340
Birch, paper (*Betula papyrifera*)	65	.48	3,000	6,700	1170	.45	16.2	12,400	49	1640	2420	340	840	470	560
	12	.55	6,900	12,300	1590	1.80	16.0		34	3610	5690	740	1210	890	910
Butternut (*Juglans cinerea*)	104	.36	2,900	5,400	970	.52	8.2	7,300	24	2020	2420	270	760	410	390
	12	.38	5,700	8,100	1180	1.59	8.2	11,200	24	4200	5110	570	1170	540	470
Cedar, Alaska (*Chamaecyparis nootkatensis*)	38	.42	3,800	6,400	1140	.77	9.2	9,100	27	2500	3050	430	840	790	440
	12	.44		11,000	1420	2.06	10.4	12,200	29	5210	6310	770	1130	760	580
Cedar, eastern red (*Juniperus virginiana*)	35	.44	3,400	7,000	650	1.08	15.0	7,000	35	2540	3570	860	1010	900	650
	12	.47	3,800	8,800	880	1.01	8.3	8,500	22			1140		570	
Cedar, incense (*Libocedrus decurrens*)	108	.35	3,900	6,200	840	.94	6.4	7,300	17	2940	3150	460	830	830	390
	12		5,900	8,000	1040	1.67	5.4	9,600	17	4760	5200	730	880		470

Species																
Cedar, northern white (*Thuja occidentalis*)	230	320	620	290	1990	1490	15	5,300	5.7	.60	640	4,200	2,600	.29	55	12
Cedar, Port Orford (*Chamaecyparis lawsoniana*)	320	450	850	380	3960	2630	12	7,100	4.8	.72	800	6,500	4,900	.40	43	12
Cedar, southern white (*Chamaecyparis thyoides*)	400	460	830	350	3130	2770	22	9,200	7.4	.65	1420	6,400	4,000	.31	35	12
Cedar, western red (*Thuja plicata*)	560	730	1080	760	6470	5890	28	13,500	7.9	.97	730	11,300	7,700	.32	37	12
Cherry, black (*Prunus serotina*)	290	400	690	300	2390	1660	18	—	9.1	.51	750	4,700	2,500	.31	55	12
Chestnut (*Castanea dentata*)	350	520	800	500	4700	2740	13	6,000	4.1	1.46	930	6,800	4,800	.32	122	12
Cottonwood, eastern (*Populus deltoides*)	270	430	710	340	4750	2470	17	7,600	5.0	.63	920	5,100	3,300	.33	11	12
Cottonwood, northern black (*Populus trichocarpa hastata*)	350	660	860	610	5020	4360	17	8,600	5.8	.44	1120	7,700	4,200	.47	132	12
Cypress, southern (*Taxodium distichum*)	660	750	1130	440	3540	2940	33	10,200	12.8	.80	1310	8,000	9,000	.50	91	12
Douglas fir (coast region) (*Pseudotsuga taxifolia*)	950	1470	1700	850	7110	5960	29	13,600	11.1	3.11	1490	12,300	9,000	.40	36	12
Douglas fir, "Inland Empire" region (*Pseudotsuga taxifolia*)	420	530	800	760	5320	2080	24	10,700	7.0	.59	930	5,600	6,100	.43	42	12
Douglas fir (Rocky Mountain region) (*Pseudotsuga taxifolia*)	540	720	1080	380	2470	3780	19	7,900	6.5	1.78	1230	5,300	6,800	.37	38	12
Elm, American (*Ulmus americana*)	340	380	680	760	5320	1740	21	7,200	7.4	.49	1010	8,500	2,900	—	89	12
Elm, rock (*Ulmus racemosa*)	430	580	930	240	2280	3490	22	10,700	7.3	.39	1370	5,300	5,700	.32	48	12
Elm, slippery (*Ulmus fulva*)	250	280	600	470	4910	1760	20	6,800	5.0	.44	1070	4,800	2,900	.35	85	12
Fir, balsam (*Abies balsamea*)	350	540	1020	200	2160	3270	22	9,800	6.7	1.25	1260	8,300	4,200	—	117	12
Fir, commercial white [5] (*Abies* sp.)	390	660	810	370	4420	3100	25	8,800	6.6	2.15	1180	6,600	7,200	.42	108	12
Gum, black (*Nyssa sylvatica*)	510	510	930	500	3580	4740	24	10,400	8.2	.85	1440	10,600	4,800	.46	55	12
Gum, red (*Liquidambar styraciflua*)	480	760	1140	900	6360	3410	24	9,800	8.6	1.96	1550	7,600	8,100	.45	81	12
Gum, tupelo (*Nyssa aquatica*)	670	530	760	510	6390	6450	30	12,700	8.6	.55	1920	11,700	3,600	.44	97	12
Hackberry (*Celtis occidentalis*)	470	720	870	910	7420	2460	22	8,700	8.6	.91	1340	6,800	7,400	.40	65	12
Hemlock, eastern (*Tsuga canadensis*)	630	450	1190	500	3240	5520	27	9,100	6.8	.65	1610	11,300	3,600	.46	111	12
Hemlock, western (*Tsuga heterophylla*)	400	740	880	950	6700	2540	26	12,100	6.4	1.60	1400	6,400	6,300	.50	74	0.38
Hickory, pecan [7] (*Hicoria* sp.)	660	790	1070	820	3000	4660	38	—	11.8	.81	1110	7,200	3,900	.49	68	.42
Hickory, true [8] (*Hicoria* sp.)	630	1240	1340	850	6060	1920	39	—	10.8	2.53	1340	11,800	7,600	.50	57	.59

Table 131. Strength Properties of Some Commercially Important Woods Grown in the United States—*Continued*

Commercial and Botanical Name of Species	Moisture Content	Specific Gravity [3]	Static Bending — Fiber Stress at Proportional Limit	Modulus of — Rupture	Modulus of — Elasticity	Work to — Proportional Limit	Work to — Maximum Load	Impact Bending — Fiber Stress at Proportional Limit	Impact Bending — Height of Drop Causing Complete Failure (50-lb Hammer)	Compression Parallel to Grain — Fiber Stress at Proportional Limit	Compression Parallel to Grain — Maximum Crushing Strength [1]	Compression Perpendicular to Grain — Fiber Stress at Proportional Limit	Shear Parallel to Grain — Maximum Shearing Strength	Hardness — Load Required to Embed a 0.444-in. Ball to ½ Its Diameter — End	Hardness — Side
1	2	3	4	5	6	7	8	9	10	11	12	13	14	15	16
	Per Cent		Lb per sq in.	Lb per sq in.	1000 lb per sq in.	In.-lb per cu in.	In.-lb per cu in.	Lb per sq in.	Inches	Lb per sq in.	Lb per sq in.	Lb per sq in.	Lb per sq in.	Lb	Lb
Honey locust (*Gleditsia triacanthos*)	63	.60	5,600	10,200	1290	1.40	12.6	11,800	47	3320	4420	1420	1660	1440	1390
	12		8,800	14,700	1630	2.74	13.3	15,400	47	5250	7500	2280	2250	1860	1580
Larch, western (*Larix occidentalis*)	58	.48	4,600	7,500	1350	1.01	7.1	9,400	24	3250	3800	560	920	470	450
	12	.52	7,900	11,900	1710	2.46	8.0	15,100	32	5950	7490	1080	1360	1110	760
Locust, black (*Robinia pseudoacacia*)	40	.69	8,800	13,800	1850	2.36	15.4	18,300	44	6120	6800	1430	1760	1640	1570
	12		12,800	19,400	2050	4.62	10.2	21,100	57	6800	10,180	2260	2480	1580	1700
Magnolia, cucumber (*Magnolia acuminata*)	80	.44	4,200	7,400	1560	.66	12.4	14,700	30	4200	3140	410	990	600	520
	12	.48	8,000	12,300	1820	.98	12.8		35	4840	6310	710	1340	950	700
Magnolia, evergreen (*Magnolia grandiflora*)	105	.46	3,600	6,800	1110	.67	12.8	14,700	54	2160	2700	570	1040	780	740
	12	.50	6,800	11,200	1400	.90	8.7		29	3420	5460	1060	1550	1280	1020
Maple, bigleaf (*Acer macrophyllum*)	72	.44	4,400	7,400	1100	1.02	7.8	8,500	23	2510	3240	550	1110	760	620
	12	.44	6,600	10,700	1450	1.66	12.5	10,200	28	3240	5950	930	1730	1330	850
Maple, black (*Acer nigrum*)	65	.52	4,100	7,900	1330	.70	11.4	13,500	48	2800	3270	740	1130	940	840
	12	.57	8,300	13,300	1620	2.39	12.5	6,800	40	4600	6680	1250	1820	1700	1180
Maple, red (*Acer rubrum*)	63	.49	3,800	7,700	1390	.71	11.0	12,400	32	2360	3280	500	1150	780	. . . .
	12	.54	8,700	13,400	1640	2.84	8.3	12,200	32	4650	6540	1240	1850	1430	590
Maple, silver (*Acer saccharinum*)	66	.47	3,100	5,800	940	.61	13.3	20,600	29	1930	2490	460	1050	670	700
	12	.56	6,200	8,900	1140	.90	16.5	10,800	25	2850	5220	910	1480	1140	970
Maple, sugar (*Acer saccharum*)	58	.63	5,100	9,400	1550	1.03	12.6	17,000	40	4360	4020	800	1460	1070	1450
	12		9,500	15,800	1830	2.76	15.0	10,900	39	5390	7830	1810	2330	1840	1030
Oak, red [9] (*Quercus* sp.)	80	.57	4,400	8,500	1360	.85	11.3	17,400	43	4610	3520	800	1220	1050	1300
	12	.63	8,400	14,400	1810	2.30	13.3	7,200	43		6920	1260	1830	1490	1070
Oak, white [10] (*Quercus* sp.)	70	.59	4,700	8,100	1200	1.08		9,000	42	2940	3520	850	1270	1110	1330
	12	.67	7,900	13,900	1620	2.31		6,700	39	4350	7040	1410	1890	1420	330
Pine, lodgepole (*Pinus contorta*)	65	.38	3,000	5,500	1080	.49	5.6	9,500	20	2110	2610	310	680	320	480
	12	.41	6,700	9,400	1340	1.97	6.8		20	4310	5370	750	880	530	310
Pine, northern white (*Pinus strobus*)	68	.34	3,100	5,000	1020	.54	5.2		17	2060	2490	290	660	310	400
	12	.36	6,000	8,800	1280	1.59	6.7		19	3680	4840	550	860	500	

Values in each cell are shown as **green material / seasoned material (≈12% moisture)**. (See footnote 2.)

Species	Moisture, %	Sp. gr.[3]	Fiber stress at P.L. (psi)	Modulus of rupture (psi)	Modulus of elasticity (1,000 psi)	Work to P.L.	(col 7)	(col 8)	(col 9)	(col 10)	(col 11)	(col 12)	(col 13)	(col 14)	(col 15)
Pine, Norway (*Pinus resinosa*)	54 / 12	.44 / .48	3,700 / 9,400	6,400 / 12,500	1380 / 1800	.59 / 2.78	5.8 / 10.0	7,500 / 15,900	28 / 25	2410 / 5330	3080 / 7340	360 / 830	780 / 1230	360 / 670	340 / 580
Pine, ponderosa (*Pinus ponderosa*)	91 / 12	.38 / .40	3,100 / 6,300	5,000 / 9,200	970 / 1260	.59 / 1.85	5.1 / 6.6	6,800 / 9,800	20 / 17	2070 / 4060	2400 / 5270	360 / 740	680 / 1160	300 / 550	310 / 450
Pines, southern yellow:															
Loblolly (*Pinus taeda*)	81 / 12	.47 / .51	4,100 / 7,800	7,300 / 12,800	1400 / 1800	.68 / 1.92	8.2 / 8.9	8,900 / 12,100	30 / 30	2550 / 4820	3490 / 7080	480 / 980	850 / 1310	420 / 750	450 / 690
Longleaf (*Pinus palustris*)	63 / 12	.54 / .58	5,200 / 9,300	8,700 / 14,700	1600 / 1990	.95 / 2.44	10.4 / 11.0	10,100 / 15,400	35 / 34	3430 / 6150	4300 / 8440	590 / 1190	1040 / 1500	550 / 920	590 / 870
Shortleaf (*Pinus echinata*)	81 / 12	.46 / .51	3,900 / 7,700	7,300 / 12,800	1390 / 1760	.63 / 1.93	8.2 / 8.2	8,600 / 13,600	30 / 33	2500 / 5090	3430 / 7070	440 / 1000	850 / 1310	410 / 750	440 / 690
Pine, sugar (*Pinus lambertiana*)	137 / 12	.35 / .36	3,400 / 5,700	5,100 / 8,000	940 / 1200	.70 / 1.53	5.4 / 5.5	7,400 / 10,700	17 / 18	2330 / 4140	2530 / 4770	350 / 590	680 / 1050	320 / 530	310 / 380
Pine, western white (*Pinus monticola*)	54 / 12	.36 / .38	3,400 / 5,200	5,200 / 9,500	1170 / 1510	.56 / 1.47	5.0 / 5.0	7,600 / 11,900	19 / 23	2430 / 4480	2650 / 5620	290 / 540	640 /	310 / 440	310 / 370
Poplar, yellow (*Liriodendron tulipifera*)	64 / 12	.40 / .42	3,400 / 6,000	5,400 / 9,200	1090 / 1500	.62 / 1.43	5.4 / 5.4	8,600 / 13,500	18 / 20	1930 / 3550	2420 / 5290	330 / 580	740 / 1100	390 / 560	340 / 450
Redwood (virgin) (*Sequoia sempervirens*)	112 / 12	.38 / .40	4,800 / 7,500	7,500 / 10,000	1180 / 1340	1.18 / 2.04	7.4 / 6.5	8,900 / 10,200	21 / 19	3700 / 4560	4200 / 6150	520 / 860	800 / 940	570 / 700	410 / 480
Spruce, eastern[11] (*Picea sp*)	46 / 12	.38 / .40	3,300 / 6,500	5,600 / 10,100	1110 / 1440	.57 / 1.68	6.9 / 8.4	7,000 / 11,400	21 / 22	2120 / 2600	2600 / 5590	290 / 590	710 / 1070	390 / 630	340 / 490
Spruce, Engelmann (*Picea engelmannii*)	100 / 12	.31 / .33	2,500 / 4,200	4,200 / 8,500	830 / 1160	.43 / 1.64	4.9 / 5.6	5,800 / 9,000	14 / 15	1680 / 1980	1980 / 4580	290 / 640	590 / 1010	250 / 450	240 / 310
Spruce, Sitka (*Picea sitchensis*)	42 / 12	.37 / .40	3,000 / 6,000	5,700 / 10,200	1230 / 1570	.53 / 1.62	6.3 / 9.4	8,400 / 11,600	24 / 25	3580 / 4580	2670 / 5610	340 / 710	760 / 1150	430 / 760	350 / 510
Sugarberry (*Celtis laevigata*)	62 / 12	.47 / .51	3,300 / 5,700	6,600 / 9,900	810 / 1140	.78 / 2.18	12.0 / 7.5	8,800 / 10,500	33 / 26	2240 / 4780	2800 / 5620	580 / 1240	1050 / 1280	760 / 1280	740 / 960
Sycamore (*Platanus occidentalis*)	83 / 12	.46 / .49	3,200 / 6,700	6,500 / 10,000	1060 / 1420	.60 / 1.66	8.5 / 7.2	7,800 / 12,500	26 / 28	1990 / 3970	2920 / 5380	450 / 700	1050 / 1470	700 / 920	610 / 770
Tamarack (*Larix laricina*)	52 / 12	.49 / .53	4,200 / 8,000	7,200 / 11,600	1240 / 1640	.84 / 2.19	7.1 / 14.6	11,900 / 16,400	23 / 37	2930 / 4780	3480 / 7160	480 / 900	860 / 1280	400 / 670	380 / 590
Walnut, black (*Juglans nigra*)	81 / 12	.51 / .55	5,400 / 10,500	9,500 / 14,600	1420 / 1680	1.16 / 3.70	10.7 / —	— / —	34 / —	3520 / 5780	4300 / 7580	600 / 1250	1220 / 1370	960 / 1050	900 / 1010

* From *Wood Handbook*, U. S. Department of Agriculture, prepared by Forest Products Laboratory, Madison, Wis.

1 Test specimens 2 by 2 in. in section. Bending specimens 30 in. long; others shorter depending on kind of test.

2 The values in the first line for each species are from tests of green material; those in the second line are from tests of seasoned material adjusted to an average air dry condition of 12 per cent moisture.

3 Based on weight when oven dry and volume when green or at 12 per cent moisture content.

4 Average of Biltmore white ash (*Fraxinus biltmoreana*), blue ash (*F. quadrangulata*), green ash (*F. pennsylvanica lanceolata*), and white ash (*F. americana*).

5 Average of sweet birch (*Betula lenta*) and yellow birch (*B. lutea*).

6 Average of lowland white fir (*Abies grandis*) and white fir (*A concolor*).

7 Average of bitternut hickory (*Hicoria cordiformis*), nutmeg hickory (*H. myristicaeformis*), water hickory (*H. aquatica*), and pecan (*H. pecan*).

8 Average of bigleaf shagbark hickory (*Hicoria laciniosa*), mockernut hickory (*H. alba*), pignut hickory (*H. glabra*), and shagbark hickory (*H. ovata*).

9 Average of black oak (*Quercus velutina*), laurel oak (*Q. laurifolia*), pin oak (*Q. palustris*), red oak (*Q. borealis*), scarlet oak (*Q. coccinea*), southern red oak (*Q. rubra*), swamp red oak (*Q. rubra pagodaefolia*), water oak (*Q. nigra*), and willow oak (*Q. phellos*).

10 Average of bur oak (*Quercus macrocarpa*), chestnut oak (*Q. montana*), post oak (*Q. stellata*), swamp chestnut oak (*Q. prinus*), swamp white oak (*Q. bicolor*), and white oak (*Q. alba*).

11 Average of black spruce (*Picea mariana*), red spruce (*P. rubra*), and white spruce (*P. glauca*).

Table 132. Working Stresses For Standard Commercial Grades of Lumber

Species and Commercial Grade		Rules under Which Graded	Allowable Unit Stresses in pounds per sq in.				Modulus of Elasticity E
			Extreme Fiber in Bending f and Tension Parallel to Grain t	Horizontal Shear H	Compression Perpendicular to Grain $c\perp$	Compression Parallel to Grain c	
Ash, white:		National Hardwood Lumber Association			600		1,500,000
2150 f grade	J.&P.		2150	145		1700	
1900 f grade	J.&P.–B.&S.		1900	145		1500	
1700 f grade	J.&P.–B.&S.		1700	145		1325	
1450 f grade	J.&P.–B.&S.		1450	120		1150	
1300 f grade	B.&S.		1300	120		1050	
1450 c grade	P.&T.					1450	
1200 c grade	P.&T.					1200	
1075 c grade	P.&T.					1075	
Beech:		National Hardwood Lumber Association			600		1,600,000
2150 f grade	J.&P.		2150	145		1750	
1900 f grade	J.&P.–B.&S.		1900	145		1525	
1700 f grade	J.&P.–B.&S.		1700	145		1350	
1450 f grade	J.&P.–B.&S.		1450	120		1150	
1550 c grade	P.&T.					1550	
1450 c grade	P.&T.					1450	
1200 c grade	P.&T.					1200	
Birch:		National Hardwood Lumber Association			600		1,600,000
2150 f grade	J.&P.		2150	145		1750	
1900 f grade	J.&P.–B.&S.		1900	145		1525	
1700 f grade	J.&P.–B.&S.		1700	145		1350	
1450 f grade	J.&P.–B.&S.		1450	120		1150	
1550 c grade	P.&T.					1550	
1450 c grade	P.&T.					1450	
1200 c grade	P.&T.					1200	
Chestnut:		National Hardwood Lumber Association			360		1,000,000
1450 f grade	J.&P.		1450	120		1200	
1200 f grade	J.&P.–B.&S.		1200	120		950	
1075 c grade	P.&T.					1075	
Cypress, southern:		National Hardwood Lumber Association			360		1,200,000
1700 f grade	J.&P.–B.&S.		1700	145		1425	
1300 f grade	J.&P.–B.&S.		1300	120		1125	
1450 c grade	P.&T.					1450	
1200 c grade	P.&T.					1200	
Cypress, tidewater red:		Southern Cypress Manufacturers Association			360		1,200,000
1700 f grade	J.&P.–B.&S.		1700	145		1425	
1300 f grade	J.&P.–B.&S.		1300	120		1125	
1450 c grade	P.&T.					1450	
1200 c grade	P.&T.					1200	
Douglas fir, coast region:		West Coast Bureau of Lumber Grades and Inspection					1,600,000
2150 f dense select structural	J.&P.		2150	145	455	1550	
2150 f dense select structural	B.&S.		2150	145	455	1550	
1900 f select structural	J.&P.		1900	120	415	1450	
1900 f select structural	B.&S.		1900	120	415	1450	
1700 f dense no. 1	J.&P.		1700	145	455	1325	
1700 f dense no. 1	B.&S.		1700	145	455	1325	
1450 f no. 1	J.&P.		1450	120	390	1200	
1450 f no. 1	B.&S.		1450	120	390	1200	
1100 f no. 2	J.&P.		1100	110	390	1075	
1550 c dense select structural	P.&T.				455	1550	
1450 c select structural	P.&T.				415	1450	
1400 c dense no. 1	P.&T.				455	1400	
1200 c no. 1	P.&T.				390	1200	
Douglas fir, inland empire:		Western Pine Association					
Select structural	J.&P.		2150	145	455	1750	1,600,000
Structural	J.&P.		1900	100	400	1400	1,500,000
Common structural	J.&P.		1450	95	380	1250	1,500,000
Select structural	P.&T.				455	1750	1,600,000
Structural	P.&T.				400	1400	1,500,000
Common structural	P.&T.				380	1250	1,500,000

Table 132. Working Stresses For Standard Commercial Grades of Lumber—*Continued*

Species and Commercial Grade		Rules under Which Graded	Allowable Unit Stresses in pounds per sq in.				Modulus of Elasticity E
			Extreme Fiber in Bending f and Tension Parallel to Grain t	Horizontal Shear H	Compression Perpendicular to Grain $c_\perp$	Compression Parallel to Grain c	
Elm rock:		National Hardwood Lumber Association			600		1,300,000
2150 f grade	J.&P.		2150	145		1750	
1900 f grade	J.&P.–B.&S.		1900	145		1525	
1700 f grade	J.&P.–B.&S.		1700	145		1350	
1450 f grade	J.&P.–B.&S.		1450	120		1150	
1550 c grade	P.&T.		....	......		1550	
1450 c grade	P.&T.		....	......		1450	
1200 c grade	P.&T.		....	......		1200	
Elm, soft:		National Hardwood Lumber Association			300		1,200,000
1700 f grade	J.&P.		1700	120		1225	
1450 f grade	J.&P.–B.&S.		1450	120		1050	
1200 f grade	J.&P.–B.&S.		1200	120		875	
1075 c grade	P.&T.		....	......		1075	
Gum, black and red:		National Hardwood Lumber Association			360		1,200,000
1700 f grade	J.&P.		1700	120		1225	
1450 f grade	J.&P.–B.&S.		1450	120		1050	
1200 f grade	J.&P.–B.&S.		1200	120		875	
1075 c grade	P.&T.		....	......		1075	
Hemlock, eastern:		Northern Hemlock and Hardwood Manufacturers Association			360		1,100,000
Select structural	J.&P.–B.&S.		1300	85		850	
Prime structural	J.&P.		1200	60		775	
Common structural	J.&P.		1100	60		650	
Utility structural	J.&P.		950	60		600	
Select structural	P.&T.		....	......		850	
Hemlock, west coast:		West Coast Bureau of Lumber Grades and Inspection			360		1,400,000
1600 f select structural	J.&P.		1600	100		1100	
1450 f no. 1	B.&S.		1450	100		1075	
1450 f no. 1	J.&P.		1450	100		1075	
1100 f no. 2	J.&P.		1100	90		850	
1075 c no. 1	P.&T.		....	......		1075	
Hickory:		National Hardwood Lumber Association			720		1,800,000
2150 f grade	J.&P.–B.&S.		2150	145		1725	
1900 f grade	J.&P.–B.&S.		1900	145		1550	
1700 f grade	J.&P.–B.&S.		1700	145		1350	
1550 c grade	P.&T.		....	......		1550	
1450 c grade	P.&T.		....	......		1450	
1325 c grade	P.&T.		....	......		1325	
Larch:		Western Pine Association					1,300,000
Select structural	J.&P.		2150	145	455	1750	
Structural	J.&P.		1900	120	415	1450	
Common structural	J.&P.		1450	120	390	1325	
Select structural	P.&T.		....	......	455	1750	
Structural	P.&T.		....	......	415	1450	
Common structural	P.&T.		....	......	390	1325	
Maple, hard:		National Hardwood Lumber Association			600		1,600,000
2150 f grade	J.&P.		2150	145		1750	
1900 f grade	J.&P.–B.&S.		1900	145		1525	
1700 f grade	J.&P.–B.&S.		1700	145		1425	
1450 f grade	J.&P.–B.&S.		1450	120		1150	
1550 c grade	P.&T.		....	......		1550	
1450 c grade	P.&T.		....	......		1450	
1200 c grade	P.&T.		....	......		1200	
Oak, red and white:		National Hardwood Lumber Association			600		1,500,000
2150 f grade	J.&P.		2150	145		1550	
1900 f grade	J.&P.–B.&S.		1900	145		1375	
1700 f grade	J.&P.–B.&S.		1700	145		1200	
1450 f grade	J.&P.–B.&S.		1450	120		1050	
1300 f grade	B.&S.		1300	120		950	
1325 c grade	P.&T.		....	......		1325	
1200 c grade	P.&T.		....	......		1200	
1075 c grade	P.&T.		....	......		1075	

Table 132. Working Stresses For Standard Commercial Grades of Lumber—*Continued*

Species and Commercial Grade	Rules under Which Graded	Extreme Fiber in Bending f and Tension Parallel to Grain t	Horizontal Shear H	Compression Perpendicular to Grain $c\perp$	Compression Parallel to Grain c	Modulus of Elasticity E
Pecan:		National Hardwood Lumber Association		720		1,800,000
2150 f grade	J.&P.–B.&S.	2150	145		1725	
1900 f grade	J.&P.–B.&S.	1900	145		1550	
1700 f grade	J.&P.–B.&S.	1700	145		1350	
1550 c grade	P.&T.				1550	
1450 c grade	P.&T.				1450	
1325 c grade	P.&T.				1325	
Pine, Norway:		Northern Hemlock and Hardwood Manufacturers Association		360		1,200,000
Prime structural	J.&P.	1200	75		900	
Common structural	J.&P.	1100	75		775	
Utility structural	J.&P.	950	75		650	
Pine, southern longleaf:		Southern Pine Inspection Bureau of the Southern Pine Association		455		1,600,000
Select structural	J.&P.–B.&S.	2400	120		1750	
Prime structural	J.&P.–B.&S.	2000	120		1400	
Merchantable structural	J.&P.–B.&S.	1800	120		1300	
Structural S.E.&S.	J.&P.–B.&S.	1800	120		1300	
No. 1 structural	J.&P.–B.&S.	1600	120		1150	
No. 1 dimension	J.&P.	1700	150		1400	
No. 2 stress dimension	J.&P.	1250	150		1025	
Select structural	P.&T.				1750	
Prime structural	P.&T.				1400	
Merchantable structural	P.&T.				1300	
Structural S.E.&S.	P.&T.				1300	
No. 1 structural	P.&T.				1150	
Pine, southern shortleaf:		Southern Pine Inspection Bureau of the Southern Pine Association				1,600,000
Dense select structural	J.&P.–B.&S.	2400	120	455	1750	
Dense structural	J.&P.–B.&S.	2000	120	455	1400	
Dense structural S.E.&S.	J.&P.–B.&S.	1800	120	455	1300	
Dense No. 1 structural	J.&P.–B.&S.	1600	120	455	1150	
No. 1 dense dimension	J.&P.	1700	150	455	1400	
No. 1 dimension	J.&P.	1450	125	390	1200	
No. 2 dense stress dimension	J.&P.	1250	100	455	1025	
No. 2 medium grain stress dimension	J.&P.	1100	85	390	875	
Dense select structural	P.&T.			455	1750	
Dense structural	P.&T.			455	1400	
Dense structural S.E.&S.	P.&T.			455	1300	
Dense No. 1 structural	P.&T.			455	1150	
Poplar, yellow:		National Hardwood Lumber Association		300		1,100,000
1500 f grade	J.&P.	1500	110		1200	
1250 f grade	J.&P.–B.&S.	1250	110		950	
1075 c grade	P.&T.				1075	
Redwood:		California Redwood Association				1,200,000
Dense structural	J.&P.–B.&S.	1700	110	320	1450	
Heart structural	J.&P.–B.&S.	1300	95	320	1100	
Spruce, eastern:		Northeastern Lumber Manufacturers Association, Inc.		300		1,200,000
1450 f structural grade	J.&P.	1450	110		1050	
1300 f structural grade	J.&P.	1300	95		975	
1200 f structural grade	J.&P.	1200	95		900	
Tupelo:		National Hardwood Lumber Association		360		1,200,000
1700 f grade	J.&P.	1700	120		1225	
1450 f grade	J.&P.–B.&S.	1450	120		1050	
1200 f grade	J.&P.–B.&S.	1200	120		875	
1075 c grade	P.&T.				1075	

Table 133. Ranges of Physical Properties of Glasses*

Property	Value
Refractive index	1.467–2.179
Specific gravity	2.125–8.120
Elastic coefficient	6,500,000–12,500,000
Compressive strength	90,000–180,000 psi
Tensile strength	4,000–1,000,000 psi
Thermal conductivity	0.0018–0.0028 cal/cm/°C/sec
Expansion coefficient	8×10^{-7}–140×10^{-7} cm/cm/°C
Softening point	500–1510 C
Annealing point	350–890 C
Volume resistivity	10^8–10^{18} ohms/cu cm
Dielectric constant	3.7–16.5

* From C. J. Phillips, *Glass: The Miracle Maker*, Chicago, Pitman, 1941.

Table 134. Specific Heats of Several Glasses*

Material	Temperature Range, °C	Mean Specific Heat, cal/gm/°C
Bottle glass	40–800	0.280
	40–1000	0.298
Plate glass	40–800	0.274
	40–1000	0.289
22% lead glass	40–800	0.211
	40–1000	0.224

* From S. R. Scholes, *Modern Glass Practice*, Chicago, Industrial Publications, 1946.

Table 135. Properties of Some Commercial Glasses*

Glass Code	Type	Color	Principal Use	Forms Usually Available	Thermal Expansion Coeff.—1 C	Upper Working Temperatures (Mechanical Considerations Only) — Annealed Normal service, °C	Annealed Extreme limit, °C	Tempered Normal service, °C	Tempered Extreme limit, °C	Thermal Shock Resistance Plates 6 by 6 in. Annealed 1/8 in. thick, °C	1/4 in. thick, °C	1/2 in. thick, °C	Thermal Stress Resistance, °C
0010	Potash–soda–lead	Clear	Lamp tubing	T	91×10^{-7}	110	380			65	50	35	19
0041	Potash–soda–lead	Clear	Thermometers	T	84×10^{-7}	110	400			70	60	40	19
0080	Soda–lime	Clear	Lamp bulbs	BMT	92×10^{-7}	110	460	220	250	65	50	35	17
0120	Potash–soda–lead	Clear	Lamp tubing	TM	89×10^{-7}	110	380			65	50	35	17
1710	Hard lime	Clear	Cooking utensils	BP	42×10^{-7}	200	650	400	450	135	115	75	29
1770	Soda–lime	Clear	General	BP	82×10^{-7}	110	450	220	250	70	60	40	19
2405	Hard red	Red	General	BPU	43×10^{-7}	200	480			135	115	75	36
2475	Soft red	Red	Neon signs	T	91×10^{-7}	110	440			65	50	35	17
3321	Hard green sealing	Green	Sealing	T	40×10^{-7}	200	470			135	115	75	39
4407	Soft green	Green	Signal ware	BPU	90×10^{-7}	110	460			65	50	35	17
6720	Opal	White opaque	General	P	80×10^{-7}	110	480	220	275	70	60	40	19
6750	Opal	White opaque	Lighting ware	BPR	87×10^{-7}	110	420	220	220	65	50	35	18
6810	Opal	White opaque	Lighting ware	BPR	69×10^{-7}	120	470	240	270	85	70	45	23
7050	Borosilicate	Clear	Series sealing	T	46×10^{-7}	200	440	235	235	125	100	70	34
7052	Borosilicate	Clear	Kovar sealing	BMPT	46×10^{-7}	200	420	210	210	125	100	70	34
7070	Borosilicate	Clear	Low-loss electrical	BMPT	32×10^{-7}	230	430	230	230	180	150	100	70
7250	Borosilicate	Clear	Baking ware	P	36×10^{-7}	230	460	230	260	160	130	90	43
7340	Borosilicate	Clear	Gage glass	BPT	67×10^{-7}	120	510	240	310	85	70	45	20
7720	Borosilicate	Clear	Electrical	BPSTU	36×10^{-7}	230	460	260	260	160	130	90	45
7740	Borosilicate	Clear	General	BP	32×10^{-7}	230	490	260	290	180	150	100	48
7760	Borosilicate	Clear	Electrical	BPTU	34×10^{-7}	230	450	250	250	160	130	90	51
7900	96% silica	Clear	High temperature	M	8×10^{-7}	800	1090			1250	1000	750	200
7910	96% silica (multiform)	White opaque	High temperature	BTU	8×10^{-7}	800	1090			1250	1000	750	200
7911	96% silica	Clear	Ultraviolet transmission	T	8×10^{-7}	800	1090			1250	1000	750	200
8870	High lead	Clear	Sealing or electrical	MTU	91×10^{-7}	110	380	180	180	65	50	35	22
9700	....	Clear	Ultraviolet transmission	TU	37×10^{-7}	220	500			150	120	80	42
9741	....	Clear	Ultraviolet transmission	BUT	39×10^{-7}	200	390			150	120	80	40

Glass Code	Type	Strain point, °C	Annealing point, °C	Softening point, °C	Working point, °C	Impact Abrasion Resistance	Density (Sp. Gr.)	Modulus of Elasticity, psi	Log₁₀ of Volume Resistivity 25 C	250 C	350 C	Power factor, %	Dielectric constant	Loss factor, %	Refractive Index Sodium D Line (0.5893 μ)
0010	Potash-soda-lead	397	428	626	970		2.85	9.0×10^6	17+	8.9	7.0	.16	6.6	1.1	1.539
0041	Potash-soda-lead	426	460	648			2.89								1.545
0080	Soda-lime	478	510	696	1000	1.2	2.47		12.4	6.4	5.1	.9	7.2	6.5	1.512
0120	Potash-soda-lead	400	433	630	975		3.05	9.8×10^6	17+	10.1	8.0	.16	6.6	1.1	1.560
1710	Hard lime	672	712	915	1200	2.0	2.53	12.7×10^6	17+	11.4	9.4	.37	6.3	2.3	1.534
1770	Soda-lime	470	503	710			2.40								1.496
2405	Hard red	506	537	802			2.50								1.508
2475	Soft red	466	501	693			2.56								1.511
3321	Hard green sealing	497	535	780			2.27								
4407	Soft green	485	518	695			2.53								
6720	Opal	499	531	775			2.58								1.525
6750	Opal	445	475	672			2.63								1.507
6310	Opal	496	529	768			2.65								1.513
7050	Borosilicate	461	496	703	1115	4.1	2.25	6.8×10^6	16	8.8	7.2	.33	4.9	1.6	1.508
7052	Borosilicate	438	475	708	1100	3.2	2.28		17+	9.2	7.4	.26	5.1	1.3	1.479
7070	Borosilicate	455	490				2.13		15	11.2	9.1	.06			1.484
7250	Borosilicate	486	524	775			2.24		16	8.5	6.7	.28	4.7	1.3	1.469
7340	Borosilicate	538	575	785			2.43	11.5×10^6	16	8.8	6.9				1.475
7720	Borosilicate	484	518	755	1110	3.2	2.35	9.5×10^6	17	8.1	7.2	.27	4.7	1.3	1.506
7740	Borosilicate	515	555	820	1220	3.1	2.23	9.8×10^6	17	9.4	7.7	.46	4.6	2.1	1.487
7760	Borosilicate	475	515	780	1210		2.23	9.1×10^6	17	9.7	8.1	.18	4.5	.79	1.474
7900	96% silica	820	910	1500		3.5	2.18	9.7×10^6	17+	11.2	9.2	.05	3.8	.19	1.458
7910	96% silica (multiform)	820	910	1500		3.5	2.18	9.7×10^6	17+	11.2	9.7	.05	3.8	.19	1.458
7911	96% silica	820	910	1500		3.5	2.18	9.7×10^6	15	11.8	6.5	.024	3.8	.091	1.458
8370	High lead	398	429	580		.6	4.28	7.6×10^6	17+	9.4	7.6	.019	3.8	.072	1.693
9700		517	558	804	1195		2.26					.09	9.5	.85	1.478
9741		407	442	705			2.16								

Column 5: B = blown ware. M = multiform ware, P = pressed ware, R = rolled sheet, S = plate glass, T = tubing and rod, U = panels.
Column 6: From 0 to 300 C in./in./°C or cm/cm/°C.
Column 7: These data are approximate only. Freedom from excessive thermal shock is assumed. See Column 8. At extreme limits annealed glass will be very vulnerable to thermal shock. Recommendations in this range are based on mechanical considerations only. Tests should be made before final designs are adapted.
Column 8: These data are approximate only. They are based on plunging sample into cold water after oven-heating. Resistance of 100 C means no breakage if heated to 110 C and plunged into water at 10 C. Tempered samples have over twice the resistance of annealed glass. Glasses 7900, 7910, 7911 cannot be tempered.
Column 9: Resistance, in Centigrade degrees, is the temperature differential between the two surfaces of a tube or a constrained plate that will cause a tensile stress of 1000 psi on the cooler surface
Column 10: These data are subject to normal manufacturing variations.
Column 11: Data show relative resistance to sandblasting.
Column 12: Units are grams per cubic centimeter.
Column 14: Data at 25° were extrapolated from high-temperature readings and are approximate only.
Glasses 7910 and 7911: Electrical properties were measured on lamp-worked specimens.
All data are subject to normal manufacturing variations.

* From Properties of Selected Commercial Glasses, Corning, N. Y., Corning Glass Works, 1949.

Table 136.　Density of Miscellaneous Non-metallic Solids

Name	Density* g per cu cm	lb per cu ft	Name	Density* g per cu cm	lb per cu ft
Agate..........	2.5–2.7	156–168	Lava, basaltic.....	2.8–3.0	175–187
Amber..........	1.06–1.11	66–69	trachytic.......	2.0–2.7	125–168
Asbestos........	2.0–2.8	125–175	Leather.........	.86–1.02	54–64
Asphalt.........	1.1–1.5	69–94	Lime, mortar.....	1.65–1.78	103–111
Basalt..........	2.4–3.1	150–193	quick (in bulk)..	0.8–0.96	50–60
Bauxite.........	2.55	159	slaked.........	1.3–1.4	81–87
Beeswax........	0.96–0.97	60–61	Limestone........	2.00–2.9	125–181
Biotite..........	2.7–3.1	168–193	Litharge		
Borax..........	1.7–1.8	106–112	(artificial)......	9.3–9.4	580–587
Brick, soft......	1.6	100	natural........	7.8–8.0	487–499
common.......	1.79	112	Magnesia		
hard..........	2.0	125	carbonate......	2.4	150
pressed........	2.16	135	Magnetite........	4.9–5.2	306–324
fire...........	2.24–2.4	140–150	Marble..........	2.6–2.84	162–177
sand lime.....	2.18	136	Masonry, dry		
Brickwork			rubble........	2.24–2.56	140–160
mortar........	1.6	100	dressed........	2.24–2.88	140–180
cement........	1.79	112	Mica...........	2.6–3.2	162–200
Carbon, diamond..	3.52	220	Muscovite.......	2.76–3.00	172–187
graphite.......	2.25	140	Oligoclase.......	2.65–2.67	165–167
Cement, natural...	2.8–3.2	175–200	Orthoclase.......	2.58–2.61	161–163
Portland.......	3.05–3.15	190–197	Paper...........	0.7–1.15	44–72
loose..........	1.44	90	Pitch...........	1.07	67
barreled.......	1.84	115	Plaster-of-Paris...	1.5–1.8	94–112
slag...........	1.9–2.3	119–144	Porcelain........	2.3–2.5	143–156
Chalk..........	1.9–2.8	119–175	Porphyry........	2.6–2.9	162–181
Charcoal, oak.....	0.57	35	Pumice..........	0.37–0.90	23–56
pine..........	0.28–0.44	17–27	Pyrite..........	4.95–5.1	309–318
Clay...........	1.8–2.6	112–162	Quartz..........	2.65	165
Coal, anthracite..	1.4–1.8	87–112	Quartzite........	2.73	170
bituminous.....	1.2–1.6	75–100	Resin...........	1.07	67
charcoal.......	0.27–0.58	17–36	Riprap		
lignite........	1.1–1.4	69–87	limestone......	1.3–1.4	81–87
Coke...........	1.0–1.7	62–106	sandstone......	1.4	87
Concrete†		144	shale.........	1.7	106
Corundum.......	3.9–4.0	244–250	Rock salt.......	2.18	136
Dolomite........	2.84	177	Rubber		
Earth			caoutchouc.....	0.92–0.96	57–60
dry, loose......	1.2	75	manufactured...	1.0–2.0	62–125
packed....	1.5	94	Salt............	0.78–1.25	49–78
moist, loose.....	1.3	81	Sand, dry.......	1.44–1.76	90–110
packed..	1.6	100	wet.......	1.89–2.07	118–129
mud, flowing....	1.7	106	Selenium........	4.82	301
packed....	1.8	112	Serpentine.......	2.50–2.65	156–165
Emery..........	4.0	250	Shale..........	2.6–2.9	162–181
Feldspar........	2.55–2.75	159–172	Silicon..........	2.42	151
Flint...........	2.63	164	Slag, bank.......	1.1–1.2	69–75
Garnet.........	3.15–4.3	197–268	bank screenings..	1.5–1.9	94–119
Gas carbon......	1.88	117	furnace........	2.0–3.9	125–244
Gelatin.........	1.27	80	machine.......	1.5	94
Glass, common....	2.5–2.75	156–172	sand..........	0.8–0.9	50–56
crystal........	2.90–3.00	181–187	Slate............	2.6–3.3	162–205
flint..........	3.2–4.7	200–294	Soapstone........	2.6–2.3	162–175
plate..........	2.45–2.72	153–170	Starch..........	1.53	95
Glue...........	1.27	80	Stone, various.....	2.16–3.4	135–212
Gneiss..........	2.4–2.7	150–169	crushed........	1.6	100
Granite.........	2.65–2.7	165–169	Sugar...........	1.61	100
Gravel, dry, loose..	1.4–1.7	87–106	Sulfur...........	2.0–2.1	125–131
packed...	1.6–1.9	100–119	Talc............	2.7–2.8	168–175
wet..........	1.9	119	Tar, bituminous...	1.02	66
Gypsum.........	2.31–2.33	144–145	Terracotta.......	1.9	119
Hematite........	4.9–5.3	306–330	Tile............	1.76–1.92	110–120
Hornblende......	3.0	187	Tourmaline......	3.0–3.2	187–200
Iodine..........	4.94	308	Traprock........	2.72–3.4	170–212
Ivory..........	1.83–1.92	114–120	Wood...........	see Table 131	see Table 131

* Ordinary temperatures understood.

Table 137. Specific Heat of Miscellaneous Non-metallic Solids

Name	Temp, C	Specific Heat, g-cal per g or Btu per lb	Name	Temp, C	Specific Heat, g-cal per g or Btu per lb
Asbestos.............	0	0.25	Granite.............	20–100	0.20
Bakelite...............	....	0.3–0.4	Ice.................	−20	.48
Basalt...............	....	.20	....	0	.505
Carbon, graphite......	−76–0	.126	India rubber, Para....	?–100	.481
	26–76	.165	Limestone............	....	.217
diamond............	0	.147	Marble.............	18	.21
Calcspar.............	0–100	.2005	Mica...............	20–98	.2061
Cellulose................	....	.32	Paraffin.............	0–20	.694
Chalk................	20–99	.214	Porcelain	15–950	.26
Clay.................	0	.224	Quartz...............	0	.17
Coal.....................	....	0.26–0.37	Rock salt.............	13–45	.219
Coke.................	21–400	.265	Sand...............	....	.191
Concrete.............	70–312	.156	Selenium.............	20.5	.077
Ebonite...............	20–100	.40	Silicon...............	18.2–99.1	.181
Glass			Sulfur, rhombic.......	15–96	.176
normal thermometer			monoclinic........	0–52	.181
16III	19–100	.1988	Woods, general........	....	0.45–0.65
crown...............	0	0.16–0.20			
flint................	10–50	.117			

Table 138. Thermal Coefficient of Linear Expansion of Miscellaneous Non-metallic Solids[a]

Name	Temp, C	Thermal Coefficient of Expansion, per °C ($\times 10^4$)	Name	Temp, C	Thermal Coefficient of Expansion, per °C ($\times 10^4$)
Amber..............	0–30	0.50	Quartz glass........	16–1000	0.0058
Bakelite, bleached...	20–60	0.22	Rock salt	40	0.4040
Caoutchouc........	16.7–25.3	0.770	Rubber, hard.......	0	0.691
Carbon, diamond....	50	0.012	Rubber, hard......	−160	0.300
graphite..........	50	0.0786	Selenium...........	0–100	0.660
Celluloid...........	20–80	1.09	Silicon.............	−3 to +18	0.0249
Ebonite.............	25.3–35.4	0.842	Tourmaline:		
Fluorspar: CaF₂.....	0–100	0.1950	‖ to longitudinal		
Glass, tube.........	0–100	0.0833	axis..........	0–100	0.0937
plate.............	0–100	0.0891	‖ to horizontal		
crown (mean).....	0–100	0.0897	axis..........	0–100	0.0773
flint.............	50–60	0.0788	Vulcanite..........	0–18	0.6360
Jena thermometer			Wedgwood ware....	0–100	0.0890
16III			Wood ‖ to fiber		
normal........	0–100	0.081	ash.............	0–100	0.0951
Jena thermometer			beech...........	2.34	0.0257
56III	0–100	0.058	chestnut.........	2.34	0.0649
Jena thermometer			elm.............	2.34	0.0565
59III	−191 to +16	0.424	mahogany........	2.34	0.0361
Gutta percha........	20	1.983	maple............	2.34	0.0638
Ice.................	−20 to −1	0.51	oak.............	2.34	0.0492
Iceland spar:			pine.............	2.34	0.0541
‖ to axis.........	0–80	0.2631	walnut...........	2.34	0.0658
⊥ to axis........	0–80	0.0544	Wood:		
Limestone...........	25–100	0.09	⊥ to fiber		
Marble.............	15–100	0.117	beech...........	2.34	0.614
Paraffin.............	0–16	1.0662	chestnut.........	2.34	0.325
Paraffin.............	16–38	1.3030	elm.............	2.34	0.443
Porcelain............	20–790	0.0413	mahogany........	2.34	0.404
Bayeux...........	1000–1400	0.0553	maple............	2.34	0.484
Quartz:			oak.............	2.34	0.544
‖ to axis..........	0–80	0.0797	pine.............	2.34	0.341
‖ to axis..........	−190 to +16	0.0521	walnut...........	2.34	0.484
⊥ to axis..........	0–80	0.1337	Wax, white........	10–26	2.300
Quartz glass........	−190 to +16	−0.0026	Wax, white........	26–31	3.120
Quartz glass........	16–500	0.0057			

[a] The coefficient of cubical expansion may be taken as three times the linear coefficient.

Table 139. Typical Composition Ranges of Fuel Gases*

Per Cent by Volume

	Hydrogen H_2	Methane CH_4	Ethane C_2H_6	Illumi-nants	Carbon Monoxide CO	Carbon Dioxide CO_2	Oxygen O_2	Nitrogen N_2	Sp Gr Air = 1.0	Gross Heating Value, Btu per cu ft
Natural gas†	0	80.5-90.2	4.3-12.1	0	0	0-2.7	0-0.4	0.1-7.5	0.59-0.67	1047-1210
Mixed refinery gas † ‡	2.0-20.9	21.6-53.3	13.4-19.3	10-16.2	0	1.5-8.0	1.5-8.0	1.5-8.0	0.83-1.15	1380-1828
Oil gas	24-55	26-60	0	3-15	0-10	1-4	0.3-0.5	2.5-4.0	0.37-0.50	540-700
Coal gas	37-50	27-40	0-3	3-4	6-8.5	1.6-2.4	0.4-1.0	3.2-11.0	0.45-0.55	540-700
Coke-oven gas	48-53	30-35	0-1	3-4	4-6	1-3	0-1.1	4-6	0.4-0.5	550-650
Producer gas	10-19	0.3-6.3	0	0-0.4	17-30	2.5-7.3	0.3-0.7	49-58	0.82-0.87	135-170
Water gas	45-70	0-4.5	0	0-1.1	32-41	3.0-5.0	0.2-0.7	4-10	0.5-0.6	290-320
Carbureted water gas	32-37	11-14	0	8-10	26-30	4-6	0.4-0.9	4.5-12.0	0.55-0.6	528-545
Blast-furnace gas	1-4	0	0	0	26-27.5	11.5-13.0	0	57.6-60	0.95-1.0	90-93
Acetylene (comm.)	0	0	0	95-96	0	0	0-0.8	3-4	0.91-0.96	1300-1400

* From various sources, including *U.S. Bureau of Mines Repor's.*
† Mekler, L. A., and von Fredersdorff, C. G., *Petroleum Refiner,* No. 2, 81, 1947.
‡ Contains 1.3 to 4.2% propane and 0.2 to 3.5% butane.

Table 140. Requirements for Motor Gasoline*

Gasoline Volatility Grade^a	Distillation — Maximum Temperatures,^b deg F (deg C), at Percent Evaporated				Residue, percent, max	Vapor/Liquid Ratio^b — Test Temperature, deg F (deg C)	Vapor/Liquid Ratio^b — V/L, max	Reid Vapor Pressure, lb, max	Research Method Octane Number,^c min	Motor Method Octane Number,^c min	Copper Strip Corrosion, max	Gum, mg/100 ml, max	Sulfur
	10	50	90	End Point									
C (Cold)	125 (51.5)	250 (121.0)	375 (190.5)	437 (225.0)	2	105 (40.6)	20	14.5	90 or 96	82 or 88^d	No. 1	5	....
M (Mild)	130 (54.5)	250 (121.0)	375 (190.5)	437 (225.0)	2	114 (45.6)	20	13.5	90 or 96	82 to 88	No. 1	5	....
W (Warm)	140 (60.0)	250 (121.0)	375 (190.5)	437 (225.0)	2	124 (51.1)	20	11.5	90 or 96	82 or 88	No. 1	5	....
H (Hot)	150 (65.5)	250 (121.0)	375 (190.5)	437 (225.0)	2	132 (55.6)	20	10.0	90 or 96	82 or 88	No. 1	5	....
E (Extreme)	155 (68.5)	250 (121.0)	375 (190.5)	437 (225.0)	2	140 (60.0)	20	9.0	90 or 96	82 or 88	No. 1	5	....

* From ASTM Designation D 439-68T.

^a For Geographical schedule of seasonal volatility grades see ASTM Designation D 439-68T.

^b At 760 mm Hg pressure.

^c In all cases the octane number shall be agreed upon between the purchaser and the seller. See Appendix, A1.3, for discussion of Research and Motor octane numbers. The numerical values shown are minimum values currently encountered in service stations. The lower value pertains to regular-price gasolines, and the higher value to premium-price gasolines. For more detailed information on current levels for both research and motor octane numbers, as well as for other characteristics of motor gasoline, reference is made to the series of semiannual reports, issued as Mineral Industry Surveys by the U. S. Bureau of Mines and entitled "Motor Gasolines." The Motor and Research octane numbers of gasoline for use in areas where altitude is greater than 2000 ft may be reduced one (1) octane number for each succeeding 1000 ft but not more than three (3) octane numbers in total. Values reported are derived from the 1967 Survey, Petroleum Products Survey No. 53.

^d The technical data available do not afford an adequate basis for specifying maximum sulfur content. At present time (1967), gasolines containing up to 0.20 percent sulfur (ASTM Method D 1266) are distributed within the United States.

Table 141. Detailed Requirements^a For Aviation Gasolines^{"*}

	Grade 80-87	Grade 100-130	Grade 115-145	Test Method^b
Knock value, min, octane number, lean rating	80	100	isooctane plus 0.47 ml of tetraethyllead per gallon	ASTM D 2700
Knock value, min, octane number, rich rating	87	isooctane plus 1.28 ml of tetraethyllead per gallon	isooctane plus 2.8 ml of tetraethyllead per gallon	ASTM D 909
Color^c	red^d	green	purple	ASTM D 2392
Dye content:				
Permissible blue dye,^e max, mg per gal	0.5	4.7	4.7	
Permissible yellow dye,^f max, mg per gal	none	7.0	none	
Permissible red dye,^g max, mg per gal	8.95	none	3.7	
Tetraethyllead,^h max, ml per gal	0.5^e	3.0	4.6	ASTM D 526
Net heat of combustion,^i min, Btu per lb	18,720	18,720	18,800	ASTM D 1405

REQUIREMENTS FOR ALL GRADES

	Value	Test Method
Distillation temperature, deg F (deg C):		ASTM D 86
10 per cent evaporated, max	158 (70)	
50 per cent evaporated, max	221 (105)	
90 per cent evaporated, min	212 (100)	
90 per cent evaporated, max	257 (125)	
Final boiling point, max, deg F (deg C)	338 (170)	
Sum of 10 and 50 per cent evaporated temperatures, min, deg F (deg C)	307 (135)	
Distillation recovery, min, per cent	97	
Distillation residue, max, per cent	1.5	
Distillation loss, max, per cent	1.5	
Acidity of distillation residue	shall not be acid	ASTM D 1093
Vapor pressure, max, lb	7.0	ASTM D 323
Copper strip corrosion, max	No.1	ASTM D 130
Potential gum (5 hr aging gum)^j max, mg per 100 ml	6	ASTM D 873
Visible lead precipitate,^k max, mg per 100 ml	3	ASTM D 873
Sulfur, max, per cent	0.05	ASTM D 1266
Freezing point, max, deg F (deg C)	-72 (-58)	ASTM D 2386
Water tolerance	volume change not to exceed ±2 ml 4.2	ASTM D 1094
Permissible antioxidants,^l max, lb per 1000 bbl (42 gal)		

*From ASTM Designation D 910-68.

a Requirements contained herein are absolute and are not subject to correction for tolerance of the test methods. If multiple determinations are made, average results shall be used.

b The test methods indicated in this table are referred to in Section 9 of ASTM Designation D 910-68.

c These colors have been approved by the Medical Director Chief, Division of Occupational Health, U. S. Department of Health, Education and Welfare.

d If mutually agreed upon between the purchaser and the supplier, Grade 80-87 may be required to be free from tetraethyllead. In such a case the fuel shall not contain any dye and the color as determined in accordance with Method D 156, Test for Saybolt Color of Petroleum Products (Saybolt Chromometer Method) shall not be darker than +20.

e The only blue dye which shall be present in the finished gasoline shall be essentially 1,4-dialkylamino-anthraquinone.

f The only yellow dye which shall be present in the finished gasoline shall be essentially p-diethylaminoazobenzene (Color Index No. 11020).

g The only red dye which shall be present in the finished gasoline shall be essentially methyl derivatives of azobenzene-4-azo-2-naphthol (methyl derivatives of Color Index No. 26105).

h The tetraethyllead shall be added in the form of an antiknock mixture containing not less than 61 per cent by weight of tetraethyllead and sufficient ethylene dibromide to provide two bromine atoms per atom of lead. The balance shall contain no added ingredients other than kerosene, and approved inhibitor, and blue dye, as specified, herein.

i Use the value calculated from Table I in Method D 1405, Test for Estimation of Net Heat of Combustion of Aviation Fuels. Method D 2382, Test for Heat of Combustion of Hydrocarbon Fuels by Bomb Calorimeter (High-Precision Method) may be used as an alternative method. In case of dispute, Method D 2382 must be used. In this latter case, the minimum values for the net heat of combustion in Btu's per pound shall be 18,700 for Grades 80-87, 81-98, and 100-130.

j If mutually agreed upon between the purchaser and the supplier, aviation gasoline may be required to meet a 16-hr aging gum test (ASTM Method D 873, Test for Oxidation Stability of Aviation Fuels (Potential Residue Method) instead of the 5-hr aging gum test. In such a case, the gum content shall not exceed 10 mg per 100 ml and the visible lead precipitate shall not exceed 4 mg per 100 ml. In such fuel the permissible antioxidants shall not exceed 8.4 lb per 1000 bbl (42 gal).

k The visible lead precipitate requirement applies only to leaded fuels.

l Permissible antioxidants are as follows:

N,N'-diisopropyl-para-phenylenediamine
N,N'-di-secondary-butyl-para-phenylenediamine
2,4-dimethyl-6-tertiary-butylphenol
2,6-ditertiary butyl-4-methylphenol
2,6-ditertiary butylphenol
Mixed tertiary butylphenols, composition:
    75 per cent 2,6-ditertiary butylphenol
    10 to 15 per cent 2,4,6-tritertiary butylphenol
    10 to 15 per cent o-tertiary butylphenol.
    72 per cent min 2,4-dimethyl-6-tertiary butylphenol, and 28 per cent max monomethyl and dimethyl tertiary butylphenols.

These inhibitors may be added to the gasoline separately or in combination, in total concentration not to exceed 4.2 lb of inhibitor (not including weight of solvent) per 1000 bbl (42 gal).

m Listing of and requirements for Grades 91-98 and 108-135 appear in the 1967 version for these specifications.

Table 142. Detailed Requirements^a for Aviation Turbine Fuels*

Property	Type^b Jet A	Type^b Jet A-1	Type^b Jet B	ASTM Test Method
Gravity, max, deg API (min, sp gr)....	51 (0.7753)	51 (0.7753)	57 (0.7507)	D 287
Gravity, min, deg API (max, sp gr)....	39 (0.8299)	39 (0.8299)	45 (0.8017)	D 287
Distillation temperature, deg F (deg C):				D 86
10 per cent evaporated, max.........	400 (204.4)	400 (204.4)	...	
20 per cent evaporated, max.........	...	...	290 (143.3)	
50 per cent evaporated, max.........	450 (232.2)	450 (232.2)	370 (187.8)	
90 per cent evaporated, max.........	...	...	470 (243.3)	
Final boiling point, max, deg F (deg C).	550 (287.8)	550 (287.8)	...	
Distillation residue, max, per cent......	1.5	1.5	1.5	
Distillation loss, max, per cent.........	1.5	1.5	1.5	
Vapor pressure, max, lb...............	...	...	3	D 323
Flash point, min, deg F (deg C)........	101 (43.3)	110 (43.3)	...	D 56
Flash point, max, deg F (deg C).......	150 (65.6)	150 (65.6)	...	D 56
Freezing point, max, deg F (deg C).....	−36 (−38)	−54 (−48)	−56 (−49)	D 2386
Viscosity at −30 F (−34.4 C), max, cST.	15	15	...	D 445
Net heat of combustion, min, Btu per lb.	18,400^c	18,400^c	18,400^c	D 1405, or D 2382
Net heat of combustion, Btu per gal....	^{d}	^{d}	^{d}	
Copper Strip ⎰ 3 hr at 122 F (50 C), max.	No. 1	No. 1	...	D 130
Corrosion ⎱ 2 hr at 212 F (100 C), max.	...	...	No. 1	D 130
Total acidity, max, mg KOH per gram..	0.1	0.1	...	D 974
Sulfur, max, per cent.................	0.3	0.3	0.3	D 1266
Mercaptan sulfur, max, per cent ^{e}.....	0.003	0.003	0.003	D 1323 or D 1219
Water tolerance, vol change, not to exceed, ml........................	±1	±1	±1	D 1094
Total potential residue, 16 hr, max, mg per 100 ml.....................	14	14	14	D 873
Thermal stability at 300 to 400 F (148.9 to 204.4 C):^{f}				
Filter press drop, max, in. of Hg.....	12	12	12	D 1660
Preheater deposit less than.........	Code 3	Code 3	Code 3	
Combustion Properties. One of the following requirements shall be met:				
(1) Luminometer number, min.......	45	45	50	D 1740
or				
(2) Smoke point, min...............	25	25	...	D 1322
or				
(3) ⎰ Smoke point, min.............	20	20	...	D 1322
⎱ Naphthalenes, max, per cent...	3	3	...	D 1840
or				
(4) Smoke volatility index, min......	...	...	54^g	D 1322 and D 86
Aromatics, vol, max, per cent..........	20	20	20	D 1319
Olefins, vol, max, per cent............	...	...	5	D 1319
Additives, max lb (not including weight of solvent) per 1000 bbl (42 gal)				
(1) Antioxidants...................	8.4	8.4	8.4	^{h}
(2) Metal deactivator..............	2.0	2.0	2.0	^{i}
(3) Other.......................	^{j}	^{j}	^{j}	

* From ASTM Designation D 1655-68.

^a The requirements herein are absolute and are not subject to correction for tolerance of the test methods. If multiple determinations are made, average results shall be used.

^b Type Jet A.—A relatively high flash point distillate of the kerosine type.

Type Jet A-1.—A kerosine type similar to Type A but incorporating special low temperature characteristics for certain operations.

Type Jet B.—A relatively wide boiling-range volatile distillate.

^c Use for Jets A and A-1 the value calculated from Table IV or Eqs 4 and 5 in ASTM Method D 1405, Estimation of Net Heat of Combustion of Aviation Fuels. Use for Jet B the value calculated from Table II or Eqs 2 and 5 in Method D 1405. ASTM Method D 2382, Test for Heat of Combustion of Hydrocarbon Fuels by Bomb Calorimeter (High-Precision Method) may be used as an alternative. In case of dispute, Method D 2382 must be used.

^d The Btu's per gallon are not limited but shall be reported. They shall be calculated as the product of the net heat of combustion in terms of Btu's per pound as obtained from Tables II and IV of ASTM Method D 1405, and the pounds per U. S. gallon at 60 F (15.6 C) as obtained from Table 8 of the Petroleum Measurement Tables (ASTM Designation: D 1250).

^e The mercaptan sulfur determination may be waived if the fuel is considered sweet by the doctor test described in Section 4(c) of ASTM Specifications D 484, for Stoddard Solvent.

^f Thermal stability test shall be conducted for 5 hr at 300 F (148.9 C) preheater temperature 400 F (204.4 C) filter temperature, and at a flow rate of 6 lb per hr.

^g The smoke volatility index, SVI, may be used for acceptance of fuels having luminometer numbers below 50 if agreeable to the user. It is determined from the following equation:

SVI = smoke point (D 1322) + 0.42 × volume [per cent boiling under 400 F (204.4 C) (D 86)]

^h Approved antioxidants are: N,N-diisopropylparaphenylenediamine; Mixed tertiary butyl phenols, composition: 75 per cent 2,6-ditertiary butyl phenol, 10 to 15 per cent 2,4,6-tritertiary butyl phenol, 10 to 15 per cent o-tertiary butyl phenol; 72 per cent min 2,4-dimethyl-6-tertiary butyl-phenol, and 28 per cent max monomethyl and dimethyl tertiary butyl phenols.

ⁱ Approved metal deactivator is: N,N' disalicylidene, 1,2 propane diamine.

^j Other antioxidants, inhibitors, and special purpose additives are permitted under Sections 4 and 7 of ASTM Designation D-1655. The quantities and types must be declared by the manufacturer and agreed to by the purchaser.

Table 143. Limiting Requirements for Diesel Fuel Oils*

Grade of Diesel Fuel Oil	Flash Point, deg Fahr (deg C)	Pour Point deg Fahr (deg C)	Water and Sediment, per cent by volume	Carbon Residue on 10 per cent Residuum, per cent	Ash, per cent by weight	Distillation Temperatures, deg Fahr (deg C) 90 per cent Point		Viscosity at 100 F (37.8 C) Kinematic, centistokes or Saybolt Universal, sec		Sulfur, per cent by weight	Copper Strip Corrosion	Cetane Number[e]
	Min	Max	Max	Max	Max	Min	Max	Min	Max	Max	Max	Min
No. 1-D { A volatile distillate fuel oil for engines in service requiring frequent speed and load changes.	100 or legal (37.8)	b	Trace	0.15	0.01	...	550 (287.8)	1.4	2.5 (34.4)	0.50	No. 3	40[f]
No. 2-D { A distillate fuel oil of lower volatility for engines in industrial and heavy mobile service.	125 9 or legal (51.7)	b	0.05	0.35	0.01	540[c] (282.2)	640 (338)	2.0[c] (32.6)	4.3 (40.1)	0.50[d]	No. 3	40[f]
No. 4-D { A fuel oil for low and medium speed engines.	130 or legal (54.4)	b	0.50	...	0.10	...	...	5.8 (45)	26.4 (125)	2.0	...	30[f]

* From ASTM Designation D 975-68.

^a To meet special operating conditions, modifications of individual limiting requirements may be agreed upon between purchaser, seller, and supplier.

^b For cold weather operation, the pour point should be specified 10 F (5.6 C) below the ambient temperature at which the engine is to be operated except where fuel oil heating facilities are provided.

^c When pour point less than 0 F (−17.8 C) is specified, the minimum viscosity shall be 1.8 cs (32.0 sec, Saybolt Universal) and the minimum 90 per cent point shall be waived.

^d For all products outside the U.S.A. the maximum sulfur limit shall be 1.0 per cent by weight.

^e Where cetane number by Method D 613, Test for Ignition Quality of Diesel Fuels by the Cetane Method, is not available, ASTM Method D 976, Calculated Cetane Index of Distillate Fuels may be used as an approximation. Where there is disagreement, Method D 613 shall be the referee method.

^f Low-atmospheric temperatures as well as engine operation at high altitudes may require use of fuels with higher cetane ratings.

Table 144. Detailed Requirements for Fuel Oils[a][*]

Grade of Fuel Oil[b]	Flash Point, deg F (deg C) Min	Pour Point, deg F (deg C) Max	Water and Sediment, per cent by volume Max	Carbon Residue on 10 per cent Bottoms, per cent Max	Ash, per cent by weight Max	Distillation Temperatures, deg F (deg C)			Saybolt Viscosity, sec				Kinematic Viscosity, centistokes				Gravity, deg API Min	Copper Strip Corrosion Max
						10 per cent Point Max	90 per cent Point Min	90 per cent Point Max	Universal at 100 F (38 C) Min	Universal at 100 F (38 C) Max	Furol at 122 F (50 C) Min	Furol at 122 F (50 C) Max	At 100 F (38 C) Min	At 100 F (38 C) Max	At 122 F (50 C) Min	At 122 F (50 C) Max		
No. 1 { A distillate oil intended for vaporizing pot-type burners and other burners requiring this grade of fuel }	100 or legal (38)	0	trace	0.15	…	420 (215)	…	550 (288)	…	…	…	…	1.4	2.2	…	…	35	No. 3
No. 2 { A distillate oil for general purpose domestic heating for use in burners not requiring No. 1 fuel oil }	100 or legal (38)	20ᶜ (−7)	0.10	0.35	…	d	540ᵉ (282)	640 (338)	(32.6)f	(37.93)	…	…	2.0ᶜ	3.6	…	…	30	…
No. 4 { Preheating not usually required for handling or burning }	130 or legal (55)	20 (−7)	0.50	…	0.10	…	…	…	45	125	…	…	(5.8)	(26.4)	…	…	…	…
No. 5 (Light) { Preheating may be required depending on climate and equipment }	130 or legal (55)	…	1.00	…	…	…	…	…	150	300	…	…	(32)	(65)	…	…	…	…
No. 5 (Heavy) { Preheating may be required for burning and, in cold climates, may be required for handling }	130 or legal (55)	…	1.00	…	0.10	…	…	…	350	750	(23)	(40)	(75)	(162)	(42)	(81)	…	…
No. 6 { Preheating required for burning and handling }	150 (65)	…	2.00ᵉ	…	…	…	…	…	(900)	(9000)	45	300	…	…	(92)	(638)	…	…

* From ASTM Designation D 396-67.

a Recognizing the necessity for low-sulfur fuel oils used in connection with heat-treatment, nonferrous metal, glass, and ceramic furnaces and other special uses, a sulfur requirement may be specified in accordance with the following table:

Grade of Fuel Oil	Sulfur, max, per cent
No. 1...................	0.5
No. 2...................	0.7**
No. 4...................	no limit
No. 5...................	no limit
No. 6...................	no limit

Other sulfur limits may be specified only by mutual agreement between the purchaser and the seller.

** For Grade No. 2 Fuel Oils sold outside the United States, the maximum sulfur limit is 1.0 per cent by weight.

b It is the intent of these classifications that failure to meet any requirement of a given grade does not automatically place an oil in the next lower grade unless in fact it meets all requirements of the lower grade.

c Lower or higher pour points may be specified whenever required by conditions of storage or use.

d The 10 per cent distillation temperature point may be specified at 440 F (226 C) maximum for use in other than atomizing burners.

e When pour point less than 0 F is specified, the minimum viscosity shall be 1.8 cs (32.0 sec, Saybolt Universal) and the minimum 90 per cent point shall be waived.

f Viscosity values in parentheses are for information only and not necessarily limiting.

g The amount of water by distillation plus the sediment by extraction shall not exceed 2.00 per cent. The amount of sediment by extraction shall not exceed 0.50 per cent. A deduction in quantity shall be made for all water and sediment in excess of 1.0 per cent.

Table 145. Commercial Sizes of Bituminous Coal*

Designation	Size	Use
Run of mine..........	Fines to large lumps	Hand firing, domestic and industrial
Lump................	Fines to 5 in.	Hand firing, domestic and industrial
Egg.................	2–5 in.	Hand firing, domestic and industrial gas producers
Nut.................	1 1/4–2 in.	Hand firing, domestic, small industrial stokers and gas producers
Stoker coal...........	3/4–1 1/4 in.	Domestic and small industrial stokers
Slack................	0–3/4 in.	Industrial stokers and pulverizers

Commercial Sizes of Anthracite Coal *†

Designation	Size, in.	Use
Broken.......................	3 1/4 –4 3/8	Hand firing, domestic and industrial
Egg..........................	2 7/16–3 1/4	Hand firing, domestic and industrial
Stove........................	1 5/8 –2 7/16	Hand firing, domestic and industrial, gas producers
Chestnut.....................	13/16 –1 5/8	Hand firing, domestic and industrial, gas producers
Pea..........................	9/16– 13/16	Stokers, domestic and industrial
Buckwheat No. 1..............	5/16– 9/16	Domestic stokers
Buckwheat No. 2 (Rice)........	3/16– 5/16	Domestic stokers
Buckwheat No. 3 (Barley)......	3/32– 3/16	Industrial stokers, pulverizers
Buckwheat No. 4..............	3/64– 3/32	Industrial stokers, pulverizers

* Otto deLorenzi, *Combustion Engineering*, Combustion Engineering Co., Inc., New York, 1947.
† ASTM D 310—34.

Table 146. Heating Value and Available Hydrogen of Various Coals

Rank	Gross Heating Value *	Available Hydrogen †
Coke.......................	14,550	0.0
Anthracite.................	16,100	.029
Semi-bituminous............	17,400	.049
Bituminous................	17,900	.054
Sub-bituminous.............	17,600	.045
Lignite....................	17,100	.037

* Btu per pound of total carbon.
† Pounds of available hydrogen per pound of total carbon.

Table 147. Typical Ultimate Analyses for Coals*

Rank	Gross Btu per Lb		Constituents, Percent by weight						
	Moist, Mineral-matter-free^a	Moist, as Received	Oxygen	Hydrogen	Carbon	Nitrogen	Sulfur	Ash	$\frac{O_2+}{H_2+C}$
Anthracite..............	14,600	12,910	5.0	2.9	80.0	0.9	0.7	10.5	87.9
Semi-Anthracite.........	15,200	13,770	5.0	3.9	80.4	1.1	1.1	8.5	89.3
Low-Volatile Bituminous..	15,350	14,340	5.0	4.7	81.7	1.4	1.2	6.0	91.4
Medium-Volatile Bituminous..	15,200	13,840	5.0	5.0	79.0	1.4	1.5	8.1	89.0
High-Volatile Bituminous A....	14,500	13,090	9.2	5.3	73.2	1.5	2.0	8.8	87.7
High-Volatile Bituminous B....	13,500	12,130	13.8	5.5	68.0	1.4	2.1	9.2	87.3
High-Volatile Bituminous C....	12,000	10,750	21.0	5.8	60.6	1.1	2.1	9.4	87.4
Sub Bituminous B........	10,250	9,150	29.5	6.2	52.5	1.0	1.0	9.8	88.2
Sub Bituminous C........	9,000	8,940	35.8	6.5	46.7	0.8	0.6	9.6	89.0
Lignite................	7,500	6,900	44.0	6.9	40.1	0.7	1.0	7.3	91.0

* Copyright by ASHRAE. Reprinted by permission from *ASHRAE Handbook of Fundamentals*, 1967.
^a (Btu as received) $\times$ 100 $\div$ (100 − 1.1 Ash).

Table 148. Properties of Gases

(All properties are at a pressure equivalent to 760 mm of mercury unless otherwise stated)

Name	Formula	Density (0° C) G per l	Density (0° C) Lb per cu ft	Molecular Weight	Melting Point, deg cent	Latent Heat of Fusion, g-cal per g	Boiling Point, deg cent	Latent Heat of Vaporization, g-cal per g	Specific Heat — Constant Pressure, C_p	Temp, deg cent	Mean Ratio, $\frac{C_p}{C_v}$	Temp, deg cent	Thermal Coef. — Const. at Volume	Temp, deg cent	Constant Pressure	Temp, deg cent	Viscosity Poises [a] × 10⁶	Temp, deg cent
Acetylene	C₂H₂	1.1708	0.07323	26.016	−81.3		−83.6	51.0	0.3832	15	1.26	15					93.5	0
Air		1.2929	.08018	28.952					0.2377	−30−+10	1.401	20	.003665	0−100	.003671	0−100	181.2	20.2
Ammonia	NH₃	0.7710	.04813	17.031	−75	108.0	−33.5	327.1	0.5202	23−100	1.3172	0					102	20
Argon	A	1.7837	.11135	39.944	−189.2	6.71	−185.7	37.7	0.1233	20−90	1.668	15	.003668 (517 mm)				224.1	17.9
Arsine	AsH₃	3.48	.217	77.95	−113.5		−54.8	87.4			1.11	15					114.0	15
Butane-iso	C₄H₁₀	2.637	.1669	58.08	−145.0		10.2	91.5									75.5	23
Butane-n	C₄H₁₀	2.519 (710 mm)	.1525 (710 mm)	58.08	−135.0		0.6										83.3	16.0
Carbon dioxide	CO₂	1.9769	.12341	44.000	−57	45.3	−80 Subl.	137.9	0.1989	15	1.304	15	.003981 (518 mm)	0−100	.003707 (518 mm)	0−100	145.7	15.0
Carbon monoxide	CO	1.2504	.07806	28.000	−207	8.00	−191.5	50.4	0.2478	15	1.404	15	.003667		.003669		184.0	20
Carbon oxychloride (phosgene)	COCl₂	4.531	.283	98.91	−118		8.3											
Carbon oxysulfide	COS	2.72	.170	66.06	−138		48				1.336	0					119.0	15
Chlorine	Cl₂	3.214	.2006	70.914	−101.6	23.0	34.7	81	0.1125	16−343							137	20
Chlorine monoxide	Cl₂O	3.89	.243	86.91	−20		3.8 (explodes)										129	12.7
Cyanogen	C₂N₂	2.335	.146	52.02	−27.90		−21.17	258.0	0.4095	15	1.256	15					107	20
Ethane	C₂H₆	1.3566	.08469	30.05	−172.0		−88.3	92.5	0.3861	15	1.22	15					101	20
Ethyl chloride	C₂H₅Cl	2.870	.1793	64.50	−138.7		12.2		0.2750	10−170	1.19	16					105	20
Ethylene	C₂H₄	1.2604	.07868	28.031	−169.4		−103.8		0.399	15−100	1.18	100					109.0	20
Fluorine	F₂	1.696	.1059	38.00	−223		−187	40.5	0.182	0								
Helium	He	0.1785	.01114	4.002	−272		−268.94	5.97	1.25	−180	1.660	−180	.003665 (567 mm)	−180			196.9	15.3
Hydrogen	H₂	0.08988	.005611	2.0156	−259.14	14.0	−252.8	106.7	3.389	+15	1.4080	4−16	.0036504 (764 mm)		.0036600 (1000 mm)	0−100	88.9	15

Table 148. Properties of Gases—Continued

(All properties are at a pressure equivalent to 760 mm of mercury unless otherwise stated)

Name	Formula	Density (0 °C) G per l	Lb per cu ft	Molecular Weight	Melting Point, deg cent	Latent Heat of Fusion, g-cal per g	Boiling Point, deg cent	Latent Heat of Vaporization, g-cal per g	Specific Heat — Constant Pressure C_p	Specific Heat — Temp, deg cent	Mean Ratio C_p/C_v	Ratio Temp, deg cent	Thermal Coef. — Constant Volume	Temp, deg cent	Constant Pressure	Temp, deg cent	Viscosity Poises[a] × 10⁶	Viscosity Temp, deg cent
Hydrogen bromide	HBr	3.6445	.22752	80.92	-86.7	7.67	-68.7	48.7	0.082	11-100	1.42	20					181.9	18.7
Hydrogen chloride	HCl	1.6392	.10233	36.465	-111.3	13.4	-83.1	105.9	0.194	13-100	1.389	20					140	20
Hydrogen fluoride	HF	0.922	.0576	20.01	-92.3		-36.7 (755 mm)		0.343	0							185.7	20.6
Hydrogen iodide	HI	5.789	.3614	127.93	-51.3	5.68	-35.7 (755 mm)	33.92	0.06	0	1.40	20-100						
Hydrogen selenide	H₂Se	3.670	.229	81.22	-64		-42											
Hydrogen sulfide	H₂S	1.539	.09608	34.08	-86		-62	131.9	0.2451	20-206	1.324						130	20
Hydrogen telluride	H₂Te	5.81	.363	129.5	-48		-1.8											
Krypton	Kr	3.708	.2315	82.9	-169		-151.8	28									246	15
Methane	CH₄	0.717	.0448	16.0317	-182.5	14.53	-161.4	138	0.5929	18-208	1.666	19					120.1	20
Methyl chloride	CH₃Cl	2.3076	.14406	50.404	-103.6		-23.73	102.3	0.24		1.316	11-30					116	20
Methyl ether	(CH₃)₂O	2.1098	.13171	46.05	-138		-24.9				1.20							
Methyl fluoride	CH₃F	1.5452	.09646	34.02			-78.0				1.11	6-30						
Monomethylamine	CH₃NH₂	1.396	.08715	31.05	-92.5		-6.8											
Neon	Ne	0.9004	.05621	20.183	-248.67	2.84	-245.9		0.232	10-180	1.642	19					312	15
Nitric oxide	NO	1.3402	.08367	30.008	-167	18.4	-153		0.2438	0-200	1.40	15					179	0
Nitrogen	N₂	1.2506	.07807	28.016	-209.86	6.1	-195.8				1.41		.0036682				170.7	10.9
Nitrosyl chloride	NOCl	2.992	.1868	65.47	-64.5		-5.5	47.8					.003676		.003719		138	0
Nitrous oxide	N₂O	1.978	.1235	44.016	-102.4		-89.8		0.2126	26-103	1.311	0					195.7	15.4
Oxygen	O₂	1.42904	.089212	32.0000	-218.4	3.33	-183.0	51	0.2175	13-207	1.3977	5-14	.003668 (75.9 cm)					
Phosphine	PH₃	1.529	.0955	34.04	-133.5		-87.4										112.0	15
Propane	C₃H₈	2.020	.1261	44.06	-189.9		-44.5											
Silicon tetrafluoride	SiF₄	4.68	.292	104.06	-76		-68											
Sulfur dioxide	SO₂	2.9269	.18272	64.06			-10	94.8	0.1544	16-202	1.256	16-34	.003845		.003903		129	20
Xenon	Xe	5.85	.365	130.2	-140	3.73	-109.1	24.4			1.666	19					222	15

[a] Dyne sec per sq cm.

Table 149. Properties of Liquids
(Normal compounds only)

Name	Formula	Density		Melting Point, deg cent	Latent Heat of Fusion, g-cal per g	Boiling Point (760 mm), deg cent	Latent Heat of Vaporization, g-cal per g	Viscosity		Specific Heat	
		G per cu cm	Temp, deg cent					Poises[b]	Temp, deg cent	G-cal per g or Btu per lb	Temp,[a] deg cent
Acetaldehyde (aldehyde)	C_2H_4O	0.806	0	−120.0		20.8	136.0	0.00231	20	0.522	26–95
Acetic acid	$C_2H_4O_2$	1.05	20	16.7	43.2	118.5	96.8	0.01222	20	0.514	3–22.6
Acetone	C_3H_6O	0.792	20	−94.6	19.6	56.1	124.5	0.0031	20	0.386	0
Allyl alcohol	C_3H_6O	0.847	25	−129.0		97.0	163.0	0.01363	20	0.711	22–125
Amyl alcohol	$C_5H_{12}O$	0.817	20	−78.5		137.9	120.2			0.512	8–82
Aniline	$C_6H_5NH_2$	1.035	0	−6.24	0.95	183.9	110.0	0.04467	20		
Benzene (benzol)	C_6H_6	0.879	20	5.56	30.34	80.12	94.2	0.00654	20	0.419	6–60
Bromine	Br	3.187	0	−7.2	16.2	58.8	48.0	0.01005	20	0.107	13–45
Butyl alcohol	$C_4H_{10}O$	0.810	20	−89.8	30.0	117.7	141.0	0.02948	20	0.687	21–115
Butyric acid	$C_4H_8O_2$	0.954	25	−5.55	30.10	163.5	114.0	0.01540	20	0.515	20–100
Carbolic acid (phenol)	C_6H_6O	1.07	25	41	29.03	182.2		0.1274	18.3	0.561	14–26
Carbon disulfide	CS_2	1.594	0	−111.8		46.26	84.1	0.00376	20	0.240	20
Carbon tetrachloride	CCl_4	1.594	−20	−22.8	41.57	76.75	46.4	0.00975	20	0.201	20
Castor oil		0.9603	20					9.86	20	0.434	15
Chloroform	$CHCl_3$	1.489	20	−63.5		61.2	59.0	0.00571	20	0.226	0–50
Decane	$C_{10}H_{22}$	0.747	20	−32.0		174.0		0.0077	22.3	0.500	0
Di-ethyl ether	$C_4H_{10}O$	0.714	20	−116.3		34.5	83.9	0.00245	20	0.529	20
Ethyl acetate	$C_4H_8O_2$	0.899	20	−83.6		77.1		0.0045	20	0.457	0
Ethyl alcohol	C_2H_6O	0.789	20	−114.6	24.89	78.32	204.3	0.012	20	0.548	15–20
Ethyl bromide	C_2H_5Br	1.45	15	−119.0		38.4	59.9	0.00402	20	0.215	0
Ethyl chloride	C_2H_5Cl	0.918	8	−138.7		12.2	92.5			0.367	20
Ethyl iodide	C_2H_5I	1.944	14	−108.5		72.1	45.6	0.00592	20	0.161	20
Ethylene bromide	C_2H_4Br	2.17		10.01		131.7	46.2	0.01721	20	0.173	20
Ethylene chloride	C_2H_4Cl	1.246		−35.3		83.7	77.33	0.00838	20	0.299	20–100
Formic acid	CH_2O_2	1.22	20	8.4	58.89	100.8	119.9	0.01784	20	0.525	0–100
Gasoline		0.66–0.69				70.0–90.0				0.5	0–100
Glycerin	$C_3H_8O_3$	1.261	20	18.1	47.5	290.0		8.30	20.3	0.576	15–50

Table 149. Properties of Liquids—Continued

(Normal compounds only)

Name	Formula	Density G per cu cm	Density Temp,[a] deg cent	Melting Point, deg cent	Latent Heat of Fusion, g-cal per g	Boiling Point (760 mm), deg cent	Latent Heat of Vaporization, g-cal per g	Viscosity Poises[b]	Viscosity Temp, deg cent	Specific Heat G-cal per g or Btu per lb	Specific Heat Temp,[a] deg cent
Heptane	C_7H_{16}	0.684	20	−90.7	....	98.4	76.3	0.00416	20	0.490	20
Hexane	C_6H_{14}	0.660	20	−95.4	....	68.7	79.3	0.00326	20	0.600	20
Kerosene	....	0.78–0.82	....	....	....	....	....	....	....	0.5	0–100
Linseed oil	....	0.934	15.5	....	....	287.0	....	0.331	30	....	....
Methyl acetate	$C_3H_6O_2$	0.927	....	−98.1	....	57.1	98.1	0.00388	20	0.468	15
Methyl alcohol	CH_4O	0.792	20	−97.8	22.0	64.7	262.8	0.00596	20	0.601	15–20
Methyl iodide	CH_3I	2.285	15	−64.0	....	42.3	45.9	0.00500	20	....	....
Naphthalene	$C_{10}H_8$	1.152	15	80.2	35.6	218.0	75.5	0.040	20	0.396	80–85
Neatsfoot oil	....	0.913–0.917	....	....	....	....	....	....	....	....	....
Nitric acid (100%)	HNO_3	1.513	20	−47	9.53	86.0	114.9	0.021	20	0.350	14
Nitrobenzene	$C_6H_5O_2N$	1.212	7.5	5.85	22.5	210.9	79.1	0.0062	22.3	0.503	0–50
Nonane	C_9H_{20}	0.718	20	−51.0	....	150.6	....	0.00542	20	0.578	20–123
Octane	C_8H_{18}	0.707	17	−56.9	....	124.6	70.95	0.840	20	0.471	6.6
Olive oil	....	0.918	15	−20±	....	300±	....	0.024	20	....	....
Pentane	C_5H_{12}	0.631	20	−129.9	....	36.0	....	....	....	....	....
Petroleum	....	0.878	0	....	....	....	....	....	....	0.511	21–58
Propionic acid	$C_3H_6O_2$	0.99	....	−20.8	....	141.1	98.8	0.01102	20	0.560	20–137
Propyl alcohol	C_3H_8O	0.804	20	−127.0	....	97.5	164.4	0.02256	20	0.57	20
Rapeseed oil	....	0.913	15	−3.5	....	....	....	1.18	15.6	....	....
Soya bean oil	....	0.919	30	....	....	....	....	0.406	30	....	....
Sperm oil	....	0.88	25	....	....	....	....	0.420	15.6	....	....
Sulfuric acid (100%)	H_2SO_4	1.831	20	10.49	24.03	(98.3%) 330.0	122.08	0.50	20.0	0.344	20
Tallow	....	0.94	15	27–41	....	....	....	0.176	66	....	....
Toluene	C_7H_8	0.882	0	−95.0	....	110.3	86.53	0.00590	20	0.440	12–99
Turpentine	....	0.873	16	−10	....	160.0	68.6	0.01487	30	0.411	0
Water	H_2O	1.00000[c]	4	0	79.70	100.0	539.44	0.010050	20	1.0000	16
o-Xylene	$C_6H_4(CH_3)_2$	0.863	20	−27.1	....	142.0	82.9	0.00881	13.88	0.411	30

[a] Where the temperature is not given, ordinary temperature is understood. [b] Dyne sec per sq cm. [c] 8 1/3 lb per gal, 62 1/2 lb per cu ft (approx.).

INDEX

Ab- (prefix), 414
Abampere, 414-6
Abbreviations, engineering terms, 4-6
  mechanics of fluids, 572-3
  scientific terms, 4-6
Abcoulomb, 415
Abelian group, 238
Aberrations, 1062
Abfarad, 415
Abhenry, 414
Abohm, 415
Absolute dielectric constant, 412
  dimensions of, 392-3
  of free space, 415
  electrical dimension system, 394-5
  magnetic permeability, 413
  dimensions of, 392-3
  temperature, 410
  unit systems, 397
  viscosity, units of, 574
"Absolute" dimension system, 389
  electrical units, 412-6
  practical units, 402, 416-7
Absorption characteristics of a body, 1119
  coefficient, mass (radiation), 1053
  electric, 976
  of radiation, 1052
  refrigeration system, 919-20
  sound, 1106
    coefficients, 1106-8
    in fluid media, 1096-7
  ultraviolet, 1071
  of x-rays, 944, 1075
Absorptivity, 1053
Abvolt, 415
Acceleration, angular, 460, 473
    and rectilinear acceleration, relation
      between, 460
    units of, conversion of, 154
  centripetal, 472
  components of, 458
  coriolis, 473
  curvilinear motion, 458
  force, mass, relation between, 462
  linear, units of, conversion of, 154
  measurement of (control), 1228
  of a particle, 455
  of any point in plane motion, 461
  precessional, 473
  rectilinear motion, 455
  -space diagram, 457
  -time diagram, 456
  translational, 472
  uniform linear (fluids), 584
  units of, 455, 462

Accelerometers, 818
Acetaldehyde, properties of, 1506
Acetic acid, properties of, 1506
Acetone, properties of, 1506
Acetylene, properties of, 1504
Acid, 1254
  Bessemer steel, 1358
  character of a solution, 1254
  nitric, properties of, 1507
  open-hearth steel, 1358
  sulfuric, 1507
Acids, bases and salts, 1254
  properties of, 1506-7
Acoustic bibliography, 1110-1
  density, 1094
  filter, 1102
  impedance, measurement of, 1110
  insulating materials, absorption coeffi-
    cients, 1107-8
  resistance, specific, 1094, 1096
  strain gage, 507
Acoustical instruments and measurements,
    1099, 1100
  law, Ohm's, 1105
Acoustics, 1090-1111
  architectural, 1106-9
  physiological, 1103-5
  radiation and light, section 11, 1051
Actinic rays, 1071
Activation energy, 1284
Activity, 1261
  coefficient, 1261-2
Adaptive control, 1212-5
Addition, 213-6
  reaction, 1299
  symbols for, 3
Additivity of extensive state function, 1265
Adiabatic compression, multistage, 865
  single-stage, 864
  engines, turbines, etc., 833
  enthalpy drop, 889
  flow, 611-2, 833
  of a perfect gas, 656-7
  process, 826
  reversible, 832, 845, 877
Admiralty (alloy), properties of, 1412, 1414
Admittance (electric), 988
  indicial, 995
  vector, 988
Advance (alloy), properties of, 1450
Aerodynamic heating, 759-61
Aerodynamics, 430, section 7, p. 637
  of bodies, 734
  wing–body combinations, 761
  of wings, 704

moment of, about a line, 434
  about a point, 433
  units of, conversion of, 157
non-concurrent, 431
  composition of, 435-8
  resultant of, 438
non-coplanar, 431
  composition of, 433, 437
  resolution of, 433
per unit area, units of, conversion of, 157
polygon, 435-6
power of, 463
produced by a magnetic field on a coil
  carrying a current, 966
single and couple, composition and
  resolution of, 435
systems of, 431
transmissibility of, 430
units of, 397, 400, 407-8, 462
  conversion of, 156
vs. weight, 407
on wire in a magnetic field, 967
work of a, 462
Forces, on curved surfaces, 582
generalized, 477
on plane surfaces, 582
Forgings, magnesium, 1406
Form factor, 979
Format statement, 359
Formation, standard heat of, 1265
Formic acid, properties of, 1506
Forms Hermitian, 233
  quadratic, 232
Formula, parabolic, 530
Formulas, approximate, 217-8
Forsterite brick, 1474, 1476
Fortran, 353-83
  constant, 353-4
  statement, 356-82
  variable, 355
Foucault currents, 965
Four-cycle engine, 908
Four-stroke cycle, 906, 908
Fourier series, 308-9, 936-7
Fractional electrical dimension system, 393-4
  powers of numbers, table, 142
  roots, 225
Fractions, 216
  partial, 219
  rational, 215
    integration of, 315
Fracture, types of, 494
Frame, inertial, 472
  non-inertial, 472
  of reference, 472
Frames, redundant, 442
Fraunhofer lines, 1069-70
Free oscillations, 1097-8
  -radical reagents, 1300
Freedom, number of degrees of, 477
Freezing point, aluminum, 1372
  magnesium, 1404
French units and standards, 405-6
Freon, p-H diagram for, 916
Frequency, change of velocity with, 1092

cumulative, definition of, 239
curve, 239
distribution, definition, 239
distributions of one variable, 239
of errors, relative, and normal distribution, 245
mass and energy relation, 935
modulated wave, 294
of periodic quantity, 978
polygon, 239
resonance (sound), 1098-9
-response methods (control), 1158-68
simple harmonic motion, 457
of sound, 1104
Friction, 479-84
axle, 480
belt, 484
boundary, 481
coefficient of, carbon brushes, 1451
  rolling, 482
  static, 479-80
coil, 484
journal, 481
kinetic, 479-80
  coefficient of, 479-80
pivot, 482-3
rolling, 482
skin, 757-9
static, 479-80
tape, dielectric strength of, 1466
Frictionless fluid, 705
Froude's number, 422, 426
Frustum of a cone, 255
  of a pyramid, 254
Fuel-air ratio, 680-3, 688
Fuel gases, composition ranges of, 1492
oil, density, 574
  Diesel, limiting requirements for, 1498-9
  requirements for, 1500-1
  viscosity of, 577-8
Fuels, 1492-1503
  injection (Diesel engine), 911
  liquid and gaseous, 1492-1501
  turbine, aviation, requirements for, 1496-7
Fugacity, 1261, 1277
  generalized chart, 1277
Fullerboard, dielectric strength of, 1466
Function, gamma, table, 141
  moment generating, 241
  stream, 595, 706
  subprograms, 379
Functional relationships, trigonometric, 274-6
Functions, circular, of plane angles, 272-6
  values and logarithms of, 75-119
complementary, 328
of complex variables, 347-9
composite, differentiation of, 305
continuity of, 301
definition of, 300
and derivatives, 300-3
expanded in series, 310-2
gravitational, 466
hyperbolic, 279-80